Viral Diseases/Anatomic Syndromes

Disease/Syndrome	Virus	Reservoir	Pages	Disease/Syndrome	Virus	Reservoir	Pages
AIDS	human immunodeficiency virus (HIV)	humans	308, 320–321, 686–692	hepatitis (infectious)	hepatitis A		
				hepatitis (serum)	hepatitis B	humans	
aseptic meningitis	arbovirus, enteroviruses, mumps virus	humans	625	hepatitis C	hepatitis C	humans	308, 581
				hepatitis D	hepatitis delta agent	humans	581
bronchiolitis	respiratory syncytial virus	humans	546	hepatitis E	hepatitis E	humans	581
Burkitt's lymphoma	Epstein-Barr	humans	308	herpetic infections	herpes simplex	humans	607–608
chickenpox	varicella zoster	humans	308, 651–652	infectious mononucleosis	Epstein-Barr	humans	684–685
condyloma acuminata (genital warts)	human papillomavirus	humans	608–610	influenza	influenza	humans, swine	309, 321–323, 544–545
colds	coronavirus	humans	533	measles	measles (rubeola)	humans	653–655
	rhinovirus	humans	309, 532–533	mumps	mumps virus	humans	309, 559–560
conjunctivitis (pinkeye)	adenovirus	humans	308	pericarditis	enterovirus	humans	309
croup	parainfluenza	humans, some other mammals	546	pneumonia	respiratory syncytial virus	humans	309, 546
				poliomyelitis	poliovirus	humans	630–632
cytomegalic inclusion disease	cytomegalovirus	humans	614–615	rabies	rabies	mammals	309, 629–630
encephalitis	arboviruses (e.g. eastern equine encephalitis, western equine encephalitis)	mammals, birds	632, 634	rubella (German or 3-day measles)	rubella	humans	308, 655
	enteroviruses	humans	634	shingles	varicella zoster	humans	308, 651–652
	herpes simplex	humans	634	smallpox	variola	humans	656
				warts, common	human papillomavirus	humans	308, 656–657
hantavirus pulmonary syndrome	hantavirus	mammals	546	warts, genital	human papillomavirus	humans	608–610
				yellow fever	yellow fever	monkeys	308, 683–684

Fungal Diseases/Anatomic Syndromes

Disease/Anatomic Syndrome	Organism	Page	Disease/Anatomic Syndrome	Organism	Page
blastomycosis	Blastomyces dermatitidis	284, 549	mycetoma	Madurella mycetomatis	659
candidiasis	Candida albicans	612, 657–658		Phialophora jeanselmei	659
coccidioidomycosis (San Joaquin valley fever)	Coccidioides immitis	284, 547–549	Pneumocystis pneumonia	Pneumocystis carinii	549
cryptococcal meningitis	Cryptococcus neoformans	284, 625	ringworm (tinea)	Epidermophyton spp., Microsporum spp., Trichophyton spp.	284, 657
histoplasmosis	Histoplasma capsulatum	284, 547	sporotichosis	Sporothrix schenckii	569

Helminths, Protozoa, and Arthropods

Disease/Anatomic Syndrome	Organism	Type	Pages	Disease/Anatomic Syndrome	Organism	Type	Pages
African trypanosomiasis (sleeping sickness)	Trypanosoma brucei	protozoan	289, 635–636	loaiasis	Loa loa	roundworm	293, 662
amoebic dysentery	Entamoeba histolytica	protozoan	289, 290, 572	lung fluke	Paragonimus westermani	flatworm	293, 295–296
ascariasis	Ascaris lumbricoides	roundworm	293, 296, 575	malaria	Plasmodium spp.	protozoan	289, 291, 297–298, 692–695
babesiosis	Babesia spp.	protozoan	696	onchocerciasis	Onchocerca volvulus	roundworm	662–663
balantidiasis	Balantidium coli	protozoan	289, 291, 573–574	pediculosis (body lice)	Pediculus spp.	louse	659–660
body lice (pediculosis)	Pediculus humanus	louse	659–660	pinworm	Enterobius vermicularis	roundworm	293, 574–575
Chagas' disease	Trypanosoma cruzi	protozoan	289, 673–674	primary amoebic encephalitis	Naegleria fowleri	protozoan	289, 290, 635
Chinese liver fluke	Opisthorchis sinensis	flatworm	581–582	scabies	Sarcoptes scabiei	mite	659
crab lice	Phthirus pubis	louse	659–660	schistosomiasis	Schistosoma spp.	flatworm	293, 696–698
cryptosporidiosis	Cryptosporidium spp.	protozoan	574	sheep liver fluke	Fasciola hepatica	flatworm	581–582
filariasis	Brugia malayi, Wuchereria bancrofti	roundworm	293, 297, 698	strongyloidiasis	Strongyloides stercoralis	roundworm	576
giardiasis	Giardia lamblia	protozoan	289, 572–573	tapeworm	Taenia spp.	flatworm	577–578
hookworm	Ancyclostoma duodenale (Old World hookworm)	roundworm	575–576		Taenia saginata (beef tapeworm)	flatworm	293, 294–295
	Necator americanus (New World hookworm)	roundworm	293, 296–297, 575–576		Echinococcus granulosus (dog tapeworm)	flatworm	293, 294–295
				toxoplasmosis	Toxoplasma gondii	protozoan	289, 291, 695–696
				trichinosis	Trichinella spiralis	protozoan	289, 613
hydatid disease	Echinococcus granulosus	flatworm	293, 294–295, 578	whipworm	Trichuris trichiura	roundworm	293, 296, 576

Introduction to
MICROBIOLOGY

JOHN L. INGRAHAM
University of California, Davis

CATHERINE A. INGRAHAM
The Permanente Medical Group, Inc., Rancho Cordova

Developmental Editor
HARRIETT PRENTISS
Evanston, Illinois

Wadsworth Publishing Company
I(T)P **An International Thomson Publishing Company**

Belmont • Albany • Bonn • Boston • Cincinnati • Detroit • London • Madrid • Melbourne
Mexico City • New York • Paris • San Francisco • Singapore • Tokyo • Toronto • Washington

BIOLOGY EDITOR: Jack Carey
DEVELOPMENTAL EDITOR: Mary Arbogast
EDITORIAL ASSISTANT: Kristin Milotich
PRODUCTION EDITOR: Deborah Cogan
MANAGING DESIGNER: Carolyn Deacy
PRINT BUYER: Randy Hurst
ART EDITOR: Kelly Murphy; Emma Nash
PERMISSIONS EDITOR: Marion Hansen; Jeanne Bosschart
COPY EDITOR: Joan Pendleton
PHOTO RESEARCHER: Stephen Forsling
ART CONCEPT DEVELOPER FOR SPECIAL FIGURES: Dana Hawley
TECHNICAL ILLUSTRATORS: Illustrious, Inc.; Carlyn Iverson; Margaret Gerrity
COMPOSITION AND COLOR SEPARATION: TSI
COVER DESIGNER: Carolyn Deacy
COVER ILLUSTRATOR: Tomo Narashima
PRINTER: R. R. Donnelley & Sons, Willard, Ohio

*This book is printed
on recycled paper.*

For more information, contact:

Wadsworth Publishing Company
10 Davis Drive
Belmont, California 94002
USA

International Thomson Publishing Europe
Berkshire House 168-173
High Holborn
London, WC1V 7AA
England

Thomas Nelson Australia
102 Dodds Street
South Melbourne 3205
Victoria, Australia

Nelson Canada
1120 Birchmount Road
Scarborough, Ontario
Canada M1K 5G4

International Thomson Editores
Campos Eliseos 385, Piso 7
Col. Polanco
11560 México D.F. México

International Thomson Publishing GmbH
Königswinterer Strasse 418
53227 Bonn
Germany

International Thomson Publishing Asia
221 Henderson Road
#05-10 Henderson Building
Singapore 0315

International Thomson Publishing Japan
Hirakawacho Kyowa Building, 3F
2-2-1 Hirakawacho
Chiyoda-ku, Tokyo 102
Japan

2 3 4 5 6 7 8 9 10—01 00 99 98 97 96 95

Ingraham, John L.
    Introduction to microbiology / John L. Ingraham, Catherine A. Ingraham;
developmental editor, Harriett Prentiss.
        p. cm.
    Includes bibliographical references and index
    ISBN 0-534-16728-4
    1. Medical microbiology. I. Ingraham, Catherine A. II. Prentiss, Harriett.
III. Title.
QR46.I53 1994
616' .01—dc20   94-10832

We dedicate this book to

Tom V., Marge, Tom I., Anna, Lisa, and Dana, who indulged us for eight years.

CONTENTS IN BRIEF

DETAILED TABLE OF CONTENTS

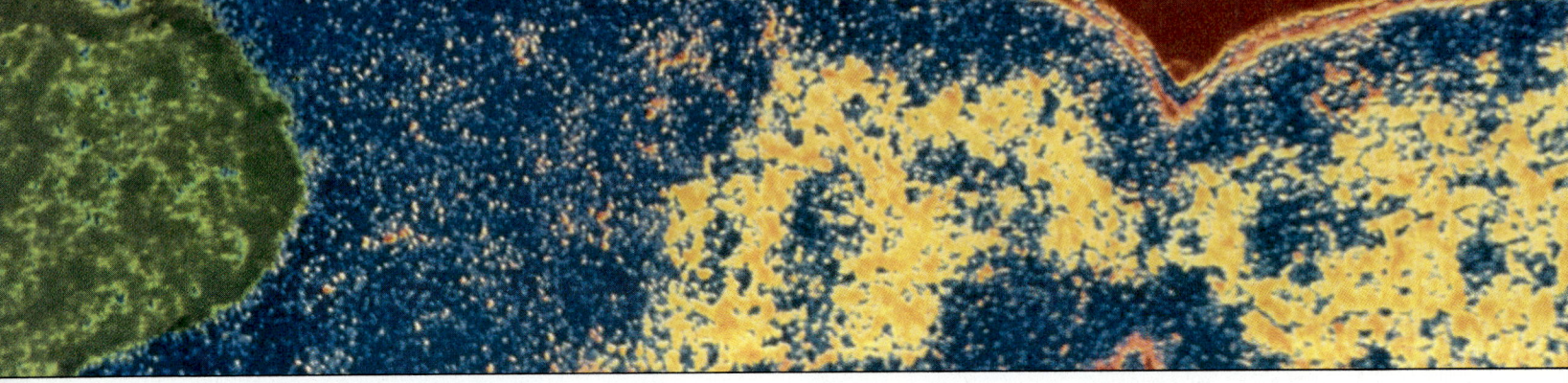

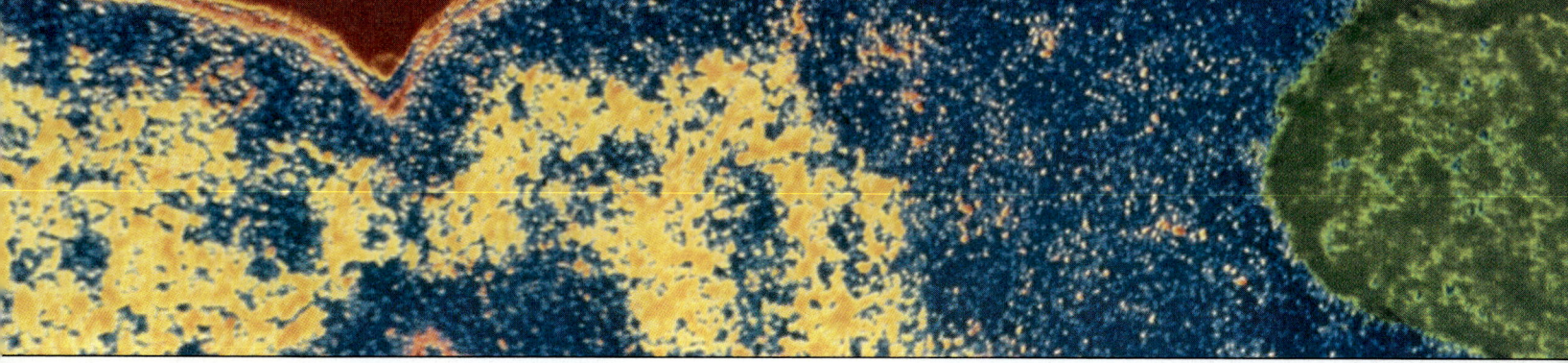

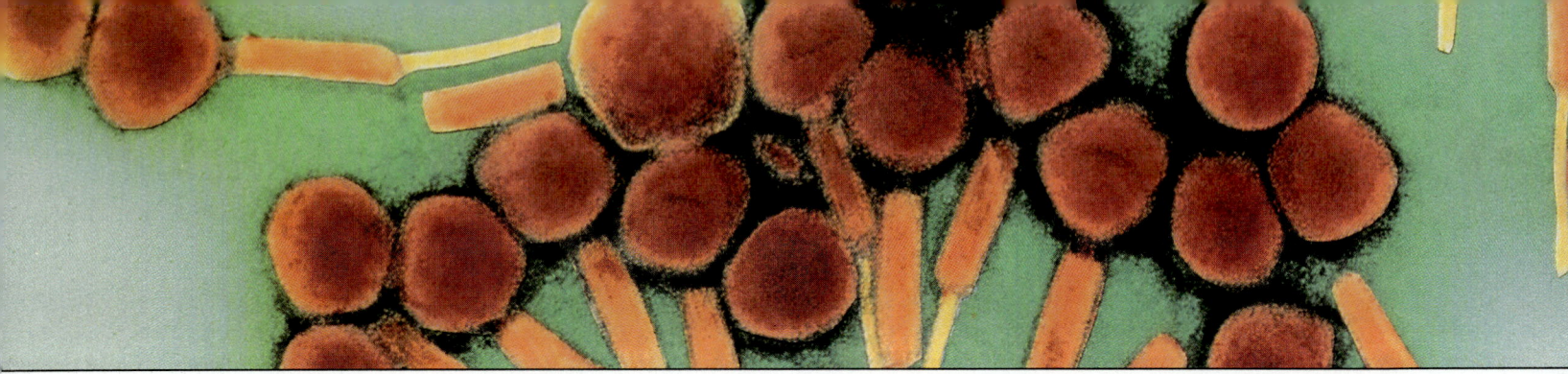

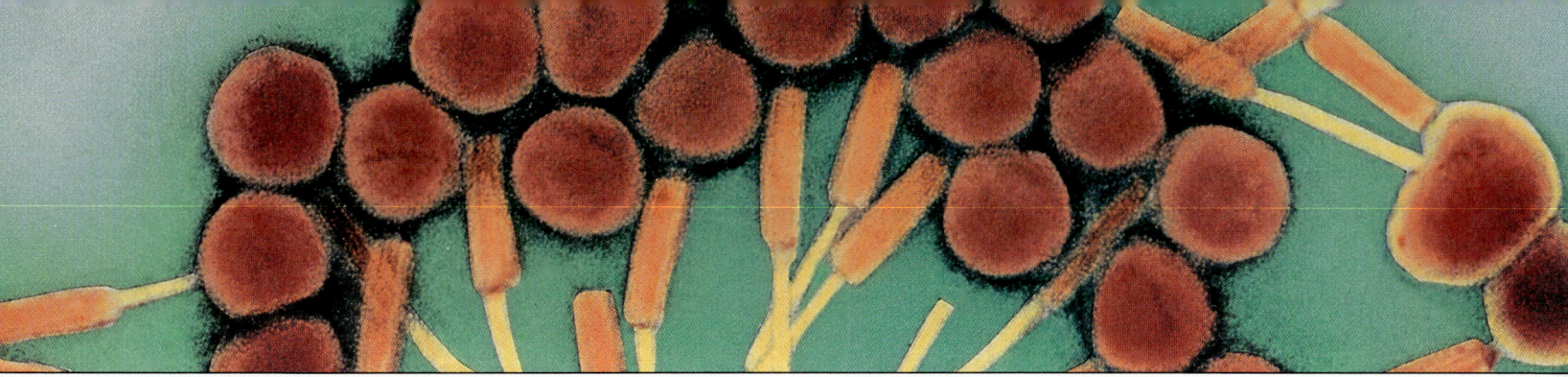

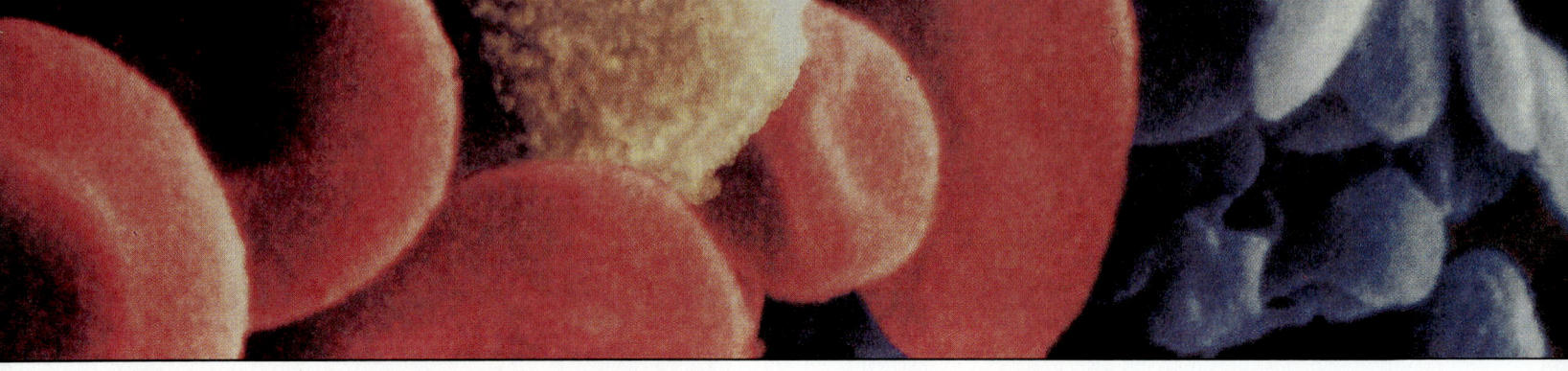

Introduction to
MICROBIOLOGY

BOOKS IN THE WADSWORTH BIOLOGY SERIES

Biology: Concepts and Applications, 2nd, Starr

Biology: The Unity and Diversity of Life, 7th, Starr and Taggart

Laboratory Manual for Biology, Perry and Morton

Human Biology, Starr and McMillan

Introduction to Microbiology, Ingraham and Ingraham

Living in the Environment, 8th, Miller

Environmental Science, 5th, Miller

Sustaining the Earth, Miller

Environment: Problems and Solutions, Miller

Resource Conservation and Management, Miller

Molecular and Cellular Biology, Wolfe

Introduction to Cell and Molecular Biology, Wolfe

Cell Ultrastructure, Wolfe

Marine Life and the Sea, Milne

Oceanography: An Invitation to Marine Science, Garrison

Essentials of Oceanography, Garrison

Oceanography: An Introduction, 5th, Ingmanson and Wallace

Plant Physiology, 4th, Salisbury and Ross

Plant Physiology Laboratory Manual, Ross

Plant Physiology, 4th, Devlin and Witham

Exercises in Plant Physiology, Witham et al.

Plants: An Evolutionary Survey, 2nd, Scagel et al.

Psychobiology: The Neuron and Behavior, Hoyenga and Hoyenga

Sex, Evolution, and Behavior, 2nd, Daly and Wilson

Dimensions of Cancer, Kupchella

Evolution: Process and Product, 3rd, Dodson and Dodson

KEY TO COVER ILLUSTRATION

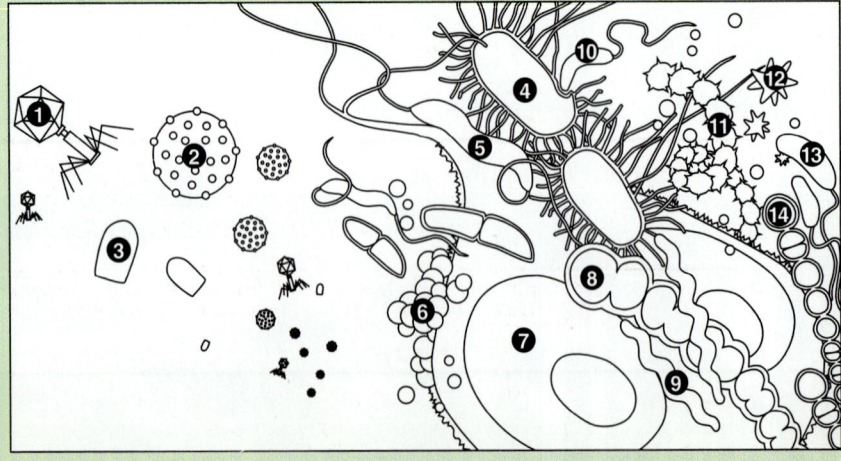

1 T4 bacteriophage
2 HIV
3 rabies virus
4 *Escherichia coli*
5 *Aquaspirillum magnetotacticum*
6 *Staphylococcus*
7 *Giardia*
8 a cyanobacterium
9 a spirochete
10 *Bdellovibrio bacteriovorans*
11 *Streptomyces* spores
12 a prosthecate bacterium
13 *Vibrio cholerae*
14 *Streptococcus*

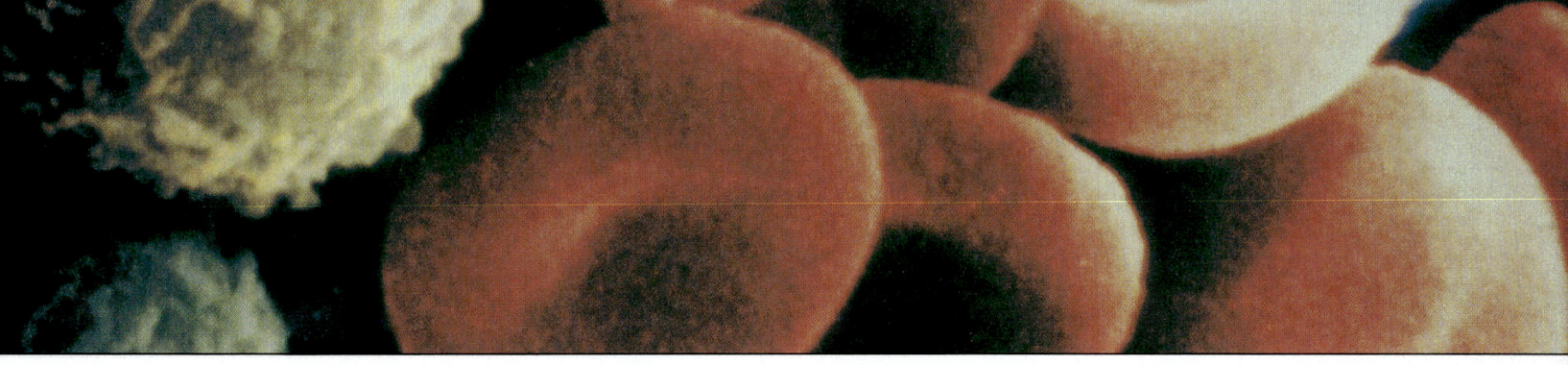

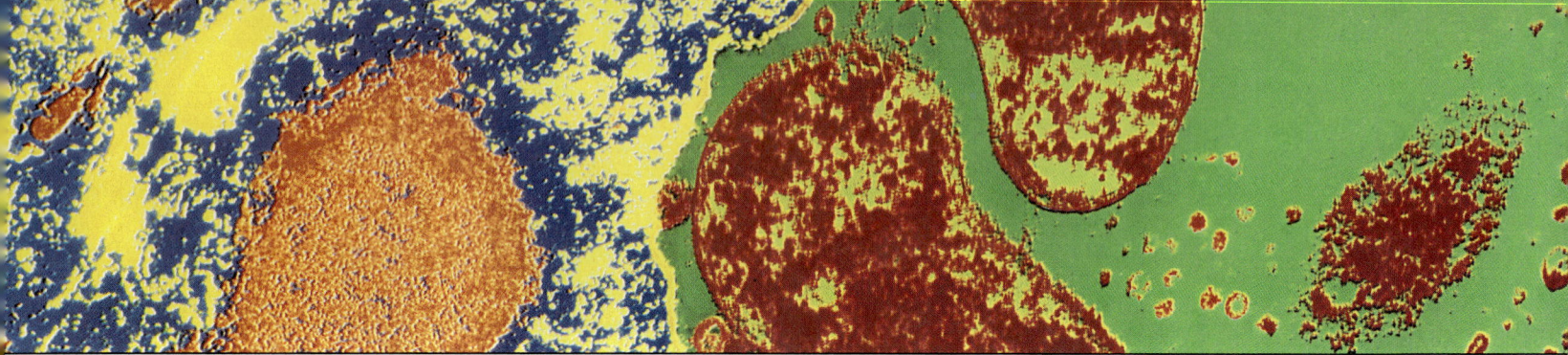

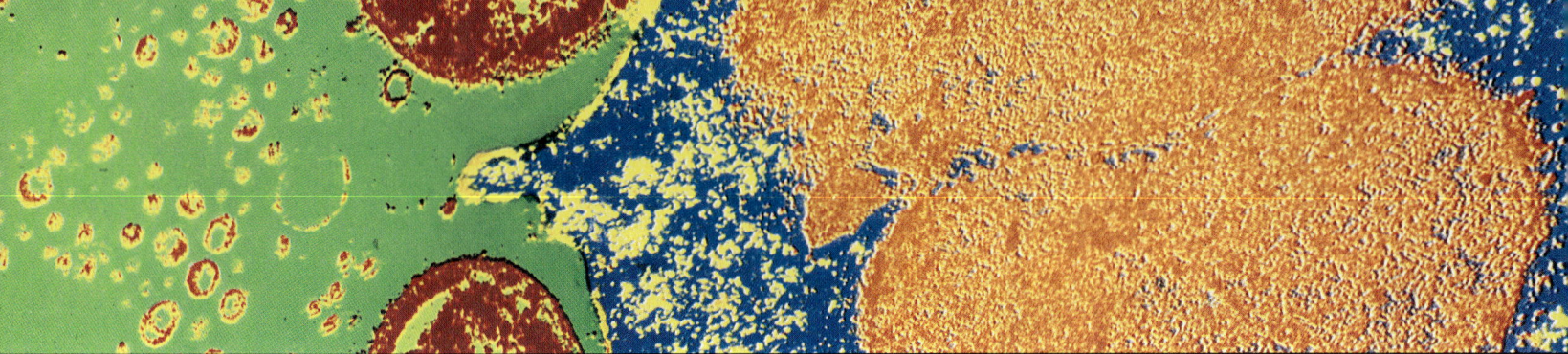

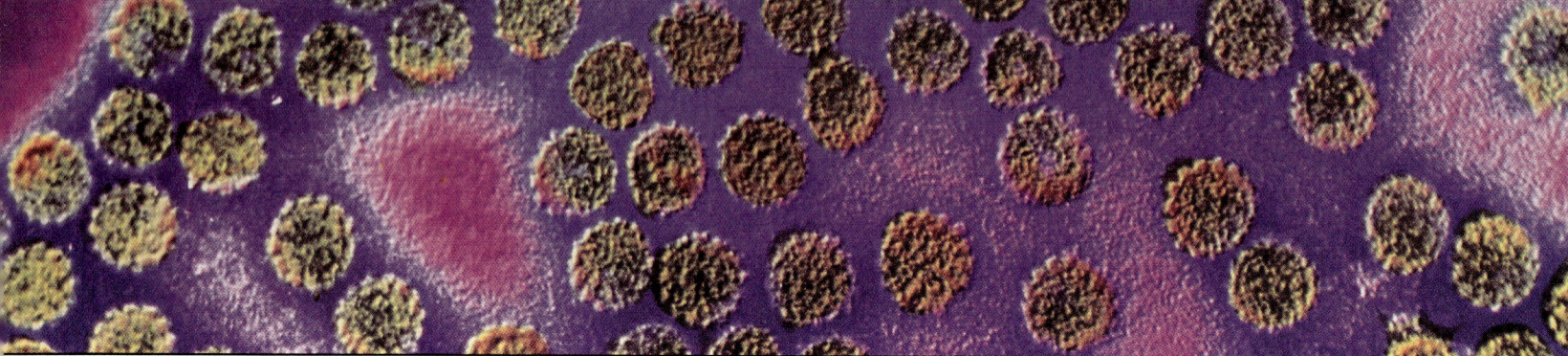

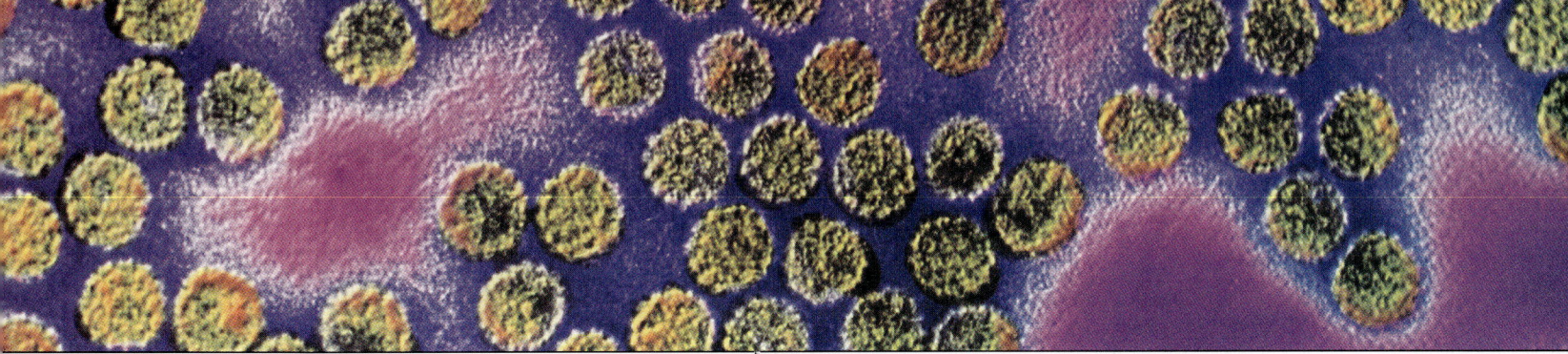

PREFACE

Among friends, favorite topics of conversation recur. They never seem to get completely talked out—especially when people with different backgrounds have similar ideas that they arrived at by different paths. Once in a while such conversations actually lead to something, as one of ours did eight years ago. It lead to this book.

We talked about the rapid expansion and generally unappreciated impact of the study of microbiology on the world. One of us would cite medical examples: the appearance of new infectious diseases, the increased difficulty treating old ones, the rapid spread of all of them by modern human travel and migration. The other would counter with examples related to environmental damage, the explosive growth of biotechnology, the unique value of microorganisms as tools to answer basic questions about all biology. Why not tell this story to the most important audience: students who were being introduced to the field for the first time? And because we represented different specialties of microbiology—one of us medical practice and the other laboratory research—but were both engaged day-by-day in practicing the science, why not present microbiology from a personal, hands-on point of view? For example, we could use medical case histories to illustrate the complexities and relevance of infectious disease and focus on the functions of metabolism and genetics rather than catalogue their complexities.

We enlisted Jack Carey's enthusiasm and book expertise, Dana Hawley's wonderful combination of artistic skills and imagination, Harriett Prentiss's presentation abilities and encouragement, and we were on our way. Along the way we had the help of many perceptive consultants who identified with our goals and clarified our path, as well as the help of the talented book-making staff at Wadsworth, especially Deborah Cogan with her organizational leadership. Now, only eight years later, we have a book that takes this approach to studying microbiology. It has some distinctive features.

Use of Medical Case Histories

To make the central theme of this book—infectious diseases of humans—more interesting and related to our lives, we took the approach of relying heavily on case histories. All these cases are real, taken from or based on the clinical experience of one of us. They help us remember microbiological facts, and we hope they also will help you. For example, when you try to recall the symptoms, progression, and treatment of erysipelas, we hope you will think about G. T. and his ordinary-looking pimple (Chapter 23).

We included two kinds of case histories: vignettes at the beginning of the chapters in Part IV (Human Diseases Caused by Microorganisms) and brief boxes within all chapters. A vignette sets the scene for a chapter and guides us through it. The signs and symptoms the patient experiences, the ups and downs of the progress of a disease, and the outcome of an infection are presented as clinical observations in the vignette. Then as the microbiological and physiological basis of the disease is discussed in the chapter, the bases of the observations are explained. We expect, and hope, you will return to the vignette from time to time as you read the chapter. Chapters in other sections of the book also begin with vignettes. Most of these are nonclinical, designed primarily to set the scene for the subject matter to follow.

The clinical boxes within chapters serve a different purpose. Each illustrates or amplifies one specific topic.

A Different Approach to Metabolism and Genetics

The 30-plus-years' research and teaching experience of one of us unrelentingly drove home two points about metabolism and genetics. First, comprehension of these interlinked topics is the *sine qua non* to understanding any other aspect of microbiology, be it medical, environmental, biotechnological, or basic biological. Second, most students consider these topics to be the most difficult, a morass of seemingly unrelated chemical reactions and processes.

To make sense of them and arrange them in a logical framework, we have taken the straightforward, practical approach that has worked well for us in the classroom: Focus on what metabolic reactions and genetic processes accomplish—the formation of two microbial cells starting with a single cell. The process is much like an assembly line in a factory. Materials are taken in, they are processed inside, and a product is assembled, all under the direction of a plan for how it should be done. We use the factory analogy to explain metabolism and genetics—how chemicals are taken into the cell, how they are processed, and how a new cell is assembled—all under a genetic plan. The myriad metabolic reactions and genetic processes are collected into groups and placed in their sequence on the metabolic assembly line.

Focus on Research

In spite of the wealth of information that makes up the subject, microbiology is still a young, rapidly expanding science driven by the real and pressing needs of our time. And research leads the way. Throughout the book, we consider aspects of microbiology that are still evolving through research. To emphasize the importance of research to the progress of microbiology, we have invited some leading contemporary microbiologists to explain their research and its significance. We have called these essays Focus on Research.

Focus on the Student

We have included several features to aid learning. Each chapter begins with Learning Goals, a brief listing of the most important ones in that chapter. Each chapter ends with a summary that has cross-referencing page numbers to the primary coverage and a series of self-study questions. Finally, each chapter contains suggested reading. To save space these references to books and scientific journals have been pruned to include only the most help-

ful and informative sources of information, most of which will lead the student to references to more highly specialized information.

At the end of the book we have included a Glossary, which includes scientific terms that are in boldface type in the text and might not be adequately defined in a dictionary.

Focus on Microorganisms

Classifying microorganisms has never been easy, and all the diverse, previously accepted schemes for accomplishing this task had a common property: They changed. We have to keep this fact in mind as we rely primarily on the scheme accepted by most microbiologists today. We emphasize how it evolved and why it, too, will almost certainly soon change. Microbiologists devise the schemes, but relationships among microorganisms, which we do not yet fully understand, are their fundamental bases. The words of probably the most famous microbiologist of all, Louis Pasteur (1822–1895), are an excellent guide for this and other topics in microbiology: "Messieurs, c'est les microbes qui auront le dernier mot." (The microbes have the last word.)

JOHN INGRAHAM
CATHERINE INGRAHAM
March 1994

REVIEWERS

David Balkwill, *Florida State University*

Cecilio R. Barrera, *New Mexico State University*

Spencer A. Benson, *University of Maryland, College Park*

Kostia Bergman, *Northeastern University*

Russell J. Centanni, *Boise State University*

William H. Coleman, *University of Hartford*

Irene M. Cotton, *Lorain County Community College*

Monica A. Devanas, *Rutgers, The State University of New Jersey*

Cindy Erwin, *City College of San Francisco*

David R. Filmer, *Purdue University*

Randy Firstman, *College of the Sequoias*

Joseph J. Gauthier, *University of Alabama at Birmingham*

Susan Harlander, *University of Minnesota*

Kevin C. Hazen, *University of Virginia Health Sciences Center*

Ted R. Johnson, *St. Olaf College*

Patricia P. Jones, *Stanford University*

Ilsa Kaattari, *College of William and Mary*

Kenneth C. Keudell, *Western Illinois University*

Juhee Kim, *California State University, Long Beach*

John Kimball

Bruno J. Kolodziej, *The Ohio State University*

Larry M. Lewis, *Bradford College*

Patricia A. Lorenz, *Penn Valley Community College*

Eleanor K. Marr, *Dutchess Community College*

Carolyn F. Mathur, *York College of Pennsylvania*

William C. Matthai, *Tarrant County Junior College*

Susan McMahon, *Pellissippi State Technical Community College*

Frank Mittermeyer, *Elmhurst College*

Henry Mulcahy, *Suffolk University*

Elinor O'Brien, *Boston College*

Judith A. Owen, *Haverford College*

Robert A. Pollack, *Nassau Community College*

Ralph J. Rascati, *Kennesaw State College*

Gordon Schrank, *St. Cloud State University*

Robert Sjogren, *University of Vermont*

Cynthia V. Sommer, *University of Wisconsin, Milwaukee*

Josephine Smith, *Montgomery County Community College*

Pamela P. Tabery, *Northampton Community College*

Ian Tizard, *Texas A&M University*

James Urban, *Kansas State University*

Robert Vinopal, *University of Connecticut*

Brian J. Wilkinson, *Illinois State University*

Michael R. Yeaman, *UCLA School of Medicine*

Shanna Yonenaka, *San Francisco State University*

LEARNING GUIDE FOR STUDENTS

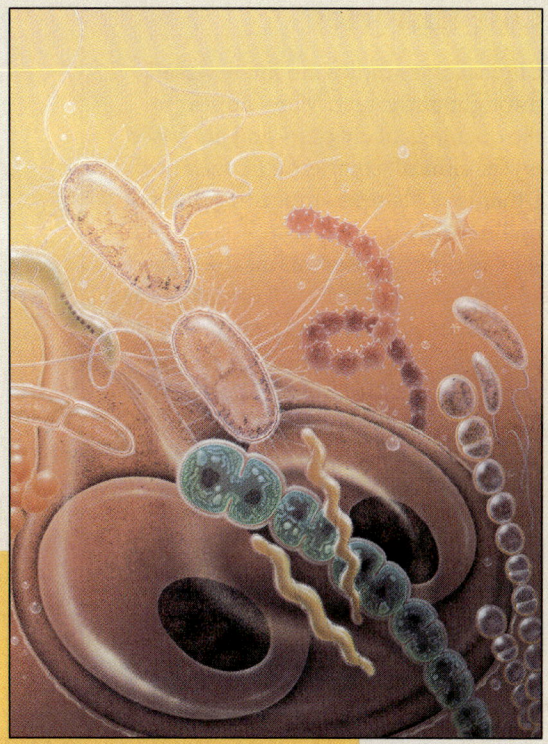

From acquired immune deficiency syndrome to the common cold, infectious diseases have had an impact on our lives since the early days of time. We've written this book to introduce you to the complex and rapidly expanding field of microbiology.

One of us is a teacher and researcher; the other a practicing physician. By combining our diverse areas of knowledge and experience and integrating our different vantage points, we hope to present microbiology as a complex yet highly relevant science. We've used the events of our own day-to-day lives and those of other practicing microbiologists to tell you a unique, personal, hands-on story. You'll find case histories and vignettes that are based on real situations we've encountered. There are research essays written by scientists actively engaged on the front lines of advancing science.

The following pages present examples of some of this material and give you a glimpse of what's to come. This visual guide will also help you use our book more effectively by pointing out some of its main features and learning aids.

Case Histories

Clinical case studies are interspersed throughout the text. Many appear in the openings of chapters and are based on the clinical experience of one of us. Other case histories appear throughout to illuminate important topics. We present as clinical observations the signs and symptoms a patient experiences as well as the progress and outcome of the disease. Then, as you read about the microbiological and physiological basis of the disease in the chapter, you'll begin to understand the scientific reasoning behind our initial clinical observations.

exchange cannot take place. Typically, the clinical presentation of a patient with pneumonia includes fever, trachyapnea, labored breathing, and a cough that may produce infected secretions. If pneumonia involves the pleura, it causes **pleurisy**, associated with painful breathing. Many different illnesses are classified as pneumonia, and the etiologic agent can be bacterial, viral, or fungal.

UPPER RESPIRATORY INFECTIONS

Bacteria and viruses are the most common agents of upper respiratory infection. Some of these infections are minor, but others can cause sudden death because of the anatomy of the respiratory tract. Because all air must pass through a single, fairly narrow airway above the spot where the trachea bifurcates (splits into the bronchi), infections that cause swelling, such as epiglottitis and diphtheria, can close off the airway and cause sudden death by suffocation.

BACTERIAL CAUSES

Bacteria are the most virulent of the upper respiratory pathogens, but most of the infections they cause can be either prevented or effectively treated.

Case History

Epiglottitis

M. L., a 4-year-old boy, arrived at the pediatric emergency room at 5 A.M. He'd been entirely well when his mother put him to bed the night before, but around 4 A.M. she was awakened by noises coming from his room. She found him sitting up in his bed, gasping loudly with each breath and warm with fever. M. L.'s mother knew immediately this was an emergency.

In the short time it took to arrive at the hospital, M. L.'s breathing grew even noisier. He was immediately taken to an examining room where the on-duty physician found an anxious youngster in acute respiratory distress with marked stridor (**Figure 22.2**). M. L. was sitting very still, holding his head forward in a sniffing position and drooling slightly. The physician quickly examined the child, taking care not to agitate him, and immediately called for an anesthesiologist and a portable x-ray machine. Blood was drawn and an intravenous line was started.

The physician's working diagnosis—based on the child's age, the sudden onset of fever, and the severe stridor—was acute epiglottitis, a life-threatening infection caused almost invariably by *Haemophilus influenzae* type b. A lateral view x ray of M. L.'s neck confirmed the working diagnosis, showing a massively enlarged epiglottis severely narrowing his airway. M. L. was holding his head forward in a sniffing position to try to keep open his nearly occluded (blocked) airway. Any agitation or sudden movement could close off the airway entirely, causing sudden death. Antibiotic treatment could not possibly act soon enough to prevent possible suffocation. An anesthesiologist was needed to pass an endotracheal tube past the swollen epiglottis into the trachea to secure an airway immediately. As the child was prepared for intubation, the ER physician administered a large dose of ceftriaxone, an antibiotic effective against *H. influenzae*.

The blood culture drawn at the time of M. L.'s admission to the hospital later grew *H. influenzae* type b. M. L. was lucky enough to reach the hospital before he became completely unable to breathe. After a brief hospitalization in the intensive care unit and additional antibiotic treatment, his epiglottis returned to normal size, his fever resolved, and he returned home in good health.

23 INFECTIONS OF THE DIGESTIVE SYSTEM

Good Reason to Be Worried

A. R., a 3-year-old boy, was brought to a pediatric emergency room in late September because of high fever and diarrhea. His mother reported that the boy had suddenly become ill the day before. His temperature rose to 105°F; he was listless, refused to eat, and vomited twice. In the morning he seemed lethargic, and at one point he trembled uncontrollably and fell unconscious. Shortly after this convulsion, A. R. passed a large watery stool. During the next few hours he passed many smaller stools streaked with blood and mucus. The history revealed that A. R. lived in a small apartment with 10 others, several of whom had recently had diarrhea.

A. R. moaned and cried quietly as the emergency room physician examined him, but he did not speak and barely put up a struggle. His temperature was 104.5°F. He seemed to be mildly dehydrated; his mouth was dry; and when he cried, there were no tears. Laboratory analysis showed only a slight elevation in the number of A. R.'s circulating white blood cells but an exceptionally high number of immature neutrophils—35 percent compared to a normal 1 percent. A mucus-containing sample of his stool stained with methylene blue and examined under the microscope revealed sheets of polymorphonuclear leukocytes, evidence of colonic inflammation. Cultures of A. R.'s blood and a stool were also sent to the microbiology laboratory.

A. R. was admitted to the hospital. His high fever, appearance, and especially the blood and mucus in his stool suggested a working diagnosis of dysentery, a colitis syndrome usually caused by a bacterial pathogen. In the United States the most common cause of dysentery is a bacterium of the genus *Shigella*. A. R. was started on a treatment course of trimethoprim-sulfamethoxazole (trimethoprim-sulfa), and an intravenous line was started for rehydration.

A. R.'s worried parents were told that with treatment their son was expected to recover, but they stayed at his bedside during most of his weeklong hospitalization.

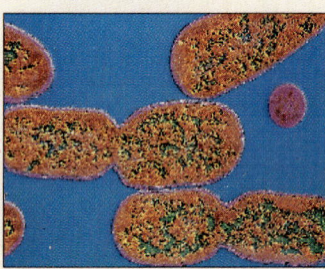

Shigella, the most common cause of dysentery in the United States.

His recovery was gradual. The day after he was admitted, A. R. was more alert and responsive, but he still had a temperature of 103°F and severe diarrhea, consisting of more than 20 small mucus-containing stools that day alone. Twenty-four hours after admission, the clinical microbiology laboratory returned its preliminary report from A. R.'s stool culture—a Gram-negative, nonlactose-fermenting rod was identified as the probable pathogen.

Over the next few days, A. R.'s temperature gradually returned to normal, and the number of stools decreased. He was discharged from the hospital thinner than before his illness, but energetic and otherwise in good health. The final report from the microbiology laboratory identified the stool culture as *Shigella sonnei*, the most common species of *Shigella* in the United States. Sensitivity studies showed this strain to be sensitive to trimethoprim-sulfa. The blood culture grew no microorganisms, indicating that A. R.'s infection was strictly intestinal.

Clinical Notes

Clinical Notes illustrate or amplify a specific topic covered in the text. Described here is the remarkable yet restricted life of David, "the boy in the bubble," who lived with SCID (severe combined immunodeficiency) longer than most people thought possible.

Clinical Notes

David—Life in a Germ-Free World

David in his germ-free environment.

Surely the best-known patient with SCID was a boy from Texas known to the public by his first name, David. David was diagnosed with SCID when he was born, and the prognosis (probable outcome) was dismal. SCID babies usually die from infection within weeks to months. But David didn't die during infancy. He was placed in a germ-free environment and protected from all contact with infection-causing agents. Pictures of David show him sitting alone inside a plastic enclosure or wearing a protective "space suit."

When he was 12, David left his germ-free enclosure to receive a bone marrow transplant from his sister that might restore his immune system. But a few months later David died. The transplant had not failed. Rather, David probably died from cancer caused by the Epstein-Barr virus (Chapter 27), an unusual complication. The virus, which usually does not cause cancer, was in the bone marrow donated by his sister. In David's immune-disordered system, however, the virus was fatal.

The technology of protective isolation allowed David to remain germ-free and therefore alive, but his life was tragically restricted.

Turning Point

Each describes a scientific or medical breakthrough that has led to new advances in the field such as the use of agar as a nutrient for growing a pure culture.

Turning Point

Frau Hesse's Pantry

Koch realized the importance of obtaining a pure culture, but doing it was another matter. He theorized that if he isolated a single bacterial cell on a solid surface, it would multiply and form a visible collection of bacterial cells (called a colony). This would be a pure culture because all the cells would be progeny of the single cell. Growth, though, requires nutrition. He tried everything from a slice of potato to gelatin. The gelatin worked well except for one thing: It melted at 37°C, body temperature, the best temperature for growing pathogens.

A neighbor of Koch's, Frau Hesse, heard of his problem and brought him a jar of agar from her pantry. Agar was a powder made from seaweed that she used to thicken jam. When mixed with nutrients such as meat broth, it was ideal—nutrient-rich and solid below 100°C. Frau Hesse's suggestion was a turning point. Koch now had a simple way to obtain a pure culture, which opened the way for rapid advances in microbiology.

Microbe Mappers

Microbe Mappers focus on actual laboratory research done by microbiologists and how it influences the field.

Microbe Mappers

Courage—and Lead Weights

Researchers working on tuberculosis have traditionally used the drug-sensitive Erdman strain of *Mycobacterium tuberculosis*. But with the possibility of a major outbreak of untreatable TB, the fastest progress is likely to come from studying the resistant strains directly. Some microbiology laboratories, called BL-3 labs, are specially equipped for the study of dangerous pathogens. But even with safeguards, researchers are uneasy dealing with bacteria that may cause untreatable tuberculosis. Ian Orme, a research microbiologist at Colorado State University's BL-3 tuberculosis lab, said in an interview with *Science* magazine, "To tell you the truth, we're fairly nervous about doing experiments on aerosolized multiple-drug-resistant strains." His lab received an aerosol machine with a 10-inch rubber gasket that should prevent leaks, but, said Orme, "I think we're going to put lead weights on top."

Focus on Research essays

Learning about significant advances in research is an important part of studying the rapidly expanding science of microbiology. To emphasize the importance of research, we've asked leading contemporary microbiologists to write about their research and its significance in special essays. These essays will help you understand how research is done and why it is vital in this field.

Focus on Research

The Discovery of Streptomycin

Albert Schatz

Albert Schatz received his B.Sc. and Ph.D. degrees from Rutgers University. His undergraduate major was soil science and his graduate work was in soil microbiology. The search for a new antibiotic challenged him because as a young boy he knew people who died from what was then called blood poisoning, ear infections, pneumonia, diphtheria, whooping cough, tuberculosis, etc. When he was in grade school, a classmate died from one disease or another almost every year. For his research that led to the discovery of streptomycin, he has received honorary degrees and medals and has been named an honorary member of medical and scientific societies. Last April, Rutgers University awarded him the Rutgers Medal at the 50th Anniversary Celebration of the discovery of streptomycin.

I began the research that led to the discovery of streptomycin in 1943 when I was working for my Ph.D. degree at Rutgers University. At that time, the discovery of penicillin had motivated a search for other antibiotics. Several had been discovered but were too toxic to be used. I was aware of that when I spent six months as a bacteriologist in army hospital with the tubercle bacillus. When I informed Dr. Waksman that I wanted to take on the TB problem as part of my Ph.D. research, he let me do that. But he transferred me to a basement laboratory and insisted that I never bring any TB cultures up to the third floor.

Dr. Waksman and others warned me that there was little likelihood of finding a cure for tuberculosis. The tubercle bacillus has a waxy coating, which is why it grows slowly and requires a special stain to make it visible through a microscope. It was assumed that no drug could get through that protective waxy coating. However, I reasoned that food had to get into the cells and waste products had to get out—otherwise neither the tubercle bacillus nor tuberculosis would exist.

Dr. Feldman provided me with a virulent strain of human tuberculosis, with which he was working. He subsequently contracted tuberculosis but recovered after two years of treatment. I feel good that no one in the building where I worked contracted tuberculosis. The laboratory I worked in was not equipped with safety equipment that is now used. It had no ultraviolet light, no special incubator, and no positive air pressure to continuously blow the air through a filter. Eventually I isolated two strains of the actinomycete *Streptomyces griseus*. Both produced a new antibiotic that I called streptomycin. One strain came from a heavily manured field soil. The other came from an agar plate that my fellow graduate student Doris Jones had streaked with a swab of a healthy chicken's throat. At about 2:00 P.M. on October 19, 1943, I knew that I had found a new antibiotic.

But would it control tuberculosis *in vivo*? Drs. Feldman and Hinshaw would find that out with guinea pigs. My job was to produce enough streptomycin for their initial tests. To do that, I had to run three stills around the clock. At night, I drew lines, with a red glass-marking pencil, on the flasks from which I was distilling and went to sleep on a wooden bench in the laboratory. The night watchman checked the flasks periodically and woke me up when the liquid in the flasks went down to the lines I had drawn. I then added more liquid, and went back to sleep. I also had to save, purify, and reuse the solvents I worked with because during World War II they were rationed. When the Mayo Clinic tests were over, I was exhausted. But I knew I would have an acceptable Ph.D. dissertation.

507

Focus on Research

A Matter of Judgment

Cynthia A. Needham

I began life in a small town in northwest Oklahoma, growing up amidst the wheat fields. I never dreamed that I would spend most of my career in Boston directing the clinical microbiology laboratory at Lahey Clinic Medical Center and teaching medical students about microbiology and infectious diseases. Thanks to some wonderful college instructors, I fell in love with science. With their encouragement, I went to graduate school, and then to a postdoctoral fellowship in medical microbiology. Working in the clinical laboratory over the last twenty years has shown me how scientific breakthroughs are translated into practical applications that benefit us all. Throughout my career, I've worked hard and put in long hours, but I've always tried to maintain the sense of adventure that first brought me to Boston. I've learned to sail and to maintain and repair my own boat. I've learned to navigate and someday I plan to sail around the world. The prairie never prepared me for becoming captain of my own sailboat, but maybe being director of a clinical microbiology laboratory did.

A clinical microbiology laboratory provides diagnostic information to help physicians treat patients who have infections. When I began my career, we had to rely heavily upon our ability to cultivate pathogens in the laboratory. Unfortunately we couldn't (and still can't) encourage the microorganisms to divide fast enough to provide information in time for the physician to make therapeutic choices. Consequently, the physician had to rely upon clinical judgment in order to begin treatment quickly. Results from the microbiology laboratory became available later; they merely confirmed or contradicted the physician's decision.

Today, by taking advantage of immunochemistry and recombinant DNA technology, we can detect and identify microorganisms in time to guide the physician's decision about treatment. But many different technologies are now available. One of the greatest challenges for the director of a clinical microbiology laboratory is to choose the best technology for a particular use while keeping the cost as low as possible. As an example, let's consider a clinical trial we did at the Lahey Clinic Medical Center in Boston to evaluate technology that aids the diagnosis, treatment, and management of pharyngitis.

Pharyngitis is among the most common out-patient complaints. It accounts for approximately 30 to 40 million physician-visits each year in the United States. The majority of these cases are caused by viruses; only 10 to 30 percent are caused by the bacterium *Streptococcus pyogenes* (commonly called "strep throat"). The distinction is important because viral and bacterial pharyngitis call for different treatment. Bacterial pharyngitis requires treatment with antibiotics to shorten symptoms, to prevent spreading the disease to others, and, most importantly, to prevent later development of a serious condition—acute rheumatic fever. In contrast, treating viral pharyngitis with antibiotics yields no benefit and exposes the patient unnecessarily to an adverse drug reaction. The decision to treat or not treat a pharyngitis patient with antibiotics is an important one.

Unfortunately, there are no reliable clinical criteria that allow a physician to differentiate between pharyngitis caused by *S. pyogenes* or something else. Typically physicians can correctly identify only about 50 percent of the cases caused by this pathogen, and they attribute the same cause to 25 to 30 percent of the cases actually resulting from something else. The laboratory can distinguish reliably between the two either by throat culture or by one of the new technologies. Culture is too slow, requiring a minimum of 24 hours. There are over 30 different devices available and licensed by the FDA for the direct and rapid detection of *S. pyogenes*. The methods, which rely upon immunoassay or genetic probes, yield results within 5 to 10 minutes. These new methods are the product of considerable basic research; they correctly identify over 90 percent of the patients infected with *S. pyogenes* and falsely identify as infected less than 5 percent of the patients who are, in fact, not infected. This level of reliability is considerably better than physician judgment.

On the other hand, most of these tests are 2 to 4 times more costly than cultures, and they are neither as sensitive nor as specific. Before introducing a new technology, it is important to know whether the timely information it yields will change the manner in which physicians manage their patients. Will the physician rely on the results and thereby decrease the number of patients receiving antibiotics from the number based only on clinical judgment? If the answer is no, the test has no value, only additional cost.

To answer this question, we set up a clinical trial. We discovered that physicians relied less on clinical judgment with the introduction of a rapid test for *S. pyogenes* and were able to choose an appropriate therapeutic strategy for a significant percentage of patients at the time of the visit. Patients who tested positive were treated. This change eliminated the need for follow-up visits for almost all patients with positive throat cultures. Physicians did choose to prescribe antibiotics for some patients when the test was negative, but the unnecessary use of antibiotics was dramatically decreased. With or without the test, all patients eventually received appropriate therapy. Still, the decision to provide regular use of the new test is a matter of judgment. Do you think the increase in direct cost for testing for this common disease was justified by improved clinical practice? If you were a laboratory director, would you provide the rapid test?

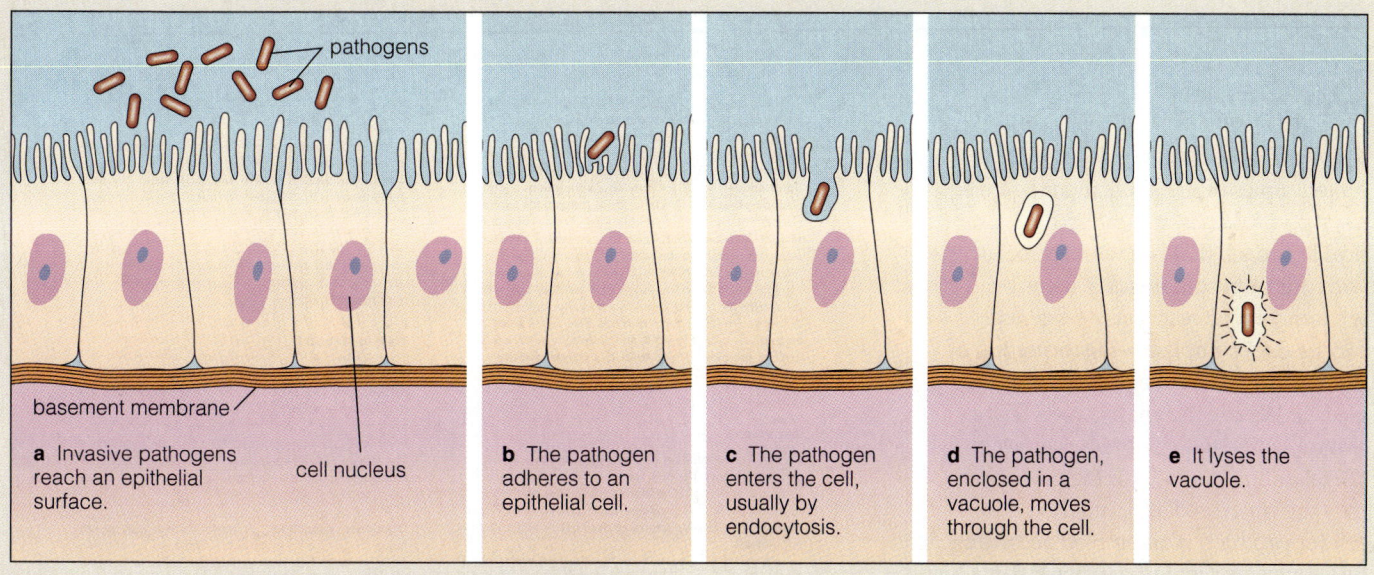

a Invasive pathogens reach an epithelial surface.

b The pathogen adheres to an epithelial cell.

c The pathogen enters the cell, usually by endocytosis.

d The pathogen, enclosed in a vacuole, moves through the cell.

e It lyses the vacuole.

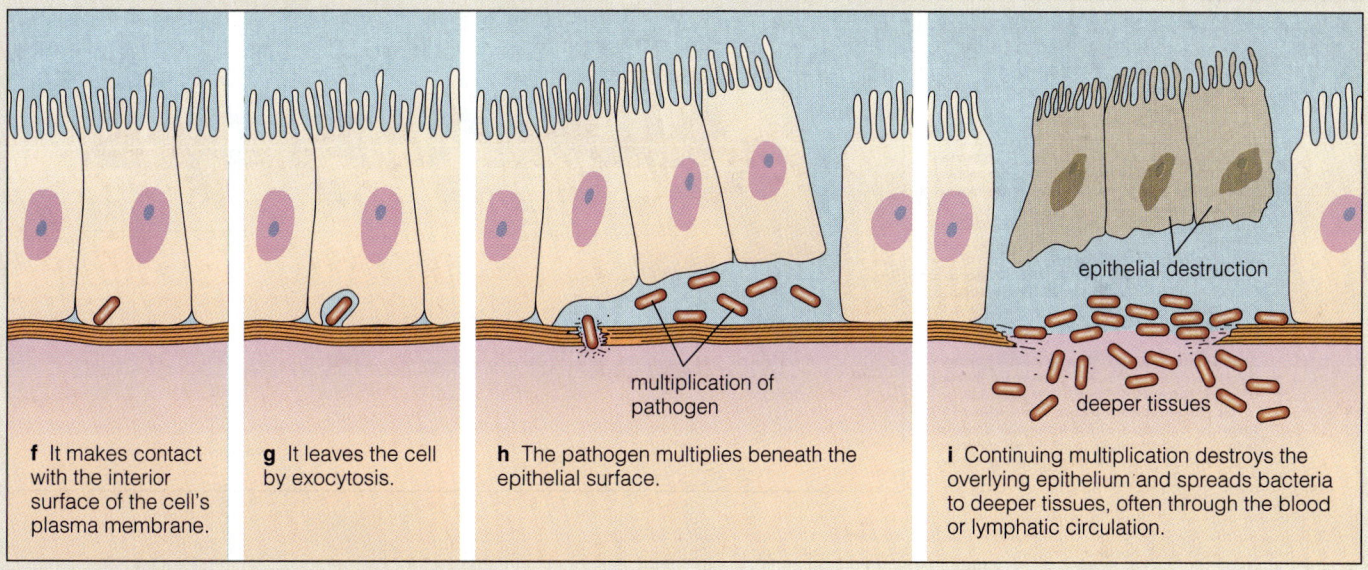

f It makes contact with the interior surface of the cell's plasma membrane.

g It leaves the cell by exocytosis.

h The pathogen multiplies beneath the epithelial surface.

i Continuing multiplication destroys the overlying epithelium and spreads bacteria to deeper tissues, often through the blood or lymphatic circulation.

Visual summaries of processes

We've created unique illustrations to help you understand key microbiological processes. Many concepts are so complex that students need additional help in learning. When this is the case, descriptions are integrated within the illustrations themselves. These "visual summaries," arranged in step-by-step sequence, allow students to create a mental image of the concept, which reinforces their understanding even further.

A logical approach to studying metabolism and genetics

To help you make sense of the inter-linked topics of metabolism and genetics, we've taken a straightforward, practical approach that has worked for us with our own students. We focus on what metabolic reactions and genetic processes accomplish—the formation of two microbial cells beginning with a single cell. We use the analogy of a factory assembly line to explain it: Chemicals are taken into the cell (or factory) where they are processed and, finally, a new cell (or product) is assembled according to a genetic (or manufacturing) plan.

a compound; **reduction** is the addition of electrons to a compound. A number of the biochemicals that *E. coli* makes to build its cellular components must be reduced and, therefore, require a supply of electrons. *E. coli* stores electrons in compounds called **nicotinamide adenine dinucleotide (NAD)** and **nicotinamide adenine dinucleotide phosphate (NADP)**. These compounds capture electrons in the form of hydrogen atoms from compounds that are being oxidized. Later they use them to reduce other compounds. In this way NAD and NADP—collectively designated **NAD(P)**—drive metabolic reductions.

ATP stores the cell's energy, and NAD(P) stores the cell's reducing power. These two critically important reserves are, in fact, interconvertible. That is, ATP can be expended to reduce NAD(P) and reduced NAD(P) can be used to produce ATP. Later in this chapter we discuss how these two driving forces are formed and how they are used.

The Plan

A plan is as essential for directing metabolism as it is for running a factory. The plan for *E. coli*'s microbial metabolism consists of about 6 million distinct pieces of information because its end product—the microbial cell—is so complex. *E. coli* manufactures approximately 1000 different proteins.

All the information necessary to direct microbial metabolism is contained in the cell's DNA. A copy of this master plan is transmitted from one generation of cells to the next, and each cell uses the plan to synthesize a new cell (Chapter 6).

METABOLISM: AN OVERVIEW

In metabolism, raw materials (substrates) from the environment are converted into the finished product. The assembly line consists of five sequential steps: (1) entry mechanisms, (2) catabolic reactions, (3) biosynthesis, (4) polymerization, and (5) assembly. First we consider the flow of materials, from substrate to new cell. Then we consider how the driving forces (energy and reducing power) are accumulated and expended during the five sequential steps. **Figure 5.2** presents an overview of the five-step metabolic assembly line.

Flow of Materials

Entry Mechanisms. Raw materials must be brought to the factory. This is the function of **entry mechanisms**—they bring substrates into the cell. Entry mechanisms must overcome barriers presented by the plasma membrane and, in Gram-negative bacteria, the outer membrane. The entry mechanisms transport substrate across membranes and maintain concentrations within the cell at sufficient levels to fuel metabolism.

Catabolic Reactions. A manufactured item is not made directly from raw materials, and a cell is not made directly from substrate molecules. **Catabolic reactions** refine substrates into the building materials needed to manufacture a new cell. The reactions are called *catabolic* from a Greek word meaning "to bring down," because the basic building materials of metabolism are, in general,

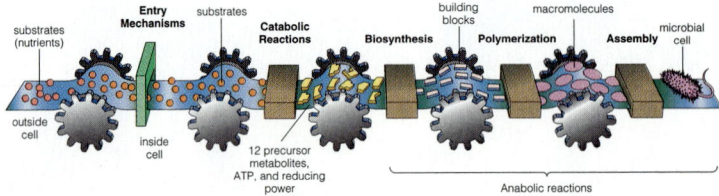

Figure 5.2 The metabolic factory. The metabolic assembly line consists of five sequential steps: entry mechanisms to bring in nutrients from outside the cell; catabolic reactions that yield 12 precursor metabolites, energy, and reducing power; biosynthesis, which produces the building blocks for macromolecules; polymerization, which produces the macromolecules; and assembly, which builds the structures that form the cell.

Acetyl CoA enters the TCA cycle by combining with the four-carbon precursor metabolite oxaloacetate to form a six-carbon intermediate, citrate. In a series of six subsequent reactions, the two added carbon atoms are released as carbon dioxide, and oxaloacetate is regenerated. In the TCA cycle, three additional precursor metabolites—succinyl CoA, alpha-ketoglutarate, and oxaloacetate—are formed.

Each turn of the TCA cycle produces one molecule of ATP by substrate level phosphorylation. In addition, considerable reducing power is stored in the form of two molecules of NADH, one as NADPH, and another as a reduced carrier called FADH$_2$. Reducing power stored as flavine adenine dinucleotide (FADH$_2$), as well as NADH and NADPH, can be converted to ATP by chemiosmosis.

Pentose Phosphate Pathway. The key metabolic intermediates of the pentose phosphate pathway and the reactions that involve NADPH are shown in **Figure 5.12**.

The pentose phosphate pathway begins when an intermediate of glycolysis, glucose-6-phosphate, enters the pathway. It passes through a complex series of reactions to produce three molecules of carbon dioxide and one

Figure 5.13 Biosynthesis. Biosynthesis pathways convert precursor metabolites into the building blocks of macromolecules: amino acids for proteins, sugars for carbohydrates, nucleotides for DNA and RNA, and so on.

Introduction to
MICROBIOLOGY

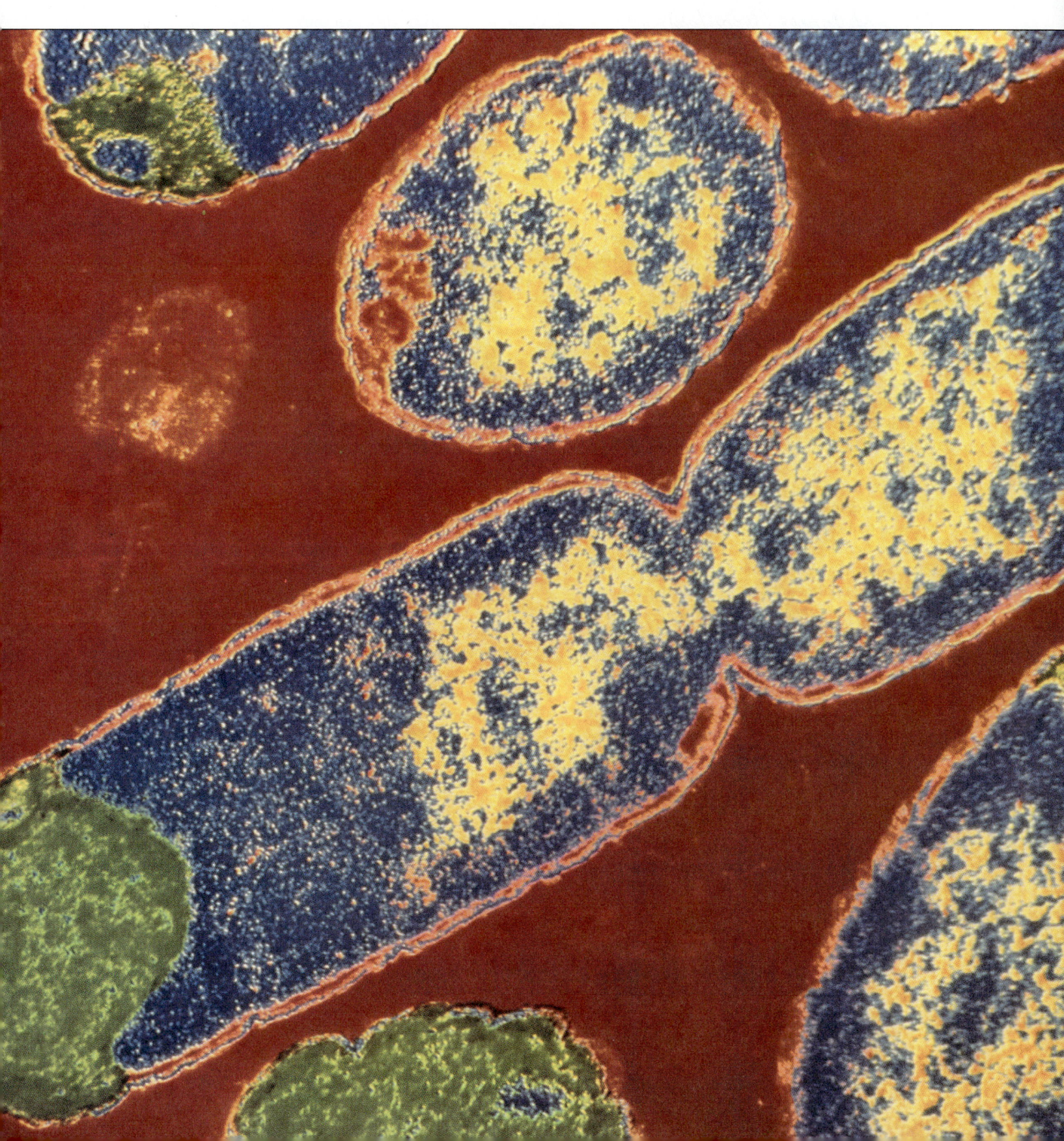

1 THE SCIENCE OF MICROBIOLOGY

The Case of James Greenlees

On August 12, 1865, an 11-year-old boy named James Greenlees was walking down a street in Glasgow when he was struck by a horse-drawn cart. A wheel ran over James's leg just below the knee. Jagged edges of bone tore through the skin.

Young James's injury was a death sentence. Compound fractures (when the bone breaks through the skin) are usually dirty wounds, and people who suffered contaminated wounds of flesh and bone during the nineteenth century almost always died from them. First, the wound exuded large quantities of pus, and then the tissue around it began to decay and stink. Without treatment, the victim developed fever, and death followed in a matter of days. The only treatment was immediate amputation. But even when the dirty wound was replaced by a comparatively clean surgical incision, nearly half the patients died.

James was taken to the Royal Infirmary, where he came under the care of a young surgeon, Joseph Lister. Lister had a theory about treating open wounds and decided to try it on James. He splinted the broken bone and dressed the wound in bandages soaked in carbolic acid. Lister believed that carbolic acid, which we use as a disinfectant in our homes today, would prevent infection. He watched James's wound closely for signs of festering. If the carbolic acid treatment didn't work, Lister would follow standard procedure and amputate.

Four days after the accident, Lister removed the carbolic acid dressing to examine James's wound. There was no evidence of infection. Encouraged, Lister reapplied bandages soaked in a more dilute solution of carbolic acid. James's wound continued to heal. Six weeks later the broken fragments of bone had become reunited by new growth, and James Greenlees was released from the hospital with two sound legs.

Lister called his carbolic acid technique *antisepsis*, meaning "against infection." It had profound practical implications, especially for the developing art of surgery. During the mid-1800s, operations performed in a hospi-

Joseph Lister.

tal were almost as likely to lead to infection and death as were contaminated fractures and open wounds.

But what made Lister think of trying carbolic acid? In fact, everything he needed to know had already been discovered by others before him. From reading scientific papers, Lister knew that microorganisms, living things too small to be seen by the unaided eye, caused wounds to fester and putrefy. He also knew that microorganisms were everywhere—in soil, water, and air. Finally, he knew that carbolic acid killed microorganisms. Thus, he concluded, if carbolic acid killed microorganisms, then applying it to the wound would help it heal by preventing decay. Lister's logical reasoning, based on established scientific fact, led him to formulate a hypothesis and test it with an experiment. His approach is the essence of the scientific method.

To understand:

- The impact of microorganisms on human affairs

- Advances and challenges in applied microbiology

- Careers available to trained microbiologists

- The scope of microbiology and why it is a separate science

- The development of microbiology as a science

- Microbiology today and where it is headed in the future

Figure 1.1 The bubonic plague that swept through Europe in the Middle Ages was also called the Black Death because internal hemorrhaging caused black patches on the skin.

THE UNSEEN WORLD AND OUR WORLD

Microbiology is the study of **microorganisms**—the unseen world of living things. Throughout history, microorganisms have had a tremendous impact on human affairs; and for most of that time, the impact was overwhelmingly negative. Before the development of microbiology as a science, when disease struck, it controlled human events. With epidemics came social and political chaos as well as human suffering. To begin our study of microbiology, let's briefly look at some major events in the long history of the unseen world and our world.

Microbes and Disease

The bubonic plague that swept through Europe during the Middle Ages killed about 25 million people—one-third of the population (**Figure 1.1**). Imagine the social and political dislocation that resulted from such mass death. Not until 500 years later, in 1890, did microbiologists identify the causative organism, a bacterium called *Yersinia pestis*. They pieced together this story: Infected fleas on rats carry the plague. Infection spreads first among the rats, which have little resistance and usually die. When rat hosts become scarce, the fleas begin to bite humans and a plague epidemic is underway. It is a painful and ugly disease, characterized by swollen lymph glands called buboes. Fever soars as the microorganism proliferates throughout the body. Though rare today, bubonic plague still occurs in parts of the world, including the western and southwestern United States (Chapter 27).

The Potato Famine. A plague of another sort—a disease of plants rather than humans—caused the great Irish migration to the United States in the nineteenth century.

The potato was a staple of the Irish diet, and so when the fungus *Phytophthora infestans* caused a deadly potato blight (rot) in Ireland, the result was devastating. By 1846 the potato harvest was so meager that starvation and hunger-based disease were widespread. An estimated 1,240,000 people had died, and 1,200,000 more had emigrated to other countries.

The Conquest of the Incas. One of the most tragic instances of disease destroying almost an entire population occurred in the Americas. When the Spanish conquistador Hernando Cortés landed in Mexico in 1519, the Native American population of the central region was 25 to 30 million. Within 50 years, the population had shrunk to 3 million—10 percent of what it had been—from the ravages of disease brought by the Spaniards.

After centuries of exposure, Europeans had developed a certain level of tolerance to smallpox and measles. But Native Americans, with no previous exposure, were particularly vulnerable.

Once established in the New World, European diseases spread with lightning speed. From Mexico, smallpox spread to Peru, killing millions, including the reigning Inca and his heir. In 1525, when the Spanish conquistador Francisco Pizarro arrived, the Inca Empire was in social and political chaos. Pizarro conquered the mighty Incas without significant military resistance.

Napoleon in Russia. Warfare and infection have always been intimately connected. Poor sanitation, movement of peoples, and malnutrition in war zones all cause outbreaks of disease. When Napoleon invaded Russia in 1812 he lost more of his troops to typhus than to all other causes, including enemy action. Soldiers injured in battle were more likely to die from wound-related infections such as tetanus and gas gangrene than from the wound itself.

Microbes and Life Today

Disease-causing microbes, called **pathogens**, are responsible for a vast spectrum of human illness. But let's not lose perspective. As the American microbiologist Otto Rahn pointed out, the fraction of microorganisms that cause disease is far less than the fraction of humans who commit first-degree murder. More important, the growth of microbiology as a science means that we now know how to control many pathogens. We have also discovered new ways to use microorganisms for our own purposes, to improve the quality of life. Let's look at some of the advances—and challenges—in the four areas of applied microbiology: medical, environmental, industrial, and agricultural.

Medical Microbiology. Each of us houses trillions of microbes. In fact, the human body contains many more microbial cells than human cells. Most of these microorganisms coexist with us harmlessly; and in some cases, as you will learn in Chapter 14, they are beneficial. We have vaccines to prevent contagion by many harmful microorganisms, and drugs and antibiotics afford treatment. Moreover, we have made tremendous progress in preventing infectious disease through public hygiene. That is not to say we control microbial disease today. Although we have vaccines, getting every child in the United States inoculated is a challenge, and a small one compared with protecting every child in the world. One major challenge for modern medical microbiology is to find a cure or effective means of prevention for AIDS (acquired immunodeficiency syndrome), which is discussed in detail in Chapter 27.

Environmental Microbiology. The study of how microorganisms affect the earth and its atmosphere is called environmental microbiology. If it were not for microorganisms, life would not be possible on earth, as you will read in Chapter 28. One of the earliest practical uses we made of our knowledge of environmental microbiology was to provide a safe and palatable drinking water supply. Among the most recent applications is the development of biodegradable materials (materials that are rapidly broken down by microorganisms normally present in water and soil). Biodegradable products are one response to a major challenge facing environmental microbiologists—how to cope with the mounting toxic waste our industries produce.

Industrial Microbiology. The first human use of microbes was to make food and ferment beverages—bread, wine and vinegar, cheeses, and olives. People also learned to preserve food for lean winters by controlling the growth of microorganisms—by drying fruit and salting meat. Today, microorganism-dependent industries are highly diverse. In addition to foodstuffs, we use microorganisms to make vitamins, antibiotics, and other pharmaceuticals, such as insulin to treat diabetes (**Table 1.1**). Undoubtedly the greatest challenges in industrial microbiology—and economic opportunities—will come from applying genetic engineering to medical, environmental, and agricultural problems.

Agricultural Microbiology. Thanks to research in agricultural microbiology, livestock and agriculturally important plants are now largely protected from microbial diseases. Microorganisms that kill insects are used as natural pesticides, and others help make soil fertile. Some of the newest research in this field examines the use of microorganisms to produce food supplements for animals and humans. Though we Americans worry about agricultural abundance, other parts of the world would benefit from a readily available, inexpensive, and palatable source of protein in particular.

Careers in Microbiology

Among the many reasons to study microbiology is that a basic knowledge of the field is often necessary to pursue careers in medical science, ecology, agriculture, or biotechnology. Some students simply want to better understand the world we live in, but others intend to make microbiology their life's work. Careers in microbiology are challenging, rewarding, and varied. The American Society for Microbiology, the national professional microbiologists' organization, has almost 40,000 members and is still growing.

Career opportunities in microbiology depend on training as well as interest (**Table 1.2**). Microbiologists

Table 1.1 Some Industrial Uses of Microorganisms

Product	Contribution of Microorganisms
Cheese	Growth of microorganisms contributes to ripening and flavor. The flavor and appearance of a particular cheese are due in large part to the microorganisms associated with it.
Alcoholic beverages	Yeast is used to convert sugar, grape juice, or malt-treated grain into alcohol. Other microorganisms may also be used; a mold converts starch into sugar to make the Japanese rice wine, sake.
Vinegar	Certain bacteria are used to convert alcohol into acetic acid, which gives vinegar its acid taste.
Citric acid	Certain fungi are used to make citric acid, a common ingredient of soft drinks and other foods.
Vitamins	Microorganisms are used to make vitamins, including C, B_2, B_{12}.
Antibiotics	With only a few exceptions, antibiotics are manufactured through the activities of microorganisms.
Amino acids	Many amino acids, including monosodium glutamate (MSG), a flavor enhancer, are manufactured using microorganisms.
Human growth hormone, insulin	These and other medically useful proteins are made by genetically engineered bacteria.

Note: Chapter 29 discusses these processes in detail.

Table 1.2 Career Opportunities for Microbiologists

Baccalaureate Degree (B.A., B.S.)	
Research associate	Does experiments along with other technical specialists under the supervision of a director.
Food, industrial, or environmental microbiologist, quality assurance technologist	Identifies microorganisms in water, food, and dairy, pharmaceutical, and environmental products. Checks quality and safety of pharmaceutical products.
Clinical or veterinary microbiologist, medical technologist	Identifies disease-causing microorganisms.
Master's Degree (M.A., M.S.)	
Supervisor or laboratory manager	Supervises activities of a laboratory.
Instructor	Teaches at a community or junior college.
Doctoral Degree (Ph.D., M.D.)	
Scientist	Conducts independent research.
University or college professor	Teaches, trains graduate students and postdoctoral fellows; does independent research.
Research director	Heads a research team.
Consultant	Advises businesses and government agencies.
Infectious disease specialist	Specializes in treating patients with infectious diseases.

with baccalaureate degrees assist in research and work in clinical laboratories or in food and environmental industries. With advanced degrees, career choices are broadened. With a master's degree, one can become a clinical or industrial laboratory supervisor or a teacher in a community or junior college. The doctor's degree (Ph.D. or M.D.) opens opportunities to do independent research,

direct the research of others, teach in a university, consult, and specialize in treating infectious diseases.

Some of the most far-reaching contributions come from microbiologists who choose a career in general microbiology. Studying microorganisms for their intrinsic interest, general microbiologists do pure research to find new microorganisms, new microbial activities, and new

Table 1.3 The Subgroups of Microorganisms

Subgroup	Cell Type	Contains Representatives That Are: Photosynthetic	Motile	Macroscopic
Bacteria	Procaryotic	Yes	Yes	No
Algae	Eucaryotic	All are	Yes	Yes
Fungi	Eucaryotic	No	No[a]	Yes
Protozoa	Eucaryotic	No	Yes	No
Viruses	Acellular	No	No	No

[a] The reproductive cells (spores) of some fungi are motile.

relationships among microorganisms. A major recent achievement in general microbiology was the discovery of archaebacteria, a group of microorganisms we discuss later in this chapter. There is a great deal of work to do in general microbiology, and when general microbiology flourishes, the applied branches of microbiology also flourish.

For information about careers in microbiology, write The Board of Education and Training, American Society for Microbiology, 1325 Massachusetts Ave., Washington, DC 20005-4171.

THE SCOPE OF MICROBIOLOGY

There are five subgroups of microorganisms: bacteria, algae, fungi, protozoa, and viruses. Except for size, these subgroups are not related. Only a single property links microorganisms—small size. In fact, the diversity of form and function found among the subgroups of microorganisms is as great as the total diversity of all living things. Bacteria, for example, are less like algae or fungi or protozoa or viruses than a shark is like a giraffe or an orchid is like an eagle.

Why, then, are such unrelated organisms grouped for common study as a subject called microbiology? The answer is a practical one. The techniques for identifying, cultivating, and studying the groups of microorganisms are similar (Chapter 3). Microbiology is a cohesive science *because of its methodology and approach to problems*, not because of the relatedness of the organisms it studies.

The most basic differences among the subgroups are in structure. Bacteria are cells, but they are **procaryotes**,[1]

1. Sometimes eucaryotic and procaryotic are spelled with a "k" rather than a "c," more accurately reflecting, it is said, their Greek roots. However, E. Chatton, who coined the terms, and C. B. van Niel and R. Y. Stanier, who brought the concept into sharp focus and the terms into wide use, used a "c." We follow their use in this book.

meaning "before a nucleus." They lack internal membrane-bound structures. The algae, fungi, and protozoa, like plants and animals, are **eucaryotes**, meaning "true nucleus." They have a membrane-bound nucleus and other membrane-defined internal structures called organelles, or tiny organs. Viruses, the fifth subgroup, are **acellular**. That is, a virus is not a cell. It is merely a small packet of nucleic acid, the chemical form of genetic information, wrapped in a coat, usually made of protein.

We will look at each group in detail in Chapters 11, 12, and 13, but here we look at their general properties. We will also look at certain helminth (worm) species that are traditionally part of microbiological study. (See **Table 1.3** for a summary of the five subgroups of microorganisms.)

Bacteria

Bacteria are distinguished from all other organisms by their procaryotic cell structure. The terms *bacteria* and *procaryotes* are synonymous. Instead of the internal membrane-bound structures that characterize eucaryotic cells, procaryotes are filled with a uniform grainy material. Most bacteria are unicellular (single cells). Another distinguishing property of procaryotes is their small size, even for microorganisms (**Figure 1.2**). A typical bacterial cell has only about one one-thousandth the volume of a typical eucaryotic cell.

It was only in the 1970s that we discovered that procaryotes consist of two extremely different groups, the **archaebacteria** (ancient bacteria) and the **eubacteria** (true bacteria). The evolutionary distance between the two is reflected in their different chemical compositions. Superficially, however, they are similar in appearance and size.

Among themselves, the eubacteria are very diverse. Most species have a characteristic cell shape (**Figure 1.3**). They can be spherical, rod-shaped, helical, comma-shaped, or even square. Some bacteria are motile; others are not. Some obtain energy by processing organic compounds (foods), as animals do. Others get energy through photosynthesis, as plants do. Still others process inor-

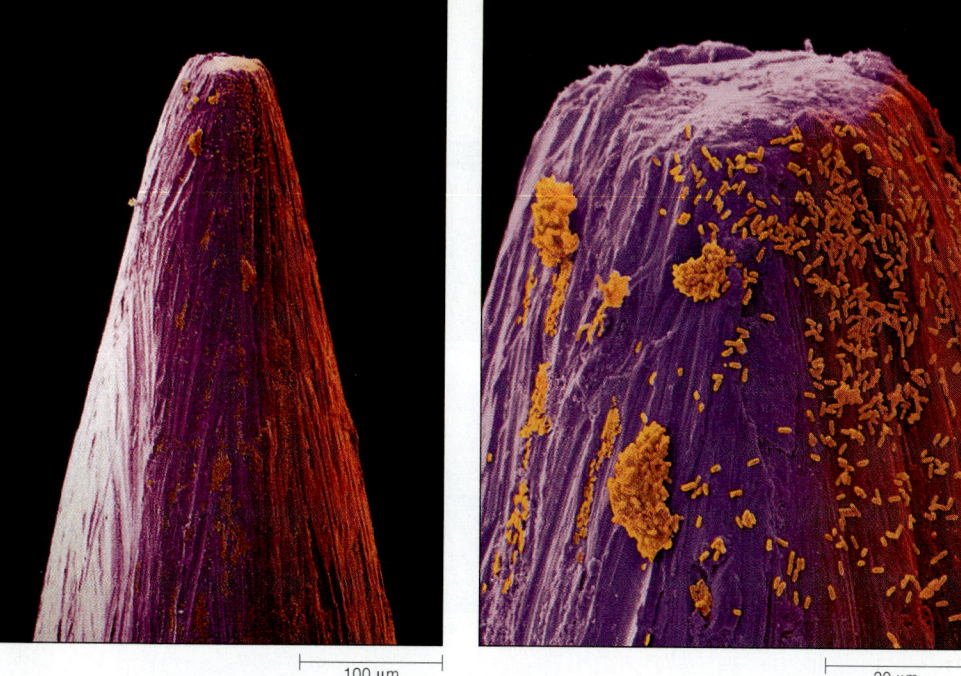
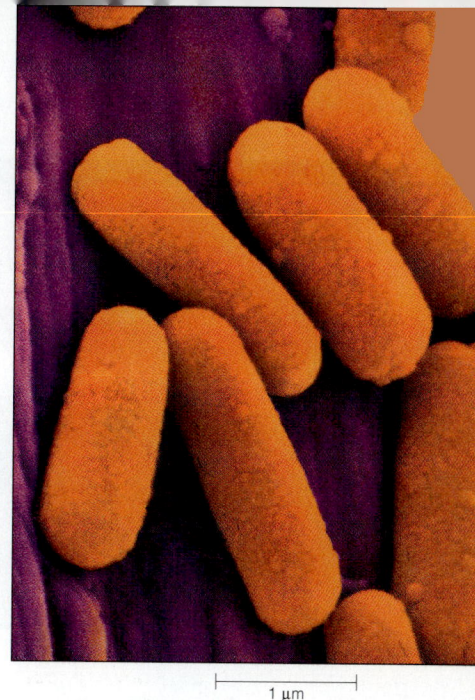

Figure 1.2 Bacteria have two distinguishing properties. They are procaryotes (no organelles in their cells), and they are extremely small. Shown here, *Bacillus* cells on the tip of a pin.

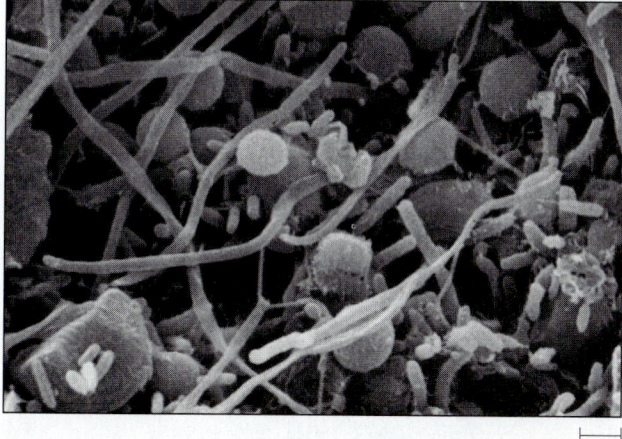

Figure 1.3 A micrograph of various eubacteria, showing some of their diverse shapes.

ganic materials, such as sulfur or iron, for energy. Some bacteria can grow in temperatures as low as –20°C, lower than the freezing point of water, while others thrive where temperatures reach 110°C, higher than the boiling point of water. Some grow best under conditions more acidic than lemon juice, while others favor environments more alkaline than household ammonia.

Bacteria come first to mind for the diseases they cause—from food poisoning and toxic shock syndrome to syphilis and typhoid fever. But while bacteria may sometimes take life, they also make plant and human as well as other animal existence possible by keeping the environment and atmosphere in life-sustaining balance (Chapter 28).

Algae

The **algae** (*sing.*, alga) are eucaryotic organisms that carry out photosynthesis. Like all eucaryotes, they have a nucleus and membrane-bound organelles. They also have chloroplasts, structures where photosynthesis takes place. Some algae are unicellular and microscopic. Certain algae, however, consist of so many cells they are **macroscopic**—they can be seen without the aid of a microscope. Kelp, the large brown seaweed that washes up on Pacific Ocean beaches, is a macroscopic alga (**Figure 1.4**). Multicellular algae may look superficially like higher plants, but they lack characteristic plant organs, such as stems, roots, and leaves.

Algae make up the mass of organisms called **phytoplankton** that are found near the surface of marine and fresh water. Phytoplankton are at the base of all aquatic food chains. Algae, thus, are critical to our global ecology, but they are of negligible medical importance.

Fungi

Fungi (*sing.*, fungus) include mushrooms, yeasts, and molds. They are eucaryotic and nonphotosynthetic. There are both microscopic and macroscopic fungi (**Figure 1.5**). Most fungi are scavengers and, thus, ecologically important because as they live off dead matter, they decompose it. A few fungi are pathogenic to animals and humans. Some cause trivial infections, such as ringworm and athlete's foot. Others cause life-threatening infections; an example is pneumocystis pneumonia, which invades the lungs of immunologically weakened individuals, such as AIDS patients. Many fungi are pathogenic to plants.

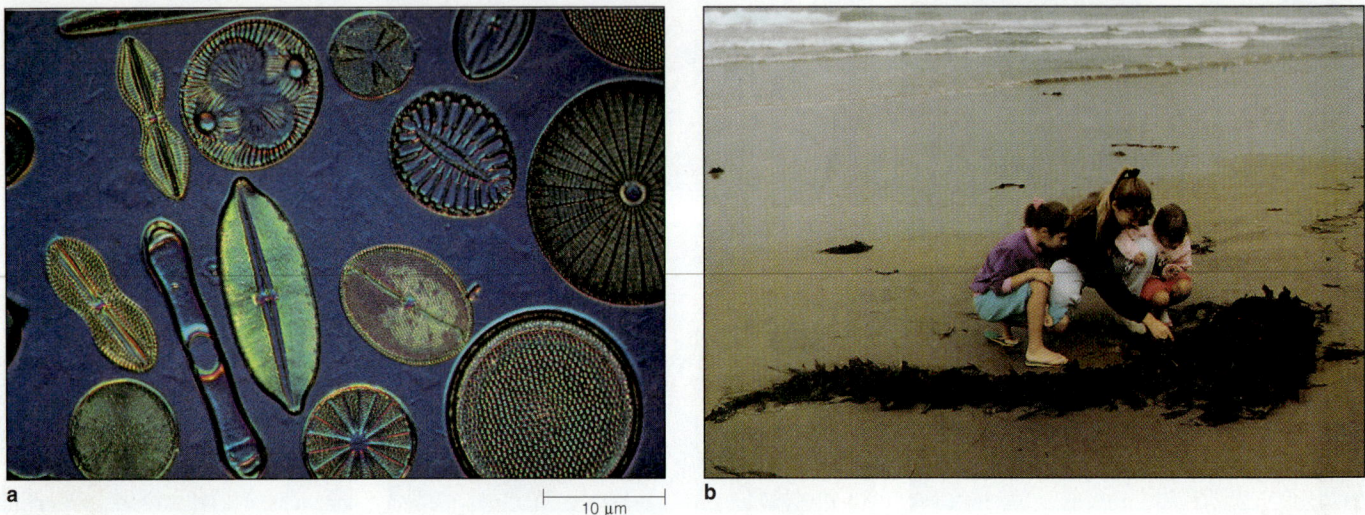

a

10 µm

b

Figure 1.4 All algae are photosynthetic. Some are microscopic and others are macroscopic. **(a)** Phytoplankton—the base of the marine food chain—are single-cell algae. They are sometimes so abundant they make water cloudy. **(b)** Children looking at kelp that has washed up on a beach. Kelp is a multicellular macroscopic brown alga that grows in the Pacific Ocean.

a

10 µm

b

Figure 1.5 **(a)** The microscopic fungus *Epidermophyton fluccosum* is one of the many associated with athlete's foot. **(b)** The mushroom *Amanita muscaria*, which causes hallucinations if eaten, is a macroscopic fungus.

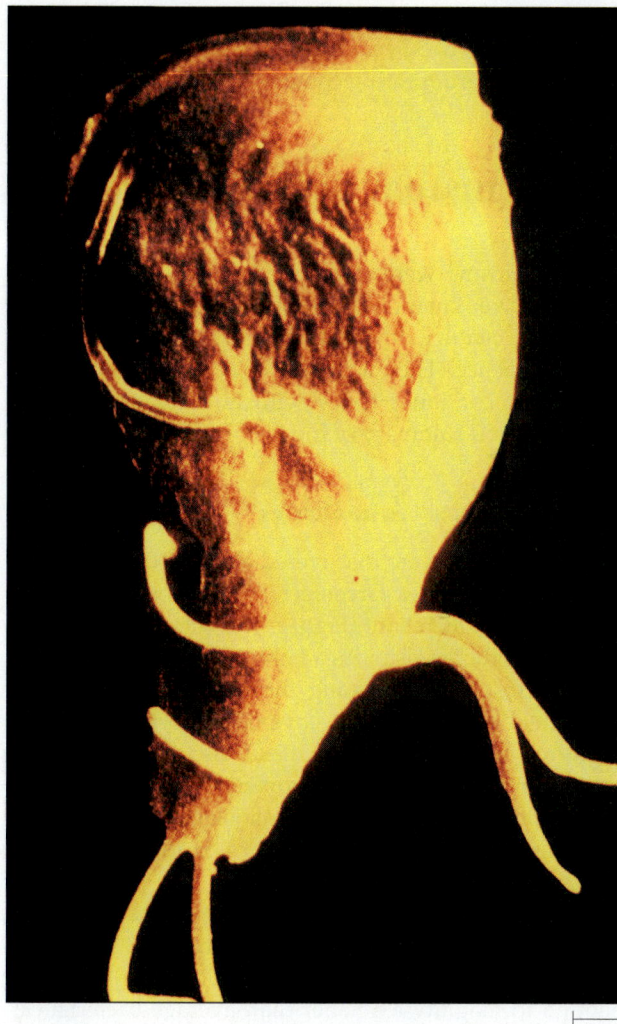

Figure 1.6 The protozoan *Giardia lamblia* is a common cause of diarrhea in humans. It was first seen by Leeuwenhoek during a bout of intestinal distress.

Diseases like corn smut and wheat rust cause great economic losses.

Most fungi grow as multibranched tubes that make up a structure called a mycelium. What we call mushrooms are the above-ground structure formed from an extensive underground mycelium. Unicellular fungi are called yeasts. Molds are lower-order fungi that infect plants, but rarely humans.

Protozoa

Protozoa means "first animals." As the name suggests, they are superficially animal-like. They are nonphotosynthetic and usually motile (**Figure 1.6**). The amoebae move by extending tubelike structures called pseudopods. The flagellates and ciliates move by long (flagella) or short

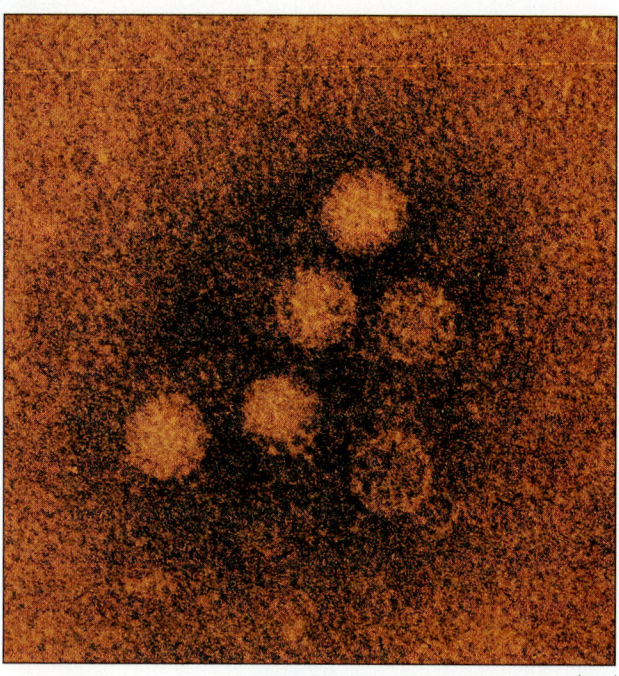

Figure 1.7 The common cold is caused by a virus, usually a rhinovirus, like the one pictured here. There is no effective vaccine.

(cilia) hairlike extensions that wave or beat. Virtually all protozoa are microscopic. The protozoa are the height of unicellular organization. Some have myriad organelles that are almost as complex in form and function as are some tissues in higher organisms.

Protozoan diseases, such as malaria and African sleeping sickness, kill millions of people every year. Protozoan diseases and helminth-caused diseases (which you will read about shortly) are cross-classified as parasitic diseases (though, technically, every infectious disease is a case of parasitism). In a **parasitic** relationship, one organism benefits at the expense of the other. The study of protozoan- and helminth-caused disease is called **parasitology**.

Viruses

Viruses are not cells. They are merely particles of nucleic acid, either RNA (ribonucleic acid) or DNA (deoxyribonucleic acid), but never both. They are usually enclosed in a protein coat. Viruses are incapable of carrying out the activities needed for them to reproduce. As a result, they can reproduce only inside a host cell. In other words, viruses are **obligate intracellular parasites**. They can infect animals, plants, and other microorganisms (**Figure 1.7**).

The second major property of viruses is their extremely small size, even compared to bacteria. The largest viruses are about one-tenth the size of a typical bacterial cell. The smallest are about one one-thousandth the size. Viruses cannot be seen through an ordinary microscope. They must be studied biochemically or with an electron microscope.

Many major diseases of plants, humans, and other animals are caused by viruses. Smallpox, yellow fever, and polio are viral diseases that have been particularly deadly in the past. The human immunodeficiency virus (HIV) that causes AIDS was discovered in the 1980s.

As chemically and structurally simple as viruses are, there are even simpler infectious agents. **Prions**, discovered in the early 1980s, are composed exclusively of protein. How they reproduce remains a mystery. Prions cause rare neurological diseases in humans, "mad cow disease" in cattle, and scrapie in sheep.

Helminths

Helminths are worms and, as such, they belong to the animal kingdom. They are also macroscopic. Why, then, do we study them in microbiology? Some helminths go through microscopic stages in their life cycles, and many cause parasitic diseases in plants, humans, and other animals. Only with microbiological techniques can we fully study these helminths in their own right and find ways to prevent and treat the diseases they cause.

The two types of helminths that concern us are flatworms and roundworms (**Figure 1.8**). Flatworms include the beef tapeworm, which can grow to lengths of 30 feet in human intestines. Roundworms include hookworms,

parasites that were common in the southern United States until about 50 years ago, and *Trichinella*, which humans acquire from eating contaminated pork.

A BRIEF HISTORY OF MICROBIOLOGY

Microorganisms were discovered more than 300 years ago, but we knew little about them until the mid-nineteenth century, when microbiology became an experimental science. Then a period of accelerating progress began that continues today with no end in sight. (See **Table 1.4** for a summary of the history of microbiology.)

Leeuwenhoek's "Animalcules"

Microorganisms were discovered about 200 years before Lister treated James Greenlees. Antony van Leeuwenhoek, a Dutch merchant (**Figure 1.9**), was the first person to see a microorganism. As a hobby he made small hand-held microscopes. Squinting through the lens at specimens held on a pin, he discovered a world of invisible creatures he called "animalcules," small animals. They were everywhere—in water droplets, particles of soil, his teeth scrapings. In 1674 Leeuwenhoek communicated his discoveries to the Royal Society of London, sending detailed drawings.

All of his drawings and nine of the estimated 500 microscopes that Leeuwenhoek made still exist. The most powerful of these has a magnification of 266×—powerful enough to magnify a smaller-than-average bacterial cell to the size of the period at the end of this sentence. But

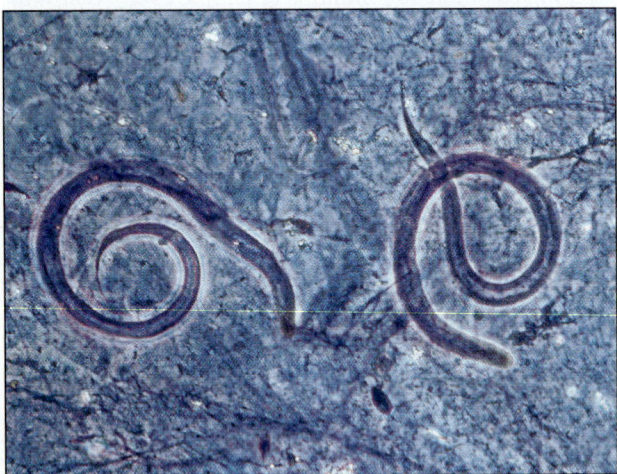

Figure 1.8 In its microscopic stage, the roundworm *Strongyloides stercoralis* invades human intestines and sometimes migrates to different parts of the body, causing painful rashes.

Table 1.4	Highlights in the History of Microbiology
Year	Event
1674	Leeuwenhoek discovers microorganisms.
1796	Jenner creates a vaccine for smallpox.
1847	Semmelweiss establishes the cause of childbed fever.
1859	Pasteur disproves spontaneous generation of microorganisms.
1865	Lister introduces antiseptic techniques.
1876	Koch proves that specific microorganisms cause specific diseases.
1881	Koch uses agar to obtain a pure culture.
1892	Iwanowski discovers viruses.
1894	Ehrlich articulates the principle of selective toxicity.
1929	Fleming discovers penicillin.

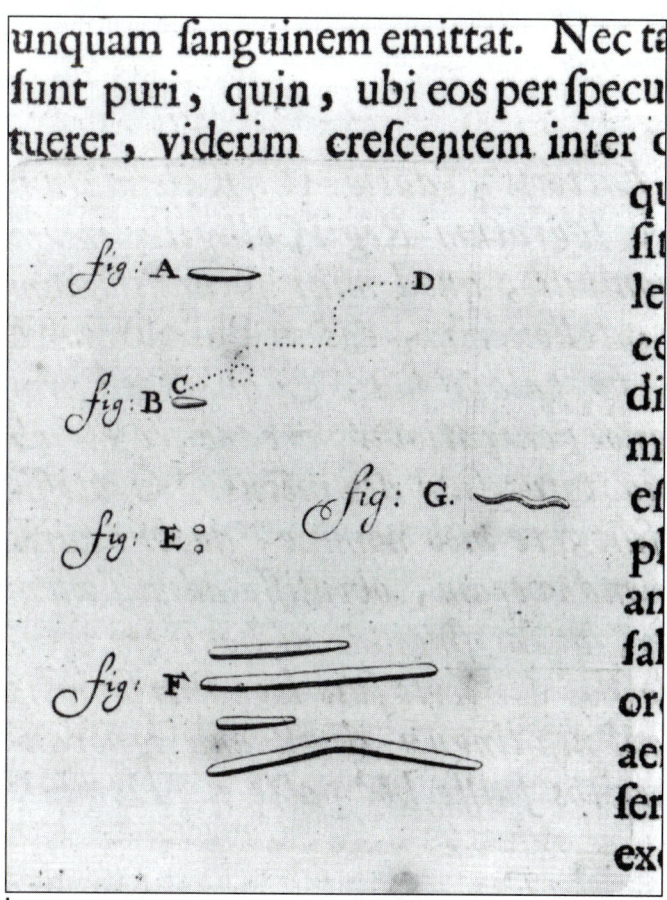

Figure 1.9 Antony van Leeuwenhoek was the first person to see microorganisms. He used a tiny hand-held microscope (**a**) and in 1684 published drawings of bacteria that he called "animalcules" (**b**).

judging from the details of his sketches, he must have made considerably more powerful microscopes that have been lost.

Leeuwenhoek was not the first to make and use microscopes, but he made better ones and used them with greater skill than anyone else. He jealously guarded his simple (single lens) microscopes as an improvement over the compound (double lens) microscopes available, refusing to sell them or teach others how to make them. Not until 200 years later were microscopes superior to Leeuwenhoek's developed.

Hooke and the Cell Theory

When Leeuwenhoek was sending his drawings to the Royal Society in London, the curator of instruments there, Robert Hooke, was experimenting with compound microscopes. Hooke's instruments could magnify 300–500×, but the images were masked by colored rings of light. He couldn't see an object as small as a bacterial cell. He did, however, make a discovery of fundamental importance. Hooke's observations of thin slices of cork showed a honeycomb of chambers. He called them *cellulae*, Latin for "small rooms." This discovery led to the formulation of the **cell theory**, which states that cells are the basic unit of organization for all living things.

Spontaneous Generation

While Leeuwenhoek was observing microorganisms in Holland, a physician in Italy was making groundbreaking discoveries of his own. In 1665, Francesco Redi proved that **spontaneous generation** of macroscopic animals does not occur; in other words, living organisms do not arise from inanimate matter. Redi did an experiment with jars of covered and uncovered meat. By showing that maggots developed only in meat that flies could reach to lay eggs on, he provided powerful additional evidence that *living things come only from preexisting living things.*

Still, maybe *microorganisms* were an exception to this rule. They always appeared, and in large numbers, soon after the death of a plant or animal. Did decomposition cause microorganisms to form or did microorganisms cause decomposition?

Needham versus Spallanzani. For 80 years the debate was conjectural. But then the spontaneous generation proponents seemed to gain ground. In 1745, an English clergyman named John Needham proposed an experiment to settle the issue. Everyone knew boiling killed microorganisms. So he would boil chicken broth, put it in a flask, and seal it. If microorganisms grew, then it could only be because of spontaneous generation. Indeed, microorganisms grew. But an Italian priest and professor named Lazzaro Spallanzani was not convinced. Maybe microorganisms entered the broth after boiling but before sealing? So Spallanzani put broth in a flask, sealed it, creating a vacuum, and then boiled it. When he tested the cooled broth, there were no microorganisms. The critics were not persuaded. Spallanzani didn't disprove spontaneous generation, they said, he just proved that spontaneous generation required air.

Pasteur's Epic Experiments. The controversy continued another 100 years. Finally, in 1859, the French Academy of Science sponsored a competition to prove or disprove the theory of spontaneous generation. A young French chemist named Louis Pasteur entered. To offset the argument that air was necessary to spontaneous generation, Pasteur would use barriers in his experiments that would allow the free passage of air but prevent the entry of microorganisms.

In his most famous experiment, Pasteur boiled meat broth in a flask and then drew out and curved the neck of the flask in a flame (**Figure 1.10**). No microorganisms developed in the flask. But when he tilted the flask so some broth flowed into the curved neck and then tilted it back so the broth was returned to the base of the flask, the broth quickly became cloudy with the growth of microbial cells. Gravity had caused the microbial cells that had entered the flask in air and dust to settle at the low point of the neck, never reaching the broth in the base until they were later washed in by the broth. Thus Pasteur proved, first, that spontaneous generation of microorganisms does not occur even in the presence of air and, second, that the growth of microorganisms causes food to spoil and, by extension, dead plant and animal matter to decompose.

Pasteur's success was due partly to good luck. Many early experiments to disprove spontaneous generation failed because the samples contained endospores, highly heat-resistant bacterial structures that are not killed by boiling. Experiments with vegetable broths were particu-larly doomed to failure, because plants often house endospore-forming species of bacteria. Meat broths like Pasteur's rarely contain endospores.

Pasteur's simple but epic experiments grounded microbiology in scientific fact: (1) no living things, including microorganisms, arise by spontaneous generation, (2) microorganisms are *everywhere*—even in the air and on dust particles, and (3) the growth of microorganisms causes dead plant and animal tissue to decompose and food to spoil.

The Germ Theory of Disease

Once spontaneous generation of microbes was disproved, the field of microbiology exploded. Microbiology changed from an observational science to an experimental science. Now scientists could address highly practical issues, including the cause of infectious diseases. The way was open for the germ theory of disease. Building on Pasteur's work, a German physician, Robert Koch (**Figure 1.11**), proved not only that microbes—germs—cause disease, but also that specific microorganisms cause specific diseases.

Koch's Postulates. In 1876 Koch was studying anthrax, a disease of cattle and sheep that also affects humans. He observed that the same microorganisms were present in all blood samples of infected animals. He cultivated these microorganisms, which today we know as the bacterium *Bacillus anthracis*, in a pure form outside the infected animal. He then injected a healthy animal with the cultured bacteria. That animal became infected with anthrax, and its blood samples showed the same microorganisms as did the originally infected animals. These four steps, generalized, constitute **Koch's postulates**, which we still use today. If fulfilled, Koch's postulates provide absolute proof that a particular microorganism causes a particular disease. (For a more detailed explanation, see the box on Koch's postulates in Chapter 15, "One Microbe, One Disease.")

In his work on anthrax, Koch made another critically important contribution to the growing science of microbiology. He developed a technique to obtain and cultivate bacteria in **pure culture**. A pure culture contains only a single kind of bacterium. This is in contrast to the mixtures of various kinds of bacteria, called **mixed cultures**, that are found in nature. It is difficult to do valid experiments using mixed cultures because they are so complicated. In fact, early studies on mixed cultures led to fanciful descriptions of the "life cycles" of microorganisms. Scientists described how microorganisms changed their shape, transforming themselves from one cell type into another, from round to elongated to spiral over time. In

a

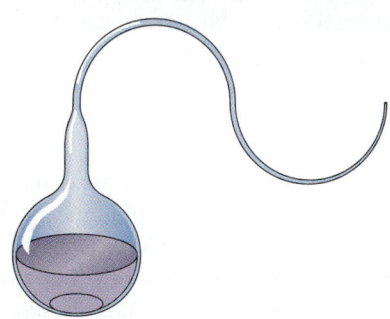

b

Figure 1.10 **(a)** Louis Pasteur made many fundamental contributions to microbiology. **(b)** In 1861 he devised this swan-necked flask to prove that microorganisms did not generate spontaneously in sterilized broth exposed to air.

fact, they were merely observing the way different species dominated a mixed culture at different times.

Koch's postulates, his technique for obtaining pure cultures, and his introduction of agar as a culture medium (see the box "Frau Hesse's Pantry") led to spectacular advances in microbiology as an experimental science. Between 1882 and 1900 the organisms that caused almost all

Figure 1.11 Robert Koch, another giant in microbiology, in his laboratory.

the bacterial diseases then prevalent in Europe were isolated, including typhus, dysentery, syphilis, gonorrhea, pneumonia, and—by Koch himself—tuberculosis.

Immunity

If specific microorganisms caused specific diseases, then it should be possible to control **infectious diseases**, those caused by microorganisms. It would be necessary only to control the microorganisms, reasoned Koch's colleagues. Thus, the new science of microbiology moved rapidly on two fronts. The first was **immunity**, stimulating the body's own ability to combat infection. The second was **public hygiene**, promoting cleanliness and reducing exposure to disease.

From ancient times, it was a recognized fact that people who suffered from certain diseases never got them again. Apparently, some protective change occurred in the body. In other words, *infection could produce immunity*. **Immunization** produces immunity by providing exposure to altered organisms that do not cause disease.

Frau Hesse's Pantry

Koch realized the importance of obtaining a pure culture, but doing it was another matter. He theorized that if he isolated a single bacterial cell on a solid surface, it would multiply and form a visible collection of bacterial cells (called a colony). This would be a pure culture because all the cells would be progeny of the single cell. Growth, though, requires nutrition. He tried everything from a slice of potato to gelatin. The gelatin worked well except for one thing: It melted at 37°C, body temperature, the best temperature for growing pathogens.

A neighbor of Koch's, Frau Hesse, heard of his problem and brought him a jar of agar from her pantry. Agar was a powder made from seaweed that she used to thicken jam. When mixed with nutrients such as meat broth, it was ideal—nutrient-rich and solid below 100°C. Frau Hesse's suggestion was a turning point. Koch now had a simple way to obtain a pure culture, which opened the way for rapid advances in microbiology.

Jenner and Smallpox. Edward Jenner, an English physician, observed that dairymaids who had naturally contracted a mild infection called cowpox seemed to be protected against smallpox, a horribly disfiguring disease and a major killer. In 1796, Jenner inoculated an 8-year-old boy with fluid from cowpox blisters on the hand of a dairymaid, Sarah Nelms (**Figure 1.12**). The boy contracted cowpox. Then Jenner inoculated him with fluid from a smallpox blister. There was no reaction. The child had been immunized by exposure to cowpox. The technique of inducing infection for protection became known as **vaccination** from *vacca*, Latin for "cow." Jenner was acclaimed and smallpox was brought under control.

But not enough was known about microbiology for Jenner to extend the principles of immunization to other diseases.

The First Vaccines. Based on Jenner's and Koch's work, Pasteur articulated the principles by which **vaccines**—agents that conferred immunity without causing disease—could be developed. Like Koch, Pasteur had been doing some work with anthrax. In the early 1880s, he was also doing experiments on fowl cholera. He found that injecting an attenuated (weakened) form of the bacterium that caused the disease into healthy chickens protected them against fowl cholera. He applied this principle to developing a vaccine for anthrax and, in 1885, for rabies. About this same time, two American microbiologists, Daniel E. Salmon and Theobald Smith, were also developing ways to create vaccines. Salmon and Smith experimentally demonstrated that, in addition to attenuated strains of microorganisms, killed microbial cells were effective as vaccines. This discovery led to the development of vaccines against many infectious diseases.

Public Hygiene

Immunization represented a tremendous advance in the prevention of infection, but even more lives have been saved by improved public hygiene. Before the germ theory of disease gained prevalence, sewage regularly mixed with drinking water. Improving sewage disposal and assuring a clean public water supply prevented mass outbreaks of cholera and typhoid fever. Similar improvements came in food preservation and, eventually, inspection. **Pasteurization**, which kills most pathogens by a brief exposure to heat, is merely one example. (Pasteur originally developed the pasteurization process to keep wine from spoiling.)

The germ theory of disease also strengthened the arguments of physicians who maintained that improved personal hygiene, especially careful hand washing, could prevent disease (see the box "Childbed Fever," p. 16). Better hygiene not only decreased the likelihood of exposure to disease-causing microorganisms, but it also improved a person's general state of health and thus the ability to fight off disease.

MICROBIOLOGY TODAY

The late nineteenth century was the golden age of microbiology because advances came so rapidly and life was so dramatically changed as a result. The advances during the twentieth century have been no less striking. Let's look at four key areas: chemotherapy, immunology, virology, and genetic engineering.

Figure 1.12 A statue honoring Edward Jenner, the English physician who struck the first telling blow against smallpox.

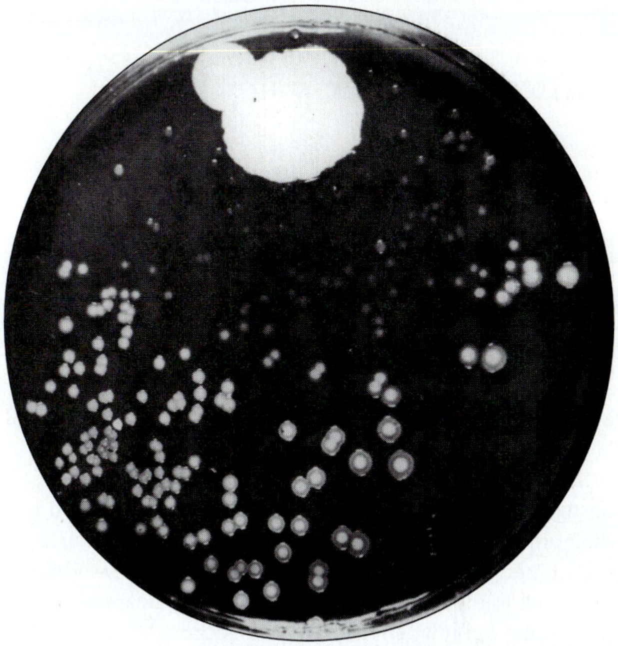

Figure 1.13 This is the actual petri dish that led to the discovery of penicillin in 1929. The plate on which the disease-causing bacterium *Staphylococcus aureus* was being cultivated accidentally became contaminated by a fungal spore that developed into a colony of *Penicillium*. Fleming observed that bacterial colonies did not develop near the fungus and those that developed closest to the fungus appeared to be damaged. He reasoned that the fungus might be producing a chemical that killed bacteria. The chemical proved to be penicillin, probably the most useful antibiotic known.

Chemotherapy

The most significant advance in medical microbiology during the 1900s was the development of **chemotherapy**, the treatment of disease with chemicals called **drugs**. Nineteenth-century microbiologists discovered ways to *prevent* many infections, but not until the twentieth century did we acquire the ability to *treat* infections.

The German physician-chemist Paul Ehrlich is called the father of chemotherapy because he articulated its guiding principle, **selective toxicity**. For a drug to be effective against infection, it must be selectively toxic, deadly to the infecting microorganism but relatively harmless to human cells. Ehrlich conceived of drugs as being "magic bullets," agents that would kill the microbe but not the host. Ehrlich gave the fledgling science of chemotherapy its first success in 1908 by discovering a drug for the treatment of syphilis. He called his drug salvarsan, from the Latin word meaning "to save." For over 20 years syphilis was the only infectious disease that could be treated by chemotherapy.

Stimulated by Ehrlich's success, research continued. The **sulfa drugs** were the first major class of drugs to come into widespread clinical use. The sulfas are synthetic drugs—organic chemicals manufactured in the laboratory. They were discovered in the 1930s when a German chemical company, I. G. Farben, began systematically testing various compounds as possible chemotherapeutic agents. The sulfas had been manufactured originally as dyes.

Antibiotics are natural chemotherapeutic agents—produced by microorganisms. They proved to be more versatile than the sulfas. The first medically useful antibiotic, **penicillin (Figure 1.13)**, was discovered in 1929 by the Scottish microbiologist Alexander Fleming (1881–1955); but it did not come into widespread clinical use for another decade because of technical problems in purifying and mass-producing it. In the 1940s, however, during World War II, funds were provided to make penicillin readily available. The dramatic effectiveness of penicillin gave it the deserved nickname "wonder drug."

Childbed Fever

Entering a hospital in the mid-nineteenth century to give birth was fraught with risk. Many women who delivered normal, healthy infants never survived to take them home because of an illness known as childbed fever, or puerperal sepsis. At that time, physicians had no idea what caused childbed fever. We now know that it is caused by the bacterium *Streptococcus pyogenes*. Infection begins in the uterus and spreads rapidly through the body. *S. pyogenes* can infect any wound and, untreated, the infection shows the identical symptoms: fever, chills, delirium, and death.

In 1847, Ignaz Semmelweiss was a physician in the obstetrics ward of a Viennese hospital. He was horrified by the number of women he saw dying. He was also struck by the strange pattern of deaths. Women in labor were admitted either to the First or Second Clinic of the hospital, depending upon when they arrived. Although these two areas were seemingly identical, almost all the childbed fever deaths occurred in the First Clinic. All Vienna knew about it. Women in advanced labor would wait in the halls, announcing their arrival at a time they hoped would allow them to enter the Second Clinic.

The only clear difference between the clinics was in staffing. Medical students attended in the First and mid-wives in the Second. But what could that have to do with it? The clue Semmelweiss needed came when a medical student cut his finger during an autopsy and died of an illness identical to childbed fever. Medical students began their days in the morgue, doing autopsies to learn anatomy. Then they went to the First Clinic. So, reasoned Semmelweiss, they must be taking something with them on their hands and, in delivering the child, contaminating the mother. Semmelweiss ordered the medical students to wash their hands in a chlorine solution before entering the clinic. Hand washing was not routine. Physicians often had blood and pus on their hands while treating patients. Semmelweiss chose a chlorine solution because it removed the characteristic odor of the morgue, but chlorine also killed the deadly streptococcal bacteria and thus prevented infection. The number of deaths from childbed fever in the First Clinic soon fell to the prevailing low levels in the Second.

Nevertheless, Semmelweiss was ridiculed. Three decades later the work of Pasteur and Koch would provide a scientific explanation for Semmelweiss's findings and lead to a revolution in controlling infection. But Semmelweiss did not live to see it. Ironically, he died of a streptococcal infection in 1865, an outcast from the medical community.

Immunology

In the days of Pasteur and Koch, immunology was a branch of microbiology devoted to developing vaccines for preventing infectious diseases. Today, immunology is an independent and fast-developing science in its own right. We have learned that the immune system is extremely complicated, delicate, and prone to defects, some causing disease directly and some leaving the person vulnerable to infection and cancer. Chapter 17 discusses the immune system, and Chapter 19 discusses the clinical applications of what we know about using immunity to combat infectious disease.

Virology

Virology, the study of viruses, began in 1892 when the Russian microbiologist Dmitri Iwanowski discovered the tobacco mosaic virus (see Chapter 13). Iwanowski was studying a disease of tobacco plants called tobacco mosaic disease. To isolate the microorganism, he forced juice from diseased plants through filters that retained the smallest bacteria. But he found the filtered juice still caused disease. Because bacteria were believed to be the smallest microorganisms, Iwanowski first thought his methodology might be flawed. But repeated experimentation convinced him that smaller disease-causing agents were passing through the filter. He called these tiny agents "filterable viruses." They could not be seen, even under the most powerful microscopes of that time. Until the electron microscope was developed in the 1930s, we knew viruses existed, but little more.

Genetic Engineering

Two facts about microorganisms have led to particularly exciting developments in the second half of the twentieth century. First, the metabolism and genetic properties of microorganisms, particularly bacteria, are remarkably similar to those of plants and animals. Thus, what we learn is often directly applicable to higher forms of life. In fact, much of our knowledge of the fundamental properties of all living things came first from studies on bacteria. Second, microorganisms are especially suitable for experimental investigation.

The suitability of microorganisms for experimental study rests on the ease with which they can be cultured and the rapidity with which they multiply. Under proper conditions certain bacteria can double their numbers every 20 minutes. The number of individual organisms in a single milliliter of a bacterial culture can exceed the number of human beings on Earth. We can thus study enormous numbers of organisms in an extremely short period of time and undertake experiments that would be impossible using larger, slower-growing organisms.

Intensive laboratory study of microorganisms has in turn led to the evolution of a remarkable set of techniques collectively called **genetic engineering**, or **recombinant DNA technology**. With this technology researchers can manipulate DNA, the cell's genetic material, outside the organism from which it was obtained and reintroduce the modified DNA into another cell where it will exert its effect. The DNA can be obtained from any organism, and fragments of DNA from different organisms can be joined together. The organism that is most frequently used as the host for manipulated DNA is the bacterium *Escherichia coli*. For example, we use *E. coli* to produce insulin for the treatment of diabetes and human growth hormone for the treatment of pituitary dwarfism.

The potential of recombinant DNA technology seems almost limitless. We explain genetic engineering techniques in Chapter 7 and describe some of its exciting uses in Chapter 29.

The Future

As an active experimental science, microbiology is little more than a hundred years old. Its pattern of accelerating progress seems likely to continue. Medical microbiology and virology will need to solve problems of resistance to antibiotics and evolution of new viruses; controlling infectious disease remains a challenge. New antibiotics and other chemotherapeutic agents may be the answer, though some researchers emphasize enhancement of the natural immune system. Whatever route researchers follow, they will depend heavily on recombinant DNA technology.

The other area that will expand rapidly in the future is environmental microbiology. We will rely increasingly on **bioremediation**, using microorganisms to clean up toxic chemicals we have added to our environment.

The unseen world of microorganisms is huge and diverse. Most have not yet even been identified or named. They benefit us in many ways, but unsolved problems remain and new challenges constantly arise. Many new microbiologists are needed to attack the problems the twenty-first century will surely place before us. You might consider the career yourself. It asks for dedication, and it demands interest. Whether this commitment is right for you or not, we hope you enjoy the study of the unseen world you are about to undertake.

Summary

THE UNSEEN WORLD AND OUR WORLD (pp. 3–6)

1. Microorganisms are living things too small to be seen by the unaided eye.

2. Lister developed antisepsis, a technique for preventing infection, by building on established scientific fact, formulating a hypothesis, and testing it with an experiment. This approach is the basis of the scientific method.

3. Microbiology is the study of microorganisms.

Microbes and Disease (pp. 3–4)

1. Throughout history, pathogens (disease-causing microorganisms) have had a tremendous negative impact on human affairs. Epidemic disease brought social and political chaos along with human suffering. Only a small fraction of microorganisms are pathogenic.

Microbes and Life Today (p. 4)

1. The development of microbiology as a science has allowed us to control microbes and use them for our benefit.

2. Medical microbiology focuses on preventing and treating infectious diseases.

3. Environmental microbiology is the study of how microorganisms affect the earth and its atmosphere.

4. Industrial microbiology deals with microorganism-dependent industries, such as those producing foodstuffs, fermented beverages, and pharmaceuticals. Industrial microbiologists will increasingly apply genetic engineering to solving medical, environmental, and agricultural problems.

5. Research in agricultural microbiology has led to healthier livestock and more disease-free crops.

Careers in Microbiology (pp. 4–6)

1. Career opportunities in microbiology depend on training as well as interest.

THE SCOPE OF MICROBIOLOGY (pp. 6–10)

1. There are five groups of microorganisms: bacteria, algae, fungi, protozoa, and viruses.

2. Microbiology is a cohesive science because of its methodology and approach to problems, not because of the relatedness of the organisms it studies.

3. Bacteria are procaryotes; they lack internal membrane-bound structures.

4. Algae, fungi, and protozoa are eucaryotes; their organelles are membrane-bound.

5. Viruses are acellular.

Bacteria (pp. 6–7)

1. The terms *bacteria* and *procaryotes* are synonymous.

2. Bacteria are extremely small, even for microorganisms.

3. The two groups of procaryotes, eubacteria (true bacteria) and archaebacteria (ancient bacteria), are similar in appearance and size but different biochemically.

4. Eubacteria vary in shape, motility, and how they get energy. There are species that can withstand freezing, boiling, and extreme acidity or alkalinity.

5. Some bacteria cause disease, but others keep our environment in life-sustaining balance.

Algae (p. 7)

1. Algae are eucaryotic organisms that carry out photosynthesis.

2. Some algae are unicellular and microscopic. Others consist of so many cells they are macroscopic.

3. Algae are not important medically, but they are critically important to global ecology.

Fungi (pp. 7–9)

1. Fungi include mushrooms, yeasts, and molds. Fungi are eucaryotic and nonphotosynthetic. They are both microscopic and macroscopic.

2. A few fungi are pathogenic to humans, and many are pathogenic to plants (corn smut, wheat rust).

Protozoa (p. 9)

1. Protozoa are eucaryotic microorganisms that are superficially animal-like, nonphotosynthetic, and usually motile. Virtually all are microscopic.

2. Examples of protozoa are the amoebae, flagellates, and ciliates.

3. The study of protozoan (and helminth-caused) diseases is called parasitology.

Viruses (pp. 9–10)

1. Viruses are particles of nucleic acid (either RNA or DNA), usually enclosed in a protein coat.

2. Viruses are obligate intracellular parasites.

3. Viruses are extremely small, even compared to bacteria, and must be studied biochemically or with an electron microscope.

4. Viruses can infect animals, plants, and other microorganisms, such as bacteria.

5. Prions are even smaller infectious agents than viruses. They are composed entirely of protein, and their method of reproduction is a mystery.

Helminths (p. 10)

1. Helminths are macroscopic worms but some go through microscopic stages in their life cycle and cause parasitic diseases in plants and animals, including humans.

2. The helminths important to health studies are flatworms and roundworms.

A BRIEF HISTORY OF MICROBIOLOGY (pp. 10–14)

1. Once microbiology became an experimental science in the mid-nineteenth century, a period of accelerating progress began.

Leeuwenhoek's "Animalcules" (pp. 10–11)

1. Antony van Leeuwenhoek, whose hobby was making microscopes, was the first to see microorganisms (about 1674). He called them "animalcules."

Hooke and the Cell Theory (p. 11)

1. Robert Hooke's observation that thin slices of cork had a honeycomb of chambers, or cellulae, led to the formulation of the cell theory: cells are the basic unit of organization for all living things.

Spontaneous Generation (pp. 11–12)

1. In 1665 Francesco Redi's experiment with covered and uncovered meat jars proved that living things come only from preexisting living things.

2. Many scientists were convinced by John Needham's 1745 experiment that spontaneous generation of microorganisms did occur. They said that Lazzaro Spallanzani's experiment with sealed flasks proved only that microorganisms needed air for spontaneous generation.

3. In 1859 Louis Pasteur finally disproved spontaneous generation. Using special swan-necked flasks he performed experiments that allowed the free passage of air but prevented the entry of microorganisms.

4. Pasteur's experiments grounded microbiology in scientific fact: No living things, including microorganisms, arise by spontaneous generation; microorganisms are everywhere—even in the air and on dust particles; the growth of microorganisms causes dead plant and animal tissue to decompose and food to spoil.

The Germ Theory of Disease (pp. 12–13)

1. Once spontaneous generation was disproved, microbiology changed from an observational to an experimental science. Robert Koch developed the germ theory of disease: microorganisms (germs) cause disease, and specific microorganisms cause specific diseases.

2. Koch developed four postulates that, if fulfilled, provide absolute proof that a particular microorganism causes a particular disease.

3. Koch also developed a technique for obtaining a pure culture.

4. Koch also discovered that agar added to broth provided a solid, nutrient-rich medium for obtaining pure cultures.

Immunity (pp. 13–14)

1. Infectious diseases are those caused by microorganisms.

2. Once the germ theory of disease was accepted, microbiology rapidly advanced. Controlling microorganisms became its focus.

3. Immunity is stimulating the body's own ability to combat infection. The idea of immunization was based on the recognized fact that people who suffered once from certain diseases did not get them again.

4. Edward Jenner used fluid from cowpox blisters to provide protection against smallpox. Inducing immunity for protection came to be known as vaccination.

5. Pasteur developed vaccines for anthrax and rabies, using attenuated forms of the disease-causing bacteria. D. E. Salmon and Theobald Smith demonstrated that killed microbial strains were also effective as vaccines.

Public Hygiene (p. 14)

1. Acceptance of the germ theory advanced the idea of public hygiene, promoting cleanliness and reducing exposure to disease, which saved even more lives than immunization did.

2. Concern for public hygiene led to clean drinking water, improvements in food preservation (such as pasteurization), and hand washing in hospitals and for personal hygiene.

MICROBIOLOGY TODAY (pp. 14–17)

1. Advances in twentieth-century microbiology have been most striking in the areas of chemotherapy, immunology, virology, and genetic engineering.

Chemotherapy (p. 15)

1. Chemotherapy is the treatment of disease with chemicals called drugs.

2. Paul Ehrlich articulated the guiding principle of chemotherapy, selective toxicity: For a drug to be effective against infection, it must kill the infecting microorganism without damaging the host.

3. The first major class of drugs to gain widespread clinical use was the sulfa drugs, which are synthetic chemicals.

4. Antibiotics are natural chemotherapeutic agents produced by microorganisms. Penicillin, the first medically useful antibiotic, was discovered in 1929 by Alexander Fleming.

Immunology (p. 16)

1. Immunology studies the immune system, whose defects can either cause disease directly or leave a person vulnerable to it.

Virology (p. 16)

1. Virology, the study of viruses, began in 1892 when Dmitri Iwanowski discovered the tobacco mosaic virus; viruses couldn't be seen until the electron microscope was developed.

Genetic Engineering (pp. 16–17)

1. Genetic engineering (recombinant DNA technology) is a group of techniques for manipulating DNA outside the organism from which it was obtained and reintroducing the recombinant DNA into another cell where it will exert its effect.

2. *Escherichia coli* is most often used as the new host for manipulated DNA.

3. Microorganisms lend themselves to experimentation because (1) the metabolism and genetics of microorganisms, particularly bacteria, are remarkably similar to those of plants and animals and (2) microorganisms are easy to culture and multiply rapidly so that enormous numbers can be studied in short times.

The Future (p. 17)

1. Genetic engineering and bioremediation, using microorganisms to clean up toxic chemicals added to the environment, promise to be areas where rapid progress will be made.

Review Questions

THE UNSEEN WORLD AND OUR WORLD

1. Define microorganisms. Give some examples of how microorganisms have affected human history.

2. Give some examples of advances in the field of medical microbiology and challenges to be faced.

3. What is environmental microbiology?

4. Name some microorganism-dependent industries. In what directions is industrial microbiology moving today?

5. Give some examples of advances and challenges still to be faced in agricultural microbiology.

THE SCOPE OF MICROBIOLOGY

1. Name the five subgroups of microorganisms.

2. What is the one property all microorganisms have in common? Explain this statement: Microbiology is a cohesive science because of its methodology.

3. Explain the difference between procaryotes and eucaryotes.

4. What are eubacteria and archaebacteria? How are they alike, and how are they different?

5. Explain this statement and give examples: Eubacteria are extremely diverse.

6. What are algae? What do they look like? Of what importance are they?

7. What are fungi? Name the different types. How are fungi important medically? Ecologically?

8. What are protozoa? Give some examples of motile protozoa and tell how they move. What is parasitology?

9. Explain this statement: Viruses are acellular. Why are viruses called obligate intracellular parasites? How are viruses studied? Why are viruses medically important?

10. What are prions?

11. Why do we study helminths in microbiology?

A BRIEF HISTORY OF MICROBIOLOGY

1. Who was the first person to see microorganisms? When did this occur?

2. What was Robert Hooke's contribution to microbiology?

3. Define spontaneous generation. What roles did Redi, Needham, and Spallanzani play in dispelling or propagating the theory?

4. When and by whom was spontaneous generation finally disproved? How did he do it?

5. Explain this statement: Once spontaneous generation was disproved, microbiology changed from an observational science to an experimental science.

6. Discuss Pasteur's contributions to the field of microbiology.

7. What is the germ theory of disease?

8. Discuss Koch's contributions to the field of microbiology.

9. Define these terms: infectious diseases, immunity, immunization, vaccination, vaccine. What is the difference between attenuated and killed vaccines?

10. Compare Jenner's approach to developing a vaccination for smallpox with Lister's development of the carbolic acid method of antisepsis.

11. Define public hygiene. Give some examples of progress made in public hygiene after acceptance of the germ theory of disease.

MICROBIOLOGY TODAY

1. Define chemotherapy. What is Ehrlich's principle of selective toxicity? What is the difference between synthetic drugs and antibiotics? Give an example of each.

2. Define immunology, and describe how the field today differs from that of Pasteur and Koch's era.

3. What is virology? How were viruses discovered and by whom?

4. What is genetic engineering? Why is it also called recombinant DNA technology? What role does *Escherichia coli* play in genetic engineering? Name some products of genetic engineering.

5. Explain the two reasons microorganisms lend themselves so well to experimentation.

Essay Questions

1. If you could meet one of the figures mentioned in the history of microbiology, who would it be and why? What questions would you ask him?

2. What do you think is the most pressing problem facing microbiologists today? Explain.

Suggested Readings

Brock, T. D. 1988. *Robert Koch, a life in medicine and bacteriology*. Madison, Wis.: Science Tech.

Dubos, R. J. 1988. *Pasteur and modern science*. Edited by Thomas Brock. Madison, Wis.: Science Tech.

McNeill, W. H. 1977. *Plagues and people*. Garden City, N.Y.: Anchor Press.

Roueché, B. 1984. *The medical detectives*. 2 vols. New York: Times Books.

2 BASIC CHEMISTRY

The Chemical Soup of Life

Living cells are chemical factories that make more chemical factories like themselves. In fact, almost all organic (carbon-containing) chemicals, such as those that compose wood, leaves, coal, and petroleum, were made by living cells. Cells also make the organic compounds that become more cells. It is a self-perpetuating cycle—cells make organic compounds that become more cells that make more organic compounds. But where did the organic compounds that formed the first cells come from? Charles Darwin, best known for his writings on evolution, might have been the first to formulate a plausible explanation. In 1871 he wrote to a friend, "If we could conceive in some warm little pond, with all sorts of ammonia and phosphoric salts, light, heat, electricity, etc., present, that a protein compound was chemically formed ready to undergo still more complex changes. . . ." In other words Darwin speculated that the organic compounds that make up cells might form spontaneously from simple inorganic compounds under the right conditions. If no living things were present to consume them, they would accumulate and might assemble themselves into a self-reproducing structure—a primitive cell.

About 80 years later, Stanley Miller, an American graduate student working with the Nobel Prize–winning chemist Harold Urey, put Darwin's speculation to an experimental test. He put a chemical mixture of methane, ammonia, and water vapor (the simple gases presumed to be present in Earth's atmosphere before life appeared) in a flask and exposed it to electrical discharges. He found, as he expected, that organic compounds formed in the flask. But quite unexpectedly he found the mixture included a group of amino acids—such as glycine, alanine, aspartate, and glutamate—compounds found in the proteins of all living things. Miller's results had a great impact on scientists who study the origin of life. His results supported Darwin's speculations with scientific fact. The organic compounds that compose cells can be made by ordinary chemical reactions from materials that were probably present on Earth before life appeared.

This is probably how the earth looked about 4 billion years ago. Only inorganic compounds were present, but they formed the "soup" of life from which the first cells would appear about 500 million years later. (The large object over the horizon is the moon. It used to be much closer to Earth.)

To understand:

- The basic building blocks of matter—subatomic particles, atoms, elements, and molecules

- The ways in which atoms bond to make molecules and compounds

- What a chemical reaction is, what determines whether a reaction occurs, and what determines the rate of a chemical reaction

- The unique chemical properties of water

- The structure of organic molecules

- The structures of the macromolecules—proteins, nucleic acids, carbohydrates, and lipids—from which all cellular microorganisms are built

THE BASIC BUILDING BLOCKS

Most microorganisms are too small to be seen by the naked eye, but we are constantly aware of their effects. We see a tomato rot or a slice of bread turn moldy, or we find that the milk we left too long in the refrigerator has turned sour. The effects microorganisms have on their environment result from the chemical processes they carry out in order to grow.

Microbiologists follow most microbial activities chemically rather than visually. Therefore, to learn about microorganisms and what they do, you need some knowledge of basic chemistry. **Chemistry** is the science that studies the composition of matter and the changes it undergoes. **Matter** is an all-inclusive term describing everything (other than a complete vacuum) that fills space. In our discussion of the chemical nature of matter, we will emphasize the chemicals and chemical reactions in living things.

Matter is composed of small particles called **atoms**. Atoms, in turn, are composed of three kinds of **subatomic particles**—**protons**, **neutrons**, and **electrons**. Different numbers of protons, neutrons, and electrons combine to produce the different atoms. A gold atom, for example, contains 79 protons, 118 neutrons, and 79 electrons, while an oxygen atom contains 8 protons, 8 neutrons, and 8 electrons.

The three kinds of subatomic particles differ in size and electrical charge (**Table 2.1**). Protons and neutrons have approximately equal mass and are enormous compared to an electron. A single proton or neutron has 1837 times more **mass** than an electron. Protons and neutrons are consequently assigned a value of 1 mass unit, or 1 atomic weight unit (awu).

A proton carries a single positive (+) charge. An electron carries a single negative (−) charge. A neutron is electrically neutral—it carries no charge. The number of protons always equals the number of electrons in an atom. Thus, the net electrical charge of an intact atom is zero because the equal positive and negative charges cancel each other out.

Atoms

All atoms have the same basic structure. A cloud of electrons orbits a **nucleus** composed of densely packed protons and neutrons (**Figure 2.1**). Electrons orbit the nucleus at different energy levels or **shells**. Inner shells must always be filled before electrons can enter more distant shells. The innermost shell can hold up to 2 electrons. Atoms that have only 1 or 2 electrons contain them all within this shell. The next shell can contain as many as 8 additional electrons. The third shell can contain another 8 electrons. Thus the first three shells can accommodate as many as 18 electrons.

Additional shells exist, but most of the atoms that make up microorganisms and other living things have fewer than 18 electrons and therefore contain all their electrons within the first three shells. The electrons in an atom's *outermost* shell are called **valence electrons**. Valence electrons have special significance, as you will see when we discuss chemical bonds.

The **atomic number** of an atom is the number of protons it contains. Every atom has its own atomic number, from 1 to 106. For example, the atomic number of carbon (and only carbon) is 6, while the atomic number of sulfur (and only sulfur) is 16.

The **atomic weight** of an atom, in atomic weight units, is the number of protons plus neutrons it contains.

Table 2.1 Properties of Subatomic Particles			
Subatomic Particle	Unit Charge	Relative Weight	Location in Atom
Proton	+1	1	Nucleus
Neutron	0	1	Nucleus
Electron	−1	1/1837	Cloud surrounding nucleus

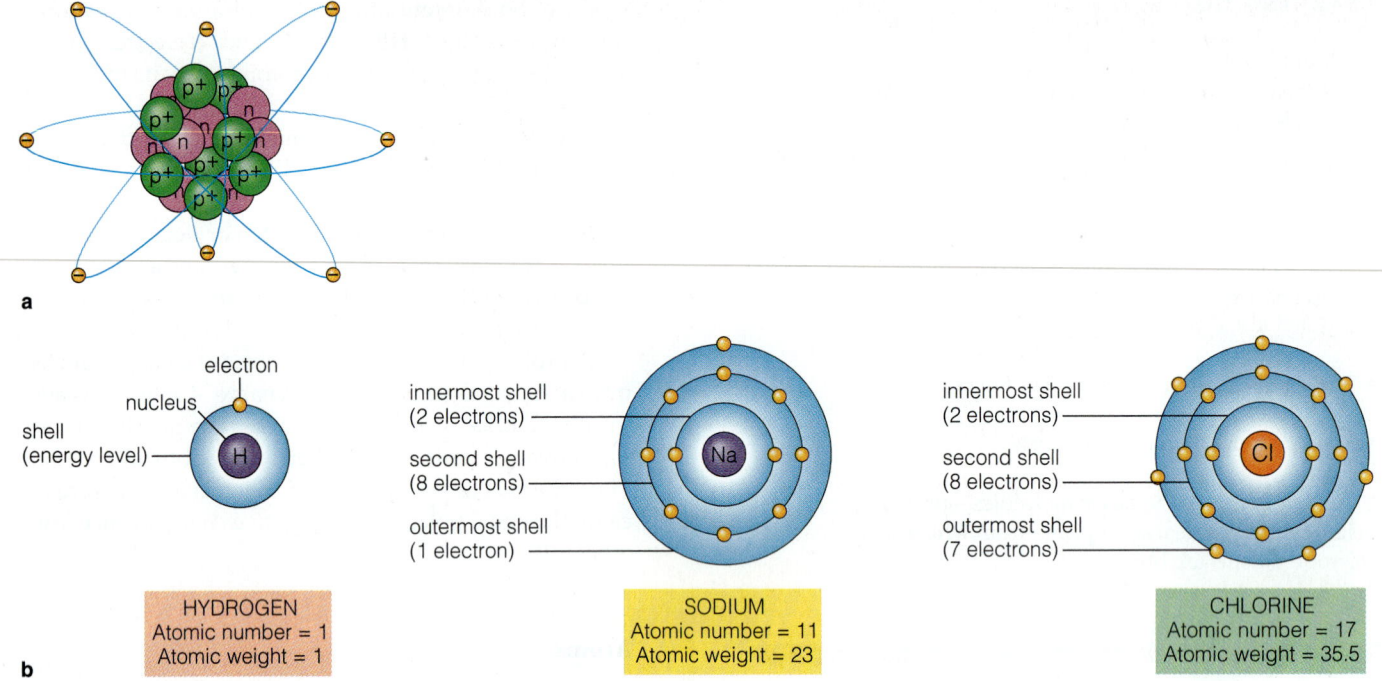

Figure 2.1 (**a**) Simplified model of an atom showing protons and neutrons in nucleus and electrons orbiting. This is an oxygen atom with 8 protons, 8 neutrons, and 8 electrons. (**b**) Structures of hydrogen, sodium, and chlorine atoms showing the arrangement of their electrons in shells and their atomic number and atomic weight.

The same element, as defined by its atomic number, *may* have more than one form that differ by their number of neutrons and therefore their atomic weights. Atoms with the same atomic number but different atomic weights are called **isotopes**. For example, chlorine has an atomic number of 17 (because it has 17 protons). Most chlorine atoms in nature have 18 neutrons, giving them an atomic weight of 35. But a small percentage of chlorine atoms have 20 neutrons, giving them an atomic weight of 37. (The average awu of chlorine is, therefore, 35.45.) Carbon atoms typically have 6 protons and 6 neutrons for an awu of 12. But there are carbon atoms with 7, 8, 9, and 10 neutrons. In other words, carbon has four isotopes. Isotopes may be stable (also called "heavier" because of the additional neutrons) or unstable. An unstable isotope is **radioactive** because it releases radiation as it decays (changes) into a stable element (see the box "Radioactive Tracers").

Elements

Matter composed of only one kind of atom is called an **element**. Because an atom is the simplest particle of matter, it necessarily follows that an element cannot be reduced chemically to a simpler form. Ninety-two elements occur naturally, and 14 more have been synthesized (created ar-

tificially) in the laboratory. Every element has distinctive physical and chemical properties. Chemists use one- or two-letter abbreviations for elements. For example, H stands for hydrogen, C for carbon, and Fe for iron.

Of all the naturally occurring elements, only about 25 are essential to life. Of these, by far the most abundant in living organisms are carbon, hydrogen, nitrogen, and oxygen, plus phosphorus and sulfur. These six elements account for more than 99 percent of the weight of most living things. The rest is composed of **trace elements**, elements that are essential to life but present in only minute amounts. Iron is one example of a trace element essential to microorganisms, as well as to plants and animals. It constitutes less than 0.2 percent of the dry weight of most living things, but life cannot exist without it. Some trace elements are listed in **Table 2.2**.

Molecules

A **molecule** consists of two or more atoms joined by linkages called **chemical bonds**. The atoms that make up a molecule may be the same or different. A molecule of oxygen gas, for example, consists of two oxygen atoms, while a molecule of water consists of one oxygen atom and two hydrogen atoms. A DNA molecule consists of carbon, hydrogen, oxygen, nitrogen, and phosphorus

Table 2.2 Elements That Occur Most Abundantly in Microorganisms

Element	Symbol	Atomic Number	Most Common Mass Number	Abundance in *Escherichia coli* (% of dry weight)
Hydrogen	H	1	1	8
Carbon	C	6	12	50
Nitrogen	N	7	14	14
Oxygen	O	8	16	20
Sodium	Na	11	23	1
Magnesium	Mg	12	24	0.5
Phosphorus	P	15	31	3
Sulfur	S	16	32	1
Chlorine	Cl	17	35	0.5
Potassium	K	19	39	1
Calcium	Ca	20	40	0.5
Iron	Fe	26	56	0.2

atoms. Molecules vary enormously in size and complexity. Some are particles as small as 2 atoms. Macromolecules, however, can be as big as the millimeter-long strand of DNA that makes up a bacterial chromosome and must be highly folded to fit inside the cell.

A **compound** is composed of molecules that contain more than one type of atom. Thus, water, which is composed of both hydrogen and oxygen, is a compound. But metallic gold, which is composed of a single type of atom, is not. **Organic** (as opposed to inorganic) compounds are the ones that most interest microbiologists. The term *organic* refers to carbon-containing compounds—including the ones that characterize living things. Organic compounds from living things are also called **biochemicals**.

Molecular Weight. A **molecular formula** tells which atoms and how many of each kind are joined to form a particular molecule. The molecular formula for oxygen gas, for example, is O_2, indicating that two oxygen atoms are joined together. The molecular formula for water is H_2O, indicating that two hydrogen atoms and one oxygen atom are joined.

A molecular formula allows us to calculate **molecular weight**, which is the total of the atomic weights of all of a molecule's atoms. For example, the molecular weight of oxygen gas, O_2, is 32.

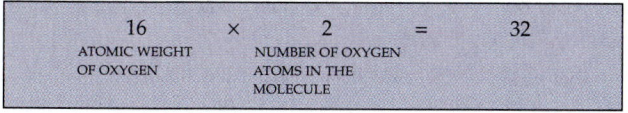

The molecular weight of water, H_2O, is 18.

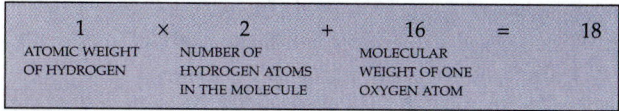

Moles. The size of a molecule (or an atom) is determined by how much a certain number of them weighs. Much as a grocer might weigh a dozen eggs to determine whether they are small, medium, or large, chemists weigh a fixed number of molecular or atomic particles. The number selected as the basis for this comparison is called **Avogadro's number**. Its value is 6.02×10^{23}. Avogadro's number of any type of molecule is called a **mole** of that substance (the equivalent term for atoms is called a **gram atom**). Avogadro's number is convenient because a mole is the molecular weight of the substance in grams. Thus, a mole of oxygen weighs 32 grams (O_2, molecular weight of 32) and contains 6.02×10^{23} molecules. Similarly, a gram atom of pure carbon weighs 12 grams (C, atomic weight 12) and contains 6.02×10^{23} atoms.

CHEMICAL BONDS AND REACTIONS

The science of chemistry seeks to understand not only the nature of matter, but also the changes it undergoes. All chemical changes in matter come about by forming or

Radioactive Tracers

Until the 1940s, we didn't have the technology to produce relatively pure isotopes. Then researchers learned how to make some and purify others from naturally occurring mixtures. Ordinary hydrogen, for example, with its one neutron could be separated from its heavier but stable two-neutron form (called deuterium) and its radioactive three-neutron form could be made (called tritium).

Isotopes enable us to label and trace biological structures. Both heavy (stable) and radioactive (unstable) isotopes can be used to label and to trace, allowing researchers to locate them at different places in an organism or a molecule. Isotopes can be used conveniently on organisms because radioactive isotopes exist for three biologically important elements: ^3H for hydrogen, ^{14}C for carbon, and ^{32}P for phosphorus. (Only heavy isotopes exist for oxygen. Nitrogen has a heavy and a radioactive isotope, but the radioactive element is so highly unstable that it is inconvenient.) Minute amounts of radiation can be detected from a radioactive isotope. An instrument called a scintillation counter can detect even a single radioactive isotope molecule releasing radiation.

Using isotopes to label and trace has led to many important discoveries in microbiology. We discovered how bacterial chromosomes replicate by labeling them with heavy isotopes. Radioactive tracers have also been used to find where individual atoms move in chemical reactions. The first step in photosynthesis was discovered this way. An alga was grown in a radioactive carbon dioxide environment. Then researchers traced the carbon dioxide through the chemical reactions by which it was incorporated into other carbon-containing molecules.

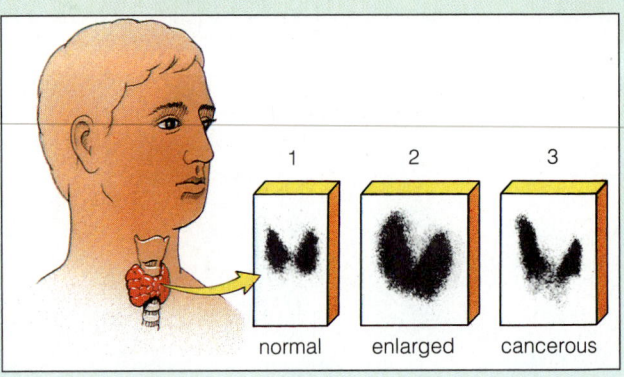

Images of thyroid glands that have taken up radioisotopes of ^{123}iodine.

Radioactive tracers play an important role in clinical medicine. Radioisotopes, for example, are routinely used to diagnose thyroid gland abnormalities. The thyroid is the only body structure to take up iodine as a trace element. So the radioisotope for iodine is intravenously injected, and then the thyroid is scanned with a radiation detector. See the illustration for examples of diagnostic scanner images. Radioisotopes are also used to do whole-body scans for cancer. The radioactive isotope ^{67}Ga of the metallic element gallium is used. Isotope ^{67}Ga has the unusual property of being taken up selectively by cancer cells. The isotope is intravenously injected, and then the entire body is scanned with positive emission tomography (PET) to detect radioactivity. The scanner image locates any cancer cells.

breaking chemical bonds, the forces that hold atoms together in molecules. Chemical bonds form if the resultant molecule will be at a lower energy state, and therefore more stable, than the original configuration of atoms. Thus, energy is released (largely as heat) when any chemical bond forms, and energy is required to break it. In spite of this, chemists call certain bonds that need small amounts of energy to break **high-energy bonds** (Chapter 5) because when they are broken, other bonds form that release larger amounts of energy.

An atom's valence electrons determine its capacity to form chemical bonds. An atom is most stable (in its lowest energy state) when its outermost shell is filled with electrons. As a result, pairs or groups of atoms share, lose, or gain valence electrons to fill their outermost shells (**Table 2.3**). The three kinds of bonds formed are covalent, ionic, and hydrogen. They are summarized in **Table 2.4**.

Covalent Bonds

A **covalent bond** is formed when atoms share electrons. Covalent bonds are extremely stable. A covalent bond may be single (two shared electrons), double (four shared electrons), or triple (six shared electrons) (**Figure 2.2**). All six of the major elements in biochemicals—carbon, hydrogen, nitrogen, oxygen, phosphorus, and sulfur—have vacancies in their valence shell of electrons and are, therefore, capable of forming covalent bonds. Because it has four vacancies in its valence shell, carbon can form four single covalent bonds, or any combination of single, double, or triple bonds that adds to four. The critical role that carbon plays in the structure of biochemicals is discussed later in this chapter.

Sometimes the shared electrons in a covalent bond are not equally spaced between the two atoms they join.

Table 2.3 Electron Shells and Covalent Bonds of Biologically Important Elements

Element	Atomic Number	Number of Electrons in:				Number of Single Covalent Bonds It Can Form
		First Shell	Second Shell	Third Shell	Valence Shell	
Hydrogen	1	1	0	0	1	1
Carbon	6	2	4	0	4	4
Nitrogen	7	2	5	0	5	3 or 5
Oxygen	8	2	6	0	6	2
Phosphorus	15	2	8	5	5	3 or 5
Sulfur	16	2	8	6	6	2

Table 2.4 Important Kinds of Chemical Bonds in Living Systems

Bond	Properties	Notes
Covalent	Atoms share a pair of electrons	Strong bonds. Can be single, double, or triple. Covalent bonds link atoms in most biochemicals.
Nonpolar	Electrons shared equally	Most covalent bonds are nonpolar.
Polar	Electrons shared unequally	Polar covalent bonds give molecules a positive and negative pole.
Ionic	Attraction between positive and negative charges	Strong but not as strong as covalent. They occur in salts and between charged parts of macro-molecules.
Hydrogen	A hydrogen atom is shared between two other atoms	Weak bonds but very important to living things. They give water its unusual properties. They link the two strands of DNA and produce the secondary and quaternary structures of proteins.

When this occurs, the part of the molecule with the greater concentration of electrons carries a more negative charge than the other part of the molecule. Such a molecule has a positive pole and a negative pole and is called a **polar molecule (Figure 2.3)**. Water, the most important molecule found in living things, is markedly polar. Due to the location of the electrons that are shared between the oxygen and the two hydrogen atoms, both hydrogens are located toward one side of the molecule. This side has a slight positive charge, and the side of the molecule near

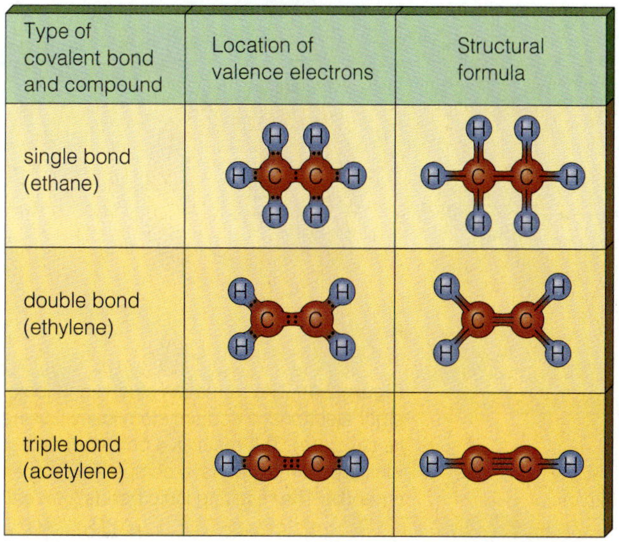

Figure 2.2 Covalent bonds. All six of the major biochemicals can form covalent bonds, but carbon can form all three types of covalent bonds—single, double, and triple—because it has four vacancies in its valence shell. All the examples here show carbon and hydrogen covalent bonds. In all cases, the valence shells of both carbon and hydrogen are filled: each hydrogen has 2; each carbon has 8.

Type of covalent bond and compound	Location of valence electrons	Structural formula
single bond (ethane)		
double bond (ethylene)		
triple bond (acetylene)		

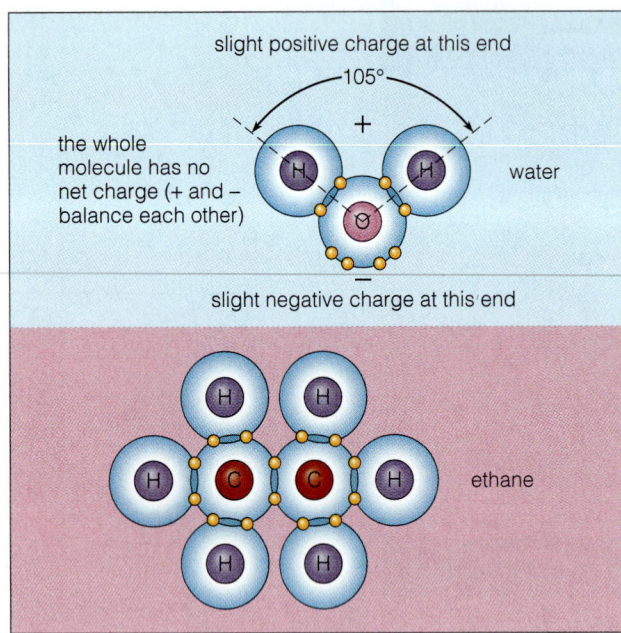

slight positive charge at this end
105°
+
the whole molecule has no net charge (+ and − balance each other)
water
slight negative charge at this end

ethane

Figure 2.3 Polar and nonpolar molecules. Water is a polar molecule because the shared electrons are pulled closer to the oxygen atom, giving it a slight negative charge and leaving the hydrogen atom with a slight positive charge. Water molecules are bent (forming a 105-degree angle), and they always have a positive and negative pole. Ethane is nonpolar because electrons are shared equally between atoms.

the oxygen atom has a slight negative charge. Molecules in which electrons are evenly distributed lack polarity and are called **nonpolar molecules**. Thus, covalent bonds can be single, double, or triple polar or single, double, or triple nonpolar.

Ionic Bonds

To understand ionic bonds, you must understand what an ion is. **Ions** are atoms that have acquired a complete valence shell by gaining or losing electrons. Unfilled valence shells are unstable and seek stability. Atoms with a nearly empty or nearly complete complement of electrons in their valence shell readily form ions. An atom that loses an electron to become stable is called an **electron donor**. When an atom takes up an electron to fill its shell it is called an **electron acceptor**. **Figure 2.4** shows how a sodium atom and a chlorine atom become ions and then form an ionic bond.

The distinctive characteristic of ions is that they are **charged**. Normally, the number of protons equals the number of electrons in an atom. But because ions have gained or lost electrons, they have an unequal number of protons and electrons. This means they have a net electrical charge equal to the number of electrons they have gained or lost. Ions formed by losing electrons have excess protons and thus a positive charge. They are called

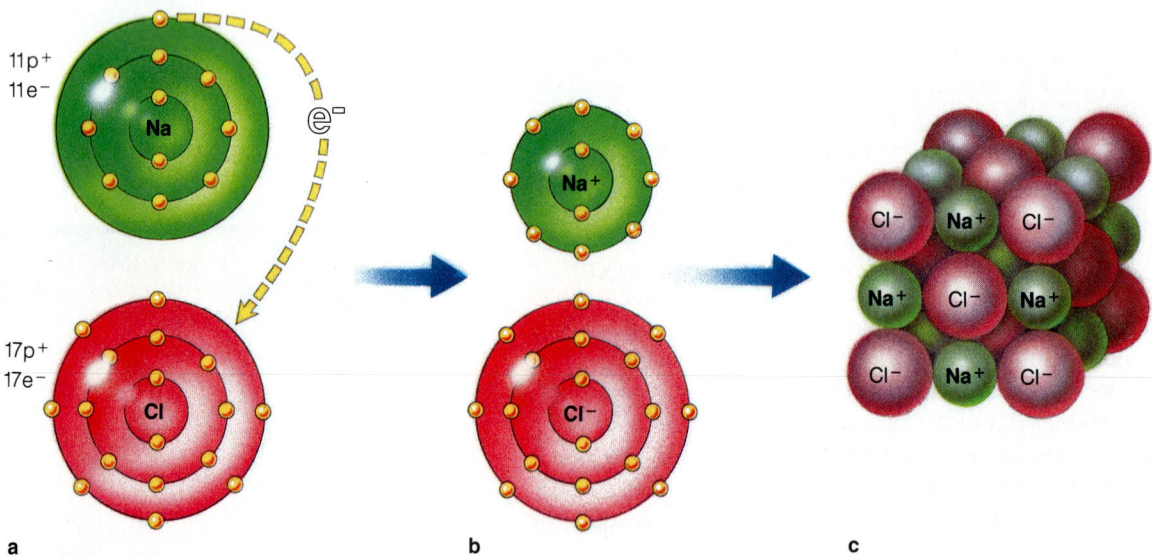

Figure 2.4 The formation of ions and ionic bonds. To attain a completely filled outer shell of eight electrons, a sodium atom loses its single valence electron and a chlorine atom adds one to its seven (**a**). The result is a positively charged sodium ion and a negatively charged chloride ion. Ions are atoms that have complete valence shells. The opposite charges of ions attract each other (**b**). The resulting ionic bonds form salt crystals (**c**).

cations. Ions formed by gaining electrons have a negative charge and are called **anions**.

Ions play important roles in living systems, including microorganisms. For example, magnesium (Mg^{2+}) activates certain enzymes, and calcium (Ca^{2+}) signals environmental changes to eucaryotic cells. Certain bacteria produce nitrate ions, which are critical to the nitrogen cycle (Chapter 28). Other ions play nonspecific roles. They neutralize charges on other molecules or increase the concentration of solutes in cells to balance the solutes outside.

Ionic bonds are formed by the attraction between oppositely charged ions or molecules. Ionic bonds unite anions and cations to form salt crystals. For example, the single positive charge of a sodium ion (Na^+) attracts the single negative charge of chloride (Cl^-), forming sodium chloride (NaCl), table salt. Ionic bonds are strong, but not so strong as covalent bonds.

Hydrogen Bonds

Hydrogen bonds form when hydrogen atoms between two molecules or between different parts of the same molecule are shared (**Figure 2.5**). Hydrogen bonds occur between molecules containing hydrogen atoms bonded to nitrogen, oxygen, or fluorine atoms. These three bonds (H—N, H—O, and H—F) are extremely polar. The hydrogen atom therefore has a slight positive charge, and the nitrogen, oxygen, or fluorine atoms have a slight negative charge.

Hydrogen bonds are weak. They cannot bind atoms into molecules. The energy required to break a hydrogen bond is only about one-twentieth as much as is required to break a covalent bond. Nevertheless, hydrogen bonds are critically important to living things. They stabilize the structure of many large biochemical molecules, including DNA. Hydrogen bonds help large molecules retain the shape they need to fulfill their biological roles. Hydrogen bonds also form between water molecules, giving liquid water the unusual properties that make life possible.

Chemical Reactions

A **chemical reaction** occurs when atoms or molecules (called **reactants**) collide and are transformed into different combinations of the same atoms or molecules (called **products**). In the process, chemical bonds break and new ones form. Energy is taken up (required or consumed) when bonds are broken and is released, mostly in the form of heat, when they form. Chemical reactions can occur only if the total amount of energy released by forming the new bonds exceeds the amount of energy required to break the old ones. How much energy is required or released depends on the strength of the bond.

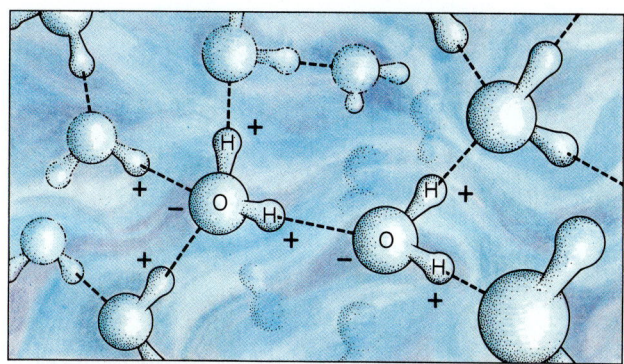

Figure 2.5 Hydrogen bonds give water its unusual properties. The tendency of oxygen to hold electrons more tightly than hydrogen makes part of the water molecule more negative than other parts. The positive and negative charges in water molecules attract each other, and hydrogen bonds form by sharing hydrogen atoms between oxygen atoms.

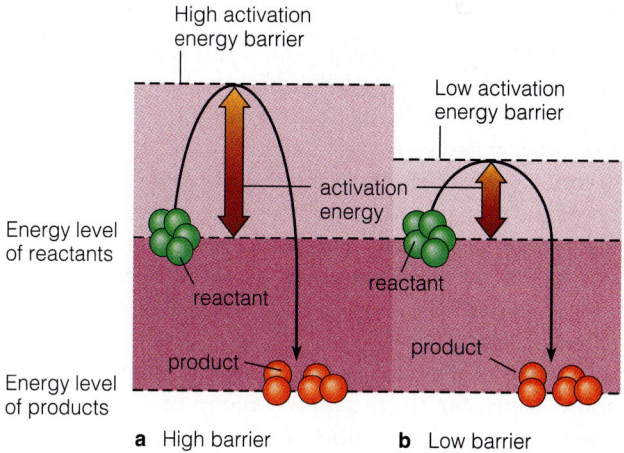

Figure 2.6 Chemical reactions occur only between molecules with sufficient energy to enter an activated state. (**a**) If there is a high activation energy barrier, reactants are less likely to reach an activated state and the reaction proceeds slowly. (**b**) On the other hand, if the activation energy barrier is low, many reactant molecules attain an activated state and the reaction proceeds rapidly. Enzymes and other catalysts lower activation energy barriers.

Reaction Rates. Even if the amount of energy released exceeds the amount taken up, chemical reactions do not occur every time reactant molecules collide. They occur only between molecules with sufficient energy to enter an **activated state** (**Figure 2.6**). If the barrier of activation energy is high, reactants are less likely to reach an activated state and enter into a reaction. Therefore, the reaction proceeds slowly. On the other hand, if the barrier is

Table 2.5	Factors That Increase the Rate of Chemical Reactions
Factor	Effect
Concentration	Chemical reactions proceed more rapidly if the concentration of reactants is increased. The greater the number of atoms or molecules in an area, the greater the likelihood that they will collide—the first essential step for a chemical reaction to proceed.
Heat	Higher temperature speeds chemical reactions because it adds energy, which pushes the reaction over the activation energy barrier.
Catalysts	Catalysts speed chemical reactions by lowering the activation energy barrier. Higher concentrations of a catalyst further speed reactions because more catalytic sites are present. The catalysts in living things are enzymes.

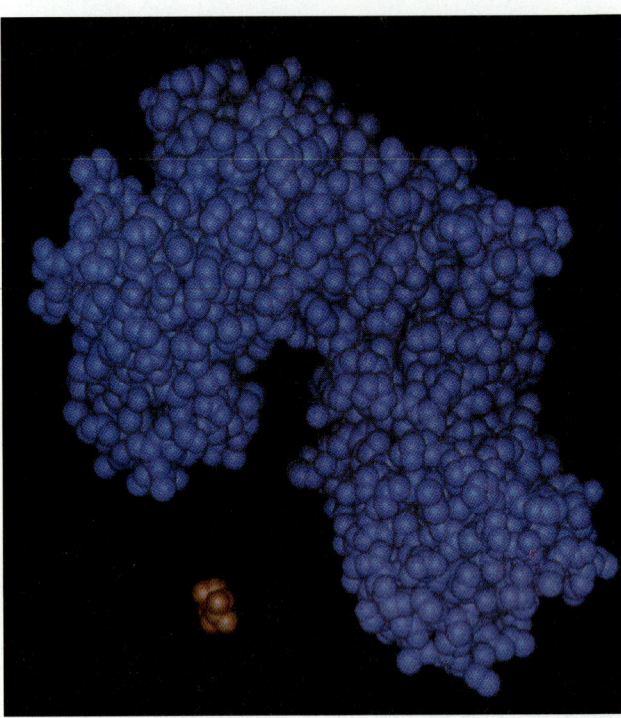

Figure 2.7 In this model, the glucose molecule (red) is about to enter the catalytic site of the enzyme hexokinase (blue). Enzymes increase the rate of a chemical reaction by lowering the activation energy barrier. Nearly every chemical reaction in a cell requires a specific enzyme.

low, many reactant molecules attain an activated state simultaneously and the reaction proceeds rapidly.

A number of other factors influence the rate of entering the activated state and, therefore, the rate of a reaction (**Table 2.5**). Higher concentrations of reacting molecules speed the rate of reaction because they increase the probability that molecules will collide with one another. Higher temperature increases the rate of reactions because molecules are raised to an activated state by absorbing thermal (heat) energy.

Catalysts also increase the rate of a reaction; they do so by decreasing the activation energy barrier. Catalysts are substances that position the reactants, bringing together the parts of the molecules that must interact for the reaction to occur. Some reactions occur at perceptible rates only if a catalyst is present. The distinctive feature of a catalyst is that it is not used up or inactivated by the reaction it promotes. Just a small amount can catalyze many chemical reactions and convert large amounts of reactants into new products. A familiar example of a catalyst is the platinum in the catalytic converters of modern automobiles. It catalyzes a reaction between unburned gasoline and oxygen, thereby decreasing smog-causing emissions. In living things, the catalysts that promote chemical reactions are **enzymes**.

Enzymes. All enzymes are proteins except for a few recently discovered RNA molecules. Unlike nonenzyme catalysts (such as platinum) that can catalyze many different reactions, most enzymes are highly specific. They catalyze only one reaction, but they do it very well. A single molecule of the enzyme acetylcholine esterase, found in nerve tissue, can catalyze a million reactions in a second.

Enzymes catalyze specifically and rapidly because of pockets in their surface called **catalytic sites**. Reactants enter catalytic sites and are precisely positioned to react with one another (**Figure 2.7**). Because catalytic sites are tailored to specific reactions, nearly every chemical reaction that occurs in a cell requires a different enzyme. Moreover, most cellular reactions have such high activation energies that they would not occur in the absence of enzymes. The bacterium *Escherichia coli*, for example, requires about 1000 different enzymes to catalyze the approximately 1000 different reactions it needs to grow and reproduce.

Enzyme names always end in "ase" and suggest the reaction they catalyze. For example, the enzyme that liquefies agar is named agarase.

WATER

Water is essential for life. About 70 percent of the weight of microorganisms consists of water molecules. Moreover, most molecules in cells participate in the chemical reactions necessary for life only if they are in an aqueous

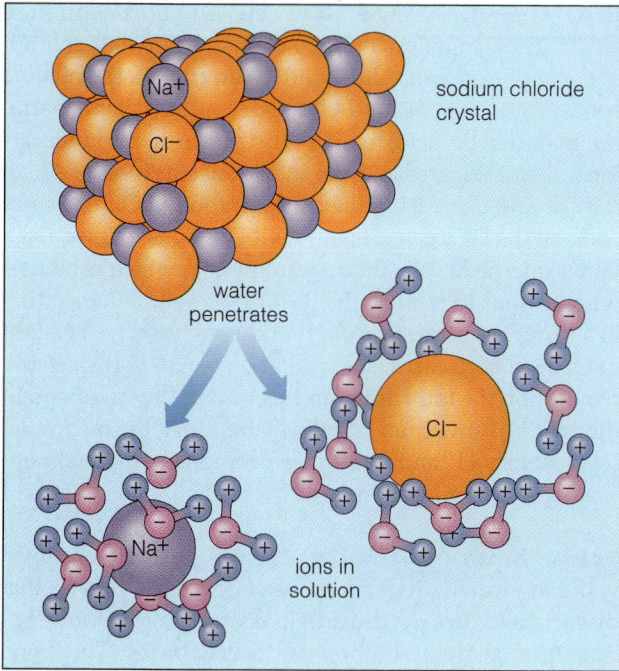

Figure 2.8 Water as a solvent. When water molecules penetrate a sodium chloride crystal, the sodium and chloride ions dissociate. Then water molecules surround the ions to form spheres of hydration that keep the ions in solution. The negative parts of water molecules are directed toward positively charged sodium ions, and the positive parts are directed toward negatively charged chloride ions.

(watery) environment. Water makes life possible for two reasons. First, water has the capacity to form hydrogen bonds with other water molecules and polar molecules. Second, water is a polar molecule. Let's look more closely at these characteristics.

Special Properties of Water

The capacity of water to form hydrogen bonds gives it properties usually associated with larger molecules. One of water's most important properties is its relatively high boiling point and relatively low freezing point. Water can exist as a liquid over a wide range of temperatures—from 0° to 100°C (32° to 212°F), the range that exists most places on earth.

Also, hydrogen bonding in water stabilizes temperature inside cells. It does this by increasing water's **specific heat** (the amount of heat necessary to raise one gram of material 1°C, measured in calories). Higher specific heat means water-filled cells can absorb considerable heat before temperature increases dangerously. Hydrogen bonds also increase water's **heat of vaporization** (the amount of heat necessary to convert a liquid to a gas). In other words, evaporation of water takes up enough heat to cool cells significantly.

Hydrogen bonds make water cohesive. Its cohesive property allows it to be drawn to the tops of trees.

Finally, hydrogen bonding between water molecules makes Earth more habitable for life. The extensive hydrogen bonding in ice gives it an open structure that makes it *less* dense than water (a highly unusual situation—the solid forms of most compounds are more dense than their liquid forms). As a result, ice floats; and ice on lakes and oceans is exposed to the sunlight and melts in the spring. Were ice denser than water, it would sink to the bottom where sunlight cannot strike it, probably keeping it frozen all year.

Water as a Solvent

Water's polarity makes it an excellent solvent. In a **solution**, substances exist together but are not joined by chemical bonds. The material in smaller concentration is the **solute**, and the material in larger concentration is the **solvent**. Inside cells, water is the solvent in which biochemicals are dissolved and undergo the chemical reactions that characterize life. To understand the chemical reactions in microbial cells, it is critical to know which kinds of molecules can dissolve in water and which cannot. Compounds that dissolve in water are called **hydrophilic** (water loving). Those that do not interact with water are called **hydrophobic** (water fearing).

Many salts dissolve in water. If you put sodium chloride (NaCl), table salt, for example, in water, individual Na^+ and Cl^- ions **dissociate** (separate) from one another and disperse. Water molecules cluster around each ion in a particular way. The negative poles of water molecules orient toward Na^+ ions, while the positive poles of water molecules orient toward Cl^- ions (**Figure 2.8**). In this way, water molecules form **spheres of hydration** around each ion, keeping it in solution.

Colloids. Not all materials that remain stably dispersed in water are dissolved solutes. Some are tiny solid particles, called **colloids**. They range in diameter from about 0.001 to 1 μm (micrometer). Ions attach to these tiny particles and attract water molecules, which form spheres of hydration, thus keeping the particles from aggregating into chunks. But unlike solutions, which are completely clear, most **colloidal dispersions** are turbid (cloudy). Furthermore, when they become concentrated or cooled, colloidal dispersions may form **gels**, open networks of interconnected colloidal particles. Although gels are semisolid and sometimes quite firm, they are mainly water. For example, gels made of the carbohydrate agar, which you read about in Chapter 1, are more than 98 percent water. Colloids are important because cells contain colloidal dispersions and gels. The shape of some parts of microbial cells—for example, the periplasm of bacteria (Chapter 4)—is maintained by gels. In addition, amoebae move by periodic formation and breakdown of gels (Chapter 12).

Hydrophobic and Hydrophilic Interactions. Nonpolar compounds, such as oils and gasoline, dissolve poorly in water because they have no charged regions to interact with water molecules. When nonpolar molecules are put in water, the hydrophobic regions clump together, further minimizing their interaction with water. For example, when we rinse out a pan used to make candy, the polar sugar dissolves but the nonpolar margarine floats as a blob to the top.

Hydrophobic and hydrophilic interactions are very important to the organization of living things. Cellular membranes, for example, are formed by polar molecules that are hydrophilic at one end and hydrophobic at the other. In water, these molecules position themselves so that their hydrophilic regions maintain contact with water molecules and their nonpolar hydrophobic regions are pointed away and shielded from the water. These interactions with water cause the molecules to aggregate spontaneously and form membranes. (The section on phospholipids later in the chapter further discusses this point.)

Hydrogen and Hydroxide Ions

Water's polarity makes it prone to **ionize**, dissociate into ions, according to the chemical equation:

$$H_2O \longleftrightarrow H^+ + OH^-$$

It dissociates into an equal number of positively charged **hydrogen ions** (H^+) and negatively charged **hydroxide ions** (OH^-). This is a reversible reaction, and when it reaches equilibrium, only a few ions of each type are present in a large amount of water. Nevertheless, hydrogen and hydroxide ions play crucial roles in the chemistry of life. *Almost every chemical reaction that occurs in living things involves the participation of one or the other of these two ions.* In addition, macromolecules, particularly proteins, exist in a stable state only if their environment has the proper concentration of hydrogen and hydroxide ions. Thus, a microbiologist studying the life processes of a microorganism must know the concentration of hydrogen and hydroxide ions in and around microbial cells to cultivate them and study their reactions.

The concentrations of H^+ and OH^- ions in any solution are inversely related to one another. That is, a high concentration of hydrogen ions means a low concentration of hydroxide ions, and vice versa. In pure water, the concentration of H^+ ions and OH^- ions is always equal because one of each is formed as water dissociates

($H_2O \longleftrightarrow H^+ + OH^-$). But cells do not contain pure water. They contain solutions of many different solutes, and the presence of other solutes can change the relative concentrations of H^+ and OH^-. For example, some bacteria make lactic acid, which dissociates to produce H^+ ions, thereby increasing the total concentration of H^+ in the cell and decreasing the concentration of OH^- because some of the H^+ ions combine with OH^- ions to form H_2O. Other microbial cells produce ammonia, which combines with H^+ ions to remove them from the solution, resulting in a lower concentration of H^+ and a higher concentration of OH^-. In both cases, survival and growth of the cell depend upon the ratio of H^+ and OH^- in its environment. In the artificial environment of the laboratory, the microbiologist adjusts H^+ and OH^- concentrations to maintain optimal growing conditions.

Acids, Bases, and Salts. Solutes that dissociate in water to form hydrogen ions are called **acids**. Solutes that dissociate to form hydroxide ions or react to remove hydrogen ions from solution are called **bases**. Therefore, solutions that contain more H^+ than OH^- are **acidic**, while solutions that contain more OH^- than H^+ are **basic** or **alkaline**.

The dissociation of any acid can be represented by the equation:

$$HA \longleftrightarrow H^+ + A^-$$

The A^- represents the anion that remains after separation from the hydrogen ion. Another way to view this reaction is to consider the acid a **proton donor** (since a hydrogen ion is a proton). Bases may either dissociate to form a hydroxide ion or accept a hydrogen ion. The dissociation of a base can be represented by the equation:

$$COH \longleftrightarrow C^+ + OH^-$$

The C^+ represents the cation that remains after separation from the base (COH). Bases are referred to as **proton acceptors** because they or the OH^- ions they release take up hydrogen ions. Salts are chemicals that dissociate into anions and cations, but not hydrogen or hydroxide ions.

The pH Scale. The **pH scale** describes how acidic or basic a solution is by assigning it a number. The scale is based on the concentration of hydrogen ions in a solution. However, because concentrations of H^+ and OH^- ions are inversely related to each other, the pH of a solution also indicates its concentration of hydroxide ions.

The pH scale numbers from 0 to 14. Zero describes highly acidic solutions, while 14 describes highly basic solutions (**Figure 2.9**). Because the possible concentration

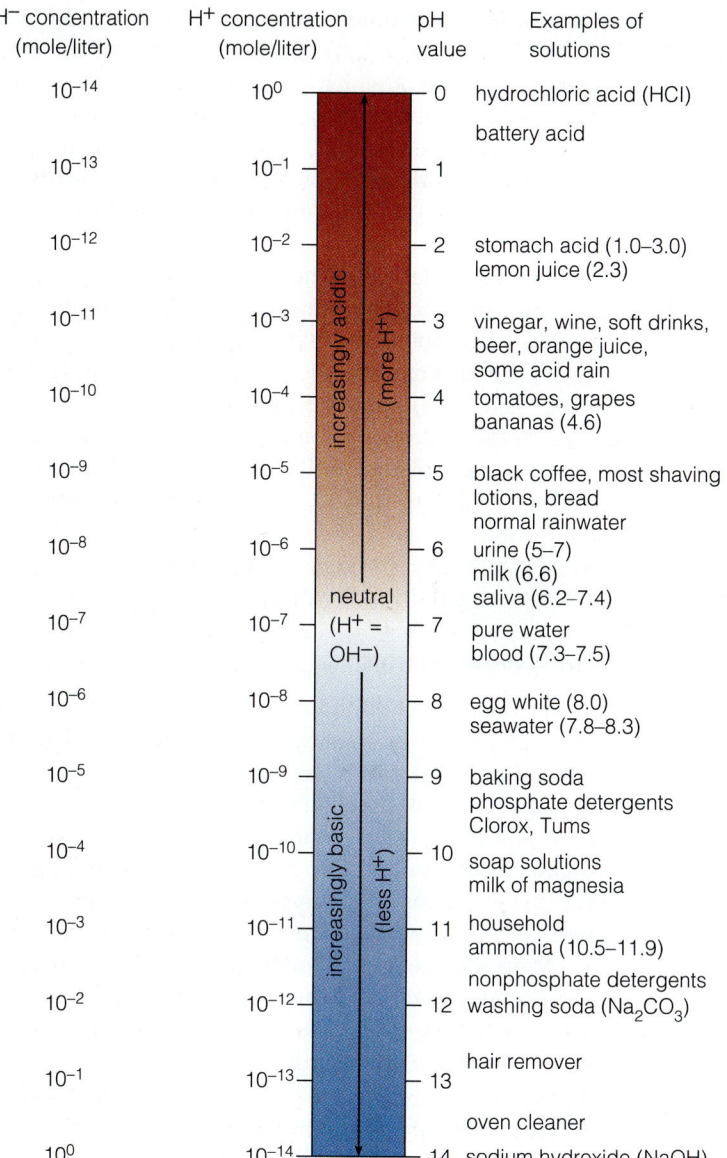

OH⁻ concentration (mole/liter)	H⁺ concentration (mole/liter)	pH value	Examples of solutions
10^{-14}	10^0	0	hydrochloric acid (HCl)
			battery acid
10^{-13}	10^{-1}	1	
10^{-12}	10^{-2}	2	stomach acid (1.0–3.0) lemon juice (2.3)
10^{-11}	10^{-3}	3	vinegar, wine, soft drinks, beer, orange juice, some acid rain
10^{-10}	10^{-4}	4	tomatoes, grapes bananas (4.6)
10^{-9}	10^{-5}	5	black coffee, most shaving lotions, bread normal rainwater
10^{-8}	10^{-6}	6	urine (5–7) milk (6.6) saliva (6.2–7.4)
10^{-7}	10^{-7}	7	pure water blood (7.3–7.5)
10^{-6}	10^{-8}	8	egg white (8.0) seawater (7.8–8.3)
10^{-5}	10^{-9}	9	baking soda phosphate detergents Clorox, Tums
10^{-4}	10^{-10}	10	soap solutions milk of magnesia
10^{-3}	10^{-11}	11	household ammonia (10.5–11.9)
10^{-2}	10^{-12}	12	nonphosphate detergents washing soda (Na_2CO_3)
10^{-1}	10^{-13}	13	hair remover
			oven cleaner
10^0	10^{-14}	14	sodium hydroxide (NaOH)

increasingly acidic (more H⁺)

neutral (H⁺ = OH⁻)

increasingly basic (less H⁺)

Figure 2.9 The pH scale. The product of the concentration of hydrogen (H⁺) and hydroxide (OH⁻) ions is always $10^{-14}M$, making hydrogen and hydroxide inversely related. A typical vinegar, for example, at pH 3.0 has a hydrogen ion concentration of $10^{-3}M$, which makes its hydroxide concentration 10^{-11} ($10^{-3} \times 10^{-11} = 10^{-14}$).

of hydrogen ions spans a vast range, the pH scale is not based on the numbers themselves, but on exponents. That is, it is **logarithmic**, with base 10. The value of pH is the value of the negative exponent of the concentration of H⁺ ions. For example, a solution that contains $10^{-3}M$ H⁺ has a pH value of 3. (M is the abbreviation for **molarity**, the moles per liter of an ion or molecule.) A change of one pH unit represents a tenfold change in hydrogen ion concentration. Thus vinegar, with a pH of 3, is ten times more acidic than tomato juice, with a pH of 4, and one hundred times more acidic than coffee, with a pH of 5 (10×10). Pure water, which is neutral because the concentration of hydrogen and hydroxide ions is equal, has a pH value of 7 (it has an H⁺ and OH⁻ concentration of $10^{-7}M$).

The key to understanding the pH scale is understanding the relationship between H⁺ and OH⁻ ions. Namely, (H⁺) × (OH⁻) = 10^{-14}. Thus, if the H⁺ = $10^{-3}M$, then OH⁻ = $10^{-11}M$ (3 + 11 = 14). The relationship is always inverse and the total value is always 14.

Buffers. Most living things can function only within a relatively limited range of pH. The optimal pH for a cell depends upon the proteins it contains. Different proteins are stable at and function best at different pH values. Human cells must be near pH 7.4 to function properly. The bacterium *Escherichia coli*, however, has proteins that require a pH of 7.6. Species of bacteria that live in acid environments like leaching gold mines have an intracellular pH as low as 6.0 because of their unusual proteins. Bacteria that live in highly basic environments, such as desert lakes, have an intracellular pH as high as 8.5.

For a cell to survive, intracellular pH must be maintained at a near-constant value, although many factors act continuously to change it. Hydrogen ions are continually being produced and consumed by the chemical reactions necessary to sustain life. In addition, changes in the external environment of microorganisms threaten the constancy of intracellular pH. Therefore, cells have a variety of mechanisms to stabilize their intracellular pH, one of which is a buffer system.

Buffers are weak acids or weak bases that stabilize or "buffer" the pH of a solution against change. They do this by absorbing hydrogen ions if their concentration rises and releasing them if their concentration falls. Although buffers do not maintain a completely constant pH, they moderate consequences of additional H^+ or OH^- ions.

Buffers help *E. coli* maintain a constant internal pH of 7.6, even if the pH of the immediate environment fluctuates as low as 5.5 or as high as 9. In blood and animal tissue, the weak acid and base, carbonic acid and bicarbonate, are important buffers. Carbonic acid dissociates to release a hydrogen ion if the pH rises, while bicarbonate accepts a hydrogen ion if pH falls.

In addition to their natural role in maintaining intracellular pH, buffers are important in the laboratory (Chapter 3). The chemical reactions that occur in a bacterial cell tend to change the pH of its external environment. In the artificial laboratory environment, buffers must be added to the medium to moderate these changes and promote cell growth. The microbiologist must choose the appropriate buffer for the microorganism being studied because different buffers maintain pH at different values.

ORGANIC MOLECULES

The term *organic* comes from the word *organism*—a living thing. Nineteenth-century chemists believed carbon-containing compounds could only be made biologically. Thus, **organic chemistry** was the study of the carbon-containing molecules that were found only in living things or their remains (inorganic chemistry studied inanimate matter). Today we know how to make organic (carbon-containing) compounds in the laboratory ("humanmade" as opposed to "natural"). We now use the term **biochemistry** to refer to the study of carbon-containing molecules made by organisms.

Carbon

A carbon atom can form four covalent bonds with other atoms (Figure 2.2). This gives it great bonding versatility.

Carbon atoms bond with other carbon atoms to form the molecular backbone structures of many different biochemicals, some of which are very complex. These structures may be linear, branched, or in the form of rings (**Figure 2.10**). The three-dimensional shape of these carbon backbones depends on the nature of the bonds that join the carbon atoms. Carbon atoms joined by single bonds can rotate around each other. Double or triple bonds fix the relative position of carbon atoms.

Some organic compounds, called **hydrocarbons**, consist of only carbon and hydrogen atoms. In these molecules each carbon atom is joined to one or more other carbons, and the remainder of its four covalent bonds connect to hydrogen atoms. Hydrocarbons are found mostly in petroleum and coal deposits (the fossilized remains of living things). Most organic compounds in living things contain combinations of other atoms—primarily nitrogen, oxygen, phosphorus, and sulfur—in addition to carbon and hydrogen.

Functional Groups

In an organic compound, atoms other than carbon and hydrogen are joined in patterns called **functional groups**. Every functional group has characteristic properties. For example, a polar functional group makes a compound more soluble in water. Functional groups determine if an organic compound is an acid or a base. In addition, the functional groups in a compound (there can be more than one) determine which other compounds it can react with.

Many different functional groups exist. Two of the most common are the **carboxyl group** (—COOH) and the **amino group** (—NH$_2$). Molecules that have a carboxyl group are acids, because they readily ionize to produce hydrogen ions. Molecules that have an amino group are bases, because they readily accept hydrogen ions.

Functional groups are the basis for classifying organic molecules into major groups that share important characteristics. Molecules that contain a carboxyl group belong to a class of organic compounds called **organic acids**. Molecules that contain an amino group belong to a class called **amines**. Examples of organic compounds and the functional groups they contain are presented in **Figure 2.11**.

MACROMOLECULES

Some organic molecules are small, containing only a few carbon atoms. Others are enormous, containing thousands or even millions of atoms. These extremely large molecules, called **macromolecules**, play a central role in biology. They determine most of a cell's structure and functions. The four main classes of macromolecules

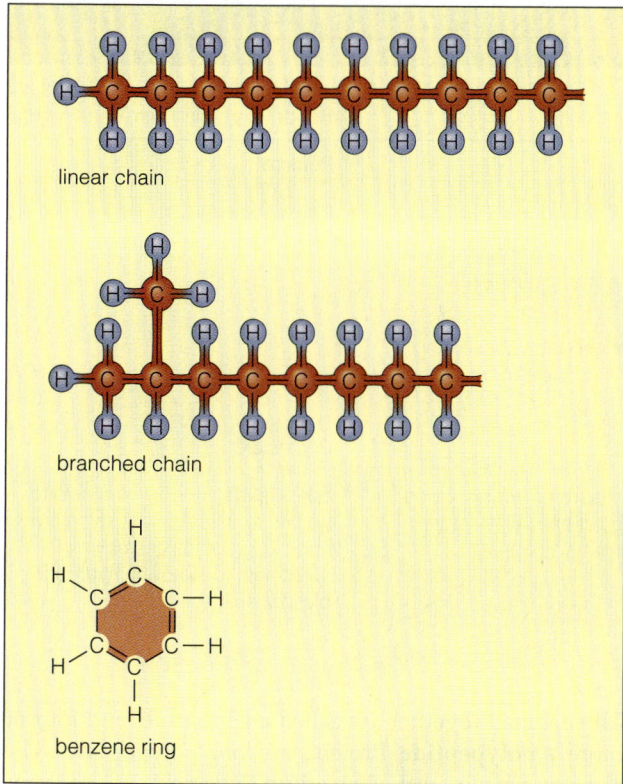

Figure 2.10 Possible arrangements of carbon atoms in organic molecules. Carbons can be linear chains, branched chains, or rings. This single ring is a benzene ring.

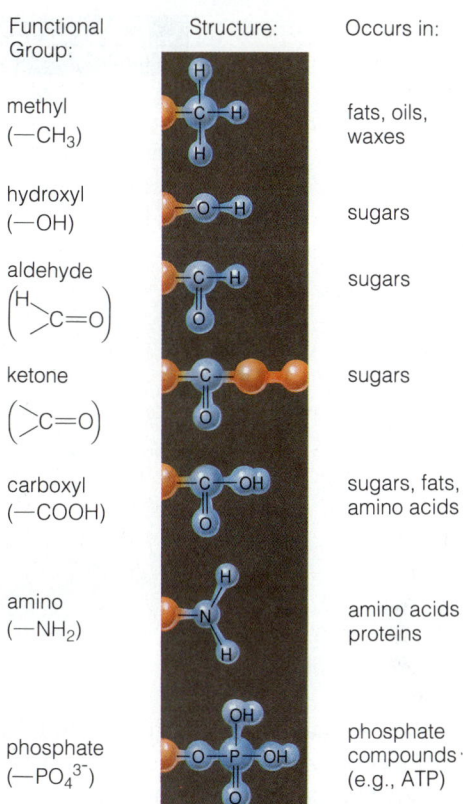

Functional Group:	Structure:	Occurs in:
methyl (—CH$_3$)		fats, oils, waxes
hydroxyl (—OH)		sugars
aldehyde		sugars
ketone		sugars
carboxyl (—COOH)		sugars, fats, amino acids
amino (—NH$_2$)		amino acids proteins
phosphate (—PO$_4^{3-}$)		phosphate compounds (e.g., ATP)

Figure 2.11 This is a sampling of the great number of functional groups that occur in biologically important molecules.

found in cells are proteins, nucleic acids, polysaccharides, and lipids. Polysaccharides are members of the larger class of compounds called carbohydrates.

Most macromolecules are **polymers** (many parts), built from smaller modular units called **monomers** (one part). Thus, monomers are the building blocks from which macromolecular polymers are made. The combination of monomers into polymers is termed **polymerization**. With the exception of lipids, macromolecules are distinguished by the building blocks from which they are made and the kinds of bonds that unite them (**Table 2.6**). Lipids are not defined by chemical structure. Instead, lipids are any biochemical molecule that dissolves in a nonpolar solvent.

Although fundamentally different reactions form polymeric macromolecules, the bonds between monomers all look as though they might have been formed the same way—by taking a hydrogen ion from one monomer and a hydroxide ion from a neighbor, a net loss of one molecule of water. For this reason, before the actual chemical reactions forming macromolecules were known, the process was called **dehydration synthesis**. As you will see in Chapter 5, monomers are not themselves polymerized into macromolecules. Instead, chemical

groups are added to the monomers before polymerization. These added groups are eliminated during polymerization.

The reverse of dehydration is **hydrolysis** (the addition of water). In hydrolysis, an organism breaks down macromolecules into monomers to obtain nutrients or to build new cell structures. As the covalent bond is broken, one part of the molecule combines with a hydrogen ion, while the other combines with a hydroxide ion. Thus, the outcome of hydrolysis is the splitting of a macromolecule and a water molecule as a hydrogen ion and a hydroxide ion are used.

Proteins

Next to water, **proteins** are the most abundant component of cells. Proteins make up over half the dry weight of microorganisms. They perform many essential biological functions. Most of a cell's proteins are enzymes, which catalyze the thousand or more different chemical reactions essential to life. Some proteins play structural roles, holding parts of the cell together or building cellular organelles. Other proteins carry molecules in and out of the

Table 2.6   The Major Macromolecules

Macromolecule (Polymer)	Building Block (Monomer)	Bonds That Join Them
Proteins	Amino acids	Peptide
Nucleic Acids		Phosphodiester
DNA	Nucleotides (a phosphate, deoxyribose, and a base—adenine, guanine, thymine, or cytosine)	
RNA	Nucleotides (a phosphate, ribose, and a base—adenine, guanine, uracil, or cytosine)	
Polysaccharides	Monosaccharides	Glycosidic
Lipids	Unlike the other macromolecules, lipids are not defined by chemical structure. Lipids are any organic nonpolar molecule.	Some lipids are polymers held together by ester bonds; some are huge aggregates of small molecules held together by hydrophobic interactions.

cell across the cell membrane. Still others fulfill highly specialized functions such as cell movement. For example, the hairlike flagella that move bacteria are composed entirely of protein.

Amino Acids. The 20 **amino acids** are the building blocks of proteins (**Figure 2.12**). Amino acids all share the same basic structure—a carbon atom attached to four different groups: a carboxyl group (—COOH), an amino group (—NH₂), a hydrogen atom (H), and an R group. (*R* stands for the rest of the molecule, whatever it may be.) The R group is different in each of the 20 naturally occurring amino acids. The R group may be polar or nonpolar, acidic, basic, or neutral, ionized or not. All amino acids contain carbon, hydrogen, oxygen and nitrogen, and two also contain sulfur.

The four different atoms or functional groups can attach to a carbon atom in two ways. These alternative arrangements produce two different forms of each amino acid, called the L-**isomer** and the D-**isomer**. L- and D-isomers are mirror images of each other (**Figure 2.13**). Only the L-isomers of amino acids are found in proteins. The D-isomers are rare in nature, but some occur in peptidoglycan, the macromolecule from which eubacterial walls are made, and in a few antibiotics (see the box "Peptidoglycan: Molecular Boilerplate").

Peptide Bonds. The bonds that join amino acids in proteins are called **peptide bonds**. They are formed between the amino group of one amino acid and the carboxyl group of another (**Figure 2.14**). Two amino acids joined together by a peptide bond are called a **dipeptide**.

Three joined together are called a **tripeptide**, and four or more, a **polypeptide**. Proteins are long polypeptides. The average protein in the bacterium *Escherichia coli* contains 270 amino acids.

Protein Structure. A protein's **primary structure** is determined by its specific amino acids and the order in which they are joined. Since most proteins are strings of hundreds of amino acids, and each position can be filled by 1 of 20 different amino acids, an almost infinite number of different primary structures is possible.

But proteins do not exist as an extended chain of amino acids. Rather, they are folded into precise three-dimensional shapes. The amino acids in the primary structure interact to create a particular protein's secondary, tertiary, and quaternary structures (**Figure 2.15**).

A protein's **secondary structure** is determined by the hydrogen bonds that form at intervals between nonadjacent amino acids. The formation of hydrogen bonds is somewhat limited because some atoms are rigidly bound in the peptide bond. Nevertheless, enough flexibility exists so that certain proteins twist around a central axis, forming an **alpha helix**. An alpha helix is held together by hydrogen bonds between every fourth amino acid. Other stretches of amino acids assume the conformation of a **beta sheet** (also called a **pleated sheet**). A beta sheet is held together by hydrogen bonds between adjacent peptide chains. Some sequences of amino acids tend to form alpha helices, while others are more likely to form beta sheets. **Hemoglobin**, the oxygen-carrying protein in blood, is composed of alpha helices. Silk proteins are composed of beta sheets. Some proteins—for example,

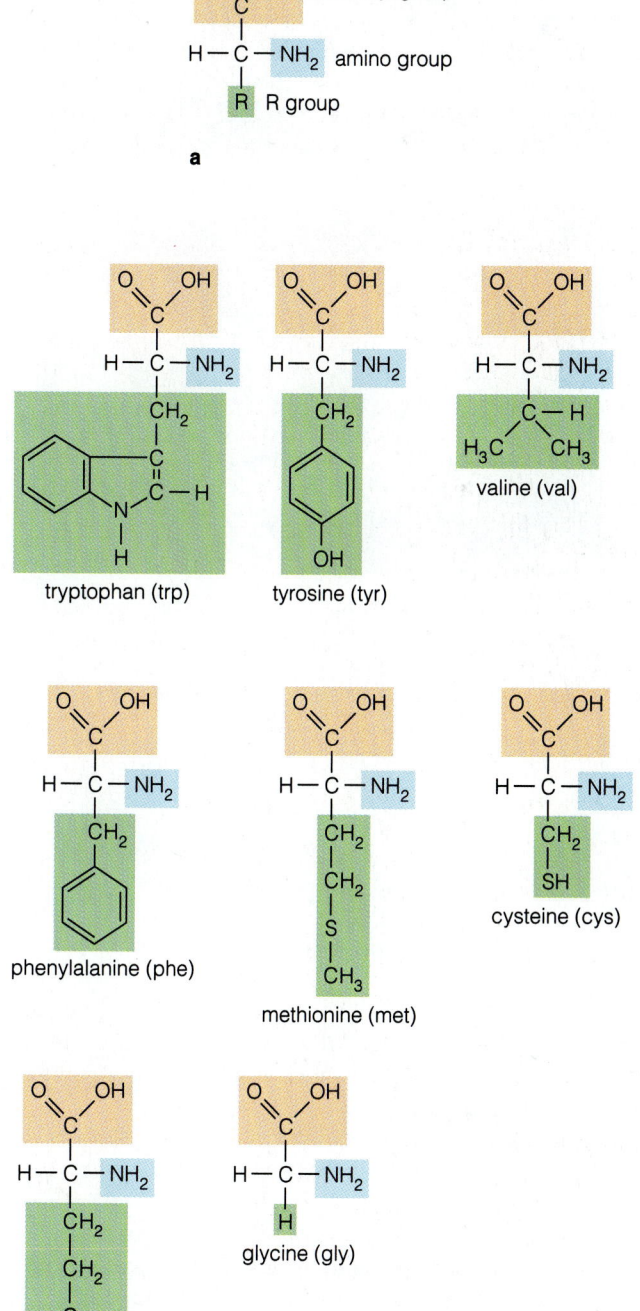

a

tryptophan (trp)

tyrosine (tyr)

valine (val)

phenylalanine (phe)

methionine (met)

cysteine (cys)

glutamate (glu)

glycine (gly)

b

Figure 2.12 Amino acids. (**a**) All 20 naturally occurring amino acids have four components: an amino group, a carboxyl group, a hydrogen atom attached to a carbon atom, and a distinctive fourth groups (R). (**b**) The structures of eight amino acids are shown here, with their distinctive R groups highlighted. Some R groups are carboxyl groups (glutamate). Some are benzene rings (phenylalanine, tyrosine, and tryptophan).

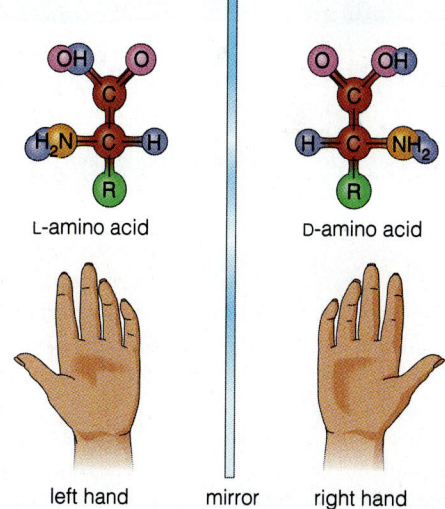

L-amino acid   D-amino acid

left hand   mirror   right hand

Figure 2.13 D- and L-isomers. The D- and L-isomers of an amino acid are mirror images of each other, differing as our left and right hands do. Amino acids have isomers because four different atoms or groups can be attached to a carbon atom in two different ways in three dimensions.

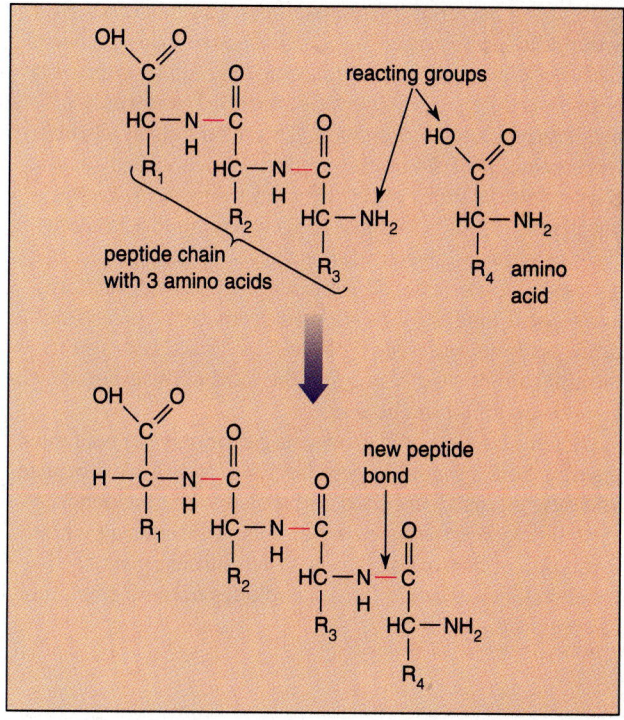

Figure 2.14 Peptide bonds. Peptide bonds form between the amino group of one amino acid and the carboxyl group of the next. The equivalent of one molecule of water is released for each peptide bond formed. This figure shows one step in the synthesis of a polypeptide chain—the addition of an amino acid to a peptide chain with three amino acids. Peptide bonds are shown in red.

Peptidoglycan: Molecular Boilerplate

A boilerplate is the steel exterior of a boiler that gives it shape and enough strength to withstand tremendous internal pressure. Eubacteria have their own boilerplate—an unusual and extremely strong protective covering found nowhere else in nature. It is a macromolecule called **peptidoglycan**. Other macromolecules are long, linear or branched chains of monomers, but peptidoglycan is an edgeless sheet forming a hollow structure that confers shape on a bacterial cell.

In many bacterial species, the peptidoglycan is one layer thick, but in others, it is several layers thick. Peptidoglycan gets its strength from being a single molecule with all its atoms linked by covalent bonds. The eubacterial wall has to be very strong to contain the pressure of the cytoplasm pushing out against it. Some bacterial cells exert more than 350 pounds of pressure per square inch (from internal osmotic pressure). This is 20 times greater than the pressure in a household pressure cooker.

As its name suggests, peptidoglycan has both protein and polysaccharide components. ("Peptido" implies peptide, a short protein, and "glycan" is a polysaccharide.) In a rod-shaped bacterial cell, for example, glycan strands run around the short axis of the cell, like hoops around a barrel. Each hooplike glycan strand is joined at regular intervals to both of its neighbors by peptide strands. The result is a mesh resembling chicken wire. Glycan strands in all eubacteria are made of alternating units of two modified sugars, *N*-acetylglucosamine and *N*-acetylmuramic acid. The linking peptide strands that join *N*-acetylmuramic acid molecules in adjacent glycan strands vary with different bacterial species. In *Escherichia coli*, for example, tetrapeptide (four-amino-acid-long) strands stick out from each *N*-acetylmuramic acid molecule. Most of these are linked between the terminal amino acid of one and the third of the other, which is meso-diaminopimelic acid.

When a bacterial cell grows, its peptidoglycan wall must enlarge. Bonds that hold peptidoglycan together are broken, and new pieces are inserted and sealed with new bonds. Penicillin kills growing bacterial cells by interfering with the formation of the sealing bonds. When this happens, the bacterial wall is weak, and the cell explodes from its great internal pressure.

1 μm

a Micrograph of a purified bacterial wall composed of a single molecule of peptidoglycan.

b The chemical structure of peptidoglycan. This is a schematic of the intact peptidoglycan wall of *Escherichia coli*. The G represents *N*-acetylglucosamine acid and the M represents *N*-acetylmuramic acid. They are joined by glycosidic bonds. The ⊥ symbol represents cross-linked tetrapeptide side chains.

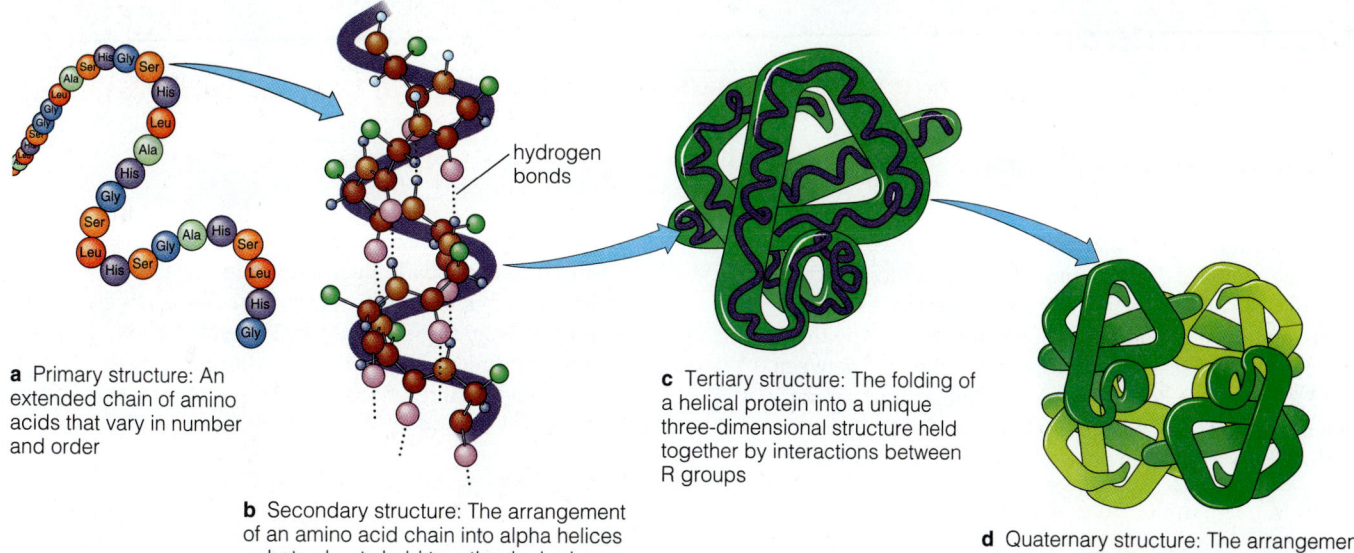

a Primary structure: An extended chain of amino acids that vary in number and order

b Secondary structure: The arrangement of an amino acid chain into alpha helices or beta sheets held together by hydrogen bonds

hydrogen bonds

c Tertiary structure: The folding of a helical protein into a unique three-dimensional structure held together by interactions between R groups

d Quaternary structure: The arrangement of polypeptide chains in a protein that has more than one chain

Figure 2.15 Protein structure. (**a**) The sequence of amino acids in a protein is its primary structure. (**b**) The secondary structure is based on hydrogen bond formation. Some sequences of amino acids form hydrogen bonds that link the strand into an alpha helix; other sequences form beta sheets. All helices and beta sheets have the same three-dimensional structure. (**c**) The tertiary structure gives a protein its *unique* three-dimensional character. The long strands of protein are folded back on themselves and held together by ionic bonds, disulfide bonds, and hydrophobic interactions. (**d**) Proteins with more than one polypeptide chain have a quaternary structure.

the hormone insulin—have regions of alpha helices and regions of beta sheets.

A protein's **tertiary structure** is determined by interactions among the R groups of its various amino acids. For example, hydrophobic R groups tend to lie next to one another to minimize their interaction with water. Oppositely charged regions of a protein are held together by ionic bonds. Some proteins that contain more than one molecule of the amino acid cysteine (for example, insulin) are held together by covalent disulfide bonds (—S—S—). The R group of each cysteine unit ends in a sulfhydryl group (—SH), which reacts to form disulfide bonds. The chemical interactions that determine tertiary structure cause protein chains to fold back on themselves, creating a uniquely shaped molecule.

Some complex proteins consist of more than one polypeptide chain. They have a **quaternary structure**, determined by the way in which their different chains fit together. The bacterial protein **RNA polymerase**, which catalyzes the formation of all of a bacterium's RNA, consists of four polypeptide chains (RNA is explained later in this chapter).

Denaturation. Secondary, tertiary, and quaternary protein structures depend upon many relatively weak molecular interactions, such as hydrogen bonds, hydrophobic bonds, and ionic attractions. These interactions are readily disrupted by heat, unfavorable pH, or high concentrations of salts. Destruction of a protein's three-dimensional structure is called **denaturation**. Disrupting a primary structure is not denaturation.

Denaturation almost always destroys a protein's biological activity. Thus, cells can function only within the range of temperature, pH, and osmotic strength that prevents denaturation. Bacteria that grow at temperatures near the boiling point of water can do so only because they have proteins that are highly resistant to denaturation by heat. Most microorganisms have proteins that are much less stable and are readily killed at temperatures as low as 55°C. Once proteins are denatured they usually cannot be returned to their original state. A familiar example of protein denaturation is cooked egg white. Heat irreversibly changes the normally clear and viscous protein albumin into a solid white mass of denatured protein.

A Tantalizing Question

Louis Pasteur observed that organisms produce only D or L forms (not both) of organic compounds. Chemical reactions that do not involve enzymes produce a **racemic mixture**, equal amounts of D and L forms. We now know why this is true. The flat surface of enzymes can discriminate between D and L forms, but ordinary chemical interactions cannot. Scientists at the National Aeronautics and Space Agency (NASA) used this fact to answer a tantalizing question.

Since amino acids are components of proteins, and proteins are produced only by living things, does the presence of amino acids in some meteors prove that life exists elsewhere in our universe? No, says NASA. The amino acids in meteors are a racemic mixture; therefore, they are formed by ordinary chemical reactions, not living things—as we know them.

You will read more about denaturation in Chapter 9, where we cover sterilization (killing all the microorganisms on an object). One of the most effective sterilization methods is heating to denature a microorganism's vital proteins.

Nucleic Acids

Nucleic acids are the third most abundant component (after water and protein) of microbial cells. There are two types of nucleic acid molecules, **deoxyribonucleic acid** (DNA) and **ribonucleic acid** (RNA). Although DNA makes up only a small percentage of the dry weight of a typical microorganism, it is of central importance, as it is in all living things. DNA is the molecule that encodes the cell's genetic information. In reproduction, it is DNA that ensures progeny are of the same species as parents. Closer resemblances, such as facial features in humans, are also encoded in DNA.

RNA plays a critical role in building proteins. It interprets the information for protein construction encoded in DNA. The organelles on which proteins are manufactured, **ribosomes**, are composed partly of RNA. RNA is much more abundant than DNA, particularly in microorganisms. It can constitute as much as a third of the dry weight of a fast-growing bacterial cell.

Nucleotides. Nucleic acids are composed of monomers called **nucleotides**. Nucleotides contain a phosphate group and a **nucleoside**, composed of a five-carbon sugar and a nitrogen-containing base (**Figure 2.16**). The sugar may be either **deoxyribose** (found in DNA) or **ribose** (found in RNA). Bases may contain either a **pyrimidine** molecule (single-ringed) or a **purine** molecule (double-ringed). The pyrimidines are **cytosine** (found in DNA and RNA), **thymine** (found only in DNA), and **uracil** (found only in RNA). The purines found in both DNA and RNA are **adenine** and **guanine**. The three components of a nucleotide are joined by covalent bonds. Nucleotides are joined in nucleic acids by **phosphodiester bonds**.

In addition to their role as the building blocks of nucleic acids, nucleotides fulfill other essential functions in cells. For example, the nucleotide **adenosine triphosphate (ATP)** (**Figure 2.17**) is the major carrier of cellular energy. It is indispensable in many chemical reactions that are essential to life, as you will read in Chapter 5. ATP contains the nucleoside adenosine, a molecule composed of adenine and ribose, and three phosphate groups. Another adenosine phosphate molecule, **cyclic AMP (cAMP)**, acts as a chemical messenger in cells. It carries different messages in different cells. In bacteria, for example, it signals the cell to shift its activities when certain nutrients are depleted. Two nucleotides function as coenzymes, nonprotein facilitators in chemical reactions. **Nicotinamide adenine dinucleotide (NAD)** and **flavine adenine dinucleotide (FAD)** facilitate biochemical reactions by transporting hydrogen atoms between molecules. You will read more about NAD and FAD when you study metabolism in Chapter 5.

DNA. DNA is built from long covalently bonded polymers of nucleotides. Sugar-phosphate groups form the molecular backbone, and bases (cytosine, thymine, adenine, and guanine) are attached to this backbone. Each molecule of DNA contains two of these strands wrapped around each other in the form of a **double helix** (**Figure 2.18**). The two strands of the DNA molecule are attached to each other by hydrogen bonds that form between pairs of bases. Adenine always bonds to thymine, and guanine always bonds to cytosine. The sugar-phosphate backbone is the same in all molecules of DNA, but the sequence of bases is unique to each molecule. Some molecules of

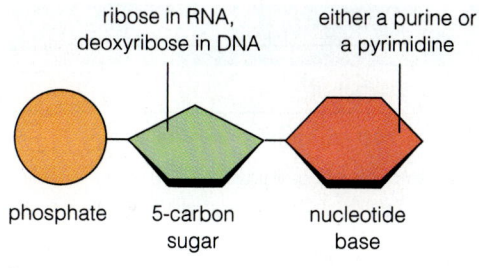

ribose in RNA, deoxyribose in DNA

either a purine or a pyrimidine

phosphate   5-carbon sugar   nucleotide base

a

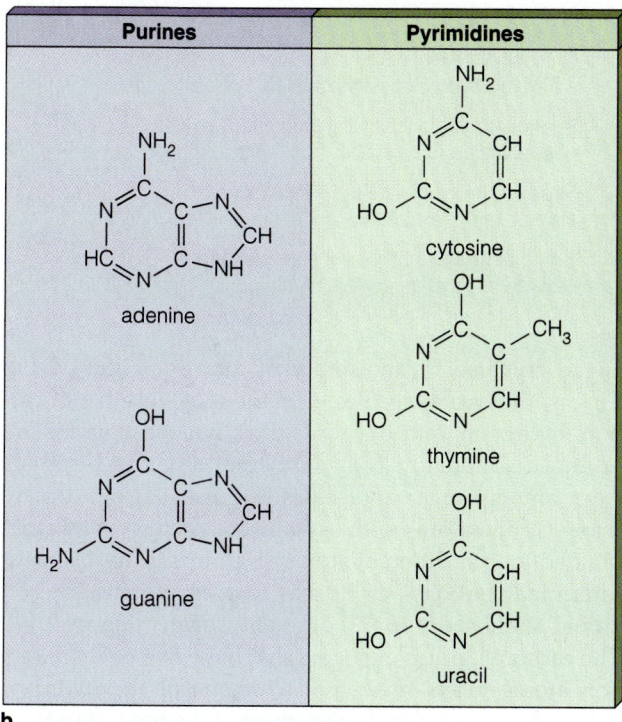

Purines	Pyrimidines

adenine

guanine

cytosine

thymine

uracil

b

Figure 2.16 Nucleotides. Nucleotides are the building blocks of nucleic acids. (**a**) They consist of a phosphate group, a sugar, and a nucleotide base. The sugar in RNA is ribose and in DNA, deoxyribose. (**b**) The nucleotide bases found in RNA are adenine, guanine, cytosine, and uracil. The bases in DNA are adenine, guanine, cytosine, and thymine.

DNA have more adenine-thymine base pairs than guanine-cytosine base pairs, while in other molecules the reverse is true.

The sequence of base pairs in a molecule of DNA encodes its genetic information. The discovery of the structure of DNA in 1953 earned the Nobel Prize for biologists James Watson and Francis Crick. Once the structure of DNA was known, it was possible to decipher the genetic code, which in turn led to enormous advances in microbiology. The way in which genetic information is encoded and preserved in DNA is discussed in detail in Chapter 6.

RNA. RNA is also a polymer of nucleotides, but they differ from those in DNA (Figure 2.16). All nucleotides in

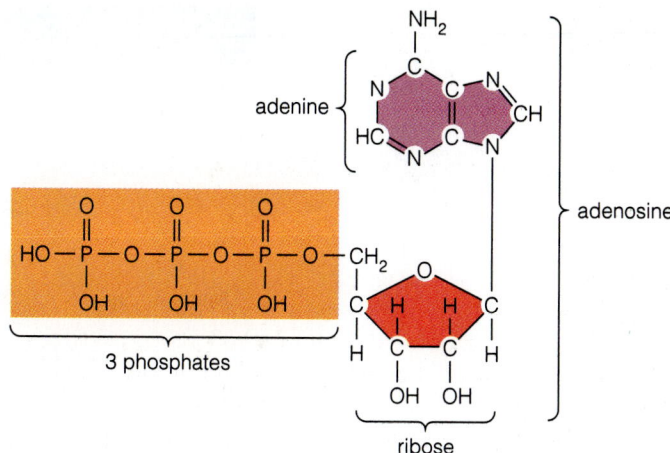

adenine

adenosine

3 phosphates

ribose

Figure 2.17 The structure of adenosine triphosphate (ATP). ATP is an extremely important nucleotide. It is made of adenine, ribose, and three phosphate groups. It is the major carrier of cellular energy. Adenine linked to ribose is called adenosine.

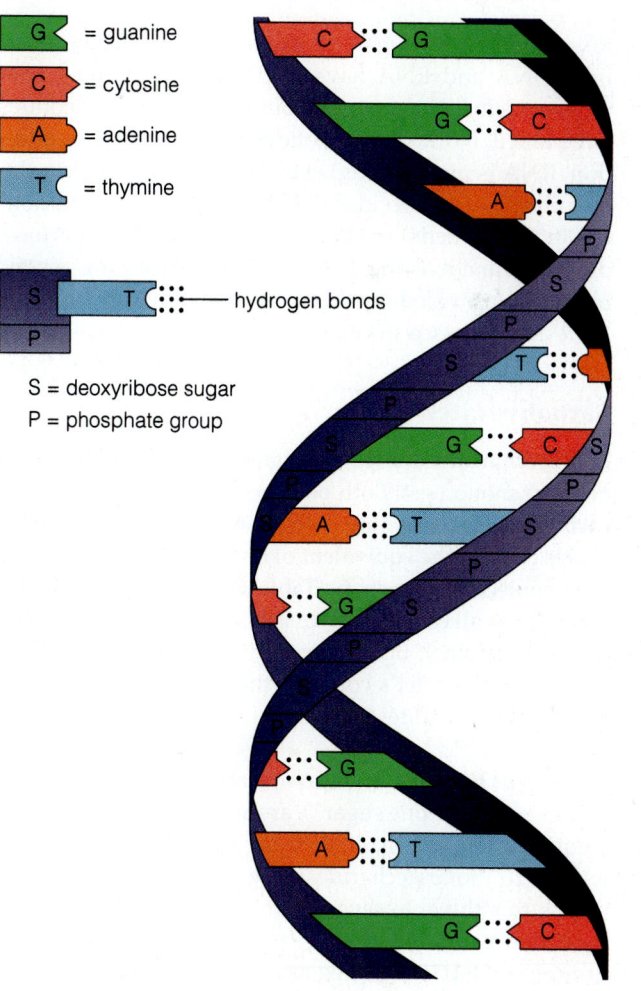

G ◁ = guanine
C ◁ = cytosine
A ◁ = adenine
T ◁ = thymine

S — T ⁞⁞⁞ — hydrogen bonds
P

S = deoxyribose sugar
P = phosphate group

Figure 2.18 The DNA double helix. DNA forms a double helix composed of two helical (spiral) strands of deoxyribose and phosphate joined by hydrogen bonds between pairs of bases. Guanine bonds with cytosine, and adenine with thymine.

Table 2.7   Classes of Carbohydrates

Class	Examples	Composition and Distribution
Monosaccharides	Glucose	A hexose; found in most organisms
	Fructose	A hexose; large amounts occur in fruit
	Ribose	A pentose; constituent of RNA
	Deoxyribose	A pentose; constituent of DNA
Disaccharides	Sucrose	Glucose-fructose; table sugar
	Lactose	Glucose-galactose; milk sugar
	Trehalose	Glucose-glucose; occurs in most organisms; protects membranes from damage caused by desiccation
Polysaccharides	Cellulose	Straight-chain polymer of alpha glucose; occurs mainly in cell walls of plants and algae
	Starch	Straight-chain polymer of beta glucose; storage material in plants and some microorganisms
	Glycogen	Branched-chain polymer of beta glucose; storage material in animals

RNA contain the sugar ribose rather than deoxyribose. Both DNA and RNA have nucleotides containing adenine, cytosine, and guanine, but some RNA nucleotides have uracil. No RNA nucleotides have thymine. In addition, RNA is usually **single stranded**—it has single polymer chains of nucleotides (DNA in cells is always a double-stranded helix). Different kinds of RNA perform different functions in building proteins from the genetic blueprint provided by DNA. These different types of RNA are discussed in Chapter 6.

Carbohydrates

Polysaccharides (meaning "many sugars") are polymers of sugar monomers. Both polysaccharides and sugars are **carbohydrates** (meaning "carbon with water"). Carbohydrates contain the equivalent of one molecule of water for each molecule of carbon (**Table 2.7**). Thus, the general formula for all carbohydrates is $(CH_2O)_n$ (n is the number of carbon atoms). Before discussing the macromolecular polysaccharides, let's consider their building blocks, the simple carbohydrates (sugars).

Monosaccharides.   The simplest sugars, **monosaccharides** (meaning "one sugar"), are small organic molecules composed of carbon, hydrogen, and oxygen. Approximately 20 monosaccharides occur in nature. Any molecule with a three- to seven-atom carbon backbone and at least one **carbonyl group** (=O) and several **hydroxyl groups** (—OH) is classified as a sugar.

Names of individual sugars and classes of sugars have the suffix *ose*. Five-carbon sugars are called **pentoses**. A common example is ribose. Six-carbon sugars are called **hexoses**. **Glucose** is a hexose. Another six-carbon

sugar, **fructose**, is an **isomer** of glucose (**Figure 2.19**). That is, the two molecules have the same number of carbon, hydrogen, and oxygen atoms, but those atoms are arranged differently. Many different structural isomers exist among sugar molecules because sugars contain many carbon atoms with asymmetric centers—one atom bonded to four different atoms or groups. A hexose with a terminal carbonyl group has four asymmetric carbon atoms, so there are 16 (2^4) different isomers (Figure 2.19). The carbonyl group of sugars that have five or more carbon atoms reacts reversibly with one of the hydroxyl groups to form a ring. In fact, most of the time, carbonyls exist as a ring and their chemical structure is usually shown as a ring. When the ring forms, the carbonyl carbon atom becomes asymmetric, and so two different ring forms are possible. They are designated **alpha (α)** and **beta (β)**. A solution of glucose and most other sugars contains all three forms of monomers—straight chain, alpha ring, and beta ring (**Figure 2.20**).

Monosaccharides are important to microorganisms as sources of energy and nutrition. Enzymes split monosaccharides into simpler forms, releasing energy that is used to fuel other cellular reactions and form the molecules that are eventually built into macromolecules. Almost all microorganisms possess the enzymes necessary to use glucose, so it is frequently used as a nutrient in laboratory cultures.

Disaccharides.   Two monosaccharides joined by **glycosidic bonds** form a **disaccharide** sugar. Disaccharides are plentiful in nature and, like monosaccharides, serve as sources of energy for metabolism. One disaccharide is the common table sugar **sucrose**, which is composed of a molecule of glucose bonded to a molecule of fructose.

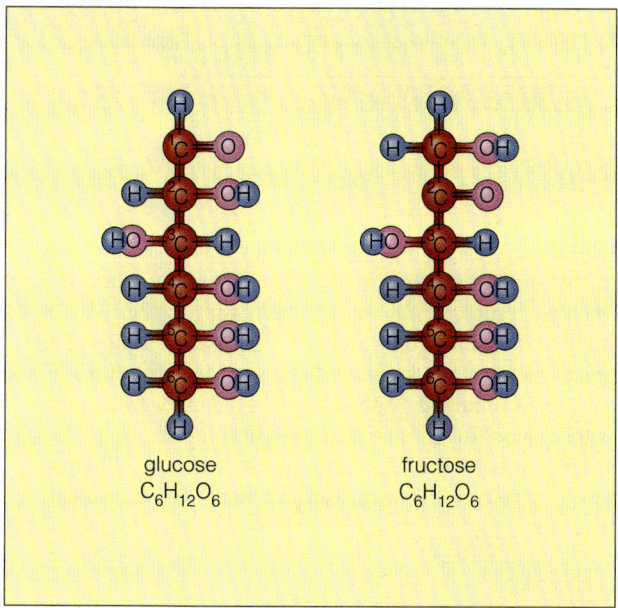

glucose
$C_6H_{12}O_6$

fructose
$C_6H_{12}O_6$

Figure 2.19 Isomers. Glucose and fructose are isomers because they have the same chemical composition, but the atoms are arranged differently. Many different isomers exist among sugars, which is why there are so many sugars. Glucose is an example of a hexose with a terminal carbonyl group (on carbon 1); it has four asymmetric carbon atoms (carbons 2, 3, 4, and 5), so there are 16 (2^4) different isomers. If it weren't for isomers, there would be one pentose, one hexose, and so forth.

Another disaccharide is **lactose**, or milk sugar, composed of one molecule of glucose and one molecule of the monosaccharide **galactose**. Microorganisms that make their home in the human intestine use lactose as a source of energy because this is one of the few natural environments where it is plentiful.

Polysaccharides. Monosaccharides are the building blocks of polysaccharides, which can be extremely complex. They can be linear or branched chains. They can be built from one or several different monosaccharide monomers. All polysaccharides have glycosidic bonds. Polysaccharides serve important structural functions in plant and bacterial cells. They also serve as storage forms of simple sugars that can be broken down for nutrients as need arises. Glucose alone is built into three different polysaccharides—cellulose, starch, and glycogen. These are distinguished by having linear or branched chains and glucose rings in either the alpha or beta form (**Figure 2.21**).

Cellulose is the most abundant polysaccharide in nature because it is the principal component of plant cell walls. It is also found in most algae and a few bacteria. Cellulose is composed of glucose rings in the beta form that are covalently linked in long linear chains. Cellulose

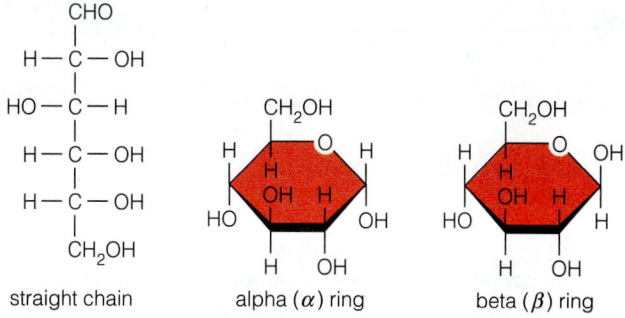

straight chain

alpha (α) ring

beta (β) ring

Figure 2.20 The structure of glucose. Three forms of glucose occur in solution—a straight chain and alpha and beta rings. Rings form when the carbonyl group (=O) reacts with a hydroxyl group (—OH).

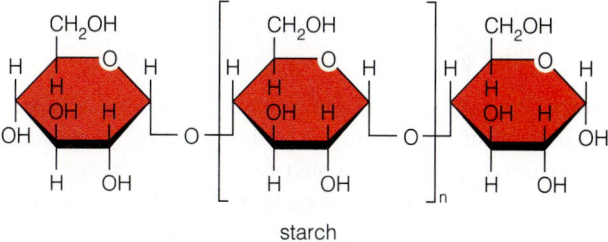

starch

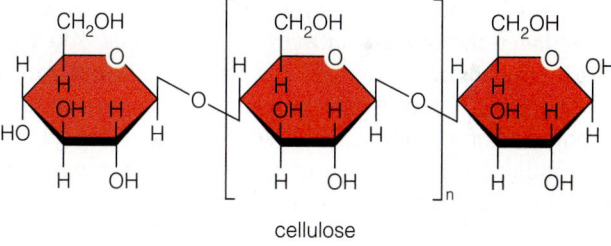

cellulose

Figure 2.21 Biologically important polysaccharides composed of glucose monomers joined by glycosidic bonds. Starch is a long chain of glucose monomers in the alpha form. Cellulose is a long chain of beta glucose monomers.

is a strong material that resists breakdown by most microorganisms. Wood is largely cellulose, and cotton fibers are almost pure cellulose.

Structurally, **starch** differs from cellulose only in the form of its glucose monomers. The glucose rings in starch are in the alpha form instead of the beta form. The properties and biological roles of starch and cellulose are very different. Starch is the reserve material that plants and some protozoa make to store excess nutrients for later use.

Glycogen, like starch, is a polymer of glucose rings in the alpha form, but its polymer strands are multiply

Bread in the Desert

During World War II, American troops in North Africa found a curious-looking powder among captured German supplies. The powder was dried yeast cells, but they were still alive and could make bread dough rise. Researchers in the American food industry had tried for years to produce active dried yeast without success. Under the gentlest drying conditions, the yeast cells died. Even with the German samples in hand, Americans couldn't duplicate the process. When the war ended, so did the mystery. The yeast had to be grown under conditions that caused cells to accumulate the disaccharide **trehalose**. Then they remained alive and active when dried and could be stored for long periods. Trehalose protects cells during desiccation (drying) by fitting precisely into molecular crevices on the surface of biological membranes, keeping these sensitive structures intact as water is removed. Trehalose plays a similar role in the water bear, a tiny animal that lives in soil and ponds. The water bear contains large amounts of trehalose, allowing it to survive complete desiccation when its environment dries up.

branched. Glycogen is the animal equivalent of starch. That is, animals, including human beings, store excess nutrients as glycogen, which can provide energy when needed. Algae, fungi, and certain bacteria, including *Escherichia coli*, also make glycogen and accumulate it as intracellular storage granules.

Finally, polysaccharides are a major structural component of bacterial cells. They form the protective outermost layer, the **capsule**, of many bacterial cells. You will read more about the capsule and its medical importance in Chapter 4. The cell walls of some fungi contain **chitin**, a polymer of a modified glucose molecule, *N*-acetylglucoseamine. The chitin in fungi is the same tough material found in the shells of crabs and lobsters.

Lipids

Unlike the other types of macromolecules, **lipids** are not classified by their building blocks. This is because some lipids are true polymeric macromolecules (for example, the poly-beta-hydroxyalkanes that some bacteria make to store excess nutrients), but not all. Some lipids are simply huge aggregates of small molecules held together by hydrophobic interactions (for example, the lipid components of most cell membranes). Therefore, lipids are defined as any nonpolar molecule. As such they are soluble in nonpolar solvents, such as gasoline and kerosene, and insoluble in water.

In microbial cells, lipids are most important as components of membranes. They also play other roles, including serving as an intracellular reserve food. Lipids make up about 10 percent of the dry weight of most microorganisms. They are categorized by the presence or absence of subunits called **fatty acids**.

Lipids Containing Fatty Acids. Fatty acids are long chains of carbon and hydrogen atoms, with a carboxyl (—COOH) group at the end. **Fatty acids** can be **saturated** or **unsaturated**. Fatty acid molecules that do not contain double bonds, and therefore have the maximum possible number of hydrogen atoms, are called saturated. On the other hand, fatty acid molecules that contain double bonds between carbon atoms, and therefore fewer hydrogen atoms, are called unsaturated.

When three fatty acids are covalently joined to the alcohol **glycerol**, they form **fats** or **oils**, depending upon whether they are solid or liquid at room temperature. Fats and oils are reserve food storage products for most organisms and cellular microorganisms, but not for bacteria. Fatty acids in fats and oils are joined to glycerol by **ester linkages**—a bond between a carboxyl group on the fatty acid and hydroxyl groups on the glycerol molecule.

Phospholipids are more complex lipids. They are composed of a glycerol backbone, two fatty acids, and a phosphate group to which another constituent (R—for "rest of molecule") is attached (**Figure 2.22**). Phospholipids are the essential components of membranes in plants, animals, and microorganisms. The fatty acid component of the molecule is hydrophobic and constitutes the nonpolar "tail" (insoluble in water). The phosphate group constitutes the small polar "head" (soluble in water). In the watery environment of a cell, phospholipids spontaneously arrange themselves by hydrophobic interactions to form membranes.

Lipids Not Containing Fatty Acids. Some lipids do not contain fatty acids. The **sterols**, for example, are lipids composed of hydrocarbon rings. Most bacteria do not have sterols, but sterols are found in the membranes

Clinical Notes

Bacterial Products That Heal

Cellulose is the tough, light polysaccharide component of the walls of plants and some algae. A few bacteria, including *Acetobacter xylinum*, the bacterium we use to make vinegar, also produce cellulose, but not for cell walls. *A. xylinum* releases its cellulose into its aqueous environment where it floats to the surface carrying *A. xylinum* cells with it. There they have access to the air they need to grow. The cellulose

produced by *A. xylinum* is not like plant cellulose. It is pure and, as such, is nontoxic and strong. It is ideal for dressing wounds because it serves as a firm matrix (form-giving base) to promote healing. *A. xylinum* is also a "hi fi" microbe. Because the cellulose it makes is strong and resilient, it is used in acoustical diaphragms for high-quality stereo speakers.

of all eucaryotes, including eucaryotic microorganisms. Sterols are clinically important because they are the target of one group of antibiotics, the **polyenes**. The polyenes bind to sterols, damaging them and weakening the microbial membrane. **Cholesterol**, which is associated with heart disease, is a sterol produced only by animals.

Summary

THE BASIC BUILDING BLOCKS   (pp. 21–23)

1. Matter is composed of small particles called atoms. Atoms are composed of protons, neutrons, and electrons.

2. Protons carry a positive charge, and electrons carry a negative charge. Neutrons have no charge.

Atoms   (pp. 21–22)

1. All atoms have the same basic structure. A cloud of electrons orbits a nucleus of densely packed neutrons and protons. Electrons orbit in shells, or energy levels.

2. The atomic number of an atom is the number of protons it contains. The atomic weight of an atom is the number of protons and neutrons it contains.

3. Atoms with the same atomic number but different atomic weights are called isotopes.

Elements   (p. 22)

1. Matter composed of only one kind of atom is called an element. The most abundant elements in living organisms are carbon, hydrogen, nitrogen, and oxygen, plus phosphorus and sulfur.

Molecules   (pp. 22–23)

1. A molecule consists of two or more atoms, of the same or different elements, joined by chemical bonds. A compound is composed of molecules containing more than one type of atom.

2. A molecular formula tells which atoms and how many of each kind form a particular molecule.

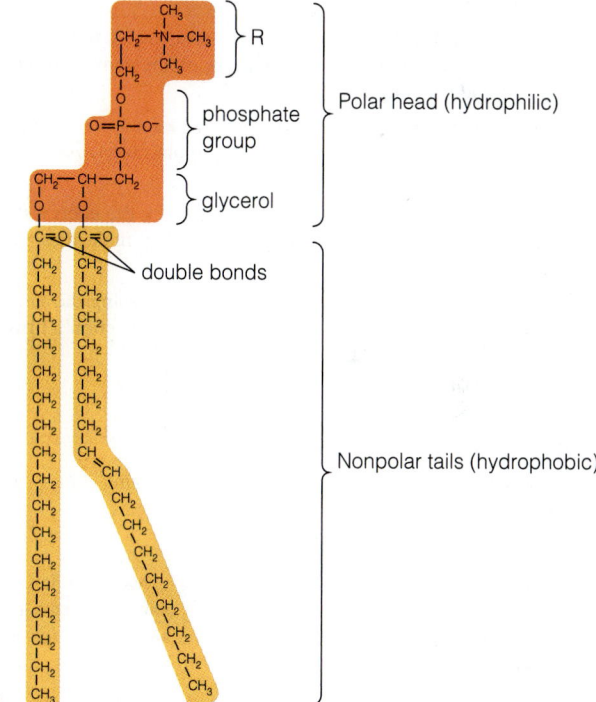

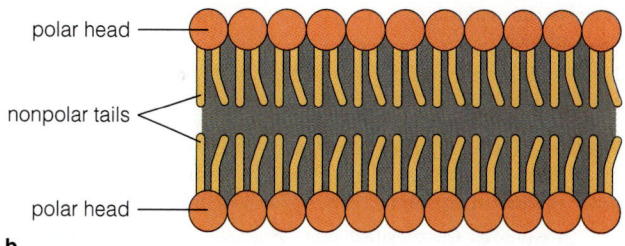

Figure 2.22 Phospholipids. (**a**) Phospholipid molecules have a polar head and nonpolar tail. (**b**) Because of their hydrophilic-hydrophobic structure, they aggregate spontaneously to form membranes in living things. Double bonds in the fatty acid tails bend the molecule, making the membrane more fluid.

3. Avogadro's number (6.02×10^{23}) of any type of molecule is called a mole of that substance. A mole is the molecular weight of the substance in grams.

CHEMICAL BONDS AND REACTIONS (pp. 23–28)

1. Chemical bonds form if the resultant molecule will be at a lower energy state and therefore more stable than the original configuration. Energy is released when a chemical bond forms, and energy is required to break a chemical bond.

2. Valence electrons determine an atom's capacity to form chemical bonds.

3. A covalent bond is formed when atoms share electrons; covalent bonds are extremely stable.

4. Covalent bonds form nonpolar molecules (shared electrons are equally spaced) or polar molecules (shared electrons are not equally spaced). Polar molecules, such as water, have a negative and positive pole.

5. Ionic bonds are formed by the attraction between oppositely charged ions or molecules. Ionic bonds are strong but not so strong as covalent bonds.

6. Hydrogen bonds form when hydrogen atoms are shared between two molecules or between different parts of the same molecule. Hydrogen bonds are weak.

7. A chemical reaction occurs when reactants are transformed into different combinations of the same atoms or molecules; these combinations are called products.

8. The rate of a chemical reaction is influenced by the energy state of the reactants, the concentrations of reacting molecules, and temperature.

9. Catalysts increase the rate of a reaction by decreasing the activation energy barrier. A catalyst is not altered by the reaction it brings about.

10. Catalysts in living things are called enzymes.

WATER (pp. 28–32)

1. About 70 percent of the weight of microorganisms comes from water molecules.

Special Properties of Water (p. 29)
1. Water's capacity to form hydrogen bonds gives it a relatively high boiling point and relatively low freezing point; stabilizes temperature inside cells; makes water cohesive; and makes ice less dense than water, which makes life possible on Earth.

Water as a Solvent (pp. 29–30)
1. Water's polarity makes it an excellent solvent. Compounds that dissolve in water are hydrophilic and those that do not are hydrophobic.

2. Colloids are tiny solid particles that create turbid liquids. Some colloidal dispersions that become concentrated or cooled form gels.

Hydrogen and Hydroxide Ions (pp. 30–32)
1. Almost every chemical reaction that occurs in living things involves the participation of either hydrogen or hydroxide ions.

2. Water ionizes (dissociates) into an equal number of positively charged hydrogen ions and negatively charged hydroxide ions.

3. Solutes that dissociate in water to form hydrogen ions are acids. Solutes that dissociate to form hydroxide ions are bases.

4. Salts dissociate into anions and cations but not hydrogen or hydroxide ions.

5. The pH scale (logarithmic with base 10) describes how acidic or basic a solution is by assigning it a number from 0 to 14.

6. Buffers are weak acids or weak bases that stabilize the pH of a solution.

ORGANIC MOLECULES (p. 32)

1. Biochemistry is the study of carbon-containing molecules made by organisms.

2. A carbon atom can form four covalent bonds with other atoms; carbon atoms bind with other carbon atoms to form the molecular backbone of biochemicals.

3. In an organic compound, atoms other than carbon and hydrogen are joined in patterns called functional groups, which give a compound its characteristic properties and determine which other compounds it will react with.

MACROMOLECULES (pp. 32–43)

1. Macromolecules are large molecules formed through polymerization, a process that removes added chemical groups from monomers, the building blocks of macromolecules.

2. In the opposite process, called hydrolysis, macromolecules are broken down into monomers to obtain nutrients or build a new cell structure.

Proteins (pp. 33–38)
1. Proteins function as enzymes, as components of cell structures, in membrane transport, and in cell movement.

2. The building blocks of proteins are amino acids, which are joined by peptide bonds.

3. The primary structure of a protein is its specific order of amino acids. The secondary structure is determined by hydrogen bonds that arrange amino acids into an alpha helix or a beta sheet. The tertiary structure is determined by interactions among R groups. Quaternary structure is the way the different chains fit together.

4. Denaturation is the destruction of a protein's three-dimensional structure by extremes of temperature, pH, or osmotic strength.

Nucleic Acids (pp. 38–40)
1. The two kinds of nucleic acid molecules are deoxyribonucleic acid (DNA) and ribonucleic acid (RNA). DNA encodes all the cell's genetic information. RNA interprets information encoded in DNA to construct proteins.

2. Nucleic acids are composed of monomers called nucleotides. The five-carbon sugar in DNA is deoxyribose. In RNA it is ribose. DNA also contains phosphate and the bases cytosine, adenine, guanine, and thymine. RNA contains phosphate and the first three bases plus uracil in place of thymine. Nucleotides are joined by phosphodiester bonds.

3. Adenosine triphosphate (ATP) is the major carrier of energy in the cell. Cyclic AMP (cAMP) is a chemical messenger in cells. Nicotinamide adenine dinucleotide (NAD) and flavine adenine dinucleotide (FAD) are important coenzymes.

4. The DNA molecule is a double helix, but RNA is usually single stranded.

Carbohydrates (pp. 40–42)
1. Polysaccharides are polymers composed of monosaccharides, small organic molecules made up of carbon, hydrogen, and oxygen.

2. Two monosaccharides joined by glycosidic bonds form a disaccharide sugar, such as lactose.

3. Polysaccharides can be linear or branched and built from one or several different monosaccharides. Glucose is built into three polysaccharides—cellulose, starch, and glycogen.

Lipids (pp. 42–43)

1. Lipids are nonpolar molecules that are soluble in nonpolar solvents such as gasoline and insoluble in water.

2. Lipids are classified according to whether they contain fatty acids. Fatty acids in fats and oils are joined to the alcohol glycerol by ester linkages. Lipids that do not contain fatty acids include the sterols.

3. Phospholipids are complex lipids that form membranes in plants, animals, and microorganisms.

Review Questions

THE BASIC BUILDING BLOCKS

1. Describe the structure of an atom. How do mass and electrical charge of the subatomic components differ?

2. What is the special significance of valence electrons?

3. Explain the difference between atomic number and atomic weight.

4. Explain this statement: An element cannot be reduced chemically to a simpler form.

5. Define these terms: molecule, macromolecule, compound, organic compound.

6. What is the molecular formula for water? How does knowing a molecular formula allow you to calculate molecular weight?

7. What is Avogadro's number?

CHEMICAL BONDS AND REACTIONS

1. What are chemical bonds? Explain this statement: Chemical bonding has to do with energy states. What determines an atom's capacity to form chemical bonds?

2. What is a covalent bond? Explain the difference between a polar molecule and a nonpolar molecule. What is the most important polar molecule in living things?

3. Define these terms: ion, electron donor, electron acceptor, cation, anion.

4. Name some functions of ions in living systems, including microorganisms.

5. What are ionic bonds? Give an example.

6. What is a hydrogen bond? How are hydrogen bonds important to living things?

7. Which type of bond is most stable? Which type is least stable?

8. Define these terms: chemical reaction, reactant, product.

9. Complete this sentence: When chemical bonds are broken, energy is _____, and when chemical bonds are formed, energy is _____.

10. Name some factors that affect chemical reaction rates.

11. Why are enzymes important?

WATER

1. What special properties does water have because of its capacity to form hydrogen bonds?

2. Explain how water's polarity makes it an excellent solvent.

3. What are hydrophobic and hydrophilic groups, and why are they vital to living things?

4. What are colloids, colloidal dispersions, and gels?

5. Why are hydrogen and hydroxide ions so crucial to the chemistry of life?

6. Distinguish between acids, bases, and salts.

7. What is the pH scale? Why do we say it is logarithmic?

8. What is a buffer and how do buffers function in living things? How are they used in the laboratory?

ORGANIC MOLECULES

1. What is biochemistry?

2. Explain the following statement: A carbon atom has great bonding versatility.

3. What are functional groups? Give an example.

MACROMOLECULES

1. Define these terms: polymers, monomers, polymerization.

2. How are macromolecules formed? What is the reverse process and what functions does it serve in a cell?

3. What are the building blocks of proteins, and what kinds of bonds hold them together? Give some examples of how proteins function in a cell.

4. What are L-isomers and D-isomers? Why is each important?

5. Explain protein structure. What is denaturation? Define and give an example.

6. Name the two kinds of nucleic acid molecules and tell what role each plays in a cell.

7. What are the basic building blocks of nucleic acids, and what kinds of bonds hold them together?

8. How are DNA and RNA alike chemically? How are DNA and RNA different?

9. What are the functions of these nucleotides: ATP, cAMP, NAD, FAD?

10. Define these terms: polysaccharide, monosaccharide, carbohydrate. Give some examples of monosaccharides and tell why they are important to living things.

11. What kinds of bonds hold polysaccharides together? Explain how each of these polysaccharides is important to plants, animals, and microorganisms: cellulose, starch, glycogen. How are polysaccharides important to bacteria?

12. How are lipids defined? How are lipids that contain fatty acids different from lipids that do not? What are ester linkages?

13. What are phospholipids and why are they important to living things?

14. How is peptidoglycan different from other macromolecules? What are its basic building blocks? How does it function?

Suggested Readings

Alberts, B.; Bray, D.; Lewis, J.; Raff, M.; Roberts, K.; and Watson, J. D. 1989. *Molecular biology of the cell*. 2d ed. New York: Garland.

Miller, G. T. 1991. *Chemistry: A contemporary approach*. Belmont, Calif.: Wadsworth.

Scientific American. 1985. Molecules of life. Special issue. October.

Stryer, L. 1989. *Biochemistry*. 3d ed. San Francisco: W. H. Freeman.

3 METHODS OF STUDYING MICROORGANISMS

Everything Was Going According to Plan, Until . . .

It was 1962 and Heinz Stolp, a young German microbiologist, was spending a year in the United States studying the *Erwinia* species of bacteria. *Erwinia* are major plant pathogens, causing such diseases as fire blight of apples and pears. Stolp was trying to figure out the relationships between various strains (subgroups) of the species. He reasoned that if he could determine which viruses attack which strain of *Erwinia* bacteria, then he could assume that strains attacked by the same virus are closely related.

Stolp proceeded in the usual way. He made pour plates and inoculated them with strains of *Erwinia*. Then he added soil inocula suspected to contain viruses that prey on *Erwinia* and incubated the plates. The *Erwinia* grew as a confluent layer or lawn covering the plate. When viruses destroyed *Erwinia* cells, they left circular clear zones, called plaques. Stolp then picked virus stocks from the plaques with an inoculation loop.

Every day Stolp prepared many plates and isolated new strains of virus from the previous day's plates. The project went well, until one day, while discarding some old plates, he noticed that some of the plaques had gotten bigger. But plaques formed by viruses do not enlarge. Their size is fixed when bacterial growth stops, which is usually the day after they are prepared. What was consuming the nongrowing bacterial cells?

Stolp took a direct approach. With an inoculating loop, he picked a region from the clear plaque, made a wet mount, and observed it under a phase-contrast microscope. He saw only a few *Erwinia* cells, nothing else. But from time to time the *Erwinia* cells moved abruptly, as though they had been struck by an unseen missile. Whatever was striking them might be destroying them and forming the plaques. If so, the predator was too small to be seen with a phase-contrast microscope. So Stolp put a sample from the plaque under a transmission electron microscope and saw *Bdellovibrio*, as he named them (*bdello*, meaning "leach," and *vibrio*, describing their commalike shape). *Bdellovibrio* are bacteria that attack Gram-negative soil bacteria by ramming into them

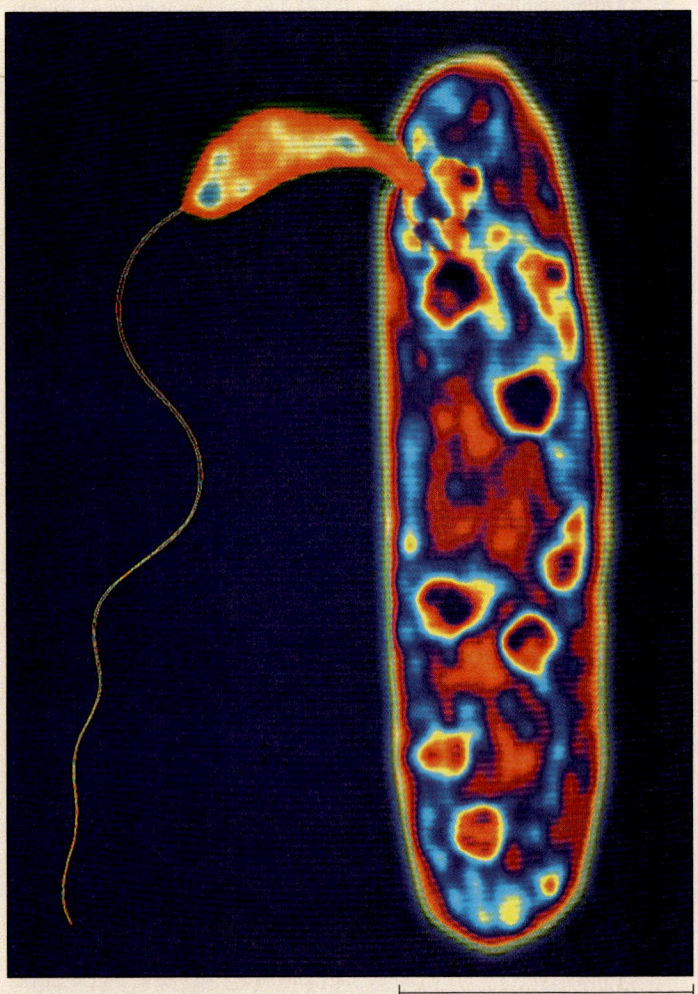

1 µm

Bdellovibrio penetrating its prey.

at such speed that the host cell recoils from the impact. Then the *Bdellovibrio* bacterium enters its new host while it is still living. Once inside, the *Bdellovibrio* consumes the host, which accounts for the enlarged plaque.

Bdellovibrio constitute a major bacterial group, and they are ecologically important. They help maintain the balance of microorganisms in soil. Stolp's discovery of these bacteria is a tribute to the power of microscopy and careful observation.

To understand:

- The fundamental properties of light: reflection, transmission, absorption, diffraction, and refraction

- The principles of microscopy: how magnification, contrast, and resolution are achieved

- The preparations necessary to view microorganisms by microscopy

- The way the compound light microscope works

- How the phase-contrast, darkfield, fluorescence, and Nomarsky modifications of the light microscope increase contrast

- How transmission and scanning electron microscopy work

- What pure cultures are, why they are important, and how they are obtained: methods of sterilization and isolation

- How to cultivate a pure culture: types of media and the laboratory environment

- How cultures are preserved

VIEWING MICROORGANISMS

Like all life scientists, microbiologists need to observe the organisms they study, which means their indispensable tool is the microscope. **Microscopy**, the construction and use of microscopes, began in the seventeenth century with Antony van Leeuwenhoek and his little hand-held microscope (Chapter 1).

Also like other life scientists, microbiologists need to perform **experiments**. Experimentation on microorganisms almost always requires cultivating them in the laboratory. To **cultivate** a microorganism means to provide suitable conditions for its growth and multiplication.

In this chapter, you will learn about the tools and techniques of the modern microbiological laboratory—the different kinds of microscopes and the methods of propagating (growing) microorganisms in the artificial environment of the laboratory.

Properties of Light

We see the visible world around us because of the lens and retina in our eyes. The lens in the front of the eye pro-

duces images on the light-sensitive retina at the back of the eye. But for the eye's lens to produce images of invisible microorganisms, it needs the help of the lenses in a microscope. And to understand how microscopes work, we must know something about the properties of light.

There is a continuous spectrum of **electromagnetic waves**. At one end are the *lowest*-frequency, *longest* waves, called **radio waves** (**Figure 3.1**). Radio waves can be as long as 2000 meters (more than a mile). At the other end of the spectrum are the *highest*-frequency, *shortest* waves, called **gamma rays**. Gamma rays, produced within the nucleus of radioactive atoms, are shorter than 0.01 nm (nanometer), smaller than a virus.

Both radio waves and gamma waves are invisible. So are most of the other waves in the electromagnetic spectrum. In fact, our eyes can detect only a very small part of the electromagnetic spectrum, the part we call **light**.

Wavelength. The properties of light waves and all electromagnetic waves are much like those of ocean waves. Light waves have peaks (high points) and troughs (low points). The distance between two peaks or two troughs is the **wavelength**. The wavelengths of visible light are between about 400 and 700 nm, depending on color. Blue light is the shortest light wavelength, and red is the longest. Yellow and green are intermediate. White light is a mixture of all wavelengths. Sunlight is white light, but colors appear when sunlight is altered by absorption or refraction (which are explained below). The **frequency** of a light wave is the number of peaks or troughs passing a particular point each second. The **intensity** of a light wave is the height of the wave.

In addition to describing light as waves, we can describe it as a stream of particles called **photons** or as a narrow beam called a **ray**.

Reflection, Transmission, and Absorption. When a ray of light strikes an object head-on, it has three possible fates: **reflection**, **transmission**, or **absorption**. If the ray bounces back, we say it is reflected. If it passes through the object, we say it is transmitted. If it transfers some of its energy to the object, we say it is absorbed (**Figure 3.2**).

Reflected rays bounce off a smooth surface as a ball would. Rays that strike straight on bounce directly back. If they strike at an angle to the surface, they bounce off at the same angle.

When a ray of light hits a clear object, such as pure water in a clear glass, it is transmitted undiminished. But if the object is not clear (suppose the glass is smoked or colored), the light is absorbed. If all wavelengths of light are absorbed equally, the intensity of the light is diminished but the color is unchanged. However, if certain wavelengths, but not others, are absorbed, the color of

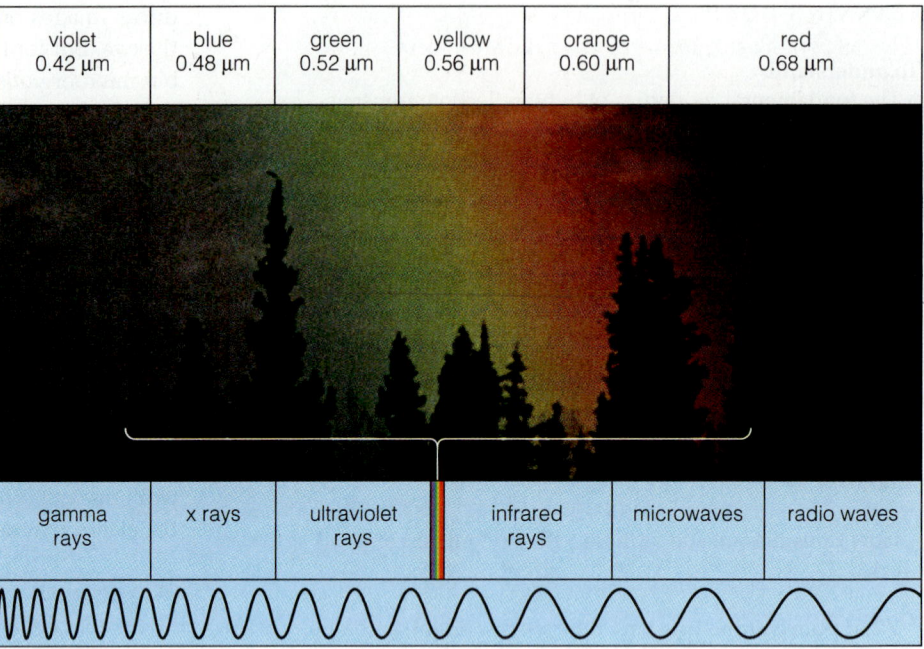

Figure 3.1 Visible light makes up only a tiny part of the electromagnetic spectrum, the part between 400 and 700 nm.

violet 0.42 μm	blue 0.48 μm	green 0.52 μm	yellow 0.56 μm	orange 0.60 μm	red 0.68 μm

gamma rays	x rays	ultraviolet rays	infrared rays	microwaves	radio waves

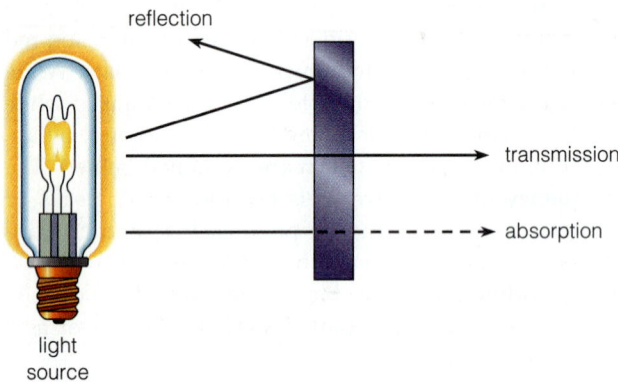

Figure 3.2 Light rays that strike an object have three possible fates: reflection, transmission, or absorption.

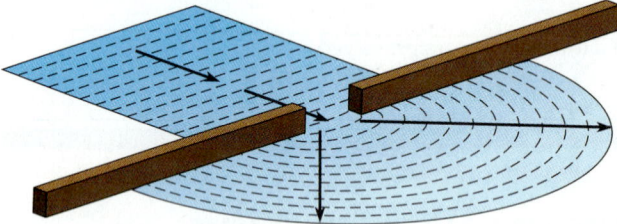

Figure 3.3 Diffraction causes light rays to bend as they pass through a small opening or go around an obstacle (any opaque object). The parallel waves of light spread out—diffract—much as ocean waves do when passing through a breakwater. The light rays (arrows), which are perpendicular to the wave front, bend.

light changes. We see the color of the transmitted or reflected light. For example, a solution of red dye is red because it absorbs the blue component of white light and allows the red to pass through. A leaf is green because it absorbs the blue and red components of sunlight and reflects the green component.

Diffraction. When rays of light pass through a small opening or pass by the edge of an opaque object (one that transmits no light), light rays bend. This kind of bending is called **diffraction**. Because light rays diffract, a sharp edge or an opaque object casts a fuzzy shadow. Diffraction occurs because of the wave nature of light. Consider an ocean wave example; picture the waves coming toward

a breakwater near shore. As the waves approach, they are parallel. But as they pass through the opening in the breakwater, or around the end of it, they spread out in semicircles (**Figure 3.3**). Light rays (which are perpendicular to the wave front) therefore bend when they go through a small opening or around an obstacle.

Refraction. Refraction also bends rays of light but by a different mechanism. **Refraction** occurs when a ray of light meets an object of a different density at an angle. There are, then, two things involved in refraction. First, when the light enters a denser medium (air to water, for example), it slows down. Second, when the light enters at an angle, the edge of the ray entering the substance first is slowed more than at the opposite edge, which enters later (**Figure 3.4**). The amount of refraction, or bending, depends upon the angle. The greater the angle, the greater the refraction. The ratio of the velocity (speed) of light traveling through a vacuum to the velocity in any particular material is termed its **refractive index**. The higher the density, the higher the refractive index. For example, the refractive index of pure water (at room temperature) is 1.33, meaning light travels 1.33 times slower in water than in a vacuum. The refractive index of air is 1.0002.

Reflection, transmission, absorption, diffraction, and refraction all play a part in creating a visible image with a microscope, as you will see in the next section.

Microscopy

We use microscopes to create a visible, detailed image of something that is otherwise too small to see. Different

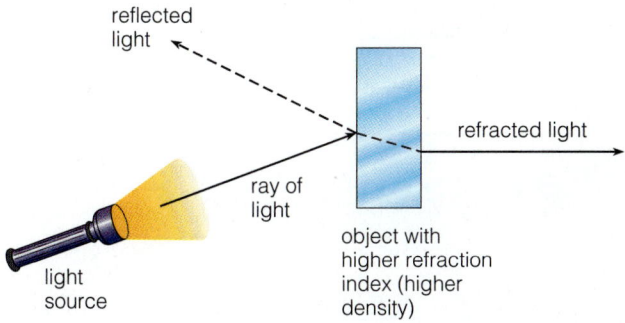

Figure 3.4 A light ray from the source at the left is reflected and refracted by the glass. Refraction occurs when light enters a medium denser than the one in which it is traveling. Here, refraction occurs because glass is a denser medium than air.

types of microscopes create a visible image in different ways, but in all cases good images depend on three factors: magnification, contrast, and resolution. All three must be adequate to produce a clear image.

Magnification. The first important function of a microscope is **magnification**, the enlargement of an image. Microorganisms must be magnified to be visible. Objects are magnified with a **convex lens**, which is thicker in the center than at the edge. A convex lens bends parallel rays of light by refraction so that they meet at a single point, the **focal point**. This forms an enlarged image of the object (**Figure 3.5**). Magnification can be increased in two ways, by making the lens more convex or by bringing the object to be magnified closer to the focal point. A single convex lens is adequate to view microorganisms as small as 1 μm, as Leeuwenhoek showed.

Magnification is essential for creating a visible image of a small object, but it is not enough. Even the largest image, if fuzzy, tells us little about the object under study. Detail is required to convey information. The two main factors that affect our ability to perceive detail in a magnified image are contrast and resolution.

Contrast. The term **contrast** refers to a difference in light intensity. Variations in light intensity allow us to see that one part of an image is different from the part nearby or from the background. As we've seen, some of the light that strikes an object is absorbed by the object, and some is transmitted through it. The image we see is created by the transmitted light. If the same amount of light is transmitted from an object and from its immediate surroundings, the image will blend into the background. Similarly, if the same amount of light is transmitted from all parts of

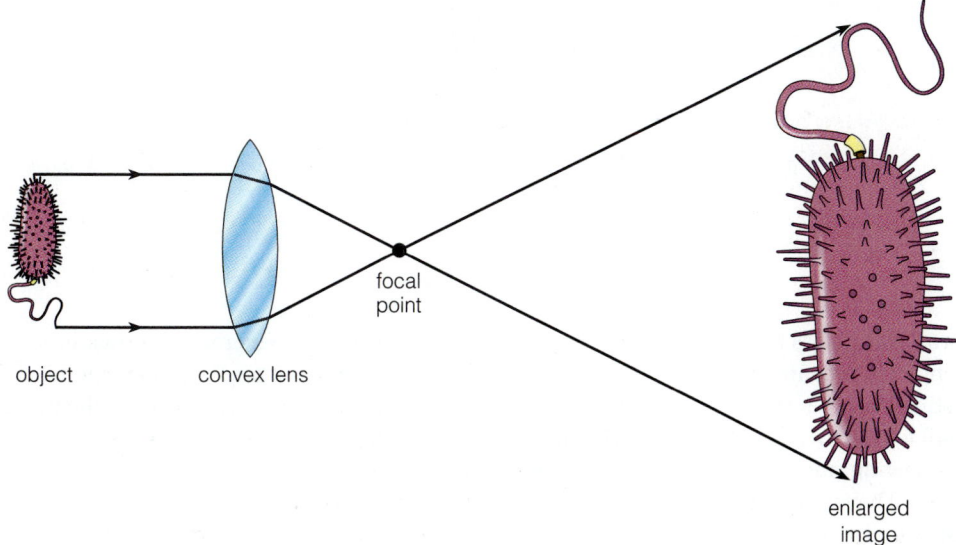

Figure 3.5 A convex lens magnifies because it bends parallel rays of light so that they meet at a focal point. This produces an enlarged image of the object behind the lens.

Measuring Microorganisms

We know microorganisms are small, but "small" is a relative term. We could say that an average-size bacterium is about a million times smaller than an average-size human, but that is not particularly helpful. How can we measure their "smallness"?

As in science generally, microbiology uses the **metric system**. The metric system consists of basic units used with a series of prefixes that differ by powers of ten. For example, **meter** is the basic unit for length, **liter** for volume, and **gram** for mass. Thus, a decimeter (dm) is one-tenth of a meter, a centimeter (cm) is one one-hundredth of a meter, and a millimeter (mm) is one one-thousandth of a meter.

The two units of measurement most often used for mea-

The Metric System (International System of Units)

Quantity	Basic Unit	Symbol	English Equivalent
Length	Meter	m	39 inches—about a yard
Mass	Gram	g	About 1/30 of an ounce
Volume	Liter	L	About 1.06 quarts

Prefix	Symbol	Means Multiply by
	basic unit	10^0 (1)
Deci	d	10^{-1} (0.1)
Centi	c	10^{-2} (0.01)
Milli	m	10^{-3} (0.001)
Micro	μ	10^{-6} (0.000001)
Nano	n	10^{-9} (0.000000001)
Pico	p	10^{-12} (0.000000000001)

Commonly Used Units

1 meter	=	10^2 cm	1 gram	=	10^3 mg	1 liter	=	10^3 ml
	=	10^3 mm		=	10^6 μg		=	10^6 μl
	=	10^6 μm		=	10^9 ng			
	=	10^9 nm		=	10^{12} pg			
	=	10^{10} Å (Angstroms)						

Note: For example, *Escherichia coli* is about 1 μm or 0.000001 m wide and weighs about 1 pg or 0.000000000001 g.

an object, the image will have a uniform, featureless appearance. In either case, no detail is visible. Thus, contrast is needed, first, to see an object as distinct from its surroundings and, second, to perceive detail within it.

Contrast is a special problem with viewing microorganisms because most are almost colorless. They transmit all wavelengths of visible light almost equally. To get a useful image of most microorganisms, contrast must be increased artificially. One way to increase contrast is with color. With microorganisms, we apply **stains**, chemicals that add color to microbial cells or parts of them, in preparation for viewing. Some kinds of microscopes can also increase contrast.

Resolution. Good **resolution** means you can perceive two adjacent points of an image as separate from each

suring microorganisms are smaller yet. **Micrometer** (μm) is one one-millionth of a meter (1/1,000,000 or 10^{-6} meter). **Nanometer** (**nm**) is one one-billionth of a meter (1/1,000,000,000 or 10^{-9} meter.) A smaller unit called the **angstrom** (Å), which is one ten-billionth of a meter (1/10,000,000,000 or 10^{-10} meter), is also used, although it is not officially part of the metric system.

Cells of bacteria, fungi, algae, and protozoa are measured in micrometers. Viruses and subcellular structures, such as ribosomes and membranes, are measured in nanometers. Atoms and molecules are measured in angstroms.

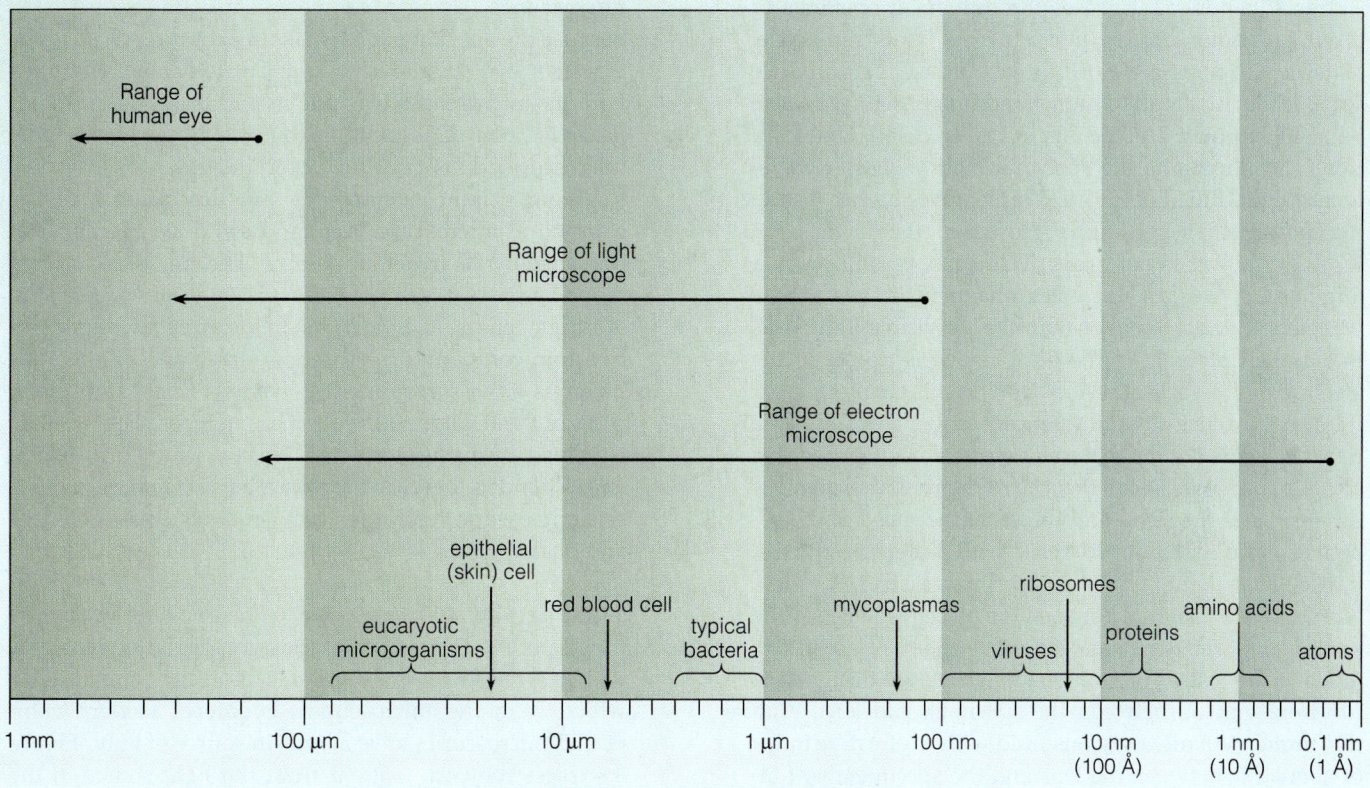

Microorganisms and other small objects are measured in metric units. Eucaryotic microorganisms and bacteria are measured in micrometers (μm), viruses in nanometers (nm), and atoms and molecules in angstroms (Å).

other. If an image has poor resolution, the two points merge and look like one. Naturally, an image with poor resolution conveys little detail. For example, a television set with high resolution has a clear picture, while a set with poor resolution has a fuzzy picture. Increasing magnification does not improve resolution. A big-screen TV, if it has poor resolution, simply has a larger fuzzy image.

Unlike contrast, which is a property of the object under study, resolution is a property of the lens system used to view it. The **resolving power** of a microscope refers to the smallest distance between two adjacent points that can still be perceived as separate from each other (see the discussion of numerical aperture, below). When more closely spaced points can be distinguished, more detail becomes visible, and smaller objects can be viewed. Resolving power is determined by three factors:

(1) the size of the first magnifying lens—called the objective lens, (2) the wavelength of the light illuminating the **specimen**, the object to be viewed, and (3) the refractive index of the material between the objective lens and the specimen. Remember that a resolving power of 4 nm is greater than one of 5 nm.

Larger lenses have greater power of resolution than do smaller lenses. This is because larger lenses allow a larger cone of light to enter. Thus, you can increase resolving power—which is always the goal—by increasing the size of the objective lens. There are, however, practical limits to the size of the lens in a microscope.

How does the wavelength of the light illuminating the specimen influence resolving power? Recall that light diffracts (bends) as it passes through small openings or around opaque objects and emerges as a semicircle. When two points are close to each other, they produce wave fronts that tend to merge and make them appear to be a single point. This results in loss of detail. One solution is to illuminate the specimen with a shorter-wavelength light. Blue light, for example, gives higher resolving power than red light. However, the increase in resolving power is not great. To increase resolution by a large factor, we can use electromagnetic waves with a wavelength much smaller than that of visible light. This is the principle behind the high resolving power of the electron microscope described later in this chapter.

The third factor that influences resolving power is the refractive index of the material between the objective lens and the specimen. Recall that the refractive index involves differences in density—and therefore the speed of light—between two materials. Material with a higher refractive index slows light, which in turn increases resolution. Normally, the material between the objective lens and the specimen is air. Air has a refractive index of about 1.0. Some lenses, called oil immersion lenses, are designed to function with oil between the specimen and lens. With this **immersion oil**, a viscous fluid with a refractive index of about 1.5, between the lens and the specimen, you increase resolution. Because immersion oil and glass have the same refractive index, no light is lost by reflection off the surface of the objective lens.

Two of the factors determining resolving power—lens size and use of immersion oil—are intrinsic properties of the lens. A measurement of these two factors, called **numerical aperture (NA)**, is stamped on the side of a microscope's objective lens. Knowing NA and the wavelength of light used to illuminate the specimen, you can calculate the distance (d) between two points that can be resolved as separate entities:

$$d = \frac{\text{wavelength}}{2NA}$$

For example, if the wavelength of blue light used to illuminate the specimen is 450 nm and the value of NA is 0.65 (typical of a high-quality oil immersion lens), the distance is 346 nm [$450/(2 \times 0.65) = 346$], or about one-third the length of an ordinary-size bacterial cell. This means that you will not be able to see many internal details in a bacterium.

The Compound Light Microscope

Most microscopes use visible light as a source of illumination. They are called **light microscopes**. Light microscopes, like Leeuwenhoek's, that have a single lens operate like a simple magnifying glass. They are called **simple microscopes**. Using a simple microscope requires skill because the specimen must be held extremely close to the viewer's eye. (Some say that Leeuwenhoek's success with his microscopes was due in part to his nearsightedness.) Because of aberrations (defects) in the lens, simple microscopes do not produce good images.

Today's light microscopes are descendants of the **compound microscope** that had been developed but not perfected by Leeuwenhoek's time. The compound microscope has two lenses, which increases magnification; but all single lenses—whether one in a simple microscope or two in a compound microscope—have aberrations. The image is often surrounded by colored rings, and not all parts of the field of observation are in focus. The problem is solved by using **corrected lenses**, which are several lenses bonded together to make a lens system. So a modern microscope really has two **lens systems**—objective and ocular.

The Parts of a Microscope. The two lens systems of a compound light microscope are the **objective lens**, which is close to the specimen, and the **ocular lens**, which is in the microscope's **eyepiece**. Modern compound microscopes have a built-in source of light. **Figure 3.6** traces the path of light from the light source in the base to the viewer's eye. The light from the lamp often passes through a blue filter (to filter out long wavelengths) and through a series of lenses called the **condenser**. The condenser directs the light through the specimen where some light is absorbed and some is diffracted. The transmitted light enters the objective lens, which forms an image within the **body tube** of the microscope. The ocular lens further magnifies that image and projects it to the last lens in the series—the one within the eye. The eye forms an image on the retina, and the brain perceives it.

An ordinary compound microscope gives **brightfield illumination**. That is, the condenser directs light *through* the specimen, and the background is brightly lit. The compound microscope can be adapted, however, to view a specimen by (1) phase-contrast, (2) darkfield, (3) fluorescence, or (4) differential interference contrast (Nomarsky) methods, all of which are discussed later in this chapter.

Total Magnifying Power. Most compound microscopes have several objective lenses, each with a different magnifying power. Typically, the low-power lens magnifies an object 10 times (10×), the high-power lens 40 times (40×), and the oil immersion lens 100 times (100×). Most ocular lenses magnify the image another 10 times (10×). You can calculate total magnification by multiplying the power of both lens systems in use. See **Table 3.1** for total magnifying power of the three lens systems. Because its field of vision is large, low power is best for scanning a specimen to get a sense of its overall appearance. Oil immersion, on the other hand, has a small field of vision but greater detail (**Figure 3.7**).

Because most microorganisms have little natural contrast, they require special preparation before being examined under an ordinary light microscope. The preparation and staining (coloring with dye) of a specimen are critical to the quality of the image. Decades of trial and error have gone into developing standard methods of preparation. We will discuss wet mounts for live specimens and various types of staining.

Wet Mounts

The simplest way to prepare a specimen for microscopic examination is to make a **wet mount**. There are two types of wet mounts. A **simple wet mount** is made by placing a drop of liquid containing microorganisms on a microscope slide and covering it with a coverslip. A **hanging**

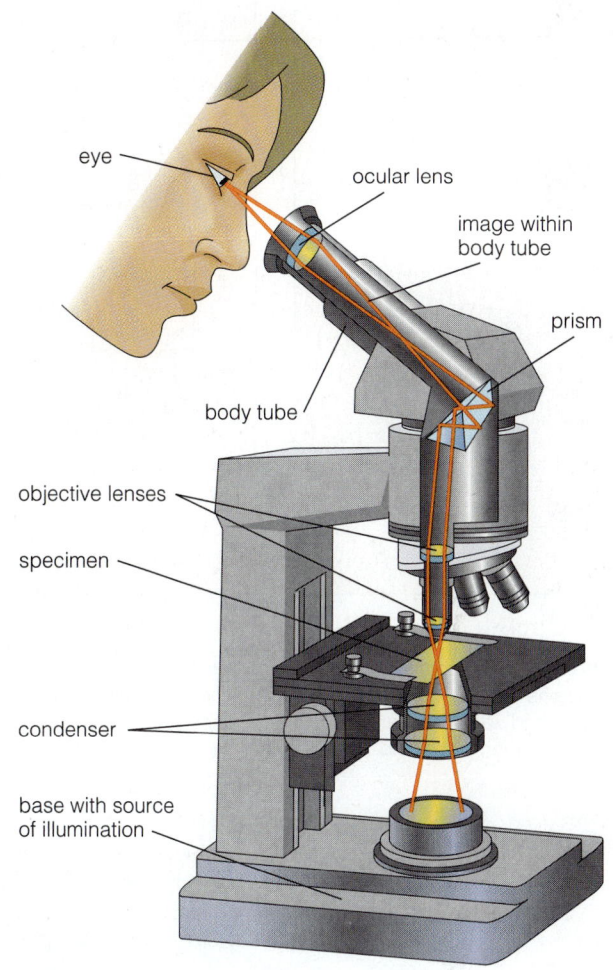

Figure 3.6 The path of light in the compound microscope. The light source is in the base. The condenser focuses the beam on the specimen, where some light is absorbed, some refracted, and some transmitted. The transmitted light enters the objective lens, which magnifies the image. The prism bends the image so that it forms in the body tube, making viewing more comfortable. The ocular lens adds magnification and focuses the image on the eye.

Table 3.1	Total Magnification			
Microscope	Objective Lens	Ocular Lens		Total Magnification
Light Microscopes				
Low power	10×	× 10×	=	100×
High power	40×	× 10×	=	400×
Oil immersion	100×	× 10×	=	1000×
Electron Microscopes				
Transmission (TEM)				~200,000×
Scanning (SEM)				~10,000×

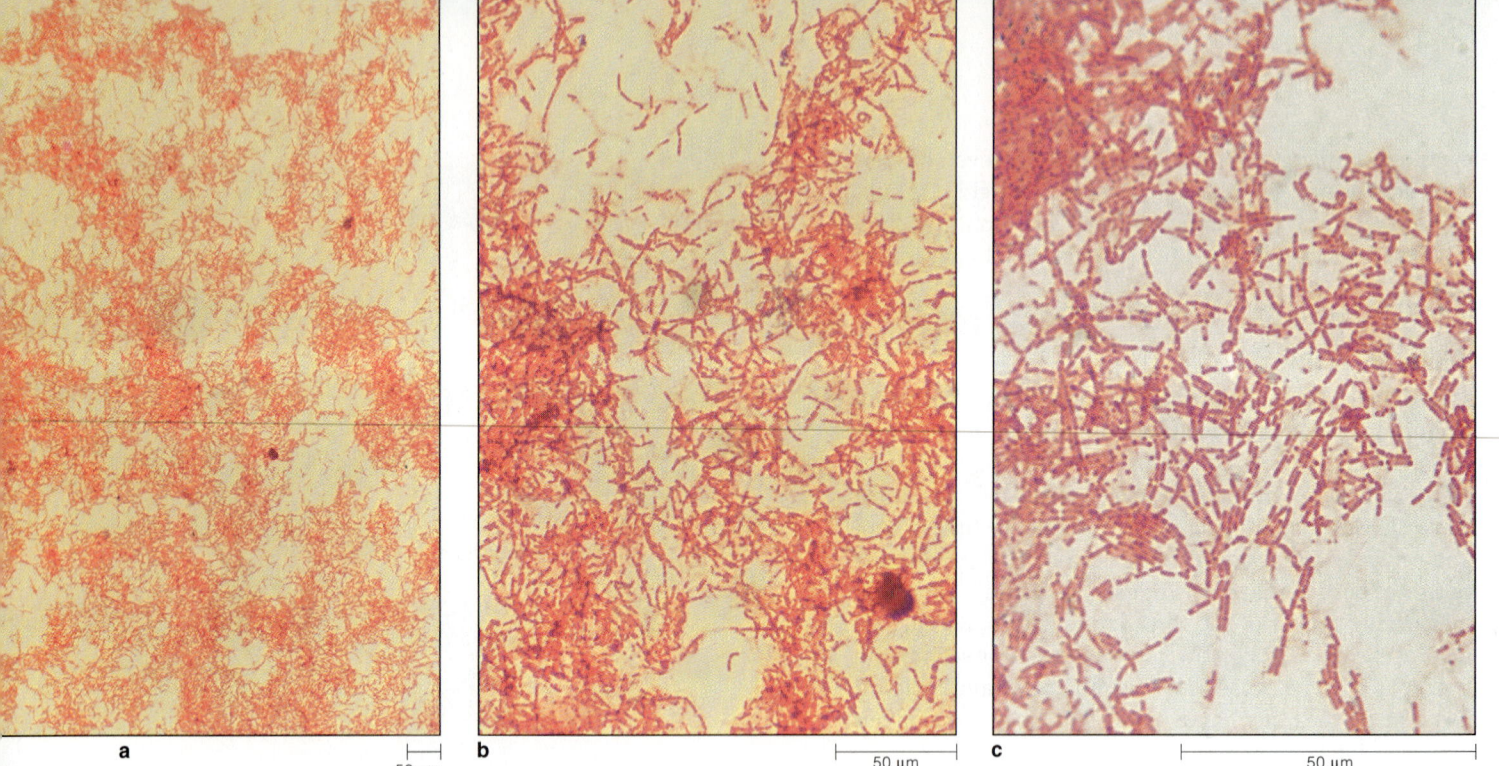

a   ⊢ 50 μm ⊣   b   ⊢ 50 μm ⊣   c   ⊢ 50 μm ⊣

Figure 3.7 The compound microscope has three objective lenses (three systems of lenses). The same field of *Bacillus subtilis* is shown here magnified at low power (100×), at high power (400×), and with oil immersion (1000×). Higher power reveals progressively more detail about a smaller portion of the field.

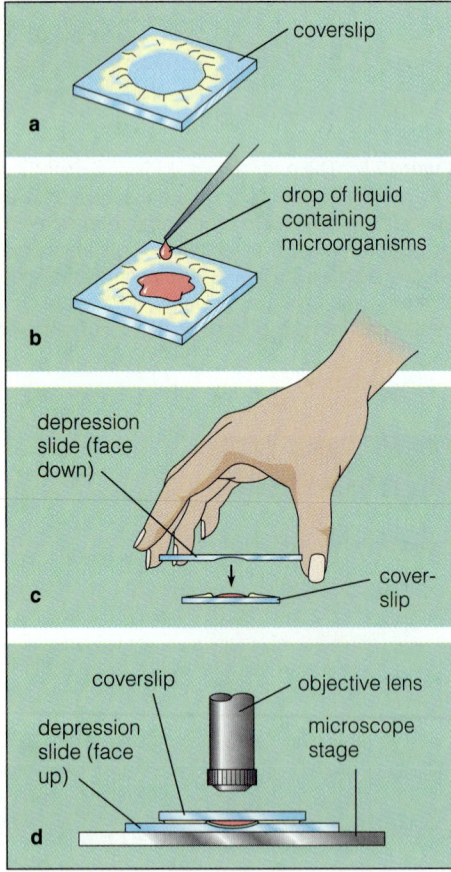

Figure 3.8 Hanging drop mount. A hanging drop mount is made by first placing a drop of liquid containing microorganisms within a ring of petroleum jelly on the coverslip (**a**). The jelly will act as a seal. (**b**) A depression slide is placed over the coverslip. (**c**) The prepared slide is inverted and placed on the microscope stage (**d**) for viewing. Hanging drop mounts are used for viewing living microorganisms.

drop mount is made by placing a drop of liquid containing microorganisms on a coverslip and suspending it in a depression slide (**Figure 3.8**). Petroleum jelly around the well of the depression slide seals the mount. The advantage of hanging drop preparations is that they resist drying, and so microorganisms can be observed for longer times than with simple wet mounts.

Wet mounts are used to observe living microorganisms. For example, wet mounts are routinely used to examine vaginal secretions for *Trichomonas vaginalis*. This highly motile protozoan causes inflammation of the vagina and urethra. If the specimen contains fish-shaped cells that move jerkily across the field, they are most probably *T. vaginalis* and a diagnosis can be made.

The disadvantage of a wet mount, however, is that it does nothing to increase the contrast lacking in live organisms. The uses of wet mounts with an ordinary brightfield light microscope are therefore limited. Usually, one of the four other types of light microscopes is used to observe living microorganisms.

Table 3.2 Common Staining Procedures for Bacteria

Stain	Use	Procedure
Simple Stains	One dye; provides contrast for better viewing an entire organism	Stain with a basic dye (methylene blue, crystal violet, or carbolfuchsin) for about 5 minutes. Rinse briefly with water. Almost all bacteria are stained; most tissues are not.
Differential Stains		
Gram stain	Two or more dyes; distinguishes between Gram-positive and Gram-negative bacteria	Flood slide of fixed bacterial cells with crystal violet and then a solution of iodine (mordant). All bacterial cells become stained a dark violet color. Flood slide with 95% acetone (decolorizing agent).
		Gram-positive cells remain violet; Gram-negative cells lose their stain. Flood with safranin stain (counterstain). Gram-positive bacteria turn darker violet; Gram-negative are counterstained pink.
Acid-fast stain (Ziehl-Neelsen)	Two dyes; distinguishes acid-fast mycobacteria and closely related bacteria from all others	Stain cells with carbolfuchsin and heat to steaming for 5 minutes. All bacteria are stained red. Treat briefly with dilute alcoholic sulfuric or hydrochloric acid (decolorizing agent). Acid-fast bacteria remain red; all others are decolored. Flood with methylene blue (counterstain). Acid-fast bacteria remain red; all others counterstain blue.
Special Stains		
Wirtz-Conklin spore stain	Selectively stains endospores	Flood slide with malachite green and heat to steaming for 60 seconds. Wash with water for 30 seconds. Stain with safranin. Spores retain the intense green of malachite green; the rest of the cell takes the pink of safranin.
Leifson flagella stain	Reveals flagella	A mixture of tannic acid (mordant) and rosaniline dye or fuchsin dye is applied to the fixed cell. The tannic acid thickens the flagella and the dye stains them.
Negative stain	Reveals presence of capsule	India ink or nigrosin dye is applied to a wet mount of the specimen. The dye particles do not penetrate the capsular region, revealing it as a clear zone around the cell.

Stains

Brightfield microscopy is most useful for observing stained specimens. Stains are chemical dyes used to increase contrast. A few stains, called **vital stains**, can be added directly to a wet mount. They stain living cells. But most stains are effective only after microorganisms are **fixed**, killed, and attached to a microscope slide. In **heat fixation**, a drop of liquid containing microorganisms is spread in a thin film (a **smear**) on a microscope slide and allowed to air dry. Then the slide is passed rapidly through an open flame. Heat from the flame kills the microbial cells by denaturing their protein. The coagulated protein attaches them to the slide. **Chemical fixation** is used with extremely delicate specimens because it does less damage than heat. Osmic acid, formaldehyde, or glutaraldehyde is most often used. A drop of the chemical agent is simply added to the liquid containing the microorganisms.

Fixation has disadvantages. It often distorts the cell's appearance, which can make identification more difficult, and movement cannot be studied in a fixed specimen. After fixation, stain is applied to the slide and left in place long enough to be absorbed by the specimen; then excess stain is washed away by flooding the slide, usually with water.

Types of Dyes. Most dyes used for staining are salts, compounds composed of charged ions. **Basic dyes** are composed of positively charged ions, and **acidic dyes** are composed of negatively charged ions. Because the surfaces of most microorganisms carry a slight negative charge, basic dyes bind best to them. Some commonly used basic dyes are safranin, carbolfuchsin, crystal violet, and methylene blue. Acidic dyes bind to negatively charged parts of cells, including proteins. They are used to stain animal tissues that microorganisms have invaded. Commonly used acidic dyes include eosin, acid fuchsin, and Congo red.

Although they are not dyes, compounds called **mordants** are important to some staining procedures. Mordants intensify staining by increasing a cell's affinity for a dye. Mordants can also be used to coat cell appendages, such as flagella, making them thicker and therefore more visible when stained.

There are three types of staining procedures—simple, differential, and special (**Table 3.2**).

Simple Stains. Only one dye is used in **simple stains**, and it is always a basic dye. Simple staining serves merely to increase contrast, staining all cells that absorb the dye the same color. Thus, simple staining enhances viewing of the whole cell. The specimen is fixed, the dye is applied, time is allowed for absorption, excess dye is removed, and the slide is ready for viewing. If a mordant is used, it is added just before the dye is applied.

Differential Stains. To differentiate between types of microorganisms, **differential stains** are used. Differential staining procedures usually involve two steps—**primary staining** (which is the same as simple staining) followed by **counterstaining**. In counterstaining, another dye is applied to stain (and therefore reveal) the cells unstained by the primary dye. Two differential stains are extensively used in microbiology, the Gram stain and the acid-fast stain. Both are for bacteria.

The Gram Stain. The **Gram-stain technique** was developed by the Danish bacteriologist Christian Gram in 1884. It divides bacteria into two groups, the **Gram-positives** and the **Gram-negatives** (**Figure 3.9**). The Gram stain involves staining, decolorizing, and counterstaining:

1. The specimen is stained with the primary stain, the purple dye crystal violet.

2. Iodine is applied, which acts as a mordant.

3. A decolorizing agent, usually an acetone solution, is added. At this stage, Gram-negative bacteria lose their violet stain, but Gram-positive bacteria do not.

4. The pink dye safranin is added as a counterstain. Safranin turns the decolorized Gram-negatives pink and the Gram-positive bacteria a deeper violet color.

Figure 3.9 Gram-staining procedure. The Gram stain divides bacteria into two groups, Gram-positive and Gram-negative. The micrograph shows a Gram-stained specimen containing both *Staphylococcus aureus* and *Escherichia coli*. The *S. aureus* bacteria are revealed as Gram-negative because they lose their violet color during decolorizing and turn pink on application of the counterstain.

The staining difference between Gram-positives and Gram-negatives reflects their different outer surface. Gram-negative bacteria have a thin cell wall with a membrane outside it. Gram-positive bacteria have a thick wall with no membrane outside. This important difference is discussed further in Chapter 4.

The Gram stain is extremely useful clinically. It is almost always the first test performed on a disease-causing bacterium recovered from an infected person. With only a microscope and a few bottles of stain we can know whether an organism is Gram-positive or Gram-negative and the shape and arrangement of its cells. This is often enough information to make a good guess at what kind of bacterium we have. Moreover, the treatment of an infection depends upon whether the causative bacterium is Gram-positive or Gram-negative. Some antibiotics work only on the former, and others only on the latter. Penicillin, for example, is most effective on Gram-positives; streptomycin and tetracycline work on Gram-negatives as well.

The Acid-Fast Stain. The **acid-fast stain**, a second type of differential stain, was developed by Paul Ehrlich in 1882. Today we use a modification of Ehrlich's technique called the **Ziehl-Neelsen stain**. The acid-fast stain colors bacteria of the genus *Mycobacterium*. All other bacteria are colored blue by counterstaining. Acid-fast staining is used to identify *M. tuberculosis* and *M. leprae*, the organisms that cause tuberculosis and Hansen's disease (**Figure 3.10**). Mycobacteria respond to the acid-fast stain because they have complex fatty acid lipids forming a waxlike material

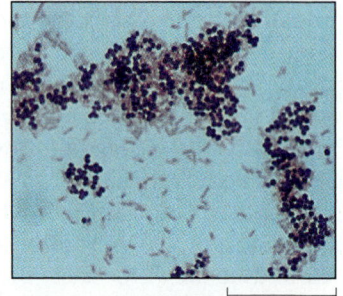

10 µm

STEP 1 Crystal violet (1 minute), drain and rinse	STEP 2 Iodine (1 minute), drain and rinse	STEP 3 Decolorize with acetone-alcohol (one quick rinse) immediately after, rinse with water	STEP 4 Safranin (30–60 seconds), drain, rinse, and blot
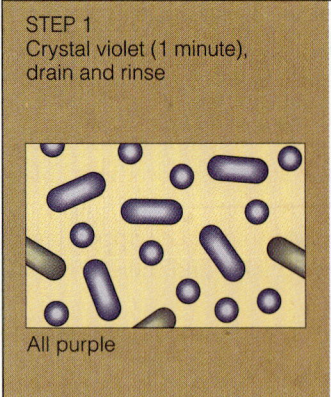			
All purple	All purple. Iodine acts as mordant to set stain	Gram + cocci = purple Gram – rods = clear	Gram + cocci = purple Gram – rods = red (pink)

Clinical Notes

First, Do a Gram Stain

A nineteen-year-old male who had been perfectly well in the morning began to behave strangely in the afternoon. His parents found him pacing around the house in his underwear, swearing and muttering. He became increasingly agitated and hostile. Finally, his father and a neighbor had to force him into the car to take him to the emergency room. Although his bizarre behavior could have been due to drugs, his fever of 104°F suggested infection. Specimens of blood, urine, and cerebrospinal fluid (obtained by a spinal tap) were taken. Within minutes, the emergency room physician had a definitive diagnosis. A Gram stain of the spinal fluid revealed numerous Gram-negative diplococci (spherical bacteria that occur in pairs). They were *Neisseria meningitidis*, a common cause of bacterial meningitis in young adults (Chapter 25). Prompt diagnosis by Gram staining allowed early treatment and complete recovery from this life-threatening infection.

---

in their cell wall that resists decolorization (Chapter 2). An acid-fast stain is prepared by the following steps:

1.   The primary stain, carbolfuchsin, which stains all cells red, is applied.

2.   The slide is heated slightly to drive the dye into the cell.

3.   A decolorizing solution of hydrochloric acid in ethanol is added. This removes the red dye from all cells except the mycobacteria, which retain it due to their waxy surface.

4.   Methylene blue is added as a counterstain. It colors all decolorized cells blue, heightening the distinction between the still-red mycobacteria and other cells in the specimen.

Special Stains.   While differential stains differentiate between whole microorganisms, **special stains** heighten contrast within microbial cells to reveal particular structures, including endospores, flagella, and capsules.

The Wirtz-Conklin Spore Stain.   Some bacteria have **endospores**, exceptionally hardy dormant structures that allow them to survive extreme environmental conditions (Chapter 4). Endospores have thick, impermeable walls that do not absorb most stains. The **Wirtz-Conklin spore stain**, however, is effective:

1.   Malachite green is applied to the specimen.

2.   It is then heated with steam for three to six minutes, allowing the stain to penetrate the spore wall.

3.   The specimen is washed with running tap water, which removes the green from all parts of the cell except the spores.

4.   The pink dye safranin is applied as a counterstain.

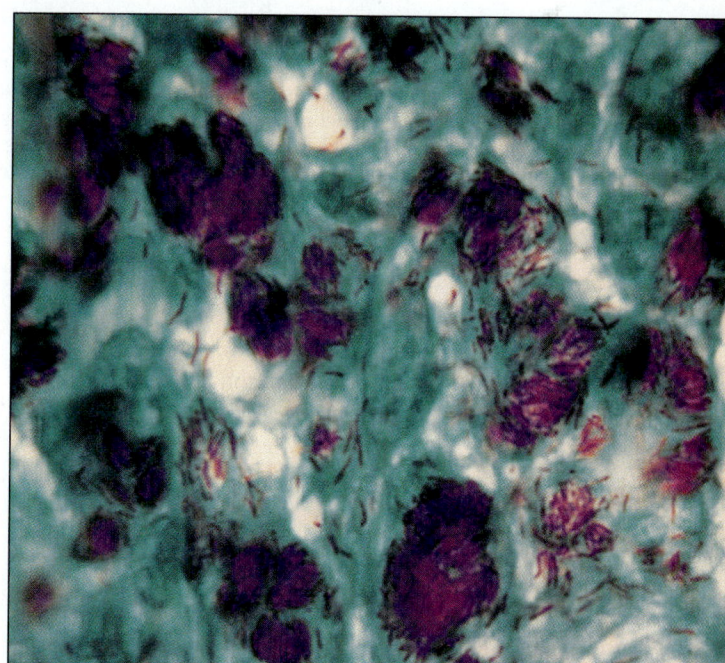

Figure 3.10   The acid-fast stain is used to reveal *Mycobacterium* and related species that have waxes in their cell walls that resist coloration. This micrograph reveals masses of *Mycobacterium leprae* cells within host cells. Mycobacteria remain red, while all other bacteria counterstain blue.

At the end of the procedure, the bacterial endospores are green and the rest of the cell is pink (**Figure 3.11a**). The bacteria *Bacillus anthracis* and *Clostridium perfringens*, which cause anthrax and gangrene, form spores and can be identified through spore staining.

The Leifson Flagella Stain.   **Flagella**, the long threadlike cellular extensions that allow some bacteria to swim

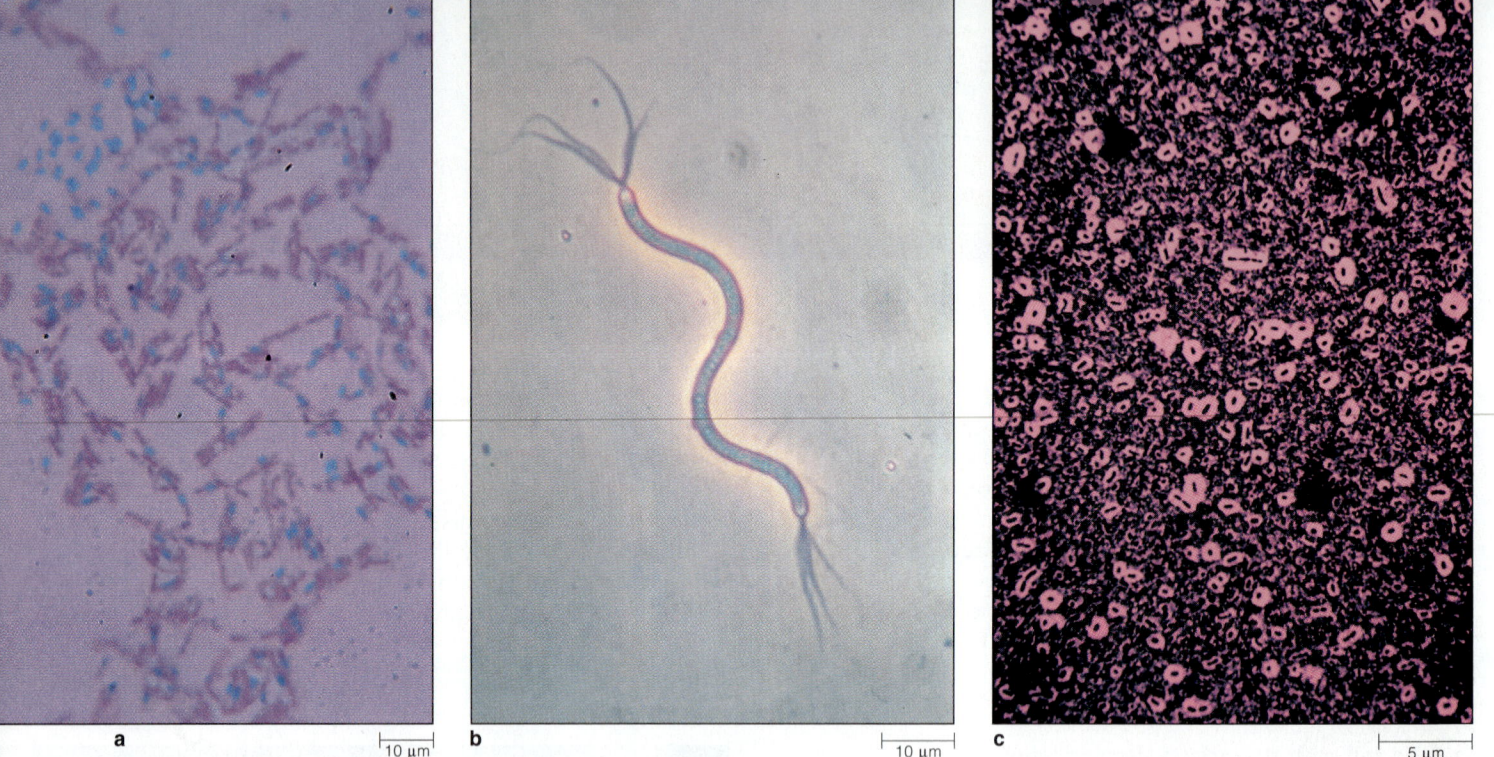

a   10 μm   b   10 μm   c   5 μm

Figure 3.11 Special stains. **(a)** The Wirtz-Conklin spore stain reveals bacteria with endospores. Here, *Bacillus cereus* are revealed as endospore bacteria because they retain the primary green stain while all other bacteria counterstain pink. **(b)** The Leifson flagella stain reveals that these bacteria have flagella, which help identify them as *Spirillum volutans*. **(c)** Negative staining with India ink reveals that these bacteria have capsules, which help identify them as *Klebsiella pneumoniae*, a cause of pneumonia.

(Chapter 4), are too thin to be visible by light microscopy without special staining techniques. Different staining techniques use various combinations of mordants and metals to thicken the flagella and dyes to stain them. The **Leifson flagella stain** involves the following steps:

1. The suspension of bacteria is fixed chemically, with formalin, and spread on a glass slide.

2. It is allowed to air dry without heating.

3. A freshly prepared mixture of tannic acid and rosaniline dye is then added to the slide. The tannic acid thickens the flagella, and the rosaniline stains them.

4. The excess stain is washed off by flooding the slide with water.

5. It is again allowed to air dry before being examined under the microscope.

Flagella staining reveals the number and arrangement of flagella on bacteria, vital information for identifying many species (**Figure 3.11b**). Flagella staining is an art that develops only with practice.

Negative Staining. Some bacteria are surrounded by a protective structure called a **capsule** (Chapter 4). A technique called **negative staining** is used to reveal the presence of a capsule:

1. A wet mount of the specimen is prepared.

2. India ink is added. The carbon particles in the ink cannot penetrate the capsule, so only the background is blackened.

3. Under the microscope, the capsule and cell within are revealed as a clear zone (**Figure 3.11c**).

4. A simple stain may then be applied to make the cell interior visible (most stains do not stain capsules). Instead of India ink, the dye nigrosin may be used.

Light Microscopy: Other Ways to Achieve Contrast

Brightfield is the preferred microscopy in clinical laboratories, and staining is the best way to increase brightfield contrast. But over the years, compound light microscopes have been developed to heighten contrast without having to stain the specimen. These microscopes are the phase-contrast, darkfield, fluorescent, and Nomarsky.

Phase-Contrast Microscopy. The **phase-contrast microscope** gives a clear and detailed image of living unstained cells. In an ordinary light microscope, contrast comes from different materials absorbing different amounts of light. But in the phase-contrast microscope,

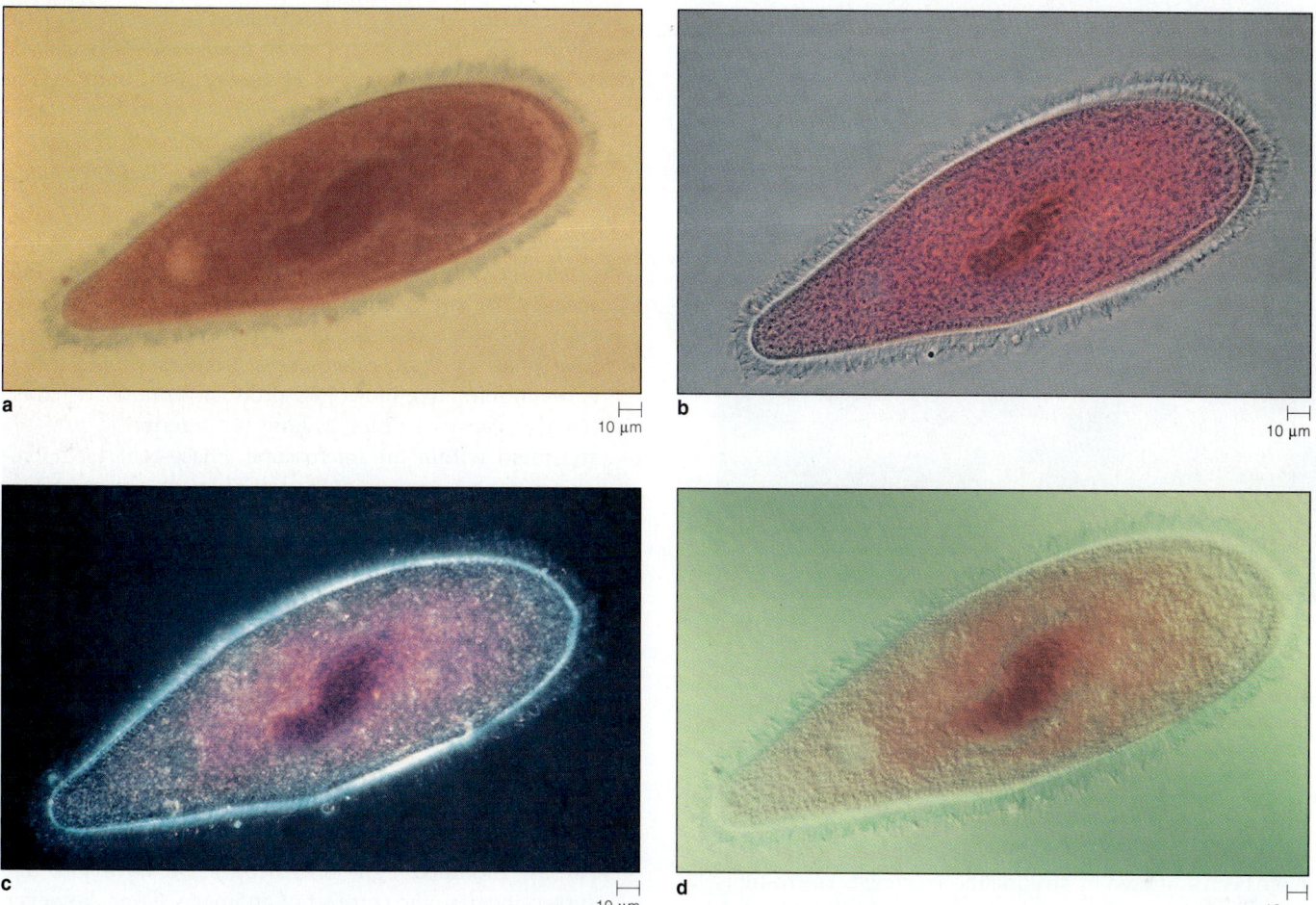

Figure 3.12 Images of *Paramecium* at the same magnification (1000×) under different microscopes. (**a**) Brightfield microscopy shows the shape and larger internal structures such as the nucleus. (**b**) The phase-contrast image shows greater internal detail and the characteristic halo. (**c**) Darkfield microscopy reveals the presence of cilia. (**d**) The Nomarsky image is almost three-dimensional.

contrast comes from the difference in refractive index between the parts of microorganisms and their background. The only structural differences between a standard light microscope and a phase-contrast microscope are in the objective lens and the condenser, which contain special opaque rings.

The phase-contrast microscope is based on the principle that waves of light move at different speeds through materials of different refractive index. So the light rays that pass through the specimen are out of phase with the rays that go around the specimen. Picture ocean waves coming to shore from different directions. They augment or cancel out each other, depending upon whether the peaks arrive at the same or different times. Similarly, the phase-contrast microscope combines the light rays from the specimen and its surroundings. They augment or can-

cel out each other to produce different light intensities and, thus, increased contrast. The consequence of combining out-of-phase light rays is called **interference**.

The major advantage of phase-contrast microscopy is that it can be used to study cellular movement and internal cell structures that have not been distorted by fixation and staining. Phase-contrast shows considerable internal detail. Its disadvantage is that the image is surrounded by a halo of light, an unavoidable consequence of generating contrast by interference (**Figure 3.12**).

Darkfield Microscopy. The **darkfield microscope** also allows viewing of living unstained cells. Darkfield microscopy operates on the principle of **scattering**, which means that a ray of light will change direction, or scatter, when it strikes a small object. A special condenser with a

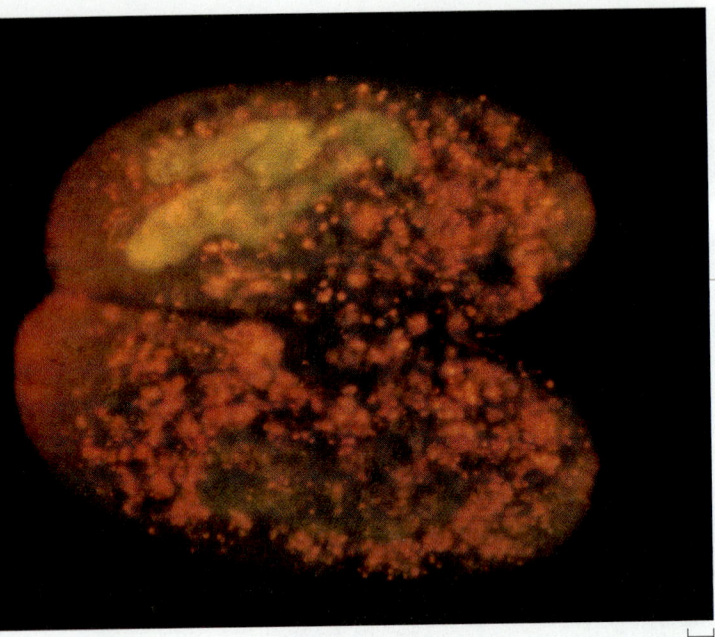

⊢——⊣
10 μm

Figure 3.13 Some microorganisms are naturally fluorescent, but fluorescence can be artificially added. When a fluorochrome is applied to these mating *Paramecium tetraurelia* cells, they glow neon orange.

disk redirects the light beam so that it misses the objective lens; the only rays that enter the lens are those that have been scattered by striking the specimen. The result is a bright image against a dark background (**Figure 3.12c**). Darkfield microscopy is most effective in external structures, especially flagella. It reveals less internal detail of cells than does phase-contrast microscopy. Darkfield technique is used to detect *Treponema pallidum*, the highly motile bacterium that causes syphilis (Chapter 24).

Fluorescence Microscopy. In **fluorescence microscopy**, contrast is increased through **fluorescence**, the property of certain materials to absorb light of one wavelength and give off light of a higher wavelength. When fluorescent materials are illuminated by short-wave (invisible) ultraviolet light, they give off visible light, glowing brightly against a dark background (**Figure 3.13**). Unlike phase-contrast and darkfield techniques, fluorescence microscopy depends on a property of the specimen, not of the microscope. Nevertheless, a microscope must be modified with special filters to illuminate the specimen with short-wave ultraviolet light.

Some microorganisms, including photosynthetic bacteria and algae, are naturally fluorescent. But most microorganisms must be rendered fluorescent by dyes called **fluorochromes**. Some fluorochromes attach readily to microorganisms. Some can be chemically attached to **antibodies**, proteins produced by the immune system that bind to specific microorganisms (Chapter 17). Thus fluorescence microscopy can be used to identify as well as observe microorganisms. **Fluorescent antibody**, or **immunofluorescence, tests** are important diagnostic uses of fluorescence microscopy. They are discussed in detail in Chapter 19.

Nomarsky Microscopy. Based on the principles laid down by Georges X. Nomarsky in the late 1950s, **Nomarsky**, or **differential interference contrast, microscopy** has come into use only during the past 20 years. Like phase-contrast microscopy, Nomarsky uses differences in refractive index and produces contrast by interference. The two differ in how the interfering rays are separated within the microscope. Phase-contrast microscopes use a complementary pair of opaque rings, one located below the condenser and the other on the back side of the objective lens. Nomarsky microscopes use a pair of prisms in similar locations. These different ways to direct the rays produce different kinds of images. Nomarsky microscopy produces an almost three-dimensional image of finer detail than phase-contrast (**Figure 3.12d**). Nomarsky microscopy is used to distinguish living microorganisms in animal tissue.

Electron Microscopy

All the modified light microscopes we have just discussed increase the contrast of an image. None, however, can significantly increase resolving power because the resolution of a light microscope—the extent to which it distinguishes individual points of detail—is limited by the wavelength of visible light. The development of **electron microscopy** in the 1930s was, therefore, a major technical breakthrough. Electron microscopes have vastly greater resolution than light microscopes and therefore produce higher *useful* magnification. There are two kinds of electron microscopes in common use, the **transmission electron microscope (TEM)** and the **scanning electron microscope (SEM)**. The TEM has higher resolving power than the SEM, but the SEM produces a remarkable three-dimensional image.

Transmission Electron Microscopy. The transmission electron microscope (TEM) has greater resolving power than does a light microscope because it uses a beam of electrons rather than visible light to define the object (**Figure 3.14**). The TEM takes advantage of the fact that a beam of electrons behaves as electromagnetic radiation with an extremely short wavelength. A TEM can resolve points as close together as 1 nm, as compared to 200 nm for a light microscope. Thus electron microscopy produces a clear, detailed image at magnifications of

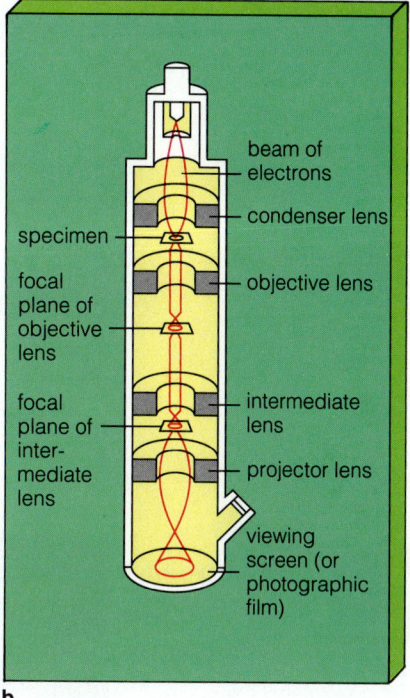

beam of electrons
condenser lens
specimen
objective lens
focal plane of objective lens
intermediate lens
focal plane of intermediate lens
projector lens
viewing screen (or photographic film)

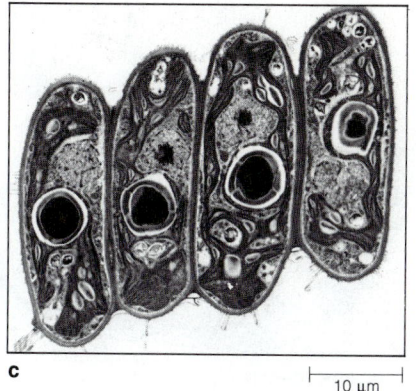

10 μm

a   b   c

Figure 3.14   The transmission electron microscope. (**a**) The TEM is a research tool, much too large and complicated for ordinary laboratory use. (**b**) It uses a beam of electrons and electromagnetic lenses to form images with great resolution and detail. (**c**) The transmission micrograph shows the green alga *Scenedesmus*. Compare this TEM image with the SEM image of the same specimen in Figure 3.17.

up to 200,000×, as compared to about 1000× for a light microscope.

In many ways, the TEM functions much as a light microscope does. Instead of a beam of light, a beam of electrons is transmitted through the specimen. Instead of glass lenses, electromagnetic lenses focus the image. Instead of producing the image directly, it is recorded on a fluorescent screen or a photographic plate (the image is called a **transmission electron micrograph**). Instead of mounting the specimen on a glass slide, it is held on a copper grid that allows electrons to pass through. Instead of using dyes to increase contrast, heavy metals, such as lead, tungsten, and uranium, that cause the specimen to absorb different quantities of electrons are used.

In other ways, the TEM presents special technical problems. The main difficulty is that a beam of electrons has very little penetrating power. As a result, a specimen requires **ultra-thin sectioning** to see any intracellular detail. The specimen must be cut with an instrument called a **microtome** into slices no thicker than 0.1 μm.

Instead of ultra-thin sectioning, specimens can also be prepared by **freeze-fracturing** and **freeze-etching** (**Figure 3.15**). These techniques are especially good for exam-

ining intracellular membranes. The specimen is frozen in ice and fractured with a sharp blow of a knife. The fracture can run through organelles and often a membrane, exposing internal structure. Details of the structure can then be enhanced by freeze-etching. The fractured sample is put in a vacuum and water is removed (a process called sublimation). This leaves solid material protruding above the surface. Then carbon is sprayed on the still-frozen specimen, forming a replica of the surface. The replica is removed and it is viewed by TEM.

Sometimes contrast is increased by **shadow-casting**, coating the specimen with a thin layer of a heavy metal, such as gold or platinum, deposited at an angle (**Figure 3.16**). Objects in the specimen produce "shadows" in the metal layer, giving the image a three-dimensional appearance that reveals their shape and, with some calculation, size.

After these exacting procedures, the specimen must be examined in a vacuum so that air does not interfere with transmission of the electron beam. This means the sample must be completely desiccated. The elaborate preparation and desiccation sometimes introduce **artifacts**, structures or details in the image that do not really

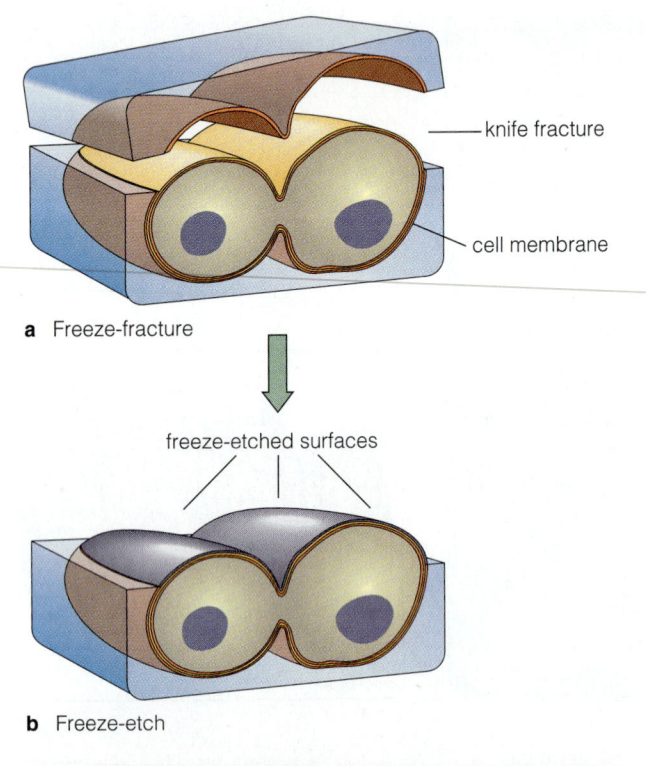

knife fracture

cell membrane

a Freeze-fracture

freeze-etched surfaces

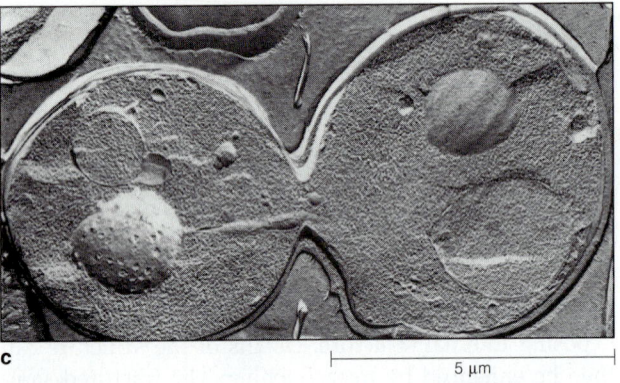

b Freeze-etch

c

5 μm

Figure 3.15 Freeze-fracturing and freeze-etching. Transmission electron microscopy requires special preparation because the electron beam has very little penetrating power. Cells may be prepared by ultra-thin sectioning or they may be freeze-fractured and etched. (**a**) In freeze-fracturing, the sample is frozen into a block of ice and fractured by a sharp blow with a knife. (**b**) The sample may then be freeze-etched for greater detail: The fractured sample is put in a vacuum and water is removed by sublimation. This causes structures to protrude. (**c**) A transmission micrograph of a freeze-fractured and freeze-etched yeast cell that is budding.

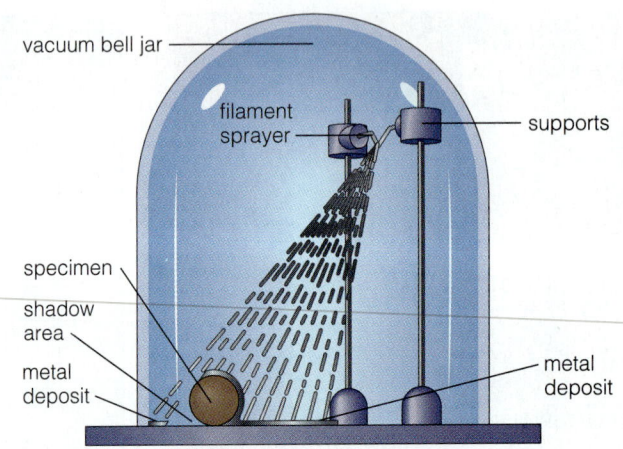

vacuum bell jar

filament sprayer

supports

specimen

shadow area

metal deposit

metal deposit

a Shadow casting

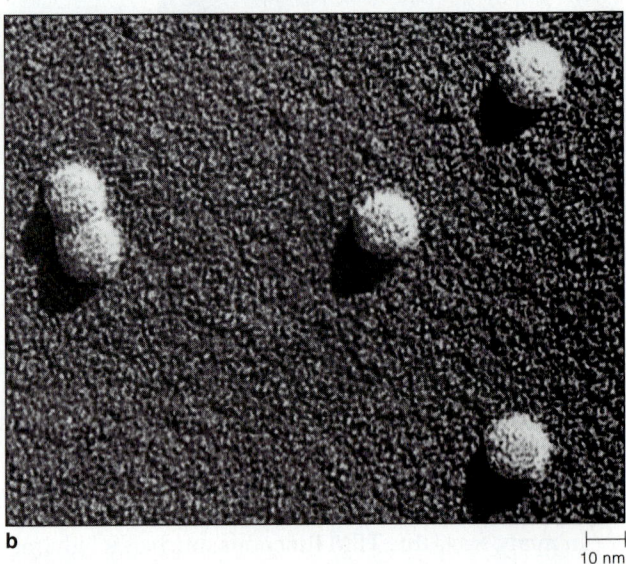

b

10 nm

Figure 3.16 Shadow-casting. The primary advantage of transmission microscopy is great magnification, which allows viewing of viruses and the ultrastructure (internal structure) of cells. However, the TEM can also be used for exterior views of intact structures if the specimen is prepared by shadow-casting. (**a**) The specimen is coated with a thin layer of heavy metal, which gives a three-dimensional appearance. (**b**) A transmission micrograph of a shadow-casted polio virus. It shows all three dimensions: length, width, and thickness.

exist in the specimen. For example, for a long time bacteria were believed to have complex convoluted membranes called **mesosomes**. We now know these are artifacts produced during preparation of the specimen for electron microscopy. Considerable experience is required to interpret electron micrographs.

Scanning Electron Microscopy. The scanning electron microscope (SEM) resembles the transmission electron microscope in its use of a beam of electrons, but in most other ways it is different (**Figure 3.17**). All the other microscopes we have discussed (with the exception of darkfield) are based on the passing of a beam of light or

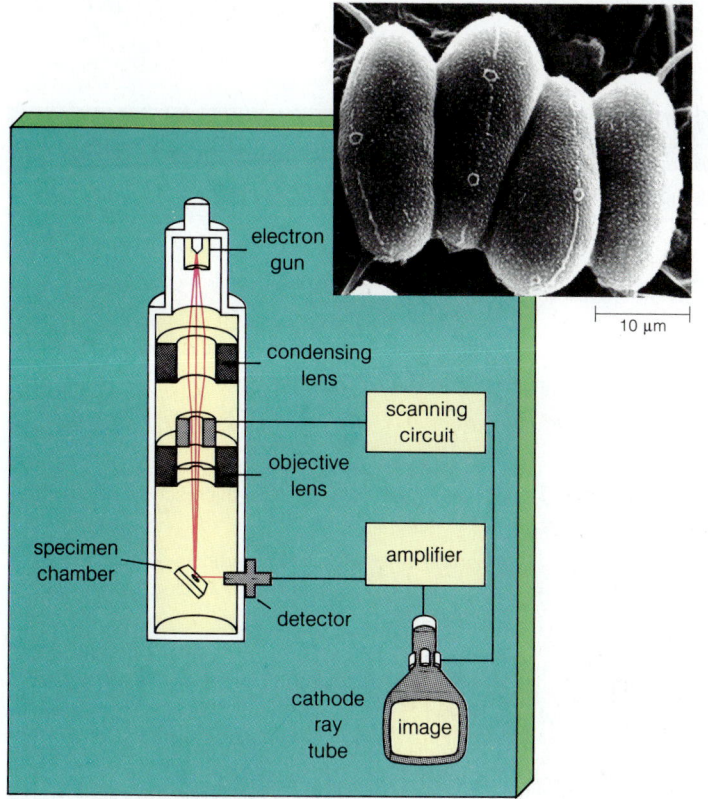

10 μm

Figure 3.17 The SEM uses a beam of electrons to scan the specimen and produce secondary electrons that are then collected to produce the image. It does not transmit light or electrons through the specimen. The SEM produces images of surfaces, therefore, with great three-dimensional quality, as in this image of *Scenedesmus*.

electrons *through* the specimen. The SEM, on the other hand, bombards the surface of the specimen with a beam of electrons. These "primary" electrons eject "secondary" electrons from the specimen. The number of secondary electrons ejected varies with the composition and conformation of different parts of the surface. The secondary electrons are collected to generate a signal that is processed electronically. The processed signals produce an image of the surface on a cathode-ray tube (like the picture tube of a TV set). Because a SEM "sees" only the object's surface, it has great depth of focus, producing striking three-dimensional images. The SEM can resolve objects as close together as 20 nm. Thus, it produces clearly defined detail at magnifications of up to 10,000×.

The scanning electron microscope is usually used with intact cells. Specimens are **freeze-dried** (desiccated in a vacuum while frozen) and then coated with a thin layer of heavy metal, such as gold or platinum. Coating prevents electrons from penetrating the specimen, thereby sharpening the image (called a **scanning electron micrograph**). The SEM also views specimens in a vacuum. Though the SEM is usually used to depict the surface of an object, the freeze-fracture technique allows this powerful microscope also to be directed at structures inside the cell (Figure 3.15).

The Uses of Microscopy

Each of the various kinds of light and electron microscopes has its particular advantages, disadvantages, and uses. In general, light microscopes are for everyday use. Specimen preparation and operation are relatively rapid and simple, and light microscopy provides valuable information about the size, shape, and general appearance of cells. Resolution is limited, however, and therefore so is useful magnification.

Electron microscopy is principally a research tool. The practical problems of working with electrons rather than visible light make these instruments large and complex compared to the compact and simple light microscopes. Electron microscopes allow us to view the detail of cells, sometimes called their **ultrastructure**. It is the only form of microscopy with sufficient resolution and useful magnification to view viruses and smaller objects such as large molecules. **Table 3.3** summarizes light and electron microscopy.

CULTIVATING MICROORGANISMS

In many ways, smallness makes microorganisms ideal experimental subjects. Billions of organisms can be subjected to study, all within a single milliliter of culture. The rapid multiplication of these tiny creatures is also an experimental advantage because many generations can be studied in a day. Moreover, the lessons learned from studying microorganisms can often be generalized to cell systems, plants, and animals, including humans. To do experiments with microorganisms it is usually necessary to cultivate them in the laboratory.

Obtaining a Pure Culture

A **pure culture** consists of a single species of microorganism derived from a single cell. Pure cultures exist rarely in nature. In their natural environment—for example, in soil, water, or the human body—we find **mixed cultures**. In a mixed culture, many different microbial species live together. A pure culture is artificially created by a laboratory scientist.

For the balance of this chapter, we will discuss methods of obtaining and cultivating pure cultures. These methods were developed for bacteria and fungi, but with

Table 3.3   The Uses of Microscopes

Type of Microscopy	Key Features	Appearance	Principal Uses in Microbiology
Light Microscopy			
Brightfield	Contrast achieved by absorption.	Clear image.	Specimens usually require staining. Shows whole organism.
Phase-contrast	Contrast achieved by interference; differences in refractive index shift phase of light from specimen.	Clear detailed images surrounded by halos.	Staining not required, so live cells can be viewed in wet mounts.
Darkfield	Only light scattered by the specimen enters the objective lens, producing high contrast.	Image is brilliantly lit against a black background.	For viewing live cells or flagella that are too thin to be seen by phase-contrast microscopy; shows little internal detail.
Fluorescence	Depends on fluorescence of specimen to give off visible light when illuminated by ultraviolet light.	Specimen is highly colored (luminous) against a dark background.	Can be used to view one particular kind of microorganism in a complex mixture using fluorescent antibodies.
Nomarsky	Uses the same principle as phase-contrast but splits interfering ray with prisms instead of rings.	Produces an almost 3-dimensional image.	Useful for viewing living microorganisms in animal tissue.
Electron Microscopy			
Transmission (TEM)	Passes a beam of electrons through specimen, producing high useful magnification.	Produces a highly magnified image with great detail.	For viewing viruses and ultrastructure of cells. Live cells cannot be viewed because the specimen must be completely desiccated. Can produce artifacts.
Scanning (SEM)	Scans specimen with beam of electrons, ejecting secondary electrons that create a signal which, in turn, generates a computer image.	Produces a 3-dimensional image.	For viewing structure of intact organisms and internal structures in realistic detail. Not for live cells; viewing also requires desiccation.

slight modifications they are used with algae and protozoa. Methods of obtaining pure cultures of viruses are discussed in Chapter 13.

Obtaining a pure culture is a two-step process. First, materials are sterilized to eliminate all microorganisms. Second, one single microorganism is isolated and cultivated to produce a clone of descendants.

Sterilization

Special precautions must be followed to obtain a pure culture because microorganisms exist everywhere in nature, in the laboratory, and on the instruments and containers used in experiments. These unwanted or **contaminating** microorganisms must be eliminated from a culture to make it pure.

Eliminating all microorganisms is called **sterilization**. The principles of sterilization and its practical applications in clinical medicine and industry are discussed in

Chapter 9. Here we will discuss sterilization only as it relates to cultivating bacteria in the laboratory.

All apparatus and materials used to obtain a pure culture must be sterilized. That includes the **medium** (*pl.*, media), the liquid or solid (gelled) material that supplies nutrients to the culture. It also includes the flasks, test tubes, and dishes that hold the media and the apparatus used to transfer a culture from one container to another, pipettes and inoculating needles. In the laboratory, heat, filtration, and chemicals are usually used to sterilize.

Heat Sterilization.   Heat is the method of choice for laboratory materials that are not damaged by high temperatures. The specific requirements for sterilization—to what temperature an object must be heated and for how long—depend on whether moist or dry heat is used.

Moist heat kills microorganisms more readily than dry heat does. Moist heat at 121°C (250°F) for 20 minutes reliably sterilizes most laboratory materials. Bulky materials and large volumes of liquid require more time to

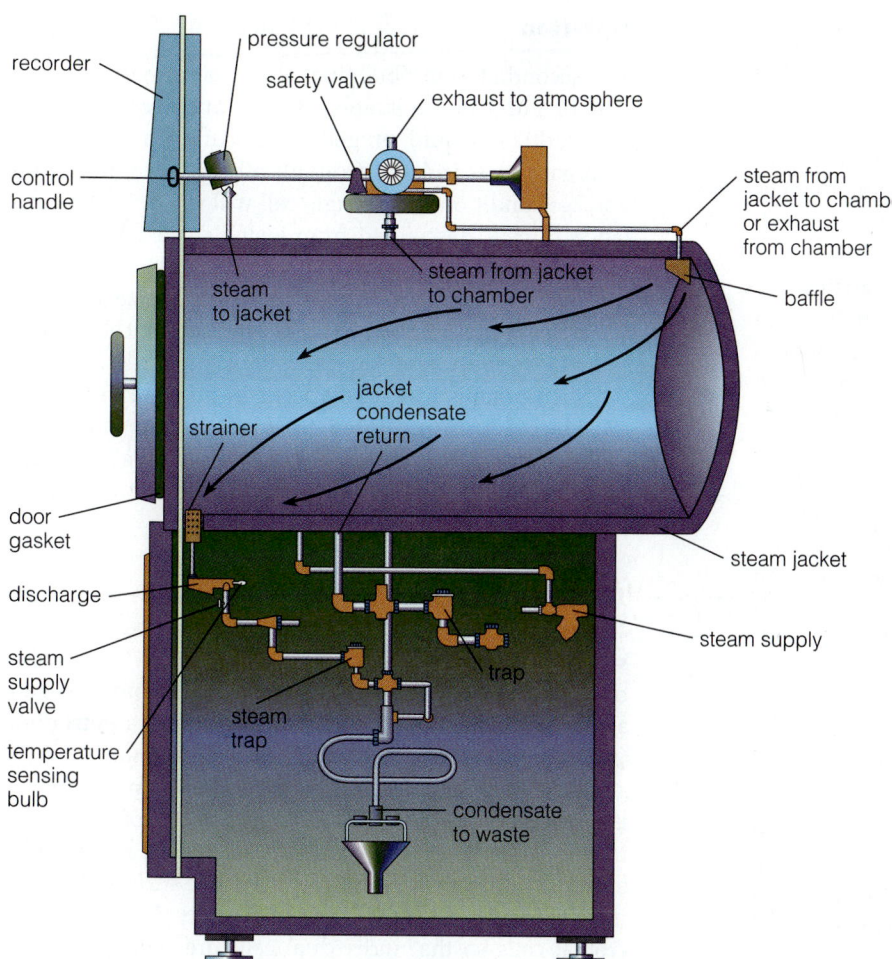

recorder

pressure regulator

safety valve

exhaust to atmosphere

control handle

steam to jacket

steam from jacket to chamber

steam from jacket to chamber or exhaust from chamber

baffle

jacket condensate return

strainer

door gasket

discharge

steam supply valve

temperature sensing bulb

steam trap

trap

steam jacket

steam supply

condensate to waste

Figure 3.18 An autoclave uses steam under pressure to sterilize objects by heat. Most autoclaves are designed so the steam sweeps air out of the chamber; otherwise, objects caught in pockets of air would not be sterilized. They are also fitted with recorders to check that sterilizing temperatures are reached and maintained for the required time.

achieve a lethal temperature at the center. For example, it takes only 15 minutes to sterilize 10 ml of liquid in a test tube, but it takes an hour and 10 minutes to sterilize a 10-L flask that is two-thirds filled with liquid. Because the temperature needed for sterilization is higher than the boiling point of water (100°C or 212°F at sea level), objects cannot be sterilized merely by boiling them in an open vessel. Instead, moist heat sterilization takes place in a pressurized chamber. Under a pressure of 15 pounds per square inch, water can be heated to a temperature of 121°C without boiling. The contents of a chamber raised to this temperature and pressure will be sterilized in 20 minutes provided all the air in the chamber is replaced by steam.

The pressurized container usually used to sterilize objects in a laboratory is an **autoclave** (**Figure 3.18**). In many ways, an autoclave resembles an ordinary home pressure cooker. In fact, in some small microbiological laboratories, pressure cookers are used instead of autoclaves. Both are closed metal containers with walls strong

enough to contain pressurized steam. Autoclaves have large steel chambers and automatically introduce pressurized steam and time the period of sterilization. Objects must be loaded so that steam comes in contact with all parts of each one and so that no air is trapped within the objects. Air that is not replaced by steam creates a dry pocket that may not be sterilized.

Microbiological media and fabrics, such as towels and lab coats, are sterilized in an autoclave. So are other dry materials that can withstand high temperature, including glass pipettes and empty flasks or test tubes, especially if the autoclave has automatic drying. Materials sterilized by moist heat become damp. Glass and metal instruments may also be sterilized by dry heat. They are placed in a hot air oven and heated at 170°C (338°F) for 90 minutes. Fabrics cannot be sterilized by dry heat because the high temperature chars them.

An open flame is also used to sterilize. Any time a flask or test tube is opened to add or remove materials, the neck is first passed through an open flame to kill any microorganisms that might have landed there. This procedure minimizes the chances of microbial cells falling into the vessel and contaminating its contents. The wire inoculating loops that microbiologists use to pick up microorganisms from one surface and move them to another, for example, are often sterilized by placing them in the flame of a Bunsen burner until the metal glows red. This method of sterilization is quick and convenient, but it is dangerous when working with disease-causing microorganisms. Sudden heating in a flame can form an **aerosol**, a suspension of tiny droplets of liquid, containing live microbial cells that might be inhaled by people in the laboratory. Using a small furnace to heat the loop avoids this hazard.

Filtration. Microbial cells can be removed from liquids or gases by filtration. Filtration is more time-consuming and expensive (a new filter must be used each time) than autoclaving, so only liquids that are destroyed by heat—and therefore cannot be autoclaved—are filtered. Solids that change chemical composition if heated (heat-labile) can also be treated by filtration if they are first dissolved in a liquid. Technically, filtration does not sterilize because viruses pass through the filters, but the process is adequate for most routine laboratory purposes. The filters that are used are called **membrane filters**. They are sheets of uniformly porous nitrocellulose, a chemically modified form of cellulose also used as smokeless gunpowder. The pores are approximately 0.45 μm in diameter, small enough to filter out all microorganisms except viruses and a few species of unusually small bacteria. A liquid is filtered by pouring it on a membrane filter fitted to a filter flask. A vacuum is applied to pull the liquid through the filter, leaving microorganisms behind on the filter's surface. Filtration is the preferred method for treating solutions of vitamins, antibiotics, and heat-sensitive compounds that will be added to a medium.

Chemicals. Most chemicals cannot be used to sterilize culture media because they remain, making the media toxic to microorganisms. One chemical agent is in common use, however, when experiments are completed. Sodium hypochlorite (household bleach) is used to destroy cultures of most disease-causing microorganisms so they are not released into the environment. It is also used to sterilize the surface of certain biological materials, such as seeds and plant tissues. Chemicals are also used to minimize contamination throughout the laboratory. Quaternary ammonium salts are used to disinfect laboratory surfaces (benches, tables, etc.). In large institutions, the toxic gas **ethylene oxide** is used to sterilize certain heat-sensitive solid materials, such as clothing and plastic containers, in autoclave-like pressurized chambers.

Isolation

The second step in obtaining a pure culture is to inoculate, or introduce, a single cell of a microorganism into the sterilized liquid or gelled medium. Thus, one microorganism is isolated from all others. Then, under favorable conditions, this single cell will multiply. A population of cells descended from a single cell constitutes a **clone**. Clones large enough to be visible on a solid medium are called a **colony**. A medium-size bacterial colony contains a few billion (10^9) individual cells, about as many as the total number of people on Earth.

Microorganisms usually occur in huge numbers that must be reduced, or **diluted**, in order to isolate individual cells. Dilution and ultimately isolating a single cell are usually done in one of three ways: by the streak plate, pour plate, or spread plate method.

The Streak Plate Method. The easiest and most commonly used method of obtaining a pure culture is the **streak plate method** (**Figure 3.19**). A sterile wire inoculation loop is dipped into a mixed culture and streaked across the surface of a solid agar medium in a **petri dish** (petri dishes are also called plates). As the loop is streaked back and forth, fewer and fewer microorganisms are deposited on the surface. Then the loop is sterilized in a flame, touched to the last-streaked region, and streaked onto a fresh, sterile region of the surface. Repeating this process several times sufficiently dilutes the microbial cells so that individual cells are separated from one another.

Then the plates are **incubated**, allowed to grow in a warm place, until the individual cells have undergone enough divisions to form visible colonies. Although each colony probably represents a clone derived from a single cell, we cannot be sure. Perhaps two cells were deposited so close together that they formed a single mixed colony. Therefore, to be certain that the culture is pure, the entire procedure is repeated starting with an isolated colony on the first streak plate. Isolated colonies that develop the second time are almost surely pure cultures.

The Pour Plate and Spread Plate Methods. In the **pour plate** and **spread plate** methods, suspensions of microbial cells are diluted in liquid *before* they are put on the plate. These two methods are used when the required dilutions are too great to be accomplished in a single step. For example, a suspension with a billion cells per milliliter must be diluted to one ten-millionth of its original concentration to obtain a suspension with a hundred cells per milliliter. Therefore, **serial dilutions** (multistep) are done, usually tenfold (one-tenth original concentration) but sometimes a hundredfold in each step (**Figure 3.20**). For tenfold dilutions, 1 ml of cells is added to 9 ml of ster-

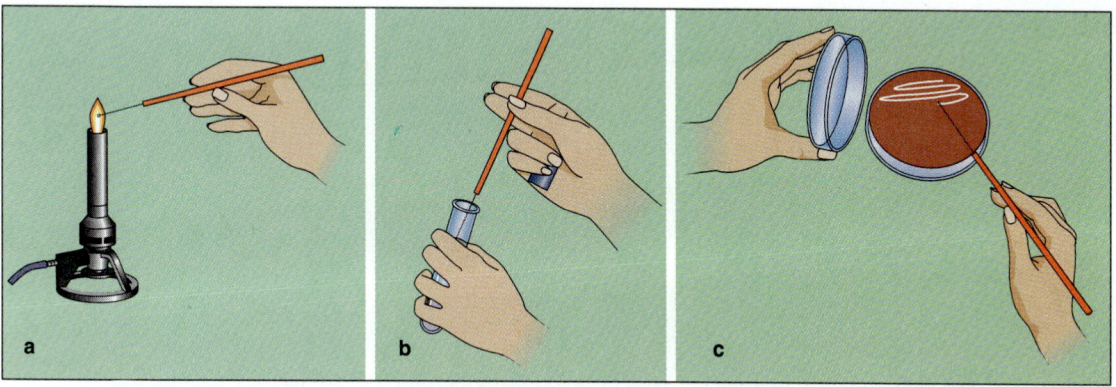

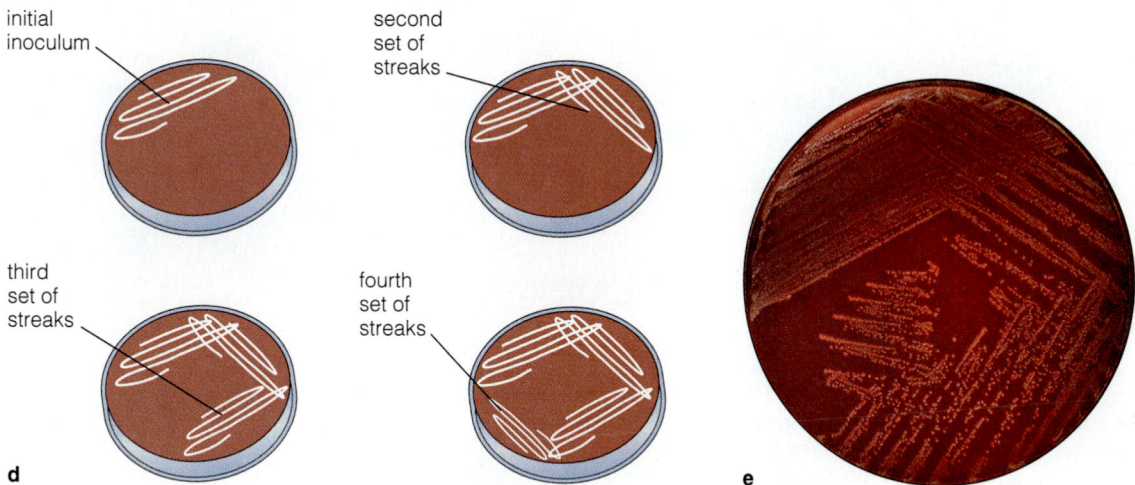

Figure 3.19 Streak plate method. To obtain pure cultures by the streak plate method, (**a**) sterilize an inoculating loop in a flame, (**b**) dip it into a suspension of microbial cells, (**c**) streak it across the solid medium of a petri dish, and (**d**) resterilize the loop, dip it in the streaked region, and create a second set of streaks in a new region. Repeat the process a third and fourth time to dilute the microbial cells until individual cells are separated out. (**e**) After incubation, isolated colonies develop. To be sure the culture is pure, the entire procedure is repeated. A second streak plate is made from an isolated colony on the first streak plate.

ile culture medium or saline (salt) solution. Then the mixture is shaken thoroughly to dilute the cells evenly, and the process is repeated until the cells have been diluted sufficiently. The dilutions are performed the same way for both the pour plate and spread plate methods. But at the point that the sample contains only a few hundred cells, the two methods diverge.

In the pour plate method, the diluted sample is added to melted agar, mixed, and poured into a petri dish. Some colonies that develop are embedded in the agar and some develop on the surface. The surface colonies spread and appear to be larger. In the spread plate method, the diluted sample is poured onto the surface of an agar plate

and spread evenly with a sterile glass rod. The liquid and dissolved material are absorbed into the agar, leaving microbial cells on the surface. In either method the plates are incubated until individual colonies appear. As with the streak plate method, there is no assurance that the colonies that develop on spread plates or within the solidified agar of poured plates are pure cultures until the process is repeated.

The pour plate and spread plate methods produce more isolated colonies than the streak plate method does, so they are preferred when isolating a particular strain from a sample that contains several different types of microorganisms.

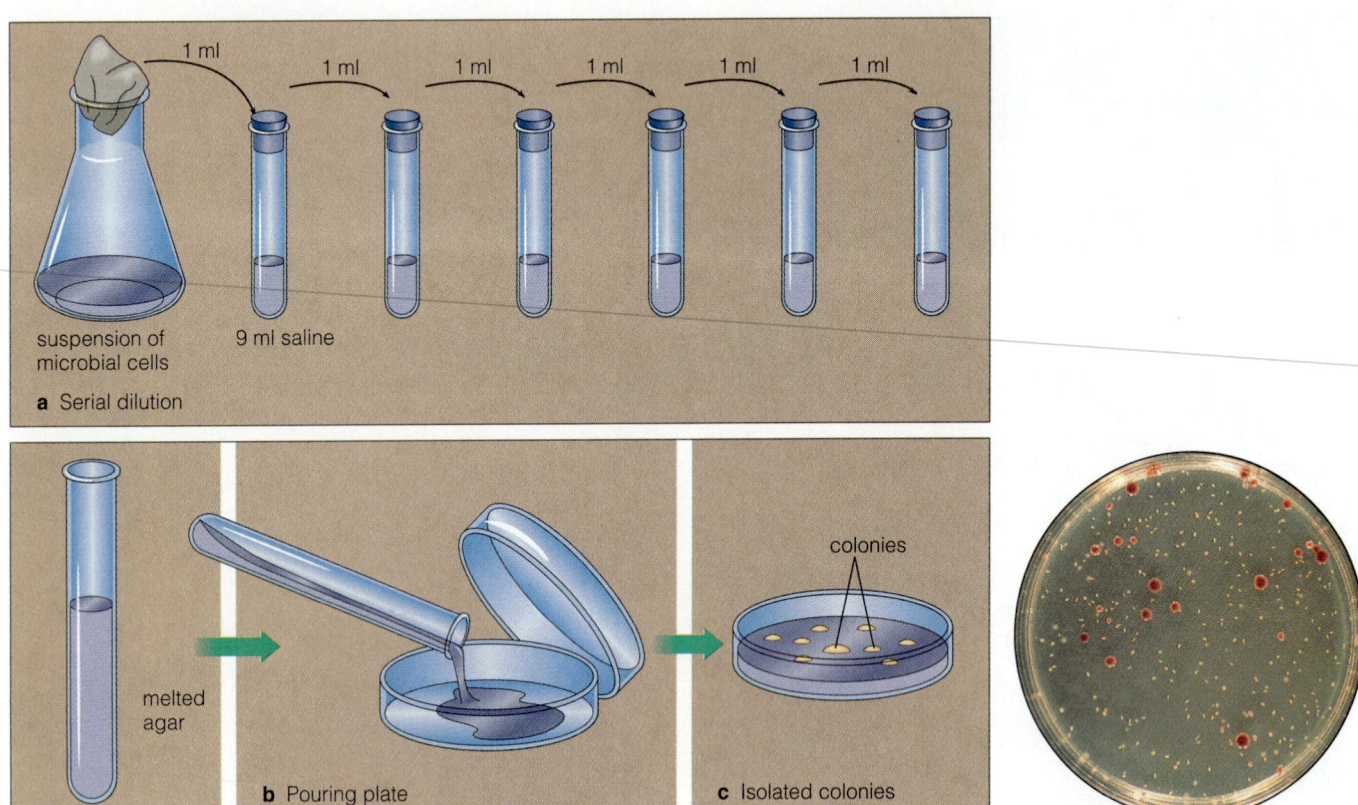

Figure 3.20 Pour plate method. To obtain a pure culture by the pour plate method, **(a)** serially dilute a suspension of microbial cells to get a sample containing only a few hundred cells. **(b)** Pour the sample into a tube of melted agar, mix it, and pour it into a petri dish. **(c)** After incubation, isolated colonies develop in the agar (small colonies) and on its surface (large colonies).

In the figure, labels include: "1 ml" (repeated across arrows), "suspension of microbial cells", "9 ml saline", **a** Serial dilution, "melted agar", **b** Pouring plate, "colonies", **c** Isolated colonies.

Growing a Pure Culture

Once a culture is obtained, it is usually propagated (grown or reproduced). At the least, a researcher wants to obtain greater quantities of cells to experiment with. To cultivate microorganisms in the laboratory, all the nutrients needed for growth must be supplied in the **culture medium**, the liquid or gel (solid) that contains nutrients. Solid media are usually made by adding agar (see the box "Frau Hesse's Pantry" in Chapter 1 to review agar's special properties).

Types of Culture Media. The medium a microbiologist uses depends upon the microorganism and why it is being cultivated. Microorganisms have vastly different nutritional requirements. A minimal medium is used to determine a microorganism's nutritional requirements, while a rich medium is used to obtain a mass of cells quickly. Media can be formulated to favor development of one microbial species or to distinguish among species or strains. Thus, media can be classified as defined, complex, selective, differential, selective-differential, or enrichment, depending upon composition and purpose.

Defined Media. A **defined medium** is one for which we know the exact chemical composition because it is prepared from pure chemicals. *Escherichia coli* can grow in a relatively simple chemically defined medium (**Table 3.4**). It requires an organic source of carbon (for example, glucose), but it can obtain all its other essential nutrients from inorganic salts. In contrast, some organisms, termed **fastidious**, require a very complex chemical medium. The bacterium *Leuconostoc citrovorum*, for example, is extremely fastidious, requiring a medium with many ingredients (**Table 3.5**). Defined media are generally used in genetic studies; but otherwise, the disadvantages often outweigh the advantages. Preparing a defined medium is time-consuming, and bacteria grow more slowly on defined media than on complex media. Also, since we do not know the nutritional requirements of all bacteria, sometimes the option isn't available.

Complex Media. The exact chemical composition of **complex media** is unknown. They are made from extracts of natural materials, such as beef, blood, casein (milk protein), yeasts, and soybeans. A liquid complex medium is

The Importance of Being Pure

All microbiologists know they must have pure cultures to do meaningful experiments. But even skilled microbiologists have been fooled. It happened to Ralph Wolfe, one of our country's most distinguished general microbiologists.

In 1960 Wolfe obtained a culture of the well-studied bacterium *Methanobacillus omelianskii*. *M. omelianskii* is a strict anaerobe that makes methane (natural gas). Wolfe wanted to research how *M. omelianskii* converts ethanol and carbon dioxide into methane gas. The procedure was clear. He had to lyse (burst) the bacterial cells and isolate the enzymes that catalyze the conversion in a test tube. It sounded simple enough. But doing it had frustrated all previous researchers and, for about a year, Wolfe as well. Then Wolfe succeeded. One solution of enzymes made methane gas in a test tube! It was a spectacular breakthrough—until a colleague wondered if the culture were pure.

Wolfe set about repurifying the culture. But instead of adding ethanol and carbon dioxide to the culture, he added hydrogen gas instead of ethanol. Colonies appeared on the new medium, but when they were transferred to a medium with ethanol and carbon dioxide, they did not grow. What had happened? A colleague, M. J. Wolin, realized what was going on. Although *M. omelianskii* formed uniform isolated colonies on a medium with ethanol and carbon dioxide, it was really a mixture of two different bacterial species. One, designated strain S, converted ethanol to hydrogen gas. The other, designated strain M, converted hydrogen gas and carbon dioxide to methane. Neither one alone could grow on ethanol and carbon dioxide. All Wolfe's experiments had been done with a mixed culture. They had to be repeated to determine which enzymes came from strain S and which from strain M.

What lessons are to be learned from Wolfe's story? First, technical problems (such as obtaining a pure culture) that are so basic they are explained in an introductory textbook are very real and plague even the most advanced research scientists. Second, as in this particular case, even a good microbiologist can be fooled. Pairs of species tend to grow together (to form a consortium) and appear to be a pure culture when they are not.

Table 3.4 Ingredients of a Defined Medium Suitable for Cultivating *Escherichia coli*		
Ingredient	Amount	Comments
KH_2PO_4	13.6 g	Source of phosphate and buffers pH changes.
$(NH_4)_2SO_4$	2.0 g	Source of nitrogen and sulfur.
$CaCl_2$	0.01 g	Source of calcium; chloride is not required by bacteria.
$FeSO_4 \cdot 7H_2O$	0.0005 g	Source of iron.
$MgSO_4 \cdot 7H_2O$	0.02 g	Source of magnesium. Since this compound is relatively impure it also serves as a source of trace elements.
Glucose	1.0 g	Source of carbon.
Distilled water	1000 ml	
Agar	15 g	Added if a solidified medium is desired.

Note: Medium is adjusted to pH 7.4 by adding NaOH.

called a **broth**. When casein or other proteins are used in media, they are usually hydrolyzed with enzymes or acid to make them more soluble and therefore nutritionally more readily available. Partial hydrolysis breaks proteins into peptides (Chapter 2). Complete hydrolysis breaks them down to amino acids. Partially hydrolyzed proteins are called **peptones**. Commercially available peptones include proteose peptone, tryptone, and tryptose. Completely hydrolyzed casein is called casein hydrolysate.

The various ingredients of complex media are commercially available as dried powders. Hundreds of mixtures already formulated into specific complex media are also available. One of these is **nutrient broth**, probably the most frequently used complex medium (**Table 3.6**).

Table 3.5	Ingredients of a Defined Medium Suitable for Cultivating *Leuconostoc citrovorum*
Water	1 L
Energy Source	
Glucose	25 g
Nitrogen Source	
NH_4Cl	3 g
Minerals	
KH_2PO_4	600 mg
K_2HPO_4	600 mg
$MgSO_4 \cdot 7H_2O$	200 mg
$FeSO_4 \cdot 7H_2O$	10 mg
$MnSO_4 \cdot 4H_2O$	20 mg
NaCl	10 mg
Organic Acid	
Sodium acetate	20 g
Amino Acids	
DL-Alanine	200 mg
L-Arginine	242 mg
L-Asparagine	400 mg
L-Aspartic acid	100 mg
L-Cysteine	50 mg
L-Glutamic acid	300 mg
Glycine	100 mg
L-Histidine-HCl	62 mg
DL-Isoleucine	250 mg
DL-Leucine	250 mg
L-Lysine·HCl	250 mg
DL-Methionine	100 mg
DL-Phenylalanine	100 mg
L-Proline	100 mg
DL-Serine	50 mg
DL-Threonine	200 mg
DL-Tryptophan	40 mg
L-Tyrosine	100 mg
DL-Valine	250 mg
Purines and Pyrimidines	
Adenine sulfate·H_2O	10 mg
Guanine·HCl·$2H_2O$	10 mg
Uracil	10 mg
Xanthine·HCl	10 mg
Vitamins	
Thiamine·HCl	0.5 mg
Pyridoxine·HCl	1.0 mg
Pyridoxamine·HCl	0.3 mg
Pyridoxal·HCl	0.3 mg
Calcium pantothenate	0.5 mg
Riboflavin	0.5 mg
Nicotinic acid	1.0 mg
p-Aminobenzoic acid	0.1 mg
Biotin	0.001 mg
Folic acid	0.01 mg

Source: H. E. Sauberlich and C. A. Baumann, "A Factor Required for the Growth of *Leuconostoc citrovorum*," *Journal of Biological Chemistry* 176 (1948):166.

Table 3.6	Ingredients of Nutrient Broth, a Complex Medium Suitable for Cultivating Many Species of Bacteria		
Ingredient	Amount	Comments	
Peptone	5 g	Casein that has been partially hydrolyzed by the enzyme trypsin	
Beef extract	3 g	Dried solids of a hot water extract of beef	
NaCl	8 g	Added to keep cells from clumping	
Distilled water	1000 ml		

Note: When nutrient broth is solidified by adding 15 g of agar per liter, it is called nutrient agar.

When solidified with agar, it is called **nutrient agar**. Complex media are generally favored because they are easy to use and they promote rapid growth.

Selective Media. A **selective medium** favors the growth of certain microorganisms while suppressing the growth of others. Selective media are used to isolate a particular species from a complex mixture. For example, a selective medium is used to isolate *Salmonella typhi*, the bacterium that causes typhoid fever, from feces, which contain hundreds of different microorganisms. Some selective media work on the basis of chemicals, such as sodium azide, potassium tellurite, or crystal violet, that inhibit development of some microorganisms but not others. SPS agar (so named for the chemicals it contains, sulfadiazine and polymyxin sulfate) is used to identify *Clostridium botulinum*, the bacterium that causes a dangerous food poisoning. SPS promotes the growth of the *botulinum* species and suppresses other *Clostridium* species. Some types of selective media employ an extreme pH value or an unusual carbon source.

Differential Media. A **differential medium** is used to identify colonies of a particular type of microorganism. *Streptococcus pyogenes*, the bacterium that causes strep throat (Chapter 22), is usually identified using the differential medium blood agar, which is agar containing red blood cells. *S. pyogenes* shows up as colonies surrounded by a clear zone because they lyse (kill by bursting) the red blood cells around them (**Figure 3.21**). *Escherichia coli* and related bacteria are identified on media containing a pH indicator because they produce acidic metabolic products that change the color of the indicator and, hence, their colonies.

Selective-Differential Media. Some media are both selective and differential. MacConkey agar is an example

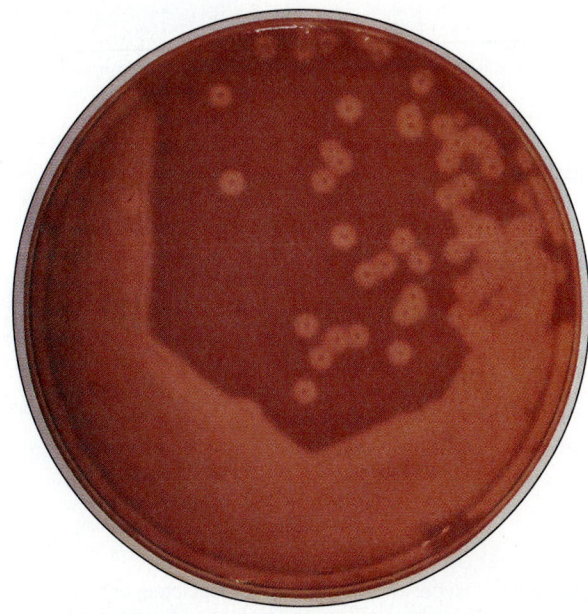

Figure 3.21 Differential medium. *Streptococcus pyogenes* differentiates itself from other bacteria on blood agar because it lyses red blood cells to create a clear zone around each colony.

of a **selective-differential medium**. It is used to detect strains of *Salmonella* and *Shigella*, enteric (intestinal) bacteria that cause dysentery (Chapter 23). Any stool sample sent to the laboratory teems with vast numbers of growing bacteria of many species. The selective component of MacConkey therefore acts as a kind of coarse screen to narrow the field of identification. Crystal violet and bile salts in MacConkey inhibit the growth of most Gram-positive bacteria (*Salmonella* and *Shigella* are Gram-negative). Now the differential component of MacConkey becomes important. The differentiating ingredient is lactose. *Salmonella* and *Shigella* do not ferment lactose, but most other enteric bacteria do. Therefore, *Salmonella* and *Shigella* are differentiated when their colonies turn red and other enteric bacteria colonies do not.

Enrichment Culture. An **enrichment culture** is used to isolate a particular microorganism or type of microorganism from a large, complex natural population. For example, endospore-forming bacteria can be isolated by boiling a sample of soil and culturing the survivors. Only endospores survive boiling temperatures and grow. Nitrogen-fixing bacteria can be isolated by culturing a soil inoculum in a nitrogen-free medium. Only nitrogen-fixing bacteria will grow because they derive nitrogen from the atmosphere. Environmental microbiologists frequently use enrichment cultures. For example, to find a microorganism that can break down a particular toxic

chemical, they might inoculate soil into a medium in which the toxic chemical is the only source of carbon. The microorganism that flourishes is the one that might be used for bioremediation (Chapter 29).

The chemical classes of nutrients that various microorganisms need to grow are discussed in Chapter 8.

Providing a Suitable Laboratory Environment

A medium is formulated to provide a microorganism with the nutrients it needs to grow, but this is not sufficient. A microorganism in the laboratory must also be provided with a suitable environment. Temperature and pH must be maintained within proper ranges, and oxygen must be provided or excluded.

Temperature. Different microbial species have characteristic temperature ranges over which they grow best (Chapter 8). But, in general, microorganisms grow more rapidly as temperature increases, until it reaches a point that damages the microorganism's proteins (Chapter 2). To foster rapid growth, then, bacteria are cultivated in as warm an environment as they can tolerate, which usually reflects their natural environment. *Escherichia coli*, the bacterium used most commonly for microbiological experimentation (Chapter 5), exists naturally in the intestine of humans and other mammals. It grows best in the laboratory at 37°C, the temperature of the human body.

Cultures are maintained at a constant temperature in the laboratory in thermostatically controlled **incubators** or **water baths**. Incubators are air-filled chambers suitable for growing cultures on solid or liquid media in any sort of container, including petri dishes, test tubes, or flasks. Liquid media in test tubes or flasks can be held in water baths, containers of warm water. Water baths are convenient because they can be kept on the laboratory bench so cultures can be readily sampled.

pH. Optimum pH varies among microbial species, but any particular species can grow only within its own relatively narrow range. Most bacteria grow best at pH values near neutrality—in the range of 6.5 to 7.5. Fungi, on the other hand, grow better in a pH range between 4.5 and 6.0.

Though the pH of a medium may be ideal for a species when the culture is first inoculated, microbial growth often changes it. This happens because some microorganisms selectively use acidic or basic components from the medium and produce acidic or basic materials as by-products of their growth. To minimize changes in pH value in defined media, buffers (Chapter 2) are usually added. Buffers are added less frequently to complex media because these natural materials act as weak buffers. The most effective buffers at near-neutral values of pH are phosphate-containing salts and calcium carbonate. Both are often added to supply nutritional requirements for phosphorus and calcium, but if they are to be used for buffering as well, larger quantities must be added.

Oxygen. The concentration of oxygen in the environment is a critical determinant of microbial growth (Chapter 8). Some microorganisms, called **strict aerobes**, cannot grow without oxygen because they need it to obtain energy. Other microorganisms, **strict anaerobes**, cannot grow in the presence of oxygen. For them, oxygen is toxic. Other types of organisms, such as **facultative anaerobes** and **aerotolerant anaerobes**, can grow either in the presence or the absence of oxygen. Facultative anaerobes use oxygen when it is available but can do without it. Aerotolerant anaerobes cannot use oxygen, but it does not adversely affect them. **Microaerophiles** need low concentrations of oxygen. Higher concentrations, as in air, are toxic to them. Thus, depending upon the microorganism being cultured, oxygen must be either provided, restricted, or excluded from the laboratory environment.

Providing oxygen to bacterial species that need it or that grow more rapidly in its presence presents technical difficulties. Aerobes growing on the surface of agar plates obtain adequate amounts of oxygen directly from the air, although the interior of each colony is completely anaerobic. Providing oxygen to organisms growing in a liquid culture is more difficult because not much oxygen dissolves in water and aerobic microorganisms use it rapidly. Liquid cultures deeper than a centimeter or two must be aerated, actively supplied with oxygen. This is usually done by forcing a stream of air through the culture or by shaking it continuously. Most microbiological laboratories have shaking machines, platforms with holders for flasks that are rotated or shaken continuously to mix oxygen into the culture. Supplying oxygen to cultures in thousands of gallons of medium, such as those used commercially to produce antibiotics, is an engineering challenge. Cultures are stirred rapidly by large propellers while huge volumes of air are forced through them.

Excluding oxygen from the environment of anaerobes also presents technical problems. Many anaerobes can be cultivated in test tubes that are exposed to air if the medium contains a chemical such as **thioglycolate**, which reacts with oxygen, keeping the lower part of the tube oxygen-free. Anaerobes can grow on the surface of petri dishes if they are placed in containers where oxygen has been completely replaced by an inert gas such as nitrogen or argon. Oxygen can also be eliminated chemically. Commercially packaged combinations of reactants will produce hydrogen gas when water is added, and the

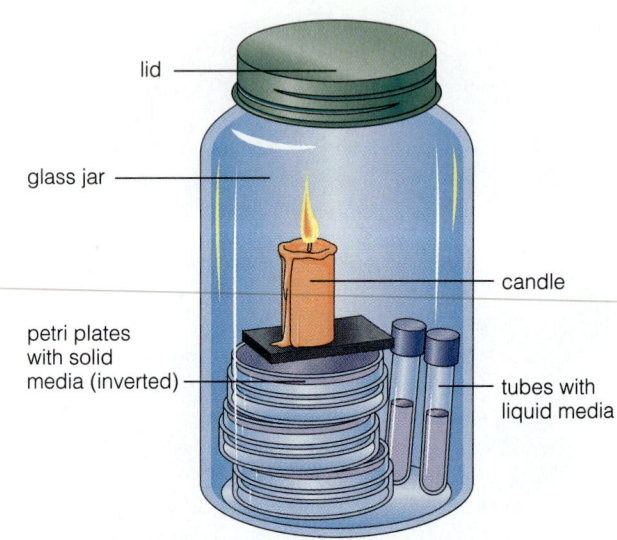

Figure 3.22 Oxygen must be provided, restricted, or totally removed from the laboratory environment, depending upon a microorganism's growth requirements. Candle jars are used for microaerophiles, which require low concentrations of oxygen, and aerotolerant anaerobes, which can grow in either the presence or absence of oxygen.

hydrogen reacts with oxygen in the presence of a catalyst. The reaction produces water and uses up the oxygen. A simple method for lowering oxygen (and simultaneously increasing the concentration of carbon dioxide, which benefits some microorganisms) is to burn a candle in a closed container until it is extinguished from having used up most of the oxygen. These **candle jars** were once commonly used to cultivate aerotolerant anaerobes, such as lactic acid bacteria. They have also been used for microaerophiles, especially when enriched with carbon dioxide (**Figure 3.22**).

Strict anaerobes, which are killed by even brief exposure to oxygen, can be grown in glass containers that have an airtight seal. When the container must be opened to add media or remove samples, a stream of nitrogen gas is used to flush air out of the culture vessel. Occasionally, more elaborate procedures are required. Sometimes anaerobic cultures are maintained within an **oxygen-free glove box**. All work is done using the gloves that extend into the box (**Figure 3.23**). Some laboratories have oxygen-free rooms where the technicians wear oxygen masks.

Preserving a Pure Culture

Once a pure culture is obtained, microbiologists often want to keep it for later experimentation or share it with other scientists. They could continue to cultivate the organism by periodic monthly or even daily transfer to a

And They Grew Happily Ever After

Douglas Nelson, a general microbiologist working today, wanted to grow strains of the bacterium *Beggiatoa*. *Beggiatoa* are interesting for the way they obtain metabolic energy—they oxidize reduced sulfur compounds. Nelson knew these bacteria grow and multiply in the presence of hydrogen sulfide and oxygen, plus a few other nutrients. But he also knew these gases were incompatible. Hydrogen sulfide and oxygen react spontaneously. They exist together for only a short time. But Nelson had an idea. He would use an agar gel. Although an agar gel is quite firm, it is composed

almost entirely of water, so dissolved substances diffuse through it readily. Why not let hydrogen sulfide diffuse to the bacteria from one direction and oxygen from another, meeting where *Beggiatoa* could use them? Nelson filled a test tube with agar and provided hydrogen sulfide at the bottom. He left the tube open to air at the top, and inoculated it with *Beggiatoa*. His plan worked. Hydrogen sulfide diffused up from the bottom, oxygen from the air diffused down from the top, and *Beggiatoa* grew in a thin line where they met—about halfway down the tube.

fresh medium. But a pure culture can also be preserved in a nongrowing state and remain viable for years. Cultures that are maintained for study and reference are called **stock cultures**. Most laboratories have stock cultures of strains they have isolated or that they have obtained from **stock culture collections**, organizations that maintain huge numbers of stock cultures as a service to microbiologists. There are many preservation techniques, but most involve either desiccation (removal of all water) or low-temperature storage.

A culture is often desiccated by **lyophilization**, freeze-drying. A liquid culture of the microorganism to be preserved is poured into a small test tube. Skim milk is added to protect the cells during the process. The mixture is frozen with dry ice and exposed to a vacuum for sublimation (removal of all water while frozen). Sublimation keeps the culture frozen during desiccation. Then, while still exposed to the vacuum, the test tube is sealed by fusing the neck shut in an intense flame. Once lyophilized, cultures are stored at room temperature.

A culture to be preserved by low-temperature storage is first mixed with a substance that minimizes freeze-killing—skim milk, glycerol, or dimethyl sulfoxide (DMSO). The test tube is then sealed and stored below –50°C, in either liquid nitrogen or an **ultradeep-freeze**, a freezer that is capable of maintaining especially low temperature.

Microorganisms That Cannot Be Cultivated in the Laboratory

Many microbial species cannot be cultured in an artificial environment. In fact, evidence suggests that the majority of microbial species probably cannot be cultured in the

laboratory. Samples have been taken from soil and the number of different microbial species visible under the microscope counted. When that same sample is cultivated, less than half the number of species are present, regardless of the technique used.

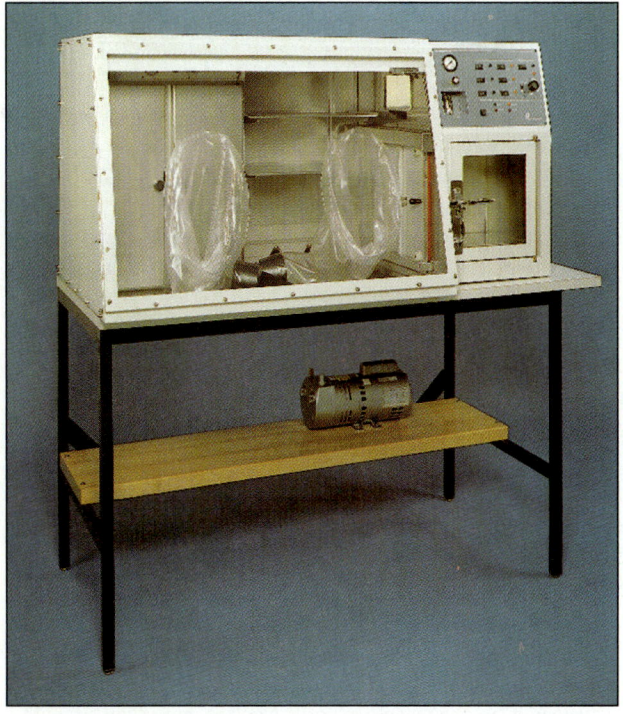

Figure 3.23 Oxygen-free glove box. Strict anaerobes are killed by even the briefest exposure to oxygen. Less elaborate procedures usually suffice.

Several extremely serious disease-causing bacteria cannot be cultivated in laboratory media, but they can be cultivated in intact animals. *Treponema pallidum*, which causes the sexually transmissible disease syphilis (Chapter 24), is cultivated by inoculating it into rabbits. *Mycobacterium leprae*, which causes Hansen's disease (leprosy), is cultivated in armadillos (Chapter 26). The rickettsia and chlamydia bacterial families (Chapter 11) and all viruses can be cultured only within living host cells because they are obligate intracellular parasites. In addition to intact animals, though, they can be grown in **tissue cultures**, cultures of animal cells.

Actually, most microorganisms that cannot be cultured in the laboratory have never been identified. We don't know how many of these mystery organisms exist or what their importance may be. We expect, however, that as genetic engineering technology advances, we will be able to identify microorganisms and even classify them according to species merely by recovering a few fragments of their genes. When this happens it will be possible to study microorganisms without culturing them.

Summary

VIEWING MICROORGANISMS (pp. 47–63)

1. Unless microorganisms can be seen and cultivated, little can be learned about them.

Properties of Light (pp. 47–49)

1. Light is the part of the spectrum of electromagnetic waves that is visible to the human eye.

2. Wavelength is the distance between two peaks or two troughs. The intensity of a light wave is the height of the wave.

3. If a ray of light hits an object and bounces back, it is reflected. If it passes through the object, it is transmitted. If it transfers some of its energy to the object, it is absorbed.

4. When light rays pass through a small opening or by the edge of an opaque object, they bend. This kind of bending is called diffraction.

5. Refraction is bending that occurs when a ray of light enters an object with a different density at an angle.

Microscopy (pp. 49–52)

1. All microscopes depend upon the same three factors to produce a clear image. Magnification is the enlargement of an object. Contrast refers to the difference in light intensity. Good resolution means being able to perceive two adjacent parts of an image as separate from each other.

2. The resolving power of a microscope can be increased by using a larger magnifying lens, by illuminating the specimen with a shorter wavelength, and by using a material (such as immersion oil) with a higher refractive index than air between the objective lens and the specimen.

3. Numerical aperture (NA) is a measurement of the objective lens size and depends on whether the lens is to be used with immersion oil.

The Compound Light Microscope (pp. 52–53)

1. A light microscope uses visible light as a source of illumination. A compound light microscope has two lens systems, the objective lens and the ocular lens. Another series of lenses called the condenser directs light through the specimen. Light transmitted through the specimen forms an image in the tube of the microscope, which the ocular lens magnifies and projects to the lens in the eye.

2. Compound light microscopes typically are provided with several objective lenses: low power (10×), high power (40×), and oil immersion (100×).

3. An ordinary compound light microscope gives brightfield illumination—light goes through the specimen and the background is brightly lit.

Wet Mounts (pp. 53–54)

1. A simple wet mount is used to observe living microorganisms. A hanging drop mount keeps the microorganisms from drying out, allowing longer viewing. Neither preparation provides much contrast.

Stains (pp. 55–58)

1. Stains are chemical dyes used to increase contrast. Most stains are effective only after microorganisms are fixed—killed and attached to a microscope slide. Basic dyes are composed of positively charged ions, and acidic dyes are composed of negatively charged ions.

2. A mordant is a compound that increases a specimen's affinity for a dye or that coats a structure to make it more visible.

3. A simple stain uses only one dye. A differential stain involves two steps, a primary stain and a counterstain.

4. The Gram stain, a type of differential stain, distinguishes between Gram-positive and Gram-negative bacteria, reflecting differences in their outer surface.

5. The acid-fast stain—or Ziehl-Neelsen stain—is a differential stain that colors mycobacteria red and all other bacteria blue during counterstaining.

6. The Wirtz-Conklin spore stain identifies endospores, dormant structures in some bacteria. The Leifson flagella stain uses stains and mordants to thicken flagella, threadlike appendages used for motility. Negative staining is used to reveal the protective capsule some bacteria have.

Light Microscopy: Other Ways to Achieve Contrast (pp. 58–60)

1. In phase-contrast microscopy, contrast comes from different materials absorbing different amounts of light. Because it can be used with living unstained cells, it can be used to study cellular movement. There is no distortion from fixing and staining, making it especially good for showing internal detail. There is an unavoidable halo of light around the image.

2. In darkfield microscopy, the image is bright against a dark background. It can be used with living unstained cells.

3. Fluorescence microscopy increases contrast through fluorescence and can be used to identify as well as observe microorganisms. The fluorescent antibody test (or immunofluorescent test) is an important diagnostic tool.

4. Nomarsky microscopy produces contrast by interference, as phase-contrast does, but gives a three-dimensional-like image with finer detail.

Electron Microscopy (pp. 60–63)

1. The electron microscope provides much greater resolving power and, therefore, greater useful magnification than a light microscope.

2. The transmission electron microscope (TEM) uses a beam of electrons rather than visible light to define the object. It produces magnifications up to 200,000× (as compared to about 1000× for a light microscope).

3. The TEM requires special preparation of specimens, such as ultra-thin sectioning, freeze-fracturing, or freeze-etching. Sometimes contrast is increased through shadow-casting, coating with a heavy metal.

4. The scanning electron microscope (SEM) bombards the surface of the specimen with a beam of electrons in such a way that it produces images with apparent three-dimensional depth.

5. SEM preparation involves freeze-drying (desiccation in a vacuum while frozen) and then coating with a heavy metal. Specimens are viewed in a vacuum. SEM is usually used with intact cells, but freeze-fracturing and freeze-etching allow interior structures to be viewed.

The Uses of Microscopy   (p. 63)

1. In general, light microscopes are for everyday use, giving valuable information about the size, shape, and general appearance of cells. But since resolution is limited, so is magnification.

2. Electron microscopes, which allow a view of the ultrastructure of cells, are used in research. Viruses and smaller objects, such as macromolecules, can only be seen with electron microscopes.

CULTIVATING MICROORGANISMS   (pp. 63–74)

1. Microorganisms are ideal laboratory subjects because billions can be studied in a single milliliter of culture, because they multiply rapidly, and because what we learn can often be generalized to cell systems, plants, and animals (including humans).

Obtaining a Pure Culture   (pp. 63–64)

1. A pure culture consists of a single species of microorganism derived from a single cell. Mixed cultures occur in nature.

Sterilization   (pp. 64–66)

1. The first step in obtaining a pure culture is sterilization—eliminating all microorganisms from an area. All instruments and materials involved in obtaining a pure culture must be sterilized, including the medium.

2. Heat sterilization kills by denaturing the proteins in microbes. Moist heat kills faster than dry heat, but the temperature required is higher than the boiling point of water; an autoclave is usually used.

3. Dry heat is used for materials that would be damaged by moisture. Hot air sterilization requires temperatures of at least 171°F for a minimum of one hour. Direct flaming is another form of dry-heat sterilization.

4. Filtration is passing a fluid through a barrier with holes too small for cellular microorganisms. Because viruses pass through filters, filtration does not remove them.

5. Chemical sterilization is used to destroy cultures after an experiment and to minimize contamination in the laboratory.

Isolation   (pp. 66–67)

1. The second step in obtaining a pure culture is to inoculate, or introduce, a single cell of a microorganism into the sterilized medium. A population of cells descended from a single cell constitutes a clone. Clones grown on a solid medium large enough to be visible are called a colony.

2. Before a single cell can be isolated, the cells in a culture must be diluted.

3. In the streak plate method, an inoculation loop is dipped into a mixed culture and streaked across a solid medium in a petri dish.

The process is repeated until individual cells are isolated. The cells are incubated until colonies form. Then the entire process is repeated.

4. In the pour plate and spread plate methods, suspensions of microbial cells are diluted *before* they are put on the plate. In the pour plate method, the diluted sample is added to melted agar, mixed, and poured into a petri dish. In the spread plate method, the diluted sample is poured onto the surface of an agar plate and spread evenly with a glass rod. Plates are incubated until colonies appear. Then the process is repeated.

Growing a Pure Culture   (pp. 68–71)

1. A defined medium is prepared from pure chemicals. Fastidious organisms require a very complex chemical medium. Defined media are time-consuming to prepare and bacteria usually grow more slowly than on complex media.

2. Complex media are made from extracts of natural materials such as beef, blood, or casein. A liquid complex medium is called a broth. Complex media are easy to prepare and promote rapid growth.

3. Selective media favor the growth of certain microorganisms while suppressing the growth of others. They are used to isolate a particular species from a complex mixture.

4. Differential media, such as blood agar, are used to identify colonies of a particular type of microorganism. Selective-differential media both narrow the field of identification and then identify a specific bacterium.

5. An enrichment culture is used to isolate a particular microorganism or type of microorganism from a large, complex natural population.

Providing a Suitable Laboratory Environment   (pp. 71–72)

1. In general, microorganisms grow more rapidly in warmer temperatures. To minimize pH changes, buffers are usually added to media.

2. Strict aerobes require oxygen to obtain energy. Strict anaerobes cannot grow in the presence of oxygen. Facultative anaerobes use oxygen if it is available but can do without it. Aerotolerant anaerobes cannot use oxygen but it does not adversely affect them. Microaerophiles need low concentrations of oxygen. Providing, restricting, or excluding oxygen from the environment all present technical problems.

Preserving a Pure Culture   (pp. 72–73)

1. Cultures that are preserved for study and reference are called stock cultures.

2. A culture can be preserved by desiccation (removing all water). Usually a culture is desiccated by lyophilization, freeze-drying. Cultures can also be preserved by low-temperature storage. After freezing, the culture is stored below –50°C, either in liquid nitrogen or in an ultradeep-freeze.

Microorganisms That Cannot Be Cultivated in the Laboratory   (pp. 73–74)

1. Many microorganisms cannot be cultivated in laboratory media, including the bacteria that cause syphilis and Hansen's disease (leprosy).

2. The rickettsia and chlamydia bacterial families and all viruses can be cultivated only in living host cells. Intact animals or tissue cultures are used.

VIEWING MICROORGANISMS

1. Why is understanding the properties of light important to the study of microorganisms? What is light?

2. Explain this statement: When a ray of light strikes an object head-on, it has three possible fates.

3. Explain these terms: reflection, diffraction, refraction, refractive index.

4. All microscopes are based on the same three factors—magnification, contrast, and resolution. Explain each of these factors.

5. What is the resolving power of a microscope? How can resolving power be increased? What is the numerical aperture (NA) of a microscope lens?

6. Sketch a compound light microscope and label these parts: objective lens, ocular lens, condenser, body tube.

7. Explain why a compound light microscope has lens systems and corrected lenses. What is brightfield illumination?

8. Explain how you would prepare a wet mount and a hanging drop mount. What are the advantages and disadvantages of this type of viewing?

9. What are the advantages and disadvantages to staining a sample?

10. Explain these terms: basic dye, acidic dye, mordant.

11. Explain the difference between these types of stains:
    a. simple stain   c. Gram stain
    b. differential stain   d. acid-fast stain (Ziehl-Neelsen stain)

12. What is a special stain? Tell how these special stains are used: Wirtz-Conklin, Leifson flagella stain, negative stain.

13. Compare and contrast the various types of light microscopy. For each, tell the underlying principle, describe the image, and give the advantages and disadvantages.
    a. brightfield illumination   d. fluorescence microscopy
    b. phase-contrast microscopy   e. Nomarsky microscopy
    c. darkfield microscopy

14. Answer question 6 for the two types of electron microscopy. Also describe the preparation techniques—ultra-thin sectioning, freeze-fracturing, freeze-etching, and freeze-drying.

15. What, in general, are the uses of light and electron microscopy?

CULTIVATING MICROORGANISMS

1. What are the advantages of smallness in cultivating microorganisms? What are the disadvantages?

2. What is a pure culture? What are the two basic steps in obtaining a pure culture?

3. Define sterilization. What are the different uses of heat sterilization and chemical sterilization in the laboratory? What are the advantages and disadvantages of moist heat and dry heat? What special precautions must be taken to assure sterilization in an autoclave?

4. Describe filtration. Does filtration sterilize? Explain. When is filtration used in the laboratory?

5. Define these terms: inoculate, clone, colony, incubate, dilute, serial dilutions.

6. Describe how you would carry out these methods to isolate a microorganism for culturing: streak plate method, pour plate method, spread plate method.

7. Compare and contrast these types of media:
    a. defined medium   d. differential medium
    b. complex medium   e. selective-differential medium
    c. selective medium   f. enrichment culture

8. What special considerations must be given to temperature, pH, and oxygen in cultivating a microorganism?

9. Define these types of microorganisms:
    strict aerobes   aerotolerant anaerobes
    strict anaerobes   microaerophiles
    facultative anaerobes

10. Tell how you would preserve a culture by desiccation and by low-temperature storage.

11. Are there microorganisms that cannot be cultivated in the laboratory? Explain and give examples.

Suggested Readings

American Society for Microbiology. 1993. *Methods of general microbiology*. Washington, D.C.

Balows, A.; Hausler, W. J.; Herrmann, K. L.; Isenberg, H. D.; and Shadomy, H. J. 1991. *Manual of clinical microbiology*. 5th ed. Washington, D.C.: American Society for Microbiology.

Block, S. S. 1991. *Disinfection, sterilization, and preservation*. Philadelphia: Lea and Febiger.

James, J., and Tanke, H. J. 1991. *Biomedical light microscopy*. Boston: Kluwer.

4 PROCARYOTIC AND EUCARYOTIC CELLS

A Matched Team

Hiroshi Nikaido.

We can usually figure out how a machine works just by looking at it, but not so with a microbial cell. Because the primary activities of a microbial cell are chemical, not mechanical, knowing how it works means finding out how its molecules are arranged. Success in this endeavor often leads to new insights about the entire cell—and sometimes to important medical advances.

A major advance in understanding microbial structure was made in 1993 by Hiroshi Nikaido, a microbiologist at the University of California, Berkeley. Nikaido discovered how mycolic acids (long-chain fatty acids with 70 or more carbon atoms) are arranged to form a membrane surrounding *Mycobacterium tuberculosis*, the bacterium that causes tuberculosis. Nikaido's discovery explains some of the unusual properties of *M. tuberculosis* and provides invaluable help to pharmaceutical researchers designing drugs to control the disease. New drugs are essential because the number of reported cases in the United States has been rising alarmingly since the mid-1980s, and many of these cases are caused by drug-resistant strains. Drugs that had been highly effective are useless in these cases. Tuberculosis is an even more serious problem in developing countries. The World Health Organization estimates 31 million people will die of tuberculosis in this decade, compared with 10 million deaths from AIDS.

Only *M. tuberculosis* and closely related bacteria have mycolic acids, which have long been assumed to form a waxy layer around the *M. tuberculosis* cell, giving it some of its unusual properties—slow growth rate, acid-fast staining, and resistance to many antibacterial drugs. A waxy layer would account for these properties by impeding passage of nutrients, dyes, and antibacterial drugs. Still, some nutrients and antibacterial compounds do enter the cell. How can they pass through the outer waxy layer of mycolic acids?

Nikaido used x-ray diffraction (a procedure that reveals molecular structure) to show that the mycolic acid molecules are arranged in two layers with their hydrophobic tails directed toward the space between them. The mycolic acids form a highly ordered membrane, not a disorganized waxy layer. The membrane completely surrounds the cell, but embedded proteins form water-filled pores through which nutrients and certain drugs pass slowly.

Nikaido's discovery not only helps the search for new drugs to fight tuberculosis, but it also illustrates an important principle—cellular structure and function are always related. Like a matched team of horses, one depends on the other.

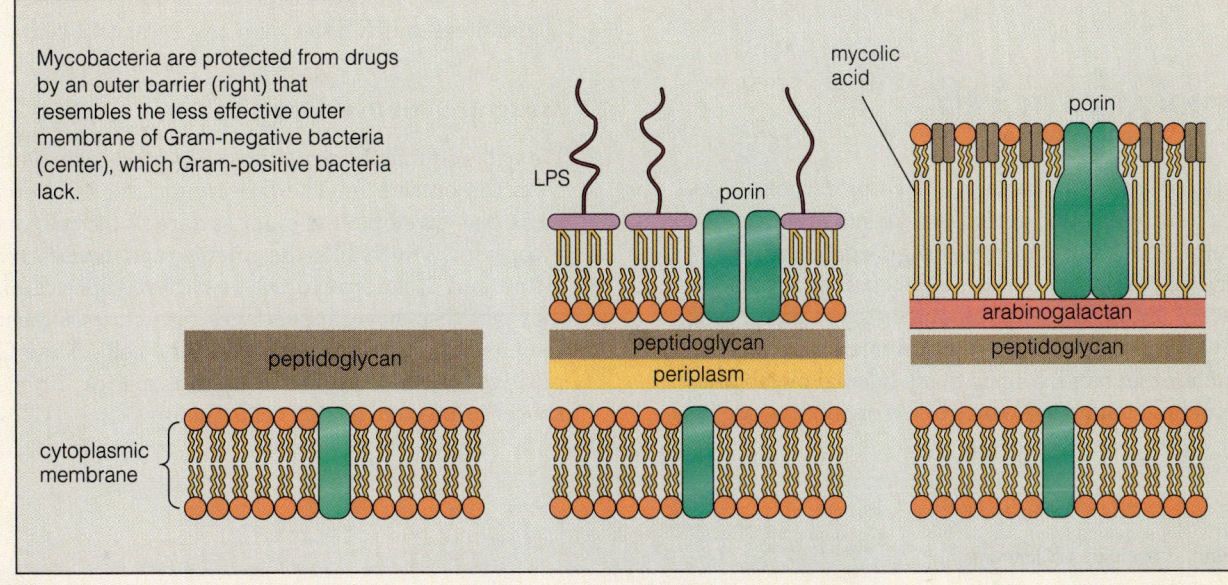

Mycobacteria are protected from drugs by an outer barrier (right) that resembles the less effective outer membrane of Gram-negative bacteria (center), which Gram-positive bacteria lack.

To understand:
- The principal differences between procaryotic and eucaryotic cells

- The structural and functional differences between Gram-positive bacteria, Gram-negative bacteria, and mycoplasmas

- The structural and chemical differences between eubacteria and archaebacteria

- The structure of eucaryotic cells and the functions of the organelles

- How molecules cross cell membranes: simple diffusion, osmosis, facilitated diffusion, active transport, group translocation, and engulfment

STRUCTURE AND FUNCTION

To understand the structure of living things, we must think in terms of function. The process of natural selection inextricably links biological structures to function. In the course of evolution, useful structures persist, less useful ones improve, and useless ones are usually lost. This principle—that structure determines function, and function determines structure—is operative at every level. Even the basic structural unit of all living things, the cell, is best understood in terms of structure and function.

Of the five groups of microorganisms, only the viruses are not organized on a cellular model. These tiny particles, which some biologists argue are not even alive, are discussed in Chapter 13. In this chapter we look at the structure and function of microbial cells. First, we consider the simpler structure of procaryotic cells, and then we contrast it with the more complicated structure of eucaryotic cells.

THE PROCARYOTIC CELL

All bacteria—and only bacteria—have procaryotic cells. The terms *bacterium* and *procaryote* are synonymous. All other cellular microorganisms—fungi, algae, and protozoa—and all plants and animals have eucaryotic cells.

Viewed under the electron microscope, procaryotic cells have a grainy but fairly uniform interior. Eucaryotic cells, on the other hand, contain many internal membranes and membrane-bound structures called **organelles**, or "lit-tle organs." Procaryotic cells also contain a few internal structures called organelles, but they are not bounded by lipid membranes (this is an important difference). In particular, procaryotic cells have a nuclear region (a mass of DNA), while eucaryotic cells have a membrane-bound nucleus. (Recall from Chapter 1 that procaryote in Greek means "before a nucleus," and eucaryote means "true nucleus.") Compare **Figures 4.1** and **4.2**, which show a procaryotic cell and a eucaryotic cell under a light microscope and a transmission electron microscope.

The first difference between the two kinds of cells, then, is structural. Another important difference is chemical. Virtually all procaryotic cells have a cell wall, and its primary, if not only, component is **peptidoglycan** (Chapter 2). A eucaryotic cell never has peptidoglycan.

A third distinguishing characteristic is size. Most procaryotic cells measure from slightly less than one to several micrometers (μm). They are the size of some of the organelles in eucaryotic cells. Whole eucaryotic cells are about 10 times larger than procaryotes. Being small may seem a trivial property, but its consequences are profound. Size affects the rate at which a cell can grow and function. The small size of procaryotic cells allows them to grow faster and multiply more rapidly than eucaryotic cells.

A cell's size affects its growth rate because of the relationship between a cell's surface area and the volume of its contents. Surface area determines how fast nutrients can enter the cell from the environment. The larger the surface area, the more rapidly nutrients enter. Volume determines the cell's need for nutrients. The larger the volume, the more rapidly nutrients are used for maintenance, repair, growth, and reproduction. Small cells have a higher **surface-to-volume ratio** than large cells. In other words, they have relatively more surface area for the same volume. The surface-to-volume ratio for a typical bacterium is 20 times greater than that for a typical human cell (**Table 4.1**). Thus, procaryotic cells, being small, meet their relatively modest nutritional needs easily and grow rapidly. For example, under ideal conditions, the bacterium *Escherichia coli* doubles about every 20 minutes—much faster than any eucaryotic cell.

Structure: An Overview

We will examine the structure of a typical procaryotic cell from the outside in. The procaryotic cell has two basic parts: the envelope (the outer surface of the cell) and the cytoplasm (which fills the interior and houses the nucleoid and other structures). A bacterial species may or may not also have appendages (structures attached to the envelope that extend beyond the cell). Some Gram-positive and a few Gram-negative species form endospores (protective resting cell structures).

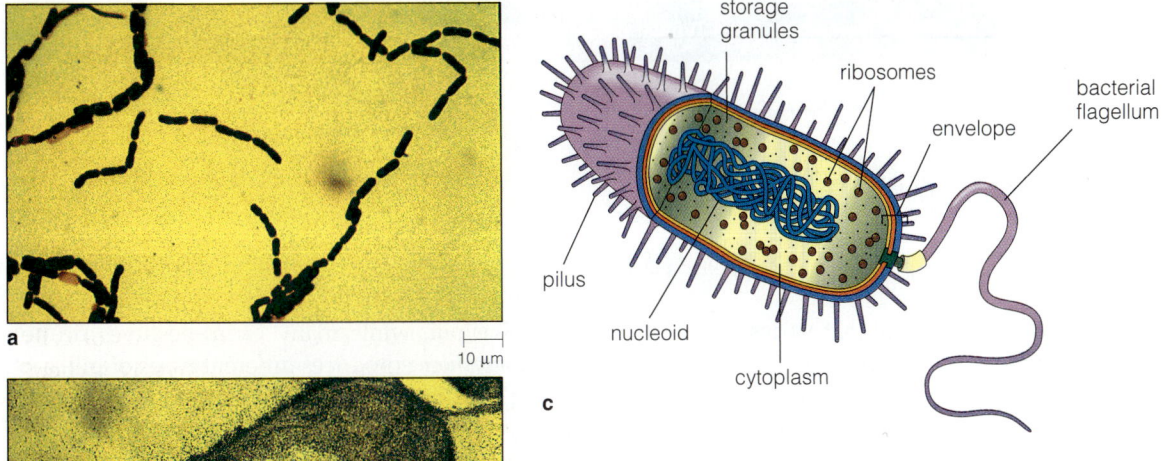

storage
granules

ribosomes

bacterial
flagellum

envelope

pilus

nucleoid

cytoplasm

c

Figure 4.1   Procaryotic cell. This is the bacterium *Bacillus megaterium* under (**a**) the light microscope and (**b**) the transmission electron microscope. In both, it looks like a typical procaryote—colorless, grainy, and undifferentiated. (**c**) Parts of a typical procaryotic cell. Compare these images with the complex eucaryotic cell in Figure 4.2.

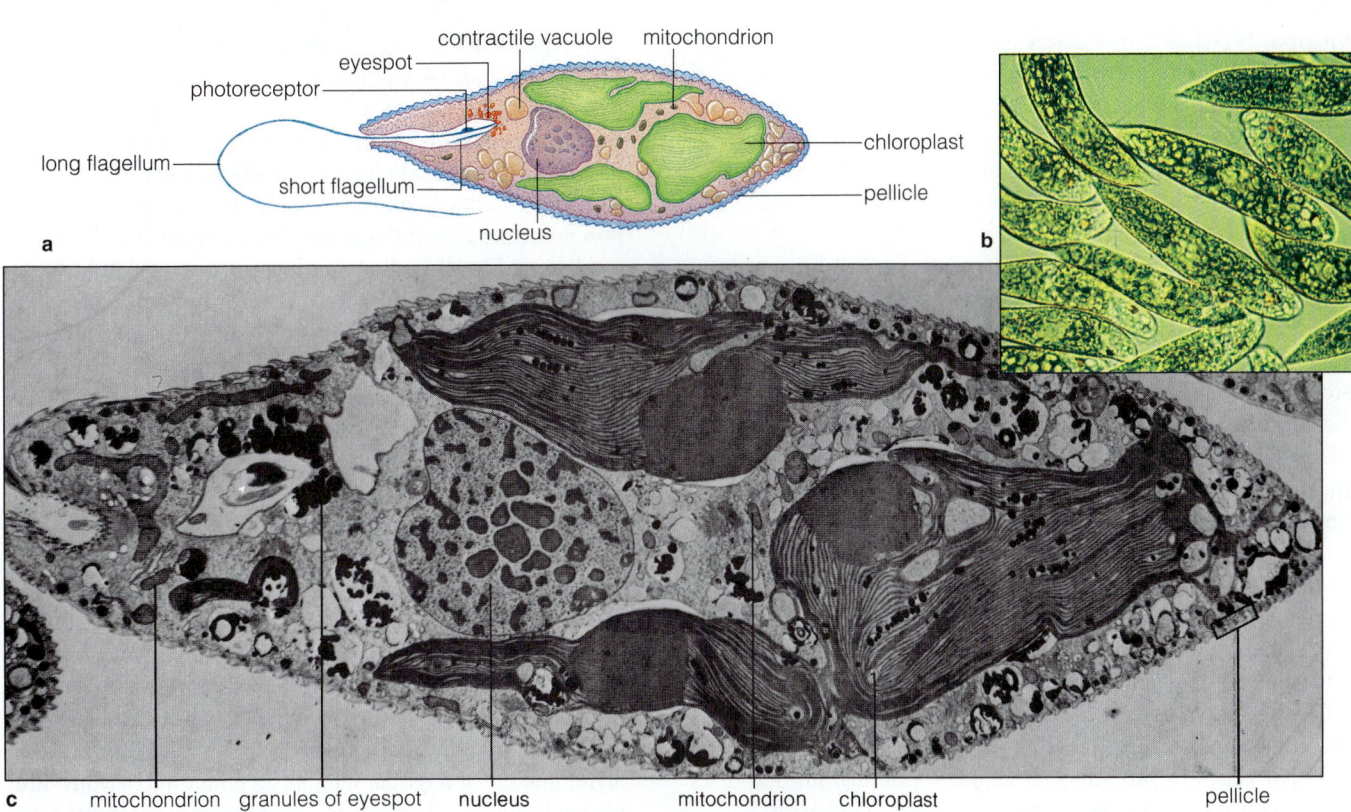

contractile vacuole   mitochondrion

eyespot

photoreceptor

long flagellum

short flagellum

nucleus

chloroplast

pellicle

a

b

c   mitochondrion   granules of eyespot   nucleus   mitochondrion   chloroplast   pellicle

Figure 4.2   Eucaryotic cell. This is the alga *Euglena*. Although only a single-celled creature, it is extremely complex compared to a procaryote. (**a**) Parts of a *Euglena* cell. (**b**) Under the light microscope, the interior appears compartmentalized. (**c**) Under the transmission electron microscope, you can clearly see its numerous organelles.

Table 4.1 Comparison of a Typical Bacterial Cell and a Typical Human Cell

	Bacterial Cell	Human Cell	Comparison
Diameter	1 μm	10 μm	Bacterium is 10 times smaller
Surface area	3.1 μm^2	1257 μm^2	Bacterium is 405 times smaller
Volume	0.52 μm^3	4190 μm^3	Bacterium is 8057 times smaller
Surface-to-volume ratio	6	0.3	Bacterium is 20 times greater

The **envelope** is a multilayered structure that varies in composition and complexity among four groups: the Gram-positive bacteria, the Gram-negative bacteria, the **mycoplasmas** (a group of bacteria that are wall-less), and the archaebacteria. The Gram-negative bacteria have the most complex envelopes and the mycoplasmas, the least complex. In between are the Gram-positive bacteria. The envelopes of archaebacteria vary, and they are unusual in other ways. We consider them separately, at the end of the procaryote section.

Appendages

Some bacterial cells have **appendages**, structures attached to the cell envelope and extending out. The two principal types of appendages are pili (*sing.*, pilus) and flagella (*sing.*, flagellum). A bacterium may have one, both, or neither of these, depending upon the species and the conditions under which the cells are cultivated. One particular group of bacteria, the spirochetes, have a modified flagellum called an axial filament, or endoflagellum.

Pili.   The **pili** are straight hairlike appendages. They are usually short but can be up to several cell lengths long. Pili are composed of protein molecules, called **pilin**, that are arranged helically around a central hollow core. Differences between pili of different bacterial species—the width of the pilus, for example, or the pitch (angle of the helix)—are due to slight differences in pilin molecules. Another word for pili is *fimbriae* (*sing.*, fimbria), a term commonly used in Britain and in medical literature.

The principal function of pili is to attach bacteria to other cells. One type of pilus, called a **sex pilus**, attaches one bacterial cell to another during mating (see Figure 6.20). This process, during which DNA is transferred from one bacterial cell to another, is described in Chapter 6. Cells of *Escherichia coli* that are capable of donating DNA in sexual exchange bear about six sex pili.

Other types of pili attach bacteria to plant or animal cells. *E. coli* has 100 to 300 of this type. Pili allow bacteria to maintain themselves in a favorable environment. In the case of disease-causing bacteria, if pili have been lost (through mutation, perhaps), the bacteria cannot establish an infection (Chapter 15). Virtually all Gram-negative

bacteria have pili, while many Gram-positives do not. *Neisseria gonorrhoeae* produces different types of pili, each enabling it to adhere to a different surface of the human body. Depending upon where *N. gonorrhoeae* is deposited—the male or female urogenital tract, the eye, the rectum, the throat—different types of pili adhere more effectively.

Flagella.   Like pili, **flagella** are thin structures that extend outward from the surface of the envelope, but that's as far as the similarity goes. The two appendages differ considerably in function and structure. Pili are used for attachment; flagella are used for locomotion. Bacteria with flagella are **motile**; they can move under their own power.

Flagella rotate and thereby propel the cell through its watery environment in much the same way as a propeller drives a ship (**Figure 4.3**). They can do this because of their three-part structure. The outermost part of the flagellum is the helical-shaped, hairlike **filament**. This part causes the cell to move by turning. It is composed of protein subunits called **flagellin**. The filament is attached to a thickened **hook**, which acts as a universal joint, allowing the filament to point in different directions. The hook is attached to an elaborate structure called the **basal body**, which has two functions. It is entirely within the envelope and thus firmly anchors the flagellum; it also causes the flagellum to rotate. It is thus the motor that turns the filament. The basal body consists of a series of rings on a rod. Gram-positive bacteria have two inner rings and Gram-negatives have two inner and two outer rings. In both cases, the inner ring is called the **M ring**, because it is embedded in the plasma membrane. It is the M ring that turns.

Bacteria can have both pili and flagella, and various species of both Gram-positive and Gram-negative bacteria have flagella. Different types of bacteria have different numbers and arrangements of flagella. The number and arrangement of flagella, which can be **monotrichous**, **amphitrichous**, **lophotrichous**, or **peritrichous**, give the cell a characteristic movement, which is helpful for identifying species in wet mounts. **Figure 4.4** presents a visual summary of bacterial types, by number and arrangement of flagella.

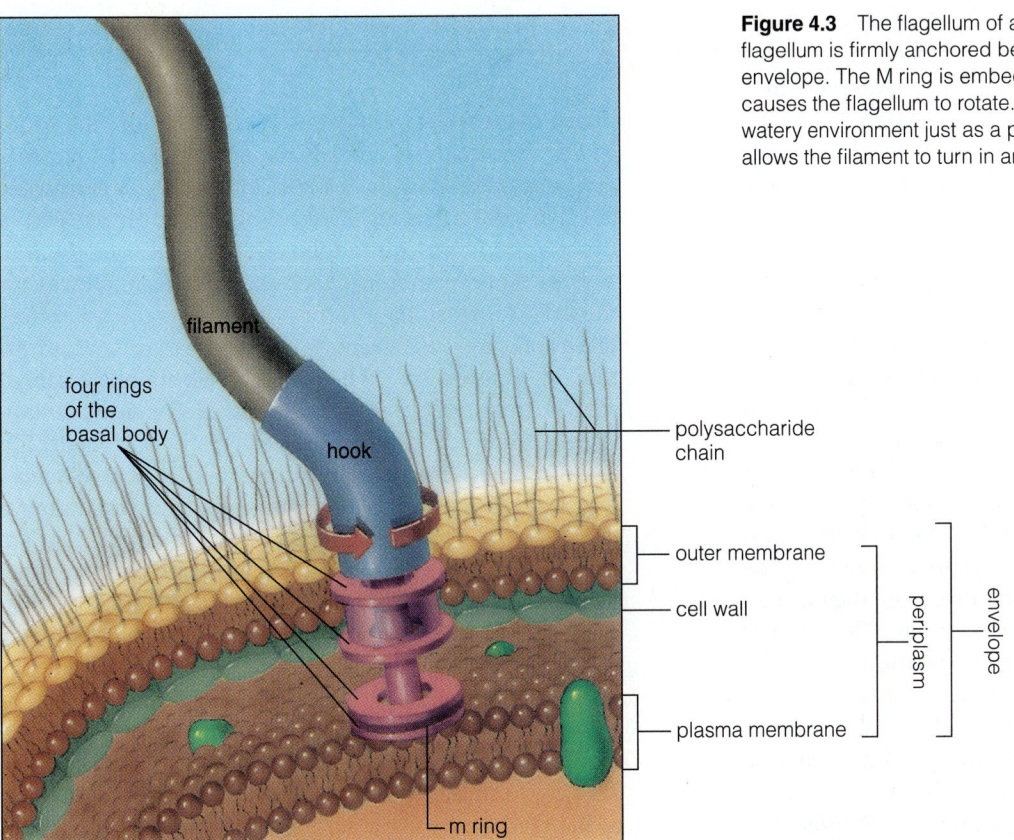

Figure 4.3 The flagellum of a Gram-negative bacterium. The flagellum is firmly anchored because it is entirely within the envelope. The M ring is embedded in the plasma membrane. It causes the flagellum to rotate. Rotation propels the cell through its watery environment just as a propeller drives a ship. The hook allows the filament to turn in any direction.

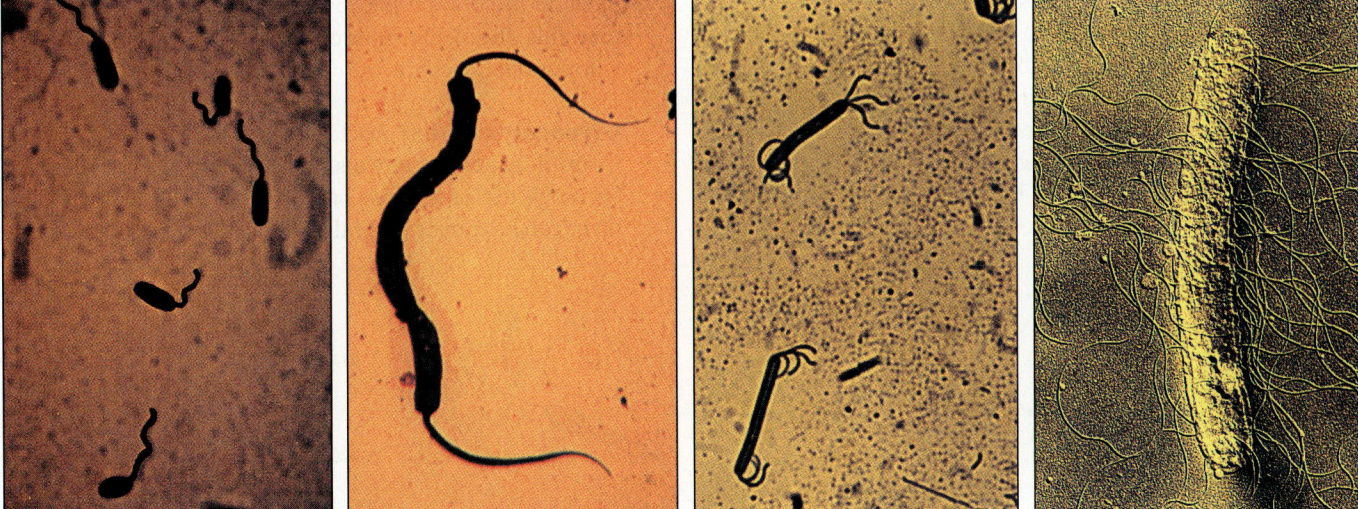

a Monotrichous: Single flagellum at one pole.

b Amphitrichous: Single flagellum at each pole.

c Lophotrichous: Two or more flagella at one or both poles.

d Peritrichous: Flagella all over the surface of the cell.

Figure 4.4 Arrangements of flagella on bacteria.

Flagella allow bacteria to seek out favorable environments and avoid harmful ones. All such behavior is called **taxis** (*pl.*, taxes), but there are many different kinds of taxes. By **chemotaxis**, bacteria sense certain chemicals and swim toward regions that contain more nutrients and away from regions with toxic materials. By **aerotaxis**, they swim to regions that contain favorable concentrations of dissolved oxygen. By **phototaxis**, photosynthetic bacteria swim to regions of optimal light intensity and quality. A few bacteria are capable of **magnetotaxis**, allowing them to travel along magnetic lines of force toward sediments at the bottom of marine or fresh waters where conditions are most favorable for them. Bacteria sense positive and negative stimuli because of receptors in their membranes.

The most thoroughly studied of the tactic responses is the chemotaxis of *Escherichia coli*. *E. coli* is peritrichous, with 8 to 10 flagella spread over its surface. When all the flagella turn in a counterclockwise direction, they form a single ropelike structure that turns as a unit. This propels the cell in a straight line through the medium—a swimming motion called a *run* (**Figure 4.5**). Periodically, flagella reverse their rotation. When flagella suddenly begin to turn in a clockwise direction, the ropelike structure flies apart and the cell is said to *tumble*. When rotation reverses again, the cell again swims in a straight line, but usually in a different direction. Normally, runs last a second or two and tumbles last only a fraction of a second. Chemotaxis (and other taxes as well) changes the period of the straight-swimming run. When swimming in a favorable direction (toward a nutrient or away from a toxic compound), the duration of the run is extended up to several minutes. In an unfavorable direction (away from a nutrient or toward a toxic compound), the run lasts even less than a second. Thus, by sensing concentrations of a nutrient or toxic substance, a bacterium swims for longer periods in a favorable direction than in a neutral or an unfavorable one.

Axial Filaments. Only one bacterial group, the spirochetes, has axial filaments (Chapter 11). **Axial filaments** are bundles of flagella covered in protein. A spirochete has two axial filaments. Each is anchored near one end of the cell and wraps around the cell body between the wall and the outer membrane. Each extends about halfway up the cell. Together they form a helical bulge that moves like a corkscrew as the entrapped flagella turn and propel the cell (**Figure 4.6**). The spirochete's unique form of movement is well suited to the viscous environments— mud and mucous membranes—where it is generally found. Some spirochetes live in mud. *Treponema pallidum*, which causes syphilis, inhabits mucous membranes (Chapter 24). Spirochetes swim fastest in liquids that are about as viscous as heavy motor oil. Ordinary flagellated bacteria are almost immobilized in such thick liquids. They swim best in less viscous material, such as blood or water.

Procaryotic Envelope

Moving from the outside toward the inside, we will look at the glycocalyx, the outer membrane and periplasm (in Gram-negatives only), the cell wall, and the plasma membrane (also called the inner membrane or the cytoplasmic membrane because it is next to the cytoplasm that fills the cell's interior). **Figure 4.7** shows the envelopes of eubacteria.

Glycocalyx. Most bacteria, both Gram-positive and Gram-negative, secrete a slimy or gummy substance that becomes the outermost layer of the cell envelope. This layer can be thin or several times as thick as the rest of the

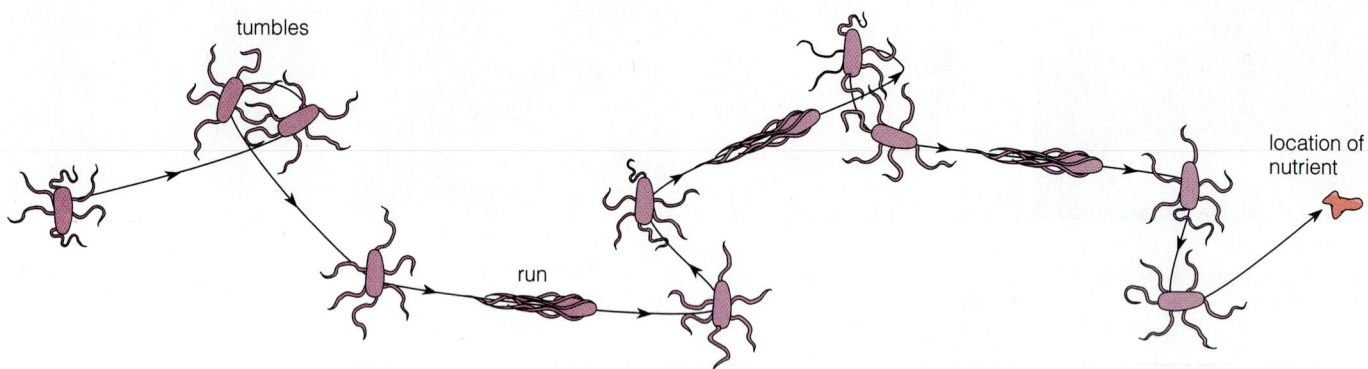

Figure 4.5 Flagella allow bacteria to seek out favorable environments and avoid unfavorable ones. Such behavior is called taxis. Though the motion may appear random, it is not. When swimming toward a nutrient or away from a toxic compound, a bacterium increases the duration of runs and takes fewer tumbles.

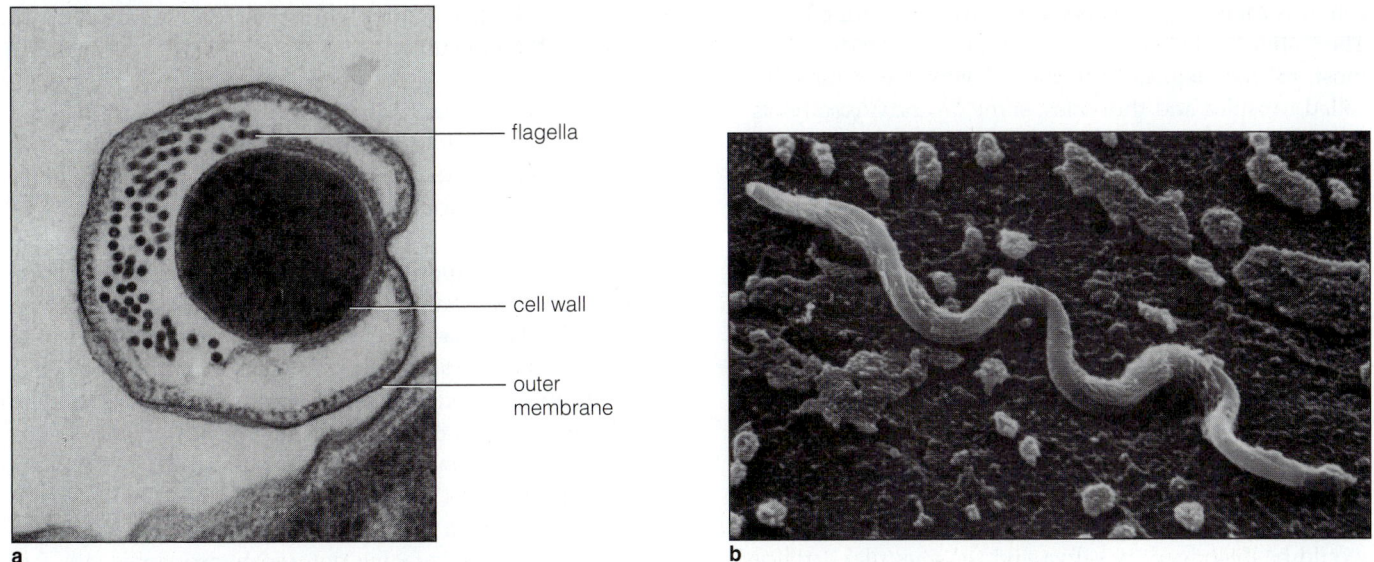

Figure 4.6 Axial filaments are modified flagella that all spirochetes—and only spirochetes—have. (**a**) A cross section through a spirochete shows the flagella lying inside the outer membrane. (**b**) The axial filaments—composed of many flagella—form a helical bulge extending the length of the spirochete. The entrapped flagella turn, propelling the cell like a corkscrew.

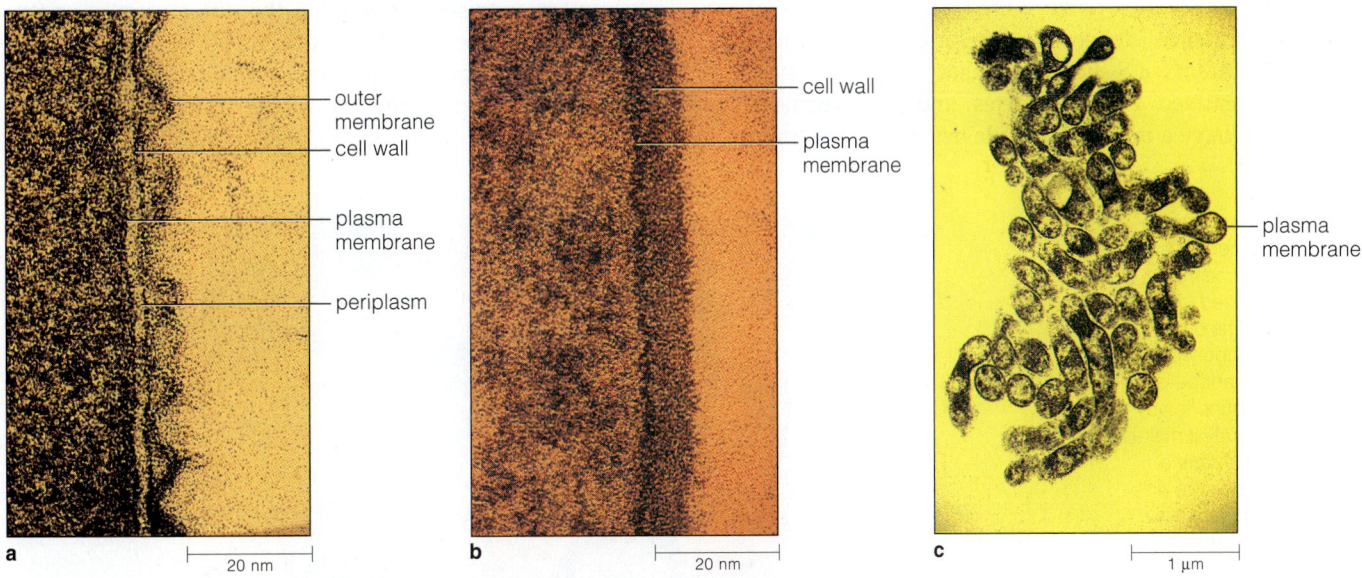

Figure 4.7 Eubacterial envelopes. (**a**) The Gram-negative envelope, the most complex, has an outer membrane and periplasm, neither of which are found in Gram-positives. The cell wall consists of a single enormous molecule of peptidoglycan. (**b**) The Gram-positive envelope does not have an outer membrane and the cell wall is thick. It also contains teichoic acids in addition to peptidoglycan. (**c**) Mycoplasmas are the only eubacteria that do not have cell walls. Note that no glycocalyx is shown in any of the photos. It is not present in all bacteria, though Gram-negatives usually have one.

cell. It is called a **glycocalyx**, a **capsule**, or a **slime layer**. The distinction between the terms is not clearly drawn by most microbiologists, but thick layers are commonly called capsules and thin ones, slime layers. *Glycocalyx* is the more general term. The glycocalyx may be composed of a single polysaccharide or a mixture of several polysaccharides and proteins (Chapter 2).

The principal function of the glycocalyx is protection. The thicker the layer, of course, the more protection it gives. Any amount of glycocalyx, however, protects the cell against drying out. The sticky nature of the material also helps a cell adhere to a surface where conditions are favorable for growth. Otherwise, it might be washed away to a nutrient-poor environment. For example, the capsule of *Streptococcus mutans*, the bacterium principally responsible for tooth decay, anchors the bacterium to the surface of the tooth. If it weren't for its capsule, *S. mutans* would be dislodged by saliva and the muscular activity of chewing and speech.

Capsules also provide protection against phagocytosis, engulfment and destruction by another cell (Chapter 16). A slippery capsule makes it difficult for a phagocyte to establish contact with the invading bacterium (a phagocyte is a type of white blood cell that seeks to destroy microorganisms). Consider *Streptococcus pneumoniae*, the bacterium that causes a life-threatening pneumonia. Its virulence depends entirely on its capsular protection. Strains of *S. pneumoniae* that lack a capsule are readily consumed by phagocytes and are harmless. Strains that have a minimal capsule usually cause a mild

pneumonia, and strains with the thickest capsules can cause a fatal pneumonia.

Outer Membrane. In Gram-negative bacteria *only*, the next layer in the envelope is the **outer membrane**. The outer membrane is a bilayer membrane (two rows of molecules with space in between) **(Figure 4.8)**. The inner layer is composed of phospholipid, and the outer layer is composed of **lipopolysaccharide (LPS)**. LPS is a compound found only in the outer membrane of Gram-negative bacteria—not in any other living thing. As the name implies, lipopolysaccharide is a complex molecule, with lipid on one end and polysaccharide on the other. The lipid end is **lipid A**. Lipid A is hydrophobic (water-fearing), and the polysaccharide tail is hydrophilic (water-loving). Thus, lipopolysaccharide has a hydrophilic head and a hydrophobic tail. In this regard (having hydrophilic heads facing out and hydrophobic tails facing in), it is like all phospholipid membranes. But unlike other biological membranes, LPS is a barrier to both polar and nonpolar molecules. As a result, only water and a few gases can diffuse across the outer membrane.

All molecules (including nutrients) that enter the Gram-negative cell must pass the outer membrane. Most enter through special proteins called **porins**. Porins create small channels—or pores—in the outer membrane that allow molecules to diffuse in. Some pores are specialized. For example, one porin in *Escherichia coli* allows vitamin B_{12} to enter and another allows the disaccharide, maltose, to enter.

Figure 4.8 Outer membrane. The inner surface is like the plasma membrane, composed of phospholipid. But the outer surface is composed of lipopolysaccharide—a material not found in any other kind of cell. It makes the outer membrane a formidable barrier to diffusion, so most materials enter through proteins called porins. The lipid end of lipopolysaccharide, lipid A, is toxic to humans.

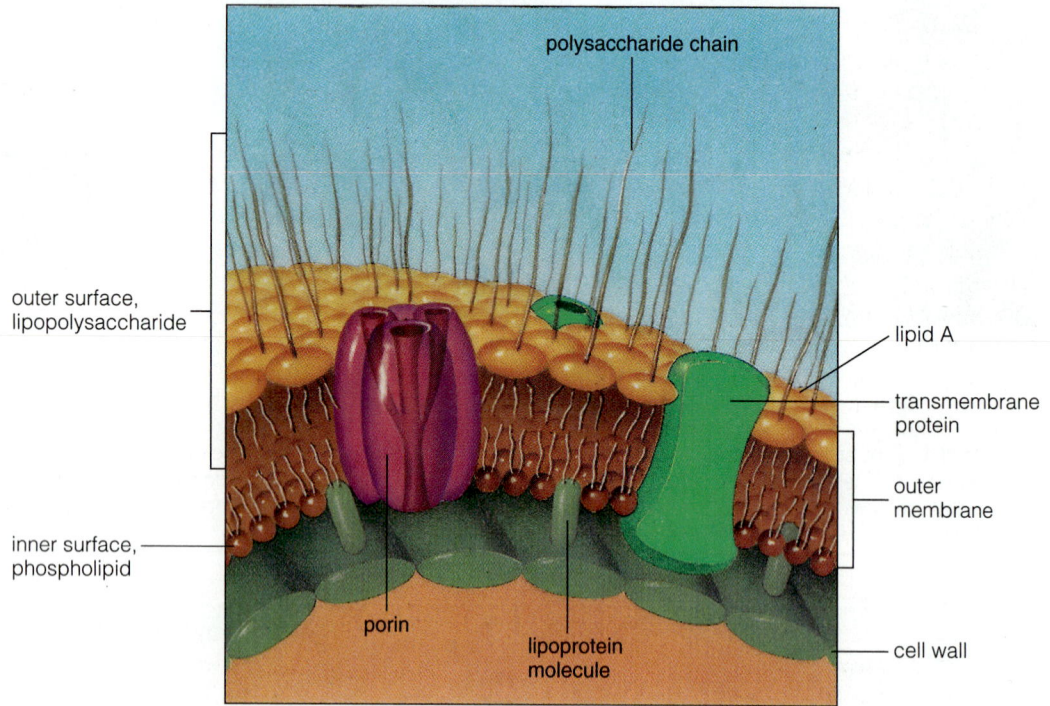

Clinical Notes

The Capsule Is the Clue

Microorganisms that produce a capsule are extraordinarily important in clinical medicine. Consider this case of a 32-year-old white male: The man, who had always enjoyed good health, gradually began to suspect that he was ill. At first he noticed unusually frequent and severe headaches. As weeks and months went by, he developed fever, fatigue, and occasional morning vomiting. When he became aware that he sometimes wasn't thinking clearly, he was sufficiently alarmed and called his doctor. His doctor did a thorough medical evaluation, including a spinal tap, which in the end provided the diagnosis. One sample of spinal fluid was negatively stained with India ink. It revealed many perfectly round cells, each surrounded by a capsule several times larger than itself. The laboratory technologist identified these encapsulated cells as the fungus *Cryptococcus neoformans*. The diagnosis of cryptococcal meningitis was later confirmed by culture. This life-threatening infection of the membranes surrounding the brain and spinal cord is extremely rare in healthy people, but rather common in people with AIDS. Unfortunately, further testing established

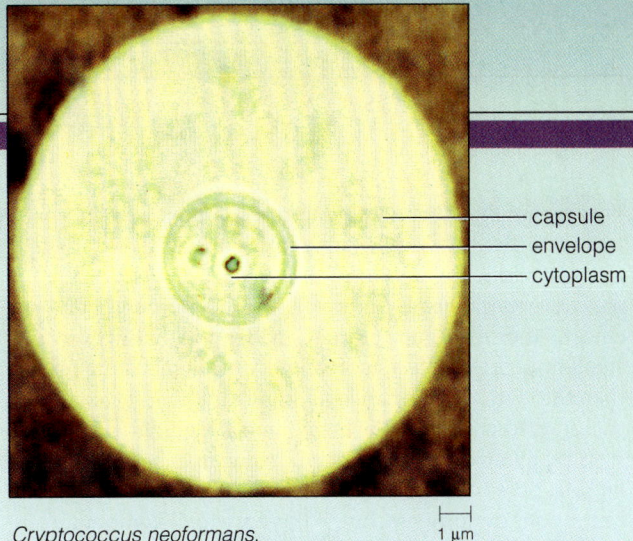

Cryptococcus neoformans.   ⊢—⊣ 1 µm

— capsule
— envelope
— cytoplasm

that this patient had AIDS, and he died four months later despite aggressive medical treatment.

It is no coincidence that many potentially fatal infections are caused by capsular microorganisms. A capsule greatly enhances a microbe's virulence (Chapter 15). Bacterial meningitis, pneumonia, and epiglottitis are other potentially fatal infections caused by encapsulated microbes. (Note that *C. neoformans* is a fungus with a capsule. As you will read later in this chapter, some eucaryotic cells also have a glycocalyx.)

The outer membrane is anchored all the way to the plasma membrane by **Bayer junctions**. Bayer junctions are very tight connections where the inner surface of the outer membrane is contiguous with the outer surface of the plasma membrane. In addition to connecting the outer membrane to the rest of the cell, Bayer junctions are the entryway for DNA—both DNA from a genetic exchange between cells (Chapter 6) and DNA that enters when a virus infects a bacterial cell (Chapter 13).

The most important function of the outer membrane is protection. Because of their outer membrane, Gram-negative bacteria are generally more resistant than Gram-positive bacteria to toxic substances in the environment, including antibiotics. Many toxic compounds are simply too large to diffuse through the outer membrane's porins. Of course, if the bacterium is a pathogen, what benefits the Gram-negative cell harms the host. The antibiotic rifampin, for example, kills both Gram-negative and Gram-positive bacteria if it reaches its target enzyme—RNA polymerase—in the cytoplasm. But Gram-positive cells are 1000 times more sensitive than Gram-negative cells to rifampin because it cannot readily pass through the Gram-negative bacteria's outer membrane or porins.

The LPS component of the outer membrane is also known as **endotoxin**, harmful to humans and other animals. Most of the toxicity is due to lipid A. Endotoxin is discussed further in Chapter 15.

Periplasm. In older microbiological literature, the periplasm was called the *periplasmic space* because electron micrographs showed it as an empty region extending between the outer membrane and the plasma membrane—containing the cell wall's peptidoglycan. Today, however, we know that **periplasm** is an organelle of Gram-negative bacteria. The periplasm is gelatinous material containing two major types of proteins. One type consists of essential enzymes that break down certain nutrients into smaller molecules that can pass through the plasma membrane. The second type consists of binding proteins that facilitate passage of nutrients across the plasma membrane and speed their entry.

The Cell Wall. A cell wall is found in all eubacteria except mycoplasmas. In Gram-negative bacteria, the cell wall lies just inside the outer membrane. In Gram-positives, it lies just inside the glycocalyx, if one exists. We

The Microbial Motto: Be Prepared

Pore size in the outer membrane is a compromise between letting nutrients in and keeping toxic substances out. Both nutrient and toxin molecules come in all sizes, but procaryotes have mechanisms for breaking down nutrient molecules so they can enter through smaller pores. However, small molecules diffuse more rapidly through a large pore than through a smaller one and more rapidly if the concentration gradient across the membrane is high.

It is advantageous for a microorganism to have pores as large as possible if it lives in an environment with low concentrations of nutrients. But, if there are toxic compounds in the environment, it is better to have smaller pores. Thus, bacteria that live in lakes or streams with low nutrient concentrations and few toxic compounds evolve larger porins than do bacteria that live with concentrated nutrients and many toxic substances.

The bacterium *Escherichia coli* encounters both kinds of environment. Normally, *E. coli* is a harmless inhabitant of the intestinal tract (though certain strains are pathogenic and some cause diarrhea). The intestinal tract contains concentrated nutrients and toxic compounds. When *E. coli* is shed in feces, however, it usually finds itself in an aqueous environment, poor in nutrients and toxic compounds. But *E. coli* makes two different porins. One (OmpC) has small pores, and the other (OmpF) has larger pores. Warm temperature and moderate osmotic strength (conditions in the intestinal tract) signal it to make OmpC. Cooler temperatures and lower osmotic strength (conditions typical of aqueous environments) signal it to make OmpF. Thus, *E. coli* is adapted to either environment.

---

will look at three aspects of the cell wall: its structure and composition, how it confers shape on a cell (one major function), and how it withstands turgor pressure (its other major function).

Structure and Composition of the Cell Wall. The chief component of both Gram-negative and Gram-positive cell walls is peptidoglycan (a kind called **murein**). Peptidoglycan is composed of long chains of polysaccharide (glycan) cross-linked by peptides (short proteins). The polysaccharide component always consists of alternating units of the two sugars *N*-acetylglucosamine (NAG) and *N*-acetylmuramic acid (NAM). The peptide component varies among bacterial species. But the most common type in both Gram-negative and Gram-positive bacteria is four amino acids long and contains L-isomers, which are normally found in proteins, and D-isomers, which are rare in nature (Chapter 2). When linked together, the glycan and peptide chains create the single rigid meshlike molecule that forms the bacterial cell wall. Review the box "Peptidoglycan: Molecular Boilerplate" in Chapter 2 for additional detail.

The cell walls of Gram-negative and Gram-positive bacteria are slightly different. A Gram-negative cell wall consists of a peptidoglycan mesh that is only one layer thick. In other words, Gram-negative bacteria are enclosed by a single enormous molecule of peptidoglycan, a thin peptidoglycan sac. Gram-positive bacteria, on the other hand, have a peptidoglycan wall that is many layers thick and also includes a second component, **teichoic**

acids. Teichoic acids consist of glycerol or ribitol (a five-carbon polyalcohol) linked by phosphate groups. Although there are many theories about the function of teichoic acids, none has yet been proved. We do know that they are vital to the structural integrity of the cell wall.

Cell Shape. One function of the cell wall is to confer shape on a bacterium. Bacteria come in innumerable shapes. Some are square, some are star-shaped, and some have unusual irregular shapes. But most bacteria fall into one of three groups: spherical, called **cocci** (*sing.*, coccus); rod-shaped, called **bacilli** (*sing.*, bacillus); or spiral-shaped, called **spirilla** (*sing.*, spirillum). There are also intermediate shapes. The most common are a short bacillus (or elongated coccus), called a **coccobacillus**, and a short spirillum that looks comma-shaped, called a **vibrio**. (The term *bacillus* can refer to a cell's shape or, if capitalized, the genus *Bacillus*, which includes *B. anthracis*, the species that causes the animal and human respiratory disease anthrax.)

Bacterial cells of some species stick together after division, forming distinctive clusters that help in identifying them. Cocci can form many patterns, depending upon whether they divide on one plane or more than one. Bacilli always divide in one plane; and if they don't separate, they form a simple pattern of rods joined end to end. Spirilla usually remain as single microorganisms. Their variability comes from having more or fewer corkscrew turns and more or less rigidity or flexibility. **Figure 4.9** shows cell types and arrangements.

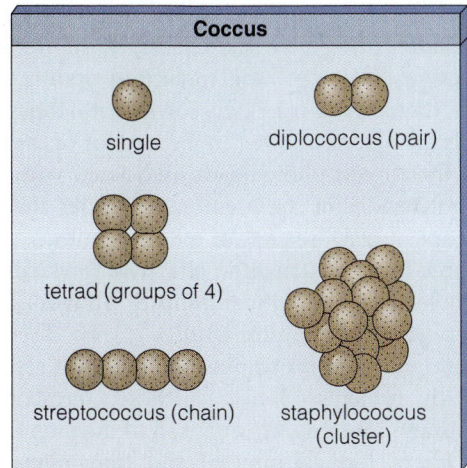
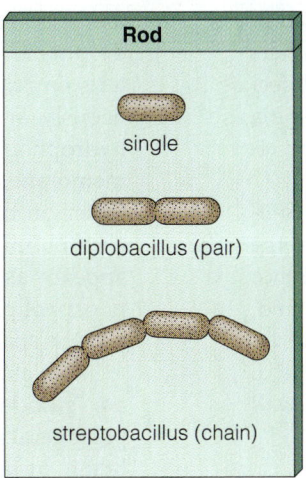
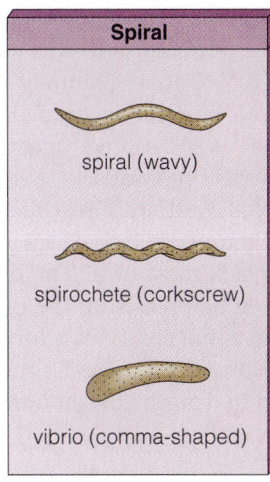
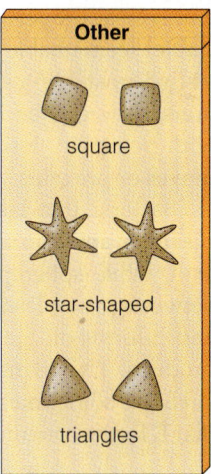

Figure 4.9 Shapes and arrangements of bacterial cells.

How does the cell wall confer shape on a bacterium? Shape is determined largely by the enzymes that cross-link the peptide chains of peptidoglycan. Most bacterial species produce several different enzymes, called **autolysins**, that break cross-linking bonds and several others, called **transpeptidases**, that reseal the links. When the links are resealed, new peptidoglycan monomers are added, enlarging the peptidoglycan sac and increasing cell size. Some transpeptidases form spheres and build cocci. Two types of transpeptidases—sphere-forming and cylindrical-forming—are involved in building rod-shaped bacteria because rods have hemispherical ends connected by a straight cylinder. Transpeptidases that form curved cylinders build spirilla and vibrios.

A simple experiment demonstrates that cell shape depends on the integrity of the peptidoglycan molecule. The enzyme lysozyme breaks glycan bonds of peptidoglycan, and so if we put bacteria in a solution with lysozyme, it will destroy the integrity of the cell wall. The cell assumes the spherical configuration that is characteristic of wall-less objects such as soap bubbles.

Turgor Pressure. The lysozyme treatment also demonstrates the second major function of the cell wall: containing **turgor pressure**. A cell's turgor is the internal pressure from its contents. A bacterial cell's turgor is extremely high because its contents are concentrated. Unless the lysozyme-treated bacterial cell's high internal concentration of solute molecules is matched by an increase in the concentration of solutes in the external environment, the spherical wall-less cell begins to enlarge as water moves into it. The swelling eventually tears the relatively weak plasma membrane, causing **lysis**, which is rupture and destruction of the cell. The cell wall withstands the bacterial cell's tremendous turgor. Some micrococci, for example, have turgor pressures of 350 pounds per square inch, which would rupture a steel water tank.

Some antibiotics work by damaging the cell wall and causing lysis. Penicillin, for example, binds to the transpeptidase enzymes that reseal the peptidoglycan molecule during cell growth. For this reason, transpeptidase enzymes are also called *penicillin-binding proteins* (Chapter 21). Binding inactivates the enzymes, and the cells cannot repair the mounting number of breaks in the cross-linking bonds of the cell wall that normally occur as the cell grows. When the wall becomes too weakened to offset turgor pressure, the bacterial cell lyses.

Mycoplasmas. The mycoplasmas lack a cell wall. They avoid lysis from turgor pressure by a completely different mechanism: They maintain nearly equal pressure between their cytoplasm and their external environment by actively pumping sodium ions out of the cell. If mycoplasmas are deprived of an energy source and therefore cannot pump out sodium ions, they swell and lyse. In addition, the membranes of mycoplasmas are slightly strengthened because they contain **sterols**, a type of lipid not found in most other bacterial membranes, but present in all eucaryotic membranes (Chapter 2).

L Forms. Strains of bacteria that have lost the ability to form walls but are still able to reproduce if provided osmotic protection are called **L forms** (or *L variants*), for the Lister Institute where they were first observed over 50 years ago. L forms from many genera of bacteria have been observed. Most microbiologists believe that they are a universal property of bacteria. Some L forms (called β-stable) seem unable to return to a normal state, but others (called unstable) can. L forms occur naturally in infections; they can be induced by treating normal bacteria with penicillin, lysozyme, or some other agent that removes or prevents synthesis of cell walls.

Bacteria that lack a cell wall are also called **protoplasts** or **spheroplasts**. Protoplasts lack all traces of the wall, while spheroplasts lack most of it. The difference between protoplasts and spheroplasts on the one hand, and L forms on the other, is the ability of L forms to reproduce. The terms *protoplast L forms* and *spheroplast L forms* are sometimes used to describe the amount of wall that L forms contain.

Plasma Membrane. The membrane that encloses the cytoplasm of *any* cell is called the **plasma (cytoplasmic) membrane**. The major functions of the plasma membrane are to contain the cytoplasm and to transport—to regulate what comes in and what goes out of the cell. The plasma membranes of bacteria are very similar to plasma membranes in eucaryotes. In fact, all membranes in procaryotic and eucaryotic cells that have the same structural and chemical characteristics, including the plasma membrane, are categorized as **unit membranes**. (The outer membrane of Gram-negative bacteria and protein membranes are not unit membranes because they are chemically different.)

Unit membranes are composed primarily of phospholipids (Chapter 2). They thus have a hydrophilic head, made of glycerol and a phosphate group, and a hydrophobic tail, made of two fatty acid chains. In the watery environment of the cell, these molecules spontaneously arrange themselves into two rows, with their hydrophobic tails facing one another in the interior of the membrane, and their hydrophilic heads associated with water on both outer sides of the membrane. Under the electron microscope, a cross-section of a unit membrane appears as a three-layered structure: the two rows of phospholipid molecules—the **phospholipid bilayer**—and the space between them (**Figure 4.10**).

Both eucaryotic and procaryotic phospholipid bilayers are studded with proteins. Some of these, termed **peripheral membrane proteins**, are attached to the membrane surface. Others, termed **integral** and **transmembrane proteins**, penetrate through the membrane. Still others are **transport proteins** (also called **carrier proteins**) that physically carry ions and molecules through the lipid **matrix** (the ground material). There are more, and more kinds of, proteins in procaryotic plasma membranes than in eucaryotic. Typically, proteins make up about half of the dry weight of a bacterial plasma membrane. Because these proteins can move within the membrane, the membrane is typically described as "dynamic." The membrane structure as a whole is referred to as fitting the **fluid mosaic model**.

The carrier proteins found in procaryotic plasma membranes are called **permeases**. Permeases are important because only water, polar molecules, and certain gases can pass through the phospholipid portion of the

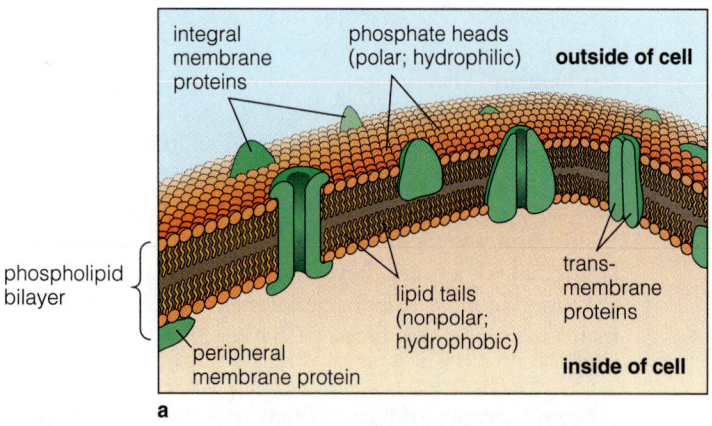
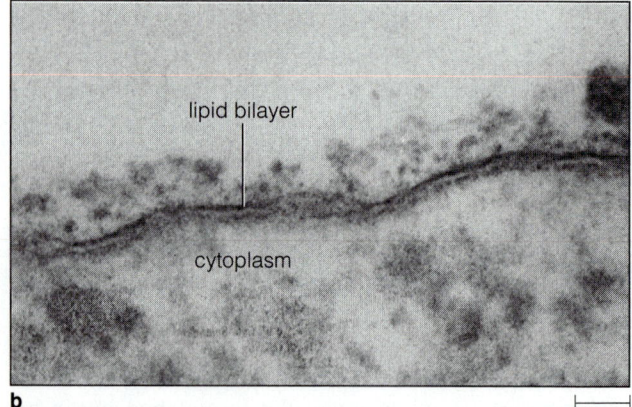

Figure 4.10 Plasma membrane. (**a**) The plasma membranes of all procaryotes (and all eucaryotes) are phospholipid bilayers, or two layers of phospholipids. The phosphate heads point outward and the lipid tails point inward. The bilayer is studded with different kinds of proteins that can move as they perform their functions. The plasma membrane is therefore described as dynamic and fits the fluid mosaic model. (**b**) Under the transmission electron microscope, the two layers appear to be three because of the space between the rows of molecules.

plasma membrane. All other substances must be physically transported by permeases. Permeases are discussed further in Chapter 5.

Two other kinds of bacterial membrane proteins are also important. There are the proteins in the electron transport chain, which helps the cell generate energy for growth and reproduction. Also, in some bacteria, certain proteins participate in photosynthesis, the process that generates energy from light. Therefore, procaryotic membranes also have a function in generating energy. (Both the electron transport chain and photosynthesis are discussed in Chapter 5.)

A cross-sectional view of a bacterial cell through the electron microscope usually shows the plasma membrane as a bag around the cytoplasm. Sometimes, though, the membrane **invaginates**, folds back on itself, forming structures that extend into the cytoplasm. Intricately folded plasma membranes are seen in cells that have an unusual need for membrane proteins, particularly the types that generate energy (**Figure 4.11**). Because the protein fraction of the membrane changes very little, the only way to get more proteins is to have more membrane surface. Bacteria that grow on low-energy-yielding substrates and those that generate energy through photosynthesis require greater amounts of energy-generating proteins. Folds of plasma membrane often extend all through the cytoplasm of these bacteria.

Cytoplasm

The **cytoplasm** is a matrix composed primarily of water (90 percent) and protein. In electron micrographs of procaryotic cells, it appears colorless and uniformly grainy. But appearances can be deceiving. The cytoplasm is where most of the cell's vital chemical reactions, called metabolism (Chapter 5), take place. Indeed, certain internal structures can be distinguished. The cytoplasm of all bacteria contains a nuclear region and ribosomes. Most bacteria also have various structures that are grouped as **inclusions**. Inclusion bodies are visible structures in the cell other than the nuclear region and ribosomes. The inclusion most commonly found is storage granules. In addition to these structures, different bacteria have various specialized structures.

Nucleoid.   The **nucleoid** or **nuclear region** is a mass of DNA. A transmission electron microscope shows that the nuclear region is relatively well defined, even though it is not surrounded by a membrane as the nucleus of a eucaryotic cell is. Bacterial DNA—like all DNA—carries the cell's genetic information. Most of a bacterium's DNA is arranged in a single circular molecule called the **bacterial chromosome**. Some bacteria also contain smaller circular DNA molecules called **plasmids**. Plasmids carry genes

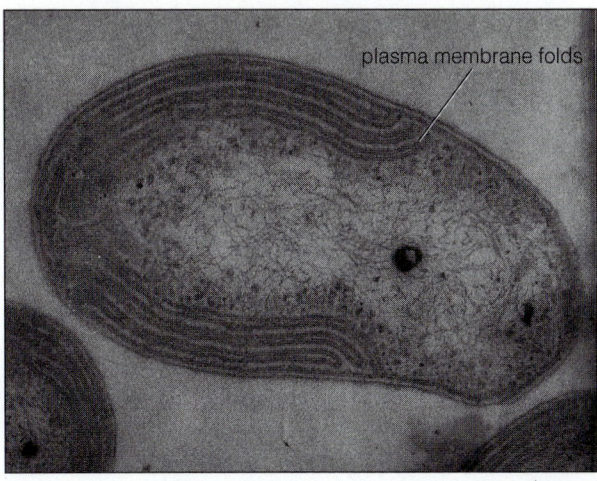

100 nm

Figure 4.11   *Plasma membrane of* Nitrobacter. *The plasma membrane of procaryotes includes proteins involved in generating energy for cell maintenance and reproduction. (In eucaryotes, organelles perform this function.) Bacteria such as* Nitrobacter *that grow on low-energy-yielding substrates and therefore have greater requirements for energy-generating proteins often have folds of plasma membrane extending into their cytoplasm.*

that encode specialized, nonessential functions, such as resistance to antibiotics, pigmentation, and the ability to use unusual nutrients. DNA is discussed in Chapters 6 and 7.

Ribosomes.   The cytoplasm of bacteria is packed with **ribosomes**, small structures that manufacture proteins. The way ribosomes synthesize proteins is described in Chapter 8. It is the small size and great number of ribosomes that give cytoplasm its grainy appearance. A single bacterial cell may contain as many as 20,000 ribosomes. Whether procaryotic or eucaryotic, ribosomes are composed of two subunits. One subunit is always smaller and one always larger, but both are composed of protein and ribonucleic acid (Chapter 2). Procaryotic ribosomes are smaller than eucaryotic. They are called **70S ribosomes** (S for Svedberg units, the way they are measured). Eucaryotic ribosomes are called **80S ribosomes**. This difference is important in antibiotic actions, which often target ribosomes (Chapter 21). Tetracycline, erythromycin, streptomycin, and chloramphenicol all act by binding to ribosomes and interfering with their function. Because of the 70S-80S difference, antibiotics can specifically target bacterial ribosomes and not harm the human or animal host's eucaryotic ribosomes.

Storage Granules.   As the name implies, **storage granules** are granular inclusions in the cytoplasm that hold reserve supplies of nutrients. Their function is similar to that of the globules of fat that some cells in the human body accumulate. Most bacterial species store the nutrients they need. Some species store carbon in granules composed of glycogen, polymers of glucose. Other species store carbon as lipid granules composed of poly-beta-hydroxyalkanes, and some don't store carbon at all. A few bacterial species store reserves of sulfur or nitrogen. Many bacterial species, as well as algae, fungi, and protozoa, store phosphate as **polyphosphate** (also called *volutin*). Polyphosphate is used for many cellular functions, including producing ATP for energy (Chapter 5). Polyphosphate granules have unusual staining characteristics—they appear red when stained with the blue dye, methylene blue. Thus they are called **metachromatic** (changing color) **granules**. Many bacterial species have several kinds of storage granules.

Other Inclusions. Some inclusions fulfill highly specialized functions. Many photosynthetic aquatic bacteria contain **gas vacuoles**. These are gas-filled regions surrounded by a protein membrane that help them float to the water level that provides the best conditions for growth. Some photosynthetic bacteria contain **chlorosomes**, structures just inside the cell membrane that house pigments necessary for photosynthesis. An unusual group of bacteria has iron-containing structures called **magnetosomes** that allow them to react to magnetic lines of force and seek the bottom of a body of water where conditions favor their growth (**Figure 4.12**).

Endospores

Endospores are extremely hardy **resting**, or nongrowing, structures that some bacteria—principally Gram-positives—produce. Endospores can survive up to hundreds of years, withstanding extreme heat, dehydration, toxic chemicals, and radiation. When favorable conditions return, endospores produce new **vegetative cells**, cells that grow and reproduce. Endospores are not an essential step in the life of the bacteria that produce them. They form only when nutrients are exhausted or other conditions become unfavorable for growth. When an endospore-forming species stops growing, it usually starts forming endospores. (Endospores are sometimes simply called *spores*. Do not confuse these with the spores fungi produce; see Chapter 12.)

Endospores can withstand harsh environmental conditions in part because they contain so little water. Eucaryotic cells consist of more than 90 percent water, and procaryotic cells are about 70 percent water. Bacterial endospores, however, are less than 15 percent water. How endospores become so dry has been studied for years, but the mechanism is still not well understood. We do know that endospores' ability to withstand intense heat is related to their low moisture content and probably to their

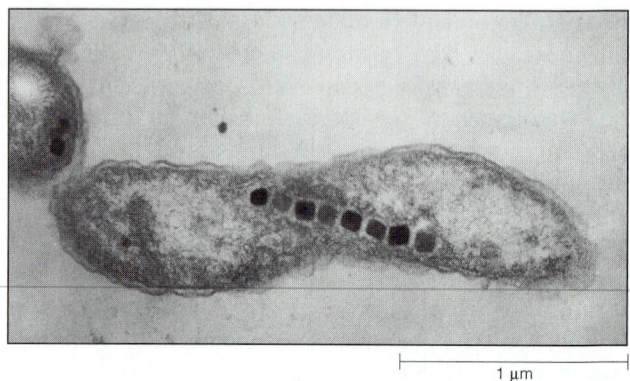

Figure 4.12 This is the bacterium *Aquaspirillum magnetotacticum*. The dark spots are magnetosomes, a type of procaryotic cell inclusion. Magnetosomes act like tiny compasses to guide *A. magnetotacticum* toward a favorable environment.

high concentrations of calcium and dipicolinic acid, a compound not found in vegetative cells.

Sporulation and Germination. When starvation or other unfavorable conditions trigger a bacterial cell to **sporulate**, or form an endospore, its cytoplasm divides into two unequal parts, with a complete complement of DNA in each part (**Figure 4.13**). The larger part then engulfs the smaller part, called the **forespore**. The forespore eventually forms the endospore, and the encircling vegetative cell provides materials for it to build a thick protective wall.

The wall consists of an inner **cortex** and an outer **spore coat**. The cortex is composed largely of peptidoglycan, but with fewer cross-linkages than in a vegetative cell. Some of the peptidoglycan backbone also contains a derivative (muramic lactam) of muramic acid. The spore coat (also sometimes divided into inner and outer) is composed of a protein called keratin-like, because it resembles the tough protein in outer layers of skin. Inside lies the spore body. It contains all the essential (inside-the-plasma-membrane) cell parts—the ribosomes and DNA—plus some spore-specific materials. In the course of maturation, the endospore dehydrates. When sporulation is complete, the surrounding vegetative cell lyses, releasing the endospore.

When conditions become favorable for growth, endospores **germinate**, grow into new vegetative cells. Most endospores cannot germinate immediately after they are formed, however. They must rest for several days or be **heat-shocked**, heated to 60°C for about 30 minutes. Then, in the proper environment, the permeability of the protective wall changes. Water enters and causes the spore to swell. Then the spore coat ruptures and a **germ tube** grows out. The germ tube is a developing vegetative cell surrounded by a peptidoglycan wall

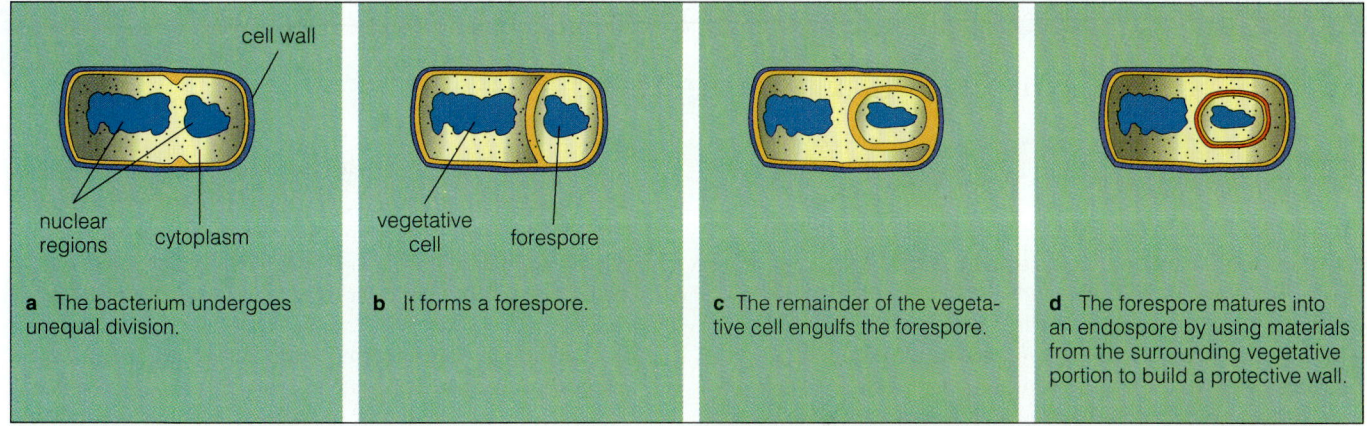

a The bacterium undergoes unequal division.

b It forms a forespore.

c The remainder of the vegetative cell engulfs the forespore.

d The forespore matures into an endospore by using materials from the surrounding vegetative portion to build a protective wall.

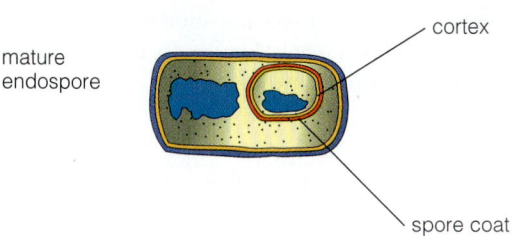

e The wall is composed of an inner cortex and an outer spore coat. The endospore also dehydrates during maturation.

f Transmission electron micrograph of *Clostridium tetani*, the bacterium that causes tetanus. The endospore has enlarged one end of the cell. Later, the cell will lyse, releasing the endospore.

Figure 4.13 Sporulation is the process whereby a vegetative procaryotic cell (one that can grow) produces an endospore (a dormant structure). It is a survival mechanism for periods of starvation.

(some coming from the cortex). It contains cell contents (some coming from the spore body).

Sporogenesis, the complete cycle of sporulation and germination, is a mechanism for survival. Metabolism ceases because conditions are unfavorable, and metabolism resumes when conditions are again favorable. Sporogenesis is not a means of reproduction. One vegetative cell produces one endospore which, in turn, produces one new vegetative cell.

Practical Implications. Some of the Gram-positive bacteria that produce endospores are pathogenic to humans. Species of *Bacillus* cause anthrax, and species of *Clostridium* cause botulism, tetanus, and gas gangrene (deep-wound infections). Endospores are the reason it is important to get a tetanus shot immediately if you cut your foot, say, on glass or a nail that was embedded in the ground. *C. tetani* can live for years in soil and then, when it enters human tissue and finds the right environment, can germinate and cause a life-threatening infection.

Endospores complicate our attempts to control microbial growth (Chapters 3 and 9).

The Archaebacterial Cell

In electron micrographs, archaebacteria and eubacteria look quite similar. But there are profound differences in the compounds that form the structures. For example, murein, the peptidoglycan that is the major component of eubacterial cell walls, is never found in archaebacterial walls. Rather, archaebacterial cell walls are composed of either protein or a peptidoglycan called **pseudomurein**, which is not found in any other living thing. In addition, some archaebacteria do not have walls.

The plasma membranes of archaebacteria also differ. Plasma membranes of eubacteria (including mycoplasmas) and of all eucaryotes are composed of phospholipids, in which unbranched lipids are joined to glycerol by ester linkages (Chapter 2). In archaebacteria, branched lipid molecules are joined to glycerol by a type of bond

No One Knows for Sure, but a Long Time

How long can an endospore survive in its state of suspended animation? That's a tough question to answer. Many experiments have shown that over half of a population of endospores will survive a year's storage at 10°C (50°F). Dried plant samples wrapped and stored without opening for over a hundred years in botanical museums have been found to contain viable endospores, and so most microbiologists are willing to say that endospores can last at least 100 years. Finding viable endospores on older, unwrapped samples is problematic. It is always possible for endospores to have entered during storage or for endospores to have germinated and produced new endospores. This is the problem with the claim that viable endospores have been recovered from Egyptian tombs that are more than 7000 years old.

How do you know that the endospores you are looking at are first generation? You don't. More convincing is a claim for 2000-year-old endospores. In 1976, the Roman fort Vindolanda, dated A.D. 90–95, had been drained and was being excavated. Viable endospores of *Thermoactinomyces vulgaris* were found in debris that was compacted in layers of clay. *T. vulgaris* is a thermophilic, aerobic bacterium. It requires warmth and oxygen to grow. But Vindolanda was sealed, cold, and anaerobic (because it was flooded), making it unlikely that *T. vulgaris* entered later or that vegetative cells formed during the nearly 2000-year-long interim. So what is the answer to the question we started out with? No one knows for sure how long endospores remain viable, but it is an extraordinarily long time.

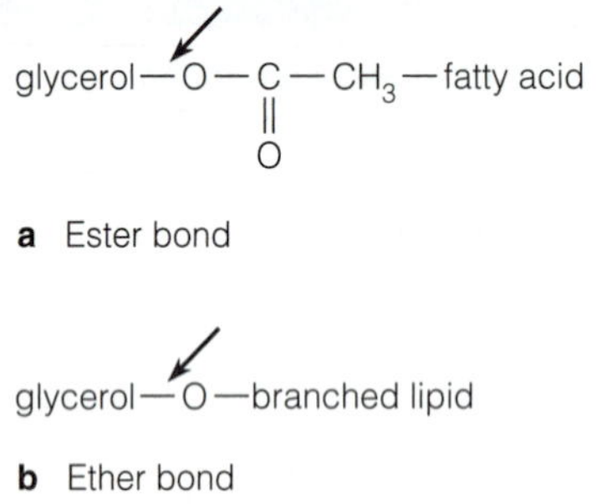

Figure 4.14 Chemical composition of archaebacterial plasma membranes. (**a**) In all organisms except archaebacteria, plasma membranes contain fatty acids joined to glycerol by ester linkages. (**b**) In archaebacteria, branched lipid molecules are joined to glycerol by ether linkages, which are stronger and may account for archaebacteria's ability to withstand environmental extremes.

the eucaryotic cell is divided into separate compartments (organelles) bounded by unit membranes. The membranes around the organelles keep different chemical activities of the cell separate and provide systems of transportation, which are necessary because eucaryotes are more complex, are larger, and have a smaller surface-to-volume ratio.

Structure: An Overview

We will look at the eucaryotic cell in the same way we looked at the procaryotic: from the outside in. Although some have appendages, a glycocalyx, and a cell wall, others don't. All, however, have a plasma membrane and membrane-bound organelles, including a nucleus. **Table 4.2** compares a procaryotic and eucaryotic cell.

Appendages

Eucaryotic cells do not have pili or axial filaments, but like procaryotes, some have flagella. The function is the same—locomotion—but their structure is vastly different. Procaryotic flagella are rigid three-part structures. Eucaryotic flagella are flexible, composed of individual fibers called **microtubules** made of the protein **tubulin**. Some eucaryotes have smaller versions of these flagella that are called **cilia** (*sing.*, cilium). Eucaryotic flagella undulate and cilia beat in waves, quite different from the procaryotic flagella that rotate. In humans, cilia are called **projections** rather than appendages, since their function

called an **ether linkage** (**Figure 4.14**). Because ether bonds are stronger than ester bonds, this unusual plasma membrane may help archaebacteria survive extreme temperature and pH.

THE EUCARYOTIC CELL

All cells except bacterial cells are eucaryotic. Cells of algae, fungi, and protozoa are therefore structurally similar to human cells. Unlike the procaryotic cell, which looks uniformly grainy under the electron microscope,

Table 4.2   Comparison of a Procaryotic Cell and a Eucaryotic Cell

Structure	Procaryote	Eucaryote
Appendages	Pili, flagella, axial filaments in spirochetes	Flagella; structurally very different from procaryotic flagella
Glycocalyx	Usually; can be a thin slime layer or thicker capsule	In some algae, fungi, protozoa, and human cells that lack walls
Outer membrane and periplasm	Gram-negatives only	Never
Cell wall	All eubacteria except mycoplasmas; basic component is peptidoglycan	Algae, fungi, protozoa, plants, but never with peptidoglycan; no cell walls in animals, including human cells
Plasma membrane	All; a phospholipid bilayer	All; a phospholipid bilayer
Cytoplasm	Undifferentiated	Cytoskeleton and cytoplasmic streaming in all; pseudopods in amoebae
Organelles	Not membrane-bound; they include nucleoid, 70S ribosomes, various inclusions, and the periplasm in Gram-negatives	Always membrane-bound; include nucleus, endoplasmic reticulum, Golgi apparatus, 80S ribosomes (except 70S in organelles), mitochondria and/or chloroplasts (algae and some protozoa have both)
DNA	Usually one circular chromosome	Paired chromosomes; associated with histones
Cell division	Binary fission or budding for both growth and reproduction	Mitosis for growth and meiosis for reproduction
Endospores	Principally in Gram-positive bacteria, but in some Gram-negative	Never
Size	Typically one or a few micrometers	Typically 10 micrometers or more

is movement of materials rather than locomotion. Cilia on epithelial (skin) cells of the nose move mucus up and out of the body—a protective mechanism that removes the microbes carried by mucus.

Two groups of protozoa have appendages. They are appropriately called the **flagellates** and the **ciliates** (Chapter 12).

Cell Wall

Although virtually all procaryotes (except mycoplasmas and some archaebacteria) have cell walls, eucaryotes show much more variation. Most animal cells do not have a cell wall, and human cells never do. Fungi and algae typically have a cell wall. Protozoa do not have a cell wall, but some have a **pellicle** outside their plasma membrane. A pellicle functions as a wall but is flexible and therefore more skinlike.

The chemical structure of eucaryotic cell walls (when they have them) varies. The cell walls of plants and of most algae and some fungi consist of the polysaccharide cellulose (Chapter 2). Some primitive fungi have cell walls composed of **chitin**, a nitrogen-containing polysaccharide. Others contain nonnitrogenous polysaccharides other than cellulose.

Cell walls in eucaryotes function in the same ways they do in procaryotes. They give shape to a cell, and they resist turgor pressure. Like the wall-less mycoplasmas, unwalled protozoal and animal cells lack turgor pressure.

Plasma Membrane

The plasma membrane of eucaryotic cells and the plasma membrane of procaryotic cells are structurally similar. Both are unit membranes composed of phospholipids with proteins embedded in them that can move within the lipid matrix. Both regulate what enters and leaves the cell. There are some differences, however.

First, the procaryotic plasma membrane contains the proteins involved in the electron energy chain and in photosynthesis. In eucaryotic cells, these proteins are located in organelles, called mitochondria (found in animals and plants) and chloroplasts (found in plants and algae). Also, the eucaryotic cell membrane contains sterol lipids (Chapter 2). In procaryotes, only the wall-less mycoplasmas have sterols.

Cytoplasm and Its Contents

Like procaryotic cytoplasm, eucaryotic cytoplasm is mostly water; and like procaryotic cytoplasm, eucaryotic cytoplasm functions as the matrix housing the cell's organelles. The eucaryotic cell, however, has a cytoskeleton, a lattice network of protein. We will look first at the cytoskeleton and then at the other contents of the cytoplasm: the nucleus, the cytomembrane system, and the mitochondria and chloroplasts.

Cytoskeleton. The **cytoskeleton** is composed of three types of threadlike protein structures: **microtubules**, **microfibrils**, and **intermediate filaments**. The microtubules are the largest and are made of the protein tubulin, as flagella and cilia are. Microfibrils are composed largely of the protein actin. Intermediate filaments are composed of a variety of proteins, including keratin, which adds rigidity to the cytoskeleton (**Figure 4.15**). Most of the cytoskeleton is unchanging, but other parts appear and reappear only as needed, for example, during cell division.

The cytoskeleton has one main function: to allow the cytoplasm of the larger, more complex eucaryotic cell to move in an orderly way. Microtubules, for example, form during cell division, ensuring that each daughter cell receives one complete copy of the genetic blueprint. Microfibrils cause a characteristic cytoplasmic movement called **cytoplasmic streaming**, another essential feature of cell division and the means by which nutrients reach all parts of the cell. Streaming also allows one type of protozoan, the amoebae, and one type of fungus, the slime molds, to move. **Amoebae** and slime molds project blobs of cytoplasm to form structures called **pseudopods**, or "false feet," when they move across a flat surface (see Figure 12.15). The cytoplasm retracts when conditions are not right.

Nucleus. Instead of the indistinct nuclear region found in procaryotic cells, the eucaryotic nucleus is defined by a double-membrane structure called the **nuclear envelope** (**Figure 4.16**). The nuclear envelope is covered with pores that allow the nucleus to communicate with the cytoplasm. The nuclear envelope contains the **nucleoplasm**, or the gelatinous matrix of the nucleus, and several **nucleoli**. Nucleoli are dense masses of RNA and protein that manufacture ribosomes.

The nucleus is the largest organelle in a eucaryotic cell. Its principal function is to house almost all the cell's DNA (a small amount of DNA is housed in mitochondria and chloroplasts). Eucaryotic DNA is chemically identical to procaryotic DNA, but structurally different. Also, unlike the DNA of procaryotic cells, eucaryotic DNA is associated with histones, basic proteins forming a DNA-protein structure called a nucleosome. Nucleosomes, in turn, are organized into paired structures called chromosomes. Chromosomes carry the cell's genes and are visible under the microscope around the time of cell division.

The chromosome arrangement allows the massive amount of DNA in a eucaryotic cell to be separated effi-

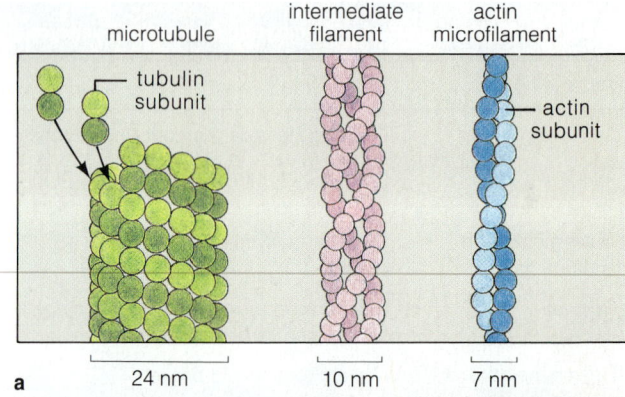

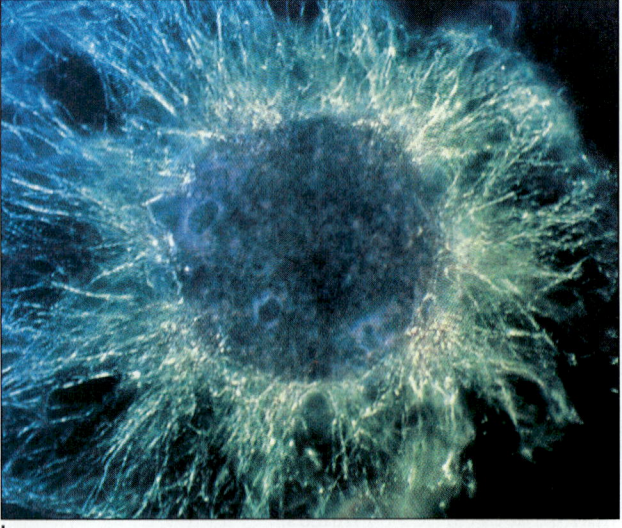

Figure 4.15 Cytoskeleton. (**a**) Eucaryotes have a cytoskeleton composed of three types of protein structures: microtubules, intermediate filaments, and microfilaments. (**b**) This is a cell of the African blood lily plant. The green structures are microtubules. The purple center is the nucleus.

ciently during cell division. Unlike procaryotic cells, which merely duplicate their single molecule of DNA and distribute one copy to each daughter cell, eucaryotic cells undergo two types of nuclear division, **mitosis** and **meiosis**. Mitosis occurs during cell division for growth. Meiosis is an essential feature of sexual reproduction of eucaryotes.

Mitosis consists of four phases: **interphase**, **prophase**, **metaphase**, and **anaphase** (**Figure 4.17**). In mitosis, the nucleus divides and each daughter cell receives one pair of each type of chromosome that the cell contains. During meiosis (**Figure 4.18**), the chromosome pairs split and each new cell receives only one copy of each chromosome. Thus, the products of meiosis, called **gametes**, are **haploid**—they contain only half as many chromosomes

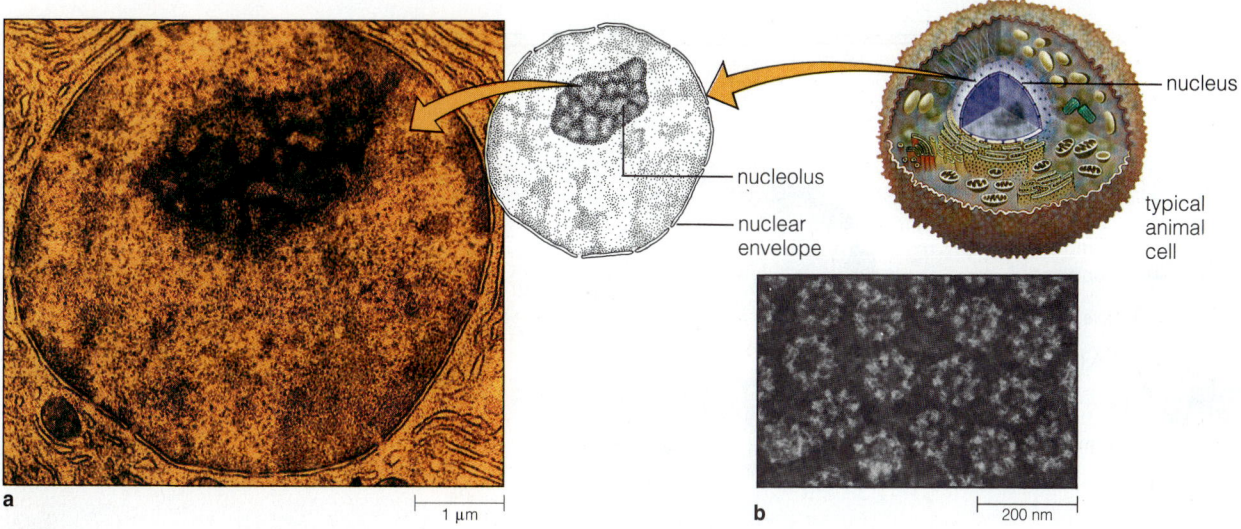

nucleus

nucleolus

nuclear envelope

typical animal cell

a

1 µm

b

200 nm

Figure 4.16 Nucleus. (**a**) This transmission electron micrograph shows a nucleus with one nucleolus, a mass of RNA and protein where ribosomes are manufactured. (**b**) A close-up of the pores that cover the nuclear envelope.

as the original **diploid** vegetative cell. Later in the life cycle, the haploid gametes fuse, producing a **zygote**, a diploid cell that develops into a new organism.

Cytomembrane System and Ribosomes. In addition to the cytoskeleton, a **cytomembrane system** runs through the eucaryotic cell. The cytomembrane system functions to sort, organize, and package the different kinds of molecules that the cell produces to maintain and reproduce itself. The cytomembrane system is connected to the nuclear envelope, but not to the plasma membrane. The two main bodies of the cytomembrane system are the endoplasmic reticulum and the Golgi apparatus.

The **endoplasmic reticulum (ER)** is a double membrane that folds back upon itself, creating a complex pattern of tubes and layered sacs. Some regions of the ER appear smooth under the electron microscope, while others appear rough. Each has a different function. The smooth regions, called the **smooth endoplasmic reticulum (SER)**, are the major site of phospholipid synthesis for the cell. Thus, the SER produces the components for building and repairing membranes. The rough regions, called the **rough endoplasmic reticulum (RER)**, owe their grainy appearance to the presence of ribosomes on their outer surface (**Figure 4.19**, p. 98). Some ribosomes are also free in the cytoplasm. Both free and attached are 80S ribosomes, larger than the 70S ribosomes found in procaryotic cells. The RER keeps some of the newly formed proteins and sends others to the Golgi apparatus.

At certain places in the eucaryotic cell, stacks of flattened membranes form an organelle called the **Golgi apparatus (Figure 4.20)**. The endoplasmic reticulum trans-

fers molecules of protein and phospholipid to the Golgi apparatus in membrane-bound packages called **vesicles**. Vesicles bud from the ER and fuse with the Golgi apparatus. Within the Golgi apparatus the molecules are modified and concentrated. Then they are repackaged in new vesicles, which leave the Golgi apparatus on their way to a final destination, either inside or outside of the cell.

One specialized type of vesicle that leaves the Golgi apparatus is called a **lysosome**. Lysosomes are small but powerful packets of enzymes that can destroy many kinds of molecules. They can also destroy microbial cells captured by phagocytes.

Mitochondria and Chloroplasts. In procaryotic cells, the proteins involved in generating ATP through respiration (Chapter 5) and photosynthesis are in the plasma membrane. In eucaryotic cells, these energy-generating proteins are located in the intracellular membranes that surround organelles called **mitochondria** (respiration) and **chloroplasts** (photosynthesis).

Mitochondria and chloroplasts have similar structures (**Figure 4.21**, p. 100). They are bounded by double membranes: An outer membrane encloses a highly folded system of internal membranes. This creates two regions, one between the outer and inner membrane and another within the folds of the internal membrane itself. The inner membrane of a chloroplast is arranged in stacks called **thylakoids**. Thylakoids contain the pigment chlorophyll, which is involved in photosynthesis. Plants and all but a very few colorless forms of algae have chlorophyll. The deeply folded inner membrane in mitochondria is called

(text continues on page 100)

Mitosis

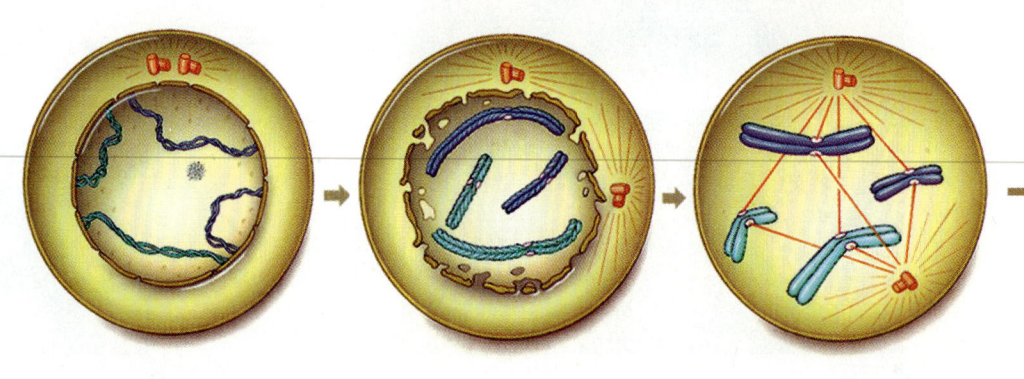

Cell at Interphase

The cell prepares for mitosis by replicating (duplicating) its DNA.

- pair of centrioles
- cytoplasm
- nucleus
- DNA (decondensed)
- nuclear envelope
- plasma membrane

Figure 4.17 Mitosis. Mitosis is division for cell growth. Each new cell receives a pair of each type of chromosome.

Early Prophase
Duplicated chromosomes start to condense. The nucleus is 2n; one set of 2 chromosomes is shaded purple and the other blue.

Late Prophase
Chromosomes condense more, spindle starts to form.

Transition to Metaphase
Condensed chromosomes attach to spindle apparatus.

Meiosis I

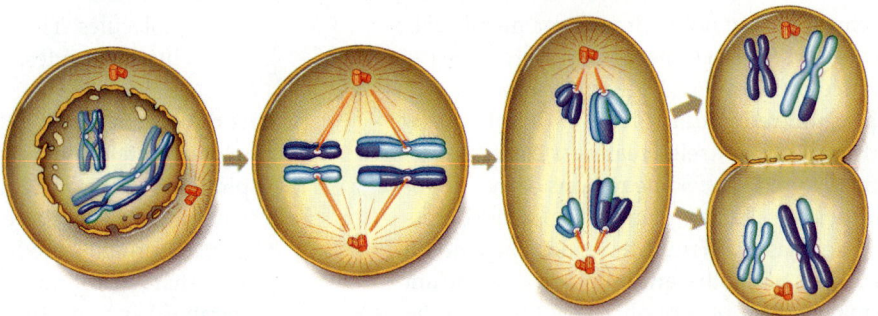

Prophase I
Duplicated chromosomes start to condense, spindle starts to form; duplicate chromosomes pair.

Metaphase I
Pairs of chromosomes align at equator; the two are attached to opposite poles.

Anaphase I
Pairs of chromosomes separate and move to opposite poles.

Telophase I
A haploid number (n) of chromosomes is clustered at each pole.

Figure 4.18 Meiosis, sexual reproduction in eucaryotic cells. Only reproductive cells undergo meiosis. Chromosomes split and each new cell receives one copy of each chromosome.

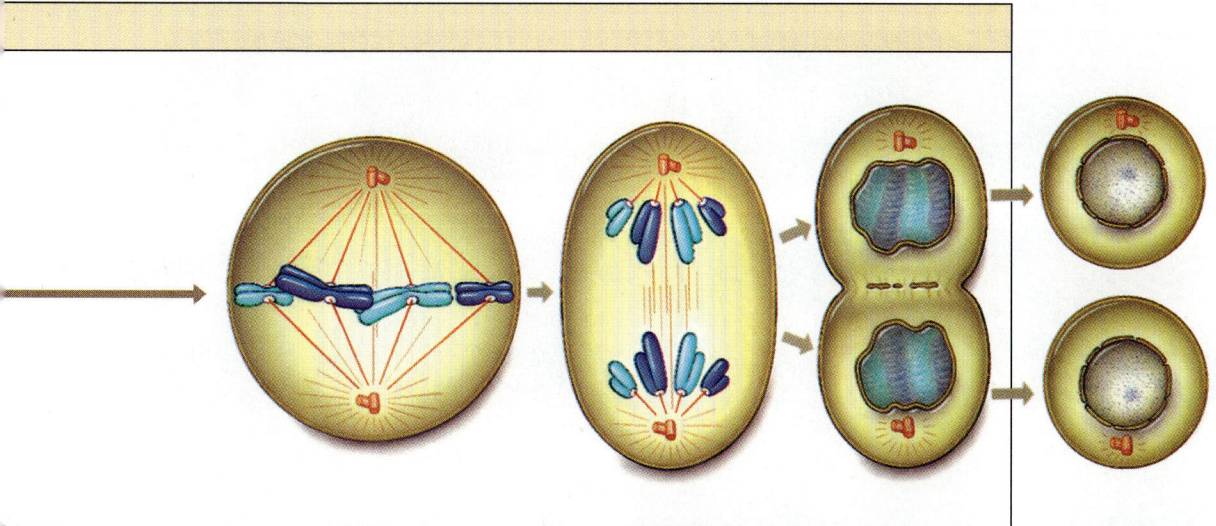

Metaphase
All chromosomes line up at spindle equator.

Anaphase
Chromosomes are separated and moved to opposite poles.

Telophase
Chromosomes decondense; nuclear membranes form and the cell starts to divide.

Interphase
The product of mitosis is two daughter cells, each of which is diploid (2*n*).

Meiosis II

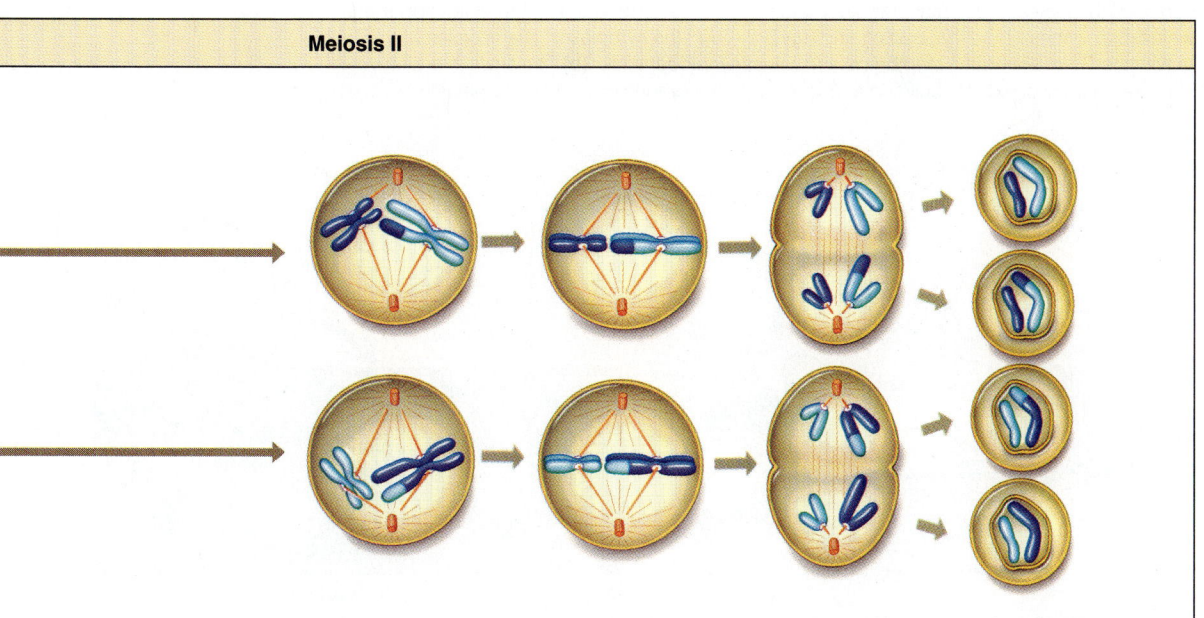

Prophase II
Brief stage; each chromosome still duplicated.

Metaphase II
Chromosomes line up at spindle equator.

Anaphase II
Chromosomes separate, move to opposite poles.

Telophase II
Each daughter nucleus is *haploid* (*n*), with one of each pair of homologous chromosomes that was present in the parent nucleus.

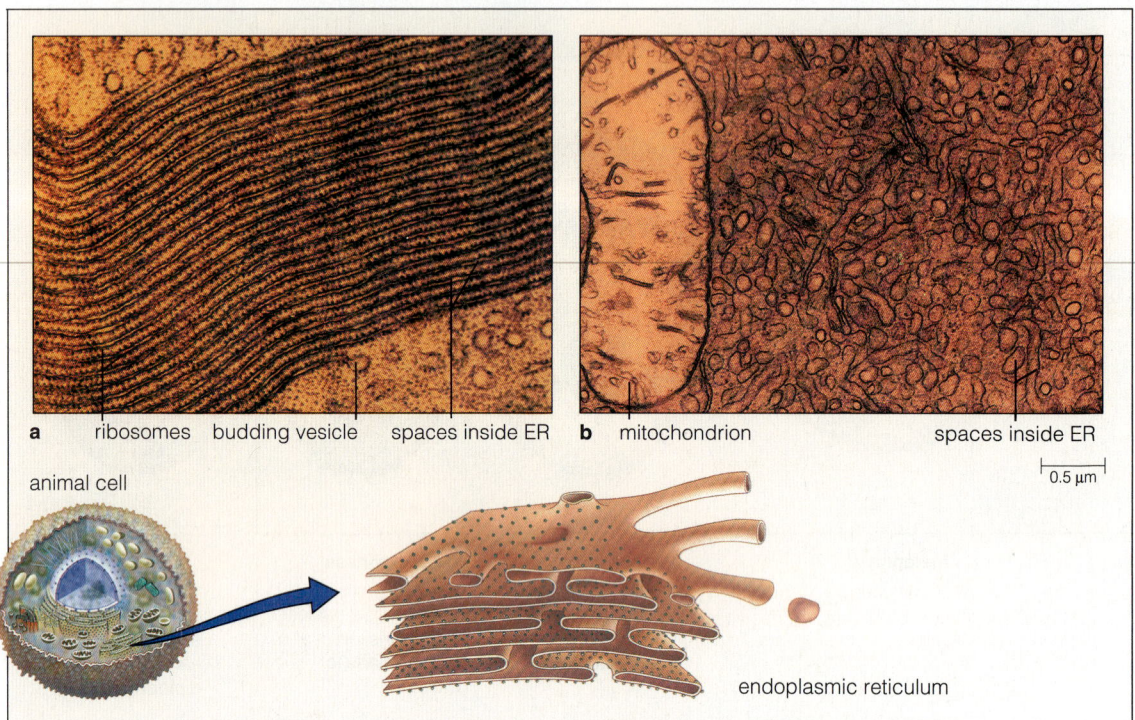

a   ribosomes   budding vesicle   spaces inside ER   b   mitochondrion   spaces inside ER

0.5 µm

animal cell

endoplasmic reticulum

Figure 4.19 Endoplasmic reticulum. The endoplasmic reticulum (ER) is a membrane system in eucaryotic cells. It has rough regions (RER) and smooth regions (SER). **(a)** The rough region is studded with ribosomes, which manufacture proteins. It sends its products to the Golgi apparatus in vesicles, bits of membrane that bud off and travel through the cytoplasm. The vesicles then merge with the Golgi apparatus membrane and release their contents. **(b)** The smooth region manufactures phospholipids for building and repairing membranes (including the plasma membrane).

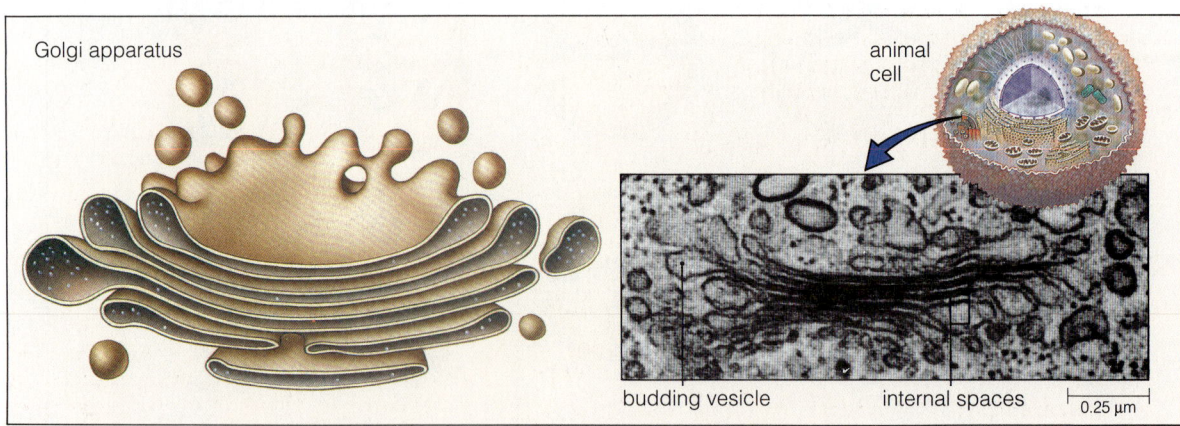

Golgi apparatus

animal cell

budding vesicle   internal spaces   0.25 µm

Figure 4.20 Golgi apparatus. The Golgi apparatus is a stack of flattened membranes that receives phospholipids and proteins from the endoplasmic reticulum. The Golgi modifies and concentrates the molecules and repackages them in vesicles to send them to their final destination within or outside of the cell.

A Different Kind of Mitochondrion?

Paul Baumann and
Linda Baumann

*Paul Baumann has a Ph.D.
in bacteriology from the
University of California,
Berkeley. Linda Baumann
has an M.A. in microbiology
from Tufts University
Medical School, Boston.*
*During the past 20 years they have worked together on such diverse
topics as the taxonomy and physiology of marine bacteria, the mos-
quitocidal toxins of* Bacillus sphaericus, *and currently the procaryotic endosymbionts of aphids.*

Associations between eucaryotic cells and intracellular procaryotic organisms are widespread in the living world. At
one extreme are *Rickettsia* and *Chlamydia*, two bacterial genera with species of intracellular human pathogens that cannot be grown outside of cells. At the other extreme are
mutually beneficial associations between intracellular procaryotic organisms and their eucaryotic hosts. Many insects
contain such associations, and one which has been studied
in some detail is that between aphids and a procaryotic
endosymbiont.

Aphids are analogous to living syringes in that each
aphid has an appendage that penetrates the surface of plants
and sucks up plant sap. Repeated feedings deplete the plant
of nutrients but, more importantly, transmit a variety of
plant viral diseases that reduce the economic value of agricultural crops. In their body cavity, aphids have specialized
cells that contain about 5×10^6 cells of a procaryotic endosymbiont. Treatment of the aphid with procaryote-specific
antibiotics eliminates the endosymbiont, and the aphid dies.
The endosymbiont cannot be grown outside the aphid host.
This is an example of a mutualistic association: Both the host
and the endosymbiont depend on each other for survival.
The endosymbiont of aphids may be a stage between a free-
living organism and an organelle such as mitochondria,
which are known to have originated from procaryotic cells.
Understanding the nature of the endosymbiont and its essential contributions to the host should point to strategies for
the control of aphid populations.

We were initially interested in finding out if the endosymbiont more closely resembles a free-living bacterium or an
organelle. Organelles originating from procaryotes undergo a
number of changes. The most important is a reduction in the
genome size and a consequent loss of genes encoding a variety of essential cellular functions. We cloned and sequenced a

variety of genes from the endosymbionts and found that they
have many genes coding for proteins involved in DNA replication, transcription, translation, protein secretion, and
energy-yielding metabolism. Thus the endosymbionts
resemble free-living bacteria and not organelles.

Then we wanted to establish the evolutionary relationships of endosymbionts to free-living bacteria. We cloned and
sequenced the 16S ribosomal RNA gene from endosymbionts
of different aphid species and compared their sequences to
one another and to 16S ribosomal RNA sequences from representative procaryotes. We found that the endosymbionts
are closely related to one another and that *Escherichia coli* is
their closest free-living relative. The data also suggest the
following scenario: About 200–250 million years ago a free-
living bacterium entered the cells of an aphid ancestor and
became an endosymbiont. With time, both the aphid and the
endosymbiont evolved, giving rise to the current species of
aphids and their endosymbionts (cospeciation). The evidence
for a single ancestor of the aphid endosymbionts suggests
that at least some of the contributions made by the endosymbionts to the aphid host may be similar or identical in all of
the associations.

How does the endosymbiont contribute to the aphid
host? The diet of aphids (plant sap) is rich in sugars but deficient in amino acids and vitamins. It is known that insects
cannot synthesize 10 essential amino acids and consequently
require them in their diet. Nutritional studies have suggested that aphids do not require these amino acids, and evidence has been presented indicating that they are made by
endosymbionts. The best available evidence is for overproduction of the amino acid tryptophan. Using recombinant
DNA techniques we have found that the endosymbiont contains all of the genes of the tryptophan biosynthetic pathway.
In free-living bacteria the first enzyme of the tryptophan
pathway is feedback inhibited by the end product (tryptophan), but usually this inhibition is not complete. In order
for the aphid endosymbiont to overproduce tryptophan, this
regulatory mechanism must be modified. We have found
that in the endosymbiont the genes coding for the first enzyme
are amplified, being arranged as four tandem repeats on a
plasmid which is present as four copies per endosymbiont
cell. The increase in gene copies assures overproduction of
the enzyme and retention of sufficient enzyme activity even
in the presence of inhibitory concentrations of tryptophan.
The adaptation of the endosymbiont to overproduction of
tryptophan by amplification of the genes for the first enzyme
of the biosynthetic pathway is similar in mechanism to the
gene amplification observed in some antibiotic-resistant
pathogens.

the **cristae**. It courses through the matrix, the gelatinous ground material. Mitochondria are involved in ATP production in all animals, including humans, and in algae, fungi, and most protozoa (Chapter 5).

PASSAGE OF MOLECULES ACROSS CELL MEMBRANES

Plasma membranes of procaryotes and eucaryotes have a critical function: They determine which molecules enter and leave the cell. The physical properties of all phospholipid bilayer (unit) membranes determine how molecules cross—or fail to cross—this barrier.

All unit membranes allow certain molecules to cross freely while blocking the passage of others. In other words, these membranes are **semipermeable**. Because cell membranes are composed primarily of lipids, which are hydrophobic (polar), hydrophobic molecules can pass directly through the membrane by dissolving in it. Hydrophilic molecules, on the other hand, such as sugars and amino acids, cannot. Neither can molecules that bear a positive or negative charge, such as potassium (K^+) and phosphate (PO_4^{3-}) ions. Water is an exception. Although it is a polar molecule, it passes cell membranes extremely rapidly through water-transmitting pores that appear and disappear as the phospholipid molecules constantly move. Biologically important molecules that can cross cell membranes freely include water, carbon dioxide, oxygen, ethanol, and medium-length fatty acids. Most other molecules cannot cross the plasma membrane without the help of carrier proteins.

We will look at six ways that molecules pass across plasma membranes. Both procaryotic and eucaryotic cells exhibit four of these ways: simple diffusion, osmosis, facilitated diffusion, and active transport. Procaryotes alone use group translocation, and only eucaryotes use engulfment.

Simple Diffusion

Any molecule that crosses a membrane freely will eventually reach equal concentrations on either side because of the process of **diffusion**. That is, all molecules are subject to random movement. They collide with molecules nearby and eventually distribute themselves evenly throughout all the space available to them (**Figure 4.22**). Any molecule that crosses a membrane freely will diffuse through it until it reaches **equilibrium**, equal distribution on both sides. Thus, a net flow of molecules into the cell by diffusion occurs only when driven by a **concentration gradient**, when the concentration of the molecule is greater outside than inside the cell. No energy is needed to drive the flow.

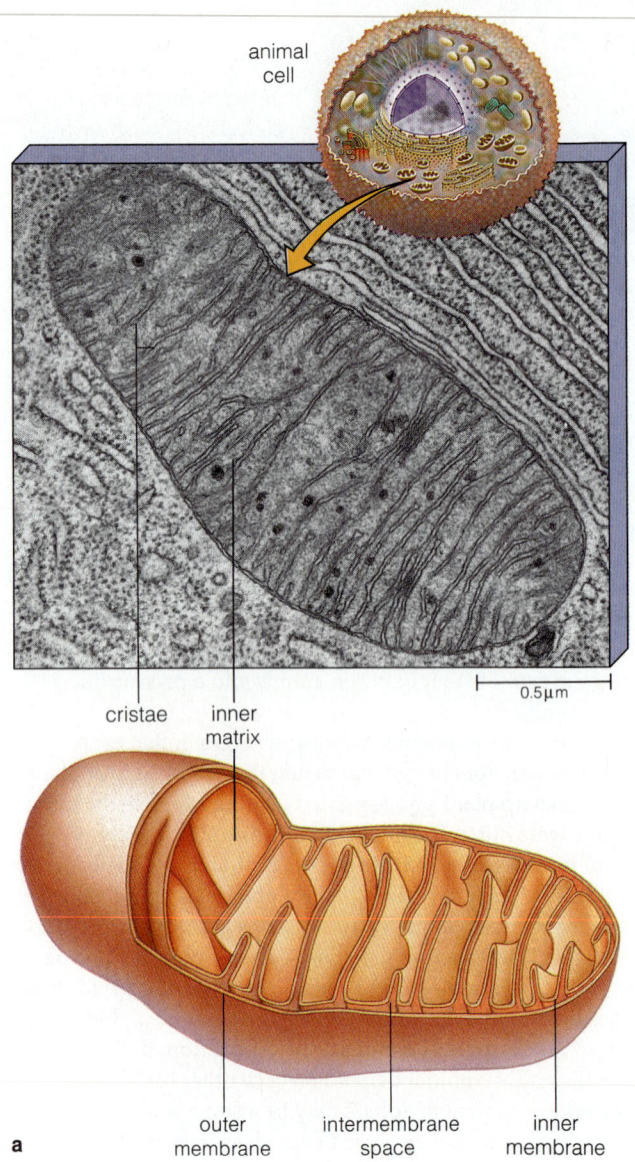

Figure 4.21 Mitochondria and chloroplasts. These organelles house the energy-generating proteins that, in procaryotes, are found in the plasma membrane. Mitochondria are found in all aerobic eucaryotes: all animals, (including humans), algae, fungi, and most protozoa. Chloroplasts are found in all plants, almost all algae, and a few protozoa. (**a**) The inner membrane of mitochondria is highly folded, forming cristae. (**b**) The equivalent inner membrane in chloroplasts is called the thylakoid. A stack of thylakoids is called a granum. The stroma is a thick ground substance (matrix).

animal cell

0.5 μm

cristae   inner matrix

a

outer membrane   intermembrane space   inner membrane

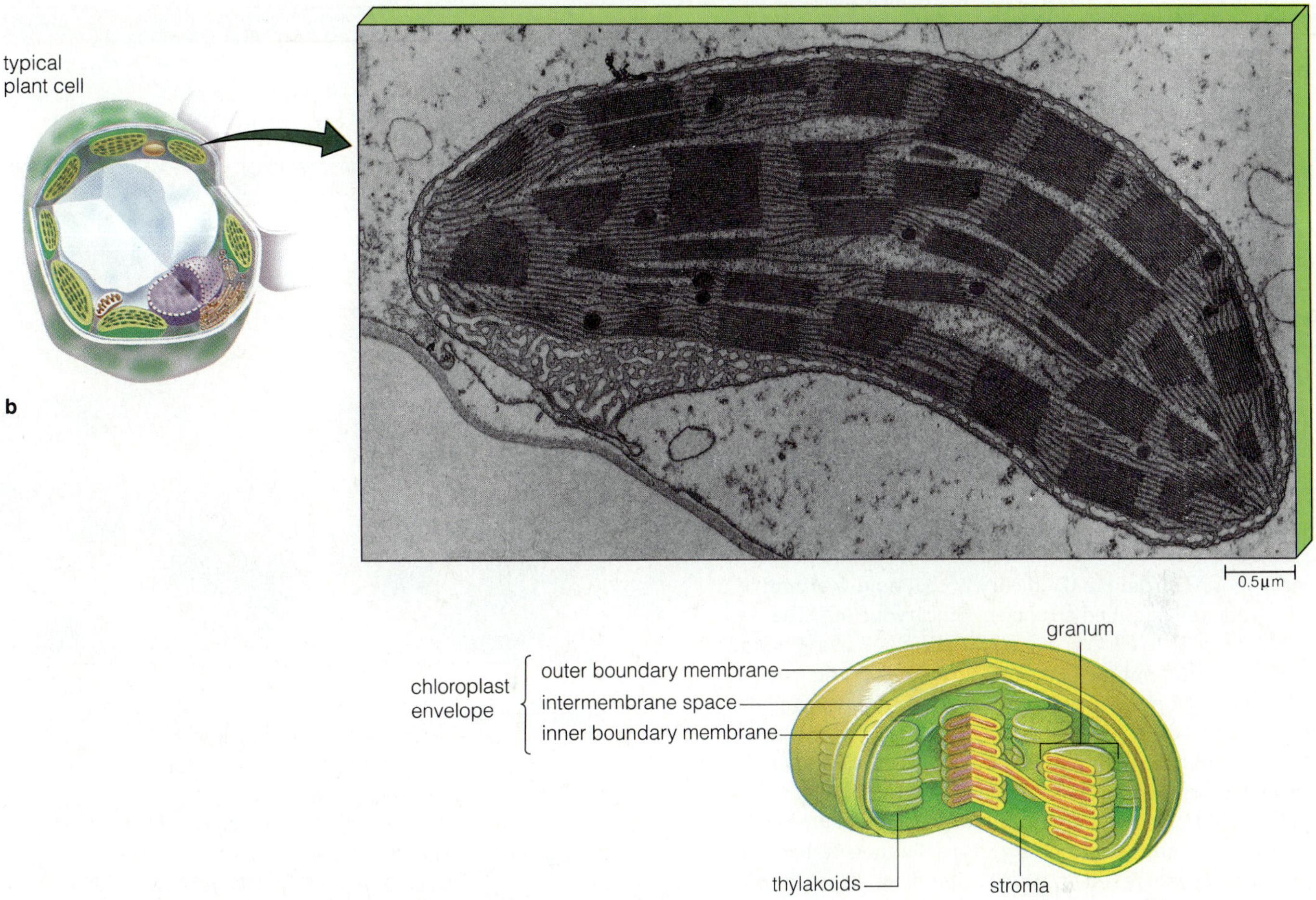

typical
plant cell

b

0.5μm

chloroplast
envelope
{
outer boundary membrane
intermembrane space
inner boundary membrane

granum

thylakoids

stroma

Osmosis

The same principles of diffusion apply to the passage of water molecules across a membrane. However, diffusion of water is referred to as **osmosis**. By osmosis, water crosses a membrane toward the side with the higher concentration of solute molecules and, therefore, lower concentration of water molecules. That is, the presence of other molecules decreases the concentration of water, causing water molecules to move in a direction that will tend to equalize their concentration on both sides of the membrane. The total concentration of solute molecules in a membrane-bound space, such as a cell, is referred to as its **osmotic strength**. As water flows into a closed space, it causes an increase in internal pressure, called **osmotic pressure**. In a cell it is called *turgor pressure*.

A cell's external environment affects osmotic movement and the integrity of the cell (**Figure 4.23**). A **hypertonic** environment has a higher concentration of solutes

Figure 4.22 Simple diffusion. Dye molecules become evenly distributed throughout a volume of water by simple diffusion. The flow is driven by a concentration gradient. No energy is required.

Endosymbiosis: Something Old, Something New

Mitochondria and chloroplasts resemble procaryotic cells in many ways. Both of these organelles are approximately the same size as a procaryotic cell. They are the only eucaryotic organelles to contain DNA; and in both, the DNA is arranged like a single circular procaryotic chromosome. Both organelles contain 70S ribosomes, like those found in procaryotic cells, rather than the 80S ribosomes found elsewhere in eucaryotic cells. Moreover, mitochondria and chloroplasts multiply the way procaryotic cells do, by binary fission.

The resemblance of mitochondria and chloroplasts to procaryotic cells reflects the origin of these eucaryotic organelles from procaryotes. Procaryotes first appeared about 3.5 billion years ago, and eucaryotes about a billion years ago. According to the theory, at that time, a primitive eucaryotic cell engulfed an ancient procaryotic cell. The two cells continued to evolve, one within the other, in a symbiotic relationship (mutually beneficial). The procaryote benefited by having a place to live and a ready source of nutrients, and the eucaryote benefited by having organelles that make ATP by respiration or photosynthesis. Eventually the modern eucaryotic cell, with its membrane-bound mitochondria and/or chloroplasts, evolved. This is called the *endosymbiotic theory* of cellular evolution because an organism that lives within another is called an **endosymbiont**.

If mitochondria and chloroplasts are the evolutionary product of a primitive procaryotic cell, it must have been a Gram-negative one. The double membranes of mitochondria and chloroplasts resemble the plasma and outer membranes of Gram-negative bacteria. The external membrane of these organelles even has pores that resemble the porins in the outer membrane of Gram-negatives.

How do we know it wasn't another procaryote that engulfed a smaller procaryote? The answer to this question is the strongest piece of evidence in favor of the endosymbiotic theory. Only eucaryotes are capable of phagocytosis, and phagocytosis is undoubtedly the way the procaryote was captured. In engulfing nutrients, the primitive eucaryote also engulfed procaryotes.

The primitive eucaryote was probably similar to the modern-day protozoan *Giardia*, which causes diarrhea in humans (Chapter 23). *Giardia* is a eucaryote, but it has no mitochondria and its distant relationship to other eucaryotes suggests that *Giardia* is evolutionarily primitive. It probably did not lose mitochondria—it never had them.

Mitochondria and chloroplasts have changed enormously during their billion-year stay inside eucaryotic cells. They lost properties that do not benefit their eucaryotic host—their cell wall, many enzymes, and much of their DNA. But

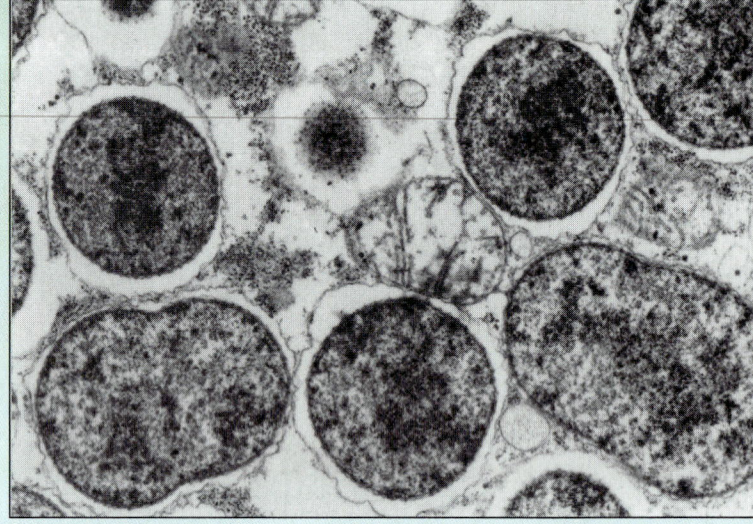

Endosymbiont in an aphid.

they kept capacities needed by their eucaryotic hosts—capturing energy through respiration or photosynthesis.

Are mitochondria and chloroplasts the only remnants of procaryotic endosymbionts in eucaryotes? Decidedly not. Eucaryotes constantly phagocytize procaryotes, and some persist as endosymbionts when both partners benefit. Aphids and related insects (whiteflies and mealybugs) have procaryotic symbionts. Since the relationship is fairly recent—only about 200 million years old—the endosymbionts have not yet lost their Gram-negative envelope. But they can no longer grow on their own outside their host. Some species are closely related to *Escherichia coli*.

Aphids need their endosymbionts. When they are eliminated by antibiotic or heat treatment, the aphid dies. The endosymbionts supply their aphid host with essential nutrients. Apparently, aphids have lost the ability to synthesize about 10 amino acids, which are not present in adequate amounts in an aphid's diet of plant juices. What evolutionary scenario could lead to aphids losing the capacity to synthesize so many amino acids, only to regain it by capturing a procaryote? Most probably, aphid ancestors lived in a more nutrient-rich environment (most insects do) and benefited by being spared from synthesizing their own amino acids. But in order to adapt to a diet of plant juices, aphids suddenly had to reacquire this ability. They did it secondhand, by capturing a procaryote. (See the Focus on Research box "A Different Kind of Mitochondrion?")

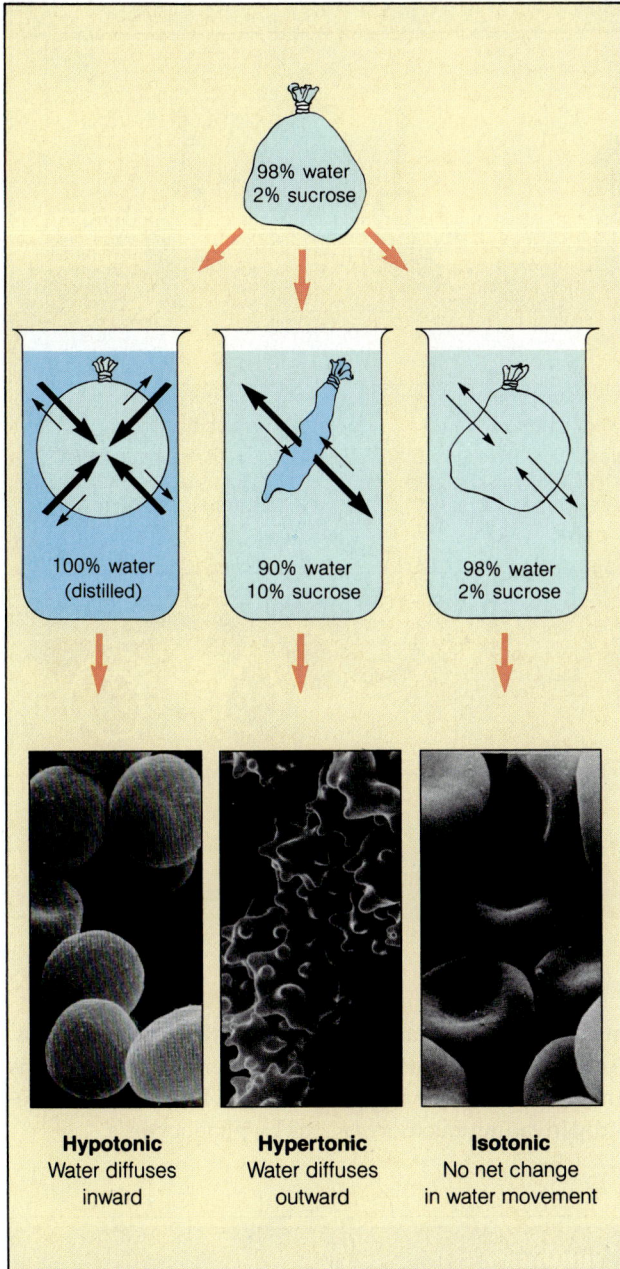

Figure 4.23 Osmosis. Simple diffusion of water down a concentration gradient is called osmosis. A cell's external environment affects osmotic movement and the fate of the cell. Imagine that the membranous bag with a solution of 98 percent water and 2 percent sucrose is a cell. In a hypotonic environment, when water enters the cell, it swells; and if the membrane isn't strong enough to counteract the increased internal turgor pressure, the cell will lyse. In a hypertonic environment, water will leave the cell, causing it to shrivel (plasmolysis). In an isotonic environment, there will be no net movement of water, and the cell keeps its normal shape. The cells shown here are red blood cells.

than does the cell's interior. It therefore draws water out of the cell by osmosis, causing the membrane-bounded volume to decrease—a condition called **plasmolysis**. A **hypotonic** environment has a lower concentration of solutes than the cell's interior. It therefore loses water to the cell's interior, which increases the cell's turgor pressure. If turgor pressure is stronger than the plasma membrane and cell wall (if the cell has one), the cell lyses, or bursts. An **isotonic** environment has the same concentration of solutes as the cell's interior and so exerts no osmotic effect on the cell.

Facilitated Diffusion

Substances necessary to sustain life that cannot diffuse across unit membranes must somehow still enter. In some cases, passage is mediated by carrier proteins embedded in the membrane. Carrier proteins that mediate **facilitated diffusion** bind a compound on one side of the membrane and release it on the other (**Figure 4.24**). However, facilitated diffusion transports molecules only from regions of higher concentration to regions of lower concentration, just as diffusion would if the membrane were permeable to that molecule.

Carrier proteins that mediate facilitated diffusion are highly specific, meaning they usually transport only one kind of molecule. Not many compounds enter procaryotes by facilitated diffusion. For example, only glycerol enters *Escherichia coli* by this mechanism. But many compounds, including most sugars, enter most eucaryotic microorganisms by facilitated diffusion.

Active Transport

Some carrier proteins transport molecules from regions of lower concentration to regions of higher concentration—the opposite of what occurs with diffusion. Such a process is necessary to bring essential nutrients into the cell when there are low concentrations outside. The movement of molecules *against* a concentration gradient requires energy. Energy can be in the form of ATP or a concentration gradient of ions, usually protons (Chapter 5), and is called **active transport**. Most nutrients, including sugars, amino acids, and vitamins, enter procaryotes by active transport because many procaryotes live in low-nutrient environments.

Group Translocation

Group translocation is a variation of active transport that occurs only in bacterial cells and only with certain molecules. In group translocation, a molecule is transported into the cell and at the same time chemically changed into a slightly different molecule. The latter process requires

Figure 4.24 Mechanisms by which substances move across cell membranes. Facilitated diffusion and active transport involve carrier proteins, but only active transport requires energy. The driving force in facilitated diffusion is still a concentration gradient, so no energy is required. The carrier protein simply binds a molecule on one side and releases it on the other side. Active transport requires energy because the solute is moving against the concentration gradient. In endocytosis, the plasma membrane folds in on itself, wholly surrounds a solid foreign particle, and brings it into the cell, where it buds off and travels in a vacuole. Exocytosis is the reverse procedure. The vacuole merges with the plasma membrane, and the contents are released outside.

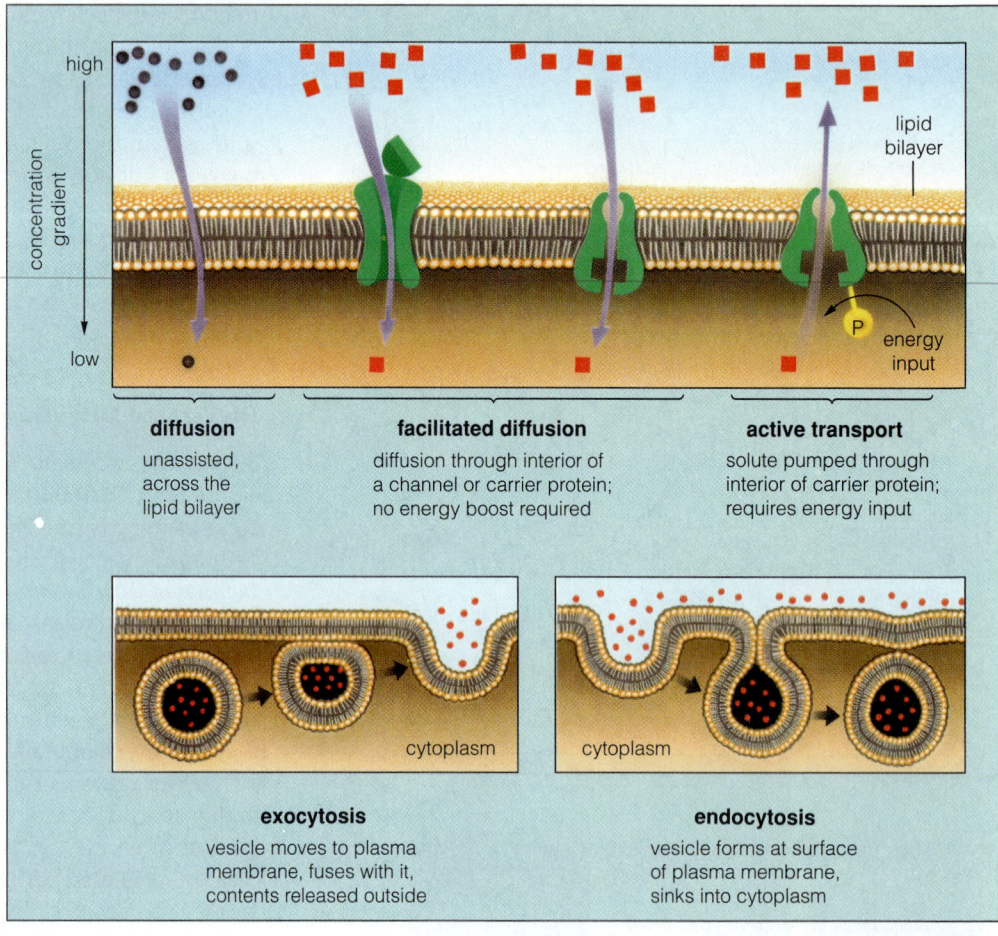

diffusion
unassisted, across the lipid bilayer

facilitated diffusion
diffusion through interior of a channel or carrier protein; no energy boost required

active transport
solute pumped through interior of carrier protein; requires energy input

exocytosis
vesicle moves to plasma membrane, fuses with it, contents released outside

endocytosis
vesicle forms at surface of plasma membrane, sinks into cytoplasm

energy. But the energy is well spent because the chemical modification prevents the molecule from leaving the cell. For example, bacteria take up the nutrient glucose by group translocation. A carrier protein moves glucose molecules from the environment into the cell, where it is **phosphorylated**, meaning a phosphate group is added to it so it cannot leave. The phosphorylated form is used to fuel the cell's metabolic reactions.

Engulfment

Some eucaryotes can also move molecules across the plasma membrane by engulfing them. Only eucaryotes without cell walls can do this. In **engulfment**, the plasma membrane folds in on itself, wholly surrounds a foreign particle, brings it into the cell, and releases it as a separate membrane-bound structure called a **vacuole**. When solid material is engulfed, the process is called **endocytosis**. If a cell is taken in by certain types of white blood cells, the process is called **phagocytosis** ("cell eating"). If liquid material is taken in, the process is called **pinocytosis** ("cell drinking"). **Exocytosis** refers to the expulsion of

material, such as waste materials, by the reverse process. Amoebae take in most of their nutrients by endocytosis and pinocytosis (Chapter 12). White cells in our blood destroy invading microorganisms by phagocytosis.

Summary

STRUCTURE AND FUNCTION (p. 78)

1. The cell is best understood in terms of structure and function, which are inextricably linked.

THE PROCARYOTIC CELL (pp. 78–92)

1. Procaryotic cells have an undifferentiated, grainy interior with a few organelles and a nuclear region or nucleoid (a mass of DNA), rather than a membrane-bound nucleus.

2. Procaryotic cells are usually much smaller than eucaryotic cells.

Structure: An Overview (pp. 78–80)

1. The procaryotic cell has two basic parts: the envelope (the outside of the cell) and the cytoplasm (which fills the interior and houses the nucleoid and other structures).

Appendages (pp. 80–82)

1. Some bacterial species have appendages (structures attached to the envelope and extending beyond the cell).

2. Pili are straight, usually short, hairlike appendages that attach bacteria to other cells. The sex pilus attaches mating bacteria.

3. Flagella have a complex three-part structure that allows them to rotate and propel the cell toward favorable environments and away from harmful ones (a behavior called taxis). In chemotaxis, bacteria swim toward nutrients and away from toxic materials.

4. There are four types of flagellar arrangements: monotrichous, amphitrichous, lophotrichous, and peritrichous.

5. The spirochetes have axial filaments, bundles of flagella wrapped around the cell body that propel the cell like a corkscrew through viscous environments.

Procaryotic Envelope (pp. 82–89)

1. The glycocalyx is the slimy or gummy outermost layer of the envelope. Thick layers are also called capsules, and thin layers may be called slime layers. The glycocalyx offers protection from phagocytosis and helps the cell adhere to surfaces.

2. The outer membrane is a bilayer composed of phospholipid on the inner surface and, on the outer surface, lipopolysaccharide (LPS). LPS is not found in any other living thing. Only Gram-negative bacteria have an outer membrane.

3. The periplasm is a gelatinous organelle containing essential enzymes and other proteins. Only Gram-negative bacteria have a periplasm.

4. All eubacteria except mycoplasmas have a cell wall that is composed primarily of peptidoglycan (murein).

5. The cell wall in Gram-negative bacteria is one layer thick. In Gram-positives, it is many layers thick and also contains teichoic acids.

6. The cell wall functions to confer shape on a bacterium. Most bacteria fall into one of three groups, spherical (cocci), rod-shaped (bacilli), or spiral-shaped (spirilla). There are also intermediate shapes (coccobacillus, vibrio).

7. The cell wall also contains a cell's turgor pressure, the internal pressure from its contents. A bacterial cell's turgor is extremely high because its contents are concentrated.

8. Mycoplasmas lack a cell wall. They avoid lysis from turgor pressure by actively pumping sodium ions out of the cell.

9. L forms are strains of bacteria that have lost the ability to form walls but can still reproduce if provided osmotic protection.

10. The membrane that encloses the cytoplasm of any cell is called the plasma (cytoplasmic) membrane. Plasma membranes contain the cytoplasm and regulate what enters and leaves the cell.

11. All membranes, including the plasma membrane, that have the same structural and chemical characteristics are called unit membranes.

12. Unit membranes are composed primarily of phospholipids that spontaneously arrange themselves into two rows with their hydrophobic tails facing one another and their hydrophilic heads associated with the water on both outer sides of the membrane. Unit membranes are thus described as a phospholipid bilayer.

13. Unit membranes are studded with proteins that can move as they function, so the membrane structure as a whole fits the fluid mosaic model.

14. Unlike eucaryotic cells, procaryotic cells also contain proteins that are members of the electron transport chain and proteins that participate in photosynthesis.

Cytoplasm (pp. 89–90)

1. The cytoplasm is a matrix composed primarily of water and protein where the cell's vital chemical reactions (metabolism) take place.

2. The nucleoid or nuclear region is a mass of DNA arranged in a single circular molecule, the bacterial chromosome. Some bacteria also have smaller circular DNA molecules called plasmids that encode specialized, nonessential functions.

3. Bacterial cytoplasm is packed with ribosomes, small structures that manufacture proteins. Procaryotic ribosomes (called 70S) are smaller than eucaryotic ribosomes (called 80S).

4. Inclusions are visible structures in the cell other than the nuclear region and ribosomes; examples are storage granules (hold the cell's reserve supplies of nutrients), gas vacuoles (gas-filled regions), or magnetosomes (iron-containing structures).

Endospores (pp. 90–91)

1. Endospores are extremely hardy, resting (nongrowing) structures that some bacteria—principally Gram-positives—produce through the process of sporulation when nutrients are exhausted. When favorable conditions return, endospores germinate to produce new vegetative cells, which grow and reproduce.

2. Sporogenesis, the complete cycle of sporulation and germination, allows survival under unfavorable conditions.

The Archaebacterial Cell (pp. 91–92)

1. Archaebacterial cell walls are composed of either protein or a peptidoglycan-like material called pseudomurein, not found in any other living things. Some archaebacteria do not have cell walls.

2. The bonds in the plasma membrane of archaebacterial cells are ether linkages, which are stronger than the ester bonds found in eubacteria.

THE EUCARYOTIC CELL (pp. 92–100)

1. Cells of algae, fungi, and protozoa are eucaryotic, as are human and plant cells.

2. Many membrane-bound organelles help keep different chemical activities separate and provide systems of transportation in the more complex eucaryotic cell.

Appendages (pp. 92–93)

1. Eucaryotic cells do not have pili or axial filaments, but some have flagella and smaller versions called cilia that also function for locomotion.

2. Humans have cilia—projections whose function is the movement of materials rather than locomotion. Two groups of protozoa—the flagellates and the ciliates—have appendages.

Cell Wall (p. 93)

1. Algae and fungi typically have a cell wall. Some protozoa have a pellicle, which functions much as a cell wall, though it is more flexible and skinlike. Human cells never have walls. The cell walls of plants and of most algae and some fungi consist of polysaccharide cellulose.

Plasma Membrane (p. 93)

1. The plasma membranes of procaryotes and eucaryotes both are unit membranes. Both also serve to regulate what enters and leaves the cell.

Cytoplasm and Its Contents (pp. 93–100)

1. The cytoskeleton is a lattice network composed of three types of protein structures that allows the cytoplasm in larger eucaryotic cells to move in an orderly way.

2. The eucaryotic nucleus is surrounded by a nuclear envelope containing nucleoplasm and nucleoli, dense masses of RNA and protein that manufacture ribosomes.

3. Eucaryotic DNA is associated with proteins called histones that form a structure called a nucleosome. Nucleosomes form paired structures called chromosomes, which carry a cell's genes.

4. Eucaryotic cells undergo two types of cell division, mitosis (for growth) and meiosis (for reproduction).

5. The cytomembrane system sorts, organizes, and packages the different molecules that the cell produces. It consists of the endoplasmic reticulum (ER) and the Golgi apparatus.

6. Some regions of the ER are smooth (SER) and others are rough (RER). The SER is the major site of phospholipid synthesis, creating the components for the cell's membranes. The RER appears rough (grainy) because of the ribosomes (the major site of protein synthesis) on its surface.

7. The Golgi apparatus is made up of stacks of flattened membranes that modify and concentrate phospholipids and proteins. They also produce lysosomes, powerful packets of enzymes that destroy microbial cells captured by phagocytes.

8. The proteins involved in generating ATP are found in two types of double-membrane-bound organelles, mitochondria (respiration) and chloroplasts (photosynthesis). Plants and nearly all algae have chloroplasts. Animals (including humans), fungi, and some protozoa have mitochondria.

PASSAGE OF MOLECULES ACROSS CELL MEMBRANES (pp. 100–104)

1. The plasma membrane is semipermeable: it allows certain molecules to cross freely but not others. Molecules that cannot diffuse must cross the plasma membrane by some other means, which may require the cell to expend energy.

2. In simple diffusion, molecules flow into a cell along a concentration gradient until the molecules reach equilibrium.

3. The diffusion of water is referred to as osmosis. As water flows into a closed space, it increases the internal pressure, or osmotic pressure. In a cell, it is called turgor pressure.

4. A cell's external environment may be hypertonic (higher concentration of solutes than the cell's interior, which may cause plasmolysis), hypotonic (lower concentration of solutes than the cell's interior, which can cause the cell to lyse), or isotonic (same concentration of solutes outside and inside the cell so there is no osmotic effect).

5. In facilitated diffusion, carrier proteins bind a compound that cannot diffuse across the membrane and release it on the other side. Facilitated diffusion transports molecules only from regions of higher concentrations to regions of lower concentration, so no energy is required.

6. In active transport, carrier proteins transport molecules from regions of lower concentration to regions of higher concentration. Energy is required.

7. Group translocation occurs only in procaryotes and only with certain molecules. At the same time a molecule is transported into the cell, it is chemically changed to prevent it from leaving the cell. Energy is required.

8. In engulfment, the plasma membrane folds in on itself, wholly surrounds a foreign particle, brings it into the cell, and releases it in a membrane-bound structure called a vacuole. Only eucaryotes without cell walls can engulf molecules.

9. In endocytosis, solid material is engulfed; in phagocytosis, a whole cell is taken in; in pinocytosis, liquid material is taken in. Exocytosis is expelling material from a cell by the reverse process.

THE PROCARYOTIC CELL

1. Describe three major differences between procaryotic and eucaryotic cells.

2. What are cellular appendages? Name the three types of procaryotic appendages.

3. Compare the functions of pili and flagella.

4. Sketch a procaryotic flagellum and describe what makes it turn. What are the four types of flagella, as determined by number and arrangement?

5. Define taxis. Describe chemotaxis in *Escherichia coli*.

6. How does the structure of axial filaments suit their function?

7. Sketch and label the parts of the procaryotic cell envelope. Which parts are found only in Gram-negative procaryotic cells?

8. Define glycocalyx, capsule, and slime layer. How are structure and function related in the procaryotic glycocalyx?

9. What is the structure of the outer membrane? What is its function? What are porins and Bayer junctions? Why is lipid A called an endotoxin?

10. What is the structure and function of periplasm?

11. Do all bacteria have a cell wall? Explain.

12. What is the chief component of all bacterial cell walls? What is its chemical structure? What are the two major functions of procaryotic cell walls?

13. Compare and contrast the cell walls in Gram-positive bacteria, Gram-negative bacteria, and mycoplasmas.

14. What are the three most commonly found shapes of bacteria?

15. How does the lysozyme experiment demonstrate that the cell wall confers shape on a procaryotic cell?

16. What is turgor pressure? Why is a cell's turgor pressure so high? How do mycoplasmas offset turgor pressure?

17. Compare and contrast L forms, protoplasts, and spheroplasts.

18. Describe the structure of the plasma membrane.

19. What are permeases?

20. Name two functions that the plasma membranes of all procaryotic and eucaryotic cells share. Name a function of the procaryotic plasma membrane that is not performed by the eucaryotic plasma membrane.

21. Describe the structure and function of procaryotic cytoplasm.

22. Describe the nucleoid. What is the bacterial chromosome? Define plasmid.

23. What is the function of ribosomes? What is the significance of the 70S-80S difference in ribosomes?

24. Define inclusion, and give some examples of inclusions.

25. What are endospores? Describe sporogenesis.

26. Name two ways the archaebacterial cell is different from the eubacterial cell.

THE EUCARYOTIC CELL

1. Give some examples of eucaryotic cells. How are eucaryotic organelles different from procaryotic organelles?

2. What types of appendages do eucaryotic cells have? How are eucaryotic flagella different from procaryotic flagella? How are they the same?

3. Which eucaryotic microorganisms have cell walls? Do human cells have cell walls? What is a pellicle?

4. What is the main difference between cell walls in procaryotes and cell walls in eucaryotes (when they have them)? How do the functions of cell walls compare in procaryotes and eucaryotes?

5. How are the plasma membranes of procaryotic and eucaryotic cells the same? Name two important differences.

6. What is the cytoskeleton? Give some examples of how the cytoskeleton functions.

7. Sketch and label the parts of a eucaryotic nucleus.

8. Compare and contrast procaryotic and eucaryotic DNA.

9. Sketch and explain mitosis.

10. What is the difference between mitosis and meiosis?

11. What is the cytomembrane system in a eucaryotic cell?

12. Describe the endoplasmic reticulum (ER). Why is some ER called smooth (SER) and some rough (RER)?

13. What is the function of SER? of RER? How do vesicles function in the cytomembrane system?

14. Describe the Golgi apparatus and how it functions. What are lysosomes?

15. How are mitochondria and chloroplasts structurally alike? How are they functionally alike?

16. How do some cellular biologists explain the similarity between mitochondria and chloroplasts and procaryotic cells?

PASSAGE OF MOLECULES ACROSS CELL MEMBRANES

1. Why are plasma (unit) membranes described as semipermeable? How does the chemical structure of unit membranes determine which molecules can pass and which cannot?

2. Explain simple diffusion.

3. What is osmosis? Define osmotic strength. Define turgor pressure. Give examples of how a cell's external environment affects osmotic movement and the integrity of the cell.

4. How is facilitated diffusion like simple diffusion and how is it different?

5. What is active transport?

6. Describe group translocation.

7. What is engulfment? Define endocytosis, phagocytosis, pinocytosis, and exocytosis.

8. Which of these processes require energy and which do not? Which occur in both procaryotic and eucaryotic cells? in only procaryotic? in only eucaryotic?
  a. simple diffusion   d. group translocation
  b. active transport   e. facilitated diffusion
  c. osmosis   f. engulfment

Suggested Readings

Fawcett, D. 1981. *The cell: An atlas of fine structure*. New York: W. H. Freeman.

Inouye, M., ed. 1986. *Bacterial outer membranes as model systems*. New York: John Wiley.

Krawiec, S., and Riley, M. 1990. Organization of the bacterial chromosome. *Microbiological Reviews* 54:502–39.

Mohan, S.; Dow, C.; and Coles, J. A., eds. *Prokaryotic structure and function: 47th Symposium of the Society for General Microbiology*. Cambridge: Cambridge University Press.

Rothman, J. 1985. The compartmental organization of the Golgi apparatus. *Scientific American* 244:57–67.

5 METABOLISM OF MICROORGANISMS

Introducing Escherichia coli

Without doubt *Escherichia coli* is the most thoroughly understood cellular organism. Why is that? First, *E. coli* is extremely easy to study in the laboratory. It grows rapidly as isolated single cells in simple media. Most strains do not cause disease, so they can be studied without taking special precautions. Furthermore, the more we know about *E. coli*, the easier it is to learn new things. In other words, it is easier to experiment on an organism if its basic properties are already known. Not surprisingly, the biotechnology industry uses *E. coli* extensively. Many biotechnology products—including insulin and growth hormone—are made by inserting human genes into strains of *E. coli* and then harvesting the hormones from the microbial cultures.

Escherichia coli and humans share a long and intimate history. *E. coli* probably first appeared between 120 and 160 million years ago—just about the time mammals did. *E. coli* typically inhabits the colon of mammals, and it probably evolved there. It was living with us when we emerged about a million years ago as human beings.

Life in the mammalian colon offers advantages and disadvantages. The temperature is a constant 37°C and nutrients are abundant, but the environment is anaerobic and contains some toxic compounds. *E. coli* prospers in this environment, but it can also grow in the very different environment it encounters when it leaves the colon in feces. This external environment is usually cooler, poorer in nutrients, and aerobic.

The metabolism of *E. coli* reflects these two habitats. It is a facultative anaerobe, meaning it can grow in either the presence or absence of air. It has the relatively rare ability to use lactose (milk sugar) to support growth, and lactose is produced only by mammals. Unlike most microorganisms, *E. coli* has a high tolerance for bile salts, which the liver produces and empties into the intestinal tract. *E. coli* can also rapidly shift its metabolism to respond optimally to nutritional and physical changes in both the colon and the external environment.

As is the case with all microorganisms, there are many strains (naturally occurring variants) of *E. coli*. Some strains—not those used routinely in the laboratory or in the biotechnology industry—can cause human diseases (see Chapter 23). Traveler's diarrhea, infant diarrhea, and infections of the urinary tract are frequently caused by *E. coli*.

In many countries, including the United States, health departments routinely test drinking water for *E. coli*. Because *E. coli* is present in the colons of mammals, finding *E. coli* in a water supply is presumptive evidence that it has been contaminated by feces and is dangerous to drink.

Escherichia coli shown in an image from a transmission electron microscope, with computer-enhanced color.

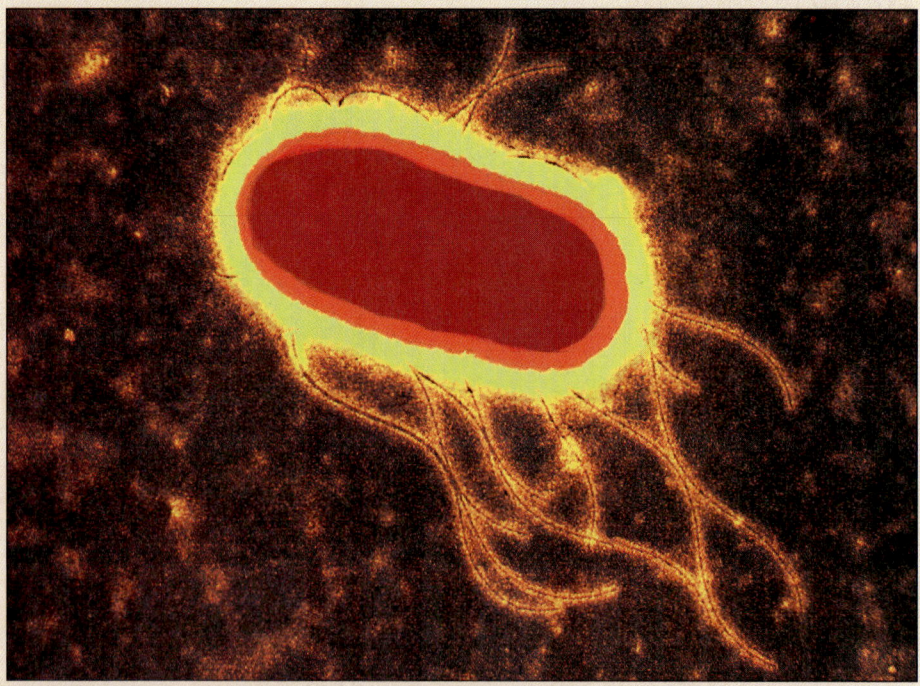

1 µm

To understand:

- The biological function and importance of metabolism

- The metabolism of *Escherichia coli* as a paradigm for aerobic and anaerobic metabolism

- How materials flow through five sequential steps in metabolism—entry mechanisms, catabolic reactions, biosynthesis, polymerization, and assembly

- How ATP is formed through substrate level phosphorylation and chemiosmosis and how it drives metabolism

- How reducing power is formed and how it drives metabolism

- How glycolysis, the TCA cycle, and the pentose phosphate cycle proceed and how they form the 12 precursor metabolites, ATP, and reducing power

- How anaerobic metabolism proceeds by way of anaerobic respiration or fermentation

- How photoautotrophs carry out metabolism through light-driven processes

- How chemoautotrophs carry out metabolism by oxidizing inorganic compounds

- How metabolism is regulated so the proper amounts and proportions of products are made efficiently

ESCHERICHIA COLI: A METABOLIC PARADIGM

As soon as a new microbial cell is formed by the process of cell division, it begins the process of synthesizing a complete set of component parts and assembling them into a new organism. The cell chemically makes all the molecules needed to build the new organism. The steps of this process are ordinary chemical reactions, but because they occur in living cells they are called biochemical reactions. Collectively, all the biochemical reactions that take place in a cell are called its **metabolism**.

The individual biochemical reactions that make up the metabolism of most living things are strikingly similar. Metabolism in any organism requires materials, a driving force, and a plan. There are some differences, however, particularly among types of microorganisms.

Much of what we know about metabolism was first discovered by studying *Escherichia coli*, so this chapter begins by looking at *E. coli* as a metabolic paradigm, a model of how metabolism is carried out in a cell. With this overview in mind, we examine the processes cells use to obtain what they need: entry mechanisms, catabolic reactions, biosynthesis, polymerization, assembly. Then we discuss aerobic metabolism and compare it to anaerobic metabolism. Finally, we turn to other forms of metabolism in bacteria and eucaryotic microorganisms.

Metabolism allows an organism to carry out all its activities, both behavioral and reproductive. But the behavioral component of microbial activity is minor. For example, moving to a more favorable environment or avoiding a toxic one (Chapter 4) places only minor metabolic demands on a microorganism. The major part of microbial metabolism is dedicated to reproduction—making a new cell. This is in startling contrast to human beings and other animals who devote most of their metabolic effort to behavioral activity and relatively little to producing new cells.

Because its primary function is reproduction, *a microbial cell is like a factory that manufactures more factories identical to itself.* Building a factory requires materials, energy to do work, and a plan. Metabolism leading to the synthesis of a new microbial cell has the same three requirements: materials, driving force (energy and reducing power), and a plan.

Materials

Raw materials are needed to manufacture a new cell, just as they are to manufacture anything (**Figure 5.1** and **Table 5.1**). A microorganism, such as *Escherichia coli*, is a complex combination of biochemicals, so molecules are the materials of metabolism. And because the primary building material of biochemicals is carbon, carbon is the material we track most closely as we follow the process of metabolic construction.

Carbon atoms are available to *E. coli* as part of naturally occurring molecules in its environment. These molecules, called **substrates** (or sometimes **nutrients** or **carbon sources**), are the raw materials of microbial metabolism. *E. coli* possesses the chemical machinery to refine substrates into basic building materials—usable compounds that can enter the assembly lines of the microbial factory. *E. coli* can produce the driving force (energy and reducing power) to run the assembly lines that transform the basic building materials into the complex components of a bacterial cell.

Driving Force

Every factory needs a driving force to do its work. In a factory, components are moved through an assembly line to produce a finished product. In *E. coli*, chemical

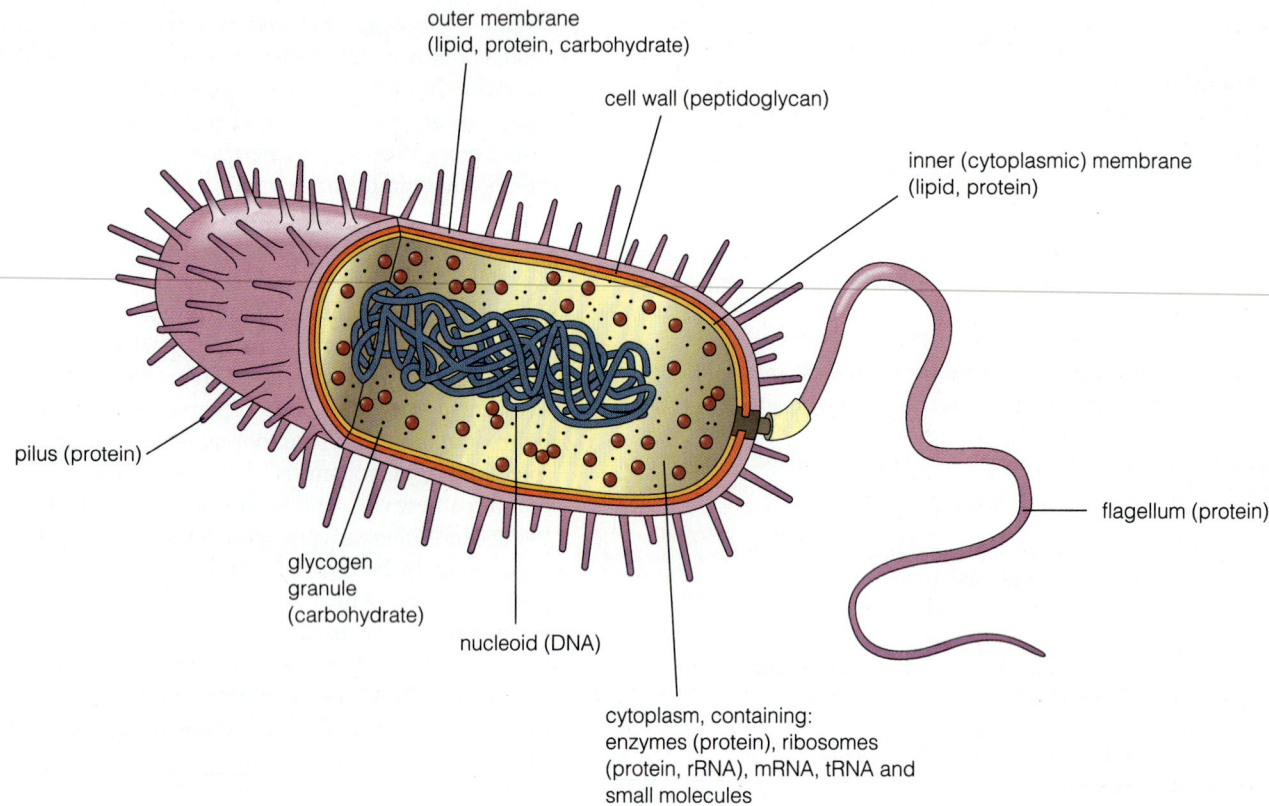

outer membrane
(lipid, protein, carbohydrate)

cell wall (peptidoglycan)

inner (cytoplasmic) membrane
(lipid, protein)

pilus (protein)

flagellum (protein)

glycogen
granule
(carbohydrate)

nucleoid (DNA)

cytoplasm, containing:
enzymes (protein), ribosomes
(protein, rRNA), mRNA, tRNA and
small molecules

Figure 5.1   The major product of microbial metabolism is a new cell.

Table 5.1 The Chemical Composition of *Escherichia coli*		
Component	Percent of Total Dry Weight[a]	Number of Different Kinds of Molecules
Protein	55.0	1050
RNA	20.5	463
DNA	3.1	1
Lipid	9.1	4
Lipopolysaccharides[b]	3.4	1
Peptidoglycan	2.5	1
Glycogen	2.5	1
Total macromolecules	96.1	
Small molecules[c]	2.9	
Ions	1.0	
TOTAL	100.0	

[a]A living cell is 70 percent water. These amounts make up the other 30 percent.
[b]A combination of lipids and polysaccharides that make up part of a cell's outer membrane.
[c]Includes building blocks, metabolic intermediates, and vitamins.

reactions synthesize a new cell. Energy and reducing power—collectively called **driving force**—drive these reactions.

Energy.   Chemical reactions proceed spontaneously only if energy is released by the reaction. Some *require* energy: by themselves these reactions would not occur. To make them occur, *E. coli* stores energy in the form of a highly reactive molecule called **adenosine triphosphate (ATP)** (Chapter 2). Some of the chemical reactions that occur during metabolism release excess energy, which is captured and stored in the form of ATP. The breakdown of ATP, which releases energy, is integrated into chemical reactions that build cellular components. These ATP-associated reactions proceed with a net release of energy and therefore make the process of building a cell energetically feasible. ATP is not the only form of stored energy that drives metabolic reactions, but it is the principal one. *All other forms of stored energy can be converted into ATP or can be formed from it.*

Reducing Power.   Many chemical reactions, including biochemical reactions of metabolism, involve oxidation and reduction. **Oxidation** is the removal of electrons from a compound; **reduction** is the addition of electrons to a compound. A number of the biochemicals that *E. coli* makes to build its cellular components must be reduced and, therefore, require a supply of electrons. *E. coli* stores electrons in compounds called **nicotinamide adenine dinucleotide (NAD)** and **nicotinamide adenine dinucleotide phosphate (NADP)**. These compounds capture electrons in the form of hydrogen atoms from compounds that are being oxidized. Later they use them to reduce other compounds. In this way NAD and NADP—collectively designated **NAD(P)**—drive metabolic reductions.

ATP stores the cell's energy, and NAD(P) stores the cell's reducing power. These two critically important reserves are, in fact, interconvertible. That is, ATP can be expended to reduce NAD(P) and reduced NAD(P) can be used to produce ATP. Later in this chapter we discuss how these two driving forces are formed and how they are used.

The Plan

A plan is as essential for directing metabolism as it is for running a factory. The plan for *E. coli*'s microbial metabolism consists of about 6 million distinct pieces of information because its end product—the microbial cell—is so complex. *E. coli* manufactures approximately 1000 different proteins.

All the information necessary to direct microbial metabolism is contained in the cell's DNA. A copy of this master plan is transmitted from one generation of cells to the next, and each cell uses the plan to synthesize a new cell (Chapter 6).

METABOLISM: AN OVERVIEW

In metabolism, raw materials (substrates) from the environment are converted into the finished product. The assembly line consists of five sequential steps: (1) entry mechanisms, (2) catabolic reactions, (3) biosynthesis, (4) polymerization, and (5) assembly. First we consider the flow of materials, from substrate to new cell. Then we consider how the driving forces (energy and reducing power) are accumulated and expended during the five sequential steps. **Figure 5.2** presents an overview of the five-step metabolic assembly line.

Flow of Materials

Entry Mechanisms. Raw materials must be brought to the factory. This is the function of **entry mechanisms**—they bring substrates into the cell. Entry mechanisms must overcome barriers presented by the plasma membrane and, in Gram-negative bacteria, the outer membrane. The entry mechanisms transport substrate across membranes and maintain concentrations within the cell at sufficient levels to fuel metabolism.

Catabolic Reactions. A manufactured item is not made directly from raw materials, and a cell is not made directly from substrate molecules. **Catabolic reactions** refine substrates into the building materials needed to manufacture a new cell. The reactions are called *catabolic* from a Greek word meaning "to bring down," because the basic building materials of metabolism are, in general,

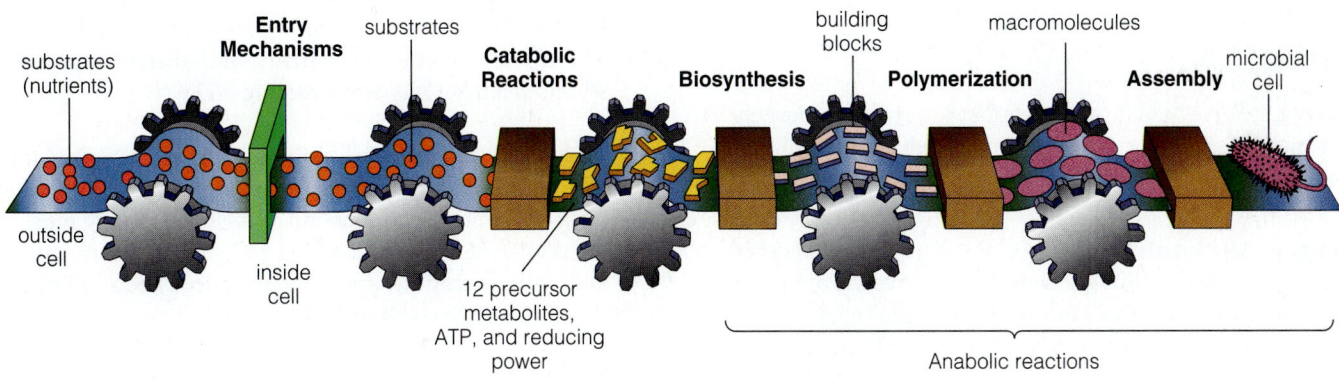

Figure 5.2 The metabolic factory. The metabolic assembly line consists of five sequential steps: entry mechanisms to bring in nutrients from outside the cell; catabolic reactions that yield 12 precursor metabolites, energy, and reducing power; biosynthesis, which produces the building blocks for macromolecules; polymerization, which produces the macromolecules; and assembly, which builds the structures that form the cell.

structurally simpler than substrate molecules. *Escherichia coli* can use many different substrates, but only 12 compounds, called **precursor metabolites**, are required to synthesize an entire microbial cell.

Biosynthesis. The basic materials for building a new factory enter assembly lines to produce the structural components, like bricks, that are needed to build a new factory. Similarly, the 12 precursor metabolites enter biochemical assembly lines called **biosynthesis pathways** to produce the building blocks of macromolecules (**Table 5.2**). For example, biosynthesis produces amino acids, the basic building materials of proteins, and deoxynucleotides, the basic material of DNA (Chapter 2). Biosynthetic reactions, along with subsequent steps in metabolism, are also called **anabolic reactions**, from the Greek word meaning "to build up."

Polymerization. Bricks can be used to build a functional subunit of a factory, such as a wall. Similarly, building blocks are *polymerized*, or joined to one another, to produce macromolecules, the functional subunits of a cell. **Polymerization** unites a series of amino acids into a protein, for example, or a series of deoxynucleotides into DNA.

Assembly. Assembling the finished subunits is the final step in a manufacturing process, and assembling macromolecules into biological structures is the final step in producing a cell. Macromolecules are complex, but they are merely chemicals, not biological structures. A mixture of all the chemical compounds contained within a cell would be just that—a mixture of chemicals—not a cell. Thus, the final step in metabolism is assembling macromolecules into structures—cell walls, flagella, or ribosomes, among others.

Driving Force

Driving force—ATP and reducing power—is largely formed in catabolic reactions and used to drive all the other processes (**Figure 5.3**).

Entry Mechanisms. Materials flow spontaneously down a **concentration gradient**—from a region of higher to lesser concentration—but moving against a concentration gradient requires energy (Chapter 4). To concentrate substrates inside the cell, ATP or another form of stored energy is needed. The amount of ATP used by entry mechanisms is a small fraction of the total needed to manufacture a new cell. Reducing power is not consumed by entry mechanisms.

Table 5.2	The Building Blocks Required to Synthesize a Cell's Macromolecules
Macromolecules	**Building Blocks**
Proteins	20 amino acids
Nucleic Acids	Nucleotides consisting of:
RNA	Adenine, guanine, cytosine, uracil, phosphate, ribose
DNA	Adenine, guanine, cytosine, thymine, phosphate, deoxyribose
Polysaccharides	Sugars
Peptidoglycan	*N*-acetylmuramic acid, *N*-acetylglucosamine, 5 amino acids[a]
Lipids	Because lipids are defined only by physical properties, they comprise many different building blocks.

[a]The number and kinds of amino acids vary by genus.

Catabolic Reactions. In general, the catabolic reactions transform substrate molecules into precursor metabolites, reducing power, and ATP. For *Escherichia coli* and similar microorganisms, the 12 precursor metabolites, ATP, and reducing power [NAD(P)] are all derived from a substrate by the same set of catabolic reactions. A substrate is converted into precursor metabolites, and in the process of forming them, NAD(P) is reduced and ATP is formed.

Biosynthesis. Many of the biosynthetic reactions that transform precursor metabolites into biochemical building blocks involve reductions. As a result, biosynthesis cannot proceed without an ample supply of reducing power, stored in the reduced form of NAD(P). In fact, most of the reducing power stored during catabolic reactions is used to fuel biosynthesis. Most biosynthetic reactions release chemical energy and therefore proceed spontaneously. Only a few require ATP.

Polymerization. The polymerization reactions that join building blocks to form macromolecules require the direct or indirect participation of ATP, consuming most of the ATP generated by catabolic reactions. Because only a few polymerizations are reductions, they require little of the reducing power stored in NAD(P).

Assembly. Some assembly reactions proceed spontaneously and do not require the expenditure of stored energy. Others require ATP, but the total amount of ATP used for assembly is far less than that used for polymerization.

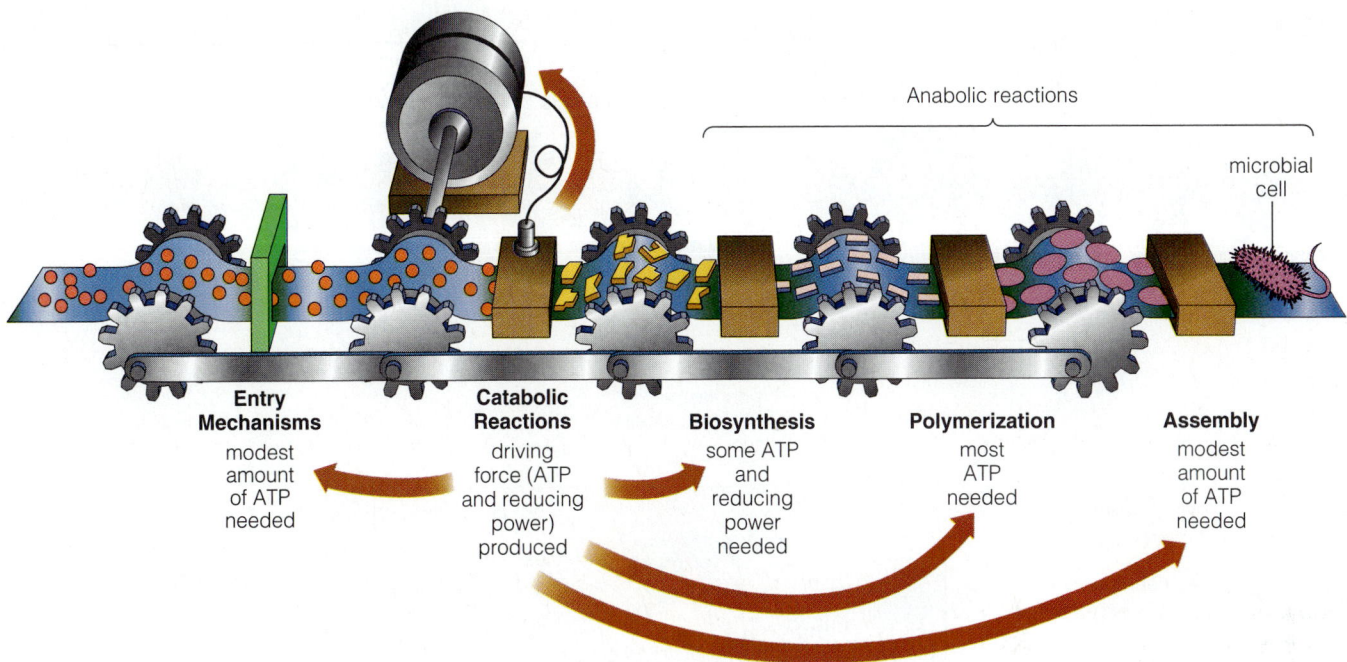

Figure 5.3 Interaction between materials and driving force. The driving force of metabolism (ATP and reducing power) is produced by catabolic reactions. In turn, it drives the other steps. Modest amounts of ATP are needed to drive entry mechanisms, biosynthesis, and assembly. Most ATP is used to drive polymerization. Reducing power is used primarily for biosynthesis.

AEROBIC METABOLISM

Keeping this overview of the materials and driving force involved in microbial metabolism in mind, we are ready to consider each step in more detail. First, let's look at aerobic metabolism, using *Escherichia coli* as an example.

Entry Mechanisms

Entry mechanisms take the raw materials of microbial metabolism through the cell envelope, which bars the passage of most molecules. Recall from Chapter 4 the structure of the procaryotic cell envelope and the ways molecules pass across cell membranes. Refer to **Figure 5.4**.

Escherichia coli is Gram-negative, so its outer membrane presents the initial barrier to the uptake of substrates. Because the outer membrane prevents the passage of both hydrophilic and hydrophobic molecules, *all* substrates enter through the tiny water-filled holes in the outer membrane called porins.

Molecules small enough to fit through the openings in porins pass through them by simple diffusion. The size and properties of porins place two limitations on the molecules that can cross the outer membrane:

1. Only molecules small enough to fit through a porin can cross the outer membrane. The maximum size is approximately that of a trisaccharide (three simple sugars joined together).

2. A substrate can cross the outer membrane only if its concentration in the environment exceeds its concentration in the periplasm. Porins cannot concentrate nutrients within the periplasm.

From the periplasm of Gram-negative bacteria or the environment of Gram-positive bacteria, substrate molecules must cross the cell wall and the plasma membrane. The peptidoglycan cell wall is a loose molecular mesh and does not present a significant barrier to the passage of substrates. The cytoplasmic membrane, on the other hand, prevents the passage of hydrophilic compounds, including hydrophilic substrates. Carrier proteins called permeases bind to a substrate and carry it across the membrane. In this way, they concentrate substrates inside the cell.

Recall from Chapter 2 that enzymes are protein catalysts with specially shaped sites that position specific molecules properly to participate in chemical reactions. Permeases act like enzymes that bring substrates into the cell. Like enzymes, permeases are specific. With a few

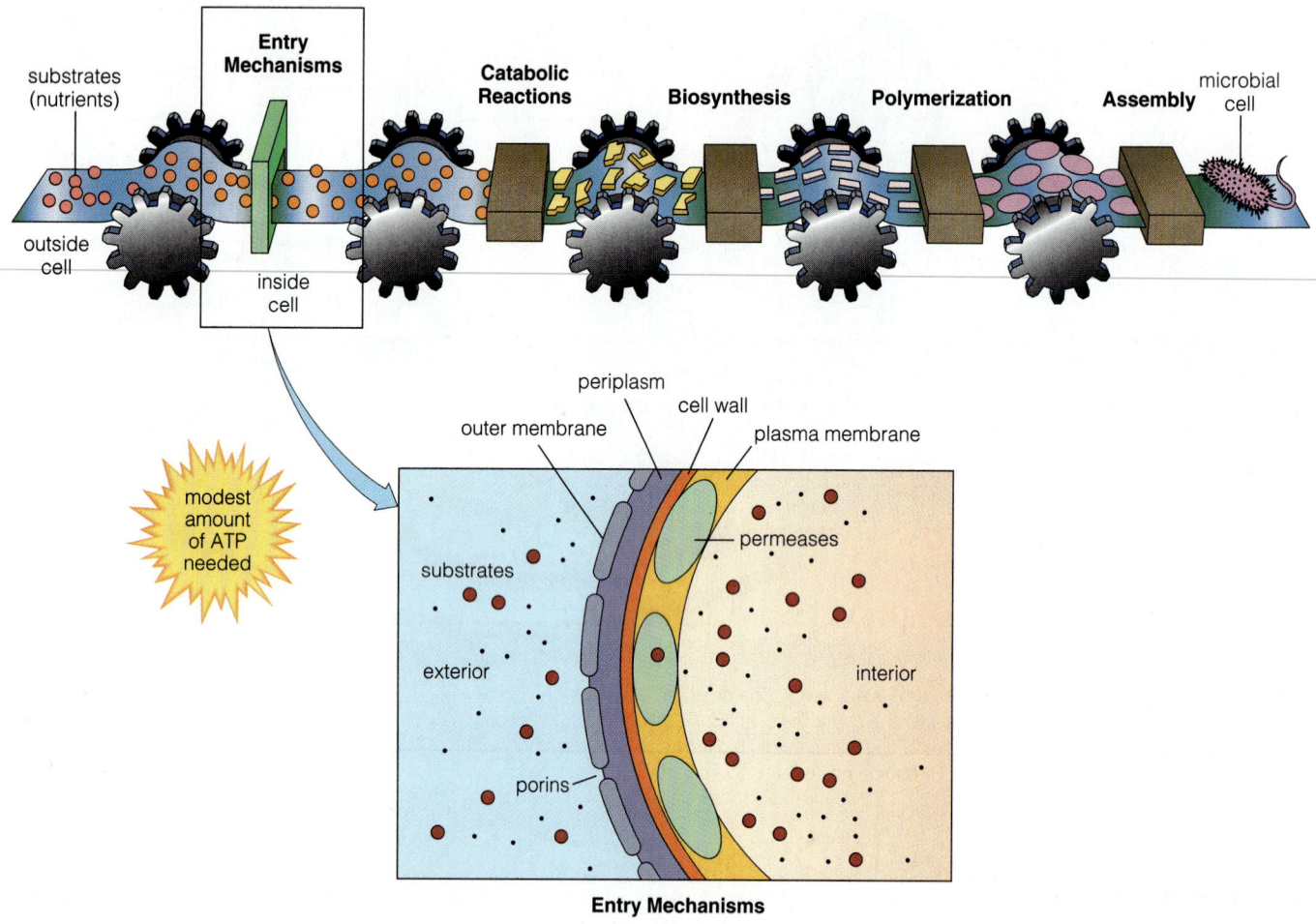

Figure 5.4 Entry mechanisms. Entry mechanisms enable substrates—nutrients—to pass through the cell envelope. Substrates first pass through porins in the outer membrane (in Gram-negative cells only) and the loose molecular mesh of the cell wall. Then permeases in the plasma membrane bring the substrate molecules into the cell.

exceptions, one permease can bring only one compound into a cell. Therefore, *E. coli* has many different permeases, each of which brings one or a few substrates into the cell.

A few nutrients cross the plasma membrane by facilitated diffusion. Most, however, enter by active transport through the action of permeases that can concentrate nutrients within the cell. Because concentration of nutrients requires metabolic energy, permeases use ATP or, as you will see later in this chapter, a **proton gradient** (or gradient of some other ion) to pump nutrients inside the cell. Active transport can achieve nutrient concentrations in the cell a thousand times higher than their concentration in the external medium. Another method that concentrates nutrients is group translocation, also an energy-requiring pro-

cess (Chapter 4). Group translocation chemically changes the substrate as it is concentrated. For example, *E. coli* uses group translocation to achieve the intracellular concentrations of the substrate glucose that it requires, concentrating it in the form of glucose-6-phosphate.

Catabolic Reactions

Catabolic reactions supply *Escherichia coli* with what it needs for growth: materials in the form of the 12 precursor metabolites and driving force in the form of stored reducing power and ATP. The many different catabolic reactions that *E. coli* performs are organized into **catabolic pathways**—sequences of chemical reactions, each catalyzed by a different enzyme, that convert substrate mol-

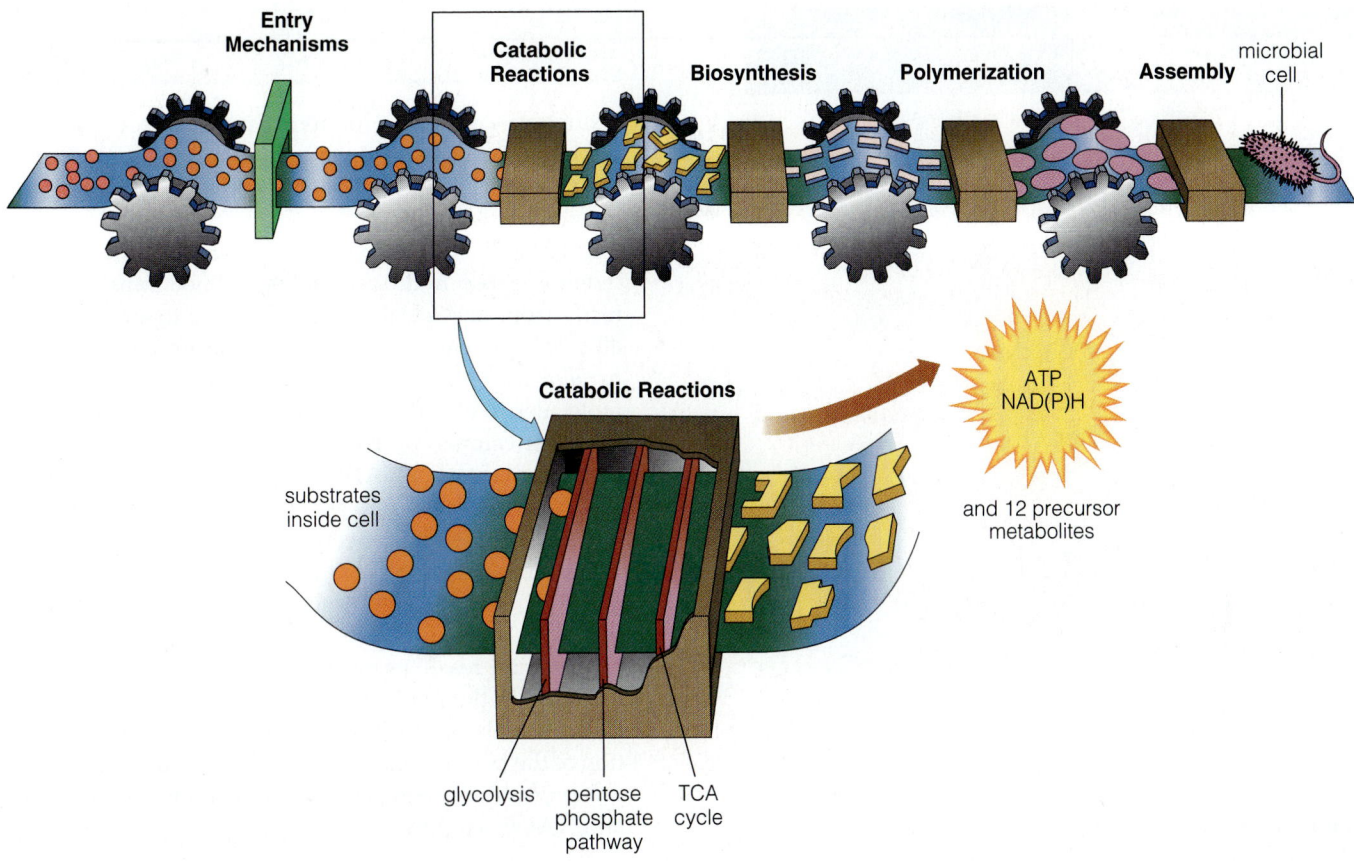

Figure 5.5 Catabolic reactions. The catabolic reactions, collectively called central metabolism, are arranged into several pathways. The most important are glycolysis, the TCA cycle, and the pentose phosphate pathway. There are three products of catabolic reactions: the 12 precursor metabolites, ATP, and reducing power.

ecules into end products. Catabolic pathways achieve this conversion through compounds called **metabolic intermediates**, some of which are precursor metabolites. In this section we consider how catabolic reactions generate precursor metabolites, NAD(P), and ATP (**Figure 5.5**).

Precursor Metabolites. *E. coli*'s various catabolic reactions produce all 12 of the precursor metabolites it and all cells need to grow. These basic building materials of metabolism are manufactured by enzyme-catalyzed reactions that chemically modify the substrate molecule. For example, the phosphorylated form of glucose, **glucose-6-phosphate**, produced by the group translocation of glucose into the cell, is a precursor metabolite. Then a reaction catalyzed by the enzyme **phosphoglucose isomerase** converts glucose-6-phosphate into a second precursor metabolite, **fructose-6-phosphate**. About 25 chemical reactions in *E. coli* are required to make the 12 precursor metabolites from glucose.

No single catabolic pathway produces all 12 precursor metabolites. A minimum of three different pathways is required: (1) **glycolysis**, which produces six precursor metabolites, (2) the **tricarboxylic acid (TCA) cycle**, which produces four more, and (3) the **pentose phosphate pathway**, which produces the final two. We look at these three essential pathways and the reactions they involve to make the precursor metabolites later in the chapter. **Table 5.3** presents the 12 precursor metabolites from which all cell structures—procaryotic and eucaryotic—are made.

Reducing Power. Many steps in a catabolic pathway involve oxidation (loss of electrons) from a metabolic intermediate, while only a few involve reduction (gain of electrons). Because only minuscule amounts of electrons can exist free in solution, oxidations and reductions are always linked. That is, as one compound loses electrons through oxidation, another compound gains them through reduction. As a result, the overall reaction, in which one

Table 5.3	The 12 Precursor Metabolites from Which All Cell Structures Are Made
Precursor Metabolite	Catabolic Pathway That Leads to Its Synthesis
Glucose-6-phosphate	Glycolysis
Fructose-6-phosphate	Glycolysis
Triose phosphate	Glycolysis
3-phosphoglycerate	Glycolysis
Phosphoenolpyruvate	Glycolysis
Pyruvate	Glycolysis
Acetyl CoA	TCA cycle
Alpha-ketoglutarate	TCA cycle
Succinyl CoA	TCA cycle
Oxaloacetate	TCA cycle
Ribose-5-phosphate	Pentose phosphate
Erythrose-4-phosphate	Pentose phosphate

compound is oxidized and the other is reduced, is called an **oxidation-reduction reaction**, often abbreviated as **redox reaction**.

The term *oxidation* implies the addition of oxygen. In inorganic chemical reactions, such as the oxidation of metallic iron to iron oxide (rust), oxygen gas often participates. This reaction illustrates the basic principle of oxidation-reduction reactions: Iron loses electrons and is oxidized, and oxygen gas gains electrons and is reduced.

As iron rusts:

$$4\,Fe + 3\,O_2 \longleftrightarrow 2\,Fe_2O_3$$
METALLIC IRON   OXYGEN GAS   IRON OXIDE, RUST

iron is oxidized, producing electrons (e^-):

$$4\,Fe \longleftrightarrow 4\,Fe^{3+} + 12\,e^-$$

that are used as oxygen is reduced:

$$3\,O_2 + 12\,e^- \longleftrightarrow 6\,O^{2-}$$

Metabolic oxidations, however, including those that occur in the catabolic pathways of *E. coli*, are different from inorganic oxidations. In metabolic oxidations, oxygen gas seldom participates in the reaction, and electrons are seldom transferred by themselves. Instead, a proton is usually transferred along with the electron. This combination of a proton and an electron is a hydrogen atom:

$$H^+ + e^- \longleftrightarrow H$$
PROTON   ELECTRON   HYDROGEN ATOM

Oxidations in which protons are removed together with electrons are called **dehydrogenation reactions**. Reductions in which they are added together are called **hydrogenation reactions**.

Metabolic oxidation-reduction reactions are also different because hydrogen atoms are not transferred directly from one metabolic intermediate to another. Usually, *hydrogen atoms are transferred to or from one of the two pyridine nucleotides*, NAD or NADP. Either of these two molecules in its reduced form stores the cell's reserves of hydrogen atoms, or reducing power (**Figure 5.6**). Reduced molecules of NAD(P) can be used to reduce metabolic intermediates or to generate new molecules of ATP (see below), thereby converting stores of reducing power into stores of chemical energy.

As *E. coli* processes substrate through its catabolic pathways to produce precursor metabolites, many dehydrogenation reactions occur. In these reactions, electrons removed in the form of hydrogen atoms are transferred to a molecule of NAD(P). For example, in the first reaction of the pentose phosphate catabolic pathway, glucose-6-phosphate is oxidized to **6-phosphogluconolactone** and NADP is reduced.

$$glucose\text{-}6\text{-}phosphate + NADP^+ \longleftrightarrow 6\text{-}phosphogluconolactone + NADPH + H^+$$

The oxidized form of NAD or NADP [NAD(P)] is designated as having a single positive charge, NAD^+ or $NADP^+$ [$NAD(P)^+$], and the reduced form as NADH or NADPH [NAD(P)H]. Oxidation or reduction of either NAD or NADP involves the exchange of two hydrogen atoms.

$$NAD(P)^+ + 2\,H \longleftrightarrow NAD(P)H + H^+$$

In biosynthesis, many hydrogenation reactions occur, reoxidizing NAD(P). For example, one step in the biosynthesis of the amino acids tyrosine and phenylalanine involves the reduction of dehydroshikimate to shikimate at the expense of reduced NADP.

$$dehydroshikimate + NADPH + H^+ \longleftrightarrow shikimate + NADP^+$$

ATP: Stored Energy. The principal compound that stores chemical energy in *E. coli* and all other cells is adenosine triphosphate, or ATP. The energy-storage capabilities of ATP depend on the two bonds that join the three phosphate groups in the molecule. To understand the role

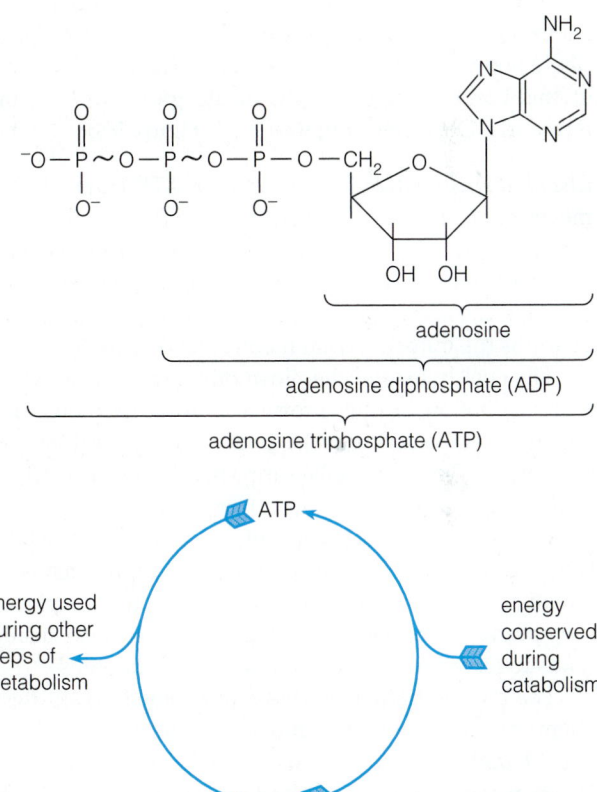

reduction (hydrogenation)

2H

NAD(P)$^+$
[oxidized NAD(P)]

2H

oxidation (dehydrogenation)

$+$ H$^+$

NAD(P)H
[reduced NAD(P)]

Figure 5.6 Reducing power. When the oxidized form of the pyridine portion of pyridine nucleotides (R represents the nucleotide portion) accepts two hydrogen atoms from a catabolic reaction, it is converted to a reduced form. The reduced form, in turn, can donate hydrogen atoms to biosynthesis reactions and be converted back to the oxidized form.

these bonds play in energy storage, we briefly review the ways in which energy is involved in breaking and forming chemical bonds.

Chemical bonds break and new ones form if the new bonds are at a lower energy state and, therefore, more stable than the old bonds (Chapter 2). The energy difference between old bonds and new bonds, which determines if a reaction will occur, is called **free energy**. The key here is *energy difference*—the tendency of a bond to break and form other bonds depends on how much energy it takes to break the original bond(s) and how much is released as the new bond(s) are formed. Certain bonds, however, are relatively unstable. The small amount of energy needed to break them is exceeded by the energy released during the formation of almost any new bonds. These bonds are said to be *highly reactive* because they participate in many different chemical reactions.

The bonds that join the three phosphate groups to ATP are among the most highly reactive bonds found in biochemicals. Because of their extraordinary reactivity, these are sometimes called **high-energy bonds** and designated by the special bond symbol ~ (**Figure 5.7**). The phosphate groups that are joined in ATP by high-energy bonds are readily donated to other compounds. These compounds that receive a phosphate group from ATP are termed **phosphorylated compounds**. They participate in chemical reactions that would not occur if the reactants were unphosphorylated. This is why we say that the energy "stored" in the bonds of ATP is used to drive other metabolic reactions.

ATP Formation. But how are high-energy phosphate bonds formed in the first place? How is energy derived from catabolic pathways stored chemically? ATP is always formed by adding a single phosphate group to **adenosine diphosphate (ADP)**. Some ADP is made by a biosynthetic pathway, while the rest is formed as a second product of phosphorylation reactions. For example, in one step of the biosynthesis of the amino acid proline,

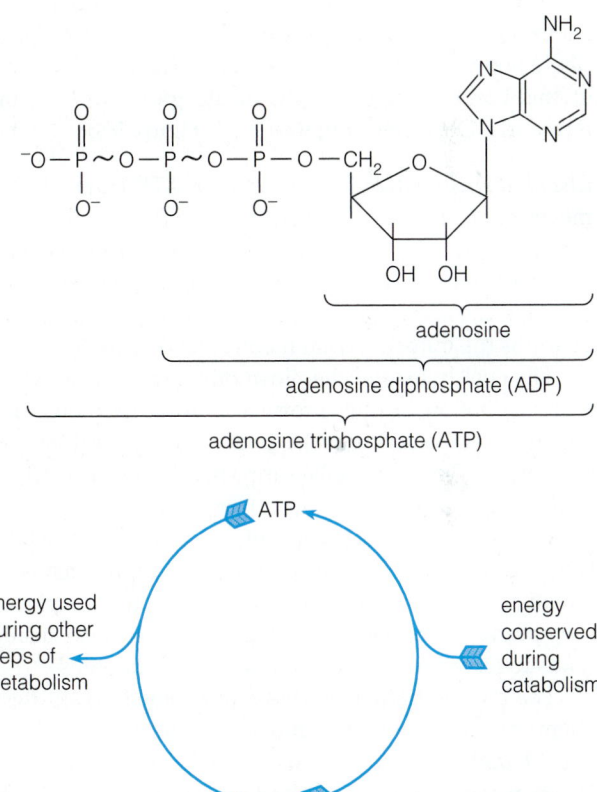

adenosine

adenosine diphosphate (ADP)

adenosine triphosphate (ATP)

ATP

energy used during other steps of metabolism

energy conserved during catabolism

ADP

Figure 5.7 Stored energy: ATP. In ATP, the two bonds (~) that join the three phosphate groups are high-energy bonds. ATP is formed by adding a single phosphate group to ADP during catabolism. The energy stored in ATP is used in other steps of metabolism as ATP is converted back to ADP.

glutamate is phosphorylated, yielding glutamylphosphate and ADP.

| glutamate + ATP $\longrightarrow$ glutamylphosphate + ADP |

Thus, ADP is formed when ATP is used to drive a reaction.

ADP can be converted to ATP in two different ways—either through **substrate level phosphorylation** or through **chemiosmosis**.

Substrate Level Phosphorylation. Substrate level phosphorylation involves a phosphate group that is already part of a metabolic intermediate. During certain catabolic reactions that would otherwise release large amounts of chemical energy in the form of heat, a preexisting phosphate bond becomes highly reactive. The high reactivity of this bond enables the phosphate group to be transferred to ADP, converting it to ATP (**Figure 5.8**).

Chemiosmosis. Chemiosmosis forms ATP from ADP by means of an enzyme called **ATPase**. ATPase catalyzes the conversion of ADP to ATP as a result of a series of chemical events that occur in and around a membrane. In procaryotes like *E. coli*, it is the plasma membrane. In eucaryotes, it is the mitochondrial membrane (**Figure 5.9**).

The energy for the chemiosmotic formation of ATP is a concentration gradient formed across the membrane. During metabolism, certain ions—usually protons, but sodium ions in some cells—are actively transported out of the cell, and so their concentration outside the plasma membrane exceeds their concentration inside the cell. Because any concentration gradient naturally tends to equalize itself, the proton gradient across the plasma membrane of procaryotes (high outside, low inside) tends to force protons back into the cell (Chapter 4).

The proton gradient across a membrane constitutes a chemical energy potential, much as water stored in an elevated tank constitutes a mechanical energy potential. When water flows out of the tank by gravity, it can be passed through a waterwheel and made to do work, such as generating electricity. When protons flow down a concentration gradient and across the membrane, they pass through a channel in the membrane that contains an ATPase enzyme (the only available channel through the otherwise proton-impermeable membrane) and do the work of generating ATP from ADP.

But creating a proton gradient, like storing water in an elevated tank, requires energy. How does *E. coli* accomplish this task chemically? The process begins during one of the energy-releasing dehydrogenation reactions of catabolism, when two hydrogen atoms (each consisting of a proton and an electron) are transferred to NAD(P)$^+$, forming NAD(P)H. NAD(P)H, in turn, transfers this pair of atoms to one of a series of compounds embedded in

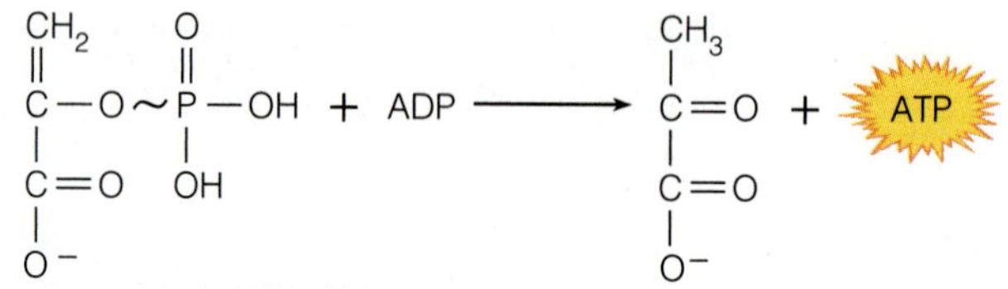

Figure 5.8 Substrate level phosphorylation. Through substrate level phosphorylation, a phosphate group in a metabolic intermediate (here, phosphoenolpyruvate) is transferred to ADP, producing pyruvate and ATP. The phosphate group in both phosphoenolpyruvate and ATP is bound to the rest of the molecule by a high-energy bond ($\sim$).

the cell membrane. These compounds, collectively called an **electron transport chain**, perform a cascade of oxidation-reduction reactions, transferring the reducing power of the original hydrogen atom from one molecule to another. Electron transport chains differ from one organism to another, but all contain certain compounds, such as quinones and flavoproteins, that accept only hydrogen atoms and other compounds, such as cytochromes, that accept only electrons (Figure 5.9).

The critical step in generating a proton gradient occurs when a hydrogen-accepting compound in the electron transport chain transfers its reducing power to an electron-accepting compound. At this point in the chain the protons and electrons of the two hydrogen atoms are separated, and the protons are released on the outer side of the membrane, contributing to the proton gradient. When this separation first occurs in the electron transport chain, the two free electrons are transported to the inner side of the membrane where they combine again with two protons, producing two new hydrogen atoms. These two new atoms reenter the chain, and the protons and electrons are separated once again. After this second separation, in which two additional protons are released outside the membrane, the electrons are transferred to a cytochrome, which reacts with oxygen at the end of the chain, reducing it to water. For this reason oxygen is called the **terminal electron acceptor**.

The net effects of passing two hydrogen atoms through the *E. coli* electron transport chain, therefore, are the movement of four protons outside the plasma membrane and the reduction of half a molecule of oxygen to one molecule of water. The process of transporting electrons through a chain ending in oxygen as a terminal electron acceptor is called **aerobic respiration**.

In summary, the *E. coli* electron transport chain pumps protons out of the cell, which creates a proton concentration gradient across the cytoplasmic membrane. The chemical potential of this gradient causes protons to flow back into the cell through a trans-membrane-bound ATPase, converting ADP to ATP. For each pair of hydrogen atoms

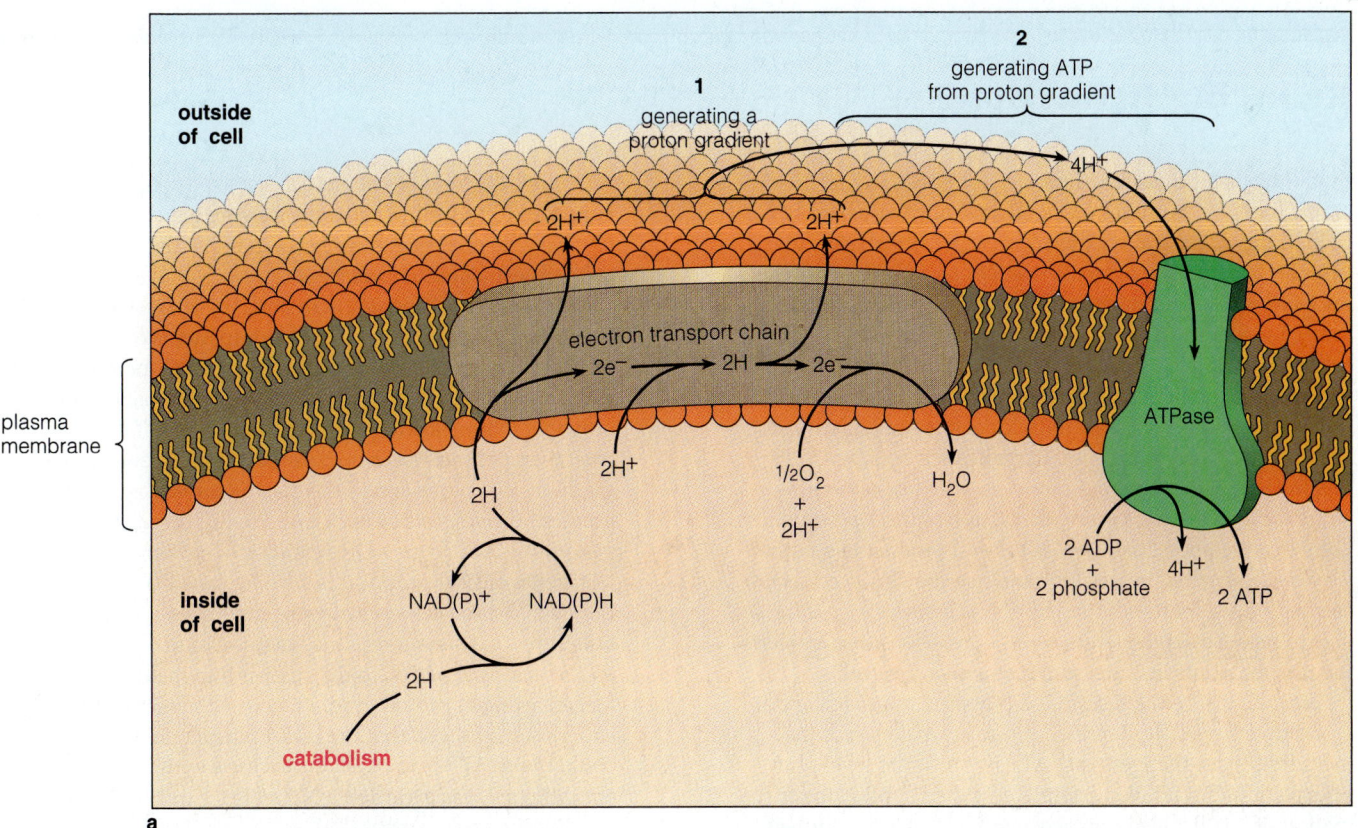

Figure 5.9 Chemiosmosis. Chemiosmosis in bacteria forms ATP in two steps: (1) generating a proton gradient and (2) using it to generate ATP. (**a**) Generating a proton gradient across the plasma membrane occurs when two hydrogen atoms from catabolism are passed through NAD(P) and an electron transport chain. (**b**) As hydrogen atoms are passed from hydrogen-carriers such as quinones to electron-carriers such as cytochromes on the outer edge of the membrane, protons leave the cell. The proton gradient, thus formed, generates ATP from ADP as it forces protons through a membrane-bound ATPase back into the cell (**a**). The energy of the gradient drives the reaction.

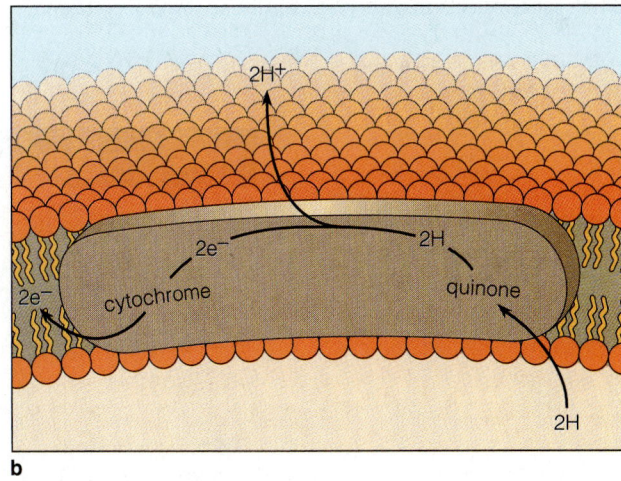

that enters the electron transport chain of *E. coli*, four protons are pumped across the cytoplasmic membrane. For each pair of protons that reenters the cell through the ATPase channel, approximately one molecule of ADP is converted to ATP. Therefore, each pair of electrons that passes through the chain generates about two molecules of ATP. Although chemiosmotic synthesis of ATP probably occurs in all cellular organisms, its efficiency varies greatly. Eucaryotic organisms generate about three molecules of ATP molecule per pair of electrons, while some autotrophic bacteria generate only one. Even *E. coli* makes only one ATP molecule per hydrogen pair if the concentration of oxygen in its environment is low.

Two aspects of ATP generation through chemiosmosis by *E. coli* growing aerobically deserve extra emphasis. First, the process requires oxygen as a terminal electron acceptor. This requirement distinguishes aerobic metabolism from anaerobic metabolism, which is discussed later in this chapter. Second, the process converts reducing power in the form of NADH into ATP.

Some organisms use ATP to generate reducing power by **reverse electron flow**. That is, they force electrons/hydrogen atoms back through an electron transport chain and reduce $NAD(P)^+$ to NAD(P)H. Thus, the two driving forces of metabolism, reducing power [NAD(P)H] and stored energy (ATP) are *interconvertible*. In a similar way, ATP and a proton gradient are interconvertible. Protons

Reducing Reductionism

By the 1960s biochemistry had developed a good set of scientific tools for studying metabolism. They were based on reductionism, the philosophical approach which says that even highly complex systems (such as metabolism) can be understood by studying each component part and then collating the accumulated bits of information to produce a picture of the whole system. In the case of biochemical reductionism, the methods were well established: break the cell, purify and characterize the individual enzymes, analyze how the reactions fit together to make metabolic sense. If the product of one enzyme-catalyzed reaction is the substrate for another, the two reactions must function sequentially in a metabolic pathway. If the end product of a pathway is the starting material for another, they interconnect. It seemed only a matter of time until biochemists would be able to reconstruct the entire series of reactions that convert the raw materials in a medium into a microbial cell.

Biochemical reductionism is a powerful approach and it worked very well. It identified most metabolic reactions, including those that generate ATP by substrate level phosphorylation. But it failed completely when applied to formation of ATP by electron transport chains. When the cell was broken, electron transport continued, but formation of ATP stopped completely. Researchers speculated that perhaps the ATP-forming reactions were extremely labile (unstable). So they sought gentler ways to break the cell. But even the gentlest yielded no success.

In 1961, Peter Mitchell, a Scottish biochemist, took a different line of reasoning. Because components of the electron transport chain are embedded in the plasma membrane, he speculated that ATP formation by electron transport occurred only when the plasma membrane was intact. If this were true, then perhaps electron transport built a concentration gradient of protons across the plasma membrane, and the gradient drove protons back into the cell through a channel generating ATP from ADP. Scientists were slow to accept Mitchell's radical proposal (which he called chemiosmosis). But even the most skeptical were convinced when an intact but empty plasma membrane was shown to convert ADP to ATP if acid was added to the suspending medium to establish an artificial proton gradient across the membrane. The experiment showed that the intact plasma membrane alone could make ATP from ADP if provided with a proton gradient. Mitchell was awarded a Nobel Prize in 1978 for his remarkable achievement.

Table 5.4 Principal Catabolic Pathways That Make Up Central Metabolism

Pathway	Yield of Precursor Metabolites	Net Yield of ATP[a]	Yield of Reducing Power	Carbon-containing End Products
Glycolytic pathway	6	$(4 - 2) = 2$	2 NADH	2 pyruvate
TCA cycle[b]	4	1	1 NADPH 4 NADH[d]	3 CO_2
Pentose phosphate pathway	2[c]	−1	2 NADPH 1 triose phosphate	3 CO_2

[a]The number of ATP molecules produced by substrate level phosphorylation minus those utilized.
[b]Three others (glucose-6-phosphate, glyceraldehyde-3-phosphate, and fructose-6-phosphate) produced by the glycolytic pathway are also produced here.
[c]Yield calculated per molecule of pyruvic acid metabolized.
[d]One of these is produced in another form; i.e., reduced flavine adenine dinucleotide.

flowing through membrane-bound ATPase back into the cell make ATP from ADP, and in the other direction, conversion of ATP to ADP by membrane-bound ATPase forces protons out of the cell, establishing a proton gradient. The capacity to interconvert a proton gradient and ATP is essential because both forms of energy are needed by the cell. ATP is needed to drive many steps in anabolism. Proton gradients are needed for other purposes, primarily to drive entry mechanisms and to turn flagella for motility.

Catabolic Pathways. Catabolic pathways are the enzyme-catalyzed reaction sequences that transform substrate molecules into precursor metabolites, store energy in the form of ATP, and store reducing power in the form of NAD(P)H. *E. coli* has enzymes that catalyze the catabolism of many different substrates, but three pathways—glycolysis, the TCA cycle, and the pentose phosphate pathway—are particularly important. Because of their importance, these three pathways are collectively called **central metabolism (Table 5.4)**.

Central metabolism begins with the sugar glucose. The components of glucose that are not used as materials in the formation of precursor metabolites are eventually excreted from the cell as carbon dioxide and water. The pathways of central metabolism are widely distributed in nature. They are found in organisms as diverse as *E. coli* and human beings, because all cellular organisms need the 12 precursor metabolites, ATP, and reducing power in order to function metabolically. Most organisms, including *E. coli*, have other catabolic pathways, as well, which use substrates other than glucose. These pathways all feed into central metabolism at various points.

Glycolysis. The metabolic intermediates of glycolysis and the reactions that evolve ATP and NADH and produce precursor metabolites are shown in **Figure 5.10**.

Glycolysis begins with the substrate glucose and leads to the formation of two molecules of pyruvate, a key metabolic intermediate that participates in many other reactions. Along the pathway from glucose to pyruvate a total of six precursor metabolites are formed: glucose-6-phosphate, fructose-6-phosphate, triose phosphate, 3-phosphoglycerate, phosphoenolpyruvate, and pyruvate.

The initial steps of glycolysis require the *expenditure* of energy in the form of two molecules of ATP. This conversion of ATP to ADP phosphorylates metabolic intermediates, making them sufficiently reactive to participate in subsequent conversions. Then the phosphorylated compounds are converted to pyruvate. During this series of reactions, four molecules of ATP are formed by substrate level phosphorylation, and two molecules of NAD^+ are reduced to NADH for each molecule of glucose that enters the pathway. The net yield of ATP by substrate level phosphorylation is only two, because of the two molecules of ATP that were spent in the early steps. The total yield of ATP can be about six, however, because the two molecules of NADH can donate two pairs of hydrogen atoms to an electron transport chain, yielding four ATPs by chemiosmosis.

The TCA Cycle. Some of the pyruvate formed by glycolysis is used in biosynthesis, and the rest is oxidized to another precursor metabolite, acetyl CoA. Some acetyl CoA enters a cyclic catabolic pathway called the tricarboxylic acid cycle (TCA cycle). The TCA cycle forms more precursor metabolites, ATP by substrate level phosphorylation, NAD(P)H, and carbon dioxide (**Figure 5.11**).

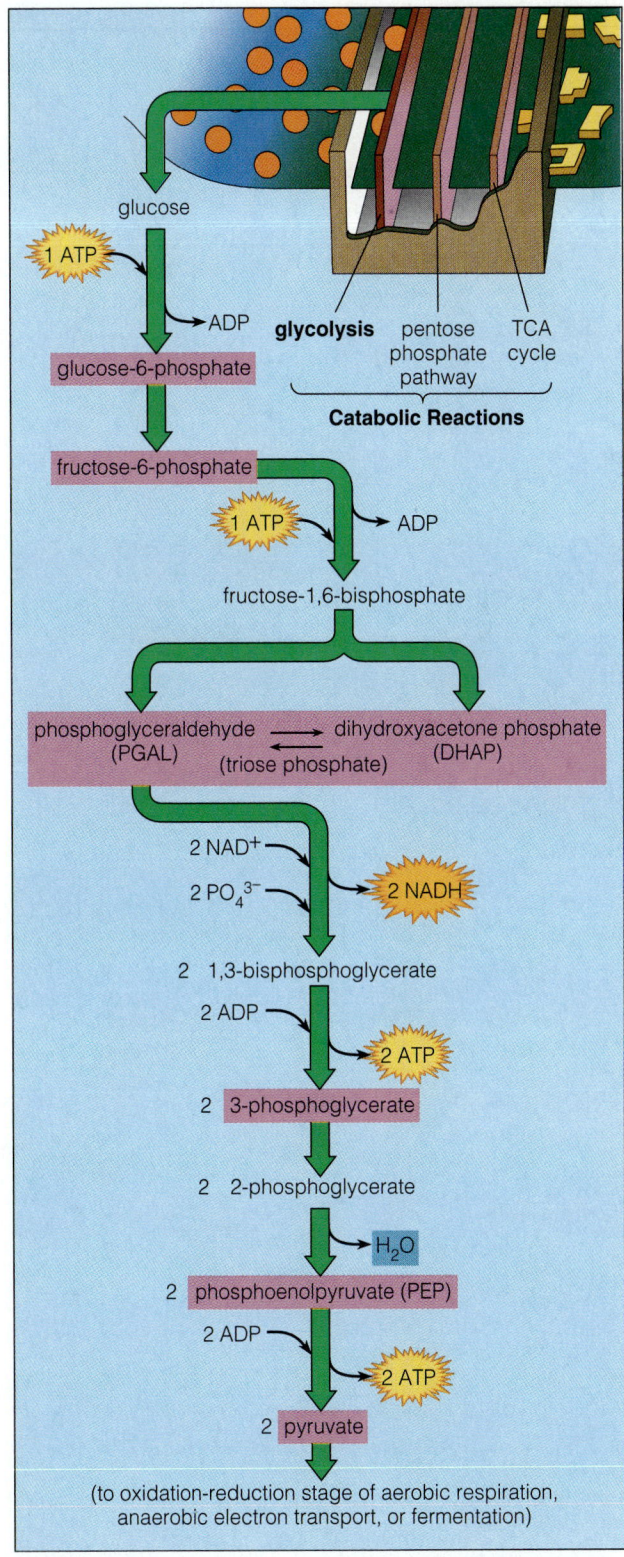

Figure 5.10 Glycolysis. Glycolysis is a pathway of central metabolism that converts a molecule of glucose into two molecules of pyruvate with a net yield of two molecules of ATP and two molecules of NADH. Six precursor metabolites, shown in boxes, are intermediates that are drawn off as needed for biosynthesis.

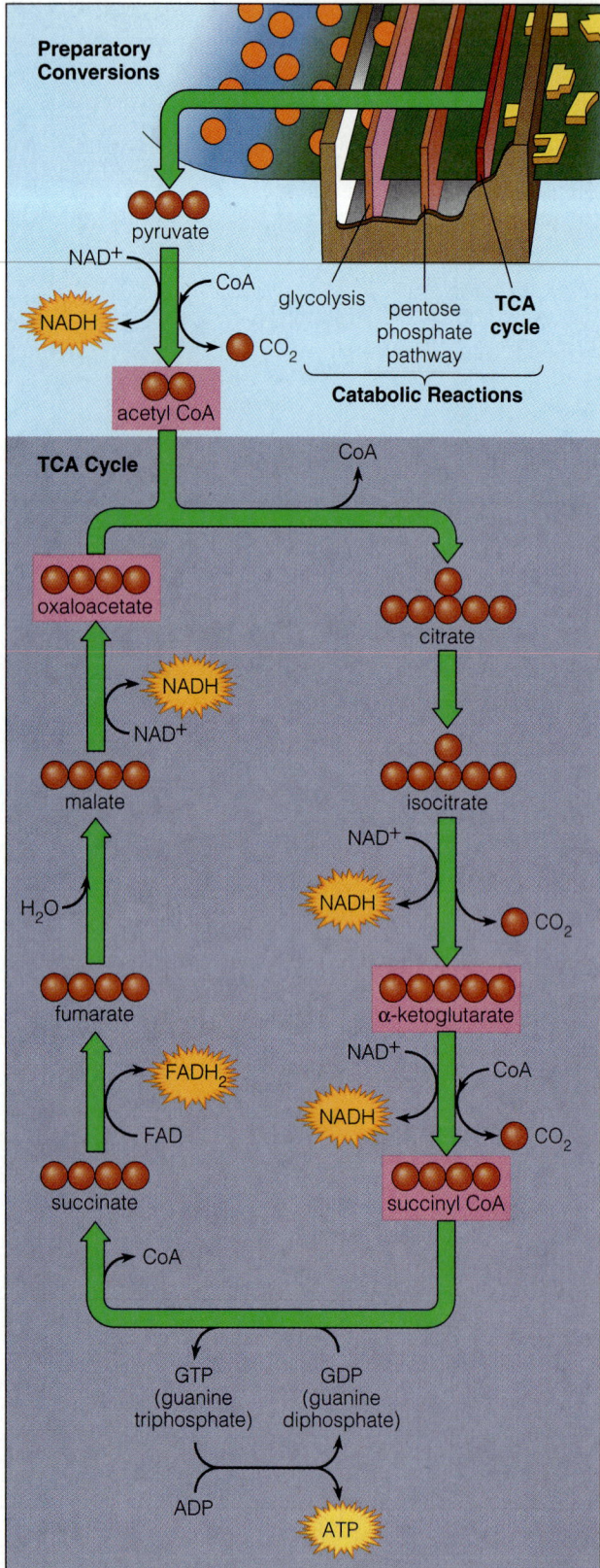

Figure 5.11 The TCA cycle. The tricarboxylic acid (TCA) cycle converts pyruvate into CO_2, reducing power, ATP (by substrate level phosphorylation), and four precursor metabolites, shown in boxes. Some of the reducing power, in the form of FADH (reduced flavine adenine dinucleotide), flows directly into an electron transport chain. The rest can flow into an electron transport chain or be used for biosynthesis.

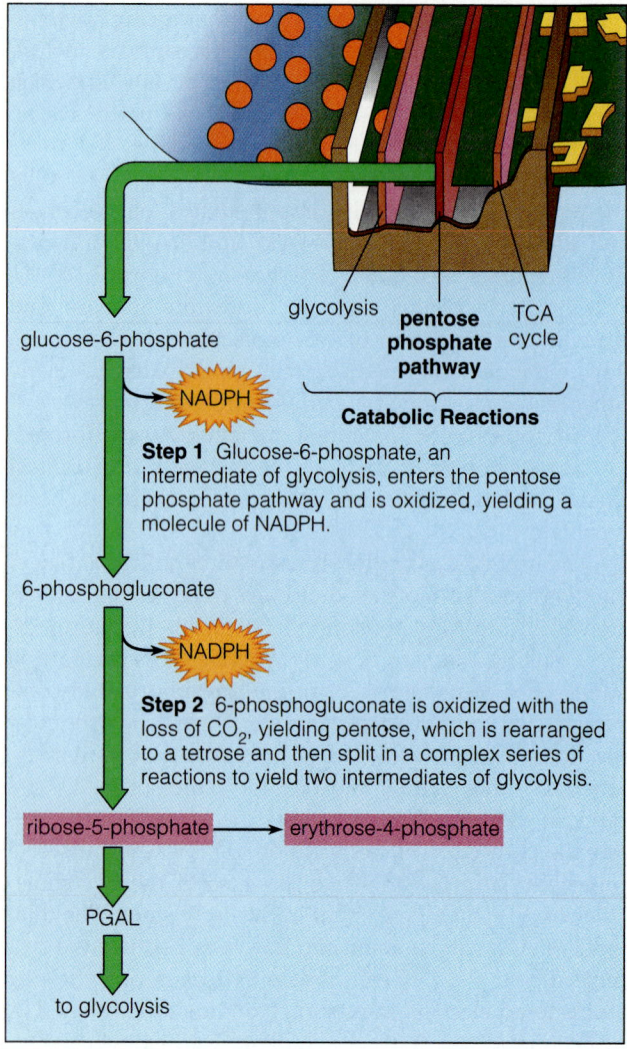

Step 1 Glucose-6-phosphate, an intermediate of glycolysis, enters the pentose phosphate pathway and is oxidized, yielding a molecule of NADPH.

Step 2 6-phosphogluconate is oxidized with the loss of CO_2, yielding pentose, which is rearranged to a tetrose and then split in a complex series of reactions to yield two intermediates of glycolysis.

Figure 5.12 The pentose phosphate pathway. The pentose phosphate pathway is a part of central metabolism that forms two precursor metabolites, shown in boxes. The pathway begins with one intermediate of glycolysis and ends with another.

Acetyl CoA enters the TCA cycle by combining with the four-carbon precursor metabolite oxaloacetate to form a six-carbon intermediate, citrate. In a series of six subsequent reactions, the two added carbon atoms are released as carbon dioxide, and oxaloacetate is regenerated. In the TCA cycle, three additional precursor metabolites—succinyl CoA, alpha-ketoglutarate, and oxaloacetate—are formed.

Each turn of the TCA cycle produces one molecule of ATP by substrate level phosphorylation. In addition, considerable reducing power is stored in the form of two molecules of NADH, one as NADPH, and another as a reduced carrier called $FADH_2$. Reducing power stored as flavine adenine dinucleotide ($FADH_2$), as well as NADH and NADPH, can be converted to ATP by chemiosmosis.

Pentose Phosphate Pathway. The key metabolic intermediates of the pentose phosphate pathway and the reactions that involve NADPH are shown in **Figure 5.12**.

The pentose phosphate pathway begins when an intermediate of glycolysis, glucose-6-phosphate, enters the pathway. It passes through a complex series of reactions to produce three molecules of carbon dioxide and one molecule of phosphoglyceraldehyde—an intermediate that reenters glycolysis. The pentose phosphate pathway is critically important because it forms the last two precursor metabolites, ribose-5-phosphate and erythrose-4-phosphate. Neither of these can be formed by glycolysis or the TCA cycle. They are the precursor metabolites needed to make three amino acids (phenylalanine, tyrosine, and tryptophan) and all nucleotides.

The pentose phosphate pathway also stores reducing power in the form of two molecules of NADPH. Although it produces no ATP by substrate level phosphorylation, the NADPH produced by the pentose phosphate pathway may generate four molecules of ATP through chemiosmosis.

Biosynthesis

Escherichia coli uses the three products of catabolism—precursor metabolites, energy, and reducing power—to construct a new *E. coli* cell. The first step in the process is *biosynthesis*, which converts precursor metabolites into the building blocks of macromolecules (**Figure 5.13**).

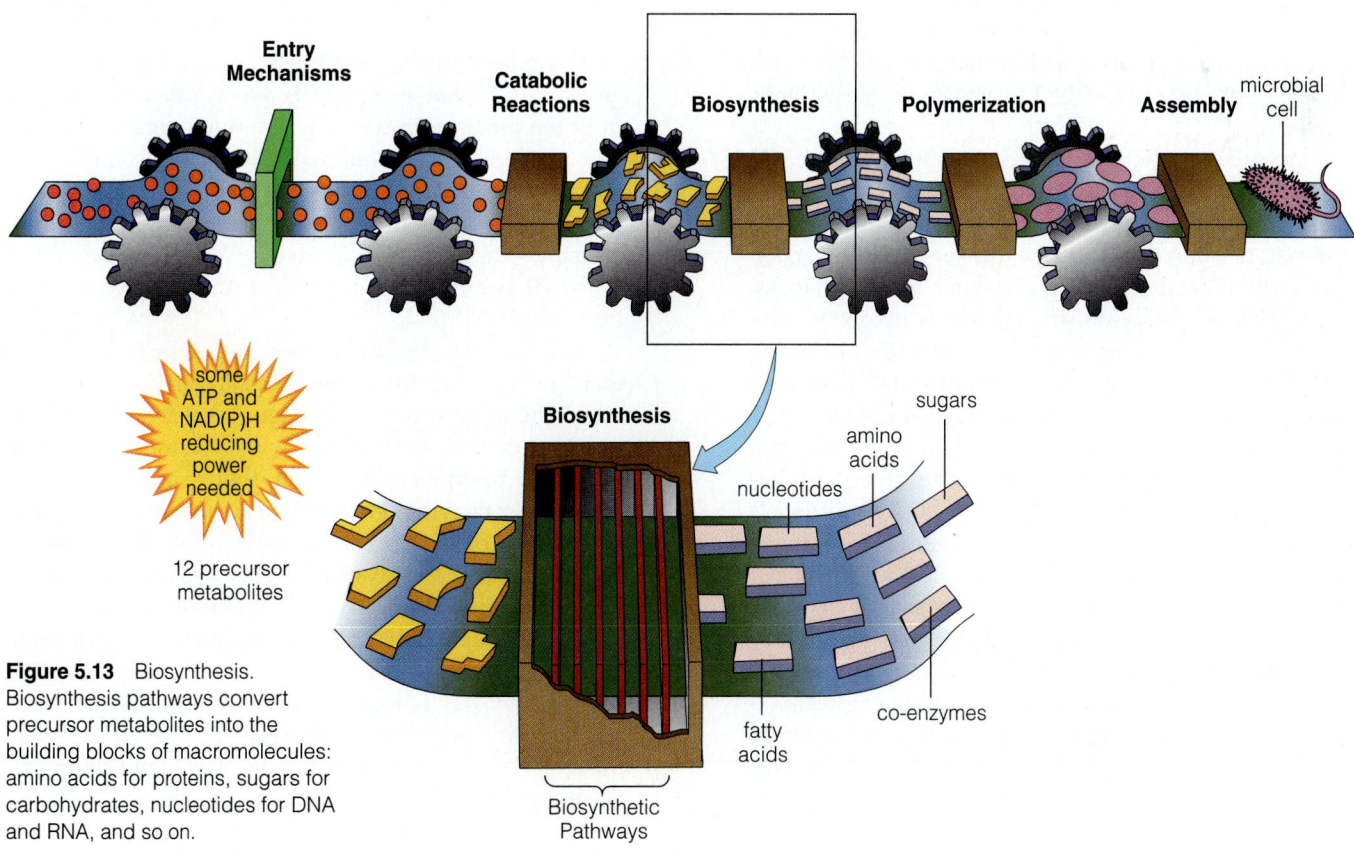

Figure 5.13 Biosynthesis. Biosynthesis pathways convert precursor metabolites into the building blocks of macromolecules: amino acids for proteins, sugars for carbohydrates, nucleotides for DNA and RNA, and so on.

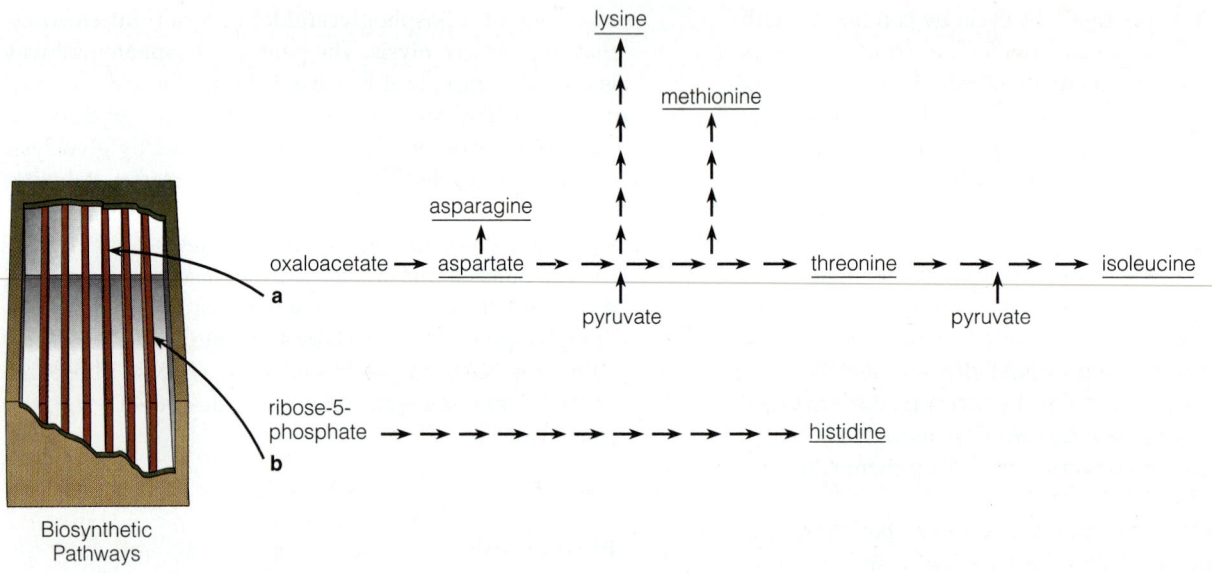

lysine

methionine

asparagine

oxaloacetate → aspartate → → → → → → threonine → → → → isoleucine

a

pyruvate

pyruvate

ribose-5-
phosphate → → → → → → → → → → → histidine

b

Biosynthetic
Pathways

Figure 5.14 Biosynthetic pathways. Some biosynthetic pathways are branched, involving several different precursor metabolites and/or producing several different building blocks. Others are linear and very simple. (**a**) A branched biosynthetic pathway converts two precursor metabolites (pyruvate and oxaloacetate) into six amino acids (underlined) by 24 enzyme-catalyzed reactions (arrows). (**b**) A much simpler biosynthetic pathway converts a single precursor metabolite (ribose-5-phosphate) to a single amino acid (histidine) by 11 reactions.

Biosynthesis is a major part of metabolism. *E. coli* produces about 1000 enzymes that catalyze about 1000 different chemical reactions within the cell. Of this total, about 200, or approximately 20 percent, are biosynthetic reactions.

Biosynthetic reactions, like catabolic reactions, are organized into reaction sequences called pathways. The **biosynthetic pathways** are composed of enzyme-catalyzed reactions that mediate the step-wise conversion of precursor metabolites into metabolic building blocks. Some of these pathways are branched, involving several different precursor metabolites and/or producing several different building blocks (**Figure 5.14**). For example, six amino acids (aspartate, asparagine, isoleucine, lysine, methionine, and threonine) are synthesized from two precursor metabolites (oxaloacetate and pyruvate) through a multiple branched pathway. Other pathways are unbranched, producing a single building block from a single precursor metabolite. For example, the amino acid histidine is synthesized through an unbranched pathway from the single precursor metabolite ribose-5-phosphate.

The biosynthetic pathways of *E. coli* are remarkably similar to biosynthetic pathways in all other organisms—bacteria, eucaryotic microorganisms, plants, and animals. Differences do arise, however, because some organisms cannot make the enzymes needed to complete certain pathways. But almost all organisms that can make a particular building block do so by the same series of reactions.

Organisms that cannot make a given building block grow only if that molecule is provided ready-made—from the medium in the case of a microorganism or the diet in the case of an animal. *E. coli*, for example, can make all 20 amino acids needed to build proteins, so it can grow in a medium that contains no amino acids. Human beings, on the other hand, are unable to make 9 of the 20 essential amino acids, so these nutrients come from our diets.

In addition to building blocks, other small molecules that are essential for a cell to grow are produced by biosynthetic pathways. Some of these are **coenzymes**, compounds that act together with enzymes to catalyze metabolic reactions. Organisms that cannot synthesize their own coenzymes grow only if small amounts of these compounds or their precursors are supplied from an external source. The ready-made molecules that act as coenzymes or their precursors are called **vitamins**.

The principal driving force that fuels biosynthesis is reducing power, stored mostly in the form of NADPH. Reducing power is required because some biosynthetic reactions are reductions. Some ATP is also required, but the major expenditure of ATP in the making of a new cell occurs in the next step—the polymerization of building blocks into macromolecules.

Not Just Proteins Anymore

As biology advances, unifying principles emerge that simplify and pull together the masses of facts. But sometimes new principles force biologists to give up long-held convictions. For example, the belief that all enzymes are proteins and only they catalyze metabolic reactions was shattered in 1987 by T. R. Cech at the University of Colorado. Cech showed that an RNA molecule had catalytic properties. It could cut a long strand of RNA in two places and rejoin two of the resulting three RNA pieces, forming a shorter RNA molecule with a missing midsection. Such RNA processing is an essential step in gene expression of eucaryotes. Soon other RNA molecules were found that had similar catalytic activities. They were named **ribozymes**, short for ribonucleic acid (RNA) enzymes. However, known ribozyme-catalyzed reactions all involved cutting and rejoining other RNA molecules. Maybe ribozymes were able to catalyze only this one highly specialized kind of reaction.

Not at all. In 1992 H. F. Noller and his colleagues at the University of California, Santa Cruz, found that a ribozyme catalyzes what is probably the cell's most important metabolic reaction, the formation of peptide bonds that link amino acids to form proteins. The ribozyme itself is a portion of the RNA component of ribosomes. Two important antibiotics, chloramphenicol and carbomycin, act by inhibiting this ribozyme-catalyzed reaction (Chapter 21).

The discovery of ribozymes has exciting implications about the origin of life. The fact that ribozymes can form the bonds that make RNA means RNA can reproduce itself. And ribozymes form the bonds that make protein. Thus, RNA can reproduce itself and make protein! An RNA molecule with both of these properties would have the fundamental characteristics of life.

Polymerization

In polymerization reactions, molecular building blocks are joined to form macromolecules. The major cellular polymerization reactions are DNA replication, RNA synthesis, protein synthesis, and polysaccharide synthesis (**Figure 5.15**).

In most macromolecules, building blocks must be joined in a specific *order*. The primary structure of a protein, for example, is determined by the order in which its particular amino acids are joined together. An almost infinite number of different proteins can be created from different arrangements of the 20 amino acids, but each cell must produce only certain proteins. For example, *E. coli* must produce precisely the approximately 1000 different proteins it needs to catalyze its metabolic reactions and build its structural components. To do so the cell must arrange amino acids in the proper order to produce the right proteins to build an *E. coli* cell.

The ordering of polymerization reactions is determined directly by the information stored in the organism's DNA. Some polymerization reactions—those that build DNA, RNA, and protein—are directly determined by the physical structure of DNA. The ordering of building blocks in these macromolecules is accomplished as information flows from DNA synthesis to RNA synthesis and then to protein synthesis. We discuss exactly how this occurs in Chapter 6.

Other polymerization reactions are indirectly determined by the information contained in DNA. For example, the polymerization reactions that build structural molecules such as polysaccharides and peptidoglycan are indirectly determined. In these cases, building blocks are ordered by the enzymes that catalyze the chemical reactions of polymerization. Because enzymes are proteins and protein structure is determined by DNA, the polymerization reactions that they catalyze are indirectly determined by DNA as well.

Not only is ordering crucial in polymerization, but so is the formation of the chemical bonds that join building blocks to create macromolecules. Often the bonds of the macromolecule are at a higher energy state than the bonds of the individual building blocks, which would seem to make polymerization reactions energetically impossible. The solution lies in expending chemical energy in the form of ATP. In many cases ATP reacts with the form of the building block that is to be polymerized. The third phosphorus atom from ATP is transferred to the building block, producing ADP and an activated building-block subunit. This activated subunit is highly reactive and so is capable of entering into a polymerization reaction, during which it loses its activating phosphate group. The many polymerization reactions needed to build a new cell consume a considerable portion of the cell's ATP.

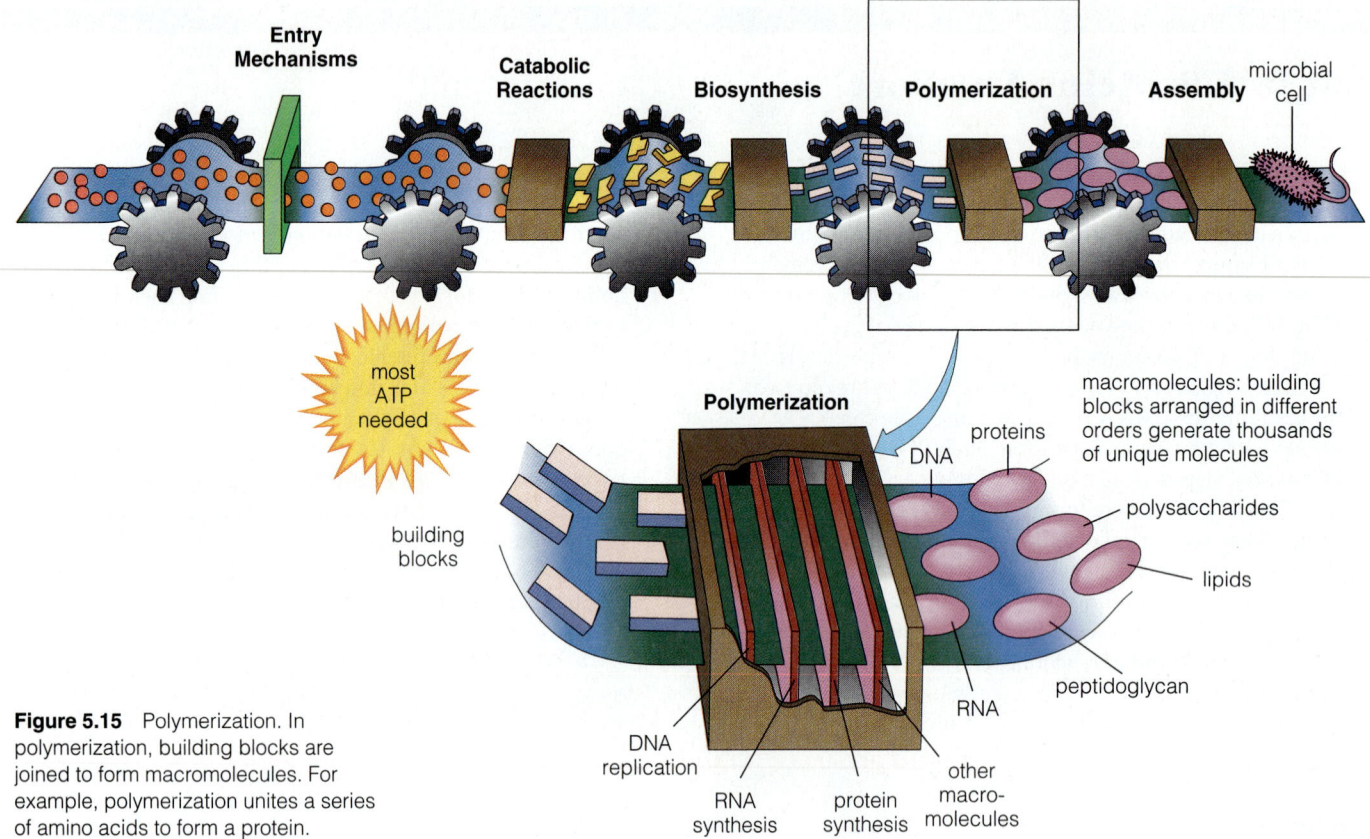

Figure 5.15 Polymerization. In polymerization, building blocks are joined to form macromolecules. For example, polymerization unites a series of amino acids to form a protein.

The synthesis of the polysaccharide glycogen is an example of enzyme-catalyzed polymerization that is indirectly determined by information contained in DNA. Glycogen is a metabolic reserve product stored by some microorganisms, including *E. coli*. Glycogen is made when the cell has an abundant source of nutrients, and it is then used to fuel catabolic pathways when nutrients are scarce. Glucose, the building block of glycogen, is converted into its activated building-block subunit, ADP-glucose, by an enzyme-catalyzed reaction that requires the expenditure of one high-energy phosphate bond in the form of ATP. ADP-glucose is polymerized into glycogen by the action of another enzyme, **glycogen synthase**.

Glycogen synthesis thus illustrates the basic principles of polymerization reactions: (1) the building block is activated by an ATP-utilizing reaction or a short pathway, and (2) activated building blocks enter into the enzyme-catalyzed polymerization, releasing the activating group to be used again later (**Figure 5.16**).

Assembly

Assembly of macromolecules into cellular structures may occur spontaneously or as a result of reactions catalyzed by enzymes (**Figure 5.17**). Spontaneous assembly, also called **self-assembly**, is an intrinsic property of certain proteins. For example, flagellin, the molecular component of flagella on bacteria like *E. coli*, self-assembles. The test of self-assembly is whether or not it will proceed spontaneously *in vitro* (outside the living organism). Under proper conditions of flagellin concentration and pH, and with the addition of a primer, a solution of flagellin will assemble spontaneously to form a flagellum with a characteristic width and helical shape. Even much more complex cellular structures like ribosomes pass the *in vitro* test of self-assembly. Ribosomes, which are composed of 54 different proteins and three different RNA molecules, can assemble from purified components under carefully controlled laboratory conditions.

Formation of the bacterial cell wall is one example of assembly that is catalyzed by enzymes. Short units of peptidoglycan are released into the periplasm, where they are assembled into an intact wall by enzyme-catalyzed reactions. These wall-building enzymes are called **penicillin-binding proteins** because the antibiotic penicillin binds to them (Chapter 21). The particular penicillin-binding proteins that a cell contains determine the shape of the bacterial cell wall and thus the shape of the cell (Chapter 4).

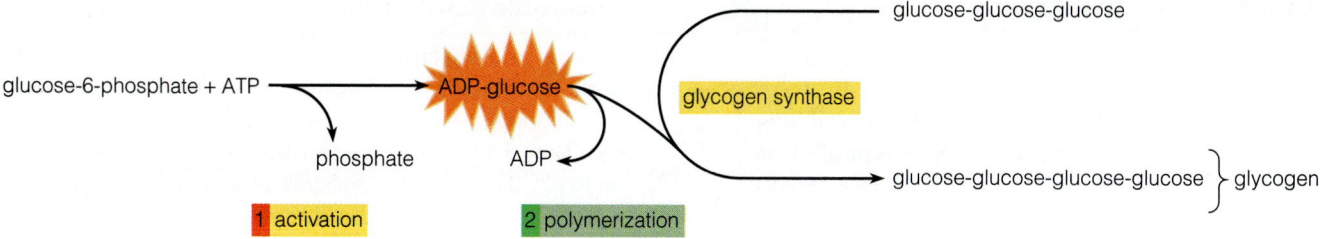

glucose-6-phosphate + ATP → ADP-glucose

phosphate

glucose-glucose-glucose

glycogen synthase

ADP

glucose-glucose-glucose-glucose } glycogen

1 activation   2 polymerization

Figure 5.16 Glycogen synthesis. Glycogen synthesis illustrates the principles of polymerization. (1) Glucose-6-phosphate, derived from the building block (glucose), is activated (into ADP-glucose). (2) It is then polymerized by an enzyme-catalyzed (glycogen synthase) reaction, thereby elongating the polymer.

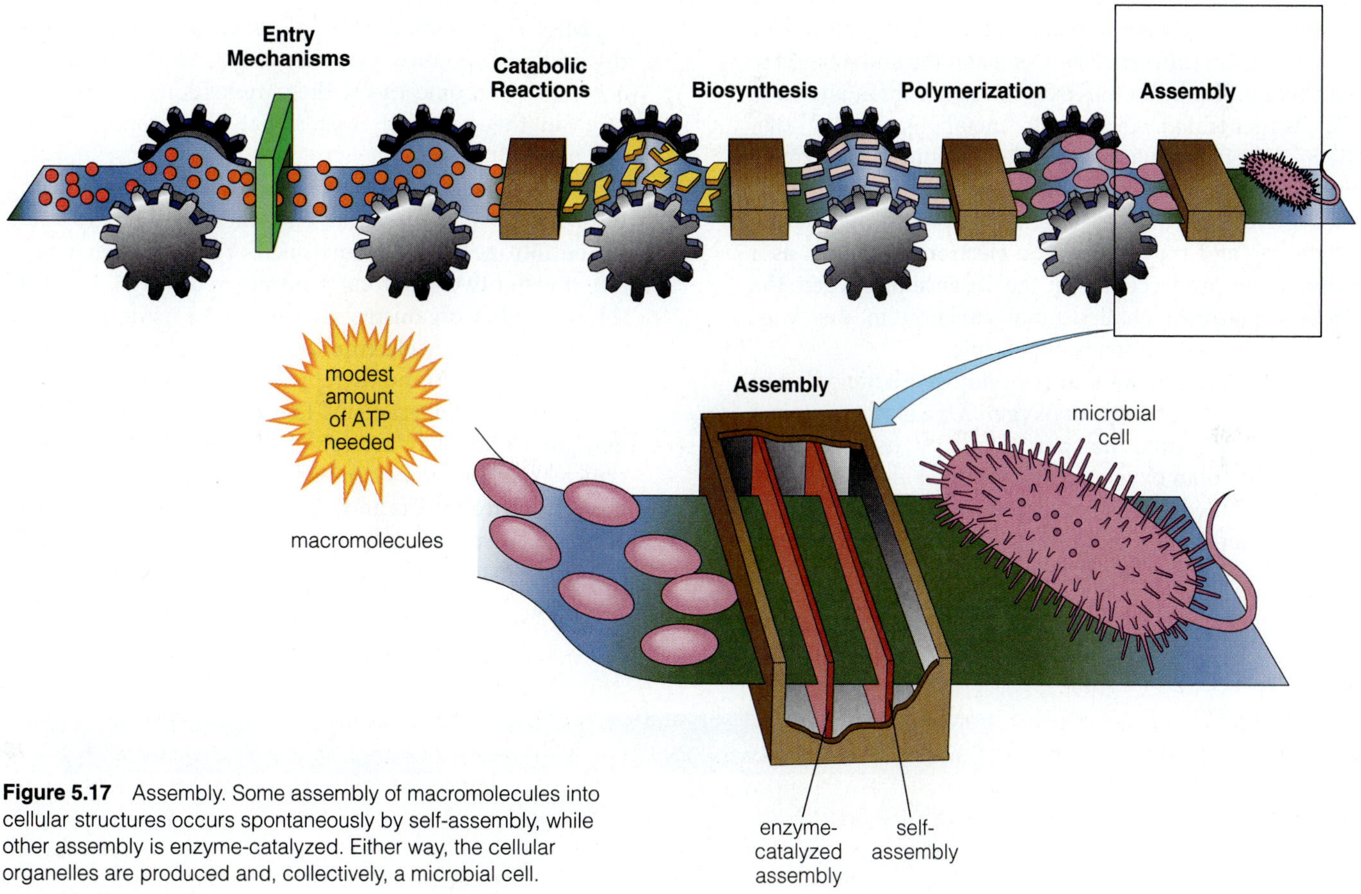

Entry Mechanisms

Catabolic Reactions

Biosynthesis

Polymerization

Assembly

modest amount of ATP needed

macromolecules

Assembly

microbial cell

enzyme-catalyzed assembly

self-assembly

Figure 5.17 Assembly. Some assembly of macromolecules into cellular structures occurs spontaneously by self-assembly, while other assembly is enzyme-catalyzed. Either way, the cellular organelles are produced and, collectively, a microbial cell.

ANAEROBIC METABOLISM

The story of how *Escherichia coli* builds another cell identical to itself when growing aerobically is typical of the metabolism of many microorganisms. But there are also other types of microbial metabolism. The differences lie almost exclusively in the reactions that produce precursor metabolites, ATP, and reducing power. In almost all cases, the other four steps—entry mechanisms, biosynthesis, polymerization, and assembly reactions—are the same.

Let's look now at **anaerobic metabolism**—the biochemical reactions that allow cells to grow in the absence of oxygen. Strict anaerobes are capable of only anaerobic metabolism, while facultative anaerobes (like *E. coli*) are capable of both aerobic and anaerobic metabolism.

The critical difference between aerobic and anaerobic metabolism lies in the electron transport chain and how ATP is generated. In aerobic metabolism of *E. coli*, chemiosmosis is driven by a proton gradient. This gradient is created by an electron transport chain that separates a hydrogen atom, releasing the proton outside the membrane and transferring the electron to oxygen as a terminal electron acceptor. In the absence of oxygen, the electron transport chain cannot function in this way. Thus, aerobic respiration is impossible.

There are two ways that nonphotosynthetic cells can make ATP in the absence of oxygen. One is to maintain a functioning electron transport chain by using a compound other than oxygen as the terminal electron acceptor. This is called **anaerobic respiration**. The second way is to generate all the cell's ATP by substrate level phosphorylation. This is called **fermentation**.

Anaerobic Respiration

Aerobic and anaerobic respiration are similar processes. Both maintain a functioning electron transport chain, which allows the cell to generate energy by chemiosmosis. The chains themselves differ somewhat and their yield of ATP varies, but the critical difference is the **terminal electron acceptor**, the compound that is reduced by accepting electrons at the end of the chain. In aerobic respiration, oxygen accepts electrons and is reduced to water. In anaerobic respiration, another compound is reduced by accepting these electrons. Compounds that can act as a terminal electron acceptor in anaerobic respiration include sulfate, nitrate, fumarate, and trimethylamine-N-oxide (**Table 5.5**). *Escherichia coli*, for example, can use nitrate or fumarate as an electron acceptor if oxygen is not available.

Microorganisms capable of anaerobic respiration reduce the compounds that they use as terminal electron acceptors. In some cases, these reductions play critical roles in the cycles of matter that are essential to life (Chapter 28).

Organisms that carry out anaerobic respiration using nitrate (NO_3^-) as a terminal electron acceptor play a role in the nitrogen cycle. Some microorganisms, such as *E. coli*, transfer two electrons to nitrate, reducing it to nitrite (NO_2^-). Other organisms are capable of using the product, nitrite, as an electron acceptor. They transfer five electrons to nitrite and thereby reduce it completely to nitrogen gas (N_2). Organisms that change nitrate to nitrogen gas in this way are called **denitrifiers**. They play a critical role in removing nitrogen from terrestrial and aquatic environments and returning it to the atmosphere. Most organisms that use nitrate as a terminal electron acceptor do so only when oxygen is not available.

Table 5.5 Some Terminal Electron Acceptors of Bacterial Electron Transport Chains		
Type of Respiration	Terminal Electron Acceptor	Reduced Product
Aerobic Respiration	Oxygen (O_2)	Water (H_2O)
Anaerobic Respiration		
Sulfate reduction	Sulfate (SO_4^{2-})	Hydrogen sulfide (H_2S)
Nitrate reduction	Nitrate (NO_3^-)	Nitrite (NO_2^-)
Fumarate reduction	Fumarate (HOOC—CH=CH—COOH)	Succinate (HOOO—CH$_2$—CH$_2$—COOH)
Denitrification	Nitrate (NO_3^-)	Nitrogen gas (N_2)
Trimethylamine-N-oxide reduction	Trimethylamine-N-oxide	Trimethylamine

Organisms that carry out anaerobic respiration using sulfate (SO_4^{2-}) as a terminal electron acceptor play a role in the sulfur cycle. These microorganisms, called **sulfate reducers**, reduce sulfate to hydrogen sulfide gas (H_2S). Sulfate reducers include members of the genus *Desulfovibrio*. Sulfate reducers typically grow in marine and river mud flats, giving these environments the rotten-egg odor of hydrogen sulfide and turning the mud black from the formation of metal sulfides.

Fermentation

Fermentation is a form of anaerobic metabolism in which all ATP is generated by substrate level phosphorylation. No ATP is generated by chemiosmosis.

The biochemistry of fermentation affects the way a microorganism generates its driving force, ATP, and reducing power. Fermentation generates fewer molecules of ATP per molecule of substrate than do aerobic and anaerobic respiration. For example, *Escherichia coli* derives about 28 ATP molecules from glucose by aerobic respiration but only about 3 by fermentation. Other microorganisms derive even fewer. For instance, lactic acid bacteria, which carry out a simpler fermentation, form only 2 molecules of ATP from each glucose molecule fermented. Because many molecules of substrate must be metabolized to supply a cell's ATP requirements, microorganisms can grow by means of fermentation only if substrate is abundant. At the same time, fermentation can support relatively rapid growth, because most bacteria can use substrate rapidly during fermentation. Some strains of *E. coli* can grow almost as rapidly by fermentation as by aerobic respiration.

The main difference between fermentation and respiration in generating reducing power has to do with the balance of oxidation-reduction reactions in the cell. Oxidation and reduction must be balanced in any cell,

regardless of its form of metabolism, because electrons removed from one compound as it is oxidized must be accepted by another compound, which is reduced. During respiration, a great deal of reducing power is consumed by reducing oxygen or some other terminal electron acceptor at the end of the electron transport chain. Reducing power is not used this way in fermentation. Because organisms growing by fermentation cannot reduce oxygen, the overall metabolic reactions of fermentation must not generate too much reducing power. This situation is possible only if substrate enters fermentation pathways at an intermediate state of oxidation—that is, neither highly oxidized nor highly reduced. Sugars, which are at an intermediate oxidation state, are therefore almost the only substrates that can be used in fermentation.

To illustrate the principles of fermentation, let's look at one specific example—**lactic acid fermentation (Figure 5.18)**. This type of fermentation is carried out by lactic acid bacteria, the organisms that cause milk to sour and are used to produce acidic dairy products such as yogurt and buttermilk. Muscle tissue of animals, including humans, also carries out a lactic acid fermentation—by the same pathway—when deprived of oxygen. Lactic acid fermentation proceeds through the glycolysis pathway, which we discussed earlier in the chapter. One molecule of glucose is metabolized to produce two molecules of pyruvate, two molecules of ATP, and two of reduced NAD. The two molecules of reduced NAD are reoxidized as pyruvate is reduced to lactic acid, thereby consuming as much reducing power as was formed. The cell cannot survive using this fermentative pathway alone because the pathway does not generate all 12 precursor metabolites. Certain reactions of the pentose phosphate pathway and the TCA cycle must also occur in order to generate them. But the overwhelming majority of substrate is metabolized through the fermentative pathway in order to meet the cell's requirement for ATP. Huge amounts of the

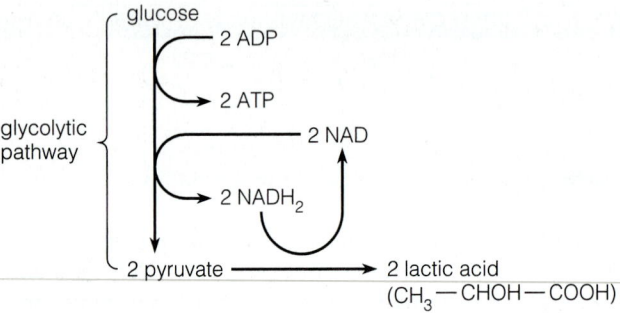

Figure 5.18 Fermentation. Lactic acid fermentation illustrates the principles of this form of anaerobic metabolism. In lactic acid fermentation, glucose is metabolized by glycolysis to pyruvate. The pyruvate is reduced to lactic acid using the reducing power produced by glycolysis.

fermentative end product, lactic acid, result, which gives the products of this fermentation their characteristic acidic taste.

There are many different kinds of fermentation, and each is characteristic of a particular kind or group of microorganisms (**Table 5.6**). Most fermentations metabolize glucose to pyruvate by means of glycolysis, but they differ in how pyruvate is reduced and NADH reoxidized. In lactic acid fermentation, pyruvate is reduced directly to lactic acid. In alcoholic fermentation, which is typical of yeast, pyruvate is converted to CO_2 and ethanol. In the mixed acid fermentation, which is characteristic of *E. coli* and related bacteria, pyruvate is converted to at least six different end products.

NUTRITIONAL CLASSES OF MICROORGANISMS

Because *Escherichia coli* can produce ATP and reducing power both in the presence of oxygen (aerobically) and in its absence (anaerobically), it has served well as a model of the types of microorganisms that use these forms of metabolism. Now we turn to organisms that are quite different from *E. coli*. We examine the vast array of starting materials that some organisms can use to produce the precursor metabolites, ATP, and reducing power that they all need for growth.

Microbiologists classify organisms according to **nutritional class**. The nutritional class to which an organism belongs depends upon two factors: (1) the source of carbon atoms it uses to make precursor metabolites and (2) how it generates ATP and reducing power. Organisms that obtain all their carbon from atmospheric carbon dioxide are called **autotrophs**, or self-feeders. Those that obtain carbon from organic compounds in their medium or diet are called **heterotrophs**, or different feeders. Organisms that generate ATP and reducing power from chemical reactions are called **chemotrophs**, or chemical feeders. Those that generate ATP and reducing power from light energy are called **phototrophs**, or light feeders. All four possible combinations of generating precursor metabolites and ATP exist among microorganisms. There are **chemoautotrophs**, **chemoheterotrophs**, **photoautotrophs**, and **photoheterotrophs** (**Table 5.7**).

In some organisms, the source of materials and the driving force are biochemically linked. That is, the same metabolic pathways generate precursor metabolites, ATP, and reducing power. Our model chemoheterotroph, *E. coli*, is an example. In other organisms, however, these processes may be completely separate. For example, some photosynthetic organisms derive carbon atoms for precursor metabolites from atmospheric carbon dioxide but generate ATP and reducing power from light energy.

Because we have discussed the pathways that chemoheterotrophs like *E. coli* use to generate precursor metabolites, ATP, and reducing power, here we look at some of the pathways that microorganisms of other nutritional classes use to supply these three needs. *E. coli*'s pathways are catabolic—they break down organic molecules like glucose into simpler precursor metabolites. In autotrophs, however, precursor metabolites are not formed by breaking down more complex molecules. Instead, more

Table 5.6	Some Types of Microbial Fermentation	
Fermentation	Some Organisms That Perform It	End Products
Alcoholic	Yeasts	Ethanol and CO_2
Lactic acid	Lactic acid bacteria	Lactic acid
Mixed acid	*Escherichia coli*	Lactic acid, acetic acid, formic acid, succinic acid, H_2, CO_2, ethanol
Butanediol	*Enterobacter aerogenes*	Lactic acid, acetic acid, formic acid, H_2, CO_2, 2,3-butanediol

Table 5.7 Nutritional Classes of Microorganisms		
Source of Carbon Atoms	Source of Energy (ATP)	
	Chemical Reactions	Light Energy
Organic compounds	Chemoheterotrophs	Photoheterotrophs
CO_2	Chemoautotrophs	Photoautotrophs

complex organic molecules are created from the simpler inorganic molecule carbon dioxide (CO_2). Pathways that *generate* precursor metabolites, ATP, and reducing power from simpler molecules, namely CO_2, are called **fueling pathways**. We look at fueling pathways in autotrophs, phototrophs, and chemoautotrophs.

Formation of Precursor Metabolites by Autotrophs

Autotrophs synthesize precursor metabolites from CO_2. In other words, autotrophic organisms, both microorganisms and plants, take molecules from an inorganic atmospheric gas and create organic molecules. Not only do these organic molecules enable the autotrophs to synthesize a new cell, but they also fuel the metabolism of all heterotrophs. That is, plants, as autotrophs, use CO_2 to create organic molecules and grow, and animals eat plants or other animals that have eaten plants and catabolize the organic molecules to grow. The transformation of inorganic carbon to organic carbon is an essential step in the earth's life-sustaining carbon cycle (Chapter 28).

Because autotrophs synthesize organic molecules from CO_2, they have special fueling pathways not found in heterotrophs. The most widespread of these pathways is the **Calvin-Benson cycle**, which incorporates CO_2 into a preexisting organic molecule (**Figure 5.19**). Gaseous CO_2 reacts with a phosphorylated sugar, **ribulosebisphosphate**, producing two molecules of 3-phosphoglyceraldehyde—which is an intermediate of glycolysis. The remainder of the Calvin-Benson cycle regenerates another molecule of ribulosebisphosphate that can react with another molecule of CO_2. Intermediates of the Calvin-Benson cycle feed into slightly modified glycolysis, TCA, and pentose phosphate pathways to generate the 12 precursor metabolites for autotrophs.

Unlike *E. coli*'s catabolic pathways, the Calvin-Benson cycle does not generate ATP or reducing power. Instead, it uses them. Thus, autotrophs must generate ATP and reducing power in other fueling pathways. These pathways derive energy from sunlight (photoautotrophs) or from the oxidation of inorganic compounds (chemoautotrophs).

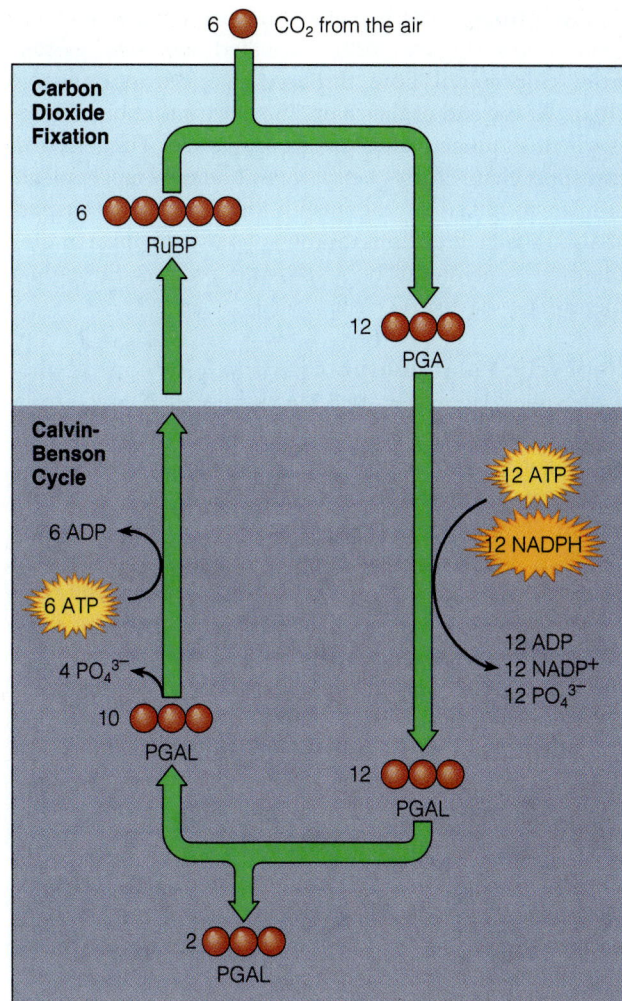

Figure 5.19 The Calvin-Benson cycle. The Calvin-Benson cycle is the way autotrophs make precursor metabolites. CO_2 reacts with ribulosebisphosphate (RuBP), eventually producing phosphoglyceraldehyde (PGAL) and regenerating more RuBP to react with more CO_2. PGAL flows into central metabolism through glycolysis to make the other precursor metabolites.

Formation of ATP and Reducing Power by Photoautotrophs

Photoautotrophs supply the ATP and reducing power needed to fuel the Calvin-Benson cycle, as well as the driving force necessary for entry mechanisms, biosynthesis, polymerization, and assembly, by means of two light-driven processes: **cyclic photophosphorylation** and **noncyclic photophosphorylation**. The green photosynthetic pigment **chlorophyll** plays a central role in both processes. When activated by **photons**, which represent light energy, electrons are ejected from chlorophyll at a high energy level. Like respiration, cyclic and noncyclic phosphorylation employ an electron transport chain embedded in a membrane, creating a proton gradient across that membrane.

Cyclic Photophosphorylation. In cyclic photophosphorylation (**Figure 5.20**), activated electrons ejected from chlorophyll flow through an electron transport chain. At the end of the chain, the electrons rejoin chlorophyll in its **unactivated**, or ground, state. This electron transport chain creates a proton gradient, by mechanisms similar to those in the respiratory electron transport chain. Like heterotrophs, autotrophs use this proton gradient to generate ATP by passing protons through a membrane-located ATPase.

Noncyclic Photophosphorylation. In noncyclic photophosphorylation (**Figure 5.21**), the ejected electron from chlorophyll in a protein complex (called **photosystem II**) flows down an electron transport chain and generates a proton gradient, just as it does in cyclic photophosphorylation. The difference occurs at the end of the chain. Instead of completing the cycle by rejoining the original chlorophyll, it joins chlorophyll in another complex of proteins (called **photosystem I**) and light energy raises it to a higher activated state. The ejected electron flows through a second electron transport chain and then joins with a proton and is transferred to NADP$^+$, forming NADPH. Thus, noncyclic photophosphorylation produces reducing power, in the form of NADPH, in addition to the ATP generated from the proton gradient.

Because noncyclic photophosphorylation does not return the electron to a chlorophyll molecule, a source of electrons other than chlorophyll is required. Green plants and some bacteria—namely, the cyanobacteria (Chapter 11)—use water as the electron source. They photosplit water to produce two electrons, two protons, and a half molecule of oxygen gas. These organisms are called **oxygenic**, or "oxygen-producing," because oxygen gas is a by-product of their NADPH-generating metabolism. Some phototrophic bacteria that cannot photosplit water rely on other sources of electrons—either reduced sulfur compounds or organic compounds—to produce reducing power. Because oxygen gas is not a by-product of their metabolism, they are termed **nonoxygenic**, or "not oxygen-producing." The by-product of nonoxygenic noncyclic photophosphorylation is either an oxidized sulfur or an organic compound. Phototrophs that use organic compounds as sources of electrons for noncyclic photophosphorylation also use them as sources of precursor metabolites. These phototrophs are photoheterotrophs, while all others are photoautotrophs.

Formation of ATP and Reducing Power by Chemoautotrophs

Chemoautotrophs derive ATP and reducing power from the oxidation of one of several inorganic compounds (**Table 5.8**). They accomplish this in much the same way

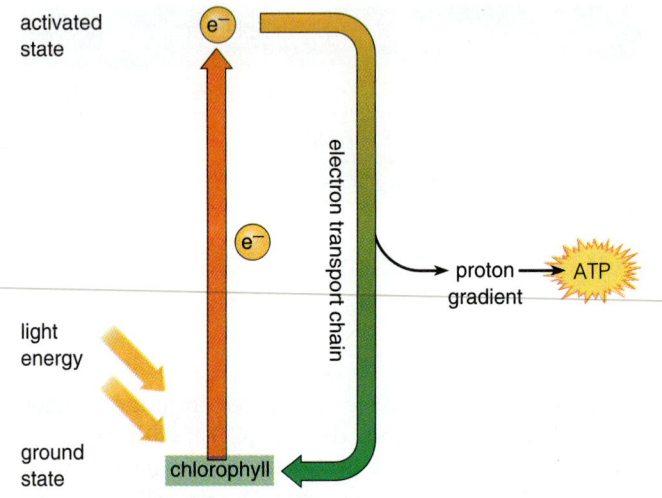

Figure 5.20 Cyclic photophosphorylation. In cyclic photophosphorylation, light energy strikes chlorophyll, ejecting an electron and raising it to an activated state. It has sufficient energy in its activated state to drive it through an electron transport chain back to chlorophyll at the ground (unactivated) state of energy. The flow through the electron transport chain forms a proton gradient that generates ATP.

as chemoheterotrophs do. They remove an electron from the inorganic substrate and pass it through an electron transport chain, thereby generating a proton gradient that is capable of producing ATP when the protons flow back into the cell through a membrane-bound ATPase. Some inorganic compounds oxidized by chemoautotrophs are reducing agents sufficiently strong to reduce NAD(P) directly. An example would be reduced sulfur compounds. But others—iron, for one—are not. Iron-oxidizing chemoautotrophs reduce NAD(P) by a process called **reverse electron transport**. In reverse electron transport, the proton gradient drives electrons back through the electron chain, reducing NAD(P).

REGULATION OF METABOLISM

At the beginning of this chapter we compared *Escherichia coli* to a metabolic factory and traced how the various cellular assembly lines produce all the components necessary to make a cell. But how are these assembly lines regulated and coordinated? How does the cell ensure that assembly lines producing components needed in large quantities will run faster than assembly lines producing components needed in small quantities? How are the various lines geared up or down as growth accelerates or slows? How is one line selectively shut down if its product becomes available from an outside source?

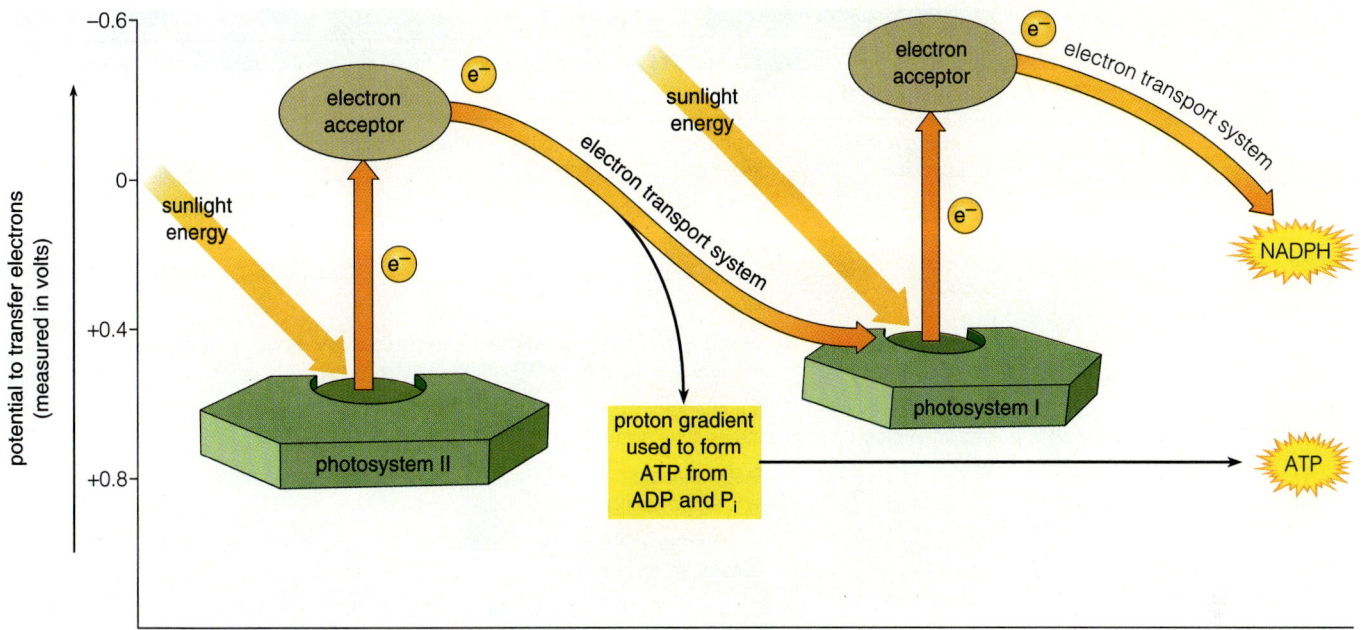

Figure 5.21 Noncyclic photophosphorylation. In noncyclic photophosphorylation, light energy striking chlorophyll in a protein complex (photosystem II) ejects an electron and raises it to an activated state. The electron has sufficient energy in its activated state to drive it through an electron transport chain, which generates a proton gradient that is used to form ATP. The electron then joins chlorophyll in another complex (photosystem I) where light energy raises it to a *higher*-energy-activated state. From there it flows through an electron transport chain to NADP, reducing it to NADPH. The original source of electrons is water, which is split by light energy in photosystem II. While cyclic photophosphorylation produces only ATP, noncyclic produces both ATP and reducing power.

Table 5.8	Classes of Chemoautotrophs and the Reactions They Use to Generate ATP	
Class of Chemoautotroph	Substance Oxidized	Product of Oxidation
Hydrogen bacteria	Hydrogen gas (H_2)	Protons (H^+)
Sulfur bacteria	Sulfide (H_2S), sulfur (S), or sulfite (SO_3^{2-})	Sulfate (SO_4^{2-})
Iron bacteria	Ferrous ion (Fe^{2+})	Ferric ion (Fe^{3+})
Ammonia oxidizers	Ammonium ion (NH_4^+)	Nitrite ion (NO_2^-)
Nitrite oxidizers	Nitrite ion (NO_2^-)	Nitrate ion (NO_3^-)

Purpose

Metabolic regulation ensures that each of the cell's different metabolic pathways operates at a rate that will supply the cell with the optimum amount of its end product. This regulatory network is complex, because different metabolic pathways operate at vastly different rates. Moreover, these rates can change suddenly and dramatically. Consider, for example, the need for the 20 amino acids that are incorporated into proteins. On the average, protein in an *E. coli* cell contains 11 times as much glycine as tryptophan. So, on the average, the biosynthesis pathway leading to glycine must operate 11 times faster than the one leading to tryptophan. But these basic rates must change in response to changing demands, both within the cell and in the environment. For example, if *E. coli* were provided with glycine in its growth medium, it

Table 5.9   Some Regulatory Mechanisms That Operate in *Escherichia coli*

Level of Operation	Effect of Regulatory Mechanism
Bringing nutrients into the cell	Produces permeases to bring certain nutrients into the cell only if its substrate is available.
Fueling pathways	Produces enzymes to metabolize a particular nutrient only if it is available.
	Produces only the enzymes to metabolize the better of two substrates if both are available.
	Metabolizes substrates just fast enough to provide optimal levels of precursor metabolites, ATP, and reducing power.
Biosynthesis of building blocks	If a building block is present in the medium, enzymes for catalyzing its synthesis are not made.
	Biosynthetic pathways operate just fast enough to maintain optimal levels of building blocks.
Polymerizations	The number of ribosomes made is just sufficient to synthesize proteins at an optimal rate.
	Chromosome is replicated just fast enough to supply daughter cells with a complete genome.
Behavioral	Flagella are not made if nutritional conditions are ideal (no need to move).

Figure 5.22 Allosteric enzymes. One of the two main types of metabolic regulation is effected by allosteric enzymes. An allosteric enzyme has two active sites. One binds substrate to catalyze a reaction. The other binds a small molecule effector that changes the enzyme's activity. In this example, the effector changes an active enzyme into an inactive enzyme.

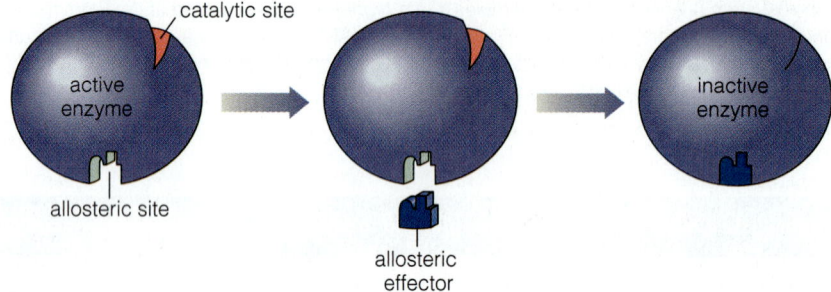

would stop making glycine but would continue to make tryptophan and the other 18 amino acids at full rate. Adequate but not excessive amounts of building blocks must be synthesized under all conditions. Metabolic regulation allows the cell to maximize growth rate and metabolic efficiency.

Metabolic regulation also increases the cell's efficiency in other ways. For example, cells usually synthesize enzymes for using substrates only when the substrate is available. Cells synthesize only the enzyme to use the best substrate—the one that supports fastest growth—when several substrates are available. Furthermore, cells synthesize only enough ribosomes to meet the demand for protein synthesis (**Table 5.9**).

Types

There are two major types of metabolic regulation. The first type regulates the *amount* of an enzyme that is synthesized. This type is called **regulation of gene expres-** sion (see Chapter 6). The second type regulates the *activity* of an enzyme once it has been synthesized. This type of metabolic regulation, which is usually effected by **allosteric enzymes**, is discussed here.

Allosteric means "different site." All enzymes have a catalytic site where reactants—also known as substrates—are positioned properly to combine with one another (Chapter 2). Allosteric enzymes, however, also have a site where small signal molecules called **effectors** bind to the enzyme and change its activity.

Effectors either increase or decrease the rate of the enzymatic reaction. Some effectors bind to an allosteric site and activate enzymes, while others bind to an allosteric site and inhibit enzymes. Both activation and inhibition occur because binding at the allosteric site changes the conformation (shape) of the enzyme. The change in conformation subtly alters the catalytic site that in turn influences the rate at which the enzyme binds its substrate and catalyzes the reaction (**Figure 5.22**).

Regulating Ribosomes

Catherine Squires and Craig Squires

Cathy and Craig Squires have been in the Department of Biological Sciences at Columbia University since 1977, where she is a Professor and he is a Research Associate. They have been working together on the expression of the transcription and translation machinery in Escherichia coli since 1976. Cathy received her bachelor's and master's degrees in bacteriology and microbiology from the University of California at Davis and her Ph.D. in molecular biology and biochemistry from the University of California at Santa Barbara. After obtaining her Ph.D., Cathy received a Helen Hay Whitney Postdoctoral Fellowship for studies in the laboratory of Charles Yanofsky at Stanford University. Craig received his bachelor's degree in agriculture at Michigan State University and did graduate work in biochemistry at both the Berkeley and Davis campuses of the University of California. Craig worked with Drs. A. G. Marr, E. Englesberg, J. Carbon, and S. Yanofsky before joining forces with Cathy.

One of the most fundamental and vexing questions in microbial metabolism is: How do cells regulate the synthesis of their ribosomes—the organelles that make protein?

The question is fundamental because cells always synthesize an optimal number of ribosomes: When growing rapidly, they make large numbers of ribosomes because they must make protein rapidly; when growing slowly, they need and make only small numbers of ribosomes. Making ribosomes is metabolically expensive; overproduction is a waste, but underproduction slows growth. Either type of metabolic sloppiness is disastrous in nature's competitive world.

The question is vexing because, in spite of intensive study and important progress, it remains unsolved. Microbiologists know the genetic and molecular details of how optimal levels of most cellular components are set; they can't say the same about ribosomes. And because the protein components of ribosomes are made in amounts that match the amount of the RNA component (rRNA), the basic question really is: How do cells regulate synthesis of rRNA? That is the question we set out to answer.

The first complication we had to overcome was the redundancy of *rrn* operons, which encode rRNA. The human genome contains hundreds of copies of the *rrn* operon, and *Escherichia coli*, which we studied, carries seven. How could we study what one *rrn* operon was making when six others were making the same thing? We used recombinant DNA technology to fuse one *rrn* operon to another gene that encodes a product, chloramphenicol acetyltransferase, that we could measure easily. By measuring chloramphenicol acetyltransferase activity, we knew how much of that particular *rrn* operon product was being made.

We used the *rrn* fusion for two types of studies. In one, we studied how changing the growth rate of a culture of *E. coli* controlled the amount of rRNA it made. This control is thought to be governed by feedback inhibition, a model first proposed by M. Nomura. When he put extra *rrn* operons in *E. coli*, it did not make more rRNA. Seemingly, the cell "knew" how much rRNA it needed and made just that amount regardless of its genetic capacity. We studied the reverse situation: How did the cell respond if 1, 2, 3, or 4 *rrn* operons were taken out of it? We found that when we removed *rrn* operons, each remaining operon increased its synthesis to make more product. Then, using electron microscopy, we found out how this happened. We saw that the *rrn* operons had more RNA polymerase molecules attached to them, and so more rRNA was being made from each remaining operon. But our calculations showed that the greater number of RNA polymerase molecules could not account for all the increase in rRNA. Another change had occurred: Each RNA polymerase molecule was moving more rapidly down the operon, making rRNA at a faster rate. We think this faster rate allowed more frequent transcription initiation, resulting in more RNA polymerase molecules per operon. More RNA polymerase molecules moving faster explained how so much rRNA was being made from the few remaining *rrn* operons. Now we are studying the genetic and metabolic mechanisms that bring this about.

We also used our *rrn* fusion to study a phenomenon called antitermination that occurs when *rrn* operons and certain others are transcribed. If, when transcribing a particular stretch of DNA, an RNA polymerase molecule encounters a sequence of DNA called a terminator, transcription stops. But if the RNA polymerase molecule encounters an antiterminator sequence first, it goes right on through the terminator. We found that *rrn* operons have antiterminators and that they are only short DNA sequences. They must be powerful regulators of rRNA expression. We are currently studying how they do this.

Figure 5.23 End-product inhibition. End-product inhibition is one of the two main types of metabolic regulation. When enough end product is produced, it shuts down, or slows, the process.

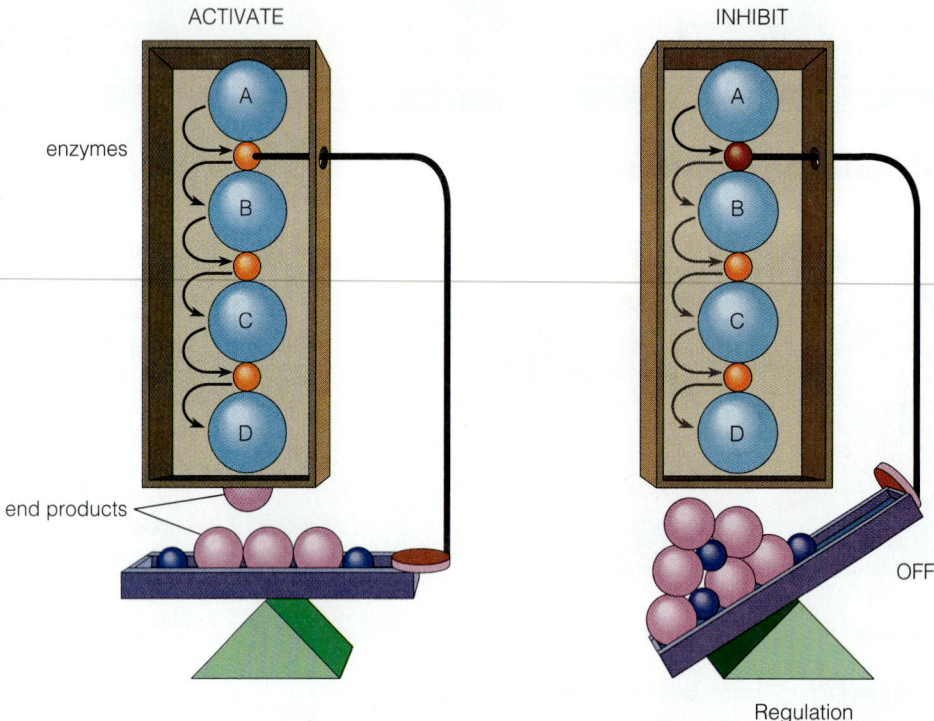

ACTIVATE

INHIBIT

enzymes

end products

OFF

Regulation

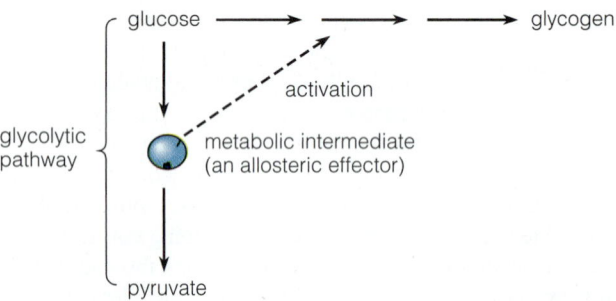

glucose → → → glycogen

activation

glycolytic pathway

metabolic intermediate (an allosteric effector)

pyruvate

Figure 5.24 Allosteric activation. Glycogen synthesis is one of the metabolic processes regulated by allosteric activation. When the concentration of an intermediate of glycolysis rises, indicating excess glucose, glycogen synthesis is activated and the excess glucose is stored as glycogen.

End-Product Inhibition. Most biosynthetic pathways are allosterically regulated by a feedback mechanism called **end-product inhibition** (**Figure 5.23**). The end product of the pathway itself is the allosteric effector. The end product (effector) binds to the enzyme that catalyzes the first reaction in that pathway and inhibits its catalytic activity. Inhibiting the first enzyme decreases the flow of materials through the entire pathway.

Consider the regulation of the biosynthetic pathway that produces the amino acid isoleucine. If a large amount of isoleucine is present in the cell, the first reac-

tion of the pathway that produces isoleucine is slowed down. This slowing prevents more isoleucine from accumulating within the cell. When growth occurs and isoleucine is used for protein synthesis, its intracellular concentration drops. The lower concentration decreases inhibition of the first enzyme and increases the flow of materials through the isoleucine pathway. When demand drops again, the opposite series of events takes place.

Allosteric Activation. Some metabolic processes are regulated by allosteric activation. This means that a rise in intracellular concentration of an effector activates an allosteric protein. Synthesis of the storage product, glycogen, is an example of a pathway regulated by allosteric activation (**Figure 5.24**). When intermediates of pathways such as glycolysis rise, indicating that more carbon is available than is needed for growth, enzymes leading to the synthesis of glycogen are activated. This enables the cell to store the excess carbon for later use.

Summary

ESCHERICHIA COLI: A METABOLIC PARADIGM (pp. 109–111)

1. Metabolism is the total of all biochemical reactions that take place in a cell.

2. Because its primary function is reproduction, a microbial cell is like a factory that manufactures more factories. Like a factory, a cell needs building materials, driving force to do work, and a plan.

3. The raw materials of metabolism are molecules and, in particular, carbon. These molecules are called substrates (nutrients).

4. Driving force is the collective term for energy and reducing power.

5. To carry out chemical reactions that do not occur spontaneously, a cell stores energy in the form of a molecule called adenosine triphosphate (ATP).

6. Reducing power is based on two related biochemical reactions: oxidation (the removal of electrons from a compound) and reduction (the addition of electrons to a compound). When compounds are oxidized, electrons are captured in the form of hydrogen atoms and stored in other compounds called nicotinamide adenine dinucleotide (NAD) or nicotinamide adenine dinucleotide phosphate (NADP). Collectively, they are termed NAD(P).

7. The plan for directing microbial metabolism is contained in the cell's DNA.

METABOLISM: AN OVERVIEW (pp. 111–112)

1. Materials flow along the metabolic assembly line in *Escherichia coli* and most other organisms in five sequential steps:
   a. Entry mechanisms transport substrates across the plasma membrane (and outer membrane in Gram-negative bacteria) so there are sufficient concentrations to fuel metabolism.
   b. Catabolic reactions break down substrates into the basic building materials needed to manufacture a new cell: 12 precursor metabolites, ATP, and reducing power.
   c. During biosynthesis, precursor metabolites are converted into the basic building blocks of macromolecules.
   d. In polymerization, the basic building blocks are activated and polymerized into macromolecules.
   e. In assembly, macromolecules are assembled into cell structures, producing a new cell.

2. ATP and reducing power are largely formed during catabolic reactions and used to drive the other steps in the metabolic assembly line.

3. Entry mechanisms use a small amount of ATP. The same set of catabolic reactions that produces the 12 precursor metabolites produces ATP and reduced NAD(P). Biosynthesis requires ample reducing power stored in the form of reduced NAD(P). Polymerization consumes most of the ATP generated by catabolic reactions. Assembly that does not occur spontaneously requires ATP, but the total amount is slight.

AEROBIC METABOLISM (pp. 113–127)

Entry Mechanisms (pp. 113–114)
1. Carrier proteins called permeases bind substrate molecules and transport them into the cell; group translocation is another entry mechanism that requires energy.

Catabolic Reactions (pp. 114–123)
1. Catabolic reactions are organized into catabolic pathways, sequences of chemical reactions, each catalyzed by a different enzyme, that convert molecules (called metabolic intermediates) step-wise from substrate to precursor metabolites.

2. Substrates are converted into the 12 precursor metabolites through enzyme-catalyzed reactions in three different catabolic pathways: the glycolysis pathway, the tricarboxylic acid (TCA) cycle, and the pentose phosphate pathway.

3. In the process of converting substrate to precursor metabolites, NAD(P) is reduced (and ATP is formed).

4. Oxidations in which protons are removed along with electrons are dehydrogenation reactions. Reductions in which they are added together are hydrogenation reactions.

5. In metabolic oxidations and reductions, hydrogen atoms are not transferred directly from one metabolic intermediate to another. Instead, they are transferred to or from NAD or NAD(P). Thus, NAD(P) stores the cell's reserves of hydrogen, or reducing power.

6. In biosynthesis, many hydrogenation reactions occur that reoxidize NAD(P)H, or reduced NAD(P).

ATP: Stored Energy (pp. 116–117)
1. ATP derives its energy-storage capabilities from the two bonds, called high-energy bonds, that join the last two of its three phosphate groups.

2. When the high-energy bonds are broken, the phosphate groups are readily donated to other compounds. Compounds that receive ATP phosphate groups are said to be phosphorylated. They participate in metabolic reactions that would not occur if reactants were unphosphorylated.

ATP Formation (pp. 117–120)
1. ATP is formed by adding a single phosphate group to adenosine diphosphate (ADP). Some ADP is formed as a product of biosynthesis, and the rest is formed as a second product of phosphorylation reactions.

2. ADP can be converted to ATP through substrate level phosphorylation. A phosphate bond that is already part of a metabolic intermediate becomes highly reactive during certain catabolic reactions, which enables the phosphate group to be transferred to ADP, converting it to ATP.

3. ADP can also be converted to ATP through chemiosmosis. The enzyme ATPase catalyzes the conversion of ADP to ATP in and around a membrane (in procaryotes, it is the plasma membrane).

4. Energy for chemiosmosis comes from a concentration gradient (usually of protons) formed across the membrane. The energy for creating the proton gradient comes from aerobic respiration, the process of passing electrons or passing hydrogen through an electron transport chain that ends in oxygen as a terminal electron acceptor.

5. Each pair of protons that reenters the cell through ATPase converts approximately one molecule of ADP to ATP. Therefore, each pair of electrons that passes through the chain generates about two molecules of ATP. Eucaryotic organisms generate three molecules of ATP per pair of electrons.

6. The two driving forces of metabolism are interconvertible. Some organisms use ATP to generate reducing power by reverse electron flow. That is, they force electrons/hydrogen back through the electron transport chain and reduce $NAD(P)^+$ to NAD(P)H.

7. Similarly, ATP and a proton gradient are interconvertible. Hydrolysis of ATP to ADP by ATPase forces protons out of the cell, establishing a proton gradient.

Catabolic Pathways (pp. 120–123)
1. The three most important catabolic pathways for all cellular organisms—glycolysis, the TCA cycle, and the pentose phosphate cycle—are collectively called central metabolism.

2. Glycolysis begins with glucose as a substrate, forms six precursor metabolites, and results in two molecules of pyruvate. The total yield of ATP can be about six molecules. The net yield by substrate level phosphorylation is two (because two ATP molecules are spent at the beginning). The two molecules of NADH can donate two

pairs of hydrogen atoms to an electron transport chain, yielding four ATPs by chemiosmosis.

3. Each turn of the TCA cycle produces four precursor metabolites, one molecule of ATP, two molecules of NADH, one of NADPH, and one of another reduced carrier called $FADH_2$.

4. The pentose phosphate pathway produces the last 2 of the 12 precursor metabolites, which are used to make three amino acids and all nucleotides. It also produces two molecules of NADPH. No ATP is directly produced, but the NADPH may generate four molecules of ATP through chemiosmosis.

Biosynthesis   (pp. 123–124)

1. Biosynthetic reactions are organized into biosynthetic pathways, enzyme-catalyzed reaction sequences that convert precursor metabolites into metabolic building blocks and coenzymes.

2. Some biosynthetic pathways are branched, involving several different precursor metabolites and/or producing several different building blocks. Others are unbranched, producing a single building block from a single precursor metabolite.

3. Coenzymes are compounds that act with enzymes in certain metabolic reactions. Ready-made molecules that act as coenzymes or their precursors are called vitamins.

Polymerization   (pp. 125–126)

1. The major polymerization reactions in cells are DNA replication, RNA synthesis, protein synthesis, and polysaccharide synthesis.

2. Ordering in polymerization is directed by the organism's DNA, either directly (DNA, RNA, and protein polymerization) or indirectly through enzymes (polysaccharides and peptidoglycan).

3. Glycogen synthesis illustrates the two basic principles of polymerization reactions: (1) the building block is activated by an ATP-utilizing reaction and (2) activated building blocks enter into the enzyme-catalyzed polymerization, releasing the activating group to be used again.

Assembly   (pp. 126–127)

1. Assembly of macromolecules into cellular structures and, collectively, a new cell may occur spontaneously (self-assembly) or as a result of enzyme-catalyzed reactions.

2. Flagella in bacteria are self-assembled, while the bacterial cell wall is assembled through enzyme-catalyzed reactions.

ANAEROBIC METABOLISM   (pp. 128–130)

1. The types of metabolism vary in their catabolic reactions, but the other four steps are essentially the same.

2. In the absence of oxygen, the electron transport chain cannot function, and chemiosmosis cannot occur unless an alternate terminal electron acceptor is present.

3. In the absence of oxygen, nonphotosynthetic cells can make ATP through anaerobic respiration or fermentation.

Anaerobic Respiration   (pp. 128–129)

1. Because anaerobic respiration, like aerobic, depends upon an electron transport chain, ATP is generated by chemiosmosis.

2. The terminal electron acceptor for anaerobic respiration can be nitrate, sulfate, fumarate, or trimethylamine-N-oxide.

Fermentation   (pp. 129–130)

1. In fermentation, all ATP is generated by substrate level phosphorylation.

2. With a given amount of substrate, fermentation generates the least amount of ATP. However, it can proceed as rapidly as aerobic respiration.

3. Sugars are almost the only substrates that can be used in fermentation because they are at an intermediate oxidation state. There are many different kinds of fermentation, such as lactic acid fermentation, which is carried out by lactic acid bacteria.

NUTRITIONAL CLASSES OF MICROORGANISMS   (pp. 130–132)

1. Autotrophs obtain their carbon from atmospheric carbon dioxide. Heterotrophs obtain organic carbon compounds. Chemotrophs generate ATP and reducing power through chemical reactions. Phototrophs generate ATP and reducing power from light energy.

2. There are four nutritional classes of microorganisms: chemoautotrophs, chemoheterotrophs, photoautotrophs, and photoheterotrophs.

3. The most widespread pathway for forming precursor metabolites in autotrophs is the Calvin-Benson cycle. Autotrophic microorganisms obtain energy from sunlight (photoautotrophs) or by oxidizing inorganic compounds (chemoautotrophs).

4. Photoautotrophs obtain ATP and reducing power for the Calvin-Benson cycle as well as for the other steps in metabolism from cyclic photophosphorylation and noncyclic photophosphorylation.

5. Both processes involve chlorophyll and an electron transport chain that creates a proton gradient. Cyclic photophosphorylation produces ATP exclusively, while noncyclic photophosphorylation produces both ATP and reducing power.

6. Chemoautotrophs derive ATP and reducing power by oxidizing an inorganic compound. Chemoautotrophs that use sulfur compounds can reduce NAD(P) directly, but iron-oxidizing chemoautotrophs cannot. They use reverse electron transport: the proton gradient drives electrons back through the electron chain, reducing NAD(P).

REGULATION OF METABOLISM   (pp. 132–136)

1. Metabolic regulation ensures that pathways operate at a rate that will supply the optimum amount of their end products and that cells synthesize only the enzyme that will use the best substrates available.

2. Regulation of gene expression controls the amount of enzyme that is produced. Allosteric enzymes regulate the activity of an enzyme once it has been produced. They have special sites to which effectors (small molecules) bind and increase or decrease enzyme activity.

3. Most biosynthetic pathways are regulated through end-product inhibition. When sufficient end product has been produced, an effector binds to the enzyme that catalyzes the first reaction in the pathway, thereby inhibiting the flow of materials through the entire pathway.

4. Allosteric activation refers to a rise in intracellular concentration of an effector that activates an allosteric protein. The glycogen synthesis pathway is regulated by allosteric activation.

Review Questions

ESCHERICHIA COLI: A METABOLIC PARADIGM

1. Define metabolism.

2. Why is E. coli such a well-understood organism?

3. What is E. coli's dual habitat and how does its metabolism reflect it?

4. How is cellular metabolism like a factory that makes more factories?

5. What are the raw materials of metabolism? What is driving force?

6. In what form does a cell store energy, and how is the energy released?

7. What is the source of a cell's reducing power? In what forms does a cell store reducing power?

8. What role does DNA play in metabolism?

METABOLISM: AN OVERVIEW

1. What are the five sequential steps in the metabolic assembly line?

2. Define these terms: catabolic reactions, precursor metabolites, polymerization.

3. At which stage of the metabolic process are ATP and reducing power largely formed?

4. At which stage of the metabolic process is most of the ATP consumed? At which stage is most of the cell's stored reducing power required?

AEROBIC METABOLISM

1. Define aerobic metabolism.

2. Explain the entry mechanisms by which raw materials enter a cell at the beginning of aerobic metabolism. What role do permeases and ATP play?

3. Define substrate, metabolic intermediates, and precursor metabolites. How do they fit together in catabolic pathways?

4. Name the three primary catabolic pathways, and tell how many precursor metabolites each produces.

5. What is an oxidation-reduction reaction? How does metabolic oxidation differ from most nonbiological oxidations?

6. How are the two types of metabolic oxidations—dehydrogenation and hydrogenation reactions—different? Where are the cell's reserves of hydrogen, or reducing power, stored?

7. What is the source of ATP's capacity to store energy?

8. Why is phosphorylation an important part of metabolism?

9. How is ATP formed?

10. Explain substrate level phosphorylation and tell why it is important.

11. What is chemiosmosis? Where does the energy for chemiosmosis come from?

12. Explain aerobic respiration and the role of the electron transport chain. What is the terminal electron acceptor in aerobic respiration?

13. Which parts of chemiosmosis are interconvertible? Why is interconvertibility important in metabolism?

14. How does glycolysis proceed? What is the total yield of ATP?

15. How does the TCA cycle proceed and what are its products?

16. Why are the two precursor metabolites made by the pentose phosphate pathway so important? How does the pentose phosphate pathway proceed?

17. Which of the five steps of metabolism are catabolic? Which are anabolic?

18. Describe the two types of biosynthetic pathways. What roles do enzymes and coenzymes play in biosynthesis? What are vitamins?

19. What are the major polymerization reactions in a cell?

20. What role does DNA play in polymerization?

21. How does glycogen synthesis illustrate the two basic principles of polymerization reactions?

22. What are the two ways assembly occurs?

ANAEROBIC METABOLISM

1. How do the catabolic reactions of aerobic and anaerobic metabolism differ?

2. What are the two ways nonphotosynthetic cells can make ATP in the absence of oxygen?

3. How are aerobic and anaerobic respiration the same? How are they different?

4. How is ATP generated in fermentation? Why are sugars virtually the only substrate for fermentation?

5. Compare the efficiency of generating ATP through aerobic respiration, anaerobic respiration, and fermentation.

6. Explain how the principles of fermentation are illustrated by lactic acid fermentation.

NUTRITIONAL CLASSES OF MICROORGANISMS

1. Upon what two factors does nutritional class depend? Define the following: chemoautotroph, chemoheterotroph, photoautotroph, and photoheterotroph.

2. Why are autotrophs so important in the chain of life? Describe the Calvin-Benson cycle.

3. Photoautotrophs obtain ATP and reducing power for metabolism from two light-driven processes—cyclic photophosphorylation and noncyclic photophosphorylation. How does each proceed? Which produces ATP exclusively?

4. How do chemoautotrophs form ATP and reducing power? Under what circumstances is reverse electron transport necessary?

REGULATION OF METABOLISM

1. How does metabolic regulation increase a cell's efficiency?

2. What are the two major types of metabolic regulation? Explain how allosteric enzymes regulate metabolism.

3. What is end-product inhibition? Explain allosteric activation.

Suggested Readings

Dawes, E. A. 1986. *Microbial energetics*. New York: Chapman and Hall.

Gottschalk, G. 1979. *Bacterial metabolism*. New York: Springer-Verlag.

Harold, F. M. 1986. *The vital force: A study of bioenergetics*. New York: W. H. Freeman.

Ingraham, J. L.; Maaløe, O.; and Neidhardt, F. C. 1983. *Growth of the bacterial cell*. Sunderland, Mass.: Sinauer.

Neidhardt, F. C. 1987. Escherichia coli *and* Salmonella typhimurium. Washington, D.C.: American Society for Microbiology.

6 THE GENETICS OF MICROORGANISMS

Serendipity and Science

In a Persian fairy tale, the three princes of Serendip are always discovering wonderful things by accident. Their adventures inspired the English word, *serendipity*. Serendipity plays an important role in science. Setting out to answer a relatively mundane question, scientists often obtain results that lead to completely unexpected discoveries, sometimes groundbreaking ones. Of course, the scientist must be astute enough to recognize the potential of unexpected experimental results. As Louis Pasteur said, "Chance favors the prepared mind."

In 1928, Frederick Griffith, a British microbiologist, set out to investigate the connection between *Streptococcus pneumoniae*'s ability to produce capsules and its ability to produce disease. But his results led to the discovery that genetic exchange occurs among bacteria and, eventually, to the discovery that bacterial genetic material is composed of DNA. Griffith's story is a prime example of the prepared mind and serendipity in science.

Griffith worked with two strains of pneumococci, as shown in the figure. One, called *smooth* (S) because it forms glistening colonies, produces lethal pneumonia when injected into mice. The other, called *rough* (R), forms nonglistening colonies and is harmless to mice. Smooth cells are surrounded by a thick capsule, while rough cells have no capsule. By mutation, smooth cultures constantly produce a few rough cells. When smooth cells lose their capsule, they lose their ability to kill mice. Griffith's question was: How is the presence of a capsule related to lethality?

Griffith found that neither rough cells nor heat-killed smooth cells killed mice when injected separately (experiments 1 and 3), but they did when injected together (experiment 4). Could it be that capsular material from any source made pneumococci lethal? Or perhaps heat treatment did not kill all the smooth cells? More experimentation was necessary. So Griffith isolated pneumococci from the dead mice that had received rough and heat-killed smooth cells. He found that the pneumococci were smooth, with the same kind of capsule as the heat-killed cells, but they had other characteristics of the rough strain as well. In other words, they had some properties of each strain he had injected: Genetic exchange had taken place. This was the serendipitous moment, and Griffith grasped it. He astutely reasoned that some substance released from the heat-killed smooth cells had permanently changed the live rough cells. The change became known as *transformation* and the substance that mediated the change was called the *transforming principle*.

The chemical nature of the transforming principle was finally identified in 1944 by three American microbiologists, Oswald T. Avery, Colin MacLeod, and Maclyn McCarthy. They purified the material capable of causing genetic change and showed it to be exclusively DNA. They established for the first time the chemical nature of genetic material.

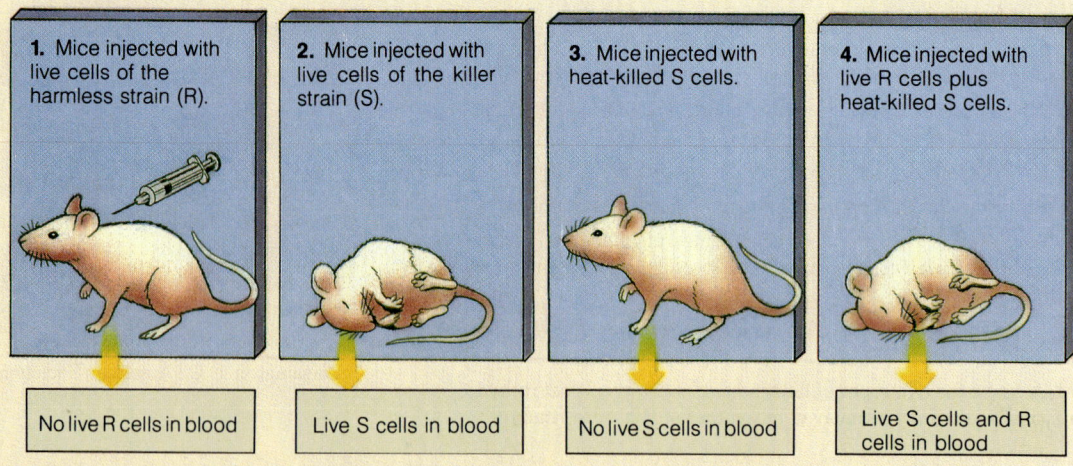

1. Mice injected with live cells of the harmless strain (R).

No live R cells in blood

2. Mice injected with live cells of the killer strain (S).

Live S cells in blood

3. Mice injected with heat-killed S cells.

No live S cells in blood

4. Mice injected with live R cells plus heat-killed S cells.

Live S cells and R cells in blood

The results of Griffith's experiments leading to the discovery of genetic exchange in bacteria.

To understand:

- The structure of DNA and how it is replicated

- The two steps in gene expression: transcription and translation

- The ways that gene expression is regulated at the levels of transcription and translation

- The microbial genome and the difference between genotype and phenotype

- How the microbial genome changes by mutation

- How genetic information is transferred between procaryotes through transformation, conjugation, and transduction

- How genetic information is exchanged between eucaryotic microorganisms

- How genetic change spreads through a population of bacteria

STRUCTURE AND FUNCTION OF GENETIC MATERIAL

Genetics, the study of heredity and variation among organisms, is new among the sciences. Its principles were set down in 1865 by Gregor Mendel, an Austrian monk, but it was not until the last half of this century that these principles were applied to microorganisms. Genetic studies of microorganisms led to rapid and dramatic advances that fundamentally changed the whole field of biology.

A cell's DNA carries the plan to make a new cell (Chapter 5). But in what form is this information stored? How is genetic information transferred between cells? And how is it transmitted from one generation to the next? How does the genetic plan direct and coordinate metabolism? Does genetic information ever change? The principles of these processes are the same in all procaryotes and eucaryotes, but the details differ.

The Structure of DNA

DNA is a macromolecule composed of deoxyribonucleotide building blocks (Chapter 2). Deoxyribonucleotides have three components: deoxyribose, phosphate, and a nucleoside base. The base may be either adenine (A), guanine (G), cytosine (C), or thymine (T). The deoxyribose and

phosphate portions of deoxyribonucleotides are linked together in DNA to form long strands. Pairs of these strands are wrapped around a central axis forming a **double helix**, which is held together by hydrogen bonds between pairs of bases (**Figure 6.1**).

The hydrogen bonds that connect single strands of DNA to form a double helix link G's to C's and A's to T's, forming G-C and A-T **base pairs**. A G in one strand is always opposite a C in the other, and the same is true for A and T. As a result, the numbers of G's and C's in any molecule of DNA, regardless of organism, are always equal, as are the numbers of A's and T's. In contrast, the proportion of G-C to A-T base pairs can vary greatly among organisms. In some organisms the G-C pairs are only 25 percent of the total, while in others they are almost 80 percent (**Table 6.1**).

The Function of DNA

DNA has one function: to store all the information required for the cell to grow, reproduce, and maintain itself. Segments of DNA called **genes** contain the information needed to manufacture a specific macromolecule. Usually the final product is a protein, but sometimes it is a molecule of RNA.

DNA enters into two kinds of reactions: **replication** and **gene expression**. Replication reproduces genetic information from generation to generation. Gene expression directs metabolism. In replication, the chromosome and plasmids are reproduced so that accurate copies of them can be passed on to daughter cells. In gene expression, the information encoded in DNA directs the synthesis of RNA and protein. There are two steps in gene expression, **transcription** and **translation**. In transcription, the information in DNA is transferred to RNA. In translation, information carried by RNA directs the synthesis of protein (**Figure 6.2**).

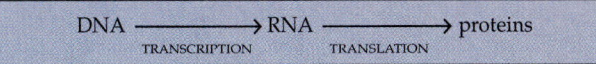

DNA ——→ RNA ——→ proteins
TRANSCRIPTION   TRANSLATION

The information contained in DNA is the master plan of metabolism (Chapter 5). Genetic information is passed from DNA through RNA to direct the synthesis of proteins. *Thus, DNA directs all of metabolism.*

Replication of DNA

Replication is the polymerization of a new macromolecule of DNA from deoxyribonucleotide building blocks. The challenges of polymerization are ordering building blocks properly and supplying the chemical energy necessary to bond them to one another. In this section, we examine how DNA replication meets these challenges.

Figure 6.1 Structure of DNA. The basic building blocks of DNA are nucleotides, which are composed of a sugar (deoxyribose), a phosphate, and a nucleoside base. The nucleotides form two helical strands held together by hydrogen bonds (the dashed lines) between bases. There are two bonds between adenine (A) and thymine (T) and three between guanine (G) and cytosine (C). The two strands are antiparallel because phosphate groups link deoxyribose units between the 3 prime (3′) of one strand and the 5 prime (5′) of the next. One strand runs in the 5′ to 3′ direction; the other runs in the 3′ to 5′ direction.

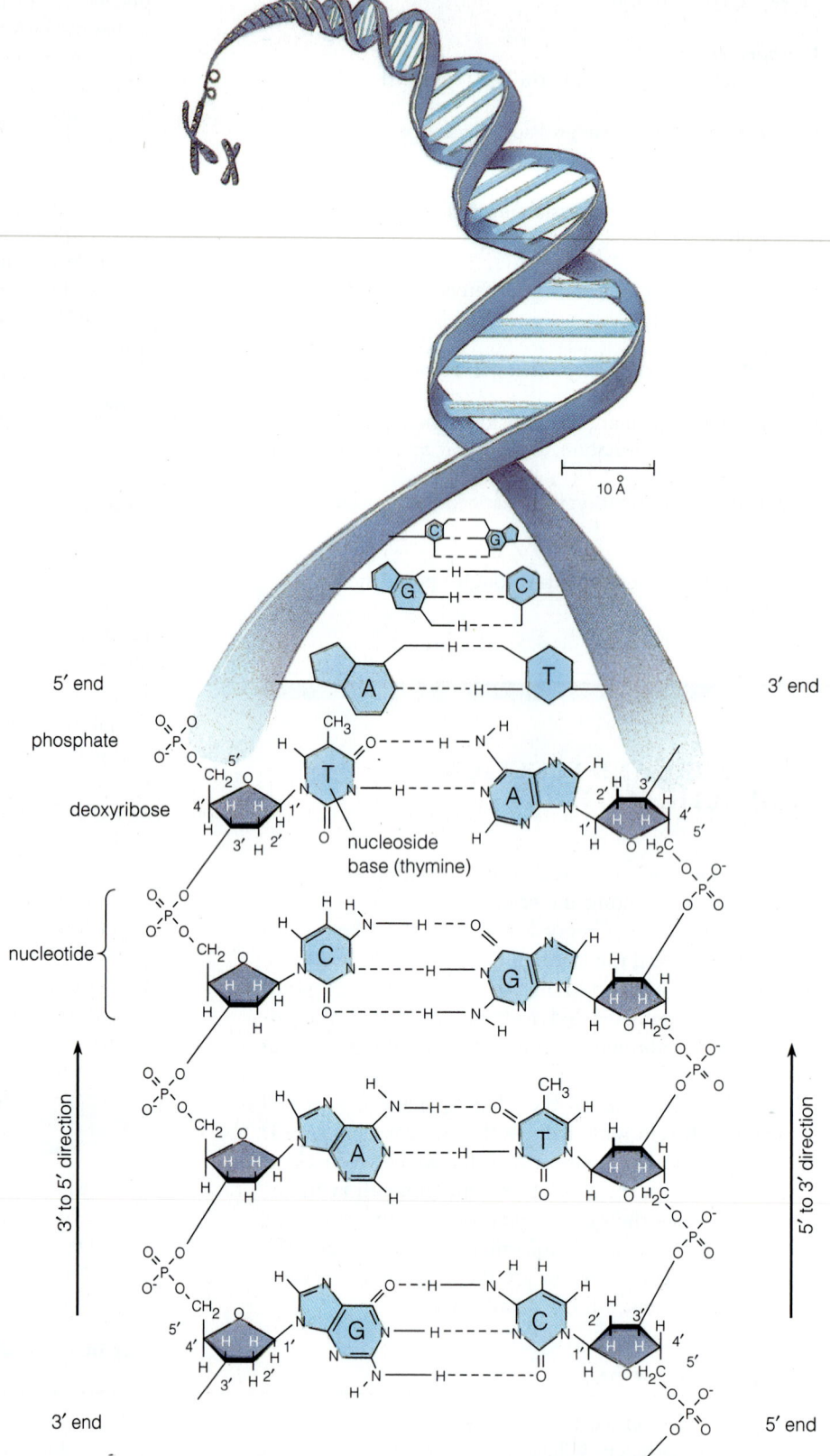

Table 6.1	Range of Percent of G-C Pairs in the DNA of Organisms
Organisms	Range of Percent of G-C Pairs in DNA
Procaryotes	
Archaebacteria	27 to 61
Eubacteria	25 to 78
Eucaryotes	
Fungi	22 to 60
Algae	37 to 68
Protozoa	23 to 67
Plants	33 to 48
Animals	
Invertebrates	32 to 50
Vertebrates	36 to 43

Replication: The Synthesis of DNA. The molecular structure of DNA suggests how it can be copied accurately. The hydrogen bonds that form the G-C and A-T pairs are the key to ordering deoxyribonucleotides during replication (**Figure 6.3**). Replication begins by breaking the A-T and G-C bonds within a short stretch of DNA. This allows the two strands to separate, forming a bubble. Exposed bases within the bubble are then free to pair with new bases in nucleotides that will be polymerized to form a new molecule of DNA. Because hydrogen bonds form only between A-T and G-C pairs, an exposed A from the old molecule will bind only to a new T, and so on. This bonding specificity between base pairs orders the new sequence of deoxyribonucleotides.

However, polymerization cannot proceed spontaneously because the bonds in DNA are at a higher energy state than are the bonds in deoxyribonucleotides. Like most polymerization reactions, DNA replication requires

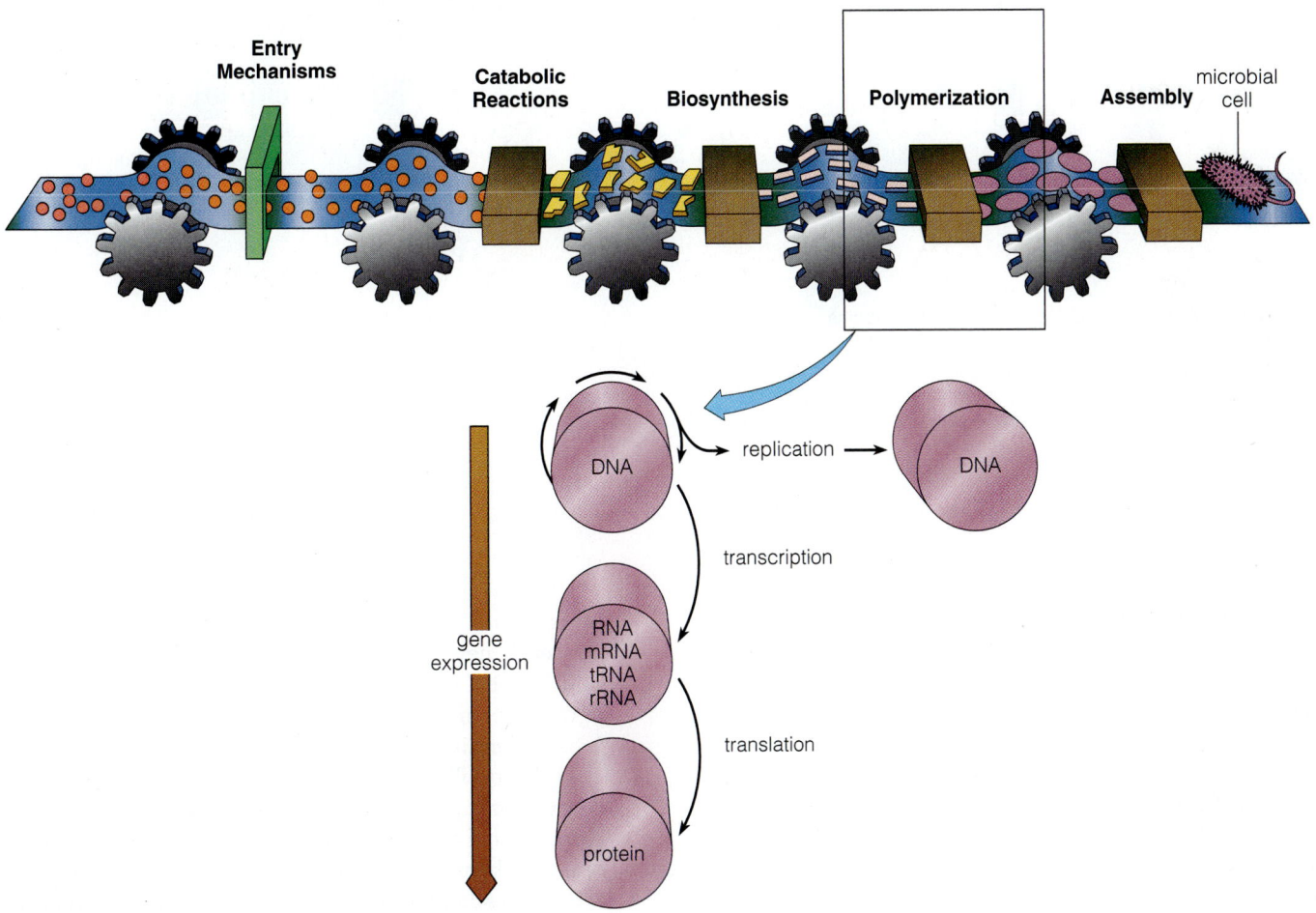

Figure 6.2 DNA carries the master plan of metabolism. DNA enters into two kinds of reactions, replication and gene expression. Replication duplicates the DNA for distribution to daughter cells. Gene expression, through transcription and translation, directs synthesis of proteins, including enzymes, which mediate cell metabolism.

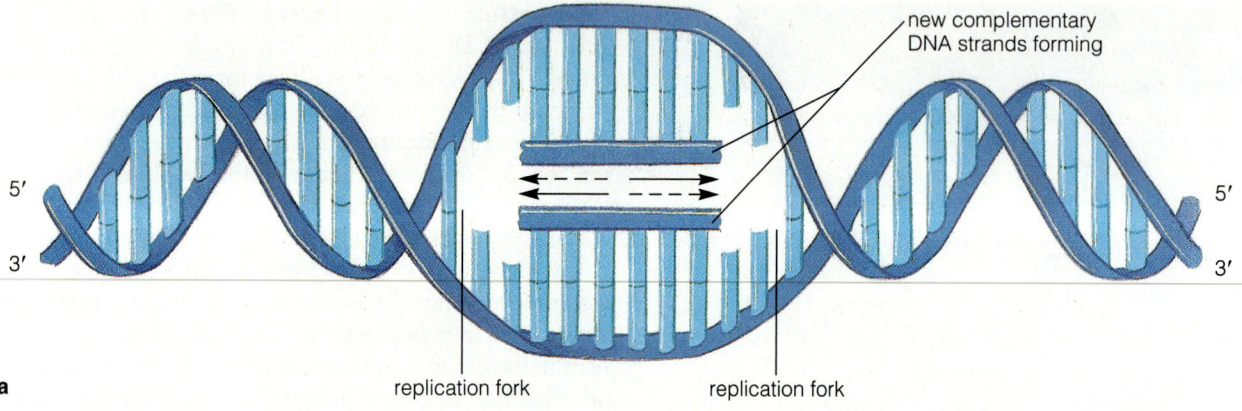

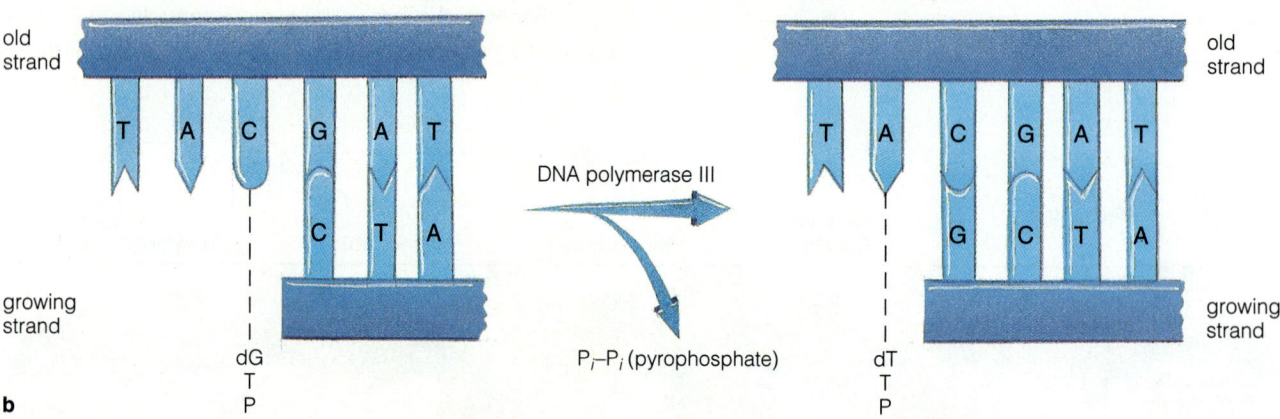

Figure 6.3 Replication. (**a**) Replication of DNA begins when hydrogen bonds in a short stretch break to form a bubble and two replication forks. (**b**) New strands grow within the bubble as nucleoside triphosphates (here, dGTP) pair with exposed bases (cytosine, C, here) on the old strand. Bases are added to the growing strand when DNA polymerase III releases pyrophosphate. The next nucleoside triphosphate (here, dTTP) pairs with the next exposed base (adenosine, A, here).

ATP (Chapter 5). Deoxyribonucleotides react with ATP to form **nucleoside triphosphates**, chemically activated forms of the nucleotide building blocks. These highly reactive nucleoside triphosphates are the form of nucleotides that pairs with exposed bases during replication. Then the enzyme **DNA polymerase III** joins the deoxyribonucleotides together to form two new strands of DNA. During polymerization, one molecule of pyrophosphate, which consists of two linked phosphates, is released for each nucleotide added.

The Mechanics of Replication. Each newly polymerized strand of DNA has an A, T, G, or C opposite to the **template** (pattern) of the old strand's T, A, C, or G.

Thus, the new strand is **complementary** to the old strand, and each new double helix is identical to the original double helix. Because each new double helix is composed of one new and one *conserved*, or old, strand, the process is called **semiconservative replication**.

The bubble that initiates replication of the bacterial chromosome forms at a genetically specified point on the chromosome called the **origin**. The two points within the bubble where the original double strand of DNA separates are called the **replication forks**. It is at the replication forks that polymerization of nucleotides into a new strand of DNA occurs. From the origin, the two replication forks travel simultaneously in *opposite* directions around the chromosome. When the two forks meet

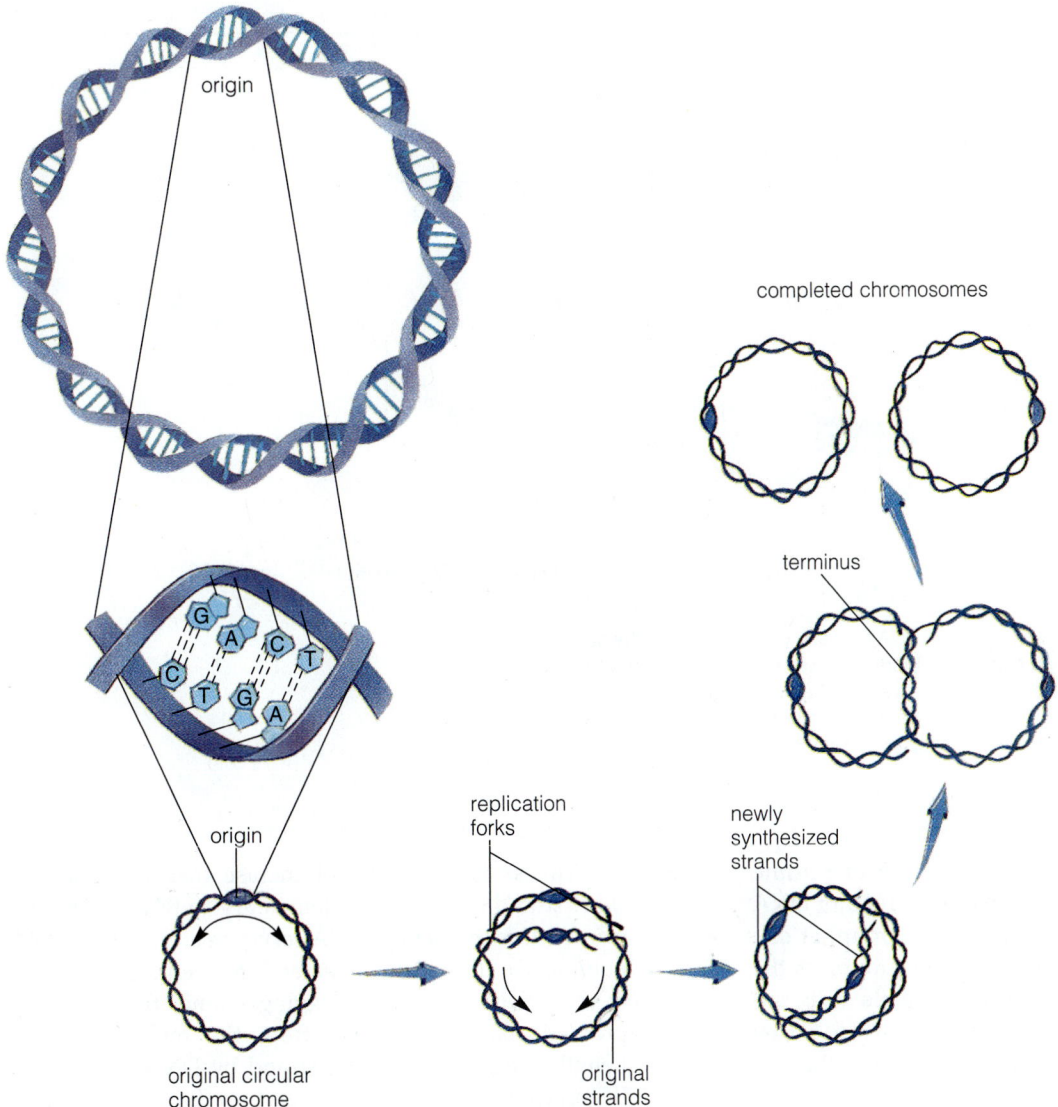

Figure 6.4 Replication of the bacterial chromosome. Replication begins when a bubble forms at the origin. This creates two replication forks at which polymerization occurs. The forks travel in opposite directions until they meet at the terminus. Then the two chromosomes separate.

halfway around the chromosome, at a point called the **terminus**, the two completed chromosomes separate, ready to be distributed to daughter cells (**Figure 6.4**).

The replication fork is biochemically complex (**Figure 6.5**). Several enzymes in addition to DNA polymerase III combine to form a loose complex called the **replication apparatus**. The replication apparatus performs two functions. It moves the replication fork and it synthesizes the two new strands of DNA. One of the enzymes, **DNA helicase**, moves the fork by unwinding and separating the strands of the old double helix. Another protein, **single-strand binding protein (SSBP)**, binds to the exposed single strands of DNA, keeping them separate during replication.

Three other enzymes in the replication apparatus are needed because of complications arising from the structure of DNA and the function of DNA polymerase III. The first complication is that the two strands of a DNA double helix are oppositely oriented, or **antiparallel**. The second complication is that DNA polymerase III can replicate a strand of DNA continuously in only one direction. Let us examine each of these complications separately.

The two strands of any double helix are antiparallel because of the way that the deoxyribose units are chemically joined and the two strands are aligned. Look again at Figure 6.1. Note that phosphate groups link deoxyribose units between the third, called the 3' or **3 prime**

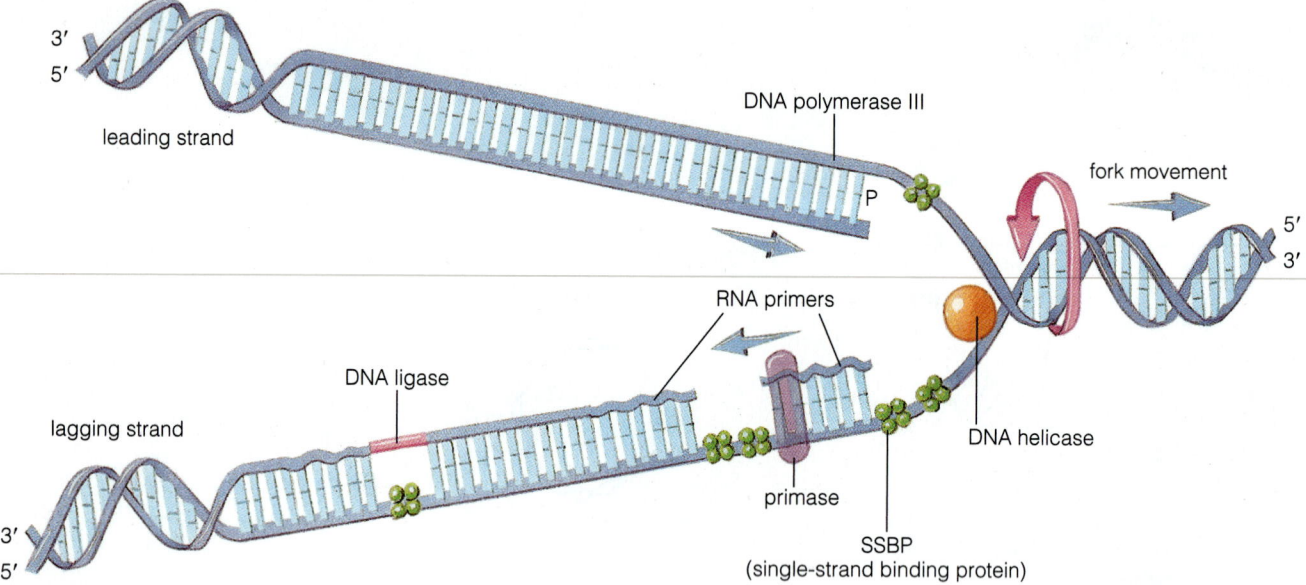

Figure 6.5 The replication fork is composed of DNA helicase (which unwinds the strand in front of the fork), single-strand binding protein (which binds to single strands and keeps them separate), DNA polymerase III (which adds nucleotides to the new strand), and primase (which adds short RNA primers to the lagging strand). The leading strand is replicated in the same direction the fork moves. The lagging strand is replicated in the opposite direction.

carbon atom, of one and the fifth, called the 5′ or **5 prime carbon atom**, of the next. Therefore, each strand of DNA has a 5′ carbon (with attached phosphate group) at one end and a 3′ carbon (with attached OH group) at the other. The two single strands in any double helix are aligned in opposite directions—one runs from 3′ to 5′ (the 3′→ 5′ direction) and the other runs from 5′ to 3′ (the 5′→ 3′ direction). As a result, the double helix has one 3′ and one 5′ carbon atom at each end.

Now let's look at the second complication, that DNA polymerase III can synthesize a continuous strand of DNA only in one direction. DNA polymerase III links new nucleotides to the 3′ OH end of an existing nucleic acid strand, lengthening it in the 5′→ 3′ direction. This preexisting strand, which is necessary for the function of DNA polymerase III, is called a **primer**. Because the DNA strands exposed at the replication fork are antiparallel, only one of the two DNA strands presents a 3′ primer that can be extended by DNA polymerase III. This strand, called the **leading strand**, is synthesized continuously in the direction that the replication fork is moving. Synthesis of the other strand, called the **lagging strand**, is more complicated because it ends in a 5′ carbon, with *no* 3′ primer available to be lengthened by DNA polymerase III. Together, three additional enzymes in the replication apparatus extend the lagging strand: **primase, DNA polymerase I**, and **DNA ligase**.

Primase is an RNA polymerase that synthesizes a short strand of RNA complementary to DNA without a primer. This segment of RNA serves as a primer for DNA polymerase III, which can extend its 3′ end. But because of the orientation of the lagging strand, this synthesis proceeds backward, away from the replication fork. Lengthening of the lagging strand by DNA polymerase III can continue only until it runs into the 5′ end of a previous RNA primer. When this collision occurs, DNA polymerase I takes over. It destroys the RNA primer and replaces it with a DNA strand. The third additional enzyme, DNA ligase, seals the gap between the newly synthesized fragment of DNA and the continuous strand in front of it.

Gene Expression: Transcription

Transcription is the first step in gene expression. In transcription, the information contained in DNA is copied into three types of RNA—**messenger RNA (mRNA)**, **transfer RNA (tRNA)**, and **ribosomal RNA (rRNA)**. When the three types of RNA have been made by transcription, the cell is biochemically equipped to make proteins.

Transcription is the polymerization of ribonucleotide building blocks into a molecule of RNA. The two challenges of polymerization—the ordering and activation of

Microbe Mappers

Jannine Zeig

In the late 1970s, Jannine Zeig, a graduate student at the University of California, San Diego, was doing research on the bacterium *Salmonella typhimurium*. Specifically, she was trying to explain **phase variation**, periodic changes in the type of flagella made by *S. typhimurium* and many other bacterial pathogens. Phase variation protects *S. typhimurium* from the host's immune system because by the time it is able to detect the kind of flagella that the invading *S. typhimurium* cells produce, some of them start to produce a second kind that the immune system cannot detect. Later some of these bacteria again start to produce the first kind of fla-

gella. Zeig knew that the change was not a mutation because it occurred too frequently. Mutations in any particular gene occur in about one out of a hundred million cells. Phase variation occurs in about one out of a thousand. After extensive experimentation, Zeig found that an enzyme caused a small piece of DNA (containing a promoter) in the *S. typhimurium* chromosome to invert. In one position, the DNA transcribes genes encoding one kind of flagella. In the other position, the DNA causes the other type to be encoded. Zeig discovered a new basis for phase variation. It has since been found that gene switching occurs in many bacteria.

---

building blocks—are met during transcription in much the same way as they are during replication. Recall from Chapter 2 that DNA and RNA are structurally similar. The differences lie in their base composition (RNA has uracil instead of the thymine found in DNA) and sugar component (deoxyribose in DNA and ribose in RNA). Ribonucleotides react with ATP to become activated **ribonucleoside triphosphates** in much the same way as deoxyribonucleotides react with ATP to become activated deoxyribonucleoside triphosphates. Like replication, transcription begins with the formation of a bubble in the DNA double helix. But unlike replication, in transcription, ribonucleoside triphosphates rather than deoxyribonucleoside triphosphates pair with exposed bases on one of the exposed single strands of DNA (**Figure 6.6**). Thus, the order of the building blocks polymerized into RNA is determined by the sequence of bases in DNA. This means the information content of the DNA must be *transcribed*, or rewritten, in the form of RNA, because RNA lacks thymine but contains uracil (U), which is structurally similar. During transcription, then, C pairs with G and A pairs with U.

In spite of these similarities, transcription differs significantly from replication. First, in transcription the point of polymerization travels only in a single direction—only one strand of the double helix is transcribed. Second, the product of transcription is a single-stranded rather than a double-stranded molecule. (Sometimes, however, the RNA product may fold back on itself to form short double-stranded regions where sequences of bases happen to be complementary to one another.) Third, the RNA produced by transcription is much shorter than the bacterial chromosome produced by

replication. Transcription copies only one or a few genes at a time. Finally, a single enzyme, **RNA polymerase** (a different enzyme from the primase that participates in DNA replication), is solely responsible for transcription in procaryotes, as opposed to the numerous enzymes needed for replication.

Although the individual products of transcription are relatively short—corresponding to only one or a few genes—at some time during the cell's life, almost its entire bacterial **genome** (chromosome plus plasmids) is transcribed. Transcription begins at many different sites, called **promoters**, on the genome and continues until it reaches a **terminator**. The sequence of bases in a promoter signals RNA polymerase to bind there. Then RNA polymerase separates the two strands, forming a bubble in the DNA, and uses only one of the separated strands, the **sense strand**, as a template. Polymerization continues in the $5' \rightarrow 3'$ direction until the sequence of bases at the terminator signals the transcription to stop. Finally, RNA polymerase and the **transcript**, the RNA product of transcription, are released, and the DNA bubble closes.

Gene Expression: Translation

Translation, the second step of gene expression, is the polymerization of amino acid building blocks into a protein macromolecule. Translation changes the information in the metabolic master plan from the language of nucleic acids into the language of protein.

Translation: The Synthesis of Protein. The products of transcription—mRNA, rRNA, and tRNA—all participate in translation: mRNA carries the information that

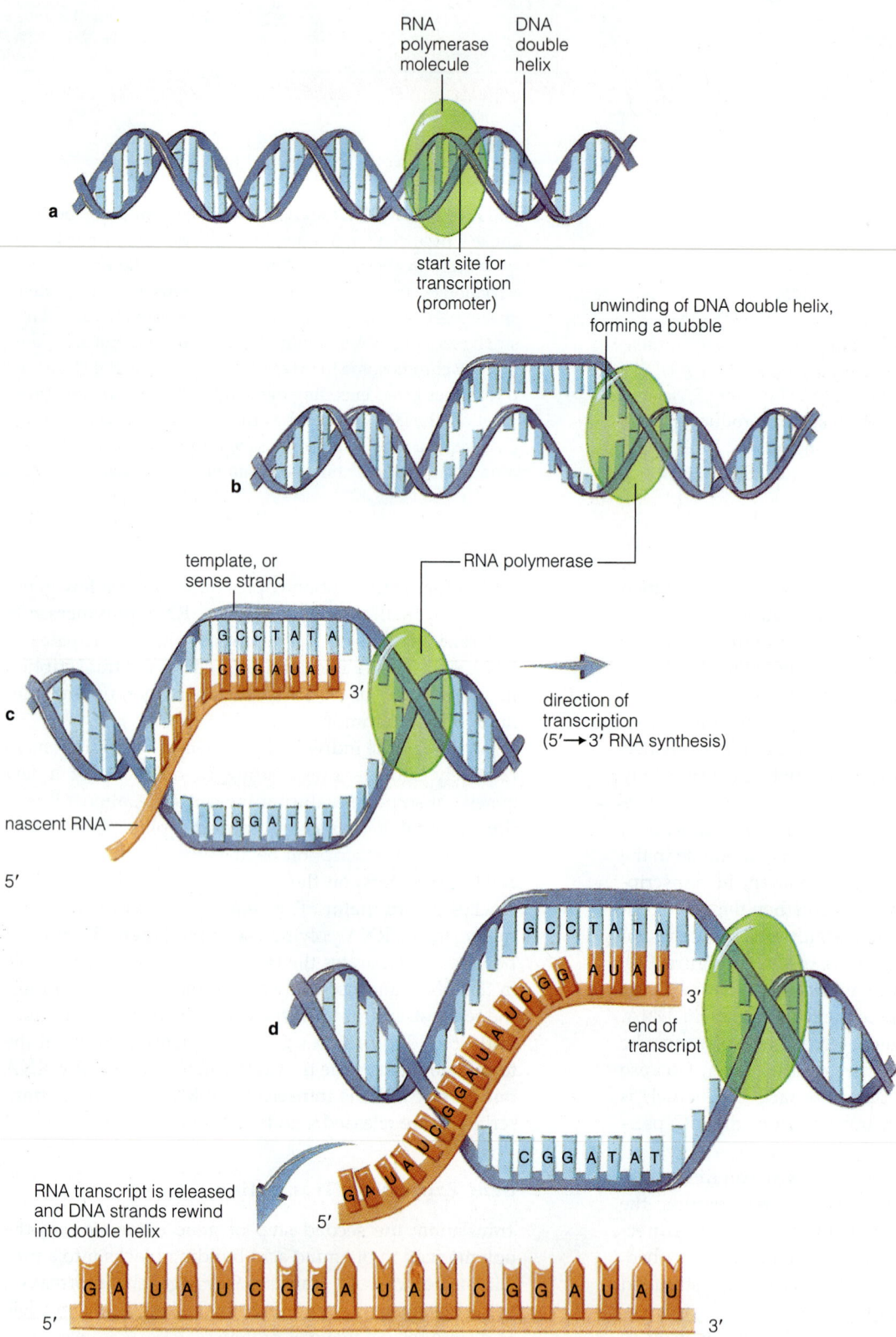

Figure 6.6 Transcription. (**a**) Transcription begins when a molecule of RNA polymerase binds to the DNA double helix at the start site. (**b**) The unwinding of the DNA helix forms a bubble. (**c**) Ribonucleoside triphosphates pair with the exposed bases and are polymerized into a nascent RNA molecule as the bubble moves in the direction of transcription. (**d**) When it reaches a terminator, the RNA transcript is released.

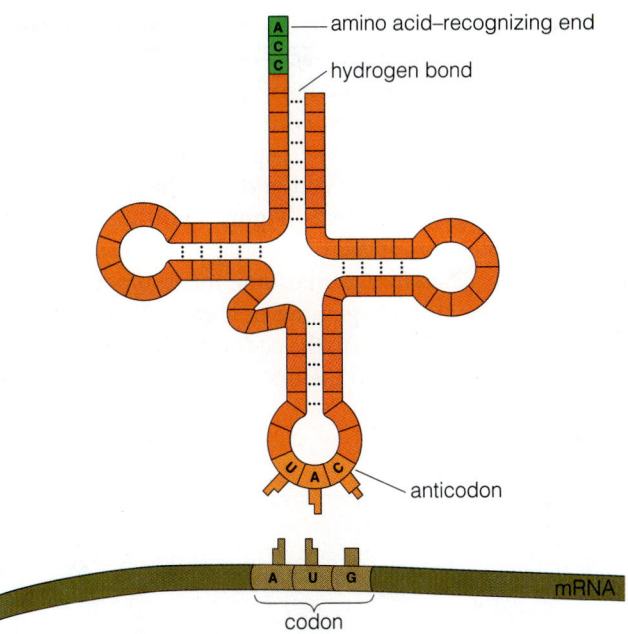

amino acid–recognizing end

hydrogen bond

anticodon

mRNA

codon

Figure 6.7 Transfer RNA has double-stranded regions that give it a cloverleaf-like shape. The amino-acid-recognizing end of the tRNA molecule becomes linked to an amino acid. The other end is the anticodon, consisting of three bases that pair with three complementary bases, the codon, on the messenger mRNA molecule.

First Base	Second Base				Third Base
	U	C	A	G	
U	phenylalanine	serine	tyrosine	cysteine	U
	phenylalanine	serine	tyrosine	cysteine	C
	leucine	serine	stop	stop	A
	leucine	serine	stop	tryptophan	G
C	leucine	proline	histidine	arginine	U
	leucine	proline	histidine	arginine	C
	leucine	proline	glutamine	arginine	A
	leucine	proline	glutamine	arginine	G
A	isoleucine	threonine	asparagine	serine	U
	isoleucine	threonine	asparagine	serine	C
	isoleucine	threonine	lysine	arginine	A
	(start) methionine	threonine	lysine	arginine	G
G	valine	alanine	aspartate	glycine	U
	valine	alanine	aspartate	glycine	C
	valine	alanine	glutamate	glycine	A
	valine	alanine	glutamate	glycine	G

Figure 6.8 The genetic code tells which amino acids are encoded by codons on mRNA molecules. For example, tryptophan is encoded by UGG. The code is redundant. For example, six codons encode leucine. Three codons are stop, or nonsense, codons. The start codon is usually an AUG methionine.

determines the order of amino acids in the protein, tRNA is the adapter that connects the information-containing bases in mRNA to the amino acids that build the protein, and rRNA (along with structural proteins) is a component of ribosomes, where translation occurs.

Transfer RNA (tRNA) plays a critical role in bridging the gap between mRNA, which carries the information of the metabolic master plan, and proteins, which determine most of the chemical reactions of metabolism. Transfer RNA fulfills its translating function because it communicates in two languages—the language of nucleic acids and the language of proteins. That is, each tRNA molecule has one site that binds to an amino acid and another site that binds to mRNA. In addition, the attachment of an amino acid to a molecule of tRNA creates an activated subunit that is energetically capable of being joined to a growing protein by a peptide bond.

This is how tRNA functions in translation: The amino-acid-recognizing end of a tRNA molecule is linked to an amino acid (**Figure 6.7**). This linkage, also called **activation**, occurs at the expense of a molecule of ATP and thus creates an energetically activated complex. The nucleic-acid-recognizing region of a tRNA molecule consists of an **anticodon** (three adjacent bases on tRNA) that pairs with a complementary **codon** (composed of three adjacent bases on mRNA). Codon-anticodon pairing properly

positions the tRNA-bound amino acid for polymerization onto a newly synthesized protein. The correspondence that tRNA imposes between mRNA codons and amino acids is quite specific. In other words, a particular kind of tRNA molecule can pair only with a certain mRNA codon and can be attached only to a particular amino acid. As a result, the sequence of bases in the mRNA molecule determines the unique sequence of amino acids in a particular protein.

The correspondence that tRNA determines between codons in mRNA and amino acids is called the **genetic code** (**Figure 6.8**). There are four different nucleic acid bases and 64 (4^3) possible three-base codons, but only 20 amino acids. As a result, the genetic code is redundant—different codons correspond to tRNA molecules that are bound to the same amino acid. In other words, most amino acids are encoded by more than a single codon. Three codons, called **nonsense codons**, do not correspond to the anticodon of any tRNA molecules and therefore do not encode any amino acid. They stop translation at the end of a gene. The other 61 codons, however, correspond to the anticodon on a molecule of tRNA and therefore encode a particular amino acid.

The Mechanics of Translation. Translation occurs on a ribosome, a biochemically complex organelle composed of protein and rRNA (Chapter 4). The ribosome moves down a molecule of mRNA, exposing successive codons in a region called the **A site** (the amino acid binding site) where an amino-acid-bearing tRNA molecule pairs with the exposed codon. This positions the amino acid to be polymerized onto the end of a growing protein at the ribosome's **P site** (the peptide binding site). Then the ribosome moves one codon down the mRNA molecule, and the process repeats (**Figure 6.9**).

The protein produced by translation corresponds to a single gene. Therefore, a ribosome must initiate translation precisely at the beginning of a gene. Molecular signals direct the ribosome to the codon on the mRNA molecule that corresponds to the first amino acid in a protein.

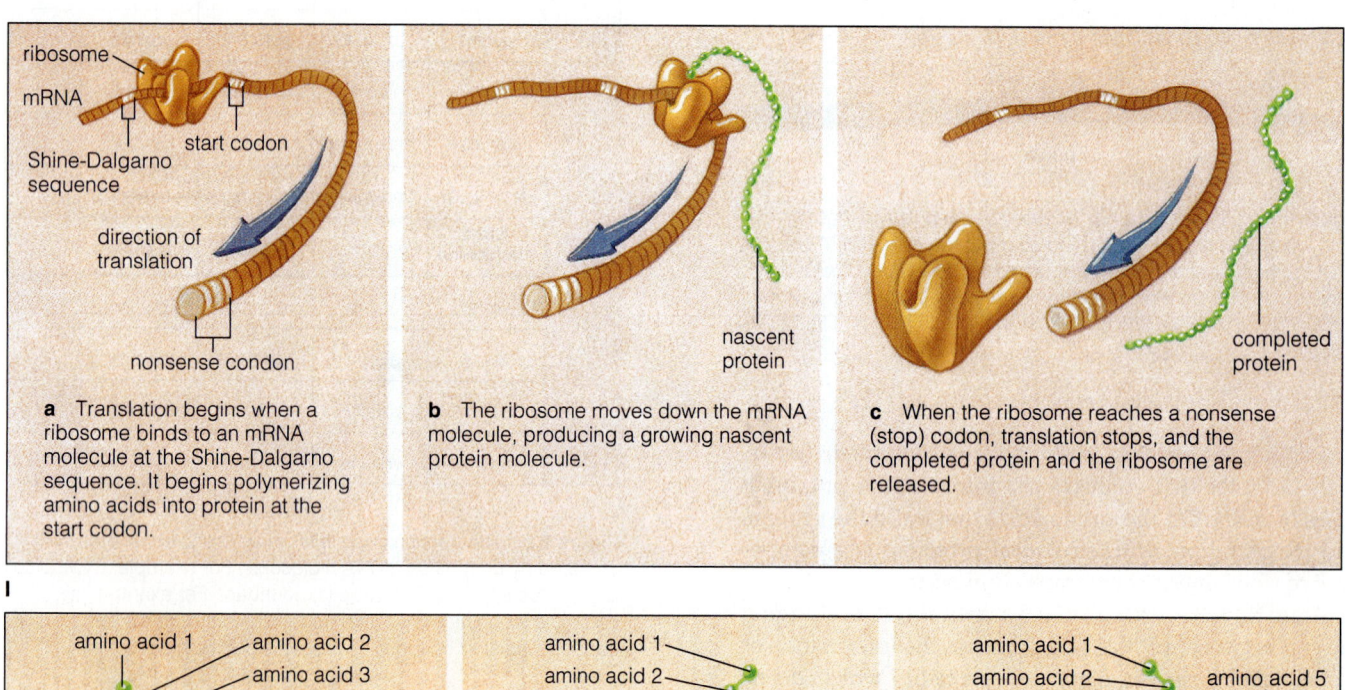

a Translation begins when a ribosome binds to an mRNA molecule at the Shine-Dalgarno sequence. It begins polymerizing amino acids into protein at the start codon.

b The ribosome moves down the mRNA molecule, producing a growing nascent protein molecule.

c When the ribosome reaches a nonsense (stop) codon, translation stops, and the completed protein and the ribosome are released.

I

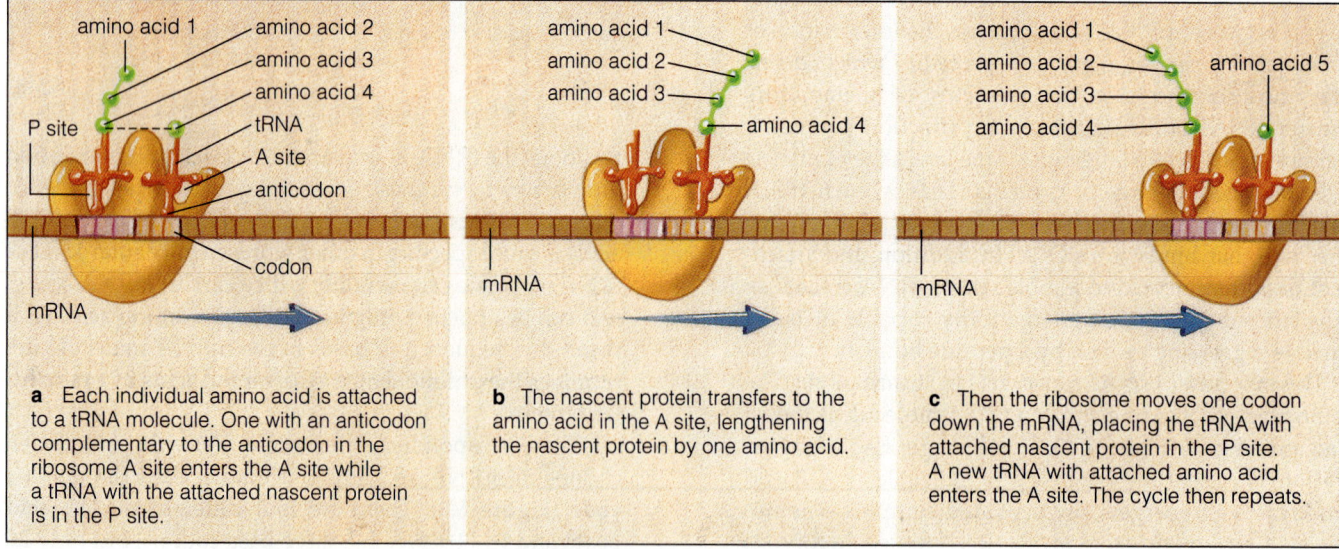

a Each individual amino acid is attached to a tRNA molecule. One with an anticodon complementary to the anticodon in the ribosome A site enters the A site while a tRNA with the attached nascent protein is in the P site.

b The nascent protein transfers to the amino acid in the A site, lengthening the nascent protein by one amino acid.

c Then the ribosome moves one codon down the mRNA, placing the tRNA with attached nascent protein in the P site. A new tRNA with attached amino acid enters the A site. The cycle then repeats.

II

Figure 6.9 Translation. (**I**) Translation of a region of mRNA corresponding to a single gene into a protein product. (**II**) The steps by which the growing nascent protein molecule is lengthened by a single amino acid.

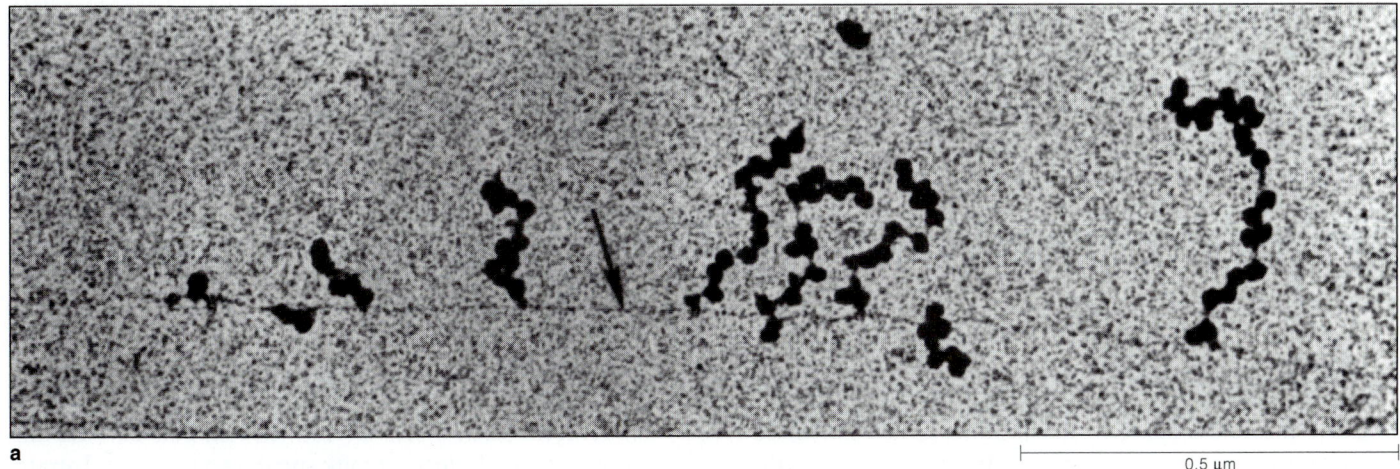

a

0.5 μm

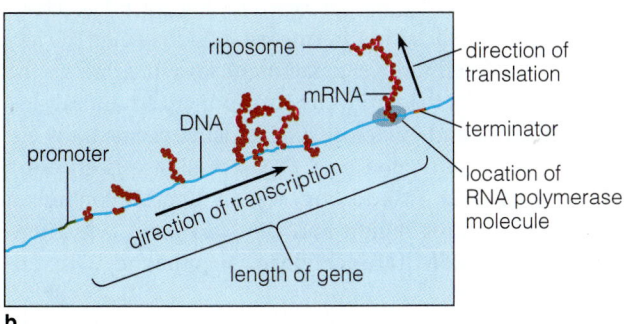

b

ribosome
direction of translation
mRNA
DNA
terminator
promoter
location of RNA polymerase molecule
direction of transcription
length of gene

Figure 6.10 Simultaneous transcription and translation. Both steps of gene expression, transcription and translation, are carried on simultaneously in procaryotes. This transmission electron micrograph shows a length of DNA being transcribed from left to right by RNA polymerase molecules (not visible in micrograph), so the RNA molecules grow larger from left to right. While being formed, mRNA molecules are being translated. Ribosomes on each mRNA molecule are making nascent proteins (not visible in micrograph). The arrow in the micrograph points to the DNA molecule at the midpoint of the gene.

One signal is that the **start**, or first, codon is usually an AUG codon for methionine, but the AUG codon appears frequently (refer again to Figure 6.8). The start codon is distinguished by its closeness to a sequence of bases called the **Shine-Dalgarno sequence**, or the **ribosome binding site**, that is complementary to a sequence of bases in the ribosome's RNA.

Thus, a protein molecule is synthesized when a ribosome binds to the start codon on a molecule of mRNA. Then the ribosome moves down the mRNA molecules one codon at a time. As a new codon is exposed, a complementary tRNA molecule with its attached amino acid pairs with it. Positioned in this way, the amino acid at the other end of the tRNA molecule is attached to the growing chain of amino acids that is building a new protein. This process continues until the ribosome reaches one or another of the three nonsense codons, to which no tRNA can bind. Then the completed protein is released from the ribosome.

Transcription and translation occur simultaneously in bacteria. **Figure 6.10** shows ribosomes at various sites on an mRNA molecule that is still in the process of being transcribed. About 20 ribosomes are simultaneously translating the average mRNA molecule, forming a structure called a **polysome**.

REGULATION OF GENE EXPRESSION

Microorganisms use substrates and synthesize macromolecules just fast enough to meet their needs. To do more would be wasteful, but to do less would starve the cell of essential components. Either extreme slows the rate of growth and makes the microorganism less able to compete with other organisms in its environment. It is, therefore, important that a cell operate efficiently. The mechanisms by which a cell maximizes its efficiency are called **metabolic regulation**.

Recall from Chapter 5 that the cellular metabolism is regulated in two ways. One way is to change the *activity* of enzymes so they act more rapidly when the need for their product increases and more slowly when it decreases. This mechanism, which involves allosteric proteins, was discussed in Chapter 5. The other way to regulate a cell's metabolism is to change the rate of *synthesis* of enzymes, that is, to produce more of them when the need for their product increases and fewer of them when the need decreases. This second type of regulation—which we discuss here—modulates gene expression.

The rate of synthesis of different enzymes is regulated according to certain patterns that increase a cell's metabolic efficiency. **Inducible enzymes** are produced only when a signal molecule is abundant in the environment. For example, many of the enzymes needed to metabolize different sugars are inducible, and the sugar itself is the signal molecule that accelerates enzyme synthesis. Enzyme induction increases metabolic efficiency because the cell produces these enzymes only when it needs them to metabolize a particular sugar. **Repressible enzymes** are produced only when a signal molecule is scarce in the environment. For example, many of the enzymes in biosynthetic pathways are repressible, and the end product of the pathway is the signal molecule that slows enzyme synthesis. This increases metabolic efficiency because the cell is spared from producing more of these enzymes when their product is already abundant. Still other enzymes are produced at a constant rate, or **constitutively**. Their synthesis is unregulated because they are always needed. Enzymes that catalyze the glycolytic pathway, for example, are constitutive.

The molecular mechanisms that cause enzymes to be inducible, repressible, or constitutive are varied and complex. Let's look a little more closely, therefore, at the types of systems that regulate transcription and translation and consider specific examples. We conclude with global regulation.

Regulation of Transcription

Any step of gene expression can be regulated, but procaryotes act principally on transcription. They usually regulate transcription by changing the interaction between DNA and RNA polymerase. Their regulatory mechanisms change the frequency, rather than the rate, at which a particular gene is transcribed. When fewer transcripts are made, less protein is made.

Regulatory Proteins That Bind to DNA. Transcription is often regulated by **regulatory proteins** that bind to DNA and change its interaction with RNA polymerase. These proteins bind to specific regions of DNA, called **operators**, that are close to a gene's promoter, where RNA polymerase normally binds. Depending upon what kind of protein is bound to the operator, transcription of the gene by RNA polymerase becomes more or less probable.

Most regulatory proteins that bind to DNA are allosteric proteins. This means that their ability to bind to DNA changes when they combine with small molecules called effectors (Chapter 5). When transcription is regulated by an allosteric DNA-binding protein, the outcome is determined by two different interactions. First, effectors bind to DNA-binding proteins and make them either more or less likely to bind to DNA. Then the regulatory proteins bind to DNA and make transcription either

more or less likely. This combination of two different events—either of which can have a positive or negative effect on transcription—generates many different molecular mechanisms for regulating inducible or repressible enzymes.

Now let's look at two examples of inducible enzyme systems that are regulated by means of DNA-binding proteins. They are the *lac* **operon** and the **arabinose operon** in *Escherichia coli*. Both of these systems are inducible, but their mechanics of regulation are different. A set of genes that is regulated and transcribed together is called an **operon**.

The Lac Operon. The lactose (*lac*) operon encodes the ability to use lactose, or milk sugar, as a growth substrate. Both of the enzymes that a cell needs to metabolize lactose are encoded by the *lac* operon (**Figure 6.11**). These enzymes are **galactoside permease**, which brings the lactose into the cell, and *β*-**galactosidase**, which splits the disaccharide lactose into its component monosaccharides, glucose and galactose. Galactoside permease is encoded by the *lacY* gene, and *β*-galactosidase is encoded by the *lacZ* gene. In addition, the *lac* operon contains a promoter-operator region and the *lacA* gene, which encodes **galactoside transacetylase**, an enzyme with unknown function.

The DNA-binding protein, called the *lac* **repressor**, that regulates expression of the *lac* operon is encoded by the regulatory gene *lacI*, which lies close to but outside the *lac* operon, has its own promoter, and is expressed constitutively.

In the absence of lactose, the *lac* repressor binds to the *lac* operator and prevents transcription of the *lac* operon. When lactose is present, a small amount of it is converted to an effector, called **allolactose**. Allolactose binds to the *lac* repressor, changing it so that it no longer binds to the *lac* operator. Thus, only when lactose is present are the genes of the *lac* operon transcribed and expressed. The result is expression of the *lac* operon only when its products are needed—when lactose is available.

The Arabinose Operon. The arabinose operon is similar to the *lac* operon because it, too, is an inducible enzyme system that encodes the ability to use a sugar (in this case, arabinose) as a growth substrate. But regulation of the arabinose operon is different from regulation of the *lac* operon. Instead of a repressor that prevents transcription when it binds to DNA, the arabinose operon is regulated by an **activator** that stimulates transcription when it is bound to DNA. In this system, the allosteric effector is arabinose. In the absence of arabinose the activator is unable to stimulate transcription, but when arabinose is present in the environment it binds to the arabinose activator, changing its conformation so that it is capable of binding to the operon's

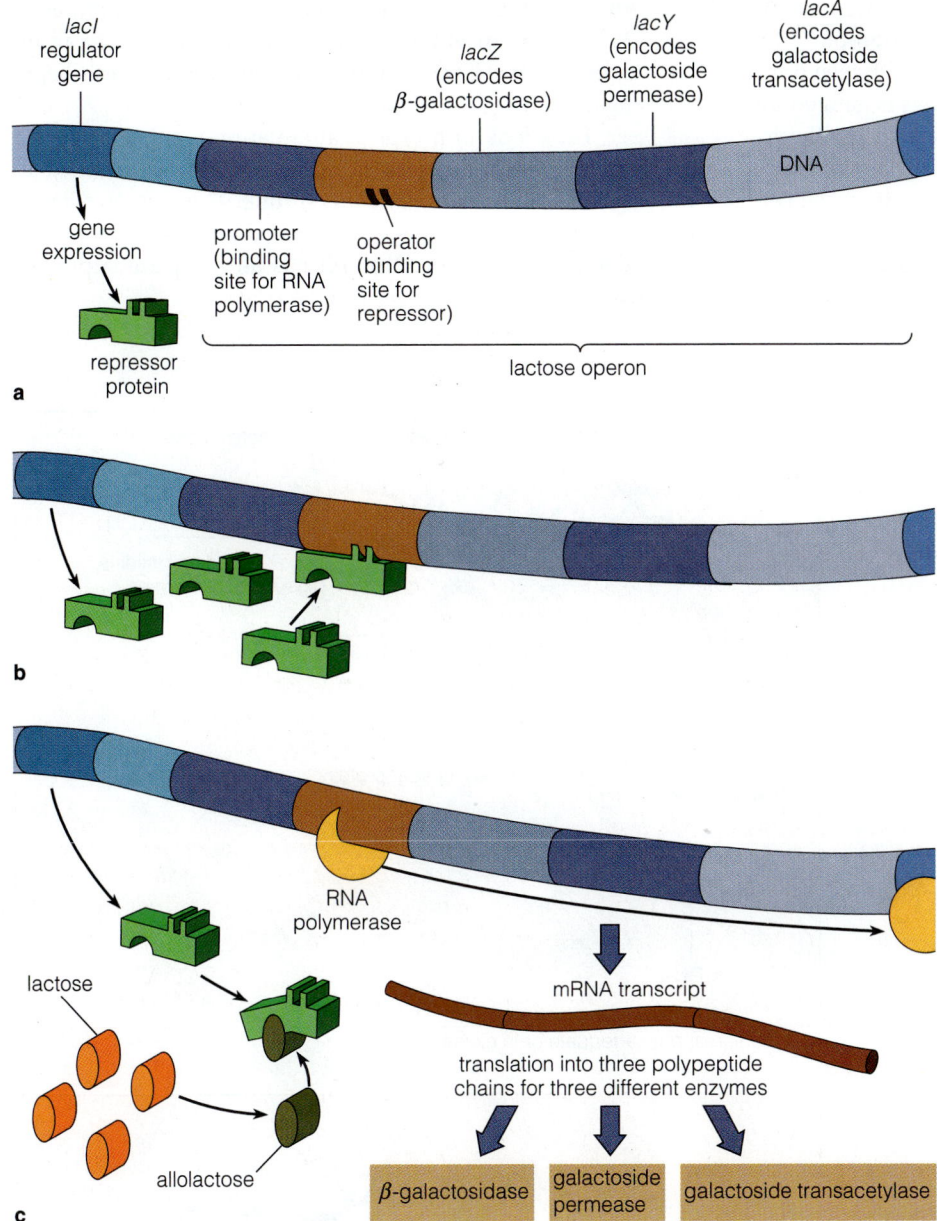

Figure 6.11 Regulation of transcription through the *lac* operon. (**a**) The lactose operon has three genes, *lacZ*, *lacY*, and *lacA*, which are regulated by the nearby regulatory gene, *lacI*. (**b**) When lactose is absent, the *lac* repressor binds to the operator, stopping transcription. (**c**) But when lactose is present, some of it is converted to allolactose, which binds to the repressor. This changes the repressor so it can no longer bind to the operator. Then the genes are expressed.

promoter. When the activator is bound to the promoter, the operon is transcribed and the enzymes necessary for the metabolism of arabinose are expressed. The result is the same as with the *lac* operon—the arabinose operon is expressed only when its products are needed, when arabinose is available.

Attenuation. Attenuation, a major means of regulating gene expression in bacteria, is a repression mechanism that regulates transcription without the participation of a DNA-binding protein. Instead, attenuation regulates gene expression based on the balance between the cell's rates of transcription and translation.

Attenuation usually regulates the production of enzymes in biosynthesis pathways—for example, the biosynthesis of an amino acid. The balance between transcription and translation in these pathways is gauged by the availability of amino acids for protein synthesis. When the intracellular concentration of a particular amino acid rises to a concentration greater than the cell needs in order for translation to occur at a rate equal to transcription, attenuation decreases transcription and thereby stops the synthesis of more of this amino acid. When the concentration of the amino acid falls to an inadequate level, transcription and enzyme synthesis begin again.

Let's look at a typical attenuation-regulated operon. The **histidine operon** encodes the enzymes needed to synthesize the amino acid histidine. This operon (**Figure 6.12**) begins with a gene that encodes a **leader protein**. It determines whether the cell has an adequate supply of

Figure 6.12 Attenuation. Regulation of the histidine operon in *Escherichia coli* is an example of attenuation. (**a**) When the histidine supply is adequate, the first ribosome closely follows the transcribing RNA polymerase molecule. An attenuator loop forms and RNA polymerase stops, so the genes encoding the enzymes that synthesize histidine are not transcribed. (**b**) When the histidine supply is inadequate, the ribosome lags behind RNA polymerase sufficiently for an antiterminator loop to form. This prevents the attenuator loop from forming. Thus, transcription continues and the histidine biosynthetic genes are expressed.

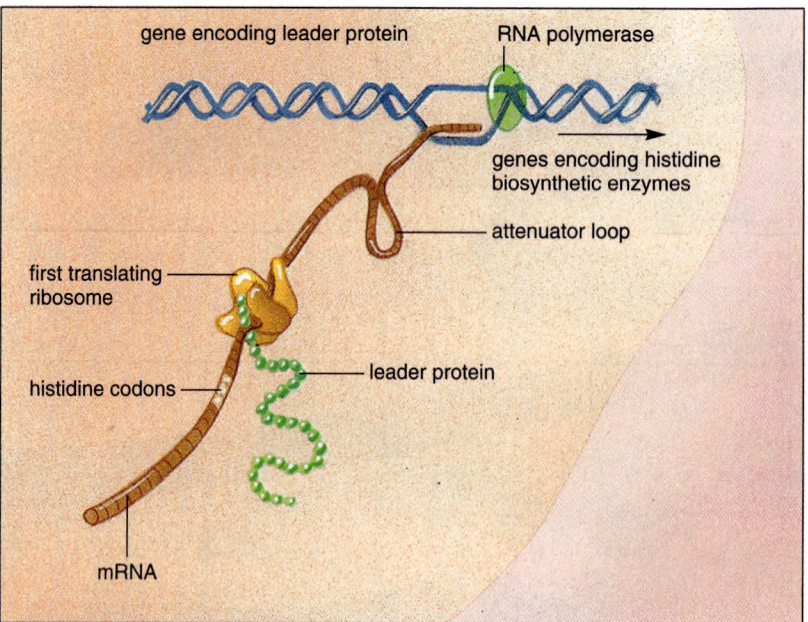

a Histidine supply adequate or in excess.

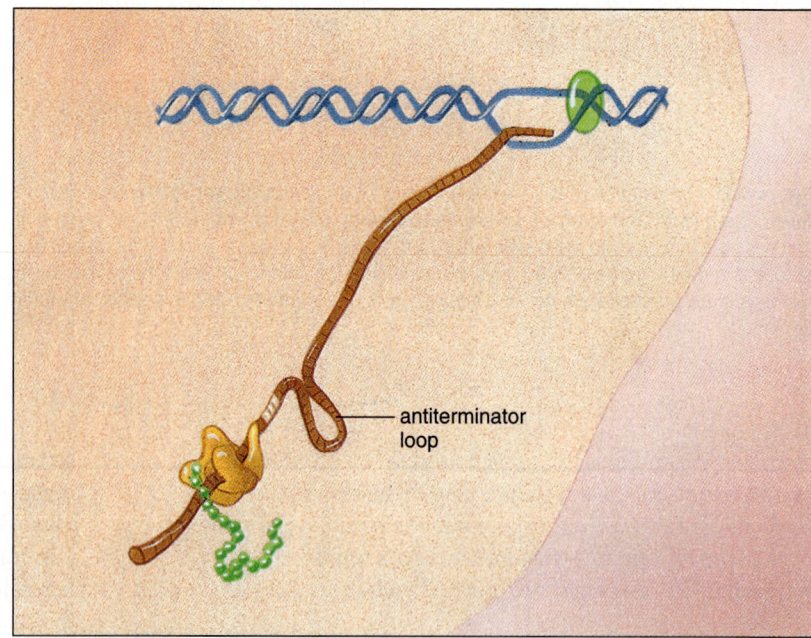

b Histidine supply inadequate.

histidine. The structure of the leader protein suits its function well. It contains seven histidine units in a row, making its rate of translation sensitive to the intracellular concentration of histidine. When the supply of histidine is adequate or excessive, translation of the leader protein occurs so rapidly that the first translating ribosome is only slightly behind RNA polymerase by the time that the end of the leader protein has been transcribed. This causes the short piece of free mRNA to fold back on itself, creating an **attenuator loop**, which displaces RNA polymerase and prevents transcription of subsequent genes in the operon. When the supply of histidine falls, the rate of translation also falls and the first translating ribosome is considerably behind RNA polymerase when the leader protein has been transcribed. This extended piece of free RNA folds back on itself, producing an **antiterminator loop** that prevents formation of the attenuator loop. Transcription of subsequent genes in the operon proceeds.

Attenuation helps adjust the supply of histidine to the cell's need for it. When more histidine is needed, the cell makes more of the enzymes that synthesize it; when excess histidine is present, lesser amounts of these enzymes are made. Histidine also regulates the activity of one of these enzymes by end-product inhibition (Chapter 5). Together, attenuation and end-product inhibition set the intracellular concentration of histidine precisely at its optimum value and spare the cell from making enzymes and histidine that are not needed.

Regulation of Translation

The synthesis of a few proteins is regulated at the level of translation rather than transcription. An example is syn-

thesis of the protein components of ribosomes, called **ribosomal proteins (Figure 6.13)**.

The genes encoding ribosomal proteins are grouped into operons. Each operon contains one gene that encodes a protein with two functions. The first function is to be incorporated into a ribosome, and the second function is to bind to its own mRNA and inhibit translation. When the supply of this protein is not excessive, all the molecules are incorporated into ribosomes. When the supply is excessive, the unused protein molecules bind to their encoding mRNA molecules and inhibit translation. This decreases further synthesis of the protein until the excess of free molecules is depleted. This regulatory system maintains the concentration of free ribosomal proteins at an optimal level for synthesis of ribosomes.

Global Regulation

So far we have discussed how expression of a gene or an operon operates in response to a specific metabolic signal. For example, the availability of a single compound such as lactose, arabinose, or histidine acts as a metabolic signal. But gene expression by microorganisms is also regulated by more general signals, such as the availability of *any* source of carbon or the shortage of *any* amino acid. This kind of regulation is called **global regulation** to reflect its broad impact (**Table 6.2**). The following examples of global regulation were obtained from studies on *Escherichia coli*, but they are thought to apply generally to all procaryotes.

Catabolite repression regulates and coordinates the expression of many genes and operons in response to the availability of carbon sources for growth. The *lac* operon

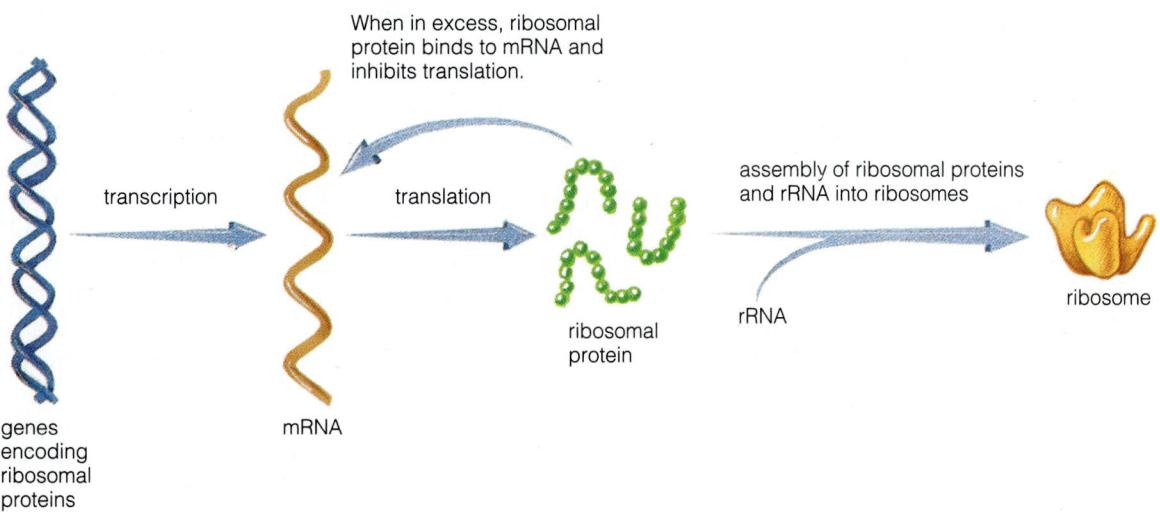

Figure 6.13 Regulation of translation. Expression of ribosomal protein genes is regulated at the level of translation. When free ribosomal proteins accumulate because they are not being used to make ribosomes, they bind to mRNA, preventing further translation.

Table 6.2 Examples of Global Regulation That Function in *Escherichia coli*

Regulatory System	Function
Catabolite repression	Prevents expression of genes encoding catabolism of carbon sources when a better carbon source is available
Nitrogen regulation	Prevents expression of genes encoding catabolism of nitrogen sources when a better nitrogen source is available
Phosphorus regulation	Prevents expression of genes encoding catabolism of phosphorus sources when a better phosphorus source is available
Stringent control	Prevents synthesis of ribosomal RNA when any amino acid is lacking
Heat shock	Stimulates synthesis of proteins that enable cells to better withstand damage from heat when temperature is increased

is one set of genes that responds to catabolite repression. We have seen that the *lac* operon is expressed only in the presence of lactose. Catabolite repression adds an additional metabolic advantage by saving the cell from expressing the *lac* operon if a "better" substrate than lactose—one that supports faster growth—is also available. If, for example, a mixture of lactose and glucose is available to *E. coli*, it uses all of the better substrate, glucose, before it makes the β-galactosidase and galactoside permease necessary to metabolize lactose.

Catabolite repression is mediated by two compounds: a small molecule, **cyclic AMP**, which belongs to a class of metabolites called **alarmones**, and a regulatory protein called **CAP** for catabolite activating protein. Alarmones serve only as metabolic signals. Cyclic AMP, for example, signals the rate at which catabolic reactions are proceeding. In the presence of a good substrate, like glucose, catabolic reactions proceed rapidly and suppress the synthesis of cyclic AMP. In contrast, catabolic reactions proceed slowly in the presence of a poor substrate such as the dicarboxylic acid succinate. This causes the intracellular concentrations of cyclic AMP to rise. When cyclic AMP combines with CAP, the complex binds to the regulatory region of the *lac* operon (and other genes and operons controlled by catabolite repression) near the site where the *lac* repressor binds. Instead of preventing transcription, however, the CAP–cyclic AMP complex stimulates transcription. In fact, very little transcription of the *lac* operon occurs unless it is stimulated by CAP–cyclic AMP.

In summary, catabolite repression enables the cell to metabolize the best substrate from the mixture that is available to it. *E. coli* would use glucose in preference to lactose and lactose in preference to succinate. Together the specific metabolic signal of enzyme induction and the global regulatory mechanism of catabolite repression prevent unnecessary protein synthesis. The proteins that metabolize lactose, for example, are made only when lactose is the best substrate available.

CHANGES IN A CELL'S GENETIC INFORMATION

Over time, a microbial cell's *genome*, its complement of genetic information, changes. Chemical changes called mutations occur in its DNA. Changes also occur by **genetic transfer** when a cell receives slightly different DNA from another microbial cell. Before discussing the two mechanisms of genetic change, let's look more closely at the genome.

The Genome

The genome is the sum total of DNA that a cell contains. The vast majority of a bacterial cell's DNA is contained in its single circular chromosome (some is in plasmids). In contrast, most of a eucaryotic cell's genome is contained in pairs of chromosomes. The chromosome in *Escherichia coli* contains enough DNA to form about 3500 genes. This is typical, though some bacteria contain about three times as much and some less than one-fourth as much. However, all procaryotic cells contain much less DNA than eucaryotic cells. The chromosome carries all the genes essential for growth and survival.

Plasmids. Most bacteria contain some DNA in structures called **plasmids**. They are 20 to 50 times smaller than a chromosome, but their structure is much the same. These circular structures are replicated the same way chromosomes are, and copies are passed on to daughter cells. Plasmids differ from chromosomes by encoding only functions that are not essential for growth and reproduction. Rarely, a cell will fail to pass a plasmid to one of its daughters. When that happens the progeny cell may be at a competitive disadvantage, but it does not die.

Functions that plasmids encode are beneficial in special environments. For example, some plasmids encode enzymes that make a bacterium resistant to antibiotics. These plasmids are called **R factors** (resistance factors). Some R factors confer resistance to only a single antibiotic. Others confer resistance to several antibiotics, so strains that carry them are said to have **multiple resistance**. Over the last few decades, R factors have proliferated to the point that some infections are difficult to cure

Pasteur Revisited

In 1881 Louis Pasteur developed a way to protect animals and humans against anthrax, a deadly respiratory disease caused by *Bacillus anthracis* (Chapter 27). He attenuated (weakened) a strain of *B. anthracis* by growing it at 43°C, a temperature just below the maximum it could tolerate, and used this strain as a vaccine (Chapter 20). Injections of these live but weakened bacteria did not cause disease. Rather, they protected against disease-causing strains of *B. anthracis*. Pasteur's discovery was a turning point because other mi-

crobiologists used his attenuation technique—growing bacteria or viruses in adverse environments or unconventional hosts to weaken them—to produce protective vaccines against many other diseases. But how attenuation of bacteria worked was not discovered until almost a hundred years later. Disease-causing information is carried in plasmids, and bacteria lose plasmids when cultivated in adverse environments. Thus, *B. anthracis* loses plasmids when grown at 43°C, and plasmid-free strains are effective, safe vaccines.

---

with antibiotics. Both chromosome and plasmid-type antibiotic resistance are discussed in detail in Chapter 21. Plasmids also encode, among other things, synthesis of the pigments that color bacterial cells, the capacity to make antibiotics, the capacity to cause plant diseases, the ability to make certain pili, the ability to use certain sources of carbon or nitrogen, and the ability to make certain disease-causing toxins. When these strains lose their toxin-encoding plasmids, they become essentially harmless (see the box "Pasteur Revisited").

Some plasmids, called **conjugative plasmids**, encode the ability to transfer a copy of themselves to another bacterial cell. This transfer may even be made to another species or genus of bacteria. Transfer occurs by the process of **conjugation**. If a plasmid confers a significant benefit on the cell, conjugation can spread plasmids quickly throughout a population of bacteria. (Conjugation is discussed later in the chapter.)

Genotype and Phenotype. The genome of a bacterium or any other organism can be considered from two different points of view, **genotype** and **phenotype**. Genotype describes the genes that a cell contains; phenotype describes the effect these genes have on the cell's appearance and function. If a cell's genome changes, its genotype changes, and usually its phenotype changes as well. For example, if an *Escherichia coli* cell loses the gene encoding the enzyme β-galactosidase, the cell's genotype is changed; its phenotype is also changed because β-galactosidase is not produced and the cell is unable to metabolize lactose.

Mutations

A **mutation** is any chemical change in a cell's genotype. Many different kinds of changes can occur (**Figure 6.14**).

A **base substitution mutation** is a change in a single pair of bases to a different pair. A **deletion mutation** is total removal of a segment of the DNA. An **inversion mutation** is the reversal in the order of a segment of DNA. A **transposition mutation** is the movement of a segment of DNA to a different position on the genome. A **duplication mutation** is the insertion of an identical new segment of DNA next to the original one. All of these mutations change the cell's genotype and, depending on the specific case, may change the phenotype.

Incidence of Mutations. Every time the chromosome is replicated, mistakes occur and mutations result. Accumulation of these mutations depends on reproduction, not time, so **mutation rate** is calculated as the *number of mutations per cell per generation*. Mutations that occur in the natural course of microbial growth are called **spontaneous mutations**; mutations caused by chemical, physical, and certain biological treatments are called **induced mutations**.

Spontaneous mutations are rare. Only about one cell in a hundred million (10^8) has a mutation in any particular gene. This frequency is equivalent to about two persons being affected in the population of the United States. Such a rare event might seem inconsequential, but it is not. Most full-grown bacterial cultures contain about 10^9 cells per milliliter (four times as many as there are people in the United States), so each milliliter of an average full-grown culture contains about 10 cells with mutations in any particular gene. Because the bacterial chromosome contains about 3500 genes, each milliliter of culture contains about 35,000 mutations that weren't present when the culture started growing from a small inoculum.

The vast majority of mutations are detrimental to the cell; some are lethal. If a mutation damages the cell, it is

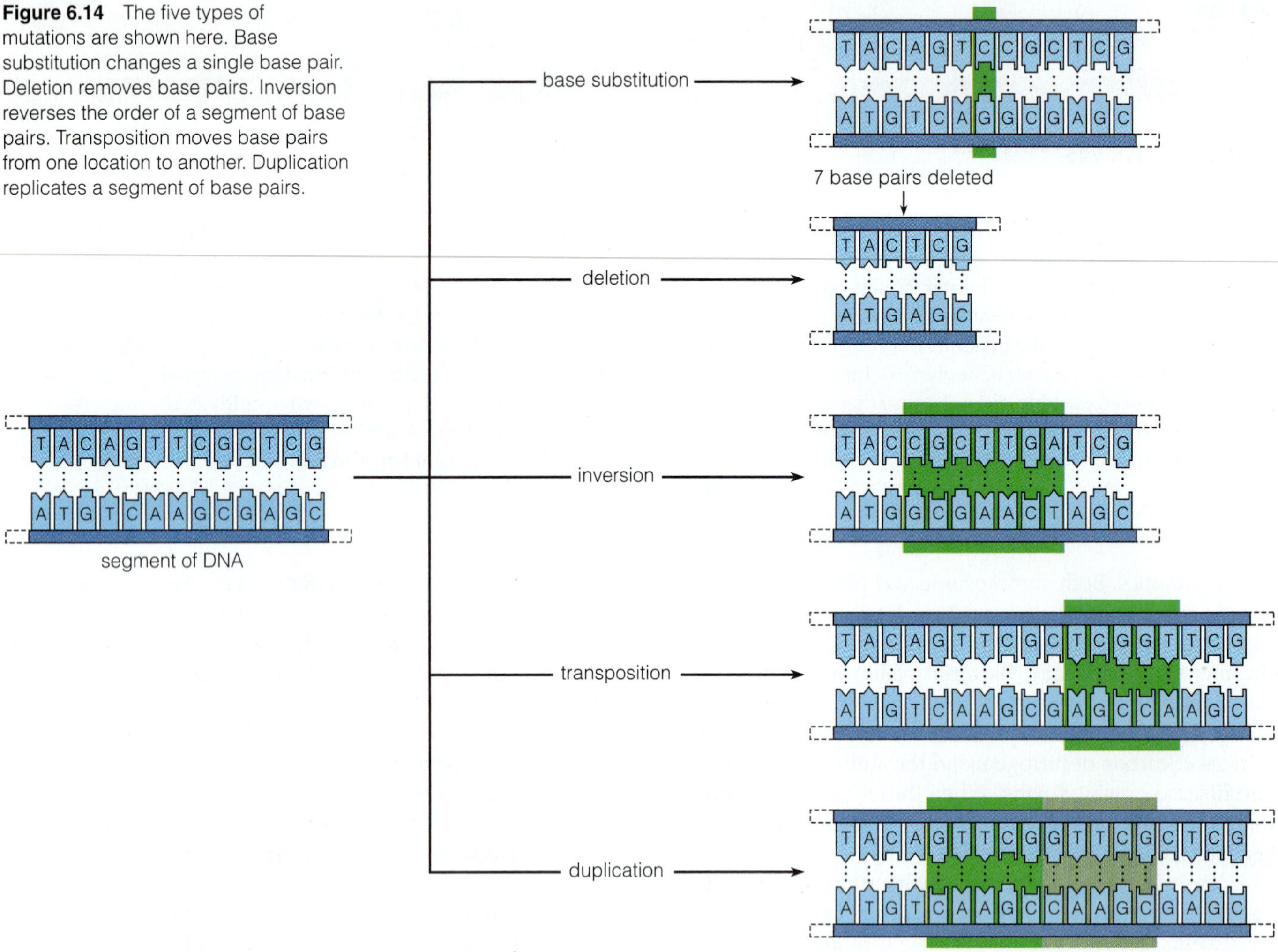

Figure 6.14 The five types of mutations are shown here. Base substitution changes a single base pair. Deletion removes base pairs. Inversion reverses the order of a segment of base pairs. Transposition moves base pairs from one location to another. Duplication replicates a segment of base pairs.

quickly lost from the population because cells that carry it are outgrown by the rest of the population. However, in the rare cases when a mutation benefits the cell—for example, a mutation that allows a cell to grow faster in a particular medium or resist a toxic agent—that cell's progeny (succeeding generations of daughter cells) soon dominate the population.

The Cause of Mutations. The frequency of mutations is increased when a culture is treated with **mutagens**, agents that induce mutations. Mutagens may be chemical, physical, or biological.

Chemical Mutagens. Certain chemicals react with DNA or the replication machinery in various and sometimes very powerful ways. For example, almost every surviving cell of a bacterial culture treated with the chemical mutagen **nitrosoguanidine** carries a new mutation somewhere in its genome.

Some chemical mutagens react with a component of DNA and change it. For example, **hydroxylamine** reacts specifically with cytosine in DNA, converting it to **hydroxylaminocytosine**. Unlike cytosine, which pairs only with guanine, hydroxylaminocytosine pairs with adenine as well. So following replication of a region of DNA where a cytosine has been converted to a hydroxylaminocytosine, the original C-G pair of bases is often converted to a T-A pair.

Other chemical mutagens—for example, 5-bromouracil (5BU)—act by being incorporated into DNA (**Figure 6.15**). Usually, 5BU, like its structural analogue, thymine, pairs with adenine. Unlike thymine, however, 5BU frequently pairs with guanine as well. So incorporation of 5BU into DNA usually changes an A-T pair of bases to a G-C pair.

Physical Mutagens. Certain physical agents also cause mutations. Physical mutagens include ultraviolet (UV)

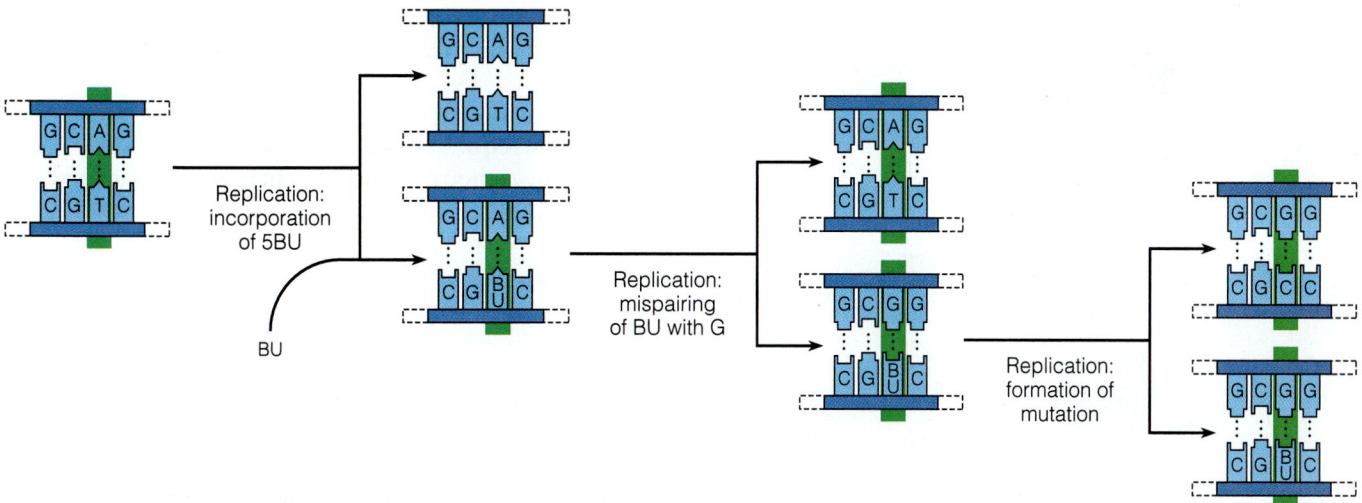

Figure 6.15 *Chemical mutagens. One of these is 5-bromouracil (5BU). Incorporating 5-bromouracil into DNA in place of thymine (T) can change an A-T pair to a G-C pair. This occurs because on replication BU pairs with G, causing it to be inserted. The next replication generates a G-C pair.*

light, x rays, gamma radiation, and decay of radioactive elements. Heat is slightly mutagenic.

UV light, which is commonly used to generate mutant strains of microorganisms, damages DNA. As the cell's DNA repair mechanisms correct this damage, mutations occur (**Figure 6.16**). UV light stimulates adjacent pyrimidine bases, usually thymines, in DNA to react with one another, becoming linked to form a **thymine dimer** (a dimer is a chemical compound composed of two identical parts). This damage activates several repair systems, one of which allows DNA replication across the damaged region at the cost of accuracy. The resulting errors are mutations.

In contrast to UV light, x rays, gamma rays, decay of radioactive elements, and heat cause mutations by changing the chemical structure of DNA. X rays and gamma rays are **ionizing radiations** (Chapter 9). That is, they expel electrons from atoms, producing ions, and these electrons react with other components of DNA to produce altered structures. Some reactions fragment the DNA backbone. Others cause mistakes to be made during subsequent replication, leading to mutations. Decay of radioactive elements can cause mutations by producing ionizing radiation or by producing a new element. When radioactive phosphorus atoms (^{32}P) decay, they become sulfur (S) atoms. If the ^{32}P atom is part of a DNA molecule, the result is a mutation.

Biological Mutagens. Microorganisms, and probably most higher organisms as well, carry biological mutagens within their genomes. Biological mutagens are sequences of DNA that themselves cause mutations by *transposing,* or moving, themselves from one part of the genome to another. Because of their ability to move from place to place within the genome, biological mutagens are called **transposable elements,** or sometimes **jumping genes.**

Two types of transposable elements exist, **insertion sequences** and **transposons.** Both are short regions of DNA, one to a dozen genes in length. Both types encode enzymes that transpose the element—simultaneously replicating the element and moving it to another place on the genome, leaving the original copy where it was. Insertion sequences encode only the ability to transpose. Transposons also carry other genes.

When the new copy of a transposable element is inserted within a gene, a mutation has occurred, and the gene's function may be destroyed. You might think that transposable elements would be a biological disaster, moving around the genome and destroying it. If they moved too frequently, they would certainly destroy the cell and themselves as well. But transposable elements also encode a regulatory protein that inhibits transposition. When a transposable element is first introduced into a cell, the inhibitory protein has not yet been produced, so the element transposes at a high frequency. Later, when the protein has been synthesized from the genes on the transposable element, further transposition occurs at a low frequency.

Some transposable elements benefit the cell by bringing it useful genes. For example, some transposons carry genes that destroy antibiotics and thereby make the cell resistant to them. But what benefits do insertion sequences confer? If they do not confer any, why have they been perpetuated during evolution? Some microbiolo-

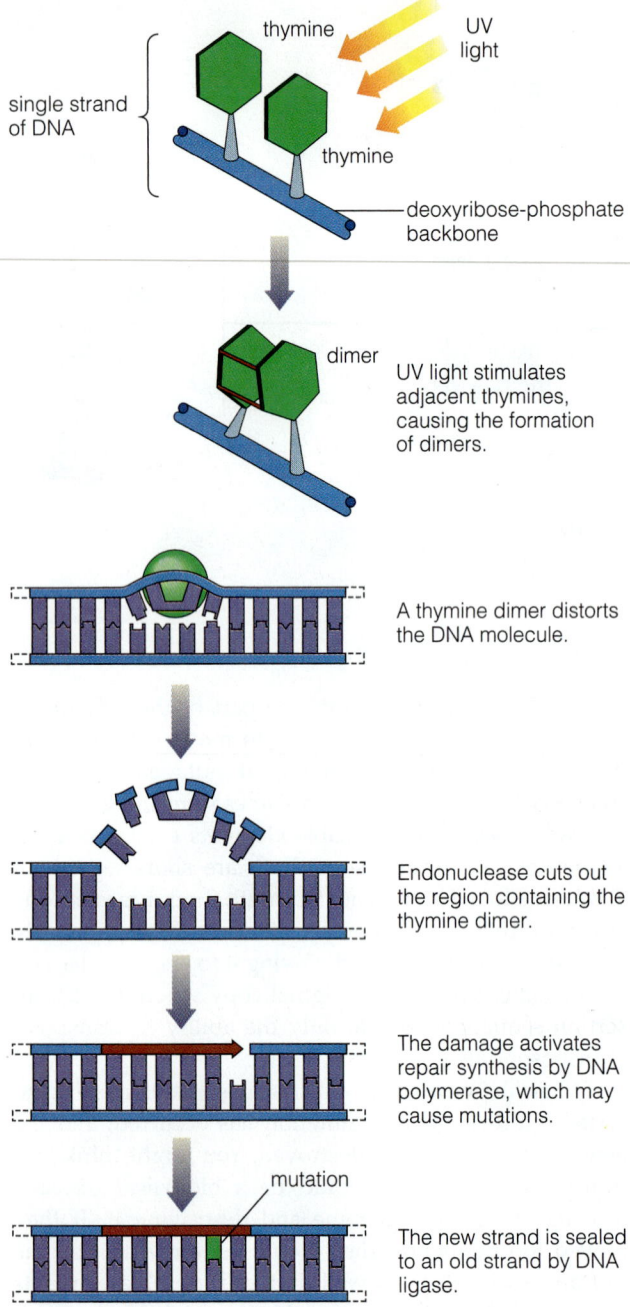

single strand of DNA

thymine

UV light

thymine

deoxyribose-phosphate backbone

dimer

UV light stimulates adjacent thymines, causing the formation of dimers.

A thymine dimer distorts the DNA molecule.

Endonuclease cuts out the region containing the thymine dimer.

The damage activates repair synthesis by DNA polymerase, which may cause mutations.

mutation

The new strand is sealed to an old strand by DNA ligase.

Figure 6.16 Physical mutagens. Ultraviolet (UV) light is mutagenic. It stimulates adjacent thymines to react chemically, forming dimers. Metabolic repair systems excise the dimers and remove neighboring bases, which are later replaced. This often causes mutations because the process is error-prone.

gists speculate that they are examples of "selfish DNA"—evolved to perpetuate themselves, not to benefit the host cell.

The Consequences of Mutations. Two factors determine how serious a mutation will be for the phenotype of a cell. The first is how much the mutation changes the gene product. The second is how important the gene product is to the cell.

Damage to the Gene Product. All mutations change the coding properties of DNA, but their effects on the gene products can vary greatly (**Table 6.3**). Some mutations may cause no change or only a minor change in the cell's phenotype. If, for example, the mutation changes the codon to another that encodes the same amino acid, the gene product remains completely unchanged. Similarly, the duplication of genes or operons might affect the quantity of gene product but usually has no effect on the cell's phenotype. A mutation that changes a codon to one that encodes a different amino acid, called a **missense mutation**, may alter the gene product only slightly if the new amino acid is similar to the original one.

However, some mutations alter the gene product profoundly or destroy it completely. A missense mutation that changes an amino acid to a very different one may seriously damage the resultant protein. If a bit of DNA is deleted, the gene product may be completely inactivated. A mutation that changes a codon to a nonsense codon, called a **nonsense mutation**, stops translation before synthesis of the protein product is completed. Transposition of a piece of DNA into a gene often destroys the gene product as well.

Surprisingly, the vast majority of mutations cause no change in the cell's phenotype that can be detected in the laboratory. But most mutations probably cause subtle changes that affect the cell's long-term competitiveness in nature.

Essential Gene Products. Some gene products are essential to a cell's survival regardless of the conditions of cultivation, while others are needed under only certain conditions. For example, if DNA polymerase is destroyed by mutation, the cell cannot polymerize nucleotides to make more DNA and therefore cannot multiply. This is called a **lethal mutation**. On the other hand, if an enzyme in a biosynthetic pathway leading to the synthesis of an amino acid is destroyed, the effect is less drastic. The mutant strain, called an **auxotroph**, can grow normally if the amino acid that it can no longer make is present in the medium. Auxotrophs are mutant strains with nutritional requirements that the parent strain lacks. Similarly, if a mutation destroys an enzyme in a pathway by which a

Death by Mutation

A 20-year-old woman who used heroin intravenously was admitted to the hospital because of a persistent fever and pain and swelling in her right leg. Her history of IV drug abuse suggested endocarditis, a life-threatening infection of the heart's lining often seen in drug addicts (Chapter 27). The diagnosis was confirmed when her blood sample grew the yeast *Candida parapsilosis*. Despite the odds, the patient's doctors hoped she might be cured because *C. parapsilosis* is highly sensitive to the drug 5-flucytosine (5-FC). 5-FC is toxic to many fungi because it is incorporated into their RNA, disrupting transcription. But it is harmless to humans because we lack one enzyme in the pathway that incorporates 5-FC into RNA. The patient was treated and improved so dramatically that after only four days her fever was gone and *C. parapsilosis* could no longer be isolated from her blood. Sadly, however, despite continuing treatment with 5-FC, her infection reappeared a month later and she died within days. An autopsy found that her heart was heavily overgrown with *C. parapsilosis*, and this time the strain was resistant to 5-FC. Thus, her treatment had failed because of a genetic mutation in the fungus. The original strain had a complete pathway for adding 5-FC to RNA, but the mutant strain had lost an essential enzyme in this pathway and therefore had become resistant to the drug. Unfortunately, this type of acquired drug resistance is common and is an enormous problem in using drugs to treat infections (Chapter 21).

Table 6.3	Effects of Mutations on Gene Products	
Mutation	Change	Probable Effect on Gene Product
Base substitution	Same amino acid	None
	Different amino acid, but similar to original	Slight loss of activity
	Different amino acid and dissimilar from original	Moderate loss of activity
	No amino acid (nonsense)	Complete loss of activity
Deletion	Loss of a segment of DNA	Complete loss of activity
Inversion	Reverses amino acid sequence	Inactivates some genes
Transposition	Interrupts gene with transposed DNA fragment	Complete loss of activity
Duplication	Doubles number of affected genes	No loss of activity

particular carbon or nitrogen source is metabolized, the mutant strain—called a **carbon source mutant** or **nitrogen source mutant**—will not be able to grow unless another carbon or nitrogen source is provided.

Sometimes a mutation renders the gene product nonfunctional only in certain environments. These are called **conditionally expressed mutations**. **Temperature-sensitive mutants**, for example, make the gene products more sensitive to extremes of temperature. **Osmotic remedial mutants** make the gene product more sensitive to the concentration of salts in the medium. A heat-sensitive mutant might be unable to grow or carry out a certain function at 42°C, but perform that function normally at 30°C. The opposite might be true for a cold-sensitive mutant. An osmotic remedial mutant might be quite normal in a high-salt medium, but completely unable to perform a certain function in a low-salt medium.

Conditionally expressed mutations allow the research microbiologist to study defects in essential genes. A temperature-sensitive mutant, for example, can first be cultivated normally at the temperature at which the gene product is functional and then grown at the temperature at which the gene product is inactive. The researcher can determine what effect the mutation has on microbial growth. Temperature-sensitive mutants were used to determine which genes and enzymes participate in transcription, translation, and replication. They are still used to probe indispensible cellular functions.

The Uses of Mutant Strains

Mutant strains are invaluable for answering fundamental questions about biology and in practical applications. In fact, the possibilities in both areas are vast. We discuss a few representative examples, but first we consider how mutant strains are identified and how a particular desired mutant strain can be isolated.

Selecting and Identifying Mutants. The number of ways to select and identify mutant strains of microorganisms is limited only by the imaginations of genetic microbiologists. But the possible ways can be grouped into three general categories: direct selection, indirect or counterselection, and brute strength. These procedures are usually applied to a culture that has been **mutagenized**, treated with a mutagen to increase its content of mutant cells, including the one being sought.

In **direct selection**, conditions are created that foster growth of only the desired mutant strain. For example, direct selection is used to identify antibiotic-resistant mutants, such as penicillin-resistant mutants. A plate containing penicillin supports the growth of penicillin-resistant strains alone, so if a dense culture of bacteria containing just a few penicillin-resistant mutant cells is plated, only the resistant cells will grow and produce colonies. Cells of the parent strain will not. Each colony that develops is a penicillin-resistant mutant strain.

Indirect selection, or **counterselection**, is the opposite of direct selection. First, conditions are created that prevent growth of the desired mutant strain. Then a condition is imposed that kills growing cells. Because the desired mutant cells are not growing, they survive the lethal treatment and constitute a larger fraction of the surviving population. They are thus easier to isolate. Counterselection is used for auxotrophic mutant strains. For example, an auxotrophic mutant that requires the amino acid histidine can be isolated using penicillin to counterselect. A population containing a few histidine auxotrophic cells is inoculated into a medium that lacks histidine. Then penicillin is added. The penicillin kills growing cells, so parental cells die, but the desired histidine auxotrophs survive because they cannot grow in a medium lacking histidine. The penicillin-treated culture is then plated on a medium containing histidine, and individual colonies are tested to see which required histidine to grow. Other agents, including radioactive substrates and certain toxic chemicals that kill only growing cells, can also be used to isolate mutant strains by counterselection.

Brute strength is the term applied to examining large numbers of clones one by one to find the desired mutant strain. There are many techniques. For example, a researcher could test to see if a clone produces a particular enzyme or if it is unable to grow in a particular environment. Following treatment with a powerful mutagen, a researcher must, on the average, examine about 10,000 surviving clones to find a particular desired mutant strain. In spite of such daunting numbers, many important mutant strains, including the first unable to produce the enzyme DNA polymerase III, have been isolated by brute strength. Using counterselection first can decrease the numbers that must be examined, and techniques have been developed to simplify examination.

One simplified brute-strength examination technique is **replica plating** (**Figure 6.17**). Replica plating allows the growth properties of many clones to be examined in a single operation. For example, replica plating can be used to detect a histidine auxotroph by testing for inability to grow on a medium lacking histidine. Replica plating is an easy way to transfer large numbers of colonies and keep track of them.

Uses of Mutant Strains in Basic Biology. Mutant strains have answered many important biological questions. For example, they were used to determine the biosynthesis pathways of amino acids from precursor metabolites (Chapter 5). A large number of auxotrophic mutants were isolated using the techniques described above and studied in various ways. If a mutant strain released a compound into the medium, that compound was probably a substrate of the enzyme that the mutant lacked. If a compound met a mutant strain's need for histidine, that compound was probably an intermediate of the histidine pathway beyond the step catalyzed by the missing enzyme.

Practical Uses of Mutant Strains. Mutant strains of microorganisms have been put to many practical uses. One is a test developed by Bruce Ames, a microbiologist at the University of California, Berkeley, to determine whether a particular chemical is carcinogenic (cancer-causing). Carcinogen tests are usually done by administering the test chemical to animals and observing whether the animals develop cancer. But, knowing that most carcinogenic chemicals are mutagens, Ames was able to develop a simpler, faster, and equally reliable method using histidine auxotrophic mutants (**Figure 6.18**). The **Ames test** determines, by direct selection, the ability of the mutagen to cause **reversions**, mutations that reverse the original mutation causing auxotrophy and enabling the strain to grow in the absence of added histidine. To increase the value of the test, the chemical to be evaluated is first mixed with macerated (mashed) rat

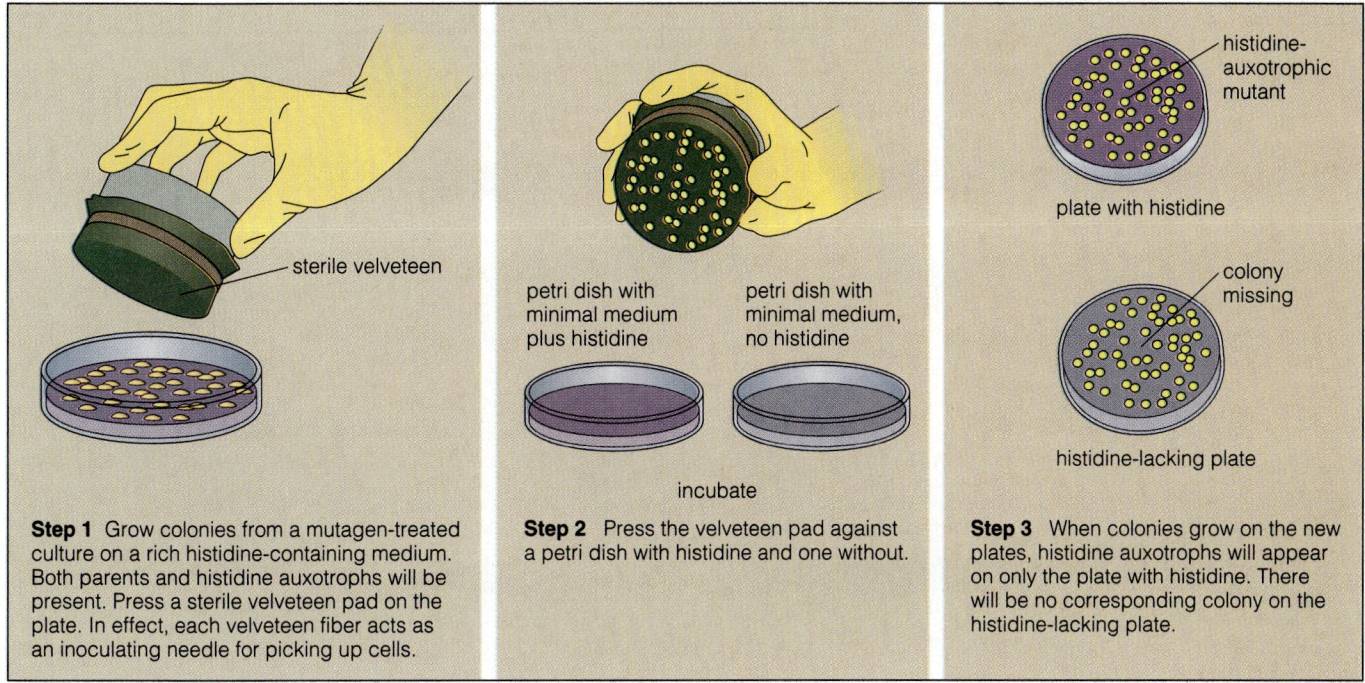

Figure 6.17 Replica plating is a technique for identifying and isolating mutant strains. This example shows how a colony of histidine auxotrophic mutant cells would be identified.

liver, a rich source of enzymes that converts some innocuous (harmless) chemicals into carcinogens. If a chemical is negative in the Ames test, it is probably innocuous. If it is positive, it is certainly mutagenic and probably carcinogenic. About 90 percent of the chemicals that test mutagenic in the Ames test are also shown to be carcinogenic in follow-up animal tests.

Genetic Exchange among Bacteria

In **genetic exchange**, genes are transferred from one cell to another. Genetic exchange among bacteria is unlike the genetic exchange that occurs during sexual reproduction of plants and animals. In most eucaryotes, genetic exchange is an essential part of the organism's life cycle. It occurs when self-replicating chromosomes from two individuals come together in the same cell, called a zygote. In bacteria, genetic exchange is not an essential step of the life cycle. When it does occur, only a portion of the genome of one cell, called the **donor cell**, is transferred to the other, the **recipient cell**. This produces a **merozygote**, or partial zygote. The transfer travels one way—the recipient transfers no DNA back to the donor.

Usually the piece of bacterial DNA donated during genetic exchange cannot replicate itself independently. In other words, it is not a **replicon**. The donated genes can be replicated only if they are incorporated into one of the cell's replicons—the chromosome or a plasmid—so that

they become a permanent part of the recipient cell's genome. The incorporation of donated genes into a replicon almost always involves the breakage and rejoining of pieces of DNA (**Figure 6.19**). This breakage and rejoining of DNA molecules is called **recombination**, or **crossing over**. Recombination is an essential feature of genetic transfer of chromosomal genes among procaryotes. It is not an essential feature of genetic plasmid transfer. If the entire plasmid is transferred, it need not recombine with DNA in the recipient cell to become a part of its genome.

There are three forms of genetic exchange in bacteria: transformation, conjugation, and transduction. Although all three processes allow DNA to leave one bacterium and enter another, there are differences.

Transformation.   During **transformation** DNA leaves one cell, exists for a time in the aqueous extracellular environment, and then is taken into another cell where it may become incorporated into the genome. Transformation can be natural or artificial.

Natural Transformation. Some bacterial species have genes in their chromosomes that enable them to absorb DNA from their environment. For example, in *Streptococcus pneumoniae*, a bacterium that causes pneumonia in humans, transforming DNA free in the extracellular environment binds to receptors on the cell surface. Then surface nucleases cut the bound DNA into fragments. One

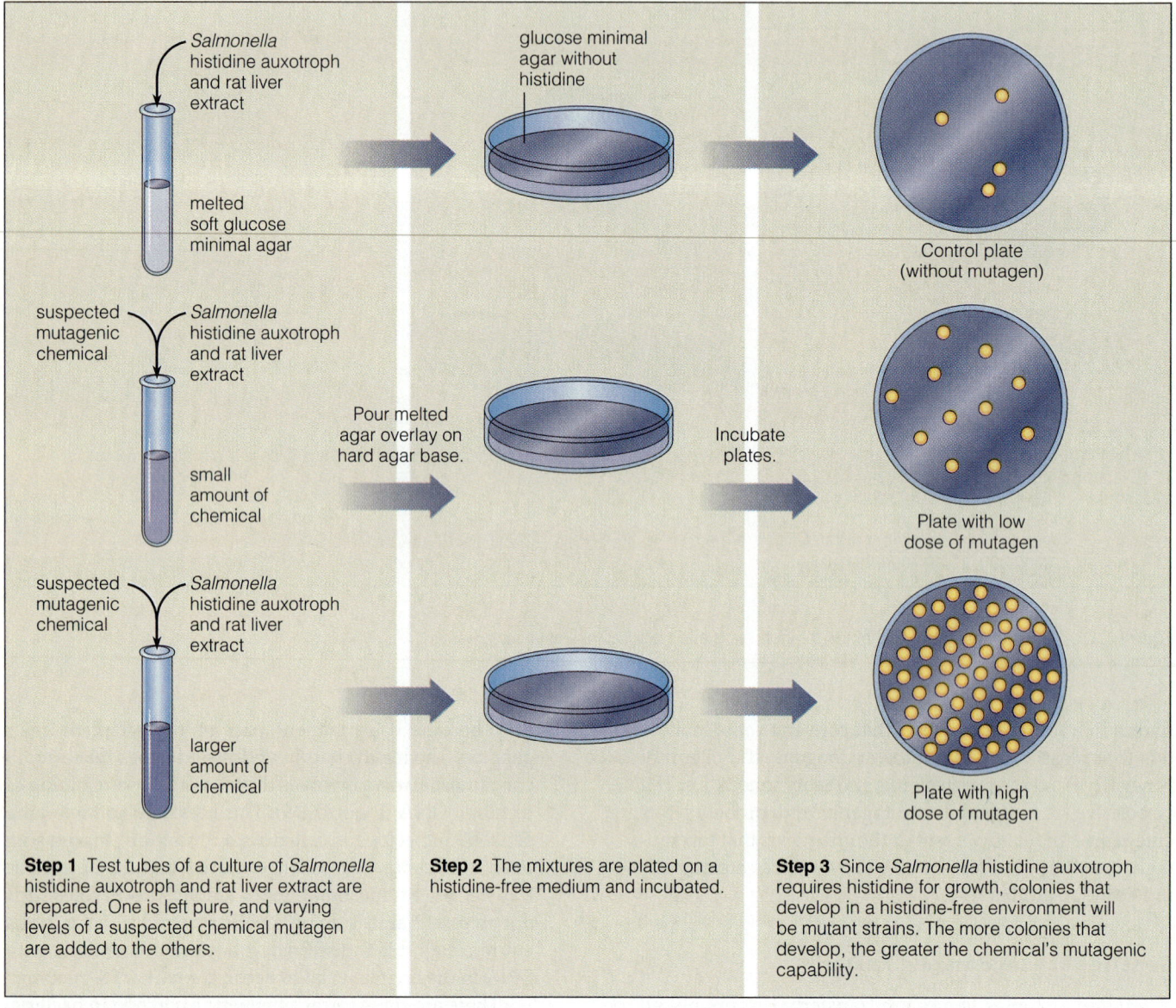

Step 1 Test tubes of a culture of *Salmonella* histidine auxotroph and rat liver extract are prepared. One is left pure, and varying levels of a suspected chemical mutagen are added to the others.

Step 2 The mixtures are plated on a histidine-free medium and incubated.

Step 3 Since *Salmonella* histidine auxotroph requires histidine for growth, colonies that develop in a histidine-free environment will be mutant strains. The more colonies that develop, the greater the chemical's mutagenic capability.

Figure 6.18 The Ames test is used to determine whether a particular chemical is mutagenic (and, therefore, probably carcinogenic). It is based on testing if a suspected mutagen will revert an auxotroph—that is, mutate again to grow without a substance it previously required for growth.

strand of each fragment is destroyed by another nuclease, while the other strand enters the recipient *S. pneumoniae* cell. Inside the cell, the donated DNA is coated with a protein that protects it from being destroyed by intracellular nucleases. When the newly absorbed DNA contacts the portion of the resident genome where matching genes are located, called the **homologous region**, one strand of the resident DNA is cut, or **nicked**, and replaced by the incoming strand. Then the ends of the incoming piece are enzymatically sealed to the rest of the resident DNA. This forms a region of **heteroduplex** DNA, where one strand of DNA from the donor cell is paired with one strand of

DNA from the recipient. When this region is replicated, one product is an exact copy of the DNA originally present in the donor cell, and the other product is a copy of the donated genes. About 12 separate genes are needed to encode the various steps of natural transformation of *S. pneumoniae*.

The details of natural transformation vary somewhat among species, but in all cases, the DNA is cut into small pieces by surface nucleases before it enters the recipient cell. As a result, the linear fragments of transformed DNA must enter a resident replicon in order to be copied and passed on to daughter cells. Chromosomal DNA that

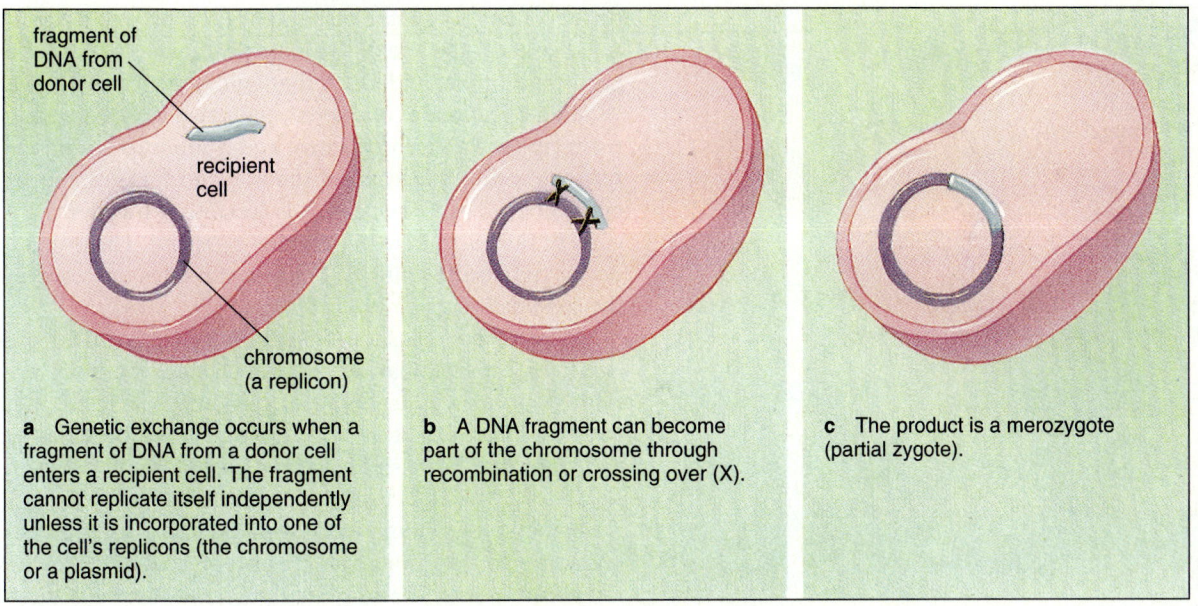

a Genetic exchange occurs when a fragment of DNA from a donor cell enters a recipient cell. The fragment cannot replicate itself independently unless it is incorporated into one of the cell's replicons (the chromosome or a plasmid).

b A DNA fragment can become part of the chromosome through recombination or crossing over (X).

c The product is a merozygote (partial zygote).

Figure 6.19 Genetic exchange in bacteria: genes (DNA) leave one bacterium and enter another.

enters a cell by transformation is sure to find a homologous part of the chromosome where it can be incorporated. If the transforming DNA comes from a plasmid, however, it can be incorporated into a functioning replicon only if the recipient cell happens to carry an identical plasmid. Consequently, plasmids are rarely transferred from one cell to another by natural transformation.

Natural transformation almost certainly evolved to make genetic exchange possible, but it remains a mystery why only a few bacterial species are capable of it. It is also an intriguing but unexplained fact that a greater percentage of pathogenic bacteria are capable of natural transformation than are nonpathogenic bacteria. Perhaps the ability to exchange genetic information provides pathogens with an additional means of acquiring the ability to overcome host cells' defense systems (Chapters 14 and 15). Examples of naturally transformable pathogenic bacteria include *Streptococcus pneumoniae*, which causes pneumonia, *Neisseria gonorrhoeae*, which causes the sexually transmissible disease gonorrhea, and *Haemophilus influenzae*, which causes meningitis, an inflammation of the membranes covering the brain and spinal cord.

Artificial Transformation. In 1972, researchers discovered how to alter the envelope of *Escherichia coli* so DNA could cross it. Based on this work, other researchers went on to develop an elaborate laboratory technique called **artificial transformation** in *E. coli* cells. In this process, *E. coli* cells are chilled and treated with a strong solution of cal-

cium chloride. Then DNA is added to the suspension, which is heated to 42°C for several minutes and quickly chilled again to 0°C. At this point, intact plasmids can be introduced into *E. coli* cells. These intact plasmids can replicate by themselves because they are not cut as they enter the cell. Modifications of this procedure (and totally different ones that have since been developed) now make it possible to introduce intact plasmids into almost any bacterium as well as many eucaryotic microorganisms, plants, and animals. The development of artificial transformation was essential to the development of recombinant DNA technology, sometimes called genetic engineering (Chapter 7).

Conjugation. Conjugation is carried out by conjugative plasmids, the type that can encode the capacity to transfer themselves to another cell. The best-studied conjugative plasmid is the **F plasmid**, which can replicate in *Escherichia coli* and closely related bacteria. Conjugation requires many genes, and the F plasmid encodes at least 13 genes. Cells that carry an F plasmid are designated F^+ and those that lack it F^-. In Gram-negative bacteria, one of these genes encodes a special pilus (Chapter 4), called a **sex pilus** or, in the case of *E. coli*, the **F pilus** (**Figure 6.20**). The F pilus allows the plasmid-bearing donor, or **F^+ cell**, to attach itself to a plasmid-free recipient, or **F^- cell**. Attachment triggers a series of events that results in the transfer of the intact plasmid from the donor to the recipient. First the pilus retracts, bringing the two cells in

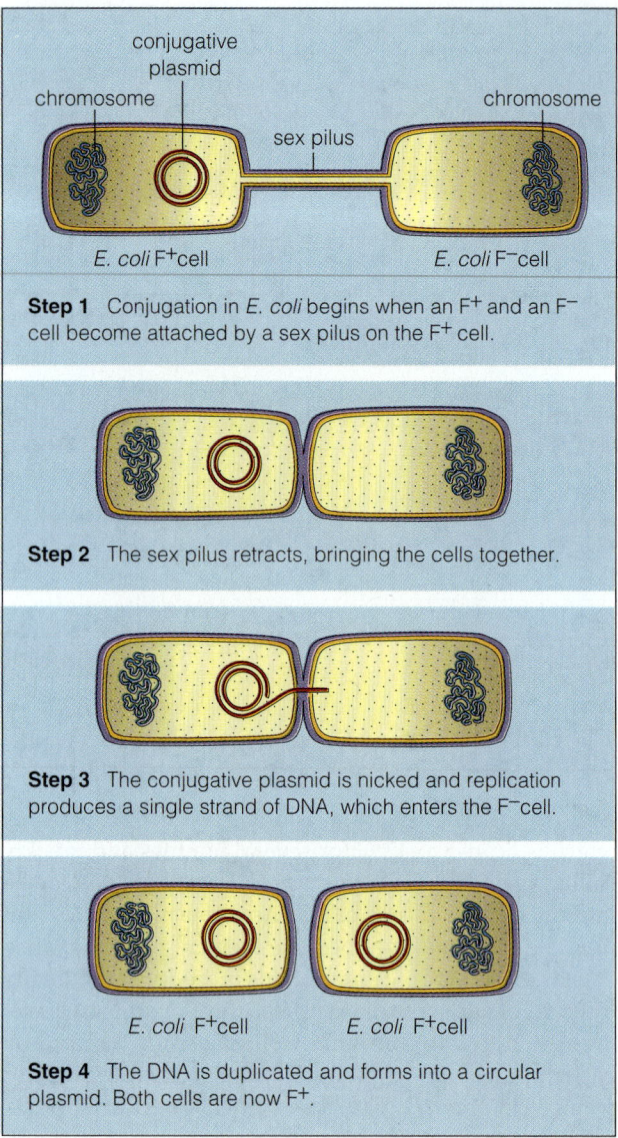

conjugative plasmid

chromosome

chromosome

sex pilus

E. coli F⁺cell

E. coli F⁻cell

Step 1 Conjugation in *E. coli* begins when an F⁺ and an F⁻ cell become attached by a sex pilus on the F⁺ cell.

Step 2 The sex pilus retracts, bringing the cells together.

Step 3 The conjugative plasmid is nicked and replication produces a single strand of DNA, which enters the F⁻cell.

E. coli F⁺cell

E. coli F⁺cell

Step 4 The DNA is duplicated and forms into a circular plasmid. Both cells are now F⁺.

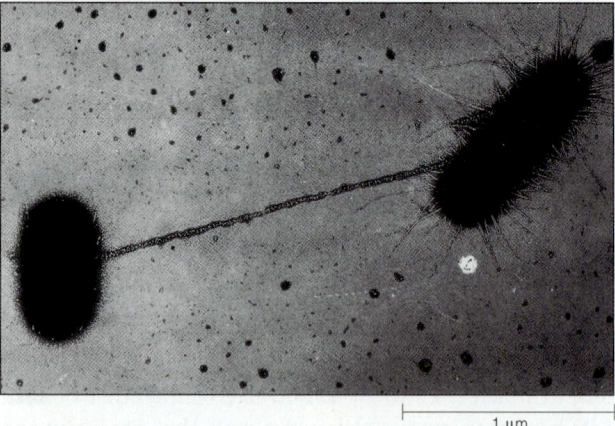

1 µm

Figure 6.20 Conjugation is a form of genetic exchange between bacterial cells mediated by conjugative plasmids. In *Escherichia coli*, the conjugative plasmid is called an F plasmid.

direct contact. Then, the plasmid DNA is nicked. Replication begins at the site of the nick, producing a linear single strand of plasmid DNA that enters the recipient cell. When replication is completed, the plasmid DNA within the recipient forms into a circle and duplicates within the recipient cell. As a result of the plasmid transfer, the recipient becomes F⁺ because it is capable of transferring a copy of the plasmid to yet another recipient. The original donor cell also remains F⁺ because it retains one copy of the conjugative plasmid.

Many other plasmids, including R plasmids, are transferred by similar mechanisms. Some of these are termed **promiscuous** because they are transferred between almost all species of Gram-negative bacteria. In contrast, the F plasmid is transferred only between strains of *E. coli* and closely related species.

Sometimes conjugative plasmids carry insertion sequences or transposons that are also present in the bacterial chromosome. If recombination occurs between these homologous regions, the plasmid becomes integrated into the chromosome, and the cell becomes an **Hfr** (**High frequency of recombination**) cell. Even though the plasmid has become part of the bacterial chromosome, it can still bring about conjugation. When a conjugative plasmid, now integrated into the chromosome, is transferred to an F⁻, some of the chromosome is transferred along with it. Usually only a portion of the chromosome is transferred because the attachment between donor and recipient cells is fragile. Most cell pairs break apart long before the 100 minutes required for the entire chromosome of *E. coli* to be transferred from an Hfr cell to an F⁻ cell. The **origin of transfer** lies within the F plasmid, so part of the F plasmid is transferred before the chromosome, and the other part is transferred after. Because the entire chromosome is rarely transferred, neither is the entire plasmid. Any population of bacteria includes a few Hfr cells that contain a conjugative plasmid. These Hfr cells can be purified, yielding pure cultures of Hfr cells that are extremely useful in studying bacterial genetics. These strains are used in matings that **map** (locate the relative position of) genes on the bacterial chromosome (**Figure 6.21**).

Transduction. Transfer of chromosomal genes during conjugation is an accident that occurs when a plasmid happens to recombine into the chromosome. Transfer of chromosomal genes during **transduction** is an accident that occurs when some of the viruses that infect bacteria, called **bacteriophages**, or **phages**, reproduce themselves.

To understand transduction, you need some understanding of how bacteriophages reproduce. Phage reproduction is discussed in detail in Chapter 13, but here we provide sufficient background to understand transduction. Based on their pattern of reproduction, there are two kinds of phages and, hence, two kinds of transduction.

Function of the Genes between 0 and 10 Units on the *Escherichia coli* Map

Gene(s)	Function
thr	A cluster of genes encoding enzymes in pathway of biosynthesis of the amino acid threonine
leu	A cluster of genes encoding enzymes in pathway of biosynthesis of the amino acid leucine
pan	A cluster of genes encoding enzymes in pathway of biosynthesis of the vitamin pantothenic acid
metD	A gene encoding an enzyme in pathway of biosynthesis of the amino acid methionine
proA	A gene encoding an enzyme in pathway of biosynthesis of the amino acid proline
lac	The *lac* operon, a cluster of genes encoding enzymes in pathway of catabolism of the sugar lactose
tsx	A gene encoding a protein in the outer membrane to which the bacteriophage T6 attaches

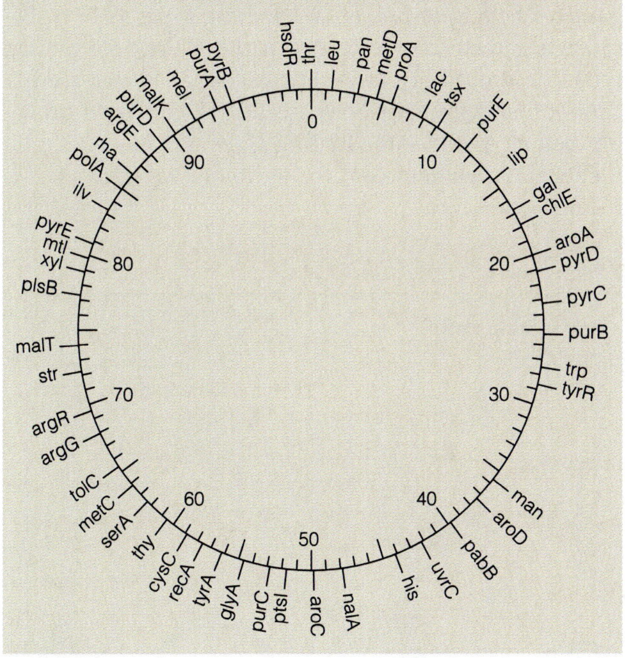

Figure 6.21 The map microbial geneticists use to locate the genes on the *Escherichia coli* chromosome is a circle divided into 100 units. Some of the 1200 genes that have been identified are shown here. *E. coli*'s chromosome has space for about 3500 genes. The table names the genes between 0 and 10 units on the *E. coli* map and indicates their functions.

Some **virulent phages** (phages that always kill their host) mediate **generalized transduction**. Some **temperate phages** (phages that can be carried passively within their host without harming it) mediate **specialized transduction**, generalized transduction, or both.

Virulent Phages and Generalized Transduction. Virulent phages infect bacteria by attaching themselves to the surface of the victim cell and injecting their DNA into it (**Figure 6.22**). The phage DNA, now inside the bacterial cell, directs the infected cell to synthesize phage components instead of proceeding with its normal metabolic functions. Phage DNA and protein made by the infected cell are assembled into phage particles that are then released. The new phage infects other cells.

Rarely—about 1 in 100,000 times—a mistake occurs in the final stages of assembly of the phage particle. Instead of assembling a phage particle that contains only phage DNA, one is assembled that contains bacterial DNA. After it is released, this abnormal phage particle, called a **transducing particle**, can attach to and inject its DNA into another bacterial cell. When this happens genetic exchange occurs. DNA from the bacterial cell where the transducing particle was formed is introduced into the next infected cell. The amount of DNA exchanged by transduction is about the same as the amount within a normal bacteriophage, which is equivalent to a few percent of the bacterial chromosome. As in transformation and conjugation, recombination is a necessary step in transduction. The DNA fragment that the phage injects persists in the host cell only if it becomes integrated by recombination into the host's genome. Transduction mediated by virulent phages is called generalized transduction because *any* bacterial gene can be transferred from one cell to another. The term *virulent* is applied to these phages because they kill their host cells, usually by lysing them.

Temperate Phages and Specialized Transduction. Temperate phages have two life cycles. One, the **lytic cycle**, is like the life cycle of virulent phages. The other life cycle, called the **lysogenic cycle**, does not kill the cell. Instead of directing the host cell to make more phage particles, phage DNA becomes part of the host cell's genome and is called a **prophage**. The prophages of some temperate phages exist as plasmids, while others become incorporated in the host cell's chromosome. The latter kind can mediate specialized transduction because prophages occasionally become reactivated and enter a lytic cycle that produces phage particles. **Specialized transducing particles** are formed when a mistake is made during reactivation. Instead of only the prophage leaving the chromosome, a few bacterial genes leave as well. When a specialized transducing particle attaches to another bacterial host and injects its DNA, it injects the bacterial genes as well. Thus, genetic exchange occurs between the two bacterial cells—the one in which the specialized transducing particle was formed and the one it later infects. The process is called specialized transduction because prophages are inserted only at a specific site on the bacterial chromosome, so only those bacterial genes adjacent to this site can be transferred by this type of transduction.

Genetic Exchange among Eucaryotic Microorganisms

Genetic exchange between eucaryotic microorganisms is similar to genetic exchange between plants and animals: Following meiosis of a diploid cell, haploid gametes develop and fuse to form a new diploid cell (Chapter 5).

To understand the principle of genetic exchange among eucaryotic microorganisms, let's look at the life cycle of baker's yeast, *Saccharomyces cerevisiae* (**Figure 6.23**). *S. cerevisiae* is capable of both asexual and sexual reproduction. A culture of baker's yeast is composed largely of diploid cells. These cells grow and reproduce asexually as long as adequate nutrients are available. Occasionally, however, conditions for growth become less favorable and some cells undergo meiosis. The products of meiosis develop into thickened **ascospores**, spores enclosed within a saclike structure called an **ascus**. When conditions become favorable, the ascospores germinate, producing haploid cells that act as gametes. Usually they undergo a few divisions before they fuse with another gamete. All gametes belong to one of two mating types, a or α. Because a gamete can fuse only with an opposite mating type, the probability of genetic exchange is increased; it cannot fuse with neighboring cells, which would probably belong to the same mating type and be genetically identical. When gametes fuse, a diploid vegetative cell is formed again.

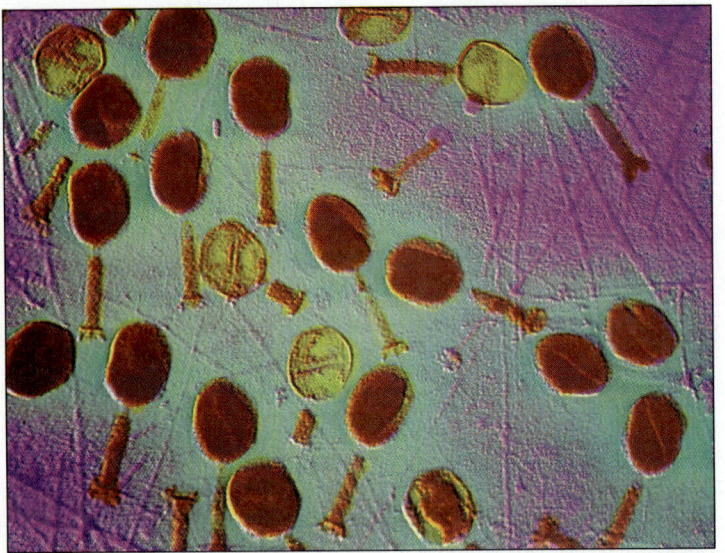

100 nm

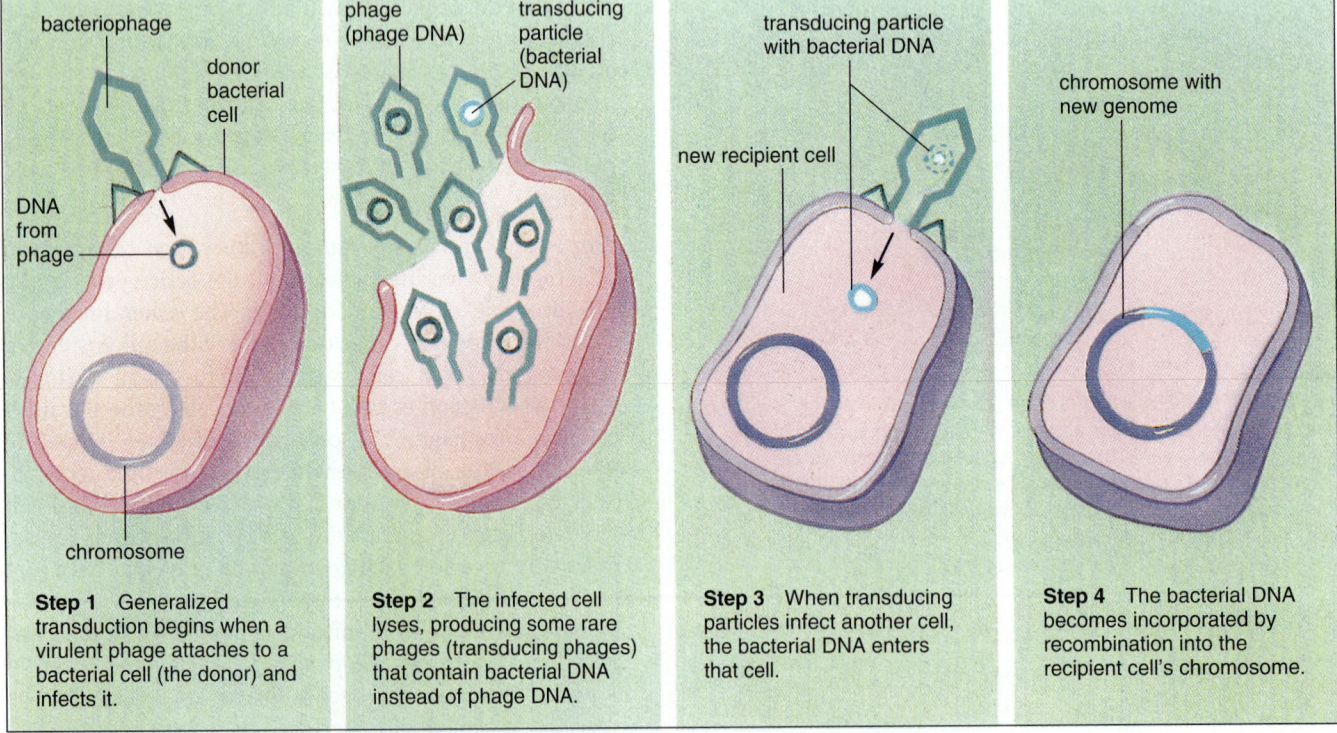

Step 1 Generalized transduction begins when a virulent phage attaches to a bacterial cell (the donor) and infects it.

Step 2 The infected cell lyses, producing some rare phages (transducing phages) that contain bacterial DNA instead of phage DNA.

Step 3 When transducing particles infect another cell, the bacterial DNA enters that cell.

Step 4 The bacterial DNA becomes incorporated by recombination into the recipient cell's chromosome.

Figure 6.22 Generalized transduction. The photo shows a mixture of phages and transducing particles; they cannot be distinguished by appearance.

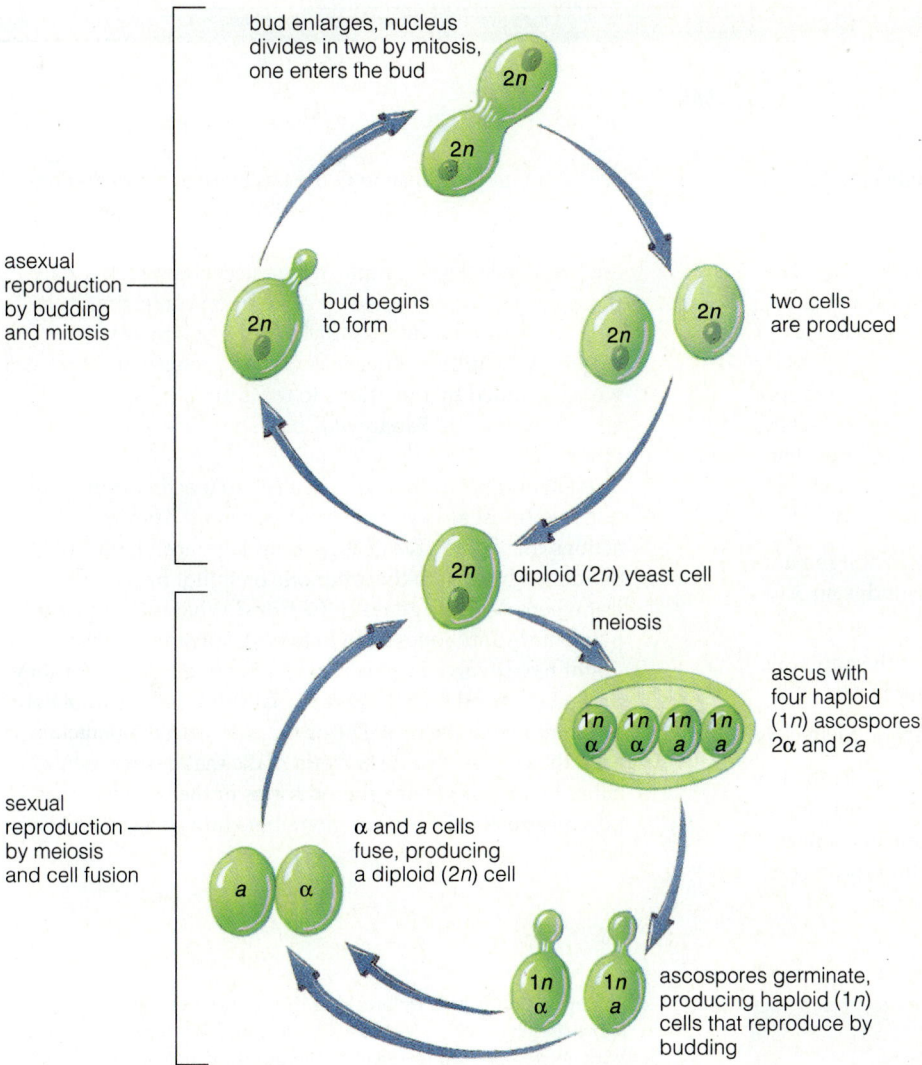

bud enlarges, nucleus divides in two by mitosis, one enters the bud

2n
2n

asexual reproduction by budding and mitosis

2n
bud begins to form

2n

2n
two cells are produced

2n
diploid (2n) yeast cell

meiosis

ascus with four haploid (1n) ascospores: 2α and 2a

1n α   1n α   1n a   1n a

sexual reproduction by meiosis and cell fusion

a   α
α and a cells fuse, producing a diploid (2n) cell

1n α   1n a
ascospores germinate, producing haploid (1n) cells that reproduce by budding

Figure 6.23 Baker's yeast is a eucaryote that can reproduce asexually or sexually. Asexual reproduction occurs by budding and mitosis. Meiosis produces two types of haploid ascospores (a and α). These germinate, producing haploid cells that fuse, producing diploid cells.

Population Dynamics

A population of microorganisms is constantly undergoing genetic change. Mutations are occurring, and favorable ones are retained while unfavorable ones are lost. If the culture is capable of transformation, conjugation, or transduction, and the conditions are favorable, genetic exchange is also occurring.

Genetic change can have a major impact on a population of bacteria. This can be seen most dramatically in a hospital setting, where antibiotic resistance can spread rapidly among bacteria. Certain genes encode antibiotic resistance, allowing bacteria to survive in the presence of antibiotics that would otherwise kill them. Genes that encode antibiotic resistance are often carried in transposons. Frequently these transposons are carried on conjugative plasmids that can be transferred from one species of bacteria to another, which might confer resistance to several antibiotics. Here, two powerful mechanisms promote genetic exchange: (1) the transposon's ability to move between the chromosome and a plasmid and (2) the plasmid's ability to move to other cells in the environment, even to cells of different species. This combination creates the potential for rapid spread of antibiotic-resistant genes through a bacterial population. In a hospital environment, where antibiotics are always present, strains of bacteria that carry genes for antibiotic resistance have a clear advantage over strains that are killed by antibiotics. As a result, a population of bacteria can be converted from antibiotic sensitivity to antibiotic resistance in a short time. The stage is set for a **nosocomial infection** (a hospital-acquired infection) that may not be treatable with antibiotics (Chapter 20).

The Genetics of Antibiotic Resistance

Microbiologists have learned about the differences between procaryotic and eucaryotic cells by studying the targets of antibiotic action (Chapter 4). For an antibiotic to be effective, the host eucaryotic cell must lack these targets or have different forms of them. The genetics of antibiotic resistance tells us even more about antibiotic action and procaryotic cells.

Antibiotic-resistant strains of bacteria first appeared soon after antibiotics came into common use. How could bacteria change genetically to become resistant to an antibiotic that strikes at a vital cellular function? Subsequent research revealed three answers.

1. The target in the bacterial cell changes so that it remains capable of fulfilling its vital cellular function and is no longer sensitive to the antibiotic.

2. The bacterial cell becomes able to exclude the antibiotic or to pump it out of the cell after the antibiotic enters.

3. The cell becomes able to destroy the antibiotic chemically.

Knowing the basis for antibiotic resistance can be the key to developing new antibiotics effective against resistant strains. Moreover, understanding the mechanisms of resistance has greatly benefited genetic research on procaryotic cells. Mutations that change the target lie in genes that encode the target, all of which are vital cellular enzymes or structures. Many such genes were first identified by antibiotic resistance. For example, the gene encoding RNA polymerase was first identified by a mutation conferring resistance to the antibiotic rifampicin (RNA polymerase is its target). Also, different genes encoding proteins in ribosomes were identified by mutations to the antibiotics streptomycin, erythromycin, kanamycin, and spectinomycin, among others.

Genetic changes that enable a cell to inactivate an antibiotic are almost always carried on plasmids. They encode reactions that add active groups (generally, acetyl, phosphate, or adenyl groups) to the antibiotic or split it by hydrolysis. But where do these genes come from? What role did they play before antibiotics came into widespread use? They might have evolved to inactivate other toxic agents. Or they might be altered forms of genes that fulfill some completely different metabolic role. This is the case with the penicillinases, the enzymes some bacteria make that destroy penicillin. Penicillinases are altered forms of the enzymes that polymerize peptidoglycan monomers into bacterial cell walls.

Summary

STRUCTURE AND FUNCTION OF GENETIC MATERIAL   (pp. 141–151)

1. Genetics is the science that studies heredity.

The Structure of DNA   (p. 141)
1. DNA is composed of the sugar deoxyribose, phosphate, and a nucleoside base. The base may be adenine (A), guanine (G), cytosine (C), or thymine (T).

2. Deoxyribose and phosphate form long strands that wrap around a central core of bases, forming a double helix. Hydrogen bonds that form between G-C and A-T base pairs hold the two strands together.

The Function of DNA   (p. 141)
1. DNA stores all the information required for a cell to grow, reproduce, and maintain itself.

2. Genes are the segments of DNA that carry the information for encoding a specific macromolecule.

3. DNA enters into two kinds of reactions: replication and gene expression.

Replication of DNA   (pp. 141–146)
1. Replication is making a copy of a DNA molecule from deoxyribonucleotides.

2. Replication begins by breaking the A-T and G-C bonds within a short stretch of DNA, forming a bubble and exposing bases to pair with nucleoside triphosphates, chemically activated forms of the nucleotide building blocks.

3. Each newly polymerized strand of DNA is complementary to and has bases that pair with the template (original strand). Each new double helix is composed of one new and one conserved strand. Thus, the process is called semiconservative replication.

4. Replication begins at a genetically specified point on the chromosome called the origin.

5. Replication forks travel simultaneously in opposite directions around the chromosome. When they meet at a point called the terminus, the two completed chromosomes separate.

6. The replication forks consist of the replication apparatus.

7. The two strands of a DNA double helix are antiparallel; one strand runs from a 3′ carbon to a 5′ carbon and the other runs from a 5′ to a 3′. Because DNA polymerase III can synthesize a continuous strand of DNA in only one direction, there is continuous synthesis on the leading strand and discontinuous synthesis on the lagging strand.

Gene Expression: Transcription   (pp. 146–147)
1. Gene expression consists of two steps, transcription and translation.

2. Transcription is the polymerization of ribonucleotide building blocks into a molecule of RNA—either messenger RNA (mRNA), transfer RNA (tRNA), or ribosomal RNA (rRNA).

3. Transcription begins when ribonucleoside triphosphates pair with the exposed bases on an exposed strand of DNA. The information in DNA is transcribed (rewritten): C pairs with G and A pairs with U (uracil).

4. Transcription begins near a site on the genome called a promoter. Here RNA polymerase separates the two strands, forming a bubble in the DNA, and uses one of them (the sense strand) as a template. Polymerization proceeds in the $5' \rightarrow 3'$ direction. It stops at a place called the terminator. The transcript is released, and the DNA bubble closes.

Gene Expression: Translation   (pp. 147–151)

1. Translation is the polymerization of amino acids into a protein.

2. All three products of transcription participate in translation: mRNA carries the information that determines the order of amino acids in the protein; rRNA is a component of ribosomes where translation takes place; and tRNA does the actual translating because each tRNA molecule has a site that binds to mRNA and a site that binds to an amino acid.

3. The nucleic-acid-recognizing end of the molecule consists of an anticodon that pairs with a codon on mRNA.

4. The specificity of codon-anticodon pairing determines the sequence of amino acids in a protein. The correspondence between codon and amino acid is called the genetic code. Any of the three nonsense codons stops translation at the end of a gene.

5. As a ribosome moves down a molecule of mRNA, exposing successive codons in a region called the A site, an amino-acid-bearing tRNA molecule attaches at each codon, positioning the amino acid to be polymerized onto a peptide at the ribosome's P site. Then the ribosome moves one codon down and the process repeats.

REGULATION OF GENE EXPRESSION   (pp. 151–156)

1. Gene expression is regulated by increasing or decreasing the rate of transcription or translation.

2. Inducible enzymes are produced only when a signal molecule is abundant, and repressible enzymes are produced only when a signal molecule is scarce in the environment. Constitutive enzymes are produced at a constant rate.

3. Transcription is often regulated by regulatory proteins that bind to DNA and change its interaction with RNA polymerase. When fewer transcripts are made, less protein is made.

4. An operon is a set of genes that is regulated and transcribed together. The *lac* operon in *Escherichia coli*, which encodes the ability to use lactose as a substrate, is regulated by a repressor. The arabinose operon in *E. coli* is regulated by an activator. In both cases the operon is expressed only when its products are needed.

5. Attenuation regulates transcription in bacteria by adjusting relative rates of transcription and translation. Regulation of the histidine operon is an example. When the supply of histidine is adequate, translation of the leader protein occurs so quickly that an attenuator loop forms, preventing transcription of the subsequent genes in the operon. When the supply of histidine is low, translation of the leader protein is slow; and an antiterminator loop is formed that prevents formation of the attenuator loop, so transcription of subsequent operon genes proceeds.

6. Translation of genes encoding ribosomal proteins is regulated by a gene that encodes a protein with two functions. One is to be incorporated into a ribosome, but the second is to inhibit translation. When concentration of the ribosomal protein is low, all the molecules are incorporated into ribosomes, but when there is enough free ribosomal protein, the second function is activated.

7. Catabolite repression is global regulation of gene expression in response to the availability of carbon. For example, catabolite repression inhibits the *lac* operon if a better substrate than lactose is available.

CHANGES IN A CELL'S GENETIC INFORMATION   (pp. 156–170)

The Genome   (pp. 156–157)

1. The genome is the sum total of a cell's genetic information (DNA).

2. The genome can change by mutation or by genetic transfer.

3. Most of a procaryote's genome is in its single circular chromosome, but some is in small circular structures called plasmids. Plasmids encode nonessential features. R factors are plasmids that encode resistance to antibiotics.

4. Most eucaryotes do not have plasmids; most of their genome is in chromosome pairs.

5. Genotype is the genes that a cell contains. Phenotype is the outward expression of a cell's genes.

Mutations   (pp. 157–162)

1. A mutation is a change in a cell's genotype. A base substitute mutation is a change in a single pair of bases to a different pair. A deletion mutation is total removal of a segment of DNA. An inversion mutation is the inversion of a segment of DNA. A transposition mutation is the movement of a segment of DNA to a different position on the genome. A duplication mutation is the addition of a new segment of DNA.

2. Every time the chromosome is replicated, mistakes occur and mutations result. Therefore, mutation rate is number of mutations per cell per generation, not time.

3. Spontaneous mutations are relatively rare but their impact is great. Induced mutations are caused by chemical, physical, or biological agents (mutagens).

4. Chemical mutagens may change a component of DNA or become incorporated into the DNA. Physical mutagens may fragment the DNA backbone or cause mistakes in replication. Biological mutations are sequences of DNA that themselves cause mutations by transposing.

5. Most mutations do not change the cell's phenotype. A missense mutation can be serious if the new amino acid that is encoded is not similar to the old. A nonsense mutation (one that changes a codon to a nonsense codon and stops translation) usually inactivates the gene product.

6. Lethal mutation results in destruction of an essential gene product. An auxotroph is a strain that has a nutritional requirement its parent did not have. Conditionally expressed mutations render a gene product nonfunctional only in certain environments.

The Uses of Mutant Strains   (pp. 162–163)

1. To isolate a mutant strain by direct selection, conditions are created that foster the growth of only the mutant strain.

2. In indirect selection (counterselection), conditions are created to prevent the growth of the desired mutant strain. Then growing cells are killed.

3. In the brute strength technique, large numbers of clones are examined one by one to find the desired mutant strain. Replica plating is a simplified brute strength technique.

4. The Ames test determines whether a particular chemical is mutagenic and therefore potentially carcinogenic.

Genetic Exchange among Bacteria (pp. 163–168)

1. Genetic exchange among bacteria is not an essential part of their life cycle. When it occurs, only a portion of the genome of the donor cell transfers to the recipient cell. A merozygote is produced.

2. A piece of transferred DNA must be incorporated into a replicon (the chromosome or a plasmid) before it is a permanent part of the recipient cell's genome. Incorporation occurs by recombination, or crossing over. A plasmid transferred intact to a recipient cell does not require recombination to become part of the recipient cell's genome.

3. Some bacteria are able to undergo natural transformation. DNA leaves the donor cell; later it is taken up by the recipient cell and incorporated into its genome.

4. In artificial transformation, the bacterial cells are treated in the laboratory to make them able to take up DNA from their environment. Artificial transformation was the basis for developing recombinant DNA technology (genetic engineering).

5. Conjugation is genetic transfer carried out by conjugative plasmids. The best-studied conjugative plasmid is the F plasmid in *Escherichia coli*. Conjugation begins when an F$^+$ cell encodes a sex pilus that attaches to an F$^-$ cell. The F plasmid is nicked and begins producing a single strand of plasmid DNA that enters the recipient cell. The recipient becomes F$^+$ and is now capable of transferring DNA to still another recipient.

6. Transduction is mediated by bacteriophages. Generalized transduction occurs during infection of a bacterial cell by a virulent phage. When a phage particle is assembled that contains the bacterial DNA, it can be introduced into a new cell where genetic exchange occurs.

7. Specialized transduction occurs during infection by a temperate phage. When a prophage leaves the cell, and carries bacterial genes as well as phage genes, it injects them into a new bacterium.

Genetic Exchange among Eucaryotic Microorganisms (p. 168)

1. Genetic exchange among microorganisms is similar to genetic exchange in other eucaryotes (plants and animals). A diploid cell undergoes meiosis; haploid gametes develop; they fuse to form a zygote or new diploid cell.

2. Baker's yeast, *Saccharomyces cerevisiae*, is a fungus that undergoes both asexual (budding and mitosis) and sexual (meiosis) production.

Population Dynamics (pp. 169–170)

1. Mutagenic resistance to antibiotics can cause nosocomial (hospital-acquired) infections.

Review Questions

STRUCTURE AND FUNCTION OF GENETIC MATERIAL

1. Define genetics.

2. How did Griffith's work set the stage for applying genetic theory to microorganisms? Why was it significant that the genetic material in bacteria was DNA?

3. What are the building blocks of DNA? Name the bases and tell which pair with which.

4. What is DNA's function? What is a gene?

5. What are the two kinds of reactions DNA enters into? What is transcription? What is translation?

6. Explain this statement: DNA is the master plan of metabolism, and enzymes are the chief mediators.

7. What is replication?

8. How does the structure of DNA lead to the correct ordering of bases during replication?

9. Why is ATP necessary for replication? Where does it enter the replication process? What role does DNA polymerase III play?

10. Explain the mechanics of replication. In your explanation, use and define these terms: template, complementary strand, semiconservative replication, replication fork, origin, terminus.

11. What is the replication apparatus and how does it function?

12. Explain this statement: The two strands of any double helix are antiparallel.

13. Why is a primer important for the function of DNA polymerase III? What are the leading and lagging strands in replication? What roles do primase, DNA polymerase I, and DNA ligase play in the final phase of replication?

14. Define transcription.

15. Complete this statement: Replication is to DNA as _____ is to RNA. How is the ordering of building blocks determined in transcription? How are replication and transcription similar? How are they different?

16. Describe how transcription occurs. Use and define these terms: promoters, terminator, sense strand, transcript.

17. What are the three products of transcription?

18. Explain this statement: Translation changes the information in the metabolic master plan from the language of nucleic acids into the language of proteins.

19. Briefly, what is the function of each of the three products of transcription in translation?

20. Explain how tRNA functions in translation. Use and define these terms: activation, anticodon, codon.

21. What is the genetic code? Why is the genetic code called "redundant"? What are nonsense codons?

22. Describe the mechanics of translation. Use and define these terms: A site, P site, start codon, Shine-Dalgarno sequence/ribosome binding site, polysome.

REGULATION OF GENE EXPRESSION

1. What is metabolic regulation and why is it important?

2. What are the two ways to regulate metabolism?

3. What roles do inducible enzymes and repressible enzymes play in metabolic regulation? Give an example of an enzyme produced constitutively.

4. How do procaryotes regulate transcription? What role do allosteric regulatory proteins play?

5. Define operon. What does the *lac* operon encode? How does the *lac* operon function in the absence of lactose? How does it function in the presence of lactose?

6. How is regulation of the arabinose operon like regulation of the *lac* operon? How is it different? How does the arabinose operon function in the presence of arabinose? How does it function in its absence? What is the end result?

7. What is attenuation?

8. Explain how expression of the histidine operon is regulated. In your explanation, use and define these terms: leader protein, attenuator loop, and antiterminator loop.

9. How is the translation of ribosomal proteins regulated?

10. How is global regulation different from regulation of the *lac* operon, for example, or different from attenuation?

11. Explain catabolite repression as an example of global regulation. What roles do cyclic AMP and CAP play?

CHANGES IN A CELL'S GENETIC INFORMATION

1. Name the two ways that changes can occur in a cell's genetic information.

2. What constitutes the genome in eucaryotes? In procaryotes?

3. Describe the appearance and function of plasmids.

4. What are R factors? Why are they medically significant?

5. What are conjugative plasmids?

6. Define genotype. Define phenotype.

7. Explain these types of mutations: base substitution, deletion, inversion, transposition, duplication.

8. Why is the rate of mutation calculated on the basis of reproduction rather than time?

9. What are spontaneous mutations? Why do they have a significant impact on laboratory cultures even though they occur relatively rarely?

10. What are induced mutations? What is a mutagen?

11. Give an example of a chemical mutagen. How does ultraviolet light act as a mutagen?

12. Why are biological mutagens called transposable elements, or jumping genes? What are transposons and how do they benefit a cell? What are insertion sequences and how do they function?

13. What two factors determine how serious a mutation will be for the phenotype of a cell? What is a missense mutation? A nonsense mutation?

14. What is a lethal mutation? An auxotroph?

15. What are conditionally expressed mutations? Give an example.

16. Explain these methods of identifying mutant strains: direct selection, indirect selection (counterselection), brute strength, replica plating.

17. How is the Ames test performed?

18. Name the three ways genetic exchange occurs in bacteria. Compare and contrast genetic exchange in procaryotes and eucaryotes.

19. What is transformation? Describe natural transformation. Describe artificial transformation.

20. Describe conjugation in *Escherichia coli*. What is an Hfr cell?

21. What are bacteriophages (phages)? Describe generalized transduction. Describe specialized transduction. Use the terms lytic cycle and lysogenic cycle in your explanation.

22. Describe the forms of genetic exchange in baker's yeast as examples of genetic exchange in eucaryotes.

23. Discuss an example of how genetic change in a population of bacteria can lead to serious hospital-acquired (nosocomial) infections.

Essay Questions

1. Mutation rate is regulated by metabolic processes that occur within a cell. Discuss why evolution has led to a low but detectable rate instead of a higher or lower one.

2. Speculate about which of the three known types of genetic exchange between bacteria is most likely to occur in soil, in a lake, or in an infected person. Justify your speculation.

Suggested Readings

Ames, B. W. 1979. Identifying environmental chemicals causing mutations and cancer. *Science* 204:587.

Drake, J. W. 1991. Spontaneous mutation. *Annual Reviews of Genetics* 25:125–46.

Freifelder, D. 1987. *Microbial genetics*. Boston: Jones and Bartlett.

Glass, R. E. 1982. *Gene function:* E. coli *and its heritable elements*. Berkeley: University of California Press.

7 RECOMBINANT DNA TECHNOLOGY

Happy Anniversary

In September 1992, two girls, ages 6 and 11, were honored guests at the National Institutes of Health (NIH) second-anniversary celebration of a medical landmark—the beginning of *gene therapy* (treating disease with new genes) for humans. Many scientists had believed gene therapy would not be possible for a long time and, even then, they doubted it would be effective. But these two children were proof that gene therapy is happening today and working astoundingly well. The girls were healthy, active, attending public school—and very much enjoying being the center of attention as the first humans to be treated by gene therapy.

Both girls were born with a defective adenosine deaminase (ADA) gene, which meant they had virtually no immune defenses against microbial disease. Their first microbial infection could be fatal. At first it seemed that the only way they could avoid infection would be to live in total isolation, receiving microorganism-free air, food, and water (see the box "David—Life in a Germ-Free World" in Chapter 18). But there might be another alternative—gene therapy. Successful gene therapy would give these children new, functional ADA genes, which would in turn give them functional immune systems and normal lives. Initially, the therapy was a treatment, not a cure; white blood cells carrying the genes had to be reinserted every few months. But a new approach—inserting normal genes into the stem cells that produce white blood cells—offers hope for a permanent cure.

The spectacular success of gene therapy in these two cases had an immediate impact. By October 1992, 18 separate gene therapy trials were underway worldwide. They were designed to treat such diverse diseases as cancer, liver disease, AIDS, and hemophilia. Almost certainly, gene therapy will eventually cure many diseases whose victims cannot be treated today.

Gene therapy was made possible by the advent of recombinant DNA technology. But it is only one of many

Molecular biologist French Anderson developed the gene therapy for ADA deficiency and led the treatment team for the two young girls. He is shown here at the opening session of the 1993 American Society for Microbiology, where he described his landmark research.

medical benefits. This powerful new technology has also enabled us to develop effective new chemotherapeutic drugs, vaccines, and diagnostic procedures. Recombinant DNA technology has found practical uses in agriculture, food processing, criminology, and basic science. Eventually it will touch all aspects of our lives. It has given new life to *biotechnology*, the practical uses of organisms and their products. In fact, biotechnology is expected to become the United States' largest industry in the twenty-first century.

To understand:
- The nature of recombinant DNA technology as a collection of techniques and the potential it has to affect every aspect of our lives

- The fundamental tool of recombinant DNA technology—gene cloning—and the five steps involved: obtaining DNA, splicing genes into a cloning vector, putting recombinant DNA into a host cell, testing, and propagating

- Methods of finding the right gene for gene cloning

- The uses of *Escherichia coli* as a host cell for recombinant DNA

- Current applications of recombinant DNA technology in medicine, industry, agriculture, and criminal investigation and applications we can expect in the future

RECOMBINANT DNA TECHNOLOGY

Recombinant DNA technology is easy to appreciate but hard to define. It is not a technique or a procedure. Rather, it is a vast collection of different procedures for taking DNA from a cell, manipulating it *in vitro*, and putting it into another cell, usually one of a different species. But the fundamental step in all these procedures is the one that gives the technology its name—*recombining* genes from different DNA molecules into a single molecule.

Recombination is a technical term geneticists use to describe the processes of forming a new combination of genes by any means, natural or artificial. For example, recombination occurs during sexual reproduction. All offspring acquire all their genes from their parents, but each offspring has a different combination of genes from those found in either of its parents. Most such recombination in eucaryotes occurs because an offspring receives half its chromosomes (and the genes they carry) from each parent. The result is a recombination of parental genes.

But sometimes (usually during meiosis in eucaryotes and in all types of genetic exchange between procaryotes) recombination has a different basis—genes on the *same* chromosome or DNA molecule recombine by crossing over (Chapter 6). In essence, chromosomes or DNA molecules break and a fragment of one joins to a fragment of another. The result is a **recombinant molecule**, part of which is derived from one chromosome (or DNA molecule) and part from another. Such recombinant DNA

molecules form frequently in living cells as a normal part of their development, but only between chromosomes or DNA molecules that are **homologous** (similar enough for a single strand of one DNA molecule to form a hydrogen-bonded double strand with a single strand of the other DNA molecule with only a few noncomplementary base pairs).

The procedures of recombinant DNA technology can also produce recombinant DNA molecules, but with two important differences. First, recombination occurs *in vitro*, not in the cell. Second, the DNA molecules that join to form a recombinant molecule do not have to be homologous—they do not even have to be similar. Pieces of DNA from distantly related organisms—humans and bacteria, for example—can be joined into a single molecule.

In this chapter we first discuss gene cloning, the fundamental tool of recombinant DNA technology. Then we look at how recombinant DNA technology is applied in basic and applied science and what applications we might expect in the future. Specific uses of recombinant DNA technology are discussed throughout the rest of this book. Its impact on **biotechnology** is discussed in Chapter 29.

GENE CLONING

Gene cloning, the basic tool of recombinant DNA technology, is the process of obtaining a set of identical copies of a gene. The process is called *cloning* because a **clone** is any identical group of progeny derived from an individual. Just as a clone of bacteria—for example, a colony on a petri dish—is a group of cells derived from a single cell (Chapter 3), a clone of genes is a group of identical genes derived from a single gene copy or molecule.

To obtain a gene clone, a single copy of the gene is inserted into a cell. The clone of cells that develops contains a clone of the inserted gene. Gene cloning is the cornerstone of recombinant DNA technology because it is the way to produce the massive numbers of gene copies needed for other recombinant DNA procedures.

Gene cloning involves five steps (**Figure 7.1**):

1. Obtaining DNA that contains the gene to be cloned

2. Splicing a piece of DNA containing the gene into a **cloning vector** (a DNA molecule that a cell will replicate)

3. Putting the recombinant DNA (the cloning vector with the desired gene spliced into it) into an appropriate host cell

4. Testing to assure that the desired gene has been inserted into the host cell

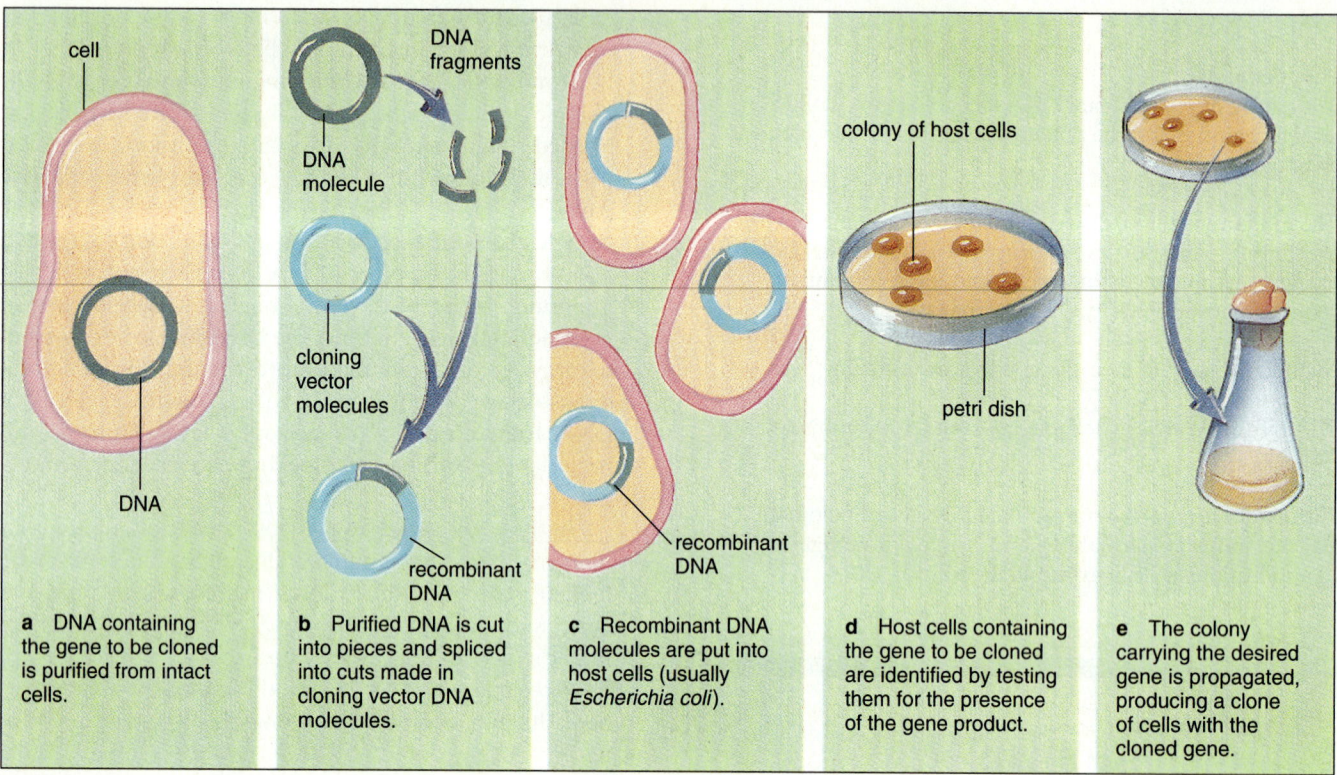

a DNA containing the gene to be cloned is purified from intact cells.

b Purified DNA is cut into pieces and spliced into cuts made in cloning vector DNA molecules.

c Recombinant DNA molecules are put into host cells (usually *Escherichia coli*).

d Host cells containing the gene to be cloned are identified by testing them for the presence of the gene product.

e The colony carrying the desired gene is propagated, producing a clone of cells with the cloned gene.

Figure 7.1 Gene cloning.

5. Propagating the host cell to produce a clone of cells with the clone of genes

Let's look at each of these steps individually.

Obtaining DNA

About 3 percent of the dry weight of a bacterial cell is DNA, and in eucaryotic cells the percentage is even smaller. But DNA is easy to separate in pure form from other cell components because it has unusual chemical and physical properties. First, DNA is an extremely large molecule. For example, in the chromosome of *Escherichia coli*, a single DNA molecule is a millimeter long. In addition, DNA is highly resistant to chemical and physical treatments that rapidly destroy other cell components. Finally, it is denser (heavier per unit volume) than most other cell components.

DNA's special properties make it relatively easy to purify DNA from a **cell extract**, the liquid content of ruptured cells (**Figure 7.2**). Because of its large size, DNA forms ropelike aggregates in the presence of alcohol; these aggregates can then be separated from the extract by wrapping them around a glass rod and pulling it out of the liquid. DNA's chemical resistance allows it to sur-

vive treatment with phenol, which denatures and precipitates proteins. Its high density allows it to be separated from the other less dense cell components by centrifugation. The first step in cloning a bacterial gene is purifying DNA, using one or a combination of such treatments.

Obtaining DNA to clone a procaryotic gene is done simply by purifying it from other components of the bacterial cell. But obtaining DNA to clone a eucaryotic gene is more complicated because eucaryotic genes contain regions called **introns**. Introns are noncoding regions that enzymes in the eucaryotic cell cut out after the gene has been transcribed into mRNA (**Figure 7.3**). But because bacterial genes lack introns, bacterial cells, the usual host cells for cloned genes, don't have the resources to cut introns out of mRNA. Thus, intron-free DNA must be used to clone eucaryotic genes into bacteria. To obtain intron-free DNA, the gene cloner purifies mRNA instead of DNA from the eucaryotic cell extract. Then the gene cloner uses reverse transcriptase to make intron-free DNA from the mRNA. (**Reverse transcriptase** is an enzyme that uses RNA as a template to make a complementary strand of DNA.) This DNA is termed **cDNA**, for **complementary DNA**. It has the coding properties of normal eucaryotic genes but lacks introns, and bacteria produce the normal eucaryotic gene product from it.

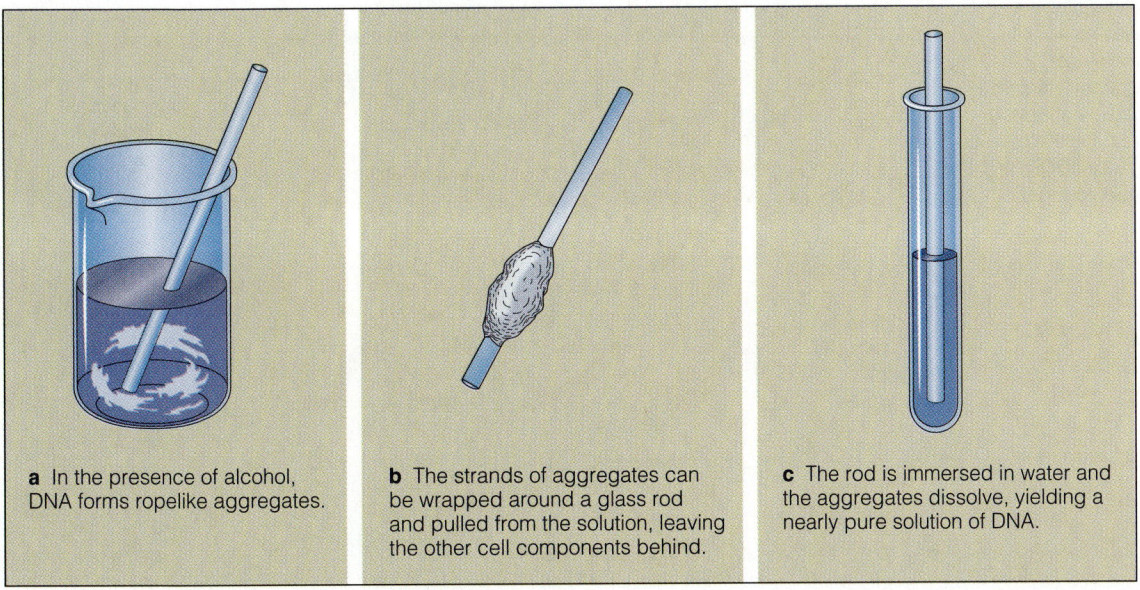

a In the presence of alcohol, DNA forms ropelike aggregates.

b The strands of aggregates can be wrapped around a glass rod and pulled from the solution, leaving the other cell components behind.

c The rod is immersed in water and the aggregates dissolve, yielding a nearly pure solution of DNA.

Figure 7.2 Obtaining DNA to clone a bacterial gene.

Splicing Genes into a Cloning Vector

After DNA containing the desired gene is obtained, it is spliced into a cloning vector. This step is necessary because only certain DNA molecules are replicated in cells. Cloning vectors are DNA molecules that contain a region called an **origin of replication** and consequently are replicated (Chapter 6). It is highly unlikely that the gene to be cloned will have an origin of replication nearby. Therefore, to assure that the fragment of DNA containing the desired gene will be replicated, it is inserted into a cloning vector. Then the gene is replicated along with the rest of the cloning vector.

Good cloning vectors need other properties in addition to an origin of replication. They must be relatively small so that their replication does not unduly tax the host cell's metabolic capacity. Also, they must carry other genes that identify host cells that contain the vector. For example, most cloning vectors carry genes encoding antibiotic resistance so that cells with the vector and the cloned gene can be easily identified by their ability to multiply in antibiotic-containing media.

Plasmids and the genomes of certain viruses are used as cloning vectors. Almost all of these cloning vectors are circular molecules of DNA. Inserting the gene to be cloned into the cloning vector involves two steps—cutting the cloning vector and sealing it back together (a process called ligation). Ligation occurs after the fragment of DNA containing the gene to be cloned is inserted between the cut ends.

Cutting DNA with Restriction Endonucleases. Cloning vehicles and DNA to be inserted into them are usually cut with enzymes called **restriction endonucleases**. These enzymes have particularly valuable properties for cloning. But before we discuss how restriction endonucleases are used to clone genes, let's consider the question, why don't restriction endonucleases cut up the DNA in the bacteria that produce them? The answer is that bacteria protect their own DNA from their own restriction endonucleases by modifying it. DNA is modified by the addition of methyl groups to certain bases within the sequences where restriction endonucleases cut (**Table 7.1**). But foreign DNA, such as DNA injected by an infecting virus, is not modified. It is rapidly cut and thereby destroyed (see the box on Werner Arber, Chapter 13). Most microbiologists believe that restriction endonucleases evolved to protect bacteria from viral attack.

Restriction endonucleases recognize specific base sequences and cut the DNA within or near them. Useful cloning vectors have only a single site at which the endonuclease to be used for cloning cuts. Because most bacterial species produce their own particular type of restriction endonuclease, these enzymes are named for the species from which they are obtained, using the initial of the genus and the two first letters of the species. For example, a restriction enzyme from *Escherichia coli* is designated *Eco*. If a species produces more than one enzyme, they are distinguished by Roman numerals. Also, if enzymes are encoded on an R factor (Chapter 6), they are

Figure 7.3 Obtaining DNA from a eucaryotic gene to clone into a eucaryote. Because many genes from eucaryotes contain introns, scientists purify the intron-free eucaryotic mRNA (instead of DNA) and use the enzyme reverse transcriptase to make cDNA, an intron-free copy of the eucaryotic gene.

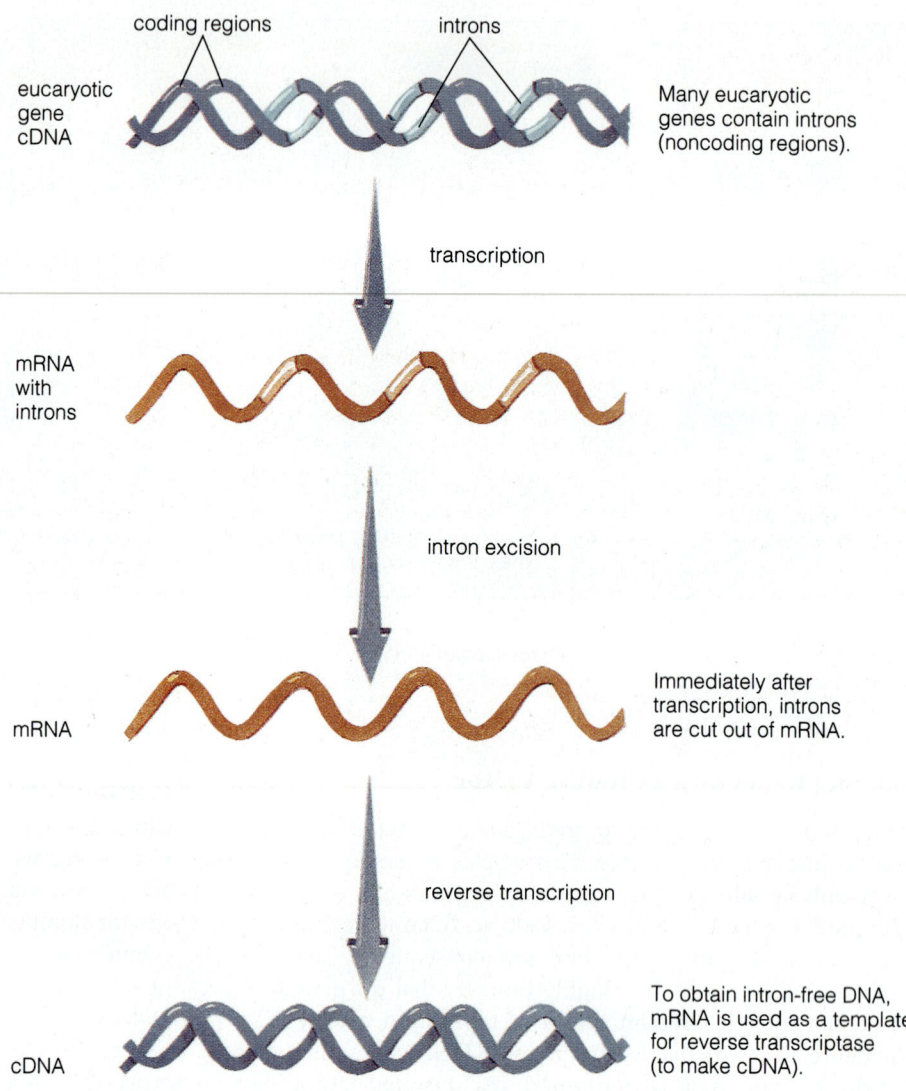

coding regions

introns

eucaryotic gene cDNA

Many eucaryotic genes contain introns (noncoding regions).

transcription

mRNA with introns

intron excision

mRNA

Immediately after transcription, introns are cut out of mRNA.

reverse transcription

cDNA

To obtain intron-free DNA, mRNA is used as a template for reverse transcriptase (to make cDNA).

designated R. Accordingly, one of the most commonly used restriction endonucleases is named *Eco* RI. It is one of several restriction enzymes produced by *E. coli*, and it is encoded by an R factor. Another useful restriction endonuclease—one obtained from the highly thermophilic bacterium *Thermus aquaticus*—is named *Taq* I.

Properties of Restriction Endonucleases. Members of a certain class of restriction endonuclease (called **type II**) are particularly valuable for cloning. Type II restriction endonucleases cut asymmetrically within base sequences that are four, five, six, or seven base pairs long and that have an **axis of rotational symmetry**. These base sequences read the same way on one strand as the opposite way on the other strand. For example, the sequence

--A-G-A-T-C-T--
--T-C-T-A-G-A--

has an axis of rotational symmetry because it reads the same backward and forward. Such sequences are termed **palindromic** because they resemble word and phrase palindromes ("Otto" and "Madam I'm Adam"). Sequences such as

--A-G-C-C-G-A--
--T-C-G-G-C-T--

are also termed palindromic, though they occur within a single strand and do not have an axis of rotational symmetry.

Many type II restriction endonucleases cut asymmetrically within the axis of rotational symmetry. Thus, they produce short, complementary, single-strand regions called **cohesive ends**, or **sticky ends**, because they readily pair with one another. For example, *Eco* RI cuts each strand between G and A (see arrows) in the sequence

Microbe Mappers

Cohen, Boyer, and Corned Beef

Beginnings in science are hard to pinpoint, but no one can argue that recombinant DNA technology didn't begin in 1972 in a Waikiki delicatessen over corned beef sandwiches.

Stanley Cohen and Herbert Boyer were attending a scientific conference in Honolulu. Cohen, who was studying bacterial plasmids, was struck by their flexibility. They replicate independently of the bacterial chromosome and can be moved readily from one bacterial cell to another even in the form of a pure solution of DNA. Cohen reasoned that he could move genes from any source just as readily if he could put them in a plasmid. In other words, he could clone a

gene. But how to do it? At the conference, Cohen attended Boyer's talk on restriction enzymes and came up with an idea.

The two scientists met for dinner and decided to collaborate. Boyer's laboratory at the University of California in San Francisco was only an hour's drive from Cohen's at Stanford University. The project went quickly—and successfully. Cohen called the recombinant DNA (the plasmid and foreign gene) a chimera, after the mythological fire-breathing beast with a lion's head, a goat's body, and a serpent's tail. The age of recombinant DNA technology was born.

Table 7.1	Properties of Some Restriction Endonucleases		Characteristics	
Restriction Endonuclease	Source (Bacterial Species)	Target Site (Cuts at Arrow)	Recognizes (No. Base Pairs)	Product
Eco RI	*Escherichia coli* R13	↓ G-A-A-T-T-C C-T-T-A-A-G ↑	6	4-base-long sticky ends
Hha I	*Haemophilus haemolyticus*	↓ G-C-G-C C-G-C-G ↑	4	2-base-long sticky ends
Sma I	*Serratia marcescens*	↓ C-C-C-G-G-G G-G-G-C-C-C ↑	6	Blunt ends
Hae III	*Haemophilus aegyptius*	↓ G-G-C-C C-C-G-G ↑	4	Blunt ends

↓
--G-A-A-T-T-C--
--C-T-T-A-A-G--
↑

producing

--G   A-A-T-T-C--
--C-T-T-A-A   and   G--

DNA from *any* organism—microorganism, plant, or animal—will produce the same sticky ends (A-A-T-T-- and

--T-T-A-A) when cut by *Eco* RI or a similar restriction endonuclease. When DNA molecules cut with this kind of restriction endonuclease are mixed and held at a low temperature (near the freezing point of water), the sticky ends pair and *anneal* (join together). They anneal because hydrogen bonds form between complementary bases. Sticky ends cut with *Eco* RI would anneal to form

--G A-A-T-T-C--
--C-T-T-A-A G--

The ends anneal, but **gaps** (missing phosphodiester bonds) exist between the G and A on both strands.

Thus, using a restriction endonuclease, any two pieces of DNA can be joined together. It is necessary only to cut both of them with the same restriction endonuclease, mix them, and allow them to anneal.

Monitoring the Reaction. Restriction endonucleases catalyze a precise chemical reaction—cutting DNA at all copies of the sequences that a particular restriction endonuclease recognizes. How do we monitor such a reaction? That is, how do we determine when all the vulnerable sites have been cut? Usually **gel electrophoresis** is used: Samples of the reaction mixture are placed on a slab of gel (usually made of **agarose**, a derivative of agar, or **polyacrylamide**, a synthetic polymer similar to the polymer used to make acrylic fabric). Then the gel is exposed to an electrical field (**Figure 7.4**). Because DNA is negatively charged, it moves toward the positive pole of the field. Different DNA becomes separated on the gel because small molecules move faster than large ones. The location on the gel of the various-sized pieces of DNA is revealed by staining the gel with **ethidium bromide** (or a similar compound) and illuminating the gel with ultraviolet light. Ethidium bromide binds tightly to DNA and fluoresces. The locations of the DNA molecules appear as red-orange lines or bands on a light pink background. When cutting by restriction endonuclease is complete, no new (smaller) bands appear. When cut with a particular restriction endonuclease, each DNA molecule produces a characteristic pattern of bands after gel electrophoresis.

This occurs because each DNA molecule has a definite number and definite location of sites at which the restriction endonuclease cuts.

Because specific patterns of bands are produced, cutting DNA samples with a restriction endonuclease and separating the fragments by electrophoresis determine if two samples of DNA are different or—with high probability—identical. Comparing gel patterns of DNA cut by restriction endonuclease is the way DNA samples are analyzed in criminal investigations. If a suspect's DNA pattern differs from that of an incriminating sample recovered at the crime scene, that individual may be ruled out as a suspect.

Ligation. After the sticky ends produced by a restriction endonuclease anneal, the gaps that remain at sites where the original cuts were made must be sealed. Otherwise, the molecule would easily fall apart. The gaps are sealed, or *ligated*, by forming a covalent bond between the adjacent bases through the action of an enzyme called **DNA ligase**. Using our earlier example, the intact structure would thus be

--G-A-A-T-T-C--
--C-T-T-A-A-G--

All organisms produce DNA ligases, but the one most commonly used in cloning is **T4 ligase**. T4 ligase is an enzyme formed by *Escherichia coli* when it is infected by bacteriophage T4.

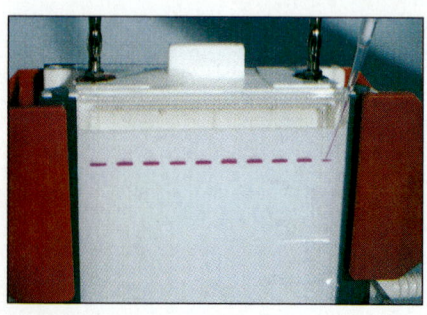

a Each sample to be examined is placed in one of the wells in the top of a slab of gel held between two pieces of glass.

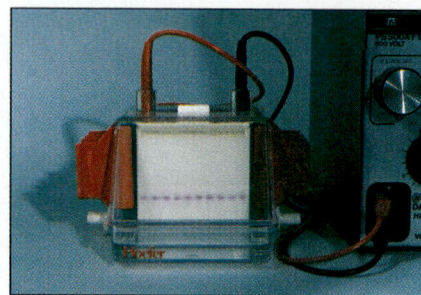

b The gel is exposed to an electric field (positive on bottom, negative on top) causing the negatively charged fragments of DNA to move down the gel. Small fragments move faster than large ones.

c The gel is stained with ethidium bromide and illuminated with ultraviolet light. Regions to which DNA molecules have moved appear as orange bands.

Figure 7.4 Gel electrophoresis is a way to monitor the cutting of DNA with restriction endonucleases. As the reaction proceeds, bands corresponding to larger molecules disappear and those corresponding to smaller molecules appear. When no more small-molecule bands appear, the reaction is complete.

Tree Testimony

Law enforcement was quick to exploit the power of recombinant DNA technology. DNA from bits of dried human blood, semen, or tissue collected at the scene of a crime can be compared with DNA of a suspect to provide powerful proof of guilt or innocence. Recently, an Arizona judge, Susan Bolton, extended the forensic authority of DNA analysis when she admitted analysis of plant DNA as evidence in a murder trial. On May 2, 1992, the body of a woman was found in the Arizona desert lying near a Palo Verde tree, a leafless tree typical of that region. Seeds from a Palo Verde tree were found in the back of a suspect's truck, but who could say from exactly which tree they had come? Recombinant DNA technology provided the answer. The pattern of DNA obtained by the polymerase chain reaction (PCR) from the seeds precisely matched the pattern obtained from DNA from the tree at the crime scene but from no others tested. The seeds almost certainly came from that particular tree.

Blunt-End Ligation. Although annealed sticky ends facilitate **ligation**, they are not absolutely necessary. Pieces of DNA without an extending single strand, which are termed *blunt-ended*, can also be ligated. The process is called **blunt-end ligation**. Blunt-end ligation is less efficient than ligation of gaps in annealed sticky ends because blunt ends contact one another only transiently. Nevertheless, it has certain advantages and is often used in cloning.

One advantage of blunt-end ligation is that the DNA fragment bearing the cloned gene and the vector need not be cut by the same restriction endonuclease. Another advantage is that the DNA to be cloned can be cut mechanically by **shearing**. In shearing, the solution of DNA is vigorously stirred or rapidly drawn through a small orifice such as the opening of a pipette. This process tends to break the DNA molecule near its midpoint and then break the two fragments near their midpoints again. To ligate sheared DNA, we have to assure that the ends are blunt by treating the solution of DNA fragments with an enzyme that removes any single strands on the ends of the molecule. Then the blunt-ended DNA fragment can be ligated into a vector that has also been cut to generate blunt ends. Some restriction endonucleases—for example, *Sma* I, from *Serratia marcescens*—form blunt ends directly. *Sma* I cuts the DNA sequence

$$\text{--C-C-C-G-G-G--}$$
$$\text{--G-G-G-C-C-C--}$$

between G and C, forming two blunt ends

$$\begin{array}{ccc} \text{--C-C-C} & & \text{G-G-G--} \\ \text{--G-G-G} & \text{and} & \text{C-C-C--} \end{array}$$

Blunt ends spontaneously anneal, though not as tightly as sticky ends because they are not held together by hydrogen bonds between complementary bases. Still, blunt ends anneal tightly enough to be ligated by the same enzymes (usually T4 ligase) under conditions used to ligate the gaps in sticky ends.

Putting Recombinant DNA into a Host Cell

Regardless of how the desired gene is inserted into a cloning vector and regardless of which cloning vector is used, the gene cannot be easily replicated (a clone will not be formed) until the recombinant molecule is put back into a cell. It is possible to replicate DNA *in vitro* using purified enzymes, but the only practical way to replicate recombinant DNA quickly, inexpensively, and in large amounts is to insert it into a living cell. A number of procedures are available to put DNA molecules into a host cell. They include transformation, transfection, microinjection, and electroporation.

Transformation is the process by which an intact cell takes up DNA from solution (Chapter 6). Some bacteria have an innate ability to be transformed by DNA from solution, a process called natural transformation. Most bacteria, however, can be transformed only after they have been specially treated, a process called **artificial transformation**. For example, *Escherichia coli* will take up DNA from solution in its environment after it has been exposed to high concentrations of calcium chloride ($CaCl_2$), chilled to 0°C, and suddenly heated to 42°C. Other treatments have been developed to transform other bacterial species as well as plant and animal cells that do not undergo natural transformation.

If DNA from a virus is used as the cloning vector, the process is called **transfection**. A fundamental property of viruses is their ability to introduce their genome into a host cell (Chapter 13). So if recombinant DNA is incorporated into a virus particle, it can be inserted into a host

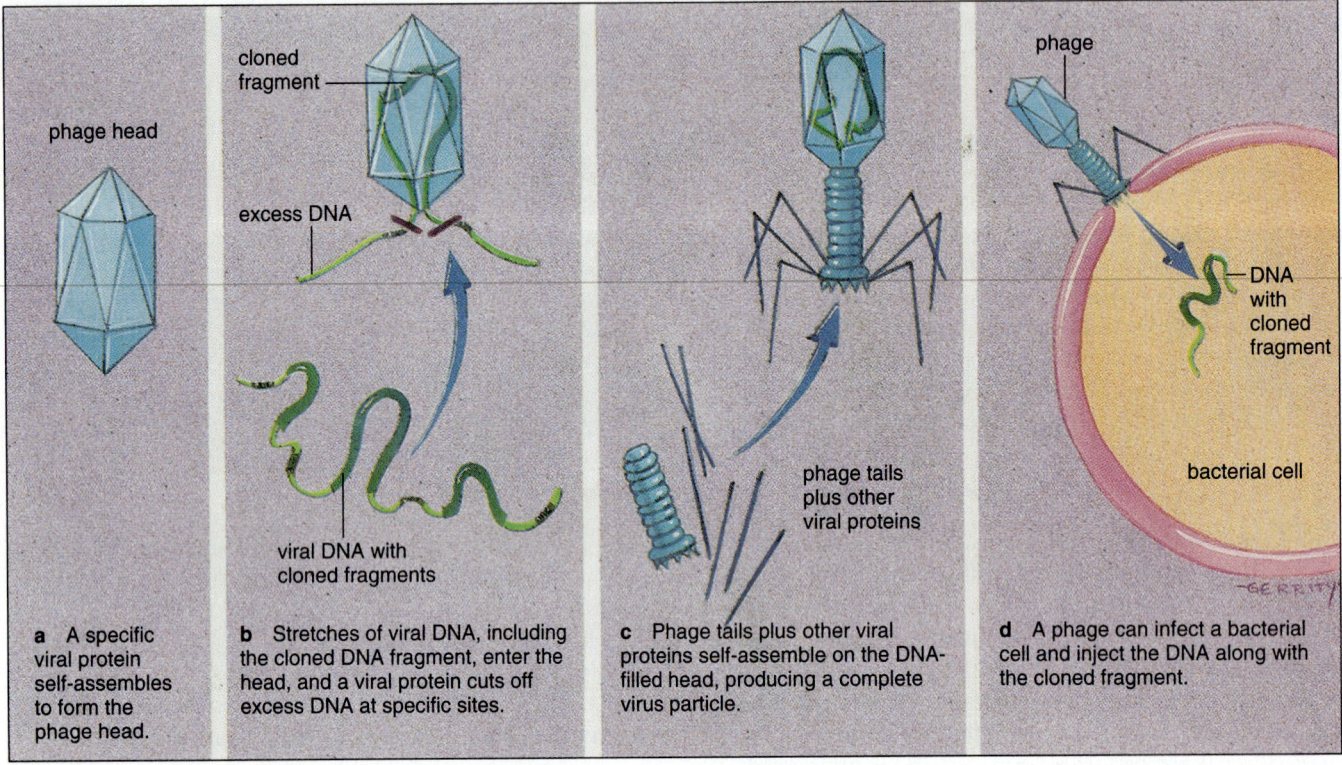

Figure 7.5 Assembling a virus capable of injecting cloned DNA into a bacterium. In the case of the bacteriophage lambda, the vector (with cloned DNA) is mixed with viral protein which assembles into a phage particle capable of injecting its DNA into an *Escherichia coli* cell.

a A specific viral protein self-assembles to form the phage head.

b Stretches of viral DNA, including the cloned DNA fragment, enter the head, and a viral protein cuts off excess DNA at specific sites.

c Phage tails plus other viral proteins self-assemble on the DNA-filled head, producing a complete virus particle.

d A phage can infect a bacterial cell and inject the DNA along with the cloned fragment.

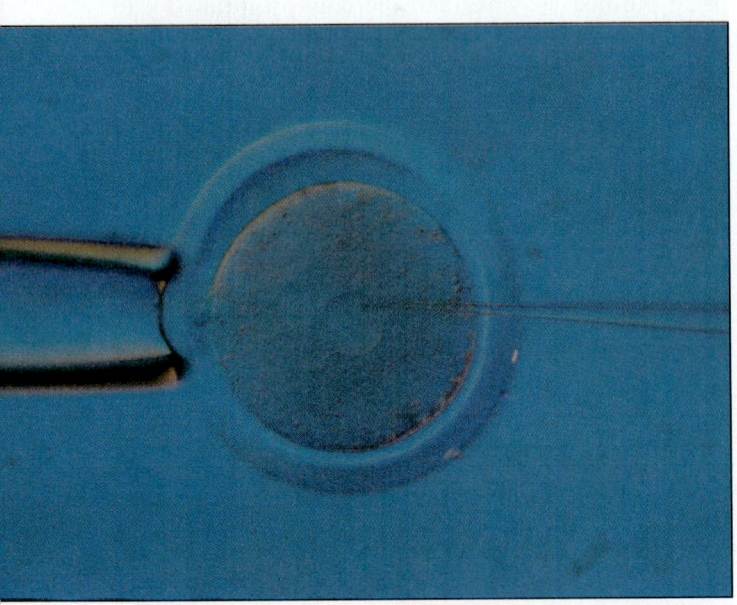

Figure 7.6 Microinjection is one way recombinant DNA can be introduced into a host animal cell.

cell the same way the virus inserts its own DNA. For example, when DNA from the bacteriophage lambda (λ) is used as the cloning vector, it can be incorporated into an infectious viral particle by adding purified viral proteins to a solution of the recombinant DNA (**Figure 7.5**). Then the artificially constructed virus can insert the recombinant molecule into a susceptible *E. coli* cell.

DNA can be inserted into some animal cells directly by **microinjection** (**Figure 7.6**). The cell is held by a slight vacuum to the end of a holding pipette, and DNA is injected into it from a tiny (diameter about 1.5 µm) pipette. Microinjection cannot be used with bacterial cells because they are too small to be injected even by such a tiny pipette.

DNA can be introduced into many bacterial, animal, and plant cells by **electroporation**. In this process, the cells are suspended in a solution of DNA and exposed to high-voltage electrical impulses. The electrical impulses destabilize the plasma membrane, making it permeable so that the DNA in solution can enter the cell.

Finding the Right Gene

The basic step in cloning is cutting a cloning vector and splicing into the cut a fragment of DNA carrying the gene to be cloned. But how do we know the fragment being spliced in carries the gene we want to clone? There are two ways—prior purification and subsequent identification.

Prior Purification.   In **prior purification**, a fragment carrying the gene to be cloned is purified and spliced into the cloning vector. To purify a gene or mRNA made from it, we have to be able to distinguish it from other similar molecules. We can do this by taking advantage of the specific hydrogen bonding that occurs between complementary single strands of DNA or between a strand of DNA and an mRNA made from it. Pairing between such complementary molecules occurs spontaneously—a process called **hybridization** (Chapter 6). So to identify a particular gene or its mRNA product, we need only a complementary DNA strand appropriately tagged (usually with a radioactive isotope). A short DNA molecule that is complementary to its corresponding mRNA is called a **probe**.

But how do we obtain an appropriate probe? As an example, let's look at how scientists cloned the gene encoding human growth hormone (hGH)—the hormone produced by the pituitary gland that stimulates our growth and, when in short supply, causes dwarfism in children (**Figure 7.7**). HGH had been studied for a long time, so they knew the sequence of amino acids that form it. With this information and knowledge of the genetic code (Chapter 6), they deduced the possible sequences of the encoding DNA. More than one sequence is possible because most amino acids are encoded by more than one codon.

They then synthesized in radioactive form all possible DNA fragments corresponding to a sequence of six amino acids in the protein. (Most modern research laboratories have machines that synthesize short pieces of DNA.) Such a DNA fragment [18 bases long (6 amino acids × 3 bases per codon)] is an adequate probe to identify the hgh-encoding mRNA molecule. They obtained mRNA from extracts of cadaver pituitary glands. Using

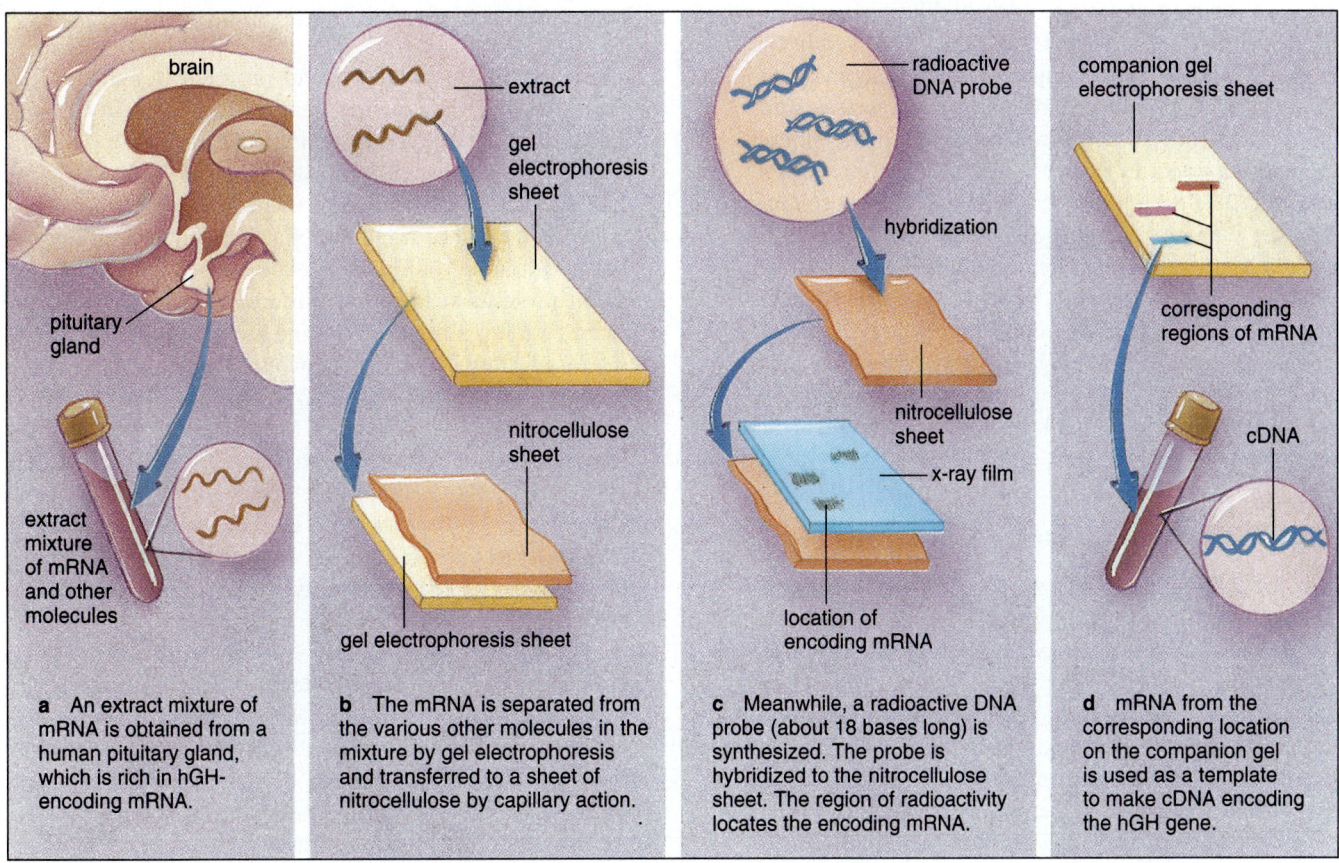

brain

pituitary gland

extract mixture of mRNA and other molecules

a An extract mixture of mRNA is obtained from a human pituitary gland, which is rich in hGH-encoding mRNA.

extract

gel electrophoresis sheet

nitrocellulose sheet

gel electrophoresis sheet

b The mRNA is separated from the various other molecules in the mixture by gel electrophoresis and transferred to a sheet of nitrocellulose by capillary action.

radioactive DNA probe

hybridization

nitrocellulose sheet

x-ray film

location of encoding mRNA

c Meanwhile, a radioactive DNA probe (about 18 bases long) is synthesized. The probe is hybridized to the nitrocellulose sheet. The region of radioactivity locates the encoding mRNA.

companion gel electrophoresis sheet

corresponding regions of mRNA

cDNA

d mRNA from the corresponding location on the companion gel is used as a template to make cDNA encoding the hGH gene.

Figure 7.7   Prior purification is one way to assure that the fragment of DNA being spliced into the cloning vector carries the desired gene. This figure shows prior purification of a eucaryotic gene encoding human growth hormone (hGH).

the probes, they isolated the mRNA. Then, using reverse transcriptase, they synthesized cDNA, which they cloned into an appropriate vector.

Subsequent Identification. In **subsequent identification**, random fragments of DNA are spliced into the cloning vector and inserted into cells that are then plated to develop into colonies. Colonies that produce the product of the gene to be cloned are selected because they must carry that gene.

We usually do not purify procaryote genes before cloning because they have relatively small genomes. Instead, the entire genome is cut into pieces, each of which is spliced into molecules of the cloning vector and then transformed into bacterial cells. By this process, called **shot gun cloning**, we produce a gene bank. A **gene bank** is a set of bacterial clones, each carrying a clone of DNA, that collectively constitutes the entire genome. Cloned DNA fragments are commonly about 30 kilobases (30,000 base pairs) long, so a complete gene bank of *Escherichia coli* could be carried in 150 bacterial clones [4500 (genome size of *E. coli*) ÷ 30]. Once made, a gene bank is saved and used later as a source of cloned genes. We only have to determine which bacterial clone carries the gene clone we want. This can be done in a number of different ways.

The gene we are looking for may be expressed in the host cell, if the host cell's gene expression machinery, including transcription and translation machinery, is similar to the machinery of the cell from which the cloned DNA was obtained. In that case, we look directly for the product of the cloned gene. We can see if the host cell has gained the characteristic encoded by the gene, such as an ability to synthesize a particular amino acid or break down a particular substrate. Even if the gene is not expressed by the host, we can make an appropriate DNA probe, as we did when purifying the eucaryotic gene prior to cloning. Then, by hybridization, we can test bacterial clones in the gene bank to find which one carries the gene we want.

Hosts for Recombinant DNA

Escherichia coli was the first organism used as a host for recombinant DNA, and it is still the most commonly used today. But with advances in recombinant DNA—particularly in transformation and other ways of putting recombinant DNA into cells—other microorganisms, insects, plants, and animals are also now being used as hosts. The organism selected depends on the reason for cloning. In basic biological studies on genes and their products, *E. coli* is almost always used. It is easy to maintain and cultivate, it grows rapidly, and we know more about it than any other organism.

Along with the advantages of *E. coli*, however, come some difficulties. For example, *E. coli* was the host used to

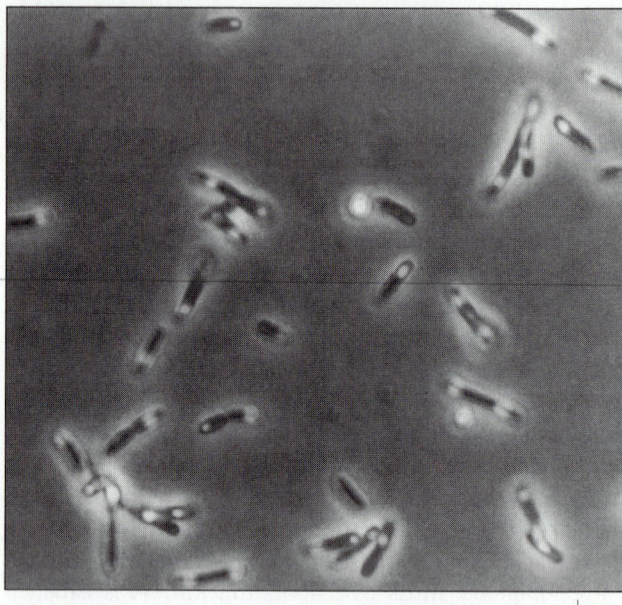

Figure 7.8 Inclusion bodies. In some cases, *Escherichia coli* expresses cloned genes so effectively that the protein product, such as human growth hormone (shown here), precipitates, forming inclusion bodies.

make the first medically useful proteins, human growth hormone and insulin (for treating diabetes). But *E. coli* makes so much of these proteins that they precipitate (solidify) within the cell, forming **inclusion bodies** (**Figure 7.8**). Before the protein from inclusion bodies can be recovered, it must be treated to make it soluble and then renatured (refolded). This is a relatively easy procedure for small proteins such as insulin and human growth hormone, but it is a major challenge for large proteins.

In addition, *E. coli* recognizes some human proteins as foreign and destroys them with proteases (protein-destroying enzymes). Finally, *E. coli* lacks certain abilities to modify proteins after they have been made, something eucaryotic cells normally do. For example, human cells **glycosylate** (add sugar molecules to) many of their proteins. (Glycosylation does not change a protein's activity, but it does change its solubility, an important characteristic of some medically useful proteins.)

Researchers have tried using other microorganisms, both bacteria and fungi, instead of *E. coli*, but none is superior. However, cultured animal cells overcame two of *E. coli*'s drawbacks. Animal cells don't destroy human proteins as readily, and it is not necessary to renature the proteins they produce. Also, animal cells can glycosylate. At the same time, using animal cells is expensive. They grow more slowly, produce less protein from recombinant DNA, and require more expensive media than do bacteria. Chinese hamster ovary (CHO) cells are being used to produce medically useful proteins, including tis-

Cloning Dinosaurs

The movie *Jurassic Park* is about a theme park populated by cloned dinosaurs and the disaster that results. The basic premise is that DNA from dinosaur bones and the guts of insects (whose last meal before they were preserved in amber was dinosaur blood) was cloned and then used to produce a dinosaur.

How plausible is this scenario? The second step stretches credulity to the limit. No one knows enough about developmental genetics even to consider creating a dinosaur from cloned DNA. But what about the first step—cloning DNA from dinosaurs? If even a minuscule amount of DNA were found, it could be amplified (by polymerase chain reaction) and cloned. So the question becomes, could any DNA survive the 60 million years since dinosaurs became extinct?

Most scientists think it could not. They cite biochemical studies on the rate of decomposition of DNA that indicate DNA lasts only 40,000 to 50,000 years. Other scientists, however, point out that DNA completely protected from water and oxygen would last much longer. Moreover, water-free, oxygen-free conditions might exist inside some dinosaur bones. In fact, DNA *has* been obtained by amplification from dinosaur bones. The only question is whether it is dinosaur DNA or comes from some other source. It might be from a human who touched the bone or from a fungus that grew on it. In this situation the enormous amplifying power of PCR leads to confusion rather than resolution. We simply do not know if any dinosaur DNA has survived.

Recovering authentic DNA from more recently deceased creatures is another story, however. The procedure is well accepted by scientists, and a whole new field, **molecular paleontology**, has emerged. For example, DNA from bones of a human who lived 2700 years ago in Oceania shows that some of the earliest settlers must have spread from Melanesia. DNA from 800-year-old skeletons of early Americans is providing evidence of their relationship to current Native Americans. DNA from pelts of wolflike creatures that were hunted to extinction in the Falkland Islands will reveal whether they were actually wolves, large foxes, or domestic dogs gone feral. Molecular paleontology probably won't produce a Jurassic Park, but without doubt it will tell us a great deal about recent evolutionary history.

---

sue plasminogen activator (to treat heart attacks) and human DNase (to treat symptoms of cystic fibrosis).

For further discussion of what makes a good host and how biotechnology industries are using intact insects and other hosts, see the box "A Good Host" in Chapter 29.

APPLICATIONS OF RECOMBINANT DNA TECHNOLOGY

Cloning is fundamental to recombinant DNA technology, and its applications are myriad. In the years to come, basic biology, medicine, agriculture, industry, and criminal investigation will all benefit (**Table 7.2**). Most applications fall into one of four classes—producing DNA, making proteins, amplifying genes, and engineering organisms.

Producing DNA

Cloned genes can be produced in massive amounts by cultures of host cells. For example, there are 20 to 40 copies of certain *Escherichia coli* plasmids (pBR type) per cell. When used as a cloning vehicle, an *E. coli* cell can be caused to produce as much cloned DNA as is present in all the rest of its genome. Such a culture can produce about 50 mg of cloned DNA per liter, a huge amount when you consider that most recombinant DNA procedures require only microgram or nanogram amounts. The cloned DNA that is produced is used in other recombinant DNA procedures, including use as a probe to detect similar DNA. DNA probes are used increasingly to identify microorganisms (see the beginning of Chapter 10).

Cloned DNA is also used as the starting material for the chemical reactions used to determine the sequence of bases in DNA. The procedure is called **sequencing** (Chapter 10). Sequencing is routinely done in microbiological laboratories today. On a much larger scale, the federal government is funding the Human Genome Project, which is sequencing the entire human genome. It is an enormous undertaking because the human genome consists of 3 billion (3×10^9) base pairs; and for acceptable accuracy, sequences will be done 10 times. In other words, the sequence of a total of 30 billion base pairs (3×10^{10}) will be determined. The project, which is on schedule for completion by the year 2005, will reveal the genetic basis of many inherited diseases and will undoubtedly provide other valuable information we can only guess about now.

Already, sequencing portions of the genomes of other organisms, some with much smaller genomes (the *Escherichia coli* genome has only about 4.5 million base pairs, about half of which were sequenced by the summer of 1993), has produced much valuable information about

Table 7.2 Some Applications of Recombinant DNA Technology

Field	Application	Importance
Basic biology	DNA sequencing	Answers questions about gene structure, gene function, and relatedness of genes and organisms
	Directed mutagenesis	Answers questions about gene function
Medicine	Therapeutic proteins	Makes human proteins for treating diseases such as diabetes, pituitary dwarfism, hemophilia
	Gene therapy	Treats genetic diseases such as cystic fibrosis
	Improved vaccines	Produces more effective vaccines with fewer side effects
	Diagnosis	Allows rapid, accurate diagnosis of infections and other diseases
	Veterinary medicine	Better diagnosis, prevention, and treatment of disease
Industry	Altering microorganisms	Improved production of antibiotics, amino acids, vitamins, and enzymes; also improved disposal of waste, including persistent toxic chemicals
Agriculture	Altering plants	More rapid breeding of disease-resistant and improved plants (e.g., tomatoes that stay fresh-tasting longer)
	Altering farm animals	More rapid development of superior breeds
Criminal investigation	DNA fingerprinting	Can determine if a biological sample such as blood, semen, or tissue is from a particular person

the relationships among organisms (Chapter 10) and the structure of genes (Chapter 6). Sequencing has become a fundamental tool of modern biology.

Making Proteins

We have already discussed how organisms carrying cloned genes are used to make medically useful proteins. This industry is already approaching a billion dollars a year. Industrial enzymes are also made from cloned DNA; the process is discussed in detail in Chapter 29.

Amplifying Genes

An amazing procedure called the polymerase chain reaction (PCR) was developed in 1985. PCR makes it possible to take a complex mixture of DNA, zero in on a single gene, and then multiply it to produce an essentially pure solution for study. PCR is so powerful it can amplify (increase) the amount of the gene more than a million times in a few hours. Because the rest of the DNA in the sample is unchanged, the product of PCR is an almost pure solution of the multiplied gene.

PCR is catalyzed by DNA polymerase, so understanding the procedure requires a basic understanding of the enzyme's properties. Recall from the discussion of replication in Chapter 6 that DNA polymerase requires a single-stranded DNA template and a free 3' OH end of a primer molecule. During PCR, the DNA sample to be analyzed is mixed with DNA polymerase, large amounts of two short primers, and the four deoxyribonucleoside triphosphates (dATP, dGTP, dCTP, and dTTP) that are the building blocks of DNA. Then the mixture is repeatedly heated and cooled to a moderate temperature. Heating melts the DNA, breaking the hydrogen bonds between base pairs and converting it into single strands. During the moderate temperature periods, the primer molecules pair with complementary regions on the single-stranded DNA. Then the DNA polymerase lays down a complementary strand from the 3' OH end of the primer. This cycle is usually repeated about 20 times (**Figure 7.9**).

The sources of two components of the PCR reaction—DNA polymerase and the primer molecule—are critical. The DNA polymerase is obtained from a thermophilic (heat-loving) bacterium, *Thermus aquaticus*, because the enzyme (*taq* polymerase) can withstand repeated heating. The two primer molecules must be complementary to re-

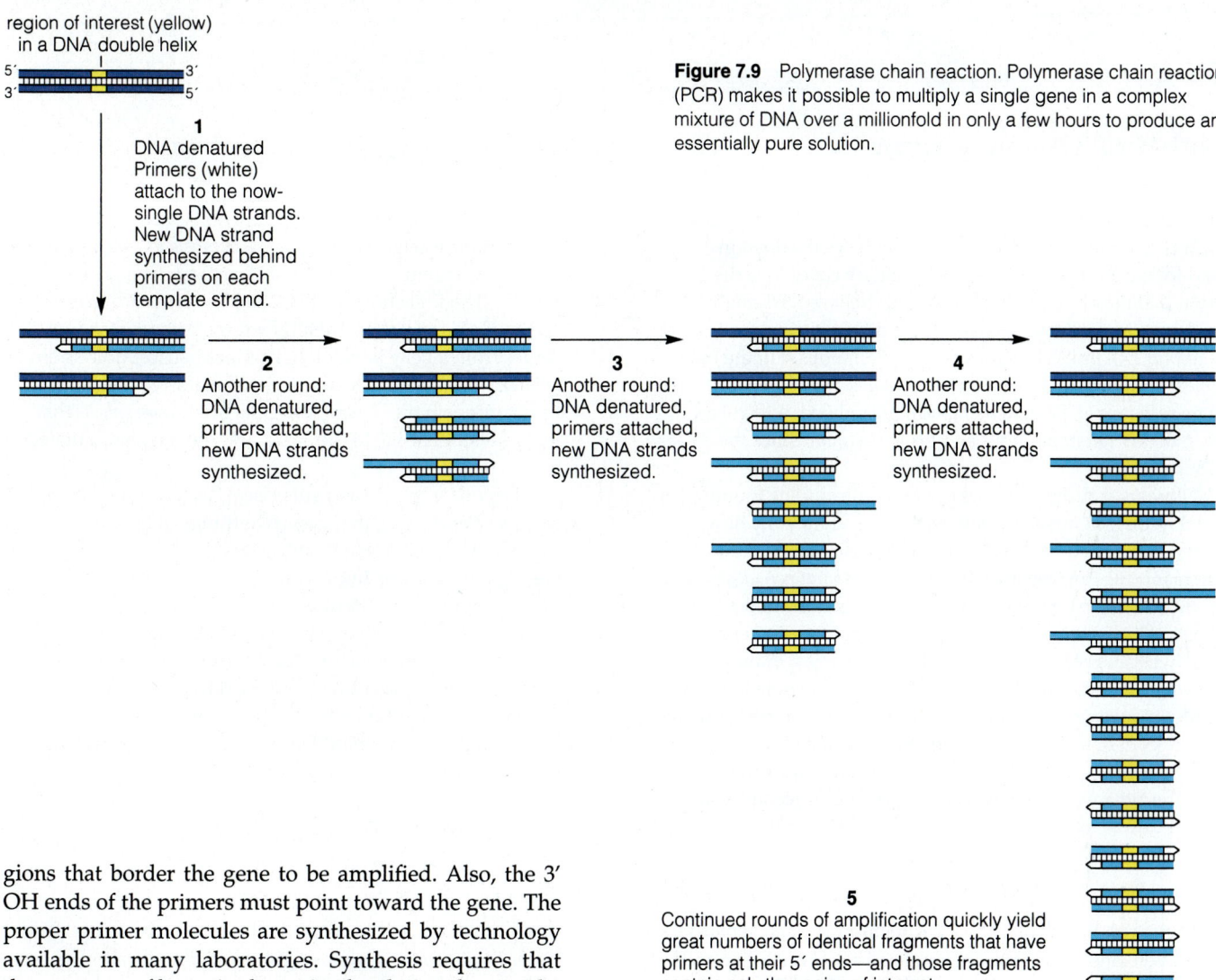

region of interest (yellow) in a DNA double helix

1
DNA denatured
Primers (white) attach to the now-single DNA strands. New DNA strand synthesized behind primers on each template strand.

2
Another round: DNA denatured, primers attached, new DNA strands synthesized.

3
Another round: DNA denatured, primers attached, new DNA strands synthesized.

4
Another round: DNA denatured, primers attached, new DNA strands synthesized.

5
Continued rounds of amplification quickly yield great numbers of identical fragments that have primers at their 5′ ends—and those fragments contain only the region of interest.

Figure 7.9 Polymerase chain reaction. Polymerase chain reaction (PCR) makes it possible to multiply a single gene in a complex mixture of DNA over a millionfold in only a few hours to produce an essentially pure solution.

gions that border the gene to be amplified. Also, the 3′ OH ends of the primers must point toward the gene. The proper primer molecules are synthesized by technology available in many laboratories. Synthesis requires that the sequence of bases in the region bordering the gene be known, but this information is usually available for a gene of interest.

Consider now the first few heat-cool cycles of PCR. Heating melts the DNA sample, producing single strands of DNA. On cooling, the primers pair with complementary regions of DNA in the sample, and the *taq* polymerase lays down a complementary strand in the 3′ direction. Then, on reheating, the newly synthesized double helix is melted, and its single strands pair with more primers, which are extended during the next cool period. Because the amount of the primer-bracketed gene doubles during each heat-cool cycle, 20 such cycles increase the amount about a millionfold ($2^{20} \cong 1,050,000$). Note also that after about three cycles almost all the new DNA is no longer than the bracketed region. Large amounts of the primers must be added because the requirement for them doubles after each cycle. That is, the requirement increases exponentially as do cells in a growing population of bacteria (Chapter 9).

The potential applications of PCR are almost limitless. In research laboratories it is a tool for studying genes. Forensic experts use PCR in criminal investigations, to match blood, hair, or sperm. It has been used to analyze the genes of ancient mummies.

PCR may even revolutionize medical diagnosis. Diagnosis often involves identifying small quantities of a specific biological substance—a task for which PCR is ideally suited. PCR could, for example, lead to preventing the transmission of AIDS through blood transfusions (Chapter 27). Most contaminated blood can be identified because an infected person produces antibodies against the virus. But in some cases antibodies are absent. It is thus difficult to be absolutely sure whether a particular sample of blood contains the AIDS virus. PCR, however, makes it possible to take a sample of blood and search for

Clinical Notes

Fighting TB—New Hope

Tuberculosis was the scourge of the nineteenth century. But with chemotherapeutic drugs, the incidence declined in developed countries during the twentieth century to the point that many public health officials believed TB might be the next infectious disease to be eradicated worldwide (Chapter 22). But the picture changed ominously in the mid-1980s. Strains of *Mycobacterium tuberculosis* appeared that were resistant to TB drugs, including isoniazid, which had been the cornerstone of tuberculosis treatment since the 1950s.

Obviously, drugs to replace isoniazid had to be found quickly, but scientists did not even know at the time how isoniazid worked (see the beginning of Chapter 6). How did isoniazid stop the growth of sensitive *M. tuberculosis* strains, and what genetic changes caused strains to become resistant? Answers to these questions would come very slowly because *M. tuberculosis* grows at an exasperatingly slow pace. It takes between 2 and 8 weeks to isolate a strain of *M. tuberculosis* from a clinical sample and then from 1 to 13 weeks to determine its sensitivity to isoniazid.

But in 1992, a team of microbiologists in London and Paris found a way to speed the research using recombinant DNA technology. They would transfer isoniazid-resistant genes from slowly growing *M. tuberculosis* to a closely related nonpathogenic species, *M. smegmatis*, which grows rapidly. The first step was to isolate an isoniazid-resistant strain of *M. smegmatis*. Then they transferred individual clones from a gene bank of an isoniazid-susceptible strain of *M. tuberculosis* into it. The clone that made *M. smegmatis* sensitive again must carry the gene in *M. tuberculosis* that makes it susceptible to isoniazid and the one that mutated to resistance.

The culprit gene was a surprise. It encoded an enzyme, catalase-peroxidase, that does two things. It breaks down hydrogen peroxide (H_2O_2) into water (H_2O) and oxygen gas (O_2). It also uses hydrogen peroxide to oxidize certain organic compounds. How could one of these reactions make *M. tuberculosis* sensitive to isoniazid? Did one of them convert isoniazid into an active form? If so, maybe the active product could be identified and used to treat patients infected with isoniazid-resistant strains of *M. tuberculosis*. Increasingly, recombinant DNA technology is providing hope and solutions in clinical research.

---

the genes of the virus. The PCR technique is so sensitive that even one gene from a single virus can be detected and identify contaminated blood. Because of this extraordinary power, PCR may soon become the diagnostic standard not only for AIDS, but also for many other infectious and noninfectious diseases.

Engineering Organisms

In addition to producing quantities of specific genes by gene cloning, we can also insert those genes into organisms to change them genetically. We have already discussed **gene therapy** of humans—introducing good genes to replace or overcome the effects of disease-causing genes. In similar ways, microorganisms, plants, and animals can be genetically altered or *engineered* for specific purposes. Microorganisms can be engineered to produce greater quantities of their products, such as antibiotics, amino acids, or enzymes. Or they can be engineered to break down toxic or polluting compounds in our environment (Chapter 28). Plants can be engineered to pro-

duce more or better crops or to resist disease or insect damage. Animals can be engineered for similar purposes. The physicist Freeman Dyson predicts that recombinant DNA technology will benefit humanity more than the industrial revolution did.

Summary

RECOMBINANT DNA TECHNOLOGY (p. 175)

1. Recombinant DNA technology is a collection of procedures for taking DNA from a cell, manipulating it *in vitro*, and putting it into another cell, usually one of a different species.

2. Recombination refers to the combination of genes, whether natural (as in sexual reproduction) or artificial (*in vitro* in recombinant DNA technology).

3. When genes on the same chromosome or DNA molecule recombine by crossing over, the result is a recombinant molecule. This happens frequently between homologous DNA molecules in living

things. In recombinant DNA technology, the DNA molecules do not even have to be similar, let alone homologous.

GENE CLONING   (pp. 175–185)

1. Gene cloning is the basic tool of DNA technology; it is the process of obtaining a set of identical copies (clones) of a gene.

2. Gene cloning involves five steps:
    a. obtaining DNA that contains the gene to be cloned
    b. splicing a piece of DNA containing the gene into a cloning vector (a DNA molecule that a cell will replicate)
    c. putting the recombinant DNA (the cloning vector with the desired gene spliced into it) into an appropriate host cell
    d. testing to assure the desired gene has been inserted into the host cell
    e. propagating the host cell to produce a clone of cells with the clone of genes

3. DNA is purified from cell extract, the liquid content of ruptured cells. The large size of DNA causes it to form ropelike aggregates in the presence of alcohol. DNA used to clone a eucaryotic gene must be intron-free, which requires an extra step.

4. Good cloning vectors have an origin of replication, are relatively small so their replication does not tax the host cell's metabolic capacity, and carry other genes that identify host cells containing the vector. Plasmids and the genomes of certain viruses are used as cloning vectors.

5. Cloning vehicles and DNA to be inserted into them are usually cut with enzymes called type II restriction endonucleases, which produce palindromic sequences and single-strand regions called sticky ends because they readily pair with another complementary strand. Sticky ends pair and anneal at low temperature. Restriction endonuclease reactions are monitored by gel electrophoresis.

6. Gaps that exist where the original cuts were made must be ligated (sealed) to stabilize the molecule; this is done with an enzyme called DNA ligase. Blunt-end ligation produces a less stable molecule, but it has certain advantages; the DNA fragment bearing the cloned gene and the vector need not be cut by the same restriction endonuclease, and the DNA to be cloned can be cut mechanically by shearing. Blunt ends spontaneously anneal.

7. DNA molecules can be inserted into a living cell several ways. In transformation, an intact bacterial cell takes up DNA from solution. Most bacteria must be specially treated, a process called artificial transformation. In transfection, DNA from a virus is used as the cloning vector. DNA can be inserted into some animal cells directly by microinjection. Finally, DNA can be introduced into many bacterial, plant, and animal cells by electroporation.

8. Testing to be certain the gene to be cloned, and not some other, has been spliced in can be done by prior purification, which involves using a probe, a short DNA molecule that is complementary to its corresponding mRNA. Testing can also be done by subsequent identification, which involves developing colonies that produce the products of the gene to be cloned.

9. Subsequent identification is usually done in cloning procaryotes. The procedure is based on a process called shot gun cloning, which produces a gene bank, a set of bacterial clones, each carrying a clone of DNA, that collectively constitutes the entire genome. Bacterial clones in the gene bank are tested by hybridization to find which carries the desired gene.

10. *Escherichia coli* is the most common host for recombinant DNA because it is easy to maintain and cultivate, it grows rapidly, and we know more about it than about any other organism. However, *E. coli* can make so much of a protein that inclusion bodies form, a problem with large proteins. Also, *E. coli* destroys certain human proteins as foreign, and it cannot modify proteins as eucaryotic cells normally do.

11. Cultured animal cells overcome the last two problems, but they grow more slowly, produce less protein, and require more expensive media. Chinese hamster ovary (CHO) cells are usually used.

APPLICATIONS OF RECOMBINANT DNA TECHNOLOGY   (pp. 185–188)

1. Recombinant DNA technology produces huge amounts of DNA, used in sequencing. DNA sequencing answers questions about gene structure, gene function, and relatedness of genes and organisms. The Human Genome Project is sequencing the entire human genome.

2. Recombinant DNA technology produces medically useful proteins for treating diabetes, pituitary dwarfism, and hemophilia. It also produces industrial enzymes.

3. Polymerase chain reaction (PCR) is an amazing procedure that makes it possible to zero in on a single gene in a complex mixture of DNA and then amplify (multiply) it to produce an essentially pure solution for study in an extremely short time. PCR is used in research laboratories, in criminal investigations to match blood, hair, or sperm, and in anthropology to analyze genes of ancient mummies. PCR may even revolutionize medical diagnosis.

4. Recombinant DNA technology allows us to engineer (genetically alter) microorganisms, plants, and animals. For example, plants can be engineered to produce better crops or resist disease or insect damage.

Review Questions

RECOMBINANT DNA TECHNOLOGY

1. Explain this statement: Recombinant DNA technology is easy to appreciate but hard to define.

2. Explain the differences between recombinant molecules produced naturally in living things and those produced by recombinant DNA technology. How do the processes differ?

GENE CLONING

1. Define these terms: homologous, gene cloning, clone, cloning vector.

2. What are the five steps in gene cloning?

3. How is DNA obtained in procaryotes and in eucaryotes?

4. What are the properties of a good cloning vector? How are cloning vectors cut? Define these terms: restriction endonuclease, palindromic sequence, sticky ends, anneal.

5. What is gel electrophoresis and how is it used in gene cloning?

6. How is ligation done and why is it necessary? What is blunt-end ligation and what are its advantages?

7. Describe these ways of inserting recombinant DNA into a host cell: transformation, transfection, microinjection, and electroporation.

8. Discuss the two ways of testing that the DNA fragment to be spliced in carries the desired gene and not some other.

9. What roles do shot gun cloning and a gene bank play in cloning genes from procaryotes?

10. What are the characteristics of a good recombinant DNA host? What are the advantages and disadvantages of using *E. coli* as a host? What are the advantages and disadvantages of using CHO cells?

APPLICATIONS OF RECOMBINANT DNA TECHNOLOGY

1. What is sequencing and why is the Human Genome Project important?

2. What are some of the medically important proteins produced by recombinant DNA technology?

3. What is PCR and why is it considered such an amazing procedure? How is PCR being used today? Explain this statement: The potential applications of PCR are almost limitless.

4. What does it mean to engineer an organism? What is gene therapy? Discuss some other uses of genetic engineering.

Suggested Readings

Bloom, B. 1992. Back to a frightening future. *Nature* 358:538–39.

DeWitt, P. E. 1994. The genetic revolution. *Time*, January 10, pp. 46–56.

Persing, D. H.; Smith, T. F.; Tenover, F. C.; and White, T. J. 1993. *Diagnostic molecular microbiology: Principles and applications.* New York: ASM Press.

Williams, R. C. 1991. *Molecular biology in clinical medicine.* New York: Elsevier Science.

8 THE GROWTH OF MICROORGANISMS

A Walking Chemostat

The dream of industrial microbiology is to convert an abundant cheap substrate into a valuable product. Microbiologists might differ about the ideal product but not the ideal substrate. They all agree that cellulose—the major component of plant cell walls—is the world's most abundant organic compound and probably the cheapest. But only microorganisms can use cellulose as a substrate and, therefore, convert it into something useful.

Asked if a commercially successful microbial conversion of cellulose would ever be discovered, Robert Hungate, a leading American microbiologist, replied that it already had—it is the cow. The cow and other ruminant animals are mobile factories that use microorganisms to convert cellulose into meat and milk. They gather cellulose (along with lesser amounts of other substrates) as grass and other forage and chew it into small bits that flow to the rumen, the large chamber of a cow's four-chambered stomach. A cow's rumen teems with bacteria and protozoa. They—not the cow itself—metabolize cellulose. Neither does the cow derive nutrients from cellulose. Rather, it derives most of its nutrients from the fermentation products produced by microorganisms in its

rumen and from the microbial cells that constantly flow out of the cow's rumen into its intestines, where they are digested and absorbed.

Hungate and his colleagues have shown that the cow's rumen functions like a chemostat, a device people use to cultivate microorganisms. The rumen is continuously supplied with nutrients, microorganisms continuously grow in it, and some of its content continuously flows out into the digestive system. That doesn't mean cows eat continuously. Rather, ruminants provide a steady supply of nutrients to the rumen by chewing their cud, regurgitated forage. As they chew, bits of forage suspended in saliva flow into the rumen. The cow does not chew its cud every moment it isn't eating. But the supply of nutrients to the rumen is essentially continuous because the solid bits of cellulose act as reservoirs of nutrients for the microorganisms.

Cows and other ruminants have achieved the goal that still tantalizes industrial microbiologists—converting cellulose into valuable products. Moreover, they do it efficiently, more efficiently than any industrial process known.

The rumen of a cow functions like a chemostat.

To understand:

- How microorganisms grow: doubling time and exponential growth

- The phases of growth that microbial cultures pass through

- The methods that can be used to keep microorganisms growing continuously

- What kinds of nutrients microorganisms require for growth

- The environmental conditions that permit microbial growth: temperature, hydrostatic pressure, pH, and osmotic strength

- How to measure a microbial population using indirect methods—measuring turbidity, dry weight, or metabolic activity

- How to measure a microbial population by direct count, plate count, most probable number, and filtration

POPULATIONS

In studying the metabolism and genetics of microorganisms, we have focused on the individual—how a microbial cell maintains and reproduces itself. In this chapter and the next, we turn our attention to large groups—how **populations** of microorganisms increase through growth and decrease or disappear as a result of death.

Focusing on populations requires that we view microorganisms from a new perspective. In studying individuals, we *describe*. We describe the parts of the cell involved in a particular process, how the cell produces the molecules it needs to build a new cell, and how it preserves its genetic information from one generation to the next. In studying populations, however, we *count*. With populations, we want to know "How many?" "How many more?" or "How many fewer?" The study of populations, whether microbial or human, requires a mathematical point of view.

Unless we mention otherwise, we use the term **microbial growth** to refer to the growth of a population, not to an increase in the size of the individual cell. Individual microbial cells grow, but most increase their size only about twofold before the cell cleaves, or divides in two. Cell division leads to the *growth*, or *increase in the number, of cells* in the population (**Figure 8.1**).

THE WAY MICROORGANISMS GROW

In a favorable environment, a microbial cell enlarges; and when it doubles in size, it divides. Most bacteria elongate and divide by binary fission, or cleavage, near the midpoint to form two daughter cells of approximately equal size. Some unicellular microorganisms, including a few bacteria, replicate by budding, forming a bubble-like growth that enlarges and separates from the parent cell. Although we discuss growth in terms of a bacterial population dividing by binary fission, the same principles apply as well to microorganisms that reproduce by budding (Chapter 5). The growth of filamentous microorganisms (those that form long tubes) and microorganisms with complex life cycles follows more complicated rules.

Doubling Time and Growth Rate

Doubling time (formerly referred to as **generation time**) is the period required for cells in a microbial population to grow, divide, and produce two new cells for each one that existed before. Doubling time is approximately the same for all cells in a given population. It does not change until nutrients become depleted or toxic metabolic products begin to accumulate. As conditions for growth become less favorable, doubling time increases before growth stops.

Doubling times vary depending on the species of microorganism and the growth conditions of the population. The doubling time of *Escherichia coli*, for example, is about 18 minutes in a rich laboratory medium. But in the intestinal tract of vertebrates, where nutrients are less abundant, doubling time is about 12 hours. A few bacteria can grow a little faster than *E. coli*; but even under the most favorable conditions, many microorganisms grow much more slowly, requiring hours or days to double.

The doubling-time figure tells us how fast the population is growing, but the relationship is inverse. That is, populations with a low value of doubling time are growing rapidly, while populations with a higher value of doubling time are growing slowly. For this reason, the same information is often expressed as **growth rate**, or doublings per hour, which is high for rapidly growing populations and low for slowly growing ones. Thus, the growth rate of the culture of *E. coli* growing with a doubling time of 18 minutes would be 3.3 (60/18) doublings per hour.

Exponential Growth

During a period of unchanging doubling time, a microbial population undergoes **exponential growth**. This

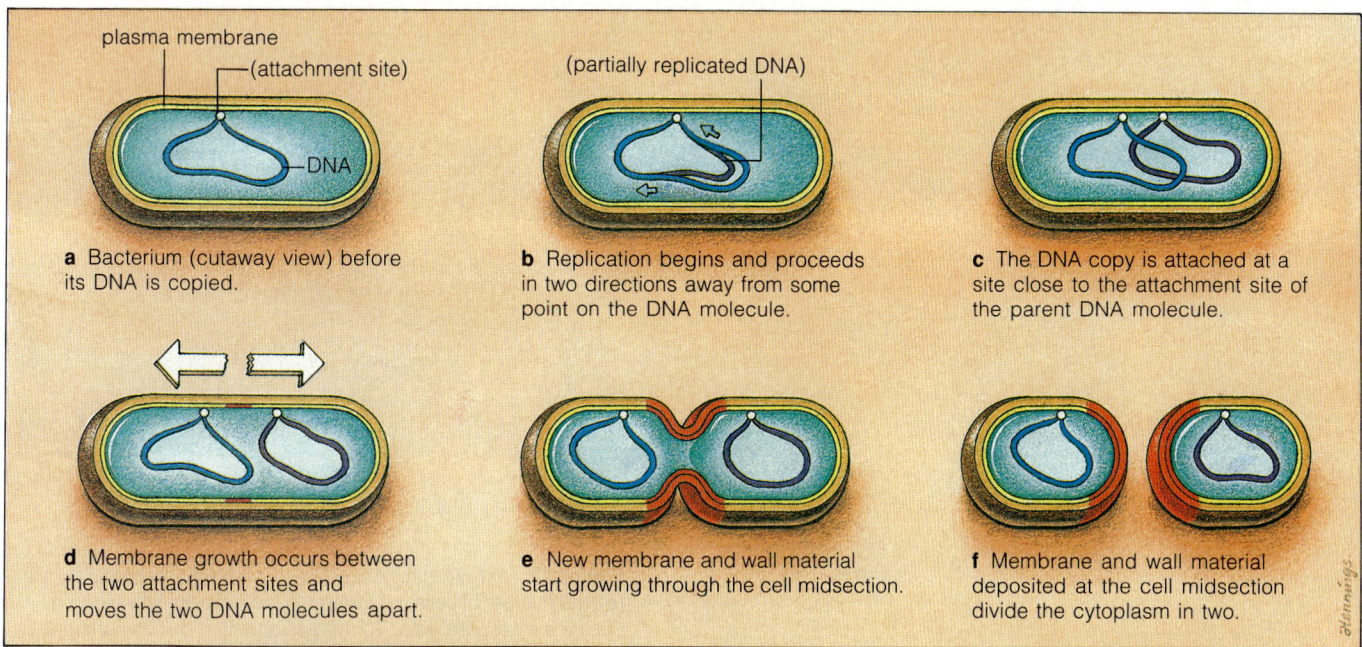

plasma membrane
— (attachment site)

— DNA

a Bacterium (cutaway view) before its DNA is copied.

(partially replicated DNA)

b Replication begins and proceeds in two directions away from some point on the DNA molecule.

c The DNA copy is attached at a site close to the attachment site of the parent DNA molecule.

d Membrane growth occurs between the two attachment sites and moves the two DNA molecules apart.

e New membrane and wall material start growing through the cell midsection.

f Membrane and wall material deposited at the cell midsection divide the cytoplasm in two.

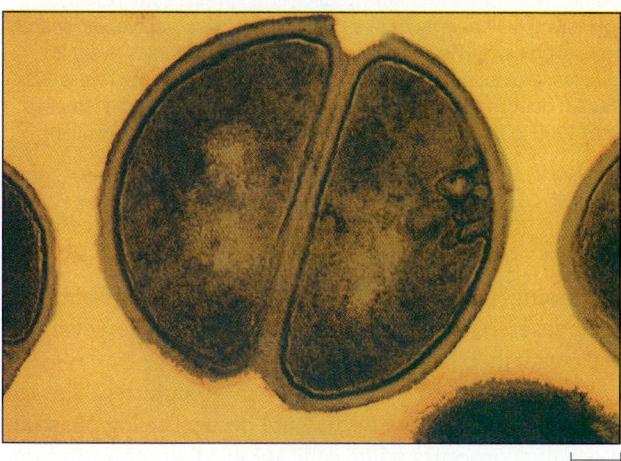

100 nm

Figure 8.1 Most microbial cells divide by binary fission, as shown in this scanning electron micrograph of the bacterium *Staphylococcus aureus*. In binary fission, the cell simply replicates its DNA and gives each daughter cell one copy.

could be recorded using an arithmetic, or simple numerical, graph, the exponential graph has significant advantages. The number of microorganisms in an exponentially increasing population increases slowly at first, then extremely rapidly. For example, during the first three doubling times of a single cell, only seven new cells are produced. During the seventh doubling, 64 cells are produced, and during the twenty-first doubling, over a million. If you plotted this curve on an arithmetic scale, the initial increase could barely be appreciated, while the later growth phases would shoot off the graph. Using an exponential scale, the curve becomes a straight line, with its slope directly related to the growth rate of the population. This is logical if you remember that with each doubling time the population increases by a factor of two. In other words, the total number of cells is equal to two raised to an exponent, where the exponent is the number of generations that have elapsed since the population began to increase. Thus, the population begins as one cell (2^0), increases to two cells (2^1), then to four cells (2^2), eight cells (2^3), 16 cells (2^4), and so on. The general formula is $N = 2^n$ where N is the number of cells in the culture after n doubling times have passed.

Why is the graph line straight and not in a stair-step pattern? If a microbial population arises from a single cell and has a fixed doubling time, wouldn't the population remain unchanged during each doubling time and then

means that during each doubling time, the number of cells in the population increases by a factor of two.

The concept of exponential growth is often illustrated with a mathematical riddle about water lilies. A water lily produces one new leaf each day for each leaf it had the day before. Starting with a single leaf, it will cover a 1-acre pond in 30 days. How long will it take to cover a 2-acre pond? The correct answer is 31 days, not 60, because the doubling time is one day. Under exponential growth conditions, living things multiply with startling rapidity. During each doubling time, as many new cells are produced as were produced cumulatively before.

Because microbial growth is exponential, we graph the growth of a microbial population using an exponential scale (**Figure 8.2**). Although the same information

| Time | | Cells/milliliter | | |
Min.	Hours	Numbers	Scientific Notation	Logarithm
0	0	1000	10^3	3.0
20	0.33	2000	2×10^3	3.301
40	0.66	4000	4×10^3	3.602
60	1.00	8000	8×10^3	3.903
80	1.33	16,000	1.6×10^4	4.204
100	1.66	32,000	3.2×10^4	4.505
120	2.00	64,000	6.4×10^4	4.806
140	2.33	128,000	1.28×10^5	5.107
160	2.66	256,000	2.56×10^5	5.408
180	3.00	512,000	5.12×10^5	5.709
200	3.33	1,024,000	1.02×10^6	6.010

a

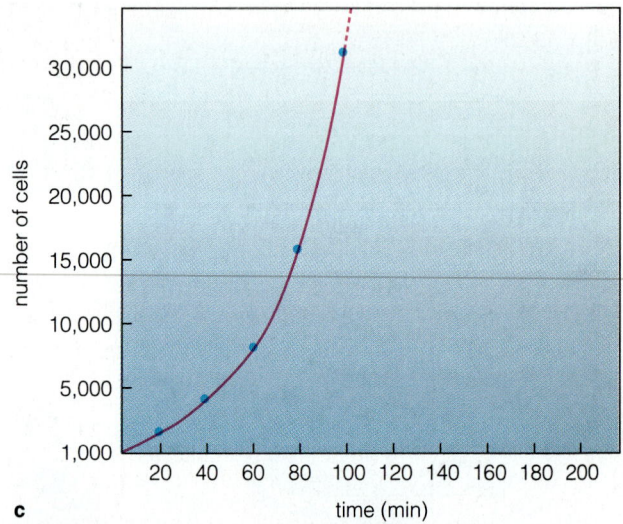

c

Figure 8.2 Following the exponential growth of a bacterial culture. (**a**) The number of cells per milliliter in the exponentially growing culture is measured every 20 minutes and recorded as numbers, by scientific designation, and by logarithm of the numbers. (**b**) Plotting the logarithm of the numbers against time gives a straight line. (**c**) Plotting the actual numbers against time gives a curve with sharply increasing slope. Because the number of cells doubles every 20 minutes, the doubling time is 20 minutes and the growth rate is 3.0 (60/20) doublings per hour. A culture growing at this rate increases from 1000 to more than 1 million cells per milliliter in 200 minutes.

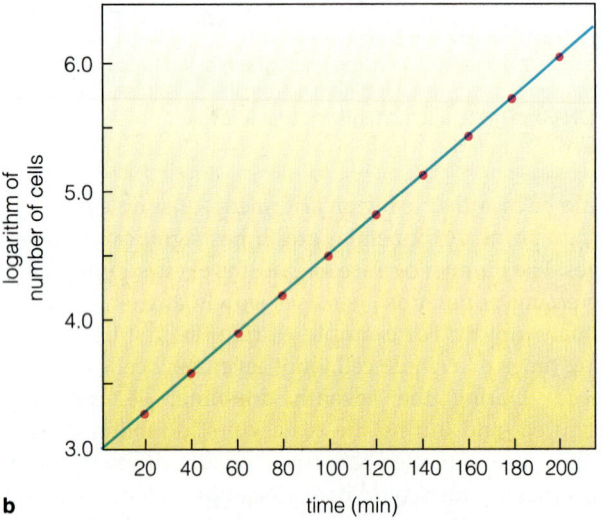

b

increase abruptly—the stair-step pattern? In fact, this pattern, called **synchronous growth**, does not occur under ordinary circumstances. Instead, individual cells in the population are likely to divide before or after the population's doubling time, in a pattern of **nonsynchronous growth**. Actual division times are distributed around the doubling time, so the population increase is smooth and the graph rises as a straight line.

Phases of Growth

A microbial culture typically passes through four distinct and sequential phases of growth: the **lag phase**, the **log phase** (also called the **logarithmic** or **exponential phase**), the **stationary phase**, and the **death phase** (Figure 8.3). In

discussing exponential growth, we were discussing the log phase, so let us pick up our story there. However, the cycle usually begins with the lag phase.

Not all growth phases occur in all cultures. The lag and death phases may or may not occur. The log phase almost always occurs when cells are in a favorable environment, but it cannot continue indefinitely. As suggested by the water lily riddle, a microbial culture that continued to undergo exponential growth would soon overrun the planet. But something always intervenes. An essential nutrient is depleted, a toxic product accumulates, or the pH becomes unfavorable. Then the culture stops growing and enters the stationary phase of growth.

Although no net increase in the mass of the culture occurs during the stationary phase, cell composition changes as the culture undergoes the transition from exponential to stationary phase. Cells become smaller and they begin to synthesize components to help them survive longer periods without growing. For example, when *Escherichia coli* enters the stationary phase, it synthesizes about 30 proteins not found in log-phase cells and it changes the composition of some of the fatty acids in its membranes. Mutant strains that cannot make these stationary-phase proteins die quickly when they stop grow-

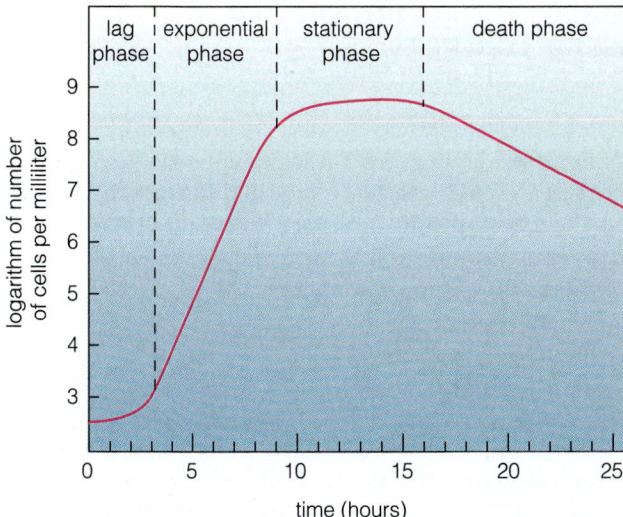

Figure 8.3 Phases of growth in a bacterial culture. (This graph is based on the culture described in Figure 8.2.) The death phase is an exponential decline, like the exponential phase, and is therefore a straight line, but it occurs at a much slower rate.

ing. Some bacterial species have evolved elaborate mechanisms such as forming endospores to survive the stationary phase (Chapter 4).

After a day or so in the stationary phase, the death phase begins as cells in the culture start to die. During the death phase, cells of most microbial species die exponentially but at a low rate—much lower than the rate of increase of cells during the log phase. Death usually occurs because the cells have depleted their intracellular reserves of ATP to repair cellular components. No new ATP can be generated by nongrowing cells, so death occurs when there is not enough energy to continue cellular repair or to reinitiate growth when nutrients become available.

The lag phase is also a no-growth period. It usually follows the inoculation of stationary-phase or death-phase cells into a fresh culture medium. Although no net growth occurs during the lag phase, considerable metabolic activity takes place as cells prepare to grow. Preparation is necessary because metabolic damage suffered during the stationary or death phase must be repaired completely before growth can begin again. If the cells used to **inoculate** (seed) a new culture are in the log phase, no lag occurs, provided the new culture medium is the same as the old one and other conditions remain the same.

The length of time a culture remains in the lag phase depends on how long the previous culture was held in the stationary or death phase, or it may depend on a change in the medium. If the cells were in the stationary phase only briefly, the ensuing lag phase may be as short as a few minutes. But if they were in the stationary phase for months, the ensuing lag phase of the surviving cells may last many hours. If an exponential culture is moved from a poor medium to a rich one, then no lag occurs. But a move from a rich to a poor medium results in a several-hour lag in most bacterial cultures.

Continuous Culture of Microorganisms

In the laboratory, most microorganisms are cultivated in **closed containers**, which means nutrients are not supplied or toxic products removed. Consequently, the cultures usually pass through all four stages of growth. A stationary-phase culture is inoculated into a fresh culture medium that contains adequate concentrations of all essential nutrients and passes sequentially through a lag phase, a log phase, a stationary phase, and a death phase. Microorganisms do not grow this way in nature. Under most natural conditions, nutrients continuously enter the cell's environment at low concentrations, and populations grow continually at a low but steady rate. The rate of growth is set by the *concentration* of the scarcest or **limiting nutrient**. Rarely is growth rate in nature set by the accumulation of metabolic by-products, because some other microorganism can usually use them as fast as they are formed.

It is possible to achieve a continuous culture in the laboratory under conditions that mimic natural conditions—by keeping the concentration of one essential nutrient low enough that it limits the rate of growth. Such concentration-limited growth is most frequently accomplished with continuous culture apparatus called the **chemostat** (**Figure 8.4**). A chemostat consists of a metering pump, a reservoir, a growth vessel, and an overflow. The metering pump adds a controlled amount of fresh medium from the reservoir continuously to the culture in the growth vessel. At the same time, an equal amount of culture leaves continuously through the overflow. Under these conditions, the number of cells in the growth vessel doesn't change. The culture grows just fast enough to replace the cells lost through the overflow.

The concentration of the limiting nutrient in the growth vessel sets the growth rate of the culture. The culture grows as fast as the limiting nutrient is supplied, which is also as fast as partially spent culture medium and cells leave through the overflow. Cells in the growth vessel of the chemostat remain in an exponential phase of growth indefinitely.

Continuous culture has important industrial applications (Chapter 29). It is an efficient way to produce microbial cells or their products (such as antibiotics and vitamins) because a dense culture is constantly produced from the overflow of the growth vessel.

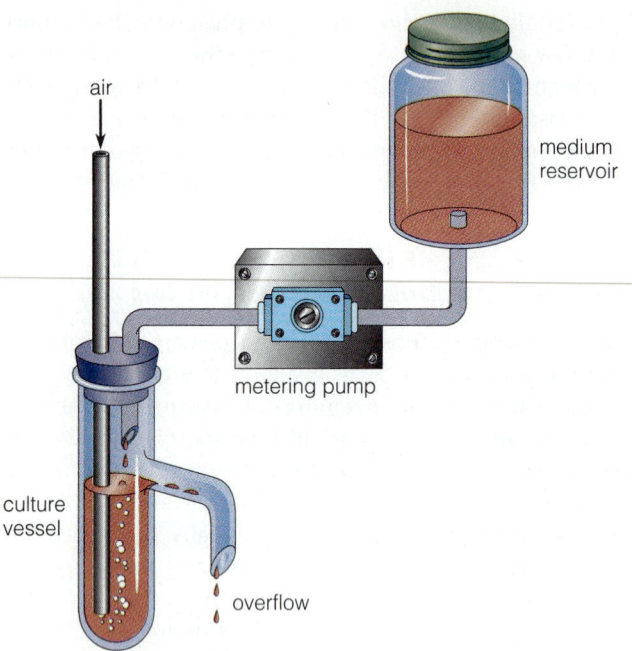

air

medium reservoir

metering pump

culture vessel

overflow

Figure 8.4 A chemostat is used to maintain a microbial population in a constantly growing state. A metering pump adds sterile medium from a reservoir to the culture at a constant rate. The culture—containing cells and partially spent medium—leaves at the same rate. An actual chemostat has many electronic controls.

Growth of a Colony

When a population of microbial cells develops from a single cell on a solid surface—for example, on an agar-solidified medium in a petri dish—they form a solid mass of cells called a **colony**. Different microorganisms form colonies with different shapes and textures. Some colonies have such distinct appearances that microbiologists can identify a microorganism by a glance at the colony it forms.

Cells in a colony are in different phases of growth, depending on their location (unlike cells in a liquid medium, which are all in the same phase of growth). Cells in the center of the colony have stopped growing and are in the stationary or death phase. They no longer have access to nutrients. On the other hand, cells on the edge are in contact with nutrients and are actively growing in the log phase. No cells in a visible colony are in the lag phase because once cells in the center of a colony stop growing, they do not start again until transferred to a fresh medium.

Location within a colony determines a cell's access to nutrients in another way, too. In a colony growing in air, for example, the cells on the surface are fully aerobic. Cells embedded in the center of the colony are fully anaerobic because surface cells of an air-grown colony remove all the oxygen in the air that diffuses into the colony.

WHAT MICROORGANISMS NEED TO GROW

Every species of microorganism requires particular conditions in order to grow. These requirements vary greatly because the environments to which different species have adapted vary greatly. But every species has two kinds of requirements, nutritional and environmental. Let's look first at the nutritional requirements.

Nutrition

All living things use chemicals called **nutrients** from the environment to build the molecules needed to build new cells. We know that nutrition affects the growth of all organisms. A well-nourished plant, animal, or human grows more rapidly than a poorly nourished one. For microorganisms, particularly bacteria, the effect of nutrition on growth rate is dramatic. *Escherichia coli*, for example, grows up to 10 times faster in a rich nutritional environment, such as meat extract, than it does in a poor nutritional environment, such as one composed of succinate and salts. The succinate-salts medium meets all of *E. coli*'s essential needs, but growth in such a medium requires that the cell carry out more biosyntheses. The result is slower growth.

A higher growth rate means a culture reaches its maximum total mass and numbers faster than in a nutrient-poor environment. However, a culture can still reach the same density, given sufficient time, provided the same total amount of nutrient is present. The *yield*, or maximum total mass, of a culture is directly proportional to the amount of nutrient.

The nutrients that various microorganisms need to grow include a source of carbon, oxygen, nitrogen, phosphorus, sulfur, trace elements, and organic growth factors. Cells use these nutrients to generate driving force or precursor metabolites (Chapter 5). Hydrogen is also required for growth, but it is never a limiting factor in a medium that otherwise supports growth.

Carbon. Carbon is the structural basis of biochemicals. Autotrophic microorganisms obtain their carbon from carbon dioxide (CO_2) in the atmosphere, while heterotrophic microorganisms obtain carbon from organic compounds in their environment (Chapter 5).

Heterotrophic microorganisms use many different organic molecules as carbon sources. In fact, probably every naturally occurring organic compound is used as a carbon source by some bacterial species. One of the most commonly used sources among microorganisms is the hexose sugar, glucose, which plays a central role in metabolism. Consequently, it is a common ingredient of microbiological media. But other sugars, as well as polysac-

The Survivors

How can nonendospore-forming bacteria survive after they stop growing? A few hours after a culture of *Escherichia coli* (and other nonendospore-forming bacteria) enters the stationary phase, the death phase begins. Though exponential decline may be slow, it will eventually kill all cells in the culture. Wouldn't it be only a matter of time before whole species died out, one population at a time? In fact, this does not happen because not all cells die during the death phase. Even after a year's storage, a stationary-phase culture of

E. coli contains over a million live cells per milliliter. Cells in the culture die exponentially for four or five days, but the cells that remain survive for long periods. Roberto Kolter, a microbiologist at Harvard University, found that nonendospore-forming bacteria have evolved their own mechanisms to survive prolonged periods of starvation. Most cells die, but a few are programmed to survive, enough to preserve the species.

charides, organic acids, alcohols, and amino acids, are often added to media as a carbon source.

For chemoheterotrophs, carbon-containing organic molecules such as glucose provide both a source of energy to generate ATP and the carbon atoms necessary to build biochemicals (Chapter 5). Carbon sources that are metabolized quickly support relatively rapid growth, while carbon sources that are metabolized slowly support only slow growth. For example, *Escherichia coli* grows twice as fast if glucose instead of the amino acid lysine is added as a carbon source to otherwise identical media.

Oxygen. All microorganisms require elemental oxygen to build their biochemical components, but not all microorganisms require *atmospheric* oxygen. Most heterotrophic microorganisms obtain oxygen from the same molecule that serves as a carbon source. Recall that the chemical formula for carbohydrate (a common carbon source for heterotrophs) is CH_2O, which means that every carbohydrate molecule gives the cell one atom of oxygen for each atom of carbon. Autotrophs that generate energy from light obtain much of their oxygen from the CO_2 fixed during photosynthesis. Most aerobic microorganisms have enzymes called **oxygenases** that can directly add atmospheric oxygen to organic molecules, but this is a minor source of oxygen. A few enzymes allow microorganisms to utilize the oxygen present in water by means of hydration reactions.

In addition to using oxygen as a nutrient, microorganisms that are capable of aerobic respiration use oxygen to generate energy. But in spite of being required by some organisms, atmospheric oxygen is toxic. Some enzymes are rapidly and irreversibly destroyed by exposure to atmospheric oxygen. An extreme example is **nitrogenase**, the enzyme that enables nitrogen-fixing bacteria to use

atmospheric nitrogen. As a result, microorganisms have evolved various mechanisms to protect even highly oxygen-sensitive enzymes and thus survive in the presence of oxygen. Some nitrogen-fixing cyanobacteria—which are aerobic and produce oxygen gas by photosynthesis—protect nitrogenase by segregating it in specialized cells, called **heterocysts**. Heterocysts do not produce oxygen and are impermeable to it. The nitrogen-fixing Azotobacters carry out aerobic respiration at a rate high enough to keep the center of the cell, where nitrogenase is located, anaerobic.

In addition to oxygen itself, organisms must deal with toxic oxygen-containing compounds that are produced as by-products of aerobic metabolism. These toxic agents include hydrogen peroxide (H_2O_2) and the even more toxic free radicals, **superoxide** (O_2^-) and **hydroxyl radical** ($\cdot OH$), which is formed from superoxide. A **free radical** is a compound with an unpaired electron, making it highly reactive. If superoxide accumulates, biochemicals are rapidly oxidized and the cell dies.

Living things that survive in the presence of air have evolved ways to detoxify free radicals. All aerobes produce the enzyme **superoxide dismutase**. It destroys superoxide by converting it into oxygen and hydrogen peroxide.

$$2O_2^- + 2H^+ \longrightarrow O_2 + H_2O_2$$

Hydrogen peroxide, which is the product of this and other enzyme-catalyzed reactions, is itself a toxic oxidant. It is converted to water and oxygen by the enzyme **catalase**.

$$2H_2O_2 \longrightarrow 2H_2O + O_2$$

Table 8.1   Relationship of Various Bacteria to Oxygen

Microbial Class	Response to Oxygen	Presence of		Example
		Catalase	Superoxide Dismutase	
Obligate aerobes	Require oxygen	Present	Present	*Pseudomonas aeruginosa*
Facultative anaerobes	Can grow with or without oxygen	Present	Present	*Escherichia coli*
Microaerophiles	Grow best with low oxygen	Present	Present	*Campylobacter jejuni*
Aerotolerant anaerobes	Grow without oxygen, but not killed by it	Absent	Present	*Streptococcus pneumoniae*
Obligate anaerobes[a]	Killed by oxygen	Absent	Absent	*Methanococcus vannielii*

[a]Some contain small amounts of superoxide dismutase.

Hydrogen peroxide is also removed by **peroxidases**, enzymes that use hydrogen peroxide to oxidize certain organic compounds. Organisms that can tolerate exposure to oxygen contain superoxide dismutase, and almost all of them also contain catalase. Organisms that lack both of these enzymes quickly die in an oxygen-containing environment.

The relationship between oxygen and growth varies enormously from one species of microorganism to another. **Obligate aerobes** can grow only in the presence of oxygen because they require oxygen as a terminal electron acceptor in order to generate energy. **Microaerophiles** tolerate less oxygen (or require greater concentrations of CO_2). They can grow only at reduced concentrations of oxygen (2 to 10 percent oxygen, compared to the 21 percent oxygen present in the air). **Facultative anaerobes** can grow in either the presence or absence of oxygen because they use oxygen as a terminal electron acceptor when it is available, but employ other biochemical pathways to generate energy when oxygen is absent. **Aerotolerant anaerobes** can generate energy without oxygen but are not killed by exposure to air. **Obligate anaerobes** die in the presence of oxygen because they lack the enzymes necessary to detoxify superoxide (**Table 8.1**).

Nitrogen.   Nitrogen makes up about 14 percent of the dry weight of most microorganisms. It is a constituent of proteins and nucleic acids as well as certain essential metabolites, so all living things require a source of this element. The form of nitrogen that microorganisms use depends upon their metabolic capacity and environment. Probably all microorganisms can use ammonia (NH_3) as a source of nitrogen because this form is incorporated in biosynthesis. Most microorganisms can derive ammonia from a variety of organic compounds, including amino acids, and some can use inorganic forms, including nitrate ion (NO_3^-). Some can fix atmospheric nitrogen gas (N_2). Nitrogen-fixing organisms prosper in natural environments where nitrogen is the limiting nutrient.

Phosphorus.   Phosphorus occurs in cells exclusively in the form of phosphate ion (PO_4^{3-}) or phosphate-containing organic compounds. It makes up about 3 percent of the dry weight of microorganisms. It is a constituent of nucleic acids, phospholipids, and certain essential metabolites. Phosphorus enters the cell in the form of a phosphate ion because most phosphate-containing organic compounds cannot pass through the plasma membrane. When microorganisms use a phosphate-containing organic compound as a source of phosphorus, phosphate is usually split off from the compound outside the cell or in the periplasm by phosphate-splitting enzymes the cell releases. Then the phosphate ion enters the cell. Phosphate is the limiting nutrient in many lakes and streams, so when phosphate from detergents or sewage is released into these bodies of water, it causes **eutrophication**, or overgrowth of algae and other microorganisms.

Sulfur.   Sulfur is a minor but essential constituent of cells, amounting to about 1 percent of the dry weight of most cells. It is a component of two amino acids, a few species of transfer RNA, and certain essential metabolites. Although sulfur enters biosynthesis as sulfide (S^{2-}), the most common source of sulfur in laboratory media and in nature is sulfate ion (SO_4^{2-}). Sulfate is abundant in the ocean but relatively scarce in some lakes and soils. The major source of sulfur in many soils is organic sulfate-

Table 8.2	Trace Elements in Microbial Cells	
Element	Principal Ion Form	Function
Potassium	K^+	Maintains turgor pressure, cofactor for certain enzymes
Magnesium	Mg^{2+}	Cofactor for many enzymes
Calcium	Ca^{2+}	Cofactor for certain enzymes
Iron	Fe^{2+}/Fe^{3+}	In cytochromes[a] and certain enzymes
Manganese	Mn^{2+}	Cofactor for some enzymes
Molybdenum	Mo^{2+}	Present in coenzyme for several enzymes
Cobalt	Co^{2+}	Present in vitamin B_{12} and related coenzymes
Copper	Cu^{2+}	Present in several enzymes
Zinc	Zn^{2+}	Present in several enzymes

[a]Cytochromes are components of electron transport chains (Chapter 5).

containing compounds that soil microorganisms take up and convert to sulfide.

Trace Elements. In addition to the major elements needed to build biochemicals, microorganisms need small but essential quantities of certain inorganic chemicals, **trace elements**, to grow. Trace elements include potassium and certain metal ions, including copper, cobalt, iron, and zinc. Trace elements fulfill a variety of essential functions in the cell (**Table 8.2**). Some trace elements are **cofactors**, *inorganic ions* that must be present for an enzyme to be active. Other trace elements are part of **coenzymes**, *organic molecules* that must be present for an enzyme to be active.

Availability of trace elements rarely limits microbial growth because they are required in such minute amounts, but it can happen. For example, although iron is abundant in nature, it occurs predominantly in the form of highly insoluble iron (ferric) hydroxide, unavailable to microorganisms unless they release iron-solubilizing compounds called **siderophores**. Living tissue is a particularly poor source of available iron. Thus, most pathogenic bacteria produce siderophores.

Organic Growth Factors. Many microorganisms can synthesize all the building blocks (amino acids, nucleotides, monosaccharides, or disaccharides) needed to make macromolecules from single sources of the elements we have just discussed. But if a microorganism cannot synthesize a particular building block, it must be supplied preformed in the environment. Such essential preformed building blocks are called **organic growth factors**. They can be any of the various building blocks discussed in Chapter 5: amino acids, purines, pyrimidines,

fatty acids, and so on. **Vitamins**, which are nutrients required in minute quantities, primarily as precursors of enzyme cofactors, are organic growth factors for some microorganisms.

Nutritional diversity refers to the different requirements among microorganisms for organic growth factors. There is a considerable range. For example, *Escherichia coli* needs no organic growth factors. In contrast, *Leuconostoc mesenteroides*, a lactic acid bacteria that can cause milk to sour, must be provided with all 20 amino acids, several purines and pyrimidines, and 10 vitamins (**Table 8.3**).

If nonessential preformed building blocks are present in the environment, most microorganisms use them to enhance their growth. Regulatory mechanisms allow microorganisms to exploit a **rich growth environment**, one with abundant preformed building blocks. Then the organism expends metabolic capacity to synthesize only those building blocks that are not available in the environment. They use the metabolic capacity they save to grow more rapidly. To accomplish this they produce more ribosomes. For example, when the bacterium *Salmonella typhimurium* is growing rapidly in a rich environment, it has over 10 times as many ribosomes per cell as when it is growing slowly in a poor medium because it must synthesize proteins faster to grow faster. This relationship between richness of a medium, growth rate, and number of ribosomes per cell is shown in **Table 8.4**. In a poor medium such as one composed of salts and lysine (a carbon source that *S. typhimurium* uses slowly), doubling time is long and cells in the medium contain only the small number of ribosomes they need to grow slowly. In richer media that contain more rapidly utilized carbon sources (such as glucose) or are enriched with building blocks such as amino acids and materials found in meat

Nutrient Source	Lactic Acid Bacterium (*Leuconostoc mesenteroides*)		Escherichia coli		Cyanobacterium (*Gleobacter violaceus*)	
Carbon	Glucose	50 g[a]	Glucose	2 g	None	
	Na acetate	40 g				
Nitrogen source	NH_4Cl	6 g	NH_4Cl	2 g	None	
Inorganic salts	KH_2PO_4	1.2 g	Na_2HPO_4	6 g	$K_2HPO_4 \cdot H_2O$	40 mg
	K_2HPO_4	1.2 g	KH_2PO_4	3 g	$MgSO_4 \cdot 7H_2O$	75 mg
	$MgSO_4 \cdot 7H_2O$	0.4 g	NaCl	3 g	$CaCl_2 \cdot H_2O$	36 mg
	$FeSO_4$	20 mg	Na_2SO_4	11 g	Ferric ammonium citrate	6 mg
	$MnSO_4$	40 mg	$MgCl_2 \cdot 6H_2O$	0.4 mg	Na_2CO_3	20 mg
	NaCl	20 mg	$CaCl_2$	11 mg	Trace metals solution[b]	1 ml
			$FeCl_3 \cdot 6H_2O$	0.8 mg		
Growth factors (amino acids)	Alanine	400 mg	None		None	
	Arginine · HCl	484 mg				
	Asparagine	800 mg				
	Aspartate	200 mg				
	Cystine	100 mg				
	Glutamate	600 mg				
	Glycine	200 mg				
	Histidine · HCl	124 mg				
	Isoleucine	500 mg				
	Leucine	500 mg				
	Lysine	500 mg				
	Methionine	500 mg				
	Phenylalanine	200 mg				
	Proline	200 mg				
	Serine	100 mg				
	Threonine	400 mg				
	Tryptophan	80 mg				
	Tyrosine	200 mg				
	Valine	50 mg				
Growth factors (bases)	Adenine · SO_4	20 mg	None		None	
	Guanine · HCl	20 mg				
	Uracil	20 mg				
	Xanthine	20 mg				
Growth factors (vitamins)	Thiamine · HCl	1 mg	None		None	
	Pyridoxine · HCl	2 mg				
	Pyridoxamine · HCl	0.6 mg				
	Pyridoxal · HCl	0.6 mg				
	Pantothenate	1.0 mg				
	Riboflavin	1.0 mg				
	Nicotinate	2.0 mg				
	p-aminobenzoate	0.2 mg				
	Biotin	2.0 μg				
	Folate	20 μg				

Source: Data taken from B. F. Steel, H. E. Sauberlich, H. S. Reynolds, and C. A. Baumann, 1949, "Media for *Leuconostoc mesenteroides* D-60," *Journal of Biological Chemistry* 177:533–551 (*Leuconostoc mesenteroides*); D. J. Clark and O. Maaløe, 1967, DNA replication and the division cycle in *Escherichia coli*, *Molecular Biology* 23:99–112 (*Escherichia coli*); and R. Rippke, J. Deruelles, J. B. Waterbury, H. Herdman, and R. Y. Stanier, 1979, Genetic assignment, strain histories, and properties of pure cultures of cyanobacteria, *General Microbiology* 111:1–61 (*Gleobacter violaceus*).
[a]Quantities shown after each ingredient are those added to 1 liter of medium.
[b]2.86 g H_3BO_3, 1181 g $MnCl_2 \cdot 4 H_2O$, 0.22 g $ZnSO_4 \cdot 7H_2O$, 0.39 g $Na_2MoO_4 \cdot 2H_2O$, 0.079 g $CuSO_4 \cdot 5H_2O$, 0.049 g $Co(NO_3)_2 \cdot H_2O$ per liter.

Selenium: Feast or Poison?

Selenium is an element that closely resembles sulfur. When selenium is available in high concentrations, some enzymes that normally catalyze reactions of sulfur compounds use it instead. The results can be disastrous. Selenium is incorporated into the organism's macromolecules, which then don't function normally. The organism might die, or it might develop into a misshapen creature. Many can be seen at a lake near Kesterton, California, which is fed with runoff irrigation water from soils that contain more selenium than most. Some animals at the lake are crippled, and some birds have twisted beaks and deformed wings.

Selenium is a dangerous toxic material, but some bacteria obtained from the lake generate ATP by oxidizing selenium, much as some bacteria derive energy by oxidizing sulfur (Chapter 5). And other bacteria, including *Escherichia coli*, incorporate selenium into certain of their proteins in the form of selenocysteine. Selenocysteine is an amino acid identical to cysteine but with a selenium atom in place of the sulfur atom in ordinary cysteine. *E. coli* contains one selenocysteine monomer in an enzyme, formic dehydrogenase, that participates in central metabolism. If the selenium atom is replaced by a sulfur atom, the enzyme becomes almost completely inactive. If selenium were to replace sulfur in the other cysteines in the enzyme or any other enzyme, presumably they too would become inactive.

How can selenium replace sulfur in this cysteine and only this one? It occurs in a highly unusual way. Selenocysteine is made on a special kind of tRNA molecule that serves only this purpose, and its incorporation into protein is encoded by a nonsense codon (UGA). Why does only this one particular nonsense codon signal the incorporation of selenocysteine? Presumably, the structure of the mRNA surrounding this codon is in some way unique. But without doubt, selenium, which can be so toxic, can also be essential.

Table 8.4   Rate of Growth of the Bacterium *Salmonella typhimurium* in Various Media

Medium	Composition	Doubling Time (minutes)	Number of Ribosomes per Cell
Lysine minimal	The amino acid lysine + salts	97	7000
Glucose minimal	Glucose + salts	50	17,000
20 amino acids	20 amino acids + salts	32	42,000
Brain-heart	An extract of beef brain and heart	21	83,000

---

extract, doubling time decreases and the number of ribosomes cells contain increases to accommodate the more rapid growth of the culture.

The Nonnutritive Environment

Factors other than nutrition influence the growth of microbial populations. They include physical factors—temperature and hydrostatic pressure—and chemical factors—pH and osmotic strength.

Temperature. Every species of microorganism grows over a range of temperature, from the **minimum temperature of growth**, the lowest that will support growth, to the **maximum temperature of growth**, the highest that will support growth. Most bacteria can grow in a temperature range spanning approximately 40 Celsius degrees. Most eucaryotic microorganisms have a narrower range. A species' **optimum temperature** is that at which the mi-

croorganism grows most rapidly. Optimum temperature is usually only a few degrees below the maximum temperature of growth. Growth rate increases from the minimum temperature to the optimum temperature because chemical reactions, including enzyme-catalyzed reactions, proceed more rapidly as temperature rises (**Figure 8.5**). Above the optimum temperature, however, growth rate declines rapidly because certain cellular components, especially proteins, are inactivated more rapidly than they can be replaced.

Microorganisms grow at extreme temperatures where most other living things cannot. In general, procaryotic microorganisms tolerate more extreme temperatures than do eucaryotes. No eucaryotes can grow at the boiling point of water, although some fungi can grow at subfreezing temperatures—as low as those tolerated by any species of bacteria—and some algae grow near the freezing point of water. The growth of some red algae, called snow algae, makes huge red patches on mountain snow

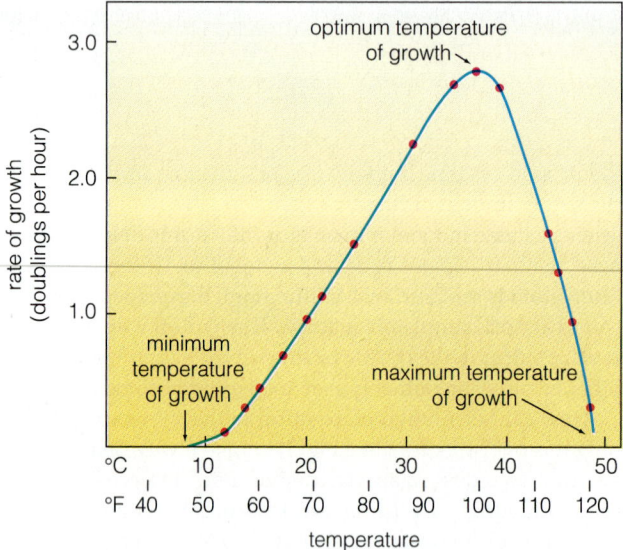

Figure 8.5 Rate of growth of *Escherichia coli* in a rich medium at various temperatures. *E. coli* grows from 8° to 48°C. The rate of growth increases rapidly above 8°C. At 37°C, *E. coli*'s optimum growth temperature, growth rate is 2.8 doublings per hour, a doubling time of about 21.4 minutes. At higher temperatures, growth rate slows, and growth stops at 48°C, its maximum temperature of growth.

fields, where snow persists for long periods and sunlight is intense.

Microorganisms are divided into three major classes on the basis of the range of temperature over which they can grow. **Thermophiles** (heat lovers) grow at high temperatures. **Mesophiles** (moderate temperature lovers) grow at intermediate temperatures, and **psychrophiles** (cold lovers) grow at cold temperatures. Although psychrophiles are capable of growth at low temperatures, these microorganisms, like all others, grow most rapidly near their maximum temperature for growth.

Thermophilic bacteria grow above 50°C—a temperature at which water causes pain if you immerse your hand. The extremely thermophilic *Pyrodictium occultum* can grow at 110°C—ten Celsius degrees above the boiling point of pure water at sea level. *P. occultum* is found in hot springs.

Mesophilic bacteria grow best near 37°C—the temperature of the human body. *Escherichia coli*, which grows naturally in the intestines of humans and other animals, is a mesophile, as are most disease-causing bacteria.

Psychrophilic bacteria grow at a significant rate at or below 5°C—the temperature of a properly functioning refrigerator. When milk spoils in a refrigerator it often has the fruity odor of *Pseudomonas* spp. or the foul odor of *Archromobacter* spp. because these bacteria are psychrophiles. In contrast, milk that sours at room tempera-

ture has the more pleasant, though sour, taste of yogurt or buttermilk because at this temperature mesophilic lactic acid bacteria predominate.

Psychrophiles are subdivided into **obligate** and **facultative** groups. Thermophiles are subdivided into **stenothermophiles**, **facultative thermophiles**, and **extreme thermophiles** (Table 8.5). Facultative psychrophiles are also called **psychrotrophs** ("cold-growing") to indicate that they tolerate rather than benefit from decreasing temperature.

What determines the maximum or minimum temperature at which an organism is able to grow? Without doubt, the maximum temperature of growth is determined by the heat stability of an organism's proteins. Proteins are usually the most heat-sensitive macromolecules in a cell, and growth cannot occur at temperatures that denature the cell's proteins (Chapter 2). Thermophiles grow at high temperatures because their proteins are exceptionally heat-stable; psychrophiles can grow only at low temperatures because some of their proteins are unusually heat-sensitive. Heat stability reflects total protein structure. That is, a mutation that changes any amino acid in a heat-stable protein is likely to decrease its heat stability.

Causes of the minimum temperature of growth are more complex, but most have the same basis. Hydrophobic interactions within proteins become weaker as temperature decreases, and the shape of proteins changes slightly. The function of some proteins, including those that regulate metabolism, is particularly sensitive to such changes, and so at low temperatures, metabolic regulatory mechanisms become distorted and stop growth. The regulatory mechanisms of some thermophiles begin to break down at temperatures as high as 37°C. For these organisms, including some soil-dwelling bacteria such as *Bacillus stearothermophilus*, human body temperature is too cold for growth. Surprisingly, these bacteria are found in ordinary soils in temperate climates, which indicates how hot the soil surface can become on a sunny day.

Hydrostatic Pressure. **Hydrostatic pressure** is pressure applied to a liquid. Hydrostatic pressure is commonly measured in **atmospheres**. One atmosphere is 14.7 pounds per square inch, the pressure we are exposed to at the earth's surface. Ordinary bacteria, such as *Escherichia coli*, thrive at pressures as great as 300 atmospheres, or 4410 pounds per square inch. Bacteria found in the deep ocean tolerate up to 1500 atmospheres, or 22,050 pounds per square inch, enough to crush all but the strongest steel vessels. In contrast, most yeasts cannot grow at pressures over 8 atmospheres, a low pressure. Some bacteria, the **barophiles** (pressure lovers), grow more rapidly at pressures greater than 1 atmosphere, and **obligate barophiles** grow *only* at pressures greater than 1 atmosphere.

Table 8.5 Growth Responses to Temperature among Various Groups of Bacteria

Class	Properties	Typical Environment
Psychrophiles (also called psychrotrophs)	Grow at appreciable rates below 5°C	
Obligate psychrophiles	Cannot grow at or above 20°C	Cold ocean water
Facultative psychrophiles	Can grow above 20°C	Soil and water
Mesophiles	Grow best at moderate temperature, around 37°C	Animals
Thermophiles	Grow above 50°C	
Facultative thermophiles	Can grow below 37°C	Soil
Stenothermophiles	Cannot grow below 37°C	Compost
Extreme thermophiles	Grow above 80°C (some above 100°C)	Hot springs

Elevated pressure does not crush the microbial cell as it would a human being because water passes readily through the cell membrane. Thus, the pressure inside and outside the cell is equalized. Increased pressure can stop microbial growth, but the effect is biochemical. Molecular volume changes in the course of most chemical reactions. High pressure inhibits any chemical reaction that undergoes an increase in molecular volume and favors any chemical reaction that undergoes a decrease in molecular volume. Some biochemical reactions increase molecular volume and are slowed or virtually stopped by increasing pressure. Essential chemical reactions of barophiles, however, decrease molecular volume and proceed more rapidly as pressure increases.

pH. The **pH scale** measures the concentration of hydrogen and hydroxide ions in a solution and indicates whether the environment is acidic, below pH 7; alkaline, above pH 7; or neutral, near pH 7 (Chapter 2). In general, bacteria grow best at a slightly alkaline (basic) pH. Fungi grow best at a slightly acid pH, and protozoa and algae at a neutral pH. Exceptions occur, however, especially among bacteria. **Acidophiles** (acid lovers) thrive in environments of extremely low pH. Certain acidophiles, for example, grow in the acid leachings of mine waste, where pH values are as low as 1.0—the acidity of sulfuric acid. **Alkaliphiles** (base lovers) thrive in environments of extremely high pH. Certain alkaliphiles, for example, grow in soda lakes, common in deserts of the American West. Here pH values are as high as 12.0—the alkalinity of hair remover.

Most bacteria survive in environments with a relatively wide pH range by adjusting their intracellular pH. By various mechanisms they pump hydrogen ions out of or into the cell. *Escherichia coli*, for example, can grow in environments that range from pH 5.0 to 8.0. But regardless of external pH, internal pH is maintained at fairly close to 7.6—the optimum value for its metabolism. *E. coli*, like most other bacteria, can perform vital metabolic reactions only within a limited pH range because many of its enzymes function properly only over a rather narrow pH range (Chapter 2).

Osmotic Strength. All microorganisms need liquid water to grow. For this reason, they cannot grow at temperatures below the freezing point of their medium or above its boiling point. High osmotic strength also deprives a cell of water. Cell membranes are highly permeable to water, so water enters or leaves a cell depending on the relative osmotic strength, or concentration, of dissolved solutes in the cell and its environment. Bacteria maintain a positive turgor pressure because the osmotic strength of their cell contents is greater than the osmotic strength of the environment (Chapter 4). Turgor pressure provides the force for the cell to grow. Instead of turgor pressure, the cytoskeleton provides the force to enlarge the cell in eucaryotic microorganisms.

If the concentration of solutes in the external environment increases, the bacterium must maintain positive turgor pressure. To do so, it pumps potassium ions (K^+) and/or **osmoprotectants**, such as the amino acid proline, into the cell and synthesizes the disaccharide **trehalose**. These solutes maintain a higher osmotic pressure inside the cell than the osmotic pressure outside. Eventually, however, this increase in internal osmotic pressure damages essential enzymes and interferes with the cell's metabolism. Thus, environments with a high osmotic strength inhibit bacterial growth. A traditional way of preserving food is to add sugar or salt, which increases osmotic strength and thereby prevents microbial growth.

Although high osmotic strength prevents the growth of most bacteria, some species, called **halophiles** (salt lovers), can withstand extremely high salt concentrations. Red archaebacteria called *Halobacteria* flourish in water

Fighting Off a Chill

Because microorganisms grow over a wide range of temperatures, they face environmental challenges that warm-blooded humans never do. One of the biggest is maintaining proper membrane fluidity. Membranes are composed of phospholipids; and so, like most lipids, they solidify in the cold and become liquid in heat. To function properly, membranes must have the right degree of fluidity over a range of temperature. How do they do this? One way: the fatty acid composition of the phospholipids in their membranes changes with the surrounding temperature.

The effect of fatty acid composition on the fluidity of fatty-acid-containing lipids is dramatic. In the refrigerator olive oil solidifies, but corn oil remains fluid. This is because corn oil is rich in unsaturated fatty acids (contain double bonds) and polyunsaturated fatty acids (contain more than one double bond), and olive oil has more saturated fatty acids (contain no double bonds). At room temperature corn oil is more fluid than olive oil. But by mixing corn oil and olive oil in the proper proportions, an intermediate level of fluidity can be attained at room temperature, in the refrigerator, or at any intermediate temperature.

This same process occurs in microorganisms. As growth temperature declines, an increasingly high proportion of the fatty acids in their phospholipid membranes are unsaturated, so their fluidity is optimal at all growth temperatures. In *Escherichia coli* this precise adjustment is made quite simply. Fatty acids are synthesized by a branched pathway. One branch leads to saturated fatty acids, and the other to unsaturated fatty acids. One enzyme near the branch point, on the unsaturated side, is sensitive to higher temperature. As temperature rises, its activity declines, and more fatty acids are made by the "saturated" branch. As temperature declines, the reverse happens. The result is constant fluidity at all temperatures.

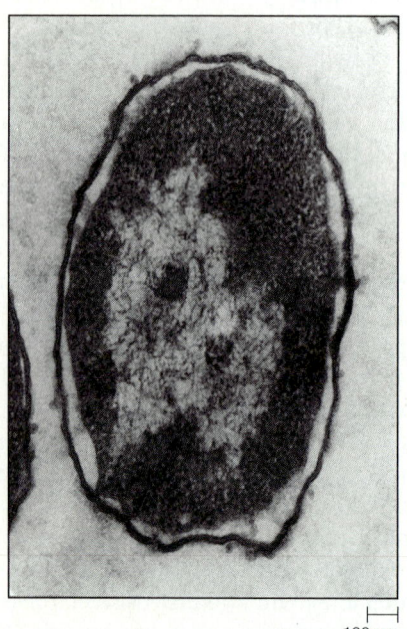

$\vdash\!\!\dashv$
100 μm

Figure 8.6 Plasmolysis of a bacterial cell pulls the plasma membrane away from the cell wall except at the site of Bayer junctions, places where the plasma membrane is attached through the cell wall to the outer membrane (Chapter 4). This transmission electron micrograph shows plasmolysis of *Escherichia coli*.

that is saturated with salt. Salt flats, bodies of sea water that are dammed off and allowed to evaporate to produce table salt, and some desert lakes are bright red from their presence. These bacteria not only tolerate high intracellular concentrations of salt but also depend on it for membrane stability. Placed in distilled or ordinary tap water, they lyse immediately.

Plasmolysis occurs when an increase in external osmotic strength causes water to leave a cell, shrinking the cytoplasm (Chapter 4). The consequences of plasmolysis differ by cell type. In bacteria, the plasma membrane pulls away from the rigid wall, but a cell can usually recover from plasmolysis (unless the osmotic pressure is extreme) by increasing its internal osmotic strength (**Figure 8.6**). Eucaryotic microorganisms, particularly protozoa, are more sensitive to plasmolysis because they lack a rigid wall.

MEASURING NUMBERS OF MICROORGANISMS

Measuring the growth and death of microorganisms in the laboratory requires special techniques. Some techniques are indirect measurements. They measure a property—turbidity, dry weight, or metabolic activity—of the mass of cells in a population. Other techniques are direct. Direct measurement may be by **direct microscopic** or **elec-**

Why Does Champagne Come in Such a Heavy Bottle?

Most microorganisms can withstand high hydrostatic pressures. But yeasts are a striking exception. They stop growing and fermenting at only 8 atmospheres of pressure. Although no one knows why yeast is so sensitive to pressure, it is an advantage in winemaking. German winemakers exploit the pressure sensitivity of yeasts to slow the rate of the yeast fermentation that converts grape juice into wine. This prevents heat from building up and damaging the taste of wine. They carry out the fermentation in closed steel tanks and let the pressure from carbon dioxide production rise until it moderates fermentation rate.

The pressure sensitivity of yeasts is even more critical in making champagne. The carbon dioxide bubbles in champagne are formed by a second fermentation usually carried out in the bottle. A calculated amount of sugar is added to the wine to assure that the proper amount of carbon dioxide is produced. But there is little risk that too much carbon dioxide will be produced and the bottle will explode. Fermentation stops when pressure in the bottle reaches 8 atmospheres, a pressure the heavy champagne bottle can withstand.

---

tronic count, plate count, or most probable number (MPN). (Sometimes filtration is a preliminary step in direct measurement.)

Direct microscopic or electronic count yields a total count, a count of all cells, living and dead. A viable count measures only the living cells in a population, those capable of reproduction (*viable* means "live"). Plate counts and MPN are viable counts. When filtration is involved, it yields a viable count. All of these techniques have advantages and disadvantages.

Turbidity

Because the cells in a pure culture are all approximately the same size, their number is proportional to their weight. In other words, two cells weigh twice as much as one cell, four cells weigh twice as much again, and so on. Thus, we can estimate the size of a microbial population from its total weight. Usually microbiologists do not laboriously collect and weigh cells. Instead, they use an optical instrument called a spectrophotometer (Figure 8.7). A spectrophotometer measures how much light a solution or a liquid culture of microbial cells transmits. The greater the mass of cells in the culture, the greater its turbidity (cloudiness); less light will be transmitted and the reading on the spectrophotometer will be higher. Because a spectrophotometer precisely measures what we see when we examine a liquid culture visually, it is not very sensitive in terms of numbers of bacterial cells. About 10 million average-size bacterial cells must be present per milliliter of liquid culture before the liquid becomes visibly turbid, so a spectrophotometer is not useful for detecting minor contamination, the presence of a small number of unwanted microbial cells. The spectrophotometer may also be used on yeast cultures as well as most bacteria cultures.

The spectrophotometer is usually used to estimate the mass of a relatively dense culture in order to chart the growth of a growing population of microorganisms (Figure 8.7). To measure mass or number of cells with a spectrophotometer, it is first necessary to prepare a standard curve, a graph relating spectrophotometer readings to cell mass in a particular culture. The same number of different kinds of bacteria usually produce different turbidity. Then the cell mass corresponding to any spectrophotometer reading can be determined from the graph. Standard curves can also be prepared to relate spectrophotometer readings to number of cells in a culture.

Like all instruments, the spectrophotometer has advantages and disadvantages. Its measurements are rapid and reproducible. However, it can be used only on relatively dense cultures, and it cannot distinguish between dead and live cells. Nor can it be used with cells that tend to aggregate, because they rapidly settle out of suspension and cloudiness disappears.

Dry Weight

We can measure the number of cells in a culture by dividing the dry weight of the culture by the dry weight of an individual cell. The dry weight of the cells in a culture is determined by separating them from the medium, drying them, and weighing them. To obtain an accurate measurement, about 25 ml of a relatively dense culture is required. The cells are removed by filtration (Chapter 3) or centrifugation. A centrifuge is an instrument that rotates vessels called centrifuge tubes at a high rate of speed, generating a centrifugal force that pushes small particles, including microbial cells, to the bottom of the tube. Then the cells are washed, resuspended in distilled water, and filtered or centrifuged again. After the second washing, cells are dried in an oven at 105°C for about 24 hours and

Figure 8.7 The spectrophotometer. (**a**) A spectrophotometer measures turbidity, the amount of light that a culture transmits in units of absorbance. The bacterial cells scatter light so it does not strike the light-sensitive detector. (**b**) The tube on the left has no bacterial cells in it and is clear. The one on the right contains about a billion (10^9) cells per milliliter and therefore appears turbid (cloudy). (**c**) A standard curve relates absorbance to the number of a particular kind of bacterial cells per milliliter. Once prepared, the curve can be used to convert any absorbance reading to cell number. For example, from this curve a reading of 1.6 absorbance units means the culture contains a hundred million (10^8) cells per milliliter.

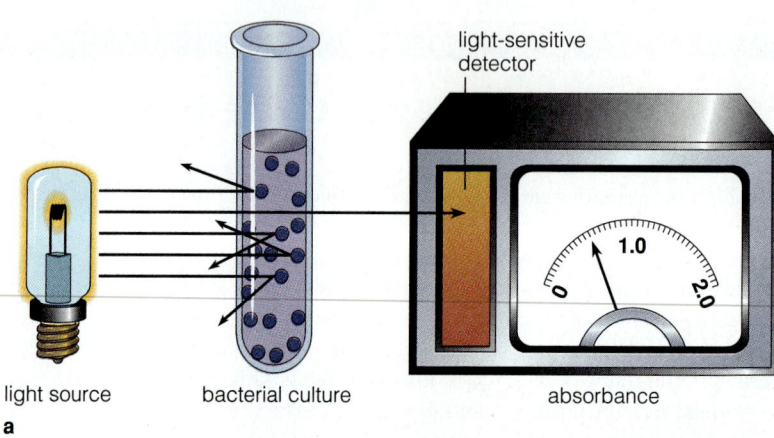

light source   bacterial culture   absorbance

a

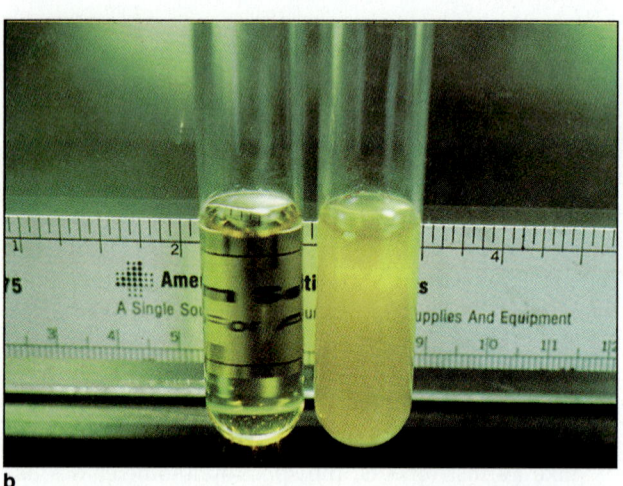

b

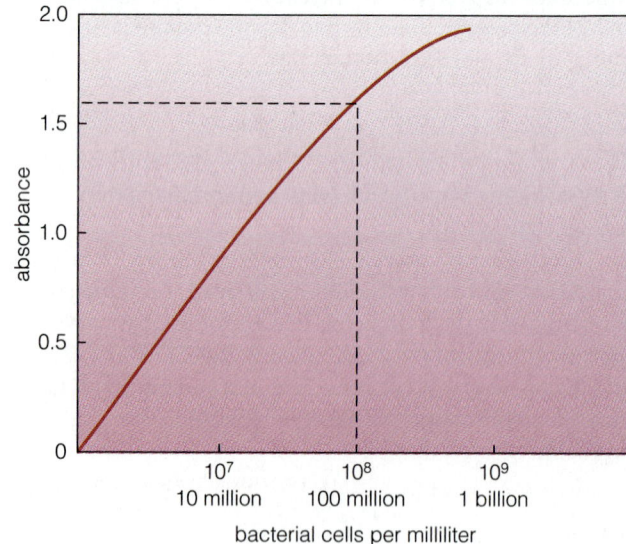

c

then cooled in a **desiccator** to prevent the cells from taking up moisture from the air. A desiccator is a chamber that is maintained chemically or mechanically at low humidity to dry materials or keep them dry. Finally, the cells are weighed.

Measurement by dry weight is tedious and time-consuming, and the sample must contain more than about 10 million cells. Nevertheless, it has certain uses. It must be used, for example, to construct a standard curve relating spectrophotometer measurement to cell mass.

Metabolic Activity

There are several ways that metabolic activity can be used to indirectly measure a quantity of microbial cells. The rate of formation of metabolic products, such as gases or acids, that a culture produces reflects the mass of cells present. Also, the rate of utilization of a substrate, such as oxygen or glucose, reflects cell mass.

Finally, the rate of reduction of certain dyes is another way to estimate microbial mass. For example, methylene

blue becomes colorless when reduced by components of microbial electron transport chains. Consequently, in the absence of oxygen, which oxidizes methylene blue, the rate of decolorization of this dye is an index of microbial mass. Rate of dye reduction is highly indirect and therefore not an accurate measurement of microbial mass. However, the technique can be used with complex materials, such as soil or milk, and no instruments.

Direct Count

A **direct count** is a count of individual cells in a microbial population. Direct counts can be done visually, using a microscope with a special microscope slide called a **counting chamber**. There are many styles of counting chambers, but the **Petroff-Hauser** chamber, named for its developers, is perhaps the most commonly used in microbiology laboratories (**Figure 8.8**). All counting chambers have known depths and are marked on the bottom by squares of known dimensions. Thus, each square that we see under the microscope marks off a known volume

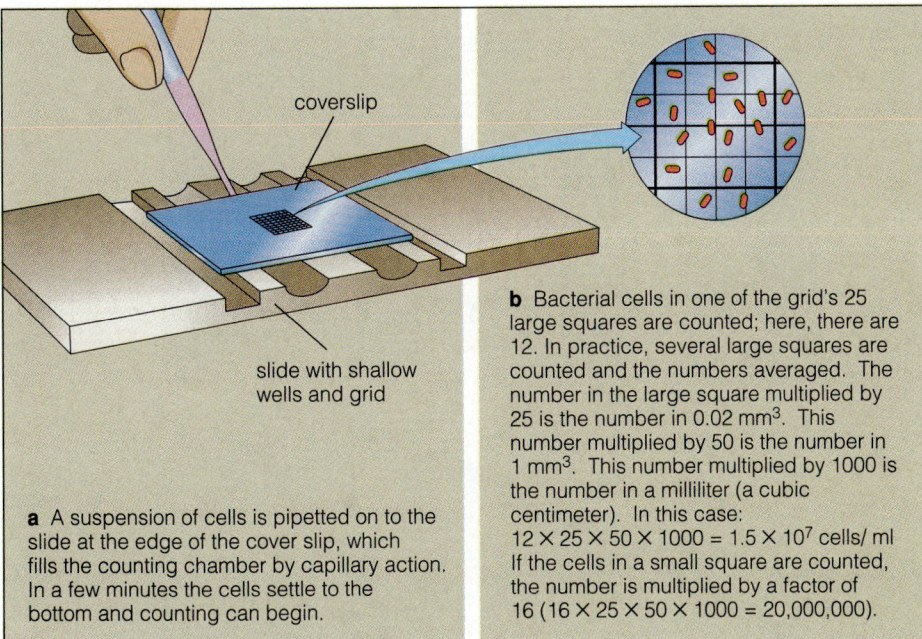

coverslip

slide with shallow
wells and grid

a A suspension of cells is pipetted on to the slide at the edge of the cover slip, which fills the counting chamber by capillary action. In a few minutes the cells settle to the bottom and counting can begin.

b Bacterial cells in one of the grid's 25 large squares are counted; here, there are 12. In practice, several large squares are counted and the numbers averaged. The number in the large square multiplied by 25 is the number in 0.02 mm³. This number multiplied by 50 is the number in 1 mm³. This number multiplied by 1000 is the number in a milliliter (a cubic centimeter). In this case:
$12 \times 25 \times 50 \times 1000 = 1.5 \times 10^7$ cells/ ml
If the cells in a small square are counted, the number is multiplied by a factor of 16 ($16 \times 25 \times 50 \times 1000 = 20,000,000$).

Figure 8.8 Direct microscopic count. The Petroff-Hauser counting chamber is used with a microscope to do a direct count of cells. It consists of a slide with shallow wells (0.02 mm deep). On the bottom of the slide, 1 mm² is inscribed with a grid dividing it into 25 large squares, each of which is divided into 16 smaller squares. So the volume over the entire grid is 0.02 (one-fiftieth) mm³, and each large square is one-twenty-fifth of that amount.

of liquid, for example, one microliter (0.001 µl). By counting the cells within several squares and averaging the results, we can calculate how many cells are present in 1 ml or any other volume of the culture. Because live and dead cells look much the same under a microscope, direct counts are total counts.

Direct counts can also be done electronically, with an instrument called a **Coulter counter**. Electronic counting is possible because microbial cells do not conduct electricity as well as the medium in which they are suspended. A measured volume of the culture to be counted is placed in one chamber of the counter and passed through a small pore (about 15 µm in diameter) into another chamber by a vacuum pump. The pore is part of an electric circuit, so each time a cell passes through, the conductivity of the circuit decreases and the presence of that cell is tallied electronically on the counter.

The two principal methods of direct counts—microscopic and electronic counting—give similar results, but they are used in quite different situations. Microscopic counting requires no expensive equipment, only a relatively inexpensive counting chamber in addition to a light microscope, which any microbiology laboratory would have. But it is a slow, tedious process. In contrast, electronic counting requires expensive equipment but is extremely rapid and accurate if cells are the only particles present. (Electronic counting is also used in clinical laboratories to count blood cells.) Microscopic counting is necessary if the sample contains foreign particles, such as blood cells or bits of soil. A person doing a microscopic count can discriminate between microbial cells and foreign particles, but the electronic counter will count them as cells if they are approximately the same size.

Plate Count

The technique for performing a **plate count** is the same as the spread and pour plate methods of culturing microorganisms described in Chapter 3. A sample of the culture is serially diluted, usually tenfold at each dilution, and then a small amount of each dilution is spread on a plate or mixed with melted medium and incubated under appropriate growth conditions. After visible colonies have formed, usually a day or so later, plates with clearly separated colonies are selected and counted. Choosing plates with between 30 and 300 colonies offers a good compromise between speed and accuracy. The number of colonies on a plate, along with the dilution of the sample that was spread on that plate, allows us to calculate the concentration of cells present in the original sample (**Figure 8.9**). Plate count gives a viable count. It is based on the principle that a single viable cell will develop into a visible colony when incubated on an agar plate.

The advantage of the plate count method is its extreme sensitivity. Even a single live cell can be detected with the appropriate medium and incubation conditions. Moreover, a plate count does not require complicated equipment. On the negative side, doing plate counts is slow and tedious and not very accurate because accuracy depends upon counting large numbers of colonies. Accuracy increases with the numbers of colonies counted because **sampling error** is decreased. Sampling error is the inevitable inaccuracy that results, no matter how carefully dilutions and plating are done, because all samples are not completely representative of the total population. Some samples of a culture will contain more cells than others, but 95 percent of the time, the true number of viable cells does not differ from the number counted by

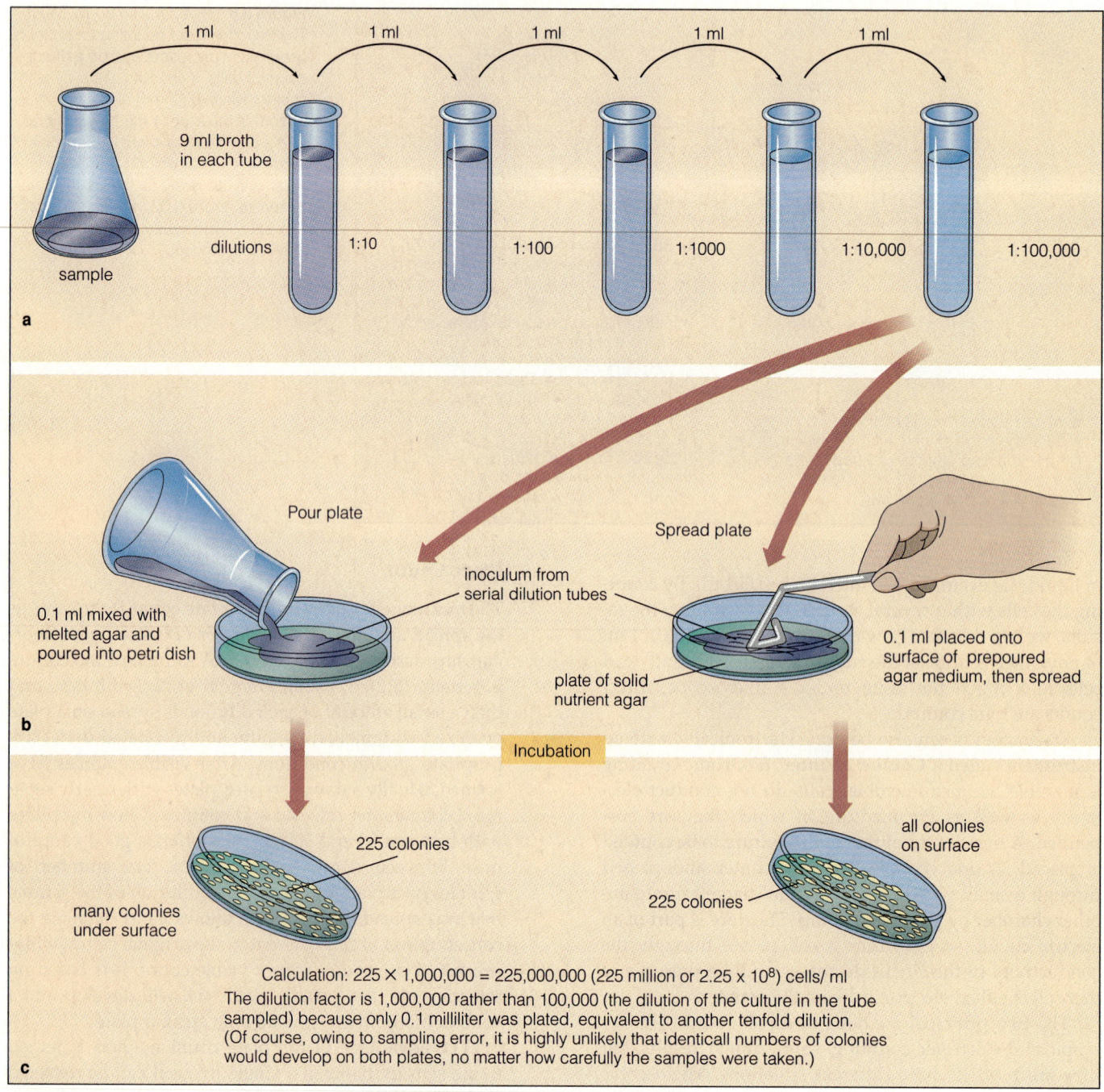

Figure 8.9 A plate count is a direct measurement and gives a viable count. First the sample is serially diluted by transferring 1 ml of the sample to 9 ml of sterile medium and mixing thoroughly. This process is repeated until an appropriate dilution is obtained—in this example 1:100,000 (10^{-5}). Then 0.1 ml is added to a plate of nutrient agar either by spreading it on the surface (spread plate method) or by pouring it with the medium (pour plate method). The plates are incubated. The colonies that develop are counted. For statistical reasons, plates should contain between 30 and 300 colonies. In the example, the plate has 225 colonies.

Text within the figure:

1 ml   1 ml   1 ml   1 ml   1 ml

9 ml broth in each tube

dilutions   1:10   1:100   1:1000   1:10,000   1:100,000

sample

a

Pour plate

Spread plate

inoculum from serial dilution tubes

0.1 ml mixed with melted agar and poured into petri dish

plate of solid nutrient agar

0.1 ml placed onto surface of prepoured agar medium, then spread

b

Incubation

225 colonies

all colonies on surface

many colonies under surface

225 colonies

Calculation: 225 × 1,000,000 = 225,000,000 (225 million or 2.25×10^8) cells/ ml
The dilution factor is 1,000,000 rather than 100,000 (the dilution of the culture in the tube sampled) because only 0.1 milliliter was plated, equivalent to another tenfold dilution. (Of course, owing to sampling error, it is highly unlikely that identicall numbers of colonies would develop on both plates, no matter how carefully the samples were taken.)

c

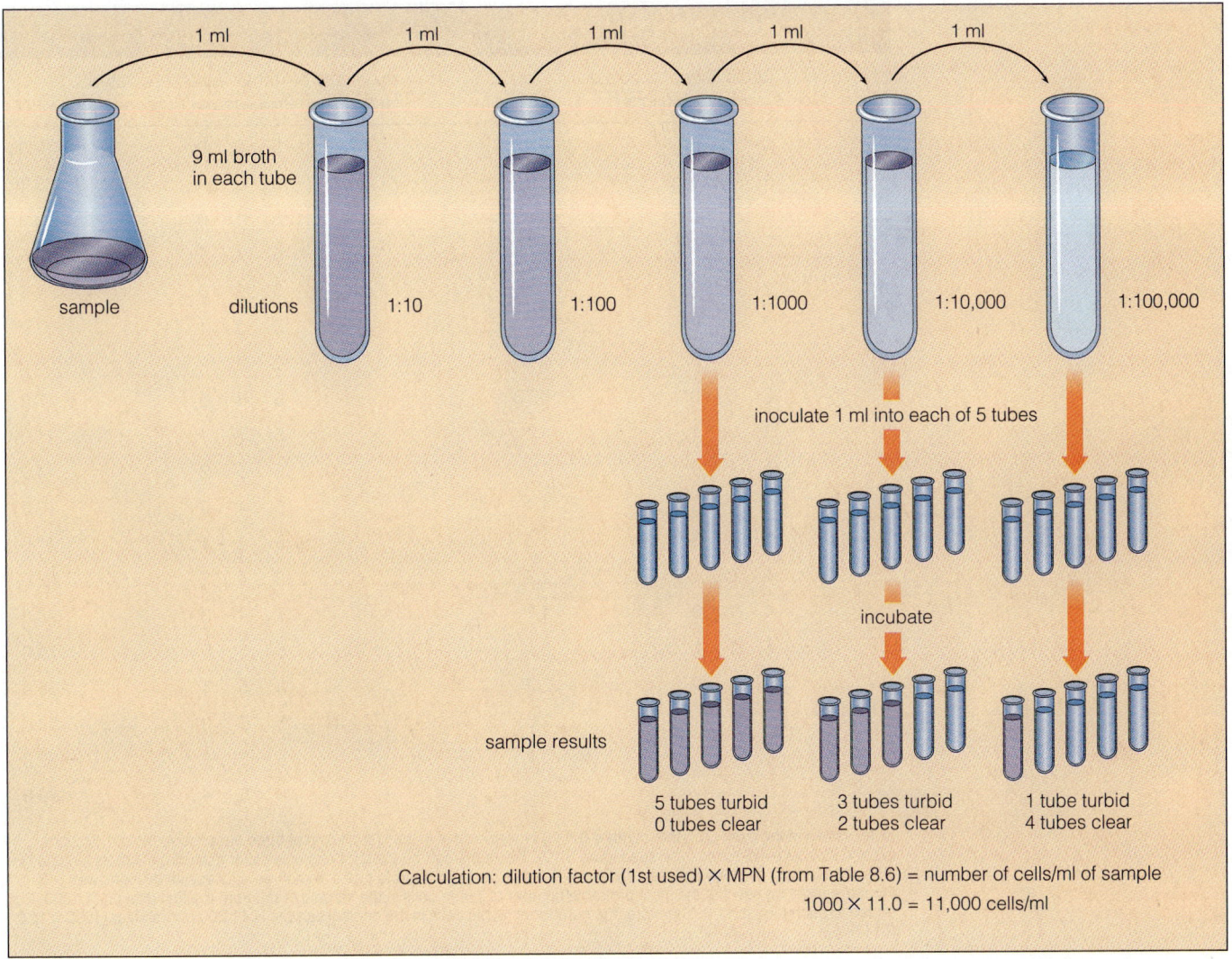

1 ml　　**1 ml**　　**1 ml**　　**1 ml**　　**1 ml**

9 ml broth
in each tube

sample　　dilutions　　1:10　　1:100　　1:1000　　1:10,000　　1:100,000

inoculate 1 ml into each of 5 tubes

incubate

sample results

5 tubes turbid　　3 tubes turbid　　1 tube turbid
0 tubes clear　　2 tubes clear　　4 tubes clear

Calculation: dilution factor (1st used) × MPN (from Table 8.6) = number of cells/ml of sample
1000 × 11.0 = 11,000 cells/ml

Figure 8.10　The most probable number (MPN) method gives a viable count. As the dilution factor increases, a point is reached at which some tubes contain only a single organism and the others, none. From the pattern of the number of turbid tubes in three successive dilutions (here 5, 3, 1), the statistically most probable number of cells (11.0) in the first of the three dilutions can be determined by referring to a table of most probable numbers (Table 8.6). This value multiplied by the dilution factor gives the number of viable cells in the sample.

more than an amount equal to twice the square root of the number of colonies counted.

Most Probable Number

The **most probable number (MPN)** method, like the plate count method, gives a viable count. It is based on the principle that a single live cell can develop into a turbid culture. MPN involves making a series of tenfold dilutions of the culture sample in a liquid medium suitable for the growth of that organism (**Figure 8.10**). Then samples from these test tubes are incubated. After enough time has passed to allow for microbial growth, the test tubes are examined. Tubes that received one or more microbial cells from the sample become cloudy, while tubes that didn't receive any cells remain clear. As the dilution factor increases, a point is reached at which some tubes contain only a single organism and others, none. By determining the probability that tubes did not receive cells, the number of microorganisms that were most probably present in the original sample can be determined using a statistically derived table. **Table 8.6** presents an MPN table.

Accuracy of the most probable number increases with the number of tubes used, but five tubes per dilution is regarded as a practical compromise between accuracy

Table 8.6 Most Probable Number Table

Numbers of Turbid Tubes Inoculated from Three Successive Dilutions			MPN	Numbers of Turbid Tubes Inoculated from Three Successive Dilutions			MPN
0	1	0	0.18	5	0	0	2.3
1	0	0	0.20	5	0	1	3.1
1	1	0	0.40	5	1	0	3.3
2	0	0	0.45	5	1	1	4.6
2	0	1	0.68	5	2	0	4.9
2	1	0	0.68	5	2	1	7.0
2	2	0	0.93	5	2	2	9.5
3	0	0	0.78	5	3	0	7.9
3	0	1	1.1	5	3	1	11.0
3	1	0	1.1	5	3	2	14.0
3	2	0	1.4	5	4	0	13.0
4	0	0	1.3	5	4	1	17.0
4	0	1	1.7	5	4	2	22.0
4	1	0	1.7	5	4	3	28.0
4	1	1	2.1	5	5	0	24.0
4	2	0	2.2	5	5	1	35.0
4	2	1	2.6	5	5	2	54.0
4	3	0	2.7	5	5	3	92.0
				5	5	4	160.0

Note: Values of the most probable number (MPN) of viable cells in the first dilution of three successive tenfold dilutions used to inoculate five tubes (also see Figure 8.10). This table can be used for any unicellular microorganism as long as five tubes and tenfold dilutions are used. In Figure 8.10 the number of turbid tubes from three successive dilutions was 5, 3, 1. From this table we see that the most probable number of viable cells in the sample (1 ml) from the first dilution (1:1000) added to each tube is 11.0. Therefore, the number of viable cells in the original sample is 11,000 per milliliter (1000 × 11.0 = 11,000).

and economy. The MPN method is used to count microorganisms that are difficult to culture on solid media. It is also used to determine the number of cells in a mixed culture that can grow in a particular liquid medium. For example, it can be used to test for contamination in drinking water by determining the number of bacteria that can grow in a lactose-containing medium. Such bacteria are probably *Escherichia coli* from contaminating sewage, and presence of *E. coli* in drinking water is presumptive evidence of contamination.

Filtration

Methods that depend on forming colonies (plate count) or developing turbid cultures (most probable number) can detect a single viable microbial cell, but for practical reasons, only about a milliliter of sample is used as the inoculum for the plate or the tube. Populations with fewer than one cell per milliliter therefore go undetected. Suppose you want to measure bacteria in a swimming pool. The sample must be concentrated before it is counted. Concentrating samples of microorganisms is usually done by filtration, the same method described in Chapter 3. A known volume of liquid or air is drawn through a membrane filter by vacuum. The pores in the filter are too small for microbial cells to pass through. Then the filter is placed on an appropriate solid medium and incubated. The number of colonies that develop is the number of viable microbial cells in the volume of fluid that was filtered. Results are usually calculated as organisms per liter (**Figure 8.11**).

The various ways of measuring numbers of microorganisms are summarized in **Table 8.7**.

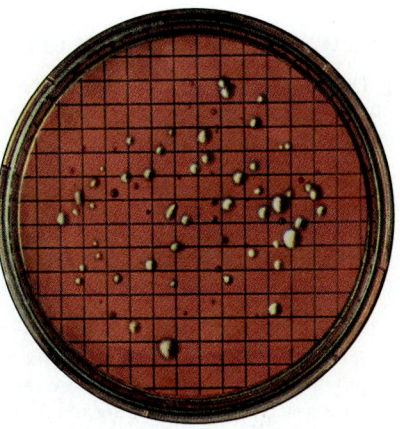

Step 2 Millipore filter is placed on a plate with agar medium and incubated.

Step 1 A known volume of sample (for example, 10 liters) is filtered through a millipore filter.

Step 3 The number of colonies on the plate is counted (here, 26 colonies).

Figure 8.11 Filtration.

Table 8.7 Methods of Measuring Numbers of Microorganisms

Method	Microorganisms It Is Used On	Type of Count	Uses/Limitations
Indirect			
Turbidity	Most bacteria and yeasts	Total	Measures turbidity with spectrophotometer; rapid and reproducible; suspension must contain more than about 10 million cells per ml; standard curve needed
Dry weight	Any microorganism	Total	Tedious and time-consuming, but accurate and reproducible
Metabolic activity	Any live microorganism	Viable	Time-consuming and can be inaccurate; can be used with complex materials
Direct			
Direct microscopic count	Any unicellular microorganism	Total	Most useful for counting one kind of cell in a mixture; time-consuming; not useful for dilute cultures
Direct electronic count	Any unicellular microorganism	Total	Most useful for counting cells in a pure culture; rapid and accurate; no foreign material can be present
Plate count	Any live unicellular microorganism	Viable	Very sensitive—can detect even a single cell; slow and tedious; large numbers of colonies must be counted to avoid sampling error
Most probable number (MPN)	Any live microorganism	Viable	Used for microorganisms that are difficult to culture on solid medium and to test for *Escherichia coli* contamination in water; time-consuming
Filtration	Any live microorganism	Viable	Concentrates a sample so a count of a small number of cells can be obtained from large volumes of liquid or gas

Summary

POPULATIONS   (p. 192)

1. Microbial growth refers to growth of a population—an increase in the number of cells, not an increase in the size of an individual cell.

THE WAY MICROORGANISMS GROW   (pp. 192–196)

1. Doubling time is the period required for cells in a microbial population to grow, divide, and produce two new cells for each one that existed before.

2. Growth rate is usually measured as doubling times per hour.

3. Exponential growth means that during each doubling time, the number of cells in the population increases by a factor of two. The formula is $N = 2^n$ where N is the number of cells in the culture after n doubling times have passed.

4. Microbial growth is graphed on a logarithmic scale.

5. A microbial culture typically passes through four distinct phases of growth: the lag phase, the log phase, the stationary phase, and the death phase. Occasionally the lag and death phases do not occur in a particular culture.

6. The lag phase usually follows the inoculation of stationary-phase or death-phase cells into a fresh culture medium. It is a no-growth period, but there is considerable metabolic activity as cells prepare to grow.

7. The log phase does not continue indefinitely because of intervening factors, such as depleted nutrients or an accumulation of toxic products.

8. No net increase of mass occurs during the stationary phase, but cells become smaller and synthesize components to help them survive nongrowth periods.

9. During the death phase, cells die exponentially but at a much lower rate than they grow during the log phase. Death usually occurs because a cell has depleted its intracellular reserve of ATP to repair cellular components.

10. Under natural conditions, nutrients continuously enter a cell's environment at low concentrations, so growth rate is set by the concentration of the scarcest, or limiting, nutrient.

11. In the laboratory, a continuous culture is achieved with a chemostat. The concentration of the limiting nutrient in the growth vessel sets the growth rate.

12. Cells in a colony are in different phases of growth depending upon their location. Cells growing on the surface of a colony are aerobic and those in the center are anaerobic.

WHAT MICROORGANISMS NEED TO GROW   (pp. 196–204)

Nutrition   (pp. 196–201)

1. Nutrients are essential chemicals from the environment that a cell uses to synthesize the molecules needed to build new cells.

2. All microorganisms require a source of carbon to grow because carbon is the structural basis of biochemicals.

3. Oxygen plays a complex role in microbial metabolism—as nutrient, electron acceptor, and generator of toxic by-products. Microorganisms capable of aerobic respiration use oxygen gas to generate energy.

4. Toxic oxygen-containing compounds are by-products of aerobic metabolism. They include hydrogen peroxide and the free radicals superoxide and hydroxyl radical.

5. Obligate aerobes can grow only in the presence of oxygen. Microaerophiles tolerate less oxygen. Facultative anaerobes can grow in either the presence or absence of oxygen. Aerotolerant anaerobes do not require oxygen but are not killed by it. Obligate anaerobes die in the presence of oxygen.

6. All living things require some form of nitrogen (a constituent of proteins and nucleic acids as well as certain essential metabolites), phosphorus (a constituent of nucleic acids, phospholipids, and certain essential metabolites), and sulfur (a component of two amino acids, a few types of tRNA, and certain essential metabolites).

7. Trace elements are small but essential inorganic chemicals, such as potassium, iron, and zinc, that microorganisms need to grow. Some trace elements act as cofactors.

8. Organic growth factors are essential preformed building blocks. If a microorganism cannot synthesize a particular building block, it must be supplied preformed in the environment. Organic growth factors can be amino acids, purines, pyrimidines, fatty acids, or vitamins.

9. Nutritional diversity refers to the different requirements among microorganisms for organic growth factors.

10. Microorganisms will exploit a rich growth environment, using preformed building blocks even if they are capable of making them.

The Nonnutritive Environment   (pp. 201–204)

1. Every species of microorganism grows over a range of temperatures, from a minimum temperature of growth to a maximum temperature of growth. A species' optimum temperature is that at which it grows most rapidly.

2. Thermophiles grow best at temperatures above 50°C. Mesophiles grow best near 37°C, the temperature of the human body. Psychrophiles grow at or below 5°C, refrigerator temperature.

3. The heat stability of a microorganism's proteins determines the maximum temperature at which it can grow. Minimum temperature is determined by weakening of hydrophobic interactions of proteins.

4. Hydrostatic pressure is pressure applied to a liquid. Most microorganisms tolerate high hydrostatic pressures. Barophiles grow better (facultative) or only (obligate) at pressures greater than 1 atmosphere.

5. In general, bacteria grow best at a slightly alkaline pH, and fungi grow best at a slightly acid pH. Acidophiles thrive in low-pH environments, while alkaliphiles thrive in high-pH environments.

6. Bacteria maintain a positive turgor pressure: Their internal cell osmotic strength is greater than the osmotic strength of their external environment. Certain bacterial species, the halophiles, require high external osmotic strength.

7. Plasmolysis occurs when an increase in external osmotic strength draws water out of the cell, shrinking the cytoplasm.

MEASURING NUMBERS OF MICROORGANISMS   (pp. 204–211)

1. Some measurement methods are indirect, measuring a property—turbidity, dry weight, or metabolic activity—of the mass of cells in a population. Other methods are direct, counting individual cells.

2. Microscopic or electronic counts yield a total count—a count of all cells, living and dead. A viable count is a count of only living cells, those capable of reproducing. Plate counts and MPN yield viable counts.

3. Turbidity is measured with a spectrophotometer. The spectrophotometer measures how much light a liquid culture of microbial cells

transmits. A standard curve is used to relate spectrophotometer readings to cell mass or numbers in a particular culture.

4. Determining dry weight depends upon separating cells from their medium, drying them, and weighing them.

5. There are several ways metabolic activity can be used to measure numbers: the rate of formation of metabolic products, such as gases or acids, that a culture produces; rate of utilization of a substrate, such as oxygen or glucose; rate of reduction of certain dyes.

6. A microscope count is done with a special microscope slide called a counting chamber, such as the Petroff-Hauser chamber. Based on the depth of the well and the grid, the number of cells in a set volume of the sample can be computed.

7. An electronic count is done with a Coulter counter. Electronic counting is possible because microbial cells do not conduct electricity as well as their medium.

8. In a plate count, a sample of the culture is diluted tenfold or a hundredfold several times, and then a small amount is spread or poured on a plate and incubated. Plates with between 30 and 300 colonies are selected and counted.

9. Most probable number (MPN) is based on the principle that a single live cell can develop into a turbid culture. After several tenfold dilutions, test tubes are incubated. Tubes that become cloudy received one or more cells. By determining the average dilution at which tubes did not receive cells, the number of microorganisms that were most probably present in the original sample can be computed using an MPN table.

10. Filtration is done prior to measuring small numbers of microorganisms in large volumes of liquid or gas.

Review Questions

POPULATIONS

1. To what does the term *microbial growth* refer?

THE WAY MICROORGANISMS GROW

1. What is doubling time? Why is growth rate often used instead of doubling time?

2. Explain this statement: Microorganisms grow exponentially.

3. Name, in sequence, the phases of growth in a culture. Explain each one.

4. How do you maintain a continuous culture in the laboratory?

5. Under natural conditions, what usually sets the growth rate?

6. Define the term *colony*. How does location in a colony affect a cell's phase of growth? How does location affect its access to oxygen?

WHAT MICROORGANISMS NEED TO GROW

1. What are nutrients? Explain the function of each of these nutrients: carbon, oxygen, nitrogen, sulfur, trace elements, organic growth factors.

2. Explain this statement: Oxygen plays a complex role in microbial metabolism. Use the term *free radical* in your explanation.

3. Explain the different relationships these types of microorganisms have with oxygen: obligate aerobes, microaerophiles, facultative anaerobes, aerotolerant anaerobes, and obligate anaerobes.

4. To what does nutritional diversity refer?

5. What is a rich growth environment? How do microorganisms exploit a rich growth environment?

6. Name the significant physical factors in a microorganism's non-nutrient environment. Name the significant chemical factors.

7. Explain how different species grow over a range of temperatures. Use these terms in your discussion: minimum temperature of growth, maximum temperature of growth, optimum temperature. Explain what determines the maximum and minimum temperatures.

8. Define these terms: thermophiles, mesophiles, psychrophiles.

9. What is hydrostatic pressure? Why can microorganisms withstand high hydrostatic pressures? How is hydrostatic pressure harmful to microorganisms?

10. What pH levels do bacteria prefer? Fungi? What are acidophiles and alkaliphiles? If metabolic functions require a relatively narrow pH, how can bacteria exist in environments with a wide range of pH values?

11. How are halophiles unlike most other bacteria? How do most bacteria maintain turgor pressure? What is plasmolysis?

MEASURING NUMBERS OF MICROORGANISMS

1. How do direct and indirect methods of measuring populations of microorganisms differ? What is the difference between a total count and a viable count?

2. Tell whether each of the following measurement methods is direct or indirect and whether it gives a total or viable count; describe the procedure, and give its advantages and disadvantages.

a. turbidity
b. dry weight
c. metabolic activity
d. direct microscopic
e. direct electronic count
f. plate count
g. most probable number (MPN)
h. filtration

Essay Questions

1. Discuss the advantages and disadvantages of using continuous culture to make antibiotics.

2. Discuss the effect of richness of a medium and concentration of a nutrient on growth rate.

Suggested Readings

Ingraham, J. L.; Maaløe, O.; and Niedhardt, F. C. 1983. *Growth of the bacterial cell*. Sunderland, Mass.: Sinauer.

Kolter, R. 1992. Life and death in stationary phase. *ASM News* 58:75–79.

Marr, A. G. 1991. Growth rate of *Escherichia coli*. *Microbiological Reviews* 55:316–33.

Mynell, G. G., and Mynell, E. 1970. *Theory and practice in experimental bacteriology*. New York: Cambridge University Press.

9 CONTROLLING MICROORGANISMS

Protecting the Tiniest Patients

Disinfectants safeguard all patients in a hospital in many ways, but they play a special role in the first days of a newborn's life. Newborns in a hospital are especially vulnerable to infection—their immune defenses are not yet fully developed and the microorganisms that help protect body surfaces against disease-causing pathogens have not yet been established. And a hospital has many antibiotic-resistant pathogens. For all these reasons, a newborn nursery is a hot spot for outbreaks of infection.

Disinfectants can play a key role in stopping these outbreaks. Within minutes of birth the newborn's eyes are treated with either an antibiotic or the disinfectant silver nitrate to kill *Neisseria gonorrhoeae*—a potentially blinding

bacterium that the newborn may have acquired from the mother during birth. When a newborn is first bathed, the umbilical cord stump is painted with triple dye—a deep-purple-colored bactericidal agent—to reduce colonization by the highly pathogenic *Staphylococcus aureus* and help prevent a life-threatening infection of the umbilicus. All hands that touch babies in the hospital nursery must be recently washed with chlorhexidine or iodophor-containing soaps. Lapses in hand washing probably cause occasional nursery epidemics of *S. aureus* infections. If infection has been a problem in the nursery recently, the baby may receive a single chlorhexidine bath.

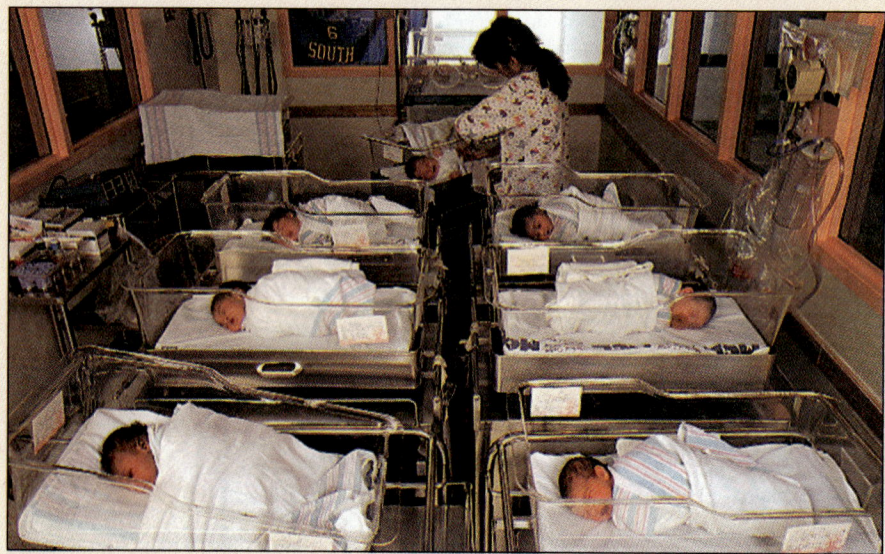

Disinfectants must be used in hospital nurseries, which have the potential for spreading serious infections among newborn infants.

To understand:

- How microorganisms die: rate of microbial death and D-value

- The physical treatments used to control microorganisms

- The chemical treatments used to control microorganisms

THE WAY MICROORGANISMS DIE

Microbial death occurs when a cell can no longer divide to form new cells. As with the growth of microorganisms, we are seldom interested in the fate of an individual microbial cell. Instead, we want to know how cell death affects the population. In particular, we want to know if a specific treatment kills all the cells in a microbial population, reduces the numbers of microorganisms to safe levels, or simply stops growth for a period of time.

Control may mean **sterilization**, treatment to destroy all microbial life; **disinfection** (or **sanitation**) to reduce the number of pathogens to a level at which they pose no danger of disease; **decontamination** to render an instrument or surface that has been heavily exposed to microorganisms safe to handle; or **antisepsis** to kill microorganisms on skin or other living tissue. Some of these terms can be used interchangeably. For example, sterilizing or disinfecting an instrument also decontaminates it.

A microbial cell is considered to be dead when it fails to form a colony on a solid medium or produce a turbid culture in a liquid medium. However, some treatments only inhibit microbial growth. That is, treated cells are temporarily unable to form a colony and therefore appear to be dead. They are able to grow again when the treatment is withdrawn or time has passed. Treatments that inhibit rather than kill are termed **microbiostatic**. For example, refrigerating food preserves it by preventing microbial growth, not by killing microorganisms. **Microbiocidal** treatments kill microbial cells. Certain treatments are microbiostatic, others microbiocidal; but the distinction is not absolute. Some microbiostatic chemicals become microbiocidal if treatment is prolonged or their concentration is increased.

Microbial populations are controlled by many different agents—heat, cold, radiation, filtration, drying, high osmotic strength, and certain chemicals. But regardless of the treatment, a population of microorganisms always declines according to the same logarithmic pattern. The time course of microbial death, like a graph of microbial growth, is a straight line when plotted on a logarithmic scale (**Figure 9.1**).

The **rate of microbial death** can be calculated even though it is impossible to predict when an individual cell will die. After exposure to the lethal agent, some die almost immediately while others die much later. Nevertheless, we can predict accurately how the population as a whole will decline because a fixed percentage of the survivors die in any given time interval. In other words, if it takes 10 minutes to kill 90 percent of the cells, 90 percent of the survivors will be killed during the next 10 minutes. Because of this pattern, the effectiveness of a particular way of killing microorganisms can be described by the **decimal reduction time** (**D-value**). D-value is the time in minutes it takes to kill 90 percent of the cells in a population. For thermal killing, the temperature of treatment is sometimes indicated by a subscript. For example, the D-value at 80°C is written $D_{80°C}$.

For most treatments the D-value depends upon four factors: temperature, the type and physiological state of the microorganisms being treated, and the presence of other substances. (1) Temperature affects the rate because lethal chemical reactions, like all others, proceed more rapidly at higher temperatures. (2) Different types of microorganisms vary considerably in their resistance to killing. For example,

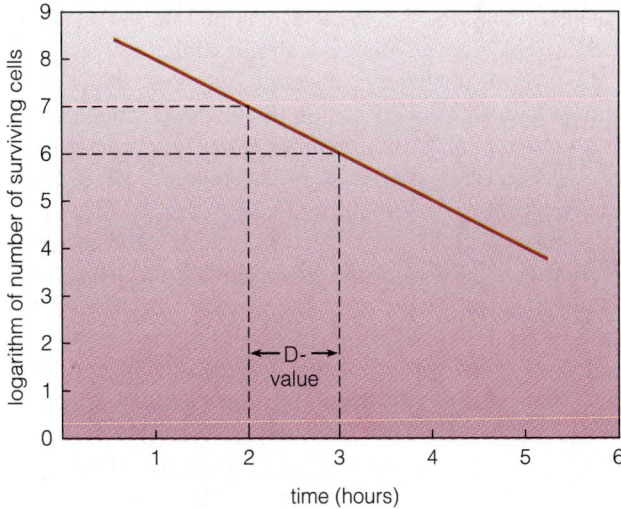

Figure 9.1 Microbial death. A graph of microbial death, like a graph of microbial growth (Figure 8.2), is a straight line when plotted on a logarithmic scale. When exposed to lethal conditions, the number of surviving live cells in a microbial population declines 90 percent (tenfold or one logarithm value) in the same period of time, called decimal time, or D-value. In this plot, the D-value is one hour.

the mycoplasmas (Chapter 4) are highly susceptible to treatment; endospores are extremely resistant. (3) Log-phase cells are in a physiological state that makes them generally more susceptible than stationary-phase cells. Sometimes such differences are dramatic. For example, over 99 percent of log-phase cells of *Escherichia coli* can be killed by cold shock (rapid chilling to 5°C or less), but no stationary-phase cells are killed by the same treatment. (4) The presence of other substances, especially proteins, protects microorganisms from many lethal treatments. Milk, for example, is quite protective. When added to a bacterial culture, it decreases killing by heat, cold, or chemicals because of the protein it contains. Microorganisms in vomit or feces are also hard to kill because of protective proteins.

To sterilize an object means to kill *all* the microorganisms, including viruses, on or in it. But how is it possible to calculate when all the microorganisms will be killed? Consider this mathematical riddle: How long will it take for a snail to reach its goal if every day it travels half the remaining distance? Of course, it would never reach its goal. But it would get very close. The same reasoning applies to the death of a bacterial culture. If the D-value remains the same throughout the period of treatment, as it usually does, we know that during each D time 90 percent of the remaining cells will die—but not all of them. However, what does it mean that 90 percent of the population will die when only one cell remains? Of course, with high probability, one D time later the material will be sterile—absolutely free of microorganisms. But because calculating microbial sterilization is based on statistics, we calculate *probability* of sterilization.

Given an unchanging D-value, we must know three things to design a sterilization treatment: (1) the D-value of the treatment, (2) the number of cells present, and (3) the desired degree of certainty that no cells will be alive at the end of the treatment.

These calculations are vital in industry and research laboratories. For example, the canning industry must know how long food must be heated to be free of live cells. If the treatment is too short, endospores of the bacterium *Clostridium botulinum*, which causes botulism, may survive and germinate, producing a potent neurological toxin that can paralyze and kill. On the other hand, excessive heating degrades the quality of the food. These calculations assure that materials are in fact sterilized—that all microorganisms are eliminated.

In addition to D-value, two other terms are useful to describe the susceptibility of a particular microorganism to killing by heat. **Thermal death point (TDP)** is the lowest temperature required to kill all microorganisms in a particular liquid suspension in 10 minutes. **Thermal death time (TDT)** is the minimal time required to kill all microorganisms in a particular liquid suspension at a given temperature.

PHYSICAL CONTROLS ON MICROORGANISMS

In Chapter 3 we discussed sterilization treatments used as the first step in preparing a bacterial culture. Now we look at principles of physical and chemical control and their applications in clinical medicine and industry. **Table 9.1** summarizes both types of control.

Physical treatments to control microorganisms include heat, cold, radiation, filtration, drying, and osmotic strength.

Heat

Heat treatment is simple, inexpensive, and effective. It is probably the best method if the material being treated is not damaged by heat. The advantage of heat is that it penetrates to kill microorganisms throughout the object. Some sterilization treatments, such as ultraviolet light and chemicals, sterilize only the surfaces they touch. After extensive and sophisticated research, the National Aeronautics and Space Agency (NASA) chose to sterilize the Viking spacecraft sent to Mars with heat (there are strict international standards to protect other planets from microbial contamination).

Heat sterilization can be dry or moist. Pasteurization is a special heat treatment that limits microbial growth but does not sterilize. All heat treatments kill by denaturing proteins (Chapter 2). But dry heat kills largely by speeding oxidations, which inactivate macromolecules. Moist heat, on the other hand, denatures proteins by breaking the bonds that confer secondary and higher protein structure.

In a microbiological laboratory, dry-heat sterilization is commonly used in the form of flaming and hot-air ovens (Chapter 3). Quickly passing a pipette or the opening of a flask through a flame is considered reliable sterilization. Reliable hot-air sterilization requires severe treatments—171°C for one hour, 160°C for two hours, or 121°C for 16 hours. Large objects require even more time.

Moist heat in the microbiological laboratory is applied either by boiling or in autoclaves (Chapter 3). Moist heat is effective at a lower temperature than dry heat and it penetrates more quickly. Boiling water kills most bacterial and fungal cells and inactivates many viruses in just a few minutes. However, vegetative cells of some thermophilic bacteria can withstand prolonged boiling. In fact, some grow vigorously at boiling temperature (100°C) because their proteins are not denatured. Endospores also survive boiling. The proteins in bacterial endospores resist inactivation by heat because the interior is dry and water cannot penetrate their thick walls. The autoclave is more effective than boiling because it uses pressure to raise the temperature considerably above that of boiling water. At a

Table 9.1 Treatments to Control Microorganisms

Treatment	Effect	Mode of Action	Uses
PHYSICAL METHODS			
Heat			
Dry heat	Sterilizes	Denatures protein	In the laboratory, used to sterilize dry materials that can withstand high temperature and any materials damaged by moisture.
Moist heat	Sterilizes	Denatures protein	In the laboratory, used to sterilize liquids and material easily charred. Used in food canning.
Pasteurization	Kills certain microorganisms	Denatures protein	Eliminates pathogens and slows spoilage of milk and dairy products, wine, beer. (Canned evaporated or condensed milk is sterilized.)
Cold	Slows or stops microbial growth	Slows chemical reactions	Preserves perishable materials, including food and microorganisms.
Radiation			
UV light	Sterilizes	Damages DNA	In the laboratory, sterilizes surfaces.
X rays and gamma rays	Sterilizes	Strips electrons from atoms	Used to sterilize plastic equipment and surface of fresh fruits and vegetables.
Filtration	Removes cellular microorganisms	Physically removes cells	In the laboratory, used with media, antibiotics, and other heat-sensitive materials. Used to preserve certain beverages.
Drying			
By evaporation or heat	Kills many microorganisms	Distorts membranes	Used to preserve foods.
By sublimation (freeze-drying)	Stops microbial growth	Stops most chemical reactions	In the laboratory, used to preserve cultures, proteins, blood. Used to make instant coffee, lightweight food.
CHEMICALS			
Phenols	Kill most microorganisms	Denature proteins	Germicides
Phenolics	Kill most microorganisms	Denature proteins and disrupt plasma membrane	Disinfectants, antiseptics.
Alcohols	Kill most microorganisms	Denature proteins and disrupt plasma membrane	Disinfect surfaces, including skin and thermometers.
Halogens	Kill microorganisms	Oxidize vital biochemicals	Disinfect surfaces, including skin and water.
Hydrogen peroxide	Kills many microorganisms	Oxidizes vital biochemicals	Mild skin disinfectant.
Heavy metals	Kill many microorganisms	React with sulfhydryl groups of proteins	Skin disinfectants.
Surfactants			
Soap, detergent	Wash away microorganisms	Physically remove microbes	Disinfect surfaces, including skin, bench tops.
Quaternary ammonium salts	Kill microorganisms	Disrupt membranes	Widely used sterilizing agents.
Alkylating agents			
Formaldehyde and glutaraldehyde	Kill microorganisms	Inactivate enzymes by adding alkyl groups	Preserve tissues, prepare vaccines. Sterilize surgical instruments.
Ethylene oxide	Kills microorganisms	Inactivates enzymes by adding alkyl groups	Gas used to sterilize heat-sensitive materials and unwieldy objects in hospitals.

pressure of 15 pounds per square inch (normal operation), the temperature in an autoclave is 121°C. Autoclaving requires certain precautions and a test that sterilization temperature was reached (Chapter 3).

Cold

Low temperature by itself does not kill microorganisms. In fact, it may be used to preserve them for limited periods of time. If held at a temperature below their growth range, most microorganisms die slowly; and the lower the temperature, the more slowly they die. An exception to this rule is the phenomenon of **cold shock**—when a growing culture is suddenly chilled, many microorganisms are killed immediately. The lethal effect of cold shock varies with the microorganism, the medium it is growing in, and the rate at which chilling occurs. For example, if a culture of *Escherichia coli* growing in a rich medium at 37°C is suddenly chilled to 5°C, more than 90 percent of the cells will be killed. In spite of this, cold or freezing does not sterilize.

Although cold does not kill microbial cells, it is an effective and widely used microbiostatic treatment. Refrigeration at 5°C preserves food because it stops the growth of most species of microorganisms. Most disease-causing microorganisms are mesophiles, not psychrophiles, so refrigeration largely prevents foodborne infection. There are some exceptions. *Listeria monocytogenes*, which causes **listeriosis**, a foodborne disease of the nervous system (Chapter 24), grows under refrigeration. But most psychrophiles multiply slowly in the refrigerator and are not a health risk, though they eventually cause food to spoil.

Freezing kills most bacteria, but the survivors remain alive for long periods in the frozen state. Because bacteria are too small for ice crystals to form within them, they are not killed by the physical destruction of essential intracellular structures, as eucaryotic cells are. Instead, bacteria are killed by the high osmotic strength that develops as water in the external environment freezes.

Bacterial cultures can be preserved by rapid freezing, sometimes with the addition of DMSO (dimethylsulfoxide), glycerol, or milk to protect proteins from denaturation and thereby decrease the number of cells killed by freezing (Chapter 3).

Radiation

Some forms of electromagnetic radiation kill living things, including microorganisms. Radiation is classified by wavelength, with cosmic rays at the short-wavelength end, through visible light in the middle, to radio waves at the long-wavelength end (Chapter 3). In general, visible light and longer wavelength radiation are not significantly lethal. High doses of radio waves will kill microorganisms but they do this by heating in much the same way a microwave oven uses radio waves for heating. Very intense visible light can kill in the presence of oxygen and pigments called photosensitizers. For example, when the dye methylene blue, a photosensitizer, is added to microbial cultures that are then exposed to intense light, oxygen is converted to a highly lethal form called singlet-state oxygen. However, this simple method of killing microorganisms has not yet found practical applications.

The types of radiation that kill bacteria directly are all of shorter wavelength than visible light. They are **ultraviolet (UV) light**, radiation with a wavelength of 10 to 400 nm (nanometers), and **ionizing radiation**, which has a shorter wavelength, extending down to about 0.001 nm.

The UV wavelength that is most lethal to microorganisms is about 265 nm. UV light kills principally by damaging DNA, and 265-nm UV is maximally effective because DNA maximally absorbs UV at this wavelength. Because sunlight is rich in UV light, microorganisms have evolved various DNA repair mechanisms to correct the damage done by UV. If they are not effective, the cell dies (Chapter 6). One of these mechanisms is **photoreactivation**. Photoreactivation depends on an enzyme that is active in the presence of visible light between 420 and 540 nanometers. Therefore, controlled experiments on the killing effects of UV must be done in a room with dim light.

Mercury vapor lamps, called **germicidal lamps**, that emit maximally at 253.7 nm are commercially available. They are simple to use, and as soon as the lamp is extinguished, the UV light disappears. However, UV light kills microorganisms only on surfaces because it cannot even penetrate glass. Also, UV light presents a health hazard. If it reaches the eyes, either directly or by reflection, it damages the cornea. Moreover, longer UV wavelengths are visible to humans, but not the shorter ones that kill microorganisms, so damage could occur without an individual's being aware of it. For these reasons, there are few practical applications of UV light. Some microbiological laboratories turn on UV lights at night, when no one is present, to decrease the microbial population and thereby minimize the possibility of contaminating pure cultures that are being studied. In the home, drying clothes in sunlight is a simple application of UV sterilization. The high UV content kills all the microorganisms directly exposed to the light.

Two forms of ionizing radiation are used to kill microorganisms: **x rays**, wavelengths from 0.1 to 40 nm, and **gamma rays**, wavelengths from 0.001 to 0.1 nm. Both forms cause a chain of ionizations by stripping electrons from atoms. The result is usually cell death. Ionizing radiation is rarely used in microbiological laboratories because it is so technically complex. It does, however, have considerable commercial application. It is used to sterilize plastic containers such as petri dishes and bottles used for tissue culture. The radioactive isotope of cobalt,

^{60}C, used as a source of gamma radiation, has great potential as a safe and cheap way to preserve food. But even though it has been approved by the FDA, public suspicion of irradiated food has prevented significant use in the United States.

Filtration

Microorganisms other than viruses can be removed from liquids by filtration (Chapter 3). Thus, it is not a sterilization technique. In the microbiological laboratory, filtration is used for certain media, vitamin solutions, and antibiotics that are heat-sensitive. In industry, filtration is replacing pasteurization in some cases because filtration causes even less damage than the slight amount caused by pasteurization. Some sweet wines are now preserved with filtration, as are beers called "draft beer" or "cold-filtered beer."

Drying

Drying, the removal of water, can be accomplished through evaporation or sublimation (Chapter 3). Drying by evaporation is almost never used in the laboratory because it causes chemical changes. But it is widely used in the food industry.

Lyophilization, or freeze-drying, removes water by **sublimation**, direct conversion from a solid state to a gaseous state. Materials to be lyophilized are frozen and placed in a chamber to which a partial vacuum is applied. Sublimation avoids the chemical changes caused by heat drying (Chapter 3). As a result, this method is frequently used in the microbiological laboratory to preserve perishable materials such as proteins, blood products, and reference cultures of microorganisms. Because it is expensive, lyophilization has limited industrial use.

Osmotic Strength

High concentrations of salt or sugar are used to preserve certain foods for the same reason drying is—microorganisms cannot grow if they are deprived of water. The high osmotic strength of salt and sugar solutions can also damage cells by plasmolysis. Osmotic strength is not used to control microbial growth in the microbiological laboratory because once added, solutes—such as salt—cannot be easily removed.

CHEMICAL CONTROLS ON MICROORGANISMS

In addition to the physical methods of controlling microorganisms, there are chemical treatments. Chemicals, including antibiotics, that are used to treat disease are called **chemotherapeutic agents**. They are discussed in Chapter 21. Here we discuss all other antimicrobial chemical agents. Chemicals that kill microorganisms are called **germicides**. Chemicals that inhibit microbial growth are called **germistats**. Germicides used on inanimate objects are also called **disinfectants**. When applied to living tissue, germicides are called **antiseptics**. Depending upon how it is used, the same germicide can sterilize a bacterial culture, disinfect the surface of a laboratory bench, or decontaminate a floor on which infectious material was spilled; it can also cause antisepsis when used as Joseph Lister did to prevent sepsis of wound infections (Chapter 1). For this reason, the general term *germicide* is more useful than *disinfectant* or *antiseptic* in referring to antimicrobial chemical agents.

Germicides are critically important in preventing infections, particularly in germ-laden environments such as hospitals and day-care centers. The use and formulation of germicides are extremely complex and standards are set by the government. The same germicide may be formulated differently by the manufacturer for particular uses—for example, as a disinfectant or an antiseptic. To register a formulation with the Environmental Protection Agency (EPA), the manufacturer must supply test results that establish its effectiveness and safety. Also, directions for use must be provided on the product label.

Many different compounds are used as germicides, and the number continues to increase. In 1973 the American Society for Microbiology surveyed 16 hospitals in the United States and found that each used an average of 14.5 formulations of 224 different products. Currently, approximately 14,000 germicidal products exist with approximately 300 active ingredients registered with the EPA. In general, the action of germicides is nonselective. That is, at a sufficient concentration they kill any microorganism.

In this section, we first discuss how to select a germicide for a particular use. Then we look at how germicides are tested for effectiveness. Finally, we survey types of germicides and the way they kill.

Selecting a Germicide

Choosing the appropriate germicide depends on answers to these questions: Will the chemical damage or destroy the tissue or object being treated? Will the chemical control the target microorganism(s)? What is the purpose of the treatment? The answers to the first two questions come from standard classifications for germicides. Every registered germicide is classified as having **high**, **intermediate**, or **low germicidal activity**. Microorganisms are then grouped according to their sensitivity to these three classes (**Table 9.2**). As for purpose, more powerful germicides or higher concentrations must be used for sterilization than for disinfection. For example, a 3- to 6-percent

Table 9.2 Susceptibility of Some Types of Microorganisms to Germicides

Microorganism	Germicidal Activity		
	High	Intermediate	Low
Bacteria			
Endospores	Killed	Not killed	Not killed
Vegetative cells[a]	Killed	Killed	Killed
Mycobacterium tuberculosis[b]	Killed	Killed	Not killed
Fungi	Killed	Killed	Some killed
Viruses			
Nonlipid and small	Killed	Some killed	Some killed
Lipid and medium-size	Killed	Killed	Killed

Source: After M. S. Favero and W. W. Bond, 1991, Sterilization, disinfection, and antisepsis. In *Manual of Clinical Microbiology*, 5th ed., ed. A. Balows (Washington, D.C.: American Society of Microbiology).
[a]Vegetative cells of most bacteria.
[b]Owing to its resistance and special importance, *M. tuberculosis* is considered separately.

solution of hydrogen peroxide is used for disinfection, but a 6- to 30-percent solution is needed for sterilization. If large amounts of protective substances, such as blood or feces, are present, a more powerful germicide is required. The price of a germicide may also be a consideration.

Testing Germicides

Germicides are tested by comparing their effectiveness to phenol, a traditional germicide. It was phenol, in fact, that Lister used to treat James Greenlees (Chapter 1). First, the test germicide is serially diluted. Then a standard volume of a culture of a test bacterium is added, usually the intestinal pathogen *Salmonella typhi* or *Staphylococcus aureus*, a bacterium that infects wounds. After 5 and 10 minutes the tubes are sampled and tested to see if any live cells remain. The samples are inoculated into test tubes containing a germicide-free medium to see if they become turbid after 48 hours' incubation. The highest dilution of the germicide that kills in 10 but not 5 minutes is the **end point**. The ratio of end points is the **phenol coefficient**. For example, a germicide that kills at a dilution of 1:1000 under the same conditions in which phenol kills at 1:100 has a phenol coefficient of 10 (1000 ÷ 100 = 10). In other words, it is 10 times more powerful than phenol.

The effectiveness of a germicide against specific microorganisms is determined by the **paper disc method** or the **use-dilution test**. In the paper disc method, a filter-paper disc is impregnated with a dilution of the test germicide. It is then placed on the surface of an agar medium in a petri dish that was previously seeded with the test organism. After an appropriate period of incubation, uniform bacterial growth develops with a clear, bacteria-free zone around the disc. This method is not precise or reproducible, but the size of the bacteria-free zone is a useful index of the germicide's effectiveness (**Figure 9.2**).

The use-dilution test is a more accurate way of determining the effectiveness of a germicide against a particular organism. The test organism is added to dilutions of the germicide, much as is done in determining a phenol coefficient. But rather than sampling the tubes, the tubes are incubated. The highest dilution of the germicide that remains clear after incubation indicates the effectiveness of the germicide.

Types of Germicides and the Way They Kill

Germicides are customarily classified according to chemical structure and activity. Thus, the six major classes are phenol and phenolics, alcohols, halogens and hydrogen peroxide, heavy metals, surfactants, and alkylating agents. Germicides are bacteriostatic or bactericidal, depending upon strength.

Phenol and Phenolics. **Phenol**, which Lister called carbolic acid, has been used since before his time as a germicide. **Phenolics** are compounds that resemble phenol. Both have hydroxyl (—OH) groups attached to a benzene ring (**Figure 9.3**). Phenol and phenolics are powerful denaturing agents, with phenol itself the most powerful. They inactivate vital cellular proteins, including enzymes. Phenolics (but not phenol) also act on lipids, and because both lipids and proteins compose the plasma membrane, the first noticeable effect of phenolics on microbial cells is disruption of the cell membrane. Phenol is used in the laboratory to purify DNA when it is mixed with protein. Phenol denatures the protein, leaving a purified solution of DNA. Phenol itself is not often used clinically today because it is so powerful. When applied to skin it denatures enough protein to turn the area white. Some throat sprays do contain low concentrations of phenol.

One class of phenolics, the **cresols**, is a common ingredient of household and hospital disinfectants. *O*-**phenylphenol** is the most commonly used representative of the class, and the most familiar brand name is probably Lysol. Cresols are effective disinfectants because they persist. They remain active even in the presence of organic materials, including proteins found in blood and feces.

Hexachlorophene is a chlorinated phenolic that is very effective as an antiseptic. It persists on skin, provid-

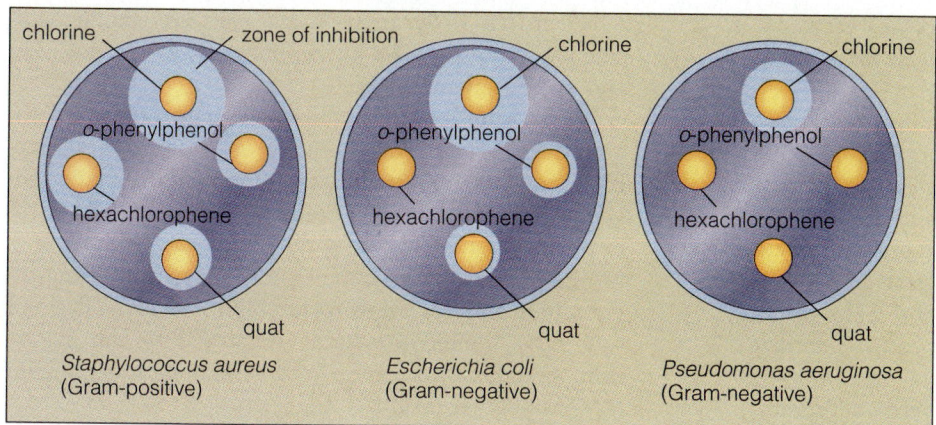

Figure 9.2 Paper disc method of testing germicides. These petri dishes show four germicides being tested: sodium hypochlorite (generates chlorine), o-phenylphenol, benzalkonium chloride, and hexachlorophene. Paper discs saturated with solutions of these germicides are placed on plates seeded with different bacteria and incubated. The size of the clear zone indicates the susceptibility of the bacterium to the germicide. For example, at the concentrations tested, *Escherichia coli* is sensitive to o-phenylphenol but resistant to hexachlorophene; *Staphylococcus aureus* is sensitive to both and *Pseudomonas aeruginosa* is sensitive to neither.

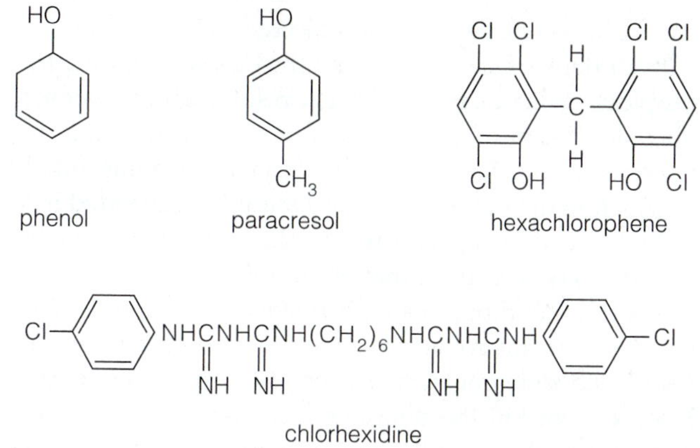

Figure 9.3 Phenol, phenolics, and chlorhexidine. Phenol is a hydroxyl (—OH) group connected to a benzene ring. Phenolics have this same basic structure but contain other groups as well. Paracresol has a methyl (—CH₃) group in the *para* position (opposite the hydroxyl group). Hexachlorophene has two connected benzene rings with chlorine (Cl) groups. Chlorhexidine has many of the same properties as hexachlorophene but it is not a phenolic because it lacks a hydroxyl group.

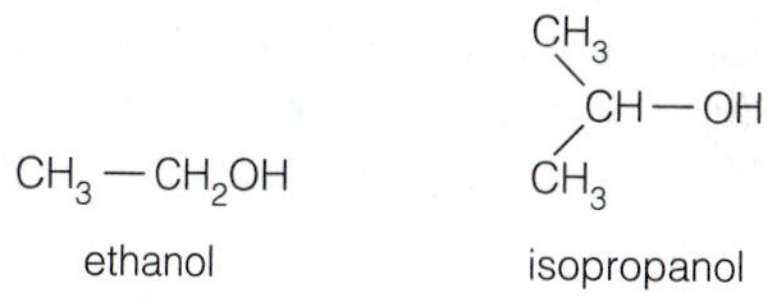

Figure 9.4 Ethanol and isopropanol are widely used as skin disinfectants.

ing long-lasting antimicrobial protection. It was once widely used as an ingredient in soaps and lotions, particularly in hospitals, where it helped control staphylococcal infections in newborns. But in the early 1970s we learned that hexachlorophene is absorbed through the skin. Premature babies who had received daily baths in 3-percent hexachlorophene for more than three days were found to be at substantially increased risk of brain damage. In 1972, the Food and Drug Administration (FDA) directed hospitals to restrict 3-percent hexachlorophene to prescription use for difficult skin infections.

Hexachlorophene has been replaced by **chlorhexidine** for most hospital uses. Chlorhexidine is not a phenolic, but it has many of the same properties and is less toxic to humans. It persists on skin and is highly effective. Concentrations as low as 10 µg (micrograms) per ml inhibit the growth of Gram-positive bacteria, though the lowest recommended concentration is 50 µg per ml. Like the phenolics, chlorhexidine acts first to disrupt the cytoplasmic membrane.

Alcohols. **Alcohols** are compounds with a hydroxyl (—OH) group. They kill by disrupting the lipids in cell membranes and by denaturing proteins. The carbon chain in alcohols penetrates the hydrophobic region of the membrane, disorganizing it and allowing the cell's contents to leak out. Two alcohols, **ethanol** and **isopropanol**, are widely used as skin antiseptics (**Figure 9.4**). They are routinely used to disinfect sites for injection or for drawing blood because alcohols dissolve lipids on the skin, thereby exposing microorganisms embedded in it and killing them. Ethanol at a 50- to 70-percent solution in water, its most effective concentration, is also used to disinfect thermometers, although isopropanol is somewhat more effective and only slightly more toxic. The main disadvantage of alcohols is that they do not kill endospores.

Halogens and Hydrogen Peroxide. The **halogens**, iodine and chlorine, are oxidizing agents. They inactivate enzymes by oxidizing certain functional groups, particularly sulfhydryl (—SH) groups. The more active agents also attack amino (—NH$_2$) and hydroxyl (—OH) groups. Iodine is used as an antiseptic, and chlorine is used as a disinfectant.

Iodine is used in two forms, as an aqueous solution and as a **tincture**, a dilute alcohol solution. Tincture of iodine contains iodide ion (I$^-$), elemental iodine (I$_2$), and triiodide ion (I$_3^-$); the active form of iodine is I$_2$. Iodine, like alcohols, disinfects the skin without damaging tissue. **Iodophors** are mixtures of iodine and **surfactants** (detergent-like agents). Iodophors are powerful, safe germicides used to disinfect skin before surgery and to sterilize dairy equipment.

Chlorine is used in many forms as a disinfectant—chlorine, **hypochlorites**, **inorganic chloramines**, and **organic chloramines**. The disinfection properties of all these compounds depend on their ability to generate chlorine (Cl$_2$). The microbiocidal activity of chlorine is markedly decreased by organic matter, so it is used to treat substances like water, which contain low amounts of organic material. Enough chlorine should be added to water supplies (Chapter 28) or swimming pools to overcome the presence of organic material and maintain enough **free chlorine**—that is, chlorine not bound to organic materials—to kill microorganisms quickly. A concentration of 0.6 to 1.0 part per million free chlorine will kill all microorganisms in 15 to 30 seconds. Hypochlorite is the active ingredient in household bleaches such as Clorox. It is also a widely used disinfectant in the dairy industry, the food industry, and hospitals.

Although **hydrogen peroxide** (H$_2$O$_2$) is not a halogen, we include it with this group because it is also an oxidizing agent and it kills microorganisms the way halogens do. It is used in a 3-percent solution as a weak antiseptic for cleaning wounds. It is also used to disinfect medical instruments and soft contact lenses because it does minimal damage to these fragile objects. When hydrogen peroxide comes in contact with tissue, it bubbles vigorously, producing oxygen gas. This is because all aerobes, including humans, contain the enzyme catalase, which decomposes hydrogen peroxide into oxygen and water. Hydrogen peroxide kills before it is destroyed by catalase or another enzyme, peroxidase. Some bacteria—for example, species of *Staphylococcus*—are relatively resistant to hydrogen peroxide because they produce large amounts of peroxidase.

Heavy Metals. Salts (Chapter 2) of heavy metals react with the sulfhydryl (—SH) groups of proteins, "poisoning" enzymes and thereby killing microbial cells. Salts of **mercury** and **silver** have been used in medicine for years. The mercury salt **mercuric chloride** was once widely used as an antiseptic but is not used today because it is highly toxic. Instead, organic compounds that contain mercury, including Merthiolate and Mercurochrome,

which are less toxic, are used. Both are basic home medicine chest and first aid kit supplies for disinfecting skin and mucous membranes.

Silver salts and **colloidal silver** (finely dispersed silver particles) were once widely used antiseptics. State laws uniformly required that a solution of the soluble silver salt **silver nitrate** be applied to the eyes of newborns to prevent ophthalmic gonorrhea and the blindness that results. Today, however, the trend is away from using silver nitrate and toward using antibiotics. In general, silver salts have limited use today because they can irritate skin, though they continue to be helpful in preventing infections of burned skin.

Surfactants. **Surfactants** are compounds that have hydrophilic (water-loving) and hydrophobic (water-fearing) parts (Chapter 2). They penetrate oily substances in water and break them apart into small droplets that become coated with surfactant molecules. The hydrophobic ends of the surfactant stick into the droplets and the hydrophilic ends into the water. The result is an **emulsion**, a fine suspension of oily droplets in water. Soaps and detergents are surfactants. They clean by forming emulsions, allowing oily material and material embedded in it to be washed away with water. Skin glands continually produce oily secretions that accumulate as a layer on the skin surface and become embedded with dead cells, dirt, dried sweat, microbial cells, and viruses (Chapter 26). Soaps and detergents, which emulsify and remove this layer, do not kill the microorganisms in it, but they remove them. As a result, frequent hand washing controls the spread of many infections. It is probably the best way to avoid colds (Chapter 22). Soaps with antibacterial agents are also available.

Some surfactants are germicidal because they penetrate and destroy phospholipid membranes, including the plasma membrane of microbial cells. **Quaternary ammonium salts** are surfactant germicides that are **cationic agents**, meaning positively charged. They have a charged nitrogen atom (an ammonium group) that is hydrophilic, with four (hence, "quaternary") hydrophobic organic groups attached to it (**Figure 9.5**). These compounds kill all classes of cellular microorganisms and viruses that have membranes, although higher concentrations are required to kill Gram-negative bacteria because of their protective hydrophilic outer membranes. Quaternary ammonium salts are nontoxic and widely used disinfectants in the home, industry, laboratories, and hospitals. Two popular quaternary ammonium salts are Cepacol (cetylpyridinium chloride) and Zephiran (benzalkonium chloride). Some mouthwashes contain quaternary ammonium salts.

Anionic surfactants act like quaternary ammonium salts but are less effective. They are widely used as mild

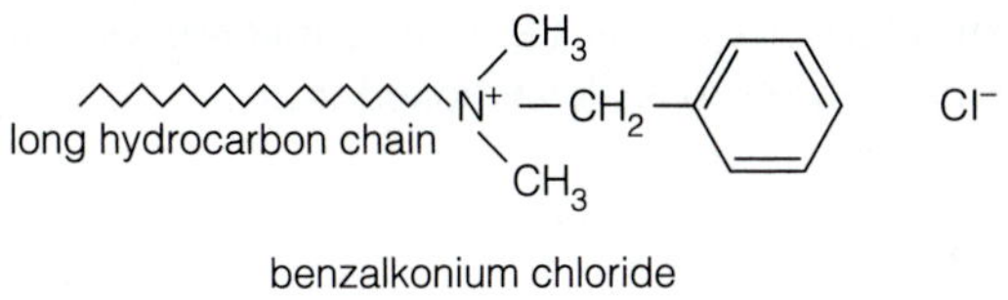

Figure 9.5 Benzalkonium chloride (commercially known as Zephiran) is a quaternary ammonium salt. Quaternary ammonium salts have four different groups attached to a charged nitrogen atom (N^+). They are powerful surfactants and form emulsions because the charged nitrogen, being hydrophilic, is attracted to water and the hydrophobic groups, such as the long hydrocarbon chain, penetrate oily substances.

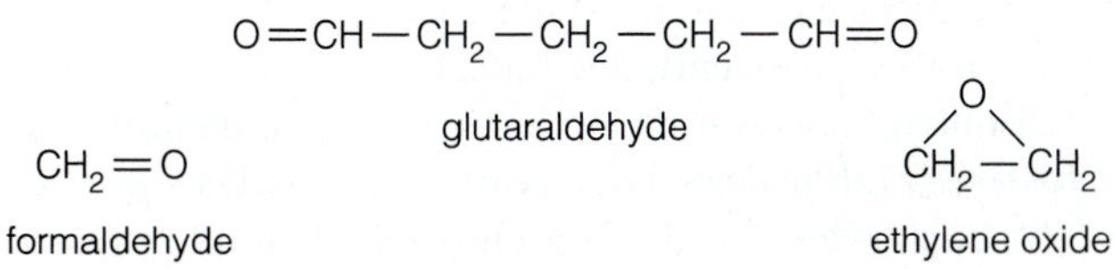

Figure 9.6 Alkylating agents kill microorganisms by adding alkyl groups (short chains of carbon atoms) to proteins, which inactivates them. Formaldehyde adds a single carbon atom, ethylene oxide adds two, and glutaraldehyde adds five.

disinfectants of skin and other surfaces, such as bench tops in microbiological laboratories. If used together, anionic and cationic agents inactivate each other.

Alkylating Agents. **Formaldehyde**, **glutaraldehyde**, and **ethylene oxide** are alkylating agents, meaning they **alkylate** or attach short chains of carbon atoms (alkyl groups) to enzymes, inactivating them and killing the cell (**Figure 9.6**).

Formalin is a 37-percent solution of formaldehyde used to preserve tissues and to embalm. It kills all microorganisms, including endospores. Lower concentrations of formaldehyde—0.2 to 0.4 percent—are used to inactivate microorganisms for killed vaccines (Chapter 20). Glutaraldehyde is used to sterilize surgical instruments if equipment for heat sterilization is not readily available.

Ethylene oxide has special advantages as a sterilizing agent because it is a gas and therefore disappears from the object after antimicrobial treatment is completed. However, ethylene oxide is extremely toxic to humans and so must be used in a sealed chamber. Airing out the chamber can take hours. Ethylene oxide kills all bacteria, including endospores, so it is used to sterilize materials that would be destroyed by heat, such as plastic medical equipment and animal feed. It is also used in hospitals to sterilize unwieldy objects such as mattresses and telephones.

Table 9.3 summarizes the antimicrobial chemical agents discussed here.

Table 9.3 Uses of Some Common Germicides

Action	Example	Concentration
Sterilization	Ethylene oxide	450–500 mg/liter
	Glutaraldehyde	Variable
	Hydrogen peroxide	6–30%
	Formaldehyde	6–8%
	Chlorine dioxide	Variable
	Peracetic acid	Variable
Disinfection	Glutaraldehyde	Variable
	Hydrogen peroxide	3–6%
	Formaldehyde	1–8%
	Chlorine dioxide	Variable
	Peracetic acid	Variable
	Chlorine compounds	500–5000 mg/liter
	Alcohols (ethanol, isopropanol)	70%
	Phenolic compounds	0.5–3%
	Iodophor compounds	30–50 mg free iodine/liter
	Quaternary ammonium compounds	0.1–0.2%
Antisepsis	Alcohols (ethanol, isopropanol)	70%
	Iodophors	1–2 mg free iodine/liter
	Chlorhexidine	0.75–4.0%
	Hexachlorophene	1–3%
	Parachlorometaxylenol	0.5–4.0%

PRESERVING FOOD

Most foods are chemically quite stable. They spoil only when microorganisms grow on them, so preserving food is a matter of eliminating microorganisms or slowing their growth. But effective and safe food preservation is a relatively recent human accomplishment. In the nineteenth century, most American farm families were malnourished during the winter months, despite an abundant fall harvest, because they could not preserve the food.

Humans have manipulated the nonnutrient environment of microorganisms—pH, osmotic strength, and temperature—to control microbial growth and preserve food. Even high hydrostatic pressure is used to slow the growth of wine yeast.

Temperature

Temperature is the environmental factor most often used to preserve food. Low temperature slows or completely arrests the growth of microorganisms, and high temperature kills them.

The temperature inside a refrigerator (about 5°C) is low enough to stop the growth of many microorganisms, including most pathogens. Psychrophilic microorganisms are the exception. They continue to grow, some relatively rapidly. When food does eventually spoil in a refrigerator, the psychrophilic population is responsible.

In contrast, almost no microorganisms can grow at the temperature inside a deep freeze (about −10°C), and those that can grow extremely slowly. As a result, food does not spoil in a properly functioning deep freeze. However, the quality of food, particularly vegetables, will deteriorate over time because of enzymes in the food itself. To minimize this deterioration, most vegetables are blanched (surface-heated by brief exposure to boiling water) before they are frozen. **Blanching** inactivates many of the destructive enzymes in vegetables.

Canning is the oldest and most widespread method of preserving food by temperature. At the beginning of the nineteenth century in France, Nicolas Appert discovered that food could be kept almost indefinitely if enclosed in an airtight container and heat-sterilized. Heat treatments for canning are designed to kill bacterial endospores because they are the most heat-resistant forms of life and because in an anaerobic environment (the airtight container) some spore-forming bacteria produce lethal toxins. For example, if spores of *Clostridium botulinum* survive the heat treatment, they germinate and produce a lethal toxin. Although the food looks, smells, and (presumably) tastes normal, even a minute amount can be lethal. Most canning procedures are designed to eliminate, with a high degree of assurance, all spores of *C. botulinum*.

Killing microorganisms follows statistical rules. The D-value, or heat treatment that kills 90 percent of a particular microbial population (decreases it by one log value), can be determined accurately; but treatment necessary to sterilize can be calculated only for a certain probability. For canning, a 12D treatment for *C. botulinum* spores is usually applied, meaning that the treatment reduces the population of *C. botulinum* by 12 log values. Such a treatment would decrease a population of 10^{12} *C. botulinum* spores—a population many times denser than ever encountered in food—to a single viable spore.

Two factors—time and temperature—determine safe heat treatments for canning. If the temperature is higher, the time can be shorter. **High-temperature, short-time (HTST)** treatments hold great promise for improving the quality of canned foods because they are as effective as conventional treatments for killing microorganisms but do not give the food an overcooked taste.

Pasteurization

Pasteurization is special heat treatment to control microorganisms in milk, dairy products, wine, and beer. Pasteurization does not sterilize because it heats at lower temperatures. But although it does not kill all the microorganisms present, it does kill most pathogens plus microorganisms that cause rapid spoilage. Moreover, it causes minimal damage to the product.

Louis Pasteur developed the process to keep wine from spoiling. Pasteurization kills the lactic acid bacteria that are most likely to spoil wine (and beer). Pasteurization of milk also extends shelf life, but it is used primarily to eliminate disease-causing bacteria. These include *Mycobacterium tuberculosis*, which causes tuberculosis; species of *Brucella*, which cause brucellosis (a serious systemic disease); *Salmonella* spp., which cause intestinal disease; rickettsiae, which cause Q fever; and *Listeria* spp., which cause food poisoning. The standard treatment for pasteurizing milk in the United States is heating to 63°C for 30 minutes or 72°C for 15 seconds (HTST). HTST treatment causes less flavor change. Milk can also be sterilized by **ultrahigh temperature (UHT) treatments** so that it can be kept without refrigeration. UHT sterilized milk is popular in Europe, but not in the United States because it has a slightly different taste. Canned evaporated or condensed milk is sterilized, not pasteurized.

pH

As the earlier discussion of lactic acid bacteria pointed out, acidity—low pH—prevents the growth of most microorganisms, especially in an anaerobic environment. Adding vinegar to foods accomplishes much the same result. For example, some cucumber pickles are made by a lactic acid fermentation, while others are made by adding vinegar. Low pH also increases the effectiveness of heat treatments. Acidic foods like tomatoes and fruits can be canned safely merely by boiling.

Water

Drying and salting food do not sterilize but preserve food by making it unable to support microbial growth for lack of water, an essential nutrient. Both treatments have been used for centuries to preserve foods. Drying is used most often for fruits and meat. Many fruits—apricots, peaches, prunes, pears, and grapes—are dried commercially in the sun in California. Most are first exposed to sulfur dioxide (SO_2) supplied as compressed gas or as the fumes of burning sulfur, but that step of the process is not essential

Figure 9.7 Field of apricots drying in the sun near Winters, California.

Chemical	Use and Activity
Calcium propionate	Antifungal agent added to bread to prevent mildew and *Bacillus* spp. that cause "ropy bread"
Sorbic acid (potassium sorbate)	Antifungal agent added to acid foods such as soft drinks, salad dressings, and cheeses
Sodium benzoate	Antifungal agent added to acid foods such as soft drinks, salad dressings, and cheeses
Sodium nitrate (nitrite)	By anaerobic respiration, nitrate is reduced to nitrite, an antibacterial agent that prevents germination of *Clostridium botulinum* spores when added to bacon, ham, hot dogs

Table 9.4 Some Chemicals Used to Preserve Food

for preservation. It is used principally to prevent the fruit from darkening in the sun (**Figure 9.7**). Other foods—for example, yeast and some prunes—are dried by heating. Smoking meat dries it and also adds chemicals that enhance flavor and inhibit microbial growth.

In freeze-drying, food is frozen and then exposed to a vacuum to dehydrate it. The heat of sublimation keeps the food frozen until it is dry. Because the drying occurs at low temperature, the food suffers very little loss of quality. Unfortunately, freeze-drying is energy-intensive and therefore costly, which restricts its use. Coffee is freeze-dried, as are specialty meals carried by backpackers.

Salting is most often used for fish and meat. Salt pork, a staple of the American frontier, would last through an entire winter because the microorganisms that normally cause meat to spoil cannot withstand the high osmotic strength of a concentrated salt solution (Chapter 2). Today, salt is still used to preserve pickles, olives, and meats (they are called *cured*). Similarly, jellies and jams are preserved by sugar, which also increases osmotic strength.

Chemicals

Various chemical preservatives are added to commercially prepared foods (**Table 9.4**). They are safe (the Food and Drug Administration regulates their use), and some are remarkably effective. For example, calcium propionate is now routinely added to bread. Before its use, bread would grow moldy in a week or so. Now it lasts almost indefinitely.

Summary

THE WAY MICROORGANISMS DIE (pp. 215–216)

1. Sterilization completely eliminates all microorganisms, including viruses. Disinfection (sanitation) reduces the number of pathogens to a safe level. Decontamination renders an instrument or surface that has been heavily exposed to microorganisms safe. Antisepsis treats living tissue to destroy or inhibit microbial growth.

2. Microbial death occurs when a cell can no longer divide to form new cells. As with microbial growth, microbial death is defined in terms of populations rather than individual cells.

3. Microbiostatic treatments inhibit (prevent) growth. Microbiocidal treatments kill microbial cells. Some microbiostatic treatments become microbiocidal if prolonged.

4. A graph of microbial death and time of treatment is a straight line when plotted on a logarithmic scale.

5. Rates of microbial death from any particular treatment can be calculated according to decimal time (D-value). D-value is the time in minutes it takes to kill 90 percent of the cells in a population.

6. The value of D depends upon temperature, the type of microorganism, its physiological state, and the presence of other substances that might be protective.

7. Given an unchanging D-value, to design a sterilization treatment you must know the D-value of the treatment, the number of cells to be eliminated, and the desired degree of certainty that no cells will be alive after treatment.

8. Thermal death point (TDP) is the lowest temperature required to kill all microorganisms in a particular liquid suspension in 10 minutes. Thermal death time (TDT) is the minimal time required to kill all microorganisms in a particular liquid suspension at a given temperature.

PHYSICAL CONTROLS ON MICROORGANISMS
(pp. 216–219)

1. The advantage of heat is that it penetrates to kill microorganisms throughout an object; some treatments sterilize only the surfaces they touch. Heat treatments may be dry (direct flaming, hot-air oven) or moist (boiling or autoclaving). Heat kills by denaturing protein.

2. Low temperature is a microbiostatic treatment; it does not sterilize. Cold shock, suddenly chilling to 5°C, can kill most cells in a growing culture.

3. Freezing kills most bacteria, but survivors remain alive for long periods in a frozen state. Rapid freezing is a way to preserve bacterial cultures.

4. Two forms of radiation kill bacteria, ultraviolet (UV) light and ionizing radiation (x rays and gamma rays). UV kills by damaging DNA, while ionizing radiation kills by stripping electrons from atoms. Only exposed surfaces are sterilized.

5. Filtration eliminates microbes by physically removing them. It is not a sterilization procedure because viruses pass through filter pores.

6. Drying is the removal of water by evaporation or sublimation (removing water vapor from ice by vacuum). Drying does not sterilize, but it controls microbes by removing an essential nutrient, water.

7. Increasing osmotic strength by adding salt or sugar is used to preserve food.

CHEMICAL CONTROLS ON MICROORGANISMS
(pp. 219–223)

1. Chemicals that kill microorganisms are called germicides. Germicides used on inanimate objects are also called disinfectants. Germicides applied to living tissue are called antiseptics. Germistats are chemicals that inhibit microbial growth.

Selecting a Germicide (pp. 219–220)

1. In choosing a germicide, consider whether it will damage the tissue or object being treated, whether it will control the target microorganism, and whether the purpose of treatment is to control or eliminate all microorganisms.

2. Germicides are classified as having high, intermediate, or low germicidal activity. Choice of germicide may also depend upon the presence of other substances, such as blood or feces, because these materials have protective proteins. Cost may also be a factor.

Testing Germicides (p. 220)

1. Germicides are tested by comparing their effectiveness to phenol. The highest dilution of the germicide that kills in 10 but not 5 minutes is the end point. The ratio of end points is the phenol coefficient.

2. The effectiveness of a germicide against a specific microorganism is determined by the paper disc method or the use-dilution test. The presence and size of a bacteria-free zone around a disc impregnated with the germicide indicate its effectiveness. In the use-dilution test, the test organism is added to dilutions of the germicide, and the tubes are incubated. The highest dilution of the germicide that remains clear indicates effectiveness.

Types of Germicides and the Way They Kill (pp. 220–223)

1. Phenol consists of a hydroxyl group attached to a benzene ring. Phenolics have this structure and additional components. Both act by denaturing protein. Phenolics also act on lipids. Hexachlorophene is a phenolic once used as an antiseptic but now replaced with chlorhexidine, which is less toxic to humans.

2. Alcohols are compounds with a hydroxyl group. They kill microorganisms by denaturing proteins and disrupting lipids in the plasma membrane. They do not kill endospores. Ethanol and isopropanol are commonly used clinical disinfectants.

3. Halogens are oxidizing agents. They inactivate enzymes by oxidizing certain functional groups. Iodine is an antiseptic and chlorine is a disinfectant.

4. Hydrogen peroxide is not a halogen but it acts in the same way. It is used as a weak antiseptic for cleaning wounds and to disinfect fragile medical instruments and contact lenses.

5. Salts of heavy metals react with the sulfhydryl groups of proteins to "poison" enzymes and thereby kill microorganisms. Merthiolate and Mercurochrome, organic compounds that contain mercury, are skin disinfectants.

6. Surfactants are compounds with hydrophilic and hydrophobic parts that penetrate oily substances in water and form an emulsion. Soaps and detergents are surfactants. They do not kill microorganisms but control them by washing them away.

7. Quaternary ammonium salts are powerful surfactant germicides. They are cationic agents that kill all classes of cellular microorganisms and viruses that have membranes, though higher concentrations are required for Gram-negative bacteria.

8. Anionic agents are mild disinfectants commonly used to disinfect bench tops in microbiological laboratories.

9. Alkylating agents attach short chains of carbon atoms to enzymes, which inactivates them and kills the cell. Formaldehyde, formalin, and glutaraldehyde are examples.

10. Ethylene oxide is a gas and a sterilizing agent used for heat-sensitive materials and unwieldy objects. However, it is extremely toxic to humans.

PRESERVING FOOD (pp. 223–225)

1. Temperature is the environmental factor most often used to preserve food. Canning is the oldest method. Two factors—time and temperature—determine safe heat treatments for canning. Refrigeration (about 5°C) is low enough to stop the growth of most microorganisms. Psychrophilic microorganisms are the exception.

2. Pasteurization is heat treatment that controls microorganisms without sterilization and with minimal effect on the food.

3. Low pH prevents the growth of most microorganisms. Adding vinegar is one method of lowering pH.

4. Drying and salting food have been used for centuries to preserve meat, fish, and fruit. Both are means of removing water, which is essential for microbial survival.

5. Chemical preservatives are routinely used. For example, calcium propionate is routinely added to bread, making it last almost indefinitely.

Review Questions

THE WAY MICROORGANISMS DIE

1. Define these terms: sterilization, disinfection, sanitation, decontamination, antisepsis.

2. How do you determine whether a cell is dead or not?

3. Explain how a rate of microbial death is determined. Use the term *D-value* in your explanation.

4. Define these terms: microbiostatic, microbiocidal, sterilization, thermal death point, thermal death time.

PHYSICAL CONTROLS ON MICROORGANISMS

1. How does heat kill? What is the advantage of heat over most other sterilization treatments?

2. Explain this statement: Cold is a microbiostatic treatment.

3. How does freezing kill bacteria as opposed to eucaryotic cells? What is cold shock?

4. Name the two forms of radiation used to sterilize and tell how each acts to kill microorganisms.

5. How does filtration control microorganisms? Does filtration sterilize? Explain.

6. Discuss the laboratory advantages and disadvantages of the two forms of drying treatments.

7. How does osmotic strength control microorganisms?

8. Give an industrial use for each of these physical antimicrobial treatments: heat, cold, radiation, filtration, drying, osmotic strength.

CHEMICAL CONTROLS ON MICROORGANISMS

1. What guidelines should be observed in choosing a germicide?

2. Discuss the importance of the phenol coefficient.

3. Explain the paper disc method and the use-dilution test for determining the effectiveness of a germicide against a specific microorganism.

4. How are germicides classified?

5. For each of the following, tell how the germicide acts chemically to kill, and give an example of its use:
   a. phenol   e. hydrogen peroxide
   b. phenolics   f. heavy metals
   c. alcohols   g. surfactants
   d. halogens   h. alkylating agents

PRESERVING FOOD

1. Explain why food sometimes spoils despite refrigeration.

2. What is the benefit of pasteurization if it does not sterilize?

3. What determines safe heat treatments for canning?

4. Explain why drying and salting food are effective preservation methods.

Suggested Readings

Banwart, G. J. 1989. *Basic food microbiology.* 2d ed. New York: Van Nostrand Reinhold.

Block, S. C. 1992. *Disinfection, sterilization, and preservation.* 4th ed. Baltimore: Lea and Febiger.

Mynell, G. C., and Mynell, E. 1970. *Theory and practice in experimental bacteriology.* New York: Cambridge University Press.

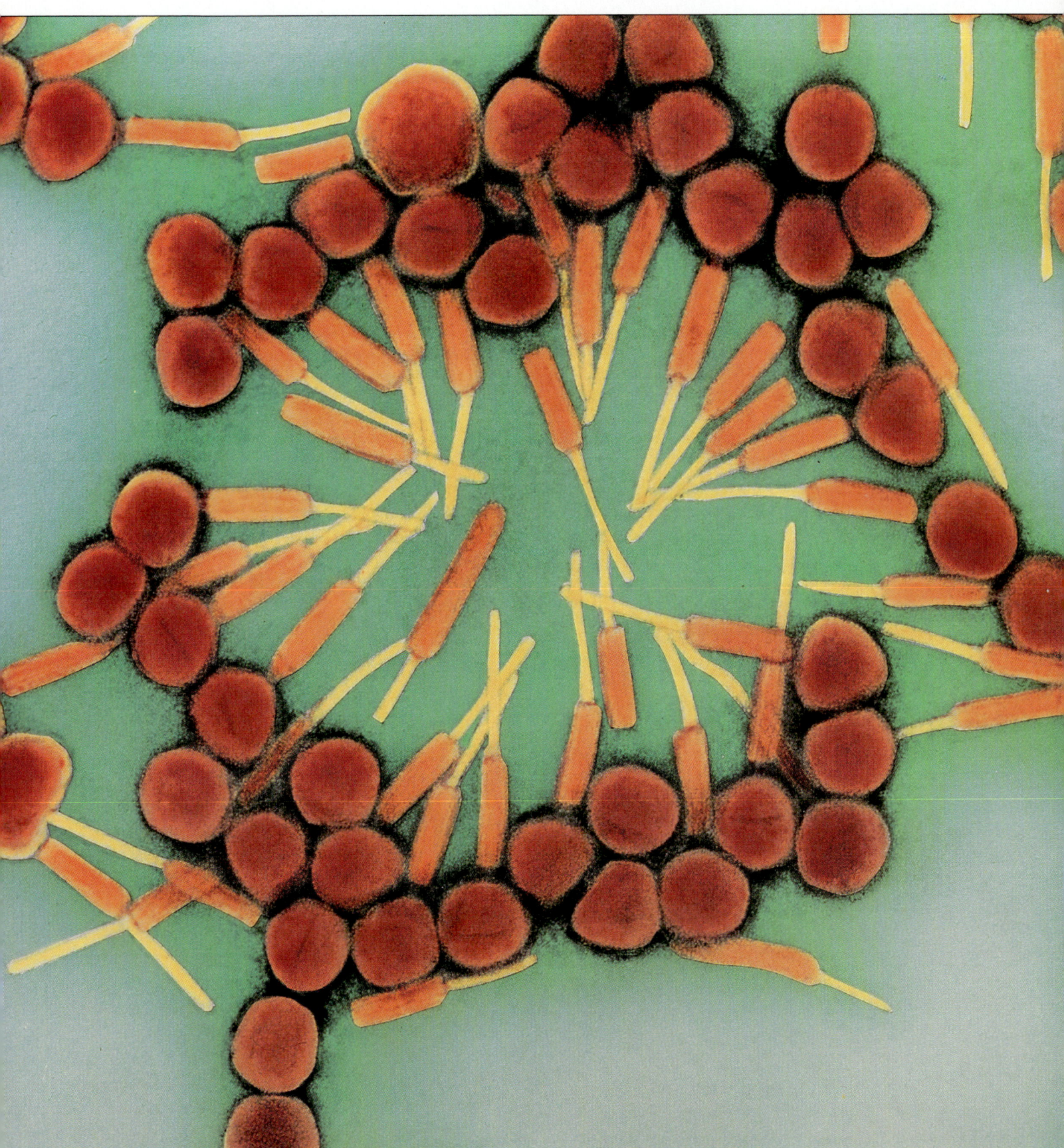

10 CLASSIFICATION

We're All Kin

In 1857, Charles Darwin wrote his friend T. H. Huxley, "The time will come I believe . . . when we shall have fairly true genealogical trees of each great kingdom of nature." The time has come, and we know much more than Darwin had hoped for. We know the relatedness of the kingdoms themselves and many of their species. We also know relative *evolutionary distance*—how closely various species are related.

Over the past decade, scientists have been determining the sequence of bases in the ribosomal RNA of various species. Then they note the number of base differences between pairs of species and relate this number to evolutionary distance, constructing a map of relatedness. At first there was some question whether the number of sequence differences was a legitimate measure of evolutionary distance, but no longer. The reliability of the proposed relationship passes the test of accuracy we put to any good map: Distances must add up. The evolutionary distance (measured in number of base differences) between one species and two others is just about the same

as the distance (measured in number of base differences) between the two others.

We now have a *tree*, or map, like the one shown below, which shows evolutionary distances between all groups of cellular organisms. Some information confirms what we know. For example, Gram-positive and Gram-negative bacteria are distinct, separate groups. But there are surprises. For example, the protozoa are not a tight evolutionary cluster. Ciliate protozoa are more closely related to animals, including humans, than they are to flagellate protozoa. The tree confirms that all living things are related and fall into three distinct groups—the eubacteria, the archaebacteria, and the eucaryotes. So far, it is not possible to locate the tree's root or beginning, because the tree can only relate organisms now alive (the information comes from rRNA of living organisms). The tree cannot tell us which organisms are the most primitive or the most closely related to our universal ancestor. That question involves other considerations and speculation.

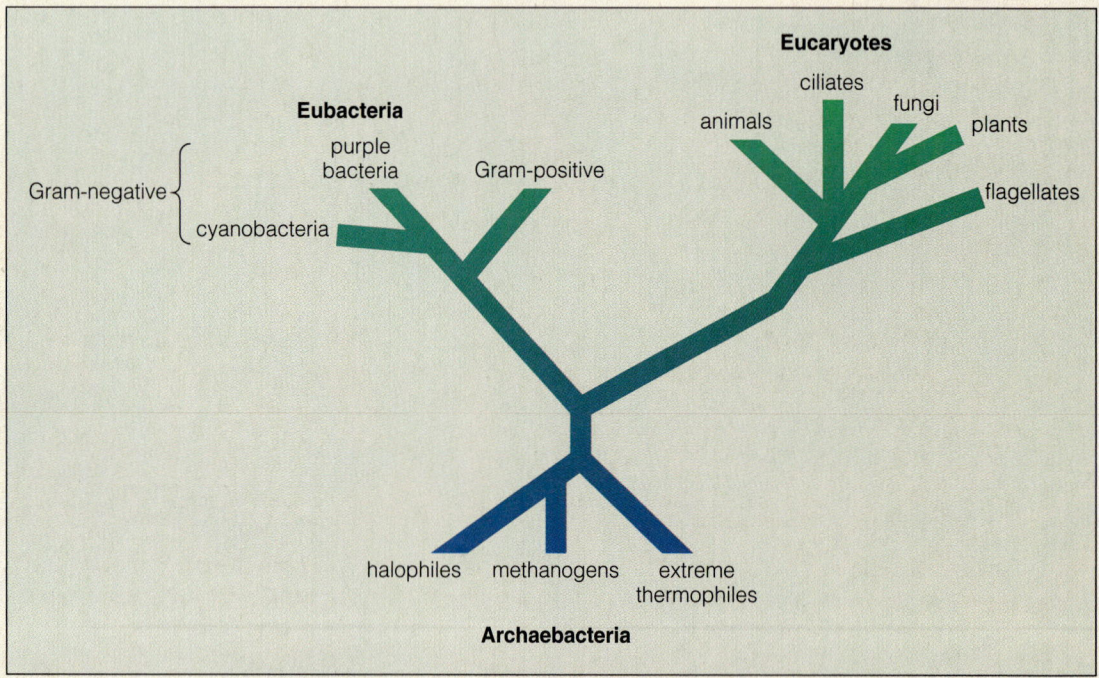

This unrooted tree shows the evolutionary distance between the three major groups of cellular organisms—eubacteria, archaebacteria, and eucaryotes.

To understand:

- The principles by which organisms are classified

- The difference between artificial and natural systems of classification

- The concept of species and how it applies to micro-organisms

- How microbial species are named

- The evolutionary relationship among microorganisms, plants, and animals

- The methods of microbial classification—numerical taxonomy, characters, and dichotomous keys

- The characters used to classify bacteria and viruses

PRINCIPLES OF BIOLOGICAL CLASSIFICATION

Faced with the seemingly infinite variety of life, it is easy to see why the study of biology is so compelling. But how do we make sense out of such diversity? The instinctive human and scientific response is to *classify*—to group similar things together. Then by studying one member of a group, we know something about all its members. As knowledge expands, we can put similar groups into larger groups, again on the basis of similarities. A **classification** system based on collecting individuals into groups and groups into progressively more inclusive and broader groups is called a **hierarchical scheme of classification**.

This chapter is about **taxonomy**, the science of classifying organisms. First we cover the basic principles of biological classification. Then we look at how microorganisms are classified.

The first recorded attempt to classify all living things into one scheme was by the Greek philosopher Aristotle, in the fourth century B.C. After Aristotle, others tried, but it wasn't until 2000 years later, in the eighteenth century, that a Swedish biologist named Carolus Linnaeus devised a scheme that was both practical and adaptable to expanding information. The **Linnaean scheme** remains the basis for biological classification today in two regards—we continue to group organisms hierarchically and we use Linnaeus's **nomenclature** (terms).

Linnaeus's scheme grouped individuals into **species** and then proceeded this way:

> species
>
> genera (*sing.*, genus)
>
> families
>
> classes
>
> orders
>
> phyla (*sing.*, phylum) or divisions
>
> kingdoms

One modern scheme groups kingdoms into **domains** (**Figure 10.1**).

As the fundamental **taxon** (*pl.*, taxa), or taxonomic group, species is the basis on which the higher taxa are ultimately determined. Therefore, all the members of a species must be very much alike. Progressively higher levels accommodate progressively greater diversity among individual members. Members of **kingdoms**, the largest grouping in most hierarchical schemes, may be quite different from one another. For example, a coral and an elephant are both members of the animal kingdom. Still, organisms grouped together within the same kingdom have more in common with one another than do organisms belonging to different kingdoms. Thus, a coral resembles *all* animals more closely than it resembles *any* plant.

Scientific Nomenclature

Linnaeus introduced a **binomial nomenclature**. That is, each organism is designated by two names. The first name is the organism's **genus** designation and the second its **specific epithet**. Together, the two constitute the species name. The species name is always Latinized and underlined or written in italics. The genus designation is capitalized, but the specific epithet is not. Thus, the proper designation for the best-studied bacterium is *Escherichia coli* and that for humans, *Homo sapiens*. By convention, the genus designation can be replaced with an initial if the complete genus name has been used recently enough to avoid possible confusion. The bacterium becomes *E. coli* and humans, *H. sapiens*. All eucaryotes and procaryotes are named this way. Viruses are not, as you will see later in this chapter.

Most species names tell us something about the organism—its appearance, its source, a characteristic property, or the scientist who discovered, described, or was connected with it. For example, *Escherichia coli* is named for microbiologist Theodor Escherich and its usual habitat, the colon. *Staphylococcus aureus* is named for the characteristic way its cells aggregate to resemble a bunch of

Category (taxon)	Corn	Housefly	Human	Microbe
Kingdom	Plantae	Animalia	Animalia	Monera
Phylum (or Division, botanical schemes)	Anthophyta (flowering plants)	Arthropoda	Chordata	Gracilicute
Class	Monocotyledonae (monocots)	Insecta	Mammalia	Scotobacteria
Order	Commelinales	Diptera	Primates	Spirochaetales
Family	Poaceae	Muscidae	Hominidae	Spirochaetaceae
Genus	*Zea*	*Musca*	*Homo*	*Treponema*
Species	*mays*	*domestica*	*saplens*	*pallidum*

Figure 10.1 Biological classification. We use the taxonomic categories established by Linnaeus to classify living things. *Treponema pallidum*, the microbe example, causes the sexually transmissible disease syphilis.

grapes (*staphyle* means "cluster") and its yellow colonies (*aureus* means "golden"). Some names are more obscure. For example, *Shigella etousae* is named for the Japanese microbiologist Kiyoshi Shiga, but the specific epithet comes from being isolated in the European Theater of Operations of the United States Army (the terminal "e" gives the proper Latin ending).

Artificial and Natural Systems of Classification

Linnaeus's system is what biologists now call an **artificial scheme** of classification because he grouped organisms on the basis of visible similarities rather than of **phylogenetic** (evolutionary) relatedness. A system based on phylogeny is called a **natural scheme** of classification. The Linnaean scheme was necessarily artificial because Linnaeus devised it about a hundred years before Charles Darwin developed his theory of evolution. With an understanding of evolution, the meaning and importance of biological classification vastly changed. Modern classification does two things—it names organisms and it indicates how closely they are related to one another. Evolution explains how and why organisms are related to one another and is the basis of natural schemes of classification.

The Fossil Record

The course of evolution can be reconstructed by studying the **fossil record**. **Fossils** are the mineralized remains of organisms from previous geologic periods. They provide a record of Earth's biological history. Nearly 200,000 species of fossil organisms have been discovered and accurately dated because geologists can accurately date the rocks in which they are embedded. There is a substantial fossil record covering about 600 million years. Older fossils exist, but they are not complete enough to provide a picture of the entire 3.5 billion years life has existed on Earth. Most of the record is therefore limited largely to evolutionarily advanced forms—plants and animals.

Only in the past 25 years have fossils of microorganisms been discovered by examining thin slices of ancient rocks with electron microscopes (**Figure 10.2**). Microbial fossils have been found in finely layered rocks called **stromatolites**, which are fossilized photosynthetic procaryotes that grew as masses of cells, or **microbial mats** (**Figure 10.3**). Stromatolites still form today and are relatively common in lagoons and hot springs. Microbial fossils do not preserve enough detail to reveal phylogenetic relationships, but they do establish how long microorganisms have been on Earth. The oldest known microbial fossils are about 3.5 billion years old, almost as old as the

No Job Too Big

Carolus Linnaeus (1707–1778) was ambitious and hard-working and had a passion for classification. He organized the biological information of his time and set up a system of classification that serves as the basis for all modern systems.

Linnaeus set his lifetime goal as a youth. He wanted to classify the three kingdoms of nature—plants, animals, and minerals. He went to medical school in Sweden and then on an expedition to Lapland for the Swedish Academy of Science. Next he traveled to Germany, England, France, and Holland. At age 28 he published three books. One was on a system of classifying plants, animals, and minerals. The second compared and analyzed all existing schemes for classifying plants, and the third described all known plant genera (935 at the time). He returned to Sweden, practiced medicine for a while, and became a professor at the University of Uppsala. There he made his major contribution—publishing books on the binomial designation of species and the classification of species into hierarchical schemes. He even classified human diseases by genera, families, and orders. Carolus Linnaeus's life work was a turning point in biology.

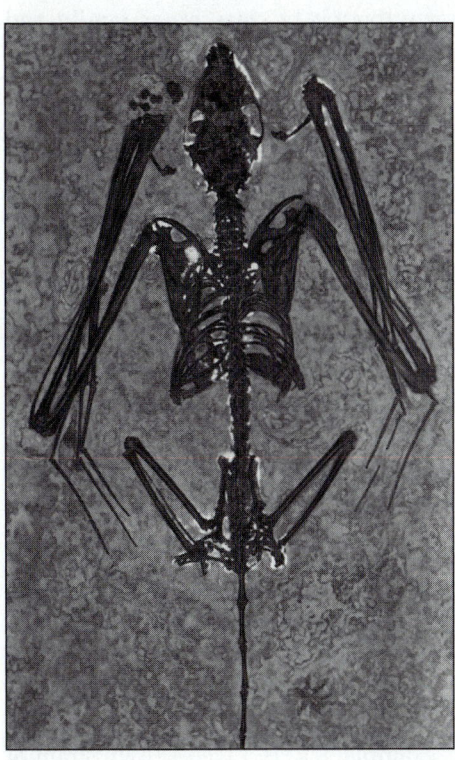
a

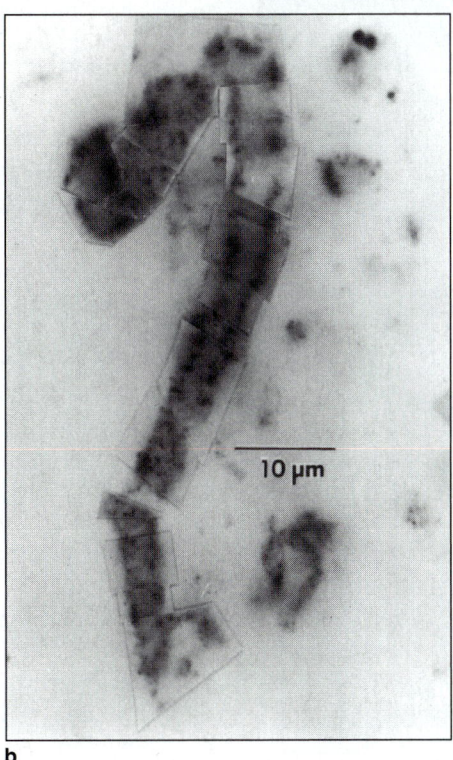
b

Figure 10.2 The fossil record. **(a)** A 50-million-year-old fossil of a bat reveals enough detail to identify its species and tells us about evolution of other species. **(b)** The oldest known fossil, a 3.5-billion-year-old filamentous microorganism, superficially resembles modern cyanobacteria, but cannot be identified.

oldest rocks (3.8 billion years) and only about 1 billion years younger than Earth itself.

The Concept of Species

Identifying a species and determining its limits are the most difficult parts of biological classification. After all, the individual members of a species are not identical. The key question is, How different can individuals be and still belong to the same species?

The Definition of Species. Before evolution was recognized as the biological process that produces new organisms, each species was considered an individual act of divine creation. The theory of evolution changed this definition but did not replace it with an equally simple one.

Possibly the clearest and most useful modern definition of *species* is a group of organisms that share and can exchange a common pool of genes. In other words, a species is a group of organisms capable of interbreeding and thereby interchanging their genes. When some mem-

Figure 10.3 Stromatolites in shallow sea water in western Australia. They are 1000–2000 years old. The oldest known fossils were found in stromatolites just like these that formed more than 3 billion years ago.

bers of a species change or become geographically isolated so they no longer have the chance to breed with the rest of the group, they may continue to change and eventually become a new species. The new species, in turn, may or may not survive to spawn other new species.

Bacterial Species.   The modern definition of species is applicable to most eucaryotes, including algae, fungi, and protozoa. It does not apply to procaryotes because breeding and the genetic exchange that goes with it are not essential in their life cycle. Genetic exchange among bacteria does occur in nature; and it can occur between distantly related organisms (Chapter 6). Viruses are again an exception of a different sort.

Defining bacterial species, then, presents special problems. We might even ask if the concept of species applies at all to organisms like bacteria that reproduce al-

most exclusively by asexual means. Because bacteria probably evolve principally by accumulating mutational changes, it might be correct to assume that individual bacteria are all just slightly different, forming a continuous variation of types. But this is not the case. Groups of bacteria with similar properties do, indeed, exist; and so the term *species* does apply. A **bacterial species** is defined by the similarities of its members.

The species of bacteria that most microbiologists recognize are catalogued in *Bergey's Manual of Systematic Bacteriology* (Chapter 11). The *Bergey's Manual* classification scheme divides bacteria into four divisions (**phyla**) on the basis of cell wall (Gram-positive or Gram-negative bacteria), lack of a wall (the mycoplasmas), and walls lacking peptidoglycan (archaebacteria). Then bacteria are grouped into 29 **sections**, a term that has no taxonomic standing and does not, therefore, imply an evolutionary relationship. Thereafter, a particular bacterium is identified by species within a section and, depending upon how much is known about it, by **class**, **order**, **family**, and genus. Finally, if strains have been identified, they are given.

Bergey's Manual disregards evolutionary relationships because they often group bacteria into assemblages that cannot be easily identified by standard laboratory procedures. Instead, *Bergey's Manual* takes a strictly practical approach so that it can be used as a comprehensive and quick reference when accuracy and speed are important, as is often the case in diagnostic laboratories.

Clones and Strains.   Microorganisms that belong to the same species are described in terms of clones and strains. A **clone** is a group of cells that results from the division of an individual cell by asexual reproduction. In other words, all cells in the same clone are presumed to be genetically identical. Clones that are presumed or known to be genetically different are called **strains**. For example, if two clones of *Escherichia coli* are isolated at different places or at different times, they are different strains. Or if a strain of *E. coli* is altered by mutation or genetic exchange, the product will be a new strain. By convention, strains are designated by capital letters followed by numbers. For example, K12 is a well-known strain of *E. coli*.

Microbiology laboratories accumulate large numbers of strains. A microbial genetics laboratory might have several thousand strains of the same species, each carrying a particular mutation or combination of mutations. An infectious disease laboratory would have strains of pathogenic bacteria isolated from different patients or outbreaks of disease. Identifying strains is an important step in tracing the spread of disease (Chapter 20). People suffering from disease caused by the same bacterial strain probably acquired it from the same source.

Viral Species. Classifying viruses presents even greater problems than classifying bacteria because we have no evolutionary history of this group of organisms. Still, an agreed-upon classification scheme is essential, so in 1966 the International Committee on Nomenclature of Viruses was formed. In 1973 it changed its name to the International Committee on Taxonomy of Viruses (ICTV). The ICTV does not use a Linnaean scheme. Rather than Latinized names, viral species are given ordinary English names, which are not translated in most countries. Thus, the virus that causes mumps is called the mumps virus, and most other virus names are similarly simple and direct. Species are grouped into genera and genera into families. No higher taxa are used because higher levels of relatedness among viruses are not yet understood (Chapter 13).

Virus species also form strains, some of them quite different. For example, the Bangladesh strain of smallpox virus killed about 40 percent of the people it infected, while other strains of the same species typically killed only a few percent of those infected.

MICROORGANISMS AND HIGHER LEVELS OF CLASSIFICATION

Fitting microorganisms into the higher hierarchical levels of biological classification is almost as difficult as defining *species*. The earliest schemes put microorganisms into one or the other of Linnaeus's two categories of living things—plants and animals (**Figure 10.4**). Protozoa, because they are largely motile, were assigned to the animal kingdom. Algae, because they are photosynthetic, and fungi, because they are nonmotile, were assigned to the plant kingdom. Bacteria were also put in the plant kingdom because there was no more logical place for them.

In the nineteenth century, scientists began to realize how artificial these assignments were. Clearly, microorganisms are neither plants nor animals. Nearly every group has representatives with plantlike properties and others with animal-like properties. Some bacteria are photosynthetic, but many are motile. All algae are photosynthetic, but some are motile. Certain algae readily lose their ability to carry out photosynthesis and thereby become indistinguishable from certain protozoa. Some fungi are motile, but none is photosynthetic.

The obvious solution was to admit that microorganisms are neither plants nor animals. Thus, the German biologist Ernst Haeckel proposed in 1866 that living things be divided into three kingdoms—plants, animals, and microorganisms, which he called **protists**. This scheme served biology well for nearly 60 years. In 1892 Dmitri Iwanowski discovered viruses. Where did they fit? To this day, viruses are not included in any classification scheme. Some scientists still argue that they should not be considered living because they are not cellular.

By the 1930s, there was additional reason to revise Haeckel's classification scheme. Using the new electron

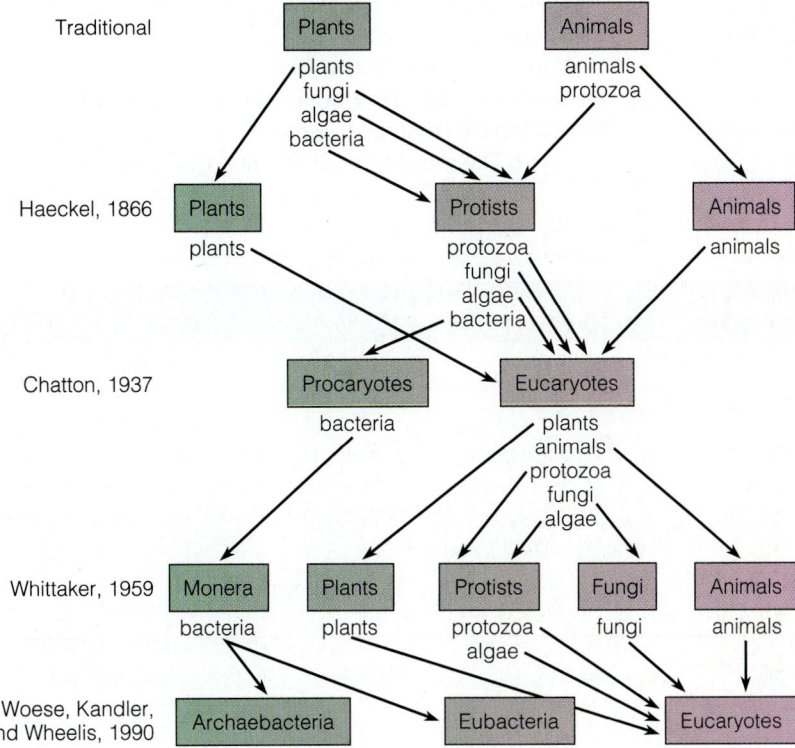

Figure 10.4 Evolution of the major schemes for classifying organisms. Biological classification changes as technology advances and scientists make new discoveries.

microscope, scientists discovered that all cellular organisms are either procaryotic or eucaryotic. A French microbiologist, Edouard Chatton, considered the differences between the two cell types so fundamental that in 1937 he proposed all living things be divided into two groups, eucaryotes and procaryotes. Under Chatton's scheme, bacteria were separated from other subgroups of microorganisms, as well as from plants and animals. In the 1940s, two American microbiologists, Roger Y. Stanier and Cornelius B. Van Niel, marshalled extensive evidence to justify the logic of this scheme, making it generally accepted by biologists.

In 1959, Robert H. Whittaker, an American taxonomist, proposed that their unique life cycle and continuous cytoplasm (Chapter 12) made fungi different enough from other organisms to justify calling them a separate kingdom. This led to the five-kingdom scheme of classification—animals, plants, fungi, protists (containing protozoa and algae), and **monera** (bacteria) (**Table 10.1**). Whittaker's scheme is still widely accepted as a logical classification of living things.

Biological classification continues to change with new discoveries. In the 1970s molecular biologists discovered that some procaryotes were chemically different from the rest. They designated the two groups the **archaebacteria**, or ancient bacteria, and the **eubacteria**, or true bacteria. Archaebacteria are as unrelated to eubacteria as they are to eucaryotes. Consequently, in 1977, two American microbiologists, C. Woese and G. Fox, suggested that living things fell logically into three groups—the archaebacteria, the eubacteria, and the eucaryotes. The new scheme was formalized in 1990 by C. Woese, O. Kandler, and M. L. Wheelis, who proposed that these three groups be designated domains, which are then divided into about 34 kingdoms, although this aspect of the proposal has not yet been completed.

Bergey's Manual has not incorporated this latest scheme. In spite of the fundamental evolutionary distance between eubacteria and archaebacteria (which *Bergey's* spells archaeobacteria), the two groups are superficially similar in appearance and metabolism, and *Bergey's* is designed to be a practical laboratory reference. Most microbiology textbooks, including this one, follow *Bergey's Manual* in considering archaebacteria and eubacteria together.

THE METHODS OF MICROBIAL CLASSIFICATION

Out of the early taxonomic studies came the concept that some **characters**, or properties of organisms, are more important than others in taxonomy. For example, flowers are more important than leaves in classifying plants, and means of reproduction is particularly important in classifying animals. Determining the most important characters allows us to classify an organism by observing only these particular characters. An alternative approach is taken by the numerical taxonomists.

Numerical Taxonomy

Michel Adanson, a contemporary of Linnaeus, believed that if enough characters of an organism were observed, regardless of their importance, the organism could be accurately classified. His approach is called **numerical taxonomy**. Numerical taxonomy gives equal weight to all characters.

Numerical taxonomy has an additional goal, to express the relatedness of organisms—or more properly, the **evolutionary distance** between them—in terms of a number. The numerical taxonomist measures many char-

Table 10.1	Major Differences among Kingdoms in the Five-Kingdom Scheme of Classification				
Property	Plantae	Animalia	Protista	Fungi	Monera
Cell type	Eucaryotic	Eucaryotic	Eucaryotic	Eucaryotic	Procaryotic
Cell organization	Mostly multicellular	Mostly multicellular	Mostly unicellular	Multicellular and unicellular	Mostly unicellular
Cell wall	Present	Absent	Present in some; absent in others	Present	Present in most
Nutritional class	Phototrophic	Heterotrophic	Heterotrophic and phototrophic	Heterotrophic	Phototrophic, heterotrophic, or chemoautotrophic
Mode of nutrition	Mostly absorptive	Mostly ingestive	Absorptive or ingestive	Absorptive	Absorptive
Motility	Mostly nonmotile	Mostly motile	Motile or nonmotile	Nonmotile	Motile or nonmotile

acters for a number of strains and then calculates the percentage of characters that various pairs of strains share. In this way relatedness between all pairs of strains is reduced to a number, the **similarity coefficient (S_J)**.

$$S_J = \frac{\text{the number of characters that two organisms share}}{\text{the total number of characters measured}}$$

If two organisms are examined for 100 characters and they share 93 of them, the similarity coefficient for the pair would be 0.93.

If a group of organisms is compared, similarity coefficients for all the possible pairs can be determined. With this information and the help of a computer, it's possible to construct a **dendogram**, a diagram rather like a family tree that illustrates the relationship among all the organisms examined. A dendogram is constructed by grouping organisms with the highest similarity coefficients. Lines that divide the graph along different similarity coefficients indicate various levels of relatedness. For example, in **Figure 10.5**, a similarity coefficient of 0.85 might define a species and 0.65 might define a genus. These values vary depending on the organisms being studied and the taxonomists applying them. Sometimes the similarity coefficient for members of a species is greater than 0.85 and sometimes it is less.

Characters Used to Classify Bacteria

The characters used to classify bacteria fall into two groups—traditional and new. Traditionally, taxonomists have used morphology, biochemistry and physiology, serology, and phage typing to make decisions about bacterial species. Today, though, they can also use percentage of base pairs that are G-C, sequences of DNA bases, DNA hybridization, sequences of mRNA bases, sequences of amino acids in proteins, and protein profiles (**Table 10.2**).

Morphology. Looking at bacteria through a microscope reveals the shape of individual cells, whether the cells are in a characteristic arrangement (chains, clusters, or regular packets), and how flagella are arranged, but not much more. Moreover, similar appearance does not necessarily indicate a close relationship. For example, many quite different bacteria are rod-shaped with peritrichous flagella (Chapter 4). Thus, morphological characters help describe bacteria, but they do not provide much information about relatedness.

There is a major exception to this statement. The Gram stain reveals significant morphological character. It tells whether a bacterium does or does not possess an outer membrane and whether it has a thick or thin peptidoglycan wall. This simple test distinguishes two divisions of bacteria—Gram-positive and Gram-negative. The other two divisions, mycoplasmas and archaebacteria, are difficult to distinguish morphologically.

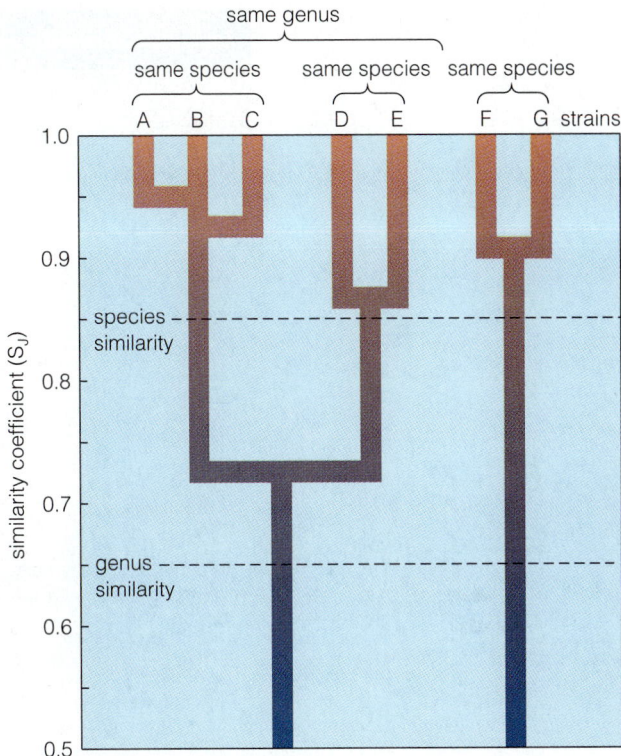

Figure 10.5   This dendogram relates seven bacterial strains. Lines from strains are joined at their similarity coefficient (for example, A and B at 0.95, and A, B, C, D, and E at 0.725). Setting species similarity at 0.85 and genus similarity at 0.65 places strains A, B, and C in one species, D and E in another, and F and G in a third. Strains A, B, C, D, and E are in one genus, and F and G in another.

Biochemistry and Physiology. Some of the biochemical and physiological characters used to classify bacteria are based on conditions that support growth. Do the bacteria grow aerobically, anaerobically, or both? What incubation temperature is most favorable? Over what range of pH do they grow? Are they able to withstand high osmotic strength? What end products and enzymes do they form?

Carbon sources that support growth are a particularly useful set of characters because most bacteria can use many different carbon sources. Whether or not a bacterium can use each particular carbon source becomes a character that can be used to calculate a similarity coefficient. In fact, it is possible to identify the species of any Gram-negative pathogen by determining which of 96 different organic compounds it can use as carbon sources. Modern techniques make this a relatively easy task. A

Table 10.2   Characters Used to Classify Bacteria

Character	Examples of Properties Observed or Tests Applied	Uses
Morphology	Cell shape, presence of flagella, arrangement of flagella, Gram stain	Principally distinguishes genera, but sometimes species; Gram stain separates eubacteria into divisions
Biochemistry and physiology	Conditions required for growth (pH, temperature, oxygen), carbon sources used, fermentation end products (acids and gases), specific enzymes or other end products	Distinguish species, genera, and higher groups
Serology	Slide agglutination, fluorescent-labeled antibodies	Distinguishes strains and some species
Phage typing	Susceptibility to a group of bacterio-phages	Identifies and distinguishes strains
Percentage of base pairs (mole percent G + C)	Melting point of DNA	Members of same genus vary within a relatively narrow range
Sequence of bases in DNA	DNA sequencing of homologous regions	Determines relatedness within genera and families
DNA hybridization	Amount of annealing between single-stranded DNA from pairs of organisms	Determines relatedness within genera
Sequence of bases in rRNA	rRNA sequencing	Determines relatedness among all living things
Sequence of amino acids in proteins	Protein sequencing	Same as DNA sequencing but much more difficult to do
Protein profiles	Separate proteins by two-dimensional PAGE	Distinguish strains

plastic plate with 96 depressions is prepared, each containing a colorless form of a **tetrazolium dye** (an oxidation/reduction indicator that is colorless when oxidized and colored when reduced) and a different carbon source. A suspension of bacterial cells is added. If the organism can use the carbon source, the dye is reduced and becomes colored. The color pattern is read by a machine and results are fed directly into a computer. In a few minutes, the computer displays the name of the bacterium (**Figure 10.6**). Before this technology was available, identification was a laborious process of comparing test results with the known properties of various species.

Fermentation properties of bacteria are also useful characters. Are acids and/or gases produced during fermentation? Both are easily detected. We can trap gases in an inverted tube or in an agar-solidified medium. When acid products of fermentation accumulate, they lower the pH of the medium. The lowered pH can be detected in various ways, though the most common is to incorporate an indicator into the growth medium that will change color with increased acidity. For example, *Esch-*

erichia coli can be distinguished from the closely related *Enterobacter aerogenes* by the **methyl red test** (**Figure 10.7**). *Escherichia coli* produces enough acid when it ferments sugars to change the color of methyl red from yellow to red, but *Enterobacter aerogenes* does not. The distinction is important in water bacteriology because *Escherichia coli* in a water supply indicates it has been contaminated with sewage. *Enterobacter aerogenes* could have come from soil.

Production of both acid and gas from a particular carbon source can be very informative. For example, an organism that produces both acid and gas as fermentation products of lactose is likely to be *Escherichia coli*. Forming acid and gas from lactose is only a presumptive identification, however, and confirmation requires additional testing.

Serology.   Another way to classify bacteria is through **serology**, the science that studies the properties and action of **serum** (*pl.*, sera), the noncellular part of blood (Chapter 19). Serum contains **antibodies**, protein mole-

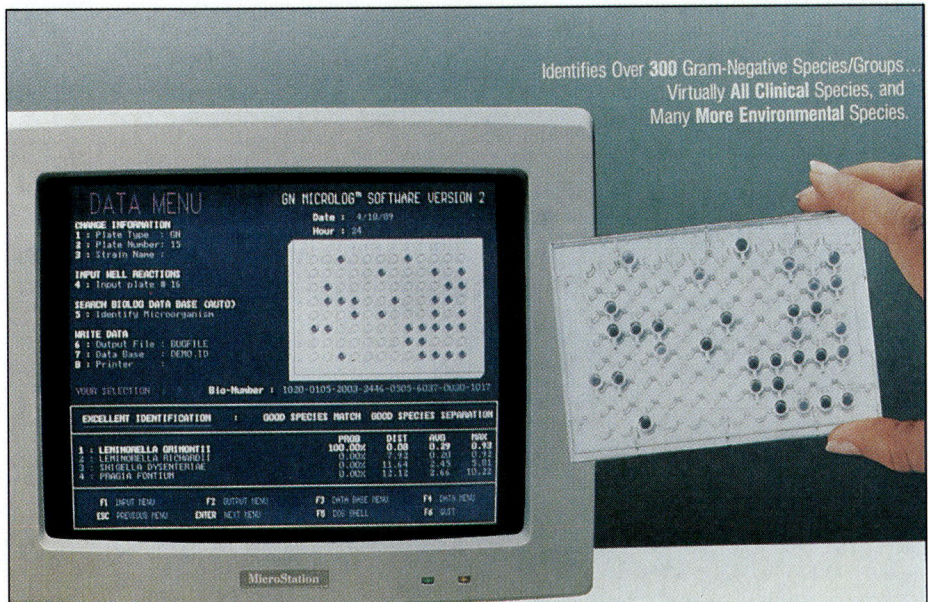

Figure 10.6 Classifying bacteria by their use of carbon sources. A suspension of the unknown bacterium is added to 96 wells in a plastic dish. Each well contains a different carbon source and an oxidation/reduction indicator. If the carbon source is metabolized, the indicator is reduced and becomes colored. A machine detects the pattern of colors and feeds the information to a computer, which displays the name of the bacterium.

Figure 10.7 Methyl red test. The methyl red test classifies bacteria by fermentation end products. *Escherichia coli* produces sufficient acid to change the color of methyl red indicator from yellow to red. *E. coli* also produces gases (carbon dioxide and hydrogen) as products of fermentation; these accumulate within inverted vials.

cules made in response to infection that attack the invading microorganism. Antibodies are highly specific, targeting specific microbes. The fact that antibodies distinguish between closely related microorganisms and even between strains makes serology a reliable method of classifying bacteria.

Serology is widely used in the diagnostic laboratory to identify pathogens and to identify nonpathogens. Nevertheless, serology can be overused. The most striking instance occurred with the genus *Salmonella*, bacteria that cause intestinal infections (Chapter 23). If serum from an animal exposed to one strain of *Salmonella* did not inacti-

vate another strain, the two strains were considered different species. This led to dividing the genus into hundreds of species when other evidence indicates there are only five species and hundreds of strains.

Preparations of sera from animals that inactivate particular bacteria are called **antisera**. Antisera are used in the **slide agglutination test** to identify species and even strains within species. **Fluorescent-labeled antibodies** are also used to identify microorganisms in a procedure similar to but more sensitive than the slide agglutination test (see Chapter 19 for complete descriptions of both tests).

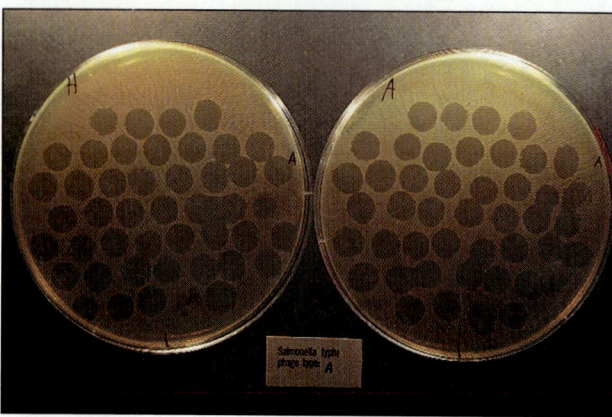

Figure 10.8 Petri dish used for phage typing. Suspensions of various phages are dropped on a lawn of bacteria in an array that identifies the phages. Clearing at the area where a phage was placed indicates susceptibility of the bacterial strain to that particular phage. The pattern of clearing shown here identifies the strain of *Salmonella typhi* being tested as belonging to phage type A. (Courtesy of the Microbial Diseases Laboratory, Berkeley, CA.)

Table 10.3 The Range of % G + C in DNA from Various Organisms

Organisms	Range of % G + C
MICROBIAL GROUPS	
Procaryotes	
Eubacteria	25 to 78
Archaebacteria	27 to 62
Eucaryotes	
Algae	37 to 68
Protozoa	21 to 65
Fungi	22 to 62
NONMICROBIAL GROUPS	
Eucaryotes	
Plants	33 to 48
Animals	32 to 50

Phage Typing. Bacteriophages, or **phages**, are viruses that attack bacteria (Chapter 13). The pattern of a bacterial strain's susceptibility to a set of bacteriophages is called its **phage type**. The number of bacterial strains that one bacteriophage will attack (termed the **host range**) is quite narrow, so most bacterial strains that are attacked by the same phages are closely related.

Determining which phages attack which bacteria is called **phage typing**. In phage typing, a thin layer of the bacterial strain to be tested is spread on the surface of an agar plate and small drops of suspensions of various phages are placed on the surface. If the bacteria are susceptible to the phages in one or more of the drops, they will lyse, producing a clear zone in the **lawn** (confluent layer of growth) that forms (**Figure 10.8**).

Strains with identical phage types are identical, while strains with similar phage types are closely related. Phage type has no taxonomic standing, but is an index of close relatedness.

Because phage typing can identify specific strains of bacteria, it is used to determine if a cluster of bacterial infections is caused by the same strain. Such information is useful in determining the source of a nosocomial, or hospital-acquired, infection and controlling its spread (Chapter 20).

Percentage of G + C. The fraction of total base pairs in DNA (Chapter 6) that are guanine-cytosine (G-C) pairs as opposed to adenine-thymine (A-T) pairs can vary widely, depending on the source of the DNA. By convention, the fraction of G-C pairs is expressed as **mole percent guanine plus cytosine** or **% G + C**, referring to the percentage of the total number of base pairs that are G-C. For example, if % G + C for a sample of DNA is 40, 40 percent of the base pairs are G-C pairs and the other 60 percent are A-T pairs.

The % G + C of DNA affects its physical properties. Because three hydrogen bonds join G-C pairs and only two join A-T pairs, higher values of % G + C mean the two strands of DNA are joined by more hydrogen bonds, making them more difficult to melt, or separate by heating. The **DNA melting point**, or temperature at which the strands separate, can therefore be used to determine the % G + C. Measurements of density can also be used, because the density of a DNA sample is proportional to its % G + C.

The % G + C values can be used to determine the relatedness of organisms. These values vary even more widely in microorganisms than in higher eucaryotes (**Table 10.3**). But the variations in values of % G + C within a species are slight, and variations within a genus are only a little greater. Thus, closely related organisms have similar % G + C values. However, similar % G + C values do not prove that two organisms are closely related because % G + C is unrelated to the coding properties of DNA. For example, humans and *Bacillus subtilis* have almost identical % G + C values.

Sequences of Bases in DNA. The sequence of bases in DN3A determines its coding properties. Thus, an identical sequence means identical organisms, and a similar sequence means closely related organisms. Comparing DNA base sequences, then, is the ultimate tool of taxonomy. In spite of recent technological advances, however,

Good Question

The mole percentage of guanine plus cytosine (% G + C) in the DNA of various species has a considerable range. Some ciliate protozoa have a % G + C as low as 21 and some eubacteria, as high as 79. But in any one species the % G + C is essentially the same throughout the genome. If it varies among species, why doesn't it vary within a genome?

The answer is that over a wide range, one value of % G + C is as good as another because the genetic code is redundant. For example, the UUC and UUA codons both specify the same amino acid, leucine. Frequent use of the UUC codon in place of the UUA codon raises the % G + C without affecting the protein being encoded. But the protein synthesizing machinery evolves to match the DNA's % G + C. In organisms with high % G + C, more tRNA molecules recognize UUC than recognize UUA. For this reason, % G + C tends not to change throughout one organism's genome.

This knowledge is important to biotechnology because when genes from high % G + C organisms are cloned into low % G + C organisms and vice versa, the genes are poorly expressed (protein is made slowly). The host cell's protein synthesizing machinery is ill-suited to express DNA with a different % G + C than its own.

---

sequencing, determining the sequence of bases for all of an organism's DNA, is still a formidable task. The number of base pairs is enormous—4.5 million in *Escherichia coli* and 3 billion in a human, for example. The complete sequence of bases is not yet known for any cellular organism, although several viruses have been sequenced because their genomes are relatively small. Within a decade, the complete sequence for *Escherichia coli* and for humans will probably be known.

A more manageable task that nonetheless yields helpful information for classifying microorganisms is determining the sequence of homologous (equivalent) regions of DNA from different organisms—for example, regions that encode the same enzyme. Many homologous studies have focused on DNA sequences encoding cytochrome oxidase because the enzyme, which occurs in all aerobic organisms, is relatively small and makes sequencing more manageable.

Comparing DNA sequences of the same enzyme from different organisms and using this information to determine relatedness between organisms is based on the assumption that all forms of the enzyme have a common origin: they were originally identical. As one species evolves into another, the DNA sequences encoding the enzyme change, but the enzyme continues to catalyze the same reaction. The similarity of DNA sequences encoding an enzyme such as cytochrome oxidase is therefore a direct measure of the relatedness of the organisms that produced those enzymes. Conversely, the number of differences in their DNA sequences is a measure of evolutionary distance between a pair of organisms.

Taxonomists have attempted to extract more information from comparisons of DNA sequences by assuming that changes occur at a constant known rate. This allows them to calculate the time in the past that two organisms began to evolve from a common ancestor.

DNA Hybridization. Sequencing the DNA encoding for a particular enzyme is more manageable than sequencing an entire genome, but it is still time-consuming. Other, quicker procedures have been developed. They yield less precise but still useful information about the relatedness of organisms. One of these techniques is DNA hybridization (**Figure 10.9**).

DNA hybridization is based on a number of principles discussed in this chapter and Chapter 2. First, heating DNA causes it to melt as the hydrogen bonds break, allowing the two strands of the double helix to separate. Second, if the DNA is cooled, hydrogen bonds re-form and the two single strands will **anneal**, or come together again. Finally, if single strands of DNA from different sources are mixed, they will anneal in regions where the sequences are the same or similar, creating a **hybrid**. Therefore, the extent of annealing is a quantitative index of the similarity of base sequences in the DNA from the two sources.

DNA hybridization is simple and precise, but it has one major drawback. For any hybridization to occur, the two samples of DNA—and hence the organisms from which they were obtained—must be similar. As a general rule, DNA hybridization gives useful information only about organisms in the same genus. Determining more distant relationships among phyla and kingdoms requires knowing the exact base sequences of DNA or RNA.

DNA hybridization can also be used to identify a particular organism by determining whether a specific short piece of DNA, called a **probe**, will hybridize to any

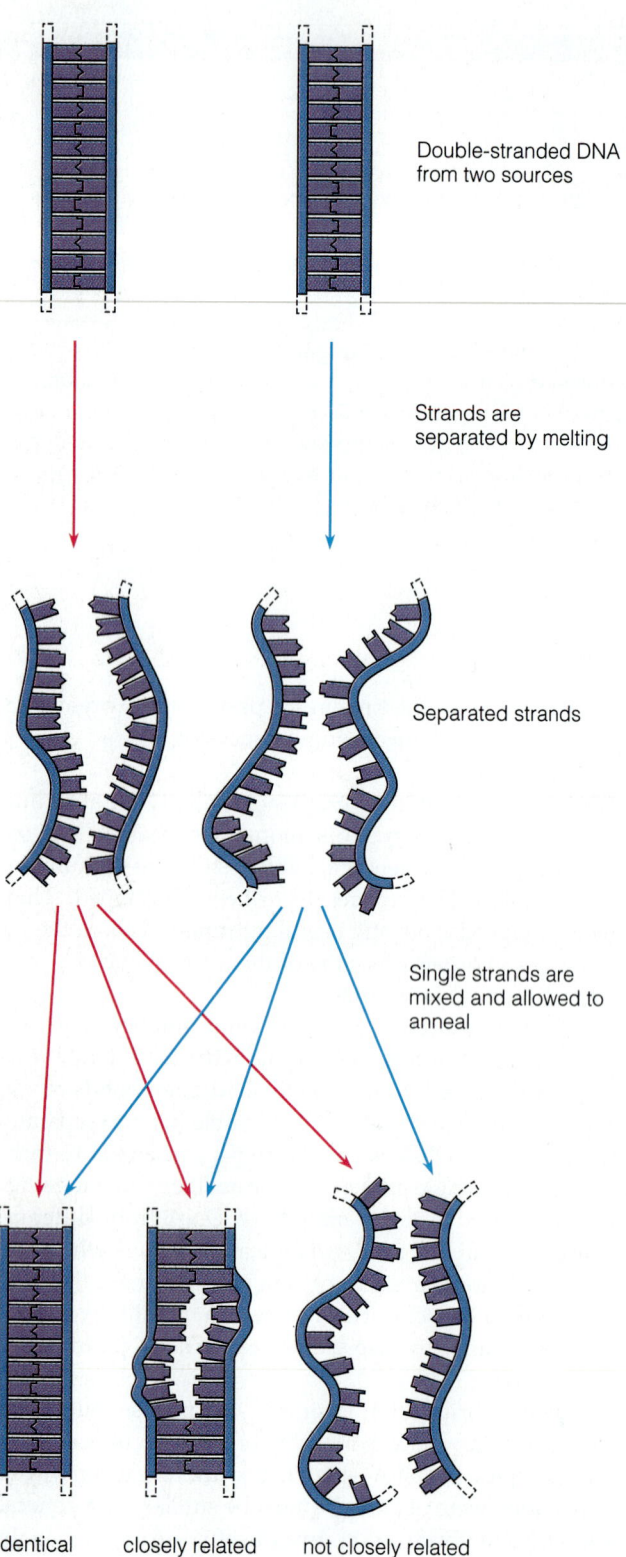

Double-stranded DNA from two sources

Strands are separated by melting

Separated strands

Single strands are mixed and allowed to anneal

identical   closely related   not closely related

Figure 10.9 DNA hybridization. DNA strands from two organisms are separated into single strands and mixed. The degree of annealing indicates relatedness of the organisms. If the strands anneal completely, the organisms are identical. The extent of partial annealing indicates the extent of similarity. If there is no annealing, the organisms are not closely related.

stretch of the unknown organism's DNA. DNA probes are chosen to hybridize only with the DNA of specific organisms (**Figure 10.10**). Probes are made by genetic engineering (Chapter 7) and tagged with a fluorescent dye, a chemiluminescent molecule, or radioactive atoms so hybridization can be detected easily. The method can be used to identify colonies on a petri dish by lysing cells and adding the probe. The probe hybridizes only to DNA from colonies it is designed to identify.

Sequences of Bases in rRNA.   DNA hybridization indicates similarity in base sequences in DNA and, therefore, similarity of closely related organisms. Comparing the DNA sequences encoding particular enzymes gives information about the similarities of more distantly related organisms—for example, groups of eubacteria. In contrast, the sequence of bases in ribosomal RNA reveals relationships among all organisms—eubacteria, archaebacteria, and eucaryotes. In fact, studies on the sequences of bases in rRNA first revealed that archaebacteria are a separate kingdom, as different from eubacteria as they are from eucaryotes.

How can ribosomal RNA show relationships over great evolutionary distances? During evolution, rRNA changes much more slowly than most macromolecules. Because the structure of the ribosome cannot tolerate much change and still remain functional, ribosomal RNA is highly **conserved**. Therefore, the rRNA of even distantly related organisms is similar enough to count the number of differences and thus calculate the evolutionary distance between them. There is no easy way to sequence RNA directly. First, the RNA is converted into DNA and the DNA is sequenced.

Over the past decade and a half, studies on the sequence of bases in rRNA have revolutionized our understanding of the relationships among microorganisms, as you will learn in the next three chapters.

The Sequence of Amino Acids in Proteins.   The relatedness of organisms can also be determined through the science of **molecular taxonomy**, which compares the sequence of monomers in macromolecules. When the field first emerged, more than 20 years ago, molecular taxonomists compared the sequence of amino acids in homologous proteins from different organisms. It is a technically complex and time-consuming task that diminished in use once it became possible to sequence DNA in 1977. Sequencing DNA is much easier and quicker, and it yields equivalent information. Knowing the genetic code and the sequence of bases in a segment of DNA means we can determine the sequence of amino acids in the protein it encodes (Chapter 6).

Escherichia coli's Family Tree

The unrooted **trees** that scientists have developed from sequencing rRNA tell us which organisms evolved most recently from a common ancestor, but they don't tell us when. Until recently, evolutionary events could be dated only by the fossil record. Fossils are dated by the age of the rocks they are found in, and evolutionary events are deduced by comparing fossils of different ages with organisms alive today.

But could the sequence of bases in rRNA tell us dates in evolutionary history? It could if evolutionary changes occur at a constant rate. Then changes in rRNA would be a molecular clock ticking off evolutionary time and recording it. To test the possibility that changes in rRNA occur at a constant rate and, if so, set the molecular clock, some key evolutionary events had to be dated. Some were dated by the fossil record, and other, older events by relating them to geological change. For example, the time that oxygen appeared in the earth's atmosphere (which is recorded by changes in rocks) must have been when aerobes evolved.

Applying these tests showed that the rate of change in the sequence of rRNA is an excellent molecular clock. By setting the molecular clock this way, base differences between a pair of organisms can be converted into years since they had a common ancestor. The method applies even to microorganisms, providing some fascinating insights. For example, it establishes that *Escherichia coli* separated from its closest relative, *Salmonella typhimurium*, about 120 million years ago, when mammals first appeared. In short, *E. coli* evolved with mammals. It is well-adapted to grow in the intestines of warm-blooded animals. It is resistant to bile, a fat-emulsifying liquid produced by the liver and toxic to many microorganisms. Finally, it can use lactose, a sugar produced only by mammals, as a carbon source in milk. The capacity to use lactose is relatively rare in bacteria, but in view of *E. coli's* evolutionary history, we should not be surprised.

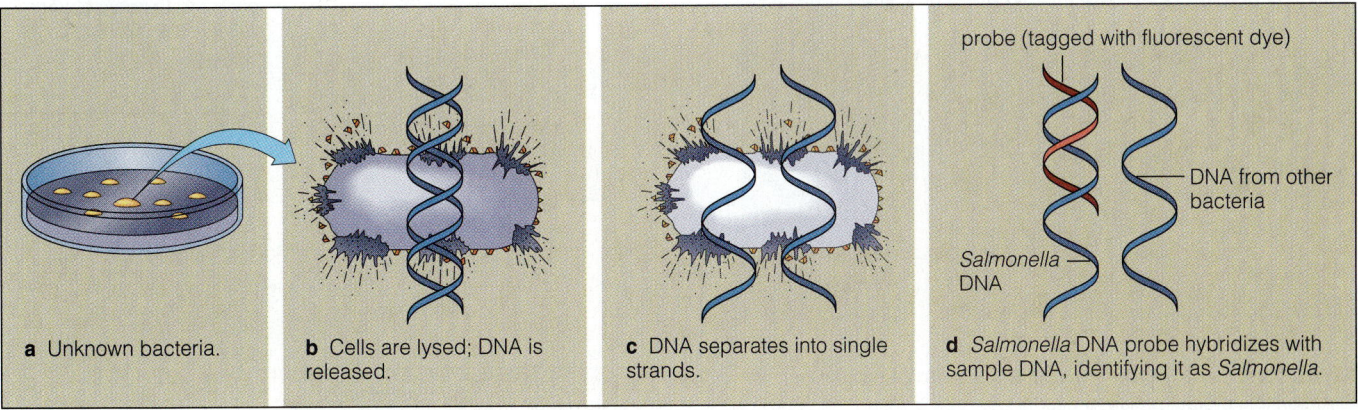

a Unknown bacteria.　**b** Cells are lysed; DNA is released.　**c** DNA separates into single strands.　**d** *Salmonella* DNA probe hybridizes with sample DNA, identifying it as *Salmonella*.

probe (tagged with fluorescent dye)

DNA from other bacteria

Salmonella DNA

Figure 10.10　Using a DNA probe to identify bacteria.

Protein Profiles.　A profile of the proteins that organisms produce tells their number, size, charge, and relative abundance. It therefore tells us about the differences among strains of the same species but reveals nothing about the relatedness of species and higher taxons.

Protein profiles are obtained by two-dimensional **polyacrylamide gel electrophoresis (PAGE)**, which separates proteins. First, cells are lysed to release the proteins. The solution of proteins is placed on a thin slab of polyacrylamide gel and electric current is applied. Because protein molecules are charged, they move with the current. But because different proteins have different charges and sizes, they move at different rates. After a few hours the gel is rotated 90 degrees so that current passes across it in a second dimension. When the current is stopped, the gel is stained. Staining reveals where various proteins have moved and allows them to be counted and analyzed.

With two-dimensional PAGE it is possible to separate virtually all of a cell's proteins. Stained protein spots corresponding to as many as 2000 proteins can be seen on a gel (**Figure 10.11**). The size of the spot reflects the

Looks Can Be Deceiving

The adage "Things are not always what they appear to be" certainly applies to *Pneumocystis carinii*—and in more ways than one. Until recently, this pathogen was known for causing mild, often asymptomatic respiratory infections (Chapter 22). But today it is known for the deadly pneumonia it causes in AIDS patients whose weakened immune systems cannot fight off *P. carinii* (Chapter 27).

From its morphology and life cycle, *P. carinii* appeared to be a protozoan closely related to *Plasmodium,* the protozoan that causes malaria (Chapter 27). Sequencing the bases in its rRNA revealed that the sequence is 90 percent identical to the sequence in certain fungi and only 40 percent identical to the sequence of *Plasmodium* and related protozoa. *P. carinii* is a fungus, not a protozoan (see Figure 22.16).

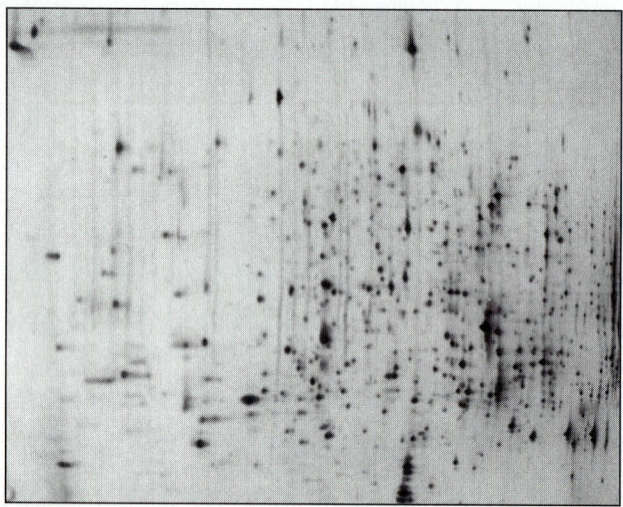

Figure 10.11 The two-dimensional PAGE technique was applied to obtain this protein profile of *Escherichia coli*. Each of the 2000-plus spots is a different protein. To compare strains, they must be grown under identical conditions, because growth in a different medium or at a different temperature changes the pattern.

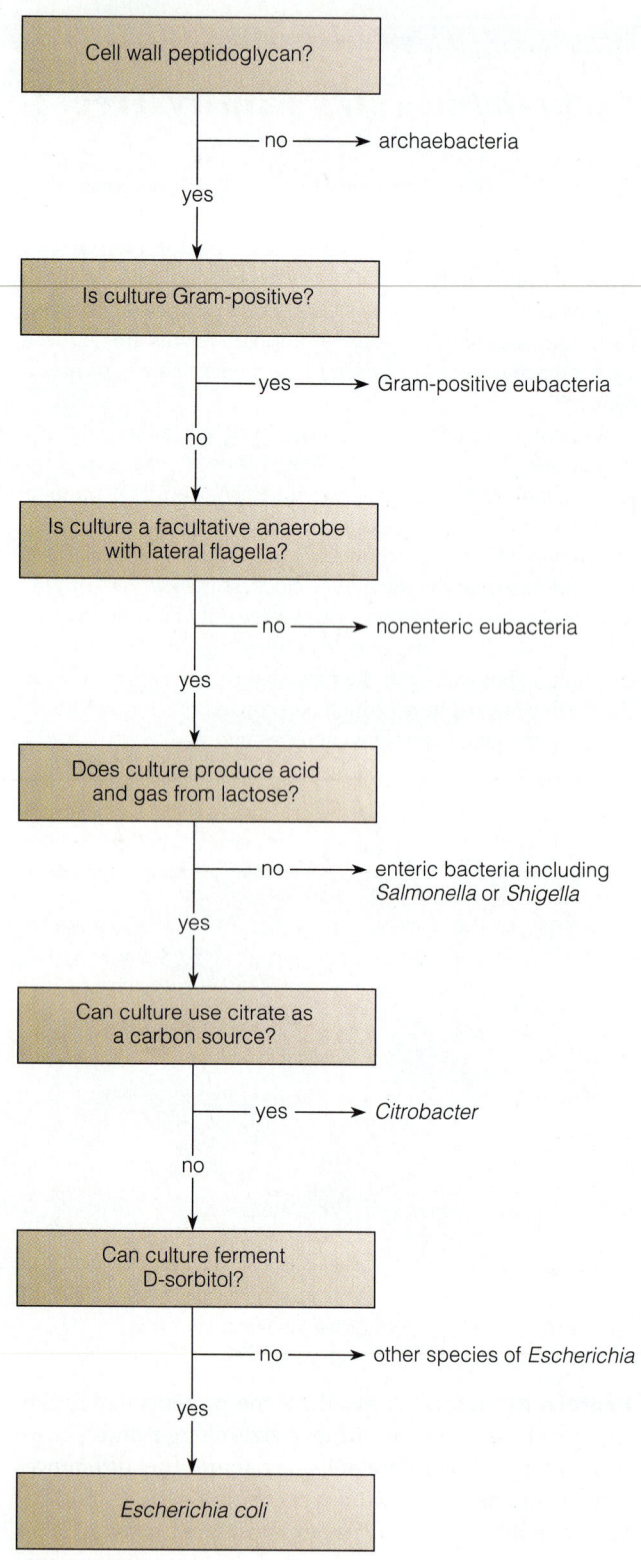

Figure 10.12 Dichotomous key. This dichotomous key divides procaryotes into smaller groups. The answers shown lead to identifying an unknown procaryote as *Escherichia coli.*

Kits Come to the Micro Lab

It has been said that taxonomy is an art, but identification is a science. Identification is also a rapidly evolving technology. Until about 15 years ago microbiologists made many kinds of media and used pipettes, inoculating loops, and racks of test tubes to identify microorganisms in clinical samples. Today microbiologists often use commercial kits.

Kits usually have small plastic chambers (in place of test tubes) filled with test media (eliminating most media preparation and pipetting). A system for simultaneous inoculation is provided (saving the time of multiple transfers with an inoculating needle). The test media contain indicators and sometimes films to collect gas. After inoculation and appropriate incubation (18 to 24 hours), the technician records color changes and gas accumulation and compares the results to charts. The charts relate patterns of results to specific microorganisms, thus completing the identification.

Some kits are designed to identify members of specific microbial groups, such as enteric bacteria (those related to *Escherichia coli*). One such kit provides a plastic tube divided into 12 chambers. All the chambers can be inoculated simultaneously from a single colony by touching the colony with a wire and drawing the wire up through the tube. Color changes and gas accumulation in the 12 chambers provide 15 test results, enough to identify most enteric bacteria. Color changes indicate whether the bacteria use certain substrates and produce specific metabolic products and enzymes. A wealth of information comes from one small plastic tube.

A technician observing results from a test kit and entering them into a computer.

Some kits are designed to give especially rapid results. They use heavier inocula and depend on the activity of enzymes already present in the cell mass. It's not necessary to wait overnight—let alone a longer period—for growth. Positive or negative tests for one particular organism are also commercially available. They are usually based on immunological or molecular reactions.

---

abundance of the protein, and its location on the gel identifies it. The two-dimensional PAGE technique can distinguish between strains differing by only a single protein.

Characters Used to Classify Viruses

The primary character used to classify viruses is the kind of nucleic acid that constitutes the virus genome—whether it is DNA or RNA, single-stranded or double-stranded. Other characters are appearance in an electron micrograph, whether the virus is surrounded by a membrane, host range, and serology. For example, the **rhabdoviruses**, which include the rabies virus, are composed of single-stranded RNA. They are bullet-shaped and surrounded by a membrane. They attack many animal species, and serum from an animal that survives an infection will protect other animals. Chapter 13 discusses the characters used to classify viruses in detail.

Dichotomous Keys

Taxonomists sometimes construct **dichotomous keys** to simplify identification of unknown organisms. Dichotomous keys consist of a series of questions about important characters that offer only two alternatives. They are arranged somewhat like a game of Twenty Questions. First, general questions are asked to divide organisms into two large groups. Then more specific questions divide organisms into smaller groups, and finally highly specific questions distinguish between similar species. **Figure 10.12** gives an example of a dichotomous key that divides procaryotes into smaller groups and eventually identifies *Escherichia coli*.

Dichotomous keys are helpful, but not perfect. Relying on only a small number of characters can lead to incorrect or equivocal conclusions because rarely is any single character the same in all members of a species. For example, the last step of the key to distinguish *E. coli* in

Figure 10.12 from other species of *Escherichia* depends on a character that is present in 94 percent of *E. coli* strains. In other words, 6 percent of the time the key will give an incorrect answer.

Summary

PRINCIPLES OF BIOLOGICAL CLASSIFICATION   (pp. 231–235)

1. Taxonomy is the science of classifying organisms. Biological taxonomists use a hierarchical scheme of classification: They collect individuals into groups and groups into progressively more inclusive and broader groups.

2. Biologists use the hierarchical classification scheme devised by Linnaeus: species, genera, families, classes, orders, phyla or divisions, and kingdoms. Kingdoms may be grouped into domains.

Scientific Nomenclature   (pp. 231–232)

1. Linnaean classification uses binomial nomenclature. The first name is the organism's genus, and the second is its specific epithet. Together the two constitute the species name.

Artificial and Natural Systems of Classification   (p. 232)

1. Linnaeus's system was an artificial system of classification because it grouped organisms on the basis of visible similarities. A natural scheme of classification is based on phylogenetic (evolutionary) relatedness.

The Fossil Record   (pp. 232–233)

1. Most natural systems of classification depend on information gained from the fossil record. Only recently have scientists found microbial fossils. Stromatolites are rocks composed of fossilized photosynthetic procaryotes that grew as masses of cells (microbial mats). This finding establishes that microorganisms have been on the earth about 3.5 billion years.

The Concept of Species   (pp. 233–235)

1. A eucaryotic species is a group of organisms that share and can exchange a common pool of genes. A bacterial species is defined by the similarities of its members.

2. *Bergey's Manual of Systematic Bacteriology* divides bacteria into four divisions on the basis of cell wall (Gram-negative or Gram-positive bacteria), lack of a wall (the mycoplasmas), and walls lacking peptidoglycan (archaebacteria). *Bergey's Manual* disregards evolutionary relationships in favor of practicality.

3. A microbial clone is a group of organisms derived from an individual cell by asexual reproduction. Clones that are presumed or known to be genetically different are called strains.

4. Viral species are grouped into genera and genera into families. No higher taxa are used. They are given ordinary English names. Viral species are also divided into strains.

MICROORGANISMS AND HIGHER LEVELS OF CLASSIFICATION   (pp. 235–236)

1. Initially, microorganisms were grouped with animals or plants. In 1866, Ernst Haeckel proposed three kingdoms—plants, animals, and microorganisms (or protists).

2. In 1937, Chatton divided all living things into eucaryotes and procaryotes.

3. In 1969, Whittaker proposed the five-kingdom scheme—animals, plants, fungi, protists, and monera.

4. In 1990, Woese, Kandler, and Wheelis proposed three domains—archaebacteria, eubacteria, and eucaryotes.

THE METHODS OF MICROBIAL CLASSIFICATION   (pp. 236–246)

Numerical Taxonomy   (pp. 236–237)

1. Numerical taxonomy gives equal weight to all characters (properties) of organisms. It aims to express the evolutionary distance between organisms in a number—the similarity coefficient (S_J).

2. Numerical taxonomists construct charts called dendograms to group organisms with the highest similarity coefficients.

Characters Used to Classify Bacteria   (pp. 237–245)

1. Morphology (the shape of bacterial cells) is a traditional character. The Gram stain divides bacteria into two groups, those with and those without an outer membrane.

2. Some of the biochemical and physiological characters used to classify bacteria are based on conditions that support growth.

3. Serology (the study of the properties of serum, the liquid component of blood) is another traditional method of classifying bacteria. Antibodies contained in serum are highly specific and thereby distinguish between closely related strains. Also, antisera (preparations from animals that inactivate certain bacteria) identify bacterial species.

4. Phage typing (determining which phages attack which bacteria) can be used to identify strains of bacteria.

5. The mole percent guanine plus cytosine (% G + C) expresses the percentage of total base pairs in DNA that are G-C, as opposed to A-T. Higher % G + C means a higher DNA melting point and density. Closely related organisms have similar % G + C values.

6. The ultimate tool of taxonomy is DNA sequencing—determining the sequence of bases in DNA. Identical sequences mean identical organisms, and similar sequences mean closely related organisms.

7. DNA hybridization involves melting DNA strands from separate sources, mixing them, and cooling so the two strands anneal, creating a hybrid. The extent of annealing indicates the extent of similarity in DNA bases. Sometimes a probe (a specific short piece of DNA) is used.

8. Sequencing bases in rRNA reveals relationships between all organisms. This is possible because during evolution rRNA changes much more slowly than DNA does.

9. It is also possible to determine relatedness between organisms by sequencing the amino acids in homologous proteins, but it is much easier and faster to sequence DNA.

10. Protein profiles tell the number, size, charge, and relative abundance of proteins that organisms produce. They are useful for distinguishing between strains of a species but do not distinguish between species or higher taxa.

Characters Used to Classify Viruses   (p. 245)

1. Viruses are classified according to the kind of nucleic acid that constitutes their genome—DNA or RNA, single- or double-stranded.

2. Viruses are also classified on the basis of morphology, host range, and serology.

Dichotomous Keys   (pp. 245–246)

1. Dichotomous keys are a series of questions about characters that offer only two alternatives.

2. Dichotomous keys can lead to incorrect conclusions because a single character is rarely the same in all members of a species.

Review Questions

PRINCIPLES OF BIOLOGICAL CLASSIFICATION

1. Define these terms: classification, hierarchical scheme of classification, taxonomy.

2. Arrange these taxa from the smallest group to the largest:
   a. order
   b. kingdom
   c. genus
   d. phylum or division
   e. family
   f. species
   g. class
   h. domain

3. Explain the scientific nomenclature of a species name. Give an example.

4. What is the difference between an artificial and a natural system of biological classification? What is the fossil record? What is the significance of stromatolites?

5. How are eucaryotic species defined? How are bacterial species defined? What is the basis for classification in *Bergey's Manual*?

6. What is the difference between a clone and a strain?

7. On what basis are viral species defined?

MICROORGANISMS AND HIGHER LEVELS OF CLASSIFICATION

1. Explain this statement: Biological classification schemes change as we get new information. Use schemes developed by Haeckel, Chatton, Whittaker, and Woese/Kandler/Wheelis as examples.

2. Why are viruses not included in any biological classification schemes?

3. Why isn't *Bergey's Manual* organized according to the most up-to-date classification scheme?

THE METHODS OF MICROBIAL CLASSIFICATION

1. How is an organism classified according to numerical taxonomy?

2. Explain how each of these traditional characters is used to classify bacteria:
   a. morphology
   b. biochemistry and physiology
   c. serology
   d. phage typing

3. Explain how each of these characters is used to classify bacteria:
   a. % G + C
   b. sequencing DNA bases
   c. DNA hybridization
   d. sequencing bases in rRNA
   e. protein profiles

4. Describe the characters used to identify viruses.

5. What is a dichotomous key? What are its advantages and disadvantages?

Suggested Readings

Margulis, L., and Schwartz, K. V. 1988. *Five kingdoms: An illustrated guide to the phyla of life on earth.* San Francisco: W. H. Freeman.

Sneath, P. H. A., and Sokal, R. R. 1973. *Numerical taxonomy: The principles and practice of numerical classification.* San Francisco: W. H. Freeman.

Woese, C. R. 1987. Bacterial evolution. *Microbiological Reviews* 51: 221–71.

Woese, C. R.; Kandler, O.; and Wheelis, M. L. 1990. Towards a natural system of organisms: Proposal for the domains archaea, bacteria, and eucarya. *Proceedings of the National Academy of Science* 87: 4576–79.

11 THE BACTERIA

An Exception Proves the Rule

Many generalizations—including "bacteria are small"—may have more exceptions than we know. We tend to accept generalizations and not look beyond them. For example, when we see a large unknown microbial cell we automatically assume it isn't a bacterium. But today's new technology is making microbiologists question old generalizations. In one new procedure a cell can be identified as bacterial or not with fluorescent-tagged DNA probes that hybridize selectively to rRNA-encoding segments of procaryotic DNA. Bacterial cells glow brightly under ultraviolet light.

One of the biggest surprises from this new identification procedure is the discovery that not all bacteria are small. In fact, some are quite large. The current record holder was found in 1992 in the intestines of a surgeon fish caught off the Great Barrier Reef in Australia. The new bacterium has been tentatively named *Epulopiscium* (meaning "guest at a banquet of a fish"). The largest specimens of *Epulopiscium* are about 0.57 mm long and 0.06 mm thick, easily visible to the naked eye. In fact, *Epulopiscium* is so large that more than a million ordinary-size bacteria could fit inside.

Aside from its enormous size, *Epulopiscium* is a typical bacterium. It has no membrane-bound nucleus or organelles. Its rRNA sequence relates it closely to the Gram-positive genus *Clostridium*. *Epulopiscium* is a symbiont that digests the algae on which its host fish grazes. In other words, it plays the same role in surgeon fish that other bacteria play in a cow's rumen (Chapter 8). Microscopic identification of bacteria using DNA probes will no doubt produce other surprises.

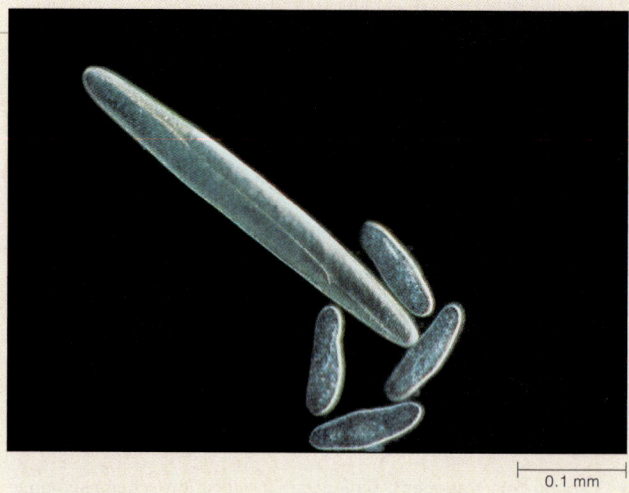

0.1 mm

The bacterium *Epulopiscium* is pictured with four smaller cells of the protozoan *Paramecium* to show its relative size.

To understand:
- The *Bergey's Manual* scheme of classifying bacteria and the significance of sections in *Bergey's Manual*

- The relation between the *Bergey's* scheme and natural schemes of classifying bacteria

To become familiar with:
- The better-known bacterial genera and species in *Bergey's Manual*

- The properties and activities of these organisms

- The impact of these organisms on our lives and our environment

IDENTIFYING BACTERIA

Some bacteria have distinctive cell shapes, but many look alike under the microscope. Consequently, in identifying bacteria, microbiologists begin with morphological characters but rely primarily on physiological characters. Physiological characters include conditions that support growth, metabolic end products, how growth and metabolism change the environment, the ability to grow on a selective medium, and appearance of a colony on a particular medium.

Usually, a number of properties must be examined to identify a genus or species. But that does not mean all properties must be examined every time a bacterial sample is presented for identification. Simply knowing the source of the bacterium narrows the field to a few species. Moreover, commercial kits and DNA probes make identification of some species fast and easy (Chapter 10).

BACTERIAL TAXONOMY

Bacterial taxonomists today find themselves in an awkward position. Sophisticated new molecular biology techniques have revealed a great deal about the evolutionary relatedness of bacteria, which means it's possible to devise natural schemes of bacterial taxonomy. But natural schemes group bacteria into assemblages that cannot be easily identified by standard laboratory procedures. As a result, the more practical artificial schemes are still generally used in bacterial taxonomy.

The most widely used artificial scheme is published under the auspices of the American Society for Microbiology (ASM). Soon after the turn of the century, the ASM realized that bacteriologists needed an authoritative manual to help them identify bacterial species. A committee chaired by bacteriologist David H. Bergey undertook the task and published the first edition of *Bergey's Manual of Determinative Bacteriology* in 1923. After the third edition was published in 1930, the ASM set up a nonprofit trust funded by royalties for revision and publication of the manual.

With successive editions, *Bergey's Manual* grew in scope. To reflect the changes, in 1984 the trust changed the name to *Bergey's Manual of Systematic Bacteriology* (first edition). The new *Bergey's Manual* covers a bacterium's physiology, ecology, cultivation, and preservation as well as its identification and classification (which is what *determinative* indicates). It is published in four volumes. The first was published in 1984, the second in 1986, and the third and fourth in 1989. In 1994, the trust published a brief single volume, the ninth edition of *Bergey's Manual of Determinative Bacteriology*, returning to the original title because the summary follows the earlier style.

Bergey's Manual takes a practical approach to classifying bacteria, which means it sometimes ignores known evolutionary relationships. For example, it groups together all phototrophic bacteria, even though molecular studies on their ribosomal RNA show that members of this group are distantly related evolutionarily. Similarly, the *Manual* considers all bacteria a single kingdom, including the archaebacteria, although the archaebacteria are evolutionarily as distant from the eubacteria as they are from eucaryotes (Chapter 10).

In this chapter we consider some clinically and ecologically important groups of bacteria. We also cover groups of bacteria that have unusual properties and are not discussed elsewhere in the text. We use *Bergey's Manual* as a framework for our discussion.

THE *BERGEY'S MANUAL* SCHEME OF BACTERIAL TAXONOMY

Bergey's Manual divides bacteria into four divisions based on cell wall composition. The Gram-negative bacteria have a thin peptidoglycan wall. The Gram-positive bacteria have a thick peptidoglycan wall. The mycoplasmas lack a rigid cell wall. The archaebacteria have walls composed of materials other than peptidoglycan. Some archaebacteria have walls composed of a peptidoglycan-like material that lacks muramic acid (NAM), while others have walls made of protein.

The main feature of *Bergey's Manual* is its highly pragmatic subdivision called a section. Bacterial species in each division are assigned to one or, in a few cases, two of 33 sections. For example, *Halobacterium*, a halophilic archaebacterium, appears in Sections 4 and 25 because halophilic (salt-loving) bacteria are considered in Section 4 and archaebacteria are considered in Section 25. Sections are based on a few readily identifiable properties, usually indicated in a nonlatinized name. For example, Section 13 is "Endospore-forming Gram-Positive Rods and Cocci." **Table 11.1** summarizes the sections in *Bergey's Manual*.[1]

Sections have no taxonomic standing—they do not fit into a hierarchical scheme of classification. They are simply groups of organisms that share certain easily identifiable properties. Sections cut across usual taxonomic boundaries. For example, Section 10 (Mycoplasmas) contains a complete division. However, Section 1 (Spirochetes) contains only one order and Section 8 (Anaerobic Gram-Negative Cocci) contains only a family. Some sections contain more than one order or family, while some contain only a few genera.

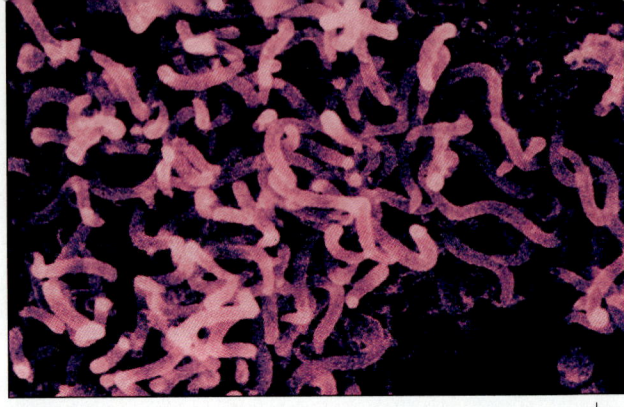

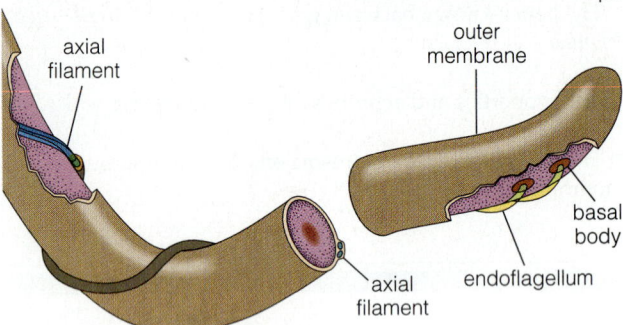

Figure 11.1 The spirochetes are distinctive for their shape and means of motility.

The Spirochetes (Section 1)

The spirochetes are Gram-negative bacteria distinguished by their unusual shape and unusual mechanism for motility. Spirochetes are comparatively long and helical—they look like a corkscrew or coil (**Figure 11.1**). They have typical bacterial flagella, but these are uniquely arranged. Unlike other bacterial flagella, which extend from the cell body, spirochete flagella, called **endoflagella**, lie within the periplasm, the region between the cell wall and the outer membrane. (The outer membrane of spirochetes is sometimes called the **outer sheath**, an older term that persists.) Endoflagella are bundled into a structure called an axial filament. Spirochetes have two axial filaments. One originates at each pole, or end of the cell, and wraps helically around the **protoplasmic cylinder**, or cell body, to meet or overlap the other axial filament at the midpoint of the cell. The number of endoflagella in an axial filament varies considerably—from 2 to more than 200.

The axial filament accounts for the spirochetes' unusual motility. As the endoflagella turn, they cause the helical ridge they form on the outer membrane to move down the cell and propel it through the medium. Spirochetes move so rapidly that when viewed under the microscope they seem to disappear from one place and suddenly reappear in another. They are motile in viscous environments that immobilize ordinary flagellated bacteria. Thus, spirochetes are often found in thick muds and on mucous membranes of animals. They are also able to move on the surface of solid media. Presumably by chang-

ing the direction of rotation of their flagella, they twitch and writhe, moving like an inchworm along a surface.

Spirochetes are metabolically diverse. Some are aerobes, others are facultative anaerobes, and still others are anaerobes. They divide by binary fission. The process begins when two new axial filaments appear near the midpoint of the cell. Then, cell division occurs between them, producing two daughter cells, each with one old and one new axial filament.

Some spirochetes live harmlessly in our mouths (Chapter 14), but others cause devastating diseases. *Treponema pallidum* causes the sexually transmissible disease syphilis (Chapter 24), *Leptospira interrogans* causes leptospirosis, a kidney infection of animals and humans who work with animals (Chapter 24), and *Borrelia* spp. cause relapsing fever, a debilitating systemic infection (Chapter 27). One species of *Borrelia*, *B. burgdorferi*, causes Lyme disease, a systemic disease that was first recognized in the 1970s (Chapter 27).

Aerobic/Microaerophilic, Motile, Helical/Vibrioid Gram-Negative Bacteria (Section 2)

All members of Section 2 are either aerobic or microaerophilic (grow best at an oxygen concentration lower than that of air) and either helical or vibrioid. The helical members are corkscrew-shaped like the spirochetes, but their flagella are ordinary, extending from one or both poles. The vibrioid members are comma-shaped.

[1]The spellings of *archaebacteria*, *rickettsiae*, and *chlamydiae* used in this book don't agree with the spellings in *Bergey's*. In this chapter, however, we follow *Bergey's* when referring to its section titles.

Table 11.1 Sections in *Bergey's Manual of Systematic Bacteriology*

Section	Bacterial Group (Group Properties)	Representative Genera/Species	Features of Species (Chapter)
1*	The spirochetes (Gram-negative spiral-shaped cells with endoflagella)	*Treponema pallidum*	Causes syphilis (24)
		Leptospira interrogans	Causes leptospirosis (24)
		Borrelia burgdorferi	Causes Lyme disease (27)
		Borrelia spp.	Cause relapsing fever (27)
2*	Aerobic/microaerophilic, motile, helical/vibrioid Gram-negative bacteria	*Campylobacter jejuni*	Causes gastroenteritis (23)
		Azospirillum spp.	Fix atmospheric nitrogen for grasses (28)
		Bdellovibrio bacteriovorus	Predator that consumes other bacteria
		Aquaspirillum magnetotacticum	Contains magnetosomes
3	Nonmotile (or rarely motile) Gram-negative curved bacteria	*Brachyarcus* spp.	Arc-shaped cells found in lakes; cannot be cultivated
		Pelosigma spp.	Bundles of S-shaped cells found in lakes; cannot be cultivated
4*	Gram-negative aerobic rods and cocci	*Bordetella pertussis*	Causes whooping cough (22)
		Brucella abortus	Causes brucellosis (27)
		Francisella tularensis	Causes tularemia (27)
		Legionella pneumophila	Causes legionellosis (22)
		Neisseria meningitidis	Causes meningococcal meningitis (25)
		N. gonorrhoeae	Causes gonorrhea (24)
		Xanthomonas spp.	Cause plant diseases
		Acetobacter spp.	Make acetic acid
		Agrobacterium spp.	Cause tumors on plants used in biotechnology
		Rhizobium and *Bradyrhizobium* spp.	Symbiotic nitrogen-fixers
		Azotobacter and *Beijerinckia* spp.	Free-living nitrogen-fixers
		Pseudomonas spp.	Abundant and versatile oxidizers of organic compounds
		P. aeruginosa	Important opportunistic pathogen (26)
		Halobacterium spp.	See Section 25
5*	Facultatively anaerobic Gram-negative rods	*Salmonella typhi*	Causes typhoid fever (23)
		Shigella spp.	Cause shigellosis (23)
		Vibrio cholerae	Causes cholera (23)
		Yersinia pestis	Causes plague (27)
		Erwinia spp.	Cause plant diseases
		Escherichia coli	Metabolic paradigm (5), some strains cause diarrhea (23)
		Photobacterium spp.	Luminescent
		Haemophilus spp.	Cause serious infections of the eye, brain, and meninges (25)
		H. influenzae	Causes upper respiratory infections (22)
		Zymomonas spp.	Ferment sugars, forming alcohol
		Klebsiella spp.	Cause pneumonia (22)
		Pasteurella multocida	Causes fowl cholera
		Gardnerella vaginalis	Causes vaginitis (24)
6*	Anaerobic Gram-negative straight, curved, and helical rods	*Bacteroides* spp.	Abundant in human intestines
		Fusobacterium spp.	Cause dental abscesses, infections of head and neck
		Selenomonas spp.	Abundant in the rumen
		Wolinella spp.	Abundant in the rumen
7*	Dissimilatory sulfate- or sulfur-reducing bacteria	*Desulfovibrio* spp.	Produce hydrogen sulfide in damp soil, corrode iron in buried pipe
8*	Anaerobic Gram-negative cocci	*Veillonella* spp.	Abundant in the rumen; contribute to periodontal disease (23)

Note: An asterisk (*) indicates sections discussed in this chapter.

Table 11.1 (continued)

Section	Bacterial Group (Group Properties)	Representative Genera/Species	Features of Species (Chapter)
9*	The rickettsias and chlamydias (tiny, Gram-negative, intracellular parasites)	*Rickettsia rickettsii*	Causes Rocky Mountain spotted fever (27)
		R. prowazekii	Causes epidemic (louseborne) typhus (27)
		R. typhi	Causes endemic (fleaborne) typhus (27)
		Coxiella burnetii	Causes Q fever (22)
		Chlamydia trachomatis	Causes trachoma (26) and NGU (24)
		C. psittaci	Causes psittacosis (22)
10*	Mycoplasmas (lack defined cell wall; all are parasites of plants, animals, or humans)	*Mycoplasma pneumoniae*	Causes primary atypical pneumonia (22)
		Ureaplasma urealyticum	Causes NGU (24)
		Thermoplasma spp.	See Section 25
11	Endosymbionts (bacteria that live inside cells of other organisms)	*Holospora* spp.	Rod-shaped cells found in nucleus of proto-zoa; cannot be cultivated
		Blattabacterium spp.	Live in cells of insects; cannot be cultivated
12*	Gram-positive cocci	*Sarcina* spp.	Anaerobes; cells arranged in cubes
		Staphylococcus spp.	Facultative anaerobes; cells arranged in grapelike clusters
		S. aureus	Causes impetigo, boils, abscesses (26), toxic shock syndrome (24)
		Streptococcus pneumoniae	Causes pneumococcal pneumonia (22)
		S. mutans	Causes dental caries (23)
		S. pyogenes	Causes strep throat (22), endocarditis (27), and other life-threatening infections
		Micrococcus spp.	Aerobes; live on skin as part of human flora
		Streptococcus, Leuconostoc, Pediococcus spp.	Lactic acid bacteria; used to preserve food and fodder
13*	Endospore-forming Gram-positive rods and cocci	*Bacillus* spp.	Aerobic rods
		B. thuringiensis	Produces proteins that kill caterpillars and mosquitoes
		B. anthracis	Causes anthrax (27)
		Clostridium spp.	Anaerobic rods
		C. tetani	Causes tetanus (25)
		C. botulinum	Causes botulism (25)
		C. perfringens	Causes gas gangrene (26)
14*	Regular nonsporing, Gram-positive rods	*Caryophanon latum*	Very large bacterium
		Lactobacillus spp.	Lactic acid bacteria
		Erysipelothrix rhusiopathiae	Animal pathogen; causes skin lesions in humans
		Listeria monocytogenes	Causes listeriosis (24)
15*	Irregular nonsporing, Gram-positive rods	*Arthrobacter* spp.	Abundant bacteria in soil
		Propionibacterium spp.	Ripen Swiss cheese
		P. acnes	Associated with acne (26)
		Bifidobacterium spp.	Abundant in intestines of breastfed babies (14)
		Corynebacterium diphtheriae	Causes diphtheria (22)
		Actinomyces israelii	Causes abscesses (26)
		Gardnerella vaginalis	See Section 5
16*	Mycobacteria	*Mycobacterium leprae*	Causes leprosy (26)
		M. tuberculosis	Causes tuberculosis (22)
17*	Nocardioforms (acid-fast bacteria)	*Nocardia* spp.	Many species abundant in soil; some pathogenic for humans
18*	Anoxygenic phototrophic bacteria	*Rhodobacter* spp.	Well-studied genus of photoheterotrophic bacteria
		Chlorobium spp.	Well-studied genus of photoautotrophic bacteria
19*	Oxygenic photosynthetic bacteria (cyanobacteria)	*Anabaena* spp.	Well-studied genus of nitrogen-fixing cyanobacteria
		A. azolla	Used with azolla ferns to increase rice yields

Table 11.1 (continued)

Section	Bacterial Group (Group Properties)	Representative Genera/Species	Features of Species (Chapter)
20*	Aerobic chemolithotrophic bacteria and associated organisms	*Nitrosomonas, Nitrosococcus, Nitrosolobus* spp.	Oxidize ammonia to nitrite ion
		Nitrobacter, Nitrospina, Nitrococcus, Nitrospira spp.	Oxidize nitrite ion to nitrate ion
		Thiobacillus spp.	Oxidize reduced sulfur compounds to sulfuric acid
		Hydrogenobacter spp.	Oxidize hydrogen to water
21*	Budding and/or appendaged bacteria	*Caulobacter* spp.	Paradigm for study of bacterial differentiation
		Hyphomicrobium spp.	Prosthecate bacteria that multiply by budding
22*	Sheathed bacteria	*Sphaerotilus* spp.	Common inhabitants of polluted streams and sewage treatment plants
23*	Nonphotosynthetic, nonfruiting gliding bacteria	*Beggiatoa* spp.	Lithotrophic oxidizers of hydrogen sulfide
		Cytophaga spp.	Metabolize cellulose
		Lysobacter spp.	Prey on algae and cyanobacteria
24*	Gliding fruiting bacteria	*Myxococcus* spp.	Form small domelike fruiting bodies
		Stigmatella spp.	Form elaborate treelike fruiting bodies
25*	Archaeobacteria	*Methanococcus* spp.	Methanogens
		Halobacterium halobium	Halophile that carries out primitive photosynthesis
		Sulfolobus spp.	Oddly shaped thermoacidophiles
		Thermoplasma spp.	Wall-less thermoacidophiles (see Section 10)
26	Nocardioform actinomycetes	*Nocardia asteroides*	Causes a rare lung infection
27*	Actinomycetes with multilocular sporangia	*Frankia* spp.	Fix nitrogen in association with roots of certain trees
		Geodermatophilus spp.	Soil organisms
		Dermatophilus spp.	Cause skin infections (26)
28	Actinoplanetes	*Actinoplanes* spp.	Filamentous organisms which produce sporangia that release motile spores
		Micromonospora spp.	Filamentous organisms which produce single nonmotile spores
29*	*Streptomyces* and related genera	*Streptomyces* spp.	Abundant in soil; source of most antibiotics
30	Maduromycetes	*Actinomadura madurae*	Common in soil; can cause mycetoma, a tumorlike swelling
31	*Thermomonospora* and related genera	*Thermomonospora* spp.	Thermophilic actinomycetes abundant in compost and decaying vegetable material
32	Thermoactinomycetes	*Thermoactinomyces* spp.	Thermophilic actinomycetes; can cause allergic alveolitis
33	Other genera (unknown relationships to other bacteria)	*Pasteuria* spp.	Oval-shaped cells in aquatic environments, often joined by their polar holdfasts forming rosettes

This group contains several interesting and important organisms. *Campylobacter jejuni* is a major cause of diarrheal illness in the United States (Chapter 23). A related species, *C. fetus*, causes a sporadic abortion in cattle and sheep, but rarely infects humans. *Campylobacter* spp. are microaerophilic and curved, spiral, or S-shaped. The closely related genus *Helicobacter* contains a species, *H. pylori*, that causes gastric ulcers in humans.

Azospirillum species, which are aerobic and helical, are important to maintaining the fertility of tropic soils. They live in close association with roots of grasses, fixing nitrogen from the atmosphere and making it available to the plants (Chapter 28). One species from a closely related genus of microaerophilic vibrioid bacteria, *Aquaspirillum magnetotacticum*, contains magnetosomes (Chapter 4), inclusions that allow the bacteria to follow magnetic lines of force toward the bottoms of bodies of water, their optimum environment. They are found in both fresh and salt water.

An aerobic vibrioid member of the group, *Bdellovibrio bacteriovorus*, preys on other bacteria. *B. bacteriovorus* is so small (0.3 μm × 1.4 μm) and fast it is easily missed in the field of a light microscope (see the beginning of Chapter 3). *Bdellovibrio* move about 100 cell lengths per second, equivalent to a person moving at about 400 miles per hour. When a *Bdellovibrio* strikes its bacterial prey, it bores through the outer membrane and establishes itself in the host's periplasm, where it grows in length, using host cytoplasm for nutrients (**Figure 11.2**). Three to four hours later, the long cell divides and the host cell lyses, releasing three or four *Bdellovibrio* cells.

Strains of *B. bacteriovorus* are **host-specific**; each attacks some species of Gram-negative bacteria but not others. Gram-positive bacteria are rarely attacked. Various strains are widespread in soil, fresh water, and marine environments, indicating that they consume large numbers of bacteria.

Gram-Negative Aerobic Rods and Cocci (Section 4)

The large and diverse group of bacteria in Section 4 is undistinguished morphologically—most are ordinary-appearing rods and cocci. But the group includes species that greatly affect human health and our environment.

Among the human pathogens from this group are *Bordetella pertussis*, which causes the childhood disease whooping cough (Chapter 22), *Francisella tularensis*, which causes tularemia, a systemic infection spread primarily by infected rabbits (Chapter 27), *Legionella pneu-*

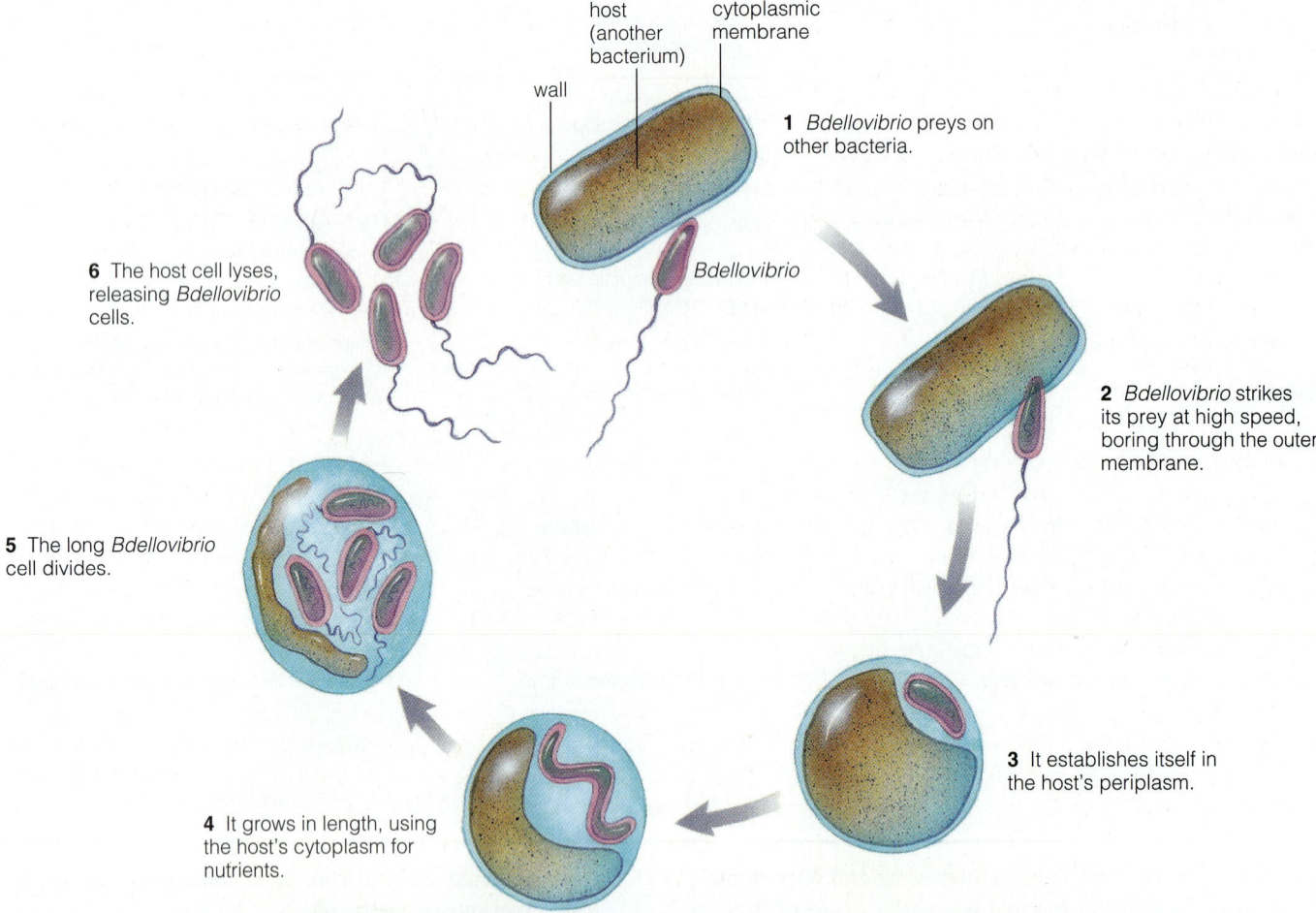

1 *Bdellovibrio* preys on other bacteria.

2 *Bdellovibrio* strikes its prey at high speed, boring through the outer membrane.

3 It establishes itself in the host's periplasm.

4 It grows in length, using the host's cytoplasm for nutrients.

5 The long *Bdellovibrio* cell divides.

6 The host cell lyses, releasing *Bdellovibrio* cells.

host (another bacterium)

cytoplasmic membrane

wall

Bdellovibrio

Figure 11.2 Life cycle of *Bdellovibrio* (Section 2). *Bdellovibrio* preys on other bacteria, usually Gram-negatives.

mophila, which causes legionellosis, a pneumonia (Chapter 22), *Neisseria gonorrhoeae*, which causes the sexually transmissible disease gonorrhea (Chapter 24), and *Neisseria meningitidis*, which causes meningococcal meningitis, an infection of the meninges (lining) of the brain and spinal cord (Chapter 25).

Section 4 also includes commercially important genera. The genus *Xanthomonas* includes one species cultivated for its capsule, which contains **xanthan gum**, a thickener for foods and paints. Many labels on foods such as salad dressing, cottage cheese, and yogurt list xanthan gum as an ingredient. Members of the genus *Acetobacter* produce **acetic acid**, the acid component of vinegar. Only microbially produced vinegar can be sold legally in the United States and many other countries. However, some species of *Acetobacter* are a bane of the food industry. They cause beer and wine to sour when they convert alcohol to acetic acid. Because *Acetobacter* spp. are aerobes, beer and wine are always protected from air.

Members of another genus, *Agrobacterium*, cause cancerlike diseases of plants. One of these, **crown gall**, is caused by *A. tumefaciens* (**Figure 11.3**). *A. tumefaciens* enters the plant through a wound at the crown (where the root and stem meet). It carries genes called **T-DNA** on a plasmid called the **Ti**, or **tumor-inducing**, **plasmid**. In the plant, these genes leave the plasmid to become incorporated into a plant cell's chromosome, transforming it into a tumor cell. Then the transformed plant cells grow and divide, eventually producing a tumorlike growth, or gall, at the crown of the plant. Inside this gall, the bacteria flourish because the transformed cells, unlike normal plant cells, produce **opines**. Opines are unusual amino acids that *A. tumefaciens* uses as nutrients. Thus, genes from *A. tumefaciens* direct the plant to make the nutrients the bacterium needs.

Plant molecular biologists are exploring practical uses of the Ti plasmid's ability to direct the transfer of some of its genes into the genome of a plant. Using recombinant DNA technology, almost any gene can be placed within the T-DNA region of the Ti plasmid and can therefore enter the plant genome along with the T-DNA. The product is a **transgenic plant**, one that contains genes from another organism. The process holds great promise for improving plant varieties and possibly for using plants to make valuable products. For example, genes from humans or animals could enable plants to make medically useful proteins.

Two other genera in this group are important in the earth's nitrogen cycle (Chapter 28). *Rhizobium* and *Bradyrhizobium* are symbiotic nitrogen-fixing bacteria. They enter the roots of legumes such as beans, peas, clover, and alfalfa and produce tumorlike nodules where nitrogen is fixed (reduced from atmospheric nitrogen gas to ammonia). The relationship between bacterium and plant is

Figure 11.3 The genus *Agrobacterium* (Section 4) causes cancerlike growths called galls. Inside this crown gall on an oak tree, bacteria flourish.

symbiotic: the plant supplies nutrients and favorable conditions of low-oxygen supply; the bacterium supplies the plant with usable nitrogen.

This group also contains two genera, *Azotobacter* and *Beijerinckia*, with species that can fix nitrogen in a free-living state—not associated with plants. Nitrogen fixation in global ecology is discussed in Chapter 28.

Other species in this section that have great impact on soil ecology are members of the genus *Pseudomonas*. *Pseudomonas* is a huge genus of species that are not closely related. All pseudomonads, however, are aerobic rods with polar flagella and the ability to metabolize many carbon sources. It is their ability to break down many organic compounds into carbon dioxide and water that makes the pseudomonads interesting to environmentally conscious industries today. For example, species that degrade petroleum have been used to clean up oil spills. *Pseudomonas aeruginosa* is a common cause of infections in weakened hosts, especially burn victims and cystic fibrosis patients (Chapter 26). Some pseudomonads produce fluorescent, water-soluble pigments that are distinctive enough to identify colonies on sight. *P. aeruginosa* produces a blue-green pigment, and related species produce green or gold pigments (**Figure 11.4**).

Facultatively Anaerobic Gram-Negative Rods (Section 5)

The bacteria in Section 5 are grouped into three families. Their members have considerable impact on humans—

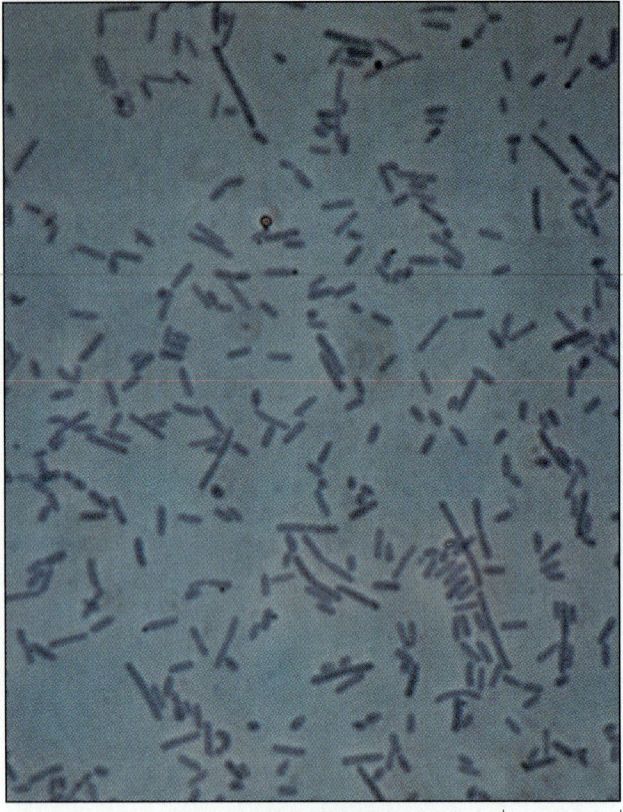

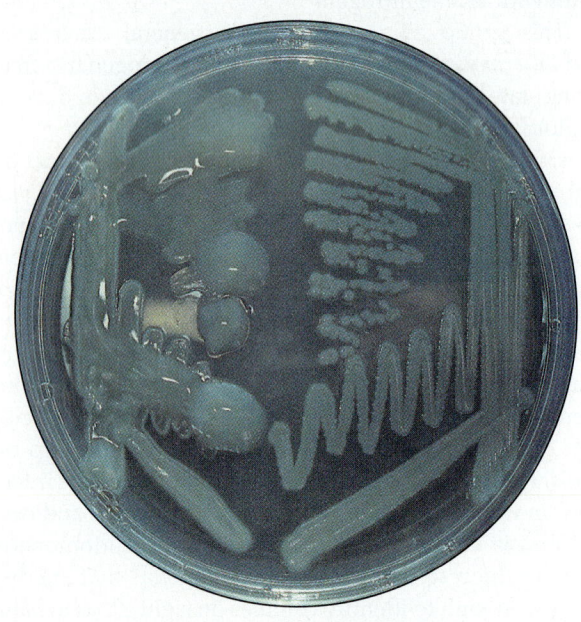

Figure 11.4 *Pseudomonas aeruginosa* (Section 4), which causes deadly infections in burn victims, has rod-shaped cells and polar flagella, like other pseudomonads. Colonies of *P. aeruginosa* are distinguished by their distinctive blue-green color. The strain on the left (derived from the parent strain on the right) is a mutant that makes large quantities of extracellular slime. Such slime-producing strains frequently infect lungs of persons who have cystic fibrosis.

the enterics (Enterobacteriaceae), the vibrios (Vibrionaceae), and the pasteurellas (Pasteurellaceae).

The Enterics (Enterobacteriaceae). The **enterics** include human pathogens, most of which infect the digestive system. *Salmonella typhi*, which causes typhoid fever, several species of the genus *Shigella*, which causes shigellosis, a form of dysentery, and *Vibrio cholerae*, which causes cholera, are all discussed in Chapter 23. The enteric family also includes *Yersinia pestis*, which causes bubonic plague (Chapter 27), and *Escherichia coli*, which has fostered research in all the life sciences and is now the centerpiece of recombinant DNA technology.

Enteric means "intestinal," but not all members of the enteric group are found in digestive tracts of animals. The enterics are linked by similar metabolism, not habitat. Some, including members of the genus *Erwinia*, cause diseases of plants. *E. amylovora*, for example, causes **fire blight** of pears and related fruits. Diseased plant leaves look as if they have been burned. *E. carotovora* causes **soft rots** of many plants. Solid carrot roots, for example, turn to mush.

Some members of the enteric group are motile by means of flagella, and some are nonmotile. All are rod-shaped. As facultative anaerobes, enteric bacteria metabolize sugars by fermentation when oxygen is not available. Two different kinds of complex fermentations, mixed-acid fermentation and butanediol fermentation, are typical of enterics. **Mixed-acid fermentation** produces a mixture of relatively strong organic acids. Products of **butanediol fermentation** are less acidic; and butanediol, a four-carbon dihydroxy alcohol, is always formed. The ability of different enteric species to carry out one or the other of these fermentations is one criterion for subdividing the group (**Table 11.2**).

It is fairly easy to determine what kind of fermentation an unknown enteric organism can perform. Though fermentation occurs only under anaerobic conditions, the interior of a bacterial colony growing on the surface of a petri plate or liquid at the bottom of an unshaken test tube is anaerobic enough to stimulate a shift to fermentative metabolism.

Three key features differentiate mixed-acid fermentation from butanediol fermentation. Mixed-acid fermentation produces large quantities of acid. Also, some mixed-acid fermenters can convert formic acid into CO_2 and H_2, producing gas, often in easily detectable quantities. Finally, butanediol fermentation produces butanediol and small amounts of acid. Mixed-acid fermentation can be detected by including an acid-base indicator—for example, neutral red—in either an agar or liquid medium. The indicator changes color if the test organism carries out mixed-acid fermentation but not if it carries out butanediol fermentation. Gas production can be detected visibly by trapping

Table 11.2	Fermentation Patterns of Key Genera of Enteric Bacteria (Section 5)			
	Mixed-Acid Fermentation		Butanediol Fermentation	
Produce H_2 and CO_2	No Gas Produced	Produce H_2 and CO_2	Produce only CO_2
---	---	---	---
Escherichia	Shigella	Enterobacter	Serratia
Proteus	Salmonella typhi		Erwinia
Salmonella (most spp.)	Yersinia		

gas bubbles within the medium. Butanediol fermentation can be detected by a simple chemical test for the presence of acetoin, the immediate precursor to butanediol, in the growth medium.

Besides carrying out mixed-acid fermentation, *Escherichia coli* exhibits three other distinctive properties. It is able to ferment lactose (milk sugar). It is unable to use citric acid as a carbon source, and it converts the amino acid tryptophan to indole, the compound that gives cultures of *E. coli* and, therefore, feces their characteristic odor. These characteristics are the basis for the tests used to distinguish *E. coli* from other enteric bacteria (**Table 11.3**, **Figure 11.5**). Such tests are commonly performed to determine if a water supply has been contaminated by sewage (Chapter 28).

The Vibrios (Vibrionaceae). The **vibrios** are curved rods. Most are motile by polar flagella and produce hydrogen and CO_2 as products of fermentation. The family includes *Vibrio cholerae*, the species that causes cholera, a deadly disease throughout the world and now at epidemic levels in the Western Hemisphere (Chapter 23).

The genera *Vibrio* and *Photobacterium* both have **luminescent** species, meaning they give off light as a metabolic product. Most, possibly all, luminescent bacteria are associated with marine life. Some of these associations are highly evolved. For example, the **flashlight fish**, *Photoblepharon palpebratus*, in the eastern Mediterranean Sea has a special light-generating organ below each eye that contains a dense culture of luminous bacteria (**Figure 11.6**). This highly evolved organ secretes nutrients to support the bacteria, has a layer of reflective tissue under it to direct the light outward, and has a closeable flap to turn off the light. It probably functions as a defense mechanism at night. When the fish changes direction, it closes the flap, and a pursuing predator is disoriented because the fish seems to disappear and then reappear somewhere else.

The Pasteurellas (Pasteurellaceae). The **pasteurellas** are a group of small nonmotile rods. They include two genera of devastating pathogens—*Pasteurella* and *Haemophilus*. The genus *Pasteurella*, named for Louis Pasteur,

Table 11.3	Key Properties That Distinguish *Escherichia coli* from Other Enteric Bacteria		
Genus	Ferments Lactose	Uses Citric Acid as Carbon Source	Makes Indole from Tryptophan
Escherichia	Yes	No	Yes
Salmonella	No	Yes	No
Shigella	No	No	Variable

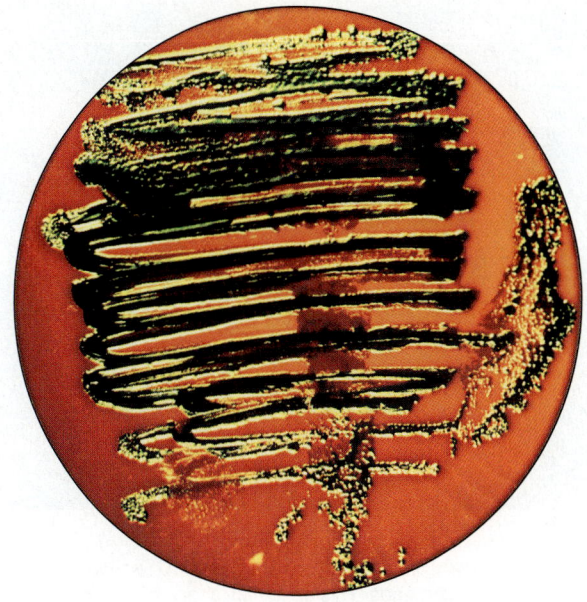

Figure 11.5 *Escherichia coli* (Section 5, enteric family) forms colonies with a distinctive metallic green sheen on a lactose-EMB plate because it ferments lactose. Fermentation produces enough acid to cause the two dyes in the medium—eosin (E) and methylene blue (MB)—to interact, creating the unusual color.

who identified *P. multocida* as the causative agent of fowl cholera, contains species that cause various diseases of animals. Bacteria of the genus *Haemophilus* are obligate parasites that cause many human illnesses, from minor eye infections to potentially fatal meningitis (Chapter 25).

a b c

Figure 11.6 The cells of *Vibrio* spp. (Section 5) are curved rods. The species shown in (**a**) is *V. cholerae*, which causes cholera. (**b**) Like the fish *Photoblepharon*, the small (approx. 4 cm) sepiolid squid *Euprymna scolopes* is symbiotically bioluminescent. The squid is a common nocturnal inhabitant of the shallow waters above reef flats in the Hawaiian Islands. (**c**) A ventral dissection of *E. scolopes* revealing the light-emitting organ as a complex, bilobed structure (approx. 0.8 cm in length) located in the center of the mantle cavity. An adult animal maintains about 100 million symbiotic luminous bacteria of the species *Vibrio fischeri* in the central tissue of the organ.

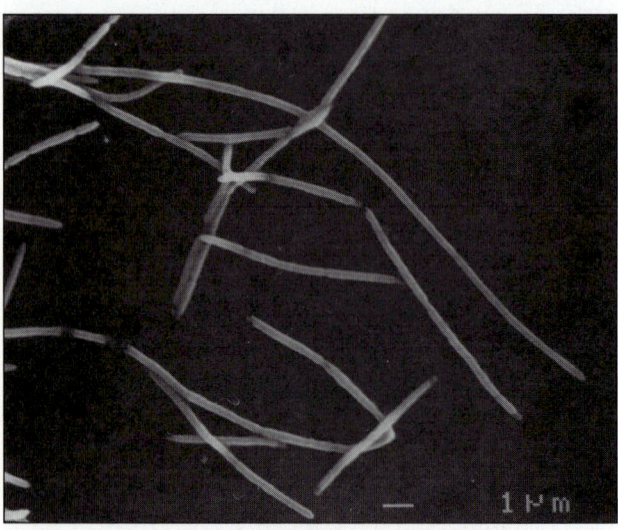

Figure 11.7 Cells of *Fusobacterium nucleatum* (Section 6) are slender with pointed ends. This obligately anaerobic bacterium inhabits the mucosus of the oropharynx, urogenital tract, and intestinal tract; it can infect the respiratory system.

Anaerobic Gram-Negative Straight, Curved, and Helical Rods (Section 6)

The bacteria in this section, including the genera *Bacteroides*, *Fusobacterium*, *Wolinella*, and *Selenomonas*, are among the most abundant microorganisms in the mouth and intestinal tract of animals, including humans. Species of *Bacteroides*, for example, can be found in concentrations of 10 billion to 100 billion (10^{10} to 10^{11}) cells per gram of intestinal contents. These organisms usually coexist harmlessly with their human hosts, but occasionally they cause disease. *Bacteroides* spp. are particularly likely to cause infections of the abdominal and pelvic cavities in weakened hosts or after surgery. *Fusobacterium periodonticum* occasionally causes dental abscesses, and *F. nucleatum* can infect the respiratory system (**Figure 11.7**). *Wolinella* and *Selenomonas* occur in large numbers in the rumen of cattle and other grazing animals. These bacteria break down cellulose into nutrients for the animal (see the beginning of Chapter 8).

Dissimilatory Sulfate- or Sulfur-Reducing Bacteria (Section 7)

Sulfur-reducing anaerobes play an important role in the cyclic interchanges of sulfur-containing compounds found in nature (Chapter 28). Through anaerobic respiration they reduce sulfate (SO_4^{2-}) and elemental sulfur to hydrogen sulfide gas, H_2S, which some bacterial phototrophs and the sulfur-oxidizing bacteria require for growth. Members of this group are united by a common metabolism, not shape. Various genera are rods, curved rods, spirals, cocci, or packets of cocci.

Another genus in this group, *Zymomonas*, produces large quantities of ethanol as an end product of fermentation. This is common among yeasts, but unusual among bacteria. *Zymomonas* can produce five times its dry weight in ethanol each hour. Pulque, an alcoholic drink in Mexico, is the juice of the cactuslike agave plant, fermented by *Zymomonas*. Tequila is the liquor made by distilling pulque.

The rotten egg odor of H_2S usually signals the presence of sulfur-reducing bacteria. Sulfur-reducing bacteria live in mud flats, usually bordering salt water or brackish water in a bay or estuary because salt water is rich in sulfate and the mud is anaerobic. The characteristic black color of mud flats results from the metal sulfides that form when H_2S reacts with metal ions, especially iron, which is present in all soils.

Desulfovibrio spp. are the most thoroughly studied of the sulfate-reducing bacteria. They are responsible for the color and hence the name of the Black Sea. The H_2S these bacteria produce in its depths forms black metal sulfides in the water. *Desulfovibrio* and other sulfate-reducing bacteria cause considerable economic loss because they accelerate the corrosion of iron pipes buried in damp soil. These bacteria convert metallic iron to iron sulfide and iron hydroxide, eventually destroying the pipe.

Anaerobic Gram-Negative Cocci (Section 8)

Section 8 is a group of fermentative cocci. The largest genus, *Veillonella*, is found in the rumen of grazing animals and in the mouth of humans. *Veillonella* is the most abundant anaerobe in human saliva, and one species, *V. parvula*, predominates in dental plaque. *Veillonella* is studied by biochemists because it ferments organic acids to propionate, a somewhat rare metabolic end product, by an unusual pathway. The group contains two other genera, *Megasphaera*, a contaminant of packaged beer, and *Acidaminococcus*, found in soil.

The Rickettsias and Chlamydias (Section 9)

Two groups of Gram-negative bacteria—the **rickettsiae** and the **chlamydiae**—were once thought to be viruses because they are smaller than most bacteria. Chlamydiae range from 0.2 to 0.7 μm in diameter, while rickettsiae are slightly larger, ranging from 0.3 to 1.0 μm. They otherwise look like other nonmotile Gram-negative rods or coccobacilli, with thin peptidoglycan walls and an outer membrane. Most species are **obligate intracellular parasites**, which cannot be cultivated outside a living host cell.

Rickettsiae and chlamydiae cause serious human infections. *Rickettsia prowazekii* causes epidemic typhus during war and times of social and natural disaster. It is transmitted by body lice. *R. typhi* causes endemic typhus, transmitted by rat fleas. *R. rickettsii* causes Rocky Mountain spotted fever, named for the area where the pathogen was first identified. These rickettsial diseases are all systemic (Chapter 27). *Coxiella burnetii* is a rickettsial pathogen that causes Q fever (Chapter 22). *Chlamydia trachomatis* causes trachoma, the most common cause of blindness in the world (Chapter 26), and the sex-

ually transmissible infection nongonococcal urethritis (NGU) (Chapter 24). *C. psittaci* is primarily a bird pathogen, but it can cause a pneumonia in humans called human **psittacosis**, or parrot fever (Chapter 22).

In general, rickettsial pathogens are carried from one human host to another by arthropods (ticks, lice, mites), while chlamydiae are spread directly from one infected human to another. Not all *Chlamydia* spp. are pathogens. Some are harmless intracellular symbionts that may even confer an advantage to their host cells.

Rickettsiae reproduce by simple division, but chlamydiae undergo complex morphological changes (**Figure 11.8**). They alternate between two cell types, the **chlamydiospore**, or **elementary body**, and the **vegetative cell**, or **reticulate body**. Elementary bodies are tiny, round structures released when an infected host cell lyses. When phagocytized, they differentiate into rod-shaped reticulate bodies that multiply within the host cell. Then they dedifferentiate into elementary bodies again before the host cell lyses.

Rickettsiae and chlamydiae can grow only within a host cell because they are energy parasites. Instead of making ATP, they take it from their host in an exchange reaction. When ADP from the bacterial cell exits, ATP from the host cell enters.

The Mycoplasmas (Section 10)

The eubacterial mycoplasmas were traditionally grouped together because all lack a defined cell wall. Now molecular biological studies confirm that these bacteria are closely related, forming a tight evolutionary cluster. They share a number of properties. All are parasites of humans, animals, or plants. Almost all are **obligate fermenters**; that is, they ferment even in the presence of oxygen. Finally, their colonies have a distinctive fried egg appearance (**Figure 11.9**).

Without a cell wall, how do the mycoplasmas contain the turgor pressure that is characteristic of bacteria? First, like eucaryotes, but unlike other procaryotes, mycoplasmas have sterols in their cytoplasmic membrane, which confer some slight added strength. More important, mycoplasmas maintain their cytoplasm at nearly the same pressure as their external environment by actively pumping sodium ions (Na^+) out of the cell. Blocking this mechanism, by depriving mycoplasmas of an energy source, causes their cells to swell and lyse in a solution of NaCl.

Mycoplasmas range from long spirals to perfectly round cocci. That they have a shape other than spherical is probably due to the layer of carbohydrate typically found on the outside of the cytoplasmic membrane. When growth conditions are suboptimal, mycoplasma cells become distorted, forming long strands that resemble fungi (thus accounting for their name; *myco* means

Figure 11.8 (**a**) Vegetative cells of *Chlamydia trachomatis* (Section 9) are shown within a human cell. This species causes trachoma, a sight-threatening infection. (**b**) The life cycle of chlamydiae alternates between two cell types—the elementary body and the reticulate body. Elementary bodies are phagocytized. They grow and divide as vegetative cells. Before the host cell lyses, they become elementary bodies again.

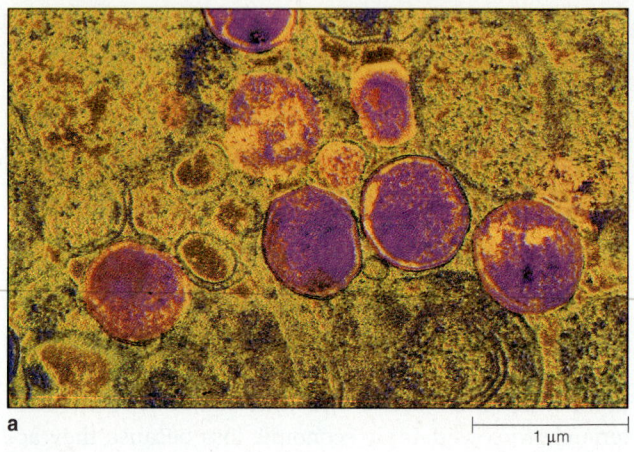

a

1 µm

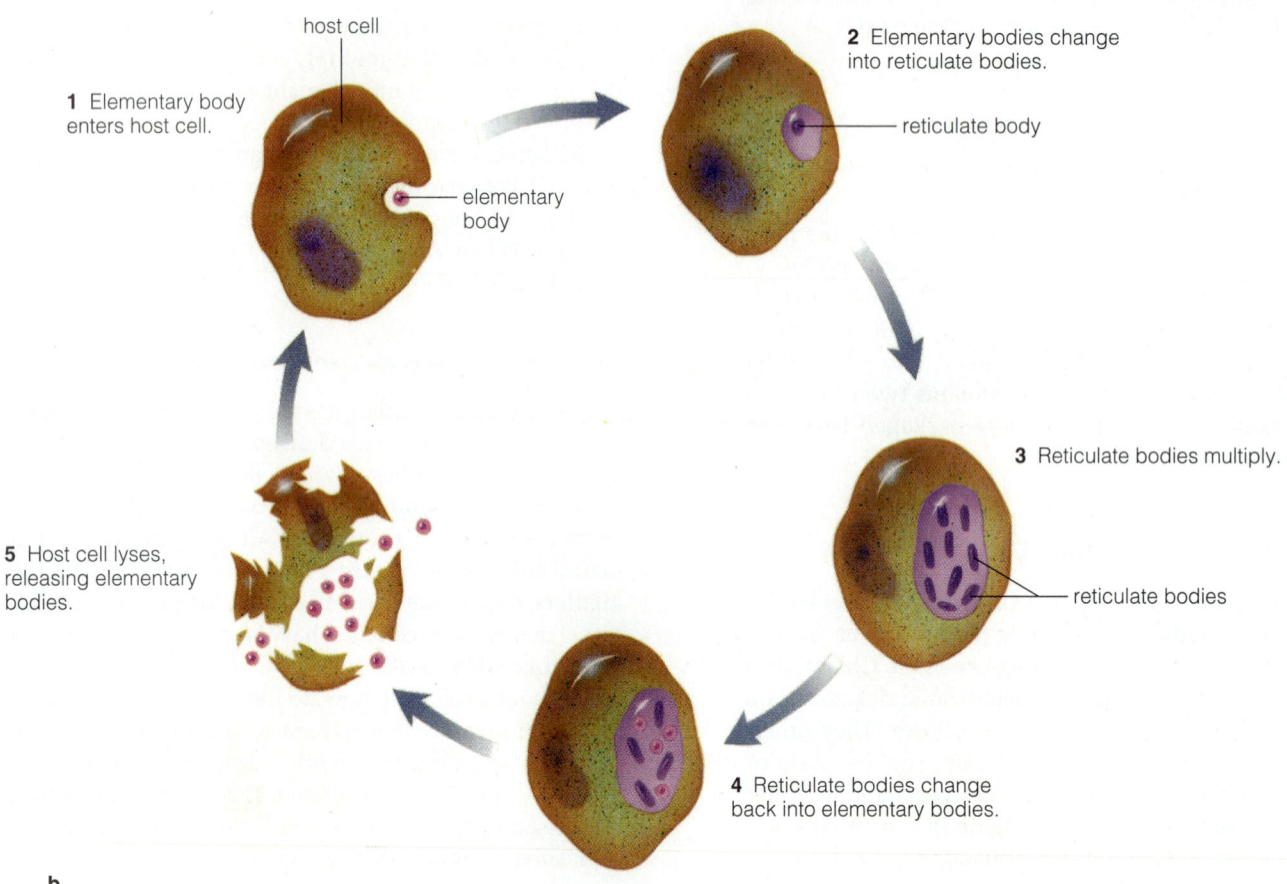

host cell

1 Elementary body enters host cell.

elementary body

2 Elementary bodies change into reticulate bodies.

reticulate body

3 Reticulate bodies multiply.

reticulate bodies

4 Reticulate bodies change back into elementary bodies.

5 Host cell lyses, releasing elementary bodies.

b

"fungus"). Their wall-less structure permits mycoplasmas to squeeze through small holes, even the pores in membrane filters used to sterilize liquids. As a result, tissue cultures of animal cells cannot be sterilized by filtration; antibiotics, usually penicillin and streptomycin, are usually added to the media to suppress growth of contaminating mycoplasmas.

Mycoplasma pneumoniae causes a mild type of pneumonia called **primary atypical pneumonia** (Chapter 22).

Many mycoplasmas cause diseases of animals, such as **rhinitis**, an inflammation of the nose, in chickens and turkeys. Another mycoplasma, *Ureaplasma urealyticum*, is a harmless inhabitant of the vaginal tract of 60 percent of normal women, but it occasionally enters the bloodstream during delivery, causing a mild postpartum fever.

This section also contains *Thermoplasma*, a genus of thermophilic, wall-less archaebacteria that is only distantly related to other mycoplasmas. *Thermoplasma* is

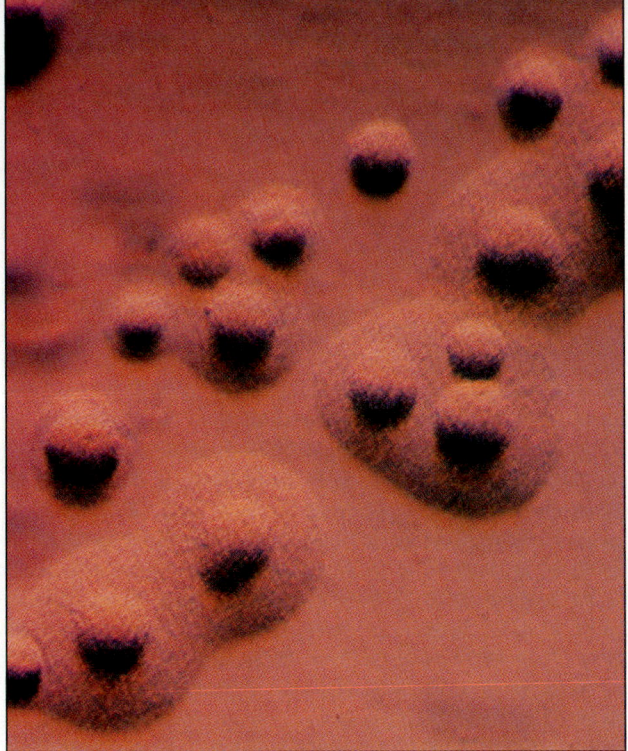

Figure 11.9 Mycoplasmas (Section 10) form distinctive colonies with a fried egg appearance.

cross-listed in Section 25. Two species, *T. acidophilum* and *T. volcanium*, have been distinguished from each other by DNA hybridization. They are found in smoldering coal-refuse piles, in hot springs, in regions surrounding volcanoes, and near deep-sea hydrothermal vents.

Gram-Positive Cocci (Section 12)

The Gram-positive cocci are a large group of organisms with similar morphology but loosely related otherwise. All are spherical or nearly so, but they vary considerably in size from species to species. Some genera are identifiable by the way cells are attached to one another in packets, chains, or grapelike clusters (**Figure 11.10**). These arrangements reflect patterns of cell division and whether or not cells remain attached to one another after division. *Sarcina* cells, for example, are arranged in cubical packets that form because cell division alternates regularly among the three perpendicular planes, and the cells tend to remain together after division. Some species of *Streptococcus* resemble a string of beads because division always occurs in the same plane and cells remain attached to one another for some time. The chains of other streptococci, for example, *S. pneumoniae*, tend to break into pairs called **diplococci**. Species of *Staphylococcus* have no regular plane of division and so form grapelike clusters.

The species of Gram-positive cocci are different physiologically and are found in many different habitats. Species of *Micrococcus*, for example, are obligate aerobes, and their habitat is human skin. Working microbiologists,

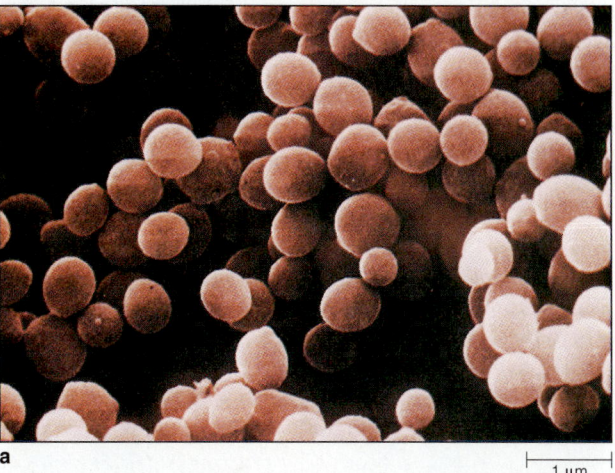

a   1 µm

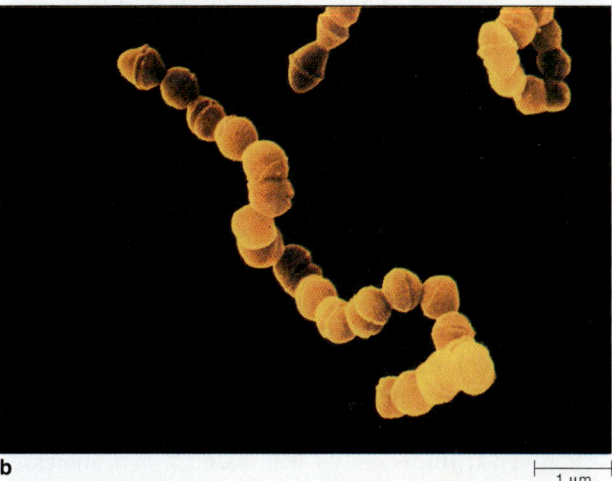

b   1 µm

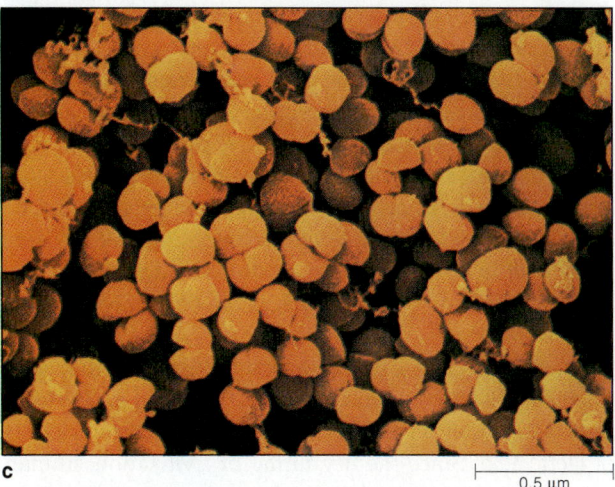

c   0.5 µm

Figure 11.10 The Gram-positive cocci (Section 12) are grouped by morphology. Some genera are identified by the way cells remain attached after division. (**a**) *Staphylococcus* cells form grapelike clusters. (**b**) *Streptococcus* forms long chains. (**c**) *Sarcina* forms cubical packets.

Enter *Staphylococcus aureus*

The patient, a 10-year-old girl, seemed well except for a limp and a mild but nagging pain just above her ankle. At first her parents thought she must have twisted her ankle or otherwise injured her foot, but rest didn't help and the pain seemed to be growing worse. The x ray her doctor took revealed that part of the bone appeared to have been eaten away and its surface was raised and distorted. Because such an abnormality can sometimes mean cancer, the child's doctor biopsied the diseased bone (surgically removed a tiny sample for laboratory examination). There were no cancer cells—only bacteria. The diagnosis was osteomyelitis, bone infection caused by the Gram-positive coccus *Staphylococcus aureus* that entered through a break in the skin. After several weeks of antibiotic therapy she recovered completely.

who like all of us carry these microorganisms as part of their normal flora, occasionally deposit them on the materials they are working with, so these bacteria turn up as contaminants on agar plates.

Staphylococcus species are also part of the skin's normal flora, but unlike the *Micrococcus*, they are facultative anaerobes. In the absence of oxygen, **staphylococci** ferment sugars, producing lactic acid as an end product. Many species also produce carotenoid pigments, which give staphylococcal colonies their characteristic yellow and orange colors. *Staphylococcus aureus* is a major human pathogen that can infect almost any tissue in the body. It causes a number of skin infections, including a weepy infection called **impetigo**, as well as boils and abscesses (Chapter 26). *S. aureus* also causes many **nosocomial** (hospital-acquired) infections (Chapter 20).

Three genera of this section—*Streptococcus*, *Leuconostoc*, and *Pediococcus*—along with one genus from Section 14, *Lactobacillus*, are often grouped as **lactic acid bacteria** because they share a set of distinctive physiological properties. The most significant is that they ferment sugars and produce lactic acid as the major end product. Lactic acid bacteria are aerotolerant anaerobes. That is, they grow well in the presence of oxygen, but do not use it in their metabolism. Fermentation is the only way they can generate ATP.

Lactic acid bacteria are resistant to the acid they produce, and they are often the sole survivors in an environment where they have grown and produced large amounts of lactic acid. Since the beginning of civilization, human beings have used lactic acid bacteria to preserve and make foods and dairy products (Chapter 29).

Some lactic acid bacteria are human pathogens. The genus *Streptococcus* contains a versatile group of pathogens that are divided into species on the basis of their effect on blood agar and the antigens they produce (Chapter 22). *S. pyogenes* causes a number of diseases, including strep throat (Chapter 22), rheumatic fever (Chapter 22), and endocarditis (Chapter 27). *S. pneumoniae* causes a life-threatening pneumonia (Chapter 22), and *S. mutans* is the principal cause of dental caries (Chapter 23).

Endospore-Forming Gram-Positive Rods and Cocci (Section 13)

Section 13 contains seven genera, all of which produce resting structures called endospores (Chapter 4). Two genera, *Bacillus* and *Clostridium*, are widespread and important clinically. Both are rods, and some species are motile by means of peritrichous flagella. All species of *Clostridium* are strict anaerobes, inhabiting soil and mud. All *Bacillus* species are capable of aerobic respiration; but some species are facultative anaerobes, able also to ferment and/or carry out anaerobic respiration. *Bacillus* inhabits both soil and aquatic environments.

Sterilizing heat treatments are designed with *Bacillus* and *Clostridium* in mind. Because they are the most heat-resistant living things, they are used as an index of sterilization. If they are killed, other microorganisms are certainly dead (Chapter 9).

Because of their low water content, unstained endospores are easily visible under the microscope. Often endospores have a greater diameter than the cells in which they are formed, so they appear as bulges. Their location within the mother cell—in the center, at the end, or near the end—is genetically determined, which means it can be used to distinguish species (**Figure 11.11**). Certain species of *Bacillus* produce protein crystals near their spores that are highly toxic to insects (see Figure 29.7). The protein produced by *B. thuringiensis* kills the caterpillar stage of butterflies and moths as well as mosquitoes but is completely harmless to vertebrates, including humans. It is commercially available for garden and agricultural use.

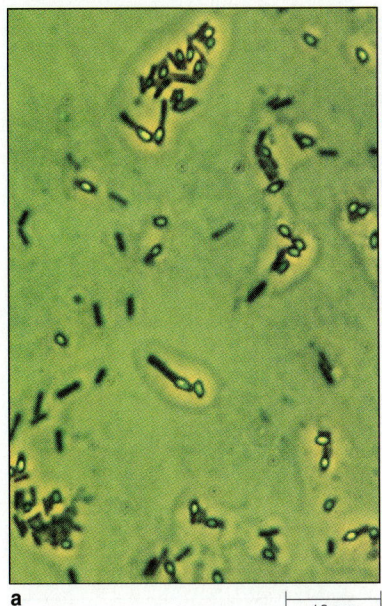

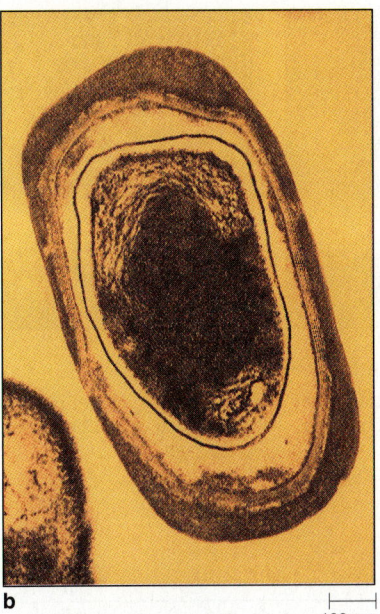

a   10 μm   b   100 nm

Figure 11.11 All members of Section 13 form endospores. In some species, endospores form near the end of the cell (**a**); in others they form near the middle (**b**). In some species, they swell the cell (**a**), but in others they don't (**b**).

Both genera contain significant pathogens. *Bacillus anthracis*, which reproduces in the blood, causes anthrax, a disease of animals that also infects humans (Chapter 27). *Clostridium tetani* causes tetanus, characterized by severe muscle spasms and a fatal rigid paralysis (Chapter 25). *C. botulinum* causes botulism, a potentially lethal food poisoning (Chapter 25). *C. perfringens* causes gas gangrene, a tissue-destroying wound infection (Chapter 26), and, occasionally, food poisoning (Chapter 23). *C. difficile* causes diarrhea when antibiotics upset the normal balance of intestinal microorganisms (Chapter 23). Other species of *Bacillus* and *Clostridium* are harmless inhabitants of soil and water.

Regular Nonsporing, Gram-Positive Rods (Section 14)

Of the seven genera in Section 14, four are of special interest—*Lactobacillus*, *Caryophanon*, *Erysipelothrix*, and *Listeria*. Along with three genera from Section 12, *Lactobacillus* spp. are lactic acid bacteria.

Until recently, *Caryophanon latum* was notable because its rod-shaped cells are exceptionally large, typically about 3 × 15 μm. However, several species of much larger bacteria have recently been discovered. The largest to date is the symbiont (not yet classified in *Bergey's Manual*) from the gut of surgeon fish in Australia described at the beginning of the chapter. Its cells are more than 500 μm long.

The group contains two pathogens, *Erysipelothrix rhusiopathiae* and *Listeria monocytogenes*. *Erysipelothrix rhusiopathiae* primarily affects swine but also causes red, painful skin lesions in humans. *Listeria monocytogenes* causes the foodborne disease listeriosis. In healthy adults,

listeriosis is a mild disease, but in the young, the old, and immunocompromised patients, it can cause a form of meningitis (Chapter 25).

Irregular Nonsporing, Gram-Positive Rods (Section 15)

The species in Section 15 have unusual shapes. Cells may be branched, club-shaped, or variable in width, and sometimes cell shape changes with the growth phase of the culture. They may be aerobic or anaerobic.

The aerobic genus *Arthrobacter* changes shape in different growth phases (**Figure 11.12**). *Arthrobacter* are abundant in soil and responsible for much of the mineralization of organic material that takes place. Stationary phase cultures of *Arthrobacter* are composed of cocci. When one of these spherical cells begins to grow, it elongates, producing a rod that divides to produce more rods, some of them branched. As the culture ages, the cells become increasingly irregular in width. Eventually the beadlike regions look more like cocci held together by a thin thread. Finally these break apart, producing a culture that appears to be made up almost entirely of uniform cocci.

During the exponential growth phase of its growth cycle, *Arthrobacter* exhibits a curious sort of movement called **snapping post-fission movement** that is also characteristic of several other members of this group. Immediately after the wall forms between two daughter cells, they appear to snap apart, forming a V-shaped pair of cells. Snapping post-fission movement occurs because the cell wall is made up of two layers (**Figure 11.13**). The inner layer grows in to form the cross wall, while the outer layer holds the two daughter cells together. When

Figure 11.12 Members of Section 15 have unusual shapes. Some—such as *Arthrobacter*—change shape in different growth phases.

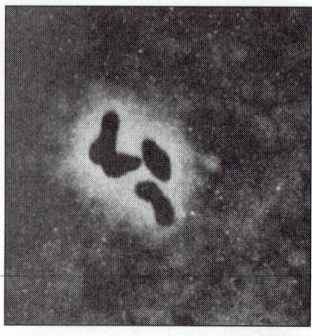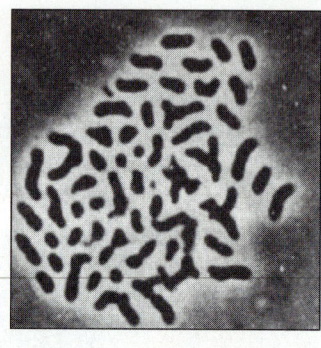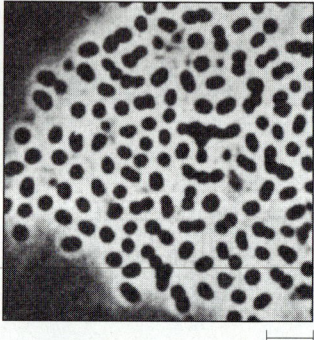

5 µm

a Although the stationary phase cells of *Arthrobacter globiformis* are cocci, in a suitable environment they elongate, becoming rods.

b The rods divide, producing more rods, which become branched.

c As the culture approaches stationary phase, the rods become irregular in width, eventually becoming cocci again.

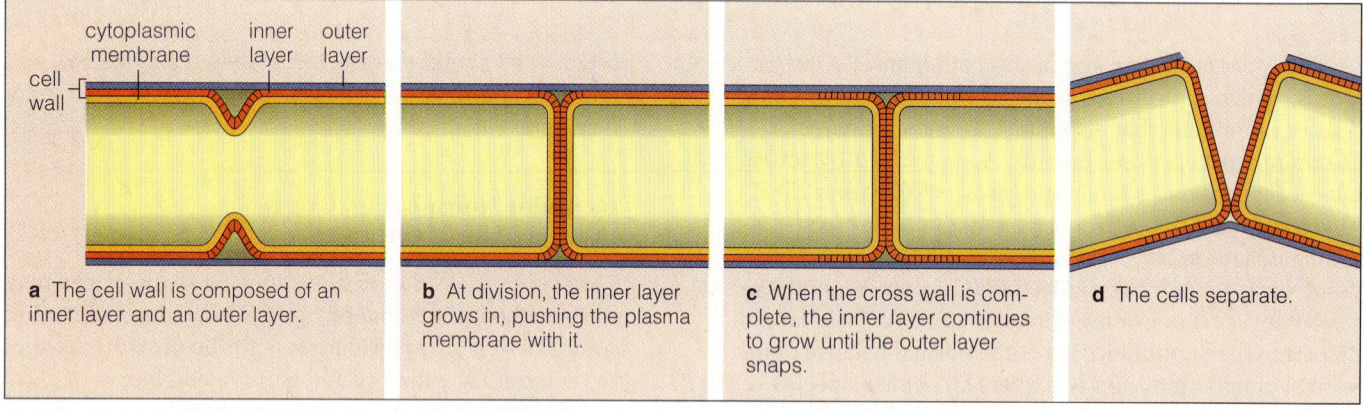

cytoplasmic membrane — inner layer — outer layer
cell wall

a The cell wall is composed of an inner layer and an outer layer.

b At division, the inner layer grows in, pushing the plasma membrane with it.

c When the cross wall is complete, the inner layer continues to grow until the outer layer snaps.

d The cells separate.

Figure 11.13 Snapping post-fission movement. *Arthrobacter* and some other members of Section 15 appear to snap apart after division.

the pressure of cell enlargement causes the outer layer to break at one side, the cells snap into their characteristic V shape.

A closely related genus, *Bifidobacterium*, has irregular, often swollen and branched cells. It is part of the intestinal microflora of breastfed babies (Chapter 14). Bifidobacteria are anaerobes that ferment lactose to lactic and acetic acids. They require *N*-acetylglucosamine to build their cell wall, which is abundant in breast milk. Because the bifidobacteria produce significant quantities of acid as they ferment lactose, their growth suppresses the growth of many diarrhea-causing pathogens.

The human pathogens that belong to this group include *Propionibacterium acnes*, which causes the skin infection acne (Chapter 26), *Corynebacterium diphtheriae*, which causes the respiratory disease diphtheria (Chapter 22; *coryne* means "club-shaped"), and *Actinomyces israelii*,

which causes severe draining abscesses, usually of the head and neck (Chapter 26). Corynebacteria exhibit snapping fission, and *Actinomyces* have irregularly shaped branching cells.

The Mycobacteria and Nocardioforms (Sections 16 and 17)

Members of the mycobacterial and nocardioform groups have an unusual wall structure. The usual peptidoglycan wall is attached by chemical bonds to a polysaccharide layer. The polysaccharide layer, in turn, is attached by ester bonds to mycolic acids, branched chains of fatty acids that bear hydroxyl (—OH) groups.

Two genera, *Mycobacterium* and *Nocardia*, have mycolic acids with especially long carbon chains. As a result, their cells have a waxy outer surface that protects against

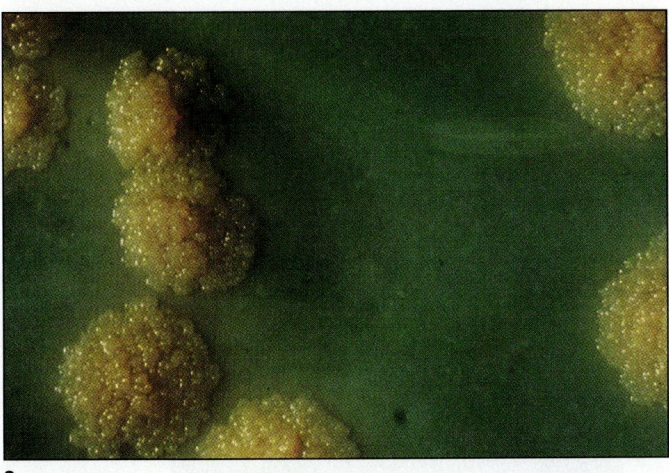
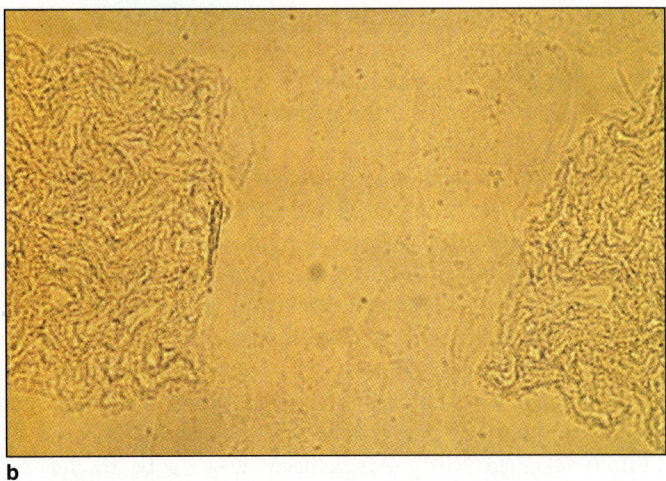

a

b

Figure 11.14 Mycobacteria (Section 16) have mycolic acids in their cell envelopes, which give them a waxy appearance and make them difficult to stain. (**a**) These cause *M. tuberculosis* colonies on solid media to have a waxy, warty appearance. (**b**) In liquid media, cells form ropelike aggregates.

hostile environments and also affects staining (see the beginning of Chapter 4). The waxy surface decreases permeability to a number of compounds, including most dyes. Cells must be heated to about 75°C before they can be stained and, once stained, they are difficult to destain, even with acid. These unusual characteristics allow *Mycobacterium* and *Nocardia* to be identified by the acid-fast stain procedure, and they are known as acid-fast bacteria (Chapter 3).

Mycobacterium and *Nocardia* are closely related, differing mainly in cell shape. *Mycobacterium* cells are mainly rods with some branched cells, while most *Nocardia* are elongated and branched. *Nocardia* are abundant in the soil and include several species that are pathogenic to humans, causing transient and chronic infections that can spread throughout the body. Nineteen species of *Mycobacterium* cause human disease, including *M. tuberculosis*, which causes the lung disease tuberculosis. Antibiotic-resistant strains of *M. tuberculosis* have contributed to the resurgence of this disease in the United States (Chapter 22). Another species, *M. leprae*, causes Hansen's disease, or leprosy, a chronic disease characterized by loss of sensation and deformed extremities (Chapter 26).

Diagnosis and basic research on mycobacterial diseases is hampered because it is extremely difficult to culture mycobacteria in the laboratory. Mycobacteria grow slowly, probably because their waxy surface impedes entry of nutrients into the cell. Even species classified as rapid growers require several days' incubation under optimum conditions before a colony can be detected. Most disease-associated mycobacteria, including tuberculosis, require two to six weeks of incubation. *M. leprae* has not yet been cultivated in vitro.

Colonies of *M. tuberculosis* have a waxy, warty appearance; and when grown in liquid culture, cells aggregate into characteristic ropelike structures (**Figure 11.14**). These distinctive growth habits reflect the presence of the cell's waxy outer layer.

Anoxygenic Phototrophic Bacteria (Section 18)

Photosynthesis is the process by which phototrophs generate ATP (Chapter 5). In oxygenic photosynthesis, the process carried out by green plants, algae, and cyanobacteria, water is the source of electrons, and oxygen (oxidized water) is one of the products. **Anoxygenic** phototrophs carry out a slightly different form of photosynthesis. These bacteria use organic compounds or a reduced sulfur compound, usually hydrogen sulfide (H_2S), as a source of electrons. As a result, anoxygenic photosynthesis produces an oxidized organic compound or an oxidized form of sulfur—usually elemental sulfur (S)—rather than oxygen.

Anoxygenic phototrophs are divided into two groups, the **purple bacteria** and the **green bacteria**, based on the kind of chlorophyll and other pigments they produce and their interior membrane arrangement (Chapter 4). *Rhodobacter* is an abundant and well-studied genus of purple bacteria. *Chlorobium* is a well-known genus of green bacteria. Section 18 further subdivides out **purple nonsulfur bacteria** and divides green bacteria into **green sulfur** and **multicellular filamentous green bacteria**.

Anoxygenic phototrophic bacteria are found exclusively in anaerobic environments; but because they are photosynthetic, they must have light. As a result, these bacteria are found in deep, clear bodies of water. They also thrive near the surface of ponds or mud flats that are rich in organic matter, because any watery environment containing high concentrations of organic compounds soon becomes anaerobic from microorganisms using oxygen in their metabolism.

Their environment is so restricted that these bacteria do not contribute significantly to the total photosynthesis on the earth. Nevertheless, research into anoxygenic phototrophic bacteria has taught biologists a great deal about the biochemistry of photosynthesis and its evolution on Earth. In producing elemental sulfur as a by-product of their photosynthesis, anoxygenic phototrophic bacteria are probably responsible for the huge deposits of elemental sulfur in various parts of the world, especially Texas. Certain lakes in North Africa seem now to be in the process of forming new sulfur deposits. Anoxygenic phototrophs flourish in them, and their bottoms are covered with elemental sulfur.

Often these bacteria are highly colored—red, orange, purple, and bright green—because of the chlorophyll and other photosynthetic pigments they contain. Water samples from deep regions of lakes with an abundance of these organisms are sometimes as unusually and intensely colored.

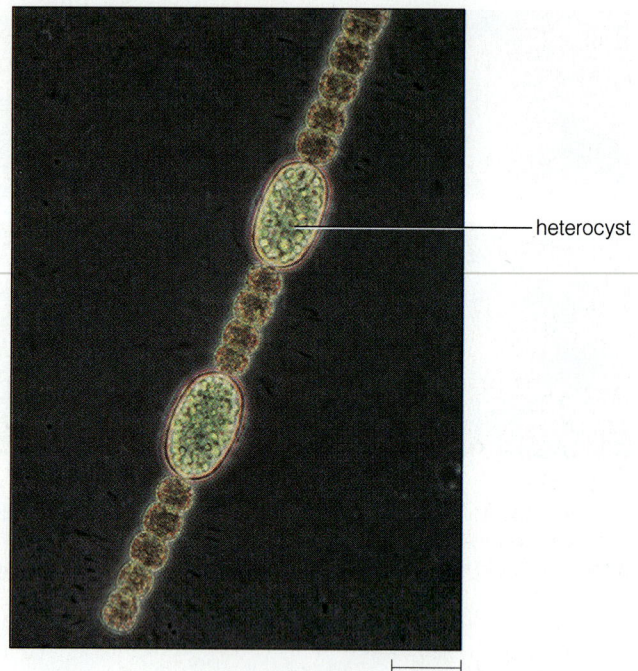

Figure 11.15 Cyanobacteria (Section 19) fix nitrogen in oxygen-impermeable heterocysts.

Oxygenic Photosynthetic Bacteria (Section 19)

Oxygenic photosynthetic bacteria are called **cyanobacteria**. They were once called the **blue-green algae**: some of them are blue-green, they are about the same size and shape as filamentous or unicellular algae, and they produce oxygen as algae do. They are classified as cyanobacteria because they are procaryotes and algae are eucaryotes. They have an outer membrane and a thin peptidoglycan wall typical of Gram-negative bacteria.

Some cyanobacteria are able to fix atmospheric nitrogen, which only procaryotes can do. As a result, these organisms, which obtain both their carbon (CO_2) and nitrogen (N_2) from the atmosphere, have the simplest possible nutrition. They need only a few minerals, which are present in almost any natural body of water, and a light source to grow. Cyanobacteria are major nitrogen-fixers in nature (Chapter 28).

Nitrogen fixation and oxygenic photosynthesis might appear to be incompatible processes because all nitrogen-fixing systems, including the cyanobacterial system, are extremely sensitive to oxygen (O_2). Many cyanobacteria solve this problem by physically separating the two processes. Nitrogen fixation is carried out in specialized cells called **heterocysts**, while photosynthesis occurs in all other cells (**Figure 11.15**). The intracellular environment of heterocysts is anaerobic and therefore suitable for nitrogen fixation. Moreover, because heterocysts do not photosynthesize, they do not produce O_2 intracellularly, and their thickened cell walls exclude extracellular O_2.

In addition to their simple nutritional requirements, cyanobacteria resist desiccation and high temperatures. They are thus found in almost all environments—oceans, lakes, streams, and soil. They are particularly abundant in nutrient-poor tropical soils.

Cyanobacteria form many symbiotic associations with plants—the bacterium contributing fixed nitrogen and the plant contributing a protective environment. One example has been important agriculturally. Rice farmers in Southeast Asia, Africa, China, and, to a limited extent, California add the tiny floating fern *Azolla* with its symbiotic cyanobacterium, *Anabaena azolla*, to a rice paddy when rice plants are young. The fern grows to cover the surface of the paddy completely, limiting the growth of weeds and supplying enough fixed nitrogen to support a crop of rice.

Aerobic Chemolithotrophic Bacteria and Associated Organisms (Section 20)

Chemolithotrophic ("chemical-stone-feeding") bacteria obtain energy by oxidizing inorganic compounds. The word *lithotrophic* is synonymous with *autotrophic*. These bacteria derive all their carbon from carbon dioxide (CO_2) gas (Chapter 5). Section 20 is subdivided on the basis of the inorganic compound oxidized (**Table 11.4**). Some of these bacteria oxidize a nitrogen compound, either ammonia (NH_3) or nitrite ion (NO_2^-), while others oxidize either a reduced sulfur compound, hydrogen gas, or a metal ion, usually iron.

Table 11.4 Classes of Chemolithotrophic Bacteria

Class	Representative Genus	Inorganic Compound Oxidized
Nitrifying bacteria	*Nitrosomonas* *Nitrobacter*	Ammonia Nitrite ion
Sulfur bacteria	*Thiobacillus*	Reduced sulfur compounds
Hydrogen bacteria	*Hydrogenobacter*	Hydrogen gas

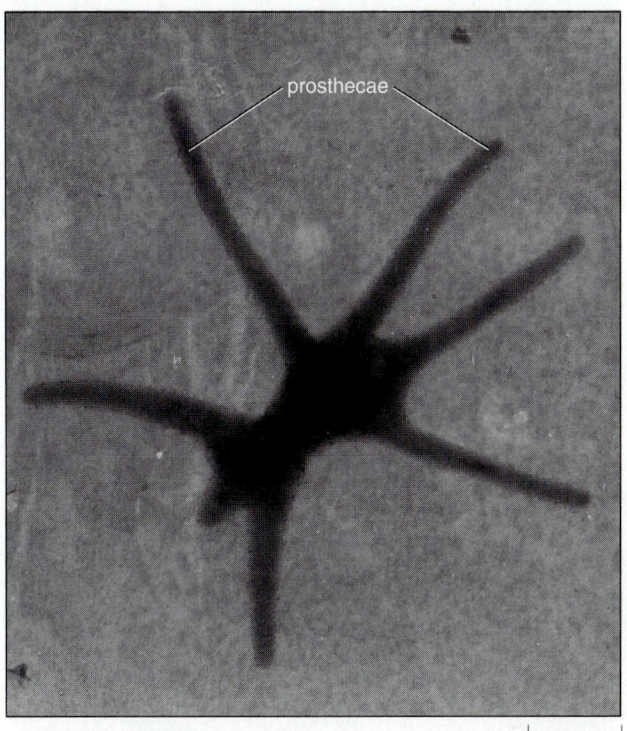

1 μm

Figure 11.16 Prosthecae (appendages) give the bacteria in Section 21 a spidery shape. This micrograph shows *Ancalomicrobium adetum*.

Most members of this group oxidize one particular inorganic compound. For example, the **nitrifiers** oxidize nitrogen compounds. There are two groups of nitrifiers. One group, which includes the genera *Nitrosomonas*, *Nitrosococcus*, and *Nitrosolobus*, oxidizes ammonia to nitrite ion:

$$NH_3 + 1.5\,O_2 \longrightarrow NO_2^- + H_2O + H^+$$

The other group, which includes the genera *Nitrobacter*, *Nitrospina*, *Nitrococcus*, and *Nitrospira*, oxidizes nitrite to nitrate ion:

$$NO_2^- + 0.5\,O_2 \longrightarrow NO_3^-$$

Together the nitrifiers oxidize NH_3 to NO_3^- and thereby play an essential role in the nitrogen cycle (Chapter 28). Without them, the ammonia produced by the decay of dead plants and animals would not be converted into the nitrate that plants need to grow and some bacteria can convert to nitrogen gas (N_2). Nitrifiers cannot oxidize any other compounds as a source of metabolic energy. In fact, many organic compounds that are readily oxidized as energy sources by most bacteria are toxic to the nitrifiers.

The colorless sulfur bacteria, which include the genus *Thiobacillus*, obtain energy by oxidizing reduced sulfur compounds, including hydrogen sulfide (H_2S) and elemental sulfur (S), to sulfuric acid (H_2SO_4). They grow in sulfur-containing wastes from mines, producing highly acidic runoff water. Gardeners spread sulfur on soil to make it suitable for acid-loving plants; the sulfuric acid produced by the bacteria acidifies the soil.

Obligate hydrogen oxidizers, including members of the genus *Hydrogenobacter*, also belong in this section. They obtain energy by oxidizing hydrogen gas to water.

Budding and/or Appendaged Bacteria (Section 21)

Most bacteria are simple geometric shapes—spheres, rods, or helices—but there are exceptions, and most of them are included in this section. Also included here are bacteria that divide by budding, as most yeasts do, rather than by fission, as most bacteria do. Some budding bacteria also have unusual shapes.

Possibly the most unusually shaped bacteria are the appendaged or **prosthecate** bacteria found in aquatic environments. Prosthecates have one or several filamentous or conical extensions of the cell, called prosthecae, which give it a spidery appearance (**Figure 11.16**). The prosthecae are part of the cell because their contents lie within the cytoplasmic membrane. Their physiological function is not established, but there are two possibilities. First, these are nonmotile aerobic species; the prosthecae may help suspend them in water by slowing their rate of settling and allowing even a gentle current to raise them, thus keeping them in a more aerobic environment. Second, prosthecae dramatically increase a cell's surface-to-volume ratio, facilitating the entry of nutrients. This is beneficial because these organisms live in extremely nutrient-poor environments. This second possibility is supported by observation. Prosthecae become longer as the growth medium becomes more dilute.

One prosthecate bacterium, *Caulobacter*, has an elaborate life cycle by bacterial standards (**Figure 11.17**). *Caulobacter* prospers in nutrient-poor aquatic environments where it attaches itself to solid surfaces (including

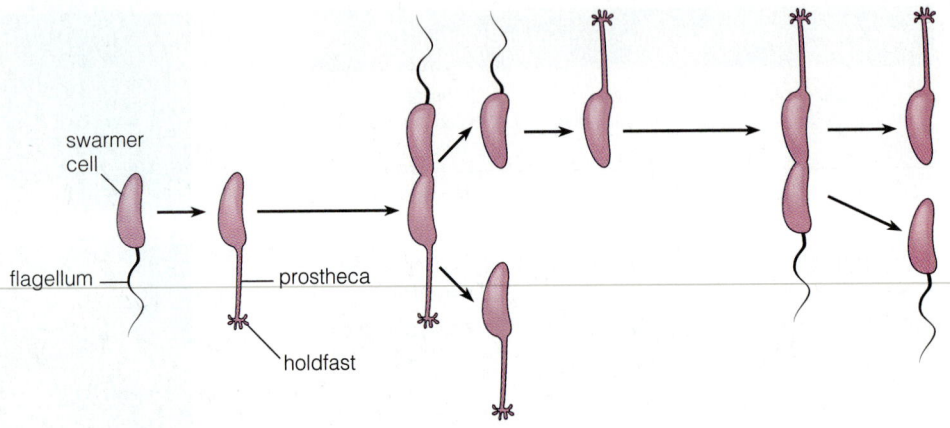

Figure 11.17 The prosthecate bacterium *Caulobacter* (Section 21) has an elaborate life cycle by bacterial standards. The single flagellum of a swarmer cell develops into a prostheca with a holdfast. When the prosthecate divides, it produces another swarmer that repeats the cycle.

swarmer cell

flagellum

prostheca

holdfast

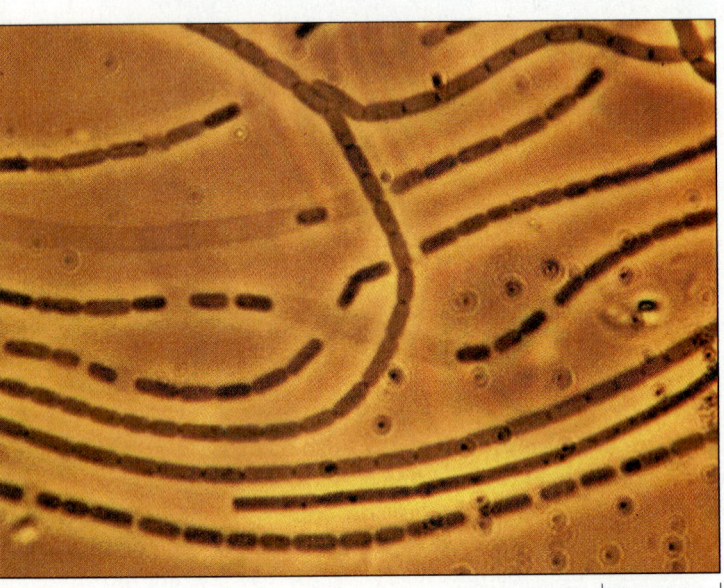

10 μm

Figure 11.18 Cells of the sheathed bacterium *Sphaerotilus* (Section 22) develop within long transparent tubes. Like other sheathed bacteria, it is found in polluted water and sewage treatment plants.

other microorganisms) by a sticky pad called a **holdfast**. The holdfast is located at the end of a single prostheca extending from the end of the kidney-shaped cell. During growth, the prosthecate cell elongates, eventually budding off a motile cell, called a **swarmer**, with a single flagellum. The swarmer may divide or, after a brief period of swimming, the flagellum may be replaced by a prostheca with a terminal holdfast. This new prosthecate cell then settles to produce more swarmer cells.

Caulobacters have been intensively studied as a model system to learn how cells differentiate—possibly the major unsolved biological problem. The molecular mechanisms that allow a swarmer to differentiate into a prosthecate cell might resemble the mechanisms that direct the differentiation of more complex organisms—for example, the differentiation of a fertilized egg into a human being. Caulobacters are easy to find in nature. A microscope slide immersed in unsterilized tap water for several days will frequently have *Caulobacter* cells attached to it.

In certain species of budding bacteria—for example, *Hyphomicrobium*—budding occurs at the tips of the prosthecae. The result is a string of cells linked by prosthecae.

Sheathed Bacteria (Section 22)

Bacteria in Section 22 develop within **sheaths**, long transparent polysaccharide tubes (**Figure 11.18**). Under favorable conditions, cells grow and divide within the sheath. Cells pushed out of the sheath form new sheath material to elongate the tube. If growth conditions become unfavorable, perhaps because an essential nutrient is depleted, cells swim out, leaving an empty sheath behind. When the swimming cells enter a favorable environment, they multiply, producing a new cell-packed sheath.

Sheathed bacteria are found in contaminated streams and sewage treatment ponds. Some are capable of oxidizing iron and manganese. In environments with elevated concentrations of these minerals, the bacterial sheaths may become heavily encrusted with insoluble iron and manganese oxides. The encrustations make tufts of these bacteria easily visible in polluted water. Tufts of *Sphaerotilus* spp. growing in sewage treatment plants sometimes become so abundant they cause **bulking**. They increase the volume of the sludge so that it does not settle properly and interferes with clarification of the treated water.

Nonphotosynthetic, Nonfruiting Gliding Bacteria (Section 23)

The bacteria in Section 23 are grouped together because of their unusual form of motility, known as **gliding**, although a few other sections contain some gliders. Gliders that form fruiting bodies are in Section 24, and gliding cyanobacteria are in Section 19. Gliders can move only when they are in contact with a solid surface. Some gliders move relatively rapidly, at rates as high as 600 μm per minute. In terms of body length, this rate for a 1-μm bacterial cell is equivalent to a 6-foot human moving about 40 miles per hour. In contrast, other gliders move as slowly as 1 μm per minute—about the rate of continental drift.

Microbiologists have intensively studied the mechanism of gliding motility, but it remains a mystery. A popular but unproven hypothesis is that these bacteria have a fluid outer membrane that flows back and forth in grooves in the peptidoglycan wall. When the moving outer membrane attaches to a solid surface, the cell glides.

Gliding motility, like all forms of motility, offers an organism a selective advantage in nature. Combined with tactic responses, it allows an organism to seek out favorable environments and avoid unfavorable ones. For example, *Beggiatoa* has a coordinated set of tactic responses that enable it to survive in a fairly special situation. *Beggiatoa* lives by metabolizing hydrogen sulfide (H_2S) through aerobic respiration, so it must find an environment where both H_2S and oxygen (O_2) are present. *Beggiatoa* plays an essential role in the sulfur cycle by oxidizing H_2S.

Other members of the gliding nonfruiting group fulfill very different ecological roles. Some, like *Cytophaga*, help break down cellulose in plants. They can be found in almost all soils. The *Lysobacter* are microbial predators. They lyse living algae and cyanobacteria and consume their nutrients.

Gliding Fruiting Bacteria (Section 24)

The gliding fruiting bacteria, also called **myxobacteria**, are soil organisms that undergo an unusual degree of differentiation for procaryotes. They glide in masses over surfaces, consuming living and dead microorganisms in their path. When nutrients are depleted they aggregate and differentiate into **fruiting bodies** that are covered with a slime layer and contain spores called **myxospores**. These are easily visible to the naked eye. Some fruiting bodies, such as those formed by species of *Myxococcus*, are simple dome-shaped structures. Others, such as those produced by species of *Stigmatella*, are treelike, branched, and quite beautiful (**Figure 11.19**). The surface of animal

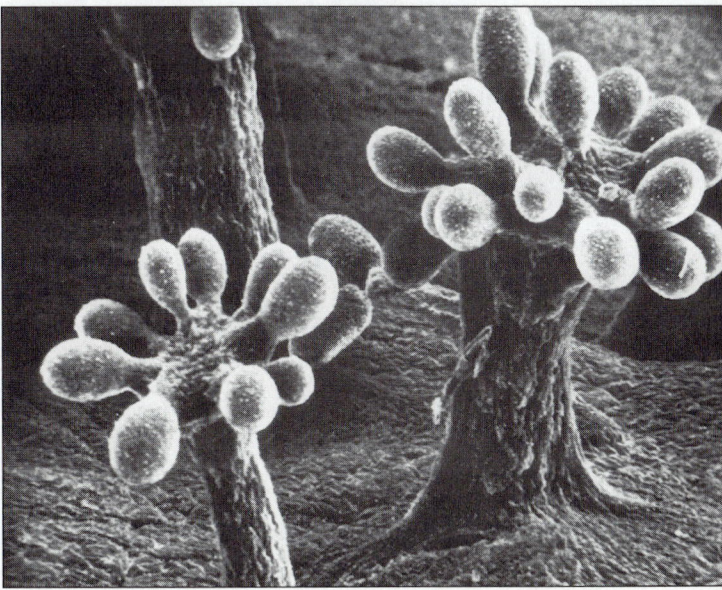

Figure 11.19 Gliding fruiting bacteria (Section 24) are found in the soil. When nutrients are scarce, they form fruiting bodies filled with spores called myxospores. The fruiting bodies of *Stigmatella*, shown here, are treelike, branched, and quite beautiful.

dung in a damp area is a good place to find myxobacteria because it contains large numbers of bacteria, the food of myxobacteria.

Archaeobacteria (Section 25)

Although the archaebacteria are evolutionarily different enough from all other bacteria to be considered a distinct biological kingdom (Chapter 10), *Bergey's Manual*, consistent with its practical approach, merely assigns them to a separate section called Archaeobacteria.

Superficially, archaebacteria resemble eubacteria. They are about the same size and their shapes are typical of eubacteria. Viewed with an ordinary microscope, it is impossible to tell whether a cell is archaebacterial or eubacterial. But physiologically and biochemically, archaebacteria are extremely different from eubacteria.

Archaebacteria are divided into three subgroups—the methanogens, the halophiles, and the thermoacidophiles—each of which favors different extremes of habitat.

Methanogens. The **methanogens**, or methane formers, make the natural gas methane (CH_4) from hydrogen gas (H_2) and either carbon dioxide (CO_2) or acetate (CH_3—COO^-). This conversion occurs only in anaerobic environments where H_2, CO_2, and acetate are produced by other species of bacteria.

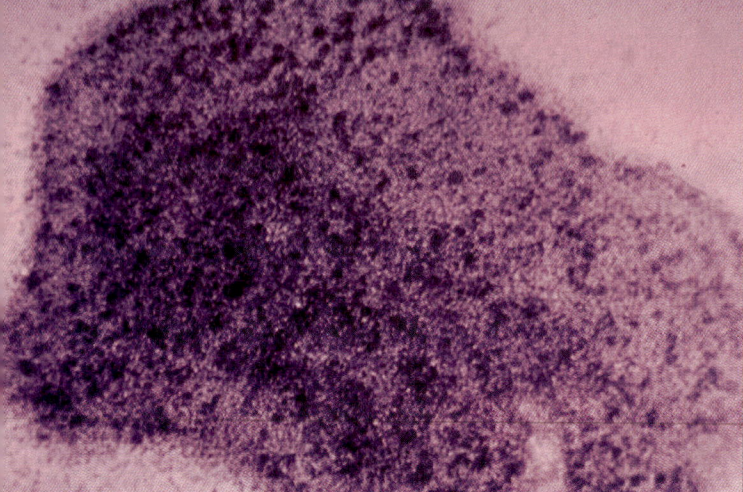

Figure 11.20 *Sulfolobus* is an archaebacterium (Section 25) found in highly acidic geothermal springs; it has irregular lobed cells.

sporangium

1 μm

Figure 11.21 *Frankia*, a member of Section 27, forms multilocular sporangia. *Frankia* is important ecologically because it fixes nitrogen.

The methanogens include many genera—*Methanococcus* is the best studied—and they are everywhere. If organic material is put in an anaerobic environment, methane will almost certainly form. This is particularly likely in water, where bacteria that metabolize organic compounds quickly create anaerobic conditions. Bubbles rising from quiet ponds usually contain methane formed by methanogens in the anaerobic mud at the bottom. Cows and other ruminants produce prodigious amounts of methane in their rumen which they belch into the atmosphere. Metabolism by methanogens is also put to practical use to produce methane from sewage (Chapter 28).

Halophiles. Many bacterial species are **halophiles** (salt lovers), but the halophilic archaebacteria, such as *Halobacterium halobium*, set the record for tolerating high concentrations of salt. These organisms are unable to grow in media containing less than 10 percent sodium chloride (NaCl), and they grow optimally in media containing more than twice that amount. These bright red bacteria flourish even in saturated salt solutions.

The halophilic archaebacteria are aerobes that generate most of their ATP by respiration, but they can also carry out a primitive form of photosynthesis. The cytoplasmic membrane of *Halobacterium* is differentiated into patches called **purple membranes** that contain the pigment **bacteriorhodopsin**. (Bacteriorhodopsin is similar to rhodopsin, the light-sensitive pigment in the eyes of vertebrates.) Bacteriorhodopsin, acting with a cofactor, retinal, converts light into chemical energy. Bacteriorhodopsin uses light energy to expel a proton from the cell, thus creating a proton gradient that can be used to generate ATP.

Thermoacidophiles. The **thermoacidophiles** (heat- and acid-loving bacteria) thrive in extremely hostile environments. They grow at temperatures of 85°C—near the boiling point of water—and pH values of 1.0—as acid as the contents of the human stomach, which is strong enough to kill most living things. They are found, for example, in geothermal springs containing sulfuric acid.

Although all members of the group are adapted to these extreme environments, they are otherwise quite diverse. Members of one genus, *Sulfolobus*, have oddly irregular lobed cells (**Figure 11.20**). Members of another genus, *Thermoplasma*, appear to lack a cell wall completely, resembling the mycoplasmas of Section 10.

Actinomycetes with Multilocular Sporangia (Section 27)

Bergey's Manual lists several sections of **actinomycetes**, bacteria that grow as **mycelia**, masses of branching filamentous cells that resemble a mycelial fungus. The bacteria in Section 27 are characterized by forming spores within a **multilocular sporangium**, a many-chambered swelling at the end of a filament (*loculus* means "small chamber"; **Figure 11.21**). The group contains three genera. *Geodermatophilus* is found in soil, *Dermatophilus* infects animals and sometimes human skin (Chapter 26), and *Frankia* fixes nitrogen. *Frankia* spp. are ecologically important. They live in nodules on the roots of 17 different genera of plants, including alder trees, which typically grow on the banks of streams. When infected with *Frankia*, alders can grow in the complete absence of fixed nitrogen, getting all they need from atmospheric nitrogen fixed by *Frankia*.

Can Bacteria Grow Anywhere? What about Mars?

Some cyanobacteria (Section 19) need no organic nutrients to live, only inorganic ions and light. Chemolithotrophs (Section 20) don't even need light. They oxidize inorganic compounds to get energy. Some bacteria can grow at temperatures above the boiling point of water—others below its freezing point. Some bacteria thrive in the crushing pressures of the deepest oceans, and others at a pH as low as 1 or as high as 12. Some bacteria can even survive high doses of radiation or starvation for hundreds of years. With such wide-ranging capabilities, it's not surprising that bacteria can survive almost everywhere on Earth. But what about other planets?

The reality of space travel has made this a critically important question in two regards. First, extraterrestrial microorganisms might wreak ecological havoc on Earth or start a deadly epidemic. Conversely, microorganisms from Earth might materially alter other planets. Certainly their presence would confound future scientific investigations because scientists could not know if they were native to the planet. For these reasons, the United States and other nations engaged in space exploration agreed in the 1970s to minimize the possibility of contaminating other planets with terrestrial microorganisms. By international agreement, the highest acceptable probability of contaminating a planet is to be one chance in a thousand. This limit is a compromise between safety and practicality.

Many factors affect the probability of contaminating other planets. They include the number of microorganisms on spacecraft, the likelihood that they can survive the trip, and the number of space missions to the planet. But merely introducing microorganisms to a planet does not permanently contaminate it. Eventually the microorganisms will die out—unless they multiply. The crucial factor is the *probability* of a terrestrial microorganism's being able to multiply there and persist indefinitely. Microbiologists estimate the probability of microbial growth by comparing the known or suspected general conditions on planets with the requirements for microbial growth. Of the planets in our solar system, our neighbor Mars is by far the most hospitable. It has the elements and temperature necessary for life and, most importantly, it has water. But the Martian environment is severe. The temperature rarely rises above the freezing point of water, and most regions are extremely dry. Also, the planet is exposed to intense ultraviolet radiation.

In 1977 the United States sent the Viking space probe to Mars with two landers to test for presence of life. At that time, the probability of a terrestrial organism growing on Mars was considered high enough to require that the landers be heat-sterilized to decrease the population of microorganisms. The landers tested a few square meters of surface and found no evidence of microorganisms.

Data collected by the landers also revealed how truly hostile the Martian environment is. Powerful unknown oxidants that are probably lethal to terrestrial microorganisms were found on the surface. Also, the low temperature combined with tiny amounts of water suggests that liquid water—an absolute requirement for life as we know it—might not exist on Mars. The chance of microorganisms growing on Mars is now considered so low that the next planned space probe probably will not be sterilized before launch. Current scientific opinion says Earth is the only place life exists in our solar system.

Streptomyces and Related Genera (Section 29)

The bacteria in Section 29 are also actinomycetes. Members of the actinomycete genus *Streptomyces* are abundant in most soils. In fact, the pleasant odor of freshly turned soil comes from the volatile compounds produced by these organisms. When an agar plate is inoculated with soil, numerous *Streptomyces* colonies develop. The colonies are easy to recognize by their pastel colors, characteristic texture—a hard mass that extends into the agar and sticks together tightly—and soillike odor. These properties reflect the growth pattern of *Streptomyces* (**Figure 11.22**). Some of the *Streptomyces* mycelium grows into the agar, anchoring the colony, while the rest extends up from the agar, forming chains of spores at the tips. The spores give the colony its characteristic pastel color.

Although the general growth pattern of *Streptomyces* is mycelial, the bacteria are distinctly procaryotic in size and structure. In cross-section an actinomycete mycelium looks like a typical Gram-positive bacterial cell.

Streptomyces and many other actinomycetes are important ecologically and industrially. Their abundant presence in soil accounts for much of the breakdown of organic matter that occurs there. But the greatest impact actinomycetes have had on our lives comes from the ability of some to produce antibiotics. Most antibiotics in current use are produced by members of the genus *Streptomyces* (Chapter 21).

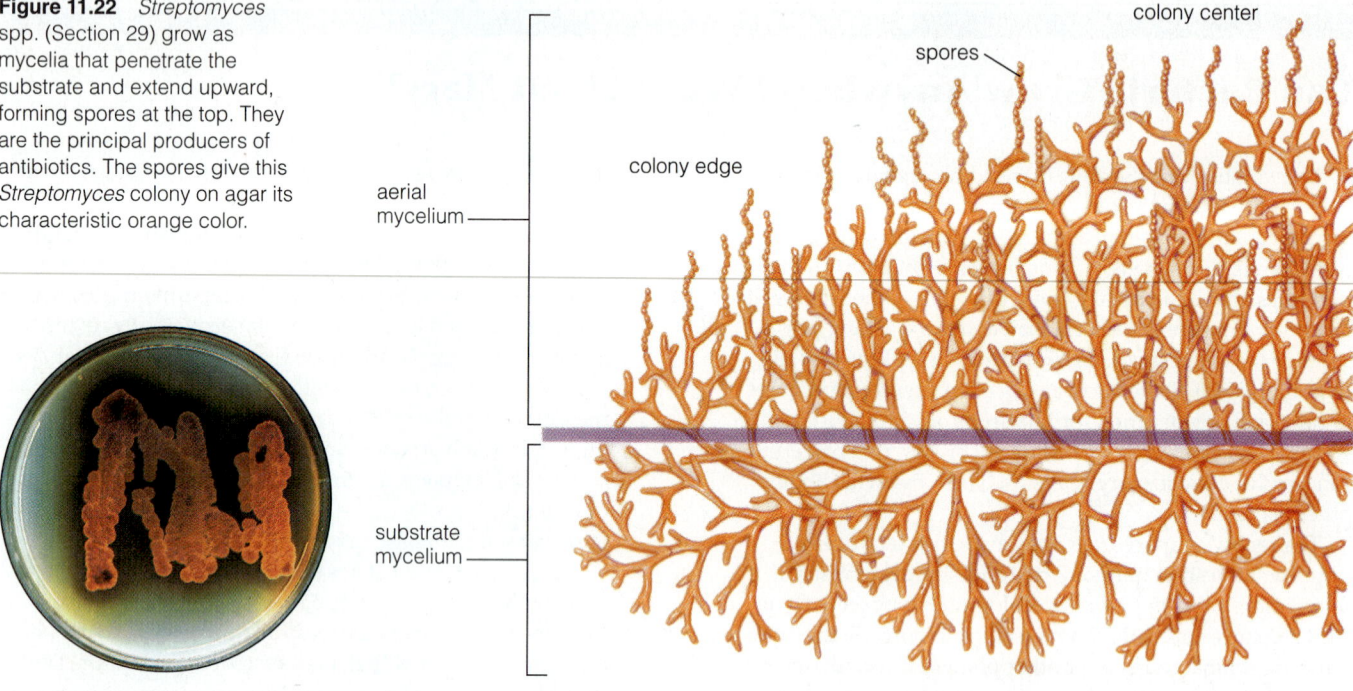

Figure 11.22 *Streptomyces* spp. (Section 29) grow as mycelia that penetrate the substrate and extend upward, forming spores at the top. They are the principal producers of antibiotics. The spores give this *Streptomyces* colony on agar its characteristic orange color.

aerial mycelium

colony edge

spores

colony center

substrate mycelium

Summary

IDENTIFYING BACTERIA (p. 249)

1. When identifying bacteria, microbiologists begin with morphological characters but rely principally on physiological characters.

BACTERIAL TAXONOMY (p. 249)

1. Though we know a great deal about the evolutionary relatedness of bacteria, a natural classification scheme is not yet practical. The artificial scheme in *Bergey's Manual* is widely used.

THE *BERGEY'S MANUAL* SCHEME OF BACTERIAL TAXONOMY (pp. 249–271)

1. *Bergey's Manual* divides bacteria into four divisions based on cell wall composition—the Gram-negative bacteria (thin peptidoglycan wall), Gram-positive bacteria (thick peptidoglycan wall), mycoplasmas (no wall), and archaebacteria (walls composed of materials other than peptidoglycan).

2. Bacterial species in each division are placed in one or more sections according to similar properties. Sections have no taxonomic standing.

3. Some clinically and ecologically important groups of bacteria are covered in this chapter.

   a. The spirochetes are Gram-negative bacteria distinguished by their corkscrew shape and unusual mechanism for motility (axial filaments).

   b. The Gram-negative aerobic rods and cocci are a diverse group that includes major human pathogens (for whooping cough, gonorrhea), plant pathogens (for crown gall), commercially important species (such as one that produces acetic acid for vinegar), and ecologically important species (*Rhizobium*, nitrogen-fixing bacteria).

   c. The facultatively anaerobic Gram-negative rods are grouped into three families that have great impact on humans—the enterics, the vibrios, and the pasteurellas.

   d. The dissimilatory sulfate- or sulfur-reducing bacteria are ecologically important to nature's sulfur cycle.

   e. The rickettsiae and chlamydiae are so small they were once thought to be viruses. They cause human infections.

   f. The mycoplasmas have no cell walls.

   g. The Gram-positive cocci are a large group with only their shape in common. Some aggregate in characteristic ways, such as *Streptococcus* (in chains or pairs) and *Staphylococcus* (grapelike clusters).

   h. The endospore-forming Gram-positive rods and cocci include human pathogens in the *Bacillus* and *Clostridium* genera.

   i. The irregular nonsporing, Gram-positive rod bacteria include the genus *Arthrobacter*, which breaks down organic matter in soil and which exhibits a curious movement called snapping post-fission movement.

   j. The mycobacteria and nocardioforms have mycolic acids in their cell walls that give them a waxy surface and stain in a distinctive way. Species of *Mycobacterium* cause tuberculosis and Hansen's disease (leprosy).

   k. The oxygenic photosynthetic bacteria are called cyanobacteria. They were called blue-green algae.

   l. The aerobic chemolithotrophic bacteria oxidize inorganic compounds and thus play an important role in global ecology.

   m. The budding and/or appendaged bacteria are a collection of unusually shaped bacteria and bacteria that divide by budding rather than fission.

   n. The archaebacteria are subdivided into three groups, based on extremes of habitat—the methanogens (methane formers), halophiles (salt lovers), and thermoacidophiles (heat- and acid-loving bacteria).

   o. The *Streptomyces* produce most of the antibiotics in current use; they are also important in the ecology of soil.

IDENTIFYING BACTERIA

1. Describe factors involved in identification of microorganisms.

BACTERIAL TAXONOMY

1. What kind of classification scheme does *Bergey's Manual* use and why?

2. What kinds of information do you find in *Bergey's Manual*?

THE *BERGEY'S MANUAL* SCHEME OF BACTERIAL TAXONOMY

1. What are the four divisions of *Bergey's Manual*?

2. On what basis are bacteria subdivided into sections? Why are some bacteria placed in more than one section?

3. *The Spirochetes (Section 1):* Describe how spirochetes move. What is their shape? Name some diseases they cause.

4. *Aerobic/Microaerophilic, Motile, Helical/Vibrioid Gram-Negative Bacteria (Section 2):* Define these terms: microaerophilic, helical, vibrioid. One of the members of this group is *Bdellovibrio bacteriovorus*, which preys on other bacteria. What does it mean that *B. bacteriovorus* is host-specific?

5. *Gram-Negative Aerobic Rods and Cocci (Section 4):* Describe the process by which *Agrobacterium tumefaciens*, a member of this group, causes crown gall disease in plants. How are xanthan gum and acetic acid, two products of this group, used industrially? What is the significance of the symbiotic relationship *Rhizobium* and *Bradyrhizobium* have with legumes?

6. *Facultatively Anaerobic Gram-Negative Rods (Section 5):* Name some human diseases caused by members of this group. What four properties of *Escherichia coli* distinguish it from other enterics?

7. *Anaerobic Gram-Negative Straight, Curved, and Helical Rods (Section 6):* Give an example of a disease caused by a member of this group.

8. *Dissimilatory Sulfate- or Sulfur-Reducing Bacteria (Section 7):* How do members of this group get their energy and why is this ecologically significant? Where are these bacteria found?

9. *Anaerobic Gram-Negative Cocci (Section 8):* What functions do these bacteria perform?

10. *The Rickettsias and Chlamydias (Section 9):* Why were these bacteria once thought to be viruses? What does it mean that most are obligate intracellular parasites? Name some of the human infections they cause. Describe chlamydial reproduction.

11. *The Mycoplasmas (Section 10):* What characteristic unites this group? What does it mean that they are obligate fermenters? How do you account for their name?

12. *Gram-Positive Cocci (Section 12):* Tell how cells in these genera are arranged: *Sarcina, Streptococcus, Staphylococcus*. Some members of this group are called lactic acid bacteria. What does this mean?

13. *Endospore-Forming Gram-Positive Rods and Cocci (Section 13):* What are endospores? Why are *Bacillus* and *Clostridium* used as an index of sterilization? Name some of the serious human diseases these two genera cause.

14. *Regular Nonsporing, Gram-Positive Rods (Section 14):* The largest known bacterium, recently discovered, belongs to this group. How large is it, and how many times larger is it than rickettsiae and chlamydiae (Section 9)?

15. *Irregular Nonsporing, Gram-Positive Rods (Section 15):* What are some of the "irregular" shapes included in this group? Some members of this group, including *Arthrobacter*, exhibit snapping post-fission movement. Describe this. *Bifidobacterium* is also a member of this group. Why is it important?

16. *The Mycobacteria and Nocardioforms (Sections 16 and 17):* What is unusual about the wall structure of these groups? Name some of the human diseases caused by mycobacteria.

17. *Anoxygenic Phototrophic Bacteria (Section 18):* How is the photosynthesis carried out by these bacteria different from that carried out by green plants, algae, and cyanobacteria? Why are these bacteria highly colored?

18. *Oxygenic Photosynthetic Bacteria (Section 19):* These bacteria were called the blue-green algae. What are they called now? Why were they originally classified as algae? Why were they reclassified? What role do heterocysts play?

19. *Aerobic Chemolithotrophic Bacteria (Section 20):* How do chemolithotrophs get their energy? On what basis are members of this group subdivided?

20. *Budding and/or Appendaged Bacteria (Section 21):* What are prosthecate bacteria? How do most bacteria reproduce if not by budding?

21. *Sheathed Bacteria (Section 22):* Explain where sheathed bacteria are found and how they grow and multiply. What is bulking?

22. *Nonphotosynthetic, Nonfruiting Gliding Bacteria (Section 23):* Describe gliding motility.

23. *Gliding Fruiting Bacteria (Section 24):* What is the other name for gliding fruiting bacteria? Under what conditions do fruiting bodies and myxospores appear?

24. *Archaeobacteria (Section 25):* How are archaebacteria like and unlike eubacteria? What characterizes the methanogens and where are they found? What are halophiles and where are they found? What are thermoacidophiles and where are they found?

25. *Actinomycetes with Multilocular Sporangia (Section 27):* Define actinomycetes. What is a multilocular sporangium? Why is *Frankia* ecologically important?

26. *Streptomyces (Section 29):* Describe the appearance of *Streptomyces* spp. colonies in a petri dish. What is their characteristic odor? What gives the colony its pastel color? How is *Streptomyces* important ecologically? medically?

Suggested Readings

Balows, A.; Trüper, H.; Dworkin, M.; Harder, W.; and Schleifer, K-H. 1991. *The prokaryotes: A handbook on the biology of bacteria: Ecophysiology, isolation, identification, applications.* 2d ed. New York: Springer-Verlag.

Clements, K. D., and Bullivant, S. 1991. An unusual symbiont from the gut of surgeonfishes may be the largest known prokaryote. *Journal of Bacteriology* 173:5359–62.

Holt, J. G., ed. *Bergey's manual of systematic bacteriology.* 1st ed. Vol. 1, 1984; vol. 2, 1986; vols. 3 and 4, 1989. Baltimore: Williams and Wilkins.

Stanier, R. Y.; Ingraham, J. L.; Wheelis, M. L.; and Painter, P. R. 1986. *The microbial world.* 5th ed. Englewood Cliffs, N.J.: Prentice-Hall.

12 EUCARYOTIC MICROORGANISMS, HELMINTHS, AND ARTHROPOD VECTORS

Parasitology: A Can of Worms?

In clinical medicine, *parasitology* is the study of diseases caused by pathogenic protozoa and helminths, two of the eucaryotic groups discussed in this chapter. Pathogenic bacteria, viruses, and fungi are also parasites, but certain features set protozoal and worm infections apart from other diseases, making parasitology a separate science and one of increasing importance.

One feature that distinguishes protozoal and helminthic infections is their immense importance in tropical countries and relatively minor importance in temperate countries such as the United States. About 1 billion people are infected by the worm *Ascaris*, 600 million by malarial protozoa, and 300 million by larval roundworms called *filaria*—almost all in the developing world. In fact, parasitology is sometimes called *tropical medicine*. But it would be a mistake to think these diseases are of no concern to us in the United States. Parasitic diseases are becoming more common in this country, in part because more infected people are moving to the United States and in part because immune deficiencies such as AIDS make more people susceptible to certain parasites.

Less is known about protozoal and helminthic parasites than about bacteria and viruses—perhaps because they are less well researched, perhaps because they are such complex organisms. For example, it is a mystery how the immune system responds to these organisms. Evidence suggests that protozoa and worms activate immune responses, but the immune system is seldom able to rid the body of them. Infections typically last for years. The chronic nature of parasitic diseases, coupled with a generally high number of infecting parasites, makes these infections especially debilitating. Parasitic infections lower the quality as well as the length of life for millions of people.

Parasitic infections are also set apart from other kinds of infections because of the complex life cycles and multiple hosts involved. Many protozoal and helminthic species change body type during their life cycle. Moreover, parasites that exist in different forms as larvae and adults often have different hosts at each stage. Creatures as diverse as mosquitoes, snails, and cattle may be essential to transmit infection.

Ascaris lumbricoides is the largest of the roundworms that infect human beings.

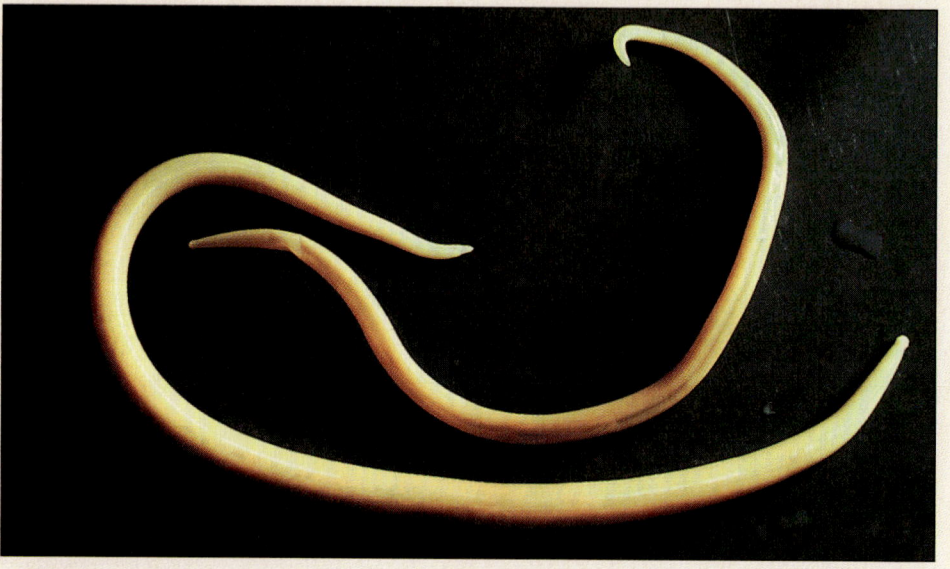

10 mm

To understand:

- The characteristics, ecological roles, and classification of the major groups of eucaryotic microorganisms—the fungi, algae, and protozoa

- The nature of lichens as stable mutualistic associations between algae and fungi

- The properties of slime molds, a small group of eucaryotic microorganisms distinct from the three major groups

- The major groups of helminths—appearance, life cycles, and the diseases they cause

- The characteristics of arthropods, including insects, ticks, and mites that cause disease or act as vectors

EUCARYOTIC MICROORGANISMS

Because microorganisms are defined by size, which is relative, the perimeters of the science of microbiology are necessarily fuzzy. Some microbiologists include groups of organisms that other microbiologists exclude. In this chapter we look at some of these groups at the perimeter. First we consider the eucaryotic microorganisms, consisting of fungi, algae, protozoa, and slime molds. Fungi, protozoa, and slime molds are included in any definition of microorganisms. Some definitions include all algae as microorganisms, but others consider some algae as protists, and therefore microorganisms, and others as plants.

We also consider helminths and certain arthropods in this chapter. Flatworms and roundworms are helminths. They are not microorganisms, but some cause infectious disease. Arthropods—animals with jointed legs, such as crabs, lobsters, insects, and arachnids—are not microorganisms. We examine the insects and arachnids because some cause infectious disease directly (Chapter 26) and others transmit it (Chapter 15).

In the five-kingdom scheme of classification, plants (including multicellular algae) and animals constitute two kingdoms (Chapter 10). The other three kingdoms consist of microorganisms, though some groups have macroscopic members (Chapter 1). Of these, all procaryotes belong to the kingdom Monera. All remaining eucaryotic microorganisms are divided between two kingdoms—the Fungi, which constitute a kingdom of their own, and the Protista, which include protozoa, unicellular algae, and slime molds. For the sake of simplicity, we will consider the fungi, algae, **lichens** (symbiotic associa-

tions of fungi and algae or cyanobacteria), protozoa, and slime molds in that order, without special regard for their kingdom classification.

Fungi

The **fungi** are a large and diverse group. With more than 250,000 named species, they constitute a kingdom that includes single-celled organisms called **yeasts**; filamentous organisms called **molds**; and organisms that form fleshy, macroscopic, fruiting structures called **mushrooms, puffballs**, or **shelf fungi**, depending on their shape (**Figure 12.1**; see also Figure 1.5b). The branch of microbiology that studies this complex group of organisms is **mycology** (*mykes* is Greek for "mushroom").

Like all animals, plants, and protists, fungi are **eucaryotic** (Chapter 4). In addition, fungi are **heterotrophic** (they use organic compounds as a carbon source) and **nonphototrophic** (they do not use light as an energy

a

b

Figure 12.1 Fungi include organisms as diverse as (**a**) shelf fungi growing from a tree trunk and (**b**) the mold *Plasmospora viticola*, which causes powdery mildew of grapes.

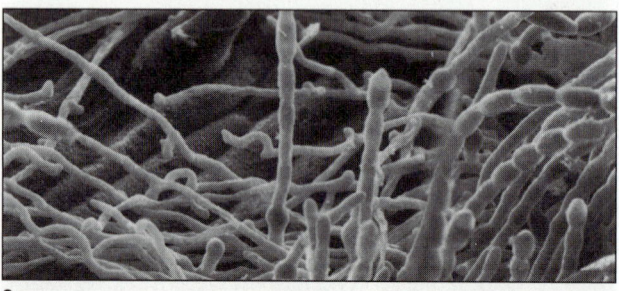

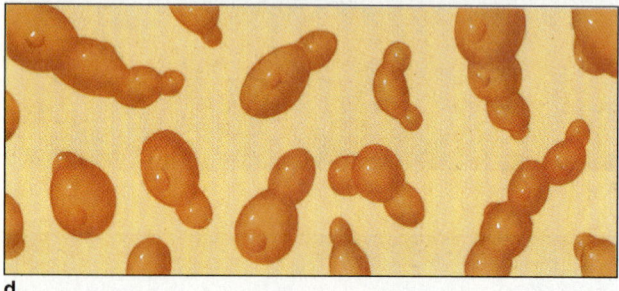

Figure 12.2 Fungal structure. (**a**) In higher fungi, the hyphae have septa (cross walls) at regular intervals, though these walls are incomplete. (**b**) Lower fungi have no cross walls. (**c**) This micrograph shows that mycelia are a mesh of hyphae. (**d**) Some yeasts that reproduce by budding form pseudohyphae. Despite the resemblance, they are not true hyphae.

source). Fungi are also **absorptive**. That is, they take in nutrients, as bacteria do: molecules in solution pass through the plasma membrane.

Most fungi are also **saprophytes**. They obtain nutrients by decomposing dead and decaying organic matter. Some are **parasites**, existing in or on living plants, animals, and humans, causing disease. Most of the fungal

pine needles

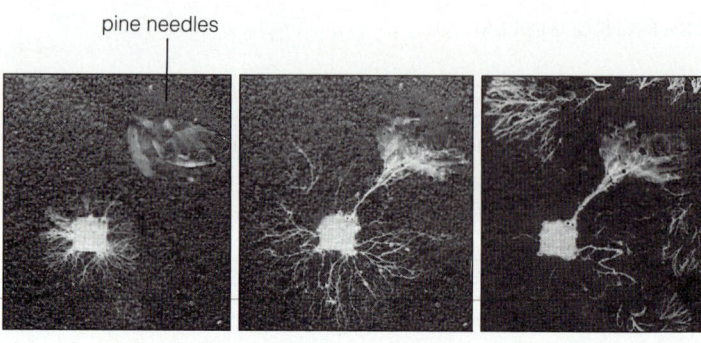

Figure 12.3 Molds grow as mycelial hyphae elongate at their tips. When some of them encounter a source of nutrients (here, a cluster of pine needles), growth is channeled there and curtailed elsewhere.

species that cause human disease are saprophytic organisms adapted to survive in human tissues.

Morphology. Most fungi, with the exception of unicellular species, have a vegetative structure known as a mycelium. A **mycelium** is a multinucleate mass of cytoplasm enclosed within a system of rigid, branched, tubelike filaments called **hyphae** (**Figure 12.2**). In the lower fungi, the mycelium is **coenocytic**, an undivided network of branching tubes. The higher fungi have **septa** (cross walls) that appear to divide the mycelium into compartments, but these walls are incomplete. They have a central opening that cytoplasm can flow through. In this sense the mycelium of a fungus can be viewed as a single giant cell. Parts of a mycelium, termed **vegetative mycelia**, are specialized to absorb nutrients and produce more cytoplasm. Others, termed **reproductive mycelia**, are specialized to produce spores (reproductive structures).

In molds, the **thallus** (body of a fungus) is composed of highly branched and loosely intertwined hyphae, forming an open mycelium. The mycelia of molds often form woolly growth on damp and decaying material in nature and fuzzy spreading colonies on solid media in culture (**Figure 12.3**).

The thallus of a fungus that forms mushrooms or other fleshy fruiting structures consists of two parts—an extensive, usually underground, moldlike mycelium and an above-ground mushroom. The mushroom is composed of tightly intertwined hyphae, forming a solid mass that contains sexual spores. Mushrooms form quickly—sometimes overnight—because cytoplasm from the underground mycelium flows into the developing mushroom.

The remarkably tough hyphae are composed of cellulose (Chapter 2), **chitin** (a polymer of acetylglucosamine, also found in the hard exoskeleton of crabs, lobsters, and some insects), or a combination of the two.

Most yeasts are oval-shaped cells that reproduce by **budding**. In budding, a bubble forms on the cell surface, grows, and pinches off to separate. Sometimes buds remain attached to the mother cell, forming a chain of cells called a **pseudohypha** (Figure 12.2). A few yeasts divide, as most bacteria do, by forming cross walls. In culture, yeasts form round, pasty, or mucoid colonies on solid media.

Ecology. Fungi survive and prosper in nearly every environment on Earth, both aquatic and terrestrial. They grow in highly acidic environments, such as acidic fruit, and in the presence of high concentrations of salts, sugars, and other nutrients, such as on the surface of home-canned jams and jellies. They grow at temperatures well below the freezing point of water (below those that support the growth of bacteria), but they cannot tolerate temperatures as high as bacteria do. Fungi can grow with minuscule concentrations of nutrients. In the laboratory, for example, they often develop in "nutrient-free" reagents, deriving nutrition from airborne organic compounds that dissolve in these liquids.

Most fungi are aerobes. A few species are anaerobes, and some species, including many species of yeast, are facultative anaerobes.

Along with bacteria, fungi are the prime decomposers of organic materials, including food. They convert dead plant and animal materials into forms that plants can use (Chapter 28). Some fungi are even better than bacteria at decomposing decay-resistant organic materials such as **lignin**, the dark-colored component of wood. Fungi such as *Polystictus* spp. and *Armillaria* spp. cause **white rot** by attacking the dark-colored lignin and leaving white cellulose. Other species of fungi (and bacteria as well) cause **brown rot**. By selectively removing cellulose and leaving lignin, they convert wood into a pulpy reddish-brown mass. Other fungi, such as *Pleurotus* spp. and *Polyporus* spp., attack both cellulose and lignin. Fungi can break down just about anything organic—including clothing, paper, leather, and paint. Some fungi even grow in jet fuel, causing clogged fuel lines. Fungi are abundant in garden compost piles, converting plant materials and other organic waste into rich friable soil.

Reproduction. Fungi are classified by how they reproduce, sexually or asexually.

Asexual reproduction occurs by elongation of hyphae, division or budding of single cells, or the production of **asexual spores**, specialized cells that are dispersed and germinate in a favorable environment to produce a new thallus. Asexual spores are products of mitosis. They are of two general types—**sporangiospores** and

conidia (also called **conidiospores**). Sporangiospores are produced within a **sporangium** (a spore-containing structure). They can be motile or nonmotile. Conidia are borne naked on the tips of specialized hyphae called **conidiophores**.

Sexual reproduction occurs by producing **sexual spores**, which form following sexual fusion of gametes. Usually, but not always, meiosis occurs before spores are formed. **Homothallic fungi** produce both male and female gametes on the same thallus. **Heterothallic fungi** produce male and female gametes on different thalli. Usually, but not always, sexual spores are more resistant than asexual spores and vegetative cells to heat, desiccation, and other hostile conditions.

Each group of fungi—Chytridiomycetes, Oomycetes, Zygomycetes, Ascomycetes, and Basidiomycetes—produces characteristic spores.

Classification of the Lower Fungi. The fungi are divided into two broad groups, called the **lower fungi** and the **higher fungi**, based on evolutionary advancement (**Table 12.1**).

The lower fungi are all coenocytic. They are divided into five classes according to the structure of their spores and gametes. The most important of these are the Chytridiomycetes and Oomycetes, which are water molds, and the Zygomycetes, which are terrestrial. The asexual spores of these three groups are sporangiospores.

The **Chytridiomycetes** are water molds, aquatic fungi that produce fuzzy, whitish growth on organic materials that fall into streams or ponds. Within a few days after being immersed in water, a piece of fruit or a dead fish, for example, becomes covered with a growth of Chytridiomycetes and other water molds. At some stage of their life cycle, Chytridiomycetes produce motile sporangiospores and motile gametes that have single posterior flagella. Some classification schemes place the Chytridiomycetes among the Protista. *Allomyces* is a well-studied Chytridiomycete that grows in tropical ponds and streams.

The **Oomycetes** are water molds that produce motile, **biflagellate** (two flagella) sporangiospores and nonmotile sexual spores formed after fusion of nonmotile female gametes with motile or nonmotile male gametes. The thin film of water on damp soil particles is adequate for these organisms to complete their life cycle. Oomycetes infect plants. Very few infect humans. Some species, such as those belonging to the genus *Saprolegnia*, commonly found in streams and ponds, are parasitic on fish (**Figure 12.4**).

The **Zygomycetes** do not produce swimming cells. They can complete their life cycle in an environment with no liquid water. The black mold *Rhizopus nigricans*, which develops on stale bread and other cereal foods, is a Zygomycete (**Figure 12.5**). Zygomycete sporangiospores are

Table 12.1   Properties of Major Groups of Fungi

Class	Distinguishing Characters	Examples
The Lower Fungi	Mycelia (if present) are coenocytic and lack septa.	
Chytridiomycetes	Water molds with uniflagellated sporangiospores and gametes.	*Allomyces* (seen as fuzzy growth in tropical ponds and streams)
Oomycetes	Water molds with biflagellate sporangiospores and male gametes, nonmotile female gametes and sexual spores.	*Saprolegnia* (seen as fuzzy growth on fish in a stream)
Zygomycetes	Terrestrial fungi with nonmotile sporangiospores and zygospores. Several genera parasitize weakened human patients.	*Rhizopus* (black bread mold)
The Higher Fungi	Mycelia (if present) are septate; all species are terrestrial; conidia and gametes (if present) are nonmotile.	
Ascomycetes	Ascospores formed in a sac (ascus); some parasitic on plants or animals.	*Peziza* (an orange cup fungus seen on soil in damp woods)
Basidiomycetes	Basidiospores formed on a club-shaped basidium; some are parasitic on plants or animals.	*Amanita* (red or white mushrooms seen in forests)
Deuteromycetes (Fungi Imperfecti)	Higher fungi that lack sexual spores; some are parasitic on plants or animals.	*Penicillium* (blue or green mold seen on fruit; used to make penicillins)
Yeasts	Unicellular, oval cells, mostly Ascomycetes and Deuteromycetes but some Basidiomycetes and Zygomycetes; some parasitic on plants or animals.	*Saccharomyces cerevisiae* (yeast used in baking, brewing, and wine making)

Figure 12.4 The Oomycete *Saprolegnia* parasitizes fish. The skin peeled and there is a fuzzy growth on the tail.

unflagellated. When released from the sporangia, they are dispersed by air currents. The gametes of Zygomycetes are ordinary-appearing short hyphae that swell and fuse when they come close to one another, forming a zygote that undergoes meiosis to produce a resistant **zygospore**. Gametes of Zygomycetes cannot be distinguished as being male or female, but they belong to one of two **mating types**, designated plus (+) and minus (−). Plus gametes fuse only with minus gametes and vice versa. Several genera of Zygomycetes, including *Rhizopus*, contain species that produce diseases called **zygomycoses** in individuals who are immunologically or otherwise weakened.

Classification of the Higher Fungi. The mycelia of all higher fungi have septa perforated by a central pore. The higher fungi are divided into Ascomycetes, Basid-

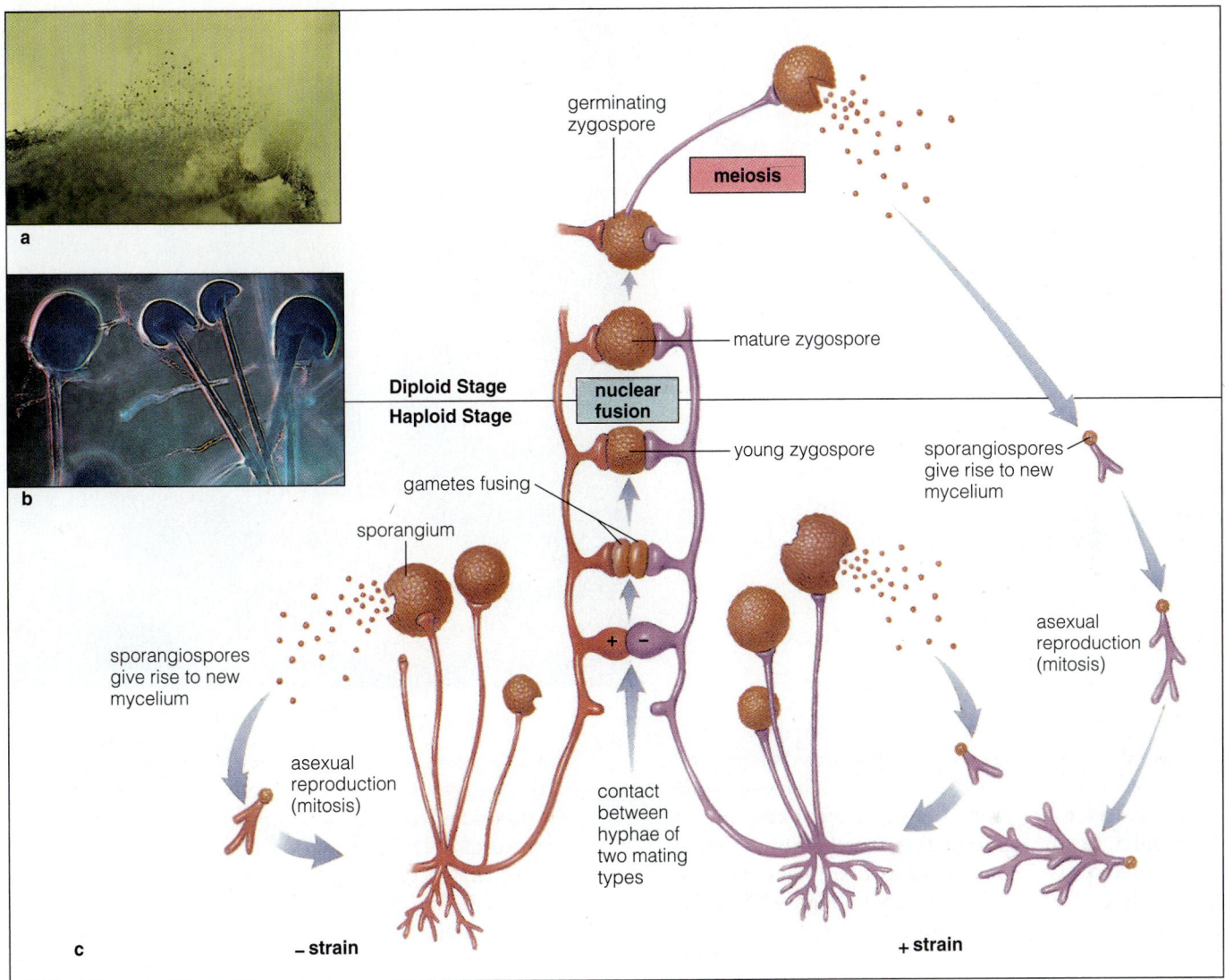

Figure 12.5 Life cycle of a Zygomycete. (**a**) The zygomycete *Rhizopus* is commonly called the black bread mold because it covers and blackens bread. (**b**) *Rhizopus* produces many sporangia filled with black sporangiospores. (**c**) Zygospores form when two hyphae (+ and –) swell and fuse.

iomycetes, and Deuteromycetes, depending on whether they form sexual spores and, if so, what kind. The asexual spores of all three groups are conidia.

The sexual reproduction of all **Ascomycetes** is quite similar—hyphal tips on the same or different thalli (homothallic or heterothallic, respectively) fuse. The nucleus within the resulting zygote undergoes meiosis, and sometimes the four nuclear products divide again. Then they develop into sexual spores called **ascospores** within the **ascus** (a sac that forms in the wall of the zygote). When released from the ascus, ascospores germinate to produce a vegetative mycelium, completing the cycle of

sexual reproduction. Asexual reproduction occurs by formation and germination of conidia.

The major difference among species of Ascomycetes is the way that asci are arranged. Some species produce individual asci. They are moldlike throughout their life cycle. Other species produce arrays of asci arranged in a structure called an **ascocarp**. The shapes of ascocarps are quite varied (**Figure 12.6**). Some, like those of *Peziza* species, are cup-shaped. They are the bright orange or red structures seen on forest floors. Others, like those of *Morchella* (morels), are mushrooms, with a globular head on a stalk. Morels are desired for their delicious taste.

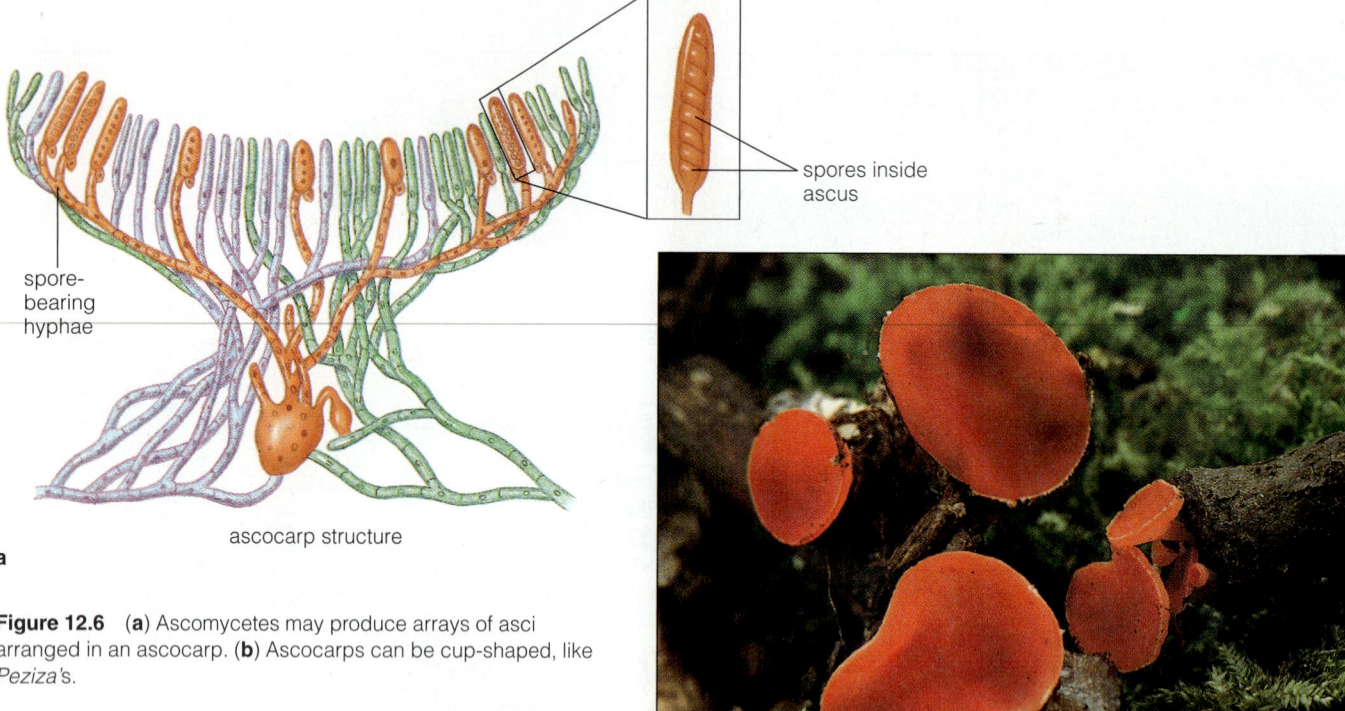

a

ascocarp structure

spore-bearing hyphae

spores inside ascus

b

Figure 12.6 (**a**) Ascomycetes may produce arrays of asci arranged in an ascocarp. (**b**) Ascocarps can be cup-shaped, like *Peziza*'s.

Most **Basidiomycetes** form **basidiocarps** (mushrooms, puffballs, or shelflike bodies on trees), but some are molds, and a few are yeasts. Asexual reproduction occurs by conidia. Sexual reproduction of mycelium-forming Basidiomycetes is unique because at one stage they undergo extensive growth as a **dikaryon** (**Figure 12.7**). The mycelium contains nuclei from the two different mating types whose hyphae have fused. The cytoplasms of the two mycelia mix but the nuclei do not fuse. In mushroom- and puffball-forming Basidiomycetes, the dikaryon phase develops underground. When conditions are right, much of the cytoplasm flows through the mycelium to the soil surface where it quickly forms the basidiocarp, or fruiting structure, which is the mushroom or puffball. Within the basidiocarp, pairs of nuclei from different mating types fuse, undergo meiosis, and develop into sexual spores called **basidiospores**. These basidiospores form on **basidia**, club-shaped cells, on the gills of gilled mushrooms or the pores of **pored mushrooms**, or inside puffballs. When mature basidiospores are shot from the surface of gills (or pores) into the space between them, they fall out of the mushroom to the ground where they germinate.

Deuteromycetes produce conidia but not sexual spores. Thus, they do not form mushrooms or puffballs. They grow as molds or yeasts. Deuteromycetes are also called **Fungi Imperfecti** because two nineteenth-century

mycologists, Charles and Louis Tulasne, termed the sexual stage *perfect* and fungi that lacked it, *imperfect*.

Deuteromycete is a classification of convenience. A culture of a higher fungus that does not form sexual spores cannot be assigned to either the Ascomycetes or the Basidiomycetes, so it is assigned, sometimes temporarily, to the Deuteromycetes. For example, a culture of one mating type of a heterothallic Ascomycete or Basidiomycete could not form sexual spores in the absence of its opposite mating type. It would be assigned to the Deuteromycetes. But if the opposite mating type became available, it would form sexual spores and could be classified as an Ascomycete or Basidiomycete. For this reason some fungal cultures originally classified as Deuteromycetes have later been reassigned to an Ascomycete or Basidiomycete genus (**Table 12.2**). Other fungal cultures that have lost through mutation the capacity to produce sexual spores remain permanently classified as Deuteromycetes.

Because they lack sexual spores, Deuteromycetes are identified principally by the shape and arrangement of their conidia (**Figure 12.8**). For example, the conidia of the well-known imperfect genus *Penicillium* form long chains on branching conidiophores, creating a brushlike structure (*penicillus* means "brush" in Latin). The conidia of a closely related genus, *Aspergillus*, form long chains on a globelike conidiophore. Some species of *Penicillium*

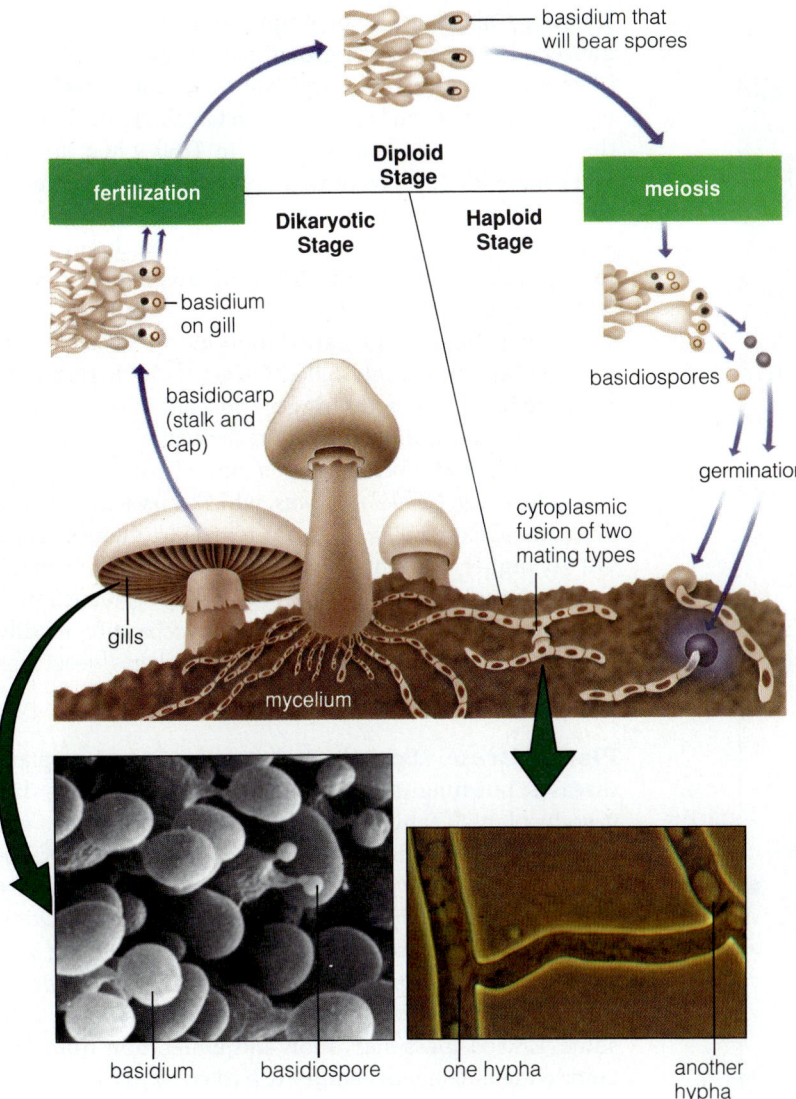

basidium that will bear spores

Diploid Stage

fertilization

Dikaryotic Stage

meiosis

Haploid Stage

basidium on gill

basidiospores

basidiocarp (stalk and cap)

germination

cytoplasmic fusion of two mating types

gills

mycelium

basidium   basidiospore

one hypha

another hypha

Figure 12.7 Life cycle of a Basidiomycete gilled mushroom. Basidiospores formed after meiosis germinate to produce haploid mycelia. The haploid hyphae fuse, forming dikaryon mycelia. When conditions are right, they form a mushroom. Then, on the gills, basidia form, the nuclei fuse (fertilization), and basidiospores develop.

produce the antibiotic penicillin. Some species of *Aspergillus* cause aspergillosis, a disease of animals, including humans. Most fungi that cause human disease—including *Blastomyces*, *Cryptococcus*, *Histoplasma*, and *Candida*—are Deuteromycetes.

Yeasts. *Yeast* is a descriptive term, not a taxonomic one. Originally, only strains of *Saccharomyces cerevisiae* (or baker's yeast) used to make bread, beer, and wine were referred to as yeasts. *S. cerevisiae* is a facultative anaerobe (Chapter 5). In the absence of oxygen, it ferments sugars to ethanol and carbon dioxide. The carbon dioxide gas makes bread dough rise and fermenting beer or wine appear to boil (*yeast* comes from the Greek word *zestos*, meaning "to boil").

Table 12.2 How Deuteromycetes Are Reclassified If Sexual Spores Form

Deuteromycete Genus (no sexual spores)	Reclassification If Sexual Spores Form	
	Class	Genus
Aspergillus	Ascomycete	*Eurotium*
Blastomyces	Ascomycete	*Ajellomyces*
Candida	Ascomycete	*Pichia*
Cryptococcus	Basidiomycete	*Filobasidiella*
Histoplasma	Ascomycete	*Gymnoascus*
Penicillium	Ascomycete	*Talaromyces*
Trichophyton	Ascomycete	*Arthroderma*

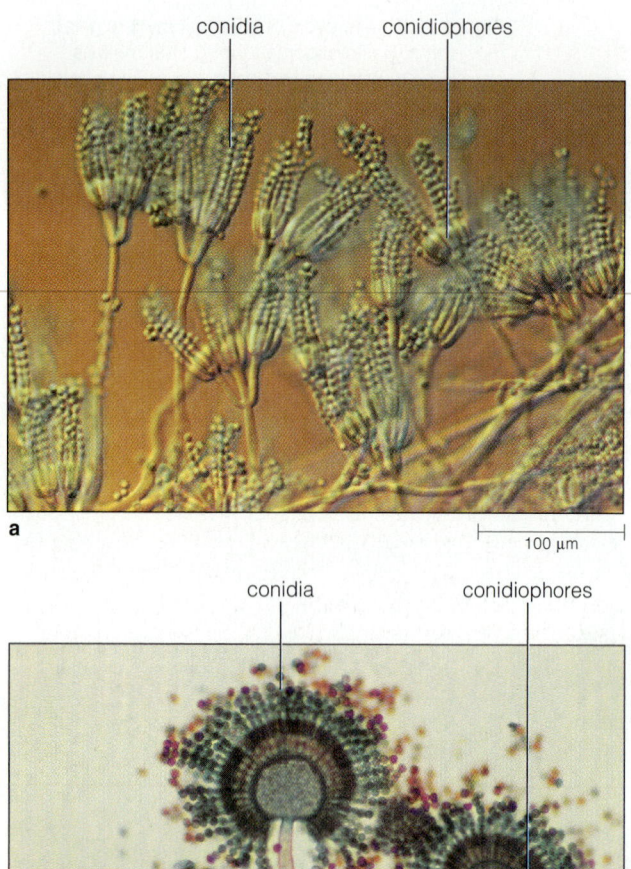

conidia   conidiophores

a   100 µm

conidia   conidiophores

b   100 µm

Figure 12.8 Deuteromycetes are identified by the shape and arrangement of conidia. (**a**) On *Penicillium*, spherical conidia form chains on branching conidiophores. (**b**) On *Aspergillus*, spherical conidia form chains on globelike conidiophores.

Today microbiologists use the term for any of the hundreds of species of nonfilamentous, single-celled, round or oval-shaped fungi. Most yeasts are Ascomycetes or Deuteromycetes, some are Basidiomycetes, and a few are Zygomycetes. Yeasts, which include aerobes as well as facultative anaerobes, are widely distributed in nature. They are found on leaves, fruits, and cured meats such as bacon and ham; in the nectar of flowers; in soil; and on our bodies as normal flora (Chapter 14). Most yeast cells multiply by budding, and a few by fission.

Dimorphic Fungi. Some fungi switch between a single-celled yeast phase of growth and a mycelial phase, a phenomenon called **dimorphism** (*di* means "two" and *morphe* means "form" or "shape" in Greek). Louis Pasteur discovered dimorphism in fungi in 1860 when he was working with the Zygomycete *Mucor rouxii*. *M. rouxii* switches from a mycelial to a yeast form if the oxygen supply decreases. More often, fungi exhibit dimorphism in response to temperature. They are mycelial at 25°C to 30°C and yeastlike at 37°C.

Pathogenic fungi that are dimorphic are mycelial outside the host and single-celled inside it. The higher temperature inside the body triggers the shift in most pathogenic fungi, including *Histoplasma capsulatum*, *Blastomyces dermatitidis*, and *Coccidioides immitis*. There are exceptions, however. *Candida albicans*, which causes thrush and vaginitis (Chapter 24), does not respond to temperature but to higher concentrations of nutrients in the body. Forming single cells is essential to causing systemic infection because single cells, but not mycelia, are readily spread in the bloodstream. Recently, baker's yeast has been found to be dimorphic.

Plant Disease. Some bacteria and viruses cause plant diseases, but fungi are the major cause of infectious disease in plants (**Table 12.3**). Fungal plant pathogens are mainly molds but include a few yeasts. Two genera of Oomycetes, *Pythium* and *Phytophthora*, are particularly devastating. *Pythium* spp. kill young plants by attacking the stem near the ground, blackening it and causing it to collapse, a disease process called **damping off**. *Phytophthora infestans*, which turns potato tubers into a dark slime, caused mass starvation and emigration from Ireland in the nineteenth century (Chapter 1).

Higher fungi cause **apple scab**, which disfigures apples (**Figure 12.9**); **corn smut**, which turns ears of corn into swollen, black, powdery masses; and **wheat rust**, which forms rusty red spots on the plant. Corn smut and wheat rust have destroyed vast quantities of these cereal grains. Another fungus causes **Dutch elm disease**, which has virtually eliminated that graceful shade tree from the United States.

Many plant diseases are successfully controlled by applying antifungal sprays or dusts. Others are controlled by breeding fungus-resistant plant varieties, such as rust-resistant wheat and smut-resistant corn. Often, plant geneticists keep barely one step ahead of the microorganisms, developing a new rust-resistant strain of wheat just before the previously developed variety becomes vulnerable to a new strain of fungus.

Fungi and Medicine. Of the 250,000 known fungal species, fewer than a hundred cause human disease. Any

Microbe Mappers

Gerald Fink

In 1992, at the Whitehead Institute in Boston, Gerald Fink discovered that *Saccharomyces cerevisiae* exhibited dimorphism—it could switch from being unicellular to being mycelial. The discovery came 132 years after Louis Pasteur discovered the phenomenon and in a microorganism that has been intensively studied by hundreds of microbiologists (baker's yeast is the *Escherichia coli* of eucaryotic microbiology) and used by bakers and brewers for thousands of years.

How could dimorphism in this organism be overlooked all this time by so many people? Fink himself was suspicious of his discovery, believing the mycelial form might be a contaminant blown in from the ventilation system. He became convinced only when one of his graduate students showed that the mycelial form could mate with unicellular forms. Fink discovered dimorphism in baker's yeast in the course of studying how *S. cerevisiae* responds to the harsh

conditions it undoubtedly encounters in nature. He found that near-starvation for a nitrogen source triggers the change. Fink speculates that by providing a way to forage for food, dimorphism offers a selective advantage. Being nonmotile, baker's yeast cannot swim to a new source of nutrients, and budding cells form a compact colony on a surface. Filaments, however, spread over a surface, enabling the organism to obtain nutrients from a larger area.

Finding that baker's yeast is dimorphic has significance beyond learning more about this one organism. The ability of most fungal pathogens to infect humans depends on their being dimorphic. But most pathogenic fungi have not been intensively studied, and we badly need antifungal drugs. A drug that stops dimorphic change might cure fungal infection and not be toxic to the human host.

Table 12.3	Some Important Plant Diseases Caused by Fungi	
Disease	**Fungus and Class**	**Impact**
Powdery mildew	*Plasmopara viticola* (Oomycete)	Farmers must spray several times a year to protect grapes from this infection, which destroys fruit
Potato blight	*Phytophthora infestans* (Oomycete)	Caused famine in Ireland (1845–1860)
Wheat rust	*Puccinia graminis* (Basidiomycete)	Sporadic epidemic that costs North American farmers billions of dollars
Corn smut	*Ustilago mayidis* (Basidiomycete)	Major destroyer of corn, turns ears black
Dutch elm disease	*Ceratocystis ulmi* (Ascomycete)	Almost eliminated elm trees from North America
Apple scab	*Venturia inaequalis* (Ascomycete)	Disfigures apple fruit

fungal infection is called a **mycosis**. The various kinds of mycoses and the organisms that cause them are discussed in Part IV (**Table 12.4**). Fungal infections are not highly contagious. Humans usually acquire fungal disease from nature, where the organisms exist as saprophytes. When fungi do cause disease, they can be difficult to treat and even deadly because most antifungal agents

are toxic to humans and many fungal pathogens infect tissues such as skin that are difficult for drugs to enter.

Some molds and mushrooms produce toxins that are hallucinogenic or highly poisonous to human beings when ingested. For example, **muscarin**, produced by the mushroom *Amanita muscaria*, is highly hallucinogenic. It is consumed as part of the religious rites of peoples

Figure 12.9 The Ascomycete *Venturia inaequalis* causes apple scab.

Table 12.4 Some Higher Fungi That Infect Humans

Fungus	Infection (Chapter)
Blastomyces dermatitidis	Blastomycosis (22)
Coccidioides immitis	San Joaquin valley fever (22)
Histoplasma capsulatum	Histoplasmosis (22)
Pneumocystis carinii	*Pneumocystis* pneumonia (22)
Candida albicans	Thrush, vaginitis, candidiasis (24, 26)
Cryptococcus neoformans	Cryptococcosis (25)
Dermatophytes (*Trichophyton*, *Microsporum*, *Epidermophyton*)	Tinea (ringworm, athlete's foot) (26)

native to the northern rim of the Pacific Ocean. The toxins **phalloidin** and **amanitin**, produced by the closely related *A. phalloides* (also known as the death cap), are highly poisonous. Ingesting even a small quantity of this mushroom can cause death from irreversible liver failure. But another closely related species, *A. caesarae* (Caesar's mushroom), is not only nontoxic but also delicious, proving that collecting and eating mushrooms is dangerous unless you know them well.

Ergot, a toxin produced by *Claviceps purpurea*, a mold that grows on rye and rarely in other grains, is highly toxic. It was once a significant source of human poisoning (see the box "St. Anthony's Fire"). **Aflatoxin**, a fungal toxin more commonly found today, is produced by certain species of *Aspergillus*, principally *A. flavus*, which grows in many kinds of plant materials. Crops such as peanuts and cereal grains, if not properly dried, can contain enough aflatoxin to cause severe liver damage and death. Low levels of aflatoxin may be carcinogenic. In the United States, rigid standards of food processing and testing have effectively eliminated disease caused by aflatoxin, but in parts of Asia and Africa it remains a serious problem.

Some fungi help in the battle against disease. The penicillins are produced by the Deuteromycete species *Penicillium notatum* and *P. chrysogenum*. Some members of the closely related family of antibiotics, the **cephalosporins**, are produced by closely related fungi, *Cephalosporium* spp. In small amounts, ergot is used to treat migraine headaches and to aid childbirth.

Algae

Metabolically, all algae resemble higher plants. They generate ATP by oxygenic photosynthesis using the same kind of chlorophyll (chlorophyll a) that plants use, and they make precursor metabolites from carbon dioxide (CO_2) through the Calvin-Benson cycle (Chapter 5). But the structure of algae and their life cycles are quite different from those of higher plants. Some algae are microscopic single-celled organisms. Others, commonly called seaweeds, form huge multicelled structures up to 50 meters long. But all algae, even the large, morphologically complex varieties, lack the tissue differentiation of higher plants. In other words, all the cells within the thallus (body) of an alga are similar. Algae also differ from plants in their means of reproduction. They do not produce **embryos**, miniature organisms, like those in the seeds of plants. Rather, they reproduce as fungi do, by producing asexual spores and gametes that fuse to form zygotes. Most, but not all, asexual spores and gametes are motile by means of flagella.

Ecology and Uses. Algae are aquatic organisms. A few, however, inhabit moist terrestrial environments such as the surface of soil, rocks, and tree trunks in damp areas. Algae live in both fresh and salt water, and their metabolic activities profoundly affect the earth's ecology. They are the earth's major fixers of CO_2. It is estimated that 80 percent of Earth's photosynthesis is carried out by **phytoplankton**, microscopic algal species floating in the open ocean (*planktos* means "drifting" in Greek). Phytoplankton are at the base of a food chain leading to fish and aquatic animals.

St. Anthony's Fire

In August 1951, 300 inhabitants of Point-Saint-Esprit, a small village in France, became ill and 5 died of a bizarre illness. People hallucinated that they were being chased by tigers or saw death walk into the room. They suffered convulsions. They jumped from rooftops. The streets were filled with screaming people. Not just humans were affected. A cat writhed, twisted, and tried to climb a wall. A dog leaped in the air, snapping viciously, and crushed a rock with its teeth until blood dripped from its mouth. Ducks strutted like penguins, flapped their wings, quacked to a crescendo, and died.

Laboratory tests on the victims were too late to detect the cause, but experts agreed that this was an outbreak of St. Anthony's fire, or ergot poisoning. It was almost certainly caused by illegally distributed ergot-contaminated rye flour that was made into bread at the village bakery. It was verified that the sick animals had also eaten the contaminated bread. John J. Fuller, a science writer who published a definitive account of the episode, concluded, "There is one and only one cause of the tragedy: some form of ergot, and that form has logically got to be akin to LSD."

Ergot poisoning is caused by the fungus *Claviceps purpurea*, an ascomycete that infects rye, a major cereal grain. The infection does little damage to the plant. Its body and most of the grain it produces look normal—only a few grains are replaced by a hardened purple-black mass of *Claviceps* mycelium called **ergot**. But when even the tiniest amount of ergot is consumed by an animal or human, it can cause disorientation, hallucinations, gangrene, abortion, or death.

Ergot poisoning was known to the Assyrians as early as 600 B.C. There were few occurrences during Roman times because Romans didn't care for rye; but during the Middle Ages, when Roman influence ended, rye became popular throughout Western Europe and major outbreaks of ergot poisoning occurred. Entire villages suffered hallucinations. In 994, 40,000 people died in the Limoges district of France. About that time ergot poisoning became known as St. Anthony's fire because the Pope authorized the Order of St. Anthony to treat its victims, some of whom complained of fiery pain. At first, theories of what caused St. Anthony's fire were as bizarre as the symptoms; they included witchcraft and poisonous air. Gradually, however, people came to associate the symptoms with infected rye, and ergot poisoning became rare.

What of the physiological symptoms, abortion and gan-

ergotamine

LSD (lysergic acid diethylamide)

Structures of ergotamine and LSD.

grene? Ergot contains an alkaloid, ergotamine, that makes smooth muscles contract. Contractions can be severe enough to induce abortion and restrict blood supply so that gangrene results. Today purified ergotamine is used clinically to induce birth, control postpartum bleeding, and control migraine headaches.

The psychological symptoms of ergotism are not as well understood. Ergotamine does not cause hallucinations, and it is not mind-altering. It is, however, structurally similar to lysergic acid diethylamide (LSD), an extremely powerful hallucinogen. A minuscule amount of LSD causes pigeons to "strut like penguins" and people to jump from high places. The threshold dose of LSD is 25 µg, and about 100 µg produces a reaction resembling a psychotic state. For comparison, an ordinary aspirin tablet is 300,000 µg. What, then, is the connection between ergot poisoning and its LSD-like psychological effects? In damp climates, other molds grow on ergot, and circumstantial evidence suggests that they convert a small amount of ergotamine to LSD or a similar mind-altering chemical. The outbreak in Point-Saint-Esprit was relatively small; St. Anthony's fire must have caused mass chaos and terror during the Middle Ages.

Indirectly, algae profoundly affect our lives—without them the planet would not be habitable by humans—but they have little direct effect on us. They do not cause infectious disease; they are photosynthetic and cannot live inside the body. The colorless alga *Prototheca* is an exception. *Prototheca* is a green alga that has lost its ability to photosynthesize and is associated with bursitis, an inflammation of the joints.

In Asian countries certain marine algae, or seaweeds, are eaten as food. **Alginic acid** (**algin**), a polysaccharide, is obtained from **kelp**, the huge brown seaweed that grows profusely off the Pacific Coast (Chapter 1). It is used as a thickener in ice cream, salad dressings, prepared sauces, and other foods. Another polysaccharide thickener, **carrageenan**, which is used in milk products, including ice cream, custards, and evaporated milk, comes from the red alga *Chondrus crispus*, also known as Irish moss. It takes its name from Carragheen, Ireland, where its properties were discovered. Other red algae, species of *Gelidium* and *Gracilaria*, that grow mainly in the western Pacific Ocean are the total source of agar, which microbiologists use to solidify media. **Diatomaceous earth**, deposits of shells from **diatoms**, an algal group, is used in industry. The siliceous (glasslike) shells, mined from regions that were once the floors of ancient seas, are used for polishing and insulating and as an additive to speed the rate of filtering liquids.

Members of the algal group **dinoflagellates**, such as *Gymnodinium* and *Gonyaulax*, produce a potent neurotoxin that can kill humans. These dinoflagellates proliferate in shallow seas during spring and summer months, forming **blooms**, dense accumulations of cells. They are called **red tides** because their red pigments color the water. Shellfish that feed on them are not harmed, but fish and humans who eat the shellfish can suffer life-threatening paralysis (**Figure 12.10**).

Classification.   The algae are divided into six groups according to the form of their thalli (whether they are unicellular, coenocytic, filamentous, or plantlike), the structure of their walls (whether they lack walls or have walls made of cellulose, algin, or silica), and the pigments they produce (**Table 12.5**).

All algae contain chlorophyll a, the kind found in plants, and they may or may not contain two other modified chlorophylls, chlorophyll b and chlorophyll c. Chlorophylls are the pigments that mediate the basic reactions of photosynthesis by converting light energy into chemical energy (Chapter 5). Besides chlorophyll, algae contain **accessory photosynthetic pigments**, either **carotenoids** (such as the pigments that give carrots and tomatoes their characteristic color) or **phycobilins** (similar in structure to cytochromes; Chapter 5). Both accessory pigments collect light energy and pass it on to the chlorophylls. The

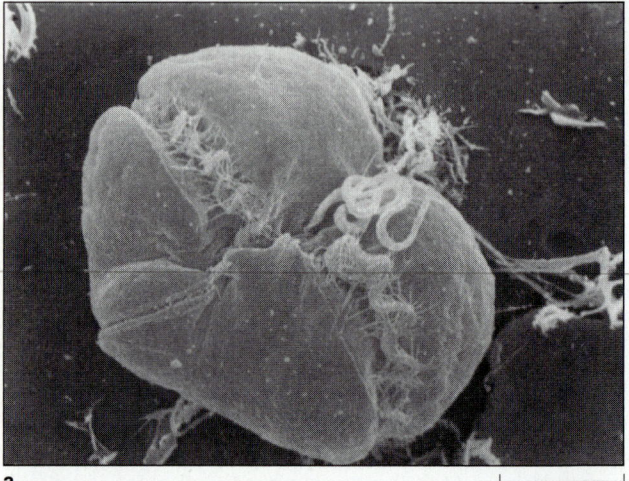

a |—— 10 μm ——|

b

Figure 12.10   Red tide. (**a**) Blooms of red dinoflagellates such as *Gymnodinium breve* color the water red from their sheer numbers. (**b**) Large numbers of fish killed by red tide.

pigments an alga contains dictate its habitat because it grows best in an environment rich in the wavelengths of light that its pigments absorb. The mixtures of pigments also give the various algal groups their characteristic colors.

Pigments are also the basis for the common names of three groups—the **green algae**, the **brown algae**, and the **red algae**. The other three groups—the euglenoids, the dinoflagellates, and diatoms—are distinguished principally by morphology. **Euglenoids** are single motile cells with two flagella of unequal length. Dinoflagellates are single cells usually with two flagella, one wrapped in a groove around the middle of the cell and the other extending from the groove. Diatoms are single cells enclosed in an elaborately sculptured, rigid silica shell that resembles the two parts of a petri dish nested together (**Figure 12.11**).

The red, brown, and green algae also differ significantly in complexity. The thalli of brown and red algae

Table 12.5 Properties of Major Groups of Algae

Phylum	Thallus Structure	Wall Composition	Pigments Chlorophylls	Others
Euglenophyta: euglenoids	Unicellular	No wall	a, b	
Pyrrophyta: dinoflagellates	Unicellular	Cellulose	a, c	Carotenoids
Chrysophyta: diatoms	Unicellular, coenocytic, filamentous	Silica	a, c	Carotenoids
Chlorophyta: green algae	Unicellular, coenocytic, filamentous, or plantlike	Cellulose	a, b	
Phaeophyta: brown algae	Plantlike	Cellulose and algin	a, c	Carotenoids
Rhodophyta: red algae	Plantlike	Cellulose	a	Phycobilins

are all plantlike and multicellular. The green algae, on the other hand, span the full spectrum from unicellular to filamentous to plantlike multicellular organisms.

Phycology is the study of algae, which are studied primarily in natural environments by specialists called **algologists**. Some algae are also studied in culture, notably *Chlamydomonas* (a green alga), *Euglena* (a euglenoid), and *Navicula pelliculosa* (a diatom).

Chlamydomonas is a unicellular green alga that swims actively by means of a posterior flagellum. It can be propagated easily in the laboratory and has been intensively researched in studies on genetics, photosynthesis, and herbicide action. Herbicides that act by inhibiting photosynthesis kill *Chlamydomonas* as well as plants, but it is easier to study cultures of *Chlamydomonas* than fields of plants. *Chlamydomonas* undergoes asexual and sexual reproduction, both of which have unusual features.

Euglena are single-celled organisms, motile by means of flagella (**Figure 12.12**). They have a bright red **eyespot** that senses direction and intensity of light, allowing the cell to swim to bright regions where it can photosynthesize more rapidly. Studies on this response mechanism reveal how eucaryotes sense and respond to visible light.

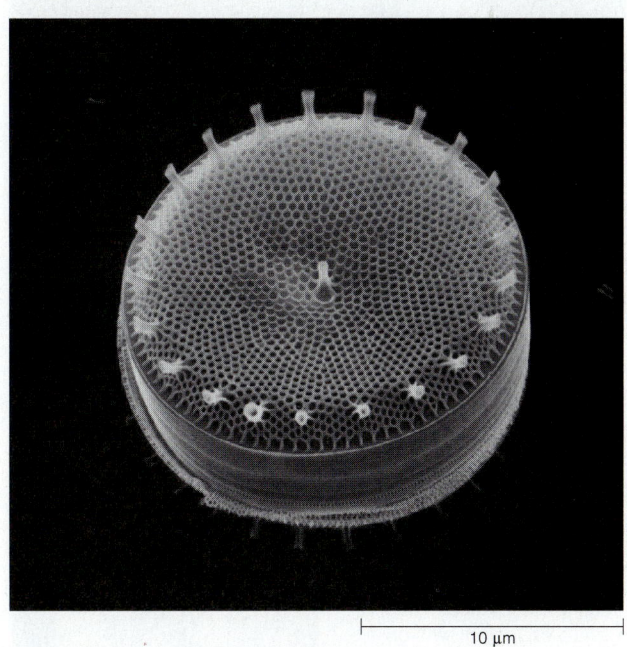

Figure 12.11 Diatoms, one of the six major groups of algae, are single-celled algae enclosed in a silica shell.

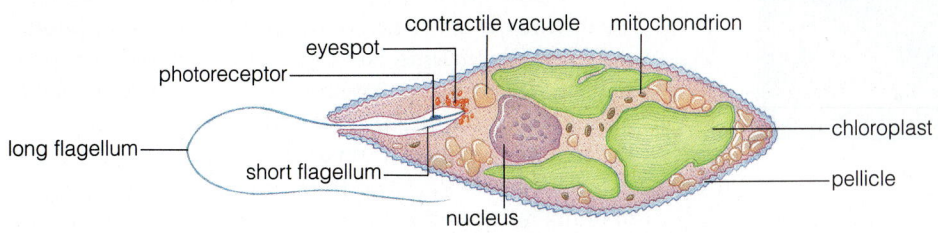

Figure 12.12 Euglenoids have two flagella of unequal length. *Euglena* is a euglenoid alga with a bright eyespot that allows it to swim toward light so that it can photosynthesize more rapidly.

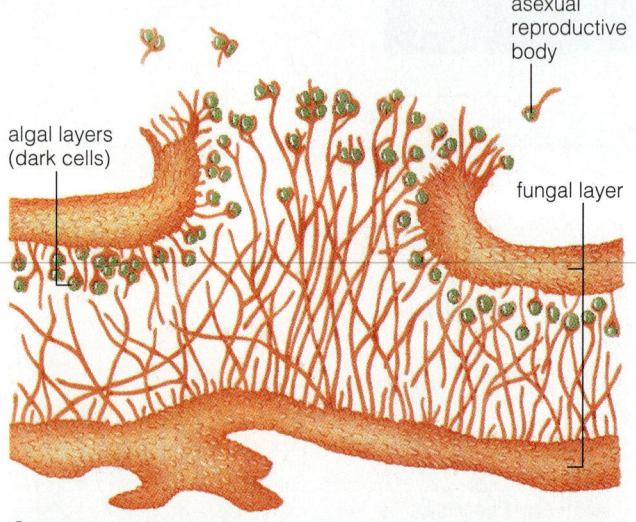

algal layers
(dark cells)

asexual
reproductive
body

fungal layer

a

b

Figure 12.13 Lichens. (**a**) A cross section through the thallus of a typical lichen shows it is composed of two layers of fungal mycelium with algal cells in between. Lichens occur in a vast variety of forms. (**b**) *Cladonia rangifernia*, reindeer moss, forms a spongy growth on soil. (**c**) *Usnea*, old man's beard, looks like moss growing off the limbs of a tree.

c

The diatom *Navicula pelliculosa* has been studied in laboratory culture to determine how it metabolizes silicon to make its silica shells and how an organism with such a wall structure divides. Division is lengthwise, which means each daughter cell retains half of the old wall and synthesizes a new half.

Lichens

Lichens are the product of a **mutualistic association** (one in which both partners benefit) between a fungus, usually an Ascomycete, and a phototroph, usually a green alga or a cyanobacterium. The two partners are so well adapted that lichens *appear* to be separate species that can be identified by morphology and distribution. More than 16,000 species of lichens have been described.

Lichens vary considerably in appearance and structure (**Figure 12.13**). The fungus is the dominant partner. It makes up most of the mass and gives the thallus its shape. The thallus may be upright (**fruticose**), leaflike and flat (**foliose**), or encrusted and flat (**crustose**). Usually some fungal mycelia extend below the lichen thallus to form **rhizoids**, rootlike structures that attach the lichen to its substrate. Lichens that hang from trees are often mistaken for mosses, while others are thin, colored patches on rocks. In all cases, they grow where nutrients are scarce and other organisms cannot survive. The manna that the Bible describes as food miraculously supplied to the Israelites in the desert wilderness is thought to have been lichens. **Litmus**, a dye used as a pH indicator, is extracted from a lichen, *Roccella tinctoria*. Lichens grow very slowly, only about 0.1 to 4 cm per year, and they are very long lived. Some lichens in the Arctic are estimated to be 1000 to 4000 years old.

When the thallus of a lichen is teased apart, both partners can be cultivated separately. The fungus thallus then assumes a shape quite different from its shape in the lichen. In contrast, the alga, which usually exists as individual cells in the lichen, continues growing as individual cells. It is possible to re-form a lichen experimentally by restricting the supply of nutrients. Lichens are extremely efficient at concentrating available nutrients within the thallus to use when needed. They can also scavenge nutrients from the air, which contributes to their hardiness, except in urban areas where pollution kills them.

The benefits the fungus derives from the association are obvious. It uses some of the alga's products of photosynthesis as nutrients. Some cyanobacterial partners provide both nitrogen compounds (by fixing atmospheric nitrogen) and products of photosynthesis. The benefits to the alga or cyanobacterium are less clear. At the least, the association protects them from desiccation and extremes of temperature and light.

Protozoa

Protozoa are nonphotosynthetic, unicellular eucaryotes. Some cause devastating human diseases, including malaria, an infection of red blood cells (Chapter 27), leishmaniasis, an infection of white blood cells, and African sleeping sickness, an infection of the nervous system that leads to coma and death (Chapter 25). Malaria affects more people than does any other infectious disease. African sleeping sickness kept parts of Africa uninhabitable for centuries, and even today at least a million people are infected at any time.

Protozoa are the pinnacle of complexity in unicellular differentiation. They have elaborate organelles almost as complex as the organs of multicellular organisms. Some protozoa are differentiated into a **cytostome**, or mouthlike organ, that takes in particulate food; **cilia** that sweep food into the mouth and move the cell; **food vacuoles** that digest the food; and a **cytoproct**, or anuslike structure, through which undigested food is expelled. A **contractile vacuole** keeps the cell from bursting from turgor pressure. Water, which constantly flows into the cell by osmosis, collects in small vacuoles that coalesce to form the contractile vacuole. When the contractile vacuole attains a critical size, it expels its contents outside the cell through the **contractile vacuole pore**. Many protozoa have a cellulose or similar polysaccharide shell outside their cell membrane. Others have a **pellicle** located just inside the cell membrane. The pellicle is a stiff layer composed of protein fibers or calcium-like or silica-like structures.

Protozoa reproduce asexually, by fission and budding. Some protozoa, including *Plasmodium*, the organism that causes malaria, undergo **schizogony**, multiple fission. Either the nucleus divides repeatedly and then some cytoplasm gathers around each of the nuclei to form daughter cells, or a giant cell undergoes many fissions without growth, producing many small cells. Sexual reproduction occurs by **conjugation**, the fusion of vegetative cells, or by the fusion of specialized gametes called **gametocytes**.

During their life cycle, some protozoa produce highly resistant, nongrowing cells called **cysts**. Cysts are surrounded by a polysaccharide capsule that protects them from desiccation, temperature extremes (though they are not as heat-resistant as bacterial endospores), and toxic materials. They can withstand long periods of starvation. Other protozoa have much more elaborate life cycles. Some sporozoa, including pathogenic species, require more than one host and change their body structure and basic physiology.

The protozoa are divided into four groups, each made up of several phyla, on the basis of their means of locomotion (**Table 12.6**). The **flagellates** move by means of one or more flagella. The **amoeboids** move by extending **pseudopods**, long lobes that form on the cell (Chapter 4). The **sporozoa** are nonmotile, and the **ciliates** move by means of many cilia.

Mastigophora. The **flagellate protozoa** (also called **Mastigophora**) resemble the euglenoid algae, differing principally by not being photosynthetic. Algologists and protozoologists sometimes quarrel over ownership of the two groups. Algologists claim flagellate protozoa are algae that have lost photosynthesis, while protozoologists claim flagellate algae are photosynthetic protozoa.

Flagella have a whiplike motion, though the arrangement of flagella and how they propel the protozoa vary considerably. Most flagellated protozoa have only two flagella. In some species one or both flagella are anterior, pulling the cell. In others, one or both flagella are posterior, driving the cell. Some flagella lie close to the cell surface but do not extend beyond the cell. They are covered by an expanded membrane called an **undulating membrane** that propels the cell by undulating as the flagellum inside moves.

Table 12.6 Major Groups of Protozoa

Group	Means of Motility	Parasitic Species	Disease (Chapter)
Mastigophora (flagellates)	One or more flagella	*Trypanosoma brucei* *Trypanosoma cruzi* *Giardia lamblia* *Trichomonas vaginalis*	Sleeping sickness (25) Chagas' disease (27) Giardiasis (23) Trichomoniasis (24)
Sarcodina (amoeboids)	Pseudopods	*Entamoeba histolytica* *Naegleria fowleri*	Amoebic dysentery (23) Primary amoebic encephalitis (25)
Sporozoa	Nonmotile	*Plasmodium* *Toxoplasma gondii*	Malaria (27) Toxoplasmosis (27)
Ciliophora (ciliates)	Numerous cilia	*Balantidium coli*	Balantidiasis (23)

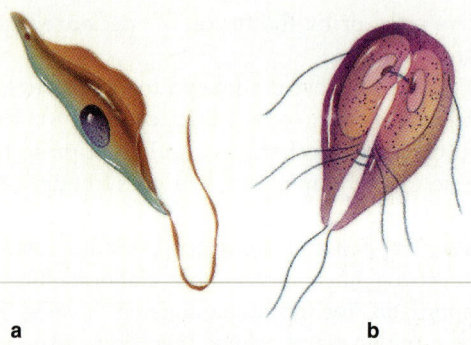

Figure 12.14 Flagellate protozoa. (**a**) One group of flagellates, the trypanosomes, has a leaflike appearance, a single flagellum, and an undulating membrane. (**b**) Members of another group, the diplomonads, have a double body containing two sets of most organelles and two to six flagella.

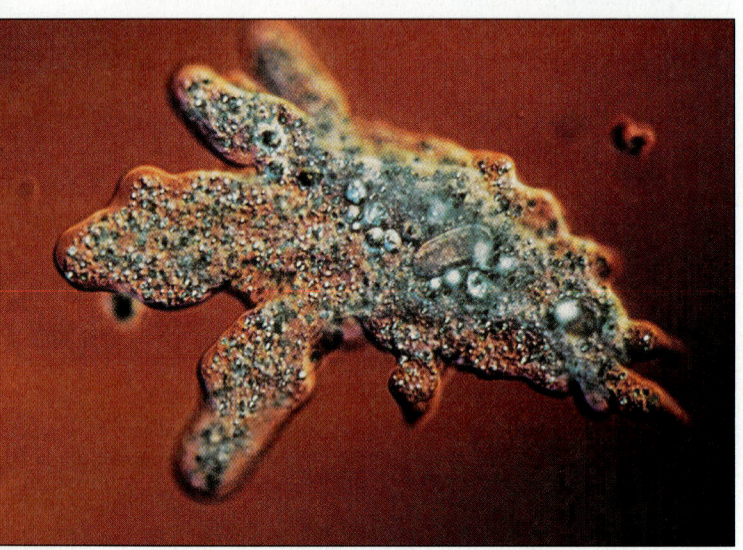

Figure 12.15 Amoeboid protozoa, such as *Amoeba proteus*, have pseudopods for motility and gathering food.

Members of one phylum of flagellate protozoa, the **trypanosomes** (including the species that cause African sleeping sickness and leishmaniasis), are characterized by a leaflike appearance, single flagellum, and undulating membrane at one stage of their life cycle (**Figure 12.14**). Members of another phylum, the **diplomonads**, have a double body containing two sets of most organelles, giving the cell the appearance of a human face. Diplomonads have two to six flagella. *Giardia lamblia*, which causes a waterborne dysentery, is a diplomonad (Chapter 23). It can survive as a cyst for extended periods outside a host.

merozoite

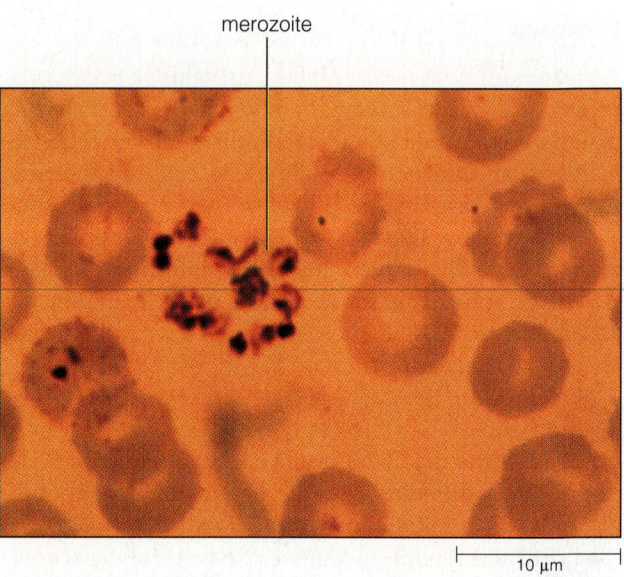

10 µm

Figure 12.16 This light micrograph of blood taken from a person suffering from malaria caused by *Plasmodium vivax* shows many red blood cells that appear normal and one that contains the merozoite stage of the parasite.

Sarcodina. Amoeboids (also called **Sarcodina**) are found in marine, freshwater, and soil (including desert) environments and in association with animals. This is a huge group of microorganisms—some protozoologists divide it into 12 phyla, one of which contains 48,000 described species in eight classes. All members, however, have the ability to produce pseudopods at some stage of their life cycle. At other stages some are not amoeboid, and a few even produce flagellated gametes. Because they can extend and retract their pseudopods, the amoeboids look very different in their resting states and motile forms. The pseudopods vary, too. Some are thick and rounded, while others are thin and pointed.

Pseudopods are organelles of locomotion and feeding. They extend to pull the cell along a solid surface and they engulf bits of food, including cells (**Figure 12.15**). When conditions are right, the protein **actin** (related to the protein in the muscles of animals) aggregates to form microtubules. These microtubules push against the cell membrane, causing bulges that the cytoplasm streams into, forming a pseudopod. Then the actin molecules disaggregate and the pseudopod retracts, pulling the cell.

Very few amoeboids cause human disease. One species, *Entamoeba histolytica*, causes amoebic dysentery, acquired by consuming fecally contaminated water or food containing cysts (Chapter 23). Another, *Naegleria fowleri*, causes a rare and usually fatal form of encephalitis, acquired by swimming in water containing this amoeba (Chapter 25).

Figure 12.17 Scanning electron micrograph showing one ciliate protozoan, *Didinium*, that has captured another, *Paramecium*, and is in the process of consuming it.

100 µm   100 µm

Sporozoa. All members of the sporozoa are parasitic. The group constitutes a single phylum with more than 300 genera and 4000 species. Some sporozoa have an elaborate life cycle, changing body form. At the actively multiplying vegetative stage, the sporozoan is called a **trophozoite**. At the stage when it is infectious to its animal or human host, it is called a **sporozoite**, and at the stage it enters and infects red blood cells, it is called a **merozoite**. More than one host is involved. *Plasmodium* spp., which cause malaria in humans, lower primates, birds, and reptiles, exist in these different forms (**Figure 12.16**). Its life cycle takes place in the human body and the female *Anopheles* mosquito (see Figure 27.14). A typical life cycle also has episodes of schizogony, resulting in sudden, large increases in the number of merozoite cells in someone with malaria and accounting for the characteristic attacks of shaking chills and high fever (Chapter 27).

Toxoplasma gondii, which causes toxoplasmosis, a systemic disease of animals, particularly cats, and humans, has a less complicated life cycle (Chapter 27). Humans acquire the disease by consuming cysts in the meat of infected animals or ingesting material contaminated by cat feces containing *Toxoplasma* cells.

Ciliophora. The **ciliate protozoa** also constitute a single phylum, the **Ciliophora**. Most ciliates have two nuclei called the **macronucleus** and the **micronucleus**. The larger macronucleus is **polyploid** (has several copies of its complement of chromosomes) and directs vegetative growth and cell division. The smaller micronucleus is ap-

parently necessary only for sexual reproduction. Strains that lack a micronucleus are able to reproduce indefinitely asexually but cannot undergo sexual reproduction.

Cilia are organs of locomotion. By coordinated beating they can move the cell rapidly, turn it sharply, and stop it abruptly—abilities ciliates need because some prey on bacteria, fungi, and other protozoa. Predation may be highly specific. For example, *Didinium nosutum* attacks only species of *Paramecium* (another ciliate), which it swallows whole (**Figure 12.17**). Because *Didinium* is not much larger than *Paramecium*, swallowing enlarges its cytostome enormously, and *Didinium* cannot eat again for at least two hours.

Cilia are also accessory feeding organs. Their beating movement sweeps small bits of food into the cytostome. Some nonmotile species that are attached to surfaces by stalks use their cilia only for feeding.

Many ciliates are associated with animals. The rumen of cattle and other grazers teems with ciliates that digest cellulose and become food for the animal (see the beginning of Chapter 8). A few ciliates are animal parasites, but only one, *Balantidium coli*, is a human pathogen. It causes balantidiasis, a severe diarrheal infection of the large intestine (Chapter 23).

The Slime Molds

The slime molds do not fit into any of the three major groups of eucaryotic microorganisms—algae, fungi, or protozoa. Some microbiologists group slime molds with

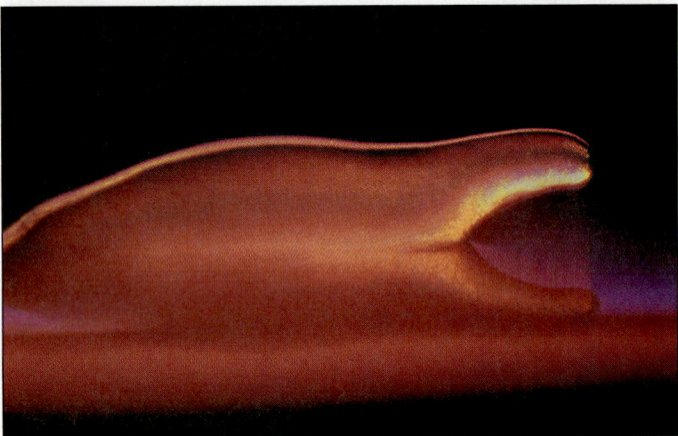

Figure 12.18 Slime molds are divided into two groups. (**a**) Myxogastria, or true slime molds, such as *Physarum*, are commonly seen on decaying logs. Acrasieae, or cellular slime molds, such as *Dictyostelium discoideum*, spend part of their life cycle as individual amoeboid cells. Then they aggregate into a slug-shaped grex (**b**) that migrates toward light and develops into a fruiting body (**c**).

Not Gone and Not Forgotten

The mitochondria (and chloroplasts) in eucaryotes are remnants of procaryotic cells engulfed by some primitive eucaryote. Microbiologists have suspected this for decades, and now the evidence is overwhelming. The biochemical and morphological structure of these organelles is clearly procaryotic. Also, antibiotics active against procaryotes inhibit much of their metabolism without affecting other parts of eucaryotes. Certainly, a eucaryote engulfed the procaryotic precursor of mitochondria. But what about its mitochondria-less cohorts? Did they persist? Or did they disappear because they couldn't compete with mitochondria-bearing eucaryotes? Now the answer is known.

Mitochondria-less eucaryotes persisted, and their descendants are alive today. They include *Giardia* spp. and a closely related group, the Microsporidia. Microbiologists have known for some time that these protozoa lack mitochondria and consequently are anaerobes. Until recently, they thought these organisms lost their mitochondria because they were unnecessary in anaerobic environments. However, sequencing of ribosomal RNA (Chapter 10) shows that *Giardia* spp. and the Microsporidia are primitive eucaryotes, more closely related than other eucaryotes to archaebacteria and eubacteria. They must be descendants of eucaryotes that never acquired mitochondria.

the fungi because at certain stages of their life cycle some produce multinucleate masses of cells. Others group them with the amoeboid protozoa because at other stages slime molds produce amoeba-like cells. We will consider this small group of about 700 species on their own. The slime molds are divided into two groups, the true slime molds and the cellular slime molds (**Figure 12.18**). None is a pathogen or is put to any industrial use.

Myxogastria. The **true slime molds** (also called **Myxogastria**) are like fungi in being multinucleate at one stage of their life cycle. True slime molds are commonly seen on decaying logs or stumps as an amorphous slimy mass called a **plasmodium**. The plasmodium, which is composed of multinucleate cytoplasm without a rigid wall, flows over a surface, such as leaves or a damp tree trunk, consuming microorganisms and bits of plant material. As long as nutrients are available, it continues to expand. It can attain a mass of half a pound or so. When a plasmodium reaches a relatively dry region it develops **fruiting bodies**, raised structures that contain spores. Meiosis occurs within the fruiting body. The spores,

Table 12.7   Major Groups of Helminths

Group	Principal Characteristics	Examples
Platyhelminthes: Flatworms		
Cestoda: tapeworms	Scolex; flattened, segmented body	*Taenia saginata* (beef tapeworm) *Echinococcus granulosus* (dog tapeworm)
Trematoda: flukes	Shaped like flattened, pointed ovals; unsegmented	*Paragonimus westermani* (lung fluke) *Schistosoma* spp. (cause schistosomiasis)
Nemathelminthes: Roundworms	Long cylindrical, unsegmented bodies, pointed ends	*Trichuris trichiura* (whipworm) *Necator americanus* (hookworm) *Trichinella spiralis* (causes trichinosis) *Wuchereria bancrofti* (causes elephantiasis) *Enterobius vermicularis* (pinworm) *Ascaris lumbricoides* (causes ascariasis) *Onchocerca volvulus* (causes river blindness) *Loa loa* (causes loaiasis)

which can stand prolonged starvation, are haploid. When conditions become favorable, the spores germinate, producing amoeboid cells that feed on bacteria. These cells can fuse to produce a zygote or, if water is present, they develop into flagellated gametes that fuse. The zygote develops into a new plasmodium.

Acrasieae. At one point in their life cycle, the **cellular slime molds** (also called **Acrasieae**) produce cells that resemble amoeboid protozoa. One species, *Dictyostelium discoideum*, is a favorite of microbiologists who study morphogenesis because it undergoes startling morphological changes during its life cycle. The vegetative stage is an amoeboid cell with a single nucleus. It feeds mainly on bacteria, ingesting them by phagocytosis. When this source of nutrients is depleted, the cells aggregate by responding chemotactically to cyclic AMP, a chemical signal produced by some cells. The aggregated mass of cells, called a **grex**, takes on the shape of a slug and moves toward light. When it reaches an area of light, it stops and differentiates into a fruiting body that has a bulbous sac filled with spores at the end. When conditions become favorable again, the spores germinate, forming amoeboid cells and completing the life cycle.

HELMINTHS

Unlike the microorganisms we have considered so far, **helminths** are animals—flatworms and roundworms, to

be specific. These complex, multicellular creatures are included in this microbiology text because certain species go through microscopic stages in their life cycles and cause parasitic diseases. Later chapters on human health and disease discuss helminthic infections. Here we discuss the worms' biology.

We focus on helminths that are human parasites, worms that have evolved to take advantage of an intimate relationship with a human host for their survival. Helminths also parasitize other animals and plants, but not all helminths are parasites. In fact, helminths are an extremely diverse group and most are free-living.

The helminths include two phyla—the **Platyhelminthes**, or **flatworms**, and the **Nemathelminthes**, or **roundworms**. The Platyhelminthes include two main types of human parasites—the tapeworms and flukes. The Nemathelminthes include many parasitic species (**Table 12.7**). Most helminths have life cycles that include larval as well as adult worms.

Platyhelminthes

The Platyhelminthes, or flatworms, are named for the flattened bodies of the adult worms. They have a head and bilaterally symmetrical bodies, as well as specialized organ systems, including a nervous system, an excretory system, and a reproductive system. Most species have a digestive system with a single opening through which food enters and waste leaves. But some parasitic species have no digestive tract and simply absorb nutrients from their host through their outer covering.

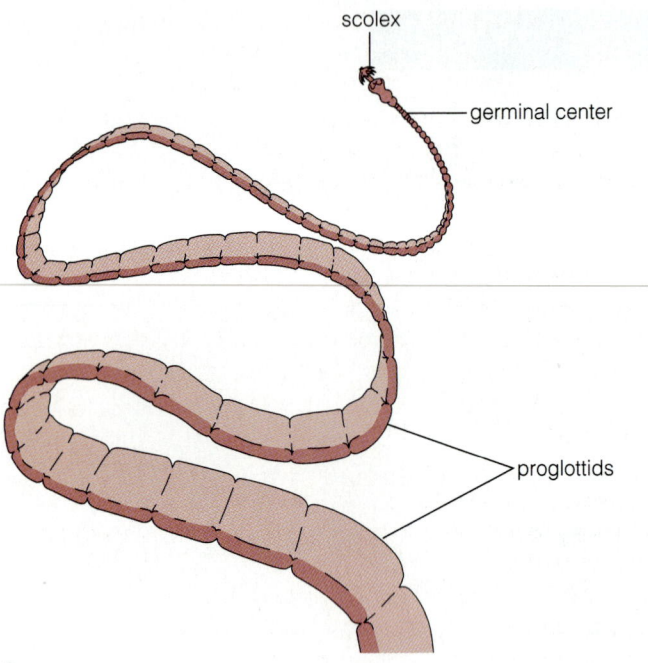

a

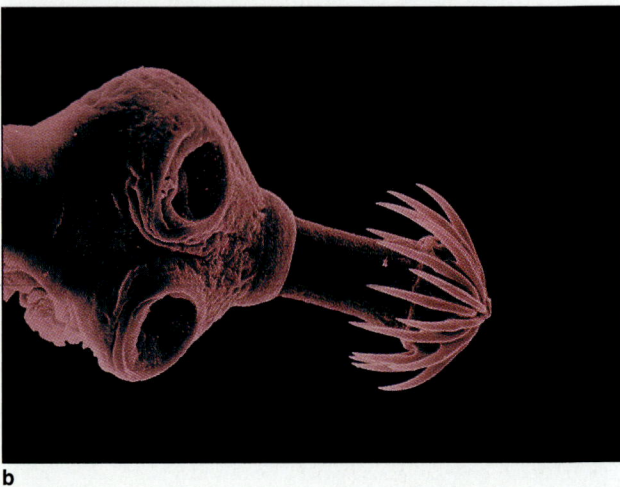

b

Figure 12.19 Cestodes. (**a**) Cestodes (tapeworms) have flat, segmented bodies consisting of a scolex, a germinal center, and proglottids. (**b**) In some, the scolex has sharp hooks (others have suckers) to attach the worm to the intestinal lining of the host.

Cestoda. The **cestodes** are also called **tapeworms** because the flat, segmented bodies of the adult worms look like a piece of tape. Tapeworms consist of three parts. A **scolex** bearing hooks or suckers attaches the worm to the intestinal lining of its host. Immediately behind the scolex is a neck, or **germinal center**, where new segments are formed. The segments themselves are called **proglottids (Figure 12.19)**. Most tapeworms are **hermaphroditic**, meaning they have both male and female reproductive organs. Fertilization occurs within mature segments near the middle of the worm, and larger and older segments near the end of the worm are packed with fertilized eggs. Proglottids are egg-making machines. They have no digestive or excretory system; instead, they absorb nutrients and excrete wastes across their surface.

Cestodes have life cycles with larval as well as adult forms. In some species it is the adult that causes human infection, while in other species it is the larvae. Adult and larval tapeworm infections are quite different. In general, adult cestodes inhabit the intestine, with the scolex attached to the intestinal wall and proglottids extending into the intestinal space. Adult tapeworms can grow up to 30 feet long and cause extreme anxiety when proglottids are passed in the stool. Nevertheless, these are relatively harmless infections. Most people with an adult tapeworm never even know they have one. Larval tapeworms, on the other hand, generally develop in muscle, brain, eye, liver, or heart tissue. As the cestode larvae develop and replace surrounding tissue, they cause serious or even life-threatening damage.

People acquire cestode infections in various ways, depending upon how humans fit into the worm's life cycle. As examples, let's consider two different tapeworm infections—*Taenia saginata*, the beef tapeworm, in which humans are parasitized by the adult worm, and *Echinococcus granulosus*, in which humans harbor the worm's larvae.

Cattle can carry *Taenia saginata* larvae, which become encysted in their muscle as **cystecerci**. The larvae are killed if meat is thoroughly cooked, but people become infected by eating undercooked infected meat (**Figure 12.20**). Larvae hatch in the intestine, the scolex attaches to the intestinal wall, and the person now has a tapeworm infection, annoying but not life-threatening. Proglottids containing infective *T. saginata* eggs are excreted in human feces. If untreated human waste is deposited on grazing land and cattle consume the eggs, larvae hatch in the cow's intestine, penetrate the intestinal wall, and encyst in muscle, thus completing the cycle.

People acquire *Echinococcus granulosus* from close contact with infected dogs. In this infection humans are hosts for the worm's larval stage. The relatively harmless adult tapeworm infects dogs, which pass proglottids filled with infective eggs in their feces. Grazing animals, particularly sheep, are the usual hosts for the *E. granulosus* larvae, which become encysted in their muscle after they ingest infective eggs. However, the same fate can befall humans who accidentally ingest traces of dog feces. In humans, larvae can encyst in various tissues, though liver and lung are the most common. These infections can cause serious disease. *E. granulosus* completes its life cycle when larvae encysted in sheep muscle are ingested by dogs. Human infection is a dead end for *E. granulosus* since animals do not eat human flesh.

The most important tapeworms that infect humans are discussed in Chapter 23.

Trematoda. Like tapeworms, **trematodes**, or **flukes**, have a flattened body, but it is not segmented. Adults are shaped like flat, pointed ovals, ranging in size from a few millimeters to a few centimeters, depending on the species. They have a digestive system, including an oral sucker and usually a ventral sucker that attaches them to their host. They also have a well-developed reproductive system. Most are hermaphroditic, although some species are **dioecious**, with separate male and female forms.

Flukes have complex life cycles with both larval and adult forms. Humans are the **definitive hosts**, harboring the sexually mature adults. Some fluke species parasitize human tissues such as lung or liver, while others live in blood vessels. Aquatic animals or plants are the **intermediate hosts**, harboring larvae. All species require at least one intermediate host—a snail or clam—and some require a second—a fish, crab, or water plant.

As an example of the flukes' complex life cycle, we consider here *Paragonimus westermani*, the lung fluke (see the life cycle of the liver fluke, Figure 23.20). Adult worms develop in the human lung. Patients suffering from this infection have fever and a cough that produces blood-tinged sputum. Eventually lung destruction may be so extensive that breathing is impaired and the patient may die. The hermaphroditic worms produce eggs that are coughed up, swallowed, and passed in the feces. Development can continue only if the eggs reach water, which happens when raw sewage is dumped in lakes or streams. Then the eggs hatch to release free-swimming larvae called **miracidia**. The miracidia penetrate the body of a snail and multiply asexually to produce numerous **redia**, each of which eventually develops into another free-swimming form called a **cercaria**. Cercariae leave the snail's body to enter the water, where they penetrate the body of crabs or crayfish, forming cysts called **metacercariae** in their muscles or other internal organs. The cycle is completed if people eat the crabs or crayfish without cooking them completely. Then the larval worm hatches from its cyst and burrows through the human intestine, the abdominal cavity, and the diaphragm to get to the lung.

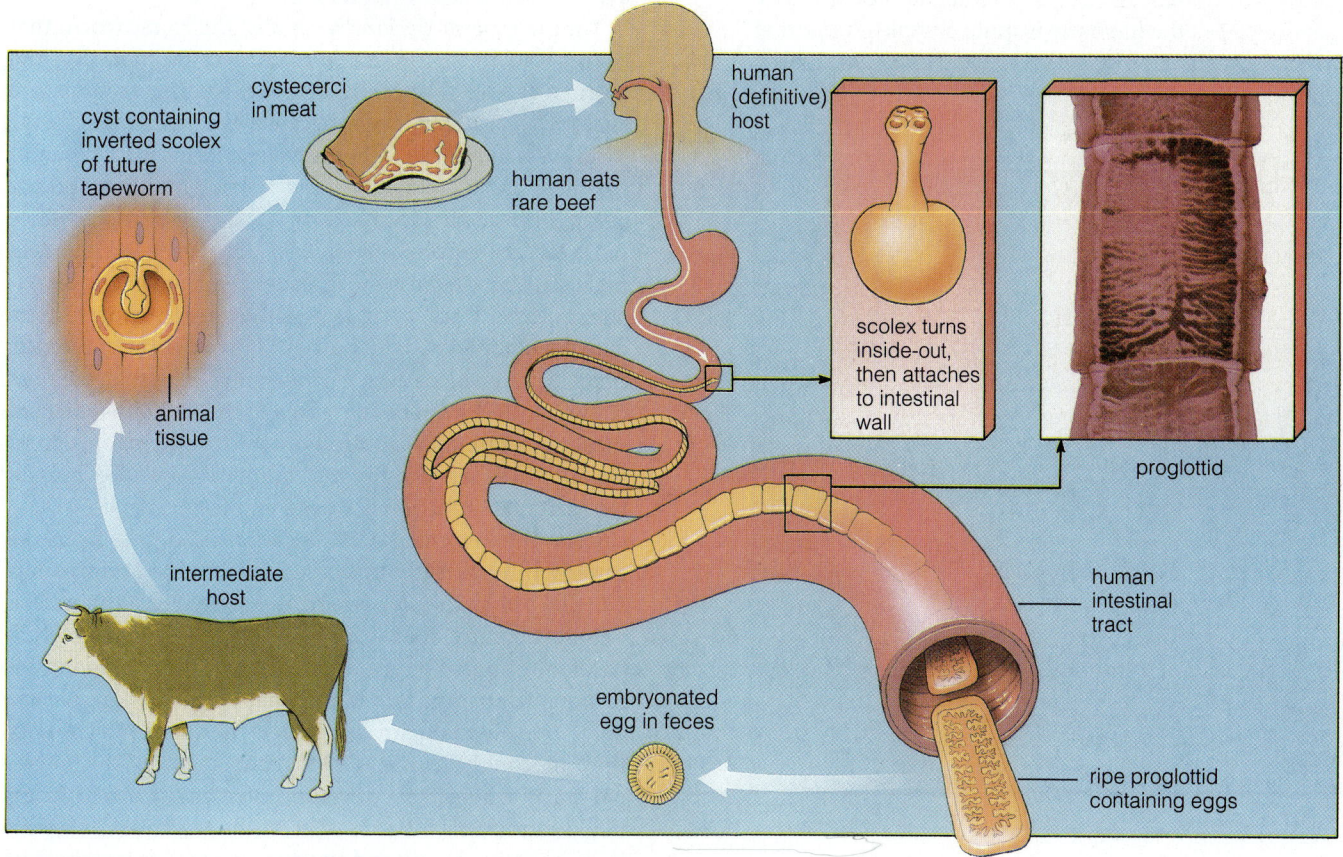

Figure 12.20 Life cycle of *Taenia saginata*, the beef tapeworm.

Paragonimus westermani, like other parasitic human flukes, is limited to areas where the proper species of intermediate hosts are found and where untreated human sewage enters water. *P. westermani* infections are acquired in Asia, South America, and occasionally Africa, but different fluke species have different geographic distributions.

Other fluke infections discussed in detail in this text include the liver fluke (Chapter 23) and the schistosome or blood fluke (Chapter 27). These diseases do not occur in the United States, but they cause enormous suffering worldwide. Approximately 200 million people are debilitated by chronic schistosome infestations.

Nemathelminthes

As their name suggests, the Nemathelminthes (also called **nematodes**), or roundworms, have cylindrically shaped bodies. These worms are bilaterally symmetrical and usually tapered at both ends. They are more highly developed than flatworms, with a complete digestive system including mouth and anus, a well-developed reproductive system, and a primitive nervous system. Most are dioecious. Roundworms are extremely diverse and found in almost every environment. About 30 species are parasitic for humans.

Adult roundworms develop from larvae. In some parasitic species the life cycle is quite complex, requiring more than one host or including a free-living larval form.

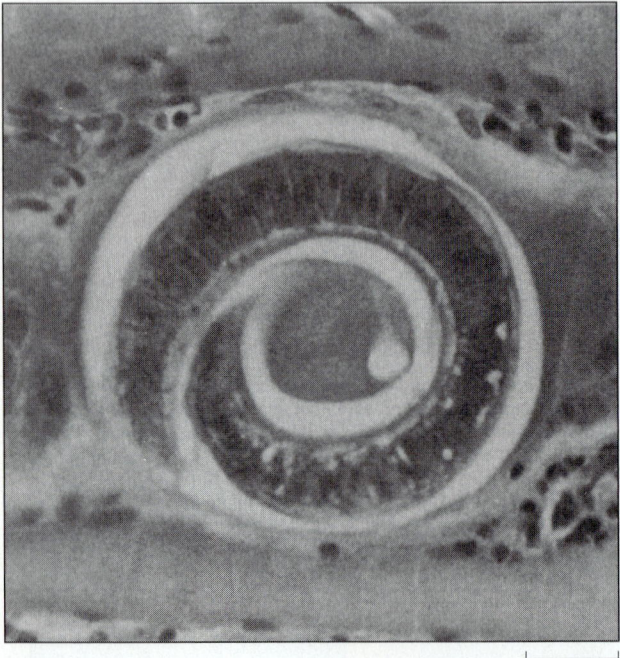

Figure 12.21 Larvae of the roundworm *Trichinella spiralis*, which causes trichinosis.

In other species, the cycle of adult to egg to larva to adult is relatively straightforward. We consider here the life cycles of several roundworms that are parasitic for humans. Diagnosis, signs, symptoms, and treatment of these helminthic infections are covered in Part IV.

The whipworm, *Trichuris trichiura*, has a relatively simple life cycle. People ingest eggs in material contaminated by human feces, and the eggs hatch in the small intestine, releasing larvae that are carried to the large intestine. There they burrow into the intestinal lining and develop into male and female adult worms, which begin to produce eggs three months later. Fertilized eggs are passed in the feces. They mature in the soil, and the cycle is completed when they are ingested by another human (Chapter 23).

Infection by *Ascaris lumbricoides* is also by ingesting eggs. Its life cycle is slightly more complex than the whipworm's. Instead of remaining in the intestines, the hatched larvae burrow through the intestinal wall and enter the blood, where they travel to the lungs and grow. Eventually they become large enough that they are coughed up and swallowed, thus returning to the intestinal tract. The adult male and female worms live in the large intestine, and fertilized eggs are passed in the feces. When mature eggs are ingested by another human, the cycle is completed (Chapter 23).

In the case of *Trichinella spiralis*, the roundworm that causes trichinosis, infection is caused by ingesting the worm's larvae (**Figure 12.21**). Meat-eating animals are susceptible to trichinosis because the infectious larvae form cysts in muscle. When a new host eats meat containing these cysts, larvae hatch in the small intestine and develop into adult worms. The adults, in turn, produce more larvae—approximately 1500 from each fertilized female. The larvae cross the intestinal wall, enter the bloodstream, and form cysts when they reach muscle. The cycle is completed when a new host eats the infectious cysts. The usual pattern of transmission is that pigs become infected from eating garbage that contains infected meat, and humans acquire the infection by eating pork. Because neither people nor animals consume human flesh, human infection is a dead end for *T. spiralis*.

The hookworm, *Necator americanus*, differs from the roundworms already discussed because it produces larvae that live in the soil, free of any animal or human host. People acquire hookworm infection when they walk barefoot on infected ground, allowing the free-living infectious **filariform larvae** to penetrate their skin. Larvae enter the bloodstream and reach the lungs. Eventually they are coughed up and swallowed, entering the intestinal tract, where they develop into adult worms. Adult hookworms attach themselves to the intestinal lining by their hooklike heads and produce enormous numbers of eggs—one female can lay up to 20,000 eggs per day.

These eggs are passed in the feces and hatch into noninfectious **rhabditiform larvae** if they reach the soil. Rhabditiform larvae develop into filariform larvae, and the cycle is completed if they encounter the bare feet of a new human host (Chapter 23).

Wuchereria bancrofti, the roundworm that causes elephantiasis (Chapter 27), requires a mosquito host as well as a human host to complete its life cycle. The mosquito is the intermediate host, and three larval stages—a first stage, second stage, and third stage—develop inside the insect's body. When a mosquito carrying third-stage larvae in its saliva bites a human being, that person becomes infected. Larvae are introduced under the skin and migrate into the lymphatic circulation where they develop into adult worms. The worms block lymphatic vessels, causing the elephant-like swellings that give this infection its name (see Figure 27.18). Nearly a year after the infecting mosquito bite, the worms are mature enough to begin producing larvae, called **microfilariae**. Microfilariae enter the bloodstream, and the cycle is completed when a susceptible mosquito consumes them during a blood meal. In this complex life cycle, the human is the definitive host, and the mosquito is both the intermediate host and a biological vector of infection (see the discussion of arthropods).

ARTHROPOD VECTORS

The arthropods are invertebrate animals with jointed legs (*arthro* means "joint," *pod* means "foot"). They are not microorganisms, but they play an important role in human disease (Chapters 15 and 20). Arthropods are **vectors**— they transmit disease between two hosts. Some are also **reservoirs** of infection, repositories for maintaining microorganisms between hosts.

The Organisms

Arthropods are a diverse group, ranging from crabs and lobsters to centipedes, honeybees, and spiders. More than a million species of arthropods have been described, and new ones are still being discovered. All arthropods have jointed appendages, segmented bodies, and a hardened exoskeleton. They have organ systems that fulfill specialized functions, including a digestive system, a respiratory system, and a well-developed nervous system with complex sensory organs. Many arthropods have complex life cycles with very different looking adult, larval, and resting forms—an adaptation that gives them great reproductive efficiency. Of the three main arthropod groups—**chelicerates**, **crustaceans**, and **insects**—we are concerned principally with certain chelicerates (ticks and mites) and insects (mosquitoes, flies, lice, fleas, and bugs).

The most medically important chelicerates are ticks and mites, sometimes called **arachnids**. Ticks and mites hatch from eggs as six-legged larvae and metamorphosize into eight-legged adults with two body segments. Although ticks and mites are closely related, ticks tend to be larger and have nearly hairless, leathery bodies. Ticks and some mites have mouthparts that allow them to suck their host's blood. Some species of ticks can pass infectious microorganisms into their eggs in a process called **transovarial transmission**. This makes ticks reservoirs for disease-causing microorganisms. Disease reservoirs are critically important in maintaining certain human infections (Chapter 15).

More than 800,000 species of insects have been described. All have three body segments—a head, thorax, and abdomen—and six legs attached to the thorax. Most also have two pairs of wings. Some insect species are of great medical importance. Female mosquitoes and many flies have mouthparts adapted for puncturing the skin and sucking blood. Lice (which are wingless) have mouthparts adapted either for sucking blood or for chewing tissues of the body's surface. They live on the body of a larger host, either animal or human. Fleas are jumping insects without wings that live on the bodies of animals or humans and suck their host's blood. Some true bugs, notably the reduviid bug found in North and South America, can also suck blood from their animal or human hosts.

Arthropods and Human Health

Arthropods affect human health in various ways. In this textbook, we consider how arthropods transmit diseases caused by microorganisms or helminths as vectors of infection (**Figure 12.22**). Arthropod vectors can be mechanical or biological.

A **mechanical vector** picks up a pathogen on its body and carries it from one place to another. For example, when a housefly lands on decaying matter such as garbage or feces, it can pick up disease-causing microorganisms on its feet and deposit them on people or their food. Mechanical arthropod vectors play their most important role in spreading diarrheal diseases, including cholera and typhoid fever (Chapter 23).

Though mechanical vectors may be a significant factor in disease transmission, they are never essential to perpetuating an infection. A **biological vector**, on the other hand, is an *essential* link in the transmission of a disease because the arthropod is a stage in the microorganism's life cycle. For example, the protozoal species of *Plasmodium* that spread malaria require a mosquito host as well as a human host to survive (Chapter 27). *Plasmodium* spp. are passed from mosquito to human and back to the mosquito when the insect takes its blood meal.

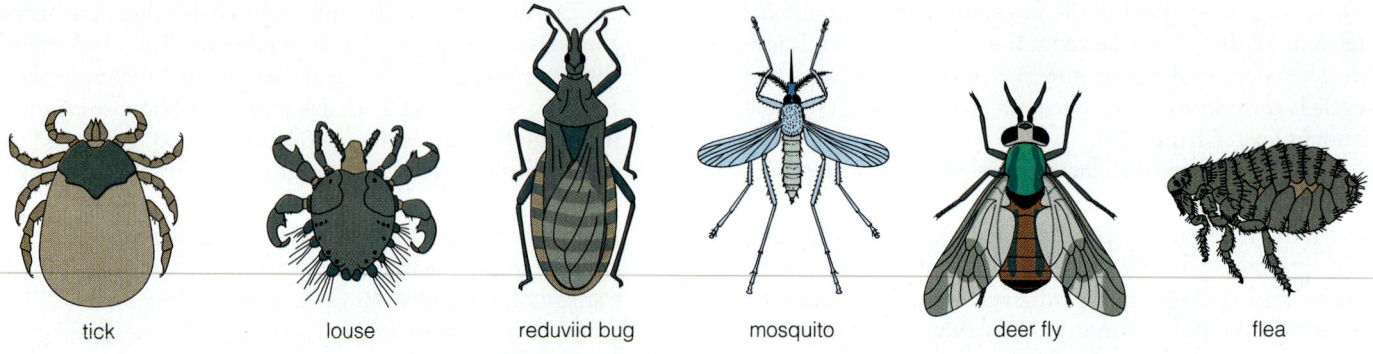

| tick | louse | reduviid bug | mosquito | deer fly | flea |

Figure 12.22 Arthropod vectors transmit microbial infection from one host to another. Some species of ticks are also reservoirs of infection.

Table 12.8 Some Microbial Infections Transmitted by Arthropod Vectors		
Microbial Pathogen	Arthropod Vector	Human Disease (Chapter)
Viruses		
Encephalitis virus (Eastern, Western, St. Louis, Japanese B)	*Culex* mosquito	Viral encephalitis (25)
Yellow fever virus	*Aedes* mosquito	Yellow fever (27)
Dengue fever virus	*Aedes* mosquito	Dengue fever (27)
Bacteria		
Rickettsia rickettsii	*Dermacentor* tick	Rocky Mountain spotted fever (27)
Rickettsia prowazekii	*Pediculus* body louse	Epidemic typhus (27)
Rickettsia typhi	*Xenopsylla* flea	Endemic (murine) typhus (27)
Francisella tularensis	Ticks, deer flies, mosquitoes	Tularemia (27)
Yersinia pestis	*Xenopsylla* flea	Plague (27)
Borrelia burgdorferi	*Ixodes* tick	Lyme disease (27)
Borrelia spp.	*Ornithodorus* tick	Relapsing fever (27)
Protozoa		
Plasmodium spp.	*Anopheles* mosquito	Malaria (27)
Trypanosoma cruzi	*Triatoma* reduviid bug	Chagas' disease (American trypanosomiasis) (27)
Trypanosoma brucei gambiense and *rhodesiense*	*Glossina* tsetse fly	African sleeping sickness (African trypanosomiasis) (25)
Helminths		
Onchocerca volvulus	*Simulium* black fly	River blindness (26)
Loa loa	*Chrysops* mango fly	Loaiasis (26)
Wuchereria bancrofti	Mosquitoes	Filariasis (elephantiasis) (27)

Microbe Mappers

One Disease, One Bacterium, Three Ticks

Lyme disease is caused by the spirochete *Borrelia burgdorferi*. This fairly new disease was first recognized in 1975, and the number of cases is increasing. Its early symptoms are mild; however, untreated, it causes painful and serious illness (Chapter 27).

Lyme disease is most common in the northeastern United States, but it also occurs in the Midwest and the West. In the Northeast, its reservoir is the white-footed mouse, and it is transmitted to humans by the bite of a tick, *Ixodes dammini*. Between 25 and 50 percent of *I. dammini* ticks in the Northeast are infected with *Borrelia burgdorferi*, a level high enough to maintain the infection in the mice. In Northern California, however, the situation is quite different. The bite of another tick, *Ixodes pacificus*, infects several hundred people each year, but only 1 to 5 percent of *I. pacificus* ticks are infected with *Borrelia burgdorferi*. This is too small a fraction to maintain the disease in an animal reservoir. So, then, what is the reservoir, and how is it maintained?

In 1992 R. N. Brown and R. S. Lane, parasitologists at the University of California, solved the double riddle. They found that the animal reservoir was the dusky-footed wood rat, and the disease was maintained in the reservoir by another tick, *Ixodes neotomae*. Fifteen percent of these ticks are infected, an adequate fraction to maintain the reservoir. *I. neotomae* doesn't bite humans, so both ticks are required for humans to be infected—*Ixodes pacificus* to transmit *Borrelia burgdorferi* to humans, and *Ixodes neotomae* to maintain the infection in wood rats.

Infected wood rats and infected ticks have been found in the mountains near Los Angeles, which could lead to an epidemic of Lyme disease in a major metropolitan center. Fortunately, we know how the infection, the microorganism that causes it, and the arthropod that transmits it are interconnected. Brown and Lane's research provides epidemiologists with the information they need to control an outbreak. (See the box "Tick Check" in Chapter 27 for photos of the ticks.)

Because the mosquito is an essential link in the chain of infection, mosquito abatement has controlled malaria in many parts of the world.

Specific species spread specific diseases (**Table 12.8**). For example, the only mosquitoes that transmit malaria are a few species of one genus, *Anopheles*.

Other arthropods that do not carry disease-causing microorganisms can live on the human body. These arthropod infections, also called **infestations**, are usually not serious. The itchy skin condition called scabies is caused by a microscopic mite (*Sarcoptes scabiei*) that lives in the skin's outer layers (Chapter 26). Infestations by the head louse (*Pediculus capitis*), which lives on hairs in the scalp, and by the pubic or crab louse (*Phthirus pubis*), which lives on hairs in the genital area (Chapter 26), can cause significant skin irritation, but they are not biological vectors of infection. However, the body louse (*Pediculus corporis*), which lives in unwashed clothing, does spread microbial disease, including typhus.

Still other arthropods threaten human health in different ways. Some biting arthropods inject a venom. These include the black widow spider, whose bite can cause painful or even fatal muscle contractions, and the brown recluse spider, whose bite can kill nearby skin and muscle. Certain ticks can also cause a severe neurological syndrome called tick paralysis if they remain attached for several days. This is a rare disorder and can be cured by removing the tick.

Honeybee and wasp stings produce a relatively harmless venom, but in rare instances they cause sudden death from a malfunction of the immune system (Chapter 18).

Summary

EUCARYOTIC MICROORGANISMS (pp. 275–293)

1. The perimeters of the study of microbiology are necessarily fuzzy because microorganisms are defined by size, which is relative. Fungi, protozoa, and slime molds are included in all definitions of microorganisms.

Fungi (pp. 275–284)

1. Fungi include single-celled organisms (yeasts), filamentous organisms (molds), and organisms that form fleshy fruiting structures (mushrooms, puffballs, shelf fungi).

2. All fungi are eucaryotic, heterotrophic, nonphototrophic, and absorptive. Most fungi are saprophytes; some are parasites.

3. Fungi are composed of a mycelium, a multinucleate mass of cytoplasm enclosed within a system of tubelike filaments called hyphae. In the lower fungi, the mycelium is coenocytic. In higher fungi, the mycelia have septa, incomplete cross walls. The reproductive structures in fungi are called spores.

4. The body of a fungus is called the thallus. Fungi that have fleshy fruiting structures have a two-part thallus, an extensive mycelium underground and a solid mass containing sexual spores above ground (such as a mushroom).

5. Most yeasts reproduce by budding. A bubble forms on the cell surface, grows, and pinches off to separate. Pseudohyphae are buds that remain attached to the mother cell in long chains.

6. Most fungi are aerobes. Many yeasts are facultative anaerobes. Fungi can break down most organic matter.

7. Fungi are classified by how they reproduce. Asexual reproduction occurs by elongation of hyphae, dividing or budding of single cells, or the production of asexual spores. Sporangiospores are produced within a special structure called a sporangium, and conidiospores, or conidia, are produced naked on tips of special hyphae called conidiophores. Sexual reproduction occurs by producing sexual spores.

8. Important lower fungi include Chytridiomycetes and Oomycetes (water molds) and the Zygomycetes (such as *Rhizopus nigricans*, the black mold on stale bread).

9. The higher fungi are divided into three groups. Ascomycetes develop ascospores within a sac called the ascus. Most Basidiomycetes form basidiospores on basidia, club-shaped cells; some have a dikaryon stage. Deuteromycetes do not form sexual spores; they are also called Fungi Imperfecti.

10. *Yeast* is a descriptive term applied to any nonfilamentous, single-celled, round or oval-shaped fungi.

11. Dimorphic fungi switch between a single-celled yeast phase and a mycelial phase.

12. Fungi are the major cause of infectious disease in plants. Only a few fungi are pathogenic for humans, but fungal infections can be deadly and difficult to treat. Fungi produce penicillin and the cephalosporins.

Algae (pp. 284–288)

1. Metabolically, algae resemble higher plants, but they reproduce as fungi do, producing asexual spores and gametes that fuse to form zygotes.

2. Most algae are aquatic organisms. They are the major fixers of CO_2; phytoplankton carry out about 80 percent of the earth's photosynthesis.

3. Algae do not cause disease because they are photosynthetic and cannot live inside the body. An exception is *Prototheca*, which has lost its ability to photosynthesize.

4. Alginic acid, derived from kelp, and carrageenan from the red alga *Chondrus crispus* are food thickeners. Other red algae are the source of agar.

5. Algae are divided into six groups. Red algae and brown algae are plantlike and multicellular. Green algae range from unicellular to filamentous to plantlike and multicellular. Euglenoids are single motile cells with two flagella of unequal length. Dinoflagellates are single cells usually with two flagella, one wrapped around the cell and the other extending. Diatoms are single cells enclosed in a silica shell.

Lichens (p. 288)

1. Lichens are the product of a mutualistic association between a fungus (usually an Ascomycete) and a green alga or a cyanobacterium. The fungus is dominant, making up most of the mass and giving the lichen its shape.

2. Lichens grow under harsh conditions and very slowly.

3. Both partners can be cultivated separately. The alga continues growing as individual cells, but the fungus assumes a quite different shape.

Protozoa (pp. 289–291)

1. Protozoa are nonphotosynthetic, unicellular eucaryotes. They have elaborate organelles and are the pinnacle of complexity in unicellular differentiation.

2. Protozoa reproduce asexually by fission and budding. The species that causes malaria undergoes schizogony, multiple fission. Protozoa reproduce sexually by conjugation or by the fusion of gametocytes.

3. Some protozoa produce cysts.

4. Protozoa are divided into four groups on the basis of locomotion. The flagellates (Mastigophora) move by means of flagella. The amoeboids (Sarcodina) move by extending pseudopods. The sporozoa (Sporozoa) are nonmotile, and the ciliates (Ciliophora) move by means of cilia.

5. Protozoa cause such serious diseases as malaria, amoebic dysentery, and African sleeping sickness.

The Slime Molds (pp. 291–293)

1. The true slime molds (Myxogastria) are multinucleate at one stage of their life cycle. They are commonly seen on decaying logs as a slimy mass called a plasmodium. When a plasmodium reaches a relatively dry region, it develops fruiting bodies, which contain spores.

2. The cellular slime molds (Acrasieae) produce cells that resemble amoeboid protozoa.

3. Slime molds are not pathogenic or put to any industrial use.

HELMINTHS (pp. 293–297)

1. Helminths are animals, but they are studied in microbiology because some pass through microscopic stages and cause infectious diseases.

2. Platyhelminthes (flatworms) include two main groups of human parasites, Cestoda (tapeworms) and Trematoda (flukes). Nemathelminthes (roundworms) include numerous parasitic species.

3. Cestodes have segmented bodies (segments are called proglottids) and life cycles with larval as well as adult forms. Adult cestodes attach to the intestines of the host with the scolex, and proglottids are passed in the stool. Larval tapeworms generally develop in tissue (such as heart or muscle).

4. Humans can acquire tapeworm infections by eating cystecerci, larvae encysted in undercooked meat.

5. Trematoda are not segmented; adults are shaped like flat, pointed ovals. Some fluke species parasitize human tissue, such as the liver or lung, while others live in blood vessels.

6. Flukes have complex life cycles. Parasitic human species are limited to areas where the proper intermediate hosts are found and where untreated human sewage enters water.

7. Roundworms are more highly developed than flatworms. Life cycles can be simple or complex.

8. People acquire roundworm infections in various ways. They can ingest eggs in fecal-contaminated material (whipworm and *Ascaris lumbricoides* infections), ingest the worm's larvae (trichinosis), become infected by walking barefoot on infected ground (hookworm), or have free-living filariform larvae penetrate their skin.

ARTHROPOD VECTORS (pp. 297–299)

1. Arthropods are invertebrate animals with jointed legs. They are studied in microbiology because some are vectors and some are also reservoirs of infection.

2. The most medically important arthropods are certain chelicerates (ticks and mites) and insects (mosquitoes, flies, lice, fleas, and bugs).

3. Ticks that pass infectious organisms into their eggs (transovarial transmission) are also reservoirs.

4. A mechanical vector picks up a pathogen on its body and carries it from one place to another.

5. A biological vector is an *essential* link in the transmission of a disease.

6. Some biting arthropods inject a venom that can be painful or even fatal.

Review Questions

EUCARYOTIC MICROORGANISMS

Fungi

1. What three characteristics define all fungi? Distinguish between yeasts, molds, and mushrooms.

2. Describe the typical structure of fungi. How are higher fungi structurally different from lower fungi?

3. What is the difference between a saprophyte and a parasite?

4. How do yeasts reproduce?

5. Discuss how lower fungi reproduce and give some examples of lower fungi.

6. Discuss how each of the following types of higher fungi reproduces:
   a. Ascomycetes
   b. Basidiomycetes
   c. Deuteromycetes

7. What are dimorphic fungi?

8. Give some examples of how fungi benefit us and how they harm living things.

Algae

1. How are algae like higher plants and how are they like fungi? What are their chief characteristics?

2. Why don't algae cause diseases?

3. Name the six types of algae and distinguish between them.

Lichens

1. What are lichens? Where are they found?

2. What happens when you tease apart the thallus of a lichen?

Protozoa

1. What are protozoa? How do they reproduce? What are cysts?

2. On what basis are protozoa classified? Name the four types.

3. Give some examples of diseases caused by protozoa.

The Slime Molds

1. Distinguish between true slime molds and cellular slime molds.

2. What is a plasmodium? What is a fruiting body?

HELMINTHS

1. What are helminths and why do microbiologists study them?

2. Distinguish between the following:
   a. Platyhelminthes
   b. Cestoda
   c. Trematoda
   d. Nemathelminthes

3. How do humans acquire a tapeworm infection? What are the symptoms?

4. Describe the life cycle of a fluke.

5. How is each of the following roundworm infections acquired?
   a. whipworm
   b. trichinosis
   c. hookworm

ARTHROPOD VECTORS

1. What are arthropods and why do microbiologists study them?

2. What are the most medically important arthropods?

3. Distinguish between a mechanical vector and a biological vector and give an example of each.

Suggested Readings

Garcia, L. S., and Bruckner, D. A. 1993. *Diagnostic medical parasitology.* 2d ed. Washington, D.C.: American Society for Microbiology.

Jeffrey, H. C., and Leach, R. M. 1975. *Atlas of medical helminthology and protozoology.* New York: Churchill Livingstone.

Large, E. C. 1940. *The advance of the fungi.* New York: Henry Holt.

Lee, J. J.; Hutner, S. H.; and Bovee, E. C. 1985. *An illustrated guide to the protozoa.* Lawrence, Kan.: Society of Protozoologists.

Ross, I. K. 1979. *Biology of the fungi.* New York: McGraw-Hill.

Sleigh, M. 1989. *Protozoa and other protists.* New York: Hodder and Stoughton.

Sze, P. 1986. *A biology of the algae.* Dubuque: Wm. C. Brown.

Tanner, J. R. 1987. St. Anthony's fire, then and now: A case report and historical review. *The Canadian Journal of Surgery* 30:291–93.

13 THE VIRUSES

New Alliances with Old Enemies

Viruses cause the common cold and they cause AIDS. They kill our livestock and ruin our crops. Can anyone argue with Nobel Prize winner Sir Peter Medawar's comment that a virus is "a piece of bad news wrapped in protein"? Who wouldn't second Nobel Prize winner David Baltimore's complaint, "If they weren't here, we wouldn't miss them"? No one—until recently.

Viruses have become important tools of genetic engineering. They are used to clone genes and to introduce beneficial genes into plants and animals. Scientists using them as vehicles of human gene therapy offer hope that genetic diseases, such as cystic fibrosis and diabetes, and even some forms of cancer can be cured.

Scientists are beginning to use certain viruses in the battle against insect pests. Baculoviruses are rod-shaped, double-stranded DNA viruses that infect and kill invertebrate animals (*baculum* means "rod" in Latin). They belong to the family Baculoviridae and the genus Baculovirus. The majority of baculovirus species—more than 600—infect insects.

Their potential as effective, safe insecticides has been recognized since the 1950s, but until recently it hasn't been possible to turn this knowledge into commercial success. Baculoviruses are expensive to produce, each species attacks only a limited range of insects, and insect death takes four to eight days. Worst of all, infection causes insects to binge-eat so that they cause much destruction before they die. In comparison, synthetic chemicals are inexpensive, lethal to a wide variety of insects, and fast-acting.

Today, however, we recognize that chemical insecticides can be health risks and damage the environment. Baculoviruses, as biopesticides—natural agents—avoid these problems. Moreover, through genetic engineering, scientists are improving the way baculoviruses work as pesticides. A viral gene, *egt*, has been identified as the cause of the destructive eating binge; deleting it from the viral genome reduces feeding damage. Faster-killing insect-selective toxins, such as scorpion toxin, have been introduced into the viral genome. The virus has been modified to infect more kinds of insects. At the same time, targeting may be more of an advantage than was imagined years ago—it ensures that only the pests and not helpful insects are killed. Finally, large quantities of baculovirus can now be produced inexpensively.

Five baculovirus biopesticides have been registered with the U.S. Environmental Protection Agency (EPA), and safety tests are underway. The U.S. Forest Service already uses a baculovirus biopesticide to control Douglas fir tussock moths in the Northwest. This new alliance with viruses should benefit both humans and the environment.

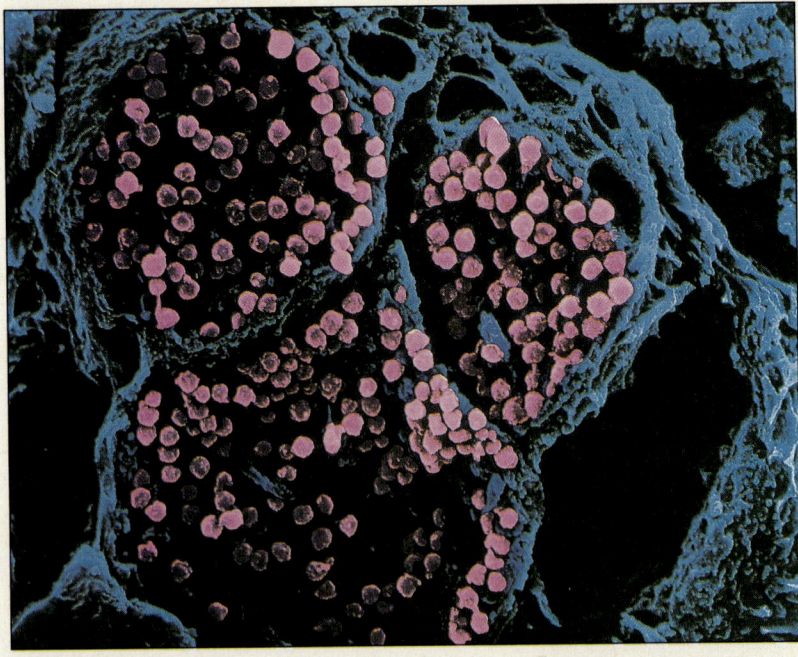

Baculovirus.

To understand:

▪ The nature of viruses—their host range, size, structure, and life cycle—and how they are classified and named

▪ The life cycles of virulent and temperate bacteriophages

▪ The general properties of animal viruses and, in more detail, the properties of retroviruses, influenza viruses, and tumor viruses

▪ The general properties of plant viruses

▪ The nature of viroids and prions

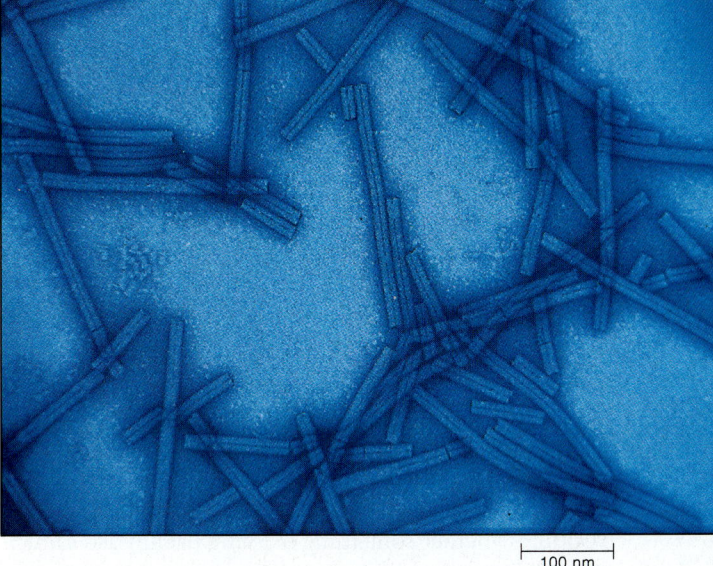

100 nm

Figure 13.1 The mottled lesions on a tobacco leaf give tobacco mosaic disease its name. Iwanowski discovered the virus in 1892, but scientists could not see it until the electron microscope was developed almost 50 years later.

THE ULTIMATE PARASITES

Viruses are parasites. They are not cells, but packages of genetic information—nucleic acids in a protein coat—that insert themselves into a host cell and direct *its* metabolic machinery to make more virus. The virus supplies information only. With few exceptions, all other metabolic requirements—both materials and driving force—are supplied by the host cell (Chapter 5). In some cases only a small portion of a cell's metabolic capacity is diverted to make viruses, and most functions of the host cell continue normally. In other cases, however, all host cell functions are diverted and the cell dies.

All cellular organisms can be attacked by viruses. A particular virus usually parasitizes only a small group of closely related organisms, but there are many different kinds of viruses. All bacteria, eucaryotic microorganisms, and animals, including humans, are vulnerable to many different viruses. Each of us has been infected by viruses many times during our lives; and even when we feel completely healthy, we harbor numerous viruses that are in a quiescent state.

The Discovery of Viruses

Virus is the Latin word for "poison" or "slime." Before the germ theory of disease was developed, the mysterious agents thought to poison a person's body and cause disease were called viruses. But with the development of medical microbiology in the 1800s and the study of disease-causing bacteria, the stage was set for the discovery of viruses as we know them today.

When scientists realized that bacteria could cause disease, they developed ways to remove bacteria from infected materials. One method was to pass a liquid through an unglazed porcelain filter with pores small enough to retain bacterial cells. Because the smallest infectious agents known were bacteria, investigators assumed that filtration would sterilize all liquids. Of course, sterilization by filtration was not always successful. Sometimes the porcelain filter was faulty, and sometimes the investigator used it carelessly or incorrectly. But scientists, like other humans, are not eager to publicize their mistakes. It was therefore an act of courage when a young Russian scientist named Dmitri Iwanowski reported in 1892 that filtration of tobacco plant extracts did not remove the agent that caused mosaic disease (**Figure 13.1**). Though he believed an infectious agent smaller than a bacterium must exist, in his report to the St. Petersburg Academy of Science, Iwanowski expressed some concern that the filter was faulty.

Six years later the Dutch microbiologist Martinus W. Beijerinck made the same observation as Iwanowski, but he had no doubts about his methods and he provided a plausible explanation for his results. He suggested that the infectious agent of tobacco mosaic disease might be an unknown organism, one small enough to pass through the pores in the filter. He called these organisms *filterable viruses* because they could pass through a filter and there was no other way to characterize them. They were too small to be visible under a light microscope. Beijerinck predicted that other viruses would be discovered, and they soon were. In the next several years, the plant diseases peach yellows and peach rosette as well as foot-and-mouth disease in animals were found to be caused by viruses. Gradually, the modifier *filterable* was dropped; we now refer to this group of microorganisms simply as **viruses**.

In 1935 Wendell M. Stanley, an American biochemist, purified and crystallized tobacco mosaic virus. Now a virus could be studied as a chemist would study any matter. With the development of the electron microscope viruses were finally identified as objects with definite sizes and shapes. But are viruses alive? Beijerinck had speculated that they might be "self-supporting molecules," and early evidence suggested that he might be right.

Are Viruses Alive?

Deciding whether something is living usually isn't much of a problem. Most living things have a number of properties that inanimate objects lack. They reproduce and carry on metabolism. They are organized as cells. They contain enzymes, nucleic acids, carbohydrates, and lipids. They evolve and adapt to changing environments. An entity that has these properties is considered living; but what if it has most, but not all, of them?

Viruses can evolve, they contain some macromolecules found in cellular organisms, and they can direct their own reproduction. In these respects they are like other living things. But viruses are not cells, and they lack a metabolism of their own. A purified preparation of **virions**—intact, nonreplicating virus particles—shows no obvious signs of life.

Most biologists consider viruses to be living organisms, albeit very simple ones, because they resemble other living things in so many respects and appear to have evolved from them. Still, some biologists insist that the concept of being alive should be restricted to cellular organisms. In this textbook, we regard viruses as one group of microorganisms.

CLASSIFICATION OF VIRUSES

Viruses are classified according to (1) host range, (2) size, (3) structure, and (4) life cycle.

Host Range

The primary classification of viruses is based on **host range**, the spectrum of organisms a virus attacks. There are animal viruses, plant viruses, viruses of eucaryotic microorganisms, and bacterial viruses. Bacterial viruses are called **bacteriophages** or simply **phages** (*phage* is the Greek root for "to eat") because they usually lyse bacterial cells.

Viruses exhibit considerable specificity for hosts, usually attacking only one species and sometimes only certain cultivars (of plants), races (of animals), or strains (of microorganisms) within a species. (Cultivars, *cultivated varieties*, are selected varieties of cultivated plants.) In humans, some viruses attack only particular types of cells.

Host specificity is determined largely by the presence of appropriate **receptors**—usually proteins—on the cell surface, to which virions attach. Cells that lack the receptor for a particular virus or lose it by mutation are not attacked by that virus. (See the box "Virus Receptors" on p. 318.)

Size

Smallness and structural simplicity are the essential features of viruses. Viruses range in size from about 25 nm to over 300 nm, about one-tenth to one-third the size of a small bacterial cell. They lack all the cellular structures described in Chapter 4—cytoplasm, a cytoplasmic membrane, ribosomes, a nucleus or nucleoid. Most are merely nucleic acid wrapped in a protein coat. As a result, most viruses are smaller than even the smallest procaryotic cell (**Figure 13.2**). Even one of the largest viruses, T4, which contains at least 77 genes, has about 50-fold fewer genes than *Escherichia coli*. Some viruses are far smaller. For example, the viruses $Q\beta$ and MS2 contain only three genes, yet they are highly successful and lethal bacteriophages.

Structure

The basic structure of a virion is a nucleic acid core surrounded by protein and sometimes also a membrane (**Figure 13.3**).

Nucleic Acid. All cellular organisms store their genetic information in the form of double-stranded (ds) DNA. They also contain RNA, but it serves other purposes and never stores genetic information. Some viruses also store their genetic information in dsDNA, but others have

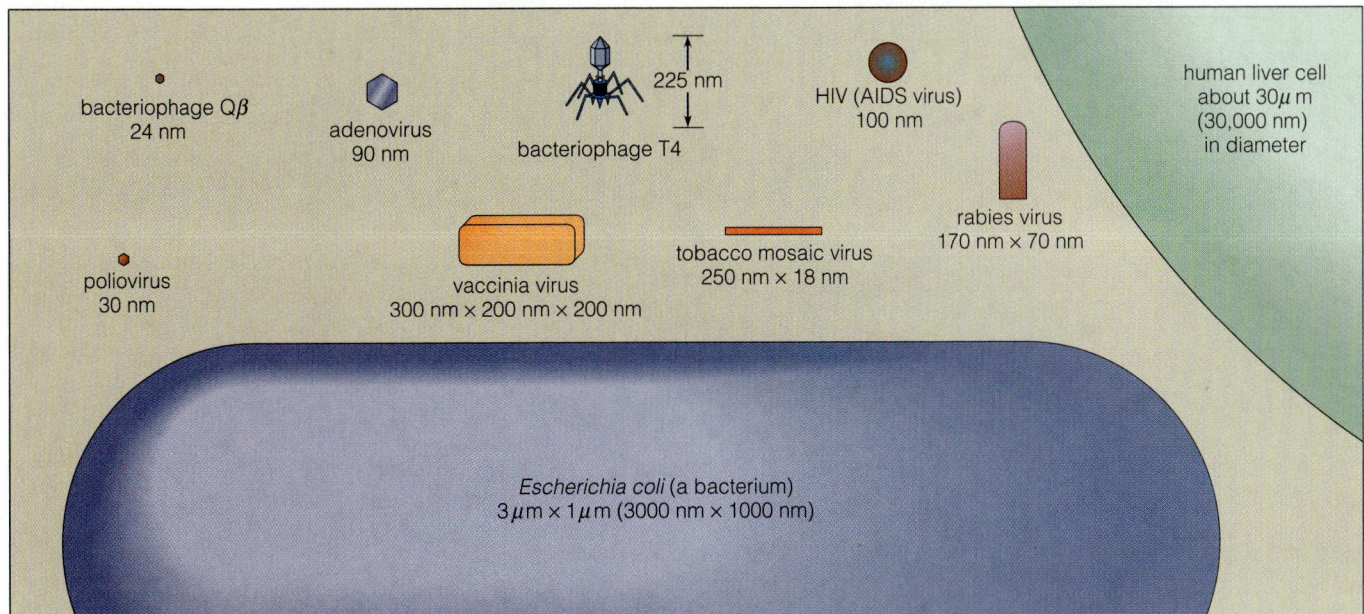

Figure 13.2 Relative size of viruses, bacteria, and human cells.

In the figure, the following labels appear:

bacteriophage Qβ
24 nm

adenovirus
90 nm

225 nm

bacteriophage T4

HIV (AIDS virus)
100 nm

human liver cell
about 30μm
(30,000 nm)
in diameter

rabies virus
170 nm × 70 nm

poliovirus
30 nm

vaccinia virus
300 nm × 200 nm × 200 nm

tobacco mosaic virus
250 nm × 18 nm

Escherichia coli (a bacterium)
3μm × 1μm (3000 nm × 1000 nm)

more space-saving or energy-saving ways of storing genetic information. Some viruses contain single-stranded (ss) DNA, which is converted into a usable double-stranded form only after it enters its host. Other viruses short-circuit the flow of information from DNA to RNA to protein (Chapter 6) by storing their genetic information in RNA—sometimes double-stranded and sometimes single-stranded. Some of these viruses store ssRNA that is translated directly by the host ribosomes. Such RNA, called the **plus-strand**, is, in fact, messenger RNA (mRNA). Other viruses store the complementary ssRNA called the **minus-strand**, which is converted into mRNA (a plus-strand) after it enters the host cell. In summary, viruses can store their genetic information in five different types of nucleic acid—dsDNA, ssDNA, dsRNA, or ssRNA (either a plus- or minus-strand)—but each virus has only one type. No known virus contains more than one kind of nucleic acid. Some nucleic acid molecules are circular, and some are linear. Most viruses have a single nucleic acid molecule, but a few have more than one.

Viral Capsids. The protein coat that surrounds the nucleic acid core of viruses is called a **capsid** (Figure 13.3). Capsids come in three basic shapes—**helical**, **polyhedral**, and **complex**—depending upon how the constituent protein molecules, called **capsomeres**, are arranged (**Figure 13.4**). Some viruses build their capsid from only one type of capsomere, while others use more than one type.

The capsomeres of helical viruses fit together as a spiral to form a rod-shaped structure. The tobacco mosaic virus is helical.

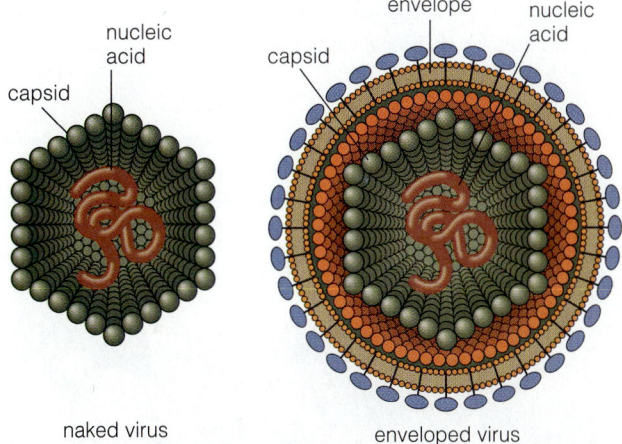

Figure 13.3 Basic structure of virions. A virion, or virus particle, consists of nucleic acid surrounded by a protein capsid and sometimes also a membrane. Virions that have a membrane are called enveloped and those without, naked.

In the figure the labels appear:

nucleic acid

capsid

envelope

nucleic acid

capsid

naked virus

enveloped virus

The capsomeres of polyhedral viruses are usually arranged in equilateral triangles that fit together to form a structure resembling a geodesic dome (**Figure 13.5**). The most common polyhedral virus structure has 20 triangular surfaces, so it is **icosahedral** (20-sided). Viruses with this many faces appear almost spherical.

Complex viruses are combination viruses, with a helical portion, called a **tail**, attached to a polyhedral portion, called a **head**. Many bacteriophages are complex. Some have additional structures, including a **tail sheath**, a

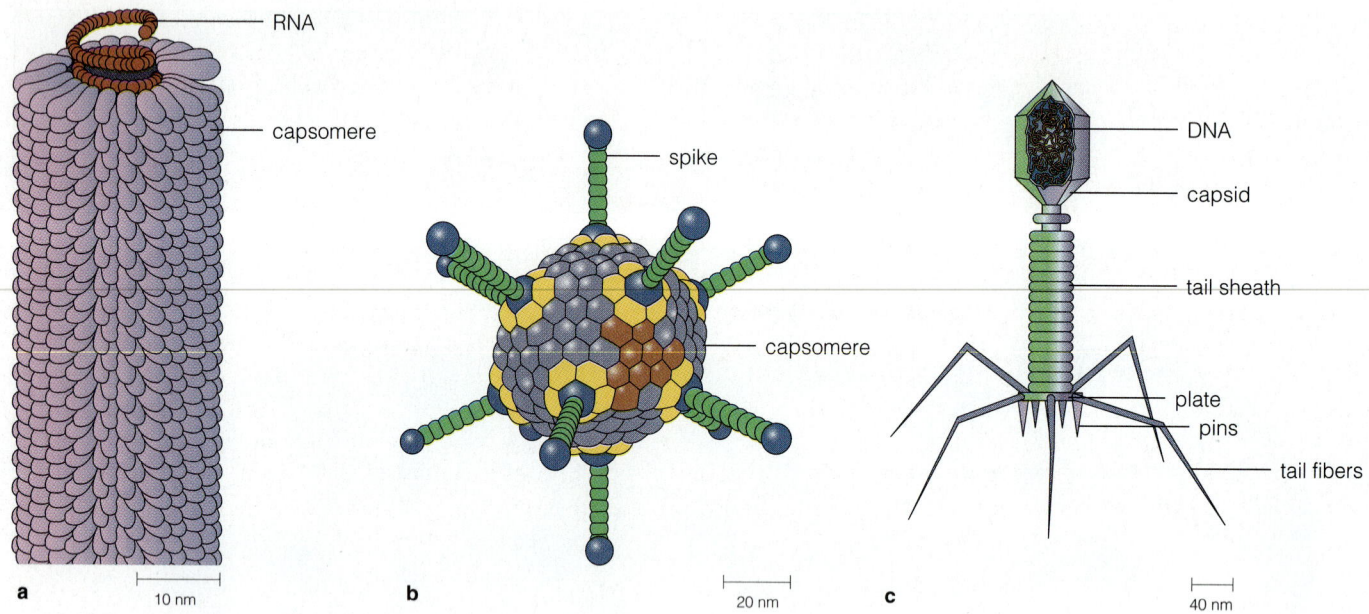

Figure 13.4 Three basic shapes of viral capsids. (**a**) In a helical virus, the capsomeres are helically arranged to form a rod. (**b**) In a polyhedral virus, the capsomeres are usually arranged in equilateral triangles: The most common polyhedral virus is icosahedral (20-sided). This icosahedral virus is an adenovirus, which has protruding spikes. (**c**) A complex virus has a helical tail attached to a polyhedral head. Many bacteriophages, including T4 shown here, are complex. They usually have additional structures, such as a tail sheath, plate, pins, and tail fibers.

Figure 13.5 Virologists got the idea of how capsomeres fit together to form the triangular faces of icosahedral and other polyhedral viruses from the way the American engineer Buckminster Fuller used triangles to create geodesic domes.

plate, **pins**, and **tail fibers**. The plate, pins, and tail fibers help the virion attach to a host cell. The tail sheath participates in injecting viral DNA into the host cell.

Viral Envelopes. Viruses surrounded by a membrane are called **enveloped viruses**. Viral **envelopes** are pieces of the host cell's plasma membrane that the virion acquires as it emerges from its host cell. Like plasma membranes, viral envelopes contain both lipids and proteins. The lipid portion of a viral envelope is encoded by host cell genes, but the proteins are viral. Some of these proteins, termed **spikes** because they stick out of the membrane, are **glycoproteins**—they have sugar molecules attached to them. Enveloped viruses are particularly sensitive to nonpolar solvents such as chloroform and ether, which destroy the membrane and thereby the virus's ability to infect. Viruses that lack envelopes are called **naked viruses**. They may have glycoprotein spikes. They are more resistant to nonpolar solvents. In practice, a quick and easy way to determine if a virus is enveloped or not is to test whether it is inactivated by ether. Enveloped viruses are, and naked viruses are not.

Life Cycle

All viruses have the same basic life cycle, which is radically different from the reproduction of any cellular organism. Outside a host cell, viruses exist only in the form of virions, which cannot replicate themselves. To initiate an infection, the virion attaches itself to the host cell, a process called **adsorption**. Then the viral genome enters the host cell, a process called **penetration**. During penetration, some types of viruses open and disassemble so that only the nucleic acid enters the host cell. The process of removing the capsid and envelope is called **uncoating**. In other cases, the entire virion is taken into the host cell and uncoating occurs later. After uncoating, infection develops and more viral components (nucleic acid and protein) are manufactured by the host cell, a process termed **viral synthesis**. Intact new virions reappear only as these components are reassembled in the process of **maturation**. Soon afterward the virus particles exit the infected cell during the process of **release**, often—but not always—killing the cell.

Later in this chapter, we look at the replication of various types of viruses in more detail.

Taxonomy

At first viruses were named according to the organ system they affected or their host range. By the 1960s, however, virologists realized that some viruses can affect more than a single organ system or host. Classifying a virus as a brain virus or a skin virus or one that infects a particular organism was no longer adequate.

A new classification system was needed. A committee, later named the International Committee on Taxonomy of Viruses (ICTV), set about that task at the International Congress of Microbiology in Moscow in 1966. The system they proposed is not the traditional Linnaean scheme used for all cellular organisms. The ICTV scheme has only three hierarchical levels—family (including some subfamilies), genus, and species. Family names all end in *viridae*. Genus names end in *virus*. Species names are English words.

In their latest report the ICTV describes more than 61 families and almost 2000 species of viruses. Using the ICTV system, the virus that causes AIDS is classified as:

family: Retroviridae

genus: Lentivirus

species: HIV (Human Immunodeficiency Virus)

Family names are often converted into English. Thus, the Retroviridae are called retroviruses. This practice removes the distinction between family and genus.

The names of viral families generally indicate something about their members. The family of hepatitis viruses that contain DNA is the Hepadnaviridae (hepa + DNA). The family containing smallpox, monkey pox, and cowpox viruses is the Poxviridae. The family with viruses that contain two molecules of RNA is the Birnaviridae (bi + RNA).

A number of human diseases discussed in Part IV are caused by viruses. The properties of the families to which they belong are summarized in **Table 13.1**.

BACTERIOPHAGES

About 40 years ago a small group of microbiologists, led by two Americans, Max Delbrück and Salvador Luria, initiated studies on seven bacteriophages (called T1 through T7) that attack *Escherichia coli*. Their results were groundbreaking for two reasons. First, they developed the methods needed to grow and enumerate bacteriophages. Second, they revealed the basic model of bacteriophage replication that proved to be applicable to all other types of viruses.

Phage Counts and Phage Growth

The methods used to count and study bacteriophages (or other viruses) are based on their mode of infecting and destroying host cells. The two commonly used methods are plaque count and one-step growth curve.

Plaque Count. To determine how many bacteriophages are present in a sample, we add a drop of a culture of the host bacterium to the virus sample or an appropriate dilution of it. Then we spread the mixture, or pour it in melted agar, over the surface of agar-solidified medium in a petri dish. After an appropriate incubation period—from a few hours up to a day or so—the surface of the petri plate becomes covered with a lawn of the host bacteria (**Figure 13.6**). Within the lawn are clear circular **plaques**, regions that contain only a few or no bacteria. In most cases, one plaque develops for each phage that was present in the suspension.

The plaque is a result of an epidemic started by a single virion. That virion infected a single host bacterial cell, which produced many more virions; these infected surrounding cells, producing still more virions. The epidemic continued until the bacteria stopped growing, preventing further phage development. By that time all the cells in the region surrounding the original site of infection had been eliminated, producing a plaque.

From the number of plaques that develop on the plate and how much the original sample was diluted, we can calculate the number of bacteriophages present in the sample, the same way we would calculate the results of a viable count of bacterial cells (Chapter 8). This procedure is termed a **plaque count**. Virions cause a plaque to form, but so does an infected bacterial cell that is destined to produce virions. Both are present in any culture infected with bacteriophages. Virions and infected cells together are called **plaque-forming units (PFUs)**.

Table 13.1 Animal Viruses Discussed in Part IV

Family	Properties	Virus (Species or Genus)	Diseases
DNA Viruses			
Herpesviridae	Enveloped dsDNA	Herpes simplex types 1 and 2 (HSV-1, HSV-2)	Cold sores, genital herpes
		Varicella zoster virus	Chickenpox/shingles
		Epstein-Barr virus (EBV)	Mononucleosis, Burkitt's lymphoma
		Cytomegalovirus (CMV)	Birth defects
Poxviridae	Enveloped dsDNA	Smallpox virus	Smallpox (variola)
Hepadnaviridae	Enveloped dsDNA	Hepatitis B virus	Hepatitis B
Papovaviridae	Naked dsDNA	Human papillomaviruses	Warts
Adenoviridae	Naked dsDNA	Human adenovirus	Respiratory disease
			Enteric diseases
			Infectious pinkeye
RNA Viruses			
Retroviridae	Enveloped plus-strand RNA	Human immunodeficiency viruses (HIV-1 and HIV-2)	AIDS
		Human T-cell leukemia viruses (HTLV-1 and HTLV-2)	T-cell leukemia
Togaviridae	Enveloped plus-strand RNA	Alphavirus	Some forms of encephalitis
		Rubella virus	Rubella (German measles)
Flaviviridae	Enveloped plus-strand RNA	Yellow fever virus	Yellow fever
		Dengue fever virus	Dengue fever
		Hepatitis C virus	Hepatitis C

Table 13.1 (continued)

Family	Properties	Virus (Species or Genus)	Diseases
RNA Viruses			
Coronaviridae	Enveloped plus-strand RNA	Coronavirus	Upper respiratory infections
Picornaviridae	Naked plus-strand RNA	Enteroviruses (includes poliovirus, coxsackie-virus, echovirus)	Polio
			Myocarditis, pericarditis
			Gastroenteritis, meningoencephalitis
		Rhinovirus	Common cold
		Hepatitis A virus	Hepatitis A
Calciviridae	Naked plus-strand RNA	Norwalk agents	Gastroenteritis
Orthomyxoviridae	Enveloped minus-strand RNA	Influenza virus	Influenza
Rhabdoviridae	Enveloped minus-strand RNA	Rabies virus	Rabies
Paramyxoviridae	Enveloped minus-strand RNA	Mumps virus	Mumps
		Measles virus	Measles (rubeola)
		Parainfluenza virus	Croup
		Respiratory syncytial virus	Bronchiolitis
Bunyaviridae	Enveloped minus-strand RNA	Hantavirus	Hantavirus-associated respiratory distress syndrome
Reoviridae	Naked dsRNA	Rotavirus	Infant diarrhea

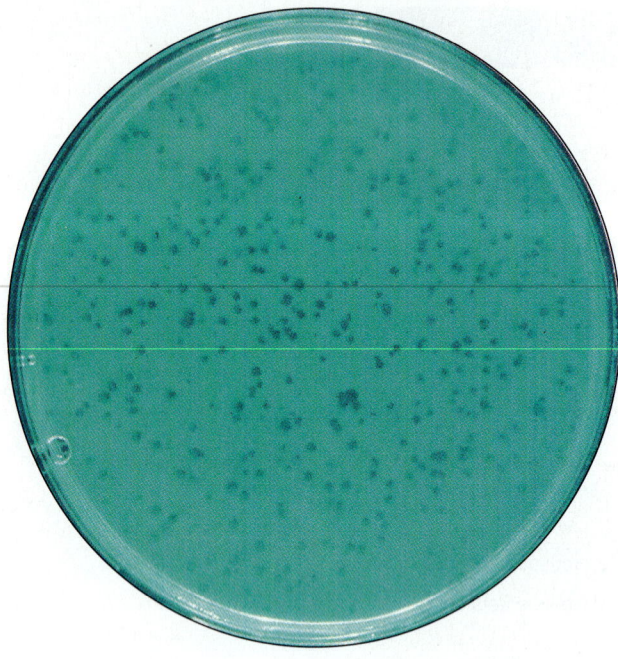

Figure 13.6 Plaques formed by bacteriophage lambda (λ). When a drop of bacterial culture is poured in melted agar over the surface of an agar plate and incubated, a continuous lawn of bacterial growth develops (light areas). If phages are present, plaques develop (dark circles), usually one plaque per phage.

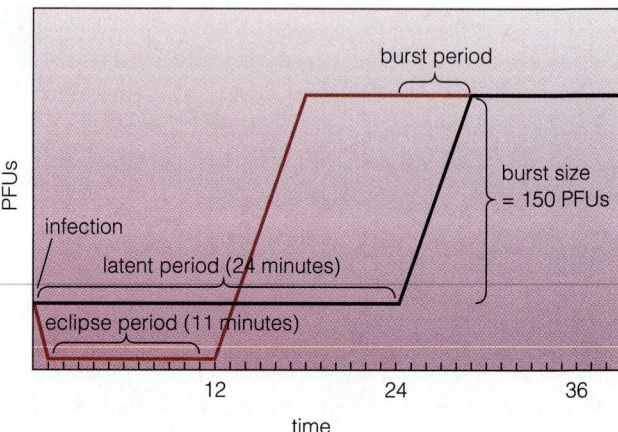

Figure 13.7 One-step growth curve for phage T4. The one-step growth curve experiment charts the change in plaque-forming units (PFUs) in a bacterial culture that has been infected with bacteriophage. The number of PFUs in the sample remains constant (black curve) during the latent period when phage components are being synthesized but no new virions have yet formed. If samples are treated with chloroform (causing premature lysis) before doing plaque counts (red curve), PFUs fall to zero during the eclipse period because no infected cells or intact virions are present. During the burst period virions are assembled. PFUs in chloroform-treated samples rise first because chloroform treatment releases mature virions from cells that have not yet lysed.

The One-Step Growth Curve. Plaque counts allow microbiologists to do quantitative experiments with phages, including the **one-step growth curve** experiment, which tells a great deal about viral replication. Bacteriophage virions are added to a bacterial culture. After a few minutes, during which the virions become attached to bacterial cells and infect them (adsorption and penetration), the mixture is diluted several-hundred-fold, making further contact between virions and host cells unlikely. Then, at intervals, the number of PFUs in the infected culture is determined by doing plaque counts on samples taken from it. Plotting the number of PFUs in the sample against the time the sample was taken produces the one-step growth curve (**Figure 13.7**).

A phage growth curve looks quite unlike the growth curve of cellular organisms, which increases steadily with time. Following infection, the number of PFUs in the culture does not increase at all for a period of time, termed the **latent period**. Then, in a **burst period**, the number of PFUs abruptly increases several-hundred-fold. This one-step growth curve tells us that during the latent period when no new PFUs are produced, the host cell, under the direction of viral nucleic acid, is producing components of new virions. Just before and during the burst period, the components are assembled into virions that are released from the cell, which is destroyed in the process. The fold increase in PFUs that occurs during the burst pe-riod—which can be a hundredfold increase or more—is called the **burst size**. It represents the average number of new virions released from each infected cell.

By counting the number of virions (instead of total PFUs) in the infected culture, we can demonstrate that the virion disassembles during infection. To count virions, each sample taken from the infected culture is treated with chloroform. This treatment kills the cells in the sample, including infected ones, but does not harm intact phage virions. This plaque count stays at zero until the burst period begins. This tells us that the virion disassembles during infection and ceases to exist as an entity until new virions are assembled just before and during the burst period. The period when no intact virions are present is called the **eclipse period**.

Replication Pathways

In broad outline the replication cycle of all phages is the same. There are variations, however—for example, in how the viral genome enters the host cell and how the mature virions leave. In addition, there are different developmental pathways. The infecting phage either enters the **lytic pathway**, leading to production of more virions, or it enters the **lysogenic pathway**, leading to a prolonged quiescent state, termed **lysogeny**. Phages that follow only the lytic pathway are termed **virulent;**

phages that can follow either pathway are termed **temperate**.

Virulent Phages. The steps in the replication cycle of a typical virulent bacteriophage such as T4 are described below and summarized in **Figure 13.8**. T4 is a complex, dsDNA bacteriophage.

First, T4 adsorbs (attaches) by its tail to a specific receptor site, the core region of a lipopolysaccharide molecule in the outer membrane of its host, *Escherichia coli*. Each kind of phage adsorbs to a specific kind of molecule on the cell surface—a lipopolysaccharide or protein component of the outer membrane of Gram-negative bacteria, teichoic acid of a Gram-positive bacterium, a protein of the flagellum, or a pilus. For example, some phages adsorb only to proteins at the end of F-pili, others only to proteins that make up the bulk of the F-pili.

In the penetration or infection stage, the tail sheath of T4 contracts, forcing its dsDNA into the cell, probably through a Bayer junction (Chapter 4). This initiates the eclipse period. The protein component of the phage, which plays no further role in replication, remains attached to the cell's surface. Thus, uncoating of T4 occurs outside the host cell.

During the latent period that follows infection, the phage DNA directs the host cell to synthesize viral components, including more viral DNA and protein components.

At the end of the latent period, the component parts are assembled into new virions (maturation). Finally, the intact virions are released during the burst period as the cell is destroyed. A phage-encoded lysozyme (Chapter 4) dissolves the cell wall, causing the cell to lyse and release the hundreds of phage virions inside it.

Virulent phage infections are so swift, deadly, and productive of phage virions that it is surprising that any bacterial cells in a phage-infected culture survive. But some always do. They are mutant cells, resistant to phage

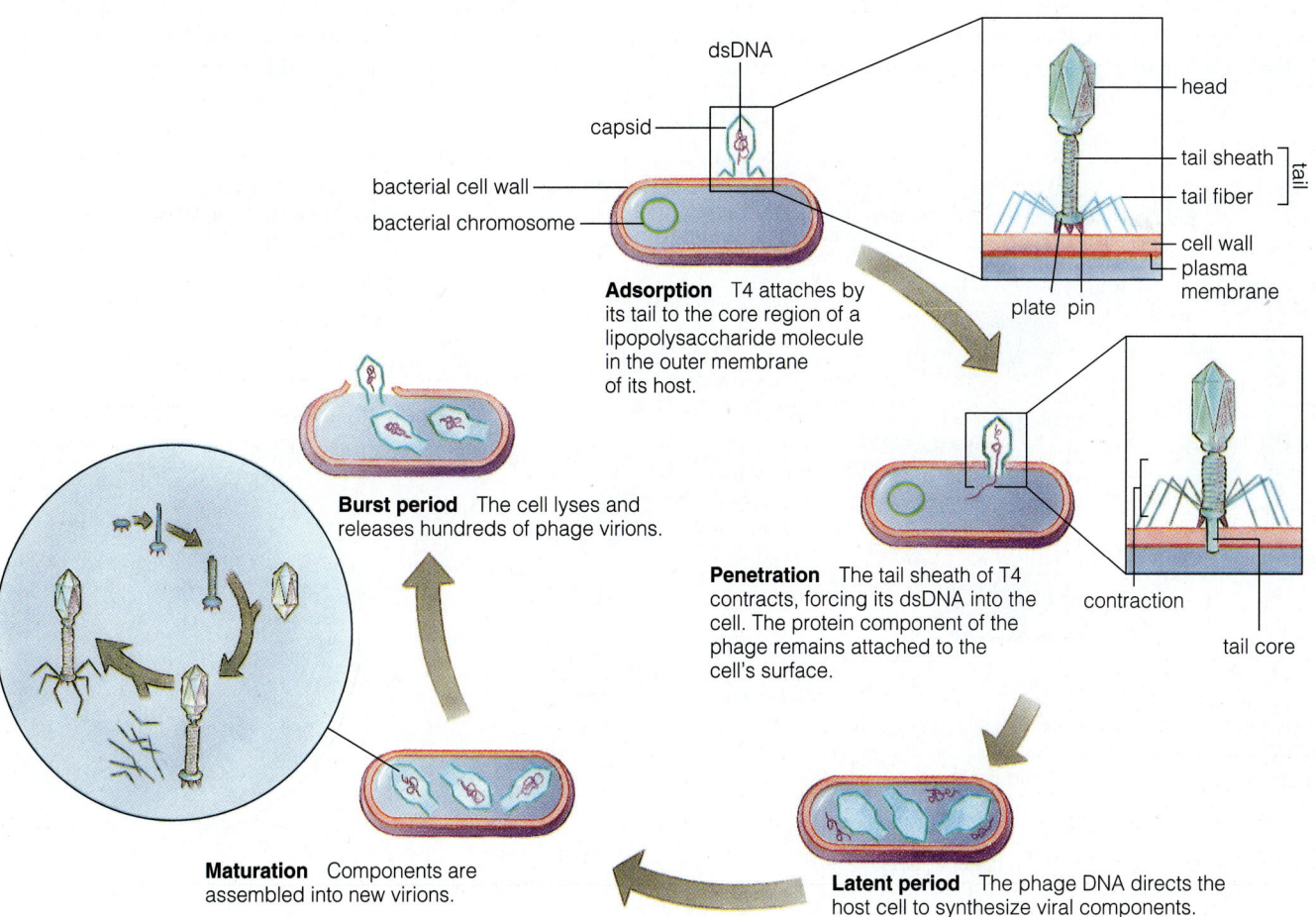

Figure 13.8 Life cycle of a virulent phage (T4).

attack. In most cases, the specific receptor site on their surface is altered so phage virions can no longer bind to them. Such mutant cells are generally present at a frequency of about one in a hundred million (10^{-8}). An ordinary bacterial culture with more than a hundred million (10^8) cells in each milliliter will almost certainly contain cells that survive the infection.

Temperate Phages. The life cycle of the temperate phage **lambda** (λ), which also infects *Escherichia coli*, is shown in **Figure 13.9**. Like phage T4, phage λ is a dsDNA complex virion.

The initial stages of infection by λ and T4 are quite similar. The λ virion adsorbs by its tail to a specific receptor protein in the bacterium's outer membrane—a porin that brings maltose across the outer membrane—and injects its linear dsDNA into the host cell, where its cohesive ends join, forming a circle.

Then lambda takes either the lytic or the lysogenic pathway. The lytic pathway, as with T4, leads to lysis of the host cell after the phage components are synthesized and assembled into virions. In contrast, the lysogenic pathway leads to a quiescent state. No phage components are synthesized, and the host cell is not damaged. Instead, the circular phage genome recombines with the host cell's chromosome and becomes integrated into it at a specific location, forming a **prophage**. A repressor protein encoded by the phage maintains the phage genome in a quiescent state. Because the phage genome becomes a part of the bacterial chromosome in lysogeny, the phage genome is replicated and distributed to progeny cells along with bacterial genes. Such a cell is termed **lysogenic**, meaning that it has the potential to cause lysis if the genes in the prophage are reactivated to initiate an infection that kills the host cell.

Even in the prophage state, however, some phage genes are expressed, which slightly changes the host cell's phenotype. For example, colonies may have a slightly different appearance. An interesting example of this phenomenon occurs in *Corynebacterium diphtheriae*, the bacterium that causes the respiratory disease diphtheria (Chapter 22). Only lysogenic strains of *C. diphtheriae* cause disease because the disease-causing toxin is encoded in the prophage of the infecting virus.

A lysogenic bacterial cell can grow, divide, and produce lysogenic progeny almost indefinitely. In fact, most bacterial cultures probably carry one or more prophages. But on rare occasions, in one infected cell out of 10,000 or 100,000, the prophage leaves the bacterial chromosome by another recombination and enters the lytic pathway.

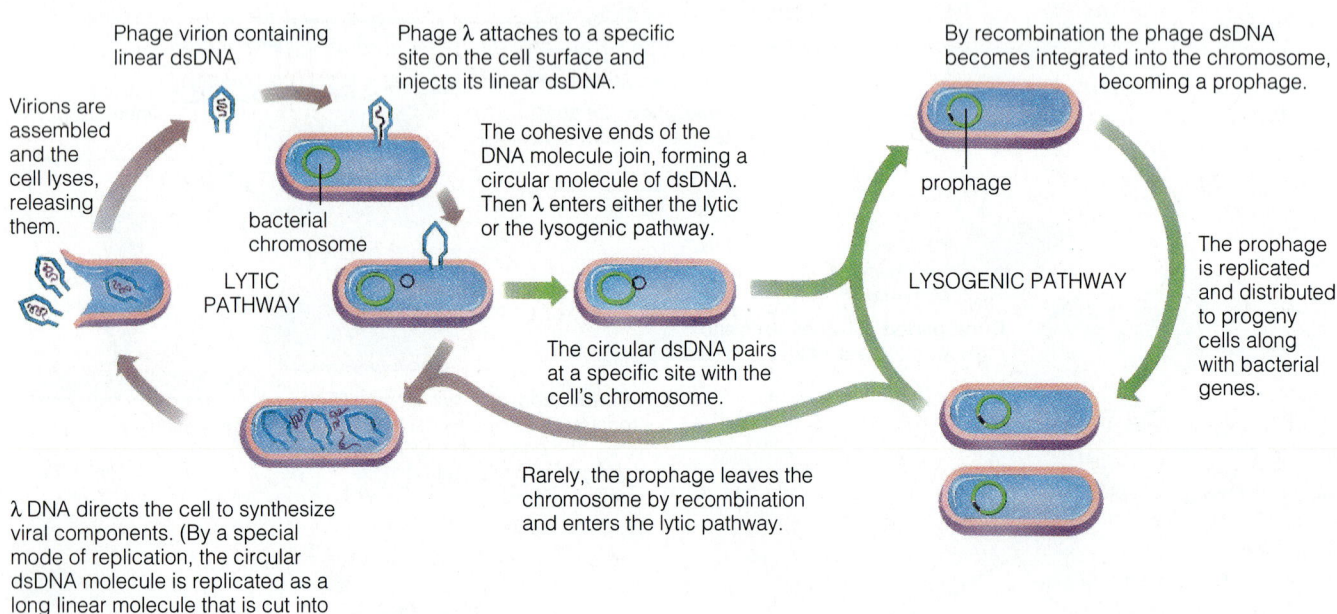

Figure 13.9 Life cycle of a temperate phage (λ).

Microbe Mappers

Werner Arber

The history of science is filled with stories of basic research leading to surprising practical applications. In the 1960s, a Swiss microbiologist, Werner Arber, was working on an esoteric problem of phage replication called host-induced modification. Arber's work laid the groundwork for what promises to be the most important scientific tool of the twenty-first century—genetic engineering.

In host-induced modification, phage λ (and other phages) changes its properties depending upon the strain of *Escherichia coli* on which it is grown. When phage λ is grown on strain B, each λ virion forms a plaque when plated on a lawn of strain B, but only 1 in 10,000 virions forms a plaque on a lawn of strain K. The opposite occurs with λ virions grown on strain K—all form plaques on a lawn of strain K but only 1 in 10,000 forms a plaque on strain B.

On investigating, Arber found that DNA from strain-B-grown phage λ was rapidly degraded when it entered a strain-K cell but not when it entered a strain-B cell. On the other hand, K-grown phage λ was rapidly degraded in strain-B cells but not in strain-K cells. Apparently these two strains of *E. coli* (and other strains as well) treated λ DNA like their own if it was made in a cell of their own strain but like foreign DNA if it was made in a cell of a different strain. They could recognize foreign DNA and destroy it.

But how could a bacterium distinguish between its own and foreign DNA? Arber found that each strain of bacteria produces its own **restriction endonuclease** that recognizes a particular sequence of four to six bases and cuts DNA there (Chapter 7). Then other enzymes destroy the DNA beginning at the site of the cut. To avoid destroying its own DNA, each strain of bacteria also has a **modification enzyme** that adds a methyl group to one of the bases in the sequence that the restriction endonuclease recognizes. The restriction endonuclease cannot cut the methylated DNA. Thus, DNA in strain-K-grown λ, for example, is modified so that it cannot be cut by the K-strain restriction endonuclease, but it can be cut by B-strain endonuclease.

Arber's studies on an obscure biological phenomenon led to an application that will eventually touch all of our lives. Restriction endonucleases are the fundamental tools of recombinant DNA technology (Chapters 7 and 29), which is the basis of genetic engineering.

Then that host cell lyses and mature virions are released into the medium.

The virions that are released do not affect other cells in the lysogenic culture because all these cells carry the same prophage in their genome. The presence of a prophage makes the cell immune to attack by a virion of the same phage, though it is vulnerable to others. The same repressor protein that keeps the prophage in a quiescent state binds to DNA from an attacking virion as it enters the cell, neutralizing its ability to cause infection and thereby making the cell immune.

Because of this immunity, temperate phages make plaques only when plated on a nonlysogenic strain. These plaques have a distinctive appearance. Instead of being clear, like the plaques formed by virulent phages, they are slightly cloudy. The plaques are visible because of cell lysis by the lytic pathway. The turbidity within the plaque is caused by the lysogenic cells that are produced by the lysogenic pathway and grow within the plaque.

Diversity

The two bacteriophages that we have discussed, T4 and λ, are structurally similar. They are large (about 225 nm long), complex, naked phages that contain dsDNA. But there are many other kinds of bacteriophages. Some bacteriophages, for example, Qβ, are as small as 24 nm. Some phages are helical; others are polyhedral. A few phages are enveloped. All the types of nucleic acid—dsDNA, ssDNA, dsRNA, and ssRNA—are found among phages. The nucleic acid content of most phages is in the form of a single molecule, but some contain more. For example, f6 contains three different molecules of dsRNA. Having multiple molecules of nucleic acid, termed a **segmented genome**, is more typical of plant and animal viruses.

ANIMAL VIRUSES

Viruses that infect animals are as diverse in size, shape, and structure as are the phages that infect bacteria. For the most part, animal-virus life cycles resemble those of the bacteriophages, but there are some differences. The methods of studying animal viruses are modeled after the methods used to study phages.

Cell Cultures

For a long time, research on animal viruses lagged behind research on bacteriophages. Bacteriophages can be cultivated by infecting a bacterial culture and counted by plaque assay—both simple procedures. Animal viruses had to be cultivated by infecting animals and counted by

Table 13.2 Cell Cultures Used to Culture Certain Animal Viruses

Virus Cultivated	Source of Cultured Tissue Used to Cultivate Virus	Cell Type	Culture Type
Human immunodeficiency virus	Human lymphocytes	White blood cells	Primary cell culture
Adenovirus	Human kidney	Human embryonic kidney	Primary cell culture
Rhinovirus, enterovirus	Human kidney, lung, foreskin	Human fetal fibroblast	Diploid cell line
Herpes simplex virus	Human cervical cancer	HeLa[a]	Continuous cell line
Influenza virus	Canine kidney	HDCK	Continuous cell line
Enterovirus	Monkey kidney	MGMK	Continuous cell line

[a]Continuous cell lines are usually designated by four letters.

dilution end points (Chapter 8), which required infecting many animals, a time-consuming and expensive process.

In the 1930s, however, virologists discovered that some animal viruses could be grown in embryonated chicken eggs. An incubator with eggs can contain as many individuals as a roomful of mice or rats—a more efficient and economical approach. The embryonated eggs are inoculated; and after the virus has multiplied, virus-containing fluids are withdrawn with a syringe. To do dilution end points, embryonated eggs are inoculated with dilutions of a virus sample to determine the highest dilution that kills the embryo. Although time-consuming, this procedure is still used today to study viruses and to produce viruses for vaccines.

In the 1950s the **cell culture** and **tissue culture** methods of cultivating animal cells and tissues were developed. These involve cultivating plant or animal cells or tissues in culture. In 1952, Renato Dulbecco, an Italian microbiologist trained in the United States and working in Sweden, used cell culture in virus research. After that, animal virology began to advance swiftly. Cell culture made it possible to count animal viruses quickly and inexpensively by plaque count. Dulbecco grew a **monolayer**, a confluent growth of animal cells, one layer thick (much like a lawn of bacteria), in a petri dish. He then infected them with a **cytopathic** animal virus, one that damages or kills animal cells. Each virion caused a region of the cell layer to die or degenerate, forming a plaque.

Cell culture also solved another problem that had plagued virus research—viral specificity. Before cell culture, a virus could be studied only in the particular animals it infected, making it almost impossible to study viruses that infect only humans, such as human immunodeficiency virus (HIV). With cell culture methods, cells from the target animal can be cultivated in the laboratory (**Table 13.2**). For example, HIV taken from lymphocytes of an infected person can be cultivated in cultured human white blood cells.

Cells are cultured in ordinary plastic glassware—flasks, bottles, or petri dishes—provided with a complex medium supplemented with serum (**Figure 13.10**). The cells can be kept suspended by stirring or they can be allowed to settle and grow as a monolayer. In order to subculture, monolayers can be separated into individual cells without damaging them by treating them briefly with the proteolytic enzyme trypsin, which dissolves the proteins that hold the cells together.

Cultured cells are categorized by their source, their chromosome number, and how long they continue to multiply in culture. **Primary cell cultures** are started from normal tissues taken directly from humans or other animals. After several subcultures, the cells that grow best come to dominate the culture. They are called a **cell line** because they are derived from cells better suited to grow in culture. If the cell line contains the same number of chromosomes as somatic cells of the species from which it was derived, it is called a **diploid cell line**. If it has a different number, it is called a **heteroploid cell line**.

Most diploid cell lines, including those from humans, grow increasingly slowly after 20 to 30 subcultures and eventually lose their ability to support viral replication. In contrast, some cell lines, usually derived from cancerous tissue, grow indefinitely in culture. They are called **continuous cell lines**. Probably the most famous continuous cell line is the HeLa cell line (for *He*len *La*ck, the donor), which has been cultivated continuously in various laboratories throughout the world since it was started in 1951 from human cervical cancer tissue. Cell culture has vastly simplified virus research. It has largely replaced the use of laboratory animals to detect and identify animal viruses.

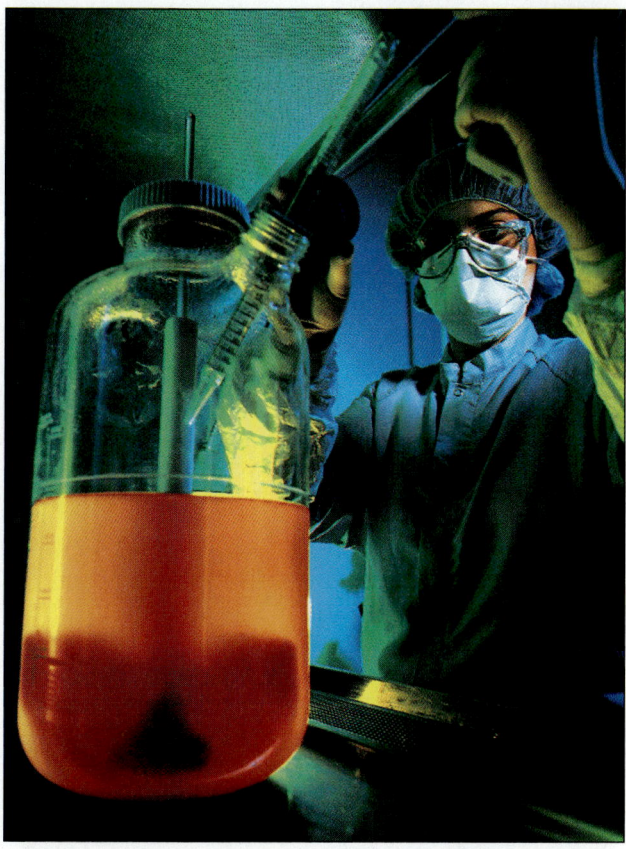

Figure 13.10 Cell culture. In this flask, the colored liquid is the medium that supplies nutrients to the cultured cells.

In spite of the dramatic advances made in research in the last 30 years, methods for diagnosing viral disease have not changed much. Cell cultures take too long for the results to be useful in planning treatment. Most virus-caused illnesses can be diagnosed if the patient develops antibodies against the virus (Chapter 17), but antibodies usually develop only after the patient has recovered, so this approach is not practical either. (AIDS is an exception—antibodies develop before the patient is seriously ill.) Most physicians therefore do not culture viruses or draw samples of blood to look for antibodies. They still rely on symptoms to diagnose viral disease.

Replication

Animal virus replication proceeds through the same stages as phage replication—adsorption, penetration, uncoating, viral synthesis, maturation, and release. As with some phages, uncoating may precede penetration.

Adsorption. Because animal cells lack walls, the plasma membrane is their outermost layer. Proteins em-

bedded in it act as receptors for a virus. During adsorption, receptor-binding proteins on the virus surface bind to cell receptors. The receptor-binding proteins are usually on the glycoprotein spikes of enveloped viruses or are a capsid protein of naked viruses. Adsorption is largely responsible for the tissue specificity of animal viruses—only cells with a complementary receptor are attacked by a particular virus.

Penetration. The attachment between a viral protein and a host cell receptor triggers a reaction that brings the virion, or its nucleic acid and a few proteins, into the cell.

Viruses penetrate animal cells in three ways (**Figure 13.11**). Some enveloped viruses exploit the physical properties of membranes to enter. The viral envelope fuses with the cell's plasma membrane, emptying the rest of the virion inside the cell. Other enveloped viruses enter the cell through phagocytosis (Chapter 4). The host cell membrane engulfs the virion, bringing it into the cell intact within a vesicle of the cell membrane. The mechanism by which the virus escapes from the vesicle and enters the cytoplasm depends upon the type of cell infected. Most naked viruses enter a cell as most bacteriophages do—the capsid adsorbs to the cell surface and remains there as the nucleic acid component of the virion enters the cell.

Uncoating. Viral genes are not expressed (translated into proteins) until the virus is uncoated. The membranes of some enveloped viruses and the capsids of most naked viruses are removed in the process of penetration. Virions that enter the cell partially or completely intact are uncoated inside the cell by the cell's own hydrolytic enzymes, sometimes those in the cell's lysosomes (Chapter 4).

Viral Synthesis. Once uncoated, the virus's nucleic acid can direct the host cell's metabolic machinery to synthesize virus components—capsid proteins, any enzymes the virion may contain, and more nucleic acid. The steps in the process are determined largely by the type of nucleic acid that the virion contains—dsDNA, ssDNA, plus-strand RNA, minus-strand RNA, or dsRNA (**Table 13.3**).

If the virion contains dsDNA, viral synthesis occurs much like the synthesis of any host-cell component (**Figure 13.12**). Viral genes are transcribed into messenger RNA (mRNA), which is translated into viral proteins, and the viral DNA is replicated. The host cell has all the enzymes to carry out these syntheses, but sometimes for special reasons virus-encoded enzymes catalyze certain steps. Poxviruses, for example, encode their own transcriptase as a means of timing the synthesis of viral components. Certain viral genes, called **early genes**, can be

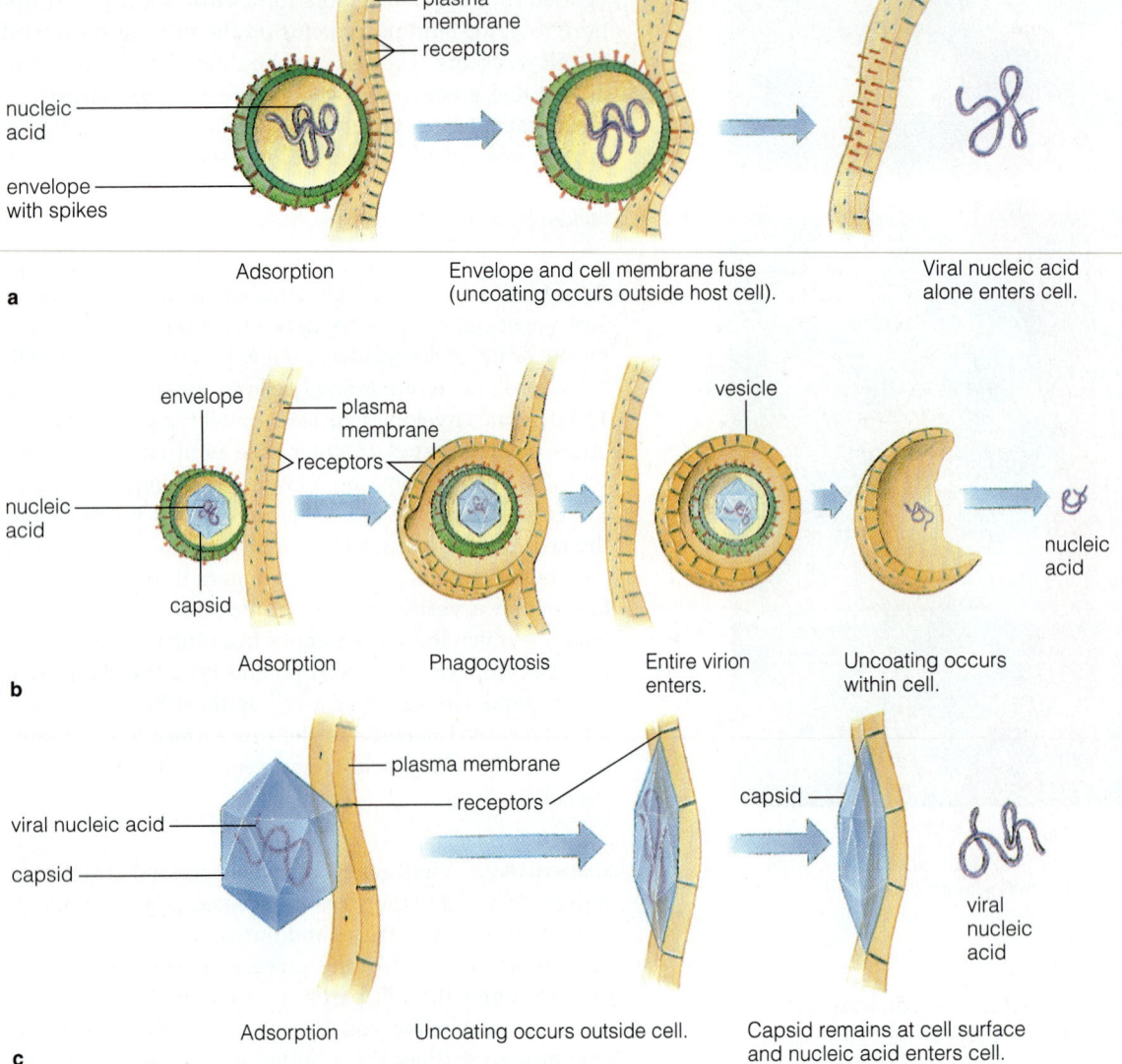

plasma
membrane
receptors

nucleic
acid

envelope
with spikes

a

Adsorption | Envelope and cell membrane fuse (uncoating occurs outside host cell). | Viral nucleic acid alone enters cell.

envelope

plasma
membrane

receptors

nucleic
acid

capsid

vesicle

nucleic
acid

b

Adsorption | Phagocytosis | Entire virion enters. | Uncoating occurs within cell.

plasma membrane

receptors

viral nucleic acid

capsid

capsid

viral
nucleic
acid

c

Adsorption | Uncoating occurs outside cell. | Capsid remains at cell surface and nucleic acid enters cell.

Figure 13.11 Viruses can penetrate animal cells in one of three ways. (**a**) The viral envelope can fuse with the cell's plasma membrane, uncoating the virus. The nucleic acid alone is emptied into the cell. (**b**) An enveloped virus can be phagocytized, in which case the entire virion enters and is uncoated within the cell. (**c**) Most naked viruses adsorb to the cell surface. The capsid remains, uncoating the virus. The nucleic acid alone enters the cell.

transcribed by the host's own transcriptase. Among these is the gene encoding the viral transcriptase. **Late genes** can be transcribed only by virus-encoded transcriptase and, therefore, only late in the infection after the virus-encoded transcriptase has been synthesized.

Viral synthesis in cells infected with ssDNA virus goes through an additional step. A complementary strand to the ssDNA is made, converting it to dsDNA. Then the dsDNA is transcribed, forming mRNA, which is translated, forming viral proteins.

All three classes of RNA viruses—dsRNA viruses, plus-strand RNA viruses, and minus-strand RNA viruses—encode a special enzyme called **RNA-dependent RNA polymerase**, which the host cell lacks, to catalyze certain steps in the expression or replication of their genomes. RNA-dependent RNA polymerase uses ssRNA as a template to synthesize a complementary ssRNA. The enzyme plays a different role in each class of RNA viruses (**Figure 13.13**).

In plus-strand RNA viruses, RNA-dependent RNA polymerase is used to replicate the genome. The plus-strand RNA in the virion can be translated into viral pro-

Table 13.3 Patterns of Animal Virus Replication

Nucleic Acid in Virion	Replication of Nucleic Acid	Transcription/Translation	Examples
dsDNA	Replication of DNA begins after some viral protein has been made.	dsDNA is transcribed by the same mechanism as chromosomal DNA.	Herpesviruses, poxviruses
ssDNA	In host cell nucleus, ssDNA is converted to double-stranded replicative form, which is copied and generates ssDNA.	Host-cell RNA polymerase transcribes double-stranded replicative-form DNA.	Parvoviruses
Plus-strand RNA	Virus-encoded RNA-dependent RNA polymerase makes minus-strand RNA, which serves as a template for more plus-strand.	Translation of plus-strand RNA begins immediately, forming capsid protein and RNA-dependent RNA polymerase.	Picornaviruses, togaviruses
Minus-strand RNA	RNA-dependent RNA polymerase makes plus-strand RNA, then more minus-strand.	Transcription is catalyzed by RNA-dependent RNA polymerase in virion.	Orthomyxoviruses, rhabdoviruses
dsRNA	RNA is replicated by virus-encoded RNA polymerase.	dsRNA is converted to mRNA by RNA-dependent RNA polymerase.	Reoviruses

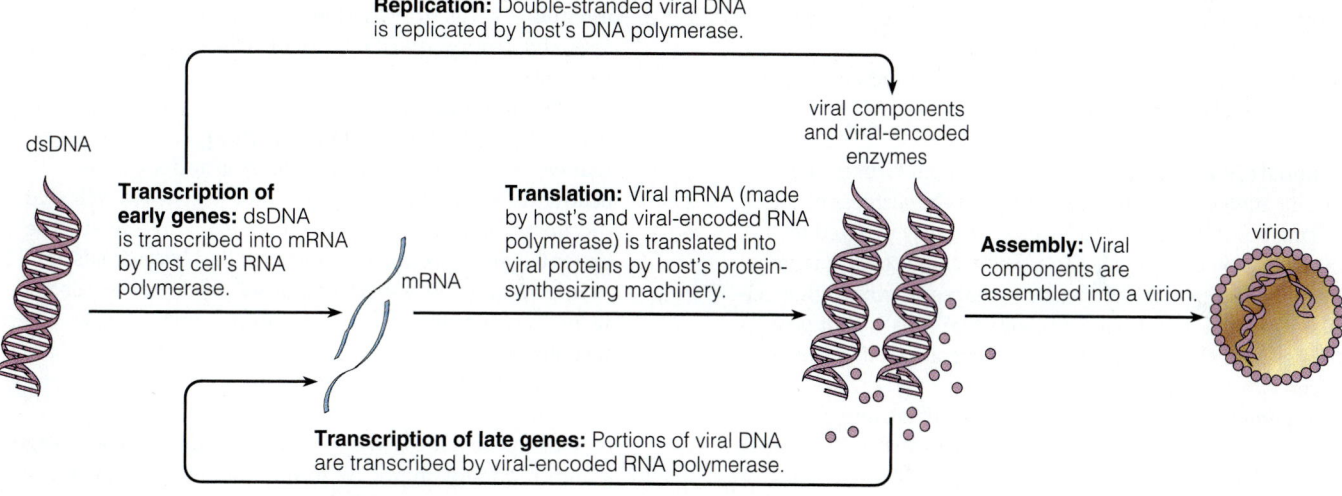

Figure 13.12 Replication and expression of viral DNA. There is an additional step in ssDNA synthesis that does not occur in dsDNA. In ssDNA, a complementary strand to the ssDNA is made, converting it to dsDNA.

teins by the host cell's enzymes as soon as it enters the host cell. But the cell lacks the machinery to make more plus-strand RNA for progeny virions. RNA-dependent RNA polymerase, encoded by the virus and made by the host, uses the plus-strand as a template to make minus-strand RNA and then uses the minus-strand to make more plus-strand for progeny virions. The newly synthesized plus-strand RNA also increases the rate at which viral proteins are synthesized. Picornaviruses (such as poliovirus) and togaviruses (including rubella or German measles) are synthesized by this route, but synthesis of

retroviruses (like HIV) occurs by a completely different pathway in which DNA is an intermediate of replication (see the discussion of retroviruses, below).

In minus-strand RNA viruses, including rhabdoviruses such as rabies virus, RNA-dependent RNA polymerase is used for gene expression and replication. This enzyme uses the minus-strand RNA as a template to make plus-strand, which serves as mRNA for viral proteins and as a template to make more minus-strand RNA for progeny viruses. Some RNA-dependent RNA polymerase is present in the virions of minus-strand RNA

Virus Receptors

Richard L. Crowell

Dr. Richard L. Crowell has been a faculty member of Hahnemann University School of Medicine in Philadelphia since 1960 and as Chairperson has led its Department of Microbiology and Immunology since 1979. He has studied human enteroviruses in mammalian cell cultures with emphasis on chronic infection as well as viral interference, factors that make cells susceptible to viral infection, and the role of coxsackieviruses in heart disease, diabetes, and myositis. He is known best for his studies on viral receptors. Dr. Crowell has been President of the Association of Medical School Microbiology Chairs and President of the American Society for Microbiology.

To initiate infection and replication, human viruses attach to specific molecules, called receptors, on the surfaces of the plasma membrane of cells. Viruses probably evolved to attach to molecules—cellular adhesion molecules, enzymes, receptors for growth factors, etc.—that serve some other cellular function, because it is unlikely that cells would produce special molecules just to invite viruses in to destroy them.

Together with my many graduate students and postdoctoral fellows, I have devoted over 35 years to the study of receptors for the group B coxsackieviruses. These viruses cause severe diseases, including viral meningitis, pleurodynia, heart and skeletal muscle diseases, pancreatitis (sometimes thought to cause insulin-dependent diabetes), and hepatitis. The major questions to be answered in our studies concerned whether different viral receptors on different types of cells might account for the fact that certain closely related viruses cause completely different types of diseases.

Early in our studies, we made a discovery that allowed us to answer some of these questions. We found that large amounts of a purified virus would bind to *all* its corresponding receptor molecules on the surface of cells, despite published research by others indicating this to be impossible. We were then able to show that a single type of receptor molecule was shared by the three serotypes of polioviruses. We did this by adding enough poliovirus type 1 (PV1) to bind to all its receptor molecules and then testing to see if the other two types of poliovirus (PV2 and PV3) could attach to these virus-saturated cells. We found that they could not; thus, all three types of poliovirus must bind to the same type of receptor molecule.

By doing similar experiments we found that the six group B coxsackieviruses bound to a different receptor from the one the polioviruses bound to. Then we extended these studies to other species of picornaviruses (the group of small RNA-containing viruses that includes coxsackieviruses, polioviruses, and human rhinoviruses, among others) and found that their receptor specificity corresponded to the type of disease they caused: Polioviruses caused destruction of neurons in the spinal cord and brain stem to produce poliomyelitis, whereas the group B coxsackieviruses caused a different spectrum of diseases, and 90 serotypes of the human rhinoviruses, which are the leading agents causing the common cold, share a common receptor that is distinct from those used by the other picornaviruses. These findings led us to propose that cellular receptors are the major determinants of virus tropism (the reason why certain viruses seek and destroy specific types of cells—for example, neurons).

Other scientists did experiments that convincingly validate our proposal. They cloned and sequenced the genes that encode the receptor for poliovirus and for some of the echoviruses, and they transferred these genes to cells that did not have these receptors. The cells that received these genes began to produce the receptor and became susceptible to infection by the respective viruses. Also, transgenic mice were prepared that produce the human protein that is the receptor for poliovirus; these mice (unlike other mice) are susceptible to paralytic disease caused by wild type poliovirus.

One particular discovery that we made many years ago stands out in my mind. We found that the receptors for the group B coxsackieviruses could be digested off cells by exposing them to the enzyme chymotrypsin, and the cells did not die in the process. Because the enzyme cannot enter the cell, we could conclude that these receptors were located on the outer surface of the plasma membrane. We disrupted the cells and did not find receptor molecules on any of the cell's internal structures, despite a publication to the contrary from another laboratory. Evidently, the other researchers had found receptors on the cells' internal structures because in the absence of prior enzyme treatment, the disrupted plasma membrane pieces, which contained receptors, became attached to small particles in the cytoplasm.

One of the powerful forces that drives scientific discovery is reading a publication that you cannot accept, because your experience in the field leads you to disbelieve the author's conclusions. This prompts you to return to the laboratory, to plan additional experiments, and to prove your beliefs. By this process of testing ideas, new findings are made, and the science is corrected. The life of a scientist is fully rewarded by learning new things and being able to teach these findings to others.

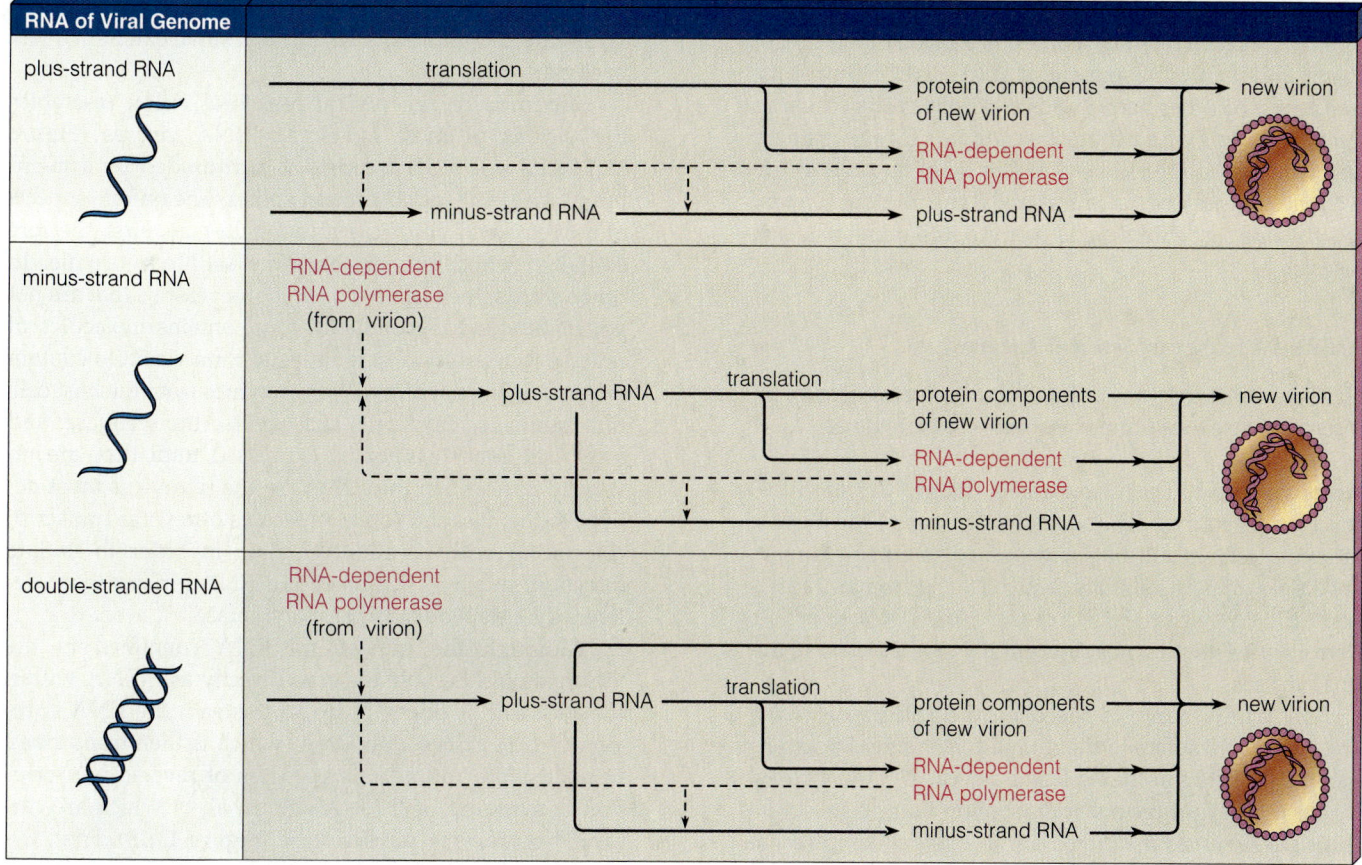

RNA of Viral Genome		
plus-strand RNA	translation → protein components of new virion; RNA-dependent RNA polymerase; minus-strand RNA → plus-strand RNA	new virion
minus-strand RNA	RNA-dependent RNA polymerase (from virion) → plus-strand RNA → translation → protein components of new virion; RNA-dependent RNA polymerase; minus-strand RNA	new virion
double-stranded RNA	RNA-dependent RNA polymerase (from virion) → plus-strand RNA → translation → protein components of new virion; RNA-dependent RNA polymerase; minus-strand RNA	new virion

Figure 13.13 Viral RNA synthesis. All three types of RNA synthesis depend upon RNA-dependent RNA polymerase, though this enzyme plays a different role in each case.

viruses because no viral protein can be synthesized until plus-strand RNA is made.

Viruses with dsRNA need RNA-dependent RNA polymerase to replicate and express the genome. In the reoviruses, such as rotavirus, which causes serious diarrhea in children, a capsid protein functions as this enzyme. From the dsRNA genome it makes plus-strand RNA, which is translated into viral proteins, one of which is another RNA-dependent RNA polymerase that makes more plus- and minus-strand RNA. The plus- and minus-strands combine, forming dsRNA for progeny virions.

Maturation. How capsids assemble around nucleic acid in animal viruses is not understood as thoroughly as maturation of phages. It seems to occur largely spontaneously. Some host cell proteins may be involved.

Release. The capsid-covered nucleic acid of an enveloped virus acquires its envelope as it leaves the host cell. Before the virus leaves the cell, viral genes direct the host cell to make viral proteins. They become embedded

in the region of the host cell membrane that will become the viral envelope. The virus pushes out the plasma membrane (or the endoplasmic reticulum), forming a bud that encloses the virus. Then the bud pinches off behind, resealing the host cell. As a result of this exit mechanism, the virion of an enveloped animal virus is a composite structure. That is, the contents of the virion and most of the proteins in its membrane are encoded by the viral genome, but the lipid membrane itself is encoded by the genome of the host cell.

Nonenveloped viruses leave by a simpler mechanism, similar to the way phage T4 is released. After completely assembled virions accumulate in the cytoplasm, the cell lyses and dies, releasing the virions.

Latency. Sometimes viral replication is arrested and the infected animal cells function normally for years, just as a lysogenic bacterium does. Such **latent viral infections** are typical of DNA viruses belonging to the herpesvirus family. One strain of herpes simplex virus (HSV-1) causes a latent symptomless infection of nerve cells principally in the lips and mouth (Chapter 26). The infection can be reactivated by a fever, a cold, too much sun, or stress; and painful blisters called **cold sores** or **fever blisters** erupt. Another herpesvirus (varicella zoster) causes a latent infection with different symptoms for the primary and reactivated infections (Chapter 26). The primary infection is expressed as chickenpox and the subsequent reactivation as shingles, a painful inflammatory skin disease.

Animal Viruses of Special Interest

Here we consider three animal viruses—retroviruses, influenza viruses, and tumor viruses—in more detail.

Retroviruses. The **retroviruses** (family Retroviridae) are a large group of RNA-containing viruses. Most cause no symptoms at all, but some cause malignant tumors and leukemias in animals; and one, the human immunodeficiency virus (HIV), causes AIDS (Chapter 27). Retroviruses are the most thoroughly studied class of animal viruses.

The family name is based on the Latin word *retro*, "backward." It refers to the unique, seemingly backward biochemical step in the replication cycle of these viruses. In the normal process of transcription that occurs in all cells, DNA serves as a template to make a complementary strand of RNA. In retroviruses, the enzyme reverse transcriptase uses RNA as a template to make a complementary molecule of DNA (Chapter 6). Going from

RNA to DNA is "backward" and occurs only in retrovirus-infected animal cells (and to a small extent in a few bacteria).

Superficially, the typical retroviral virion resembles the virions of most enveloped RNA viruses (**Figure 13.14**). Its icosahedral capsid is surrounded by a membrane with embedded protein spikes. The unique aspects of the retroviral virion are found in its core. First, the core contains two copies of the same ssRNA molecule. In other words, the virion is diploid, for reasons that are not yet understood. Second, the core contains molecules of reverse transcriptase. It is unusual for a virion to contain enzymes; they usually rely on enzymes from the host cell. But uninfected host cells lack reverse transcriptase, and retroviral genes cannot be expressed until they are reversely transcribed into DNA. So the retrovirus must deliver to the host cell some molecules of reverse transcriptase along with its RNA genome in order to initiate infection, much as minus-strand RNA viruses must deliver RNA-dependent RNA polymerase.

Although the plus-strand RNA contained in the virion should be able to serve directly as mRNA within the host cell, it does not do so. Instead, the RNA is reversely transcribed into DNA, which is then transcribed normally into mRNA. The process of reverse transcription is made up of three reactions, all of which are catalyzed by reverse transcriptase (**Figure 13.15**). First, the plus-strand of RNA in the virion is reversely transcribed into DNA. The product of this reaction is a hybrid molecule—one strand of DNA hydrogen-bonded to a strand

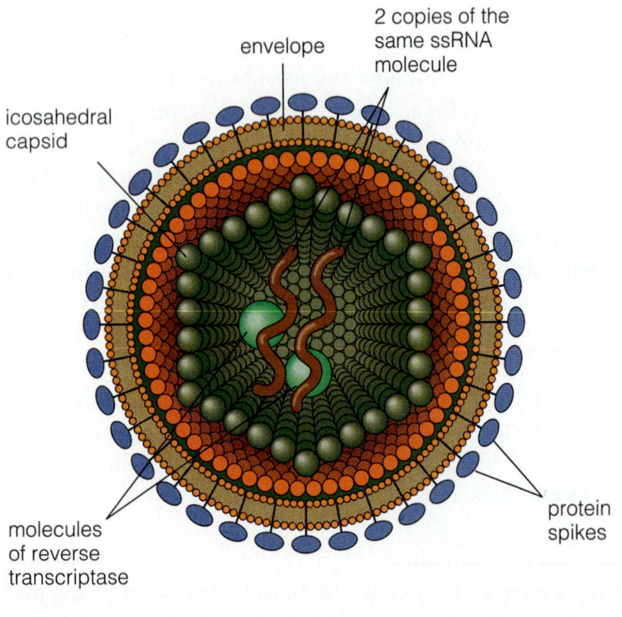

Figure 13.14 Typical retroviral virion. Superficially, a retrovirus looks like an enveloped RNA virus. However, unlike other viruses, a retrovirus is diploid (it carries two copies of the same ssRNA molecule) and it contains reverse transcriptase.

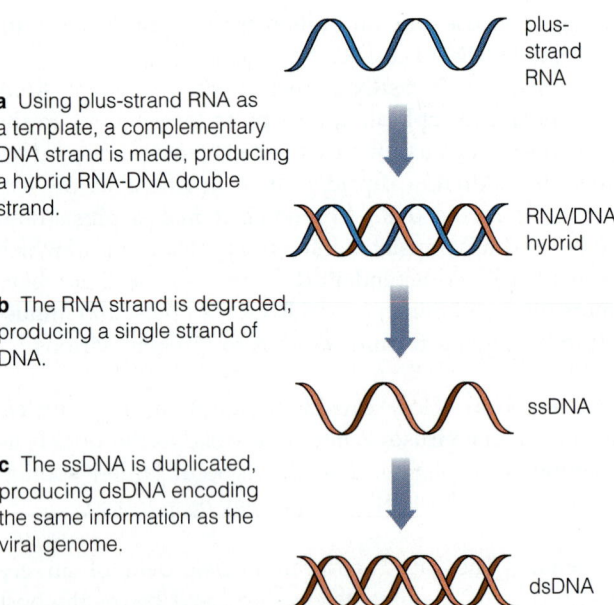

a Using plus-strand RNA as a template, a complementary DNA strand is made, producing a hybrid RNA-DNA double strand.

b The RNA strand is degraded, producing a single strand of DNA.

c The ssDNA is duplicated, producing dsDNA encoding the same information as the viral genome.

plus-strand RNA

RNA/DNA hybrid

ssDNA

dsDNA

Figure 13.15 Reverse transcriptase catalyzes three reactions.

of RNA. Second, reverse transcriptase degrades the RNA strand, producing a single strand of DNA. Third, reverse transcriptase duplicates the molecule of ssDNA producing the final product—a molecule of dsDNA with the same coding properties as the original RNA molecule found within the virion.

After the reactions catalyzed by reverse transcriptase have been completed, enzymes from the host cell act on the DNA to convert it to circular form. This circularized form of dsDNA then becomes integrated into the DNA of the host chromosome in a reaction similar to a step in the lysogenic pathway of temperate bacteriophages. Unlike the genome of temperate phages, however, the circularized dsDNA copy of the retroviral genome does not always recombine at the same position on the host chromosome. The integrated DNA copy of the retroviral genome is called a **provirus** (just as the integrated form of the genome of the temperate phage is called a prophage).

The proviral DNA is transcribed into mRNA and translated into viral proteins by the host cell. When sufficient quantities of these proteins have accumulated, they migrate to the cell surface where, along with the mRNA,

they assemble and bud through the cell membrane to produce virions. Or, as occurs with a prophage, the provirus can be replicated along with the host cell DNA, causing the cell no damage (**Figure 13.16**).

The special role reverse transcriptase plays in the replication of retroviruses offers possibilities for controlling AIDS. Chemicals such as azidothymidine (AZT) and related drugs that stop reverse transcription arrest replication of the virus. AIDS and HIV are considered in detail in Chapter 27.

Influenza Viruses. The **influenza viruses** make up the family Orthomyxoviridae. Influenza viruses are divided into types A, B, and C, groups that correspond roughly to genera. Type A is the most common. It has an extremely wide host range, infecting many animals, including humans, seals, pigs, and birds. Type A strains have been responsible for many pandemics (worldwide epidemics). The great flu epidemic of 1918, for example, caused more deaths than World War I (Chapter 20). Types B and C infect only humans and do not cause pandemics. Outbreaks of type B occur every two or three years. Type

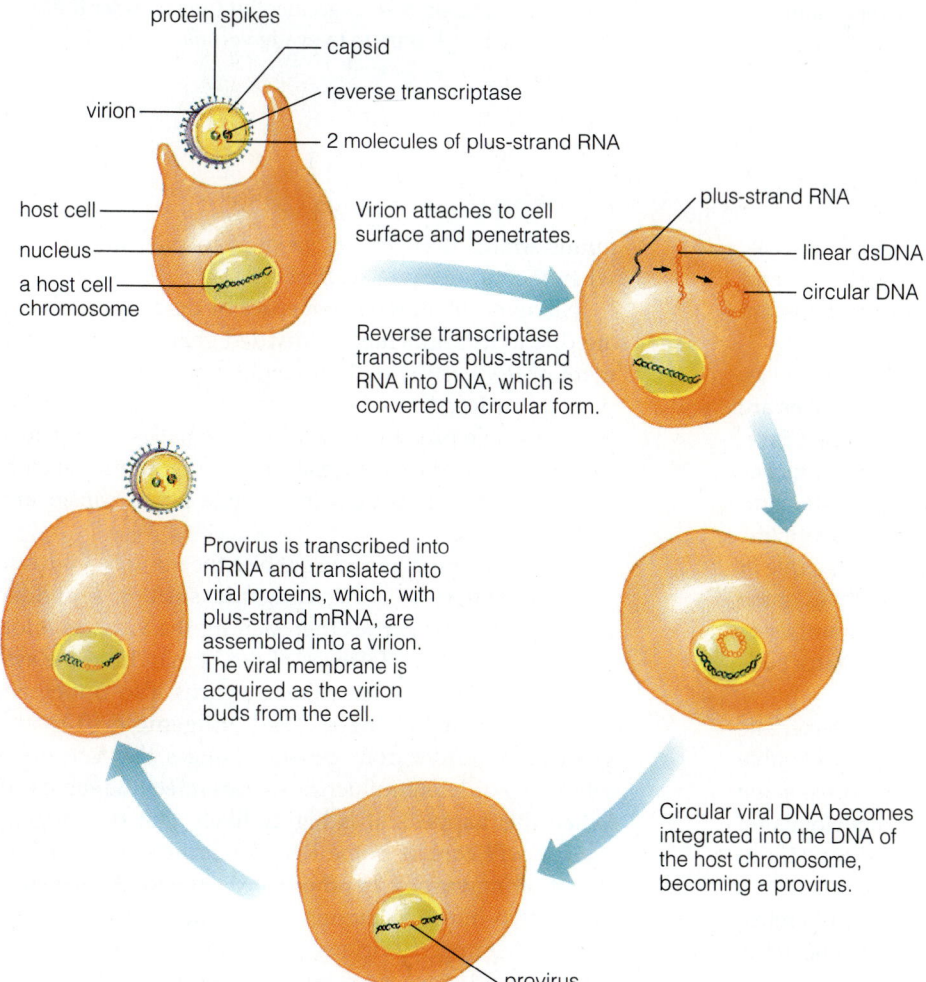

protein spikes
capsid
virion
reverse transcriptase
2 molecules of plus-strand RNA

host cell
nucleus
a host cell chromosome

Virion attaches to cell surface and penetrates.

Reverse transcriptase transcribes plus-strand RNA into DNA, which is converted to circular form.

plus-strand RNA
linear dsDNA
circular DNA

Provirus is transcribed into mRNA and translated into viral proteins, which, with plus-strand mRNA, are assembled into a virion. The viral membrane is acquired as the virion buds from the cell.

Circular viral DNA becomes integrated into the DNA of the host chromosome, becoming a provirus.

provirus

Figure 13.16 Replication of a retrovirus. Viral RNA is converted to dsDNA and integrated into the host cell's DNA before replication can proceed.

Are Viruses Smarter Than We Are?

Safe, effective antibacterial drugs are relatively easy to develop because disease-causing bacteria have structures and metabolic reactions that we, their hosts, lack. However, many once-excellent antibacterial drugs have become almost useless because mutant, drug-resistant strains of the bacteria have become widespread. Antiviral drugs, on the other hand, seemed to have the opposite profile. Although antivirals are difficult to develop, pharmacologists expected them to be useful for a long time because viruses would not become drug-resistant.

Unfortunately, this has not been the case, especially with human immunodeficiency virus (HIV), the virus that causes AIDS. HIV, being a retrovirus, offers an excellent target for chemotherapy—reverse transcriptase (RT). RT plays an essential role in replicating the virus but has no role in the human host. Azidothymidine (AZT) is an anti-RT drug. It inhibits the enzyme, effectively stopping HIV replication and therefore slowing progress of the disease. It does not cure the disease because it cannot eliminate viruses already present in the infected person. One problem with AZT that wasn't anticipated is drug resistance. Viruses *do* become drug-resistant, and HIV produces mutant strains at an especially high rate. Some of these mutant strains encode a form of RT that is subtly changed chemically and completely unaffected by AZT and other RT-blocking drugs. Resistance develops predictably and almost certainly.

Rarely can an AIDS patient be treated effectively with AZT for more than 12 to 18 months, and sometimes treatment becomes ineffective much sooner. Other RT-blocking drugs or drugs with other viral targets (for example, a protease, an enzyme like RT, essential for viral replication but not needed by the host) might then be prescribed. But at present the outlook for control of AIDS by chemotherapy is not promising because of drug resistance. Many AIDS researchers are turning their attention to other ways to control the disease, such as a vaccine. But they wonder if the "smart" virus will mutate to evade vaccine-induced immune responses as well.

C causes subclinical infections or mild coldlike illnesses. All three types of flu viruses are similar in structure and mode of replication, although their capsid proteins are quite different.

Flu viruses are enveloped minus-strand RNA viruses with protein spikes in their membranes. These spikes are of two kinds, **hemagglutinin** and **neuraminidase**. Hemagglutinin is the protein by which the virus attaches to its host cell. As its name implies, it can also attach to red blood cells, causing them to aggregate or agglutinate. Neuraminidase is an enzyme that probably plays a role in releasing the virion from its host cell. It removes some molecules from the carbohydrate (*N*-acetylneuraminic or sialic acid) attached to spike proteins. This facilitates budding of the virion through the membrane.

Unlike most viruses, which have a characteristic shape, the shape of the influenza virus is highly variable (**Figure 13.17**). Within a single strain of the virus, some virions are nearly round. Others are somewhat elongated and sometimes bent. Still others are filaments many times longer than they are wide.

Influenza viruses and other RNA viruses, like retroviruses, contain active enzymes. Attached to the RNA molecules are RNA polymerases that transcribe the minus-strand RNA of the virion into plus-strand RNA when the virion enters its host cell.

Many of the unique properties of the influenza virus are due to its segmented or divided genome. Its complement of RNA is divided into eight separate pieces. Each piece is enclosed in a helical capsid, and each piece encodes a single protein, except for the smallest two, which function as overlapping genes for two proteins. All eight pieces are then packaged in a single larger capsid and enveloped.

The segmented genome enables the influenza virus to undergo antigenic shift. **Antigenic shift** is a sudden change in the properties that identify the virus as a foreign invader to the defenses of the human immune system (Chapter 27). (The influenza virus also undergoes small mutational changes called **antigenic drift**.) Antigenic shift results from genetic changes that can occur when two different influenza viruses infect the same cell. When this happens, it is highly likely that the progeny virions will contain some RNA molecules from each of the infecting virions. In other words, the RNA molecules of the two infecting virions recombine in various ways among the progeny virions, producing a virus that is significantly different from either of the original infecting

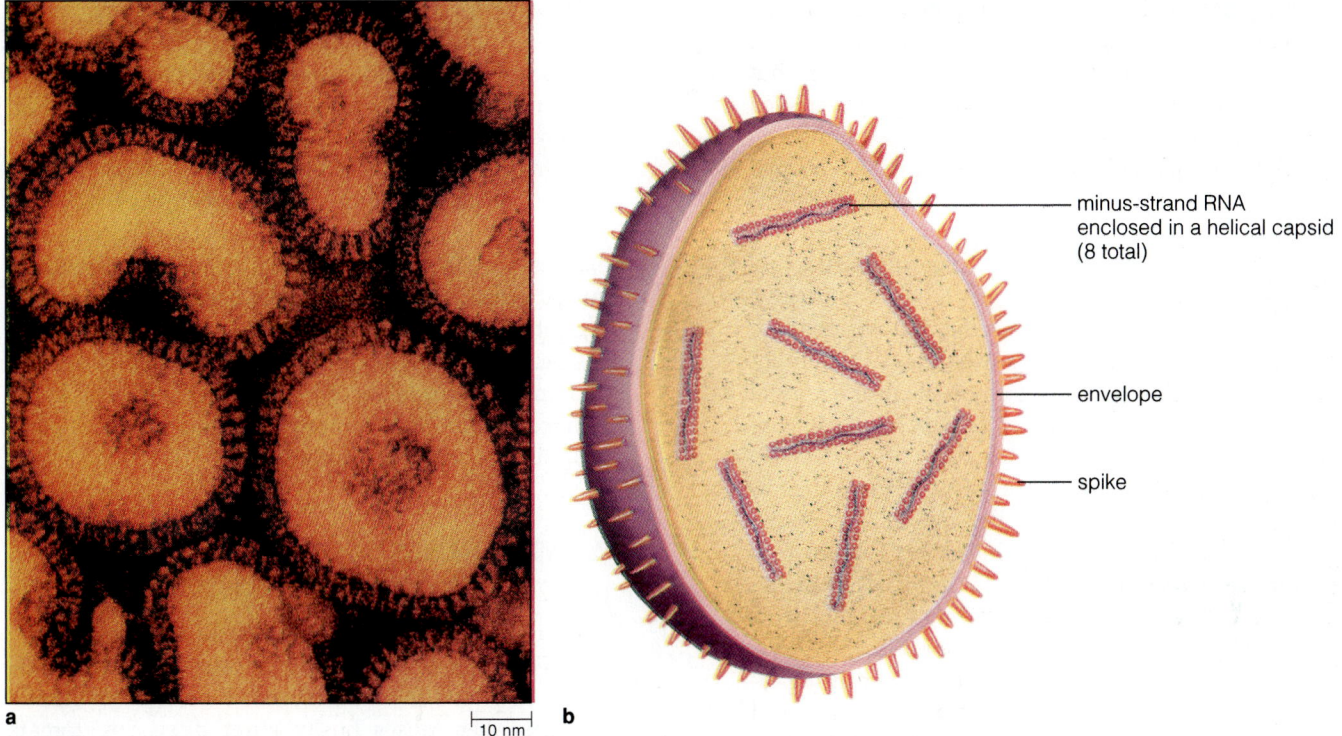

Figure 13.17 Influenza virus. (**a**) Unlike most viruses, influenza virus particles vary greatly in shape, as this electron micrograph shows. (**b**) An influenza virus has a segmented genome— eight different minus-strand RNA molecules—which enables it to undergo antigenic shift and evade the immune system. Each RNA molecule is enclosed in a helical capsid, and all are enclosed in a single capsid surrounded by an envelope with hemagglutinin or neuraminidase spikes.

minus-strand RNA enclosed in a helical capsid (8 total)

envelope

spike

a

b

10 nm

strains. While a person is protected against reinfection by the two original viruses, he or she might be totally vulnerable to the new strain. This means people become sick with flu again and again. Moreover, antigenic shift ensures that radically new viral strains emerge periodically to which no one has immunity. Influenza is discussed in greater detail in Chapter 22.

Tumor Viruses.   Several kinds of viruses cause **tumors** (uncontrolled growth of new tissue) in animals and humans. Most tumors are **benign** (non-life-threatening), but some are **invasive** (grow into other tissue) and **malignant** (deadly). Malignant tumor is another term for **cancer**. Virus-caused cancer is well-documented in experimental animals but not in humans. Because no causal relationships have been definitely established, we speak of viruses that are *associated* with human cancer. That is, a person infected by one of these viruses may develop a specific type of cancer (Chapter 15). Human T-cell leukemia virus (HTLV-1), an RNA virus, is associated with a leukemia (cancer of the blood) in humans. Three DNA viruses are associated with human cancer— Epstein-Barr virus with Burkitt's lymphoma and nasopharyngeal carcinoma, hepatitis B virus with liver can-

cer, and human papillomavirus with skin and cervical cancers (Chapter 24).

The RNA and DNA viruses that cause tumors do so by different mechanisms. All RNA tumor viruses are retroviruses. Retrovirus studies over the past two decades have established that certain genes, called **oncogenes**, cause tumors to form. Oncogenes derive from normal mammalian genes called **proto-oncogenes**, which have changed in either form or expression. In general, proto-oncogenes regulate normal cellular processes. In their oncogene form, however, normal regulation is distorted so that uncontrolled cancerous growth results.

Retroviruses convert a normal host cell to a tumor cell by introducing an oncogene into it. This happens in one of three ways. The first way is typified by Rous sarcoma virus (RSV), a virus that causes tumors in chickens. The normal genome of RSV contains an oncogene designated *src*. During the life cycle of RSV, when the provirus is inserted into the genome of the host cell, the oncogene becomes part of the host cell's genome and causes that cell to become a tumor cell.

The second way retroviruses convert a normal cell to a tumor cell involves **defective viruses**. Defective retroviruses have smaller genomes than normal. They have lost viral genes, and they have gained an oncogene. Such a defective retrovirus can insert a provirus into the genome of its host, causing it to become a tumor cell, but it cannot replicate by itself. Defective retroviruses, such as mouse sarcoma virus (MSV), which causes tumors in mice, persist in nature because the presence of a normal retrovirus in the same cell allows the defective retrovirus to replicate. The vital functions lost from the genome of the defective virus are supplied by the normal virus.

Retroviruses might cause tumors in yet a third way. Some researchers believe even normal retroviruses that do not carry oncogenes might cause tumors. Merely inserting a provirus in the host chromosome near a normal gene might alter the regulation of its expression and convert it to an oncogene.

A number of DNA viruses are associated with the formation of tumors, but the way they cause cancer to develop is not understood very well. Part of the problem is that even the DNA viruses known to cause tumors in animals do so at a very low frequency. Only 1 in 1 million to 10 million cells actually becomes a tumor cell.

One reason DNA viruses are inefficient at causing tumors is that DNA viruses lyse their host cells as part of their normal replicative cycle. Once dead, the cell cannot be transformed. The retroviruses, in contrast, transform their host in the normal course of their replication cycle. DNA tumor viruses apparently transform when unfavorable conditions suppress their normal replication. In laboratory studies, irradiation with ultraviolet light, incubation at an abnormal temperature, and exposure to certain chemicals have stimulated formation of tumor cells by DNA viruses.

How DNA viruses become tumor-forming under suppressed growth conditions can be demonstrated by introducing a virus into an animal that does not serve as a host and so does not support its growth. Human adenoviruses (*adeno* is Latin for "glands") are DNA viruses found in tonsil and adenoid glands. They cause mild respiratory infections in humans, but never tumors. When injected into newborn hamsters, however, human adenoviruses cause tumors to form. Virologists speculate that the virus becomes oncogenic because it cannot undergo its normal replication cycle and kill the hamster cell. Instead, it causes certain viral proteins to be synthesized and to accumulate. These transform the cell into a tumor cell. Some DNA tumor viruses and their oncogenic properties are shown in **Table 13.4**.

PLANT VIRUSES

Plant viruses are generally named for the disease they cause. For example, the virus that causes mosaic disease of tobacco plants is called tobacco mosaic virus (TMV). The virus that causes bushy stunt disease of tomato plants is called tomato bushy stunt virus.

Plant viruses are studied in the same way as phages—with cell cultures. But plant virology lags behind studies on phages and animal viruses.

Growth, Replication, and Control

Unlike animal cells, which have only outer membranes to maintain their integrity, plant cells have thick cell walls.

Table 13.4 Some DNA Tumor Viruses		
Virus	Host	Tumors[a]
Shope papillomavirus	Rabbit	Papillomas, carcinomas
Bovine papillomavirus	Cow Hamster	Papillomas Lymphomas
Wart papillomavirus	Human	Warts and probably cervical cancer
Simian virus 40	Monkey Hamsters, rats	Lymphomas Sarcomas
Lymphotropic papovavirus	Monkey	Sarcomas
Human adenovirus	Newborn hamsters	Various tumors
Epstein-Barr virus	Human	Burkitt's lymphoma Nasopharyngeal carcinoma
Herpesvirus	Human	Possibly cervical carcinoma

[a]Papillomas are benign wartlike overgrowths of skin. Carcinomas are malignant tumors that begin in epithelial layers of body organs, including the skin. Lymphomas are tumors of lymph tissue. Sarcomas are malignant tumors of connective tissue, bone, cartilage, or muscle.

How Safe Is Safe?

We believe that our municipal water supplies are safe. Filtration systems chemically treat water with chlorine and ozone to kill pathogenic bacteria and parasites. Cholera, typhoid fever, and shigellosis have virtually disappeared in developed countries. But what about viruses? They aren't removed by filtration, and they might survive chemical treatment. Moreover, small numbers can cause disease. It takes hundreds or thousands of bacterial *Salmonella* cells to cause enteritis, but only a few echovirus particles.

Even if there are virus particles in municipal drinking water supplies, do they cause disease? Most microbiologists and public health experts believe they do not. But in 1991, a Canadian microbiologist, Pierre Payment, studied 2400 people in suburban Montreal. Half of them used their regular tap water and half were supplied with water that had been specially processed to also remove viruses. Over the next 18 months he monitored the health of both groups. The results were surprising. Montrealers using regular tap water had a 30 to 35 percent greater chance of getting gastroenteritis than did the people drinking virus-free water. Something in the tap water was causing 20 extra illnesses for every 100 people each year.

The illnesses were relatively mild—mostly one day at home in bed—and the incidence was small—one extra case of gastroenteritis per family. But applying this incidence to the U.S. population, for example, it amounts to millions of extra cases of gastroenteritis per year with an economic loss of several billion dollars. If enteritis-causing viruses are indeed present in Montreal's water, the numbers must be extremely low. One virus per thousand liters of water would be enough to account for the incidence of illness Payment found in Montreal.

Some plant tissues, including leaves and stems, are also protected by layers of waxy material. A plant virus must break through these protective layers to enter its host cell. Some plant viruses exploit the mechanical injuries—abrasions and nicks from animals or weather—that constantly occur to plant surfaces. Other plant viruses enter with an insect bite. Aphids, leafhoppers, and other sap-consuming insects all transmit plant viruses.

Considering the ways they must enter their hosts, it is probably not surprising that plant viruses are never enveloped. But in other respects, plant viruses resemble animal viruses. Some are helical and some are polyhedral. Most contain RNA, but some contain dsDNA or ssDNA. Once inside the plant cell, viral replication proceeds much like phage and animal virus replication. Viral nucleic acid directs the host cell to make more viral components, which are then assembled into virions and released.

Most viral diseases of plants are chronic degenerative diseases, slowing plant growth and diminishing crop yield. Their economic impact can be enormous. There are two general ways to control plant viruses—control the insects that spread the virus or develop virus-resistant and virus-free plant strains. Developing resistant plants is preferable because viruses are often spread in seeds and scions used for grafting.

Ironically, virus-infected plants are often quite beautiful. In spite of their diminished crop, potato plants with **leaf roll disease** have thickened, healthy-looking leaves, and grape vines with **Pierce's disease** turn particularly beautiful shades of orange and red in the fall. Tulip plants with **tulip break disease** have beautifully variegated flowers (**Figure 13.18**).

Tobacco Mosaic Virus

Tobacco mosaic virus (TMV) was the first virus to be discovered, the first to be crystallized, and the first to be disassembled experimentally into its component parts—RNA and protein—and then reassembled into infectious virions. The most intensively studied of the plant viruses, it is *Escherichia coli*'s counterpart in the world of plant viruses.

TMV is named for the mottled mosaic-like patterns it causes on leaves of infected plants. It is a helical plus-strand RNA virus. When it infects a tobacco plant, it multiplies rapidly and accumulates in large quantities. All tobacco products—cigarettes, cigars—almost certainly contain TMV. It was because of its abundance in infected plants that virologists were attracted to it. Many fundamental principles of virology, especially relating to structure, came from studies on TMV.

Figure 13.18 Variegated tulips are the result of tulip break disease, a viral infection that changes pigment formation in different regions. The infection is passed on to new generations, which also produce variegated flowers.

VIRUSES OF EUCARYOTIC MICROORGANISMS

Studies on viruses that infect eucaryotic microorganisms have lagged in virology because these viruses do not have the economic impact of plant and animal viruses, and they are not as easy to study as phages. In fact, until about 30 years ago, microbiologists believed eucaryotic microorganisms were not infected by viruses. Viruses in protozoa were not definitely identified until 1986. Now, many viruses are known to infect these organisms. Viruses that infect fungi are called **mycoviruses**. They are mainly dsRNA viruses and may be enveloped. Viruses that infect protozoa are all RNA viruses, and most—if not all—are double-stranded.

INFECTIOUS AGENTS THAT ARE SIMPLER THAN VIRUSES

Viruses are not the simplest agents that cause disease. There are two even simpler agents, viroids and prions.

Figure 13.19 Growth stunting and leaf distortions in the tomato plant on the right are caused by the potato spindle tuber viroid.

Viroids

A **viroid** is a circular molecule of ssRNA without a capsid. Viroids cause several economically important diseases of plants. One such disease, potato spindle tuber, causes the edible part of the potato (the tuber) to become long and pointed and it stunts the growth of tomato plants (**Figure 13.19**). As few as 10 molecules of the viroid are enough to infect a potato plant—many fewer than the number of virions needed to cause most plant diseases.

A viroid is only about one-tenth the size of the smallest plant virus. Spindle tuber viroid is composed of only 359 nucleotides, about a third as many as make up an ordinary-size gene. How such a small molecule can cause disease remains a mystery. It seems unlikely that the viroid encodes a protein that causes disease. More probably it interacts in some way with the host genome, changing the expression of the host genes to cause disease. Viroids may be bits of mRNA from plant genes that have escaped during normal processing.

A number of other viroids that cause plant disease are known, and most are even smaller than potato spindle tuber viroid. One, called cadang-cadang, causes a lethal disease of coconut palms. It is an RNA molecule only 246 nucleotides long.

Prions

A **prion** is an infectious agent that seems to be composed only of protein. Six diseases of humans and animals, including scrapie, a neurological disease of sheep, are caused by prions (**Figure 13.20**). Until recently these agents were thought to be viruses. Because the prion-caused animal diseases develop slowly over a period of years, prions are also called **slow viruses**.

Good Timing

To enter a host cell, plant viruses must depend upon the wall being breached by mechanical abrasion or the bite of an insect. That gets the virus into the *first* cell, but how does it spread to other cells to cause general infection? Plant viruses encode special **movement proteins** that spread the virus from cell to cell through **plasmodesmata**, the thin protoplasmic bridges between adjoining plant cells. Tobacco mosaic virus, for example, encodes only four proteins, but one of them is a movement protein.

Plant virologists discovered movement proteins only in the last decade, but biotechnology is far enough along that commercial applications are already possible. By replacing the virus's cell-killing genes with genes for a protein of commercial value, and keeping the movement protein, biotechnologists can create virus-like packages that spread from cell to cell throughout a plant, carrying the protein-encoding gene. This technique is already being used to make tricosanthin, an experimental AIDS drug, in a tobacco field.

Prion-caused diseases affect the central nervous system. Creutzfeldt-Jakob disease, for example, causes dementia (Chapter 25). Kuru resulted in uncoordinated movements that progressively worsened and led to death (Chapter 25). Kuru infected people in a small region of Papua, New Guinea, and was transmitted through ritual cannibalism.

The existence of a self-replicating protein seems to defy current scientific dogma, but possibly prions are a type of protein that can be replicated. A more probable explanation is that the host cell itself carries the information to make more prions. Also, it is still possible that prions contain nucleic acid in amounts that are not detectable with current technology, though this possibility is becoming increasingly unlikely as more highly purified preparations become available and still no nucleic acid is found.

Although prions are found only in infected individuals, individuals in susceptible groups carry a gene that encodes a larger protein, called the **prion protein**, or PrP. The prion is a part of PrP. Apparently, when a prion infects a cell, it not only causes disease but also stimulates formation of more prion from PrP.

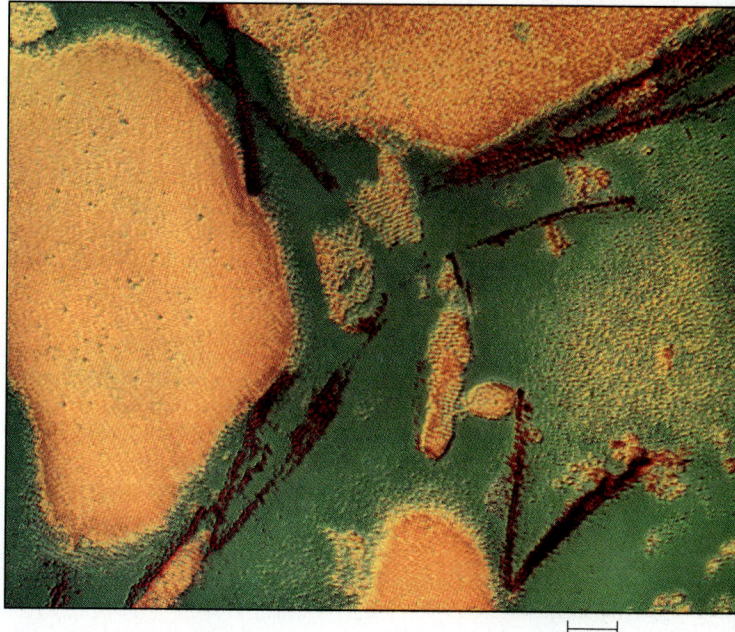

Figure 13.20 Prions are infectious agents that seem to be composed only of protein. The prion particles shown here cause scrapie, a neurological disease of sheep.

THE ORIGIN OF VIRUSES

How did viruses originate? Because their structure is so extremely simple, we might think they are very primitive and evolved early in the history of life on Earth. But because they are obligate parasites, viruses—as they are now, at least—could not have existed before their hosts evolved. More probably, viruses, like many other parasites, are examples of retrograde evolution—evolution that produces simpler rather than more complex things.

Many scientists believe that viruses evolved from genes in the very cells that they infect—that they are "genes on the loose." A hint of this evolutionary origin comes from the properties of viroids. They seem to be pieces of copies of plant genes—those pieces of mRNA called introns that are excised when plant genes are processed before translation. It is possible to imagine that pieces excised from a cellular gene could become capable of entering another cell and causing the cell to reproduce the excised gene. Such a gene would be a primitive virus. Further evolution might produce the highly successful and widespread infectious agents we call viruses. This line of reasoning suggests that viruses evolved late from cellular organisms and that viroids are primitive viruses.

New viruses continue to emerge. Some of these, such as hepatitis C virus (Chapter 23), have probably existed for a long time, but have been discovered only recently. Others, such as dengue fever virus (Chapter 27), have spread into new host populations. Still others, such as HIV and a new strain of hantavirus that caused a deadly outbreak of pneumonia in the southwestern United States in the spring of 1993, are probably genuinely new, having arisen by the interplay of mutation, cross-species transfer, and human behavior.

Summary

THE ULTIMATE PARASITES   (pp. 303–304)

1. A virus is a package of nucleic acid wrapped in a protein coat. It is acellular, lacking a metabolism of its own.

2. Virions are intact, nonreplicating virus particles. The fact that they have been crystallized associates them with highly purified chemicals rather than living things.

3. In 1892, Dmitri Iwanowski found that filtering tobacco plant extracts did not remove the disease-causing agent, which led to the discovery that infectious agents smaller than bacteria existed. Viruses were not seen until the development of the electron microscope in the 1930s.

CLASSIFICATION OF VIRUSES   (pp. 304–307)

1. The primary classification of viruses is based on host range. Viruses that attack bacteria are called bacteriophages, or simply phages. Host specificity is determined principally by receptors on the cell surface to which virions attach.

2. Viruses are also classified by size, which ranges from 25 nm to more than 300 nm. Viruses are smaller than the smallest procaryotic cell.

3. Viruses have a nucleic acid core, a protein coat, and sometimes a membrane.
   a. The nucleic acid core may contain double-stranded (ds) DNA, single-stranded (ss) DNA, dsRNA, or ssRNA (either plus-strand or minus-strand). Plus-strand is mRNA, translated directly by the host ribosome. Minus-strand must be transcribed into mRNA (a plus-strand) after it enters the host cell.
   b. The protein coat that surrounds the nucleic acid core of a virus is called a capsid. A capsid can be helical, polyhedral, or complex, depending upon the arrangement of its constituent protein molecules, called capsomeres. Complex viruses have a helical portion called a tail attached to a polyhedral portion called a head. They may have additional structures, such as a tail sheath, a plate, pins, and tail fibers.
   c. Viruses surrounded by a membrane are called enveloped viruses. Viral envelopes are pieces of the host cell's plasma membrane that the virion acquires as it emerges from its host cell. Viruses without envelopes are called naked viruses. Enveloped viruses have protruding proteins called spikes; naked viruses may not have them.

4. All viruses have the same life cycle. Outside a host cell they exist as nonreplicating virions. The virion attaches itself to a host cell in a process called adsorption. The genome enters the host cell (penetration). Uncoating, the process of removing the capsid and envelope, can occur before or after penetration. The host cell manufactures viral components in viral synthesis. During maturation, the components are assembled into intact new virions. Virus particles leave the infected cell during the process of release.

5. Viral taxonomy does not follow the Linnaean scheme. It has only three levels—family (and sometimes subfamily), ending in *viridae*; genus, ending in *virus*; and species names, which are English words.

BACTERIOPHAGES   (pp. 307–313)

1. Bacteriophages are studied by plaque count and one-step growth curve. A plaque count determines the number of bacteriophages in a sample by counting the plaques on a lawn of bacteria. Virions and infected cells together are called plaque-forming units (PFUs).

2. In the one-step growth curve experiment, plaque counts are done on an infected culture to determine the number of PFUs present. The phage growth curve shows a latent period, a burst period, and an eclipse period. The burst size is the fold increase in PFUs that occurs during the burst period.

3. Bacteriophages may immediately enter the lytic pathway, leading to production of more virions, or they enter the lysogenic pathway, leading to a prolonged quiescent state termed lysogeny. Phages that follow only the lytic pathway are termed virulent. Phages that can follow either pathway are termed temperate.

4. Virulent phages, such as T4, lyse the host cell during release. Lambda, a typical temperate phage, lyses the host cell if it takes the lytic pathway. If it takes the lysogenic pathway, no phage components are synthesized. Instead the phage genome becomes part of the host cell chromosome, forming a prophage. The cell is termed lysogenic; it has the potential to cause lysis if the genes in the prophage become reactivated. Lysogenic cells are immune to further attack by the same virus.

ANIMAL VIRUSES   (pp. 313–324)

1. Animal viruses are often studied in cell or tissue cultures. A monolayer of cells is infected with virus, causing plaques that can be counted. Cells that dominate a culture are called a cell line. Cell lines that grow indefinitely in culture are called continuous cell lines.

2. There are three modes of penetration: The viral envelope fuses with the cell's plasma membrane; the host cell engulfs the virion (phagocytosis); most naked viruses adsorb to the cell surface while the nucleic acid component enters.

3. Single-stranded DNA must be converted to dsDNA, which is transcribed into mRNA and translated into viral proteins. RNA viruses encode RNA-dependent RNA polymerase, an enzyme that plays a different role in viral synthesis of each class of RNA virus.

4. In latent viral infections, such as those caused by the herpesvirus family, viral replication is arrested until infection is reactivated.

5. Retroviruses use RNA as a template to make DNA. Influenza viruses undergo antigenic shift and antigenic drift. Some viruses are associated with human cancer; no direct causal relationship has been established.

PLANT VIRUSES   (pp. 324–325)

1. Plant viruses are usually named for the disease they cause. All plant viruses are naked.

2. Plant viruses must penetrate cell walls and sometimes protective waxy outer coverings on stems and leaves. Some plant viruses enter

through mechanical injuries to plants, while others enter through an insect bite.

3. A plant virus—tobacco mosaic virus (TMV)—was the first to be discovered, the first to be crystallized, and the first to be disassembled into its component parts and then reassembled into infectious virions.

VIRUSES OF EUCARYOTIC MICROORGANISMS (p. 326)

1. The study of viruses that infect eucaryotic organisms is in its infancy. Viruses in protozoa were not definitely identified until 1986.

2. Viruses that infect fungi are called mycoviruses.

INFECTIOUS AGENTS THAT ARE SIMPLER THAN VIRUSES (pp. 326–327)

1. A viroid is a circular molecule of ssRNA without a capsid. Viroids cause several plant diseases, including potato spindle tuber disease. Viroids are about one-tenth the size of the smallest plant virus.

2. A prion is an infectious agent that seems to be composed only of protein. Prion-caused diseases affect the nervous system.

THE ORIGIN OF VIRUSES (pp. 327–328)

1. Viruses may have evolved from genes in the cells they infect.

2. Some viruses, such as the hepatitis C virus, have existed for a long time but have been discovered only recently. Others, such as HIV, are probably new, arising from an interplay of mutation, cross-species transfer, and human behavior.

Review Questions

THE ULTIMATE PARASITES

1. Define these terms: virus, virion, acellular.

2. Briefly discuss the key events in the history of virology.

CLASSIFICATION OF VIRUSES

1. On what four bases are viruses classified?

2. What is host range? How is host specificity determined?

3. What is the basic structure of a virus? Name the types of nucleic acid that may be found in viruses. What is the difference between plus-strand and minus-strand RNA?

4. Define these terms: capsid, capsomere, tail, head. What purpose is served by such additional viral structures as a tail sheath or plate?

5. What is the difference between an enveloped virus and a naked virus? What are spikes, and how do viruses acquire them?

6. Explain these steps in the viral life cycle: adsorption, penetration, uncoating, viral synthesis, maturation, release.

7. How is viral taxonomy different from bacterial?

BACTERIOPHAGES

1. Explain plaque count and the one-step growth curve. What is a PFU?

2. Trace the two pathways bacteriophages may take on entering a host cell.

3. Explain these terms: virulent phage, temperate phage, prophage, a lysogenic cell.

ANIMAL VIRUSES

1. How are animal viruses studied? What is a continuous cell line?

2. How is animal virus replication the same as phage replication? How is it different?

3. What is a latent viral infection? Give an example.

4. What is a retrovirus?

5. Explain the difference between antigenic shift and antigenic drift.

6. Explain this statement: Some viruses are *associated* with human cancer.

PLANT VIRUSES

1. How are plant viruses studied? How are they named?

2. Explain this statement: It is not surprising that all plant viruses are naked viruses.

3. Why do we speak of tobacco mosaic virus as the plant-world equivalent of *Escherichia coli* in the history of microbiology?

VIRUSES OF EUCARYOTIC MICROORGANISMS

1. Explain this statement: The study of viruses in eucaryotic microorganisms is in its infancy.

2. What are mycoviruses?

INFECTIOUS AGENTS THAT ARE SIMPLER THAN VIRUSES

1. What is a viroid? What kinds of diseases do viroids cause?

2. What is a prion? Why are prions so puzzling? What kinds of diseases do they cause?

THE ORIGIN OF VIRUSES

1. Why are viruses probably an example of retrograde evolution?

2. Explain this statement: Viruses probably evolved from a gene on the loose.

3. How do new viruses arise?

Suggested Readings

Fields, B. N., and Knipe, D. M. 1990. *Fields' virology*. 2d ed. New York: Raven Press.

Fraenkel-Conrat, H.; Kimball, P. C.; and Levy, J. A. 1988. *Virology*. 2d ed. Englewood Cliffs, N.J.: Prentice-Hall.

Levine, A. J. 1992. *Viruses*. New York: Scientific American Library.

Prusiner, S. 1989. Scrapie prions. *Annual Reviews of Microbiology* 43:345–76.

Wang, A. L., and Wang, C. C. 1991. Viruses of protozoa. *Annual Reviews of Microbiology* 45:251–64.

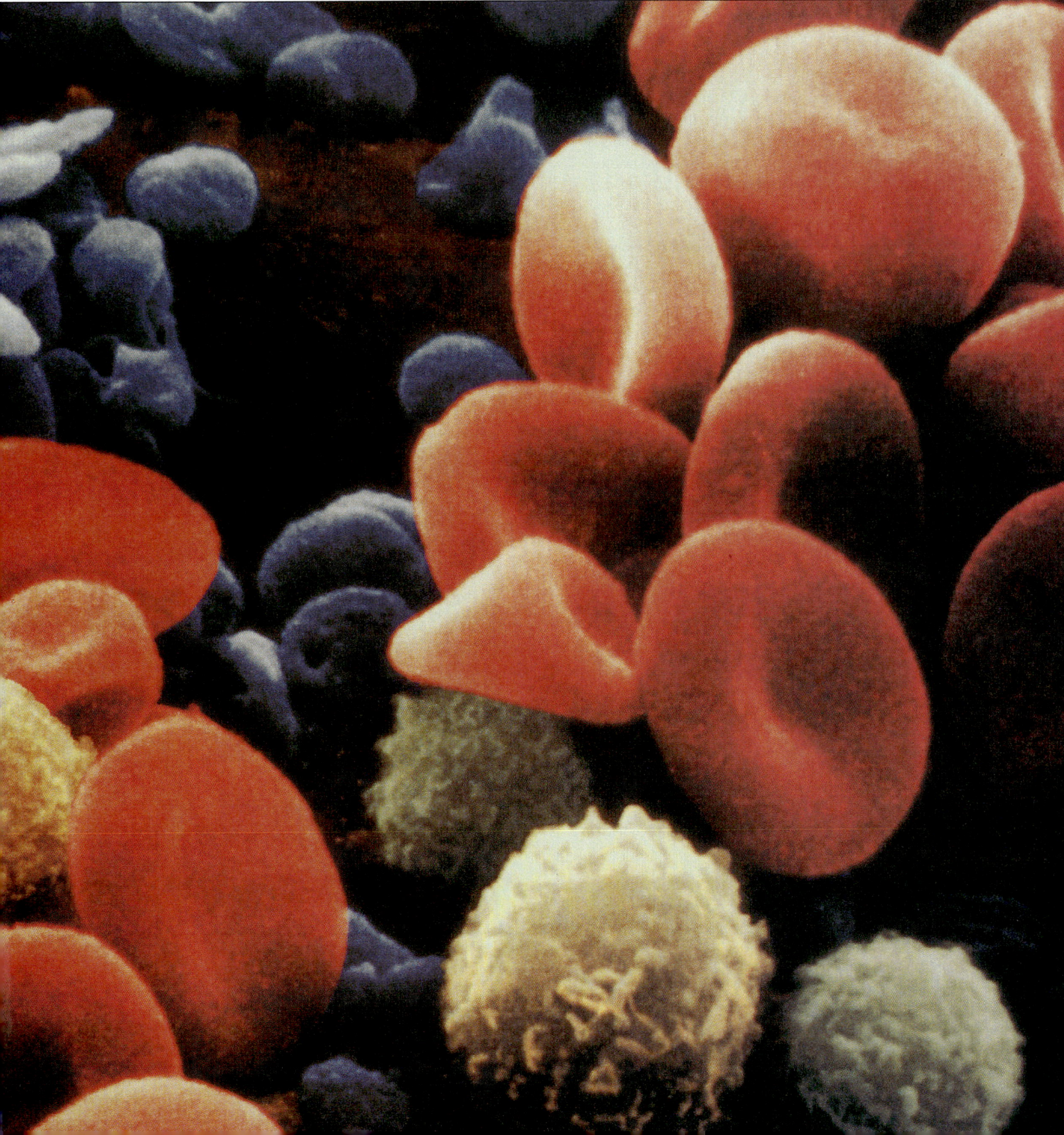

14 MICROORGANISMS AND HUMAN HEALTH

A Baby Enters the World of Microorganisms

Early in Baby Girl A's embryonic development, about 100 cells that arose from the original union of egg and sperm burrow into her mother's uterine lining. Completely surrounded by maternal tissue, the embryo develops in a sterile environment. In time, fetal membranes that are impermeable to microorganisms form a sac around her to help maintain sterility. Thus, her life begins in a germ-free environment.

She grows and develops in her protected environment until a few hours before birth. At this time, the fetal membranes rupture and her sterile barrier is broken. Now, microorganisms from her mother's vagina can ascend into the uterus. If the membranes rupture more than 24 hours before birth, infection may threaten the health or life of the mother and soon-to-be-born baby.

During delivery, Baby Girl A passes through the vagina, where microorganisms are abundant. By the time she takes her first breath, bacteria are already in her mouth, in her external ear canals, and on her skin. During her first few days of life, Baby Girl A comes in contact with a multitude of microorganisms in the air, on her clothes and bedding, and on the bodies of other people. Many microorganisms are transmitted from the skin of mother to infant. Other microorganisms enter her digestive system from food and objects she takes into her mouth. Still more enter her respiratory tract from the air she inhales. Baby Girl A's entry into the world of microorganisms is one of the most momentous transitions she will experience in her life.

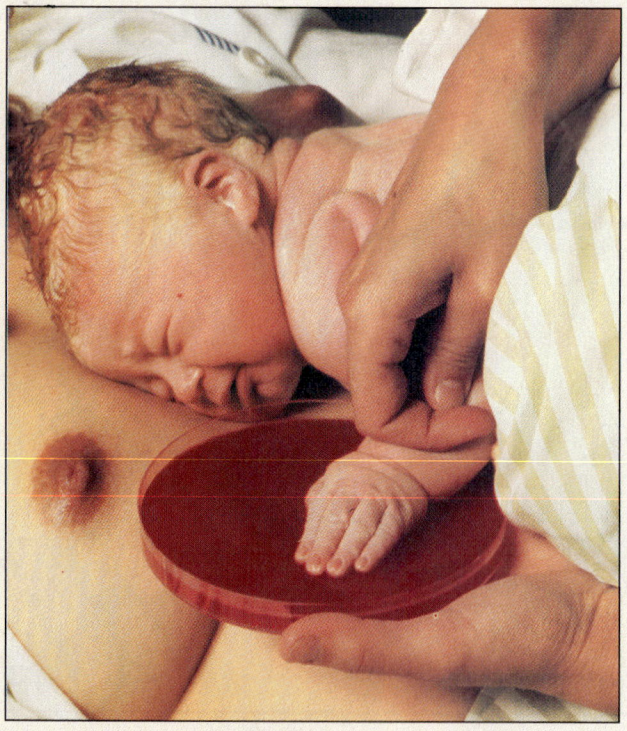
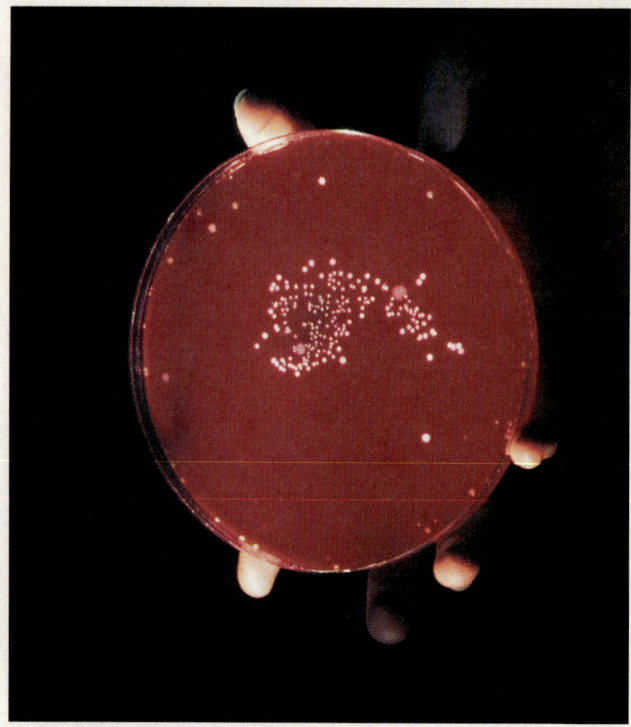

After skin-to-skin contact with the mother, a newborn's hand imprint on a blood agar plate will show colonization by microorganisms.

To understand:

- The nature of the normal human microflora and how it can change

- The three types of human-microbe symbioses: commensalism, mutualism, and parasitism

- The human factors that determine the normal microflora: structural defenses, mechanical defenses, biochemical defenses

- The microbial factors that determine the normal microflora: physical, nutritional, and special adaptations to life on living tissue

- The nature of the microflora on the skin and conjunctivae; in the nasal cavity and nasopharynx, mouth, intestinal tract, vagina, and urethra

- The impact of the normal microbial flora on its human host

- The disease-causing ability of microorganisms as it exists along a continuum, from highly virulent to almost harmless

NORMAL FLORA

Over the many millennia of evolution we humans have developed an intimate and complex relationship with the world of microorganisms. From birth until death our bodies are inhabited by hundreds of species of bacteria and fungi. Some make the human body their permanent home, and others come to rest there temporarily, but harmlessly. *Ten times more microbial cells than human cells are found in and on most healthy people.* In fact, our good health is partly due to our relationships with certain microorganisms.

In this chapter, you will read about relationships between microorganisms and human beings in a state of health. We all have an intuitive understanding of health, but in this textbook, we define **health** as a state of relative equilibrium in which the body's many organ systems function adequately. Health is the stable condition in which most of us exist most of the time. **Disease** is a state of functional disequilibrium that may be resolved by recovery or death.

The microorganisms that coexist with humans in a stable relationship constitute our **normal flora**. They thrive and multiply because they are adapted to life on the human body. Under most circumstances, these organisms do not cause disease.

Normal flora inhabit only the *surfaces* of the body. But, anatomically, there are both external and internal surfaces. External surfaces, such as the skin, come in direct contact with the environment and therefore with microorganisms. Internal surfaces are exposed to the environment indirectly. The intestines, for example, encounter microorganisms that enter the body through the mouth.

External body surfaces that support a normal flora of microorganisms include the skin and the outer covering of the eye. Normal flora also inhabit the interior surfaces of the nose, mouth, intestinal tract, vagina, and urethra (through which urine leaves the body) (**Figure 14.1**). Microorganisms found in any other tissues of the body—such as the brain, heart, muscle, or bone—are not normal flora and can be expected to cause disease.

Resident Flora

The first microorganisms on the surfaces of Baby Girl A's body are those that just happened to land there. But by the time she is 2 weeks old, the constellation of microorganisms on her body is strikingly similar to the constellation on all other humans. These microbes constitute her **resident flora**, microbial species that are present on the human body throughout life. Characteristic species will populate her skin, sparsely in cool, dry, exposed areas such as her arms and legs and more heavily in the warm, moist areas around her mouth, nose, and rectum. The exact combination of microbial species and the size of the populations present in any area—external or internal—may vary somewhat, but these microorganisms are permanent residents.

Transient Flora

Transient flora are microbial species that can be cultured from body surfaces under certain circumstances but are not there as permanent residents. They are not well enough adapted to life on the human body to persist indefinitely and are not part of the normal flora. Most species of the transient flora are harmless (like the vast majority of all microbial species), but some species may be pathogenic.

The skin is particularly likely to acquire a varied and complex transient microbial flora because it is constantly and directly exposed to the environment. Transient microorganisms land on exposed areas, particularly the hands, and are loosely attached by oil or dirt. Unlike the skin's resident flora, which can be reduced but not eliminated by washing or applying disinfectants, transient flora *can* be removed from the skin by careful cleansing.

Figure 14.1 Normal flora sites. Only the body's external and internal surfaces support a normal flora.

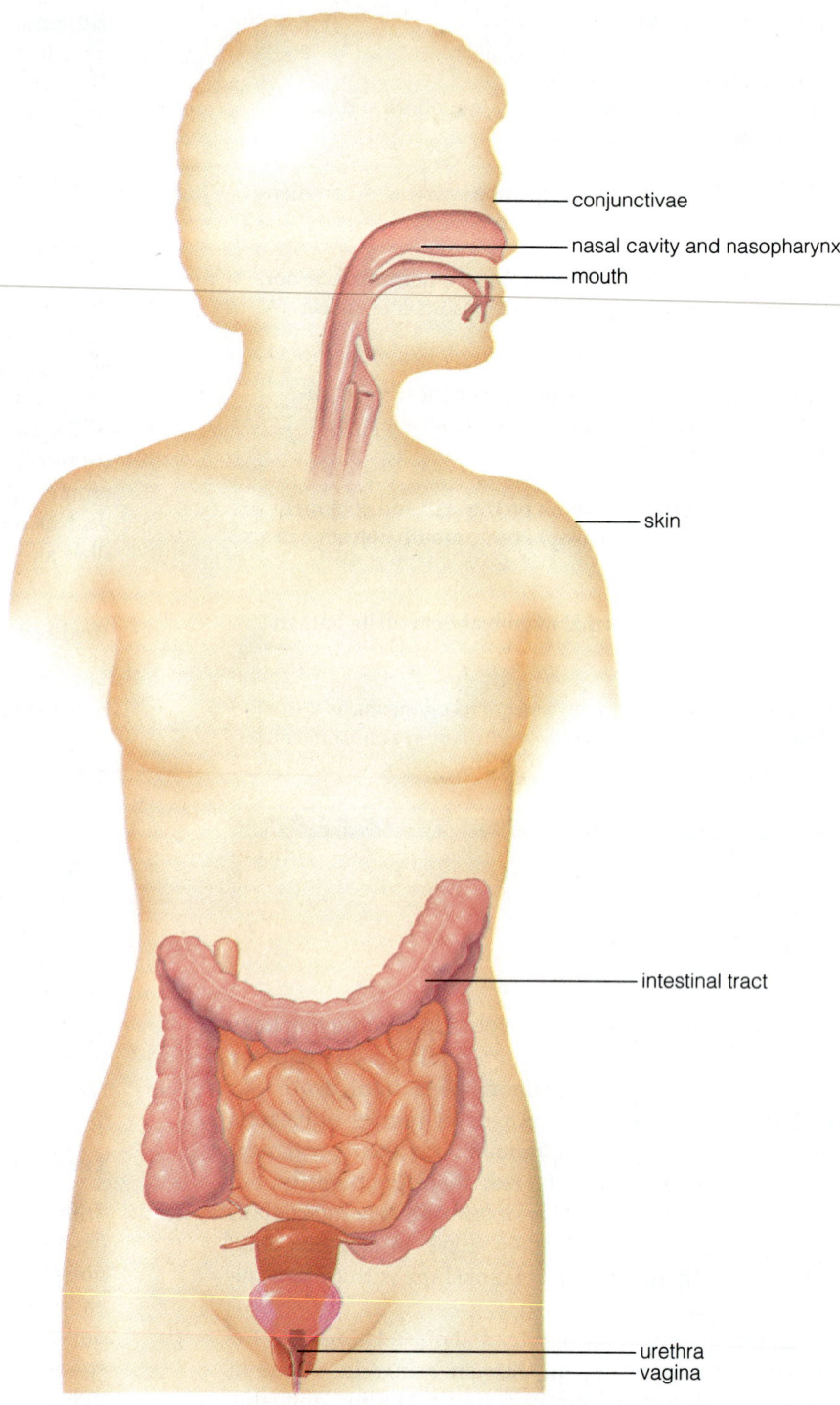

conjunctivae

nasal cavity and nasopharynx

mouth

skin

intestinal tract

urethra

vagina

Most hospital workers have a large transient microflora of pathogens because they are exposed to large numbers of pathogens daily. In particular, they often carry pathogenic *Staphylococcus aureus* (see Figure 11.10). These pathogens are harmless unless transferred to a patient whose resistance is lowered by disease or open wounds or, in the case of *S. aureus*, transferred to an infant. Even the healthiest of babies, like Baby Girl A, is extremely vulnerable to staphylococcal skin and eye infections. Because transient flora can be dangerous, good habits of hand washing are crucial in a hospital.

Opportunists

Opportunists are microorganisms that cause disease when the proper "opportunity" arises. Usually the opportunity is a breakdown in the immune system, the body's normal defense against infection. Or it may be a result of medical treatment, such as occurs with the use of **broad-spectrum antibiotics** (antibiotics that act against a wide variety of bacteria). The bioimplantation of artificial devices such as plastic catheters or metal joint replacements also allows opportunistic infections. Furthermore, different organisms become opportunists under different circumstances. For example, the yeast *Candida albicans* is a member of the normal resident flora that can become an opportunist during antibiotic therapy (see Figure 24.14). The water-borne bacterium *Pseudomonas aeruginosa* causes disease only in people whose health is already weakened by other factors.

Changing Flora

The normal flora of the human body is permanent and relatively stable, but it does change over time. As Baby Girl A grows and develops, her body changes in ways that alter the environment for her microbial flora. Some species disappear, while new species are established.

For example, the appearance of teeth alters the population of microbial flora. When Baby Girl A's first teeth erupt about 6 months after birth, the environment in her mouth changes dramatically. The teeth themselves provide new surfaces for microbial growth, especially of streptococcal species. In addition, anaerobic pockets are created in the crevices between teeth and gums. This allows strict anaerobes, such as members of the genera *Fusobacterium* and *Bacteroides*, to colonize the baby's mouth.

Another predictable change in normal flora occurs when Baby Girl A is weaned. While she is fed only breast milk, she has an intestinal flora composed almost exclusively (over 90 percent) of a bacterium called *Bifidobacterium* (**Figure 14.2**). But when she begins to drink infant formula or eat solid food, she loses her large population of *Bifidobacterium* and acquires the widely varied intestinal microflora typical of adults. It is thought that nursing infants harbor large numbers of *Bifidobacterium* because human milk contains a carbohydrate called **bifidus factor** that favors their growth.

Bifidobacterium is important to infant health. Being a lactic acid bacterium, it metabolizes milk sugars into acetic and lactic acids (Chapter 5). The result is an acid environment in the intestinal tract inhospitable to many disease-causing microorganisms, including those that cause diarrhea. Because infant diarrhea is a common cause of illness and death in many parts of the world, programs encouraging breastfeeding have a profound effect on public health.

Because of the female hormone **estrogen**, the constellation of microorganisms that colonize Baby Girl A's vagina also changes dramatically during her life. When

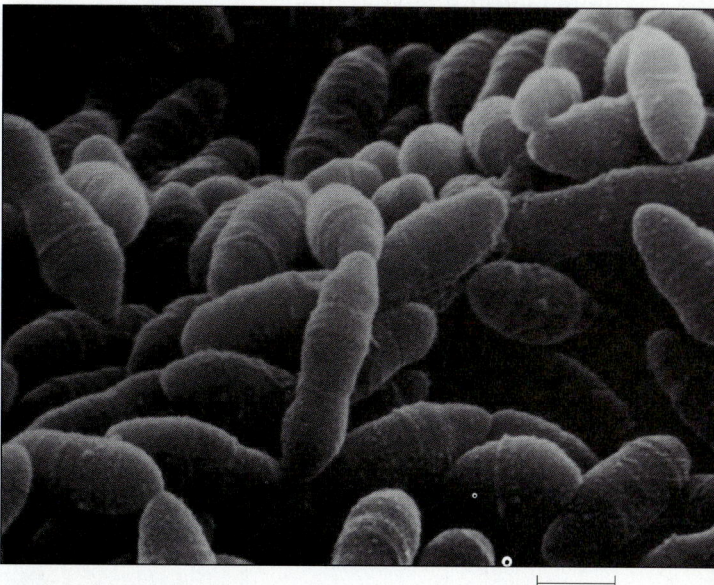

Figure 14.2 Babies who drink nothing but breast milk have an intestinal microflora dominated by the genus *Bifidobacterium*. These normal commensals protect against dangerous diarrheal infections. Upon weaning, *Bifidobacterium* disappears and the widely varied intestinal flora typical of adult humans appears.

Baby Girl A is born, estrogen that crossed the placenta from her mother encourages the growth of lactobacilli, creating an acidic vaginal environment. Over the next 2 to 3 weeks, the influence of maternal estrogen declines. As a result, the epithelial lining becomes thinner, lactobacilli nearly disappear, and the pH becomes more alkaline. These conditions, which persist through infancy and girlhood, favor colonization by only a few microorganisms, and these do not compete well against potential pathogens. Fortunately, prepubertal girls who are not sexually active are not exposed to many vaginal pathogens. However, occasional vaginal infection may occur—caused by normal bowel inhabitants, such as *Escherichia coli*. During puberty, estrogen stimulates vaginal cells to produce glycogen. When these cells are sloughed as part of normal epithelial growth, the glycogen molecules become a nutrient substrate for lactobacilli. The lactobacilli reestablish the acidic environment of the vagina and compete effectively against potential pathogens. This crucial protection of the acidic environment during Baby Girl A's sexually active and reproductive years persists until estrogen levels fall again after the menopause. Then an alkaline vaginal environment, with its resultant sparse microflora, returns.

In addition to changes in the normal flora that occur with time, there is some variation among the normal flora of different individuals. We discuss this variation later in the chapter.

Table 14.1	Types of Symbioses		
Type	Host	Symbiotic Partner	Relationship
Mutualism	Beneficial	Beneficial	Stable
Commensalism	Neither beneficial nor harmful	Beneficial	Stable
Parasitism	Harmful	Beneficial	Unstable

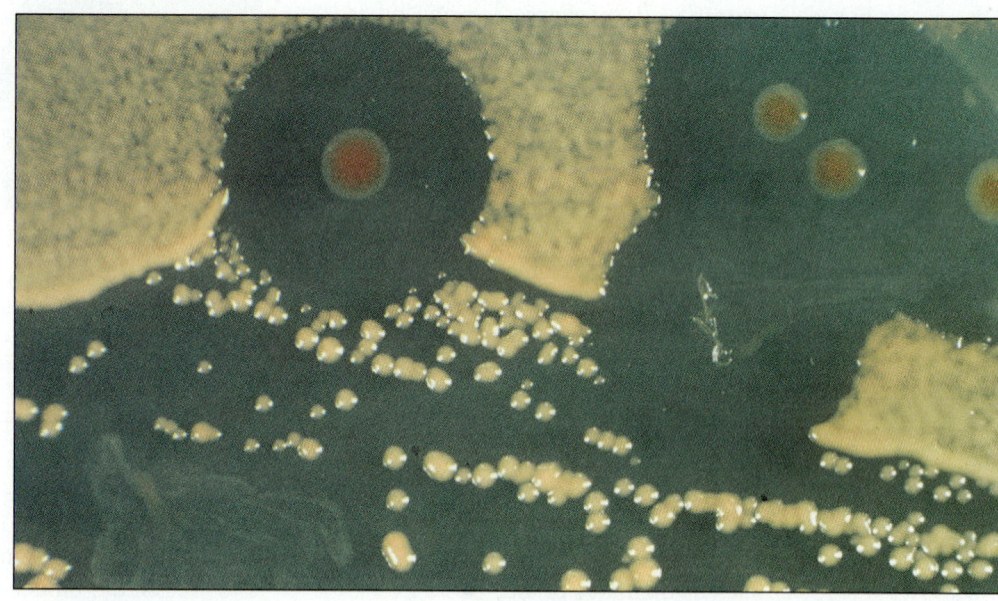

Figure 14.3 Here, the growth of the Gram-negative bacteria *Serratia* on a laboratory petri plate inhibits the growth of the Gram-positive bacteria *Micrococcus* immediately nearby. This is microbial antagonism. Similar microbial interactions on the surfaces of the human body allow the normal flora to edge out competing pathogens.

NORMAL FLORA AND SYMBIOSIS

Baby Girl A and her normal flora live in symbiotic relationships. **Symbiosis** refers to two different kinds of organisms living together. Baby Girl A, as the larger organism whose body provides a habitat for its symbiotic partners, is the **host**. The three major types of symbioses—commensalism, mutualism, and parasitism—are distinguished by whether they harm or benefit the host (**Table 14.1**).

Commensalism

Commensalism is a symbiotic relationship in which one partner is neither benefited nor harmed and the other benefits. In this case, the human host is the first partner and the commensal organism—the one that benefits—is the microorganism. Commensal symbioses tend to be long-lasting and stable. They are usually the product of extensive evolution. The stability of commensal relationships makes them compatible with health, which is a state of equilibrium. In commensal relationships between microorganisms and humans, the microorganism benefits by being provided with a suitable habitat and nutrients. *The vast majority of our normal flora are commensals.* Among them is *Escherichia coli.*

Individual species of commensal organisms do not provide any direct benefit to their host. However, collectively, the multitude of different species that constitute Baby Girl A's normal flora provides many benefits. The most important benefit is that the great numbers of harmless microorganisms successfully compete with—in effect, edge out—many pathogens that might otherwise take hold. This protective effect of the normal flora, in which one type of microorganism interferes with the growth of another, is called **microbial antagonism (Figure 14.3).**

Mutualism

Mutualism is a symbiotic relationship in which both partners benefit. In most cases, the benefits of mutualism are essential to both partners—neither can survive without the other. Like commensalism, mutualism is a highly evolved and extremely stable type of symbiosis. Ruminants (cows and similar cud-chewing mammals) have

Clinical Notes

TORCH

Baby Girl A had no contact with microorganisms before her birth, but not all babies are so fortunate. An unborn baby's lifeline is the placenta, the organ that communicates with the mother's bloodstream. The placenta delivers oxygen and nutrients to the fetus and carries away its metabolic waste. But in about 2 percent of pregnancies, the placenta also provides access for unwelcome pathogens.

Transplacental infection occurs when microorganisms from an infected mother's bloodstream cross the placenta and infect the fetus. Transplacental infections can interfere with the complex processes that shape a baby's organs and tissues. Even an infection so mild that the mother is unaware of any illness can kill the fetus. The most common infections that cause severe prenatal damage and malformation are abbreviated TORCH. The initials stand for Toxoplasmosis (Chapter 27), Others (such as syphilis, Chapter 24), Rubella (Chapter 26), Cytomegalovirus (Chapter 24), and Herpes simplex (Chapter 24).

The stage of pregnancy at which infection occurs is critical. Tissues that are undergoing active development are the most likely to be harmed. For example, if rubella (German measles) infection occurs before the twelfth week of development, heart defects are likely. But if infection occurs between the twelfth and sixteenth week, the result is more likely to be deafness. Infections that are transmitted from mother to infant during pregnancy are discussed further in Chapter 24.

TORCH Infections

Infection in Mother	Causative Microorganism	Congenital Damage to Fetus
Toxoplasmosis	*Toxoplasma gondii*	Small head, calcified areas of the brain, eye defects, deafness
Others—e.g., syphilis	*Treponema pallidum*	Rash, jaundice with enlarged liver, bone damage, severe nasal congestion
Rubella (German measles)	Rubella virus	Heart malformations, eye defects, deafness, small head due to abnormal brain development
Cytomegalovirus infection	Cytomegalovirus (a type of herpesvirus)	Enlarged liver and spleen, lack of platelets for clotting, eye and brain defects; however, many infants are completely normal in spite of congenital infection by this virus
Herpes simplex infection	Herpes simplex viruses types I and II	Skin blisters, fever, pneumonia, brain infection

mutualistic relationships with microorganisms in their digestive tract that ferment cellulose. The ruminant is able to use cellulose as a primary source of food, and the microorganism is provided with nutrients and a suitable habitat.

Using the strict definition of *essential benefit*, there are no true mutualistic relationships between humans and microorganisms. No single species of microorganism is essential to human survival. Some relationships provide nonessential benefits, and it is a matter of definition whether to call these examples "true mutualism" or not.

Parasitism

In **parasitism**, the host is harmed by its symbiotic partner and the parasite benefits. In clinical medicine, microbial parasites are called **pathogens**, and the term **parasite** is reserved for eucaryotic pathogens other than fungi—that is, protozoa, worms, and insects. Parasitism is usually an unstable symbiosis (though exceptions to this rule are discussed in Chapter 15) in which the host either dies or successfully defends itself by eliminating the pathogen from its body. This instability indicates that the two organisms are poorly adapted to live together and have probably evolved together for only a relatively short time.

The commensals that make up the vast majority of Baby Girl A's normal flora are potential pathogens because, under certain circumstances, they can become opportunists. Aside from opportunistic infections, though, the parasitic relationships our baby will experience will come in the form of at least 150 different infections over a lifetime, including the common cold.

FACTORS THAT DETERMINE THE NORMAL FLORA

The microorganisms that inhabit the human body, like all microorganisms, survive because they are adapted to their environment. But what determines which microorganisms are found in a given location on Baby A's body? This determination is made by two sets of factors: the environment itself (living tissue) and how microbes adapt to that environment.

Living Tissues as an Environment for Microbial Growth

In Chapter 8 you read about the environmental factors that influence microbial growth—nutrients and nonnutrient features, such as temperature, pH, and oxygen supply. But Baby Girl A's body surfaces are more complex than is the environment of the lab. In addition to their unique physical features—such as exposure or lack of exposure to oxygen—each bodily environment has features that are unique to *living* surfaces. Some of these features are **structural** (tissues on body surfaces are composed of tightly knit sheets of cells). Others are **mechanical** (some tissues move). Still others are **biochemical** (some tissues produce substances that affect microbial growth).

The features that prevent the growth of many microbes are called **nonspecific surface defenses (Figure 14.4).** They are called nonspecific because they act universally and uniformly; that is, they act against all microorganisms and always act in the same way. Body surface defenses are our first line of defense against infection. We will discuss the body's second and third lines of defense in Chapters 16 and 17.

Structural Defenses. All the microbially colonized surfaces of Baby Girl A's body share one essential struc-

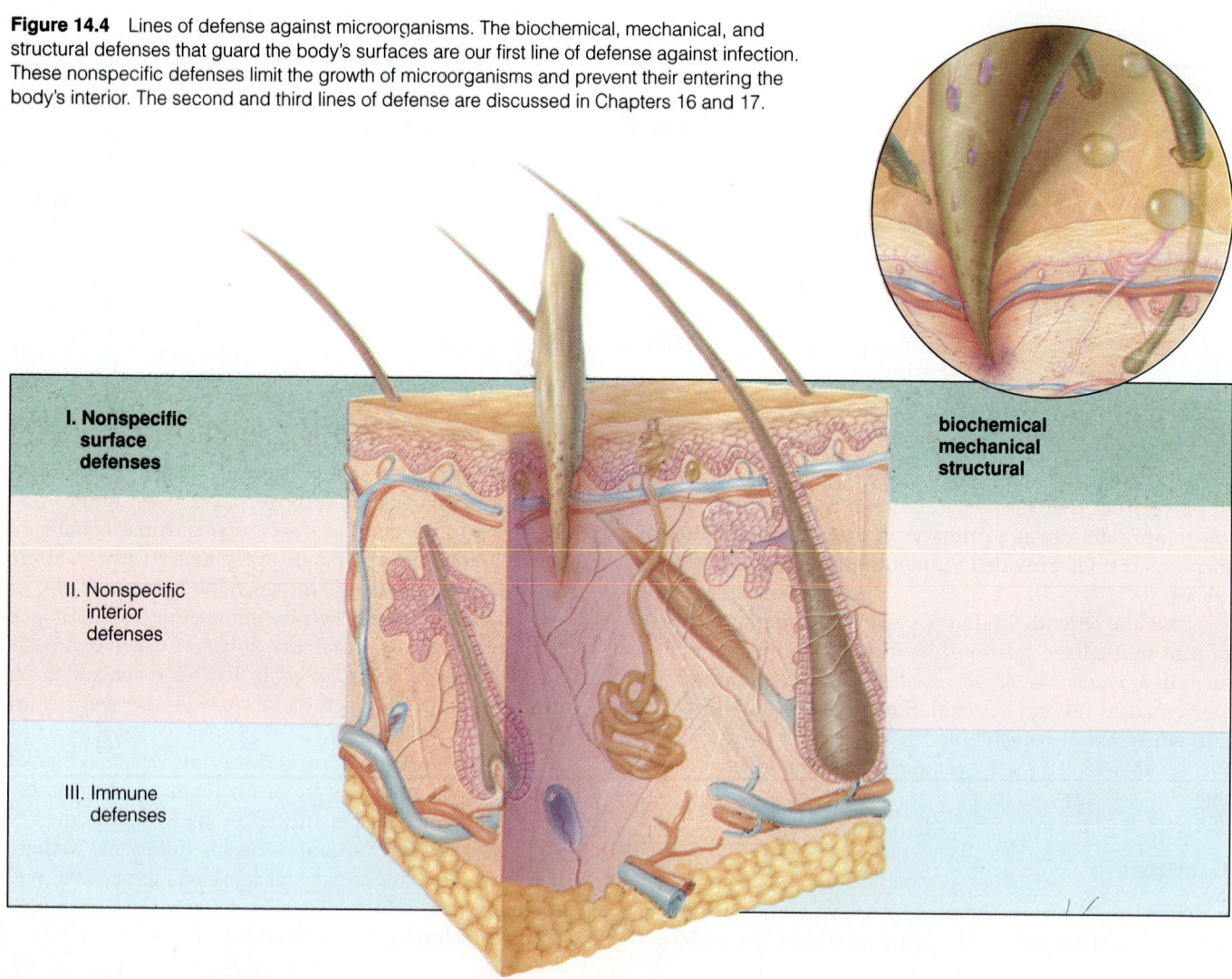

Figure 14.4 Lines of defense against microorganisms. The biochemical, mechanical, and structural defenses that guard the body's surfaces are our first line of defense against infection. These nonspecific defenses limit the growth of microorganisms and prevent their entering the body's interior. The second and third lines of defense are discussed in Chapters 16 and 17.

I. Nonspecific surface defenses

II. Nonspecific interior defenses

III. Immune defenses

biochemical
mechanical
structural

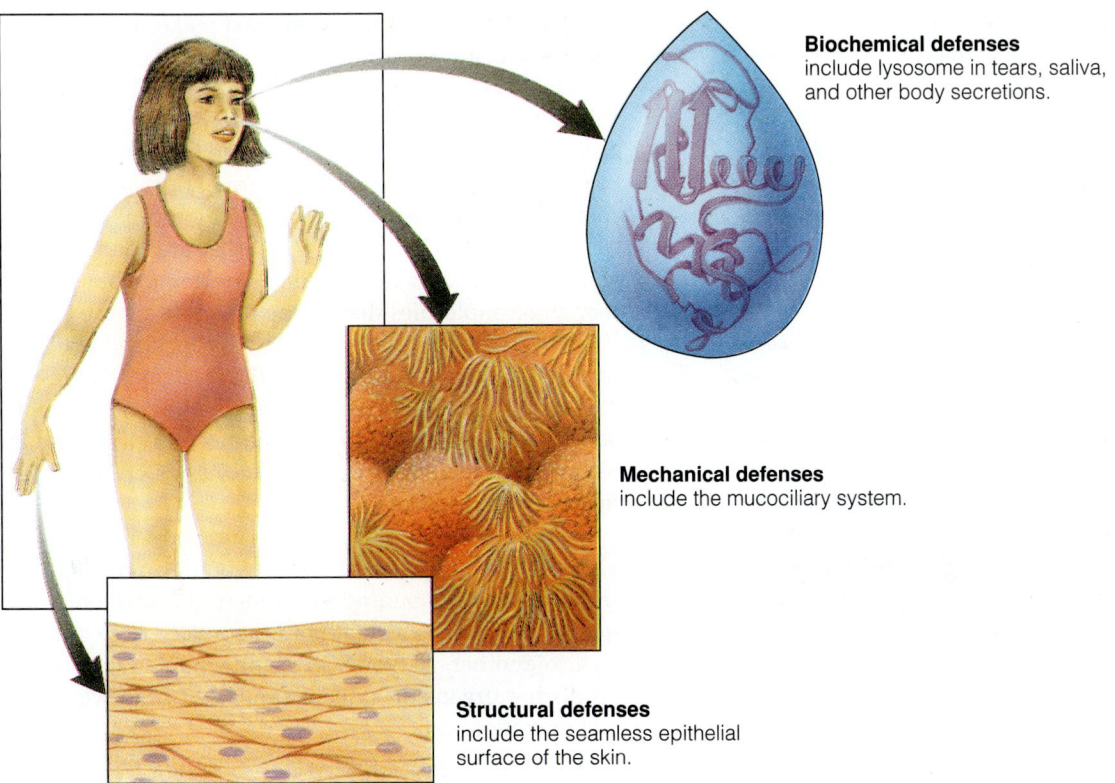

Biochemical defenses
include lysosome in tears, saliva, and other body secretions.

Mechanical defenses
include the mucociliary system.

Structural defenses
include the seamless epithelial surface of the skin.

Figure 14.5 Surface defenses—biochemical, mechanical, and structural.

tural feature—they are **epithelial surfaces**. These surfaces are composed of cells that are tightly joined to one another, creating a seamless and relatively impermeable barrier. Even the harmless commensal organisms that make up most of her normal microbial flora do not penetrate deeper tissues, where their growth would do significant damage.

There are two main types of epithelial coverings. The first type is skin, which encloses all of the body's exterior with the exception of the eye. Skin is relatively thick because it is composed of many layers of epithelial cells. The second type is mucous membrane, which covers the surface of the eye (where it is called the **conjunctiva**) and all of the body's interior surfaces. Compared to skin, mucous membranes are relatively thin and provide less protection; nevertheless, they are significant structural barriers.

Epithelium serves as a structural defense in a second way. As living tissue, epithelium grows—cells are sloughed (shed) and replaced by new ones. Moreover, epithelial cells divide rapidly and have a relatively short life span, from a few days to weeks. When dead cells are lost from the surface, the microorganisms growing on those cells are also lost. Sloughing prevents the microbial

population from continually increasing and restricts the resident flora to those species that can multiply more rapidly than they are lost (**Figure 14.5**).

Mechanical Defenses. Movements that occur along many of Baby Girl A's body surfaces eliminate most microbes that happen to land there. Some surfaces move because of the action of underlying muscles. In the intestines, for example, there is almost constant vigorous muscular activity as food and waste are moved through the digestive system.

The **mucociliary system** is another mechanical defense (Figure 14.5). The epithelial surfaces that line the nasal cavity and throat are protected by a combination of mucus production and ciliary movement. Because mucus is so viscous, microorganisms adhere to it. Epithelial cells with cilia constantly move the mucous layer toward the mouth, where it—along with the trapped microorganisms—is swallowed and eliminated.

Other surfaces are bathed by bodily fluids that dislodge microbes. In the urethra, for example, the rapid flow of urine washes away most microorganisms. Tears that wash over the conjunctivae perform a similar defensive function.

Biochemical Defenses. Biochemical defenses refer to substances produced by the body that inhibit microbial growth.

Some biochemical defenses are inside cells. For example, the skin protein **keratin** is produced in such large quantities that it essentially fills the cells that form the outermost layer of the skin. These protein-packed cells contain very little water, so the skin is very dry and therefore inhospitable to most species of microorganisms.

Other biochemical defenses are secreted onto body surfaces and act by lowering pH. Both fatty acids produced by the skin and hydrochloric acid produced by the stomach inhibit the growth of microorganisms this way.

Other biochemical secretions have specific microbiocidal effects. Bile, which is produced in the liver and secreted into the intestines, kills many microorganisms by disrupting their cellular envelope. Lysozyme, which is found in many secretions, including tears and saliva, breaks specific chemical linkages in peptidoglycan. Lysozyme is particularly destructive to Gram-positive bacteria because the peptidoglycan wall is the outermost layer of their cell envelope (Figure 14.5).

Microbial Adaptations to Life on Body Surfaces

What adaptations allow our normal flora to survive the body's defenses when most microorganisms cannot?

First, the normal flora are adapted to their specific environment on the human body—as any microorganism on Earth must be adapted to its specific physical conditions of temperature, pH, moisture, and oxygen. The environment of the skin is very different from the environment of the intestine, and each environment suits specific types of microorganisms. Anaerobic organisms that would die on the exposed outer skin prosper in airless pockets of the intestines. Species that can survive extreme dryness colonize the skin. Other species prefer the moist mucous membranes.

Microorganisms that colonize the human body must be able to tolerate any toxic chemicals produced there. Bile salts, for example, are toxic to many microbes, but not to *Escherichia coli* or the other organisms that colonize the human intestine. The fatty acids that kill many microorganisms are used by certain species of skin fungi as a substrate for growth.

Second, not only the physical environment, but also the nutritional environment determines which microbes can survive. Most microbial nutrients are readily available within the human body, with the significant exception of iron. Human secretions contain a variety of proteins that bind to the iron ion, making it unavailable to microorganisms. Thus an ability to obtain and use iron efficiently is critical to organisms that make their home on the internal surfaces of the human body.

Third, perhaps the most critical adaptation of human microflora is their ability to adhere to human cells. Otherwise, they would be overcome by the powerful mechanical defenses of surface tissues. Normal flora firmly attach themselves to the epithelial surface by means of protein molecules called **adhesins**. Adhesins are found on the microbial cell surface or on pili. They bind to complementary sugar molecules that happen to occur on the epithelial cell surface. Because they play this incidental but critical role in microbial adhesion, these sugar residues are sometimes called **receptors**. *Escherichia coli*, for example, adhere to **D-mannose** sugar residues on cells of the intestinal epithelium by means of adhesins called **type 1 fimbriae (Figure 14.6)**.

Adhesins are remarkably specific. Most bind only to a single type of epithelial cell in a single animal species. In some areas, including the human mouth, cell receptors and the adhesins that bind them vary from one tiny area to another. As you will see in Chapter 15, disease-producing organisms adhere similarly to the ways normal microflora do.

Fourth, some microflora persist on the human body because they produce compounds that are toxic to competing microorganisms. Certain bacteria in the human intestine, for example, produce proteins called **bacteriocins** that kill bacteria competing for the same space and nutrients.

SITES OF NORMAL FLORA

Only certain surfaces of the body—the skin, conjunctivae, nasal cavity, mouth, intestinal tract, vagina, and urethra—support a normal microbial flora. Let's consider each of these sites individually. We first discuss the particular environment each site presents to a potential colonizer. Then we look at the species of microorganisms that establish themselves there (**Table 14.2**).

The Skin

The skin is our most visible organ and our largest, constituting about 15 percent of our total body weight. Its two functions are to protect the body's vulnerable interior tissues from the outside environment and to help regulate body temperature. Further information about the skin is presented in Chapter 26.

Microbial Environment of the Skin. The predominant environmental factor limiting microbial growth on the skin is dryness (**Figure 14.7**). Microorganisms are

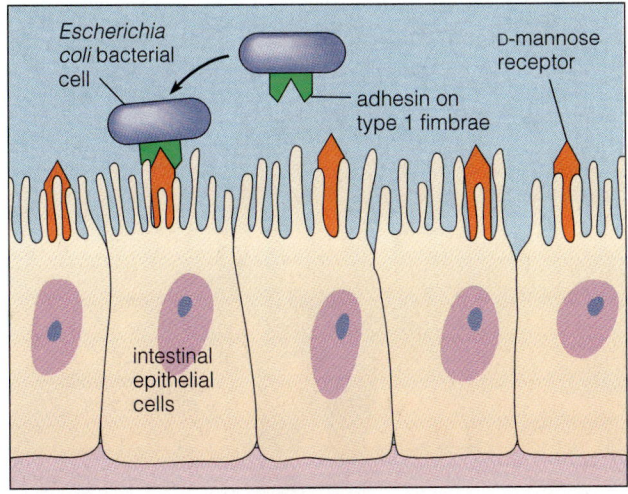

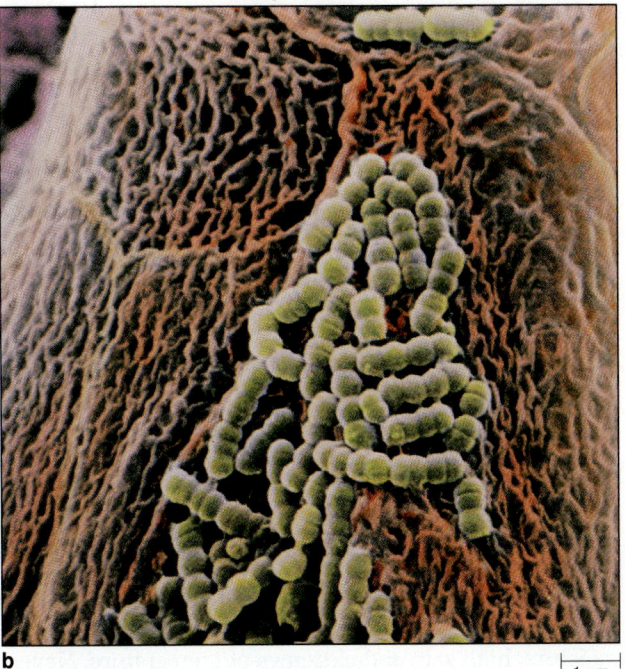

Figure 14.6 Microbial adaptations to life on body surfaces. Microorganisms adhere to body surfaces with adhesins, which attach to receptors on epithelial cells. (**a**) *Escherichia coli* attaches itself to D-mannose receptors on the surface of intestinal epithelium by means of its adhesins on type 1 fimbriae. (**b**) Bacteria adhering to the tongue's surface.

Table 14.2	Microorganisms Commonly Found on the Bodies of Healthy Human Beings
Organ	**Microorganisms (bacteria unless indicated as fungus or mite)**
Skin	*Staphylococcus epidermidis* and *aureus* *Corynebacterium* spp. *Propionibacterium acnes* *Pityrosporum ovale* and *orbiculare* (fungus) *Demodex folliculorum* (mite)
Conjunctivae	*Staphylococcus* spp. *Corynebacterium* spp.
Nasal cavity and nasopharynx	*Staphylococcus epidermidis* and *aureus* *Streptococcus* spp. *Moraxella catarrhalis* *Lactobacillus* spp. *Corynebacterium* spp. *Haemophilus* spp.
Mouth	*Streptococcus salivarius, mitis,* and alpha and gamma streptococci *Staphylococcus epidermidis* and *aureus* *Moraxella catarrhalis* *Lactobacillus* spp. *Corynebacterium* spp. *Haemophilus* spp. *Bacteroides* spp. *Fusobacterium* spp. Enterobacterial spp. *Candida albicans* (fungus)
Intestinal Tract	*Bacteroides* spp. *Bifidobacterium* spp. *Clostridium* spp. *Escherichia coli* *Fusobacterium* spp. *Klebsiella* spp. *Proteus* spp. Enterobacterial spp. *Lactobacillus* spp. Group D *Streptococcus* *Staphylococcus aureus*
Vagina	*Lactobacillus* spp. *Clostridium* spp. *Bacteroides* spp. *Fusobacterium* spp. *Candida albicans* (fungus)
Urethra	*Staphylococcus epidermidis* Enterococci

found on all normal skin, but large numbers are present only in relatively moist areas. Much of the skin's moisture comes from its own normal secretions, which include oil and perspiration. The skin is also moistened where it joins mucous membrane surfaces, near the nose, mouth, urethra, and anus.

Skin secretions may have lethal as well as beneficial effects on microorganisms, however. The oils on human skin, for example, contain microbiocidal fatty acids. Perspiration contains lysozyme and high concentrations of salt that increase osmotic pressure—an environmental limitation tolerated only by certain microorganisms.

The importance of the skin as a continuous impermeable defense against infection is dramatized when large

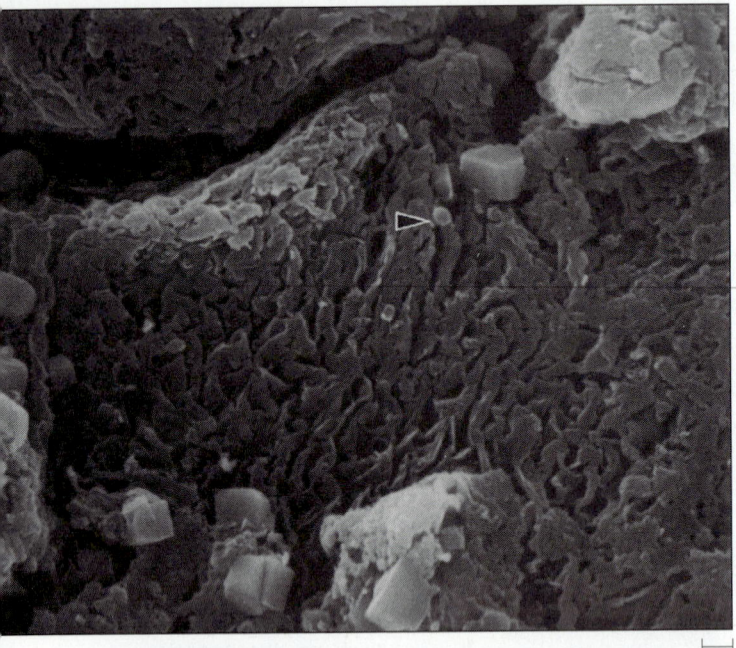

Figure 14.7 The predominant environmental factor limiting microbial growth on the skin is dryness. This electron micrograph shows a single staphylococcal cell (arrow) on the surface of the skin. Nearby is a crystal, which shows that the skin is inhospitably salty as well as dry.

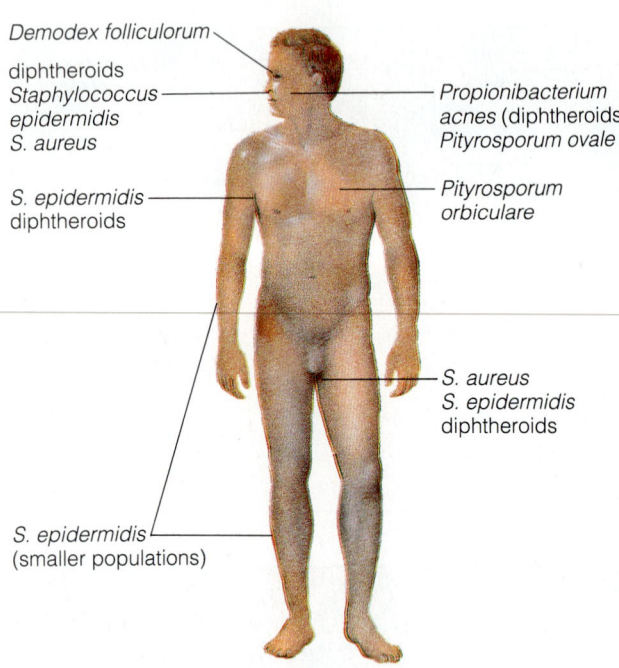

Figure 14.8 Flora of the skin. Our skin flora fall into three major groups: staphylococci, diphtheroids, and fungi. We also harbor the mite *Demodex folliculorum* on our faces.

areas are lost through burns. If a burn victim survives the initial crisis, the most common cause of death is infection, often due to the opportunistic pathogen *Pseudomonas aeruginosa*. Even intensive care in a hospital burn unit, with antibiotic therapy and scrupulous attention to cleanliness, cannot protect against deadly infection.

Although the skin is a relatively inhospitable environment for microorganisms, many manage to survive there. Baby Girl A, like other humans, will have up to 10,000 microorganisms growing on each square centimeter of dry skin, and up to 1 million per square centimeter in moist areas. Most live on the outermost skin layers and can be removed by washing. Thorough cleansing will decrease the bacterial count of the skin by as much as 90 percent. However, near-normal numbers of microbes are found again within 8 hours of washing because microorganisms that persist in deeper skin layers continue to multiply.

Flora of the Skin. Like all normal flora, the vast majority of skin flora are commensals. No species, therefore, directly benefits the host, but collectively they constitute a complex constellation that keeps the skin in healthy equilibrium and generally prevents the growth of pathogens. Furthermore, many of the fatty acids pro-duced by normal skin flora help lower the pH of the skin, thereby slowing the growth of pathogenic species.

The microorganisms that grow on the skin fall into three major groups—staphylococci, diphtheroids, and fungi (**Figure 14.8**).

Gram-positive cocci belonging to the genus *Staphylococcus* grow aerobically on the skin's surface and anaerobically in its airless pores. They are hardy enough to withstand the skin's harsh conditions, including low moisture, high salt, and extremes of temperature. Nevertheless, they are found in greatest numbers in relatively moist areas such as under the arms, around the nose, and near the anus.

Until fairly recently, most of the staphylococcal species that normally inhabit the skin were dismissed as harmless commensals. Clinical scientists now realize, however, that each staphylococcal species has disease-causing potential, but severity varies. *Staphylococcus aureus*, long known to be a dangerous pathogen, causes a wide variety of serious infections. It is found on the intact skin of only 5 to 10 percent of the population, usually around the anus (Chapter 26). Other species found on everyone's skin can become opportunists that cause problems for people with special vulnerabilities. Opportunistic staphylococci, for example, are the most common

Gonococcal Blindness

Until the 1920s, the most common cause of blindness in the United States was newborn infection with *Neisseria gonorrhoeae*. As a newborn passed through the birth canal of a mother with gonorrhea, bacteria from the mother's genital tract infected the infant's eyes. In 2 to 5 days, an intense inflammation appeared, often causing permanent eye damage. Gonococcal blindness was effectively eliminated in this country by making silver nitrate eyedrops a routine part of newborn treatment. In the 1980s, most nurseries began using the less irritating antibiotic erythromycin. Some type of treatment is required by law in most states (Chapter 26).

cause of infections in artificially implanted materials, such as plastic catheters and artificial joints. *S. epidermidis* is the umbrella term that encompasses all these strains of nonpathogenic and weakly pathogenic staphylococci.

The bacteria known as **diphtheroids** constitute the second major group of skin flora. They are Gram-positive rods related to *Corynebacterium diphtheriae*, the species that causes the respiratory disease diphtheria. Some anaerobic diphtheroids grow only in the airless regions deep within hair follicles. Here they metabolize sebaceous (fatty) secretions. One anaerobic species, *Propionibacterium acnes*, changes sebaceous secretions into the irritating fatty acids that initiate the inflammation called **acne** (Chapter 26). *P. acnes* might be considered an opportunistic pathogen because it causes disease primarily when the production of fatty skin secretions increases during adolescence. Aerobic diphtheroids grow nearer the skin surface. Most diphtheroids are commensals.

Staphylococci and diphtheroids are found on everyone's skin, but their relative numbers vary among individuals. On axillary (underarm) skin of some people, staphylococci predominate, while diphtheroids predominate in others. The skin flora of "coccal people" and "diphtheroid people" differ in other respects as well. For example, coccal people support a much larger population of staphylococci on their faces than do diphtheroid people.

The third major group of microorganisms that inhabit normal skin is fungi. Most are yeasts belonging to the genus *Pityrosporum*. These organisms use fats as a substrate for their growth and are consequently found where fatty sebaceous secretions are plentiful. *P. ovale* predominates on the face and scalp, while *P. orbiculare* is commonly found on the chest and back. Pityrospora are commensals. Other yeasts that may inhabit the skin include the commensal *Torulopsis glabrata* and the opportunistic pathogen *Candida albicans*.

Another microscopic inhabitant of human skin is the mite *Demodex folliculorum*, the only arachnid that normally inhabits the human body. It lives on the face, within hair follicles and in the openings of sebaceous glands. Baby Girl A will not acquire this organism during her infancy, but, like other humans, she is likely to provide a home by the time she reaches adulthood. Although the relationship is usually harmless to the human host, it may result in an irritation of the eyelid margins called **blepharitis**.

The Conjunctivae

The conjunctiva (*pl.*, conjunctivae) is the mucous membrane that covers the exposed surface of the eye and the interior surface of the eyelid. This epithelial barrier protects the delicate surface of the eye, which would otherwise be exposed directly to the environment.

Microbial Environment of the Conjunctivae. The conjunctivae provide a structural defense for the eyes because their continuous surface is relatively impermeable to microorganisms. They in turn are protected by the mechanical and biochemical defenses of tears. Tears are produced by the lacrimal glands beneath each eyelid. Small amounts of tears are produced constantly to bathe the conjunctivae. They are spread by movement of the eyelids and accumulate at the inner corner of the eye. From there, they pass through the nasolacrimal duct into the nose. Tears are a mechanical defense because of their flushing action and a biochemical defense because they contain such antimicrobial substances as lysozyme.

Flora of the Conjunctivae. The tear-washed conjunctivae present an inhospitable environment for microorganisms, and few species are normally found there. Those that do persist are usually microflora of the skin.

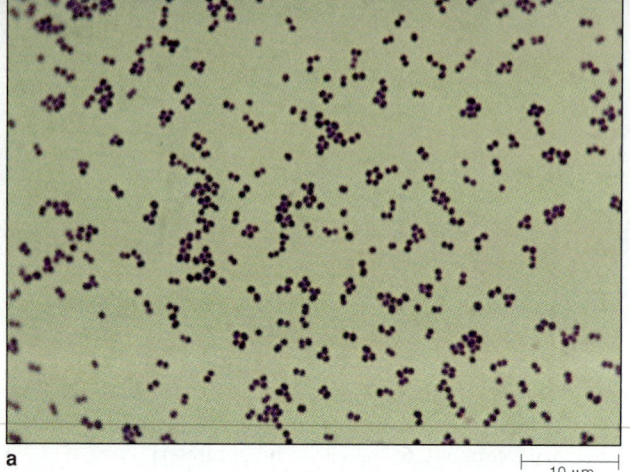

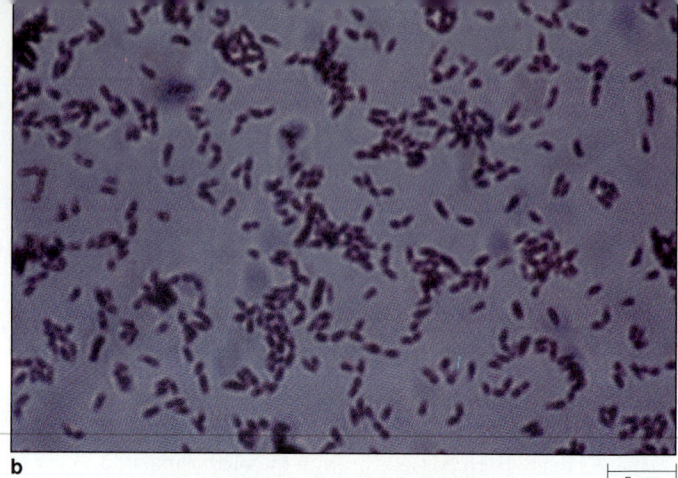

a   10 µm   b   5 µm

Figure 14.9 Flora of the conjunctivae. Only a few microorganisms, most of which are normal skin flora, manage to survive on the well-defended surface of the eye. (**a**) Micrograph of *Staphylococcus epidermidis*. (**b**) Micrograph of *Corynebacterium pseudo-diphtheriae*. (**c**) Eye and lacrimal apparatus.

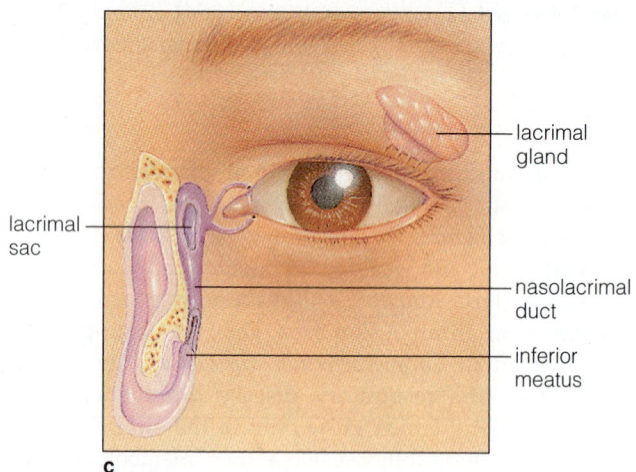

lacrimal sac

lacrimal gland

nasolacrimal duct

inferior meatus

c

The two most prominent groups are staphylococcal species, found in more than 90 percent of people, and diphtheroids, found in as many as 80 percent of people (**Figure 14.9**). Even these microorganisms are present in small numbers compared to their populations on the skin. Most microorganisms on the conjunctivae are commensals, although opportunistic pathogens may emerge if the normal flow of tears is interrupted.

The Nasal Cavity and Nasopharynx

The nasal cavity and the space behind it, called the nasopharynx, are the outermost respiratory passages that help deliver air to the lungs. Air that enters the body brings with it dust, debris, and microorganisms. The average adult inhales about 8 microorganisms per minute, or 10,000 each day. Baby Girl A's nasal cavity and nasopharynx, like the rest of her body, is sterile until the time of birth. But 24 hours later, many microorganisms can be found here.

Microbial Environment of the Nasal Cavity and Nasopharynx. Unlike the skin, the nasal cavity and nasopharynx provide microorganisms with a moist, warm environment in which to grow. Many microorganisms would survive if it were not for mechanical defenses.

The epithelial surfaces that line the nasal cavity and nasopharynx are mechanically protected by the mucociliary system described earlier in the chapter. If this system does not function normally, however, repeated and potentially life-threatening respiratory infections occur. For example, cystic fibrosis patients produce abnormally viscous (thick) mucus and therefore have trouble clearing it—and the microorganisms it contains—from their respiratory tract. The mucociliary system can be weakened by toxic substances in the environment, such as cigarette smoke. If Baby Girl A's parents smoke, it will signifi-

cantly increase her risk (and theirs) of frequent respiratory infections.

These especially powerful mechanical defenses of the nasal cavity and nasopharynx mean the microflora in these sites must be well adapted in their ability to adhere.

Flora of the Nasal Cavity and Nasopharynx. The nasal cavity and nasopharynx are densely colonized. Many of the microorganisms are the same ones found on the skin (**Figure 14.10**). *Staphylococcus epidermidis*, for example, can be cultured from the nasal cavity of more than 90 percent of human beings. Some studies indicate that as many as 80 percent also harbor the pathogen *S. aureus*, although other studies report much lower numbers. Diphtheroids are found in the nasal cavity of up to 80 percent of human beings.

The nasal cavity and nasopharynx are also colonized by microorganisms not usually found on the skin, including various species of streptococci, the vast majority of which are harmless commensals. *Lactobacillus* spp. are commonly found in the nasal cavity. The nasopharynx is inhabited by Gram-negative bacilli such as *Moraxella catarrhalis*; it has traditionally been considered a commensal, but recent studies show it to be responsible for a significant number of middle ear infections in young children. *Haemophilus influenzae* is another resident of the nasopharynx. Some strains of this organism are highly

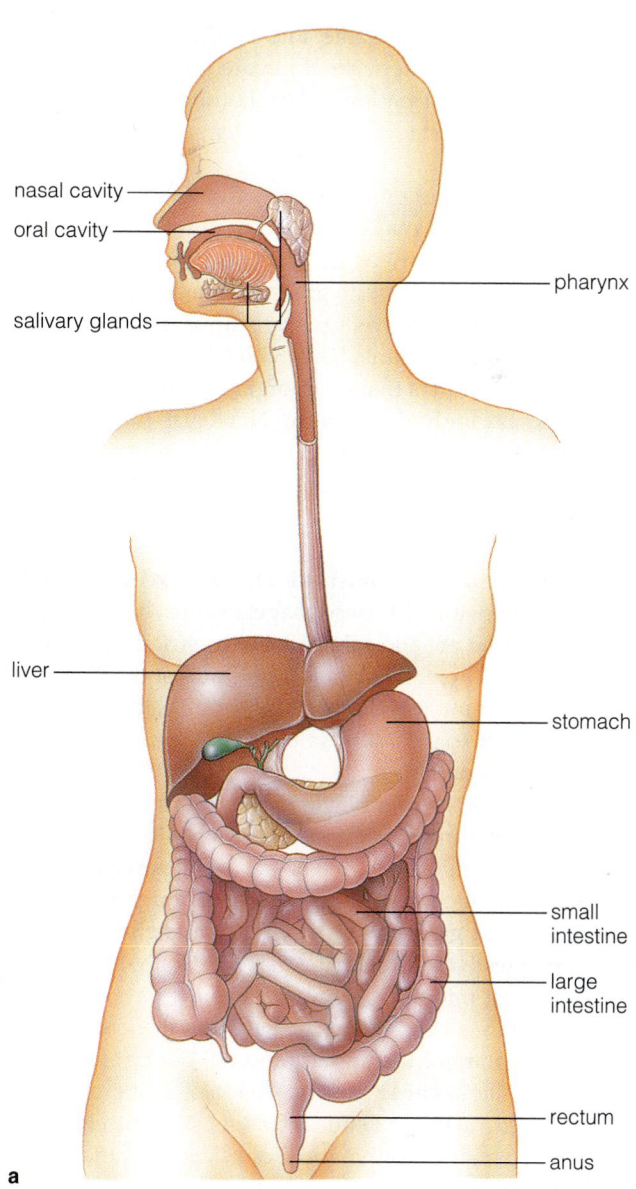

nasal cavity

oral cavity

pharynx

salivary glands

liver

stomach

small intestine

large intestine

rectum

anus

a

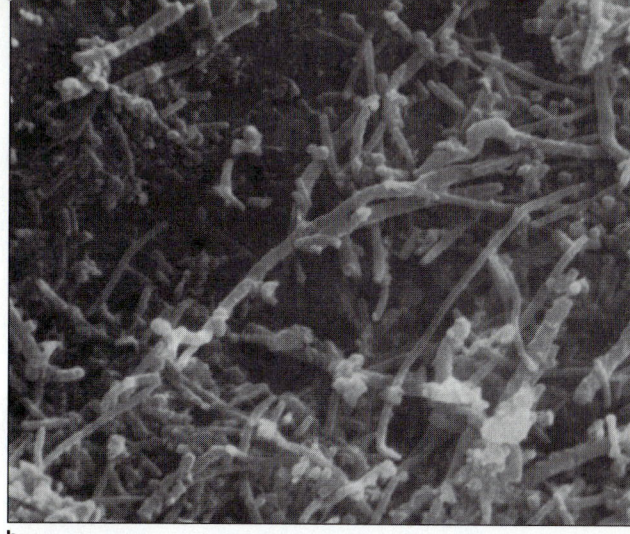

b

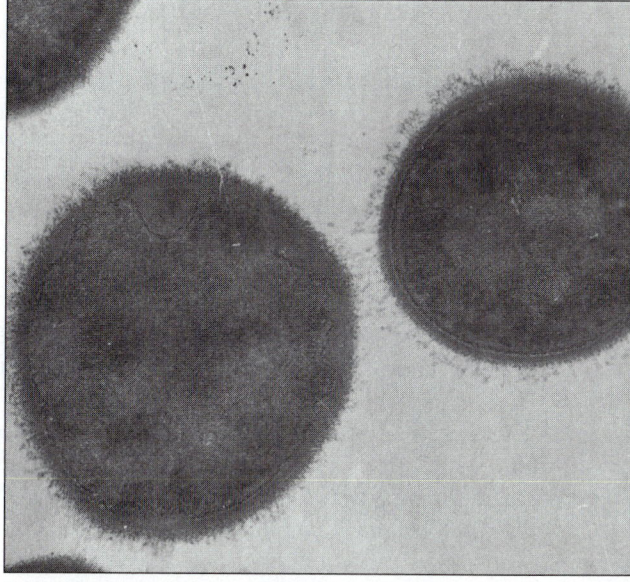

c

⊢—⊣
100 nm

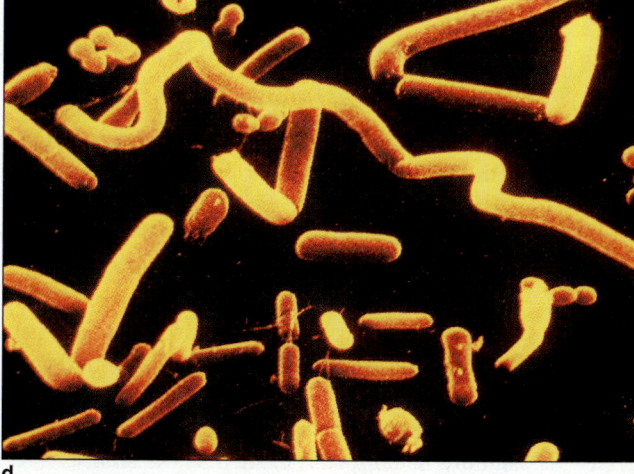

d

pathogenic (Chapter 22), but the strains that colonize the nasopharynx are usually relatively harmless.

The Mouth

The mouth is the entrance to the digestive system, the point where food is taken into the body. Inside the mouth, food is chewed and mixed with saliva.

During Baby Girl A's passage through the birth canal, microbes from her mother's vagina enter her mouth. During her first meal, microorganisms from the skin around her mother's breast enter her mouth. And during her first few months of life, as she explores her environment—mouthing blankets, her own hands, and numerous other objects—she will introduce huge numbers and wide varieties of microorganisms into her mouth.

Figure 14.10 Flora of the nasal cavity, nasopharynx, mouth, and intestinal tract. (**a**) Sites for growth of flora. (**b**) Plaque on human tooth enamel. (**c**) *Streptococcus salivarius*, one of the species of streptococci that grow in the mouth. (**d**) Intestinal bacteria.

Microbial Environment of the Mouth. Many features of the mouth make it an ideal environment for microbial growth—it is moist, warm, and abundantly supplied with nutrients. Mechanical and biochemical factors, however, limit the types of microorganisms that can survive here. Talking, chewing, and swallowing all dislodge microorganisms that are not well adapted to adhere.

In addition, the mouth is bathed with saliva, which is constantly being produced by the salivary glands and removed from the mouth by swallowing. Like tears, saliva flushes and offers biochemical defense because it contains lysozyme. People who do not produce normal amounts of saliva have difficulty maintaining a healthy microbial environment in their mouth, and many suffer from excessive tooth decay.

Flora of the Mouth. The human mouth is densely populated with more than 80 different species of microorganisms (Figure 14.10). Many of these are bacteria belonging to the genus *Streptococcus*. Streptococci are the principal microorganisms that inhabit Baby Girl A's mouth before her teeth erupt. Thereafter, streptococci will continue to be present but in smaller numbers. Streptococcal mouth inhabitants include *S. salivarius*, *S. mitis*, and numerous other species. Even some species of enterococci are found in the mouth, although they are found predominantly in the intestines. Various streptococcal species are limited to the surface of the teeth, roof of the mouth, or tongue. Regional differences are due to the different adhesins on different species.

In addition to streptococci, many other microorganisms occur in the mouth. Staphylococci are widespread. *Staphylococcus epidermidis* is in the mouths of nearly all humans, and *S. aureus* is fairly common. Also nearly universally present in the mouth are the Gram-negative coccus *Moraxella catarrhalis*, lactobacilli, diphtheroids, and the Gram-negative rod *Haemophilus influenzae*. Anaerobes such as *Bacteroides* spp. and *Fusobacterium* spp. are also common. Almost half of us also harbor small populations of the fungus *Candida albicans*.

The complex microbial population that exists within the mouth is generally quite stable. Most are commensals that maintain a healthy mouth environment. However, some species that are normally present do cause problems. One example is *Streptococcus mutans*, a mouth inhabitant that is the primary cause of tooth decay (Chapter 23). Other mouth flora, such as *Haemophilus influenzae* and *Moraxella catarrhalis*, can cause disease in nearby sites that are usually sterile, such as the sinuses and the middle ear.

The Intestinal Tract

From the mouth, food and microorganisms move into the intestinal tract. They pass through the esophagus, the stomach, the small intestine, and finally the large intestine. Undigested material is eliminated from the body as feces. Further information about the intestinal tract and other organs of the digestive system is presented in Chapter 23.

The first evacuations from Baby Girl A's digestive system, a sticky green material called meconium, are free of microorganisms. Only as the microorganisms that Baby Girl A takes into her mouth gradually pass into her lower intestinal tract do microorganisms appear in her feces. Clinical studies of infants born with a blockage of the intestinal tract confirm that microorganisms enter this body site only through the mouth. Newborn infants with bowel obstruction have a normal population of microorganisms above the blockage, but the region below remains sterile.

Microbial Environment of the Intestinal Tract. The environments of the esophagus, stomach, and upper small intestine are too inhospitable to sustain a normal flora. For one thing, biochemical products of the digestive system inhibit microbial growth. Stomach acid lowers pH in the stomach to pH 1. Bile salts destroy many organisms in the upper small intestine.

Biochemical factors play a role in limiting organisms from the upper intestinal tract, but mechanical factors are more important. The muscular action of peristalsis (the wavelike contractions that move food through the digestive tract) and gastric churning are extremely powerful, and the digestive contents flow past the epithelial surface rapidly. Most microorganisms are swept away. We know the importance of mechanical defenses in determining the normal flora of the upper intestinal tract from autopsies. Within hours after death, when intestinal movement has stopped but the chemical content has not changed, the upper small intestine acquires a large population of microorganisms.

In contrast to the upper regions, the large intestine houses many different species of normal flora. Although some toxic biochemical agents, such as bile salts, are present, intestinal movement is much slower, and so enormous populations accumulate. Approximately a third of the dry weight of the feces is made up of microorganisms. Most people eliminate between one hundred billion and one hundred trillion bacteria in their feces every day.

Flora of the Intestinal Tract. The large intestine contains an extremely complex microbial community (Figure 14.10). Hundreds of species of bacteria normally coexist there, but by far the most abundant are strict anaerobes. And the majority of these belong to the genera *Bacteroides*, *Bifidobacterium*, *Fusobacterium*, and *Clostridium*. These four genera account for more than 90 percent of all the bacteria in the bowels.

Present in smaller numbers are facultative anaerobes.

Once a Hydrogen-Producer, Always a Hydrogen-Producer

Curiously, not everyone produces the same type of intestinal gas. While most of us are hydrogen-producers, a significant minority produce methane in their large intestine. The type of gas we produce depends upon our normal intestinal flora. Everyone is colonized by eubacteria, such as *Escherichia coli*, that produce hydrogen when they ferment sugars under the anaerobic conditions of the large intestine. But some people are also colonized by a type of archaebacteria, the methogens, which convert hydrogen to methane.

Whether you are a hydrogen-producer or a methane-producer is determined early in childhood when the intestinal flora becomes established. Moreover, the type of intestinal gas we produce is fixed for life. So far, the production of methane as opposed to hydrogen has not been associated with any clinically significant difference between the two groups.

Although the type of gas we produce is fixed, the volume produced varies greatly, depending on our diet. Most of the sugars we eat are absorbed as nutrients in the small intestine, but some sugars cannot be used by humans. These sugars pass directly into the large intestine, where they are metabolized by intestinal bacteria. Beans contain two non-absorbable sugars, raffinose and stachyose, which accounts for our increased gas production after a meal with beans.

These include Gram-negative rods that belong to the Enterobacteriaceae, such as *Escherichia coli*, *Proteus* spp., *Enterobacter* spp., and *Klebsiella* spp. Gram-positive organisms include a group of streptococci called enterococci (*enteron* means "intestines" in Greek), *Staphylococcus aureus*, and *Lactobacillus* spp.

Microorganisms in the intestine interact in complex ways. Some produce bacteriocins to kill competing organisms. Other intestinal bacteria control the growth of their neighbors by producing acids. Still others change the oxidation state. All these changes help limit the growth of competing microorganisms, including pathogens.

Genetic exchange occurs among the various members of the dense intestinal microflora, sometimes with important consequences. For example, when plasmids are exchanged that encode resistance to antibiotics, a large portion of the intestinal population can become drug-resistant in a short time.

The Vagina

The vagina lies at the outlet of the female reproductive tract. It receives semen during sexual intercourse and is the birth canal through which babies pass as they leave the womb and enter the outside world. It was there that Baby Girl A came in contact with her first microorganisms. The vagina is the only site in the female reproductive tract that sustains a normal flora.

Microbial Environment of the Vagina. The vagina is a warm, moist, protected environment with few mechanical features. Only the availability of nutrients and pH limit microbial growth. As you read earlier in the chapter, when the vagina is acidic because of the influence of estrogen, the growth of lactobacilli helps protect against pathogens. During other periods, however, when the vagina is alkaline, microbial flora are sparse and the vagina is less protected.

Flora of the Vagina. The lactic acid bacteria that are responsible for the low pH of the vagina are its primary inhabitants (**Figure 14.11**) along with small populations of some other microbial species. These include aerobic and anaerobic streptococci, *Staphylococcus* spp., and anaerobes such as *Bacteroides*, *Clostridium*, and *Fusobacterium* spp. Many women also support a small population of the fungus *Candida albicans*, which can cause opportunistic vaginal infections.

The Urethra

The urethra lies at the very end of the urinary tract in both males and females. The outermost part of the urethra, near the point at which the mucous membranes of the urinary tract meet the skin, is the only part of the urinary tract that supports microflora. The upper part of the urethra, which joins the bladder, is normally free of microorganisms.

Microbial Environment of the Urethra. The structure of the urethra limits microbial growth because it is composed of tightly joined epithelial cells that resist attachment of microorganisms. In addition, the mechanical forces of urine flowing rapidly over the epithelial surface flush out microorganisms. The fluid that bathes the urethra does not provide biochemical protection (urine is an excellent culture medium), but cells of the urethral epithelium may secrete antibacterial substances.

Flora of the Urethra. The normal flora of the urethra is scant. Organisms commonly cultured from the end of the urethra include enterococci and *Staphylococcus epidermidis* (Figure 14.11).

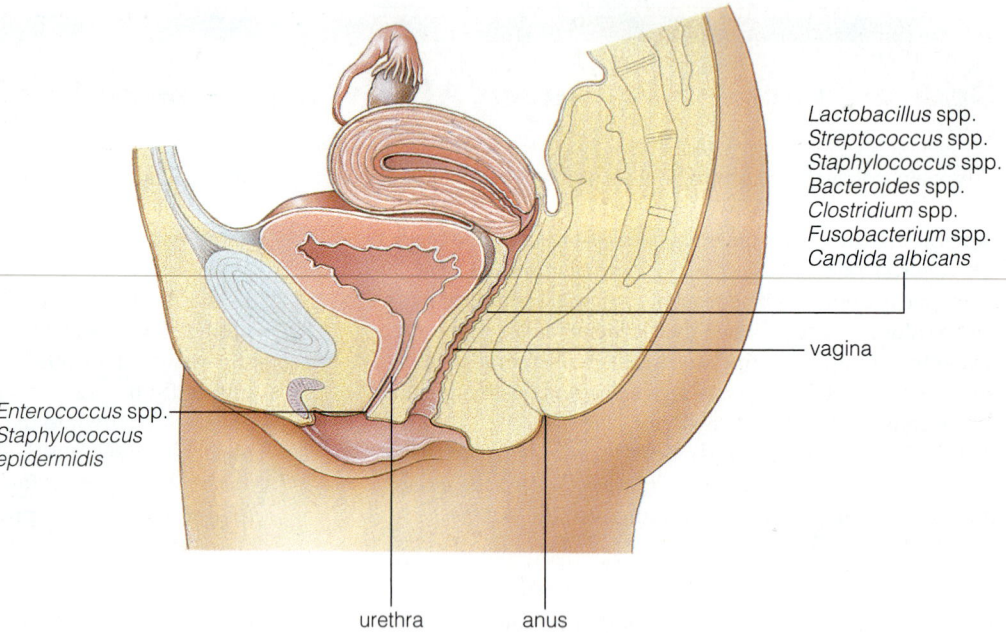

Figure 14.11 Flora of the vagina and urethra. In males and females, the outermost part of the urethra is colonized by a few microorganisms. The vagina, in contrast, is heavily colonized by a variety of different bacterial and fungal species.

Lactobacillus spp.
Streptococcus spp.
Staphylococcus spp.
Bacteroides spp.
Clostridium spp.
Fusobacterium spp.
Candida albicans

vagina

Enterococcus spp.
Staphylococcus epidermidis

urethra   anus

IS THE NORMAL FLORA HELPFUL OR HARMFUL?

The relationship between humans and their normal microbial flora is compatible with health. We might ask, however, do the many microorganisms that inhabit Baby Girl A's body exert any *subtle* beneficial or harmful effects on her? This question has fascinated microbiologists for years.

Conflicting Theories

Louis Pasteur believed that normal microbial flora was not only helpful but also essential to the life of higher animals. Pasteur's associate, the Nobel Prize–winner Elie Metchnikoff, believed that all microorganisms compete with humans for factors essential to life and therefore harm their hosts to varying extents.

Germ-free animal studies have proved Pasteur wrong. It is now possible to maintain animals germ-free by transferring them from the sterile prenatal environment directly to a sterile external environment. Raised in sterilized chambers, fed sterilized food, and handled only with sterilized gloves, these animals remain germ-free throughout their lives (**Figure 14.12**). When they reproduce, their offspring are healthy and germ-free and have the same life spans as normally colonized animals. Germ-free animals do differ from their natural counterparts in subtle ways, including differences in acquired immunity, but germ-free animals suffer no ill effects from their lack of a normal microbial flora.

Just because Pasteur has been proved wrong does not mean Metchnikoff was right. Metchnikoff's theory is difficult to prove or disprove, but we have no evidence that microorganisms on our bodies sap our strength in any way. Probably, in a natural environment, the benefits of normal flora outweigh possible harmful effects.

Harmful Effects

If there is no proof that microflora are outright harmful, are there subtle effects? One area of research has been microbial use of vitamins. Some bowel microorganisms, for example, use vitamin C as a substrate for their growth and thereby destroy it, but there is no evidence that this leads to clinically significant vitamin deficiencies.

Another area of research arose from experiments on germ-free animals. Investigators found that some chemicals are carcinogenic in normal animals but not in germ-free animals. These chemicals are converted into carcinogens by bacterial metabolism. This discovery led to the theory that our normal microflora might promote the development of cancer. Further studies showed, however, that germ-free and normal animals develop cancers at similar rates. Therefore, it is unlikely that microbially produced carcinogens are a significant cause of cancer in humans.

Beneficial Effects

Normal microflora benefit humans in several ways, but the most significant derives from microbial antagonism. Normal flora, for example, produce bacteriocins, compete

Figure 14.12 Technology now allows us to compare germ-free animals with normal animals. We know for certain that microflora are not essential to life and that a lack of normal flora is not harmful.

for receptors on epithelial cells, and, by influencing the pH of an environment, kill pathogens outright or inhibit their growth. Certain microflora in the bowel inactivate toxins produced by some species of pathogenic bacteria.

Microflora also benefit us by stimulating our immune system. The stimulation is *nonspecific*; that is, it does not provide immunity against any particular disease. It improves the entire defensive system. Studies show that normal animals are more capable than germ-free animals of mounting an immune defense against pathogens, even pathogens with which they have had no previous contact.

Finally, though most of our requirement for vitamins is satisfied by our diet, the metabolic products of some intestinal bacteria provide supplemental sources of others. Two of these are vitamin K (important for blood clotting) and vitamin B_{12} (important in the production of red blood cells and normal functioning of the nervous system).

Loss of Normal Flora

Anything that alters the equilibrium of our relationship with microorganisms may have undesirable effects. When the balance is upset, pathogens are able to establish themselves in the underpopulated regions and disease may result. Or a normally harmless resident microflora may proliferate and cause an opportunistic infection. Let's look at how this can happen as a result of antibiotic therapy.

By the age of 2 to 3 weeks, Baby Girl A will have acquired most of the normal flora typically found on the adult human body. Suppose that some time later she develops an infection of the middle ear. She will probably be treated with a broad-spectrum antibiotic to assure that the causative bacterium—which may be Gram-positive or Gram-negative—is eliminated. But, in addition to killing the bacteria infecting her ear, the antibiotic will also kill many bacteria in other parts of her body. These changes in bacterial microflora may stimulate growth of the yeast *Candida albicans*. As a fungus, *C. albicans* is unaffected by the antibiotics that eliminate many bacteria and may rapidly proliferate when competing bacteria are depleted. For our baby—and baby boys as well—the result is likely to be a severe and persistent *Candida* diaper rash. Antibiotic therapy in adult women often causes a *Candida* **vaginitis**, characterized by redness, itching, and a yeast-laden vaginal discharge. Antibiotic therapy can also cause *Candida* infections of the mouth, called **thrush**.

A Spectrum of Disease-Causing Abilities

Categorizing microorganisms as pathogens or opportunists or commensals is useful, but it implies rigid distinctions. In reality, the distinctions are blurred because the host's state of health is critical in determining whether or not disease will occur. It is more helpful, therefore, to think of microorganisms existing along a continuous spectrum of disease-causing potential (**Figure 14.13**). At one end of the spectrum are highly virulent pathogens that cause disease in almost all cases. Other

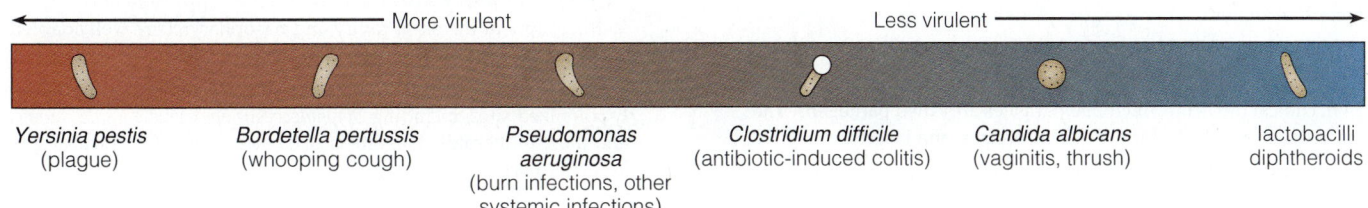

More virulent →			Less virulent →

Yersinia pestis (plague)   *Bordetella pertussis* (whooping cough)   *Pseudomonas aeruginosa* (burn infections, other systemic infections)   *Clostridium difficile* (antibiotic-induced colitis)   *Candida albicans* (vaginitis, thrush)   lactobacilli diphtheroids

Figure 14.13 The disease-causing ability of microorganisms is a continuous spectrum ranging from highly virulent pathogens that cause disease in almost all cases, to opportunists that cause disease only when the human host is weakened, to nearly harmless commensals that rarely, if ever, cause disease.

microorganisms are less virulent but often cause disease. Others are opportunistic to greater or lesser degrees, causing disease in more or less weakened hosts. Most microorganisms never cause disease, though any microbe that can sustain itself in the human body is probably capable of harming the host if debilitation is extreme.

With our advanced medical technology, people with serious diseases and badly weakened defenses now survive much longer. Today, in industrialized societies more deaths are caused by opportunistic infections than by "true" pathogens.

Summary

NORMAL FLORA (pp. 333–335)

1. Health is a state of relative equilibrium in which the body's organ systems function adequately. Disease is a state of functional disequilibrium resolved by recovery or death.

2. The microorganisms that coexist with humans in a stable relationship constitute our normal flora. Normal flora inhabit only the surfaces of our bodies: the skin and the conjunctivae, the nasal cavity and nasopharynx, mouth, intestinal tract, vagina, and urethra.

3. Resident flora inhabit the human body throughout life. They can be temporarily reduced but never eliminated.

4. Transient flora can be cultured from the human body under certain circumstances. They are not part of the normal flora and can be removed by careful cleansing.

5. Opportunists are microorganisms that cause disease when the host's normal defenses are weakened or when the normal flora is altered.

6. The normal flora change over time.

NORMAL FLORA AND SYMBIOSIS (pp. 336–337)

1. Symbiosis refers to two different kinds of organisms living together. The larger organism that provides a habitat for its symbiotic partners is the host.

2. Symbiotic relationships are classified according to whether they harm or benefit the host. In commensalism, one partner benefits (the commensal organism) and the other is neither benefited nor harmed.

3. Most normal flora are commensals. Though individual species do not benefit the human host, they do collectively because of microbial antagonism, when one type of microorganism interferes with the growth of another.

4. In mutualism, both partners benefit. In parasitism, the host is harmed by its symbiotic partner and the parasite benefits.

5. In clinical medicine, microbial parasites are called pathogens. The term *parasite* is reserved for protozoa, worms, and insects.

6. Many commensals are potential pathogens. That is, resident flora can become opportunists and cause infection.

FACTORS THAT DETERMINE THE NORMAL FLORA (pp. 338–340)

1. Our microflora survive because they are uniquely adapted to life on living tissue.

2. Structural, mechanical, and biochemical features of body surfaces constitute nonspecific surface defenses, the body's first line of defense against infection. They are nonspecific because they act against all pathogens in the same way.

3. Our structural defenses are our epithelial surfaces, the skin and mucous membranes, which are thin and cover the surface of the eyes (called the conjunctivae) and the interior surfaces. The normal sloughing (shedding) of epithelium as it grows creates another structural defense. When dead cells are lost, so are the microbes growing on them.

4. Mechanical defenses—movements—eliminate many transient microorganisms when they land. Some surfaces move because of the action of underlying muscles. The mucociliary system protects through mucus and ciliary movement. The rapid flow of urine washes microorganisms out of the urethra, and tears wash microorganisms off the conjunctivae.

5. Biochemicals that inhibit microbial growth include keratin, which keeps the skin surface dry, stomach acid and fatty acids, which lower pH, and lysozyme and bile, which have specific microbiocidal effects.

6. The normal flora are adapted to growth at a particular body site.

7. Adhesins are protein molecules by which normal flora adhere to human cells. Adhesins are very specific. Some microflora produce bacteriocins, proteins that kill other microorganisms.

SITES OF NORMAL FLORA (pp. 340–347)

The Skin (pp. 340–343)
1. Skin flora fall into three main categories: staphylococci, diphtheroids, and fungi.

2. Staphylococci, which grow aerobically and anaerobically, include the pathogen *Staphylococcus aureus* and the opportunist *S. epidermidis*.

3. Diphtheroids include the anaerobic *Propionibacterium acnes*, which causes acne.

4. Fungi on the skin include yeasts belonging to the genus *Pityrosporum*, which use fats as a substrate for growth and are therefore found in oily areas on the face, scalp, chest, and back.

5. The mite *Demodex folliculorum* lives on the face, within hair follicles and in the openings to oil glands.

The Conjunctivae (pp. 343–344)
1. The conjunctivae are defended by their continuous and relatively impermeable surface and by tears.

2. Among the few species found at this site are staphylococci and diphtheroids.

The Nasal Cavity and Nasopharynx (pp. 344–345)
1. Microflora of the nasal cavity and nasopharynx are well adapted to adhere because of the mucociliary system.

2. The same species that colonize the skin are found in these densely colonized sites, including *Staphylococcus epidermidis* and *S. aureus* and the diphtheroids. This site is also colonized by *Lactobacillus* spp. and *Moraxella catarrhalis*.

The Mouth (pp. 345–346)
1. The mouth is a warm, moist environment with abundant nutrients and is densely populated with microorganisms. Streptococci predominate before teeth erupt and thereafter are present in smaller numbers.

2. Other species include aerobes and anaerobes. The complex microbial population is stable and consists largely of commensals. One exception is *Streptococcus mutans*, which causes tooth decay.

The Intestinal Tract (pp. 346–347)

1. The first intestinal tract evacuations in a newborn are germ-free.

2. The esophagus, stomach, and upper intestine are too inhospitable to sustain a normal flora. Peristalsis and stomach churning keep this region nearly germ-free.

3. In the lower intestine, on the other hand, intestinal movement is much less vigorous, allowing a complex microbial community to develop.

4. The majority of species in the lower intestine are strict anaerobes belonging to the genera *Bacteroides*, *Bifidobacterium*, *Fusobacterium*, and *Clostridium*. Facultative anaerobes include the Enterobacteriaceae (such as *Escherichia coli*) and *Lactobacillus* spp.

The Vagina (p. 347)

1. The vagina is a warm, moist, protected environment. When influenced by estrogen, it becomes acidic with the growth of lactobacilli. When estrogen is not being produced, it is alkaline and more prone to infections.

2. Aerobic and anaerobic species colonize this site, including the fungus *Candida albicans*, which can cause opportunistic infections.

The Urethra (pp. 347–348)

1. Only the outermost part of the urethra, where the mucous membranes meet the skin, supports microflora.

2. The normal floral is scant, but usually enterococci and *Staphylococcus epidermidis* can be cultured from the outer part of the urethra.

IS THE NORMAL FLORA HELPFUL OR HARMFUL? (pp. 348–350)

1. Today scientists believe the benefits of normal flora outweigh possible harmful effects. Research has yielded no proof of even subtle harm.

2. The most significant beneficial effect of microflora derives from microbial antagonism. Some microflora produce bacteriocins; compete for receptors on epithelial cells; and, by influencing pH, kill pathogens or inhibit their growth.

3. Microflora also stimulate our immune system in a nonspecific way. Some intestinal bacteria provide supplemental sources of vitamins K and B_{12}.

4. If our relationship with microorganisms is altered, pathogens can establish themselves in underpopulated areas or normally harmless commensals can proliferate and cause opportunistic infections.

5. Categorizing microorganisms as harmful or harmless is not always clear-cut because the host's state of health frequently determines whether disease will occur. Microorganisms exist along a continuous spectrum of disease-causing potential.

6. Today, because of our advanced medical technology, more people die from infection by opportunists than by true pathogens.

Review Questions

NORMAL FLORA

1. What are normal flora and where are they found?

2. How do we acquire our normal flora?

3. Describe, as fully as you can, these three types of flora: resident flora, transient flora, and opportunists.

4. How do normal flora in the mouth and intestines change during infancy?

5. Describe the influence of estrogen on vaginal flora.

NORMAL FLORA AND SYMBIOSIS

1. Define these terms: symbiosis, host. How are symbiotic relationships classified?

2. Explain the difference between commensalism, mutualism, and parasitism. Which are stable, highly evolved relationships and which are not? Explain.

3. Explain this statement: There are no true mutualistic relationships between humans and microorganisms.

4. What is microbial antagonism?

5. How do commensals benefit us? How can they harm us?

6. How are the terms *parasite* and *pathogen* used in clinical medicine?

FACTORS THAT DETERMINE THE NORMAL FLORA

1. In addition to the physical and nutritional factors, what unique factors influence microbial growth on human tissue?

2. What are our nonspecific surface defenses?

3. Name our major structural defense against microbial invasion. What are the two types of epithelium and where are they found? Name two ways the skin defends against microbes.

4. What are mechanical defenses? Give some examples.

5. What are biochemical defenses? Give some examples.

6. Describe the ways microorganisms must adapt if they are to live on body surfaces. What are adhesins, and what role do they play in adaptation?

7. What are bacteriocins, and why are they important?

SITES OF NORMAL FLORA

1. For each of the following sites, describe the environment and name some of the representative microflora found there:
   a. skin
   b. conjunctivae
   c. nasal cavity and nasopharynx
   d. mouth
   e. intestinal tract
   f. vagina
   g. urethra

2. What is *Demodex folliculorum*, and why is it significant?

3. Why are the first evacuations in a newborn free of microorganisms?

4. Explain this statement: The intestinal tract is a complex microbial community.

5. Describe some of the complex interactions in the intestinal ecosystem.

IS THE NORMAL FLORA HELPFUL OR HARMFUL?

1. Are microflora essential to human life? Are they inevitably harmful? How would most microbiologists answer the question of whether normal flora are helpful or harmful?

2. Describe some of the research that has been done into potentially harmful subtle effects of normal flora and explain the conclusions.

3. What are the major beneficial effects of normal flora?

4. Why is the loss of normal flora dangerous? Why does the loss of normal flora from broad-spectrum antibiotics affect the population of *Candida albicans*?

5. What is the single most important reason why categorizing microorganisms as commensals, opportunists, or pathogens is not always clear-cut?

6. Explain this statement: Microorganisms exist along a continuous spectrum of disease-causing potential.

Essays

1. Discuss the human and microbial factors that lead to a continuing relationship between a microorganism and humans.

2. What have germ-free animal studies taught us about the relationship between humans and their normal flora?

Suggested Readings

Drasar, B. S., and Barrowk, P. A. 1985. *Intestinal microbiology*. Washington, D.C.: American Society for Microbiology.

Mackowiak, P. A. 1982. The normal microbial flora. *New England Journal of Medicine* 307:83.

Marsh, P., and Martin, M. 1984. *Oral microbiology*. 2d ed. Washington, D.C.: American Society for Microbiology.

Roth, R., and Jenner, W. 1988. Microbial ecology of the skin. *Annual Reviews of Microbiology* 42:441.

van der Waaij, D. 1989. The ecology of the human intestine and its consequences for overgrowth by pathogens such as *Clostridium difficile*. *Annual Reviews of Microbiology* 43:69–87.

15 MICROORGANISMS AND HUMAN DISEASE

The Infection of Baby Girl B

Baby Girl B was a healthy 3-month-old. She weighed nearly 13 pounds, could smile, laugh, and roll over by herself. Her development was normal; but because of a minor illness, she had missed her first set of baby shots at the 2-month medical checkup.

One day Baby Girl B got a runny nose and began sneezing. Her mother called the doctor and was reassured that infants often catch colds and recover from them uneventfully. But Baby Girl B did not improve. After 10 days of cold symptoms, she developed a cough that produced thick, sticky mucus. She began suffering from fits of repeated coughing that were so bad she couldn't catch her breath. After one of these fits or paroxysms, Baby Girl B would gasp so loudly it sounded like a "whoop." Often, too, her coughing episodes ended in vomiting. Baby Girl B's mother took her to the doctor.

The pediatrician made a provisional diagnosis of pertussis, or whooping cough. It was confirmed when the small Gram-negative bacterium *Bordetella pertussis* was cultured from the back of Baby Girl B's nose. As is usually the case when infants contract pertussis, the illness was serious. The doctor immediately admitted Baby Girl B to a pediatric intensive care unit. Her coughing paroxysms were almost constant, and she was unable to eat or rest. During her worst coughing spells she couldn't get enough oxygen and occasionally stopped breathing for short periods. Subtle abnormalities also developed. Her blood contained an excess of lymphocytes, a particular type of white blood cell. Cells in her tissues became unusually sensitive to histamine—a chemical released at times of stress or injury that can cause many physiologic abnormalities. Excessive production of insulin interfered with normal regulation of the glucose in her blood.

In the intensive care unit, nurses frequently suctioned the sticky mucus that clogged Baby Girl B's nose and throat. She received supplemental oxygen and intravenous fluids and nutrition. Treatment with antibiotics prevented her from spreading the infection to others but did not hasten her recovery. The supportive therapy she received allowed her to survive a month-long ordeal of almost continuous coughing, however.

By the time of her discharge from the hospital, Baby Girl B had lost weight and could no longer roll over or lift her head as well as she used to. She still suffered occasional coughing paroxysms over the next few months, but eventually her recovery was complete. By her first birthday, Baby Girl B was chubby and playful and trying to take her first wobbly steps.

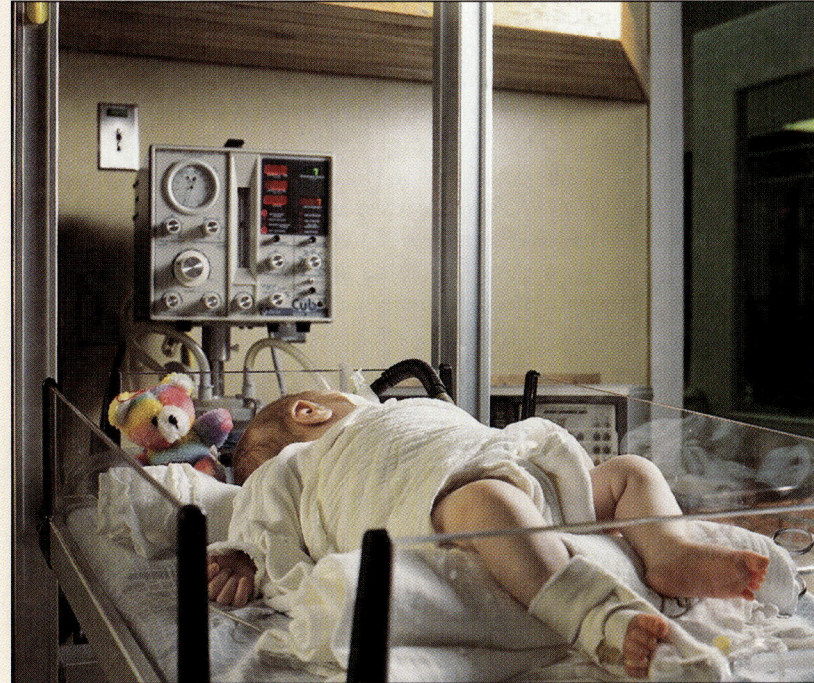

Infants with whooping cough frequently require supplemental oxygen in a pediatric intensive care unit.

**To understand how microbial pathogens
that cause human infection:**

- Survive between hosts

- Come in contact with the body

- Adhere to body surface tissues

- Gain access to deeper body tissues

- Avoid host defenses and multiply in the body

- Cause disease

- Leave to infect another host

SEVEN CHALLENGES THAT FACE PATHOGENIC MICROORGANISMS

In Chapter 14, you saw how Baby Girl A benefits from the hundreds of different species of microorganisms that colonize her body surfaces. Here, Baby Girl B suffers an acute life-threatening illness when a single bacterial species, *Bordetella pertussis*, infects her respiratory system. In this chapter we look at the relationship between humans and pathogens.

A pathogen must overcome certain challenges if it is to survive on or in a human host and cause disease. It must be highly adapted in order to

1. Maintain a reservoir, a place in which it can survive before and after infection

2. Leave its reservoir and enter the body of a human host

3. Adhere firmly to the surface of the host's body and thereby colonize it

4. Invade the body in order to enter cells or deeper tissues

5. Evade the body's elaborate defenses against microbial invaders

6. Multiply within the body, perhaps producing toxic products or stimulating host reactions that cause disease

7. Leave the body and return to the reservoir and/or enter a new host

Pathogenesis, a microorganism's ability to cause disease, depends upon its meeting all seven of these challenges. **Infection** refers to the growth of microorganisms

in the body (although the multiplication of bacteria from the normal flora is not usually considered an infection). Infection is one cause of the disequilibrium we call disease, though as you will see, infection does not *always* cause disease. The two terms—*infection* and *disease*—are often used interchangeably.

In the following sections we look at how *Bordetella pertussis* overcame each of these challenges in the process of causing Baby Girl B's illness. We also compare the adaptations made by *B. pertussis* to those made by other pathogens, many of which overcome the same challenges in different ways.

THE FIRST CHALLENGE: MAINTAINING RESERVOIRS OF INFECTION

A **disease reservoir** is a place in which a pathogenic microorganism is maintained between infections. If a pathogen's reservoirs are eliminated, it ceases to exist. The most common reservoirs for human pathogens are other humans, animals, and the environment.

Human Reservoirs

For many microorganisms, including those that cause pertussis, measles, gonorrhea, and the common cold, humans are their only reservoir. They are relatively fragile microorganisms that find the moisture, temperature, and pH of any other environment too hostile. They cannot survive for long outside the human body.

An individual, such as Baby Girl B, who is ill from an infection is obviously a reservoir. But apparently healthy people can also be **carriers**, reservoirs of infection. These apparently healthy individuals may be in the very earliest symptomless stages of their illness (**incubatory carriers**). People who are a reservoir for the HIV virus but have not yet developed any form of AIDS are incubatory carriers. A **chronic carrier** is a person who harbors a pathogen for an extended period of time (months to years) without becoming ill. People who recover from an illness may continue to harbor the pathogen within their bodies and become chronic carriers. The cook who became known as Typhoid Mary, for example, was a chronic carrier of typhoid. After she contracted the disease and recovered from it, the typhoid pathogen, *Salmonella typhi*, continued to multiply in her body and be excreted in her feces (Chapter 23). Other chronic carriers harbor a pathogen for variable periods of time and never appear to be or feel ill. Some people, for example, have *Streptococcus pyogenes*, which causes strep throat, growing in their throats for years without developing the disease.

Animal Reservoirs

Some pathogens that cause disease in humans have an animal reservoir. Like us, most animal reservoirs are warm-blooded creatures, so they provide an environment quite similar to that of the human body. Both domestic and wild animals are reservoirs for certain potential human pathogens. For example, cats, dogs, skunks, and bats are all reservoirs for rabies, a fatal nervous system infection (Chapter 25). Human infection occurs when a person comes into contact with the animal reservoir. In the case of rabies, contact usually means an animal bite. More often, pathogens pass from an animal reservoir to human beings when people consume or handle contaminated animals or animal products or when they are bitten by insects that have previously bitten an infected animal.

A human disease caused by a pathogen that maintains an animal reservoir is called a **zoonosis**. More than 150 zoonoses have been described, although fewer than half are of much importance clinically (**Table 15.1**). New zoonoses continue to be discovered. Lyme disease, for example, was only recognized as a distinct clinical entity in the late 1970s (Chapter 27). Wild mammals such as deer and mice are the primary reservoir. Another example surfaced in the spring of 1993 when a deadly outbreak of acute respiratory disease occurred in the southwestern United States. The disease was caused by a previously unknown strain (or species) of hantavirus. The new strain differed genetically from known strains and produced severe respiratory disease, whereas known strains primarily affect the kidneys. At least eight rodent species, but principally the long-tailed deer mouse, are the reservoir of the new strain (Chapter 22).

The existence of an animal reservoir can profoundly affect patterns of human disease. The multiplication of microorganisms in an animal reservoir can cause new disease-causing strains to emerge or make their control more difficult. Influenza and yellow fever illustrate these principles.

Influenza is a respiratory disease of humans, but closely related strains of the same virus infect birds, horses, and pigs. As these viral strains multiply in their different animal hosts, mutations occur and genetic variability develops. When viruses from these diverse strains happen to infect a single animal, genetic rearrangement can suddenly produce a "new" virus. If the newly created pathogen is transferred from its animal host to a human, the virus can spread rapidly from person to person.

Yellow fever will never be completely eradicated because of its large animal reservoir. Yellow fever is caused by a virus, and human infection results in serious—often life-threatening—damage to the liver. At one time, yellow fever ravaged North America, South America, and Africa. With the discovery that mosquitoes spread the disease Walter Reed and other clinician-scientists eradi-

cated yellow fever from Havana, Cuba, by controlling the mosquito population (**Figure 15.1**). It could not be eliminated in Panama, however, where, along with other tropical infections, it was slowing work on the Panama Canal. In Cuba, there was no significant animal reservoir, but in Panama, monkeys in the surrounding jungle created an immense reservoir, making eradication impossible. Instead of trying to eradicate yellow fever, public health officials sought to control its spread—a more modest but realistic goal and one they achieved.

Environmental Reservoirs

Some pathogens can survive for long periods in nonliving reservoirs, principally soil, water, and house dust. Because growth conditions in these inanimate reservoirs are quite unlike those in the human body, these pathogens must be adapted to survive in two different environments.

The soil is a reservoir for many pathogens, including *Clostridium tetani*, the bacterium that causes tetanus (Chapter 25). In the moist, warm environment of human tissues, vegetative cells of *C. tetani* multiply and produce the deadly toxin that causes agonizing muscle contractions. In the soil, however, where the environment is dry and subject to extremes of temperature, *C. tetani* survives because of its ability to produce endospores (Chapter 4). *C. tetani* endospores—like most endospores—remain viable in the soil for hundreds of years.

Many of the pathogens that have water as a reservoir infect the gastrointestinal tract when they are consumed in contaminated drinking water. One of these is *Vibrio cholerae*, which causes cholera, and another is *Salmonella typhi*, which causes typhoid fever. Both pathogens adjust quickly to the physical and nutritional conditions of the human intestine, surviving in spite of the highly acidic conditions in the stomach and the destructive effects of bile salts and digestive enzymes in the small intestine. When they reenter lakes and streams in the feces of an infected person, their metabolism readjusts to the considerably lower temperature and relatively scarce nutrients.

THE SECOND CHALLENGE: GAINING ACCESS TO A NEW HOST

Disease transmission takes place when a pathogen leaves a reservoir and *enters* the body of a host. Different pathogens are transmitted in different ways, but most have a preferred **portal of entry**, the anatomic site through which the pathogen enters the body of the host. *Bordetella pertussis*, for example, is inhaled into the respiratory system, so its portal of entry is the nose. The virus

Table 15.1 Some Important Zoonoses: Human Diseases with Animal Reservoirs

Disease	Microorganism	Animal Reservoir	Transmission from Reservoir	Chapter Reference
Bacterial				
Plague	*Yersinia pestis*	Rodents, including wild burrowing species and rats	Flea bite	27
Anthrax	*Bacillus anthracis*	Cattle, goats, horses, pigs, sheep	Inhaling spores from infected soil or animal products	27
Brucellosis	*Brucella* spp.	Cattle, goats, horses, pigs, sheep	Consuming dairy products from infected animals or direct contact with infected tissue	27
Salmonellosis	*Salmonella* spp.	Many, including poultry, turtles, rats	Eating contaminated food, drinking contaminated water	23
Tularemia	*Francisella tularensis*	Wild animals, especially rabbits	Insect bite or direct contact with infected tissue	27
Leptospirosis	*Leptospira interrogans*	Mammals, including wild species and domestic dogs and cats	Contact with infected urine or water contaminated by infected urine	24
Psittacosis	*Chlamydia psittaci*	Many kinds of birds	Inhaling the bacteria	22
Lyme disease	*Borrelia burgdorferi*	Wild mammals, especially deer and mice	Tick bite	27
Fungal				
Ringworm	*Epidermophyton* spp. *Microsporum* spp. *Trichophyton* spp.	Dogs and cats	Direct contact	26
Histoplasmosis	*Histoplasma capsulatum*	Many kinds of birds	Inhaling spores from feces	22
Protozoal				
Giardiasis	*Giardia lamblia*	Beaver, marmot, muskrat	Drinking water contaminated by feces	23
Toxoplasmosis	*Toxoplasma gondii*	Domestic cats, wild carnivores, birds, rodents	Eating undercooked meat from infected animals or contact with cat feces	27
Trypanosomiasis (African sleeping sickness)	*Trypanosoma* spp.	Wild game animals	Bite of tsetse fly	25
Viral				
Influenza	Influenza virus	Swine, birds, horses	Inhaling the virus	22
Yellow fever	Yellow fever virus	Monkeys	Mosquito bite	27
Rabies	Rabies virus	Most animals, especially carnivores	Animal bites	25
Viral encephalitis	Encephalitis virus spp.	Birds, horses	Mosquito bite	25
Helminthic				
Tapeworm	*Taenia* spp.	Cattle, pigs	Eating undercooked contaminated beef or pork	23
Trichinosis	*Trichinella spiralis*	Pigs	Eating undercooked contaminated pork	23

Critical Cues

When some pathogenic microorganisms encounter a human host, they pick up environmental cues that turn on their disease-causing capabilities. For some microbes the critical cue is a low concentration of available iron—an essential nutrient that is much scarcer inside the human body than elsewhere in the environment. When pathogenic members of the bacterial genus *Shigella*, for example, leave their watery reservoir and

enter cells of the intestinal tract, they sense the drop in iron concentration and begin to produce a cell-damaging toxin. For other microbes, the critical cue is body temperature. Intestinal pathogens of the genus *Yersinia*, for example, turn on disease-causing functions only at 37°C—the temperature inside the human body.

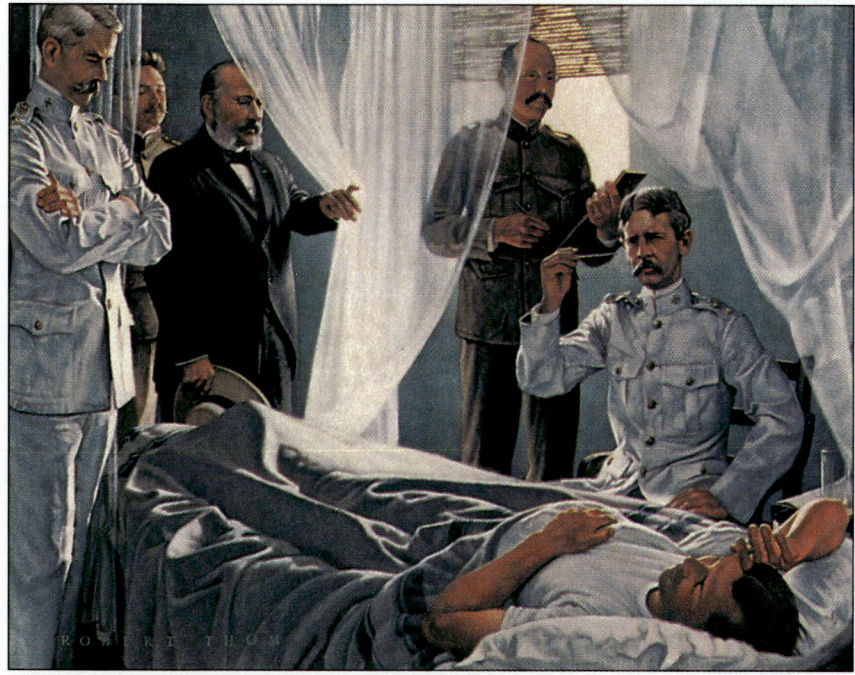

Figure 15.1 Walter Reed and his Yellow Fever Commission. Reed could eliminate yellow fever from Havana, Cuba, because the animal reservoir was limited. In Panama, however, the monkeys in the surrounding jungle provided an unlimited reservoir.

that causes yellow fever must be injected into the circulation by a mosquito bite, so its portal of entry is the bloodstream. Some pathogens enter the body through different portals of entry under different circumstances. The usual portal of entry for most sexually transmissible pathogens, for example, is the urethra in males and the vagina in females, although it may be the throat or the rectum if these are points of sexual contact (**Table 15.2**).

The likelihood of successful disease transmission is influenced by the number of pathogens that reach the portal of entry. If many infectious microorganisms gain access, infection is likely. The quantitative relationship between numbers of microorganisms transmitted experimentally to an animal host and the probability of infec-

tion can be measured. It is expressed as the ID_{50} (**infectious dose**)—the number of microorganisms that must enter the body to establish infection in 50 percent of test animals. A similar measurement, the LD_{50} (**lethal dose**), measures fatal infections—the number of microorganisms that must enter the body to cause death in 50 percent of test animals. Highly virulent organisms have a lower ID_{50} and a lower LD_{50} than do weakly virulent ones (**Figure 15.2**).

The most common portals of entry for disease-causing microorganisms are the anatomic surfaces, external and internal, of the body—the same sites normally colonized by microflora. These surfaces are the skin, conjunctivae, nasal cavity and nasopharynx, mouth, intestinal

Table 15.2 Portals of Entry for Some Microbial Pathogens

Portal of Entry	Disease	Pathogen	Chapter Reference
Skin	Abscesses (through the blood)	Many organisms, including *Staphylococcus aureus*	26
	Tetanus (through wounds)	*Clostridium tetani*	25
	Plague (through insect bites)	*Yersinia pestis*	27
	AIDS (through injections)	HIV virus	27
Nose (primarily respiratory pathogens)	Pertussis	*Bordetella pertussis*	22
	Influenza	Influenza virus	22
	Common cold	Many different viruses	22
	Pneumococcal pneumonia	*Streptococcus pneumoniae*	22
	Measles	Measles virus	26
	Diphtheria	*Corynebacterium diphtheriae*	22
	Smallpox	Variola virus	26
	Chickenpox and shingles	Varicella-zoster virus	26
	San Joaquin valley fever	*Coccidioides immitis*	22
Conjunctivae	Ophthalmia neonatorum (gonococcal eye infection of the newborn)	*Neisseria gonorrhoeae*	24
	Trachoma	*Chlamydia trachomatis*	26
Mouth (primarily gastrointestinal pathogens)	Cholera	*Vibrio cholerae*	23
	Typhoid	*Salmonella typhi*	23
	Hepatitis A	Hepatitis A virus	23
	Salmonellosis	*Salmonella* spp.	23
	Poliomyelitis	Poliovirus	25
	Giardiasis	*Giardia lamblia*	23
Urethra	Urinary tract infections (males and females)	Many organisms, especially *Escherichia coli* and other enterobacteria	24
	Gonorrhea (males)	*Neisseria gonorrhoeae*	24
	Syphilis (males)	*Treponema pallidum*	24
	AIDS (males)	HIV virus	27
Vagina	Vaginitis	Many organisms, including *Gardnerella vaginalis*	24
	Gonorrhea (females)	*Neisseria gonorrhoeae*	24
	Syphilis (females)	*Treponema pallidum*	24
	AIDS (females)	HIV virus	27
Placenta	Rubella (German measles)	Rubella virus	26
	Syphilis	*Treponema pallidum*	24
	AIDS	HIV virus	27

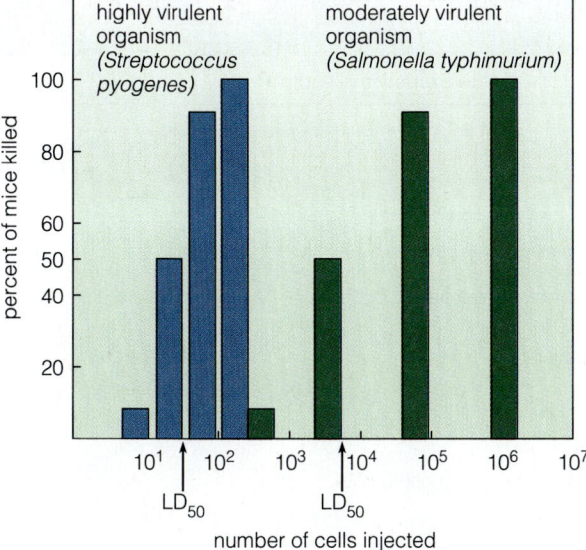

Figure 15.2 Lethal dose (LD). The LD$_{50}$ is the number of microorganisms that must enter the body in order to cause death from infection. Highly virulent organisms have a lower LD$_{50}$ than do less virulent ones—fewer microorganisms are required to cause a fatal infection.

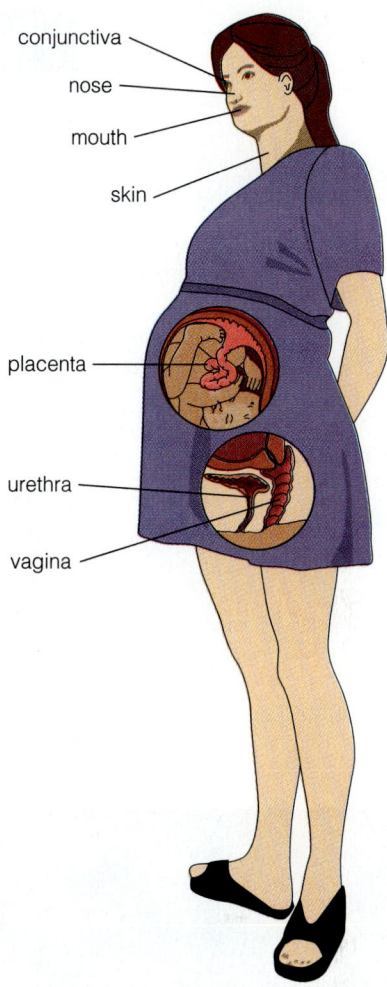

Figure 15.3 Portals of entry. Pathogens can enter the body in innumerable ways, but the most common portals of entry are the same entry sites normally colonized by microflora: skin, conjunctivae, nasal cavity and nasopharynx, mouth, vagina, and urethra; a pregnant woman may transmit pathogens to a fetus through the placenta if she is ill and has microorganisms in her bloodstream.

tract, vagina, and urethra (Chapter 14). Microorganisms may reach other portals of entry under unusual or abnormal circumstances. For example, microorganisms may be inhaled directly into the lung, or they may enter tissues below the skin through an open wound. If a pregnant woman is ill and has microorganisms in her bloodstream, the placenta may be a portal of entry into the body of the unborn baby (**Figure 15.3**).

Transmission of pathogenic microorganisms can occur in several ways. Some modes of transmission involve **contact** between human beings, which can be either direct or indirect. Other modes involve **vehicles**, inanimate objects such as food, water, eating utensils, and bedding that carry microorganisms from one person to another. Still other modes involve **vectors**, living transmitters, usually arthropods, that spread pathogens from person to person. Some pathogens can be transmitted more than one way. **Table 15.3** summarizes the modes of transmission.

Transmission by Respiratory Droplets

Pathogens are transmitted directly from one host to another in droplets of respiratory secretions that are expelled by coughing, sneezing, laughing, or talking (**Figure 15.4**). *Bordetella pertussis* is one of these. About a week before her first symptoms appeared, Baby Girl B inhaled invisible droplets of infected mucus that someone exhaled. She must have been reasonably close to the person

because the warm, moist mucus droplets last only a brief time. They can travel only about a meter before they dry and the bacteria die.

Whooping cough, like many other infectious diseases, is **communicable**, which means that it can be transmitted from one person to another. In fact, whooping cough has extremely high communicability. Over 80 percent of people exposed to *B. pertussis* become infected. Children can contract whooping cough from a person they meet once, briefly, in a public place. Baby Girl B, for example, came down with whooping cough even though no one in her family or her immediate environment had the disease.

More human diseases are transmitted by respiratory transmission than by any other method; it is an extreme-

Table 15.3 Modes of Transmission	
Mode of Transmission	Disease Examples (Chapter)
Respiratory Droplets	Pertussis, pneumonia (22); measles (26)
Fomites	
Facial tissues, household surfaces	Common cold (22)
Eating utensils	Typhoid (23)
Contaminated needles	HIV infection (27)
Direct Contact	Gonorrhea, herpesvirus infections, syphilis (24); AIDS (27)
Fecal-oral	Cholera, viral gastroenteritis, hepatitis A, giardiasis (23)
Vectors	
Mechanical	Cholera, typhoid (23)
Biological	Malaria, yellow fever (27)
Airborne	Tuberculosis, San Joaquin valley fever (22)
Parenteral	Tetanus (25); gas gangrene (26)
By injection	Hepatitis B (23); HIV infection (27)

Figure 15.4 More human diseases are transmitted by respiratory droplets than by any other method. Sneezing, coughing, and even laughing and talking communicate pathogens directly from one host to another.

ly efficient mode of transmission. One person exhaling infected mucus droplets can transmit a respiratory pathogen to every susceptible person in the immediate environment. This is especially true in crowded conditions or in a household where people live intimately together.

Transmission by Fomites

Some pathogens are hardy enough to remain infectious for brief periods on objects such as cups, towels, bedding, and handkerchiefs. Inanimate objects such as these that transmit disease are called **fomites**. The type of object that serves as a fomite for a pathogen depends on the pathogen's preferred portal of entry. Eating utensils, for example, are likely fomites for pathogens of the intestinal tract. Contaminated hypodermic needles transmit bloodborne pathogens like those that cause AIDS or hepatitis B.

Fomites are not a continuing reservoir for pathogens (as soil and water are) because they sustain live microorganisms for only a short time. As a result, they usually transmit disease between family members or other people who come into fairly close contact with one another.

The viruses that cause the common cold are usually transmitted by fomites. A cold-sufferer produces unusually large quantities of mucus, containing huge numbers of viruses. If the sufferer rubs her nose or eyes, her hands carry the virus particles to nearby fomites, such as facial tissues, tabletops, or eating utensils. When an uninfected person touches these fomites soon afterward and then touches his own nose or eyes, the cold virus completes its journey to a new host. Disinfecting fomites can decrease this type of transmission. One study, for example, showed that using facial tissues containing a virus-killing agent decreased the spread of colds in a household. The most effective way to decrease the spread of colds, however, is frequent hand washing, which prevents transferring the virus to and from fomites.

Transmission by Direct Body Contact

Some pathogens are spread by **direct contact** between the body of one person and another, such as by touching, kissing, or sexual intercourse. This type of transmission is also called **person-to-person** transmission, or sometimes **horizontal** transmission (to distinguish it from vertical transmission, discussed later).

Pathogens exchanged by direct contact are so fragile that they cannot survive outside a human for even a brief period. They must be deposited directly onto the body of a new host. Most contact this intimate occurs during sexual relations, and certain pathogens are transmitted primarily or exclusively by sexual means. Gonorrhea is a **sexually transmissible disease (STD)**. It is discussed in Chapter 24, along with others. STDs spread when the mucous membranes of an infected person come into direct contact with the mucous membranes of an uninfected person. Sexual transmission is an extremely effective way of transmitting fragile pathogens, but new cases spread through a population only if an infected person has sexual contact with multiple partners.

Some pathogens transmitted by direct body contact are spread by nonsexual means. Herpes simplex virus type 1, which usually infects the mouth or lips, is most commonly spread by kissing or exchange of saliva. Infants and toddlers who mouth each other and share toys that they put into their mouths commonly exchange saliva. As a result, most cases of primary herpes simplex type 1 infections are seen in pediatric patients. The virus can also be transmitted from one place to another on an individual's body. Children with oral herpes lesions who suck their fingers, for example, may develop the typical blistering sores of herpes on their hands, a condition known as **herpetic whitlow**.

Another nonsexual form of disease transmission through direct contact is **vertical transmission**, transmission of pathogens from mother to infant. When it occurs across the placenta before birth, it is called **prenatal transmission**. When it occurs during passage through the birth canal or during the extremely close contact between mother and infant in the first few days of life, it is called **perinatal transmission**. Many sexually transmissible infections can also be passed vertically from mother to infant; these include syphilis, gonorrhea, hepatitis B, and HIV infection.

Transmission by the Fecal-Oral Route

In the **fecal-oral route**, pathogens are transmitted from infected feces to the mouth of a new host. This route overlaps with other modes of transmission. Fecal-oral transmission can be by direct hand-to-hand or mouth-to-mouth contact, by vehicles such as water, food, and fomites, or by vectors. Most of the gastrointestinal pathogens that cause diarrhea are transmitted this way.

One pathogen that causes diarrheal disease is *Vibrio cholerae*, the bacterium responsible for the life-threatening infection cholera. The transmission of *V. cholerae* is a case study in the efficiency of the fecal-oral route. Cholera spreads rapidly when a public water supply is contaminated and large numbers of uninfected people drink the pathogen. Most major epidemics of cholera occur in this way (Chapter 20). But cholera can also be transmitted within a household if only one family member is infected. People ill with cholera pass several liters of watery feces each day, and each drop contains millions of *V. cholerae* organisms. Traces of feces are certain to get on the hands of whoever is caring for the sick; and unless hand washing is scrupulous, the *V. cholerae* organisms will eventually be transmitted to the caretaker's mouth. If the family nurse also prepares the family's food, these organisms are likely to be consumed in meals, potentially infecting everyone in the household. Fomites may also play a role. Anyone who shares or touches fecally contaminated bedding, for example, is also at risk of infection. Flies that land on feces or other infected material can act as vectors that carry the organisms to the food or bodies of uninfected people—perhaps the family next door.

Transmission by Arthropod Vectors

Most vectors, as you learned in Chapter 12, are arthropods, such as flies, mosquitoes, ticks, fleas, and lice. Some arthropods are **mechanical vectors**. They pick up pathogens on their bodies and carry them from one place to another, acting as living fomites. The flies that transmitted *Vibrio cholerae* in the preceding example were mechanical vectors of infection. Mechanical vectors are not essential to the transmission of a pathogen. **Biological vectors**, however, are an *essential* link in the chain of transmission for certain diseases because they play a role in the life cycle of the microorganism. The microorganism develops and multiplies part of the time in the body of an arthropod host and part of the time in the body of a human host (**Table 15.4**). Biological vectors usually transmit infection by biting, introducing the pathogen through

Table 15.4 Biological Vectors of Infection

Disease	Pathogen	Vector	Chapter Reference
Plague	*Yersinia pestis*	Flea	27
Malaria	*Plasmodium* spp.	Anopheles mosquito	27
Lyme disease	*Borrelia burgdorferi*	Tick	27
Yellow fever	Yellow fever virus	Aedes mosquito	27
Rocky Mountain spotted fever	*Rickettsia rickettsii*	Tick	27
Viral encephalitis	Various arboviruses	Mosquitoes	25
Epidemic typhus	*Rickettsia prowazekii*	Body louse	27
African trypanosomiasis (African sleeping sickness)	*Trypanosoma brucei*	Tsetse fly	25
American trypanosomiasis (Chagas' disease)	*Trypanosoma cruzi*	Reduviid bug	27
Filariasis (elephantiasis)	*Wuchereria bancrofti*	Mosquitoes	27

the host's skin. The flea that transmits plague is a biological vector, as are the mosquito that transmits malaria and the tick that transmits Lyme disease (**Figure 15.5**).

Airborne Transmission

Airborne transmission occurs when microorganisms that can survive for prolonged periods of time in air are inhaled into the respiratory system. Unlike microorganisms that are transmitted by means of respiratory droplets, these pathogens are hardy enough to withstand prolonged drying. Airborne pathogens, therefore, can be transmitted across long distances (greater than a meter) and between people who have not had close contact. Moreover, airborne pathogens remain viable in dust and reenter the air months or years later to cause infection.

Microorganisms transmitted by an airborne route include *Mycobacterium tuberculosis*, the bacterium that causes tuberculosis (Chapter 22). This unusually hardy bacterium can survive desiccation for months, infecting new human hosts from dried respiratory secretions. Some fungal pathogens produce spores that can remain viable in the soil for years and become airborne whenever their reservoir is disturbed. *Coccidioides immitis*, which causes San Joaquin valley fever (Chapter 22), is one example. In the 1970s, a major dust storm in California infected people who lived hundreds of miles away from the original site of the *C. immitis* spores.

Parenteral Transmission

In **parenteral transmission**, pathogenic microorganisms are deposited directly into blood vessels or tissues below the skin or into mucous membranes. The viruses that cause hepatitis B and AIDS, the protozoan that causes malaria, and the helminth that causes elephantiasis are a few examples of pathogens that can be transmitted by a parenteral route.

Parenteral transmission occurs when a biological arthropod vector introduces pathogens during a skin-penetrating bite or when breaks in the skin or mucous membranes provide microorganisms with access to deeper tissues. Parenteral transmission can also occur from deliberate penetration of the skin with a hypodermic needle. Both blood transfusions and sharing of hypodermic needles by intravenous drug users provide opportunities for parenteral transmission of the HIV and hepatitis viruses. Although blood products for transfusions are screened for certain pathogens, absolute safety cannot be ensured. (Intravenous fluids, manufactured under sterile conditions, do not contain pathogenic microorganisms.)

A special type of parenteral transmission occurs when people sustain significant wounds. Any small break in the skin can lead to a localized infection caused by normal skin flora, such as staphylococci. But wounds that are deep and surrounded by dead tissue create an

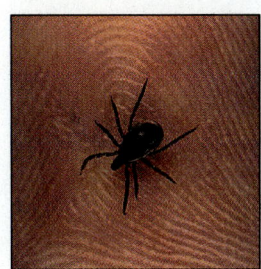

Figure 15.5 Transmission by arthropod vectors. Lyme disease is spread by an arthropod vector called the deer tick. A deer tick infected by *Borrelia burgdorferi* bites a deer, thereby infecting it. When another deer tick bites this same deer, it acquires the infection. If the newly infected tick comes in contact with a human and bites it, it spreads the infection to that person.

environment in which certain anaerobic pathogens thrive. These include the spore-forming organisms *Clostridium tetani*, which causes tetanus, and *C. perfringens*, which causes a rapidly spreading and often fatal infection called gas gangrene (Chapter 26).

THE THIRD CHALLENGE: MAINTAINING A FOOTHOLD— ADHERING TO BODY SURFACES

All microorganisms, whether normal flora or pathogens, that come to rest on internal or external surfaces of the human body are confronted by defenses that tend to dislodge them. The structural, mechanical, and biochemical surface defenses are the body's first line of defense against microbial invasion (Chapter 14). To remain on a body surface and colonize it, pathogens, like normal flora, attach to specific types of target cells by means of adhesins (see Figure 14.6). Thus, pathogens differ from normal flora not in attachment, which is necessary for both, but in the disease-causing steps that occur *after* attachment.

The *Bordetella pertussis* cells that entered Baby Girl B's body attached to cilia on epithelial cells lining her respiratory tract by means of the adhesin **filamentous hemagglutinin**. Unlike many adhesins, filamentous hemagglu-

tinin is not part of the bacterial cell envelope. It is a rod-shaped molecule that is secreted into the extracellular environment. In other words, it is a free molecule. Filamentous hemagglutinin must therefore make two different attachments—one to a molecule on the bacterial cell surface and the other to a carbohydrate receptor on a human cilium (**Figure 15.6**). This firm, bridgelike attachment withstands the mucociliary defenses of the respiratory tract. As a result, *B. pertussis* was able to colonize the baby's respiratory epithelium and begin to multiply there.

Filamentous hemagglutinin is a relatively unusual type of adhesin. Most adhesins, including those on pathogenic strains of *Escherichia coli*, other pathogenic Enterobacteriaceae, and species of the genus *Neisseria*, are molecular components of the microorganism's pili. The presence of pili is essential for the virulence of these organisms. Strains without pili cannot adhere to human tissue and therefore cannot colonize a host and initiate infection. Strains that lose their ability to produce pili become avirulent (harmless).

Some pathogenic bacteria produce different adhesins at different times. Having a repertoire of adhesins helps a pathogen in several ways. Changing adhesins helps a pathogen evade the body's immune defenses (discussed in the Fifth Challenge) and also allows the pathogen to attack more than one cell type. *Neisseria gonorrhoeae*, for example, can produce specific adhesins that bind to genital, rectal, pharyngeal, or conjunctival surfaces. Such versatility makes *N. gonorrhoeae* a highly virulent pathogen (Chapter 24).

THE FOURTH CHALLENGE: GOING DEEPER—INVADING THE BODY'S INTERIOR

Normal flora live on the inner and outer surfaces of the human body without causing disease. In contrast, most pathogens are **invasive**, meaning that they penetrate the

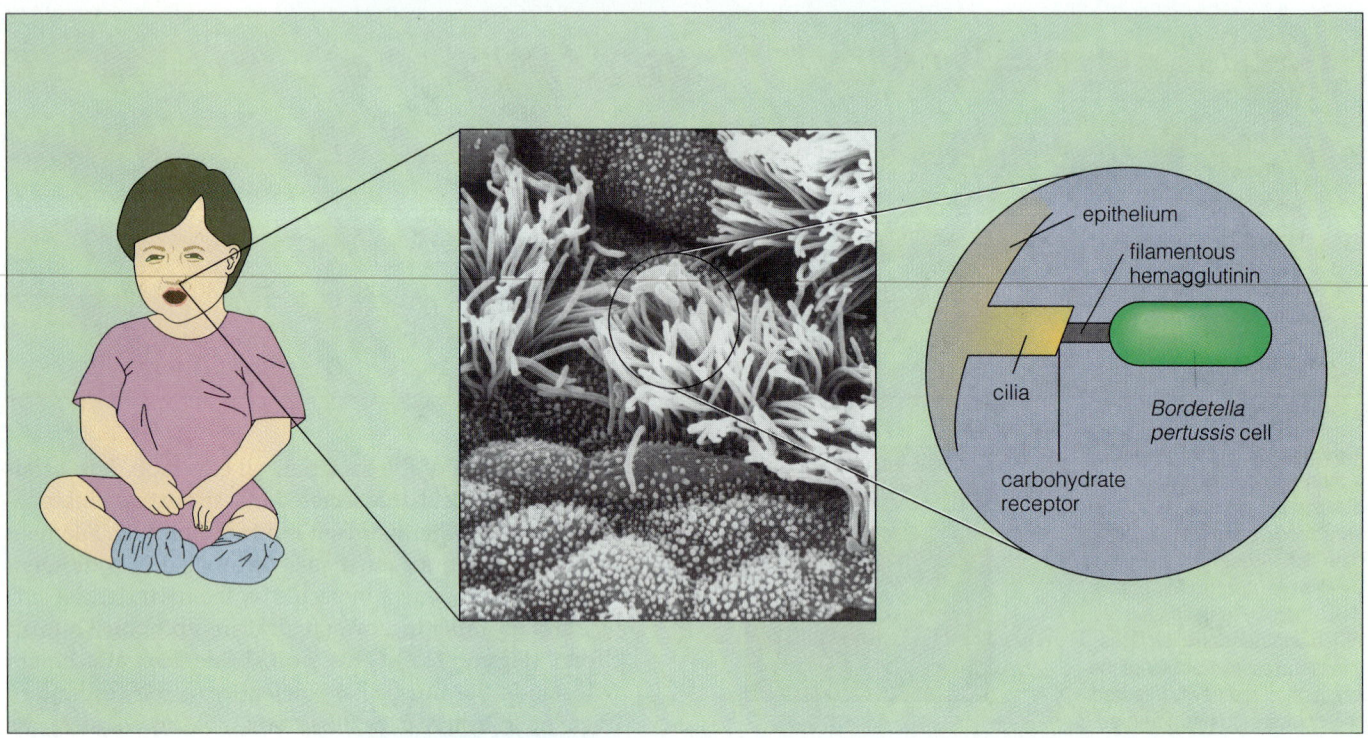

Figure 15.6 Maintaining a foothold. After *Bordetella pertussis* is inhaled, it must adhere to the respiratory epithelium if it is to colonize the site. It succeeds in adhering by means of filamentous hemagglutinin, which is secreted into the extracellular space by *B. pertussis*. It makes two connections—to a carbohydrate receptor on the cilia and to the surface of the pathogen—in effect, building a bridge. This firm attachment resists the mucociliary defenses.

body's surface to enter cells or deeper tissues (**Figure 15.7**). There are exceptions, however. Some pathogens, such as *Bordetella pertussis*, are **noninvasive**, meaning they cause disease without invading body surfaces or entering human cells. The bacteria that caused Baby Girl B's life-threatening illness remained on the surface of her respiratory tract. Another noninvasive pathogen, *Streptococcus pneumoniae*, can cause a life-threatening pneumonia.

Invasiveness is an adaptation for survival. The ability to penetrate human cells and tissues allows a pathogen both to escape certain host defenses and to gain access to a nutrient-rich environment that is free of competing microorganisms. Invasive pathogens that enter cells to live and multiply inside them are called **intracellular pathogens**. They pass through the plasma membrane and take up residence within the cytoplasm of the host cell. Other invasive pathogens pass through epithelial cells on their way to deeper tissues, such as the blood or lymphatic circulations.

Most invasive bacteria exploit the mechanisms that bring nutrients or other essential molecules into the host cell. For example, some gain entry by adhering to surface receptors that fold into the cell during endocytosis (Chapter 4). The host cell is tricked into taking in a harmful bacterium. Entering bacteria may become trapped within a vacuole formed by the infolding of the host cell membrane that occurs during endocytosis, but many produce enzymes that allow them to lyse this vacuole and escape.

Though we have a general understanding of how bacteria enter human cells, it was not until recently that we began to understand adaptations for invasiveness at a molecular level. We now know, for example, that the bacterium that causes gonorrhea triggers its uptake into epithelial cells by means of a bacterial surface protein called Protein I. Strains that lack this protein are noninvasive. In one research project, a single gene was transferred from an invasive intestinal pathogen into a noninvasive strain of *Escherichia coli*. The gene enabled the *E. coli* to invade host cells.

Chapter 13 described the three ways viruses enter human cells. Viruses with envelopes fuse with the cell's plasma membrane or enter through endocytosis. Most naked viruses adsorb to the cell's surface (see Figure 13.11).

We know much more about bacterial and viral pathogenesis than we do about eucaryotic pathogenesis. Eucaryotic pathogens—from fungi to protozoa to worms (algae are not clinically significant)—are incredibly varied, having much less in common with one another than

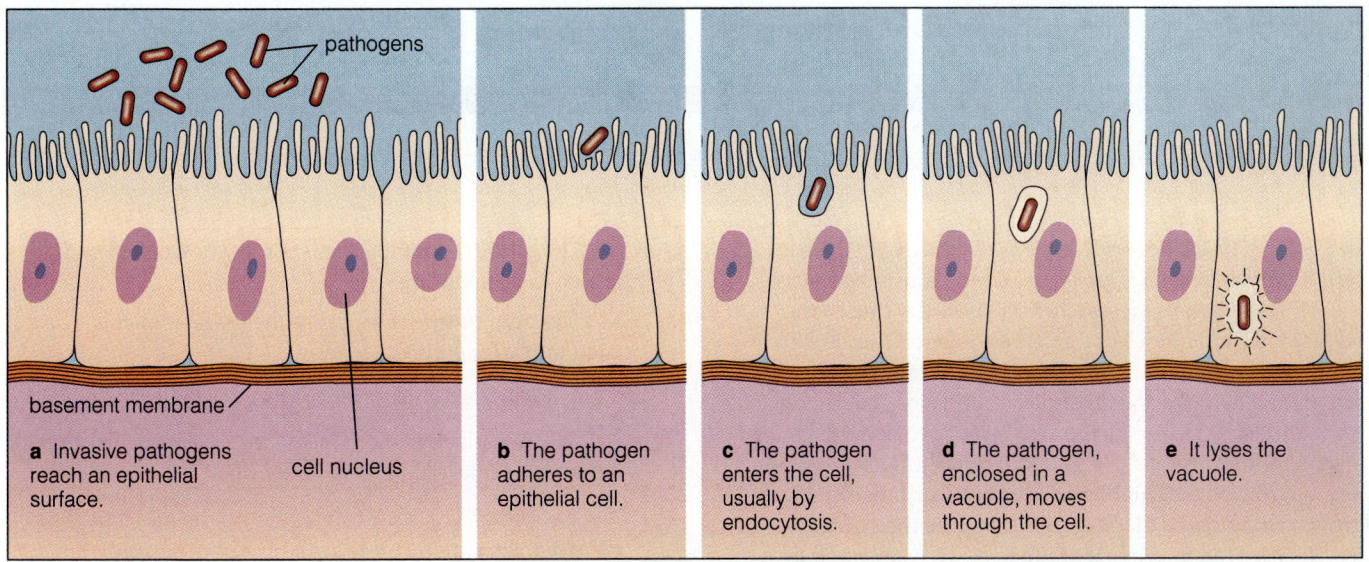

a Invasive pathogens reach an epithelial surface.

pathogens

basement membrane

cell nucleus

b The pathogen adheres to an epithelial cell.

c The pathogen enters the cell, usually by endocytosis.

d The pathogen, enclosed in a vacuole, moves through the cell.

e It lyses the vacuole.

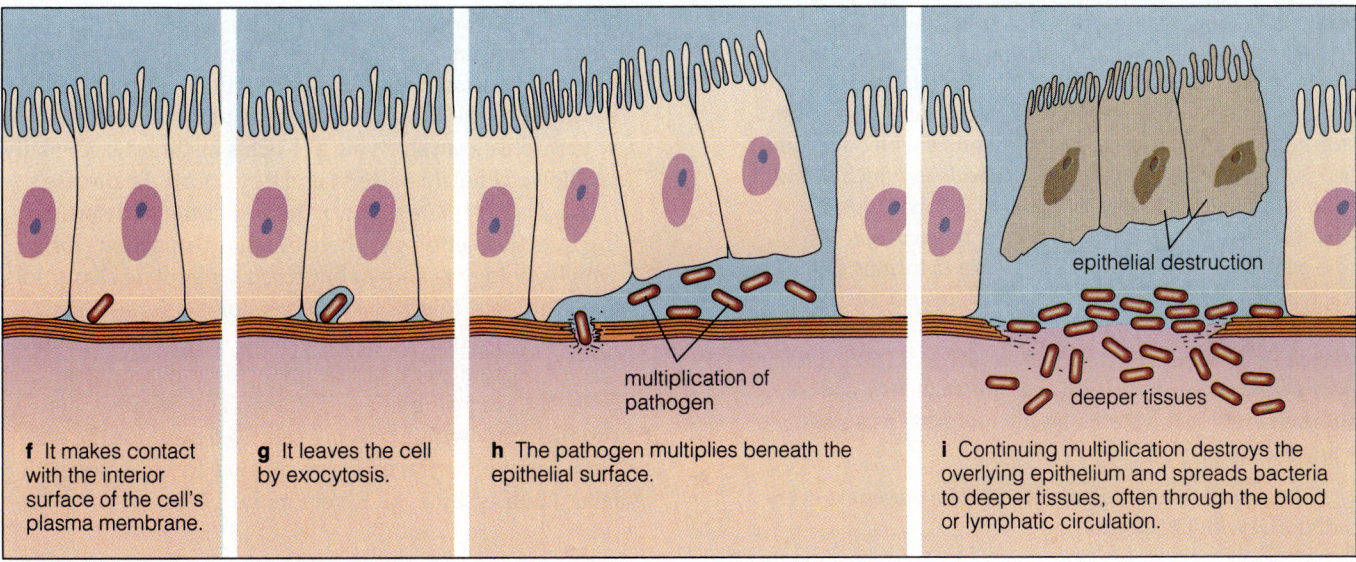

f It makes contact with the interior surface of the cell's plasma membrane.

g It leaves the cell by exocytosis.

h The pathogen multiplies beneath the epithelial surface.

multiplication of pathogen

epithelial destruction

deeper tissues

i Continuing multiplication destroys the overlying epithelium and spreads bacteria to deeper tissues, often through the blood or lymphatic circulation.

Figure 15.7 Invading the body's interior. Most pathogens are invasive—they penetrate the body's epithelial surface and enter deeper tissues, such as blood, lymph, liver, or lung.

do bacteria with other bacteria or viruses with other viruses. Most eucaryotic pathogens do not invade cells, but some do; for example, the protozoa that cause malaria (*Plasmodium* spp.) invade red blood cells.

To give you an idea of the diversity in eucaryotic pathogenesis, let's look at some random examples. The fungi that cause ringworm (members of the genus *Dermatophyte*) destroy with enzymes the keratin that makes the epidermis inhospitable to pathogens and live off skin cells (Chapter 26). The fungi that cause respiratory infections (*Blastomyces dermatitidis* and *Histoplasma capsulatum*)

cause damage through hypersensitivity, an immune system malfunction (Chapters 18 and 22). The protozoan that causes amoebic dysentery (*Entamoeba histolytica*) directly attacks and kills the cells that line the colon (Chapter 23). The helminth that causes hookworm (*Necator americanus*) attacks the intestinal lining and sucks blood from the host, causing anemia (Chapter 23). Finally, the helminth that causes schistosomiases (*Schistosoma* spp.) enters the bloodstream (but not blood cells) and causes extensive tissue damage mainly through hypersensitivity (Chapters 18 and 27).

One Microbe, One Disease

In the early 1880s, Robert Koch was working with the bacterium *Mycobacterium tuberculosis*. The organism was of great interest because researchers suspected it caused the widespread and often lethal infection tuberculosis. Koch made two important discoveries. He found a way of staining human tissue for microscopic examination that showed *M. tuberculosis* cells as thin blue rods on a brown background of human cells. He also found that *M. tuberculosis*—a slow-growing and highly fastidious bacterium—would grow on coagulated blood serum. Based on these discoveries, Koch set out to prove that tuberculosis was caused by *M. tuberculosis*. In the 1880s, no connection between the two had been proved. For that matter, there was no proof that a particular microorganism caused a particular disease.

Koch began by examining tuberculosis patients for the presence of *M. tuberculosis* cells. He found the bacterium in every patient—blue rods against brown tissue. Then Koch cultured the tuberculosis cells on coagulated blood serum and isolated pure cultures of *M. tuberculosis*, which he injected into guinea pigs. They succumbed to tuberculosis. Unequivocally, *M. tuberculosis* caused tuberculosis.

Koch's work with *M. tuberculosis* was a turning point in the science of microbiology. He provided absolute proof of the microbial **etiology**, or cause, of an important infectious disease of humans. Moreover, he enunciated a principle that is valuable today. Fulfilling **Koch's postulates** provides absolute proof that a particular microorganism causes a particular disease.

1. The causative microorganism must be present in every individual with the disease.

2. The causative microorganism must be isolated and grown in pure culture.

3. The pure culture must cause the disease when inoculated into an experimental animal.

Most modern textbooks add a fourth postulate.

4. The causative microorganism must be reisolated from the experimental animal and reidentified in pure culture.

Of course, Koch's postulates cannot be met if there is no way to culture the pathogen or if it infects only humans. Koch himself faced this dilemma later in his career when he studied cholera. He discovered that a bacterium, *Vibrio cholerae*, was present in the intestines of all patients he examined, and he was able to culture the organism. But he could not find an experimental animal susceptible to the disease. The third postulate was ultimately fulfilled when a physician studying at Koch's institute accidentally swallowed cholera bacteria and developed the disease.

Koch's postulates are not the sole route to determining infectious etiology. There are many infections for which etiology is known but Koch's postulates have not been fulfilled. *Treponema pallidum*, for example, is the cause of syphilis; but because it has never been cultured, Koch's postulates cannot be met. No viral pathogens can be tested according to Koch's postulates because they reproduce only within a living cell. To determine the etiologic agent of a viral infection, we use Rivers's postulates, codified by T. M. Rivers in the 1930s.

THE FIFTH CHALLENGE: ESTABLISHING INFECTION— EVADING HOST DEFENSES

So far we have discussed the ways pathogens reach the part of the human body where they are best adapted to survive. The real success of a pathogen, however, is measured by how well it can multiply after reaching its preferred site. To multiply—and therefore establish an infection—it must evade the body's defenses against infection. We will discuss the defenses in Chapters 16 and 17. Here we look at the pathogen's arsenal against those defenses.

Protection against Phagocytosis

Certain types of white blood cells called **phagocytes** consume and destroy pathogenic microorganisms by a process called **phagocytosis** (see Chapters 4 and 16). Only intracellular pathogens are protected from phagocytosis. Consequently, pathogens that live on body surfaces or in deep tissues have evolved intricate adaptations, such as capsules and surface proteins, to evade phagocytosis. Other pathogens evade destruction by taking up residence within the phagocyte.

Extracellular Capsules. Before a phagocyte can phagocytize (consume and destroy) a bacterium, it must

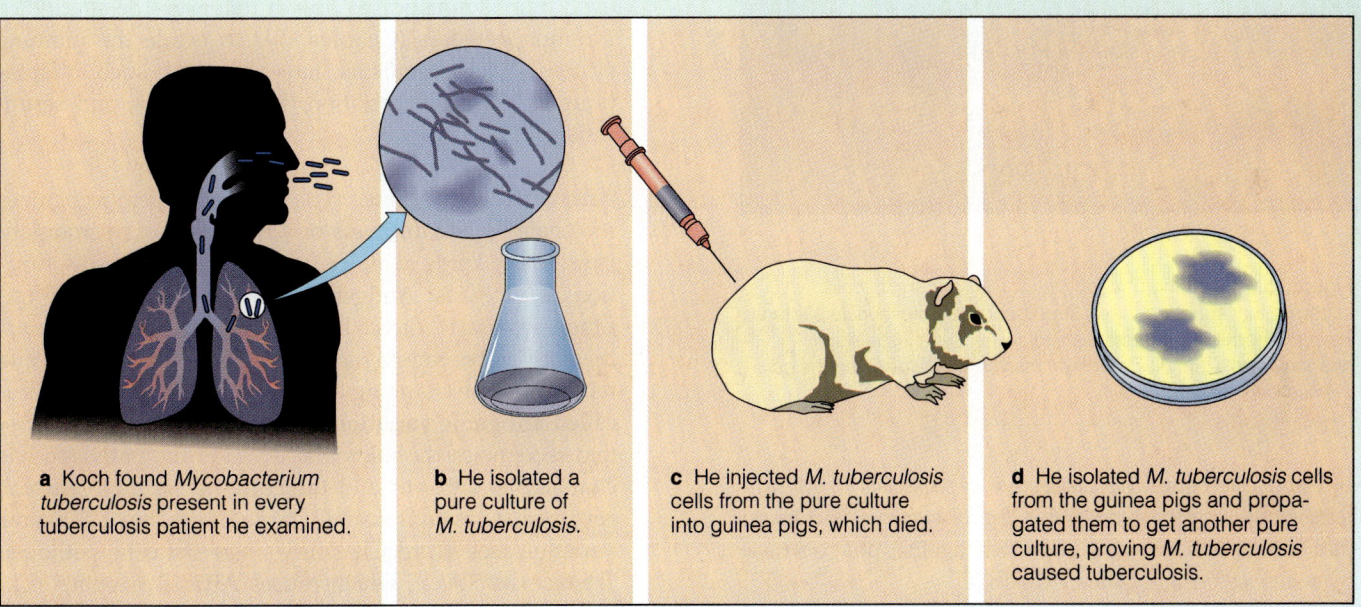

a Koch found *Mycobacterium tuberculosis* present in every tuberculosis patient he examined.

b He isolated a pure culture of *M. tuberculosis*.

c He injected *M. tuberculosis* cells from the pure culture into guinea pigs, which died.

d He isolated *M. tuberculosis* cells from the guinea pigs and propagated them to get another pure culture, proving *M. tuberculosis* caused tuberculosis.

Koch's postulates. If fulfilled, Koch's postulates provide absolute proof of the etiology, or cause, of an infection.

1. The viral agent must be found either in the host's body fluids at the time of the disease or in the infected cells.

2. The viral agent obtained from the infected host must produce the disease in a healthy animal or plant or must induce production of antibodies (proteins produced by the host in response to infection).

3. Viral agents from the newly infected animal or plant must in turn transmit the disease to another host.

establish direct contact with its prey (Chapter 4). But a slippery mucoid capsule prevents such contact (**Figure 15.8**). *Bordetella pertussis* protects itself by encapsulation. The phagocytes that normally defend Baby Girl B's airways failed to establish direct contact because of the protective capsule.

For many pathogens, producing an extracellular capsule is essential to their pathogenicity (disease-causing ability). For example, *Streptococcus pneumoniae* (Chapter 22) is virulent *only* if it produces a capsule, and strains that produce the thickest and most highly protective capsules are the most virulent. Almost all the bacteria and fungi that infect the central nervous system, including *Neisseria meningitidis*, *Haemophilus influenzae*, *Streptococcus pneumoniae*, and *Cryptococcus neoformans*, must produce a capsule in order to be virulent.

Surface Proteins. Some pathogens defend themselves against phagocytosis by means of specialized surface proteins. Like capsules, these antiphagocytic proteins interfere with the cell-to-cell contact that is essential for phagocytosis. *Streptococcus pyogenes*, the Gram-positive bacterium that causes strep throat, produces a surface protein called **M protein**. M protein appears as hairlike projections on the surface of the streptococcal cell wall (**Figure 15.9**). Pathogenic streptococci can be readily phagocytized only if antibodies (proteins produced by the host in response to infection) bind to the *S. pyogenes*

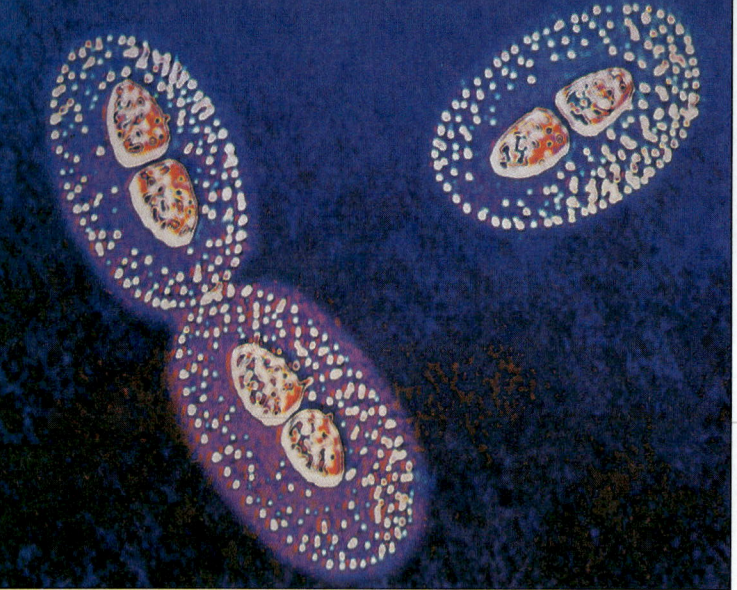

Figure 15.8 Evading phagocytosis. The slippery mucoid capsules that surround these *Streptococcus pneumoniae* bacteria keep phagocytes from establishing contact, the first step in phagocytosis. Capsules are often many times larger than the cells themselves, as these are.

cell and overcome the protective capabilities of M protein. *Neisseria gonorrhoeae*, the Gram-negative bacterium that causes gonorrhea, is protected against phagocytosis by a surface protein called Protein II.

Surviving Phagocytosis. A few pathogens avoid destruction by being readily consumed and then existing inside phagocytes. The environment within a phagocyte is very hostile, however, because phagocytes destroy most microorganisms they consume by producing lethal enzymes. Only pathogens that possess special adaptations survive; these adaptations vary. *Mycobacterium tuberculosis*, for example, survives because the destructive enzymes that kill most phagocytized microbes never get through the *M. tuberculosis* waxy cell wall (Chapters 4 and 22). *M. tuberculosis* must take up residence within

normally lethal phagocytic cells in order to be distributed throughout the body and establish a successful infection.

Adaptations to Evade the Immune System

Microorganisms that survive phagocytes and other non-specific body defenses (Chapter 16) are confronted by an even more sophisticated line of defense—the specific or immune defenses (Chapter 17). To evade the immune system, microorganisms have evolved such adaptations as antigenic variation, IgA proteases, and serum resistance.

Antigenic Variation. The immune system recognizes molecules called **antigens** on the surface of microorganisms. Many kinds of molecules, including adhesins, function as antigens. Antigenic recognition allows the immune system to direct highly specific defenses against the antigen-bearing invaders. But some pathogens evade recognition by changing their surface antigens, a process called **antigenic variation**. By the time the immune system recognizes the microbial invader, the pathogen has changed its antigens and therefore its identity. *Neisseria gonorrhoeae* has an unusually large repertoire of antigenic variation (see Third Challenge). The parasitic pathogen *Trypanosoma brucei*, which causes African sleeping sickness, also evades host defenses by antigenic variation (Chapter 25).

IgA Proteases. Some pathogens attack antibodies directly with antibody-destroying enzymes. The **IgA proteases**, for example, destroy the IgA class of antibodies (Chapter 17). IgA is present on epithelial surfaces and normally prevents infection by interfering with the microorganism's ability to adhere to a host cell. *Neisseria gonorrhoeae* uses IgA proteases, as do *Neisseria meningitidis* and most other pathogens that attack the central nervous system.

Figure 15.9 Evading phagocytosis. Like capsules, surface proteins interfere with the direct contact that must be established before a phagocyte can consume an invader. *Streptococcus pyogenes* produces a surface protein called M protein, hairlike projections on the surface of the cell.

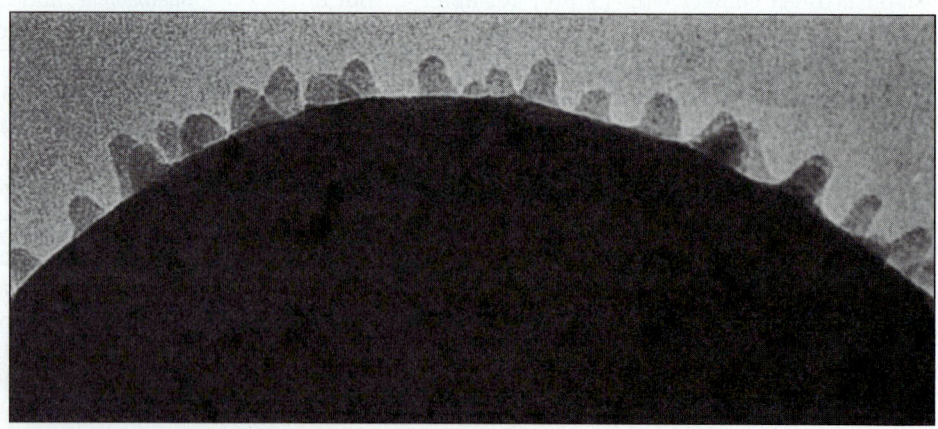

No Harm, No Foul

Pathogens face a dilemma. To survive in their human host, they inevitably cause some damage; but if the damage is too great the host will die, leaving them homeless. Ideally, a pathogen should modulate its activities so that it can grow, while damaging the host only slightly. Microbiologist John Mekalanos calls this survival strategy "no harm, no foul." Consider how this mechanism works with toxin-producing pathogens that kill host cells in order to obtain iron. When iron is scarce, the pathogen produces toxin, host cells die and release iron, and the local iron concentration rises, which signals the microbe to decrease toxin production. This feedback loop assures that the pathogen gets enough iron to continue to grow, but host damage is minimized.

Serum Resistance. Some microorganisms use a biochemical weapon called **serum resistance**. Serum resistance acts against the **complement system**, a group of host proteins with many defensive functions (Chapter 16). Complement proteins coalesce, forming a complex on the surface of a Gram-negative bacterium that makes holes, causing the bacterium to lyse. Serum resistance refers to features on the bacterial surface that interfere with the formation of the destructive protein complex. Serum resistance gets its name from the ability to confer resistance to destruction by human serum, which contains complement. Serum resistance can be a powerful disease-producing adaptation. For example, most strains of *Neisseria gonorrhoeae* that spread throughout the body and cause life-threatening disease manifest serum resistance, while strains that are localized in the genital tract and cause less serious disease usually lack serum resistance.

Adaptations to Obtain Iron

Iron is a relatively scarce but essential nutrient. It is a component of many biomolecules, including cytochromes, which are very important to both human and microbial metabolism (Chapter 5). Very little iron is available to microorganisms that grow within the human body because most extracellular iron in humans is tightly bound to iron-transport proteins such as **transferrin**, **lactoferrin**, and **ferritin**. As a result, microorganisms that infect humans are engaged in a constant struggle for iron.

Some pathogens obtain iron by producing their own iron-binding proteins called **siderophores**. Siderophores bind iron more tightly than human binding proteins, thereby making iron available to the pathogen. *Neisseria meningitidis*, a pathogen that causes meningitis, uses a different strategy. It produces receptors on its cell surface that bind to the human iron-transport proteins transferrin and lactoferrin. Then the bacterium absorbs these proteins along with the iron atoms they contain (Chapter 25).

Some interesting research has been done on how human iron metabolism affects our susceptibility to disease. Some studies, for example, suggest that giving infants supplementary iron may make them more prone to infection. However, iron is relatively scarce in the infant diet and iron deficiency in infancy can cause not only anemia but abnormal development as well. Many physicians, therefore, recommend iron supplementation if there is any indication of deficiency. Other clinical studies suggest that the excess free iron present in the blood of patients with sickle cell anemia contributes to their unusual vulnerability to infection.

THE SIXTH CHALLENGE: MULTIPLYING IN HOST TISSUES— ADAPTATIONS THAT CAUSE DISEASE

In this section we discuss the "how" of bacterial disease causation—the mechanisms of which bacteria damage tissue and cause human illness. We will discuss the two most common forms of bacterial pathogenesis—production of **toxins**, poisonous products that harm human cells and tissues, and damage caused by stimulation of the body's defenses. Certain eucaryotic microorganisms also produce toxins that are highly poisonous to human beings. Toxic mushrooms, ergot (a substance produced by a fungus that infects grain), and the algal toxins that cause red tides were discussed in Chapter 12. (Protozoa do not produce toxins.) The systemic infections caused by most fungi and helminths result from an exaggerated immune response called **hypersensitivity**, a form of microbial pathogenesis discussed in Chapter 18. Viral pathogenesis, which is quite different from bacterial and eucaryotic pathogenesis, is discussed later in the chapter.

Toxic Products

The two major types of bacterial toxins—exotoxins and endotoxins—are biochemically different, and each interacts with host cells in many different ways. Bacterial

Table 15.5 Exotoxins

Exotoxin	Bacterium	Disease	Mechanism	Clinical Effect
Cholera (enterotoxin)	*Vibrio cholerae*	Cholera	Activates adenyl cyclase, which disrupts cellular regulation by increasing concentration of cyclic AMP	Stimulates intestinal epithelium to secrete fluids, causing profuse diarrhea
Heat-stable *E. coli* (enterotoxin)	*Escherichia coli*	Traveler's diarrhea	Same as cholera	Same as cholera
Tetanus (neurotoxin)	*Clostridium tetani*	Tetanus	Binds to nerve cells, blocking inhibitory messages to muscles	Rigid contraction of skeletal muscles throughout body
Botulinum (neurotoxin)	*Clostridium botulinum*	Botulism	Blocks release of acetyl choline from nerve endings that supply muscles	Flaccid (limp) muscle paralysis
Diphtheria (cytotoxin)	*Corynebacterium diphtheriae*	Diphtheria	Interferes with protein synthesis by blocking chain elongation	Kills cells in throat where bacteria usually grow and damages distant cells, e.g., heart muscle
Shigella (enterotoxin)	*Shigella* spp.	Shigellosis	Interferes with protein synthesis by blocking chain elongation (by a different mechanism than diphtheria toxin)	Kills cells in the intestinal epithelium, causing bloody diarrhea (dysentery)

pathogens also produce various other proteins (neither exotoxins nor endotoxins) that are toxic to cells and tissues; these include extracellular enzymes.

Exotoxins. **Exotoxins** are highly destructive proteins produced by certain Gram-positive and Gram-negative pathogens (**Table 15.5**). Most exotoxins are composed of two subunits, a B, or **b**inding, component and an A, or **a**ctive, component. The B component attaches the exotoxin to molecular receptors on certain types of host cells. It is the B component that accounts for the specificity of exotoxins. A given exotoxin affects only tissues where matching receptors exist. Thus, **neurotoxins**, such as tetanus and botulinum toxin, exert their effect only on cells of the nervous system. **Enterotoxins**, on the other hand, such as cholera toxin and shigella toxin, exert their effect only on epithelial cells lining the intestinal tract. After binding, the A component enters the cell and disrupts its function, usually by inhibiting one specific metabolic reaction (**Figure 15.10**). This additional activity may account for the almost unbelievable potency of some exotoxins. For example, as little as 130 μg (micrograms) of tetanus toxin—an amount about equal to the size of the period at the end of this sentence—can kill an adult.

Some exotoxins enter the bloodstream, exerting harmful effects far from their site of production. This explains how pathogens such as *Bordetella pertussis* can cause **systemic disease**, affecting tissues throughout the body, rather than merely **local disease**, which affects only the particular region in which the bacteria are multiplying.

The **pertussis toxin** acts by disrupting the **adenylate cyclase system**, which mediates intracellular communication. Because it enters the circulation and reaches distant tissues, many cellular functions in the body are disrupted. This produces the nonrespiratory symptoms observed in Baby Girl B (oversensitivity to histamine, abnormally high levels of insulin, and a vast excess of lymphocytes). Also, the course of the infection indicates that it was probably not the bacteria themselves but the toxin that caused the most severe part of Baby Girl B's illness. The weeks of disabling coughing occurred relatively late, by which time probably only a few microorganisms were actively dividing in her respiratory tract. The unremitting and paroxysmal cough, therefore, was probably due to toxin molecules that had previously entered cells in the respiratory system and brain center where coughing is controlled, disrupting intracellular communication.

In spite of their potency, exotoxins can be inactivated. As proteins, most exotoxins are readily denatured and rendered inactive by heat. Furthermore, the immune system can neutralize exotoxins with special antibodies called **antitoxins** that bind to them. Certain exotoxins can be modified in the laboratory by treatment with heat or

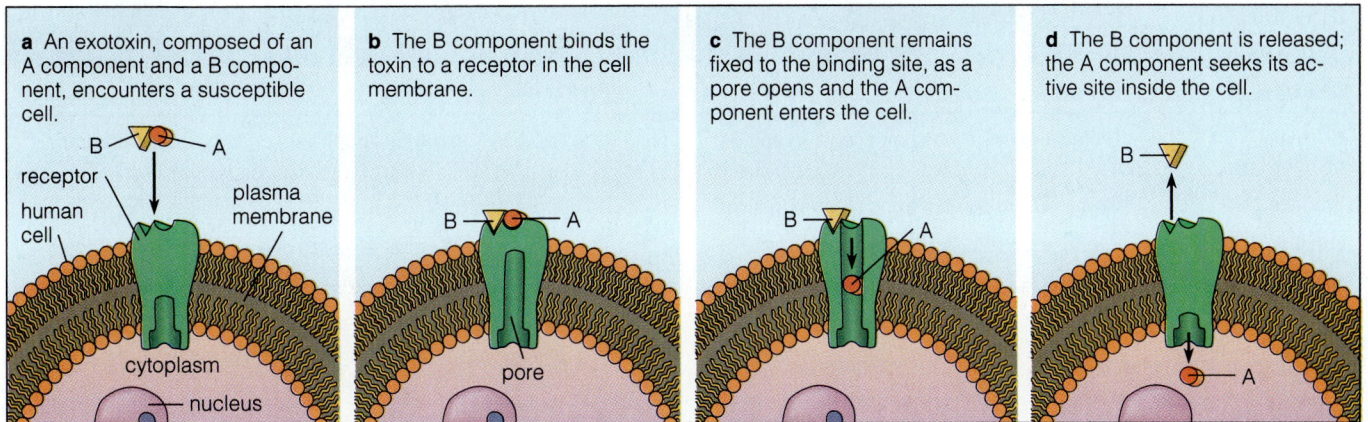

a An exotoxin, composed of an A component and a B component, encounters a susceptible cell.

b The B component binds the toxin to a receptor in the cell membrane.

c The B component remains fixed to the binding site, as a pore opens and the A component enters the cell.

d The B component is released; the A component seeks its active site inside the cell.

Figure 15.10 Exotoxins. Exotoxins are highly destructive proteins produced by certain Gram-positive and Gram-negative cells. Their two-part structure makes them very specific.

chemicals such as formaldehyde to produce **toxoids**, molecules that have lost their disease-causing properties but still stimulate the immune system to produce antitoxin. From manufactured toxoids, we can prepare vaccines to protect against diseases caused primarily by the action of an exotoxin. The vaccine against diphtheria, for example, is a toxoid (Chapter 20).

Exotoxin production is encoded by genes carried on bacterial plasmids or temperate bacteriophages (Chapter 13). Diphtheria toxin, for example, is produced only by strains of bacteria infected by a bacteriophage that carries the **tox gene**. Strains not infected by this bacteriophage do not produce toxin and are avirulent.

Microbiologists do not fully understand what value exotoxins offer most microorganisms that produce them, but there are some hints, such as the case of *Corynebacterium diphtheriae*. When *C. diphtheriae* produces toxin in a tissue such as the human throat, many nearby cells are killed. Because *C. diphtheriae* grows better than most bacteria in this environment of dead and dying cells, the toxin may give *C. diphtheriae* a selective advantage, allowing it to outgrow its competitors (Chapter 22).

Most exotoxin-caused diseases, such as cholera, certain *Escherichia coli*–caused diarrheas, shigellosis, pertussis, and tetanus, occur only if toxin-producing bacteria actively multiply in the body. However, in some diseases, toxin is produced outside the body and taken in with contaminated food, producing **food poisoning**. Botulism is a deadly food poisoning caused by the bacterium *Clostridium botulinum* (Chapter 25). A less serious but still significant food poisoning is caused by staphylococcal enterotoxin (Chapter 23). These illnesses are not infections; they are a type of poisoning.

Endotoxins. Endotoxins are released only by Gram-negative bacteria. Recall from Chapter 4 that all Gram-negative bacteria have an outer membrane that is partly composed of lipopolysaccharide (LPS), a hybrid molecule with lipid (lipid A) on one end and polysaccharide on the other. This hybrid molecule is also called **endotoxin**. Most of the toxicity is mediated (brought about) by lipid A. All Gram-negative bacteria produce endotoxin, even those that produce other toxins.

Endotoxin exerts a truly astonishing variety of toxic effects on the human body—fever, increased or decreased numbers of white blood cells, diarrhea, shock, prostration (extreme weakness), and death. These toxic effects are expressed when endotoxin stimulates human cells to secrete particular messenger proteins. For example, fever occurs when endotoxin stimulates white blood cells to produce a protein called **interleukin-1** (Chapter 16). **Endotoxic** or **septic shock** (one type of **shock**, any life-threatening loss of blood pressure) occurs when endotoxin causes phagocytes to secrete a protein called **tumor necrotizing factor**. Tumor necrotizing factor causes loss of fluid from the circulation, which means insufficient amounts of blood reach vital organs; as a result, they shut down, causing a potentially life-threatening condition.

Endotoxins differ from exotoxins in several ways (**Table 15.6**). Endotoxin is a normal structural component of the Gram-negative cell. Thus, it is not secreted by pathogens, but is released into the environment when the bacterial cell dies. Because endotoxins, unlike exotoxins, are not proteins, they are relatively heat stable. Also unlike exotoxins, endotoxins don't stimulate the immune system to produce protective antibodies. Perhaps most important, endotoxins are not specific. Endotoxin is the same from one Gram-negative pathogen to another, and its action is nonspecific—it affects a wide variety of human cells.

Endotoxin is not very potent. The endotoxins produced in Gram-negative infections—including Baby Girl

Table 15.6	Differences between Exotoxins and Endotoxins	
Characteristic	Exotoxin	Endotoxin
Chemical composition	Proteins, usually two components (A and B)	Lipid portion (Lipid A) of lipopolysaccharide outer membrane
Produced by	Certain Gram-positives and Gram-negatives	All Gram-negatives
Location	Excreted into extracellular space	Part of outer membrane, released only on death of bacterium
Heat stability	Most are heat labile (sensitive); inactivated at 60–80°C (except staphylococcal enterotoxin)	Relatively heat stable; many withstand heating above 100°C
Toxicity	High toxicity; often fatal	Low toxicity; fatal only under conditions of overwhelming infection
Effect on host	Highly variable from one toxin to another	Similar for all endotoxins
Fever producing	No	Yes
Immune response	Active, antitoxins provide host immunity	Poor, no antibodies to endotoxin produced
Toxoid production	Chemically treated toxin (toxoid) used as vaccine	Toxoid cannot be made
Diseases	Many, including tetanus, botulism, diphtheria (see Table 15.5)	Meningococcemia and overwhelming Gram-negative infections

B's whooping cough—generally play minor roles in the bacterium's pathogenesis. One exception is **meningococcemia**, a condition in which the virulent pathogen *Neisseria meningitidis* invades the bloodstream. Many of the clinical findings that typify meningococcemia are caused by the effect of endotoxin on blood vessels. Blood vessel damage in the skin causes a rash. Blood vessel damage also diminishes the blood supply to vital organs, such as the adrenalin-producing adrenal glands, causing prostration and shock. When meningococcemia causes death, as it usually does, septic shock is the terminal event.

Normally endotoxins become clinically significant only when large numbers of dying Gram-negative bacteria are circulating in the bloodstream, and this is usually caused by an overwhelming Gram-negative infection. Paradoxically, agents that kill Gram-negative bacteria, such as antibiotics or complement, may actually increase endotoxin-mediated damage. If death occurs, the immediate reason is septic shock from endotoxin.

Other Bacterial Toxins. In addition to exotoxins and endotoxins, pathogenic bacteria produce various proteins that have toxic properties. *Bordetella pertussis*, for example, produces several agents, besides pertussis exotoxin, that have toxic properties. **Tracheal cytotoxin** is a small protein in the cell wall that selectively kills ciliated respiratory cells by interfering with their ability to make DNA. **Adenylate cyclase toxin** is a bacterial enzyme that enters eucaryotic cells and acts in the same way as the human adenylate cyclase, catalyzing the conversion of ATP to cyclic AMP (a different effect on the adenylate cyclase system from that of pertussis exotoxin). **Dermonecrotic**

toxin is a two-component protein that causes tissue death when low doses are injected into mice and kills the experimental animals in higher doses. Its exact mechanisms are not well understood.

Certain species of bacteria produce damaging proteins called **extracellular enzymes**, or **exoenzymes**, enzymes secreted outside of the cell (**Table 15.7**). Species of *Streptococcus*, *Staphylococcus*, and *Clostridium* manufacture numerous kinds of toxic extracellular enzymes. Extracellular enzymes fall into three major groups. First, there are extracellular enzymes that lyse cells. The **cytolysins** attack cell membranes, causing host cells to lyse. **Hemolysins** lyse red blood cells. **Leukocidins** lyse **leukocytes**, white blood cells that are among the body's most powerful defenders against microbial infection (**Figure 15.11**).

The second type of extracellular enzyme breaks down the materials that hold cells together to form tissues. **Hyaluronidase** degrades the polysaccharide hyaluronic acid that cements cells together in many different tissues. **Collagenase** degrades the protein collagen, a major structural component of connective tissue. The third type of extracellular enzyme affects the delicate balance between formation and destruction of blood clots. **Coagulases** split the serum protein **fibrinogen** to form **fibrin**, thus creating blood clots. In contrast, **kinases** split fibrin, dissolving clots. (One of these bacterial kinases, **streptokinase**, is used clinically to dissolve the blood clots that block coronary arteries during a heart attack.)

Extracellular enzymes and other toxic proteins are capable of harming human cells, but this does not necessarily mean they play a role in causing the signs and symp-

Table 15.7 Actions of Extracellular Enzymes

Type of Extracellular Enzyme	Example of Producing Pathogen	Action
Cytolysins (lyse cells)		
Hemolysin	*Staphylococcus aureus, Streptococcus pyogenes*	Lyses red blood cells
Leukocidin	*Staphylococcus aureus*	Lyses white blood cells
Lecithinase	*Clostridium perfringens*	Destroys cell membranes
Breaks Down Materials That Hold Cells Together		
Hyaluronidase	*Streptococcus pyogenes*	Breaks down hyaluronic acid, an essential component of connective tissue
Collagenase	*Clostridium perfringens*	Breaks down connective tissue
Protease	*Clostridium perfringens*	Destroys protein in muscle tissue
Disturbs Normal Blood Clotting		
Coagulase	*Staphylococcus aureus*	Causes plasma to coagulate
Streptokinase	*Streptococcus pyogenes*	Dissolves clots

toms that we recognize as disease. That is, although they are toxins, we do not know for certain that all of them are **virulence factors**, substances that help cause disease. However, intuition suggests that they might be. For example, hyaluronidase and collagenase, by destroying the structure of tissues, could help pathogens invade them. Interfering with the blood clotting system could benefit invading microorganisms by forming protective clots around them or by freeing them from clots the body has produced in self-defense. In spite of intensive research we have no proof that these proteins are necessary for pathogenesis. In fact, evidence suggests that some may not be pathogenic adaptations. Tracheal cytotoxin, for example, is produced by avirulent as well as virulent strains of *Bordetella pertussis*. We do not completely understand the true function of these proteins and the selective advantage they may give a pathogen.

Damage Caused by Host Response to Microbial Multiplication

When pathogenic microorganisms multiply in or on human tissues, the body fights back. These defenses are essential for human survival, but they may also disrupt normal body functions in ways that produce disease. Thus, the defensive response that some pathogens provoke in the human body is a form of pathogenesis. Pathogens that damage the body in this way need not produce any toxins to be virulent.

Streptococcus pneumoniae, for example, produces no known clinically significant toxins, but it does cause a deadly pneumonia (Chapter 22). When *S. pneumoniae*

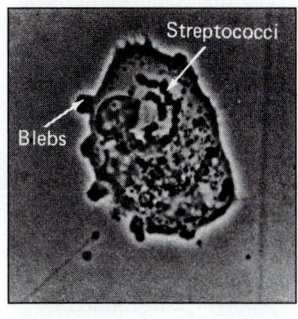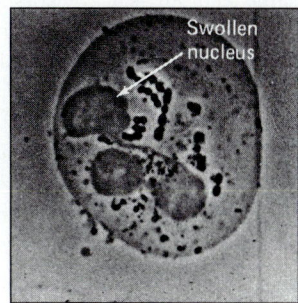

Figure 15.11 Leukocidin is a toxin produced by some pathogens that lyses phagocytes. (**a**) Chains of streptococcal cells attach to the surface of a phagocyte; the phagocyte consumes the bacterial cells. (**b**) The phagocyte is damaged; it stops moving and swells. As damage to the cell progresses, most internal structures disappear.

multiplies in the lungs, phagocytic white blood cells come to combat the infection. *S. pneumoniae* is protected by a capsule, however, so more and more phagocytes arrive to help. Dead cells of both kinds accumulate in the lungs, impairing normal gas exchange and making breathing difficult.

Much of the damage of *Bordetella pertussis* is also caused by host defenses. When *B. pertussis* adheres to respiratory epithelial surfaces, the cilia stop beating. Without normal mucociliary action, masses of bacteria begin to accumulate, and phagocytic white blood cells arrive on the scene to try to control them. These enormous numbers of white blood cells—some still active, some already dead—produced the sticky respiratory secretions that

Another Golden Age?

The study of microbial pathogenesis—the relationship between pathogenic microorganisms and the diseases that they cause—began little more than 100 years ago, when Robert Koch conclusively proved that a specific species of bacterium caused a specific human disease (see the box "One Microbe, One Disease" earlier in this chapter).

Koch's success, along with Louis Pasteur's at about the same time, initiated a period of intense research. This period—from the late 1800s through the early 1900s—became known as the golden age of medical bacteriology. Most of the major bacterial pathogens were isolated during this time. The techniques were fairly basic: Isolate the microorganism, grow it in pure culture, and examine human and microbial cells under the microscope.

Until relatively recently, most of our knowledge about microbial virulence and pathogenesis was based on the same techniques. Clinicians observed the manifestations of a disease. Pathologists examined diseased tissue, grossly and under the microscope. Microbiologists used Koch's postulates to prove that a given microorganism was the etiologic agent of a given disease.

Thus, basic microbiology explained which microorganisms caused which diseases, but—except for identifying a few toxins—little more. The molecular and biochemical mechanisms of pathogenesis were a mystery.

Today the story is different. In the last 20 years, advances in molecular biology and recombinant DNA technology have provided microbiologists with new and amazingly powerful tools. The technological sophistication of biomedical research, including recombinant DNA technology (Chapter 7), has led to major breakthroughs in our understanding of microbial pathogenicity. Geneticists have also discovered ways in which disease-causing genes are regulated. In some organisms, such as *Bordetella pertussis*, genes that encode a set of interacting virulence factors are turned on or off as a group. All the *B. pertussis* virulence factors discussed in this chapter, with the exception of tracheal cytotoxin, are controlled by a single regulatory gene called *vir*.

Knowing the molecular basis of a pathogenesis helps us develop new ways to prevent and cure illness. For one thing, we can develop better vaccines. For example, the traditional pertussis vaccine was made of whole pertussis cells—a crude preparation that provided good immunologic protection against pertussis but also caused side effects such as fever, soreness at the vaccination site, and rare neurological complications, including seizures. An increased understanding of the key pertussis virulence factors allowed researchers to design a more highly purified vaccine that contains only the few antigens critical to protection against infection. This new acellular vaccine, which has few side effects, is now used for pertussis immunization in all children over age 1. It is likely to be approved soon for use in infants, making uncomfortable pertussis vaccine reactions largely a thing of the past.

The First Golden Age of Medical Microbiology

Date	Disease	Bacterium	Date	Disease	Bacterium
1876	Anthrax[a]	*Bacillus anthracis*	1886	Pneumonia	*Streptococcus pneumoniae*
1879	Gonorrhea	*Neisseria gonorrhoeae*	1887	Meningitis	*Neisseria meningitidis*
1880	Typhoid fever	*Salmonella typhi*	1887	Brucellosis	*Brucella* spp.
1880	Malaria	*Plasmodium* spp.	1892	Gas gangrene	*Clostridium perfringens*
1881	Wound sepsis	*Staphylococcus aureus*	1894	Plague	*Yersinia pestis*
1882	Tuberculosis[a]	*Mycobacterium tuberculosis*	1896	Botulism	*Clostridium botulinum*
1883	Cholera[a]	*Vibrio cholerae*	1898	Dysentery	*Shigella dysenteriae*
1883–84	Diphtheria	*Corynebacterium diphtheriae*	1905	Syphilis	*Treponema pallidum*
1885	Tetanus	*Clostridium tetani*	1906	Whooping cough	*Bordetella pertussis*
1885	Diarrhea	*Escherichia coli*	1909	Rocky Mountain spotted fever	*Rickettsia rickettsii*

[a]Koch discovered these.

caused Baby Girl B to choke and cough. These defenses, and their protective as well as damaging effects, are further discussed under "Inflammation" in Chapter 16.

Often the sophisticated defenses of the immune system control an infection and prevent the host from dying, but they can overreact and cause more harm than benefit. This exaggerated immune response, called hypersensitivity, accounts for the major damage that some infections cause. Most of the tissue damage that occurs in chronic tuberculosis results from hypersensitivity (Chapter 18).

VIRAL PATHOGENESIS

Viruses damage human cells in ways very different from those of bacteria. Recall from Chapter 13 that viruses depend completely on the metabolism of their host cell to reproduce themselves and that they multiply only within cells. Beyond this, however, viral infections can harm human cells in many ways. Some viral infections are **cytocidal** (killing the host cell), while others are **cytopathic**

(damaging but not killing the host cell). **Figure 15.12** shows four different effects that viral infection can have on the host.

Some viral infections are called **lytic infections** because they kill the host cell by lysing it. When the virus takes over the cell's metabolic machinery, the host cell cannot carry on essential maintenance functions. Because the cell cannot synthesize lipids, membranes deteriorate. When the plasma membrane disintegrates, lysis is immediate. When the membranes that surround lysosomes disintegrate, they release destructive enzymes into the cytoplasm and destroy the cell from within, a process termed **autolysis**. When an infected cell lyses, new viruses are released that infect more host cells. An active herpesvirus infection, for example, such as a fever blister on the lip, kills host cells in this way.

On the other hand, **persistent viral infections** can last for years, producing new virus particles without killing the infected cell. Rather than lysing the infected cell, viral particles are released by budding out through the cell membrane. This process can cause little damage to the host. For example, infants who suffer a prenatal infection by cytomegalovirus, a member of the herpes family, may

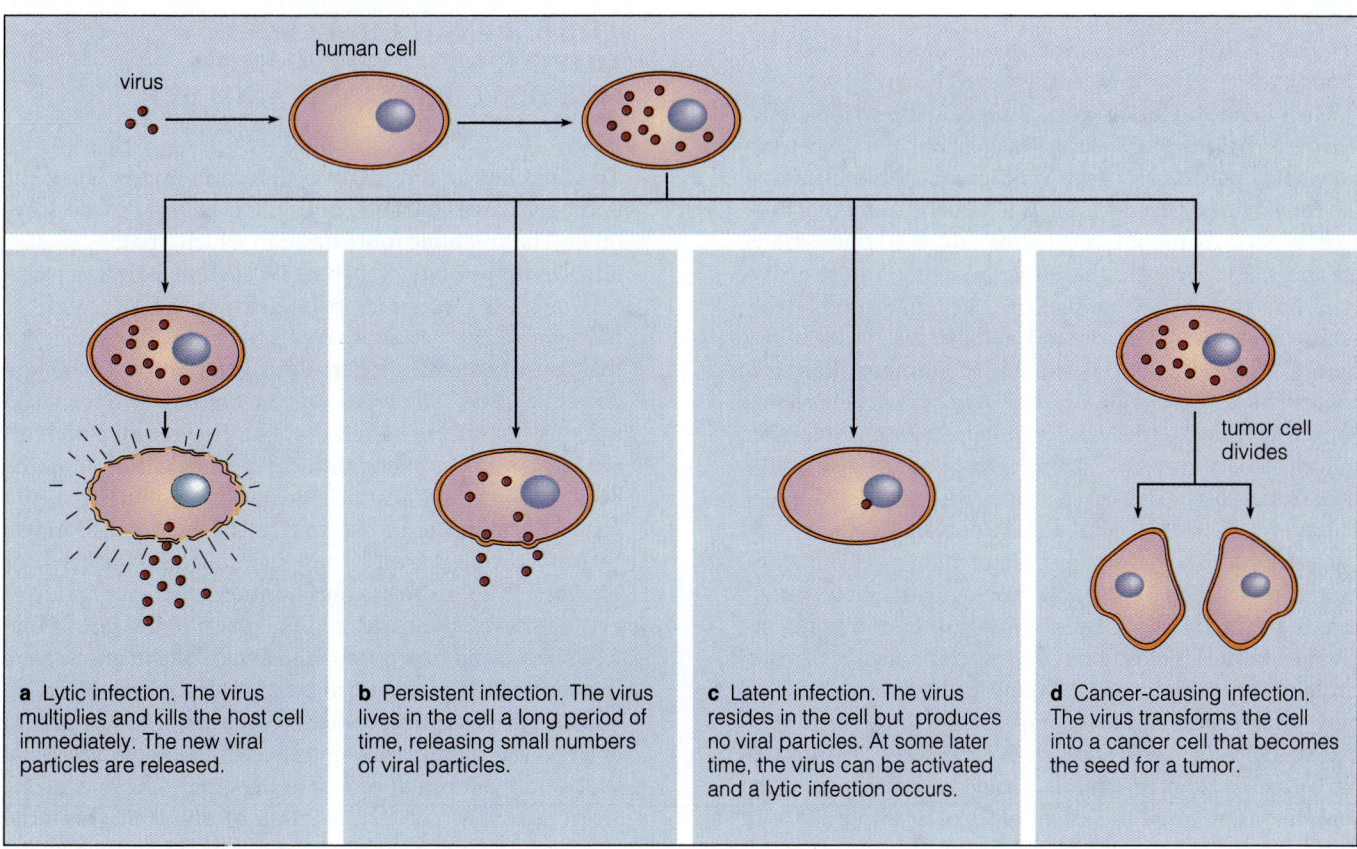

a Lytic infection. The virus multiplies and kills the host cell immediately. The new viral particles are released.

b Persistent infection. The virus lives in the cell a long period of time, releasing small numbers of viral particles.

c Latent infection. The virus resides in the cell but produces no viral particles. At some later time, the virus can be activated and a lytic infection occurs.

d Cancer-causing infection. The virus transforms the cell into a cancer cell that becomes the seed for a tumor.

Figure 15.12 Viral pathogenesis. When a virus infects a human cell, there are four possible outcomes.

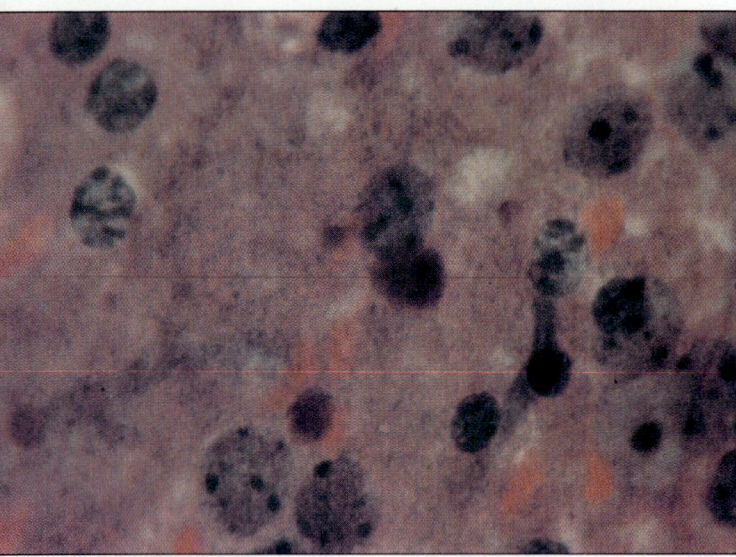

Figure 15.13 Negribodies, the magenta-colored inclusions in this stained tissue sample from a dog's brain, indicate the presence of the rabies virus.

shed the virus in their urine throughout childhood although they appear to be entirely well (Chapter 24).

In a **latent infection**, yet a different pattern develops. The virus lies dormant within the host cell, not producing new viral particles. Latent infections may be sustained for the rest of a person's life. No damage to the host occurs as long as the infection remains latent, but various stimuli may reactivate the virus, initiating a destructive lytic infection. For example, the varicella zoster virus, which causes chickenpox and shingles, is typically acquired in childhood. The initial, or **primary**, infection causes chickenpox. Although the child appears to recover completely, the herpesvirus maintains a latent infection in nearby nerve cells. When triggering stimuli, such as illness or psychological stress, reactivate the virus, the secondary lytic infection causes shingles (Chapter 26).

Viral infections sometimes affect human cells in ways that can be seen under the microscope. Some virus-infected cells can be identified because they contain **inclusion bodies**, collections of viral components such as protein and nucleic acid, waiting to be assembled into new viral particles. For example, the rabies virus produces inclusion bodies called **Negribodies** in infected nerve cells (**Figure 15.13**). Other viruses, including measles virus, cause the membranes of neighboring cells to fuse, creating giant, multinucleated cells.

Viral infections can also affect cells in ways that are not visible under the microscope, but that can have important consequences. Some viruses, for example, cause changes in the antigens that mark host cells, setting them up as targets for destruction by the body's own immunologic defenses (Chapter 17).

Certain viruses establish latent infections in human cells that can later cause fatal damage by transforming the infected cells into cancer cells. These viruses are termed **oncogenic** (cancer-causing). When an oncogenic virus infects a cell, its genetic material does not immediately direct the production of new viruses. Instead, if it is a DNA-containing virus, DNA enters the host genome and remains there, being copied along with the host's DNA each time the infected cell divides. In the case of RNA-containing oncogenic viruses, RNA is first converted into DNA by reverse transcriptase (Chapter 13). In either case, the viral-derived DNA and the genes it contains may be spliced into the human chromosomes or may replicate independently as a plasmid. Years later these viral genes may cause the cell to become **transformed**, meaning to escape the controls that normally regulate cell division and become the seed for a **tumor**, or cancerous growth. Only a few viruses have been shown to be definitely associated with human cancer (as discussed in Chapter 13).

THE SEVENTH CHALLENGE: MOVING ON—EXITING FROM ONE HOST, ENTERING ANOTHER

The final link in any chain of infection occurs when the pathogen leaves the body of one host in order to reach another. The anatomic route through which a pathogen usually leaves the body of its host is called its **portal of exit**.

For most respiratory pathogens, the portal of exit is the same as the portal of entry—the nose. The pathogens that cause influenza, tuberculosis, strep throat, meningococcemia, and all types of pneumonia are expelled through respiratory secretions. As a respiratory pathogen, *Bordetella pertussis* exited this way in Baby Girl B. After the bacterium reached a sufficient concentration in her respiratory tract, infectious bacterial cells were expelled in respiratory droplets every time she coughed, sneezed, or breathed. The greatest concentration of microorganisms existed during the early coldlike phase of her illness (before she was admitted to the hospital), so she might have transmitted the pathogen to other susceptible children.

For most gastrointestinal pathogens, the portal of entry is the mouth—the entrance to the gastrointestinal tract—and the portal of exit is the anus—the end of the gastrointestinal tract. Examples are the bacterial pathogens that cause cholera and typhoid, as well as the many viruses that cause diarrhea. Protozoal causes of diarrhea, including *Giardia lamblia* and *Entamoeba histolytica* also exit the body through this portal.

Most sexually transmitted pathogens exit the body in the same way they entered—across the mucous membrane surfaces of the genital tract. The pathogens that cause gonorrhea, syphilis, genital herpes, and AIDS all exit this way most of the time.

Pathogens that are transmitted parenterally by arthropod vectors also exit the same way they entered—in a drop of blood. They include the pathogens that cause malaria, yellow fever, and viral encephalitis.

Microbial pathogens must meet all seven challenges in order to survive. The full details of how pathogens meet these challenges are known for only the handful of microorganisms that cause disease by producing a single toxic product; examples include *Clostridium botulinum*, which causes botulism, and *C. tetani*, which causes tetanus (Chapter 25). The pathogenesis of most microbial infections (including pertussis) is so complex we understand it only in broad outline.

Summary

SEVEN CHALLENGES THAT FACE PATHOGENIC MICROORGANISMS (p. 354)

1. For a pathogen to survive in a human host and cause disease, it must be adapted to
   a. survive between hosts
   b. come in contact with the human body
   c. adhere to body surface tissues
   d. gain access to deeper body tissues
   e. avoid host defenses and multiply in the body
   f. cause disease
   g. leave to infect another host

2. Pathogenesis refers to the process by which a microorganism causes disease. Infection is the growth of a disease-causing microorganism in the body.

THE FIRST CHALLENGE: MAINTAINING RESERVOIRS OF INFECTION (pp. 354–355)

1. A disease reservoir is a place where pathogenic microorganisms are maintained between infections.

2. A human serving as a reservoir of infection is a carrier. An incubatory carrier is an apparently healthy individual who may be in the earliest, symptomless stages of the disease. A chronic carrier is someone who harbors a pathogen for a long time (months or years).

3. Most animal reservoirs provide an environment similar to that of the human host. A human disease caused by a pathogen that maintains an animal reservoir is called a zoonosis.

4. Animal reservoirs can profoundly affect the pattern of human disease, leading in some cases to a pandemic.

5. Some pathogens survive in water, soil, and house dust. *Clostridium tetani*, which causes tetanus, can survive as endospores in soil for hundreds of years.

THE SECOND CHALLENGE: GAINING ACCESS TO A NEW HOST (pp. 355–363)

1. Disease transmission takes place when a pathogen leaves a reservoir and enters the body of a host.

2. The portal of entry for a pathogen is the anatomic site through which it enters the host's body. The most common portals of entry are the same anatomic surfaces colonized by microflora.

3. The ID_{50} (infectious dose) is the number of microorganisms that must enter the body to establish infection in 50 percent of test animals. The LD_{50} (lethal dose) is the number of microorganisms that must enter the body to cause death in 50 percent of test animals.

4. Modes of transmission are categorized as contact between humans (direct or indirect), vehicles (inanimate objects), and vectors (living transmitters, usually arthropods).

5. Droplets of respiratory secretions are transmitted directly from one host to another through sneezing, coughing, laughing, or speaking. More human diseases are transmitted this way than by any other.

6. Fomites are inanimate objects such as eating utensils, towels, bedding, and handkerchiefs that transmit disease. Hand washing can break the transmission cycle.

7. Direct body contact, also called horizontal transmission, takes place by touching, kissing, or sexual intercourse. Sexually transmissible disease (STD) is spread by direct mucous membrane contact.

8. Vertical transmission is transmission of pathogens from mother to infant. It can be prenatal or perinatal (occurring in the birth canal or immediately after birth).

9. In the fecal-oral route, pathogens are transmitted from infected feces to the mouth of a new host. Fecal-oral transmission can be direct hand-to-hand or hand-to-mouth, by vehicles such as water, food, and fomites, or by vectors.

10. Arthropods can be mechanical vectors, carrying pathogens on their bodies, or biological vectors, an essential link in the transmission of a disease because the microorganism spends part of its life cycle in the arthropod host.

11. Airborne transmission occurs when microorganisms that can survive in air are inhaled.

12. In parenteral transmission, pathogenic microorganisms are deposited directly into blood vessels or deep tissues. This occurs when a biological vector bites through the skin or when intravenous drug users share needles.

13. Deep wounds can allow anaerobic pathogens such as *Clostridium tetani* (which causes tetanus) to enter.

THE THIRD CHALLENGE: MAINTAINING A FOOTHOLD—ADHERING TO BODY SURFACES (p. 363)

1. Pathogens—like the body's normal floral—adhere to a body surface by means of adhesins on their pili or surface.

THE FOURTH CHALLENGE: GOING DEEPER—INVADING THE BODY'S INTERIOR (pp. 363–366)

1. Most pathogens are invasive; they enter host cells or tissues. A few are noninvasive, remaining on the surface.

2. Invasive pathogens that enter host cells to live are called intracellular pathogens. The cell provides a nutrient-rich environment safe from body defenses.

3. In the 1880s, Robert Koch proved the germ theory of disease, that a particular microorganism causes a particular infection.

4. We still use Koch's postulates in certain cases today to prove the etiology (cause) of an infectious disease.
   a. The causative microorganism must be present in every individual with the disease.

b. The causative microorganism must be isolated and grown in pure culture.

c. The pure culture must cause the disease when inoculated into an experimental animal.

d. The causative microorganism must be reisolated from the experimental animal and reidentified in pure culture.

5. We use Rivers's postulates to determine the etiology of viral infections.

a. The viral agent must be found either in the host's body fluids at the time of disease or in infected cells.

b. The viral agent obtained by the host must produce the disease in a healthy animal or plant or must produce antibodies.

c. Viral agents from the newly infected animal or plant must in turn transmit disease to another host.

THE FIFTH CHALLENGE: ESTABLISHING INFECTION—EVADING HOST DEFENSES
(pp. 366–369)

1. Some pathogens have capsules that keep a phagocyte from establishing direct contact. Strains of *Streptococcus pneumoniae* without capsules are avirulent (harmless), while strains with the thickest capsules are the most virulent.

2. Surface proteins on pathogens also keep phagocytes from establishing contact. *Streptococcus pyogenes*, which causes strep throat, produces M protein.

3. Some pathogens, such as *Mycobacterium tuberculosis*, survive inside the phagocyte.

4. Pathogens that evade nonspecific body defenses encounter the body's immune defenses. The immune system recognizes pathogens by means of markers on their surface called antigens.

5. Some pathogens change their surface antigens to avoid recognition (antigenic variation). Some attack antibodies directly with enzymes called IgA proteases. Some use serum resistance, features on the bacterial surface that interfere with the host's defensive complement system.

6. Some pathogens obtain iron by producing iron-binding proteins called siderophores.

THE SIXTH CHALLENGE: MULTIPLYING IN HOST TISSUES—ADAPTATIONS THAT CAUSE DISEASE   (pp. 369–375)

1. The two most common forms of bacterial pathogenesis are production of toxins and damage caused by stimulation of the body's defenses.

2. Exotoxins are highly destructive proteins produced by both Gram-positive and Gram-negative bacteria. Most exotoxins are composed of two units, the A (active) unit and the B (binding) unit, and are highly specific. Usually genes on bacterial plasmids or temperate bacteriophages encode exotoxins.

3. Endotoxin is the lipopolysaccharide (LPS) component of the outer membrane of Gram-negative bacteria. It acts by stimulating human cells to secrete particular messenger proteins. For example, fever results when endotoxin stimulates white blood cells to secrete the protein interleukin 1. Endotoxin is generally not very potent.

4. Some pathogens produce extracellular enzymes. There are three types: cytolysins attack cell membranes; hemolysins lyse red blood cells; leukocidins lyse leukocytes.

5. The body defenses that fight off pathogens can disrupt body function in ways that cause disease. *Streptococcus pneumoniae*, for example, multiplies in the lungs and summons great numbers of phagocytes. As dead cells of both kinds accumulate, normal gas exchange is impaired and breathing becomes difficult.

6. Some pathogens stimulate hypersensitivity, an exaggerated immune response that causes damage.

VIRAL PATHOGENESIS   (pp. 375–376)

1. Some viral infections are cytocidal (they kill cells), while others are cytopathic (they damage but do not kill the cell).

2. Lytic infections kill the host cell by lysing it.

3. A persistent viral infection can last for years, producing new virus particles without killing the infected cell. In a latent viral infection, the virus lies dormant within the host cell, not producing new viral particles. Latent infections can last a lifetime and not be damaging unless the virus is reactivated.

4. Inclusion bodies are collections of viral components such as protein and nucleic acid.

5. Oncogenic viruses establish latent infections in human cells that transform the infected cells into cancer cells. Only a few virally caused cancers have been definitely established.

THE SEVENTH CHALLENGE: MOVING ON—EXITING FROM ONE HOST, ENTERING ANOTHER   (pp. 376–377)

1. The anatomic route through which a pathogen leaves the body of its host is called its portal of exit.

2. For most respiratory pathogens, the portal of exit is the same as the portal of entry, the nose. For most gastrointestinal pathogens, the portal of exit is the anus. Most sexually transmissible diseases exit the same way they entered, through the genital mucous membranes. Pathogens transmitted parenterally by arthropod vectors exit the same way, in a small amount of blood.

Review Questions

SEVEN CHALLENGES THAT FACE PATHOGENIC MICROORGANISMS

1. What are the seven challenges that face pathogens?

2. Define these terms: pathogenesis, infection.

THE FIRST CHALLENGE: MAINTAINING RESERVOIRS OF INFECTION

1. What is a disease reservoir? Name three disease reservoirs and describe the kinds of pathogens that exist in each.

2. Distinguish between a carrier, an incubatory carrier, and a chronic carrier.

3. What is a zoonosis? Give an example.

4. Give an example of how an animal reservoir can profoundly affect the pattern of human disease.

THE SECOND CHALLENGE: GAINING ACCESS TO A NEW HOST

1. Define disease transmission. What are the three major modes of transmission?

2. What is a portal of entry? Name the most common portals of entry for pathogens.

3. Explain ID_{50} and LD_{50}.

4. Describe how a pathogen is transmitted by each of the following means:

a. respiratory droplets
b. fomites
c. direct body contact
d. the fecal-oral route
e. arthropod vectors
f. air
g. parenteral means

5. Why are more diseases transmitted by respiratory droplets than by any other mode?

6. Compare and contrast the types of pathogens transmitted by respiratory droplets, fomites, direct body contact, and air.

7. What is an STD? Give an example. Are STDs communicable diseases? Explain.

8. Explain this statement: The fecal-oral route overlaps with other modes of transmission.

9. How are horizontal transmission and vertical transmission alike? How are they different? Name the two types of vertical transmission.

10. What is the difference between a mechanical vector and a biological vector?

11. Why are deep wounds more likely than surface breaks in the skin to lead to lethal infections?

THE THIRD CHALLENGE: MAINTAINING A FOOTHOLD—ADHERING TO BODY SURFACES

1. Why do pathogens face the same challenge as normal microflora in maintaining a foothold on body surfaces?

2. Describe how filamentous hemagglutinin functions as an adhesin for *Bordetella pertussis*. Why is it not a typical adhesin?

THE FOURTH CHALLENGE: GOING DEEPER— INVADING THE BODY'S INTERIOR

1. What is the difference between an invasive pathogen and a non-invasive pathogen?

2. What is an intracellular pathogen, and why is that an adaptive advantage?

3. Explain the process by which a bacterium invades the body's interior. How does a virus enter? Give some examples showing the variety of eucaryotic modes.

4. Define etiology.

5. What are Koch's postulates and what is their significance?

6. Can Koch's postulates be applied to all bacteria? Explain. Give an example.

7. What are Rivers's postulates?

THE FIFTH CHALLENGE: ESTABLISHING INFECTION—EVADING HOST DEFENSES

1. Define phagocytosis. What kinds of cells are phagocytes?

2. Describe two ways pathogens evade phagocytosis.

3. Instead of evading phagocytosis, some pathogens are readily consumed and then survive within the phagocyte. Explain how *Mycobacterium tuberculosis* does this.

4. Pathogens overcome the body's more sophisticated immune defenses by means of antigenic variation, IgA proteases, and serum resistance. Explain each of these.

5. Why must pathogens constantly compete for iron in the human body? Give an example of how some pathogens succeed.

THE SIXTH CHALLENGE: MULTIPLYING IN HOST TISSUES—ADAPTATIONS THAT CAUSE DISEASE

1. What are the two most common ways pathogens cause disease?

2. What are exotoxins? Describe their structure. What determines if a particular cell will produce an exotoxin?

3. In what ways are exotoxins very specific? Define neurotoxin and enterotoxin, and name a disease caused by each.

4. How is botulism toxin different from pertussis toxin? How are the diseases they cause different?

5. Compare and contrast endotoxin with exotoxin.

6. How does endotoxin act? What is interleukin 1? What is tumor necrotizing factor?

7. What are extracellular enzymes? Name the three types. Give an example of each.

8. Explain this statement: The response that pathogens evoke in the body is a form of pathogenesis. Give an example.

VIRAL PATHOGENESIS

1. What is the difference between cytocidal and cytopathic viral pathogenesis?

2. Classify each of the following as cytocidal or cytopathic and then explain the process: lytic infection, persistent infection, latent infection, cancer-causing infection.

3. Define these terms: autolysis, oncogenic virus, inclusion bodies, a transformed cell.

THE SEVENTH CHALLENGE: MOVING ON—EXITING FROM ONE HOST, ENTERING ANOTHER

1. Define portal of exit. Name the typical portals of exit for respiratory, gastrointestinal, sexually transmissible, and arthropod-transmitted pathogens.

Essay Questions

1. Trace *Bordetella pertussis*—or another pathogen of your choice—through the seven challenges of infecting a human host.

2. From time to time, there is an outbreak of disease of unknown origin. In 1976, for example, large numbers of war veterans attending a convention in Philadelphia came down with a mysterious illness. Twenty-nine died. How would you have applied Koch's postulates to identify the causative microorganism for what we now call Legionnaires' disease?

Suggested Readings

Finlay, B. B., and Falkow, S. 1989. Common themes in microbial pathogenicity. *Microbiological Reviews* 53:210–230.

Mekalanos, John J. 1992. Environmental signals controlling expression of virulence determinants in bacteria. *Journal of Bacteriology* 174:1–7.

Sharpe, A. H., and Fields, B. N. 1985. Pathogenesis of viral infections. *New England Journal of Medicine* 312:486–497.

zur Hausen, H. 1991. Viruses in human cancers. *Science* 254:1167–1172.

16 DEFENDING THE BODY'S INTERIOR: NONSPECIFIC DEFENSES

T. L.'s Close Call: Part One

T. L., a college student, was backpacking in a remote wilderness region. While pitching a tent he tripped and fell toward the branch of a dead tree. In an attempt to break his fall, he extended his arm and sustained a puncture wound to his right palm. Although the wound was painful and bled for a short time, it did not appear to be serious, and he fell asleep that night unconcerned about his condition.

By the next morning, however, T. L. noticed that the tissues immediately surrounding his wound were red, swollen, and warm. A round area about 1 inch in diameter clearly looked abnormal when compared to the rest of his hand. The affected part of T. L.'s hand was also painful, especially when he touched or bumped it. After eight hours of hiking the sore hand was even more painful, and a thick yellow discharge oozed from the open wound. T. L. felt unusually tired, his body ached, and a brief chill made him aware of a developing fever. His companions helped to elevate his arm and apply warm compresses to his palm, hoping that he would feel well enough by the next day to continue their trip. (Continued in Chapter 17.)

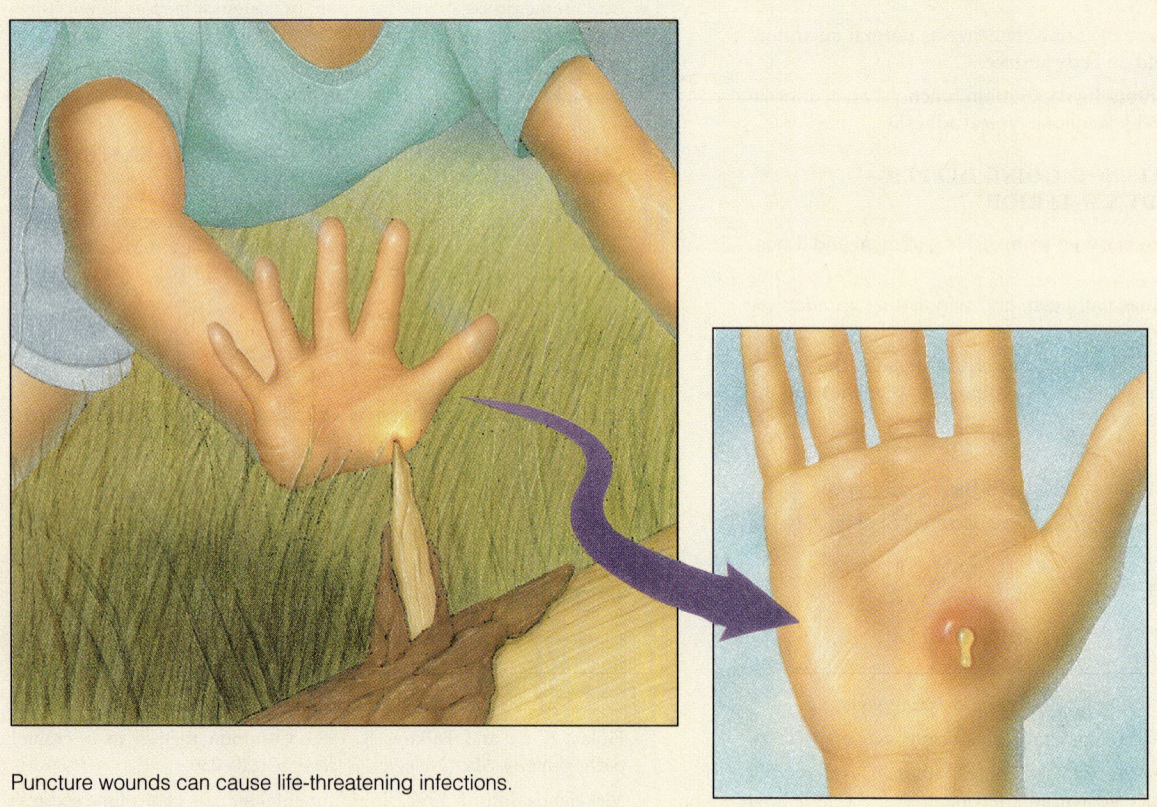

Puncture wounds can cause life-threatening infections.

To understand:

- The three lines of defense against microbial infection

- The steps in the inflammatory response and how inflammation activates and coordinates the body's nonspecific defenses

- The different types of leukocytes (white blood cells) and how they contribute to the body's defenses

- The steps in phagocytosis and its central role in the body's nonspecific internal defenses

- The complement system, including the complement cascade and the classical, alternate, and terminal pathways

- Interferon, particularly its role in defending against viral infection

THE BODY'S THREE LINES OF DEFENSE AGAINST INFECTION

Human beings come into daily contact with a multitude of microorganisms that can thrive on or in the body. Some are harmless or even beneficial (Chapter 14), but microbial infections can be life-threatening (Chapter 15). In this chapter and the next, you will read about the human body's complex, interconnected, and overlapping systems of defense against microbial infection.

The body has three lines of defense. If the first line of defense is overwhelmed, a second and even a third line are activated (**Figure 16.1**). You read about the first line of defense in Chapter 14, the nonspecific surface defenses—structural, mechanical, and biochemical. They are *nonspecific* because they function in the same way against all microorganisms. Surface defenses are considered separately because they play their role *before* microorganisms establish an infection, preventing pathogens from entering deeper tissues.

In T. L.'s case, first-line surface defenses were overcome instantly when he sustained the penetrating wound to his skin. The skin is a formidable barrier to infection and rarely allows the entry of microorganisms while it remains intact, but even a minor wound like T. L.'s provides microorganisms with access to the highly vulnerable tissues that lie below. Any break in the skin sets the stage for a potentially serious infection.

When first-line surface defenses fail and pathogens enter the body, second-line defenses are activated imme-

diately—within hours in T. L.'s case. Like surface defenses, the second-line defenses are nonspecific. They combat all invaders with the same weapons. The four major **nonspecific interior defenses** of the body that make the second line are inflammation, phagocytosis, complement, and interferon.

When T. L.'s hand became infected, certain molecules formed in response to the tissue damage, and infection provoked the process of inflammation. Inflammation increased the blood supply to his hand, bringing **leukocytes** (white blood cells) and complement proteins to the site. **Phagocytes** (certain kinds of leukocytes) attacked the invading bacteria, consuming and destroying them in a process called phagocytosis. The complement proteins helped phagocytes eliminate bacteria and also attacked microorganisms directly.

These second-line defenses caused many of the visible dramatic changes in T. L.'s hand and also caused much of his discomfort. The redness, swelling, warmth, pain, and loss of function were the result of inflammation in which phagocytosis and complement both played a crucial role. Yet these different defenses, working together, helped bring T. L.'s infection under control. Interferon did not play a role; it defends only against viral infection. The power of nonspecific defenses lies in the speed with which they act and their ability to act against many different pathogens. Sometimes—though not in T. L.'s case—the nonspecific interior defenses can eliminate microbial invaders even before illness becomes apparent.

In spite of their quick activation and wide-ranging action, nonspecific interior defenses cannot always control infection. When they are not sufficient, the body's third line of defense—**specific immune defenses**—is activated. These defenses are *specific* because they deploy cellular and chemical weapons targeted against one particular type of microorganism. The power of these defenses lies in their specificity. Individually tailored defenses eliminate microorganisms that can evade nonspecific defenses. Although immune defenses are slower to act than nonspecific defenses—sometimes taking days or weeks to become fully effective—they "remember" previous encounters with a pathogen. Upon subsequent exposure to the same pathogen, immune defenses are effective almost immediately and usually prevent reinfection. Pathogens that evade the third line of defense cause fatal or chronic infection. The immune system, its specific defenses, and the role they played in T. L.'s infection are discussed in Chapter 17.

All three lines of defense are essential parts of our response to the threat of microbial infection. Although we consider them in different chapters, these systems are cooperative and interrelated. You will see that many of the body's nonspecific defenses operate most effectively only

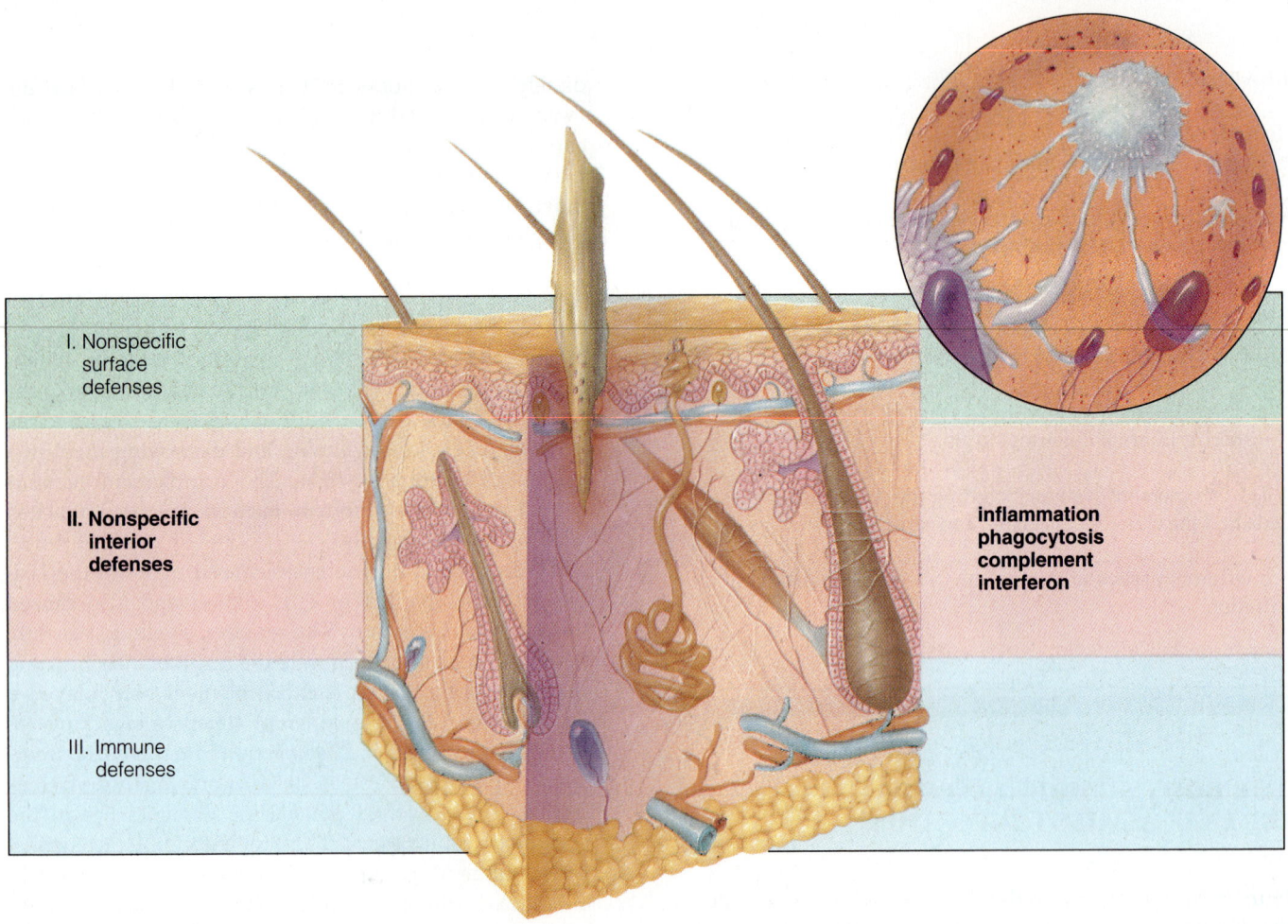

I. Nonspecific surface defenses

II. Nonspecific interior defenses

III. Immune defenses

inflammation
phagocytosis
complement
interferon

Figure 16.1 Lines of defense against infection. Although the three lines of defense are covered in separate chapters—14, 16, and 17—they are highly interrelated. Many of the body's nonspecific defenses operate most effectively only when supported by specific defenses, and vice versa. The second line of defense includes four nonspecific interior defenses that immediately attack any invader that reaches the body's interior—inflammation, phagocytosis, complement, and, in the case of viruses, interferon. These defenses often stop infection before symptoms appear.

when supported by specific defenses, and vice versa. In the next few sections, we discuss in more detail how each of the nonspecific interior defenses participated in controlling T. L.'s bacterial infection. We discuss viral infection and interferon at the end of the chapter.

INFLAMMATION

When bacteria were introduced into T. L.'s wound and began to multiply, producing millions of actively growing cells, an inflammatory response was activated. **Inflammation** is the body's reaction to injury or infection. Inflammation can occur anywhere in the body, but it is

easiest to recognize when the inflamed tissue is exposed, as it was in T. L.'s hand. Inflammation is the cornerstone of the body's nonspecific interior defenses. It activates and coordinates other interior defenses, including phagocytosis and complement. Clinically, inflammation is manifested by pain, redness, heat, and swelling. Severe inflammation can interfere with function of the affected body part, as it did in T. L.'s case. "Inflammation" is an apt description—inflamed tissue may look and feel as if it were on fire.

Inflammation can be broken into a series of steps (**Figure 16.2**). Injury or infection stimulates the release of **inflammatory mediators**, molecular messengers that mediate (bring about) inflammation. Inflammatory mediators exert their effect on **capillaries**, the microscopic blood vessels that supply tissues, causing **vasodilation** (enlargement) and increased **permeability** (leakiness). At the same time, the inflammatory mediators activate the phagocytes and attract them to the tissue. Changes in the capillaries and phagocytosis produce redness, swelling, warmth, and pain. During this phase, phagocytosis and complement destroy infecting microorganisms, but host tissues also suffer some damage. As the inflammation subsides, the damage caused by inflammation is repaired. Now let's consider each step in detail.

Inflammatory Stimulus

Any stimulus that damages human tissue can initiate inflammation. The most common inflammatory stimuli are injury and infection. T. L. suffered both—injury and infection. The stimulation of normal tissue by injury or infection produces **acute inflammation**, the common physiologic process described in the following sections. In rare cases, inflammation may result from other types of damage, such as disordered immune responses or other disease processes. The result is a clinically less dramatic but more destructive process called **chronic inflammation**, discussed in Chapter 18.

The inflammatory stimulus releases inflammatory mediators into damaged tissues. There are many types of inflammatory mediators from many sources. Among the most potent and significant are bacterial by-products, complement fragments, kinins, histamine, prostaglandins, and leukotrienes (**Table 16.1**).

Bacterial By-Products. F-met-leu-phe (a tripeptide, formyl-methionyl-leucyl-phenylalanine) is an example of a bacterial by-product that *Escherichia coli* produces. This molecule activates and attracts leukocytes to inflamed tissue.

Complement Fragments. Certain members of the complement family of serum proteins are powerful mediators of inflammation. Complement proteins are normally inactive in human serum; activation takes place when a complement protein splits into fragments. Some complement fragments act primarily on blood vessels; others activate and attract leukocytes. The complement system is discussed in detail later in this chapter.

Kinins. The **kinins** are a family of short, biologically active peptides that are produced when larger, inactive proteins are split enzymatically. The sequence of chemical events that causes blood to clot also produces kinins. **Bradykinin** has dramatic effects on blood vessels, causing vasodilation and increased permeability. It also contributes significantly to the pain of inflammation. The kinins produced as blood clotted around T. L.'s wound were partly responsible for the pain he experienced. Like

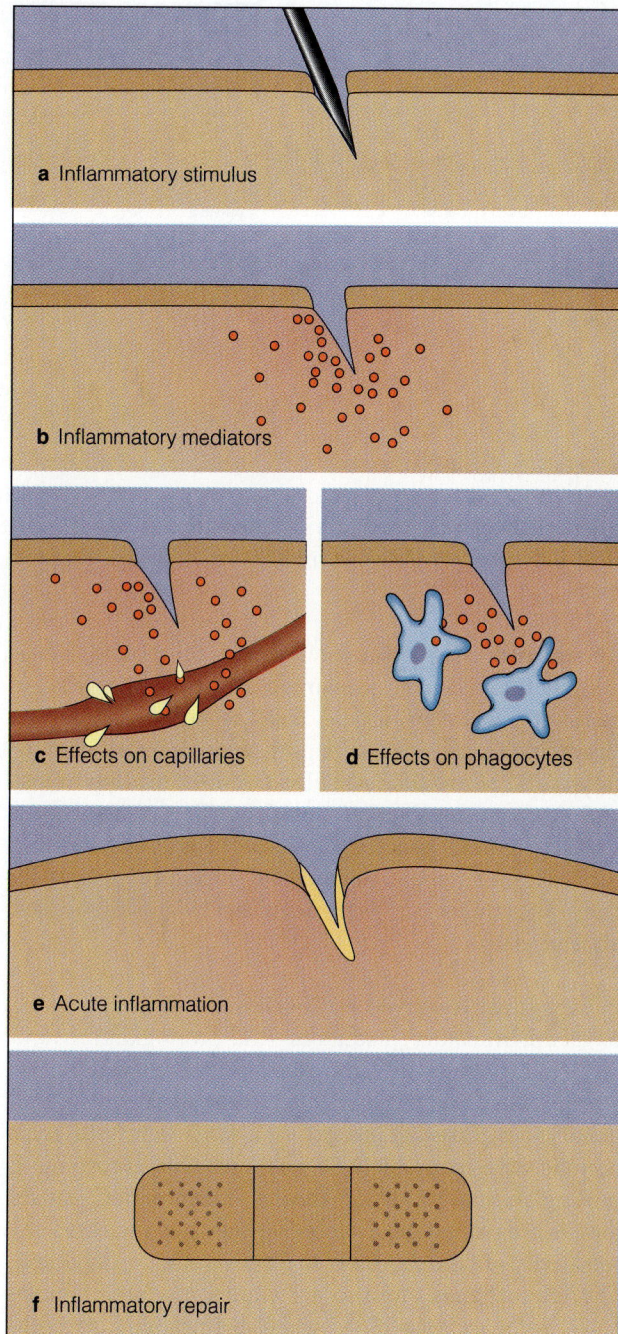

Figure 16.2 Six steps in inflammation. Inflammation begins with a stimulus (**a**), usually injury or infection, causing the formation and release of compounds known as inflammatory mediators (**b**). These mediators affect capillaries (**c**), causing vasodilation and increased permeability. They also affect phagocytes (**d**), causing activation and chemotaxis. The result is acute inflammation (**e**), manifested as redness (due to capillary dilation), swelling (due to the increased volume of fluid and cells in tissues), heat (due to increased blood flow), and pain (due to stimulation of nerve endings by swelling and pain mediators). The final stage is inflammatory repair (**f**), during which dead cells (both host and microbial) are cleaned up by macrophages and damaged tissue regenerates.

Table 16.1 Inflammatory Mediators

Inflammatory Mediator	Function
Bacterial by-products	Attract and activate leukocytes to inflamed tissue
Complement fragments	Vasodilation; activate and attract leukocytes
Kinins	Vasodilation and increased blood vessel permeability, leading to swelling; bradykinin stimulates nerve endings, causing pain
Histamine	Vasodilation and increased blood vessel permeability, leading to swelling; stimulates nerve endings, causing pain
Prostaglandins	Long-lasting dilation of blood vessels; increase sensitivity of pain receptors to bradykinin and histamine; attract and activate leukocytes
Leukotrienes	Attract and activate leukocytes; increase blood vessel permeability

most inflammatory mediators, kinins are destroyed soon after they are formed. Enzymes called **kinases** degrade them.

Histamine. The relatively simple compound **histamine** is derived from the amino acid histidine. It is released by **mast cells**, a type of cell related to leukocytes. Histamine causes vasodilation and dramatically increases the permeability of blood vessels, leading to loss of fluid from the circulation and swelling of nearby tissues. Histamine also stimulates nerve endings, causing pain. This extremely potent, and therefore potentially dangerous, agent is rapidly degraded by enzymes called **histaminases**.

Prostaglandins and Leukotrienes. Two other classes of inflammatory mediators—the prostaglandins and the leukotrienes—are produced when enzymes act on arachidonic acid, a major fatty acid component of cell membranes. **Prostaglandins** are a large family of compounds with many biological actions. One group, the E prostaglandins, produced during inflammation, causes long-lasting dilation of blood vessels and increases the sensitivity of pain receptors to histamine and bradykinin. The **leukotrienes** are also a complex mixture of molecules. Some increase blood vessel permeability, and others attract leukocytes to the inflammation site.

Effects on Capillaries

Capillaries respond to inflammatory mediators in two ways—vasodilation and increased permeability. When capillaries **dilate**, or increase in diameter, they increase the blood supply to inflamed tissues. The increased volume of blood brings infection-fighting weapons such as leukocytes and serum proteins (including complement and antibodies) to the affected region. Increased blood flow also causes **erythema** (redness) and warmth of inflamed tissue. Erythema may be the most obvious diagnostic clue to inflammation of the skin, but tissue temperature is also a useful clinical sign. To diagnose inflammation, clinicians often touch the affected part of the body in order to compare its temperature to that of normal skin. Vasodilation is caused chiefly by complement fragments, histamine, kinins, and prostaglandins.

The traditional home remedy for infection, warm compresses, also augments blood supply. Like inflammatory mediators, warmth enlarges blood vessels. Although there was little that T. L.'s friends could do to help him in their remote mountain camp, the warm compresses they applied to his hand augmented his protective inflammatory response.

Capillaries also respond to inflammatory mediators by becoming more permeable—they become leaky. Normally, the **endothelial cells** that line blood vessels make near-leakproof contact with each other, but during inflammation they separate. This allows leukocytes, proteins, and some fluid to leave the capillaries and enter the inflamed tissue. Increased vascular permeability is caused primarily by complement fragments, kinins, histamine, and leukotrienes. The fluid that enters tissues when blood vessels become more permeable produces **edema** (swelling) in the inflamed area. Edema increases pressure within the tissue, which stimulates nerve endings and causes pain. When T. L.'s friends elevated his arm they kept his inflamed hand from becoming excessively swollen and thereby decreased his discomfort.

Effects on Phagocytes

Phagocytic cells also respond to inflammatory mediators in two ways—activation and chemotaxis. Phagocytes normally exist in a resting state. **Activation** occurs when the phagocyte's cell membrane comes in contact with activating compounds, such as inflammatory mediators. Then a series of biochemical changes turns phagocytes into agents of microbial destruction. The inflammatory mediators that act on phagocytes are bacterial by-products, complement fragments, and prostaglandins.

Chemotaxis—in which chemical agents stimulate cells to move along a concentration gradient (Chapter 4)—attracts phagocytes to regions where the concentration of inflammatory mediators is high. Three types of phagocytes—neutrophils, eosinophils, and macrophages—are chemotactically attracted to inflamed tissues (see the box "Blood: A Complex Body Tissue," p. 388). Neutrophils and eosinophils travel more rapidly, arrive first, and participate primarily in acute inflammation. Macrophages travel more slowly, arrive later, and participate primarily in inflammatory repair.

Many inflammatory mediators, including f-met-leu-phe, complement fragments, and leukotrienes, chemotactically attract phagocytes to an inflamed region. First, the phagocytes migrate to the capillary wall and adhere to it, a process called **margination**. Then the phagocytes squeeze through gaps in the blood vessel's endothelial lining, dissolve the basement membrane layer that normally surrounds all blood vessels, and enter tissues, a process called **diapedesis**. Free in the tissue space, the phagocytes continue to follow the trail of increasing concentrations of chemotactic compounds.

Acute Inflammation

When capillaries and phagocytes are stimulated by inflammatory mediators, we see the clinical manifestations of acute inflammation that occurred in T. L.'s hand. Erythema results from capillary dilation. Edema results from increased fluid in tissues. Warmth results from increased blood flow. Pain results from nerve stimulation due to edema and the direct action of pain-inducing inflammatory mediators. When all of these manifestations are severe, the function of the affected part is limited (**Figure 16.3**).

Acute inflammation is beneficial to the host. Activated phagocytes, drawn to the inflamed tissue by chemoattractants, destroy many of the invading microorganisms. In addition, serum proteins that accumulate in inflamed tissue, including complement and antibodies, contribute to microbial destruction. Inflammation activates and coordinates many different weapons of tissue defense against infection.

But inflammation also damages host tissue. Pain, which we experience when our body is damaged, is one of the hallmarks of inflammation. In addition to killing microorganisms, phagocytes and complement also kill human cells. **Pus**, a mixture of dead leukocytes, microorganisms, and tissue cells, is one manifestation of host damage. The yellow fluid that began to ooze from T. L.'s wound when the inflammation reached its peak was pus.

Inflammatory Repair

In the final stage of inflammation, the tissue damage that occurred during the acute phase is repaired. The first step in **repair** is clean-up. As inflammation wanes, numerous phagocytic scavenger cells called **macrophages** arrive at the infection site and consume dead microorganisms, dead and dying host cells, and any foreign particles that may have found their way into the infected wound. If the inflammation was severe—for example, if large quantities of pus accumulated—this stage of repair may take a long time. As we will see in Chapter 17, T. L.'s hand did not return to normal for two weeks. When the debris of inflammation has been cleared away, cells in the affected tissues regenerate and finally heal. In many cases, like T. L.'s, repair is complete. Permanent damage may result, however, if tissue damage is extensive enough to cause scarring or if infection occurs in a tissue that cannot regenerate. For example, infection of the nervous system can lead to permanent brain damage.

LEUKOCYTES: DEFENDERS OF THE BODY'S INTERIOR

A large and diverse collection of leukocytes is central to the body's defense against infection. There are five types of leukocytes—**neutrophils**, **eosinophils**, **basophils**, **lymphocytes**, and **monocytes**—each with a different function in the immune system. A **leukocyte differential count** determines the number of each type of leukocyte, computed as a percentage of the total number of leukocytes (**Figure 16.4**). Some infections elevate, depress, or alter the composition of the leukocyte population in characteristic ways. Most acute infections, for example, increase the total leukocyte count by dramatically increasing the number of neutrophils. If T. L. had had a **complete blood count** (CBC)—a total count of both leukocytes and erythrocytes—during the early days of his infection, his total white count would probably have been grossly elevated, possibly as high as 20,000 cells/ml. Pertussis, however, dramatically increases the number of lymphocytes (Chapter 15), and some infections, such as shigellosis (Chapter 23), decrease the total white count. Neutrophils are sometimes subdivided into **segs** (segmented or mature neutrophils) and **bands** (band form or immature neutrophils, which have an incompletely segmented nucleus). This distinction is helpful clinically because patients with active infections are producing large numbers of new neutrophils and will therefore have a higher proportion of band forms. T. L. would probably have had a high percentage of bands on his leukocyte differential.

Figure 16.3 Acute inflammation is initiated by a stimulus such as injury or infection. Inflammatory mediators are produced at the site of the stimulus. They cause blood vessels to dilate and increase their permeability; they also attract phagocytes to the site of inflammation and activate them.

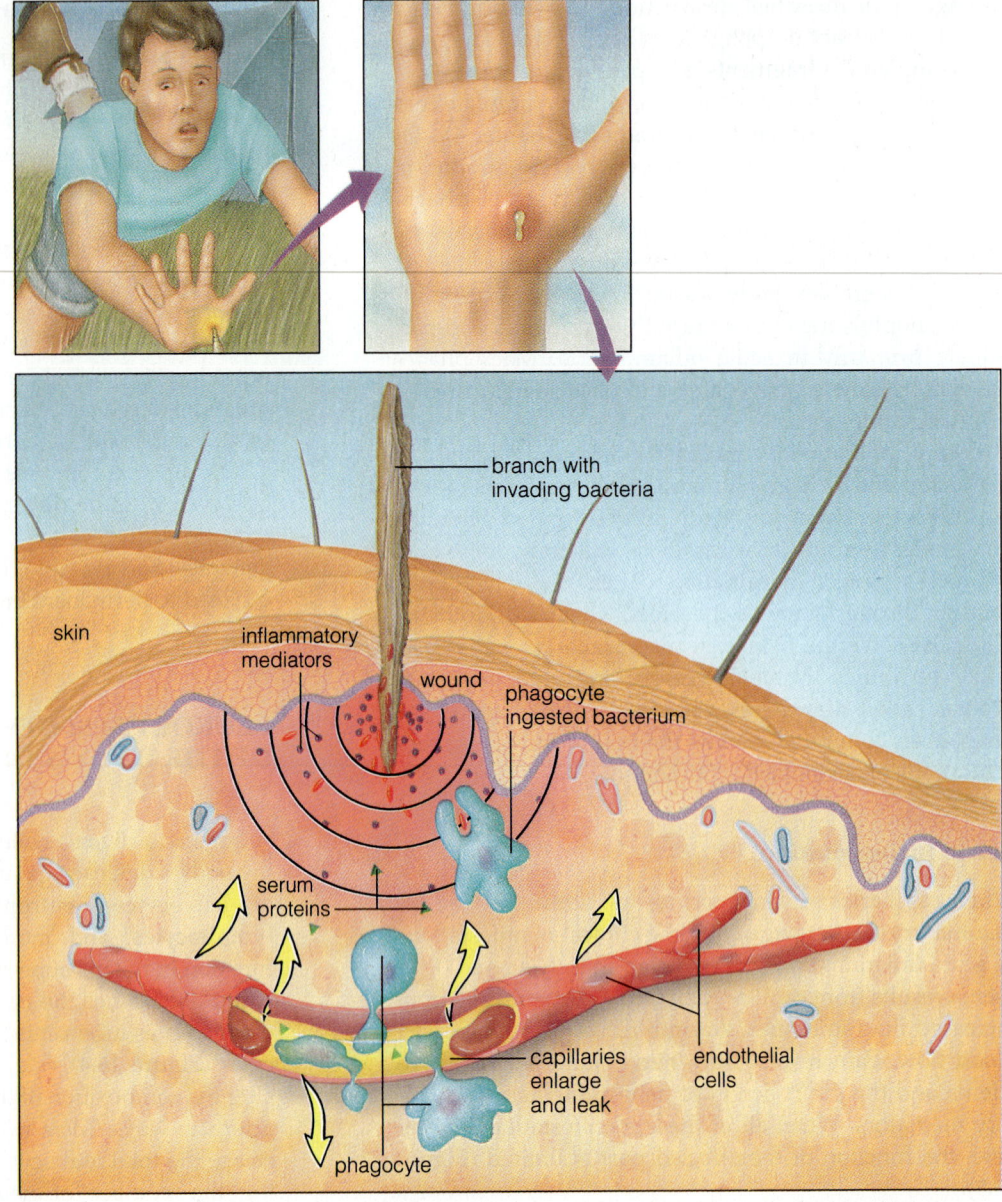

branch with invading bacteria

skin

inflammatory mediators

wound

phagocyte ingested bacterium

serum proteins

capillaries enlarge and leak

endothelial cells

phagocyte

Leukocytes, along with erythrocytes and platelets, are produced in the bone marrow, found in the spaces within certain bones. Active bone marrow contains large **hematopoietic** (blood-forming) **stem cells**, which divide continuously, producing millions of other cells. During these multiple divisions, the stem cells **differentiate**, changing from uniform-appearing cells into the distinctive highly specialized cells that we recognize as erythrocytes and leukocytes.

Early in the course of stem cell differentiation into leukocytes, a split occurs. One branch leads to the **phagocytic family** of cells and the other to the **lymphoid family**

(**Figure 16.5**, p. 390). Both branches defend against infection—the phagocytic family mainly in nonspecific defenses and the lymphoid family mainly in specific defenses of the immune system. Cells of the phagocytic family (not all of which are capable of phagocytosis) are discussed here. Lymphoid cells are discussed in Chapter 17.

The Phagocytic Family

The phagocytic family is composed of two groups, **the polymorphonuclear leukocytes** and the **mononuclear leukocytes**. The various types of member cells perform

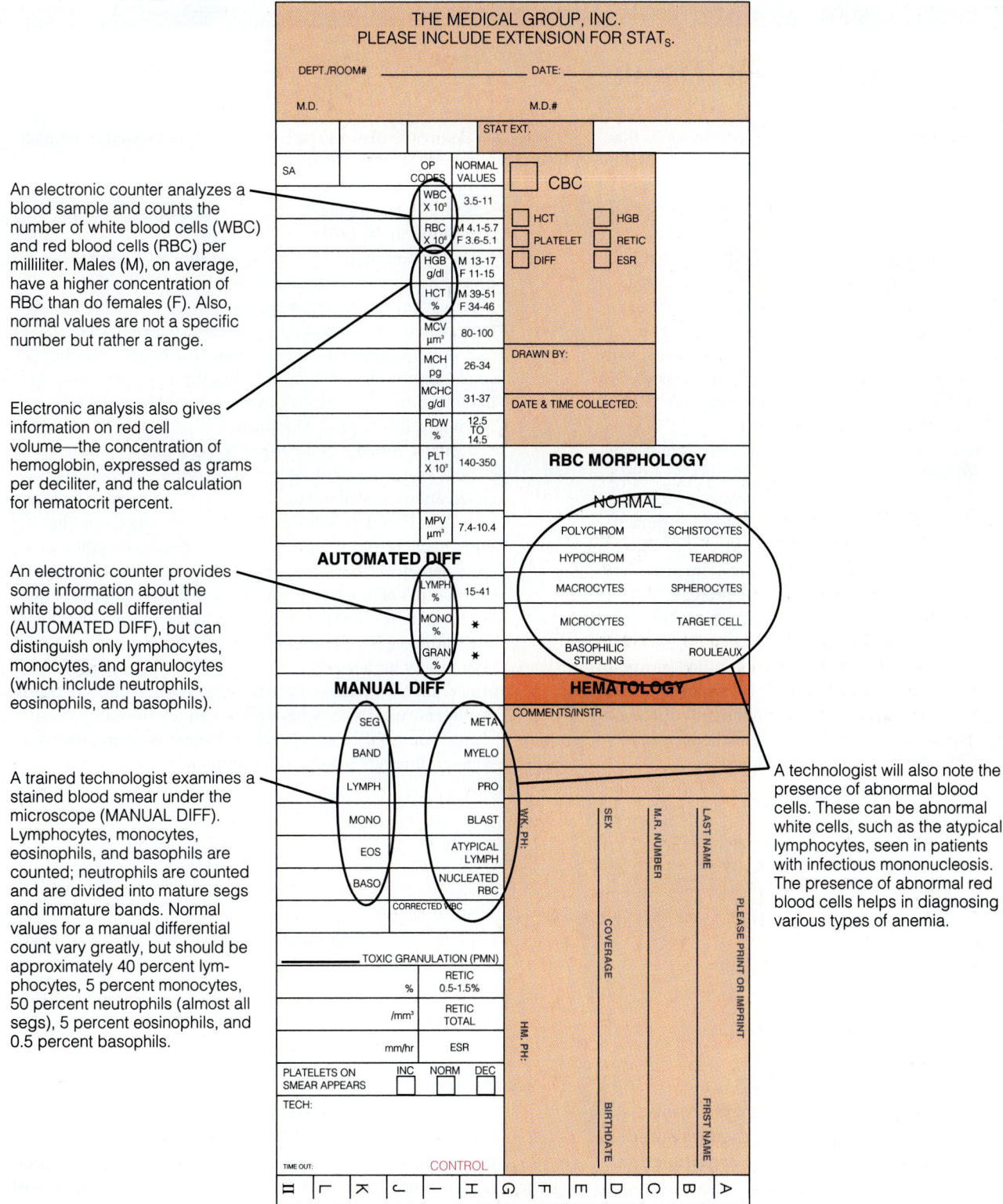

An electronic counter analyzes a blood sample and counts the number of white blood cells (WBC) and red blood cells (RBC) per milliliter. Males (M), on average, have a higher concentration of RBC than do females (F). Also, normal values are not a specific number but rather a range.

Electronic analysis also gives information on red cell volume—the concentration of hemoglobin, expressed as grams per deciliter, and the calculation for hematocrit percent.

An electronic counter provides some information about the white blood cell differential (AUTOMATED DIFF), but can distinguish only lymphocytes, monocytes, and granulocytes (which include neutrophils, eosinophils, and basophils).

A trained technologist examines a stained blood smear under the microscope (MANUAL DIFF). Lymphocytes, monocytes, eosinophils, and basophils are counted; neutrophils are counted and are divided into mature segs and immature bands. Normal values for a manual differential count vary greatly, but should be approximately 40 percent lymphocytes, 5 percent monocytes, 50 percent neutrophils (almost all segs), 5 percent eosinophils, and 0.5 percent basophils.

A technologist will also note the presence of abnormal blood cells. These can be abnormal white cells, such as the atypical lymphocytes, seen in patients with infectious mononucleosis. The presence of abnormal red blood cells helps in diagnosing various types of anemia.

Figure 16.4 Laboratory report of complete blood count and leukocyte differential. Some of the information comes from running a blood sample through an electronic counter. Other information must be supplied by a trained technologist examining a blood smear.

Blood: A Complex Body Tissue

We are all familiar with blood—the brilliant red liquid that appears at even the tiniest wound. What is not immediately obvious is that blood is a complex body tissue, composed of many different types of cells and molecules, including water.

We can see that blood is composed of different elements when it is spun in a centrifuge—an instrument that separates materials by density, size, and shape. Centrifuging a sample of blood is a routine part of diagnosing many illnesses. The procedure is quick and simple. The patient's finger is punctured with a small lance and blood is collected in a tube that has been treated to prevent clotting. The tube is centrifuged for five minutes and examined.

The upper half of the tube is filled with clear liquid. This is **plasma**, the fluid component of blood. Plasma contains no cells, but it does contain many different proteins, including complement (which is discussed in this chapter), **albumin** (which increases the osmotic strength of blood and helps keep it in the circulatory system), **globulins** (which include the antibodies), and **fibrinogen** (one of the proteins that allows blood to clot). Because plasma contains clotting proteins, it will clot if allowed to sit outside the body. The fluid left after the clot is removed from plasma is called **serum**. In addition to proteins, plasma also contains various ions, sugars, lipids, amino acids, hormones, vitamins, and dissolved gases. Many of the substances in plasma are being transported from one part of the body to another.

Below the plasma is a barely discernible band of whitish material called the **buffy coat**. This small but important component of blood is made up of leukocytes. **Platelets**, subcellular fragments that participate in blood clotting, are also found in the buffy coat.

At the bottom of the centrifuged tube is a long red column made up of **erythrocytes** (red blood cells). They contain the iron-containing pigment hemoglobin, which binds oxygen and gives blood its color. Erythrocytes transport oxygen and carbon dioxide between the lungs and other tissues. The erythrocyte volume, measured as a percentage of the total blood volume, is called the blood **hematocrit**. Normally, it is around 45 percent.

Usually, blood is centrifuged to get a quick reading of the patient's hematocrit. A decrease in erythrocyte volume, called **anemia**, is characteristic of many different disorders, including iron deficiency and certain deficiencies of vitamins, such as foliate and vitamin B_{12}. But examination of a spun hematocrit tube can also provide clues to other diagnoses, often before the results of more complicated laboratory tests can be obtained. For example, the plasma of patients with jaundice is yellow, while the plasma of people with abnormally high levels of fat in their blood has a milky appearance. Patients with leukemia, whose leukocyte count is often very high, may have an unusually thick buffy coat layer. People suffering from dehydration have an unusually wide erythrocyte band and a diminished plasma layer.

Spun hematocrit tube and components of blood.

two critical functions. They destroy microorganisms by phagocytosis and they release toxic chemicals and inflammatory mediators (a process called **degranulation**).

Polymorphonuclear leukocytes are named for their distinctive irregularly shaped nuclei with several arms or lobes. They are also called **granulocytes** because their cytoplasm is filled with granules containing antimicrobial chemicals and enzymes. Polymorphonuclear leukocytes are subdivided into neutrophils, eosinophils, and basophils, based on the staining characteristics of their granules. Neutrophils stain poorly (neutral); eosinophils stain red with acidic dyes such as eosin; and basophils stain blue with basic dyes such as methylene blue. All polymorphonuclear leukocytes are capable of degranulation, but only neutrophils and eosinophils are capable of phagocytosis.

Mononuclear leukocytes have a compact nucleus and no visible cytoplasmic granules. They are sometimes called **agranulocytes**, although we now know that they contain granules. When mononuclear leukocytes are released from the bone marrow into the bloodstream, they are called *monocytes*. When they migrate into tissues and

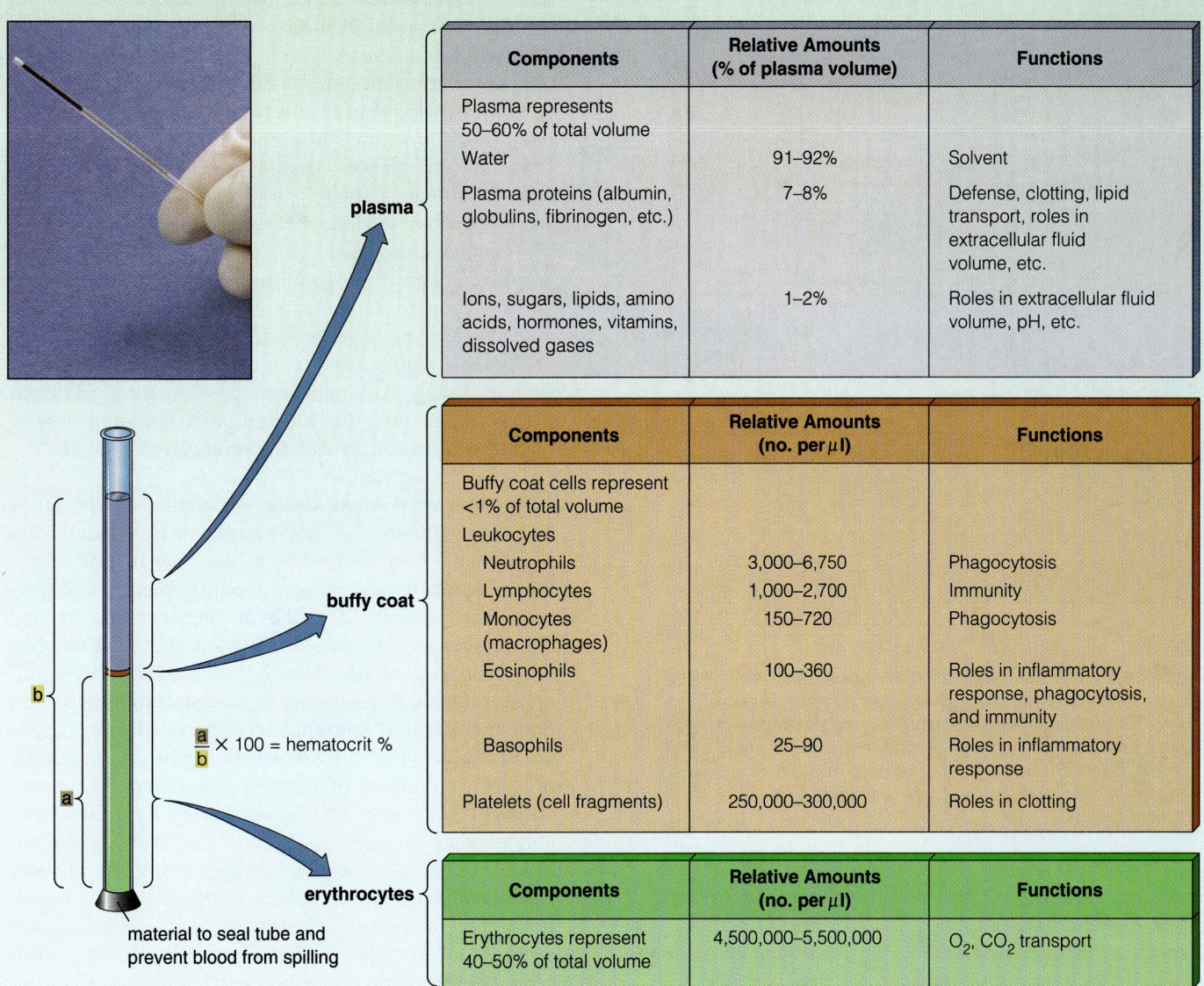

Components	Relative Amounts (% of plasma volume)	Functions
Plasma represents 50–60% of total volume		
Water	91–92%	Solvent
Plasma proteins (albumin, globulins, fibrinogen, etc.)	7–8%	Defense, clotting, lipid transport, roles in extracellular fluid volume, etc.
Ions, sugars, lipids, amino acids, hormones, vitamins, dissolved gases	1–2%	Roles in extracellular fluid volume, pH, etc.

Components	Relative Amounts (no. per μl)	Functions
Buffy coat cells represent <1% of total volume		
Leukocytes		
Neutrophils	3,000–6,750	Phagocytosis
Lymphocytes	1,000–2,700	Immunity
Monocytes (macrophages)	150–720	Phagocytosis
Eosinophils	100–360	Roles in inflammatory response, phagocytosis, and immunity
Basophils	25–90	Roles in inflammatory response
Platelets (cell fragments)	250,000–300,000	Roles in clotting

Components	Relative Amounts (no. per μl)	Functions
Erythrocytes represent 40–50% of total volume	4,500,000–5,500,000	O_2, CO_2 transport

$\dfrac{a}{b} \times 100$ = hematocrit %

material to seal tube and prevent blood from spilling

plasma

buffy coat

erythrocytes

undergo further differentiation, they become *macrophages*. Mononuclear cells are extremely effective at phagocytosis and degranulation.

Neutrophils. Over half the leukocytes found in the circulation are neutrophils. Adults normally have about 5000 neutrophils in each milliliter of blood, but during certain infections this number may increase two- to threefold. Neutrophils live for only a few days, but they are produced in the bone marrow at the astounding rate of nearly 80 million per minute. Neutrophils normally reside in the bloodstream, but when attracted by inflammatory mediators, they migrate to infected tissues.

The granules within neutrophils contain various potent chemicals, some of which attack only bacteria. Lysozyme, for example, lyses Gram-positive bacteria by breaking peptidoglycan bonds. Lactoferrin binds an essential nutrient, iron, rendering it unavailable to bacteria (Chapter 15).

Other chemicals in neutrophilic granules damage both microorganisms and human cells. **Acid hydrolase**

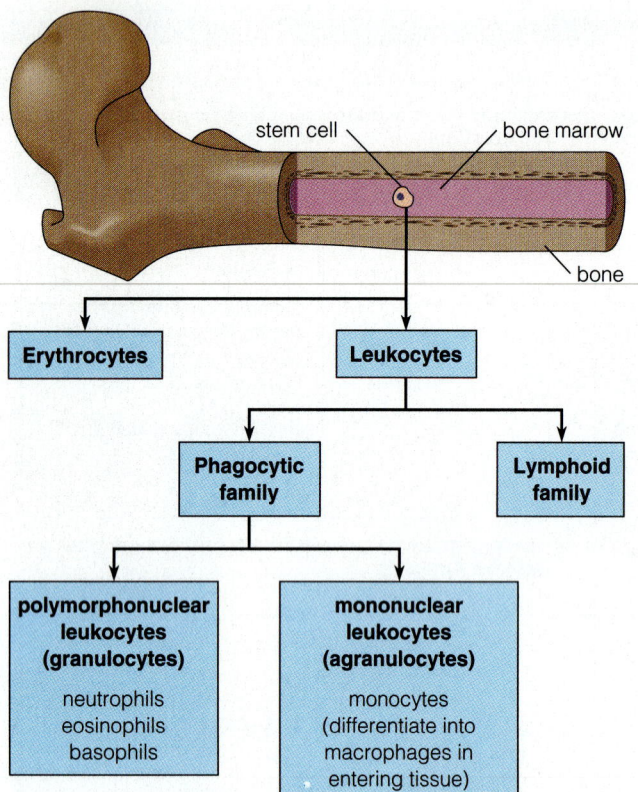

stem cell

bone marrow

bone

Erythrocytes

Leukocytes

Phagocytic family

Lymphoid family

polymorphonuclear leukocytes (granulocytes)

neutrophils
eosinophils
basophils

mononuclear leukocytes (agranulocytes)

monocytes (differentiate into macrophages in entering tissue)

Figure 16.5 Stem cell differentiation. Hematopoietic stem cells in the bone marrow divide continuously. During these multiple divisions, they differentiate, changing from uniform-appearing cells into the cells we recognize as leukocytes and erythrocytes. Early in stem cell differentiation into leukocytes, a split occurs. One branch leads to the phagocytic family of cells and the other to the lymphoid family.

degrades cellular proteins, and **myeloperoxidase** catalyzes reactions that produce lethal oxidants, including hypochlorous acid and free chlorine. Hydrogen peroxide and hydroxyl radicals are themselves lethal oxidants produced when neutrophils are activated by infection or other stresses.

These enzymes and lethal oxidants are harmless as long as they remain inside the neutrophilic granules. However, when released into intracellular vacuoles they kill microbial cells that have been consumed during phagocytosis. Released into the external environment by degranulation, they act as inflammatory mediators, help activate other phagocytes, and attack any microorganisms that might be nearby. Degranulation of T. L.'s neutrophils helped to destroy many invading bacteria, but it also contributed to the redness and pain of his infected hand.

Eosinophils. Most people have only about 250 eosinophils in each microliter of blood, about 5 percent of

their total number of leukocytes. Eosinophilic granules, which are larger than neutrophilic granules, contain a dense core of eosin-staining basic protein and many destructive enzymes, including acid phosphatase, peroxidases, and proteinases.

Like neutrophils, eosinophils are capable of phagocytosis, but their primary function is releasing their contents into the extracellular environment by degranulation. The substances they release defend primarily against pathogens, such as protozoal parasites, fungi, and worms, that are too large to be consumed by phagocytes. The eosin-staining basic protein that is released, for example, damages the cuticle (outer covering) of parasitic worms. People who suffer from parasitic infection usually have high numbers of eosinophils in their blood.

But eosinophils also play a role in acute inflammation such as T. L.'s. Although their precise role is not completely understood, they are thought to moderate inflammatory destruction, preventing serious tissue damage.

Basophils and Mast Cells. Basophils are the rarest and least understood polymorphonuclear leukocytes. Most people have fewer than 40 basophils in each microliter of blood; this constitutes about 0.5 percent of leukocytes. Basophils are incapable of phagocytosis, but they degranulate in response to different stimuli. Basophils are closely related to **mast cells**, but mast cells differentiate and remain in tissues while basophils differentiate in bone marrow and migrate to the blood. Mast cell granules contain potent inflammatory mediators, including high concentrations of histamine and **heparin**, a substance that interferes with blood clotting. Basophils probably play a critical role in defending against parasitic infection, but most of our knowledge about mast cells and basophils relates to their role in **allergy**. Allergy is a malfunction of the body's defenses in which harmless agents such as pollen cause degranulation (Chapter 18). Mast cells also degranulate in response to infection or tissue injury, such as the puncture in T. L.'s palm. The degranulation releases inflammatory mediators that help activate the body's inflammatory defenses.

Monocytes and Macrophages. Monocytes and macrophages are the mononuclear members of the phagocytic cell family. They are extraordinarily active as phagocytes, consuming foreign material within minutes after it enters the circulation.

Monocytes from the bone marrow enter the bloodstream, where they accumulate at concentrations of approximately 400 cells per microliter. Monocytes become macrophages when they migrate into tissues, where they can survive for weeks to months. **Wandering macrophages** travel to tissues that are under attack by microorganisms. **Fixed macrophages** remain permanently in a particular tissue. Recent research indicates that activated macrophages produce a highly toxic gas, nitric oxide (NO), from the amino acid arginine. Nitric oxide plays a key role in the ability of macrophages to destroy microorganisms and tumor cells.

Macrophages no doubt helped eliminate the bacteria in T. L.'s infected hand during early stages of his infection. They also played the essential role of scavenging dead cells and other debris during inflammatory repair. Macrophages also participate in the body's specific immune defenses (Chapter 17).

Phagocytosis

Phagocytosis is the process by which leukocytes contact, surround, consume, and destroy microorganisms and other foreign particles (**Figure 16.6**). It is indispensable to controlling bacterial infections. Even when sophisticated immune defenses are activated, almost all the invading bacteria are killed by phagocytosis. Phagocytosis of microorganisms (or other particles) can be divided into six steps (**Figure 16.7**).

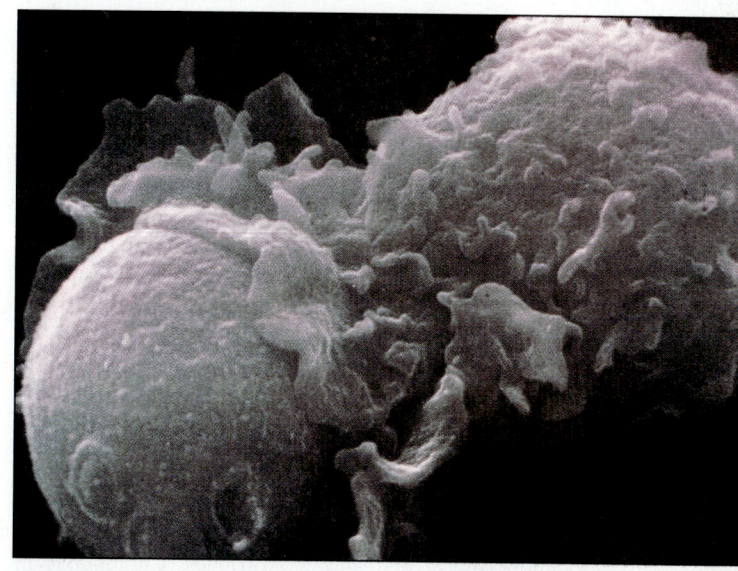

Figure 16.6 This scanning electron micrograph shows a macrophage ingesting a yeast cell during phagocytosis.

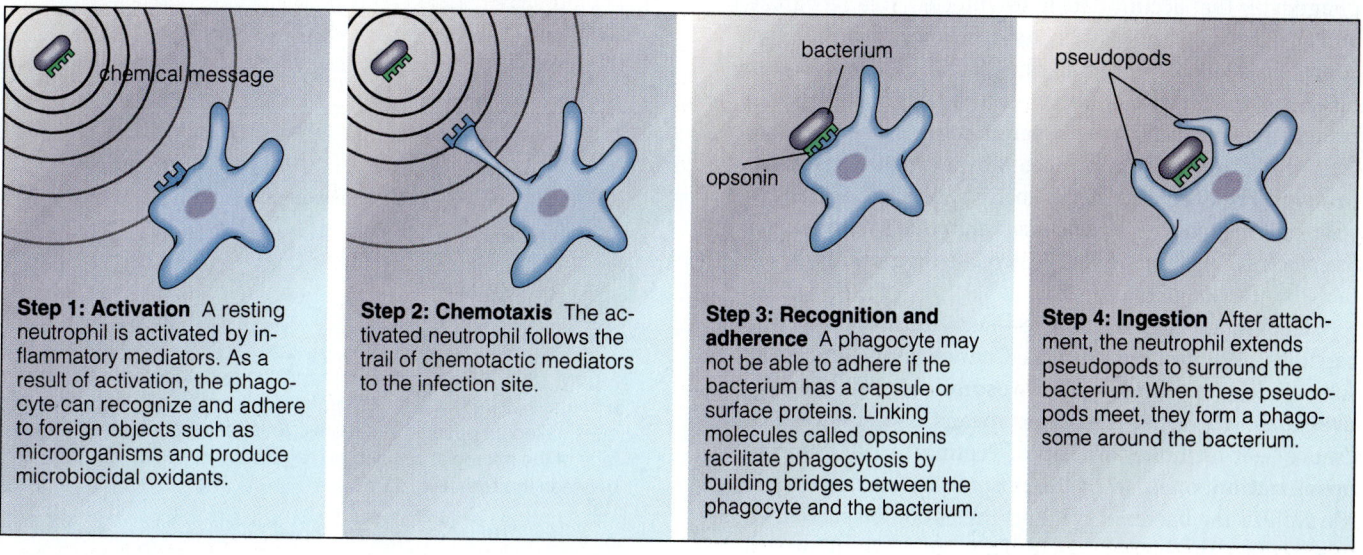

Step 1: Activation A resting neutrophil is activated by inflammatory mediators. As a result of activation, the phagocyte can recognize and adhere to foreign objects such as microorganisms and produce microbiocidal oxidants.

Step 2: Chemotaxis The activated neutrophil follows the trail of chemotactic mediators to the infection site.

Step 3: Recognition and adherence A phagocyte may not be able to adhere if the bacterium has a capsule or surface proteins. Linking molecules called opsonins facilitate phagocytosis by building bridges between the phagocyte and the bacterium.

Step 4: Ingestion After attachment, the neutrophil extends pseudopods to surround the bacterium. When these pseudopods meet, they form a phagosome around the bacterium.

Step 5: Killing and digestion The phagosome fuses with a lysosome to create a phagolysosome where the bacterium is killed and then digested.

Step 6: Expulsion The phagolysosome fuses with the cell membrane to expel the undigestible parts of the bacterium.

Figure 16.7 Steps in phagocytosis.

Step 1: Activation. Leukocytes must be activated by inflammatory mediators to become effective phagocytes. Activation stimulates the production of glycoprotein receptors on the leukocyte's cell membrane, which increase the cell's ability to adhere to surfaces and to recognize foreign objects such as bacteria. Activation also triggers a **respiratory burst**, when the phagocyte's granules produce lethal oxidants as its metabolism switches from anaerobic glycolysis to aerobic respiration. Inflammatory mediators that activate leukocytes include bacterial by-products, complement fragments, leukotrienes, and prostaglandins.

Step 2: Chemotaxis. Phagocytes migrate along a chemical gradient toward tissues where inflammatory mediators are present in high concentrations. Inflammatory mediators that stimulate this chemotaxis include bacterial by-products, complement fragments, and leukotrienes. Chemotaxis of phagocytes to infected tissue is rapid and effective. When chemotactic mediators are released, large numbers of phagocytes begin to accumulate within an hour at the infection site.

Step 3: Recognition and Adherence. The activated phagocytes that accumulate in an infected area recognize and adhere to a variety of bacterial cells by means of glycoprotein receptors on the phagocytic cell membrane. However, the attachment between the phagocyte's glycoprotein receptor and the bacterial cell surface may not form if the bacterium is protected by a capsule or surface proteins. For example, the protective polysaccharide capsule of *Streptococcus pneumoniae* and the M protein of *Streptococcus pyogenes* greatly increase their ability to establish infection (Chapter 15).

Recognition and adherence—even in the presence of capsules and proteins—proceed much more efficiently when linking molecules called **opsonins** are present (**Figure 16.8**). (*Opsonin* in Greek means "seasoning," or "sauce"—it facilitates the job of "eating" a cell.) During **opsonization**, one part of the opsonin molecule binds to the wall of the bacterial cell and then another part binds to a receptor on the phagocyte. Thus, an opsonin forms a bridge between bacterium and phagocyte to facilitate phagocytosis. The two most important opsonins are complement and antibodies. A bacterium that has bound enough opsonins to coat its surface is said to be *opsonized* and is readily phagocytized as soon as the bridge is formed to a leukocyte.

Step 4: Ingestion. After it attaches to a bacterial cell, the phagocyte extends pseudopods that engulf the bacterium. When the tips of the pseudopods meet, they fuse, forming a **phagosome**, an internal vacuole of cell membrane that contains the bacterium. The phagosome separates from the cell membrane and becomes an internal cell structure.

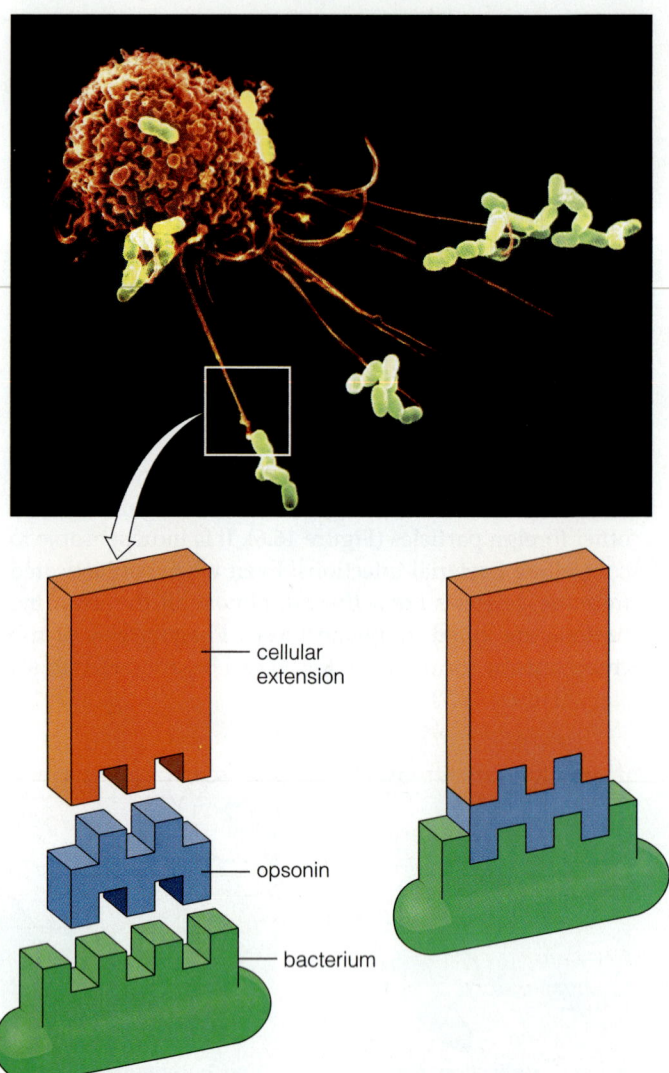

cellular extension

opsonin

bacterium

Figure 16.8 Opsonization. Recognition and adherence between a phagocyte and its prey are greatly facilitated if a linking molecule, called an opsonin, is present. An opsonin attaches to the surface of the microbial cell and to the phagocyte, building a bridge between the two cells.

Step 5: Killing and Digestion. When a phagosome fuses with lysosomes in the cell cytoplasm, they form a **phagolysosome**. The bacterium inside is killed by microbiocidal chemicals—including lysozyme, lactic acid, and other enzymes and oxidants—in the lysosome. Working together, these agents kill the bacterium in 30 minutes or less. Then enzymes also present in the lysosome digest almost all the bacterial remains.

Although phagocytosis is extremely effective in destroying most pathogens, including those that infected T. L.'s hand, some microorganisms have evolved ways to survive, and even multiply, within phagocytes. Growth within phagocytes is one type of intracellular infection. One bacterium that sustains an infection within phago-

Clinical Notes

When Phagocytes Don't Work

Chronic granulomatous disease (CGD) is an inherited disorder in which phagocytes don't kill normally. The patient's phagocytes are normal-looking, but they can't mount a normal respiratory burst and therefore cannot kill the microorganisms that they ingest. Patients with CGD suffer greatly from this failure of phagocytosis. Most develop serious and recurrent bacterial infections before the age of 2. They often develop multiple abscesses and large, infected lymph nodes that must be opened surgically before they will heal. They suffer repeated episodes of pneumonia, often caused by pathogens that are not usually virulent. The average life expectancy of a child with CGD is 10 years. Good health depends on phagocytosis.

---

cytes is *Mycobacterium tuberculosis. M. tuberculosis* must establish an infection within phagocytes in order to cause tuberculosis (Chapters 15 and 22).

Step 6: Expulsion. After the bacterium is killed and digested, the phagolysosome again fuses with the cell membrane and expels indigestible debris.

COMPLEMENT

Complement, or the complement system, refers to a family of more than 30 different proteins in serum that function together as a nonspecific defense against infection. The complement system has three major defensive functions—to produce inflammatory mediators, to bind opsonins to microbial cells, and to form a membrane attack complex that lyses cells, including many Gram-negative bacteria. Still other complement proteins regulate the complement system and prevent excessive damage to host tissues.

Many different proteins play a part in this sophisticated system, but we focus on the proteins associated with the key events in fighting microbial infection. These include a series of proteins designated C1–C9. Fragments produced when these proteins split are designated by letters (for example, C3 splits into C3a and C3b).

Complement was named for its ability to "complement," or complete, the action of antibodies (Chapter 17). Thus, the complement system operates most efficiently in conjunction with antibodies, but it plays a significant, though lesser, role early in the course of infections like T. L.'s—before antibodies are produced. Complement is a nonspecific defense because it acts against many different microorganisms and it initiates some defensive actions as soon as microorganisms enter the body.

In the following sections, we discuss the complement cascade and then its three arms—the classical pathway, the alternate pathway, and the terminal pathway.

The Complement Cascade

Complement proteins must be activated before they can function. Activation requires two events. First, a complement protein is split into smaller fragments, some of which split again and activate other proteins. Second, some fragments bind to a membrane surface. (Some also act as inflammatory mediators, but this is not part of activation.) Complement activation occurs as a **cascade**, a series of reactions in which the product of one reaction activates the next reaction. As the cascade proceeds it expands because *each* product molecule activates many subsequent conversions.

The complement cascade can be triggered in two different ways, termed the **classical pathway** and the **alternate pathway**. Both of these pathways lead to the activation of C3, the central reaction of the cascade, and then to the **terminal pathway** (Figure 16.9).

The Classical Pathway

In the classical complement pathway the proteins C1, C4, and C2 are activated, in that order (**Figure 16.10**). C1 is a particularly large and complex molecule—large enough to be seen under the electron microscope. C1 is composed of three subunits—C1q, C1r, and C1s. C1q, the largest subunit, fulfills the special function of recognizing and binding to an **antigen-antibody complex**—the molecular combination that sets off the classical pathway. (An antigen-antibody complex is a physical combination of antibody and the **antigen** it recognizes; see Chapter 17.) C1r and C1s activate each other and, in turn, C4 and C2, splitting each into two fragments. One of the C4 fragments

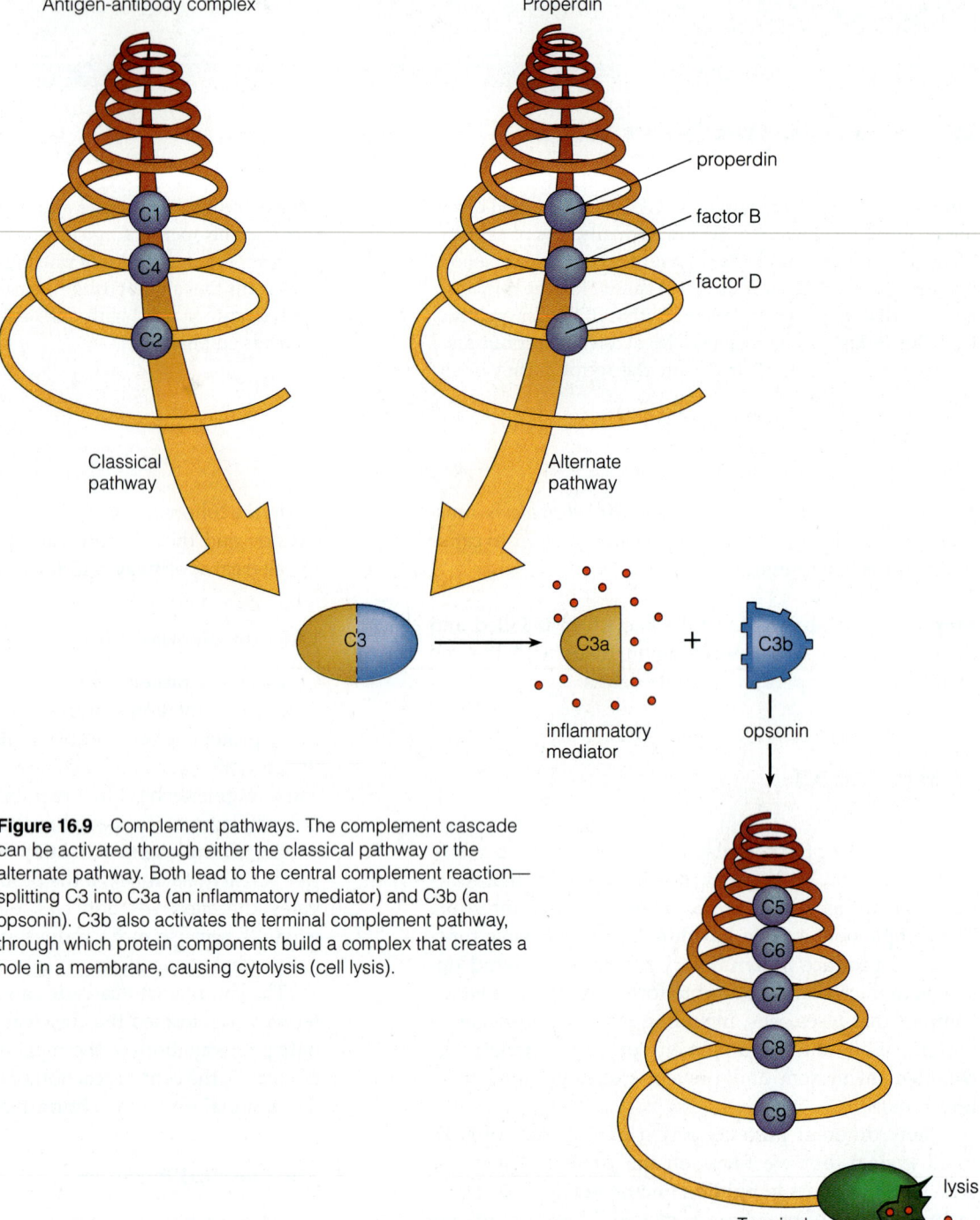

Figure 16.9 Complement pathways. The complement cascade can be activated through either the classical pathway or the alternate pathway. Both lead to the central complement reaction—splitting C3 into C3a (an inflammatory mediator) and C3b (an opsonin). C3b also activates the terminal complement pathway, through which protein components build a complex that creates a hole in a membrane, causing cytolysis (cell lysis).

and one of the C2 fragments combine to activate C3, cleaving it into fragments called C3a, which is released, and C3b, which is bound to the pathogen's membrane.

C3b is a powerful opsonin. It binds to the surface of invading bacterial cells, creating protein structures that match receptors on neutrophils, eosinophils, monocytes, and macrophages. Bacteria that have been opsonized by C3b are coated with these proteins, allowing phagocytes

to recognize and adhere to them. C3b is a critical component of the defense system because it enables phagocytes to consume and destroy even highly virulent pathogens that are protected by capsules or antiphagocytic proteins. People born without the ability to form C3b opsonins suffer from repeated and life-threatening infections by highly virulent bacteria such as *Streptococcus pneumoniae* (Chapter 22).

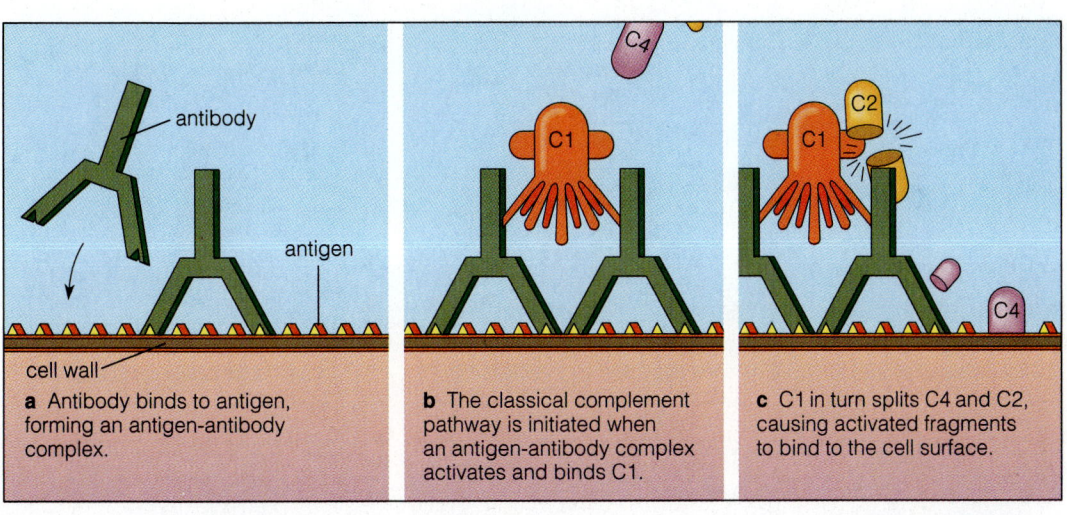

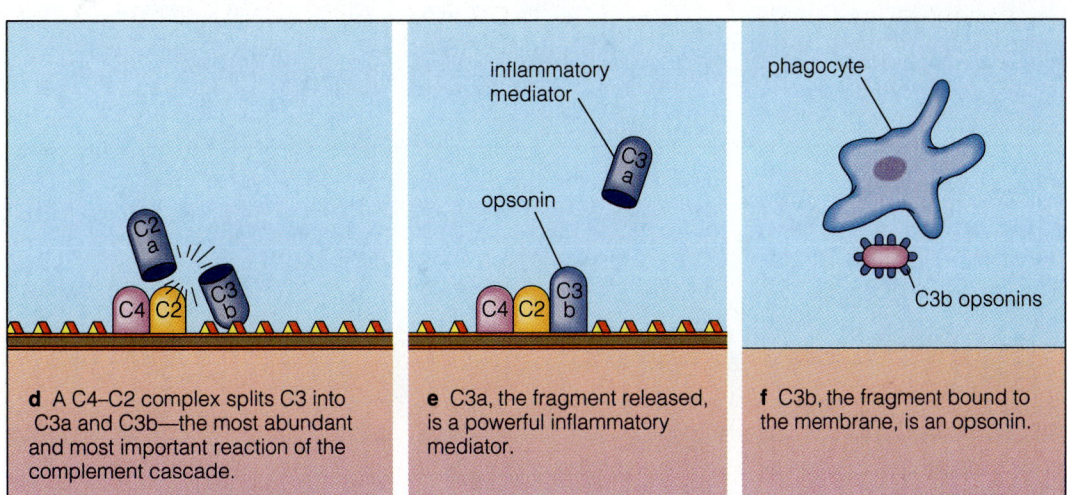

Figure 16.10 Classical complement pathway.

a Antibody binds to antigen, forming an antigen-antibody complex.

b The classical complement pathway is initiated when an antigen-antibody complex activates and binds C1.

c C1 in turn splits C4 and C2, causing activated fragments to bind to the cell surface.

d A C4–C2 complex splits C3 into C3a and C3b—the most abundant and most important reaction of the complement cascade.

e C3a, the fragment released, is a powerful inflammatory mediator.

f C3b, the fragment bound to the membrane, is an opsonin.

C3a, the other activated C3 fragment, is a powerful inflammatory mediator. It brings about both vasodilation and increased vascular permeability. Human skin exposed to even minute amounts of C3a immediately becomes red and swollen. C3a also stimulates basophils and mast cells to degranulate, releasing additional inflammatory mediators, such as histamine, into nearby tissues.

The Alternate Pathway

The alternate complement pathway activates C3 by means of three additional complement proteins called factor B, factor D, and properdin (**Figure 16.11**). Unlike the classical pathway, the alternate pathway does not require the participation of antibody. Scientists learned about this pathway long after complement was first discovered. When they realized the classical pathway was not the only route to activating complement, they named the new route the "alternate" pathway.

Activation of the alternate pathway requires three steps: (1) breaking an internal bond in the C3 molecule, which allows it to bind to factor B, (2) addition of factor D to form an active molecule that binds to properdin, forming a stable molecular complex, and (3) cleavage of C3 into C3a and C3b by the stable complex. C3a acts as an inflammatory mediator, and C3b as an opsonin, just as they do when formed in the classical pathway. C3b is a less effective opsonin when formed by the alternate pathway rather than by the classical pathway, probably because antibody helps position it optimally on the microbial cell.

Activation of the alternate pathway can take place on any surface where the appropriate four molecules come together, but the surface of bacterial cells is particularly effective because bacterial surfaces bear repeating polysaccharide molecules that dramatically increase the rate

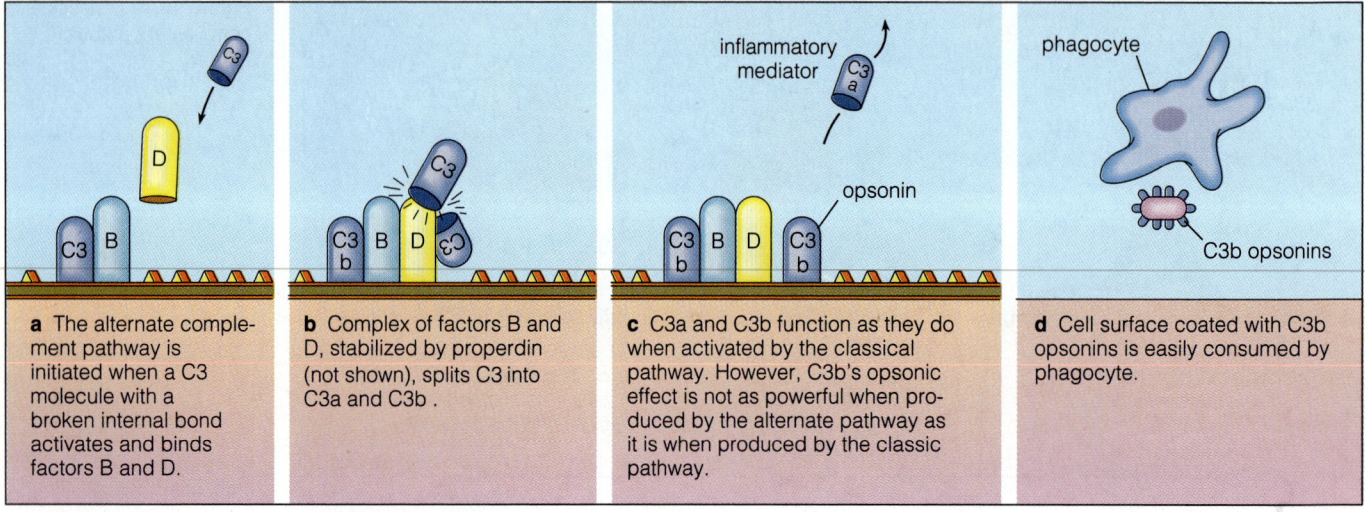

a The alternate complement pathway is initiated when a C3 molecule with a broken internal bond activates and binds factors B and D.

b Complex of factors B and D, stabilized by properdin (not shown), splits C3 into C3a and C3b.

c C3a and C3b function as they do when activated by the classical pathway. However, C3b's opsonic effect is not as powerful when produced by the alternate pathway as it is when produced by the classic pathway.

d Cell surface coated with C3b opsonins is easily consumed by phagocyte.

Figure 16.11 Alternate complement pathway.

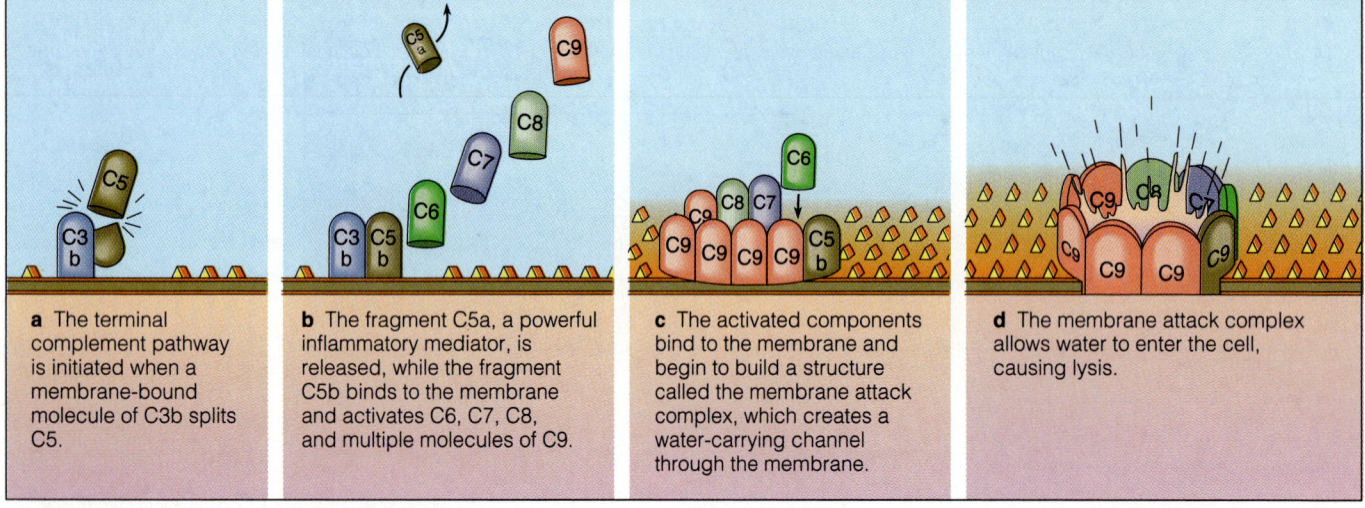

a The terminal complement pathway is initiated when a membrane-bound molecule of C3b splits C5.

b The fragment C5a, a powerful inflammatory mediator, is released, while the fragment C5b binds to the membrane and activates C6, C7, C8, and multiple molecules of C9.

c The activated components bind to the membrane and begin to build a structure called the membrane attack complex, which creates a water-carrying channel through the membrane.

d The membrane attack complex allows water to enter the cell, causing lysis.

Figure 16.12 Terminal complement pathway.

of activation. Thus, invading microorganisms stimulate the activation of complement, making the alternate pathway an important early, nonspecific line of host defense. The infection in T. L.'s hand provides a good example: During the first few days after his injury, his body had not yet produced antibodies, as we shall see in Chapter 17, so the classical complement pathway was not activated. But the alternate pathway was activated as soon as microbial cells entered his tissues.

The Terminal Complement Pathway

Cleavage of C3 to produce C3a, an inflammatory mediator, and C3b, an opsonin, is the culmination of the classical and alternate pathways. But C3b does more than act as an opsonin. It also combines with C5 to initiate the terminal pathway (**Figure 16.12**).

C3b splits C5, releasing C5a, another inflammatory mediator. C5b remains attached to C3b. C5b activates C6, C7, C8, and multiple molecules of C9, all of which form themselves into a growing membrane-bound complex. As the complex enlarges with the addition of more and more C9 molecules, it becomes increasingly hydrophobic and begins to insert itself through the membrane. Eventually, it forms the **membrane attack complex**, a cylinder-like channel through the membrane that allows water to enter the cell. This leads to **cytolysis**.

Gram-negative bacteria, which have the outer mem-

Complement Out of Control

Complement out of control can be nearly as serious as an inadequate complement defense. Hereditary angioedema, in which patients lack the ability to produce a normal C1 inhibitor, is one such disorder. Without this inhibitor, uncontrolled C1 activity leads to the splitting of C2 and the release of a protein fragment that dilates nearby blood vessels and causes tissue swelling. When C1 is activated by injury, vigorous exercise, or emotional stress, swelling occurs very

rapidly. In some areas, such as the arms or legs, edema may not be much of a problem, but swelling of the intestines can cause severe cramps, and swelling near the throat can cause death from suffocation. Interestingly, patients with hereditary angioedema rarely experience problems until they are older children or teenagers. The attacks last two or three days when they do occur.

---

brane exposed, tend to be vulnerable to cytolysis by the terminal complement pathway. (Some species have evolved mechanisms of serum resistance that protect them from complement, as discussed in Chapter 15.) In contrast, Gram-positive bacteria, which have the cell wall rather than a membrane on their outer surface, are generally not vulnerable. Although complement-mediated cytolysis is a significant defense against bacterial infection, it is not as important clinically as opsonization. People who cannot form a normal membrane attack complex remain relatively healthy, but they are especially sensitive to infections by pathogens of the genus *Neisseria* (Chapters 24 and 25).

Human cells, which are also enclosed by a membrane, can be lysed by complement under certain abnormal circumstances. In Chapter 18 we discuss the complications that occur when complement lyses red blood cells after a blood transfusion or when a mother and unborn infant have incompatible blood types. And in Chapter 19, we discuss how complement-mediated lysis of human cells can be used in the laboratory.

INTERFERON

The nonspecific interior defenses just discussed help overcome bacterial infections like the one T. L. suffered. Now, let's turn to a nonspecific interior defense that helps us overcome viral infections—**interferon**.

Interferons were discovered when scientists found that animals infected by viruses produced small proteins that helped protect against spread of the infection. These proteins were named interferons because of their ability to "interfere" with viral replication. A remarkable feature of these natural antiviral agents is their startling potency. They impair viral replication when present in concentra-

tions as low as $3 \times 10^{-14} M$—or approximately 1 molecule for every 2 thousand trillion water molecules.

Three main classes of interferons have been identified—**alpha interferon**, produced by T cells (immune system cells); **beta interferon**, produced by fibroblasts (a type of tissue cell); and **gamma interferon**, also produced by T cells. There are 13 different alpha interferons, 5 different betas, and an unknown number of gammas. Interferons are classified on the basis of chemistry and genetics, rather than function, which overlaps among the three groups. In general, the most powerful antivirals belong to the alpha and beta groups, while many of the most powerful immune regulators (Chapter 17) belong to the gamma group.

Cells that have been infected by a virus are stimulated to produce and release interferon molecules, which are taken up by healthy neighboring cells. These healthy cells produce **antiviral proteins (AVPs)**—enzymes that interfere with viral protein synthesis and stem the spread of infection (**Figure 16.13**). Cells are stimulated to produce interferon by most major types of viruses and by some species of bacteria and protozoa, double-stranded RNA, and endotoxins.

Interferons are produced by many different types of vertebrate hosts, from reptiles to human beings. Each interferon is **host-specific**, meaning that it defends only against infection in other animals of the same species. However, interferons are **virus-nonspecific**, meaning that the same interferon is active against many different viral pathogens. Thus, interferons are a nonspecific host defense—the same weapon combats many types of invaders.

At one time, scientists hoped that interferon might be used to treat some of our most serious diseases. Their antiviral effect made them logical possibilities for treating viral infections, and their ability to slow cell division made them possible agents for treating cancer. Enthusi-

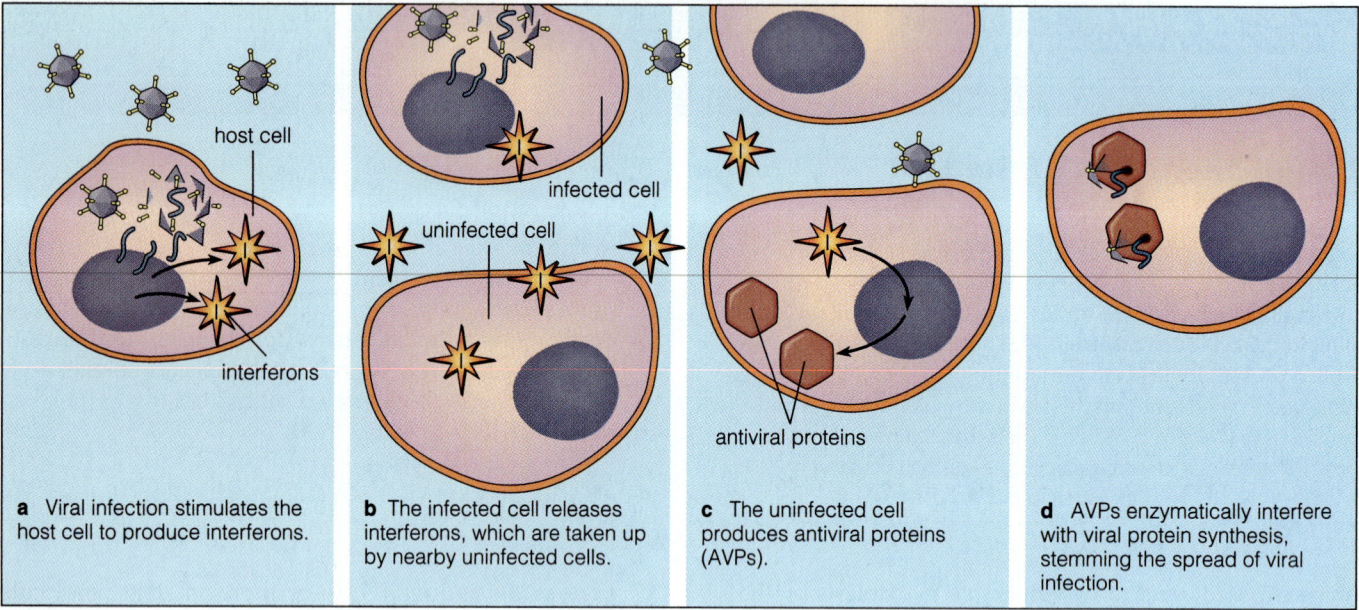

a Viral infection stimulates the host cell to produce interferons.

b The infected cell releases interferons, which are taken up by nearby uninfected cells.

c The uninfected cell produces antiviral proteins (AVPs).

d AVPs enzymatically interfere with viral protein synthesis, stemming the spread of viral infection.

Figure 16.13 Antiviral action of interferon.

asm soared when it became possible to produce large quantities of pure human interferon using recombinant DNA technology. As is often the case with new therapeutic agents, however, the results of interferon therapy were not as spectacular as hoped. Interferons proved ineffective against many cancers and viral diseases; and, in spite of the fact that these are natural agents, interferon therapy was found to be highly toxic. Interferons are used to treat certain rare leukemias (cancers of leukocytes) and some chronic viral infections, including hepatitis and herpes. As clinical research proceeds, interferons will probably find other applications, either alone or in combination with other agents.

Interferons do more than defend against viral infection. They also help regulate many cell functions, including cell motility, cell division, activation of macrophages, and transplant tissue rejection.

In this chapter we saw how T. L.'s nonspecific interior defenses fought to control a serious wound infection. In the next chapter we consider the defenses contributed by the immune system.

Summary

THE BODY'S THREE LINES OF DEFENSE AGAINST INFECTION   (pp. 381–382)

1. The body's first line of defense consists of the nonspecific surface defenses—structural, mechanical, and biochemical features that prevent pathogens from entering the body.

2. Its second line consists of the nonspecific interior defenses—inflammation, phagocytosis, complement, and interferon.

3. The first and second lines of defense are nonspecific because they function in the same way against all microorganisms.

4. When first- and second-line defenses cannot control infection, the body's third line—the specific immune defenses—is activated. They are specific because they act against one particular type of microorganism.

INFLAMMATION   (pp. 382–385)

1. Inflammation is the body's reaction to injury or infection. It is manifested as redness, heat, swelling, pain, and sometimes loss of function.

2. Inflammation can be broken down into a series of steps—an inflammatory stimulus causes the release of inflammatory mediators that affect capillaries and leukocytes. At the end, tissue damage is repaired.

3. The most common inflammatory stimuli are injury and infection, which produce acute inflammation. The inflammatory stimulus releases inflammatory mediators into the damaged tissue, including:

   a. Bacterial by-products, chemicals released by bacteria that provoke inflammation

   b. Complement fragments, which cause vasodilation and increased permeability of blood vessels and activate leukocytes

   c. Kinins, a family of peptides that includes bradykinin, which causes vasodilation, increased vessel permeability, and pain

   d. Histamine, which causes pain, vasodilation, and especially increased vessel permeability

   e. Prostaglandins, which cause long-lasting vessel dilation and increase the sensitivity of pain receptors to bradykinin and histamine

   f. Leukotrienes, which increase vessel permeability or attract leukocytes to the inflammation site

4. Inflammatory mediators chemotactically attract phagocytes. In margination, phagocytes migrate to the capillary wall and adhere to it. In diapedesis, they squeeze through gaps in the endothelial lining and enter tissues.

Fever

Within hours of his accident, T. L. began to run a fever. Normally, the human body is maintained at a relatively constant 37°C (98.6°F) by a regulatory center in the brain called the anterior hypothalamus. This region functions like a thermostat, registering body temperature and modifying the production or loss of heat in order to maintain temperature at the normal set point. Body temperature can be regulated by simple behavioral responses like putting on a sweater or by complex physiological processes like sweating (which dissipates heat) or shivering (which produces heat). Another physiologic determinant of body temperature is the amount of blood flowing through surface skin vessels. When vasodilation enlarges blood vessels and shunts large amounts of blood through surface vessels, heat is lost from the body and the skin appears flushed. When vasoconstriction (the narrowing of blood vessels) shunts blood away from the surface, heat is conserved within the body and the skin feels cold and clammy.

During fever, the body's normal temperature set point is adjusted slightly upward. At the onset of fever, shivering and vasoconstriction elevate body temperature. While fever persists, an increased rate of metabolism maintains the elevated temperature. As fever resolves, vasodilation and sweating allow body temperature to return to normal.

What causes the abnormal elevation of the temperature set point? Certain microorganisms, including bacteria and viruses, produce substances that can cause fever. Collectively, these substances are called **exogenous pyrogens** (*exogenous*, "from the outside"; *pyrogens*, "fever-generating"). One well-known exogenous pyrogen is lipid A, a component of endotoxin (Chapter 15). Exogenous pyrogens cause phagocytes to produce a protein called **endogenous pyrogen** (*endogenous*, "from the inside")—now known to be interleukin-1 (see Chapter 15). Endogenous pyrogen stimulates the anterior hypothalamus to produce prostaglandins, causing the change in body temperature that we recognize as fever.

Some of the most uncomfortable sensations of illness—chills, shivering, sweating, and a feeling of being overheated—are attributable to fever. Antipyretic medications such as aspirin and acetaminophen reduce fever by interfering with the last step in the fever-generating process—the synthesis of prostaglandins. The ill person usually feels much better, although the medication has no effect on the underlying infection. Medication to control fever is an example of treating the symptom rather than the disease.

People sometimes worry that fever will harm them—that the brain or some other part of the body will be permanently damaged by becoming too hot. Greatly elevated body temperature can, indeed, be quite harmful. Temperatures above 41°C or 105.8°F can cause permanent tissue damage, and people rarely survive temperatures above 45°C or 113°F. But such high temperatures virtually never occur as a result of fever. They usually occur from **heatstroke**, environmental conditions (high heat, high humidity, and exercise) that overwhelm the body's normal temperature-regulating mechanisms. It is exceedingly rare for fever to elevate body temperature above 40°C. If you have a high fever, you should get medical attention to rule out the possibility of a potentially serious infection, such as meningitis or pneumonia. But the fever itself is seldom dangerous.

While fever is rarely harmful, is it useful? Does fever help fight infection? No experimental evidence definitely establishes that people with fever survive infection better than people without it, but several factors suggest that fever may serve a useful purpose. Certain defensive cells, including phagocytes and cells of the specific immune system, are more active at higher temperature. Human leukocytes display maximum phagocytic activity at 38° to 40°C. Fever also decreases the concentration of iron circulating in the blood, which may restrict microbial growth. Some microorganisms grow more slowly at temperatures above 37°C, and fever may help bring these infections under control. Finally, many lower vertebrates, including reptiles and fish, have elevated body temperatures when infected, suggesting that the elevation of body temperature during infection is an ancient evolutionary adaptation; fever in humans probably confers a selective advantage.

---

5. Inflammation benefits the host because it activates and coordinates the body's weapons against infection, but it also damages host tissue. Pus is a mixture of dead leukocytes, microorganisms, and tissue cells.

6. The first step in inflammatory repair is clean-up. Macrophages consume dead and dying microorganisms, host cells, and foreign particles. This stage can take weeks.

7. After clean-up, the tissue regenerates and heals. If tissue damage was severe, scarring may occur. Some tissue—such as nervous system tissue—does not regenerate.

LEUKOCYTES: DEFENDERS OF THE BODY'S INTERIOR (pp. 385–393)

1. The five types of leukocytes—neutrophils, eosinophils, basophils, lymphocytes, and monocytes—have different functions in the immune system.

2. A complete blood cell count (CBC) is a total count of both leukocytes and erythrocytes. A leukocyte differential count determines the number of each type of leukocyte, computed as a percentage of the total number of leukocytes.

3. Most acute infections increase the total leukocyte count by increasing the number of neutrophils, but some infections depress or otherwise alter the differential count.

4. Early in leukocyte differentiation there is a split into the phagocytic family (active in nonspecific interior defenses) and the lymphoid family (active in the specific immune system).

5. The phagocytic family is composed of two groups. The polymorphonuclear leukocytes include the neutrophils, eosinophils, and basophils. All polymorphonuclear leukocytes are capable of degranulation (releasing toxic chemicals and inflammatory mediators), but only neutrophils and eosinophils are capable of phagocytosis.

6. The other subgroup consists of the mononuclear leukocytes. When they are released from bone marrow into the bloodstream they are called monocytes. When they leave the bloodstream and enter tissue, they are called macrophages. They are extremely active phagocytes.

7. The number of neutrophils may double or triple during infection. They have granules that are lethal only to bacteria, such as lysozyme and lactoferrin, and others that damage both microorganisms and human cells, such as acid hydrolase (degrades cellular proteins).

8. Eosinophils act primarily on pathogens that are too large to be consumed by phagocytes, such as protozoal parasites and worms. They may also prevent the body's defenses from causing serious tissue damage.

9. Basophils are not well understood. They are closely related to mast cells and probably help defend against parasitic infections.

10. Monocytes may become wandering macrophages, which travel to tissues under attack by microorganisms, or fixed macrophages, remaining permanently in one tissue. Macrophages scavenge dead cells during inflammatory repair and participate in the body's specific immune defenses.

11. Phagocytosis is the process by which leukocytes contact, surround, consume, and destroy microorganisms and other foreign particles.
   a. Step 1: Activation. Inflammatory mediators activate leukocytes so they recognize foreign objects, adhere to surfaces, and change their metabolism in order to produce lethal oxidants.
   b. Step 2: Chemotaxis. Activated phagocytes migrate along a chemical gradient toward tissues where inflammatory mediators are present in high concentrations.
   c. Step 3: Recognition and adherence. During opsonization, one part of a linking molecule called an opsonin binds to the bacterium and one part binds to an opsonin-recognizing receptor on the phagocyte.
   d. Step 4: Ingestion. The phagocyte extends pseudopods to engulf the bacterium. The pseudopods meet and fuse, forming a new structure called a phagosome, which contains the bacterium.
   e. Step 5: Killing and digestion. When a phagosome fuses with lysosomes in the cytoplasm, they form a phagolysosome. Microbiocidal chemicals kill the bacterium inside and enzymes digest it.
   f. Step 6: Expulsion. After digestion, the phagolysosome fuses with the cell membrane to expel indigestible debris.

COMPLEMENT (pp. 393–397)

1. Complement, or the complement system, refers to a family of over 30 different proteins in serum that function together as a nonspecific defense.

2. Complement has three major functions—to produce inflammatory mediators, to bind opsonins to microbial cells, and to form a membrane attack complex that lyses cells.

3. The key proteins involved in these functions are designated C1–C9. Fragments produced when these proteins split are designated by letters (e.g., C3a, C3b).

4. Complement plays its major role in conjunction with antibodies, but it also plays an important role before antibodies are produced.

5. The complement cascade—a series of reactions in which the product of one reaction activates the next reaction—can be triggered by the classical pathway or the alternate pathway. Both lead to the activation of C3 and then to the terminal pathway.

6. In the classical pathway, C1, C4, and C2 are activated in that order. C1 has three subunits—C1q, C1r, and C1s. C1q recognizes and binds to an antigen-antibody complex—the molecular combination that sets off the classical pathway. C1r and C1s activate each other and, in turn, C4 and C2, splitting each into two fragments. One of the C4 fragments and one of the C2 fragments combine to activate C3, splitting it into fragments called C3a and C3b. C3a (a powerful inflammatory mediator) is released and C3b (a powerful opsonin) binds to the activating membrane.

7. The alternate pathway is initiated when an internal bond in the C3 molecule breaks, allowing it to bind to factor B. The addition of factor D forms an active molecule that is stabilized by properdin. This molecular complex splits C3 into C3a and C3b, which act just as they do in the classical pathway. C3b is not as effective an opsonin because no antibodies are involved in this pathway.

8. C3b initiates the terminal pathway. C3b combines with C5, which causes C5 to split. C5a, an inflammatory mediator, is released. C5b activates C6, C7, C8, and multiple molecules of C9, which attach themselves to the membrane. The membrane-bound complex grows until it eventually forms a membrane attack complex, a channel through the membrane that allows water to enter, weakening the cell membrane and eventually causing cytolysis.

INTERFERON (pp. 397–398)

1. Interferon is the nonspecific interior defense against viral infections. There are three main classes—alpha interferon, beta interferon, and gamma interferon.

2. Cells infected by a virus produce interferon and release it. Healthy neighboring cells take up the interferon molecules and produce antiviral proteins (AVPs), enzymes that interfere with viral protein synthesis.

3. In addition to defending against viral infection, interferons also help regulate many cell functions, including cell motility, cell division, activation of macrophages, and transplant tissue rejection.

Review Questions

THE BODY'S THREE LINES OF DEFENSE AGAINST INFECTION

1. What are the body's three lines of defense? What is the difference between nonspecific and specific defenses?

2. Name the four nonspecific interior defenses.

3. What happens if a pathogen evades the third line of defense?

INFLAMMATION

1. What is inflammation? Why is it called the cornerstone of the nonspecific interior defenses?

2. How does inflammation manifest itself clinically?

3. Briefly describe the steps in inflammation.

4. What is an inflammatory stimulus? What is the difference between acute and chronic inflammation?

5. What are inflammatory mediators? Describe each of these inflammatory mediators and tell what some of their effects are:
   a. bacterial by-products   d. histamine
   b. complement fragments   e. prostaglandins
   c. kinins/bradykinin   f. leukotrienes

6. Describe the two effects that inflammatory mediators have on capillaries.

7. Define endothelial cells, edema, erythema.

8. Describe the two effects that inflammatory mediators have on phagocytes.

9. Explain margination. Explain diapedesis.

10. What are the benefits and drawbacks of inflammation to the host body? What is pus?

11. What are the two steps in inflammatory repair? Which leukocytes play the major role in repair?

12. What happens if the tissue that suffered infection is not a type that can regenerate?

LEUKOCYTES: DEFENDERS OF THE BODY'S INTERIOR

1. Name the five kinds of leukocytes.

2. What is a complete blood count? What is a leukocyte differential count? Why might someone with an infection have one or both of these tests done?

3. Define segs and bands. What is their clinical significance?

4. How and where are leukocytes produced? What are the two families of leukocytes?

5. What are the two main functions of the phagocytic family of cells?

6. What is degranulation?

7. Name the two groups of cells within the phagocytic family. Why is one group called granular and the other agranular?

8. To which of the two groups does each of these types of cells belong: neutrophils, monocytes, eosinophils, macrophages, basophils?

9. What is the difference between a monocyte and a macrophage?

10. Why does it benefit the human host that neutrophils exist in such large numbers?

11. What function do the granules in neutrophils fulfill in phagocytosis? What function do they fulfill in degranulation?

12. How do the contents of granules both benefit and harm the human host?

13. What is the key defensive function of eosinophils?

14. How are basophils and mast cells alike? How are they different?

15. What is the key defensive function of mast cells? What is probably the key defensive function of basophils?

16. What is the difference between fixed macrophages and wandering macrophages?

17. What roles do macrophages play in the body's defense?

18. Define phagocytosis. Name the six steps in phagocytosis.

19. How is a leukocyte activated? What is a respiratory burst?

20. Explain opsonization. What are the two most important opsonins?

21. Explain the difference between a phagocyte, phagosome, and phagolysosome.

COMPLEMENT

1. What is complement or the complement system? What are its three main functions?

2. What are the key events in the complement system? What are the key proteins involved?

3. Explain this sentence: Activation of complement proteins occurs as a cascade.

4. Why does the cascade expand as it proceeds?

5. How is the complement cascade triggered?

6. Explain what happens in the classical pathway.

7. What are the functions of C3a and C3b?

8. Explain what happens in the alternate pathway.

9. Aside from the different molecules involved, what is the critical difference between the classical and alternate pathways? How does this difference affect C3b as an opsonin?

10. Explain how many Gram-positive and Gram-negative bacteria stimulate the activation of the alternate pathway.

11. Which protein initiates the terminal pathway? Explain what happens in the terminal pathway. Include the terms *membrane-attack complex* and *cytolysis* in your discussion.

12. Why are Gram-negative bacteria, but not Gram-positives, vulnerable to cytolysis by the terminal pathway?

INTERFERON

1. What is interferon? How does this defense differ from the other three nonspecific interior defenses?

2. How does interferon work? What are AVPs?

3. Explain this statement: Interferon is host-specific but virus-nonspecific.

4. What are the three main classes of interferon and their functions?

Essays
1. You are examining a patient who has recently sustained a minor injury that penetrated the skin. Discuss how you might know whether treatment is necessary, what it might be, and what favorable and unfavorable consequences might result.

2. People with advanced cases of diabetes often have changes in their small blood vessels that interfere with the normal circulation of blood, especially to their feet and toes. Discuss how this condition might affect vulnerability to infection.

Suggested Readings

Kiester, E. 1984. A little fever is good for you. *Science 84*, November, 168–173.

Nilsson, L. 1987. *The body victorious.* New York: Dell. Spectacular photomicrographs and drawings of the cells that make up the body's defenses.

Rietshel, E. T., and Brader, H. 1992. Bacterial endotoxins. *Scientific American*, August, 54–61.

Roitt, I.; Brostoff, J.; and Male, D. 1993. *Immunology.* 3d ed. St. Louis: C. V. Mosby. A textbook that considers the body's defenses, with excellent diagrams and illustrations.

17 THE IMMUNE SYSTEM: SPECIFIC DEFENSES OF THE BODY'S INTERIOR

T. L.'s Close Call: Part Two

When T. L. awoke two days after sustaining a puncture wound to his palm, his condition was clearly deteriorating. The first thing he noticed was that the small area of redness and swelling had extended to include his entire hand, with the margin of redness now visible just above his wrist. Faint red streaks had appeared along the inner aspect of his arm, and tender lumps were noticeable behind his elbow and under his arm. His fever seemed higher than the night before, and he felt too weak to walk.

T. L. was clearly too sick to travel, and his companions calculated that it would take several days to return with medical help. Reluctantly, they decided to stay in camp and allow T. L. to rest. Over the next few days T. L. and his friends were relieved to find him gradually improving. The red area did not advance past his forearm. Although his hand remained painful and swollen, he felt less tired and his fever was gone. Four days later, when he was already recovering, the party turned back and took T. L. to the nearest emergency room.

Doctors diagnosed a presumed bacterial infection of T. L.'s hand and prescribed a broad-spectrum antibiotic that was likely to be effective against the unknown infecting organism. The medication was taken by mouth, and T. L. was sent home with instructions to return if his condition failed to improve. The redness and swelling of his hand subsided gradually over the next two weeks, and he had no further problems related to his infection.

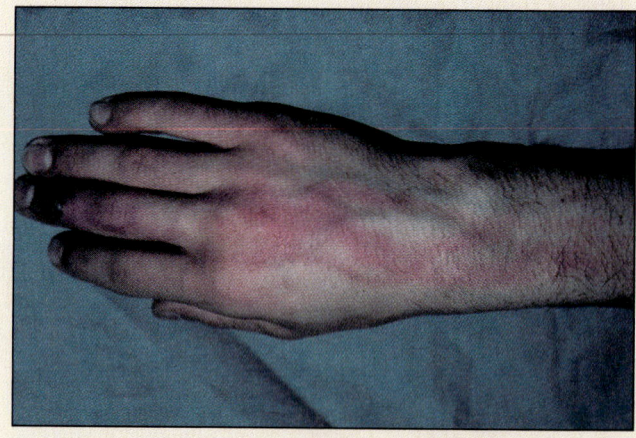

Red streaks indicate the development of lymphangitis as the infection progresses.

To understand:

- The nature of the lymphatic system—its cells, organs, and circulation

- The types of naturally and artificially acquired immunity

- The three elements of a successful immune defense—recognition, activation, and response

- Humoral immunity and the roles of antigen-presenting cells, effector B cells, memory B cells, cytokines and lymphokines, and the primary and secondary immune responses

- The structure of antibodies and the roles of the five classes of antibodies

- Cell-mediated immunity and the actions of the different types of T cells—helper T cells, cytotoxic T cells, and suppressor T cells

- Immunologic tolerance and the difference between class I and class II MHC antigens

- The roles of cytokines and lymphokines in T-cell activation and the two types of T-cell response—cytotoxicity and immunoregulation

- How non-B non-T lymphocytes differ from other lymphocytes and how they function in the immune system

CELLS AND ORGANS OF THE IMMUNE SYSTEM

This chapter discusses the complex and powerful defenses of the immune system—the third and final line of defense that fortunately eliminated the bacteria from T. L.'s body and allowed him to survive the infection.

The **immune system** is a network of cells and organs that extends throughout the body and functions as the last line of defense against infection (**Figure 17.1**). The immune system has been recognized as a separate body system only in the last 40 years. It is also known as the **lymphoid system** because its principal cells are **lymphocytes**, one of the five types of leukocytes. Lymphocytes are smooth and round, lack visible granules, and are relatively small (see the box "Blood: A Complex Body Tissue" in Chapter 16).

There are two major types of lymphocytes—**B lymphocytes (B cells)** and **T lymphocytes (T cells)**. To understand the immune system, you must understand the

difference between these two cell types. Both types recognize **antigens**—molecular markers on the surfaces of all cells and viruses that provoke a response from lymphocytes. Every microorganism has its own unique constellation of antigens. The uniqueness of antigens and the ability of lymphocytes to recognize them is what makes immune-system defenses specific. B cells respond to antigens by producing defensive proteins called **antibodies**. T cells respond to antigens by killing the marked cells or, indirectly, by helping to regulate the immune system. Thus, B cells are the agents of **antibody-mediated**, or **humoral**, **immunity**, and T cells are the agents of **cell-mediated immunity**. A third group of lymphocytes, called **non-B non-T lymphocytes**, differs fundamentally from T and B cells because they function without recognizing antigens. **Table 17.1** summarizes the types and subtypes of lymphocytes that are discussed in this chapter.

The organs of primary immunologic importance, called **primary lymphoid organs**, are the **bone marrow** and the **thymus**. Bone marrow is tissue composed primarily of fat and blood cells located within the vertebrae, ribs, sternum, long bones of arms and legs, and pelvis. The thymus is a gland in front of the heart. All blood cells—leukocytes and erythrocytes—are produced by hematopoietic stem cells in the bone marrow (Chapter 16). B cells remain there to differentiate. T cells migrate to the thymus to differentiate (**Figure 17.2**, p. 406).

The organs of secondary immunologic importance, called **secondary lymphoid organs**, are the **lymph nodes** and the **spleen**. The lymph nodes are small bean-shaped structures located throughout the body. The spleen is a fist-size, blood-filled organ located high in the abdomen. Other lymphoid tissues are the **tonsils** in the throat; the **adenoids** in the nose; the **appendix**, an outpouching of the intestinal tract; and **Peyer's patches**, small pockets of lymphoid cells within the intestinal wall (**Figure 17.3**, p. 407). Mature B and T lymphocytes congregate and interact within these organs and tissues.

Lymphocytes housed in the secondary lymphoid organs and tissues must travel to interact with other cells of the immune system or to reach the site of an active infection. Besides the bloodstream, lymphocytes are transported in the **lymphatic circulation**, a system of vessels that collects excess fluid from the body's tissues. This excess fluid originates in the blood, but escapes into the tissues across thin capillary walls. The lymphatic circulation returns the fluid—called **lymph**—and many lymphocytes to the bloodstream when the largest lymphatic vessel, the **thoracic duct**, empties into the heart.

Lymph nodes, which house huge numbers of lymphocytes, are interspersed along lymphatic vessels (**Figure 17.4**, p. 407). Blood vessels also pass through lymph nodes, so lymphocytes can pass from blood to lymph or vice versa within the nodes. Lymphatic vessels and

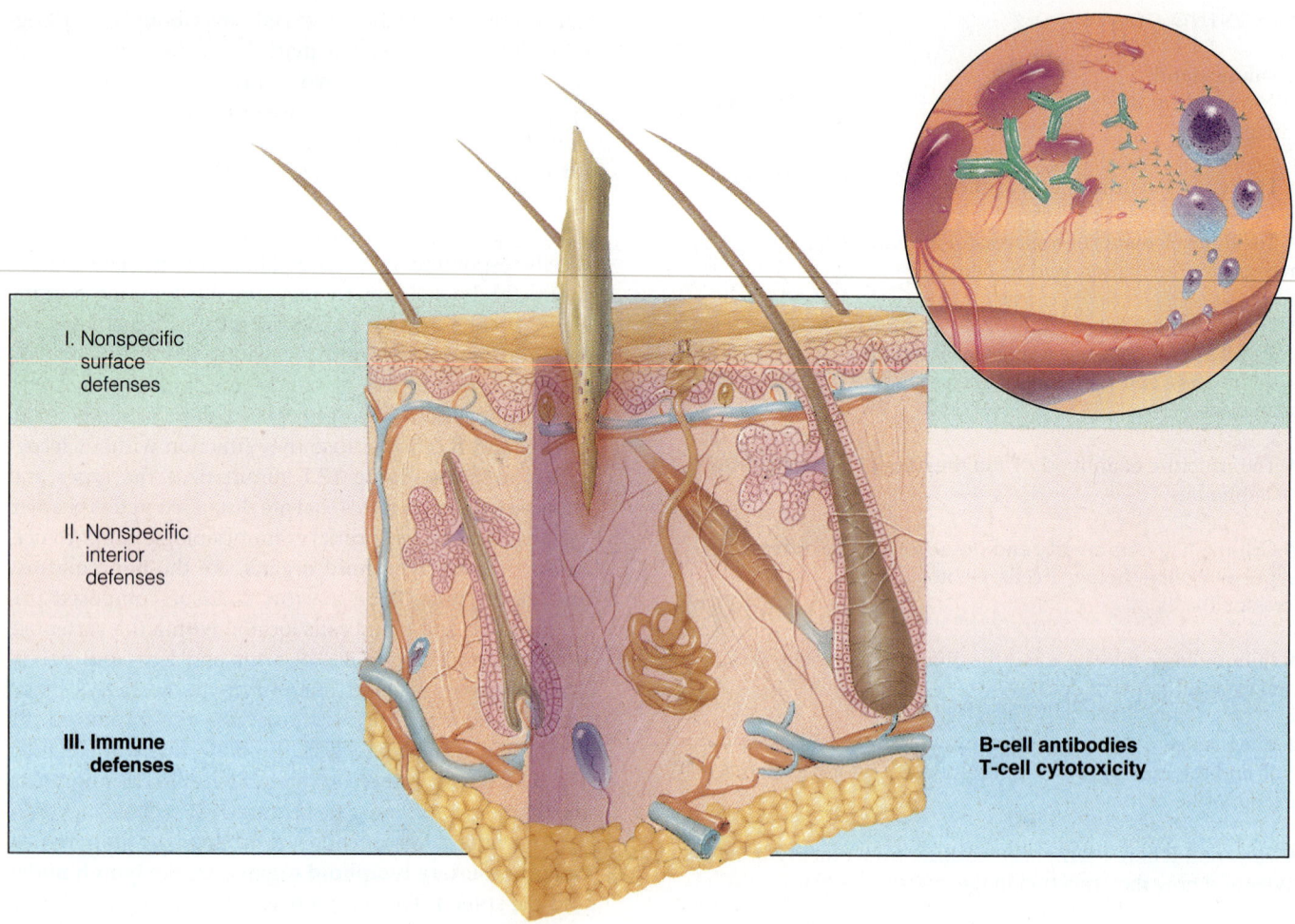

I. Nonspecific
surface
defenses

II. Nonspecific
interior
defenses

III. Immune
defenses

B-cell antibodies
T-cell cytotoxicity

Figure 17.1   Lines of defense against infection. The body's third and last line of defense against invading microorganisms is the immune system. Immune defenses are activated when nonspecific defenses are insufficient to control invading organisms and infection becomes established. Unlike the first and second lines, immune-system defenses are specific, meaning B and T cells target specific pathogens; each microbe provokes a unique response directed against it.

lymph nodes usually become swollen and inflamed during infections that activate the specific immune defenses. The red streaks that T. L. noticed in his arm were inflamed lymphatic vessels. The tender swellings behind his elbow and under his arm were lymph nodes that became enlarged as lymphocytes proliferated during the immune response. **Table 17.2** (p. 408) summarizes the organs of the immune system.

TYPES OF IMMUNITY

Immunity can be **active** (a product of the individual's own immune system) or **passive** (antibodies produced elsewhere are given to the individual). Each of these

types can in turn be **naturally acquired** or **artificially acquired**. Refer to **Table 17.3** (p. 408) as you read this section.

Naturally acquired active immunity comes from infections encountered in daily life. For example, T. L.'s injury stimulated active immunity. Naturally acquired active immunity controls infection and sometimes prevents reinfection (how the immune system does this is the subject of this chapter). **Artificially acquired active immunity** is stimulated by **vaccines**, antigen-containing preparations administered to prevent infection. DPT immunization, for example, confers artificial immunity against diphtheria, pertussis, and tetanus (Chapter 20).

Naturally acquired passive immunity refers to the antibodies transferred from mother to fetus across the placenta and to the newborn in colostrum and breast

Table 17.1   Types of Lymphocytes and Their Roles in Specific Interior Defense

Lymphocyte Type	Functions
B CELLS	Agents of humoral immunity
Effector B cells	Antibody production
Plasma cells	Differentiated B cells; mass production of antibodies to fight current infection
Memory B cells	Antigen recognition: produce accelerated and amplified antibody response if infection with an organism bearing the same antigen occurs again
T CELLS	Agents of cell-mediated immunity
T_4 Type	
Helper T cells (T_H)	Immunoregulatory cells: make possible an immune response by stimulating B cells and cytotoxic T_C cells
T_8 Types	
Cytotoxic T cells (T_C)	Killer cells: target viral-infected cells, cancerous cells, and donor-transplant cells and destroy them by secreting perforins that lyse the cell membrane
Suppressor T cells (T_S)	Immunoregulatory cells: dampen the immune response
NON-B NON-T CELLS	Function without recognizing antigen
Natural killer (NK) cells	Destruction of human cells—viral-infected cells, donor-transplant cells, and especially cancerous cells—by secreting perforins that lyse the cell membrane
Antibody-dependent killer (K) cells	Lysis: target cells coated by any antibody

milk during the first few months of life. **Artificially acquired passive immunity** consists of antibodies formed by an animal or a human and administered to an individual to prevent or treat infection. Examples include giving immune globulin obtained from many human donors to prevent hepatitis A (a viral disease that damages the liver) and administering antibodies against diphtheria toxin produced in animals to someone with diphtheria.

THE IMMUNE RESPONSE: AN OVERVIEW

The essential function of the immune system is to defend against microbial infection. To succeed, it must accomplish three tasks (**Figure 17.5**). The first, called **immune recognition**, is to perceive a threat and identify the *specific* microbial invader. The second, called **immune activation**, is to marshal all appropriate forces against the particular infecting microorganism. The final task, called **immune response**, is to counterattack to destroy or contain the invaders.

Immune Recognition

All microorganisms that invade the human body are marked by antigens. Antigens are complex molecules present in the capsules, cell walls, or flagella of bacteria, on fungal and helminthic cells, and in the capsids or envelopes of viruses. Antigens may be composed of protein; protein in combination with carbohydrates, lipids, or nucleic acids; or complex polysaccharides.

Immune-system recognition of an invader requires direct contact between the macromolecules. Lymphocytes have **antigenic receptors** (antigen-recognizing molecules). The antigenic receptors on B cells are antibody molecules. The antigenic receptors on T cells are called **T-cell antigen receptors**.

An antigenic receptor does not bind to the entire antigen molecule. Most antigen molecules are quite large (they have molecular weights of 100,000 or more). The part of the molecule that fits with its matching receptor on a lymphocyte is relatively small (usually they have a molecular weight of 1000 or less). This small region of the antigen molecule, which is the part that a lymphocyte really recognizes, is called an **epitope**, or **antigenic determinant**. Most antigen molecules have many epitopes, which may be identical to or different from one another.

Differently shaped epitopes are recognized by lymphocyte receptors with complementary shapes. In other words, there is a shape-to-shape recognition (**Figure 17.6**). The two molecules fit together the way two jigsaw puzzle pieces fit together. This recognition is similar to the interaction between an enzyme and its substrate or a microbial adhesin and its epithelial receptor. The immune

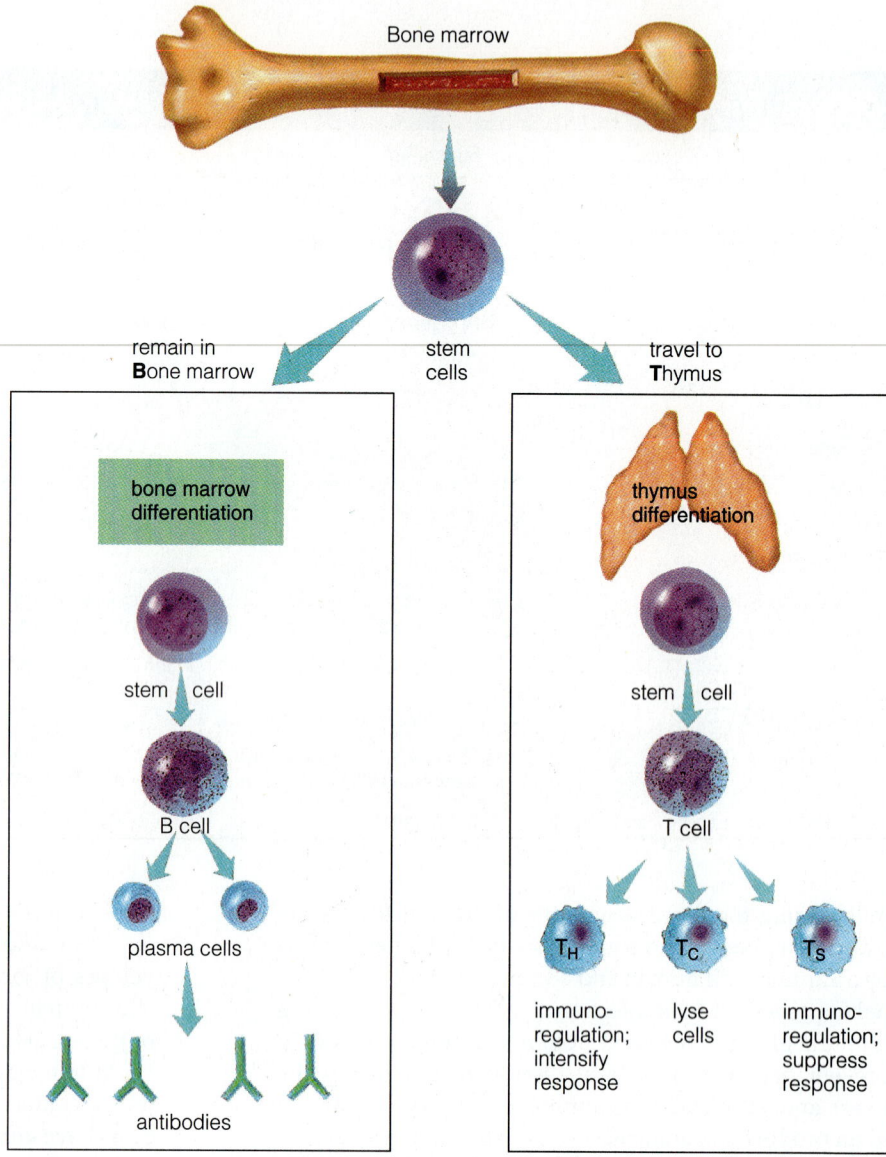

Figure 17.2 Origins of B cells and T cells. Lymphocytes, like all blood cells, arise from stem cells in the bone marrow. B cells differentiate there. Later, as mature B cells, some differentiate further into plasma cells if the body is under microbial attack. T cells, on the other hand, travel to the thymus where they differentiate into helper T cells (T_H), cytotoxic T cells (T_C), and suppressor T cells (T_S).

Bone marrow

remain in **B**one marrow

stem cells

travel to **T**hymus

bone marrow differentiation

thymus differentiation

stem cell

stem cell

B cell

T cell

plasma cells

T_H

T_C

T_S

immuno-regulation; intensify response

lyse cells

immuno-regulation; suppress response

antibodies

system's remarkable ability to recognize individual epitopes is the key to its specificity.

Antigen size is also a factor in immune recognition. A molecule must be large to stimulate a response from lymphocytes. But a small serum protein molecule called a **hapten** may become an epitope if it combines with a larger carrier protein. The penicillin molecule, for example, is too small itself to stimulate lymphocyte recognition, but when it binds to proteins in the body it creates a penicillin-protein complex, which is large enough to be an antigen in which penicillin is the epitope. Unfortunately, recognition in this case results in penicillin allergy—a harmful, rather than helpful, immune response.

To initiate a defense against his rapidly progressing infection, T. L.'s lymphocytes had to recognize the antigens of the invading bacteria. Recognition was facilitated by **antigen-presenting cells**, specific cell types, including

macrophages, that phagocytize bacterial cells and display the antigens in such a way that they can be more easily recognized by lymphocytes. Actual recognition occurred when direct contact was established between antigen molecules and antigen receptor molecules on the surface of T. L.'s B and T cells.

Immune Activation

When the immune system is not actively fighting an infection, only a few lymphocytes exist that produce each type of antigenic receptor. This enables the body to maintain a vast reserve of different antigenic receptors but a manageable number of lymphocytes. Lymphocytes become activated when they come in contact with their corresponding antigen. On activation, T. L.'s lymphocytes with matching receptors to antigens on invading bacteria

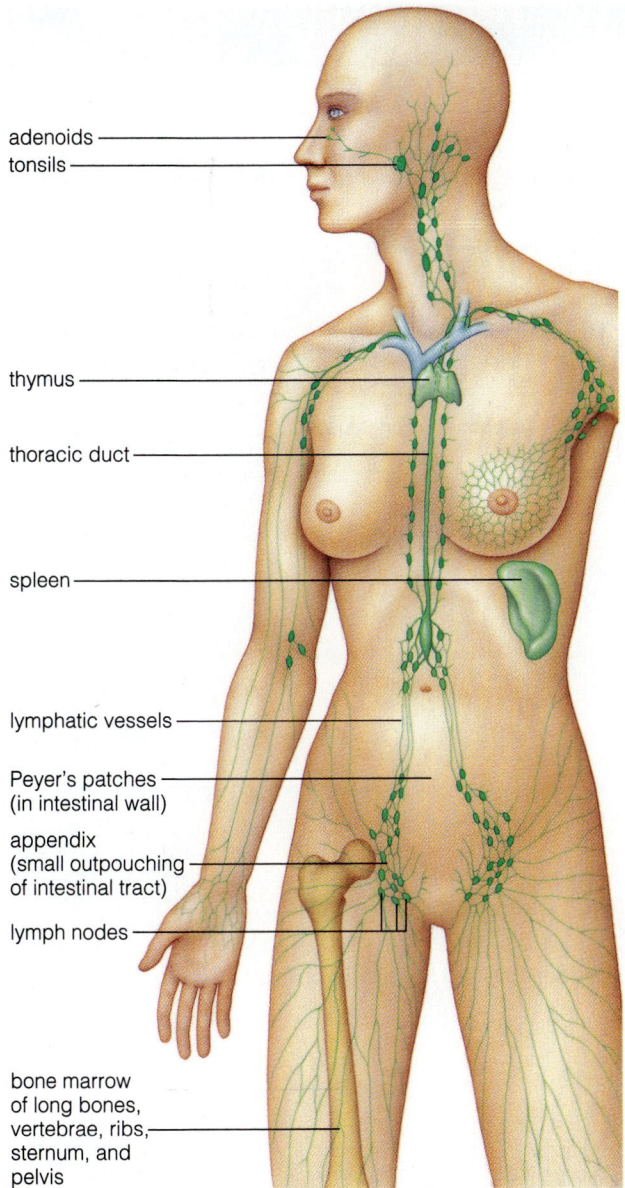

adenoids
tonsils

thymus

thoracic duct

spleen

lymphatic vessels

Peyer's patches
(in intestinal wall)

appendix
(small outpouching
of intestinal tract)

lymph nodes

bone marrow
of long bones,
vertebrae, ribs,
sternum, and
pelvis

Figure 17.3 Organs of the immune system. The primary organs of the immune system are the bone marrow and thymus. The secondary organs are the lymph nodes, spleen, and collections of lymphoid tissue—the tonsils, adenoids, appendix, and Peyer's patches. Lymphocytes congregate and interact in all of the secondary organs. Huge numbers are also found in the lymph that flows through the lymphatic vessels.

shifted their metabolism from a relatively quiescent state to one of active division. This antigen-driven activation produced many new lymphocytes to combat his infection. Lymphocyte activation generally requires the interaction of B cells, T cells, and antigen-presenting cells, as you will see.

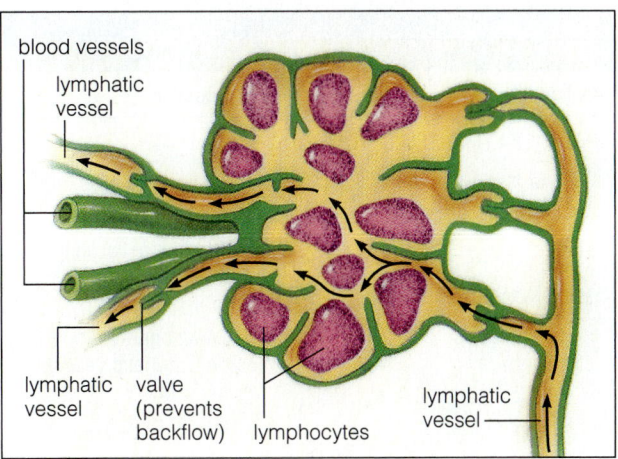

blood vessels
lymphatic vessel

lymphatic vessel
valve (prevents backflow)
lymphocytes
lymphatic vessel

Figure 17.4 A lymph node. Both blood and lymphatic vessels enter and leave a lymph node, and so lymphocytes can pass from lymph to blood and vice versa.

Immune Response

Activated lymphocytes deploy their weapons to eliminate microorganisms from the human body—this deployment constitutes the immune response. Some activated B cells differentiate into **plasma cells**, short-lived factories that mass-produce antibodies. Both B cells and plasma cells produce antibodies, but plasma cells produce vastly greater quantities. Mass production of antibodies that could bind to antigens on the bacteria invading T. L.'s hand played a central role in his recovery. Activated T cells respond in two ways—they attack certain types of infected human cells, and they help regulate the immune response, either positively or negatively. Activated T cells were also essential for T. L. to overcome his infection.

We consider in more detail how each of these events—immune recognition, activation, and response—occurs in B cells and T cells. Finally, we examine the special role of non-B non-T lymphocytes in the body's defense.

B LYMPHOCYTES

We can defend ourselves against the microorganisms that threaten us daily because we are born with millions of B lymphocytes, each producing a different antibody that can recognize a different epitope. B cells differentiate in the bone marrow extremely early in development, before we are exposed to foreign antigens. In other words, during bone marrow differentiation, each B cell becomes **immunocompetent**—it acquires the capability to produce an antibody that is a unique antigenic receptor.

Table 17.2 Organs and Vessels of the Immune System

Organ	Location	Role in Immune System
PRIMARY LYMPHOID ORGANS		
Bone marrow	Within vertebrae, ribs, sternum, long bones, pelvis	Produces all leukocytes, including lymphocytes; site of B-cell differentiation
Thymus	Gland that lies in front of the heart	Site of T-cell differentiation
SECONDARY LYMPHOID ORGANS		
Lymph nodes	Small bean-shaped organs located on lymphatic vessels throughout the body	Mature lymphocytes congregate here and interact with one another; also move between blood and lymph vessels within nodes
Spleen	Upper abdomen	Mature lymphocytes congregate and interact (also filters blood)
Collections of Lymphoid Tissue		In all, mature lymphocytes congregate and interact
Tonsils	Form a ring of lymphoid tissue in the throat	
Adenoids	Form a ring of lymphoid tissue in the nose	
Appendix	Small outpouching of lymphoid tissue in the intestinal tract	
Peyer's patches	Small collections of lymphoid tissue within the intestinal wall	
LYMPHATIC CIRCULATION	System of vessels throughout the body that collects excess fluid from body tissues and returns it to the bloodstream	Transports lymphocytes between secondary lymphatic organs and to infection sites
Lymph	Lymphocyte-filled fluid that circulates through lymphatic vessels	Houses and transports lymphocytes
Thoracic duct	Largest vessel in lymphatic circulation	Empties into heart to return lymph to bloodstream

Table 17.3 Types of Immunity

Type	Natural	Artificial
Active	B cells and T cells. Immunity developed from a naturally acquired infection	B cells and T cells. Immunity developed in response to a vaccine
Passive	Antibodies only. Antibodies transferred from mother to child across the placenta and in breast milk	Antibodies only. Antibodies from a donor are injected into an individual

I. Immune recognition

II. Immune activation

III. Immune response

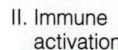

Figure 17.5 Recognition—activation—response. The immune system must accomplish three essential tasks to successfully defend the body against microbial attack. Task I is recognition—the foreign invader is identified. Task II is activation—the immune system prepares to fight back. Task III is response—the immune system acts to eliminate or contain the invading microorganism.

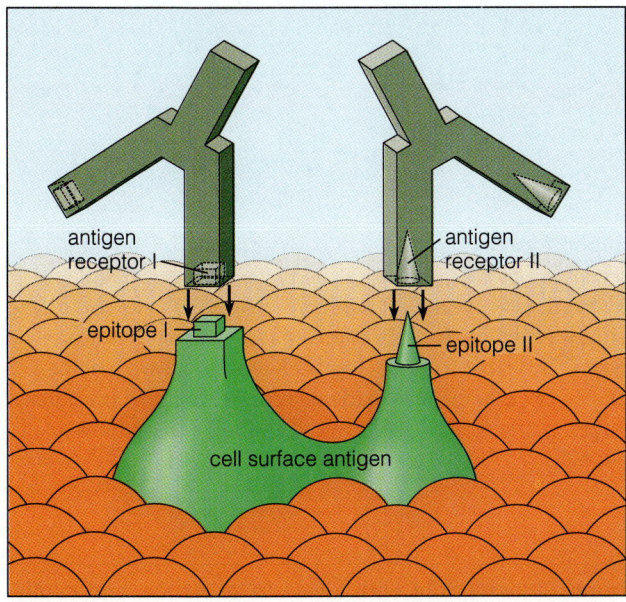

Figure 17.6 Shape-to-shape recognition. Most antigen molecules have many epitopes. An antigen receptor is able to recognize a particular epitope because of its complementary shape.

Antigen Recognition

What occurs in differentiation that generates such B-cell diversity? How can there be an antibody for every antigen? Normally, each protein (and an antibody is a protein) is encoded by a separate gene (Chapter 6). But the number of different antibody proteins is so vast that all the genes would not fit in the human genome. Some other extraordinary process must be at work.

B-cell diversity is generated by a process that involves a type of genetic recombination (Chapter 7; **Figure 17.7**). Instead of having a separate gene to encode each antibody, individual B lymphocytes *create*, during differentiation, their own genes to make a unique antibody. Lymphocytes do this by genetic cutting and pasting. Undifferentiated lymphocytes in the bone marrow contain an extensive library of modular bits of DNA that can be pieced together to create complete antibody genes. As differentiation occurs, one copy of each necessary piece of information is chosen at random from the library. Because the number of possible combinations of these small bits of genetic information is enormous, literally millions of genetically different B lymphocytes are produced from a relatively limited amount of DNA. A similar process occurs as T lymphocytes differentiate in the thymus. Thus, T. L.'s immune system created in advance the exact antibodies he would need to combat the infection in his hand. It also created antibodies that could recognize all the antigens he might encounter in his life.

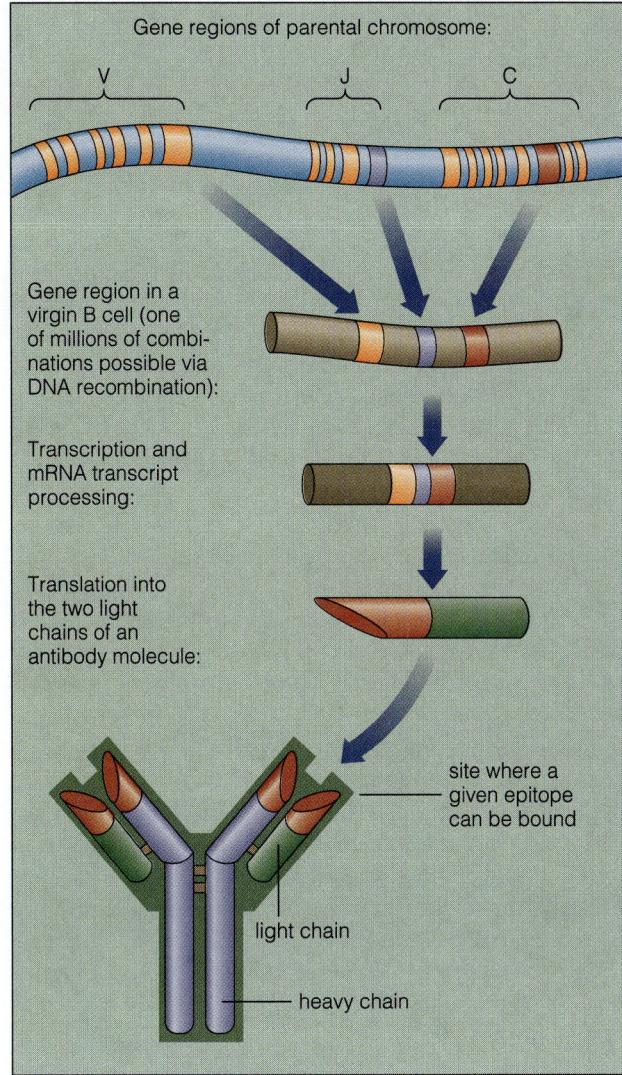

Figure 17.7 Generating antibody diversity. Antibody genes are created by a process of genetic cutting and pasting. Modular segments of DNA are selected at random from an extensive gene library. Short regions of DNA on the parental chromosome recombine to create a unique combination of variable (V), constant (C), and joining (J) regions. All these are transcribed and translated to produce a unique antibody.

Antigen-Presenting Cells. In order to recognize antigens, B lymphocytes usually require the cooperation of antigen-presenting cells (as well as T cells, as you will soon see). The most important antigen-presenting cells are the macrophages, which present antigens after phagocytosis. **Dendritic cells**, named for their branching shape, are important antigen-presenting cells found primarily in the skin, lymph nodes, spleen, and thymus. **Langerhans cells** are dendritic cells that phagocytize antigens in the skin and transport them to nearby lymph

The Missing Puzzle Piece

We have known since the late 1970s that antibody genes were pieced together from smaller pieces of DNA. But the details of the process remained hazy until recently because there was a missing piece in the puzzle. We hadn't identified the gene that encodes antibody recombinase, the enzyme responsible for splicing the gene fragments into a single functioning antibody gene. After intense effort in research laboratories across the country, the gene was finally found in the late 1980s. The importance of this gene, which recombines T-cell antigen receptor genes as well as antibody genes, is impossible to overestimate. Without it neither B cells nor T cells could recognize antigens, and the immune system would be paralyzed. Now that the gene has been identified, we may be able to tell whether any patients born with global immune failure suffer from a defect in the antibody recombinase gene.

nodes where they present the antigens to lymphocytes. Because T. L.'s infection originated in the skin, Langerhans cells were involved in his body defense.

Antigen-presenting cells phagocytize antigens as part of intact microorganisms or antigen-containing fragments. The antigen is processed within the phagocytic cell, and the epitopes are transported to the cell surface and displayed there so they can be recognized by lymphocytes. Once the antigen is displayed, the cells travel to lymph nodes or other lymphoid organs where large numbers of lymphocytes are concentrated. Here, the odds are increased that the antigen will encounter a B cell with a matching receptor. Thus, secondary lymphoid organs usually contain many antigen-presenting cells, as well as many lymphocytes.

Antigen-Antibody Binding. Approximately 100,000 identical antibody molecules are displayed on the surface of each B cell. When one binds to a matching antigen, recognition has occurred.

When an antibody binds to a matching antigen, the two molecules form an **antigen-antibody complex** because they have complementary molecular conformations. A uniquely shaped cavity in the antigen-binding region of the antibody molecule is filled by an epitope of complementary shape. Unlike the precise fit of enzyme-substrate binding, however, the fit between antigen and antibody is somewhat variable. Some antigen-antibody matches are precise, but others are poorer. Poorer fit (termed lower **binding affinity**) leads to weaker antigen-antibody interaction and, eventually, a poorer immune response.

Variable binding affinity highlights a fundamental difference between antigen-antibody interactions and the interactions between most other macromolecules. Enzymes interact only with a certain substrate, and evolution has produced enzymes that fit their substrate molecules nearly perfectly. Antibody production, on the other hand, is driven by the evolutionary necessity to have an antibody that can react with any possible antigen. Thus, unlike an enzyme, antibodies cannot "know" in advance what they will have to recognize. In order to cover all possibilities, millions of different antibodies are produced and while some happen to match their antigen well, other matches are poorer.

B-Cell Activation

Before T. L.'s hand became infected, his immune system included some B lymphocytes that could recognize (to a better or poorer extent) antigens on the invading bacteria. Their specificity, generated by random differentiation in the bone marrow, did not require previous exposure to antigens. There were only a few of these B cells, however, and they were in a resting state, incapable of mounting a defense against T. L.'s rapidly spreading infection. (T. L.'s immune system may also have included memory B cells, lymphocytes that existed precisely because of previous infection by the invading bacteria.) So before T. L.'s specific immune defense could be mounted, two things had to happen. The specific B cells that recognize the epitopes on the invading bacteria had to be found among the millions of other B cells. And then those B cells (and only those B cells) had to be activated. B cells bearing antibodies that matched the foreign antigens bound to those antigens and thereby marked these cells for activation. Activation caused the selected cells to divide, producing a clone. Activation also caused some B cells to differentiate further into plasma cells, which produced the enormous quantity of antibody needed to fight T. L.'s infection. This

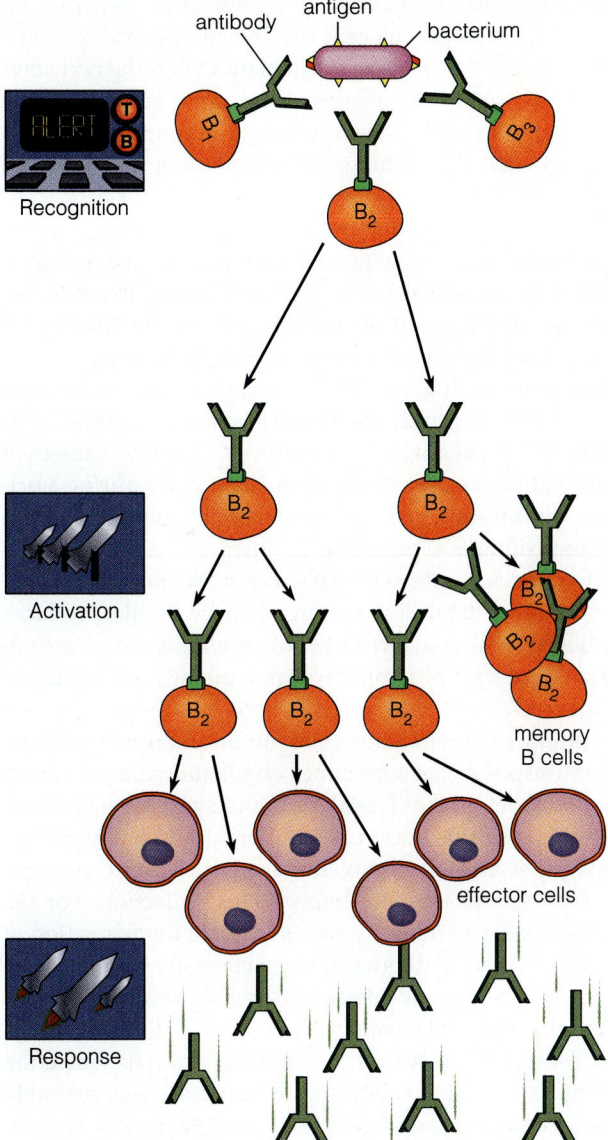

Recognition

Activation

memory
B cells

effector cells

Response

Figure 17.8 Clonal selection. *Clonal selection* refers to the multiplication of only those B cells that recognize the epitopes on the invading bacteria. Antigen recognition stimulates activation—the production of a clone of identical B cells. Some cells will be effector cells that fight the present infection. Others will be memory cells, held in reserve to fight a future infection.

selective lymphocyte stimulation is called **clonal selection (Figure 17.8)**.

Clonal selection produces a group of identical B cells capable of fighting an infection. Some members of this clone are **effector cells**, B cells and plasma cells that actively fight the current infection, while others are **memory B cells**, held in reserve to fight future infections by the same bacterial strain.

Activating Factors: Cell Contact and Lymphokines. In addition to the participation of antigen-presenting cells, activation of B lymphocytes usually requires the participation of **helper T cells** (T_H), one of several types of T lymphocytes. Some large antigens with many identical epitopes—for example, the polysaccharide capsule of pneumococci—can activate B cells without any help. These **T-independent antigens** stimulate a relatively weak immune response. Most antigens, however, are **T-dependent antigens**, which have a slightly more complicated activation sequence. It begins when antigen-presenting cells reach the secondary lymphoid organs and encounter T_H cells with receptors that match the epitopes the antigen-presenting cells are displaying. Activation of these T_H cells in turn activates B cells that recognize the same antigen.

B-cell activation is an incremental process, fostered by many stimuli. One component is antigen recognition itself. In the case of T-dependent antigens, this occurs when antigens displayed on some cells bind to antibodies on others. A network of these cellular contacts grows. Both B and T cells bind to antigens displayed on antigen-presenting cells. In addition, B cells take up, process, and display antigen, thus binding activating T_H cells to themselves. Antigen binding causes the lymphocytes to acquire new cell surface receptors, enlarge, and become more metabolically active.

Activation of lymphocytes is also mediated by free-floating messenger proteins called **cytokines**, or **lymphokines** if they are produced by lymphocytes. Lymphokines are signal molecules produced by lymphocytes. Cell-to-cell binding is critical to the function of lymphokines because these messenger molecules are destroyed soon after they are produced and they can exert their effect only on other cells that are immediately nearby.

Many cytokines and lymphokines are critical to B-lymphocyte activation. Macrophages and other antigen-presenting cells secrete a cytokine called **interleukin-1**, which activates both B and T_H cells. Activated T_H cells, in turn, secrete other lymphokines that activate additional B cells and increase their defense capabilities in various ways (**Figure 17.9**). These include **interleukin-2**, which is also a potent T-cell stimulator, **interleukin-4**, **interleukin-5**, **interleukin-6**, and gamma interferon (see Chapter 16). **Table 17.4** lists the specific role each lymphokine plays in activation.

In additon to participating in B-cell activation, lymphokines amplify the immune response in a wide variety of ways (Table 17.4). Interleukin-1 plays a role in fever production (Chapter 16) and has other actions as well. Gamma interferon activates cytotoxic lymphocytes. A closely related group of cytokines includes **tumor necro-**

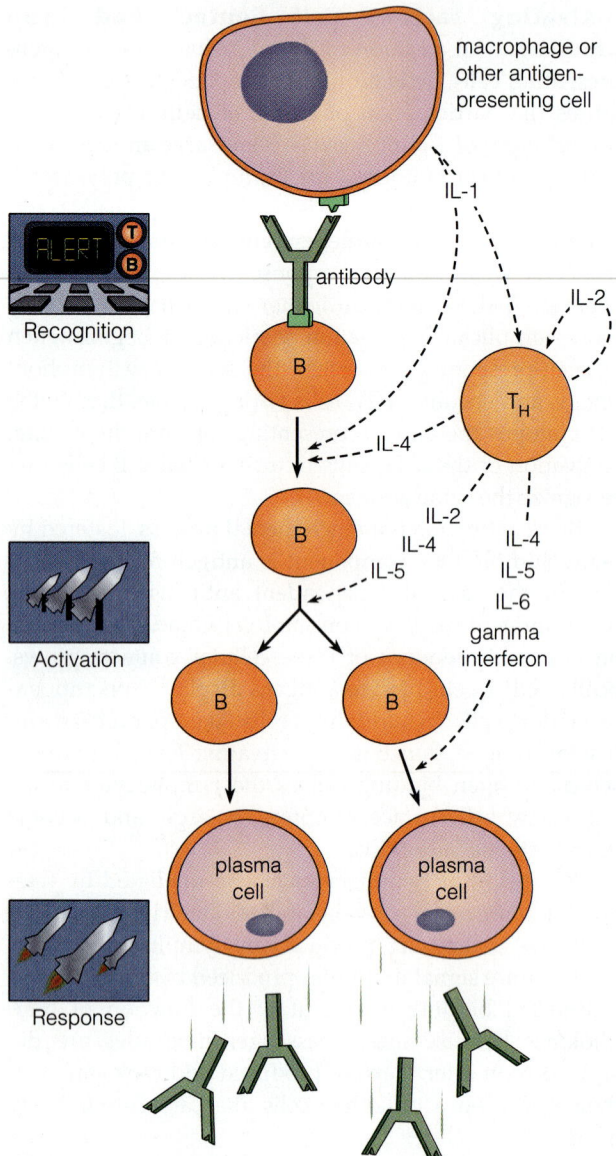

Figure 17.9 The role of lymphokines in B-cell activation. Macrophages and other antigen-presenting cells secrete interleukin-1 (IL-1), which activates B and T_H cells. Interleukin-2 (IL-2) activates B and T cells. Other interleukins stimulate B-cell clones and B-cell differentiation into plasma cells—antibody factories.

Reactivation: Immunologic Memory. When an infection is over, plasma cells die and disappear. However, memory B cells continue to circulate within the body and retain the ability to recognize the same antigen if they encounter it again. They function in **immunologic memory**, greatly accelerating and amplifying the immune response.

Memory B cells that recognize a given antigen are far more numerous than the never-stimulated or **virgin B cells** that recognize a comparable antigen, and they are able to mount an immune response sooner. Memory B cells differ from virgin lymphocytes in other ways, too. They produce higher-affinity (and therefore more effective) antibodies, and their plasma cells manufacture a different profile of antibody classes (see "The Five Classes of Antibodies" below). Production of antibody during a **primary immune response**, which occurs when a person first encounters a particular antigen, may take from one to two weeks to develop fully. But a **secondary immune response**, which is initiated by memory B cells, produces antibody within a few days. Moreover, a secondary response generates a much greater quantity of antibody (**Figure 17.10**).

Immunologic memory has profound implications. Its rapid response can sometimes eliminate pathogens from the body before the clinical signs and symptoms of illness develop. For this reason people who have contracted infections such as smallpox, measles, chickenpox, mumps, or German measles are immune to reinfection. For the same reason, **vaccines**, which stimulate the formation of memory B cells without causing a clinically significant infection, can prevent infectious disease. Vaccines are discussed in detail in Chapter 20.

Because T. L. began to recover from his infection within a few days, his immune response was probably the secondary type initiated by memory B cells. His immune system must have been exposed previously to the same (or very similar) bacterial antigens. T. L. and his friends made a serious error in judgment, however, when they decided to wait a day and see how his infection progressed, because if memory B cells had not been present to accelerate his immune response, his infection could have been fatal.

B-Cell Response

Plasma cells produce antibodies at the rate of 1000 molecules per cell per minute. They fight microbial infection both directly and indirectly.

Antibody molecules sometimes act directly by binding to antigens. For example, some antibodies, called **antitoxins**, bind to epitopes on bacterial toxins (Chapter 15), inactivating the toxin so it cannot exert its harmful effect.

sis factor (actually a lymphokine), which is directly toxic to some tumor cells and virus-infected cells.

In summary, exposure to antigen fosters direct cellular contacts and production of lymphokines. This stimulates B lymphocytes with the appropriate antibody specificity to generate an activated clone and to differentiate into plasma cells. The activated cells produce the great numbers of antibody molecules needed to fight invading microorganisms.

Clinical Notes

The Indispensable Macrophage

Body structures that we cannot live without always hold a special fascination for clinicians and researchers. Macrophages fall into this category. No one has been found without any macrophages. Macrophages are essential not only for their voracious phagocytic activity—for which they were named—but also for their ability to secrete more than 100 different substances. Among macrophage secretions are cytokines, includ-ing interleukin-1, tumor necrosis factor, and interferons (when produced by macrophages and other mononuclear cells, these cytokines are also called **monokines**). Other macrophage secretions are enzymes, such as lysozome; still others, such as prostaglandins and leukotrienes, mediate inflammation and intercellular communication. Macrophages even secrete components of the complement system.

Table 17.4 Properties of Important Lymphokines

Lymphokine	Source	Role in Lymphatic Activation	Other Actions
Interleukin-1	Macrophages and other antigen-presenting cells; B cells	Activates B cells; activates T cells and stimulates them to produce lymphokines	Inflammatory mediator; induces fever, sleep; stimulates synthesis of erythrocytes
Interleukin-2	T cells; NK cells	Activates T_C cells, lymphokine-producing T cells, and T_S cells	Growth factor for activated T cells; activates monocytes
Interleukin-3	T cells	—	Stimulates production of all types of blood cells by bone marrow
Interleukin-4	T cells	Growth factor for activated B cells; activates T cells	Mast cell growth factor
Interleukin-5	T cells	Stimulates B cells to form a clone and to differentiate into plasma cells	Stimulates differentiation of eosinophils
Interleukin-6	T cells; macrophages	Stimulates B cells to differentiate into plasma cells	Stimulates differentiation of thymus cells
Gamma interferon	T cells; NK cells	Stimulates B cells to differentiate into plasma cells; activates T_C cells	Activates macrophages; antiviral agent; activates T_C cells; increases activity of NK cells
Tumor necrosis factor/Lymphotoxin	T_C cells	—	Toxic to some tumors and virus-infected cells; inflammatory mediator; induces fever, sleep

Other antibodies act directly on viruses. Antibodies bind to the antigens that attach the virus to the host cell, thereby blocking an essential step in viral infection—the virus cannot enter the host cell.

Antibacterial antibodies act indirectly by *linking* a bacterial cell to a phagocyte, which can then destroy the bacterium by phagocytosis, or to complement, which can bring about cytolysis (Chapter 16). Linking molecules must have more than one binding site, however. The sites that bind to antigens on bacterial cells are different from both the site that binds to phagocytes and the site that binds to the activating complement protein, C1. Anti-body binding to the pathogenic bacterium usually occurs first, marking the invading microorganism for destruc-tion. Then the phagocyte or complement protein binds to its site and is poised to destroy the bacterium (**Figure 17.11**).

When antibodies link bacteria to phagocytes, they act as opsonins—molecules that facilitate phagocytosis. When they link bacteria to C1, they activate the comple-ment cascade through the classical pathway (Chapter 16). Antibodies acting indirectly as linking molecules are the

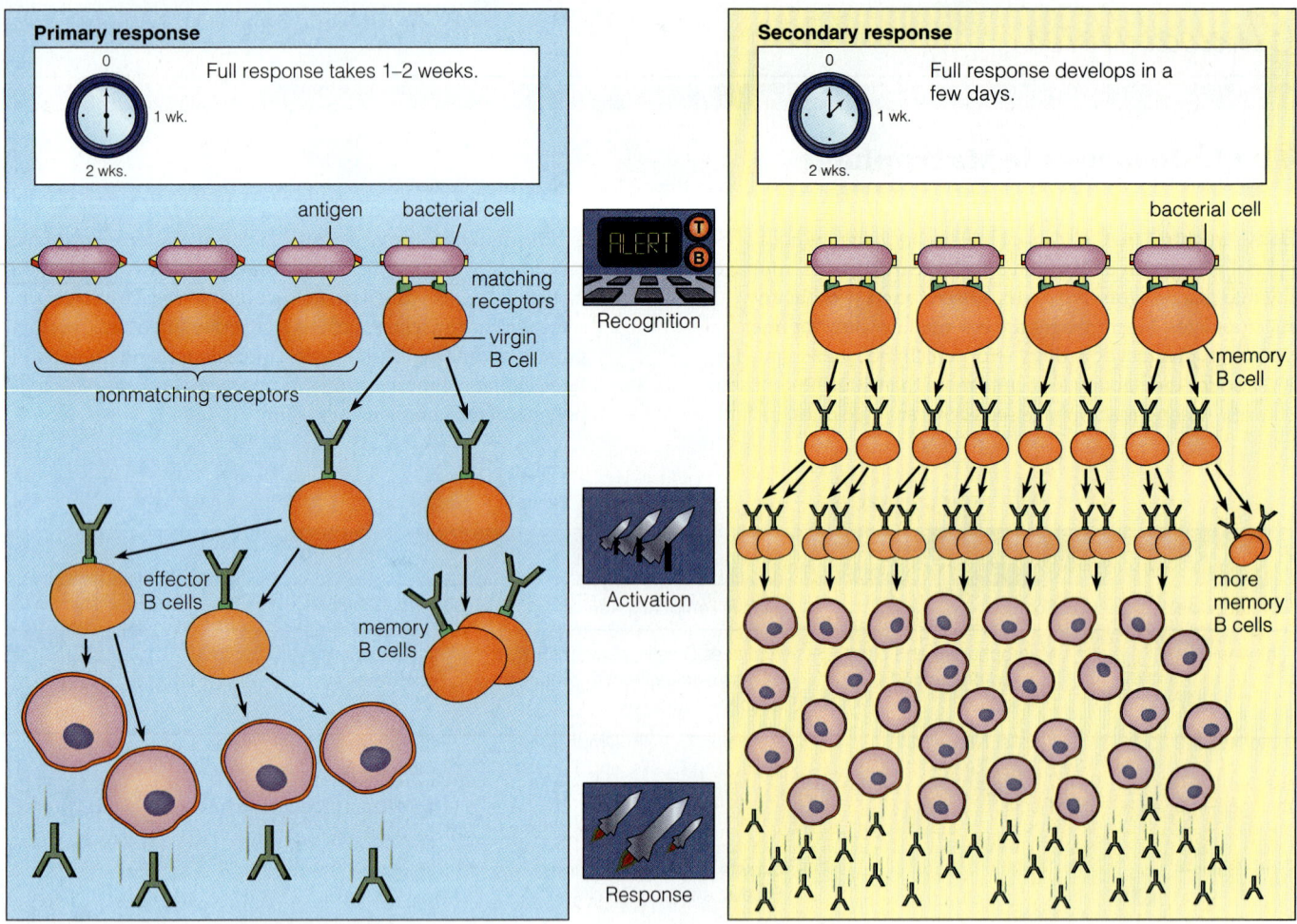

Figure 17.10 Primary versus secondary immune response. The first time a person encounters an antigen, virgin B cells mount a primary immune response by producing antibodies and some memory B cells. If a person encounters the same antigen again, the memory B cells initiate a secondary immune response, which is more rapid and effective than a primary response.

main defenders against bacterial infections and were therefore critical in helping T. L. survive his infection.

The Structure of Antibodies. Antibodies are complex proteins made up of polypeptide chains. On the basis of their chain structure, antibodies are classified into five groups. All five groups share a basic Y-shaped structure called an **antibody monomer**, but the number of component polypeptide chains varies among the five (**Figure 17.12**).

The Y-shaped antibody monomer consists of four polypeptide chains, two identical **heavy chains** and two identical **light chains** linked together by disulfide bonds. (The heavy chains have greater molecular weight than the light chains.) The bottom halves of two heavy chains join to form the stem of the Y, and the top halves form the

inside of the two arms of the Y. The light chains form the outside of the Y arms.

An antibody binds to its complementary antigen at the two **antigen-binding sites** on the ends of the monomer's Y arms. Antigen-binding sites are in the **variable regions** of the antibody, which contain the unique protein sequences created during lymphocyte differentiation by DNA recombination. The rest of the arms and the stem of the Y are called the **constant region**. In any given antibody class the constant region is composed of identical, or nearly identical, sequences of amino acids. Sites for binding to phagocytes and complement (C1) are found in the antibody's constant region. Antibodies can be split by certain proteolytic (protein-splitting) enzymes near the point where the arms of the Y join the stem. One such enzyme, papain, cuts antibodies producing two F_{ab} (anti-

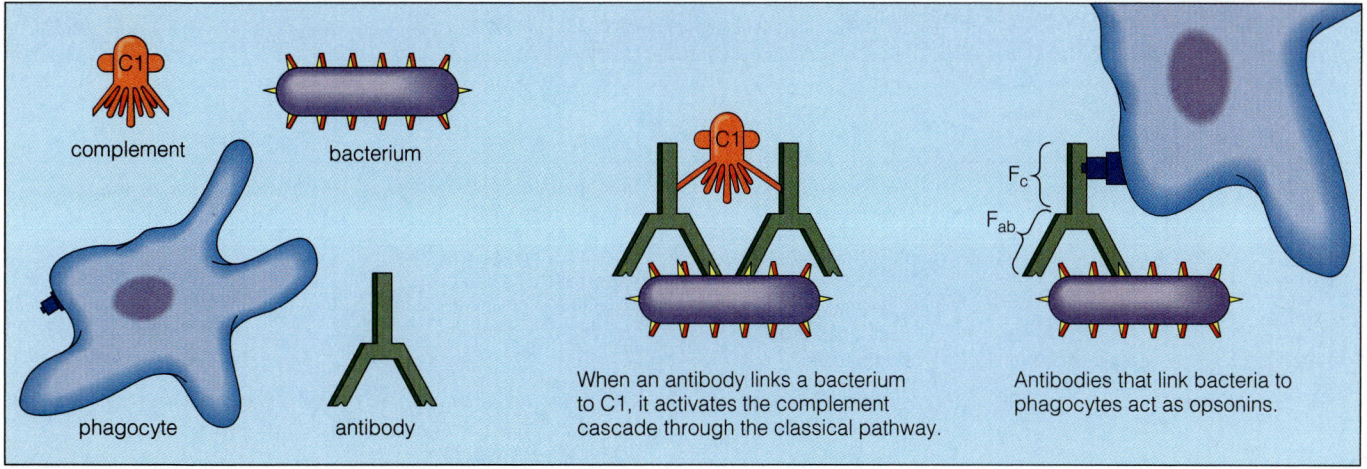

Figure 17.11 Linking function of antibodies. Antibodies themselves do not destroy or inactivate bacteria; rather, they link the bacterial cells to agents of destruction—phagocytes or complement.

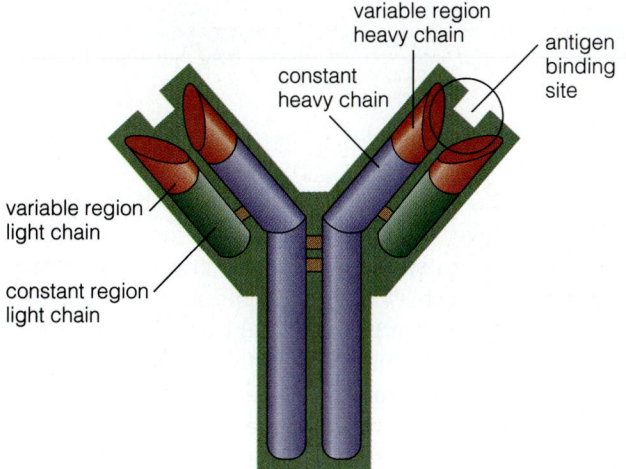

Figure 17.12 Antibody monomer. All antibodies have a basic Y-shaped structure called a monomer. The tail and the inside of the Y's arms are made up of heavy chains, and the outside of the Y's arms are made up of light chains. At the end of each arm, in the variable regions, the monomer has an antigen-binding site. Sites for binding to phagocytes and complement (C1) are in the constant regions.

body-binding) **fragments** and one F_c (**constant**) **fragment**. The terms F_{ab} and F_c are used to describe parts of antibody molecules.

Because there is an antigen-binding site at the end of each arm of the Y, an antibody monomer is said to have a **valence** of 2, meaning that it can combine with two epitopes. Members of antibody classes with more than one monomer have more antigen-binding sites and therefore higher valences. Each antigen also has a valence, the number of antibody molecules with which it can com-

bine. Some large antigens with many epitopes that can combine with many antibody molecules at once have very high valences. An antigen must have a valence of at least 2 in order to be **immunogenic**, or to stimulate the production of antibody.

The Five Classes of Antibodies. Antibodies are part of the **immunoglobulin** family of proteins. Immunoglobulins are immunologically active molecules found in the globulin protein fraction of serum. In fact, antibodies are also called immunoglobulins. Antibodies can be grouped into five major classes distinguished by the composition of their heavy chains and the arrangement of their monomer subunits (**Table 17.5**). The different antibody classes are named immunoglobulin (abbreviated Ig) and followed by a letter—G, A, M, D, or E—that is related to its heavy-chain designation.

T. L. produced all five classes of antibodies. Some were indispensable to his recovery, while others had limited usefulness during his particular struggle, though they are known to be useful in other circumstances. The function of still others is not completely understood.

IgG. The **IgG** (**immunoglobulin G**) **antibodies** form the largest class. About 75 percent of all antibodies in the human body are IgG. IgG is an antibody, or immunoglobulin, monomer. When bound to an epitope, IgG activates the complement cascade through the classical pathway. Many phagocytes bear receptors that bind to the constant region of the IgG molecule, making IgG an effective opsonin. IgG is the only immunoglobulin class that can cross the placenta. Maternal IgG, directed against antigens to which the mother is immune, enters the fetal circulation to protect the newborn for several months after

Table 17.5   The Five Classes of Antibodies

Class	Structure	Percent of Total Antibodies	Location	Primary Function	Additional Functions
IgG	Monomer	75	Blood and extracellular fluids	Predominates in secondary immune response	All antitoxins belong to this class
IgA	Monomer/dimer	15–20	A monomer in the blood; a dimer with a secretory component attached in body secretions	Protects mucosal surfaces, especially by preventing attachment of viruses	—
IgM	Pentamer	5–10	Blood and extracellular fluids	The "early" antibody produced during a primary immune response; fixes complement	Causes precipitation of high-valence antigens
IgD	Monomer	less than 1	Principally on the surface of B cells	Poorly understood; may help stimulate lymphocyte proliferation in response to antigens	—
IgE	Monomer	less than 0.01	On the surface of mast cells and basophils	Stimulates cell to degranulate when it binds to antigen	Defends against protozoa and parasitic worms; may cause harmful allergic reactions

birth. For example, if a mother is immune to measles, her newborn baby will also be immune as long as maternal IgG persists in the child's circulation. IgG antibodies are the most abundant antibody class produced during a secondary immune response.

IgG was the most important class of antibody produced by T. L.'s secondary immune response against his bacterial infection. As an opsonin, IgG helped T. L.'s phagocytes recognize and attach to the invading bacteria. As an activator of the complement system, it produced more opsonins in the form of C3b, and it generated inflammatory mediators that chemotactically attracted and activated phagocytes in the area of infection. However, skin infections like T. L.'s are usually caused by Gram-positive bacteria, which are not generally vulnerable to the lytic effect of complement, so complement-mediated bacterial lysis probably played a very limited role in T. L.'s recovery. Thus, the critical function of IgG in T. L.'s defense was primarily to enhance phagocytosis, demonstrating that the body's nonspecific and specific defenses are really one highly interrelated system.

IgA. The **IgA antibodies** are the second largest class, representing 15 to 20 percent of the body's total antibodies. IgA occurs as both a monomer and a **dimer** (two monomers joined together). The monomer is found in the blood and extracellular fluid, while the dimer enters secretions such as saliva, milk, and mucus. An antibody dimer has four antigen-binding sites and a valence of 4.

A **J chain**, or joining peptide, produced by plasma cells, holds two monomers together to form the dimer. IgA molecules in secretions have an attached **secretory component**, which is a protein produced by epithelial cells. As an IgA dimer passes through an epithelial cell to enter human secretions, it binds to a secretory component, becoming **secretory IgA**. The secretory component not only facilitates secretion of the molecule onto a mucosal surface, but it also protects IgA from being destroyed by the proteolytic enzymes in human secretions.

Secretory IgA is found in the protective mucous layer that lines organs such as the lungs, intestines, and bladder. It protects these surfaces by inactivating microorganisms, particularly viruses, before they can adhere to the mucous membrane surface. Secretory IgA is also present in tears and saliva. Its presence in breast milk and **colostrum**, the breast secretion that precedes the production of true milk, defends the gastrointestinal tract of newborn humans against infection.

The primary role of IgA is to protect the body's mucous membrane surfaces against microbial attack. Because T. L.'s infection occurred across his skin and not a mucous membrane, IgA played only a marginal role in his defense.

IgM. Approximately 5 to 10 percent of the antibody total is made up of **IgM antibodies**. The IgM antibody class is extremely effective in fixing complement. IgM molecules consist of five monomers joined together by a J chain into

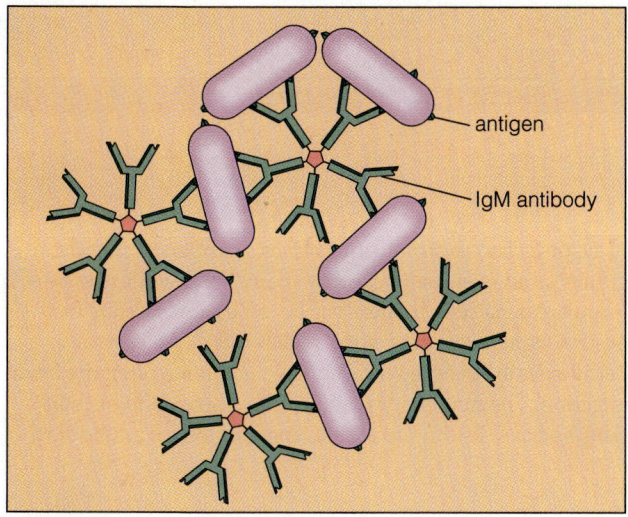

Figure 17.13 Antigen-antibody precipitation. Large numbers of antibody and antigen molecules may clump together into a lattice structure that becomes so large it precipitates—it can no longer stay in solution. IgM antibodies are particularly prone to precipitation because of their pentamer structure (five monomers) and high valence (10). Precipitation is the basis of many diagnostic laboratory tests.

Labels in figure: antigen; IgM antibody

a **pentamer**. It has a valence of 10, the highest of any antibody class. IgM is sometimes called *early antibody* because it is the first antibody class to form during a primary immune response. Its high valence allows IgM to bind firmly to antigens even though the antibodies produced during a primary immune response tend to have a weaker affinity for the target antigen than do antibodies produced during a secondary response.

IgM's pentamer structure also allows it to build complex antigen-antibody lattices that sometimes clump, forming a visible precipitate, or molecular complex, so large it can no longer remain in solution (**Figure 17.13**). The clumping is the basis of a number of laboratory assays (tests; see Chapter 19). Small quantities of IgM are also produced during a secondary immune response, so this antibody class undoubtedly contributed to T. L.'s defense against his infection. IgM is one of the antibody classes displayed on the surface of B cells.

IgD. The monomer **IgD antibody** constitutes less than 1 percent of the total amount of antibody. It is the main type of antibody found on the cell membranes of antibody-producing B cells. IgD may play a role in regulating clonal selection, but little is known for certain about this class.

IgE. The **IgE antibody** is also a monomer. It occurs in the body in minute quantities, constituting less than 0.01 percent of the antibody total. IgE is fixed to the surface of

basophils, mast cells, and eosinophils. When molecules of IgE bind to an antigen, it stimulates the microbe to degranulate, releasing powerful inflammatory mediators such as histamine. The degranulation contributes to the body's defense against disease-producing parasites, such as worms, that are too large to be eliminated by phagocytosis. Unfortunately, IgE degranulation also contributes to **allergy**, a malfunction of the immune system (Chapter 18). IgE did not contribute to T. L.'s defense because it does not play a significant role in defending against bacterial infection.

T LYMPHOCYTES

In this section, you will read how T-cell function differs from B-cell function. T cells are divided into two main classes—T_4 **cells** and T_8 **cells**, distinguished by cell markers called T_4 and T_8. T_4 cells are the T_H (**helper**) **cells**. They regulate the immune system by increasing immune responsiveness. T_8 cells are further differentiated into T_C (**cytotoxic**) **cells** and T_S (**suppressor**) **cells**. T_C cells attack and kill invading microorganisms. T_S cells regulate the immune system by decreasing (suppressing) the immune response.

Antigen Recognition

All T cells produce antigen receptor proteins with which they bind foreign antigens (thereby recognizing them). In many respects, T-cell antigen receptors resemble the antibody molecules that act as B-cell antigen receptors. Just as during B-cell differentiation, millions of genetically unique T cells are produced by recombination, each capable of producing a unique antigen receptor protein with a unique antigen-binding specificity. Also just like antibodies, each T-cell antigen receptor has a variable region that binds to and recognizes antigen and a constant region that is attached to the T-cell surface.

But T-cell antigen receptors differ from antibody molecules in one important respect—in order to recognize a foreign antigen, they must simultaneously recognize a particular type of **self antigen**. Self antigens are structural components of the surface of human cells much like foreign antigens are structural components of the surface of invading microorganisms. Unlike foreign antigens, however, self antigens do not normally stimulate a destructive immune response. The immune system's lack of response to self antigens is called **immunologic tolerance**. Tolerance is an essential safety feature of the immune system; it prevents destruction of the body's tissues by its own immunologic weapons. The ability to distinguish self antigens from foreign antigens means the immune system can distinguish self from nonself.

Runaway Antibodies

For most of us, antibodies help protect against infection. But victims of a plasma cell cancer, called multiple myeloma, suffer runaway antibody production that can be fatal. In these patients one particular clone of plasma cells turns cancerous and begins to multiply out of control, producing immense quantities of antibody in the process. Because the cancerous transformation occurs within a single clone, the excess antibody molecules are of a single type—either IgG,

A, D, or E. The enormous quantities of antibody clog the kidneys and eventually destroy them. Meanwhile, the cancerous plasma cells themselves become so numerous they destroy bone, crowding normal cells out of the marrow. Paradoxically, people with multiple myeloma often die from infection. The malignant plasma cells overwhelm normal lymphocytes, leading to a deficiency of normal antibodies.

How exactly immunologic tolerance occurs is an area of active research. One answer is clonal deletion—any clone of T lymphocytes that happens to react against self antigens is deleted, or destroyed, soon after differentiation in the thymus. In other words, the immune system itself recognizes and eliminates potentially harmful T cells before they have a chance to proliferate. Suppressor T cells may also be crucial to the development and maintenance of immune tolerance, but we know little about the precise role T_S cells play. When immune tolerance fails, the result is **autoimmune disease**; *auto* means "self" (Chapter 18). In effect, the body tries to destroy itself.

The combination of foreign and self antigens essential to T-cell antigen recognition works this way: A foreign antigen is taken up and processed by a human cell—sometimes an antigen-presenting cell or a lymphocyte or sometimes a normal body tissue cell. Then the processed antigen is combined physically with a self antigen normally present on the cell surface and displayed there. Next, the T-cell antigen receptor binds to the foreign-self antigen complex (**Figure 17.14**). Thus, the antigen-binding site of a T-cell antigen receptor must simultaneously recognize and bind both a foreign antigen and a self antigen, unlike the antigen-binding site of an antibody, which binds only a foreign antigen. Also unlike B cells, which can sometimes recognize free antigens, T cells can recognize *only* those antigens on cell surfaces that bear the appropriate self antigens. The self antigens that participate in T-cell antigen recognition are also called **restriction elements** because they restrict the recognition of foreign antigens to surfaces on which they are present.

Self antigens vary from person to person due to differences in each individual's genetic makeup. The DNA that encodes self antigens is called the **major histocompatibility complex (MHC)**. It is located on chromosome

number 6. Two major classes of MHC-encoded antigens mark cells. **Class I MHC antigens** are present on all nucleated cells in the body. **Class II MHC antigens** are present only on macrophages, other antigen-presenting cells, and certain lymphocytes.

Different classes of T cells require different MHC antigens in order to recognize foreign antigens. T_H cells recognize only antigens that are presented with class II MHC self antigens, and T_C cells recognize only antigens that are presented with class I MHC self antigens. The difference in MHC restriction is logical in view of the different cell functions. T_H cells, which help initiate an immune response, recognize foreign antigens on the surface of antigen-presenting cells (which bear class II MHC antigens), while T_C cells recognize antigens on the surface of virus-infected or cancerous tissue cells (which bear class I MHC antigens). We think MHC restriction sharpens the focus of T-cell antigen recognition, possibly preventing errors that could lead to a misdirected and harmful autoimmune response.

T-Cell Activation

All T lymphocytes exist in a resting state until their antigenic receptors encounter the appropriate combination of foreign antigen and MHC self antigen. The first cells to be activated in any immune response are the T_H cells. These were the cells that initiated T. L.'s immune response. T_H cell activation begins when a T_H cell binds to a matching foreign antigen–class II MHC complex on the surface of an antigen-presenting cell, often a macrophage. Activation is enhanced when the T_H cell, now in close contact with the macrophage, is stimulated by macrophage-produced interleukin-1. The stimulated T_H cell *stimulates itself* still further by producing interleukin-2. Responsive-

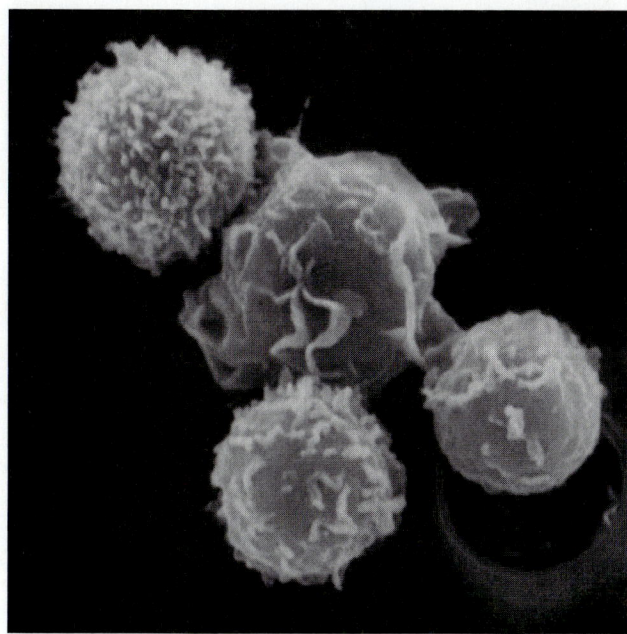

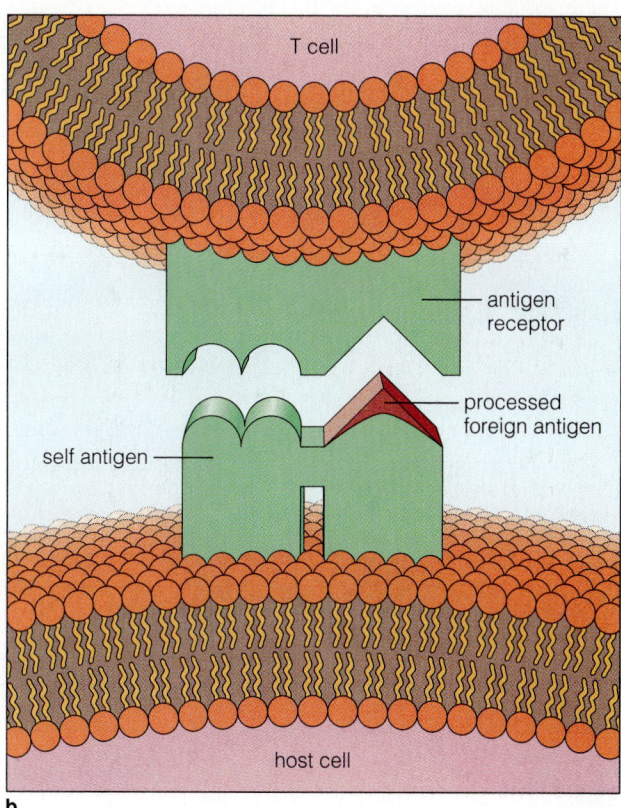

Figure 17.14 T-cell antigen recognition. T cells recognize foreign antigen only if it is processed by a host cell and then displayed in physical combination with a self antigen on the host cell's surface. The T-cell antigen receptor binds to both the foreign antigen and the self antigen. (**a**) Antigen-presenting B cell surrounded by three T cells. (**b**) Cell antigen and processed foreign antigen on a host cell bind to antigen receptor on a T cell.

ness to interleukin-2 increases as the T_H cell acquires new cell-surface receptors that recognize interleukin-2. Resting T cells have approximately 100 receptors that can recognize interleukin-2, but activated T cells develop as many as 12,000. Binding of interleukin-2 to receptors on the T-cell surface is the most powerful stimulus to T_H cell activation, but interleukin-4 also helps activate T_H cells.

The production of interleukin-2 by clones of activated T_H cells also helps to activate T_C and T_S cells. When these cells recognize antigen, presented in combination with the class I MHC restriction elements that they recognize, they respond to stimulation by interleukin-2. The combination of antigen recognition and interleukin-2 causes the T_C and T_S cells to divide, forming clones of activated cells that can participate in the immune response. This clonal selection is similar to B-cell clonal selection.

T-Cell Response

Once activated, T cells respond in one of two ways—with direct T_C cell killing, called **cytotoxicity**, or with **immunoregulation**, intensifying or suppressing the immune response (carried out by T_H or T_S cells).

Cytotoxicity. T_C cells are a subgroup of T_8 lymphocytes that recognize foreign antigens presented in physical combination with the class I MHC self antigens present on all nucleated cells in the human body. When self antigens combine with foreign antigens, the *human* cell is marked as a target for T_C cell killing.

This works to the body's advantage in controlling viral infections (**Figure 17.15**). When a virus infects a person, viral particles—which bear many surface antigens—enter the host cells. Host cells process these viral antigens and transport them to the cell surface where they combine with class I MHC antigens. The infected cell that displays class I MHC antigens in combination with foreign viral antigens is marked as a target. Destruction begins when a T_C cell with an appropriate antigenic receptor encounters and binds to the infected host cell, bringing the two cells into intimate contact for about 10 minutes. During this time, the T_C cell secretes toxic proteins called **perforins** from its lymphocytic granules. Perforins create holes in the target cell membrane that resemble the holes produced by the membrane attack complex in complement. Approximately an hour after the two cells separate, the target cell swells and lyses from membrane damage.

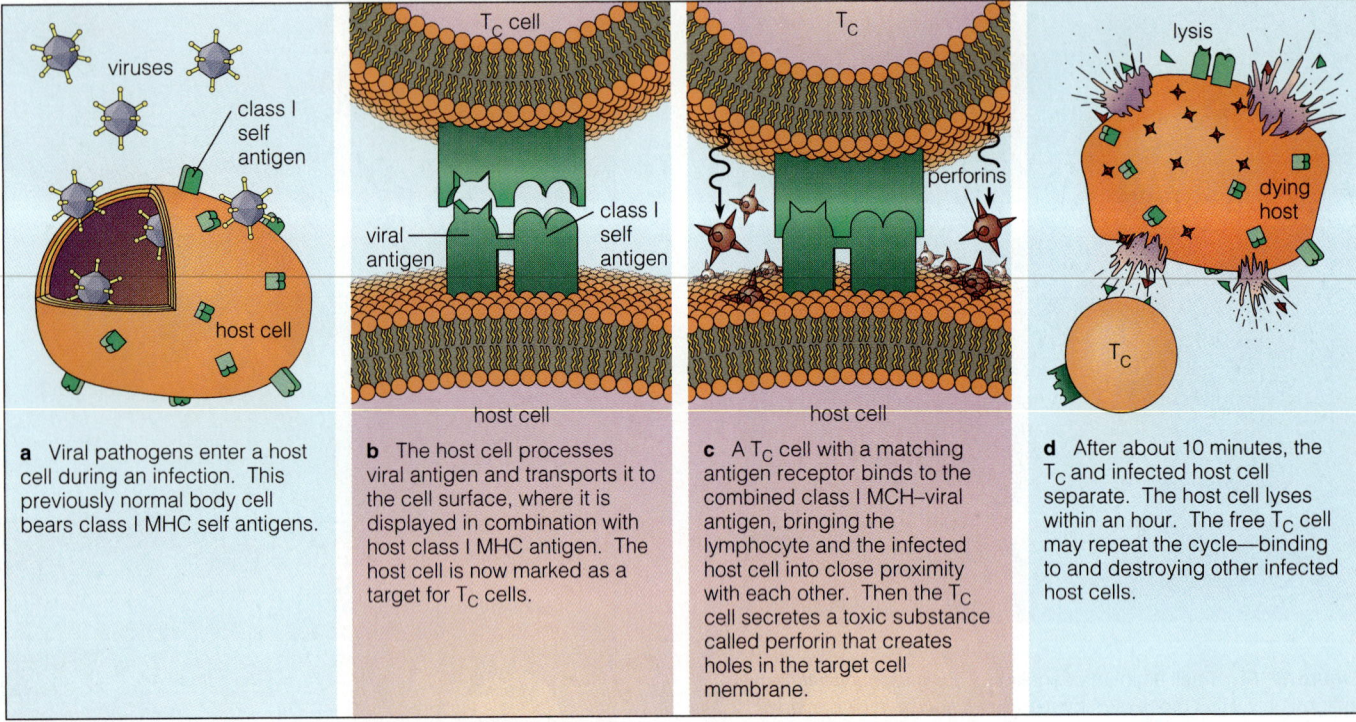

a Viral pathogens enter a host cell during an infection. This previously normal body cell bears class I MHC self antigens.

b The host cell processes viral antigen and transports it to the cell surface, where it is displayed in combination with host class I MHC antigen. The host cell is now marked as a target for T_C cells.

c A T_C cell with a matching antigen receptor binds to the combined class I MCH–viral antigen, bringing the lymphocyte and the infected host cell into close proximity with each other. Then the T_C cell secretes a toxic substance called perforin that creates holes in the target cell membrane.

d After about 10 minutes, the T_C and infected host cell separate. The host cell lyses within an hour. The free T_C cell may repeat the cycle—binding to and destroying other infected host cells.

Figure 17.15 Cytotoxicity. Once activated, T cells can destroy a virus-infected host cell.

After separating from its first victim, the T_C cell binds to another infected cell and then another. In this way a single T_C cell can eliminate many targets. By killing cells that are already infected by the virus, T_C cells help prevent the production of new viral particles and thus stem the spread of infection.

Cells infected by viruses are not the only targets for T_C cell killing. Any host cell that bears class I MHC antigens in combination with foreign or abnormal antigens may become a T_C cell victim. Cancer cells often begin to display abnormal antigens when they undergo transformation and frequently become T_C cell targets. T_C cells can also attack certain eucaryotic pathogens, such as fungal cells. But, on the negative side of the body's immune system ledger, organ transplants bearing the donor's antigens are perceived as foreign by the recipient's T_C cells. T_C cell killing is the basis of transplant rejection (Chapter 18).

Immunoregulation. Two classes of T cells act as **immunoregulatory cells**, helping to adjust the immune system for optimum effectiveness. T_H cells intensify the immune response while T_S cells dampen it. As you have read in this chapter, T_H cells are vital components of both humoral and cell-mediated immunity. In most cases, they are essential participants in the chain of events that acti-

vates B cells to produce antibody, and they are essential in activating T_C cells.

T_S cells play the opposing role—they lower immune responsiveness when it is no longer needed or might be dangerous. Although the details of their action are still poorly understood, we do know that T_S cells turn off activated lymphocytes and return the immune system to a resting state when infection has been controlled. They interact both with B cells and with other classes of T cells; and they are thought to act by secreting proteins called **suppressor factors**, which signal the immune system to decrease its activity. T_S cells probably also help develop immune tolerance because experiments that stimulate immune tolerance are known to activate T_S cells.

NON-B NON-T LYMPHOCYTES

The third group of lymphocytes, non-B non-T lymphocytes, makes up less than 3 percent of the total lymphocyte population. These cells are fundamentally different from B and T lymphocytes because they function without recognizing antigen. Non-B non-T cells also lack the markers that identify B and T cells (antigens in the case of B cells, and T_4 and T_8 markers in the case of T cells). Non-

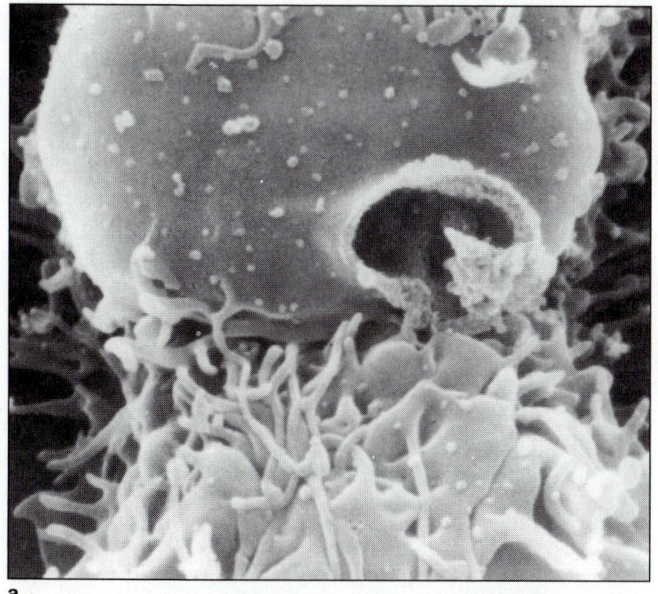

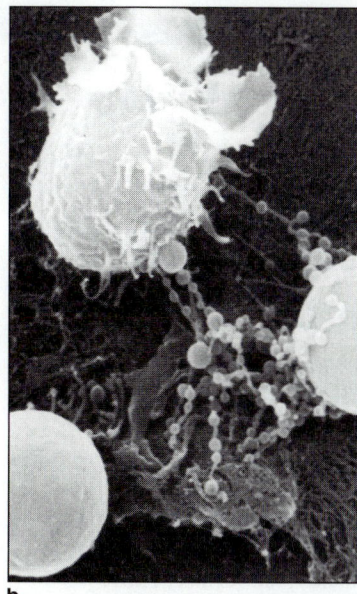

Figure 17.16 Natural killer cells. These scanning electron micrographs show an NK cell (a type of non-B non-T cell) destroying a tumor cell. (**a**) After contact between the NK cell and the cancer cell is established, the NK cell secretes perforins, which create holes in the cancer cell's membrane. The cancer cell becomes swollen and leaky. (**b**) All that remains of the cancer cell after lysis are remnants of its nucleus and other debris.

B non-T lymphocytes are sometimes called **large granular lymphocytes** because of their slightly larger size and the presence of granules in their cytoplasm.

One group of non-B non-T cells, called **natural killer (NK) cells**, lyse target human cells in much the same way T_C cells do, by secreting perforins. Soon after an NK cell contacts a target, the target cell lyses. NK cells destroy cells infected by viruses and cells in transplanted organs. They are thought to be particularly effective against cancer cells (**Figure 17.16**). Actually, because NK cells do not recognize antigen, they resemble the nonspecific defenses described in Chapter 16 more closely than they resemble the specific immune defenses described in this chapter.

A second group of non-B non-T lymphocytes called **killer (K) cells** acts slightly differently from NK cells. K cells destroy only target cells that have been coated by antibody. Their type of destruction is called **antibody-dependent cellular toxicity**. In spite of their dependence on the presence of antibody, K cells are also nonspecific because they cannot distinguish among antibodies. They attack cells marked by any of the five antibody classes.

THE ROLE OF THE IMMUNE SYSTEM IN T. L.'S RECOVERY

Let's return to the case of T. L. and trace the steps in his body's immune response—from recognition to activation and response (**Figure 17.17**).

When T. L.'s skin was punctured, large numbers of bacteria, marked by their characteristic antigen mole-cules, were introduced into his subcutaneous tissues. The bacteria multiplied rapidly, and local inflammatory defenses in his hand were overwhelmed by the developing infection. Without help from his immune system, the infection would have spread throughout T. L.'s body, probably causing his death.

Immune Recognition

T. L.'s immune system prepared to recognize the invading bacteria when antigen-presenting cells such as macrophages phagocytized some bacteria, destroying most of the bacterial cell but preserving the bacterial antigens. The antigens were processed into fragments of the right size to combine with lymphocyte receptors and displayed on the macrophage surface. Very few responsive lymphocytes were present near T. L.'s wound, and the likelihood that the displayed antigens would happen to encounter lymphocytes with matching receptors at the site was extremely remote. But macrophages transported the processed antigen to T. L.'s lymph nodes, which housed millions of immunocompetent lymphocytes with different antigen receptors.

Within the lymph nodes, antigen-presenting macrophages encountered T_H and B cells with antigen receptors that matched the displayed bacterial antigens. Recognition occurred when the lymphocyte antigen receptors bound to the displayed antigens. Because he had been exposed previously to organisms with these same antigens, T. L. had a relatively large number of memory lymphocytes with receptors that matched the particular bacterial antigens. This significantly accelerated his immune response.

Anti-immune Trickery: Various Strategies

The immune system makes life difficult for pathogenic microorganisms. Yet, as you know, people get infections and sometimes die from them. How do different species of microorganisms resist the immunologic onslaught?

Many pathogens establish chronic or fatal infections because they have evolved ways to trick the immune system—to evade immune recognition or slip away, unharmed, from a powerful immune response. Let's look at a few examples of microorganisms that have evolved strategies to help ensure their survival.

Some viral pathogens evade immune defenses by hiding their most critical antigens. All viruses must somehow enter the cells they infect. Most enter by means of specialized proteins that bind to receptors on the host cell surface. These viral host-binding proteins are especially vulnerable targets for antibody attack because an antibody molecule that combines with them and occupies the binding site prevents infection. Rhinovirus-14, one of many viruses that cause the common cold, circumvents this problem by hiding its host cell binding protein at the bottom of a deep canyon on the surface of the viral particle. Its binding target—the host cell receptor protein—is shaped so that it can reach into the canyon and establish the necessary link for viral attachment and survival. But antibody proteins are too large to enter the canyon and therefore cannot inactivate the binding protein. As a result, the rhinovirus is protected against antibody attack. The viruses that cause influenza, poliomyelitis, and AIDS similarly conceal their host-binding proteins on the floor of narrow surface canyons.

Other pathogens confuse immune defenses by changing their surface antigens. Trypanosomes, the protozoal pathogens that cause African sleeping sickness (Chapter 25), are masters at this antigenic switching. Mounting a primary immune defense takes time, so significant amounts of antibody are not produced for almost two weeks after the body detects a foreign antigen. When a trypanosome first enters a new human host, it displays a variable surface glycoprotein that activates lymphocytes and stimulates an immune response. But by the time significant quantities of antibody are produced, the variable surface glycoprotein has changed and the new, differently shaped antigen cannot be bound and inactivated by the already-produced antibody. The new antigen, in turn, stimulates a primary immune response and the cycle begins again. By the time new antibody is produced—nearly two weeks later—the trypanosome's antigens have changed yet again. The trypanosome is capable of producing

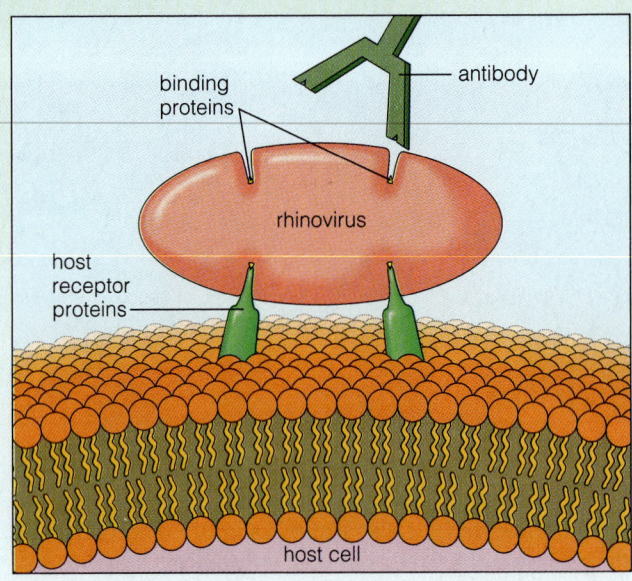

a Rhinovirus 14. The binding protein that allows the rhinovirus to attach itself to and enter a human cell is located at the bottom of a molecular canyon. The host-cell receptor is narrow enough to enter the canyon, which allows the virus to initiate an infection, but antibodies are too fat to enter and inactivate the virus.

more than 1000 different antigens, and by continually changing them it remains one jump ahead of the immune system. As a result, humans cannot eliminate the pathogen from their bodies, and untreated African sleeping sickness is a chronic, usually fatal infection. Other pathogens that rely on antigenic switching for survival are the bacteria *Neisseria gonorrhoeae* and *Haemophilus influenzae*.

T-cell defenses, too, can be sabotaged. One example is the adenovirus, which infects the human respiratory and gastrointestinal systems. T_C cells, the usual defenders against viral infections, can recognize viral antigens only if they are displayed in complex with class I MHC antigens on a host cell surface. But adenoviruses produce proteins that bind to class I MHC antigens and trap them within the cell, preventing them from being transported to the cell surface and displayed. Thus, although viral antigens are present within the host cell, they cannot be appropriately displayed, and so immune recognition is thwarted. The immune response never gets under way.

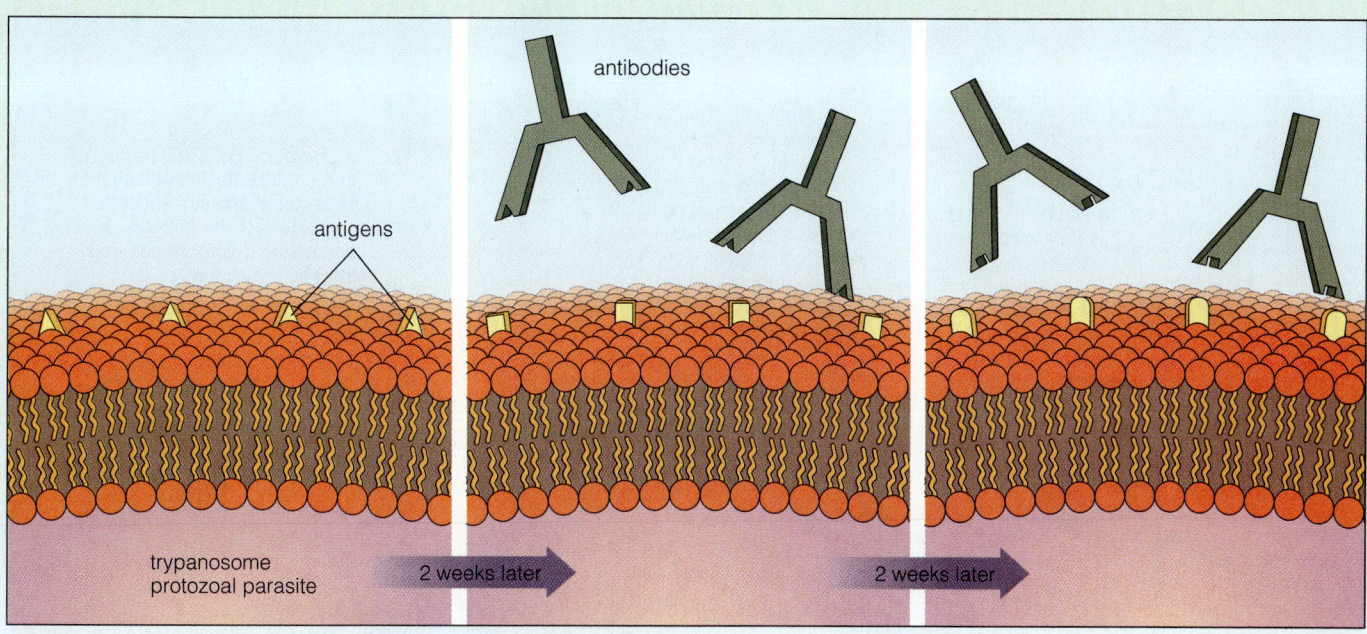

b Trypanosomes. Antigens on the surface of trypanosomes change so frequently that the antibodies produced by a primary immune response cannot keep up. Consequently, untreated African sleeping sickness is a chronic and often fatal disease.

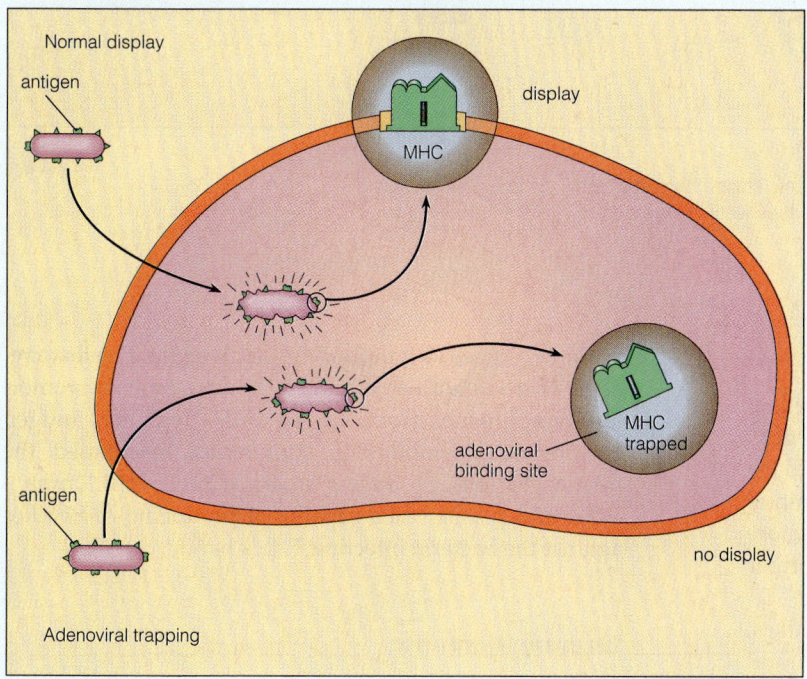

c Adenovirus. The adenovirus produces a protein that binds to the class I MHC antigen, but not at the normal binding site for foreign antigens. Instead, the adenoviral protein traps the MHC antigen within the host cell so it can't display and trigger the immune response.

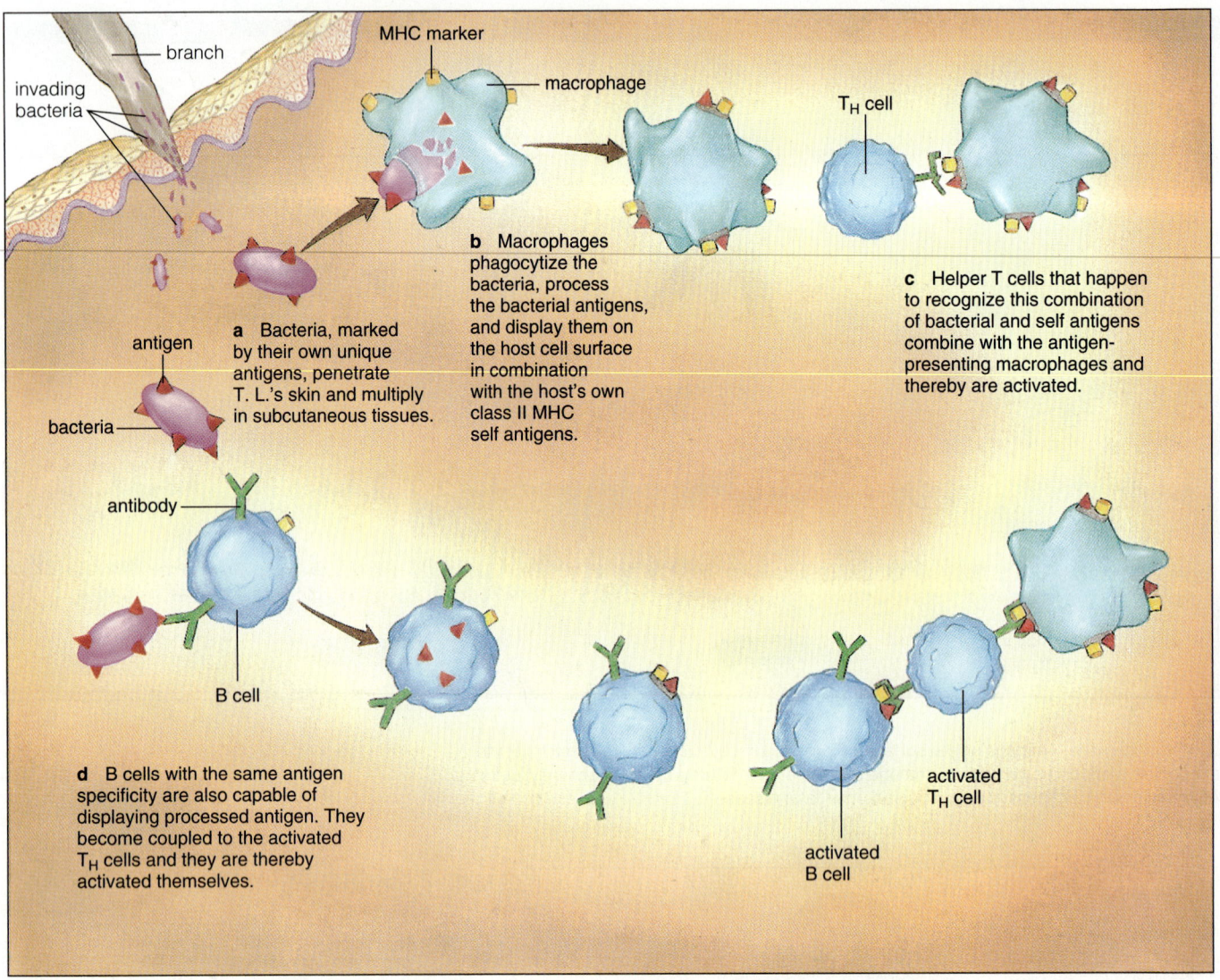

branch

invading bacteria

MHC marker

macrophage

T_H cell

antigen

bacteria

a Bacteria, marked by their own unique antigens, penetrate T. L.'s skin and multiply in subcutaneous tissues.

b Macrophages phagocytize the bacteria, process the bacterial antigens, and display them on the host cell surface in combination with the host's own class II MHC self antigens.

c Helper T cells that happen to recognize this combination of bacterial and self antigens combine with the antigen-presenting macrophages and thereby are activated.

antibody

B cell

d B cells with the same antigen specificity are also capable of displaying processed antigen. They become coupled to the activated T_H cells and they are thereby activated themselves.

activated B cell

activated T_H cell

Figure 17.17 T. L.'s immune response. The immunologic events shown here enabled T. L. to recover from his infection when his first- and second-line defenses were overwhelmed.

Immune Activation

Some of the processed bacterial antigen was displayed on the macrophage surface in a complex with class II MHC antigens. These antigen complexes were recognized by T_H cells in T. L.'s lymph nodes, and antigen recognition stimulated the lymphocytes to enlarge and become more metabolically active. The T_H cells were stimulated further by macrophage-produced interleukin-1 molecules. T_H cells, in turn, produced self-stimulating lymphokines, particularly interleukin-2, that caused cell division and formation of an active T_H cell clone.

At the same time, memory B cells were taking up and presenting bacterial antigens, causing some B and T_H cells to become coupled together. The close proximity of these two types of lymphocytes allowed the B cells to receive large quantities of activating and growth-promoting lymphokines from the T_H cells. Together, antigen binding and lymphokine stimulation reactivated the memory B cells, producing an army of active B lymphocytes and plasma cells capable of producing antibodies against the bacteria infecting T. L.'s hand.

Immune Response

Mass production of antibody began when B-cell clones that could produce antibody against the bacteria in T. L.'s hand began to proliferate. Production increased when some activated B cells were stimulated by lymphokines

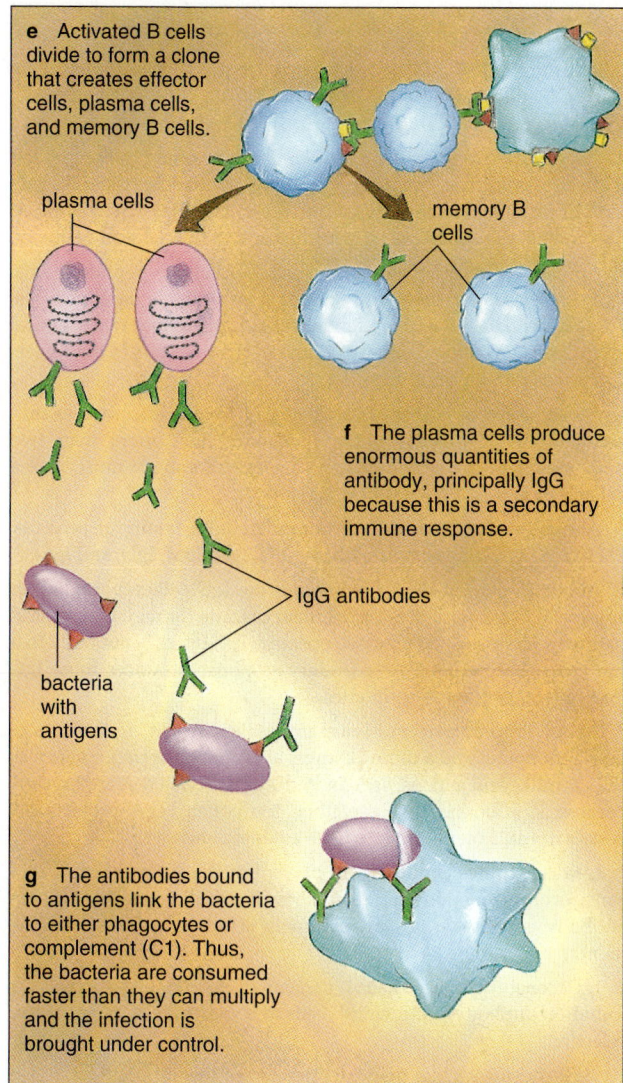

e Activated B cells divide to form a clone that creates effector cells, plasma cells, and memory B cells.

plasma cells

memory B cells

f The plasma cells produce enormous quantities of antibody, principally IgG because this is a secondary immune response.

IgG antibodies

bacteria with antigens

g The antibodies bound to antigens link the bacteria to either phagocytes or complement (C1). Thus, the bacteria are consumed faster than they can multiply and the infection is brought under control.

to differentiate further and become short-lived plasma cells that produce huge quantities of antibody. Because this was a secondary immune response, most of the antibody produced was IgG.

Millions of IgG molecules entered T. L.'s circulation. Wherever they encountered an invading bacterium, their variable region bound to antigens on the bacterial cell surface. At sites within their constant region, these same antibody molecules bound to complement or phagocytes, activating complement and causing opsonization. Opsonization made the invading bacteria easy prey for neutrophils and other phagocytes. Before antibody levels rose, the bacteria multiplied faster than T. L.'s phagocytes could eliminate them. But after antibody became plentiful, phagocytosis outpaced bacterial multiplication and T. L.'s immune system won the battle against the invading bacteria.

As T. L.'s infection subsided, suppressor T cells returned the activated lymphocytes to a resting state. Some memory lymphocytes, however, persisted in his body to defend him against yet another infection by the same type of bacteria.

B cells played the major role in controlling T. L.'s infection. Another type of infection might have been controlled primarily by cell-mediated immunity.

Summary

CELLS AND ORGANS OF THE IMMUNE SYSTEM (pp. 403–404)

1. The immune system is a network of cells and organs that extends throughout the body.

2. Its principal cells are lymphocytes. The two main types of lymphocytes are B cells and T cells.

3. B cells are agents of humoral immunity; they respond to antigens by producing defensive proteins called antibodies. T cells are agents of cell-mediated immunity; they respond to antigens directly by killing the marked cells or indirectly by regulating immune system function.

4. A third type of lymphocyte, the non-B non-T cells, functions without recognizing antigens.

5. The primary lymphoid organs are the bone marrow and the thymus. The secondary lymphoid organs are the lymph nodes and spleen. Other lymphatic tissues are the adenoids, tonsils, appendix, and Peyer's patches. Lymphatic circulation is a system of vessels that collects excess fluid and lymphocytes (lymph) from the body's tissues.

TYPES OF IMMUNITY (pp. 404–405)

1. Immunity can be active (a product of the individual's own immune system) or passive (antibodies produced elsewhere are given to the individual). Both types can also be naturally or artificially acquired.

2. Naturally acquired active immunity results from infections. Artificially acquired active immunity comes from vaccines. Naturally acquired passive immunity refers to the antibodies a child receives from the mother across the placenta and in breast milk. Artificially acquired passive immunity consists of antibodies formed by an animal or human and administered to an individual.

THE IMMUNE RESPONSE: AN OVERVIEW (pp. 405–407)

1. Immune recognition requires direct contact between the invader and a lymphocyte. Lymphocytes have antigen-recognizing molecules called antigenic receptors. B-cell antigenic receptors are antibodies; on T cells they are called T-cell antigen receptors.

2. Antigenic receptors bind to a small region of the antigen called the epitope. There is shape-to-shape recognition.

3. Most small molecules cannot stimulate an immune response. A hapten, a small molecule, may combine with a larger carrier protein to create an antigen in which the hapten is the epitope.

4. Recognition is facilitated by antigen-presenting cells, cells that phagocytize the microorganism, process the antigen, and then display the antigen on their surface.

5. When the immune system is not fighting infection, only a few lymphocytes that produce each type of antigenic receptor exist.

6. On activation, a lymphocyte multiplies rapidly to produce enough cells to fight an infection.

7. Some activated B cells differentiate into plasma cells, short-lived factories that mass-produce antibodies. Activated T cells kill invaders or regulate the immune response.

B LYMPHOCYTES (pp. 407–417)

1. When B cells differentiate in the bone marrow, they become immunocompetent, capable of producing an antibody that is a unique antigenic receptor. There is a B cell with an antibody for every antigen we encounter in our lives.

Antigen Recognition (pp. 409–410)

1. B-cell diversity is a result of genetic recombination. As differentiation occurs, undifferentiated lymphocytes randomly assemble modular bits of DNA, producing millions of genetically different B cells from a relatively limited amount of DNA.

2. A similar process occurs with T cells in the thymus.

3. Macrophages are the main antigen-presenting cells. They phagocytize pathogens and process the antigens so the epitopes are displayed; then they travel to nearby lymph nodes where odds are increased that they will encounter a lymphocyte with a matching receptor.

4. When an antibody binds to an antigen, the two molecules form an antigen-antibody complex. The fit may be good or poor. A poor fit is termed lower binding affinity and results in a poor immune response.

B-Cell Activation (pp. 410–412)

1. Only the B cells bearing antibodies that match the invaders' antigens are activated. The selected cells divide to produce a clone (clonal selection). Some members of the clone are effector cells that fight the current infection. Others are memory B cells that persist to fight future infections.

2. B cells can be activated directly by T-independent antigens, large antigens with many identical epitopes; they stimulate a relatively weak immune response.

3. Most antigens are T-dependent antigens, which involve antigen-presenting cells and helper T cells (T_H). In this case, B-cell activation takes several steps. Antigen-presenting cells travel to lymphoid organs and encounter T_H cells with receptors that match the epitope being displayed. Activation of the T_H cells activates B cells that recognize the same antigen.

4. Macrophages and other antigen-presenting cells secrete a cytokine called interleukin-1 that activates B and T_H cells.

5. Activated T_H cells secrete lymphokines: interleukin-2 activates B and T cells. Interleukin-4 and interleukin-5 stimulate B cells to form clones and to differentiate into plasma cells. Interleukin-6 and gamma interferon stimulate B cells to differentiate into plasma cells.

6. After an infection is over, memory B cells function as immunologic memory. Memory B cells mount a stronger response and do it faster than do the never-stimulated virgin B cells that recognize the same antigen.

7. A primary immune response takes one to two weeks to produce antibody. A secondary immune response is initiated by memory B cells and produces antibody in a few days.

8. Vaccines stimulate the formation of memory B cells without causing clinically significant infection.

B-Cell Response (pp. 412–417)

1. Antibodies fight infection both directly and indirectly. Some—called antitoxins—bind to epitopes on bacterial toxins and inactivate them. Other antibodies act directly on viruses.

2. Antibacterial antibodies act indirectly by linking a bacterial cell to a phagocyte or to complement (C1). One site binds to the epitope and one site binds to the phagocyte or C1. Antibodies that link bacteria to phagocytes act as opsonins. When they link bacteria to C1, they initiate the complement cascade through the classical pathway.

3. Antibodies are complex proteins made up of polypeptide chains. They have a basic Y shape and are called an antibody monomer. The Y monomer consists of two identical heavy chains and two identical light chains linked by disulfide bonds. At the two ends of the Y arms are the antigen-binding sites.

4. An antibody monomer has a valence of 2 because it has two antigen-binding sites. Some antibody classes have more than two monomers and therefore higher valences. An antigen must have a valence of at least 2 to be immunogenic.

5. Antibodies are part of the immunoglobulin family of proteins. There are five classes of antibodies—IgG, IgA, IgM, IgD, and IgE.

6. About 75 percent of our antibodies are IgG. When bound to an epitope, IgG can initiate a complement cascade through the classical pathway. IgG is also an effective opsonin. It is the only antibody that can cross the placenta. A secondary response produces more IgG than any other class of antibodies.

7. About 15 to 20 percent of our antibodies are IgA. IgA occurs as both a monomer and a dimer. Dimers are found in secretions such as saliva, milk, and mucus. IgA in secretions is called secretory IgA because it is bound to a secretory component. IgA protects the body's mucous membrane surfaces against attack.

8. IgM constitutes about 5 to 10 percent of the antibody total. IgM consists of five monomers, which allow it to build complex antigen-antibody complexes. IgM is extremely effective in fixing complement.

9. IgD constitutes less than 1 percent of the total antibody. It is found on antibody-producing B cells and may play a role in clonal selection.

10. IgE constitutes less than 0.01 percent of the antibody total. It is fixed to the surfaces of basophils, mast cells, and eosinophils. IgE stimulates degranulation, releasing inflammatory mediators. IgE degranulation is an important defense against pathogens too large to be phagocytized, but it also contributes to allergy.

T LYMPHOCYTES (pp. 417–420)

1. There are two main types of T lymphocytes—T_4 cells (T_H, or helper, cells) and T_8 cells (T_C, or cytotoxic, cells and T_S, or suppressor, cells).

2. T_H cells increase immune responsiveness. T_C cells attack and kill invading microorganisms. T_S cells decrease (suppress) immune responsiveness.

Antigen Recognition (pp. 417–418)

1. T-cell antigen receptors resemble and function the same as B-cell antigen receptors (antibodies), except to recognize a foreign antigen T-cell antigen receptors must simultaneously recognize a particular type of self antigen, a structural component of the surface of human cells. Self antigens do not normally provoke a destructive immune response.

2. In T-cell recognition, a foreign antigen is taken up and processed by a human cell. The processed antigen is combined physically with a self antigen and displayed on the human cell surface. Then a T-cell antigen receptor binds to the foreign-self antigen complex.

3. The DNA that encodes self antigens is called the major histocompatibility complex (MHC). There are class I MHC antigens (present on all nucleated cells in the body) and class II MHC antigens (present on macrophages, other antigen-presenting cells, and certain lymphocytes).

4. T_H cells require class II antigens to recognize foreign antigens, while T_C cells require class I.

T-Cell Activation   (pp. 418–419)

1. The first T cells to be activated are the T_H cells. A T_H cell binds to a matching antigen–class II MHC complex on the surface of an antigen-presenting cell, often a macrophage. The macrophage produces interleukin-1, which stimulates the T_H cell to stimulate itself still further by producing interleukin-2. Interleukin-4 also helps activate T_H cells.

2. Interleukin-2 also activates T_C and T_S cells to divide in a process similar to B-cell clonal selection.

T-Cell Response   (pp. 419–420)

1. T cells respond in one of two ways—cytotoxicity (direct T_C cell killing) or immunoregulation (T_H cells intensify immune response and T_S cells suppress it).

2. T_C cells recognize foreign antigens in physical combination with class I MHC self antigens. Thus, the human cell is a marked target for T_C cell killing.

3. In viral infections the host cell processes the viral antigens and displays them in combination with class I MHC antigens; this process targets the human cell for destruction. A T_C cell with a matching antigenic receptor binds to the infected host cell and secretes perforins that create holes in the target cell membrane. About an hour after the two cells separate, the target cell swells and lyses.

4. T_C cells also attack cancer cells and certain eucaryotic pathogens, such as fungal cells.

5. T_H cells intensify humoral immunity by helping to activate B cells; they intensify cell-mediated immunity because they activate T_C cells.

6. T_S cells interact with both B and T cells to reduce immune responsiveness, but the details of their action are not well understood. They might secrete proteins called suppressor factors that slow the immune system.

NON-B NON-T LYMPHOCYTES   (pp. 420–421)

1. Non-B non-T cells make up less than 3 percent of the total lymphocyte population. Unlike B and T cells, they function without recognizing antigen and they do not have markers that other lymphocytes bear (antibodies for B cells, and T_4 and T_8 markers for T cells).

2. One group of non-B non-T cells, the natural killer (NK) cells, lyse human cells as T_C cells do, by secreting perforins. NK cells act against virus-infected cells, cancerous cells, and transplanted organs. Because they do not recognize antigen, they are more like nonspecific interior defenses than specific immune defenses.

3. The other group of non-B non-T cells, the killer (K) cells, kills only cells coated with antibody. Their type of action is called antibody-dependent cellular toxicity. They also act nonspecifically.

CELLS AND ORGANS OF THE IMMUNE SYSTEM

1. What is the immune system? Name the three kinds of lymphocytes.

2. What are the two types of immunity?

3. Name the primary lymphoid organs and describe their functions. Do the same for the secondary lymphoid organs.

4. Describe lymphatic circulation. Why are lymph nodes critical to immune-system functioning?

5. What is an antigen? an antibody?

TYPES OF IMMUNITY

1. What distinguishes active from passive immunity?

2. Define these types of immunity and give an example of each: naturally acquired active immunity, naturally acquired passive immunity, artificially acquired active immunity, artificially acquired passive immunity.

THE IMMUNE RESPONSE: AN OVERVIEW

1. What are the three tasks the immune system must accomplish to successfully defend against infection?

2. Explain how antigen shape and size are important in immune recognition.

3. Define these terms: antigenic receptors, epitope, hapten, antigen-presenting cell.

4. What are plasma cells?

B LYMPHOCYTES

1. Explain the process of genetic recombination that leads to B-cell diversity.

2. Describe the main types of antigen-presenting cells and how they function.

3. How is an antigen-antibody formed?

4. Explain this statement: Unlike the lock-and-key fit between enzyme and substrate, antigen-antibody binding is variable.

5. What is clonal selection in B-cell activation? How do effector B cells differ from memory B cells?

6. How do T_H cells function in B-cell activation?

7. What are lymphokines? Which lymphokines are involved in B-cell function? How are they involved?

8. What is immunologic memory? What is the difference between a primary immune response and a secondary immune response? What is a vaccine?

9. What is antitoxin? Explain the B-cell response to virus-infected cells. Explain the linking response of antibodies.

10. Describe the structure of antibodies. Use these terms in your explanation—antibody monomer, heavy chains, light chains, antigen-binding sites, variable regions, constant regions, valence, immunogenic.

11. Name the five classes of antibodies. For each, tell what percent of the total antibody population they constitute and what their main functions and features are.

T LYMPHOCYTES

1. Name the three types of T cells and describe the function of each.

2. Describe how T-cell antigen recognition occurs. Why are the self antigens that participate in T-cell recognition also called restriction elements?

3. What is the difference between class I MHC antigens and class II MHC antigens? Why does it make sense that T_H cells would require class II MHC antigens and T_C cells would require class I?

4. What role do lymphokines play in T-cell activation?

5. What are the two kinds of responses T cells make?

6. Describe how a T_C cell kills.

7. How do T_H cells regulate the immune system? Be specific. How do we think T_S cells regulate it?

NON-B NON-T LYMPHOCYTES

1. Name the ways non-B non-T cells are different from B and T cells.

2. How do NK cells kill? How do K cells kill?

3. Explain this statement: Non-B non-T cells are more like nonspecific defenses than like specific defenses.

Essay Questions

1. Discuss the similarities and differences between B cells and T cells.

2. Trace your body's response to an infection you recently suffered. How did the microorganisms evade surface defenses? How did your nonspecific interior defenses function? What did your immune system do to overcome the infection? Diagram the three lines of defense your body used.

Suggested Readings

Ada, G., and Nossal, Sir G. 1987. The clonal-selection theory. *Scientific American*, August, 62–69.

Schindler, L. W. 1988. *Understanding the immune system*. Bethesda, Md.: National Institutes of Health, NIH Publication No. 88-529.

Scientific American. 1993. Life, death, and the immune system. Special issue, September.

18 IMMUNOLOGIC DISORDERS

"It's Nothing—Just a Bee Sting"

Mr. C. was a 31-year-old operator of farm machinery. One day, while cultivating an almond orchard that was being pollinated by honeybees, he was stung on his hand. Within five minutes his hand was red and swollen. Having experienced similar reactions to bee stings, he was unconcerned and continued his work. During the next 45 minutes, however, Mr. C. became alarmed as the redness and swelling extended up his forearm almost to his elbow. His entire body began to itch and red blotchy patches appeared on his face, chest, and left arm. His lips became swollen and he felt nauseated. The itching became particularly intense under his chin, and when he called to a fellow worker for help, he was hoarse. He could barely talk because of a dry, uncontrollable cough.

Mr. C. was rushed to a nearby emergency room and immediately treated with a subcutaneous injection of epinephrine (adrenaline). Within a few minutes he improved markedly. Fifteen minutes later he felt back to normal. The emergency room doctor kept Mr. C. several hours for observation. She told him he had suffered an anaphylactic reaction due to a bee sting allergy. When he was discharged, she gave him an injection of a long-acting adrenaline-like medication and a prescription for an antihistamine to take orally at home. She also gave him a prescription for an adrenaline syringe to use at once if he suffered similar symptoms from another bee sting. Finally, considering that Mr. C.'s work might put him at similar risk in the future, she advised him to consult an allergist to see if he could undergo immunotherapy to prevent similar reactions.

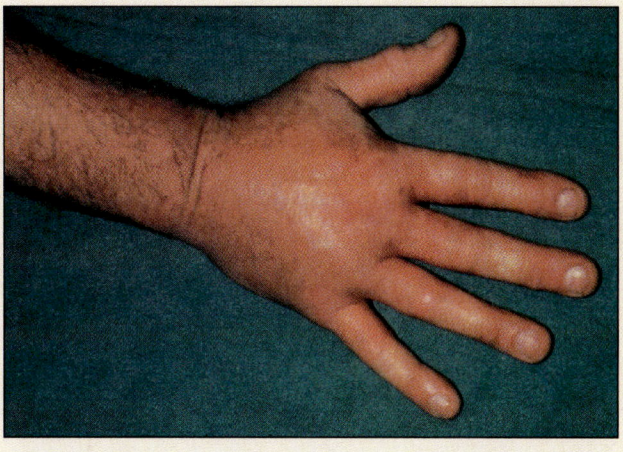

a Typical manifestations of an allergic reaction as a result of a bee sting.

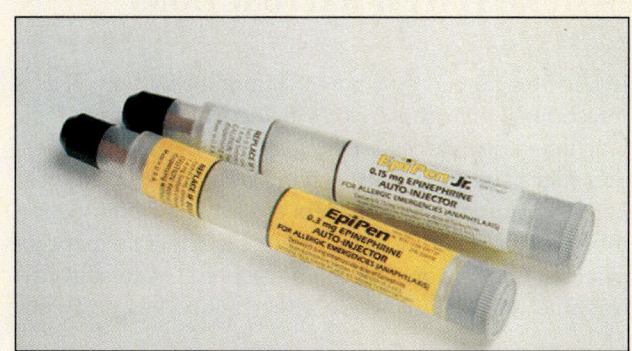

b The EpiPen Epinephrine Auto Injector is a device used for emergency treatment of an allergic reaction such as that following a bee sting. A prescription anaphylaxis emergency kit may include an alcohol pad for cleaning the injection site; a syringe containing epinephrine, which counteracts the most immediately life-threatening manifestations of anaphylaxis; antihistamine pills that extend the epinephrine effect; and a tourniquet to tie above the site of the sting if it occurs on an arm or leg, helping to prevent more venom from entering the circulation.

To understand:

▪ How abnormal or misdirected immune responses, called hypersensitivity, can harm the body

▪ The basis of type I hypersensitivity and its relationship to allergy

▪ The basis of type II hypersensitivity and its relationship to blood transfusion and certain autoimmune diseases

▪ The basis of type III hypersensitivity and its relationship to certain autoimmune diseases

▪ The basis of type IV hypersensitivity and its relationship to tissue transplantation, contact dermatitis, and granulomatous infections

▪ Congenital and acquired immunodeficiency disorders

▪ How failure of immune surveillance might lead to cancer

IMMUNE SYSTEM MALFUNCTIONS

In Chapter 17, you read how the immune system *combats* infection. In this chapter you will learn how immunologic reactions *cause* disease. Paradoxically, the immune system that is essential to surviving infection can do great harm. **Immunologic disorders** are caused when the immune system malfunctions, producing either an inappropriate or an inadequate immune response. A misdirected immune response in which either antibodies or T cells cause significant damage to human tissues is called **hypersensitivity**. Failure to mount an adequate immune response, on the other hand, causes an **immunodeficiency disorder**, which is characterized by recurrent infection. Immunodeficiencies can be **congenital**, present from birth, or **acquired**, developed during a person's life. Some cancers may result from a more subtle breakdown of immune function, but they are not considered immunodeficiency disorders.

HYPERSENSITIVITY

There are four types of hypersensitivity reactions:

1. Anaphylactic (type I) hypersensitivity causes insect sting reactions, hay fever, and other allergies.

2. Cytotoxic (type II) hypersensitivity causes transfusion reactions, hemolytic disease of the newborn, and some autoimmune diseases.

3. Immune-complex (type III) hypersensitivity causes many different autoimmune disorders.

4. Cell-mediated or delayed (type IV) hypersensitivity causes transplant rejection, poison ivy and poison oak rashes, and the tissue damage associated with certain types of infections.

The first three types of hypersensitivity are mediated by antibodies and the fourth type by T cells (**Table 18.1**).

Type I: Anaphylactic Hypersensitivity (Allergy)

Type I hypersensitivity—the type that caused Mr. C.'s reaction to the bee sting—is also called **immediate hyper-**

Table 18.1	Hypersensitivity		
Type	Names	Examples	Reaction Time
Type I	Anaphylactic hypersensitivity Immediate hypersensitivity Allergy	Insect sting allergy, hay fever, asthma, food allergy	Immediate, within 20 minutes
Type II	Cytotoxic hypersensitivity	Transfusion reactions, Rh disease of the newborn, Goodpasture's syndrome, Graves' disease	Variable
Type III	Immune-complex hypersensitivity	Systemic lupus erythematosus (SLE), rheumatoid arthritis, farmer's lung disease, serum sickness	Variable
Type IV	Cell-mediated hypersensitivity Delayed hypersensitivity	Organ transplantation rejection, poison ivy/oak/sumac, clothing and jewelry allergies, granulomatous reaction	12 hours to several weeks

sensitivity because its symptoms develop rapidly, usually within 10 to 20 minutes. You probably know type I hypersensitivity by its more common name, **allergy**. Antigens that stimulate type I reactions are also called **allergens**.

Type I hypersensitivity reactions are also called **anaphylaxis** (*ana*, "away from"; *phylaxis*, "protection"). An anaphylactic reaction like Mr. C.'s begins when the immune system is exposed to and produces antibodies against a foreign antigen. But rather than protecting the individual against future threat, the first, or **sensitizing**, exposure to the allergen sets the stage for a harmful reaction the next time the person encounters the same antigen.

Allergic reactions may be **local**, affecting only a limited part of the body, such as the tissues immediately around the site of a bee sting or the nasal passages in the case of hay fever. In some cases, though, life-threatening **systemic** reactions like Mr. C.'s occur. These dramatic events are the result of allergic responses that affect the bronchial tree (air passages that supply the lungs) and blood vessels throughout the body. People who are allergic to a particular substance may experience a localized reaction after one exposure and a systemic reaction after another exposure.

Immunologic Basis of Type I Hypersensitivity. How does a sensitizing dose of allergen lead to an anaphylactic response on subsequent exposure to the same allergen? In Mr. C.'s case the sensitizing agent was **phospholipase A**—the most powerful allergen in bee venom. During one of his previous bee stings, phospholipase A had stimulated B and T lymphocytes to cooperate in producing IgE antibody against this antigen (Chapter 17). Like all antibodies, IgE molecules have antigen-binding sites, but the IgE class also has a binding site in the molecule's constant region that allows it to attach to the surface of mast cells or basophils, which contain histamine and other powerful inflammatory mediators (Chapter 16). So Mr. C.'s sensitizing exposure to phospholipase A produced IgE antibodies with antigen-binding sites specific for phospholipase A; these antibodies bound to basophils and mast cells. Then, when Mr. C. was reexposed to phospholipase A by a bee sting, the allergen bound to IgE molecules on these cells. Some of the phospholipase A antigens—large molecules with many epitopes—bound two adjacent IgE molecules (**Figure 18.1**), triggering the mast cell or basophil to degranulate.

A variety of preformed and newly formed inflammatory mediators were released by the degranulating cells, including histamine, prostaglandins, and leukotrienes (Chapter 16). Histamine, a preformed mediator that is abundant in mast cell granules, had the most immediate effect, dilating blood vessels and constricting the bronchial passages that deliver air to the lungs. The leukotrienes LTC_4 and LTD_4 are formed at the time of degranulation. These two—jointly called **SRS-A**, the **slow-reacting substance of anaphylaxis**—have a more delayed effect than histamine and cause prolonged respiratory difficulty.

Mediators from mast cells or basophils produced all the **signs** (objective indications of illness on physical exam) and **symptoms** (subjective reports of illness) of Mr. C.'s allergic reaction—local inflammation, pain, redness, warmth, and swelling (edema) at the site of his bee sting. Constriction of bronchial smooth muscle, one effect of SRS-A, caused the dry spasmodic cough. Increased vascular permeability caused edema in nearby tissues. Edema of his skin caused the raised itchy rash known commonly as **hives** and medically as **urticaria**. Edema of his lips caused them to become large and pale. Edema of his intestinal tract produced nausea. **Laryngeal**

Antibodies Involved	Action	Damage Caused by Antibody Action	Treatment
IgE	IgE antibodies link mast cells and allergen	Mast cells and basophils release chemicals (e.g., histamine) during degranulation	Epinephrine, antihistamines, immunotherapy (desensitization)
IgG, IgM	IgG and IgM mark normal human cells as targets for immune effectors	Complement and phagocytosis	Varies; supportive treatment only for transfusion reactions, exchange transfusion for Rh disease; insulin for diabetes
IgG, IgM	Antigen-antibody complexes are distributed throughout the body	Complement and phagocytosis stimulated by antigen-antibody complexes	Varies; often steroids to suppress further antibody production
None	Lymphokines and macrophages interact, causing inflammation	T cells activated	Varies; often steroids to suppress T-cell function

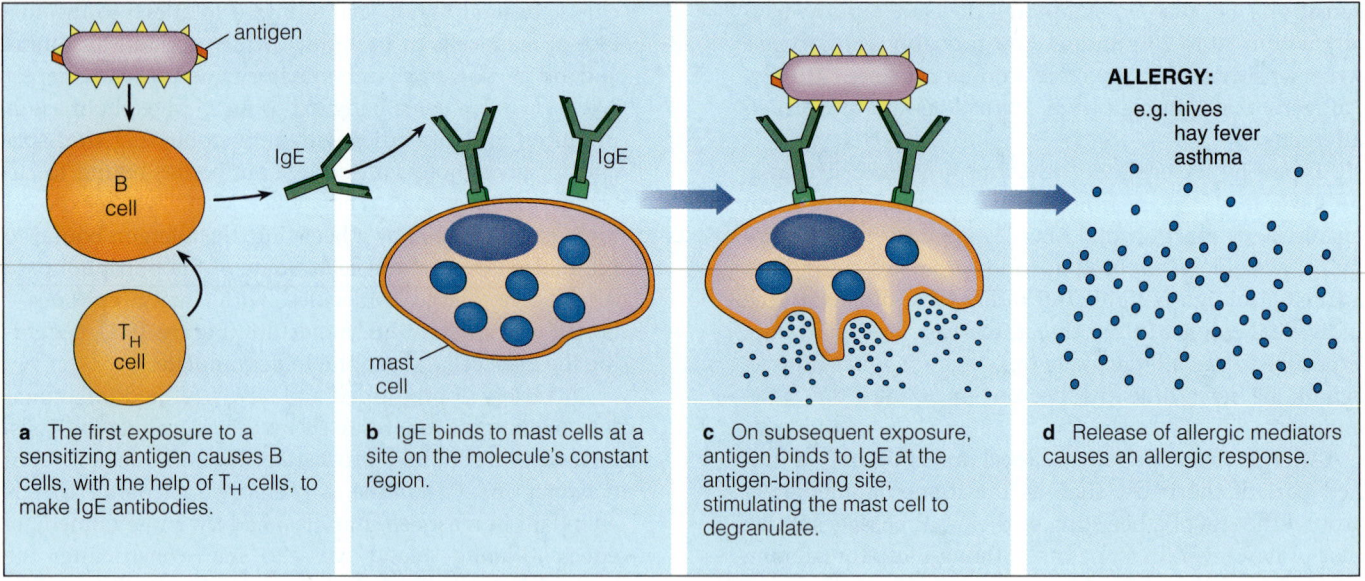

Figure 18.1 Type I (anaphylactic) hypersensitivity, commonly known as allergy.

Within the figure:

- antigen
- IgE
- B cell
- T$_H$ cell
- IgE
- mast cell
- IgE
- **ALLERGY:** e.g. hives hay fever asthma

a The first exposure to a sensitizing antigen causes B cells, with the help of T$_H$ cells, to make IgE antibodies.

b IgE binds to mast cells at a site on the molecule's constant region.

c On subsequent exposure, antigen binds to IgE at the antigen-binding site, stimulating the mast cell to degranulate.

d Release of allergic mediators causes an allergic response.

---

edema narrowed the airway within Mr. C.'s larynx (voice box), causing hoarseness and the itching under his chin.

Mr. C.'s reaction to these inflammatory mediators was potentially life-threatening. Laryngeal edema sometimes completely obstructs the airway, causing immediate death from suffocation. Sometimes, too, rapid loss of fluid from the circulation causes **anaphylactic shock**, a precipitous fall in blood pressure that can result in sudden death. About 40 Americans die each year from anaphylactic reactions due to bee stings.

Administering epinephrine reversed the manifestations of Mr. C.'s anaphylaxis almost immediately by constricting blood vessels throughout his body and relaxing bronchial muscle. Antihistamine medication blocked the effects of histamine by competing for its binding sites on target cells. Mr. C.'s chances of suffering an anaphylactic reaction if he suffers another bee sting are about 60 percent. The preloaded syringe of epinephrine prescribed for him would counteract anaphylaxis if used immediately.

Because Mr. C.'s outdoor agricultural job places him at high and continuing risk for bee stings, he might decide to undertake **immunotherapy** designed to **desensitize** him to bee venom. In immunotherapy, increasing amounts of bee venom are injected into allergic patients. When patients are able to tolerate substantial doses of injected venom, they are considered desensitized to bee venom and therefore protected from bee sting anaphylaxis.

The immunologic basis of this seemingly paradoxical form of treatment is thought to be the production of IgG **blocking antibodies** against bee venom. They bind to and inactivate molecules of venom, thereby preventing them from interacting with IgE on mast cells. Desensitization also seems to increase the number of T$_S$ cells and thereby limit production of IgE, although the exact details of how this occurs are not well understood. Immunotherapy is effective in preventing adverse reactions to bee sting in 97 to 100 percent of patients.

Other Examples of Type I Hypersensitivity. Type I hypersensitivity reactions are very diverse and very common, affecting up to 20 percent of the population of the United States. A tendency toward allergies is strong in certain families, perhaps as a result of a genetic predisposition to excessive production of IgE antibodies (**Figure 18.2**).

Allergens that provoke type I hypersensitivity reactions enter the body in a variety of ways. Some, like insect venoms and penicillin or other medications, are injected into tissues beneath the skin. Injected allergens, which readily enter the circulation, are particularly likely to cause life-threatening reactions of generalized anaphylaxis. Many allergens that are small enough to be inhaled, such as pollen, mold spores, animal dander, and feces of microscopic house dust mites, provoke localized respiratory allergies (**Figure 18.3**). Ingested allergens, including medications, foods such as shellfish and peanuts, and food additives, can provoke localized or generalized allergic reactions.

Inhaled allergens that primarily affect the upper respiratory tract cause **allergic rhinitis** (hay fever), the most common allergic disorder. Type I hypersensitivity reac-

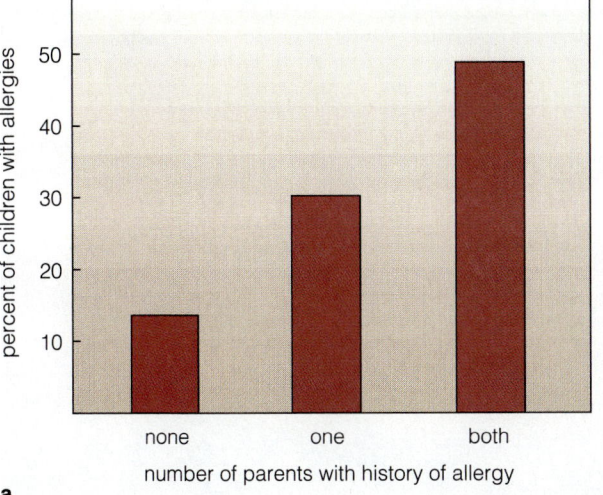

a

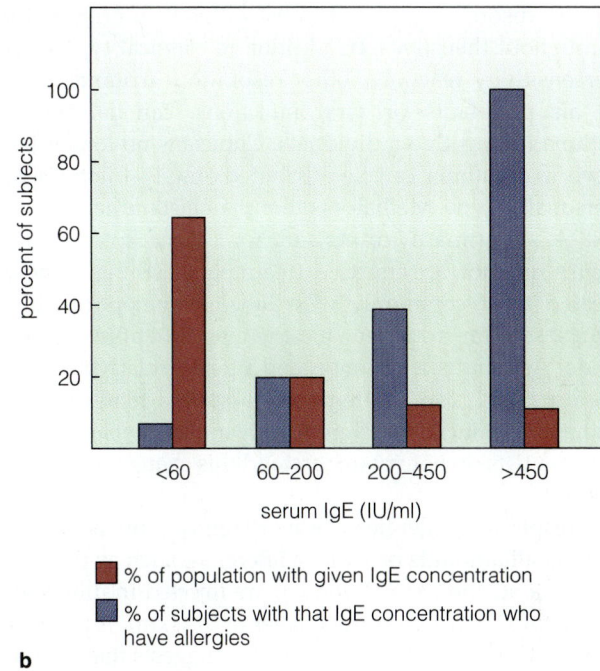

b

Figure 18.2 Genetic predisposition to allergies. Allergic disorders such as hay fever and asthma run in some families. (**a**) A child's risk of becoming allergic rises sharply if one parent is allergic and even more dramatically if both are allergic. (**b**) A genetic predisposition may be related to excessive production of IgE.

■ % of population with given IgE concentration

■ % of subjects with that IgE concentration who have allergies

a   10 μm   b   10 μm   c   10 μm

Figure 18.3 Some of the inhaled allergens that cause hay fever, the most common allergic disorder. Scanning electron micrographs show (**a**) grass pollen, found almost everywhere, (**b**) *Ambrosia*, or ragweed pollen, the most common hay fever culprit in the American East and Midwest, and (**c**) the house dust mite. The mite's fecal pellets, which are about the same size as pollen grains, become airborne and cause hay fever.

tions trigger mast cells in the mucous membranes of the nose to degranulate, releasing inflammatory mediators that cause tissue swelling and excess mucus production. Allergic rhinitis is characterized by nasal congestion, itching, and mucus discharge. If the offending allergen is a type of pollen that is airborne only at certain times of year, hay fever may be **seasonal**. If the allergen is present year-round, symptoms are **perennial**. Because hay fever is mediated primarily by histamine, antihistamines provide effective treatment. Severe cases can also be treated with nasally administered steroids that suppress the immune response or sodium cromolyn, which prevents mast cell degranulation by stabilizing the cell membrane.

Inhaled allergens that affect the lower respiratory tract can cause **asthma**, a breathing disorder that results from constriction of the smooth muscle surrounding bronchial passages combined with edema of the airways themselves. The small, collapsible bronchi narrow, making it difficult for air to leave the lungs. Asthma is a chronic disorder that affects most sufferers intermittently throughout their lives. In addition to classical type I hypersensitivity reactions, other respiratory irritants, such as air pollutants or viral infections, can precipitate asthma in sensitive individuals. Clinicians no longer believe that asthma can be attributed simply to stress or personality type. Mediators other than histamine, such as SRS-A, are primarily responsible for asthma, so antihistamines are not an effective treatment. Asthma is now treated most commonly with inhaled epinephrine-like drugs, sodium cromolyn, and the intermittent use of steroids. Asthma is a common and often a relatively mild disease; but it can be life-threatening, and fatal cases of asthma have increased significantly in recent years. Clinical scientists do not yet understand this change in disease pattern.

Respiratory allergies can be diagnosed by **skin testing**. Small amounts of various allergens, such as pollens, house dust, and animal dander, are injected in dilute solution into the dermis (lower layer of the skin). A positive test, indicated by a hivelike reaction, suggests that the patient is allergic to that allergen (**Figure 18.4**). But knowing what someone is allergic to is not always helpful in treatment. If a person is allergic only to dog or cat dander, finding a new home for a pet may improve symptoms. Usually, though, the offending allergens include dust or pollen—substances present everywhere. Therefore, skin testing is usually done on patients who wish to follow up with desensitization. Unfortunately, desensitization to respiratory allergens is much less effective than desensitization to insect venom. Only about half of hay fever patients improve significantly, and desensitization seems ineffective for preventing asthma. The procedure is uncomfortable, time-consuming, expensive, and not without risk: The skin testing takes several hours and produces itchy hivelike swelling at the site of every positive test, and the desensitization shots must be administered twice a week for an indefinite period of time. Each injection also carries a small risk of precipitating a life-threatening generalized anaphylaxis. As a result, many physicians believe that respiratory allergies are better treated by medication.

Ingested allergens can also cause serious reactions. Most so-called food allergies are not allergies at all, but the result of irritation of the intestinal tract or problems with digestion. Some adverse reactions to food, however, *are* true examples of type I hypersensitivity. Highly allergenic foods such as peanuts or shellfish can cause hives or systemic anaphylactic reactions identical to Mr. C.'s. Food allergies cannot be diagnosed by skin testing, so diagnosis depends upon a careful medical history associating ingestion of the offending food with symptoms typical of type I hypersensitivity.

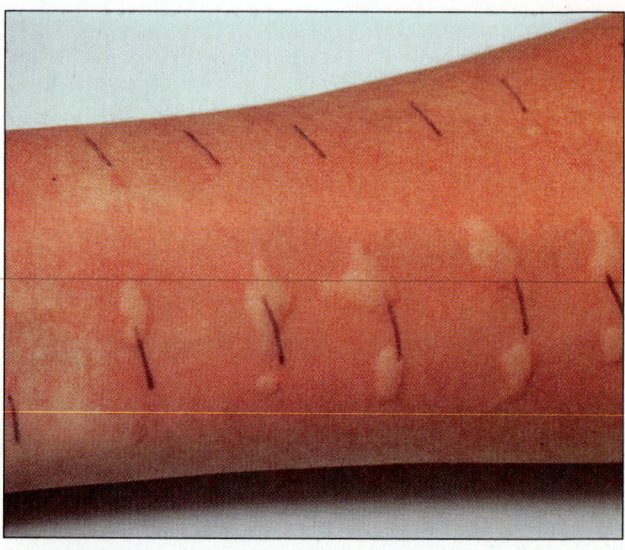

Figure 18.4 Allergy skin tests are usually done on the forearm (as in this photo) or the back. In the test shown here, the hivelike reaction increases as the strength of the diluted allergen increases.

Type I hypersensitivity reactions to ingested medications are more common than food allergies. Drug allergy is a frequent cause of **iatrogenic** (medically induced) disease. Penicillin, for example, is a hapten that becomes a potent allergen if it binds to the carrier protein human albumin (Chapter 17). If IgE antibodies against the penicillin-albumin complex are produced and bind to mast cells or basophils, a person becomes sensitized to the drug and may experience adverse reactions if exposed again. Allergic drug reactions usually cause urticaria alone; but systemic anaphylaxis, including laryngeal edema and shock, may occur, particularly if the drug is injected rather than ingested—the reason patients who receive a penicillin shot are asked to remain in the doctor's office for 20 minutes of observation. Each exposure to penicillin carries a risk of provoking allergy, even in people who have taken the drug safely many times before. Preventing drug sensitization is one reason that penicillin should not be prescribed needlessly—for example, as a treatment for colds or other viral illnesses. A patient exhibiting type I hypersensitivity to penicillin should never again be given penicillin or any other drug from the penicillin family, such as ampicillin or dicloxacillin.

Type II: Cytotoxic Hypersensitivity

Unlike type I hypersensitivity reactions, in which IgE antibodies stimulate degranulation of mast cells and basophils, type II reactions occur when antibodies bind directly to antigens on the surface of human cells, marking them. Type II hypersensitivity is also called cytotoxic

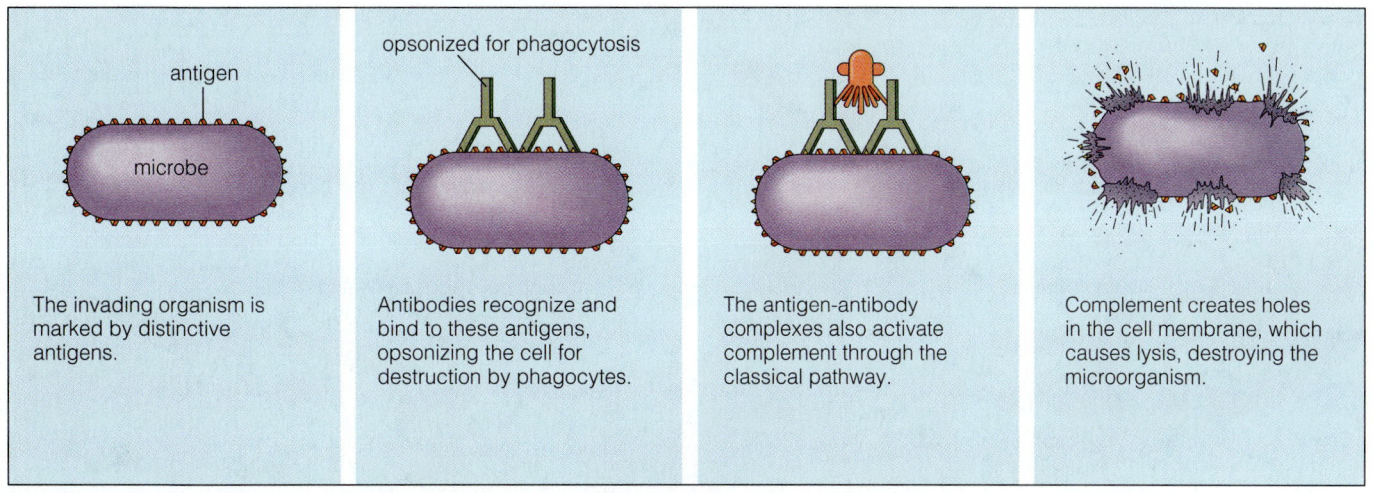

The invading organism is marked by distinctive antigens.

opsonized for phagocytosis

Antibodies recognize and bind to these antigens, opsonizing the cell for destruction by phagocytes.

The antigen-antibody complexes also activate complement through the classical pathway.

Complement creates holes in the cell membrane, which causes lysis, destroying the microorganism.

a Normal antimicrobial action

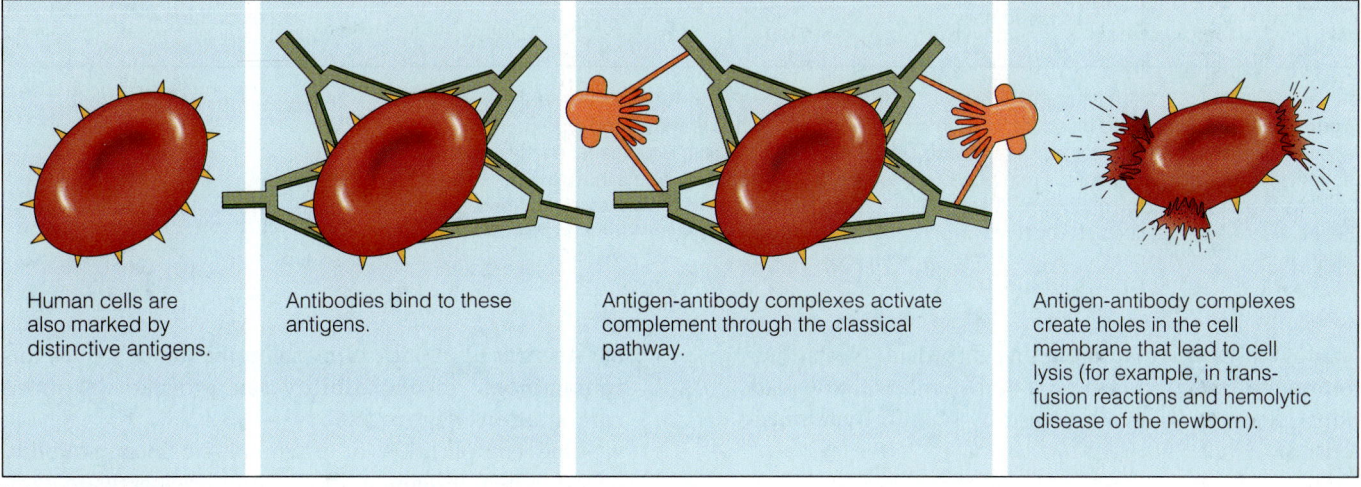

Human cells are also marked by distinctive antigens.

Antibodies bind to these antigens.

Antigen-antibody complexes activate complement through the classical pathway.

Antigen-antibody complexes create holes in the cell membrane that lead to cell lysis (for example, in transfusion reactions and hemolytic disease of the newborn).

b Hypersensitivity

Figure 18.5 Type II (cytotoxic) hypersensitivity. The destructive mechanisms of type II hypersensitivity and normal immunological events that allow the body to rid itself of pathogenic microorganisms are very similar. (**a**) Normal immunologic defense against microbes. (**b**) The same immunologic mechanisms directed against human cells and the damage they cause.

hypersensitivity because the antibody-marked cell usually dies.

The Immunologic Basis of Type II Hypersensitivity. Type II hypersensitivity occurs when IgM or IgG antibodies bind abnormally to antigens on the surface of human cells, just as they bind normally to the surface of a pathogenic bacterium. When they bind to human cells, they sometimes activate complement and lyse the target cell through the terminal complement pathway (**Figure 18.5**; Chapter 16). This is the mechanism of cell destruction in transfusion reactions and hemolytic disease of the newborn (discussed below). In other cases, antibodies

and activated complement attract and activate human leukocytes, including phagocytes and killer cells, that attack and damage the marked human cells. This mechanism occurs, for example, in **Goodpasture's syndrome**, a kidney disease. In a few cases, the binding of antibodies to human cells alters their function and causes disease without killing them. Graves' disease, in which the thyroid gland becomes hyperactive, is an example.

Immune tolerance (Chapter 17) normally prevents immunologic attack against normal body tissues, but when tolerance fails and the immune system launches an attack against its own self antigens, the result is an **autoimmune disorder**. Goodpasture's syndrome and

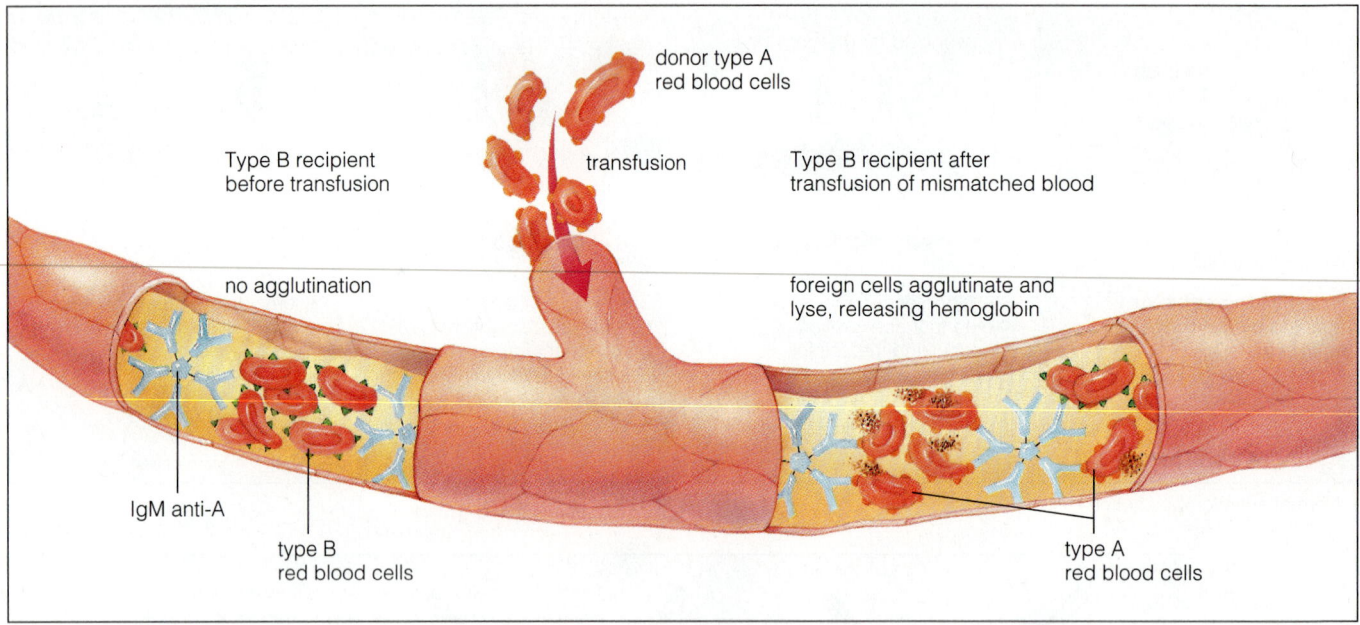

donor type A
red blood cells

Type B recipient
before transfusion

transfusion

Type B recipient after
transfusion of mismatched blood

no agglutination

foreign cells agglutinate and
lyse, releasing hemoglobin

IgM anti-A

type B
red blood cells

type A
red blood cells

Figure 18.6 A mismatched blood transfusion: what happens if someone with blood type B receives blood type A. The recipient's red blood cells are marked only with B antigens, so they recognize A as a nonself antigen. From previous encounters with microorganisms that carry similar antigens, the recipient has anti-A antibodies, produced long before the transfusion. When the recipient's anti-A antibodies encounter A antigens, they lyse and agglutinate the transfused cells.

Graves' disease are autoimmune disorders. Not all autoimmune disorders are type II hypersensitivity reactions, however. Some are caused by type III hypersensitivity.

The ABO Antigen System and Transfusion Reactions. Erythrocytes (red blood cells) fulfill the essential function of carrying oxygen to human tissues. Erythrocytes bear a variety of self antigens that vary according to a person's **blood type**. The best-known, most basic group of blood antigens constitutes the **ABO antigen system**, described by Austrian-born Karl Landsteiner in the early 1900s. Genes inherited from both parents determine which of two antigens—designated A and B—is present on a person's erythrocytes. People who produce only A antigens are said to have blood type A. Those who produce only B antigens have blood type B. People who produce both A and B antigens have the blood type AB, and those who produce neither have blood type O.

Erythrocytes can be damaged by antibodies that bind to their antigens. Immunologic tolerance usually prevents the immune system from producing antibodies against antigens on a person's own cells, but complications may occur if erythrocytes are taken from one person and administered to someone with a different blood type in a **transfusion**. If antibodies in the transfusion recipi-

ent's serum react with antigens on the donor's cells, this **immunologic incompatibility** can provoke disastrous complications (**Figure 18.6**).

One complication of immunologic incompatibility occurs when antigen-antibody complexes activate complement to form membrane attack complexes. These membrane-perforating complexes lyse erythrocytes, decreasing the number of circulating red cells available to carry oxygen and releasing large amounts of hemoglobin and other potentially damaging cell contents into the circulation. A second complication is the agglutination (clumping) of erythrocytes, which occurs when antigen-antibody binding forms interconnecting networks of many cells. Red cells that become enmeshed in these networks cannot fulfill their oxygen-carrying function and may clog the circulation. Such complications cause a **transfusion reaction**—a clinical response that includes the rapid onset of fever, vomiting, and shock. Landsteiner's identification of A and B antigens made safe blood transfusions possible.

People who lack A or B antigens on their erythrocytes produce antibodies (primarily of the IgM class) against these antigens even if they have never been exposed to another person's blood, probably because many microorganisms carry antigens similar to A and B. Because of these "naturally occurring" anti-A and anti-B antibodies,

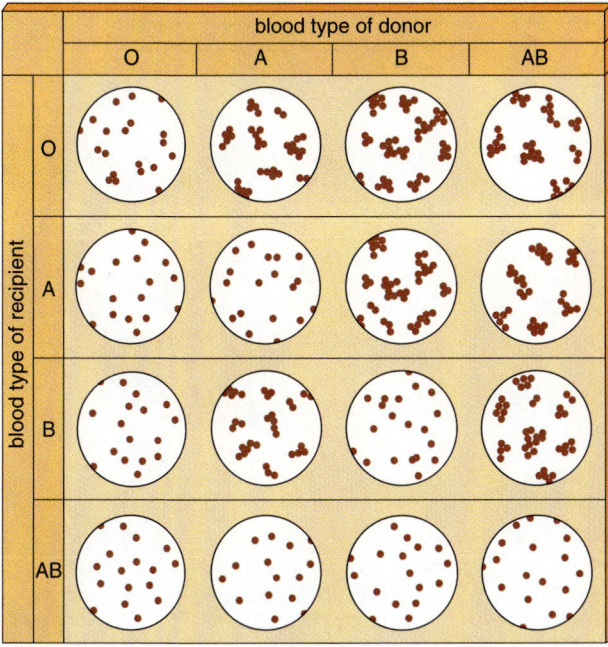

Figure 18.7 A major blood crossmatch is performed before every transfusion. This graph shows transfusion compatibility based on the ABO antigen system. There is no agglutination in the O donor column because type O individuals are universal donors, and no agglutination in the type AB recipient row because type AB individuals are universal recipients.

a patient who receives erythrocytes with nonself A or B antigens is likely to experience a transfusion reaction.

Let's consider a few examples of compatible and incompatible blood types for transfusion. A person with type AB blood recognizes A and B as self antigens and will not produce antibodies against either of them. Thus, such a person can receive any type of blood—O, A, B, or AB. People with type AB blood are called **universal recipients**, but they can donate blood only to others with type AB blood. In contrast, a person with type O blood recognizes both A and B as foreign antigens and will have anti-A and anti-B IgM antibodies in his or her serum. A type O person, therefore, can receive only antigen-free type O blood but can donate antigen-free cells to anyone. People with type O blood are called **universal donors**.

Besides ABO antigens, erythrocytes have many other antigens (organized into systems other than ABO) that can trigger transfusion reactions. However, most are comparatively weak and thus less likely to cause severe transfusion reactions. Because it would be impractical to type donor and recipient blood for every antigen system, a **major crossmatch** is performed for each transfusion. The donor's cells and recipient's serum are mixed and examined for adverse reactions. Any crossmatch that demonstrates agglutination indicates the transfusion

would be unsafe (**Figure 18.7**). A **minor crossmatch** checks for agglutination on mixing donor serum and recipient cells—a type of reaction less likely to cause significant clinical problems. It is also done before a transfusion.

Rh Antigens and Hemolytic Disease of the Newborn. More than 30 years after his trail-blazing work on the ABO antigen system, Landsteiner identified another human erythrocyte antigen. He named it **Rh factor** because the same antigen is found on the red blood cells of **rh**esus monkeys. Blood type is termed **Rh-positive** if red blood cells carry this antigen and **Rh-negative** if they do not. The Rh antigen system is independent of the ABO system. That is, a person of any ABO blood type is also either Rh-positive or Rh-negative—for example, O⁺, or O⁻. The Rh system is also used to determine compatibility for blood transfusion. In addition, Rh antigens cause the most severe form of **hemolytic disease of the newborn** (*hemo*, "blood"; *lytic*, "related to lysis"), a potentially fatal blood disorder of infants.

When anti-Rh antibodies bind to an Rh-positive red blood cell, the antigen-antibody complex activates complement, forming a membrane attack complex and lysing the cell. Unlike anti-A and anti-B antibodies, most of which are of the IgM class and therefore incapable of crossing the placenta, anti-Rh antibodies belong primarily to the IgG class, which can cross the placenta readily. Moreover, anti-Rh antibodies do not occur naturally. They are produced only by people who have been previously exposed to Rh-positive blood. A person who has been exposed to the Rh antigen and produced antibodies against it is said to be **Rh-immunized**. Rh immunization may be the result of a mismatched transfusion, but it usually occurs when blood from an Rh-positive baby (who inherited an Rh gene from the father) enters the circulation of an Rh-negative mother during pregnancy or around the time of birth (**Figure 18.8**). During a first pregnancy, therefore, the mother is seldom Rh-immunized and the baby is not at risk for hemolytic disease.

If a previously immunized Rh-negative mother carries another Rh-positive baby, however, hemolytic disease of the newborn may be a life-threatening problem. IgG antibodies against the Rh antigen cross the placenta, enter the fetal circulation, bind to the baby's Rh-positive cells, and mark them for destruction by complement. The more fetal red cells that are destroyed, the more severe the consequences to the developing baby and the more likely it is to die from a profound anemia. Severely affected babies who survive to birth may die of shock soon afterward or suffer a type of permanent brain damage called **kernicterus** from lysis-produced toxins. Severely affected newborns are treated by an **exchange transfusion**—the baby's Rh-positive red cells are removed and

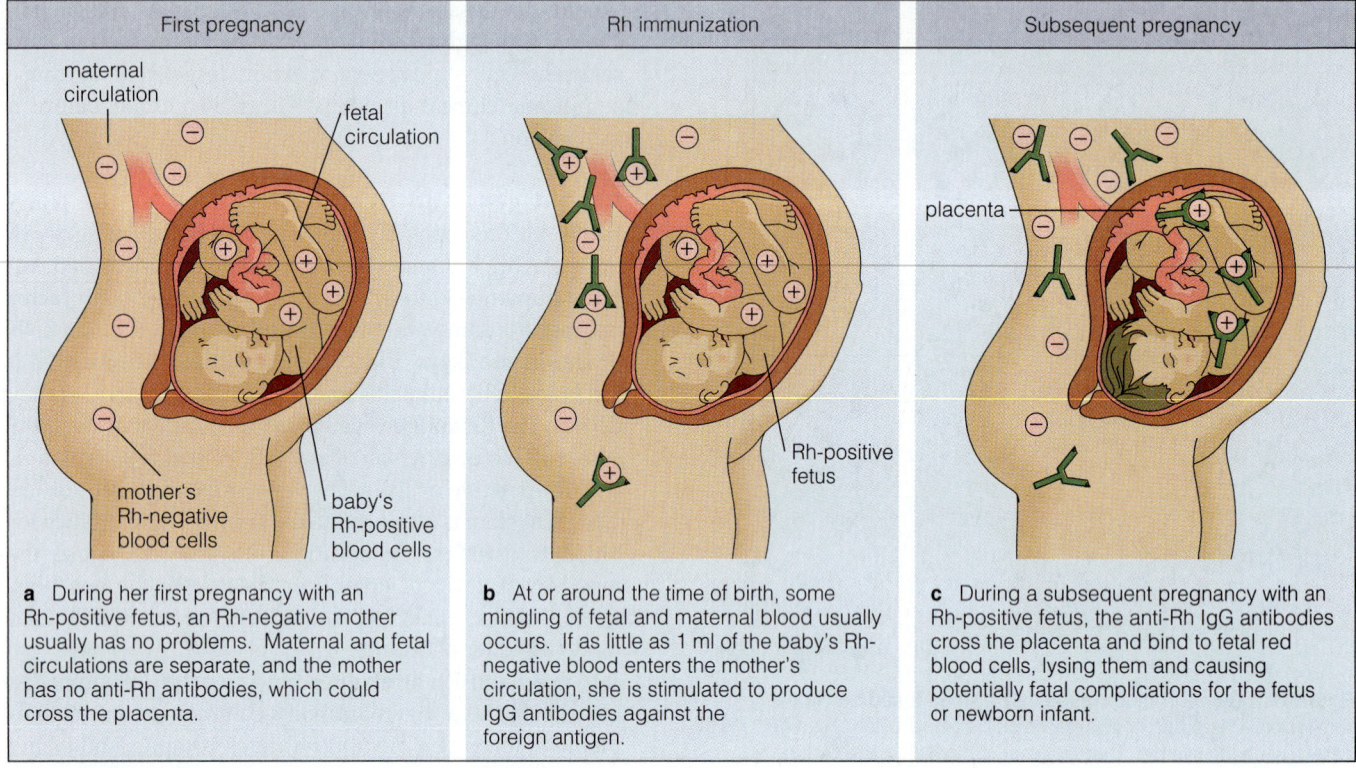

First pregnancy	Rh immunization	Subsequent pregnancy

maternal circulation

fetal circulation

mother's Rh-negative blood cells

baby's Rh-positive blood cells

Rh-positive fetus

placenta

a During her first pregnancy with an Rh-positive fetus, an Rh-negative mother usually has no problems. Maternal and fetal circulations are separate, and the mother has no anti-Rh antibodies, which could cross the placenta.

b At or around the time of birth, some mingling of fetal and maternal blood usually occurs. If as little as 1 ml of the baby's Rh-negative blood enters the mother's circulation, she is stimulated to produce IgG antibodies against the foreign antigen.

c During a subsequent pregnancy with an Rh-positive fetus, the anti-Rh IgG antibodies cross the placenta and bind to fetal red blood cells, lysing them and causing potentially fatal complications for the fetus or newborn infant.

Figure 18.8 Hemolytic disease of the newborn, a type II hypersensitivity reaction.

replaced by Rh-negative cells that cannot be damaged by the maternal antibodies in the baby's circulation (Chapter 17).

Hemolytic disease of the newborn can usually be prevented by keeping Rh-negative mothers from becoming Rh-immunized. If the mother is injected with anti-Rh antibodies, antibody and complement will destroy most of the baby's Rh-positive cells before immunization can occur. The routine use of anti-Rh immune globulin has dramatically decreased the incidence of Rh-mediated hemolytic disease.

Other Examples of Type II Hypersensitivity. The other diseases caused by type II hypersensitivity reactions involve the binding of antibodies to human cells or tissues. Many are autoimmune diseases, in which immune tolerance fails and autoantibodies are produced against the body's own antigens.

One such disease is **Goodpasture's syndrome**, in which the body begins to produce antibodies against antigens in the kidneys. As antibodies bind to the kidney antigens, they fix complement and attract activated leukocytes, initiating a destructive inflammatory reaction that may persist for months or years. Many patients die of renal failure unless treated with dialysis or a kidney transplant.

In **Graves' disease**, autoantibodies bind to the cells that produce thyroid hormone. Unlike other type II reactions, Graves' disease is not cytotoxic. Rather than destroying the target cells, the antibodies stimulate hormone production. Graves' disease is a serious problem because abnormally high levels of thyroid hormone cause many physiologic abnormalities, including weight loss and protruding eyes. Because the autoantibodies of Graves' tissue belong to the IgG class, they can cross the placenta, and babies born to mothers with Graves' disease are affected with the disorder for the first few weeks of life while maternal antibodies are in the baby's bloodstream.

Type III: Immune-Complex Hypersensitivity

Type III hypersensitivity resembles type II hypersensitivity in that IgG and IgM antibodies react with a person's own antigens within the body. The critical difference is that type II reactions involve antigens bound to cells or tissues, while type III reactions involve antigens that are free in the circulation. When antibodies bind to circulating antigens, they form antigen-antibody complexes that remain **soluble**, or free in body fluids. Type III hypersensitivity is also called **immune-complex hypersensitivity**.

Clinical Notes

Blood Transfusion

The idea of replacing blood lost as the result of injury arose centuries ago. It seemed the obvious way to prevent the death that nearly always occurred with massive blood loss. As early as the 1600s, physicians tried giving animal blood to bleeding humans. But their efforts at transfusion, along with most others, were disastrous failures. Although there were occasional lifesaving miracles, early transfusions resulted in so many deaths that they were banned in many European countries. Only the Incas seem to have performed routinely successful blood transfusions—a result of the fact that almost all native South Americans belong to blood group O and can therefore donate blood to one another without immunologic complications.

Early in the 1900s, Karl Landsteiner made meticulous studies that led to his description of the ABO blood antigen system. His work set the stage for a rational approach to matching blood for transfusions, but no one immediately put Landsteiner's discovery to practical use. Then, during World War I, when more than 20 million soldiers were seriously wounded, physicians began to collect blood, type it according to the ABO system, and use it to transfuse men who were bleeding to death. The discovery that sodium citrate prevented blood from clotting, which allowed it to be stored for prolonged periods of time in blood banks, was another breakthrough.

Now, more than 3 million blood transfusions are performed each year in the United States. In addition to the transfusion of whole blood to injury victims, many other types of transfusions are done routinely. Because blood can be separated into its various components—including oxygen-carrying erythrocytes, blood-clotting platelets, and serum—proteins such as immunoglobulins or clotting factors can be extracted. These blood products can then be administered selectively. Patients who lose blood during surgery, for example, usually receive packed red blood cells to replenish their lost oxygen-carrying capacity. Patients who begin to bleed because of lack of platelets, which is often a result of leukemia treatment, receive a platelet transfusion. Patients with the genetically determined bleeding disorder hemophilia receive an infusion of the clotting proteins that they are unable to produce for themselves.

But blood transfusion is a risky business. To prevent the potentially fatal mistake of giving a patient the wrong blood, checks and double-checks are required. Before any transfusion, two health-care professionals must check the identification code on the unit of blood and compare it to the patient's hospital identification number. Then both must sign to verify their identification. The patient is also observed closely during a transfusion so that it can be stopped if signs of a transfusion reaction develop.

Another risk of transfusion is transmission of infection. Many viral pathogens survive in banked blood and can be spread from donor to recipient. For years, hepatitis B was the most feared infectious complication of blood transfusion, but today HIV is a greater concern. To make transfusions safer, blood donors are carefully questioned for their risk of infection and donated blood is screened for infection-related antibodies. Still, no blood can ever be certified pathogen-free.

Blood transfusion today remains a potentially lethal but also lifesaving form of therapy. As with all medical treatments, the likely benefits must be weighed against the known risks.

The Immunologic Basis of Type III Hypersensitivity. A type III hypersensitivity reaction is initiated when antibodies bind to circulating antigens and form soluble antigen-antibody complexes. These complexes are ordinarily removed from the circulation by macrophages in the spleen and liver, but some complexes persist in the circulation, especially if the amount of antigen slightly exceeds the amount of antibody available for binding or if huge numbers of complexes overwhelm the macrophage system. Then the complexes lodge in the capillaries of normal tissues, activating complement and provoking an inflammatory response that attracts activated leukocytes. These leukocytes degranulate, releasing destructive enzymes that damage nearby tissues (**Figure 18.9**).

Systemic Lupus Erythematosus. The autoimmune disorder **systemic lupus erythematosus (SLE)** is a type III hypersensitivity in which autoantibodies are produced against antigens found in the cell nucleus. Presence of these **antinuclear antibodies (ANAs)** is the basis for diagnosing SLE.

Like many immune-complex diseases, SLE manifests itself clinically in a wide variety of ways. Antigen-antibody complexes that enter the circulation can be deposited anywhere. Wherever they lodge, they activate

Figure 18.9 Type III (immune-complex) hypersensitivity reactions involve antigens that are free in the circulation.

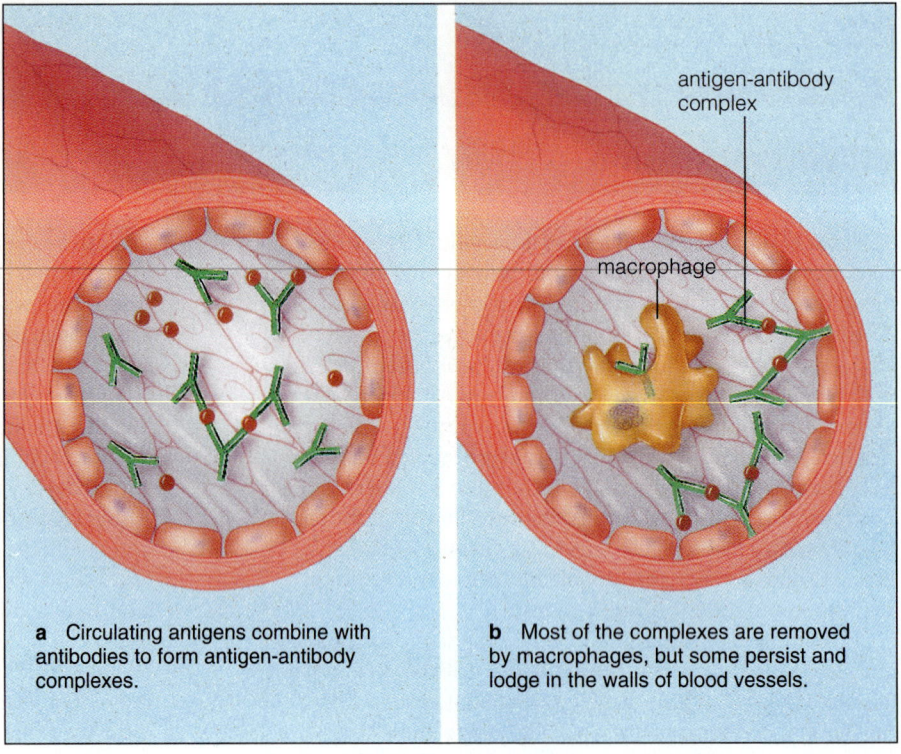

a Circulating antigens combine with antibodies to form antigen-antibody complexes.

b Most of the complexes are removed by macrophages, but some persist and lodge in the walls of blood vessels.

complement and leukocytes, stimulating a chronic, destructive inflammatory reaction. Immune complexes are most likely to end up in tissues with a rich blood supply, such as the kidneys, the skin, and the joints. But any organ, including the brain, can be affected. Chronic inflammation of the kidney is one of the most common causes of death in SLE. Inflammation in the skin often causes a facial "butterfly rash" (**Figure 18.10**). Inflammation of the joints causes arthritis, and inflammation of brain tissue causes disorders in thinking that may be mistaken for mental illness.

Treatment of SLE or any other autoimmune disorder presents the same problem: How do we control the destructive reaction of autoimmunity without also short-circuiting the essential immune responses? Unfortunately, we have no completely satisfactory solution. Patients with SLE are treated with immunosuppressive drugs, including steroids. However, blocking immune function with drugs makes SLE patients vulnerable to infection.

Other Examples of Type III Hypersensitivity. Many kinds of antigens provoke type III hypersensitivity reactions—self antigens, microbial antigens, and antigens from plants and animals.

As with SLE, self antigens stimulate the production of autoantibodies in **rheumatoid arthritis**. Joints throughout the body may be damaged or completely destroyed. Many patients with rheumatoid arthritis produce an autoantibody called **rheumatoid factor** that is directed against IgG antibodies. The presence of rheumatoid factor in a patient's serum helps in diagnosing rheumatoid arthritis.

Chronic infection from long-term exposure to microbial antigens may cause type III hypersensitivity by stimulating production of antibodies that form immune complexes. In some cases of chronic viral hepatitis, for example, antibodies against the virus are produced, but they do not rid the body of infection. They continue to circulate, and some form complexes with viral antigens that are likely to be deposited in blood vessels, where they initiate a destructive inflammatory reaction.

Nonmicrobial antigens from outside the body may also set off a type III hypersensitivity reaction. In some cases, these antigens enter the body in the course of daily life. For example, "farmer's lung" is a lung disease contracted by people who continually inhale certain antigens from molds, plants, or animals. In other cases, these antigens enter the body in the course of medical treatment. **Serum sickness**, for example, occurs when proteins from

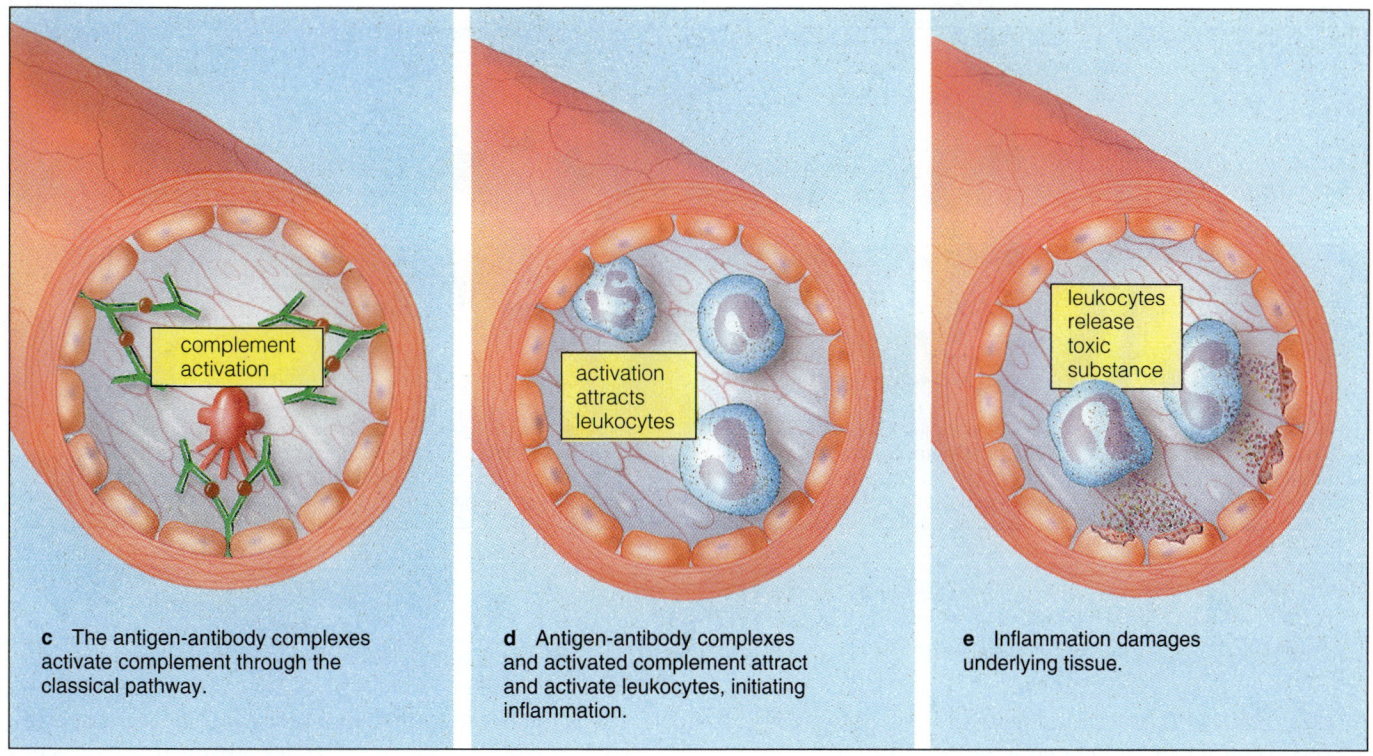

c The antigen-antibody complexes activate complement through the classical pathway.

d Antigen-antibody complexes and activated complement attract and activate leukocytes, initiating inflammation.

e Inflammation damages underlying tissue.

animal serum are used in medical therapy. It produces a dramatic hivelike rash and joint swelling and tenderness. Horse antiserum, for example, was once used in the treatment of diphtheria and is still used in the treatment of venomous snake bites.

Type IV: Cell-Mediated Hypersensitivity (Delayed)

Type IV hypersensitivity—also called cell-mediated hypersensitivity—differs from types I, II, and III because it does not involve antibodies. Instead, these reactions are initiated by certain CD4 lymphocytes sometimes called T_D cells because of their role in delayed (type IV) hypersensitivity. T_D cells produce macrophage-stimulating lymphokines, which are responsible for much of the tissue damage in this type of hypersensitivity. Because a type IV reaction occurs from 12 hours to several weeks after exposure, it is also known as **delayed hypersensitivity**.

Immunologic Basis of Type IV Hypersensitivity.

Type IV is the least-understood type of hypersensitivity. We know that T_D cells mediate all type IV reactions and

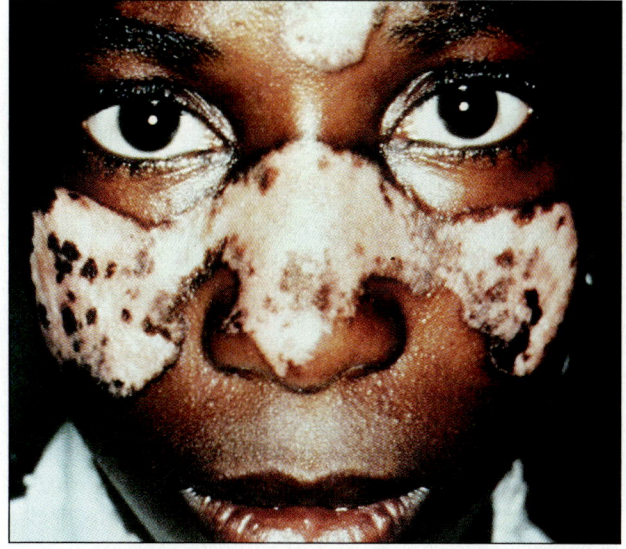

Figure 18.10 Systemic lupus erythematosus (SLE) is an autoimmune disorder mediated by type III hypersensitivity. The butterfly-shaped rash results when immune complexes are deposited in the small blood vessels of the skin.

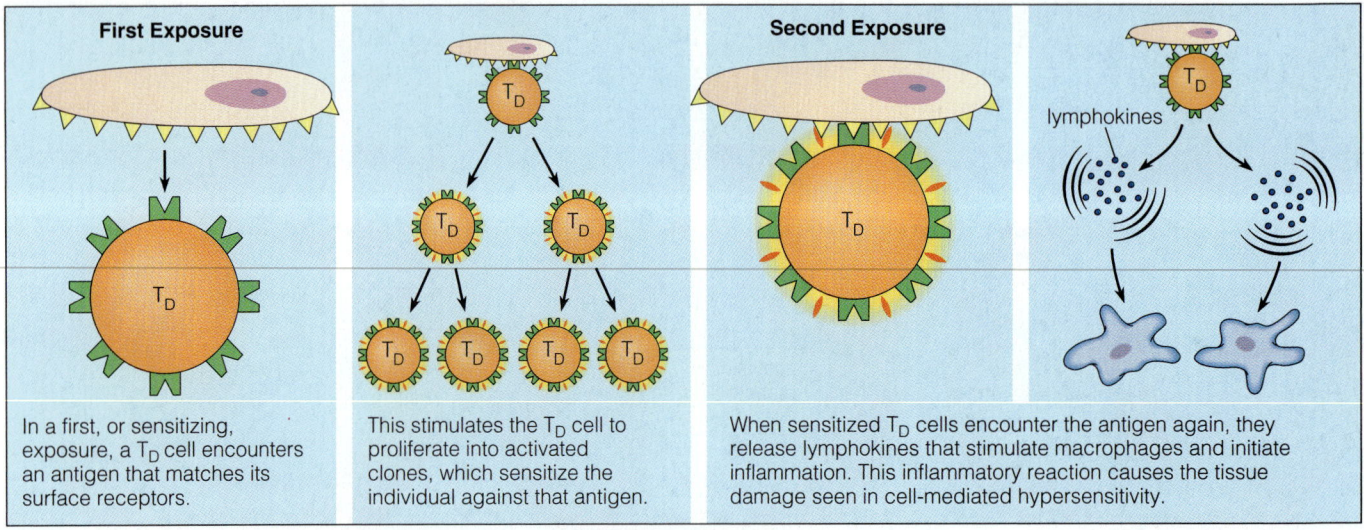

T_D

T_D

T_D

T_D

T_D

T_D

T_D

T_D

T_D

In a first, or sensitizing, exposure, a T_D cell encounters an antigen that matches its surface receptors.

This stimulates the T_D cell to proliferate into activated clones, which sensitize the individual against that antigen.

T_D

T_D

lymphokines

When sensitized T_D cells encounter the antigen again, they release lymphokines that stimulate macrophages and initiate inflammation. This inflammatory reaction causes the tissue damage seen in cell-mediated hypersensitivity.

Figure 18.11   The basic mechanism of type IV (cell-mediated) hypersensitivity.

that T_H cells and T_C cells participate in some. Type IV reactions begin when an antigen is presented to a T_D cell that bears a matching antibody receptor. Contact with the antigen stimulates the T_D cell to proliferate into an activated clone, sensitizing the individual to that particular antigen. If the person encounters the antigen again, the T_D cells respond by releasing lymphokines that stimulate macrophages and provoke inflammation (**Figure 18.11**).

Cell-mediated hypersensitivity reactions range from the life-threatening rejection of transplanted organs to the everyday annoyance of poison ivy and poison oak, a form of contact hypersensitivity. The tuberculin skin test, used to diagnose tuberculosis, is also an example of cell-mediated hypersensitivity. Probably the most significant form of type IV hypersensitivity in clinical medicine is the granulomatous reaction, which is directly responsible for the tissue damage seen in many chronic infections, including tuberculosis.

Organ Transplantation.   In **organ transplants**, also called **tissue grafts**, tissues are transferred from one part of the body to another or from one individual to another. Skin can be removed from a healthy area and grafted to another part of the same person's body to help heal a burn or extensive injury. A kidney, heart, lung, or liver, as well as bone marrow, may be removed from one person and given to another to sustain life.

There are four types of grafts (**Table 18.2**). An **autograft** is a transplant from one part of a person's body to another. An **isograft** is a transplant from a genetically identical individual, such as when one identical twin donates a kidney to another. An **allograft** is a transplant be-

tween genetically different members of the same species, such as when an organ from a recently dead person is donated to a matched recipient. A **xenograft** is a transplant from a nonhuman primate to a human, such as when a baboon heart was once given experimentally to an infant with a fatal heart defect.

Transplant organs can be rejected because of immunologic reactions against foreign antigens on the grafted tissue. Because autografts and isografts do not involve foreign antigens, the transplant is not rejected. Allografts and xenografts, however, do introduce foreign antigens, making rejection likely.

Organ rejection sometimes occurs within a few minutes to 48 hours of the transplant, because of an antibody-mediated type II hypersensitivity reaction called **hyperacute rejection**. Usually, though, rejection occurs weeks to months later, after T_D cells become sensitized from prolonged contact with the foreign graft antigens.

Surprisingly, transplantation of the cornea, the transparent covering of the eye through which we see, does not provoke an immunological response or lead to rejection. The eye is an **immunologically privileged site** that lacks normal lymphatic drainage, so rejection does not occur unless blood vessels grow into the transplanted cornea. Then the transplant becomes accessible to the recipient's immune system and the graft may be rejected.

An unusual type of reverse tissue rejection, called **graft-versus-host disease**, occurs fairly frequently after bone marrow transplantation. In preparation for a bone marrow transplant, the recipient's immune system is completely obliterated by treatment with whole-body radiation. With the recipient's immune system destroyed,

Table 18.2	Types of Tissue Grafts				
Name	Source of Tissue	Example	Foreign Antigens	Outcome
Autograft (*auto* = self)	Transplant from one part of a person's body to another	Skin graft for burns	None	No rejection
Isograft (*iso* = same)	Transplant from a genetically identical individual, such as clinical grafts between identical twins	Usually bone marrow; rarely, kidney	None	No rejection
Allograft (*allo* = different)	Transplant from a genetically different member of the same species, such as a nonidentical relative or a recently dead person	Bone marrow, kidney, heart, liver, lung, cornea	Variable; depends upon the closeness of tissue match between donor and recipient	Rejection would normally occur but is frequently prevented with immunosuppressive medication
Xenograft (*xeno* = foreign)	Transplant from nonhuman primate to human	Baboon heart to infant with fatal heart defect (experimental)	A great many foreign antigens	Rejection appears inevitable

the only functioning T cells are from the donor's marrow. Unless the donor and recipient are genetically identical, the donor T cells consider the recipient's cells the bearers of foreign antigens and launch an immune attack against them. Once graft-versus-host disease becomes established, it is usually fatal.

Tissue Typing and Matching for Transplantation. One way to avoid the rejection of transplanted organs is to match the tissue antigens of the transplant donor and recipient as closely as possible. If the transplanted organ bears only a few antigens that the recipient recognizes as foreign, the graft stands a reasonable chance of survival.

The primary determinants of tissue compatibility are the MHC antigens (Chapter 17), just as the ABO and Rh systems are the major determinants of blood compatibility. Each animal species has particular antigens that determine **histocompatibility** (*histo* means "tissue"). In humans these antigens are encoded by a group of genes called the **HLA**, or **human leukocyte antigen complex**. These genes encode class I MHC antigens (designated HLA A, B, and C) and class II antigens (designated HLA DP, DQ, and DR). HLA antigens were first discovered because of their role in transplant rejection, but we now know that these antigens normally help present foreign antigens to T lymphocytes—T_H lymphocytes can identify only foreign antigens presented in a complex with class II MHC antigens, and T_C lymphocytes can identify only foreign antigens presented in a complex with class I MHC antigens (Chapter 17).

A person's **tissue type**, or constellation of HLA antigens, is determined by mixing antigen-bearing lymphocytes with complement and antisera that contain monoclonal antibodies (clone cells produced by B lymphocytes that have identical antibodies against one specific epitope) against different HLA antigens (*sera* is the plural of serum). When lymphocytes are mixed with antibodies that happen to match their HLA antigens, the antigen-antibody complex binds complement and leads to cell lysis. (See the boxes on monoclonal antibodies in Chapter 19.)

Unfortunately for transplant technology, each gene in the HLA complex exists in many different forms that can specify many different antigens. Unlike the ABO blood antigen system with only three possible manifestations (A, B, or O), the HLA B gene can occur in nearly 50 different forms. Other HLA genes have many forms as well. There are more than 10,000 possible combinations of HLA antigens, each of which determines a distinct tissue type. Thus, perfect donor-recipient tissue match is all but impossible except between identical twins. Nevertheless, partial tissue type matching increases the probability of transplant success. In addition to the best possible HLA matching, the donor and recipient for an organ transplant must also have the same ABO blood type.

Immunosuppression. Transplant survival is also fostered by **immunosuppressive drugs**, which inhibit the immune response and thus block organ rejection. Unfortunately, most immunosuppressive drugs also leave the patient vulnerable to life-threatening infections. However, a drug called **cyclosporine**, the natural product of a fungus, is an improvement. Cyclosporine came into clinical use during the early 1980s. It interferes selectively with T-cell function, leaving B-cell function nearly

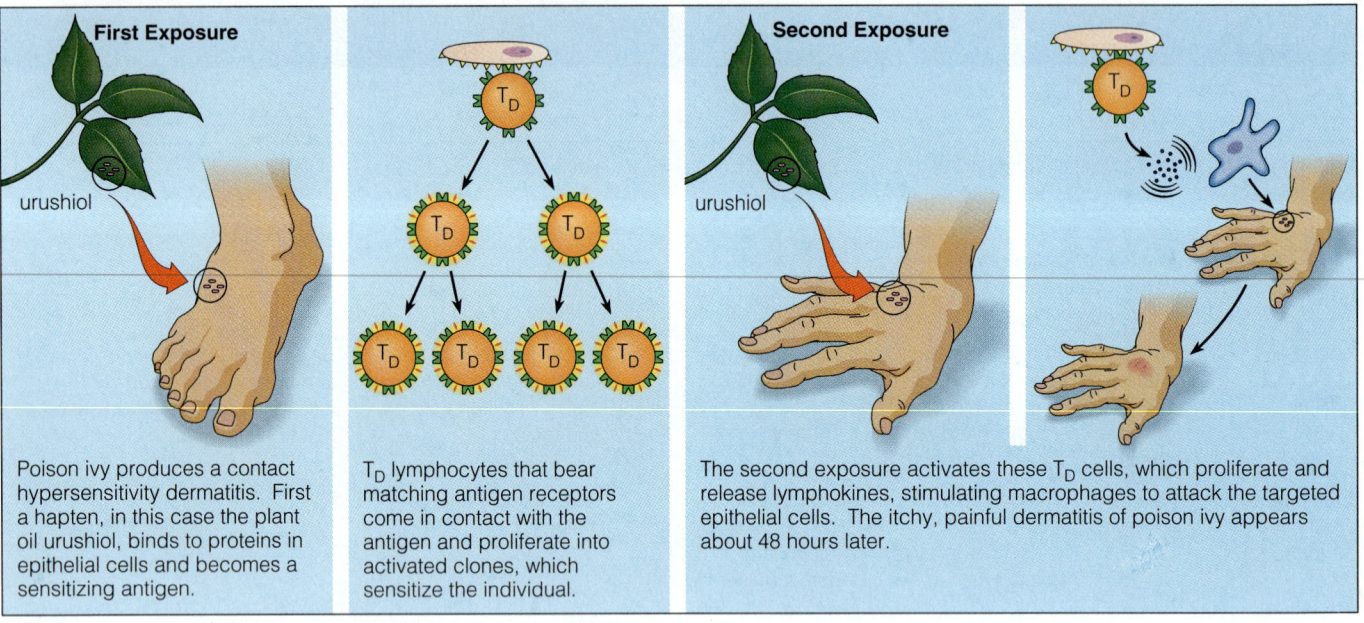

First Exposure

Poison ivy produces a contact hypersensitivity dermatitis. First a hapten, in this case the plant oil urushiol, binds to proteins in epithelial cells and becomes a sensitizing antigen.

T_D lymphocytes that bear matching antigen receptors come in contact with the antigen and proliferate into activated clones, which sensitize the individual.

Second Exposure

The second exposure activates these T_D cells, which proliferate and release lymphokines, stimulating macrophages to attack the targeted epithelial cells. The itchy, painful dermatitis of poison ivy appears about 48 hours later.

Figure 18.12 Contact hypersensitivity, a kind of type IV hypersensitivity that produces an itchy dermatitis (rash).

normal. In other words, cyclosporine blocks the cell-mediated processes that lead to organ rejection while sparing the body's nonspecific and antigen-mediated defenses against infection. The availability of cyclosporine during the last decade has led to a marked increase in the success rate of organ transplants. Cyclosporine therapy is not without toxicity, however. High doses can cause liver and kidney damage and may increase the risk of cancer. Research to find new and more selective immunosuppressive agents continues. A new fungus-derived immunosuppressive agent called FK506 looks promising in clinical trials.

Besides drugs, antibodies directed against the lymphocytes that mediate graft rejection can also be used to block rejection of transplanted tissue. The earliest antibody preparations—a mixture of antilymphocyte antibodies collected from laboratory animals that had been injected with human cells—had complications. Patients developed fever, rash, and a reaction similar to serum sickness. But monoclonal antibody technology (see boxes in Chapter 19) has led to the production of a pure antibody preparation targeted specifically against T lymphocytes. This new preparation, called OKT3, has been somewhat effective, but because it reacts with all T cells it is not a selective immunosuppressant. Research to obtain antibodies directed against the specific lymphocytes that participate in graft rejection continues.

Other Forms of Type IV Hypersensitivity. A common form of type IV hypersensitivity, called **contact hypersensitivity**, produces a painful, itchy **dermatitis**, or skin rash, from exposure to various compounds. Some of these irritating compounds are produced by toxic plants of the genus *Toxicodendron*, which includes poison ivy, poison oak, and poison sumac (**Figure 18.12**). Other compounds are present in certain types of clothing or jewelry. The irritating agents are haptens that become complete antigens when they attach to skin proteins (Chapter 17). The hapten-protein complexes sensitize T_D cells. Upon subsequent exposure to the same agent, the haptens and proteins once again combine to form a potent antigen, and T_D cells initiate an inflammatory reaction that produces a characteristic rash 48 hours later.

Poison ivy is one of the most common forms of contact dermatitis. The sensitizing hapten produced in the sap of poison ivy is **urushiol**. Although the plant leaves, stems, or roots must be slightly damaged to release the sap, the contact that leads to a poison ivy rash is usually so brief it goes unnoticed. Anything that has touched poison ivy—clothes, garden tools, or even dogs—can transfer urushiol to the skin. Washing with detergent within 15 minutes of contact will remove urushiol. After that time, the urushiol has bound to skin proteins and created the offending antigen. The red and itchy rash of poison ivy, which becomes blistery or weepy in severe cases, no longer contains any free urushiol, so poison ivy does not spread from one part of the body to another or from person to person. Although some people claim to be immune to poison ivy, almost anyone with a functioning immune system will become sensitized and react with a rash if exposed to urushiol on several occasions.

To identify antigens causing a contact sensitivity, a person can have **patch tests** done. A healthy part of the skin is gently rubbed and then an agent that commonly

Table 18.3 Immunodeficiency Disorders: Selected Examples

Disorder	Immune Defect	Congenital vs. Acquired	Clinical Manifestations
Severe combined immunodeficiency (SCID)	Deficiency of B and T cells	Congenital	Overwhelming infections that cause death in early infancy unless patient is maintained in germ-free environment or receives a bone marrow transplant
X-linked agammaglobulinemia	Severe deficiency to near absence of B cells	Congenital (inherited on the X chromosome, so it affects males only)	Frequent, often life-threatening bacterial infections that begin in infancy
IgA deficiency	Selective inability to produce IgA	Congenital	Repeated bacterial infections of the respiratory and gastrointestinal tracts
DiGeorge's syndrome	Abnormal development of T cells associated with abnormal thymus	Congenital	Severe immunodeficiency associated with other developmental abnormalities, often leading to death in infancy
AIDS	Severe deficiency of T_H lymphocytes	Acquired through infection by HIV	Recurrent bacterial, viral, fungal, and protozoal infections beginning when T_H cell count falls below 500 per microliter of blood
Chronic granulomatous disease	Deficiency in phagocyte function—neutrophils unable to kill ingested pathogens	Congenital; inherited on the X chromosome	Recurrent bacterial infections, both bloodborne and soft-tissue, beginning during the first year of life
C5 dysfunction	Abnormal function of the C5 complement protein	Congenital	Recurrent bacterial infections

causes contact sensitivity is taped to the skin and sealed with an airtight covering. After 48 hours the covering is removed. If a red rash that resembles poison ivy has developed, the patient has a contact sensitivity to the test agent. Although everyone is sensitive to highly potent agents such as urushiol, only certain people react to the synthetic materials in clothing, elastic, or the metals in costume jewelry.

The tuberculin skin test used to identify people infected by the tuberculosis bacterium is another manifestation of delayed hypersensitivity. In this test, an antigen extracted from tuberculosis bacilli is injected into the skin. If sensitized T_D cells are present from a previous tuberculosis infection, they initiate an inflammatory response that causes large numbers of lymphocytes and macrophages to accumulate near the blood vessels of the affected skin. This produces a red lump at the injection site after two to three days. The test has been invaluable in controlling the spread of tuberculosis (Chapter 22).

The **granulomatous reaction** is probably the most important form of delayed hypersensitivity because it causes serious disease in the greatest number of people. Most of the tissue damage of tuberculosis, leprosy, and schistosomiasis (a disease caused by parasitic worms) is due to granulomatous reactions. Granulomatous reactions occur when microbial antigens that a macrophage cannot completely destroy persist within the cell. These persistent antigens cause T_D cells to release lymphokines that stimulate the production of a **granuloma**, a nodule of

activated monocytes and macrophages, some of which differentiate further. The continuing inflammation that goes on within and around granulomas displaces and destroys normal cells, often leading to significant tissue damage. Granulomatous reactions take at least two weeks to develop—the longest of any type of delayed hypersensitivity.

IMMUNODEFICIENCY DISORDERS

Immunodeficiency disorders occur because of a breakdown, or deficiency, of immune function. In some cases, the immune system is completely paralyzed. In other cases, the defect is more selective, involving, for example, an inability to produce a certain type of antibody or an abnormality in the function of a particular type of T cell. Immunodeficiencies can be categorized as defects in T-cell function, B-cell function, both T- and B-cell function, phagocyte function, or complement (**Table 18.3**).

Regardless of mechanism, immunodeficiencies result in recurrent infection. Immunodeficiency disorders can appear at any age, from infancy to old age. **Congenital immunodeficiencies** are inherited or develop before birth and usually appear early in life. **Acquired immunodeficiencies** develop later; they are caused by infection, cancer, or the side effects of immunosuppressive medications.

David—Life in a Germ-Free World

David in his germ-free environment.

Surely the best-known patient with SCID was a boy from Texas known to the public by his first name, David. David was diagnosed with SCID when he was born, and the prognosis (probable outcome) was dismal. SCID babies usually die from infection within weeks to months. But David didn't die during infancy. He was placed in a germ-free environment and protected from all contact with infection-causing agents. Pictures of David show him sitting alone inside a plastic enclosure or wearing a protective "space suit."

When he was 12, David left his germ-free enclosure to receive a bone marrow transplant from his sister that might restore his immune system. But a few months later David died. The transplant had not failed. Rather, David probably died from cancer caused by the Epstein-Barr virus (Chapter 27), an unusual complication. The virus, which usually does not cause cancer, was in the bone marrow donated by his sister. In David's immune-disordered system, however, the virus was fatal.

The technology of protective isolation allowed David to remain germ-free and therefore alive, but his life was tragically restricted.

Congenital Immunodeficiencies

Congenital immunodeficiencies often lead to death from infection early in life. The most profound congenital immunodeficiencies are called **severe combined immunodeficiencies (SCID)**. Several disorders that disable both B-cell and T-cell immunity are grouped together under this heading. Patients with SCID die from infection during early infancy unless extreme measures are taken to restore immune function (see beginning of Chapter 7) or protect them from all contact with microorganisms (see the box "David—Life in a Germ-Free World").

Some immunodeficiency disorders affect only B lymphocyte function and antibody production. Patients with **X-linked agammaglobulinemia**, for example, are unable to produce antibodies although they have fairly normal cell-mediated immunity. They are severely threatened by bacterial infections, against which antibody defenses are crucial, but they recover normally from viral diseases such as measles, against which cytotoxic T cells are the main defenders. Agammaglobulinemia is a sex-linked inherited disease that shows up only in males. They die from bacterial infection during childhood unless they receive antibody-containing injections throughout their lives. Other people have more limited antibody deficiencies, such as a selective defect in IgA production. These patients suffer from an unusual number of bacterial infections, but many survive to adulthood without special treatment.

Other congenital immunodeficiency disorders affect only T-lymphocyte function. In **DiGeorge's syndrome**, the patient lacks a normal thymus gland, which means normal T-lymphocyte differentiation cannot take place. These patients have other associated defects and usually die in infancy.

Some immunodeficiency disorders involve defects in nonspecific defenses—phagocytosis or complement. For example, in **chronic granulomatous disease**, neutrophils lack the ability to produce the respiratory burst that normally leads to phagocytic killing (Chapter 16). These patients frequently get severe bacterial infections. Other disorders involve the absence of a particular protein of the complement cascade. C5 deficiency, for example, causes an increased susceptibility to infections by species of bacteria that are normally killed by complement's membrane attack complex.

Congenital immunodeficiencies are rare. Many doctors may not see even one patient with a congenital immunodeficiency during their career. Still, these diseases are extremely important in clinical research because they help scientists understand immune function. For example, patients with agammaglobulinemia helped clinical scientists define which aspects of immunity are attributable to antibody function and which are T-cell mediated.

Acquired Immunodeficiencies

Acquired immunodeficiencies are far more common than congenital immunodeficiencies. Patients receiving long-term therapy with systemic steroids and patients with leukemias (blood cancers) and lymphomas (lymph node cancer) are at particular risk. Although we do not understand the mechanisms at work, immune defenses seem to decline when a person becomes gravely ill for any reason. As a result, many people with serious illnesses die from infectious complications. When the serious illness involves cells critical to immune system function, the risk of life-threatening infection is greatly increased.

Even comparatively mild illnesses, such as the flu, can depress immune function and make the ill person susceptible to additional infections. Immunoregulation is extremely complex and sophisticated. We do not yet understand any of the mechanisms at a molecular level.

Acquired immunodeficiency syndrome (AIDS) is a special example of an acquired immunodeficiency. AIDS is a syndrome of immune failure caused by infection with HIV, a human retrovirus. This new disorder, first described in 1981, is discussed in detail in Chapter 27. Because AIDS irreversibly affects the survival of T_H cells, eventually affecting both humoral and cell-mediated immunity, most patients die from infection.

CANCER AND THE IMMUNE SYSTEM

The immune system may play a crucial role in preventing cancer. Cells become cancerous when they undergo **transformation**, a process leading to uncontrolled multiplication that produces large numbers of undifferentiated cells. These primitive, rapidly dividing cells form large growths called **tumors** that crowd and eventually kill normal neighbor cells. Transformed cells often express new surface antigens that are not found on normal cells. These **tumor-specific antigens** mark the cancer cells for destruction. T cells, NK cells, and macrophages are all known to destroy antigen-marked transformed cells.

The **immune surveillance theory** proposes that cell transformation occurs frequently, but that the immune system eliminates most malignant cells before they cause cancer. The fact that people with defective immune systems are at high risk of developing cancer supports the theory. AIDS patients, for example, develop otherwise unusual types of cancer, such as Kaposi's sarcoma. This theory is also consistent with increased incidence of cancer in the elderly as immune defenses weaken with advanced age.

But the immune surveillance theory must be reconciled with the fact that tumors carrying cancer antigens also cause fatal disease in otherwise healthy people. This phenomenon is called **immunologic escape**. Cancer immunologists propose four theories to explain the apparent contradiction:

1. Small numbers of cancer cells may fail to stimulate an effective immune response until the tumor is too large to be controlled.

2. Other molecules may mask the cancer cells' new antigens, making them invisible to defending lymphocytes.

3. Cancer cells may shed such large quantities of antigen that they occupy all antigen-binding sites on nearby lymphocytes, thereby overwhelming immunologic defenses.

4. Cancer cells may secrete chemical messengers that interfere with immune function—for example, stimulating suppressor lymphocytes at the expense of cytotoxic lymphocytes.

Many cancer immunologists believe that the immune system may be the ideal tool to treat widespread cancers. Most cancer treatments are highly toxic because they are nonspecific. That is, they cannot be directed against cancer cells alone. For example, most anticancer drugs are designed to kill rapidly dividing cells. These drugs are highly effective in killing cancer cells, but they also kill cells that normally divide rapidly, such as hair cells (causing baldness), cells of the intestinal lining (causing pain and digestive problems), and bone marrow stem cells (causing a scarcity of white blood cells and vulnerability to infection). A treatment that is highly specific for cancer cells would be very effective.

Some attempts at **cancer immunotherapy**—using products of the immune system to treat cancer—have been successful. Interferons (Chapter 16) are now standard therapy for a rare blood cancer called **hairy-cell leukemia**. These lymphokine messengers activate macrophages and NK cells to attack cancerous cells. In a still-experimental therapy for **melanoma**, a highly malignant skin cancer, the patient's own lymphocytes are recovered from the tumor, cultured in the laboratory, and artificially stimulated by lymphokines. Then the **lymphokine-activated killer cells (LAK cells)** are injected into the patient where they aggressively attack tumor cells. This approach has occasionally brought about a complete cure of this otherwise fatal disease.

Advances in basic immunology and biotechnology will extend cancer immunotherapy and make it even more effective. In an intriguing new approach, pure antibodies are directed against tumor-specific antigens. Such antibodies, bound to radioactive tracer substances, have already been used to reveal the location of tumor cells. This immunologic imaging can detect even a few cancer cells, unlike standard diagnostic techniques, such as x rays or magnetic resonance imaging (MRI), which cannot locate tumors until they become fairly large. Related research is in progress to design and test **immunotoxins**—tumor-specific antibodies bound to lethal drugs or radioactive compounds. These injected agents, which would seek out and kill cancer cells selectively, would be a long-awaited breakthrough in cancer therapy.

Summary

IMMUNE SYSTEM MALFUNCTIONS   (p. 430)

1. Immunologic disorders are caused by a malfunction of the immune system that produces either an inappropriate or an inadequate immune response.

2. Hypersensitivity is a misdirected immune response in which either antibodies or T cells damage tissue.

3. Failure to mount an adequate immune response causes immunodeficiency disorder. Immunodeficiencies can be congenital, or they can be acquired during a person's lifetime.

HYPERSENSITIVITY   (pp. 430–445)

1. There are four types of hypersensitivity reactions: anaphylactic (immediate), cytotoxic, immune-complex, and cell-mediated (delayed).

Type I: Anaphylactic Hypersensitivity (Allergy)   (pp. 430–434)

1. Type I reactions affect about 20 percent of the population. Type I is also called immediate hypersensitivity because the symptoms develop within 10 to 20 minutes. The common name is allergy. Antigens that stimulate type I reactions are called allergens.

2. The first exposure to a sensitizing antigen causes B cells, with the help of T_H cells, to make IgE antibodies. IgE binds to mast cells; and, on subsequent exposure, antigen binds to IgE, which stimulates the mast cell to degranulate. The inflammatory mediators that are released, including histamine, cause an allergic response.

3. The leukotrienes LTC_4 and LTD_4 are also formed during degranulation. They are jointly called SRS-A (the slow-reacting substance of anaphylaxis) because they take effect later and cause prolonged respiratory difficulty.

4. Signs and symptoms of type I anaphylactic reactions usually include local inflammation, pain, redness, warmth, and swelling. There may also be bronchial constriction, a dry spasmodic cough, and different types of edema, causing hives (urticaria), nausea, and hoarseness. Complete laryngeal edema can cause suffocation. Anaphylactic shock is a precipitous fall in blood pressure that can result in sudden death.

5. Treatment consists of epinephrine and antihistamine medication. Individuals who are at frequent risk of bee sting may undergo immunotherapy.

6. Allergens can be injected into the body (insect venom, medications), inhaled (pollen, animal dander, feces of microscopic dust mites), or ingested (medications, foods, food additives).

7. Inhaled allergens that affect the upper respiratory tract cause allergic rhinitis (hay fever), the most common type I reaction. Mast cells in the mucous membranes of the nose degranulate, releasing mediators that cause nasal congestion, itching, and mucus discharge. Treatment is usually with antihistamines.

8. Inhaled allergens that affect the lower respiratory tract can cause asthma, a breathing disorder resulting from constricted bronchial passages. Treatment is usually with inhaled epinephrine-like drugs, cromolyn, and—occasionally—steroids. Asthma is a chronic condition that can be life-threatening.

9. Many food "allergies" are really the result of an irritated intestinal tract or digestive problems. True food allergies produce hives or systemic anaphylactic reactions. Shellfish and peanuts are highly allergenic foods.

10. Penicillin can be a powerful allergen.

Type II: Cytotoxic Hypersensitivity   (pp. 434–438)

1. Cytotoxic hypersensitivity occurs when IgM or IgG antibodies bind abnormally to antigens on the surface of human cells.

2. One outcome is that the cell-bound antibodies activate complement and lyse the target cell through the terminal complement pathway. This is what occurs in transfusion reactions and hemolytic disease of the newborn.

3. Another outcome is that antibodies and activated complement attract and activate phagocytes and killer cells that attack the target cell. This occurs in Goodpasture's syndrome, an autoimmune disorder that affects the kidneys.

4. Occasionally, the bound antibodies alter cell function and cause disease without killing the target cell. This happens in Graves' disease, an autoimmune disorder in which the thyroid gland becomes hyperactive.

5. If antigens on the erythrocytes in a transfusion react with antibodies in the recipient (immunologic incompatibility), antigen-antibody complexes lyse the transfused erythrocytes and/or the erythrocytes agglutinate (clump together).

6. People with type A blood can receive only type A, and people with type B can receive only type B. People with type AB are universal recipients, but they can donate blood only to others with type AB blood. People with type O blood are universal donors, but they can receive only antigen-free type O blood.

7. A transfusion reaction produces rapid onset of fever, vomiting, and shock. Before every transfusion, a major crossmatch is performed; agglutination indicates a bad match.

8. A person who has the Rh antigen is said to be Rh-positive and one who doesn't is Rh-negative. Rh antigens cause the most severe form of hemolytic disease of the newborn, a potentially fatal blood disorder.

9. During her first pregnancy, an Rh-negative mother usually has no problems with an Rh-positive fetus. At or around the time of birth, some mingling of blood between mother and child typically occurs,

and the mother produces anti-Rh antibodies. During a subsequent pregnancy with an Rh-positive fetus, the anti-Rh antibodies cross the placenta, bind to the fetal erythrocytes, and lyse them. Severely affected babies are given exchange transfusions.

Type III: Immune-Complex Hypersensitivity (pp. 438–441)

1. A type III hypersensitivity reaction is initiated when antibodies bind to circulating antigens and form antigen-antibody complexes that remain soluble. Macrophages normally remove these complexes, but if they persist they lodge in capillaries and activate complement. The inflammatory response attracts leukocytes, which degranulate, releasing enzymes that destroy the tissue.

2. Type III hypersensitivity is the basis of systemic lupus erythematosus (SLE), an autoimmune disorder. Wherever the circulating antigen-antibody complexes lodge, they cause a destructive inflammatory response (typical sites are the skin, joints, and kidneys).

3. Currently, SLE patients are treated with immunosuppressive drugs, including steroids, but this is not a totally satisfactory solution.

4. Many kinds of antigens provoke type III hypersensitivity reactions, including self antigens, microbial antigens, and antigens from plants and animals.

5. Chronic viral hepatitis is an example of a microbial antigen that causes type III hypersensitivity. Serum sickness is caused when antigens in animal serum provoke a type III response.

Type IV: Cell-Mediated Hypersensitivity (Delayed) (pp. 441–445)

1. Type IV hypersensitivity does not involve antibodies. It is initiated by T_D cells and is called cell-mediated hypersensitivity. It is also known as delayed hypersensitivity because the reaction occurs from 12 hours to several weeks after exposure.

2. Type IV reactions begin when an antigen is presented to a T_D cell with a matching antibody receptor. The T_D cell is stimulated to proliferate into an activated clone, which sensitizes the individual to that particular antigen. When the person encounters the antigen again, the T_D cells release lymphokines that stimulate macrophages and provoke inflammation.

3. Rejection of organ transplants (tissue grafts) is due to type IV hypersensitivity. There are four types of grafts: autografts, isografts, allografts, and xenografts. An unusual type of reverse tissue rejection called graft-versus-host disease often occurs after bone marrow transplantation.

4. MHC antigens are used to determine tissue compatibility (histocompatibility). Immunosuppressive drugs, such as cyclosporine, and antibodies directed against T cells help block immunologic rejection.

5. Contact hypersensitivity is a common form of type IV reaction. It produces an itchy dermatitis (skin rash) from exposure to certain compounds in plants (poison ivy, poison oak), clothing, and jewelry.

6. The irritating agents are haptens that attach to skin proteins. The hapten-protein complexes sensitize T_D cells, and on subsequent exposure the T_D cells initiate an inflammatory response.

7. The tuberculin skin test is a delayed hypersensitivity reaction.

8. The granulomatous reaction causes most of the tissue damage in tuberculosis, leprosy, and schistosomiasis. When microbial antigens are not eliminated and persist within a cell, they cause T_D cells to release lymphokines that stimulate the production of a granuloma. The continuing inflammation in the area destroys normal cells.

IMMUNODEFICIENCY DISORDERS (pp. 445–447)

1. Immunodeficiency disorders occur because of a breakdown, or deficiency, of immune function. In general, immunodeficiencies can be categorized as defects in T-cell function, B-cell function, both T- and B-cell function, phagocyte function, or complement. Immunodeficiencies are characterized by recurrent infection.

2. Congenital immunodeficiencies are inherited or develop before birth. Several profound congenital immunodeficiencies that disable both T and B cells are grouped under the name severe combined immunodeficiencies (SCID). These patients usually die early in infancy unless extreme measures are taken.

3. Agammaglobulinemia is a sex-linked immunodeficiency of B cells. Because these patients cannot produce antibodies, they are prone to bacterial infections.

4. DiGeorge's syndrome is an immunodeficiency of T cells. These patients lack a normal thymus gland, which means normal T-lymphocyte differentiation does not take place; patients usually die in infancy.

5. In chronic granulomatous disease, neutrophils lack the ability to produce the respiratory burst that normally leads to phagocytic killing. A C5 deficiency means an individual lacks that protein in the complement cascade.

6. Acquired immunodeficiencies develop after birth as a result of infection, cancer, or the side effects of immunosuppressive medications. Patients receiving long-term therapy with systemic steroids and patients with leukemia and lymphomas are at particular risk. AIDS is an acquired immunodeficiency caused by a virus.

CANCER AND THE IMMUNE SYSTEM (pp. 447–448)

1. Cells that undergo transformation (a process leading to uncontrolled multiplication that produces large numbers of undifferentiated cells) form large growths called tumors. Tumor-specific antigens appear on the cancer cells and mark them for destruction by T cells, NK cells, and macrophages.

2. The immune surveillance theory proposes that cell transformation occurs frequently, but the immune system eliminates most malignant cells before they cause cancer. This theory must be reconciled with immunologic escape, the fact that people who are otherwise healthy develop cancer.

3. Cancer immunotherapy—using products of the immune system to treat cancer—may be the ideal treatment. Interferons are standard therapy for a rare blood cancer called hairy-cell leukemia, and lymphokine-activated killer cells (LAK cells) are occasionally used to treat melanoma.

Review Questions

IMMUNE SYSTEM MALFUNCTIONS

1. Explain the difference between an immunologic disorder and an immunodeficiency disorder.

HYPERSENSITIVITY

1. What is hypersensitivity? Name the four types.

2. Why is anaphylactic hypersensitivity also called immediate? What is the common name by which people know anaphylactic hypersensitivity?

3. Explain the immunologic basis of anaphylactic hypersensitivity. Which antibodies are involved?

4. What are the signs and symptoms of anaphylactic hypersensitivity? What role does SRS-A play? What causes anaphylactic shock? Describe the treatment for anaphylactic hypersensitivity caused by insect venom.

5. Give some examples of injected, inhaled, and ingested allergens.

6. Explain the difference in cause, signs and symptoms, and treatment of hay fever and asthma.

7. Explain this sentence: Penicillin should be used judiciously to prevent drug sensitization.

8. Explain the immunologic basis of cytotoxic hypersensitivity, including the three possible outcomes. Which antibodies are involved?

9. Describe the ABO antigen system and blood typing. Define universal recipient and universal donor.

10. What happens if there is immunologic incompatibility in a blood transfusion? What is a transfusion reaction? How and why is a major crossmatch done?

11. What is the Rh system of blood typing and why is it important? How does hemolytic disease of the newborn occur? Why is an exchange transfusion the best treatment for a severely affected newborn?

12. What is an autoimmune disorder? Describe Goodpasture's syndrome and the type II mechanism that causes it. Do the same for Graves' disease.

13. What is the basic difference between type II hypersensitivity and type III? Describe the immunologic basis of type III hypersensitivity. Which antibodies are involved?

14. Why is systemic lupus erythematosus (SLE) classified as an autoimmune disorder? What happens at the cellular level to cause the signs and symptoms of SLE?

15. Explain this statement: The treatment of SLE presents the same issues raised by all autoimmune disorders.

16. What kinds of antigens besides self antigens cause type III reactions? Give some examples of type III hypersensitivity (other than SLE).

17. How is type IV hypersensitivity different from types I, II, and III? Why is type IV also called delayed hypersensitivity?

18. What is the immunologic basis of type IV reactions?

19. Explain how organ transplants are rejected because of cell-mediated hypersensitivity. Explain the type of rejection called graft-versus-host disease.

20. Define these terms and give an example of each: autograft, isograft, allograft, xenograft.

21. Define histocompatibility. What is a person's tissue type? How is it determined?

22. Explain the two major ways of blocking organ rejection.

23. What is contact hypersensitivity? Give some examples.

24. Explain why someone may not be affected at first by exposure to poison ivy but inevitably will. Describe the events at a cellular level.

25. Explain this statement: The tuberculin skin test is a delayed hypersensitivity reaction.

26. What is a granulomatous reaction?

IMMUNODEFICIENCY DISORDERS

1. How are immunodeficiency disorders usually categorized?

2. For each of the following congenital immunodeficiency disorders, tell what the breakdown is in immune function:
   a   severe combined immunodeficiencies (SCID)
   b. X-linked agammaglobulinemia
   c. DiGeorge's syndrome
   d. chronic granulomatous disease
   e. C5 deficiency

3. Who is at most risk for acquired immunodeficiencies? What is the breakdown in immune function in AIDS?

CANCER AND THE IMMUNE SYSTEM

1. Define the following terms: transformation, tumor, tumor-specific antigens.

2. What is the immune surveillance theory? What evidence is in its favor? What is immunologic escape? Give an example.

3. What is cancer immunotherapy? Give an example.

Essay Questions

1. Write a case history of yourself or someone you know who suffers from a type I hypersensitivity reaction. Describe the agent, the signs and symptoms, and treatment. Also describe the cellular events at each stage.

2. Research the procedures for organ donation in your state. Driver's licensing facilities usually have informative brochures. Your library has additional information.

Suggested Readings

Buisserit, P. D. 1982. Allergy. *Scientific American*, August, 86.

Carpenter, C. B. 1989. Immunosuppression in organ transplantation. *New England Journal of Medicine* 322:1224–26.

Cohen, I. R. 1988. The self, the world, and autoimmunity. *Scientific American*, April, 52.

19 DIAGNOSTIC IMMUNOLOGY

Diagnostic Immunology at Home: Pregnancy Tests

C. G. was a 22-year-old woman who thought she might be pregnant. Though her menstrual periods were often irregular, this one was several weeks late. She called her doctor for an appointment but none was available for two weeks. C. G. made the appointment but decided to try a home pregnancy test to find out sooner whether she was pregnant.

At the drugstore, C. G. found more than a dozen different brands of home pregnancy tests. The package information revealed that all the kits were basically alike. They contained a plastic dipstick to be immersed in urine; an indicator dot changed color if the woman was pregnant. C. G. bought the least expensive kit, took it home, and followed the simple directions carefully. There was no color change—indicating C. G. was not pregnant.

Although she was confident she had used the test exactly according to directions, C. G. decided to keep her doctor's appointment anyway. Perhaps a more sensitive test would show an early pregnancy her home kit missed. The doctor sent C. G. to a small clinical laboratory where she was asked to collect another urine specimen. Curious about the whole process, C. G. asked the lab technologist how she planned to test the specimen.

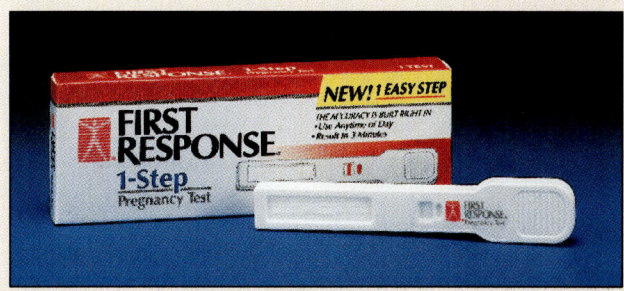

Home pregnancy kit.

The technologist explained that pregnancy tests are based on the presence or absence of a hormone called human chorionic gonadotropin (HCG), which is produced by the placenta and therefore found only in pregnant women. The test, she explained, contains a monoclonal antibody linked to an enzyme and a dye system. If HCG is in the urine, it binds to the antibody and the enzyme causes the dye to change color. In fact, she told C. G., the clinical lab test is just like the home pregnancy tests sold in the drugstore. All can detect even early pregnancies. The lab test confirmed C. G.'s home result—no color change, no pregnancy.

To understand:

• How antigen-antibody reactions can be used to diagnose infectious and noninfectious diseases

• The procedures for performing the basic serological assays—precipitation, agglutination, complement fixation, immunofluorescence, radioimmunoassays, and enzyme-linked immunosorbent assays (ELISA)

• The different types of assays: qualitative, quantitative, direct, and indirect

SEROLOGY

You read in Chapter 17 how immunologic reactions defend against microbial invaders and in Chapter 18 how some immunologic reactions cause disease. In this chapter, you will learn how clinicians use these reactions to diagnose many immunologic and nonimmunologic disorders. Most immunologically based diagnostic tests employ antigen-antibody reactions. Some of the most recent—like the pregnancy test—use monoclonal antibodies (identical antibodies against one specific epitope; see the boxes in this chapter).

Antigens and antibodies are far too small to be seen, but the reactions they cause are easily detectable. In general, immunologic testing is designed to detect a particular antigen, such as a bacterial antigen present during an infection, or a particular antibody, such as an abnormal autoantibody or an antibody produced during a previous infection. Diagnostic clinical immunology is also called **serology** because many of the tests detect antibodies in **serum**—the cell-free liquid component of blood.

Serology is the branch of immunology that studies antigen-antibody interactions for medical diagnosis. Serologic tests, or **assays**, are used to identify unknown agents in clinical specimens. In these assays a specimen, such as blood or tissue, is recovered from a patient and combined with test reagents that contain known concentrations of antigens or antibodies (a reagent is something added to a solution that participates in a chemical reaction). A reagent containing antibodies detects antigens; a reagent containing antigens detects antibodies in the specimen. Identifying the antigens or antibodies allows a clinician to diagnose a present or past infection. For example, hepatitis A (infectious hepatitis) is usually diagnosed by identifying *antibodies* against the causative virus. Hepatitis B (serum hepatitis), on the other hand, is usually diagnosed by identifying *antigens* from the viral coat. Many noninfectious disorders, such as autoimmune disorders, are diagnosed by identifying abnormal antibodies.

Some serologic tests are **qualitative**; they reveal only the presence or absence of an antigen or antibody. Others are **quantitative**; they measure the concentration of a particular antigen or antibody. Quantitative tests help monitor the progress of a disease in which antigen or antibody

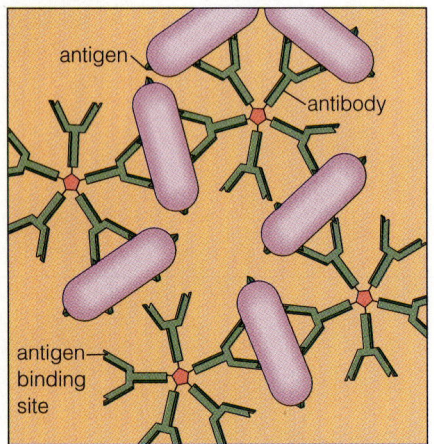
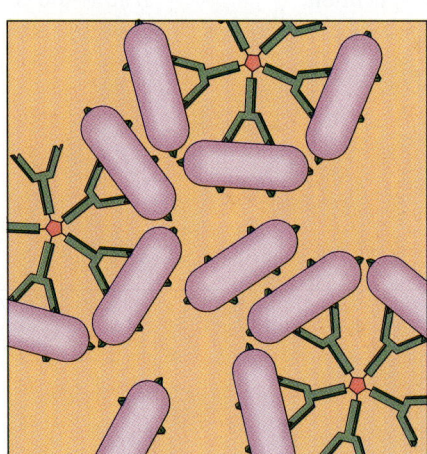
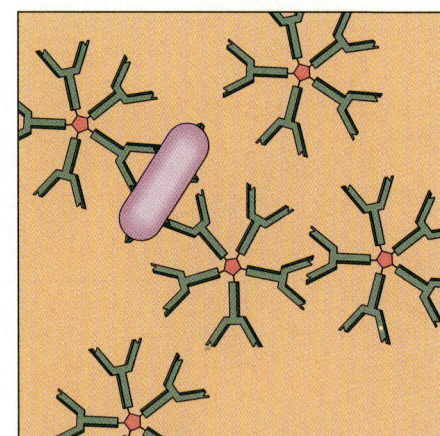

a Lattice formation at optimal proportions   **b** Antigen excess   **c** Antibody excess

Figure 19.1 Precipitation reactions occur when molecules of antigen and antibody combine to form visible lattices that precipitate. The relative concentrations of antigen and antibody are critical to the formation of precipitates. (**a**) If antibody and antigen are present in optimal proportions—approximately equal concentrations—a lattice forms and precipitates from solution. But if antigen is in excess (**b**) or antibody is in excess (**c**), a visible lattice is not likely to form.

levels rise or fall. Both qualitative and quantitative tests have advantages and disadvantages. For example, qualitative serologic tests are the most specific and reliable for making a definitive diagnosis of syphilis, but only quantitative serologic tests, which show a fall in antibody concentration as the result of antibiotic treatment, can document a cure (Chapter 24).

SEROLOGIC TESTS

There are six basic types of serologic tests—precipitation reactions, agglutination reactions, complement fixation reactions, immunofluorescence reactions, radioimmunoassays, and enzyme-linked immunosorbent assays (ELISA).

Precipitation Reactions

Free antigen and antibody molecules sometimes combine to form **lattices**, huge interlocking macromolecular webs. The formation of these lattices is the basis for a type of serologic assay called a **precipitation reaction**. Precipitation reactions are simple to perform and have been in common use for decades.

Antigens and antibodies that are not attached to cells or large particles dissolve in almost any type of aqueous fluid. But because of their size, lattices become visible as they **precipitate** or leave the solution. These precipitates usually remain suspended in the fluid in which they were formed and are detectable as cloudy regions in the test mixture. Significant quantities of precipitate are most likely to form when the reacting antigen and antibody are present in **optimal proportions**, or approximately equal concentrations (**Figure 19.1**).

Precipitation reactions can be performed in a liquid or a gel. A **precipitin ring test** is a type of precipitation reaction commonly performed in a liquid. When an antibody-containing liquid is mixed in a test tube with an antigen-containing liquid, a cloudy zone of precipitation forms where antigen and antibody meet each other in optimal proportions (**Figure 19.2**). The initial antigen-antibody binding occurs within minutes, but it may take hours to form lattices large enough to precipitate out of solution. The precipitin ring test requires a minimum of equipment, but layering the different liquids into the tube can be difficult, which limits its convenience. This test is used more for research than clinical diagnosis.

An **immunodiffusion assay** is a type of precipitation reaction performed in a gel—in a petri plate of agar, for example. Preparations of antibodies and antigens are placed in wells, slight depressions in the agar medium. The molecules diffuse through the gel until they meet and form a precipitate. Areas of precipitation are visible as cloudy lines in the otherwise clear gel.

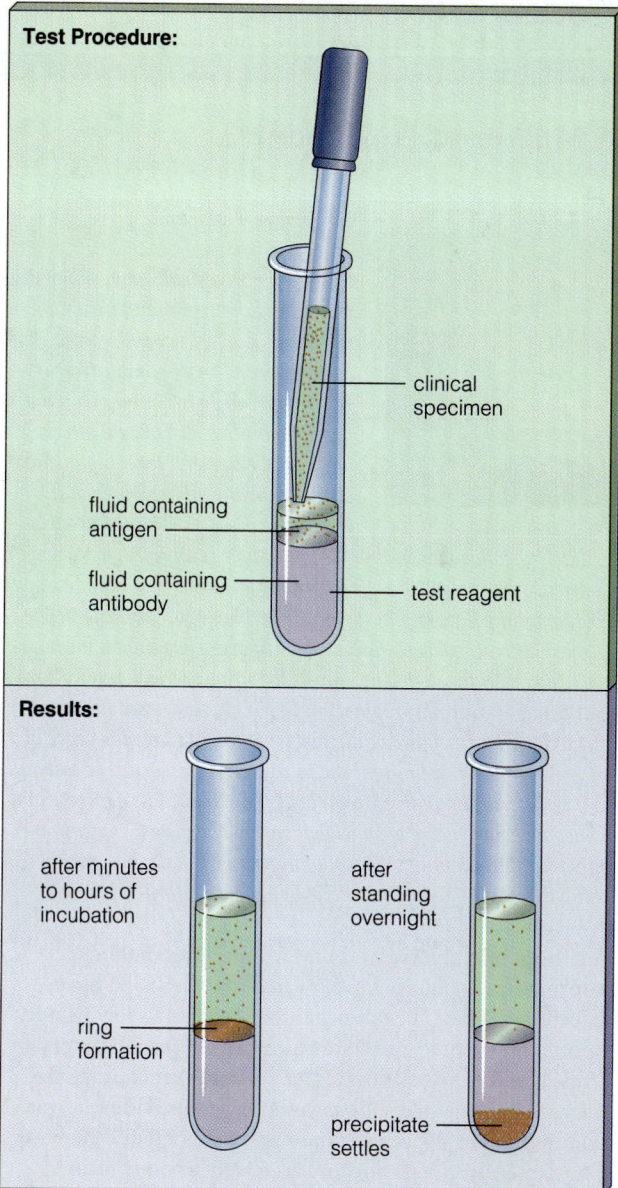

PRECIPITIN RING TEST

Test Procedure:

clinical specimen

fluid containing antigen

fluid containing antibody

test reagent

Results:

after minutes to hours of incubation

after standing overnight

ring formation

precipitate settles

Figure 19.2 Precipitin ring test. In the precipitin ring test, antibody- and antigen-containing liquids are mixed in a test tube. If the clinical specimen contains an antigen or antibody that combines with the antibody or antigen in the test reagent, antigen-antibody lattices form a visible ring where antigen and antibody meet in optimal concentrations. After standing overnight, the precipitate settles to the bottom of the tube.

Immunodiffusion techniques range from simple and straightforward to complicated. In the simplest of these tests, called **single immunodiffusion tests**, the test gel contains a single antibody, and clinical specimens suspected to contain a matching antigen are tested against

A Matter of Judgment

Cynthia A. Needham

I began life in a small town in north-west Oklahoma, growing up amidst the wheat fields. I never dreamed that I would spend most of my career in Boston directing the clinical microbiology laboratory at Lahey Clinic Medical Center and teaching medical students about microbiology and infectious diseases. Thanks to some wonderful college instructors, I fell in love with science. With their encouragement, I went to graduate school, and then to a postdoctoral fellowship in medical microbiology. Working in the clinical laboratory over the last twenty years has shown me how scientific breakthroughs are translated into practical applications that benefit us all. Throughout my career, I've worked hard and put in long hours, but I've always tried to maintain the sense of adventure that first brought me to Boston. I've learned to sail and to maintain and repair my own boat. I've learned to navigate and someday I plan to sail around the world. The prairie never prepared me for becoming captain of my own sailboat, but maybe being director of a clinical microbiology laboratory did.

A clinical microbiology laboratory provides diagnostic information to help physicians treat patients who have infections. When I began my career, we had to rely heavily upon our ability to cultivate pathogens in the laboratory. Unfortunately we couldn't (and still can't) encourage the microorganisms to divide fast enough to provide information in time for the physician to make therapeutic choices. Consequently, the physician had to rely upon clinical judgment in order to begin treatment quickly. Results from the microbiology laboratory became available later; they merely confirmed or contradicted the physician's decision.

Today, by taking advantage of immunochemistry and recombinant DNA technology, we can detect and identify microorganisms in time to guide the physician's decision about treatment. But many different technologies are now available. One of the greatest challenges for the director of a clinical microbiology laboratory is to choose the best technology for a particular use while keeping the cost as low as possible. As an example, let's consider a clinical trial we did at the Lahey Clinic Medical Center in Boston to evaluate technology that aids the diagnosis, treatment, and management of pharyngitis.

Pharyngitis is among the most common out-patient complaints. It accounts for approximately 30 to 40 million physician-visits each year in the United States. The majority of these cases are caused by viruses; only 10 to 30 percent are caused by the bacterium *Streptococcus pyogenes* (commonly called "strep throat"). The distinction is important because

viral and bacterial pharyngitis call for different treatment. Bacterial pharyngitis requires treatment with antibiotics to shorten symptoms, to prevent spreading the disease to others, and, most importantly, to prevent later development of a serious condition—acute rheumatic fever. In contrast, treating viral pharyngitis with antibiotics yields no benefit and exposes the patient unnecessarily to an adverse drug reaction. The decision to treat or not treat a pharyngitis patient with antibiotics is an important one.

Unfortunately, there are no reliable clinical criteria that allow a physician to differentiate between pharyngitis caused by *S. pyogenes* or something else. Typically physicians can correctly identify only about 50 percent of the cases caused by this pathogen, and they attribute the same cause to 25 to 30 percent of the cases actually resulting from something else. The laboratory can distinguish reliably between the two either by throat culture or by one of the new technologies. Culture is too slow, requiring a minimum of 24 hours. There are over 30 different devices available and licensed by the FDA for the direct and rapid detection of *S. pyogenes*. The methods, which rely upon immunoassay or genetic probes, yield results within 5 to 10 minutes. These new methods are the product of considerable basic research; they correctly identify over 90 percent of the patients infected with *S. pyogenes* and falsely identify as infected less than 5 percent of the patients who are, in fact, not infected. This level of reliability is considerably better than physician judgment.

On the other hand, most of these tests are 2 to 4 times more costly than cultures, and they are neither as sensitive nor as specific. Before introducing a new technology, it is important to know whether the timely information it yields will change the manner in which physicians manage their patients. Will the physician rely on the results and thereby decrease the number of patients receiving antibiotics from the number based only on clinical judgment? If the answer is no, the test has no value, only additional cost.

To answer this question, we set up a clinical trial. We discovered that physicians relied less on clinical judgment with the introduction of a rapid test for *S. pyogenes* and were able to choose an appropriate therapeutic strategy for a significant percentage of patients at the time of the visit. Patients who tested positive were treated. This change eliminated the need for follow-up visits for almost all patients with positive throat cultures. Physicians did choose to prescribe antibiotics for some patients when the test was negative, but the unnecessary use of antibiotics was dramatically decreased. With or without the test, all patients eventually received appropriate therapy. Still, the decision to provide regular use of the new test is a matter of judgment. Do you think the increase in direct cost for testing for this common disease was justified by improved clinical practice? If you were a laboratory director, would you provide the rapid test?

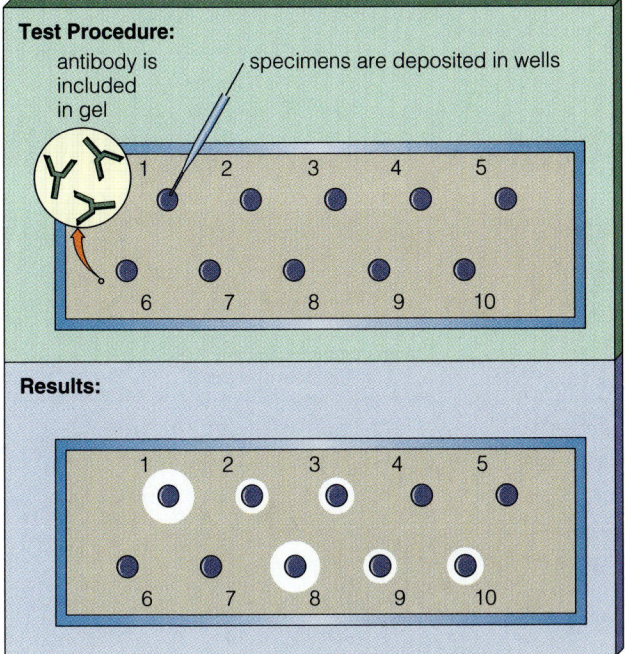

Figure 19.3 In this gel-based single immunodiffusion precipitation assay, a single type of antibody is included in the gel and multiple wells are filled with clinical specimens that might contain matching antigen. A circular band of visible precipitate will form around each well that contains matching antigen. The more antigen present, the larger the band. In this assay, for example, samples 1 and 8 had relatively large concentrations of antigen, samples 2, 3, 9, and 10 had lower concentrations, and samples 4, 5, 6, and 7 had no detectable antigen.

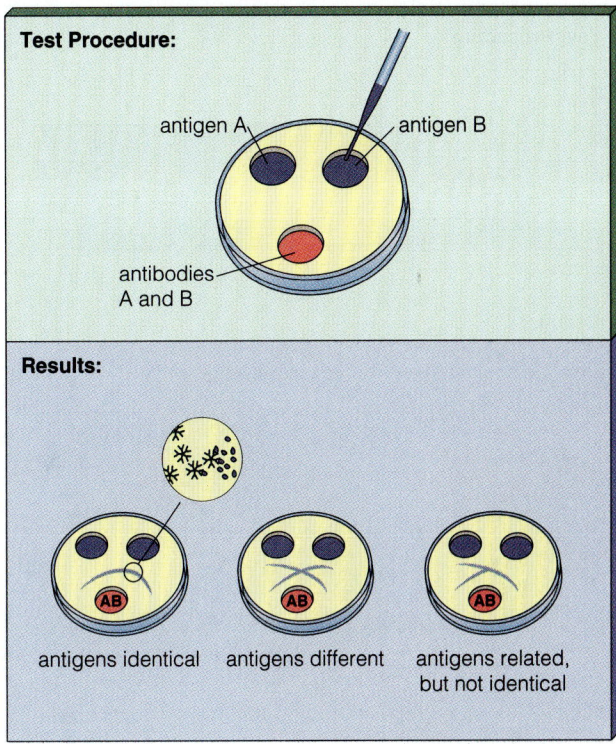

Figure 19.4 In this gel-based immunodiffusion precipitation assay, two types of antigens and antibodies are placed in nearby wells (holes cut in agar) and allowed to diffuse toward each other. In this case, the two antibodies were put in one well and the two antigens to be tested were put in separate wells. Antigens and antibodies diffuse toward each other, forming a line of precipitation where they meet in optimal proportions. Identical antigens form a single continuous line; completely different antigens form two crossing lines; related but not identical antigens form a line with a "spur" or branch.

that antibody. Circular, or radial (not a complete circle), lines of precipitate form around the wells that contain matching antigens (**Figure 19.3**). The advantage of single immunodiffusion tests is that they can be interpreted quantitatively. The larger the ring or the greater the number of radial lines, the higher the concentration of antigen in the test specimen. The more complicated tests are called **double immunodiffusion tests**. Different combinations of antigens and antibodies are placed in separate wells. Visible lines of precipitate form where matching antigen and antibody meet. Characteristic patterns reveal the number and identity of antigens or antibodies in a specimen (**Figure 19.4**). The advantage of these tests is that they can be used to differentiate between closely related antigens. For example, immunodiffusion assays can be used to diagnose histoplasmosis, a respiratory infection caused by a fungus (Chapter 22), and determine whether it is an active or a chronic infection. An immunodiffusion assay of a patient's serum shows whether it contains antibodies against one or both of two similar antigens that the fungus produces. The presence of both

antibodies indicates an active infection; the presence of only one of them indicates a chronic infection or a very early active infection.

A slightly different type of immunodiffusion assay also performed in gels is the **immunoelectrophoresis assay**. In these assays, the movement of antigen and antibody through the gel is accelerated by an electric current. Electrophoresis has two advantages. It speeds up movement, producing test results more rapidly than is possible with simple diffusion, and a complex mixture of antigen or antibody that may not separate completely into distinct precipitin bands with simple diffusion will do so with immunoelectrophoresis.

Countercurrent immunoelectrophoresis (CIE) rapidly tests a specimen for the presence of a single antigen or antibody (**Figure 19.5**). It is often used in the early diagnosis of life-threatening bacterial infections such as bacterial

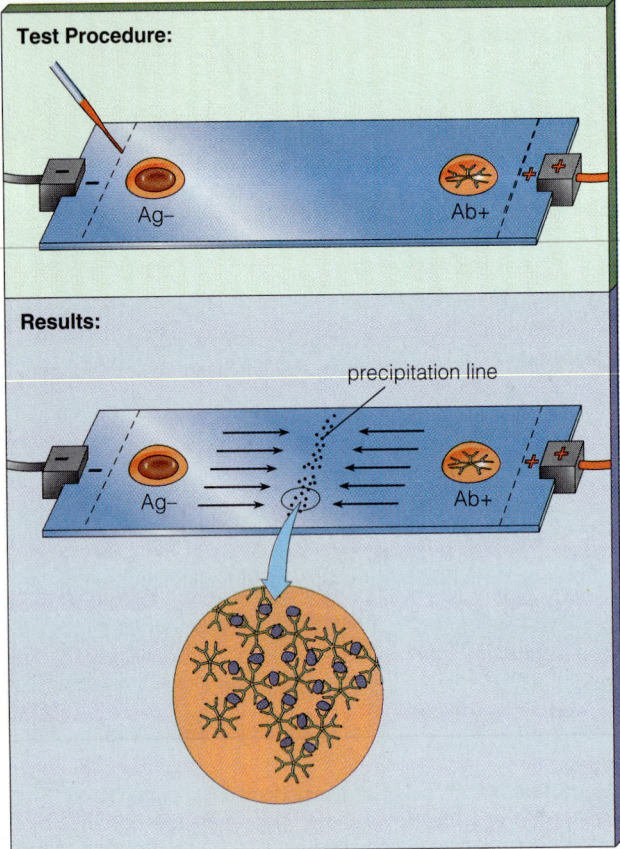

Test Procedure:

Results:

precipitation line

Ag−   Ab+

Figure 19.5   In a countercurrent immunoelectrophoresis (CIE) assay, antigen (Ag) and antibody (Ab) are placed at opposite ends of an agar slab. An electric current is applied, and antigen and antibody move toward each other. If a precipitation line forms somewhere in the middle of the slab, the test is positive and the test specimen must contain the antigen or antibody in question. This test is valuable in clinical diagnosis because results are available within an hour.

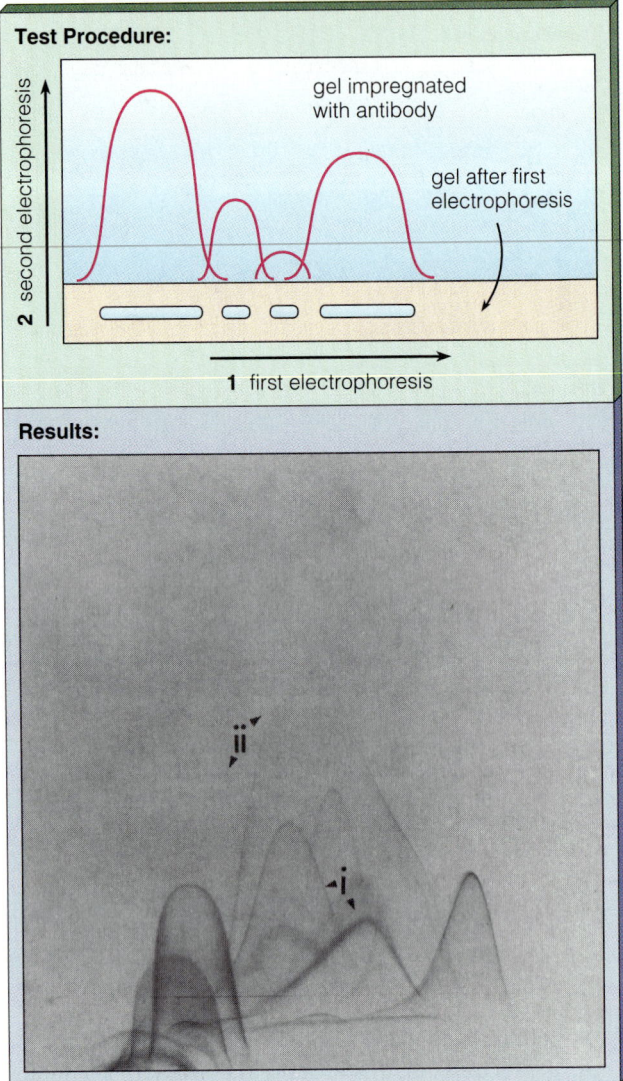

Test Procedure:

gel impregnated with antibody

gel after first electrophoresis

2 second electrophoresis

1 first electrophoresis

Results:

Figure 19.6   In two-dimensional immunoelectrophoresis assays, an electric current is applied to separate the different proteins in a complex mixture (direction 1 in the drawing). After separation has been achieved, the orientation of the current is changed. Now proteins are separated along a second axis (direction 2). The photo shows a precipitin gel obtained as the result of two-dimensional immunoelectrophoresis. Each of the precipitin arcs represents a different protein present in the original mixture.

meningitis (Chapter 25). A specimen of cerebrospinal fluid or urine, suspected to contain bacterial antigens, is placed on one end of a gel slab. Antibody against a known bacterial antigen is placed on the other end. When a current is applied, antigen and antibody each migrate to the opposite ends of the slab. If antigen-antibody binding occurs, a zone of precipitation becomes visible somewhere in the middle of the slab, indicating a positive test. If precipitation occurs when the specimen is combined with a known preparation of antibodies against *Haemophilus influenzae*, for example, the patient probably has an *H. influenzae* infection.

Two-dimensional electrophoresis is particularly effective in separating complex mixtures of antigen or antibody. The mixture is electrophoretically separated in one direction, and then the orientation of the current is changed to separate it in the opposite direction. The number and location of precipitin arcs tell the number and identity of the different proteins present in the mixture (**Figure 19.6**). This relatively complex test has more applications in research than in clinical medicine.

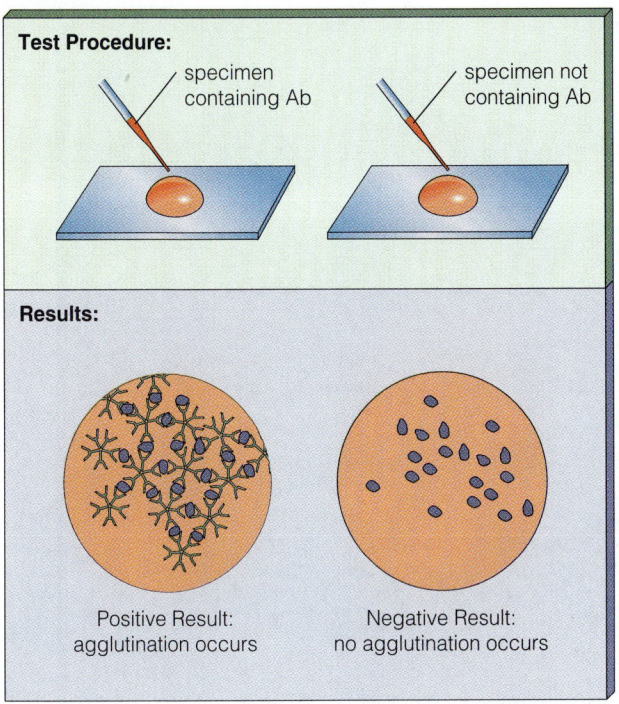

Figure 19.7 Qualitative agglutination assays determine the presence or absence of antibody (Ab) or antigen (Ag) in a test specimen. If agglutination occurs, the mixture of specimen and test reagent forms clumps and the test is positive. If no agglutination occurs, the test is negative.

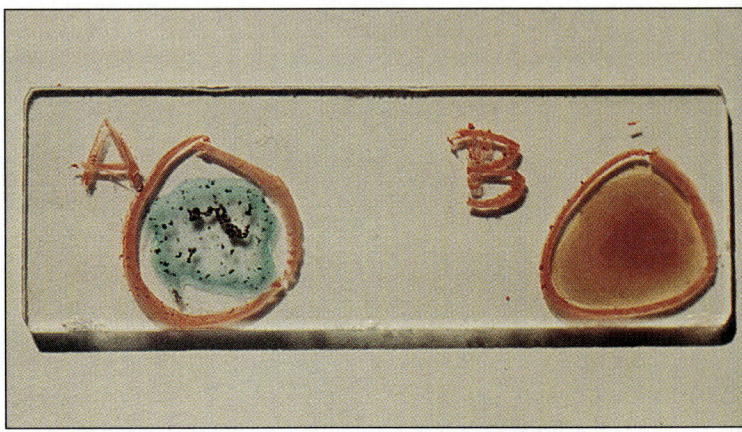

Figure 19.8 Hemagglutination reactions are used to type blood. In this example, the patient's antigen-bearing red blood cells were combined with two test reagents, one containing antibodies against antigen A, and the other containing antibodies against antigen B. Agglutination occurred with the anti-A reagent, but not anti-B, which means the patient has type A blood.

Agglutination Reactions

Agglutination reactions are similar to precipitation reactions because they combine antigen and antibody molecules into a visible, easily detectable network of molecules. In **agglutination reactions**, the antigen or antibody is attached to a large particle, such as a cell or a latex bead. Therefore, agglutination reactions produce a visible clump composed of antigen, antibody, and attached particles. Agglutination reactions are a classic serologic technique, some of these assays having been in use since the late 1800s.

Agglutination reactions can be either direct or indirect. A **direct agglutination reaction** involves antigens or antibodies that are naturally a part of a larger particle, such as a microorganism or an erythrocyte. If the particle is an erythrocyte, the reaction is called **hemagglutination**. An **indirect agglutination reaction** involves antigens or antibodies that are artificially adsorbed onto a particle such as a latex bead.

Agglutination assays are extremely sensitive, detecting antibody at concentrations as low as 1 microgram/ml. Many are designed to diagnose present or past microbial infection by identifying serum antibodies produced in response to that infection. The test reagent contains known microbial antigens. If the antigens in the test reagent agglutinate with antibodies in the patient's serum, the test is positive—the person has probably been infected by the microorganism in question, either recently or sometime in the past.

Qualitative agglutination reactions are usually performed on a slide and are easy to do. A drop of the clinical specimen is combined with a drop of the test reagent. If agglutination occurs, the mixture develops clumps and the test is positive, indicating presence of antibodies. If no agglutination occurs, clumps fail to form and the test is negative (**Figure 19.7**).

Qualitative slide agglutination tests are widely used in clinical medicine. For example, an indirect latex agglutination slide test is standard for diagnosing strep throat. Material swabbed from the patient's throat is obtained and mixed with a test reagent containing particle-bound antibodies against the strep bacterium. Before monoclonal antibody technology became cost-effective, the test for pregnancy was based on agglutination. Blood type is also determined by a slide agglutination test. Test reagents containing antibodies against the A, B, and Rh red blood cell antigens are combined with a patient's blood cells. A positive result is indicated by hemagglutination (**Figure 19.8**).

Quantitative agglutination tests usually involve progressive dilutions of a clinical sample until agglutination with the test reagent no longer occurs. Results are usually expressed as a **titer**—the highest dilution of a test serum

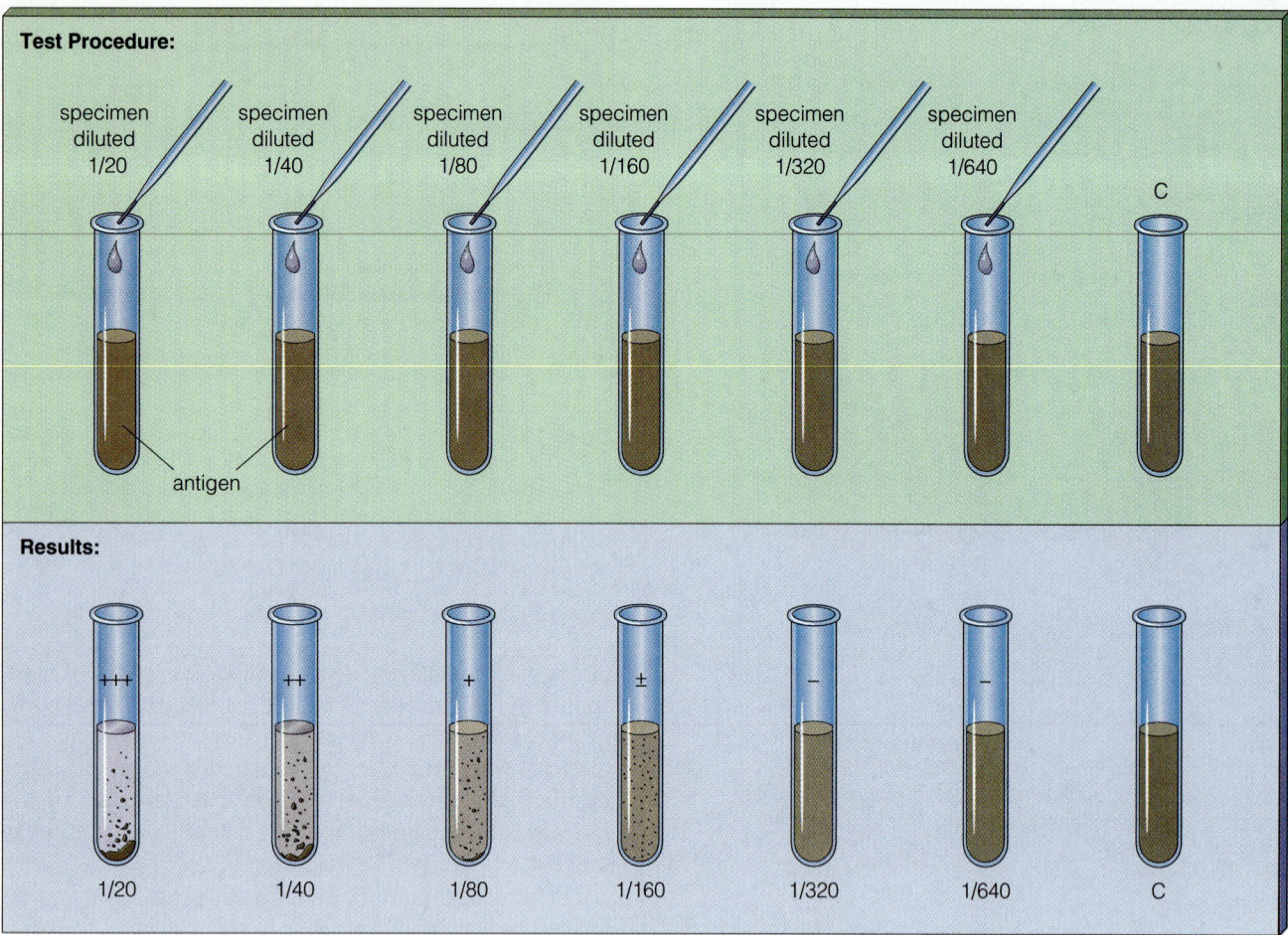

Figure 19.9 Quantitative agglutination reactions to determine antibody titer. Each tube contains an equal amount of a certain type of microbial antigen bound to a carrier particle and suspended in a salt solution. A serum specimen taken from a patient suspected of being infected by that particular microorganism is prepared in progressively more dilute suspensions and added to each test tube. If antigen-antibody binding occurs, the tube will show agglutinated clumps of antigen, antibody, and carrier particles, and the test is scored as positive. If no clumps appear, it is assumed that no agglutination has taken place, and the test is scored as negative. (A control, or C, tube is included to which no specimen is added; it is a known negative.) The antibody titer, a quantitative measure of antibody concentration, is the highest dilution at which a positive reaction occurs. Strengths of reactions are scored by pluses and minuses: +++, very strong; ++, strong; +, weak; ±, probably positive; –, negative. Here the antibody titer is 1:160.

that gives a positive agglutination response. In a typical quantitative agglutination test, serial dilutions of an antibody-containing specimen are combined with a standard test antigen. The dilutions are usually twofold, meaning that each tube contains half as much of the original specimen as the previous one. The last dilution at which agglutination still occurs, for example 1:160, is the antibody titer (**Figure 19.9**). Direct and indirect agglutination tests are used to diagnose many different infections.

Agglutination reactions to detect antibodies produced in response to microbial infection have many applications. For example, if no antibodies are detected at any dilution of the specimen, that person has never been infected by the microorganism in question and remains susceptible to infection. A negative antibody test against rubella (German measles) in a woman planning to become pregnant, for example, indicates that she is susceptible to infection by the virus and should be immunized.

Harvesting Antibodies

Every B lymphocyte and its clone cells produce antibody against only one epitope (Chapter 17). Any single B-cell clone, therefore, produces a uniform **monoclonal antibody**, antibody molecules with identical antigen specificities. When the immune system reacts to an infection, however, several thousand different B-cell clones produce **polyclonal antibodies**. These antibodies react with many different epitopes on invading microorganisms, providing the immune system with a wide array of infection-fighting weapons.

A complex mixture of polyclonal antibodies is essential for combating microbial infection, but pure monoclonal antibodies with known specificities have immense practical value to clinicians and scientists. Unfortunately, normal B cells do not multiply or produce antibody under laboratory conditions, making production of monoclonal antibodies a technological challenge.

The first to meet this challenge were César Milstein and Georges Köhler of the Medical Research Council Laboratory in Cambridge, England. They received the Nobel Prize in 1984 for their accomplishment. They fused one line of antibody-producing B cells from a mouse with a second line of cancer cells to produce a **hybridoma cell**, a cell that divides indefinitely in tissue culture.

The hybridoma procedure has almost limitless applica-

tions, but it is complex and cumbersome. Hybridoma cells have to be grown for several months in vitro and in mice. Then cells that produce the desired monoclonal antibody must be selected from a mixture that includes many cells producing other antibodies. Moreover, mouse cells cannot produce the human antibodies that are essential for certain therapeutic applications.

Recently, recombinant DNA technology (Chapter 7) has been used to produce monoclonal antibodies. Human antibody genes are packaged in bacteriophages and introduced into *Escherichia coli*, which then produces human monoclonal antibodies. The recombinant DNA procedure has many advantages over the hybridoma procedure. It is simpler, faster, and less expensive. A person's or animal's entire antibody repertoire can be genetically engineered into a culture of bacteria. Then the particular bacterial cells that produce the desired monoclonal antibody can be identified and isolated. The antibodies produced by this procedure can be improved: the cloned genes can be extracted, modified by mutation, and reintroduced into the antibody-producing bacteria to create an antibody with an even greater specificity. This promising new technique will probably replace the hybridoma technique in the near future.

But if the test shows a positive antibody titer, the virus cannot reinfect her and will not harm her fetus (Chapter 26). Finally, a series of tests over time showing a rising antibody titer or conversion from a negative to a positive titer indicates a recent infection. Being able to diagnose a recent or ongoing infection by means of a change in antibody titer is helpful where the causative microorganism is dangerous to cultivate in the clinical laboratory, for example, tularemia (Chapter 27). However, it takes up to six weeks for antibody titers to rise, and so these tests may not help in making decisions about treatment.

Complement Fixation Reactions

Antibodies bound to antigen can activate complement and initiate the formation of a membrane attack complex that will lyse a target cell (Chapter 16). **Complement fixation assays** take advantage of this phenomenon to detect antibody present in a clinical specimen. Moreover, complement fixation assays are quantitative, which means

that the antibody concentration can also be determined. Like agglutination assays, complement fixation assays are extremely sensitive.

Complement fixation assays entail a number of steps. First, a clinical specimen suspected to contain a certain antibody is combined with a test reagent that contains the corresponding antigen. If antibody is present, antigen-antibody complexes form. Next, a measured amount of complement is added. If antigen-antibody complexes are present, they fix and consume the complement. But if no complexes are formed, free complement remains in the mixture. Finally, this mixture of antigen, antibody, and complement is combined with an indicator solution containing erythrocytes coated with an unrelated antibody. If all the free complement has been fixed in the first two steps of the assay, no complement is available to bind to the target cells, and they fail to lyse. If free complement remains, however, it binds to the antibody-coated target cells and lyses them. This destruction of red cells can be scored visually. The assay is quantitative because the

COMPLEMENT FIXATION ASSAY

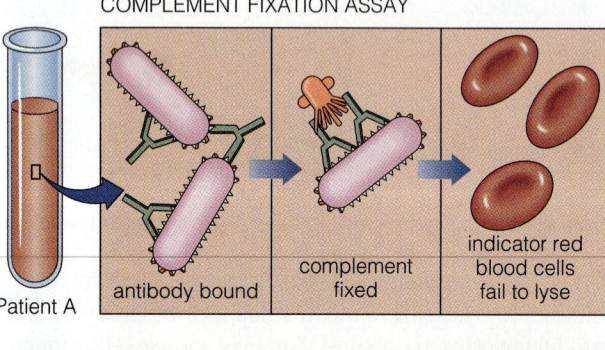

Patient A — antibody bound | complement fixed | indicator red blood cells fail to lyse

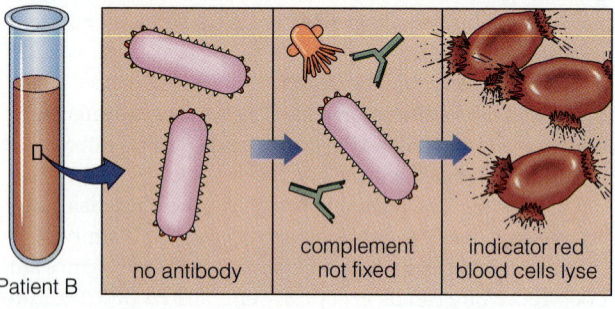

Patient B — no antibody | complement not fixed | indicator red blood cells lyse

Figure 19.10 Complement fixation assay. Here, patient A has a positive complement fixation test, indicating that her serum contains antibodies against the microorganism being tested for. Patient B, on the other hand, has a negative test, indicating the absence of antibodies against the test microorganism.

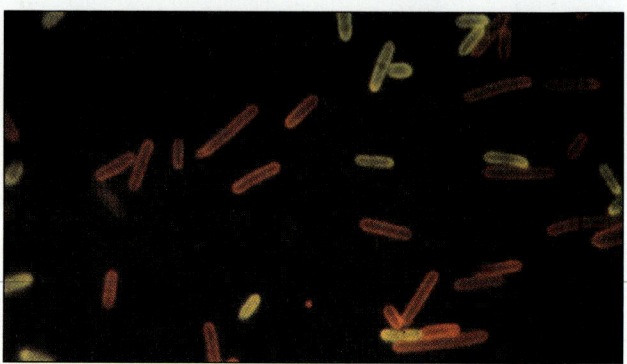

Figure 19.11 In immunofluorescence assays, antibodies are tagged with fluorescent dyes that make them easy to identify visually. Two species of rod-shaped bacteria are present in this sample. Antibodies against one species are tagged with a dye called isothiocyanate, which fluoresces green. Antibodies against the other species are tagged with a dye called rhodamine B, which fluoresces red. As a result, all the cells belonging to one species glow green, while cells belonging to the other species glow red.

amount of complement added to the test reagent is carefully measured and the degree of lysis can also be measured, which means the concentration of antibody in the original specimen can be determined (**Figure 19.10**).

Because several steps are involved, complement fixation assays are tedious and complicated to perform and can take at least 48 hours to complete. Clinicians have simpler and quicker tests. However, complement fixation is still used to diagnose respiratory syncytial virus, and the test has taken on increased importance now that antiviral therapy is available to treat this life-threatening respiratory infection of infants (Chapter 22). Complement fixation assays are also used to diagnose influenza A and B, though rarely for clinical purposes. Rather, epidemiology labs that chart the yearly progress of this common infection use it (Chapter 22). Other rarer infections diagnosed by means of complement fixation assays include Q fever (Chapter 22) and fungal infections such as histoplasmosis, blastomycosis, and coccidioidomycosis (Chapter 22).

Immunofluorescence Reactions

In **immunofluorescence assays** antibodies are tagged with fluorescent dyes that make them easy to identify visually. The color of the tagged antibodies is determined by the dye used. Fluorescent antibodies can be used to detect antigens or antibodies within tissues or on cells.

Immunofluorescence tests can be direct or indirect. Direct tests employ fluorescent dye-tagged antibodies that bind to the antigens under study and make them visible under ultraviolet light. **Figure 19.11** shows how two bacterial species are identified by a direct immunofluorescence test.

Indirect tests are more complicated because they are designed to detect antibody rather than antigen. A fluorescent antibody must be prepared that binds to the antibody under study. An antibody that binds to another antibody—called an **anti-antibody**—may seem paradoxical, but it is not. Antibodies are complex proteins, and they possess epitopes that stimulate the formation of antibodies against themselves. Test reagents for indirect immunofluorescence assays, therefore, contain a fluorescent antibody that binds to, and illuminates, the type of antibody under identification in a clinical specimen.

Indirect immunofluorescence assays are widely used and extremely valuable in clinical medicine. They are used for the serologic diagnosis of toxoplasmosis (Chapter 27), Legionnaires' disease (Chapter 22), varicella zoster (chickenpox and shingles, Chapter 26), syphilis (Chapter 24), and many other infections.

Radioimmunoassays

In a **radioimmunoassay**, the test reagent contains an antibody that is tagged with a radioactive molecule rather than a fluorescent one. Radioimmunoassays can detect extremely small quantities of antigen or antibody because the decay of even a single radioactive atom can be measured by modern instruments. These tests are performed

Clinical Notes

Putting the Immune System in a Bottle

Monoclonal antibody preparations are useful because of what antibodies can do—recognize and bind to one specific type of antigen molecule in a sea of closely related antigens. This remarkable specificity has many clinical applications. Monoclonal antibodies tagged with radioactive or fluorescent markers can detect extremely small quantities of a substance in the body, such as an antibody, microorganism, or drug. For example, digitalis, a heart medication, is used in such extremely small doses that, until recently, it was impossible to monitor serum drug levels. Now, with the help of a monoclonal-based drug assay, doctors can check digitalis levels of heart patients and fine-tune dosage of this potentially toxic medication.

Monoclonal antibodies are also used in medical research, which also often depends upon identifying and precisely measuring small quantities of a specific biological substance. For example, monoclonal antibodies were the key to unraveling the complexity of the MHC antigen system (Chapter 17). Human populations carry many different MHC genes, which code for similar but significantly different MHC antigens. Determining which of these antigens were identical to one another and which were ever-so-slightly different required tests based on the specificity of monoclonal antibodies.

Monoclonal antibodies may someday play a significant role in medical treatment, especially of cancer. The problem with most cancer treatments is lack of specificity—normal cells are destroyed along with the cancer cells. Monoclonal antibodies directed against tumor antigens could be coupled with lethal drugs or radioactive substances, possibly creating the magic bullet that cancer therapy has sought for so long. This ideal form of cancer treatment remains a dream of the future, but the first clinical trials have already begun.

The ease with which monoclonal antibodies can be produced through genetic engineering may further extend their use in research, diagnosis, and treatment. According to Richard Lerner from Scripps Clinic in La Jolla, California, one of the developers of the new recombinant DNA technique for manufacturing monoclonal antibodies, the new technology "opens the way to doing in a bottle exactly what the immune system would do."

---

by machine and often in a chemistry laboratory rather than a serology laboratory.

Direct radioimmunoassays are usually used to detect nonmicrobial antigens, such as hormones, in human serum. Indirect radioimmunoassays are sometimes used to detect antimicrobial antibodies in human serum. For example, radioimmunoassays are occasionally used to detect antibodies against the hepatitis B virus in the serum of people who have had this infection. Indirect radioimmunoassays employ anti-antibodies tagged with a radioactive compound. Most indirect assays, including immunofluorescence assays, radioimmunoassays, and enzyme-linked immunosorbent assays (see below), are performed by means of a sandwich technique. The antibody under study is sandwiched between its antigen and a radioactively labeled anti-antibody. The technique is described in **Figure 19.12**.

Enzyme-Linked Immunosorbent Assays (ELISA)

Enzyme-linked immunosorbent assays (ELISA) are similar to immunofluorescence and radioimmunoassays, except that in these assays the indicator antibodies are linked to an enzyme instead of a fluorescent or radioactive molecule. The enzymes serve as markers because, when combined with their substrate, they produce a visible color change. Enzyme-linked assays are preferable to radioactive assays because they can be scored visually and do not require the elaborate instruments necessary to count radioactive emissions. ELISA reagents are usually fixed to a surface and combined in a sandwich technique similar to an indirect radioimmunoassay. A direct ELISA detects antigens, while an indirect ELISA detects antibodies. The principles behind performing both are the same.

The ELISA is widely used in clinical diagnosis, and its applications are increasing. The advantage of ELISA is that it can be performed easily on many serum samples at the same time. The disadvantage is that assays are not quantitative. The pregnancy test discussed in the opening of this chapter is an ELISA. ELISA is also used to diagnose viral hepatitis (Chapter 23), herpes simplex infections (Chapters 24 and 26), and rubella virus (Chapter 26), among others. ELISA is also used to diagnose rotavirus infection, a common cause of childhood diarrhea (Chapter 23). But since there is no treatment for this infection, tests are seldom done to diagnose individual children. Rather, they are used in epidemiologic studies.

INDIRECT RADIOIMMUNOASSAY TEST

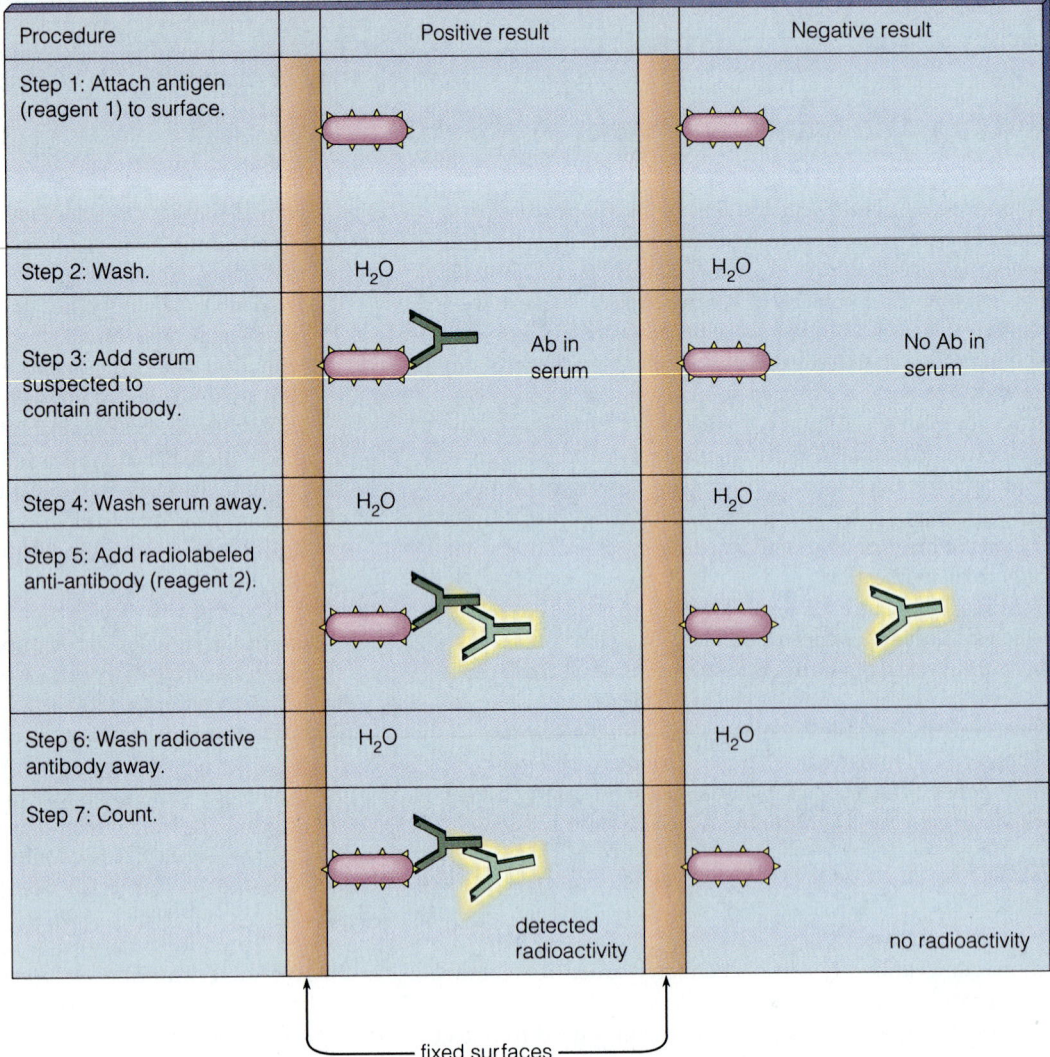

Procedure		Positive result		Negative result
Step 1: Attach antigen (reagent 1) to surface.				
Step 2: Wash.		H₂O		H₂O
Step 3: Add serum suspected to contain antibody.		Ab in serum		No Ab in serum
Step 4: Wash serum away.		H₂O		H₂O
Step 5: Add radiolabeled anti-antibody (reagent 2).				
Step 6: Wash radioactive antibody away.		H₂O		H₂O
Step 7: Count.		detected radioactivity		no radioactivity

fixed surfaces

Figure 19.12 Indirect radioimmunoassays test for the presence of a particular antimicrobial antibody. They are usually performed by attaching one of the test reagents to a fixed surface. This technique is sometimes called a sandwich technique because the antibody under study is sandwiched between its antigen and a radioactively labeled anti-antibody. In a series of seven steps, two different reagents and a clinical specimen are added. Between each step, the surface is washed to remove any unbound reagents. In step 1, an antigen known to bind to the antibody under study—for example, a microbial antigen—is fixed to a surface. In the third step, the serum specimen is added to this surface. If the patient has been infected by that microorganism and has antibodies against it, the antibodies will bind to their antigen. In the fifth step, radioactively labeled anti-antibodies directed against human antibodies are added as an indicator. If antimicrobial antigens have bound to the test surface, the radioactive anti-antibodies will also bind to the surface, where they can be detected by a radioactive counter.

Summary

SEROLOGY   (pp. 452–453)

1. The immunologic reactions that defend against infection can be used to diagnose many disorders.

2. Serology is the branch of immunology that applies antigen-antibody interactions to medical diagnosis. The term *serology* comes from *serum*—the cell-free liquid component of blood—because many serologic tests detect antibodies in serum.

3. Serologic tests (assays) identify unknown agents in clinical specimens usually by mixing a specimen (such as blood or tissue) with a test reagent that contains known concentrations of antigens or antibodies.

4. Serologic assays detect a particular type of antigen or antibody, allowing the clinician to diagnose a present or past infection or an autoimmune disorder.

5. Qualitative serologic tests reveal the presence or absence of an antigen or antibody; quantitative tests measure the concentration of an antigen or antibody. Quantitative tests allow a clinician to monitor the progress of a disease.

SEROLOGIC TESTS (pp. 453–462)

1. There are six types of serologic tests—precipitation, agglutination, complement fixation, immunofluorescence, radioimmunoassays, and enzyme-linked immunosorbent assays (ELISA).

Precipitation Reactions (pp. 453–456)

1. Precipitation reactions depend on the tendency of some antigens and antibodies to form lattices, huge interlocking macromolecular webs that become visible as they precipitate, forming cloudy regions in the test mixture. For a precipitate to form, antigen and antibody must be present in approximately equal concentrations.

2. A precipitin ring test, used more in research than in the clinical laboratory, is performed in a liquid. An antibody-containing liquid is added to a test tube with an antigen-containing liquid. A cloudy zone of precipitation forms where the two meet in optimal proportions.

3. An immunodiffusion assay is performed in a gel. Antibodies and antigens are placed in wells and molecules diffuse toward one another, forming a precipitate where they meet.

4. In a single immunodiffusion test, one well contains an antibody and others contain clinical specimens. Circles form around the wells that contain matching antigens. The thicker the ring, the higher the concentration of antigen. Thus, the results are quantitative.

5. In a double immunodiffusion test, different antigens and combinations of antibodies are placed in separate wells. They diffuse, and the patterns of precipitate reveal the number and identity of antigens or antibodies in the specimen.

6. In immunoelectrophoresis assays, electric current is applied to speed diffusion through the gel.

7. In countercurrent immunoelectrophoresis (CIE), antibody and antigen are placed on opposite ends of a slab of gel. When current is applied, they migrate toward the opposite ends of the slab, forming a precipitate where complementary pairs meet.

8. Two-dimensional electrophoresis separates complex mixtures of antigen or antibody. An electric current is applied in one direction, then in another. The resulting precipitin arcs indicate the number and identity of proteins present. This test is used more in research than in the clinical laboratory.

Agglutination Reactions (pp. 456–459)

1. Agglutination reactions produce a visible clump of antigen, antibody, and attached particles. Direct agglutination reactions involve antigens or antibodies that are part of a particle, such as a microorganism or erythrocyte. If the particle is an erythrocyte, the reaction is called hemagglutination.

2. Indirect agglutination reactions involve antigens or antibodies adsorbed onto a particle such as a latex bead.

3. Agglutination assays are very sensitive.

4. Qualitative agglutination tests are performed on a slide. A drop of the test specimen is added to a drop of reagent containing microbial antigens. If clumping occurs, the test is positive.

5. The qualitative slide agglutination test is used widely for diagnosing strep throat, determining pregnancy, and typing blood.

6. Quantitative agglutination tests involve progressive dilutions of a clinical sample until agglutination with the reagent no longer occurs. Results are expressed as a titer—the highest dilution of a test serum that gives a positive agglutination response.

7. Agglutination reactions to detect antibodies in response to bacterial infection have many applications. A pregnant woman with a negative antibody test against rubella would be at great risk without immunization.

Complement Fixation Reactions (pp. 459–460)

1. Complement fixation assays are based on antibodies binding to antigens and activating complement to initiate the formation of a membrane attack complex that will lyse a target cell.

2. Complement fixation assays are quantitative: they can measure antibody concentration.

3. Complement fixation is complicated and takes longer than other serologic tests, but it is used to diagnose certain infections, including infant respiratory syncytial virus, Q fever, and certain fungal infections.

Immunofluorescence Reactions (p. 460)

1. In immunofluorescence assays, antibodies are tagged with fluorescent dyes that make them visible under ultraviolet light.

2. Direct immunofluorescence tests tag antibodies.

3. Indirect immunofluorescence tests require an anti-antibody—an antibody that binds to another antibody. Indirect tests are widely used clinically. This is the assay for syphilis and varicella zoster (chickenpox and shingles), among many other infections.

Radioimmunoassays (pp. 460–461)

1. In these assays, the test reagent antibody is tagged with a radioactive molecule. They are extremely sensitive tests.

2. Direct radioimmunoassays are usually used to detect nonmicrobial agents, such as hormones, in serum. Indirect radioimmunoassays are used to detect antimicrobial antibodies.

3. In indirect radioimmunoassays—and most other indirect techniques—the antibody is sandwiched between the antigen and an anti-antibody. The anti-antibody is radioactively labeled. The radioactive emissions are counted to give a quantitative result.

Enzyme-Linked Immunosorbent Assays (ELISA) (pp. 461–462)

1. In ELISA the anti-antibody is linked to an enzyme instead of a radioactive molecule. After the immunologic reaction, an enzyme substrate is added; the enzyme catalyzes a change in color.

2. ELISA can be scored visually and do not require complicated instruments, but they are not quantitative.

3. Clinical use of ELISA includes diagnosis of herpes simplex infections and rubella virus.

Review Questions

SEROLOGY

1. What is serology, and what is the basis of the term?

2. Define: serum, assays, test reagent.

3. What is the difference between a qualitative and quantitative serologic test?

SEROLOGIC TESTS

1. What is the basis of precipitation reactions?

2. Describe how you would perform each of these assays:
   a. precipitin ring test
   b. single immunodiffusion test
   c. double immunodiffusion test
   d. immunoelectrophoresis assay
   e. countercurrent immunoelectrophoresis (CIE)
   f. two-dimensional electrophoresis

3. What is the basis for agglutination reaction assays?

4. Describe the difference between direct and an indirect agglutination. Which type is hemagglutination?

5. Describe how you would do a qualitative slide agglutination test, and give some examples of its use.

6. How would you perform a quantitative agglutination test? Define titer.

7. Give an example of a practical use for agglutination reactions to detect antibodies in response to microbial infections.

8. What is the basis for a complement fixation reaction? Why is this a quantitative test?

9. Describe how you would perform a complement fixation assay and when you might want to use one.

10. What is the difference between a direct and an indirect immunofluorescence test?

11. What is an anti-antibody?

12. Give some examples of infections diagnosed by immunofluorescence reactions.

13. How is a radioimmunoassay like an immunofluorescence assay?

14. How are direct radioimmunoassays used? Describe the sandwich technique used in indirect radioimmunoassays.

15. Are radioimmunoassays quantitative or qualitative tests? Explain.

16. How is the enzyme-linked immunosorbent assay like immunofluorescence and radioimmunoassays? What is the advantage of ELISA over radioactive assays?

17. How would you perform an ELISA? Which infections might you use it to diagnose?

18. Is ELISA quantitative or qualitative? Explain.

Essay Questions

1. Some serologic tests have been in use a very long time while others are relatively new. Which of the tests described in this chapter fall into each category? Explain.

2. Explain Richard Lerner's comment that the new DNA technology for producing monoclonal antibodies is like putting the immune system in a bottle. Give some examples of applications that would improve human health.

Suggested Readings

Herrman, J. E. 1986. Enzyme-linked immunoassays for the detection of microbial antigens and their antibodies. *Advances in Applied Microbiology* 31:271.

Marx, J. 1989. Learning how to bottle the immune system. *Science* 246: 1250–51.

Rose, N. R.; E. C. deMacario; J. L. Fahey; H. Friedman; and G. M. Penn. 1992. *Manual of clinical laboratory immunology.* 4th ed. Washington, D.C.: American Society for Microbiology.

20 PREVENTING DISEASE: EPIDEMIOLOGY AND PUBLIC HEALTH

Cholera Epidemic on Broad Street

For centuries cholera has been a devastating epidemic disease. Its special horror is its suddenness. A healthy person can go to work in the morning, fall victim to the profuse, watery diarrhea of *Vibrio cholerae*, and be dead from dehydration and shock by nightfall.

Cholera was widespread in Europe in the 1800s and couldn't be controlled because no one knew what caused the outbreaks. So John Snow, an English physician, decided to take a more direct and practical approach. Instead of looking for the *cause* of the disease, he reasoned, why not find its *source* and eliminate it?

In 1854, cholera broke out in the London neighborhood where Broad Street joined Cambridge Street. Snow called it "the most terrible outbreak of cholera which ever occurred in this kingdom." During the first 10 days of September, more than 500 people died in an area 250 yards by 250 yards. Had people not begun fleeing the neighborhood, the number of deaths would have been even greater.

As the epidemic raged, Snow began a systematic investigation. He reviewed all the records of deaths registered to cholera, compiled information about each victim, and interviewed all the survivors who remained in the neighborhood. From what he learned, Snow made a map, which revealed that nearly every cholera death occurred in a house near the Broad Street water pump. Interviews with survivors further strengthened his theory that the Broad Street pump was the critical factor. Victims from houses nearer other pumps had used the Broad Street pump because they attended school nearby or because they preferred its water.

In some cases, information about those who survived provided equally valuable information. A workhouse in the heart of the Broad Street neighborhood reported only 5 cholera deaths among more than 500 inmates, and the employees of a nearby brewery were entirely spared. It turned out that the workhouse had its own water supply and did not take water from Broad Street, and the brewery employees never went to the pump for water because the proprietor allowed them to drink beer. The source of this epidemic, concluded Snow, was the water from the Broad Street pump.

From his investigations of previous epidemics, Snow had suspected that water contaminated by fecal matter

1866 sketch showing King Cholera pumping drinking water.

was the source of cholera. Now, even though the water from the Broad Street pump ran clear, Snow went to the local parish church Board of Guardians and convinced them to remove the handle of the Broad Street pump.

This ended the cholera epidemic at Broad Street, though London was visited by many similar outbreaks over the next decade. Even knowing that contaminated water was a source of cholera did not solve the problems caused by lack of public sanitation. Snow's work was a turning point, however. His method of drawing generalizations based on the study of many individual cases was a new approach to medical research and the beginning of the science of epidemiology.

To understand:

- How the science of epidemiology contributes to our understanding of disease

- How epidemiologists collect information and why statistics are central to epidemiology

- The types and uses of epidemiology—descriptive epidemiology, surveillance epidemiology, field epidemiology, and hospital epidemiology

- How public health organizations help prevent disease

- How diminishing reservoirs and interfering with disease transmission help prevent disease

- The types of vaccines and how active immunization helps prevent disease

THE SCIENCE OF EPIDEMIOLOGY

Epidemiology is the study of when and where diseases occur and how they are transmitted in human populations. The term **public health** refers to the development and implementation of plans to prevent and control disease. In other words, epidemiology generates the information needed to carry out effective public health programs.

Prevention of disease eliminates the suffering of human illness—treatment only decreases it. Prevention is also more economical. A single sewer system or immunization program can prevent thousands of illnesses. Programs of disease prevention, based on epidemiological studies, are largely responsible for the good health that most people in industrialized countries enjoy today.

The distinctive feature of an epidemiologic approach to disease control is its focus on large groups of people, or **populations**, rather than individuals. John Snow studied the population of the Broad Street neighborhood to learn how cholera victims were like, and different from, other members of the population. Snow's approach (studying a population) and method of reasoning (generalizing from large numbers of individual cases) characterize the science of epidemiology today. John Snow's conclusion about the source of the Broad Street epidemic seems obvious from his map, but it would have been almost impossible to draw this conclusion by studying patients one by one.

Epidemics are a pattern of disease transmission that affects many members of a population within a short time. Cholera, diphtheria, and polio are epidemic diseases. An epidemic that spreads worldwide is a **pandemic**. Today the world is experiencing an AIDS pandemic. Some diseases are **endemic**, always present in a population at about the same level; examples include chickenpox and gonorrhea. Other diseases, such as tetanus and trichinosis, are **sporadic**, occurring only occasionally in a population. Mass death due to infection is much less common today than in the past, but epidemics still occur. In 1992, South America experienced a cholera epidemic.

The word *epidemiology* was first used to describe the study of epidemic disease. But modern epidemiology addresses a wide variety of health-related questions. Epidemiological studies helped prove that pellagra, a fatal disease of neurological deterioration, is due to a vitamin deficiency. Epidemiology provided irrefutable evidence for the harmful effects of smoking and established that fluoride can prevent tooth decay. Epidemiological studies even help determine which screening tests should be included in routine medical checkups. In short, epidemiological research has led to major advances in medical knowledge (**Table 20.1**).

Epidemiology remains critical to understanding infectious disease. Epidemiologists helped solve the mystery of three "new" infectious diseases—legionellosis (Chapter 22), Lyme disease (Chapter 27), and AIDS (Chapter 27). In all three cases, a disease outbreak was identified but no infectious cause was obvious. Epidemiological investigations, however, identified the mode of transmission—inhaling contaminated air for legionellosis, a tick bite for Lyme disease, and sexual transmission or exposure to infected blood for AIDS. As John Snow demonstrated, knowing the sources showed how steps could be taken to prevent transmission even while the causes were still unknown. Eventually, microbiological research identified the infectious agents—bacteria for legionellosis and Lyme disease and a retrovirus for AIDS.

THE METHODS OF EPIDEMIOLOGY

Instead of test tubes and microscopes, epidemiologists use pencils, paper, and calculators to collect information about populations and to interpret it.

Sources of Information

Much useful information for epidemiologists is part of the public record—vital statistics, census data, and disease reports to public health authorities. Epidemiologists obtain additional data from questionnaires, surveys, and hospital records.

Vital statistics—records of births, deaths, and other human events such as marriages and divorces—are collected by almost all governments. Before the 1800s, when vital statistics were first systematically and accurately recorded, an epidemiologic approach to health problems was almost impossible. Snow's cholera study, for example, relied on death certificates listing cholera as the cause of death.

A **census** also generates useful information for epidemiologists. Knowing the number of people living in an area and their distribution by age, race, sex, occupation, national origin, marital status, and income is invaluable because diseases are often unevenly distributed among subpopulations. Tuberculosis, for example, is most common among the elderly, the poor, and recent immigrants, while coccidioidomycosis and blastomycosis are limited to people who live in certain geographic areas (Chapter 22). If Snow conducted his study today, census data could provide an accurate estimate and description of the Broad Street residents even before he began his interviews.

Government agencies also record all cases of certain diseases. Physicians who treat patients with these **notifiable diseases (reportable diseases)** are required to notify their local department of public health when the diagnosis is made (**Table 20.2**). Some of these diseases are of immediate public health concern and must be reported immediately by telephone. Others must be reported within one working day by phone or mail, and a third category must be reported within seven calendar days. A public health department needs to know about a case of diphtheria to stop a potential epidemic. Typhoid carriers must be registered to prohibit them from working as food handlers, which would spread the disease. Public health departments forward names of patients with Alzheimer's disease or disorders characterized by lapse of consciousness to state authorities responsible for issuing driver's licenses.

Regulations for disease reporting vary from state to state. Local public health statistics are forwarded to state agencies and the **Centers for Disease Control and Prevention (CDC)** in Atlanta, Georgia. The CDC maintains and compares reports from year to year, allowing epidemiologists to track both regional and seasonal disease patterns.

Epidemiologists at the CDC prepare the *Morbidity and Mortality Weekly Report (MMWR)*. This publication is a compilation of weekly and cumulative annual statistics on reportable diseases from around the United States. **Morbidity** refers to illness and disability, while **mortality** refers to death. In addition to disease statistics, each issue of *MMWR* contains three or four short articles on morbidity and mortality trends in the population of the United States. **Figure 20.1** shows the table of contents from a recent issue of *MMWR*, indicating the kinds of topics covered. Not all relate to infection, but all are based on an epidemiologic analysis of disease and death. Anyone can

Table 20.1	Significant Medical Advances Made by Epidemiology
Year	Discovery
1700–1713	Bernardino Ramazzini (Italian physician) described the first occupational diseases by studying the diseases specific to different tradesmen—painters, gilders, and so on.
	Similar discoveries were made later that painters suffered illness from exposure to lead-based paint and surgeons and pharmacists experienced other illnesses from exposure to mercury.
1847	Oliver Wendell Holmes (American physician) and Ignaz Semmelweiss (Austrian physician) discovered at the same time that puerperal fever occurs when birth attendants do not wash their hands after examining infected patients or doing autopsies.
1854	John Snow (English physician) showed that cholera epidemics are related to a contaminated water supply.
1900	Walter Reed (American surgeon) discovered that yellow fever is transmitted by mosquitoes.
1926	Joseph Goldberger (American physician and microbiologist) discovered that pellagra, a potentially fatal disorder manifested by rash, intestinal upset, and mental deterioration, is caused by a dietary deficiency of niacin.
1955	Evarts Ambrose Graham (American surgeon) and E. Cuyler Hammon (American statistician) proved that smoking is a significant risk factor for premature death from lung cancer.

Table 20.2   Reportable Diseases in California

Requirements

Report Immediately by Phone to the County Health Department

- Anthrax
- Botulism
- Cholera
- Dengue fever
- Diarrhea of the newborn, outbreaks
- Diphtheria
- Rabies, human or animal
- Yellow fever

Report within One Working Day by Phone or Mail

- Amoebiasis
- Campylobacteriosis
- Conjunctivitis, acute, infectious of the newborn
- Encephalitis, viral, bacterial, fungal, parasitic (specify)
- *Haemophilus influenzae*, invasive disease
- Hepatitis A
- Listeriosis
- Malaria
- Measles
- Meningitis, viral, bacterial, fungal, parasitic (specify)
- Meningococcal infections
- Pertussis
- Q fever
- Relapsing fever
- Salmonellosis
- Shigellosis
- Streptococcal infections (outbreaks and food handlers only)
- Syphilis
- Trichinosis
- Typhoid fever, cases and carriers

Report within Seven Calendar Days

- Acquired immunodeficiency syndrome (AIDS)
- Alzheimer's disease and related conditions
- Brucellosis
- Chancroid
- Chlamydial infection (*Chlamydia trachomatis*)
- Coccidioidomycosis
- Cryptosporidiosis
- Cysticercosis
- Disorders characterized by lapses of consciousness
- Food poisoning (report by telephone outbreaks of 2 or more cases)
- Giardiasis
- Gonococcal infections
- Granuloma inguinale
- Hepatitis, non-A, non-B (including hepatitis C)
- Hepatitis, unspecified
- Hepatitis B, cases and carriers
- Hepatitis delta (D)
- Kawasaki syndrome
- Legionellosis
- Leprosy (Hansen's disease)
- Leptospirosis
- Lyme disease
- Lymphogranuloma venereum
- Mumps
- Nongonococcal urethritis (not chlamydial)
- Pelvic inflammatory disease
- Reye's syndrome
- Rheumatic fever, acute
- Rocky Mountain spotted fever
- Rubella (German measles)
- Tetanus
- Toxic shock syndrome
- Tuberculosis
- Tularemia
- Typhus fever

Note: According to Title 17, Section 2500, California Code of Regulations. Each state makes its own regulations.

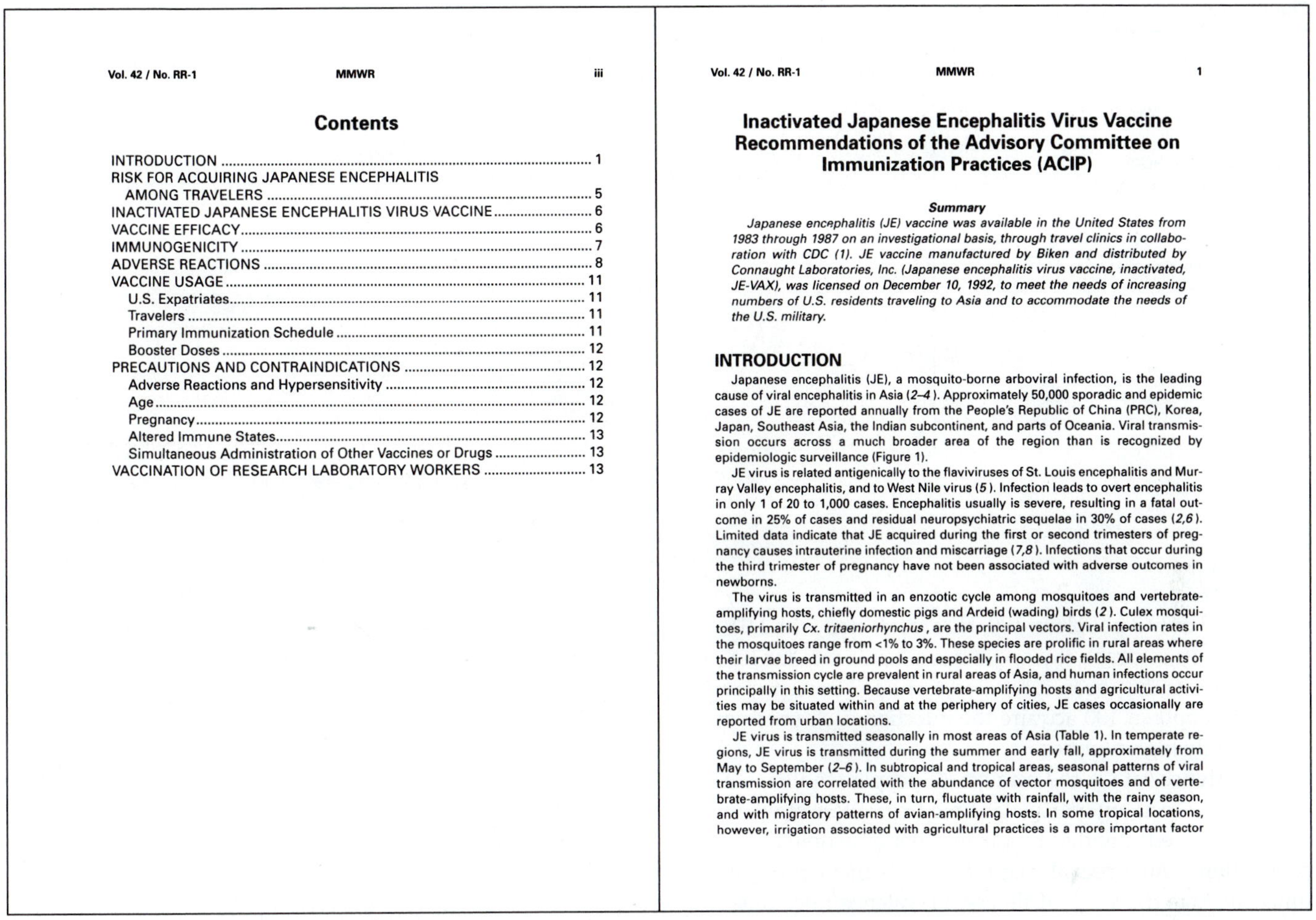

Contents

Inactivated Japanese Encephalitis Virus Vaccine Recommendations of the Advisory Committee on Immunization Practices (ACIP)

Summary

Japanese encephalitis (JE) vaccine was available in the United States from 1983 through 1987 on an investigational basis, through travel clinics in collaboration with CDC (1). JE vaccine manufactured by Biken and distributed by Connaught Laboratories, Inc. (Japanese encephalitis virus vaccine, inactivated, JE-VAX), was licensed on December 10, 1992, to meet the needs of increasing numbers of U.S. residents traveling to Asia and to accommodate the needs of the U.S. military.

INTRODUCTION

Japanese encephalitis (JE), a mosquito-borne arboviral infection, is the leading cause of viral encephalitis in Asia (2–4). Approximately 50,000 sporadic and epidemic cases of JE are reported annually from the People's Republic of China (PRC), Korea, Japan, Southeast Asia, the Indian subcontinent, and parts of Oceania. Viral transmission occurs across a much broader area of the region than is recognized by epidemiologic surveillance (Figure 1).

JE virus is related antigenically to the flaviviruses of St. Louis encephalitis and Murray Valley encephalitis, and to West Nile virus (5). Infection leads to overt encephalitis in only 1 of 20 to 1,000 cases. Encephalitis usually is severe, resulting in a fatal outcome in 25% of cases and residual neuropsychiatric sequelae in 30% of cases (2,6). Limited data indicate that JE acquired during the first or second trimesters of pregnancy causes intrauterine infection and miscarriage (7,8). Infections that occur during the third trimester of pregnancy have not been associated with adverse outcomes in newborns.

The virus is transmitted in an enzootic cycle among mosquitoes and vertebrate-amplifying hosts, chiefly domestic pigs and Ardeid (wading) birds (2). Culex mosquitoes, primarily Cx. tritaeniorhynchus, are the principal vectors. Viral infection rates in the mosquitoes range from <1% to 3%. These species are prolific in rural areas where their larvae breed in ground pools and especially in flooded rice fields. All elements of the transmission cycle are prevalent in rural areas of Asia, and human infections occur principally in this setting. Because vertebrate-amplifying hosts and agricultural activities may be situated within and at the periphery of cities, JE cases occasionally are reported from urban locations.

JE virus is transmitted seasonally in most areas of Asia (Table 1). In temperate regions, JE virus is transmitted during the summer and early fall, approximately from May to September (2–6). In subtropical and tropical areas, seasonal patterns of viral transmission are correlated with the abundance of vector mosquitoes and of vertebrate-amplifying hosts. These, in turn, fluctuate with rainfall, with the rainy season, and with migratory patterns of avian-amplifying hosts. In some tropical locations, however, irrigation associated with agricultural practices is a more important factor

Figure 20.1 Epidemiologists at the Centers for Disease Control and Prevention (CDC) in Atlanta publish the *Morbidity and Mortality Weekly Report* (*MMWR*) to provide up-to-date information on trends in diseases.

subscribe to *MMWR*, which is published by the Massachusetts Medical Society. *MMWR* can also be found in any science or biomedical library.

Other sources of information available to epidemiologists include surveys, questionnaires, interviews, and hospital records. The disadvantage of these sources is that the information must be gathered specially for each project—an expensive undertaking—while vital statistics, census information, and reportable disease statistics are available as a part of the public record. The advantage of these sources, however, is that information is collected to answer specific questions.

Uses of Statistics

Epidemiologists usually express health information about a population as a **rate**—a ratio of the number of people in a particular category (such as people who de-velop a certain disease, people who smoke, people who live in New York City) to the total number of people in the population being studied. Rates are preferable to simple numbers because a comparison can be made. For example, it is more meaningful to say that less than 1 percent of the residents of the Broad Street workhouse died from cholera than to say that five deaths occurred there.

Two of the most commonly used rates in epidemiology are incidence rate and prevalence rate. **Incidence rate** is the number of people who develop a disease or condition during a certain period of time (for example, one year) divided by the total number of people in the population. Incidence rates measure the growth of an epidemic. In the AIDS epidemic, for example, the incidence of new HIV infections tells us how rapidly the virus is spreading through the population. The crude incidence rate, gauged by new infections detected in repeat blood donors, has been estimated at 0.003 percent per year. This

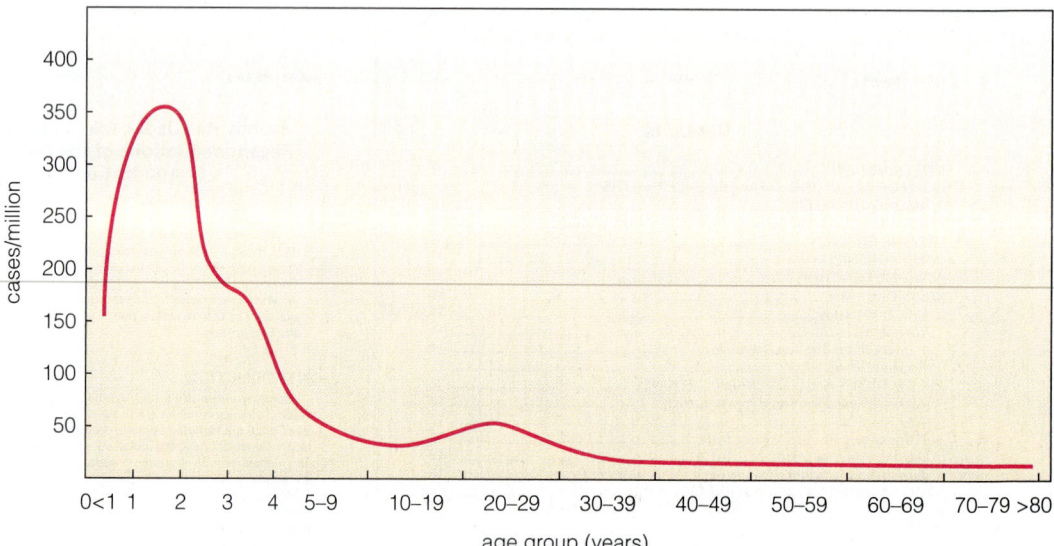

Figure 20.2 Age-adjusted incidence rate for shigellosis. Raw statistics suggest that the disease is relatively rare—fewer than 50 cases per million population—but age-adjusted data reveal that it is a significant illness among children—over 350 cases per million.

means that only 3 people in 100,000 are infected with HIV each year. In contrast, the specific incidence rate for homosexual men has been estimated to be 1–3 percent per year—3 people in 100 acquire the infection each year—a staggeringly high number.

Prevalence rate is the number of people who have a certain disease at any particular time (regardless of when it first appeared) divided by the number of people in the population. Thus, prevalence rate equals incidence rate times average duration of illness. Prevalence rates measure the magnitude of an epidemic. Although the prevalence of HIV infection cannot be measured exactly (because it would require that every person in the United States undergo an HIV blood test), estimates from various studies in 1991 led to a rough estimate of 1 to 1.5 million infected Americans. This translates to a prevalence rate of approximately 0.5 percent—much higher than the crude incidence rate of 0.003 percent. This difference is typical of AIDS and similar disorders that people live with for many years after they are infected. For a disease like cholera, however, which results in recovery or death within a few days, incidence and prevalence rates are nearly the same.

Births and deaths are also expressed as rates—the number of births or deaths in a certain time period divided by the total number in the population. Death rates are usually subdivided by age group. An **age-adjusted death rate** is calculated for a particular age group. This age breakdown provides useful information. We expect a high death rate among the elderly, but not among infants or young adults. A rising death rate among the young calls for epidemiologic investigation.

Disease incidence and prevalence rates may also be age-adjusted to highlight clinically important informa-tion about a disease under study—for example, age-adjusted data show that shigellosis is largely a disease of children (**Figure 20.2**).

TYPES AND USES OF EPIDEMIOLOGICAL STUDIES

All types of epidemiological studies share certain common elements:

1.   They focus on groups rather than on individuals.

2.   Results are expressed as rates and are compared statistically.

3.   They draw generalizations about the cause and pre-vention of a certain disease.

Epidemiologic studies differ widely, however, in their methods and immediate goals. For example, **retrospective studies** analyze events that have already occurred, while **prospective studies** record events as they happen and analyze them at the end of the study. Most studies, like John Snow's, deal with naturally occurring events, but a few are **experimental**, meaning that events are deliberately influenced by the investigator—an approach often limited by ethical considerations. Some studies are undertaken with an urgent goal in mind, such as finding the source of an epidemic. Others provide long-range background information about a disease.

In this section, we will look at four of the many different types of epidemiological studies—descriptive epidemiology, surveillance epidemiology, field epidemiology, and hospital epidemiology.

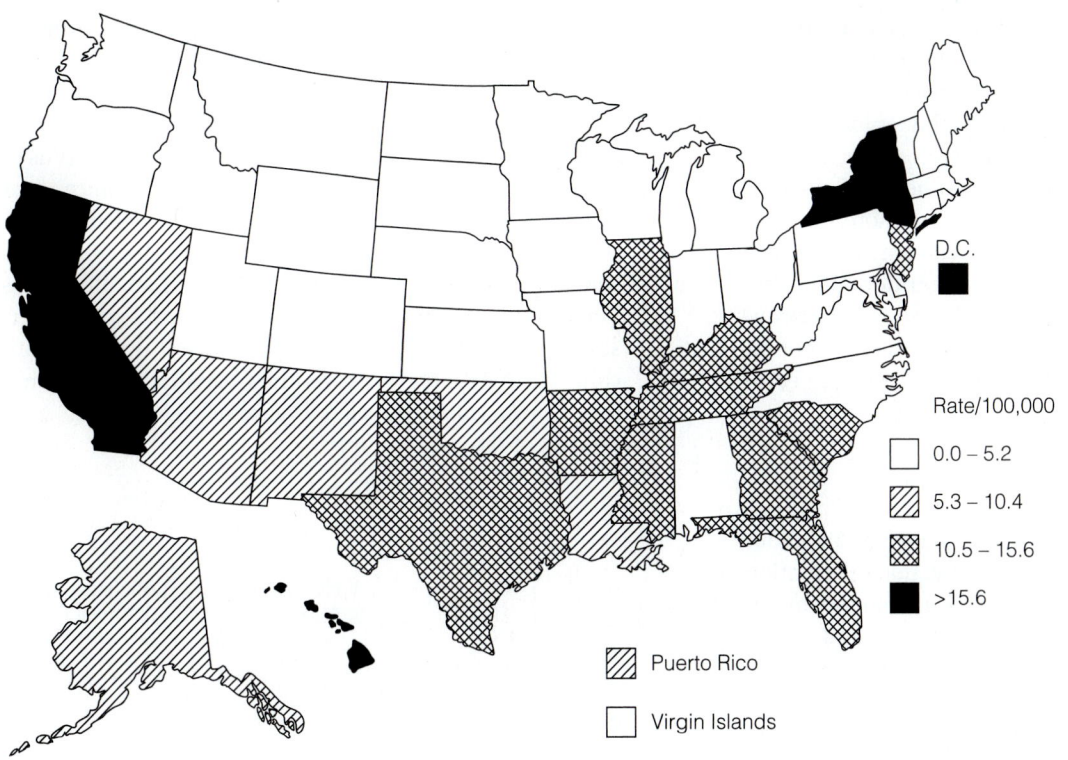

Rate/100,000

0.0 – 5.2

5.3 – 10.4

10.5 – 15.6

>15.6

Puerto Rico

Virgin Islands

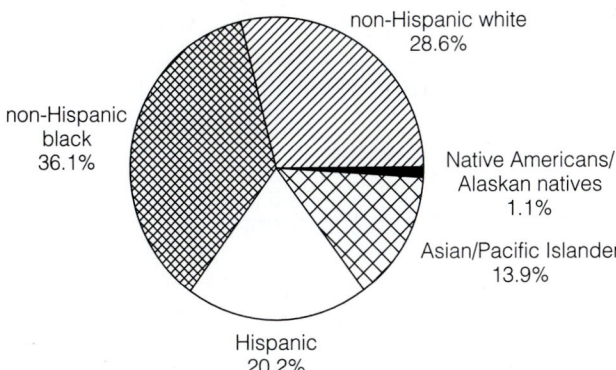

non-Hispanic white
28.6%

non-Hispanic black
36.1%

Native Americans/
Alaskan natives
1.1%

Asian/Pacific Islander
13.9%

Hispanic
20.2%

Descriptive Epidemiology

Descriptive epidemiology provides general information about a disease. For example, what are the incidence and prevalence rates of tuberculosis? In what parts of the country is tuberculosis most common? What is the death rate from tuberculosis in the general population and in selected subpopulations, such as different ethnic groups (**Figure 20.3**)? What is the ratio of fatal to nonfatal cases? How do factors such as age, geographical distribution, race, economic status, and sex affect the likelihood that a person will contract tuberculosis? A descriptive study

might also consider where victims seek medical treatment and what type they receive. Thorough descriptive studies often take years to complete, so they usually are retrospective. From this information, physicians and public health officials might expose a need for better availability and delivery of health care. In this way, descriptive epidemiologic studies have practical benefits.

Descriptive studies can provide a perspective on whether a disease is increasing or declining and allow epidemiologists to draw conclusions as to why this may be the case. Tuberculosis, for example, was a common and often fatal infection during the 1700s. During the early 1800s, however, before the causative microorganism was even identified, the disease began to decline significantly—a phenomenon that epidemiologists still cannot explain. With the development of effective antitubercular chemotherapy in the 1950s, tuberculosis rates in industrialized countries plummeted. Tuberculosis was confined to high-risk groups such as the poor, alcoholics, and recent immigrants from developing countries—it was considered under control. But the 1990s have seen a marked resurgence because of the AIDS epidemic (immunosuppressed individuals are vulnerable to TB), the emergence of new drug-resistant strains of *Mycobacterium tuberculosis*, and social changes. Tuberculosis once again presents a formidable challenge to public health workers (Chapter 22).

Descriptive studies can be particularly useful in demonstrating how nonmicrobial factors contribute to infectious disease. For example, Koch proved that *Mycobacterium tuberculosis* causes tuberculosis, but epidemiologic research showed that people who live in overcrowded conditions, whose nutrition or general health is poor, or who suffer from alcoholism or AIDS are more likely to contract tuberculosis than are those who are affluent and otherwise healthy. Thus, poverty, malnutrition, ill health, alcoholism, and AIDs also "cause" tuberculosis. Epidemiologists view infectious diseases as being the result of interactions between three factors: the pathogen, the environment, and the host.

Surveillance Epidemiology

Descriptive epidemiology provides information about diseases. In **surveillance epidemiology**, this information is used to track epidemic diseases. Epidemiologists monitor situations that might lead to an epidemic and follow the progress of an epidemic when one begins.

Surveillance means close observation. Traditionally, epidemiologic surveillance tracked individuals who were a risk to public health. For example, people who entered a smallpox-free country from one where smallpox still occurred were monitored carefully by the public health department for the first few weeks after their arrival to make sure they had not brought the deadly virus with them.

Epidemiologic surveillance today includes more sophisticated techniques. These new techniques played a crucial role in eradicating smallpox worldwide. An effective vaccine for smallpox had been available for over a hundred years, but the financial and organizational problems of vaccinating the entire world population were overwhelming, particularly in developing countries where smallpox was most prevalent.

In the 1970s, the World Health Organization's smallpox eradication team developed reliable ways of reporting all new cases, or **index cases**, of smallpox, and thus of maintaining effective epidemiologic surveillance over potential outbreaks. In some countries they offered rewards to citizens who reported smallpox. Workers were sent to investigate each reported case. If smallpox was verified, affected persons were isolated and exposed persons vaccinated, confining the outbreak. By systematically following these procedures, smallpox was eliminated from many areas by vaccinating as little as 6 percent of the population.

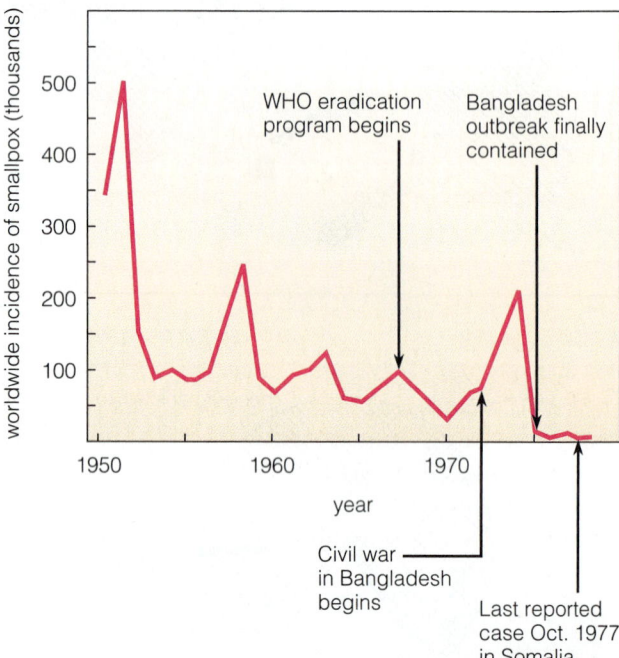

Figure 20.4 Conquering smallpox. The World Health Organization (WHO) was making great progress toward eradicating smallpox worldwide until civil war broke out in Bangladesh in 1971. Ten million people, some infected with smallpox, fled to India and were housed in refugee camps, where infection spread. When infected refugees returned to Bangladesh, an epidemic sparked a marked upturn in the incidence of smallpox. Intensive epidemiological surveillance and immunization eradicated smallpox six years later. The photo shows Bangladeshi refugees with smallpox in a Bihari relief camp.

Microbe Mappers

Broad Street Revisited

In July 1984 the Minnesota Department of Public Health became aware of an outbreak of chronic diarrhea in the city of Brainerd. Physicians were reporting an unusual number of patients with severe diarrhea lasting more than four weeks. In the month of May, for example, more than 30 cases were identified. To trace the source of the outbreak, the health department initiated a **case-control study**. Twenty-three **cases**, people with the disease, were studied and compared to 46 **controls**, people of the same age and sex as the cases but without the disease. The study revealed one critical difference—the cases had consumed unpasteurized milk from a local dairy; the controls had not. When the dairy voluntarily stopped selling raw milk, no further cases occurred.

The health department's investigation of Brainerd diarrhea illustrates the advantages of an epidemiologic approach to disease control. Although physicians strongly suspected that the diarrhea was infectious, no causative microorganism had been identified. Several antibiotics were prescribed, but none was effective. Like Broad Street during the cholera epidemic, the community of Brainerd was affected by an epidemic for which there was no known cause or treatment. The solution was the same—an epidemiologic investigation to identify the "cause" and eliminate it.

Although the procedures sound simple, the logistical complications were immense. Eradicating smallpox from Bangladesh, the last country in the world to suffer a major series of smallpox epidemics, required the cooperation of 100 Bangladeshi epidemiologists, 12,000 local workers, 13,000 village recruits, and 75 international health workers. Success was finally declared in 1977 (**Figure 20.4**). Surveillance techniques have also been used to control (but not eliminate) malaria, diphtheria, and polio.

Field Epidemiology

Surveillance epidemiologists keep an eye on diseases that are known to be a problem. **Field epidemiology** involves investigating unexpected outbreaks of disease. At the beginning of a field epidemiologist's investigation, the source of the problem is usually a mystery. John Snow's work during the Broad Street epidemic is an example of field epidemiology.

There are many ways that public health authorities become aware of a new epidemic. Occasionally a health department will be informed of an unusual number of cases of disease in a certain area (see the box "Broad Street Revisited"). AIDS was first noticed when many cases of *Pneumocystis carinii* pneumonia, previously a rare disease, began to be reported among young men (Chapter 27). Investigation revealed that all the patients had other manifestations of immunodeficiency and that all were homosexual. From the ensuing investigation, epidemiologists concluded that some infectious agent, transmitted by sexual contact, was destroying the pa-

tients' immune system—an accurate conclusion reached before virologists identified the HIV retrovirus that causes AIDS.

Sometimes large numbers of people who have been together at some gathering will simultaneously become ill, an epidemiologic phenomenon known as a **common source epidemic**. Many common source epidemics turn out to be cases of food poisoning. A typical outbreak may involve a large gathering at which incompletely cooked or inadequately refrigerated food is contaminated by *Staphylococcus aureus* enterotoxin. Dozens or even hundreds of people who eat the tainted food suffer from diarrhea and vomiting within a few hours (Chapter 23). The outbreak of legionellosis in a Philadelphia hotel was a case of an airborne common source epidemic (Chapter 22).

Citizens' complaints to a health department may draw attention to a developing epidemic. Residents of Old Lyme, Connecticut, complained to the local health department and to physicians at the Yale University medical center that there seemed to be an extraordinary number of cases of arthritis in their community. At first, medical experts were skeptical, but in 1977 epidemiologic investigation confirmed a statistically significant increase in serious cases of arthritis. The infection was eventually traced to a tick bite (Chapter 27). As with AIDS, much of what is known today about Lyme disease was inferred by epidemiologists before the causative microorganism was identified.

Sometimes an epidemic is identified when a hospital reports an unusual number of cases of a certain illness.

Table 20.3   Ten Basic Questions Asked in an Epidemiologic Interview

Question	How It Might Help the Investigator
1. Have you had recent contact with a person who had a similar illness?	"Yes" would suggest an infectious disease.
2. Where do you live?	Some diseases are found only in certain geographic areas, such as coccidioidomycosis (San Joaquin valley fever in the San Joaquin valley of California).
3. Who lives at home with you?	An adult with young children might have contracted one of the diseases commonly spread in day-care centers, such as hepatitis or giardiasis.
4. Do you have any pets?	"Yes" might suggest a zoonotic disease, such as toxoplasmosis among cat owners.
5. What is your work or daily routine?	This might provide clues to occupational illnesses, such as cancer among workers exposed to asbestos.
6. Where do you work or study?	This might provide a connection to others with the same disease, such as children attending a school where a measles outbreak is occurring.
7. What are your hobbies?	Leisure activities might involve disease exposure; hunters can contract tularemia through contact with wild animals.
8. Have you traveled away from home recently?	International travelers may contract a variety of diarrheal illnesses, such as traveler's diarrhea or shigellosis.
9. Are your immunizations up-to-date?	Unimmunized people are at risk for diseases such as measles that immunized people are unlikely to contract.
10. Do you use illegal drugs?	People who inject drugs illegally often use contaminated needles, putting them at risk for many bloodborne infections, including HIV and hepatitis B.

This happened in Los Angeles County during the 1980s when physicians at Los Angeles County Women's Hospital noticed an unusually high number of cases of listeriosis among new mothers and their infants (see the box "Listeriosis Outbreak" in Chapter 24). Epidemiologic investigation revealed that all the women had recently eaten a particular brand of unpasteurized soft cheese. Containers of cheese still on grocery store shelves cultured positive for *Listeria monocytogenes*, and when further cheese sales were stopped, the epidemic ended.

Field epidemiologists are like detectives. They compile as much information as possible about each victim, as well as the victim's family and neighbors (**Table 20.3**). They piece together the information to determine what caused the epidemic and what can be done to control it (**Figure 20.5**). Field epidemiologists deal with practical matters rather than research, but new information about disease transmission does come from field studies. For example, while investigating an outbreak of infectious hepatitis, field epidemiologists found that all victims had recently eaten shellfish. This observation led to the discovery that shellfish harvested from contaminated waters can spread viral hepatitis.

Hospital Epidemiology

People in a hospital—patients and staff—are a population of special epidemiologic interest. Bringing large numbers of sick people together to one place greatly facilitates the transmission of infection. Approximately 8 percent of all inpatients acquire a new infection while in the hospital, and at least 20,000 Americans die of hospital-acquired or **nosocomial infections** each year. Hospital personnel are also at risk. The direct cost of nosocomial infections in the United States each year is in the billions of dollars.

Factors That Foster Nosocomial Infections.   With our thorough understanding of the modes of disease transmission and our armamentarium of antibiotics, shouldn't nosocomial infections be preventable or at least treatable? Unfortunately, nosocomial infections are inevitable because of three factors—immunocompromised patients, invasive medical procedures, and antibiotic resistance.

Many patients in a modern hospital would surely not have survived for so long in the past. These people—victims of advanced cancer, recipients of organ transplants, patients in kidney failure, the frail elderly—are extremely infection-prone. In fact, all gravely ill people have weakened immune defenses and are vulnerable to infection (Chapter 18). Moreover, the treatments for some illnesses, including anticancer medications and steroids, depress immune function still further. For these patients, even microorganisms from their own normal flora may cause life-threatening infections (Chapter 15).

a

c

b

Figure 20.5 Field epidemiologists are scientists whose work ranges from (**a**) searching a disease-ridden area for infected insect vectors to (**b**) interviewing people around the world. (**c**) The final task is interpreting the data—deciding what it all means.

Invasive medical procedures foster infection because they bypass the normal structural, mechanical, and biochemical defenses of the body's surfaces (Chapter 14). Blood drawing, intravenous lines, and urinary catheterization are routine procedures. Some patients also undergo specialized invasive procedures. In endoscopy, instruments are introduced into the gastrointestinal tract (through either the mouth or anus). In bronchoscopy, instruments are introduced deep into the respiratory tract; and in laparoscopy, instruments are introduced into the abdominal cavity. Major surgical procedures like coronary bypass surgery are also routine in many hospitals. Any invasive procedure carries some risk of infection.

Antibiotics help control hospital epidemics, but many bacteria found in hospitals have developed resistance to some of the most useful drugs. When these resistant organisms are transmitted to immunocompromised patients, antibiotics may not be effective in treating their infection, leading to fatality.

Types of Nosocomial Infection. A patient can acquire a nosocomial infection in almost any body part, but certain nosocomial infections are particularly common. Infection of the urinary tract accounts for nearly half of hospital-acquired infections. The most common causative microorganisms are *Escherichia coli* and other enteric organisms, including *Proteus* spp., *Klebsiella* spp., *Enterobacter* spp., and enterococci. Many urinary tract infections are the result of urinary catheterization—a procedure in which a **catheter**, or plastic tube, is passed through the urethra to drain urine from the bladder.

Surgical wound infections are the second most frequent type of nosocomial infection. *Staphylococcus aureus* is a common cause, and Gram-negative enterics of the same species that cause urinary tract infections often infect abdominal wounds. All operating rooms practice strict aseptic techniques, including **scrubbing**—a long, ritualized procedure of hand washing. Operating room personnel wear sterile masks and gowns, all instruments

Microbe Mappers

MMWR Alert

The *Morbidity and Mortality Weekly Report* contained the following story of a hospital epidemic in Wisconsin. From October 1986 through June 1988, nearly 9 percent of patients who underwent endoscopy of the upper gastrointestinal tract became infected with a particular strain of *Pseudomonas aeruginosa*—a much higher infection rate than normal. *P. aeruginosa* is a notorious nosocomial pathogen, capable of surviving for extended periods in almost any liquid environment, even some disinfectant solutions. Although not highly virulent, *P. aeruginosa* can cause serious, even fatal, infections in immunocompromised patients. An epidemiologic investigation revealed that the apparatus for sterilizing the endoscopes was faulty. (Endoscopes are delicate tubular instruments made of flexible light-emitting fibers that enable a physician to look inside the body—at an ulcer in the stomach or a tumor in the intestines.) After use, endoscopes were placed in an apparatus where they were flushed with a detergent solution, disinfected with a chemical germicide, and rinsed with tap water. But in the Wisconsin hospital, the equipment had become contaminated. A thick film of *P. aeruginosa* was growing in the detergent holding tank, inlet water hose, and air vents. Only when hospital personnel began to rinse and dry the endoscopes by hand was the outbreak brought under control.

---

are sterilized, and the air is filtered to remove microorganisms. Still, at least 10 percent of patients who undergo surgery develop an infection afterward. The infection rate is highest in cases where bacterial contamination is inevitable, such as in operations that open the microbe-laden intestinal tract and in surgery to repair dirty wounds.

Pneumonias rank close behind wound infections as the third most common type of nosocomial infection. Many different microorganisms can cause nosocomial pneumonias, including *Streptococcus* and *Staphylococcus* spp., *Pseudomonas aeruginosa*, and, occasionally, enterics. Some pneumonias occur as the direct result of medical intervention. For example, mechanical respirators and intubation to help a patient breathe introduce microorganisms and provide a continuous route of entry just as catheters do. Pneumonias also occur in severely ill patients who **aspirate**, or inhale, microbe-containing secretions from their upper respiratory tract directly into their lungs.

Skin infections can also be acquired in the hospital. Extremely vulnerable patients are most at risk—babies in newborn nurseries and burn victims. Nursery epidemics are usually due to *Staphylococcus aureus*. These infections can be troublesome and unsightly, causing blistering and peeling, but they are seldom fatal. Many nosocomial skin infections do not develop until the baby has gone home, and so it is difficult to know how common these infections really are.

Nosocomial skin infections of burn victims, however, can be life-threatening because they often progress to systemic bloodborne infections. Although almost any microorganism can infect a burn victim, the most common is *Pseudomonas aeruginosa*. The loss of skin, which provides our first and most crucial line of defense, presents an almost insurmountable infection-control problem. Despite specialized care and scrupulously observed aseptic technique, infection is the most common cause of death among burn victims who survive the initial injury.

Hospital Epidemiologists and Infection Control. Because nosocomial infections are a constant threat, most hospitals have staff epidemiologists, often nurses, who work with doctors on an infection-control committee. Hospital epidemiologists are specially trained to recognize and interrupt a potential hospital epidemic.

Once an epidemic is recognized, the infection-control team takes specific action that depends upon the situation, but it usually involves increasing staff awareness of the problem and enforcing infection-control policies. If there is a *Staphylococcus aureus* outbreak in a nursery, for example, improving staff hand washing is critical. Cultures may be taken from workers' hands to see if they are carrying the offending bacterium, and nasal cultures may be obtained to determine which staff members are *S. aureus* carriers.

Other infection-control measures include isolating infected patients, restricting the transfer of patients from one part of the hospital to another, or even closing certain areas until the outbreak is contained. **Isolation**, the physical separation of individuals, locates an infected person where he or she is unlikely to spread disease to healthy

Clinical Notes

Respiratory Isolation

It was a typical busy January evening in the pediatric emergency department and T. D., a young doctor, picked up the chart of yet another baby with a cough. As soon as he saw this baby, T. D. knew she was sicker than most. The baby was struggling to breathe. She had to be admitted to the hospital immediately.

T. D. called the pediatric ward for a bed—often a problem during the winter when the hospital is full-to-overflowing with children suffering from various infections. The nurse's first question was predictable: Would the baby need isolation? Yes, he told her, the baby probably had RSV (respiratory syncytial virus). She would require respiratory isolation to protect other patients from this highly contagious infection. This need made the bed situation even worse because it meant the baby would have to be housed alone in a room

designed to hold four cribs. The nurse told T. D. she would see if she could move children around the ward, putting patients who did not require isolation together to make room for his new patient. If there was no way to clear a room, the infant would have to be transferred to another hospital.

Twenty minutes later the nurse phoned T. D. to say she had managed to clear space for his baby. Two days later, when the baby's nasal smear came back positive, confirming the RSV diagnosis, the infant was moved into a room with other known RSV patients. Since they all had the same infection, they did not need to be isolated from one another. This opened space for more patients who might require respiratory isolation.

---

people. **Reverse isolation** separates infection-prone patients from sources of infection. David's germ-free environment, described in the box in Chapter 18, is an extreme example of reverse isolation.

In addition to monitoring potential epidemics, hospital epidemiologists enforce the CDC program of Universal Blood and Body Fluid Precautions, initiated in 1983 in response to the threat the AIDS epidemic posed to health-care workers. Workers are instructed to treat all patients as though they were infected. Whenever direct physical contact with blood or other secretions that might contain the AIDS virus may occur, hospital workers should wear protective gloves, gowns, and goggles. Despite the universal precautions, a few cases of occupationally acquired AIDS are inevitable because of accidents such as needle sticks or inadvertent cuts during surgery. Members of the infection-control team play an indispensable role in hospital safety, but they have a challenging and often unpopular job as the enforcers of strict standards of hygiene.

PUBLIC HEALTH: PREVENTING DISEASE

Public health deals with disease **prophylaxis**, or prevention. There are two principal methods of prophylaxis. One is to limit people's exposure to pathogenic microorganisms by decreasing or eliminating the pathogen's reservoir (Chapter 15) or by interrupting disease trans-

mission. The other method is **immunization**, which artificially augments the body's natural immune defenses.

Public Health Organizations

Many agencies have been established at local, state, national, and global levels to safeguard the public health. City and county departments of public health help citizens and practicing physicians in numerous ways—from inspecting food distributors to providing free immunizations. They track reportable diseases and make information about current disease trends available to health-care providers. For example, if a doctor is treating a patient with an animal bite and wants to know about recent rabies cases in the area, he or she can telephone the local health department.

Local health departments report to state public health agencies. Because state laws govern disease reporting, state health departments play a critical role in tracking reportable diseases. Most state health departments publish a bulletin of epidemiological trends similar to *MMWR*. News of the Brainerd diarrhea outbreak, for example, first appeared in the Minnesota Department of Health's *Disease Control Newsletter*. State health departments also maintain laboratories that do diagnostic testing for certain diseases that affect the public health. In California, for example, Lyme disease serology (Chapter 27) and rabies pathology (Chapter 25) tests are among those performed by state labs.

The United States Public Health Service (USPHS) is our national public health agency. The CDC is one of its agencies. In addition to monitoring infectious and other preventable diseases and disseminating the information through the *MMWR*, the CDC conducts research. It was the CDC, for example, that isolated the bacterium that causes legionellosis (Chapter 22). In addition, the CDC issues recommendations on the use of antibiotics and vaccines, plans and executes national public health education campaigns, and trains epidemiologists.

The international organization most concerned with improving public health is the United Nations–sponsored World Health Organization (WHO). WHO has undertaken ambitious programs to improve the health of people in the world's poorest nations. Its smallpox eradication program was history-making. Some of its current programs to improve world health include campaigns to increase childhood immunization and breastfeeding and educational programs to improve infant nutrition.

Diminishing Reservoirs and Interfering with Transmission

Infection can spread in an astonishing number of ways (Chapter 15). A drink of water, a mosquito bite, even a conversation with a friend can cause disease. Ways of preventing exposure to pathogenic microorganisms are equally varied, ranging from closing the screen door to filtering the water supply of a city. Public health practices that have most dramatically decreased the incidence of disease include clean water, clean food, personal cleanliness, insect control, prevention of sexually transmissible diseases, and prevention of respiratory diseases.

Clean Water. Many diseases, including cholera, typhoid fever, and viral and bacterial diarrheas, can be spread when human sewage carrying pathogens contaminates the water supply. Waterborne disease transmission can be prevented by adequate sewage treatment systems and a clean public water supply.

The history of public health reforms to ensure water purity makes interesting—and, at times, horrifying—reading. Until the late nineteenth century, cities were not safe places to live because raw sewage sat in open cesspools and clean drinking water was unavailable. Infection was so widespread that more people died in cities than were born in them; urban populations were maintained only by constant immigration from the countryside. The invention of the toilet was a public health disaster because sewage disposal systems did not exist (see the box "One Step Forward, Two Steps Back"). Many people initially opposed public programs for sewage disposal because they were expensive and were considered a government intrusion into people's private lives. However, reforms in public sanitation made modern urban life possible.

Most of us take clean drinking water for granted, but in many parts of the world contaminated water and waterborne infections are still common. In these places cholera and typhoid fever still exist, and diarrhea is a leading cause of death among infants and young children. When societies are unable to ensure clean water, individuals can protect themselves against most infections, except for some viral diseases like hepatitis A, by boiling their drinking water.

Clean Food. Sometimes microorganisms merely spoil food; other times they cause disease. Many food-preservation techniques, including canning, were developed before people understood why they worked, but methods of controlling specific foodborne diseases were devised only after the germ theory of disease was accepted.

Milk is particularly likely to carry infections, including tuberculosis, typhoid fever, scarlet fever, brucellosis, and diphtheria. Milkborne illnesses were rampant before **pasteurization**, a process that kills bacteria by briefly heating a liquid. However, because Pasteur developed pasteurization to kill the microorganisms that caused spoilage in wine, it was years before the process was applied to milk. In 1908 Chicago passed the first law requiring milk to be pasteurized. Today in the United States pasteurization is almost universal and milkborne disease is rare.

Most microorganisms that cause foodborne disease can be killed by heat. If food is adequately cooked, diseases such as trichinosis, salmonellosis, and tapeworm infection can all be prevented (Chapter 23). Refrigeration can also prevent foodborne diseases, even though cold only slows the growth of microorganisms. Staphylococcal food poisoning, for example, can be prevented if food is kept cool enough that the bacteria cannot grow and produce their toxins (Chapter 23).

Personal Cleanliness. Many disease-causing microorganisms are transmitted from person to person by dirty hands (direct-contact transmission) or contaminated objects (fomite transmission) (Chapter 15). Probably the single most important habit of personal cleanliness that can prevent the spread of disease is the simple act of hand washing. Even though we know invisible organisms on our own hands transmit disease, studies show that even nurses and doctors in hospitals often fail to practice good habits of hand washing unless they can see that their hands are dirty. The importance of hand washing in preventing illness in the home, in day-care centers, and in hospitals cannot be overemphasized.

Insect Control. Diseases that are transmitted by biting insects, such as malaria and yellow fever, can be controlled by decreasing the population of the insect vector (Chapter 15). Early public health programs concentrated

One Step Forward, Two Steps Back

We tend to think of toilets as an improvement over chamber pots and outhouses, but the early ones caused the death of thousands of people. The toilet, or water closet, was patented and first manufactured in England during the 1770s. Before this time, human excrement was collected and disposed of as night soil, so the waste and bacteria remained in relatively contained areas. Toilets, on the other hand, drained into sewers, which emptied into rivers. Drinking water was supplied by private companies that collected river water—now contaminated—and transported it back to cities. When the toilet came into European homes, so did higher risk of cholera and typhoid fever.

As early as the 1820s, some physicians and sanitary engineers suspected that drinking water contaminated by human waste was a threat to health. As one London engineer put it, "a stream which receives daily the evacuations of a million human beings with all the filth and refuse of various offensive manufacturers cannot require to be analyzed, except by a lunatic, to determine whether it ought to be pumped as a beverage for the inhabitants of the Metropolis

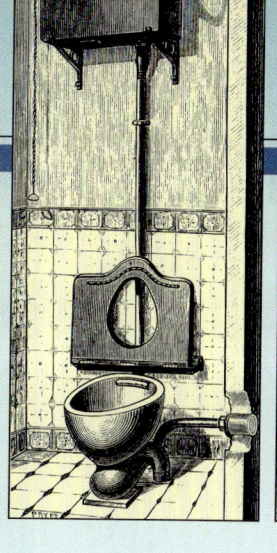

Early water closets.

of the British empire." Still, many were unconvinced. Even when microscopic examination revealed that London's water supply was teeming with "animalcules," some experts declared this was "no more harmful than eating fish." Sanitary reform was slow, advancing only as the germ theory of disease gained acceptance in the late 1800s. At that time, European cities began to develop sewage-treatment and water-filtering systems to furnish safe water.

on draining swamps to destroy breeding grounds, screening living areas, and using mosquito netting.

When insecticides such as dichlorodiphenyltrichloroethane (DDT) became available in the 1940s, insect control programs expanded. At first DDT seemed miraculous. Not only did it eliminate huge insect populations, but it was also inexpensive. Unfortunately, pesticide-resistant insects soon emerged along with serious environmental problems. It is now generally accepted that insecticides will not provide the final answer to controlling mosquito-borne disease. In 1987, after having spent more than $3 billion on insecticides over 25 years, public health planners in India announced that they would abandon the use of pesticides in their antimalaria campaign. Instead, they would concentrate on environmental measures such as removing stagnant water and developing new programs for biological insect control.

Prevention of Sexually Transmissible Disease. Syphilis and gonorrhea have caused human suffering and death for centuries, and now AIDS has joined them. All sexually transmissible diseases are completely preventable by interrupting the chain of transmission. This interruption must be done on an individual basis by limiting sexual exposure or using condoms. Unfortunately, disease prevention through ongoing individual effort is always less effective than control through public health

projects, such as a new water treatment plant. Consequently, sexually transmissible infections continue to be a major problem in developed as well as developing countries. In fact, the incidence of sexually transmissible disease rose during the 1980s among poor American city dwellers, particularly those addicted to drugs.

Control of syphilis and gonorrhea is based on a combination of early diagnosis and treatment for cure. For AIDS, however, prevention is crucial because no cure exists. So far, AIDS prevention programs have relied on public education, encouraging people to limit their sexual exposure and to use condoms.

Prevention of Respiratory Diseases. Pathogenic microorganisms can be excluded from the water we drink and the food we eat, but it is almost impossible to exclude them from the air we breathe (**Figure 20.6**). Everyone who goes outdoors or has contact with other human beings runs the risk of inhaling pathogenic microorganisms that have been aerosolized from the soil or exhaled by someone nearby.

Spread of some respiratory diseases can be minimized by isolating infected individuals. Before childhood immunization for diphtheria became common, isolation was one of the major ways the disease was controlled. Children with diphtheria were taken out of their homes and isolated in a hospital diphtheria ward to decrease the

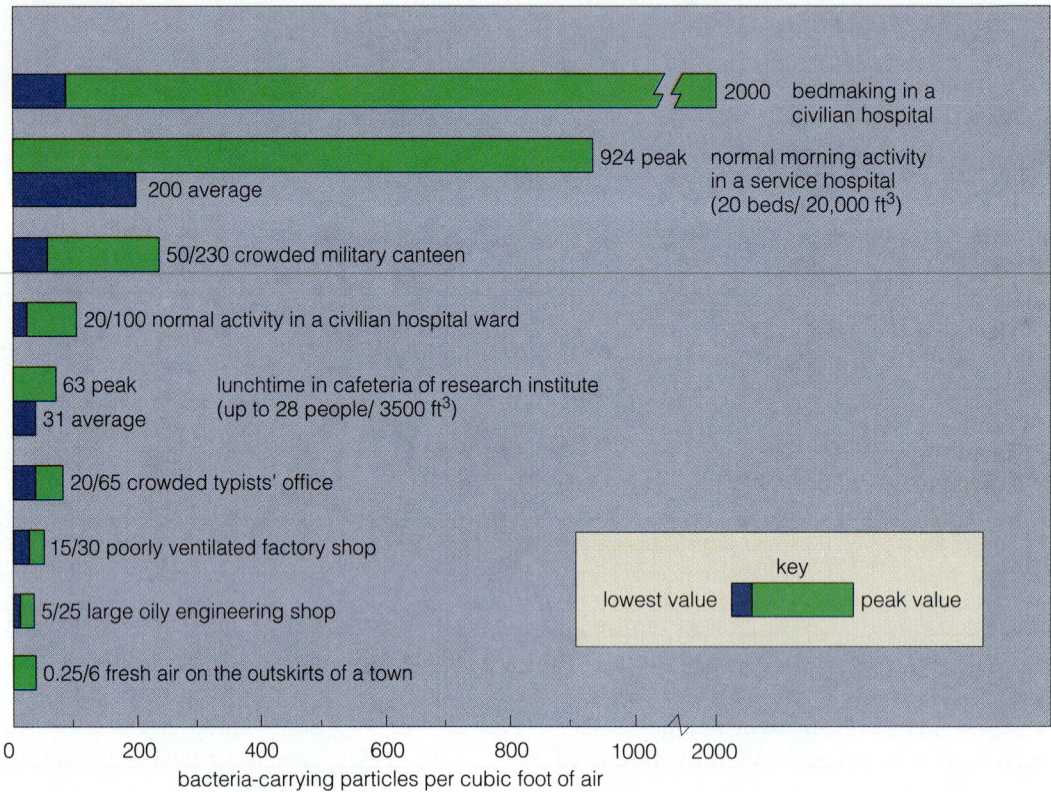

	2000 bedmaking in a civilian hospital
924 peak	normal morning activity in a service hospital (20 beds/ 20,000 ft^3)
200 average	
50/230 crowded military canteen	
20/100 normal activity in a civilian hospital ward	
63 peak 31 average	lunchtime in cafeteria of research institute (up to 28 people/ 3500 ft^3)
20/65 crowded typists' office	
15/30 poorly ventilated factory shop	
5/25 large oily engineering shop	
0.25/6 fresh air on the outskirts of a town	

key
lowest value | peak value

bacteria-carrying particles per cubic foot of air

Figure 20.6 Bacterial content of the air in different environments: results of a study done by the American armed forces on the concentration of bacteria-carrying particles in the air in different environments. The highest bacterial concentration was recovered during bed making in a hospital, an activity that involves shaking tiny particles into the air. In general, more airborne bacteria are found in hospitals than in even poorly ventilated offices and factories.

likelihood that other family members would be infected. But isolation cannot eliminate diphtheria because patients often begin to exhale pathogens before they show signs of illness. Furthermore, healthy carriers transmit the infecting microorganism.

In hospitals, face masks slow the spread of some respiratory diseases, but the only effective method for preventing transmission of respiratory pathogens is immunization.

IMMUNIZATION

As part of our natural defenses, antibody-producing B cells and different types of T cells cooperate to eliminate microorganisms from the body (Chapter 17). Immune defenses can be manipulated artificially to *prevent* infectious diseases through active or passive **immunization**. In **active immunization**, a person's own immune system is stimulated. In **passive immunization**, antibodies from an immune person or animal are transferred to a recipient.

Active Immunization

Certain infectious diseases afflict a person only once during a lifetime because memory B lymphocytes produced during the first infection mount an immune response so quickly a second infection never occurs (Chapter 17). Active immunization artificially stimulates this protective immunity, usually without making a person significantly ill. Active immunization, usually referred to simply as *immunization*, or *vaccination*, has done more to reduce the burden of infectious disease on human beings than has anything else except public sanitation.

A **vaccine** is an agent capable of inducing active immunity without causing disease. All good vaccines must be safe and relatively free of uncomfortable side effects. However, no vaccine is completely safe. The safety of a vaccine must be weighed against the likelihood and hazard of contracting the natural disease. For example, variolation, an early method of protecting against smallpox (see the box "The First Vaccination"), had a 1 to 2 percent fatality rate. Even Jenner's smallpox vaccination led to fatal complications in approximately 1 out of every 1 mil-

Turning Point

The First Vaccination

Smallpox, a disfiguring and often fatal disease, was the first disease against which a safe and effective immunizing agent was developed. It is also the first and only disease to be eliminated from Earth as the result of public health efforts.

In Asia and Africa, active immunization against smallpox was practiced for centuries as folk medicine. In a practice called **variolation**, crusts from a smallpox blister were inoculated into a healthy person through a tiny cut in the skin or into the lining of the nose, to induce smallpox. Usually, smallpox acquired from variolation was mild—only a few blisters appeared—but 1 to 2 cases in every hundred died. On the other hand, half the people who were naturally infected died, and variolation conferred lifelong immunity. When variolation was introduced in Europe in the late 1700s, Catherine the Great of Russia and Louis XVI of France thought the risk worthwhile.

About this time, Edward Jenner was a boy in England. He underwent variolation and suffered an unusually severe case of smallpox. He never forgot the experience. He became a country physician. Like other country people, he was familiar with the folk wisdom that dairymaids who had recovered from a mild disease called cowpox were protected for life against smallpox. Most physicians considered the cowpox-smallpox connection an old wives' tale. But Jenner was intrigued, and in 1796 he performed an experiment that showed cowpox infection could prevent smallpox.

Jenner took crusts from the hand of a dairymaid, Sarah Nelms, who was suffering from cowpox and inoculated them into the skin of James Phipps, a healthy 8-year-old who had never had smallpox. James became mildly ill for one day. Then James underwent variolation. He failed to contract even a mild case of smallpox. Jenner inoculated James again—and again James suffered no effect. Jenner concluded, correctly, that cowpox had made young James immune to smallpox.

Jenner's experiment was a turning point in medicine. The cowpox virus became the first safe vaccine, and vacci-

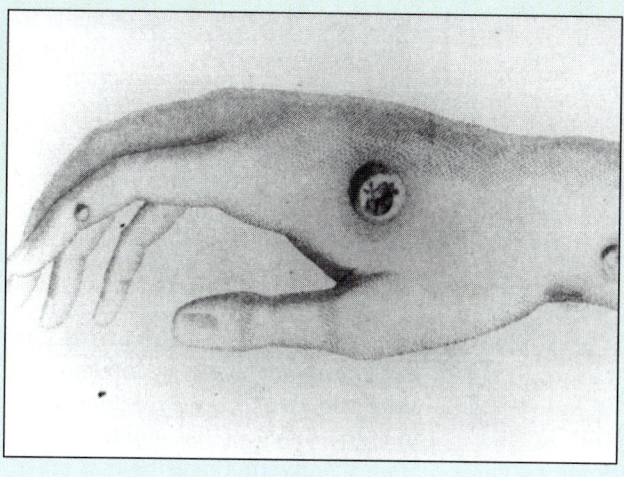

Cowpox crusts on Sarah Nelms's hand.

nation replaced variolation as a much safer way to prevent smallpox. We know now that cowpox infection provides immunity against smallpox because cowpox and smallpox are closely related viruses that share certain antigens.

We no longer vaccinate against smallpox because the disease no longer exists, but interest in the vaccinia virus that was long used for smallpox vaccination has revived. Using genetic engineering, researchers have identified genes in many pathogens that encode antigens which might be used for vaccines. Some of these genes have been transferred into the vaccinia virus and expressed there. Because we have a great deal of practical information about immunization with vaccinia—accumulated during the worldwide smallpox eradication program—vaccinia might become an ideal vehicle for immunizing against diseases other than smallpox. This is only one of the ways recombinant DNA technology may soon be used to develop new vaccines (Chapter 7).

lion people immunized. Vaccines in use today must meet high safety standards, but risk can never be completely eliminated. Oral polio vaccine, for example, causes paralytic polio in approximately 1 out of 7 million people.

Besides being safe, vaccines must be highly **immunogenic**, meaning they should stimulate an immune response strong enough to confer protection against natural infection. No vaccine provides complete protection

to every person immunized. The measles vaccine, for example, protects 90 to 95 percent of those immunized. This level of protection was considered sufficient to keep measles under control by means of **herd immunity**, the prevention of epidemics due to the scarcity of new susceptible hosts (Chapter 26). But in the late 1980s, numerous measles outbreaks occurred in schools. Epidemiologists concluded that the 5 to 10 percent of students who were immunized but unprotected were sufficient to fuel an epidemic. As as result, a second dose of measles vaccine was recommended. The second round of immunization can be expected to protect 90 to 95 percent of those who remained unprotected after their first immunization, leaving a negligible number of people unprotected and thus ensuring herd immunity.

To be effective, vaccines must be appropriately administered. Every immunizing agent has a preferred route—either **orally** (by mouth), **subcutaneously** (below the skin), or **intramuscularly** (into muscle). The Sabin polio vaccine, for example, is administered orally, while the Salk polio vaccine is administered subcutaneously. To maintain their effectiveness, vaccines must also be properly stored and administered within a given time of their production.

Types of Vaccines for Active Immunization. All vaccines contain one or more microbial antigens that stimulate an immune response and protect against infection, but the nature and source of the antigens differ. Some vaccines contain live microorganisms that are attenuated, or weakened. Other vaccines contain intact (whole cell) microorganisms that are inactivated, or killed. The newest vaccines are composed of purified, genetically engineered antigens. Vaccines against toxins are made of modified toxins called toxoids. Some vaccines—such as the one for measles—stimulate both antibody and cell-mediated defenses. Others—such as the tetanus and diphtheria toxoids—stimulate primarily antibody-mediated immunity. **Table 20.4** lists the vaccines licensed by the FDA.

Attenuated vaccines cause a limited infection, usually without serious illness. Before the development of molecular genetics, we did not understand the mechanisms behind attenuating highly virulent microorganisms; they were sometimes not weakened sufficiently, which increased the risk of serious infection. The pathogen was usually cultivated under abnormal conditions until it lost virulence. For example, Pasteur attenuated his anthrax vaccine by growing *Bacillus anthracis* at the highest temperature it could tolerate. Now we know that high temperature causes the anthrax bacterium to lose the plasmid that encodes anthrax toxin. With attenuated vaccines, the organism multiplies within the host's body, producing antigens and in turn stimulating a powerful protective immune response. Attenuated vaccines tend to produce strong and long-lasting immunity. Most attenuated vaccines now used are for viral illnesses, including polio, mumps, measles, and rubella (German measles).

Some microorganisms cannot be used safely in vaccines unless they have been inactivated (killed) either by heat or by chemical agents such as formalin, phenol, or acetone. But **inactivated vaccines** have significant drawbacks. First, the process of inactivation can destroy the antigens that stimulate immunity, particularly when the microorganism is heat-inactivated. Also, because the microorganisms in inactivated vaccines cannot multiply in the host, the vaccine dose must contain enough antigen to produce a protective immunologic reaction. Most inactivated vaccines must be administered more than once, in **booster** doses. A booster dose stimulates production of much higher antibody levels, within only a day or two. This booster dose activates memory B cells produced during the initial response (Chapter 17). Inactivated vaccines are used to protect against many bacterial illnesses, including pertussis and typhoid fever, and some viral illnesses, including rabies and hepatitis B.

Research continues to develop new, safer, and more effective vaccines. The whole-cell vaccine for pertussis, for example, contains many antigens in addition to those that stimulate protective immunity. These other cellular constituents probably contribute to the frequent undesirable side effects of pertussis vaccination (Chapter 22). An acellular pertussis vaccine was recently licensed in the United States. It contains only the antigens that stimulate protective immunity. This genetically engineered vaccine has far fewer side effects than the traditional whole-cell vaccine, but there is some concern about its ability to stimulate vigorous immunity. Thus, the acellular vaccine is restricted to children over age one; infants still receive the whole-cell vaccine at two, four, and six months. Further clinical trials are underway, however, and the acellular vaccine may soon replace whole-cell DPT (the combined diphtheria, pertussis, tetanus inoculation) entirely.

A genetically engineered vaccine has already replaced an inactivated vaccine for immunization against hepatitis B, which causes a potentially chronic and life-threatening liver disease (Chapter 23). The first vaccine against hepatitis B, released in the early 1980s, was a major breakthrough in disease prevention; but many patients were uneasy because the viral particles used to prepare the vaccine were recovered from the serum of infected patients, many of whom also had other infections, including AIDS. In fact, the vaccine was safe and licensed by the Food and Drug Administration (FDA)—it protected against hepatitis B and did not transmit other infectious agents. Nevertheless, when a laboratory-produced purified vaccine was developed (involving no blood products), the original was replaced and immunization against hepatitis B became routine.

A genetically engineered vaccine may someday be produced for the sexually transmissible disease syphilis (Chapter 24). A syphilis vaccine has not been developed because the infecting microorganism cannot be cultured in the laboratory, and so adequate amounts of immunizing antigen cannot be produced. However, if a suitable immunizing antigen were identified, it could be cloned and produced in an easily cultivated organism such as *Escherichia coli*.

A better understanding of the immune response has also allowed researchers to develop new inactivated vaccines, such as the one against *Haemophilus influenzae* type b (Chapter 22). Antibodies against the polysaccharide capsule of *H. influenzae* type b protect against infection, but polysaccharides are weak stimulants of antibody production (Chapter 17). This problem was solved by *conjugating*, or combining, the polysaccharide antigen with a protein, such as diphtheria toxoid, making it a much more powerful antigen. Several different *H. influenzae* **protein conjugate** vaccines have since been developed that can be used to immunize babies as young as 2 months, protecting them against the many life-threatening forms of *H. influenzae* infection, including meningitis (Chapter 25) and epiglottitis (Chapter 22).

Because tetanus and diphtheria are caused by exotoxins rather than by the microorganisms themselves (Chapter 15), these diseases can be prevented simply by conferring immunity to the toxin. This is done with vaccines made of **toxoids**, toxins that have been modified by heat or chemical agents such as formalin so that they remain immunogenic but are harmless (**Figure 20.7**). Toxoids stimulate the production of antibodies called **antitoxins**, which combine with toxin molecules and inactivate them.

Active Immunization and Public Health Policy. Global immunization programs sponsored by the World Health Organization (WHO) have made spectacular progress in recent decades. Eradicating smallpox is its most stunning success, but the Expanded Program on Immunization has made significant progress against the six preventable diseases that cause the greatest worldwide childhood death and disability—measles, diphtheria, pertussis, tetanus, polio, and tuberculosis. In 1974 only 5 percent of the children in the developing world were immunized against all six of these diseases; by 1988 almost 50 percent were, saving 1.4 million lives per year, at a cost of less than 50 cents per child. Still, about 2 million children in developing countries die each year from measles, for example.

In the United States, the U.S. Public Health Service has established immunization goals for *all* Americans, because younger American children are the most adequately immunized and American adults the least (**Table 20.5**). For example, 77 percent of 2-year-olds are completely immunized; 97 percent of kindergarteners have received basic immunizations, but only 20 percent of adults who need

Table 20.4	Vaccines Licensed by the U.S. Food and Drug Administration
Vaccine	**Type**
Bacterial Diseases	
Acellular DPT	Protein antigens (pertussis) and toxoids (diphtheria and tetanus)
BCG (against tuberculosis)	Attenuated bacteria
Cholera	Inactivated bacteria
DPT (diphtheria, pertussis, tetanus)	Inactivated bacteria (pertussis) and toxoids (diphtheria and tetanus)
Haemophilus influenzae type b (against meningitis)	Polysaccharide-protein conjugate
HDPT (*Haemophilus influenzae* type b, diphtheria, pertussis, tetanus)	Combination of *Haemophilus influenzae* type b and DPT vaccines
Meningococcal meningitis	Polysaccharide
Plague	Inactivated bacteria
Pneumococcal pneumonia	Polysaccharide
Td (diphtheria, tetanus for adults)	Toxoids
Typhoid fever	Inactivated bacteria (injected) or attenuated bacteria (oral)
Viral Diseases	
Hepatitis B	Inactivated viral antigen derived from recombinant yeast
Influenza	Inactivated virus
Measles	Attenuated virus
MMR (measles, mumps, rubella)	Attenuated virus
Poliovirus	Inactivated virus (injected—IPV) or attenuated virus (oral—OPV)
Rabies	Inactivated virus
Yellow fever	Attenuated virus

protection against influenza and pneumonia have been immunized.

Passive Immunization

Active immunization stimulates the recipient's own immune system to mount a protective response that prevents later infection. Passive immunization is a kind of

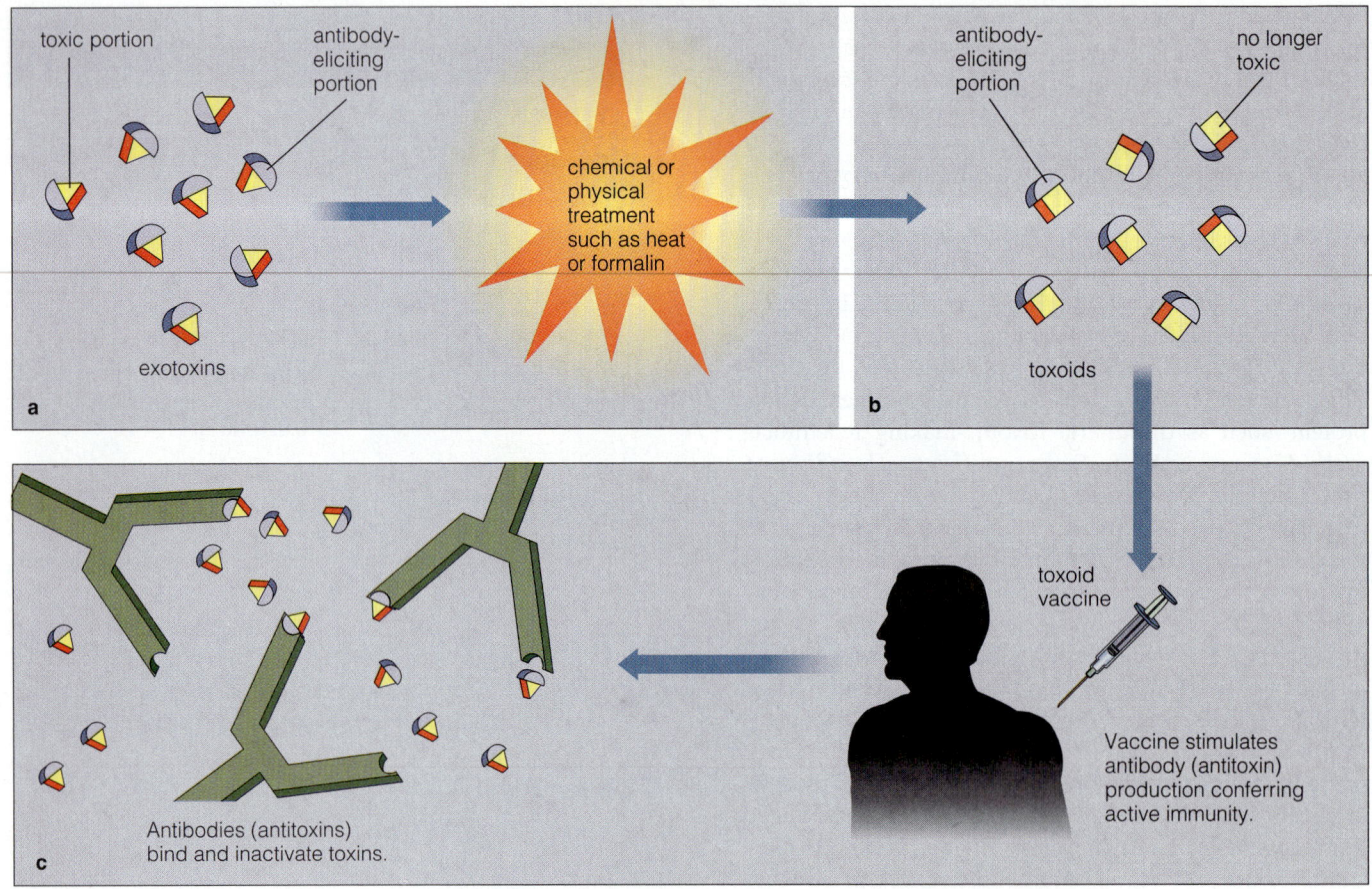

Figure 20.7 Preparation and action of toxoid vaccines. (**a**) Exotoxin is treated with heat or a chemical compound such as formalin. (**b**) Treatment produces a toxoid, which is harmless but still immunogenic. (**c**) When injected as a vaccine, the toxoid stimulates the formation of antibodies called antitoxins. The antitoxins react with natural exotoxins to inactivate them and thus confer active immunity.

antibody transfusion. Antibodies produced by a human or animal donor are collected, purified, and administered to the recipient. This type of immunization is called *passive* because protection does not require participation of the recipient's immune system. However, the protection lasts only as long as the antibody molecules survive in the recipient—months with antibodies from another human, but only weeks with antibodies from animals.

The antibody preparations used may be **gamma globulin**—a collection of antibodies from the pooled serum of many different donors—or special preparations that contain high titers of specific antibodies. For example, **tetanus immune globulin** contains high concentrations of antibodies against tetanus toxin, and **varicella zoster immune globulin** contains high concentrations of antibodies against the virus that causes chickenpox and shingles.

Passive immunization has certain advantages. Even severely immunosuppressed patients can be protected,

and protection is immediate. But passive immunization also has disadvantages. Protection lasts only a short time, and if animal antibodies are used, there is a risk of serum sickness (Chapter 18). In general, other forms of prevention or treatment usually work better than passive immunization, but for certain diseases such as hepatitis A, it is used extensively because no vaccine is yet available.

Passive immunization can occasionally be useful in preventing tetanus (Chapter 25). Most of us who have received the tetanus toxoid vaccine only need a booster dose of the toxoid after receiving a tetanus-prone wound (such as stepping on a dirty nail). Because the immune system has responded to the vaccine once, the booster dose stimulates a memory response immediately—in plenty of time to protect against tetanus. A never-immunized person, however, who will mount a primary rather than a memory immune response when receiving the vaccine, needs immediate protection against the poten-

Table 20.5	Recommended Immunization Schedule for Children
Age	Immunization
2 months	DPT (diphtheria, tetanus, whole-cell pertussis vaccine)
	Haemophilus influenzae type b conjugate vaccine (against meningitis)
	OPV (oral polio vaccine)
4 months	Same as 2 months
6 months	Same as 2 months
Within first year	Three doses of hepatitis B vaccine with the second dose given one month after and the third dose three months after the original dose
15 months	MMR (measles, mumps, rubella)
	Haemophilus influenzae type b conjugate vaccine booster
15–18 months	Acellular DPT and OPV booster
4–6 years (school entry)	Acellular DPT and OPV booster
	Second MMR
14–16 years	Td (tetanus and diphtheria) booster

tially fatal tetanus toxin. Thus, unimmunized people who suffer a dirty wound should receive tetanus immune globulin (passive immunization) at the same time they receive their first dose of tetanus toxoid vaccine (active immunization). They should then go on to receive the full series of tetanus boosters so they will be adequately protected in the future.

Summary

THE SCIENCE OF EPIDEMIOLOGY (p. 466)

1. Epidemiology is the study of when and where diseases occur and how they are transmitted in populations. Public health refers to the development and implementation of plans to prevent and control disease.

2. Epidemiology focuses on populations rather than individuals. Disease can be transmitted in an epidemic, affecting many members of a population in a short time, or in a pandemic, which is an epidemic that spreads worldwide. Some diseases are endemic, always present in a population at about the same level; others are sporadic, occurring only occasionally in a population.

THE METHODS OF EPIDEMIOLOGY (pp. 466–470)

1. Many sources of epidemiological information are part of the public record, including vital statistics and census data. Government agencies track notifiable, or reportable, diseases, which are of immediate public health concern.

2. The Centers for Disease Control and Prevention (CDC) is the national agency that maintains and reports epidemiological trends in disease. It publishes the *Morbidity and Mortality Weekly Report* (*MMWR*).

3. Epidemiologists use surveys, questionnaires, interviews, and hospital records to gather specific information.

4. Epidemiologists usually express health information about a population as a rate—the ratio of the number of people in a particular category to the total number of people in the population being studied. Unlike raw data, rates allow comparisons to be made.

5. Incidence rate, which measures the growth of an epidemic, is the number of people who develop a disease or condition during a certain period of time divided by the total number of people in the population.

6. Prevalence rate, which measures the magnitude of an epidemic, is the number of people who have a certain disease at any particular time divided by the number of people in the population.

7. Age-adjusted rates are usually more informative than raw data.

TYPES AND USES OF EPIDEMIOLOGICAL STUDIES (pp. 470–477)

1. All epidemiological studies focus on groups rather than individuals, express results as rates, compare them statistically, and draw generalizations about the cause and prevention of disease.

2. Retrospective studies analyze events that have already occurred. Prospective studies record events as they are happening and analyze them at the end of a certain period. Experimental studies deliberately influence events to see what the outcome will be.

Descriptive Epidemiology (pp. 471–472)

1. Descriptive epidemiology provides general information about a disease, including how nonmicrobial factors contribute to infectious disease.

Surveillance Epidemiology (pp. 472–473)

1. Surveillance epidemiologists track epidemic diseases by monitoring situations that might lead to epidemics and following the progress of an epidemic if one begins. Surveillance epidemiology played an important role in global eradication of smallpox.

Field Epidemiology (pp. 473–474)

1. Field epidemiologists investigate unexpected outbreaks of disease by compiling information and piecing it together to determine the source of the epidemic and how it can be controlled. Field epidemiologists discovered the sources of AIDS, Lyme disease, and legionellosis.

2. When large numbers of people at one gathering become ill simultaneously the phenomenon is called a common source epidemic.

Hospital Epidemiology (pp. 474–477)

1. Nosocomial infections are hospital-acquired infections. Gravely ill patients are susceptible to infection because of weakened immune systems. Invasive procedures bypass the body's first-line surface defenses.

2. Urinary tract infections, usually caused by catheterization, account for nearly half of all nosocomial infections. The most common causative organisms are *Escherichia coli* and other enteric organisms.

3. Despite scrupulous precautions, surgical wound infections are the second most frequent type of nosocomial infection. The most

common causative organisms are *Staphylococcus aureus* and Gram-negative enterics.

4. Pneumonias rank third among nosocomials, caused by *Streptococcus* and *Staphylococcus* spp., *Pseudomonas aeruginosa*, and enterics.

5. Skin infections, especially among newborns and burn victims, are a close fourth among nosocomials. The most common causative organism is *Pseudomonas aeruginosa*.

6. Hospital staff epidemiologists are specially trained to recognize and interrupt a potential hospital epidemic. The measures taken depend upon the nature of the outbreak. Staff epidemiologists communicate the problem and enforce routine infection-control policies such as hand washing.

7. Isolation is a physical separation of individuals to prevent the transmission of infection. Reverse isolation separates infection-prone patients from sources of disease.

8. The Universal Blood and Body Fluid Precautions, issued by the CDC to protect hospital workers, apply to all patients because it is impossible to tell which patients are HIV-infected.

PUBLIC HEALTH: PREVENTING DISEASE (pp. 477–480)

1. Public health deals with prophylaxis, prevention of disease. The two major methods are limiting exposure to a pathogen, by decreasing or eliminating its reservoir or interrupting transmission, and immunization, artificially augmenting the body's natural immune defenses.

2. Agencies to protect the public health exist at the local, state, national, and global levels. The United States Public Health Service (USPHS) is our national public health agency. The CDC is one of its arms. The World Health Organization (WHO) is the international organization concerned with improving health.

3. Waterborne diseases can be prevented by clean water supplies and adequate sewage treatment. Contaminated water is the cause of cholera, typhoid fever, and deadly infant diarrheas.

4. Heat kills most foodborne diseases; refrigeration slows the growth of microorganisms. Pasteurization makes milk—which is particularly likely to carry infection—safe to drink.

5. Probably the single most important habit of personal cleanliness that can prevent the spread of disease is hand washing.

6. Diseases transmitted by insect bites can be controlled by decreasing or eliminating the reservoirs; insecticides are not the final answer.

7. All sexually transmissible diseases are preventable by interrupting the chain of transmission. This depends upon ongoing individual effort. AIDS prevention is imperative because there is no cure.

8. Immunization is the best way to prevent transmission of respiratory diseases because it is virtually impossible to eliminate pathogens from the air we breathe.

IMMUNIZATION (pp. 480–485)

1. Immune defenses can be manipulated artificially to prevent infection.

Active Immunization (pp. 480–483)

1. In active immunization, or vaccination, a person's own immune system is artificially stimulated to ward off disease. A vaccine is an agent capable of inducing active immunity without causing disease.

2. Good vaccines are safe, relatively free of side effects, and highly immunogenic.

3. Herd immunity refers to prevention of epidemics because of scarcity of susceptible hosts.

4. Every vaccine has a preferred route of administration, either orally, subcutaneously, or intramuscularly.

5. Attenuated vaccines contain live microorganisms that are weakened. They cause a limited infection, usually without serious illness, and stimulate a powerful and long-lasting immune response because the live organisms multiply in the body, producing antigens. Most attenuated vaccines are used for viral illnesses, including polio, mumps, and measles.

6. Inactivated vaccines contain killed microorganisms. However, the process of inactivation can destroy the antigens that stimulate immunity. The dose must contain enough antigen to produce a protective immunological response because the cells cannot multiply to produce more. Most inactivated vaccines thus require a booster shot. Inactivated vaccines are used against pertussis, typhoid fever, rabies, and hepatitis B.

7. Genetic engineering is changing the way vaccines are made. For example, a genetically engineered vaccine for hepatitis B contains no contaminated blood products, making hepatitis B immunization more acceptable to many patients.

8. The new vaccine for *Haemophilus influenzae* type b is a protein conjugate vaccine. The polysaccharide antigen is conjugated, or combined, with a protein, such as diphtheria toxoid, which transforms it into a much more powerful antigen.

9. Toxoid vaccines are made of toxins modified by heat or chemicals so they remain immunogenic but are harmless. Toxoids stimulate the production of antibodies called antitoxins. These vaccines are used against tetanus and diphtheria.

10. WHO has made significant progress against the six preventable diseases that cause the greatest worldwide childhood death and disability: measles, diphtheria, pertussis, tetanus, polio, and tuberculosis.

11. The USPHS has established immunization goals for all Americans. Even in the United States, more widespread active immunization would improve public health.

Passive Immunization (pp. 483–485)

1. In passive immunization, antibodies produced in a donor human or animal are collected, purified, and administered to the recipient. The recipient's immune system is not involved.

2. Passive protection lasts only as long as the antibody molecules survive, weeks to months. Gamma globulin, a collection of various antibodies from the pooled serum of many donors, or special preparations with high titers of specific antibodies are used. For example, tetanus immune globulin contains high concentrations of antibodies against tetanus toxin.

3. Passive immunization can be used with immunosuppressed patients and protection is immediate. Because protection lasts only a short time, antibodies must be administered after every known exposure. With animal antibodies, there is a risk of serum sickness.

4. For some diseases, passive immunization is best. An injection of gamma globulin for someone exposed to hepatitis A is almost certain to provide needed protection. Someone who has never been immunized against tetanus and receives a tetanus-prone wound should receive a shot of tetanus immune globulin and be actively immunized.

THE SCIENCE OF EPIDEMIOLOGY

1. What is epidemiology? How does it differ from the field of public health?

2. Distinguish between these patterns of disease transmission in a population—epidemic, pandemic, endemic, sporadic.

THE METHODS OF EPIDEMIOLOGY

1. Why are reliable sources of information important in epidemiology? Why are vital statistics and the census useful? Name some other sources of information epidemiologists use.

2. What are notifiable (reportable) diseases? What is the CDC? the *MMWR*?

3. Explain this statement: Statistics is the language of epidemiology.

4. Define these terms: rate, incidence rate, prevalence rate, age-adjusted rate. Give an example of how knowing a rate would be more useful than having raw data.

TYPES AND USES OF EPIDEMIOLOGICAL STUDIES

1. What are the three things that all epidemiologic studies have in common?

2. What is the difference between a retrospective study and a prospective study? What is an experimental study?

3. What is descriptive epidemiology? Explain how a descriptive epidemiologist might come to the conclusion that poverty causes tuberculosis as much as *Mycobacterium tuberculosis* does.

4. What is surveillance epidemiology? Why was effective surveillance epidemiology critical in eradicating smallpox worldwide?

5. Explain this statement: Field epidemiologists are like detectives.

6. What are nosocomial infections? What three factors make nosocomial infections almost inevitable?

7. Describe the most common kinds of nosocomial infections and, in each case, the microorganisms involved.

8. Describe the work of a hospital epidemiologist. What are universal precautions?

PUBLIC HEALTH: PREVENTING DISEASE

1. Give some examples of how local, state, national, and global agencies help scientists and clinicians improve public health.

2. Explain this statement: Public health deals with prophylaxis.

3. Discuss some ways public health has diminished reservoirs and interrupted transmission in each of these areas: clean water, clean food, personal cleanliness, insect control, prevention of sexually transmissible diseases, and prevention of respiratory diseases.

IMMUNIZATION

1. Define these terms: immunization, vaccine, immunogenic, herd immunity.

2. What is the difference between active immunization and passive immunization?

3. Explain this statement: All good vaccines are *relatively* safe and *relatively* effective.

4. What is an attenuated vaccine? What is an inactivated vaccine? What are the advantages and disadvantages of each?

5. Give an example of how genetic engineering has made a vaccine safer. Give an example of how genetic engineering has made a vaccine more effective. Why do we look to genetic engineering to provide a vaccine for syphilis?

6. What are toxoid vaccines and how are they produced? Give some examples of toxoid vaccines.

7. Explain this statement: Even in the United States, more widespread immunization would improve public health.

8. What is passive immunization? What are its advantages and disadvantages compared to active immunization? Give an example of when passive immunization is best.

9. What is gamma globulin? What is tetanus immune globulin?

Essay Questions

1. Discuss John Snow's study of the Broad Street cholera epidemic from the standpoint of how it defines the field of epidemiology.

2. Sketch some scenarios illustrating how age-adjusted death rates might be essential to reaching a valid conclusion in an epidemiological study.

Suggested Readings

Robbins, A., and Freeman, P. 1988. Obstacles to developing vaccines for the third world. *Scientific American*, November, 126–33.

Rose, G., and Barker, D. J. P. 1986. *Epidemiology for the uninitiated*. 2d ed. London: British Medical Journal.

Snow, J. 1965. *Snow on cholera*. New York: Hafner.

21 ANTIMICROBIAL PHARMACOLOGY

Scarlet Fever—Then and Now

T. M., a 6-year-old kindergartner, came home from school tired and listless, with a fever of 102°F. His mother put him to bed and gave him the prescribed dose of children's acetaminophen to relieve the fever. That night T. M. refused dinner, complaining that he wasn't hungry and it hurt to swallow. In the morning his father noticed he had a rash over most of his body and his tongue was bright red and swollen. He called the pediatrician for an appointment that day.

When T. M.'s pediatrician examined him, he still had a temperature and appeared tired and uncomfortable. His throat was bright red, with white patches on his tonsils and tiny red spots on the soft palate. His tongue was inflamed, swollen, and bumpy. Lymph nodes in his neck were enlarged and tender, and the bright red rash covering most of his body felt like coarse sandpaper. From these signs and symptoms, T. M.'s pediatrician made a probable diagnosis of scarlet fever—a clinical syndrome caused by the highly pathogenic bacterium *Streptococcus pyogenes*.

Her diagnosis was confirmed by a latex agglutination test on cells swabbed from T. M.'s throat. Since T. M. had no history of drug allergies, the physician prescribed penicillin, the preferred drug for *S. pyogenes* infections. T. M. received a liquid preparation of penicillin VK, a potassium salt of the orally absorbed penicillin V, three times a day. Twelve hours after the penicillin therapy began, T. M. was feeling much better. His recovery was complete and uneventful and soon he was back at school.

Fifty years ago, a child who contracted scarlet fever was extremely ill and often died. Today most children

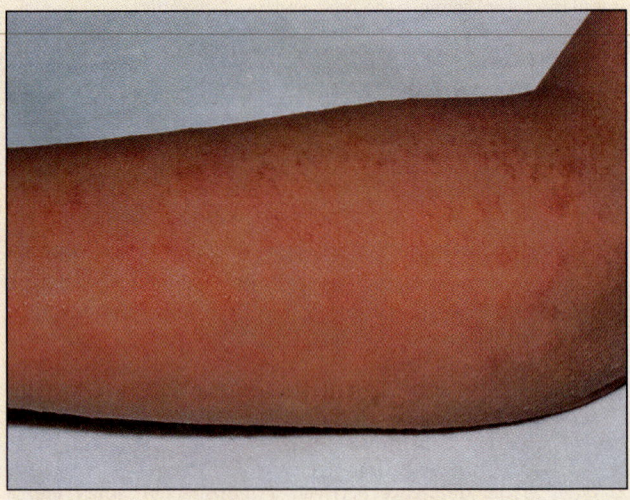

The rash typical of scarlet fever.

are only mildly ill and recover quickly with antibiotic treatment. In fact, to avoid the frightening associations that a diagnosis of scarlet fever can still hold for a parent, physicians sometimes use the term *scarletina*. As much as antibiotics help, though, the dramatic difference between scarlet fever today and yesterday is due to the diminished virulence of streptococcus. But streptococcus virulence is rising again. Will antibiotics continue to be helpful, or will antibiotic resistance become prevalent?

To understand:

- How drugs are chosen and administered

- How drugs are distributed in the body and eliminated

- How drugs act and how microorganisms acquire resistance to them

- How antimicrobial susceptibility is tested and standard dosage determined

- The characteristics of the drugs most frequently used to combat bacterial, mycobacterial, fungal, parasitic, and viral infections

PRINCIPLES OF PHARMACOLOGY

T. M.'s pediatrician used her knowledge of pharmacology to choose the safest, most effective drug treatment for T. M.'s infection. Today, most bacterial infections and some viral, fungal, and parasitic infections can be treated with antimicrobial drugs, most of which were developed only within the last 50 years. To get a perspective on the impact these drugs have had on our lives, consider this: Finding a cure for all cancers would probably add 2 years to the average life expectancy of Americans; antimicrobial drugs added 10 years. Many readers of this textbook survived a childhood infection thanks to treatment with one of the drugs in this chapter.

Earlier chapters have described how drugs disrupt microbial metabolism. In this chapter we focus on the ways drugs interact with microorganisms in the human body. First, we consider the basic principles of pharmacology. Next, we consider how antimicrobials act on microorganisms. Finally, we discuss some of the most useful drugs available today.

Pharmacology is the study of **drugs**—any chemicals that have a physiological effect on living things; alcohol, nicotine, and caffeine are drugs. Drugs used to treat disease are called **chemotherapeutic agents**, and specific chemotherapeutic agents used to treat infections are called **antimicrobial agents**.

The wide variety of antimicrobial agents available today can be categorized in several ways. One is by source (**Table 21.1**). **Antibiotics** are metabolic products of one microorganism that kill or inhibit the growth of other microbes (**Figure 21.1**). **Synthetic drugs** are chemicals produced in the laboratory, and **semisynthetic drugs** are antibiotics that have been chemically modified in the lab-

Table 21.1	Microorganisms That Produce Antibiotics
Microorganism	**Antibiotic**
BACTERIA	
Streptomyces spp.	
S. griseus	Streptomycin
S. kanomyceticus	Kanamycin
S. fradiae	Neomycin
S. noursei	Nystatin
S. mediterranei	Rifampin
S. aureofaciens	Tetracyclines
S. nodosus	Amphotericin B
S. venezuelae	Chloramphenicol (also synthetic)
S. orientalis	Vancomycin
S. erythreus	Erythromycin
Bacillus spp.	
B. subtilis Tracy-I	Bacitracin
B. polymyxa	Polymyxin B
Micromonospora spp.	
M. purpurea	Gentamicin
FUNGI	
Penicillium spp.	
P. notatum, P. chrysogenum	Penicillin
P. griseofulvum	Griseofulvin
Cephalosporium spp.	
C. acremonium	Cephalosporins

oratory. Antimicrobial agents are also categorized by clinical usefulness. **Antibacterial agents** are used for bacterial infection, **antimycobacterial agents** for tuberculosis and related mycobacterial infections, **antiviral agents** for viral infections, **antifungal agents** for fungal infections, and **antiparasitic agents** for protozoal and helminthic infections.

Drug Administration

To start a patient on drug therapy, the clinician must decide on a route of administration, which may be external or internal. **External therapy**—also called **local** or **topical**—is the application of an antimicrobial drug directly to the infected area. Minor infections of the body surface, such as impetigo or yeast infections (Chapter 26), are treated with local therapy. Most infections, however,

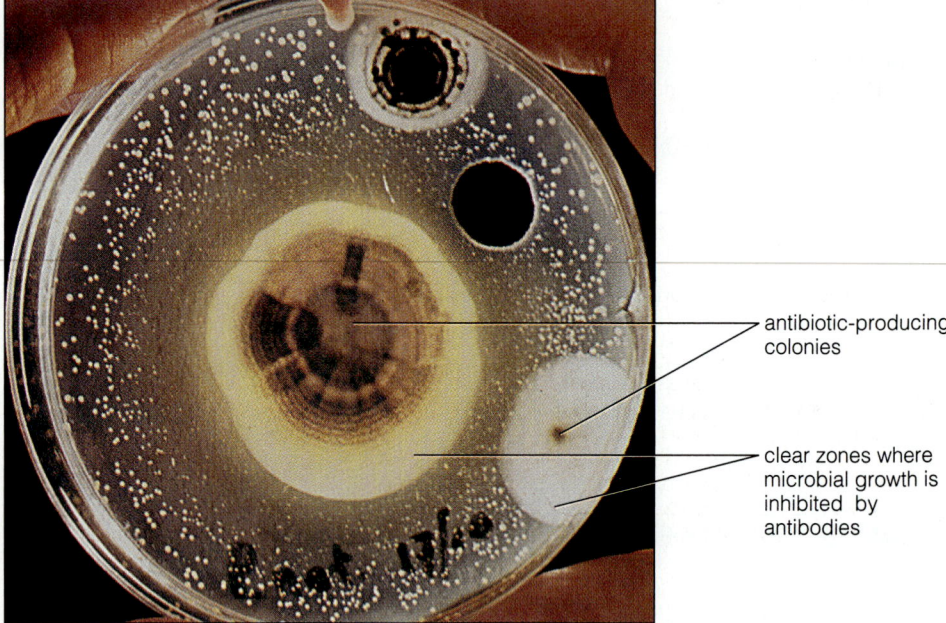

Figure 21.1 Antibiotic-producing microorganisms. A plate inoculated with soil often has colonies, like these, that produce antimicrobial compounds, a few of which are useful clinically as antibiotics.

antibiotic-producing colonies

clear zones where microbial growth is inhibited by antibodies

require **systemic therapy**, meaning that the drug enters the patient's bloodstream.

Administration of systemic therapy may be intravenous, intramuscular, or oral. In **intravenous (IV) administration**, the drug is injected directly into a vein. This is the fastest way to get high levels of a drug into the bloodstream. However, it is technically difficult and painful; and a needle or plastic catheter must remain in the patient's vein throughout therapy, risking opportunistic infection. Intravenous drug therapy is therefore reserved for the treatment of serious infections in hospitalized patients.

In **intramuscular (IM) administration**, the drug is injected into a muscle, usually the deltoid or gluteus maximus. This mode allows a drug to reach peak levels in the circulation within 15 minutes because muscle is richly supplied by blood vessels. However, IM administration is painful, and the injection must be administered by a trained professional.

In **oral administration** (or **PO**, from the Latin *per os*, "by mouth"), the drug is swallowed and absorbed into the bloodstream from the gastrointestinal tract. This is the simplest and most common route of drug administration and is painless. But oral administration can mean slow and possibly inefficient absorption because only a fraction of the dose is absorbed and reaches the bloodstream (**Figure 21.2**). Moreover, most orally administered drugs must be taken several times a day on consecutive days, which can result in dosage errors or failure to complete the full course of treatment.

T. M.'s physician decided on oral administration of penicillin because even though T. M. was uncomfortable,

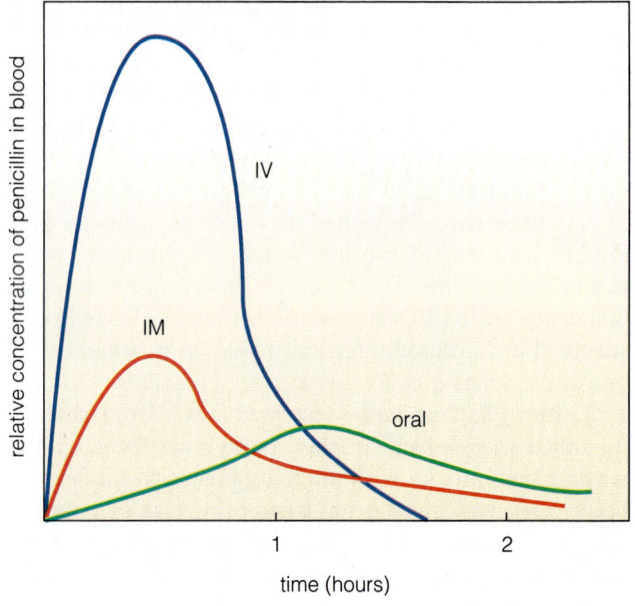

relative concentration of penicillin in blood

IV

IM

oral

1   2

time (hours)

Figure 21.2 Route of administration and blood levels of penicillin. Intravenous administration produces the highest blood levels within the shortest period of time. Intramuscular administration produces relatively high levels in less than an hour. Orally administered penicillin takes more than an hour to reach its maximum blood level, which is much lower than that reached by the other methods.

he did not require hospitalization and intravenous drug therapy. A single IM injection of a benzathine salt of penicillin G would also have been a good choice because it dissolves slowly and provides a continuous source of penicillin during the entire 10-day treatment period. This

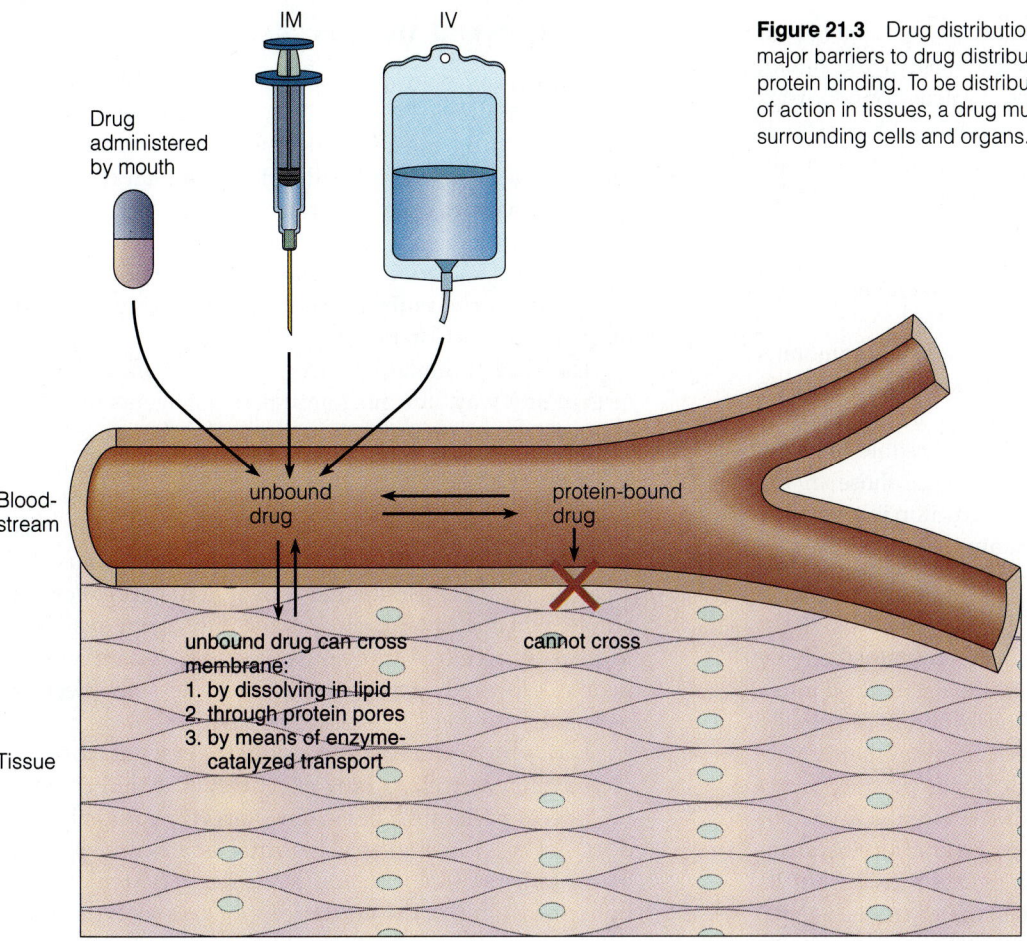

Figure 21.3 Drug distribution in body tissues. There are two major barriers to drug distribution in the body—membranes and protein binding. To be distributed from the bloodstream to the sites of action in tissues, a drug must pass through membranes surrounding cells and organs.

route of administration was rejected only because T. M.'s parents did not want him to receive the painful shot. Oral therapy with penicillin VK is the most common form of treatment for scarlet fever as well as for the more common streptococcal throat infection. Penicillin VK is preferred to oral penicillin G because penicillin G is so readily destroyed by stomach acid that only one-third of the ingested dose is absorbed into the bloodstream. In contrast, about 65 percent of ingested penicillin VK is absorbed. As long as T. M. swallowed the medication and his parents remembered to give it to him three times a day during the prescribed 10-day course, oral treatment would be fully effective.

Drug Distribution

The **distribution** of a drug refers to where in the body it is found after systemic administration (**Figure 21.3**). The membranes surrounding cells and organs present significant barriers to drug distribution. Drug delivery to the brain and eye tissue is particularly difficult because they are surrounded by nearly impermeable membranes. As a result, infections of the central nervous system and the eye are relatively difficult to treat with antimicrobial agents. High dosages must be prescribed to be sure the drug is well distributed to these organs.

Membranes in the human body, like the cell membranes of microorganisms, are composed of a phospholipid bilayer with embedded proteins (Chapter 4). Drugs cross human membranes by passing through one or the other of these two components. Drugs that dissolve readily in the phospholipid component pass through the membrane by diffusion, eventually reaching equal concentrations on both sides. However, most drugs, including lipid-soluble ones, exist within body fluids in partly ionized form, and only uncharged molecules can dissolve in phospholipids and diffuse through them. Drugs that are not lipid-soluble can only pass through the protein part of a membrane, either through pores or by means of special enzyme-catalyzed transport systems. Some of these enzyme-catalyzed systems mediate active transport (Chapter 4).

Another barrier to distribution is the tendency of many drugs to bind to plasma proteins in the blood, forming protein-drug complexes. Protein-bound drugs cannot cross membranes. Protein binding, however, is partial and reversible. As the unbound fraction is metabolized or excreted from the body, previously bound molecules are released. Thus, if a drug is 50 percent protein-bound, half of the drug molecules in the body will be bound and half will be available to cross membranes, regardless of the total amount of drug present.

T. M.'s physician was confident that adequate amounts of penicillin would travel to his throat tissues because penicillin is widely distributed throughout the body. It reaches concentrations that can kill susceptible microorganisms within almost all tissues except those in the brain and eye. About 65 percent of penicillin is reversibly bound to the plasma protein **albumin**.

Eliminating Drugs from the Body

Sooner or later all drugs are eliminated from the body by one means or another. Some drugs are metabolized into a different compound. Drug metabolism occurs principally in the liver and usually produces a chemical that is inactive as a drug. Sometimes, though, it may produce an active compound or even one with enhanced activity. For example, the drug prontosil, one of the first successful antimicrobial agents, has no antimicrobial activity, but its metabolic product, sulfa, is a potent antimicrobial agent.

Most drugs are eliminated by excretion, usually in the urine, which is formed as the kidneys filter blood to remove impurities. A few drugs are excreted by liver cells into the bile and eliminated from the body in the feces. It is critically important to know how a drug is excreted, because a drug administered to someone whose liver or kidneys are not functioning normally might rise to toxic levels.

Before prescribing penicillin, then, T. M.'s physician considered how the drug would leave his body. Penicillin is excreted rapidly through the kidneys because it is actively transported from the blood into the urine. This means penicillin levels in the bloodstream decline rapidly (Figure 21.2). Within an hour and a half after a person receives penicillin intramuscularly, most of the dose can be found in the urine. T. M., a healthy child with normally functioning kidneys, could be expected to excrete his oral dose of penicillin almost as rapidly. Thus, to maintain adequate levels of penicillin to combat the infection, he would have to take his medication regularly and often. In the early days of penicillin, when the drug was scarce and expensive, penicillin was recovered from patients' urine so that the precious material could be reused. Today, probenecid, a drug that slows the rate of excretion, is sometimes administered along with penicillin to prolong its action.

Side Effects and Allergies

A drug that affected microbial and human cells equally would be too toxic to administer. To be effective, an antimicrobial drug must be **selectively toxic**—it must inhibit or kill microorganisms but leave human cells relatively unharmed. Antimicrobial drugs do this by attacking biochemical targets, such as the peptidoglycan cell wall or bacterial ribosomes, that do not exist in human cells or are substantially different in humans and microorganisms (Chapter 4).

The ideal antimicrobial drug would not affect human cells in any way, but most antimicrobial agents do have **side effects**, or undesirable toxicity. Some drug side effects are explainable in terms of similarities between the drug target in microorganisms and structures in human cells, but the exact mechanism of most side effects is poorly understood. The danger of side effects must be weighed against the potential benefit of the drug. For example, when streptomycin was the only agent available to treat tuberculosis (previously a lethal disease), some people knowingly accepted the disabling side effect of permanent deafness in order to be cured.

T. M.'s physician had to consider the potential toxicity of penicillin as well as how effectively it would combat his infection. Because penicillin's target, peptidoglycan, does not exist in humans, the drug is a nearly perfect magic bullet, killing bacteria while leaving human cells unharmed. Only in the most exceptional cases does penicillin cause side effects. However, drugs can be dangerous—even life-threatening—if they stimulate allergic reactions (type I hypersensitivity reactions; Chapter 18). Penicillin is a relatively potent hapten, and because of its extensive use many people have developed allergies to it (**Figure 21.4**). This was the reason T. M.'s physician inquired carefully about a history of penicillin allergy before prescribing it. Once an allergy to penicillin develops, no drug from the penicillin family can be administered without risking anaphylaxis. Because every exposure to an antibiotic carries a small risk of stimulating drug allergy, penicillin, like other antimicrobials, should never be administered needlessly. For example, using penicillin to treat a viral infection such as a cold is not only completely ineffective but also increases the risk of allergy.

Drug Resistance

Drug resistance means a microorganism can grow and reproduce in the presence of a particular drug. Antimicrobial resistance can be either natural or acquired. When pharmacologists first test a new antimicrobial agent, some species of microorganisms are found to be sensitive to its action while other species are unaffected. Thus, **natural drug resistance** is a property of an entire microbial species. Over time, certain strains of a drug-sensitive

Turning Point

The Original Wonder Drug

When penicillin became available for widespread clinical use in the 1940s, the treatment of infectious diseases was transformed. People with potentially fatal infections were cured within hours. Penicillin was called a "wonder drug," and justly so. Penicillin remains the most widely used of all antibiotics. It is the drug of first choice for treatment of infections caused by *Streptococcus pyogenes*, including streptococcal pharyngitis (strep throat), streptococcal wound infections, and scarlet fever.

The antibiotic penicillin is produced by fungi belonging to the genus *Penicillium*. Penicillin kills bacterial cells by interfering with the process by which they incorporate new peptidoglycan units into the cell wall. Normally, as bacterial cells grow, chemical bonds in the peptidoglycan wall break and new peptidoglycan units are inserted. But penicillin binds and inactivates the enzymes, called **penicillin-binding proteins**, that catalyze the resealing process. When the breaks cannot be resealed, the wall is weakened and the cell lyses.

The penicillin molecule consists of two fused rings and a changeable R group. One of these rings, the **beta-lactam ring**, can be destroyed by **beta-lactamase** (also called **penicillinase**), an enzyme produced by certain strains of penicillin-resistant bacteria. Destruction of the beta-lactam ring inactivates the penicillin molecule. The different R groups (chemical side chains attached to the beta-lactam ring) modify penicillin's antimicrobial action. The penicillin that was first isolated (Chapter 1), now termed penicillin G, has a benzyl ring R group, but *Penicillium* can add different R groups and they can be chemically substituted at this site to create other types of penicillin. The R group in penicillin V makes the penicillin molecule more resistant to stomach acid and therefore more suitable to oral administration. The R group in amoxicillin makes penicillin effective against Gram-negative bacteria, which are naturally resistant to penicillin G. Still other side chains—like the one methicillin has—make the beta-lactam ring more resistant to penicillinases.

The second ring of penicillin is the **thiazolidine ring**. A chemical substitution at the carboxyl group attached to this ring creates a charged molecule that forms salts with compounds such as procaine, benzathine, or potassium. These salts have special therapeutic purposes. The procaine salt of penicillin G, for example, dissolves rapidly and releases

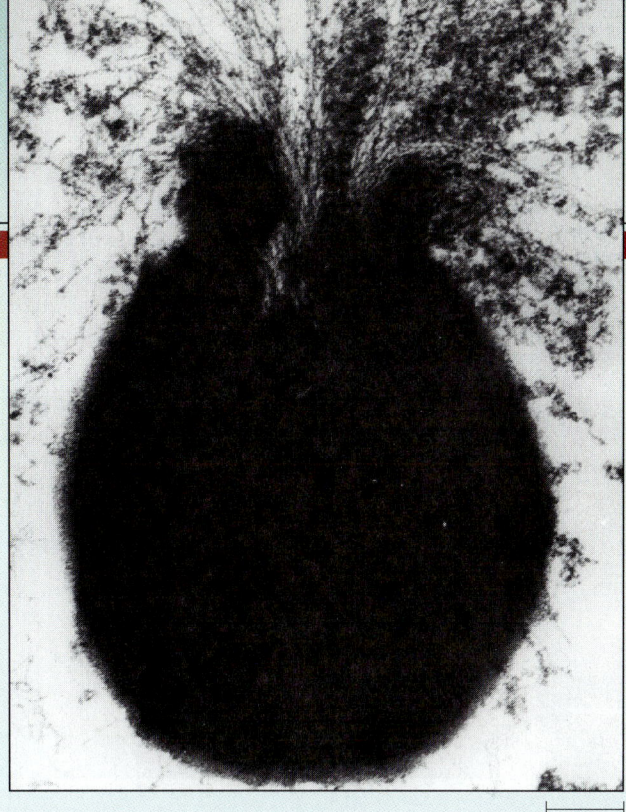

a Penicillin acts by interfering with the process by which bacterial cells incorporate new peptidoglycan units into cell walls. This causes cells to lyse, like this *Staphylococcus aureus* cell, after treatment.

b The penicillin G molecule contains a beta-lactam ring, a thiazolidine ring, and a benzyl ring R group.

high concentrations of penicillin immediately after injection. The benzathine salt dissolves slowly and releases low levels of penicillin for several weeks. The potassium salt of penicillin V, called penicillin VK (K is the chemical notation for potassium), is one of the most common forms of penicillin administered orally.

493

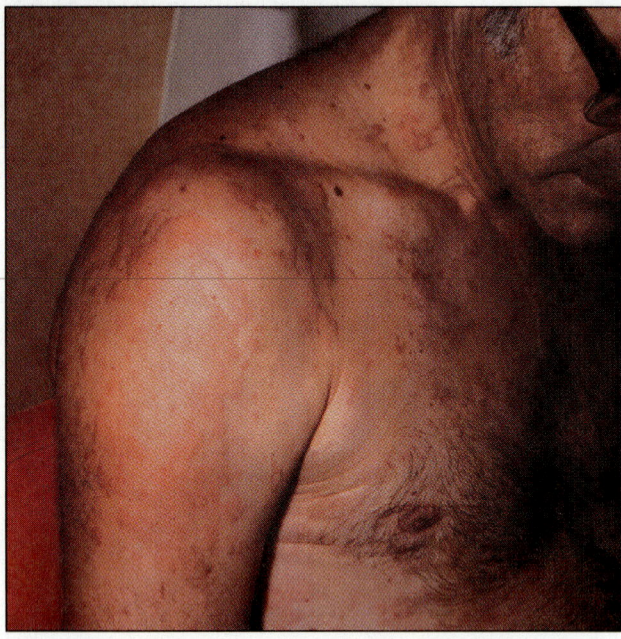

Figure 21.4 This patient's rash is a type I hypersensitivity reaction to penicillin.

species may become resistant. **Acquired drug resistance** is a property of individual strains within a species. Acquired resistance is due to a genetic alteration that is passed from one generation of a microbial strain to the next.

Natural Drug Resistance and Drug Spectrum. Natural resistance to antimicrobial agents occurs for various reasons. Some species lack the target that the antimicrobial drug attacks. For example, penicillin interferes with peptidoglycan cell wall synthesis, so any organism that lacks a peptidoglycan cell wall is naturally resistant. Penicillin is therefore not prescribed for infections caused by viruses, fungi, protozoa, or wall-less bacteria, such as the mycoplasmas. Some species are naturally resistant because the antimicrobial drug is not able to enter the cell and exert its damaging effect. Gram-negative bacteria are protected from many drugs, including most penicillins, by their relatively impermeable outer membrane. Although they enlarge their cell wall with the same penicillin-sensitive enzymes as do the Gram-positives, the outer membrane prevents penicillin molecules from reaching penicillin-binding proteins.

Based on their pattern of natural sensitivity or resistance, antibacterial agents are classified as being either narrow-spectrum or broad-spectrum (**Figure 21.5**). **Narrow-spectrum** antibacterial drugs affect only a single microbial group and are the best choice when the causative microorganism of an infection is known. They target the

pathogen fairly specifically and leave most of the body's normal protective microbial flora unharmed. T. M. was treated with penicillin VK, a narrow-spectrum drug that kills Gram-positive bacteria, because scarlet fever is caused only by *Streptococcus pyogenes*. Penicillin G is a narrow-spectrum drug also effective almost exclusively against Gram-positive bacteria. Polymixin B, on the other hand, is effective only against Gram-negatives. Isoniazid, another narrow-spectrum drug, is effective only against the mycobacteria that cause tuberculosis.

Broad-spectrum antibacterial drugs are effective against two or more microbial groups. Ampicillin, a semisynthetic derivative of penicillin, affects both Gram-positive and Gram-negative bacteria. Chloramphenicol and tetracycline have an even broader spectrum, affecting chlamydiae and rickettsiae in addition to Gram-positives and Gram-negatives. Streptomycin has a particularly broad spectrum, encompassing chlamydiae and rickettsiae, mycobacteria, and Gram-negatives (though not Gram-positives).

When treating an infection with an unknown cause, broad-spectrum agents have a definite advantage. Ear infections, for example, can be caused by either Gram-positive or Gram-negative species; the broad-spectrum ampicillin is, then, a good choice, while the narrow-spectrum penicillin VK is not. However, because broad-spectrum drugs are effective against so many kinds of microorganisms, they significantly alter the body's normal microbial flora (Chapter 14), which may cause unpleasant symptoms or allow other microorganisms to gain a foothold and cause a harmful **superinfection**—an infection that occurs in addition to the original infection. For example, children treated with ampicillin for ear infections frequently suffer from diarrhea caused by disruption of the normal intestinal flora or from rashes caused by a yeast superinfection when the skin flora is altered.

Mechanisms of Acquired Drug Resistance. Bacterial strains become drug-resistant because of changes in their genetic makeup. These genetic changes are not caused by exposure to antibiotics, but rather occur naturally, like all mutations and genetic exchange (Chapter 6). When antibiotics are plentiful in the environment, however, antibiotic-sensitive strains will be eliminated while strains that happen to be antibiotic-resistant will persist, multiply, and become dominant. During the 1940s, nearly all strains of the virulent pathogen *Staphylococcus aureus* were sensitive to penicillin. Today, over 90 percent of strains are penicillin-resistant.

Genes for drug resistance encode proteins that protect the microorganism. Many different mechanisms of drug resistance are already well understood, and still others will doubtless be described in the future (**Table 21.2**). We

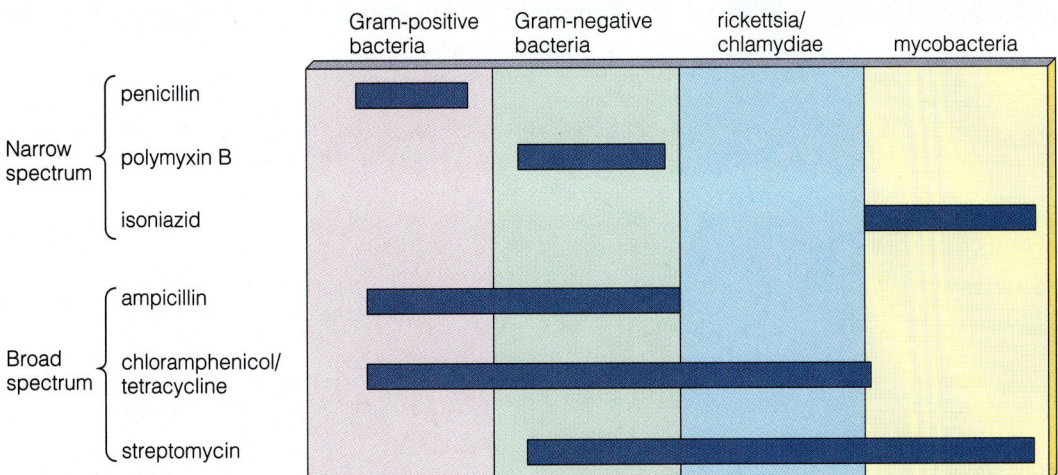

Effective against:

Figure 21.5 Antibacterial drug spectrum. Narrow-spectrum antibacterials act on only one bacterial group, while broad-spectrum drugs act on two or more groups.

will discuss three of the most common mechanisms of drug resistance—destroying the drug molecule, changing the target of drug action, and keeping the antibiotic agent out of the cell or removing it once it enters.

A common mechanism of drug resistance is the production of bacterial enzymes that destroy the drug. For example, the bacterial enzyme beta-lactamase splits a bond in the beta-lactam ring of the penicillin molecule, thereby inactivating the drug (see the box "The Original Wonder Drug"). Producing beta-lactamase is the mechanism by which most staphylococci have become resistant to penicillin. Detoxifying enzymes also confer resistance to other antibiotics, such as the cephalosporins.

Some bacterial strains become drug-resistant by changing the target of drug action. For example, some strains of the respiratory pathogen *Streptococcus pneumoniae* acquired penicillin resistance by undergoing a series of mutations in their penicillin-binding proteins (the enzymes that catalyze formation of the peptidoglycan cell wall) that make them less able to bind penicillin.

A third mechanism of acquired resistance is based on adaptations that make it more difficult for antibiotics to enter the bacterial cell or that actively remove the drug from the cell after entry. Some strains of *Neisseria gonorrhoeae*, for example, have become penicillin-resistant because changes in the outer membrane porin proteins bar the drug from entering the cell. In contrast, some microorganisms become resistant to tetracycline when their membrane transport system develops the ability to pump antibiotic molecules out of the cell.

Genetics of Acquired Drug Resistance. The mechanisms of drug resistance can be acquired either through genes on the bacterial chromosome or through genes car-

ried extra-chromosomally on plasmids (Chapter 6). Chromosomally mediated drug resistance takes place through mutation. Although most mutations result in weakened organisms that cannot survive as well as their normal competitors, mutations that happen to confer drug resistance produce microorganisms with an increased likelihood of survival in an antibiotic-laden environment.

At one time, it was thought that drug resistance due to chromosomal mutation was rare, but we now know it is fairly common. Some chromosomally drug-resistant strains have emerged gradually, over decades of antibiotic use. A series of mutations renders a bacterial strain gradually more and more drug-resistant. Higher and higher doses of antibiotics are required to cure infection until resistance becomes so great that the drug is no longer effective. Penicillin-resistant *Streptococcus pneumoniae* with altered penicillin-binding proteins and penicillin-resistant *Neisseria gonorrhoeae* with altered outer-membrane porins are both examples of gradual chromosomally mediated resistance.

Drug resistance due to plasmid-borne genes can be passed exceedingly rapidly from one strain of microorganism to another, and even from one species to another, by genetic conjugation (Chapter 6). Plasmids that carry genes encoding drug resistance are called **R factors** (R for resistance). A single R factor may carry genes that encode resistance to several different antibiotics. Penicillin resistance from beta-lactamase is an example of plasmid-borne resistance. Plasmid-borne drug resistance was recognized soon after the antibiotic era began in the 1940s. Unlike chromosomally mediated resistance, plasmid-borne changes in antimicrobial sensitivity and resistance can be dramatically sudden. See the box "A Game We May Never Win" in Chapter 24.

Table 21.2 Mechanisms of Acquired Drug Resistance

Mechanism	Antimicrobial Agent	Drug Action	Mechanism of Resistance
Destroys drug	Aminoglycosides	Binds to 30S ribosome subunit, inhibiting protein synthesis	Plasmids encode enzymes that chemically alter the drug (e.g., by acetylation or phosphorylation), thereby inactivating it.
	Beta-lactam antibiotics (penicillins and cephalosporins)	Binds to penicillin-binding proteins, inhibiting peptido-glycan synthesis	Plasmids encode beta-lactamase, which opens the beta-lactam ring, inactivating drug.
	Chloramphenicol	Binds to 50S ribosome subunit, inhibiting formation of peptide bonds	Plasmids encode an enzyme that acetylates the drug, thereby inactivating it.
Alters drug target	Aminoglycosides	Binds to 30S ribosome subunit, inhibiting protein synthesis	Bacteria make an altered 30S ribosome that does not bind to the drug.
	Beta-lactam antibiotics (penicillins and cephalosporins)	Binds to penicillin-binding proteins, inhibiting peptido-glycan synthesis	Bacteria make altered penicillin-binding proteins that do not bind to the drug.
	Erythromycin	Binds to 50S subunit, inhibiting protein synthesis	Bacteria make a form of 50S ribosome that does not bind to the drug.
	Quinolones	Binds to DNA topoisomerase, an enzyme essential for DNA synthesis	Bacteria make an altered DNA topoisomerase that does not bind to drug.
	Rifampin	Binds to RNA polymerase, inhibiting initiation of RNA synthesis	Bacteria make an altered polymerase that does not bind to drug.
	Trimethoprim	Inhibits the enzyme dihydro-folate reductase, blocking the folic acid pathway	Bacteria make an altered enzyme that does not bind to the drug.
Inhibits drug entry or removes drug	Penicillin	Binds to penicillin-binding proteins, inhibiting peptido-glycan synthesis	Bacteria change shape of outer membrane porin proteins, preventing drug from entering cell.
	Erythromycin	Binds to 50S ribosome subunit, inhibiting protein synthesis	New membrane transport system prevents drug from entering cell.
	Tetracyclines	Binds to 30S ribosome subunit, inhibiting protein synthesis by blocking tRNA	New membrane transport system transports drug out of cell.

Slowing Acquired Drug Resistance. Each year, huge quantities of penicillin and other antibiotics are prescribed. At least a third of all hospital patients receive one or more courses of antibiotic therapy during their stay. In addition, antimicrobials are widely used as an additive to animal feed. The result of such wide distribution of drugs is a rising tide of drug resistance that makes it difficult to treat certain infections (**Table 21.3**). Worse, the more antibiotics we use, the more rapidly drug-resistant strains will come to predominate over sensitive strains. Thus, our best opportunity to slow the emergence of drug resis-

tance and preserve the efficacy of antimicrobial drugs—a precious resource—is to limit their use.

One frequently discussed possibility is to limit non-medical uses of antibiotics, including animal feed additives. About half the antibiotics manufactured in the United States are fed in low doses to livestock to keep them healthy and make them grow faster. The antibiotic-sensitive organisms in their intestines are killed, but the antibiotic-resistant organisms survive and multiply. Epidemiologic studies have shown that the transmission of drug-resistant organisms from animals to humans is

Table 21.3	Rising Number of Penicillin G–Resistant Strains of *Staphylococcus aureus*	
Year	Percentage of Resistance
1940[a]	Less than 3
1946	14
1948	58
1960	83
1990	More than 90

[a]First use of penicillin after clinical trials.

more than a theoretical concern. An outbreak of drug-resistant *Salmonella newport* infection that caused 17 severe illnesses and 1 death was traced to contaminated hamburger meat from cattle that received antibiotics in their feed.

It is widely acknowledged that the medical use of antibiotics ought to be more selective and thoughtful. Probably the greatest misuse is administering antibacterial agents, such as penicillin, for untreatable viral infections, such as colds. In many countries these agents can be bought without prescription and taken by anyone who feels ill. Even in countries like the United States where antibiotic use is regulated, physicians too often prescribe indiscriminately or give in to the demands of patients who want an antibiotic for every runny nose.

In addition, a course of antibiotics ought to be prescribed at a high enough dosage and taken long enough to eradicate the infection. This prevents the spread of the drug-resistant organisms that survive partial treatment. Another approach to slowing drug resistance in the clinical setting is **combined therapy**, administering two different drugs at the same time. It is extraordinarily improbable that a pathogen would be simultaneously resistant to two drugs because the probability of mutation in two genes is the product of their individual chances of sustaining a mutation. In other words, if there is a one in a billion (10^9) probability of a pathogen's being resistant to either of two drugs, there would be a one in a billion billion (10^{18}) probability of its being resistant to both. When dealing with bacterial cultures, the first is highly probable, while the second is extraordinarily improbable. The recent emergence of highly drug-resistant strains of tuberculosis-causing mycobacteria is largely the result of patients not following instructions to take both drugs during combined therapy.

Dosage and Antimicrobial Susceptibility

Drug dosage—the quantity of a drug to be administered—depends upon many factors, including feasible routes of administration, drug distribution, and drug elimination. Together, these factors determine the level of a drug achievable in a patient's bloodstream. For antimicrobial agents, dosage depends upon another critical factor—the antimicrobial susceptibility of the infecting organism.

Antimicrobial susceptibility is a measure of how much of a drug is required to kill or stop the growth of a pathogen. Information about antimicrobial susceptibility and pharmacologic data on achievable blood levels of different agents are both considered in choosing the most effective agent for treating an infection.

Sometimes a microorganism is more susceptible to two agents administered together than to either administered alone. This enhanced effect, termed **drug synergism**, is not well understood, but it seems to make sense. A drug that damages the microbial cell envelope, for example, may allow another drug to enter the cell more readily, thereby increasing its effectiveness. Other drug combinations, however, are less effective than either agent administered singly. This is **drug antagonism**. A drug that slows cell growth, for example, may diminish the effectiveness of another agent that can damage only growing cells.

Antimicrobial susceptibility (sensitivity) can be determined in the microbiology laboratory through **susceptibility tests**. Test methods include disc diffusion, broth dilution, and serum killing-power, a variation of the broth-dilution method. These methods apply almost exclusively to bacterial pathogens and antibacterial drugs.

Disc-Diffusion Method. The **disc-diffusion method**, sometimes called the **Kirby-Bauer method**, tests drug susceptibility using filter-paper discs impregnated with known quantities of antimicrobial agents. A petri plate is inoculated with a bacterium; and discs, each impregnated with a different antimicrobial agent, are placed on its surface (**Figure 21.6**). The drug diffuses from the disc through the gel so the highest concentrations exist near the disc and progressively lower concentrations exist farther away.

When the plate has been incubated long enough for bacteria to produce a confluent lawn of growth, results can be interpreted. Some of the drug discs will be surrounded by a clear halo, indicating antimicrobial inhibition of bacterial growth. If the microorganism's growth is inhibited by a low drug concentration, the halo will be large. If growth is inhibited only by a high drug concentration, the halo will be small. The size of each halo is measured and compared to susceptibility standards for each drug. Based on these comparisons, the microorganism is judged to be either *sensitive*, meaning highly susceptible, *intermediate*, or *resistant* to each drug.

Figure 21.6 Disc-diffusion susceptibility test. (**a**) A culture of the microorganism to be tested is spread with a cotton swab over the surface of an agar-solidified medium in a petri dish. (**b**) Discs impregnated with antimicrobial agents are placed on the agar surface, sometimes with a device like the one shown here. The petri dish is incubated and the drugs diffuse out from the disc. (**c**, **d**) When the microorganism forms a confluent lawn of growth, clear zones in which growth did not occur are visible around the discs impregnated with agents to which the test organism is susceptible; larger zones reflect greater susceptibility.

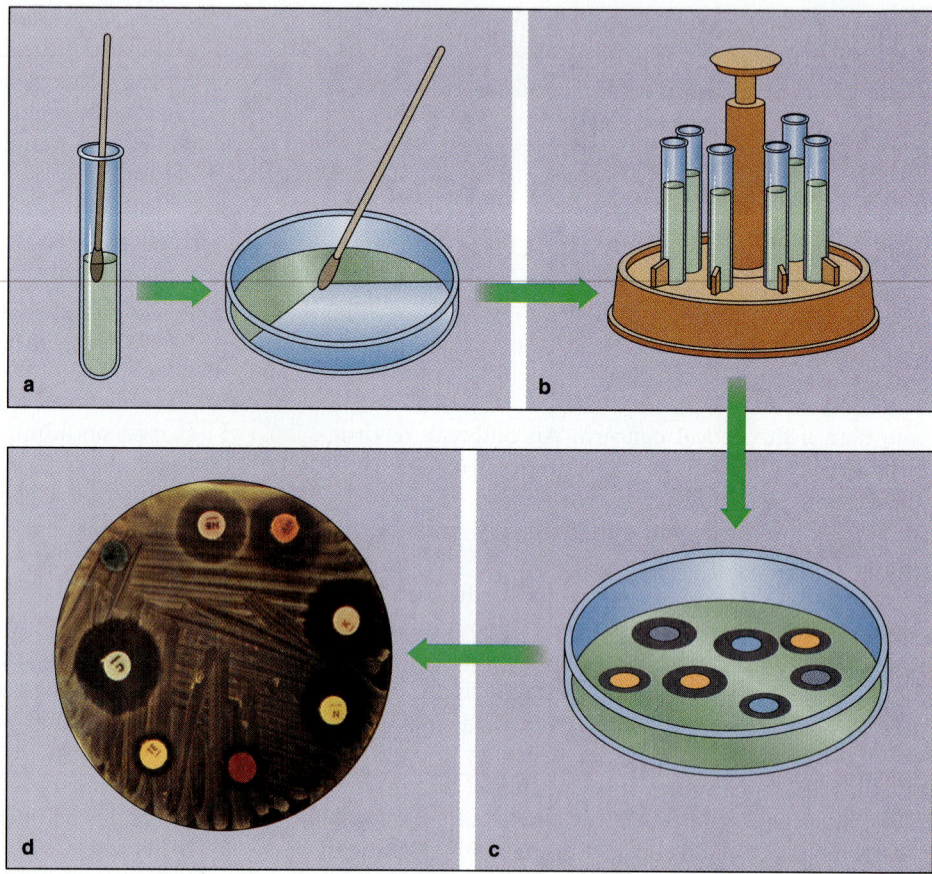

Results of a disc-diffusion susceptibility test can be extremely useful. Knowing if an organism is sensitive, intermediate, or resistant to a given antimicrobial agent is almost always enough information on which to base a treatment decision. Moreover, disc-diffusion tests are technically easy and relatively inexpensive to perform. Disc diffusion is by far the most common type of susceptibility test used in clinical medicine.

Broth-Dilution Method. The **broth-dilution method** tests drug susceptibility by cultivating bacteria in liquid cultures with progressively higher concentrations of an antimicrobial agent. A series of tubes containing decreasing concentrations of a drug in solution is prepared and inoculated with the bacterium to be tested.

Results can be interpreted when growth has occurred, making some tubes turbid (cloudy). Growth occurs in tubes that contain no drug or low drug concentrations, but not in tubes with higher concentrations. The clear test tube with the lowest drug concentration contains the **minimum inhibitory concentration (MIC)** of the drug.

More information can be obtained by taking bacteria from the growth-inhibited tubes and attempting to grow them on a petri plate (**Figure 21.7**). If organisms did not grow in the broth culture but do grow on a drug-free petri plate, the antimicrobial agent inhibited the organism's growth but did not kill it. This is a **bacteriostatic** drug concentration. If organisms do not grow on a drug-free petri plate, the organism has been killed. This is a **bactericidal** drug concentration. The tube with the lowest drug concentration that proves to be bactericidal contains the **minimum bactericidal concentration (MBC)** of that drug.

Knowing the MIC and MBC gives the clinician valuable information. The MIC and MBC can be compared to serum drug levels to see if inhibitory or bactericidal concentrations are being achieved in the patient's body. In addition, MIC and MBC provide information about the antimicrobial agent itself. If the MBC is much higher than the MIC, the antimicrobial agent is bacteriostatic, but if the MIC and MBC are nearly the same, the agent is bactericidal. Many infections can be treated effectively by bacteriostatic drugs because they tip the balance in favor of the patient's immune system. When a person's immune system is weak, however, bactericidal agents may be essential.

Broth-dilution susceptibility testing is more complicated and expensive than disc-diffusion testing, but may be helpful in certain clinical situations. If a patient is critically ill or does not respond to what ought to be appro-

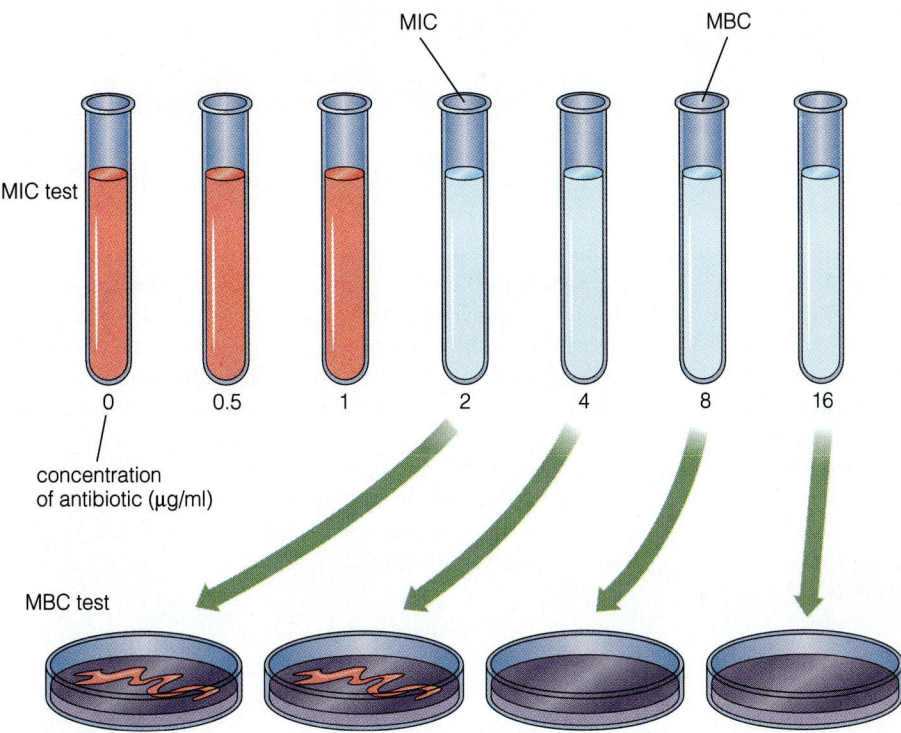

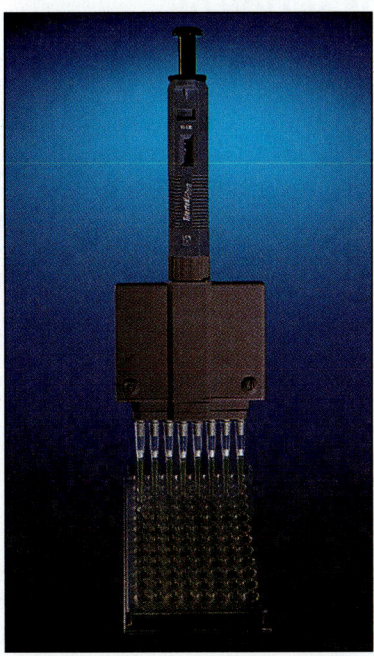

Figure 21.7 MIC and MBC testing. The MIC (minimum inhibitory concentration) is the concentration of a drug that stops microbial growth. The MBC (minimum bactericidal concentration) is the concentration that kills cells. Plating broth from growth-free dilution tubes onto petri plates determines whether a drug is bacteriostatic (merely stops growth) or bactericidal (kills microorganisms). In this illustration, the MIC is 2 µg/ml and the MBC is 8 µg/ml.

priate antimicrobial therapy, comparing the organism's MIC and MBC to the patient's serum drug levels may help a clinician decide whether to continue with that drug or change therapy. In particularly difficult cases, a **serum killing-power test** may be done. Some of the patient's own drug-containing serum is withdrawn and tested to see if it kills the infecting microorganism in the laboratory.

The trend today is toward automated versions of broth-dilution tests, though the degree of automation varies (**Figure 21.8**). In some systems, only the reading of final test results is automated, but others—spinoffs of space research designed to test for microbes on other planets—make dilutions and inoculations, incubate the cultures, and print out results. Automation saves labor; however, a fully automated machine costs as much as $100,000. A 1991 survey indicated that about 20 percent of clinical laboratories were using some type of automated susceptibility testing.

All strains of *Streptococcus pyogenes*, the bacterium that caused T. M.'s illness, are uniformly and highly susceptible to penicillin. The MBC of *S. pyogenes* is as low as 1 ng/ml, and safe drug levels in the bloodstream reach 100 µg/ml, or 100,000 times the MBC. Thus, although bacteria were swabbed from T. M.'s throat to confirm the

Figure 21.8 The broth-dilution test can be automated by using multitipped pipettes to fill depressions in plastic plates (microtiter plates). After incubation, MIC is determined on the same basis as the manual method—turbidity.

The Serum Killing-Power Test

A simple determination of a microorganism's susceptibility to an antibiotic is usually all a clinician needs to guide therapy, but sometimes more information is called for. Such was the case of a young woman with endocarditis, an infection of the lining of the heart (Chapter 27). In many ways she was a typical endocarditis patient—she had a mild heart defect that made her prone to infection, and she was infected by *Streptococcus viridans*, one of the most common microbial causes of endocarditis. But her case was unusual because the strain of *S. viridans* was relatively resistant to penicillin—the first-line drug for treating endocarditis. Because endocarditis infections are difficult to eradicate

under the best of circumstances, her doctors were concerned. They therefore requested a serum killing-power test. A sample of the young woman's serum was obtained shortly after she received a dose of penicillin and added to cells of *S. viridans* that had been isolated from her blood. The results indicated that her undiluted serum was capable of killing the bacteria, but even a twofold dilution of the serum caused the drug concentration to fall to an ineffective level. Because the margin of safety was so narrow, treatment was continued for longer than usual. The young woman recovered completely.

diagnosis, it was not necessary to perform drug susceptibility tests on them. The standard dose of penicillin to treat patients with scarlet fever is well established.

MODES OF ACTION OF ANTIMICROBIAL DRUGS

All antimicrobial drugs act by damaging a vital cell structure or inhibiting a vital metabolic function, but different drugs have different targets (**Figure 21.9**). There are five principal targets of antimicrobial drugs. They interfere with synthesis of the bacterial cell wall, disrupt the cell membrane, interfere with protein synthesis, interfere with nucleic acid synthesis, or interfere with synthesis of folic acid, an essential cofactor. The drugs given as examples are discussed in the last section of this chapter.

Drugs That Inhibit Synthesis of the Cell Wall

The cell wall—a rigid peptidoglycan mesh that withstands the procaryotic cell's turgor pressure—is unique to bacteria and a perfect target for antimicrobial agents. Drugs that block peptidoglycan synthesis kill bacteria, but have no effect on eucaryotic cells, including human cells. Because growing cells that can no longer maintain their cell wall lyse and die, cell-wall inhibitors tend to be bactericidal. Microbes that lack peptidoglycan are naturally resistant to these drugs, as are microbial species

with structures that prevent the drug from reaching its peptidoglycan target.

The penicillins and many other drugs fall into this category. Among the most useful clinically are the cephalosporins, a drug family that has come into wider use as acquired resistance to older agents has grown more widespread. Both the penicillins and cephalosporins block peptidoglycan synthesis at the final step of cross-linking the mesh. Other drugs in this category, such as vancomycin and bacitracin, interfere with earlier steps of synthesizing linear peptidoglycan strands.

Drugs That Disrupt the Cell Membrane

Some drugs inhibit microbial growth by damaging components of the plasma membrane, destroying its selective permeability, and allowing vital molecules to leak out. Because all cells, both microbial and human, have a plasma membrane, selective toxicity is much poorer for these agents than for inhibitors of peptidoglycan synthesis. Many of these drugs, when administered systemically, are highly toxic.

The antimicrobial spectrum of different members of this drug group depends upon the specific type of molecule damaged. Some agents are detergents that damage membrane phospholipids; they kill most Gram-negative bacteria because they destroy the outer membrane. These drugs are administered topically to control infections of the body surfaces; polymyxin B is an example. Other agents selectively damage sterols in the plasma membrane of eucaryotic cells. These drugs are selectively toxic

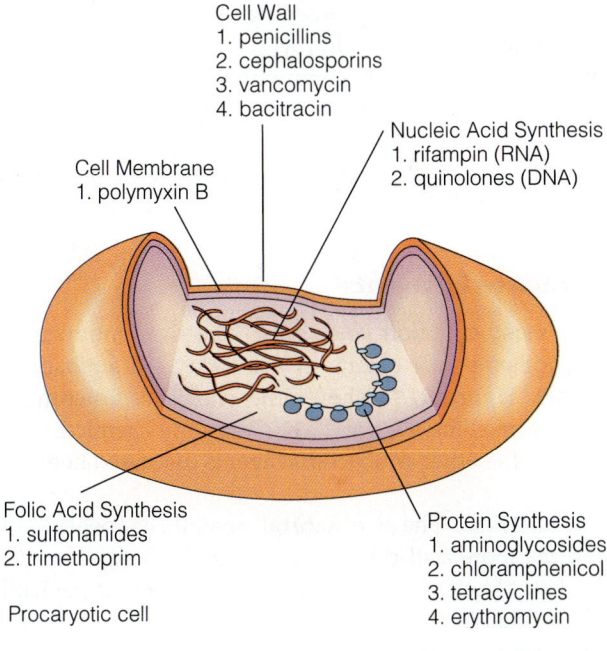

Cell Wall
1. penicillins
2. cephalosporins
3. vancomycin
4. bacitracin

Cell Membrane
1. polymyxin B

Nucleic Acid Synthesis
1. rifampin (RNA)
2. quinolones (DNA)

Folic Acid Synthesis
1. sulfonamides
2. trimethoprim

Protein Synthesis
1. aminoglycosides
2. chloramphenicol
3. tetracyclines
4. erythromycin

a Procaryotic cell

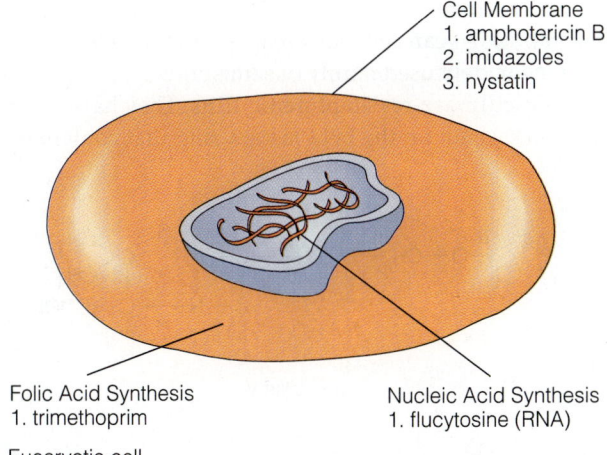

Cell Membrane
1. amphotericin B
2. imidazoles
3. nystatin

Folic Acid Synthesis
1. trimethoprim

Nucleic Acid Synthesis
1. flucytosine (RNA)

b Eucaryotic cell

Figure 21.9 Targets of antimicrobial drugs. The primary targets of drug action for both procaryotic and eucaryotic microorganisms are the cell wall (peptidoglycan synthesis), cell membrane structure, protein synthesis, nucleic acid synthesis, and folic acid synthesis. In general, different drugs act on procaryotic and eucaryotic targets.

to fungi because they do greater damage to ergosterol, which is found in fungal cell membranes, than to cholesterol, which is found in human cell membranes. Even so, the selective toxicity of these drugs is relatively poor. Some of these agents are used topically, but others are administered systemically to treat life-threatening fungal infections. Amphotericin B, for example, has many serious side effects, including fever, chills, vomiting, and kidney failure.

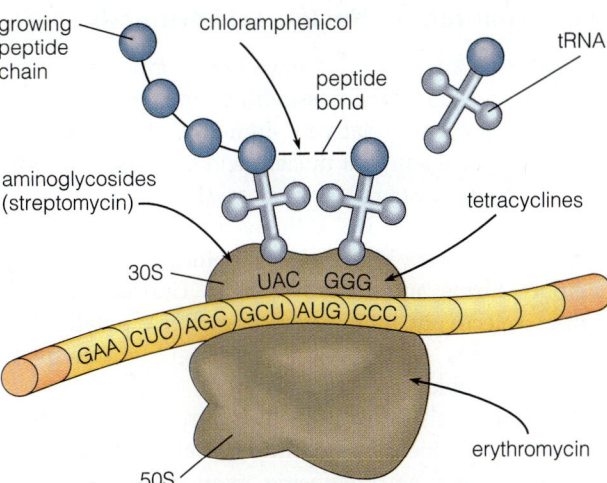

growing peptide chain

chloramphenicol

peptide bond

tRNA

aminoglycosides (streptomycin)

tetracyclines

30S   UAC   GGG

GAA CUC AGC GCU AUG CCC

erythromycin

50S

Figure 21.10 Some antimicrobials that inhibit protein synthesis work by binding to 70S procaryotic ribosomes. The arrows indicate the binding site of four commonly used drugs: Aminoglycosides, which are bactericidal, bind to the 30S ribosomal subunit. Tetracyclines, which are bacteriostatic, bind reversibly to the 30S subunit. Erythromycin, a bacteriostatic agent, binds reversibly to the 50S ribosomal subunit. Also bacteriostatic, chloramphenicol binds reversibly to the 50S subunit, preventing peptide-bond formation.

Drugs That Inhibit Protein Synthesis

Antimicrobial drugs that interfere with protein synthesis affect the bacterial ribosome. Differences between the procaryotic 70S ribosomes, made up of a 50S and a 30S subunit, and the eucaryotic 80S ribosomes (Chapter 4) are the basis for selective toxicity. But eucaryotic mitochondria, like bacteria, synthesize protein on 70S ribosomes, which accounts for some of the toxicity of these drugs. When ribosomal function is disturbed, the synthesis of protein may stop entirely or may be slowed so significantly that normal growth cannot proceed.

Many widely used antimicrobial drugs belong to this group, and different agents affect ribosomal function in different ways (**Figure 21.10**). Agents of the aminoglycoside group, including streptomycin, bind irreversibly to the 30S subunit of the ribosome, slowing protein synthesis and causing the cell to make faulty proteins. As a result, aminoglycosides are bactericidal—the only drugs in this group that cause cell death. Their selective toxicity is increased by the fact that bacteria actively transport aminoglycoside molecules into the cells; however, they enter human cells with great difficulty. The tetracyclines also bind to the 30S ribosome subunit. They interfere with the attachment of the amino-acid-carrying tRNA molecule to the growing chain. The effect of tetracycline is reversible, however, and protein synthesis begins again when tetracycline levels fall. Erythromycin and chloramphenicol both inhibit protein synthesis by reversible binding to the 50S ribosome subunit.

Drugs That Inhibit Nucleic Acid Synthesis

All cells—procaryotic and eucaryotic—must manufacture nucleic acids, so the possibilities for selective toxicity in this category are relatively limited. Only a few clinically useful drugs inhibit nucleic acid synthesis. In some cases, though, essential enzymes—the topoisomerases that unwind existing DNA chains and the polymerases that extend the new chains—are significantly different in microorganisms and human beings. The antibiotic rifampin selectively inhibits bacterial RNA polymerase. The quinolones, a relatively new drug group, selectively inhibit a microbial topoisomerase.

Differences also exist in the pathways through which nucleosides are incorporated into nucleic acids, and certain drugs, such as flucytosine, exert their effect here. Fungi, which are sensitive to flucytosine, have an enzyme that converts the drug to 5-fluorouracil, a compound that inhibits both RNA synthesis. Since humans lack this enzyme, they are unaffected by flucytosine. Flucytosine is considered an **antimetabolite** because it blocks a vital step in microbial metabolism as a result of its chemical similarity to a normal metabolic precursor. Other antimetabolites lead to synthesis of faulty nucleotides and hence nonfunctional nucleic acid.

Drugs That Inhibit Folic Acid Synthesis

A major group of antimicrobials acts by interfering with the production of the enzyme cofactor folic acid, which has many functions, including synthesis of the nitrogenous bases that build DNA. These drugs are selectively toxic to microorganisms because humans do not synthesize folic acid. Instead, we obtain it as a vitamin in our diet. Since drugs in this category block steps in microbial metabolism, they are also considered antimetabolites.

The sulfonamides are competitive inhibitors of a key enzyme in the folic acid pathway. A **competitive inhibitor** is a molecule similar enough to an enzyme's normal substrate to bind to its active site, but incapable of being converted to the next intermediate. As a result, it blocks the pathway at that point. The sulfonamides closely resemble the enzyme substrate para aminobenzoic acid (PABA), one of the intermediates in the pathway that produces folic acid. By blocking conversion of PABA to the next intermediate, the bacteriostatic sulfonamides stop production of folic acid and thereby stop microbial growth. Another drug, trimethoprim, blocks the same pathway at a different point.

ANTIMICROBIAL DRUGS

Many new antimicrobial agents are developed each year, while other agents pass out of common use because of emerging microbial resistance or the availability of better agents. Understandably, keeping up-to-date is a constant challenge for clinicians. In this section we classify the antimicrobial drugs most commonly used in clinical medicine today according to the type of infection they are usually used to treat—bacterial, mycobacterial, fungal, parasitic (helminthic and protozoal), and viral (**Table 21.4**, p. 504).

Antibacterial Agents

The most common and most effective antimicrobial chemotherapeutic agents in use today are antibacterial agents. **Figure 21.11** shows the chemical structures of three penicillins. **Figure 21.12** shows the chemical structures of the other antibacterial agents discussed here.

Penicillins. The two **natural penicillins**, penicillin G and the more acid-resistant penicillin V, are discussed in the box "The Original Wonder Drug." Natural penicillin is modified in the laboratory to produce several types of **semisynthetic penicillins**, which are also commonly prescribed. All the penicillins, both natural and semisynthetic, are bactericidal—they kill growing cells by lysis as the peptidoglycan cell wall weakens and breaks.

One widely used family of semisynthetic penicillins is the **penicillinase-resistant penicillins**. They have a modified side chain on the beta-lactam ring that confers rela-

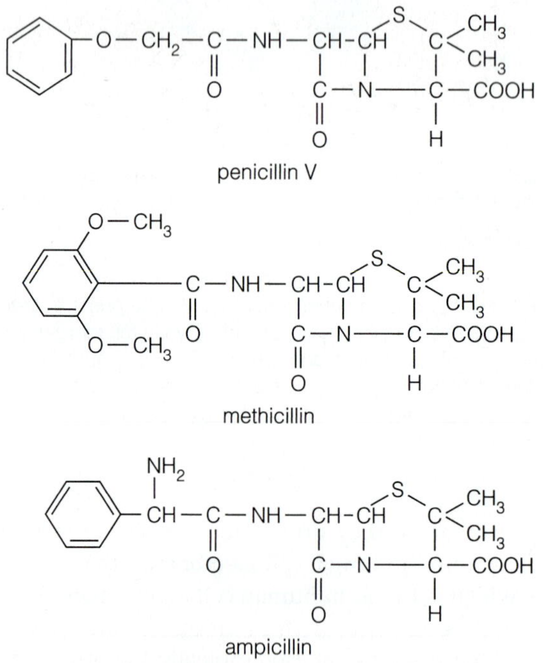

Figure 21.11 Chemical structures of some types of penicillin. Penicillin V is a natural drug. Methicillin is a penicillinase-resistant semisynthetic, and ampicillin is a broad-spectrum semisynthetic. (The structure of penicillin G is shown in the box "The Original Wonder Drug.")

tive resistance to this penicillin-destroying enzyme. These drugs include methicillin, nafcillin, and the isoxazolyl penicillins (oxacillin, cloxacillin, and dicloxacillin). They are used to treat infections by penicillinase-producing strains of *Staphylococcus aureus*, including toxic shock syndrome and wound and skin infections (Chapters 24 and 26). Infections by penicillinase-producing bacterial strains are also treated by coadministering clavulanic acid, a drug that has no antimicrobial activity of its own but can bind to penicillinase and inactivate it.

Also widely used are the **broad-spectrum penicillins**, semisynthetic drugs that are effective against Gram-negative bacteria. Ampicillin and the closely related amoxicillin are active against all the bacteria killed by penicillin G plus many Gram-negative rods, including *Haemophilus influenzae, Escherichia coli, Salmonella* spp., and *Shigella* spp. Carbenicillin resembles ampicillin but has the added advantage of being effective against many species of *Proteus* as well as *Pseudomonas aeruginosa*, which is notoriously resistant to antibiotics and difficult to eradicate

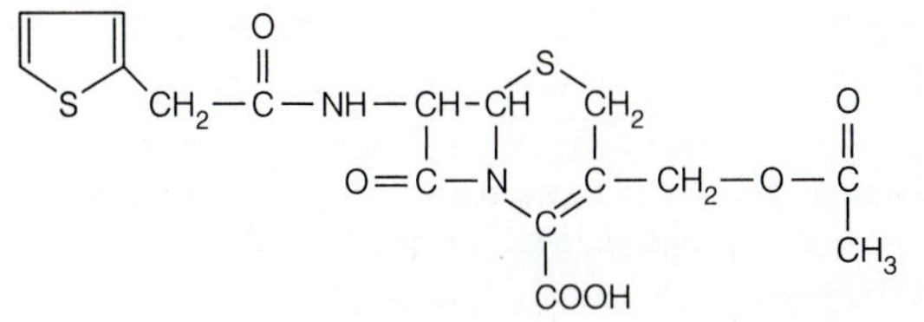

cephalothin (a cephalosporin)

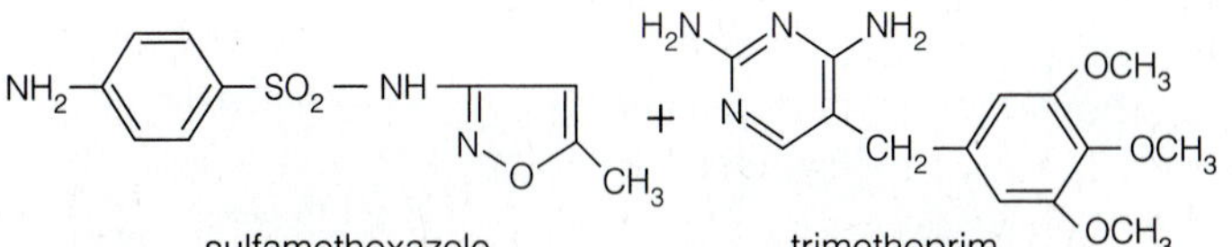

sulfamethoxazole

trimethoprim

cotrimoxazole

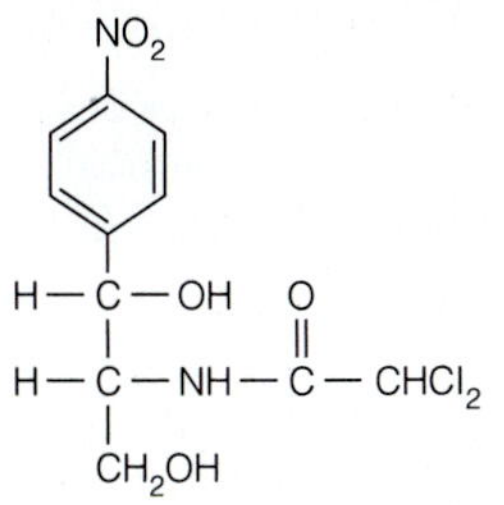

chloramphenicol

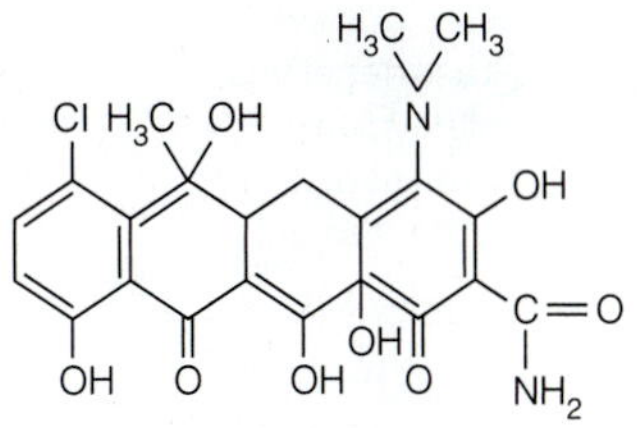

chlortetracycline

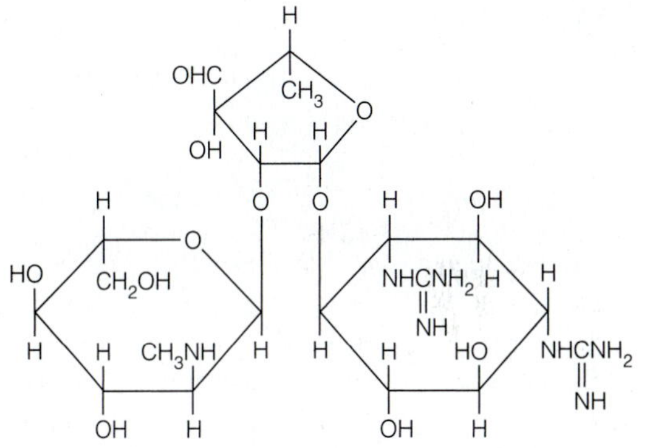

streptomycin

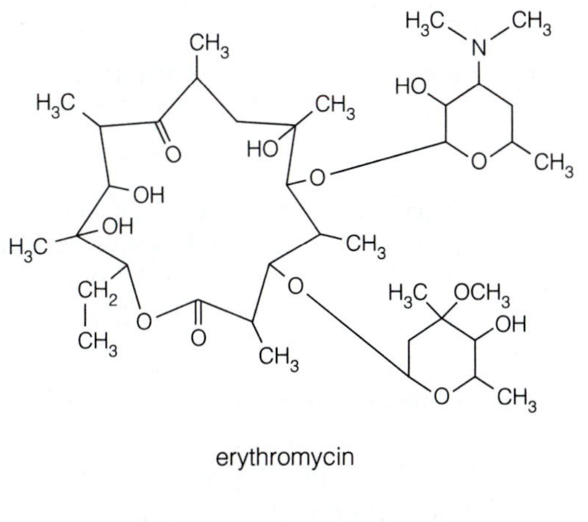

erythromycin

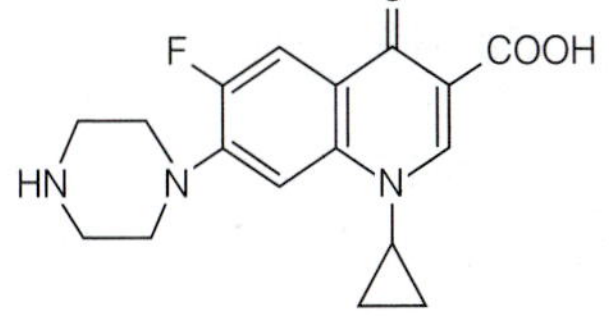

ciprofloxacin
(a quinolone)

Figure 21.12 Chemical structures of major antibacterial agents. Compare the structure of cephalosporins with that of the closely related penicillins in Figure 21.11. Cotrimoxazole is a combination drug.

Table 21.4 Antimicrobial Drugs

Drug	Natural vs. Synthetic	Effect	Spectrum	Side Effects and Allergy
Antibacterial Agents				
Penicillins	Natural and semisynthetic	Bactericidal	Natural agents cover mostly Gram-positives; semisynthetics cover some Gram-negatives.	Side effects are negligible, but allergy is common.
Cephalosporins	Natural and semisynthetic	Bactericidal	Natural agents cover mostly Gram-positives; semisynthetics cover many Gram-negatives.	Side effects are negligible, but allergy is common.
Sulfonamides	Synthetic	Bacteriostatic	Gram-positives, Gram-negatives, and chlamydia, but all greatly diminished by acquired resistance.	Many side effects, including crystals in urine and depressed stem cell production in bone marrow; allergy is also common.
Aminoglycosides	Most natural, a few semi-synthetic	Bactericidal	Aerobic Gram-negatives.	Major side effects, including loss of hearing and kidney function; allergy is rare.
Chloramphenicol	Natural, now artificially synthesized	Bacteriostatic	Gram-positives, Gram-negatives, chlamydiae, rickettsiae, and mycoplasmas.	Rare but life-threatening toxicity includes aplastic anemia and gray baby syndrome; allergy is rare.
Tetracyclines	Natural and semisynthetic	Bacteriostatic	Gram-positives, Gram-negatives, chlamydiae, rickettsiae, and mycoplasmas.	Abdominal pain and vomiting, rashes on exposure to sunlight, permanent stains in developing teeth; allergy is rare.
Erythromycin	Natural	Bacteriostatic or bactericidal	Gram-positives.	Nausea, vomiting, abdominal pain, rare liver damage; allergy is rare.
Quinolones	Synthetic	Bactericidal	Gram-positives, Gram-negatives, mycoplasmas, mycobacteria.	Side effects are uncommon but include abdominal pain and hearing loss; allergy is rare.
Antimycobacterial Agents				
Isoniazid	Synthetic	Bacteriostatic for resting cells; bactericidal for rapidly growing cells	Mycobacteria.	Side effects are uncommon but include rash, fever, liver damage.
Rifampin	Semisynthetic	Bacteriostatic	Mycobacteria.	Side effects are uncommon but may include rash, fever, nausea, and vomiting.

from infected tissue. Other broad-spectrum penicillins used principally for the treatment of serious infections due to *P. aeruginosa* are ticarcillin, piperacillin, mezlocillin, and azlocillin.

Cephalosporins. The **cephalosporins** are one of the fastest-growing drug families of the last few years—both in numbers of new drugs and in clinical usefulness. The natural cephalosporins were first discovered as products of *Cephalosporium acremonium*, a fungus isolated in 1948 from a sewer outlet near the Sardinian coast.

Cephalosporins are closely related to penicillins. Chemically, they have a penicillin-like beta-lactam ring structure. In addition, their mode of action is identical to

Table 21.4 (continued)

Drug	Natural vs. Synthetic	Effect	Spectrum	Side Effects and Allergy
Ethambutol	Synthetic	Bacteriostatic	Mycobacteria.	Side effects are uncommon butmay include rash, fever, decreased vision.
Antifungal Agents				
Nystatin	Natural	Fungicidal	*Candida albicans.*	Topical use only; negligible side effects.
Amphotericin B	Natural	Fungicidal	Many fungal species, including *Candida*, *Cryptococcus neoformans*, *Blastomyces dermatitidis, Coccidioides immitis.*	Highly toxic with systemic administration, usually causing fever, chills, and kidney damage.
Imidazoles and triazoles	Synthetic	Fungistatic	Many fungi, including *Candida.*	Few side effects when used topically; nausea and vomiting when administered systemically.
Flucytosine	Synthetic	Fungistatic	Many fungi, including *Candida* and *Cryptococcus neoformans.*	Little toxicity.
Griseofulvin	Natural	Fungistatic	Dermatophytes.	Occasionally headaches.
Antihelminthic and Antiprotozoal Agents				
Mebendazole	Synthetic	—	Many roundworms.	Negligible side effects.
Metronidazole	Synthetic	—	Many protozoa and Gram-negative bacteria.	Many side effects, including headache, nausea, vomiting, and abdominal pain; has caused cancer in animal studies—bacterial mutations.
Chloroquine	Semisynthetic	—	Malarial parasites.	Little toxicity.
Antiviral Agents				
Amantadine	Synthetic	—	Influenza A.	Nervousness, headache, insomnia.
Acyclovir	Synthetic	—	Herpesviruses.	Negligible side effects.
Ribavirin	Synthetic	—	Respiratory syncytial virus, influenza viruses.	Damages growing tissue, possibly causes birth defects.
Azidothymidine (AZT)	Synthetic	—	Human immunodeficiency virus.	Depresses stem cell production in bone marrow.
Dideoxyinosine (ddI)	Synthetic	—	Human immunodeficiency virus.	Depresses stem cell production in bone marrow.
Dideoxycytidine (DDC)	Synthetic	—	Human immunodeficiency virus.	Depresses stem cell production in bone marrow.
Interferons	Natural; mass-produced by genetic engineering	—	Human papillomavirus, hepatitis viruses.	Flulike syndrome, including fever and chills.

that of the penicillins—they interfere with the cross-linking of bacterial cell walls in exactly the same way. Also like penicillins, cephalosporins are bactericidal, lysing growing cells.

Cephalosporins are broad-spectrum drugs, but are particularly effective against the Gram-positives. Chemical modification of natural cephalosporins has produced numerous semisynthetic antibiotics that have evolved as successive generations, each more active against Gram-negative organisms than its predecessor. Third-generation cephalosporins are widely used to treat many life-threatening infections, including meningitis, pneumonia, and sepsis. Cephalosporins are more resistant than penicillins to destruction by beta-lactamases, making them much more effective against most strains of *Staphylococcus aureus* that are resistant to penicillin G. Some of the newer cephalosporins have additional advantages, such as good penetration into the central nervous system or unusually long persistence in the bloodstream; as a result they need not be administered frequently.

Side effects of the cephalosporins, like the penicillins, are minimal; but they are highly allergenic. Because of the chemical similarity between cephalosporins and penicillins, some patients who are allergic to penicillin are allergic to cephalosporins on first exposure. Moreover, the likelihood of this type of allergy crossover is hard to predict. Immunologic studies indicate that about 20 percent of penicillin-allergic patients will also react to cephalosporins, though clinical experience indicates the figure is much lower.

Sulfonamides. The **sulfonamides**, or sulfa drugs, were the first antimicrobial agents used to cure infections in humans, after salvarsan. These drugs are synthetic organic chemicals first produced by the German chemical company I. G. Farben as dyes.

Sulfonamides are a large family of drugs that act by interfering with the bacterial cell's ability to synthesize folic acid. They have a broad spectrum, which includes Gram-positives, some Gram-negatives, and chlamydiae. All are bacteriostatic.

The history of sulfonamides shows how antimicrobial drugs can become less useful with time. When sulfonamides first became available in the 1930s, their use dramatically reduced death rates from many bacterial infections—gonorrhea, meningococcal meningitis, shigella dysentery, and all infections caused by staphylococci and streptococci. But sulfonamides were used so extensively that microbial resistance emerged and spread rapidly. Today, none of these infections is routinely treated with sulfonamides because so many strains of the causative agents have become sulfonamide-resistant. Moreover, when sulfonamides were the only antimicrobial agents available, their allergies and side effects were acceptable. When safer drugs became available, they were of course preferred. Today, sulfonamides by themselves are generally used only to treat uncomplicated infections of the urinary tract.

Sulfonamides are, however, commonly used in combination with another antimicrobial drug, **trimethoprim**. The mixture, called cotrimoxazole, is many times more active against sulfa-sensitive bacteria than is either drug alone. Administering the two drugs in combination achieves drug synergism because both drugs inhibit synthesis of folic acid. Sulfonamides inhibit the conversion of PABA to the next compound in the folic acid pathway, and trimethoprim inhibits the following step in the same pathway. Administering two agents also decreases the likelihood of drug resistance. Sulfa-trimethoprim combinations are commonly used to treat urinary tract infections, minor respiratory infections, and infections caused by the opportunistic pathogen *Pneumocystis carinii*. Pneumonia caused by *P. carinii* is often a fatal infection in AIDS patients.

Aminoglycosides. The **aminoglycosides** were discovered in a deliberate, intensive search for antimicrobials effective against Gram-negative organisms. Microbiologists Selman Waksman and Albert Schatz at Rutgers University systematically examined soil actinomycetes to see if they produced antimicrobial compounds. In 1943 a strain of *Streptomyces griseus* was isolated that produced the drug streptomycin (See Focus on Research, "The Discovery of Streptomycin.") Related drugs—including neomycin, gentamicin, tobramycin, kanamycin, and amikacin—were later additions to this drug family.

The aminoglycosides are composed of amino sugar molecules connected to one another by glycoside bonds. All aminoglycosides must be injected into a vein or muscle. They are not well absorbed when taken orally because the molecules are positively charged and therefore dissolve poorly in lipid membranes. For the same reason they do not penetrate membrane-protected spaces such as the central nervous system.

Aminoglycoside antibiotics are used only to treat serious infections in hospitalized patients. They are extremely effective against certain infections, such as kidney infections caused by Gram-negative bacteria that grow under aerobic conditions. Their use is limited primarily by their toxicity. All aminoglycoside antibiotics can irreversibly damage the inner ear and the kidneys if blood levels rise too high and persist too long. Another drawback of the aminoglycosides is the rapid emergence of drug-resistant strains. Resistance to streptomycin develops so frequently that the drug must be used in combination with another antimicrobial agent.

Chloramphenicol. The natural antibiotic **chloramphenicol** was first extracted from a culture of *Streptomyces venezuelae* isolated from a soil sample in Venezuela. When the molecule was found to have a relatively simple chemical structure, drug companies began to produce it synthetically.

When chloramphenicol was discovered, it seemed the ideal drug. It readily crosses lipid membranes, which means it can be taken orally, and it penetrates well into protected areas such as the brain and the interior of the eye. It is chemically quite stable and can be stored for long periods without refrigeration. It has an extremely broad spectrum that includes Gram-positive and Gram-negative bacteria, rickettsiae, chlamydiae, and mycoplasmas. Drug allergy is rare, and in the vast majority of patients, it has

The Discovery of Streptomycin

Albert Schatz

Albert Schatz received his B.Sc. and Ph.D. degrees from Rutgers University. His undergraduate major was soil science and his graduate work was in soil microbiology. The search for a new antibiotic challenged him because as a young boy he knew people who died from what was then called blood poisoning, ear infections, pneumonia, diphtheria, whooping cough, tuberculosis, etc. When he was in grade school, a classmate died from one disease or another almost every year. For his research that led to the discovery of streptomycin, he has received honorary degrees and medals and has been named an honorary member of medical and scientific societies. Last April, Rutgers University awarded him the Rutgers Medal at the 50th Anniversary Celebration of the discovery of streptomycin.

I began the research that led to the discovery of streptomycin in 1943 when I was working for my Ph.D. degree at Rutgers University. At that time, the discovery of penicillin had motivated a search for other antibiotics. Several had been discovered but were too toxic to be used. I was aware of that when I spent six months as a bacteriologist in army hospitals, where I saw deaths caused by Gram-negative bacteria against which there were no effective drugs. Therefore, during off-duty hours, I isolated and tested soil fungi and actinomycetes to see which ones inhibited the growth of Gram-negative bacteria on agar plates. When I was discharged and resumed graduate work, I continued the search that I had begun in army hospital laboratories.

Shortly after I returned to Rutgers, Drs. William Feldman and H. Corwin Hinshaw at the Mayo Clinic asked Dr. S. A. Waksman, chairman of the Department of Soil Microbiology, to look for an antibiotic to control tuberculosis. Dr. Waksman was reluctant to take on that project because he was afraid of tuberculosis, which was also known as The Great White Plague. He did not want the tubercle bacillus, *Mycobacterium tuberculosis*, *hominis*, which had killed more than a billion people, in his third-floor laboratory, right next to his office. Like Gram-negative bacteria, the tubercle bacillus was resistant to all chemotherapy. But I had handled pathogenic bacteria in the army and was confident I could safely work with the tubercle bacillus. When I informed Dr. Waksman that I wanted to take on the TB problem as part of my Ph.D. research, he let me do that. But he transferred me to a basement laboratory and insisted that I never bring any TB cultures up to the third floor.

Dr. Waksman and others warned me that there was little likelihood of finding a cure for tuberculosis. The tubercle bacillus has a waxy coating, which is why it grows slowly and requires a special stain to make it visible through a microscope. It was assumed that no drug could get through that protective waxy coating. However, I reasoned that food had to get into the cells and waste products had to get out—otherwise neither the tubercle bacillus nor tuberculosis would exist.

Dr. Feldman provided me with a virulent strain of human tuberculosis, with which he was working. He subsequently contracted tuberculosis but recovered after two years of treatment. I feel good that no one in the building where I worked contracted tuberculosis. The laboratory I worked in was not equipped with safety equipment that is now used. It had no ultraviolet light, no special incubator, and no positive air pressure to continuously blow the air through a filter. Eventually I isolated two strains of the actinomycete *Streptomyces griseus*. Both produced a new antibiotic that I called streptomycin. One strain came from a heavily manured field soil. The other came from an agar plate that my fellow graduate student Doris Jones had streaked with a swab of a healthy chicken's throat. At about 2:00 P.M. on October 19, 1943, I knew that I had found a new antibiotic.

But would it control tuberculosis *in vivo*? Drs. Feldman and Hinshaw would find that out with guinea pigs. My job was to produce enough streptomycin for their initial tests. To do that, I had to run three stills around the clock. At night, I drew lines, with a red glass-marking pencil, on the flasks from which I was distilling and went to sleep on a wooden bench in the laboratory. The night watchman checked the flasks periodically and woke me up when the liquid in the flasks went down to the lines I had drawn. I then added more liquid, and went back to sleep. I also had to save, purify, and reuse the solvents I worked with because during World War II they were rationed. When the Mayo Clinic tests were over, I was exhausted. But I knew I would have an acceptable Ph.D. dissertation.

few side effects. But then chloramphenicol's reputation crashed.

In rare cases chloramphenicol can be fatally toxic. The most serious toxic manifestation is aplastic anemia, in which the bone marrow completely stops producing blood cells. Although this reaction occurs in fewer than 1 out of every 30,000 people who receive the drug, the condition is inevitably fatal. Another life-threatening form of chloramphenicol toxicity is the **gray baby syndrome**. Newborns who fail to metabolize the drug at a normal rate die from multiple toxic complications. Because of its potentially lethal side effects, chloramphenicol is used in the United States only to treat seriously ill hospitalized patients; it is no longer used with infants. It is still the preferred drug for treating typhoid fever and may be used for life-threatening infections, such as peritonitis or pelvic infections caused by certain anaerobic species, especially the *Bacteroides*.

Tetracyclines. The **tetracyclines** were discovered in a worldwide search for new antibiotics in the 1950s. They were the first drugs designated broad-spectrum because they were effective against so many different types of bacteria. They control Gram-positive and Gram-negative bacteria and chlamydiae, rickettsiae, and mycoplasmas as well. The natural tetracyclines, including chlortetracycline and oxytetracycline, are produced by the genus *Streptomyces*. Other drugs of this family, including tetracycline itself, are produced semisynthetically.

The tetracyclines interfere with bacterial protein synthesis by blocking the ribosomes' acceptance of tRNA. Tetracyclines are well absorbed orally and widely distributed throughout the body, although they enter the brain poorly. Side effects include gastrointestinal pain and diarrhea caused partly by direct intestinal irritation and partly because of dramatic alteration of the normal flora of the bowel. Tetracyclines also cause increased sensitivity to sunlight and permanently stain developing teeth brown, which prevents their use in children and pregnant women (**Figure 21.13**). Allergy, however, is rare.

Because they are easy to administer and have an extremely broad spectrum, tetracyclines were widely used clinically. They were also added to animal feed to promote the growth of livestock. Consequently, tetracycline-resistant strains of pathogens emerged and the tetracyclines are used much less often than they were 20 years ago. Today, tetracyclines are used primarily to treat acne and infections caused by chlamydiae, rickettsiae, and mycoplasmas.

Erythromycin. The antibiotic **erythromycin** has been a mainstay of antimicrobial therapy ever since its discovery in 1952. It is a member of the macrolide antibiotic family produced by *Streptomyces erythreus*. *S. erythreus*

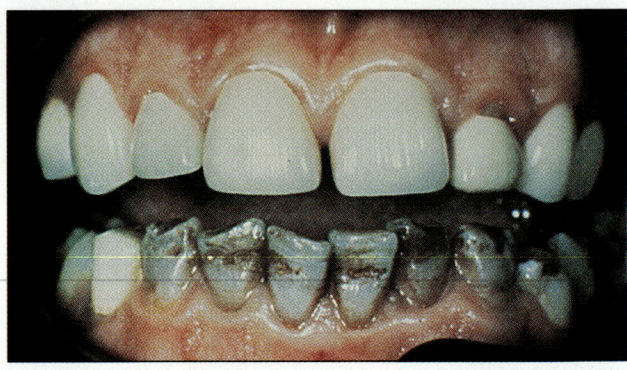

Figure 21.13 Tetracycline should not be prescribed for children or pregnant women because it causes permanent staining in still-forming teeth.

was originally obtained from a soil specimen collected in the Philippines.

Erythromycin is effective only against Gram-positive bacteria. It is widely used to treat strep throat and other common respiratory infections, particularly in patients who are allergic to penicillin. If T. M. had been allergic to penicillin, he would probably have been given erythromycin for his scarlet fever. Erythromycin is also the preferred drug for treating mycoplasma pneumonia, legionellosis, and carriers of diphtheria and whooping cough.

Erythromycin is absorbed well when administered orally, but often causes stomach pain, nausea, and vomiting—side effects that may sound minor but in fact often limit the drug's use. Serious side effects, which include liver damage, are rare. Allergic reactions are also uncommon.

Although erythromycin is still the most widely used macrolide drug, closely related drugs are now on the market. Clarithromycin is a semisynthetic macrolide used increasingly for minor respiratory infections such as ear infections. It causes less nausea and vomiting than does erythromycin.

Quinolones. The **quinolones** are a new and extremely promising family of synthetic antimicrobial agents. They are broad-spectrum agents that block DNA replication in bacteria by inhibiting the unwinding enzyme DNA topoisomerase. They were derived from nalidixic acid, a narrow-spectrum antibiotic (see below). The quinolones are extremely versatile, acting against Gram-positives, many Gram-negatives, and intracellular pathogens such as *Mycoplasma*, *Legionella*, *Brucella*, and *Mycobacterium*. They are easily administered orally and have relatively few side effects. Moreover, drug resistance does not emerge readily. Drugs from this family, which include ciprofloxacin, are already in wide use and likely to increase in the near future.

The Medical Letter

Other Antibacterials. Some antibacterial agents that are used rarely or under limited circumstances are still clinically important. They include nalidixic acid and nitrofurantoin, vancomycin, bacitracin, and polymixin B.

Nalidixic acid and **nitrofurantoin** are synthetic antimicrobials used only for the treatment of urinary tract infections. Administering a dose high enough to treat a lung infection, for example, would not be possible because therapeutic blood levels would be difficult to attain and toxic. But these drugs are concentrated by the kidneys, making them suitable for urinary tract infections. They are, in fact, called *urinary antiseptics* because they are effective against most of the Gram-negative organisms that infect the urinary tract. Both drugs inhibit DNA synthesis.

Vancomycin is an antibiotic produced by the actinomycete *Streptomyces orientalis*. It acts on Gram-positive bacteria by interfering with peptidoglycan synthesis. Vancomycin is not related chemically to any other antimicrobial agent. Because it is not well absorbed orally, it must be administered intravenously. It also has a fairly high incidence of toxic side effects, including damage to the ears and kidneys. Nevertheless, vancomycin is extremely useful for treating life-threatening infections caused by multiple-drug-resistant strains of *Staphylococcus aureus*. In many of these cases, it is the only effective drug. Vancomycin is also useful in other special circumstances, such as treatment of antibiotic-associated colitis caused by the anaerobe *Clostridium difficile*.

Bacitracin and **polymyxin B** are polypeptide antibiotics that destroy lipid membranes. Both agents are highly toxic if used systemically, but safe and effective when used topically for the treatment or prevention of minor skin infections. Polymyxin B is produced by a spore-forming soil bacterium of the genus *Bacillus*. It kills Gram-negative bacteria by interacting with phospholipids in the outer membrane, thus destroying the cell envelope. Bacitracin is produced by the Tracy-I strain of *Bacillus subtilis*, so-named because the strain was isolated from dirt cleaned out of a compound fracture in a young girl named Tracy. Bacitracin is active against a variety of Gram-positive cocci and bacilli. Ointments containing one or a mixture of both of these drugs are available without prescription.

Antimycobacterial Agents

The *Mycobacterium* genus includes the species that cause tuberculosis and leprosy (Chapters 22 and 26). Tuberculosis is the most widespread mycobacterial infection in this country, and its incidence is rising as the AIDS epidemic grows. Treating tuberculosis with antimicrobial agents is difficult for several reasons:

1. Mycobacteria are resistant to most antimicrobial drugs, in part because the mycolic acids that compose the mycobacterial cell wall are nearly impermeable.

2. Mycobacteria grow very slowly, so antimicrobial therapy must continue for months or years.

3. Some antibiotic-resistant mycobacteria exist in every infection, so at least two antituberculosis drugs must usually be used simultaneously to prevent the emergence of drug-resistant strains.

4. *Mycobacterium tuberculosis* is an intracellular pathogen, so effective antituberculosis drugs must be able to enter human cells.

Despite these difficulties, many antituberculosis drugs exist, and most cases of tuberculosis can be treated effectively, although the increasing number of drug-resistant strains causes great concern. Research for new antituberculosis drugs is very active now. **Figure 21.14** shows the chemical structures for the antituberculosis drugs discussed in the following paragraphs.

Isoniazid. The synthetic antimicrobial **isoniazid** is the most useful drug currently available for treating tuberculosis. It is related to nicotinamide, a compound found, by chance, to inhibit the growth of mycobacteria. Isoniazid is activated by a mycobacterial enzyme, a peroxidase, to destroy several cellular targets. At least 95 percent of strains of *M. tuberculosis* are isoniazid-sensitive. Isoniazid is bacteriostatic for resting cells and bactericidal for rapidly growing cells.

Isoniazid is well absorbed orally and readily penetrates cell membranes—an essential feature for a drug aimed against an intracellular pathogen. Isoniazid is inactivated in the liver at a rate that varies somewhat from person to person and then is excreted by the kidneys. Occasionally isoniazid causes liver damage, which can be fatal in extreme cases; people taking isoniazid are therefore checked regularly for signs of liver toxicity.

Isoniazid is also effective for prophylaxis (prevention) of tuberculosis in people who have been infected by the tubercle bacillus but have not yet suffered any illness. Because the number of bacteria in these healthy patients is relatively small, isoniazid alone is sufficient to prevent development of active tuberculosis.

Rifampin. The semisynthetic antibiotic **rifampin** is the drug most often administered in combination with isoniazid. Its natural precursor, rifamycin B, is produced by *Streptomyces mediterranei*. Rifampin acts against many Gram-positive and Gram-negative bacteria and so is sometimes used for infections other than tuberculosis. It is well absorbed orally and penetrates almost all parts of the body. Like isoniazid, rifampin can enter human cells and kill intracellular tubercle bacilli. Unlike isoniazid, rifampin has relatively few toxic side effects. Like isoniazid, it is almost always used as part of a multiple-drug therapy.

Ethambutol. The synthetic drug **ethambutol** is effective only against mycobacteria, but almost all strains of *Mycobacterium tuberculosis* are sensitive to it. Ethambutol's mechanism of action is not clearly understood, but it has been shown to inhibit the incorporation of mycolic acid into the mycobacterial cell wall. It is well absorbed orally, and toxic side effects are rare. When three drugs are used to treat tuberculosis—the preferred treatment for an infected person who enters the United States from a country where drug-resistant strains of *M. tuberculosis* are common—they are usually isoniazid, rifampin, and ethambutol.

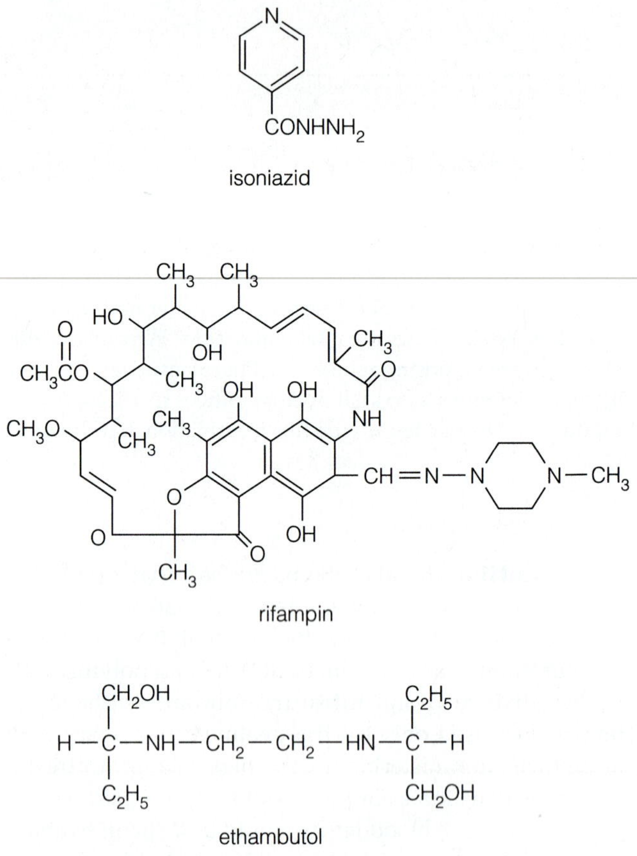

Antifungal Agents

Fungi are eucaryotes, so fungal cells and human cells have much in common. Their similarities make it difficult

Figure 21.14 Chemical structures of major antimycobacterial agents.

to find agents that damage fungal cells while sparing human cells. Systemic fungal infections are particularly difficult to cure. Therapy of serious fungal infections is further complicated by the fact that there is no really good way to test the antimicrobial susceptibility of pathogenic fungi in the laboratory, as it is possible to test antibacterial susceptibility. Nevertheless, fungal infections of the skin and mucous membranes can be treated safely and effectively with various drugs. The chemical structures for the drugs discussed below are shown in **Figure 21.15**.

Nystatin. A member of the polyene family of antibiotics produced by *Streptomyces noursei*, **nystatin** was named after its discovery in the New York State Health Laboratory. Nystatin acts by damaging the plasma membrane and allowing cell contents to leak out. It is selectively toxic for fungi because it combines primarily with ergosterol, a membrane sterol produced by fungi but not humans. Nystatin is used mainly against *Candida albicans*, to treat vaginal and skin infections, for example, and intestinal infections in very ill, immunocompromised patients. For gastrointestinal infections, it is administered

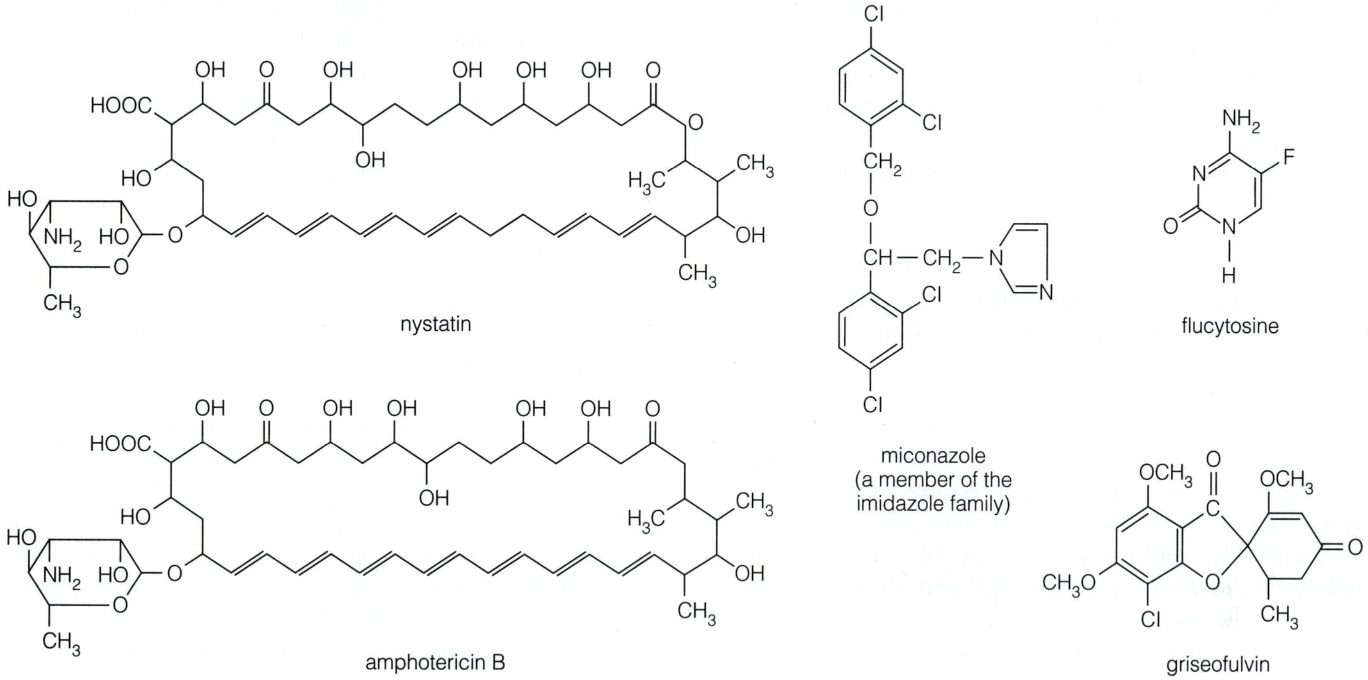

Figure 21.15 Chemical structures of major antifungal agents. Finding agents that damage fungal cells without harming human cells is challenging because both are eucaryotes.

orally. It is not absorbed across the intestinal wall, which means it never enters the blood and remains in the intestine to act locally on the intestinal lining.

Amphotericin B. Like nystatin, **amphotericin B** belongs to the polyene family of antibiotics and acts by disrupting fungal cell membranes. It was isolated from a strain of *Streptomyces nodosus* found in the Orinoco River valley of Venezuela. Unlike nystatin, amphotericin B is used to treat systemic infections. Fungal cells are more susceptible to the drug's destructive action than human cells, but human cells suffer substantial damage. About half the patients who receive a first intravenous dose of amphotericin B experience chills, vomiting, and fever, and many continue to do so during weeks or months of treatment. Over 80 percent also sustain some permanent kidney damage. Despite what would seem to be an unacceptable level of toxicity, amphotericin B is the most effective drug available for treating life-threatening systemic fungal infections such as cryptococcosis and mucormycosis. It is also the drug of first choice in patients with coccidioidomycosis, histoplasmosis, and blastomycosis if the patient is seriously ill or immunocompromised (Chapter 22).

Imidazoles and Triazoles. In recent years, the use of **imidazoles** and **triazoles**, related families of antifungals,

has been increasing. These agents—which include ketoconazole, miconazole, and the newly developed terconazole—interfere with fungal growth by inhibiting the synthesis of plasma membrane sterols. Some of these agents are used topically to treat local infections, but others are used systemically. Ketoconazole, in particular, is used instead of the far more toxic amphotericin B for many systemic mycoses. Miconazole, on the other hand, is used only for fungal infections of the skin, such as athlete's foot and ringworm. Terconazole is used to treat vaginal yeast infections that do not respond to nystatin.

Flucytosine. Another synthetic drug used to treat systemic fungal infections is **flucytosine**. Flucytosine is much less toxic than amphotericin B, but it is also much less effective. Many fungal species are naturally resistant to its action, and drug-resistant strains emerge rapidly among sensitive species, which include *Candida* species and *Cryptococcus neoformans*. Flucytosine is usually administered in combination with amphotericin B.

Griseofulvin. When **griseofulvin** was isolated from *Penicillium griseofulvum* in the 1930s, it was generally ignored because it didn't kill bacteria. Only later was it found to be an effective antifungal agent. Griseofulvin inhibits the growth of fungal cells by interfering with cell division. The only fungi susceptible to griseofulvin are the dermatophytes, which invade the keratin-producing cells in human skin, hair, and nails. They cause the infections commonly known as ringworm. Most ringworm infections are easily cured by topical creams (such as miconazole from the imidazole group), but others can be eradicated only by prolonged oral treatment with griseofulvin. The drug is especially effective against ringworm fungi because it is deposited in keratin-precursor cells. This is the reason newly produced skin or hair cells are the first to become ringworm-free. Griseofulvin does not have serious toxic side effects or drug resistance.

Antiparasitic Agents

Human diseases caused by protozoa or helminths are traditionally known as **parasitic infections**, although, of course, all infections involve a parasitic relationship between humans and an infectious agent. Protozoal and helminthic infections are now relatively uncommon in the United States because of improved public sanitation, but parasitic infections cause illness and death throughout the rest of the world. Finding drugs that are selectively toxic to these eucaryotic organisms is difficult, but some effective chemotherapeutic agents are available. The chemical structures for the drugs discussed below are shown in **Figure 21.16**.

Mebendazole.   The antihelminthic agent **mebendazole** was developed during the 1970s. It is effective against many types of roundworms, including whipworm, hookworm, pinworm, and ascaris (Chapter 23). Mebendazole is used for veterinary as well as human infection. Mebendazole interferes with the helminth's—but not the human cell's—ability to take up glucose. Less than 10 percent of the drug is absorbed from the gastrointestinal tract after oral administration, so most remains in the intestines where these parasitic worms live. Mebendazole is often used in the United States to treat pinworms, a minor infection seen frequently in children. Because so many different worm species are sensitive to mebendazole, it is a good drug for patients simultaneously infected by several types of roundworm, a common condition in developing countries.

Metronidazole.   The antimicrobial **metronidazole** has an unusual spectrum. It is effective against obligate anaerobic bacteria as well as many protozoal parasites. The mode of action of metronidazole is not completely understood, but it is known that microorganisms expend electrons in reducing this drug, which may lead to a critical shortage of reducing power and failure of normal energy-producing pathways (Chapter 5). In addition, reduced forms of the drug form cytotoxic products that destroy the cell.

Metronidazole is used to treat infection by several species of protozoa, including *Trichomonas vaginalis*, a protozoan transmitted by sexual contact; *Entamoeba histolytica*, the agent of amoebic dysentery; and *Giardia lamblia*, a common cause of diarrhea. It is also used to treat certain bacterial infections. Metronidazole is well absorbed orally and has few side effects, but it does cause cancer in animals and an unusually high rate of mutations in certain bacteria. Although no direct evidence exists of risk to humans, these laboratory results dictate caution, and metronidazole is not used during pregnancy.

Chloroquine.   Malaria, which kills more than a million people each year, is still the world's most devastating infection (Chapter 27), and chloroquine is still the most effective antimalarial drug. **Chloroquine** is a synthetic compound derived from the natural antimalarial substance quinine, an extract from the bark of the South American cinchona tree discovered hundreds of years ago. Chloroquine is selectively toxic for protozoa because it is highly concentrated within red blood cells, the site of malarial parasite infection. Its exact mode of action remains unknown.

Many characteristics of chloroquine make it the mainstay of antimalarial therapy around the world. It is readily absorbed from the gastrointestinal tract, and its side effects, such as headache and abdominal discomfort, are usually mild and disappear when treatment is stopped. It can cure or prevent the most virulent form of malaria caused by strains of *Plasmodium falciparum*. It is less effective, but still useful, against the usually milder form of malaria caused by *P. vivax* (Chapter 27). Recently, however, chloroquine-resistant strains of falciparum malaria have emerged, dealing a severe blow to international efforts to control malaria. New antimalarials, such as the recently released quinine-related drug mefloquine, offer some hope for the near future.

Antiviral Agents

A virus and the cell it infects are so intimately associated that it is extremely difficult to develop selectively toxic chemotherapeutic agents. Some of the more serious viral infections are controlled by vaccines, and a few antiviral drugs are now available. Their success raises hopes that we may soon enter an era of antiviral chemotherapy. The chemical structures for some of the drugs discussed below are shown in **Figure 21.17**.

Amantadine.   The antiviral **amantadine** is an orally administered, synthetic drug long known to be clinically effective against influenza A virus, which causes the illness most people refer to as "the flu" (Chapter 22).

Amantadine interferes with viral replication at an early stage in the process, probably inhibiting viral uncoating. It prevents the development of influenza A in 80 percent of people who have recently been exposed to the virus, but it is much less effective once symptoms appear.

The fact that amantadine helps the most before people become ill makes it difficult to use the drug effectively. It is obviously impractical to have everyone take amantadine every day during the winter flu season. Under special circumstances, however, amantadine can be used effectively for influenza prevention. In a closed population of highly susceptible individuals such as nursing home residents, for example, giving amantadine to everyone may make sense after one case of influenza has been diagnosed.

Even though benefits are limited after symptoms appear, some people feel amantadine is still worthwhile if started during the first two days of influenza symptoms—fever disappears sooner and days of illness may be shortened by as much as 50 percent. The problem is that it is very difficult to know whether a patient has true influenza A, which will respond to amantadine, or a sim-ilar viral illness against which the drug is useless. Another deterrent to treating without a firm diagnosis is that amantadine has some unpleasant side effects, which can include anxiety, headache, and insomnia.

Acyclovir. Because **acyclovir** closely resembles the nucleoside guanine (Chapter 6), it acts like flucytocine and other drugs that closely resemble the bases of DNA or RNA, interfering with the synthesis of nucleic acids. Acyclovir's selective toxicity is due to an enzyme found only in virus-infected human cells. The enzyme transforms acyclovir into a compound that interferes with the synthesis of new viral DNA, but not human DNA.

Acyclovir is effective against viruses of the herpes family. The herpes simplex viruses (Chapter 24) are particularly sensitive, and this drug is widely used to treat genital herpes infections. Administered orally and applied directly to infected skin, acyclovir significantly shortens the period of painful genital blisters during an initial infection, but unfortunately does not prevent recurrences. Administered intravenously, acyclovir is also used to treat life-threatening herpes infections such as

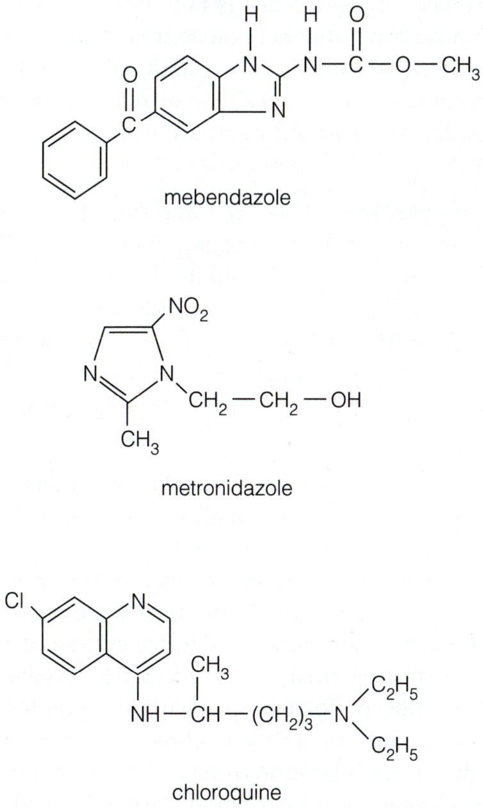

Figure 21.16 Chemical structures of antihelminthic and antiprotozoal agents. Mebendazole is used for human and veterinary helminthic infections. Metronidazole is effective against many protozoal parasites and anaerobes. Chloroquine is used to treat malaria.

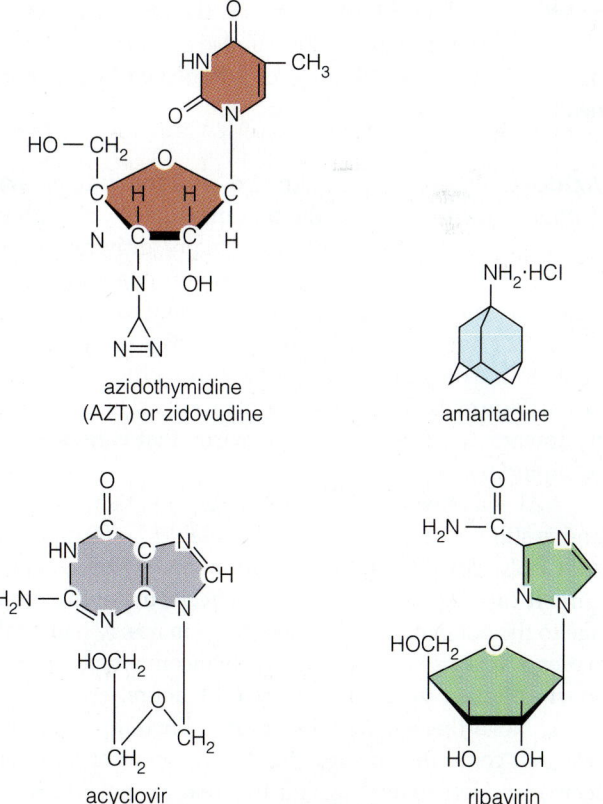

Figure 21.17 Chemical structures of antiviral agents. Because a virus and the cell it infects are so intimately associated, it is extremely difficult to develop selectively toxic antiviral agents.

herpes simplex encephalitis. Unfortunately, herpesvirus strains that are resistant to acyclovir have already been identified as the cause of severe disease in immunocompromised people such as AIDS patients.

Acyclovir is also somewhat effective against the varicella zoster herpesvirus that causes chickenpox and shingles (Chapter 26). Children with chickenpox treated with acyclovir are not cured, but they seem to have a slightly shorter period of fever and fewer blisters than do untreated children. Some physicians are now recommending acyclovir for selected high-risk patients.

Ribavirin.   Like acyclovir, **ribavirin** closely resembles the nucleoside guanine and also interferes with viral nucleic acid replication. The exact mechanism is unknown, but ribavirin probably interferes with viral nucleic acid polymerases, and it may become incorporated into viral mRNA. So far, ribavirin has been used mainly to treat infections caused by respiratory syncytial virus, an RNA virus that can cause serious lung infections in infants (Chapter 22). The drug is made into a mist and inhaled directly into the lungs. Although toxic side effects of ribavirin have been minimal, there continues to be concern about possible long-term effects on the rapidly developing lungs of these babies. There is also concern that it may be harmful to the developing fetus if inhaled by pregnant health-care workers.

Azidothymidine.   Also known as zidovudine, **azidothymidine (AZT)**, was the first major breakthrough in the drug treatment of AIDS. Like many antiviral drugs, the molecule is a close chemical analogue of a normal nucleic acid precursor, in this case the nucleoside thymidine. AZT interferes with reverse transcriptase, an enzyme found only in retroviruses that copy their RNA into DNA (**Figure 21.18**). This accounts for its selective toxicity toward the deadly HIV retrovirus that causes AIDS (Chapter 27).

AZT was first administered to AIDS patients in 1985 and is still the most useful drug available. Unfortunately, AZT only slows the pace of immune failure—it is not a cure. It can also have severe side effects, including damage to the bone marrow, which occurs in nearly half of all patients. It is also an extraordinarily inconvenient drug. It must be taken by mouth every four hours, around the clock. Nevertheless, AZT has been a breakthrough, and research continues toward the development of more effective and less toxic drugs for the treatment of AIDS.

Dideoxyinosine.   Another antiviral agent effective in treating AIDS, **dideoxyinosine (ddI)**, was developed after AZT and acts the same way, by inhibiting reverse

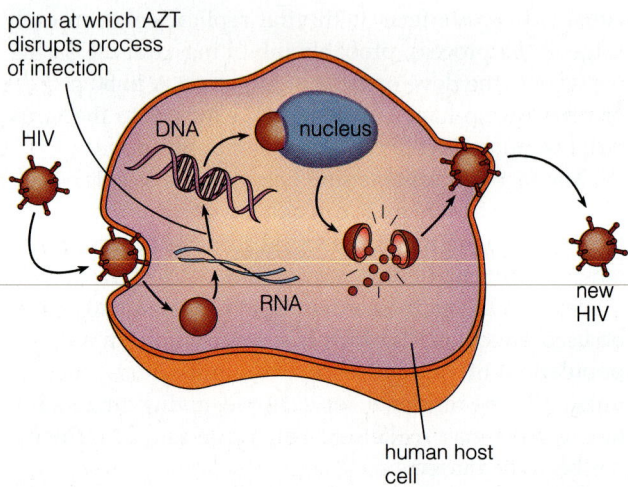

point at which AZT disrupts process of infection

HIV

DNA

nucleus

RNA

new HIV

human host cell

Figure 21.18   HIV infection and AZT. Normally, when HIV enters a human host cell, the enzyme reverse transcriptase copies the RNA of the viral genome into the DNA that is incorporated into the host's genome. AZT blocks this enzyme. Because normal human cells do not have reverse transcriptase, AZT is selectively toxic for the virus.

transcriptase. Although ddI is not superior to AZT, it is useful in certain patients because it is less toxic to bone marrow. Furthermore, it may inhibit strains of HIV that have become resistant to AZT. Research to evaluate combination therapy with AZT and ddI is underway.

Dideoxycytidine.   Like AZT and ddI, this newest antiviral agent approved for use in patients with AIDS is a nucleoside analogue that inhibits the reverse transcriptase enzyme of the retrovirus. Molecule for molecule, **dideoxycytidine (DDC)** seems to be the most potent of the three currently available drugs. Clinical trials to evaluate DDC, as well as the other drugs, are in progress.

Interferons.   In response to viral infections, human cells produce **interferons**, small glycoproteins that have an inhibitory effect against many types of viruses (Chapter 16). Genetic engineering has made it possible to produce pure interferons in amounts large enough for chemotherapy. Unfortunately, interferons have been ineffective in treating many viral infections and have also shown significant toxicity, particularly symptoms that mimic viral infection. Some clinical uses have been found, however. Interferons are used in the United States to treat chronic viral hepatitis (Chapter 23) and genital warts caused by the human papillomavirus (Chapter 24). They have also been found helpful in patients suffering from the AIDS-related cancer Kaposi's sarcoma (see Chapter 18).

Turning Point

Antimicrobial Drugs in American History

In the 1920s, when Calvin Coolidge was president of the United States, 16-year-old Calvin Jr. became ill. He had been playing tennis and developed a blister on his toe. The blister became infected and the infection spread. In spite of the best medical attention, the boy died. The diagnosis was streptococcal septicemia, an overwhelming infection caused by the bacterium *Streptococcus pyogenes*. Ten years later, in 1936, another president's son suffered from an infection caused by the same bacterium. Franklin Roosevelt Jr., then a student at Harvard, was admitted to the hospital with a streptococcal throat infection. Newspapers around the country carried the story and public concern grew—the memory of young Coolidge's death was fresh. Soon, however, the press reported that FDR Jr. was being treated with a new drug from Germany that could kill streptococcal bacteria. The drug was Prontosil, an antimicrobial of the sulfonamide family, and its use to cure FDR Jr.'s strep throat marked the beginning of the antimicrobial era in the United States.

It's hard to imagine what life was like before the development of antimicrobial drugs—only 50 years ago. Anyone—young or old—could suddenly be taken ill and die from infection. Medical science could do little, if anything, to save a person. Today when people become ill they believe that if they consult a doctor they can be cured, and often they are right. This confidence is largely the result of antimicrobial drugs. Walsh McDermott, a doctor who lived through the transformation of medical practice that resulted from antimicrobial therapy, described it as a "historic watershed": "One day we could not save lives, or hardly any lives; on the very next day we could do so across a wide spectrum of diseases. This was an awesome acquisition of power. . . ."

Obviously, antimicrobial drugs help people hospitalized with pneumonia or meningitis, but patients suffering from diseases as different as cancer, diabetes, and stroke often depend just as much on antimicrobial drugs. People who are seriously ill for *any* reason are prone to develop infections. Without antimicrobial drugs, modern surgery would be severely limited, and organ transplantation could never have been attempted.

Antibiotics have also played an important role in other medical breakthroughs. In developing a new live viral vaccine, for example, the first step is to grow a tissue culture of individual animal cells in the laboratory. But until antibiotics were added to the culture medium, contaminating microorganisms almost always overwhelmed the animal cells. Thus, antibiotics made the perfection of tissue-culture techniques possible, leading to the development and mass production of live viral vaccines. Indirectly then, antibiotics are responsible for the virtual disappearance of polio, measles, and mumps.

Walsh McDermott and physicians of his generation were awestruck by the power of antibiotics, but we now know that there are serious limitations to the use of this power. Antimicrobial chemotherapy is highly effective only against infections caused by bacteria. In spite of recent advances, most patients with life-threatening fungal and viral infections still die from them, just as young Coolidge died from his streptococcal infection in the 1920s. Even more distressing, physicians today can see the power to cure bacterial infections beginning to ebb against mounting antibiotic resistance. Antibiotics that were once effective in treating potentially fatal infections have now become almost useless. The sulfa drugs that miraculously saved FDR Jr.'s life in 1936, for example, are seldom used today because almost all the bacteria against which they were once effective are now resistant.

New drugs have been developed, however, and most bacterial infections can still be treated successfully by one antimicrobial agent or another. But drug selection must be made carefully and practitioners must stay up-to-date on constantly changing patterns of antibiotic resistance. It is inevitable that microbes will eventually develop resistance to today's drugs. How medical technology will keep pace with this growing problem is a challenging and urgent question.

Summary

PRINCIPLES OF PHARMACOLOGY   (pp. 489–500)

1. Pharmacology is the study of drugs, any chemicals that have a physiological effect. Drugs used to treat disease are called chemotherapeutic agents. Chemotherapeutic agents used to treat infections are called antimicrobial agents.

2. Antibiotics are metabolic products of one microorganism that kill or inhibit the growth of other microbes. Synthetic drugs are chemicals produced in the laboratory. Semisynthetic antibiotics have been chemically modified in the laboratory.

3. Antimicrobial agents can be classified according to their targeted microorganism. Thus, there are antibacterials, antimycobacterials, antivirals, antifungals, and antiparasitics.

Drug Administration (pp. 489–491)

1. External (local, topical) therapy is the application of an antimicrobial drug directly to the infected area. Most infections require systemic therapy; administration can be intravenous (IV), intramuscular (IM), or oral (PO).

Drug Distribution (pp. 491–492)

1. There are two major barriers to drug distribution—membranes and protein binding.

Eliminating Drugs from the Body (p. 492)

1. Eventually all drugs are eliminated from the body, usually in the urine.

Side Effects and Allergies (p. 492)

1. An effective drug is selectively toxic. Most antimicrobial drugs have side effects, or undesirable toxicity.

Drug Resistance (pp. 492–497)

1. A microorganism is drug-resistant if it can grow in the presence of a drug. Natural drug resistance is an intrinsic property of a microbial species. Acquired drug resistance is a property gained by individual strains.

2. Narrow-spectrum antimicrobial drugs affect only a single microbial group. Broad-spectrum antimicrobials affect two or more microbial groups. Broad-spectrum drugs may alter the normal flora, causing a superinfection.

3. Microbial strains acquire drug resistance through genetic change.

4. There are three mechanisms of acquired drug resistance: the microorganism produces an enzyme that destroys the drug, the target of drug action changes, and adaptations make it difficult for the antibiotic to enter the microbial cells or the cell actively expels it after it enters.

5. Drug resistance can be encoded by genes on the microbial chromosomes or on plasmids. Chromosomally encoded resistance occurs by mutation.

6. Plasmid-encoded resistance can be spread rapidly from one strain to another strain or species by genetic conjugation. Plasmids that carry genes encoding drug resistance are called R factors.

7. Acquired drug resistance can be slowed by limiting nonmedical uses of antibiotics, being more selective in the medical use of antibiotics, and, under some conditions, administering two drugs at the same time (combination therapy).

Dosage and Antimicrobial Susceptibility (pp. 497–500)

1. Drug dosage depends upon route of administration, drug distribution, drug elimination, and antimicrobial susceptibility (a measure of how much of a drug is required to kill or stop the growth of a pathogen).

2. The simplest and most commonly used antimicrobial susceptibility test is the disc-diffusion susceptibility test (or Kirby-Bauer test). In the broth-dilution susceptibility test, the clear test tube with the lowest drug concentration contains the minimum inhibitory concentration (MIC) of that drug. The tube with the lowest drug concentration that is bactericidal contains the minimum bactericidal concentration (MBC) of that drug. In the serum killing-power test, some of the patient's serum is tested in the laboratory to see if it kills the infecting microorganism.

MODES OF ACTION OF ANTIMICROBIAL DRUGS (pp. 500–502)

1. Drugs that inhibit peptidoglycan synthesis have high selective toxicity and are bactericidal; penicillins, cephalosporins, vancomycin, and bacitracin are examples.

2. Drugs that damage the cell membrane, causing increased permeability and allowing vital molecules to leak out, have limited selective toxicity and are microbiocidal; polymyxin B and amphotericin B are examples.

3. Drugs that inhibit protein synthesis by affecting the 70S bacterial ribosome are usually bacteriostatic. The aminoglycosides bind to the 30S subunit and are bactericidal. Tetracyclines bind to the 30S subunit and are bacteriostatic. Erythromycin and chloramphenicol bind to the 50S subunit and are bacteriostatic.

4. Drugs that inhibit nucleic acid synthesis include the quinolones, which inhibit topoisomerase; rifampin, which inhibits bacterial RNA polymerase; and flucytosine and other antimetabolites, which block a step in biosynthesis of nucleotides.

5. Drugs that inhibit folic acid synthesis are antimetabolites. The sulfonamides are competitive inhibitors of the enzyme substrate PABA; trimethoprim is a competitive inhibitor of another intermediate of the folic acid pathway.

ANTIMICROBIAL DRUGS (pp. 502–515)

1. Antimicrobial drugs can be classified by the infection they are usually used to treat.

Antibacterial Agents (pp. 502–509)

1. Penicillins interfere with cross-linking peptidoglycan. Natural penicillins (G and V) are most effective against Gram-positives. Ampicillin and amoxicillin are broad-spectrum semisynthetics that act against both Gram-positives and many Gram-negatives. Side effects are negligible, but allergies are common.

2. Cephalosporins also interfere with cross-linking peptidoglycan. They are broad-spectrum and can be natural or semisynthetic. The third generation is particularly effective against meningitis, pneumonia, and sepsis; some penetrate the central nervous system well, and others persist in the bloodstream. Side effects are negligible, but people allergic to penicillin can be allergic on first exposure.

3. Sulfonamides are broad-spectrum synthetics that act on Gram-positives, some Gram-negatives, and chlamydiae. They interfere with folic acid synthesis. Today they are generally used by themselves or in combination with trimethoprim (cotrimoxazole) to treat urinary tract infections. Side effects include crystals in urine and depressed stem cell production; allergy is common.

4. Aminoglycosides, such as streptomycin and neomycin, developed to control Gram-negatives, must be administered intravenously or intramuscularly. Major side effects include loss of hearing and kidney function. Drug resistance emerges rapidly.

5. Chloramphenicol is an antibiotic that can be manufactured synthetically. It can be taken orally, penetrates the brain and eye, can be stored for long periods with refrigeration, is extremely broad-spectrum, has few side effects, and rarely causes allergy; but rarely it can be fatally toxic. In the United States, it is used only with hospitalized patients and never with infants.

6. Tetracyclines, the first antibiotics designated broad-spectrum, are used to treat acne and infections caused by chlamydiae, rickettsiae, and mycoplasmas. They can be natural or semisynthetic. They cause increased sensitivity to sunlight. Because they stain developing teeth, they cannot be used with children or pregnant women. Resistant strains have emerged.

7. Erythromycin is a narrow-spectrum antibiotic effective against only Gram-positives. It is often used for patients allergic to penicillin and is prescribed for strep throat and other common respiratory infections. Side effects include stomach upset and, rarely, liver damage. Allergies are rare.

8. The quinolones are broad-spectrum drugs derived from a narrow-spectrum antibiotic, nalidixic acid. They act against Gram-positives, many Gram-negatives, and intracellular pathogens. They are administered orally and have relatively few side effects; allergy is rare, and drug resistance is slow to emerge.

9. Nalidixic acid and the synthetic drug nitrofurantoin are narrow-spectrum drugs that tend to concentrate in the kidneys, which makes them most suitable for treating urinary tract infections. They inhibit DNA synthesis.

10. Vancomycin acts on Gram-positives. It must be administered intravenously and has serious side effects, including damage to the ears and kidneys, but it can be the only treatment for life-threatening infections caused by multiple-drug-resistant strains of *Staphylococcus aureus*.

11. Bacitracin and polymyxin B, which are highly toxic if used systemically, are effective topically for minor skin infections. Ointments containing one or both of these are available without prescription.

Antimycobacterial Agents   (pp. 509–510)

1. Isoniazid, a synthetic, is the main drug used in treating tuberculosis. It is bacteriostatic for resting cells and bactericidal for rapidly growing cells. It can cause liver damage, which is fatal in extreme cases.

2. Rifampin, usually used in combination with isoniazid, is semisynthetic and acts by inhibiting the synthesis of RNA. It is also used to treat many Gram-positive and Gram-negative bacteria. Side effects are negligible.

3. Ethambutol, a synthetic, is effective only against mycobacteria. It is used in combination with isoniazid and rifampin. Side effects are negligible.

Antifungal Agents   (pp. 510–512)

1. Nystatin belongs to the polyene family of antibiotics. It disrupts the plasma membrane and is selectively toxic. It is used locally, primarily to treat *Candida albicans* infections.

2. Amphotericin B, also a polyene, also disrupts cell membranes. Side effects include chills, vomiting, fever, and, in 80 percent of patients, kidney damage. Nevertheless, it is the preferred treatment for certain systemic infections, such as cryptococcosis and mucormycosis.

3. The imidazoles and triazoles both inhibit the synthesis of plasma membrane sterols. Some agents are used topically (miconazole), while others are used systemically (ketoconazole).

4. Flucytosine is a synthetic drug used to treat systemic infections. Because drug resistance emerges rapidly, it is usually used in combination with amphotericin B.

5. Griseofulvin acts only on dermatophytes, which cause ringworm infections. It is administered orally when topical preparations do not work; it interferes with cell division. Side effects are negligible, and there is no serious drug resistance.

Antiparasitic Agents   (p. 512)

1. Mebendazole is used for many roundworm infections. In the United States, it is used mostly to treat pinworms. It interferes with the helminth's ability to take up glucose.

2. Metronidazole is effective against protozoal parasites and is used to treat trichomoniasis, amoebic dysentery, and *Giardia lamblia* diarrhea. It is also effective against obligate anaerobic bacteria. Side effects are few, though it can cause cancer in laboratory animals and possibly birth defects in humans.

3. Chloroquine is the most effective drug for treating malaria, especially the most virulent form. It is synthetic and selectively toxic

because it concentrates in red blood cells. Side effects include mild headache and abdominal discomfort. Resistant strains have emerged.

Antiviral Agents   (pp. 512–515)

1. Amantadine, a synthetic drug, interferes with viral replication. It is most useful in preventing influenza A, especially if administered soon after exposure. Side effects include anxiety, headache, and insomnia.

2. Acyclovir interferes with the synthesis of viral DNA. It is used to treat herpes infections orally and topically and, in serious systemic cases, intravenously. It is somewhat effective against the varicella zoster herpesvirus.

3. Ribavirin also interferes with viral nucleic acid replication. It is inhaled as a mist for respiratory syncytial virus infections. Immediate side effects are minimal, but long-term effects are not known.

4. Azidothymidine (AZT) is used to treat AIDS. It interferes with reverse transcriptase, making it selectively toxic. It must be taken orally every four hours and side effects, including damage to bone marrow, can be severe.

5. Dideoxyinosine (ddI), another AIDS drug, inhibits strains of HIV resistant to AZT. It acts as AZT does and is less toxic to bone marrow than AZT.

6. Dideoxycytidine (DDC), the newest drug for AIDS, acts as AZT and ddI do but is more potent.

7. Interferons can be made by genetic engineering. In the United States, interferons are used to treat chronic viral hepatitis and genital warts caused by human papillomavirus. They have significant side effects.

Review Questions

PRINCIPLES OF PHARMACOLOGY

1. What is pharmacology? What is the difference between chemotherapeutic agents and antimicrobial agents? between antibiotics, synthetic drugs, and semisynthetic drugs?

2. What is the difference between local, or topical, therapy and systemic? Describe the three ways of administering drugs systemically and the advantages and disadvantages of each.

3. Discuss the barriers to drug distribution in the body.

4. Why is it important to know how and in what period of time drugs are eliminated from the body?

5. What are drug side effects? Give an example of how side effects must be weighed against drug benefits. Discuss an example of drug allergy.

6. What determines if a microorganism is drug resistant? What is the difference between natural and acquired resistance? Why are some species naturally resistant? How do species acquire resistance?

7. What is the difference between narrow- and broad-spectrum drugs?

8. Compare and contrast chromosomally mediated resistance and resistance due to plasmid-borne genes.

9. Discuss ways drug resistance can be slowed and why it is important to do so.

10. Define these terms: drug dosage, antimicrobial susceptibility, drug synergism.

11. How would you perform a disc-diffusion susceptibility test? A broth-dilution susceptibility test? What is the difference between a bacteriostatic and a bactericidal drug?

12. Discuss the significance of the MIC and the MBC. What information does a serum killing-power test give a clinician?

MODES OF ACTION OF ANTIMICROBIAL DRUGS

1. Discuss the details of each mode of action and give some examples of drugs that act this way:
   a. drugs that inhibit cell-wall synthesis
   b. drugs that disrupt the cell membrane
   c. drugs that inhibit protein synthesis
   d. drugs that inhibit nucleic acid synthesis
   e. drugs that inhibit folic acid synthesis

ANTIMICROBIAL DRUGS

1. Tell whether each of the following antibiotics is narrow- or broad-spectrum, what the mode of action is, how the drug is used, and whether allergy or side effects are a concern:

   a. penicillins
   b. cephalosporins
   c. sulfonamides
   d. aminoglycosides
   e. chloramphenicol
   f. tetracyclines
   g. erythromycin
   h. quinolones
   i. nalidixic acid and nitrofurantoin
   j. vancomycin
   k. bacitracin and polymyxin B

2. Why is tuberculosis difficult to treat with antimycobacterials? How does each of these antimycobacterials act—isoniazid, rifampin, ethambutol? Why are these drugs usually used in combination?

3. Why is it difficult to develop effective antifungal agents? Discuss the modes of action, use, and side effects for each of the following antifungals:

   a. nystatin
   b. amphotericin B
   c. imidazoles and triazoles
   d. flucytosine
   e. griseofulvin

4. What are the modes of action for these antiparasitics: mebendazole, metronidazole, chloroquine? How is each used? What are the side effects?

5. Why is it difficult to develop selectively toxic antiviral agents? For each of the following antivirals, discuss mode of action, use, and side effects:

   a. amantadine
   b. acyclovir
   c. ribavirin
   d. interferons
   e. azidothymidine (AZT)
   f. dideoxyinosine (ddI)
   g. dideoxycytidine (DDC)

Essay Questions

1. "It is inevitable that microbial resistance will eventually develop to today's drugs." Discuss this statement from a historical point of view and from an epidemiologist's point of view. Why is drug resistance a major and urgent problem?

2. Do you support or oppose the use of antibiotics in livestock? Explain your position.

Suggested Readings

Dixon, B. 1986. Overdosing on wonder drugs. *Science 86*, 40–43.

Hirsch, M. S., and Kaplan, J. C. 1987. Antiviral chemotherapy. *Scientific American*, April, 76–85.

Hirsch, M. S., and Schooley, R. T. 1989. Resistance to antiviral drugs: The end of innocence. *New England Journal of Medicine* 320:313–14.

Lietman, P. S. 1986. What is an antibiotic? *Journal of Pediatrics* 108:824–29.

Thomas, L. 1984. Medicine's second revolution. *Science 84*, 93–95.

FACING PAGE: Helicobacter pylori *is a Gram-negative curved bacterium that is able to colonize the human stomach and cause peptic ulcers.*

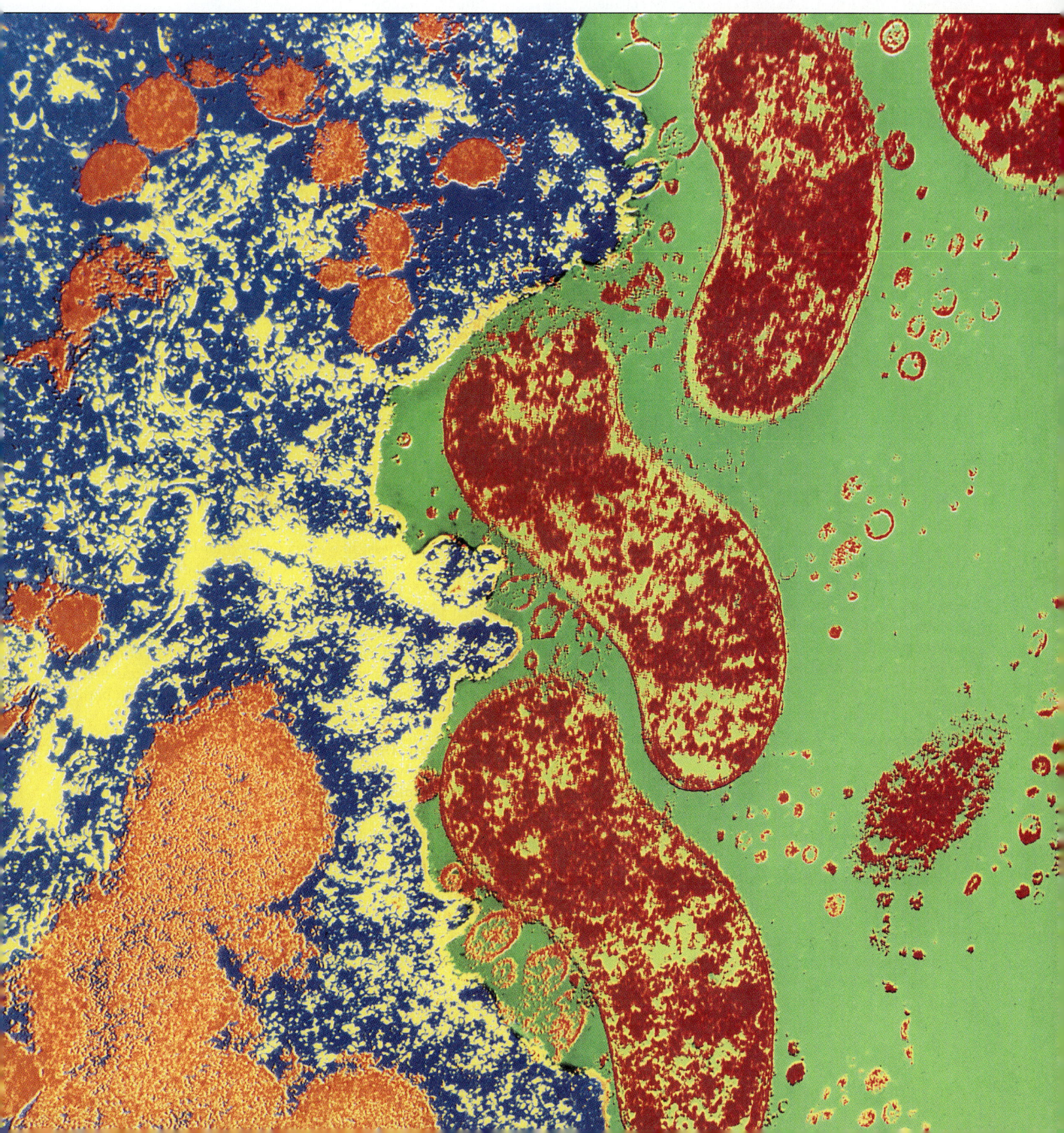

INTRODUCTION TO PART IV

Both microorganisms and human beings are creatures of amazing complexity. Understanding either is a challenge, and understanding how the two interact to produce human disease is a greater challenge.

CLINICAL SCIENCE

The study of infectious disease in humans is one practical application of the science of microbiology. It is a clinical science based on observation and day-to-day experience treating patients. Clinical study can be frustrating because it is imprecise. Many factors influence the manifestation and outcome of an illness, so there is no such thing as a *typical* illness. No two cases are exactly the same. Nevertheless, the pressing need to help people who are suffering from infectious diseases makes clinical science one of the most compelling fields in applied microbiology.

A medical **clinician** treats people who are ill. To help someone seeking medical attention, the clinician must address three fundamental questions:

1. What is wrong with this person?

2. What is likely to happen next?

3. What can be done to help?

The answers to these three questions provide, respectively, a diagnosis, a prognosis, and a treatment plan. A **diagnosis** is a determination of the nature and cause of a disease. A **prognosis** describes the probable course of the disease and its possible outcome. The treatment plan describes the measures that will be taken to help the patient recover.

The Diagnostic Process

The first of the clinician's three fundamental questions is the most crucial because prognosis and treatment are based on accurate diagnosis. Until the problem is identified, it is impossible to guess what will happen or make a plan to help.

In making a diagnosis the clinician attempts to determine the **etiology**, or cause, of the patient's illness. This approach is based on the assumption of a cause-and-effect relationship—some event (the cause) produces an illness (the effect). The clinician, who sees only the effect, must rely on knowledge and previous experience to infer the probable cause. Working backward from what one

sees to what one supposes to be the cause is the process of clinical diagnosis.

Diagnosis begins with the **clinical presentation**, information about the ill person that provides the facts upon which to base a diagnosis. Information is obtained by eliciting a medical history and performing a physical examination. To elicit a medical history, the clinician asks questions such as "Where is your pain?" and "When did you become ill?" The history tells about a person's **symptoms**, the subjective experience of the illness. In the physical examination, the clinician takes objective measurements (such as measuring body temperature) and applies clinical skills (such as listening to the heart). The physical examination reveals **clinical signs**, the clinician's objective assessment of the patient's illness. In some cases, the clinician obtains further information through laboratory or x-ray studies.

In the process of collecting information, the clinician may accumulate items that seem unrelated or even contradictory. The task of diagnosis is to fit all the pieces together into a coherent whole. In most cases, the clinician formulates a **differential diagnosis**, a list of alternative diagnostic possibilities ranked in order of probability. As additional information becomes available, the list is refined. In the meantime, the clinician must make decisions about prognosis and treatment based on a best guess, a **working diagnosis**.

Every illness, like every individual human being, is unique. To make a diagnosis, however, the clinician must generalize, comparing the illness with others that have been seen and described before. A diagnosis usually has two components—an **anatomic diagnosis**, which identifies the part of the body that is affected, and an **etiologic diagnosis**, which identifies the fundamental cause of the problem. A **conclusive diagnosis** identifies both components.

Anatomic Diagnosis. Often the clinical presentation immediately suggests an anatomic diagnosis. A patient's **chief complaint** (the first statement to the clinician) may lead directly to an anatomic diagnosis. For example, the chief complaint "I'm having trouble breathing" draws immediate attention to the respiratory system. The anatomic diagnosis is refined as the clinician determines more precisely which part of the respiratory system is affected. The specific anatomic structures implicated depend upon other clinical patterns. For example, difficult breathing along with abnormal breath sounds heard through a stethoscope implicate lung damage, regardless of whether the damage is caused by infection, allergy,

Table 1 Categories of Human Disease

Category	Definition	Example
Infection	Illness caused by multiplication of microorganisms in the body	Streptococcal pharyngitis
Trauma	Injury inflicted by a physical agent	Broken bone
Malignancy	Uncontrolled multiplication of the body's own cells	Brain tumor
Metabolic defect	Illness caused by inability to perform normal biochemical transformations	Diabetes
Inherited or congenital defect	Abnormality caused by abnormal genes (inherited) or problem in development before birth (congenital)	Cystic fibrosis; spina bifida
Immunologic disorder	Illness caused by weakness or malfunction of the immune system	Allergy
Nutritional disorder	Illness caused by poor nutrition	Obesity, vitamin deficiency
Degenerative disorder	Decrease in function associated with advancing age	Osteoarthritis
Toxicity	Debility caused by poisoning	Lead poisoning
Iatrogenic disease	Disease that results from medical intervention	Nosocomial infection
Idiopathic disease	Unknown cause	Nephrotic syndrome (childhood kidney disease)

poisoning, or shock. Constellations of signs and symptoms that implicate specific structures are called **anatomic syndromes**, or **clinical syndromes**.

Etiologic Diagnosis. Etiologic diagnosis usually follows anatomic diagnosis. The first attempt at etiologic diagnosis may merely assign the cause of illness to a broad category, such as infection (illness caused by microorganisms), trauma (injury), malignancy (cancer), malnutrition (poor nutrition), or toxicity (poisoning), among others (**Table 1**). Sometimes the etiology is obvious, as it is after an accident (trauma). Other times the cause may be elusive, as is often the case with environmental poisonings (toxicity). This textbook considers only those human illnesses caused by infection. Clinical clues that suggest infection may include recent exposure to a communicable disease, fever, and inflammation.

After a diagnosis of infection, questions remain. A conclusive diagnosis requires that the **etiologic agent** (the causative microorganism) be identified, usually by tests done in a clinical laboratory. This may answer critical questions regarding prognosis and treatment. But identification of the causative microorganisms may be impossible. Suppose, for example, that microorganisms are multiplying deep within human tissues and cannot easily be recovered—often the case with pneumonia. Or suppose the microorganism is one of the many viruses

for which there is no readily available laboratory test—often the case with the common cold. In these situations, knowing which microorganisms usually infect different parts of the body, along with information about the patient's age and previous state of health, helps the clinician formulate a fairly reliable working diagnosis. For example, if signs and symptoms suggest pneumonia, the clinician knows that *Streptococcus pneumoniae* must be considered. If the patient is a young child, *Haemophilus influenzae* is also a candidate. If the patient has AIDS and poor immune defenses, *Pneumocystis carinii* is a strong possibility.

Correlating Anatomic and Etiologic Diagnoses. A few anatomic syndromes (for example, chickenpox and measles) have only a single etiologic agent. For this class of infection a conclusive diagnosis can be made without laboratory confirmation. Usually, however, correlating anatomic and etiologic diagnoses of infection is not so clear-cut. A given anatomic syndrome such as pharyngitis (sore throat) can be caused by many different etiologic agents. Conversely, an etiologic agent such as the group A *Streptococcus* bacterium can produce many different anatomic syndromes (**Figure 1**).

The variable relationship between anatomic and etiologic diagnoses makes it difficult to organize a discussion of human infection. Grouping infections caused by the

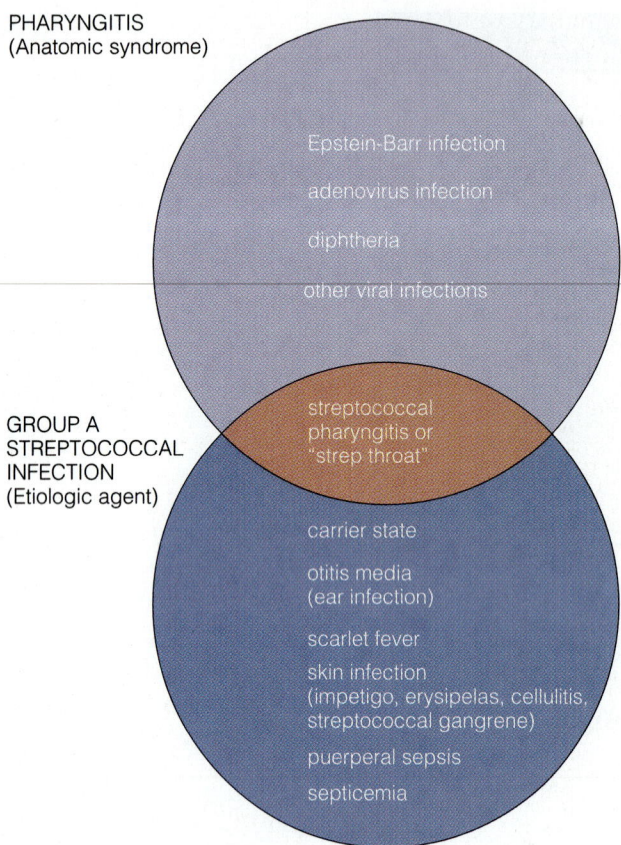

PHARYNGITIS
(Anatomic syndrome)

Epstein-Barr infection

adenovirus infection

diphtheria

other viral infections

GROUP A
STREPTOCOCCAL
INFECTION
(Etiologic agent)

streptococcal pharyngitis or "strep throat"

carrier state

otitis media
(ear infection)

scarlet fever

skin infection
(impetigo, erysipelas, cellulitis, streptococcal gangrene)

puerperal sepsis

septicemia

Figure 1 Correlating anatomic and etiologic diagnoses.

same or closely related etiologic agents results in discussing a wide range of illnesses that are otherwise unrelated. For example, if we discussed all the illnesses caused by *Streptococcus pyogenes*, we would cover ailments ranging from ear infections to life-threatening infections of the bloodstream. If we organize our discussion of infection around similar anatomic syndromes, then we must consider quite different and unrelated microorganisms. This last approach, which we will follow, has the advantage of following the usual sequence of clinical diagnosis—first identify the site of infection, then identify the causative agent.

Prognosis and Treatment

The importance of accurate diagnosis becomes clear when we consider prognosis and treatment—the two principal ways clinicians can help their patients. As an example, consider strep throat and infectious mononucleosis, two common illnesses that can be difficult to distinguish clinically. A patient with either of these illnesses, particularly a teenager or young adult, is likely to come to a clinician complaining of sore throat. On physical examination both are likely to have bright red throats with white patches of pus on their tonsils and large, tender,

swollen lymph nodes in the neck. But these two ailments differ in their prognosis and treatment—only by making the correct diagnosis can the clinician know what to do.

Streptococcal pharyngitis is an infection for which the prognosis is guarded but treatment is excellent. It is fairly easy to diagnose strep throat after a throat swab, either by means of a latex agglutination test for streptococcal antigens (results available immediately) or by culture (results available within 48 hours; Chapter 19). Knowing the diagnosis, the clinician can tell the patient that without treatment, he or she is likely to be very uncomfortable with fever and sore throat for only a few days, but that in some cases significant problems occur. Rare patients may develop a widespread streptococcal infection, while others may recover only to become more seriously ill with rheumatic fever or acute glomerulonephritis. Treatment with an antibiotic—either penicillin or erythromycin (for penicillin-allergic patients)—eliminates most symptoms within a day and prevents serious complications.

For infectious mononucleosis ("mono") the prognosis is good, but treatment is extremely limited. Like strep throat, "mono" is easy to diagnose in the laboratory by means of a serologic blood test. Knowing the diagnosis, the clinician can advise the patient that he or she will get better, but that it may take time. Sore throat and fatigue may come and go for weeks or even months. Knowing the course of the disease reassures patients who might otherwise be worried as the illness drags on. Mononucleosis is a viral infection, and so no antimicrobial drug can cure it. In the most extreme cases, when sore throat impairs swallowing or breathing, steroids may be used to relieve symptoms temporarily.

Clinical Description

In addition to anatomic and etiologic diagnoses, clinicians describe infections in many ways. Some terms, such as **acute** and **chronic**, tell about the suddenness with which symptoms appear and how long they may last. Others, such as **asymptomatic** and **persistent**, describe the association between infection and symptoms. Still other terms, such as **local** and **systemic**, tell whether infection occurs in one part of the body or everywhere. Finally, terms like **primary**, **secondary**, and **mixed** describe relationships between one infection and another (**Table 2**).

Other terms are used to describe the progress of an infection as symptoms come and go. The **incubation period** occurs after infection but before the first symptoms. The **prodromal period** is the time of early, mild symptoms. During the **active period** of illness, symptoms reach their peak. During the **convalescent period**, symptoms diminish and resolve (**Figure 2**). Some infections, such as pertussis (see beginning of Chapter 15), demonstrate all these stages clearly, while others may demonstrate only some.

Table 2 Some Types of Infection		
Type of Infection	Properties	Examples
Acute	Infection with a short, well-defined course	Pertussis, measles, common cold
Chronic	Infection that progresses slowly over a long or indeterminate time	Tuberculosis, schistosomiasis
Persistent	Infection in which the infectious agent persists in the body for an extended period	Herpes simplex virus, varicella zoster virus (the cause of chickenpox and shingles)
Asymptomatic	Infection with no apparent symptoms	Many cases of mumps and hepatitis A (particularly in children)
Local	Infection confined to a limited area of the body	Bladder infection, abscess
Systemic	Infection that affects the entire body	Measles, chickenpox
Primary, secondary	Series of infections in which one pathogen initiates the first (primary) infection, weakening the body's defenses and thereby making possible another (secondary) infection caused by another pathogen	Influenza (primary) and bacterial pneumonia (secondary); common cold (primary) and middle-ear infection (secondary)
Mixed	Simultaneous infection by several pathogenic species	Wound infections, urinary-tract infection in individual with abnormal bladder function

ORGANIZATION OF PART IV

Perhaps the best way to learn about the diseases caused by microbial infection is to adopt the perspective of a clinician. Accordingly, this section of the text is organized the way a clinician thinks about disease—by anatomic syndrome. Chapters are devoted to each of the major human organ systems: respiratory, digestive, genitourinary, nervous, skin (and conjunctivae), and cardiovascular and lymphatic. Each chapter begins with a brief review of the organ system's anatomy, physiology, and natural defenses. Then we describe the common clinical syndromes of infection. Finally, we discuss the microorganisms and diseases that are most likely to cause these syndromes. Throughout, we use case studies to illustrate discussions.

Organizing our discussion of infectious diseases by organ systems is a useful but not a perfect plan because some infections affect more than one system. Infections that are easiest to classify have their portal of entry in a given system, affect that system almost exclusively, and have their portal of exit through the same system. For example, cholera (Chapter 23) enters the body through the digestive system, affects the function of the intestines, and exits through the digestive system. For other infections, however, the pattern is less clear. Typhoid fever (Chapter 23), for example, like cholera, has the digestive system as its portal of entry, but typhoid bacteria leave that system and enter the bloodstream, affecting many other parts of the body, before reentering the digestive system to exit. Some infections have marked effects in

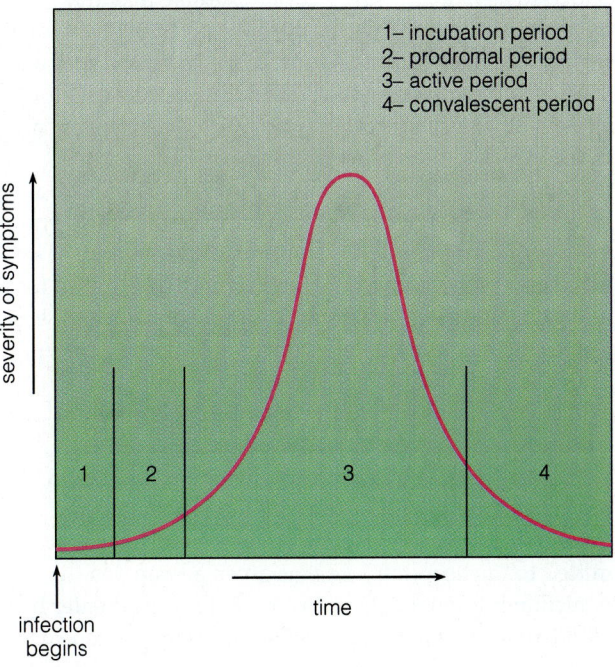

Figure 2 The progress of an infection.

more than one system; leprosy, for example (Chapter 26), affects both the skin and the nervous system. Some infections affect the body so diffusely that an anatomic syndrome is difficult to pin down and diagnosis is difficult. Many of these are discussed in Chapter 27 on the cardiovascular and lymphatic systems.

22 INFECTIONS OF THE RESPIRATORY SYSTEM

Fighting for Every Breath

J. G., a 10-year-old boy, had had a mild cold for several days when his temperature suddenly rose to 103°F and he developed a shaking chill. The next day he said the right side of his chest hurt, and coughing or taking a deep breath caused sharp pain. His breathing had become rapid and shallow and he looked anxious and much sicker than on the day before.

J. G.'s mother realized that he needed immediate attention and brought him to a hospital emergency room. J. G. was so weak and uncomfortable that he had to be taken to the examining room in a wheelchair. His temperature was 104°F, and although his rapid respirations were shallow, he was working hard to breathe.

The examining physician tapped on his chest. Instead of the normal hollow thump expected over air-filled lungs, she heard a dull note over his lower right chest. She used a stethoscope to further examine that area and—instead of hearing the normal soft flow of air entering and leaving the lungs—she heard no breath sounds at all in some places and wet crackling sounds called rales in other places. An x ray of J. G.'s chest showed a black left lung, the normal appearance when x rays travel mainly through air. But the lower lobe of his right lung was white, the characteristic appearance of x rays passing through fluid. J. G. continued to complain of severe chest pain and was unable to cough up sputum for microbiological examination. A sample of his blood was sent to the hematology laboratory, and another, drawn under sterile conditions to prevent contamination by skin bacteria, was sent to the microbiology laboratory for culture.

Based on J. G.'s history and physical examination, the physician attending him felt confident in diagnosing pneumonia. Because his illness was severe, he was admitted to the hospital. J. G. was given a large dose of penicillin intravenously to assure that it would enter his bloodstream immediately, and medication was repeated every six hours. By the next day, he was much better and

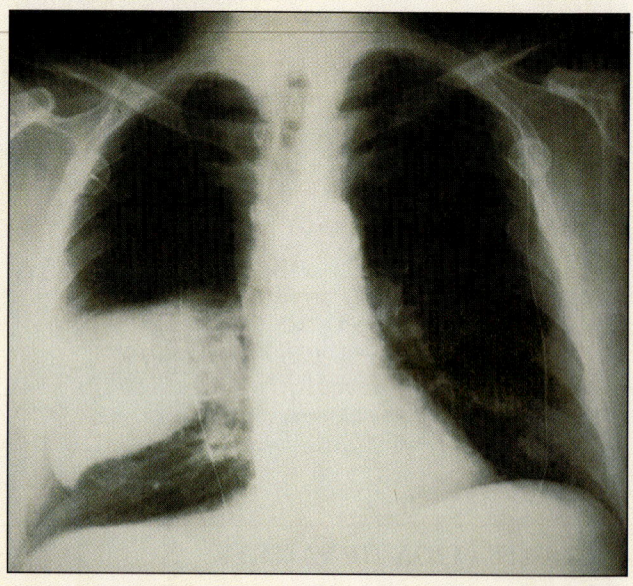

This chest x ray from a patient with pneumonia shows an abnormal, white right lower lobe where the lung is infected and fluid-filled. Both lungs would be clear in a healthy patient.

his temperature had returned to normal. J. G.'s hematology test results showed an unusually high number of neutrophils, many of which looked immature. Later the microbiology laboratory reported that his blood culture grew a Gram-positive coccus, *Streptococcus pneumoniae*.

Within two days, J. G. was breathing comfortably and had no more episodes of fever, so he was released to continue recuperating at home. He was given a prescription for oral penicillin, and a respiratory therapist instructed his mother in percussion and postural drainage, physical techniques to help J. G. loosen and cough up the fluid from his infected lung. J. G. returned to school the following week and had no further problems from his brief but serious illness.

To understand:

▪ The anatomy and function of the respiratory system and its defenses against microorganisms

▪ The clinical syndromes that characterize respiratory infections

▪ The bacterial and viral causes of upper respiratory infections; their diagnosis, prevention, and treatment

▪ The bacterial, viral, and fungal causes of lower respiratory infections; their diagnosis, prevention, and treatment

THE RESPIRATORY SYSTEM

The respiratory system extends from the exposed mucous membranes of the nose to the millions of alveolar air sacs that lie deep within the chest. Continuous operation of this organ system—which exchanges oxygen from the air for carbon dioxide produced in the tissues—is essential to our survival. If respiratory gas exchange is interrupted for more than a few minutes, permanent brain damage or death is inevitable. The entire respiratory tract is constantly exposed to air and the microorganisms it contains, which means this organ system is infected more frequently than any other.

Structure and Function

The respiratory system is divided into two regions, upper and lower (**Figure 22.1**). The upper respiratory system consists of the nasal cavity and nasopharynx. The **nasal cavity** is the external chamber that warms and filters incoming air. The **nasopharynx** is the space behind the nasal cavity. It includes the uppermost part of the throat, or **pharynx**. The nasopharynx is directly connected to

Figure 22.1 Anatomy of the upper and lower respiratory systems.

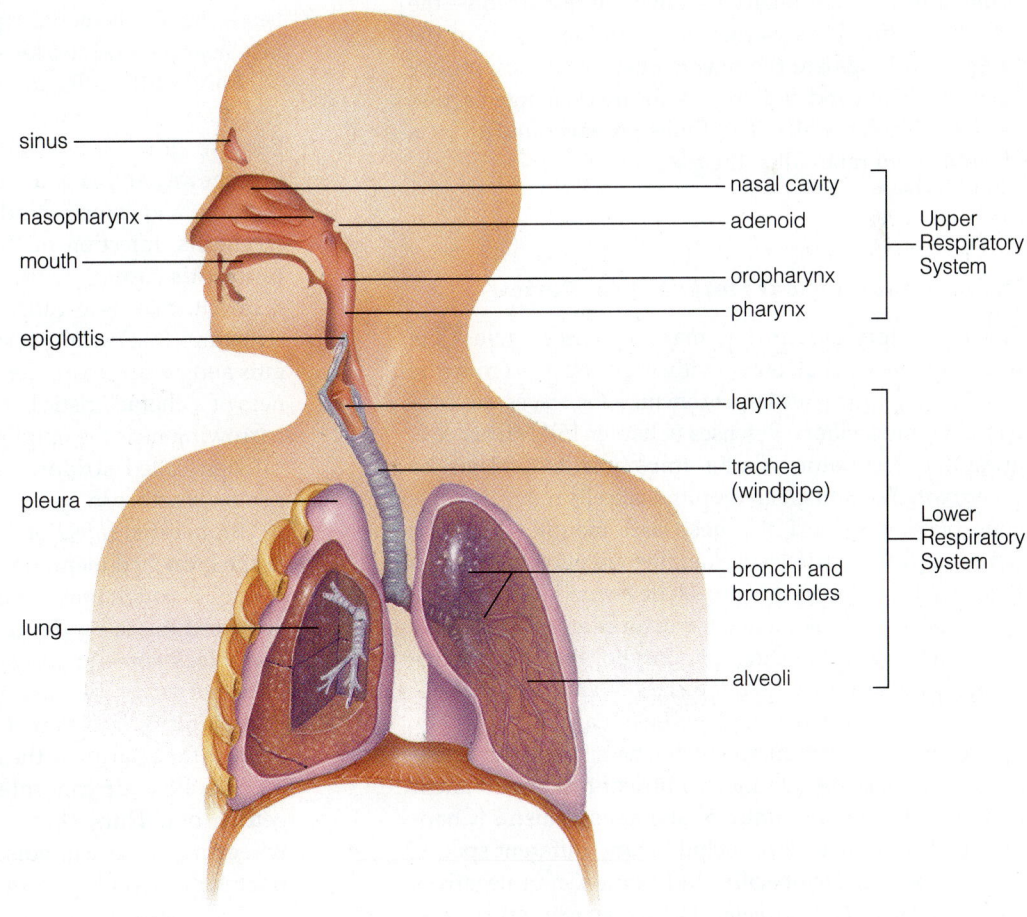

other air-containing spaces in the head, including the **sinuses** (cavities within the bones of the skull) and the middle ear (a small air-filled compartment between the eardrum and the inner ear). Although these structures are outside the main airflow, they open onto the respiratory system and therefore may become infected. Also prone to infection are the **adenoids**, collections of lymphoid tissue in the nasopharynx. The mouth and the **oropharynx** (the space below the nasopharynx) are shared by the respiratory and digestive systems. The tonsils are frequently infected lymphoid tissues of the oropharynx. Both air and food pass through the oropharynx until they reach the **epiglottis**, a flaplike structure that blocks the entrance to the lower respiratory system during swallowing, diverting food into the digestive tract. At other times the epiglottis is open, allowing air to enter the trachea.

The lower respiratory system begins at the **larynx** (voice box) and includes the trachea, bronchi, bronchioles, and alveoli. The **trachea**, or windpipe, is a single large airway that branches into two smaller airways, the right and left **primary bronchi**. The primary bronchi, in turn, branch into smaller and smaller **secondary** and **tertiary bronchi**. The tertiary bronchi branch further into **bronchioles**, which finally terminate in the alveoli—the millions of tiny air sacs that make up the tissues of the **lungs**. The lungs are the major organ of the respiratory system; oxygen and carbon dioxide are exchanged across its thin alveolar walls. The lungs are surrounded by a smooth membrane called the **pleura**.

Defenses and Normal Flora: A Brief Review

The respiratory system has many defenses against the microorganisms that enter with each breath. From the nasal cavity to the tertiary bronchi, its surfaces are protected by mucociliary defenses (Chapter 16). Mucus covering the respiratory lining traps microorganisms or other particles, and ciliated epithelial cells move the mucous blanket toward the nose and mouth where the mucus and the particles it contains are eliminated from the body. Bronchioles and alveoli lie beyond this mucociliary escalator. These smaller structures are protected by other defenses, including phagocytic **alveolar macrophages** and secretory IgA, the class of antibody that defends many of the body's mucous membranes (Chapter 17).

In spite of its antimicrobial defenses, the upper respiratory system is a warm, moist, nutrient-rich environment densely colonized by commensal microorganisms (Chapter 14). The normal flora include many different species of streptococci, lactobacilli, and some Gram-negatives such as *Moraxella catarrhalis*. The lower respiratory system (as well as the sinuses and middle ear) is normally sterile.

Clinical Syndromes

Infections of the anatomic structures in and around the respiratory tract manifest themselves in a remarkable number of ways (**Table 22.1**, p. 528).

Rhinitis (nasal inflammation) causes swelling of the membranes that line the nose and stimulates excess mucus production. Rhinitis is the most common of all respiratory syndromes. We experience it as a cold, usually at least once a year. **Adenoiditis** is infection of the adenoids, which occurs frequently in certain children. **Pharyngitis** is infection of the throat, called **tonsillitis** if the tonsils are primarily infected. Pharyngitis symptoms include sore throat and sometimes fever. An infected throat usually appears red and may be covered by a milky white exudate (film), ulcers, blisters, or even a grayish membrane. The appearance of the throat provides diagnostic clues about which microorganism is causing the pharyngitis.

Sinusitis and **otitis media** occur when the sinus or middle ear, respectively, fills with fluid and becomes infected; these infections frequently result when rhinitis closes off these areas from the nasopharynx. **Purulent** (pus-producing) sinusitis typically causes fever and headache. Otitis media typically causes ear pain, temporary hearing loss, and fever.

Epiglottitis, infection of the epiglottis, can cause this structure to swell suddenly to many times its normal size. Because all respiratory airflow must go through a single passageway at this point, acute epiglottitis can cut off respiration and cause sudden death. On rare occasions, **laryngitis**, infection of the larynx, and **laryngotracheobronchitis** (croup), which involves the larynx and tissues below it, may also cause such severe swelling in these structures that breathing stops. Typically, however, laryngitis and croup are milder infections that produce hoarseness or a characteristic barking cough. Any severe airway narrowing near the epiglottis or the larynx causes a clinical sign called **stridor**, a whistling sound heard as the person breathes in.

Bronchitis, infection of the bronchi, causes swelling of the bronchial membranes and produces thick, infected mucus. A cough that brings up infected phlegm is typical of bronchitis, as is fever. Infected bronchial secretions may block some air passages, but complete obstruction does not occur because the bronchi are so numerous. **Bronchiolitis**, infection of the bronchioles, causes inflammation that narrows these tiny collapsible airways. In bronchiolitis, air can enter the lungs but has difficulty getting out. Thus, clinical signs of bronchiolitis include **wheezing**, a musical noise heard during expiration, and **trachyapnea**, rapid breathing.

The complex syndrome **pneumonia** signifies infection of the lungs, with fluid and microorganisms replacing the air that normally fills the alveoli. Normal gas exchange cannot take place. Typically, the clinical presentation of a patient with pneumonia includes fever, trachyapnea, labored breathing, and a cough that may produce infected secretions. If pneumonia involves the pleura, it causes **pleurisy**, associated with painful breathing. Many different illnesses are classified as pneumonia, and the etiologic agent can be bacterial, viral, or fungal.

UPPER RESPIRATORY INFECTIONS

Bacteria and viruses are the most common agents of upper respiratory infection. Some of these infections are minor, but others can cause sudden death because of the anatomy of the respiratory tract. Because all air must pass through a single, fairly narrow airway above the spot where the trachea bifurcates (splits into the bronchi), infections that cause swelling, such as epiglottitis and diphtheria, can close off the airway and cause sudden death by suffocation.

BACTERIAL CAUSES

Bacteria are the most virulent of the upper respiratory pathogens, but most of the infections they cause can be either prevented or effectively treated.

Figure 22.2 A child with epiglottitis is in a typical sniffing position to try to keep his nearly blocked airway open.

Case History

Epiglottitis

M. L., a 4-year-old boy, arrived at the pediatric emergency room at 5 A.M. He'd been entirely well when his mother put him to bed the night before, but around 4 A.M. she was awakened by noises coming from his room. She found him sitting up in his bed, gasping loudly with each breath and warm with fever. M. L.'s mother knew immediately this was an emergency.

In the short time it took to arrive at the hospital, M. L.'s breathing grew even noisier. He was immediately taken to an examining room where the on-duty physician found an anxious youngster in acute respiratory distress with marked stridor (**Figure 22.2**). M. L. was sitting very still, holding his head forward in a sniffing position and drooling slightly. The physician quickly examined the child, taking care not to agitate him, and immediately called for an anesthesiologist and a portable x-ray machine. Blood was drawn and an intravenous line was started.

The physician's working diagnosis—based on the child's age, the sudden onset of fever, and the severe stridor—was acute epiglottitis, a life-threatening infection caused almost invariably by *Haemophilus influenzae* type b. A lateral view x ray of M. L.'s neck confirmed the working diagnosis, showing a massively enlarged epiglottis severely narrowing his airway. M. L. was holding his head forward in a sniffing position to try to keep open his nearly occluded (blocked) airway. Any agitation or sudden movement could close off the airway entirely, causing sudden death. Antibiotic treatment could not possibly act soon enough to prevent possible suffocation. An anesthesiologist was needed to pass an endotracheal tube past the swollen epiglottis into the trachea to secure an airway immediately. As the child was prepared for intubation, the ER physician administered a large dose of ceftriaxone, an antibiotic effective against *H. influenzae*.

The blood culture drawn at the time of M. L.'s admission to the hospital later grew *H. influenzae* type b. M. L. was lucky enough to reach the hospital before he became completely unable to breathe. After a brief hospitalization in the intensive care unit and additional antibiotic treatment, his epiglottis returned to normal size, his fever resolved, and he returned home in good health.

Table 22.1 Clinical Syndromes of Respiratory Infections

Syndrome	Region Affected	Signs and Symptoms	Causative Agents
Rhinitis	Nasal cavity	Nasal discharge, sneezing	Rhinovirus, coronavirus, and others
Adenoiditis	Adenoids	Nasal discharge, obstruction of nasal passages	Bacterial pathogens including *Haemophilus influenzae* and *Streptococcus pneumoniae*
Pharyngitis	Pharynx	Sore throat, fever	*S. pyogenes* and many viruses
Tonsillitis	Tonsils	Sore throat, fever	*S. pyogenes* and many viruses
Sinusitis	Sinuses	Fever, headache	Bacterial pathogens including *H. influenzae*, *S. pneumoniae*, *Moraxella catarrhalis*
Otitis media	Middle ear	Ear pain, fever, temporary hearing loss	Bacterial pathogens including *H. influenzae*, *S. pneumoniae*, *M. catarrhalis* and many viruses
Epiglottitis	Epiglottis	Fever, sudden obstruction of upper airway	*H. influenzae*
Laryngitis	Larynx	Hoarse voice	Many viruses
Laryngotracheo-bronchitis (croup)	Larynx, trachea, and primary bronchi	Hoarse, barking cough	Parainfluenza virus
Bronchitis	Bronchi	Cough that produces infected mucus, fever	Many bacterial pathogens
Bronchiolitis	Bronchioles	Wheezing, rapid breathing	Respiratory syncytial virus (RSV)
Pneumonia	Lungs	Cough, fever, difficult breathing	Many bacterial, viral, and fungal pathogens
Pleurisy	Lungs and pleura	Cough, fever, labored and painful breathing	Many bacterial, viral, and fungal pathogens

Haemophilus influenzae: Epiglottitis, Bacterial Meningitis

Haemophilus influenzae is a Gram-negative, largely rod-shaped, but somewhat pleomorphic (variable-shaped) bacterium. It is nonmotile and does not form endospores. *Haemophilus* cells cannot produce two cofactors they need for growth: **hemin**, an iron-containing group found in the cytochromes, and NAD (or NADP; Chapter 5). These growth factors are present in human red blood cells, leading to the genus name *Haemophilus* ("blood-loving"). The species name *influenzae* was assigned because this bacterium was thought to cause influenza, but it does not. Influenza is caused by a virus. Humans are the sole reservoir for *H. influenzae*. The organism is difficult to culture in the laboratory; it grows best on a chocolate agar medium containing red blood cells broken down by heat to make the growth factors readily available.

Some strains of *Haemophilus influenzae* produce a polysaccharide capsule, the major cause of virulence with this microorganism. Encapsulated strains are relatively resistant to phagocytosis and lysis by complement and they can invade human tissues. The encapsulated strains are designated types a through f, according to their capsular antigen, but type b causes over 95 percent of the serious infections, including epiglottitis and meningitis. Strains of *H. influenzae* that do not produce a capsule are less virulent, are noninvasive, and infect only mucous membrane surfaces.

Epiglottitis is probably the most dramatic clinical syndrome caused by *H. influenzae* type b infection, but hardly the most common. The organism also causes meningitis, infection of the membranes that surround the brain and spinal cord (Chapter 25), and cellulitis, a rapidly progressive and life-threatening infection of the skin and subcutaneous tissues (Chapter 26). Both syndromes are life-threatening, and some children who recover from *H. influenzae* meningitis are permanently disabled. The nonencapsulated strains of *H. influenzae* cause mild upper respiratory infections, particularly otitis media and sinusitis.

Antibody against the polysaccharide capsule can neutralize the capsule's protective effect and prevent infection. Most adults have antibodies against *H. influenzae* type b, but children between 6 months and 5 years who have lost their high levels of transplacental maternal antibodies and not yet fully developed their own antibodies are extremely vulnerable (older children and adults are occasionally affected). Thus, until very recently—the last five years—*H. influenzae* was the scourge of pediatric wards, causing death and disability comparable to that caused by polio in the years before the vaccine. The year-round nature of *Haemophilus* infections, however, compared to the seasonal, epidemic nature of polio meant the toll went relatively unnoticed. Older drugs, such as ampicillin, are almost useless because of plasmid-borne antibiotic drug resistance, but most strains of *H. influenzae* remain sensitive to chloramphenicol and the new cephalosporins, such as the ceftriaxone used to treat M. L. Today, however, a vaccine against *H. influenzae* infections means that cases like M. L.'s are historical rather than real.

The vaccine—called Hib for *Haemophilus influenzae* type b, the only strain against which it protects—is now available. Hib was difficult to develop for two reasons. First, the *H. influenzae* capsular antigen, like most polysaccharides, is only weakly immunogenic (Chapter 17). Second, young children, the population most at risk for *Haemophilus* infections, have a comparatively weak antibody response. The breakthrough came in 1990 with the development of a conjugate vaccine that combined the weakly immunogenic polysaccharide antigen with a highly immunogenic protein, thereby eliciting a protective immune response.

Today babies are routinely immunized against *H. influenzae* type b with a series of injections beginning at age 2 months. Protection is good and side effects are minimal. Pediatric wards are already empty of patients with epiglottitis and *Haemophilus* meningitis. In contrast to our setbacks with tuberculosis, AIDS, and antibiotic drug resistance, the Hib vaccine is a major step forward in the control of infectious disease.

Streptococcus pyogenes: Streptococcal Pharyngitis

The genus *Streptococcus* includes many species of Gram-positive cocci, several of which are significant human pathogens. *Streptococcus pyogenes* (*pyogenes* means "pus forming"), also known as **group A beta-hemolytic streptococcus**, is highly virulent and causes many different clinical syndromes of infection (Chapters 24 and 26).

Streptococci are facultative anaerobes with a fermentative metabolism. They do not produce the enzyme **catalase**, which distinguishes them from staphylococci, another group of Gram-positive cocci that are also virulent pathogens (Chapter 26). Because they divide in a single plane and do not separate after cell division, *S. pyogenes* typically form long chains of cells (**Figure 22.3**). Medically important streptococci are classified by their hemolytic and serologic properties. They are divided into groups designated A through S on the basis of the polysaccharide antigens present in their cell walls.

When unknown streptococci are cultured for diagnosis, they are grown on a blood agar medium. The effect microbial growth has on red cells in the culture medium helps identify the pathogen (see Figure 3.21). If colonies are surrounded by a clear halo (caused by lysis of nearby red blood cells), the pattern is called **beta hemolysis**. If

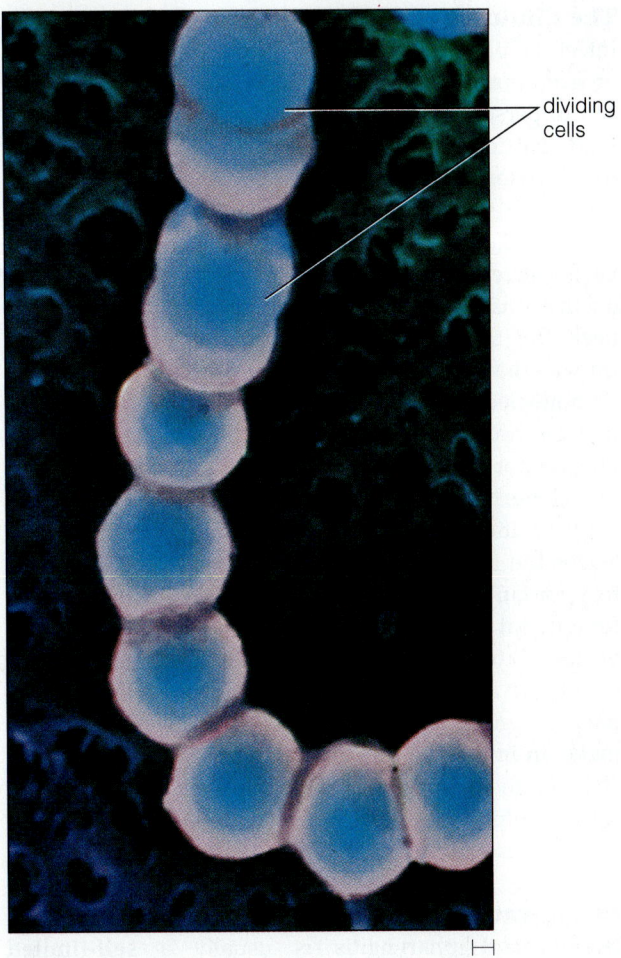

dividing cells

0.1 μm

Figure 22.3 *Streptococcus pyogenes.* Scanning electron micrograph of dividing *S. pyogenes* cells.

colonies are surrounded by a greenish halo (caused by chemical transformation of nearby red blood cells), the pattern is called **alpha hemolysis**. If microbial growth has no visible effect on nearby red blood cells, the pattern is called **gamma hemolysis**. Bacteria that demonstrate beta hemolysis and are also sensitive to the antibiotic bacitracin are likely to be *Streptococcus pyogenes*. Identification is confirmed serologically by demonstrating the presence of group A polysaccharide antigens in the cell wall.

Humans are the only natural reservoir for group A streptococci. Healthy people can be asymptomatic carriers of the pathogen, but this is uncommon. Most people with group A streptococci suffer clinically significant infections. *S. pyogenes* is a common cause of pharyngitis and can cause life-threatening infections of the female genital tract after childbirth or abortion (Chapter 24). Group A streptococcal infections can also involve the skin and underlying soft tissues (Chapter 26).

The Clinical Syndrome. When group A streptococci infect the pharynx, they cause **streptococcal pharyngitis**, or **strep throat**. Infection is usually transmitted from person to person by respiratory droplets, but contaminated food, particularly unpasteurized milk, can also spread the disease. Strep throat is common in school-aged children. The typical history is severe sore throat, fever, chills, and headache. Physical examination usually reveals a severely inflamed pharynx with a whitish exudate on the tonsils and swollen, tender lymph nodes in the neck. But this clinical syndrome is not unique to strep throat—the Epstein-Barr virus, which causes infectious mononucleosis, is one of many other microorganisms that can cause an identical illness. Streptococcal pharyngitis cannot be definitively diagnosed on clinical examination alone; laboratory tests are needed for confirmation.

If a patient appears to have strep throat, the clinician swabs the throat and sends the specimen to the laboratory for culturing. However, it may take up to two days for a throat culture to reveal bacterial growth. Newly developed **latex agglutination kits** (Chapter 19) that detect streptococcal antigens in a throat swab provide a preliminary diagnosis within minutes, helping the clinician make an immediate decision about antibiotic treatment. A laboratory culture is usually done, nevertheless, because agglutination tests occasionally fail to detect antigens in some patients.

Complications and Changing Virulence. Today, streptococcal pharyngitis is usually a **self-limited illness**—most people recover within a few days without medical treatment. But serious complications can occur. Certain strains of *S. pyogenes* produce an **erythrogenic** (*erythro*, "red"; *genic*, "producing") toxin that causes scarlet fever, a clinically identifiable syndrome of rash and fever (Chapter 21). Erythrogenic toxin kills cells and causes intense inflammation. It generally increases the strain's virulence. At the turn of the century, scarlet fever was a life-threatening illness, but today's cases are relatively mild. The change seems to be part of an unexplained decrease in the virulence of *S. pyogenes* during the last few decades (Chapter 26).

Other complications occur when bacteria spread from the throat into the sinuses, the middle ear, the lungs, or even into the bloodstream, causing a life-threatening septicemia. Like scarlet fever, fatal systemic infections are far less common than they were 50 years ago, but some recent deaths, including that of Jim Henson, creator of the Muppets, may be a sign that killer strains of streptococcus are returning. (Henson's illness is called **TSLS** for **toxic shock–like syndrome**.)

Researchers theorize that one aspect of this changing streptococcal virulence has to do with which of three types of erythrogenic toxins, designated A, B, and C, a particular strain produces. Some strains produce none of these (and therefore do not produce scarlet fever or other toxin-associated complications), some produce one, and some produce another. Type A seems to be the most virulent toxin; types B and C are less likely to cause lethal complications. Apparently, strains that produce type A toxin were common early in the century and became relatively uncommon by mid-century. It's possible that these strains are on the rise again. The highly lethal toxin in the strain that killed Jim Henson was genetically mapped and determined to be identical to toxins in strains found in the United States and Europe during the mid-1980s. Because the genes for the lethal toxin can be transferred from one bacterial strain to another by bacteriophages, these highly lethal strains of streptococcus could spread rapidly.

A third category of life-threatening complications consists of those that occur *after* the streptococcal infection is over. These postinfectious complications, which are immunologically mediated (Chapter 18), include rheumatic fever and acute poststreptococcal glomerulonephritis failure. Both are childhood diseases.

Rheumatic fever causes inflammation in many organs of the body, including the joints, skin, and brain, but permanent and life-threatening damage to heart valves is the most serious complication. Rheumatic fever is the leading cause of heart disease among children in developing countries, affecting over 6 million children in India alone. If strep throat is treated with penicillin within the first 10 days of illness, rheumatic fever can be prevented. Someone who has already suffered an episode of rheumatic fever, however, is at special risk because another streptococcal infection can reactivate the disease and cause fatal heart damage. As prophylaxis against streptococcal infection, rheumatic fever patients should receive a monthly injection of benzathine penicillin.

Acute poststreptococcal glomerulonephritis causes sudden kidney failure. The urine becomes scant and dark colored, body parts swell from retained fluid, and blood pressure rises alarmingly. The condition is life-threatening and often requires hospitalization and intensive medical care. Fortunately, poststreptococcal glomerulonephritis affects the kidneys only temporarily, so children who survive the initial shutdown do not suffer lasting kidney damage.

Because strep throat can lead to serious complications and because the risk of killer strains is a frightening possibility today, any child with fever and sore throat should have a throat swab done. If results are positive for strep, antibiotic treatment can prevent all complications. Penicillin is the drug of choice, with erythromycin used for penicillin-allergic patients.

Corynebacterium diphtheriae: Diphtheria

Corynebacterium diphtheriae is a Gram-positive, rod-shaped bacterium, though cell shape may be slightly irregular. Groups of cells sometimes look like small stacks of coins or Chinese characters (**Figure 22.4**). *C. diphtheriae* grows best aerobically. It does not form endospores but is resistant to drying, light, and cold. It is cultivated in the laboratory on a special medium called **Loeffler's medium**, named after Frederick Loeffler, who proved in 1884 that this bacterium causes the disease **diphtheria**. This complex medium of coagulated serum is particularly useful for cultivating *C. diphtheriae* because it does not support the growth of streptococci, which, on most media, overwhelm the slower-growing diphtheria organisms. A medium containing cysteine and potassium tellurite will also grow *C. diphtheriae* from a throat swab, in which case the organisms appear as dark gray-black colonies.

Virulent strains of *C. diphtheriae* produce **diphtheria toxin**, a two-subunit protein exotoxin that causes almost all the clinical manifestations of diphtheria. The gene that encodes diphtheria toxin, called *tox*, is carried by a temperate bacteriophage, so only strains of *C. diphtheriae* infected by this phage can produce toxin and cause diphtheria. Strains of *C. diphtheriae* can be tested for toxin production in several ways. Bacteria can be injected into the skin of a live guinea pig to assess the toxic effect. The toxin-producing organisms can be grown as an overlay of tissue culture (toxicity kills the vulnerable animal cells). Or the toxin itself can be identified by an immunodiffusion assay (Chapter 19).

We understand the molecular action of diphtheria toxin quite well. One of the toxin's two protein subunits (fragment B) binds to receptors on human cells, allowing the toxin to enter. Inside the cell, the other subunit (fragment A) interferes with protein synthesis by inactivating a ribosomal protein factor necessary for polypeptide elongation. This factor is found only in eucaryotic cells, so protein synthesis within *C. diphtheriae* itself is unaffected. Thus, diphtheria toxin is an example of selective toxicity in reverse.

Diphtheria toxin is an antigen against which humans and animals produce neutralizing antibodies called **diphtheria antitoxin**. Binding between toxin and antitoxin prevents toxin molecules from entering susceptible cells. The production of antitoxin that occurs during an infection allows some people to recover from diphtheria without treatment, but the risk involved in letting the disease go untreated cannot be overestimated. The recommended therapy for human diphtheria is antitoxin produced artificially in horses. Serum sickness is a potential but acceptable risk because the disease is life-threatening (Chapter 18). If antitoxin is administered before a signifi-

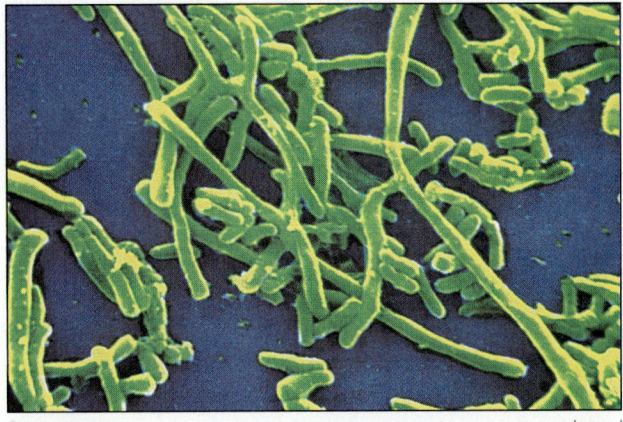

a

1 μm

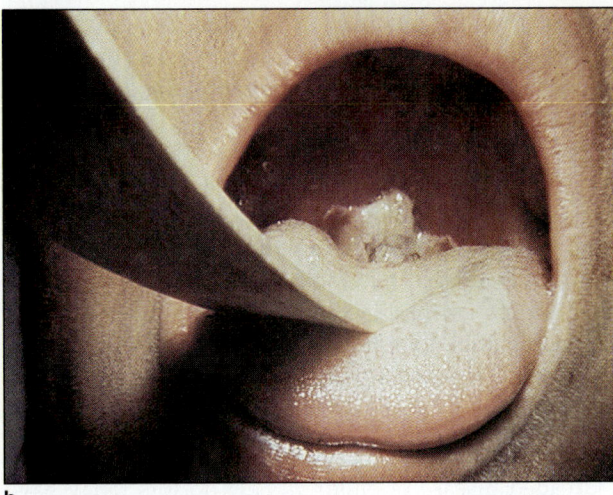

b

Figure 22.4 *Corynebacterium diphtheria.* (**a**) *C. diphtheria* cells arrange themselves like stacked coins or Chinese characters. (**b**) Infection with *C. diphtheria* produces a tough gray covering of dead cells and microorganisms called a pseudomembrane. Here a patient has such a pseudomembrane on the tonsils.

cant amount of toxin has entered host cells, recovery is almost certain. But as the number of cells poisoned by diphtheria toxin grows, treatment becomes less and less effective. With every day that antitoxin therapy is delayed, risk of death increases. Erythromycin is also administered to stop the production of diphtheria toxin and prevent transmission of the disease to others, but it does not neutralize the toxin already formed.

The Clinical Syndrome. Human beings are the only natural hosts for *Corynebacterium diphtheriae*. The organism is present in the throats of healthy carriers as well as people ill with diphtheria. It is transmitted by respiratory droplets. *C. diphtheriae* is noninvasive, infecting only the mucous membranes of the pharynx, though deeper tissues are affected because they absorb the diphtheria toxin. A toxin-damaged throat swells and becomes covered by a tough, grayish pseudomembrane composed of dead human cells and microorganisms, a clinical finding called **membranous pharyngitis** (Figure 22.4). Lymph nodes in the neck also swell markedly. A clinical finding of membranous pharyngitis strongly suggests diphtheria, but a conclusive diagnosis requires that a throat swab grow a toxin-producing strain of *C. diphtheriae* in a laboratory culture.

The swollen throat tissue and pseudomembrane of diphtherial pharyngitis can obstruct the airway and lead to death by suffocation, which was the leading cause of infant death from diphtheria. Fatal complications can also occur if toxin enters the bloodstream and damages distant organs such as the heart, kidneys, brain, or nerves. These organs may not be affected until several weeks after the original throat infection, when the disease is in an advanced stage and little can be done to help the patient.

Diphtheria can also infect the skin. **Cutaneous diphtheria** usually occurs in the tropics but is sometimes seen in colder climates among indigent people whose general health is poor. Cutaneous diphtheria usually begins in a wound already infected by other bacteria. The wound develops a gray membrane and fails to heal unless treated with antibiotics. Diphtherial skin infections do not usually cause toxin-mediated damage of distant organs.

Epidemiology and Prevention. Diphtheria was once a dreaded disease of children. Few doctors today have ever seen a case. Treatment with horse antitoxin became possible in the 1890s, but the disease was not conquered until diphtheria toxoid made widescale immunization possible in 1923. Before 1920, over 100,000 American children contracted diphtheria every year, and more than 10,000 of them died. Between 1980 and 1988, 18 cases of diphtheria and 2 deaths occurred in the United States. In 1993, there were none. In the parts of the world where immunization is not available, however, diphtheria remains a major public health problem. The World Health Organization estimates 50,000 reported cases annually, with many cases going unrecorded.

Diphtheria toxoid is produced by treating diphtheria toxin with formaldehyde, which destroys the protein's toxic properties but leaves its potent antigenic capabilities intact. Diphtheria toxoid stimulates the production of diphtheria antitoxin in the immunized host. In the United States most infants are immunized against diphtheria during their first 6 months as part of the DPT (diphtheria-pertussin-tetanus) vaccine. Toxin-producing strains of *C. diphtheriae* are now so rare among immunized children that the entire population benefits from herd immunity. Herd immunity against diphtheria probably sets in when more than 75 percent of children are immunized.

However, we should not take herd immunity for granted. During a recent diphtheria outbreak in Sweden, Shick testing for susceptibility to diphtheria revealed that immunity levels varied greatly. (In the **Shick test**, a tiny amount of diphtheria toxin is injected into the skin; if an inflammatory reaction occurs, the person does not have a sufficient level of protective antibody.) Children made up the most highly immunized group (a 95-percent immunity rate), while elderly women formed the most poorly immunized group (less than a 20-percent immunity rate). In the early 1990s, there was an outbreak of fatal diphtheria in Manchester, England. Many people without up-to-date immunization were infected with *C. diphtheriae*, but it was a non-toxin-producing strain so they didn't become ill. But when a child from Africa who had diphtheria arrived, the *tox* gene and the phage that carried it entered the community. The harmless English strain began to produce the dangerous toxin and an outbreak was underway. Because immunization with diphtheria toxoid does not confer lifelong immunity, adults should have a booster dose at least every 10 years, especially if they are traveling to developing countries where diphtheria still commonly occurs.

VIRAL CAUSES

It is safe to say that every reader of this textbook has had a viral upper respiratory infection at some time—and probably many more than one. Such infections are extraordinarily common and much less serious than upper respiratory bacterial infections. Nevertheless, prevention and treatment are major challenges.

Rhinoviruses

Rhinoviruses are single-stranded RNA viruses belonging to the picornavirus group. They are extremely small—only 25 to 30 nm in diameter—with icosahedral symmetry (see Figure 1.7). There are probably about 100 different serotypes of rhinoviruses, each with a different antigen in its protein capsid. New types continue to be identified.

Rhinoviruses are named for their portal of entry, the nose (*rhino*, "nose"). They were identified in 1956, after years of searching for the agent of the common cold. However, although rhinoviruses cause from one-fourth to one-half of all colds, more than 200 other types of microorganisms also cause the common cold.

The clinical syndrome of a **cold**, or **upper respiratory infection** (**URI**), includes sneezing, **rhinorrhea** (the production of excess nasal mucus), and nasal congestion, and sometimes sore throat, fever, and headache. The *malaise*, or feeling of general discomfort, that accompanies most colds is partly due to interferons produced to combat the infection (Chapter 16). Typically, a cold lasts about one week. Although colds are usually considered trivial illnesses, they cause the loss of more school and work days than does any other illness. Economically, then, rhinoviruses are extremely important pathogens.

Rhinoviruses occur in all populations and usually from October through April in north temperate climates. Their only reservoir is human beings. Most rhinovirus infections are transmitted by direct hand-to-hand contact or by fomites. Transmission by respiratory droplets (sneezing or coughing) also occurs, but it is secondary. As a result, rhinovirus transmission can be interrupted by frequent hand washing and disinfecting fomites.

Type-specific antibodies are produced in response to rhinovirus infections, and IgA antibodies in the nasal mucosa can inactivate virus particles before they initiate infection. Thus, people who recover from a cold are protected against becoming infected again with that type of rhinovirus, though this antibody-mediated immunity probably begins to wane after about 18 months. Complete immunity against rhinoviral infection would require protective antibodies against every rhinovirus serotype. Children, who have had the least opportunity to develop protective antibodies, are the most susceptible to rhinovirus infection. Infants and young children typically contract several colds a year, each of which helps increase antibody levels. Children under 2 are usually the ones who bring a cold into the household, often causing half of the other family members to become ill.

Colds cannot be cured—recovery depends on the individual's immune system. Antibiotics are useless because the infectious agent is a virus. Nonprescription remedies can help alleviate cold symptoms but do not shorten the illness.

No prevention or cure for the common cold has been developed, and none is on the horizon. The large number of rhinoviral serotypes and the relatively short duration of natural immunity makes a vaccine impractical. Hand washing and the use of antiviral chemicals to decrease fomite transmission help, but not enough to have a major effect on frequency of colds. Interferons seemed promising because they are natural antiviral agents, and large quantities are now available through recombinant DNA technology. Clinical studies found that alpha interferon administered intranasally prevents rhinoviral infections in nearly 80 percent of household contact after one family member comes down with a cold. Unfortunately, however, colds caused by other viruses occur in spite of interferon, so only about 40 percent of all colds are prevented by this expensive treatment.

Other Causes of the Common Cold

Coronaviruses also cause the common cold (**Figure 22.5**). These medium-size RNA viruses—about 100 nm in di-

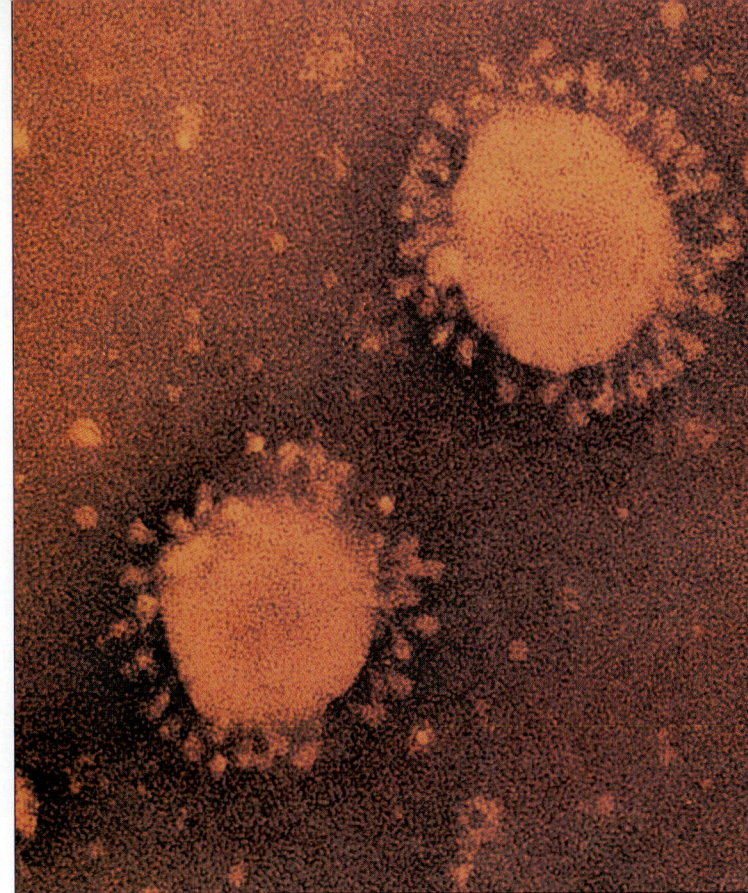

10 nm

Figure 22.5 Coronaviruses, which cause colds, are named for the glycoprotein spikes on their outer surface that resemble the sun's corona.

ameter—are named for the prominent glycoprotein spikes on their outer surface that resemble the solar corona. Coronaviruses are extremely difficult to isolate in cell culture, but serologic studies indicate that they probably cause 10 to 15 percent of upper respiratory infections in adults. They also cause pneumonia and intestinal infections.

Many other microorganisms cause the common cold, including coxsackieviruses, echoviruses, adenoviruses, myxoviruses, and even some bacteria such as *Mycoplasma pneumoniae* and *Coxiella burnetii*. These microorganisms, which usually cause other clinical syndromes, are discussed later in this chapter. In almost half of cold-sufferers, no microorganism can be isolated, so it is likely that more cold-causing microorganisms remain to be identified.

LOWER RESPIRATORY INFECTIONS

Unlike the upper respiratory tract, which has a single narrow airway, the lower respiratory tract has numerous, highly branched airways. Lower respiratory tract infections involve dangers such as obstruction from swelling, but breathing cannot be cut off abruptly. Infections of the lower respiratory tract have bacterial, viral, and fungal causes.

BACTERIAL CAUSES

Bacterial infections of the lower respiratory tract, many of which are pneumonias, are potentially serious but vary widely in their clinical presentation and severity.

Streptococcus pneumoniae: Pneumococcal Pneumonia

Streptococcus pneumoniae, also known as the **pneumococcus**, is a Gram-positive bacterium. Because its nearly round cells (with slightly pointed ends) usually occur in pairs, these organisms are described as **lancet-shaped diplococci (Figure 22.6)**. As a member of the genus *Streptococcus*, pneumococci do not produce the enzyme catalase. Unlike other streptococci, however, pneumococci are very sensitive to a chemical called optochin and lyse when exposed to a disc impregnated with it (Figure 22.6b). Colonies grown on a blood agar medium exhibit the green halo characteristic of alpha hemolysis. *S. pneumoniae* grows best anaerobically but is usually cultured aerobically at a pH of 7.4 and a temperature of 37°C—conditions identical to those of the human lung.

Humans are the sole reservoir for *S. pneumoniae*. Because about 10 percent of healthy adults harbor it in their throats, the pneumococcus is sometimes considered part of the normal bacterial flora of the human pharynx. However, its polysaccharide capsule can be a critical factor in virulence. An electrochemical charge on the capsule repels phagocytes, preventing direct contact—the essential first step in phagocytosis. The polysaccharide capsule appears as an unstained halo around pneumococcal cells surrounded by India ink (Figure 22.6c) and gives streptococcal colonies their characteristic shiny, mucoid appearance. Streptococcal strains that do not produce a capsule are avirulent.

The polysaccharides in the capsule are antigenic and vary from one strain of bacteria to another. Strains of pneumococci with the same capsular antigen belong to the same serotype. More than 80 pneumococcal serotypes have been isolated and identified by number. Serotypes with the thickest and most mucoid capsules tend to be the most virulent. When type-specific antibody reacts with the pneumococcal capsular antigen, the capsule appears to swell. This phenomenon is the basis of the serologic test (called the **quellung reaction**) used to distinguish the pneumococcus from other Gram-positive cocci (Figure 22.6d). Individual pneumococcal serotypes are identified by immunologic tests, either countercurrent immunoelectrophoresis (CIE) or latex agglutination.

The Clinical Syndrome. In the case history at the opening of the chapter, the emergency room physician immediately suspected pneumococcal pneumonia, even though viruses are the most common cause of pneumonia. J. G. had a severe illness, high fever, and chest pain; his x-ray pattern indicated that the entire lobe of his right lung was involved. His signs and symptoms were typical of bacterial pneumonia, and the pneumococcus is by far the most common cause of bacterial pneumonia serious enough to require hospitalization.

It was unusual that J. G.'s physician could confirm the diagnosis by isolating *S. pneumoniae* from his blood culture. Significant numbers of bacteria circulate in the blood of only a small percentage of critically ill pneumonia patients. In most cases, bacteria are found only in the lungs and are difficult to recover. **Sputum**, mucus coughed up for microbiologic examination, may contain the etiologic agent of a pneumonia but is frequently contaminated by normal mouth flora. Invasive procedures, such as aspirating the lungs or trachea (inserting a needle to draw out fluid), are painful, potentially dangerous, and very seldom attempted. As a result, a definitive microbiological diagnosis of pneumonia is rarely possible.

S. pneumoniae enters the body when a host inhales infected respiratory droplets that have been exhaled by someone nearby. Most infected droplets come to rest on epithelial surfaces protected by mucociliary defenses, but occasionally these defenses fail. J. G.'s defenses, for example, were probably weakened by his recent cold, allowing pneumococci to multiply within his respiratory tract and infect his alveoli. Occasionally, *S. pneumoniae* bacteria drift directly into the lower respiratory tract, which is not protected by mucociliary defenses. If they escape phagocytosis by alveolar macrophages and are not inactivated by secretory IgA, they cause pneumonia.

Pneumococci multiplying within the lung trigger an intense inflammatory response (Chapter 16). Leukocyte production is stimulated, as reflected in J. G.'s blood count. Leakage of small blood vessels in the inflamed lungs allowed fluid, blood cells, and serum proteins to flow into the affected alveoli, filling them.

Inflammation of the lung accounts for almost all the clinical findings of pneumococcal pneumonia. The affected region of the lung becomes **consolidated**, giving the impression of a solid organ when tapped, listened to, or penetrated by x rays. Normal gas exchange cannot occur in a consolidated lung, causing difficult and labored breathing, the hallmark of pneumococcal pneumonia. Inflammation, however, rarely stops the infection. Even the huge numbers of phagocytes activated by inflammatory defenses cannot eliminate pneumococci, which are protected by a thick polysaccharide capsule. They multiply, actively spread through the lung to local

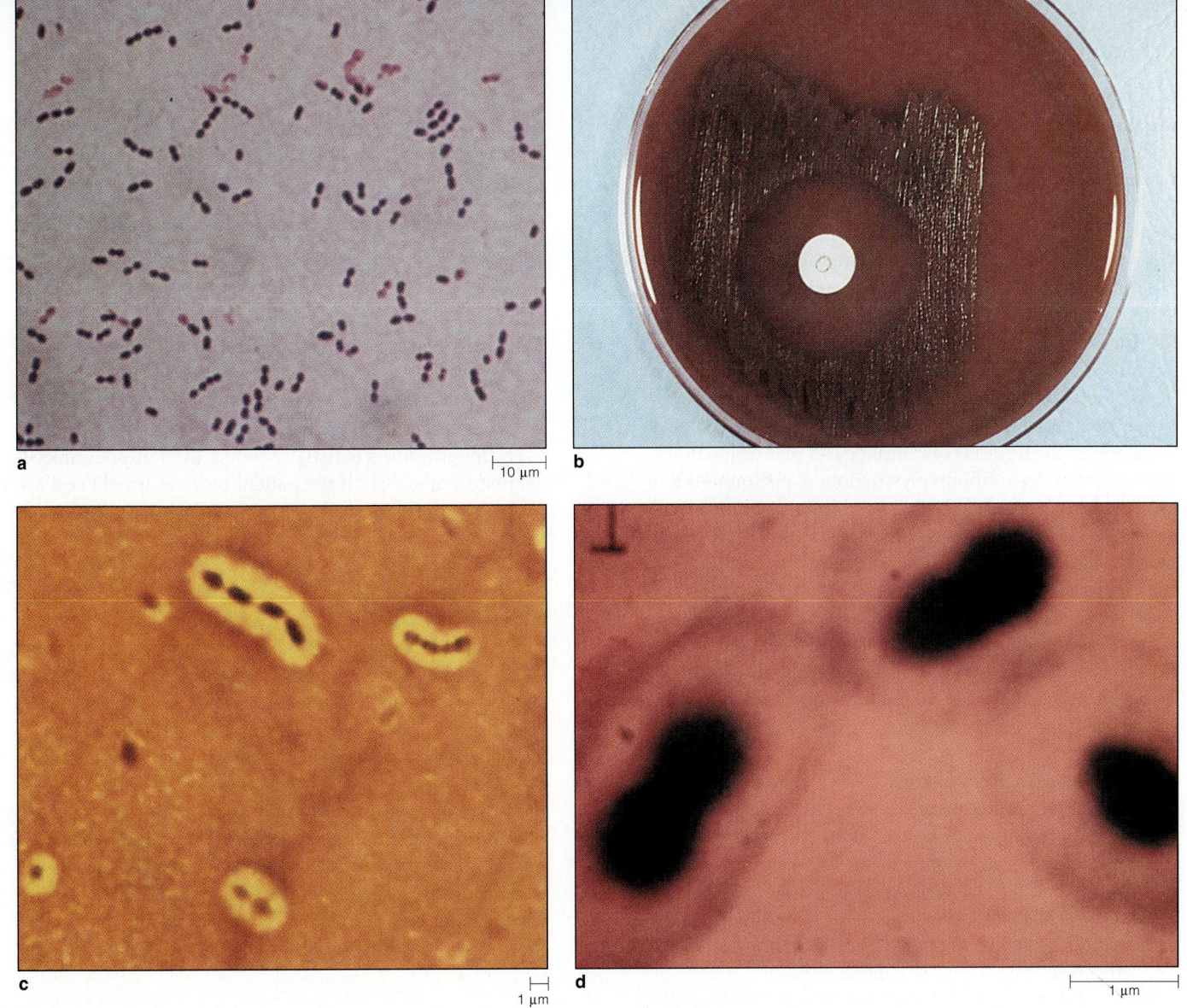

Figure 22.6 Features of *Streptococcus pneumoniae*, also called the pneumococcus, help identify it in laboratory specimens. (**a**) *S. pneumoniae* are lancet-shaped diplococci—round but with slightly pointed ends—that usually occur in pairs. (**b**) These organisms are extremely sensitive to a chemical called optochin, so growth on an agar plate is inhibited by an optochin disc (optochin lyses cells). (**c**) With negative staining, the antiphagocytic capsule appears as a clear halo around cells. (**d**) The pneumococcal capsule appears to swell when polysaccharide capsular antigens combine with type-specific antibodies. This is the basis of a serologic test called the quellung reaction.

lymph nodes, and may enter the bloodstream, as they did in J. G.'s case.

About 30 percent of untreated patients with pneumococcal pneumonia die from their infection. They suffer from unrelenting fever and progressively worsening respiratory problems until breathing becomes so labored they struggle for each breath. Eventually, for reasons not fully understood, they suffer a sudden decrease in blood pressure, sometimes accompanied by heart failure. Death can occur within days, but patients are often sick for a week or more before they die.

Despite the virulence of *S. pneumoniae*, immune defenses allow most patients to recover. Before antibiotics, pneumonia patients usually experienced a crisis 6 to 10 days into their illness—a drenching sweat and the sudden disappearance of fever indicated the beginning of recovery. This phenomenon coincided with the appearance of type-specific antibody against the pneumococcal capsule.

Anticapsular antibody forms an opsonizing bridge between bacterium and phagocyte that greatly facilitates phagocytosis (Chapter 16). It also activates complement through the classical pathway, creating more opsonizing bridges. This overcomes the invasive advantage of the pneumococcus and allows phagocytic white blood cells to eliminate bacteria from the body. In the course of recovery, macrophages migrate to the consolidated lung and clear away the debris. Healing is complete after pneumococcal pneumonia, whereas other types of pneumonia can cause permanent lung damage.

The spleen plays a vital part in the body's immune defense against the pneumococcus. It filters bacteria out of the blood and presents them to splenic macrophages. These macrophages process pneumococcal antigen and present it to the millions of lymphocytes that reside in the spleen, initiating antibody production. A person without a normal spleen is highly susceptible to **pneumococcal sepsis**, a life-threatening infection in which bacteria rapidly multiply in the circulation to concentrations as high as 1 million organisms per milliliter of blood.

Patients who survive a pneumococcal infection are permanently immune to reinfection by *that* pneumococcal serotype; but because a different opsonizing antibody is required for each of the 80 pneumococcal serotypes, repeated pneumococcal infections are common.

Treatment and Prevention.   Fifty years ago, before we had antibiotic treatment for pneumococcal pneumonia, severely ill patients like J. G. with bacteria multiplying in their bloodstreams almost surely would have died. Today, less than 5 percent of patients who receive timely and appropriate antibiotic treatment die, and most of these deaths occur among people weakened by advanced age, serious illness, or immune deficiency.

Penicillin has been the standard treatment since the 1940s. In the 1970s, however, a few penicillin-resistant pneumococci were reported in patients who had recently been treated with antibiotics or had become infected in the hospital. During the 1980s, reports of community-acquired infection with penicillin-resistant pneumococci became more frequent. Many of these strains are highly virulent and resistant to multiple antibiotics. Most infections by these organisms have been fatal, and although such cases remain rare, their incidence is alarming.

Highly susceptible people, including anyone over the age of 65 or suffering from a chronic illness or immune deficiency, should be immunized against *S. pneumoniae*. The vaccine in use (Pneumovax) contains polysaccharide antigens that stimulate the production of opsonizing antibodies against the *S. pneumoniae* capsule. It is a **polyvalent**, or multiple-antigen, vaccine, containing antigens from the 23 most common pneumococcal serotypes. Most adults who receive the vaccine show a significant rise in levels of protective antibody, and side effects are negligible (redness and pain at the immunization site). Pneumococcal polysaccharides are relatively weak immunizing

agents, however, so immunosuppressed patients and infants cannot be protected effectively by this vaccine.

Other Causes of Acute Bacterial Pneumonia

Although approximately 90 percent of acute bacterial pneumonias are caused by the pneumococcus, other bacteria occasionally cause an identical clinical syndrome. These organisms include *Haemophilus influenzae*, *Klebsiella* spp., *Staphylococcus aureus*, *Streptococcus pyogenes*, *Escherichia coli* (Chapter 23), and *Proteus* spp. (Chapter 24). When pneumonia is caused by these organisms, however, the prognosis is usually worse than in pneumococcal pneumonia even if the patient receives timely and appropriate antibiotic treatment. Permanent lung damage often results.

Various *Klebsiella* species cause less than 5 percent of bacterial pneumonias. They are virulent because of their antiphagocytic capsule, and many strains are resistant to multiple antibiotics. *Klebsiella* pneumonia occurs most commonly in people whose health is already poor, such as alcoholics, causing widespread lung destruction. Even with optimal medical treatment, most patients die.

Staphylococcus aureus also causes a small percentage of bacterial pneumonias. This infection is so damaging to the lungs that a chest x ray may show multiple abscesses. Staphylococcal pneumonia is most common in weakened hosts—the very young, the very old, those recovering from acute infections such as influenza, and those with chronic illnesses. Mortality is high. *S. aureus* is a much more common cause of other clinical syndromes (Chapters 23 and 26).

Mycoplasma pneumoniae: Mycoplasmal Pneumonia

Mycoplasmas are tiny, extremely simple procaryotes that lack a cell wall (**Figure 22.7**). Mycoplasmas are unable to synthesize many of their own building blocks, so they must remain in close contact with a host cell. They adhere to human epithelial cells, but do not invade deeper tissues. To be cultured in the laboratory, mycoplasmas require a complex medium that provides many growth factors. Even under optimal conditions, most mycoplasmas grow so slowly that it takes as long as two weeks to detect their typical "fried egg" colonies, even under magnification (see Figure 11.9).

Mycoplasma pneumoniae usually grows in the trachea and is transmitted in respiratory droplets. It causes pneumonia in less than 10 percent of people infected, but the pneumonia is different from the bacterial pneumonias described earlier. Mycoplasmal pneumonia usually comes on gradually, is mild, produces few definite signs and symptoms, and shows a diffuse patchy pattern in

Clinical Notes

Pneumonia—A Clinical Grab Bag

From a microbiologist's point of view, the logical way to think about pneumonia is in terms of the causative microorganism. This leads to a neat organizational scheme like the one presented in this chapter—there are bacterial pneumonias, viral pneumonias, and fungal pneumonias. But it is difficult, if not impossible, to recover samples from the lungs of patients to culture the causative microorganism. Moreover, pneumonia is not one disease. Clinicians have developed various ways to characterize pneumonias.

One way clinicians categorize pneumonias is by the pattern of infection that shows up in a chest x ray. Some pneumonia patients, like J. G. in our case history, have **lobar pneumonia**, a dense x-ray whiteout of one or more lobes of the lung. Others have **bronchopneumonia**, a patchy consolidation of small regions throughout the lung.

Another clinical view of pneumonia is based on patient risk factors. Categories include pneumonia in infants, pneumonia in the elderly, pneumonia in the alcoholic, and pneumonia in the immunocompromised host. These categories may

seem obvious, but they are very useful in prognosis and treatment.

Yet another categorization of pneumonia is based on clinical severity. We instinctively recognize differences of this type in our use of the nonscientific terms *walking pneumonia* (which describes a mycoplasma infection, but is commonly used by laypersons to describe a pneumonia in which the patient is still walking around) and *double pneumonia*. Clinicians describe severity with the terms **typical** (or **classical**) **bacterial pneumonia** (the severe form) and **atypical pneumonia** (the less severe form).

Taken together, these different schemes of clinical classification often help clinicians make a good guess about the etiologic agent of pneumonia. J. G., for example, had a lobar pneumonia and was a normal, healthy child with no special pneumonia risk factors, and the sudden onset and severity of his illness were characteristic of typical bacterial pneumonia. This clinical profile usually indicates *Streptococcus pneumoniae* infection—exactly what J. G. turned out to have.

chest x rays rather than dense consolidation of an entire lobe. It is rarely fatal, though unusually severe cases do occur. Because the clinical pattern is so different from pneumococcal pneumonia, mycoplasmal pneumonia is called *atypical* or **primary atypical pneumonia (PAP)** (or sometimes walking pneumonia). Other infectious agents, including viruses, chlamydiae, and rickettsiae, cause similar atypical pneumonias. *M. pneumoniae* may cause as many as 60 percent or as few as 10 percent of community-acquired pneumonias, depending upon the epidemiologic pattern at the time.

Mycoplasmal pneumonia occurs most frequently in school-age children and teenagers. It typically causes headache, low-grade fever, and a persistent dry cough. Illness may last as long as three weeks but can be shortened by antibiotic therapy. As wall-less organisms, mycoplasmas are not sensitive to the penicillins or cephalosporins. Tetracycline is the treatment of choice except for pregnant women and young children, for whom erythromycin is a good alternative.

Chlamydia psittaci: Ornithosis, Psittacosis

Chlamydiae are extremely simple bacteria that cause disease in birds and mammals. They resemble Gram-negatives because the cell envelope contains an outer and

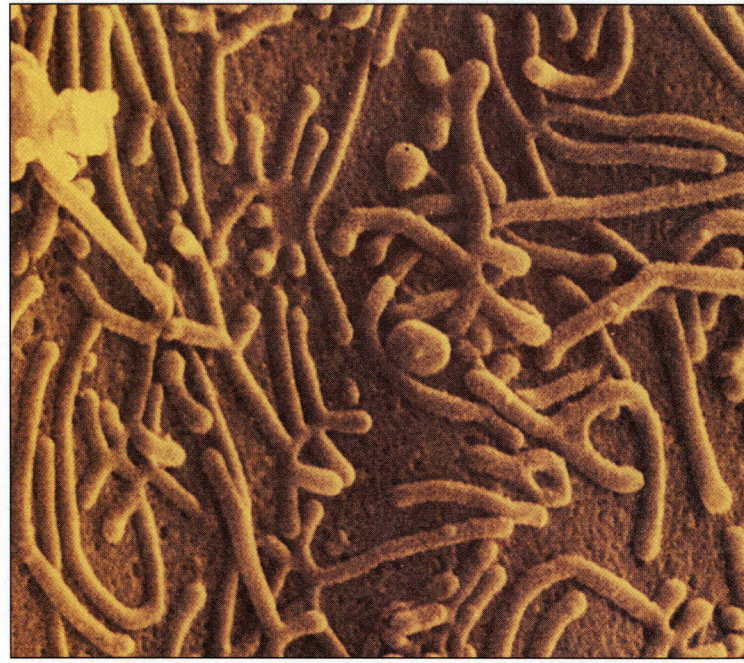

0.2 μm

Figure 22.7 *Mycoplasma pneumoniae.* Scanning electron micrograph showing the irregular shapes of these wall-less microorganisms.

inner membrane, but they lack peptidoglycan. These tiny microorganisms are incapable of generating their own energy, which makes them **obligate intracellular parasites**, capable of growing and reproducing only inside another cell.

As intracellular parasites, chlamydiae cannot be cultured in the laboratory using standard bacteriological techniques. They can be propagated only in the yolk sacs of chick embryos or in cultures of human cells. Culturing is slow, expensive, and tedious, and therefore seldom attempted. Infections by *Chlamydia psittaci* are usually diagnosed serologically, testing the patient's blood to detect a rise in antibody titer after infection.

Members of the genus *Chlamydia* exist in two forms during their life cycle—elementary bodies and reticulate bodies (Chapter 11). The microorganism is transmitted as an **elementary body**, which is resistant to drying. When an elementary body enters a host cell, it changes into a **reticulate body**, which is nonvirulent but capable of multiplying. The reticulate bodies turn back into elementary bodies and the host cell lyses. The released elementary bodies infect other cells.

C. psittaci commonly infects all types of birds—wild birds, farm birds (chickens and turkeys), and pets—which may not appear to be ill. When infected birds become stressed, as they do during shipment for sale, the disease spreads. Chlamydiae invade various organs and are excreted in the bird's droppings. If a human inhales microorganisms from the droppings, pneumonia can result. Occasionally, a person with *C. psittaci* pneumonia will transmit the infection to other people in respiratory droplets. The illness is called either **ornithosis**, because of its association with birds, or **psittacosis**, because it was first described in parrots (*psittakos* means "parrot" in Greek). Sometimes the illness is called parrot fever.

People with *C. psittaci* pneumonia develop fever, headache, chills, and cough. If illness progresses to persistent high fever, mental confusion, and marked shortness of breath, it can be life-threatening. A history of exposure to sick birds or employment in the poultry industry can be critical clues in diagnosing patients with pneumonia. In fact, the disease is an occupational hazard for bird handlers and workers in poultry slaughterhouses. There is no vaccine, but tetracycline and erythromycin are effective treatment. Ornithosis is a distinctly uncommon pneumonia (fewer than 100 cases a year are reported in the United States). Many infections are probably prevented by antibiotic supplements in bird feed and antibiotic treatment of imported parrots. In a bad ornithosis epidemic, up to 20 percent of patients die.

Coxiella burnetii: Q Fever

Coxiella burnetii belongs to the rickettsia family, a group of small intracellular parasites with a Gram-negative outer envelope. (Other rickettsiae that cause human disease are discussed in Chapter 27.) *C. burnetii* was named after an American named Cox and an Australian named Burnet who made significant contributions to the study of this microorganism.

Rickettsiae have complex life cycles, most of which require both insect and vertebrate hosts. Wild and domestic animals normally become infected from a tick bite. Then, because *C. burnetii* is hardy enough to survive drying, humans usually become infected by inhaling the microorganism from infected animal placentas, feces, or amniotic fluid. Infection can also be acquired from the milk of infected animals. *C. burnetii* can survive the pasteurization that kills tuberculosis organisms.

Human infection with *C. burnetii* is usually asymptomatic but may cause **Q fever**, a disease first described in the late 1930s as an ailment of meat-packinghouse workers. The mysterious illness was named Q (for query) fever because its etiologic agent was unknown. Q fever is an atypical pneumonia that cannot be distinguished clinically from mycoplasmal pneumonia or ornithosis. Definitive diagnosis can be made by specific serologic tests, but these tests are usually not readily available. Although *C. burnetii* infection is probably fairly common, Q fever is rare and almost never causes death. It is responsible, however, for occasional epidemics among people who raise cattle and sheep, so a history of exposure to livestock is critical to diagnosing Q fever. Tetracycline is used to treat serious infections, but antibiotic treatment is not as effective against Q fever as it is against most rickettsial infections.

Legionella pneumophila: Legionellosis

In July 1976, 182 American Legion conventioneers in Philadelphia became ill with a mysterious form of pneumonia. When the outbreak was over, 29 people were dead. No pathogen could be identified, and the public was alarmed, fearing further fatal epidemics. After six months of intensive investigation, microbiologists at the Centers for Disease Control and Prevention (CDC) isolated a previously unknown bacterium that they named *Legionella pneumophila* for the Legionnaires' epidemic that led to its discovery.

Why was *L. pneumophila* not discovered until almost 100 years after most other bacterial pathogens? First, there is the little-appreciated fact that a definitive microbiologic diagnosis is seldom made in patients with pneumonia. *Legionella* infections before 1976 were misdiagnosed clinically as other types of pneumonia. If the **legionellosis** (or **Legionnaires' disease**) epidemic in Philadelphia had not been so spectacular, we probably still would not have identified the pathogen. The second factor is that *L. pneumophila* is unusually difficult to identify in the laboratory. It stains poorly or not at all by the

Clinical Notes

Deal Me Out

Playing poker with friends doesn't seem a likely way to catch pneumonia, but the *New England Journal of Medicine* reported that's what happened to 12 residents of Halifax, Nova Scotia, in the winter of 1987. A group of friends regularly got together two or three times a week to play poker at one player's house. On Valentine's Day, 1987, the family cat chose a corner of the card room to give birth to three kittens, one of whom died shortly after. A few weeks later, everyone there that night came down with headache, fever, chills, and cough. One man with a history of heart disease died of his infection. Investigation revealed this was an epidemic of Q fever. *Coxiella burnetii* was recovered from the cat's uterus and identified in the lungs of the man who died. Blood samples from the other patients showed serologic evidence of *C. burnetii* infection. Q fever, usually a disease of farmers and animal handlers, can affect city dwellers, too.

Gram technique and can be seen in clinical specimens only by means of special stains, such as silver-impregnation stains, or by direct immunofluorescence (Chapters 3 and 19). The organism is so fastidious in its growth requirements that it was finally isolated by infecting embryonated hens' eggs and guinea pigs, techniques usually used to cultivate rickettsia.

L. pneumophila is a small, aerobic, rod-shaped bacterium that moves by means of flagella. Macrophages consume the bacterium but are unable to kill it, so *L. pneumophila* multiplies as an intracellular pathogen (**Figure 22.8**). Cell-mediated immune defenses, as well as antibody production, contribute to recovery from legionellosis. Various studies show a range of incidence of legionellosis as low as 1 and as high as 25 percent among community-acquired pneumonias.

L. pneumophila lives in natural and artificial water supplies and can survive both heat and chlorination. Infections occur when waterborne microorganisms become aerosolized and are inhaled. Several epidemics have been traced to *L. pneumophila* growing in water-cooled air conditioning systems, the probable source of the Philadelphia epidemic.

Most people infected by *L. pneumophila* experience only minor symptoms, but some develop a virulent pneumonia, like the one that hit the Legionnaires. Onset is sudden, with weakness, headache, high fever, cough, and shaking chills. X rays show consolidation of an entire lobe. The elderly, and especially those with another serious illness or a weakened immune system, are most often affected. Smokers and alcoholics are particularly susceptible. *L. pneumophila* is resistant to penicillin and cephalosporin, the drugs usually used to treat severe pneumonia when the causative organism is unknown. Thus, the clinician must suspect the diagnosis of Legionnaires' disease early in order to begin effective antimicrobial therapy with erythromycin. Legionellosis can be diagnosed by antigen detection tests, such as radioimmunoassay or enzyme-linked immunoassays; by direct fluorescent antibody stains that detect the organism in clinical specimens; or by culture in a buffered charcoal-yeast-extract agar.

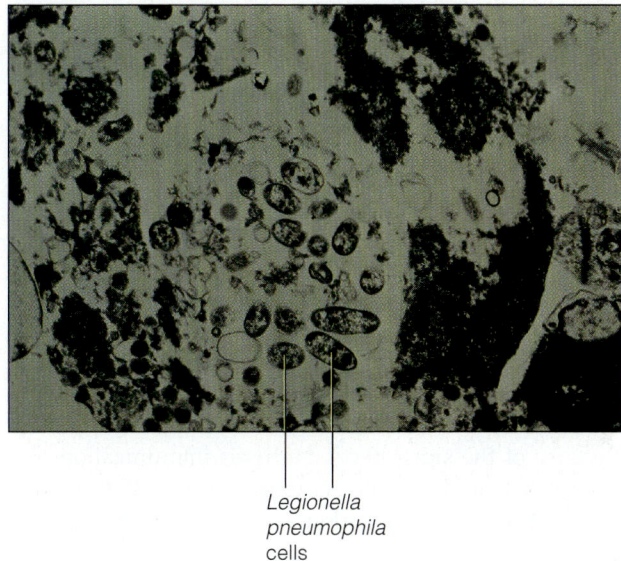

Legionella pneumophila cells

Figure 22.8 This section of human lung shows an alveolar macrophage—a phagocyte that patrols the lower respiratory system—filled with cells of *Legionella pneumophila*. Unlike most bacteria, these intracellular pathogens thrive inside macrophages.

Bordetella pertussis: Pertussis (Whooping Cough)

Bordetella pertussis, the microorganism that causes **pertussis**, or whooping cough, is a small Gram-negative coccobacillus. It is nonmotile and does not form spores. It has a capsule, requires oxygen for growth, and is slow-growing. *B. pertussis* is difficult to culture in the laboratory because of its highly fastidious growth requirements. It is usually isolated on the complex **Bordet-Gengou medium** (potato-glycerol-blood agar), named after the microbiologists who first identified the microorganism in 1906. Identification of *B. pertussis* can be confirmed by fluorescent antibody tests.

B. pertussis produces several powerful virulence factors, including a cell-attachment protein called filamentous hemagglutinin, a protein exotoxin called pertussis toxin, and a tracheal cytotoxin. The microorganism and its pathogenesis are discussed in detail in Chapter 15. Humans are the sole reservoir of *B. pertussis*.

Pertussis, which means "intensive cough," is a tracheobronchitis of infants and young children that produces prolonged and uncontrollable fits of coughing (see Chapter 15). It is highly contagious and can be fatal, especially in infants. By the time pertussis is diagnosed, pertussis toxin has already caused considerable damage, so antibiotic treatment does little to shorten the illness. The only treatment is supportive nursing care. Erythromycin is administered, but only to prevent transmission of the disease.

Once a common childhood illness, now only about 3000 cases of pertussis are reported in the United States each year. Most infants are protected during the first 6 months of life by the DPT immunization series. They receive booster doses in their second year and when they enter school. In spite of its effectiveness, the vaccine is criticized for its side effects. Most children develop a slight fever as well as pain and redness at the injection site, but a small number of infants suffer more serious complications, including convulsions. Extremely rare unconfirmed cases of brain damage have also been reported. Because of the side effects, pertussis immunization was suspended or drastically curtailed in Sweden and the United Kingdom, but within a few years both countries experienced an alarming increase in the number of pertussis cases.

Abandoning pertussis vaccination is clearly not the answer. It is likely that the DPT vaccine will be replaced by ADPT, an acellular vaccine composed of purified pertussis antigens. Several have been developed and tested in Japan, and one of these is licensed in the United States for use as a booster dose for children more than a year old. Otherwise, the traditional whole-cell vaccine is still recommended, although clinical trials are in progress that may establish the new vaccine as effective for infants.

Mycobacterium tuberculosis: Tuberculosis

Mycobacterium tuberculosis, which causes tuberculosis, is—and has been throughout history—one of the most widely distributed pathogens afflicting human beings. It is the leading killer among all infectious diseases today, responsible for an estimated 10 million new cases of active tuberculosis a year with 3 million deaths. Worldwide, TB accounts for 26 percent of avoidable adult deaths. Until the last few years, tuberculosis seemed to be on the decline in the United States, but that has changed.

M. tuberculosis, sometimes called the **tubercle bacillus**, is a rod-shaped obligate aerobe. It was identified as the causative agent of tuberculosis by Robert Koch (see the box "One Microbe, One Disease," Chapter 15). Members of this genus have unusual waxy cell walls containing complex lipids called **mycolic acids**. Because of its distinctive cell wall, special staining techniques must be used to identify *M. tuberculosis*. For clinical specimens, the **Ziehl-Neelson technique** is used; and fluorochrome stains can also be used (Chapter 3).

The high lipid content in its cell wall allows *M. tuberculosis* to survive prolonged drying, which, in turn, facilitates the transmission of tuberculosis. Bacterial cells that enter the air when a tuberculosis patient coughs can remain viable and infectious for as long as eight months if they are not in direct sunlight. Some of the lipids also add to the microorganism's disease-causing potential—protecting them against the lysosomal enzymes and oxidants within the phagocytes they enter. One of the mycolic acids, called **cord factor**, causes the strains that produce it to form cordlike, or parallel, rows of cells. Cord formation is found only in virulent strains of *M. tuberculosis* (see Figure 11.14).

M. tuberculosis requires oxygen and grows best where oxygen concentrations are high. In humans, oxygen concentrations are highest in the upper lobes of the lungs. Even under the most favorable conditions, however, *M. tuberculosis* grows very slowly. Typical doubling time is approximately 20 hours (compared to 20 minutes for *Escherichia coli*). In practical terms, the slow growth means laboratory identification can take up to six weeks. Disease usually develops slowly as well, but occasionally can progress extremely rapidly, as it does in AIDS patients.

The Clinical Syndrome. Tuberculosis is discussed in this chapter because it is usually transmitted by respiratory droplets and the lungs are most often affected, but *M. tuberculosis* can also affect the lymphatic system, the genitourinary system, the skeletal system, and the nervous system. The clinical syndromes produced by *M. tuberculosis*, then, are extremely diverse, and the infection may take many courses after the tubercle bacillus enters the body (**Figure 22.9**).

Primary infection begins when someone inhales *M. tuberculosis* from an infected person's respiratory secretions. Microorganisms that reach the lungs are phagocytized by alveolar macrophages but are not killed, and bacteria continue to multiply slowly inside and outside host cells. Some microorganisms enter the bloodstream and lymphatic circulation and spread throughout the body.

Several weeks after infection a critical change occurs. T lymphocytes, stimulated by mycobacterial antigens, begin to proliferate and secrete lymphokines, which activate macrophages, the body's principal defenders against *M. tuberculosis*. This immune response produces cellular immunity against tuberculosis and **tuberculin hypersensitivity**, a form of type IV delayed hypersensitivity (Chapter 18) and the basis for the tuberculin skin test. Cellular immunity helps control the infection; hypersensitivity causes most of the tissue damage associated with severe cases.

More than 90 percent of the people who contract a primary tuberculosis infection do not become ill. Activated macrophages engulf the bacteria and isolate them within nodules called **tubercles** or **granulomas**. A tubercle is an inflammatory lesion. It contains a few phagocytized mycobacteria, but it is mostly a dense collection of activated macrophages and lymphocytes. If host cells in the center of a tubercle die, the dead tissue looks dry and crumbly, like cheese, a phenomenon called **caseation necrosis**, cheeselike death. A few bacteria survive in the caseous center and persist in an inactive state for many years.

People whose immune defenses control a primary infection and who do not become ill usually establish a stable symbiosis with the walled-off tubercle bacilli that persists indefinitely. Sometimes calcified caseous tubercles called **Ghon complexes** are noticed on a healthy person's routine chest x ray, indicating he or she overcame a primary tuberculosis infection. However, sometimes **secondary infection**, or **reactivation infection**, occurs.

People who control their primary tuberculosis infection may develop disease months or years later when walled-off bacilli inside old tubercles escape. The caseous center of a pulmonary tubercle may liquefy and open, forming a tuberculous cavity. The immune system, activated during the primary infection, prevents bacteria from entering the blood and lymphatic circulation, but the associated hypersensitivity can destroy the lungs **(Figure 22.10)**.

Patients with reactivation tuberculosis commonly suffer fever, fatigue, weight loss, and a chronic cough—a clinical syndrome once known as **consumption**. Symptoms may develop over many years, but without treatment a debilitating and ultimately fatal illness usually results. Malnutrition, old age, and infections such as AIDS

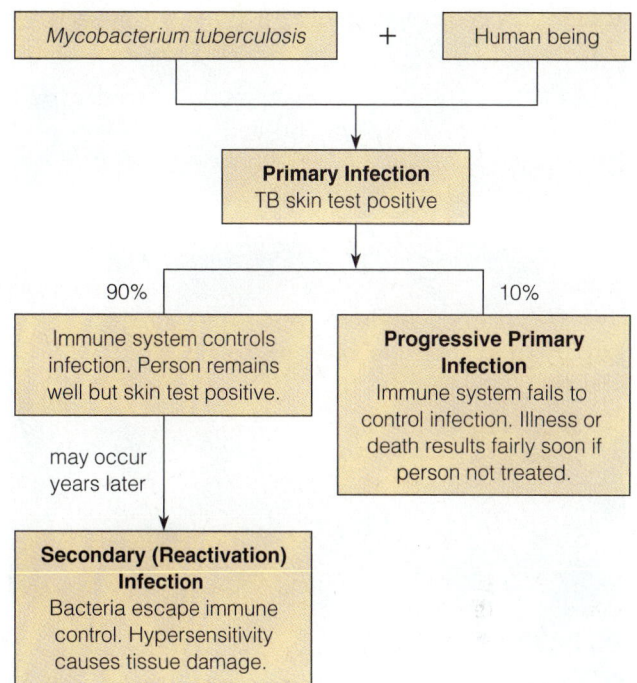

Figure 22.9 Possible outcomes of tuberculosis infection.

make a person particularly susceptible to reactivation tuberculosis.

In less than 10 percent of cases—usually young children or adults with a weakened immune system—cellular immunity fails to control primary tuberculosis, and the infection turns into **progressive primary tuberculosis**. In many cases, infection spreads directly through the lungs or into the nearby pleura, causing extensive tissue destruction and sometimes creating large bacteria-containing cavities in the lungs. In other cases, primary tuberculosis spreads outside the respiratory system, to the bones and kidneys. If infection spreads to the membranes in the central nervous system, the result is tuberculous meningitis (Chapter 25). Sometimes, particularly in children, bacteria spread through the blood and lymph and establish tiny infected areas. This is a life-threatening form of the disease called **miliary tuberculosis** because the infection sites are so numerous and tiny they look as if seeds of millet (grain) have been scattered through the body.

Without treatment, patients with progressive primary tuberculosis often die. Death from meningitis or miliary tuberculosis occurs relatively quickly, within weeks to months. Patients with progressive pulmonary disease usually become ill more gradually, but eventually suffer from fever and increasing fatigue. Those with large pulmonary cavities cough, often producing sputum tinged with blood. The air-loving mycobacteria thrive in open lung cavities; and when these patients cough, they spew

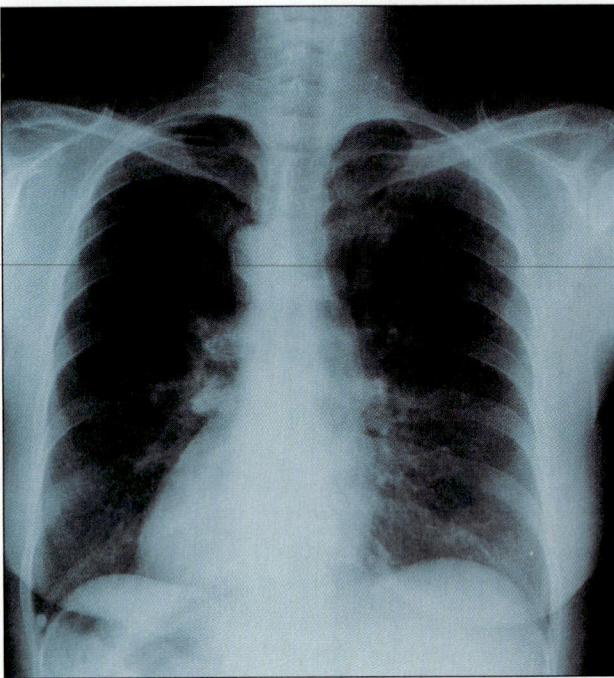

a

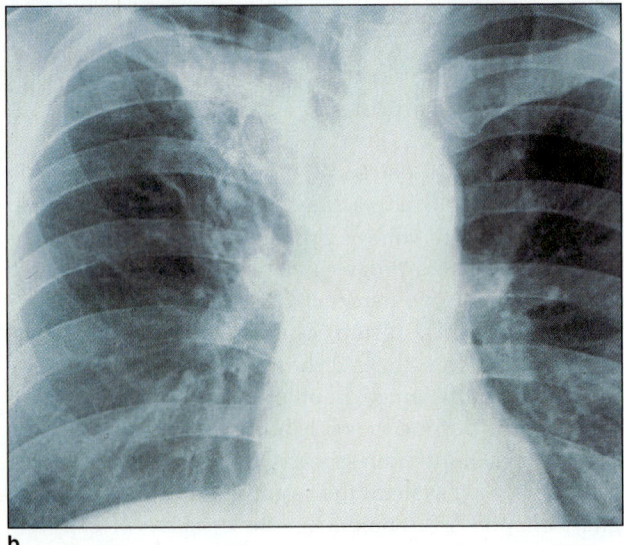

b

Figure 22.10 Tuberculosis x rays. (**a**) A calcified tubercle, or Ghon complex, can sometimes be seen in the lungs of a person who has successfully controlled tuberculosis infection. (**b**) Patients ill with reactivation tuberculosis often have cavities, or large areas of tissue destruction, in the upper lobes of their lungs.

enormous numbers of infectious microorganisms into the air, transmitting tuberculosis to others.

Prevention and Treatment. A live vaccine to prevent infection with *Mycobacterium tuberculosis*, called BCG for **Bacillus Calmette-Guerin**, is made from an attenuated strain of the closely related bacterium *Mycobacterium bovis*. BCG is used widely around the world, but it

is not entirely reliable. Clinical studies show that immunity develops in between 0 and 80 percent of the recipients. There is good evidence, though, that it reliably protects children from miliary TB and TB meningitis. Moreover, widespread immunization with BCG makes sense in places where enormous numbers of people die from tuberculosis each year. It is one of the six vaccines the World Health Organization wants all children in developing countries to receive.

BCG vaccination is not routine in the United States. Public health authorities rely instead on a program of tuberculin skin testing. A purified mycobacterial antigen is injected into the skin with a needle or with **tines**, or prongs. Anyone previously infected with tuberculosis has sensitized lymphocytes that mount a delayed hypersensitivity reaction, and within 48 hours a firm, raised area appears at the injection site. A positive skin test is followed up with a chest x ray or culture to see if the disease is active. About 5 to 10 percent of Americans test positive.

Skin-test screening is an effective program of tuberculosis control if the disease can be cured with antimycobacterial drugs. Individuals who test positive but do not have active tuberculosis are treated with a single drug, usually isoniazid, for one year (Chapter 21). This is prophylactic treatment because it *prevents* reactivation infection. People with active infections receive multiple drug therapy for as long as two years. This treatment cures the infection and protects the community because it keeps people with active disease from spreading it to others.

TB in the 1990s. Until recently, tuberculosis seemed to be a vanishing problem in the United States. Effective screening and treatment were readily available and, every year, the number of new cases dropped. Public health officials projected that tuberculosis would be eliminated from the United States by 2010. Research funding was cut back, the government stopped monitoring the appearance of drug-resistant strains, and pharmaceutical companies curtailed research into antimycobacterial drugs.

But tuberculosis case rates have been increasing by 3 to 6 percent each year since the late 1980s, and 21,479 new cases were reported to the CDC in 1993. In 1991 several prison inmates in New York City died after infection with a multiple-drug-resistant strain of *M. tuberculosis*. So far, most of these new TB patients are members of identifiable high-risk groups—AIDS patients, recent immigrants from countries where tuberculosis rates are high, nursing home residents, prison inmates, and the homeless (**Figure 22.11**). The disease burden of tuberculosis falls disproportionately on the poor, the malnourished, chronic drug and alcohol abusers, and people with limited access

Figure 22.11 *Tuberculosis in the 1990s.* This 700-bed shelter for the homeless in New York City is a perfect setting for transmitting tuberculosis.

to medical care. Epidemiologists associate the resurgence of tuberculosis with failure to provide adequate medical care for many citizens.

Making matters worse, new strains of *M. tuberculosis* are emerging that are resistant to most currently available drugs. Most drug-resistant organisms have emerged in patients who did not finish their full course of medication and are lost to medical follow-up. Because they are not cured, these patients are likely to become ill again and this time from organisms that are resistant to the drugs they were taking before. There are few new antimycobacterial drugs, and strains resistant to all the old ones cause potentially fatal, untreatable infections. These killer strains could be transmitted from high-risk patients into the general population. In fact, this epidemic scenario is far more likely than is the wildfire spread of AIDS. Any of us can contract tuberculosis from a stranger in a public place.

The final conquest of tuberculosis may depend more than anything else on microbiological research. We know amazingly little about *M. tuberculosis*. Mycobacterial genetics, including the genetics of drug resistance, is poorly understood, but the structure of this organism's unique cell wall is just now being understood (see the opening to Chapter 4, "A Matched Team"). We don't know which are the most important antigens or how an effective immune response gets under way. These slow-growing organisms take so long to culture in the laboratory that clinical diagnosis is extremely difficult—AIDS patients with rapidly progressive tuberculosis can die before culture results are available. Advances in basic research could lead to rapid diagnosis based on DNA probes, an improved vaccine, and the development of new drugs. The most

encouraging recent development was the discovery of the mechanism by which some strains of *M. tuberculosis* become resistant to isoniazid (see the box "Fighting TB—New Hope," Chapter 7).

Mycobacterium bovis and the Atypical Mycobacteria

Mycobacterium tuberculosis is the major cause of tuberculous infection but not the only one. *Mycobacterium bovis*, closely related to *M. tuberculosis*, usually infects cows, but it also causes the disease **bovine tuberculosis** in people who drink unpasteurized milk from infected cows. Thus, *M. bovis* usually enters the human body through the gastrointestinal system and causes tuberculosis of the lymph nodes and bone. Bovine tuberculosis has declined dramatically as a result of pasteurization of milk and the monitoring of commercial dairy herds.

The atypical mycobacteria, including species such as *M. avium*, *M. intracellulare*, and *M. kansasii*, are more distantly related to *M. tuberculosis*. The reservoirs for these organisms are soil, water, or other animals. Infection is not transmitted from person to person. These organisms cause various clinical syndromes, including lymph node infections in children, chronic ulcers of the skin, and a disease indistinguishable from pulmonary tuberculosis. All of these infections are rarer and less severe than tuberculosis, but in AIDS patients, atypical mycobacterial infections can be rapidly progressive and life-threatening. The atypical mycobacteria, along with other uncommon pathogens, are receiving renewed attention because of the AIDS epidemic.

Courage—and Lead Weights

Researchers working on tuberculosis have traditionally used the drug-sensitive Erdman strain of *Mycobacterium tuberculosis*. But with the possibility of a major outbreak of untreatable TB, the fastest progress is likely to come from studying the resistant strains directly. Some microbiology laboratories, called BL-3 labs, are specially equipped for the study of dangerous pathogens. But even with safeguards, researchers are uneasy dealing with bacteria that may cause untreatable tuberculosis. Ian Orme, a research microbiologist at Colorado State University's BL-3 tuberculosis lab, said in an interview with *Science* magazine, "To tell you the truth, we're fairly nervous about doing experiments on aerosolized multiple-drug-resistant strains." His lab received an aerosol machine with a 10-inch rubber gasket that should prevent leaks, but, said Orme, "I think we're going to put lead weights on top."

VIRAL CAUSES

Viral lower respiratory infections encompass a wide spectrum of clinical disease. These infections are usually relatively minor, but some can be life-threatening. In general, we have few drugs with which to treat them.

Influenza Virus: Influenza (Flu)

Almost everyone on Earth has been infected by the influenza virus, and related viruses cause disease in many animal species, including pigs, dogs, and birds, particularly ducks. Influenza got its name in the 1400s when people in Italy attributed an outbreak to the *influenza*—the "influence"—of the stars. We often forget that influenza can kill, but thousands of people die each year from flu. It is a virulent pathogen.

The influenza virus is a spherical RNA virus, about 100 nm in diameter, and a member of the orthomyxovirus family. Some of its unusual structural features help explain the pathogenesis of influenza (see Figure 13.17). The viral core is composed of eight separate pieces of single-stranded RNA, which makes the influenza virus particularly likely to undergo genetic recombination when two different viruses infect the same host. This phenomenon greatly contributes to the virus's genetic variability and potential to cause epidemics.

The helical capsid surrounding the RNA core contains antigens that distinguish the three main groups of influenza viruses—A, B, and C. Influenza A causes the most severe disease and is responsible for all pandemics (worldwide epidemics). Influenza B is a common cause of influenza in children. Influenza C is not important clinically.

The influenza virus is enveloped by a lipid membrane through which two types of protein spikes protrude.

Hemagglutinin, or the **H spike**, allows the virus to adhere to the membranes of host epithelial cells and enter them. IgA antibodies against H antigens prevent viral attachment and, thus, provide immunity against influenza. **Neuraminidase**, or the **N spike**, breaks the bond between hemagglutinin and the host cell, allowing the virus to leave one cell and pass to others nearby, thereby spreading infection. Antibodies against neuraminidase do not prevent infection, but they do help limit its spread.

Epidemiology. Infection with the influenza virus stimulates the production of antibodies that prevent reinfection, yet people suffer from influenza again and again. Recurrence is possible because the H and N antigens change frequently, creating new strains that are not inactivated by older antibodies. Minor genetic changes, called **antigenic drift**, are caused by spontaneous point mutations that affect only one amino acid in the H or N protein. A more radical change, called **antigenic shift**, occurs when recombination between different viral strains affects many amino acids, creating a brand-new H or N protein and therefore a new viral strain (Chapter 13).

Strains of influenza are identified by a shorthand, in which H stands for hemagglutinin, N for neuraminidase, and subscripts record the number of antigenic shifts in each. For example, the strain that caused most cases of influenza A in the winter of 1989–90 was H_3N_2. New strains are also named for the place they are first identified, for example, Hong Kong or Beijing.

When influenza antigens change slightly by antigenic drift, existing antibodies offer limited protection against the new strain, but radical changes by antigenic shift make existing antibodies useless. After a major antigenic shift—which occurs about every 10 years—everyone in the world is susceptible to the new strain of influenza. The stage is set for an influenza pandemic.

By far the most disastrous pandemic occurred in 1918, just before the end of World War I. It may have been the most devastating epidemic ever to afflict the human race. At least 20 million people died in only 120 days, including 12.5 million in India and one-half million in the United States. In Alaska, more than half the population was lost. In major American cities, morgues overflowed and coffins became scarce. The 21,000 deaths reported during the week of October 23, 1918, are the highest weekly mortality rate ever recorded in the United States (**Figure 22.12**).

Animal reservoirs are critical to the epidemiology of human influenza. Influenza A is widespread in birds; and although birds cannot transmit the virus directly to people, both birds and humans can exchange the virus with pigs. Recombination between human and avian strains probably occurs in pigs, creating the major antigenic shifts that lead to fatal pandemics. Because the exceptionally virulent 1918 strain probably arose this way, it is sometimes referred to as swine flu. When epidemiologic surveillance indicated that a swine flu might occur in 1976, an extensive vaccination campaign was mounted, but the outbreak never materialized.

Even during nonepidemic years, influenza outbreaks are major economic and public health problems. Approximately 20,000 Americans, most of whom are elderly or in ill health, die of influenza every year. The direct cost of influenza in absence from work, visits to the doctor, and admissions to the hospital exceeds $5 billion annually in the United States alone.

Clinical Syndromes. Illness from influenza is severe and disabling but usually fairly brief. Symptoms appear abruptly—fever, headache, muscle aches, and cough. The most common clinical syndrome is a tracheobronchitis, in which the ciliated epithelial cells lining both airways are infected and killed by the virus. This weakens the mucociliary defenses of the respiratory tract, making the person prone to secondary bacterial infection. Other viruses sometimes cause an identical respiratory illness, and definitive diagnosis of influenza requires culture of the virus or demonstration of viral antigens by immunofluorescent stains. Clinicians seldom order these expensive tests, however, because a definitive diagnosis of these flu-like illnesses is not particularly useful. These viral infections are self-limited, and specific treatment is seldom prescribed.

The influenza virus can also cause pneumonia, which, although uncommon, is life-threatening. Damage to the alveoli causes fluid to accumulate within the lungs, severely impairing gas exchange and sometimes causing death within hours. During the 1918 pandemic 20 percent of the victims developed viral pneumonia, accounting for the large number of deaths among young healthy adults.

Figure 22.12 Hoping to escape death from influenza during the 1918 pandemic, this telephone operator wears a gauze mask.

Although most influenza deaths today are due to secondary *bacterial* pneumonias, the majority of deaths in 1918 resulted from influenza alone. Thus, if a similarly virulent strain of influenza were to appear again, antibiotics would be of little value in saving lives.

Prevention and Treatment. Influenza can be prevented by active immunization, but only 70 percent of those vaccinated are protected. The changing antigens of the influenza virus require that a new vaccine be produced each year before the flu season begins. Manufacturers use epidemiologic information from around the world to design the next year's product. The recent success of genetic engineers in cloning the influenza hemagglutinin gene may facilitate vaccine production in the near future. Because the capacity for producing vaccine is limited, influenza immunization is recommended only for high-risk individuals—the elderly, chronically ill, and institutionalized as well as health-care professionals. Most vaccine side effects, such as pain at the injection site, are relatively minor. The 1976 swine flu immunization program, however, which was recommended for everyone, was associated with a significant number of cases of temporary paralysis, called **Guillain-Barré syndrome**. The antiviral agent amantadine is as effective as immunization in preventing influenza A and can speed recovery if it is administered during the first two days of illness (Chapter 21).

Parainfluenza Virus: Croup

Parainfluenza viruses belong to the paramyxovirus family, along with the viruses that cause measles and mumps. Parainfluenza viruses contain a single piece of single-stranded RNA surrounded by a helical capsid and a lipid envelope. Spike proteins protruding from the lipid layer allow the virus to adhere to human cells. The types of parainfluenza virus are numbered 1 through 4.

Parainfluenza viruses are transmitted by respiratory droplets. Although some of these viruses cause the common cold, parainfluenza type 1 is the most common cause of childhood **croup**, a laryngotracheobronchitis that causes the airway to narrow at and below the vocal cords. Extremely severe cases of croup can resemble epiglottitis, but the illness is usually milder and more gradual in onset. The typical croup patient is a toddler who becomes hoarse and develops a loud, barking cough. Symptoms are worst at night when mucus accumulates in the swollen airways, narrowing them still further. Moisture in the air thins the secretions and makes it easier to clear them with a cough. Thus, home treatment consists of a humidifier in the child's room or a brief visit to a steamy bathroom.

Croup epidemics occur during the fall. There are no effective antiviral agents against parainfluenza virus, so treatment is limited to supportive nursing care and agents that minimize the swelling of affected tissues. Natural infection does not produce immunity, so children may suffer recurrent episodes of croup.

Respiratory Syncytial Virus: Bronchiolitis

Respiratory syncytial virus (RSV), a paramyxovirus like parainfluenza, is the most common cause of fatal lower respiratory infection in young children and infants. It was named because infected respiratory tissues develop **syncytia**, large, abnormal cells with multiple nuclei.

Outbreaks of RSV infection occur yearly during the late winter and early spring, just as the flu season is ending. People of all ages are infected by the virus (almost everyone has been infected by the age of 4), but most older children and adults suffer only coldlike symptoms. Infants under 6 months suffer most from RSV. They usually contract a lower respiratory infection such as bronchiolitis or viral pneumonia. The typical RSV patient is a wheezy young infant who begins to breathe faster and faster. These babies can be cared for at home unless their lungs are so severely affected that they need extra oxygen or they are breathing too fast to eat.

The virus is transmitted by hand contact and to some extent by respiratory droplets. It spreads quickly among members of a family and causes nosocomial outbreaks in pediatric wards and nurseries. Epidemics in intensive care nurseries, where babies are small and sick, are devastating—almost every infant will be infected and a few are sure to die. Transmission can be decreased by careful hand washing.

Respiratory syncytial virus infection is a significant public health problem. Although only 1 percent of infected babies are ill enough to need hospitalization, infections are so common that these patients account for nearly half of all admissions to pediatric units during the peak season. Natural infection does not produce lasting immunity, and attempts to develop a vaccine against respiratory syncytial virus have been unsuccessful. Treatment consists mainly of supportive care, including administering oxygen, though the antiviral agent ribavirin sometimes helps in life-threatening situations.

Hantavirus: Hantavirus Pulmonary Syndrome

In the early 1990s, a mysterious respiratory illness appeared among otherwise healthy young people in the Four Corners area of the American Southwest. The initial symptoms—fever, muscle aches, and respiratory distress—are not alarming, but about 70 percent of the cases result in death within five to six days, often within hours of taking a dramatic turn for the worse. Death is caused by catastrophic lung failure. Capillaries leak profusely and fluid fills the air spaces. By November 1993, 45 cases had been reported in 12 states, with 27 deaths.

Researchers identified the cause as a new hantavirus strain, the Muerto Canyon hantavirus, that occurs principally in the long-tailed deer mouse (**Figure 22.13**). The mice shed the virus in their urine, feces, and saliva, and people contract the disease by inhaling aerosolized particles of the virus. Finding that a hantavirus was the cause came as a surprise because hantaviruses had previously been associated with kidney disease. Hantaviruses are a group of arboviruses named for the Hantaan River in North Korea. Westerners first encountered hantaviruses during the Korean War when about 3000 soldiers were infected and almost 300 died of a hemorrhaging kidney disease. Other strains discovered since have also attacked the kidneys (though some lung involvement has recently been discovered in the Seoul strain).

Virologists do not yet know whether the hantavirus pulmonary syndrome is an emerging disease or has been with us a long time but has gone unrecognized. Every year, thousands of Americans die of acute respiratory disease syndrome (ARDS). Some of these cases may be caused by hantavirus. In the meantime, the virus has been isolated, and epidemiologists are evaluating how great a public health risk it is and how it can be prevented and treated.

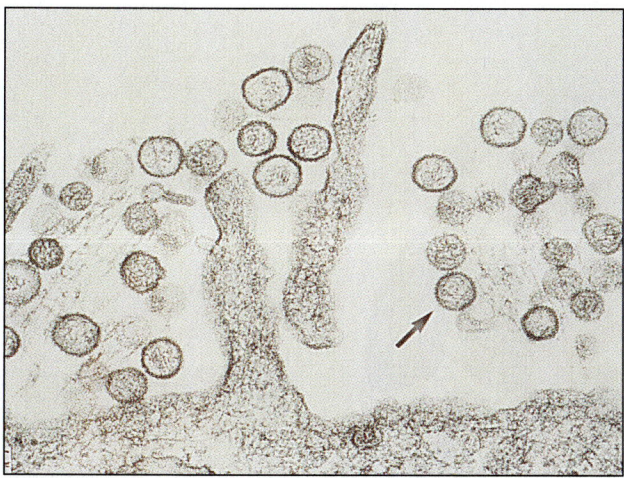

Figure 22.13 The Muerto Canyon hantavirus was first reported among Navajo Indians in New Mexico. In 1993, researchers were able to isolate and grow the virus in a laboratory, as shown in this section of cultured cells.

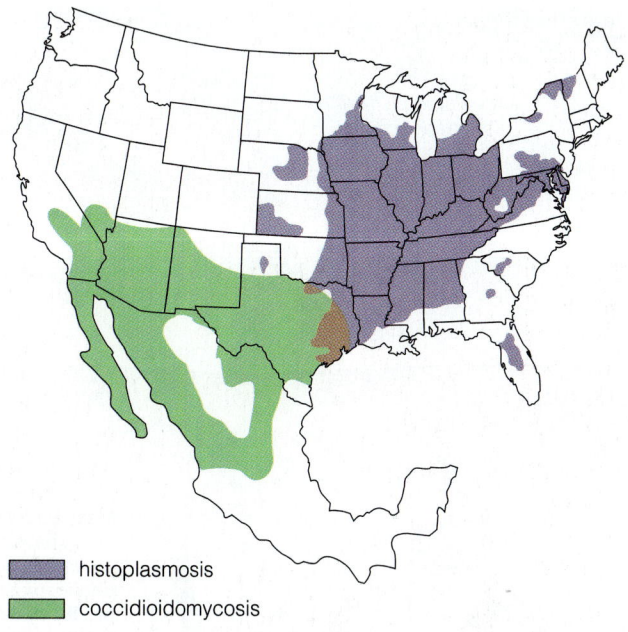

■ histoplasmosis
■ coccidioidomycosis

Figure 22.14 Geographic distribution of histoplasmosis and coccidioidomycosis in North America. Although the majority of people who live in endemic areas are infected at some time during their lives, very few of these infections cause symptoms.

FUNGAL CAUSES

Lower respiratory infections caused by fungi are quite common and rarely cause serious illness. Still, some infections that begin in the respiratory system have the potential to spread to tissues throughout the body. Consequently, these infections—histoplasmosis, coccidioidomycosis, and blastomycosis—are called **systemic mycoses** (*mycosis* means "fungal infection").

Histoplasma capsulatum: **Histoplasmosis**

Histoplasma capsulatum causes **histoplasmosis**, the most common respiratory disease caused by a fungus. *H. capsulatum* is dimorphic, growing in two different forms— below 35°C, it produces mycelia, and above that, it is a yeast. Conidia (asexual spores) initiate infection when inhaled by humans. *H. capsulatum* is found in the soil, particularly where bat or bird droppings are plentiful. Bats are hosts for *H. capsulatum*, but birds are not; their excrement merely adds nutrients to the soil that encourage fungal growth. People are likely to inhale spores and become infected by *H. capsulatum* when they are exploring caves, cleaning chicken coops, or entering attics, barns, or other bird-roosting areas.

Histoplasmosis occurs worldwide but is much more common in certain areas. In the United States it is endemic in the Ohio–Mississippi River valley and relatively uncommon elsewhere (**Figure 22.14**). We can track the epidemiology of histoplasmosis with skin tests that are like the tuberculin skin test except that *Histoplasma* rather than *Mycobacterium* antigens are used. Skin tests show that 80 percent of long-term residents of the Ohio–Missis-

sippi River valley test positive, indicating past infection, while the national rate is only about 20 percent.

Although many people are infected by *H. capsulatum*, only about 1 percent ever become ill. These people develop a flulike illness, but most recover without treatment because their immune system contains the infection. However, in people with a weakened immune system—patients with AIDS or cancer or on immunosuppressive medication—the disease progresses. Some histoplasmosis patients develop a chronic respiratory illness similar to pulmonary tuberculosis. Others develop a **disseminated** infection, meaning that the infection has spread. Diagnosis rests on serology, or finding the organism in infected tissue and culture. Histoplasmosis is treated with either amphotericin B or ketoconazole, but cure is difficult in immunosuppressed people with widespread disease.

Coccidioides immitis: **Coccidioidomycosis**

Coccidioides immitis is a dimorphic fungal pathogen that causes the disease **coccidioidomycosis**, also called **San Joaquin valley fever**. It occurs in the San Joaquin Valley of California, the American Southwest, and parts of Mexico and South America that have a semiarid climate and alkaline soils (Figure 22.14). *C. immitis* usually grows in the soil as a mycelium. The mycelia give rise to arthrospores that remain viable in the soil for hundreds of years

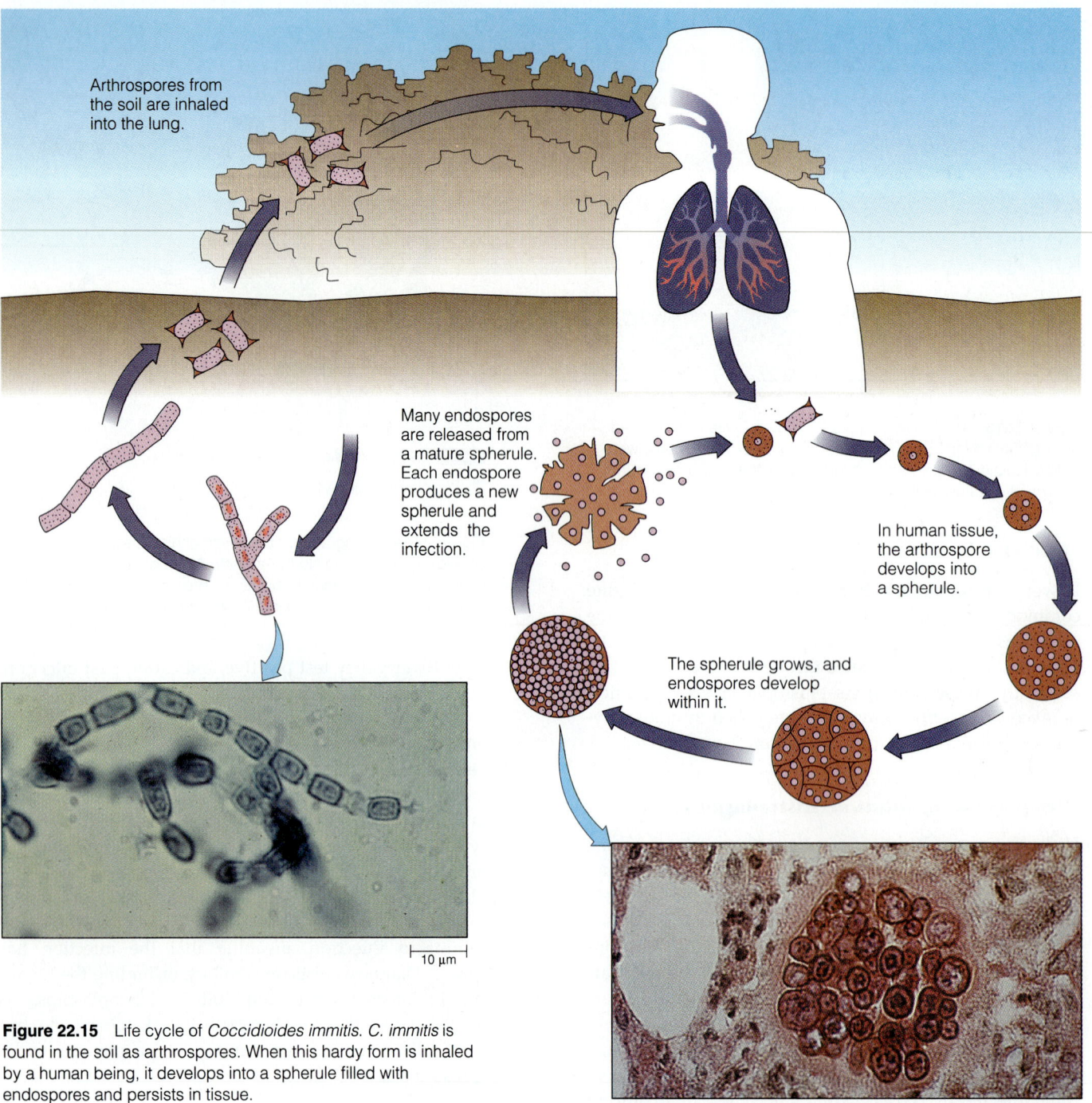

Arthrospores from the soil are inhaled into the lung.

Many endospores are released from a mature spherule. Each endospore produces a new spherule and extends the infection.

In human tissue, the arthrospore develops into a spherule.

The spherule grows, and endospores develop within it.

10 μm

10 μm

Figure 22.15 Life cycle of *Coccidioides immitis*. *C. immitis* is found in the soil as arthrospores. When this hardy form is inhaled by a human being, it develops into a spherule filled with endospores and persists in tissue.

and are small enough to be inhaled into the alveoli. In human tissue, *C. immitis* develops into thick-walled spherules, structures that contain endospores, which, when released, develop into new spherules.

Coccidioidomycosis occurs when arthrospores from infected soil become airborne and are inhaled by susceptible human beings (**Figure 22.15**). Skin-test surveys in endemic areas indicate that the majority of residents are

infected by *C. immitis* at some time during their lives. Nearly all, however, remain well or suffer only a mild respiratory illness. A small number develop a chronic respiratory illness similar to tuberculosis, and an even smaller number develop a disseminated disease, which can be manifested as a fatal form of chronic meningitis (Chapter 25). African Americans and Filipinos seem more prone to develop meningitis and other forms of widespread disease than do whites and Hispanics. Disseminated coccidioidomycosis is treated with amphotericin B but is usually fatal. No vaccine exists yet, so the only method of prevention is avoiding exposure to the infectious arthrospores.

Blastomyces dermatitidis: **Blastomycosis**

Blastomyces dermatitidis, another dimorphic fungus, is very closely related to *Histoplasma capsulatum*. It causes the disease **blastomycosis** in humans and other animals, especially dogs and horses. In the soil, *Blastomyces dermatitidis* grows as a mycelium, producing conidia that cause infection when they are inhaled by humans. In tissues it grows as a yeast. *B. dermatitidis* appears to be limited to North America and Africa, but its exact geographic distribution is unknown because it is very difficult to cultivate this organism from environmental sources.

In humans, infection begins in the lungs and may cause a clinical syndrome resembling pulmonary tuberculosis. More often, however, blastomycosis manifests itself when it affects distant organs—especially skin, bone, and testes. Cutaneous blastomycosis causes raised, wart-like lesions, usually on exposed areas such as the face or hands. Bone and testicular lesions cause pain, swelling, and sometimes drainage. Diagnosis is by culture or microscopic identification of the fungus in affected tissue. Amphotericin B is the treatment of choice. Without treatment, systemic blastomycosis is rapidly fatal.

Pneumocystis carinii: **Pneumocystis Pneumonia**

The taxonomic classification of *Pneumocystis carinii* has been a subject of debate. Traditionally it was classified as a protozoan, so its cellular forms have been given names from protozoology. Therefore, we speak of *P. carinii* as a small, unicellular trophozoite that attaches to pulmonary cells, initiates infection, and develops into a resistant cyst containing eight sporozoites, each able to produce a new trophozoite (**Figure 22.16**). But recent sequencing of the bases that compose *P. carinii* ribosomal RNA—an excellent indicator of evolutionary relatedness—shows that this organism is more closely related to the fungi.

P. carinii is an opportunistic pathogen that causes a life-threatening pneumonia in immunocompromised patients. Very little is known about it, including its reservoir, its life cycle in nature, and its mode of transmission. It is, however, found in the lungs of many human beings—between 75 and 90 percent of healthy children under the age of 4 have evidence of past infection by *P. carinii*. This suggests it is inhaled during infancy or childhood without causing disease. Then, if the immune system is significantly weakened, the infection may be reactivated, leading to a severe lung infection manifested by

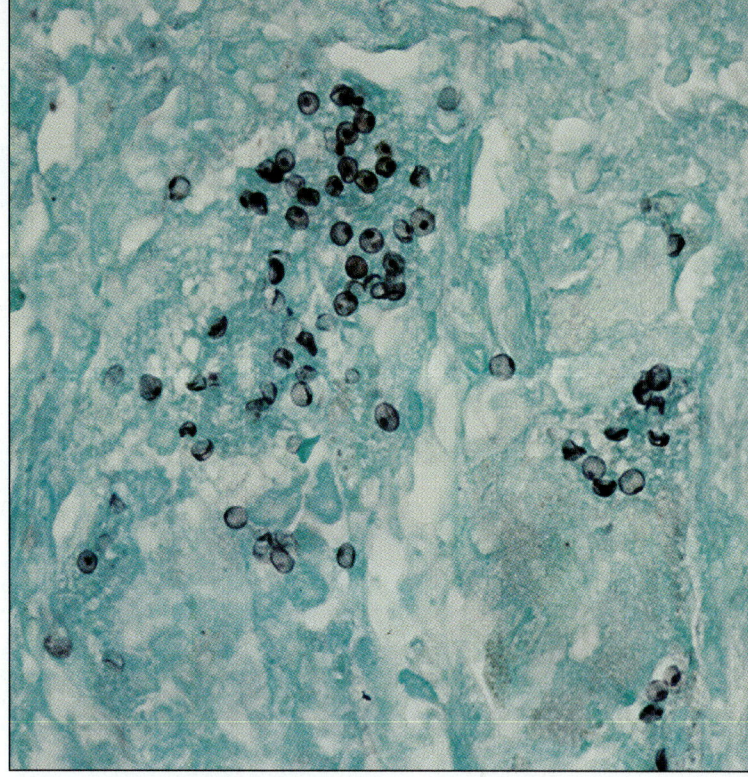

Figure 22.16 Tissue biopsy of a patient with *Pneumocystis* pneumonia, often the only way to make a diagnosis. The purple structures are *Pneumocystis carinii* cysts stained with Gomori methenamine silver.

fever, cough, tachypnea (rapid breathing), and a potentially fatal disturbance of alveolar gas exchange.

Pneumocystis pneumonia used to be a rare disorder seen on cancer wards, in patients receiving immunosuppressive medications, and occasionally in severely malnourished infants. Since the beginning of the HIV/AIDS epidemic, however, it has become a major killer, causing more deaths in the United States than do "traditional" infections such as tuberculosis. Moreover, AIDS patients present a slightly different clinical syndrome than do others with *Pneumocystis* pneumonia. They seem to get sick more slowly and survive longer without treatment. One thing is clear, however: People who develop *Pneumocystis* pneumonia often die from it, and it is the most common cause of death among people with AIDS.

To prevent *Pneumocystis* infection, AIDS patients are routinely treated with trimethoprim-sulfamethoxazole—one of the few times antimicrobial therapy is routinely used before an infection develops. If prevention fails, trimethoprim-sulfamethoxazole and pentamidine isethionate, another antimicrobial agent, are effective treatment. Unfortunately, many AIDS patients cannot tolerate these medications because of their serious side effects, including a life-threatening decrease in white blood cells.

Table 22.2 summarizes the infections of the upper and lower respiratory tracts.

Table 22.2 Infections of the Respiratory System

Infection	Causative Agent	Mode of Transmission	Symptoms	Prevention and Treatment
UPPER RESPIRATORY INFECTIONS				
Bacterial				
Epiglottitis	*Haemophilus influenzae* type b	Respiratory droplets	Fever, sudden and severe respiratory distress	Vaccination; treatment with antibiotics and surgical intervention to secure airway
Streptococcal pharyngitis	*Streptococcus pyogenes*	Respiratory droplets	Sore throat, fever, chills, headache; may develop into scarlet fever, rheumatic fever, or acute poststreptococcal glomerulonephritis	Treatment with penicillin or erythromycin
Diphtheria	*Corynebacterium diphtheriae*	Respiratory droplets	Swollen throat covered with pseudomembrane which obstructs airway; toxin may damage heart, kidneys, brain, or nerves	Toxoid vaccination; treatment with antitoxin effective if administered early enough
Viral				
Common cold	Rhinovirus, coronavirus, others	Primarily hand-to-nose	Sneezing, nasal discharge	Prevention by hand washing
LOWER RESPIRATORY INFECTIONS				
Bacterial				
Pneumococcal pneumonia	*Streptococcus pneumoniae*	Respiratory droplets	Fever, painful breathing, often life-threatening	Pneumovax; treatment with penicillin or other antibiotics
Other typical bacterial pneumonias	Many, including *Haemophilus influenzae*, *Staphylococcus aureus*, *Klebsiella* spp., *Streptococcus pyogenes*	Respiratory droplets	Fever, difficulty breathing, often life-threatening	Treatment with antibiotics appropriate to the infecting organism
Mycoplasma pneumonia (primary atypical pneumonia)	*Mycoplasma pneumoniae*	Respiratory droplets	Headache, low fever, persistent dry cough, not life-threatening	Treatment with erythromycin or tetracycline
Ornithosis (psittacosis)	*Chlamydia psittaci*	Inhaling organisms shed by birds; respiratory droplets	Headache, fever, chills, cough, occasionally life-threatening	Prevention by controlling infection in birds; treatment with erythromycin or tetracycline
Q fever	*Coxiella burnetii*	Inhalation	Headache, fever, cough, chills, rarely life-threatening	Treatment with tetracycline

Summary

THE RESPIRATORY SYSTEM (pp. 525–527)

1. The upper respiratory system consists of the nasal cavity and the nasopharynx, the space behind the nasal cavity.

2. The lower respiratory system begins at the larynx. The trachea branches into two smaller airways, the bronchi, which branch into secondary and tertiary bronchi and then bronchioles. The bronchioles terminate in the alveoli, tiny air sacs that make up the lungs. The lungs are surrounded by a membrane called the pleura.

3. Clinical syndromes of the respiratory system include rhinitis, pharyngitis, epiglottitis, bronchitis, and pneumonia (infection of the lungs with fluid and microorganisms replacing the air that normally fills the alveoli).

4. Clinical syndromes can be caused by many different microorganisms. Pneumonia, for example, can be bacterial, viral, or fungal.

Table 22.2 (continued)

Infection	Causative Agent	Mode of Transmission	Symptoms	Prevention and Treatment
Legionellosis	*Legionella pneumophila*	Inhaling water droplets contaminated with bacteria	Weakness, headache, cough, high fever, chills; life-threatening in elderly and weakened individuals	Treatment with erythromycin
Pertussis (whooping cough)	*Bordetella pertussis*	Respiratory droplets	Coughing paroxysms; life-threatening in infants	DPT vaccination; treatment with supportive care
Tuberculosis	*Mycobacterium tuberculosis*	Inhalation	Pulmonary tuberculosis characterized by fatigue, cough, fever, weight loss; many other syndromes associated with dissemination to other organ systems	BCG vaccination or tuberculin skin test; asymptomatic individuals treated with isoniazid; actively infected individuals treated with multiple antibiotics
Viral				
Influenza	Influenza virus	Respiratory droplets	Fever, headache, muscle aches, cough; pneumonia uncommon but life-threatening	Vaccination; treatment with amantadine
Croup (laryngotracheobronchitis)	Usually parainfluenza virus	Respiratory droplets	Hoarseness, barking cough, stridor in severe cases	Treatment with supportive care
Bronchiolitis	Usually respiratory syncytial virus (RSV)	Close personal contact	Wheezing, rapid breathing	Treatment with supportive care; ribavirin in life-threatening cases
Fungal				
Histoplasmosis	*Histoplasma capsulatum*	Inhalation	Tuberculosis-like	Treatment with amphotericin B, ketoconazole
Coccidioidomycosis (San Joaquin valley fever)	*Coccidioides immitis*	Inhalation	Tuberculosis-like; occasionally a fatal chronic meningitis	Treatment with amphotericin B
Blastomycosis	*Blastomyces dermatitidis*	Inhalation	Tuberculosis-like; often disseminates to skin, bone, testes, causing painful lesions and, without treatment, death	Treatment with amphotericin B
Pneumocystis pneumonia	*Pneumocystis carinii*	Unknown; perhaps inhaled in infancy and reactivated with weakened immune system	Fever, cough, difficulty breathing; frequently life-threatening	Treatment with trimethoprim-sulfamethoxazole, pentamidine isethionate

UPPER RESPIRATORY INFECTIONS (pp. 527–533)

Bacterial Causes (pp. 527–532)

1. The encapsulated strains of *Haemophilus influenzae* cause epiglottitis and bacterial meningitis, life-threatening infections. Nonencapsulated strains cause mild infections such as sinusitis. *H. influenzae* was the scourge of pediatric wards until 1990 when the Hib (*Haemophilus influenzae* type b) vaccine was developed.

2. *Streptococcus pyogenes* (also called group A beta-hemolytic streptococcus) is highly virulent and causes many clinical syndromes, including streptococcal pharyngitis (strep throat). Complications include scarlet fever and TSLS (toxic shock–like syndrome). Rheumatic fever can develop after the streptococcal infection is over.

3. *Corynebacterium diphtheriae* is noninvasive, but it produces diphtheria toxin, which infects the mucous membranes of the pharynx and causes diphtheria. The throat swells and becomes covered with a grayish pseudomembrane of dead cells. Treatment is with diphtheria antitoxin, produced artificially in horses. Infants can be immunized as part of the DPT (diphtheria, pertussis, tetanus) vaccine.

Viral Causes (pp. 532–533)

1. Rhinoviruses cause the clinical syndrome of a cold or upper respiratory infection (URI). Their only reservoir is human beings. Transmission is usually by direct contact or fomites, which means transmission can be interrupted by frequent hand washing. Recovery depends upon the immune system; antibiotics are useless because the infectious agent is a virus.

2. The common cold is also caused by other viruses, primarily coronaviruses.

LOWER RESPIRATORY INFECTIONS (pp. 533–550)

Bacterial Causes (pp. 534–543)

1. *Streptococcus pneumoniae* (the pneumococcus) causes pneumococcal pneumonia. Penicillin is the drug of choice, and recovery is complete; however, penicillin-resistant strains seem to be increasing. A vaccine (Pneumovax) is available.

2. Acute bacterial pneumonia caused by other bacteria, such as *Klebsiella* spp. and *Haemophilus influenzae*, often leaves permanent lung damage.

3. *Mycoplasma pneumoniae* causes a different clinical syndrome (gradual onset, few definite signs and symptoms, patchy x-ray pattern rather than dense consolidation of an entire lobe) and is often called atypical pneumonia. Treatment is with tetracycline.

4. Humans can contract a pneumonia by inhaling *Chlamydia psittaci* from bird droppings. The illness is also called ornithosis and psittacosis.

5. *Coxiella burnetii* causes Q fever, which is clinically indistinguishable from mycoplasmal pneumonia or ornithosis. Humans become infected by inhaling microorganisms from infected animal placentas or feces. Like all rickettsiae, *C. burnetii* has a complex life cycle that requires both an insect and vertebrate host.

6. *Legionella pneumophila* causes legionellosis, a pneumonia named for a virulent outbreak at an American Legion convention. *L. pneumophila* lives in water supplies and can be inhaled when aerosolized. The elderly, individuals with weakened immune systems, smokers, and alcoholics are most susceptible.

7. *Bordetella pertussis* causes pertussis (whooping cough), a highly contagious and frequently fatal childhood infection. The pertussis vaccine (whole cell) is criticized for its side effects and may be replaced by ADPT (acellular diphtheria, pertussis, tetanus).

8. *Mycobacterium tuberculosis* causes tuberculosis. *M. tuberculosis* has unusual waxy cell walls that resist drying and remain viable for up to eight months, which facilitates transmission. Primary infection produces tubercles or granulomas that become calcified (Ghon complexes) in patients who overcome infection. Or the bacilli in old tubercles may escape to cause a secondary, or reactivation, infection. The clinical syndrome includes fever, fatigue, weight loss, and a chronic cough. There is a vaccine (BCG), but it is not reliable. The United States has a tuberculin skin-testing program that was effective until the late 1980s, when TB case rates began to rapidly increase. New drug-resistant strains of *M. tuberculosis* are emerging.

9. *Mycobacterium bovis* causes the human disease bovine tuberculosis in people who drink unpasteurized milk from infected cows. It is not a pulmonary tuberculosis, though the symptoms are indistinguishable.

Viral Causes (pp. 544–546)

1. Almost everyone has been infected by the influenza virus. Illness may be severe (fever, headache, muscle aches, cough), but it is usually brief. The virus's structure makes it prone to minor genetic changes (antigenic drift) and radical changes (antigenic shift) that

can result in new strains. The worst antigenic shift occurred in 1918, causing a pandemic. A vaccine is available for high-risk individuals.

2. Some parainfluenza viruses cause the common cold, but parainfluenza type 1 is the most common cause of childhood croup, which causes upper airways to narrow and produces a hoarse cough. Severe cases resemble epiglottitis. Treatment is limited to supportive care.

3. Respiratory syncytial virus (RSV) is the most frequent cause of fatal lower respiratory infection in infants and young children (though all ages can be infected). Outbreaks occur yearly, as the flu season is ending. It is transmitted primarily by hand contact, so it spreads quickly within a family and in pediatric wards and intensive care nurseries. Treatment consists of supportive care, including administering oxygen.

4. A new hantavirus strain was identified as the cause of an outbreak of respiratory illness in the American Southwest. By November 1993, 45 cases had been reported in 12 states, with 27 deaths. Symptoms include fever, muscle aches, and respiratory distress, and a death rate of 70 percent due to lung failure. People become infected by inhaling particles of the virus, which is shed by certain rodents.

Fungal Causes (pp. 547–550)

1. Histoplasmosis, caused by *Histoplasma capsulatum*, is the most common fungal respiratory disease. People become infected when they inhale spores where bat and bird droppings are plentiful. Only about 1 percent of infected people become ill (with flulike symptoms) and, in the United States, that is largely in the Ohio–Mississippi River valley. Immunocompromised patients develop chronic or disseminated infection.

2. *Coccidioides immitis* causes coccidioidomycosis (also called San Joaquin valley fever). Humans become infected when they inhale airborne arthrospores. Most people in endemic areas get a mild case at some time, but a few develop a chronic infection similar to tuberculosis or a disseminated infection that can be fatal.

3. *Blastomyces dermatitidis* causes blastomycosis. Humans and other animals become infected when they inhale conidia from infected soil. The clinical syndrome resembles tuberculosis, though the skin, bone, and testes may also be affected. Without treatment with amphotericin B, systemic blastomycosis is fatal.

4. *Pneumocystis carinii* has traditionally been classified as a protozoan, but RNA sequencing shows it is more closely related to the fungi. *P. carinii* is an opportunistic pathogen that causes *Pneumocystis* pneumonia, a life-threatening infection in immunocompromised patients. Once a rare disease, it is now the most common cause of death among people with AIDS.

Review Questions

THE RESPIRATORY SYSTEM

1. Sketch the respiratory system, and label these parts:

   a. nasal cavity   f. trachea   k. alveoli
   b. nasopharynx   g. primary bronchi   l. lungs
   c. pharynx   h. secondary bronchi   m. pleura
   d. epiglottis   i. tertiary bronchi
   e. larynx   j. bronchioles

2. Why is the respiratory system infected more often than any other?

3. What is the function of the respiratory system?

4. Briefly review the normal flora and defenses of the respiratory system.

5. What is a clinical syndrome? Describe these clinical syndromes of the respiratory system:
   a. pharyngitis   c. epiglottitis   e. laryngitis
   b. bronchitis   d. bronchiolitis   f. pneumonia

6. Explain this statement and give an example: Any of the respiratory clinical syndromes can be caused by many different microorganisms.

UPPER RESPIRATORY INFECTIONS

1. Why is the new *Haemophilus influenzae* type b (Hib) vaccine such a breakthrough? What does the "type B" designation mean?

2. Why should strep throat never be ignored? How is it diagnosed? What is the causative agent? Why do researchers think streptococcal virulence may be changing in the United States?

3. Describe how diphtheria toxin causes the various clinical manifestations of diphtheria. What is the treatment? What is cutaneous diphtheria?

4. Why would a prescription for penicillin be useless in treating the common cold? What does interrupt the transmission of a cold? Name the most common causative agents.

5. Describe the primary causative agent, the clinical manifestations, and the treatment for acute bacterial pneumonia.

6. Why is mycoplasmal pneumonia also called atypical pneumonia? Why is penicillin not the drug of choice?

7. Name the microorganism that causes ornithosis and describe its life cycle. Name the microorganism that causes Q fever and describe its life cycle. In what ways are both of these infections occupational hazards?

8. Why was *Legionella pneumophila* discovered about 100 years after most other bacterial pathogens?

9. Describe the clinical syndrome of pertussis and its treatment. Why is the pertussis vaccine controversial, and what is being done to resolve the controversy?

10. List reasons why tuberculosis is the leading killer among all infections today. Draw on everything you know about the causative microorganism, the clinical syndrome, and public health issues.

11. Why is bovine tuberculosis an atypical tuberculosis, and why is it getting renewed attention?

LOWER RESPIRATORY INFECTIONS

1. Why is the influenza virus prone to antigenic drift and antigenic shift? Could the flu pandemic of 1918 happen again? Explain.

2. Name the causative agent of croup and describe the symptoms and treatment.

3. Why is prevention so important with respiratory syncytial virus (RSV)?

4. When a mysterious respiratory illness with a 70-percent fatality rate broke out in the American Southwest in the early 1990s, epidemiologists were called in. What have they—and virologists—since learned?

5. How are histoplasmosis, coccidioidomycosis (San Joaquin valley fever), and blastomycosis transmitted? Describe these infections and their treatment.

6. Why has *Pneumocystis carinii* been reclassified as a fungus? Why is it not surprising that *Pneumocystis* pneumonia is the most common cause of death among people with AIDS?

Suggested Readings

Austrian, R. 1985. Life with the pneumococcus—notes from the bed side. Philadelphia: University of Pennsylvania Press.

Edman, J. C.; Kovacs, J. A.; Masur, H.; Santi, D. V.; Elwood, H. J.; and Sogin, M. L. 1988. Ribosomal RNA sequence shows *Pneumocystis carinii* to be a member of the fungi. *Nature* 334:519–22.

Finegold, S. M. 1988. Legionnaire's disease—still with us. *New England Journal of Medicine* 318:571–73.

Gordan, R. D. 1990. Prophylaxis and treatment of influenza. *New England Journal of Medicine* 322:443–50.

Karzin, D. T., and Edwards, K. M. 1988. Diphtheria outbreaks in immunized populations. *New England Journal of Medicine* 318:41–43.

Musher, D. M. 1983. *Haemophilus influenza* infections. *Hospital Practice, August,* 158–73.

Tuberculosis: new views of an old disease (Editorial). 1985. *New England Journal of Medicine* 312:1514–15.

23 INFECTIONS OF THE DIGESTIVE SYSTEM

Good Reason to Be Worried

A. R., a 3-year-old boy, was brought to a pediatric emergency room in late September because of high fever and diarrhea. His mother reported that the boy had suddenly become ill the day before. His temperature rose to 105°F; he was listless, refused to eat, and vomited twice. In the morning he seemed lethargic, and at one point he trembled uncontrollably and fell unconscious. Shortly after this convulsion, A. R. passed a large watery stool. During the next few hours he passed many smaller stools streaked with blood and mucus. The history revealed that A. R. lived in a small apartment with 10 others, several of whom had recently had diarrhea.

A. R. moaned and cried quietly as the emergency room physician examined him, but he did not speak and barely put up a struggle. His temperature was 104.5°F. He seemed to be mildly dehydrated; his mouth was dry; and when he cried, there were no tears. Laboratory analysis showed only a slight elevation in the number of A. R.'s circulating white blood cells but an exceptionally high number of immature neutrophils—35 percent compared to a normal 1 percent. A mucus-containing sample of his stool stained with methylene blue and examined under the microscope revealed sheets of polymorphonuclear leukocytes, evidence of colonic inflammation. Cultures of A. R.'s blood and a stool were also sent to the microbiology laboratory.

A. R. was admitted to the hospital. His high fever, appearance, and especially the blood and mucus in his stool suggested a working diagnosis of dysentery, a colitis syndrome usually caused by a bacterial pathogen. In the United States the most common cause of dysentery is a bacterium of the genus *Shigella*. A. R. was started on a treatment course of trimethoprim-sulfamethoxazole (trimethoprim-sulfa), and an intravenous line was started for rehydration.

A. R.'s worried parents were told that with treatment their son was expected to recover, but they stayed at his bedside during most of his weeklong hospitalization.

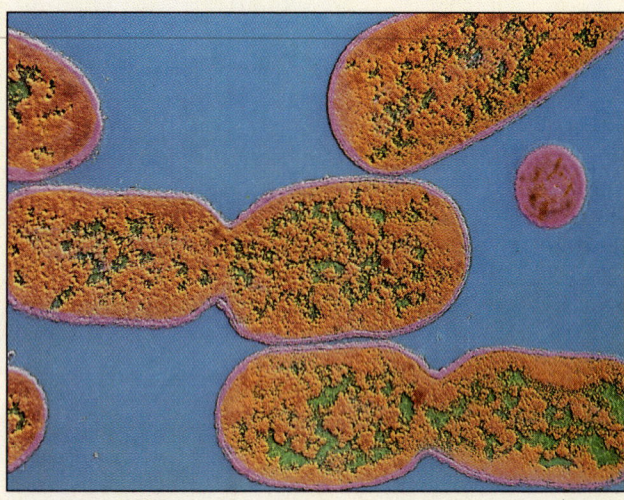

	200 nm

Shigella, the most common cause of dysentery in the United States.

His recovery was gradual. The day after he was admitted, A. R. was more alert and responsive, but he still had a temperature of 103°F and severe diarrhea, consisting of more than 20 small mucus-containing stools that day alone. Twenty-four hours after admission, the clinical microbiology laboratory returned its preliminary report from A. R.'s stool culture—a Gram-negative, nonlactose-fermenting rod was identified as the probable pathogen.

Over the next few days, A. R.'s temperature gradually returned to normal, and the number of stools decreased. He was discharged from the hospital thinner than before his illness, but energetic and otherwise in good health. The final report from the microbiology laboratory identified the stool culture as *Shigella sonnei*, the most common species of *Shigella* in the United States. Sensitivity studies showed this strain to be sensitive to trimethoprim-sulfa. The blood culture grew no microorganisms, indicating that A. R.'s infection was strictly intestinal.

To understand:
- The anatomy and function of the digestive system and its defenses against microorganisms

- The clinical syndromes that characterize digestive system infections

- The bacterial and viral causes of oral cavity and salivary gland infections; their diagnosis, prevention, and treatment

- The bacterial, viral, protozoal, and helminthic causes of intestinal infections; their diagnosis, prevention, and treatment

- The viral and helminthic causes of liver infections and their clinical syndromes; their diagnosis, prevention, and treatment

THE DIGESTIVE SYSTEM

The function of the digestive system—also called the gastrointestinal system (*gaster* is Greek for "stomach")—is to supply the body with essential nutrients. Because the digestive system takes in food from the environment, it is continually exposed to huge numbers of microorganisms.

Structure and Function

The digestive system is composed of two parts (**Figure 23.1**). The first part is a long tubelike tract—about six times longer than a human being—running between the mouth and the anus. It includes the oral cavity (mouth), pharynx (throat), esophagus, stomach, small intestine, and large intestine. The other part consists of six associated structures—teeth, tongue, salivary glands, liver, gallbladder, and pancreas. Together, these organs consume food, digest it, absorb digested nutrients, and eliminate unabsorbable waste.

Part of the digestive process is mechanical. Chewing decreases the size of food particles; and **peristalsis**, the movement that propels food through the intestinal tract, breaks the particles into even smaller pieces. The other part of the digestive process is chemical. Food, which is composed largely of macromolecules, is **hydrolyzed**, or split into smaller molecules, by enzymes. Other compounds also contribute to digestion, including stomach acid and **bile**, a biological detergent that breaks up fatty lumps of food. Cells lining the intestines absorb these small molecules.

An easy way to understand the structure and function of the digestive system is to follow a mouthful of food traveling through it. In the mouth, food is chewed and exposed to digestive enzymes in **saliva**, a secretion of the salivary glands. Then the food is swallowed into the pharynx and passes into the **esophagus**, where peristaltic action transports it to the highly muscular stomach. In the **stomach**, hydrochloric acid, enzymes, and strong mixing waves reduce the food to a liquid called **chyme**. The partially digested food leaves the stomach through a narrow muscular neck at its outlet to enter the **duodenum**, the first part of the small intestine. In the duodenum, nutrients are further digested by powerful enzymes manufactured in the **pancreas** and by bile, produced by the **liver** and stored in the **gallbladder**. Contents of the duodenum pass farther along in the **small intestine** to the regions called the **jejunum** and the **ileum**. Here digestion is completed by enzymes produced within the intestinal lining. Most products of digestion, such as simple sugars and amino acids, are absorbed in the small intestine. Indigestible components of food pass into the **large intestine**, or **colon**, where they move slowly while water and certain vitamins are absorbed. The residual waste is stored in the **rectum** until it is eliminated from the body as **feces**.

The liver performs an astonishing number of chemical reactions necessary for our survival. It produces bile for digestion, **detoxifies** (inactivates) many harmful compounds (including drugs), excretes a toxic yellow pigment called **bilirubin**, and manufactures essential proteins, including the factors that allow blood to clot.

Defenses and Normal Flora: A Brief Review

Like other body surfaces, the lining of the intestinal tract is protected by structural, mechanical, and biochemical defenses (Chapter 14). These include a seamless surface of epithelial cells, the muscular motions of chewing and peristalsis, and highly bactericidal chemicals, such as hydrochloric acid and bile. In addition, secretory IgA protects the intestinal epithelium (Chapter 17).

But constant exposure to the outside environment allows millions of microorganisms to enter the digestive system, and huge populations of commensal microorganisms become established there. Streptococci, anaerobes, and enterobacteria inhabit the mouth and large intestine, creating a complex microbial ecosystem important to good health (Chapter 14).

In contrast, the liver, gallbladder, and pancreas, which are separate from the microbe-laden digestive tract, are normally free of microorganisms. Like other sterile body tissues, these organs are protected by both nonspecific and specific immune defenses (Chapters 16 and 17).

Figure 23.1 Anatomy of the digestive system.

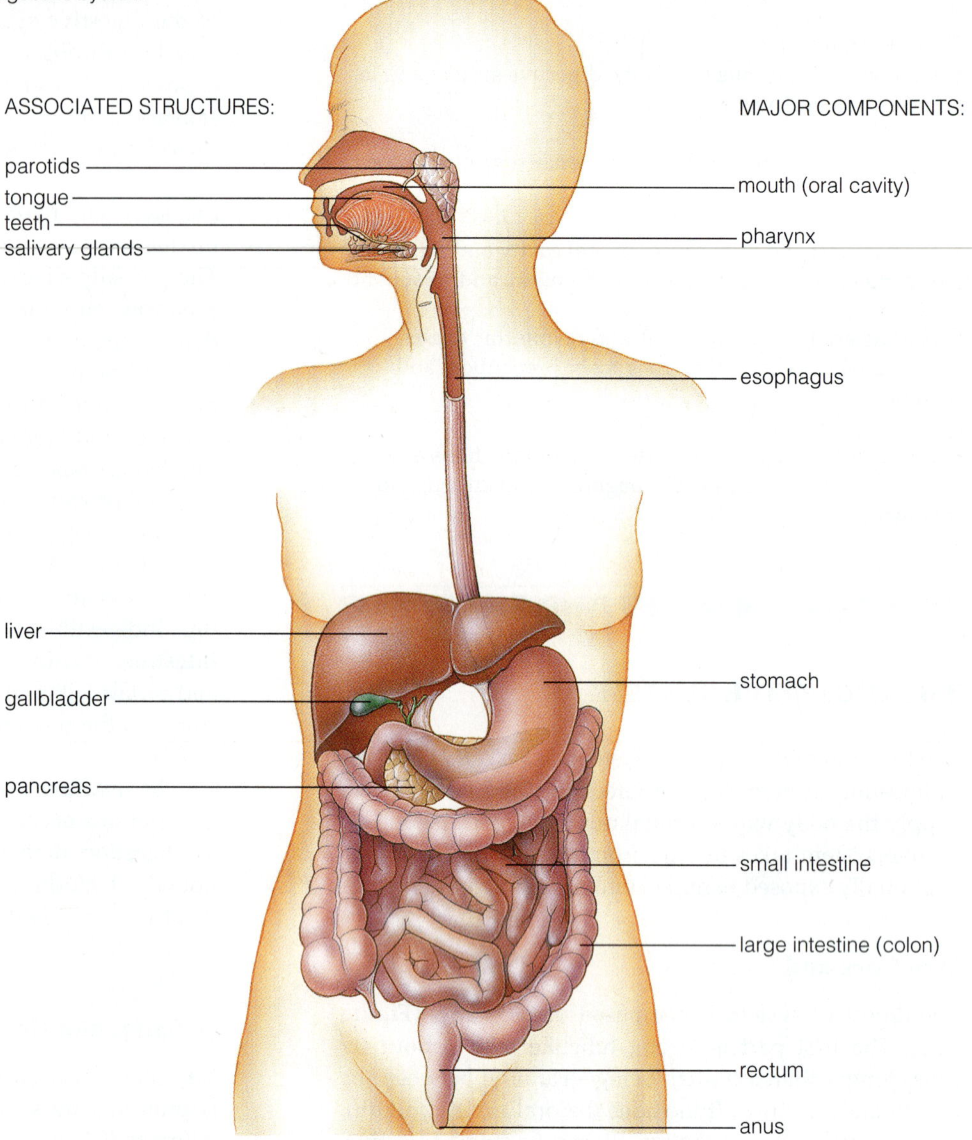

ASSOCIATED STRUCTURES:

MAJOR COMPONENTS:

parotids

tongue

teeth

salivary glands

mouth (oral cavity)

pharynx

esophagus

liver

gallbladder

pancreas

stomach

small intestine

large intestine (colon)

rectum

anus

Clinical Syndromes

Pathogens as well as commensals take advantage of the ready access to the body's interior that the digestive system offers. The mouth is the portal of entry for numerous disease-causing bacteria, viruses, protozoa, and helminths. Some of these organisms penetrate the epithelial barriers of the gastrointestinal system and spread to other organ systems. Pathogens that cause **enteric fevers**, like typhoid, use the gastrointestinal system as a portal of entry but eventually enter the bloodstream and affect the entire body. Most infections, like A. R.'s, remain in the digestive tract, however.

Infections that affect only organs of the digestive tract cause various clinical syndromes (**Table 23.1**). **Gastritis**, infection or inflammation of the stomach, causes pain in the upper abdomen and occasionally bleeding. **Gastroenteritis** (*enteron*, Greek for "intestine") is a poorly defined but extremely common syndrome characterized principally by diarrhea and sometimes nausea, vomiting, and crampy abdominal pain. The syndrome is associated with a variety of viral and bacterial pathogens. Many of the bacteria produce enterotoxins that affect the intestinal tract. Gastroenteritis can also be caused by bacteria that produce toxins outside the body that are then ingested in food. These illnesses, however, are really **intoxications**, or food poisonings, rather than infections (Chapter 15). Although gastroenteritis can be intensely uncomfortable, it is seldom life-threatening to adults.

Colitis is an intestinal syndrome that primarily involves the colon, or large intestine. Because colitis may involve some of the lower small intestine as well, it is also called **enterocolitis** or sometimes **dysentery**. Unlike gastroenteritis, colitis typically involves significant cellular

Syndrome	Region Affected	Signs and Symptoms	Example
Enteric fever	Digestive system is the portal of entry, but entire body becomes infected	Fever, other systemic symptoms	Typhoid fever
Gastritis	Stomach	Upper abdominal pain, occasional bleeding	Associated with *Helicobacter pylori* colonization
Gastroenteritis	Stomach and intestine	Diarrhea, nausea, vomiting, abdominal pain	Salmonellosis, rotavirus diarrhea
Colitis (dysentery, enterocolitis)	Colon	Diarrhea with blood and/or mucus	Shigellosis, amoebiasis
Dental caries	Teeth	Enamel destruction leading to pain and tooth loss	Associated with *Streptococcus mutans* colonization
Periodontal disease	Gums and bone supporting teeth	Loosening and eventual loss of teeth	Associated with *Bacteroides gingivalis* colonization
Parotitis	Parotid (salivary) glands	Swelling and pain below ear and over angle of the jaw	Mumps
Hepatitis	Liver	Jaundice, loss of appetite, dark urine	Hepatitis A

damage to the intestinal wall, so diarrhea in these patients often contains blood and mucus.

Structures associated with the digestive tract also become infected, producing typical clinical syndromes. Microorganisms in the mouth produce acidic metabolic byproducts that erode and destroy the teeth, causing **dental caries** (cavities, or tooth decay). Anaerobic microorganisms that thrive in the crevices around the teeth can cause destruction of gum and bone tissue called **periodontal disease** (*peri*, "around"; *dontal*, "teeth"), or **periodontitis**. Salivary glands may also become infected. The **parotids**, a pair of salivary glands located over the jaw just below the ear, produce a particularly distinctive syndrome of facial swelling when they become infected (called **parotitis**).

Liver damage produces a clinical syndrome called **hepatitis** (*hepar*, Greek for "liver"). Causes include infectious agents, toxic chemicals, physical blockage of the bile drainage system, and even hereditary disease. Patients with hepatitis become *jaundiced* (yellow) because bilirubin builds up in their bodies; and their urine, colored by excessive bilirubin, becomes dark (**Figure 23.2**). Hepatitis patients typically lose their appetite, experience changes in their sense of taste and smell, and, if damage is severe, have uncontrollable bleeding.

INFECTIONS OF THE ORAL CAVITY AND SALIVARY GLANDS

The oral cavity, which normally sustains a dense microflora, nevertheless suffers considerable damage as the

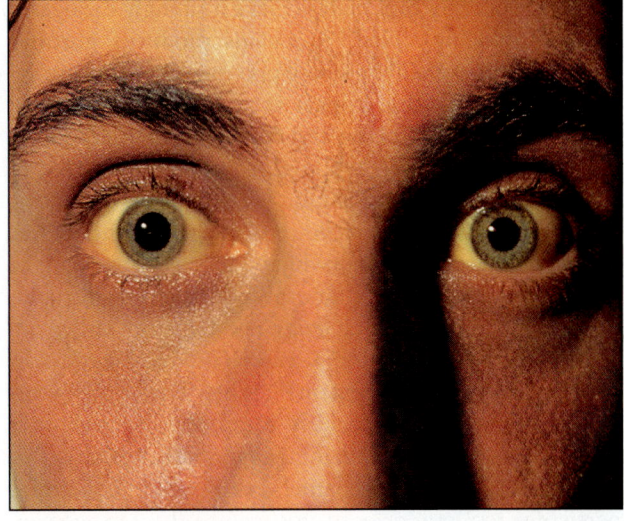

Figure 23.2   Liver damage produces a clinical syndrome called hepatitis in which patients become jaundiced, or yellow, because of the buildup of bilirubin.

result of bacterial infection. Most of us are victims of bacterial tooth or gum destruction at some time in our lives. The most common infection of the salivary glands is mumps.

BACTERIAL CAUSES

Each microbial species in the mouth occupies its own ecological niche. *Streptococcus salivarius* lives on the tongue,

S. mutans and S. sanguis on the teeth, S. mitor on the inner surface of the cheek, and Bacillus melaninogenicus and other anaerobic bacteria in the crevices between teeth and gums. Most are harmless commensals, but some are links in the chain of events that leads to the two most common diseases of the human mouth—dental caries and periodontal disease.

Streptococcus mutans: Dental Caries

Dental caries are an almost universal problem today. Over 6 percent of all current health expenditures in the United States—about $42 billion in 1990—are to restore or replace teeth damaged by dental caries. In contrast, the fossil record shows that tooth decay was almost unknown among early humans. Until recently, caries were even uncommon among people in developing countries.

Dental caries are caused primarily by *Streptococcus mutans*, which possesses adhesins on its pili that allow it to cling firmly to **tooth enamel**, the hard material that covers exposed surfaces of the teeth (see Figure 14.10). Once *S. mutans* colonizes a tooth's surface, it is difficult to dislodge.

Two properties of *S. mutans* make it strongly **cariogenic** (caries-producing). First, *S. mutans* produces **glucans**, high-molecular-weight glucose polymers, by splitting sucrose (common table sugar) into its two component monosaccharides, glucose and fructose. Fructose is used as a growth substrate, while glucose is polymerized to form a glucan mesh. This glucan mesh, together with a huge population of *S. mutans*, other bacteria, and debris, makes up **dental plaque**, which adheres so strongly to the tooth surface that the mouth's normal cleansing solution, saliva, cannot wash it away. Second, as an end product of glucose fermentation, *S. mutans* produces lactic acid, which damages dental enamel. When bacteria exhaust the supply of sucrose, acidity diminishes because it is neutralized by saliva. The minerals in saliva can also repair minor damage to enamel, but if acidic conditions persist for too long or occur too frequently, damage cannot be reversed. If decay completely penetrates an enamel surface, the tooth will die if not protected by a filling.

A person's susceptibility to tooth decay is largely determined by the presence of *S. mutans*. Almost everyone's mouth contains this bacterium, but some people have more and acquire it earlier than others. Studies show that children who acquire *S. mutans* before the age of 2 have eight times as many caries as children who acquire it after age 4. Children usually acquire *S. mutans* from their intimate caretakers, particularly their mothers, so repair of active caries in new mothers protects their infants.

Another factor that affects susceptibility to tooth decay is the consumption of sucrose. Except for sugar cane and sugar beets, most naturally occurring sweet foods owe their sweetness to fructose. A near-absence of sucrose-containing foods in the diets of early humans protected them from tooth decay, and a low-sucrose diet still promotes dental health. The constant presence of even small quantities of sucrose, however—which occurs with the frequent consumption of sucrose-containing snacks—creates a constantly acidic environment that promotes dental caries.

Yet a third factor is the condition of the teeth themselves. Some people are more caries-prone than others because of genetic or early environmental factors. Certain external factors also affect tooth susceptibility to decay, especially the mineral **fluoride**. People who take in adequate amounts of fluoride, either in drinking water or as part of dental care, have fluoride-containing enamel surfaces that are stronger and more caries-resistant. Fluoridation of public water supplies has been found to decrease the incidence of caries by 60 percent on a community-wide basis and is extremely cost-effective. Dental procedures, such as sealing tooth surfaces with caries-resistant polymers, are also effective.

Microbiological research on *S. mutans* may soon lead to the development of a vaccine. Oral administration of immunogenic proteins from *S. mutans* could stimulate the production of IgA salivary antibodies against the bacterium and possibly eliminate it from the mouth. Alternatively, antibodies against specific bacterial components—for example, the enzyme that makes glucan mesh and creates plaque—could eliminate *S. mutans'* ability to form caries. Yet another approach would be to modify the mouth's normal flora by establishing populations of non-caries-producing mutants to compete with the usual acid-producing strains of *S. mutans*.

Bacteroides gingivalis: Periodontal Disease

Periodontal disease is the leading cause of tooth loss in adults. It has been known for centuries—ancient Egyptian and Chinese scholars describe it in detail—and until recently was considered an unavoidable problem of aging. However, microbiological research suggests that periodontal disease is caused by bacterial infection that can be prevented and treated.

Unlike caries, which damage the teeth themselves, periodontal disease damages the tissues that surround and support the teeth. The **periodontum**, or tissues that support the teeth, include three structures—the **tooth root**, which is covered by cementum rather than enamel, the **periodontal ligament**, which consists of collagen fibers that connect tooth to bone, and the **bone socket** in which the tooth is embedded. The **gingiva**, or gums, which cover the tooth-supporting bone and come in contact with the base of the teeth, are also intimately involved in periodontal disease.

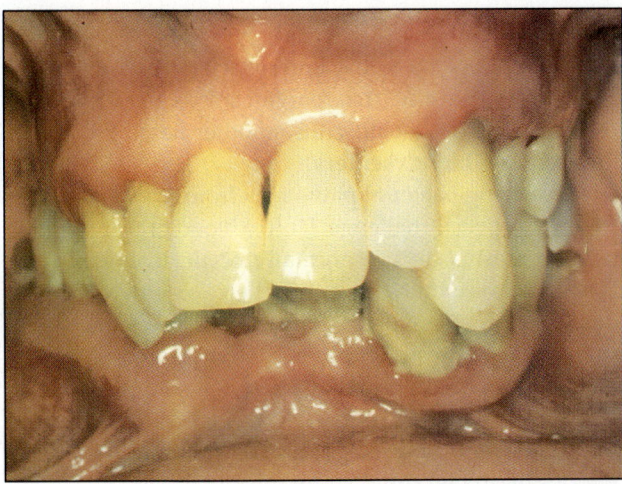

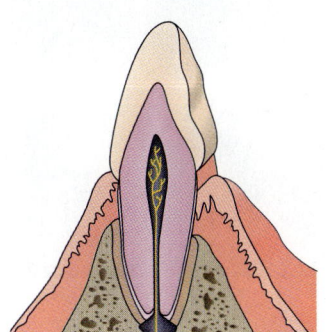

Figure 23.3 This person with advanced periodontal disease has lost most of the gum and bone tissue that support the teeth. The teeth with exposed roots are probably already loose. The accompanying sketch shows the loss of bone and gum tissue around the tooth's root.

Most periodontal infections affect the gingiva. The process begins painlessly and insidiously enough when plaque is allowed to accumulate at the gum margin. Inflammation of the gums, called **gingivitis**, may cause redness, swelling, and easy bleeding. If the infection proceeds to the periodontal structures, it is called **periodontitis**. Deep pockets of damaged tissue form at the gum line. Here, anaerobic bacteria multiply, causing further inflammation that eventually damages the periodontal ligament and surrounding bone. As the process advances, otherwise healthy teeth become loose and eventually fall out (**Figure 23.3**).

Dental microbiologists have long known that bacterial multiplication was associated with periodontal disease, but it was extremely difficult to determine which bacterial species, if any, were *uniquely* responsible for these infections. They speculated that the sheer mass of dental plaque and oral bacteria caused periodontal disease. But further research showed the microflora surrounding healthy teeth (usually streptococci and actinomyces) differed significantly from the microflora in areas of periodontal destruction. This allowed researchers to identify the handful of bacterial species—out of the more than 300 found in the mouth—that cause periodontal disease. The primary causative agent is the anaerobe *Bacteroides gingivalis*, although *B. intermedius*, other Gramnegative anaerobes, and spirochetes probably play a role.

The best way to prevent or slow the progress of periodontal disease is to remove plaque from gum margins and periodontal pockets with thorough brushing and flossing. A growing understanding of the microbiology of periodontal disease may soon provide new approaches to diagnosis and treatment. Tests based on enzymelinked or immunofluorescent antibodies or DNA probes may allow early detection of *B. gingivalis* and other pathogens in periodontal tissues. There is also considerable research into using antibiotics to control periodontal infections. In one clinical trial, placing synthetic fibers filled with tetracycline into the periodontal pockets greatly reduced bacterial populations and improved gingivitis.

VIRAL CAUSES

Many viruses that cause skin rashes, called **exanthems**, also cause redness or blisters inside the mouth, called **enanthems**. Measles is one example. **Stomatitis** (*stoma*, "mouth"; *itis*, "inflammation") is a painful inflammation of the mouth usually characterized by blisters or ulcers. The herpes simplex virus is a common cause of stomatitis in children. Because measles virus and herpes simplex virus primarily cause skin infections, however, we discuss them in Chapter 26. Here, we discuss the most common viral infection of the salivary glands, mumps.

Mumps Virus: Mumps

The **mumps virus** infects only humans. It is a member of the paramyxovirus family, like the measles, parainfluenza, and respiratory syncytial viruses. Twenty years ago, nearly every child in the United States was infected with this virus before age 15. Today, routine pediatric immunization makes mumps unusual.

The mumps virus, carried by the saliva or respiratory secretions of an infected person, enters a new host through the respiratory system. The virus replicates in the upper respiratory epithelium and probably spreads directly to the salivary glands. Twenty-five percent of infections with the mumps virus cause no symptoms at all. Typical clinical evidence of mumps is parotitis, swelling of one or both of the parotid glands below the ear and over the angle of the jaw (**Figure 23.4**), along with mild pain and sometimes fever. Salivary glands beneath the tongue or the jaw may also be inflamed, even if the parotids are unaffected.

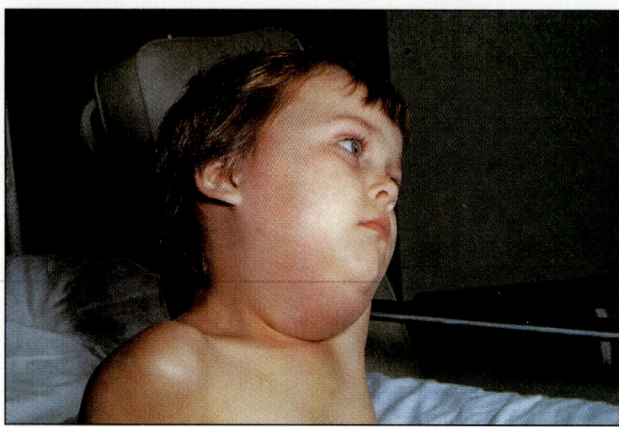

Figure 23.4 This youngster with mumps shows typical swelling below her ear and over the margin of her jaw where the parotid salivary gland lies.

After a few days of local multiplication, the mumps virus enters the bloodstream and occasionally spreads to tissues throughout the body, causing serious complications. In about 10 percent of mumps patients, the virus affects the central nervous system. Most of these patients develop only a severe headache, but a few experience a temporary loss of balance, and a very few suffer permanent brain damage. In adult males the testes may be infected, producing a painful testicular inflammation called **orchitis**, which, contrary to popular belief, almost never results in sterility. The pancreas and thyroid are sometimes affected and, in rare cases, mumps destroys the inner ear, causing deafness.

Mumps vaccine, a live attenuated strain of mumps virus, is part of the MMR (measles, mumps, rubella) vaccine given at age 15 months. Since the MMR vaccine was released in 1968, the incidence of mumps in the United States has decreased 97 percent.

INFECTIONS OF THE INTESTINAL TRACT

Infections of the intestinal tract range from mild one-day food poisoning to deadly cholera and typhoid. Intestinal infections can be bacterial, viral, protozoal, or helminthic, but in nearly all cases, transmission is by the fecal-oral route.

BACTERIAL CAUSES

Bacteria cause most lethal infections of the intestinal tract. With improved sanitation and clean drinking water, many have become uncommon in developed countries, and others are treatable with antimicrobial agents.

Shigella spp.: Shigellosis

Shigella spp. belong to the Enterobacteriaceae, which include many intestinal microorganisms. They are Gramnegative rods, or bacilli. Thus, another name for A. R.'s illness is **bacillary dysentery**.

Identifying bacterial pathogens from stool specimens presents a special challenge. Unlike blood or urine, where microorganisms do not normally occur, intestinal samples always contain large numbers of enterobacteria. Identification, therefore, is usually based on the organism's ability to grow on differential media or on its distinctive biochemical characteristics. For example, because the bacteria cultured from A. R.'s stool did not ferment lactose, did not produce gas from carbohydrate, did not produce hydrogen sulfide from thiosulfate, and were nonmotile, they were identified as members of the genus *Shigella*. O antigens on the *Shigella* cell surface divide the genus into four species—*S. dysenteriae, S. flexneri, S. boydii,* and *S. sonnei*.

Shigella spp. can be highly virulent, as in A. R.'s case. The mechanisms of its pathogenesis are adherence, invasiveness, and toxin production (Chapter 15). Like most Enterobacteriaceae, shigellae have pili with adhesin proteins that bind to carbohydrate receptors on cells of the human colon. When a shigella bacterium binds to a receptor on the host cell, the cell engulfs the bacterium, even though colon epithelial cells are not normally capable of phagocytosis. The bacterium enters in a vacuole, which it lyses to enter the cytoplasm (**Figure 23.5**). Free in the cell, it multiplies rapidly and inhibits host protein synthesis. Approximately six hours after invasion, the host cell lyses, releasing many bacterial cells to infect neighboring epithelial cells. This pattern of invasion and multiplication produces characteristic patchy areas of destruction and inflammation, called **microabscesses**, in the colon of a shigellosis patient. Shigellae do not leave the intestinal wall, which explains why blood samples of dysentery patients like A. R. are usually free of bacteria.

Most *Shigella* spp. produce a potent toxin. The toxin produced by *S. dysenteriae* is **shiga toxin**, and similar molecules produced by other species are called **shiga-like toxins**. Shiga toxin is a two-subunit protein toxin (Chapter 15). Strong evidence indicates that this toxin contributes to the development of disease, including damage to blood vessels in the intestinal wall and intense inflammation of the intestine. Bleeding and inflammation cause the stools to be streaked with blood and contain strings of mucus composed of many polymorphonuclear leukocytes. Shiga toxin may also contribute to the watery diarrhea sometimes seen in these infections. Though the bacteria themselves do not disseminate, shiga toxin probably does affect other organs. The toxin's effect on the nervous system probably contributes to the convulsions, like A. R.'s, that are a common feature of shigellosis in children.

Transmission and Epidemiology. *Shigella*, which lives only in the intestinal tract of human beings who are suffering or recovering from bacillary dysentery, is transmitted through the fecal-oral route. Flies, fingers, and food are common fecal-oral vehicles, and shigellae can survive for a considerable period in contaminated water or on fomites. People who live in overcrowded conditions where cleanliness is difficult are particularly likely to contract *Shigella* infections. A. R.'s infection probably could have been prevented by the most basic form of personal hygiene—good hand washing.

Children are affected by shigellosis far more often than adults. About half the reported cases occur in children under the age of 5. For one thing, children are too young to follow good hygiene habits; but they may also be intrinsically more susceptible to *Shigella* infection. For unknown reasons, *Shigella* infections usually occur in the late summer and early fall.

Many fecal-oral infections, including cholera and typhoid fever, have been nearly eradicated in industrialized countries because of good sanitation, but not *Shigella* dysentery. In the United States, thousands of cases of bacillary dysentery are reported annually to the Centers for Disease Control and Prevention (CDC) in Atlanta. Shigellosis is difficult to eradicate partly because it is so highly communicable. To contract typhoid fever or cholera a person must ingest thousands or millions of bacteria, but as few as 200 *Shigella* organisms can cause disease. In homes with an infected family member, large numbers of live bacteria can be recovered from the floor around the toilet, so even a minor lapse in hand washing can result in a new infection. Before the causative bacterium was identified, shigellosis was known as *asylum dysentery* because of the mass outbreaks in mental institutions. Today, it spreads rapidly through day-care centers. Unless medical microbiologists perfect an oral vaccine, which is under development, *Shigella* infections may never be entirely preventable.

Clinical Syndrome and Treatment. A. R. manifested the classical symptoms of **shigellosis**—fever, severe illness (sometimes called **toxicity**), diarrhea, and colitis, as indicated by blood and mucus in the stools. However, many victims, particularly adults in the United States, have only a watery diarrhea with little fever or toxicity. Because the symptoms are so variable, shigellosis can be diagnosed with certainty only by culturing the organism from an infected person's stool or a rectal swab.

Shigellosis is usually a self-limited disease, even in children, but as A. R.'s case showed, it can be life-threatening. Probably the greatest threat to A. R.'s survival was **dehydration**, depletion of the body's water reserves when more water is lost (through diarrhea, in this case)

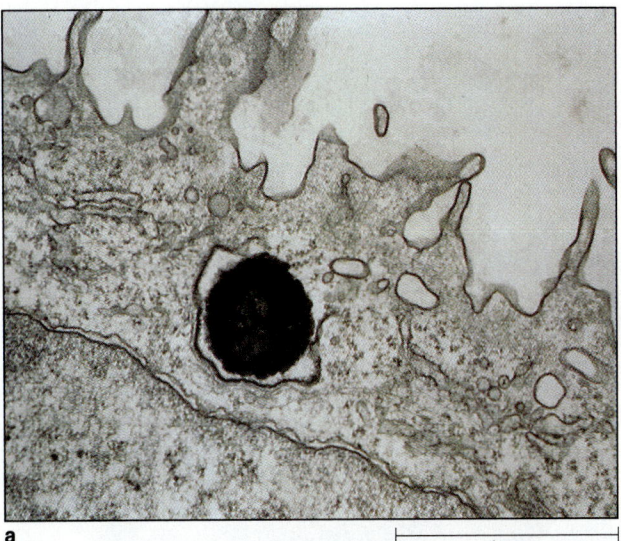

a

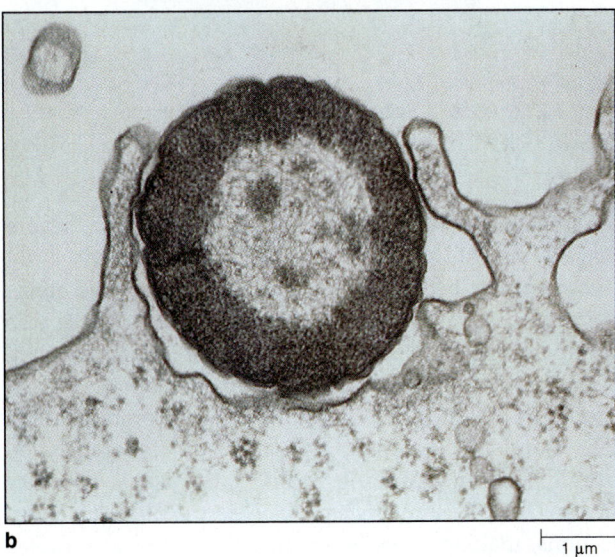

b

Figure 23.5 (a) *Shigella flexneri* inducing entry into an epithelial cell; note the filopodial structures that engulf the bacterial body. (b) *S. flexneri* escaping into the cell's cytoplasm by inducing lysis of the host-cell phagocytic vacuole.

than is taken in (through food and drink). If dehydration causes more than a 15-percent loss in body weight, it is life-threatening. A decreased volume of fluid in the circulatory system lowers blood pressure, limits the blood supply to the body's vital organs, causes shock and, ultimately, death. When A. R. was admitted to the hospital, the physician estimated that dehydration had decreased his body weight by 5 to 10 percent. If his dehydration had worsened, A. R. could have died. Children suffering from malnutrition are at especially high risk because immune defenses are compromised by prolonged hunger.

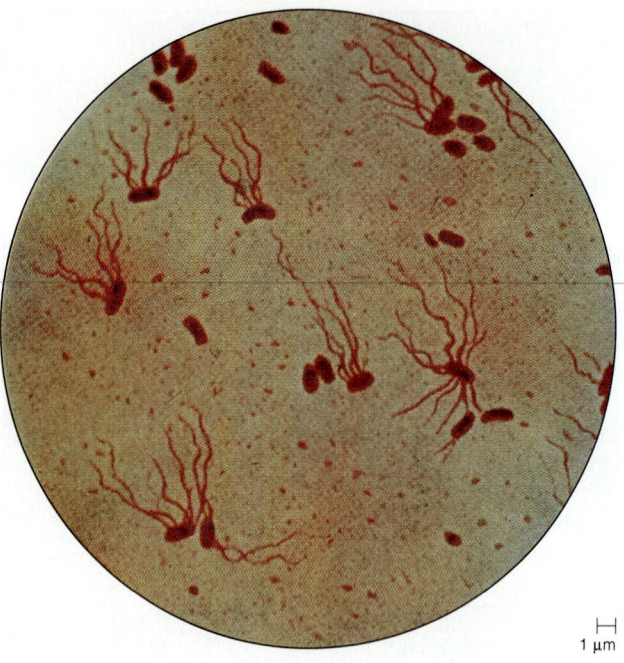

Figure 23.6 One of *Salmonella typhi*'s identifying properties is motility. These cells are stained to show their flagella.

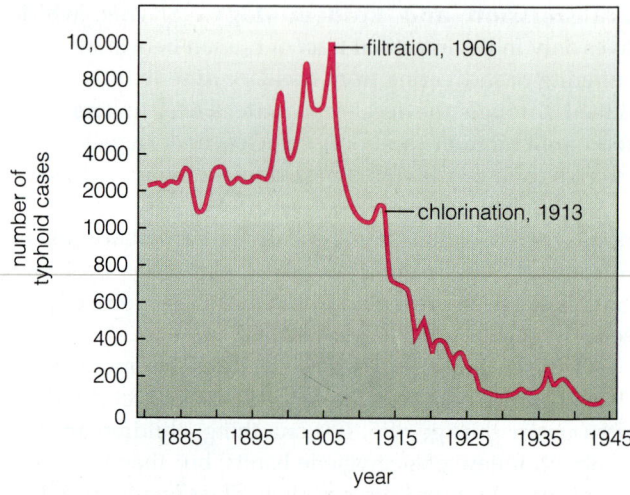

Figure 23.7 The number of typhoid cases in the city of Philadelphia decreased after public water supplies began to be filtered and, later, chlorinated. After 1945, the graph would be a flat line near zero.

In addition to fluid therapy for dehydration, severe *Shigella* infections require antimicrobial treatment. Drug treatment has been complicated, however, by the emergence of drug-resistant strains. At one time, *Shigella* was routinely treated with sulfa drugs. Now, sulfas alone are almost useless; trimethoprim-sulfamethoxazole is often used, as with A. R., but resistance to this combination is also emerging. Nalidixic acid and quinolones may be useful when the infecting strain is resistant to all other drugs. Drug resistance is a problem in treating *Shigella* infections, making it necessary to test all clinical samples for susceptibility to antimicrobial agents.

Salmonella spp.: Typhoid Fever, Salmonellosis

Salmonella spp., like *Shigella* spp., are Gram-negative rods that belong to the Enterobacteriaceae. Unlike *Shigella*, however, *Salmonella* infects many animals as well as humans, and some are capable of invading tissues outside the intestinal tract. Members of the genus *Salmonella* are identified in the laboratory by their distinctive biochemical properties—they do not ferment lactose, they produce hydrogen sulfide from thiosulfate and visible gas when they ferment sugars, and most are motile (**Figure 23.6**).

Salmonella is a large genus, which has more than 2000 distinct strains that can be distinguished serologically on the basis of two major surface antigens, the O (lipopolysaccharide) and H (flagellar) antigens. Many of these serotypes, or **serovars**, were given species names when

they were isolated. For example, *S. dublin* was named for the place where it was isolated and *S. typhimurium* was named for the typhoidlike disease it causes in mice. Taxonomists have grouped these serovars into a few species—*S. typhi*, *S. choleraesuis*, and *S. enteritidis*—but species names of serovars continue to be used in the medical literature. A more recent scheme divides the serovars among five subgroups based on reservoir and potential for pathogenicity. Serovars that cause almost all human disease belong to subgroup 1. The other four subgroups include organisms found primarily in the environment or in cold-blooded animals.

Serologic classification is not essential for clinical identification, but it is useful epidemiologically. People with *Salmonella* organisms belonging to the same serotype are likely to have been infected from the same source. Knowing the serotype often allows field epidemiologists to track down the source of an outbreak (Chapter 20). About 95 percent of reported cases are caused by about 40 of the 2000 serotypes.

Salmonella typhi. The species *Salmonella typhi* causes *typhoid fever*, a potentially fatal enteric fever. These organisms are unusual among *Salmonella* spp. because they infect humans only. Virulence factors of *S. typhi* include invasiveness and an endotoxin that may account for the high fevers typical of typhoid.

Feces from people infected by *S. typhi* often contaminate food or water, which become vehicles for infecting new victims. Direct person-to-person transmission rarely occurs. Thus, when communities started treating public

Typhoid Mary

In 1900, Mary Mallon, an Irish immigrant, was working as a cook for a wealthy Mamaroneck, New York, family. A young visitor to the home contracted typhoid fever, and shortly after, Mary left to take a job with a family in New York City. Just before Christmas 1901, the New York City family's laundress died of typhoid; and after her death, Mary took another position. In Dark Harbor, Maine, Mary was the cook for a lawyer's family. Within two weeks of her arrival, seven of the eight people in the household became ill with typhoid fever. Mary stayed to nurse them, and the lawyer rewarded her with an extra $50 before she left. There was also a waterborne outbreak of typhoid in Ithaca, New York, and another cluster of cases among servants in a house in Sands Point, Long Island, where Mary was working in 1904. Finally, in 1906, an outbreak of six cases of typhoid in Oyster Bay, New York, prompted an epidemiologic investigation.

When the health authorities in Oyster Bay could not find the source of typhoid fever, they contacted George Soper, a sanitary engineer from the New York City Health Department who was known for his talents as an investigative epidemiologist. Dr. Soper checked all the likely sources of typhoid—from the family's milk, water, well, and cesspool to clams from the local bay—with no better success than the local authorities. Eventually, his investigation led him to Mary Mallon. Six months later he traced her back to New York City, where she was cooking for a family on Park Avenue under an assumed name. Her employer's daughter had recently died of typhoid. When Soper presented Mary with evidence of her past association with typhoid, she attacked him with a cleaver and ran to a nearby outhouse to hide. Police finally took her—kicking, screaming, and biting—to Riverside Hospital for Communicable Disease on North Brother Island in the East River. Microbiologic examination of her stools confirmed that Mary was indeed a typhoid carrier. Her story created a sensation in the local press and among wealthy New Yorkers who worried that their Irish help might be exposing them to disease and death.

At Riverside Hospital, Dr. Soper explained to Mary that she could probably be cured by having her gallbladder removed, but she refused surgery. She also refused to stop working as a cook, which was not only her sole means of support but also a great source of enjoyment. As a result, Mary was kept at Riverside as a prisoner-patient for three years. She was trained as a hospital laundress and released in 1910 on the promise that she would not return to cooking and would report regularly to the health department. But Mary disappeared and returned to working as a cook under the name Mrs. Brown. A typhoid epidemic—25 infections and 2 deaths—at the Sloane Hospital for Women led authorities to Mary three years later.

Mary was returned to Riverside, where she was quarantined for the remainder of her life—25 years. She worked as a laboratory technician and lived in her own cottage near the hospital until paralyzed by a stroke in 1932. Mary Mallon died in November 1938. In all, 1300 cases of typhoid were linked to her.

water supplies to eliminate pathogens, typhoid declined dramatically in developed countries (**Figure 23.7**). In the United States only a few hundred cases of typhoid are now reported annually, and most of these are acquired by people traveling overseas.

Symptoms of typhoid fever typically begin a week or two after infection, when bacterial cells enter the bloodstream. The most dramatic clinical manifestation is a high fever (usually over 104°F) that continues for days or weeks. Typhoid patients are tired, confused, and lose their appetite. They suffer from headache and other aches and pains. Some develop **rose spots**, a characteristic faint rash. Late in the illness, bacteria that have infected the liver are excreted in the bile, reenter the gastrointestinal tract, and are shed in the feces.

Most people with typhoid fever are seriously ill for about a month. Although the majority recover, as many as 10 percent die from complications such as intestinal bleeding or perforation (rupture) of the intestinal tract. Antibiotic treatment shortens the illness and improves the chances of survival; but like other enteric microorganisms, many *S. typhi* strains are resistant to several antimicrobial drugs. In spite of its toxicity, chloramphenicol is currently the drug of choice (Chapter 21).

Most typhoid fever patients stop excreting *S. typhi* in their feces when they recover, but about 3 percent develop a chronic gallbladder infection that makes them persistent carriers. They may excrete huge numbers of deadly *S. typhi* bacteria in their feces for years. Typhoid carriers must register with the local public health department and report at intervals to have their stool tested for the continued presence of *S. typhi*. As long as they remain carriers they must inform the public health department of their whereabouts and may not work at certain occupations, including any involving food or children. This infringement of personal liberty is considered justified in order to prevent epidemics of typhoid fever (see the box "Typhoid Mary"). Usually, though, typhoid carriers stop excreting bacteria if they are treated with antibiotics and have their gallbladder removed.

Table 23.2 Food Poisoning by Bacteria		
Organism	Infection or Intoxication	Foods
Salmonella spp. (*Salmonella enteritidis*)	Infection—bacteria multiply in the small intestine	Meat, poultry, eggs
Staphylococcus aureus	Intoxication—preformed toxin ingested in food	Whipped cream, eggs, mayonnaise, meat (especially ham)
Bacillus cereus	Intoxication—preformed toxin ingested in food	Rice, meat, vegetables
Clostridium perfringens	Intoxication—bacteria are ingested in food and toxin released when spores form in the intestine	Meat
Clostridium botulinum (Chapter 25)	Intoxication—preformed toxin ingested in food	Nonacidic canned foods, especially green beans
Vibrio parahaemolyticus	Infection—bacteria multiply in the stomach	Fish

Salmonellosis. Though usually classified as a food poisoning, **salmonellosis** is a true infection—caused by bacteria multiplying in the bowel (**Table 23.2**). It is caused by many serovars of *Salmonella* traditionally grouped as *S. enteritidis* or *S. choleraesuis*.

Infection begins when a person ingests large numbers—usually millions—of *Salmonella* organisms. Bacteria invade the epithelium of the small intestine and multiply there; but unlike typhoid bacilli, they usually do not spread further. About a day after consuming contaminated food, gastrointestinal symptoms begin. Almost all salmonellosis victims have diarrhea, possibly caused by a *Salmonella* enterotoxin, and many also have abdominal cramps, fever, nausea, and vomiting. Most healthy adults recover without treatment within a few days. Complications usually occur only in very young, very old, or immunosuppressed hosts, although otherwise healthy people may die if enough organisms are ingested. In these few cases, the salmonellosis develops into severe dehydration or a systemic bloodborne infection.

Salmonellosis is a major public health concern. About 40,000 of the most serious cases are reported yearly to public health departments, but millions of cases doubtless go unreported. Most of us have had salmonellosis at one time or another. Though few people die, many workdays are lost and the cumulative burden of human suffering is immense.

Salmonellosis is widespread because we can become infected almost any time we eat. Many *Salmonella* serovars infect domestic livestock. Meat sold in supermarkets, unpasteurized milk, and even Grade A shell eggs often harbor the pathogen. Of the 4 billion chickens eaten in the United States each year, at least 1.4 billion are contaminated with *Salmonella* (**Figure 23.8**). Thorough

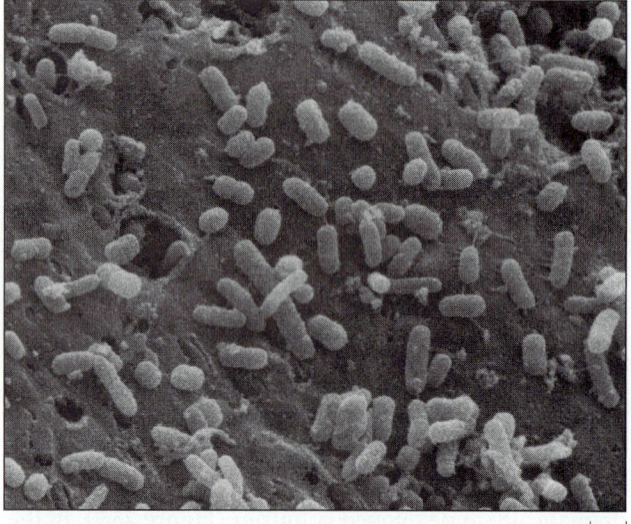

├──────┤
2 μm

Figure 23.8 A scanning electron micrograph shows chicken meat completely covered with *Salmonella* bacteria.

cooking to 145°F kills the bacteria, but poultry in particular may not be cooked long enough to kill bacteria inside the body cavity. Cutting boards or utensils can spread bacteria from uncooked meat to other foods. Salmonellosis can be prevented by using a nonporous cutting board that can be disinfected, by thoroughly cooking all foods containing meat or eggs, and by washing hands frequently.

Salmonella gastroenteritis is usually a self-limited illness—most people never consult a doctor and don't need to. In fact, antibiotic treatment of uncomplicated salmonellosis is **contraindicated** (medically inadvisable). Treat-

Clinical Notes

Salmonellosis Outbreak

In July 1989, about nine hours after they attended a baby shower in New Jersey, 21 of the 24 guests came down with severe diarrhea, vomiting, fever, and abdominal cramps. Twenty of the victims sought medical care, and 18 were hospitalized, among them a woman who delivered her baby shortly afterward. Although the baby was critically ill and required prolonged hospitalization, it survived. *Salmonella enteritidis* was isolated from the stools of all 21 patients and from the baby's bloodstream.

The epidemiologists who set out to find the source of contamination began by interviewing the shower guests.

They discovered that all 21 who became ill—but none who didn't—had eaten a homemade ziti pasta dish. One of the ingredients was an egg. The ziti was baked for one hour and served right from the oven, but those who ate it remembered that the center was cold. *S. enteritidis* was isolated from samples of the leftover ziti and from all seven eggs left in the carton. Chickens from the egg producer that supplied the supermarket were also found to be infected.

---

ing mildly ill patients with antibiotics may cause them to become chronic carriers of *Salmonella*. But in serious illnesses, particularly when bacteria enter the bloodstream, antimicrobial treatment may be essential for recovery. Unfortunately, there is growing drug resistance among *Salmonella* strains. More than one-fourth of the *Salmonella* strains isolated from hospitalized patients are resistant to two or more antimicrobial agents. The increasingly serious drug resistance in enteric pathogens may be partly due to the widespread use of antibiotics in animal feed (Chapter 21).

Escherichia coli: Diarrhea, Dysentery

Escherichia coli, a Gram-negative rod, is the most abundant facultative anaerobe in the large intestine of humans. Most strains are harmless commensals that help protect us against infection by competing with enteric pathogens such as *Shigella* and *Salmonella* (Chapter 14), but some *E. coli* strains are virulent pathogens. They are a major cause of infant diarrhea in developing countries, killing untold numbers of children, and they are a leading cause of **traveler's diarrhea**, which affects most Westerners traveling in the developing world. Food- and waterborne outbreaks of *E. coli* diarrhea have also occurred in the United States. *E. coli* is also an important pathogen of the urinary tract (Chapter 24).

Pathogenic strains of *E. coli* have virulence factors that nonpathogenic strains lack. They can adhere to human cells, invade human tissue, and produce toxins. Depending upon its particular constellation of virulence factors, a pathogenic strain that infects the intestinal tract is classified as enterotoxigenic, enteroinvasive, or enteropathogenic.

Enterotoxigenic strains of *E. coli* are the primary cause of traveler's diarrhea and infant diarrhea in developing countries. These strains have adhesins that bind to intestinal epithelium and contain plasmids encoding enterotoxins that cause diarrhea (Chapter 15). Two *E. coli* enterotoxins are known. One, termed **heat labile**, is destroyed by heat; the other, termed **heat stable**, is not. The heat-labile toxin is similar to the toxin that causes cholera, although it is not as powerful.

Infants in countries with poor sanitation are susceptible to enterotoxigenic *E. coli* because they have not yet developed immunity, and travelers are prone to infection for the same reason. During the Gulf War in 1990, some American units had rates of diarrhea as high as 10 percent per week—more than 10 times higher than rates of heat-related illness. As with all types of diarrhea, patients who are significantly dehydrated, usually children, should receive fluid replacement. Prevention and treatment of traveler's diarrhea are controversial. Most authorities recommend against medication to attempt to prevent infection; but treatment, either with bismuth salicylate or an antimicrobial agent such as trimethoprim-sulfamethoxazole or ciprofloxacin, should be started as soon as diarrhea is noted. With early treatment, illness that would otherwise last three to five days may be over within hours.

Enteroinvasive strains of *E. coli* cause a dysentery syndrome almost identical to shigellosis. In addition to adhesins and toxins, enteroinvasive *E. coli* produce plasmid-encoded proteins that allow the bacteria to invade human cells. *E. coli* toxin resembles *Shigella* toxin. It inhibits protein synthesis and eventually kills cells in the intestinal lining, which leads to dysentery and colitis. Enteroinvasive *E. coli* are not as virulent as *Shigella*, and many more bacterial cells are required to initiate an infection.

Enteropathogenic strains of *E. coli* cause diarrhea in newborn infants, sometimes causing epidemics in hospital nurseries. This type of pathogenic *E. coli* is less well understood than the other two, but probably produces a toxin that destroys the microvilli in the intestinal epithelium (tiny projections that increase the surface available to absorb nutrients). The adhesins of enteropathogenic *E. coli* are part of the outer membrane, not pilus proteins like most adhesins.

Research on *E. coli* suggests that we may be able to prevent *E. coli* diarrhea with vaccines that stimulate immunity to *E. coli*'s adhesion factors or toxoids that protect against *E. coli* toxins.

Vibrio cholerae: Cholera

The bacterium *Vibrio cholerae* causes Asiatic **cholera**, one of the world's most dreaded diseases. During the 1800s a series of cholera pandemics started in India and swept across the globe. The panic in Europe was as great as that during epidemics of black plague in the Middle Ages. People were finally frightened enough to adopt modern systems of sewage disposal and public sanitation (Chapter 20). In 1947 an epidemic in Egypt of 33,000 cases left 20,000 people dead. In endemic regions of India and Bangladesh, cholera still claims many victims annually, most of them children. Recently cholera has become endemic in the Western Hemisphere.

Humans are a major reservoir for *V. cholerae*. Bacteria from the feces of an infected person usually enter the water supply, multiply there, and infect people who drink the contaminated water. As a result, most cases of cholera occur where public sanitation is inadequate. But infected shellfish are also a significant reservoir. The few cases of cholera that have been reported in the United States in recent years were caused by inadequately cooked shellfish caught in estuaries that contained *V. cholerae*.

V. cholerae is a short, curved, Gram-negative rod that moves by means of a single polar flagellum (see Figure 11.6a). *V. cholerae* is subdivided into six different serotypes based on their O antigens. Two **biotypes** (strains) of the O1 group—cholerae and El Tor—cause classical cholera. The most significant characteristic of virulent strains is their ability to produce the cholera exotoxin.

The cholera exotoxin, which causes all the symptoms of cholera, is composed of two peptides. One peptide binds to the surface of the host cell and facilitates entry of the second peptide, which raises the intracellular level of the metabolic messenger cyclic AMP (Chapter 5). The increased level of cAMP stimulates epithelial cells to secrete large quantities of chloride into the intestine, causing water, sodium, and other electrolytes to follow and leave the body as diarrhea. The loss of water can reach astounding proportions—as much as a liter an hour. Diarrhea can be so profuse that the stools lose the appearance of feces and resemble water flecked with small particles of mucus. These are the characteristic **rice water stools** of cholera.

Pathophysiology and Treatment. The pathophysiology (disturbances of body function) of cholera is due to dehydration. Although dehydration for any reason results in the clinical syndrome described in the box "The Medical Advance of the Century," the dehydration of cholera is sudden and dramatic. A person who loses body water at the rate of 1 liter—2.2 pounds—an hour has no hope of surviving until the end of the day. This rate of fluid loss explains sudden death from cholera. A person who is well in the morning can become a shriveled, dehydrated corpse by evening. Because they have a smaller reserve of body water, children are even more vulnerable to death from dehydration. Loss of as little as 2 liters can be lethal for a 1-year-old who has only about 6 liters of total body water. The death rate from cholera is 60 percent without treatment.

Once the infection begins, the only effective treatment is replacing lost fluid. Traditionally, water and electrolytes were administered intravenously. But in most developing countries this kind of treatment, which requires sterile supplies and trained personnel, is impossible to offer all cholera victims. Research in the 1960s established that oral rehydration—a much cheaper and easier form of therapy—could also save cholera victims.

The fluids used for oral rehydration, called oral rehydration solution (ORS), contain prescribed concentrations of glucose and sodium so they are absorbed by the large intestine more rapidly than enterotoxins cause fluid to be secreted by the small intestine. Oral rehydration therapy (ORT) can drop the death rate from cholera to under 1 percent. Cholera patients should also receive an antibiotic, preferably tetracycline, to decrease the number of pathogens multiplying in the intestine. Having fewer pathogens means reduced production of new cholera toxin, shortening the illness and decreasing the likelihood of infecting others. Prevention, which is preferable to the best treatment, can be achieved by adequate sanitation and clean drinking water. Until that is possible, preventing cholera depends on developing a cheap and effective vaccine. The cholera vaccine currently available must be administered by injection and is not very effective.

Recent Epidemiology. Cholera has always been endemic in many areas of India and Asia. Pandemics begun in these regions would spread as far as the Americas, but all of these epidemics would wane within a few years and leave areas outside Asia completely cholera-free between times. In 1961, however, cholera epidemiology

The Medical Advance of the Century

Human beings are approximately 60 percent water, so the body of a person weighing 70 kilograms (154 pounds) contains about 42 liters, or 93 pounds, of water. Most of this water is inside cells and the spaces between them. Only about 3.5 liters circulate in the bloodstream. In a day, the average person takes in about 3 liters of fluid to replace the amount lost in urine, sweat, feces, and respiration. If more water is lost from the body than can be replaced, problems soon arise. The 70-kilogram person can lose 3 or 4 liters of water and suffer only from thirst and symptoms of mild dehydration—decrease in urine volume, dry mouth, and perhaps some lightheadedness on standing. When more than 8 or 9 liters of fluid are lost, however, life-threatening symptoms develop. Not enough fluid remains to maintain an adequate volume of circulating blood, and blood pressure falls. As a result, vital organs do not receive the oxygen and other nutrients they need. The heart beats faster in an attempt to compensate for falling blood pressure. If fluid losses exceed about 12 liters, the person goes into shock (the circulatory system shuts down). Death is the inevitable result.

Dehydration is a risk in all diarrheal illnesses, and the essential treatment is rehydration. It was, therefore, a medical breakthrough to find that oral rehydration therapy (ORT) could effectively replace traditional intravenous treatment for even the most serious cholera cases. The British medical journal *Lancet* announced that ORT was "potentially the most important medical advance in this century." Oral rehydration has saved not only the lives of many cholera victims, but also the lives of children dehydrated by other diarrheal illnesses. Making oral rehydration solutions available is one of the highest public health priorities of the World Health Organization (WHO); but in spite of all efforts, diarrheal dehydration still kills about 3 million children each year.

changed. A series of epidemics began that were caused by a genetically new biotype of *V. cholerae* called the **El Tor biotype**, or *V. cholerae* O1. This strain spread through the Middle East, reached Africa in 1970 (where it invaded 29 countries), and did not go away. Cases are still occurring in these countries.

In January 1991, cholera cases caused by the El Tor biotype were found in Peru; and by mid-February, more than 10,000 cases were being reported each week. The epidemic spread to a new country almost every month. By the end of 1992, over 700,000 cholera cases had been reported in 21 countries in the Western Hemisphere and over 6000 people had died (**Figure 23.9**). Cases have been transmitted by contaminated water, raw vegetables that have been irrigated with sewage, and shellfish from contaminated waters. Cholera is probably in the Americas to stay now that reservoirs have been established—plankton, shellfish, water, and humans.

In many countries hit by the El Tor biotype, health officials do not even hope to stem the epidemic by ensuring clean water for everyone—it is far too expensive. Costs are estimated at $200 billion over the next 12 years. Short-term, cost-effective solutions include repairing existing water systems, chlorinating water, and educating people to boil household water and cook vegetables. Just as important is making oral rehydration therapy available to victims. Oral rehydration is the reason relatively few people have died so far.

Vibrio parahaemolyticus: Gastroenteritis

Vibrio parahaemolyticus, related to the more highly pathogenic *Vibrio cholerae*, causes a gastroenteritis that may produce a mild, self-limited diarrhea or an explosive cholera-like illness. Usually the watery diarrhea is mild and accompanied by low-grade fever, abdominal cramps, nausea, and vomiting. Recovery without treatment occurs within a few days. Occasionally, however, dehydration occurs, requiring fluid replacement and antibiotic therapy. Only strains that produce a hemolysin (an enzyme that lyses erythrocytes) are pathogenic.

V. parahaemolyticus is a salt-requiring Gram-negative rod that lives in coastal waters and estuaries around the world. In 1951, Japanese researchers discovered that *V. parahaemolyticus* is a common cause of illness among people who eat fish caught in bacterially contaminated waters. Since then, it has been found to cause about a quarter of all reported cases of diarrhea in Japan, where fish is often eaten raw. Outbreaks have also been reported around the world, including almost all coastal states of the United States. *V. parahaemolyticus* is usually referred to as a food poisoning, but it is a true infection.

Figure 23.9 Cholera epidemic in Latin America, 1991–1992. The epidemic began in Peru in 1991. Outbreaks spread up and down the coast of South America, reaching Mexico by December 1992.

Legend:
- Initial epidemics January 1991
- .... August 1991
- – – – February 1992
- —— December 1992

Preventing *V. parahaemolyticus* gastroenteritis is difficult because these organisms multiply extremely rapidly. Their doubling time under ideal conditions is said to be as short as nine minutes. As a result, contaminated fish may come to contain huge numbers of microorganisms within a short time.

Yersinia enterocolitica: Enterocolitis

Yersinia enterocolitica is a Gram-negative bacterium belonging to the same genus as the microorganism that causes plague (Chapter 27). These microorganisms are commonly found in animals, which may be a significant disease reservoir, but they also infect humans. Infection can come from wild or domestic animals, raw milk, oysters, and water. Chitterlings are a common food source.

Y. enterocolitica has a number of virulence factors—it can adhere to the intestinal epithelium, invade cells, and produce an enterotoxin. It is unusual because it grows best at room temperature (between 22°C and 25°C—significantly below body temperature) yet can survive and multiply at refrigerator temperatures. This may be why *Y. enterocolitica* infections seem to be more common in cold climates, such as Northern Europe and the colder parts of North America.

Y. enterocolitica causes a true infection of the gastrointestinal tract, where bacteria multiply and cause disease. It usually causes an enterocolitis. Abdominal pain is also common—the pain may be so intense that appendicitis is suspected and patients undergo surgery. In some cases, bacteria may enter the bloodstream and produce a systemic infection. *Y. enterocolitica* infection is most common in young children.

Campylobacter jejuni: Campylobacteriosis

Campylobacter spp. are slightly curved, Gram-negative, rod-shaped bacteria (**Figure 23.10**). They were not recognized as human pathogens until the 1970s because they are so difficult to cultivate in the laboratory. Cultivation requires microaerophilic conditions, unusually high incubation temperatures (42°C), and selective media. *Campylobacter* spp. cause over 2 million illnesses in the United States each year—more than either *Salmonella* or *Shigella*. *C. jejuni* is the major cause of diarrheal illness, while the closely related *C. fetus* more often causes a bloodborne infection in elderly or immunocompromised patients.

C. jejuni grows in the intestinal tract of healthy cattle, sheep, dogs, cats, and poultry. Human infection probably occurs from ingesting contaminated meat or milk, although person-to-person transmission may also occur. People with *Campylobacter* diarrhea have frequent episodes of bloody diarrhea, abdominal pain, and sometimes fever. The epithelial lining of the intestinal tract is usually significantly damaged, probably from the invasion of these cells and production of a cytotoxin. *Campylobacter* diarrhea is usually a non-life-threatening, self-limited illness that lasts less than a week; treatment with erythromycin or tetracycline may help eradicate the infection.

Helicobacter pylori: Gastritis, Peptic Ulcers

Helicobacter pylori is a Gram-negative rod that used to be called *Campylobacter pylori* until RNA sequencing, a reliable indicator of evolutionary relatedness, caused it to be reclassified in 1989. *Helicobacter pylori* is found in the stomach, where it survives the inhospitably acidic conditions by producing an enzyme that converts urea to ammonia, raising the pH in its vicinity.

H. pylori has been the subject of intense interest recently because of its association with ulcer disease and gastritis, painful stomach ailments that have never been thought of as infections (**Figure 23.11**). *H. pylori* can be cultured from the stomach lining of 95 percent of patients with duodenal ulcers, but from only 20 percent of healthy volunteers.

This association was clarified in 1993 when clinical studies firmly established that *H. pylori* is, indeed, a cause of gastritis and stomach ulcers, which can eventually lead

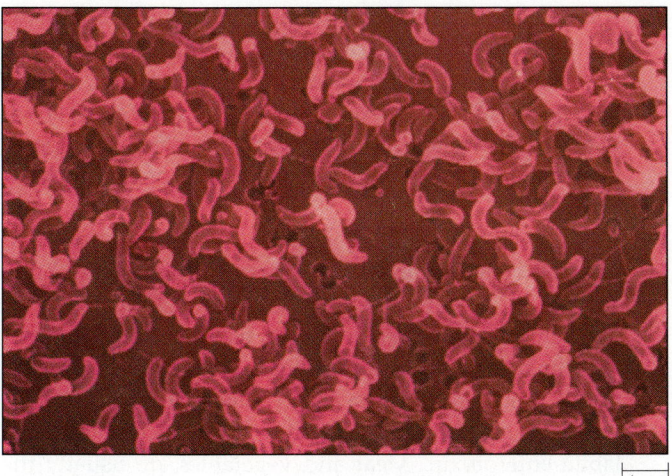

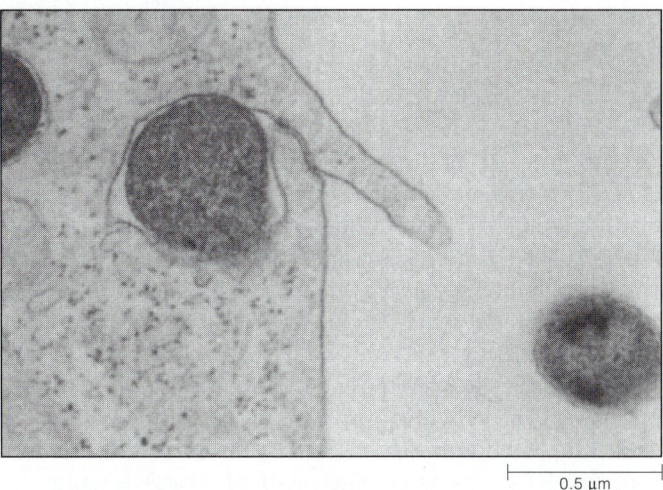

| 1 μm | 0.5 μm |

Figure 23.10 *Campylobacter.* (**a**) Scanning electron micrograph showing the slightly curved bacterial cells of *C. jejuni.* (**b**) Transmission electron micrograph showing a single cell of *C. fetus* invading the intestinal epithelium of an infected chicken.

to gastric carcinoma, one of the most common forms of cancer in humans. *H. pylori* penetrates the mucosal layer that protects the gastric epithelium from stomach acid and then binds to epithelial cells. The next step is not completely understood, but the immune system mounts an attack that damages the epithelium, allowing *H. pylori* to become established and proliferate. Over a period of years, the damaged region may grow, resulting in an ulcer, or remain small, resulting in gastritis. The steps leading to gastric carcinoma are not yet clear.

The critical question of how *H. pylori* binds to the stomach epithelium was also clarified in 1993. *H. pylori* can bind either to certain carbohydrate residues (sialic acid) on epithelial cells or to a blood group antigen that they, like red blood cells, have on their surface. *H. pylori* binds to the Lewis[b] (Le[b]) antigen, which is exposed only in people with type O blood. This discovery explains the long-known fact that people with type O blood are 1.5 to 2 times more likely to develop ulcers and gastric carcinomas than are people with type A or B blood. Understanding the cause and development of ulcers suggests strategies for developing drugs to prevent and treat them.

Clostridium difficile: C. difficile Diarrhea

Clostridium difficile is a toxin-producing bacterium that causes one type of **iatrogenic** (medically induced) diarrhea—the unintended result of antibiotic therapy. Small populations of this organism normally inhabit the human bowel (Chapter 14). During antibiotic therapy, many other bacteria may be eliminated from the intestinal tract, while the relatively drug-resistant *C. difficile* survives and flourishes, producing both an enterotoxin and a cyto-

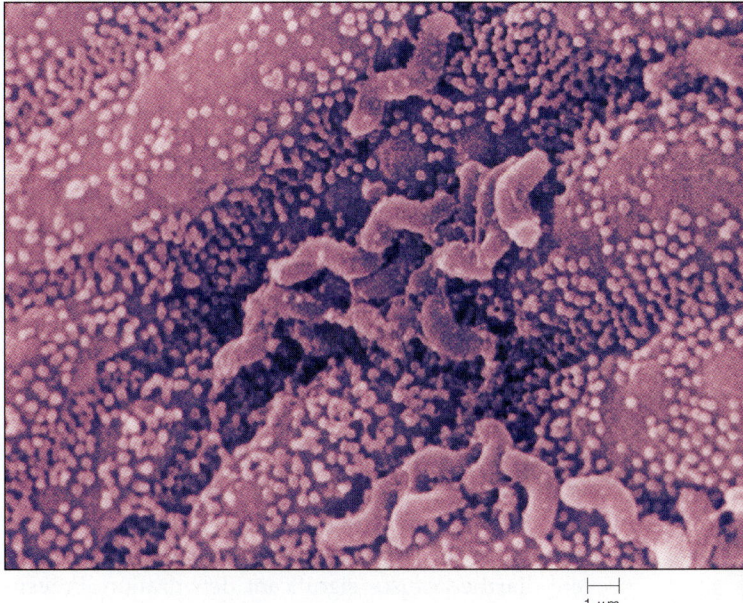

| 1 μm |

Figure 23.11 Peptic ulcer disease. A scanning electron micrograph of *Helicobacter pylori* cells on stomach epithelium. *H. pylori* has recently been shown to cause peptic ulcer disease.

toxin. The result is usually a relatively mild diarrhea that resolves when antibiotic therapy is discontinued. But sometimes *C. difficile* causes a life-threatening enterocolitis, or severe diarrhea persists. In these cases, the usual treatment is vancomycin, which eradicates the *C. difficile* and controls the diarrhea. *C. difficile* has also recently been found to be a significant cause of nosocomial infection (Chapter 20).

Staphylococcus aureus: Food Poisoning

Staphylococcus aureus is a major pathogen of human beings that causes various clinical syndromes (Chapter 26). Strains that produce a heat-stable protein enterotoxin cause an intoxication—a true example of food poisoning. *S. aureus* is a Gram-positive coccus that usually appears under the microscope as grapelike clusters of cells (see Figure 11.10a). Staphylococci are unusually resistant to environmental stresses such as heat, drying, and increased osmotic pressure, which allows them to survive on the skin and in many foods with a high sugar or salt content.

The gastroenteritis caused by staphylococcal food poisoning is usually accompanied by vomiting, diarrhea, and crampy abdominal pain. Because the toxin is formed before the food is eaten, the onset of illness is quite rapid—two to six hours, compared to one to two days for a true infection like salmonellosis. Illness is usually brief and self-limited. Staphylococcal food poisoning is the most frequently reported food poisoning in the United States, probably because it often occurs in large outbreaks—at picnics or other social gatherings—and is therefore relatively easy to identify.

Most staphylococcal food poisoning occurs because of lapses in food preparation or storage. Staphylococci are usually introduced into food (often whipped cream, eggs, mayonnaise, or meat, especially ham) from the body of the person preparing it. (Enterotoxin-producing staphylococci are found in many healthy people, in the mucous membranes of the nose or in tiny cuts in the skin; see Chapter 14.) Staphylococci multiply rapidly in warm food, producing enterotoxin as they grow. Staphylococcal enterotoxin is odorless and tasteless, so there is no immediate evidence of contamination. Once formed, the toxin cannot be destroyed by refrigeration or cooking, except in a pressure cooker.

Prevention is the key to controlling staphylococcal food poisoning. Antibiotic treatment is useless because symptoms are caused by a preformed toxin rather than bacterial growth. Fluid replacement is necessary only if severe diarrhea causes significant dehydration. Prevention, on the other hand, is relatively easy. Careful hand washing can prevent staphylococci from getting into foods that support their growth. And even if foods become contaminated, refrigeration at 5°C prevents the bacteria from multiplying and producing significant amounts of toxin.

Bacillus cereus: Food Poisoning

Like *Staphylococcus aureus*, *Bacillus cereus* can cause a foodborne intoxication. *B. cereus* is a Gram-positive facultative anaerobe that forms spores. Its ability to form heat-resistant spores explains the pattern of *B. cereus* food poisoning. The organism is present in soil, water, and the gastrointestinal tract of humans and animals, so it is often found in human food. When the food is cooked, vegetative cells of *B. cereus* are killed, but the spores survive. If the food is left unrefrigerated after cooking, the spores germinate and the new vegetative cells produce an enterotoxin.

B. cereus food poisoning is relatively uncommon, and the gastroenteritis it produces is usually mild and brief. There are two forms of this illness, associated with two different enterotoxins. One form, which primarily causes vomiting, is produced by a heat-stable enterotoxin, and rice is usually the contaminated food. The other form, which primarily causes diarrhea, is produced by a heat-labile enterotoxin. Meat and vegetables are usually the contaminated foods.

Clostridium perfringens: Food Poisoning

Clostridium perfringens is another spore-forming bacterium that can cause a foodborne intoxication. It is a Gram-positive anaerobe that produces a toxin. Gastroenteritis is only one of its many clinical manifestations (Chapter 26). *C. perfringens* lives in the gastrointestinal tract of animals and humans and is common in feces-rich soils. It usually contaminates meat. Vegetative cells are killed by cooking; but if meat or gravy is kept warm, spores can germinate and produce new vegetative cells that are ingested along with the food. When these vegetative cells once again form spores, they release an enterotoxin in the intestinal tract.

C. perfringens food poisoning usually causes a relatively mild gastroenteritis, characterized principally by diarrhea. Illness lasts less than a day and seldom causes clinically significant dehydration. Serologic studies show that many American adults have been exposed to the *C. perfringens* enterotoxin; but this type of food poisoning is seldom reported, perhaps because noticeable illness occurs only if enormous numbers of microorganisms are ingested—probably a billion or more.

Clostridium botulinum, a related spore-forming anaerobe, also causes a foodborne intoxication. But *C. botulinum* produces a neurotoxin rather than an enterotoxin. The disease it causes—botulism—primarily affects the nervous system (Chapter 25).

VIRAL CAUSES

Viral gastroenteritis is familiar to all of us as "stomach flu," and only the common cold occurs more frequently. The two major causes of viral diarrhea are rotavirus and Norwalk agents; but because a definitive microbiologic diagnosis is seldom made, other significant viral pathogens may yet be discovered.

Rotavirus: Gastroenteritis

In spite of its clinical importance, rotavirus was not identified until 1973. It is a reovirus. Electron micrographs (**Figure 23.12**) show a wheel-shaped double capsid surrounding the virion (*rota* means "wheel" in Latin). It is relatively resistant to environmental damage, including changes in temperature and exposure to solvents such as ether.

Rotavirus is transmitted by the fecal-oral route. It is highly infectious, and diarrhea often spreads rapidly through a family. Epidemics also occur in day-care centers and hospitals. Infants and young children are most frequently affected, partly because they are the most susceptible (previous infection provides limited immunity) and partly because their inability to maintain good hygiene makes them likely victims for any fecal-oral disease. Outbreaks of rotavirus infection are most common in the winter.

Symptoms of rotavirus gastroenteritis begin about 48 hours after infection. Watery diarrhea is the primary manifestation, often accompanied by fever, nausea, and vomiting. We do not yet fully understand exactly how the virus acts, but intestinal epithelial cells are damaged during infection. Severity depends upon how much fluid is lost from the body. Although the vast majority of American children with rotavirus diarrhea recover uneventfully at home, about half of all children hospitalized for dehydration are infected by rotavirus. Children in developing countries who may be malnourished or infected by parasites frequently suffer severe dehydration and die from a rotavirus infection.

Rotavirus can now be diagnosed fairly readily and inexpensively by means of an ELISA immunologic assay (Chapter 19). Though the test may be helpful epidemiologically, it is seldom used clinically because definitive diagnosis does not help with prognosis or treatment. *All* acute gastroenteritis syndromes improve within a few days, and *all* must be treated with rehydration therapy if dehydration is severe. There is no specific antiviral therapy for rotavirus infection.

Prevention is the only way to control rotavirus infections. Breastfeeding probably helps prevent disease in nursing infants, because both breast milk and colostrum contain IgA antibodies that neutralize rotavirus virions (Chapter 17). Standard habits of good hygiene—as always, hand washing—help prevent all infections transmitted by the fecal-oral route. One promising development is an experimental vaccine, consisting of an attenuated virus administered orally, that shows excellent protection in

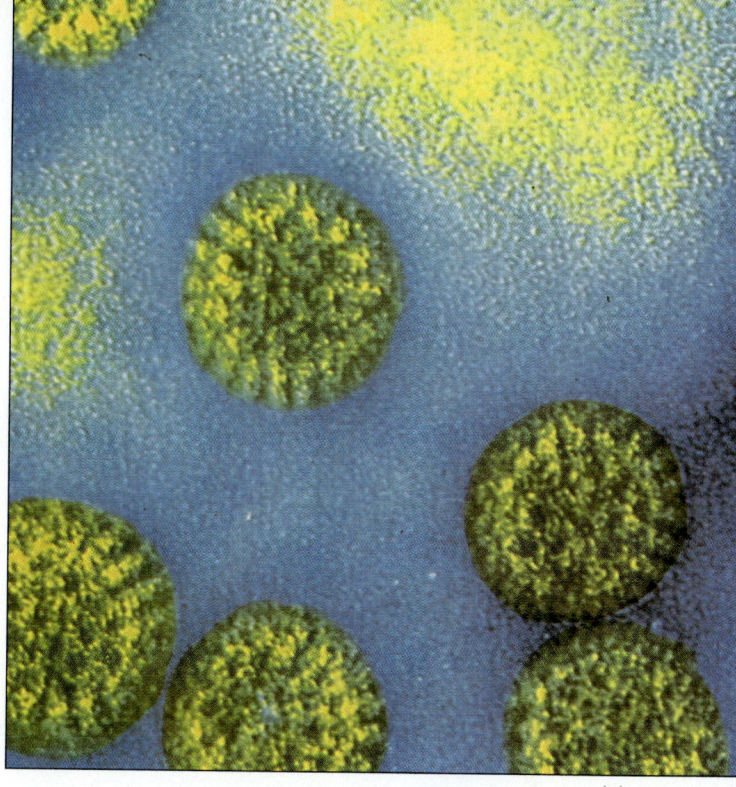

Figure 23.12 This electron micrograph shows rotavirus's wheellike appearance.

some people; however, results are inconsistent. An effective rotavirus vaccine would bring a major childhood illness under control.

Norwalk Agents: Gastroenteritis

Norwalk agents are named for an epidemic of gastroenteritis that occurred in an elementary school in Norwalk, Ohio, in 1969. They are particularly small viruses classified as **calci-like agents** because they resemble the calcivirus family of animal pathogens. We do not yet know whether Norwalk agents are DNA or RNA viruses. The virions appear to contain a single structural protein.

Probably a third of all gastroenteritis outbreaks are caused by Norwalk agents. The *Morbidity and Mortality Weekly Report* reported a typical Norwalk agent epidemic in 1987. More than 100 students attending a college football game became ill with gastroenteritis after consuming fecally contaminated ice. The virus is transmitted by the fecal-oral route and causes the usual clinical picture of gastroenteritis—nausea, vomiting, crampy abdominal pain, and diarrhea. Unlike rotavirus, however, Norwalk agents mostly infect older children and adults, rather than infants. The incubation period is about 48 hours, and the illness is self-limited and usually brief. Fluid-replacement therapy is rarely needed. Immunity after infection is not long lasting. Outbreaks can be prevented only by careful attention to hand washing and general hygiene.

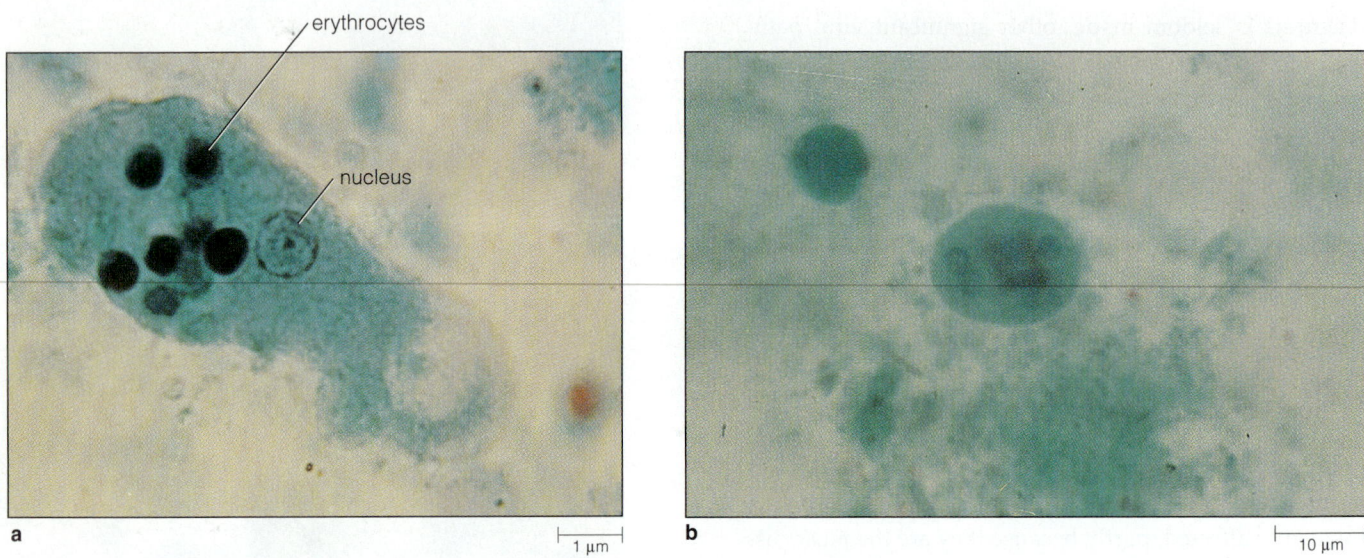

Figure 23.13 Stained preparations of *Entamoeba histolytica* seen under the light microscope. (**a**) A motile amoeba-like trophozoite that has ingested erythrocytes. (**b**) A hardy, resting cyst.

PROTOZOAL CAUSES

Protozoal infections of the intestinal tract, also transmitted by the fecal-oral route, cause a range of clinical syndromes, from mild diarrhea to life-threatening systemic infections. Some occur primarily in developing countries, but others are common causes of disease in the United States.

Entamoeba histolytica: Amoebic Dysentery

Entamoeba histolytica is an amoeba that lives in the intestine of human beings. Most people who harbor this microorganism are entirely well or suffer only mild and occasional problems such as abdominal distention, loose stools, or constipation. But *E. histolytica* has the potential for significant virulence because it can kill human cells, causing either an invasive colitis (**amoebic dysentery**) or widespread infection (extraintestinal **amoebiasis**). Amoebic dysentery is characterized by bloody mucoid stools, fever, and abdominal pain. When parasites enter the circulation and reach distant organs, the most common life-threatening complication is amoebic liver abscess.

E. histolytica exists in two forms—a motile amoebic form called a trophozoite and a hardier resting form called a cyst (**Figure 23.13**). Cysts are passed in the feces of infected people and enter the digestive system of a new host through the usual fecal-oral routes. The cysts survive the journey through the stomach and release the invasive trophozoites.

Like most fecal-oral infections, *E. histolytica* occurs

where sanitation is poor. On average, 10 to 15 percent of the population in developing countries, and particularly tropical countries, is infected by *E. histolytica*, although the figure can be as high as 50 percent. Most people have asymptomatic infections but continue to pass infectious amoebic cysts in their feces, thus perpetuating the infection. In the United States the infection rate is probably less than 2 percent, although institutionalized patients, who typically have poor hygiene, and individuals who practice oral-anal sex have higher rates of infection.

Because *E. histolytica* infection is relatively rare in the United States, it often goes undiagnosed. Serological tests are useful; but definitive diagnosis requires identification of the microorganism in stool samples, and identification can be difficult. Trophozoites disintegrate and become unrecognizable within an hour after the stool is passed, and many nonprescription medications used to treat diarrhea destroy the amoebae and make identification in the stool impossible. Moreover, *E. histolytica* may be confused with other harmless amoebae that commonly inhabit the intestinal tract. At the same time, prompt diagnosis is extremely important because even extensive infections involving potentially fatal liver abscesses can be controlled with metronidazole in combination with iodoquinol.

Giardia lamblia: Giardiasis

When Antony van Leeuwenhoek suffered an acute case of diarrhea in 1681, he examined a drop of his stool with a

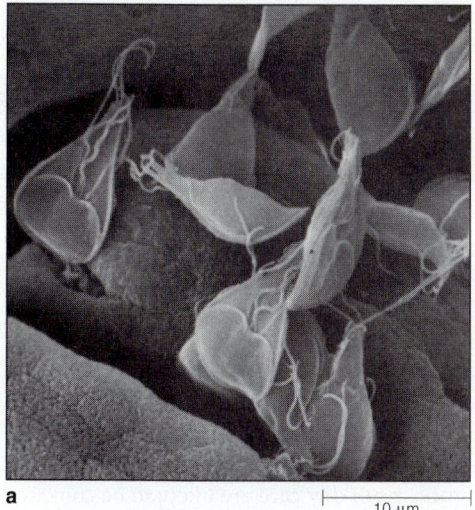 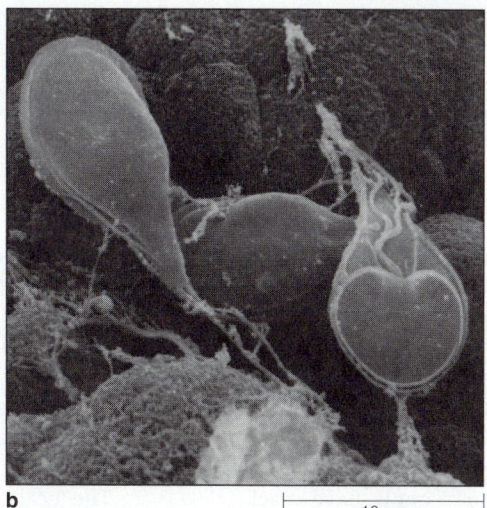

Figure 23.14 Scanning electron micrographs of *Giardia lamblia*. (**a**) A mass of trophozoites on the intestinal epithelium. (**b**) The disc-shaped ventral sucker seen on the right allows *Giardia* to remain attached to the nutrient-rich upper small intestine.

simple microscope and found it swarming with *Giardia lamblia* (Chapter 1). Until the last few decades, it was considered a harmless commensal; but it is a significant and common intestinal pathogen. Today, whenever doctors in the United States encounter a case of diarrhea that lasts longer than a week, they consider giardiasis.

G. lamblia is a distinctive-looking flagellated protozoan found in two forms—a trophozoite that causes infection and a cyst that is passed from host to host (**Figure 23.14**). The trophozoite has two nuclei, multiple flagellae, and a disc-shaped **ventral sucker** on its underside that binds the parasite firmly to the intestinal epithelium. Trophozoites multiply by binary fission in the intestine. The cyst is oval-shaped and contains several nuclei. Trophozoites form cysts as the stool becomes dry in the large intestine, and the infective cysts are shed in feces. Cysts, in turn, are ingested by a new host and release trophozoites after they pass through the stomach.

Many people infected by *G. lamblia* remain symptom-free while continuing to pass infective cysts in their stools. Others, however, suffer a sudden, unpleasant illness that begins about two weeks after infection. The first sign of giardiasis is usually an explosive, foul-smelling, watery diarrhea followed by a bloated abdomen and copious amounts of foul-smelling intestinal gas. The acute symptoms usually last only a few days and are then replaced by more subtle long-term symptoms, including abdominal pain, nausea, and occasional episodes of diarrhea. In rare cases severe diarrhea can persist for months, interfering with normal absorption of food and leading to malnutrition.

Giardiasis is transmitted by the fecal-oral route. There are several epidemiologic patterns, which put certain groups at high risk. Campers are a high-risk group be-

cause of a **sylvatic cycle** (*sylvatic*, "wilderness") of *G. lamblia*. The parasite is found in mountain streams contaminated with human feces or with the feces of *Giardia*-carrying wild animals, especially beavers. Waterborne epidemics of giardiasis that affect entire cities are a potential problem because *Giardia* cysts are resistant to chlorine, which is often used to purify municipal water. But probably the most common way to contract giardiasis is through direct person-to-person contact, usually in a family or day-care setting. In the United States, children are the most likely to get *Giardia* infections. Also, as with other fecal-oral infections, individuals who practice oral-anal sex are at particular risk. Giardiasis was traditionally diagnosed by examining stool samples for *G. lamblia* cysts, but now most labs use an ELISA test that detects *Giardia* antigens.

There are several approaches to preventing giardiasis. Campers should boil or chemically treat water before drinking it, and municipal water districts must filter drinking water to remove the *Giardia* cysts that are not killed by chlorination. Strict hygiene, especially hand washing, helps reduce the spread of *Giardia* in homes and day-care centers. High-risk sexual behavior should be avoided. Giardiasis can be cured by the antimicrobial drugs metronidazole and quinacrine.

Balantidium coli: **Balantidiasis**

Balantidium coli is the only ciliate and the largest protozoan known to infect humans. Like *Entamoeba histolytica* and *Giardia lamblia*, it has two forms—a metabolically active trophozoite and a resting cyst. The trophozoite moves by means of cilia that cover its surface and has many recognizable organelles, including two nuclei, a

Clinical Notes

Hygiene vs. Intimacy

It should come as no surprise to anyone who has cared for children at home that day-care centers, where a limited number of adults care for many diapered infants and messy toddlers, are ideal settings for the spread of fecal-oral infections. Diarrheal illness is a special problem.

Shigellosis, hepatitis A, rotavirus, *Campylobacter*, and *Giardia* in particular are easily and rapidly transmitted from person to person by a relatively small number of organisms. Day-care-based *Giardia* epidemics, for example, have **attack rates** (the number of exposed people who became ill) as high as 90 percent, and the attack rate for some rotavirus outbreaks has been 100 percent. The **secondary attack rate**

(the number of cases in the infected children's families) ranges from 12 to 80 percent. Interestingly, rotavirus outbreaks usually affect infants, while *Giardia* is more common among older toddlers.

Although good hygiene, especially hand washing, can decrease the frequency and extent of these outbreaks, complete prevention is not practical. Day-care contact, like family contact, is just too intimate ever to be absolutely hygienic. The fecal-oral problems of day care are likely to be controlled only when safe and effective vaccines become available, as is now the case for hepatitis A.

---

mouthlike organ called a **cytosome**, and numerous vacuoles. The cyst survives passage to a new host and thereby spreads infection.

Balantidium coli invades the epithelium, the colon, and the lower small intestine, causing a colitis called **balantidiasis**. Diarrheal stools typically contain blood and pus, and patients often suffer from abdominal pain, nausea, and loss of appetite. Patients may suffer intermittently, having long symptom-free periods. Some people who continue to pass infective cysts in their feces may be entirely healthy carriers of *B. coli*.

B. coli is found throughout the world, but symptomatic infection is usually restricted to tropical countries. Unlike giardiasis, this disease is rarely seen in North America. Animal reservoirs, especially pigs and monkeys, perpetuate balantidiasis. Humans are usually infected when they ingest food or water contaminated by pig feces, but human-to-human transmission also occurs. Good hygiene is generally sufficient for prevention, and tetracycline or metronidazole is an effective cure.

Cryptosporidium: Cryptosporidiosis

Because **cryptosporidiosis** causes severe problems in immunocompromised people—especially AIDS patients—it is receiving increased attention. *Cryptosporidium*, the protozoan that causes the infection, inhabits the intestinal tract of many kinds of animals, including fish, reptiles, and mammals (humans among them). People become infected by fecal-oral contamination from animal reservoirs or other humans. High-risk groups include male homosexuals and people who work with animals.

Cryptosporidium invades the intestinal epithelium and multiplies there, causing a mild enterocolitis that usually

produces abdominal pain and a watery, bloodless diarrhea. People with a normal immune system are mildly ill for 10 days or less, but AIDS patients can develop a life-threatening diarrhea, consisting of 50 or more stools a day, that may last for months. No known antimicrobial agent cures cryptosporidiosis, and treatment is limited to rehydration therapy when fluid loss becomes dangerous. Prevention—including good hygiene and avoiding high-risk sexual behavior—is crucial.

HELMINTHIC CAUSES

When people talk about "having parasites," they usually mean parasitic worms—helminths (Chapter 12). Most helminthic infections begin when the infective form enters the human mouth in feces-contaminated material or infected animal tissues (meat). In the digestive tract, different species pursue their own unique patterns of infection, remaining in the intestinal tract or traveling to the lungs, liver, muscles, or brain.

Helminthic infections are most common in countries with poor sanitation, though some also occur in the United States. The worldwide burden of helminthic gastrointestinal disease is enormous. In some parts of the world, up to 90 percent of the population suffers from debilitating or life-threatening worm infestations. Most helminthic infections are treatable by one or another of the many new antiparasitic drugs.

Enterobius vermicularis: Pinworm

Enterobius vermicularis causes **pinworm infection**, the most common helminthic infection in the United States.

E. vermicularis is a small white nematode (roundworm) less than 1 cm long. People become infected when they ingest pinworm eggs. The eggs hatch and the worms mate in the intestinal tract. Then the females crawl out, usually at night, to lay their eggs on skin near the anus. Parents who look for them may see these worms crawling across the skin of their infected child (**Figure 23.15**). Eggs perpetuate the infection by fecal-oral transmission. Diagnosis is made by recovering the eggs from the perianal skin with a sticky, transparent material such as cellophane tape.

Pinworms usually infect young children. It is extremely easy to acquire this infection because eggs remain infective for more than a week on fomites such as toys, furniture, and countertops. As a result, pinworm infections spread not only in family and day-care settings, but also in schools and other public places. Adults who care for infected children are also at high risk of infection. Fortunately, pinworm infection is not serious. Many infected children suffer no symptoms at all, while others complain of perianal itching due to worm migration.

Pinworm infections are easily treated with pyrantel pamoate or mebendazole, but eliminating eggs from the home and thus preventing reinfection is much more difficult. When one child in a family is diagnosed with pinworms, the rest of the family is usually presumed to be infected as well and is treated.

Ascaris lumbricoides: Ascariasis

Ascaris lumbricoides is another roundworm parasite. It is less common than *Enterobius vermicularis* in the United States, and has a more complex life cycle. Ascariasis begins when a new host ingests *Ascaris* eggs shed in the feces of an infected person. Larvae hatch and burrow through the wall of the small intestine, entering the bloodstream and eventually reaching the lungs. There they break into the alveoli and continue to grow. After about three weeks of maturation the *Ascaris* larvae are coughed out of the lungs and swallowed, thus reentering the digestive tract, where they often grow to 12 inches or more and produce infective eggs. Infection is diagnosed by finding the characteristic eggs in a fecal specimen.

Ascaris is extremely common in parts of the world with poor sanitation, particularly where human waste is used for fertilizer. Probably more than a billion humans suffer from ascariasis. Surprisingly, people can harbor large populations of these worms in their intestines without suffering any symptoms at all. However, infection can be uncomfortable or dangerous. Large numbers of larvae migrating through the lungs can cause an asthma-like cough. Mature worms can crawl into the bile duct or liver and block the normal flow of bile; and, occasionally, an enormous tangled mass of worms causes intestinal obstruction.

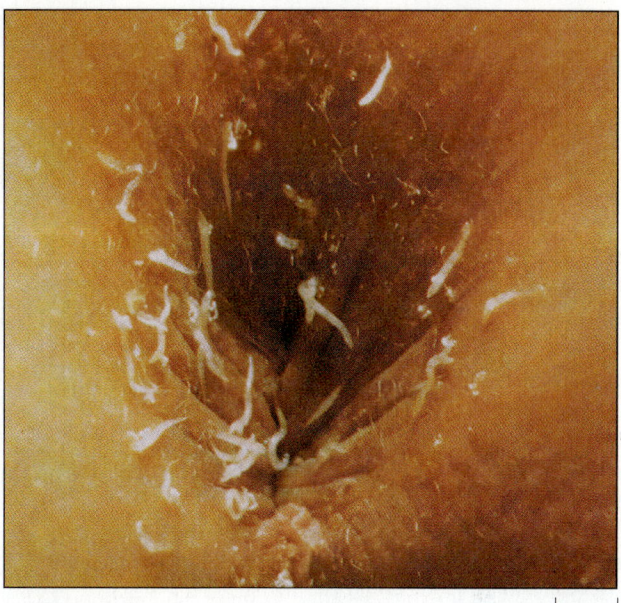

| 10 mm |

Figure 23.15 Female pinworms leaving the anus of a 5-year-old child to deposit their eggs.

Ascaris infections are treated with various antiparasitic drugs, including mebendazole, pyrantel pamoate, and piperazine. When heavily infested patients are treated, they pass masses of dead worms (see opening of Chapter 12). Control of ascariasis, however, depends upon prevention rather than treatment.

Ancyclostoma duodenale and *Necator americanus:* Hookworm

Ancyclostoma duodenale and *Necator americanus* are two species of roundworms known as hookworms. The two species are almost identical but are found in different parts of the world. *A. duodenale* is called Old World hookworm, and *N. americanus*, New World hookworm. Hookworm eggs from human feces hatch in the soil and become infective **filariform** larvae that can penetrate human skin. People usually acquire **hookworm infections** by walking barefoot in dirt that contains larvae, which remain infective for several months. After entering the body, the hookworm larvae make their way through the bloodstream to the alveoli, where they are coughed up and swallowed into the intestinal tract. Mature worms live and mate in the small intestine, producing eggs that are passed in the feces. Diagnosis depends on finding hookworm eggs in the feces.

Infection by hookworms is more serious than that by pinworms or *Ascaris* because adult worms have mouthparts called **biting plates** that allow them to suck blood from the intestines of their host (**Figure 23.16**). Heavy infestations, therefore, can cause profound anemia, with resultant weakness and fatigue. In addition, hookworm

Figure 23.16 Scanning electron micrograph of the hookworm *Necator americanus* showing the internal biting plates by which it attaches to the intestinal lining so it can suck blood from its host.

victims may suffer from malnutrition, nausea, vomiting, and diarrhea.

Hookworm infection frequently occurs in warm climates where people often go barefoot. *N. americanus* was once common in the southern United States, but improved sanitation has significantly decreased infection. Infection can be treated by mebendazole or pyrantel pamoate. Patients may also need iron supplementation to recover from anemia. Prevention depends on disposing of human feces so they do not contaminate the soil and wearing shoes in soil that could be contaminated.

Strongyloides stercoralis: Strongyloidiasis

Strongyloides stercoralis, like hookworm, is a roundworm with a larval form that can penetrate human skin (see Figure 1.8). Also like hookworm, infection begins in the blood, spreads to the lungs, and eventually produces adult worms in the intestinal tract. But **strongyloidiasis** differs from hookworm infection in several important respects. *S. stercoralis* can reproduce without a human host (it has a **free-living life cycle**), and the eggs hatch into infective larvae within the intestine. As a result, a person can suffer continuous reinfection, or **autoinfection**, as infective larvae are passed, come in contact with that person's skin, and reenter the body. This pattern leads to extremely heavy infestations. Diagnosis of *S. stercoralis* infection is made by finding the larvae in concentrated stool specimens.

Strongyloidiasis usually occurs in warm, moist climates but is sometimes seen in the United States and Canada. Infected patients may have respiratory complaints when the larvae are migrating through the lungs, and invasion of the bowel wall can cause profuse bloody diarrhea and malnutrition. People with compromised immune systems suffer extremely debilitating infections that can be fatal. Prevention is best—not contaminating soil with feces and avoiding known contaminated soil. Treatment is with mebendazole or pyrantel pamoate.

Trichuris trichiura: Whipworm

The nematode *Trichuris trichiura* is called whipworm because its body resembles a whip. *T. trichiura* is similar to *Ascaris* in many ways but has a simpler life cycle. **Whipworm infection** begins with ingesting eggs, which hatch in the small intestine and penetrate the wall, where they mature into adults. Without leaving the intestinal tract, adult whipworms mate and lay eggs, which are passed in the feces. Diagnosis depends on finding whipworm eggs in feces. The eggs require three weeks of maturation in the soil before they become infective.

Infections with a small number of whipworms are usually asymptomatic, but heavier infestations can cause irritation of the bowel, bloody diarrhea, abdominal pain, and weight loss. Extremely large numbers of whipworms can cause intestinal obstruction. Treatment with mebendazole is effective. Prevention depends upon good hygiene and properly disposing of human feces.

Trichinella spiralis: Trichinosis

Trichinella spiralis is a nematode that infects meat-eating animals. A new host is infected by consuming muscle that contains encysted larvae of *T. spiralis* (**Figure 23.17**). Larvae develop into worms in the small intestine, where they live for several months, producing hundreds or even thousands of new larvae. These larvae burrow through the intestinal wall, enter the bloodstream, and form cysts when they are deposited in muscle. Infection is perpetuated when muscle of the infected animal is eaten. Humans are infected when they eat infected meat—usually pork but occasionally bear. Unless an infected person is eaten by another animal, human infection is a dead end for *T. spiralis*. **Trichinosis** is usually diagnosed when a single source of contaminated meat causes an outbreak involving many people. Diagnosis can be confirmed by serological tests or by a muscle biopsy showing encysted larvae.

The symptoms of human trichinosis—fever, muscle pain, and malaise—are caused by larval migration through the tissues. Symptoms are mild if only a few larvae are ingested, but may be severe if large numbers are involved. In severe cases—when larval migration damages the heart, lungs, diaphragm, and brain—trichinosis can be fatal within a month or two of infection.

Clinical Notes

Reason to Squirm?

The idea of worms burrowing through our internal organs is unpleasant, yet this is what happens in **visceral larva migrans**, a clinical syndrome most commonly caused by *Toxocara canis*, the canine roundworm. *T. canis*–infected dogs excrete the eggs in their feces. After maturing in soil for about three weeks, the eggs can infect humans or any other animals that happen to ingest them. In humans, a dead end for these parasites, the eggs hatch in the small intestine, burrow their way through the intestinal wall, and then travel through tissues, including the lung and liver.

Most clinically diagnosed cases of visceral larva migrans occur in young children, particularly those who eat dirt. Rarely do children become seriously ill with visceral larva migrans, but a milder, symptomless infection may be fairly common. Though we carefully dispose of human excrement—dramatically decreasing infestations with human worms such as *Ascaris*—dog feces are everywhere. Studies show that up to 30 percent of soil samples in public playgrounds and parks are contaminated with eggs of *T. canis*. This is not surprising, since *T. canis* infects up to 80 percent of puppies and at least 20 percent of adult dogs kept as pets in the United States. Studies show that many people who never noticed symptoms have detectable antibodies against *T. canis*—evidence that they have had visceral larva migrans. The extent of the problem is unknown.

Treatment with thiabendazole, mebendazole, and possibly steroids to decrease inflammation is only moderately effective. Trichinosis, like other helminthic infections, must be controlled through prevention—in particular, through making sure that slaughtered pigs are not infected by *T. spiralis*. Trichinosis declined dramatically in the United States after World War II because pork producers stopped using raw garbage (which often contained infected meat) as feed for their livestock; it is now a rare issue in this country. Trichinosis can also be prevented by freezing or thoroughly cooking pork to kill the encysted larvae.

Taenia spp.: Tapeworm

Taenia spp. are among the most common cestodes, or tapeworms, of the flatworm (platyhelminthes) family. *T. solium* is the pork tapeworm, and *T. saginata* is the beef tapeworm. The life cycle of these worms is similar. Humans become infected when they eat pork or beef muscle that contains encysted larvae. The larvae mature in the intestine, and the adult worm attaches itself to the intestinal lining by means of suckers and hooklets on its head, or **scolex** (**Figure 23.18**). The tapeworm grows, adding individual segments called **proglottids**—becoming, in some cases, several yards long. Mature proglottids, which contain infective eggs, are passed in the feces, eventually reinfecting pigs or cattle and thereby perpetuating the infection. Tapeworm infection is diagnosed by finding proglottids in the stool. Differences in the appearance of the proglottids distinguish *T. solium* from *T. saginata*.

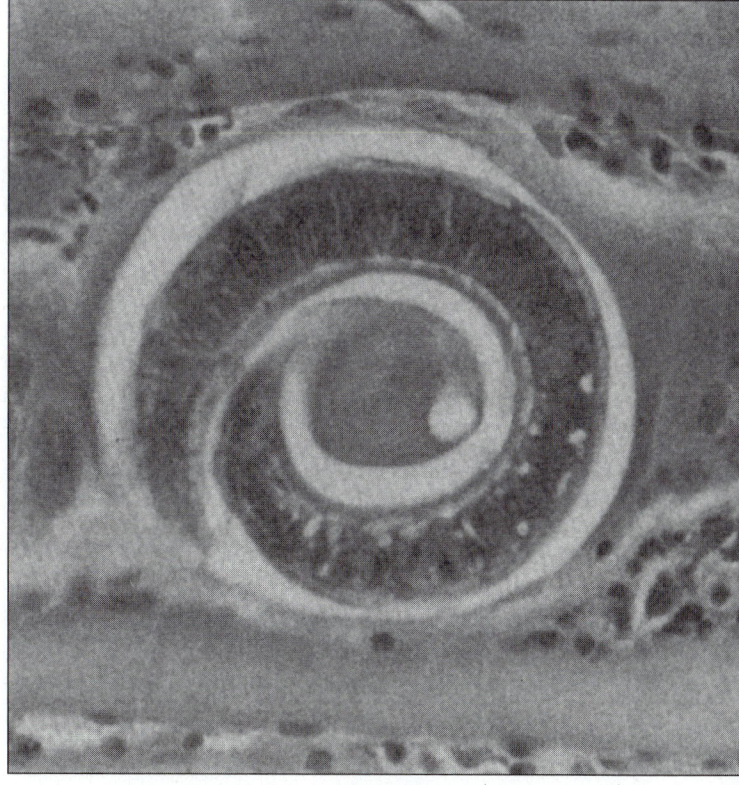

Figure 23.17 *Trichinella spiralis* larva coiled in a cyst in a still-living human muscle cell.

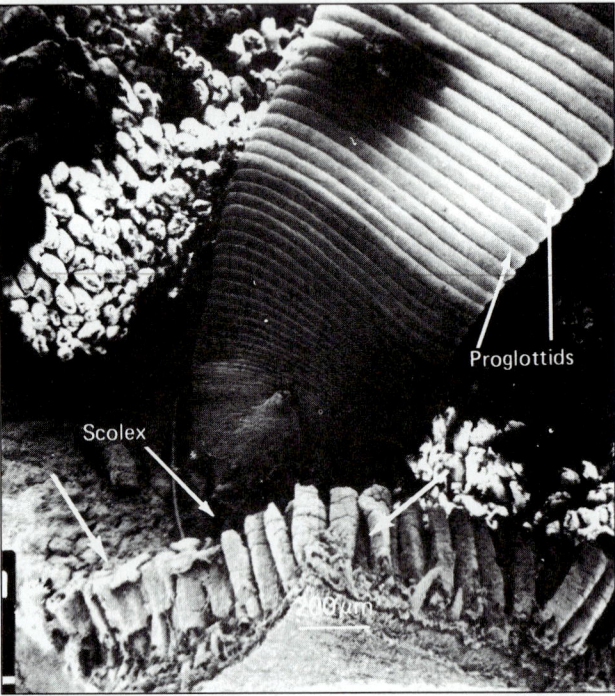

Figure 23.18 Scanning electron micrograph showing a tapeworm anchored to intestinal epithelium by its scolex.

Tapeworm infections cause surprisingly few symptoms. Even patients who harbor enormous tapeworms are seldom aware of their infection unless they see proglottids in their stool, although some abdominal pain, diarrhea, and indigestion may occur. A much more serious syndrome occurs, however, when people ingest *T. solium* eggs as the result of human fecal contamination and fulfill the pig's usual role in the worm's life cycle. In this infection, called **cysticercosis**, larvae become encysted in human tissues. If the cysts develop in vital tissues such as the brain or the eye, infection can be disabling or life-threatening.

Tapeworm infections are treated with niclosamide or praziquantel. Cysticercosis may require surgery. Preventing these infections—which are rare in the United States—consists of avoiding or thoroughly cooking infected beef or pork, or, in the case of cysticercosis, avoiding human fecal contamination of food and water.

Echinococcus granulosus: Hydatid Disease

Echinococcus granulosus is a tapeworm that ordinarily infects carnivores and herbivores. Canines, such as dogs or wolves, eat infected meat and harbor the adult tapeworms, while grazing animals, such as sheep or cattle, in-

gest the infective proglottids and develop encysted larvae in their flesh.

Humans are infected only incidentally, when they ingest eggs from dog feces and encysted larvae, called **hydatid cysts**, develop in their tissues. These cysts can be as large as 12 inches in diameter, occupying so much space that they do significant damage to vital organs. **Echinococcosis**, or **hydatid disease**, usually affects sheepherders who ingest food or water contaminated by the feces of their infected sheepdogs. This infection is therefore limited mainly to sheep-raising areas. Symptoms depend upon the location of the hydatid cysts, which is usually the lung or liver but sometimes the brain or bone. The only treatment is surgical removal of the cyst. Prevention depends upon educating at-risk individuals to protect themselves with good hygiene.

INFECTIONS OF THE LIVER

Viruses and helminths that infect the liver damage this vital organ in several ways. Sometimes the damage is minor; in other cases, it can be life-threatening.

VIRAL CAUSES

Many different viral pathogens—some identified only within the last few years—infect the liver and cause hepatitis. The hepatitis viruses cause similar signs and symptoms of liver damage, but there are some important differences.

Hepatitis A Virus (HAV): Type A Hepatitis

Hepatitis A virus (HAV) is a small nonenveloped virus that contains single-stranded RNA. It belongs to the picornavirus family and the genus Enterovirus, which includes other viruses that enter the body through the gastrointestinal tract—such as poliovirus, the coxsackie viruses, and the echoviruses.

Hepatitis A virus is spread readily from person to person by the fecal-oral route. Huge numbers of virions are excreted in the feces of infected persons, even before they become ill. Hepatitis A, also called **infectious hepatitis**, is the most common cause of hepatitis epidemics, although most cases occur sporadically rather than as part of a large-scale outbreak. Transmission is primarily through close personal contact, for example, from an infected family member. Families with young children in day care are at particularly high risk. Significant epidemics may occur when people contract hepatitis A from a common source of contaminated food or water. Hepatitis A can also be contracted after oral-anal sexual contact.

One study in Seattle, before the safe-sex campaign of the AIDS era, showed that 22 percent of susceptible gay men contracted hepatitis A each year.

After ingesting the hepatitis A virus, people feel well during an incubation period that lasts, on average, 25 days. During this period, virus is absorbed from the gastrointestinal tract, passes briefly through the bloodstream, and becomes concentrated in the liver. Virus multiplies there and, in the process, kills many **hepatocytes** (liver cells). Damage to the liver causes the typical symptoms of hepatitis—nausea, vomiting, loss of appetite, fatigue, disorders of taste and smell, dark urine, and **jaundice**, a yellow discoloration of the skin and whites of the eyes. The liver is enlarged and tender, and serum levels of bilirubin and liver enzymes are high.

The disorders of taste and smell that typically cause hepatitis patients to lose their appetite remain mysterious. The other signs and symptoms can be attributed to loss of normal liver function. If the liver is damaged and unable to excrete bilirubin at an adequate rate, it accumulates in body tissues and urine. Serum levels of liver enzymes rise because they are released into the blood as hepatocytes die and disintegrate. If the liver is too profoundly damaged to produce normal quantities of blood clotting factors, life-threatening bleeding can occur. Hepatitis A is usually a benign and self-limiting illness, but fatal complications sometimes occur.

Many people who are infected by the hepatitis A virus remain asymptomatic. This type of **anicteric** (jaundice-free) hepatitis is particularly common in young children. Eighty percent of children under the age of 3 suffer few or no symptoms, but they can transmit the virus to adults. About 75 percent of infected adults become jaundiced. Hepatitis A epidemics are quite common in infant day-care centers, where many diapered infants are cared for together, but an outbreak is usually recognized only when the adult day-care workers and parents become jaundiced.

Hepatitis A is distinguished from other types of hepatitis by its viral markers. Usually diagnosis rests on detecting **anti-HA** (antibody directed against the major hepatitis A virus antigen). When symptoms of hepatitis appear, levels of early IgM antibodies against HA rise. By the time a patient becomes ill enough to consult a doctor, extremely high levels of anti-HA IgM are usually present. Within three or four months after infection IgM levels fall, but low titers of IgG antibodies against the same viral antigens persist indefinitely, providing lifelong immunity against reinfection.

After recovery, hepatitis A virus is eliminated from the body. Unlike other hepatitis viruses, the A virus does not cause chronic infections that predispose a person to life-threatening liver disease or liver cancer.

Because antibodies against the virus prevent hepatitis A infection, immunization is possible. Until recently, the only method available was passive immunization (administration of anti-HA antibodies). People given **immune serum globulin**, an antibody-rich preparation concentrated from the serum of many donors, are protected against hepatitis A until most of those antibodies are degraded—about six months. Receiving immune serum globulin within two weeks of exposure to the virus gives a 90-percent chance of escaping infection. Family members of hepatitis patients are usually advised to receive immune serum globulin, as are day-care workers when a case occurs. Casual contacts of hepatitis patients, such as fellow workers, are not at a significantly increased risk and do not need prophylaxis. People traveling to countries where sanitation is poor and hepatitis A is common may also receive a preventive injection.

The Food and Drug Administration is evaluating an experimental vaccine that would replace immune serum globulin for preventing hepatitis A. Active immunization has important advantages over passive—protection may last as long as seven years, antibody levels are 200 times higher, and as a result, the vaccine is closer to 100 percent effective. Furthermore, the injection is less painful than is the immune serum globulin injection.

Hepatitis B Virus (HBV): Type B Hepatitis

Hepatitis B virus (HBV) causes a clinical syndrome that often cannot be distinguished from hepatitis A, but the viruses themselves, the circumstances, and the consequences of infection are quite different. HAV causes a relatively harmless infection, while HBV is second only to tobacco as a known cause of human cancer.

Hepatitis B virus is a DNA-containing virus belonging to the hepadnavirus family, which includes other viruses that infect the livers of animals. Unlike the hepatitis A virus, which has only one significant viral antigen, hepatitis B virions (sometimes called **Dane particles**) have three—a surface antigen (**HBsAg**), a core antigen (**HBcAg**), and an antigen called **e** (**HBeAg**). Excess surface antigen is produced and forms empty shells that resemble spheres or tubules, which are found in the serum of patients infected by HBV (**Figure 23.19**).

HBV has a very long incubation period, an average of 75 days. It occurs in high concentrations in the blood, body fluids, tears, saliva, and breast milk. Because it is bloodborne, hepatitis B is sometimes called **serum hepatitis**. Transmission is often through contaminated blood or blood products—via transfusions, shared needles among drug users, and accidental needle sticks in hospitals. Health-care workers are at increased risk, and illicit drug users are a very high risk group. HBV can also be

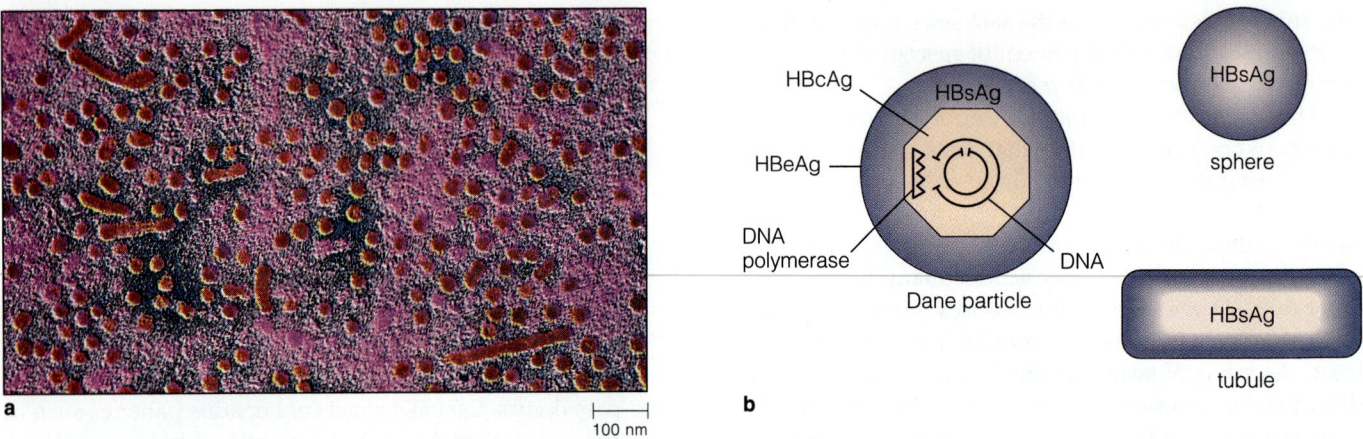

Figure 23.19 The hepatitis B virus is a DNA-containing virus with three antigens—surface (HBsAg), core (HBcAg), and e (HBeAg). An HBV virion is also called a Dane particle. Excess surface antigen that is produced forms into spheres and tubules.

transmitted by intimate contact, including sexual contact, birth, and mother-infant contact. The virus is highly infectious. The chance of contracting hepatitis B after an accidental needle stick involving HBV-contaminated blood is about 40 percent, compared to a risk of approximately 0.5 percent of contracting AIDS after a similar accident. In other respects, however, the transmission of HBV is so similar to that of HIV that hepatitis B has been called AIDS's twin.

About half the people infected with HBV never show symptoms, and more than 99 percent of patients who develop a symptomatic infection survive the initial illness. A few patients develop **fulminant hepatitis**, a disastrous condition of total liver failure that causes death within a few days. Hepatitis B may also become a lifelong, life-threatening problem if chronic infection occurs. This is a complication in about 10 percent of infected adults and as many as 80 percent of infected infants. In chronic hepatitis B infection, the virus persists and may initiate a slow but ultimately fatal form of **cirrhosis** (liver destruction). It also increases a person's risk of developing liver cancer by as much as 300 times. In parts of the developing world where HBV is widespread, liver cancer is a common cause of death.

Hepatitis B infection is diagnosed by identifying the viral antigens, or antibodies against them, in a patient's serum. Most people who become ill with type B hepatitis already have detectable levels of HBsAg in their serum. These levels fall, and levels of **anti-HBs** (antibody against hepatitis B surface antigen) rise as the person recovers. If chronic infection occurs, antibody levels are not detectable, but antigen levels remain high. Serum levels of HBeAg, **anti-HBe** (antibody against e antigen), and **anti-HBc** (antibody against the core antigen) can also be used to diagnose hepatitis B.

Hepatitis B is a major public health concern in the United States and around the world. Worldwide, there are more than 300 million hepatitis B carriers. In the United States, infection rates soared during the 1980s, and by the early 1990s hepatitis B was the second most common reportable infection after gonorrhea. Clinical studies suggest that steroids and alpha interferon may have some value in treating hepatitis B, probably because they inhibit the host defenses that contribute to hepatitis tissue damage. But these treatments are still experimental, and the focus remains on prevention.

Several new and effective approaches to preventing hepatitis B are encouraging. Sensitive serological tests for hepatitis B antigens and antibodies mean most HBV-contaminated blood can now be identified and discarded. Passive immunization with hepatitis B immune globulin can prevent infection of a person who has recently been exposed to HBV, for example, a hospital worker who receives an accidental needle stick with infected blood. Passive immunization has been particularly valuable in preventing hepatitis B among infants born to chronically infected mothers, because the overwhelming majority of babies infected at the time of birth become chronic carriers of HBV themselves.

But the biggest step in preventing hepatitis B was the release of an effective hepatitis B vaccine in the early 1980s. An even newer vaccine, which is currently in use, was the first to be produced by recombinant DNA technology. Vaccination, which is safe and effective, still remains relatively expensive. Until recently, it was recommended only for high-risk groups, but in 1992 the Immunization Practices Advisory Committee (IPAC) of the CDC and the American Academy of Pediatrics recommended universal immunization for newborns. HBV vaccination is now part of the routine infant vaccination schedule. Physicians are also vaccinating older children, particularly adolescents, who are at increased risk if they are sexually active.

In theory, the new HBV vaccine should virtually eradicate hepatitis B within the next generation. In reality, immunization rates are apparently lagging because the vaccine is expensive and because parents feel babies get so many shots (2-month-old infants, for example, may get three shots and an oral polio vaccine). Moreover, three shots of HBV vaccine are required for full protection and, unlike other immunizations, a child does not need proof of HBV immunization to enter school. Progress toward controlling the rising incidence of hepatitis B infection in the United States depends upon how many children will get the full course of immunization.

Hepatitis Delta Agent: Type D Hepatitis

In 1977, researchers found a new antigen—**delta antigen**—and antibodies against it in patients with HBV. Delta antigen is part of a distinct RNA-containing virus called **hepatitis delta**. The delta virus can only infect someone already infected by HBV because the delta agent depends on the hepatitis B virus for its transmission and/or replication. Type D hepatitis is a severe acute and chronic liver disease. The acute infection is frequently fulminant, and most patients with chronic hepatitis delta progress to cirrhosis and fatal liver disease. No treatment or prevention currently exists, and any chronic carrier of HBV is at risk. Uninfected people can be protected by vaccination against hepatitis B.

Hepatitis C Virus (HCV): Type C Hepatitis

Even after most blood contaminated with the hepatitis B virus began to be identified and removed from blood banks, 5 to 10 percent of patients receiving blood transfusions still developed posttransfusion hepatitis. Because the infecting agent was unknown, and because these patients were not infected by either the hepatitis A or the hepatitis B virus, the illness was usually referred to as *non-A non-B hepatitis*. However, recombinant DNA technology identified the agent by molecular cloning of the single-stranded RNA genome, and in 1989, the newly identified virus was named **hepatitis C virus (HCV)**. HCV is thought to belong to the togavirus or flavivirus family.

Like HBV, HCV is transmitted by blood or sexual contact. Of the approximately 150,000 hepatitis C infections that occur in the United States each year, about 10 percent occur after transfusion, 50 percent are associated with intravenous drug abuse, and 10 percent are sexually transmitted. How the remaining community-acquired cases are transmitted is not known. The mystery is complicated by the unusually long interval—up to six months—between exposure to HCV and the onset of illness or the detection of antibody against hepatitis C virus in a patient's blood.

Only about half the patients infected by HCV develop clinical evidence of hepatitis, but as many as half of those develop chronic infection and liver disease. Though much less common than hepatitis B, hepatitis C is a serious disease that merits careful public health attention. Detecting antibodies against the HCV antigen and eliminating infected blood from the transfusion pool should prevent transfusion-acquired hepatitis C, but since most cases are acquired by other means, we must develop other strategies if the disease is to be controlled. Research has accelerated since HCV was identified, and there is some preliminary evidence that alpha interferon may be useful in treating chronic liver disease due to HCV.

Hepatitis E Virus (HEV): Type E Hepatitis

Tens of thousands of people in parts of Asia, Africa, and, most commonly, India contract an infectious hepatitis but have no detectable antibodies against hepatitis A virus, the usual agent. These cases—all enterically transmitted—often occur in epidemics that can be traced to a fecally contaminated water supply. Infection is not usually transmitted to household contacts. The only known cases in Western countries were acquired during travel to the affected areas. Until recently, the agent was classified as yet another type of non-A non-B hepatitis. But the RNA-containing virus that causes this infection has been identified by DNA technology and named **hepatitis E virus (HEV)**. HEV is thought to be related to the calcivirus family.

Like hepatitis A, hepatitis E is usually benign and self-limited. It is not known to cause chronic infection or persistent liver disease. However, hepatitis E is an extremely serious disease for pregnant women—about 20 percent of pregnant HEV victims die from acute hepatitis. Like hepatitis C, hepatitis E has been identified so recently that much remains to be learned about it.

HELMINTHIC CAUSES

Certain species of trematodes, or flukes, of the flatworm (platyhelminthes) family cause infections that do their primary damage to the liver.

Fasciola hepatica, Opisthorchis sinensis: Liver Fluke Infection

Fasciola hepatica, the sheep liver fluke, and *Opisthorchis sinensis*, the Chinese liver fluke, are two of the most frequent causes of liver fluke infection. These worms enter a

Figure 23.20 Life cycle of *Fasciola hepatica*, the sheep liver fluke. For liver fluke infection to occur, infected human feces must be deposited in fresh water, a certain species of water snail must be present to serve as an intermediate host, and water plants or fish containing the infectious metacercaria (larvae) must be eaten.

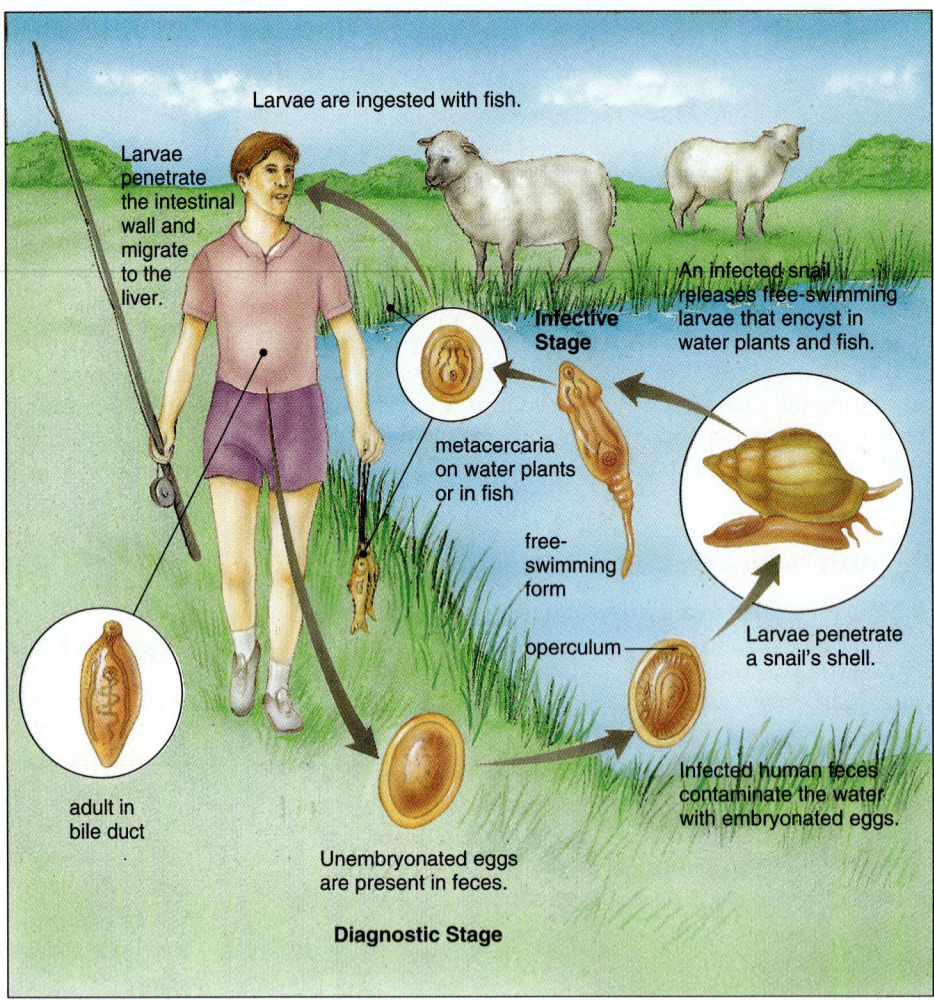

Larvae are ingested with fish.

Larvae penetrate the intestinal wall and migrate to the liver.

Infective Stage

An infected snail releases free-swimming larvae that encyst in water plants and fish.

metacercaria on water plants or in fish

free-swimming form

Larvae penetrate a snail's shell.

operculum

adult in bile duct

Infected human feces contaminate the water with embryonated eggs.

Unembryonated eggs are present in feces.

Diagnostic Stage

new human host when an infective larval form, called a **metacercaria**, is ingested with water plants (*F. hepatica*) or freshwater fish (*O. sinensis*). Larvae penetrate the intestinal wall and migrate to the liver, where they enter the bile ducts and develop into egg-producing adult flukes. Eggs, which are diagnostic of infection, are passed in the feces. Fluke eggs are distinctive because they have an **operculum**, or lid, that opens to release the larva. If egg-containing feces contaminate water, the larvae penetrate the shell of an aquatic snail, where the worm completes its life cycle. Infected snails release free-swimming larvae that encyst in water plants or fish, completing the cycle of infection when people eat them (**Figure 23.20**).

Because these flukes invade the liver, infections can cause signs and symptoms of liver damage: liver enlarge-

ment, tenderness, and jaundice. Liver fluke infection may cause extensive cell death, called **liver rot**, or predispose the infected person to the development of bile duct cancer. *F. hepatica*, which infects sheep as well as humans, causes infection in sheep-raising areas of Egypt, Japan, Latin America, and countries of the former Soviet Union. *O. sinensis* occurs in China, Japan, Korea, and Vietnam. Both infections occur in the United States only among people who have traveled to the affected areas. The drug of choice for both infections is praziquantel. Prevention depends on sanitation to prevent fecal contamination of water.

Table 23.3 summarizes the infections of the digestive system.

Table 23.3 Infections of the Digestive System

Infection	Causative Agent	Mode of Transmission	Symptoms	Prevention and Treatment
ORAL CAVITY AND SALIVARY GLAND INFECTIONS				
Bacterial				
Dental caries	*Streptococcus mutans* and others	Early colonization	Tooth enamel decay, eventual tooth loss	Prevent by limiting sucrose, cleaning frequently, adding fluoride to drinking water; treat by repairing cavity with filling
Periodontal disease	*Bacteroides gingivalis* and others	Early colonization	Destruction of tissues that support teeth	Prevent by cleaning carefully around gum margin
Viral				
Mumps	Mumps virus	Respiratory droplets	Swelling of parotids and other salivary glands; occasionally other organs involved, such as brain or testes	Prevent with vaccination
INTESTINAL TRACT INFECTIONS				
Bacterial				
Shigellosis	*Shigella* spp.	Fecal-oral route	Dysentery; fever, abdominal pain, possibly seizures in children	Prevent with good hygiene; treat with rehydration and antibiotics
Typhoid fever	*Salmonella typhi*	Fecal-oral route	Enteric fever with headache, nausea, fatigue, confusion; complications include intestinal bleeding or perforation	Prevent with good hygiene and clean water supplies; treat with antibiotics
Salmonellosis	*Salmonella enteritidis*	Contaminated foods	Watery diarrhea, vomiting, abdominal pain; complications include dehydration and blood-borne infection	Treat with rehydration and antibiotics
Traveler's diarrhea	*Escherichia coli* and other Gram-negative rods	Fecal-oral route	Watery diarrhea, nausea, vomiting; complications include dehydration	Prevent with good hygiene, boiling water, and cleaning or cooking food; treat with rehydration and antibiotics for severe cases
Cholera	*Vibrio cholerae*	Fecal-oral route	Profuse watery diarrhea, often life-threatening dehydration	Prevent with good hygiene and clean water supply; treat with rehydration
V. parahaemolyticus gastroenteritis	*Vibrio parahaemolyticus*	Eating undercooked, contaminated fish	Watery diarrhea, nausea, vomiting; complications include dehydration	Prevent by refrigerating and cooking fish
Y. enterocolitica diarrhea	*Yersinia enterocolitica*	Fecal-oral route	Bloody diarrhea, fever, abdominal pain	Prevent with good hygiene; treat with antibiotics
Campylobacteriosis	*Campylobacter jejuni*	Fecal-oral route	Bloody diarrhea, fever, abdominal pain	Prevent with good hygiene; treat with antibiotics in severe cases
Gastritis, peptic ulcer	*Helicobacter pylori*	Unknown	Colonization of the gastric lining by *H. pylori* causes ulcer disease	Treatment remains controversial
C. difficile diarrhea	*Clostridium difficile*	Opportunistic pathogen—iatrogenic after antibiotic treatment; also nosocomial transmission	Colitis with possibly severe diarrhea	Treat with the antibiotic vancomycin in severe cases

Table 23.3   (continued)

Infection	Causative Agent	Mode of Transmission	Symptoms	Prevention and Treatment
Staphylococcal food poisoning	*Staphylococcus aureus*	Eating food that contains preformed toxin	Intoxication that causes early onset of watery diarrhea and vomiting	Prevent by properly preparing and refrigerating food; treat by rehydration in severe cases
B. cereus food poisoning	*Bacillus cereus*	Eating contaminated food	Intoxication, usually mild; one form causes principally diarrhea and another, principally vomiting	Prevent by properly refrigerating food
C. perfringens food poisoning	*Clostridium perfringens*	Eating contaminated food	Usually a mild diarrhea	Prevent by properly refrigerating food
Botulism	*Clostridium botulinum*	Eating contaminated food	Food poisoning with neurological manifestations (Chapter 25)	Prevent by properly canning and cooking food
Viral				
Rotavirus diarrhea	Rotavirus	Fecal-oral route	Profuse watery diarrhea and vomiting; often severe dehydration; usually affects children	Prevent with good hygiene
Norwalk diarrhea	Norwalk agents	Fecal-oral route	Watery diarrhea and vomiting; often occurs in epidemics	Prevent with good hygiene
Protozoal				
Amoebiasis	*Entamoeba histolytica*	Fecal-oral route	Colitis, with bloody diarrhea; may spread to distant organs, may cause liver abscesses	Prevent with good hygiene; treat with metronidazole and iodoquinol
Giardiasis	*Giardia lamblia*	Fecal-oral route	Diarrhea, abdominal pain, foul-smelling intestinal gas	Prevent with good hygiene; treat with metronidazole
Balantidiasis	*Balantidium coli*	Fecal-oral route	Diarrhea with blood and pus, abdominal pain, loss of appetite	Prevent with good hygiene; treat with tetracycline or metronidazole
Cryptosporidiosis	*Cryptosporidium* spp.	Fecal-oral from other humans or animal reservoirs	Diarrhea, often profuse and chronic in AIDS patients	Prevent with good hygiene; treatment limited to rehydration
Helminthic				
Pinworms	*Enterobius vermicularis*	Fecal-oral route	Asymptomatic or anal itching; common in American children	Prevent with good hygiene; treat with pyrantel pamoate or mebendazole
Ascariasis	*Ascaris lumbricoides*	Fecal-oral route	Often asymptomatic but can cause cough during lung migration, bile duct or intestinal obstruction	Prevent with good hygiene; treat with mebendazole, pyrantel pamoate, or piperazine
Hookworm	*Ancyclostoma duodenale* or *Necator americanus*	Worm larvae penetrate skin	Nausea, vomiting, abdominal pain; significant anemia can be a complication	Prevent by not walking barefoot on contaminated ground and not contaminating soil with infected feces; treat with mebendazole or pyrantel pamoate
Strongyloidiasis	*Strongyloides stercoralis*	Worm larvae penetrate skin	Abdominal pain, diarrhea, malnutrition, coughing, or wheezing due to lung migration	Prevention same as for hookworm

Table 23.3 (continued)

Infection	Causative Agent	Mode of Transmission	Symptoms	Prevention and Treatment
Whipworm	*Trichuris trichiura*	Fecal-oral route	Usually asymptomatic, but may cause bloody diarrhea, weakness, weight loss	Prevent with good hygiene; treat with mebendazole
Trichinosis	*Trichinella spiralis*	Eating infected meat that is incompletely cooked	Fever, muscle pain	Prevent by cooking meat well and not feeding pigs garbage; treatment with thiabendazole, mebendazole, and steroids only moderately effective
Tapeworm infections	*Taenia saginata* (beef tapeworm) and *T. solium* (pork tapeworm)	Eating incompletely cooked, infected meat	Often asymptomatic except for passing worm proglottids; may cause malnutrition and weight loss	Prevent by eating well-cooked meat; treat with niclosamide or praziquantel
Echinococcosis (hydatid disease)	*Echinococcus granulosus*	Fecal-oral route, from dogs	Tissue destruction due to growth of cysts	Prevent by good hygiene; treat by surgically removing cysts

INFECTIONS OF THE LIVER

Viral

Infection	Causative Agent	Mode of Transmission	Symptoms	Prevention and Treatment
Hepatitis A	Hepatitis A virus (HAV)	Fecal-oral route	Jaundice, nausea, vomiting, loss of appetite; usually not life-threatening, no chronic infection or late complications	Prevent by good hygiene and new vaccine; only treatment is supportive care
Hepatitis B	Hepatitis B virus (HBV)	Contaminated blood or blood products, sexual contact, mother-infant contact during pregnancy or at time of birth	Jaundice, fulminant infection with liver failure may be fatal; late complications include chronic infection that may lead to cirrhosis or liver cancer	Prevent by vaccine and avoiding high-risk behaviors such as illicit IV drugs and unprotected sex; treatment is supportive only
Hepatitis delta	Hepatitis delta agent	Unknown	Coinfection with HBV required; may cause fulminant acute hepatitis or chronic infection	No prevention or treatment aside from avoiding HBV infection
Hepatitis C	Hepatitis C virus (HCV)	Contaminated blood or blood products, sexual contact	Acute hepatitis like type A or B and a chronic infection that may lead to liver damage like type B	Avoid high-risk behaviors (see hepatitis B)
Hepatitis E	Hepatitis E virus (HEV)	Fecal-oral route	Usually asymptomatic or mild infection like type A, but often fatal in pregnant women	Prevent by good hygiene (see hepatitis A)

Helminthic

Infection	Causative Agent	Mode of Transmission	Symptoms	Prevention and Treatment
Liver fluke infections	*Fasciola hepatica* and *Opisthorchis sinensis*	Eating larval worms in fish or water plants	Enlarged, tender liver; jaundice	Prevent by avoiding fecal contamination of water; treat with praziquantel

Summary

THE DIGESTIVE SYSTEM (pp. 555–557)

1. The digestive system has two parts—the first is a tubelike tract that includes the oral cavity, pharynx, esophagus, stomach, small intestine, and large intestine, and the second consists of six associated structures (teeth, tongue, salivary glands, liver, gallbladder, and pancreas).

2. Together, these organs consume food, digest it, absorb nutrients, and eliminate unabsorbable waste. Part of the process is mechanical and part is chemical.

3. Clinical syndromes of the intestinal tract include gastritis (upper abdominal pain), gastroenteritis (diarrhea, sometimes vomiting and cramping pain), and colitis (also called dysentery, which is characterized by diarrhea containing blood and mucus).

4. Gastroenteritis can also be caused by bacteria that produce toxins outside the intestinal tract, which are then ingested in food. These are intoxications (food poisoning), not infections.

5. Clinical syndromes of the associated structures include dental caries, periodontal disease, parotitis, and hepatitis.

INFECTIONS OF THE ORAL CAVITY AND SALIVARY GLANDS (pp. 557–560)

1. *Streptococcus mutans*, the primary cause of dental caries, produces glucans (the basis of dental plaque) and lactic acid (which can penetrate tooth enamel).

2. *Bacteroides gingivalis* is the primary cause of periodontal disease. Infection begins when plaque accumulates at the gum margin, causing gingivitis. Eventually, deep pockets form in the infected tissue where anaerobic bacteria multiply (periodontitis). Teeth eventually become loose and fall out.

3. The mumps virus is transmitted through saliva or respiratory secretions of an infected person. It causes parotitis (swelling of the parotid glands), usually with mild pain and fever. After local multiplication, the virus enters the bloodstream, occasionally causing complications. In adult males, the testes may be painfully infected. Immunization for mumps is part of the MMR (measles, mumps, rubella) vaccine.

INFECTIONS OF THE INTESTINAL TRACT (pp. 560–578)

1. Transmission of nearly all intestinal tract infections is by the fecal-oral route; the infections range from mild to deadly.

Bacterial Causes (pp. 560–570)

1. *Shigella* spp. cause shigellosis, or bacillary dysentery. Over half the reported cases are in children under age 5. Transmission is by the oral-fecal route, and it is highly communicable, spreading rapidly through a family or day-care center. The bacteria produce a potent shiga (or shiga-like) toxin. An oral vaccine is under development, but at the same time, more drug-resistant strains are emerging. Sulfa, once the drug of choice, is useless unless used in combination.

2. *Salmonella typhi* causes typhoid fever, which begins with a high fever that lasts for days or weeks and develops into headache, other aches and pains, and sometimes a rash called rose spots. Patients shed the bacteria in feces and thereby infect others through the oral-fecal route. A small number become carriers. Many *S. typhi* strains are drug-resistant.

3. Salmonellosis is most often caused by *S. enteritidis* or *S. cholerae-suis*. It is often mistakenly classified as food poisoning, but it is a true intoxication. Symptoms include diarrhea, cramps, fever, and vomiting. The infection is usually self-limiting, and antibiotic treatment is contraindicated unless there are complications.

4. Enterotoxigenic strains of *Escherichia coli* are the major cause of traveler's diarrhea and infant diarrhea in developing countries. Enteroinvasive strains of *E. coli* cause an illness almost identical to shigellosis. Enteropathogenic strains of *E. coli* cause diarrhea in newborns.

5. *Vibrio cholerae* causes classical Asiatic cholera, with its characteristic rice water stools. Diarrhea is so profuse that dehydration is sudden, and death often follows. Treatment consists of oral rehydration and tetracycline. In January 1991, the El Tor strain was found in Peru; cholera is probably in the Americas to stay for the first time in history. Most cases occur because of an infected water supply, but shellfish now also appear to be a reservoir.

6. *Vibrio parahaemolyticus* causes either a mild, self-limiting diarrhea or a cholera-like illness. Because it comes from eating contaminated fish, it is often mistakenly considered a food poisoning; it is a true intoxication. About a quarter of all reported diarrhea cases in Japan are caused by *V. parahaemolyticus*.

7. *Yersinia enterocolitica* causes an enterocolitis; abdominal pain can be so severe that appendicitis is suspected. Infection can come from wild or domestic animals, raw milk, oysters, water, or chitterlings.

8. *Campylobacter jejuni* was not considered a pathogen until the 1970s, and since then, scientists have found that it probably causes more diarrheal illness in the United States than either *Salmonella* or *Shigella*. Campylobacteriosis is characterized by bloody diarrhea, abdominal pain, and sometimes fever. It is usually self-limiting.

9. *Helicobacter pylori* (formerly *Campylobacter pylori*) has been established as a cause of gastritis and stomach ulcers, which can lead to gastric carcinoma.

10. *Clostridium difficile* causes iatrogenic diarrhea as a result of antibiotic therapy. When therapy stops, the problem is usually resolved. For severe or chronic diarrhea, treatment is with vancomycin.

11. *Staphylococcus aureus* causes the most frequently reported food poisoning in the United States. Onset is sudden, with vomiting, diarrhea, and crampy abdominal pain. Staphylococci are accidentally introduced into food (most often whipped cream, eggs, mayonnaise, ham) by someone carrying an enterotoxin-producing strain of *S. aureus*. The bacteria multiply in warm food. Prevention is possible through careful hand washing during food preparation and through refrigeration.

12. *Bacillus cereus* causes a mild, brief food poisoning. Spores of *B. cereus* not killed during cooking multiply if the food (usually meat, vegetables, rice) is not refrigerated.

13. *Clostridium perfringens* can cause a food poisoning so brief and mild that it is seldom reported. Spores in meat or gravy that is not refrigerated germinate and, when ingested, produce vegetative cells that produce an enterotoxin.

Viral Causes (pp. 570–571)

1. Rotavirus causes a gastroenteritis characterized by watery diarrhea, fever, and vomiting—commonly called "the stomach flu." It is highly infective by the fecal-oral route; the best prevention is hand

washing. Malnourished children in developing countries become dehydrated and die from rotavirus infection.

2. Norwalk agents are extremely small viruses named for an epidemic of gastroenteritis that occurred in Norwalk, Ohio. Norwalk agents affect adults more often than children; usually the illness is brief and self-limiting.

Protozoal Causes   (pp. 572–574)

1. *Entamoeba histolytica* is an amoeba normally found in the intestines. Where sanitation is poor, it can cause amoebic dysentery (bloody mucoid stools, fever, abdominal pain) and amoebiasis (the parasites enter the bloodstream and can cause fatal liver abscesses). *E. histolytica* exists as a motile trophozoite and as a cyst, which is passed in the feces. Treatment is with metronidazole in combination with iodoquinol.

2. *Giardia lamblia*, a flagellated protozoan, causes giardiasis, which begins with an explosive diarrhea and copious amounts of foul-smelling intestinal gas. The first few such days are followed by a week or more of abdominal pain, nausea, and occasional diarrhea. *G. lamblia* exists as a trophozoite and cyst; infected individuals shed cysts. *G. lamblia* has a sylvatic cycle, putting campers at special risk. When it occurs in the United States, it is most often in children. Treatment is with metronidazole or quinacrine.

3. *Balantidium coli* is a ciliate protozoan that exists as a trophozoite and as a cyst. *B. coli* causes balantidiasis, most often seen in tropical countries. Infected individuals may have long symptom-free periods. Other carriers may be entirely healthy. Metronidazole or tetracycline is effective.

4. *Cryptosporidium* is a protozoan that causes cryptosporidiosis, normally a mild enterocolitis. However, AIDS patients and other immunocompromised individuals can develop life-threatening diarrhea that lasts for months. Treatment is limited to rehydration.

Helminthic Causes   (pp. 574–578)

1. *Enterobius vermicularis*, a nematode, causes pinworm, the most common helminthic infection in the United States. Infection comes from ingesting pinworm eggs, which can live for up to a week on fomites. Children are most commonly infected, though adults who handle children are at risk. Perianal itching is the primary symptom. Treatment is with pyrantel pamoate or mebendazole.

2. *Ascaris lumbricoides* is a nematode. Infection comes from ingesting eggs, which hatch and burrow through the wall of the intestine to enter the bloodstream. They travel to the lungs, where they mature in three weeks. Larvae are coughed out of the lungs, swallowed, and reenter the digestive tract. Worms that mature in the intestines grow to 12 inches or more and produce eggs that are passed in the feces. Treatment is with various antiparasitic drugs.

3. *Ancyclostoma duodenale* and *Necator americanus* are nematodes known collectively as hookworms. Hookworm eggs hatch in soil, and the filariform larvae penetrate human skin. Larvae travel to the lungs, where they get coughed up and swallowed into the intestinal tract; mature worms in the intestines produce eggs that are passed in the feces. Hookworms attach to the intestines and suck blood, so heavy infestations can cause diarrhea, vomiting, and even malnutrition and anemia. Mebendazole and pyrantel pamoate are effective.

4. *Strongyloides stercoralis*, a nematode that causes strongyloidiasis, penetrates the skin, travels to the lungs, and produces adult worms in the intestinal tract. It has a free-living life cycle and the eggs hatch into infective larvae in the intestine, so the person can suffer continuous reinfection (autoinfection). Immunocompromised individuals can suffer a fatal diarrhea. Treatment is with mebendazole or pyrantel pamoate.

5. *Trichuris trichiura*, a nematode, causes whipworm. Infection begins by ingesting eggs, which hatch in the small intestine and burrow into the wall. They never leave the intestine. Heavy infestations can cause bloody diarrhea, abdominal pain, and weight loss. Treatment is with mebendazole.

6. *Trichinella spiralis*, a nematode, causes trichinosis. Infection is by ingesting encysted larvae in meat, usually pork. The larvae develop into worms in the small intestine and produce more larvae that enter the bloodstream and travel to muscle, where they form cysts. Symptoms are fever, muscle pain, and malaise. In severe cases, larvae enter the heart, lungs, and brain. Trichinosis can be prevented with thorough cooking or freezing.

7. *Taenia* spp. are cestodes (tapeworms) of the flatworm family. Infection begins with eating encysted larvae in pork (*T. solium*) or beef (*T. saginata*). The larvae mature in the intestines, where the adult attaches itself by suckers on its scolex and grows by adding proglottids. Mature proglottids, which contain infective eggs, are passed in the feces. Symptoms may include abdominal pain, diarrhea, and indigestion. A life-threatening infection called cysticercosis may develop if ingestion of eggs occurs as a result of human fecal contamination. Thoroughly cooking meat is the best prevention.

8. *Echinococcus granulosus* is a cestode that causes hydatid disease, which is endemic to sheep-raising areas. A person ingests eggs from dog feces. Hydatid cysts develop in their tissues, sometimes to 12 inches in diameter. Symptoms depend on the location. Surgical removal of the cyst is the only treatment.

INFECTIONS OF THE LIVER   (pp. 578–582)

1. Hepatitis A virus (HAV) causes type A hepatitis (infectious hepatitis). Transmission is usually by the fecal-oral route. Symptoms include jaundice, fatigue, loss of appetite, vomiting, and disorders of taste and smell. Illness is usually self-limiting and the virus is eliminated from the body. Treatment is with immune serum globulin. An experimental vaccine is being evaluated by the FDA.

2. Hepatitis B virus (HBV) causes hepatitis B (serum hepatitis). Transmission is through contaminated blood or sexual contact. The clinical syndrome is indistinguishable from that of hepatitis A, but HBV causes cancer, fulminant hepatitis (total liver failure), and chronic infection leading to death. HBV vaccination is part of the routine infant inoculation schedule.

3. Hepatitis D affects only those already infected with hepatitis B; it is frequently fatal.

4. Hepatitis C virus (HCV) causes hepatitis C (called non-A non-B hepatitis until DNA technology identified it). Transmission in 70 percent of the cases is by blood or sexual contact, but how the other 30 percent are transmitted is unknown. About one-fourth of infected individuals develop chronic liver disease, but symptoms do not appear for six months after infection. Alpha interferon may be useful in treatment.

5. Hepatitis E (HEV) virus was only recently identified by DNA technology. It causes hepatitis E. Transmission is by the fecal-oral route; hepatitis E is self-limiting and benign except for pregnant women, about 20 percent of whom die from acute hepatitis.

6. *Fasciola hepatica* (sheep liver fluke) and *Opisthorchis sinensis* (Chinese liver fluke) invade the liver and cause jaundice and liver rot or predispose the person to bile duct cancer. Infection comes from eating the larval form (metacercaria) with water plants or fish. The worm requires a particular aquatic snail to complete its life cycle.

THE DIGESTIVE SYSTEM

1. Describe the two parts of the digestive system, naming the six associated structures.

2. What is the function of the digestive system? Explain this statement: Part of the digestive process is mechanical and part is chemical.

3. Describe the following clinical syndromes:
   a. gastritis   d. dental caries   f. parotitis
   b. gastroenteritis   e. periodontal disease   g. hepatitis
   c. colitis (dysentery)

4. What is the difference between an intoxication and food poisoning?

INFECTIONS OF THE ORAL CAVITY AND SALIVARY GLANDS

1. Explain how *Streptococcus mutans* causes dental caries. What new treatments are being researched?

2. What is periodontal disease and how does it develop? How is it treated?

3. What causes the mumps? Is this infection serious? Explain.

INFECTIONS OF THE INTESTINAL TRACT

Bacterial Causes

1. Describe the cause, transmission, pathogenic mechanisms, and symptoms of shigellosis. How and why has treatment changed, and what are the prospects for the future?

2. Discuss the prevention of typhoid fever, including problems with carriers. What problems are arising with treatment today?

3. Is salmonellosis a food poisoning or an intoxication? Why might there be any confusion? Why is antibiotic treatment normally contraindicated?

4. Distinguish between enterotoxigenic strains of *Escherichia coli*, enteroinvasive strains, and enteropathogenic strains. Mention the illnesses each causes.

5. How does the *Vibrio cholerae* toxin work? How can cholera be prevented? What new breakthroughs are there in treatment? Discuss the recent epidemiology of cholera.

6. Describe the clinical syndrome caused by *Vibrio parahaemolyticus* infection. Who is at risk?

7. What is distinctive about the enterocolitis caused by *Yersinia enterocolitica*? What are the sources of infection?

8. Is infection by *Campylobacter jejuni* a major public health concern? Explain.

9. How did *Campylobacter pylori* come to be reclassified as *Helicobacter pylori*? Does *H. pylori* cause gastritis and ulcers? Explain your answer.

10. What is an iatrogenic illness? How does *Clostridium difficile* cause an iatrogenic diarrhea? How is the illness usually resolved?

11. Describe how *Staphylococcus aureus* can come to contaminate food and what circumstances would then lead to food poisoning. What are the symptoms? How do you prevent staphylococcal food poisoning?

12. How does *Bacillus cereus* cause food poisoning? How is this mechanism the same as or different from the one that causes *Clostridium perfringens* food poisoning?

Viral Causes

1. Describe the cause, symptoms, prevention, and treatment of the gastroenteritis we commonly call the stomach flu.

2. What are Norwalk agents? How is the illness they cause like and unlike rotavirus infection?

Protozoal Causes

1. What are the symptoms of amoebic dysentery? How is it diagnosed, why is diagnosis difficult, and why, at the same time, is prompt diagnosis important? What is the treatment?

2. Name the two forms of *Giardia lamblia* and their roles in infection. Explain this statement: *G. lamblia* has a sylvatic cycle. Describe symptoms, prevention, and treatment of giardiasis.

3. How does *Balantidium coli* cause infection? Where is balantidiasis most often seen?

4. Why are there more cases of cryptosporidiosis today than in the past? What preventive measures can high-risk individuals take?

Helminthic Causes

1. What is the most common helminthic infection in the United States? How serious is it (explain)? Why is it highly infectious? Name the microorganism that causes it.

2. Describe the life cycle of *Ascaris lumbricoides*. How does the life cycle cause symptoms in an infected individual? What kind of helminth is *A. lumbricoides*?

3. How are the two species of hookworm different? How are they alike? Trace the life cycle, from filariform larvae to egg-producing adult. How is hookworm diagnosed and treated? Why might iron supplements be required?

4. How is *Strongyloides stercoralis* like hookworms? How is it different? What is autoinfection?

5. Describe a whipworm infection—from transmission through treatment.

6. How does *Trichinella spiralis* cause illness in its human host? What are the symptoms? Can trichinosis be fatal? Explain. What kind of helminth is *T. spiralis*? Why is trichinosis relatively rare in the United States today?

7. What kind of helminths belong to the *Taenia* spp.? How do *T. solium* and *T. saginata* cause illness? How is infection diagnosed? What is cysticercosis?

8. How does *Echinococcus granulosus* cause hydatid disease? Who is most at risk? What is the treatment? What kind of helminth is *E. granulosus*?

INFECTIONS OF THE LIVER

1. For each type of hepatitis (A, B, C, D, and E), give the following:
   a. the causative agent
   b. mode of transmission
   c. individuals most at risk
   d. clinical syndrome and complications (if any)
   e. treatment
   f. prevention

2. Explain this statement: Theoretically, the HBV vaccine ought to eliminate hepatitis B within a generation. Discuss the specific obstacles.

3. What kind of helminths are *Fasciola hepatica* and *Opisthorchis sinensis*? Describe their life cycle. What are the symptoms and treatment? Why do infections in the United States occur only among people who have traveled to the infected areas?

Suggested Readings

Alter, M. J. 1989. Hepatitis C: And miles to go before we sleep. *New England Journal of Medicine* 321:1538–39.

Cover, T. L., and Aber, R. C. 1989. Medical progress: *Yersinia enterocolitica*. *New England Journal of Medicine* 321:16–24.

Gordon, R. 1989. Tales of typhoid Mary. *Hippocrates*, March/April, 102–3.

Grady, G. F. 1986. The here and now of hepatitis B immunization. *New England Journal of Medicine* 315:250–51.

Mahmoud, A. A. F. 1989. Parasitic protozoa and helminths: Biological and immunological challenges. *Science* 246:1015–22.

Shaw, J. H. 1987. Medical progress: Causes and control of dental caries. *New England Journal of Medicine* 317:996–1004.

Williams, R. C. 1990. Medical progress: Periodontal disease. *New England Journal of Medicine* 322:373–82.

Case Study: You Can't Be Too Careful

A.V., a 25-year-old married woman, noticed some discomfort on urination one morning. Over the next few hours, her discomfort became a burning sensation, and she found it necessary to urinate often and suddenly although she passed very little urine each time. These symptoms continued to worsen and, by the next morning, she found she had a fever of 101°F. She felt nauseated, vomited after breakfast, and noticed pain on her right side. A.V. called her doctor.

A.V.'s physician recorded her history of fever, dysuria (painful urination), urinary frequency, and urinary urgency. When he examined her, she had a fever of 102°F and appeared quite ill. He noted tenderness when he pressed on the part of the abdomen overlying the bladder. When he tapped on the right side of her back near her kidney—where the ribs meet the spine—A.V. winced in pain. Her physician asked A.V. to collect a urine specimen for immediate urinalysis. Under the microscope the specimen showed large numbers of bacteria and leukocytes, with the leukocytes clumped together in long masses called casts. All of these findings were abnormal. The rest of the urine sample was sent to a microbiology laboratory for culture.

A.V.'s physician made a preliminary diagnosis of urinary tract infection (UTI). The fever and tenderness around her right kidney further suggested the infection was probably a pyelonephritis, kidney infection. He prescribed trimethoprim-sulfamethoxazole, an oral antimicrobial agent effective against most enterobacteria as well as rarer causes of UTI, such as enterococci. He asked A.V. to call him that evening to report whether she was able to take the medication without vomiting. He also wanted to see her again the next morning. Twenty-four hours later A.V. felt and looked much better. Her temperature had returned to normal, and her urinary symptoms were improving, though not gone. At her doctor's office, a urine specimen showed very few white blood cells and no bacteria. A.V. was told to continue taking the medicine for 10 days and return once more. Urine samples from her second and third visits were also sent to the laboratory for culturing.

The laboratory results on the urine sample collected during A.V.'s first visit reported a pure culture of *Escherichia coli*, containing more than 100,000 bacteria per milliliter of urine. Fortunately, this strain of *E. coli* was sensitive to trimethoprim-sulfamethoxazole, and the sample taken 24 hours later produced no bacterial growth. The urine sample taken at the end of her course of treatment was also sterile.

ROUTINE URINALYSIS

THE MEDICAL GROUP, INC.

M.D. Vigran, C.   EXT. 303

SEND REPORT TO _____ FACILITY
☐ MENSES
DATE 8/2/94   TIME COLLECTED 8:00 (AM/PM)   ☐ PRE-OP
COMMENTS   ☐ OP DATE _____
COMMENTS

pH	6.0
PRO	–
GLU	–
KET	+
BIL	–
BLO	–
NIT	+
URO	–
S.G.	1.025

INTERP.	TRACE	1+	2+	3+
GLU	100	250	500	1000
KET	TRACE	15	40	80
PRO	TRACE	30	100	300

Microscopic Performed When BLD, PRO, NIT or LEU Positive.

☐ Microscopic Not remarkable (Refer to Local Policy)

CULTURE PER FACILITY PROTOCOL
☐ ☐ MICRO EVEN IF DIPSTICK NORMAL
TIME OUT:
TIME IN:

ROUTINE URINE

>100	/hpf	WBC
0	/hpf	RBC
0	/lpf	Hyaline Casts
10–20	/lpf	Fine Granular Casts
5–10	/lpf	Course Granular Casts
0	/lpf	Other Casts
0	/lpf	Epith. Cells
4+		Bacteria
0		Crystals
neg.		Abnormal Color/Appearance

Specimen submitted for culture
☑ YES   ☐ NO

COMMENTS

TECH: LAB

PT. PHONE NO./ROOM NO.
SEX F
M.R. NUMBER
LAST NAME V.
COVERAGE
BIRTHDATE 7/7/69
FIRST NAME A.
PLEASE PRINT OR IMPRINT

Laboratory slip for urinalysis.

To understand:
- The anatomy and function of the urinary and reproductive systems and their defenses against microorganisms

- The clinical syndromes that characterize genitourinary infections

- The causes of urinary tract infections and their clinical syndromes; the prevention and treatment of these infections

- The bacterial and viral causes of sexually transmissible diseases (STDs) and their clinical syndromes; prevention and treatment of these infections

- The bacterial, fungal, and protozoal causes of female reproductive tract infections and their clinical syndromes; the prevention and treatment of these infections

- The causes of male reproductive tract infections and their clinical syndromes; the prevention and treatment of these infections

- The bacterial and viral causes of infections transmitted from mother to infant and their clinical syndromes; prevention and treatment of these infections

THE GENITOURINARY SYSTEM

Together, the **genital**, or reproductive, and urinary organs form the **genitourinary system (Figure 24.1)**. These organs are closely related anatomically but fulfill two entirely different functions—reproduction and excretion of metabolic wastes. The genitourinary system is the only organ system in which anatomy differs significantly between males and females. In females, the reproductive and urinary tracts are completely separate. In males, the reproductive and urinary tracts have separate origins, but join in a common outflow tract.

Structure and Function of the Urinary System

The urinary system of both males and females includes two **kidneys** that clear metabolic waste from the body. The kidneys filter the blood, removing impurities that are concentrated in the **urine**. Urine formation begins in a cluster of blood vessels called the **glomerulus** that filters blood through a porous membrane under high pressure. The filtrate passes through a series of microscopic tubules in the kidney where certain molecules are added to the urine and others are reabsorbed into the bloodstream. Water is also reabsorbed, concentrating the urine as it passes through the tubules. Urine leaves the kidneys through two conduits called **ureters**. The kidneys and the ureters constitute the **upper urinary tract**.

The ureters transport the urine to the saclike **bladder**, where it is stored until urine leaves the body through a tube called the **urethra**. In females the urethra is only a few centimeters long and transports only urine. The male urethra is longer, leading to the head of the penis. The male urethra is a conduit for both urine and the sperm-containing fluid, **semen**. The bladder and urethra constitute the **lower urinary tract**.

Structure and Function of the Reproductive System

The reproductive systems of males and females produce, store, and transport the **gametes**, reproductive cells, that fuse to create a new human being. The female reproductive tract also provides a place for the fetus to develop before birth.

In the male reproductive system, two organs called **testes**, contained within the saclike **scrotum**, produce **sperm**, the male gametes. Sperm is conducted from the testes to the urethra through a series of ducts. Along this pathway several glands, including the **prostate gland**, contribute components to the semen. Semen leaves the body through the urethra by ejaculation. Because it transports both urine and semen, the male urethra belongs to both the urinary and genital systems.

In females, two **ovaries**, organs contained within the abdominal cavity, produce **ova**, the female gametes. Ova are transported from the ovaries through the **fallopian tubes** to the **uterus**, which ends in the **cervix**, a necklike extension that projects into the **vagina**. If the ovum unites with a sperm cell, the result is an **embryo**, or human being in its earliest stages of development. The embryo develops within the uterus. Contents of the uterus, including menstrual secretions as well as products of conception, leave the body by passing through the vagina. Sperm enter the female body through the vagina on their way to fertilize the ovum.

Defenses and Normal Flora: A Brief Review

The genitourinary system transports urine and reproductive products from the inside to the outside of the body. In this respect the genitourinary system differs fundamentally from the respiratory and gastrointestinal systems, which transport material (air and food) from

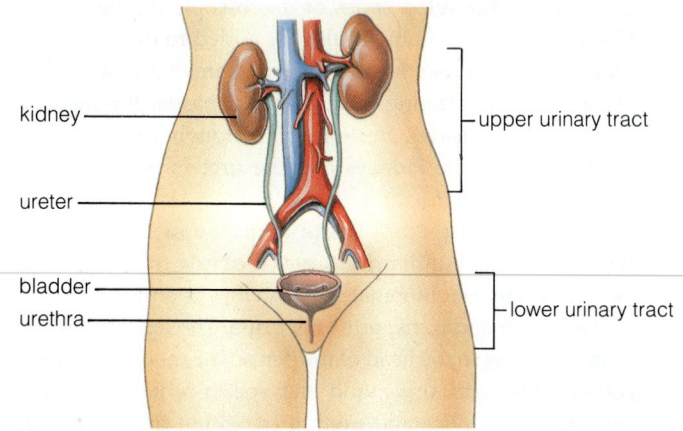

a

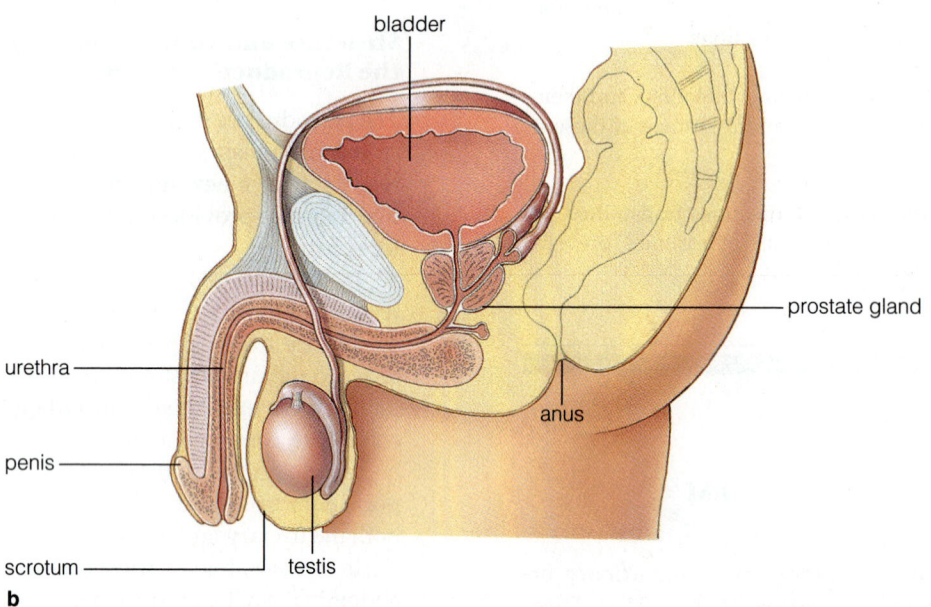

b

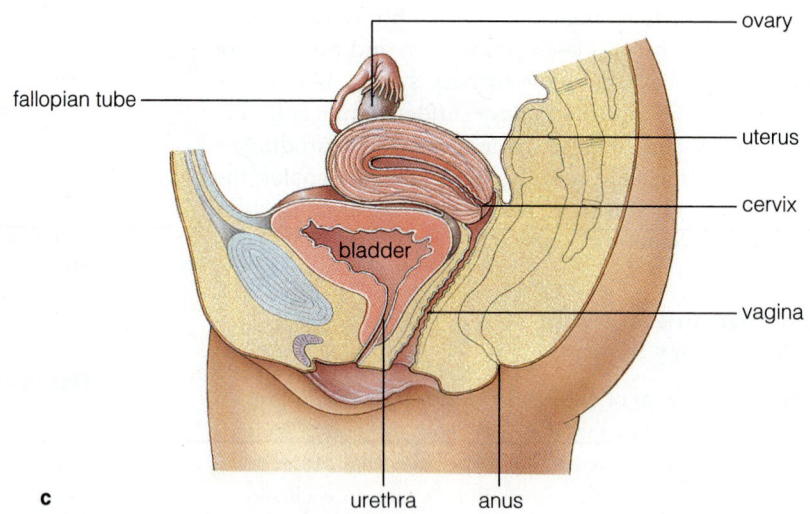

c

Figure 24.1 (**a**) Organs of the urinary system. (**b**) Components of the male reproductive system. (**c**) Components of the female reproductive system.

outside the body to the inside. The respiratory and gastrointestinal systems are constantly exposed to the microorganisms in the environment and are heavily colonized by commensal species (Chapter 14). In contrast, the genitourinary system comes in contact with microorganisms only where its outflow meets the skin and so is largely free of microorganisms.

A significant exception to this generalization is the female genital tract. The vagina, which accepts semen and the male reproductive cells that it contains, is the only region of the genitourinary tract that is heavily colonized by microorganisms (Chapter 14).

Although most of the genitourinary system is normally sterile, pathogens invade this system and cause disease. Some pathogens enter at the point where the genitourinary system meets the skin, making their way against the flow. Pathogens readily enter the female genital tract during childbirth or abortion. Many pathogens are adapted to exploit the close person-to-person contact that occurs during sexual intercourse. These can cause the **sexually transmissible diseases**, or **STDs**. Pathogens that make their way into the female reproductive tract can, in turn, be transmitted from mother to child during birth.

The urethra is defended by tightly joined epithelial cells and the mechanical force of urine that flushes out microorganisms. Low pH limits growth of microorganisms in the vagina. The other urinary and reproductive organs depend upon the body's immune defenses (Chapter 17).

Clinical Syndromes

The urinary tract is infected frequently, usually by bacterial pathogens. **Cystitis**, infection of the bladder, is associated with **urethritis**, inflammation of the urethra that causes frequent and painful urination. **Pyelonephritis** is infection of the kidneys, associated with fever and flank pain. It is often difficult to determine the exact extent of a urinary infection, in which case it is simply called a **urinary tract infection** (**UTI**).

The female genital tract can be infected by bacteria, viruses, fungi, and protozoa. When the vagina is infected, the syndrome is called **vaginitis**, an inflammation characterized by vaginal irritation and discharge. Vaginitis is uncomfortable but not life-threatening. Infection of the lining of the uterus is **endometritis**, infection of the fallopian tubes is **salpingitis**, and infection of the ovaries is **oophoritis**. Upper genital tract infections often involve all these organs and the abdominal cavity itself and are commonly referred to as **pelvic inflammatory disease** (**PID**). PID is often associated with fever and abdominal pain and may be life-threatening. **Table 24.1** summarizes these clinical syndromes.

Many infections that produce quite different clinical syndromes are grouped together as sexually transmissible diseases. They have only one thing in common—their mode of transmission. Some STDs are systemic. Others affect only the female genital tract, causing vaginitis or PID, or the male genital tract, causing urethritis and a discharge from the penis.

Table 24.1 Clinical Syndromes of Infection of the Genitourinary System

Syndrome	Region Affected	Signs and Symptoms	Causative Agents
Cystitis	Bladder	Painful and frequent urination	Many organisms, including *Escherichia coli*
Urethritis	Urethra	Painful and frequent urination; often a discharge in males	Many organisms, including *Neisseria gonorrhoeae* in males
Pyelonephritis	Kidneys	Fever, flank pain	Many organisms, including *E. coli*
Urinary tract infection (UTI)	Any or all parts of the urinary system (bladder, urethra, kidneys)	Any or all of the symptoms of cystitis, urethritis, pyelonephritis	Many organisms, including *E. coli*, *Pseudomonas* spp., *Proteus* spp., *Klebsiella* spp.
Vaginitis	Vagina	Vaginal irritation, often a discharge	Many organisms, including *Candida albicans*, *Trichomonas vaginalis*, *Gardnerella vaginalis*
Endometritis	Uterus	Fever, tender and enlarged uterus, abdominal pain	Many organisms, including *E. coli*, *N. gonorrhoeae*, group B streptococci
Salpingitis/oophoritis	Fallopian tubes/ovaries	Fever, abdominal pain; complications include scarring and infertility	Many organisms, including *N. gonorrhoeae*, *Chlamydia trachomatis*
Pelvic inflammatory disease (PID)	Vagina, uterus, fallopian tubes, ovaries, abdominal cavity itself	Fever, abdominal pain; can be life-threatening	Many organisms, including *N. gonorrhoeae*, *C. trachomatis*

In this chapter we also discuss infections that are frequently transmitted from mother to infant at or around the time of birth (although they may also be transmitted in other ways). These **perinatal** (around birth) infections may be transmitted across the placenta, during passage through the birth canal, or after birth through breast milk or intimate mother-infant contact.

URINARY TRACT INFECTIONS

All major pathogens of the urinary tract are bacteria. They enter by passing up the urinary tract or through the bloodstream. Infections that do permanent damage to the kidneys or spread beyond the urinary tract to cause a septicemia can be life-threatening.

Urinary tract infections are extremely common, and A.V.'s case was typical in many respects. Most people who contract UTIs are women or girls who become infected by bacteria that inhabit their own gastrointestinal tract. *Escherichia coli* is the most common cause of urinary tract infections. Bacteria are introduced into the urethra from fecal contamination of the surrounding skin. Sometimes trauma to the female urethra during sexual intercourse forces bacteria into the urinary tract. Infection is frequently confined to the bladder, in part because the ureters enter the bladder at such an angle that backflow of urine into the upper urinary tract is largely prevented. Occasionally, however, bacteria pass from the bladder into the ureters and kidneys, causing pyelonephritis.

The female urethra is shorter than the male urethra and located relatively close to the anus, which partly accounts for the higher incidence of urinary tract infections in women than in men. Any type of interference with the normal flow of urine predisposes a person to infection. A partial obstruction of the urinary tract (from an enlarged prostate gland, for example, or a congenital anatomic abnormality) or an inability to completely empty the bladder (from a spinal cord injury, for example) often causes recurrent UTIs. Pregnant women and people with diabetes mellitus are also at high risk for UTIs, though the reasons are not clear. Hospitalized patients who have urinary catheterization are particularly likely to contract UTIs—they are the most common type of nosocomial infection (Chapter 20).

Diagnosis

A preliminary diagnosis of urinary tract infection is often based on a clinical history, physical examination, and urinalysis. A.V.'s physician, for example, initiated treatment on this basis. Urine collected as a midstream specimen after thorough cleaning of the urethral opening should contain very few bacteria or blood cells of any kind. When the urinary tract is inflamed, the urine typically contains large numbers of bacteria, as well as white and red blood cells, visible under light microscopy. The presence of white blood cell **casts**, as in A.V.'s urinalysis, strongly suggests a diagnosis of pyelonephritis because casts are formed in the small collecting tubules of the kidney.

For a definitive diagnosis of UTI, bacteria must be cultured from the infected urine. But since urine provides an ideal culture medium for bacteria, a specimen need only stand at room temperature for a few minutes for the small number of skin contaminants in any urine sample to begin multiplying rapidly. Therefore, a urine culture must be collected carefully to minimize contamination, be refrigerated or plated on a culture medium immediately, and produce a pure culture of more than 100,000 microorganisms per milliliter. A lower concentration of microorganisms or a mixed culture generally indicates contamination of the specimen.

Lower versus Upper Urinary Tract Infections

The seriousness of a urinary tract infection depends upon how much of the urinary tract is affected. Infections that involve only the lower urinary tract—the urethra and the bladder—are common and do not have serious long-term consequences, although they can be quite uncomfortable. The typical symptoms of lower urinary tract infection—**dysuria**, urgency, and frequency—are a natural defense mechanism because each time the bladder is emptied, bacteria are eliminated. When a normal bladder empties completely, leaving less than 1 ml of infected urine behind, sterile urine from the ureters enters, which may dilute the concentration of bacteria left in the bladder enough to control the infection. Thus, patients with urinary tract infections should drink large quantities of fluid.

Infections like A.V.'s that also involve the kidneys and ureters are less common and more worrisome than lower tract infections. Patients with upper urinary tract infections tend to be sicker—with fever, vomiting, and flank pain. Bacteria in the kidneys may also enter the bloodstream, causing a life-threatening infection. This concern prompted A.V.'s physician to follow her progress closely after diagnosing pyelonephritis. Some patients with pyelonephritis are admitted to the hospital for treatment with intravenous antibiotics. Repeated episodes of pyelonephritis may contribute to lasting kidney damage (**chronic pyelonephritis**), which can eventually cause kidney failure, necessitating dialysis or a kidney transplant.

Distinguishing a lower from an upper urinary tract infection is often difficult. A.V.'s history of fever and flank pain, coupled with finding white blood cell casts in her urinalysis, made the diagnosis of pyelonephritis almost certain. But white blood cell casts are a relatively rare finding on urinalysis, even in cases of pyelonephritis. Consequently, clinicians often make the less specific diagnosis of UTI. Fortunately, the treatment for upper and lower urinary tract infections is the same, except in seriously ill patients who require hospitalization.

Treatment

Because diagnosis of a UTI is usually made before the results of the urine culture are available, treatment is begun before the identity and antibiotic sensitivity of the infecting microorganism are known. Antibiotic sensitivity testing is particularly critical, because many strains of intestinal bacteria that cause UTIs carry R factors that make them resistant to several antibiotics (Chapter 21). So even if a physician knew the *species* of the infecting bacterium, it would be impossible to know for sure which antibiotic to prescribe. In some hospital studies, for example, about 90 percent of *Escherichia coli* strains are sensitive to trimethoprim-sulfamethoxazole (trimethoprim-sulfa), while 10 percent are resistant, making prescription of this drug for an infection like A.V.'s a good bet, but far from a sure one.

Unfortunately, no antibiotic is effective against all strains of common UTI pathogens. As a result, the physician prescribes on the basis of known patterns of antibiotic resistance in her or his community and changes drugs if symptoms fail to improve or if the culture reveals that the pathogen is resistant to the antibiotic chosen. Agents commonly used to treat urinary tract infections include the broad-spectrum penicillins, such as ampicillin, the sulfa drugs, and the so-called urinary antiseptics, nalidixic acid and nitrofurantoin (Chapter 21). Sulfa drugs are often effective even against relatively resistant microorganisms because they are concentrated by the kidneys and reach extremely high levels in the urine. A.V. received a combination sulfa drug.

Escherichia coli: Urinary Tract Infections

Escherichia coli is discussed in detail in Chapter 5. As the most abundant commensal in the large intestine, *E. coli* competes with enteric pathogens to help prevent disease (Chapter 14). Paradoxically, *E. coli* is the most common Gram-negative pathogen isolated in clinical laboratories in the United States. It causes serious and even life-threatening infections of the gastrointestinal system (Chapter 23), nervous system (Chapter 25), and urinary tract. Disease-causing strains of *E. coli* possess powerful virulence factors that nonpathogenic strains lack. (The

special virulence factors that allow some strains of *E. coli* to become significant gastrointestinal pathogens are discussed in Chapter 23.)

E. coli is by far the most common pathogen of the urinary tract. Ninety percent of healthy people who acquire their first urinary tract infection are infected by *E. coli*, and most nosocomial urinary tract infections are caused by this bacterium. One critical virulence factor in **uropathogenic** strains of *E. coli* (strains capable of causing infection of the urinary system) is adherence. Each *E. coli* bacterium that initiates a urinary tract infection has from 10 to 200 adhesin-bearing pili that bind firmly to matching cell receptors on the urinary epithelium, allowing the bacterium to withstand the normal flow of urine. Recent research suggests that some women have more numerous or more accessible epithelial receptors in their urinary tract than do other women and so are vulnerable to repeated urinary tract infections.

Uropathogenic strains of *E. coli* have additional virulence factors. Many produce hemolysins that can damage human cells and colicins that are toxic to other competing strains of *E. coli*. Some are unusually resistant to the bactericidal action of normal human serum, and many carry R factors that encode antibiotic resistance. Nosocomial infections are particularly likely to be caused by multiple-drug-resistant strains because the large quantities of antibiotics used in a hospital select for drug resistance (Chapter 20).

Certain other bacterial species cause urinary tract infections that are clinically indistinguishable from A.V.'s *E. coli* infection. Thus, clinical laboratory identification of *E. coli* depends upon differentiating it from other closely related enteric bacteria and *Pseudomonas*. *Pseudomonas* and the enteric bacteria *Proteus* and *Klebsiella* most often infect nosocomial cases or individuals prone to recurrent UTIs. Because *Proteus* and *Klebsiella* spp. break down urea and create an alkaline urine, they may also foster the formation of kidney stones (collections of solid material that can block urinary flow through the ureters). Some Gram-positive bacteria, particularly enteric streptococci, also cause urinary tract infections. *Staphylococcus saprophyticus* was recently recognized as a cause of UTI in young, sexually active women.

Leptospira interrogans: Leptospirosis

Leptospirosis is a systemic infection caused by *Leptospira interrogans*, a member of the spirochete family. All *Leptospira* cells are tightly coiled with a characteristic hook at one or both ends. They move by two periplasmic flagellae, one anchored at each end of the cell. The *Leptospira* that cause disease in humans or animals belong to the species *interrogans*, named for the single hook that makes the cell resemble a question mark (**Figure 24.2**). *L. interrogans*

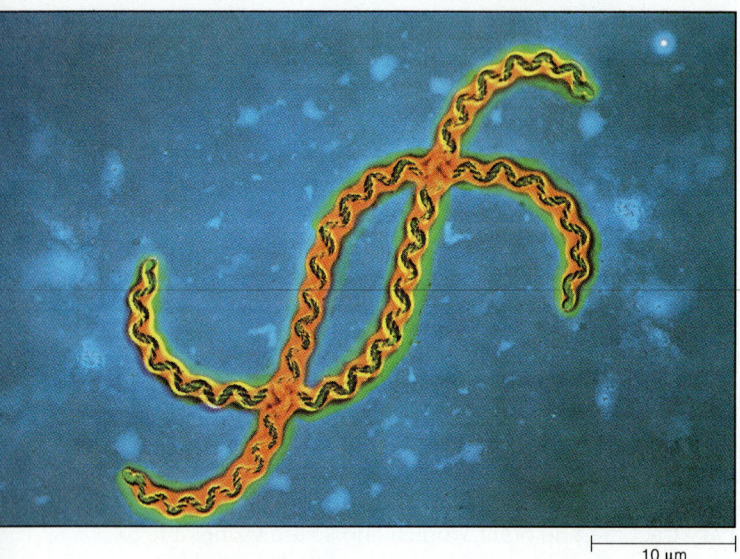

Figure 24.2 *Leptospira interrogans* gets its species name from the hooked end that resembles a question mark. Leptospirosis is transmitted by infected urine.

10 µm

is subdivided into many different serovars, and the pathogenicity of serovars varies greatly.

Leptospira interrogans is transmitted in urine from animals to humans, either directly, when urine from an infected animal enters the body through a break in the skin, or indirectly, in contaminated water and soil. Thus, animals—including dogs, cats, cattle, rodents, and wild animals—and water or soil with their urine are reservoirs for leptospirosis. The highly motile organisms penetrate intact mucous membranes or slightly damaged skin, enter the bloodstream, and are distributed to tissues throughout the body, including the central nervous system. Organisms that reach the kidneys multiply there and are excreted in the urine.

Probably most leptospirosis infections are asymptomatic, and most of those who become ill suffer only from a flulike syndrome of fever, headache, muscle pain, and reddened eyes that is usually misdiagnosed as a viral syndrome. These people usually recover within a week of their first symptoms. But some leptospirosis patients progress to a life-threatening illness characterized by liver damage, kidney damage (which may progress to renal failure), rash, bleeding, and shock. Severe cases of leptospirosis are usually marked by jaundice, and about 10 percent of these patients die. In patients who recover, however, liver and kidney function returns to normal.

Laboratory diagnosis of leptospirosis is difficult. The organisms are so small they are barely visible with light microscopy, and culture of *L. interrogans* requires special media not available in most clinical laboratories. As a result, diagnosis usually rests on serologic tests that show

an increasing titer of antibodies against *L. interrogans*. Treatment with penicillin or tetracycline, if started early, can shorten the illness and prevent serious complications.

Under 100 cases of leptospirosis are reported in the United States each year, but these are probably only the most severe. The vast majority of asymptomatic or mild infections probably go undiagnosed. A history of contact with animals may be a clue to diagnosis. At one time leptospirosis was seen mainly among packinghouse workers, sewer workers, and people who lived in rat-infested housing. Today most cases occur in young people who swim or wade in infected water. Leptospirosis is a common disease in many tropical countries. In Thailand and Vietnam, for example, as much as one quarter of the population shows serologic evidence of previous infection with *Leptospira*.

Streptococcus pyogenes: Acute Poststreptococcal Glomerulonephritis

Though *Streptococcus pyogenes* is a pathogen of the upper respiratory tract and the skin (Chapters 22 and 26), not the urinary tract, some strains are *nephritogenic*. That is, after the initial throat or skin infection has resolved, damage to the kidneys may occur as an autoimmune complication. This is not an infection, but a postinfectious complication. It is termed *acute poststreptococcal glomerulonephritis* and is discussed in Chapter 22.

SEXUALLY TRANSMISSIBLE DISEASES (STDs)

Many different infections are grouped together because of their common mode of transmission and are called STDs, or sexually transmissible diseases. They used to be known as **venereal diseases**, after Venus, the goddess of love. Until the 1970s many clinicians believed that STDs would soon be a problem of the past, wiped out by antibiotics. Sadly, new sexually transmissible pathogens have been identified, and many are viral, which means they are not treatable by antibiotics (**Table 24.2**). Moreover, the "traditional" bacterial STDs have made a stunning comeback. In the United States today, more than 12 million sexually transmitted infections are diagnosed each year, at an estimated cost to society of $3.5 billion. The majority of Americans are likely to contract at least one STD before the age of 35. At highest risk for repeated infections are people with multiple sexual partners who do not use condoms.

The long-term consequences of sexually transmissible infections primarily affect women and children. STD complications include blocked fallopian tubes, causing

Table 24.2 Some Microorganisms That Cause Sexually Transmissible Infection

Causative Agent	Infection	Comments
Bacteria		
Neisseria gonorrhoeae	Gonorrhea	Most common reportable STD in the United States; usually symptomatic in men and asymptomatic in women; new antibiotic-resistant strains appearing.
Treponema pallidum	Syphilis	Manifests many clinical syndromes; disease progresses through stages over several decades; treatable with penicillin.
Chlamydia trachomatis	PID in women and NGU in men; lymphogranuloma venereum (LGV)	Serovars D–K cause the most commonly transmitted STD in the United States; mild illness and difficult to diagnose, though treatable. LGV is rare, painful, and treatable.
Ureaplasma urealyticum, Mycoplasma hominis	Urethritis, vaginitis	Widespread, often asymptomatic, but can cause PID in women and NGU in men.
Haemophilus ducreyi	Chancroid	Open sores on genitals can lead to scarring without treatment; on the rise in U.S. inner cities; treatable.
Calymmatobacterium granulomatis	Granuloma inguinale	Draining ulcers that can persist for years; treatable; rare in U.S.
Viruses		
Herpes simplex virus	Genital herpes simplex	Painful blisters; enters latent stage, with reactivation due to stress; also oral, pharyngeal, and rectal herpes from sexual contact. No cure. Extremely prevalent in U.S.
Human papillomavirus	Condyloma acuminata; probably predisposes to cervical cancer	Genital warts, though often asymptomatic. No cure. Very common in U.S.

infertility and life-threatening tubal pregnancies, cancer of the female genital tract, fetal death, birth defects, and newborn blindness. Women's vulnerability is increased because transmission of these infections is more efficient from men to women than vice versa. Also, early diagnosis of most STDs is more difficult in women than in men. Current epidemiologic patterns show, too, that the women and children most likely to die or suffer permanent handicaps from STDs are poor and members of ethnic minority groups.

AIDS—the fatal and incurable STD that erupted in the 1980s (Chapter 27)—has had a profound effect on the epidemiology of sexually transmissible infections. Because of the common mode of transmission, people with one STD are likely to have another. AIDS patients coinfected with syphilis, for example, suffer from unusually virulent infections. Also, STDs that cause genital ulcers—such as syphilis, chancroid, and herpes—facilitate the sexual transmission of AIDS. On the other hand, the

AIDS epidemic focused long-overdue attention on safer sex practices that would limit the transmission of all STDs. Safer sex means reducing the number of sexual contacts—when people engage in sexual contact with multiple partners, each infected person can infect many others (**Figure 24.3**)—and using condoms to help prevent infection with each contact.

BACTERIAL CAUSES

Many of the best-known and most common STDs, including gonorrhea and syphilis, are caused by bacteria. Although some of these infections remain significant public health problems, all are curable by antibiotics.

Neisseria gonorrhoeae: Gonorrhea

Neisseria gonorrhoeae, a Gram-negative diplococcus, causes **gonorrhea**—the most common reportable disease in the

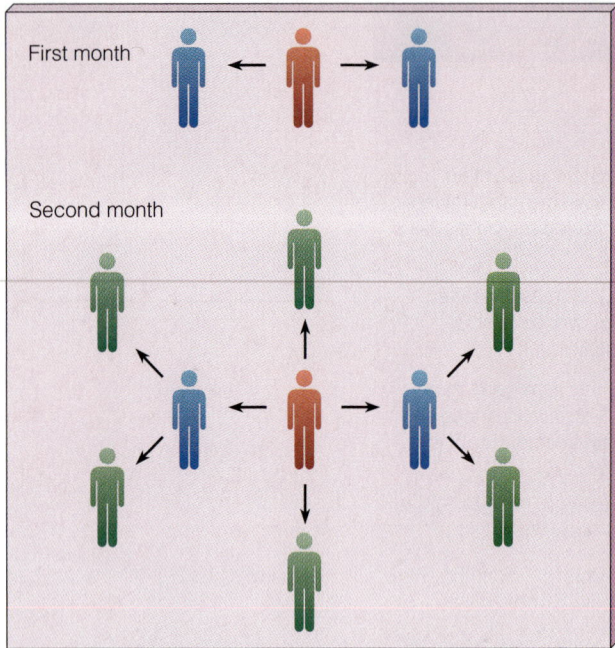

a Suppose people have two different sexual contacts every month. After only two months, one infected person has exposed eight others.

Figure 24.3 Sexually transmissible diseases spread more rapidly as the number of sexual contacts increases.

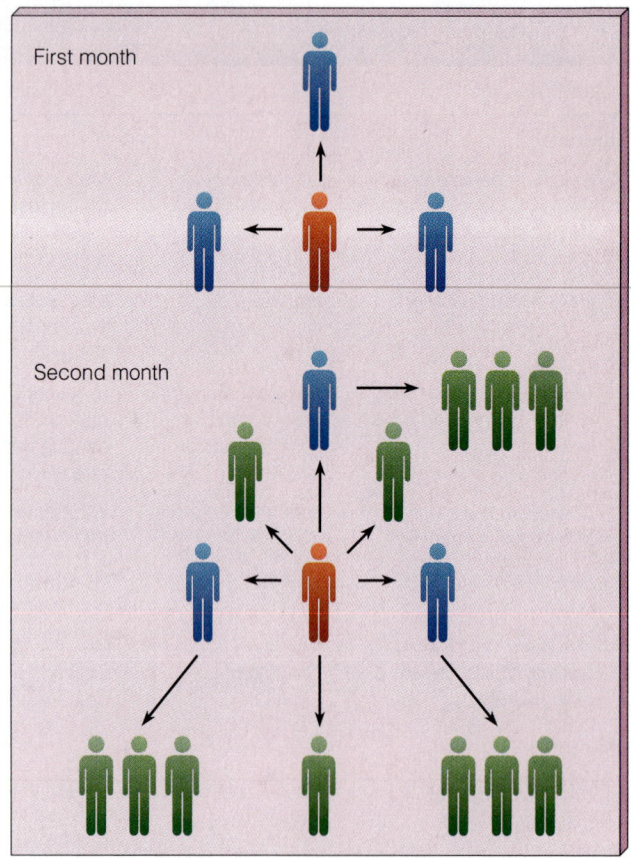

b If people have three sexual contacts every month, at the end of two months, one infected person has exposed 15 others. Of course, as the number of contacts rises, the infection spreads even more rapidly.

United States. In the 1970s, the incidence of gonorrhea was almost epidemic. Increased sexual freedom, especially among young women, and decreased use of condoms because of the availability of birth control pills led to over 450 cases being reported to the Centers for Disease Control for every 100,000 Americans. Rates remain high in the early 1990s, approximately 280 cases per 100,000 population. At highest risk are young adults between the ages of 16 and 25 and members of ethnic minority groups, who suffer gonorrhea rates up to 25 times higher than white Americans.

N. gonorrhoeae, like many other sexually transmissible pathogens, is fragile—easily inactivated by drying, room temperature, or exposure to sunlight. It is difficult to culture in the clinical laboratory because it requires complex media to meet its multiple nutritional requirements. It is usually isolated on a selective medium called **Thayer-Martin medium**, which contains antibiotics to suppress the growth of other bacteria. It grows best in the presence of 2–8 percent carbon dioxide. Diagnosing gonorrhea in

females usually requires culturing material obtained from a cervical swab during a pelvic examination. In males diagnosis is reliable if Gram-negative diplococci are found in a urethral discharge (**Figure 24.4**).

N. gonorrhoeae is a highly adapted pathogen with many powerful virulence determinants, including adhesin-bearing pili. Strains that lack pili are nonvirulent. Receptors for *N. gonorrhoeae* adhesins are present on epithelial cells in the male and female genital tracts, the eye, throat, and rectum and are present on sperm, which can transport the pathogen into the uterus and fallopian tubes. Although the gonococcus produces only one type of pilus at a time, it can rapidly switch the type being produced. This allows it to adhere more strongly to different types of epithelial cells—in the throat or rectum, for example. Switching also helps the gonococcus escape recognition and destruction by the body's immune defenses.

The gonococcus has an outer membrane protein called Protein II that also promotes adherence to epithe-

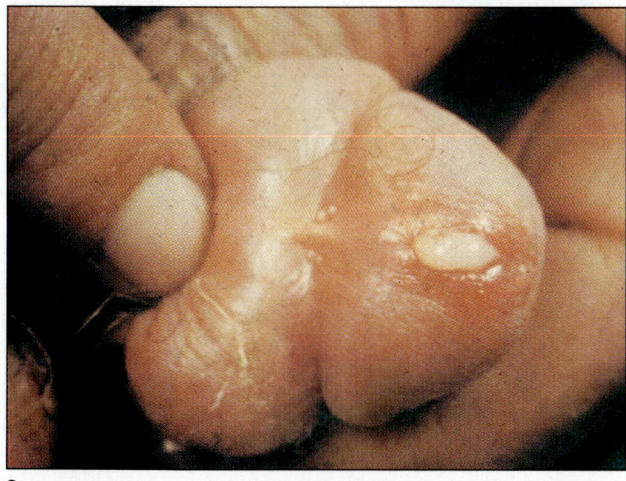

a

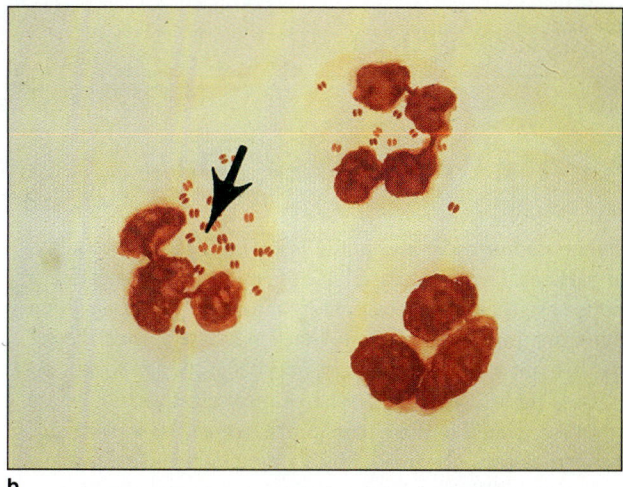

b

Figure 24.4 As with many STDs, gonorrhea is easier to recognize and diagnose in males than in females. (**a**) Most men with gonorrhea develop a pus-laden urethral discharge, like this patient's. (**b**) A Gram stain of the discharge, which takes only a few minutes, is diagnostic if *Neisseria gonorrhoeae* diplococci are seen in polymorphonuclear leukocytes, as in this photo. Women, on the other hand, may have no symptoms of their infection, and diagnosis requires a pelvic examination and culture.

lial cells. As with pili, the type of Protein II produced can switch, promoting adherence to different types of epithelial cells. Protein II also causes gonococcal cells to adhere to one another, forming clumps. When one bacterium adheres to a human cell, many others stick to it to create an **infectious unit**. In addition, Protein II helps the gonococcus escape phagocytosis.

Toxic components of the gonococcal outer membrane, particularly lipid A, are a major virulence factor for *N. gonorrhoeae*. Gonorrheal endotoxin destroys cells in tissue culture and is probably responsible for the cell death and inflammation that occur during infection. The outer membrane toxins can paralyze the cilia lining the fallopian tubes, making it easier for the gonococcus, as well as other bacteria, to multiply there. Mixed infections involving *N. gonorrhoeae* and other bacterial species are common in the fallopian tubes.

Clinical Syndromes. *N. gonorrhoeae* infects only human beings and must be transmitted by direct body contact, usually sexual intercourse or birth. About 20 percent of men contract gonorrhea after a single exposure compared to 50 percent of women. Infection in adults usually begins in the lower genital tract—the urethra in males and the cervix in females—although the throat (**pharyngeal gonorrhea**) and rectum (**anal gonorrhea**) are also vulnerable if they are points of sexual contact.

Most infections in males are symptomatic. Two to seven days after the infecting sexual contact, most men experience painful urination and a discharge of pus from the urethra. If antibiotics are administered, symptoms resolve rapidly, but without treatment they may go on for weeks or months. Women with gonorrhea may or may not notice an abnormal vaginal discharge. People with asymptomatic infections are not likely to seek the medical treatment that would allow cure. As a result, they spread the pathogen to their sexual partners and are at risk for serious complications.

Some complications of gonorrhea involve only the reproductive organs. In women, pelvic inflammatory disease due to *N. gonorrhoeae* typically causes fever, dysuria, and severe abdominal pain. Inflammation of the fallopian tubes during PID can cause scarring and infertility, and microorganisms can spread into the abdominal cavity, causing a life-threatening inflammation. In men, infection of the passages that transport sperm from the testes to the urethra can also cause scarring and infertility.

Occasionally *N. gonorrhoeae* causes systemic illness, entering the bloodstream and causing purulent (pus producing) arthritis and a rash in skin and joints. In some rare cases, the heart, liver, or meninges become infected. Disseminated disease affects only 3 percent of people with gonorrhea, usually women, but it is life-threatening. This form of gonorrhea is difficult to diagnose because there is often no sign of genital tract infection.

Infants passing through an infected birth canal can also become victims. *N. gonorrhoeae* can cause blindness or a systemic life-threatening infection (Chapter 26).

A Game We May Never Win

Overall, gonorrhea rates in the 1990s have fallen, probably in part because of increased use of condoms and more cautious sexual behavior fostered by the AIDS epidemic. But at the same time, antibiotic resistance in *Neisseria gonorrhoeae* has become a serious problem, and it's growing worse.

Less than 10 years ago, when almost all strains of *N. gonorrhoeae* isolated in the United States were drug-sensitive, gonorrhea could be treated by a single intramuscular injection of penicillin or a brief course of oral tetracycline for penicillin-allergic patients. Now, neither of these drugs can be counted on to cure a high enough percentage of patients to justify their use as a first-line agent. This might seem like an overreaction because only a little over 8 percent of gonorrhea cases diagnosed in 1990 were caused by drug-resistant strains. But research shows that treatment failures in 1 patient out of 50 are sufficient to spread a strain of resistant bacteria throughout a community. The treated patient who has not been cured will pass the resistant strain to his or her next sexual contact, who will, in turn, spread the drug-resistant infection to someone else. Thus, even a low failure rate in treatment is unacceptable from a public health standpoint.

There is a little more good news, though. We have new drugs for treating penicillin- and tetracycline-resistant strains of *N. gonorrhoeae*. Cure is virtually certain after a single—if extremely painful—injection of ceftriaxone, from the cephalosporin family, or oral administration of the closely related cefixime. Newer drugs of the fluroquinolone family are also effective, but are considered second-line agents because they do not cure syphilis, which also infects many gonorrhea patients.

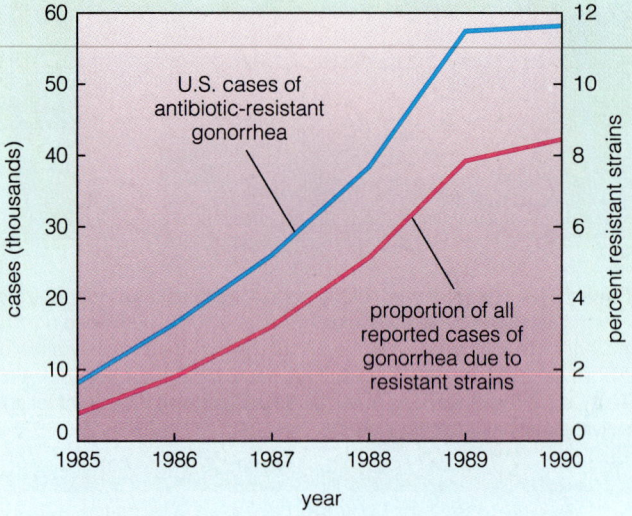

Drugs that cured gonorrhea 10 years ago are no longer first-line agents as drug-resistant strains emerge—a serious problem that is growing worse.

These new drugs, however, are much more expensive than penicillin. Curing a single patient with ceftriaxone may cost as much as $80 compared to less than $1 for penicillin. This is a major problem in publicly funded clinics. Moreover, no one knows how long these new drugs will remain effective as drug-resistant strains continue to emerge. In trying to control gonorrhea with antimicrobial treatment, we may be playing a game of catch-up that we can never win.

Prevention and Treatment. Gonorrhea has no reservoir outside infected human beings; and reliable, inexpensive diagnosis and treatment have been available for years. In theory, therefore, it should be easy to control or even eliminate the disease. In fact, no one believes it will be controlled soon, let alone eliminated. Many factors contribute to the perpetuation of this infection.

First, *N. gonorrhoeae* is perpetuated by the existence of many antigenically different strains with variations in their pili and outer membrane proteins. This tremendous genetic variability makes natural immunity poor (many people suffer repeated gonorrheal infections) and prospects for a vaccine limited.

The discovery of penicillin was a major step forward in gonorrhea control. An untreatable disease became curable with a single injection. Recently, though, antibiotic-resistant strains of *N. gonorrhoeae* have been proliferating (see the box "A Game We May Never Win"). Gonococci can become penicillin-resistant by either one of two genetic mechanisms (Chapter 21). Strains that undergo mutations on the bacterial chromosome experience small, incremental changes in resistance. If only a few mutations have occurred, a strain may remain sensitive to high concentrations of penicillin, but strains that have accumulated many mutations may be completely resistant. Resistance can also be caused by plasmids that encode a beta-lactamase that

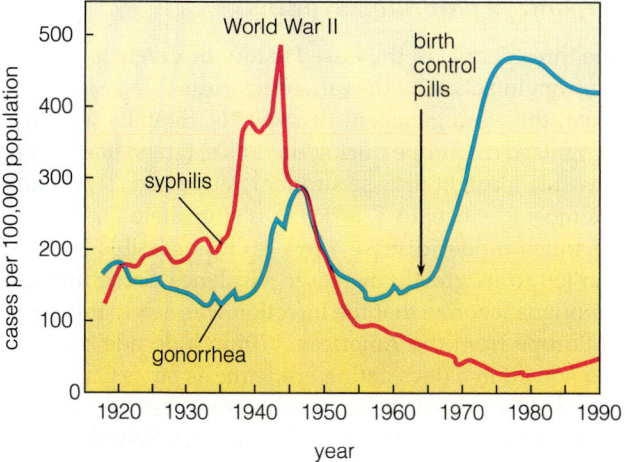

Figure 24.5 Gonorrhea and syphilis. During the late 1960s and 1970s, after the introduction of birth control pills, the incidence of gonorrhea in the United States rose alarmingly. In contrast, syphilis has never again become as common as it was during the early 1940s, before the introduction of penicillin. However, syphilis incidence began to rise in the 1980s, in large part because of the crack cocaine epidemic and the exchange of sex for drugs.

destroys the penicillin molecule. These strains immediately become highly penicillin-resistant.

Controlling gonorrhea is also complicated by the fact that it is a sexually transmissible disease. Even intelligent and well-educated people sometimes find it difficult to deal objectively and openly with diagnosis and treatment. Until recently, for example, college textbooks stated that children could acquire gonorrhea by nonsexual routes, though the mechanism was unknown. Today we acknowledge these cases are the result of sexual abuse. Similar emotional reactions led to the myth that gonorrhea could be acquired from fomites such as toilet seats.

Social and cultural attitudes toward sex often undermine public health efforts to monitor and control the spread of gonorrhea. When a case of gonorrhea is identified, it must be reported to the public health department. The infected person is asked to name his or her recent sexual contacts so these people can be cultured and, if necessary, treated for gonorrhea. However, neither patients nor physicians always cooperate. Physicians often do not report the diagnosis, hoping to spare their patients embarrassment. Also out of embarrassment, patients may not reveal all their sexual contacts. About 1 million cases of gonorrhea are reported to public health departments each year, but the total number of cases in the United States is estimated to be as high as 3 or 4 million. Gonorrhea and most other STDs can usually be prevented with condoms (**Figure 24.5**).

Case History

The Stages of Syphilis

Patient Number 1

A 22-year-old student was examined at a county-sponsored health clinic. He had noticed an open sore on his penis several days earlier, but it caused him no discomfort. The examining physician noted the sore was about 1 cm in diameter with a weepy bright red base and a raised border (**Figure 24.6**). The doctor also found one enlarged lymph node in the patient's groin. The young man reported sexual contact with a new partner about three weeks earlier.

Patient Number 2

A 53-year-old businessman made an appointment to see a private dermatologist. He was developing a red bumpy rash all over his body, even on the palms of his hands and the soles of his feet (**Figure 24.7**). He had also noticed a few irregular bald patches on his scalp, and recently he had been feeling unusually tired and occasionally feverish. The physician who examined him found enlarged lymph nodes throughout his body and several white patches inside his mouth. After being pressed by the physician, the man admitted to sexual contact with a prostitute about three months earlier.

Patient Number 3

A 60-year-old housewife was taken by her sister to see a psychiatrist. Over the last several years, the sister explained, the patient had become increasingly irritable and her personality had changed. She made inappropriate comments at social gatherings and was becoming careless about her dress and grooming. She was forgetful and could no longer do simple tasks like balancing her checkbook. Sometimes she seemed depressed, but at other times she was extremely expansive, boasting of grandiose plans. The sister stated that the patient's health had always been quite good and she seldom consulted a doctor. When asked specifically, she did remember some sort of treatment for venereal disease about 40 years before.

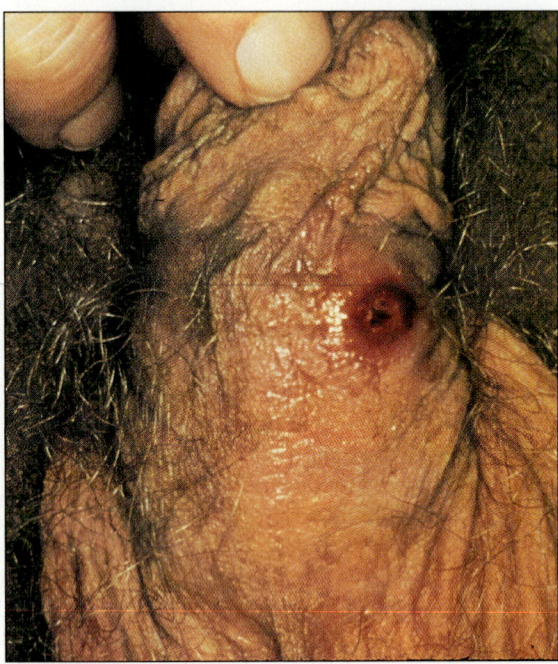

Figure 24.6 Typical chancre of primary syphilis.

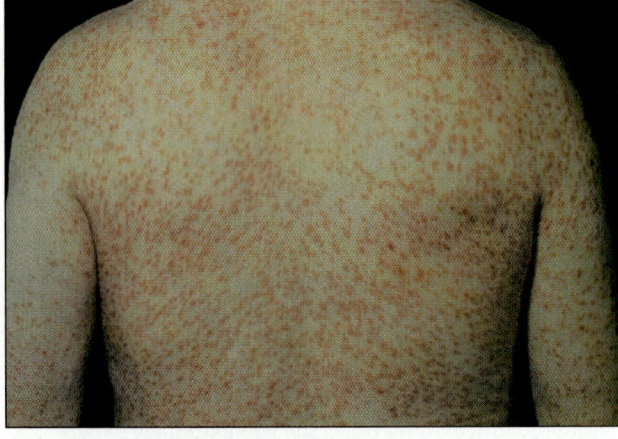

Figure 24.7 Disseminated rash of secondary syphilis.

Treponema pallidum: Syphilis

The three people in the Case History box were all suffering from infection by the same bacterium—*Treponema pallidum*, the etiologic agent of **syphilis**. Syphilis was first recognized in Europe during the 1490s, but no one knows why this virulent disease suddenly appeared. *T. pallidum* is almost identical to *T. pertenue*, a spirochete that causes the much milder disease **yaws**, so it is possible a mutation led to its greatly increased virulence. Some medical historians theorize that the infection was newly imported to Europe from the Americas. Within a decade, syphilis had spread in a devastating epidemic as far as China. Because of the disfiguring sores that appeared all over the body, it was called *great pox* to distinguish it from smallpox.

After penicillin made it possible to cure syphilis, the disease lost much of its terror. Even so, syphilis continues to be a major public health problem in the United States. During the 1970s and early 1980s, syphilis rates rose—mainly among homosexual men. The 1990s have seen an upsurge in the number of cases among heterosexuals, particularly the poor and members of ethnic groups. This new pattern is attributed partly to the epidemic of crack cocaine addiction, which results in addicts exchanging sex for drugs. Other new patterns have also emerged in the 1990s. Congenital syphilis (syphilis present at the time of birth)—an infection virtually never seen in the 1980s—is now being reported from all parts of the United States. In addition, syphilis patients coinfected with AIDS are much harder to cure than other syphilis patients. Finally, although syphilis, with an overall rate of under 20 cases per 100,000 people, is much less common than gonorrhea, the rates in some subgroups are much higher. Black men, for example, have a syphilis rate of nearly 160 cases per 100,000.

Treponema pallidum is a motile spirochete (*treponema* means "turning thread" in Greek). Motility, which can be observed under darkfield microscopy is one way of diagnosing syphilis. A smear from Patient Number 1's open lesion would have shown motile spirochetes (**Figure 24.8**), providing a diagnosis.

T. pallidum infects only human beings. It is an extremely fragile microorganism, and until quite recently could not be grown in the laboratory. These bacteria are still not cultivated routinely in diagnostic microbiology labs. If diagnosis cannot be made by darkfield microscopy, serologic testing is required.

There are two kinds of serologic tests for syphilis—nontreponemal tests and treponemal tests. **Nontreponemal tests**, also called **reagin tests**, detect the presence of reagin antibodies in syphilis patients. The major antigen for nontreponemal tests is **cardiolipin**, a substance derived from beef heart. We do not entirely understand the origin of reagin antibodies. They may be produced in response to lipids released from human cells damaged in the early stages of syphilis, but there is also evidence that similar lipids exist on the bacterial cell surface. The two most common nontreponemal serologic tests are the **VDRL (Venereal Disease Research Laboratory)** and the **RPR (rapid plasma reagin)** tests. Nontreponemal tests are inexpensive and fairly reliable and can be used to test for cure of syphilis (they become negative after adequate antibiotic treatment). They are also the only tests that reliably detect syphilis in cerebrospinal fluid.

Treponemal tests, on the other hand, use antigens derived from treponemal organisms to detect antibodies specifically directed against spirochetes. Antigens are derived from treponemes easier to cultivate than *T. pallidum*. The two most common treponemal tests for syphilis are the **FTA (fluorescent treponemal antibody)** and **MHA (microhemagglutination)** tests. Treponemal tests are used to confirm the diagnosis of syphilis in patients who have positive serology with a nontreponemal test because the VDRL and RPR can sometimes give false positive results (they yield positive results in patients who do not have syphilis). False positive syphilis serology most often occurs in patients with immunologic disorders.

Clinical Syndromes. Patients with syphilis manifest a variety of clinical syndromes—the patients described in the Case History box are only three examples. Syphilis can affect anyone, from newborn babies to the elderly, and can damage almost any organ in the body. At one time entire departments in medical schools were devoted to *syphilology*, the study of the many manifestations of syphilis.

Syphilis as a clinical syndrome is distinguished by a characteristic progression of signs and symptoms that develops over many years. The disease is divided into four stages—primary, secondary, latent, and tertiary (**Table 24.3**). Not all syphilis patients enter the tertiary stage, and the disease can be interrupted at any of the first three stages by treatment with penicillin or another appropriate antibiotic.

Infection occurs when *T. pallidum* enters the body through a break in the skin or across an unbroken mucous membrane. Because it is an extremely fragile microorganism, *T. pallidum* is transmitted only by means of direct body contact, usually sexual.

Primary syphilis begins a few weeks to a month after infection when the **chancre**, a weepy ulcer with raised borders, appears at the entry site. This was the painless sore that led Patient Number 1 to visit the health clinic. Fluid from the center of the chancre contains enormous numbers of spirochetes, and diagnosis of primary syphilis is conclusive if *T. pallidum* is found by darkfield examination. A characteristic chancre along with a positive sero-

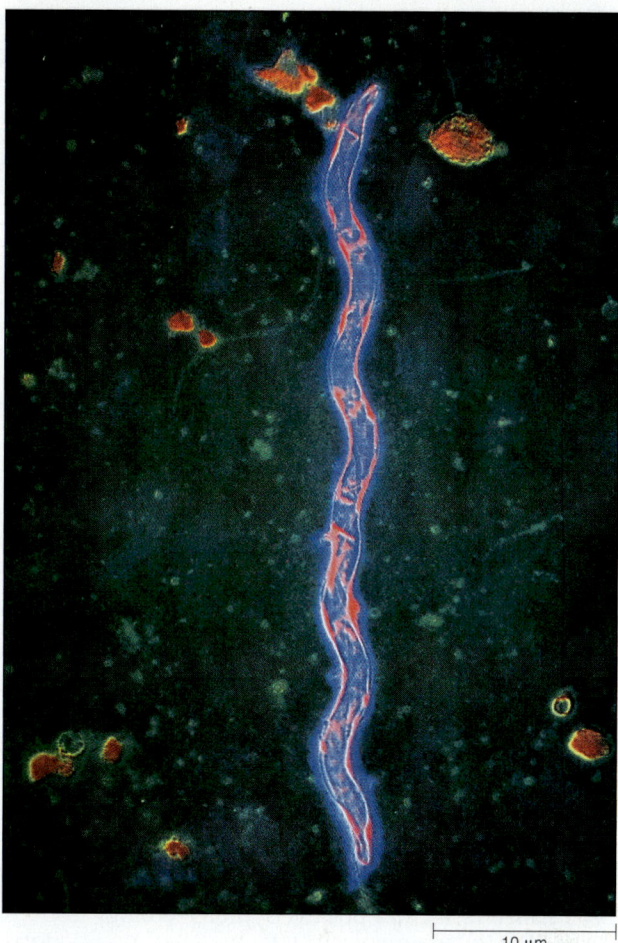

Figure 24.8 This darkfield micrograph shows the coiled cell of *Treponema pallidum*, the bacterium that causes syphilis.

logical test for syphilis is also diagnostic. With or without treatment, a syphilitic chancre will heal within a few weeks. A chancre on the penis is so noticeable that men usually seek medical attention. Women, however, often do not notice the chancre because it typically occurs on the cervix or within the vagina, so they are therefore less likely to receive treatment.

Secondary syphilis usually begins 6 to 8 weeks after the chancre appears. Signs and symptoms of this stage involve the skin and mucous membranes, so people sometimes consult a dermatologist, as Patient Number 2 did. But the skin lesions of secondary syphilis are extremely variable, making clinical diagnosis difficult. At the same time, live spirochetes are present in skin lesions, and contact with mucous membrane lesions like those in Patient Number 2's mouth can spread syphilis. Systemic symptoms include fatigue, fever, and enlarged lymph nodes. Bacteria spread throughout the body, and damage to blood vessels causes a rash.

Table 24.3 Stages of Syphilis

Stage	Time after Infection	Manifestation	Diagnosis	Treatment	Infectious
Primary	2 to 3 weeks	Chancre at site of infection	Observation of spirochetes from a chancre under dark-field microscopy or presence of a chancre with positive syphilis serology	Penicillin	Yes
Secondary	8 to 11 weeks	Skin rash, fatigue, fever, enlarged lymph nodes	Fluorescent antibody assays, serology	Penicillin	Yes
Latent	After secondary	None	Serology	Penicillin	For first 4 years; not thereafter
Tertiary	5 to 40 years	Neurological symptoms, blood vessel damage, gummas	Serology, but may not be reliable	Penicillin, but only gummas respond to treatment	No

At the secondary stage, diagnosis can be made by identifying motile spirochetes from mucous membrane lesions outside the mouth. This kind of diagnosis is possible only in patients like Number 2 with mouth lesions if special fluorescent antibody markers are used. Otherwise, spirochetes that normally occur in the mouth cannot be distinguished from *T. pallidum* under routine dark-field microscopy. Diagnosis can also be confirmed by the typical skin rash and a positive syphilis serology test.

If the patient is not treated during secondary syphilis, the disease enters the stage called **latent syphilis**. During this time, the only indication of infection is a positive serologic test and the ability to infect others, at least for a few years. Otherwise, the person looks and feels entirely well. After about four years into the latent stage, most patients are no longer infectious. Most people remain in the latent stage forever and suffer no further signs or symptoms of disease.

However, people who progress from latent syphilis to **tertiary**, or **late**, **syphilis** may suffer severe illness or death. Progression to this final stage occurs from 5 to 40 years after secondary syphilis has subsided. Like secondary syphilis, late syphilis can be manifested in an astonishing number of ways. About 8 percent of people with latent syphilis eventually develop some form of **neurosyphilis**, in which the disease affects the meninges or central nervous system. Neurosyphilis may resemble almost any neurologic disease. Patient Number 3, for.example, had a form of late syphilis called **general paresis**. (In spite of what her sister remembered, she obviously had not received adequate treatment for her syphilis.) Another 10 percent of patients suffer from **cardiovascular syphilis**, in which large blood vessels, such as the aorta and coronary arteries, are damaged. This is a fatal form of syphilis. The most common form of late syphilis is characterized by **gummas**—soft granulomas that usually replace skin or bone but may occur in any organ (**Figure 24.9**). Gummas are not life-threatening unless they destroy vital organs, so this form is sometimes called **late benign syphilis**. Gummas disappear with adequate antibiotic therapy. Penicillin cannot reverse other damage already done by late syphilis, although further damage can be prevented.

Late syphilis is usually not infectious because only a few live spirochetes remain in these patients. Sometimes, serologic tests are not even positive. The damage of late syphilis is caused by reactions of the immune system. Like tuberculosis, late syphilis is a disease of hypersensitivity (Chapter 18), though the relationship between *T. pallidum* and the immune system is not as well understood. Both humoral and cell-mediated immunity against *T. pallidum* are vigorous, yet the microorganism establishes an infection that lasts for decades.

Still another form of syphilis exists—mothers who have *T. pallidum* in their bloodstream can infect their babies before birth. Bacteria can cross the placenta during the middle months of pregnancy. In **prenatal** (or **congenital**) syphilis, about half of infected fetuses die before or at birth. Some live-born infants have birth defects, such as tooth and bone deformities or deafness. Some develop mucous membrane lesions such as **snuffles**, a thick nasal discharge containing many *T. pallidum* cells. Some infants appear normal initially, but at age 1 or 2 develop painful bone lesions among multiple other problems.

Prevention and Treatment. Preventing syphilis, like all other STDs, requires limiting sexual exposure and using condoms. The gay community's campaign for safer sex to prevent AIDS has led to a decrease in new cases of

The Neapolitan-French-West Indian-Spanish-Canton-Chinese Disease

Europeans did not know the cause of the devastating disease that swept across the continent in the 1490s or even what to call it. When Charles VIII of France attacked the city of Naples in 1495, the Neapolitans sent infected prostitutes out to entertain the besieging troops. Naples nevertheless fell quickly—as did Charles's victorious troops on their return home. They became acutely ill with disfiguring skin lesions, and many died of what Charles called the *Neapolitan disease*. When Charles's mercenaries returned to their homes in England, Poland, Russia, and Scandinavia, the infection they brought with them came to be called the *French disease*. No country wanted to claim this disaster as its own. Spaniards called it the *West Indian disease*, while the French called it the *Spanish disease*. The Chinese called it the *Canton disease*, and the Japanese, the *Chinese disease*. In 1530, an Italian physician wrote a poem about this infection as a curse from the Sun God. The literary victim—a shepherd named Syphilis—gave the disease its modern name, syphilis.

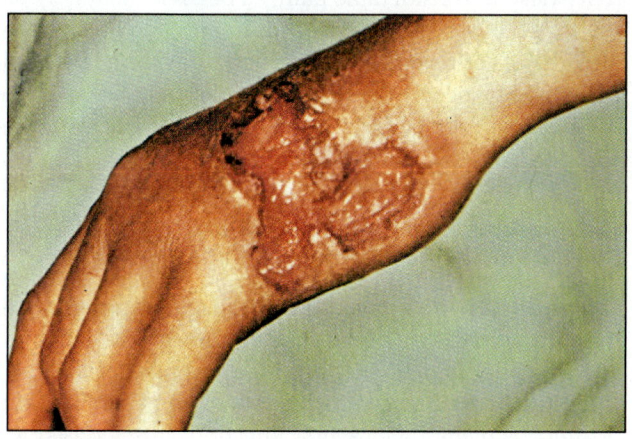

Figure 24.9 Late syphilis (tertiary stage) is most often characterized by the development of gummas—soft granulomas that usually replace skin or bone but may occur in any organ.

syphilis among members of this high-risk group. Early diagnosis and treatment of syphilis patients also help interrupt transmission and control the disease. Congenital syphilis is preventable by prenatal testing of mothers with a VDRL in the first trimester and early treatment (before the second trimester, when syphilis crosses the placenta). However, the growing number of poor, medically underserved populations where women do not get prenatal care has contributed to the resurgence of congenital syphilis in the 1990s

Natural immunity to *T. pallidum* is poor, and a vaccine seems unlikely. But *T. pallidum* is one of the few microorganisms that has remained uniformly and exquisitely sensitive to penicillin. Most infections can be cured with a single injection of benzathine penicillin. Patients with neurosyphilis, particularly people who also have AIDS, require hospitalization for a 10-day course of intravenous penicillin.

Chlamydia trachomatis: Chlamydia

Chlamydia trachomatis is subdivided into more than 15 different serovars. Some serovars cause an extremely common genital infection (called "chlamydia") among sexually active adults; others cause a relatively rare genital infection called **lymphogranuloma venereum (LGV)**.

C. trachomatis serovars D through K probably account for between 3 and 5 million new infections in the United States each year. This infection—the most prevalent bacterial STD in the United States today—does not have a specific name and is usually referred to simply as "chlamydia." Symptoms in men and women appear gradually one to three weeks after unprotected contact with an infected sexual partner. The illness is mild, consisting of dysuria and a small amount of mucoid discharge from the urethra or the vagina. Many people notice no symptoms at all. But many women suffer infections of the uterus and fallopian tubes, and chlamydial infection is the leading cause of female infertility and ectopic pregnancy in this country. Moreover, infected women transmit the pathogen to their infants at the time of birth, causing neonatal eye infections and pneumonia.

Because it was so difficult to diagnose, chlamydial infection was not recognized as an important STD until recently. *Chlamydia* cannot be identified by routine bacterial culture; it must be propagated in tissue culture. With the advent of rapid antigen detection tests in the mid-1980s (Chapter 19), we began to understand the microorganism, the infection, and the importance of controlling it. Because symptoms are subtle (only half of women with chlamydia-caused infertility remember any illness), ideally, everyone who is sexually active should be regularly screened. Since this is impractical, screening must focus on groups at highest risk. These include women under the age of 25 who have had a new sexual partner in the preceding two months or whose partners do not consistently use condoms, all pregnant women, and anyone with a vaginal or urethral discharge. Infection can be cured with doxycycline, erythromycin, or the new drug azithromycin.

In contrast to "chlamydia," lymphogranuloma venereum (LGV) is uncommon—fewer than 500 cases are reported in the United States each year. LGV is caused by *C. trachomatis* serotypes $L_1–L_3$. It produces a classical clinical syndrome. One to four weeks after sexually transmitted infection, a small sore usually appears on the genitals and the person experiences fever and headache. The primary lesion heals without treatment. In the second stage, lymph nodes in the groin enlarge and may become painful. Left untreated, the nodes may turn into draining ulcers. Scarring and lymphatic obstruction damage the external genitalia. Diagnosis is made by culturing organisms from the infected nodes. Treatment with doxycycline or erythromycin is effective.

Ureaplasma urealyticum and *Mycoplasma hominis*: Nongonococcal Urethritis

Ureaplasma urealyticum and *Mycoplasma hominis* are mycoplasmas that are common residents of the urethra or vagina of sexually active men and women. More than half of all people with five or more lifetime sexual partners are colonized by these bacteria, and most have no symptoms. Some men, however, suffer from a urethral discharge that is believed to be due to *U. urealyticum*, and *M. hominis* may cause some cases of symptomatic infection in women. But because these organisms are so widespread, it is not practical to screen for them or treat them routinely. When men seek care for urethral discharge, however, the drug of choice is usually doxycycline, which is effective against the mycoplasmas as well as many other sexually transmissible pathogens.

Haemophilus ducreyi: Chancroid

Haemophilus ducreyi causes the sexually transmissible disease **chancroid**. This rod-shaped bacterium enters the body only through a break in the skin or mucous membranes, so it is less communicable than *Neisseria gonorrhoeae* and *Treponema pallidum*, which can infect intact epithelial surfaces. The first manifestation of chancroid is the **soft chancre**, an open sore on the genitals that resembles the chancre of primary syphilis but is softer and often quite painful. Chancroid is also distinguished from primary syphilis because no motile spirochetes can be recovered from the open lesion and syphilis serology is negative. Also unlike syphilis, chancroid is a local infection that does not affect distant organs, although nearby lymph nodes may become swollen and painful. The sores of chancroid eventually heal by themselves, but antibiotic treatment speeds recovery. Without treatment, significant scarring and damage to the genitals sometimes occur.

Chancroid is fairly common in tropical countries, but it was rare in the United States until the mid-1980s. In 1990 over 5000 cases of chancroid were reported, compared to less than 1000 per year in the early 1980s. Chancroid appears to be more closely linked to the crack cocaine epidemic than any other STD; most patients are prostitutes and cocaine addicts. Furthermore, since asymptomatic infections are rare and ulcers are painful, transmitting chancroid means people continue sexual activity despite considerable pain—a behavior that may be motivated by a desperate need for money or drugs. A single dose of ceftriaxone cures chancroid in the majority of patients, but people with AIDS and chancroid may require more prolonged treatment.

Calymmatobacterium granulomatis: Granuloma Inguinale

Calymmatobacterium granulomatis causes **granuloma inguinale**, an infection of the skin and mucous membranes of the genital organs. It is not highly communicable because the bacteria must enter through a break in the skin or mucous membrane. A first infection causes raised lesions at the site of entry. Eventually the lesions develop into open draining ulcers that may persist for years. Sometimes they spread far enough to involve the legs or abdomen. Diagnosis is made on the basis of characteristic intracellular inclusion bodies called **Donovan bodies** that are visible under a light microscope in scrapings from the ulcers. Donovan bodies contain *C. granulomatis*. Granuloma inguinale can be cured by several antimicrobial agents, including ampicillin, tetracycline, and trimethoprim-sulfamethoxazole. The infection is extremely rare in the United States, but it is seen in Asia, Africa, and South America.

VIRAL CAUSES

Viral pathogens that cause sexually transmissible diseases infect sexually active adults, cause devastating infections of the newborn, and can predispose a person to cancer. Unlike bacterial infections, virally caused STDs are not curable.

Herpes Simplex Virus (HSV): Herpes

Herpes simplex virus is by far the most common sexually transmitted pathogen in industrialized countries. At least 20 percent of Americans between ages 16 and 40 have serologic evidence of past infection. Before 1970 little attention was paid to this pathogen, and clinicians were not even sure it was sexually transmissible. But during the 1970s the incidence of genital herpes simplex in the United States increased dramatically. Today, the 20 million symptomatic episodes of genital herpes that occur annually in adults are a serious public health concern, causing considerable pain and disability. But the real horror of herpes is infection of newborns.

Herpes simplex virus (HSV) is a DNA virus belonging to the herpes family. It causes both active and latent infections in every infected person (Chapter 15). During **active infections** viral genes produce 50,000 to 200,000 new virions from each infected cell. The host cell's metabolism in inhibited, its DNA is degraded, and the cell dies. Active infections may be asymptomatic, or painful blisters may form in the infected tissue. During **latent infections** the viral genome is not expressed, though at least one copy of the viral DNA remains in the nucleus of the host cell. Latent infections do not interfere with the cell's normal function but may be activated later, causing an active infection in nearby cells.

Herpes simplex virus affects epithelial cells and neurons. Infection begins when virus is introduced directly into a break in the skin or mucous membrane. At the same time, viruses enter nearby nerve endings, spreading infection along the nerve fiber to the body of the neuron, where a latent infection is established. Once there, it can return to skin. Infection of a nerve cell does not kill it, but it does affect the skin or mucous membrane that the nerve supplies. This phenomenon accounts for one of the most distinctive features of HSV infection—its tendency to recur at the same site again and again.

There are two types of herpes simplex virus, **HSV-1** and **HSV-2**. These subtypes share about 50 percent of their DNA sequences. Some evidence suggests that the two types bind to different host cell receptors. HSV-1 typically causes infection of the mouth and face (Chapter 26), while HSV-2 typically infects the genital tract, although the reverse can also occur. Furthermore, the lesions caused by each subtype—cold sores on the lip and genital blisters—are clinically indistinguishable. If a person engaged in oral sex, for example, and contracted an HSV-2 mouth infection or an HSV-1 genital infection, this would not be apparent on clinical examination. Genital recurrences, however, occur much more often when the infection is caused by HSV-2.

The human immune response to infection by herpes simplex virus is not well understood. It is known that

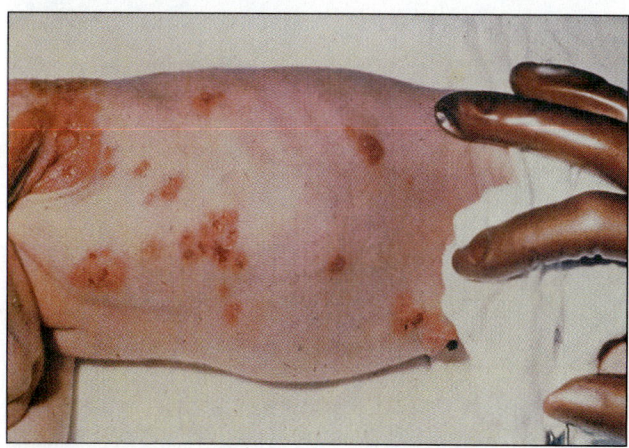

Figure 24.10 Herpes simplex type 2. Vesicles, small fluid-filled blisters, are the hallmark of herpes simplex rashes. This 10-day-old baby is likely to die or be permanently handicapped from neonatal herpes.

both antibody-mediated and cell-mediated reactions help control the severity of infection and reactivation of latent infections. People with an impaired immune system have a markedly decreased ability to control herpes simplex. In some immunocompromised patients, the virus can be fatal, affecting almost every organ of the body.

Clinical Syndromes. HSV enters the body of a new host during sexual contact and infects the external genitalia, the urethra, and the cervix. Rectal and pharyngeal herpes are also caused by sexual contact.

Genital herpes is characterized by a painful rash of tiny fluid-filled blisters called **vesicles** (**Figure 24.10**). The fluid contains infectious viruses. A first infection with HSV is called a **primary infection** and usually causes fever, headache, and muscle aches in addition to the rash. Reactivation of a latent infection usually produces rash alone. Stress, fever, or trauma—all of which can temporarily depress the immune response—seems to precipitate reactivation of HSV. Reactivation is extremely common. Typically, a person who suffers a genital infection with HSV-2 will suffer four recurrences during the next year. Recurrences are generally less severe and briefer than the primary infection, but they can be extremely painful. Both a primary infection and reactivation can occur without symptoms so that apparently healthy people who are shedding the virus unknowingly transmit herpes to their sexual partners or their newborns.

Neonatal herpes is a devastating disease. Most infected newborns develop widespread infections involving the brain and other internal organs. If they don't die, most suffer serious lifelong disabilities—blindness, deafness, profound retardation. Although babies can be infected after birth by close contact with infected adults,

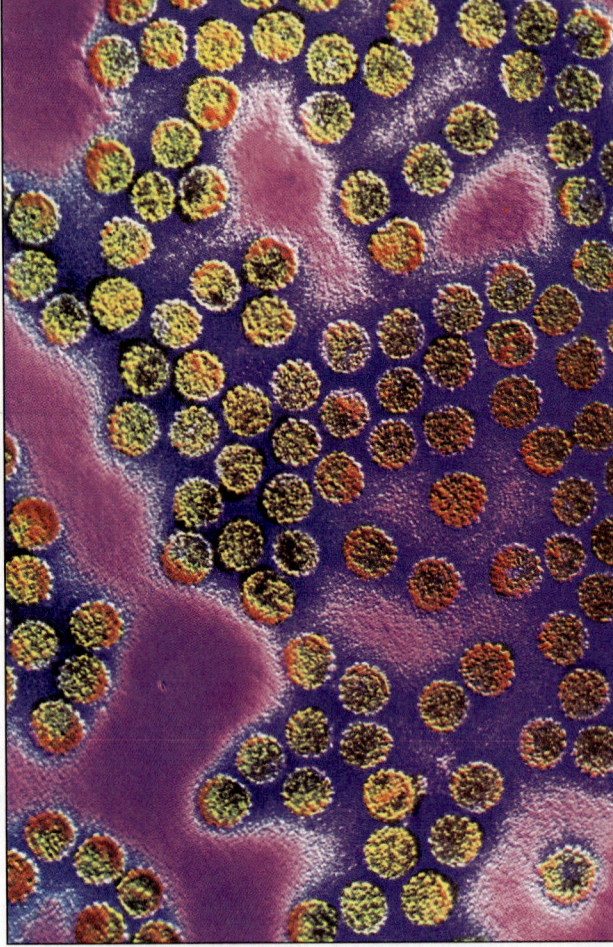

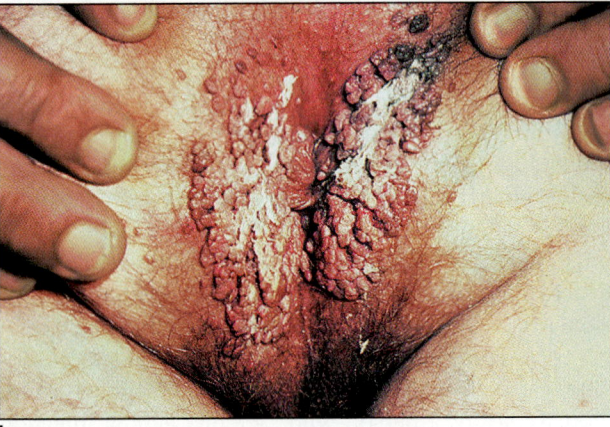

b

Figure 24.11 (**a**) Human papillomavirus (HPV). Different strains of HPV cause warts on different parts of the body, and some strains are sexually transmissible. (**b**) The genital warts shown here are perianal and vulvar.

most infections are acquired during passage through the birth canal. Delivery by cesarean section would prevent these tragic infections, but most affected babies are born to asymptomatic women who are unaware of their infection. Any woman who has active genital herpes lesions at the time of delivery should be delivered by cesarean section.

Prevention and Treatment. Genital herpes, like other STDs, can be prevented by interrupting transmission—avoiding sexual contact with an infected person. This is more difficult for herpes than for many other sexually transmissible infections, however, because many people who have the infection are not aware of it; and once infected, a person has the potential to transmit infection for the rest of his or her life. Further complicating the problem, condoms are not always protective because a woman's external genitalia or parts of the penis not covered by a condom may spread the virus. Still, using a condom provides far more protection against herpes than not using one.

HSV is not curable and, like most viral infections, once was not even treatable. Today, acyclovir, administered orally, significantly shortens the duration of a primary infection but does not prevent the establishment of a latent infection. In recurrent infections, it shortens the symptoms so slightly that it is rarely prescribed. Acyclovir is used on a daily basis, however, for the few people who experience such frequent recurrences that they are in almost constant pain. Therapy must continue indefinitely in order to be effective. As soon as the person stops taking acyclovir, the latent infection will be reactivated.

Acyclovir and another antiviral agent, **vidarabine**, are also used to treat life-threatening herpes simplex infections, those that involve the brain or other internal organs or that occur in immunocompromised patients. For these critically ill people, acyclovir is administered intravenously. In neonatal herpes, acyclovir has decreased the mortality rate from 65 percent to 25 percent; but, tragically, most of the survivors have brain damage. Recently, acyclovir-resistant strains of HSV have been identified in AIDS patients—a major setback in antiviral chemotherapy.

Human Papillomavirus (HPV): Condyloma Acuminata (Genital Warts)

Human papillomavirus (HPV) is a DNA virus of the papovavirus family that infects the skin and mucous membranes of human beings, causing warts (**Figure 24.11**). This virus cannot be cultivated in the laboratory, so strains of HPV were not identified until the 1980s when DNA technology became available (Chapter 7). If a papillomavirus was found to share less than 50 percent of its

Clinical Notes

Safer Sex—What Does It Really Mean?

Safer sex has become a familiar phrase during the AIDS epidemic, but what does it mean? In fact, safer sex refers to behaviors that keep people safer from all STDs, not just AIDS. There are only two sure ways to avoid sexually transmitted infection—either abstain from sex altogether or establish a permanent relationship in which both partners have sex only with each other. The key to this second option is *permanence*. Many people have relationships with a single partner, but the partner changes over weeks, months, or years. This serial monogamy puts people at extremely high STD risk—maybe even at higher risk than those who practice promiscuous, multiple-partner sex if the total number of contacts over a period of time is high. Even in committed relationships, therefore, condoms must be used for all sexual contact that involves penile penetration of any body orifice. Latex condoms with spermicides are the best, but the most important message of the safer sex campaign is to use *some* type of condom because *any* condom is far better than none.

Safer sex campaign poster.

DNA sequence with known strains, it was considered distinct. By this criterion, there are nearly 50 different strains, which are designated by number, for example, HPV-11. Different strains of HPV cause warts on different parts of the body—either the external genitalia, the upper airways, the soles of the feet, or other skin areas.

HPV types 6, 11, and 16 in particular are transmitted during sexual contact and cause genital warts, or **condyloma acuminata**. Genital warts occur on the skin or mucous membranes of the external genitalia, including the penis, vulva, cervix, and perianal skin. The same strains also cause wartlike growths on the vocal cords, called **laryngeal papillomatosis**, which usually occur in children and can cause hoarseness or even respiratory obstruction. HPV is transmitted during passage through the birth canal.

Genital HPV infection is not a reportable disease, but clinical data indicate that infection is quite common.

Screening tests show that up to 70 percent of female patients at STD clinics and 20 percent at family planning clinics carry HPV in their genital tract. Many of these infections are asymptomatic.

Genital warts used to be considered only a cosmetic problem, but HPV may cause cancer. Cervical and vulvar carcinoma in women and squamous cell carcinoma of the penis and rectum in men have been clearly associated with HPV, particularly HPV-16, and current research suggests a causal relationship. The association of HPV with cervical cancer is especially significant because this malignancy is so common, mostly affecting sexually active women between the ages of 18 and 30. Smoking, which is also a risk factor for cervical cancer, may interact with HPV to increase susceptibility.

Like all STDs, HPV infection can be largely prevented by safer sex practices. Condoms are effective in preventing transmission. Genital warts can usually be controlled or eliminated by freezing them with liquid nitrogen or painting them with liquid podophyllin or trichloroacetic acid. Although treatment solves the cosmetic problem, it does not eliminate the infection, and patients probably remain at risk for cancer. It is extremely important, therefore, that women with any history of genital warts have annual pap smears to detect early cervical cancer.

Other Sexually Transmissible Viruses

Two other viruses are also often sexually transmitted—hepatitis B and HIV—but both are considered in other chapters (hepatitis in Chapter 23 and AIDS in Chapter 27) for two reasons. First, they principally affect organ systems other than the urogenital, and, second, both can also be directly transmitted by blood transfusions and syringes used for illegal drugs. As sexually transmissible diseases, however, both can also be transmitted to newborns from infected mothers.

FEMALE REPRODUCTIVE TRACT INFECTIONS

Infections of the female genital tract can be caused by bacterial, fungal, or protozoal pathogens. Some are sexually transmissible, but others are not.

BACTERIAL CAUSES

Bacteria can cause infections of both the lower and upper female reproductive tracts. Clinical syndromes range from a relatively harmless vaginitis, to pelvic inflammatory disease, which can cause serious illness and sterility, to toxic shock syndrome, which can cause sudden death.

Gardnerella vaginalis: Gardnerella **Vaginitis**

Gardnerella vaginalis is a motile, pleomorphic bacterium that has recently been reclassified. It used to be considered a member of the Gram-negative *Haemophilus* genus, but it is now known to have the cell-wall structure of a Gram-positive.

G. vaginalis can be cultured from the vaginal secretions of up to 40 percent of sexually active, asymptomatic women. In women with vaginitis, it is present in much larger populations. Vaginal overgrowth of *G. vaginalis* is accompanied by unusually large numbers of anaerobic bacteria, a diminished population of normal lactobacilli, and an abnormally high vaginal pH. Though these abnormalities are typically associated with one another, the cause and effect are not clear—one reason the clinical syndrome is also called **bacterial vaginosis**. *G. vaginalis* infection causes a fishy-smelling vaginal discharge that contains **putrescine** and **cadaverine**—by-products of anaerobic metabolism that are also found in rotting fish.

Many women visit the doctor complaining of vaginal irritation and discharge, a clinical syndrome that can be caused by many different pathogens. The vaginal discharge is examined under a microscope, allowing the clinician to identify the causative agent and thereby prescribe appropriate treatment. This diagnostic procedure, called a **vaginal wet mount**, can be performed in a clinical laboratory or in the clinician's office. Microscopic diagnosis is based, in part, on the absence of other more easily visible vaginal pathogens, such as *Trichomonas vaginalis* and *Candida albicans*. Further evidence of *G. vaginalis* infection is the presence of **clue cells**—sloughed vaginal epithelial cells with many *G. vaginalis* cells clustered around them (**Figure 24.12**). *Gardnerella* vaginitis is treated with metronidazole, which decreases the population of vaginal anaerobes and helps reestablish a normal vaginal flora.

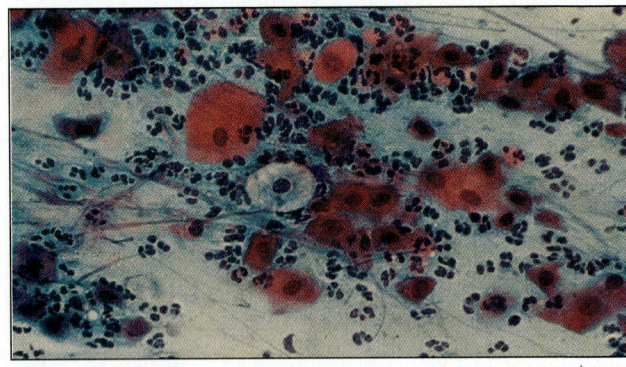

10 μm

Figure 24.12 Finding clue cells in a vaginal wet mount is the key to diagnosing *Gardnerella* vaginitis. A clue cell is a single cell sloughed from the vaginal epithelium that has many *Gardnerella* bacteria adhering to its surface.

Case History

Toxic Shock Syndrome

L. L. was a 19-year-old woman who came to an emergency clinic complaining of a high fever. She had been perfectly well until a few hours earlier when she started to think she might be coming down with the flu because she ached all over and felt feverish. Her mother, concerned about the sudden change, insisted she see a doctor. As part of the history, the admitting physician noted that L. L. had begun her menstrual period three days before and was using vaginal tampons.

The physician noted that L. L. had a fever of 105°F, but he was more concerned by the bright red rash that looked like sunburn over most of her body (**Figure 24.13**). Her blood pressure was slightly low. Because of her history of tampon use, the physician considered the possibility that L. L. was suffering from **toxic shock syndrome**. He admitted her to the hospital for observation—a decision that saved L. L.'s life. Within two hours she became unresponsive and her blood pressure was too low to be measured with a standard blood pressure cuff. L. L. was in shock.

She was transferred to the intensive care unit where heroic efforts maintained her vital functions. Intravenous fluids and medications sustained her blood pressure. A mechanical respirator supported her breathing. She was treated with nafcillin, an antimicrobial effective against *Staphylococcus aureus*. After about a week, L. L.'s condition stabilized. At this time her skin began to peel, particularly on her palms and soles. Eventually she was discharged in good condition. The only permanent reminder of her near-fatal illness was the loss of two toes because of poor blood supply while she was in shock.

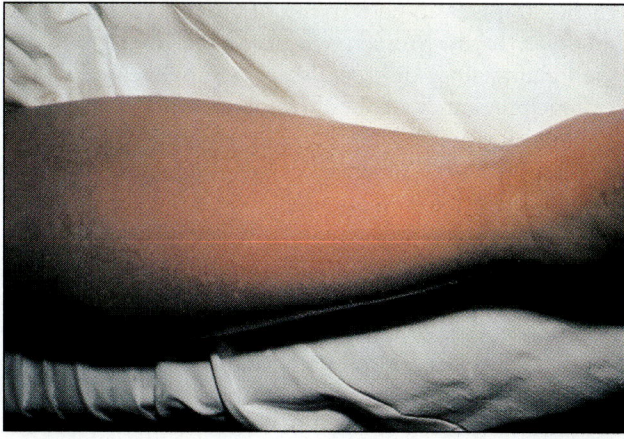

Figure 24.13 A bright red rash may be a symptom of toxic shock syndrome.

Toxin-producing strains of *S. aureus* are part of the normal vaginal flora of some women but in populations small enough that they do not cause disease. Tampons, however, set the stage for problems. Tampon insertion may cause vaginal abrasions, which speed the uptake of toxin, and the blood-soaked tampons themselves create an ideal environment for rapid bacterial multiplication. In general, the longer a tampon is retained in the vagina, the more likely it is to create a site for staphylococcal infection.

Illnesses like L. L.'s were first described in the medical literature in 1978, although they must have occurred before. Epidemiologic investigators determined that toxic shock was most likely to strike women who used a particular brand of super-absorbent tampon, probably because of the long retention time in the vagina. Taking these tampons off the market and publicizing the importance of changing tampons regularly brought this life-threatening syndrome under control. Today, the majority of cases are still related to tampons, but the disease is rare even among those at highest risk. Fewer than 3 in 100,000 menstruating women develop toxic shock syndrome each year.

Staphylococcus aureus: Toxic Shock Syndrome

Toxic shock syndrome is caused by strains of *Staphylococcus aureus* that produce a potent toxin called **toxic shock syndrome associated toxin (TSST)**. As bacteria multiply in the body, the disease-producing toxin is absorbed into the bloodstream, causing an illness like L. L.'s. Although toxin-producing staphylococci can multiply in many parts of the body and cause toxic shock syndrome in men, women, or children, the typical toxic shock patient is a menstruating woman like L. L. who is using vaginal tampons.

Pelvic Inflammatory Disease (PID) Pathogens

Sometimes bacteria from the vagina travel upward and infect the uterus, the fallopian tubes, and the ovaries, causing pelvic inflammatory disease (PID). Infection may be asymptomatic, but often fever, abnormal vaginal bleeding, and severe lower abdominal pain occur. These infections occur commonly in young sexually active women. More than 1 million women are affected in the United States each year. PID may be fatal if untreated, and even with treatment, nonfatal complications are significant. Each episode of pelvic inflammatory disease carries at least a 15-percent risk of infertility from scarred fallopian tubes. Women who have suffered from pelvic inflammatory disease also have an increased risk of **ectopic pregnancy**—a life-threatening condition in which pregnancy develops in a partially blocked tube, eventually rupturing the tube and causing massive abdominal bleeding.

PID is caused by various pathogens, most sexually transmissible. *Chlamydia trachomatis* is the most common cause of PID in the United States, followed by *Neisseria gonorrhoeae* and *Mycoplasma hominis*. Women who have multiple sexual partners are at highest risk. The contraceptive intrauterine device (IUD) was withdrawn from the market largely because of its association with PID. Cervical cultures are taken at the time of diagnosis, but results are not available for several days. Treatment—which must be begun immediately—is therefore designed to eliminate all of the most likely infecting microorganisms. An intramuscular injection of ceftriaxone, to cover *Neisseria gonorrhoeae*, followed by a 10-day course of oral tetracycline, to eliminate *Chlamydia* and *Mycoplasma*, is standard.

Infections after Childbirth or Abortion

Endometritis is particularly likely to occur after childbirth or abortion, when the epithelial lining of the uterus is disrupted. It is characterized by fever and a tender, enlarged uterus. If a bloodborne infection develops—which can progress to septicemia and shock—it is called **puerperal sepsis**. Although *Streptococcus pyogenes* was once the most common nosocomial cause of endometritis (see the box "Childbed Fever," Chapter 1), the organisms usually responsible today are *Escherichia coli, Neisseria gonorrhoeae, Chlamydia trachomatis, Clostridium* spp., and anaerobic streptococci. Thus, antimicrobial treatment must provide broad-spectrum coverage for Gram-positives, Gram-negatives, and anaerobes. Cephalosporins are the most commonly prescribed.

Microorganisms can be introduced into the uterus during the pelvic examinations performed during labor or when instruments such as forceps are inserted to help deliver the baby. Risk of infection increases greatly if part of the placenta—foreign tissue—remains in the uterus after delivery. These medical disasters are much less common today than in Semmelweis's time, but they still occur, and young, healthy mothers may die despite antimicrobial treatment.

Abortions, either spontaneous or induced, also provide an opportunity for microorganisms to enter the uterus. Infection after an induced abortion occurred frequently when terminating a pregnancy was illegal. Many illegal abortions were not performed under sterile conditions, did not completely empty the uterus, and left women afraid to seek medical treatment if infection occurred.

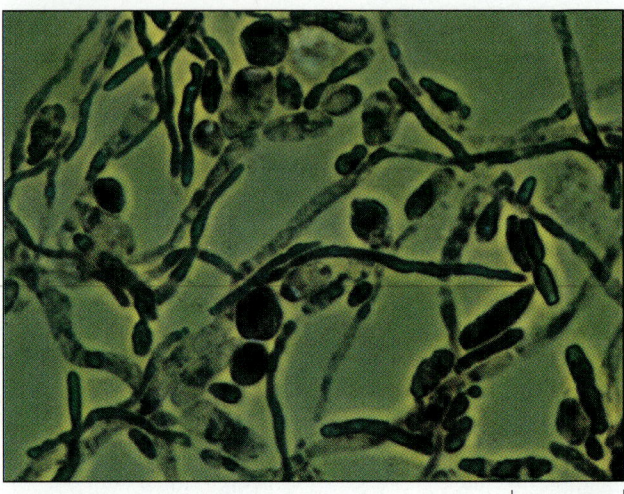

|———————| 10 µm

Figure 24.14   This wet mount was taken from a sample obtained during a pelvic examination of a woman complaining of vaginal itching. It shows the fungal hyphae and budding yeast of *Candida albicans*. *Candida* is part of the normal vaginal flora in many women, but several factors can lead to an overgrowth, resulting in candidiasis.

FUNGAL CAUSES

The yeast *Candida albicans* is one of the most common causes of vaginitis.

Candida albicans: Candidiasis

Small populations of *Candida albicans* are normally harmless inhabitants of the vagina, but if the host becomes unusually susceptible, they can proliferate and cause vaginitis. Overgrowth of *C. albicans* causes a thick white vaginal discharge and severe vaginal itching. When infected vaginal secretions are mixed with potassium hydroxide (KOH), which dissolves the keratin-containing epithelial cells that often obscure the *Candida*, fungal hyphae can be seen under the microscope (**Figure 24.14**). Local treatment with antifungal medications such as nystatin or the newer terconazole is usually effective.

Because *Candida* is an opportunistic pathogen, certain factors predispose a woman to vaginal candidiasis, including a change in hormone levels due to pregnancy or the use of oral contraceptives. Taking broad-spectrum antibiotics causes a change in the normal microbial flora of the vagina, which may allow an overgrowth of *C. albicans* (Chapter 14). Diabetics are particularly vulnerable to candidiasis, including the vaginal form.

PROTOZOAL CAUSES

Protozoal pathogens cause vaginitis, but not infections of the upper female reproductive tract.

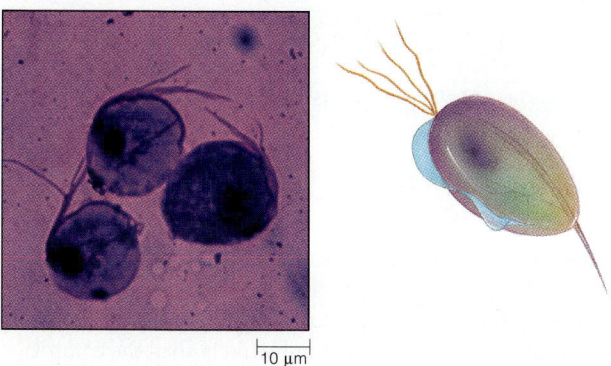

Figure 24.15 Viewing a wet mount slide of *Trichomonas vaginalis* is always interesting because these protozoa are very active and move in a characteristic jerky way. The illustration highlights the flagella that give it its characteristic movement.

Trichomonas vaginalis: Trichomoniasis

Trichomonas vaginalis, a flagellate protozoan, causes trichomoniasis, which is sexually transmissible. We discuss it here instead of with other STDs, though, because its primary clinical manifestation is a painful vaginitis characterized by a copious, frothy vaginal discharge. Some estimates put infection by *T. vaginalis* at 25 percent of women in the United States. Because infection is so widespread and often asymptomatic, *T. vaginalis* was once thought to be part of the normal vaginal flora. In males, *Trichomonas* usually infects the urethra but causes no symptoms unless it is accompanied by a bacterial infection.

Trichomoniasis is usually diagnosed by examining infected vaginal secretions under a microscope. The protozoa can be readily spotted because they move in a characteristic jerky way (**Figure 24.15**). *T. vaginalis* is easy to eradicate with oral metronidazole. To prevent reinfection, a woman's asymptomatic male sexual partner must also be treated.

MALE REPRODUCTIVE TRACT INFECTIONS

Men who seek medical attention for a reproductive tract infection usually complain of urethritis. They describe dysuria, urethral itching, and/or urethral discharge. Many pathogens can cause urethritis, and most are sexually transmissible. Urethritis is usually not a serious condition but is important in the transmission of STDs.

The classical cause of male urethritis is *Neisseria gonorrhoeae*. Gonococcal urethritis typically causes a copious, purulent urethral discharge loaded with intracellular Gram-negative diplococci. Most cases of urethritis, however, are caused by other pathogens. Until fairly recently, these others were lumped together under the heading

nongonococcal urethritis (NGU). If the infection wasn't gonorrhea, it usually wasn't possible to determine exactly what it might be.

Today, most pathogens that cause nongonococcal urethritis are identified. By far the most common is *Chlamydia trachomatis*, which causes from 30 to 50 percent of NGU. *Ureaplasma urealyticum* is a close second, causing 20 to 30 percent of cases. Other mycoplasmas may also cause NGU. A small number of cases, probably less than 5 percent, are attributable to nonbacterial pathogens—herpes simplex virus and *Trichomonas vaginalis*.

INFECTIONS TRANSMITTED FROM MOTHER TO INFANT

Some pathogens with very different clinical syndromes are grouped for study because they can be transmitted to the fetus or newborn, causing a much more serious infection than in the mother. Some of these pathogens are sexually transmissible, but others are not. They can be bacterial or viral.

BACTERIAL CAUSES

Some of the most devastating bacterial infections in newborns are acquired from the mother during pregnancy or at the time of birth.

Listeria monocytogenes: Listeriosis

Listeria monocytogenes is abundant in the environment. Its reservoirs are soil, animals, and asymptomatic human carriers. We do not know exactly how *L. monocytogenes* is transmitted to humans. Foodborne outbreaks are well-documented, but it may also be transmitted by inhalation or sexual contact. It is very difficult to eliminate *Listeria* from foods because it thrives at low temperatures, growing well in the refrigerator.

The disease is called **listeriosis** and takes many forms. *L. monocytogenes* infects the meninges, causing meningitis (Chapter 25); the bloodstream, causing septicemia (Chapter 27); and the heart, causing endocarditis (Chapter 27). It is an opportunistic pathogen, rarely causing serious disease in healthy, nonpregnant adults. People with compromised immune function are at high risk, and the number of cases has increased dramatically with the AIDS epidemic.

Women who are pregnant are also at increased risk of contracting listeriosis. The illness is mild, but bacteria enter the bloodstream and infect the placenta. Bacteria can cross the placenta to infect the fetus before birth or infect the newborn when membranes rupture during delivery. Fetal infections often cause abortion or stillbirth. Babies with listeriosis can suffer from widespread abscesses,

Listeriosis Outbreak

In 1985, obstetricians at a county hospital in Los Angeles noticed a striking number of mother-infant pairs diagnosed with listeriosis—a fairly unusual infection in healthy people. In all, 142 cases were identified between January and August. Sixty-five percent were pregnant mothers and their newborns (counted as a single case), and 35 percent were nonpregnant adults, almost all with suppressed immune-system function from drugs, illness, or advanced age. There were 48 deaths, 30 among unborn or newborn babies.

Epidemiologists attacked the problem with a **case-control study**. They matched infected mothers with uninfected women who gave birth the same day and interviewed both groups. They were looking for factors that might predispose the person to listeriosis. The epidemiologists found that the listeriosis victims were much more likely than the control group to have eaten a particular brand of soft, Mexican-style cheese. Further investigation proved that this cheese was contaminated with unpasteurized milk and grew *Listeria monocytogenes*. The contaminated cheese was recalled, the factory closed, and the epidemic brought under control.

---

meningitis, or septicemia, and many die, even with early treatment.

Group B Streptococci: Group B *Streptococcus* Infection

Streptococci classified as group B (also known as *Streptococcus agalactiae*) are significant pathogens of the female reproductive tract and newborn infants. Originally, group B streptococci were recognized as a cause of puerperal sepsis, but today they are more important as causes of life-threatening neonatal infections.

Group B streptococci are commonly found in the throat, gastrointestinal tract, and vagina of adults. Fifteen to 20 percent of pregnant women probably carry the bacterium in their reproductive tract. Babies are infected when fetal membranes rupture during delivery. Infants are at special risk if they are premature, making them more vulnerable to all infections, or if fetal membranes are ruptured for more than 12 hours before delivery, placing them in prolonged contact with the pathogen.

Pneumonia, meningitis, and sepsis are the most common forms of neonatal group B strep infection. Two of every 1000 babies are born infected by this organism—a high number—and nearly a quarter of these infants die in spite of antibiotic treatment. Many others are left permanently handicapped after surviving neonatal meningitis. Group B strep is also a significant pathogen of new mothers. It causes one out of five cases of postpartum endometritis and is the second most common cause of systemic infection following cesarean section. Recent studies suggest that high-risk mothers and their infants may be protected by receiving ampicillin intravenously during labor and delivery.

VIRAL CAUSES

Many viruses—including herpes simplex (discussed earlier), rubella (Chapter 26), and cytomegalovirus—cause devastating infections and birth defects when transmitted from mother to fetus or newborn.

Cytomegalovirus (CMV): Cytomegalic Inclusion Disease

Cytomegalovirus (CMV), a member of the herpesvirus group, is an extremely common cause of human infection around the world. Nearly 100 percent of people in developing countries and 75 percent of people in industrialized countries show evidence of CMV infection. Like other herpesviruses, cytomegalovirus establishes a latent infection that can be reactivated. Cells infected with the virus swell and develop inclusion bodies in their nuclei and cytoplasm, which accounts for the virus's name, cytomegalovirus (*cyto*, cell; *megalo*, large: large cell virus), and the disease it causes, **cytomegalic inclusion disease**.

CMV is transmitted sexually, by close contact such as the exchange of saliva, and in infected blood. Infection in healthy children and adults is usually asymptomatic, although it may cause a brief, mononucleosis-like illness. If a first infection occurs during pregnancy, however, and the virus crosses the placenta to infect the fetus, it can cause spontaneous abortion, stillbirth, or birth defects such as blindness, deafness, and mental retardation. Approximately 10 percent of infected infants show clinical evidence of disease (**Figure 24.16**). Both handicapped infants and those born with asymptomatic CMV may excrete the virus in their urine and saliva for years. Toddlers often transmit the virus to their caretakers or to one another. Day-care centers are hotbeds of CMV transmission, and pregnant women without serologic evidence of previous infection should probably avoid employment there.

Immunocompromised patients may also suffer life-threatening systemic infections from a new CMV infection or a reactivated infection. Many AIDS patients, for example, suffer from CMV infections of the eyes, lungs, or gastrointestinal tract. The only antiviral agent that appears to be at all useful in treatment is ganciclovir.

Nosocomial transmission of CMV in blood transfusions presents special ethical problems. Like other viral pathogens, CMV can be detected in banked blood. But because the pathogen is so widespread, it is not practical to discard all contaminated blood. It's been suggested that particularly susceptible patients—for example, severely ill newborns—should receive only CMV-negative blood. But this raises a difficult question. Should healthcare providers knowingly administer infected blood to some people, reserving noninfected blood for others?

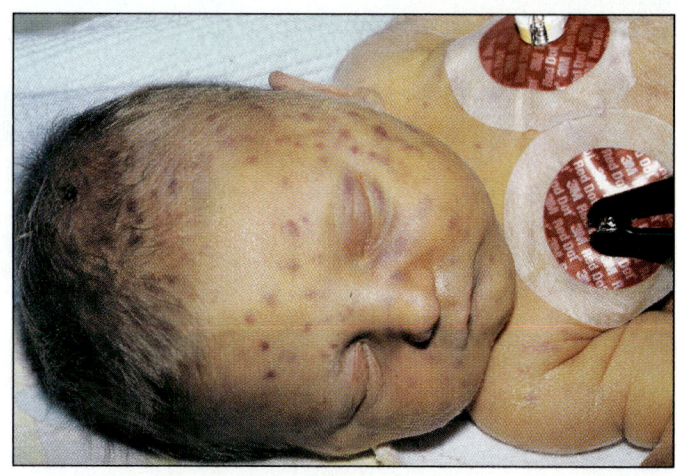

Figure 24.16 Ten percent of infants who acquire cytomegalovirus when it crosses the placenta are born ill. Note the baby's rash.

Table 24.4 summarizes infections of the genitourinary system.

Table 24.4 Infections of the Genitourinary System

Infection	Causative Agent	Mode of Transmission	Symptoms	Prevention and Treatment
URINARY SYSTEM				
Urinary tract infection (UTI)	*Escherichia coli, Proteus* spp., *Klebsiella* spp., *Pseudomonas* spp., enterococci, *Staphylococcus saprophyticus*	Fecal skin contaminants travel upward	Urinary frequency, urgency, dysuria; fever and flank pain in cases of pyelonephritis	Antibiotic choice depends upon strain sensitivity; common drugs include ampicillin, trimethoprim-sulfamethoxazole, nalidixic acid
Leptospirosis	*Leptospira interrogans*	Exposure to animal urine	Flulike illness; jaundice and kidney failure in severe cases	Penicillin, tetracycline
SEXUALLY TRANSMISSIBLE INFECTIONS				
Bacterial				
Gonorrhea	*Neisseria gonorrhoeae*	Sexual or perinatal	Urethral discharge; vaginal discharge and pelvic pain; rare systemic infections produce arthritis or rash	Ceftriaxone, cefixime
Syphilis	*Treponema pallidum*	Sexual or perinatal	Enormous variety of clinical manifestations depending on disease stage; no symptoms during latent stage; prenatal infection causes birth defects, snuffles, bone pain	Penicillin
Chlamydia	*Chlamydia trachomatis* serovars D–K	Sexual or perinatal	Mild dysuria, mucoid discharge leading to PID, NGU; high likelihood of female infertility and infection of newborns	Doxycycline, erythromycin, azithromycin
Lymphogranuloma venereum (LVG)	*Chlamydia trachomatis* serovars L_1–L_3	Sexual	Genital sore, enlarged groin lymph nodes	Doxycycline, erythromycin
Genital mycoplasma	*Ureaplasma urealyticum, Mycoplasma hominis*	Sexual	Usually asymptomatic; may lead to PID, NGU	Doxycycline

Table 24.4 (continued)

Infection	Causative Agent	Mode of Transmission	Symptoms	Prevention and Treatment
Chancroid	*Haemophilus ducreyi*	Sexual	Painful genital ulcer	Ceftriaxone
Granuloma inguinale	*Calymmatobacterium granulomatis*	Sexual	Persistent draining ulcers	Ampicillin, tetracycline, trimethoprim-sulfameth-oxazole

Viral

Genital herpes	Herpes simplex virus (usually HSV-2)	Sexual or perinatal	Painful genital blisters; systemic infection in newborns and immunocompromised patients	Acyclovir lessens symptoms, but does not cure infection or prevent recurrence
Genital warts (condyloma acuminata)	Human papillomavirus (HPV), particularly types 6, 11, 16	Sexual	Warts on external genitalia; predisposition to cervical and other cancers	Liquid nitrogen or podophyllin decreases warts, but no cure for infection and doesn't reduce risk of cancer

INFECTIONS OF THE FEMALE REPRODUCTIVE SYSTEM

Bacterial

Bacterial vaginosis	Overgrowth of *Gardnerella vaginalis*, associated with increase in vaginal anaerobes and decrease in lactobacilli	Sexual	Foul-smelling vaginal discharge; vaginal irritation	Metronidazole
Toxic shock syndrome	Toxin-producing *Staphylococcus aureus*, usually growing in vagina	Overgrowth of normal flora, usually associated with tampon use	Flulike illness, rash, shock	Nafcillin or other penicillinase-resistant penicillin; aggressive supportive care
Endometritis	*Escherichia coli*, *Neisseria gonorrhoeae*, *Chlamydia trachomatis*, *Clostridium* spp., anaerobic streptococci, group B streptococci	Bacterial contamination of uterus after childbirth or abortion	Uterine tenderness, fever, life-threatening bloodborne infection	Cephalosporin

Fungal

Vaginal candidiasis	*Candida albicans*	Overgrowth of normal vaginal flora	Vaginal itching, thick white discharge	Nystatin, terconazole

Protozoal

Trichomoniasis	*Trichomonas vaginalis*	Sexual	Vaginal itching, copious vaginal discharge	Metronidazole

PERINATALLY TRANSMITTED INFECTIONS

Bacterial

Listeriosis	*Listeria monocytogenes*	Foodborne, and possibly sexual; also perinatal	Meningitis, endocarditis, septicemia in newborns and immunosuppressed adults	Penicillin, gentamicin
Group B streptococcal disease	Group B streptococcus	Perinatal	Septicemia, meningitis, pneumonia of newborns	Ampicillin

Viral

Cytomegalic inclusion disease	Cytomegalovirus (CMV)	Close contact (exchange of saliva); blood transfusion; perinatal	Mononucleosis-like illness; stillbirth, birth defects	Ganciclovir

Summary

THE GENITOURINARY SYSTEM   (pp. 591–594)

1. The genital and urinary tracts are closely related anatomically but fulfill different functions—reproduction and excretion of metabolic wastes.

2. The upper urinary tract includes the kidneys and ureters; the bladder and urethra constitute the lower urinary tract.

3. The male reproductive system includes the testes (within the scrotum), the urethra (considered part of both the urinary and reproductive systems because it transports urine and semen), and a series of ducts and glands, including the prostate gland.

4. The female reproductive system includes the ovaries, the fallopian tubes, the uterus, and the vagina.

5. Clinical syndromes of the urinary tract include cystitis, urethritis, pyelonephritis, and urinary tract infection (UTI). Clinical syndromes of the female genital tract include vaginitis, endometritis, salpingitis, oophoritis, and pelvic inflammatory disease (PID).

URINARY TRACT INFECTIONS   (pp. 594–596)

1. All major pathogens of the urinary tract are bacteria. Diagnosis is based on a clinical history, physical examination, and urinalysis. Upper urinary tract infections are much more serious than lower urinary tract infections.

2. It can sometimes be difficult to tell how much of the urinary tract is affected, but treatment of both lower and upper urinary tract infections is the same: broad-spectrum penicillins, sulfa drugs, and certain drugs called urinary antiseptics, nalidixic acid and nitrofurantoin.

3. *Escherichia coli* is by far the most common cause of UTIs, including nosocomial urinary tract infections. Virulence factors of uropathogenic strains include adherence, hemolysins and colicins, serum resistance, and R factors.

4. Other bacterial species cause infections that are indistinguishable from *E. coli* infection. They include *Proteus, Pseudomonas,* and *Klebsiella* spp., and certain enteric streptococci.

5. *Leptospira interrogans* causes a systemic infection called leptospirosis. It is transmitted when urine from an infected animal enters the body directly through a break in the skin or indirectly through soil or water. Most infections are asymptomatic and self-limiting, but they can be fatal when liver or kidney damage occurs. Diagnosis depends on serologic tests; treatment is with penicillin or tetracycline.

SEXUALLY TRANSMISSIBLE DISEASES (STDs)   (pp. 596–610)

1. The epidemiology of STDs in the United States has changed recently with the emergence of new viral infections and increased incidence of others. People with one STD are likely to become coinfected.

2. Bacterial STDs are curable; viral STDs are not.

3. Women are more vulnerable to infection by STDs and their consequences for women are much greater (infertility, life-threatening pregnancies, fetal death, cancer). Children can suffer birth defects and newborn blindness.

4. Safer sex means reducing the number of sexual contacts and using condoms to help prevent infection with each contact.

Bacterial Causes   (pp. 597–606)

1. *Neisseria gonorrhoeae* causes gonorrhea, the most common reportable disease in the United States. It is transmitted only by direct contact. Virulence factors include adhesin-bearing pili, Protein II, and endotoxin. Symptoms in men—painful urination and a urethral discharge—appear two to seven days after exposure. Infections in women are usually asymptomatic. *N. gonorrhoeae* causes blindness in newborns. Treatment is with penicillin or tetracycline; drug-resistant strains are becoming more common.

2. *Treponema pallidum* causes syphilis. Diagnosis is by darkfield microscopy or serologic testing. As a clinical syndrome, syphilis has four stages: primary, secondary, latent, and tertiary. Treatment with penicillin is effective during the first three stages. Tertiary syphilis may be fatal or benign, characterized by gummas, soft granulomas in nonvital organs. Prenatal (congenital) syphilis results in fetal death or birth deformities, such as snuffles. Drug-resistant strains have not emerged.

3. *Chlamydia trachomatis* causes "chlamydia," the leading cause of infertility in women. Symptoms are subtle in both males and females (mild dysuria and a small discharge), which makes screening critical. Women under 25 and pregnant women are most at risk (*C. trachomatis* causes neonatal blindness and pneumonia). Treatment is with doxycycline, erythromycin, or azithromycin. Certain serotypes cause lymphogranuloma venereum (LGV), which can lead to painful ulcers.

4. *Ureaplasma urealyticum* and *Mycoplasma hominis* are part of the normal urethral and vaginal flora of most sexually active men and women. They sometimes cause a nongonococcal urethritis. Treatment for men with a discharge is doxycycline.

5. *Haemophilus ducreyi* causes chancroid. It is characterized by a soft chancre, which eventually heals; but without treatment (with ceftriaxone), it can leave scarring. In the United States chancroid is closely linked to the crack cocaine epidemic.

6. *Calymmatobacterium granulomatis* causes granuloma inguinale. Infection begins with lesions that develop into spreading ulcers. Diagnosis is on the basis of Donovan bodies. Treatment is with antibiotics, including ampicillin. Granuloma inguinale is rare in the United States.

Viral Causes   (pp. 606–610)

1. Herpes simplex virus (HSV) causes herpes, a highly communicable STD. Both epithelial and nerve cells are affected, and infection is both active (characterized by painful blisters) and latent. Acyclovir and vidarabine may be prescribed in severe cases. Neonatal herpes causes death, blindness, deafness, and profound retardation. Most affected infants are born to women unaware of their infection.

2. Human papillomavirus (HPV) causes condyloma acuminata (genital warts). The cosmetic problem is minor compared to the association that apparently exists between HPV and cervical cancer. Because there is no cure, lifelong monitoring through pap smears is critical.

FEMALE REPRODUCTIVE TRACT INFECTIONS   (pp. 610–613)

1. *Gardnerella vaginalis* causes a common vaginitis. The discharge is odorous because it contains by-products of anaerobic metabolism (an overgrowth of *G. vaginalis* is accompanied by an unusually large anaerobic bacterial population). Diagnosis is by the presence of clue cells in a vaginal wet mount. Treatment is with metronidazole to reestablish the normal vaginal flora.

2. Toxic shock syndrome—a life-threatening infection—is caused by strains of *Staphylococcus aureus* that produce toxic shock syndrome

associated toxin (TSST). Most at risk are menstruating women using vaginal tampons; there are few cases today.

3. Pelvic inflammatory disease (PID) occurs when bacteria from the vagina travel up to infect the uterus, fallopian tubes, and ovaries. Pathogens include *Chlamydia trachomatis*, *Neisseria gonorrhoeae*, and *Mycoplasma hominis*. Diagnosis is by cervical culture, but treatment must begin immediately with ceftriaxone and oral tetracycline.

4. Endometritis is an infection of the uterus caused most often today by *Escherichia coli*, *Neisseria gonorrhoeae*, *Chlamydia trachomatis*, *Clostridium* spp., and anaerobic streptococci. Treatment is with cephalosporins. Uterine infections occur after birth (when the uterine lining is disturbed) or after abortions not performed under sterile conditions.

5. The major fungal cause of vaginitis is *Candida albicans*, an opportunistic pathogen. Candidiasis is characterized by a thick white vaginal discharge and severe vaginal itching. A hormonal imbalance, antibiotic therapy, and especially diabetes put a woman at risk.

6. *Trichomonas vaginalis* is a protozoan that causes trichomoniasis. It may be asymptomatic or characterized by a painful vaginitis and copious, frothy discharge. In males, infection is asymptomatic. Treatment is with metronidazole.

MALE REPRODUCTIVE TRACT INFECTIONS (p. 613)

1. The classical male reproductive tract infection is caused by *Neisseria gonorrhoeae* and characterized by a copious, purulent urethral discharge.

2. Most infections, however, are nongonococcal urethritis (NGU). Most cases are caused by *Chlamydia trachomatis* and *Ureaplasma urealyticum*. Occasionally the cause is nonbacterial—*Trichomonas vaginalis* or herpes simplex virus.

INFECTIONS TRANSMITTED FROM MOTHER TO INFANT (pp. 613–615)

1. Pathogens with very different clinical syndromes are grouped because they affect the newborn more than the mother. They are transmitted across the placenta, during passage through the birth canal, or after birth through breast milk or intimate contact. (Many are also discussed under STDs.)

2. *Listeria monocytogenes* causes listeriosis. We do not know how it is transmitted, but foodborne outbreaks are well documented (*Listeria* thrives at low temperatures). Pregnant women are most at risk. The illness is mild, but it causes abortion, stillbirth, or—in the newborn—septicemia, meningitis, and often death, despite treatment. Immunocompromised individuals are also at high risk.

3. Group B streptococci cause neonatal group B strep infections when fetal membranes rupture during delivery. The most common forms are pneumonia, meningitis, and sepsis, causing death despite treatment in one-fourth of the cases and permanent handicaps in survivors. Ampicillin administered intravenously to high-risk mothers and infants during delivery appears to be helpful.

4. Cytomegalovirus (CMV), a member of the herpes group, causes cytomegalic inclusion disease. CMV is transmitted sexually, by close contact (exchanging saliva), and in blood. Infection is usually asymptomatic, but if first infection occurs during pregnancy, it can cause abortion, stillbirth, or birth defects. Handicapped and asymptomatic children shed the virus for years, infecting adult caretakers and other children. Blood-bank blood can contain CMV.

THE GENITOURINARY SYSTEM

1. Name the organs in the lower urinary tract, the upper urinary tract, the male reproductive system, and the female reproductive system. What are the functions of the urinary and reproductive systems?

2. Describe the following clinical syndromes:
   a. cystitis   d. urinary tract infection (UTI)
   b. urethritis   e. pelvic inflammatory disease (PID)
   c. pyelonephritis   f. vaginitis

3. What are perinatal infections?

URINARY TRACT INFECTIONS

1. Why are women anatomically more prone to urinary tract infections than men? Name some causes of recurrent UTIs. What precautions are taken to be sure a simple urine sample does not get contaminated?

2. Why should patients with a UTI drink large quantities of water? What are the general symptoms of lower urinary tract infections? of upper urinary tract infections?

3. Why are sulfa drugs particularly effective against urinary tract pathogens?

4. Describe the virulence factors in uropathogenic strains of *Escherichia coli*.

5. Why do so many cases of leptospirosis go undiagnosed?

6. Explain this statement: Until the 1970s, many clinicians believed STDs would be a problem of the past. What is safer sex?

7. Why are women more vulnerable to infection by an STD than men? Why are the long-term consequences of infection greater for a woman? Explain this statement: Children are victimized by STDs.

SEXUALLY TRANSMISSIBLE DISEASES (STDs)

1. Describe the virulence factors of the gonococcus. How is gonorrhea transmitted? Describe the symptoms in men and in women. Is gonorrhea treatable? Explain.

2. How is syphilis diagnosed? Describe primary, secondary, latent, and tertiary syphilis. What is the drug of choice for treating syphilis? Describe congenital syphilis.

3. What group is at highest risk for "chlamydia"? What are the symptoms and possible long-range complications for women? for men? Describe prevention and treatment.

4. Describe the clinical syndromes caused by *Ureaplasma urealyticum* and *Mycoplasma hominis*.

5. Why is chancroid a less communicable STD than gonorrhea and syphilis? Explain how the increasing rate of chancroid is linked to the crack epidemic in the United States.

6. How is granuloma inguinale diagnosed? What are the symptoms and treatment?

7. How prevalent are genital herpes infections in the United States? Explain how the herpes simplex virus causes both active and latent infections. Describe the clinical syndrome of primary infection and recurrent infections. Why are condoms not as helpful in preventing herpes infections as they are in preventing some other STDs? Should they still be used? Explain. How can neonatal herpes be prevented? What is the treatment and how effective is it for infants? for adults?

8. What is the treatment for genital warts and how effective is it?

FEMALE REPRODUCTIVE TRACT INFECTIONS

1. How does *Gardnerella vaginalis* cause vaginitis? How is it diagnosed? What is the treatment?

2. Describe the pathogenesis of toxic shock syndrome. Who is most at risk? What are the symptoms and treatment?

3. What significant nonfatal complications can result from pelvic inflammatory disease (PID)? How is PID transmitted? Discuss the prevention and treatment.

4. Name some situations that frequently lead to uterine infections. What is the treatment?

5. In what way is *Candida albicans* an opportunistic pathogen? What are the symptoms of candidiasis? How is it treated?

6. Describe the microorganism *Trichomonas vaginalis*. How is trichomoniasis transmitted? What are the symptoms and treatment?

MALE REPRODUCTIVE TRACT INFECTIONS

1. Discuss the symptoms of male urethritis. What is the most frequent cause of male urethritis? Name the pathogens that most frequently cause nongonococcal urethritis (NGU).

INFECTIONS TRANSMITTED FROM MOTHER TO INFANT

1. Why is *Listeria monocytogenes* easily foodborne? How is it an opportunistic pathogen and who is at highest risk? What forms does listeriosis take? How is a fetus or newborn affected?

2. How do infants acquire group B streptococcus infection? What are the risks to new mothers? How can infection be prevented?

3. How widespread is cytomegalic inclusion disease? Discuss the transmission, symptoms, prevention, and treatment.

Suggested Readings

Cates, W., Jr., and Hinman, A. R. 1991. Sexually transmitted diseases in the 1990s. *New England Journal of Medicine* 325:1368–70.

Handsfield, H. H. 1991. Recent developments in STDs: I. Bacterial diseases. *Hospital Practice* 47–56.

Handsfield, H. H. 1991. Recent developments in STDs: II. Viral and other syndromes. *Hospital Practice* 175–200.

Luby, J. 1982. Therapy in genital herpes. *New England Journal of Medicine* 306:1356–57.

Morbidity and Mortality Weekly Report. 1990. Progress toward achieving the 1990 objectives for the nation for sexually transmitted diseases. 39:53–57.

Schaechter, J. 1989. Why we need a program for the control of *Chlamydia trachomatis. New England Journal of Medicine* 320:802–4.

Witook, E., III, and Marra, C. M. 1992. Acquired syphilis in adults (review article). *New England Journal of Medicine* 327:1060–69.

25 INFECTIONS OF THE NERVOUS SYSTEM

Case History: Twenty-Four Hours

C. H. was a healthy 19-year-old male college student who suddenly developed a severe pounding headache one evening. Within an hour he had chills and a fever. Feeling sicker than ever before in his life, he asked his roommate to drive him to an emergency clinic. There the examining physician found no abnormality except a temperature of 105°F. C. H. was diagnosed as having a flu-like illness and advised to return if he didn't feel better in the next day or so.

The next day C. H. was much worse. Red spots appeared on his arms and legs in the morning. During the day, they grew and changed to a purplish color. He felt nauseous and gagged for no apparent reason. He could not eat or drink. His headache grew even more painful, and by evening his back hurt and his neck felt stiff. He became extremely weak. Within 24 hours, C. H. had gone from completely well to critically ill. Again his roommate drove him to the emergency clinic.

C. H. was barely able to sit up during the examination. His temperature was down to 101°F, but he looked extremely ill. His skin was pale with slightly raised purple spots—some nearly 1 cm in diameter—covering his trunk and extremities. His mouth was dry, and he appeared to be slightly dehydrated. He could move his head from side to side easily, but forward movement was so painful it was nearly impossible. The stiff neck suggested *meningmus*, irritation of the meninges.

C. H.'s physician recommended lumbar puncture, the only way to diagnose what she thought might be a life-threatening infection. Her suspicions were confirmed. Cerebrospinal fluid (CSF) is normally clear, but C. H.'s was cloudy—a sure sign of meningitis. The pressure of his CSF was also much higher than normal, causing his headache. Samples of C. H.'s CSF, blood, and urine were sent to a clinical laboratory for analysis.

C. H.'s blood showed a significant excess of leukocytes, suggesting a serious infection. Moreover, although CSF doesn't normally contain leukocytes, C. H.'s sample was packed with them—more than 15,000 per cubic millimeter (mm^3) of fluid. C. H.'s physician diagnosed *purulent* meningitis, usually caused by bacterial infection. (Purulent meningitis is diagnosed if the CSF contains

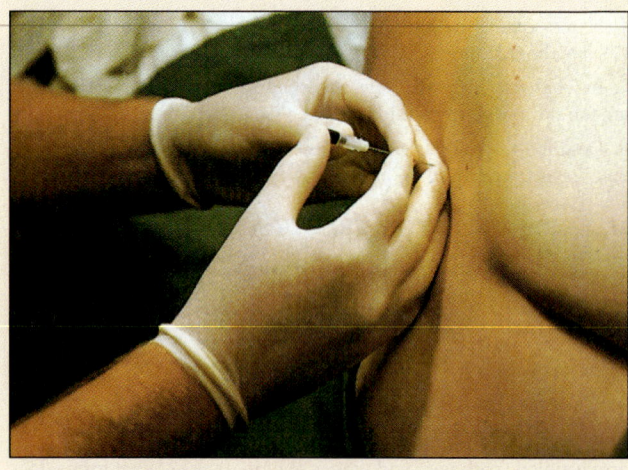

Lumbar puncture, a clinical procedure for recovering cerebrospinal fluid, is the only way to definitively diagnose meningitis. The patient's back is bent to widen the spaces between the vertebrae and allow the needle to enter the spinal canal. Because the needle is inserted below the end of the spinal cord, the procedure is safe.

more than 1000 leukocytes per mm^3. Nonpurulent meningitis, which has a lower leukocyte count, is usually nonbacterial.) A Gram stain of the CSF showed many Gram-negative diplococci among the leukocytes, making a diagnosis of meningitis caused by *Neisseria meningitidis* almost certain. Confirmation came from culturing the CSF.

As soon as C. H.'s physician received results of the CSF cell count and Gram stain, she started him on intravenous penicillin and a cephalosporin antibiotic. She admitted C. H. to the hospital and placed him in respiratory isolation. Four hours after his first dose of antibiotics, C. H. was feeling well enough to watch television, though he still complained of a headache and stiff neck.

When the culture and antibiotic sensitivity reports from his CSF became available a few days later, the cephalosporin was discontinued. C. H. continued on intravenous penicillin therapy for 10 days, during which all the signs and symptoms of his illness disappeared. He was discharged from the hospital in his normal state of good health.

To understand:

- The anatomy and function of the nervous system and its defenses against microorganisms

- The clinical syndromes that characterize nervous system infections

- The bacterial, viral, and fungal causes of infections of the meninges; their diagnosis, prevention, and treatment

- The bacterial, viral, prion-associated, and protozoal causes of neural tissue infections; their diagnosis, prevention, and treatment

THE NERVOUS SYSTEM

The nervous system integrates and coordinates all sensory and motor activities. Its largest organ, the brain, performs higher functions that allow us to appreciate life through consciousness, thought, and memory. Brain function is so central to our concept of humanity that it is the basis for our definition of death. Even if "vegetative functions" such as respiration and heartbeat can be maintained, a person is considered dead when the brain has ceased to function.

Structure and Function

The nervous system is composed of **neurons**, nerve cells, and **neuroglia**, cells that link neurons and perform accessory functions. Unlike the cells that compose many other body systems, neurons do not undergo a cycle of death and replacement. The neurons present at birth constitute the total pool available for old age. Damage to the nervous system is therefore likely to cause permanent disability.

The nervous system has two divisions. The **central nervous system (CNS)** includes the **brain** and **spinal cord**. The **peripheral nervous system (PNS)** includes the **nerves** (bundles of neurons and supporting tissue) that transmit messages of sensation inward from sense organs to the brain and commands for movement and other functions outward from the brain and spinal cord to the body (**Figure 25.1**).

The central nervous system is well protected against infection and injury. The brain is encased within the bony skull, and the spinal cord lies within the **spinal canal**, protected by many small interconnected bones called **vertebrae**. Further protection is provided by three continuous membranes called the **meninges**. The thinnest and innermost layer, which lies immediately over the brain and spinal cord, is called the **pia mater**. The middle layer is called the **arachnoid**, and the thickest and outermost layer, next to the bone, is called the **dura mater**.

A clear liquid called **cerebrospinal fluid (CSF)** circulates around the brain and spinal cord in the **subarachnoid space** between the pia mater and the arachnoid. Lying between nerve tissue and bone, CSF helps absorb the shock of injury. Samples of CSF can be removed from the spinal canal, below the point at which the spinal cord ends, by a procedure called a **lumbar puncture**, or **spinal tap**. The concentration of cells and molecules in CSF helps clinicians diagnose infections or other disorders of the CNS.

The CNS is also protected by the **blood-brain barrier**. Unlike the capillaries (smallest blood vessels) in most tissues, capillaries in the central nervous system have a thickened outer layer that severely limits their permeability. Some molecules—such as oxygen, carbon dioxide, glucose, and the essential amino acids—cross the blood-brain barrier freely, but many toxins and microorganisms cannot. Unfortunately, many antibiotics cannot cross the blood-brain barrier either, which limits the treatment of CNS infections when they do occur.

No commensal microorganisms live in the central nervous system. Infection can occur from massive trauma that opens the skull and from the bloodstream or the cerebrospinal fluid; it can spread from nearby sites, such as the sinuses or middle ear.

Clinical Syndromes

Most infections of the nervous system affect the CNS—the brain, the spinal cord, and the meninges. The most common clinical syndromes of infection are meningitis, encephalitis, and myelitis (**Table 25.1**).

Meningitis is inflammation of the meninges—the pia mater, arachnoid, and dura mater. The normally clear CSF fluid becomes cloudy with white blood cells attracted to the site of inflammation, and swelling around the brain increases pressure inside the skull, causing intense headache. Inflammation and irritation of the spinal meninges (**meningmus**) affect nearby muscles, causing a stiff neck. Fever is almost always present. Brain function is usually affected, because meningeal inflammation can diminish the flow of blood to the brain.

Encephalitis is inflammation of brain tissue. Encephalitis affects brain function, causing changes in a patient's state of consciousness or behavior. However, many disorders, even a high fever, can disturb brain function and cause a clinical syndrome that resembles infectious encephalitis. When no other cause of disturbed consciousness can be identified, infection is a likely possibility.

Myelitis is inflammation of the spinal cord. It has many causes, one of which is infection. Because the

Figure 25.1 Organs of the nervous system.

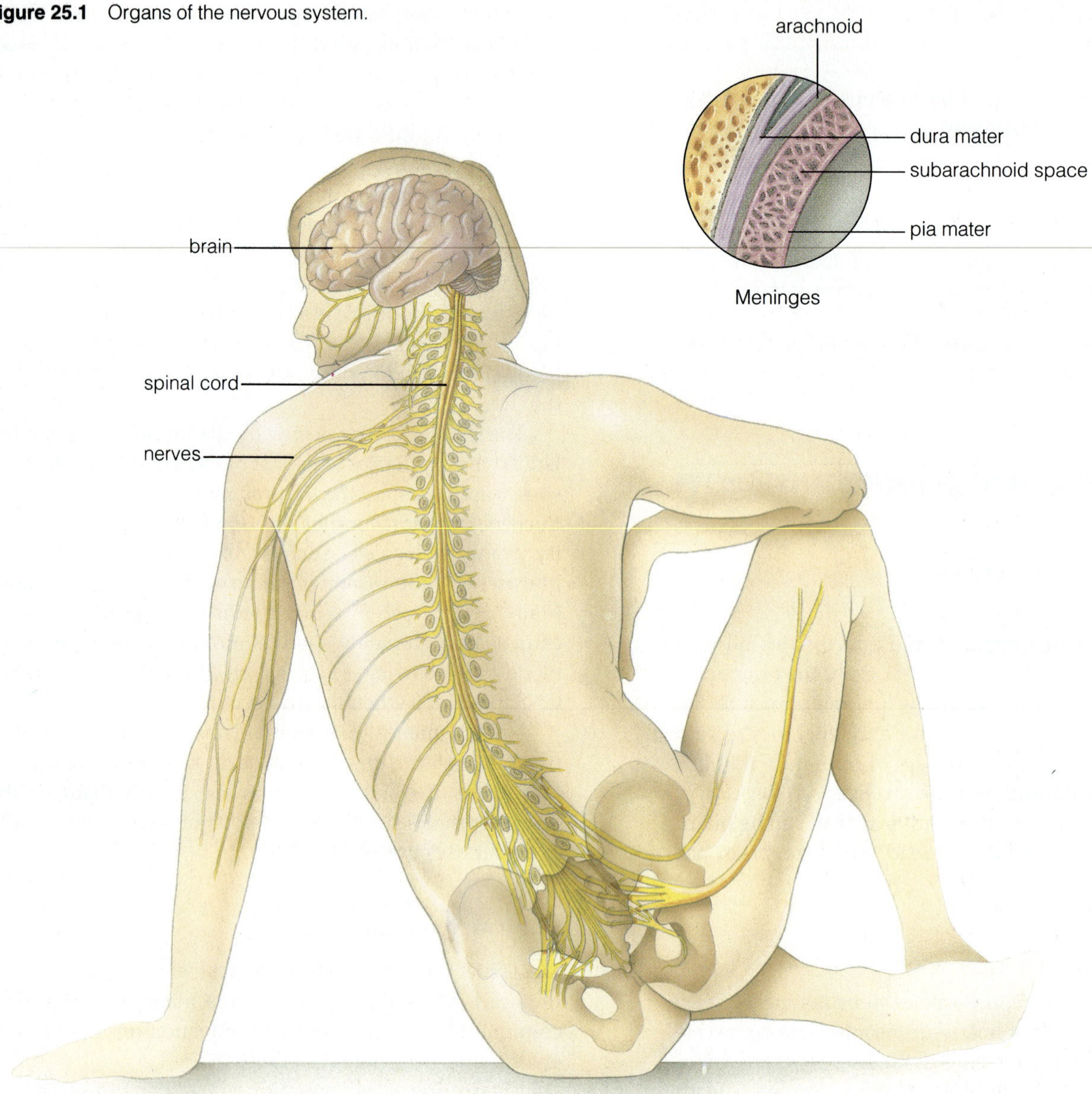

arachnoid

dura mater

subarachnoid space

pia mater

Meninges

brain

spinal cord

nerves

Table 25.1	Clinical Syndromes of Nervous System Infections		
Syndrome	Region Affected	Signs and Symptoms	Causative Agents
Meningitis	Meninges and surrounding spaces	Fever, headache, stiff neck, usually some disturbance in brain function	*Neisseria meningitidis, Haemophilus influenzae, Streptococcus pneumoniae,* various viruses, *Cryptococcus neoformans*
Encephalitis	Brain	Disturbance in brain function, usually fever	Rabies virus, arboviruses, *Naegleria fowleri, Trypanosoma brucei gambiense, Trypanosoma brucei rhodesiense,* prions
Myelitis	Spinal cord	Interrupted spinal impulses	Poliovirus
Bloodborne neurotoxins (not a nervous system infection)	Central or peripheral nerve function	Paralysis, rigid (tetanus) or flaccid (botulism)	*Clostridium tetani, Clostridium botulinum*

nerves in the spinal cord carry all messages for sensory and motor function, the particular clinical syndrome that results from a myelitis depends on which region of the cord is affected and which nerve connections are interrupted.

Some bacterial pathogens do not infect the nervous system but produce **neurotoxins**, highly destructive proteins that are transported through the blood or lymphatic circulations and affect nerve function (Chapter 15). These intoxications do not produce the typical syndromes of CNS infections described above. Their clinical manifestations depend entirely upon the action of the neurotoxin.

INFECTIONS OF THE MENINGES

Infections of the meninges are fairly common. Bacterial, viral, and fungal pathogens can cause meningitis.

BACTERIAL CAUSES

Various bacteria can infect the meninges and produce the typical clinical picture of **bacterial meningitis (Table 25.2)**.

Neisseria meningitidis: Meningococcal Meningitis

C. H.'s illness was caused by the Gram-negative pathogen *Neisseria meningitidis*, also called the **meningococcus**. Its only natural reservoir is the human body, and it often causes life-threatening diseases. C. H.'s illness was mild compared to many meningococcal infections.

N. meningitidis is usually grown in the clinical laboratory on chocolate agar medium. The colonies that develop are nonhemolytic and grow best in an atmosphere enriched with 5–10 percent carbon dioxide. *N. meningitidis* is distinguished from its close relative *N. gonorrhoeae* (Chapter 24) by its ability to ferment maltose.

The meningococcus is a superbly adapted pathogen. First, it is readily transmitted from person to person in respiratory droplets (Chapter 15). C. H.'s respiratory tract was probably first infected by someone who carried the meningococcus in his or her nose or throat. That person was probably not sick—in some populations, as many as a third of all healthy people are asymptomatic carriers of *N. meningitidis*.

Some meningococci that reached C. H.'s respiratory epithelium were immediately eliminated by IgA antibodies (Chapter 17), but others survived because *N. meningitidis* (and most other bacteria that cause meningitis) produces an enzyme called **IgA protease** that inactivates IgA. Also, meningococci bear pili with adhesins, making it difficult for mucociliary defenses to remove them (Chapter 15).

Meningococci might have lived harmlessly in C. H.'s nose and throat, making him just another asymptomatic carrier, but instead they displayed another capability of well-adapted bacterial pathogens—invasiveness. In the deeper tissues the meningococci had to secure nutrients, including iron, the nutrient in shortest supply in human tissues. *N. meningitidis* obtains iron by binding to transferrin and lactoferrin (iron-carrying proteins), releasing bound iron, and bringing it into the cell (Chapter 15).

The meningococci are protected by a polysaccharide capsule (Chapter 15). The immune system can produce type-specific antibodies against the capsule that promote phagocytosis of the invaders (Chapter 16). But C. H.'s immune system did not have enough time to mount a primary immune response, and a rapid secondary immune response was not possible because his immune system had never before been exposed to this particular antigen.

From C. H.'s bloodstream, bacteria were carried to all parts of his body. Some multiplied in the capillaries of his skin, releasing endotoxins when bacteria died. Endotoxic damage, mediated by the release of substances such as tumor necrotizing factor (Chapter 15), affected blood vessels and allowed blood to seep into his tissues, producing the purplish spots (**petechiae**).

When the meningococci entered C. H.'s central nervous system, they rapidly multiplied in his meninges,

Table 25.2 Bacterial Causes of Meningitis			
Bacterium	Percent of Cases	Case Fatality Rate with Treatment (percent)	Susceptible Groups
Haemophilus influenzae	46	5	Children age 6 months to 2 years
Neisseria meningitidis	27	13	Military recruits or others housed in crowded, stressful conditions
Streptococcus pneumoniae	11	29	Patients with weakened immune systems, including infants, sickle cell patients, alcoholics
Escherichia coli	Less than 5	50	Newborns and neurosurgical patients
Others	About 10	—	—

causing inflammation. This was a life-threatening event. If his infection had not been interrupted by antimicrobial drugs, C. H. would almost surely have been among the 70–90 percent of patients who die from untreated meningococcal meningitis.

Acute Purulent Meningitis. C. H.'s illness was typical of acute **purulent** meningitis in its sudden onset, high fever, and rapid appearance of central nervous system symptoms, including a stiff neck and headache. His deterioration within 24 hours was also typical. Not typical of bacterial meningitis was the fact that C. H.'s state of consciousness was never altered. Although meningitis primarily involves the meninges, brain function is usually affected, because inflammation disrupts cerebral blood flow. Meningitis patients may suffer from convulsions, extreme drowsiness (which may progress to coma), or agitated and bizarre behavior similar to that caused by drug overdose. Also atypical was C. H.'s age. Meningitis occurs most often in children between 6 months and 2 years who are no longer protected by maternal antibodies but have not yet fully developed their own antibodies.

One manifestation of C. H.'s infection—his purplish skin rash—is typical only of meningococcal infection. This rash is so characteristic that its presence alone allows a working diagnosis of meningococcal disease. The rash is also seen in **meningococcemia**, an illness in which *N. meningitidis* invades the bloodstream but not the central nervous system.

Epidemiology. C. H.'s infection was an isolated case, but meningococcal infection sometimes occurs in devastating epidemics. Outbreaks are particularly likely where large numbers of people are housed under crowded and stressful conditions, such as military barracks. Entire cities may be affected by epidemics. An outbreak in São Paulo, Brazil, during the early 1970s lasted for three years. The last nationwide epidemic of meningococcal meningitis in the United States occurred in the mid-1940s.

Epidemic disease may be unusually severe, progressing to death within hours despite antibiotic therapy or causing permanent brain damage in a high percentage of survivors. When epidemics are in progress, as many as 90 percent of healthy people in the community may become asymptomatic carriers. Family members of meningitis patients are almost certain to be carriers.

Various measures can stop an epidemic. One is to decrease the number of asymptomatic carriers—who are primarily responsible for spreading the infection—by treating them with the antibiotic rifampin. Rifampin is not, however, adequate treatment for systemic meningococcal infection (penicillin is the drug of choice), so these individuals must continue to be observed in case systemic illness develops. Eliminating the carrier state by treatment with rifampin is called **meningococcal prophy-**laxis. Even when an epidemic is not in progress, close contacts of patients with meningococcal infection are sometimes given rifampin prophylaxis. Because C. H. lived with only two roommates, physicians decided simply to observe them for early signs of illness. If he had lived in a large college dormitory, all his dormitory contacts might have been given rifampin.

Eventually, epidemics may be stopped with a comprehensive meningococcal vaccine. Various strains of *N. meningitidis* produce distinct capsular antigens. The major types are A, B, C, Y, and W135. Effective vaccines against *N. meningitidis* types A, C, Y, and W135 have already been developed, but one for type B—the most common cause of significant meningococcal disease—is still being researched.

Haemophilus influenzae: H. influenzae Meningitis

Haemophilus influenzae, like *Neisseria meningitidis*, is an invasive bacterial pathogen with a polysaccharide capsule (**Figure 25.2**). It also infects the respiratory tract and is discussed in detail in Chapter 22.

Like all types of meningitis, *H. influenzae* meningitis can cause permanent brain damage without prompt antibiotic therapy, so early diagnosis is essential. Diagnosis is more difficult in infants under 6 months, who seldom develop the stiff neck typical of older patients. Consequently, if there is any suspicion of meningitis with an infant, a lumbar puncture must be performed. Like meningococcal disease, *H. influenzae* meningitis is often transmitted by asymptomatic carriers, and cases sometimes occur in clusters. Rifampin prophylaxis is commonly prescribed for household and day-care contacts.

Until recently, *H. influenzae* caused about 75 percent of all meningitis in infants and young children. But the epidemiology of *H. influenzae* infection changed in late 1990 with the licensing of *H. influenzae* type b vaccine (Hib), a polysaccharide-protein conjugate vaccine. This vaccine is the first to stimulate effective immunity in infants—the group at highest risk. Within the first year of Hib licensure, *H. influenzae* type b diseases in children under 5 fell from 38 per 100,000 to under 10 per 100,000. Rates continue to fall as more children are immunized, and herd immunity helps protect children who missed their immunization. (Also see the discussion of Hib and epiglottitis in Chapter 22.)

Streptococcus pneumoniae: Pneumococcal Meningitis

Streptococcus pneumoniae is another encapsulated bacterial pathogen that can cause meningitis. Though *S. pneumoniae* more often causes pneumonia (Chapter 22), it also causes about half the cases of meningitis in adults over the age of 40. Pneumococcal meningitis often strikes people who are in a weakened state of health, such as newborns, people with sickle cell disease, and alcoholics. Bacteria usually enter the central nervous system from the bloodstream, although infection can also spread directly from respiratory sites such as the middle ear or sinuses. In spite of treatment with penicillin, about 25 percent of patients die. Permanent brain damage occurs much more frequently after *S. pneumoniae* than after *Neisseria meningitidis* or *Haemophilus influenzae* meningitis.

Escherichia coli: *E. coli* Meningitis

Escherichia coli causes less than 5 percent of bacterial meningitis cases, but newborn infants and patients who have undergone neurosurgery are particularly prone to this deadly infection. Even with appropriate treatment with ampicillin and gentamicin, half the patients die.

VIRAL CAUSES: ASEPTIC MENINGITIS

Sometimes patients with meningitis symptoms—severe headache, fever, and stiff neck—have no bacteria in their cerebrospinal fluid and, typically, fewer than 500 leukocytes per mm^3 of CSF. The diagnosis in these cases is nonpurulent, or aseptic, meningitis presumed to be caused by a virus. **Aseptic meningitis** is fairly common, especially among children and young adults, accounting for about 40 percent of all cases of meningitis.

Many different viruses can cause meningitis. The most common are the mumps virus (Chapter 23), members of the enterovirus group (including echoviruses, coxsackie viruses, and polioviruses), and arboviruses. The meningitis syndromes produced by different viruses can be indistinguishable, and, in fact, the causative virus is not usually identified.

Fortunately, it is not important to know which virus is the cause because antimicrobial treatment is not effective and people usually recover within a few days or weeks without medical intervention. It is extremely important, however, to distinguish between bacterial meningitis, which has a poor prognosis and requires immediate treatment, and viral meningitis, which has a good prognosis and requires no treatment. Many of the viruses that cause meningitis also cause other infections of the nervous system.

FUNGAL CAUSES

Fungi that cause meningitis produce a very different clinical syndrome from those caused by bacteria or viruses. The two primary fungal agents—*Coccidioides immitis* and *Cryptococcus neoformans*—cause chronic meningitis (*Coccidioides immitis* was discussed in Chapter 22 as the cause of coccidioidomycosis). A very few patients with coccidioidomycosis develop the same syndrome described below for cryptococcal meningitis.

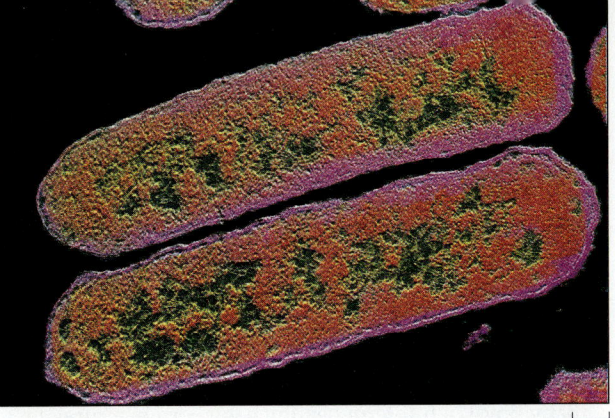

Figure 25.2 *Haemophilus influenzae* is an invasive bacterial pathogen that causes meningitis.

Cryptococcus neoformans: Cryptococcal Meningitis

Cryptococcus neoformans is a yeast with a polysaccharide capsule that can be two to three times as thick as the cell itself (see the box in Chapter 4, "The Capsule Is the Clue"). Unlike most other fungal pathogens, it is not dimorphic. *C. neoformans* is plentiful in the soil and can reach astounding concentrations in bird droppings, especially from pigeons.

C. neoformans is inhaled into the lungs. It may establish a respiratory infection with few if any symptoms. Serious disease results, though, if *C. neoformans* spreads in the bloodstream to the central nervous system. This most commonly occurs in patients with weakened immune systems. Cryptococcal meningitis is rare, affecting only a few hundred people in the United States every year, but its incidence is increasing because of the AIDS epidemic.

The chronic meningitis syndrome is quite different from the purulent meningitis syndrome seen in C. H. *C. neoformans* meningitis comes on gradually and progresses slowly. Over days or weeks the patient experiences headaches that may be accompanied by confusion, weakness, or loss of coordination. Sometimes a stiff neck is evident, sometimes not. Lumbar puncture reveals only a few hundred leukocytes per mm^3, most of which are mononuclear. The presence of *C. neoformans* cells in the spinal fluid confirms the diagnosis of cryptococcal meningitis. Without treatment, 90 percent of patients die within a year from complications caused by brain swelling and increased pressure within the skull. Treatment with amphotericin B, sometimes in combination with 5-flucytosine, decreases the death rate to only about 40 percent.

NEURAL TISSUE INFECTIONS

Bacterial, viral, and protozoal pathogens can all damage neural tissue. Some infect the brain, causing encephalitis, or the spinal cord, causing myelitis. Others do not infect the nervous system but produce neurotoxins that interfere with the function of neurons. Infectious agents called prions cause two rare nerve infections.

Case History

A Stiff Baby and a Limp Baby

The Stiff Baby

Baby Boy T, a full-term healthy baby born in a rural region of a developing country, was delivered at home after a short and uneventful labor. The woman assisting the delivery cut the baby's umbilical cord with an unsterilized instrument and, as was customary, dressed the cord with a paste made of clay. Baby Boy T was vigorous and nursed well. About the sixth day, however, he was unusually irritable, cried a lot, and had trouble sucking from the breast and swallowing milk. Over the next few days, he seemed to be growing stiff and sometimes suffered spasms that contracted muscles all over his body. The alarmed parents took Baby Boy T to the nearest health clinic, but by then his body was entirely rigid. His back was arched, and he was so stiff he could be supported with one hand under his head and another under his feet (**Figure 25.3**). His grieving parents were told that nothing could be done. Soon, Baby Boy T stopped breathing and died.

The Limp Baby

Baby Boy B was a normal newborn delivered in a large hospital in the United States. A day later he was sent home with his parents. He nursed well and grew rapidly, but at 4 months his mother began to notice subtle changes. He seemed listless and less active than usual. He lost his developing ability to hold his head up. He became constipated and began to breastfeed poorly. Even his cry was weak. The parents took Baby Boy B to an emergency clinic where the examining physician noted that he was limp—his muscles lacked normal tone. When the doctor raised Baby Boy B's arm or leg and dropped it, it flopped back down on the examining table as if he were asleep. His breathing was weak and his gag reflex was absent. The doctor immediately admitted him to the hospital.

Baby Boy B's weakness worsened and he became unable to breathe. He was transferred to the intensive care unit, put on a mechanical respirator, and fed intravenously (**Figure 25.4**). After several weeks of supportive care, his muscle strength spontaneously began to return. Improvement was gradual, but eventually he was discharged from the hospital. Baby Boy B continued to develop normally and had no lasting effects from his illness.

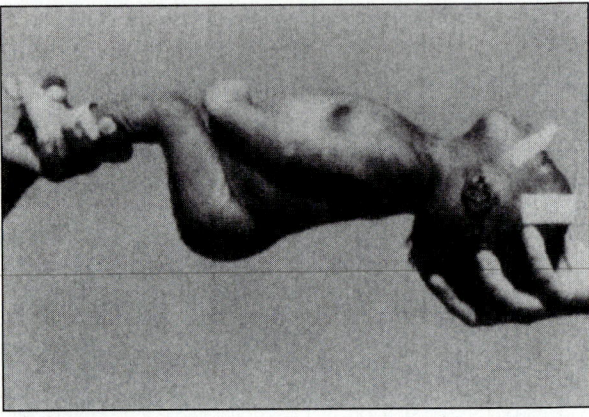

Figure 25.3 A baby with tetanus is rigid from uncontrollable contractions of the back muscles.

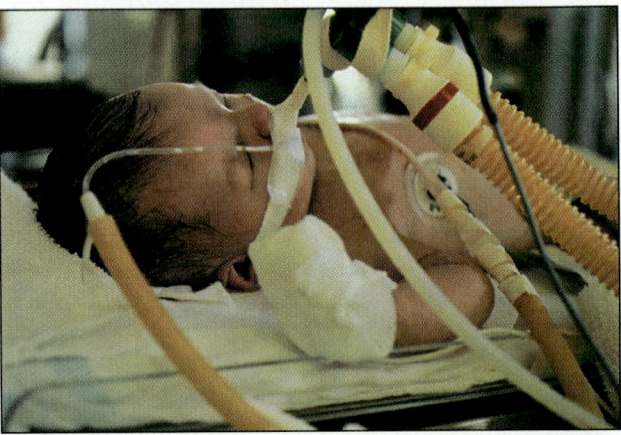

Figure 25.4 A baby with botulism is on a mechanical respirator because of flaccid paralysis of respiratory muscles.

BACTERIAL CAUSES

Bacteria that affect nervous tissue do so by means of neurotoxins. One of these—botulinum toxin—is the most poisonous natural substance known.

Clostridium tetani: Tetanus

The microorganism that caused Baby Boy T's horrible illness and death was *Clostridium tetani*—a Gram-positive rod that grows only under anaerobic conditions and moves by means of peritrichous flagella. It grows best at 37°C and a pH of 7.4, conditions that exist in the human body.

C. tetani is part of the normal intestinal flora of horses and cattle and some humans. *C. tetani* produces spores

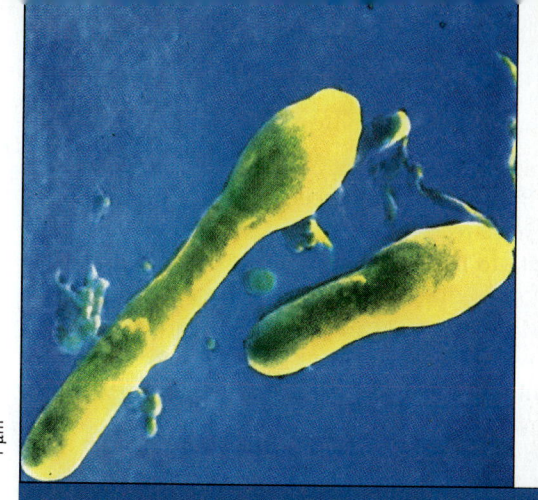
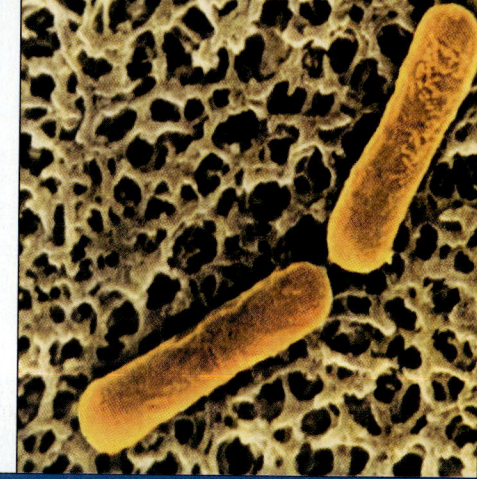

Figure 25.5 Comparison of *Clostridium tetani* and *Clostridium botulinum*.

Clostridium tetani	Clostridium botulinum
Bacterium	
Anaerobic, Gram-positive rod; motile; the endospore appears as a swelling at one end of the cell.	Anaerobic, Gram-positive rod, with slightly curved ends; motile; the endospores are more heat-resistant than *C. tetani*'s and appear as spherical bodies within the cell.
Infection	
Infection occurs when spores contaminate a deep, anaerobic wound or as a result of nonsterile surgical procedures.	Infection occurs when spores contaminate improperly canned food or are ingested by babies (for example, in honey). Rarely, infection occurs when spores contaminate a wound.
Toxin	
Tetanospasmin interferes with inhibitory nerve messages, and so muscles never relax but are in a state of constant contraction that causes rigid paralysis.	Botulinum toxin interferes with nerve message telling muscles to contract, resulting in flaccid paralysis.
Immunization	
Immunization with tetanus toxoid effectively prevents both neonatal and adult infection. Tetanus immune globulin can be administered after toxin production begins but before clinical symptoms appear.	No immunization exists. Food poisoning (an intoxication) can be prevented by careful canning. Infant cases are difficult to prevent.
Treatment	
Usually ineffective, even with the most sophisticated life-support systems.	Sophisticated supportive treatment, including respirator support, usually allows the patient to survive.

when conditions are inhospitable for growth. A spore-containing *C. tetani* cell looks like a drumstick because the spore swells one end of the rod-shaped bacterium (**Figure 25.5**). Endospores are found almost everywhere, but especially in cultivated fields where animal feces (used as fertilizer) are present.

C. tetani infection—**tetanus**—develops when endospore-containing dirt enters a deep, anaerobic wound. A wound becomes anaerobic when the blood supply to tissues is cut off and aerobic bacteria deplete existing oxygen. In the case of Baby Boy T, endospores germinated in the tissues of his decaying umbilical cord; this is the way that newborns typically acquire tetanus. Adults usually become infected through wounds or nonsterile surgical procedures. In the past, tetanus often occurred as a fatal complication of illegal abortions, and in wartime, it often killed more soldiers than did battle-inflicted injuries. By World War II, however, as a result of immunization, only 12 American soldiers contracted tetanus.

Tetanospasmin—The Tetanus Toxin. *C. tetani* has almost no invasive ability. It multiplies only within the infected wound. The toxin it produces, however, enters the bloodstream and is carried to the central nervous system, where it exerts its deadly effect. Baby Boy T's death and all his signs and symptoms of tetanus were caused by the neurotoxin **tetanospasmin**, one of the most potent toxins known. As little as 130 micrograms of tetanospasmin can kill an adult.

Normally, neurons send two kinds of messages—to stimulate muscle contraction and to inhibit it. Tetanospasmin affects neurons of the CNS that stimulate muscle contraction. We do not know the exact molecular action, but somehow the toxin interferes with the inhibitory messages. As a result, muscles throughout the body receive only stimulatory messages and contract uncontrollably. The poison strychnine produces a similar effect.

When all the muscles in Baby Boy T's body began to stiffen, strong muscles overpowered the weak ones, causing his back to arch—a clinical finding called **opisthotonos**. Adult tetanus victims also manifest **trismus**, contraction of the jaw muscles (**lockjaw**). Survivors of tetanus say these muscle spasms are excruciatingly painful. Spasms can be strong enough to break bones. When spasms prevent breathing, as they did for Baby Boy T, death occurs.

Prevention and Treatment. By the time clinical tetanus is recognized, it is difficult to treat. After tetanospasmin has bound to nerve cells, there is no way to interfere with its action. Tetanus toxin does begin to wear off after a week, and recovery will be complete if the victim survives that long. Even the most sophisticated intensive medical treatment, however—including drug-induced muscle paralysis and artificial ventilation—often fails to help. Survival in young adults, the population most likely to recover from tetanus, rarely exceeds 50 percent. Newborns are doomed. Even with the best treatment available, for example, Baby Boy T almost certainly would have died. Worse, cases like Baby Boy T's are quite common. In some parts of the developing world up to 10 percent of neonatal deaths are from tetanus.

Treatment—the administration of tetanus immune globulin—is possible after toxin production begins but before clinical signs appear. The immune globulin contains high concentrations of tetanus antitoxin (antibodies against tetanus toxin) that binds to and neutralizes tetanospasmin before it attaches to nerve cells. Preferable, of course, is prevention with tetanus toxoid (Chapter 20).

Universal tetanus immunization as part of the DPT inoculation has made tetanus exceedingly rare in the United States—the incidence rate is about 0.05 per 100,000 people. People may contract tetanus when they suffer a tetanus-prone wound—which is likely to happen to all of us from time to time—if they do not maintain their immunity with regular tetanus immunization and do not get a booster when the accident occurs.

Is there any way Baby Boy T's death could have been avoided? If his mother had been adequately immunized he would have received enough of her antibody through the placenta to be protected. Immunization of pregnant women could eliminate neonatal tetanus. His death might also have been prevented if his umbilical cord had been cut with a sterile instrument and kept clean or if tetanus antitoxin had been administered after infection but before he became ill.

Clostridium botulinum: Botulism

The bacterium that caused Baby Boy B's illness was similar in many ways to the one that killed Baby Boy T. Both are species of the genus *Clostridium*, both grow only under anaerobic conditions, and both are motile with peritrichous flagella. Also, both are rod-shaped, but *C. botulinum* cells are slightly curved with rounded ends and its endospores are even more heat-resistant than *C. tetani*'s (Figure 25.5).

Clostridium botulinum is found virtually everywhere in the environment. Because it is so prevalent in soil, spores are often present in food, especially raw agricultural products such as vegetables and honey. Spores in ocean sediment can contaminate fish and seafood. Epidemiologic investigators have even found *C. botulinum* spores inside home vacuum cleaners.

Botulinum Toxin. The neurotoxin produced by *C. botulinum*—**botulinum toxin**—is the most poisonous natural substance known, even more toxic than tetanospasmin. A lethal dose of toxin for a human is measured in nanograms, an amount that consists of only a few million molecules. The genes for toxin production are carried by a bacterial prophage. Only phage-carrying strains produce toxin, and the type of phage determines the type of toxin produced.

There are several antigenically distinct forms of botulinum toxin designated A through G. Only types A, B, and E cause human disease. Type A toxin, the most potent of all, is the principal type produced by strains of *C. botulinum* west of the Mississippi River. Type B is more common in the eastern United States and Europe. Type E toxin is the most common type to contaminate preserved fish and seafood.

As the illnesses suffered by Baby Boy T and Baby Boy B show, tetanospasmin and botulinum toxin have vastly different actions. Botulinum toxin affects the peripheral rather than the central nervous system. Toxin that enters the circulatory system binds directly to the neurons at the site where they join the muscle. When toxin enters the neuron, it prevents the release of acetylcholine—the molecular messenger that delivers the signal for muscle contraction. The paralysis that results is a limp, or **flaccid**,

paralysis, unlike the rigid paralysis of tetanus victims. In adults the nerves that control the head are affected first, typically causing double vision and difficulty with speaking and swallowing. Eventually other muscles, including the ones for breathing, may become paralyzed.

Clinical Syndromes. Botulism is always caused by the paralytic neurotoxin of *C. botulinum*, but people can be exposed to the toxin in different ways. Baby Boy B became ill because he swallowed *C. botulinum* cells that colonized his intestinal tract. The bacteria produced neurotoxin inside his body, a tiny fraction of which was absorbed into his bloodstream, causing his illness. This syndrome, in which bacteria produce the toxin while multiplying inside the body, occurs in babies and is called **infant botulism**.

Infant botulism was unrecognized before 1976, but now it is the most common form of botulism reported in the United States. Several hundred cases are diagnosed each year. No one knows why this type of botulism affects only infants under 6 months of age, because older people also ingest these bacteria. Age-related differences in the normal bacterial flora of the intestines probably make infants especially vulnerable.

Most adults acquire botulism as a form of food poisoning. The word *botulism* comes from the Latin word for "sausage," because the disease was first recognized during the 1700s when it was associated with eating blood sausage. Like staphylococcal food poisoning (Chapter 23), foodborne botulism is an intoxication rather than an infection—bacteria do not multiply inside the body.

Botulism is usually caused by food improperly canned at home. Spores of *C. botulinum* survive and germinate during storage, forming vegetative cells that multiply and produce toxin. The toxin is heat-labile, but if people eat the contaminated food without reheating, they ingest the potent toxin. Only a minute amount of the ingested toxin is absorbed into the blood (the rest is excreted), but it is enough to cause death. Botulism is usually caused by home-canned vegetables like green beans, peppers, and mushrooms. Tomatoes and other acidic foods, such as pickles, rarely transmit botulism because *C. botulinum* cannot grow in an acidic environment.

The least common way to contract botulism is through a wound. When this occurs, the mechanism is the same as with tetanus. Spores enter an anaerobic wound and produce toxin that enters the bloodstream. Of course, botulinum toxin causes the same flaccid paralysis whether it is produced inside the body or out.

Baby Boy B survived because his vital functions were maintained until the toxin wore off. With good supportive care, including mechanical ventilation, only 3 percent of infant botulism victims die. Mechanical respirators are also crucial in treating adult botulism because the usual

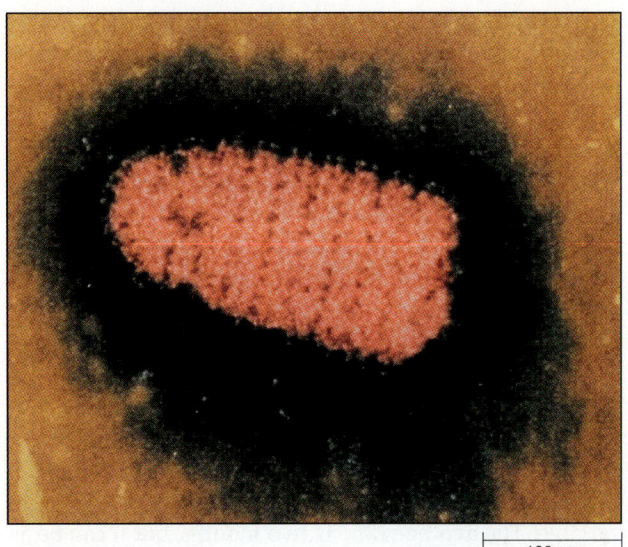

Figure 25.6 Electron micrograph showing a single bullet-shaped rabies virus.

cause of death is respiratory paralysis. The fatality rate for adults with botulism is a little higher than for babies—about 10 percent. For foodborne and wound botulism, a polyvalent botulism antitoxin containing antibodies against types A, B, and E toxin is recommended. The antitoxin is derived from horse serum, and allergic reactions are common (Chapter 18).

Most cases of foodborne botulism are preventable by carefully following canning procedures and boiling canned food before eating. Infant botulism is more difficult to prevent because *C. botulinum* spores are everywhere, and most of us consume them fairly regularly. Epidemiologists have identified honey as a risk factor for babies. As a raw agricultural product, honey often contains large numbers of *C. botulinum* spores and so is not recommended for children under the age of 1.

VIRAL CAUSES

Many viruses are **neurotropic**—they can infect neurons. Some, like herpes simplex (Chapters 24 and 26) and varicella zoster (Chapter 26), also infect other types of cells. Those that predominantly affect the nervous system cause some of the most devastating neural infections.

Rabies Virus: Rabies

Rabies virus belongs to the rhabdovirus family of bullet-shaped viruses (**Figure 25.6**). It contains a single strand of RNA surrounded by a lipid envelope with projecting antigens that stimulate the production of protective antibodies.

The rabies virus infects many different mammals, but humans only rarely. Infection usually occurs after a bite, when rabies virus in the animal's saliva enters the wound. Other modes of transmission can occur, though. Infection may spread across unbroken mucous membranes when an infected mother animal licks her young. Infection by inhalation occurs in caves with dense populations of infected bats. One case of human rabies was traced to a transplanted cornea that was infected.

Clinical Syndrome. When the rabies virus enters a wound, it multiplies in the surrounding muscle tissue. Eventually the virus infects nearby nerves and travels along nerve fibers until it reaches the brain, where it exerts its deadly effect. The time between the bite and the appearance of central nervous system symptoms is quite variable. The average time is two months, but it can be as short as nine days or as long as several years.

The clinical syndrome of rabies is horrifying. The disease usually begins with flulike symptoms, but even this **prodromal phase** (earliest symptoms) is often associated with strange sensations. The region around the bite may tingle or burn, and the victim may feel nervous or depressed or sense impending death. During the **excitation phase**, severe neurological abnormalities appear. The person begins to lose control of muscle function, speech and vision are impaired, and anxiety and apprehension may become unbearable. The classic symptom is **hydrophobia** (fear of water). The muscles that control swallowing contract abnormally, so the person can't drink. When a mouthful of water is taken, the rabies victim either chokes violently or spits out the liquid in an uncontrollable spasm. After this painful experience, the mere thought of drinking may provoke a similar attack. Throughout the excitation phase, the victim remains awake and entirely alert. Eventually the **paralytic phase** begins. Muscles weaken, consciousness fades, and death occurs. The fatality rate for rabies approaches 100 percent even with the best available supportive treatment.

Rabies can often be diagnosed clinically and then confirmed by an immunofluorescent antibody test that detects the presence of rabies virus in infected nerve cells. The sample may come from a sacrificed animal or be obtained at autopsy. Before this test became available in the late 1950s, the diagnosis was based on finding characteristic inclusion bodies, called **Negri bodies**, in infected nerve cells. Negri bodies develop where new virus particles are being assembled.

Epidemiology and Prevention. Because humans are rarely infected by the rabies virus, perpetuation of the disease depends on a reservoir of infected animals. In rural areas, commonly infected animals include skunks, foxes, raccoons, coyotes, and bats. In urban areas, dogs and cats may be infected. An infected animal can be rec-

ognized by its abnormal behavior. Nocturnal animals appear in the daytime, wild animals approach humans, and gentle pets suddenly become aggressive.

In the United States, only 15 cases were reported in the 1960s, 23 in the 1970s, and 10 in the 1980s; and some of these infections were not acquired in this country. Mass immunization of pets is one reason for the low incidence. Still, an enormous, ineradicable reservoir exists among wild animals.

Rabies can be prevented by active immunization even after infection because it has such a long incubation period. The first rabies vaccine was developed by Louis Pasteur in 1885. A young boy bitten by a rabid dog was brought to Pasteur. He vaccinated the child with weakened rabies virus from the brain and spinal cord of infected rabbits, saving the boy's life. The **Pasteur treatment** consisted of 14 to 21 injections into the abdominal wall. Not only were the injections painful, but hypersensitivity reactions were also common.

The Pasteur treatment was used into the mid-twentieth century, when better immunizing agents were developed. The current vaccine is prepared in human diploid cell culture. This human diploid cell vaccine (HDCV) is reasonably free from side effects, can be administered intramuscularly, and requires only six injections. If the risk of rabies is considered real, prophylaxis—both passive immunization with human rabies immune globulin (a serum preparation rich in antibodies against the rabies virus) and active immunization with HDCV—is begun.

Poliovirus: Poliomyelitis

Polioviruses belong to the picornavirus family. They are closely related to the hepatitis A virus and coxsackie enteroviruses (Chapter 23).

Poliovirus occurs in three different serotypes—1, 2, and 3. Antibodies against poliovirus are type-specific, so each type of antibody can neutralize only one type of virus. To be completely protected, a person must have antibodies against all three viral types. Because only humans are naturally infected by poliovirus, polio research depends on tissue culture. Not until this complex technique was perfected was development of a polio vaccine possible.

Clinical Syndrome. Polioviruses, like other enteroviruses, are transmitted by the fecal-oral route. They survive for long periods outside the body, making polio readily transmissible by water or food contaminated with feces. In parts of the world where sanitation is poor, almost everyone is exposed to poliovirus during infancy. Where sanitation is relatively good, many people are not exposed until childhood or even adulthood.

Ninety percent of those infected by polioviruses show no signs of illness, and most who do become sick experi-

Clinical Notes

A Bizarre and Deadly Encounter

Americans who stay close to home hardly ever die of rabies. Only four domestically acquired cases were reported to the Centers for Disease Control (CDC) between 1980 and 1990. So it was a surprise when the *MMWR* reported three such cases within a three-month period in 1991, one each in Texas, Arkansas, and Georgia. Even more unsettling, none of the victims had a clear history of animal bite.

The Arkansas case was typical. On August 17, a lifelong resident of Arkansas came down with a sore throat and had trouble swallowing. Two days later the man was treated with antibiotics, but that same night he began pacing and spitting frequently. His facial muscles twitched uncontrollably and he acted anxious and fearful. He was hospitalized with possible diagnoses of drug overdose or viral encephalitis. Rabies was thought unlikely because there was no history of animal bite. Although he remained alert and oriented

for several days, he was increasingly agitated, complained of itching all over his body, and began to vomit repeatedly. His temperature rose at times to 106°F. He died on August 25.

Brain samples taken at autopsy were positive by monoclonal-antibody typing for a strain of rabies commonly found in the silver-haired bat. The man lived in a previously abandoned rural house and had certainly had close contact with bats. A friend reported that one night in early July a bat had landed on the man's mouth. He killed it but may have been scratched during the encounter.

After reviewing this case and the two others CDC epidemiologists reaffirmed their recommendation that rabies prophylaxis is essential for anyone who is bitten, scratched, or has mucous membrane contact with any wild animal of a species known to carry rabies.

---

ence only a flu-like illness characterized by fever, headache, sore throat, and gastrointestinal complaints. A few people (less than 1 percent) also complain of a stiff neck. Only 0.01 percent of infected humans suffer the dreaded clinical syndrome of paralytic **poliomyelitis**. The paralysis of polio is flaccid like that of botulism.

Poliovirus produces a lytic infection that kills the infected cell (Chapter 15). First the virus replicates inside the epithelial cells lining the nose, throat, and intestine, where it causes little damage. It spreads to nearby lymph nodes and, after further multiplication, enters the bloodstream. If bloodborne virus enters the central nervous system, the stage is set for serious infection.

Polio is classified as a myelitis because the **anterior horn cells** of the spinal cord are particularly susceptible to the poliovirus. **Motor neurons**, which carry messages from the spinal cord to the muscles, originate in the anterior horn cells; if they are damaged or killed, paralysis results. Neurons in the brain that control muscle function can also be affected.

Poliovirus produces varying degrees of paralysis. Weakness usually progresses during the first few days of illness and then gradually improves as surviving neurons resume normal function. In many patients, however, some degree of paralysis is permanent because dead neurons do not regenerate. When respiratory muscles are affected, polio can be fatal. Today, polio patients with respiratory failure can be kept alive on a mechanical ventilator. During the polio epidemics of the 1950s, the iron lung was used.

The Polio Vaccine. Today we consider polio one of the many diseases against which children are routinely vaccinated. Less than 50 years ago, however, polio held terror for every parent (**Figure 25.7**). During the warm months when enteroviruses are prevalent, polio epidemics spread through the country. At the peak of the epidemics in the early 1950s, 39,000 cases of polio and 1900 deaths were reported in a single year. Worried families kept their children away from crowds and public swimming pools. Adults were also vulnerable—as President Franklin D. Roosevelt made everyone aware. In an effort to find a cure, the Mothers' March of Dimes raised millions of dollars for medical research.

In 1955 the medical miracle occurred. Jonas Salk, an American microbiologist, developed a vaccine against polio, and polio in the United States essentially disappeared. The last small outbreak was reported in 1979 in a group of people who had declined immunization. Since then, there have been no naturally occurring cases of polio inside the United States. The Salk vaccine is an **inactivated poliovirus vaccine** (IPV). Virus particles are treated with formalin so they can no longer replicate, but the antigens can still stimulate the immune system to produce protective antibodies (Chapter 20). It is a **trivalent vaccine** containing antigens of all three poliovirus serotypes.

Shortly after the Salk vaccine was released, a few batches were incompletely inactivated and some children contracted paralytic polio from their immunization. Confidence was shaken. In fact, safety is not a problem when

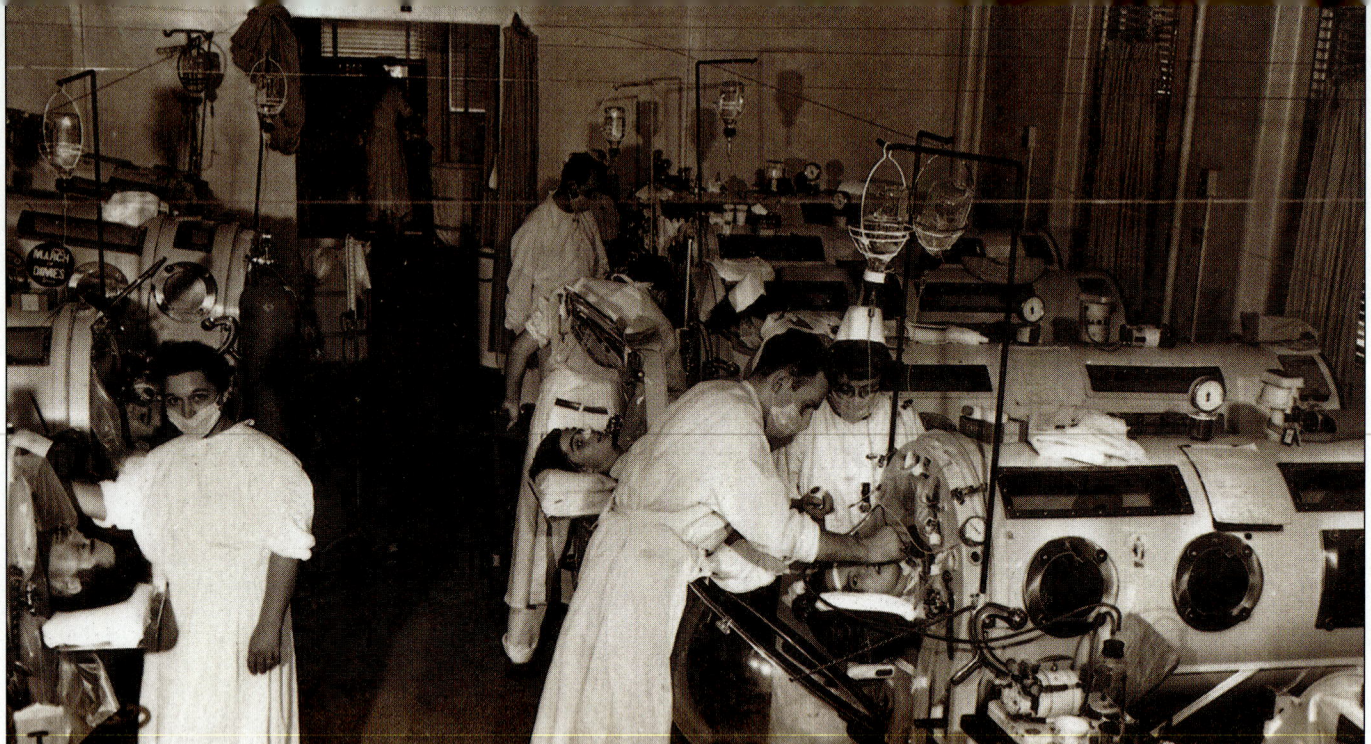

Figure 25.7 During the polio epidemics of the 1940s and 1950s, patients whose respiratory muscles were affected were kept alive with a mechanical respirator called the iron lung.

IPV is properly prepared, but as with other inactivated vaccines, continued booster immunizations are necessary (Chapter 20). In 1961 an oral trivalent vaccine was adopted for mass immunization in the United States. **Oral polio vaccine (OPV)**, also called the Sabin vaccine after its discoverer, is an attenuated vaccine that confers a vigorous, long-lasting immunity and is easy to administer (Chapter 20). Ironically, the only drawback to OPV is safety. In about 1 in 8 million cases, the oral vaccine causes paralytic poliomyelitis. More than 80 cases of vaccine-related paralytic polio have been reported in the United States since 1980. The fact that there have been no naturally occurring cases during the same period makes the vaccine-related cases especially tragic.

Physician advisory groups in the United States maintain that OPV is best for mass immunization against polio, though patients with a weakened immune system and previously unimmunized adults should not receive OPV. There is growing discussion in the United States of a modified immunization program. An improved inactivated vaccine called **E-IPV (enhanced inactivated polio vaccine)** would be administered first, followed by live vaccine after immunity has been established. This should significantly reduce, or even eliminate, the small risk of vaccine-related paralysis.

Arboviruses: Encephalitis

Arboviruses are a diverse group of RNA viruses that share a common mode of transmission—they are all *ar*thropod-*bor*ne. They include alphaviruses (of the to-gavirus group), flaviviruses, and bunyaviruses. Over 400 different arboviruses have been identified that affect many vertebrate species and cause devastating diseases in humans, including yellow fever and dengue fever (Chapter 27).

Arboviruses also cause encephalitis. In the United States, the prevalent encephalitic arboviruses are the **eastern equine encephalitis (EEE)**, **western equine encephalitis (WEE)**, **California encephalitis (CE)**, and **St. Louis encephalitis (SLE)** viruses. In other parts of the world different encephalitic arboviruses predominate. **Japanese B encephalitis**, for example, is common in many Asian countries. Animals, especially horses and birds, are the main reservoir of infection. The types of encephalitis discussed here are all transmitted by mosquitoes, though some types are also transmitted by ticks. Encephalitis usually occurs in epidemics during the summer months, when mosquitoes are plentiful.

Each type of encephalitis is caused by a different virus and has its own distinctive clinical characteristics. Some, such as eastern equine encephalitis, are so severe that death or permanent brain damage is almost inevitable. Others are much milder. Most of the human types are **subclinical** (without symptoms). The clinical syndrome of **infectious encephalitis** is characterized by disturbed brain function—confusion, paralysis, convulsions, and sleepiness that may progress to coma. The first signs are fever, headache, and usually a stiff neck, indicating irritation of the meninges. If significant numbers of neurons are killed by the virus, brain function does not return to normal when the infection is over.

Polio—The Next Conquest?

Smallpox was the first disease to be permanently eradicated (Chapters 20 and 26). Will polio be the next? The World Health Organization's Expanded Programme on Immunization (EPI) is committed to eradicating poliomyelitis by the year 2000, and great progress has already been made. Worldwide polio immunization increased from less than 5 percent in 1974 to 80 percent in 1991. But the task is still daunting. An estimated 116,000 cases of paralytic polio occurred in 1990, mostly in Africa and Asia.

Polio will be more difficult to eliminate than smallpox, for several reasons. Smallpox could be recognized easily, but most cases of polio are asymptomatic. This makes it harder to know where polio is occurring and when it has been eliminated. Also, the smallpox vaccine was nearly ideal for use in developing countries because it was heat-stable, one dose provided protection for several years, and vaccine recipients developed a scar that made them easy to identify. OPV, on the other hand, must be refrigerated, requires several doses, and leaves no identifying scar. Moreover, the OPV program has not received the political, financial, and social support it needs from many countries around the world.

But backers of the global effort are encouraged by the results of a polio eradication campaign begun in 1985 in the Americas. The American program couples immunization with sophisticated epidemiological surveillance. In addition to routine immunization, in areas where polio remains endemic, national immunization days are scheduled twice a year for all children under age 5 to receive OPV, regardless of their previous immunization status. Surveillance is carried out by over 20,000 regional health facilities that report weekly on new cases of polio. When a case is confirmed, public health workers initiate intensive "mopping up" immunization programs. As a result of these efforts, the number of polio-positive stool specimens has declined steadily—from 38 in 1988, to 24 in 1989, to 15 in 1990. In the first three quarters of 1991, only seven cases were confirmed.

Eradicating polio will be difficult, but WHO says it is achievable with current technology. It would be far more expensive than smallpox eradication, but the potential savings are enormous. The annual savings in vaccination costs in the United States alone would be $114 million. Most important, though, the conquest of polio would be a precious gift to the children of the twenty-first century.

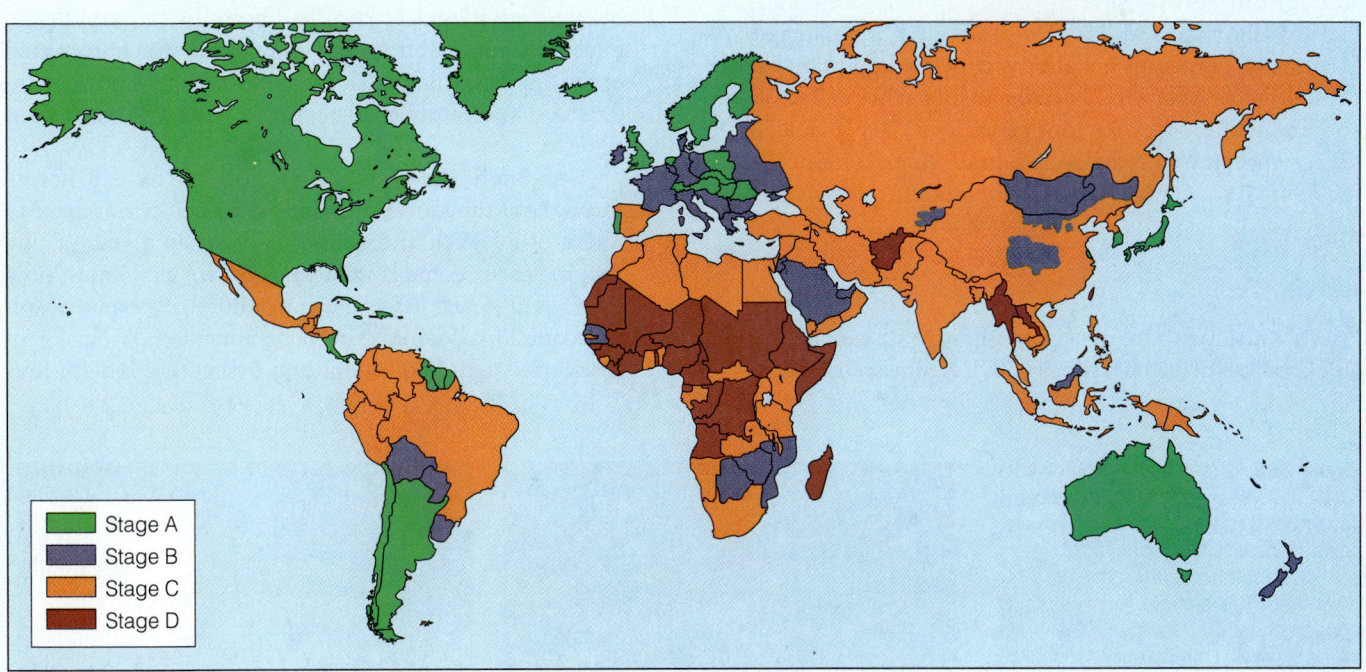

Map showing global progress toward eradicating polio as of December 1990. Stage A indicates immunization rates of at least 80 percent and no reported cases of naturally occurring polio. Stage B indicates immunization rates of at least 50 percent and fewer than 10 reported cases of naturally occurring polio each year. Stage C refers to immunization rates of 50 percent and more than 10 reported cases of naturally occurring polio each year. Stage D indicates immunization rates less than 50 percent or unknown and more than 10 or an unknown number of cases of polio each year.

Diagnosis is difficult. The lumbar puncture usually shows a few hundred leukocytes per mm³ of CSF, but no infectious organisms. When epidemics are known to be in progress, diagnosis may be made from case history and physical examination alone. Only occasionally are antibodies against arbovirus detected during or after the illness. Evidence of infections, such as viral inclusion bodies, can sometimes be detected in autopsy specimens by electron microscope. Only one arbovirus encephalitis, Japanese B, can be prevented by vaccination. But once it or any other arbovirus infection is underway, no treatment is available. Only supportive care can be offered.

Other Viral Encephalitic Agents

Many other viral pathogens produce a clinical syndrome that is indistinguishable from arbovirus encephalitis. Very rarely, each of the common viral pathogens of childhood—measles, mumps, chickenpox, and rubella—causes encephalitis. Taken together, however, these viruses infect so many people that they account for a significant number of cases. Enteroviruses, such as coxsackie and echoviruses, also cause encephalitis. It is impossible to say which viral agents cause the greatest number of cases because a definitive etiologic diagnosis is rare with encephalitis. Seasonal epidemiology, however, suggests that most cases are due to arboviruses or enteroviruses.

Herpes simplex virus types 1 and 2 (Chapters 24 and 26) can also cause a devastating viral encephalitis that usually results in death or profound brain damage. Diagnosing herpes encephalitis is critically important because it is the only viral encephalitis for which effective antiviral therapy (acyclovir or ganciclovir) is available. If herpes encephalitis is suspected, brain biopsy (surgically removing a small piece of tissue for study) is usually recommended.

PRION CAUSES

Prions cause two rare neurodegenerative diseases, kuru and Creutzfeldt-Jakob disease (CJD). Neither resembles a typical infection. There is no fever, no inflammation, and no immune response. Nevertheless, since both are transmitted from person to person and from humans to animals, they are considered infectious. Kuru and CJD, along with similar infections of animals such as the sheep disease **scrapie**, are called **transmissible spongiform encephalopathies** because infected brain tissue has a characteristic spongy appearance (**Figure 25.8**).

Initially, spongiform encephalopathies were thought to be caused by a type of virus with a long incubation period (a **slow virus**), but no viral particles could be found in purified preparations of the infectious agents. Purified preparations contained only protein, no nucleic acids. These tiny proteinaceous infectious particles were named prions (Chapter 13). Some researchers believe prions are the etiologic agents, but others believe they are only associated with the infections, not their cause.

Kuru was discovered among the Fore people of Papua, New Guinea—a community of 35,000 whose lives have changed little since the Stone Age. The word *kuru* in Fore language means "to tremble with fear." It is used to describe an illness that begins with clumsiness and progresses to total incapacitation and death within a year or two. Until 1957 the disease killed an incredible 2 to 3 percent of the population yearly. Epidemiologic research revealed an amazing story. The Fore practiced ritual cannibalism. When someone died, members of the immediate family prepared and ate the brain of the dead person for the wisdom it held. If kuru had been the cause of death, infectious material from the brain tissue was transmitted to the family participants. Once the transmission was understood and cannibalism stopped, no new cases of kuru occurred.

Creutzfeldt-Jakob disease is also rare but still occurs throughout the world. It begins with **dementia** (deterioration of intellectual function) and within a year or two progresses to coma and death. We do not know how Creutzfeldt-Jakob disease is transmitted, but some hospital-acquired cases have been documented. A cluster of cases was identified in children treated for growth hor-

Figure 25.8 Creutzfeldt-Jakob disease (CJD). CJD, kuru, and other diseases caused by prion-associated agents are called transmissible spongiform encephalopathies because infected brain tissue has a characteristic spongy appearance. Compare the normal brain tissue in (**a**) and the porous tissue in (**b**), which was taken from a CJD patient.

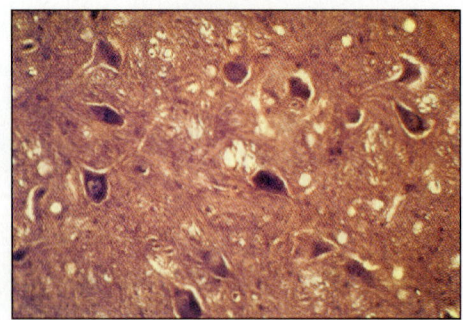
a

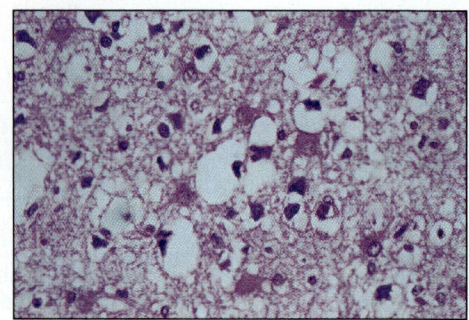
b

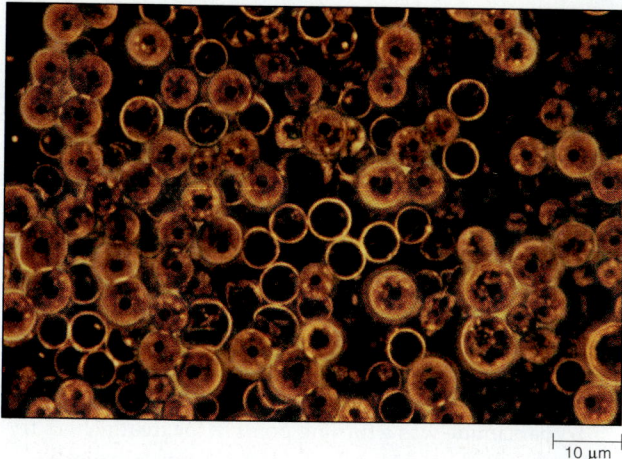

Figure 25.9 Cysts of *Naegleria fowleri*.

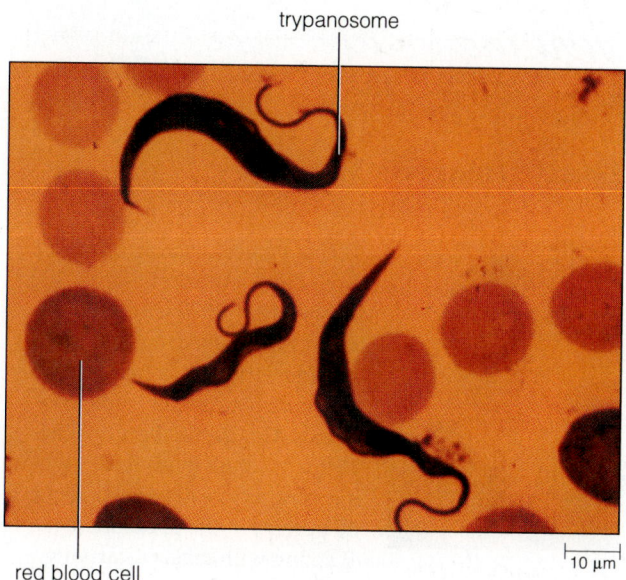

Figure 25.10 Trypanosomes in blood.

mone deficiency with human brain extracts that were presumably contaminated by the Creutzfeldt-Jakob agent. (Human growth hormone is now synthesized in *Escherichia coli* from genes inserted by means of recombinant DNA techniques, so there is no longer risk of CJD infection.) Another case was traced to a contaminated cornea transplant.

PROTOZOAL CAUSES

Protozoal pathogens also cause devastating neurological diseases.

Naegleria fowleri: Primary Amoebic Encephalitis

Naegleria fowleri is a free-living amoeba that causes a rare form of encephalitis (**Figure 25.9**). It is a common inhabitant of freshwater ponds in the United States and many other parts of the world. The usual victims are children and young adults who become infected by swimming in stagnant, infested ponds.

N. fowleri enters the body when water is forced up the nose in diving or swimming underwater. Then the amoebae pass through tiny openings in the skull where nerves from the nose join the brain. When *N. fowleri* enters the brain, the person develops severe headaches and fever, quickly progressing to **primary amoebic encephalitis**, an infection that causes death within a week or two. No effective treatment exists, though there are a few reports of survival after early treatment with amphotericin B.

Trypanosoma brucei: African Trypanosomiasis (Sleeping Sickness)

Trypanosomes are flagellated protozoa with an undulating membrane (**Figure 25.10**). Two species—*Trypanosoma brucei gambiense* and *Trypanosoma brucei rhodesiense*—cause an encephalitis of humans known as **African trypanosomiasis** or **sleeping sickness**. Both are transmitted from one human to another by the large and aggressive tsetse fly, which is found only in equatorial Africa (**Figure 25.11**). *T. b. rhodesiense* also sustains a reservoir among game animals and domestic cattle and sheep.

Trypanosomes multiply in the tsetse fly and become concentrated in its saliva. When an infected fly bites a human, a hard, red nodule appears at the site. Trypanosomes enter the circulatory system and incubate for a few days to weeks before disseminating to the spleen, lymph nodes, and liver. There they multiply, causing fever, malaise, and enlarged lymph nodes.

Eventually the trypanosomes invade the central nervous system, where they damage blood vessels and brain tissue. The first signs of neurological disease are changes in personality or a decreased ability to concentrate. As damage progresses, so does neurologic impairment, and soon the infected person cannot speak or walk. Eventually, the symptom for which the disease was named appears—the person becomes exceptionally tired and apathetic and sleeps almost continuously. Coma and death follow. Disease due to *T. b. gambiense* can last for several years, but *T. b. rhodesiense* infection is usually fatal within a few months.

Louise Pearce's Mission

Louise Pearce.

Louise Pearce, a young scientist from the Rockefeller Institute, left New York in 1920 with a small supply of a new drug—tryparsamide—that she hoped would be a cure for African trypanosomiasis. Her destination was the Belgian Congo (now Zaire), where sleeping sickness infected about 10 percent of the population. Buzzing with malaria-carrying mosquitoes and tsetse flies, Leopoldville, the colonial capital, was a frontier town inhabited by few European or American women. Undeterred, Pearce began work at the local hospital in a crowded ward reserved for blacks. She wrote her clinical notes at bedside, recording the inexorable progression of sleeping sickness. She was deeply affected by the suffering around her, and she had only enough tryparsamide to treat 77 patients. She carried out her testing program, though, and amazingly, 80 percent of her patients recovered, including many with advanced brain involvement.

Tryparsamide was a turning point in the treatment of trypanosomiasis, but not the final answer. Some patients became blind, and though tryparsamide was highly effective against *T. b. gambiense*—the species found in most of Africa—it was useless against *T. b. rhodesiense*. Tryparsamide is still used and newer drugs have been developed. In 1991, an agent called O-11 that replaces a critical fatty acid in the trypanosomal cell membrane was announced and may turn out to be the perfect cure Louise Pearce set out to find 75 years ago.

Figure 25.11 The tsetse fly, *Glossina moritans*, is the vector for African trypanosomiasis.

African trypanosomiasis is diagnosed by identifying parasites in the blood. Treatment is possible before the central nervous system is involved. Afterward, only compounds that cross the blood-brain barrier can stop progression of the disease, and these agents can be highly toxic. Europeans were eager to find a cure for sleeping sickness because it was an obstacle to colonial development in Africa. Paul Ehrlich was looking for such a cure when he discovered salvarsan, which was effective against syphilis (Chapter 1). Although several effective drugs have been developed in this century—including melarsoprol, suramin, and tryparsamide—the search for a better antitrypanosomal drug continues (see the box "Louise Pearce's Mission").

Sleeping sickness is a major cause of suffering and death in parts of Africa—at least a million people are infected at any time. Related diseases in domestic animals cause major economic losses. Preventing sleeping sickness by eliminating the tsetse fly is difficult because it inhabits millions of square miles. Moreover, because of antigenic variation, humans develop no long-lasting natural immunity to the disease and stimulating artificial immunity by vaccination seems unlikely (see the box "Anti-immune Trickery," Chapter 17).

Table 25.3 summarizes the infections of the meninges and neural tissue.

Table 25.3 Infections of the Nervous System

Infection	Causative Agent	Mode of Transmission	Symptoms	Prevention and Treatment
INFECTIONS OF THE MENINGES				
Bacterial				
Bacterial meningitis	*Neisseria meningitidis, Haemophilus influenzae, Streptococcus pneumoniae, Escherichia coli*	Usually respiratory	Acute onset of fever, headache, stiff neck, disturbed brain function (usually)	Vaccines for *N. meningitidis* and *H. influenzae*; all treatable with antibiotics
Viral				
Viral (aseptic) meningitis	Mumps virus, arboviruses, enteroviruses, and others	Respiratory or fecal-oral	Acute onset of fever, headache, stiff neck	Prevention through good hygiene; no treatment
Fungal				
Cryptococcal meningitis	*Cryptococcus neoformans*	Inhaling fungus, particularly from bird droppings	Gradual onset of headache, confusion, weakness	Treatment with amphotericin B and 5-flucytosine
NEURAL INFECTIONS				
Bacterial				
Tetanus	Toxin of *Clostridium tetani*	Wound infection	Rigid paralysis	Prevention by toxoid immunization; supportive treatment
Botulism	Toxin of *Clostridium botulinum*	Usually food poisoning or intestinal infection; rarely, wound infection	Flaccid paralysis	Prevention by proper food preparation and storage; treatment by antitoxin and supportive care
Viral				
Rabies	Rabies virus	Usually animal bite	Anxiety, impaired swallowing, coma, death	Prevention by immunizing pets and post-exposure immunization of humans; no effective treatment
Polio	Poliovirus	Fecal-oral	Most infections asymptomatic; flaccid paralysis	Prevention by immunization; supportive treatment only
Arbovirus encephalitis	Various alphaviruses, flaviviruses, bunyaviruses	Insect vector, usually mosquito	Fever, headache, stiff neck progressing to coma and death	Avoiding mosquito bites is best prevention; supportive treatment only
Prion				
Kuru	Prion-associated agent	Ritual cannibalism of infected brains	Tremor and weakness progressing to coma and death	Stopping cannibalism; supportive treatment only
Creutzfeldt-Jakob disease	Prion-associated agent	Unknown; some cases documented through exposure to extracts of infected brains	Dementia leading to coma and death	No systematic program for prevention; supportive treatment only
Protozoal				
Primary amoebic encephalitis	*Naegleria fowleri*	Swimming in contaminated water	Headache and fever rapidly progressing to death	Prevention by avoiding contaminated water; no effective treatment
African trypanosomiasis (sleeping sickness)	*Trypanosoma brucei gambiense* and *T. b. rhodesiense*	Bite of tsetse fly	Fever, enlarged lymph nodes progressing to somnolence, coma, and death	Prevention by avoiding fly bites; treatment with melarsoprol, suramin, tryparsamide

Summary

THE NERVOUS SYSTEM (pp. 621–623)

1. The nervous system is composed of the central nervous system (CNS) and the peripheral nervous system (PNS). The CNS includes the brain and spinal cord, which are covered by continuous membranes called the meninges. Cerebrospinal fluid (CSF) circulates around the brain and spinal cord. The PNS is made up of nerves (bundles of neurons and supporting tissue).

2. The outer layer of capillaries in the CNS is thickened, which limits permeability. This extra protection is referred to as the blood-brain barrier. Many toxins and microorganisms cannot enter the brain, but neither can many antibiotics, which limits treatment of CNS infections.

3. Clinical syndromes include meningitis, encephalitis, and myelitis. Some bacteria produce neurotoxins that affect nerve function in specific ways.

4. Diagnosis of nervous system infections often requires a sample of CSF, which is withdrawn from the spinal cord in a procedure called the lumbar puncture, or spinal tap.

INFECTIONS OF THE MENINGES (pp. 623–625)

1. *Neisseria meningitidis* causes meningococcal meningitis. Transmission is by respiratory droplets; many people are asymptomatic carriers. Acute purulent meningitis is sudden in onset, characterized by a high fever, stiff neck, headache, distinctive purplish skin rash, and often disturbances in brain function. Children between 6 months and 2 years are at highest risk. Epidemic disease is unusually severe, causing death or permanent brain damage. Epidemics can be interrupted by meningococcal prophylaxis with rifampin. Vaccines exist for some strains, but not yet for type B, which causes most cases.

2. *Haemophilus influenzae* causes *H. influenzae* meningitis. Like the meningococcus, *H. influenzae* is invasive and has a polysaccharide capsule. *H. influenzae* once caused about 75 percent of all meningitis in infants and young children, but in 1990, *H. influenzae* type b vaccine (Hib) was released and infection rates dropped dramatically.

3. *Streptococcus pneumoniae* causes pneumococcal meningitis. This type of bacterial meningitis strikes mainly adults, especially those in a weakened state of health. Brain damage occurs more frequently from this type of meningitis than from others.

4. *Escherichia coli* causes few cases of *E. coli* meningitis; it is a deadly disease for newborns and patients who have undergone neurosurgery.

5. Patients with meningitis symptoms and no bacteria in their CSF are diagnosed with nonpurulent or aseptic meningitis, presumably caused by a virus. The most frequent causative viruses are the mumps virus, members of the enterovirus group, and arboviruses. Children and young adults are most at risk. Recovery occurs within a few weeks without medical intervention. However, it is vitally important to distinguish between bacterial and viral meningitis because, without immediate treatment, bacterial meningitis can be fatal.

6. Two fungi—*Coccidioides immitis* and *Cryptococcus neoformans*—cause chronic meningitis. A few coccidioidomycosis patients go on to develop chronic meningitis. *C. neoformans* is inhaled and is serious if it spreads through the bloodstream to the CNS. The clinical syndrome is the same for both agents. Onset is gradual; there may or may not be a stiff neck; headaches and other neurological symptoms such as confusion and loss of coordination appear. Treatment is with amphotericin B. The AIDS epidemic has led to an increase in the incidence of cryptococcal meningitis.

NEURAL TISSUE INFECTIONS (pp. 625–637)

1. *Clostridium tetani* causes tetanus when endospore-containing dirt enters a deep anaerobic wound. *C. tetani* produces the neurotoxin tetanospasmin, which causes a rigid paralysis and excruciating muscle spasms. Before clinical signs appear, treatment with tetanus immune globulin is effective. Immunization is part of the DPT (diphtheria, pertussis, tetanus) inoculation. Immunity can be maintained with regular tetanus immunization.

2. *Clostridium botulinum* spores produce botulinum toxin, the most poisonous natural substance known. The toxin causes a flaccid (limp) paralysis that can be fatal if breathing muscles are affected. Infant botulism is the most frequently reported type. Adults usually acquire botulism as a food poisoning. Spores can also enter an anaerobic wound and produce toxin. Treatment may require a mechanical respirator. A botulism antitoxin is available. Preventive measures include not giving honey to infants under age 1, following home-canning procedures carefully, and boiling home-canned food.

3. The rabies virus causes rabies. Infection is usually from the bite of an infected animal. The incubation period can be brief or up to several years. The prodromal phase is characterized by flu-like symptoms and unusual sensations (a sense of dread, tingling around the wound). During the excitation phase, severe neurological abnormalities appear, including hydrophobia. In the paralytic phase, the person goes into a coma and dies. If rabies is suspected, prophylaxis is begun with human rabies immune globulin and six shots of human diploid cell vaccine (HDCV).

4. Poliovirus causes poliomyelitis, a flaccid paralysis. If respiratory muscles are affected, it can be fatal. Poliovirus is transmitted by the fecal-oral route. The peak polio epidemics of the early 1950s were halted by the Salk vaccine, an inactivated poliovirus vaccine (IPV). It was replaced in 1961 with the Sabin vaccine, an oral polio vaccine (OPV). Because there have been no cases of naturally occurring polio in the United States since 1979, but more than 80 vaccine-related cases, the U.S. immunization program may change to E-IPV (enhanced inactivated polio vaccine) followed by a live vaccine.

5. Arboviruses cause various types of encephalitis: EEE (eastern equine encephalitis), WEE (western equine encephalitis), Japanese B encephalitis, and others. Each type has its own characteristics. Transmission is usually by a mosquito bite, and epidemics occur during the summer months. Infectious encephalitis is characterized by disturbed brain function. A vaccine is available for Japanese B type, but no others. The only treatment is supportive care.

6. Enteroviruses and some common viral pathogens of childhood—measles, mumps, chickenpox, and rubella—can cause encephalitis. Herpes simplex virus can cause an encephalitis that usually results in death or profound brain damage. Prompt diagnosis by brain biopsy is important because treatment with acyclovir or ganciclovir is effective.

7. Prion-associated agents cause two spongiform encephalopathies, kuru and Creutzfeldt-Jakob disease. Kuru was discovered in New Guinea; symptoms progress from clumsiness to total incapacitation and death. The disease was eradicated when epidemiologists found transmission was by ritual eating of infected brain tissue.

8. Creutzfeldt-Jakob disease begins with dementia and progresses to coma and death. The mode of transmission is not understood, but cases around the world have been documented.

9. *Naegleria fowleri*—a free-living amoeba found in freshwater ponds in the United States—causes primary amoebic encephalitis. Infection occurs when water is forced up the nose in diving or swimming underwater. Symptoms include headaches and fever progressing to death within a week or two. Amphotericin B has been successful in a few cases.

10. *Trypanosoma brucei gambiense* and *Trypanosoma brucei rhodesiense*, flagellated protozoa, cause African trypanosomiasis or sleeping sickness. Transmission is by the tsetse fly. Early symptoms include fever, malaise, and enlarged lymph nodes; when pathogens reach the brain, the person loses the ability to concentrate, becomes exceptionally tired and apathetic, and sleeps almost continuously. Coma and death follow. Treatment with tryparsamide or a new agent called O-11 is possible before the brain is involved. Sleeping sickness is a major cause of suffering and death in parts of Africa.

Review Questions

THE NERVOUS SYSTEM

1. Name the two divisions of the nervous system, their main component parts, and their functions. What is the blood-brain barrier, and how does it function to our advantage and disadvantage?

2. Describe these clinical syndromes: meningitis, encephalitis, myelitis. What are neurotoxins? How and why is a lumbar puncture performed?

INFECTIONS OF THE MENINGES

1. Compare and contrast the types of meningitis caused by each of the bacteria listed below. Name the infection each causes and discuss transmission, symptoms, diagnosis, groups at highest risk, prevention, and treatment.
   a. *Neisseria meningitidis*   c. *Streptococcus pneumoniae*
   b. *Haemophilus influenzae*   d. *Escherichia coli*

2. Discuss the virulence factors that make *Neisseria meningitidis* pathogenic.

3. How is aseptic meningitis different from bacterial meningitis? Why is it crucial to distinguish between the two in a diagnosis?

4. How is cryptococcal meningitis different from bacterial and viral meningitis? How is the epidemiology of cryptococcal meningitis changing?

NEURAL TISSUE INFECTIONS

1. Compare and contrast the infections caused by *Clostridium tetani* and *Clostridium botulinum*. Discuss the pathogenic mechanism that causes illness, transmission, symptoms, diagnosis, prevention, and treatment.

2. Describe the clinical syndrome of rabies. How is rabies diagnosed and treated? How can it be prevented? Can rabies ever be totally eradicated?

3. Describe the clinical syndrome of poliomyelitis. Why was the Salk vaccine replaced with the Sabin, and why might the Sabin be replaced? What kinds of vaccines are the Salk, Sabin, and E-IPV?

4. Name some of the encephalitic arboviruses prevalent in the United States. Which is the most severe? Describe the syndrome of infectious encephalitis. How is it transmitted and treated?

5. What other viral pathogens can cause encephalitis? Why is prompt diagnosis of herpes simplex encephalitis crucial?

6. What is the epidemiology of kuru? What do we know about Creutzfeldt-Jakob disease? Why are these infections called prion-*associated*?

7. How is primary amoebic encephalitis transmitted? What is the causative agent?

8. How is the trypanosomiasis caused by *Trypanosoma brucei rhodesiense* like that caused by *T. b. gambiense*, and how is it different in regard to transmission, symptoms, prevention, and treatment?

Suggested Readings

Bartlett, J. C. 1986. Infant botulism in adults. *New England Journal of Medicine* 315:254–55.

Bingham, R. 1981. Outrageous ardor. *Science 81* 2:54–61.

Monath, T. P. 1988. Japanese encephalitis: A plague of the orient. *New England Journal of Medicine* 319:641–43.

Quegiliarillo, W., and Schield, W. M. 1992. Bacterial meningitis: Pathogenesis, pathophysiology, and progress. *New England Journal of Medicine* 326:864–72.

26 INFECTIONS OF THE BODY SURFACES: SKIN AND EYE

Clinical Sketches: A Twice-Told Tale

R. C. was a 7-year-old boy whose parents took him to his pediatrician because of fever and an itchy, blistering rash. R. C. appeared well despite his slight fever, but the physician observed a few dozen thin-walled vesicles on his trunk. Her immediate diagnosis was *varicella*—chickenpox. She told R. C.'s parents what to expect. He would continue to have a fever and develop small blisters for the next five days. The blisters would come in crops, or showers, with many appearing seemingly at once and another set coming on hours or days later. As a result, R. C. would have spots in many different stages of healing. Blisters a few days old would begin to crust while new wet ones were still appearing. The earliest blisters were on his trunk but newer lesions would spread outward, covering the entire surface of his skin and perhaps the lining of his mouth and throat.

The fluid in R. C.'s blisters contained live chickenpox virus, so during the week his sores remained wet he could have transmitted the infection to others and had to stay home from school. If any sores failed to heal or drained pus, he would need to return to the doctor. No treatment was prescribed, but he was given medication for itching. The physician clearly instructed R. C.'s parents not to administer aspirin for the boy's fever, although non-aspirin-containing medication was all right. R. C.'s younger sister might also come down with chickenpox in two to three weeks. By the end of the week, R. C. was covered with blisters and scabs, looking much sicker than he really was. He recovered uneventfully and happily returned to school.

K. V. was a healthy 72-year-old woman who noticed tingling in her skin along a narrow strip that extended from just below her left shoulder blade around her left side to her abdomen. By the next day the tingling had become a searing pain, and the day after that tiny blisters appeared in the painful area. K. V.'s physician diagnosed *zoster*—shingles—and prescribed medication for the pain, which had become almost unbearable. The blisters healed in about a week but the pain persisted. Six months later K. V. continued to complain of pain in that same strip of skin, though it was diminishing. Laboratory tests done at the time of K. V.'s illness were all normal, and her health otherwise remained good.

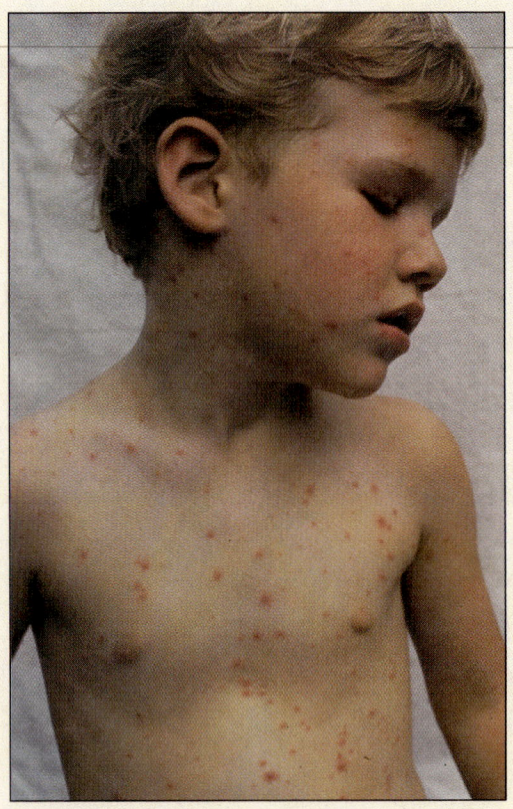

a Typical chickenpox rash.

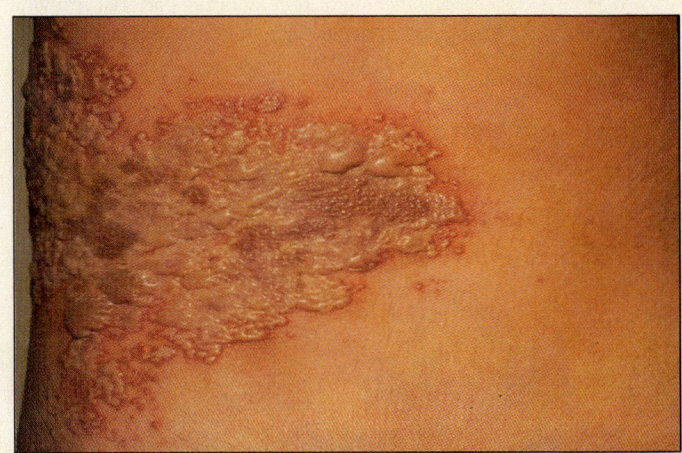

b Zoster rash occurring in characteristic stripes.

LEARNING GOALS

To understand:
- The anatomy and function of the skin and eye and their defenses against microorganisms

- The clinical syndromes that characterize infections of the skin and eye

- The bacterial, viral, fungal, and arthropod causes of skin infections; their diagnosis, prevention, and treatment

- The bacterial, viral, and helminthic causes of eye infections; their diagnosis, prevention, and treatment

THE BODY'S SURFACES

The body's outermost surfaces—the skin and exposed surfaces of the eye—defend the body's internal tissues against microorganisms. Without a tough, intact external covering, overwhelming microbial infection would be inevitable and fatal.

Structure and Function

The skin is our largest and most visible organ (**Figure 26.1**). The outer portion, the **epidermis**, is composed of epithelial cells tightly cemented together and arranged in layers, one on top of the next. In parts of the body where skin is thin, such as the face, there are only a few epidermal layers, but where skin is tough and thick, such as the palms of the hands and the soles of the feet, many epidermal layers are stacked on top of one another.

The innermost layer of the epidermis, the **stratum germinativum**, is made up of living cells that divide and move upward to replace skin as it is sloughed off. The outermost layer, the **stratum corneum**, is composed of dead cells packed with **keratin**, a protein that waterproofs the skin and creates a formidable barrier to microorganisms.

Beneath the epidermis but firmly attached to it is the **dermis**, a layer of connective tissue. Here, cells are loosely joined together. Many structures pass through the dermis, including blood vessels, lymphatic vessels, nerve endings, sweat glands, sebaceous (oil) glands, and hair follicles. Sebaceous glands produce **sebum**, a fatty substance that keeps the hair and skin from drying out. Hair and sebaceous secretions do not fulfill essential functions, but sweat glands play a critical role in regulating body temperature.

Sweat glands and hair follicles form channels through the epidermis to the surface of the skin, creating openings in the otherwise continuous keratin layer. These tiny pores are vulnerable to the entry of microorganisms. Most skin infections begin here. Sebum and perspiration foster the growth of microorganisms by providing moisture, but they also contain antimicrobial compounds. Sweat, for example, contains the enzyme **lysozyme**, which lyses Gram-positive bacteria by splitting the peptidoglycan backbone of their cell wall. Sebum contains fatty acids that are toxic to many microorganisms.

Figure 26.1 Anatomy of the skin.

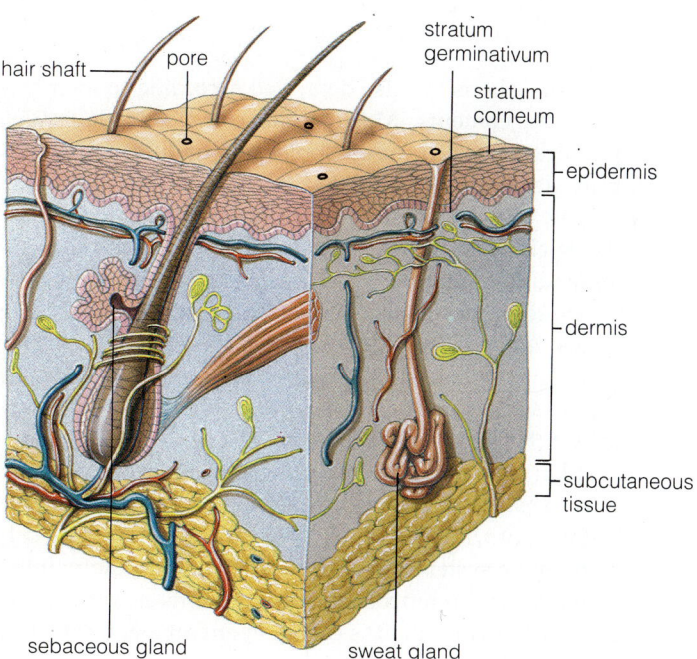

hair shaft — pore
stratum germinativum
stratum corneum
epidermis
dermis
subcutaneous tissue
sebaceous gland
sweat gland

Beneath the dermis lies **subcutaneous tissue**. Infections that begin in the skin may extend to and spread through this loosely organized layer of connective tissue.

The eye is a delicate, anatomically complex, and highly vulnerable organ with direct connections to the brain. We are concerned only with the exposed surface of the eye that comes into direct contact with the environment (**Figure 26.2**). Most of the eye's surface, including the white portion of the eyeball and the inner surface of the eyelid, is protected by a loose layer of connective tissue covered with an epithelial membrane, collectively called the **conjunctiva** (*pl.,* **conjunctivae**). Like the skin, the conjunctivae form a barrier to microorganisms. The only external surface of the eye not covered by conjunctivae is the **cornea**. This transparent window through which we see is covered by its own dense layer of protective epithelium. Infections of the cornea are much more serious than infections of the conjunctivae because clouding or scarring of this normally clear structure can cause blindness.

Defenses and Normal Flora: A Brief Review

Microorganisms have ready access to the skin and eye surfaces because they are constantly exposed to the environment. Natural defenses—primarily the dryness of the skin and tears in the eye—keep these surfaces from being overwhelmed by microbial growth. Nevertheless, both sustain a normal flora. The microbial flora of the skin consists of numerous commensal species and opportunistic pathogens (Chapter 14). The greatest number and variety of microorganisms are found in relatively moist areas, such as under the arms, in the groin, and around the nose and mouth. The normal flora of the conjunctiva is much sparser and is made up primarily of skin flora, such as staphylococcal species and diphtheroids.

Clinical Syndromes

Unlike infections of other organ systems, skin infections are not usually described by anatomic syndromes. Rather, they are described by their appearance—a logical approach because the infected tissue can be examined directly (**Table 26.1**). Clinicians use many different terms to describe the appearance of skin lesions. A **lesion** is any abnormality. It may or may not be caused by an infection.

Skin lesions are often described by color. **Erythema** is a reddening of the skin, usually a result of inflammation that enlarges blood vessels and increases skin temperature. **Petechiae** are tiny purplish discolorations of the skin, and **purpurae** are large ones.

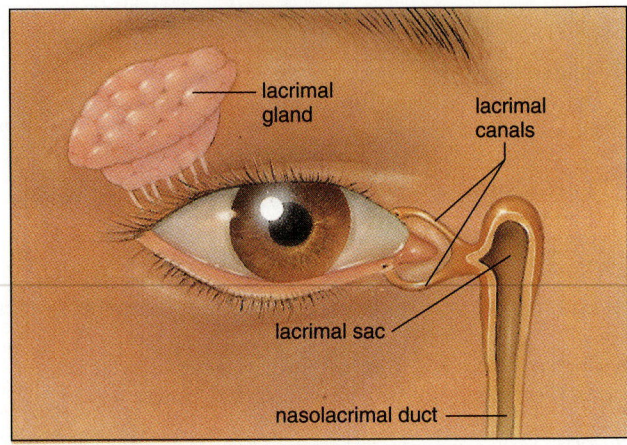

a

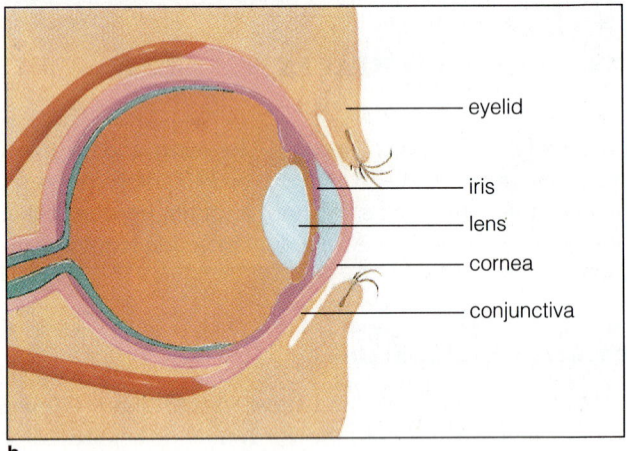

b

Figure 26.2 Anatomy of the eye.

Skin lesions are also described according to their texture and contents. **Macules** are completely flat spots, **papules** are small raised spots, and **nodules** are larger, firm elevations. Rashes that have some flat and some slightly bumpy regions are described as **maculopapular**. A **vesicle** is a small water-filled blister, and a **bulla** is a larger one. A **pustule** is a blister that contains pus.

In some cases, the skin surface breaks and a sore forms. A superficial loss of skin is an **erosion**. A deeper hole that may bleed or leave a scar is called an **ulcer**. A visible accumulation of dried blood or serum on the surface of the skin is called a **crust**.

Skin lesions can be important in diagnosing systemic infections. In these cases, skin lesions are only one manifestation of widespread tissue disease. A skin rash associated with a systemic disease is called an **exanthem**. If the rash appears on mucous membrane surfaces, such as the inside of the mouth, it is called an **enanthem**. Character-

Table 26.1   Classification of Skin Lesions

Basis of Classification	Name of Lesion	Characteristics
Color of lesion	Erythema	Red
	Petechiae	Purple (tiny)
	Purpura	Purple (large)
Texture of unbroken skin	Macules	Flat spots
	Papules	Small raised spots
	Maculopapular	Flat and bumpy regions
	Nodules	Larger, firm, round elevations
Contents of lesion	Vesicle	Small, water-filled
	Bulla	Large, water-filled
	Pustule	Pus-filled
Properties of broken skin	Erosion	Superficial skin loss
	Ulcer	Deeper skin loss
	Crusts	Lesions covered with dried blood or serum
Site of lesion	Exanthem	Skin rash caused by systemic infection
	Enanthem	Rash on mucous membrane caused by systemic infection

Table 26.2   Clinical Syndromes of Eye Infections

Syndrome	Region Affected	Signs and Symptoms	Causative Agents
Conjunctivitis (pinkeye)	Conjunctivae	Inflammation of conjunctivae only; often reddening, discharge, discomfort but no threat to vision	Many agents, including *Neisseria gonorrhoeae*, strains of *Chlamydia trachomatis* that cause trachoma, herpes simplex virus, adenoviruses, enteroviruses, and other viruses.
Keratitis	Cornea	Inflammation; may or may not be painful; deep lesions cause scarring and threaten vision	Same as for conjunctivitis.
Keratoconjunctivitis	Conjunctivae and the cornea	Characteristics of both conjunctivitis and keratitis	Same as for conjunctivitis.

istic exanthems and enanthems are the key to diagnosing measles, chickenpox, and scarlet fever.

Infections of the conjunctivae and cornea produce typical clinical syndromes (**Table 26.2**). Infection of the conjunctivae causes **conjunctivitis**. Because superficial blood vessels are usually dilated, conjunctivitis is sometimes called **pinkeye**. It is extremely common, particularly in children. Infections of the cornea are called **keratitis**. Keratitis is much less common than conjunctivitis and more serious. **Keratoconjunctivitis** is infection of both the conjunctiva and cornea.

SKIN INFECTIONS

Bacteria, viruses, fungi, and arthropods all cause infections of the skin.

BACTERIAL CAUSES

Most bacterial skin infections are localized, but some may spread to become life-threatening systemic illnesses. In general, bacterial infections can be treated effectively with antibiotics.

Case History

The Little Red Spot

G. T. was a 45-year-old businessman; one morning while shaving, he noticed a small red bump near the right side of his nose. It looked like an ordinary pimple, so he was not concerned and went to work as usual. During the day, however, the redness began to spread. By the time G. T. returned from work he had a bright red patch the size of a dime on his cheek. It was hard, warm, slightly raised, and tender. He noticed, too, that the area had a very distinct margin. It looked almost as if a line had been drawn—on one side the skin was uniformly affected, while on the other side the skin appeared entirely normal (**Figure 26.3**). G. T. went to bed hoping the annoying spot would be gone in the morning; but when he awoke, it had, in fact, grown alarmingly. Now his entire cheek and right side of his face were red and swollen. He felt sick, too. He took his temperature and found it was 102°F. He tried to eat breakfast, but felt nauseated and vomited. Instead of going to work, G. T. visited his doctor.

After listening to the history and performing a physical examination, G. T.'s physician arranged to have him admitted to the hospital. The doctor asked about a history of allergy to antibiotics and then wrote orders for G. T. to receive a cephalosporin antibiotic intravenously. On admission, a sample of G. T.'s blood was obtained for culture. After the first dose of antibiotic, the margin of G. T.'s skin infection stopped advancing and the affected area never became any larger. Within 24 hours his fever was gone and the redness was slowly beginning to fade. Three days later G. T. was discharged with a prescription for oral penicillin. At the time of his discharge, his blood culture was reported to be growing a *Streptococcus pyogenes* (group A streptococcus). His discharge diagnosis was **erysipelas**.

Streptococcus pyogenes: Impetigo, Erysipelas

Streptococcus pyogenes, or the group A streptococcus, is a Gram-positive coccus that often forms long chains of cells. This virulent pathogen causes streptococcal pharyngitis (Chapter 22), scarlet fever (Chapter 21), puerperal sepsis (Chapters 1 and 24), and streptococcal septicemia (Chapter 27). It also causes many infections of the skin and underlying soft tissues.

Group A streptococcal infections vary greatly in severity, depending upon the depth of the infection and the virulence of the infecting strain. The most common is

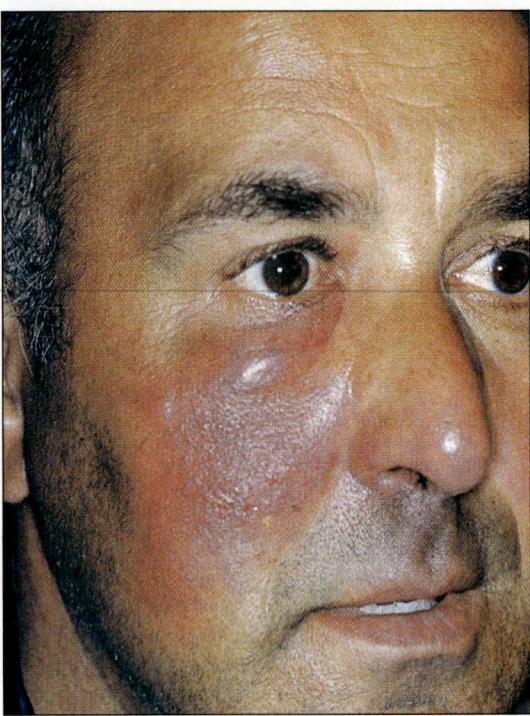

Figure 26.3 Erysipelas affects the epidermis and the underlying dermis and can spread quickly to cause a dangerous systemic infection.

a superficial epidermal infection called **impetigo**, which is usually seen in children (**Figure 26.4**). Typically, a small break in the skin fails to heal. It gradually grows larger and clear fluid oozes from the eroded skin. The fluid dries into a honey-colored crust that breaks when the lesion begins to ooze again. Often, both *S. pyogenes* and *Staphylococcus aureus* can be cultured from these lesions. *Streptococcus pyogenes* is usually considered the primary pathogen and *Staphylococcus aureus* the secondary, although the reverse may be the case, particularly in blistered or **bullous impetigo**.

Impetigo is not life-threatening, but it can become extensive and troublesome. Occasionally it leads to poststreptococcal glomerulonephritis, a serious complication (discussed below). Open lesions contain huge numbers of bacteria, and impetigo is readily transmitted by close contact or by fomites such as toys or clothing. It is most prevalent in warm, damp climates and among toddlers and young children in families and day-care centers. It is usually treated with oral or topical antibiotics.

G. T.'s streptococcal skin infection, erysipelas, was deeper than impetigo and much more dangerous. Erysipelas affects both the epidermis and the underlying dermis. Erysipelas spreads quickly because bacteria can travel freely through the lymphatic vessels in the dermis. Bacteria in the dermis may also enter the bloodstream, as

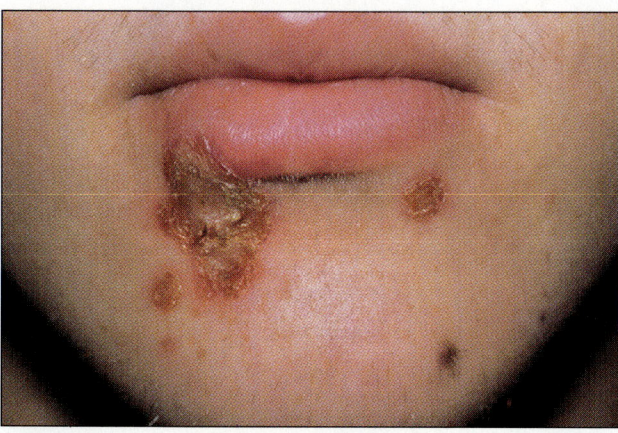

Figure 26.4 The honey-colored, crusted sores around this young man's mouth are typical of impetigo, a common bacterial skin infection.

in G. T.'s case, initiating a systemic infection characterized by fever and malaise. Untreated systemic infections caused by group A streptococci are frequently fatal.

Deeper streptococcal skin infections can progress even more rapidly than G. T.'s erysipelas. If infection extends into the subcutaneous layer, bacteria can spread through the spaces between tissue planes at an astonishing rate. In **streptococcal gangrene**, inflammation can become so intense that blood vessels are destroyed and the overlying skin dies. Bacteria from streptococcal soft tissue infections may also enter the bloodstream to initiate a life-threatening systemic infection.

Virulence and Pathogenesis. Group A streptococci have a vast and varied repertoire of mechanisms for causing human disease. Their adaptations for pathogenesis include a surface protein that protects the bacterium from phagocytosis, numerous toxins and destructive enzymes, and the ability of some strains to provoke immunologically mediated tissue destruction (**Table 26.3**).

Group A streptococci have an antiphagocytic protein, the **M protein**, that appears in electron micrographs as a hairlike projection from the cell wall (see Figure 15.9). M protein can be neutralized by antibody, so a person who has been infected by a given streptococcal serotype is immune from reinfection with the same type. However, because there are more than 80 different antigenic types of M protein, few people are completely immune to streptococcal infection. Streptococci that do not produce an M protein do not cause disease because they are easily phagocytized.

M protein played a key role in the development of G. T.'s infection. When the group A streptococci entered the dermal layer of his skin—probably through a pore, a hair follicle, or an epidermal cut—they initiated an intense inflammatory response. But phagocytes could not destroy the bacteria because they were surrounded by M protein. Phagocytosis would have been successful if antibodies against the M protein had been produced, but G. T. had never before been exposed to this serotype and so could not mount a rapid secondary antibody response. He probably would have died before the 7 to 10 days required to produce antibody through a primary immune response.

Streptococci produce many toxic substances that damage human tissues. One of these is erythrogenic toxin, produced by strains of group A streptococci that are lysogenic for a bacteriophage (meaning they carry phage DNA in their genome). Erythrogenic toxin causes scarlet fever (see the beginning of Chapter 21). Usually the toxin-producing bacteria that cause scarlet fever multiply in the throat, but toxin can be produced anywhere in the body. If bacteria multiply in an infected wound or a surgical incision, the syndrome is called **surgical scarlet fever**. People produce antitoxin against erythrogenic toxin as the result of infection. Therefore, most people contract scarlet fever only once, while repeated streptococcal infections are common.

Other toxic products of group A streptococci include **leukocidins** (which destroy leukocytes), **streptolysins** (which lyse red blood cells), **streptokinase** (which dissolves blood clots), and **hyaluronidase** (which dissolves hyaluronic acid, a polysaccharide that cements cells together). No one of these extracellular enzymes is essential for streptococci to cause disease; but collectively they contribute to the tissue damage that occurs during a streptococcal infection.

Another pathogenic consequence of group A streptococci is their ability to stimulate rheumatic fever or acute poststreptococcal glomerulonephritis, autoimmune reactions that can occur days to weeks after the streptococcal infection has subsided (Chapter 18). We do not completely understand the immunologic basis of these diseases, but research suggests that M protein is the cause. The antigenic structure of M protein resembles antigens found in human tissues. Consequently, when antibodies produced against the streptococcal M protein react with human tissue, a destructive inflammatory reaction is initiated. It is not clear what adaptive value, if any, this antigenic mimicry may have for the streptococci.

Treatment and Prevention. Thanks to antibiotic treatment, G. T. missed less than a week of work. Without treatment he probably would have died, as many otherwise healthy young adults did in the era before antibiotics (see the box "Antimicrobial Drugs in American History," Chapter 21).

Streptococcus pyogenes is one of the few Gram-positive cocci that has remained uniformly susceptible to small doses of penicillin. G. T. was initially treated with a

Table 26.3 Some Properties of Skin-Infecting Bacteria That Cause Disease

Organism	Property or Complication	Action
Group A streptococci	M protein	Allows streptococci to evade phagocytosis
	Erythrogenic toxin	Causes signs and symptoms of scarlet fever
	Leukocidin	Destroys leukocytes
	Streptolysins	Destroy red blood cells
	Streptokinase	Dissolves blood clots
	Hyaluronidase	Dissolves hyaluronic acid, which cements cells together
	Rheumatic fever	Inflames many organs, including heart valves
	Poststreptococcal glomerulonephritis	Inflames kidney tissue
Staphylococcus aureus	Alpha toxin	Damages cell membranes
	Delta toxin	Damages cell membranes
	Leukocidin	Destroys leukocytes
	Exfoliative toxin	Causes outer layer of skin to peel
	Coagulase	Causes plasma to clot
	TSST	Causes rash, falling blood pressure, and other manifestations of toxic shock syndrome
	Enterotoxin	Causes diarrhea and vomiting
Pseudomonas aeruginosa	Thick capsule	Inhibits phagocytosis; occludes respiratory passages in cystic fibrosis disease
Clostridium perfringens	Alpha toxin	Kills human cells
	Enterotoxin	Causes diarrhea and vomiting
Propionibacterium acnes	Fatty acids	Sebum trapped in pores is metabolized to irritating fatty acids, which initiate acne inflammation

cephalosporin antibiotic only because the physician could not definitely rule out infection by a penicillin-resistant strain of *Staphylococcus aureus*. When his blood culture confirmed the diagnosis, treatment was continued with penicillin. Erythromycin is generally used in cases of penicillin allergy, but the emergence of *Streptococcus pyogenes* strains resistant to erythromycin may someday make treatment of streptococcal infections more difficult in penicillin-allergic patients.

Developing a vaccine against *S. pyogenes* is complicated by the large number of M protein antigens and their ability to undergo rapid change. Still, molecular biologists are making remarkable progress toward understanding the structure and antigenic variation of the M protein, and eventually this may lead to a vaccine.

Data collected by the CDC indicate that group A streptococci virulence is on the rise in the United States. See the discussion on changing virulence in Chapter 22.

Staphylococcus aureus: Impetigo, Boils, Abscesses

Staphylococcus aureus is one of the most frequently isolated pathogenic bacteria in hospital laboratories in the United States. It causes diseases as diverse as staphylococcal food poisoning (Chapter 23) and toxic shock syndrome (Chapter 24). It also causes many skin infections.

The genus *Staphylococcus* is divided into two broad groups. The virulent *S. aureus* strains produce the enzyme **coagulase**, ferment mannitol, produce hemolysins, and grow as yellowish colonies. The avirulent or weakly virulent *S. epidermidis* strains do not produce coagulase, ferment mannitol, or produce hemolysins; and they grow as whitish colonies.

Microbiologists use phage typing (Chapter 10) to subdivide *S. aureus*. Phage typing is useful epidemiologically, but it does not include all strains. Many antibiotic-resistant strains of *S. aureus* and all strains of *S. epi-*

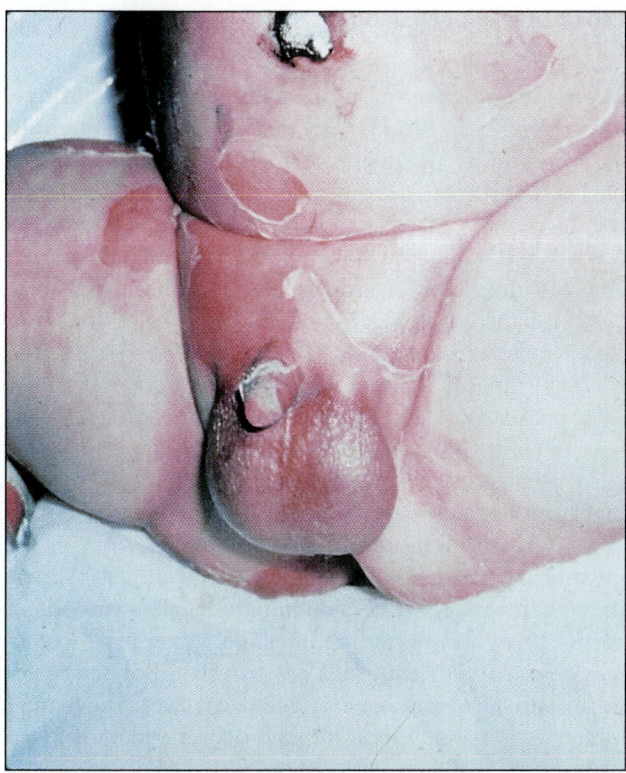

Figure 26.5 Staphylococcal scalded skin syndrome occurs when the exfoliative toxin of *Staphylococcus aureus* causes the outer layers of skin to separate and peel away.

dermidis are not susceptible to phage infection. These strains are identified by the patterns of plasmids that they carry, a technique called **plasmid pattern analysis**.

Like streptococci, staphylococci cause an intense inflammatory reaction, and the toxins and damaging enzymes they produce play a major role in their ability to cause disease. Two substances, **alpha** and **delta toxin**, damage cell membranes, thereby lysing human cells. Two other substances, collectively called **leukocidin**, destroy leukocytes, including phagocytes. **Exfoliative toxin**, which only bacteriophage-infected strains of bacteria produce, causes the outer layers of human skin to separate and peel away. Large quantities of exfoliative toxin produce a horrifying clinical syndrome called **staphylococcal scalded skin syndrome (Figure 26.5)**. Whole sections of skin surface peel away, leaving weepy areas that look like burns. It is a life-threatening condition usually seen in infants and the elderly; but, fortunately, it is rare. Another staphylococcal toxin, TSST, produces toxic shock syndrome, and staphylococcal enterotoxin causes food poisoning.

Pathogenesis. Staphylococci are part of the normal flora of the human skin. *Staphylococcus epidermidis* is found on almost everyone's skin and nasal mucous membranes. Many healthy people also carry virulent strains of *S. aureus* in their nasal passages.

Staphylococci usually infect the skin and its underlying soft tissues, but clinical case reports document staphylococcal infections of every organ in the body. The bones, joints, lungs, and heart are affected fairly frequently.

Like streptococcal skin infections, staphylococcal skin infections are characterized according to their depth. Impetigo is extremely superficial, but deeper staphylococcal infections often produce pus. In fact, a significant accumulation of pus suggests the infection is more likely caused by staphylococci than by streptococci. Staphylococcal coagulases that cause blood to clot help wall off these collections of pus and protect the bacteria from bodily defenses. **Folliculitis** is a rash composed of tiny, pus-containing lesions within hair follicles. **Boils**, or **carbuncles**, are somewhat larger pus-containing lesions within the dermis. **Abscesses** are significant accumulations of pus that may penetrate into deeper tissues. **Cellulitis** is a more diffuse and extensive infection that spreads through the skin and underlying soft tissues. Unlike boils and abscesses, staphylococcal cellulitis often introduces bacteria into the bloodstream, leading to systemic and life-threatening infections.

Staphylococcal infections often become established when a foreign body such as a surgical suture or a splinter is present in a wound or if the victim's immune system is weakened. Infections are therefore common in infants and the elderly, who are relatively immunosuppressed because of their age. Opsonization and phagocytosis can usually control early staphylococcal infections.

Treatment. Staphylococcal infections are treatable with antibiotics, although large pus-containing lesions must be *lanced*, opened surgically, for drug treatment to be effective. The biggest problem in *S. aureus* treatment, however, is the emergence of antibiotic-resistant strains. At the beginning of the antibiotic era, all strains of *S. aureus* were uniformly sensitive to small doses of penicillin, but penicillinase-producing strains of staphylococci soon began to appear. Today, most staphylococcal infections are treated with semisynthetic penicillins, such as methicillin, or with cephalosporins. Recently, however, strains of **methicillin-resistant staphylococci** have emerged, causing life-threatening infections and occasional hospital epidemics. Epidemiologists worry that *S. aureus* may become an increasingly common cause of fatal nosocomial infections as more strains become resistant to multiple antibiotics.

Pseudomonas aeruginosa: Folliculitis, Pseudomonas Infection

Bacteria of the genus *Pseudomonas* are found almost everywhere—in soil, water, and the gastrointestinal tract of humans and animals. They have minimal nutritional requirements and can tolerate a wide temperature range, which means they can be isolated from almost any moist environment.

No species is highly virulent for humans, though one, *Pseudomonas aeruginosa*, is a significant opportunistic pathogen. This is partly because it has pili for adherence and an extracellular slime layer that interferes with phagocytosis. **Hot-tub folliculitis**, a pustular infection of multiple hair follicles, is fairly common among people who bathe in hot tubs or spas contaminated by *P. aeruginosa*. The rash that results can be unsightly but is not serious and resolves without antibiotic treatment. *P. aeruginosa* can also be cultured from patients with **otitis externa**, swimmer's ear, which is common in otherwise healthy people.

Most victims of *P. aeruginosa* infection are people already ill or injured. Burn patients are at particularly high risk because they have lost their skin's protection. Eighty percent of all burn fatalities are due to infection, and *P. aeruginosa* is the most common infecting organism. It is so prevalent in the hospital environment that it is extremely difficult to eradicate from a burn unit. *P. aeruginosa* can survive in bedpans, respiratory equipment, whirlpool baths, and even disinfectant solutions. Burn patients with *Pseudomonas* infections may have wounds that exude blue-green pus. If the infection becomes bloodborne, these patients are in danger of dying from Gram-negative shock. Preventing these infections depends on scrupulous cleanliness and on proper burn **debridement** to remove dead or dying tissue that can become a focus for infection.

Patients with cancer or diabetes are also vulnerable to *Pseudomonas* infections, and the airways of children with cystic fibrosis (a genetic disease) are almost universally colonized by *P. aeruginosa*. These strains produce an especially thick capsule that occludes the respiratory passages. *P. aeruginosa* causes 10 to 20 percent of nosocomial infections.

P. aeruginosa infections are extremely difficult to treat with antibiotics. A combination of antimicrobial compounds, including an aminoglycoside and a beta-lactam agent, usually must be used, and even then *P. aeruginosa* may become more drug-resistant during treatment. Because most *Pseudomonas* victims are already seriously ill, the infection is frequently fatal. Infection-control teams in hospitals closely monitor *Pseudomonas* infections to prevent hospital epidemics.

Clostridium perfringens: Gas Gangrene

Clostridium perfringens is an anaerobic, Gram-positive, rod-shaped bacterium widely distributed in the environment. It can be the cause of infections following childbirth or abortion (Chapter 24) and is closely related to the organism that causes tetanus (Chapter 25). *C. perfringens* owes its pathogenicity to its numerous protein toxins. An enterotoxin causes clostridial food poisoning (Chapter 23). **Alpha toxin**, which is cytotoxic, is primarily responsible for tissue damage.

C. perfringens enters the body through a break in the skin. Because it is abundant in soil, wounds that are contaminated by dirt are often infected by *C. perfringens*. Also, because *C. perfringens* multiplies only in an environment with a low oxygen concentration—which occurs around **necrotic** (dead and dying) tissue—*C. perfringens* infections usually occur in deep wounds with significant tissue damage. This infection initiates a chain reaction of tissue injury and spreading infection. Toxin produced in wound-damaged tissue kills nearby cells, allowing bacteria to spread to the newly devitalized area. Preventing infection depends upon timely debridement of deep wounds.

C. perfringens has various clinical manifestations, ranging from completely asymptomatic colonization to **gas gangrene** (**Figure 26.6**). **Gangrene** is tissue death as a result of impaired blood supply. It can be caused by various agents, including streptococci. The gangrene produced by massive *C. perfringens* infections, which often destroy large amounts of muscle as well as skin, is distinguished by the gas (hydrogen and carbon dioxide) produced as a by-product of its rapid metabolism. Gas bubbles may be seen in the skin or in fluid oozing from the wound, and pressure over the affected area causes **crepitance**, a crackling sound produced by gas moving through tissue. The overlying skin turns black and dies. Gas gangrene progresses rapidly. About a week after infection, symptoms abruptly appear—extensive tissue death, renal failure, and shock. Death can occur within two days of the first signs of illness.

Patients are treated with high-dose penicillin. They may also be given clostridial antitoxin and placed in an oxygen-enriched chamber to prevent further growth of the anaerobe, though these measures have questionable value. The only truly effective way to control a gas gangrene infection is to remove all dead tissue, which usually means amputation. If amputation is impossible because of the wound's location (for example, the trunk), infection is often lethal in spite of medical therapy. Gas gangrene is occasionally caused by other toxin-producing species of *Clostridium*, such as *C. septicum* and *C. novyi*.

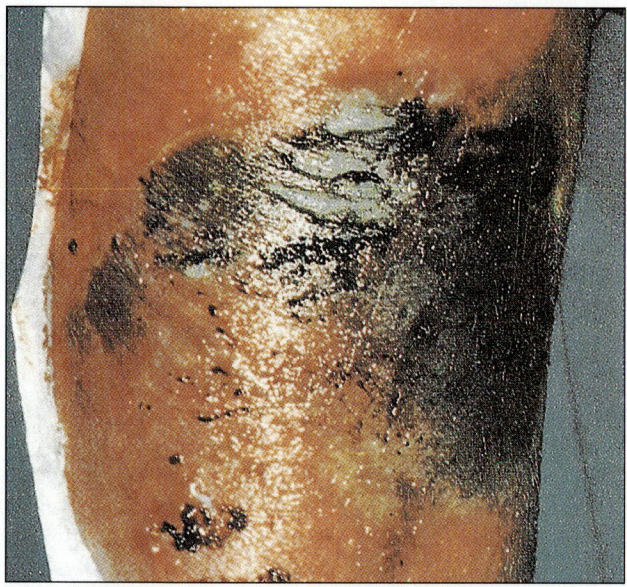

Figure 26.6 This patient had a bone-exposing fracture of his leg that became infected by *Clostridium perfringens* and developed into gas gangrene.

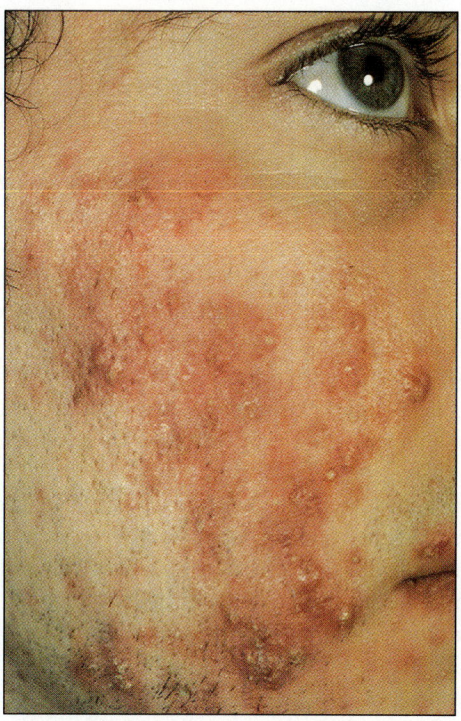

Figure 26.7 Acne is usually a minor annoyance, but for patients with deep, cystic lesions it can be extremely painful and permanently disfiguring.

Propionibacterium acnes: Acne

At some time in their lives, most people suffer from **acne**. Usually it is a relatively minor annoyance—small inflamed papules and pustules appear on the face and upper part of the body and heal spontaneously within a few days. But acne can be a painful and disfiguring disorder. People with **cystic acne** develop large tender nodules deep in the skin that leave lifelong scars (**Figure 26.7**).

Acne is an inflammatory disorder, not an infection. Onset is associated with an increased output of sebum stimulated by steroid hormones produced during adolescence. Acne lesions develop when sebum becomes trapped in pores. Instead of being discharged normally to the skin surface, sebaceous secretions are forced out into subsurface tissue. Here, commensals that normally live in the pores break down the sebum, producing extremely irritating fatty acids that initiate the inflammatory response. The most significant of these commensals is the anaerobic Gram-positive rod *Propionibacterium acnes*.

Most cases of acne respond readily to treatment. Creams that peel away the outermost epidermal layers decrease the likelihood of pores becoming clogged. Antibiotics applied directly to the skin or taken orally decrease the population of *P. acnes*. The tetracyclines are especially effective because they penetrate sebum and are secreted in sweat, thus achieving high concentrations in the skin.

The severest cases of cystic acne that do not respond to standard therapy are treated with a new drug called isotretinoin (Accutane). This compound, closely related to vitamin A, produces dramatic improvement. One major problem exists, however. Isotretinoin is a potent teratogen. This side effect was anticipated because vitamin A is also teratogenic in large doses, and women were advised not to become pregnant while taking the drug. Still, a number of birth defects have occurred and the drug manufacturer has established extremely stringent guidelines for selecting and educating patients.

Mycobacterium leprae: Hansen's Disease (Leprosy)

Leprosy is the term used in scientific literature for the disease caused by *Mycobacterium leprae*, but clinicians use the term **Hansen's disease**, after the Norwegian scientist Gerhard Hansen, who identified the causative microorganism in 1878. *M. leprae* infects both the skin and the peripheral nerves. When a patient has a chronic skin rash with loss of sensation, such as a decreased ability to perceive touch or temperature, the clinician must suspect Hansen's disease. Patients with Hansen's disease are usually treated by dermatologists, physicians who specialize in diseases of the skin.

Although *M. leprae* was the first microorganism to be identified as a cause of human disease, surprisingly little is known about it after more than a hundred years. It is closely related to *M. tuberculosis*, the organism that causes tuberculosis (Chapter 22). Both have a complex cell wall that can be stained only by an acid-fast technique. *M. leprae* grows slowly, with a doubling time of about 12 days. It grows best slightly below body temperature, which accounts for its tendency to infect cooler body parts, such as the nose, ears, toes, and fingers.

Initially, research on *M. leprae* was difficult because it could not be grown in the laboratory, even in human tissue culture. In 1971, however, researchers discovered that fairly large numbers of *M. leprae* could be recovered from artificially infected armadillos (armadillos are prone to a natural infection very similar to leprosy). This led to the development of an experimental vaccine. Recombinant DNA technology may eventually provide a cheap and plentiful supply of *M. leprae* antigens, speeding leprosy research.

Clinical Syndrome.   Leprosy is an ancient disease described in the Bible and feared throughout history as contagious and disfiguring. Infected people were required to wear a bell to warn others of their approach—an early isolation technique. Although not as contagious as once believed, leprosy does spread fairly readily. Family members of a leprosy patient have about a 10-percent chance of contracting the disease, making it about as communicable as active tuberculosis. The disease is probably transmitted from person to person by infected nasal secretions and direct skin-to-skin contact.

About 3 to 5 years after infection, patients develop **indeterminate leprosy**, manifested by a few innocent-looking skin lesions—usually depigmented macules. Nerve function is normal, although a skin biopsy will show early evidence of nerve damage and confirm a diagnosis at this stage. Untreated, the infection usually progresses to either tuberculoid or lepromatous leprosy.

Patients with **tuberculoid leprosy** mount a vigorous cell-mediated immune response and hold their infection in check. Skin biopsy shows tuberculoid granulomas, areas of activated macrophages and lymphocytes similar to those seen in tuberculosis, but no actively multiplying bacteria. A **lepromin** (leprosy skin test) is positive, indicating that the cell-mediated immune system is responding against *M. leprae* antigens. Recent immunologic studies show that T_4 helper cells, releasing primarily gamma

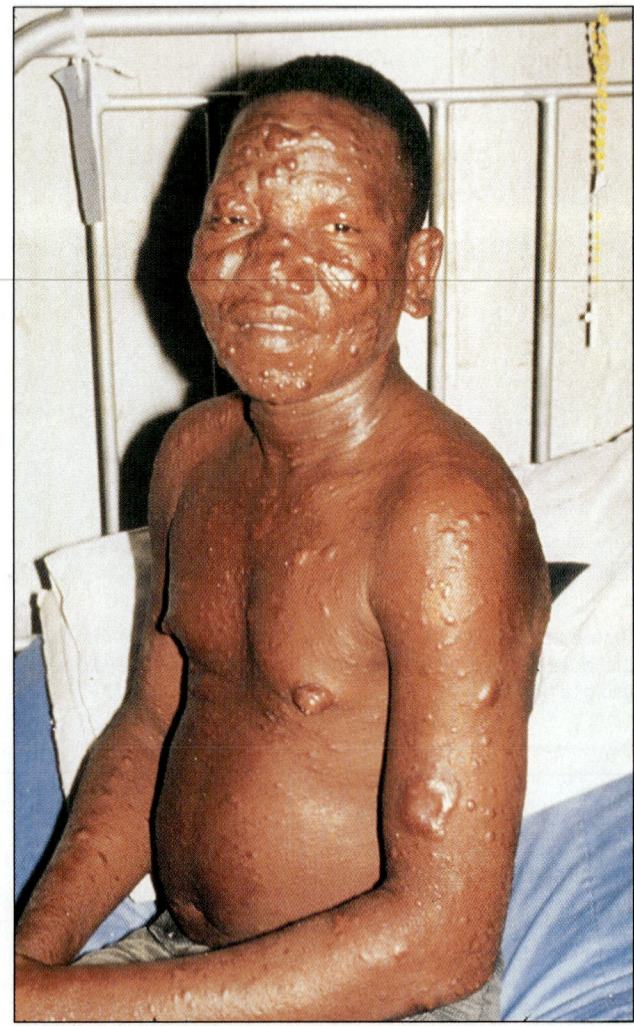

Figure 26.8   A man with lepromatous leprosy.

interferon and interleukin-2 (Chapter 16), predominate in tuberculoid leprosy lesions. Skin damage is usually limited, consisting of a few erythematous or depigmented lesions, but sensation can be completely lost in the affected area.

Patients with **lepromatous leprosy**, whose immune systems fail to mount an effective cell-mediated defense, may become grossly disfigured by their infection (**Figure 26.8**). Skin biopsy shows many actively proliferating *M. leprae* cells, and infective bacteria in nasal secretions mean these patients can transmit their infection to others. The lepromin test is negative, indicating a lack of cell-mediated immune response against *M. leprae* antigens. Interestingly, patients with lepromatous leprosy have extremely high antibody titers against *M. leprae*, but these antibodies have no protective value. T_8 suppressor cells, releasing primarily interleukin-4, predominate in the skin lesions of lepromatous leprosy. Skin damage is extensive, with many flat or raised lesions and underlying bone or cartilage often destroyed. Facial features become thickened, taking on the classical leonine (lionlike) appearance of leprosy, and the nose may collapse from extensive tissue destruction. Fingers or toes may be lost entirely. Loss of sensation, however, is usually patchy.

No one knows why some people develop tuberculoid leprosy and others lepromatous leprosy. Some specialists believe that tuberculoid disease inevitably drifts toward lepromatous disease as cell-mediated defenses are exhausted. Others believe the patient's genetic makeup determines the type he or she develops. Certain genetic markers that predispose to tuberculoid leprosy have been identified.

Epidemiology and Treatment. Hansen's disease is a major public health problem in tropical countries and parts of Asia. About 12 million people are infected worldwide. In the United States, it is rare, but on the rise because of infected people who have entered the country. About 250 new cases are reported to the CDC annually. The U.S. Public Health Service maintains a National Hansen's Disease Center at Carville, Louisiana. Anyone is entitled to receive free treatment for leprosy there or at six outlying centers in cities around the country. Treating imported cases should prevent transmission of Hansen's disease within the United States.

Leprosy has been treatable with antimicrobial drugs since the 1940s. **Dapsone**, the original antimicrobial agent used, is a sulfone drug chemically related to the sulfonamides. Drug resistance tends to develop during the prolonged treatment necessary to cure leprosy, however, and many dapsone-resistant strains of *M. leprae* have emerged. As a result, some cases must be treated with a combination of drugs—usually dapsone, rifampin, and clofazimine.

VIRAL CAUSES

Systemic viral infections include many of the classical diseases of childhood—chickenpox, measles, rubella—as well as life-threatening infections such as smallpox. Viruses also cause warts, the most common benign tumors of the skin.

Varicella Zoster: Chickenpox, Shingles

The two very different clinical syndromes described in this chapter's opening clinical sketches are caused by the same infectious agent—the **varicella zoster virus (VZV)**, a member of the herpesvirus family. **Varicella (chickenpox)** and **zoster (shingles)** affect only human beings. Most people have been infected by this virus by the time they reach adulthood.

Clinical Syndromes. Like the herpes simplex virus (Chapter 24), varicella zoster establishes a latent infection in nerve cells that can be reactivated later. People infected for the first time develop a generalized infection called varicella, or chickenpox, that produces blisters all over the body. Recovery is complete, but the virus remains latent in **spinal ganglia** (large masses of neurons lying near the spinal cord). If infection is reactivated in one of these ganglia, it travels down the neurons to produce new chickenpox-like blisters on the skin that the nerve supplies. Association with the nerve supply accounts for the stripelike distribution of shingles. Usually only one or two **dermatomes** (the area of skin supplied by a single sensory nerve) are involved.

In children, varicella is usually a mild disease. Blisters occur in the outer skin layer, and significant scarring is uncommon. Because the skin as a protective covering is disturbed, varicella sometimes allows more serious streptococcal or staphylococcal infections to become established (this is what R. C.'s parents were advised to watch for), but serious complications are rare. Shortly after recovery from chickenpox, a few children develop a life-threatening illness called **Reye's syndrome**, in which liver and brain functions rapidly deteriorate. This complication is largely preventable. Epidemiologists have found that most children who develop Reye's syndrome have taken aspirin during a preceding viral illness, usually either chickenpox or influenza. It was for this reason R. C.'s physician warned so firmly against giving him aspirin.

In adults, chickenpox is more serious. Patients are much more uncomfortable and may contract a life-threatening viral pneumonia. Adults with chickenpox often require hospitalization. Chickenpox during pregnancy is especially serious. Women infected during the first three months may suffer abortion or deliver a malformed baby. Women infected at term may deliver babies with active chickenpox, and infected newborns have a 1 in 3 chance of dying.

Reactivation of a latent varicella zoster infection occurs most often in patients like K. V., who are over the age of 40, though anyone who has had chickenpox may develop shingles. For people who are otherwise well, shingles usually occurs once, lasts only briefly, and is non-life-threatening, though it can be very painful. The most significant complication is persistent pain in the affected nerves, a syndrome called **post-herpetic neuralgia** that can be disabling for elderly victims.

Cell-mediated immunity is crucial for normal recovery from varicella zoster infections. For people with impaired T-lymphocyte function—cancer patients receiving immunosuppressive chemotherapy or AIDS patients—VZV infections can be fatal. The virus spreads to the lungs, liver, brain, and other organs. Many older cancer patients develop shingles as a painful complication. Children with leukemia are also at special risk. From 100 to 200 people die of chickenpox or shingles each year in the United States.

Prevention and Treatment of Chickenpox. Preventing chickenpox in vulnerable hosts is critical. Passive immunization with **varicella zoster immune globulin (VZIG)** significantly decreases the severity of chickenpox if administered within three days of exposure. It is obtained from recovering zoster patients who have extremely high titers of antiviral antibodies.

Preventing chickenpox in the 3 million healthy American children is really more of a financial than medical consideration. Chickenpox is a far less serious illness than measles or pertussis, for example, but parental time off work while children must stay home from school or day care costs over $380 million annually. As a result, the pressure to control this relatively benign illness is enormous.

Acyclovir, which is highly effective against herpes simplex, is less effective against VZV. Very early treatment—on day 1—may shorten the course of illness by a day or two and reduce the number of spots, but these modest effects do not seem to justify routine treatment. In fact, treating every chickenpox-infected child in the United States with acyclovir would cost at least $128 million a year, thus *increasing* the financial burden. Because acyclovir is a safe drug, however, it is recommended for individuals who may become seriously ill from chickenpox, including people over age 13 and adults on oral steroids.

Preventing chickenpox through active immunization sounds like an attractive option, but clinicians are cautious about immunizing with live attenuated herpesviruses. The herpes simplex and Epstein-Barr viruses have been associated with human cancers. Also, since varicella zoster establishes latent infections, weak immunity against chickenpox might make a person vulnerable to repeated episodes of shingles. In spite of these concerns, a live attenuated varicella zoster vaccine called the **Oka strain** has been developed in Japan. The vaccine has been used extensively there and seems to protect against chickenpox without increasing the risk of shingles. It is expected that the Oka strain will soon be approved in the United States.

Herpes Simplex Type 1: Fever Blisters, Gingivostomatitis

Like herpes simplex type 2 (HSV-2), type 1 (HSV-1) is a large, double-stranded DNA virus that infects the skin and mucous membranes and establishes a latent infection in nerves. The two viral types share many antigens but are distinguished by specific glycoproteins. HSV-1 usually causes infections above the waist, while HSV-2 causes most infections below the waist (Chapter 24).

HSV causes vesicles on infected skin. When the vesicles occur on mucous membrane surfaces such as the inside of the mouth, they rapidly evolve into ulcers. Recurrent lesions near the lips are sometimes called **cold sores** or **fever blisters**. Vesicles are painful when they first appear, but they crust and heal within about a week. Fluid in the vesicles contains live virus, and herpes spreads when this infectious material comes in contact with the mucous membranes or broken skin of a susceptible host.

The most common clinical syndrome of primary HSV-1 infection is **gingivostomatitis** (*gingiva*, gums; *stoma*, mouth). It is a moderately severe illness characterized by fever and painful blisters on the mouth and gums. Toddlers, who exchange saliva as they play closely together or mouth each other's toys or bottles, are frequently affected, especially if they live in crowded conditions. Epidemiologists estimate that 80 percent of low-income children are infected compared to only 40 percent of children from more affluent families. Adults who have never been infected remain susceptible to herpes gingivostomatitis. Some studies show that half of college students are susceptible.

Acyclovir, which is used routinely for primary genital herpes, is probably effective in shortening oral herpes infections as well, but extensive clinical studies have not been undertaken and the drug is not used routinely in children with gingivostomatitis.

After a primary oral HSV infection, the virus remains latent in nearby nerve cells. Fever, sunlight, emotional stress, or other stimuli can trigger reactivation at the same site any number of times. A cluster of tiny thin-walled blisters on a bright red base appears on the lips (**Figure 26.9**). Like primary lesions, reactivation vesicles contain live virus, and patients with cold sores can spread infection. Viral reactivation can also be asymptomatic so that apparently healthy people shed contagious viral particles.

HSV-1 can infect any part of the body if virus-containing secretions come in contact with a break in the skin. For example, **herpetic whitlow** is a finger infection seen in medical personnel caring for herpes patients and in children with oral herpes who suck their fingers. **Herpes gladiatorum** is seen in wrestlers who sustain prolonged skin-to-skin contact with one another. The most serious

HSV-1 infection, a syndrome called **herpes keratitis**, affects the surface of the eye.

Measles (Rubeola) Virus: Measles

The measles or rubeola virus belongs to the paramyxovirus family. It is a member of the same genus that causes canine distemper. The measles virus is a coiled RNA nucleocapsid surrounded by a lipid envelope bearing many short projections. The envelope enables the measles virus to destroy and agglutinate red blood cells and to fuse human cells to one another. Cell fusion creates the giant multinucleated cells seen in measles-infected tissue under the microscope. The measles virus is extremely fragile and can survive only a short time outside the human body, its only reservoir.

Clinical **measles** follows a typical pattern. A susceptible person (usually a child) inhales the virus. During the next few days, the virus multiplies in the respiratory tract and then spreads throughout the body, multiplying in lymphoid tissue. The first symptoms—fever, cough, **coryza** (runny nose), and conjunctivitis—usually appear on day 11 of the infection. In these first days **Koplik spots** also begin to appear inside the mouth. They look like tiny white flecks of sand on a bright red base. If this enanthem is noticed, the diagnosis of measles can be made even before the rash appears on day 14. The erythematous maculopapular rash of measles appears first on the face and gradually spreads downward to cover the entire body. As the rash progresses, the individual spots become so numerous they merge, so that at its peak the entire surface of the skin is deep reddish-purple and puffy. By the time the rash reaches the legs, the child is usually beginning to feel better. The rash fades in the same order as it appeared, from head to foot.

Measles is often a serious illness. The patient feels miserable and complications are common, especially ear infections and pneumonia caused by secondary bacterial invaders. Bacterial pneumonia is the leading cause of death among children with measles. Permanently handicapping complications also occur. During recent epidemics in the United States, 2 to 3 children per 1000 suffered brain damage from measles encephalitis.

Intact cell-mediated immunity is essential for normal recovery. Sensitized T lymphocytes recognize antigens present on the surface of measles-infected cells, and cytotoxic lymphocytes destroy these cells, limiting viral replication and bringing the infection under control. Cell-mediated immunity is depressed by cancer chemotherapy and by the general weakness of malnutrition. As a result, measles is a terrible killer of children in developing countries. The World Health Organization estimates that every minute three children die of measles and three more are permanently disabled.

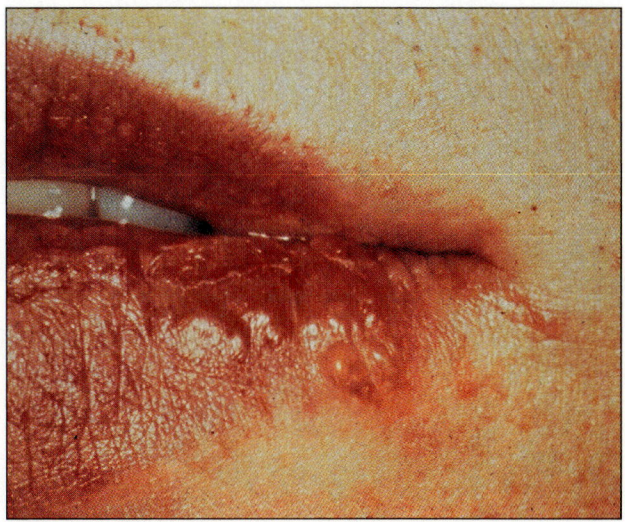

Figure 26.9 Herpetic vesicles on the lips are commonly called cold sores or fever blisters.

Epidemiology. To survive, the measles virus must sustain an unbroken chain of infection in human beings. At first, it might appear that the odds are against the virus. It can live within a host for only a few weeks, and an infected person can spread the virus to others during only about a week (from the time the first symptoms develop until about four days after the rash appears). During this brief period, the virus must be transmitted directly to a new host—it cannot survive on fomites or be transmitted by an uninfected carrier. Moreover, the pool of susceptible people is severely limited. Anyone previously infected is immune for life. If this chain of infection were ever broken, measles would be eradicated.

But the measles virus sustains the chain of infection because it is one of the most highly communicable diseases known. Over 90 percent of susceptible people who are exposed become infected. Contact does not have to be intimate. For example, a person can become infected by being on the same airplane as someone who has measles. Because it is so communicable, measles in a natural setting is almost exclusively a disease of children. Virtually everyone has measles by adulthood, so there is no risk of adult infection, as there is with chickenpox.

The only factor limiting the spread of measles virus is the availability of new hosts. Epidemiologists calculate that a population of 300,000 to 500,000 is probably necessary to sustain the measles virus. In smaller communities all the susceptible people may be infected at once, depriving the virus of new susceptible hosts. But when the measles virus is reintroduced after a period of many births, an epidemic will spread through the new susceptible population with unbelievable rapidity. Such an importation epidemic was documented in Greenland in

Figure 26.10 (a) Reported cases of measles have declined dramatically in the United States since immunization began in 1963. But a series of epidemics in the late 1980s led to a recommendation that every child receive two doses of measles vaccine rather than one. (b) The photo shows a toddler with a typical measles rash.

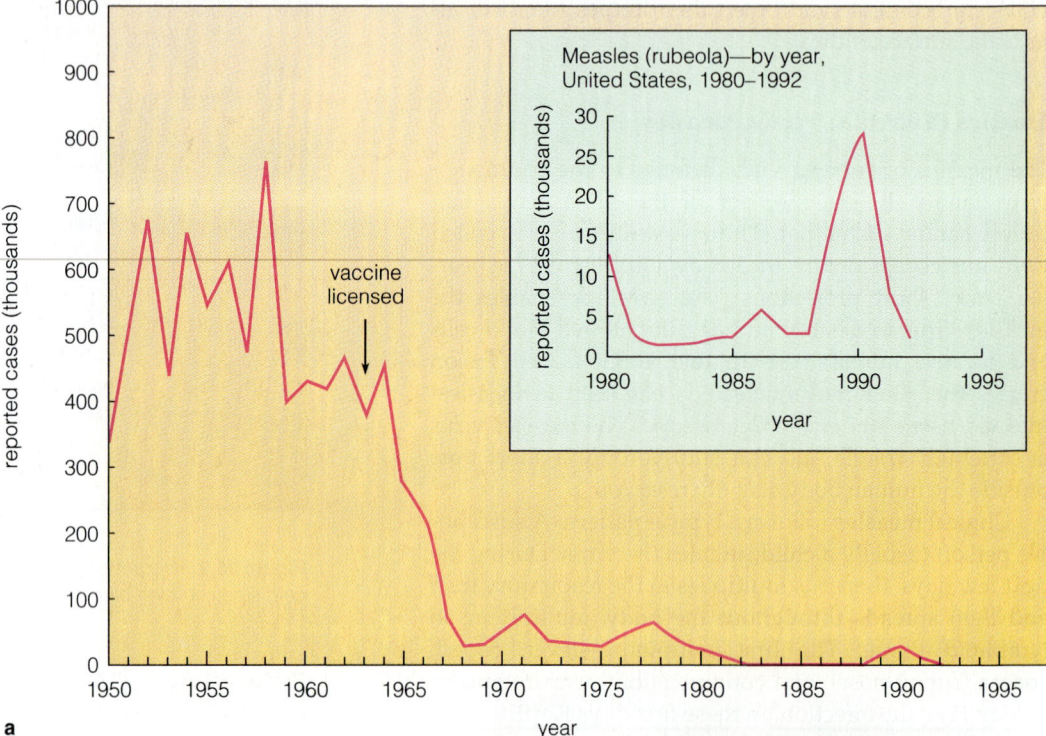

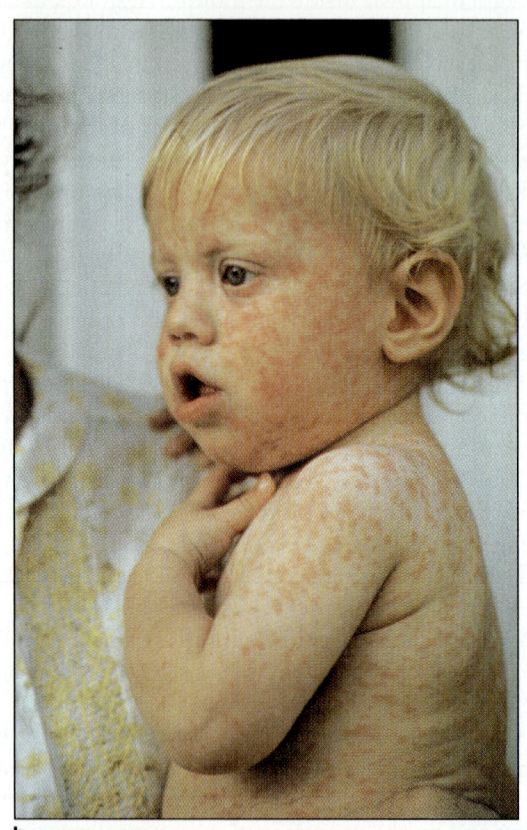

a

b

1951. Of 4600 susceptible people, all but 5 contracted measles within six weeks. Similar epidemics devastated native populations in the Americas when the measles virus was first introduced from Europe.

The epidemiology of measles was radically altered in 1963 when a vaccine with live attenuated measles virus was developed (**Figure 26.10**). Widespread immunization of 15-month-old children as part of the MMR (measles, mumps, rubella) vaccine has reduced the incidence of measles in the United States by more than 99 percent. As a result, many young parents and health professionals in industrialized countries are unfamiliar with measles, and some mistakenly think of it as a mild illness like chickenpox.

The late 1980s saw a rise in the number of reported measles cases among children and young adults who had not been adequately immunized. These were probably the 5 percent of people who fail to produce protective antibodies after a single immunization. The CDC now recommends immunization with MMR at 15 months and reimmunization upon entering school, at age 5. Because many children already in school at the time the recommendations were changed may have missed their second dose of MMR, many colleges today require proof of reimmunization for enrollment.

A major obstacle to measles control exists, however, in inner cities where many preschool children are still not immunized against measles. Because the law requires proof of measles immunization for school, children over 5 are immunized, but their younger brothers and sisters often remain unvaccinated if basic health care is expensive or inaccessible. Wherever a significant population of children without immunization exists, measles outbreaks will continue to occur. Until all children receive timely immunization, measles will not be eradicated from the United States.

The struggle to control measles in the developing world is more desperate. Malnourished children are often infected before their first birthday, when they are too young to be adequately protected by the present vaccine. (The attenuated virus often fails to replicate in children less than 1 year old who still have relatively high levels of transplacental maternal antibodies against measles.) The answer lies partly in developing a new vaccine administered directly into the respiratory tract (like natural measles virus) or a modified version of the current vaccine to protect infants as young as 4 to 6 months. In theory, measles, like smallpox, could be eradicated by a worldwide vaccination program. In reality, public health experts believe that until people around the world are provided with basic necessities—clean water, adequate food, and basic health care—measles will continue to kill children.

Subacute Sclerosing Panencephalitis (SSPE). Subacute sclerosing panencephalitis (SSPE) is an extremely rare but fatal form of measles that affects the nervous system. It may be caused by mutant strains of the measles virus, because the frequency of SSPE—about one per million cases of measles—is nearly the same as the frequency of viral mutations.

SSPE typically occurs in a child who contracts measles at an unusually young age. The measles virus replicates in brain cells, but the virions are not assembled and released into the bloodstream as they are during a normal infection. The patient appears to recover completely and continues to develop normally. Five or more years later, however, symptoms of severe brain damage begin to appear—intellectual deterioration, inappropriate behavior, and abnormal speech and movement. The disease progresses to seizures, paralysis, and, eventually, coma. Diagnosis is suggested by extremely high levels of measles antibodies in the patient's blood and cerebrospinal fluid and confirmed by cultivating measles virus from the victim's brain at autopsy. There is no treatment for SSPE and it is almost always fatal. Like measles itself, SSPE has decreased markedly where measles immunization is widespread. In the United States it has virtually disappeared.

Rubella Virus: Rubella (German Measles)

Rubella—also called **German measles** and **three-day measles**—is not related to rubeola, or true measles, at all. Rubella virus was isolated in tissue culture in 1962, finally allowing identification of the infectious agent. It is classified as a togavirus, although it has little in common clinically with the arboviruses that chiefly constitute the group (Chapter 25).

Rubella is one of many mild, rash-producing illnesses of childhood. The virus is inhaled, and after an incubation period of about two weeks a mild fever and maculopapular rash appear. Lymph nodes are often enlarged and tender. The rash consists of faint spots that appear first on the face and gradually spread downward over the rest of the body. The rash lasts less than three days. Most people with rubella are not very ill, although adult women may complain of stiffness and swelling in their joints. Complications are rare. Because rubella is not as highly communicable as measles or chickenpox, a significant number of people reach adulthood without contracting the disease. People who have recovered from rubella are immune for life.

The rubella virus is not to be underestimated, though—it is a powerful teratogen. Infection during the first three months of pregnancy is highly likely to cause abortion or birth defects of the eye, heart, and brain. A major rubella epidemic swept the United States in 1964, and more than 20,000 babies were born with congenital rubella syndrome while thousands more died before birth (**Figure 26.11**). The disaster spurred interest in developing a rubella vaccine.

The goal of vaccination—to prevent congenital rubella rather than childhood rubella—presented researchers with special problems. Immunizing women of childbearing age seemed dangerous because the live attenuated virus itself might cause birth defects. Immunizing children might produce a weak immunity that would wear off by adulthood and actually increase the frequency of the disease in pregnant women. In spite of these fears, a rubella vaccine was introduced in 1969, and it has worked well. Childhood rubella and congenital rubella have declined substantially.

In the United States, children are vaccinated at 15 months with MMR vaccine (against measles, mumps, and rubella). Immunization is also recommended for all adults except pregnant women or women who may become pregnant during the next few months. Although the vaccine virus has not been shown to cause the congenital rubella syndrome, as a precaution it is nevertheless viewed as a potential teratogen. A blood test to check for the presence of antibodies against rubella is often part of a premarital exam. If a woman does not have detectable antibodies she is susceptible and should be immunized before becoming pregnant. Rubella serology is also performed at the first prenatal exam. Although pregnant women cannot be immunized, a negative test signals maternal susceptibility and raises the possibility of fetal infection if the mother contracts rubella during the early months of pregnancy.

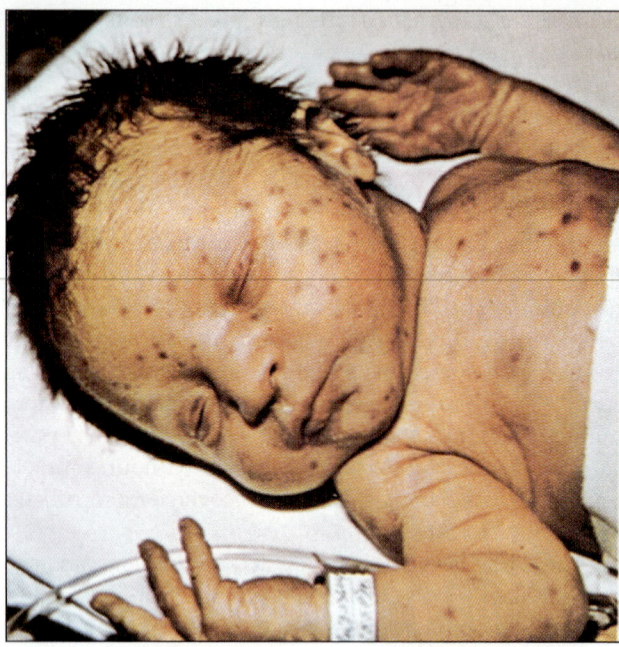

Figure 26.11 The rubella virus, which causes rubella (German measles), is a powerful teratogen. This congenitally infected infant has a typical "blueberry muffin" rash and probably other less obvious but more serious birth defects.

Variola (Smallpox Virus): Smallpox

Variola (smallpox virus) is a large double-stranded DNA virus that belongs to the poxvirus family. A closely related poxvirus, vaccinia, is used for smallpox immunization. Other poxviruses cause animal diseases, such as monkeypox, cowpox, and even camelpox.

The severity of clinical **smallpox** depended upon the viral strain causing the infection. The most virulent strain, **variola major**, produced a high fever and severe blistering rash, killing about half its victims. The pustules of smallpox originated deep in the skin, and most survivors were permanently scarred (**Figure 26.12**). In fatal cases, the blisters were often so numerous they touched one another, covering the entire body. A much milder form, **variola minor**, had a mortality rate of less than 1 percent. Survival seemed to depend entirely on the virulence of the infecting strain. It was never established that any medical treatment helped people recover.

Smallpox had nearly the same epidemiology as measles. The virus was inhaled and an unbroken chain of human infection had to be maintained to perpetuate the virus. A significant difference, however, is that smallpox was much less communicable. Close personal contact was necessary for transmission. On the other hand, relatively low communicability was offset by increased hardiness. The virus could survive on fomites and in the environment for days to weeks. Lower communicability

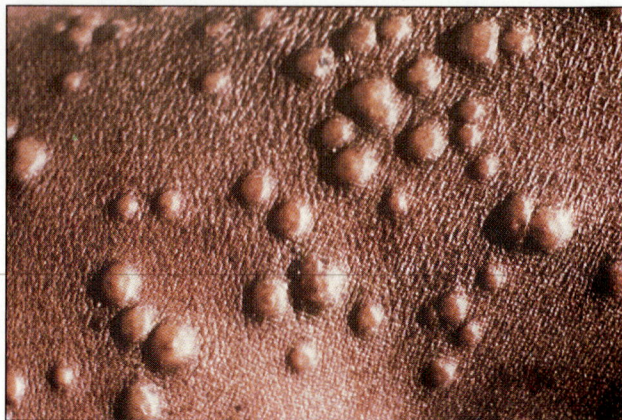

Figure 26.12 Smallpox blisters originate deep in the skin and often leave permanent scars. The disease was eradicated in 1979.

also meant that epidemics progressed less rapidly but lasted longer than measles epidemics. Because smallpox depended on a continuing chain of human transmission, eradication was possible, and the successful effort is described in Chapter 20.

Human Papillomavirus (HPV): Common Warts

Human papillomaviruses (HPV) are DNA viruses of the papovavirus family that cause warts on body surfaces. Some strains cause genital warts (condyloma acuminata) and laryngeal papillomatosis (Chapter 24). Other strains—principally HPV-1, 2, 3, and 4—cause the **common warts** that occur on most skin surfaces and **plantar warts** on the soles of the feet. Genital infection is sexually transmitted, but no one knows how other strains of HPV are transmitted. Warts on the skin are not highly contagious but may spread if HPV-infected skin cells are directly inoculated into a break in the skin of another person. This association with trauma may explain why warts are most common on the hands and feet.

The association between human papillomaviruses and cancer has stimulated interest in these organisms. The first link between HPV and cancer was established in people with a rare syndrome of persistent warts (not common warts, but a particular type of warty growth). Thirty percent of these patients eventually develop skin cancers, and viral DNA can be found in the malignant cells.

Warts on the skin appear with no warning, usually before puberty, and disappear without treatment, usually after a few years. Sometimes a single wart will develop into clusters of warts, probably due to viral autoinoculation. Although skin warts are harmless and *not* associated with an increased risk of cancer, people often want them

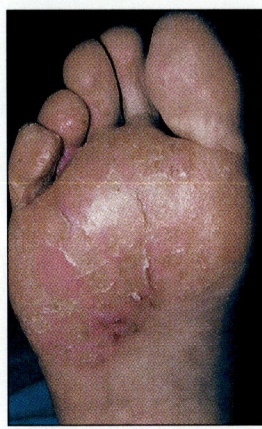

Figure 26.13 Athlete's foot is caused by a dermatophyte, a ringworm-causing fungus (see Figure 1.5 for a photo of *Epidermophyton fluccosum*).

removed for cosmetic reasons. Warts can be eliminated by applying liquid nitrogen (which freezes and thereby kills infected cells), by **keratolytic** chemicals (which destroy keratin and thereby dissolve the infected cells), or by surgery. Wart removal is relatively easy and leaves no scar because warts occur in the epidermis—the outermost skin layer that is regularly replaced by skin growth.

FUNGAL CAUSES

A variety of fungi survive in or on the surfaces of our body. Though annoying, most of these infections are not serious in otherwise healthy people.

The Dermatophytes: Ringworm

The dermatophytes are a group of fungi that infect the body's outermost surfaces causing athlete's foot, jock itch, and ringworm. The medical names for these infections are **tinea** (meaning **ringworm**) followed by the affected skin surface. Thus, **tinea pedis** is ringworm infection of the foot; **tinea cruris**, infection of the groin; **tinea capitis**, infection of the scalp; **tinea unguinum**, infection of the nails; and **tinea corporis**, infection of other parts of the body. Despite its name, ringworm has nothing to do with worms. It is estimated that as many as 90 percent of men have been infected at least once by the time they reach middle life. We do not know why women are not affected as often. Dermatophyte infections can be annoying and uncomfortable; but because they never penetrate into deeper tissues or affect vital organs, they are not disabling or life-threatening.

Dermatophytes belong to three genera—*Trichophyton*, *Microsporum*, and *Epidermophyton* (**Figure 26.13**). All produce enzymes that digest keratin, and all infect only ker-

atin-containing surfaces—the outermost layer of the epidermis (the stratum corneum), the hair, and the nails. There are more than 40 different species of dermatophytes, though fewer than 10 are found in the United States. Some live only on human beings and are transmitted directly from person to person, while others have animal reservoirs or survive in the soil. Infection is more prevalent in tropical climates and under crowded living conditions.

The body's immune response influences both the symptoms of dermatophyte infections and the chances for complete recovery. Cell-mediated immunity is activated, but exactly how the long filamentous hyphae of these fungi (too large to be phagocytized) are actually destroyed is a mystery. Some infections, such as ringworm contracted from animals, stimulate a powerful immune reaction. The infected areas are highly inflamed—becoming bright red, weepy, and unsightly—but they tend to heal completely within a week or two. Other infections, such as athlete's foot, stimulate a comparatively weak immune reaction with little inflammation—not becoming red at all and showing only mild cracking and peeling—but they can persist for years. The long-lasting but relatively mild infections indicate an extremely stable and well-adapted symbiosis between humans and microorganisms. Curiously, acute infections are usually seen in children while athlete's foot does not usually develop until after puberty.

Because dermatophyte infections are superficial, most can be cured with topical antifungal agents such as miconazole or clotrimazole. Infections that affect thickened areas such as the scalp or nails often fail to respond to local therapy and are usually treated orally with griseofulvin. Griseofulvin is especially effective against dermatophytes because it concentrates in the stratum corneum. Although it has some toxic side effects, including headache, it can usually be used quite safely for the weeks to months required to eliminate dermatophyte infections.

Candida albicans: Candidiasis

Candida albicans and other closely related *Candida* species are commensal fungi that colonize the skin, mucous membranes, and gastrointestinal tract of almost every human being (Chapter 14). As a harmless commensal, *Candida* grows as a budding yeast, but a hyphal form also exists and is associated with infection.

Candida is also an opportunistic pathogen. Many minor disruptions of the body's equilibrium (such as pregnancy, antibiotic therapy, oral contraceptives) predispose a person to **candidiasis**, symptomatic infections of body surfaces. Infants and the elderly are especially susceptible, as are people with immunodeficiency disorders.

Absolutely Astonishing

What is the connection between warts and the immune system? Most warts disappear spontaneously, probably because the immune system finally identifies the wart and rejects it as foreign tissue. But what triggers this sudden immune activity? Perhaps the virus is hidden from immune recognition by its location in the outermost layers of the body, and rejection occurs only when viral antigen escapes its protected site. Some well-documented scientific studies suggest that the connection between warts and the immune system is less concrete—that mental suggestion is enough. In one study people with many warts were hypnotized and told that the warts on one side of their body would disappear, and within a few weeks the warts on that side were gone.

The late writer-physician Lewis Thomas described the possibility that hypnosis can eliminate warts as "absolutely astonishing, more of a surprise than cloning or recombinant DNA or endorphins or acupuncture." As links between the brain and the immune system are established, we may learn more about the appearance and disappearance of warts and, more importantly, more about the deeper mysteries of immune function.

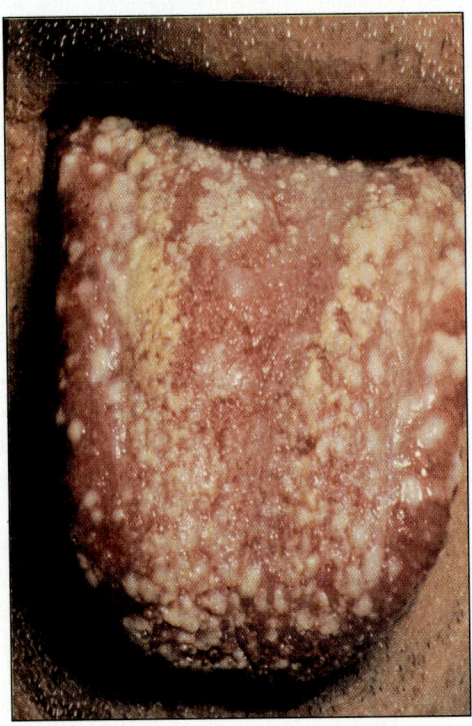

Figure 26.14 *Candida albicans* infection of the tongue.

The most common sites of infection are the vagina (*Candida* vaginitis, Chapter 24), the mouth (thrush), and the diaper area of infants. **Thrush**, which often occurs in healthy infants, causes white plaques on the mucous membrane. The lesions look like milk but cannot be wiped off. *Candida* **diaper dermatitis** is a fiery red, raised rash, often with small pimples, or **satellite lesions**, just beyond its border. Because *Candida* thrives in a moist, warm environment, any part of the skin that remains wet may be infected. *Candida* of the hands, for example, is often seen in people who have jobs washing dishes. Superficial candidiasis is treated with antifungal creams containing nystatin or clotrimazole. Thrush infections are treated with a topically applied liquid (**Figure 26.14**).

Another type of local infection, called **chronic mucocutaneous candidiasis**, occurs only in people with defective cell-mediated immunity, and even then, rarely. In this disorder, the yeast grows as an intracellular parasite rather than in the spaces between cells. Chronic mucocutaneous candidiasis affects only the skin and mucous membranes, so it is not life-threatening; but it can be long-lasting and the overgrown, warty lesions grossly disfiguring.

Occasionally, *Candida* invades deeper tissues of patients with serious immunodeficiency diseases, causing systemic and life-threatening infections. Under these conditions, *Candida* can affect any organ and cause an almost limitless number of clinical syndromes. Systemic candidiasis may be the fatal event for people with terminal illnesses such as AIDS or disseminated cancer. Systemic antifungal agents, such as amphotericin B, may be prescribed but treatment is seldom effective.

Subcutaneous Mycoses

When fungi infect the skin and its underlying connective tissue and lymphatic vessels, the disorders are called **subcutaneous mycoses**. One of the most common subcutaneous mycoses is **sporotrichosis**, caused by the dimorphic fungus *Sporothrix schenckii*. This organism is found in soil and plant material around the world, especially in Mexico, and enters the body through a break in the skin. In the United States sporotrichosis is most often acquired while gardening. Most infections are minor and progress slowly, producing swelling and ulceration in the skin and nearby lymph nodes. Occasionally muscle and bone are destroyed, and very rarely the organism reaches deeper organs such as the lungs or heart. All but the most widespread cases of sporotrichosis can be cured by oral potassium iodide.

Chromomycosis is a subcutaneous mycosis that can be caused by several species of fungi. Like sporotrichosis, the fungus enters the body through a minor wound. As the fungus grows, it produces large wartlike deformities on the overlying skin. Infection progresses slowly and is not fatal, but may become disfiguring and disabling. Chromomycosis is usually seen in tropical countries.

Mycetoma is a chronic infection caused by various actinomycete bacteria and fungi such as *Madurella mycetomatis* and *Phialophora jeanselmei*. These local infections are not life-threatening but can cause grotesque swelling with draining pus that contains granules of the causative microorganism. The foot is most commonly affected, a condition called **Madura foot** because it is especially common in the Madura province of India. The infection can be so severe it is crippling.

ARTHROPOD CAUSES

We have discussed biting arthropods in other chapters as vectors of infection. A few species, however, cause true skin infections, living on body surfaces for prolonged periods and being transmitted person to person.

Sarcoptes scabiei: Scabies

Scabies is a skin infection caused by the mite *Sarcoptes scabiei*, a parasitic arthropod. It is transferred from person to person by close personal contact and fomites such as clothing or bedding. Scabies is common around the world, including the United States, where the incidence rises and falls. Epidemics last about 15 years, with another 15 years passing before a new outbreak. A major U.S. epidemic that began during the mid-1970s has just about ended.

S. scabiei lives in the epidermis of human skin. The female mite burrows into the dead keratin-packed stratum corneum and lays 15 to 20 eggs that hatch a few days later. In a few more days the newly hatched larvae become mature mites, capable of digging their own skin burrows and carrying on the life cycle. These creatures travel across the human body at the rapid rate of about an inch per minute. They prefer the wrists and the spaces between the fingers but may be found anywhere. Surprisingly, most people with scabies harbor only a few adult female mites—usually about 11. Scabies can resemble many other itchy skin diseases, so a definitive diagnosis can be made only if the mites or their eggs are recovered and identified under the microscope.

Scabies is a harmless infection limited to the very surface of the skin, but the itching it produces—which is worst at night—can be almost intolerable. Untreated, the infestation may continue for years. For this reason it is sometimes referred to as the seven-year itch. When itching leads to such severe scratching that skin breakdown occurs, secondary bacterial infections develop.

The mites themselves do not harm the skin; a hypersensitivity reaction does the damage. When a person is first infected, the mites multiply without causing any symptoms. Once the immune system is stimulated, however, the mites and their feces become a powerful irritant. Treatment with the arachnicide gamma benzene (lindane or Kwell) kills the mites, but dead mites and their fecal pellets continue to cause irritation until the infected layer of skin is completely replaced several weeks later.

Pediculus spp.: Pediculosis (Lice)

Pediculosis is an infestation with lice, blood-sucking insects that live on human skin. The three main types of human lice are *Pediculus humanus capitis* (the head louse), *Pediculus humanus corporis* (the body louse), and *Phthirus pubis* (the pubic or crab louse) (**Figure 26.15**). All types cause itching that can lead to skin breakdown and bacterial superinfection, but only body lice can transmit microorganisms that cause epidemic diseases. Body lice and the diseases they spread, such as typhus (Chapter 27), are

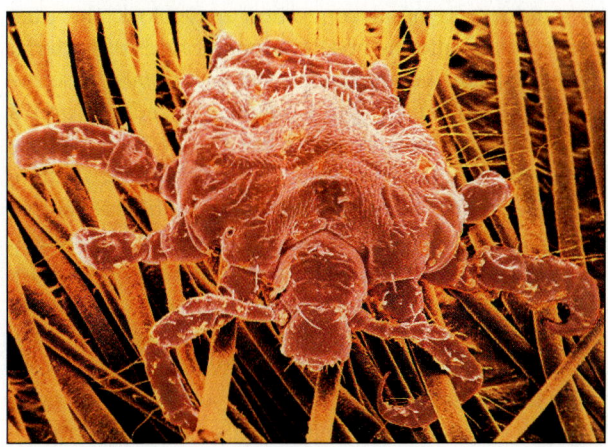

Figure 26.15 Scanning electron micrograph showing a single pubic louse, *Phthirus pubis*, clinging to pubic hair.

most often associated with the social dislocations of wartime.

Lice are usually transferred by direct body contact or by fomites such as clothes, hats, and bedding. Pediculosis is most commonly found among people living in unsanitary and crowded conditions. Head lice among children who play together, sharing hats and toys, have reached epidemic proportions in the United States in recent years. Once a head louse infestation has developed at a school, only careful and frequent checks that infected children are being treated can control the epidemic. Pubic lice are usually transmitted during sexual contact, although fomites such as bedding and even toilet seats are vehicles.

Lice are large enough to be seen with the unaided eye, so diagnosis can be made by careful examination. Head lice and pubic lice can be diagnosed by finding tiny eggs called nits firmly cemented to hair shafts. Body lice are usually easiest to find in the seams of clothing. All lice can be killed by insecticides, such as the prescription agent gamma benzene (lindane, Kwell), and pyrethrins, which are available without prescription.

EYE INFECTIONS

Bacteria and viruses that infect the eye's surfaces, the conjunctivae and cornea, are discussed here, along with helminthic infections that affect the skin and the eyes.

BACTERIAL CAUSES

Bacteria regularly come in contact with the surface of the eye, but the cleansing, antibacterial action of tears usually keeps infection from developing. Thus, any condition that blocks the tear ducts increases the likelihood of bacterial conjunctivitis. In early infancy, the problem may be congenitally blocked ducts (Chapter 14). Later, colds or allergies resulting in nasal congestion may block the duct where it opens into the nose. Common causes of bacterial conjunctivitis include *Streptococcus* spp., *Staphylococcus aureus*, and *Haemophilus* spp. Conjunctival redness and a pus-containing discharge occur, but these infections usually do not affect the cornea and so do not threaten vision. These infections are easily cured with antibacterial eye drops, especially sulfacetamide, neosporin, or gentamicin.

As foreign bodies that can interfere with normal defenses, contact lenses predispose a person to bacterial conjunctivitis, especially if lenses are not properly cleaned or are kept in the eyes for extended periods. In these cases, the infecting agent is often *Pseudomonas aeruginosa*. Contact lens conjunctivitis can sometimes affect the cornea and threaten vision. Treatment involves using antibacterial eye drops and disinfecting contact lenses to prevent further infection.

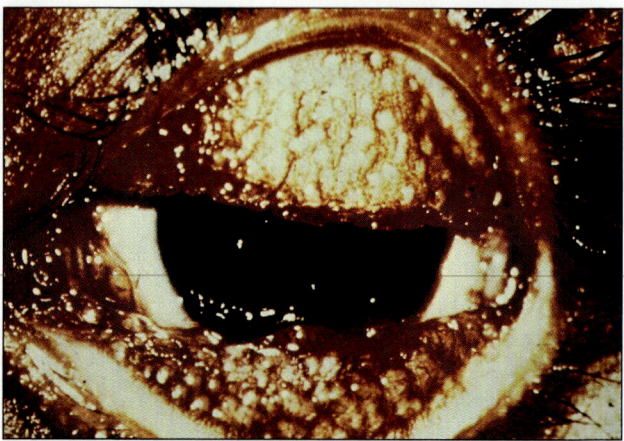

Figure 26.16 Trachoma, the most common preventable cause of blindness in the world, is caused by a chronic infection by *Chlamydia trachomatis.*

Chlamydia trachomatis: Inclusion Conjunctivitis, Trachoma

Chlamydia trachomatis is an intracellular bacterial parasite that produces several clinical syndromes. Strains that affect the genital tract can be transmitted to infants at birth (Chapter 24). Although the most serious complication of neonatal infection is chlamydial pneumonia, eye infection is far more common. The infection is called **inclusion conjunctivitis** in reference to intracellular chlamydial inclusion bodies. The eyelid is swollen, and there may be a thick purulent discharge. Infection can be prevented by treating the mother for chlamydia during pregnancy; silver nitrate or antibiotic eye drops administered at birth do *not* keep inclusion conjunctivitis from developing. Infants may be given oral erythromycin for two weeks, but the infection usually resolves spontaneously and the cornea is not damaged. Rarely, sexually active adults may contract inclusion conjunctivitis from infected genital secretions.

Certain other *C. trachomatis* strains cause the most common preventable blindness in the world—trachoma. **Trachoma** is a keratoconjunctivitis (**Figure 26.16**). About 2 million people, of all ages and mostly living in developing countries, are blind from trachoma.

Trachoma begins when *C. trachomatis* comes in contact with the surface of the eye and multiplies in the conjunctivae. As chlamydiae multiply, the eye becomes inflamed and bumps appear on the normally smooth conjunctival surface. (The word *trachoma* means "rough" in Greek.) Inflammation progresses gradually over many years. Eventually the conjunctivae become scarred, and if inflammation damages the cornea or the eyelids become too distorted to close, blindness results.

Infectious *C. trachomatis* organisms are shed from the eye. Transmission occurs by direct contact with infected

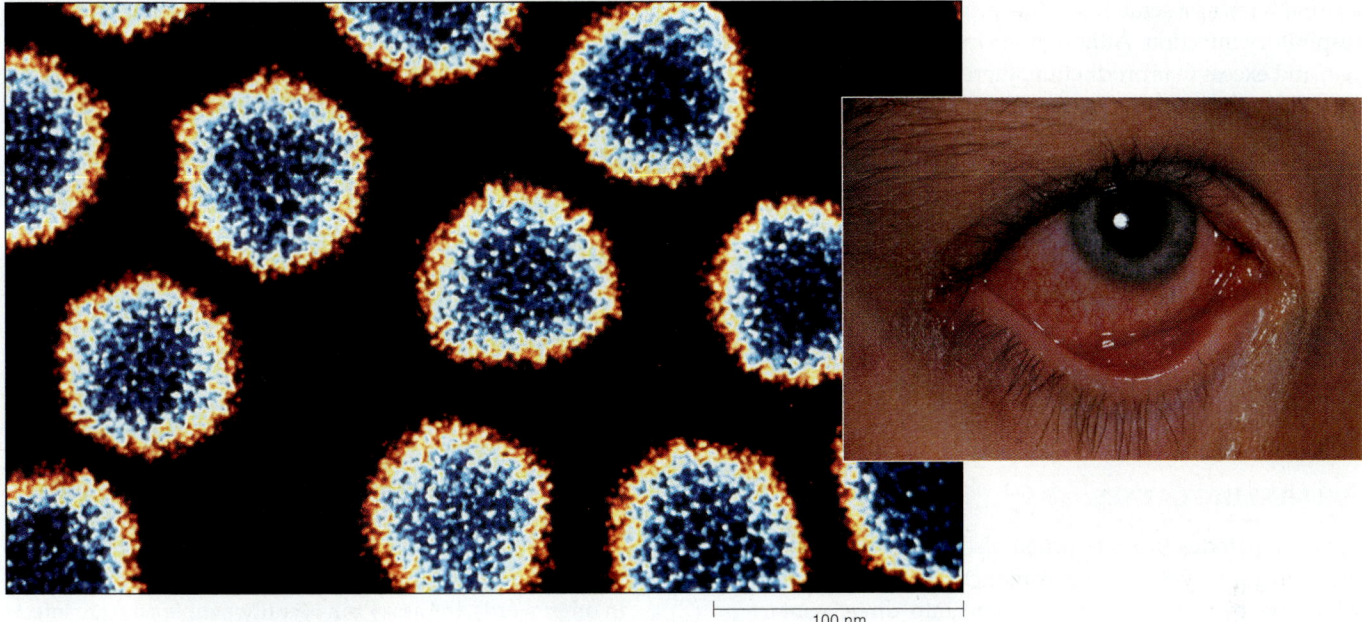

Figure 26.17 Epidemic viral conjunctivitis. The adenovirus, shown in an electron micrograph, is one of the most common causes of epidemic conjunctivitis, or pinkeye. The infection usually resolves itself in a few days and does not threaten sight.

ocular secretions, by fomites such as towels or water, or by flies. Trachoma is common where living conditions are crowded and unsanitary. It is found most widely in the Middle East, Africa, and Asia. It is also seen fairly frequently among Native Americans in the southwestern United States.

Chlamydiae are sensitive to many different antibiotics, including erythromycin and tetracycline, but unfortunately these agents cannot eradicate *C. trachomatis* from the eye. Antibiotics can only control bacterial multiplication and moderate the resultant inflammation. This does mean that when antibiotic ointments are used early they can help preserve sight, so treatment is sometimes recommended for entire populations that are at high risk. Once corneal scarring has occurred, however, only a corneal transplant can restore vision.

Neisseria gonorrhoeae: Neonatal Gonorrheal Ophthalmia

Neisseria gonorrhoeae can be transmitted from an infected mother to the eyes of her newborn as it passes through the birth canal (Chapter 24). This infection, called **neonatal gonorrheal ophthalmia**, was once a common cause of blindness. It typically appears two to five days after birth, with a thick, purulent discharge and swollen eyelids. Scarring and, occasionally, perforation of the cornea cause blindness.

Today, routine prenatal care includes a cervical culture for *N. gonorrhoeae*, allowing diagnosis and cure before delivery. Still, the law in most states requires that every newborn be treated with antibacterial eye drops (formerly silver nitrate, but now usually erythromycin) to prevent gonococcal blindness. This prophylaxis prevents infection in industrialized countries. Infants diagnosed with this infection must be treated with systemic antibiotics such as ceftriaxone.

VIRAL CAUSES

Viral infections of the conjunctiva are extremely common and usually not serious. Agents include the adenoviruses and enteroviruses. The common and highly contagious pinkeye epidemics that race through schools and daycare centers are usually viral (**Figure 26.17**). Viral conjunctivitis cannot be reliably distinguished from bacterial conjunctivitis by clinical examination alone, but viral infections tend to cause more intense redness and less purulent discharge than bacterial. No treatment is effective, but the infection resolves spontaneously within a few days; and because the cornea is not involved, viral conjunctivitis does not threaten sight.

A sight-threatening viral infection can be caused by the herpes simplex virus (HSV) if it happens to be introduced into the eye. **Herpetic keratitis** is an ulceration of the cornea. It usually begins after some sort of minor trauma, such as a scratch to the eye, sunburn, or an upper respiratory infection. Although the infection causes irritation and excess tear production, there is surprisingly little pain. Usually, only one eye is affected. Keratitis often resolves spontaneously, though scarring and visual loss may occur if deeper layers of the cornea are involved. Eye drops containing the antiviral agent iododeoxyuridine shorten the infection but do not prevent the recurrences that are typical of HSV. If an incorrect diagnosis leads to prescription of steroid-containing eye drops, which suppress immune function, the infection may progress rapidly, leading to corneal destruction and blindness. As in all cases of permanent damage to the cornea, sight can be restored only by a transplant.

HELMINTHIC CAUSES

Some nematodes (roundworms) also cause infections of the skin and eyes. Unlike conjunctivitis and keratitis, which begin when pathogens come into direct contact with exposed eye surfaces, these are systemic diseases in which pathogens reach the skin and eye through blood and lymphatic channels.

Onchocerca volvulus: Onchocerciasis (River Blindness)

Onchocerciasis, river blindness, is a serious disease of the skin and eyes that affects more than 18 million people in Africa and Latin America. It frequently causes blind-

Hope at Last

If onchocerciasis, river blindness, is diagnosed early, it can often be cured. But usually, onchocerciasis is diagnosed only by finding microfilariae in the skin or eyes or by identifying adult worms in subcutaneous nodules—methods that require minor surgery and may not be positive for several months to a year after infection. A blood test would be much more practical, but traditional serologic methods have not worked. The answer seems to lie in genetic engineering.

Researchers have identified a critical onchocercal antigen, OV-16, and cloned its DNA. This allows them to mass-produce the antigen and design an immunoassay to detect the low levels of antibody produced during early infection. In other words, the assay can identify people who are infected by the river blindness pathogen but do not yet have visible manifestations of disease. Field trials in Mali, where onchocerciasis is common, show that the test is very specific for *Onchocerca volvulus* and can diagnose infection up to a year earlier than any other method. This raises hopes that screening programs, along with drug treatment, may at last bring this infection under control.

Figure 26.18 This man has been blinded by the filarial parasite *Onchocerca volvulus*. The boy leading him has a greater than 50 percent chance of suffering the same fate—onchocerciasis, or river blindness.

ness, and in some areas of Africa, over half the male population becomes totally blind by age 50 (**Figure 26.18**).

The pathogen is *Onchocerca volvulus*, transmitted by black flies and buffalo gnats of the genus *Simulium*. The insects leave larval worms under the skin when they bite. The larvae grow into mature nematodes that produce millions of motile larvae called **microfilariae** about a year later. The microfilariae migrate through subcutaneous tissues, stimulating an inflammatory reaction that causes extensive damage. Microfilariae in the skin can cause itching, rash, or marked thickening of the inflamed skin. Those that migrate to the eyes and invade the cornea can cause blindness. The infection is perpetuated when a biting black fly ingests larvae and then bites someone else.

Diagnosis is difficult; but, if recognized in time, onchocerciasis is treatable (see the box "Hope at Last"). The adult worms create clearly visible subcutaneous nodules that can be removed surgically, thus preventing production of more microfilariae. Antihelminthic drugs—including suramin, diethylcarbamazine, and mebendazole—will also kill the adult worms. The drug ivermectin, originally used for veterinary infections, also shows great promise because it kills migrating microfilariae.

Loa loa: Loaiasis

Loa loa is another nematode that infects the skin and eyes. The infection—called **loaiasis**—occurs only in Africa. Transmission is by an insect vector, the mango fly. Larvae are deposited under the skin and grow into adult worms that migrate through subcutaneous tissues and begin to produce microfilariae. The adult worms can produce microfilariae for as long as 17 years. Skin and eyes may be irritated or itchy. Sometimes migrating adult worms create painful lumps under the skin called **Calabar swellings**, or worms migrate through the conjunctivae where they are easily visible. But many infections are asymptomatic, and blindness or other serious tissue damage does not occur. The infection can be treated with diethylcarbamazine, and worms visible in the eye can be removed surgically.

Table 26.4 summarizes skin and eye infections.

Summary

THE BODY'S SURFACES (pp. 641–643)

1. The body's surfaces are the skin, the largest organ, and the surface of the eye, or conjunctiva. The cornea is not covered by the conjunctiva. The skin consists of an upper layer (epidermis) and lower layer (dermis). Beneath is the subcutaneous tissue, composed of connective tissue and lymphatic vessels.

2. Skin infections are described by appearance. A lesion (any abnormality) can be described by color (erythema, petechiae, purpura) or texture and content (macules, papules, nodules, vesicles, bullas, pustules). Breaks in the skin include an erosion, ulcer, and crust.

3. Clinical syndromes of the eye include conjunctivitis (pinkeye), keratitis, and keratoconjunctivitis.

SKIN INFECTIONS (pp. 643–660)

Bacterial Causes (pp. 643–651)

1. *Streptococcus pyogenes* (group A streptococcus) causes impetigo, erysipelas, and streptococcal gangrene. Transmission is by close contact or fomites. Virulent group A streptococci have M protein and produce toxins and destructive enzymes. Treatment is with penicillin or erythromycin, but drug-resistant strains are emerging.

2. *Staphylococcus aureus* causes folliculitis, impetigo, boils (carbuncles), and abscesses. Strains that produce exfoliative toxin cause staphylococcal scalded skin syndrome. Staphylococcal cellulitis is a systemic infection that can be life-threatening. Infection usually results from the presence of a foreign body in a wound or a person's being in a weakened condition. Treatment is with methicillin or cephalosporins; however, methicillin-resistant staphylococci have emerged. Pus-containing lesions must be lanced.

3. *Pseudomonas aeruginosa* is an opportunistic pathogen that causes most of the infections fatal to burn victims and 10 to 20 percent of nosocomial infections. Cancer and diabetes patients are at special risk of *Pseudomonas* infections, and children with cystic fibrosis carry a strain that clogs airways. Treatment is with aminoglycoside and a beta-lactam antibiotic, but it is rarely effective because patients are usually already seriously ill.

4. *Clostridium perfringens* causes gas gangrene, characterized by gas bubbles, crepitance, and dead, black skin. Death can occur within days from shock and renal failure. The only effective treatment is amputation. If location makes surgical removal impossible, treatment is with high-dose penicillin, clostridial antitoxin, and a hyperbaric oxygen chamber, but usually the infection is fatal.

5. *Propionibacterium acnes* can cause a minor acne or cystic acne, large tender nodules deep in the skin that leave scars. Acne is an inflammatory disorder brought on by increased production of sebum in adolescence. Commensals break down the sebum, producing fatty acids, which initiate inflammation. Treatment is with tetracycline and a new drug called isotretinoin.

6. *Mycobacterium leprae* causes Hansen's disease, an infection of the skin and peripheral nerves. Transmission is probably by infected nasal secretions and direct skin contact. Incubation is three to five years. The first stage is indeterminate leprosy, characterized by a few depigmented macules. If the immune system mounts an effective response, the person develops tuberculoid leprosy, with minor lesions and often complete loss of sensation in the affected area. If the immune system is overwhelmed, the patient develops lepromatous leprosy, with extensive disfigurement and patchy loss of sensation. Treatment is with dapsone. Hansen's disease is a major public health problem in tropical countries and parts of Asia.

Viral Causes (pp. 651–657)

1. Varicella zoster virus causes an active infection—varicella (chickenpox)—and a latent infection—zoster (shingles). Varicella is a mild disease in children, characterized by epidermal blisters; Reye's syndrome, a life-threatening complication, is preventable by not giving aspirin. Varicella in adults can cause viral pneumonia; in pregnant women, it can cause abortion, birth defects, or a fatally infected

Table 26.4 Infections of the Body's Surfaces: Skin and Eye

Infection	Causative Agent	Mode of Transmission	Symptoms	Prevention and Treatment
SKIN INFECTIONS				
Bacterial				
Impetigo, erysipelas, streptococcal gangrene	Group A streptococci	Close contact, fomites	Range from annoying superficial skin lesions to life-threatening systemic infections	Some strains preventable with good hygiene; all strains treatable with penicillin
Impetigo, boils, abscesses	*Staphylococcus aureus*	Close contact, fomites, foreign body in a wound	Range from superficial lesions to systemic infections	Some preventable with good hygiene; treat with cephalosporins or penicillinase-resistant penicillins
Folliculitis, otitis externa	*Pseudomonas aeruginosa*	Fomites	Range from superficial rash to fatal systemic infections in compromised hosts, especially burn patients	Scrupulous hygiene to prevent nosocomial infections; treatment with combined drugs often ineffective in serious cases
Gas gangrene	*Clostridium perfringens*	Wound contamination	Tissue crepitance and death	Prevent with prompt debridement; treat with penicillin, amputation
Acne	*Propionibacterium acnes*	Acquired as a component of normal flora	Inflamed papules and pustules; painful cysts in severe cases	No effective prevention; treat with topical or oral antibiotics, agents to peel the skin surface, and isotretinoin in severe cases
Hansen's disease (leprosy)	*Mycobacterium leprae*	Direct skin-to-skin contact and nasal secretions	Chronic skin rash and loss of sensation leading to tuberculoid or lepromatous leprosy	Prevent by avoiding contact; treat with combined dapsone, rifampin, and clofazimine
Viral				
Varicella (chickenpox)	Varicella zoster virus	Respiratory	Crops of blisters over entire body; mild in children but risk of Reye's syndrome if child takes aspirin; can be life-threatening in adults	Prevent in high-risk hosts by immunizing with varicella zoster immune globulin (VZIG); acyclovir marginally effective
Zoster (shingles)	Varicella zoster virus	Reactivation of earlier infection	Blisters appear in striped patterns; can be extremely painful but usually not life-threatening	No effective prevention; treat for relief of symptoms only
Fever blisters, gingivostomatitis, whitlow	Herpes simplex type 1 virus (HSV-1)	Direct contact of infectious materials across mucous membranes	Fluid-filled vesicles; recurrent infection at same site	Resolve spontaneously; rarely, acyclovir
Measles	Measles virus	Respiratory, highly infectious	Cough, coryza, conjunctivitis, Koplik spots followed by erythematous maculopapular rash; life-threatening in malnourished children but rarely disabling or fatal in normal children	Effective vaccine since 1963 (MMR—measles, mumps, rubella)

Table 26.4 (continued)

Infection	Causative Agent	Mode of Transmission	Symptoms	Prevention and Treatment
SKIN INFECTIONS				
Viral				
Subacute sclerosing panencephalitis	Measles virus	Late complication of measles	Intellectual deterioration, coma	Extremely rare; no treatment, eventually fatal
Rubella (German measles)	Rubella virus	Respiratory	3-day rash with mild illness, but a powerful teratogen	Effective vaccine since 1969 (MMR—measles, mumps, rubella)
Variola (smallpox)	Variola (smallpox) virus	Respiratory	Variola major produced high fever, severe blistering, scars, and often death; variola minor was the mild form	Worldwide eradication confirmed in 1979
Warts	Human papillomavirus (HPV)	Unknown	Usually a cosmetic problem	No prevention; treat with liquid nitrogen, keratolytic chemicals, surgery
Fungal				
Ringworm (athlete's foot)	Dermatophytes	Person-to-person, animal-to-person, fomites	Highly inflamed lesions in children, mild but often chronic inflammation after puberty	Prevent by avoiding contact; treat with antifungals such as miconazole or, in resistant cases, oral griseofulvin
Candidiasis (thrush, vaginitis, chronic mucocutaneous candidiasis)	*Candida albicans*	Opportunistic	White plaques on mucous membranes; rash and warty skin growths in chronic mucocutaneous type; can be life-threatening in immunocompromised hosts	Treat with topical antifungals containing nystatin or clotrimazole; amphotericin B used for systemic infections but seldom effective
Subcutaneous mycoses (Madura foot)	Various actinomycete bacteria and fungi	Break in the skin	Swelling and disabling or disfiguring but not life-threatening lesions	Some treatable with oral potassium iodide
Arthropod-caused				
Scabies	*Sarcoptes scabiei*	Close contact, fomites	Intense itching and rash	Treat with arachnicide gamma benzene (Kwell or lindane)
Pediculosis (lice)	*Pediculus* spp., *Phthirus pubis*	Close contact, fomites	Itching and skin irritation; body lice can be vectors for typhus	Treat with gamma benzene or pyrethrins
EYE INFECTIONS				
Bacterial				
Inclusion conjunctivitis	*Chlamydia trachomatis*	Usually occurs in infants who acquire the infection at birth	Eyelid swelling, purulent discharge	Preventable by treating mother; treat with oral erythromycin
Trachoma	*Chlamydia trachomatis*	Direct contact with ocular secretions, fomites	Inflamed conjunctivae, scarring and damaged cornea, distorted eyelid, and often blindness	Erythromycin and tetracycline control infection and may prevent blindness before corneal involvement

Table 26.4 (continued)

Infection	Causative Agent	Mode of Transmission	Symptoms	Prevention and Treatment
EYE INFECTIONS				
Bacterial				
Neonatal gonorrheal ophthalmia	*Neisseria gonorrhoeae*	Infant acquires in passing through birth canal	Swelling and purulent discharge	Preventable by treating mother; silver nitrate or antibiotic eyedrop prophylaxis; infected newborns treated with ceftriaxone
Viral				
Viral conjunctivitis (pinkeye)	Many, including adeno-viruses and enteroviruses	Direct contact, highly contagious; often occurs in school epidemics	Redness, discomfort; not serious	Resolves spontaneously without treatment
Herpes keratitis	Herpes simplex virus (HSV)	Begins with introduction of HSV into the eye; may be associated with minor trauma to the eye (scratch, sunburn)	Irritation, tearing, little pain; can threaten sight	Usually self-limited; treat with iododeoxyuridine
Helminthic				
Onchocerciasis (river blindness)	*Onchocerca volvulus*	Black flies and buffalo gnats	Migrating microfilariae cause skin irritation, itching, and thickening; microfilariae that reach cornea cause blindness	If diagnosed early, treatable with various antihelminthics, such as suramin and ivermectin
Loaiasis	*Loa loa*	Mango fly	Painful subcutaneous lumps; worms may be visible in conjunctivae; does not threaten vision	Treat with diethylcarbamazine; remove visible worms surgically

infant. Zoster usually occurs in only one or two dermatomes, but it can produce a painful post-herpetic neuralgia. Immunization with varicella zoster immune globulin (VZIG) within three days of exposure is effective. Treatment with acyclovir on the first day is minimally effective. The FDA may soon approve a varicella zoster vaccine called the Oka strain.

2. Herpes simplex type 1 virus causes fever blisters and gingivostomatitis, which is characterized by fever and painful blisters on the mouth and gums. Gingivostomatitis spreads rapidly among young children. When HSV-1 infects the fingers, the infection is called herpetic whitlow. The active infection can be asymptomatic, so apparently healthy people shed the virus. Acyclovir can probably shorten the duration in adults, but it is not routinely used for children.

3. The measles (rubeola) virus causes measles, a serious and highly communicable disease. The virus is inhaled. Symptoms, including Koplik spots, usually appear on day 11; by day 14 the rash appears and begins to move from the face down the body. Complications include bacterial pneumonia and brain damage. Immunization in the United States is with the MMR (measles, mumps, rubella) vac-

cine at 15 months and at age 5 upon entering school. A new vaccine is needed for children under 1 year who are not protected by the available vaccine. Theoretically, because humans are the only reservoir, it should be possible to eradicate measles.

4. Subacute sclerosing panencephalitis (SSPE) is a rare but usually fatal form of measles. The child appears to recover completely from a normal case of the measles, but measles virus replicates in brain cells. Five or more years later, symptoms of severe brain damage appear. There is no treatment.

5. Rubella virus causes rubella (German measles), which is not related to rubeola. In children, rubella is a mild illness; the rash lasts three days. Infection during the first three months of pregnancy, however, can cause abortion or birth defects. Vaccination is with MMR (measles, mumps, rubella) at 15 months. Rubella serology is routinely performed at the first prenatal exam.

6. Variola (smallpox) virus caused smallpox, which was eradicated in 1979. Variola major killed about half its victims and in survivors caused permanent scarring from deep pustules; variola minor was mild. There was no effective treatment.

7. Certain strains of human papillomavirus (HPV) cause common warts and plantar warts. Mode of transmission is unknown. Common warts can be removed by surgery or keratolytic chemicals, though they usually disappear without treatment after a few years. A rare syndrome of persistent warts is associated with skin cancer.

Fungal Causes   (pp. 657–659)

1. Dermatophytes are fungi that cause tinea (ringworm), an annoying but superficial infection. Dermatophytes digest keratin. Men are more commonly affected than women. Treatment is with topical antifungal agents such as miconazole or clotrimazole, or orally with griseofulvin.

2. *Candida albicans* is an opportunistic pathogen that causes candidiasis. The most common sites of infection are the vagina (*Candida* vaginitis), the mouth (thrush), and the diaper area (*Candida* diaper dermatitis). Treatment is with antifungal creams; thrush is treated with a topically applied liquid. In chronic mucocutaneous candidiasis, the yeast form grows as an intracellular parasite. Systemic candidiasis strikes immunodeficient patients; treatment with amphotericin B is rarely effective.

3. Subcutaneous mycoses involve the skin and underlying subcutaneous tissue. *Sporothrix schenckii* causes sporotrichosis, producing swelling and ulceration. Gardeners are vulnerable; treatment with oral potassium iodide is effective. Chromomycosis is characterized by large wartlike growths; it is most common in tropical countries. *Madurella mycetomi* causes local infections with grotesque swelling and draining pus; the foot is most commonly affected (Madura foot).

Arthropod Causes   (pp. 659–660)

1. *Sarcoptes scabiei* is a mite that causes scabies, characterized by severe itching, especially at night. The disease is a hypersensitivity reaction to the mite and its feces. Secondary infections develop. Epidemics occur at 15-year intervals and last about 15 years. Treatment is with gamma benzene (lindane or Kwell).

2. Three types of human lice cause pediculosis: *Pediculus humanus capitis* (the head louse), *P. humanus corporis* (the body louse), and *Phthirus pubis* (the pubic or crab louse). All types cause itching and can lead to bacterial superinfections. Head lice among children has reached epidemic proportions in the United States. Body lice are the only ones that can transmit microorganisms that cause epidemic disease. Pubic lice are transmitted by sexual contact or sometimes fomites. Lice can be killed by gamma benzene or pyrethrins.

EYE INFECTIONS   (pp. 660–663)

1. *Streptococcus* spp., *Staphylococcus aureus*, and *Haemophilus* spp. commonly cause bacterial conjunctivitis. Symptoms include conjunctival redness and a pus-containing discharge; vision is not threatened. Treatment is with sulfacetamide, neosporin, or gentamicin. Contact lens conjunctivitis is usually caused by *Pseudomonas aeruginosa*; vision can be threatened.

2. Strains of *Chlamydia trachomatis* that infect the genital tract cause inclusion conjunctivitis in newborns. Infection can be prevented by treating the mother during pregnancy. Other strains cause trachoma, a preventable blindness found in Africa, Asia, the Middle East, and among Native Americans in the United States. Chlamydiae multiply in the conjunctivae, causing inflammation, bumps on the surface, scarring, and blindness if the cornea is affected or the eyelids cannot close. Organisms are shed from the eye; transmission is by direct contact or fomites. Erythromycin and tetracycline can control infection but cannot eradicate the organism. High-risk populations can use antibiotic ointments prophylactically.

3. *Neisseria gonorrhoeae* causes neonatal gonorrheal ophthalmia, once a frequent cause of blindness. The infection is prophylactically prevented in industrialized countries by routinely treating newborns with erythromycin or silver nitrate.

4. Viral infections of the conjunctivae are common and rarely serious. Viral conjunctivitis tends to cause more redness and a less purulent discharge than bacterial conjunctivitis. The infection resolves by itself in a few days; eyesight is not threatened.

5. Herpes simplex virus type 1 causes herpetic keratitis, which can be serious because the cornea is involved. Usually a minor trauma produces an infection that resolves spontaneously but recurs; eye drops with iododeoxyuridine can shorten the time of infection. If deeper layers of the cornea are involved, scarring and blindness can occur. If a misdiagnosis leads to prescribing steroid eye drops, the infection can spread rapidly and destroy the cornea. Sight can be restored only by a transplant.

6. *Onchocerca volvulus* is a nematode that causes onchocerciasis (river blindness) in Africa and Latin America. Infection is transmitted by black flies and buffalo gnats that deposit larval worms when they bite. The larvae mature into nematodes, which produce microfilariae that migrate through subcutaneous tissue, causing itching, rash, or thickening of the skin. Microfilariae that reach the cornea cause blindness. If diagnosed in time, treatment with suramin, diethylcarbamazine, or mebendazole will kill the adult worms; ivermectin kills the microfilariae.

7. *Loa loa* is a nematode that causes loaiasis in Africa. Transmission is by the mango fly. Infection can be asymptomatic, or migrating adult worms can cause painful lumps in the skin (Calabar swellings); worms that migrate through the conjunctivae are visible. Treatment is with diethylcarbamazine; worms in the eye can be removed surgically.

Review Questions

THE BODY'S SURFACES

1. Name and describe the layers and sublayers of the skin. What parts of the eye does the conjunctiva cover?

2. How are infections of the skin described? Define these terms:
   a. lesion
   b. erythema
   c. petechiae
   d. purpura
   e. macules
   f. papules
   g. nodules
   h. maculopapular
   i. vesicle
   j. bulla
   k. pustule
   l. erosion
   m. ulcer
   n. crust
   o. exanthem
   p. enanthem

3. Describe these clinical syndromes of the eye: conjunctivitis, keratitis, keratoconjunctivitis.

SKIN INFECTIONS

Bacterial Causes

1. Describe the virulence factors of group A streptococcus. Name and describe streptococcal skin infections, their transmission, prevention, and treatment.

2. How is phage typing useful in identifying nosocomial staphylococcal infections? What are the virulence factors of *Staphylococcus aureus*? Name and describe staphylococcal skin infections, their transmission, prevention, and treatment.

3. Why are *Pseudomonas* infections difficult to control, especially in a hospital? Which groups are at highest risk of *Pseudomonas* infection? Describe the treatment.

4. How does *Clostridium perfringens* cause infection? Describe the symptoms and treatment of gas gangrene.

5. Explain this statement: Acne is not an infection. What dangers are involved with using the new drug isotretinoin?

6. Describe *Mycobacterium leprae*. Describe the development of Hansen's disease, its treatment, and its epidemiology in the United States.

Viral Causes

1. Compare and contrast the seriousness of varicella infection in children, adults, and pregnant women. Describe the pathogenesis of shingles. Discuss the prevention and treatment of chickenpox, including the risks and benefits of the new Oka strain vaccine.

2. Describe the infections caused by herpes simplex type 1. How is HSV-1 transmitted?

3. Describe the transmission, development, basis for recovery, and potential complications of rubeola (measles). Why is measles a childhood killer in developing countries and what are some solutions? How is measles controlled in the United States? Why is measles eradicable?

4. Describe the pathogenesis of subacute sclerosing panencephalitis (SSPE).

5. Compare and contrast rubella (German measles) and rubeola (true measles). What can be done to prevent congenital rubella?

6. What significant difference between the epidemiology of smallpox and measles made smallpox easier to eradicate? What additional obstacles are there to eradicating measles?

7. Describe the infections caused by human papillomavirus and their treatment. Why is there new research interest in HPV?

Fungal Causes

1. What are dermatophytes and what infections do they cause? Explain this statement and give some examples: The body's immune response influences both the symptoms of dermatophyte infections and the chances for complete recovery.

2. Name and describe the major skin infections caused by *Candida albicans*. What is the treatment? Who is most at risk for chronic mucocutaneous candidiasis? for systemic candidiasis?

3. Describe these subcutaneous mycoses and tell how they are treated: sporotrichosis, chromomycosis, mycetoma.

Arthropod Causes

1. Name the microorganism that causes scabies and describe the pathogenesis. Explain how it damages the skin.

2. What are the most common forms of pediculosis and how is each transmitted and treated?

EYE INFECTIONS

1. What are the most common causes of bacterial conjunctivitis? What is the treatment? How does *Pseudomonas aeruginosa* cause eye infections and how serious are they?

2. Describe the cause, prevention, and treatment of inclusion conjunctivitis. Discuss the transmission, pathogenesis, prevention, and treatment of trachoma. In what parts of the world is trachoma most prevalent?

3. Describe the cause, prevention, and treatment of neonatal gonorrheal ophthalmia.

4. Discuss the transmission, pathogenesis, clinical syndrome, and treatment of onchocerciasis.

5. What are the symptoms of loaiasis? What kind of organism causes this infection? What is the treatment?

Suggested Readings

Advisory Committee. 1989. Measles prevention: Recommendations of the immunization practices. *Morbidity and Mortality Weekly Report* 38:S–9.

Bhbehani, A. M. 1991. The smallpox story: Historical perspective. *American Society of Microbiology News* 57(11):571–76.

Brumfitt, W., and Hamilton-Miller, J. 1989. Methicillin-resistant *Staphylococcus aureus. New England Journal of Medicine* 320:1188–96.

Chickenpox: Reexamining our options. 1991. *New England Journal of Medicine* 325:1577–79.

Howley, P. M. 1986. On human papillomaviruses. *New England Journal of Medicine* 315:1089–90.

Lobess, E.; Weiss, N.; Karanis, M.; Taylor, H.; Ottesen, E.; and Nutman, T. 1991. The immunogenic *Onchocerca volvulus* antigen: A specific and early marker of infection. *Science* 251:1603–5.

Shepard, C. C. 1982. Leprosy today. *New England Journal of Medicine* 397:1640–41.

Weller, T. H. 1983. Varicella and herpes zoster: Changing concepts of the natural history, control and importance of a not-so-benign virus. *New England Journal of Medicine* 309:1362–68; 1434–40.

INFECTIONS OF THE CARDIOVASCULAR AND LYMPHATIC SYSTEMS

Case History: Risky Business

C. D. was a 24-year-old woman who had been using heroin intravenously for several years. She felt well until about three weeks before her admission to the hospital. First, she noticed she was running a fever but decided to ignore it. She became increasingly weak, tired, and pale. Then she noticed red spots, some slightly raised, on her palms and soles. Numerous splinterlike discolorations appeared under her fingernails. She began to cough, and breathing became difficult. When she became so weak she could hardly walk, her friends insisted she go to the emergency room.

The attending physician was concerned when he learned of C. D.'s intravenous drug abuse and continuing fever. In taking a history, he questioned her carefully about a history of heart disease but discovered nothing. She had a temperature of 102°F and was very pale (laboratory tests later confirmed anemia). The discolorations on her palms, on her soles, and under her fingernails suggested that *emboli* (tiny clots) had entered her bloodstream and were clogging small vessels. Most worrisome, she had a significant heart murmur, suggesting damaged heart valves, and she showed signs of heart failure. C. D. was admitted to the intensive care unit with a diagnosis of probable bacterial endocarditis.

Her physician drew several blood samples for bacterial culture and ordered an echocardiogram (a soundwave image of the heart). The echocardiogram showed *vegetations* (abnormal growths) on several severely damaged heart valves. Though the culture results would not be available for several days, the clinical presentation so

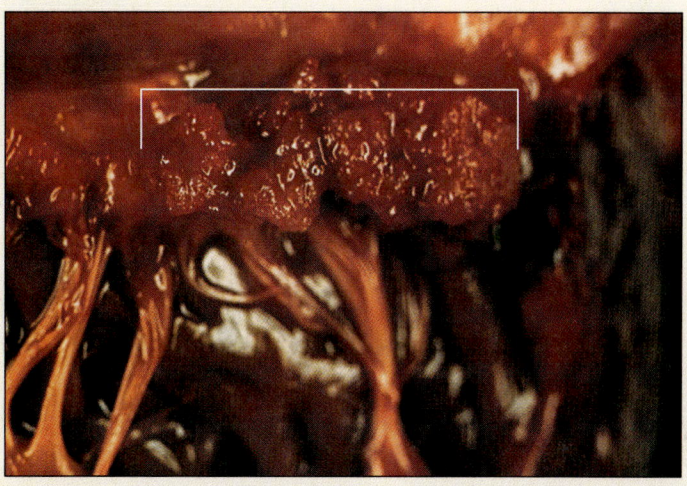

A postmortem photo of heart valves from a patient with endocarditis shows a large vegetation adhering to a valve (white bracket).

strongly suggested endocarditis that C. D. was immediately treated with a combination of drugs, including methicillin. The tentative diagnosis of bacterial endocarditis was confirmed later when *Staphylococcus aureus* was isolated from C. D.'s blood sample.

C. D.'s physician hoped to control her infection with antibiotics and stabilize her condition. She would need surgery to replace her damaged heart valves. But C. D. died three days later.

To understand:

- The anatomy and function of the cardiovascular and lymphatic systems and their defenses against microorganisms

- The clinical syndromes that characterize cardiovascular and lymphatic infections

- The principal microbial infections of the heart; their diagnosis, prevention, and treatment

- The bacterial, viral, protozoal, and helminthic causes of systemic cardiovascular and lymphatic infections; their diagnosis, prevention, and treatment

- What is known about the human immunodeficiency virus (HIV), the clinical syndromes of AIDS, its transmission, prevention, and treatment; how the epidemiology of AIDS is changing and what future prospects might be

THE CARDIOVASCULAR AND LYMPHATIC SYSTEMS

The cardiovascular and lymphatic systems transport fluids through the body. These two systems unite the organs and tissues of the body—delivering, exchanging, and removing cells and molecules.

Structure and Function

The cardiovascular system moves blood from the heart to the tissues and back again. The lymphatic system has two functions. It recovers fluid lost from the blood and returns it through lymphatic vessels to the bloodstream, and it helps fight microbial infection (**Figure 27.1**).

The Cardiovascular System. The **cardiovascular system** consists of the heart, blood vessels, and blood. The **heart** is a four-chambered muscle-driven pump that propels blood through the blood vessels. The upper chamber on the right and left sides of the heart is an expandable thin-walled **atrium**, which receives blood returning to the heart from the tissues; the lower chamber is the powerful thick-walled **ventricle**, which ejects blood into outflow vessels. **Valves** at the outlet of each chamber prevent the blood from flowing backward when the heart relaxes.

The right side of the heart sends blood at low pressure to the lungs, where oxygen is absorbed and carbon dioxide (CO_2) is removed (**pulmonary circulation**). The left side of the heart sends blood at much higher pressure to all other tissues of the body, delivering oxygen and nutri-

ents and removing carbon dioxide and metabolic wastes (**systemic circulation**).

Oxygen-rich blood leaves the left ventricle to enter systemic circulation through the **aorta**, the largest of the thick-walled **arteries** that carry blood from the heart to the tissues. As blood travels farther from the heart, arteries branch and narrow, finally becoming so small that only one blood cell at a time can pass through. These microscopic blood vessels, the **capillaries**, bathe all the tissues of the body. Their thin, highly permeable walls allow nutrients and oxygen to enter the tissues and carbon dioxide and other metabolic wastes to pass into the capillaries. Capillaries carrying waste products branch into larger and larger **veins**, which carry oxygen-poor, CO_2-enriched blood back to the heart, where it enters pulmonary circulation and the cycle begins again. In its transit through the body the blood passes through the kidneys, where metabolic wastes are removed, and through networks of capillaries in the walls of the gastrointestinal system, where nutrients are absorbed.

Blood is a complex tissue composed of erythrocytes, leukocytes, platelets, and proteins (such as complement). It also transports molecules, such as hormones and nutrients. See the box "Blood: A Complex Body Tissue" in Chapter 16.

The Lymphatic System. The lymphatic system, also called the immune system, was described in detail in Chapter 17. The location and function of lymph vessels, nodes, organs, and tissues are shown in Figure 17.3 and Table 17.2.

Defenses and Normal Flora: A Brief Review

The circulatory and lymphatic systems do not support a normal flora of microorganisms but are occasionally contaminated by pathogens. One function of lymphatic circulation is to pick up microbes that have invaded tissue and deliver them to a lymph node where they can be phagocytized.

Even microbes that enter the blood are usually eliminated before serious infection begins. **Transient bacteremias** are brief, asymptomatic periods in which bacteria are recovered from the blood. The sustained presence of microorganisms in either the cardiovascular or lymphatic system, however, indicates a true infection.

Clinical Syndromes

Infections of the heart are easily categorized according to anatomic syndromes (**Table 27.1**). **Endocarditis** is infection of the heart's inner lining, the **endocardium**. **Myocarditis** is infection of the heart muscle itself, the **myocardium**. **Pericarditis** is infection of the heart's surrounding membrane, the **pericardium**. Endocarditis is

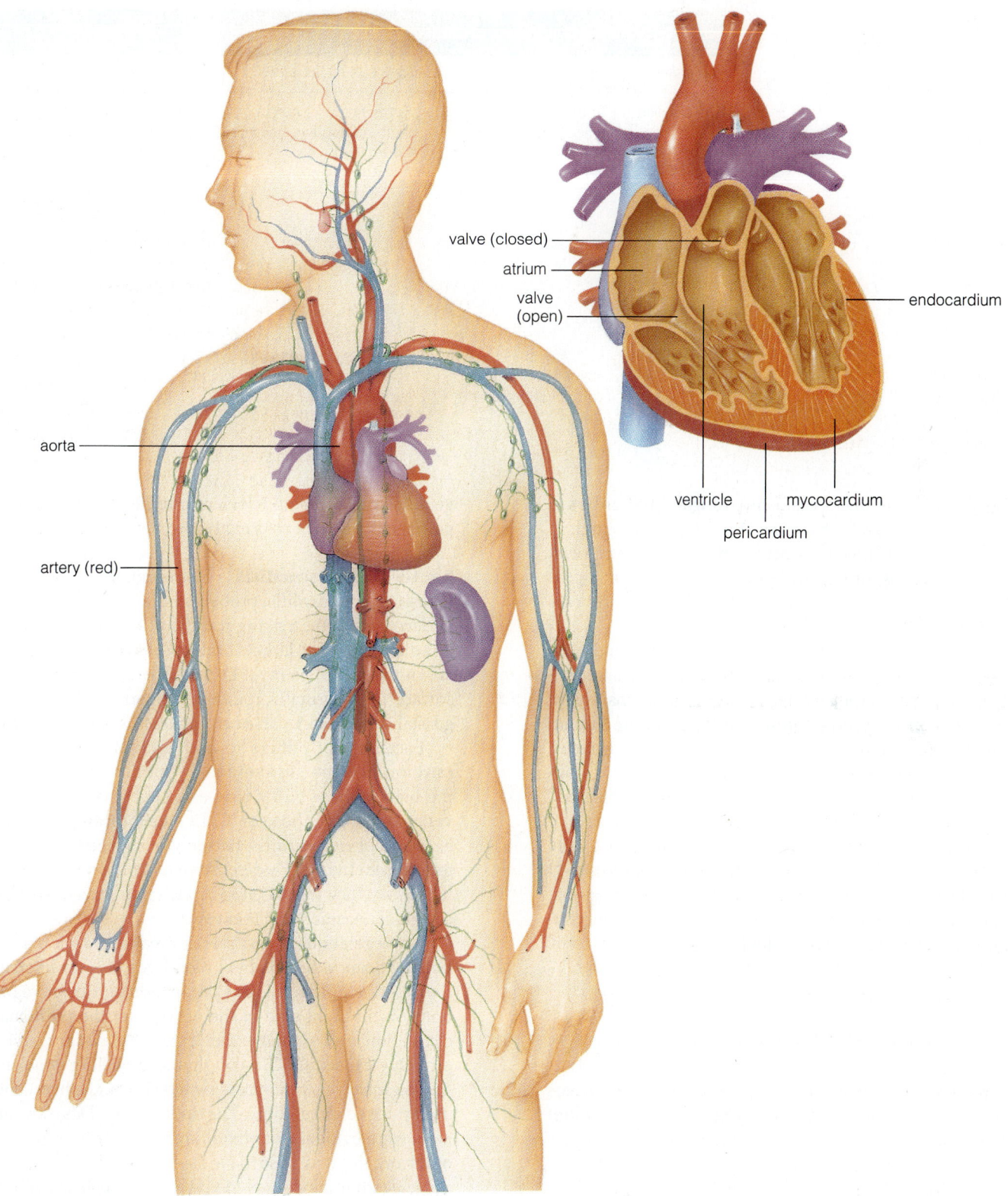

Figure 27.1 Organs of the cardiovascular and lymphatic systems.

Table 27.1 Clinical Syndromes of Cardiac Infection

Syndrome	Region Affected	Signs and Symptoms	Causative Agent
Endocarditis	Endocardium—the smooth inner lining of the heart, including valves	Fever, fatigue, anemia, heart murmurs or heart failure from valve damage	Acute bacterial infections, usually from highly virulent pathogens; subacute bacterial or fungal infections from less virulent or opportunistic pathogens
Myocarditis	Myocardium—the heart muscle	Irregular heartbeat, chest pain, heart failure	Usually viral, particularly coxsackie virus, but may be fungal, bacterial (e.g., diphtheria toxin myocarditis), or protozoal (Chagas' disease)
Pericarditis	Pericardium—the smooth membrane surrounding the heart	Few; in severe cases, scarring or pericardial tamponade may interfere with heart function	Usually viral but may be bacterial or fungal

usually bacterial, but occasionally fungal. Myocarditis is usually viral (rarely bacterial or fungal), although a unique form of infectious myocarditis called American trypanosomiasis, or Chagas' disease, is caused by a protozoan. Pericarditis is usually viral, but occasionally bacterial or fungal.

Noncardiac infections of the cardiovascular and lymphatic systems, however, are difficult to classify anatomically. Because blood and lymphatic circulations are in contact with every organ in the body, the entire human being is affected by infections involving these systems. These bodywide **systemic infections** are difficult to diagnose. For example, a patient who reports "I can't breathe" probably has a respiratory infection, and one who says "I have diarrhea" probably has a gastrointestinal infection. But a patient with a systemic infection is likely to say "I don't feel well," giving the clinician much less to go on. However, every systemic infection has some specific clinical manifestations, which, along with laboratory tests, make it unique and ultimately identifiable.

INFECTIONS OF THE HEART

Infections of the heart are both uncommon and serious. Any disease that affects this vital organ is potentially life-threatening.

Endocarditis

Most endocarditis is caused by bacteria, although fungi, including *Candida albicans*, may be responsible. The **bacterial endocarditis** that killed C. D. is one of many serious infections that can occur as a complication of intra-venous drug abuse, in which nonsterile needles introduce pathogenic organisms directly into the bloodstream.

Bacterial Endocarditis. Microorganisms seldom establish an endocarditis, probably because the endocardial surface is smooth, making it difficult for microorganisms to adhere. Only if large numbers of virulent pathogens are in the blood, as in C. D.'s case, or if the endocardial surface is abnormally roughened, perhaps from a congenital heart defect, can microorganisms get a foothold.

Bacteria that adhere to and multiply on the endocardium become surrounded by clotlike material from the blood. These bulky masses of bacteria and clots, or **vegetations**, like those on C. D.'s **echocardiogram**, are a hallmark of endocarditis. Bacteria in a vegetation are protected from the body's defenses and from bloodborne antibiotics, which means endocarditis requires a prolonged course of antimicrobial therapy.

As vegetations grow, pieces break off, forming **emboli** that can lodge in small blood vessels and interrupt the blood supply to distant organs. Emboli from vegetations on the right side of the heart enter the pulmonary circulation and damage the lungs. Pulmonary emboli probably caused C. D.'s cough and breathing problems. Emboli from vegetations on the left side of the heart enter the systemic circulation. The spots on C. D.'s skin and nails were evidence of emboli in her surface tissues, but other organs may also have been affected. Emboli can cause death if they travel to vital organs such as the brain or kidneys.

Vegetations also damage the endocardium itself. Heart valves—which are composed of endocardium—are particularly vulnerable. If valves do not function properly, the heart cannot maintain adequate blood pressure to supply vital organs. This condition, called **heart failure** (which can result from any significant heart disease), was the immediate cause of C. D.'s death.

Bacterial (or infective) endocarditis can be difficult to diagnose because clinical signs and symptoms depend upon which heart valves are affected and where emboli lodge. Fever and significant anemia occur in almost all patients. Definitive diagnosis depends on isolating the infecting microorganism from the patient's blood.

Acute and Subacute Endocarditis. Bacterial endocarditis is divided into two broad clinical syndromes, acute and subacute.

Acute bacterial endocarditis is sudden in onset and progresses rapidly. C. D.'s illness was typical. Her heart was normal until large numbers of virulent staphylococci (probably contaminants from her skin) were introduced into her bloodstream during an intravenous injection of drugs. *Staphylococcus aureus* causes more than half the cases of bacterial endocarditis in intravenous drug addicts. Non–drug users are at risk for bacterial endocarditis only if significant numbers of a virulent pathogen enter the bloodstream—in a serious postoperative infection, for example.

Acute bacterial endocarditis is usually treated with cephalosporins and penicillinase-resistant penicillins, followed by surgically replacing damaged heart valves. But extremely ill patients like C. D. may die a few weeks after symptoms appear.

Subacute bacterial endocarditis usually occurs in people who have an abnormal heart before the infection begins. Even minor abnormalities allow bacteria to adhere. Subacute infection starts more gradually, progresses more slowly, and lasts longer than acute disease. Patients may complain of fever and fatigue for months, and multiple blood cultures are sometimes necessary to isolate the infecting organism. It is usually a weakly pathogenic bacterium of the respiratory or gastrointestinal flora, such as viridans group *Streptococcus* spp. Small numbers of these bacteria enter the blood during routine activities of daily living. For example, about one-third of people who have a tooth extracted have detectable viridans streptococci in their bloodstream afterward. Normally these bacteria are rapidly cleared from the circulation, but people with heart defects are at risk of developing bacterial endocarditis and are therefore advised to take penicillin before dental or other minor surgery. Subacute bacterial endocarditis is usually curable with early diagnosis. Without treatment, it is invariably fatal.

Myocarditis

Myocarditis in the United States is usually viral, but in South America it is usually protozoal. Fungi and bacteria occasionally cause myocarditis. Diphtheria toxin, for example, is highly poisonous to the heart, and untreated diphtheria patients often die of myocardial failure.

Viral Myocarditis. The specific virus that causes a case of myocarditis is not always identified; it is often coxsackie virus, although many other viruses cause an identical clinical syndrome. Patients with viral myocarditis generally have mild symptoms, experiencing only a rapid or slightly irregular heartbeat. Chest pain or signs of heart failure develop rarely. The only treatment is rest, and most patients do well, although viral myocarditis can be life-threatening (probably accounting for the sudden deaths that can occur with minor viral infections).

***Trypanosoma cruzi:* American Trypanosomiasis (Chagas' Disease).** The flagellated protozoan *Trypanosoma cruzi* causes an infection called **American trypanosomiasis,** or **Chagas' disease** after Carlos Chagas, who first described it in Brazil in 1909. *T. cruzi* exists in a motile form in blood and a nonmotile form in skeletal muscles, the central nervous system, and the heart (**Figure 27.2**).

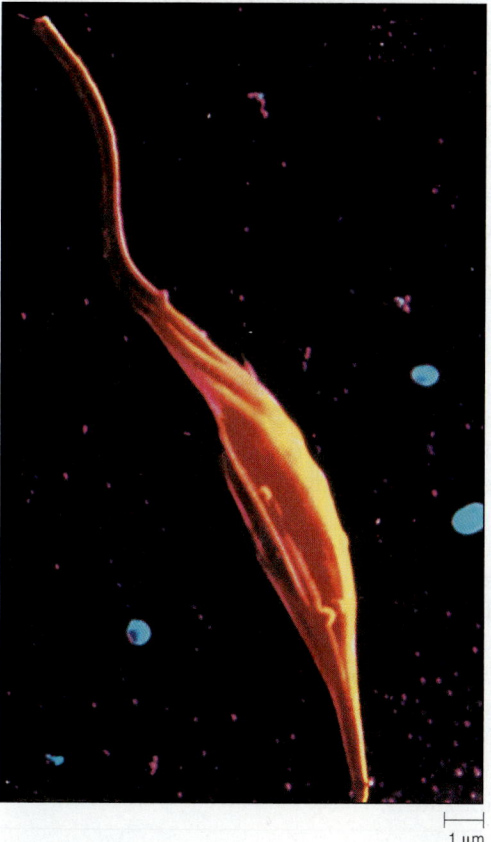

Figure 27.2 Scanning electron micrograph of the flagellated, bloodborne form of *Trypanosoma cruzi* that causes American trypanosomiasis.

This protozoan has a complicated life cycle requiring two different hosts—a mammal and a blood-sucking insect. Over 100 different mammals, including human beings and domesticated animals such as dogs and cats, can be infected by *T. cruzi*. The most common insect hosts belong to the genus *Triatoma*. They are commonly called **reduviid bugs**, or **kissing bugs**, because of their tendency to bite humans on the face. Trypanosomes pass from an infected mammal to an insect during a blood meal. A new host is infected when a trypanosome-carrying insect bites and at the same time defecates on the victim's skin. The insect feces carry the infectious protozoa, which enter the bite when the person rubs or scratches it. Infection can also enter through the eye.

Trypanosoma cruzi exists in many parts of the Western Hemisphere, including the southern United States. But almost all cases of Chagas' disease occur in rural Latin America where people live in wood and mud houses infested by reduviid bugs. In impoverished, highly endemic areas of South America, nearly half the population is infected by *T. cruzi*, and Chagas' heart disease is the most common cause of death among young adults.

Humans with Chagas' disease suffer both an acute and a chronic illness. The acute illness appears one to two weeks after the infecting insect bite, as protozoa enter, multiply within, and kill brain cells, liver cells, and myocardial cells. Clinical manifestations include fever, fatigue, a tender swelling at the site of the bite, and a characteristic swelling of the face and eyelids. Children are the most vulnerable to acute infection, and as many as 10 percent die.

The early stage of infection can be successfully treated with antiparasitic medications. The drug of choice is nifurtimox. Victims who survive the acute stage without treatment usually suffer chronic disease when parasites enter the lymphatic system and infection spreads. Many organs can be damaged, including the esophagus (causing difficulty swallowing) and the brain (causing paralysis), but myocardial damage most often kills young adults. The only treatment is medication to sustain the failing heart.

In its early stages, when protozoa are fairly numerous in the blood, Chagas' disease can often be diagnosed by blood smears. During later stages of infection, when most parasites are inside cells, diagnosis is more difficult. No serologic test is completely satisfactory. Sometimes a suspected patient is allowed to be bitten by an uninfected reduviid bug, and the insect is examined several weeks later to see if it has acquired the infection. This is called **xenodiagnosis**.

Pericarditis

Pericarditis is usually viral but can be bacterial or protozoal. Most viral pericarditis is a complication of an every-day viral infection. Enteroviruses are the most common agents, though coxsackie viruses and echoviruses probably also cause infection. Like viral myocarditis, viral pericarditis is usually benign and self-limited.

Bacterial pericarditis is much more serious. It is usually caused by *Staphylococcus* or *Streptococcus* spp. spreading from a lung infection or through the bloodstream. Pus can fill the space between the pericardium and the myocardium, putting such pressure on the heart that it cannot beat normally. This physical compression of the heart, called **pericardial tamponade**, can cause sudden death.

Chronic fungal or mycobacterial infections can also lead to pericarditis. Immunosuppressed patients suffering from opportunistic fungal infections and chronic tuberculosis patients are especially vulnerable. Chronic pericardial inflammation produces scarring, which eventually causes the heart to constrict until it fails. In such cases a **pericardectomy**, surgical removal of the pericardium, may be lifesaving.

SYSTEMIC INFECTIONS OF THE CARDIOVASCULAR AND LYMPHATIC SYSTEMS

The microorganisms discussed in this section either enter the body through the blood or lymphatic circulations, or these circulations help them spread. Keep in mind, however, that *any* bacterial infection—no matter where it begins—can eventually spread to the blood if it is not controlled.

If the bacterial invasion of the bloodstream is relatively brief and harmless, it is called a **bacteremia**. If it is persistent and serious, it is called a **septicemia**. In most cases septicemia is fatal, even with the best medical treatment. Immunocompromised hospital patients with nosocomial infections make up the majority of septicemia cases today.

BACTERIAL CAUSES

Bacterially caused systemic infections include one of the oldest and most dreaded human diseases—the plague—and one of the most recently discovered—Lyme disease.

Yersinia pestis: **Plague**

We tend to associate the plague with Europe in the Middle Ages, but this infection is confined to neither Europe nor the past (Chapter 1). Six million people died in India during an outbreak that began in 1891, and a pandemic was narrowly averted in the early 1900s (see the box "Steamships and the Plague"). *Yersinia pestis* is permanently established in North America west of the Rocky

Turning Point

Steamships and the Plague

The invention of the steamship put millions of Americans at risk for plague, as the historian William H. McNeill explains in his book *Plagues and Peoples*. The caravans and sailing ships that brought exotic goods from Asia to Europe during the Middle Ages also brought rats infected with the plague bacillus. But infected rats, which die quickly from the plague, could not survive the long transoceanic sail from Europe to the New World, so the Americas were safe from plague for centuries. By 1894, when plague broke out in Hong Kong, however, the steamship had dramatically shortened the trip to the Americas. Within a decade, plague appeared in San Francisco and Buenos Aires.

Fortunately, the potential pandemic occurred during the golden age of microbiology (Chapter 1), so scientists were one step ahead of disaster. They meticulously charted the progress of the epidemic and imposed seaport quarantines to prevent infection from spreading to the general population. Americans were saved from the Black Death, but ground-burrowing rodents in both continents became infected, establishing the reservoir that continues to cause occasional cases of human plague in the Americas today.

Mountains. Millions of wild rodents—including ground squirrels, prairie dogs, and chipmunks—form a huge plague reservoir that will never disappear. Fortunately, humans have little contact with these rodents and their fleas, so transmission is relatively rare. In 1993, for example, only 10 cases of plague were reported in the United States.

Y. pestis, a Gram-negative coccobacillus, is transmitted from one mammalian host to another by infected fleas. Infection blocks the flea's digestive tract, causing it to regurgitate infected material into a bite wound during its blood meal. The intestinal blockage also makes it impossible for the flea to take in a complete meal, so it bites more and more frequently and aggressively. Communities of wild rodents can be chronically infected but highly resistant to plague, so few animals die. This rural plague cycle is usually fairly stable and seldom involves human beings. But if hunters or campers carry infected fleas into human communities, an urban plague can begin. In cities, infection spreads first among rats, which have little natural resistance to the plague and usually die of the infection. When rat hosts become scarce and infected fleas begin to bite human beings, an epidemic like that in medieval Europe is underway.

Y. pestis is one of the most virulent bacterial pathogens known, and the disease is gruesome (**Figure 27.3**). Infection usually begins when an infected flea (often *Xenopsylla cheopis*, the oriental rat flea) bites a person and introduces bacteria into subcutaneous tissues where they enter the lymphatic system and become concentrated in a nearby lymph node. Two to six days after the infecting bite, the person develops a high fever and the affected lymph node becomes massively enlarged and tender. The diseased node is called a **bubo** and the fleaborne infection, **bubonic plague**. Bacteria are phagocytized within the node but continue to multiply intracellularly.

In almost all untreated cases, the infection is not contained by the immune system and organisms enter the blood—a stage called **septicemic plague**. Once bloodborne, bacteria spread to almost every organ of the body, initiating an overwhelming infection that is fatal in about 75 percent of cases. Black spots appear on the skin (the origin of the term *Black Death*), caused by subcutaneous bleeding from destruction of blood vessels. Internal organs are probably similarly affected.

If the lungs are infected by plague, bacterial cells are aerosolized when the infected person coughs. A new host can inhale the organisms; this form of infection, called **pneumonic plague**, is even more virulent than bubonic plague. Symptoms begin within two to three days and mortality is nearly 100 percent.

Today, because plague is so rare, it can be difficult to recognize. Definitive diagnosis is usually established by culturing the bacterium from the infected bubo or from the blood. Serologic tests that demonstrate a rise in antibody titer against *Y. pestis* confirm the plague. Treatment with streptomycin, tetracycline, or chloramphenicol is usually effective but must begin early. Because of delays in diagnosis, bubonic plague still kills about 20 percent of its victims. The most effective way to prevent plague is to eliminate rats from human housing. A vaccine is also available for high-risk individuals, such as biologists or geologists who work in plague-infested areas.

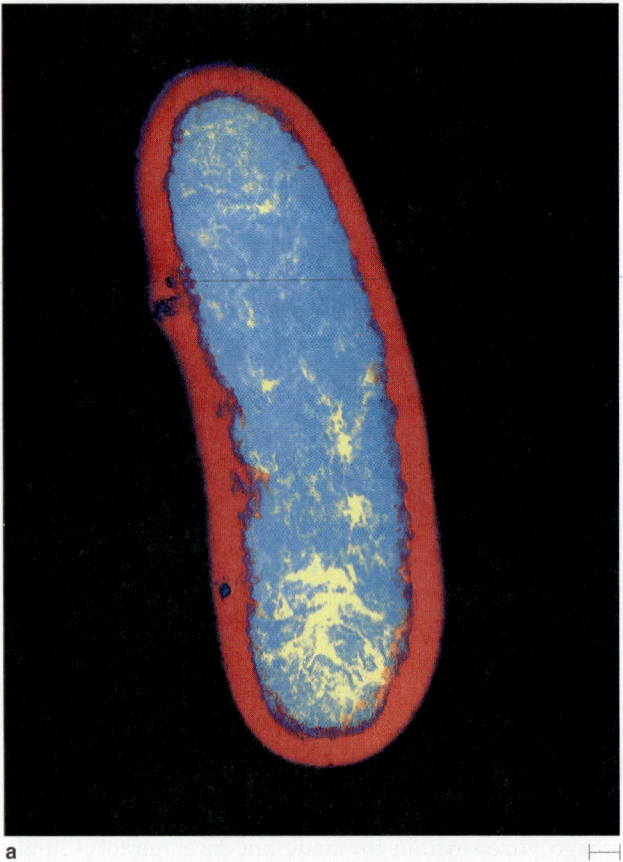

100 nm

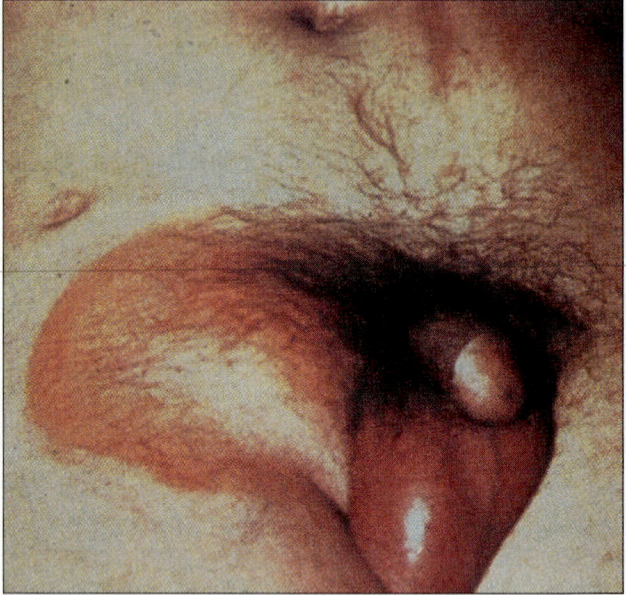

b

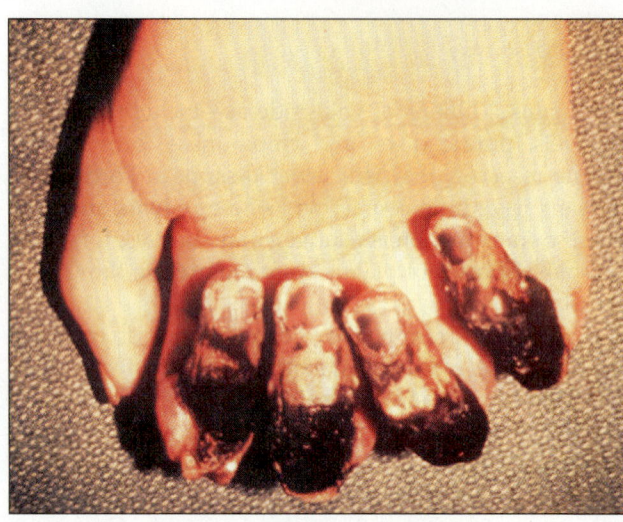

c

Figure 27.3 Pneumonic, bubonic, and septicemic plague. (**a**) A postmortem section of lung from a patient who died from pneumonic plague shows the causative microorganism, *Yersinia pestis* (the very small blue dots). (**b**) An enlarged lymph node, or bubo, in a patient who probably received his plague-infecting bite on the leg. (**c**) A patient with septicemic plague who is likely to die from the infection. Subcutaneous bleeding causes blackened spots all over the body—the origin of the term *Black Death*.

Francisella tularensis: Tularemia

Francisella tularensis, a small Gram-negative coccobacillus, was first isolated in 1911 from ground squirrels in Tulare County, California, that appeared to have died of plague. Today we recognize *F. tularensis* as the causative agent of a worldwide disease of wild mammals that occasionally affects human beings. **Tularemia** commonly infects squirrels, muskrats, and deer. Rabbits are a major reservoir. As many as 1 percent of the wild rabbits in North America harbor *F. tularensis*. Ticks, another significant reservoir, can transmit the bacterium **transovarially**

(through the eggs) from one generation to the next without an intermediate mammalian host.

Tularemia is transmitted from infected animals to humans in several ways. Bacteria can enter the body through a break in the skin or across the conjunctival surface of the eye—when a hunter skins or cleans a diseased animal, for example. Bacteria can be ingested in infected meat and can also be inhaled, a risk in clinical laboratories where the organism is being cultured. Finally, bacteria can be transmitted by vector, a tick or fly that has recently fed on an infected animal.

The most common clinical syndrome in humans is **ulceroglandular tularemia**. A small sore or ulcer appears where the bacteria enter the body, and the surrounding lymph nodes become swollen and tender. Although bacteria are phagocytized in the regional lymph nodes, they continue to multiply intracellularly and are not eliminated. Systemic symptoms, including high fever, shaking chills, and debilitating headaches, follow. Severely ill victims may develop pneumonia. Only about 10 percent of untreated patients die, but the illness can be serious and persistent. Fever often lasts for more than a month, and relapses usually occur. Less common clinical syndromes of tularemia include **glandular** disease (lymph nodes are swollen but no ulcer is visible), **typhoidal** disease (only systemic symptoms are evident), and **occuloglandular** disease (infection begins in the eye).

Diagnosing tularemia is difficult because the primary symptoms—fever and headache—are common and nonspecific. If the patient does not provide a clue when giving a history—such as hunting or a recent tick bite—the clinician has little basis for making a preliminary diagnosis of tularemia. Diagnosis cannot be made by culture because most standard media will not support the growth of *F. tularensis*. Furthermore, culturing is hazardous because of the risk of airborne infections to laboratory workers; as a result, definitive diagnosis is usually made through serology. Antibody levels against *F. tularensis* during early stages of the illness (when levels are low) are compared with antibody levels during later stages of the illness (when levels are high). Serologic diagnosis requires two to four weeks, so early diagnosis and treatment of tularemia depend on an astute clinician. Treatment is effective if begun early. Streptomycin is considered best, but tetracycline is also effective.

Brucella spp.: Brucellosis

The bacterial genus *Brucella* is named after Sir David Bruce, a British army physician who first isolated these organisms in 1887 from soldiers with an illness then known as **Malta fever**. *Brucella* are small Gram-negative coccobacilli that cause disease in domestic animals. Different *Brucella* species, which are distinguished by their surface antigens, prefer different animal hosts. The most common species that cause human disease are *B. abortus* (which preferentially infects cattle), *B. suis* (swine), *B. melitensis* (sheep and goats), and *B. canis* (dogs). *B. abortus* and *B. canis* tend to cause mild disease in humans, while infection by *B. suis* or *B. melitensis* can be fatal.

Brucellosis is transmitted to human beings by direct contact with infected animals or their secretions. People who handle diseased animals can be infected by a break in the skin, across the conjunctiva, or by inhalation.

Drinking infected milk is a particularly likely mode of transmission because *Brucella* are concentrated in the mammary glands of infected animals. Pasteurization kills *Brucella*, so milkborne infection (and brucellosis in general) is much less common in the United States than in developing countries. While over half a million cases of brucellosis are documented worldwide annually, fewer than 200 occur annually in the United States. The U.S. cases occur mainly in meat packers, farmers, and veterinarians. Texas, California, and Florida report the highest number of cases.

Like *Francisella tularensis*, *Brucella* spp. are able to establish infection because they survive as intracellular parasites. Bacteria pass from lymph to nodes to the blood and from there to organs throughout the body, including the liver, spleen, and bone marrow.

As with many other systemic infections, clinical diagnosis is difficult because the typical symptoms—weakness, fatigue, and loss of appetite—are vague and common complaints. However, at some time in their illness, brucellosis patients have a fever; and a pattern of rising fever late in the day can be diagnostic (brucellosis is also called **undulant fever**). Most patients recover without treatment within weeks, but a few develop a chronic illness that can last for months. Chronic illness can cause continuing fever and fatigue or can become localized in an organ, often the bones, endocardium, lungs, or brain.

Brucella species are a challenge to isolate in the laboratory. First, they require a complex culture medium and must be incubated in an atmosphere with an elevated concentration of CO_2. Second, laboratory technicians must take special precautions to avoid becoming infected themselves with these highly infectious organisms. Nevertheless, culturing the bacterium from the blood or tissues is the most reliable way to confirm a diagnosis of brucellosis. Documenting a rising titer of antibodies against *Brucella* can also be diagnostic. Treatment is with tetracycline and streptomycin over a course of several weeks.

Borrelia burgdorferi: Lyme Disease

In the nineteenth century, Pasteur, Koch, and other pioneering microbiologists identified almost all the bacterial pathogens known today. But the story of **Lyme disease** and *Borrelia burgdorferi*, the causative microorganism, has unfolded only during the last 20 years.

The first clues to the existence of Lyme disease came in 1975 when a woman in Old Lyme, Connecticut, notified the state health department that there were an unusual number of cases of childhood arthritis in Old Lyme. At about the same time, another woman notified the Yale medical school of an "epidemic" of arthritis in her family. The ensuing investigation eventually led to the clinical

Good Intentions

In 1986 a 67-year-old woman died in a New Jersey hospital. The unanswered questions in her case bore investigation, and eventually the following story came to light. A young friend of the woman had gone rabbit hunting. He killed two rabbits (one of which was losing its fur), cleaned them, and gave them to the victim and her husband, who skinned and froze them. A few days after dressing the rabbits the young man became ill with an ulcer on his hand, swollen nodes under his arm, and fever. He was examined at a local hospital but the diagnosis of tularemia was missed. He was treated only with medication to control his fever.

The elderly neighbor had also developed a sore on her hand, but her illness was not diagnosed until two weeks later when she and her husband were admitted to the hospital with persistent hand ulcers and a severe systemic illness. Within a few days of her admission she was put on streptomycin, but she was already too ill to recover. The young

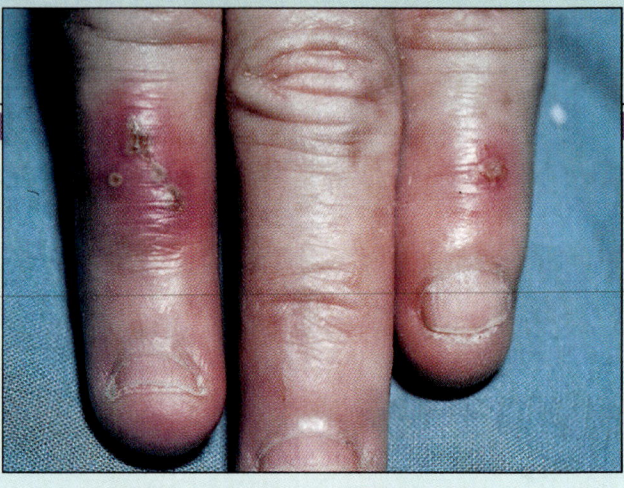

A small sore, or ulcer, appears where *Francisella tularensis* enters the body.

hunter was subsequently treated with streptomycin and recovered rapidly. All three patients showed rises in *Francisella tularensis* antibody titers, and the Centers for Disease Control cultured the organism from the bone marrow of the frozen rabbits.

---

description of Lyme disease, evidence that it was transmitted by ticks, and identification of the spirochete *B. burgdorferi* by Montana bacteriologist Willy Burgdorfer in 1982 (**Figure 27.4**). Since then, Lyme disease has gone from being a medical curiosity to being the most commonly diagnosed vector-borne disease in the United States. In 1982, 497 cases were reported in 11 states. In 1992, 9677 cases were reported in 45 states.

Lyme disease has three clinical stages. Some people manifest all three, while others manifest only one, two, or none. The first stage begins from 3 to 32 days after the infecting bite, with a circular red rash resembling a bull's eye at the site. This distinctive rash, called **erythema chronicum migrans**, is the clinical hallmark of Lyme disease, even though it occurs in only 75 percent of patients. Several weeks after the bull's-eye rash disappears, the second stage begins. Bacteria spread to nearby lymph nodes, the blood, and eventually other organs, possibly including the brain, joints, heart, liver, spleen, and kidneys. Early symptoms often include headache, stiff neck, muscle aches, and fatigue. Then patients often become quite ill. Some may develop meningitis or myocardial damage. About six months after the infecting bite, the late stage begins—chronic arthritis that may persist for years with episodes of painful joint swelling (especially in the knees) lasting months.

Transmission of Lyme disease depends upon several arthropod and mammalian hosts, all of which must be infected by *B. burgdorferi*. Ticks of the genus *Ixodes* are the primary vectors—*I. dammini* in the American Northeast and Midwest, and *I. pacificus* and *I. neotornae* on the West Coast. There are two strands in the transmission story. In the Northeast and Midwest, *I. dammini* has a two-year life cycle and must obtain three blood meals, the first two on a small mammal, preferably a mouse, and the last on a deer. Although almost any small mammal will do for the early feedings, deer are essential to perpetuation of the infection. In areas where deer populations have been eliminated, Lyme disease has virtually disappeared. Humans are involved in the infectious cycle when an adult tick attaches to a person rather than a deer. (See the life cycle of *I. dammini* in Figure 15.5.) On the West Coast, *I. neotornae* maintains the reservoir in the dusky-footed wood rat while *I. pacificus* transfers the infection to humans (see the box "One Disease, One Bacterium, Three Ticks" in Chapter 12). In all regions, ticks reach maturity in the summer, so the high-risk months for Lyme disease are May through July—the time when people are most likely to be outdoors in tick-infested areas.

Diagnosing Lyme disease is either fairly easy or very difficult, with little middle ground. Part of the problem is that no good laboratory test has yet been developed. *B.*

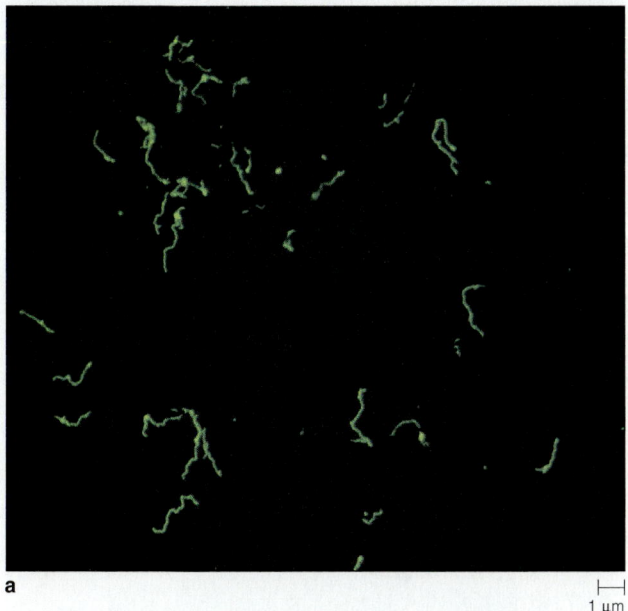

a

⊢——⊣
1 μm

Figure 27.4 **(a)** The causative agent of Lyme disease is the spirochete *Borrelia burgdorferi*, named after Willy Burgdorfer, the microbiologist who isolated it in 1982. **(b)** Most Lyme disease patients develop a characteristic bull's-eye rash where the bite occurred.

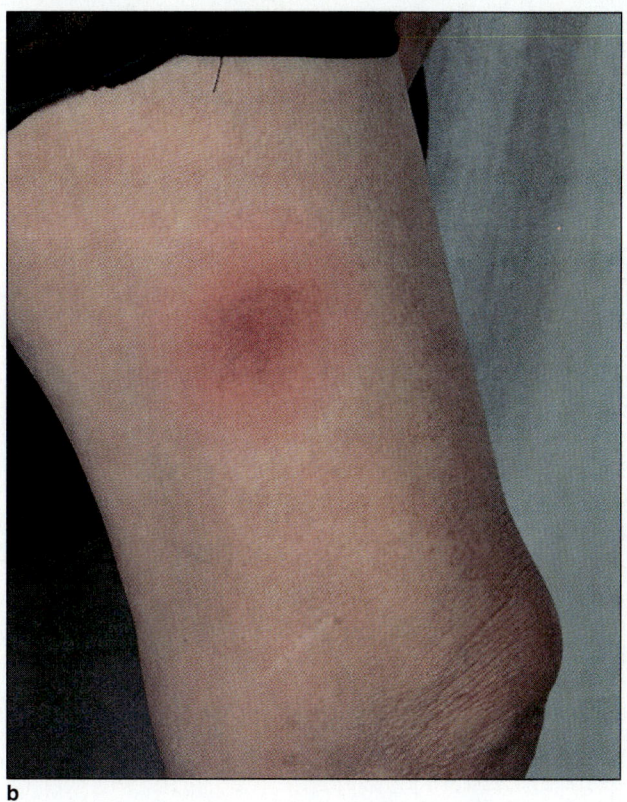

b

burgdorferi can be cultivated in the laboratory and is cultured fairly easily from infected ticks, but rarely are bacteria recovered from the blood of Lyme disease patients. Moreover, the serologic tests that exist are not entirely reliable. Both false negative and false positive results occur, even with the most reliable ELISA assay. Thus, clinical criteria are the most common basis for diagnosis. If the patient develops a bull's-eye rash, diagnosis is fairly easy, but diagnosis can be difficult otherwise. Early diagnosis is critical, however, because the infection can be cured much more easily during its earliest stages. Later, long-term intravenous therapy may be required. The drugs most commonly used include doxycycline, amoxicillin, and erythromycin.

Considering how much we know about Lyme disease, the question arises—why wasn't this relatively common infection recognized earlier? Is Lyme disease a new disease? Although *B. burgdorferi* must have existed long before its 1982 discovery, there are several reasons infection has become more common. First, the population of the *Ixodes* tick vector seems to be on the rise. Second, suburban America has grown out from cities until now it reaches wooded rural areas. In other words, more people now live where the ticks have always lived. Finally, doctors, like everyone else, are more likely to recognize something once they know what to look for.

Borrelia spp.: Relapsing Fever

Several species of the spirochete *Borrelia* cause **relapsing fever**. The infection is characterized by the sudden onset of fever, chills, headache, and muscle aches. The liver and spleen may become enlarged, and a faint red rash may appear on the trunk. After three to six days the symptoms disappear and the patient seems to have recovered. But about a week later a relapse occurs, and all the signs and symptoms reappear. Periods of illness and apparent recovery continue, though the symptomatic periods gradually become shorter and less intense. Patients may have from 2 or 3 relapses to as many as 13.

Relapsing fever appears to come and go because *Borrelia* are capable of antigenic variation. When *Borrelia* infection initially occurs, type-specific antibody is produced, bacteria are cleared from the blood, and the illness seems to resolve. But bacteria multiplying in other tissues survive. They alter their surface antigens and enter the bloodstream again, reactivating the infection. Eventually, bactericidal antibodies are produced that terminate the infection.

Relapsing fever has two distinct epidemiologic patterns. **Epidemic relapsing fever**, caused by *Borrelia recurrentis*, is transmitted by the body louse and tends to occur under crowded, unsanitary conditions, for example,

Tick Check

Lyme disease has received a great deal of publicity in recent years, but the relatively simple things we all can do to avoid infection have not. Probably the most important precaution is a daily tick check. The longer a tick is allowed to feed, the more likely it is to transmit infection. If you are in a deer- and *Ixodes*-infested area, examine yourself and your children every evening and remove the ticks with fine tweezers. If you find a deer tick, save it in a jar to show your doctor. Use tick repellents. The most effective skin repellent is N,N-diethyl-metatoluamide (DEET). It seems safe for adults, but repeated use on children can be toxic. Clothing repellents are less toxic and give good protection, which lasts about a month. The active ingredient is permethrin. A combination of DEET in low concentrations and permethrin sprayed on clothing once a month probably offers the best protection for adults and children.

The tick that transmits Lyme disease to humans in the Northeast and Midwest, *Ixodes dammini*, is the size of an apple seed and has an orange body with a black circle toward the head.

during war or natural disaster. Epidemic relapsing fever has recently been reported in Ethiopia and Sudan. The illness is serious, and mortality in untreated cases can be as high as 40 percent. **Endemic relapsing fever**, caused by *B. hermsii, B. turicatae,* and *B. parkeri,* among other species, is transmitted by ticks and occurs in many parts of the world, including the western United States. Tickborne relapsing fever is milder than louseborne. Mortality in untreated cases is about 5 percent.

Relapsing fever can sometimes be diagnosed by seeing the bacteria in blood smears of symptomatic patients. Culturing, usually done with immature laboratory mice, is a difficult procedure and is seldom undertaken. Tetracycline, chloramphenicol, penicillin, and erythromycin are all effective. Treatment should begin with a relatively low dose, however, because serious complications can arise when many bloodborne bacteria die simultaneously.

Bacillus anthracis: Anthrax

Anthrax is usually a disease of domestic and wild plant-eating animals such as cattle, goats, sheep, swine, buffalo, and deer, but humans occasionally become infected. *Bacillus anthracis* is a Gram-positive, spore-forming, bacterial rod. It is a well-adapted pathogen with a capsule and three different protein exotoxins that are critical to virulence, causing almost all the manifestations of the disease, including the respiratory failure that occurs in fatal cases. Experiments show that animals injected with pure anthrax toxin contract an illness identical to natural anthrax.

The spores that *B. anthracis* produces help sustain its chain of infection. They can remain viable for more than 40 years, which means that if spores infect a pasture, generations of animals that graze there are vulnerable.

The most common form of human infection is **cutaneous anthrax**, in which bacteria enter through a break in the skin (**Figure 27.5**). The bacteria usually come from infected animals or animal material such as wool or hides, especially goat hides. A blister forms at the site of entry, gradually developing into a blackened crater. In most cases, the ulcer heals and the person develops limited immunity. In 10 to 20 percent of untreated cases, however, bacteria invade the bloodstream, causing a sudden, severe, and fatal septicemia. Cutaneous anthrax can be treated with penicillin during its early stages; but once organisms enter the blood and produce substantial quantities of toxin, antibiotics are not effective.

A less common but more dangerous form of anthrax—**respiratory anthrax**—occurs when people inhale bacterial spores. The respiratory symptoms are mild; and by the time signs of septicemia develop, death is imminent. Almost none of these patients survive, even with

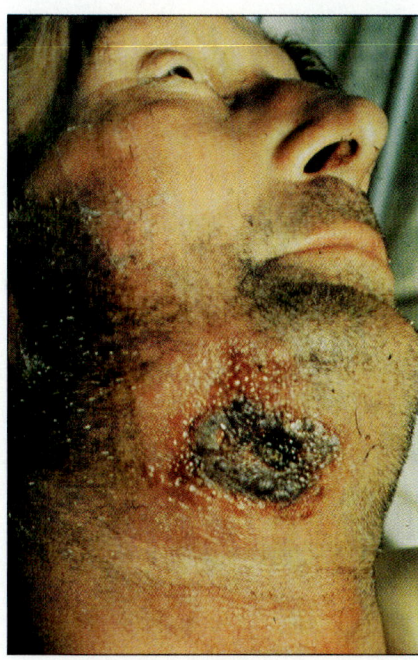

Figure 27.5 The most common type of infection by *Bacillus anthracis* is cutaneous anthrax, characterized by a blackened ulcer where the bacteria entered.

antibiotic treatment. This fatal syndrome is also called **woolsorter's disease** because it was an occupational hazard for people who worked with anthrax-contaminated wool. An extremely rare form, **gastrointestinal anthrax**, occurs in people who eat the meat of infected animals.

Controlling human anthrax depends upon controlling anthrax in domestic animals. An aggressive program of vaccinating healthy animals and burning or carefully burying infected animals has nearly eliminated the disease in the United States. The one or two cases reported each year usually occur in industrial workers exposed to anthrax-contaminated wool or hides imported from endemic countries. Rarely, agricultural workers or veterinarians are infected. It is probably impossible to completely eradicate anthrax because the spores have such a long survival time. In Iran, Turkey, Pakistan, Sudan, and some other countries the disease is highly endemic.

Anthrax can be diagnosed by culturing *B. anthracis*, which can be recovered from a cutaneous ulcer or from the blood if septicemia develops. Most deaths from anthrax are preventable by treatment with penicillin, but human infection is so rare that early diagnosis can be difficult.

Rochalimaea henselae: Cat Scratch Disease

The typical patient with cat scratch disease appears in the doctor's office with one extremely swollen lymph node,

usually under the arm or in the groin. The node is not painful and usually the patient has no fever or other symptoms. Often there are healing scratch marks on the arm or leg near the affected node and a small sore at this site where the infecting organism entered the body. The patient, most often a child, remembers having been scratched by a cat or kitten. Atypically, the node may remain swollen to the size of a tennis ball for several months, though it eventually returns to its normal size. A few patients are more seriously ill, complaining of fever, fatigue, or even symptoms of meningitis.

The clinical syndrome of cat scratch disease has been recognized for years, but only recently was the causative agent identified. It is *Rochalimaea henselae,* a pleomorphic Gram-negative rod that is probably part of the normal oral flora of domestic cats and dogs. Although the infection will resolve on its own, recent evidence suggests that treatment with trimethoprim-sulfamethoxazole or ciprofloxacin speeds recovery.

Rickettsia rickettsii: Rocky Mountain Spotted Fever

Rickettsiae are extremely small rod-shaped Gram-negative bacteria that can survive only as intracellular parasites. Rickettsiae parasitize arthropods, including ticks, lice, and fleas, and infect mammals during a blood meal. (Q fever, which is caused by rickettsiae and discussed in Chapter 22, is an exception. It is usually transmitted by inhalation.)

Rickettsia rickettsii causes **Rocky Mountain spotted fever** (**RMSF**), a severe disease characterized by fever and rash. RMSF is named for the region where it was first diagnosed but is found in many parts of the Western Hemisphere (**Figure 27.6**). RMSF is not a common disease, but it occurs much more frequently nationwide than do tularemia or brucellosis. In some years, as many as a thousand cases are reported to the Centers for Disease Control and Prevention (CDC). The disease is transmitted by a tick. In the Rocky Mountain states it is usually the wood tick, *Dermacentor andersoni*; in the southeastern states, the dog tick, *Dermacentor variabilis*; and in the south-central states, the Lone Star tick, *Amblyomma americanum*.

Rickettsia rickettsii can be passed transovarially from one generation of ticks to the next, so no other animal reservoir is needed to perpetuate RMSF. In humans, *R. rickettsii* multiplies in the smooth inner lining of blood vessels, damaging capillaries in the skin to produce the spotty rash for which the disease is named. In severe cases, internal organs suffer similar capillary damage. In 20 percent of untreated cases, overwhelming infection with abnormal blood clotting leads to shock and death.

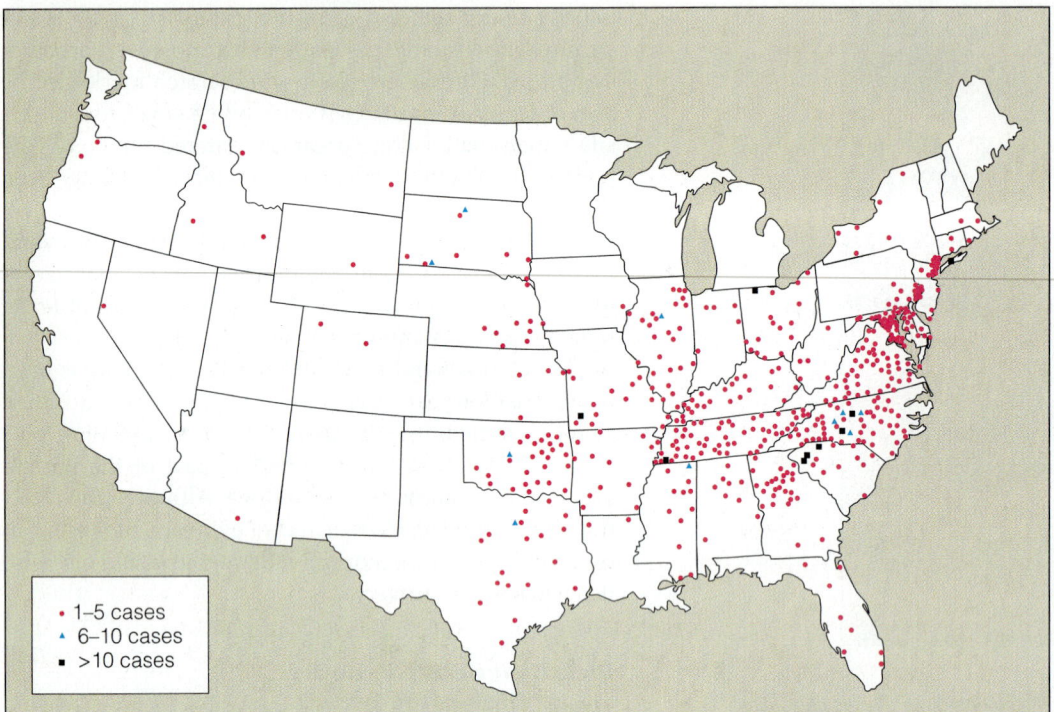

a

Figure 27.6 **(a)** The first cases of Rocky Mountain spotted fever were identified in the Rocky Mountain region, but the infection is much more common in the eastern United States. **(b)** The char–acteristic rash usually appears about a week after infection. It covers the entire body, including the palms and soles.

- 1–5 cases
- ▲ 6–10 cases
- ■ >10 cases

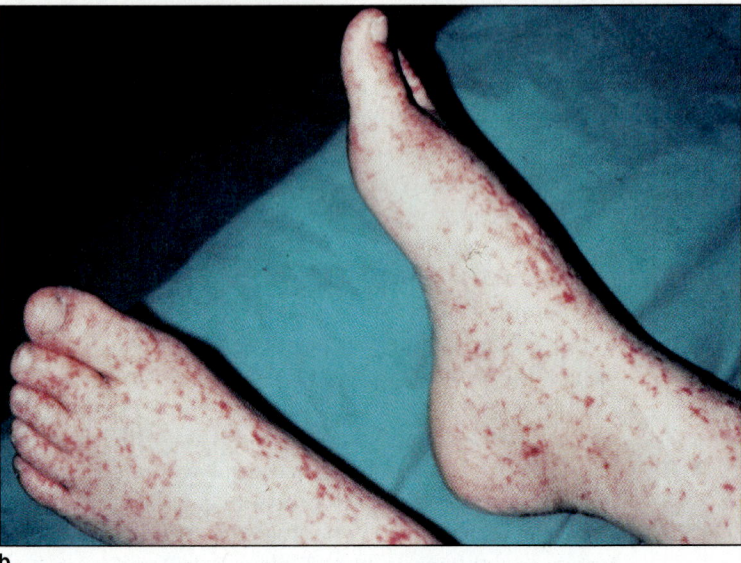

b

Tetracycline and chloramphenicol are the only antibiotics effective against *R. rickettsii*.

Early diagnosis and treatment of RMSF are difficult because no test exists for rapid identification of *R. rick-ettsii*. The organism is not cultured in the laboratory because it is so hazardous to personnel. Definitive labo-ratory diagnosis must wait until serological testing be-comes positive, which can take up to several weeks. At the same time, if antibiotic treatment is delayed, the risk of serious complications or death rises. Consequently, early treatment of Rocky Mountain spotted fever must be based on clinical and epidemiologic clues.

Infection is especially likely among children who play in tick-infested areas in the summer, so a history of expo-sure to ticks is often the key to diagnosis. The earliest signs and symptoms—fever, headache, and vomiting—are com-mon and nonspecific, but within a week a distinctive rash usually appears on the wrists and ankles. Eventually this rash covers the entire body, including the palms and soles. It becomes dark and slightly raised, sometimes re-sembling tiny bruises. Though children are more likely to

be infected, adults are more likely to die of RMSF. How-ever, the disease responds dramatically to antibiotic treat-ment, and early treatment prevents most fatalities.

Rickettsia prowazekii and *Rickettsia typhi*: Typhus

The two forms of **typhus** are caused by two closely re-lated organisms. Epidemic (louseborne) typhus is caused by *Rickettsia prowazekii* and murine (fleaborne) typhus by *Rickettsia typhi*.

Epidemic (Louseborne) Typhus.

Rickettsia prowazekii is named after two bacteriologists who died of typhus while investigating this deadly disease—Howard Ricketts and Stanislav von Prowazek. They and their successors discovered the following cycle of infection. **Epidemic typhus**, or **louseborne typhus**, is spread by infected body lice. Lice acquire the rickettsial parasite from an infected human being during a blood meal and transmit it to another human host when they bite again. All lice on a person's body are likely to be infected. When lice feed, they defecate on the skin; when the victim scratches a fresh bite, the feces enter the bite wound. *R. prowazekii* first multiply in the wound and then enter the bloodstream in large numbers; there they multiply in the lining of small blood vessels.

As with other systemic diseases, early diagnosis of typhus is difficult. However, after an 8- to 12-day incubation period, most typhus patients become suddenly ill with unbearable headache, fever, muscle aches, and shaking chills. Often they become delirious or unconscious. (In Greek, *typhus* means "smoky" or "stuporous," referring to the patient's changed consciousness.) Several days later, a rash appears in almost all patients. It resembles the rash of Rocky Mountain spotted fever but does not affect the palms, soles, or face. Tetracycline and chloramphenicol are very effective against typhus. One dose of tetracycline can cure a person who appears to be near death. Untreated, typhus may last as long as three weeks; and in about 40 percent of patients, shock leads to death. Older adults are especially vulnerable. Although most people who recover from typhus acquire long-lasting immunity, the infection can become latent, with mild recurrences, called **Brill-Zinsser disease**, years later.

Typhus is transmitted through a chain of infection from louse (the human body louse, *Pediculus humanus*) to human to louse. No other animal hosts are necessary to maintain typhus, and latent human infection sustains lice between epidemics. The key to breaking the chain of infection is eliminating human lice. These parasites, which live in unwashed human clothing, multiply when people live under crowded and unsanitary conditions. Thus, typhus epidemics occur during war and at other times when people are displaced from their homes. Typhus killed millions of civilians during World War I, but after World War II mass outbreaks were prevented by delousing refugees with the insecticide DDT.

Populations that are free of lice do not suffer from epidemic typhus. The last outbreak in the United States occurred in 1922. Typhus still exists in many parts of the developing world, particularly in Africa, where typhus is continuously present, making epidemics less apparent.

Murine (Fleaborne) Typhus.

Rickettsia typhi, which is closely related to *R. prowazekii*, causes **murine typhus**, or **fleaborne typhus**. Fleaborne typhus is similar to louseborne typhus but less severe and rarely fatal. Epidemiology and mode of transmission are also different. Murine typhus is mainly a disease of mice and rats (Muridae is the family of rodents to which rats and mice belong). It is transmitted from one animal to another by infected fleas (usually the rat flea, *Xenopsylla cheopis*) and occasionally to humans who come in contact with diseased rodents. Most cases begin with the gradual onset of headache, muscle ache, and fever, followed three to five days later by a typical rickettsial rash.

Murine typhus is endemic in the United States. People who work in mice- or rat-infested areas (such as shipyards or grain storage elevators) are most at risk. Fewer than 100 cases are reported annually, mostly in the Southeast and Texas, but many mild illnesses may go unrecognized. Murine typhus is sometimes misdiagnosed as Rocky Mountain spotted fever and, because the rickettsial microorganisms are closely related, even serologic testing does not reveal the error. Like other rickettsial diseases, murine typhus can be effectively treated with tetracycline or chloramphenicol.

VIRAL CAUSES

Systemic illnesses can also be caused by viral pathogens. This group includes a wide variety of diseases, including AIDS.

Yellow Fever Virus: Yellow Fever

Yellow fever is caused by the **yellow fever virus**, a small enveloped RNA virus belonging to the togavirus family and the flavivirus genus. The virus is inoculated under the skin of a new human host by the bite of a mosquito. The principal vector is *Aedes aegypti*. After inoculation, the virus is absorbed by lymphatic vessels and transported to nearby lymph nodes. Here it multiplies, eventually entering the bloodstream and reaching distant organs, including the liver, where it multiplies rapidly.

Most people infected by the yellow fever virus suffer a few days of mild fever, headache, and weakness, symptoms not usually recognized clinically as yellow fever. But 20 percent suffer the classical symptoms—severe fever, chills, headache, and significant liver damage. Liver damage causes the jaundice that gives the disease its name. In addition, blood clotting factors normally produced by the liver may become depleted, leading to uncontrollable bleeding. About 50 percent of severely affected people die, but those who recover suffer no lasting damage and are immune thereafter.

Yellow fever, also known as **yellow jack**, has caused devastating epidemics around the world and in cities as far north as Philadelphia in the United States. *A. aegypti* prefers to feed on human beings and has adapted to their habits. It lives in inhabited areas, breeds in water that people store for washing or drinking, and bites by day. Walter Reed and his colleagues confirmed mosquitoes as the mode of transmission of yellow fever in 1900. They eradicated yellow fever from Havana, Cuba, and instituted mosquito-control programs that allowed completion of the Panama Canal (see Chapter 15).

The endemic jungle form of yellow fever exists today in jungle areas of the Americas, including Panama, where monkeys are the reservoir. The urban form has been eliminated from the Western Hemisphere by eradicating *A. aegypti*, but it still occurs in Africa. A live attenuated vaccine developed in the 1930s could prevent most cases but is not widely used.

Flaviviruses: Dengue Fever

Dengue fever resembles yellow fever in its epidemiology and mode of transmission. Dengue fever is caused by several closely related flaviviruses of the togavirus family and is transmitted principally by the mosquito *Aedes aegypti*. Unlike yellow fever, dengue fever occurs widely in most tropical regions. Occasionally, it occurs in temperate climates, and in 1980 its transmission was documented in the United States. Closely related arboviruses cause similar tropical diseases with such exotic names as Chikungunya fever and O'nyong nyong fever.

Dengue fever is rarely life-threatening, but it can be painful. Two to seven days after being bitten by an infected mosquito, people complain of high fever and headache. Shortly after, a rash and deep pain in the limbs develop. Limb pain is so intense that the illness is sometimes called **breakbone fever**. The patient appears to recover after a few days, but soon the fever and rash return. Diseases with this pattern of apparent recovery and recurrent illness are said to have a **biphasic course**. Many patients complain of weakness and mental depression long after recovery. No vaccine exists to prevent dengue fever, and treatment is limited to supportive care. The best prevention is control of *A. aegypti* and avoiding mosquito bites.

A life-threatening syndrome called **dengue hemorrhagic fever** or **dengue shock syndrome** occasionally occurs among children who have previously had dengue fever. Patients suffer from uncontrollable bleeding and shock. This is probably an immune-complex hypersensitivity reaction that occurs when the previously stimulated immune system encounters the pathogen again. These children require intensive supportive medical care, and even then as many as 40 percent die.

Epstein-Barr Virus (EBV): Infectious Mononucleosis

The **Epstein-Barr virus (EBV)** is a large DNA-containing herpesvirus surrounded by a lipid envelope (**Figure 27.7**). Like other herpesviruses, EBV produces latent and active infections. Because latent infection occurs in B lymphocytes, EBV's pathogenesis is intimately associated with the immune system.

The Epstein-Barr virus infects most human beings at some time in their lives. In regions where standards of living and hygiene are poor, infection usually occurs in early childhood, producing a mild illness or none at all. In the United States and other countries with high standards of living and hygiene, infection typically occurs between ages 15 and 25. Either an asymptomatic infection or a clinical syndrome called **infectious mononucleosis (mono)** results. In developing countries almost all children have antibodies attributable to past EBV infections, while in industrialized countries only 60 percent of young adults have such antibodies. The 40 percent without antibodies are susceptible to infection until they finally encounter the virus. By midlife, more than 90 percent of U.S. residents have been infected by EBV.

Pathogenesis. The symptoms of infectious mononucleosis—fever, fatigue, sore throat, and swollen lymph nodes—appear one to two months after infection. By that time, blood contains an increased number of lymphocytes, many of which are enlarged and atypical in appearance. Half the patients also have an enlarged spleen. Diagnosis is confirmed by serology. IgM **heterophile antibodies** that agglutinate sheep red blood cells are detectable for three to six months after infection; IgG antibodies specific for the Epstein-Barr virus persist for life. Most patients with mononucleosis resume normal activities in four to six weeks. Major complications, involving the central nervous system or fatal rupture of the spleen, are rare.

Active, virus-producing EBV infection occurs in epithelial cells near the salivary glands. Active infection kills the host cell and transmits infection through virus-containing oral secretion. Infectious mononucleosis is therefore aptly nicknamed the kissing disease, although it can also be transmitted by saliva-carrying fomites such as drinking glasses or eating utensils. The virus establishes a latent infection in B lymphocytes. Latent infection does not produce new viral particles but allows the viral genome to be replicated along with host DNA. Infected B cells are not killed and do not change appearance, but they are profoundly altered by EBV. They become **immortalized** in tissue culture, meaning that they continue to multiply indefinitely without normal controls on cell division.

This abnormal proliferation of EBV-infected B cells initiates a chain of events that results in clinically appar-

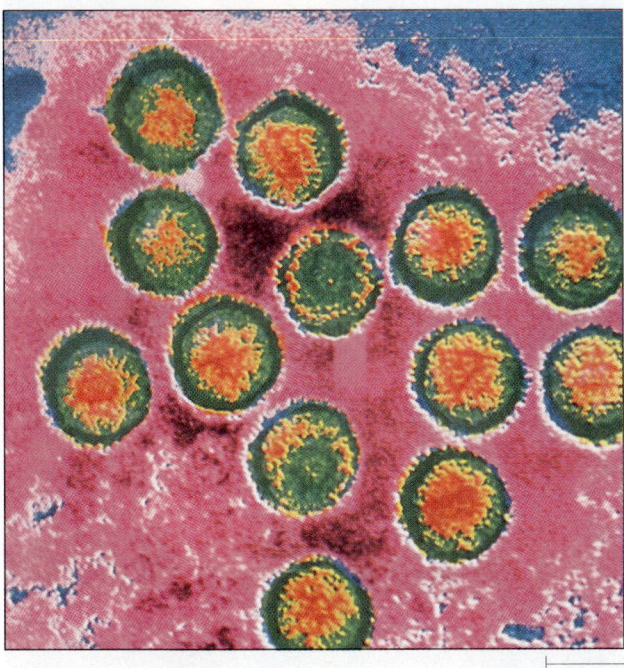

Figure 27.7 Epstein-Barr virus.

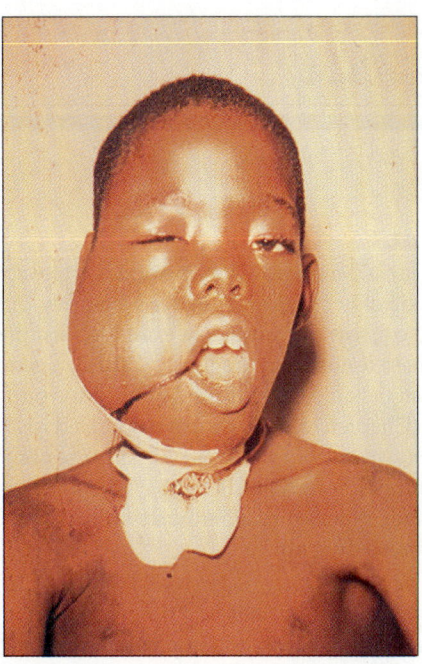

Figure 27.8 Burkitt's lymphoma, caused by the Epstein-Barr virus, is a rare—and curable—tumor of the jaw seen in children of equatorial Africa.

ent infectious mononucleosis. When infected lymphocytes enter circulation, early symptoms of fever and fatigue appear. As B cells continue to proliferate, suppressor T cells are stimulated to bring the abnormal cells back under control. These reactive T cells are the enlarged, atypical lymphocytes of mono. T-cell proliferation can also cause swelling of lymph nodes, spleen, and liver in later stages. Eventually a normal equilibrium between B and T cells is reestablished, ending the symptoms, though latent virus persists in the lymphocytes for life.

EBV-Associated Cancers. The Epstein-Barr virus is one of the few viruses that have been proved oncogenic (cancer-causing). In immunosuppressed cancer patients and organ transplant recipients, Epstein-Barr infection can cause a fatal cancer called **B-cell lymphoma**. If suppressor T cells cannot stop the proliferation of infected B cells, runaway multiplication continues, producing tumorlike growths throughout the body and EBV-induced lymphoma. David, the "bubble boy" who survived for many years in a sterile environment despite a severe congenital immunodeficiency, died from EBV-induced lymphoma (see the box "David—Life in a Germ-Free World," Chapter 18).

EBV has been definitively linked to a relatively rare cancer called **Burkitt's lymphoma**. Burkitt's is a malig-

nant tumor of the jaw in children of equatorial Africa (**Figure 27.8**). The Epstein-Barr virus was discovered during research on Burkitt's. A typical herpesvirus particle was seen in electron micrographs from Burkitt's lymphoma tumors. The link between the Epstein-Barr virus and mononucleosis was established during this same research project when one of the laboratory technicians became ill with infectious mononucleosis and simultaneously developed antibodies against EBV.

EBV is also associated with **nasopharyngeal carcinoma**, a malignant tumor of the nasopharynx that is common in China. No one is sure why this cancer and Burkitt's lymphoma are found only in certain countries when EBV infection is worldwide. Development of cancer probably requires an extra stimulus, or **cofactor**, in addition to EBV infection. Researchers speculate that the critical cofactor in Burkitt's lymphoma may be early exposure to malaria parasites, while nasopharyngeal carcinoma may depend on genetic factors.

Because almost all adults carry latent EBV, questions have arisen about a possible connection to chronic disease. Specifically, EBV was proposed as the cause of **chronic fatigue syndrome**, an ill-defined syndrome that has some similarities to infectious mononucleosis. No solid evidence was found linking EBV to chronic fatigue, though, and the theory has been largely abandoned.

Case History

AIDS

Patient 1

A 27-year-old married woman noticed white spots on the inside of her mouth. They were painless, but they didn't go away. Her physician examined scrapings from these lesions and diagnosed thrush, a common infection in babies but unusual in healthy young adults (Chapter 26). Her thrush responded to antifungal medications, but this was only the first of many opportunistic infections. Patient 1, who died two years later, reported that her husband had a history of intravenous drug abuse.

Patient 2

An 8-month-old baby suddenly became sick with a high fever and such extreme lethargy that he was hardly arousable. He was rushed to an emergency room where streptococcal meningitis was diagnosed. Treatment with antibiotics led to recovery, but he died from another overwhelming bacterial infection only two months later. His mother admitted to using illegal intravenous narcotics.

Patient 3

A 32-year-old attorney noticed a cough and fever that became progressively worse until his breathing was rapid and difficult. Because of his respiratory distress, he was admitted to the hospital, where he was diagnosed with *Pneumocystis* pneumonia. In spite of antibiotic treatment and aggressive supportive care, including a mechanical ventilator, his condition continued to worsen and he died three weeks later. The man's homosexual companion had died recently of a similar infection.

Human Immunodeficiency Virus (HIV): Acquired Immunodeficiency Syndrome (AIDS)

Patients 1, 2, and 3 each suffered from different bacterial and fungal infections. All, however, had the same underlying and much more serious infection by the **human immunodeficiency virus (HIV)**. All were diagnosed eventually with **acquired immunodeficiency syndrome (AIDS)**, which crippled their immune defenses and allowed other infectious agents to cause their deaths.

HIV was first isolated from the lymph node of a patient in 1983. It is a retrovirus, which means it has an

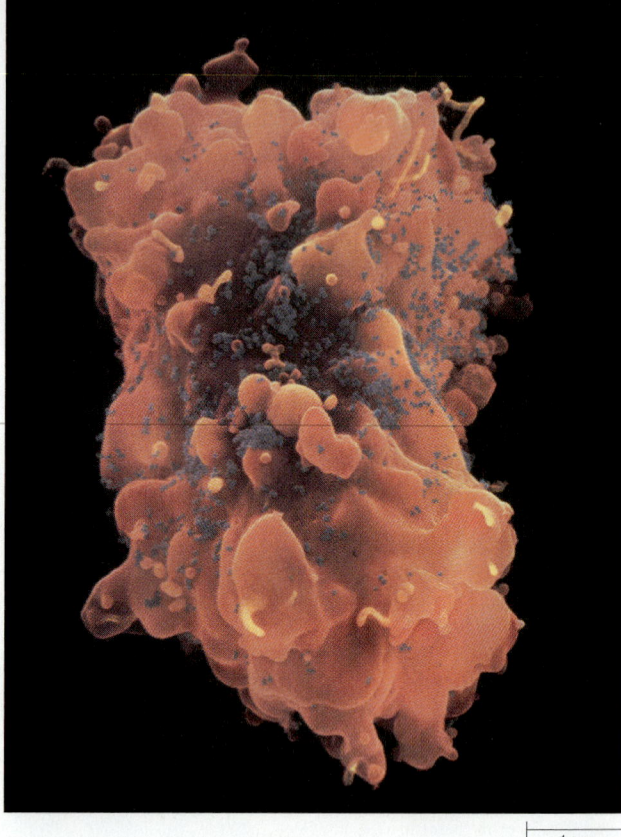

Figure 27.9 In this micrograph, a T cell is being attacked by the human immunodeficiency virus (the tiny blue particles), which causes AIDS.

RNA genome that is converted into a DNA copy of the same information inside the infected cell (see Chapter 13). A closely related virus, which also causes AIDS, was isolated from patients in Africa in 1985. This virus was named HIV-2, and the original 1983 virus was renamed HIV-1.

HIV affects several types of human cells, but its most devastating effect is on T_4 helper lymphocytes (**Figure 27.9**). These lymphocytes fulfill a critical role in the immune process—neither cell-mediated nor humoral immunity can operate effectively without them (Chapter 17). It is the depletion and eventual near disappearance of T_4 helper cells that causes the profound and ultimately fatal immunodeficiency known as AIDS.

The Infective Process. The process of infection begins when HIV enters the body of a new host and ends with death from AIDS. There are several distinct steps in between.

1.   HIV enters the circulatory-lymphatic system of a new host. There it comes into direct contact with susceptible cells and initiates an infection.

2.   A protein on the surface of the viral envelope binds to a CD4 receptor protein on the surface of the target cell (**Figure 27.10**). The CD4 protein also plays a crucial role in the ability of T_4 lymphocytes to recognize antigen

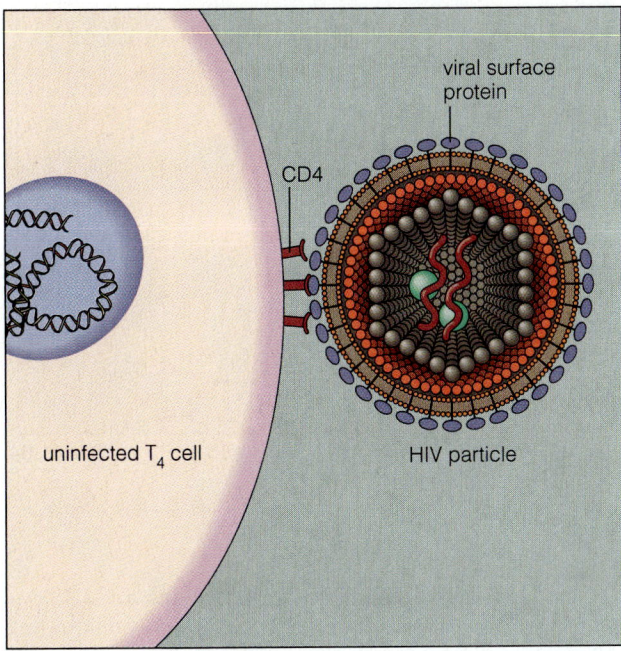

Figure 27.10 Any cells in the human body that bear the CD4 protein can be infected by HIV. HIV has surface proteins that bind to CD4. Binding allows the viral envelope to merge with the host cell membrane and enter.

(Chapter 17). Any cells, including macrophages and dendritic cells, that bear CD4 can be infected by HIV. Binding between the viral protein and CD4 allows the viral envelope to merge with the membrane of the lymphocyte, thereby bringing viral RNA and reverse transcriptase inside the host cell.

3. Viral RNA is transcribed into DNA by reverse transcriptase and incorporated into the host genome, becoming a provirus. Once the provirus enters the host's DNA, infection will persist throughout the life of that cell and will be passed on to the cell's progeny if it divides. This step seems irreversible, and scientists have not determined any way in which the provirus could ever be removed from the host cell genome. If the provirus cannot be removed, then an infected person could never be "cured" of an HIV infection.

4. HIV infection kills T_4 lymphocytes, but the exact way it kills is not clear. Some infected T_4 lymphocytes, particularly those that are stimulated to divide rapidly during an immune response, die as they produce huge numbers of new viral particles. This is relatively uncommon, however—not quite 1 in 10,000 T_4 cells is infected by the virus—and does not in itself explain why so many T cells die. One theory (of many) proposes that HIV encodes "superantigens," producing a massive stimulation of lymphocyte activity that eventually exhausts the cells

and kills them. An answer to the question of how T cells die would not only increase our understanding of the virus, but would also help researchers design new AIDS therapies (see the box "The Number One Question").

5. The total number of circulating T_4 lymphocytes declines markedly. Normal people have more than 500 T_4 helper lymphocytes in each cubic millimeter (mm^3) of blood, but patients with advanced HIV infection have very few of these cells (**Figure 27.11**). When the number of T_4 lymphocytes falls below 200 per mm^3, early signs of immune dysfunction appear and the person may contract opportunistic infections of the skin or mucous membranes, as Patient 1 did. When the number falls below 100 (on average, 10 years after infection for adults), patients suffer severe life-threatening immunodeficiency.

Clinical Syndrome. Being infected with HIV is not the same as having AIDS. HIV-positive people do not have AIDS until they develop another problem on the list of diagnostic criteria. AIDS—as the name states—is a clinical syndrome, or diagnosis based on a constellation of findings. Initially, most of these findings were unusual opportunistic infections typically found in gay men, such as *Pneumocystis* pneumonia. But in 1993, the CDC revised the case definition. It added four items to the already long list of conditions upon which AIDS can be diagnosed: a CD4 count under 200 per mm^3, pulmonary tuberculosis, recurrent pneumonia, and invasive cervical cancer. Although the time lag between HIV infection and progression to AIDS may be long, clinical studies indicate that nearly all HIV-infected individuals eventually develop AIDS.

A person newly infected by HIV may suffer from a brief illness resembling infectious mononucleosis, but most people are otherwise entirely well for many years. Only serological testing for antibodies against HIV (the AIDS blood test) can identify asymptomatic viral carriers. During the prolonged asymptomatic period the person may be unaware of the infection and unknowingly transmit the virus to others.

As HIV infection cripples the immune system, a person becomes susceptible to overwhelming and often fatal systemic infections. Thus, a secondary or complicating infection is usually the immediate cause of death in an AIDS patient. In the case of Patient 2, for example, it was overwhelming bacterial septicemia, and in Patient 3, *Pneumocystis* pneumonia—two of the most common. But *any* microorganism pathogenic to humans can be the terminal manifestation of AIDS. By the time HIV infection produces the profound immunodeficiency characteristic of AIDS, the prognosis is grim. Fifty percent of patients die within a year of diagnosis, and almost all are dead five years later.

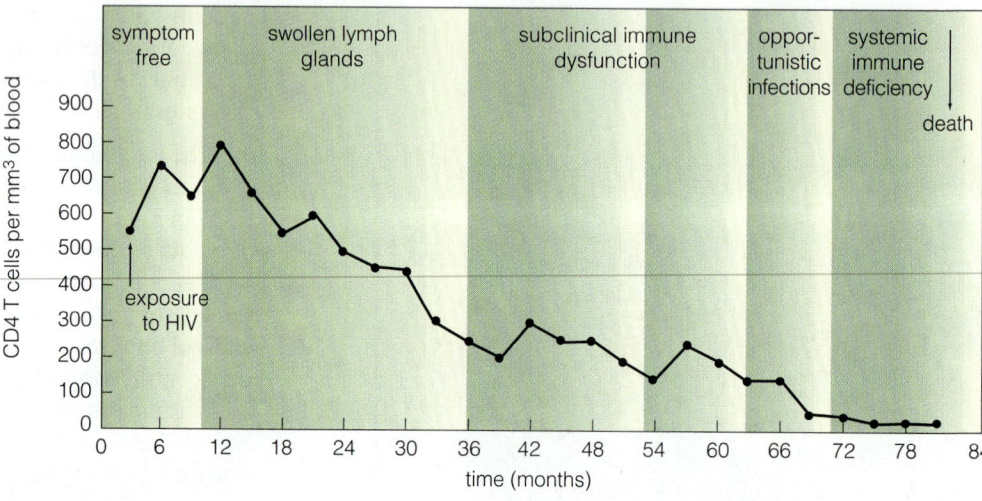

Figure 27.11 The progression of AIDS in a young male patient. This young man, like most AIDS patients, remained relatively well while his T_4 count was above 200 per mm^3. When it fell below 200, serious opportunistic infections began to occur. When the count fell below 100, he entered a state of critical immune failure and died less than a year later.

Table 27.2 Transmission of the Human Immunodeficiency Virus (HIV)

Route	Example
Known Routes of Transmission	
Sexual transmission	Homosexual intercourse between men
	Heterosexual intercourse (men to women and women to men)
Blood	Transfusion of blood or blood products (e.g., clotting factors for hemophiliacs)
	Shared needles among intravenous drug users
	Needle stick, open wound, and mucous-membrane exposure in health-care workers
	Medical injection with an unsterilized needle (common in developing countries)
Mother to infant	Across the placenta before birth
	Passage through birth canal
	Intimate contact around time of birth
Routes Investigated and Shown Not to Mediate Transmission	
Close personal contact	Household contact
	Caring for AIDS patients without exposure to blood or infected secretions
Insect vectors	Mosquitoes, flies
Fomites	Water, soil, humanmade materials

The pathogens that are most threatening to AIDS patients are the ones that can be controlled only by vigorous cell-mediated immune defenses. Among these are *Pneumocystis carinii* (Chapter 22), *Toxoplasma gondii* (discussed later in this chapter), *Cryptococcus neoformans* (Chapter 25), *Histoplasma capsulatum* (Chapter 22), and cytomegalovirus (Chapter 24). All are opportunistic pathogens that usually cause serious infections only in people with impaired immune defenses.

AIDS has also sparked a resurgence of tuberculosis, another infection normally controlled by vigorous cell-mediated immunity. The dramatic rise in TB is even more worrisome because many AIDS-related infections are caused by multiple-drug-resistant strains of *Mycobacterium tuberculosis* (Chapter 22).

Children with AIDS are vulnerable to a wider variety of pathogens than are adults, succumbing to infections in which antibody-mediated as well as cell-mediated defenses are critical. Thus, standard bacterial infections, like the streptococcal meningitis that struck Patient 2, are common in pediatric AIDS patients but unusual in adults.

Although opportunistic infections are the hallmark of HIV disease, they are not its only manifestation. Other problems, so far not directly linked to immune failure, are seen both early and late in the infection. For example, many HIV-infected people experience fevers, night sweats, lymph node enlargement, and marked weight loss. These complaints may develop gradually and persist for months or years before T_4 cell counts fall low enough to permit infection by opportunistic pathogens. In Africa, severe weight loss is so common among HIV-infected individuals that AIDS is also called slim disease.

HIV also appears to be a primary pathogen of the central nervous system. Viral particles cross the blood-brain barrier and cause a deterioration of mental function known as **AIDS dementia**. Dementia can occur early or

Microbe Mappers

The Number One Question

There is compelling evidence that *links* HIV infection to AIDS epidemiologically, but the crucial question of *how* HIV infection causes AIDS is nearly as mysterious as it was 10 years ago. We simply do not know the pathogenesis. This fact is what leads some scientific dissenters, including the famous virologist Peter Duesberg from the University of California at Berkeley, to say HIV does not cause AIDS. For anyone to say it does—and then not be able to explain the mechanism—is illogical and unscientific according to Duesberg. Moreover, says Duesberg, HIV couldn't cause AIDS. If HIV were virulent enough to kill, it would. Instead, the virus multiplies while the HIV-infected person remains healthy, usually for years, so something else must be causing AIDS. Almost no scientists agree with Duesberg (and the pathogenesis of many diseases is not well understood), but they do acknowledge that they cannot refute him. The fact is that the HIV virus is the best-studied virus of all time, yet we do not know how it causes AIDS.

In 1993, the journal *Science* sent a questionnaire to 150 AIDS researchers around the world asking what they thought were the top 10 questions that needed to be answered to come up with a cure for AIDS. The number one question was "What causes the immune system collapse seen in AIDS?" This, of course, is essentially the question of pathogenesis—Duesberg's elusive mechanism. *Science*'s summary of the AIDS situation, based on survey results, was that the more we learn, the less certain we are.

---

late in the course of HIV infection, sometimes disabling people whose T_4 counts and general health are still good. The vast majority of HIV-infected babies seem to have retarded neurological development.

Another well-known manifestation of HIV infection is the increased likelihood of developing certain types of cancer, particularly a rare blood vessel malignancy called **Kaposi's sarcoma**. Kaposi's sarcoma may be the first manifestation of HIV infection in individuals who otherwise appear to be healthy. Cancers of the lymphatic system, the rectum, and the tongue are also unusually frequent in AIDS patients. Why HIV-infected patients are vulnerable to these malignancies is not clearly understood.

Transmission. AIDS appears to be a new disease. Until 1980, when the first mysterious cases of *Pneumocystis* pneumonia in young men were reported, the acquired immunodeficiency syndrome was completely unknown to scientists and clinicians. We do not know the source of this new virus, but some think it evolved from closely related viruses that infect African monkeys. Studies on the mutation rate (which is quite rapid) suggest that HIV, in its present form, has been infecting humans for more than 20 but less than 100 years. The current pandemic probably began in the 1970s. PCR analysis of clinical specimens dates some HIV infections as far back as the 1960s, but these are rare (Chapter 7).

HIV has three main modes of transmission (**Table 27.2**):

1. By sexual contact with an infected person that transfers infected body fluids—blood, semen, or vaginal secretions

2. By receiving infected blood, usually in transfusions or from sharing intravenous needles with an infected person

3. From mother to infant during pregnancy (blood-borne transmission of HIV across the placenta) or around the time of birth (through infected vaginal secretions or infected breast milk)

Casual contact during normal social interaction in school or the workplace does not spread AIDS. Even the more intimate contact that occurs among siblings and family members has not been shown to transmit the virus, as long as infected body fluids are not exchanged. Neither insect vectors nor contaminated food or water transmits HIV. Even sexual contact with an infected person usually does not cause infection—most people who become infected with HIV as the result of sexual contact have been exposed repeatedly. In spite of fears aroused when HIV was first identified, we now know it has extremely low communicability. Nevertheless, infection *can* occur on a first and single exposure, and numerous opportunities exist because infected people appear healthy for years after infection. People with ulcer-causing STDs are at increased risk of contracting AIDS because open sores on the genitals provide an easy portal of entry for HIV (Chapter 24).

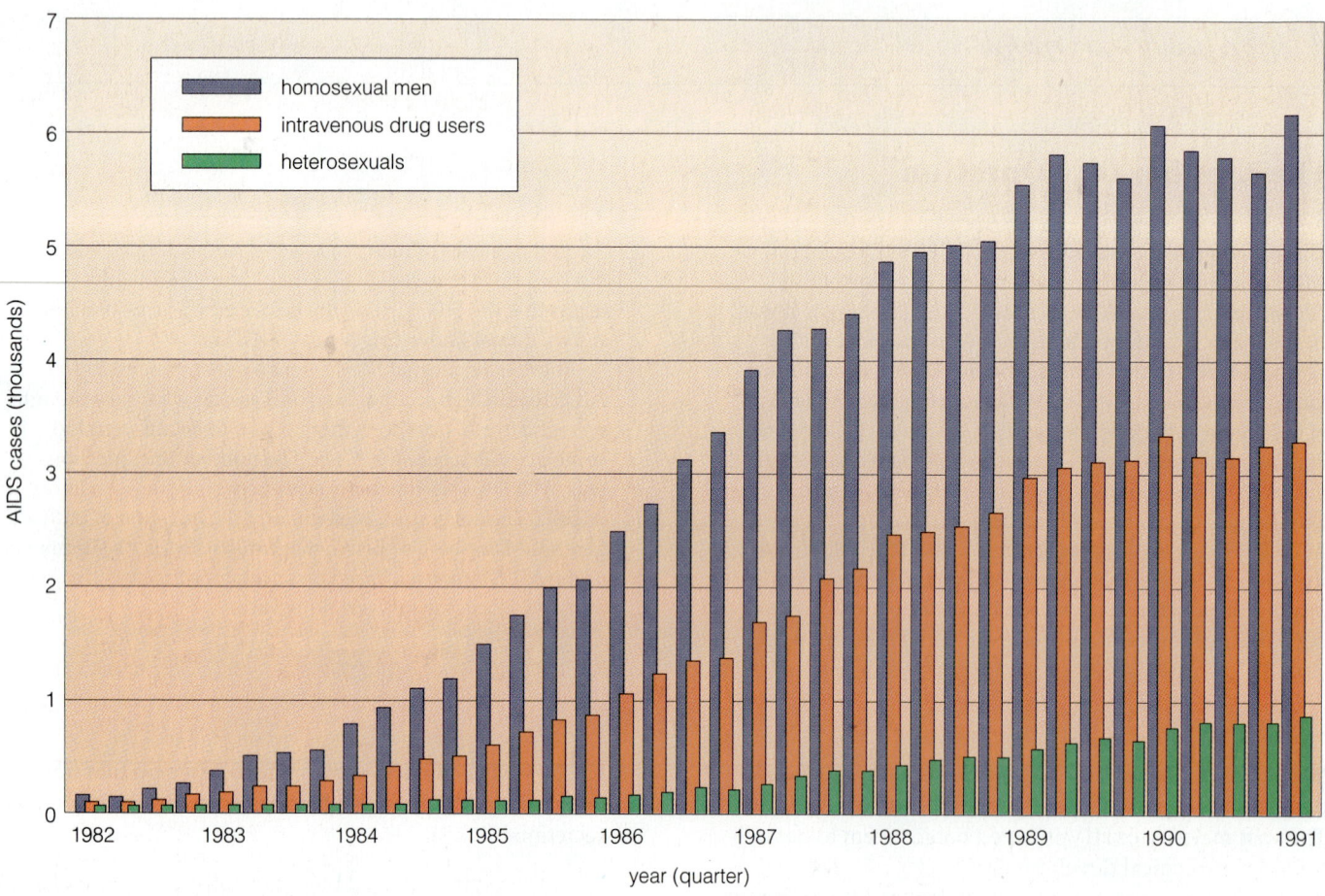

Figure 27.12 The epidemiology of AIDS in the United States is changing. During the 1980s, most cases occurred among male homosexuals. Today, the incidence is rapidly increasing among intravenous drug users and heterosexuals.

Epidemiology. When AIDS was first recognized in the United States, it occurred almost exclusively among male homosexuals, like Patient 3, and it was assumed that homosexuals were particularly susceptible to infection. Now we know that this pattern occurred because homosexual males had extremely high numbers of sexual contacts within a community in which the virus was already prevalent. Mucous membrane tears during rectal intercourse (allowing portals of entry for the virus) and a high rate of ulcer-causing STDs in the homosexual community probably fostered transmission as well. Because of this epidemiologic pattern, AIDS in the United States has primarily affected men.

However, this pattern started to change in the late 1980s. AIDS cases are increasing among drug abusers who share contaminated needles (**Figure 27.12**). This new pattern has significant implications for spread of the disease in the United States. AIDS will be seen more often

among women and children, and there will be increased sexual transmission in the heterosexual community. The majority of men and women with drug-acquired AIDS are heterosexual, and HIV can be transmitted from men to women or women to men during sexual intercourse. Thus, people like Patient 1 who acquired the disease by heterosexual exposure are already becoming more common as the pool of infected individuals in the heterosexual population increases.

In developing countries, the epidemiologic pattern of AIDS has been very different and, in some countries, the burden of disease far greater. In Africa, the continent most devastated by AIDS, infection is spread almost exclusively by heterosexual intercourse. Men and women die from the disease in approximately equal numbers. Many children are infected by HIV at birth, and many more have lost both parents. In some villages in Uganda most surviving children are orphans.

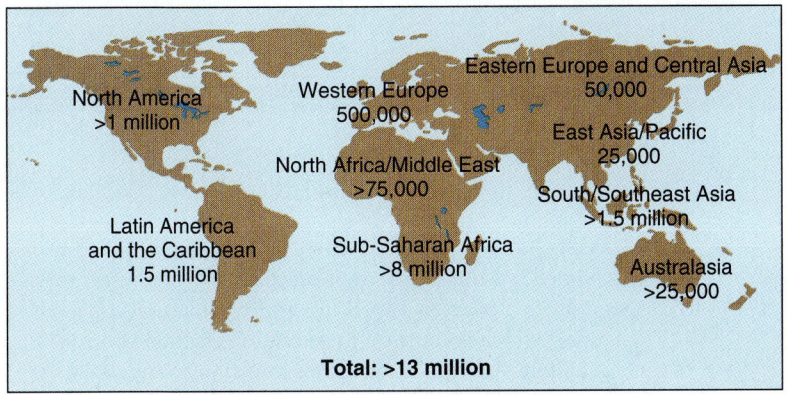

Figure 27.13 Worldwide epidemiology of AIDS. The pandemic we are experiencing has hit Africa hardest, as is illustrated by the estimated distribution of cumulative HIV infections in adults, by continent or region, as of mid-1993 shown here.

North America
>1 million

Western Europe
500,000

Eastern Europe and Central Asia
50,000

East Asia/Pacific
25,000

North Africa/Middle East
>75,000

South/Southeast Asia
>1.5 million

Latin America
and the Caribbean
1.5 million

Sub-Saharan Africa
>8 million

Australasia
>25,000

Total: >13 million

The long lag time between HIV infection and the appearance of AIDS makes tracking the pandemic difficult, but there is no question AIDS is a major threat to public health worldwide. By 1992 AIDS had claimed nearly 300,000 lives, and at least 10 million more people are estimated to be HIV-infected and certain to develop AIDS before the turn of the century. The World Health Organization estimates that in parts of the AIDS belt in Africa, as much as 80 percent of the population may already be infected (**Figure 27.13**).

In the United States the situation appears equally grim. The CDC estimates that 1.5 million Americans were infected with HIV by 1990, and epidemiologists suspect the virus is being transmitted at an alarming, though unknown, rate. In 1991, nearly 30,000 residents of the United States died of AIDS. In parts of New York City where drug abuse is especially high, AIDS was the leading cause of death among men and women in their thirties by 1991.

As concern about AIDS rises, however, it is important to remember that other threats to human health are even greater. Heart disease and cancer still kill many more Americans every year than AIDS, and nearly 7 million children worldwide die annually of measles and diarrheal dehydration—almost as many in one year as the total number of people worldwide who are likely to be infected with HIV in the next several years.

Prevention and Treatment. Great progress was made in AIDS research between 1980 and 1990. The illness was identified, the etiologic agent isolated, and a serologic test for diagnosis developed. In contrast, the two most pressing problems—prevention and cure—remain unsolved, and there is little hope on the immediate horizon.

Many forms of AIDS treatment rely on standard anti-infective therapy, such as antimicrobial drugs to combat the complicating infections that usually kill AIDS pa-

tients. Standard therapies are being applied in new ways, however. For example, people with advanced AIDS survive longer if they receive preventive daily medication for *Pneumocystis* pneumonia (which killed Patient 3). Also, many people in early stages of HIV infection are immunized against some of the pathogens most likely to cause life-threatening infections later in their illness. But no matter how effective these strategies are, they cannot cure AIDS—at best, they prolong survival.

Ten years ago researchers were confident that in a short time they would develop antiviral drugs that would kill or prevent the replication of HIV and arrest the progress of AIDS if not cure it. But the breakthroughs haven't come. With retroviruses like HIV, the viral genome becomes a part of the host cell, and eliminating the pathogen means damaging the host cell. Nevertheless, scientists are investigating ways to interfere with HIV replication at every stage of its developmental cycle—binding to the CD4 receptor, fusion of the virus with the host cell, reverse transcription of viral DNA to RNA, and integration of viral DNA into the human chromosome.

The first drug proved clinically safe and effective against HIV was azidothymidine (AZT). AZT interferes with the ability of reverse transcriptase to produce viral DNA from RNA (Chapter 21). The drug clearly prolongs the survival of most patients with advanced AIDS and seems to slow progression of the disease if administered earlier. It also seems to help patients with AIDS dementia. Currently, it is recommended that any HIV-positive patient with a T_4 cell count under 500 receive AZT, but this may change as a result of recent large studies that show early treatment may not really prolong survival. Dideoxyinosine (ddI), another anti-HIV agent, became available in 1992, and in 1993, a third, dideoxycytidine (DDC). The newest drugs are less toxic than AZT, but they still have side effects and they are not a cure (Chapter 21). Research in combining these drugs and finding more effective ones continues.

Clinical Notes

AIDS Scare

Ordinarily, a patient does not expect to be infected by a health-care provider because casual contact does not transmit AIDS. However, after several cases of AIDS were traced to a Florida dentist in 1991, Americans worried that there might be a new way to acquire the dreaded infection.

Studies were immediately undertaken to determine the exact mode of transmission from dentist to patient and whether other HIV-infected health-care workers may unknowingly have infected patients. The mode of transmission has never been clearly understood in the Florida case, and six investigations published before November 1991 failed to document additional cases of provider-to-patient transmission. Even a physician working with open lesions on his hands—which is contrary to standard infection-control guidelines—did not transmit his infection to any of the more than 300 patients on whom he performed invasive examinations or vaginal deliveries. The studies unanimously concluded that provider-to-patient transmission of AIDS is extraordinarily rare. In fact, the risk is so low it cannot be accurately measured.

Perhaps more critical than treating patients with AIDS is preventing more people from becoming infected. A vaccine would be a powerful weapon, but so far research has not been promising. No effective vaccines have ever been produced against retroviruses. Worse, HIV surface antigens mutate rapidly and continually, which would make a vaccine obsolete almost as soon as it was developed.

HIV vaccine researchers are focusing on three big questions. First, what type of human immune response could protect against HIV infection? Second, how can vaccine makers overcome the problem of viral variation? Finally, what parts of HIV should be used for a vaccine, and how should they be presented to the immune system? Basic research on the pathogenesis of HIV and its genes and antigenic variation may one day help in answering these questions.

In the absence of a scientific breakthrough to cure or prevent AIDS, efforts to contain the pandemic must rely on minimizing HIV transmission through campaigns that promote safer sex and programs to help intravenous drug users overcome their addiction. Results have been mixed because of difficulties in modifying behavior. A recent poll in New York City showed that over 80 percent of people knew condoms were protective but few used them consistently, even when engaged in behavior they knew was risky. AIDS education has decreased high-risk behavior in homosexual men, though continuing surveillance suggests that even this modest progress may not be sustained over time. Female condoms, which provide a barrier to semen entering the vagina, will soon be available in the United States, and nonirritating spermicides protective against HIV might also help slow the growing heterosexual transmission of HIV.

Only a tiny fraction of all HIV infections are acquired in the hospital, and efforts to reduce even that small number have been encouraging. Most hospital-acquired HIV infection results from a contaminated transfusion (which carries an extremely high risk of infection after a single exposure) or inadvertent exposure of a health-care worker to blood or body secretions of an infected patient (which carries a low risk of infection after a single exposure). Most infected blood has been eliminated from blood banks since 1985, when tests to identify HIV antibodies became available. Also, universal precautions designed by the CDC help protect health-care providers from infection with HIV while on the job (Chapter 20). A few cases of occupationally acquired AIDS will be inevitable, however, because of accidental needle sticks and cuts during surgery.

PROTOZOAL CAUSES

Worldwide, protozoa are among the most common pathogens of the blood and lymphatic systems.

Plasmodium spp.: Malaria

"If we take as our standard of importance the greatest harm to the greatest number, then there is no question that malaria is the most important of all infectious diseases," said Sir Macfarlane Burnet in the *Natural History of Infectious Diseases*. The media may call AIDS the most "important" infection of the 1990s, but malaria causes far more suffering and death. At any given time, approximately 100 million people are suffering symptoms of malaria, and more than 1 million people—mostly children— die from it every year. Almost all of this enormous burden of illness and death is borne by countries in the developing world.

Life Cycle of *Plasmodium*. Malaria is caused by four species of protozoa that belong to the genus *Plasmodium*. These microorganisms have a complex life cycle in mosquitoes and human beings (**Figure 27.14**). Sexual reproduction occurs in female mosquitoes belonging to the genus *Anopheles*, while asexual reproduction occurs in the liver and red blood cells of humans.

When a mosquito carrying malarial parasites bites a human, a form of the protozoan called a **sporozoite** passes from the insect's salivary gland into the person's bloodstream. Within an hour it enters the liver. In liver cells, the sporozoites multiply and produce another form, the **merozoite**. In one malarial species a single sporozoite can produce 40,000 merozoites. Eventually, infected liver cells burst, releasing huge numbers of merozoites into the bloodstream. Within minutes they invade red blood cells. Merozoites multiply within red blood cells, causing cells

to rupture and release more infectious merozoites. As merozoites infect more red blood cells and produce even more merozoites, infection escalates. A few merozoites develop into male and female gametocytes that perpetuate the life cycle if they are consumed by a mosquito during her blood meal. Infected red blood cells can be identified under the light microscope, confirming diagnosis of the disease.

All four species of malarial parasites—*Plasmodium falciparum, P. ovale, P. malariae,* and *P. vivax*—have similar life cycles, but certain minor differences are significant. *P. vivax* and *P. ovale* can remain dormant in the liver as **hypnozoites**, sleeping forms, for months or even years, leading to relapses long after the original illness seems to be over. *P. malariae* and *P. falciparum* do not cause relapsing malaria. *P. falciparum* is the most virulent species, causing almost all fatal cases of malaria. It has also developed resistance to the most effective antimalarial drugs.

Clinical Syndrome. Malarial destruction of red cells is often synchronized, releasing many new merozoites at the same time. Cell lysis typically occurs every 72 hours for *P. malariae* and every 48 hours for the other species. When this mass destruction of red cells occurs, the person

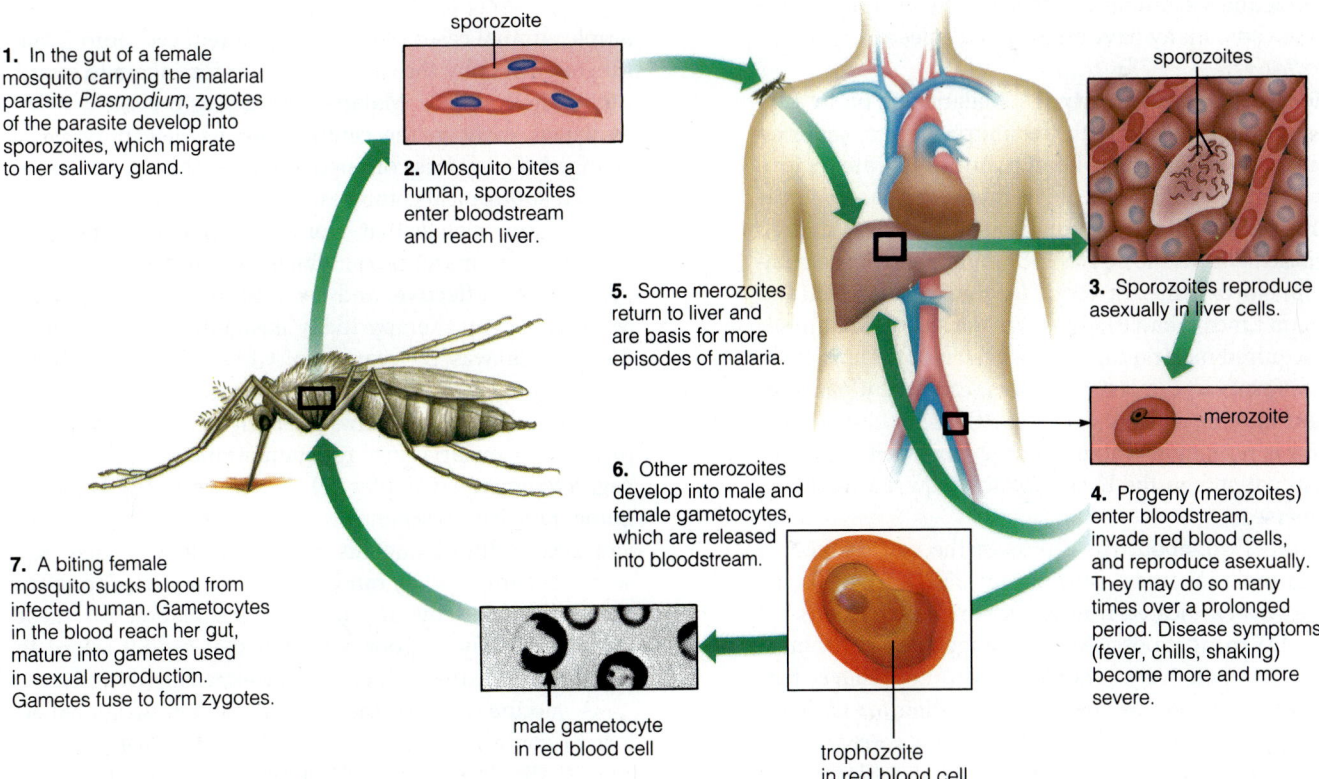

1. In the gut of a female mosquito carrying the malarial parasite *Plasmodium*, zygotes of the parasite develop into sporozoites, which migrate to her salivary gland.

sporozoite

2. Mosquito bites a human, sporozoites enter bloodstream and reach liver.

5. Some merozoites return to liver and are basis for more episodes of malaria.

sporozoites

3. Sporozoites reproduce asexually in liver cells.

merozoite

6. Other merozoites develop into male and female gametocytes, which are released into bloodstream.

4. Progeny (merozoites) enter bloodstream, invade red blood cells, and reproduce asexually. They may do so many times over a prolonged period. Disease symptoms (fever, chills, shaking) become more and more severe.

7. A biting female mosquito sucks blood from infected human. Gametocytes in the blood reach her gut, mature into gametes used in sexual reproduction. Gametes fuse to form zygotes.

male gametocyte in red blood cell

trophozoite in red blood cell

Figure 27.14 Life cycle of *Plasmodium*.

experiences a shaking chill followed by a fever as high as 104°F and a drenching sweat. Between attacks, symptoms may be minimal. Malarial antigens released in huge quantities during the red blood cell lysis probably stimulate T cells to produce cytokines, including tumor necrotizing factor, which causes the characteristic fever and chills.

Destruction of large numbers of red blood cells has other damaging consequences. If the total number of red blood cells is significantly depleted, anemia results. If masses of damaged red cells clog the circulation, blood supply to vital organs such as the kidneys or brain may be decreased. Autopsied patients who have died from **cerebral malaria** have capillaries filled with parasitized red cells and surrounded by neural tissue that has died from lack of oxygen. When massive numbers of red cells are destroyed, released hemoglobin turns the urine black. This clinical syndrome, called **blackwater fever**, usually indicates that the infection will be fatal. In chronic cases of malaria, tissues become stained by degraded hemoglobin.

The spleen, where macrophages remove damaged red cells from the circulation, often becomes greatly enlarged (**Figure 27.15**). As the burden of damaged red cells grows, circulation in the spleen becomes congested and the spleen becomes swollen. Public health workers who want a quick estimate of the prevalence of malaria in a community sometimes examine a representative sample of people to see how many have an enlarged spleen.

Epidemiology and Control. Malaria can occur anywhere humans coexist with the anopheline mosquitoes. This includes most tropical and many temperate regions of the world. In the United States there are two species of malaria vectors, *Anopheles freeborni* in the West and *A. quadrimaculatus* in the Southeast. Malaria was once common in North America (both George Washington and Abraham Lincoln had it). Although a handful of domestically acquired malaria cases have been reported in Southern California in the last few years, caused by parasites brought into the country by Latin American immigrants, malaria is no longer endemic in the United States. Most malaria patients in the United States acquired the infection overseas.

Before 1940, about two-thirds of the world's people lived in malarial areas. During the 1950s and 1960s the disease was eradicated from most temperate regions because DDT made it possible to kill huge numbers of mosquitoes inexpensively. People began to believe that malaria would soon be conquered, funding for basic research was cut, and antimalarial programs were relaxed. But DDT-resistant mosquitoes emerged and *Plasmodium falciparum* became drug-resistant, bringing about a horrifying resurgence during the 1970s. In Sri Lanka, for ex-

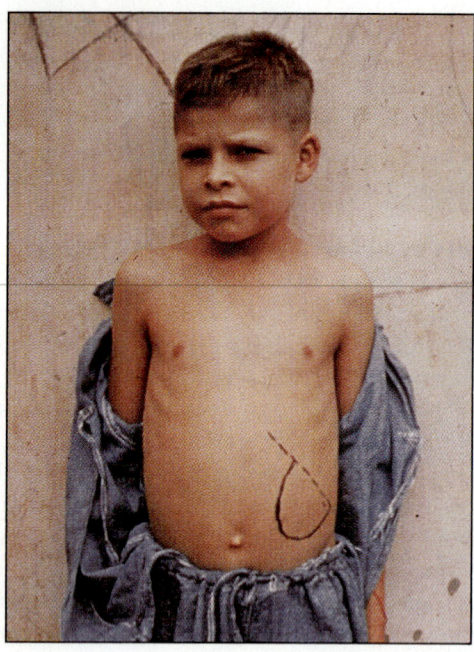

Figure 27.15 An enlarged spleen is characteristic of malaria, as shown in this young patient. His massively enlarged spleen is traced on his distended abdomen. Normally the spleen is about the size of a fist.

ample, annual cases of malaria were reduced from 3 million to only 18 by the early 1960s, but rose again to over 1 million in the 1970s. Malaria is almost as prevalent today as it was early in the century, but its distribution has changed. Today it is confined almost exclusively to tropical and subtropical countries.

We now realize that worldwide malaria cannot be controlled by insecticides, which have become more expensive, less effective, and harmful to the environment. Moreover, drug therapy today is significantly less effective than it was 25 years ago. Chloroquine, which replaced the centuries-old quinine cure, was extremely successful and inexpensive for both treatment and prophylaxis of malaria until resistant strains of *P. falciparum* began to emerge (Chapter 21). The newest drug is mefloquine, but drug resistance is a constantly evolving problem and additional agents are desperately needed. Research is hampered by our lack of basic knowledge about *Plasmodium* and by the fact that the market for these drugs is mainly in poor countries that cannot afford to support expensive research or buy expensive drugs.

A vaccine could be the turning point in the global effort to control malaria, but vaccines are most effective against diseases for which natural immunity is strong and long-lasting. Unfortunately, natural immunity to malaria is relatively weak, slow to develop, and tempo-

Malaria and Evolution

Sickle cell anemia, a potentially painful and life-threatening blood disorder seen primarily in Africans and African Americans, is a genetic disease based on inheriting two sickle genes. People who inherit only one of these genes are healthy and seem to have increased resistance to malarial infection—an extremely valuable trait in equatorial Africa, where malaria has probably been the number one health risk for thousands of years. Geneticists speculate that this protective effect against malaria accounts for the fact that so many people carry sickle genes.

rary. Furthermore, *Plasmodium* spp. frequently change their surface antigens. Nevertheless, epidemiology may be about to turn a corner.

Trials are underway to test a malaria vaccine developed by Manuel Patarroyo, a Colombian immunologist. The United States is sponsoring a trial in 1000 Thai children, and the World Health Organization, in 600 Tanzanian children. In a previous trial, in 1993 in Colombia, the vaccine was 39 percent effective in 1548 volunteers. Patarroyo's vaccine uses peptides to stimulate antibody response and targets merozoites, but exactly how the vaccine works is not understood. Nevertheless, protection against malaria for one in three infected people would be tremendous progress. If Patarroyo's vaccine is successful, it would also be a first in two other regards. This would be the first vaccine against a parasitic disease; and if it is approved, Patarroyo says he will donate the patent—worth millions of dollars—to WHO.

Until an effective vaccine becomes available, traditional "low-tech" strategies must be used to control malaria. Clinical studies show that one of the best is the use of mosquito netting impregnated with insecticides.

Toxoplasma gondii: Toxoplasmosis

Toxoplasmosis is a zoonosis that occurs in many different animal species, including reptiles, birds, and mammals. In human populations, incidence varies considerably. In the United States about one-third of adults show serological evidence of having had toxoplasmosis.

The protozoan *Toxoplasma gondii* is an obligate intracellular parasite that exists in several forms. **Tachyzoites** are the rapidly multiplying cell type that can invade all mammalian cells except mature red blood cells. During active infections, tachyzoites can either cause the host cell to rupture or turn into inactive but persistent **cysts** that remain in infected tissue indefinitely. *Toxoplasma* cysts do not harm the chronically infected animal, but can transmit the infection to other animals that eat them. A different cycle of *Toxoplasma* infection exists in cats, the only hosts in which *T. gondii* can reproduce sexually. The **gametocytes** produced are excreted as **oocysts** in cat feces, which can also transmit infection (**Figure 27.16**).

People acquire *Toxoplasma* infection in two ways—by eating meat that contains tissue cysts or by eating oocysts that have been excreted by cats. *Toxoplasma* cysts are relatively common in meat. In one study, 25 percent of pork samples and 10 percent of beef samples were contaminated. Cysts are destroyed by heating above 60°C (140°F), so thoroughly cooked meat is safe. Exposure to infective oocysts from cat feces probably occurs most often in a household setting, when—because of inadequate hand washing—people inadvertently consume traces of contaminated cat feces from litter boxes, garden soil, or sandboxes. One epidemic of toxoplasmosis among U.S. soldiers in Panama was traced to creek water that had been contaminated by wild jungle cats.

Most adults with normal immune function suffer little from toxoplasmosis. As the immune response develops, tachyzoites disappear and harmless tissue cysts take their place. Temporary enlargement of lymph nodes may be the only sign of infection. Most people who demonstrate serological evidence of toxoplasmosis were never aware of their infection.

But people who are immunosuppressed and pregnant women are at risk for serious illness. In immunosuppressed hosts, such as AIDS or cancer patients, toxoplasmosis can occur as a newly acquired infection, but more often it is a reactivation of prior infection. Most often, the brain, heart, or lungs are affected. Diagnosis can be confirmed by isolating the microorganism, by identifying it in samples of infected tissues or fluids, or by serology. Antimicrobial treatment with the sulfonamides and pyrimethamine keeps clinical manifestations from worsening. Sometimes it takes as long as a year to bring infection completely under control.

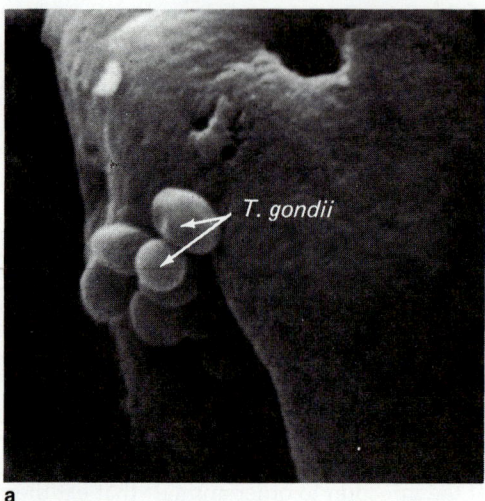 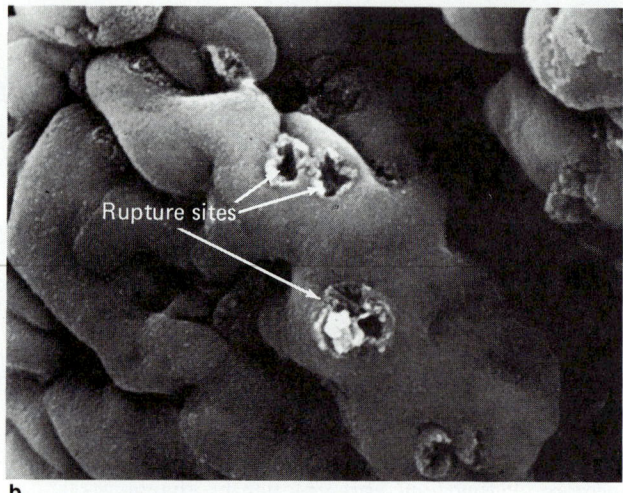

Figure 27.16 Electron micrographs showing the intestinal lining of a *Toxoplasma*-infected cat. (**a**) Protozoal parasites within cells. (**b**) Rupture site where mature oocytes have passed into the intestine to be excreted

In pregnant women toxoplasmosis infection during the early months can severely damage or kill the embryo, stimulating spontaneous abortion. If infection occurs during later months, the fetus may be born with **congenital toxoplasmosis**. Some congenitally infected babies are healthy at birth but develop signs and symptoms later. Congenital toxoplasmosis usually affects the nervous system, causing seizures, blindness, deafness, or mental retardation. Usually, infants also have a rash and enlarged liver and spleen. Congenital toxoplasmosis can be diagnosed with certainty if a newborn has IgM antibodies against *T. gondii*. Because IgM antibodies do not cross the placenta, they must have been produced by the infected fetus before birth.

No vaccine exists for toxoplasmosis. Prevention depends on thoroughly cooking potentially infected meat and avoiding cat feces. Because oocysts are not infectious for the first 24 to 48 hours after they are passed, infection can be minimized by emptying cat litter boxes daily. Cats that are always kept indoors and have no chance to acquire the infection by eating other infected animals are a low risk to humans.

Babesia spp.: Babesiosis

Babesia are protozoal parasites that resemble *Plasmodium*. Many species are found around the world—in Africa, Asia, Europe, and North America. **Babesiosis** is a zoonosis. A tick vector transmits the parasite from one mammalian host to another, usually deer, cattle, or small rodents. Humans become infected if they happen to be bitten by an infected tick. Finding *Babesia* in a patient's blood smear is diagnostic. Most human infections are relatively mild, although fever, headache, chills, fatigue, and weakness may occur. Rarely, infection can be fatal in the elderly or the immunosuppressed. Babesiosis can be treated with clindamycin and quinine administered together, but most patients recover without treatment. In the United States babesiosis is seen on the New England coast. The vector is *Ixodes dammini*, the same tick that transmits Lyme disease in that region. Small rodents are the natural reservoir.

HELMINTHIC CAUSES

Helminthic infections of the blood and lymphatic systems are seen primarily in developing countries.

Schistosoma spp.: Schistosomiasis

Blood flukes, or **schistosomes**, are a type of trematode or flatworm that causes the disease **schistosomiasis**. There are three major species of schistosomes, all with a complex life cycle that involves a freshwater snail. Sometimes the infection is called **snail fever** or **bilharzia** after the German parasitologist Theodor Bilharz, who identified the schistosomal parasites.

The schistosome life cycle begins when fresh water is contaminated with human feces or urine (**Figure 27.17**). An infected person excretes schistosome eggs, which

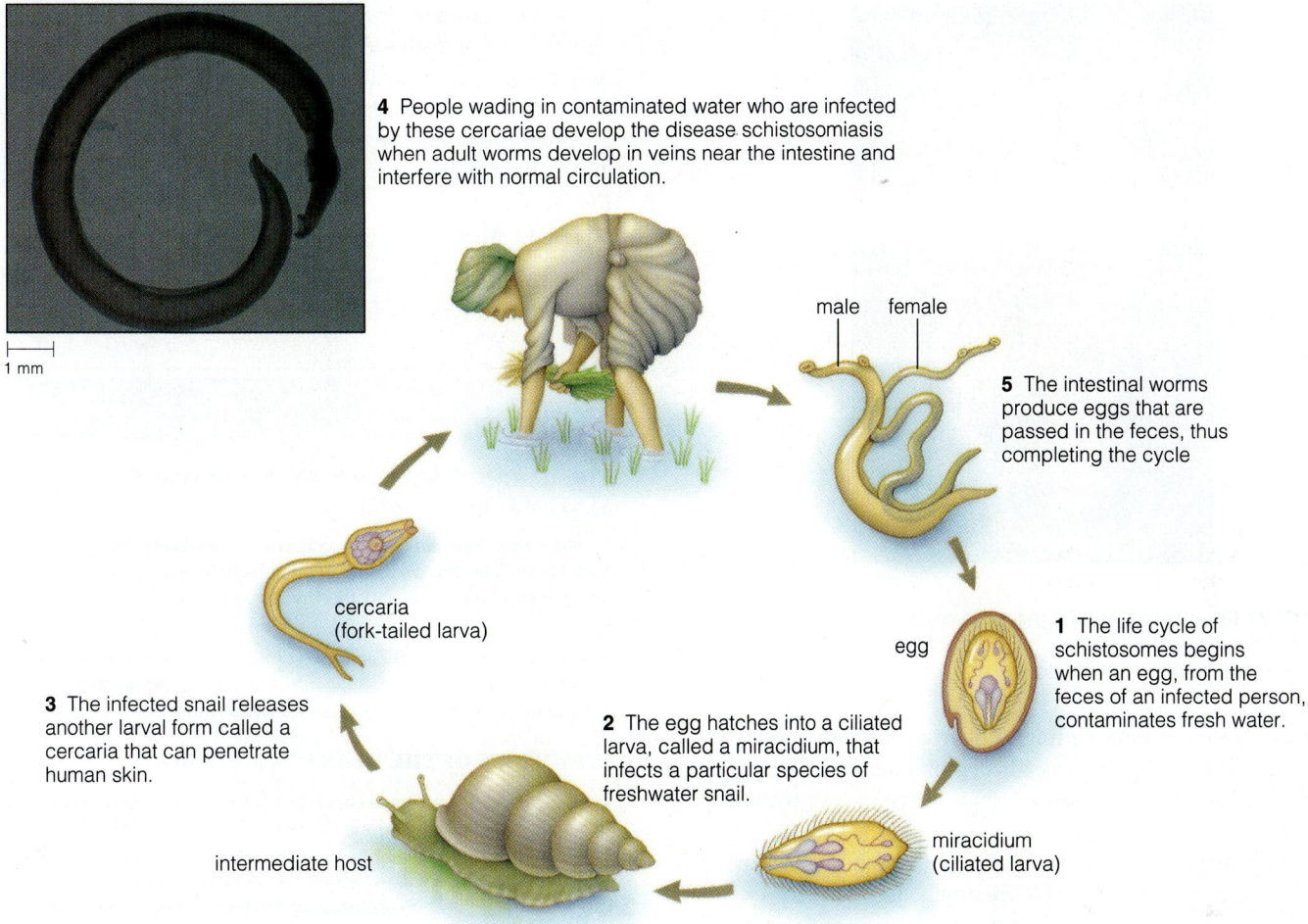

4 People wading in contaminated water who are infected by these cercariae develop the disease schistosomiasis when adult worms develop in veins near the intestine and interfere with normal circulation.

male   female

5 The intestinal worms produce eggs that are passed in the feces, thus completing the cycle

cercaria (fork-tailed larva)

1 The life cycle of schistosomes begins when an egg, from the feces of an infected person, contaminates fresh water.

egg

3 The infected snail releases another larval form called a cercaria that can penetrate human skin.

2 The egg hatches into a ciliated larva, called a miracidium, that infects a particular species of freshwater snail.

intermediate host

miracidium (ciliated larva)

1 mm

Figure 27.17   The complex life cycle of a schistosome, a type of flatworm, requires a certain freshwater snail. An adult *Schistosoma japonicum* worm is shown at the left.

hatch into infective larvae called **miracidia**, which parasitize a particular species of snail. The infected snail, in turn, releases larvae called **cercariae** that can penetrate human skin, infecting people who wade in the contaminated water. Cercariae enter the blood, where they mature into male and female worms. The female lives in a groove in the male's body, and the two persist in the bloodstream, producing enormous numbers of fertile eggs. Different schistosomal species migrate to the blood vessels of different organs: *Schistosoma mansoni* migrates to veins near the colon, *S. japonicum* to veins near the small intestine, and *S. haematobium* to veins near the bladder. Schistosome eggs pass from the blood into the nearby organ and are excreted in the stool or the urine. If these products contaminate water that contains the required snail species, the cycle begins again. Finding the eggs in a stool or urine sample is diagnostic.

Once infected with schistosomiasis, a person can remain infected for decades because of an unusual immunologic phenomenon. The schistosomes acquire on their own surface host-derived antigens that keep them from being recognized and eliminated by the immune system. As a result, enormous numbers of worms and their eggs develop in the circulation over the years. As the worms multiply, the patient experiences fever, fatigue, and pain in the affected organ. Eventually, the infected tissues become so severely irritated that the body reacts against them as a foreign body, producing granulomas. Thus, the extensive tissue damage that occurs with chronic schistosomiasis is a type of hypersensitivity.

Schistosomes, which are prevalent in Asia, Africa, and the Middle East, cause a huge burden of human disease. At any time, about 200 million people around the world suffer from schistosomal infections. Ironically, economic development has increased the incidence because dams and irrigation projects create new bodies of fresh water that are ideal habitats for schistosomes and their snail hosts. Although schistosomiasis is seen in the

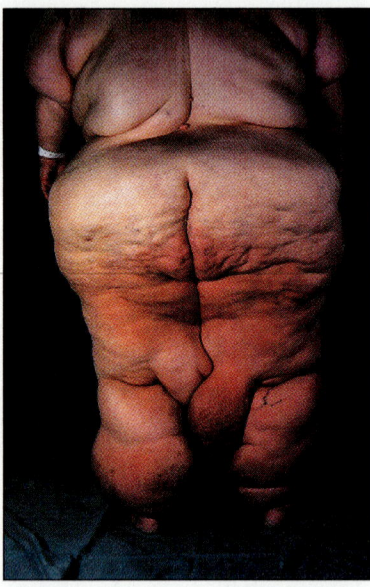

Figure 27.18 Elephantiasis, a grotesque swelling caused by worms obstructing the lymphatic circulation, is caused by two closely related species of nematodes.

United States among immigrants from endemic countries, the infection can never become endemic in North America because the essential snail hosts are not found here.

Schistosomiasis can be treated with praziquantel or oxiaminiquine (Chapter 21). Treatment is relatively expensive, though, and not a practical solution for the millions of victims around the world. Prevention through improved sanitation and controlling the vector snails would be more desirable and cost-effective. Despite the fact that the immune system is relatively ineffective in combating schistosomiasis, research is underway to develop a vaccine.

Wuchereria bancrofti and *Brugia malayi:* Filariasis

Wuchereria bancrofti and *Brugia malayi* are two closely related species of nematodes that cause an infection of the blood and lymphatic circulations called **filariasis**. It is a mosquito-borne infection transmitted by species of *Anopheles, Culex,* and *Aedes* mosquitoes in Africa, Asia, and the Pacific Islands.

People acquire filariasis when an infected mosquito introduces nematode larvae beneath the skin during a blood meal. The larvae enter the lymphatic vessels, usually in the extremities or the groin, and mature into adult worms. Nine months to a year later the worms mate and produce **microfilariae** that reenter the circulation, where they can infect a new mosquito. Presence of microfilariae in a patient's blood confirms diagnosis.

Many filariasis infections produce only fever and chills when parasites enter the blood. The most obvious manifestation of infection is **elephantiasis**, swelling of a limb or the scrotum that results when adult worms block lymphatic circulation over the course of years (**Figure 27.18**). Treatment with diethylcarbamazine is effective if the diagnosis is made before elephantiasis becomes established. Mosquito control is the best prevention.

Table 27.3 summarizes the infections of the cardiovascular and lymphatic systems.

Summary

THE CARDIOVASCULAR AND LYMPHATIC SYSTEMS   (pp. 670–672)

1. The cardiovascular system consists of the heart, blood vessels, and blood. The lymphatic system, which includes lymph vessels and nodes, the spleen, and the thymus, is also called the immune system.

2. Clinical syndromes of heart infections are endocarditis, myocarditis, and pericarditis. Noncardiac infections of the cardiovascular and lymphatic systems are called systemic infections.

INFECTIONS OF THE HEART   (pp. 672–674)

1. Bacterial endocarditis is characterized by vegetations (masses of bacteria and blood clots). Emboli are pieces of vegetations that break off, enter the circulation, and interrupt blood flow, causing death if they enter a vital organ. Vegetations that damage heart valves can cause heart failure. Symptoms of bacterial endocarditis include fever and anemia.

2. Acute bacterial endocarditis is sudden in onset and progresses rapidly. *Staphylococcus aureus* causes most of these infections in intravenous drug users. Treatment is with cephalosporins and penicillinase-resistant penicillins; heart valves may be surgically replaced.

3. Subacute bacterial endocarditis starts gradually, progresses slowly, and lasts longer than acute disease. Symptoms are fever and fatigue. People with heart defects are at risk; and without treatment, the infection is fatal. Prophylactic treatment with penicillin is routine before oral or other surgery.

4. Viral myocarditis is often caused by coxsackie virus. The only symptom may be an irregular or rapid heartbeat. Treatment consists of rest. The prognosis is good, though rarely the condition can be life-threatening.

5. *Trypanosoma cruzi* causes American trypanosomiasis (Chagas' disease). It is transmitted by the reduviid bug. The acute stage is characterized by fever, fatigue, and swelling around the bite. During the chronic stage the heart and other organs become damaged. The acute stage is treated with nifurtimox; the only treatment during the chronic stage is supportive care.

6. Viral pericarditis is usually a complication of a viral infection; it is usually benign and self-limited. The most common agents are enteroviruses.

7. Bacterial pericarditis is usually caused by *Staphylococcus* or *Streptococcus* spp., spreading from a lung or systemic infection. Pericardial tamponade can cause sudden death.

8. Chronic fungal and mycobacterial infections can also cause pericarditis. The resulting inflammation can cause scarring, leading to heart failure; a pericardectomy is usually performed.

Table 27.3 Infections of the Cardiovascular and Lymphatic Systems

Infection	Causative Agent	Mode of Transmission	Symptoms	Prevention and Treatment
INFECTIONS OF THE HEART				
Endocarditis	Bacteria (usually), fungi (rarely)	Acute disease most common in otherwise healthy people who abuse intravenous drugs; subacute disease common in people with previous heart abnormality	Fever, fatigue, anemia, valvular damage, possibly heart failure	Avoid drug abuse; prevent subacute disease with antibiotic prophylaxis; treat with antibiotics, sometimes surgery to replace valves
Myocarditis	Viruses (usually), rarely bacteria, fungi	Usually a complication of an everyday viral illness	Irregular heartbeat, chest pain, heart failure	Supportive care
Trypanosomiasis (Chagas' disease)	*Trypanosoma cruzi*	Bite of reduviid bug	Protozoal myocarditis; fever, fatigue, face swells; can be acute or chronic	Avoid bites; treat acute cases with nifurtimox; treat chronic heart disease with medications to reduce symptoms
Pericarditis	Viruses (usually), bacteria or fungi (occasionally)	Usually a complication of an everyday viral disease or a chronic bacterial or fungal infection	Few; may cause scarring or pericardial tamponade and thereby affect heart function	Supportive care; possibly a pericardectomy
SYSTEMIC INFECTIONS				
Bacterial				
Plague	*Yersinia pestis*	Infected flea bite, respiratory droplets	Chills, fever, swollen lymph nodes (bubos), black patches of dead skin	Rat abatement programs, quarantine; treat with streptomycin, tetracycline, chloramphenicol
Tularemia	*Francisella tularensis*	Direct contact with an infected animal (usually rabbit in U.S.); bite of infected tick or fly	Ulcer at site of infection, fever, fatigue, headache	Avoid contact with infected animals; treat with streptomycin or tetracycline
Brucellosis	*Brucella* spp.	Direct contact with infected animals; consuming infected unpasteurized dairy products	Fever (often with characteristic undulating pattern), fatigue, headache, loss of appetite	Eliminate infected animals from domestic herds; pasteurize milk; treat with tetracycline or streptomycin
Lyme disease	*Borrelia burgdorferi*	Infected tick bite	Bull's-eye rash and fever in acute form; chronic infection can damage joints, heart, nervous system; can cause chronic arthritis	Avoid or remove ticks; treat early with doxycycline, amoxycillin, or erythromycin
Relapsing fever	*Borrelia* spp.	Body louse (epidemic disease) or tick (endemic disease)	Recurrent episodes of fever, headache, muscle aches	Avoid contact with vectors; treat with tetracycline, chloramphenicol, penicillin, or erythromycin
Anthrax	*Bacillus anthracis*	Contact with infected wool or hides; rarely, respiratory droplets	Black sore at site of infection, fever, shock, severe pneumonia	Eliminate infected animals from domestic herds; avoid contact with animal products from endemic countries; treat with penicillin
Cat scratch disease	*Rochalimaea henselae*	Direct contact with infected animal	Usually one swollen lymph node and healing scratch	Avoid contact with infected animals; treat with trimethoprim-sulfamethoxazole or ciprofloxacin
Rocky Mountain spotted fever	*Rickettsia rickettsii*	Infected tick bite / Infected body louse bite	Characteristic rash over entire body, including palms and soles, fever	Avoid vector; treat with tetracycline or chloramphenicol

Table 27.3 (continued)

Infection	Causative Agent	Mode of Transmission	Symptoms	Prevention and Treatment
Bacterial				
Epidemic typhus	*Rickettsia prowazekii*	Infected fleabite	Unbearable headache, fever, muscle aches, chills; often delirium or unconsciousness; rash over body except palms and soles; shock; can be fatal	Good hygiene; treat with tetracycline or chloramphenicol
Murine typhus	*Rickettsia typhi*		Mild fever, headache, muscle aches; rickettsial rash	Avoid rat-infested areas; treat with tetracycline or chloramphenicol
Viral				
Yellow fever	Yellow fever virus	Infected mosquito bite	Fever, headache, jaundice	Avoid mosquito bites; vaccine available; supportive treatment only
Dengue fever	Dengue fever virus	Infected mosquito bite	Fever, headache, rash; rarely, dengue hemorrhagic fever	Avoid mosquito bites; supportive treatment only
Infectious mononucleosis	Epstein-Barr virus	Person-to-person	Fever, fatigue, sore throat, enlarged lymph nodes and spleen	No prevention (infection is almost universal); treat symptoms only
Burkitt's lymphoma	Epstein-Barr virus	Person-to-person but only in children in equatorial Africa	Malignant jaw tumor	No prevention; curable with chemotherapy
Acquired immunodeficiency syndrome (AIDS)	Human immunodeficiency virus (HIV)	Sexual contact, exposure to infected blood or body fluids, mother to infant during pregnancy or near time of birth	Sometimes a mononucleosis-like illness at time of infection; years later, opportunistic infections develop; eventually fatal	Prevent with safer sex practices and avoiding illicit intravenous drugs; treat opportunistic infections with antimicrobials; slow progress with AZT, ddl, DDC; no cure
Protozoal				
Malaria	*Plasmodium* spp.	Infecting mosquito bite	Cyclic chills and fever; tissue damage from red blood cell destruction	Avoid mosquito bites; treat with antimalarials chloroquine or mefloquine; vaccine in field trials
Toxoplasmosis	*Toxoplasma gondii*	Eating meat from an infected animal or ingesting organisms in traces of cat feces	None or few; very serious in immunosuppressed patients and fetus	Cook meat well; avoid infected cat feces; treat with sulfonamides or pyrimethamine
Babesiosis	*Babesia* spp.	Infected tick bite	Mild fever, headache, chills	Avoid tick bites; treat with clindamycin and quinine together
Helminthic				
Schistosomiasis	*Schistosoma* spp.	Wading in contaminated water; freshwater snails are essential intermediate host	Fever, fatigue, muscle aches, chronic damage to organs; gravely debilitating	Keep water free of human feces and urine; control snail host; treat with praziquantel or oxiaminiquine
Filariasis	*Wuchereria bancrofti* and *Brugia malayi*	Infected mosquito bite	Body parts swell where lymphatic vessels are blocked; debilitating and disfiguring	Avoid mosquito bites; treat with diethylcarbamazine

SYSTEMIC INFECTIONS OF THE CARDIOVASCULAR AND LYMPHATIC SYSTEMS (pp. 674–698)

1. Any bacterial infection can spread to the blood. Bacteremias are brief infections; septicemias are persistent and serious.

Bacterial Causes (pp. 674–683)

1. *Yersinia pestis* causes plague. In the United States, rodents west of the Rocky Mountains are a permanent reservoir. Transmission is by a flea bite. A nearby lymph node becomes massively enlarged and tender (a bubo). Bacteria multiply and enter the blood, spreading to all the organs. Black spots appear on the skin from subcutaneous bleeding. Plague spread by fleas is called bubonic plague. If the lungs become infected, bacterial cells are aerosolized and the infection is spread by the respiratory route (pneumonic plague). Early treatment with streptomycin, tetracycline, or chloramphenicol is usually effective. A vaccine is available for high-risk individuals.

2. *Francisella tularensis* causes tularemia. Rabbits and ticks are the main reservoirs. Bacteria can be ingested or inhaled; or they can enter through a break in the skin. The most common syndrome is ulceroglandular tularemia; a small sore appears where the bacteria enter the body and symptoms of systemic infection develop—fever, chills, headache. Diagnosis is difficult; untreated, the illness can be persistent and serious, with relapses. Streptomycin is the drug of choice.

3. *Brucella* spp. cause brucellosis. Transmission is by direct contact with infected animals or their secretions. Before pasteurization, brucellosis was often milkborne. Brucellosis is also called undulant fever because of characteristic recurrent fever. Recovery is usually spontaneous, but chronic illness can last for months. Diagnosis is difficult; treatment is with tetracycline and streptomycin.

4. *Borrelia burgdorferi* was discovered as the cause of Lyme disease in 1982. The first sign of infection is often a bull's-eye rash. During the next stage, bacteria spread, causing headache, muscle aches, and fatigue, often followed by serious illness, such as meningitis. The late stage may last for years, with chronic arthritis and painful joint swelling. *Ixodes* ticks are the primary vectors. Without the characteristic bull's-eye rash, diagnosis can be difficult. Treatment is with doxycycline, amoxicillin, and erythromycin.

5. *Borrelia* spp. cause relapsing fever. Onset is sudden, with fever, headache, muscle aches, enlarged liver and spleen, and a faint rash on the trunk. The patient appears to recover in three to six days, but a week later has a relapse. The cycle continues, with up to 13 relapses. Epidemic relapsing fever is transmitted by the body louse and has a high mortality; endemic relapsing fever is transmitted by ticks and is comparatively mild. Tetracycline, chloramphenicol, and penicillin are effective.

6. *Bacillus anthracis* causes anthrax, an animal disease humans can contract through a break in the skin (cutaneous anthrax) or by inhaling bacterial spores (respiratory anthrax). Both are treatable with penicillin in the early stages, but not when septicemia develops. Cases in the United States occur from anthrax-contaminated hides or wool imported from endemic countries. The main virulence factor is anthrax toxin; spores sustain the chain of infection.

7. *Rochalimaea henselae* causes cat scratch disease, characterized by one extremely swollen lymph node and usually no other symptoms. In rare cases, the patient complains of fever, fatigue, or symptoms of meningitis. The infection resolves itself, but trimethoprim-sulfamethoxazole or ciprofloxacin speeds recovery.

8. *Rickettsia rickettsii* causes Rocky Mountain spotted fever. The bacterium is passed transovarially, so no animal reservoir other than ticks is needed. Early signs include fever, headache, vomiting, and a distinctive rash on the wrists and ankles that eventually spreads over the entire body, including the palms and soles. Diagnosis is difficult; there is no test to identify *R. rickettsii*. Untreated, there is risk of overwhelming infection leading to shock and death. Treatment is with tetracycline and chloramphenicol.

9. *Rickettsia prowazekii* causes epidemic (louseborne) typhus. Infection occurs when a person scratches a bite and rubs louse feces, with their rickettsial parasites, into the skin. Onset is sudden, with severe headache, fever, muscle aches, and shaking chills, often leading to delirium and unconsciousness. A rash resembling that of Rocky Mountain spotted fever appears but does not cover the palms, soles, or face. Tetracycline and chloramphenicol are effective. Mild recurrences of epidemic typhus—called Brill-Zinsser disease—may appear years later.

10. *Rickettsia typhi* causes murine (fleaborne) typhus, which is much milder than epidemic disease. Transmission occurs when humans come in contact with infected rodents. Onset is gradual, followed by a typical rickettsial rash. Treatment is with tetracycline or chloramphenicol.

Viral Causes (pp. 683–692)

1. The yellow fever virus causes yellow fever, transmitted principally by the *Aedes aegypti* mosquito. Symptoms include high fever, severe chills and headache, liver damage causing jaundice, and sometimes uncontrollable bleeding. Mosquito-control programs have eliminated urban yellow fever, but it is endemic in jungles of the Americas. There is a vaccine but it is seldom used.

2. Several flaviviruses cause dengue fever, also transmitted by the *Aedes aegypti* mosquito. High fever and headache are followed by a rash and intense limb pain; periods of recovery and recurrent illness alternate. Treatment is limited to supportive care. Occasionally children who have recovered from dengue fever become reinfected and develop dengue hemorrhagic fever or dengue shock syndrome, which is fatal in 40 percent of cases.

3. The Epstein-Barr virus (EBV) causes infectious mononucleosis. As a member of the herpesvirus family, EBV establishes an active and latent infection. Transmission is through oral secretion, directly or via fomites. Symptoms include fever, fatigue, sore throat, and swollen lymph nodes. EBV has been proved oncogenic, causing B-cell lymphoma cancer in immunosuppressed cancer patients and organ transplant recipients; Burkitt's lymphoma, a jaw cancer in African children; and nasopharyngeal carcinoma, which occurs only in China.

4. The human immunodeficiency virus (HIV) causes acquired immunodeficiency syndrome (AIDS). HIV cripples the immune system—in particular, T_4 lymphocytes—making a patient vulnerable to other infections. When the T_4 count falls below 200 per cubic millimeter, opportunistic infections of the skin and mucous membranes appear. When the count falls below 100, life-threatening systemic illnesses occur, sometimes accompanied by mental deterioration. Death is inevitable.

   a. Transmission occurs in three ways: through sexual contact that transfers infected body fluids, through infected blood from transfusions or shared IV needles, and from mother to infant during pregnancy or around the time of birth. During the prolonged asymptomatic period, an individual may unknowingly transmit the virus to others.

   b. The epidemiology in the United States has shifted from homosexual men to drug users sharing contaminated needles; this change has led to increased sexual transmission in the heterosexual community and a rapidly growing number of infected women and children. In Africa, the disease is spread almost exclusively by heterosexual intercourse.

c. Diagnosis is by serologic testing. Prevention depends upon education in safer sex, campaigns against drug use, and—among health-care providers—following universal precautions. Because there is no cure, treatment is designed to prolong survival. Drugs include AZT, ddI, and DDC. Some standard therapies are being used in new ways.

Protozoal Causes (pp. 692–696)

1. Four species of *Plasmodium* cause malaria. Transmission is by *Anopheles* mosquitoes. Sporozoites deposited during a bite travel to the liver where they produce merozoites. As infected liver cells lyse, merozoites enter red blood cells to multiply. Red blood cells lyse in cycles, every 48 to 72 hours, producing attacks of shaking chills, high fever, and drenching sweat. Patients develop anemia and a swollen spleen; some die from cerebral malaria or blackwater fever. Treatment is with chloroquine and mefloquine, but resistance to both drugs is emerging. A promising vaccine is in field trials.

2. *Toxoplasma gondii* causes toxoplasmosis, a zoonosis. Humans acquire the infection by eating meat containing cysts or by eating oocysts excreted by cats. *T. gondii* exists as tachyzoites as well as inactive cysts. Enlarged lymph nodes may be the only symptom; but in immunosuppressed hosts, the brain, heart, or lungs are often affected. A pregnant woman risks abortion or an infant born with congenital toxoplasmosis, producing seizures, blindness, deafness, or retardation. Prevention depends upon cooking potentially infected meat and avoiding infected cat feces.

3. *Babesia* spp. cause babesiosis, another zoonosis. In the United States, babesiosis is primarily found in New England, where it is transmitted by the same tick that transmits Lyme disease, *Ixodes dammini*. Symptoms include fever, headache, chills, and fatigue, but recovery is usually spontaneous. Serious cases may be treated with clindamycin and quinine in combination.

Helminthic Causes (pp. 696–698)

1. Schistosomes (blood flukes) are trematodes that cause schistosomiasis. An infected human contaminates water with egg-carrying feces; larvae, called miracidia, hatch from the eggs. Miracidia parasitize a particular species of snail that in turn releases larvae called cercariae. Cercariae penetrate the skin of humans wading in contaminated water and enter the blood vessels of various organs. Enormous numbers of worms can develop in the circulation over decades; symptoms include fever, fatigue, and pain in the affected organs. Treatment is with praziquantel or oxiaminiquine. Prevention through sanitation and eliminating the vector snails is best. Schistosomiasis can never become endemic to the United States because the essential snail host is not found here.

2. *Wuchereria bancrofti* and *Brugia malayi* are nematodes that cause filariasis. Transmission is by species of mosquitoes found in Africa, Asia, and the Pacific Islands. Larvae enter the lymphatic vessels in the groin or extremities and mature; worms mate and produce microfilariae that reenter the circulation. Treatment with diethylcarbamazine is effective only before elephantiasis appears. Mosquito control is the best prevention.

Review Questions

THE CARDIOVASCULAR AND LYMPHATIC SYSTEMS

1. Describe the organs of the cardiovascular system and their functions. Describe the organs of the lymphatic system and their functions.

2. Define these clinical syndromes: endocarditis, myocarditis, pericarditis. What is a systemic infection, and why is it difficult to diagnose?

INFECTIONS OF THE HEART

1. Discuss the differences between acute and subacute bacterial endocarditis; include causative microorganisms, symptoms, individuals at high risk, and treatment in your discussion. What are vegetations, and what medical problems can they cause?

2. What type of myocarditis is most often found in the United States? What are the most frequent causative microorganisms, symptoms, and treatment?

3. What type of heart infection is Chagas' disease? What is the causative microorganism? How is it transmitted and what are the endemic areas? Describe the pathogenesis, diagnosis, and treatment.

4. Compare and contrast the types of pericarditis caused by bacteria, viruses, and fungi.

SYSTEMIC INFECTIONS OF THE CARDIOVASCULAR AND LYMPHATIC SYSTEMS

1. Define these terms: bacteremia, septicemia.

2. Discuss transmission, symptoms, treatment, and prevention of bubonic plague. What is pneumonic plague and how virulent is it?

3. How is tularemia transmitted? Name and describe the most common clinical syndrome and its treatment.

4. Who is at highest risk of contracting brucellosis, and what is the most indicative diagnostic sign?

5. Is Lyme disease a new disease? Explain. Describe the three clinical stages, vectors and mode of transmission in different parts of the country, diagnosis, and treatment.

6. Discuss the two epidemiological patterns of relapsing fever—endemic and epidemic. Why must antibiotic treatment be gradual?

7. Compare and contrast cutaneous anthrax and respiratory anthrax. Is anthrax a disease we can expect to eradicate one day? Explain.

8. How serious is cat scratch disease? Explain your answer.

9. Describe the clinical syndrome of Rocky Mountain spotted fever and difficulties in its diagnosis and treatment.

10. Compare and contrast the two forms of typhus. Name the causative microorganisms and describe the transmission, clinical syndromes, treatment, prevention, and prevalence in the United States.

11. Describe the principal vector, mode of transmission, and classical symptoms of yellow fever. What distinguishes the urban form from the endemic form?

12. Explain this statement: Dengue fever resembles yellow fever in its epidemiology and mode of transmission. Describe the clinical syndrome.

13. Describe the latent and active infections caused by the Epstein-Barr virus. Discuss this statement and give examples: EBV is one of the few viruses that has been proved oncogenic.

14. What type of virus is HIV and of what clinical significance is this? Which human cells are most damaged by HIV? What is their function? Describe the infective process that begins with infection and ends with death. How is AIDS transmitted? How is AIDS *not* transmitted? Describe the clinical syndrome of AIDS. Discuss the prevention and treatment of AIDS. What questions must be answered before a vaccine can be developed? Discuss AIDS in Africa. How is the epidemiology of AIDS changing in the United States and why?

15. How does the life cycle of *Plasmodium* cause the symptoms of malaria? Discuss the epidemiology and control of malaria in the United States and worldwide. What new hope is there for treating malaria?

16. How do humans acquire toxoplasmosis? How serious is this infection? Explain. How can it be prevented?

17. Compare and contrast babesiosis and Lyme disease in terms of transmission, symptoms, and virulence.

18. Describe how the schistosome life cycle causes the clinical syndrome of schistosomiasis. What is the treatment? Explain this statement: Schistosomiasis can never become endemic in North America.

19. How do the life cycles of *Wuchereria bancrofti* and *Brugia malayi* cause filariasis and its most obvious clinical manifestation, elephantiasis? Discuss prevention and treatment.

Suggested Readings

Aoun, H. 1989. When a house officer gets AIDS. *New England Journal of Medicine* 321:693–96.

Durack, D. T. 1988. Rus in Urbe: Spotted fever comes to town. *New England Journal of Medicine* 318:1388–90.

Marshall, E. 1990. Malaria research—what next? *Science* 247:399–402.

McCabe, R., and Remington, J. S. 1988. Toxoplasmosis: The time has come. *New England Journal of Medicine* 318:313–15.

McEvedy, C. 1988. The bubonic plague. *Scientific American*, February, 118–23.

Maurice, J. 1993. Controversial vaccine shows promise. *Science* 259:1689–1690.

Relman, A. S. 1984. Epstein-Barr virus—immortalization and replication. *New England Journal of Medicine* 310:1255–57.

Science. 1993. AIDS: The unanswered questions. Special issue. May 28.

Steere, A. C. 1989. Lyme disease. *New England Journal of Medicine* 321:586–97.

Zinsser, H. 1934. *Rats, lice, and history.* New York: Bantam.

28 MICROORGANISMS AND THE ENVIRONMENT

Every Link in the Chain Matters

Life is interdependent. On an organism-to-organism scale, for example, fungi and algae form mutualistic symbioses that produce lichens. On a global scale, biogeochemical transformations make Earth suitable for life. In between are *ecosystems*, complexes of organisms living interdependently in a particular environment. Ecosystems are differentiated by temperature and by the availability of water, sunlight, oxygen, and other factors. Deserts, tropical rain forests, temperate forests, Arctic tundra, coral reefs, open oceans, lakes, and rivers are examples of ecosystems.

In spite of their diversity, all ecosystems share certain features. With a few exceptions, such as hydrothermal vents, ecosystems run on energy from the sun. Photoautotrophs—including plants, algae, cyanobacteria, and nonoxygenic procaryotic phototrophs—are the *primary producers*. They use solar energy to build their constituent organic compounds from atmospheric carbon dioxide (CO_2). All other organisms in the system are *consumers*, heterotrophs that live directly or indirectly at the expense of primary producers. Herbivores eat primary producers. Carnivores eat herbivores or other animals, and omnivores eat primary producers and animals. Detritivores consume dead organic matter. Bacteria and fungi, the *decomposers* of the ecosystem, live on the remains or products of other organisms and convert them to forms usable by primary producers.

Ecosystems need a continuous input of energy, which is continuously dissipated as heat produced by organisms in the system. Most nutrients are recycled within the system, but some are lost—for example, in runoff water or when animals migrate. Some are replaced—for example, by minerals exposed through erosion or nutrients carried in the atmosphere.

The pattern of consumption in an ecosystem—who eats whom—is called a *food web*. In the Antarctic, for example, the primary producers in the food web are phytoplankton. Krill feed on phytoplankton. Blue whales, fishes, and penguins feed on krill. Some penguins also eat fish. Fish and penguins are eaten by seals, which are consumed by killer whales. And this is only part of the picture.

Ecosystems are extremely fragile. Because all organisms are interdependent, the loss of one species threatens the entire system.

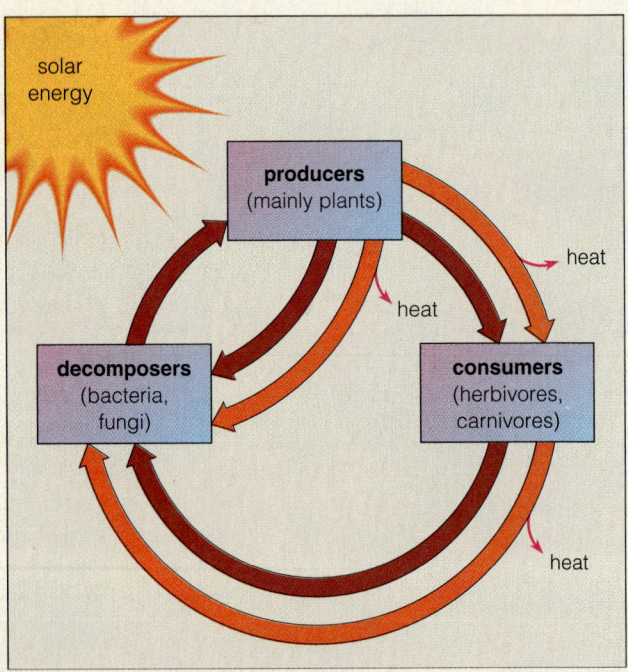

Major parts of an ecosystem.

To understand:

▪ How the evolution of organisms led to our present environment

▪ Soil and water as habitats for the microorganisms that mediate major biogeochemical transformations

▪ The major cycles of matter—the nitrogen cycle, the carbon cycle, the phosphorus cycle, and the sulfur cycle—and the microorganisms involved

▪ The ecological impact of the major cyclic transformations of matter

▪ How knowledge of the cycles of matter is applied to treating waste water and drinking water

▪ How some human-made chemicals escape from the carbon cycle and what can be done about it

LIFE AND THE EVOLUTION OF OUR ENVIRONMENT

A large part of this textbook discusses how microorganisms cause disease and how we try to stop them. But it would be a mistake to think of the microorganisms in our environment only as adversaries. The preceding description of ecosystem interactions reveals that all life, including our own, depends on the activities of microorganisms.

Earth is a rare environment in its suitability for life. The more biologists learn from space probes, the more they think life does not exist anywhere else in our solar system.

Of the many requirements of carbon-based life (life as we know it), liquid water is paramount. For a planet to support life, it must orbit at a suitable distance from its sun to maintain a temperature in the narrow range in which water is liquid. Also, it must have enough mass for its gravity to prevent water from escaping the atmosphere. A second critical requirement is usable sources of the major biological elements—carbon, hydrogen, nitrogen, oxygen, phosphorus, and sulfur (abbreviated CHNOPS).

The environment determines whether life can exist, but life changes the environment. Chemicals from the environment that enter a cell as substrate are changed during metabolism, and different chemicals leave the cell to enter the environment. Because microorganisms are so plentiful, their metabolic activities have a major im-

pact on the environment. Collectively, the biologically mediated changes in Earth's chemical composition are called **biogeochemical transformations** (*bio*, "life"; *geo*, "earth").

Life and Earth's environment evolved together, producing the near-balanced state of coexistence we see today. When Earth was formed about 4.6 billion years ago, there was no oxygen in the atmosphere. Laboratory experiments (**Figure 28.1**) show that this oxygen-free atmosphere suited the slow formation and accumulation of the building blocks of life—amino acids, purines, pyrimidines, carbohydrates, and lipids. Electrical discharges in the form of lightning provided most of the energy for the gases in the primitive atmosphere to react and form organic compounds. These compounds dissolved in surface waters and accumulated because no microorganisms were present to metabolize them (see the beginning of Chapter 2).

The first organisms on Earth were probably anaerobic, heterotrophic organisms formed from and living on

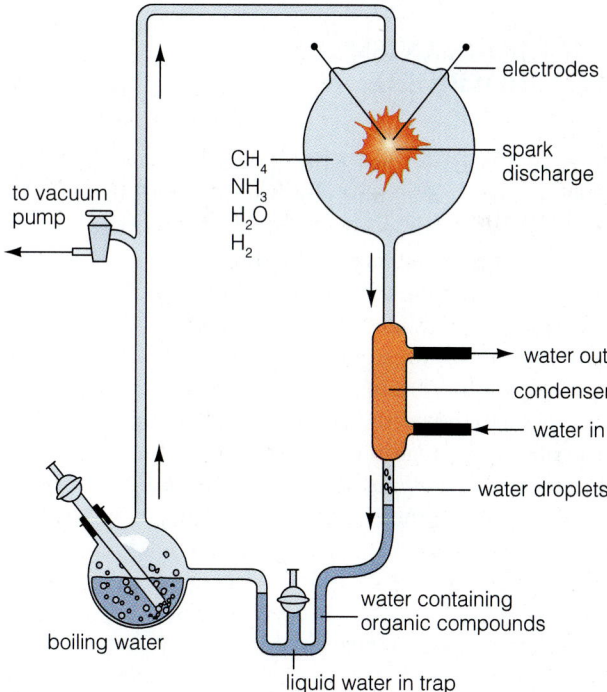

Figure 28.1 This laboratory apparatus mimics the conditions of Earth's primitive atmosphere and produces the building blocks of life. Boiling water (H_2O) at a reduced pressure created by a vacuum pump produces water vapor that flows through a reaction vessel with the gases believed to have constituted the primitive atmosphere [methane (CH_4), ammonia (NH_3), and hydrogen (H_2)]. The mixture is bombarded with electrical sparks to simulate lightning; organic compounds are synthesized. A condenser with flowing cold water condenses the vapor into droplets containing the organic compounds, and the droplets are collected in the trap for analysis.

the accumulated building blocks. But the oldest known fossils (3.5 billion years old) are of filamentous organisms that look quite like the cyanobacteria found in soil, lakes, and streams today. These ancient cyanobacteria, like modern forms, probably carried out oxygenic photosynthesis. The oxygen they produced began to accumulate in Earth's atmosphere about 2 billion years ago. Between 800 and 600 million years ago, the concentration of oxygen became high enough to support the metabolism of aerobic organisms, and plant and animal life appeared. Today, oxygen constitutes 21 percent of Earth's atmosphere. This figure represents a dynamic balance between oxygen produced through photosynthesis and oxygen used in respiration and combustion.

Over the last 300 to 600 million years, other life-sustaining elements stabilized in a similar way so that each is formed and used at approximately equal rates. These complex biogeochemical cycles keep the earth in dynamic equilibrium, both maintained by life and supporting it, and microorganisms play a particularly important role.

MICROORGANISMS IN THE BIOSPHERE

Natural materials move through a series of biological conversions. Leaves that fall are degraded by microorganisms or consumed by animals and returned to their elemental components to fulfill different roles in perpetuating life. All living things contribute to life; but microorganisms play particularly important roles. Microorganisms transform huge quantities of matter, and only they can carry out certain essential transformations. These transformations occur in various **ecosystems** within the **biosphere**, the region of the earth that can support life. Many transformations occur in the soil, others in aquatic environments or the atmosphere.

Soil

Although it appears inert, the topmost layer of soil teems with microorganisms mediating myriad chemical transformations vital to geochemical transformations and to soil fertility. **Soil** is a complex, heterogeneous medium consisting of minerals and other inorganic materials (oxides of iron, aluminum, and silicon) from the earth's crust, living organisms, and the organic residue of dead ones (**Table 28.1**). Some organic residues, such as **lignin**, the stable component of woody plants, are long-lasting because they are resistant to microbial decomposition. They accumulate and form the organic fraction of soil called **humus**, which gives soil its brown or black color.

Table 28.1	Mineral Composition of a Typical Sandy Loam Soil
Element	Percent of Element in Soil (by weight)
Silicon (Si)[a]	24.7
Aluminum (Al)	3.7
Potassium (K)	1.7
Iron (Fe)	0.8
Calcium (Ca)	0.6
Sodium (Na)	0.3
Magnesium (Mg)	0.09
Phosphorus (P)	0.06
Manganese (Mn)	0.003

Source: Martin Alexander, 1977, *Soil microbiology,* 2d ed. (New York: John Wiley).
Note: Though the composition of different soils varies, the relative amounts of these elements are about the same.
[a]Silicon dioxide (SiO_2) is the dominant component of most soils, often accounting for 70 to 90 percent of the total mass.

Mineralization. In the biogeochemical transformation known as **mineralization**, microorganisms convert organic material in soil to an inorganic form. The rate and extent of mineralization depend on the availability of oxygen. Compared to anaerobic metabolism, aerobic metabolism is more versatile—more compounds are attacked. It is also more complete, producing carbon dioxide and water instead of organic acids and alcohols (Chapter 5). Many organic materials are mineralized only if oxygen is available, but oxygen penetrates soil readily, down to a foot or so, when it is relatively dry and loose. Even then, small regions within soil particles are anaerobic because oxygen-consuming microorganisms use the oxygen faster than it diffuses in. When soils are flooded, they rapidly become completely anaerobic because water slows diffusion of oxygen to less than the rate that aerobic microorganisms use it. As a result, mineralization proceeds slowly in waterlogged soils such as swamps and bogs. Such gradual mineralization was dramatically demonstrated in the 1960s when the body of a Bronze Age man was found almost intact in a bog in Denmark (**Figure 28.2**). Waterlogged soils typically contain more than 90 percent organic material, while well-aerated agricultural soils usually contain less than 10 percent.

Soil fertility describes a soil's ability to support plant growth. Fertility depends in large measure on the amount of inorganic nitrogen, phosphorus, and potassium in the soil. Forms of these nutrients that plants can use are produced by microorganisms as they mineralize organic material. Fertilizers are added to enrich a soil's complement of these elements. Potassium is added to fertilizer largely as an inorganic salt. Nitrogen and phos-

Figure 28.2 Mineralization is the process by which microorganisms convert organic material in soil to an inorganic form. In anaerobic bogs mineralization is an extremely slow process, as this well-preserved body of a Bronze Age man (more than 4000 years old) taken from a bog in Denmark shows.

Figure 28.3 The vast majority of microorganisms living in the soil are bacteria, as shown in this micrograph of sandy soil.

phorus can be added in either organic or inorganic form because the organic forms are readily mineralized by microorganisms. Thus, through mineralization, microorganisms improve soil fertility even when commercial fertilizers are used. Some bacteria improve soil fertility in a different way, by *fixing nitrogen*—converting nitrogen gas from the atmosphere into solid form in the soil.

Organisms in the Soil. Many organisms live in the soil. Some, including insects, worms, and small vertebrates, are visible to the naked eye. But the vast majority in terms of weight and metabolic capacity are microscopic, and most of these are bacteria (**Figure 28.3**).

Bacteria. Soil bacteria are extremely diverse. They include aerobes, anaerobes, and facultative anaerobes, which continue to proliferate as their habitat cycles between aerobic and anaerobic. Most soils contain thermophilic microorganisms, reflecting the fact that the soil

surface can become extremely hot during the day. Soil temperature varies widely, however, and bacteria that inhabit the soil vary greatly in their optimum temperature for growth. Soil contains psychrophiles and mesophiles, as well as thermophiles. Similarly, soil bacteria grow over a wide range of pH.

Actinomycetes, aerobic Gram-positive bacteria that form branching mycelia (Chapter 11), are important contributors to the ecology of soil. They break down plant and animal remains and keep the soil loose and friable. Over a million actinomycete colonies, representing more than 20 genera, can be recovered from 1 g of soil. The most numerous and widely distributed is the genus *Streptomyces*. The typical odor of soil is attributable to two volatile substances, **geosmin** and **2-methyl-isoborneol**, produced by streptomycetes. *Streptomyces* colonies on laboratory media produce the same earthy odor. Availability of nutrients and oxygen determines the number and kinds of actinomycetes in soil. Actinomycetes are particularly significant degraders of complex polymers (including chitin) and hydrocarbons, which are relatively resistant to attack by other microorganisms.

A few bacteria that cause human disease are found in soil. They include *Bacillus anthracis*, which causes anthrax (Chapter 27), *Clostridium perfringens*, which causes food poisoning (Chapter 23), *Clostridium tetani*, which causes tetanus (Chapter 25), and *Clostridium botulinum*, which causes botulism (Chapter 25). *Pseudomonas aeruginosa*, which causes opportunistic infections in burn patients and immunologically weakened individuals (Chapter 26), occurs in almost all soils.

Fungi. Fungi are another group of active aerobes that degrade organic materials in the soil. They break down both simple compounds such as sugars and organic acids and complex polymers such as cellulose, starch, pectin, and lignin. Colony counts commonly underestimate the fungal population of soil because a mass of fungal hyphae may produce only a single colony when plated.

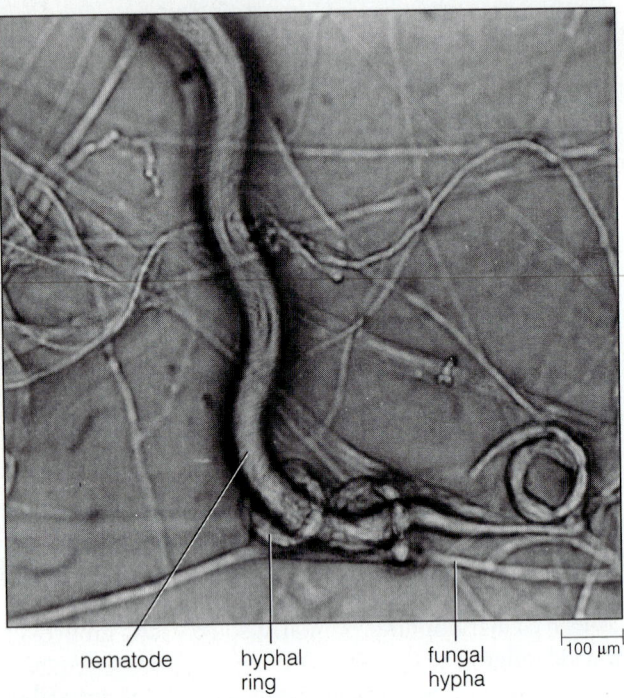

nematode | hyphal ring | fungal hypha | 100 μm

Figure 28.4 Soil fungi that are active predators limit the populations of soil protozoa and nematodes. Here, *Arthrobotrys conoides* has entrapped a nematode.

Fungal **biomass** (total weight of organisms) is a more informative estimate of their impact on soil ecology. An acre of soil contains between 500 and 5000 pounds of fungi in the complete soil layer.

Some soil fungi are predators. They produce specialized appendages or hyphal extensions that form rings to trap protozoa or nematodes (**Figure 28.4**). Then the fungal hyphae invade the captured prey and secrete enzymes that degrade it into small molecule nutrients, which they absorb. These predator fungi limit the populations of soil protozoa and nematodes.

Certain soil fungi, notably species of *Trichoderma* and *Laetisaria*, are **mycoparasites**. They attack other fungal species, including some that cause plant disease. Treating soil and seeds with mycoparasitic fungi can protect plants from disease. For example, *Trichoderma harzianum* controls damping-off (a disease that kills seedlings by blackening and shrinking their stems) of beans, peas, and radishes by *Rhizoctonia solani* or *Pythium* spp. *Trichoderma hamatum* improves survival of sugar beet seedlings, and *Laetisaria arvalis* protects seedlings of many species from fungal pathogens in the soil. For other reasons, however, commercial agriculture makes little use of mycoparasites.

Soil is also the major reservoir of some fungi that are pathogenic to humans. They include *Blastomyces dermatitidis*, which causes blastomycosis; *Histoplasma capsulatum*, which causes histoplasmosis; and *Coccidioides immitis*, which causes San Joaquin valley fever (all are discussed in Chapter 22).

Other Microorganisms. Algae are present on the surface of all soils, but usually in small numbers. A gram of soil contains 100 to 50,000 colony-forming units, amounting to between 7 and 300 pounds of algal biomass per acre. Algae and phototrophic procaryotes do not contribute significantly to soil fertility except in rice paddies, where cyanobacteria, free-living or in association with plants, fix considerable amounts of nitrogen.

The numbers of protozoa in soil are small, but there is probably no soil that lacks them completely. Protozoan cell counts in soil vary between about 10,000 and 100,000 per gram. Their direct effect on biochemical transformations in the soil is minor. Indirectly, however, they play a critical role by preying on the bacterial population and thus regulating its size and composition.

Many small animals spend part or all of their life cycle in the soil. These include earthworms, slugs, snails, centipedes, millipedes, wood lice, arachnids, insects, flatworms, and roundworms. Together, they amount to several hundred pounds of animal tissue in each acre of ordinary soil. Many of the roundworms, or nematodes (Chapter 12), attack plant roots, sapping their strength and sometimes killing them. They form characteristic knobs on roots. Tomato roots almost always have them by the end of the growing season. Larvae of the hookworm *Necator americanus*, a nematode that infects humans, are free-living in soil (Chapter 23).

Symbiosis. Many soil microorganisms live in symbiosis with other organisms. Some of these relationships are mutualistic (benefiting both partners), and most of them are with plants (though lichens are mutualistic associations between algae and fungi; Chapter 12). Two important mutualistic symbioses between microorganisms and plants are the mycorrhizae and the rhizosphere.

Mycorrhizae. Some soil fungi form **mycorrhizae**, intimate associations with roots of plants; the fungi act as additional roots, helping the plant acquire nutrients. Most plants probably have mycorrhizae. Mycorrhizal associations between basidiomycetes and beech, birch, and pine trees are particularly abundant in forest soils of temperate regions. The fungal mycelium penetrates the outermost layers of the tree root, but most remains outside, forming a sheath up to 40 μm thick. The association is usually essential to the fungus—most mycorrhiza-forming fungi cannot be cultivated free of the plant with which they normally associate. Presumably, the fungus derives some essential nutrients from its plant partner. The plant can survive, although not very well, without its

Not Just Another Soil Fungus

Many species of mycorrhizal fungi produce edible mushrooms, including *Amanita* spp. and *Boletus edulis*: some of these are prized gourmet treats. But the most desirable and certainly the most expensive mushroom is *Tuber melanosporum* and the related species *T. aestivum*, *T. brumale*, and *T. uncinatum*—better known as truffles. Truffles are ascomycetes that form mycorrhizal associations with forest trees, predominantly hazelnut and oak trees. But the edible parts of truffles (their ascocarps, Chapter 12) are produced underground, which makes them difficult to find. Female pigs are extremely sensitive to the smell of truffles because it resembles the odor of a sexually active boar. In France trained pigs smell the underground truffles and root them up with their snout. Hunting truffles with pigs is so successful and the market so good that over the years there has been a precipitous decline in production. In 1892, 2000 tons of truffles were produced in France. In 1973, only 60 tons were produced. Since then, artificial inoculation of forest soils with *Tuber* spp. has increased production.

---

fungal partner. In the laboratory, when mycorrhiza-free plants are cultivated in sterilized forest soil, they are yellow and stunted. By comparison, plants that have mycorrhizae are vigorous and deep green (**Figure 28.5**).

The fungus supplies its symbiotic plant partner with mineral nutrients that are usually in short supply in forest soils. Orchids are dependent on their mycorrhizal partners, which penetrate deep into their root tissue, to supply the plant with organic as well as mineral nutrients. Many orchids cannot be grown without mycorrhizal fungi unless supplied with certain organic nutrients, including vitamins.

The Rhizosphere. The microbial ecology of the **rhizosphere**, the region of soil immediately surrounding the roots of plants, is significantly different from the rest of the soil. The microorganisms in these regions are not as intimately associated with plants as mycorrhizae are, but they nevertheless have a profound effect on plant growth. Considerably greater numbers of specific kinds of microorganisms are found in the rhizosphere. This concentration, termed the **rhizosphere effect**, is described quantitatively by the **R:S ratio**, the ratio of microorganisms in the same amount of rhizosphere and adjacent nonrhizosphere soil. For Gram-positive bacteria, actinomycetes, protozoa, and algae, the R:S ratio is relatively small—2 or 3 to 1. But for Gram-negative bacteria—particularly species of *Pseudomonas*, *Flavobacterium*, and *Alcaligenes*—the R:S ratio can be several hundred to 1. The rhizosphere effect occurs because plant roots excrete both nutrients and antimicrobial agents that inhibit some microorganisms but not others. Thus, microorganisms in the rhizosphere benefit the plant, although the mechanism is not always clear. For example, barley associated with a normal rhizosphere microflora takes up phosphate about twice as effectively as barley grown in

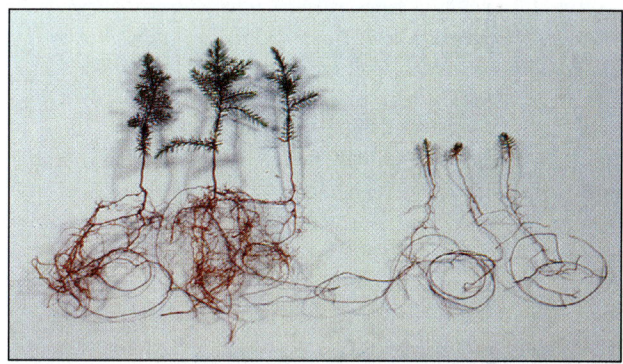

Figure 28.5 The three juniper seedlings on the left were grown in the presence of fungi that form mycorrhizae. Those on the right were grown in the same soil but without the fungus.

sterile soil, and the rhizosphere microflora is a barrier against fungal pathogens that attack plant roots.

Water

Seventy percent of Earth is covered by water, and all the major classes of microorganisms can live in aqueous environments. Microorganisms are found in boiling hot springs, in Antarctic waters that rarely rise above the freezing point of salt water, and in salt-saturated seas. They even grow near hydrothermal vents at the bottoms of oceans (see the box "Deep-Sea Hydrothermal Vents," p. 724).

Streams, rivers, ponds, lakes, and oceans are different aqueous environments, and each of these bodies of water is heterogeneous. The air-water surface, the water column itself, suspended particles in the water column, and the bottom are different environments. Covering so much of Earth's surface, the ocean is a major contributor to the cycles of matter. The biosphere is, on the average, about

Vibrio cholerae: Life in a Human, a Hemisphere, and a Habitat

Gary Schoolnik and David Thornton

Dr. Gary Schoolnik is Chief of the Division of Infectious Diseases and Geographic Medicine at Stanford Medical School. In 1984 he started a tropical medicine research center in Chiapas, Mexico, where cholera cases now occur in large numbers. Besides his laboratory research, he cares for patients on the internal medicine and infectious diseases services at Stanford University Hospital. David Thornton earned an A.B. degree in biochemical sciences from Harvard College in 1989. He is currently an M.D. and Ph.D. degree student at Stanford Medical School, pursuing graduate studies in the Department of Microbiology and Immunology.

It is the end of a long day for this man. Though he is strong and hardy, his work in the heat has left him tired. And thirsty. There is a lake nearby that pours forth a small river, and he goes there to quench his thirst. The water is a little cloudy, with a green tint, but as a cure for thirst it looks perfectly satisfying.

It is hours later—he can't tell exactly how long. The metronomes of his life have gone mad. His breathing is shallow and rapid. His heart beats frantically, but no pulse can be felt for his blood has become viscous and sluggish. His skin and eyes have collapsed inward, disguising his strength. He lies at the origin of another river. This one flows more furiously for it originates not from a lake but from the intestines of this thirsty man—a fountain of liters of his life's water.

What can compare to the intensity of a severe case of cholera? A great city is situated beside a harbor. The harbor is this city's mouth, speaking greetings to the world and ingesting the nourishment of trade. A ship from another hemisphere comes to the harbor and exchanges ballast water for cargo. The ballast carries with it an organism that hasn't been here for a century. Now conditions are right, and it takes hold. Cases of cholera spring up along the coast. The numbers soar—tens to hundreds to thousands—and medical resources quickly become strained. The disease travels inland and begins to march down highways, spreading at the rate of one new country every month. There are more cholera deaths from this outbreak in a year than occurred worldwide in the previous five years combined.

This is a true story about a new cholera epidemic in our hemisphere. Like many infectious agents, *Vibrio cholerae*, the bacterial cause of cholera, possesses not only the capacity to cause disease in individual patients but also the ability to move with astounding speed between persons in large epidemics. And, when not causing outbreaks of human disease, it can live in lakes, rivers, and marine environments, where it forms symbiotic partnerships with other members of the aquatic flora, including cyanobacteria. Viewed in this way, *V. cholerae* is a remarkable organism: It causes a fatal diarrheal illness; it spreads rapidly in human populations, across continents and oceans; and it can live innocently, in the absence of humans, in water. It follows that the control of cholera must be undertaken by an interdisciplinary task force. The physician, concerned with the diagnosis and treatment of individual patients, fights the disease with rehydration therapy and antibiotics. The epidemiologist works at the community level, searching like a detective for infected aquatic habitats and transmission mechanisms so that effective public health interventions can stop the epidemic. The molecular biologist uses the tools of genetic engineering to discover how the organism causes disease at the molecular and cellular levels with the expectation that this will lead to new vaccines and drugs. And the ecologist studies how *V. cholerae* survives as a species in nature and in particular how human-driven changes in the environment—the use of chemical fertilizers, deforestation, and the greenhouse effect—might favor its spread and persistence.

We are particularly interested in testing the hypothesis that the use of chemical fertilizers, deforestation, slash-and-burn agricultural practices, and the resulting greenhouse effect have all contributed to the ongoing cholera epidemic in Latin America by nourishing the algae with which it lives. To test this hypothesis, we have constructed microcosms in our laboratory at Stanford University that resemble a tropical pond where we can study the interaction of *V. cholerae* and cyanobacteria. We are analyzing the metabolic transactions between the two organisms in order to understand why their symbiotic partnership is mutually advantageous. And we are using molecular biology techniques to find the underlying genetic basis for this survival strategy. We anticipate that the responsible genes will be very old in evolutionary time, having arisen long before the arrival of vertebrates on earth. However, beyond satisfying our curiosity, we hope that our work will lead to the discovery of new ways of controlling cholera through the use of environmental interventions that undermine the organism's capacity to survive in nature.

38 meters thick over the land masses, but it extends throughout the entire volume of the ocean. In these terms, 99 percent of the earth's biosphere is sea water, although much of it is not a particularly favorable environment for life. Over half the water in the world's oceans is colder than 2°C and under a pressure greater than 100 atmospheres.

Nutrients. In general, aqueous environments support smaller populations of microorganisms than soil because most aqueous environments have a low concentration of nutrients.

The concentration of dissolved organic nutrients in oceans is so low that many marine microbiologists believe heterotrophic microorganisms grow only when they are attached to nutrient-containing particles. In contrast, phototrophic microorganisms flourish in the upper regions of the ocean where light penetrates and the inorganic nutrients they require are present in adequate concentrations. These organisms, called **phytoplankton**, constitute a major portion of the world's total photosynthetic capacity. In the lower regions of the ocean, where light does not penetrate, heterotrophs get most of their nutrition from dead phytoplankton drifting down from the top regions. These small particles settle so slowly—at a rate of 0.1 to 1.0 meter per day—that most decompose before they reach the bottom. The 1 percent that does reach the ocean floor supports a teeming microbial activity in the top layer of ocean sediments.

Unless they are polluted, bodies of fresh water are also nutrient-poor, which adds to their clarity and deep blue color. When aqueous environments become nutritionally enriched with nitrogen and/or phosphorus by runoff water from construction projects and municipal sewage or by industrial or agricultural waste water, large populations of microorganisms develop (**Figure 28.6**). Such bodies of water become **eutrophic**—enriched in inorganic and organic nutrients. Their color changes from blue to green.

Many eutrophic ponds and streams support huge populations of cyanobacteria that grow near the surface, collecting in unsightly masses called **blooms**. Within this rich source of nutrients, aerobic microorganisms often deplete dissolved oxygen more rapidly than it can be supplied from the atmosphere or the cyanobacteria. Then the masses of microbial cells are decomposed anaerobically by other microorganisms. The result can be unpleasant because anaerobic decomposition includes fermentation; and some products of fermentation—including fatty acids, amines, and mercaptans—have offensive odors.

If heavily polluted, even large bodies of water become anaerobic and foul-smelling. In England during the early 1800s, the city of London routinely discharged its sewage into the Thames River. A river as large as the Thames nor-

Figure 28.6 This eutrophic pond is covered with duckweed and supports a huge population of cyanobacteria.

mally has sufficient flow to dilute an immense load and mix in oxygen to remain aerobic. Eventually, however, the Thames became overwhelmed and organic pollution became anaerobic. The fermentation that followed led Londoners to refer to their situation as "the big stink." Modern methods of sewage treatment are discussed later in this chapter.

Pathogens in Aqueous Environments. Fresh water is a significant reservoir for pathogens that cause human disease. Most of these pathogens enter the water in human feces and infect new hosts when the contaminated water is used for drinking. In most cases, the diseases are gastrointestinal (**Table 28.2**), but some pathogens that live in water cause respiratory disease (see the box "When Help Becomes Harm").

Air

Air is not a habitat for microorganisms—they do not grow in air. But bacteria, algae, protozoa, fungi, and viruses are all found in air, usually as passengers on **aerosols** (tiny particles of liquid) or dust particles. We make microorganism-bearing aerosols when we cough, sneeze, or talk. Any agitation of water, such as waves breaking, rapids, and sprays, also creates aerosols. Even sudden heating of water—for example, flaming a wet inoculating loop (Chapter 3)—can produce an aerosol

Clinical Notes

When Help Becomes Harm

The patient was a 71-year-old woman with a long history of respiratory problems. After more than 50 years of smoking, her lungs were so damaged by emphysema that she had trouble breathing even during normal activity. When she was admitted to the hospital with a broken hip, breathing became more difficult. Her doctor prescribed an aerosolized medication; she would inhale the drug in a fine mist. These breathing treatments were helpful, but soon the patient became critically ill with what appeared to be pneumonia. Fortunately, legionellosis was quickly diagnosed (Chapter 22). She was

immediately treated with tetracycline and given supportive care in the intensive care unit, which allowed her to make a complete recovery. Curiously, water—rather than air or exposure to another person—was the source of this patient's infection. The bacterium that causes legionellosis, *Legionella pneumophila*, grows in water—lakes, streams, air conditioning towers, and even the water supplies of hospitals. The water used to prepare this patient's breathing treatments was contaminated. As is often the case, other patients in the hospital were also affected.

Table 28.2 Some Human Pathogens and Diseases Transmitted in Water	
Pathogen	**Disease**
Bacteria	
Salmonella typhi	Typhoid fever
Salmonella spp.	Salmonellosis (gastroenteritis)
Escherichia coli	Diarrhea
Legionella pneumophila	Legionellosis
Vibrio cholerae	Cholera
Vibrio parahaemolyticus	Gastroenteritis
Yersinia enterocolitica	Enterocolitis (diarrhea)
Campylobacter jejuni	Diarrhea
Viruses	
Hepatitis A virus	Type A (infectious) hepatitis
Hepatitis E virus	Type E hepatitis
Poliovirus	Poliomyelitis
Protozoa	
Entamoeba histolytica	Amoebic dysentery
Giardia lamblia	Giardiasis (diarrhea)
Balantidium coli	Balantidiasis (diarrhea)
Cryptosporidium	Cryptosporidiosis (diarrhea)
Helminths	
Fasciola hepatica (liver fluke)	Hepatitis (liver rot)
Opisthorchis sinensis	Hepatitis (liver rot)

Note: All of these pathogens and diseases (except legionellosis and poliovirus) are discussed in Chapter 23. Legionellosis is discussed in Chapter 22 and poliovirus in Chapter 25.

bearing live microorganisms. Aerosols and dust are the principal means of transmitting respiratory diseases.

THE CYCLES OF MATTER

Each of the major biological elements occurs in several different chemical forms. Nitrogen, for example, exists in the atmosphere as nitrogen gas (N_2) and on the earth's surface largely in organic compounds and as ammonia (NH_3) and nitrate (NO_3^-). These various forms of nitrogen are interconvertible. Nitrogen gas is converted to organic nitrogen, to ammonia, to nitrate, and back to nitrogen gas. Most of these chemical changes are carried out by living organisms, principally microorganisms. Moreover, these conversions are roughly balanced, so the net production of each form equals the net rate of utilization. Virtually all the molecules of nitrogen gas now in our atmosphere were once a part of some living thing, and they will be again. The cyclic chemical interconversion of each of the biological elements is a **biogeochemical cycle**.

The cycles of matter themselves are interrelated because most natural compounds are made of several elements. For simplicity, however, we look at each of the major biogeochemical conversions separately—nitrogen, carbon, phosphorus, and sulfur.

The Nitrogen Cycle

Of all the biological elements, nitrogen is especially important to agriculture and ecology because its concentration in most soils is low enough to limit the growth rate of plants and the yield of crops. Moreover, nitrogen undergoes an especially complex series of chemical conversions because it exists in many different forms with dif-

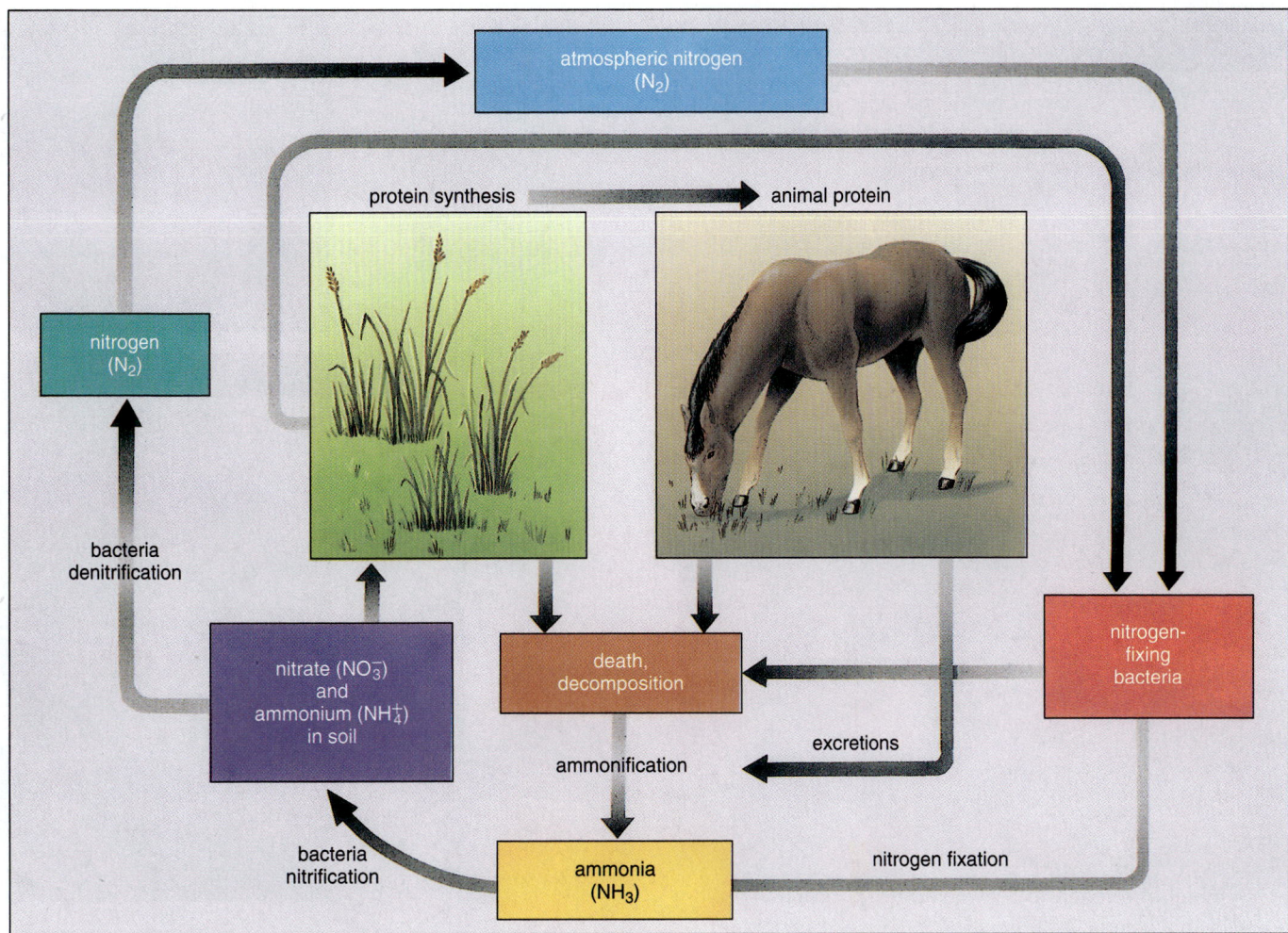

Figure 28.7 In the nitrogen cycle, nitrogen gas is converted to ammonia by nitrogen fixation. Ammonia is converted to nitrate by nitrification. And nitrate is converted to nitrogen gas—completing the cycle—by denitrification.

ferent chemical properties. The **nitrogen cycle** involves the conversion of nitrogen gas to ammonia by **nitrogen fixation**, the conversion of ammonia to nitrate by **nitrification**, and the regeneration of nitrogen gas from nitrate by **denitrification (Figure 28.7)**. Along the way intermediates, particularly ammonia, are incorporated into the organic components of organisms.

Nitrogen Fixation. Free nitrogen gas (N_2) constitutes about 80 percent of Earth's atmosphere, but it is not available to most organisms. Only a few procaryotes are able to **fix** nitrogen gas—to convert it to a usable form—and they pay a high price. They must expend more than 16 molecules of ATP to fix each molecule of N_2.

$$N_2 + 16\ ATP + \text{reducing power} \longrightarrow 2\ NH_3 + 16\ ADP + 16\ PO_4^{3-}$$
NITROGEN GAS   AMMONIUM ION   PHOSPHATE

There are two classes of nitrogen-fixing bacteria—those that form symbiotic relationships with plants and free-living (nonsymbiotic) ones that do not (**Table 28.3**). Both classes face two major metabolic problems. They must pay the high metabolic cost of fixation, and they must maintain a strictly anaerobic environment because the enzyme that catalyzes fixation, **nitrogenase**, is rapidly inactivated by oxygen.

Free-living nitrogen-fixers have evolved ways to protect their nitrogenase from oxygen. The aerobe *Azotobacter* respires oxygen at its surface so rapidly that the interior of the cell, where the nitrogenase is located, is anaerobic. The aerobic (and oxygen-producing) cyanobacteria fix nitrogen in **heterocysts**, specialized cells that do not produce oxygen inside and do not let it enter from the outside (Chapter 11). The free-living anaerobes such as *Clostridium pastorianum* need no special adaptations because they inhabit oxygen-free environments.

Table 28.3 Some Genera of Bacteria That Fix Nitrogen	
FREE-LIVING	
Aerobes	Nonoxygenic Phototrophs
Azotobacter	*Rhodobacter*
Beijerinckia	*Rhodospirillum*
Spirillum	*Rhodomicrobium*
Facultative Anaerobes	*Chlorobium*
Bacillus	Cyanobacteria
Enterobacter	*Anabaena*
Klebsiella	*Gleocapsa*
Anaerobes	*Nostoc*
Clostridium	*Oscillatoria*
SYMBIOTIC	
Rhizobium	
Bradyrhizobium	
Frankia	
Anabaena	

root hair

root nodule

Figure 28.8 The nitrogen-fixing bacteria *Rhizobium* and *Bradyrhizobium* form symbiotic relationships with legumes. When bacterial cells contact a root hair, they induce it to form an infection thread. The bacteria travel down the thread into the root tissue, penetrating cells where they multiply, forming a nodule. In the nodule, the bacteria differentiate into bacteroids and devote their metabolic energy to fixing nitrogen.

Symbiotic nitrogen-fixers, including *Rhizobium* spp., some cyanobacteria, and *Frankia* spp., depend on a plant to protect them from oxygen and to supply nutrients as well. The bacterium supplies the plant with fixed nitrogen. The complexity of these symbioses between bacteria and plants varies considerably. At one extreme, nitrogen-fixing cyanobacteria simply accumulate in small pockets on the surface of certain lower plants, such as bryophytes and ferns.

At the other extreme, *Rhizobium* spp. form complex and intimate relationships with leguminous plants. When the bacterium comes in contact with the root hair (slender extension of a surface cell) of a legume, the root undergoes a complex series of morphological changes. Eventually, a tube called an **infection thread** forms, through which the bacterium can penetrate deep into the root tissue. There it forms nodules that become filled with rhizobial cells (**Figure 28.8**). These nodules are small nitrogen-fixing factories. Along with nutrients, the plant supplies the nodule with **leghemoglobin**, an oxygen-binding protein that maintains an optimal concentration for nitrogen fixation and rhizobial metabolism. Within the nodule, rhizobial cells differentiate into **bacteroids**, nongrowing cells that devote their entire metabolic capacity to fixing nitrogen.

Until recently, almost all nitrogen on Earth was fixed by bacteria. The rest—a relatively minor amount—was fixed by volcanic activity and lightning. However, early in this century, Fritz Haber, a German chemist, discovered how to convert gaseous nitrogen and hydrogen into ammonia. He was awarded a Nobel Prize in 1918. Now about half the world's supply of nitrogen is fixed by the **Haber process**, an industrial procedure based on Haber's work. Industrially fixed nitrogen is used mainly to produce fertilizers for agriculture, which has contributed greatly to increased world food production. But fertilizer production creates its own problems. The Haber process requires huge amounts of natural gas, which makes it costly. The runoff water from fields of fertilized crops stimulates algal growth. In Chesapeake Bay, for example, algal growth has diminished the productivity of fisheries.

Increased amounts of fixed nitrogen for agriculture can also be supplied by fostering symbiotic nitrogen fixation. In Western countries, leguminous crops, which harbor symbiotic nitrogen-fixers, are part of crop rotation to replenish the soil's nitrogen content. For centuries rice paddies in Southeast Asia have been supplied with additional nitrogen by inoculating them with the small floating water fern *Azolla*, which harbors a symbiotic nitrogen-fixing cyanobacterium. This highly successful practice is now being employed for organically grown rice in California.

Nitrogen fixation is a chemical reduction—nitrogen gas, at an oxidation state of 0, is reduced to ammonia, at an oxidation state of –3. The ammonia is used without oxidation or reduction to make nitrogen-containing cell components such as amino acids, purines, and pyrimidines for the bacterium itself and for the host plant. With rare exceptions, nitrogen-fixing organisms fix only enough nitrogen for their own and their host's use. Only when these primary nitrogen recipients die or are consumed does the fixed nitrogen become available to other organisms. Organisms such as fish, snails, and other small animals that consume nitrogen-fixers usually use the complex nitrogenous building blocks (amino acids, for example) directly. But any of these nitrogen compounds that reenter the soil are converted back to ammonia (NH_3) by **ammonification**, a process carried out largely by heterotrophic bacteria.

Nitrification. Ammonia does not accumulate in the soil because it is rapidly oxidized by **nitrifying bacteria** to nitrate ion (NO_3^-) in the process of nitrification. Nitrifying bacteria are autotrophs that generate ATP from the energy released in these oxidations. Nitrification occurs in two steps, mediated by different nitrifying bacteria. The first kind, typified by *Nitrosomonas* spp., oxidizes ammonia to nitrite ion (NO_2^-).

$$NH_3 + 1\tfrac{1}{2}O_2 \longrightarrow NO_2^- + H_2O + H^+ + \text{energy}$$
$$\text{AMMONIA} \qquad\qquad\qquad \text{NITRITE ION}$$

The second kind, typified by *Nitrobacter* spp., oxidizes nitrite ion to nitrate ion (NO_3^-).

$$NO_2^- + \tfrac{1}{2}O_2 \longrightarrow NO_3^- + \text{energy}$$
$$\text{NITRITE ION} \qquad\qquad \text{NITRATE ION}$$

Nitrate is the principal form of nitrogen used by plants. But it cannot be stored in soil because it is highly soluble and does not adsorb to soil as ammonia does. Nitrate not used by plants leaches into ground water or runs off into streams. This presents two different but related problems. First, the accumulation of nitrate in ground water can contaminate drinking water, and high levels of nitrate are dangerous to humans, particularly infants. Second, the loss of nitrogen in runoff water depletes soil fertility.

Denitrification. Denitrification, the conversion of nitrate to nitrogen gas, is a cascade of anaerobic respirations (Chapter 5) mediated exclusively by bacteria. Many different kinds of bacteria can denitrify, including archaebacteria and eubacteria, Gram-positives and Gram-negatives. Most oxidize organic compounds and transfer these electrons to nitrate, reducing it stepwise to nitrogen gas. Intermediates in the reductive pathway include nitrite ion and nitrous oxide, N_2O, also known as laughing gas. When high concentrations of nitrate are present, some nitrous oxide is released along with nitrogen gas. Most denitrifying bacteria are facultative aerobes. When oxygen is available, they use it as the terminal electron acceptor of their electron transport chains. When oxygen is not available, they use nitrate.

The ecological impact of denitrification is complex. On one hand, it is essential to life on the planet. Without denitrification our atmosphere would rapidly (in geological terms) become nitrogen-free. (Half the nitrogen in our atmosphere is fixed every 20 million years.) Earth's supply of nitrogen would accumulate as nitrate in the oceans, while nitrogen starvation would stop plant growth on land. Denitrification also helps purify waste water, impeding eutrophication, and drinking water, making it safe.

On the other hand, denitrification is costly to agriculture because up to 30 percent of nitrogen added as fertilizer is returned to the atmosphere through denitrification before plants can use it. Denitrification also contributes to the destruction of the earth's protective ozone layer because it releases some nitrous oxide. (Internal combustion engines also form nitrous oxide and other nitrogen oxides.) Production of nitrous oxide by denitrification is rising because heavy use of fertilizer increases nitrate levels in soil, a condition that favors formation of nitrous oxide.

Denitrification completes the major loop of the nitrogen cycle. The individual steps of this loop—nitrogen fixation, ammonification, nitrification, and denitrification—are all mediated by microorganisms. Three of these steps—nitrogen fixation, nitrification, and denitrification—are mediated *exclusively* by bacteria (although humans now participate in fixing nitrogen by the Haber process).

The nitrogen cycle also has a minor loop. **Herbivores** obtain nitrogen by consuming plants. They in turn supply the nitrogen requirements of **carnivores**. Animal excreta and dead (decomposed) plants and animals return nitrogen to the major loop of the cycle.

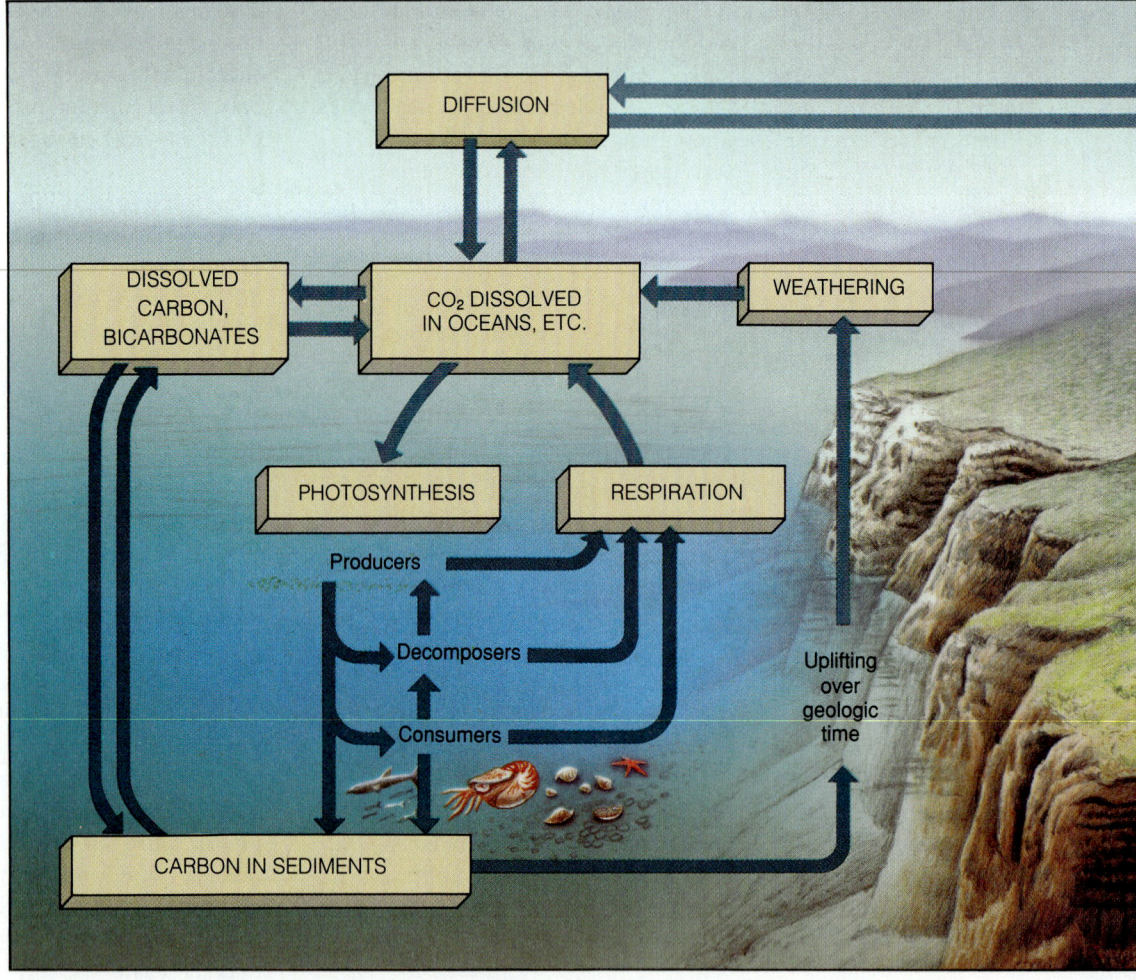

Figure 28.9 The carbon cycle. Carbon dioxide in the atmosphere moves through marine (left) and terrestrial (right) ecosystems by biological (photosynthesis and respiration) and nonbiological (combustion, volcanic action, dissolution, and weathering) processes.

The Carbon Cycle

The major loop of the **carbon cycle** is the conversion of carbon dioxide (CO_2) into the organic compounds of living organisms and their conversion back to carbon dioxide.

Like the nitrogen cycle, carbon cycling involves an atmospheric gas—in this case, carbon dioxide (**Figure 28.9**). But carbon dioxide makes up a much smaller fraction of the atmosphere (0.03 percent) than does nitrogen gas (78 percent). The turnover of carbon dioxide is thus more rapid. In fact, at the present rate of use, without resupply, the atmosphere's supply of carbon dioxide would last only about 20 years.

In the major loop of the carbon cycle, carbon dioxide is removed from the atmosphere by autotrophs, principally phytoplankton and plants, which fix atmospheric CO_2 through photosynthesis, reducing it and incorporating it into organic compounds (Chapter 5). Carbon diox-

ide is regenerated and returned to the atmosphere by respiration and combustion, both of which oxidize organic compounds to water and CO_2 (Chapter 5).

Both respiration and combustion—the major routes that return CO_2 to the atmosphere—require oxygen. As a result, organic materials deposited in anaerobic environments, such as bogs or sediments at the bottoms of bodies of water, tend to be removed from the major loop of the carbon cycle. Under natural conditions carbon in this organic material returns to it slowly, as anaerobes convert organic compounds into acetate, which methanogens convert to methane (CH_4). Methane, being a gas, can escape the anaerobic environment. When it comes in contact with air it can be oxidized to CO_2 by methane-oxidizing bacteria or it can accumulate in the atmosphere. (See the box "Methane in the Mud," p. 720.)

Over the earth's history, huge quantities of organic materials have been trapped in anaerobic environments.

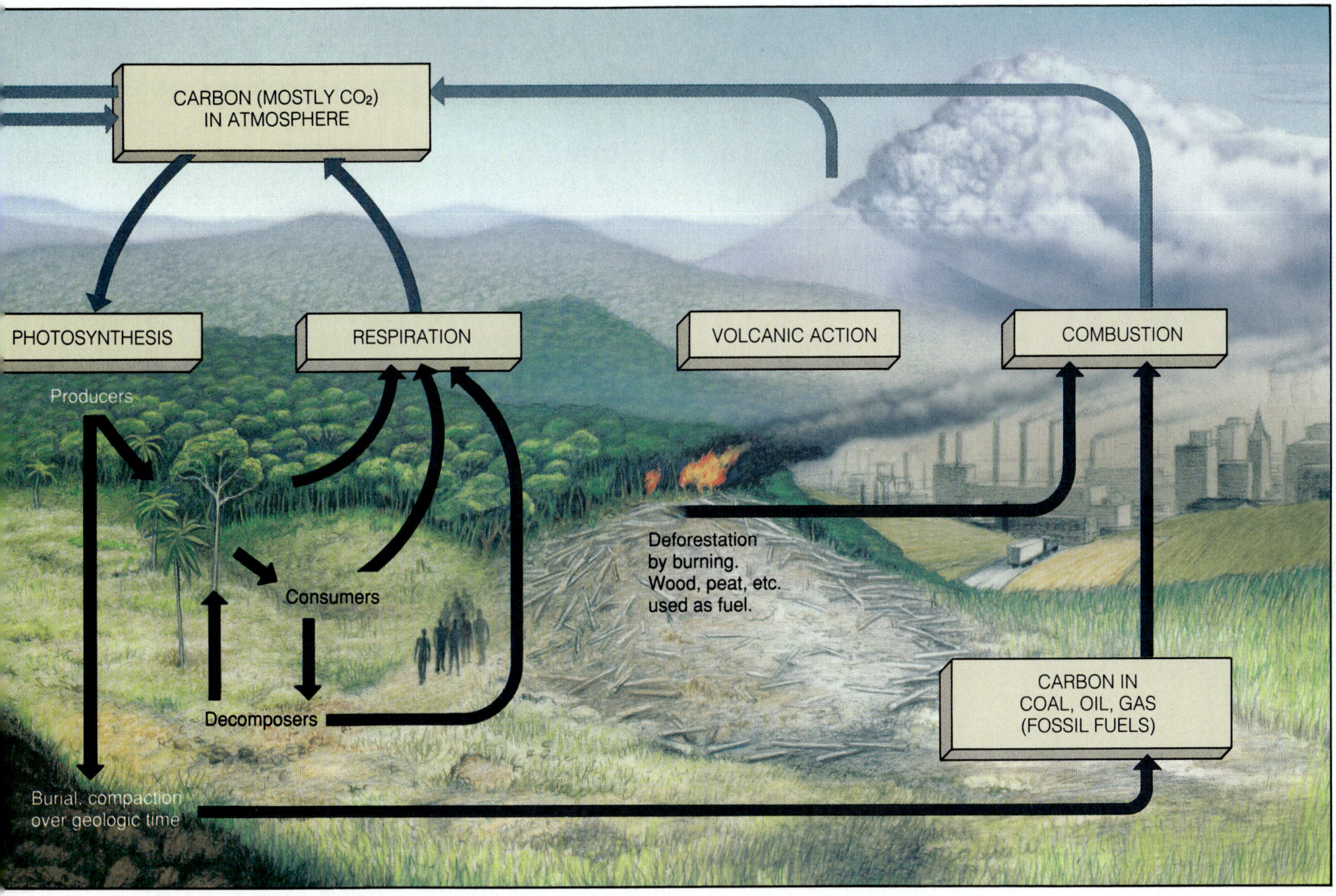

CARBON (MOSTLY CO2)
IN ATMOSPHERE
PHOTOSYNTHESIS
RESPIRATION
VOLCANIC ACTION
COMBUSTION
Producers
Consumers
Decomposers
Deforestation
by burning.
Wood, peat, etc.
used as fuel.
CARBON IN
COAL, OIL, GAS
(FOSSIL FUELS)
Burial, compaction
over geologic time

Cumulatively, they form the fossil fuels—peat, coal, and petroleum. Other global repositories of CO_2 are the oceans and sedimentary rocks. Oceans contain CO_2 in the form of dissolved bicarbonate, and sedimentary rocks such as limestone contain CO_2 in the form of carbonate (CO_3^{2-}). The rate at which the ocean takes up and releases CO_2 is one of the major unknown factors governing the rate of CO_2 accumulation in our atmosphere. The rate of incorporation and release of CO_2 from sedimentary rocks is very slow.

Since the beginning of the industrial revolution in the mid-eighteenth century, humans have burned fossil fuels at an ever-increasing rate, short-circuiting the natural carbon cycle. This, along with a decline in photosynthesis because forests are being cut down, has changed the carbon cycle, causing the CO_2 content of the atmosphere to rise sharply (**Figure 28.10**). During the three decades between 1958 and 1988, atmospheric CO_2 rose more than 11 percent. This might be serious because CO_2 is a **greenhouse gas**. That is, it acts like an atmospheric blanket that prevents some solar radiation from reradiating into space. The long-term consequence of an imbalanced car-

348
344
340
336
332
328
324
320
316
312
CO2 in atmosphere (parts per million)
1958
1962
1966
1970
1974
1978
1982
1986
year

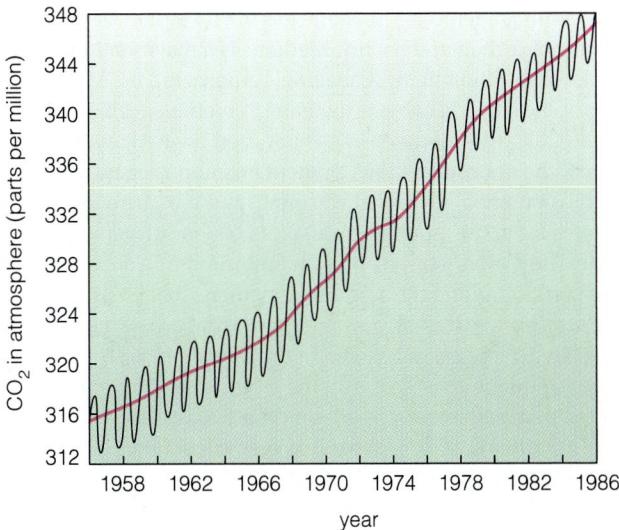

Figure 28.10 Short-circuiting the carbon cycle. Burning fossil fuels and cutting forests have caused the carbon dioxide content of the atmosphere to increase sharply. (The fluctuations in the graph line represent seasonal changes.)

Methane in the Mud: Methanotrophs in Lake Sediments

Mary Lidstrom

Mary Lidstrom grew up near Princeville, Oregon, and did her undergraduate work at Oregon State University, graduating magna cum laude in Microbiology in 1973. She did her graduate work with R. S. Hanson in Bacteriology at the University of Wisconsin–Madison as a National Science Foundation fellow and then as an American Association of University Women fellow, receiving an M.S. degree in 1975 and a Ph.D. in 1977. She spent a year with J. R. Quale at the University of Sheffield, England, as a Leverhulme postdoctoral fellow and took a position as Assistant Professor of Microbiology at the University of Washington in 1978. In 1984 she became Associate Professor with tenure and in 1985 moved to the Center for Great Lakes Studies, Milwaukee, Wisconsin, as Associate Scientist. In 1987 she moved to the Environmental Engineering Science program at the California Institute of Technology as Associate Professor of Applied Microbiology. She is currently Professor of Applied Microbiology in the same program. She lives in Altadena, California, with her husband, two children, and three dogs.

Methanotrophic (methane-utilizing) bacteria play a critical role ecologically because they can use single-carbon organic compounds including methane gas (CH_4) as their total source of carbon and energy. Methane is a major end product of decomposition in most anaerobic environments. Methanotrophic bacteria convert methane to biomass (cells) and carbon dioxide. Methane that is not used by methanotrophs enters the atmosphere. It is a potent greenhouse gas, with serious potential for global warming, and it is currently increasing in the atmosphere at a rate of about 1 percent per year. The methanotrophs play an important role on a global basis in controlling the amount of methane that escapes to the atmosphere from aquatic and terrestrial ecosystems.

I have long been interested in the role methanotrophs play in aquatic ecosystems. In the 1980s, several researchers found that methanotrophs are concentrated during the summer months at the oxic/anoxic (oxygen-containing/oxygen-free) interfaces in the water column of lakes that are anaerobic in deeper portions. This result suggested that these bacteria position themselves where they have access to both oxygen and methane, their two major required nutrients. However, the water columns of many lakes are totally aerobic, but sediments at the bottom are anaerobic. Knowing where methanotrophs occur in these lakes is important for assessing the role of different types of lakes in global methane cycling.

We chose to study Lake Washington, just off the University of Washington campus. The water column of Lake Washington remains aerobic all year, but the sediment is anaerobic just a few centimeters below the bottom of the lake. Leslie Somers, a student at the University of Washington, and I worked as a team with a group in Chemical Oceanography at the University of Washington—Professor Al Devol, and Professor Jim Murray and his graduate student Kathy Kuivila—and Dr. Claire Reimers from Scripps Oceanographic Institution in La Jolla, California, to locate methanotrophs in the lake. We used a small research vessel and a sampling apparatus called a box core to obtain a box-shaped section of the sediment 60 meters beneath us, which we brought to the surface with a winch. We used water-sampling devices called Niskin bottles to obtain water from specific depths throughout the water column. In the laboratory we analyzed these samples. Leslie Somers and I determined dissolved methane and dissolved oxygen concentrations in the water samples, and we incubated them with $^{14}CH_4$ at different concentrations of methane. Then we determined how much and how rapidly $^{14}CH_4$ was converted to carbon dioxide in water samples and in subsamples of the sediment. Dr. Reimers determined the concentration of dissolved oxygen at 0.5-mm intervals into the sediment using micro-oxygen electrodes. Kathy Kuivila determined the concentration of methane in the sediment at intervals of 0.5 cm of depth. From these data we calculated rates of methane consumption and plotted these results along with methane and oxygen concentrations versus depth.

The results showed conclusively that very little methane oxidation occurs in the water of Lake Washington, regardless of the season; methane oxidation occurs predominantly in the sediment. The maximum methane oxidation occurs in the upper 0.7 cm of the sediment, the zone in which both oxygen and methane are present, just above the oxic/anoxic interface. About half of the methane produced in the sediment was consumed by the methanotrophs. This and later studies confirmed that in freshwater and marine environments, methane oxidation is generally concentrated at oxic/anoxic interfaces, regardless of whether those interfaces occur in the water column or the sediment. Methane produced in the lower anaerobic zones diffuses to the oxic/anoxic interface where it is oxidized by methanotrophs.

When we did these experiments, it was not possible to measure populations of methanotrophs directly, due to the difficulties of enumerating methanotrophs by methods requiring growth. Therefore, we were unable to correlate numbers of methanotrophs with methane oxidation activity. However, the development of gene-probing methodologies for enumerating bacteria in unaltered natural samples provides a new tool for reexamining these habitats. We are now entering a new and exciting phase of studying methanotrophs in nature, in which we will be able to assess the relationship between natural populations of bacteria, their activities, and key environmental variables such as temperature, methane, oxygen, and trace elements.

bon cycle is global warming, which could cause more frequent droughts in the United States and flooding of coastal regions from the melting of polar ice and expansion of sea water. However, a major unknown factor in the global warming scenario is cloud formation. Certainly a warmer world will have more clouds, but will they act largely as blankets, increasing the rate of warming, or as insulators, decreasing it? Thus, the ultimate consequences of global warming are unknown, but no one doubts the world will become a different place if CO_2 continues to rise.

The Phosphorus Cycle

The **phosphorus cycle** converts phosphorus from inorganic phosphate to organic phosphate and back (**Figure 28.11**). The phosphorus cycle differs fundamentally from the nitrogen and carbon cycles because there is no significant gaseous intermediate, making global distribution through the atmosphere impossible. Distribution is fur-

ther restricted by the high solubility of many inorganic phosphates. When dissolved phosphate enters the ocean, there are only two significant routes for returning it to land. First, seabirds that feed on phosphorus-containing sea creatures return the phosphorus to shore in their feces, in some places forming huge guano deposits on land. This route redistributes only a small amount of phosphorus. The geological uplift of ocean floors to form land masses also returns phosphorus from ocean sediments to the land. This route returns enormous quantities of phosphorus, though very slowly.

The phosphorus cycle also differs from the nitrogen and carbon cycles in another important respect—phosphorus is neither oxidized nor reduced. Instead, it remains at the oxidation state of phosphate. As phosphorus passes from one organism to another, the phosphate group itself remains intact.

Phosphorus is a limiting nutrient in many soils, so most fertilizers contain this element. Rather than being commercially manufactured, like nitrate, however, phosphate is mined. Phosphate-rich deposits are the product

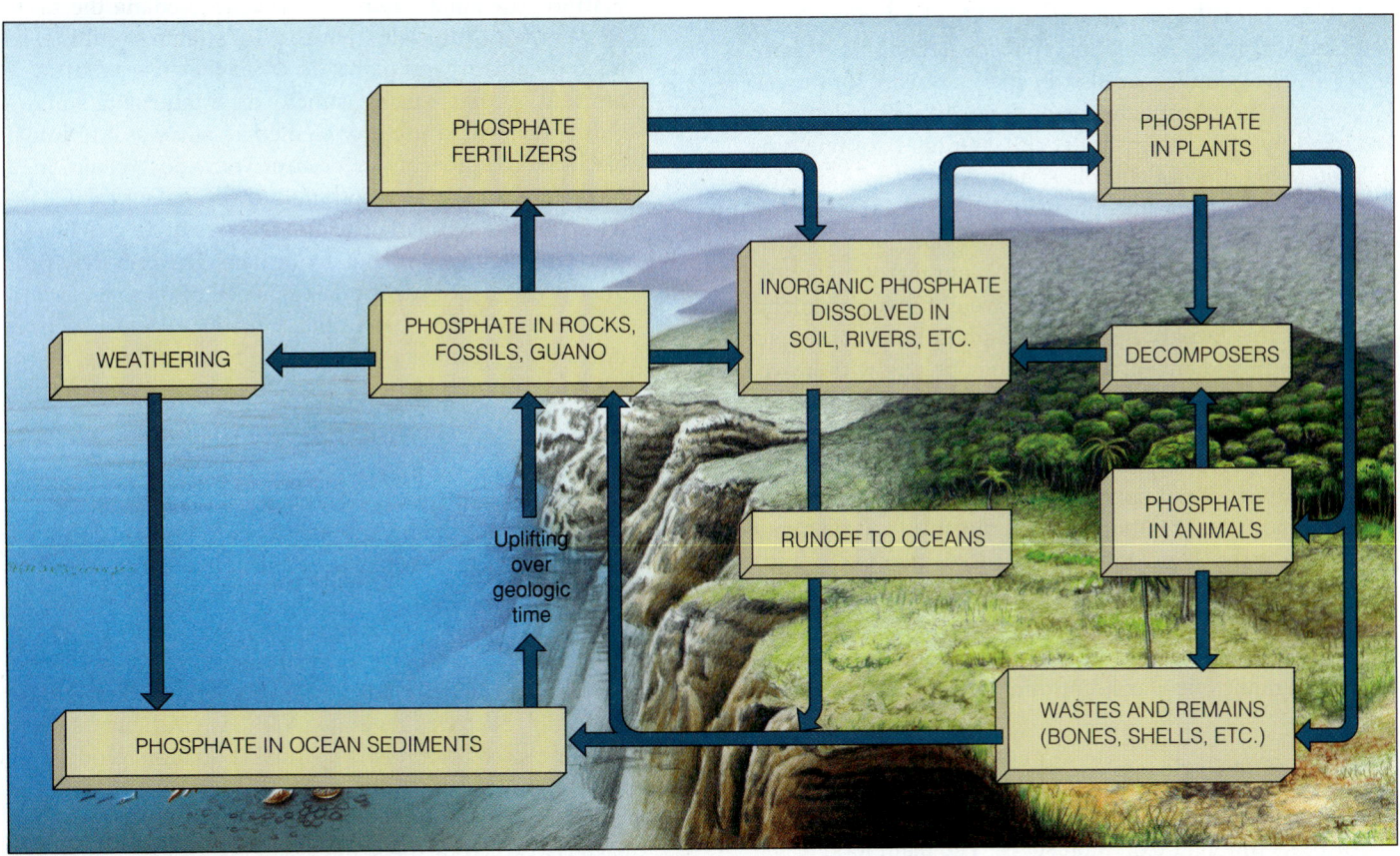

Figure 28.11 The phosphorus cycle does not have a gaseous intermediate. It cycles as organic and inorganic phosphates through plants, animals, and microorganisms. When inorganic phosphate dissolves in water that flows into the ocean, it accumulates in sediments that are returned to land by seabirds that deposit their guano on land and by the uplift of ocean floors to form new land masses.

of previous geological uplifts. Although there is no foreseeable prospect of a world phosphorus shortage, the distribution of commercially feasible phosphorus mines is restricted. North America and Africa are particularly rich in phosphate deposits.

Human intervention has altered the phosphorus cycle by adding phosphate compounds to lakes and streams, causing some ecological damage. As with soils, the productivity of many lakes is limited by availability of phosphorus. When such lakes are enriched, they become eutrophic, with all the subsequent negative consequences. Until recently, when many detergents were reformulated to be phosphate-free, sewage was dangerously rich in phosphorus. But even if sewage is completely mineralized, it can contaminate the receiving lake or stream, making it eutrophic.

Recently, an ingenious microbiological solution was developed in the Netherlands to restrict the phosphate outflow of sewage treatment plants. Mineralized sewage is diverted into an aerobic pond that is slightly enriched with raw sewage as a carbon source. It is then managed in such a way as to develop a dense culture of the bacterium *Acinetobacter calcoaceticum*. This bacterium stores phosphate in the form of polyphosphate granules (Chapter 4). The granules are also an energy reserve for the cell because the phosphate residues in polyphosphate are linked by high-energy anhydride bonds—the same type used to link the last two phosphate residues to ATP. With the rich supply of phosphate and the favorable aerobic growth conditions in the treatment pond, *Acinetobacter*, an obligate aerobe, stores massive quantities of polyphosphate—up to 30 percent of its dry weight. When the pond reaches this level of accumulated phosphate, conditions are abruptly changed. The carbon source is withdrawn, the volume is decreased, and the pond is allowed to become anaerobic. These changes cause *Acinetobacter* to degrade its polyphosphate, releasing phosphate ions into the relatively small pond. The result is a phosphate solution concentrated enough to sell directly to a detergent manufacturer. Thus, the process serves two useful purposes. It purifies sewage so it does not cause eutrophication and it produces a valuable product. At the end of the process, the *Acinetobacter* cells in the pond are ready to begin a new cycle of concentrating phosphate.

The Sulfur Cycle

The **sulfur cycle** is a series of chemical conversions of sulfur in its different oxidation states. The main loop of the cycle consists of reducing sulfate ions (SO_4^{2-}) to hydrogen sulfide gas (H_2S) and reoxidizing it to sulfate. Elemental sulfur is an intermediate of both processes.

Sulfur is the tenth most abundant element on Earth, so few if any natural environments are sulfur-deficient.

As with phosphorus, there is no atmospheric form of sulfur. (Although hydrogen sulfide is a gas, it cannot exist for long in the atmosphere because it reacts spontaneously with oxygen.) Because of its abundance, sulfur is rarely a growth-limiting nutrient in soil or water. Instead, it plays a dramatic role in metabolism because of the various ways in which bacteria process sulfur compounds to generate ATP (**Figure 28.12**).

Reduction of Sulfate. Most microorganisms and plants derive their sulfur from sulfate ions (SO_4^{2-}), which are abundant in the environment. The legendary white cliffs of Dover in England, for example, are almost pure gypsum (calcium sulfate). Only a few unusual constituents of cells contain sulfur as a sulfite or sulfate. Microbes and plants reduce sulfates to sulfides (S^{2-}) and use the sulfides to synthesize all their sulfur-containing components. Animals derive sulfide-containing compounds from their diet.

Some microorganisms convert organic sulfur compounds to hydrogen sulfide in a process called **desulfurylation**. The sulfate-reducing bacteria mediate the same step of the sulfur cycle (reduction of sulfate to sulfide) as microorganisms and plants do when they use sulfate as a nutrient. Besides using sulfate as a nutrient, sulfate-reducing bacteria use it or elemental sulfur as a terminal electron acceptor of an electron transport chain in a process of anaerobic respiration to generate ATP (Chapter 5). Sulfate-reducing bacteria live in sulfate-rich anaerobic environments such as mud flats. The H_2S they produce accounts for the rotten-egg smell of these areas and for the mud being black. Black metal sulfides, such as iron sulfide, form as a result of the reaction between H_2S produced by the bacteria and metal ions present in the mud.

Oxidation of Reduced Sulfur to Sulfate. When the remains of plants and animals are broken down by microorganisms, much of their sulfur is converted to H_2S. This foul-smelling gas is oxidized to produce sulfate, thereby completing the sulfur cycle.

The conversion of H_2S to sulfate is mediated by two different kinds of bacteria. Chemoautotrophic sulfur-oxidizing bacteria generate ATP by oxidizing sulfur-containing compounds. The final oxidation product of these organisms is sulfate, but elemental sulfur also accumulates as an intermediate. Nonoxygenic phototrophic bacteria carry out the same reactions anaerobically, also using reduced sulfur as a source of electrons. These organisms also oxidize elemental sulfur when it is available. This microbial oxidation of elemental sulfur accounts for the common gardening practice of adding sulfur to soil that is too alkaline to support vigorous plant

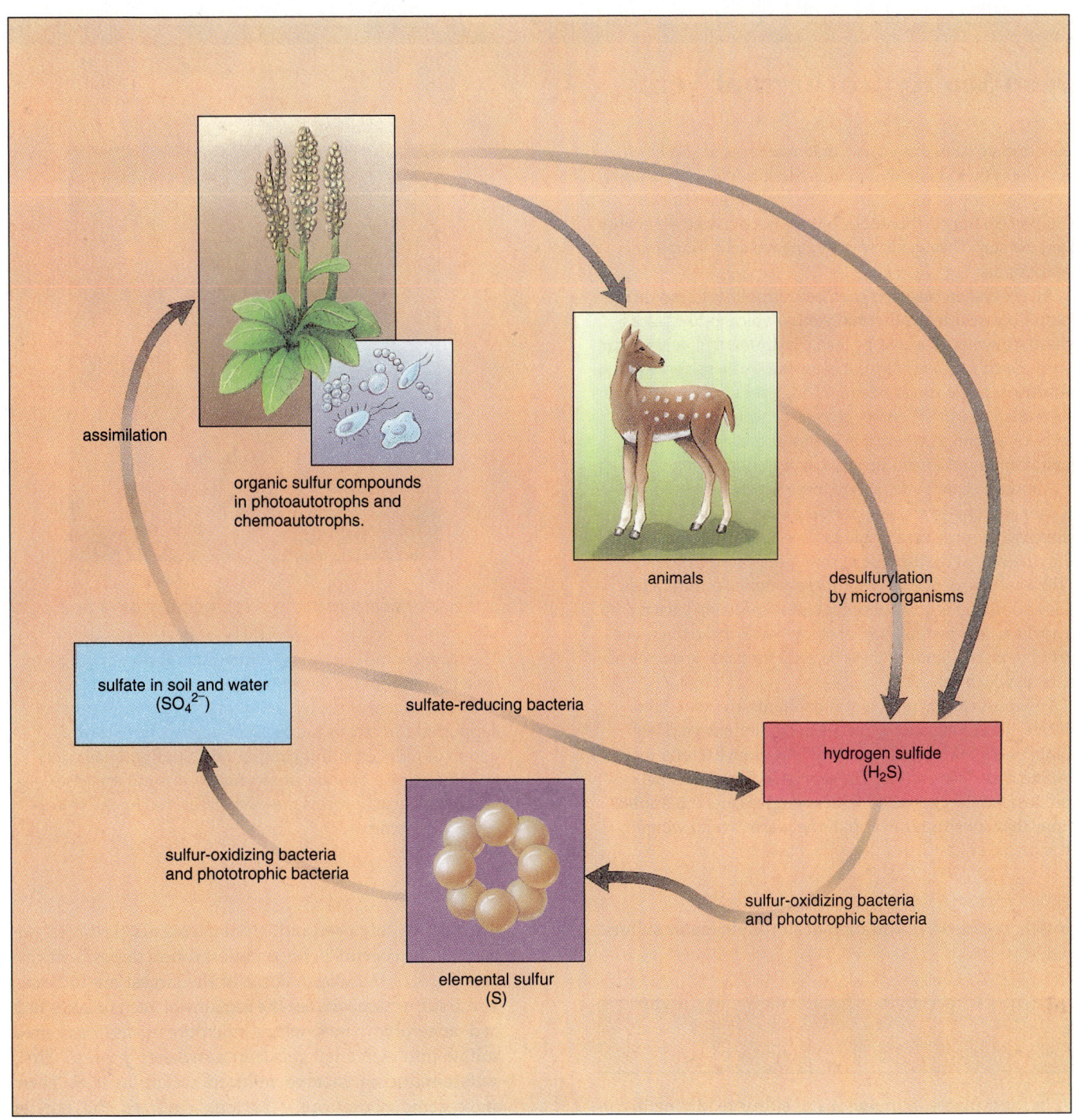

assimilation

organic sulfur compounds
in photoautotrophs and
chemoautotrophs.

animals

desulfurylation
by microorganisms

sulfate in soil and water
(SO_4^{2-})

sulfate-reducing bacteria

hydrogen sulfide
(H_2S)

sulfur-oxidizing bacteria
and phototrophic bacteria

elemental sulfur
(S)

sulfur-oxidizing bacteria
and phototrophic bacteria

Figure 28.12 The main loop of the sulfur cycle passes through sulfate and hydrogen sulfide. Plants and microorganisms reduce sulfate to form their organic sulfur-containing constituents, which animals consume. These compounds are desulfurylated to hydrogen sulfide by microorganisms. Sulfate-reducing bacteria form hydrogen sulfide directly from sulfate. Sulfur-oxidizing bacteria and phototrophic bacteria oxidize hydrogen sulfide (with elemental sulfur as an intermediate) to sulfate. The huge deposits of elemental sulfur in nature are formed from sulfate by the combined activities of sulfate-reducing bacteria and phototrophic bacteria.

Deep-Sea Hydrothermal Vents

The ultimate source of energy that maintains life on Earth and drives the cycling of matter is the sun. Without sunlight both phototrophs and heterotrophs would cease to exist (Chapter 5). But chemoautotrophs do not derive their energy from the sun. They get it by oxidizing reduced inorganic compounds.

An ecosystem based on chemoautotrophs is found on the ocean floor near **hydrothermal vents**—volcano-like places where sea water is heated by lava. Hydrothermal vents occur where Earth's geological crust is constantly being formed and re-formed. These zones extend through the Atlantic, Pacific, and Indian Oceans. In many places, sea water enters these vents, or openings, where it is heated and enriched in inorganic compounds, including hydrogen sulfide.

Almost no light penetrates to these depths, so photosynthesis cannot occur—yet dense, thriving communities of microorganisms and invertebrate animals live here. Oxidation of H_2S from the vent at the expense of the O_2 dissolved in sea water is the total source of energy for these complex communities. Huge populations of autotrophic sulfur-oxidizing bacteria, up to a billion cells per milliliter of ocean water, make the water turbid in places. Some of these bacteria form dense microbial mats on which fish graze.

The most spectacular vent inhabitants are huge tube worms, 8 feet long and 15 inches in circumference. They obtain energy in a highly unusual way from bacteria that oxidize hydrogen sulfide. The worms lack an intestinal system. Instead they are filled with spongy tissue, the **trophosome**, that constitutes over half the worm's total weight.

Tube worms inhabit deep-sea hydrothermal vents.

The trophosome consists of symbiotic sulfur-oxidizing bacteria that supply the worm with its nutrients.

All members of the hydrothermal vent community are supported directly or indirectly by energy from the oxidation of reduced sulfur compounds. Sunlight plays only an indirect role. Having supported photosynthesis elsewhere on Earth, it forms the oxygen needed by the sulfur-oxidizing bacteria.

---

growth, because the oxidation of sulfur generates sulfuric acid.

$$S + 1\tfrac{1}{2}O_2 + H_2O \longrightarrow H_2SO_4$$
$$\text{ELEMENTAL SULFUR} \qquad\qquad \text{SULFURIC ACID}$$

In most places, populations of sulfur-oxidizing bacteria are limited by the supply of reduced sulfur made by other microorganisms; but in regions of the deep sea near hydrothermal vents, the supply of reduced sulfur is copious. Huge numbers of sulfur-oxidizing bacteria develop and form the basis of a highly unusual ecosystem (see the box "Deep-Sea Hydrothermal Vents").

Products and Reactions. One intermediate of the sulfur cycle, elemental sulfur, is a valuable commodity. It is the starting material for producing sulfuric acid, one of the most widely used industrial chemicals. Sulfur in elemental form occurs in huge underground deposits in various parts of the world. Some of the largest are in Texas. These sulfur deposits are the remains of ancient lakes that had anaerobic zones where microorganisms converted sulfate into elemental sulfur in a two-step process. First, sulfate-reducing bacteria reduced sulfate to H_2S. Then, anoxygenic phototrophic bacteria oxidized the H_2S to elemental sulfur. There are lakes in North Africa where this sort of mixed bacterial metabolism is actively occurring today. The bottoms of these lakes are covered with thick layers of elemental sulfur. Underground sulfur deposits are easily mined. Hot water is pumped down holes drilled into the sulfur deposits, and a slurry of melted sulfur in water returns to the surface where it solidifies, forming yellow mountains of sulfur.

On an experimental basis, the microbiological conversions that make deposits of elemental sulfur have been induced artificially in ponds. Sewage and gypsum are added to the ponds. Sulfate-reducing bacteria use the sewage as a carbon source to reduce the gypsum to hydrogen sulfide, and anoxygenic phototrophic bacteria convert the hydrogen sulfide to elemental sulfur. The process requires only sewage, gypsum, and sunlight. Although elemental sulfur is inexpensive, a microbiological process for making it might become economically feasible for certain countries because sulfur deposits are rare in the less developed parts of the world.

The oxidative portion of the sulfur cycle, which involves the conversion of sulfide to sulfate, has already been adapted to human use. Many valuable minerals occur as insoluble metal sulfides or are trapped in a matrix of sulfides—for example, copper as copper sulfide and gold trapped in a matrix of iron pyrite (FeS_2). If these ores are stacked loosely and sprayed with water, a vigorous population of sulfur-oxidizing bacteria develops that oxidizes the sulfide to sulfate. The result is that the copper solubilizes, and the gold is released.

TREATMENT OF WASTE WATER

Our communities and industries produce huge amounts of waste water that contain human and chemical wastes. Until fairly recently, waste water was simply discharged into the nearest large body of water with the expectation that the natural cycling of matter would eventually purify it. But as population and industrialization increased, natural purification did not always occur fast enough. Instead, the receiving bodies of water became anaerobic, blackened by sulfides, and foul-smelling from the production of H_2S and products of fermentation and devoid of natural flora and fauna. Just 50 years ago, for example, large parts of San Francisco Bay became anaerobic from the discharge of raw sewage. The installation of modern sewage treatment plants has returned the bay to relative health and people can fish again.

Sewage Treatment Plants

Sewage is municipal waste water. It includes all the material that goes down household drains, from both toilets and sinks, and industrial drains. Sewage contains not only human waste and the residue of soaps and detergents, but also many other chemicals used in manufacturing. The major component of sewage is water, but dissolved materials and some solid organic waste are also present. Large cities produce enormous quantities of sewage.

The primary purpose of sewage treatment plants is to remove the organic material from sewage. Some sewage treatment plants employ a preliminary anaerobic treatment to reduce the amount of organic material present, but most of it is removed by the aerobic respiration of microorganisms.

Sewage plants are designed to add oxygen to sewage at a faster rate than it is removed by microbial respiration. Waste water is classified according to its **biochemical oxygen demand** (**BOD**), how much oxygen is needed for microorganisms to respire all the organic material in it. BOD is determined by putting a sample of the sewage (with its natural complement of microorganisms) into an oxygen-saturated bottle and measuring how much oxygen remains after five days of incubation at 20°C. The decrease in oxygen is the BOD of the sample. Knowing BOD, operators of sewage plants know how many hours or days of treatment are required.

A typical sewage plant is divided into functional units called primary and secondary treatment (**Figure 28.13**). A few plants also have tertiary treatment. **Primary treatment** is mechanical rather than biological. Solid material is usually ground up, and **primary sludge**, the portion that remains insoluble, is allowed to settle out.

The **effluent**, or liquid coming from primary treatment, is subjected to **secondary treatment**, which is biological. To decrease its BOD, the sewage is aerated, either by **forced aeration**—pumping air into a sewage-filled tank—or with a **trickling filter**—spraying a thin film of sewage onto a bed of rocks so that it readily absorbs oxygen from the atmosphere. In either case, the sewage is mineralized by a dense population of aerobic bacteria, including *Zoogloea ramigera* (**Figure 28.14**). *Z. ramigera* has a slimy extracellular matrix in which other bacteria and solid material become embedded, forming a **floc**. Floc formation is essential for proper functioning of an aerated tank because it settles rapidly, producing a relatively clear effluent. Some of the mass of floc, called **activated sludge**, is used to inoculate the next batch of sewage entering the aeration tank. The rest is added to the primary sludge.

The combination of primary sludge and excess activated sludge is held in an **anaerobic sludge digester**, a deep unaerated tank in which the mixture is partially degraded by anaerobic bacteria. **Digested sludge**, the solid portion that remains, is dried and disposed of, often as fertilizer. The anaerobic organisms in an anaerobic digester produce various metabolic end products, including acetate, hydrogen gas, and carbon dioxide. These are converted into methane (CH_4), or natural gas, by **methanogens**, methane-forming archaebacteria. Methane is used as fuel during the winter to warm the sewage during secondary treatment and thus speed the process. It can also be used to generate electricity. In some developing countries, particularly China and India, many rural households have small anaerobic digesters. They are

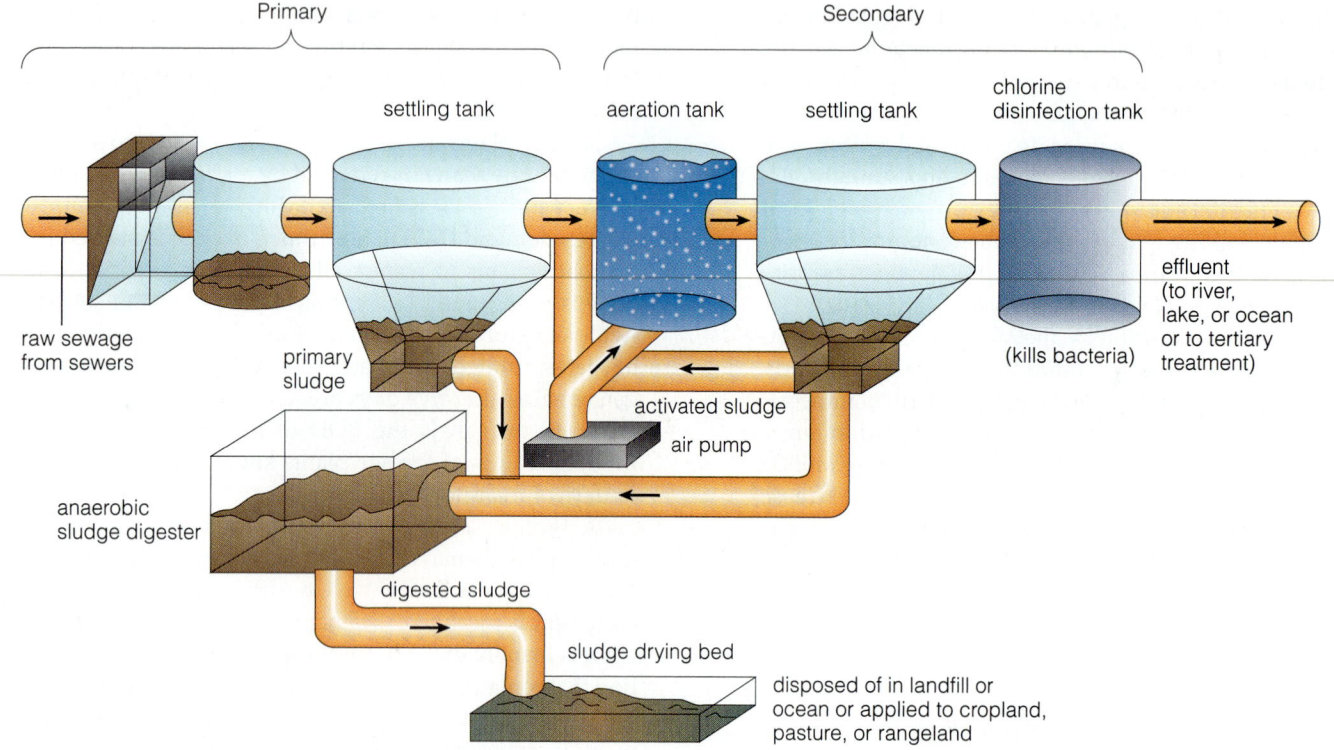

Figure 28.13 A sewage treatment plant is divided into units for primary and secondary treatment and sometimes tertiary treatment. Primary treatment mechanically separates primary sludge. Secondary treatment biologically oxidizes dissolved organic compounds, producing activated sludge and purified liquid effluent. The combined primary and activated sludge is digested in an anaerobic digester and disposed of. The effluent is chlorinated and released or subjected to tertiary treatment to reduce its content of nitrate and phosphate.

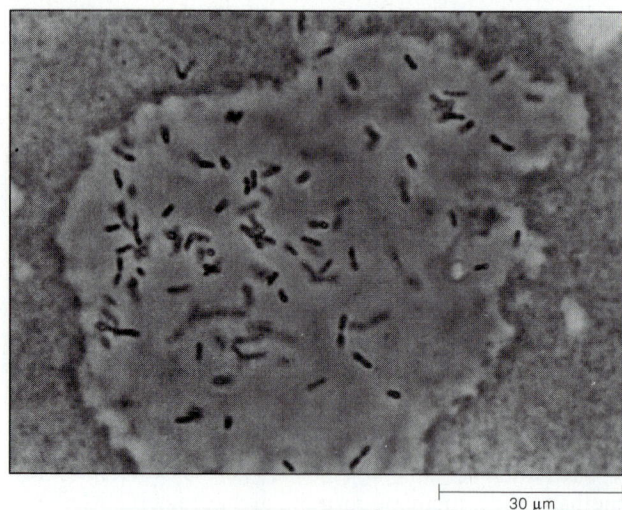

30 μm

Figure 28.14 This micrograph shows a particle of floc formed by *Zoogloea ramigera*. The black dots in the particle are embedded bacterial cells and solid matter.

charged with domestic sewage and manure from farm animals. The methane generated by these digesters is used to cook food, provide light, and heat the house.

The effluent from secondary treatment contains phosphate and nitrate ions, which can cause eutrophication. For this reason, **tertiary treatment** is sometimes used to remove or reduce the concentration of these ions. Most tertiary treatments are chemical rather than biological. For example, lime, alum (potassium/ammonium aluminum sulfate), or ferric chloride is added to remove phosphate. Biological treatments—such as the use of *Acinetobacter* to remove phosphate, described earlier in this chapter—are gaining favor, however. In many places, including the United States, denitrifying bacteria are used to remove nitrate by converting it to nitrogen gas.

The final effluent from a properly functioning sewage plant contains only a small amount of organic material and has a correspondingly low BOD. It can be added to a river or to the ocean without fear that it will make the water anaerobic. For public health reasons, the effluent is treated prior to release with chlorine gas (Cl_2) to kill pathogenic microorganisms that might be present.

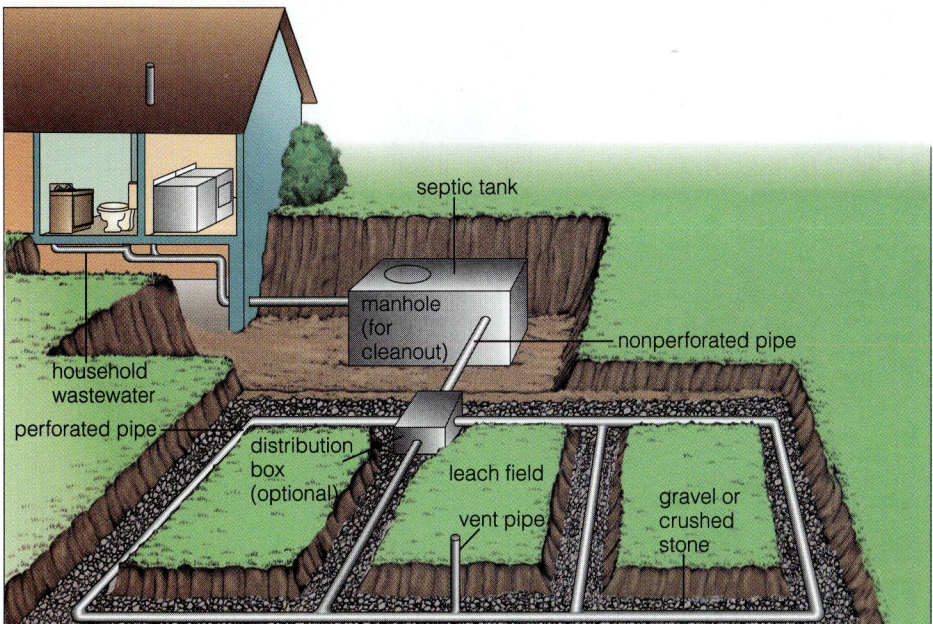

Septic Tanks and Oxidation Ponds

Many homes and farms are not linked to a municipal sewage system. Instead, they use **septic tanks**, which are essentially small anaerobic digesters (**Figure 28.15**). The sludge settles in the tank, eventually to be pumped out and disposed of. The digested effluent from the tank is distributed by a network of perforated pipes into a **leach field**, ground where the liquid's organic content is mineralized. The effectiveness of the leach field depends on the capacity of the soil's microflora to mineralize organic materials. A septic tank system can process small amounts of sewage almost indefinitely, so long as the accumulated sludge is pumped out every few years. It is impractical, however, to chlorinate the effluent, so care must be taken to prevent contamination of drinking water. The septic system must be located away from a well or other source of water.

In arid regions of the world, the soil's vigorous microbial activity can be used to process high volumes of sewage from municipalities or industries. Sewage is distributed into a series of **oxidation ponds**, earthen ponds stirred with large paddles to promote aeration. Oxygenic phototrophs develop in these ponds, which is helpful because they add more oxygen. Sometimes in arid regions, sewage is simply added to ditches and allowed to seep into the ground, a process called **sewage farming**.

TREATMENT OF DRINKING WATER

Drinking water from wells or clear mountain streams usually can be consumed safely, but surface water from most lakes and rivers must be treated to **clarify** it, to rid it of pathogenic microorganisms, and sometimes to improve its taste (**Figure 28.16**).

Processing Methods

Water to be processed for drinking is pumped into a tank where alum is added. The aluminum ions in alum neutralize charges on the colloidal particles, allowing them to aggregate into sizes that settle out in the flocculation tank. Then the water is clarified by filtering it through beds of sand or diatomaceous earth. Finally, passing water through beds of activated charcoal removes odor-producing compounds and some toxic compounds.

Filtering water through beds of sand 2 to 4 feet deep removes most bacteria that cause waterborne diseases (Chapter 20). Modern sewage treatments remove many viruses as well because they become trapped in alum-created flocs (see the box "How Safe Is Safe?" in Chapter 13). But some microorganisms still pass through the filters. Most municipal water treatment plants add chlorine or ozone either as a concentrated solution or directly as gas to kill residual microorganisms. Enough chlorine is added to react with organic material that might be present and have 0.2 to 0.6 μg/ml left over as free chlorine, which kills most microorganisms within 30 minutes.

Testing Methods

Many human diseases—including those caused by bacteria, protozoa, and viruses—are acquired from contaminated drinking water (Table 28.2). To assure safety, municipal water supplies are routinely tested. It would be

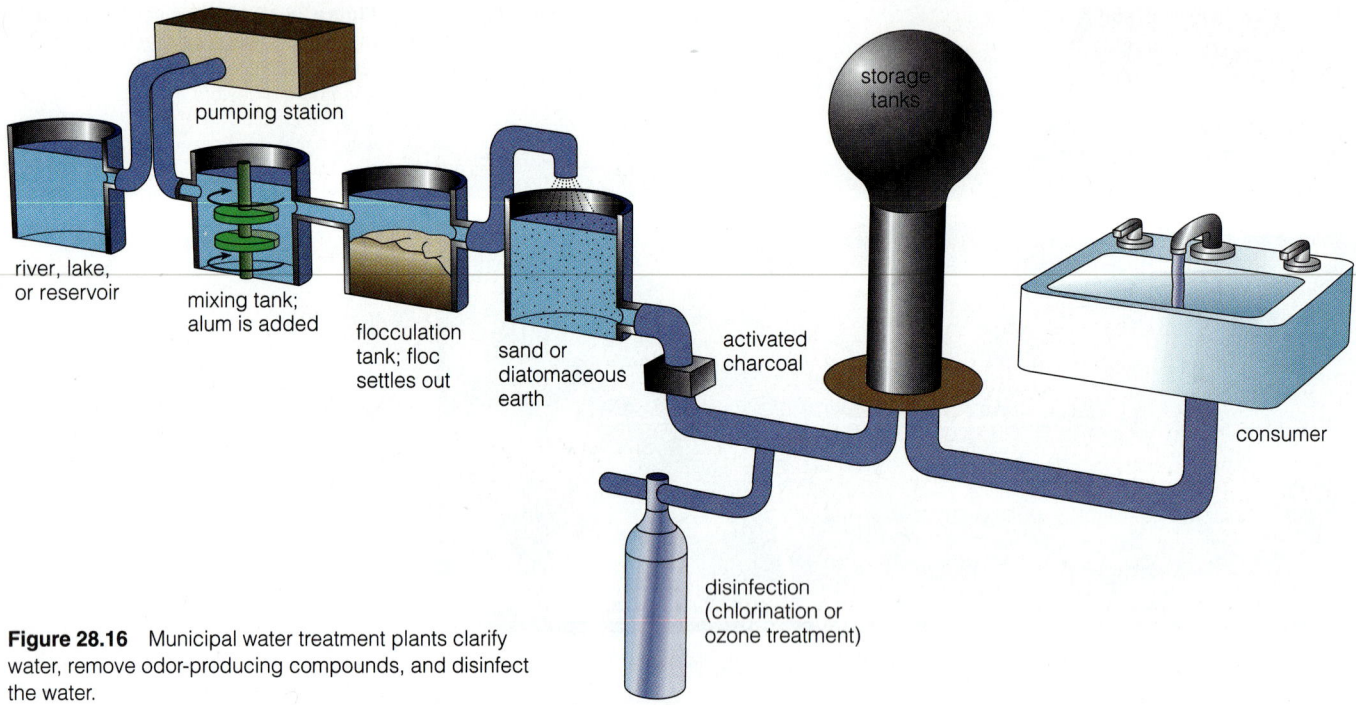

Figure 28.16 Municipal water treatment plants clarify water, remove odor-producing compounds, and disinfect the water.

impractical to test for all possible pathogens, however, so **indicator organisms** are selected. The reasoning is that if certain indicator organisms are absent, pathogens probably are also.

In many countries, including the United States, **coliform bacteria**—defined as aerobic and facultative anaerobic, Gram-negative, nonspore-forming, rod-shaped bacteria that ferment lactose with gas formation within 48 hours at 35°C—are used as indicators. This practical definition is chosen to include *Escherichia coli*, which is almost always present in feces of humans and other animals. In an aqueous environment outside the body, *E. coli* eventually dies, but not faster than other bacterial pathogens. So water that is free of coliform bacteria is most probably safe, and water with them should not be consumed.

Two methods are used to test for coliform bacteria— the traditional **most probable number (MPN) test** and the now more commonly used **membrane filter (MF) test** (**Figure 28.17**).

The MPN procedure consists of three steps—the **presumptive test**, the **confirmed test**, and the **completed test**. In the presumptive test, tubes of lactose broth are inoculated with water to be tested, incubated, and observed after 24 and 48 hours. If gas is produced, the presumptive test is positive, and the presumed number of

coliforms can be calculated from most probable number tables (Chapter 8). To confirm that the gas-producing organisms are coliforms, samples from the highest dilution showing a positive result are streaked on eosin-methylene blue (EMB) agar. On these plates, coliforms produce distinctive colonies with a metallic sheen and a dark center. If such colonies develop, the confirmed test is positive. To complete the test, these colonies are used to inoculate lactose broth and slants (test tubes with solid medium that are allowed to solidify at an angle). If gas is produced in the broth and Gram-negative, nonspore-forming, rod-shaped bacteria develop on the slants, the completed test is positive, establishing that coliform bacteria are present and the water is not safe to consume.

In the MF procedure, at least 100 ml of water is filtered and the filter is placed on the surface of a plate such as an EMB plate that identifies coliform bacteria. After incubation, typical coliform colonies are counted. Standards of drinking water safety for the United States are set by the Environmental Protection Agency (EPA). They require that the number of coliform bacteria shall not exceed any of the following: (1) an average of 1 per 100 ml in all samples examined in a month, (2) 4 per 100 ml in any one sample if fewer than 20 samples are examined each month, or (3) 4 per 100 ml in 5 percent of the samples if 20 or more samples are examined each month.

Most probable number (MPN) test

Membrane filter (MF) test

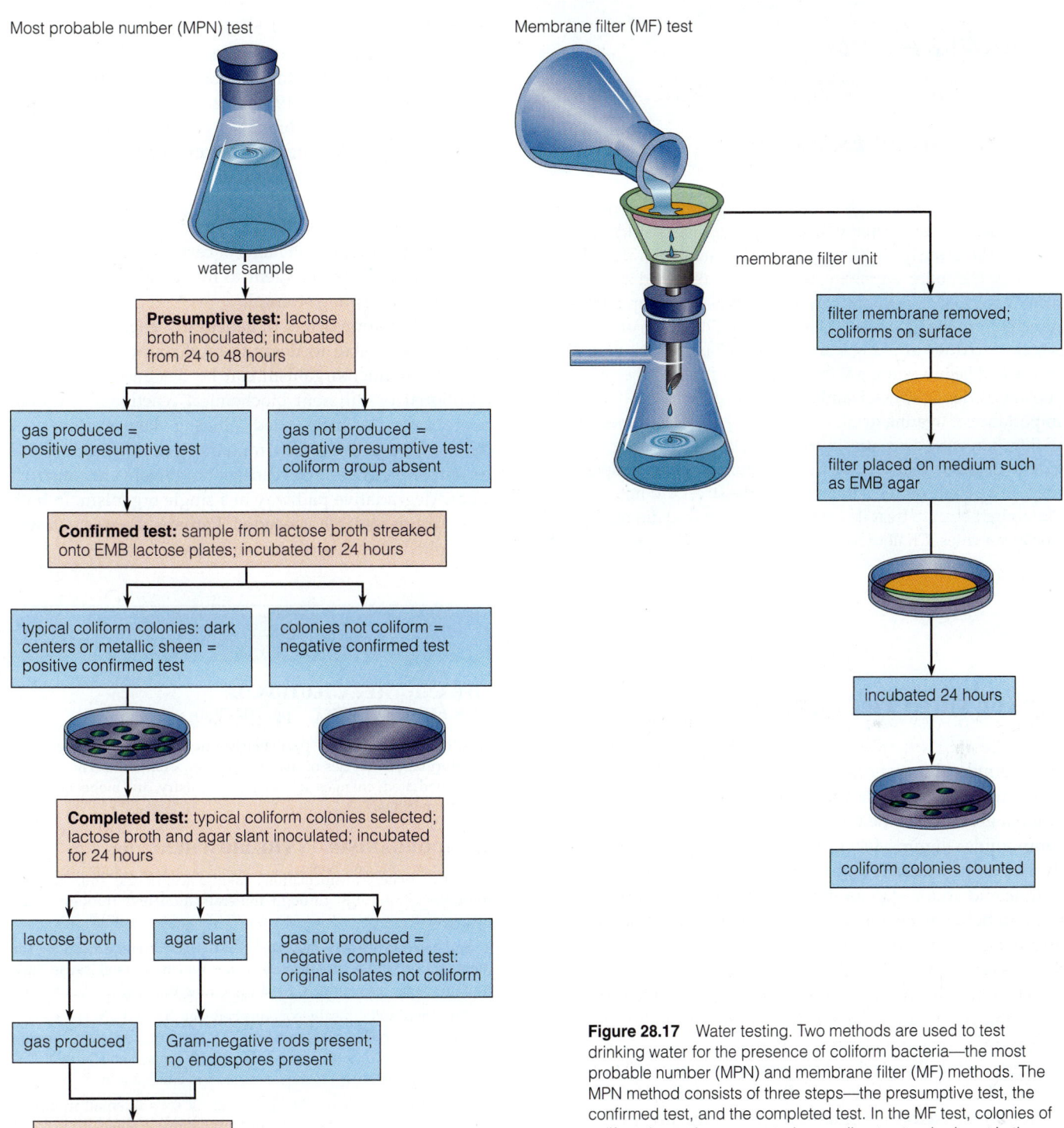

water sample

Presumptive test: lactose broth inoculated; incubated from 24 to 48 hours

gas produced = positive presumptive test

gas not produced = negative presumptive test: coliform group absent

Confirmed test: sample from lactose broth streaked onto EMB lactose plates; incubated for 24 hours

typical coliform colonies: dark centers or metallic sheen = positive confirmed test

colonies not coliform = negative confirmed test

Completed test: typical coliform colonies selected; lactose broth and agar slant inoculated; incubated for 24 hours

lactose broth

agar slant

gas not produced = negative completed test: original isolates not coliform

gas produced

Gram-negative rods present; no endospores present

coliform group present = positive completed test

membrane filter unit

filter membrane removed; coliforms on surface

filter placed on medium such as EMB agar

incubated 24 hours

coliform colonies counted

Figure 28.17   Water testing. Two methods are used to test drinking water for the presence of coliform bacteria—the most probable number (MPN) and membrane filter (MF) methods. The MPN method consists of three steps—the presumptive test, the confirmed test, and the completed test. In the MF test, colonies of coliform bacteria are counted according to standards set in the United States by the Environmental Protection Agency.

Turning Point

An Unplanned Experiment

About a hundred years ago, in 1892, an unplanned experiment took place in Germany that changed forever world opinion about the importance of treating municipal water supplies. Hamburg, a growing urban center, obtained its drinking water as inexpensively as possible—by pumping it directly from the Elbe River. The adjoining little town of Altona also drew drinking water from the Elbe, but the year before had begun filtering it through beds of sand. When a cholera epidemic struck Hamburg but not Altona, the importance of treating drinking water and the effectiveness of filtration were dramatically evident. On the Hamburg side of the street that divided the two municipalities, people died. But people on the Altona side of the street did not. Hamburg began to treat its drinking water, and so did other European cities. Cholera has never returned to Europe.

ESCAPE FROM THE CARBON CYCLE

From simple observations we know that natural organic compounds are degraded by one microorganism or another at about the same rate they are made. Some organic compounds—lignin, for example—are degraded relatively slowly. Others escape degradation for long periods of time and accumulate in the form of peat, coal, and petroleum because they were preserved in an anaerobic environment. But *all* naturally occurring organic compounds can be mineralized by microorganisms.

The equilibrium between synthesis and degradation of organic compounds that was maintained for billions of years has become unbalanced during the past fifty. Some products of the chemical industry are extremely resistant to microbial attack. They are called **recalcitrant organic compounds**. Chlorine-containing pesticides are prime examples. Other compounds, including most plastics, are impervious to microbial degradation. Recalcitrant and nonbiodegradable compounds are accumulating everywhere in the world and causing significant environmental damage.

In effect, these compounds have escaped from the carbon cycle, a major problem for which there is no single solution. But there are some encouraging approaches. The first and most obvious is to restrict the manufacture of compounds that are not susceptible to microbial attack (see the box "Bacteria That Make Plastic"). This approach has been successful in the case of the **alkylbenzene sulfate detergents**. In the 1950s this class of detergents contained molecules with branched chains of carbon atoms, which are highly resistant to microbial attack. The detergents passed unaffected through sewage plants and accumulated in fresh water supplies, causing water in streams and even from household taps to foam. But simply changing the branched chains of carbon atoms to straight chains opened the detergent to microbial attack without affecting its cleaning strength.

Another approach is to genetically engineer new microorganisms able to degrade recalcitrant organic compounds. A microorganism can be designed to perform additional or different biochemical reactions. Thus far, microbiologists have tried altering key enzymes of degradative pathways by mutating encoding genes and recombining genes from different organisms to construct a new degradative pathway in a single organism. In both cases, more research is needed. Perhaps the research will be applied to ecological problems soon.

Summary

LIFE AND THE EVOLUTION OF OUR ENVIRONMENT   (pp. 707–708)

1. Microorganisms play particularly important roles in the cyclic transformations (cycles of matter) that perpetuate life. The biologically mediated changes in Earth's chemistry are biogeochemical transformations.

MICROORGANISMS IN THE BIOSPHERE   (pp. 708–714)

1. Many cyclic transformations take place in the soil. Bacteria improve soil fertility through mineralization and through fixing nitrogen.

2. Soil bacteria are extremely diverse. Actinomycetes are among the most important. A few bacteria that live in the soil cause such human diseases as anthrax, food poisoning, tetanus, and botulism.

3. An acre of soil typically contains between 500 and 5000 pounds of fungi that degrade organic materials. Some fungi are active predators, others are mycoparasites, and others cause such human diseases as blastomycosis, histoplasmosis, and San Joaquin valley fever.

4. Algae are present on the surface of all soils in small numbers. Protozoa help control bacterial soil populations. Larvae of the hookworm *Necator americanus* are free-living in the soil.

5. Some soil fungi form mycorrhizae, intimate associations with the roots of plants in which the fungi act as additional roots to help the plant acquire nutrients. The rhizosphere effect, described by the R:S ratio, is another symbiotic association between plants and microorganisms in the region of the rhizosphere.

6. Phytoplankton in the world's oceans constitute a major portion of the world's total photosynthetic capacity. Eutrophic bodies of water support huge populations of cyanobacteria called blooms. Fresh water is a significant reservoir of human pathogens.

Bacteria That Make Plastic

Waste plastic has accumulated almost everywhere on Earth, even on the remote beaches. Plastic litter is becoming a serious ecological problem. Plastics are insoluble in water, and microorganisms have not evolved enzymes to metabolize chemicals that never existed before humans made them. Birds, fish, turtles, and marine mammals become entangled in plastic debris or eat it and die when it blocks their intestines. Plastic in landfills used for garbage disposal remains permanently. Long after other organic compounds are broken down by microorganisms, plastic will remain intact.

One solution to the accumulation of plastics in nature is to develop **biodegradable plastics**, susceptible to microbial degradation. Unfortunately, most plastic products touted as biodegradable are not. They are made of bits of undegradable plastic held together by a matrix of a biodegradable substance like starch. Microbial action dismantles the matrix but leaves the small bits of plastic intact.

Although humans have so far failed to synthesize completely biodegradable plastics, bacteria have been successful. Some bacteria store their carbon reserves as granules of poly-beta-hydroxyalkanes (Chapter 4), also called polyhydroxyalkanoate or PHA. They accumulate PHAs in huge quantities when a carbon source is available in excess and metabolize them when it is depleted. PHAs, which are polyesters, are a biodegradable plastic. The bacteria that make PHAs and other bacteria as well readily use PHAs as a carbon source. Some bacteria excrete an extracellular enzyme, a depolymerase, that degrades PHAs into their component beta-hydroxy acids, which they or other microorganisms then use as nutrients.

Bacteria make different forms of PHA, depending on the source of carbon they are fed. Both the length of the alkane chain and the length of the polymer can be varied by changing bacterial nutrition. PHAs can, therefore, be fabricated into many different kinds of plastic, suitable for making plastic bags, squeeze bottles, clear glasslike material, and even fabrics. PHAs are also versatile because they are thermoplastic—when heated, they can be shaped.

The idea of using PHAs for commercial plastics has been around for some time. Patents were filed in the United States in 1962. But the plastics were not made industrially until 1982 when Imperial Chemical Industries of Britain marketed them under the trade name of Biopol. The first consumer

Plastic debris litters the shore of a beach on one of the Hawaiian Islands.

product, bottles for biodegradable shampoos, was marketed in 1990 by Wella AG in Germany.

The hydrogen-oxidizing bacterium *Alcaligenes eutrophus* is used to make PHAs. The substrates for autotrophic growth—hydrogen gas, carbon dioxide, and air—are inexpensive but potentially explosive, so *A. eutrophus* is grown in a glucose-salts medium. When growth ceases after about 60 hours because the phosphate is exhausted, little PHA has been made. Then more glucose is added; and during the next 48 hours, massive amounts of PHA accumulate, up to 75 percent of the total biomass. PHA is extracted with hot nonpolar solvents (either chloroform or methylene chloride), and cells are filtered out. On cooling, PHA precipitates from the solvent.

On the market, PHAs cost several times more than ordinary plastics, which are inexpensive because they are made from petroleum. But when environmental costs are taken into account, these 100 percent biodegradable plastics might be a bargain.

7. Microorganisms do not grow in air, but all are found as passengers on aerosols or dust particles.

THE CYCLES OF MATTER   (pp. 714–725)

1. The major biological elements occur in several different chemical forms that are interconvertible.

2. In the nitrogen cycle, nitrogen gas is converted to ammonia by nitrogen fixation. Ammonia is converted to nitrate by nitrification, and nitrate is converted to nitrogen gas by denitrification, completing the cycle.

3. All the steps of the nitrogen cycle are mediated by microorganisms. Two classes of bacteria, symbiotic (such as *Rhizobium* spp.) and free-living (such as *Azotobacter*), fix nitrogen. About half the world's supply of nitrogen is now fixed by the Haber process. Heterotrophic bacteria carry out ammonification, the conversion of nitrogen compounds back to ammonia. Nitrifying bacteria prevent ammonia from accumulating in the soil.

4. In the carbon cycle, carbon dioxide is removed from the atmosphere by autotrophs and fixed through photosynthesis; carbon dioxide is regenerated and returned to the atmosphere through respiration and combustion. The carbon cycle is out of balance today because atmospheric carbon dioxide has risen more than 11 percent since 1958. Global warming results from this rise.

5. The phosphorus cycle has no significant gaseous intermediate, and phosphorus is neither oxidized nor reduced. The phosphorus cycle converts phosphorus from inorganic phosphate to organic phosphate and back.

6. The bacterium *Acinetobacter calcoaceticum* can be used to remove phosphate from sewage.

7. In the sulfur cycle, sulfate ions are reduced to hydrogen sulfide gas (desulfurization) and then reoxidized to sulfate.

8. Sulfate-reducing bacteria live in sulfate-rich anaerobic environments such as mud flats. The conversion of hydrogen sulfide gas to sulfate is mediated by chemoautotrophic sulfur-oxidizing bacteria and nonoxygenic phototrophic bacteria.

TREATMENT OF WASTE WATER   (pp. 725–727)

1. Sewage is municipal waste water. Microorganisms oxidize almost any organic compound in waste water until dissolved oxygen is depleted, so sewage plants add oxygen to sewage at a faster rate than it is removed by microbial respiration. Waste water is classified by its biochemical oxygen demand (BOD).

2. The primary treatment unit in a sewage plant grinds the solid material; the primary sludge settles out while the effluent proceeds to secondary treatment. The sewage is aerated and then mineralized by aerobic bacteria, including *Zoogloea ramigera*, forming a floc. The rapidly settling floc is called activated sludge. The primary sludge and the activated sludge are sent to an anaerobic sludge digester where they are partially degraded by anaerobic bacteria. Methanogens convert by-products of the anaerobic bacteria into methane, which can be used to generate electricity. The digested sludge is dried and often used as fertilizer. Effluent may be given tertiary treatment; increasingly, this is biological.

3. Septic tanks are small anaerobic digesters used where municipal sewage systems are not available. Digested effluent is distributed to leaching fields.

TREATMENT OF DRINKING WATER   (pp. 727–729)

1. After processing in flocculation tanks, drinking water is passed through sand beds that remove most bacteria and some viruses that cause waterborne diseases. Chemicals kill most remaining microorganisms.

2. Municipal drinking water is checked for safety by testing for coliform bacteria, which are indicator organisms, because it is impractical to test for all possible pathogens.

3. The two tests are the most probable number (MPN) test and the more commonly used membrane filter (MF) test. Standards for drinking water are set by the Environmental Protection Agency in the United States.

ESCAPE FROM THE CARBON CYCLE   (p. 730)

1. All naturally occurring organic compounds can be mineralized by microorganisms. However, humans have created products that are extremely resistant to microbial attack (recalcitrant organic compounds), including most plastics and chlorine-containing pesticides.

2. Solutions include using biodegradable products and genetically engineering new microorganisms able to degrade nonbiodegradable products.

Review Questions

LIFE AND THE EVOLUTION OF OUR ENVIRONMENT

1. Explain this statement: Though microorganisms cause disease, it would be a mistake to think of them only as adversaries.

2. What are biogeochemical transformations? What are the oldest known fossils and what do they tell us about the evolution of life and Earth's atmosphere?

MICROORGANISMS IN THE BIOSPHERE

1. By what processes do microorganisms improve soil fertility? Give some examples of microorganisms found in the soil and their functions. Discuss mycorrhizae and the rhizosphere as examples of symbiosis among soil microorganisms.

2. What kinds of microorganisms are found in water, and what are their ecological roles?

3. Air is not a habitat for microorganisms, but bacteria, algae, fungi, protozoa, and viruses are all found in the air. How does this occur?

THE CYCLES OF MATTER

1. How do the cycles of matter show the interdependence of living things?

2. Describe the nitrogen cycle and the role of bacteria in nitrogen fixation, nitrification, ammonification, and denitrification. What different ways have *Azotobacter* and *Rhizobium* spp. evolved to protect their nitrogenase from oxygen?

3. What is the Haber process and why is it significant? How is the water fern *Azolla* used to increase fixed nitrogen for agriculture?

4. Describe the major loop of the carbon cycle. How has the carbon cycle become unbalanced, and what are some possible ecological consequences?

5. How does the phosphorus cycle differ from the nitrogen and carbon cycles? Describe the phosphorus cycle.

6. Describe the sulfur cycle. What roles do bacteria play? What industrially valuable product is a by-product of the sulfur cycle?

TREATMENT OF WASTE WATER

1. What is sewage? What is biochemical oxygen demand (BOD) and why it is important in sewage treatment?

2. Describe the primary, secondary, and tertiary treatment units of a municipal sewage plant. What roles do *Zoogloea ramigera* and the methanogens play?

3. Discuss these alternative sewage treatments: small anaerobic digesters, septic tanks, oxidation ponds, sewage farming.

TREATMENT OF DRINKING WATER

1. How is drinking water processed? How is it tested? What are indicator organisms?

ESCAPE FROM THE CARBON CYCLE

1. What are recalcitrant organic compounds? Give some examples.

2. Discuss some possible ways we can restore the balance between synthesis and degradation of organic compounds.

Suggested Readings

Anderson, A. J., and Dawes, E. A. 1990. Occurrence, metabolism, metabolic role, and industrial uses of bacterial polyhydroxyalkanoates. *Microbiological Reviews* 54:450–72.

Atlas, R. M., and Bartha, R. 1987. *Microbial ecology: Fundamentals and applications*. Menlo Park, Calif.: Benjamin/Cummings.

Goodfellow, M., and Williams, S. T. 1983. Ecology of actinomycetes. *Annual Reviews of Microbiology* 37:189–216.

Racke, K. D., and Coats, J. R., eds. 1990. *Enhanced biodegradation of pesticides in the environment*. Washington, D.C.: American Chemical Society.

29 MICROBIAL BIOTECHNOLOGY

Say Cheese

Long before the Danish microbiologist Sigurd Orla-Jensen identified *Propionibacterium* in 1906, cheesemakers were using the microorganism to make Swiss cheese. *Propionibacterium* ferments lactic acid, forming propionic acid, acetic acid, and carbon dioxide (CO_2) as end products. (Lactic acid is always present in unripened cheeses that are made by allowing lactic acid bacteria to ferment the lactose in milk.) The two acid products of the fermentation contribute to the special flavor of Swiss cheese. The third product, CO_2, creates the holes that characterize Swiss cheese.

But making Swiss cheese is not as easy as it sounds. The balance of fermentation products must be just right. If there is too much or too little propionic or acetic acid, the cheese will not have its identifiable nutty flavor. Carbon dioxide formation must be controlled to create just enough holes, not so many that they split the cheese. In addition, the strains of *Propionibacterium* that are used must be carefully chosen. They should make enough of the amino acid proline to give the cheese a sweet taste and enough diacetyl for a buttery taste.

The fermentation products of *Propionibacterium* are also an excellent natural preservative. *P. shermanii* is cultivated in skim milk and dried to a powder. The product, called Microgard, is used to preserve cottage cheese. It inhibits Gram-negative bacteria and some yeasts and molds. Other preparations of *Propionibacterium* may soon be used to preserve grains and bread.

2 µm

Propionibacterium is the microorganism used to make Swiss cheese.

- The traditional uses of lactic acid bacteria to preserve food and make cheese and other dairy products

- How yeasts are used to make wine, beer, and bread, and, with acetic acid bacteria, vinegar

- The principal methods of slowing or preventing growth of microorganisms to preserve food

- How microorganisms are used as insecticides

- The industrial fermentations that produce solvents, amino acids, antibiotics, and enzymes

- How genetically engineered microbes are used to make new products for medicine, industry, and agriculture

TRADITIONAL USES OF MICROORGANISMS

Even before humans knew that microorganisms existed, they used them to make food (Swiss cheese is one of many examples), improve crop yields, and dispose of waste. As our knowledge of microorganisms grew, we improved and expanded their uses. Early trial-and-error processes developed into technologies we collectively called **industrial microbiology**. Managing microorganisms became a science that profoundly changed our lives by producing antibiotics, vitamins, food supplements, and industrial chemicals.

In the past 20 years, promises of greater benefits have come from the emergence of molecular biology and recombinant DNA technology. This new industry was called **biotechnology** (Chapter 7). The term now applies to all uses of microorganisms, including traditional ones. Although the term *biotechnology* describes the use of any organism for any practical purpose, microorganisms are and probably will continue to be its primary focus. In this chapter we discuss traditional uses, uses of microorganisms as chemical factories, and the use of genetically engineered microbes to make products from the genes of other organisms.

The first human use of microorganisms was probably to make and preserve foods and beverages. Lactic acid bacteria and yeasts played dominant roles.

Lactic Acid Bacteria

All lactic acid bacteria ferment various sugars into lactic acid in amounts large enough to inhibit or kill most other microorganisms (Chapter 11). But with very few exceptions, including some streptococci, lactic acid bacteria are harmless to humans, and the metabolic products of lactic acid bacteria have a pleasant taste. These properties enable us to use lactic acid bacteria to prepare and preserve food. The food must contain enough sugars for the lactic acid bacteria to produce inhibiting amounts of lactic acid (most plant materials and dairy products do). Also, air must be excluded so that aerobic microorganisms, which metabolize more rapidly, cannot use the sugar before the lactic acid bacteria have a chance to develop. Usually, it is not necessary to add lactic acid bacteria to the food because most plant materials and dairy products contain an adequate natural population.

Plant Foods. Making *sauerkraut* (meaning "acid cabbage" in German) illustrates a simple, traditional way of using lactic acid bacteria to preserve plant material for food. Cabbage is shredded and packed firmly into a nonmetal container such as a ceramic crock, and water is added to fill pockets of air. The lactic acid bacteria naturally present on the cabbage multiply and produce lactic acid from sugars in the cabbage. The sauerkraut becomes so acidic that most microorganisms are inhibited or killed. A few aerobic microorganisms can attack these high concentrations of lactic acid, but if the sauerkraut is protected from air in closed containers, it lasts almost indefinitely.

Silage—livestock feed—is prepared using the same principles. Chopped plants, principally corn, are deposited in huge airtight cylindrical silos. The lactic acid bacteria normally present, mainly *Lactobacillus plantarum*, act on the sugars in the plant material to produce enough lactic acid to preserve the plant material (**Figure 29.1**).

Spanish-style green olives are also preserved by a spontaneous lactic acid fermentation.

Cheese and Other Dairy Products. Most traditional fermented dairy products are made using the lactic acid bacteria that happen to be present in milk. Usually, the bacteria come from the tools and vessels used to make previous batches. The distinctive properties and tastes of various soured milk products—including buttermilk, acidophilus milk, and yogurt—depend on the kind of lactic acid bacterium causing the souring. Different species produce different products (**Table 29.1**). Even different strains of the same species affect the quality and flavor of the product. A great deal of research has been done to

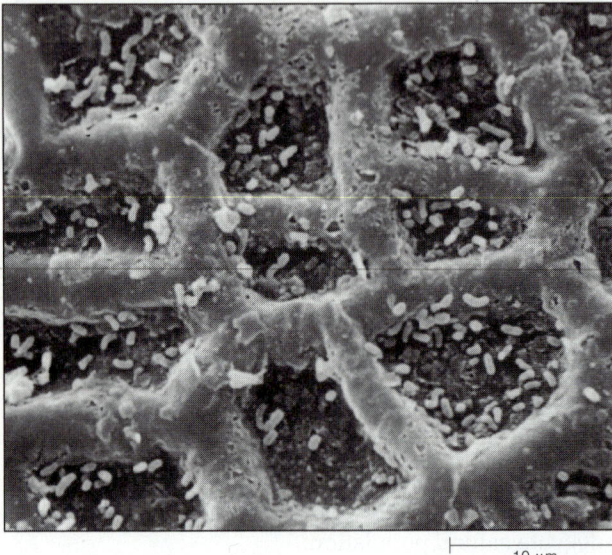

Figure 29.1 The lactic acid bacterium *Lactobacillus plantarum* on the surface of a cucumber fruit.

Table 29.1 Microbiology of Dairy Products and Cheese

Product	Microorganism(s)	Microbial Activity
Acidophilus milk	*Lactobacillus acidophilus*	Lactic acid fermentation
Butter	*Streptococcus diacetilactis, Leuconostoc cremoris*	Lactic acid fermentation
Sour cream	*Streptococcus diacetilactis, S. lactis, Leuconostoc cremoris*	Lactic acid fermentation
Buttermilk	*Lactobacillus bulgaricus*	Lactic acid fermentation
Kefir	*Streptococcus lactis, Lactobacillus bulgaricus,* lactose-fermenting yeasts	Combined lactic acid–alcoholic fermentation
Yogurt	*Lactobacillus bulgaricus, Streptococcus thermophilus*	Lactic acid fermentation
Cheese		
Cheddar	Lactic acid bacteria	Lactic acid fermentation
Roquefort	*Penicillium roquefortii*	Production of blue pigment
Swiss	Lactic acid bacteria, *Propionibacterium* spp.	Lactic/propionic acid fermentation

produce superior strains for particular products. This research led to the development of a branch of microbiology called **dairy bacteriology**.

Cheese making, largely a microbiological process, has also been studied by dairy bacteriologists (**Figure 29.2**). It consists of two steps—**curdling** and **ripening**. Curdling is the conversion of milk into a solid mass, the **curd**. The protein fraction—**casein**—is precipitated, bringing the fat with it. The liquid fraction, the **whey**, is drained off. Curdling can be accomplished either by allowing lactic bacteria to develop (the resulting lactic acid denatures the casein, causing it to precipitate) or by adding **renin**, an enzyme extracted from a calf stomach. The way that the milk is curdled depends on the type of cheese being made. For example, cottage cheese, which is not acidic, is made with renin.

After curdling, the curd is pressed into the desired shape of the final product and allowed to ripen. The species of microorganism involved depends on the type of cheese being made. In traditional cheese manufacture, the inoculation of ripening microorganisms is accidental and uncontrolled, coming from utensils or tools used to prepare previous batches. Even so, cheeses produced in the same region by the same procedures are ripened by the same strains of microorganisms and there is little difference in taste or quality from batch to batch. Swiss cheese has a buttery, nutty taste and holes because it is always ripened by *Propionibacterium*. Roquefort and other blue cheeses are always ripened by the fungus *Penicillium roquefortii*. Its blue spores color the cheese, and its metabolic products contribute flavor and modify texture.

Ripening is a biochemically and microbiologically complex process. Some of the changes come from flavor-giving compounds made by the microorganisms. Others are caused by changes in milk proteins, which are hydrolyzed into a mixture of small soluble peptides and amino acids. In general, the more hydrolysis, the softer the cheese. Only about a quarter of the protein in hard cheeses such as cheddar and Swiss is hydrolyzed to soluble products. Almost all the protein in soft cheeses like Camembert and Limburger is hydrolyzed.

Yeasts

Yeasts are found on most fruits and flowers and in the exudates of plants. These yeasts ferment sugars to produce CO_2 and ethanol. If fruits are crushed and held in a container with limited surface contact with air so that competing aerobic microorganisms do not outgrow the yeasts, alcoholic fermentation is almost certain to occur. The result is a wine. The yeasts used today to make wine, beer, whiskey, and bread are almost always strains of *Saccharomyces cerevisiae* and closely related species.

Figure 29.2 Cheese making is largely a microbiological process. It consists of two steps—curdling and ripening. Here, curdled milk is being stirred.

Figure 29.3 Freshly squeezed grape juice, or must, is fermented in large steel tanks. The finished wine is then aged in wood barrels.

Wine. The origins of **enology**, winemaking, must be as ancient as **viticulture**, grape growing, and almost as ancient as agriculture itself. Wine was a well-established article of commerce by the third century B.C., the height of Greek civilization. Although the basic procedures for making wine are simple, almost inevitable—crush the grapes and let them stand in a container with restricted air contact—enology has become a highly sophisticated science (**Figure 29.3**). The University of California at Davis and California State University at Fresno have Departments of Enology.

The price of wines depends largely on the variety and quality of the grapes used. High-quality, flavorful varieties often have low crop yield, and farming practices that enhance quality often decrease yield. The cool climates that favor flavorful grapes are not conducive to high yields.

The microbiological aspects of winemaking are also important. Traditionally, and still in some regions, the inoculum producing **primary alcoholic fermentation** in **must** (freshly squeezed grape juice) is the natural yeast flora of the grape. The natural yeast flora is a mixed culture that can add flavor complexity. Today, in many regions of the world, the natural inoculum is virtually eliminated by treating the must with sulfite and then adding a pure culture of a reliable yeast strain. Yeasts used for winemaking are usually classified as *Saccharomyces ellipsoideus*. They have a somewhat different shape but in most other respects are quite like *S. cerevisiae*.

Sometimes different kinds of yeasts and other microorganisms are used to make different kinds of wine, but the most important distinction—red or white—depends on the grapes and winemaking practice (**Table 29.2**). Only white wine can be made from white grapes, but either red or white wine can be made from red grapes. The red color is in the grape skins. If the skins are left in the fermenting must, the red color will be extracted (by alcohol) into the finished wine.

Two kinds of yeast are used to make Spanish sherry, a fortified wine. (Sherry is a British corruption of the name of the Spanish town, Jerez, where Spanish sherry is made.) After ordinary *Saccharomyces ellipsoideus* has fermented all the sugar in the must, **brandy** (distilled wine) is added to increase the alcohol content to about 15 percent (**Figure 29.4**). Then a layer of mixed yeast, including *S. beticus, S. cheresiensis,* and *S. fermenti,* develops on the surface of the wine, which is stored in half-filled barrels. This layer thickens and becomes crinkly, resembling flower petals and accounting for its Spanish name, *flor*

Table 29.2 Microbiology of Wine

Organism	Role(s) in Winemaking
Saccharomyces ellipsoideus	Primary alcoholic fermentation—glucose and fructose in must are fermented to ethanol and carbon dioxide
	Secondary alcoholic fermentation—in bottle or tank, added sugar is converted to ethanol and carbon dioxide to produce sparkling wines such as champagne
	Clouding—yeast grows in finished sweet wines, producing undesirable turbidity
Saccharomyces beticus, S. cheresiensis, S. fermenti	Flor layer on sherry—some ethanol is oxidized to acetaldehyde, producing typical sherry flavor
Botrytis cinerea	Production of naturally sweet, flavorful wines—mold raises sugar content by desiccating grapes, decreases acidity by oxidizing malic acid to carbon dioxide and water, and adds flavor and color
Pediococcus, Leuconostoc, Lactobacillus spp.	Malolactic fermentation—malic acid in wine is converted to lactic acid and carbon dioxide, decreasing acidity and increasing flavor complexity of wine
Acetobacter spp.	Aerobic spoilage—oxidizes some ethanol to acetic acid, giving wine an undesirable vinegar taste

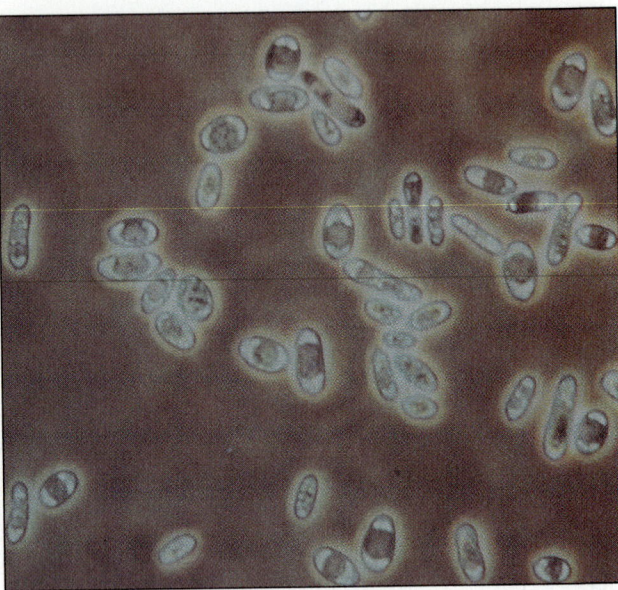

Figure 29.4 Today, in place of the natural yeast flora of the grape, reliable strains of *Saccharomyces ellipsoideus* are usually used in winemaking.

(meaning "flower"). The metabolic products of the surface yeast, including large amounts of acetaldehyde, give sherry its distinctive taste.

Certain sweet wines from the Sauternes district of France and parts of Germany are products of a more complicated microbiological process. In these relatively damp regions, the grapes become infected by the fungus *Botrytis cinerea*, which enters the grape by means of shortened hyphae called **haustoria**. The haustoria act as small wicks that dry the grape, concentrating its sugars so much that they cannot be fermented completely by the yeasts when the must is made into wine. The fungus produces a distinctive color and flavor, and some glycerol, which contributes a tactile thickness and richness to the wine. *Botrytis* does not metabolize sugar needed by the yeast. Instead it uses **malic acid** (an organic acid found in all grapes), which makes the wine less acid. This fungal infection—called "the noble rot" by winemakers—produces dessert wines that are highly valued for their distinctive soft, sweet flavor and bright golden color.

Bacteria also play roles in winemaking. In most wine regions the malic acid of premium red wine is removed by a traditionally spontaneous (now inoculated) **malolac-** tic fermentation—mediated by lactic acid bacteria such as *Pediococcus, Leuconostoc,* and *Lactobacillus* spp.—during the first winter after the wine is made. This secondary fermentation converts malic acid, which contains two acid groups, into lactic acid, which contains only a single acid group.

$$\text{HCOOH—CH}_2\text{—CHOH—COOH} \longrightarrow \text{CH}_3\text{—CHOH—COOH} + \text{CO}_2$$

MALIC ACID   LACTIC ACID   CARBON DIOXIDE

The result is a less acidic and more flavorful wine.

Microorganisms can also spoil wine. Some lactic acid bacteria grow in sweet wines, giving it a foul taste. If wine is exposed to air, acetic acid bacteria will develop, oxidizing the alcohol to acetic acid and giving the wine a vinegary taste.

Other Alcoholic Beverages. Most strains of yeast can ferment only sugars. Fruit juices, including grape juice (which contains a mixture of glucose and fructose), can be fermented without complication, but grain products contain starch, which must be **saccharified**—hydrolyzed by amylase, a starch-hydrolyzing enzyme, to its constituent monomers, glucose—before it can be fermented by yeasts. In the Western world **amylase** from **malt**, heat-dried germinated grains of barley, is used to saccharify cereal grains for making beer and whiskey.

Figure 29.5 This vat is being filled with wort, cooked and saccharified cereal grains, which will ferment into beer. Finished beer is aged in steel tanks.

Beer is made from fermented cereal grain; traditionally, barley was used, but now large amounts of rice are used. Cereal grains are cooked and saccharified with malt, producing **wort**, which is fermented in large vats (**Figure 29.5**). Two kinds of yeast are used in beer making—*Saccharomyces carlsbergensis* for beer and *S. cerevisiae* for ale. In either case, some of the CO_2 produced during the yeast fermentation is trapped. Finally, beer is flavored and to an extent preserved by adding **hops**, the flowers of the vine *Humulus lupulus*.

Whiskey is distilled, or extracted, from fermented cereal grains. Potatoes, which are used to make vodka, contain largely starch, so they, too, must be saccharified.

In Asia, amylase from fungi is used to saccharify rice before it is fermented to produce Japanese *sake* and Chinese *rice wine*. The moistened rice is heaped on the floor and inoculated with a previous batch of rice on which mixtures of fungi have grown. Within a few days the fungi grow through the pile of rice, producing enough amylase to saccharify all the starch.

Vinegar. The food product **vinegar** is a solution of acetic acid made in a two-step process. First, yeasts ferment the sugar in a fruit juice, usually apple (to make cider vinegar) or grape (to make wine vinegar), to ethanol. The yeast is usually *Saccharomyces cerevisiae*.

$$C_6H_{12}O_6 \longrightarrow CH_3—CH_2OH + CO_2$$
GLUCOSE OR   ETHANOL   CARBON
FRUCTOSE   DIOXIDE

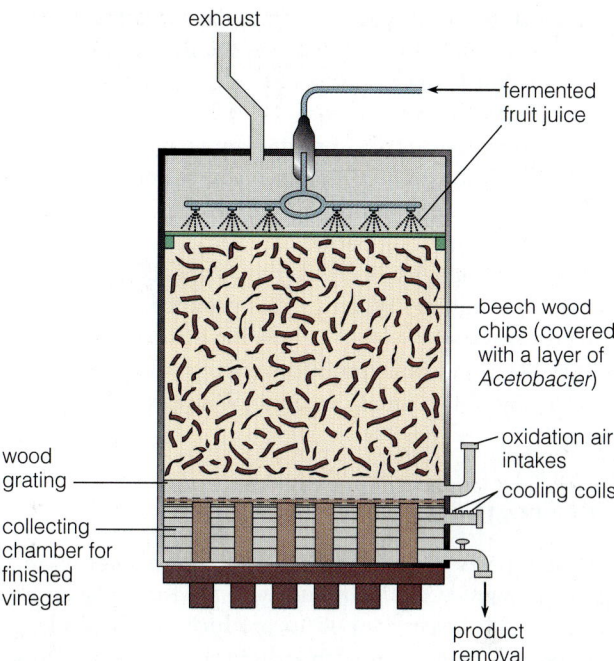

Figure 29.6 A vinegar generator, used to make large quantities of vinegar, is a large column filled with wood chips that are covered with a layer of *Acetobacter*. Fermented fruit juice is added at the top, while oxygen is forced in from the bottom.

Then species of *Acetobacter*, an obligately aerobic bacterium, from a culture or from equipment and tools used to make previous batches, oxidize the ethanol incompletely to acetic acid and water, instead of to CO_2 and water as most microorganisms would do.

$$CH_3—CH_2OH + O_2 \longrightarrow CH_3—COOH + H_2O$$
ETHANOL   ACETIC ACID

There are a number of ways to make vinegar. All are designed to provide enough oxygen to oxidize alcohol to acetic acid. The oldest and slowest—but some say best—way is to fill a barrel half full of the alcohol solution. A film of *Acetobacter* develops on the surface, oxidizing the ethanol at a rate set by the diffusion of oxygen into the solution. A more rapid way is to use a **vinegar generator**, a device that transfers oxygen to the solution at a high rate (**Figure 29.6**). Vinegar generators are large columns, usually filled with beech wood chips covered with a layer of *Acetobacter* cells. The alcohol solution is fed into the top and trickles down through the chips while air forced in at the bottom rises up through the column.

Bread. Using yeast to raise bread (or *leaven* it, which comes from an old word for yeast) before baking is a relatively recent human innovation—only a few hundred years old. Brewer's yeast, a by-product of brewing beer, was first used, but now we use specially selected strains called **baker's yeast**. Both brewer's yeast and baker's yeast are strains of *Saccharomyces cerevisiae*.

Baker's yeast is produced commercially and dried under conditions that maintain viability and enzyme activity (see the box "Bread in the Desert," Chapter 2). It can be stored for long periods at room temperature without loss of activity. CO_2 is the fermentation product needed for baking. When CO_2 is produced in dough, it creates small bubbles that lighten the texture. Usually the natural sugar in flour produces enough CO_2 to raise the dough, though a small amount of sugar is sometimes added.

Mixed Cultures

In some parts of the world, mixed cultures are used to make fermented foods. In Europe, a mixture of lactic acid bacteria and yeasts is used to produce mildly alcoholic, soured milk beverages such as **kefir**. In Asia, the production of foods such as soy sauce and miso (made from rice and soybeans) from mixed cultures of microorganisms is a huge industry (**Table 29.3**). In Japan, mixed-culture food industries produce 20 times more revenue than does the industrial alcohol industry. Usually, **starter cultures** containing complex mixtures of bacteria and fungi are used to inoculate food. Because the methods of producing starter cultures have not changed for centuries, these mixtures of microorganisms are highly adapted to coexist and produce a uniform product. Some microorganisms in these starters are found nowhere else and some do not survive in nature.

Stable mixed cultures have certain advantages over pure cultures. They have high growth rates and enhanced yields (some mixed cultures of lactic acid bacteria produce more lactic acid than does any component culture

alone). They are highly resistant to contamination. They can mediate simultaneous multistep transformations of complex mixtures of substrates.

MICROBES AS INSECTICIDES

Certain species of the endospore-forming bacterial genus *Bacillus*—including *B. larvae*, *B. popilliae*, and *B. thuringiensis*—are insect pathogens. These species form **parasporal bodies**, crystals of protein, beside their endospores (**Figure 29.7**). When a susceptible insect eats the endospore, the protein is cleaved in the insect's alkaline gut, producing a highly destructive form that ulcerates the intestinal wall. The endospore germinates, penetrates the insect tissue, multiplies, and eventually produces more endospores and lethal protein levels.

For more than 30 years, *B. thuringiensis* endospores have been produced commercially and marketed as a **bioinsecticide** called Bt. Bt is sprayed or dusted on plant leaves to control caterpillars and other insects. Within an

Table 29.3	Mixtures of Microbial Cultures Used to Make Asian Foods
Food	Microorganisms in Mixture
Soy sauce	*Aspergillus soyae*
	Aspergillus oryzae
	Saccharomyces rouxii
	Candida versitilis
	Pediococcus halophilus
	Lactobacillus delbrueckii
Miso	*Aspergillus oryzae*
	Aspergillus soyae
	Saccharomyces rouxii
	Candida etchellsii
	Pediococcus halophilus

100 nm

Figure 29.7 Some species of *Bacillus thuringiensis* are insect pathogens. These species form parasporal bodies, crystals of protein, beside their endospores. When a susceptible insect eats the endospore, the protein in the parasporal body cleaves and becomes lethal.

hour after consuming the endospores, caterpillars stop feeding; they die several days later.

B. popilliae endospores are produced commercially as a bioinsecticide called Bp to control Japanese beetles. Bp causes **milky blood disease**. Distributed over a lawn, the endospores persist in the soil and control Japanese beetles for 15 to 20 years.

Spores of the protozoan *Nosema locustae* are sold commercially as a bait to combat grasshoppers, locusts, and crickets. They act slowly, requiring four to six weeks for control. Baculoviruses are being developed as another bioinsecticide (Chapter 13).

Unlike some chemical insecticides, Bt, Bp, *N. locustae*, and baculoviruses are all ecologically safe. These bioinsecticides kill only certain insect species and have no deleterious effect on plants or animals, including humans.

MICROBES AS CHEMICAL FACTORIES

With their rapid rates of metabolism and growth, microorganisms can synthesize cells and metabolic end products at prodigious rates. Because of this, microorganisms are used industrially to manufacture many commercial products in addition to food. These include organic solvents, vitamins, amino acids, antibiotics, enzymes, proteins with medical uses, and the microorganisms themselves.

The first industrially produced microbial products were manufactured anaerobically by fermentation. Somehow the term stuck, so that today, whether the manufacturing process is aerobic or anaerobic, industrial transformations by microorganisms are called **fermentations**. The vessels in which these transformations take place are called **fermenters**.

Anaerobic Fermentations

Because fermentations (in the scientific sense) generate only small amounts of ATP (Chapter 5), anaerobic microorganisms process large quantities of substrate and produce large quantities of end products when they ferment. Some fermentative end products have commercial value, and producing them is relatively uncomplicated. Aeration of the culture is not necessary, nor is it necessary to follow rigorous aseptic procedures because contamination by aerobes is impossible and most competing anaerobes grow too slowly to displace the established population.

The economic feasibility of an anaerobically produced industrial product is determined by the relative price of the product and substrate and by the cost of recovering the product from the culture liquid. The source of substrate for the fermentation, usually a sugar, must be inexpensive. The most common substrates are blackstrap molasses (the dark residual liquid remaining after the maximum amount of sugar has been recovered from cane or beet syrup) and cereal grains (which can be saccharified, if necessary).

The price/cost margin is almost always narrow because most anaerobically produced products are relatively inexpensive industrial chemicals that can also be made by other processes—for example, chemically from petroleum. Price or cost changes often, so once successful industrial fermentations may be abandoned only to be reinitiated at a later date. Most of the food fermentations discussed earlier continue in spite of economic fluctuations. By law, beverage alcohol and acetic acid for vinegar must be microbiological products.

Ethanol. Ethanol (also called ethyl alcohol) is a constituent of alcoholic beverages and is used as a solvent and fuel. Before World War II, large amounts of ethanol were made by yeast fermentation for industrial purposes. Later, with the rise of the petrochemical industry, the price of ethanol dropped so low that the fermentation product was not economic. Within the past decade, however, the prospects of ethanol have again improved. Ethanol made by fermentation is the major source of automobile fuel in Brazil. In the United States, ethanol is being added to gasoline to make gasohol (90 percent gasoline, 10 percent ethanol). Gasohol will help extend the world's petroleum supply and is less polluting than conventional gasoline. Ethanol is also a renewable resource because it is made from saccharified cereal grains, a commodity that for the moment is in excess supply worldwide.

Acetone and Butanol. Acetone and butanol are commercially useful as organic solvents. They are fermentation products of various species of *Clostridium*. One species, *Clostridium acetobutylicum*, has been used on and off since before World War I to make a mixture of these solvents. The economic ups-and-downs of this industry, along with some of the technological innovations that it sparked, have captured the attention of industrial microbiology (see the box "Microbes versus Petroleum").

The technological advances associated with acetone-butanol fermentation stem from the fact that the process is successful only if other microorganisms can be excluded from the relatively rich medium. As a result, the technology of pure culture on a mass scale had to be developed. These advances were essential to the later development of the antibiotics industry.

Microbes versus Petroleum

Making solvents by fermentation first became an industry in England. Initially, the English made butanol, a starting material for synthetic rubber, because natural rubber was in short supply. One of the products of the butanol process was acetone, a solvent for nitrocellulose, which is an ingredient of cordite (smokeless powder). With World War I, the market for acetone grew and English solvent manufacturing expanded quickly. After the war, the industry continued to expand because butanol was an ingredient of the rapid-drying nitrocellulose paints used by the expanding automobile industry. A third product of the process was riboflavin (vitamin B_2), which met still another market need.

Yet by the early 1960s the solvent industry had virtually disappeared in Britain (and the United States). The petrochemical industry could make butanol-acetone less expensively, and the pharmaceutical industry had developed better microbiological methods for making riboflavin. Today, however, the butanol-acetone process is being intensively reexamined. It may again offer economic advantages because genetic modifications can be made in the producing organism and cheaper waste substrates such as whey are available. Moreover, the price and availability of petroleum are too often dependent on the world political situation.

Aerobic Processes

Most industrial microbiological processes—including those that make penicillin, other antibiotics, amino acids, vitamins, and therapeutically useful proteins—are aerobic. Because oxygen is only sparingly soluble in water and is used rapidly by the dense microbial cultures that industrial processes employ, keeping a mass culture aerobic is not easy or inexpensive. Moreover, to prevent contamination, the massive inflow of air must be sterilized.

Developing the technology to grow tens of thousands of gallons of microbial cultures aerobically without contamination was the major engineering prerequisite of the modern aerobic fermentation industry. The rate at which oxygen can be added to a culture broth is limited by the surface area of liquid that can be exposed to the gas. To maximize the surface area, the modern industrial fermenter is designed to be filled with small bubbles of air, which are retained in the liquid as long as possible. This is done by forcing air through a sparger (which looks somewhat like a shower head) beneath a rapidly turning impeller (a rotating shaft with sets of paddles). The impeller shatters the streams of air into minute bubbles that follow a helical (and therefore prolonged) path to the surface (**Figure 29.8**).

Microorganisms are removed from the incoming air by filtration, but the valves through which materials are added to the fermenter or taken out are also major portals of contamination. No matter how tightly they are closed, foreign microorganisms can pass through them and enter the fermenter. Therefore, most fermenter valves are bathed in steam.

Antibiotics. The antibiotic industry began with the British microbiologist Alexander Fleming in 1929 (Chapter 1). Fleming was studying wound infections caused by *Staphylococcus aureus* when he noticed that a contaminating mold colony (later identified as *Penicillium notatum*) seemed to be destroying the surrounding staphylococcal colonies. He reasoned that the mold was producing an antimicrobial substance that might have chemotherapeutic value. He named the active substance **penicillin**, after the mold, and tried to isolate it. Fleming failed, but 11 years later, two British chemists, Howard Florey and Ernst Chain, succeeded. Fleming, Florey, and Chain were awarded the Nobel Prize for their work.

Once penicillin had been isolated, chemists believed mass production was only a step behind. It was only necessary to determine the structure of penicillin and then synthesize it chemically. However, even though penicillin is a small molecule, it is extremely difficult to synthesize chemically. (In fact, it may never be economically feasible to chemically synthesize penicillin.) Procedures had to be developed to grow large amounts of *P. notatum* aerobically in rich media, using rigorously pure culture, because many contaminating bacteria produce an enzyme, **penicillinase**, that destroys penicillin. With the coming of World War II, penicillin was desperately needed. With intense international cooperation, success came quickly, and the antibiotics industry was born.

The spectacular success of penicillin, along with its limitations, stimulated a major search for new antibiotics in the years after World War II, especially among American drug companies. In particular, researchers hoped to find a cure for tuberculosis. Soil samples from all over the world were collected and examined for antibiotic-producing microorganisms. In the 1950s, streptomycin was discovered by Albert Schatz and Selman A. Waksman at Rutgers University. Streptomycin was effective

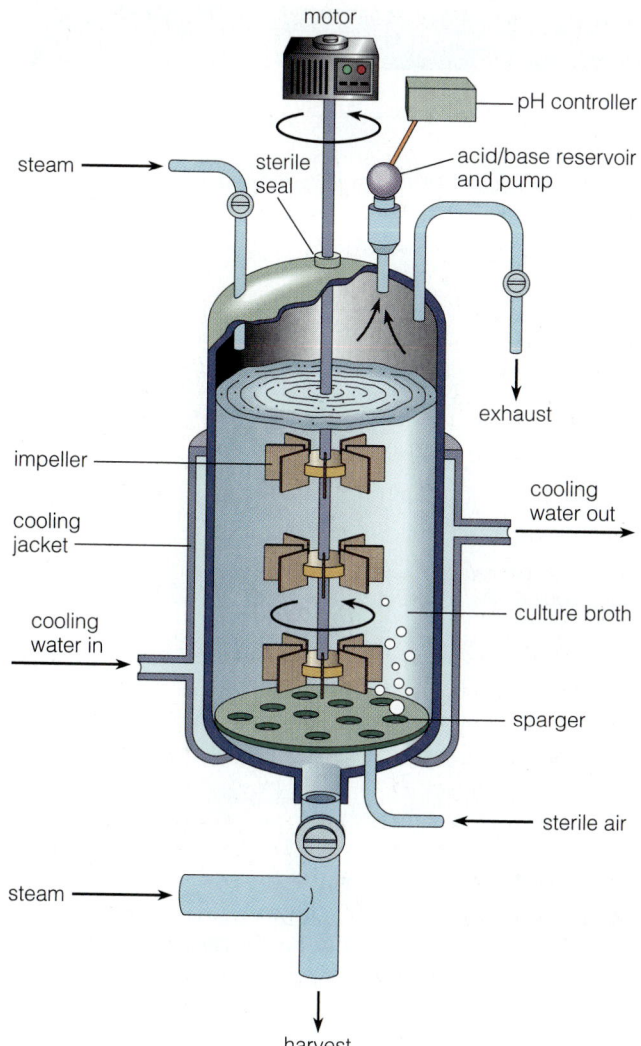

Figure 29.8 An industrial fermenter grows tens of thousands of gallons of a microbial culture aerobically without contamination. To expand the surface area of liquid exposed to oxygen, a stream of sterile air is forced through a sparger. An impeller shatters the streams of air into tiny bubbles, thus increasing the liquid surface area. This interior view of an industrial fermenter shows an impeller and cooling coils, which remove heat produced by microbial metabolism.

against tuberculosis and some Gram-negative bacteria (see the box "The Discovery of Streptomycin," in Chapter 21).

Then several broad-spectrum antibiotics, effective against a wide variety of Gram-positive and Gram-negative bacteria, were discovered at drug companies. Although chemically quite different from penicillin, these newly discovered antibiotics fit into a relatively small number of chemically distinct types, with specific targets in the bacterial cell (**Table 29.4**). With the exception of chloramphenicol, all these antibiotics are too chemically complex for commercial chemical synthesis. Instead, they are made by aerobic fermentation.

Over the years the search for antibiotics has continued, and new ones have been discovered. But the failure to find new chemical types suggests that most classes of antibiotics have already been discovered. As old antibiotics become less useful because pathogens become resistant to them, we must recognize that we won't be able to continue replacing them with newly discovered antibiotics.

Amino Acids. Several amino acids are made by aerobic fermentation for use in the food industry (**Table 29.5**). Some are used to improve or modify the flavor of a food. Others are used to enhance nutritional value. For example, lysine is added to bread in Japan because, without enrichment, bread does not contain enough of this essential amino acid to meet human nutritional requirements for protein. Lysine is not added to bread in the United States because Americans consume enough milk and meat to meet their needs for lysine. Japan also manufactures large amounts of monosodium glutamate (MSG) by fermentation for use as a flavor enhancer in many foods.

Enzymes. Microorganisms are a rich source of enzymes that have many commercial uses (**Table 29.6**). For example, in food manufacture, commercial enzymes are

Table 29.4 Some Commercially Produced Antibiotics

Antibiotic	Producing Microorganism	Class
Produced by Fungi		
Cephalosporin	*Cephalosporium acremonium*	Broad-spectrum
Griseofulvin	*Penicillium griseofulvum*	Fungi
Penicillin	*Penicillium chrysogenum*	Gram-positive bacteria
Produced by Gram-Positive, Spore-Forming Bacteria		
Bacitracin	*Bacillus subtilis*	Gram-positive bacteria
Polymyxin B	*Bacillus polymyxa*	Gram-negative bacteria
Produced by Gram-Positive Bacterium, Actinomycete		
Amphotericin B	*Streptomyces nodosus*	Fungi
Chloramphenicol	*Streptomyces venezuelae* (now chemical synthesis)	Broad-spectrum
Cycloheximide	*Streptomyces griseus*	Pathogenic yeasts
Cycloserine	*Streptomyces orchidaceus*	Broad-spectrum
Erythromycin	*Streptomyces erythreus*	Mostly Gram-positive bacteria
Kanamycin	*Streptomyces kanomyceticus*	Gram-positive bacteria
Lincomycin	*Streptomyces lincolnensis*	Gram-positive bacteria
Neomycin	*Streptomyces fradiae*	Broad-spectrum
Nystatin	*Streptomyces noursei*	Fungi
Streptomycin	*Streptomyces griseus*	Gram-negative bacteria (*Mycobacterium tuberculosis*)
Tetracycline	*Streptomyces rimosus*	Broad-spectrum

Table 29.5 Uses of Commercially Produced Amino Acids

Amino Acid	Use
Alanine	Added to fruit juice to improve taste
Aspartate	Added to fruit juice to improve taste
Cysteine	Added to bread and fruit juice to enhance flavor
Glutamate (MSG)	Added to many foods to enhance flavor
Glycine	Enhances flavor of sweetened foods
Histidine + tryptophan	Prevents rancidity in various foods
Lysine	Used in Japan to make bread a more complete protein
Methionine	Makes soybean products a more complete protein

used to keep candy from crystallizing, to clarify juices, and to curdle milk. Medically, commercial enzymes are used as digestive aids and to treat victims of heart attack by dissolving blood clots. Laundry detergents contain proteolytic (protein-destroying) enzymes that remove certain stains, including blood stains, from fabrics by dissolving the proteins that bind the stain to the fibers of the fabric. Most enzymes are produced aerobically.

Chemical Reactions

Microorganisms offer a degree of chemical specificity unmatched by ordinary chemical reactions. The action of their enzymes is specific enough to change one chemical group in a molecule and leave similar ones untouched. For this reason, microorganisms are used to carry out difficult steps in the industrial synthesis of certain compounds.

An example of this use of microorganisms is one step in the synthesis of the corticosteroids cortisone and pred-

Table 29.6 Some Commercially Useful Enzymes Produced by Microorganisms

Enzyme	Activity	Producing Microorganism	Use
Cellulase	Hydrolyzes cellulose	*Trichoderma konigi*	Digestive aid
Collagenase	Hydrolyzes collagen	*Clostridium histolyticum*	Promotes wound/burn healing
Diastase	Hydrolyzes starch	*Aspergillus oryzae*	Digestive aid
Glucose isomerase	Converts glucose to fructose	*Streptomyces phaeo-chromogenes*	Converts glucose from hydrolyzed cornstarch to a sweetener
Invertase	Hydrolyzes sucrose	*Saccharomyces cerevisiae*	Candy manufacture
Lipase	Hydrolyzes lipids	*Rhizopus* spp.	Digestive aid
Pectinase	Hydrolyzes pectin	*Sclerotina libertina*	Clarifies fruit juice
Protease	Hydrolyzes protein	*Bacillus subtilis*	Used in detergents

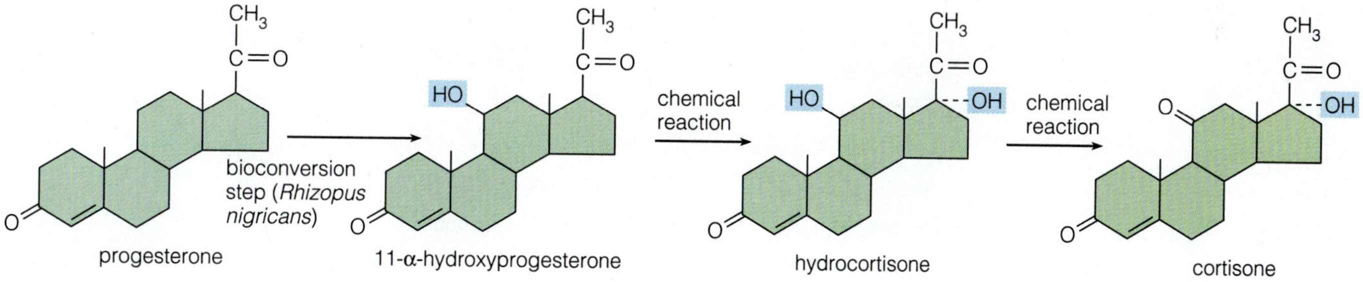

Figure 29.9 An enzyme from the fungus *Rhizopus nigricans* is used to carry out a difficult step in the industrial synthesis of cortisone, an anti-inflammatory agent. In the first step, the fungus *R. nigricans* inserts a hydroxyl group in the 11-α position to produce 11-α-hydroxyprogesterone. The two simpler reactions are accomplished chemically.

nisone, compounds that suppress inflammation. The step requires that a hydroxyl group be inserted at a position in the steroid molecule, which is protected from ordinary chemical reactions (**Figure 29.9**). An enzyme from the fungus *Rhizopus nigricans* can do this. The steroid is added to a culture of the microorganism, which carries out the desired reaction. Microorganisms are used industrially to carry out many other difficult chemical reactions.

USING GENETICALLY ENGINEERED MICROBES TO MAKE NEW PRODUCTS

Recombinant DNA technology (Chapter 7) changed the fermentation industry fundamentally. Before these procedures became available, microorganisms could produce only materials that were products of their normal metabolism. With recombinant DNA technology, microorganisms can be engineered, at least in principle, to produce any compound normally produced by any other organism. The ability to engineer organisms to make totally new compounds has tremendously expanded the fermentation industry, now called biotechnology. Scores of new companies have formed to make valuable products through genetic engineering. **Table 29.7** summarizes some of these products.

The impact of recombinant DNA technology has not been limited to engineering microorganisms to make new products. All aspects of microbial biotechnology, including the most traditional, have benefited from this powerful new technology. For example, most strains of lactic acid bacteria used to make cheese have been altered by recombinant DNA technology.

Medical Uses

New compounds manufactured by genetically engineered organisms have largely been hormones and other

Table 29.7 Some Useful Products Made by Genetically Engineered Cells

Product	Use	Status
Medical		
Human growth hormone (hgh)	To treat pituitary dwarfism	Available
Human DNase	To treat cystic fibrosis patients	Available
Insulin	To treat diabetes	Available
Factor VIII	To treat hemophilia	Under development
Relaxin	Used during childbirth	Under development
Tissue plasminogen activator (tPA)	To treat heart attacks	Available
Industrial		
Ascorbic acid (vitamin C)	Diet supplement, fruit preservative	Under development
Amino acids	Flavor enhancers	Available
Natural plant and animal products	Rubber	Under development
	Bovine growth hormone, for example	Available
Agricultural		
Bt toxin	Insecticide	Available
Ice-minus bacteria	Makes snow at ski resorts	Available
	Protects crops against frost	
Better silage	Protects against spoilage	Under development
Epidermal growth factor (EGF)	Sheep shearing	Under development

medically useful human proteins. Their value well justifies the considerable research and development costs.

Hormones. Hormones made by genetically engineered organisms can be used to supplement the needs of patients whose bodies are genetically unable to produce sufficient hormones on their own. Two human hormones, insulin and human growth hormone, are already commercially available. Both are made by strains of *Escherichia coli* containing human genes cloned into plasmids.

Some children are born with a malfunctioning pituitary that is unable to make sufficient human growth hormone (hGH). Untreated, they become pituitary dwarfs. Before hGH was produced microbially, it was purified from the pituitary glands of human cadavers. The supply was limited, and many children went untreated. Moreover, hGH obtained from cadavers carried a risk of prion infection. Batches of hGH were prepared from many pituitary glands, and a single infected gland could infect the entire batch.

The recombinant DNA methods (Chapter 7) used to clone and express the hgh gene in bacteria are typical of those used to make other human proteins. All human cells carry the gene that encodes hGH, but how does a scientist separate that gene from all others and clone it? One way is to start with mRNA instead of DNA. Human pituitary gland tissue makes hGH, so its cells must contain large amounts of the corresponding mRNA. The correct mRNA molecules can be identified by making an appropriate **probe**, a short DNA molecule corresponding to part of the hGH. The probe, being complementary to its corresponding mRNA, will **hybridize** (form hydrogen-bonded base pairs) with it, thereby identifying it. There is a second and more compelling reason for isolating mRNA rather than isolating the gene directly. Genes from eucaryotes, including humans, contain **introns**, stretches of DNA that do not code the protein because they are cut out of mRNA as it matures. Because procaryotes lack the enzymes to eliminate introns, they would synthesize an incorrect protein from a eucaryotic gene carrying introns.

At this stage, however, the scientist is presented with a new problem. How does one determine the sequence of bases needed to make the probe? First, it's necessary to know the sequence of amino acids in hGH protein. Then the correct sequence of bases for the probe can be determined by referring to the genetic code (Chapter 6). But the genetic code is redundant—most amino acids are encoded by several different codons. The genetic code states precisely which amino acid sequence is encoded by a particular sequence of bases in DNA but not the reverse. Because of redundancy of the code, all the DNA sequences that designate the desired amino acid sequence must be synthesized, and the mixture is used as a probe.

Once the correct mRNA has been isolated, **reverse transcriptase**, the enzyme from retroviruses that uses RNA as a template (Chapter 13), is used to make DNA. This DNA product is termed **cDNA** (complementary DNA) to indicate that it is a copy of mRNA, not the DNA in the gene itself. This DNA is cut with restriction endonucleases and ligated into a bacterial plasmid, which is inserted into the bacterial host by transformation (Chapter 6). In the host, the recombinant DNA molecule is replicated; and its genes, including the hGH gene, are expressed.

The method used to clone hgh is not the only way it could be done. Many variations are possible and new ones are developed almost daily. The most fundamental changes are in the **cloning vector** and the host cell. A plasmid can be used to clone hGH, but viral DNA can also be used. Introducing native or recombinant DNA into a host is termed **transfection**. Almost any kind of cell or intact organism can now be used as a host for the recombinant DNA molecule (see the box "A Good Host"). But no matter which host cell is finally selected to make the protein product, bacteria are almost always used in the cloning process.

After the hGH-producing bacterial strain is obtained, the job of producing hGH commercially has just begun. It is necessary to modify the plasmid and develop the proper cultural conditions so the bacterial strain carrying the hGH gene will produce useful amounts of hGH. Producing huge amounts of hGH and other foreign proteins is detrimental to the cell, so methods have been devised to trigger *Escherichia coli* to make hGH only after a dense culture develops. That is, the growth phase is separated from the production phase. Methods have been devised to purify hGH from the bacterial culture, to unfold the protein, and to refold it properly.

Microbiologists went through all these steps in the 1980s. Then the final product had to be extensively tested to assure that it was safe and effective. Finally, when it was proved safe to the satisfaction of the Federal Food and Drug Administration (FDA), genetically engineered hGH was released in 1985.

Insulin is a hormone secreted by the pancreas that lowers glucose levels in the blood. Insufficient insulin production creates imbalances in the body's metabolism, leading to dehydration, excessive urination, and sometimes coma and death. The disorder is called **diabetes mellitus**. The most serious type of diabetes appears in childhood and requires insulin injections throughout the person's life.

In the United States, bovine insulin (from cattle pancreas) was traditionally used for treatment; and in Europe, porcine insulin (from pig pancreas) was used. Although lifesaving, animal insulin differs from human insulin, suggesting that it is less effective than the human hormone. Moreover, the increasing use of feedlots (rather than range grazing) to fatten cattle threatened the insulin supply because it decreases the insulin content of cattle pancreases. Now an abundant supply of human insulin is produced by *Escherichia coli* carrying the human gene cloned into a plasmid.

Research is underway to develop microbial production of other human hormones. They include **factor VIII** (the hormone that most hemophiliacs need) and **relaxin** (a hormone that causes the cervix to dilate during delivery of a baby; some women do not produce an adequate amount). Along with fermentation, recombinant DNA technology holds the promise that adequate supplies of all needed human hormones will one day be available.

Other Proteins.   In addition to hormones, other medically useful human proteins are being produced by fermentation, or procedures for their production are being developed. One of the most important is **tissue plasminogen activator** (**tPA**), which binds to blood clots and dissolves them. It is extremely useful for treating victims of heart attack, in which a blocked artery cuts off the blood supply to the heart and results in damaged or dead heart tissue. If tPA is administered soon enough after the attack, the clot is dissolved and the patient may recover with no lasting damage.

Although the procedures used to clone and produce tPA are much like those used for human growth hormone, the cell used to produce tPA commercially is not a microbe. Instead, a mammalian cell line derived from Chinese hamster ovaries (CHO cells) is used because its cells are better producers. CHO cells are grown in huge fermenters much like microbial cells, using many of the techniques and procedures originally developed for penicillin production.

Streptokinase, a bacterial protease (Chapter 15) that is much less expensive than tPA, is also used to treat victims of heart attack. Some controversy exists over which treatment is better. **Human DNase**, an enzyme that degrades DNA, is produced in CHO cells to treat patients with **cystic fibrosis**, a genetic disease that causes the lungs to excrete large amounts of fluid. Affected children rarely survive to adulthood, partly because they become infected with heavily encapsulated strains of the opportunistic bacterial pathogen *Pseudomonas aeruginosa*. Macrophages, which accumulate to fight the infection by

A Good Host

Choosing the right host is one of the most important decisions a scientist must make to express a cloned gene. To manufacture medically useful human proteins, a human gene must be cloned in an appropriate cloning vector, but it must be expressed in a living host cell. Which host is best? How does a scientist decide?

When recombinant DNA technology was first developed in the late 1970s, there was no choice. Cloned genes could be inserted and expressed only in *Escherichia coli*. The technology rapidly improved, however. By developing new cloning vectors and procedures, it became possible to insert cloned genes into many species of bacteria, yeasts and other fungi, cultured plant and animal cells, and intact plants and animals. Suddenly the biotechnology industry had a wide choice of hosts.

What properties should a scientist look for when choosing a host?

1. *A good host should produce large amounts of the protein product.* Producing large amounts of a protein requires more than efficient transcription and translation. Often foreign proteins are synthesized rapidly only to be destroyed almost as rapidly by the host cell's proteolytic enzymes. Changing the growth medium or using mutant strains that produce fewer proteolytic enzymes often solves the problem. But sometimes it is necessary to use a different host.

2. *A good host should be easy to cultivate.* A good host should grow rapidly in an inexpensive medium and should require a relatively simple fermenter.

3. *The biology of a good host should be thoroughly understood.* The more that is known about a host's biology, the easier it is to improve its potential as a host and to solve problems that arise during production.

4. *A good host should produce the correct form of the protein product.* After translation in its native host, the new protein

New host candidates for genetically engineered human proteins include insect cells and intact insects. Among the most promising are silkworms, shown here with both their natural food, mulberry leaves, and a commercially prepared diet.

folds in a particular way and often is modified. Human proteins are usually glycosylated, meaning specific sugar molecules are attached to the protein at particular locations. Folding and modification occur differently in other hosts. A good host allows the protein to fold correctly and modifies it in ways not radically different from the way the native host would.

phagocytizing the bacteria, become packed with bacterial cells that are protected by their capsules. The bacteria-packed macrophages die, releasing large amounts of DNA into the lungs and forming a mucus that makes breathing difficult. The lungs must be cleared. Traditional treatment involves pounding on the back while the patient tries to cough up mucus. Inhaling human DNase breaks up the mucus painlessly. This treatment is expected to prolong the lives of cystic fibrosis patients and improve their quality of life. Nonhuman DNase cannot be used because the immune system would recognize it as foreign protein and produce antibodies. Subsequent treatments might cause anaphylactic shock (Chapter 18).

Industrial Uses

The first commercial applications of genetically engineered microorganisms to make products of foreign genes were medically useful proteins with great enough value to justify the high costs of research and production. As with most new technology, costs come down with advanced research. In time, then, genetically engineered microorganisms with foreign genes will be used to make less expensive products, such as ascorbic acid.

Large amounts of ascorbic acid—vitamin C—are used as a diet supplement and as an antioxidant to keep frozen and dried fruits from turning brown. Although vitamin C

Escherichia coli satisfies the first three requirements better than any other organism. It is the most thoroughly understood cellular organism. Also, it is easy and inexpensive to cultivate, and its growth medium and genetic constitution can be readily and predictably changed. But the fourth requirement can be a problem. Many human proteins are folded incorrectly in *E. Coli*, a troublesome but not always insurmountable problem. Sometimes the misfolded form can be purified, unfolded chemically, and refolded in vitro into the correct form. Other times the cloned gene is reengineered so its product is secreted outside the cell, where it might fold properly. Glycosylation is a more severe problem because *E. coli* lacks the metabolic machinery to glycosylate. Sometimes the unglycosylated form of a medically useful protein is as active and useful as the natural, glycosylated form. Glycosylation rarely changes a protein's activity but often changes its solubility, which is critically important for some proteins.

Largely because of folding and glycosylation, biotechnology is continuously developing new hosts. Cultured mammalian cells are now used commercially to produce certain human proteins. For example, Chinese hamster ovary (CHO) cells are used to make tissue plasminogen activator (tPA), a human protein that saves the lives of many victims of heart attack by dissolving the blood clots that cause them. CHO cells do not glycosylate exactly as human cells do, but closely enough to be effective. They also grow quite slowly, require expensive media, and are difficult to manipulate genetically.

The search for new and better hosts seems unending because none used so far is ideal. Promising new candidates include cultured insect cells and intact insects, with certain baculoviruses used as cloning vectors (Chapter 13). In nature, virions of these baculoviruses are released from infected cells in packages called **polyhedra**. Polyhedra consist of many individual virions embedded in copious amounts of a matrix composed of the protein **polyhedrin**. Polyhedra protect the virions, allowing them to survive in the wild until consumed by a susceptible insect. In the insect's digestive tract, polyhedrin is broken down, releasing the virions, which then infect the insect.

A cell infected with this kind of baculovirus is converted into a factory for making polyhedrin. To use a baculovirus as a vector, its polyhedrin gene is replaced by a cloned gene that is attached to the polyhedrin promoter—the region of the gene where transcription begins (Chapter 6). In this way, the infected cell will become a factory for making the product of the cloned gene. The productivity of cells infected with such a vector can be 10,000 times greater than the productivity of a mammalian cell culture. Moreover, host insect cells perform posttranslational modifications, including glycosylation, almost as human cells do.

Cultured cells of the fall armyworm infected with engineered baculoviruses are being used to produce proteins. An AIDS vaccine made this way is already in preliminary clinical trials. Intact insects are also being evaluated as hosts. A very promising insect host for the *Baculovirus* system is the silkworm, *Bombyx mori*. Silkworms have been cultivated in mass for thousands of years to make silk, so the methods for growing and handling them are highly developed. Some steps have even been automated. Early results suggest that silkworms are excellent hosts for *Baculovirus*-cloned genes. Larvae are infected by injecting engineered virus. Three to four days later each silkworm has accumulated about 1 mg of the protein product of the cloned gene, a large amount compared to other methods—and much more valuable than silk.

occurs naturally in many plants, none contains enough to make purification commercially feasible. Instead, vitamin C is manufactured by a mixture of chemical steps and microbial fermentations. For many years manufacturers used the **Reichstein route**, which requires one fermentation and five chemical steps. Then a process was developed in which a key intermediate (2-keto-L-gulonic acid, 2-KLG) was made from glucose tandem fermentations, the first mediated by an *Erwinia* species and the second by a *Corynebacterium* species. 2-KLG is then converted to ascorbic acid by a simple chemical transformation. Now, the gene in the *Corynebacterium* species encoding the second step has been cloned into *Erwinia herbicola*, enabling it to convert glucose to 2-KLG in a single fermentation. This process should be considerably less capital-intensive and more efficient to operate.

Many natural products made by plants or animals might also one day be made by fermentation involving genetically engineered microbes. The choice of products is limited only by the imagination.

Agricultural Uses

Microorganisms play important roles in agriculture, and genetic engineering has increased their value. Many microorganisms have been engineered to protect plants from pests or the environment, control decay after harvest, and improve animal husbandry. Four of these—Bt toxin, ice-minus bacteria, better silage makers, and epidermal growth factor—are described here.

Bt Toxin.   The capacity to produce the insect-killing Bt toxin has been transferred from bacilli into *Pseudomonas fluorescens*, a species of soil bacteria that inhabits the rhizosphere (Chapter 28). Producing Bt toxin right at the root surface is an excellent way to protect plant roots from destructive soil insects. Preliminary experiments indicate that inoculating soil with engineered strains of *Pseudomonas fluorescens* stimulates plant growth.

Ice-Minus Bacteria.   In regions of the semiarid western United States, rainfall is heavier over areas with denser vegetation. But which is cause and which is effect? Microbiologists and meteorologists working together in Montana found that certain bacteria, including *Pseudomonas syringae*, growing on the leaves of plants nucleate water and cause rain. The bacteria act much like silver iodide crystals, which are used to seed clouds, but *P. syringae* is more effective when it is swept from leaves into the clouds by wind that usually precedes a storm. Nucleation is caused by a protein on the bacterial surface that mimics the shape of ice crystals. In fact, killed *P. syringae* cells are used for snowmaking at ski resorts. The dead cells are mixed with water and sprayed into the air when the temperature is below freezing.

    P. syringae on leaves also makes frost damage more likely. Normally, pure water such as that in dew can be supercooled to 4°C below its freezing point and remain liquid. But if nucleation sites are present, including the surface protein of *P. syringae*, frost forms right at the freezing point (**Figure 29.10**). Frost, not low temperature alone, damages many plants because frost-sensitive plants cannot tolerate ice forming within their tissues. So the presence of *P. syringae* can damage frost-sensitive crops on cold nights. Frost damage to a plant is related directly to the logarithm of the number of ice-nucleation bacteria on its surface.

    Steven Lindow, a plant pathologist, has developed a method for ridding crops of frost-stimulating *P. syringae*. Using genetic engineering, Lindow deleted the gene encoding the nucleation protein from *P. syringae* to produce a strain he called **ice-minus**. Spraying young frost-sensitive crops, such as strawberries, with ice-minus bacteria establishes them on the plant, leaving no room for normal frost-inducing strains. Crops are protected up to 95 percent against frost damage. Considering that frost damage in the United States is estimated at over $1 billion annually, Lindow's contribution could have a significant economic impact.

Figure 29.10   A bean leaf with wild type *Pseudomonas syringae* causes ice-cold water to freeze (right), but one with an ice-minus strain does not (left).

Better Silage Makers.   Silage is protected against damage from anaerobic bacteria, particularly species of *Clostridium*, by lactic acid bacteria, principally *Lactobacillus plantarum*, that are naturally present in the plant material. However, sometimes whole crop cereals such as barley or wheat do not have enough sugar to generate protecting levels of lactic acid, and the silage spoils. Researchers have genetically engineered two possible ways to avoid such loss. They have cloned an α-amylase-encoding gene from *Bacillus amyloliquefaciens* into *Lactobacillus plantarum*, which enables it to convert the abundant starch in these plants into lactic acid. They are also working on cloning genes encoding compounds that are lethal to *Clostridium* spp. into *Lactobacillus plantarum*.

Sheep-Shearing Protein.   Epidermal growth factor (**EGF**), a natural animal protein, has been cloned and produced by *Escherichia coli* much as human growth hormone has. EGF can be used to shear sheep because it temporarily interrupts the function of hair follicles, producing a break in the strand of wool as it grows within the follicle. The sheep is given an injection of EGF and zipped into a coat that keeps the wool from falling off in patches. In about a week new wool growth pushes the gap out of the follicle, the coat is removed, and all the wool outside the gap falls off.

Summary

TRADITIONAL USES OF MICROORGANISMS
(pp. 735–740)

1. The term *biotechnology* refers to all uses of organisms for practical purposes, though microorganisms are its primary focus.

2. Lactic acid bacteria ferment sugars under anaerobic conditions to make sauerkraut, silage, green olives, cheese, and other dairy products. Except for some streptococci, lactic acid bacteria are harmless to humans, and their metabolic products have a pleasant taste.

3. Yeasts ferment sugars to produce carbon dioxide and ethanol; strains of *Saccharomyces* are almost always used. *S. ellipsoideus* is usually used for wines. Sauternes and other sweet wines are made from grapes infected by the fungus *Botrytis cinerea*. Lactic acid bacteria growing in sweet wines produce a foul taste, and wine exposed to air allows acetic acid bacteria to develop, giving the wine a vinegary taste.

4. Beer and whiskey are made from fermented cereal grain; *Saccharomyces carlsbergensis* and *S. cerevisiae* are most often used. Because grains contain starch, they must first be saccharified (hydrolyzed to glucose). In the United States, malt is added because it contains the starch-hydrolyzing enzyme amylase. Potatoes used to make vodka must also be saccharified. In Asia, fungi are used as a source of amylase to saccharify rice to produce sake and rice wine.

5. Vinegar is a solution of acetic acid made in a two-step process. Yeast, usually *Saccharomyces cerevisiae*, ferments sugar in juice to ethanol. Then *Acetobacter* spp. are used to oxidize the ethanol incompletely to acetic acid and water. Vinegar generators are used to add oxygen at a high rate.

6. Leavened bread was originally made by using brewer's yeast, a by-product of brewing beer. Today we use baker's yeast, specially selected strains of *Saccharomyces cerevisiae*.

7. Mixed cultures are also used to make fermented foods, including kefir, soy sauce, and miso. Starter cultures are complex mixtures of bacteria and fungi used to inoculate food.

MICROBES AS INSECTICIDES (pp. 740–741)

1. Certain species of *Bacillus* are insect pathogens; they form parasporal bodies (protein crystals) in their endospores that are lethal to insects. *B. thuringiensis* endospores are marketed as a bioinsecticide called Bt to control caterpillars and other insects. *B. popilliae* endospores produce Bp, which kills Japanese beetles.

2. The protozoan *Nosema locustae* is used as bait to kill grasshoppers and locusts. Baculoviruses are also being developed as bioinsecticides.

3. Unlike chemical pesticides, bioinsecticides are ecologically safe.

MICROBES AS CHEMICAL FACTORIES (pp. 741–745)

1. The transformations by which microorganisms produce antibiotics, vitamins, organic solvents, and so forth are called fermentations, whether they occur aerobically or anaerobically. Similarly, the vessels in which the transformations take place are called fermenters.

2. Anaerobic fermentations usually use blackstrap molasses and cereal grains (which are saccharified if necessary) as substrates.

3. Ethanol (also called ethyl alcohol) is a constituent of alcoholic beverages and is used as a solvent and fuel. It is made from saccharified cereal grains.

4. Acetone and butanol, used as solvents, are fermentation products of *Clostridium* spp. To produce large amounts during World War I, the technology of pure cultures on a mass scale had to be developed; this technology was also essential to commercial production of antibiotics.

5. Aerobic processes are used to make antibiotics, amino acids, vitamins, and therapeutically useful proteins. Fermenters with a sparger and impeller keep oxygen in the culture broth a maximum amount of time. Valves are bathed in steam to prevent foreign microorganisms from entering the fermenter.

6. The antibiotics industry began with penicillin; the search for new antibiotics continues as resistant strains emerge, but there is a good chance that most classes of antibiotics have already been discovered.

7. Amino acids are used in the food industry to improve flavor (monosodium glutamate) or enhance nutritional value (lysine).

8. Enzymes are used in food manufacture to keep candy from crystallizing, clarify juices, and curdle milk. They are used medically as digestive aids and to dissolve blood clots. Laundry detergents contain protein-destroying enzymes to remove blood and other stains.

9. Microorganisms are used to carry out difficult steps in the industrial synthesis of certain compounds. For example, an enzyme from the fungus *Rhizopus nigricans* is used in the synthesis of steroids such as cortisone and prednisone.

USING GENETICALLY ENGINEERED MICROBES TO MAKE NEW PRODUCTS   (pp. 745–750)

1. Before recombinant DNA technology, microorganisms could produce only the products of their normal metabolism. Now they can be engineered, at least in principle, to produce any compound produced by any other organism.

2. Medical uses include human growth hormone for treating pituitary dwarfism, insulin for treating diabetes, tissue plasminogen activator (tPA) and streptokinase for dissolving blood clots, and human DNase for treating cystic fibrosis.

3. Ascorbic acid (vitamin C), used as a diet supplement and food preservative, can be produced more cheaply and in large amounts, using genetically engineered species of *Erwinia* and *Corynebacterium*.

4. Through genetic engineering, Bt toxin can be transferred to soil bacteria in the rhizosphere to protect plant roots from destructive soil insects. A genetically engineered strain of *Pseudomonas syringae*, called ice-minus, is used to protect crops from frost damage. *Lactobacillus plantarum* has been genetically engineered to convert starch in crop cereals to sugar to prevent silage from spoiling. Epidermal growth factor (EGF), an animal protein that stops the growth of hair in a follicle, can be produced in *Escherichia coli* and used to shear sheep.

Review Questions

TRADITIONAL USES OF MICROORGANISMS

1. What is biotechnology?

2. Discuss some ways lactic acid bacteria have been used to preserve plant foods and make cheese.

3. What roles, positive and negative, do various microorganisms play in winemaking?

4. Why must cereal grains be saccharified for beer and whiskey making and how is it done? Which yeasts are most commonly used in beer making?

5. Describe how vinegar is made.

6. How are microorganisms used in making bread?

7. What are starter cultures? What are the benefits of mixed cultures over pure cultures?

MICROBES AS INSECTICIDES

1. Classify each of these microorganisms and tell how they act as insecticides: *Bacillus thuringiensis*, *B. popilliae*, *Nosema locustae*, baculoviruses.

2. What are the advantages of bioinsecticides over chemical insecticides?

MICROBES AS CHEMICAL FACTORIES

1. Define fermentation and fermenter.

2. Explain this statement about anaerobic fermentations: Successful industrial fermentations may be abandoned only to be reinitiated at a later date.

3. Name some products made by anaerobic fermentations.

4. Describe the technology that had to be developed before the modern aerobic fermentation industry could be successful. Give some examples of products made by aerobic fermentations.

5. How can microorganisms be used to carry out difficult chemical reactions? Give an example.

USING GENETICALLY ENGINEERED MICROBES TO MAKE NEW PRODUCTS

1. Explain this statement: Recombinant DNA technology changed the fermentation industry completely.

2. Discuss some medical products of recombinant DNA technology. How is human growth hormone produced?

3. Discuss these examples of genetic engineering in agriculture: Bt toxin, ice-minus bacteria, better silage makers, and sheep-shearing protein.

Suggested Readings

Demain, A. L., and Solomon, N. A., eds. 1986. *Manual of industrial microbiology and biotechnology.* Washington, D.C.: American Society for Microbiology.

Halvorson, H. O. 1985. *Engineered organisms in the environment: Scientific issues.* Washington, D.C.: American Society for Microbiology.

Hesseltine, C. W. 1983. Microbiology of oriental fermented foods. *Annual Reviews of Microbiology* 37: 575–601.

Jones, D. T., and Woods, D. R. 1986. Acetone-butanol fermentation revisited. *Annual Reviews of Microbiology* 50: 484–524.

Knight, P. 1991. Baculovirus vectors for making proteins in insect cells. *ASM News* 57: 567–70.

APPENDIX I
Metric Measurements and Conversions

Scientific Notation		
	Name	Prefix
$10^{-15} = 0.000000000000001$	quadrillionth	femto (f)
$10^{-12} = 0.000000000001$	trillionth	pico (p)
$10^{-9} = 0.000000001$	billionth	nano (n)
$10^{-6} = 0.000001$	millionth	micro (μ)
$10^{-3} = 0.001$	thousandth	milli (m)
$10^{-2} = 0.01$	hundredth	centi (c)
$10^{-1} = 0.1$	tenth	deci (d)
$10^{2} = 100$	hundred	hecto (h)
$10^{3} = 1,000$	thousand	kilo (k)
$10^{6} = 1,000,000$	million	mega (M)
$10^{9} = 1,000,000,000$	billion	giga (G)

Volume	
1 microliter (μl)	1 cubic millimeter (mm^3) 0.000001 liter
1 milliliter (ml)	1 cubic centimeter (cm^3, cc) 0.001 liter 0.061 cubic inch 0.03 fluid ounce
1 liter	1000 milliliters 1.06 quarts
1 cubic inch ($in.^3$)	16.39 cubic centimeters
1 fluid ounce	0.03 liter
1 quart	0.95 liters 32 fluid ounces

Mass	
1 picogram (pg)	0.000000000001 g
1 nanogram (ng)	0.000000001 g
1 microgram (μg)	0.000001 gram
1 milligram (mg)	0.001 gram
1 gram (g)	1000 milligrams 0.353 ounce
1 kilogram (kg)	1000 grams 2.2 pounds
1 ounce	28.35 grams
1 pound	453.6 grams 0.45 kilograms 16 ounces

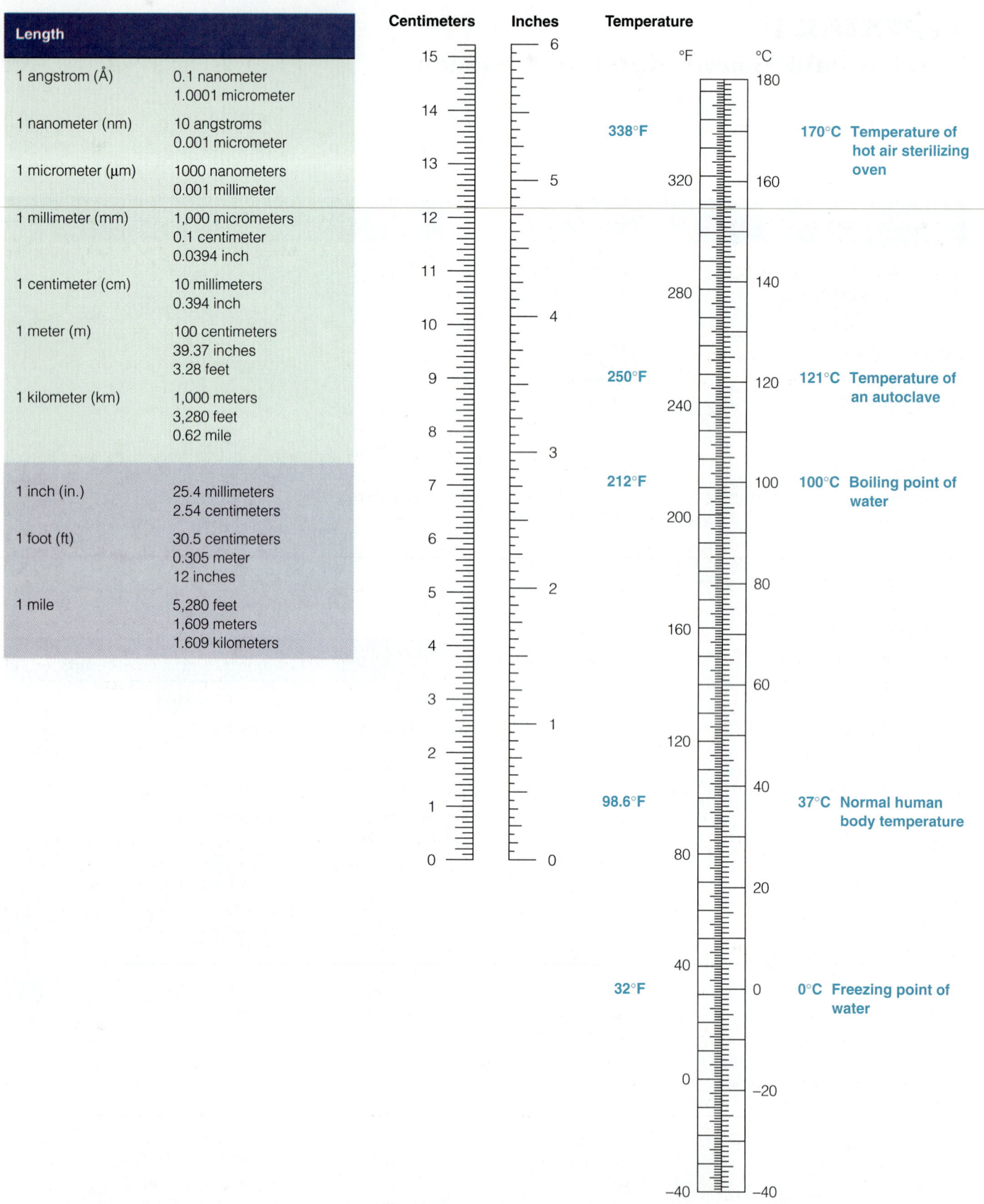

Length

1 angstrom (Å)	0.1 nanometer 1.0001 micrometer
1 nanometer (nm)	10 angstroms 0.001 micrometer
1 micrometer (μm)	1000 nanometers 0.001 millimeter
1 millimeter (mm)	1,000 micrometers 0.1 centimeter 0.0394 inch
1 centimeter (cm)	10 millimeters 0.394 inch
1 meter (m)	100 centimeters 39.37 inches 3.28 feet
1 kilometer (km)	1,000 meters 3,280 feet 0.62 mile
1 inch (in.)	25.4 millimeters 2.54 centimeters
1 foot (ft)	30.5 centimeters 0.305 meter 12 inches
1 mile	5,280 feet 1,609 meters 1.609 kilometers

Centimeters   **Inches**   **Temperature**

°F   °C

338°F   170°C **Temperature of hot air sterilizing oven**

320   160

280   140

250°F   121°C **Temperature of an autoclave**

240   120

212°F   100°C **Boiling point of water**

200

160   80

120   60

98.6°F   37°C **Normal human body temperature**

80   40

40   20

32°F   0°C **Freezing point of water**

0   −20

−40   −40

To convert temperature scales
Fahrenheit to Celsius: °C = 5/9 (°F − 32)
Celsius to Fahrenheit: °F = 9/5 (°C) + 32
Kelvin to Celsius: K = °C + 273.2

APPENDIX II
Some Word Roots Used in Microbiology

a-, an- not, without, absence. Examples: aseptic, free from infection; anaerobic, in the absence of air.

acet- pertaining to acetic acid. Example: *Acetobacter*, a bacterium that converts ethanol to acetic acid, or vinegar.

actino- having rays. Example: *Actinomyces*, a bacterium that forms star-shaped or rayed colonies.

aer- air. Example: aerobic, in the presence of air.

agglutino- clumped or glued together. Example: agglutination, clumping.

alb- white. Example: *Candida albicans*, a fungus with white colonies.

amphi- around, on both sides. Example: amphitrichous, having flagella at both ends of a cell.

amyl- starch. Example: amylase, an enzyme that degrades starch.

ana- up. Example: anabolic reaction, any synthesis reaction in an organism.

aqua-, aque- water. Examples: aquatic, taking place in water; aqueous, made with water.

archae- ancient. Example: archaebacteria, bacteria thought to resemble the most ancient life.

arthro- joint. Example: arthropod, an invertebrate having a jointed body and limbs.

asc- sac. Example: ascus, a saclike structure holding spores.

-ase enzyme. Example: polymerase, an enzyme that catalyzes the formation of the polymers DNA or RNA.

aure- gold. Example: *Staphylococcus aureus*, a bacterium with gold-pigmented colonies.

azo-, azoto- nitrogen. Example: *Azospirillum*, a nitrogen-fixing soil bacterium with helical shape.

bacill- small rod. Example: bacillus, a rod-shaped bacterium.

bacteri-, -bacter denoting bacteria. Example: *Agrobacterium*, a plant-infecting bacterium.

bio- life, living organisms. Example: biology, the study of life.

blast- bud. Example: blastomycosis, a disease caused by a yeastlike fungus.

bovi- cow. Example: *Mycobacterium bovis*, a bacterium found in cattle.

butyr- butter. Example: butyric acid, the fatty acid responsible for the odor of rancid butter.

carb- having carbon or carboxyl. Example: carbohydrate, an organic compound formed from carbon, hydrogen, and oxygen.

carcin- cancer. Example: carcinogen, a cancer-causing agent.

cardio- heart. Example: endocarditis, inflammation of the heart's inner lining.

-caryo, -karyo kernel, center. Example: procaryote, a cell without a true center or nucleus.

caseo- cheese. Example: casein, a protein produced when milk is curdled by rennet as when making cheese.

caul- stem, stalk. Example: *Caulobacter*, an appendaged or stalked bacteria.

cephalo- of the head or brain. Example: encephalitis, inflammation of the brain.

chlamyd- covered, cloaked. Example: chlamydiospore, spore formed inside a fungal hypha.

chloro- green. Example: chlorophyll, the green pigment in leaves.

-chrome colored. Example: cytochrome, an iron-containing pigment that plays a role in cellular oxidations.

chryso- gold, yellow. Example: *Penicillium chrysogenum*, a mold with yellowish colonies.

-cide killing. Example: bactericide, an agent that kills bacteria.

cili- eyelash. Example: cilia, hairlike organelles.

cocc- berry. Example: coccus, a spherical cell.

coeno- common, shared. Example: coenocytic, having many nuclei not separated by septa.

coli-, colo- colon. Example: coliform bacteria, bacteria found in the large intestine.

con- together. Examples: concentric, having a common center, together in the center.

conidio- dust. Example: conidia, dust-like spores produced by fungi.

coryne- club. Example: *Corynebacterium*, a club-shaped bacterium.

cut- skin. Example: cutaneous, relating to or affecting the skin.

cyan- blue. Example: cyanobacteria, blue-green pigmented bacteria.

cyst- bladder, sac. Example: cystitis, inflammation of the urinary bladder.

cyt-, -cyte cell. Examples: cytoplasm, the fluid within a cell; phagocyte, a cell that consumes foreign material and debris.

dermat- skin. Example: dermatophyte, a fungus parasitic on the skin.

di-, diplo- twice, double. Examples: dimorphic, having two different forms; diplococci, pairs of cocci.

dia- through, apart. Example: diagnosis, the art of distinguishing among diseases by their signs and symptoms.

dys- difficult, abnormal. Example: dysentery, severe abnormal diarrhea.

-emia condition of the blood. Example: septicemia, invasion of the bloodstream by virulent microorganisms.

en-, endo- in, within. Examples: engulf, flow over and enclose; endospore, spore formed inside a cell.

entero- intestine. Example: enterotoxin, a poison affecting the intestine.

epi- upon, over. Example: epidemic, a disease affecting an entire population at once.

erythro- red. Example: erythrocyte, red blood cell.

eu- true, proper, normal. Example: eucaryote, a cell with a true nucleus.

ex-, exo- out of, from, outside. Example: exogenous, from outside the body.

fila- thread. Example: filament, a thin threadlike process or appendage.

flagell- whip. Example: flagellum, a whiplike projection from a cell.

flav- yellow. Example: *Flavobacterium*, a bacterium that produces a yellow pigment.

gamet- to marry. Example: gamete, a reproductive cell.

gastr- stomach. Example: gastritis, inflammation of the stomach.

gen-, -gen, -genesis cause, origin, production. Example: generation, a group of individuals that originated at the same time; pathogen, a microorganism that causes disease; pathogenesis, production of disease by microorganisms.

germin- bud, sprout. Example: germinate, sprout or develop.

gingiv- gum. Example: gingivitis, inflammation of the gums.

-globulin type of protein. Example: immunoglobulin, a protein of the immune system.

glyc- sweet, sugar. Example: glycoprotein, a protein with sugars attached.

gon- reproduction. Example: gonorrhea, pus-producing infection of the reproductive system caused by the bacterium *Neisseria gonorrhoeae*.

hal- salt. Example: halophile, an organism that thrives in high salt concentrations.

haplo- one, single. Example: haploid, half the number of chromosomes or one set.

hemo-, hemat- blood. Examples: hemoglobin, a pigment in red blood cells; hematocrit, an instrument for determining the ratio of red blood cells to whole blood.

hepat- liver. Example: hepatitis, inflammation of the liver.

hetero- different, other. Example: heterotroph, an organism that obtains its nutrients from other organisms.

hist- tissue. Example: histology, the study of tissues.

homo- same, similar. Example: homologous, having the same structure.

hydr-, hydro- water. Example: hydrologic cycle, water cycle.

hyper- above, excessive. Example: hypersensitivity, excessively sensitive or susceptible.

hypo- below, deficient. Example: hypotonic, having deficient tension or osmotic pressure.

-iasis disease condition. Example: schistosomiasis, a disease condition caused by trematode worms of the genus *Schistosoma*.

immun- resistance. Example: immunity, the condition of being able to resist a particular infection.

inter- between. Example: interphase, the interval between cycles of mitosis.

intra- within. Example: intracellular, within a cell.

-ism disease or condition. Example: botulism, food poisoning caused by the toxin secreted by *Clostridium botulinum*.

iso- equal, uniform. Example: isotonic, having the same tension or osmotic pressure as another.

-itis inflammation. Example: rhinitis, inflammation of the nose.

kerato- horn. Example: keratin, the horny substance making up skin and nails.

kin- motion. Example: streptokinase, an enzyme produced by streptococci that breaks down or moves fibrin.

lact- milk. Example: *Lactobacillus*, a lactic acid–forming bacterium.

lepto- slender. Example: *Leptospira*, a slender spirochete.

leuko-, leuco- white, colorless. Example: leukocyte, white blood cell.

lip-, lipo- fatty, lipid. Example: lipopolysaccharide, a large molecule composed of fats and sugars.

-logy science, field of study. Example: microbiology, the science that studies microorganisms.

lopho- tufted. Example: lophotrichous, having a tuft of flagella on one side of a cell.

-lysis breaking down, disintegration. Example: hydrolysis, chemical breakdown of a compound into other compounds as a result of taking up water.

macro- large. Example: macromolecule, large molecule.

mening- membrane. Example: meningitis, inflammation of the membranes of the brain.

meso- middle, intermediate. Example: mesophile, an organism that thrives at middle-range temperatures.

meta- beyond, among, change. Example: metabolism, chemical changes occurring within a living organism.

-metry measure. Example: symmetry, measured to be the same or balanced.

micro- very small. Example: microorganism, a very small organism.

mono- single. Example: monotrichous, having a single flagellum.

morph- form. Example: morphology, the study of the form and structure of organisms.

multi- many. Example: multicellular, having many cells.

mur- wall. Example: murein, a polymer characteristic of bacterial cell walls.

muri- mouse. Example: murine typhus, a form of typhus endemic in mice.

mut- change. Example: mutagen, an agent that can cause genetic change.

myco-, -mycete, -myces a fungus. Examples: *Mycobacterium*, a bacterium that often forms filaments like a fungus; *Saccharomyces*, sugar fungus, a genus of yeast.

myxo- mucus, slime. Example: myxobacterium, a slime-producing bacterium.

necro- death. Example: necrosis, cell death or death of an area of tissue.

-nema thread. Example: *Treponema*, a bacterium with long, threadlike cells.

neo- new. Example: neonatal, newborn.

nigr- black. Example: *Rhizopus nigricans*, a mold with black sporangiospores.

nitro- nitrate. Example: *Nitrobacter*, a bacterium that oxidizes nitrite to nitrate.

nitroso- nitrite. Example: *Nitrosomonas*, a bacterium that oxidizes ammonium to nitrite.

noso- disease. Example: nosocomial, a disease acquired in the hospital.

ob- to, toward. Example: obligate, bound or restricted to a particular mode of life.

oculo- eye. Example: ocular lens, the lens closest to the eye in a microscope.

-oid resembling. Example: nucleoid, the region of a procaryote that resembles the eucaryotic nucleus.

-ole small. Example: bronchiole, small, thin-walled branch of a bronchus.

oligo- few, deficient. Example: oligosaccharide, a complex sugar composed of a relatively few simple sugars.

-oma tumor. Example: melanoma, a tumor containing dark pigment.

onco- cancer, tumor. Example: oncogene, a gene that causes tumor formation.

oro- mouth. Example: oropharynx, the region of the throat nearest the mouth.

ortho- straight. Example: orthomyxovirus, a virus with a straight, tubular capsid.

-ose sugar. Example: lactose, milk sugar.

-osis, -sis condition. Example: brucellosis, a condition cased by bacteria of the genus *Brucella*.

-otic relating to a condition. Example: necrotic, relating to local tissue death.

pan- all, completely. Example: pandemic, an epidemic affecting a very large region.

para- beside, abnormal. Example: parasite, an organism that lives at the expense of another.

path- abnormal, diseased. Example: cytopathic, related to abnormal changes in cells.

peri- around. Example: peritrichous, having flagella projecting from all sides.

-phage one that eats. Example: bacteriophage, a virus that digests bacteria.

philo-, -philic liking, having an affinity for. Example: hydrophilic, having an affinity for water.

-phobic fearing, having an aversion to. Example: hydrophobic, having an aversion to water.

-phore bearer of, carrier. Example: conidiophore, a fungal hypha that bears conidia.

photo- light. Example: photosynthesis, the formation of chemical compounds using light as the energy source.

-phyte plant. Example: saprophyte, a plant that obtains nutrients from rotting organic matter.

pil- hair. Example: pilus, a hairlike projection from a cell.

-plast organized granule. Example: chloroplast, an organized granule or organelle containing the green pigment chlorophyll.

pleur- membrane surrounding the lung. Example: pleurisy, inflammation of the membrane surrounding the lung.

pneumo- lung, pulmonary. Example: *Pneumocystis*, a fungus that forms cysts in the lungs.

-pod foot. Example: pseudopod, a footlike structure.

poly- many. Example: polysaccharide, a large molecule composed of many simple sugars.

pre-, pro-, proto- before, in front of. Examples: precursor, substance from which another substance is formed; protozoan, member of Phylum Protozoa, single-celled organisms that existed before animals.

psychro- cold. Example: psychrophile, an organism that thrives at low temperatures.

pyo- pus. Example: *Streptococcus pyogenes*, a pus-producing species of *Streptococcus*.

pyro- fire, heat. Example: pyrogenic, fever-producing.

rhabdo- rod. Example: rhabdovirus, an elongated, bullet-shaped virus.

rhin- nose. Example: rhinitis, inflammation of mucous membranes in the nose.

rhizo- root. Examples: mycorrhiza, the mutualism between a fungus and the roots of a plant.

rhodo- red. Example: *Rhodospirillum*, a red-pigmented, spiral-shaped bacterium.

-rrhage excessive discharge. Example: hemorrhage, excessive bleeding.

-rrhea discharge. Example: diarrhea, abnormal discharge of liquid feces.

sacchar- sugar. Example: disaccharide, a sugar composed of two simple sugars.

sapr- rotten. Example: *Saprolegnia*, a fungus that lives on dead animals.

sarco- fleshy. Example: sarcoma, a tumor of muscle or connective tissues.

schizo- split. Example: schizogony, multiple splitting producing many new cells.

-scope, -scopic see, examine. Example: microscope, an instrument used to examine small things.

semi- half. Example: semipermeable, partially permeable, to small but not large molecules.

-septic rotting. Example: antiseptic, killer of bacteria that can cause rotting.

-some body. Example: ribosome, a small, RNA-rich body in the cytoplasm of a cell.

speci- individual, particular. Example: species, the smallest taxonomic group of organisms with common attributes.

spiro- coil, helix. Example: spirochete, a bacterium with a helical cell.

sporo- spore, seed. Example: sporozoan, an immobile, parasitic protozoan.

-stasis, -static arrest, stop. Example: bacteriostatic, able to stop bacterial growth.

strepto- twined, twisted, knotted. Example: *Streptococcus*, a bacterium that forms chains of connected (knotted) spherical cells.

sub- under, beneath. Example: subcutaneous, just under the skin.

super- above, over. Example: superficial, on or just above the surface.

sym-, syn- together, with. Examples: symbiosis, living together; syndrome, a group of signs and symptoms that occur together.

-taxis, taxon- orderly arrangement, orientation. Examples: chemotaxis, orientation of an organism in relation to chemicals; taxonomy, the arrangement of organisms into natural groups.

therm- heat. Example: thermophile, an organism that thrives at high temperatures.

thio- sulfur. Example: *Thiobacillus*, a bacterium that oxidizes sulfur-containing compounds.

-tome, -tomy to cut. Examples: microtome, an instrument that cuts extremely thin slices; anatomy, cutting up a plant or animal to discover its structure and function.

-tonic tension. Example: hypertonic, having excessive tension or osmotic pressure.

tox- poison. Example: toxic, poisonous.

trans- across, through. Example: transport, movement of substances across a membrane.

trich- hair. Example: *Trichomonas*, a protozoan with several long, hairlike flagella.

-troph food, nutrition. Example: eutrophic, well-nourished or rich in dissolved nutrients.

undul- wave. Example: undulant fever, rising and falling fever caused by brucellosis.

uni- single. Example: universal, affecting the whole.

-uria pertaining to urine. Example: dysuria, difficult or painful urination.

vaccin- from cows. Example: vaccine, a preparation used to produce or increase immunity, the first of which contained matter from cows.

vacu- empty. Example: vacuole, an intracellular structure that appears empty.

vaso- pertaining to vessels. Example: cardiovascular, pertaining to the system of heart and blood vessels.

vesic- bladder, blister. Example: vesicle, a bubble.

-vore devour. Example: detritivore, an animal that eats detritus.

xantho- yellow. Example: *Xanthomonas*, a bacterium with yellow colonies.

xeno- stranger, foreigner. Example: xenograft, a graft from a different species.

zoo- animal. Example: zoonosis, a disease communicable from animals to humans.

zygo- pair, union. Example: zygospore, a spore formed from the fusion of two cells.

-zyme ferment. Example: enzyme, a complex protein that catalyzes biochemical reactions including fermentations.

APPENDIX III
Directory to Useful Facts

This directory is designed to help you quickly find information collected in tabular form at various places in the book.

APPENDIX IV
Pronunciation of Scientific Names

Because many scientific names are long and all have the odd look typical of Latinized words, they can be intimidating. But you shouldn't be afraid to use them. Pronunciation of scientific names varies with country, region, and individual scientist. Even the experts on scientific Latin take the issue casually. The authoritative reference (William T. Stearn. 1983. *Botanical Latin*. London: David and Charles) concedes, "How they [scientific names] are pronounced really matters little provided they sound pleasant and are understood by all concerned. . . . "

A few simple guidelines make Stearn's provisions easy to satisfy.

1. Divide the name carefully into syllables (it is safest to assume every vowel belongs to a different syllable) and pronounce each syllable.

   Example: *Thermoactinomyces* is Ther-mo-ac-tin-o-my-ces
   (*not* Ther-moac-tin-o-my-ces)

2. The accent usually falls on the next to the last syllable.

   Example: *Bacillus* is Ba-**cil**'-lus
   (*not* **Ba**'-cil-lus)

   In compound names the next to last syllable of both parts is sometimes accented.

   Example: *Acinetobacter* is A-ci-**ne**'-to-**bac**'-ter

   But there are exceptions. First, terminal and subterminal double vowels are usually pronounced as two syllables, but the *preceding* syllable is accented.

-eae	is pronounced -e-ae	*Example:* Enterobacteriaceae is En-ter-o-bac-ter-**ac**'-e-ae
-ei	is pronounced -e-i	*Example:* brucei is **bru**'-ce-i
-eus	is pronounced -e-us	*Example:* proteus is **pro**'-te-us
-ia	is pronounced -i-a	*Example:* Nocardia is no-**car**'-di-a
-iae	is pronounced -i-ae	*Example:* malaria is ma-**lar**'-i-ae
-iens	is pronounced -i-ens	*Example:* tumefaciens is tu-me-**fac**'-i-ens
-ii	is pronounced -i-i	*Example:* carinii is car-**in**'-i-i
-io	is pronounced -i-o	*Example:* Desulfovibrio is De-sul-fo-**vib**'-ri-o
-ium	is pronounced -i-um	*Example:* typhimurium is ty-phi-**mur**'-i-um
-ius	is pronounced -i-us	*Example:* acidocaldarius is a-**ci**'-do-cal-**dar**'-i-us

   Second, certain genus and species names are based on proper names. In these cases the original sound should be maintained.

   Example: *douglasii* is **doug**'-las-ee-eye
   (not dou-**glass**'-ee-eye)

3. Most consonants are pronounced as in English. The following pronunciation of vowels and certain consonants is preferred by most scientists:

a	It's best to be consistent with your own pronunciation of English words; use the broad **a** only if you use it in normal speech (**a** as in "hat" or **ä** as in "father").	*Example: Bacillus is either* Bacillus *or* Bäcillus
ae	Nonterminal ae usually pronounced **ē** as in "see"; terminal -ae pronounced as long **ī** as in "ice," not "ee" or "ay."	*Example: Haemophilus is* Hēmophilus *cholerae is* cholerī
i	Usually short **i** as in "sit."	*Example: utilis is* utilus
-ii	Vowel may be slightly separated, generally pronounced "ee-eye" (and count as one syllable for accent rules).	*Example: leichmannii is* leichman-**ē-ī**
-oea	Usually pronounced "ee-a."	*Example: Zoogloea is* Zooglē-ä
y	Usually pronounced like **i** as in "sit"; sometimes as **ē** as in "see."	*Example: gossypii is* gossipii *Blastomyces is* Blastomēces
ch	Generally pronounced as **k**, not as in "ouch."	*Example: Chlorobium is* klorobium

These guidelines and a little self-confidence are all you need.

APPENDIX V
Biochemical Pathways

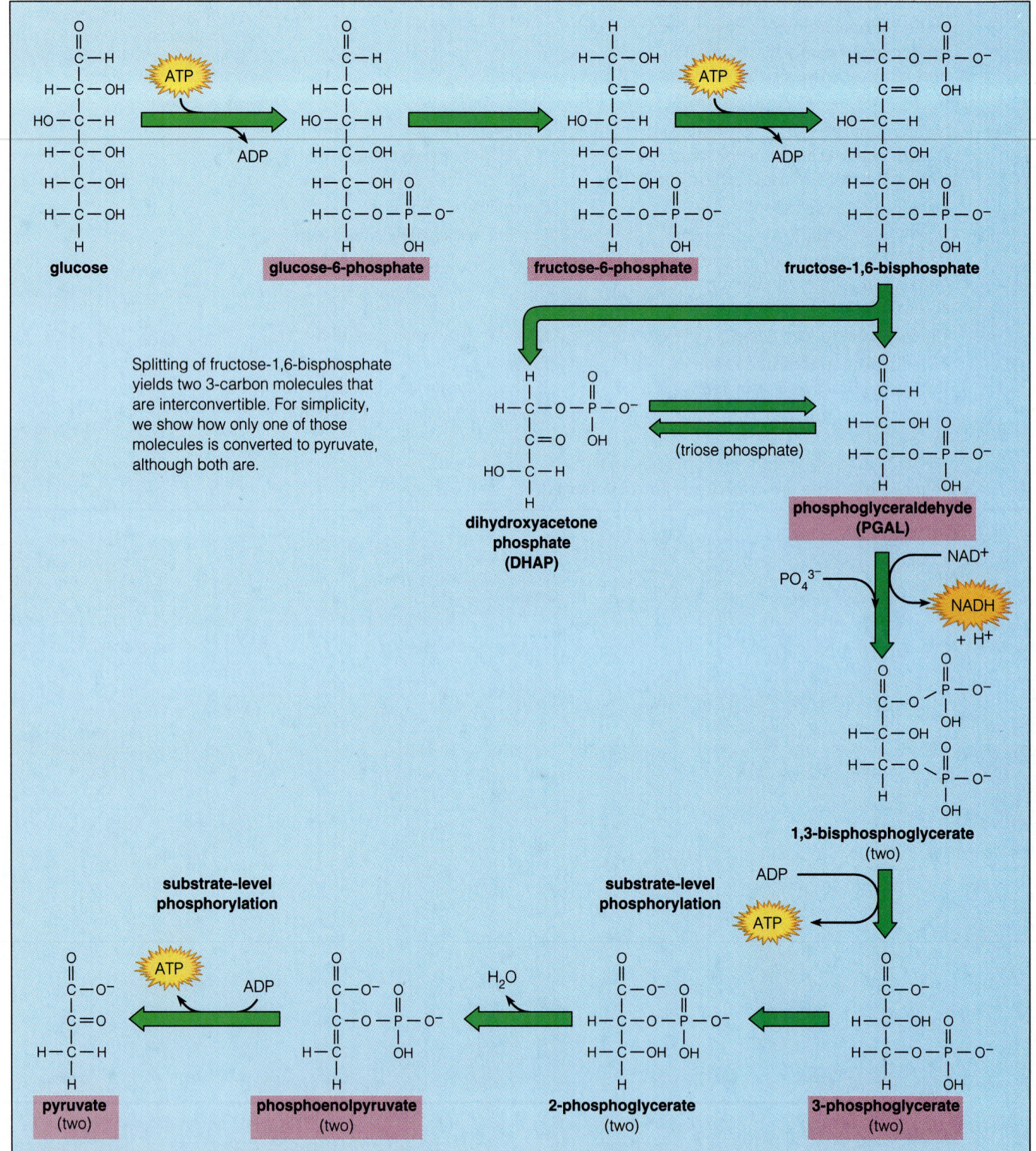

Figure a Glycolysis, ending with two pyruvate molecules for each glucose entering the pathway. The pathway produces two molecules of ATP (two are used and four are formed), two molecules of reducing power (as NADH), and six of the twelve precursor metabolites (highlighted in color).

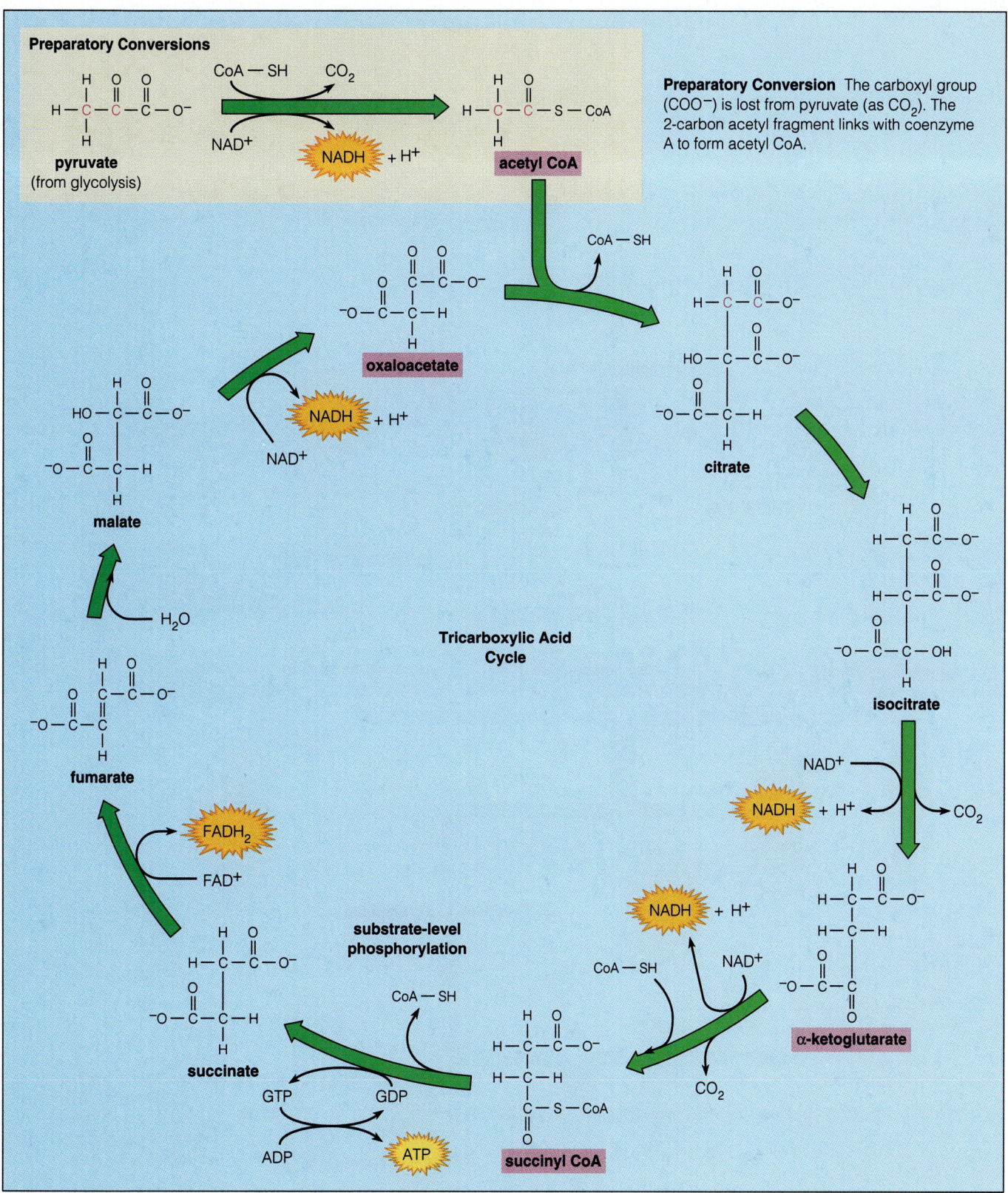

Figure b The Tricarboxylic Acid Cycle. With each turn of the cycle, two carbon atoms are added by acetyl-CoA and two are lost as CD_2. Each turn produces one molecule of ATP, four molecules of reducing power (one as $FADH_2$, three as NADH) and three precursor metabolites (shown in color). In addition, one molecule of NADH and one precursor metabolite (in color) are made in the preparatory conversion.

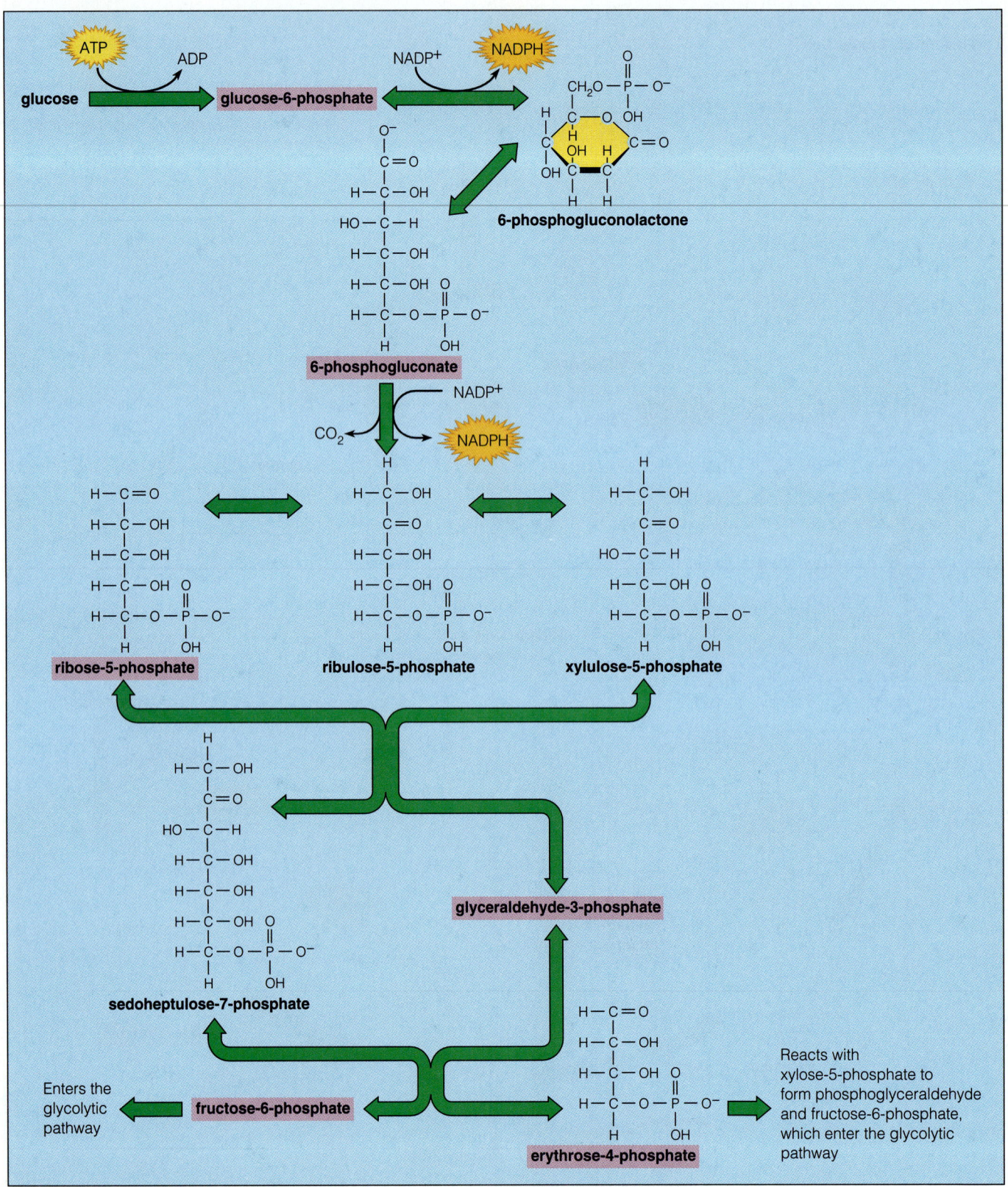

Figure c The Pentose Phosphate Pathway. The pathway does not form ATP directly, but it does generate reducing power (in the form of NADPH) in two steps and produces five precursor metabolites. Three of these precursor metabolites (glucose-6-phosphate, phosphoglyceraldehyde, and fructose-6-phosphate) are also made by the glycolytic pathway; two (ribose-5-phosphate and erythrose-4-phosphate) are made only by this pathway.

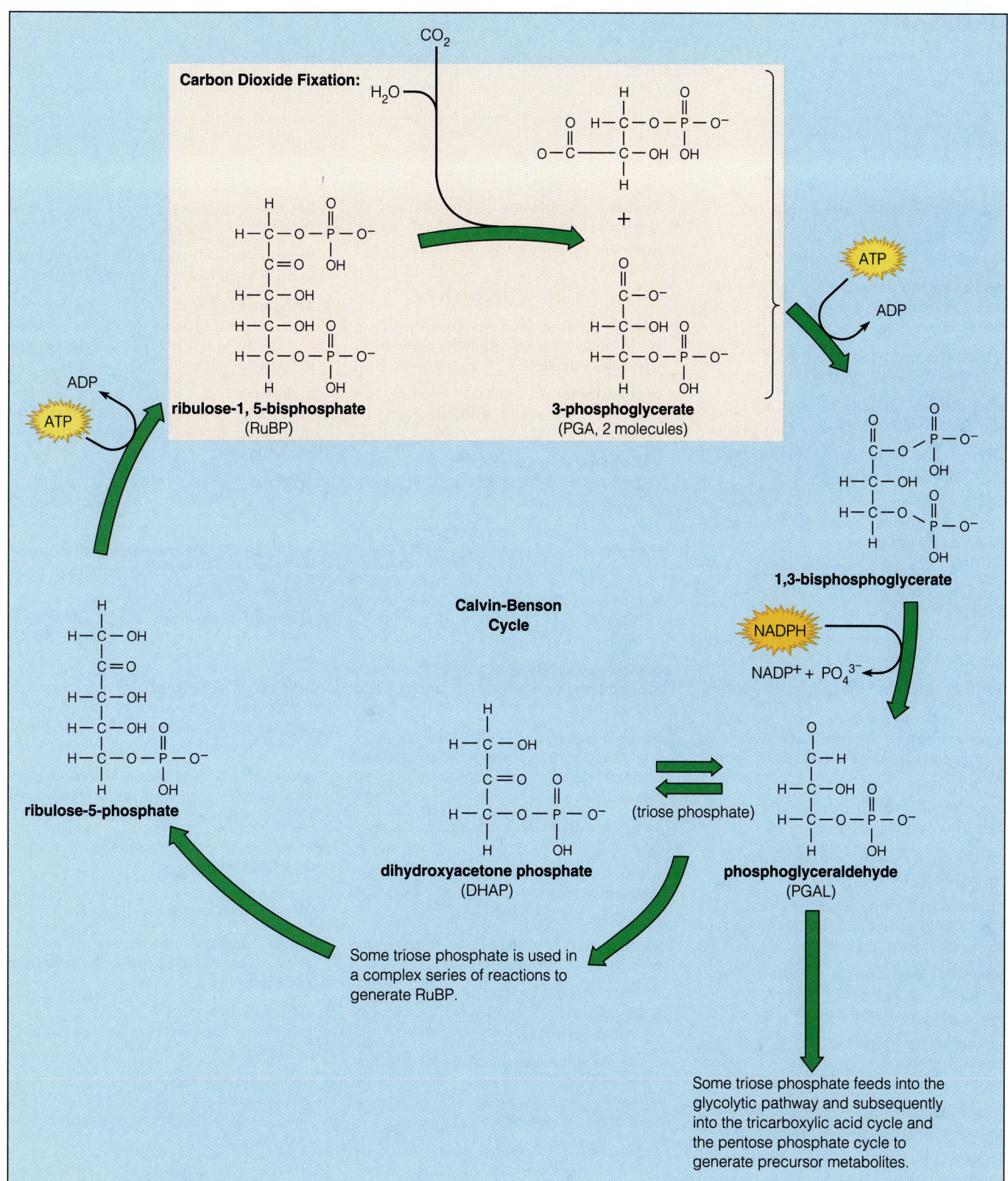

Figure d The Calvin-Benson Cycle. This cycle is used by some autotrophs to fix CO_2. It generates triose phosphate, some of which is recycled to formed RuBP to fix more CO_2; the rest flows into other pathways to generate precursor metabolites.

GLOSSARY

age-adjusted death rate Death rate in a certain age group.

agglutinate To clump together.

agglutination reaction Antigen-antibody reaction that produces a visible clump of particles.

alkaline A greater concentration of OH^- ions than H^+ ions; also called basic.

alkaliphiles Organisms that thrive in alkaline environments.

allergy The common name for type I hypersensitivity reactions.

allosteric enzymes Enzymes whose activities change when bound to small molecules, called effectors.

alternate pathway The antibody-independent complement cascade leading to activation of C3.

Ames test Method used to test for mutagens.

amino acids The building blocks of proteins.

ammonification Microbial conversion of nitrogen in organic compounds to ammonia (NH_3).

amoeba (also called Sarcodina) Protozoa that move by pseudopods.

amoeboid Cells that resemble amoebae by forming pseudopods.

anabolic reactions Reactions that build up: biosynthesis and subsequent steps of metabolic assembly line.

anaerobic metabolism Metabolism that occurs in the absence of oxygen: anaerobic respiration and fermentation.

anaerobic respiration A functioning electron transport chain with a terminal electron acceptor other than oxygen.

anaphylactic shock A precipitous fall in blood pressure caused by anaphylaxis.

anaphylaxis Systemic type I hypersensitivity reactions; sensitivity to foreign antigen resulting from previous exposure.

anatomic syndrome Constellation of signs and symptoms implicating specific anatomic structures in the diagnosis of an illness.

anions Negatively charged ions.

anneal Term used to describe formation of double-stranded nucleic acid from previously separated single strands.

anti-antibody An antibody produced in response to another antibody acting as an antigen.

antibiotics A metabolic product of an organism that kills or inhibits the growth of microorganisms.

antibody Defensive protein produced by the immune system in response to exposure to an antigen.

anticodon Three adjacent bases on a tRNA molecule that pair with a complementary codon on mRNA.

antigen-antibody complex The molecular combination between an antigen and an antibody.

antigen-presenting cells Lymphocytes that phagocytize foreign materials and display fragments of them as antigens on the cell surface.

antigenic drift Small mutational changes in a virus.

antigenic shift A sudden change in the properties of a virus resulting from exchange of nucleic acid molecules.

antigenic receptors Antigen-recognizing molecules on lymphocytes.

antigenic variation The process by which some pathogens evade recognition by changing their surface antigens.

antigens Molecular markers on the surfaces of all cells, viruses, and other large molecules such as certain drugs that provoke a response from lymphocytes.

antisepsis A treatment of living tissue that inhibits or destroys microorganisms.

antisera Preparations of sera that inactivate particular microorganisms.

antitoxins Antibodies that bind to and inactivate toxin molecules.

appendages Structures that extend beyond the envelope of microbial cells.

arabinose operon A cluster of genes, including regulatory genes that encode the ability to use a sugar, arabinose, as a growth substrate.

archaebacteria A distinct group (domain) of procaryotes as distantly related to other procaryotes (eubacteria) as they are to eucaryotes.

artificial scheme of classification A scheme not based on phylogenetic relationships.

artificial transformation Introducing DNA into a cell by laboratory manipulations.

artificially acquired active immunity Protection from disease stimulated by vaccines.

artificially acquired passive immunity Protection from disease conferred by administering antibodies formed by an animal or other human(s).

ascospores Sexual spores enclosed within a saclike structure (ascus) produced by one group of fungi (ascomycetes).

atomic number The number of protons an atom contains.

atomic weight The number of protons plus neutrons an atom contains.

atoms Smallest component of an element; composed of protons, neutrons, and electrons.

attenuated vaccine Vaccine composed of weakened, nonpathogenic, live microorganisms.

attenuation A mechanism of regulating gene expression in bacteria based on the balance between rates of transcription and translation.

autoclave A pressurized chamber for moist-heat sterilization.

autoimmune disorder A condition in which the immune system launches an attack against the body's own tissues.

autoinfection Continuous reinfection as infective larvae are passed and re-enter the body.

autolysis A process by which destructive enzymes destroy the cell from within.

autotrophs Organisms that obtain all their carbon from carbon dioxide.

axial filament (or **endoflagellum**) The modified flagella of spirochetes that lie within the periplasm; used for movement.

azidothymidine (AZT) A thymine analogue used as a drug to control HIV infection.

B lymphocytes (B cells) Lymphocytes that produce antibodies.

bacillus (plural, **bacilli**) Lower case: a rod-shaped bacterium; capitalized and italicized: a genus of rod-shaped bacteria.

bacteria Synonymous with the term *procaryote*.

bacterial chromosome The single, circular, DNA molecule that encodes all a bacterial cell's essential functions.

bactericidal Having a lethal effect on bacteria.

bacteriocins Toxic proteins produced by bacteria that kill other bacteria.

bacteriophages (or **phages**) Viruses that infect bacteria.

bacteriostatic Having an inhibitory effect on bacterial growth.

baker's yeast A specialized strain of *Saccharomyces cerevisiae* used for making bread.

barophiles Bacteria that grow more rapidly at pressures greater than 1 atmosphere.

beta sheet (also called a **pleated sheet**) A protein formation in which adjacent peptide chains are held together by hydrogen bonds between amino acids.

binary fission A mode of cell division in which a cross wall forms, producing two cells of approximately equal size.

binomial nomenclature A system of naming organisms in which each organism is identified by a genus designation and a specific epithet.

biochemistry Branch of chemistry that studies molecules made by organisms.

biological vector An organism that transports a pathogen from one host to another; the pathogen fulfills an essential part of its life cycle in the vector.

bioremediation Using organisms to detoxify or eliminate toxic materials.

biosynthesis pathways Sequences of metabolic reactions that convert the 12 precursor metabolites into building blocks for synthesis of macromolecules.

biotechnology The use of organisms for practical purposes.

blue-green algae Former designation of cyanobacteria.

blunt-end ligation Enzymatic joining of two pieces of DNA that lack extending single strands.

bradykinin A substance released during inflammation that causes vasodilation and increased blood vessel permeability.

brightfield illumination Microscopy in which visible light passes through the specimen, causing the background to be brightly lit.

broad-spectrum antibiotics Antibiotics that act against a wide variety of both Gram-positive and Gram-negative bacteria.

broth A liquid complex medium.

budding A form of asexual reproduction common in yeasts in which a bubble forms on the cell surface, grows, and pinches off, forming a new cell.

buffers Mixtures of weak acids or weak bases and their salts that resist changes in pH.

burst period The time interval of the one-step growth curve during which bacterial cells lyse and phage virions are released.

Calvin-Benson cycle The pathway through which most autotrophs incorporate CO_2 cellular constituents.

capsule The diffuse outermost layer, usually carbohydrate, of many bacterial and some other microbial cells.

carbon cycle The cyclic conversion of carbon-containing compounds that occurs in nature.

carriers Healthy-appearing individuals who are reservoirs of infection.

caseation necrosis The dead tissue that looks like cheese, at the center of tubercles caused by tuberculosis.

catabolic reactions The metabolic steps by which certain substrates are broken down into simpler compounds.

cations Positively charged ions.

cDNA (complementary DNA) Intron-free DNA synthesized from the mRNA using reverse transcriptase.

cell-mediated immunity Immune responses carried out by T cells.

Centers for Disease Control and Prevention (CDC) The national agency located in Atlanta, Georgia, that does research, collects statistics, and publishes information on infectious diseases.

cercariae The disc-shaped larvae of Trematode flukes; have a tail-like appendage.

chancre A painless weepy ulcer with raised borders that appears at the entry site in primary syphilis.

chemical bonds The forces that hold atoms together in molecules.

chemiosmosis A means of generating ATP by forming an ion gradient across a membrane that drives ATP synthesis from ADP by an enzyme (ATPase) located in the membrane.

chemoautotrophs Microorganisms that generate ATP and reducing power from inorganic chemical reactions.

chemotaxis The process by which cells sense certain chemicals and swim toward regions that contain optimal concentrations of them.

chemotherapy The treatment of disease with chemicals called drugs.

chitin A polymer of N-acetylglucosamine.

chlamydiospore A thick-walled, asexual, resting spore.

chloramphenicol A broad-spectrum antibiotic produced by *Streptomyces venezuelae*; now synthesized chemically.

chloroplasts Intracellular organelles in which photosynthesis occurs in phototrophic eucaryotes.

chloroquine Synthetic antimalarial drug.

ciliates A class of protozoa having cilia on part or all of the cell.

cirrhosis Disease of the liver characterized by replacement of functioning liver cells with fibrous scar tissue.

Class I MHC antigens Antigens present on all nucleated cells in the body.

Class II MHC antigens Antigens present only on macrophages, other antigen-presenting cells, and certain lymphocytes.

classical pathway The antibody-activated complement cascade leading to activation of C3.

clonal selection Mechanism by which the presence of an antigen causes proliferation of lymphocytes that recognize the antigen.

clone A population of organisms descended from a single individual by asexual reproduction.

cocci (sing., **coccus**) Spherical-shaped bacterial cells.

codon Three adjacent bases on an mRNA molecule that encode the addition of a particular amino acid to a growing peptide chain or signal it to stop growing.

coenocytic A multinucleate, continuous mass of cytoplasm.

coenzymes Organic molecules that certain enzymes need to be active.

cofactors Inorganic ions that certain enzymes need to be active.

colitis Inflammation of the colon.

colloids Particles ranging in diameter from about 0.001 to 1 μm that remain stably dispersed in water.

colony A clone of cells large enough to be visible on a solid medium.

commensalism A symbiotic relationship in which one partner is neither benefited nor harmed and the other benefits.

communicable disease A disease that can be transmitted from one host to another.

competitive inhibitor A molecule similar enough to an enzyme's normal substrate to bind to its active site, thereby inhibiting the enzymes activity.

complex media Extracts of natural materials used to support growth of microorganisms.

complement A family of more than 30 different proteins in serum that function together as a nonspecific defense against infection.

complement fixation assays Tests that detect antigen-antibody reactions by their utilization (fixation) of complement.

compound Matter composed of molecules that contain more than one type of atom.

conjugative plasmids Circular DNA molecules that encode the ability to transfer a copy of themselves to another bacterial cell by cell-to-cell contact.

conjunctiva (pl., **conjunctivae**) The epithelial covering of the eye.

conjunctivitis Inflammation of the conjunctiva.

contaminate To render impure; e.g., with unwanted microorganisms.

contraindicated Medically inadvisable.

cord factor A mycolic acid found only in virulent strains of *Mycobacterium tuberculosis* that causes these strains to form parallel rows of cells, called cords.

countercurrent immunoelectrophoresis (CIE) Immunologic assay in which antigen and antibody are moved in a gel slab by an electric current.

covalent bond The chemical bond formed by sharing pairs of electrons between atoms.

crepitance A crackling sound produced by gas moving through tissue.

culture medium A fluid or gelled solution, containing the nutrients needed for growth of a microorganism.

cyanobacteria The oxygen-producing, phototrophic eubacteria.

cyclic AMP (cAMP) 3′,5′-cyclic adenosine monophosphate that acts as a chemical messenger in cells.

cyclic photophosphorylation The process by which phototrophs generate ATP by passing an activated electron through a membrane-located electron transport chain to its ground state.

cystitis Inflammation of the bladder.

cytocidal Cell killing.

cytolysins Extracellular enzymes that lyse cells.

cytolysis See *lysis*.

cytomembrane system A complex of membranes that runs through eucaryotic cells.

cytopathic A damaging, nonlethal effect on a cell.

cytoplasm The matrix (ground substance) of a cell, composed primarily of water and protein.

cytosine A pyrimidine base found in DNA and RNA.

cytoskeleton The intracellular structure of eucaryotic cells, composed of microtubules, microfibrils, and intermediate filaments.

D-isomer One of the two forms in which a carbon atom attached to four different groups can exist.

Dane particle A component of the hepatitis B virions.

dapsone A sulfone drug; the first antimicrobial agent used to treat leprosy.

darkfield microscopy Method of viewing objects suspended in liquid in which the field of view is illuminated from the side, making the object brilliantly luminous against a dark background.

death phase The phase of microbial growth following the stationary phase in which cells die at an exponential rate.

decontamination A process of rendering a surface that has been heavily exposed to microorganisms safe to handle.

defined medium Culture medium for which the chemical composition is known because it is prepared from pure chemicals.

definitive host The host in which sexual reproduction of a parasite occurs; other hosts are intermediate hosts.

degranulation The process by which phagocytes release toxic chemicals and inflammatory mediators.

dehydration Medical: depletion of the body's water reserves; chemical: a reaction that removes hydrogen and oxygen atoms from a compound as water.

dehydrogenation reactions Chemical oxidations in which hydrogen atoms are removed from a compound.

delayed hypersensitivity (type IV hypersensitivity; also called cell-mediated hypersensitivity) Reactions initiated by T_D cells that produce macrophage-stimulating lymphokines and occur from 12 hours to several weeks after exposure to antigen.

denaturation Destruction of a protein's three-dimensional structure.

denitrification A bacteria-mediated cascade of anaerobic respirations that converts nitrate ion to nitrogen gas.

deoxyribonucleic acid (DNA) A macromolecule composed of four deoxynucleotides (A, G, T, and C) that encodes an organism's genetic information.

deoxyribose The five-carbon sugar found in DNA.

dermatitis Inflammation of the skin.

desulfurylation The process of converting sulfur constituents of organic compounds to hydrogen sulfide.

diapedesis The process by which phagocytes escape from blood vessels to enter tissues.

dichotomous key System of answering sequential questions with only two alternatives to identify species.

differential media Media used to identify microorganisms.

diffraction Bending of light rays when they pass through a small opening or by the edge of an opaque object.

dikaryon An organism composed of cells, each of which contains two genetically distinct nuclei.

dimorphism The switching between a yeast and a mycelial phase of growth characteristic of some fungi.

diplococci Spherical-shaped bacteria that occur in pairs.

direct microscope count Determining the number of microbial cells in a population by counting under the microscope the number of cells in a chamber of known dimensions filled with diluted sample of the population.

disaccharide A sugar composed of two monosaccharides joined by glycosidic bonds.

disc diffusion method (also called the **Kirby-Bauer method**) Determining the sensitivity of a microorganism to antimicrobial drugs by seeding a plate with the microorganism and placing filter paper discs impregnated with known quantities of different antimicrobial agents on it.

disease A state of functional disequilibrium that is resolved by recovery or death.

disease reservoir Environment where a pathogenic microorganism survives between infections.

disinfection (or **sanitation**) Treatment to reduce the number of pathogens to a level at which they pose no danger of disease.

DNA ligase An enzyme that seals gaps (missing phophodiester bonds) in DNA molecules.

DNA melting point The temperature at which double-stranded DNA separates into single strands.

DNA polymerase III The enzyme that catalyzes a reaction between a strand of DNA and a nucleoside triphosphate that produces pyrophosphate and a lengthened DNA strand.

Donovan bodies Intracellular inclusion bodies diagnostic of granuloma inguinale.

double helix Double-stranded DNA.

doubling time (formerly, **generation time**) The period required for a microbial population to produce two new cells for each one that existed before.

driving force The collective term for energy and reducing power.

drug antagonism Interaction between drug action causing the combination to be less effective than either administered alone.

drug resistance A microorganism's being able to grow and reproduce in the presence of a particular drug.

drug synergism An enhanced effect from using drugs in combination.

D-value (decimal time) The time required for a particular lethal treatment to kill 90 percent of a microbial population.

dysentery Disease characterized by diarrhea that often contains blood and mucus.

eclipse period The period following phage infection when no intact virions are present.

edema Swelling of tissue by fluid entering spaces between cells.

effector cells Cells that actively fight an infection.

electron A subatomic particle that carries a single negative charge.

electron acceptor A compound or atom that takes up electrons, thereby becoming reduced.

electron donor A compound or atom that loses electrons, thereby becoming oxidized.

electron transport chain A group of compounds embedded in a membrane that undergo sequential oxidation-reduction reactions and in so doing create a proton gradient across the membrane.

electronic count Determining the number of cells (or other particles) in a suspension by electronically scoring decreases in conductivity as the suspension is forced through a small pore.

element Matter composed of only one kind of atom.

elementary bodies (also called **chlamydiospores**) Resistant cell forms of *Chlamydia* that are released when an infected host cell lyses and that transmit infection to a new host.

emulsion A fine suspension of oily droplets in water.

enanthem Redness or blisters on mucous membrane surfaces, such as the inside of the mouth, that are caused by infection.

encephalitis Inflammation of brain tissue.

endemic Diseases that always present in a population at about the same level.

endocarditis Infection of the heart's inner lining, the endocardium.

endocytosis The process by which a cell engulfs solid material and brings it into the cell.

endoflagella Spirochete flagella that lie within the periplasm.

endometritis Inflammation of the lining of the uterus.

endoplasmic reticulum (ER) Part of the eucaryotic cell's cytomembrane system; a double membrane that folds back upon itself, creating a complex pattern of tubes and layered sacs.

endospores Exceptionally hardy dormant structures that form within the cells of certain species of bacteria.

endotoxin The lipopolysaccharide component of the outer membrane of Gram-negative bacteria that is harmful to humans and other animals; most of the toxicity is mediated by lipid A.

enrichment culture A method of cultivating microorganisms designed to isolate a particular microorganism or type of microorganism from a large, complex natural population.

enteric Intestinal.

enterotoxins Compounds that are harmful to the epithelial cells lining the intestinal tract.

enzyme-linked immunosorbent assays (ELISA) Diagnostic immunologic tests that contain an enzyme linked to the indicator antibody.

enzymes Proteins (with the exception of a few recently discovered RNA molecules) that catalyze specific metabolic reactions.

epidemic A pattern of disease transmission in which many members of a population are affected within a short time.

epidemiology The study of when and where diseases occur and how they are transmitted in human populations.

epitope (also called the **antigenic determinant**) The small region of the antigen molecule that a lymphocyte recognizes.

erythema Abnormal redness of the skin.

erythrocytes Red blood cells.

erythrogenic toxin A substance produced by *Streptococcus pyogenes* that kills cells and causes the rash of scarlet fever.

ester bonds Chemical linkages that form between carboxylic acid and alcohol groups to form esters.

etiologic agent Cause of a disease.

eubacteria One of the two groups of procaryotes (the other is archaebacteria).

eucaryotes Organisms composed of cells with internal membranes (all cellular organisms except bacteria).

euglenoids A class of protozoa; single celled; motile by two flagella of unequal length.

evolutionary distance A quantitative estimate of the phylogenetic relatedness of organisms.

exanthem A skin rash, associated with an infectious disease.

exocytosis The process of expelling material from a cell by the reverse process of endocytosis.

exotoxins Highly destructive proteins produced by certain Gram-positive and Gram-negative pathogens; most exotoxins are composed of two subunits, a B, or binding component, and an A, or active component.

exponential growth (also called the logarithmic growth) The growth phase during which the number of cells in the population continues to double during the same time interval.

facilitated diffusion Movement of molecules across a membrane from regions of higher to lower concentration mediated by proteins that permit passage only of specific molecules.

facultative anaerobes Organisms that use oxygen to grow when it is available but also grow without it.

fastidious organisms Organisms that require numerous complex nutrients to grow.

fats Esters formed between a molecule of glycerol and three fatty acid molecules that are solid at room temperature.

fatty acids Organic acids with a single carboxylate group and a chain of carbon atoms; constituents of fats, oils, phospholipids, and other biochemicals.

fecal-oral route A pattern of disease transmission by which pathogens shed in feces enter a new host through the mouth.

fermentation Scientific: anaerobic generation of ATP from organic compounds totally by substrate-level phosphorylation; industrial: all microbial transformations, either aerobic or anaerobic.

fever Elevated temperature of the body.

flagella Appendages of bacteria and eucaryotic cells that confer motility.

flagellates (Mastigophora) A class of protozoa that move by means of flagella.

flavine adenine dinucleotide (FAD) A compound that transfers reducing power by taking up and releasing hydrogen atoms.

floc Material that forms in sewage treatment systems composed largely of particles embedded in extracellular slime produced by bacteria.

fluid mosaic model The proposal that proteins embedded in a phospholipid membrane move freely.

fluorescence microscopy Increases contrast through fluorescence, the property of certain materials to absorb light of one wavelength and give off light of a higher wavelength.

fluorescent-labeled antibodies Antibodies chemically bonded to fluorochromes (fluorescent chemicals).

fomites Inanimate objects such as cups, towels, bedding, and handkerchiefs that transmit disease.

freeze-dried Desiccated in a vacuum while frozen.

fruiting bodies Structures that contain or bear spores.

fueling pathways Metabolic pathways that generate precursor metabolites, ATP, and reducing power.

functional groups Parts of organic compounds, composed of specific patterns of atoms; e.g., amino groups.

fungi A large and diverse group of nonphototrophic microorganisms that includes yeasts, molds, and mushrooms.

gamma globulin The antibody-containing globulin protein fraction of serum.

gangrene Tissue death due to impaired blood supply.

gas gangrene An infection caused by *Clostridium perfringens*.

gastritis Inflammation of the stomach.

gastroenteritis Inflammation of the stomach and intestines.

gel electrophoresis Movement of charged molecules through a gel with an electric current; molecules of different size and/or charge become separated.

gels Open networks of interconnected colloidal particles.

gene cloning The process of obtaining a set of identical copies of a gene.

gene expression The process of converting the information encoded in DNA into RNA and protein.

gene therapy Treating genetic disease by introducing new genes into the affected individual.

genetic code The correspondence between codons in mRNA and amino acids in proteins.

genetic engineering (or **recombinant DNA technology**) A group of techniques for manipulating DNA outside the organism from which it was obtained and reintroducing the recombinant or modified DNA into another cell where it will exert its effect.

genome The totality of genetic information that an organism has.

genotype The form of the genes that an organism has.

German measles Rubella, a rash-producing viral disease that resembles but is unrelated to measles, rubeola.

germicides Chemicals that kill microorganisms.

germinate To begin to grow or develop.

germistats Chemicals that inhibit microbial growth.

Ghon complexes Calcified caseous tubercles that indicate a past primary tuberculosis infection.

gingivitis Inflammation of the gums.

global regulation Response to a general signal, such as the shortage of an amino acid, that alters expression of many genes.

glycocalyx The slimy or gummy substance that constitutes the outermost layer of the envelope of some bacteria.

glycogen An alpha-linked, branched-chain polymer of glucose.

glycolysis The catabolic, metabolic pathway that converts a molecule of glucose into two molecules of pyruvate.

Golgi apparatus An organelle in eucaryotic cells that modifies molecules and sends them to their proper location inside the cell or outside it.

Gram-negative bacteria Eubacteria that have a thin cell wall surrounded by an outer membrane.

Gram-positive bacteria Eubacteria that have a thick wall and no outer membrane.

Gram-stain A technique that colors Gram-positive bacteria a deep blue and Gram-negative bacteria a light red.

group A Beta-hemolytic streptococcus Also called *Streptococcus pyogenes*, a highly virulent bacterium that causes many different clinical syndromes.

group translocation Entry of a compound into a cell and simultaneous chemical alteration.

growth rate Measurement of how rapidly a microbial population is increasing, usually expressed in doubling times per hour.

gummas Soft granulomas that usually replace skin or bone but may occur in any organ during late syphilis.

halophiles Microorganisms that grow well in environments with high salt concentrations.

hanging drop preparation A drop of liquid containing microorganisms on a coverslip suspended over a depression slide

hapten A small molecule, not itself antigenic, that becomes an epitope when attached to a protein.

health A state of relative equilibrium in which the body's organ systems function adequately.

heat labile Easily destroyed by heat.

helminths Worms.

helper T cells (TH) T lymphocytes that activate the functions of other lymphocytes.

hemagglutination Agglutination of red blood cells.

hemagglutinin Proteins that agglutinate red blood cells.

hemolysins Proteins that lyse red blood cells.

hepatitis Inflammation of the liver.

herd immunity The prevention of epidemics due to the scarcity of new susceptible hosts.

heterocysts Specialized, oxygen-impermeable cells of some cyanobacteria in which nitrogen fixation occurs.

heterotrophs Organisms that obtain carbon from organic compounds in their medium or diet.

high-energy bonds Chemical bonds that require considerably less energy to break than is released when new ones form.

higher fungi The ascomycetes, basidiomycetes, and deuteromycetes.

histamine Decarboxylation product of the amino acid histidine that causes vasodilation and increases the permeability of blood vessels.

histidine operon A group of contiguous genes that encodes the enzymes needed to synthesize the amino acid histidine.

horizontal transmission (also called **person-to-person transmission**) The spread of pathogens by direct contact between one person and another, such as touching, kissing, or sexual intercourse.

host range The spectrum of strains or species that a pathogen attacks.

humoral immunity Protection conferred by antibodies.

hybrid DNA The annealed product of mixing single stands of DNA from different sources.

hydatid cysts Encysted larvae of tapeworms of the genus *Echinococcus*.

hydrogen bonds Linkages that form when hydrogen atoms are shared between two molecules or between different parts of the same molecule.

hydrogen ion H^+, a proton.

hydrogenation reactions Addition of hydrogen atoms to a molecule.

hydrolysis Splitting of a molecule by the addition of a molecule of water.

hydrophilic compounds Compounds that dissolve in water.

hydroxide ions OH^-.

hypersensitivity An exaggerated immune response that harms the body.

hypertonic environment A cell's environment with a higher concentration of solutes than the cell's interior.

hyphae The tubelike filaments that constitute a mycelium.

hypotonic environment A cell's environment with a lower concentration of solutes than the cell's interior.

iatrogenic Medically induced disease or complication of disease.

IgA antibodies The second largest class of antibodies; it primarily protects mucous membrane surfaces.

IgD antibodies Constitutes less than 1 percent of the antibody total; it is the main type of antibody found on B cells.

IgE antibodies Constitutes less than 0.01 percent of the antibody total; it causes leukocyte degranulation, which is a primary defense against parasites too large to be eliminated by phagocytosis, such as worms.

IgG antibodies The largest class of antibodies; it activates the complement cascade through the classical pathway.

IgM antibodies Constitute approximately 5–10 percent of the antibody total; it is the first antibody class to form during a primary immune response.

imidazoles A major family of antifungal agents that act by inhibiting the synthesis of plasma membrane sterols.

immune serum globulin An antibody-rich preparation from the pooled serum of many donors.

immune system A network of cells, principally lymphocytes, and organs that extends throughout the body and functions as a defense against infection.

immunization Artificially stimulating the body's immune defenses.

immunocompetence The process by which lymphocytes acquire the capability to function fully in the body's defense.

immunodiffusion assay A type of precipitation reaction in which antigens and antibodies are diluted and mixed by diffusion through a gel.

immunoelectrophoresis assay An immunodiffusion assay in which diffusion is sped up by electrophoresis.

immunofluorescence assays Tests in which antigen-antibody reactions are detected by fluorescence because one of the reactants is tagged with a fluorescent dye.

immunogenic Capable of stimulating an immune response.

immunoglobin Synonym for antibody.

immunologic memory The ability of memory lymphocytes to recognize an antigen if they encounter it again, greatly accelerating and amplifying the immune response.

immunologic tolerance The immune system's ability not to respond to self antigens.

inactivated vaccines Vaccines containing killed microorganisms.

incidence rate The number of people who develop a disease or condition during a certain period of time divided by the total number of people in the population.

inclusion bodies In bacteria, visible structures within the cell other than the nuclear region and ribosomes.

incubate To allow to grow in a warm place.

inducible enzymes Enzymes synthesized in response to an environmental signal.

infection The growth of microorganisms in the body.

infectious disease Diseases caused by microorganisms.

infectious dose (ID) The number of microorganisms that must enter the body to establish infection in a certain percent of test animals; e.g., $ID_{50} = 50$ percent.

inflammatory response The body's nonspecific reaction to injury or infection, consisting of redness, pain, heat, swelling, and sometimes loss of function.

inflammatory mediators Molecular messengers that mediate inflammation.

interferon Small glycoproteins produced by host cells in response to viral infections.

interleukin-1 A cytokin produced by white blood cells; one of its many actions is to cause fever.

intoxication A poisoning.

introns Noncoding regions within eucaryotic genes.

invasive Ability of a pathogen to enter host cells or deeper tissues.

ionic bonds Chemical attraction between oppositely charged ions.

ionize To form ions.

ions Charged atoms or groups of atoms.

isomers Molecules with the same kind and number of atoms, but with different arrangement.

isoniazid An antimycobacterial drug.

isotonic environment A cell's environment with the same concentration of solutes as the cell's interior.

isotopes Atoms with the same atomic number but different atomic weight.

jaundice A yellow skin color caused by the buildup of bilirubin when the liver does not function properly.

keratitis Inflammation of the cornea.

killer (K) cells A group of non-B non-T lymphocytes that destroy target cells marked by any of the five antibody classes.

Kirby-Bauer method See *disc diffusion method*.

Koch's postulates Four steps developed by Robert Koch used to prove that a particular microorganism causes a particular disease.

L-forms Strains of bacteria that have lost the ability to form walls.

L isomer See *D isomer*.

***lac* operon** Group of contiguous genes that encode enzymes to metabolize lactose.

lactic acid bacteria Aerotolerant bacteria that produce lactic acid as a major product of fermentation.

lactic acid fermentation Fermentation that produces lactic acid as a major product.

lag phase Phase of microbial growth cycle in which metabolism prepares cells to grow.

latent period The period of time following viral infection during which no new virions are produced.

lawn A confluent layer of cells.

lesion An injury, hurt, wound.

lethal dose (LD) The number of microorganisms that must enter the body to kill a certain percent of test animals; e.g., $LD_{50} = 50$ percent.

leukocidins Enzymes that lyse leukocytes.

leukocytes White blood cells.

leukotrienes A class of inflammatory mediators; some increase blood vessel permeability; others attract leukocytes to the inflammation site.

lichen A symbiotic association between a fungus and an alga or cyanobacterium.

ligation In recombinant DNA technology, sealing a gap in a DNA molecule with the enzyme DNA ligase.

limiting nutrient The scarcest nutrient in a medium.

Linnaean scheme Linnaeus's hierarchical scheme of classifying organisms into species, genera, families, classes, order, phyla or divisions, and kingdoms.

lipopolysaccharide (LPS) A compound found only in the outer membrane of Gram-negative bacteria.

local therapy Applying a drug directly to the infected area.

log phase (also called the **logarithmic**, or **exponential phase**) The phase of microbial growth cycle when exponential growth occurs.

lower fungi Coenocytic fungi that are divided into five classes according to the structure of their spores and gametes.

lymph nodes Small bean-shaped organs of the lymphatic system located along lymphatic vessels throughout the body.

lymphatic circulation A system of vessels that collects lymph from tissues and returns it to the bloodstream through the thoracic duct.

lymphocytes A type of leukocyte, part of the body's defense system.

lymphokines Messenger proteins produced by T-cells.

lyophilization Freeze-drying.

lysis Rupture of the plasma membrane and destruction of the cell.

lysogenic cycle One of the two life cycles of temperate phages (along with the lytic cycle) in which phage DNA becomes part of the host cell's genome and is called a prophage.

lysogeny The state in which an infecting phage exists as a virion.

lysosome Vacuoles that contain enzymes and other chemicals that can destroy most microbial cells.

lytic cycle The life cycle of a phage that lyses the host cell as virions are released.

lytic infections Viral infections that kill the host cell by lysing it.

macrophages Phagocytic scavenger cells that consume dead microorganisms, dead and dying host cells, and foreign particles; also important as antigen-presenting cells.

macroscopic Visible without the aid of a microscope.

magnetotaxis Movement of bacterial cells along magnetic lines of force.

major histocompatibility complex (MHC) The DNA that encodes self antigens of humans.

margination The process of migration of phagocytes to the walls of capillaries.

mast cells Leukocytes that release histamine and heparin, potent inflammatory mediators.

medium See *culture medium*.

meiosis Nuclear division that converts a $2n$ nucleus into four $1n$ nuclei.

membrane attack complex A cylinder-like protein complex of the terminal complement pathway that makes a hole through the plasma membrane and causes cell lysis.

membrane filter A nitrocellulose membrane with holes too small for microbial cells to pass through.

memory B cells Residual B cells formed in response to an infection that allow more rapid response to a subsequent infection.

meningitis Inflammation of the meninges, the membranes that surround the brain and spinal cord.

meningococcus The Gram-negative pathogen *Neisseria meningitidis*.

merozoite The stage of the malarial parasite's life cycle that infects red blood cells.

mesophiles Organisms that grow best at moderate temperatures, around 37°C.

messenger RNA (mRNA) Carries the information from DNA to ribosomes that determines the order of amino acids in a protein.

metabolic intermediates Compounds formed at various steps of metabolic pathways.

metabolism All of the biochemical reactions that take place in a cell.

methanogens The methane-forming archaebacteria.

mordants Compounds that intensify staining reactions.

microaerophiles Microorganisms that need lower concentrations of oxygen than are present in air.

microbial antagonism Inhibition of microbial growth by another microorganism.

microbiostatic Ability to inhibit microbial growth.

minimum bactericidal concentration (MBC) Lowest concentration of a drug that can kill a particular microorganism.

minimum inhibitory concentration (MIC) Lowest concentration of a drug that can inhibit the growth of a particular microorganism.

minus-strand Single-stranded RNA comprising a viral genome that must be transcribed by RNA-dependent RNA polymerase to act as mRNA.

miracidia First larval state in the trematode life cycle that parasitizes a particular species of intermediate host, usually a snail.

missense mutation A mutation that changes a codon to one that encodes a different amino acid.

mitosis Cell division in which each daughter cell receives the same chromosomes that were present in the parent cell.

mixed culture A culture containing more than one kind of microorganism.

molds Filamentous fungi.

mole Avogadro's number (6.02×10^{23}) of molecules.

molecular formula Tells which atoms and how many of each kind constitute a particular molecule.

molecular weight The total of the atomic weights of all of a molecule's atoms.

molecule Two or more atoms joined by chemical bonds.

monera The bacteria.

monoclonal antibodies Antibodies produced by a single lymphocyte clone that act against a single epitope.

monocytes A large phagocytic lymphocyte with an oval or horseshoe-shaped nucleus.

monosaccharides A monomer sugar not joined to another by glycosidic bonds.

morbidity Illness and disability.

mortality Death.

most probable number (MPN) An estimate of numbers of microorganisms in a sample based on a statistical analysis of the probability of cells being present in a diluted sample.

murine typhus Flea-borne typhus.

mushrooms Fleshy, macroscopic fruiting structures produced by some higher fungi.

mutagens Agents that can induce mutations.

mutation Any chemical change in a cell's genotype.

mutation rate The number of mutations per cell per generation.

mutualism A symbiotic relationship in which both partners benefit.

mycelium A mass of hyphae produced by some fungi and actinomycetes.

mycolic acids Long-chain, complex organic acids found in the waxy cell envelope of mycobacteria.

mycoplasmas A group of small, wall-less eubacteria.

myelitis Inflammation of the spinal cord.

myocarditis Inflammation of the myocardium (heart muscle).

naked viruses Viruses not surrounded by a membrane.

narrow-spectrum antimicrobial drugs Drugs that are effective against only a limited number of similar microorganisms; e.g., Gram-negative or Gram-positive bacteria.

natural killer (NK) cells A group of non-B non-T cells that lyse target human cells by secreting perforins.

naturally acquired active immunity Immunologic protection that follows recovery from an infectious disease.

naturally acquired passive immunity Immunologic protection acquired from antibodies transferred from mother to fetus across the placenta and to the newborn in colostrum.

negative staining Use of a stain that does not penetrate cells or capsules, causing them to appear bright against a dark, stained background.

Negri bodies Inclusion bodies that develop in the brains of rabies victims.

nematodes Roundworms, a phylum of helminths.

neurotoxins Toxic proteins that specifically affect nerve function.

neutron Electrically neutral particle in the nuclei of atoms.

nicotinamide adenine dinucleotide (NAD) A compound that acts as a reservoir of reducing power by accepting and donating pairs of hydrogen atoms.

nicotinamide adenine dinucleotide phosphate (NADP) A phosphorylated form of NAD that serves a similar function.

NAD(P) Designation for NAD and/or NADP.

nitrification The conversion, by bacteria, of ammonia to nitrate.

nitrogen cycle Conversions of nitrogen compounds that occur in nature.

Nomarsky microscopy (also called **differential interference contrast microscopy**) Technique that uses differences in refractive index to produce contrast by interference.

non-B non-T lymphocytes Lymphocytes that function without recognizing antigens.

nonpolar molecules Molecules with no electrically charged regions.

nonsense codons Three codons that do not correspond to the anticodon of any tRNA molecules and therefore stop translation.

nonsense mutation A mutation that changes a codon encoding an amino acid to a nonsense codon.

nonspecific defenses Body defenses that act universally and uniformly against all microorganisms.

nonspecific interior defenses Inflammation, phagocytosis, complement, and interferon—the body's second line of defense against infection.

nonspecific surface defenses Surface features that prevent microbial growth or penetration—the body's first line of defense against infection.

normal flora The microorganisms that coexist with humans in a stable relationship on body surfaces.

nosocomial infections Infections while in the hospital.

notifiable diseases Diseases that must be reported to government agencies because they affect the public health.

nuclear envelope A double-membrane structure that defines the eucaryotic nucleus.

nucleoid (or **nuclear region**) The mass of DNA in bacterial cells—not membrane-bound.

nucleoli Dense masses of RNA and protein within the eucaryotic nucleus that manufacture ribosomes.

nucleoplasm The gelatinous matrix of the nucleus.

nucleoside triphosphate A purine or pyrimidine bonded to ribose or deoxyribose and three phosphate groups.

nucleotide A purine or pyrimidine bonded to ribose or deoxyribose and one or more phosphate groups.

numerical taxonomy Biological classification based on comparing many characters and using similarities and differences to calculate relatedness among organisms.

nutrient agar Nutrient broth solidified with agar.

nutrient broth A frequently used complex medium.

obligate aerobes Organisms that can grow only in the presence of oxygen.

obligate anaerobes Organisms that are killed by oxygen.

obligate barophiles Bacteria that grow only at pressures greater than 1 atmosphere.

obligate intracellular parasites Microorganisms that can reproduce only inside a host cell.

oncogenic Cancer-causing.

one-step growth curve A procedure for simultaneously infecting a bacterial culture with phage virions in order to study their development.

operon A set of contiguous genes that is regulated and transcribed together.

opportunistic infection Infections caused by microorganisms that can infect only debilitated hosts.

opsonin Proteins that facilitate phagocytosis.

opsonization The process by which an opsonin facilitates phagocytosis.

organic compounds Carbon-containing compounds.

organic growth factors Organic compounds that certain microorganisms need to grow.

outer membrane The lipopolysaccharide-containing membrane that surrounds Gram-negative bacteria.

oxidation Removal of electrons from an atom or molecule.

oxidation-reduction reaction (**redox reaction**) A reaction in which one atom or molecule loses electrons and another gains them.

pandemic A worldwide epidemic.

paper disc method The procedure for determining the effectiveness of a germicide: a filter paper disc impregnated with it is placed on a plate seeded with the test microorganism and incubated.

parasitism A symbiotic relationship in which the host is harmed and the parasite benefits.

parasitology The study of protozoan and helminth-caused disease.

passive immunity Immunity conferred when antibodies are administered.

pasteurization Moderate heat treatment to kill pathogens and extend the shelf life of liquid foods.

pathogens Disease-causing microbes.

pellicle Flexible covering that surrounds some protozoa.

pelvic inflammatory disease (PID) Infections of the female upper reproductive tract.

penicillin An antibiotic produced by certain species of *Penicillium*.

pentose phosphate pathway The catabolic metabolic pathway that begins with glucose and produces pentose phosphates.

peptide bonds The bonds that join amino acids in proteins.

peptidoglycan The macromolecule that forms the cell walls of eubacteria.

pericarditis Inflammation of the pericardium, the membrane surrounding the heart.

perinatal Occurring just before, during, or just after birth.

periplasm The organelle between the cytoplasmic membrane and outer membrane of Gram-negative bacteria.

pH scale Quantitative description of acidity or alkalinity.

phage typing Identifying bacterial strains by their susceptibility to phages.

phages Short for bacteriophages, viruses that infect bacteria.

phagocytosis Engulfment of one cell by another.

pharyngitis Inflammation of the throat.

phase-contrast microscope Microscope that generates contrast by interference between phase-shifted light rays that pass through the specimen and those that do not.

phenotype The outward expression of a cell's genes.

phosphodiester bonds Chemical linkages that join nucleotides in nucleic acids.

phospholipid bilayer A phospholipid membrane.

phospholipids Constituents of unit membranes; composed of a glycerol, two fatty acids, and a phosphate group to which another group is attached.

phosphorus cycle Chemical conversions of phosphorus that occur in nature.

phosphorylation Chemical addition of a phosphate group to a molecule.

phototaxis Movement of cells toward optimal intensity and quality of light.

phototrophs Organisms that generate ATP and reducing power from light energy.

phycology The study of algae.

phytoplankton Floating, microscopic algal species.

pili Straight hairlike appendages that extend out from surface of a bacterial cell.

pinocytosis Cellular engulfment of liquid.

plaque count A procedure for determining the number of bacteriophages and phage-infected cells in a sample.

plaque-forming units (PFUs) Virions and virus-infected cells.

plaques Circular clear zones on a lawn of cells.

plasma The fluid, cell-free component of blood, including clotting proteins.

plasma membrane (also called the **cyto-plasmic membrane**) The phospholipid membrane that encloses all cells.

plasmids Small circular DNA molecules found in some bacteria and other microorganisms that encode nonvital functions.

plasmodium The amorphous slimy mass that constitutes a true slime mold. *Plasmodium*: the genus of protozoa that causes malaria.

plasmolysis The drawing out of water from a cell in a hypertonic environment, decreasing the volume of the cell.

plate count Enumerating microbial cells in a sample by distributing them on an agar plate and counting the colonies that develop after incubation.

platelets Subcellular fragments that participate in blood clotting.

platyhelminths Flatworms, a phylum of helminths.

pleurisy Inflammation of the pleura, the membrane that surrounds the lungs.

plus strand Single-stranded RNA constituting a viral genome that can act directly as mRNA.

pneumococcus *Streptococcus pneumoniae*.

pneumonia Inflammation of the lungs.

polar molecule A molecule that has a positive and a negative region.

polymerization The process by which monomers are joined together to produce a macromolecule.

polymers Macromolecules built from repeating subunits.

polysaccharides Polymers built from simple sugars.

porins Proteins that form pores in the outer membrane.

portal of entry The anatomic site through which a pathogen enters a host.

portal of exit The anatomic site through which a pathogen leaves its host.

precipitation reaction An antigen-antibody reaction that forms lattices large enough to precipitate.

precipitin ring test A precipitation reaction that forms a ring of precipitate in a column of liquid.

precursor metabolites The 12 compounds from which all constituents of a cell can be synthesized.

prevalence rate The number of people who have a certain disease at any particular time divided by the number of people in the population.

primary immune response The production of antibody that occurs when a person first encounters a particular antigen.

primary structure of a protein The sequence of amino acids.

prion An infectious agent composed only of protein.

probe In recombinant DNA technology: a short, complementary, tagged molecule of nucleic acid used to detect specific pieces of DNA by hybridization.

procaryotes Bacteria.

prophage A phage genome integrated into the chromosome of a host cell.

prophylaxis Prevention of disease.

prostaglandins A large family of cytokines that are potent inflammatory mediators.

protein A macromolecule composed of polymerized amino acids.

protists Eucaryotic microorganisms in Haeckel's classification scheme.

proton A subatomic particle that carries a single positive charge; ionized hydrogen atom.

proton acceptors Bases.

proton donor Acids.

protoplast Bacterial cell from which the wall has been removed completely.

protozoa Nonphotosynthetic, unicellular eucaryotes.

provirus A viral genome integrated into the chromosome of a host cell.

pseudopods Tubelike structures that amoeboid cells project and withdraw in order to move.

psychrophiles (psychrotrophs) Organisms that grow at low temperatures.

public health A discipline that deals with the development and implementation of plans to prevent and control disease.

puerperal sepsis A bloodborne infection acquired at time of childbirth.

pure culture A culture that contains only a one kind of organism.

purulent Pus-producing.

pus A mixture of dead leukocytes, microorganisms, and host cells.

pyelonephritis Inflammation of the kidneys.

quaternary structure of proteins The way separate polypeptide chains fit together.

radioimmunoassay A test to detect antigen-antibody reactions in which one of the reactants is tagged radioactively.

recombinant DNA technology See *genetic engineering*.

recombination (genetic) Reassortment of genes.

reduction The addition of electrons to an atom or molecule.

refraction Bending that occurs when a ray of light enters an object with a different density at an angle.

refractive index The ratio of the speed of light traveling through a vacuum to the velocity in any particular material.

replica plating Inoculating a fresh plate by pressing it against a piece of velveteen that has previously been pressed onto a plate with colonies of microorganisms.

replication The biochemical process of making a copy of a DNA molecule.

replication forks The two points within the bubble form in a DNA molecule where replication occurs.

repressible enzymes Enzymes whose synthesis is inhibited in the presence of a signal molecule (repressor).

reservoirs Infectious disease: repositories for pathogens between hosts.

resistance (R) factors Plasmids that carry genes encoding drug resistance.

resolution The capacity to perceive two adjacent parts of an image as separate from each other.

respiratory burst The event that occurs when a phagocyte's granules produce lethal oxidants.

reticulate bodies Nonvirulent, reproductive cells of *Chlamydia* spp.

retroviruses (retroviridae) Family of viruses that convert their RNA genome into DNA by reverse transcriptase as part of their life cycle.

reverse transcriptase An enzyme that uses RNA as a template to make a complementary strand of DNA.

Rh factor An erythrocyte antigen found in humans and rhesus monkeys.

rhinitis Inflammation of nasal membranes.

ribonucleic acid (RNA) Macromolecular polymer of ribonucleotides.

ribosomes The organelles on which proteins are synthesized.

rifampin A semisynthetic antibiotic that inhibits eubacterial RNA polymerases.

ringworm The common name for tinea, a fungal infection of the skin.

RNA polymerase Enzyme that uses DNA as a template to make RNA.

rubella See *German measles*.

rubeola Measles, a rash-producing illness caused by the rubeola virus.

sampling error The inevitable inaccuracy that occurs because no sample is precisely representative of the total population.

sanitation Disinfection: treatment to reduce the number of pathogens to a level at which they pose no danger of disease.

saturated fat Fat composed of fatty acid molecules that do not contain double bonds.

scanning electron microscope (SEM) An electron microscope that generates an image by scanning the surface of the specimen with an electron beam.

secondary immune response Response initiated by memory lymphocytes.

secondary structure of a protein The alpha helix and beta sheet structures that result from hydrogen bonds forming between amino acids.

selective media Media that favor the growth of certain microorganisms over others.

selective toxicity Pharmacology: a drug that harms a pathogen without harming the host.

self-limited illness Illness from which most people recover without medical treatment.

semiconservative replication Refers to DNA replication because each new double helix is composed of one new and one old (conserved) strand.

semipermeable Describes properties of membranes that allow certain molecules to cross freely while blocking the passage of others.

sense strand The DNA strand used as a template for mRNA.

septa Cross walls.

septicemia Persistent and serious infection of the bloodstream.

serology Diagnostic clinical immunology.

serotype (also called a **serovar**) A taxonomic category identified by serology.

serum The cell-free liquid component of blood, not including clotting proteins.

serum killing-power test A procedure in which a patient's drug-containing serum is tested for its ability to kill the infecting microorganism.

serum resistance Inherent ability of certain bacteria to avoid destruction by serum.

serum sickness A type III hypersensitivity reaction that sometimes occurs when proteins from animal serum are used in medical therapy.

Shick test An immunological test for immunity to diphtheria.

siderophores Iron-chelating compounds released by microorganisms to obtain iron.

signs Clinical: objective indications of illness on physical exam.

similarity coefficient (S$_J$) A number that expresses the relatedness among organisms.

simple stains A single dye used to increase contrast of a specimen.

simple wet mount A drop of liquid containing microorganisms on a microscope slide covered with a cover slip.

slide agglutination test Identifying bacteria by mixing a suspension of the unknown bacterium with a known antiserum on a microscope slide and observing if agglutination occurs.

slime layer A thin slimy or gummy layer that surrounds some bacterial cells.

slime molds Two groups of microorganisms: cellular slime molds (Acrasieae) and true slime molds (Myxogastria).

smear Microscopy: a thin film spread on a microscope slide.

special stains Stains that heighten contrast within microbial cells to reveal particular structures, including endospores, flagella, or capsules.

sporadic diseases Occurring only occasionally in a population.

sporogenesis Development of a spore.

sporozoa Nonmotile protozoa; all sporozoa are parasitic.

sporozoite Form of malarial parasite that passes from the insect's salivary gland to a person's bloodstream.

sporulation The process of forming sores.

stains Microscopy: dyes used to increase contrast.

sterilization Eliminating all microorganisms from an area.

sterols Lipids composed of hydrocarbon rings.

stock cultures Microbial cultures maintained for study and reference.

storage granules Granular inclusions in the cytoplasm that hold reserve supplies of nutrients.

strains Clones that are presumed or known to be genetically different.

streak plate method Commonly used method of obtaining a pure culture.

strict aerobes Bacteria that cannot grow without oxygen.

strict anaerobes Bacteria that cannot grow in the presence of oxygen.

stridor A hoarse sound when a person inhales if the airway near the epiglottis or larynx is narrowed.

substrates Molecules used as nutrients for microorganisms or as reactants for enzyme reactions.

sulfur cycle The chemical conversions of sulfur that occur in nature.

sulfonamides (sulfa drugs) Synthetic antimicrobial agents that act by interfering with the bacterial cell's ability to synthesize folic acid.

surfactants Detergent-like agents; they penetrate oily globules in water, producing an emulsion.

symbiosis Two different kinds of organisms living together.

symptoms Subjective reports of illness by a patient.

systemic disease Body-wide infection spread through the bloodstream.

T cells (T lymphocytes) The agents of cell-mediated immunity.

T_4 cells, also called **T_H (helper) cells** Lymphocytes that increase immune responsiveness.

T_8 cells Further differentiated into T_C (cytotoxic) cells and T_S (suppressor) cells.

T_C (cytotoxic) cells T lymphocytes that attack and kill invading microorganisms.

T_S (suppressor) cells T lymphocytes that regulate the immune system by decreasing (suppressing) the immune response.

taxis Movement of cells toward favorable environments and away from harmful ones.

taxonomy The science of classifying organisms.

temperate phages Phages that can enter the lysogenic state.

terminal electron acceptor Metabolism: compound at the end of an electron transport chain; e.g., oxygen for aerobic respiration.

terminal pathway Complement action: the cascade leading from C3 to lethal antimicrobial activity.

tertiary protein structure Structure determined by interactions among the R groups of its various amino acids.

tetracyclines Broad-spectrum antibiotics that interfere with ribosome activity by binding at the A site.

thallus The body of a fungus or alga.

thermal death point (TDP) The lowest temperature required to kill all microorganisms in a particular liquid suspension in 10 minutes.

thermal death time (TDT) The minimal time required to kill all microorganisms in a particular liquid suspension at a given temperature.

thermoacidophiles A group of archaebacteria that grow at high temperature and low pH.

thermophiles Organisms that grow at high temperature.

thrush A *Candida* infection of the mouth causing white patches on mucous membranes.

tinea Ringworm; infections of the skin, hair, and nails caused by the dermatophyte fungi.

tissue culture A cultivation of eucaryotic cells or tissues in vitro.

titer The highest dilution of a test solution that is active.

toxins Poisonous proteins produced by some microorganisms.

toxoids Treated toxins that have lost their harmful properties but still stimulate the immune system to produce antitoxin.

trace elements Certain inorganic elements that are essential to life but required in only minute amounts.

transcription The first step in gene expression; formation of RNA from a DNA template.

transduction The transfer of chromosomal genes by phage particles assembled with bacterial DNA.

transformation (1) Entrance of DNA from the environment into a cell. (2) Conversion of a normal cell into a cancer cell.

transfusion reaction A clinical response to a blood transfusion with mismatched blood types.

translation The second step in gene expression; synthesis of protein directed by mRNA.

transmission electron microscopy (TEM) Use of a beam of electrons rather than visible light to form a magnified image of an object.

transovarial transmission Passage of infectious microorganisms from one generation of host to the next through their eggs.

transposable elements Short stretches of DNA that have the capacity to move from one location to another in a genome (jumping genes).

trematodes Flukes, a type of flatworm.

tricarboxylic acid (TCA) cycle A cyclic metabolic pathway that oxidizes acetate and forms four precursor metabolites.

trophozoite The actively multiplying vegetative stage of a sporozoa.

tubercles Granulomas produced by tuberculosis infection; dense nodules containing mostly activated macrophages and monocytes.

tumor An abnormal tissue growth; may be cancerous or benign.

tumor necrotizing factor A protein secreted by phagocytes that causes loss of fluid from the circulation.

turbidity Cloudiness of a liquid caused by suspended particles.

turgor pressure A cell's internal pressure resulting from a higher intracellular than extracellular osmotic strength.

ulcer An open sore.

ultrastructure Detailed microscopic cell structure.

uncoating Virology: the process by which the capsid and envelope of a virion are removed.

unit membrane A phospholipid bilayer.

universal donor A person with type O blood who can give blood to people with any blood type—O, A, B, or AB.

universal recipient A person with type AB blood who can receive any type of blood—O, A, B, or AB.

unsaturated fat Fat composed of fatty acid molecules that contain double bonds.

urethritis Inflammation of the urethra, the outlet of the urinary system.

use-dilution test A specific microorganism is added to dilutions of the germicide in a culture medium to determine which concentrations inhibit growth.

vaccination Using vaccines to produce artificial active immunity.

vaccines Agents that confer immunity without causing disease.

vacuole A membrane-bounded intracellular vesicle.

vaginitis Inflammation of the vagina.

valence electrons The electrons in an atom's outermost shell.

varicella-zoster virus (VZV) A member of the herpesvirus family that causes varicella (chickenpox) and zoster (shingles).

vasodilation Blood vessel enlargement.

vectors Agents that transmit pathogens.

vegetations Abnormal growths on the heart valves that occur in endocarditis.

vegetative cells Cells that grow and reproduce asexually.

vertical transmission Direct transmission of pathogens from mother to fetus or infant.

vesicles Tiny fluid-filled skin lesions.

viable count Measure of the living cells in a population.

virions Intact, nonreplicating virus particles.

viroids Small circular molecules of ssRNA without a capsid that cause many plant diseases.

virology The study of viruses.

virulence factors Substances or features of a microorganism that help it cause disease.

virulent phages Phages that always kill their bacterial host.

virus A microscopic packet of nucleic acid usually wrapped in a protein coat.

vital stain A stain for living cells.

vitamins Nutrients required in minute quantities, primarily as precursors of enzyme cofactors.

Wirtz-Conklin spore stain A staining technique that selectively colors endospores.

yeast A single-celled fungus.

Ziehl-Neelsen stain A special staining technique for identifying *Mycobacterium tuberculosis* and closely related bacteria.

zoonosis A human disease caused by a pathogen that maintains an animal reservoir.

CREDITS

Frontmatter

Page ii Art by Craig Hanson / **Page vii** Jan Hinsch/Science Photo Library/Photo Researchers / **Pages viii–ix** The Wellcome Centre Medical Photographic Library, London / **Pages x–xi** Biozentrum/Science Photo Library/Photo Researchers / **Pages xii–xiii** Ken Eward/Photo Researchers / **Pages xiv–xv** A. B. Dowsett/Science Photo Library/Photo Researchers / **Pages xvi–xvii** Institut Pasteur/CNRI/Phototake / **Pages xviii–xix** W. A. Banaszweski/Visuals Unlimited

Page 1 Science Photo Library/Custom Medical Stock Photos

Chapter 1

Page 2 The Wellcome Centre Medical Photographic Library, London / **1.1** Courtesy the Master and Fellows, Magdalene College, Cambridge / **1.2** (all three photos) Dr. Tony Brain and David Parker/Science Photo Library/Photo Researchers / **1.3** T. Beveridge and S. Schultze, the University of Guelph/Biological Photo Service / **1.4** (a) Jan Hinsch/Science Photo Library/Photo Researchers; (b) Courtesy of Catherine Ingraham / **1.5** (a) David M. Phillips/Visuals Unlimited; (b) Hans Reinhard/Bruce Coleman Ltd. / **1.6** Jerome Paulin/Visuals Unlimited / **1.7** Heather Davies/Science Photo Library/Photo Researchers / **1.8** Cecil H. Fox/Photo Researchers / **1.9** (a) Historical Collections, National Museum of Health and Medicine, Armed Forces Institute of Pathology; (b) Reproduced by permission of the President and Council of the Royal Society, London / **1.10** (a) The Bettmann Archive / **1.11** The Bettmann Archive / **1.12** The Granger Collection / **1.13** The Bettmann Archive

Chapter 2

Page 20 *Rivers of Molten Stone* by Chesley Bonestell/Space Art International / **page 24** From Starr and Taggart, *Biology: The Unity and Diversity of Life*, 6th Edition, © 1992 by Wadsworth Publishing Co., Inc. / **2.4** From Starr and Taggart, *Biology: The Unity and Diversity of Life*, 6th Edition, © 1992 by Wadsworth Publishing Co., Inc. / **2.5** From Starr and Taggart, *Biology: The Unity and Diversity of Life*, 6th Edition, © 1992 by Wadsworth Publishing Co., Inc. / **2.7** Thomas A. Steitz, Yale University / **2.9** From Starr and Taggart, *Biology: The Unity and Diversity of Life*, 6th Edition, © 1992 by Wadsworth Publishing Co., Inc. / **2.11** Art by Palais/Beaubois from Starr and Taggart, *Biology: The Unity and Diversity of Life*, 6th Edition, © 1992 by Wadsworth Publishing Co., Inc. / **Page 36** (a) Milton R. J. Salton, New York University Medical Center

Chapter 3

Page 46 Alfred Pasieka/Peter Arnold / **3.1** (photo) Kelly Barker/FPG International / **3.7** (a), (b), (c) Bruce Iverson / **3.9** (photo) Raymond B. Otero/Visuals Unlimited / **3.10** John Cunningham/Visuals Unlimited / **3.11** (a) Jack M. Bostrack/Visuals Unlimited; (b) Eric Grave/Phototake; (c) John Cunningham/Visuals Unlimited / **3.12** (a), (b), (c), (d) David M. Phillips/Visuals Unlimited / **3.13** J. Sonneborn/Science VU/Visuals Unlimited / **3.14** (a) George Musil/Visuals Unlimited; (b) from Starr and Taggart, *Biology: The Unity and Diversity of Life*, 6th Edition, © 1992 by Wadsworth Publishing Company, Inc.; (c) Jeremy Pickett-Heaps, School of Botany, University of Melbourne / **3.15** (c) From J. H. Willison and G. C. Johnston, *Canadian Journal of Microbiology*, 31:109–118, 1985. Photo courtesy Gerald C. Johnston, Dalhousie University, Halifax, Nova Scotia / **3.16** (b) J. J. Cardamone, Jr. and B. A. Phillips/Biological Photo Service / **3.17** (photo) Jeremy Pickett-Heaps, School of Botany, University of Melbourne; **3.17** From Starr and Taggart, *Biology: The Unity and Diversity of Life*, 6th Edition, © 1992 by Wadsworth Publishing Company, Inc. / **3.19** (e) Courtesy Becton Dickinson Microbiology Systems / **3.20** (photo) Lillian Therrien and E. C. S. Chan/Visuals Unlimited / **3.21** G. W. Willis/Biological Photo Service / **3.23** Courtesy Forma Scientific, Inc.

Chapter 4

Page 77 (photo) Courtesy of Hiroshi Nikaido; art adapted from Fig. 1, pages 382–388, H. Nikaido, *Science*, Vol. 264, April 15, 1994. Copyright 1994 by the AAAS. Reprinted by permission of the American Association for the Advancement of Science and the author / **4.1** (a) George J. Wilder/Visuals Unlimited; (b) Ralph A. Slepecky/Visuals Unlimited / **4.2** (a) Art by Palais/Beaubois from Starr and Taggart, *Biology: The Unity and Diversity of Life*, 6th Edition, © 1992 by Wadsworth Publishing Co., Inc.; (b) T. E. Adams / Visuals Unlimited; (c) From P. L. Walne and H. J. Arnott, *Planta* 77 (1967), 325–354 / **4.3** Art by Carlyn Iverson / **4.4** (a) E. C. S. Chan / Visuals Unlimited; (b) Carolina Biological Service/Visuals Unlimited; (c) E. C. S. Chan/Visuals Unlimited; (d) Fred Hossler/Visuals Unlimited / **4.6** (a) J. A. Breznak and H. J. Pankratz/Biological Photo Service; (b) John McN. Sieburth, University of Rhode Island/Biological Photo Service / **4.7** (a) T. J. Beveridge/Science VU/Visuals Unlimited; (b) T. J. Beveridge/Visuals Unlimited; (c) David M. Phillips/Visuals Unlimited / **4.8** Carlyn Iverson / **Page 85** Courtesy Edward J. Bottone, Mount Sinai Hospital, New York / **4.10** (b) H. C. Aldrich, University of Florida / **4.11** From Stanley W. Watson, *International Journal of Systematic Bacteriology*, 21:254–270, 1971. Courtesy American Society for Microbiology / **4.12** © 1982 by Annual Reviews Inc. and Richard P. Blakemore. (Richard P. Blakemore, "Magnetotactic bacteria." *Annual Review of Microbiology* 36:217–238, 1982.) Photo: D. Balkwill / **4.13** (f) T. J. Beveridge, University of Guelph/Biological Photo Service / **4.15** (a) From Starr and Taggart, *Biology: The Unity and Diversity of Life*, 6th Edition, © 1992 by Wadsworth Publishing Company, Inc.; (b) Andrew S. Bajer, University of Oregon / **4.16** (a) Don W. Fawcett/Visuals Unlimited; (b) From A. C. Faberge, *Cell and Tissue Research*, 151:403–415, 1974. Courtesy Springer-Verlag New York; art (left) by D. & V. Hennings, (right) by Leonard Morgan, both from Starr and Taggart, *Biology: The Unity and Diversity of Life*, 6th Edition, © 1992 by Wadsworth Publishing Co., Inc. / **4.17** Art by Raychel Ciemma from Starr, *Biology: Concepts and Applications*, 2nd Edition, © 1994 by Wadsworth, Inc. / **4.18** Art by Raychel Ciemma from Starr, *Biology: Concepts and Applications*, 2nd Edition, © 1994 by Wadsworth, Inc. / **4.19** (a), (b) Don W. Fawcett/Visuals Unlimited; art (left) by Leonard Morgan, (right) by Robert Demarest, both from Starr and Taggart, *Biology: The Unity and Diversity of Life*, 6th Edition, © 1992 by Wadsworth Publishing Co., Inc. / **4.20** (photo) Gary Grimes, Hofstra University; art (left) by Robert Demarest after a model by J. Kephart, (right) by Leonard Morgan, both from Starr and Taggart, *Biology: The Unity and Diversity of Life*, 6th Edition, © 1992 by Wadsworth Publishing Co., Inc. / **Page 99** Photo courtesy of Paul and Linda Baumann / **4.21** (a) (photo) Keith R. Porter, University of Pennsylvania; art from Starr and Taggart, *Biology: The Unity and Diversity of Life*, 6th Edition, © 1992 by Wadsworth Publishing Co., Inc.; (b) (photo) L. K. Shumway, College of Eastern Utah; art (left) by Leonard Morgan, (right) by Palay Beaubois, both from Starr and Taggart, *Biology: The Unity and Diversity of Life*, 6th Edition, © Wadsworth Publishing Co., Inc. / **4.22** Art by Raychel Ciemma from Starr, *Biology: Concepts and Applications*, 2nd Edition, © 1994 by Wadsworth, Inc. / **Page 102** D. McLean and M. Kinsey, photo courtesy Paul Baumann, University of California, Davis / **4.23** Micrographs M. Sheetz, R. Painter, and S. Singer, *Journal of Cell Biology*, 70:193, 1976, by copyright permission of the Rockefeller University Press; art from Starr and Taggart, *Biology: The Unity and Diversity of Life*, 6th

Edition, © 1992 by Wadsworth Publishing Co., Inc. / **4.24** by Leonard Morgan from *Biology: Concepts and Applications,* 2nd Edition, © 1994 by Wadsworth, Inc. (above) after Alberts et al., *Molecular Biology of the Cell,* 2nd Edition. Copyright 1989 by Garland Publishing, Inc., New York. Used by permission

Chapter 5

Page 108 Scott Camazine/Sharon Bilotta-Best/Photo Researchers / **Page 135** Photo courtesy of Catherine and Craig Squires

Chapter 6

Page 140 From Starr and Taggart, *Biology: The Unity and Diversity of Life,* 6th Edition, © 1992 by Wadsworth Publishing Company, Inc. / **6.1, 6.3–6.6, 6.9** Art by Margaret Gerrity / **6.10** (a) The Bettmann Archive / **6.12, 6.13, 6.19** Art by Margaret Gerrity / **6.20** (photo) C. C. Brinton, Jr., and J. Carnahan / **6.22** (photo) Institut Pasteur/CNRI Phototake; art by Margaret Gerrity / **6.23** Art by Margaret Gerrity

Chapter 7

Page 174 Photo courtesy of French Anderson; **7.1, 7.3** Art by Margaret Gerrity / **7.4** Courtesy Hoefer Scientific Instruments / **7.5** Art by Margaret Gerrity / **7.6** Courtesy Robert Hammer, Howard Hughes Medical Institute, Dallas / **7.7** Art by Margaret Gerrity / **7.8** Courtesy Norman Lin, Genentech / **7.9** Art by Jeanne Schreiber from Starr and Taggart, *Biology: The Unity and Diversity of Life,* 6th Edition, © 1992 by Wadsworth Publishing Company, Inc.

Chapter 8

Page 191 Lara Hartley / **8.1** (photo) George Musil/Visuals Unlimited; art from Starr and Taggart, *Biology: The Unity and Diversity of Life,* 6th Edition, © 1992 by Wadsworth Publishing Company, Inc. / **8.6** Courtesy Manfred E. Bayer, Fox Chase Cancer Center, Philadelphia / **8.7** (b) K. Talaro/Visuals Unlimited / **8.11** (photo) Courtesy Millipore Corporation

Chapter 9

Page 214 Hank Morgan/Science Source/Photo Researchers / **9.7** Stan Lester, Lester Farms, Winters, California

Page 229 Biozentrum/Science Photo Library/Photo Researchers

Chapter 10

10.1 (corn) Grant Heilman Photography; (fly) Runk/Schoenberger/Grant Heilman Photography; (woman) Dr. Charles Henneghien/Bruce Coleman Ltd.; (microbe) D. Chase/Phototake / **10.2** (a) Courtesy Princeton University Museum of Natural History; (b) Courtesy Stanley W. Awramik, University of California, Santa Barbara / **10.3** C. A. Henley/ Biofotos / **10.6** Courtesy Biolog, Inc. / **10.7** (left) Raymond B. Otero/Visuals Unlimited; (right) Christine Case/Visuals Unlimited / **10.8** Courtesy of the Microbial Diseases Laboratory, Berkeley, CA/

10.11 Courtesy Fred Neidhardt, University of Michigan / **Page 245** Courtesy Becton Dickinson Microbiology Systems

Chapter 11

Page 248 Esther R. Angert, Indiana University / **11.1** (photo) Veronika Burmeister/Visuals Unlimited / **11.2** Art by Carlyn Iverson / **11.3** D. A. Glawe/Biological Photo Service / **11.4** (top) Dr. K. S. Kim/Peter Arnold; (bottom) Courtesy Dennis Ohman, University of Tennessee / **11.5** Carolina Biological Service/Visuals Unlimited / **11.6** (a) Science Photo Library/Custom Medical Stock Photos; (b), (c) William Ormerod, University of Southern California. Photos courtesy E. G. Ruby and M. McFall-Ngai / **11.7** Forsyth Dental Center/Biological Photo Service / **11.8** (a) CNRI/Science Photo Library/Photo Researchers; (b) art by Carlyn Iverson / **11.9** Michael Gabridge/Visuals Unlimited / **11.10** (a), (b) David M. Phillips/Visuals Unlimited; (c) R. Kessel and G. Shih/Visuals Unlimited / **11.11** (a) George J. Wilder/Visuals Unlimited; (b) Dr. Tony Brain/Science Photo Library/Photo Researchers / **11.12** From H. Veldkamp, G. Van den Berg, and L. P. T. M. Zevenhuizen, "Glutamic Acid Production by *Arthrobacter globiformis,*" *Antonie van Leeuwenhoek* 29:35–51, 1963 / **11.14** (a) CDC/Biological Photo Service; (b) Koneman/Visuals Unlimited / **11.15** Paul W. Johnson/Biological Photo Service / **11.16** Courtesy James T. Staley, University of Washington / **11.18** Michael Richard/Visuals Unlimited / **11.19** Courtesy Patricia Grilione, San Jose State University. Photo: Patricia Grilione and J. Pangborn / **11.20** Corale Brierley/Visuals Unlimited / **11.21** R. Howard Berg/Visuals Unlimited / **11.22** (photo) C. B. C./Phototake; art by Carlyn Iverson

Chapter 12

Page 274 Larry Jensen/Visuals Unlimited / **12.1** (a) R. M. Meadows/Peter Arnold; (b) Penn State University Teaching Collection. Photo courtesy William Merrill / **12.2** (a), (b), (d) Art by Carlyn Iverson; (c) Garry T. Cole/Biological Photo Service / **12.3** From Dowson, Springham et al. *New Phytologist III,* 501–509. Courtesy Alan D. M. Rayner, University of Bath, and the New Phytologist Trust / **12.4** Heather Angel/Biofotos / **12.5** (a) John Cunningham/Visuals Unlimited; (b) David M. Phillips/Visuals Unlimited; (c) art by Carlyn Iverson / **12.6** (a) Art by Carlyn Iverson; (b) Robert Simpson/Nature Stock / **12.7** (bottom left) From *Living Images* by G. Shih and R. G. Kessel. © 1982 by Jones & Bartlett, Inc.; (bottom right) Martin Ainsworth from A. D. M. Rayner, *New Scientist,* November 19, 1988; art from Starr and Taggart, *Biology: The Unity and Diversity of Life,* 6th Edition, © 1992 by Wadsworth Publishing Co., Inc. / **12.8** (a) Courtesy G. L. Barron, University of Guelph; (b) Bruce Iverson / **12.9** Eric Crichton/Bruce Coleman Ltd. / **12.10** (a) Courtesy Florida Marine Research Institute; (b) C. C. Lockwood/Cactus Clyde Productions / **12.11** Courtesy Greta A. Fryxell, Texas A & M University / **12.12** Art by Palais/Beaubois from Starr and Taggart, *Biology: The Unity and Diversity of Life,* 6th Edition, © 1992 by

Wadsworth Publishing Company, Inc. / **12.13** (a) Art by Carlyn Iverson; adapted from Raven, Evert, and Eichhorn, *Biology of Plants,* Fourth Edition, 1986. Used by permission; (b) Don and Pat Valenti/Tom Stack & Associates; (c) Richard H. Thom/Tom Stack & Associates / **12.14** Art by Carlyn Iverson / **12.15** M. Abbey/Visuals Unlimited / **12.16** Arthur M. Siegelman/Visuals Unlimited / **12.17** Gary Grimes and Steven L'Hernault, Hofstra University / **12.18** (a) Edward S. Ross, California Academy of Sciences; (b) Carolina Biological Supply Company/Phototake; (c) Fig. 1, p. 82, from "Morphogen hunting in *Dictyostelium,*" Robert R. Kay, Mary Berks, and David Traynor, *Development* 1989 Supplement, 81–90. © The Company of Biologists 1989 / **12.19** (b) Cath Ellis, Department of Zoology, University of Hull, Science Photo Library/Photo Researchers / **12.20** (photo of proglottid) Carolina Biological Supply Company/Phototake; art by K. Kasnot from Starr and Taggart, *Biology: The Unity and Diversity of Life,* 6th Edition, © 1992 by Wadsworth Publishing Co., Inc. / **12.21** Photo by David Lacomis, University of Pittsburgh School of Medicine. Reprinted with permission from *The New England Journal of Medicine,* p. 1134, October 15, 1992

Chapter 13

Page 302 Jean Claude Revy/Phototake / **13.1** (top) Kenneth M. Corbett; (bottom) Dr. O. Bradfute/Peter Arnold / **13.5** Courtesy of John L. Ingraham / **13.6** E. C. S. Chan/Visuals Unlimited / **13.8, 13.9** Art by Carlyn Iverson / **13.10** James Holmes/Cell Tech Ltd/Science Photo Library/Photo Researchers / **13.11** Art by Carlyn Iverson / **Page 318** Photo courtesy of Richard L. Crowell / **13.16** Art by Carlyn Iverson / **13.17** (a) K. G. Murti/Visuals Unlimited; (b) art by Carlyn Iverson / **13.18** Breck's, Peoria, Ill. / **13.19** Courtesy Agricultural Research Service, USDA / **13.20** EM Unit, CVL, Weybridge/Science Photo Library/Photo Researchers

Page 331 Ken Eward/Photo Researchers

Chapter 14

Page 332 Copyright and photos Lennart Nilsson, *The Body Victorious,* Dell Publishing Company / **14.1** Art by Carlyn Iverson / **14.2** Courtesy Research Labs, Nestlé Products Technical Assistance Co., Ltd., Lausanne, Switzerland / **14.3** Larry Jensen/Visuals Unlimited / **14.4, 14.5** Art by Carlyn Iverson / **14.6** (b) Prof. P. Motta, Department of Anatomy, University "La Sapienza," Rome/Science Photo Library/Photo Researchers / **14.7** Robert P. Apkarian, Yerkes Primate Research Center, Emory University / **14.8** Art by L. Calver from Starr and Taggart, *Biology: The Unity and Diversity of Life,* 6th Edition, © 1992 by Wadsworth Publishing Co., Inc. / **14.9** (a) Paul W. Johnson/Biological Photo Service; (b) Martin M. Rotker; (c) art by Carlyn Iverson / **14.10** (a) Art by Kevin Somerville from Starr, *Biology: Concepts and Applications,* 2nd Edition, © 1994 by Wadsworth, Inc.; (b) Z. Skobe/Biological Photo Service; (c) Z. Skobe/Biological

Photo Service; (d) David M. Phillips/Visuals Unlimited / **14.11** Art by Kevin Somerville from Starr, *Biology: Concepts and Applications,* 2nd Edition, © 1994 by Wadsworth, Inc. / **14.12** Courtesy Allentown Caging Equipment Co., Inc.

Chapter 15

Page 353 Mauritius GmbH/Phototake / **15.1** Courtesy Parke-Davis, division of Warner-Lambert Co. / **15.4** Kent Wood/Photo Researchers / **15.5** (photo) Scott Camazine/Photo Researchers / **15.6** (photo) K. E. Muse, Duke University Medical Center/Biological Photo Service / **15.8** Alfred Pasieka/Peter Arnold / **15.9** From *Clinical Microbiological Review,* July 1989, p. 288. Reprinted by permission of the American Society for Microbiology. Photo courtesy Vincent A. Fischetti, Rockefeller University / **15.11** From G. W. Sullivan and G. L. Mandell, *Infection and Immunity* 30:272–280 (October 1980). Courtesy American Society for Microbiology / **15.13** Courtesy Viral and Rickettsial Zoonoses Branch, Division of Viral and Rickettsial Diseases, CDC

Chapter 16

Page 380 Art by Carlyn Iverson / **16.1, 16.3** Art by Carlyn Iverson / **Page 389** Anne Dowie / **16.6** Biology Media/Science Source/Photo Researchers / **16.8** (top) Copyright Boehringer Ingelheim International GmbH, photo Lennart Nilsson

Chapter 17

Page 402 Ken Greer / Visuals Unlimited / **17.1, 17.2** Art by Carlyn Iverson / **17.3** Art from Starr, *Biology: Concepts and Applications,* 2nd Edition, © 1994 by Wadsworth, Inc. / **17.4** Art by Carlyn Iverson / **17.14** (a) Mary Lee/Phototake / **17.16** Courtesy Gilla Kaplan, Rockefeller University / **17.17** Art by Carlyn Iverson

Chapter 18

Page 429 (a) Science Photo Library/Custom Medical Stock Photos; (b) Courtesy Center Laboratories, Port Washington, N.Y. / **18.3** (a) David M. Phillips/Visuals Unlimited; (b), (c) David Scharf/Peter Arnold / **18.7** After F. Ayala and J. Kiger, *Modern Genetics,* Second Edition, Copyright © 1984 by the Benjamin/Cummings Publishing Company. Reprinted by permission. / **18.9** Art by Carlyn Iverson / **18.10** Ken Greer/Visuals Unlimited / **Page 446** Sygma

Chapter 19

Page 451 Courtesy Carter Products, division of Carter-Wallace, Inc. / **Page 454** Photo courtesy of Cynthia A. Needham / **19.6** (photo) From C. J. Smyth, A. E. Friedmans Kien, and M. R. J. Salton, *Infection and Immunity* 13, 1273–1288, 1978. By permission of the American Society for Microbiology. Photo courtesy Milton R. J. Salton, New York University Medical Center /

19.8 George Whiteley/Photo Researchers / **19.11** Courtesy the Wellcome Research Laboratories, Beckenham, Kent, England

Chapter 20

Page 465 George John Pinwell, *Death's Dispensary,* c. 1866. Philadelphia Museum of Art: Purchased: SmithKline Beckman Corporation Fund / **20.4** Bernard Pierre Wolff/Photo Researchers / **20.5** (a) Stuart Franklin/Magnum; (b) Burt Glinn/Magnum; (c) Lara Hartley / **Page 479** Mary Evans Picture Library / **Page 481** National Library of Medicine

Chapter 21

Page 488 Science Photo Library/Photo Researchers / **21.1** C. James Webb/Phototake / **Page 493** (a) Courtesy Victor Lorian, M.D., the Bronx Lebanon Hospital, New York / **21.4** Camera M.D. Studios / **21.6** (d) Courtesy George A. Wistreich, East Los Angeles College / **21.8** Courtesy ICN Biomedicals / **Page 507** Photo courtesy of Albert Schatz / **21.13** Custom Medical Stock Photos

Page 519 A. B. Dowsett/Science Photo Library/Photo Researchers

Chapter 22

Page 524 J. Croyle/Custom Medical Stock Photos / **22.1** Art by Carlyn Iverson / **22.2** Courtesy Robert Berg, M.D., University of Arizona Health Science Center / **22.3** B.S.I.P./Custom Medical Photos / **22.4** (a) CNRI/Science Photo Library/Photo Researchers; (b), Science VU/Visuals Unlimited / **22.5** Science Photo Library/Photo Researchers / **22.6** (a) CNRI/Phototake; (b) Raymond B. Otero/Visuals Unlimited; (c) Arthur M. Siegelman/Visuals Unlimited; (d) Raymond B. Otero/Visuals Unlimited / **22.7** David M. Phillips/Visuals Unlimited / **22.8** CDC/Biological Photo Service / **22.10** (a) Courtesy Michael D. Iseman, M.D., University of Colorado School of Medicine; (b) Richard D'Amico, M.D./Custom Medical Stock Photos / **22.11** Jon Levy/Gamma-Liaison / **22.12** The Bettmann Archive / **22.13** Centers for Disease Control / **22.15** (both photos) CDC/Biological Photo Service / **22.16** Dr. F. C. Skvara/Peter Arnold

Chapter 23

Page 554 CNRI/Science Photo Library/Photo Researchers / **23.1** Art by Kevin Somerville from Starr, *Biology: Concepts and Applications,* 2nd Edition, © 1994 by Wadsworth, Inc. / **23.2** Vincent Zuber/Custom Medical Stock Photos / **23.3** (photo) Edward H. Gill/Custom Medical Stock Photos / **23.4** Centers for Disease Control / **23.5** (a), (b) Courtesy P. J. Sansonetti, M. C. Prévost, P. Gounon, personal collection, Institut Pasteur, Paris / **23.6** Centers for Disease Control / **23.8** David Scharf/Peter Arnold / **23.10** (a) D. Rollins/Science VU/Visuals Unlimited; (b) Courtesy M. E. Konkel, S. F. Hayes, and W. Cieplak, Jr., Rocky Mountain Laboratories, Hamilton, Montana / **23.11** Veronika Burmeister/Visuals Unlimited / **23.12** Science Photo Library/Custom Medical Stock Photos / **23.13** (a) Centers for Disease Control;

(b) Courtesy James W. Smith, M.D., Indiana University / **23.14** (a) From *Transactions of the Royal Society of Tropical Medicine and Hygiene,* Vol. 74, No. 4, 1980, 429–433. Photo courtesy Robert L. Owen, M.D., Veterans Administration Medical Center, San Francisco; (b) From J. R. Carlson, M. F. Heyworth, and R. L. Owen, *Survey of Digestive Diseases* 2:201–213, 1984. Used by permission of S. Karger AG, Basel, Switzerland. Photo Courtesy Robert L. Owen, M.D. / **23.15** Reprinted with permission from *The New England Journal of Medicine,* Vol. 328, No. 13, p. 927. Photo by Martin Weber, M.D., MRC-Labs, Fajara, Banjul, The Gambia, West Africa / **23.16** Marsik/Science VU/Visuals Unlimited / **23.17** Photo by David Lacomis, University of Pittsburgh School of Medicine. Reprinted with permission from *The New England Journal of Medicine,* p. 1134, October 15, 1992 / **23.18** From K. Anderson, *International Journal of Parasitology* 5:487–493, 1975 / **23.19** (a) Tektoff Rhone/Meriuex/CNRI/Science Photo Library/Photo Researchers / **23.20** Art by Carlyn Iverson

Chapter 24

24.1 (a), (b), (c) Art by Kevin Somerville from Starr, *Biology: Concepts and Applications,* 2nd Edition, © 1994 by Wadsworth, Inc. / **24.2** CNRI/Science Photo Library/Photo Researchers / **24.4** (a) Science VU/Visuals Unlimited; (b) Centers for Disease Control / **24.6** CDC/Biological Photo Service / **24.7** Custom Medical Stock Photos / **24.8** D. Chase/Phototake / **24.9** Science VU/Visuals Unlimited / **24.10** Centers for Disease Control / **24.11** (a) Institut Pasteur/CNRI/Phototake; (b) Ken Greer/Visuals Unlimited / **Page 609** Courtesy San Francisco Department of Public Health, STD Prevention and Training Unit. Photo: Anne Dowie / **24.12** Michael Davidson/Custom Medical Stock Photos / **24.13** Centers for Disease Control / **24.14** George J. Wilder/Visuals Unlimited / **24.15** (photo) Arthur M. Siegelman/Visuals Unlimited; art by Carlyn Iverson / **24.16** Courtesy Rise Maura Jampel, M.D.

Chapter 25

Page 620 From *Handbook of Thoraco-abdominal Nerve Block* by Jordan Katz and Hans Renck. Reprinted with permission by Information Consulting Medical AB, Malmo, Sweden / **25.1** (left) Art by Kevin Somerville from Starr and Taggart, *Biology: The Unity and Diversity of Life,* 6th Edition, © 1992 by Wadsworth Publishing Co., Inc.; (right) art by Carlyn Iverson / **25.2** CNRI/Science Photo Library/Photo Researchers / **25.3** From *Infectious Diseases Illustrated: An Integrated Text and Color Atlas* by Harold P. Lambert and W. Edmund Farrar. Gower Medical Publishing, London, United Kingdom, 1982. Reprinted by permission / **25.4** Melanie Carr/Custom Medical Stock Photos / **25.5** (both photos) CNRI/Science Photo Library/Photo Researchers / **25.6** CNRI/Phototake / **25.7** The Bettmann Archive / **Page 633** Map adapted from P. F. Wright et al., "Strategies for the Global Eradication of Poliomyelitis by the Year 2000," *The New England Journal of*

Medicine, Vol. 325, No. 25, p. 1776 / **25.8** (a) David M. Phillips/Visuals Unlimited; (b) Frederick C. Skvara/Peter Arnold / **25.9** Phototake / **25.10** John D. Cunningham/Visuals Unlimited / **Page 636** (top) Courtesy Rockefeller Archive Center, North Tarrytown, N.Y. / **25.11** F. Lambrecht/Visuals Unlimited

Chapter 26

Page 640 (a) Martin/Custom Medical Stock Photos; (b) AFIP/Science Source/Photo Researchers **26.1** Art by Robert Demarest from Starr and Taggart, *Biology: The Unity and Diversity of Life,* 6th Edition, © 1992 by Wadsworth Publishing Co., Inc. / **26.2** Art by Carlyn Iverson / **26.3** From *Dermatology* by O. Braun-Falco, G. Plewig, H. H. Wolff, and R. K. Winkelmann, Third Edition. Courtesy and © Springer-Verlag GmbH, Berlin-Heidelberg, 1984 / **26.4** Science Photo Library/Photo Researchers / **26.5** Ken Greer/Visuals Unlimited / **26.6** From *Infectious Diseases Illustrated: An Integrated Text and Color Atlas* by Harold P. Lambert and W. Edmund Farra. Gower Medical Publishing, London, United Kingdom, 1982. Reprinted by permission / **26.7** Biophoto Associates/Photo Researchers / **26.8** Science Photo Library/Custom Medical Stock Photos / **26.9** Centers for Disease Control / **26.10** (b) Lowell Georgia/Science Source/Photo Researchers / **26.11** Courtesy Kenneth A. Schiffer, M.D. / **26.12** CDC/Biological Photo Service / **26.13** Dr. P. Marazzi/Science Photo Library/Photo Researchers / **26.14** Biophoto Associates/Photo Researchers / **26.15** E. Gray/Science Photo Library/Photo Researchers / **26.16** AFIP/Science VU/Visuals Unlimited / **26.17** (left) Omikron/Science Source/Photo Researchers; (right) Barts Medical Library/Phototake / **26.18** Courtesy The World Bank

Chapter 27

Page 669 From *Infectious Diseases,* © J. B. Lippincott Co., Philadelphia. Photo courtesy Paul D. Hoeprich, M.D. / **27.1** Art by Carlyn Iverson / **27.2** James Dennis/Phototake / **27.3** (a) A. B. Dowsett/Science Photo Library/Photo Researchers; (b) Charles W. Stratton/Science VU/Visuals Unlimited; (c) Centers for Disease Control / **Page 678** Ken Greer/Visuals Unlimited / **27.4** (a) Courtesy Dr. Tom G. Schwan, Rocky Mountain Laboratories, Hamilton, Montana; (b) Larry Mulvehill/Photo Researchers / **Page 680** Robert Calentine/Visuals Unlimited / **27.5** Charles W. Stratton/Science VU/Visuals Unlimited / **27.6** (b) Ken Greer/Visuals Unlimited / **27.7** Custom Medical Stock Photos / **27.8** Centers for Disease Control / **27.9** Copyright Boehringer Ingelheim International GmbH, photo Lennart Nilsson / **27.11** Adapted from Brock et al., *Biology of Microorganisms,* 7th Edition. Englewood Cliffs, NJ: Prentice-Hall, 1994, p. 549 / **27.13** Adapted from Fig. 1, p. 1266, *Science,* Vol. 260, May 28, 1993. Copyright 1993 by the AAAS. Reprinted by permission / **27.14** Art by Leonard Morgan from Starr and Taggart, *Biology: The Unity and Diversity of Life,* 6th Edition, © 1992 by Wadsworth Publishing Co., Inc.; additional art by Carlyn Iverson / **27.15** Centers for Disease Control / **27.16** (a), (b) From W. M. Hutchinson, R. M. Pittilo, S. J. Ball, and J. C. Siim, *Annals of Tropical Medical Parasitology* 74:427–437, 1980 / **27.17** (photo) Courtesy Robert Calentine; Art by Raychel Ciemma from Starr, *Biology: Concepts and Applications,* 2nd Edition, © 1994 by Wadsworth, Inc. / **27.18** Dianora Niccolini

Page 705 W. A. Banaszweski/Visuals Unlimited

Chapter 28

28.2 Courtesy Nationalmuseet, Copenhagen, Denmark / **28.3** Ken Wagner/Phototake / **28.4** Courtesy David Pramer, Rutgers University / **28.5** Courtesy F. Brent Reeves, Colorado State University / **Page 712** Photo courtesy of Gary Schoolnik and David Thornton / **28.6** Doug Sokell/Visuals Unlimited / **28.7** Art by Carlyn Iverson / **28.8** (top photo) Courtesy Mark E. Dudley and Sharon R. Long; (bottom photo) Adrian Davies/Bruce Coleman Ltd; art by Jennifer Wardrip from Starr and Taggart, *Biology: The Unity and Diversity of Life,* 6th Edition, © 1992 by Wadsworth Publishing Company, Inc. / **28.9** From Starr and Taggart, *Biology: The Unity and Diversity of Life,* 6th Edition, © 1992 by Wadsworth Publishing Company, Inc. / **Page 72** Photo courtesy of Mary Lidstrom / **28.11** From Starr and Taggart, *Biology: The Unity and Diversity of Life,* 6th Edition, © 1992 by Wadsworth Publishing Company, Inc. / **28.12** Art by Carlyn Iverson / **Page 724** Fred Grassle, Woods Hole Oceanographic Institution / **28.14** From *International Journal of Systematic Bacteriology* 21:91–99, 1971. Reprinted by permission of the American Society for Microbiology. Photo courtesy Richard F. Unz, Pennsylvania State University / **Page 731** Ken Sakamoto/Black Star

Chapter 29

Page 734 (left) From M. Rüegg, U. Moor, and B. Blanc, *Milchwissenschaft 35,* 329–334, 1980. Photo courtesy Dr. M. Rüegg, Federal Dairy Research Institute, Liebefeld, Switzerland; (right) Lara Hartley / **29.1** Courtesy Henry P. Fleming, North Carolina State University / **29.2** J. P. Amet/Sygma / **29.3** Jonathan E. Pite/Gamma-Liaison / **29.4** Courtesy R. E. Kunkee, University of California, Davis. Photo: M. R. Vilas / **29.5** Courtesy Anheuser-Busch, Inc. / **29.7** J. R. Adams/Science VU/Visuals Unlimited / **29.8** Courtesy Eli Lilly and Co. / **Page 748** Courtesy Susumu Maeda, University of California, Davis / **29.10** Courtesy Steven Lindow, University of California, Berkeley

INDEX

ORGANISMS AND DISEASES/ANATOMIC SYNDROMES CAUSED BY THEM

Anatomic syndromes are set in bold type to distinguish them from diseases caused by particular organisms. For a discussion of anatomic syndromes and diagnosis, see pages 520–523 of the text.

Viruses

Virus	Group/Family	Disease/Anatomic Syndrome	Pages	Virus	Group/Family	Disease/Anatomic Syndrome	Pages
adenovirus	Adenoviridae	**conjunctivitis, enteric diseases, respiratory diseases**	308	hepatitis delta agent		type D hepatitis	581
				hepatitis E	Calciviridae?	type E hepatitis	581
arboviruses	Flaviviridae, Togaviridae	**encephalitis**	632, 634	herpes simplex	Herpesviridae	**conjunctivitis, encephalitis,** fever blisters, **gingivostomatitis,** herpes, herpetic keratitis	308, 607–608, 652–653, 661–662
coronavirus	Coronaviridae	**colds, upper respiratory infections**	309, 533				
coxsackievirus	Picornaviridae	**aseptic meningitis, colds, encephaalitis, pericarditis, pharyngitis,** systemic illness of newborn	625, 634	human immunodeficiency (HIV)	Retroviridae	AIDS	308, 686–692
				human papillomavirus	Papovaviridae	common warts, condyloma acuminata, genital warts	308, 608–610, 656–657
cytomegalovirus	Herpesviridae	**birth defects,** cytomegalic inclusion disease	308, 614–615	influenza	Orthomyxoviridae	influenza, **pneumonia**	309, 321–323, 544–545
				measles (rubeola)	Paramyxoviridae	measles	309, 653–655
flavivirus	Flaviviridae	dengue fever (breakbone fever)	308, 684	mumps	Paramyxoviridae	mumps, **aseptic meningitis**	559–560, 625
enterovirus	Picornaviridae	**conjunctivitis, encephalitis, myocarditis, pericarditis,** polio	309	parainfluenza	Paramyxoviridae	croup	309, 546
				Norwalk agents	Picornaviridae	**gastroenteritis**	571
Epstein-Barr	Herpesviridae	Burkitt's lymphoma, infectious mononucleosis, nasopharyngeal carcinoma	308, 684–685	poliovirus	Picornaviridae (enterovirus)	poliomyelitis	630–632, 633
				rabies	Rhabdoviridae	rabies	309, 629–630
hantavirus	Arboviridae	hantavirus pulmonary syndrome	546	respiratory syncytial virus	Paramyxoviridae	bronchiolitis, bronchitis	309, 546
hepatitis A	Picornaviridae	type A hepatitis (infectious hepatitis)	309, 578–579	rhinovirus	Picornaviridae	**cold, upper respiratory infection**	309, 532–533
hepatitis B	Hepadnaviridae	type B hepatitis (serum hepatitis)	308, 579–581	rotavirus	Reoviridae	**gastroenteritis, infant diarrhea**	309, 571
				rubella	Togaviridae	rubella (German or 3-day measles)	308, 655
hepatitis C	Flaviviridae, Togaviridae?	type C hepatitis	308, 581	varicella zoster	Herpesviridae	chickenpox, shingles	308, 651–652
				variola	Poxvirus	smallpox	656
				yellow fever	Flaviviridae	yellow fever	308, 683–684

Fungi

Organism	Disease/Anatomic Syndrome	Pages	Organism	Disease/Anatomic Syndrome	Pages
Blastomyces dermatitidis	blastomycosis	248, 549	*Histoplasma capsulatum*	histoplasmosis	284, 546
Candida albicans	candidiasis, endocarditis, thrush, vaginitis	282, 284, 349, 612, 657–658, 672	*Madurella mycetomatis*	mycetoma, Madura foot	659
			Microsporum spp.	ringworm (tinea)	284, 657
Cocciodioides immitis	coccidioidomycosis (San Joaquin valley fever), meningitis	284, 547–549, 625	*Phialophora jeanselmei*	mycetoma	659
			Pneumocystis carinii	*Pneumocystis* pneumonia	284, 549
Cryptococcus neoformans	cryptococcosis, meningitis	85, 284, 625	*Sporothrix schenckii*	sporotrichosis	659
Epidermophyton spp.	ringworm (tinea)	8, 284, 657	*Trichophyton* spp.	ringworm (tinea)	284, 657

ORGANISMS AND DISEASES/ANATOMIC SYNDROMES CAUSED BY THEM

Bacteria

Organism	Gram Stain	Basic Morphology	Disease/ Anatomic Syndrome	Pages
Actinomadura spp.	+	rod, some filamentous forms	mycetoma	253
Actinomyces israelii	+	filamentous, diptheroid, and coccal	actinomycosis, **head and neck abscesses**	252, 264
Bacillus anthracis	+	rod, encapsulated	anthrax	91, 157, 252, 680–681
Bacillus cereus	+	rod, encapsulated	**food poisoning**	570
Bacteroides gingivalis	−	small rod	**periodontal disease**	558–559
Bordetella pertussis	−	coccobacillus	pertussis (whooping cough)	251, 254, 353, 540
Borrelia burgdorferi	−	spiral	Lyme disease	250, 251, 677–679, 680
Borrelia spp.	−	large spiral	relapsing fever	679–680
Brucella spp.	−	coccobacillus	brucellosis (Malta fever, undulant fever)	251, 677
Calymmatobacterium granulomatis	−	rod, encapsulated	granuloma inguinale	606
Campylobacter jejuni	−	rod	campylobacteriosis, gastroenteritis	251, 253, 568
Chlamydia psittaci	NA	coccoid, very tiny	ornithosis, psittacosis	537–538
Chlamydia trachomatis	NA	coccoid, very tiny	chlamydia, **conjunctivitis,** lymphogranuloma venereum, **pelvic inflammatory disease,** psittacosis, trachoma, **urethritis**	252, 259, 605–606, 612, 613, 660–661
Clostridium botulinum	+	rod	botulism, **food poisoning**	70, 216, 252, 628–629
Clostridium difficile	+	rod	**diarrhea**	263, 569
Clostridium perfringens	+	rod	gas gangrene, **food poisoning**	252, 263, 570, 648
Clostridium tetani	+	rod	tetanus	91, 252, 263, 626–628
Corynebacterium diphtheriae	+	rod, club-shaped, pleomorphic, forms palisades	diphtheria	252, 264, 343, 531–532
Coxiella burnetii	NA	coccobacillus	colds, Q fever	252, 259, 533, 538
Escherichia coli	−	rod	**diarrheal infections, meningitis, urinary tract infections**	251, 565–566, 595, 625
Francisella tularensis	−	small rod (coccobacillus)	tularemia	251, 254, 676–677, 678
Gardnerella vaginalis	−	small rod	**vaginitis**	251, 610
Haemophilus ducreyi	−	slender rod (coccobacillus)	chancroid	606
Haemophilus influenzae	−	coccobacillus, some strains form	**meningitis, epiglottitis upper respiratory infections**	165, 251, 528–529, 623–624
Helicobacter pylori	−	curved rod	**gastritis,** peptic ulcer	253, 568–569
Klebsiella spp.	−	rod, encapsulated	**pneumonia, urinary tract infections**	251, 536, 595
Legionella pneumophila	−	coccoid rod	legionellosis (Legionnaire's disease)	251, 254–255, 538–539
Leptospira interrogans	−	spiral	leptospirosis	250, 251, 595–596
Listeria monocytogenes	+	rod	listeriosis	218, 252, 263, 613–614
Mycobacterium bovis	A-F	rod	bovine tuberculosis	543
Mycobacterium leprae	A-F	rod	Hansen's disease (leprosy)	252, 265, 649–651
Mycobacterium tuberculosis	A-F	rod, branching forms	tuberculosis	77, 188, 252, 265, 366, 540–543
Mycoplasma hominis	NA	coccoid to short filaments	**nongonococcal urethritis, pelvic inflammatory disease**	606, 612
Mycoplasma pneumoniae	NA	too small to be visualized by light microscope	mycoplasmal pneumonia	252, 260, 533, 536–537
Neisseria gonorrhoeae	−	cocci in pairs	blindness, **conjunctivitis,** gonorrhea, neonatal gonorrheal ophthalmia, **pelvic inflammatory disease, urethritis**	214, 251, 255, 343, 597–601, 612, 613, 661
Neisseria meningitidis	−	cocci in pairs; capsules formed in young cells	**meningitis**	57, 251, 623–624
Nocardia asteroides	+	rod, some filamentous forms	**lung infections**	253
Propionibacterium acnes	+	rod	acne	252, 264, 343, 649
Pseudomonas aeruginosa	−	rod	**contact lens conjunctivitis, folliculitis**	255, 648, 660
Rickettsia prowazekii	NA	coccobacillus	epidemic typhus	252, 259, 682–683
Rickettsia rickettsii	NA	coccobacillus	Rocky Mountain spotted fever	252, 259, 681–682
Rickettsia typhi	NA	coccobacillus	murine typhus	252, 259, 682–683
Rochalimaea henselae	NA	coccobacillus	cat scratch disease	681
Salmonella choleraesuis, Salmonella enteritidis	−	rod	salmonellosis (**food poisoning**)	564–565
Salmonella typhi	−	rod	typhoid fever	70, 251, 562–563
Shigella spp.	−	rod, generally single	shigellosis (bacillary dysentery)	71, 251, 560–562
Staphylococcus aureus	+	cocci in clusters	abscesses, boils, **endocarditis, food poisoning,** impetigo, osteomyelitis, **pericarditis,** pneumococcal pneumonia, toxic shock syndrome	252, 262, 536, 570, 611, 646–647